ANATOMY & PHYSIOLOGY

Elizabeth Mack Co

Australia • Brazil • Canada • Mexico • Singapore • United Kingdom • United States

***Anatomy & Physiology*, 1st Edition**
Elizabeth Mack Co

SVP, Product: Erin Joyner

VP, Product: Thais Alencar

Product Director: Maureen McLaughlin

Product Manager: Diana Baniak

Product Assistant: Emily Loreaux

Learning Designer: Kristin Meere

Senior Content Manager: Brendan Killion

Associate Digital Project Manager: Jessica Witczak

Developmental Editor: Julie Scardiglia

Senior Director, Product Marketing: Jennifer Fink

Product Marketing Manager: Adam Kiszka

Senior Product Development Researcher: Nicole Hurst

Content Acquisition Analyst: Ann Hoffman

Production Service: MPS Limited

Designer: Chris Doughman

Cover Image Source: Irina Shi/Shutterstock.com

For product information and technology assistance, contact us at
Cengage Customer & Sales Support, 1-800-354-9706
or support.cengage.com.

For permission to use material from this text or product, submit all requests online at **www.copyright.com**.

Library of Congress Control Number: 2022909327

ISBN: 978-0-357-80221-2
Loose-leaf Edition: ISBN: 978-0-357-96964-9

Cengage
200 Pier 4 Boulevard
Boston, MA 02210
USA

Cengage is a leading provider of customized learning solutions. Our employees reside in nearly 40 different countries and serve digital learners in 165 countries around the world. Find your local representative at: **www.cengage.com**.

To learn more about Cengage platforms and services, register or access your online learning solution, or purchase materials for your course, visit **www.cengage.com.**

Printed in the United States of America
Print Number: 01 Print Year: 2022

Dedication

In many ways, I have tried to package my classroom self into the pages of this book. I started teaching years ago, but answering questions, writing case studies, telling stories, and—most importantly—being inspired by the curiosity and wonder of thousands of students changed me. It was because of the students that I became the teacher that I am today. And so, I dedicate this, my first book, to them.

For my students, thank you.

Brief Contents

Unit 6 Deviations from Homeostasis

Table of Contents

Unit 6 Deviations from Homeostasis

About the Author

Dr. Elizabeth Mack Co is Assistant Clinical Professor in the Departments of Biology and Health Sciences at Boston University (BU). She teaches Gross (cadaveric) Anatomy, Human Physiology, and Physiology of Reproduction. Her teaching career spans a variety of courses, including Human Infectious Diseases, Introductory Biology, Cellular and Molecular Biology, Human Biology, Human Pathophysiology, and Histology. Dr. Co received her Ph.D. in Biomedical Sciences (with a focus on Immunology) from the University of California, San Francisco; she earned her BA with High Honors in Biology and Education at Mount Holyoke College in Massachusetts.

Outside of teaching, Dr. Co holds positions in a number of science organizations, including:

- The HAPS (Human Anatomy and Physiology Society) Learning Objectives Panel
- Principal Investigator of Assessing Student Engagement and Efficacy of Remote Learning, BU
- Independent Contractor and Presenter for Howard Hughes Medical Institute (HHMI) BioInteractive
- Society for College Science Teachers
- National Science Teachers Association
- American Association of Anatomists
- American Physiological Society

As a professor, Dr. Co is renowned for her passion—both in regard to the human body and about learning itself. In 2018 she was nominated by students and members of the faculty at BU and received the Metcalf Award for Excellence in Teaching, Boston University's highest teaching award.

Dr. Co's current research focuses on learning, particularly critical thinking skills development. She puts this research into practice by integrating an active learning and study skills curriculum in her courses. One of the focal points she investigates is how "student awareness about their learning impacts their assessment performance."

Preface

As I move around in the world from doctors' offices to dinner parties, I occasionally am asked what I do for work. When I reply "I teach Anatomy and Physiology," people tend to have very strong reactions. From "Oh! I absolutely LOVED that course in college" to "A&P almost ended my interest in medicine" to "Oh, wow, you must have a very hard job." Everyone has intense feelings about A&P. I typically reply "Actually, I have the BEST job in the world because everyone has a body and so there is a common baseline of shared curiosity." But the fact of the matter is that A&P can be intimidating for students (and for us teachers, too!) because there is a lot of information that can be taught. One of the keys as an instructor to having students leave with *positive* intense feelings is finding a balance in how much information to present and how to present it.

One of the goals I had for this book was to lighten and streamline the content. Instructors are in a perennial push-pull with the desire to reduce the burden on our students and the desire to prepare them for their future exams and courses. If we sacrifice anatomical structures by not teaching enough of them, we might leave our students underprepared; if we sacrifice the interesting connections, our students may lack the passion and excitement for the subject; and if we keep everything in, they might be overwhelmed. In the writing of this book, I wanted to give instructors and learners the choice whether to explore deeper into the interesting connections and content. I separated these topics into "Cultural Connections" and "Digging Deeper" features. **Cultural Connections** are usually ways that the topics being covered connect to everyday health, or the way that history impacts health or science today. **Digging Deeper** features often discuss relevant diseases or the science behind healthcare technology.

Increasing student persistence is a core need shared across community colleges and universities in the United States, with 30–50 percent[1,2] of students never completing the two-semester Anatomy & Physiology course. Recognizing that the course can be extraordinarily challenging due to its expansive depth and breadth of coverage, I address this lack of student persistence by improving student preparedness and helping students see themselves and the world around them reflected in what they are learning to engage them in the learning process. Increasing student confidence in both these areas will allow them to tackle the rigors of this course and lead to greater persistence in their education—and hopefully their career path.

Learner Support

As an instructor, how often are you asked "So, what will be on the test?" When we stop to think about the job of a student, they are not only trying to learn the material, but trying to learn how the instructor teaches and assesses their students. One effective mechanism for conveying the expectations of the course is through the use of learning objectives. **Learning Objectives** (LOs) can be used by an instructor to frame their pedagogy, as a list of goals for your students to make sure that you prepare them for. Sharing these same LOs with your students communicates to them what you are going to be assessing them on. In other words, it effectively answers the question "What will be on the test?" before it is even asked.

The Human Anatomy and Physiology Society (HAPS) has set out to define a set of learning outcomes (LOs) for Anatomy, Physiology, and A&P courses. As a member of the Physiology LO panel, I have participated firsthand in the careful process through which these LOs are created. The LOs are honed over months through the collaboration of some of the colleagues I respect most in our field. It is no exaggeration that some of these LOs have individually taken hours to write or improve. They are the most thorough, well-considered and well-crafted set available. I have written this text almost completely tailored to the LOs provided by HAPS. On occasion I provide additional LOs specific to the content I have created; often these are higher-order LOs.

Chapter 1 is dedicated entirely to helping each student build a learning framework so they are more prepared to engage with all the content that follows. This chapter explains the science behind learning and introduces the concept of metacognition to set the foundation for learning A&P in a systemic way. From there, each chapter orients students around the metacognitive aspects of the content to help them tie the concepts to the wider world around them.

Features in each chapter that support student learning and success in A&P include:

- **Learning Checks:** This feature provides periodic section assessments throughout the chapter to check your learning.
- **Student Study Tips:** These tips are written by *actual students* who have been successful in A&P.

[1] Gultice, Amy, Ann Witham, and Robert Kallmeyer. "Are your students ready for anatomy and physiology? Developing tools to identify students at risk for failure." *Advances in Physiology Education* (June 2015): 108-115. Doi:10.1152/advan.00112.2014, https://pubmed.ncbi.nlm.nih.gov/26031727/

[2] Vedartham, Padmaja B. *Investigating Strategies to Increase Persistence and Success Rates among Anatomy & Physiology Students: A Case Study at Austin Community College District*. 2018. National American University, Ed.D. dissertation. *ERIC*, https://eric.ed.gov/?id=ED583935

- **Learning Connections:** Chapter 1 details different ways to learn the material in A&P. Throughout the book, Learning Connections suggest learning approaches.
- **Chapter Review:** At the end of every chapter are assessment questions for each section, including a Mini Case with questions. LOs, questions, and a brief summary are provided for each section.
- **Cultural Connections:** Sometimes the things that we fall in love with in A&P are the ways that the material connects to our everyday lives. Cultural Connections are small features in each chapter that attempt to do just that. Consider these topics the ones you are most likely to share with others at the dinner table.
- **Digging Deeper:** The content in this text is streamlined, but sometimes you may want to know more about a given topic. This feature provides that deeper exploration.
- **Apply to Pathophysiology:** In most of scientific history we have learned about how the body works by studying the times when it doesn't! We did not understand blood sugar regulation until we studied diabetes; we did not understand much about how viruses affect our cells until the HIV pandemic. The Apply to Pathophysiology feature helps students strengthen their understanding of physiology by examining a disease state.
- **Anatomy of:** This feature describes different concepts graphically.

Critical Thinking

A 2017 study of Google employees[3] found that among the qualities of the most successful workers, STEM expertise comes in last. The top skills were listening well, teamwork, and critical thinking. Fostering critical thinking in the classroom is often cited as a goal of instructors, including myself. When I first began measuring my students' critical thinking skills using Bloom's taxonomy and my own exam data (Co, 2019),[4] I was shocked that, despite this being a top priority of mine, my students performed poorly on higher-order cognitive skill questions. I realized that my pedagogical approach had been to *show* critical thinking, to *tell* them how important it is, but that, as with any skill, I needed to give them supported opportunities to *practice*.

Based on these findings, I developed a new pedagogical approach in my classroom. I teach critical thinking practice through multiple choice questions that build upon each other, concept by concept, to arrive at an understanding of a complex system or pathology. Blood pressure, for example, is a factor in the body with many layers (osmosis, Starling's forces, capillary dynamics, vasoconstriction, neural control). To create a critical thinking activity on blood pressure, I start with simple concepts such as osmosis, which helps the students to remind themselves of the fundamental ideas and build confidence. From there we move on, like steppingstones in a path, to more and more complex ideas.

In this text I provide stepwise practice for students to build their critical thinking skills in each chapter. These **Apply to Pathophysiology** features first ask students to recall fundamental information and then apply it in a new situation.

To give students additional practice in critical thinking, Chapter 28 is composed entirely of case studies. The 10 cases within can be assigned at the end of the semester or at intervals as the instructor chooses.

Inclusivity and Diversity—and Accuracy

The need to include more diverse content in the Anatomy and Physiology course was one of my biggest inspirations for writing this text. An example of how this issue came into my awareness occurred about 10 years into my teaching career. I attended a science museum with an exhibit on the human body that, naturally, I gravitated toward. One feature used a size measurement to estimate the volume of blood in each museum visitor. My body, the exhibit computer told me, contained 4.6 liters of blood. I walked away shaking my head. I'd been teaching physiology for a decade and everyone knows the human body contains 5 to 5.5 liters of blood. How many times had I said that in class? All the calculations we use in class on the cardiovascular system are based on this range.

It dawned on me that I, a slightly smaller-than-average woman, may not necessarily be represented in physiological estimates. I dug into the background on this and discovered that most of the average numbers we teach came from studies of men—young men, probably mostly white men, with an average weight of 150 pounds. Since then, when I teach, I introduce the idea of "average man," a mythological 150-pound white man on whom we base our calculations. The students and I look around our classroom (which is a large auditorium) and reflect that average man is not usually among us. We are beautifully diverse and require a wider range of numbers. In this book I have tried, when possible, to research and provide more accurate and representative numbers for us. In other cases I provide the anatomical and physiological factors that influence the value range.

This text will not only prompt students to think deeper about their learning; it will also challenge them to take a broader look

---

[3] https://www.washingtonpost.com/news/answer-sheet/wp/2017/12/20/the-surprising-thing-google-learned-about-its-employees-and-what-it-means-for-todays-students

[4] **Co, E.** (2019). "The power of practice: adjusting curriculum to include emphasis on skills." *Journal of College Science Teaching* **48** (5): 22–27.

at the cultural assumptions that have an impact on anatomy and medicine. For example, Hoffman et al. (2016)[5] explored biases held by medical students and revealed deeply racist beliefs that could affect their patient care—ideas such as the one that Black Americans have higher pain tolerances and therefore require less treatment for their pain. I believe that no one holds onto racist or bias-laden beliefs because they want to; rather, racist ideas persist because they haven't been sufficiently challenged. There hasn't been, in the education of these medical students, enough accurate information about the diversity and unity in function that exists among human bodies. Anatomy and Physiology is often the threshold into health education that our future clinicians pass through, and so this subject represents especially fertile ground to facilitate critical thinking when it comes to biases within scientific information.

In developing this text, I benefited from the reviews of a number of instructors, including Inclusion and Diversity (I&D) advocate Dr. Edgar Meyer, author of "Diversity and Inclusion in Anatomy & Physiology Education, Degree Programs, and Professional Societies."[6] My goal was to create an inclusive and diverse textbook in Anatomy & Physiology. Taking into consideration Dr. Meyer's feedback, I created a seven-point I&D plan, including:

1. More diverse models and imagery throughout the text
2. "Cultural Connections" features that link science with culture
3. In-class active learning opportunities to maximize inclusive and diverse academic experiences
4. Curriculum and anatomy pertaining to transgender and gender nonconforming individuals
5. Inclusive language that is sensitive to diversity in students
6. Clinical examples detailing the predispositions of certain racial and socioeconomic groups to display health care disparities, such as sickle cell anemia
7. Defining "average" values and providing comparative data, where applicable

Organization of the Text

Chapter 1, "The Art and Science of Learning in Anatomy and Physiology," is devoted to explaining methods students can use to approach learning. Dr. Co's commitment to her students comes through as she explains the proper mindset needed to study Anatomy & Physiology. She shares study tips from her previous students to offer a peer perspective to learners.

Chapters are organized as follows:

- **Unit 1 (Level of Organization)** contains chapters that examine the levels of organization of the structures of the human body.
- **Unit 2 (Support and Movement)** includes the chapters of the musculoskeletal system.
- **Unit 3 (Regulation, Integration, and Control)** contains chapters about systems that contribute to homeostasis through control over other systems.
- **Unit 4 (Fluids and Transport)** examines the systems that regulate fluids and fluid flow throughout the body.
- **Unit 5 (Energy, Maintenance, and Environmental Exchange)** is all about exchange with the external environment.
- **Our last unit, Unit 6 (Deviations from Homeostasis),** contains two chapters that are beyond the homeostatic function of the body: reproduction and the changes that occur in our bodies in disease.

Art

Throughout my teaching career I have spent a lot of time and energy observing how students interact with instructional art. Often illustrations, symbols, or representations that seem easy to understand using our expert eyes are not intuitive for an introductory learner. In designing each figure, I drew the structures and concepts as I would do and have done while talking to students in office hours. Our amazing illustrators then took what I produced and turned it into art. This instructional art approach is infused throughout the book but especially prominent in our "Anatomy of…" features.

Years ago, I learned in an undergraduate educational psychology class that science students often had trouble developing a scientist identity because they didn't see images of scientists that looked like them. The idea of representation has wide ranging ripple effects. Clinicians see a variety of patients; students see a variety of possible selves. When we choose a limited palate of representative humans in instructional art, we limit the scope of what students are exposed to in several ways. Therefore, I asked our art team to work with me on creating and sourcing images that represented a spectrum of bodies from young to old, across different sizes, gender expressions and ethnicities.

Whether you are a student or an instructor, please know that I have written this book for *you*. I hope that it helps you learn or teach and that you find it helpful in your A&P journey. I would love to hear back from you about it, whether it is typos, constructive suggestions, ideas for enrichment or, perhaps you'd

[5] Hoffman, K.M., Trawalter, S., Axt, J.R., and Oliver, N.M. (2016). "Racial bias in pain assessment and treatment recommendations, and false beliefs about biological differences between blacks and whites" *Proceedings of the National Academy of Sciences* **113** (16): 4296–4301. Doi: 10.1073/pnas.1516047113, https://www.pnas.org/content/113/16/4296

[6] Meyer, E. R., and Cui, D. (2019). "Diversity and inclusion in anatomy & physiology education, degree programs, and professional societies." *HAPS Educator* **23** (2): 396–419. Doi: 10.21692/haps.2019.012, https://files.eric.ed.gov/fulltext/EJ1233545.pdf

like to suggest your own study tips. Please know my virtual door is always open and I would love to chat with you.

I hope you develop the most positive feelings about A&P!

Liz Co, July 2022
eco@bu.edu

MindTap

MindTap is the online learning platform that gives you complete control of your course. Craft personalized, engaging learning experiences that boost performance and deliver access to eTextbooks, study tools, assessments, and student performance analytics—whether you are teaching an in-person, online, or hybrid course.

To provide a personalized, engaging digital learning experience, you can deliver your course in MindTap integrated with your institution's Learning Management System (LMS). MindTap supports your instruction by teaching students core concepts using a prebuilt, customizable learning path that progresses from understanding to application with readings from the eTextbook, study tools, interactive media, auto-graded assessments and more, including:

- **Learn Its:** 15-minute activities that pair concise narrative and multimedia with assessments to improve students' conceptual understanding of difficult concepts.
- **Case Studies:** Engage students with clinical scenarios and challenge them to higher-level understanding with auto-graded assessments.
- **Mastery Training** (powered by Cerego): Use cognitive science principles to help students learn key terms faster and more effectively by utilizing retention, distributed learning, and retrieval practice.
- **Virtual Labs:** Includes 3D models and Institute of Human Anatomy cadaver dissection videos, with assessment content for pre- and post-lab.
- **Practice Tests:** Mimics exam questions so students can confidently prepare on their own. Students choose the chapters they want to study and generate a test with the desired number of questions.
- **Chapter Quizzes:** Measures how well students have mastered material after completing chapter readings and activities. Students see feedback explaining the correct answer after they submit the quiz.
- **Flashcards:** Students can use ready-made, mobile-friendly flashcards, or create their own, to learn key terms and concepts.

Instructor Resources

- **Test Bank:** Build your exam from a list of multiple-choice and short-answer questions.
- **Lecture PowerPoints:** Key points from the text are outlined, along with images and active-learning activities to keep students engaged.
- **Image PowerPoints:** These slides give instructors easy access to all images from the text.
- **Instructor Manual:** Each chapter's Instructor Manual includes a chapter outline, key terms list, suggested activities, discussion questions, video links to share with students, and answers to Apply to Pathophysiology questions from the text.
- **Labeling Worksheets:** Students can practice labeling anatomical structures during class, while studying, or for a grade.

Cengage Unlimited

Boost Access and Affordability

Cengage Unlimited is the cost-saving student plan that includes access to our entire library of eTextbooks, online platforms and more—in one place, for one price. For just $124.99 for four months, a student gets online and offline access to Cengage course materials across disciplines, plus hundreds of student success and career readiness skill-building activities. See www.cengage.com/unlimited/instructor for more information.

HAPS

We created both print and digital content in this product with direct ties to HAPS learning outcomes. Each piece of information was intentionally chosen to support A&P student learning. Throughout the text and MindTap, students and instructors can see which content is tied to each learning outcome. The Human Anatomy and Physiology Society includes more than 1,700 educators who work together to promote excellence in the teaching of this subject area. The HAPS A&P Learning Outcomes measure student mastery of the content typically covered in a two-semester Human A&P curriculum at the undergraduate level. The full Learning Outcomes are available at https://www.hapsweb.org.

OpenStax

Certain content in this product was developed using open-source content from OpenStax's Anatomy & Physiology product. OpenStax (www.openstax.org) is part of Rice University and is a 501(c)(3) nonprofit charitable corporation. We would like to thank the contributing authors and editorial team for their work on Openstax's Anatomy & Physiology, which can be accessed here: https://openstax.org/details/books/anatomy-and-physiology.

Acknowledgments

Writing a book is an enormous and consuming task. As I sit here writing the acknowledgments, I feel as though I could write for days because there are so many people who have helped either directly or indirectly.

I'll start with my family. In many ways, this book is my third child, I feel as though I dreamed about it, hoped for it, and poured my heart and soul into it. It also, at times, interrupted my life in the ways that a newborn can. Therefore, my first acknowledgment is to my two human children, Talia and Eliot, who had to be patient and understanding and occasionally got less of their mom's time for the duration of the writing cycle. For all the times you sat next to me writing your books as I wrote mine, drew bones on the floor of my office while I finished a paragraph, and generally cheered me on, thank you. You are the loves of my life.

Sam, thank you for describing my early morning typing as a "steady pitter patter" instead of disruptive, for all the things you took on or waited for when I needed to write, for all of your listening, but most of all, for your steadfast and contagious belief in LFC, I do not think this book could or would exist without you.

I need to thank my work wife, who patiently answered my exasperated phone calls, took walks with me when I needed to get out of my head, edited many sentences, brought clarity and insight to biological concepts, and who has always been in my corner, thank you. I wouldn't have wanted to share this work journey or all those nachos with anyone else.

Now for the book team, first, thank you, Julie Scardiglia, for being the maven of details and always patiently redirecting me to my priorities. Brendan Killion patiently kept us all on task and driving forward, and we all owe him thanks that this book is being published in 2022, instead of 2032.

Akshat Mehta, Angelina McNulty, and Emily Ackerman, you each brought your own strengths to the book and I am so very grateful. Akshat, I've always said that there is a teacher in you! Thank you for sharing that teacher with the world in the pages of this book. Angelina and Emily, many, many students will be helped by the student study tips you've shared with them. You are learning assistants through and through. Thank you for providing your unique insights the process of learning A&P. I am not only grateful for your contributions to the book, but for the privilege of getting to a part of your journey as you transitioned from my students to my colleagues. I cannot wait to see what you do next in the world.

Nanette Tomicek, thank you for your insights on the hardest of the chapters. Janet Brodsky, your precision is unparalleled, I am thankful for your partnership throughout the book.

Hilary Engebretson, your comments as a reviewer were always spot on. You corrected content, adjusted my voice, and lent a perspective that always helped me bring things back in check. I very much appreciate your contributions.

This text would not make half as much sense as it does and would have commas in all the wrong places, if not for the careful editing of Christopher Chien. Thank you.

One of the most important things to me in conceiving this book was that the images within it represent a diverse set of bodies. I dreamed that we could fill the pages of the book with beautiful images of people of different shapes, sizes, ages, and appearances. Dragonfly brought that dream into vivid color and created a book that is more magnificent than I could have dreamed. I owe the amazing artists a huge thank you.

To Kelsey Kerr, Katherine Caudill-Rios, Maureen McLaughlin, and Diana Baniak. You saw an author in me. When I believed that a book would be a dream, you picked me out, coached me along, and willed it into reality. I am a person of many, many words, but I am not sure if I will ever find all of the right ones to be able to thank you for helping me make this transition. You are the birth coaches of this book baby. I am forever thankful.

This text, MindTap content, and instructor supplements were improved thanks to the insights of many Anatomy & Physiology instructors who shared their ideas, concerns, and feedback with the product team at Cengage. Thank you to the following participants who dedicated their time to improving the product through focus groups, surveys, interviews, and product testing.

Special thanks to the following reviewers:

Janet Brodsky, *Ivy Tech Community College*
Stephen Burnett, *Clayton State University*
Mary Colon, *Seminole State College of Florida*
Hilary Engebretson, *Whatcom Community College*
Stephen Henry, *Houston Community College*
Austin Hicks, *The University of Alabama*
Nathaniel King, *Palm Beach State College*
Ann LeMaster, *Ivy Tech Community College*
Edgar Meyer, *University of Mississippi Medical Center*
Andrew Nguyen, *CUNY Queensborough Community College*
Julie Posey, *Columbus State Community College*
Rosemary Stelzer, *University of Wisconsin-Milwaukee*
Nanette Tomicek, *Thomas Jefferson University*
Padmaja Vedartham, *Lone Star College System*
Kelsha Washington, *Florida A&M University*
Theo Worrell, *Delgado Community College*

1

The Art and Science of Learning in Anatomy and Physiology

Rawpixel.com/Shutterstock.com

Chapter Introduction

In some ways learning can be discipline specific. However, in many ways learning is learning, regardless of the content. In this chapter we will explore some of the science behind how we learn in order to help you foster success in Anatomy and Physiology (A&P).

1.1 The Science of Learning

1.1a Foreign, Familiar, and Mastery-Level Understanding

Let's take an example of a person who is learning a new language. We could also use an example of a spectator trying to learn about the game of baseball, or a student studying physics, or an architect learning about building construction, but let's take a new language. Let's say our learner's name is Talia. Talia goes to Spanish class for the first time. The teacher speaks in Spanish and Talia understands none of it. She guesses as to the instructions based on the instructors' gestures and tone of voice. After a few classes, Talia begins to pick up on some of the vocabulary of the class. She is learning Spanish words in her own time, using flashcards, online quizzing apps, recognizing phrases. She eventually becomes very familiar with Spanish words. After a few years, Talia travels to Mexico. She finds she is able to recognize enough Spanish words to be able to understand signs and ads that are printed in Spanish. However, Talia now sits in a café in Mexico City. A friendly stranger approaches her and begins to make conversation in Spanish. Will Talia be able to converse with the stranger?

Probably not, because there is a difference between being familiar with a language and being able to converse in it. Conversation is a skill, a dynamic skill that involves listening, decoding, recall, comprehension, and creativity. Talia has not yet practiced these skills. She has the content she needs but will need to practice the skill of conversation before she can employ it with ease.

Learning A&P (or really most anything) can be thought of the same way: there is the content of what you are learning, and the skills involved in its mastery and application. In terms of A&P, we have vocabulary, concepts, functions, structures, and locations, and then we need to learn how to connect ideas and apply knowledge in a new context, analyze new data. For example, if you were to become a clinician and a patient had a pain in their abdomen, it would not be sufficient to list off a memorized set of organs in the abdomen. You, the clinician, must apply your knowledge of the area and an understanding of the symptoms that the patient is experiencing and analyze the data from clinical tests to diagnose and treat the patient. You must think critically about the information you know and apply it in a new circumstance, one that may be nuanced or unique. Therefore, it is insufficient to simply memorize in A&P; we must learn the life-saving skills of application, analysis, and critical thinking.

Let's go back to Talia, the Spanish language learner, for a moment. Talia needs to engage in separate processes of memorizing vocabulary and practicing the skill of conversation. She cannot dive first into practicing the skill, nor is it sufficient to stop at memorization. If we were to diagram Talia's learning it would look something like **Figure 1.1**. Talia's stage 1 of learning happened in the classroom when she was a beginning learner. She is currently at an advanced level of stage 2; she has a lot of vocabulary

Figure 1.1 **Stages of Learning**

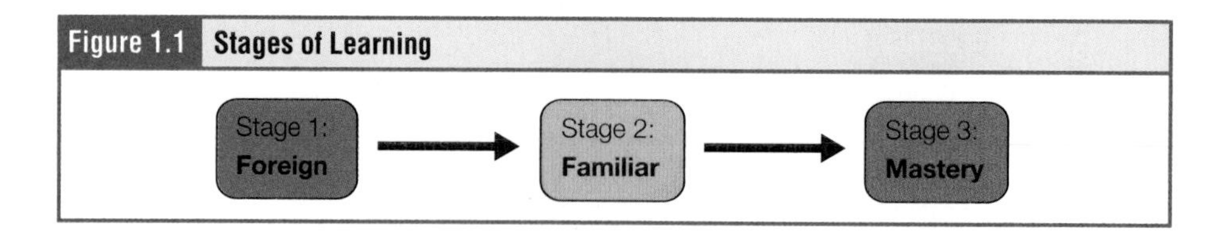
Stage 1:
Foreign
Stage 2:
Familiar
Stage 3:
Mastery

and is able to understand written text and recall meaning. If she practices the skills involved in conversation, she will reach stage 3—mastery of Spanish—and able to converse with ease.

The science of learning tells us that the transition between foreign and familiar is one that is best achieved through repetition and exposure. The methods of repeat exposure may look different from one type of learning to another, but in A&P learning, this may look like: reading/rereading your notes, copying your notes over, watching and rewatching lectures, reading and rereading your textbook, and using flashcards to memorize terminology.

The transition from familiarity to mastery, however, is a bit different. Here, the difference in our language-learning example is between being able to translate printed words or being fluid in a conversation. In A&P, mastery means not only being able to identify the deltoid muscle but being able to predict functional deficits with deltoid injury, or being able to identify, based on a set of symptoms, the location of an injury. In further mastery, you may find, as a clinician, that you need to communicate about A&P on a variety of levels. For example, you may use one set of terms to discuss a case with a colleague who is also a clinician, but you may need to translate information when conveying it to others in ways that are respectful of their level of knowledge and understanding.

In A&P, examples of practice might include: teaching or explaining the material to someone else, drawing a structure or structures, writing out an explanation of a concept, recreating a graph with explanations, completing practice problems, or answering the question *what happens if these components don't work?* (**Figure 1.2** and **Table 1.1**).

Figure 1.2 **Transitioning Between the Stages of Learning Using a Variety of Learning Techniques**

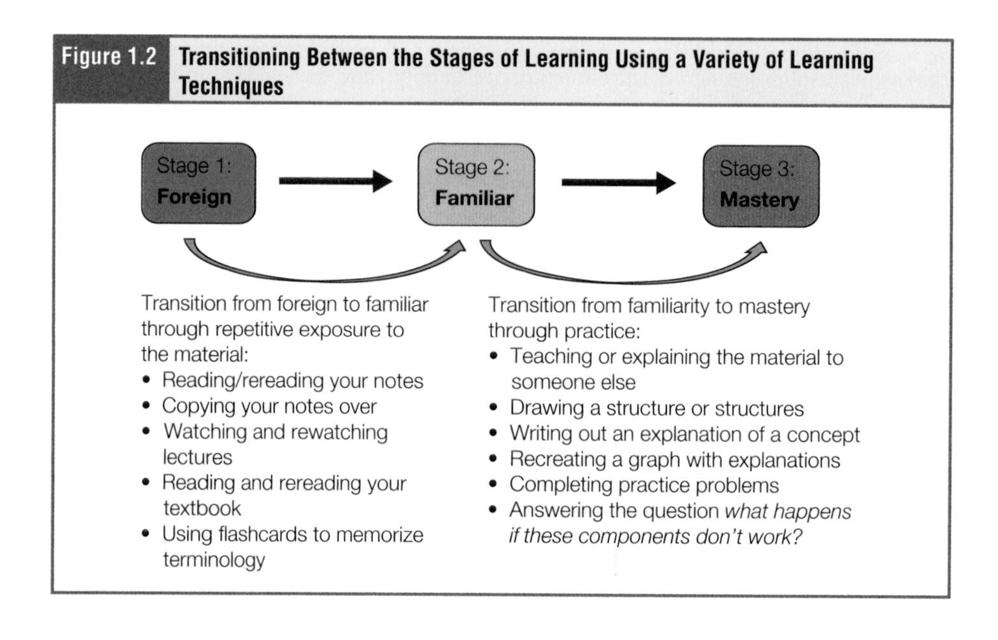
Stage 1:
Foreign
Stage 2:
Familiar
Stage 3:
Mastery
Transition from foreign to familiar through repetitive exposure to the material:
• Reading/rereading your notes
• Copying your notes over
• Watching and rewatching lectures
• Reading and rereading your textbook
• Using flashcards to memorize terminology
Transition from familiarity to mastery through practice:
• Teaching or explaining the material to someone else
• Drawing a structure or structures
• Writing out an explanation of a concept
• Recreating a graph with explanations
• Completing practice problems
• Answering the question *what happens if these components don't work?*

Table 1.1 Learning Strategies That Build Familiarity and Those That Build Mastery

Familiarity-Building Learning Strategies	Mastery-Building Learning Strategies
Reading	Drawing a representation of an anatomical structure
Looking over notes	Teaching someone else the material
Copying notes over	Quizzing yourself
Flashcards	Having a friend quiz you
Listening to/watching lectures	Explaining a process
Labeling an unlabeled drawing	Ordering the parts of a process after mixing them up
	Looking at a process you've learned, ask yourself about each component "what would happen if this was broken, how would that impact the entire process?"

1.1b Memory Formation and "Chunking"

Everyone has limits to their learning and memory. People who enter memorization competitions (yes! There are such things!) have a variety of different strategies for learning vast quantities of information quickly. The first technique is quite easy to employ. It is storytelling. Let's say that you have been asked to remember four words and recall them back four hours later. The four items are *cookie, staircase, yellow*, and *music*. Imagine you go about the rest of your day for the next four hours and then are asked to recall these four random words. Few people would be able to do so easily. But what if you paused for a moment, after getting your list of words, and created a quick story or visual for learning (*Cookie monster is sitting on the bright yellow staircase listening to his favorite music*)? If you paint this image in your mind, it is much more likely that you will be able to remember the four words a long time from now. Similarly, in A&P we can create short stories known as *mnemonic devices*. For example, there are 12 nerves that exit the brain and innervate structures mostly found in the head and neck. These 12 nerves are known as cranial nerves. Their names are: olfactory, optic, oculomotor, trochlear, trigeminal, abducens, facial, vestibulocochlear, glossopharyngeal, vagus, accessory, and hypoglossal. You got that? What was the fifth one? It's a challenge. However, what if, in order to remember them, we take just their first initials: OOOTTAFVGVAH and make a story such as *On, on, on they traveled and found Voldemort guarding very ancient horcruxes*? If you have read the Harry Potter books (or seen the movies) you will remember this mnemonic pretty easily. You still have to learn the names of the cranial nerves and that the second V, for example, stands for vagus, but it makes the process easier.

Chunking is the term for grouping items together as you learn them. It's a way to cheat your working memory into holding onto more information. Let's take an example. Please remember the following numbers—4195082637—in order, and repeat them back in five minutes. It seems like a challenging task. What if, instead, we chunk them like this: (419) 508-2637? That form of chunking is widely employed for memorizing telephone numbers. Birth dates are another example; instead of 10152004 you may remember a birthday as 10/15/2004. So how do we chunk information in A&P? One great example is learning structure and function together. Forms of chunking you might use could be:

- similarities and differences.
- structure and function.
- structures in order, for example, from top to bottom or superficial to deep.
- structures that are together in one location—for example, the three layers of protective tissue, called *meninges*, over the brain—can be learned together in one chunk.

Table 1.2 Examples of Chunking

Example 1: Chunking structures by shared functions	**Cellular Structures and Their Functions**		**Recycling Functions**	**Protein-Building Functions**
	• Lysosomes break down material through enzymes • Peroxisomes break down material through hydrogen peroxide • Smooth endoplasmic reticulum breaks down some toxins • Nucleus holds the instructions for building proteins • Ribosomes build proteins • Golgi apparatus modifies proteins	**chunked**	• Lysosomes • Peroxisomes • Smooth endoplasmic reticulum	• Nucleus • Ribosomes • Golgi apparatus
Example 2: Chunking structures by similarities	**Structures within the Abdominal Cavity**		**Organs**	**Membranes**
	• Stomach • Greater omentum • Small intestine • Large intestine • Lesser omentum • Liver • Mesocolon	**chunked**	• Stomach • Small intestine • Large intestine • Liver	• Greater omentum • Lesser omentum • Mesocolon

Two examples of chunking in A&P are provided in **Table 1.2**. Chunking is a great tool to incorporate as you move from foreign to familiar in your learning.

1.1c Retrieval Practice

Retrieval practice refers to memory strengthening. The more often you are asked to recall information, the easier it becomes to remember. Let's say that you change the passcode to your new mobile phone, email account, or favorite website. The first few times you enter this passcode it may be difficult to remember, but after the thirtieth time you've logged in, the passcode becomes rote, or habitual. The learning concept of retrieval practice is part of the philosophy behind assigning homework or quizzes or in-class questions during a course. The more times you are asked about a concept, a structure, or an idea, the more habitual it becomes to recall that piece of information. Your instructor may provide opportunities for you to practice retrieval, such as quizzes, but you can do this for yourself as a learner as well.

Examples of retrieval practice in studying can be:

- using flashcards, redrawing, or rewriting from memory.
- labeling an unlabeled diagram.
- paraphrasing a process or idea (especially effective if you are trying to tell someone else about this process or idea).
- predicting what questions could be asked of you about a structure/concept/idea.
- summarizing information at the end of a section or chapter.

Retrieval practice is a great tool to use as you move from foreign to familiar in your learning.

1.2 Bloom's Taxonomy

In our example of learning a new language, we differentiated between content (words, concepts, ideas) and skills (application, analysis, and so on). In education there is a well-accepted framework for considering the skills involved in a question or task. The framework is called **Bloom's Taxonomy** (**Figure 1.3**).

Figure 1.3 **Bloom's Taxonomy**

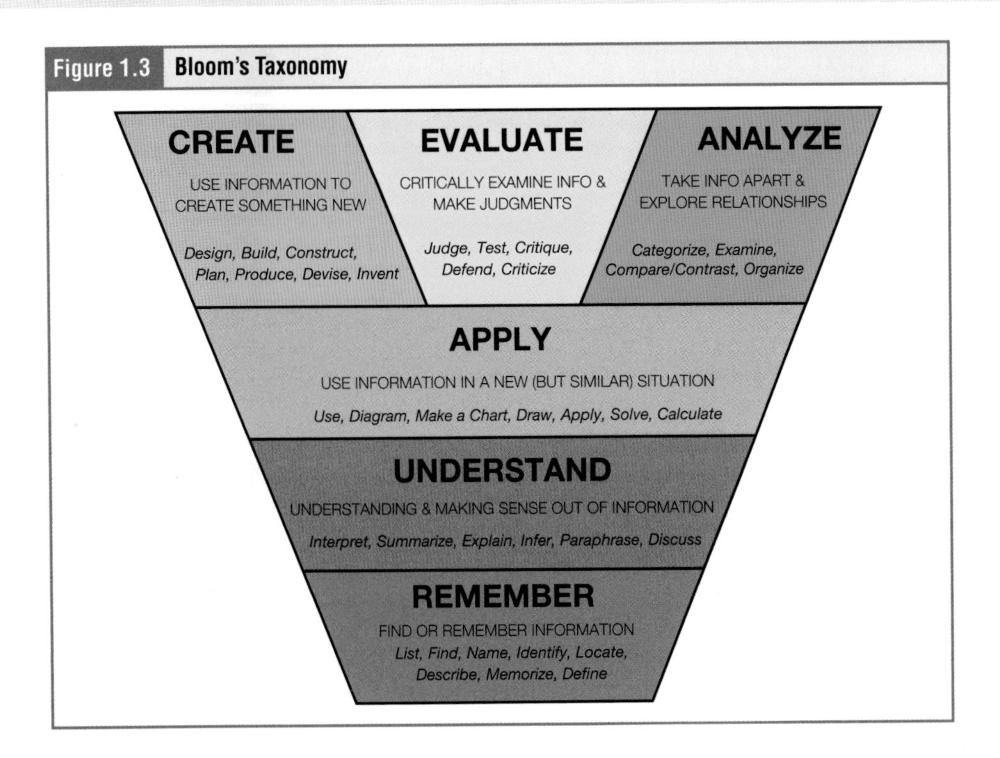

The framework conveys a few central points:

- Different assessment questions or tasks target different skills. If you are asked, "which of the following words best defines the term *tissue*?" You are being asked to *recall* the definition of tissue. If, instead, you were asked "You are looking through the microscope at a tissue that has a high degree of cellularity, with every cell contacting another cell except for a free edge at the top of your view. What type of tissue are you most likely looking at?" You are being asked to *apply* what you know about tissue types to a new situation you have never encountered before. These tasks and the skills involved are different from each other.
- These skills are iterative. You cannot *apply* information in a new context if you cannot remember it. Therefore, application depends on knowledge.
- The skills build on each other, but one is not necessarily harder than another. It could be more challenging to apply knowledge than to remember something, but if you were asked a remembering question about a tiny obscure tidbit of knowledge, or asked to apply a general concept, the application question in this case would be easier than the remembering question.
- Everyone is working with a different brain. Some students are natural memorizers; these students need to work on their skills of application and analysis. Some students struggle with memorization, but problem solving comes easily to them. To be the most successful, you need to *learn about yourself as a learner* first, then work on the skills that need the most building.

You can talk with your instructor about what to expect on your assessments in your A&P course. Within this book we label the assessment questions with the Bloom's level the question is targeting to get you familiar with your skills and growth areas. It is common for 70–80 percent of the questions on an exam to be either remembering or understanding questions, and 20–30 percent of the questions to be applying or analyzing questions. Your instructor may also include questions or tasks that target the creating Bloom's level in your course. Study strategies that target each Bloom's level are summarized in **Table 1.3**.

Table 1.3 Learning Strategies and Question Examples That Target Each Bloom's Level

Bloom's Level	Skills	Example Question	Individual Strategies	Group Strategies
Remember	• Memorize • Define • Recall details	*Which of these is the correct definition of osmolarity?*	• Practice labeling diagrams • List characteristics • Flashcard practice/quiz • Draw, classify, select, or match items • Write out the textbook definitions	• Check a drawing that another student labeled • Create lists of concepts and processes that your peers can match • Place flash cards in a bag and take turns
Understand	• Recognize a concept in a new situation • Thoroughly understand a concept	*Which of the solutions in this diagram is hyperosmotic?*	• Describe a biological process in your own words • Provide examples of a process • Write a sentence using new vocabulary words	• Discuss content with peers • Take turns quizzing each other about definitions and have your peers check your answers
Apply	• Apply information in a new situation • Use a previous example to or model to understand a new example	*What will happen to interstitial fluid volume if the concentration of glucose increased there?*	• Review each process you have learned and then ask yourself: what would happen if you increase or decrease a component? • If possible, graph a biological process and create scenarios that change the shape or slope of the graph	• Practice writing out answers to practice questions (or write your own) and have your peers check your answers • Take turns teaching your peers a biological process while the group critiques the content
Analyze	• Assess a new piece of data or graph to make a conclusion	*The blood pressure readings of four patients can be found in this table. Which one might have the highest blood osmolarity?*	• Analyze and interpret data in primary literature or a textbook without reading the author's interpretation and then compare your analysis to the author's • Compare and contrast two ideas or concepts • Create a map of the main concepts by defining the relationships of the concepts using one- or two-way arrows	• Work together to analyze and interpret data and defend your analyses to your peers • Work together to identify all of the concepts in a paper or textbook chapter, create individual maps linking the concepts together with arrows and words that relate the concepts, and then review your peer's maps and defend your own
Evaluate	• Justify a conclusion • Choose a best among a group of plausible explanations	*Which of these diseases is likely to cause an increase in osmolarity of the blood?*	• Practice justification of each of the wrong answers for complex questions	• Justify your stance to your group on each answer of case study questions
Create	• Produce a new work, idea, or writing	*Design a solution that is hyperosmotic to blood plasma. Describe the effects on the tissues (and red blood cells) if this solution was injected into the bloodstream*	• Write a paragraph summarizing a major idea or process • Draw your own physiological diagram to illustrate a concept or process	• Design your own fictional case study and ask your peers to do the same. • Exchange case studies for each other to solve

Adapted from Crowe et al., 2008.

The activities that you use to study and learn can incorporate all these ideas. As you study you want to make sure you move from familiarity-building techniques to mastery-building techniques (Table 1.1). You also want to be sure to incorporate activities that target different Bloom's levels (Table 1.3). Remember that no two learners have the same brain; therefore, the ways that you learn best may very well be different from those of a friend or even your instructor! As you progress through your education, you will discover that some learning strategies work for you and some do not! You may also find that some strategies work better for some material, while other strategies are more adapted for other courses. For example, you may find differences in how you approach even A&P material within the same course. Ultimately, as you become more in tune with your learning processes, you will have a better sense of which strategies to apply when and in what order.

1.3 What Is a Learning Objective?

Learning objectives (LOs) are used as a means of transparency between the instructor and the learner. They are, essentially, a means of answering the age-old student question: What are we supposed to learn? In this book we have included LOs at the top of every section of text. The LOs appear again at the end of each chapter along with questions that target those objectives. The goal of this design is to help you to become familiar with anticipating the types of assessment questions you want to prepare for. The text indicates higher-order and lower-order Bloom's level LOs using different icons. Remember that you may want to use different learning strategies for different levels. Your instructor may have additional LOs that they include in their lectures or syllabus. Again, LOs are tools of transparency; your instructor is communicating with you about what they feel is important for you to learn. You can also ask your instructor questions when you are unsure of your objectives. An example of an LO and how to use it in your learning can be found in **Figure 1.4**.

Figure 1.4 Anatomy of a Learning Objective (LO)

Example LO from Chapter 4 and one approach to learning the material it targets.

Anatomy of an LO

The verb: Compare and contrast

The content: chromatin, chromosomes, and chromatids.

	Drawing	Description	Timing	Function
Chromatin		Loosely wound DNA	Interphase	DNA (genes) accessible for transcription
Chromosome	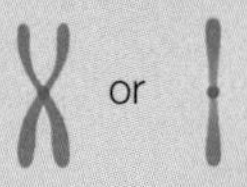 or	Compacted DNA	Mitosis Meiosis	Condense DNA for movement and sorting during mitosis and meiosis
Chromatid	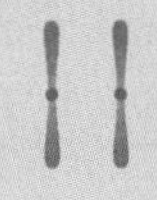	One of two replicates	Mitosis Meiosis	One of two replicated sisters separates to daughter cells

1.4 The Anatomy of Art

1.4a Image Translation in Anatomy

Learning anatomy will require a skill you have likely not had a lot of time to practice in your education: learning visually. Unlike most of your classes where the majority of your interaction with the material was through words or numbers, anatomy—and to some extent physiology—really asks that you learn images. In particular, you will be asked to transfer your visual understanding from images you have seen before to novel images. You will often find that you have studied an image from one perspective, only to need to translate to another. For example, if you have studied an image from the front, you will use what you understand about the three-dimensional relationships among the organs or structures to be able to break down and understand an image of the same area taken from above. Similarly, studying tissues through microscopes is challenging for the first few times you do it. After practice, your ability to translate what you see to the structures you have learned will grow. Beginning A&P students often find these skills challenging, until they practice them over time. Looking at a new image can be overwhelming at first. Like many anatomists, you may find it helpful to first find a landmark, a structure you know well. From that landmark, you can begin to identify the other structures around it that you recognize. In **Figure 1.5** you can walk through this process. **Figure 1.5A** illustrates a simple and straightforward diagram of the heart from an anterior (front) perspective. For simplicity, only three structures are labeled in this exercise. Now examine **Figure 1.5B**, which is an image of the same organ in the same perspective (anterior/front) but drawn by a different artist. Can you find the three structures? Now examine **Figure 1.5C**. Here you are looking at the same organ (the heart) from a different perspective (superior/from above). Find your three structures again. Repeating this strategy with new images is a great way to work toward mastery. In your study process, you can repeat this with unlabeled images as many times as is helpful to you. A last step in this process can be to draw the image for yourself. Keep in mind that the point of this is not about artistry, but that working through the task of representing the relationship among anatomical structures is a fantastic way to assess your mastery of the material.

Another example of image translation is applying what you are learning in one type of image to another type. In A&P texts we use drawings more than any other type of instructional image because medical illustrators can make the structures stand out from each other using color and lines and it is the easiest type of image for the beginning anatomist to be able to discern. However, as a clinician, the anatomical imagery you will use every day is much less straightforward. As we learn, we can begin to practice the translation of our knowledge from drawing to more relevant types of images by using photography and clinical images such as CT scans and X-rays. As with the examples in Figure 1.5, the first step in this process is to find a landmark. From there, build your understanding of the image out by finding and labeling the structures adjacent to your landmark. Compare the artist drawing (see Figure 1.5A), a photograph (**Figure 1.5D**), and a magnetic resonance image (MRI) of the heart (**Figure 1.5E**). In Figure 1.5A we have labeled the largest artery in the body, the aorta. Can you find the aorta in Figures 1.5D and 1.5E?

1.4b Comparing a Micro and Macro View

A frequently used illustration technique in A&P is presenting a zoomed-in view of a structure simultaneously with the zoomed-out view. In science, we typically use the terms *micro* and *macro* for these views, respectively. You can see an example in **Figure 1.6**. Can you find where on the macro view (**Figure 1.6A**) the structure presented

Figure 1.5 Image Translation

Anatomy requires visual literacy and image translation. Here we compare two drawings of the anterior perspective of the heart (A and B) with a superior perspective of the heart (C). Translating these diagrams to a photo (D) and finally an MRI (E) requires a strong working knowledge of the heart structures.

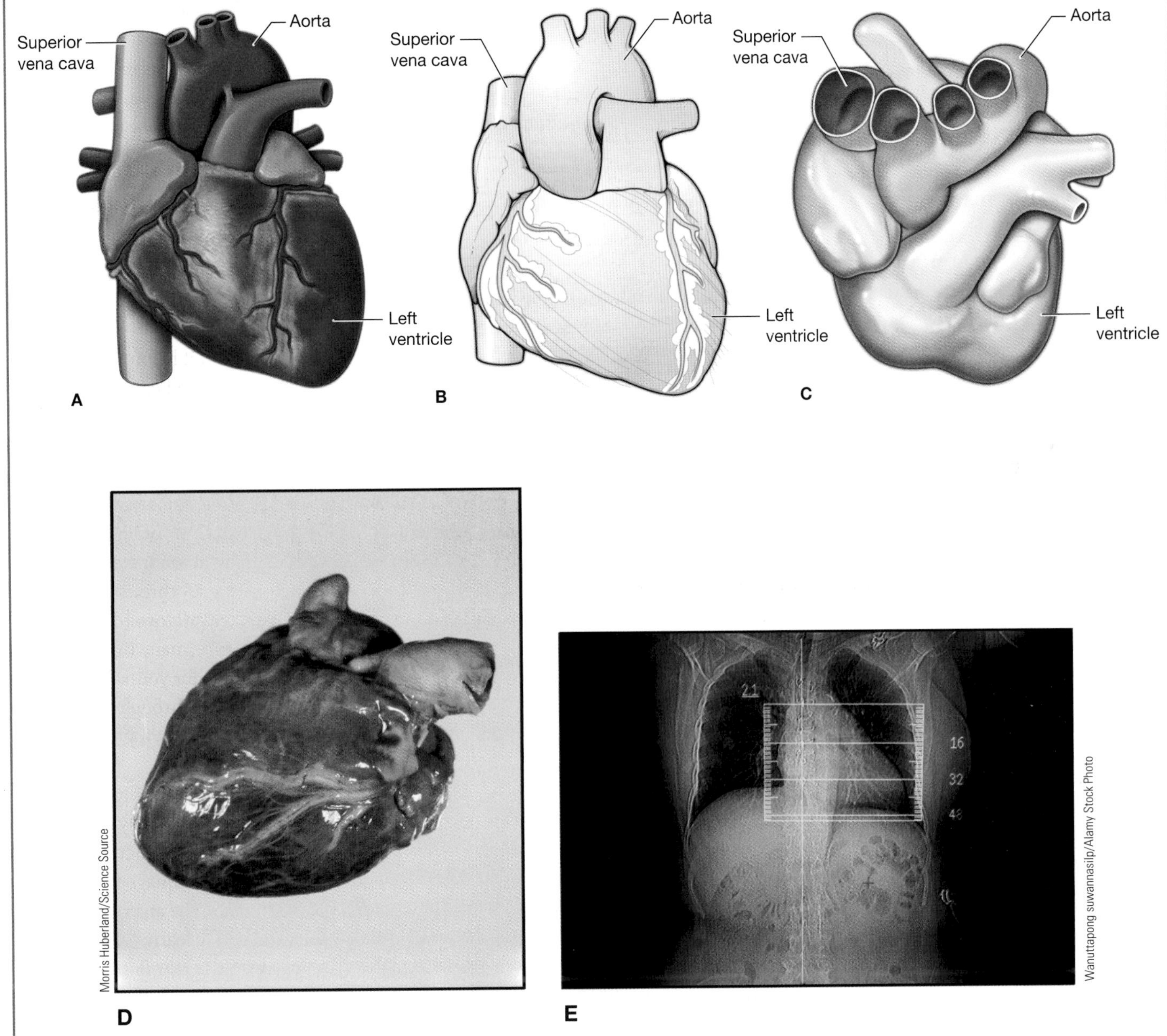

in the micro view (**Figure 1.6B**) is? Always make sure to reorient yourself when you are presented with figures such as these.

1.4c Learning from Physiology or Process Diagrams

Illustrating a dynamic process in a static image is a challenging task for our illustrators. These images, as straightforward as we try to make them, can be complicated to digest as a learner.

Figure 1.6 **Micro and Macro Views**

Comparing a (A) micro and (B) macro view of a muscle cell sarcomere.

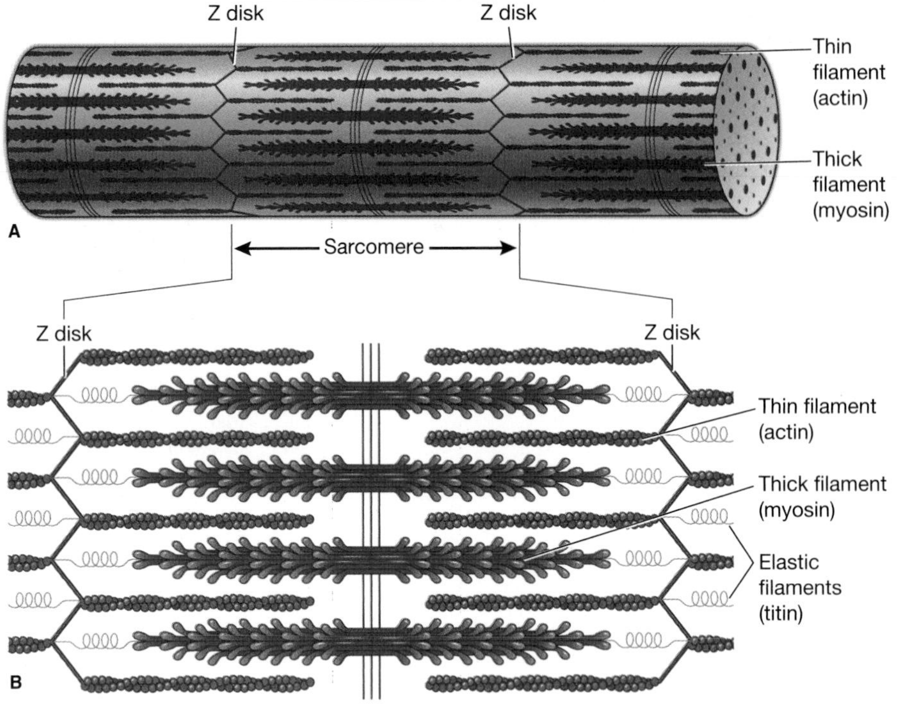

Before you set about trying to memorize or learn from a figure, make sure that you understand what the figure is trying to convey. Together, let's go through **Figure 1.7**. This is a fairly standard drawing of a foundational physiological concept (osmosis). As we break down this image, some questions you might ask yourself are:

- There are four different arrows used in the image—one orange, one green, a single-headed arrow in black, and a double-headed arrow in black. What are the arrows trying to illustrate?
- There are red dots. What do these represent? Is it significant that the red dots are on one side of the U-shaped tube and not the other?
- There is a dashed line at the bottom of this tube. What does it represent?
- Is this image trying to show two U-shaped tubes side-by-side, or is it trying to show what happens in one tube over time?

Arrows are of particular trickiness in physiological diagrams. They can be used to show the direction of movement in a dynamic process, used to convey time, or used to illustrate differences. In this image they are doing all three things! Which arrow do you think is trying to convey the passage of time? If you answered the green arrow, you are correct! The orange arrow is trying to convey the direction of movement, in this case of water. In the U-shaped tube on the right, time has passed, water has moved, and the black arrow is pointing out to you that an increase in water volume can be observed.

Asking yourself questions about a figure you are learning from not only ensures that you are beginning your learning with the correct understanding but also works

Figure 1.7 **An Example Figure Illustrating the Movement of Water in Response to an Osmotic Gradient**

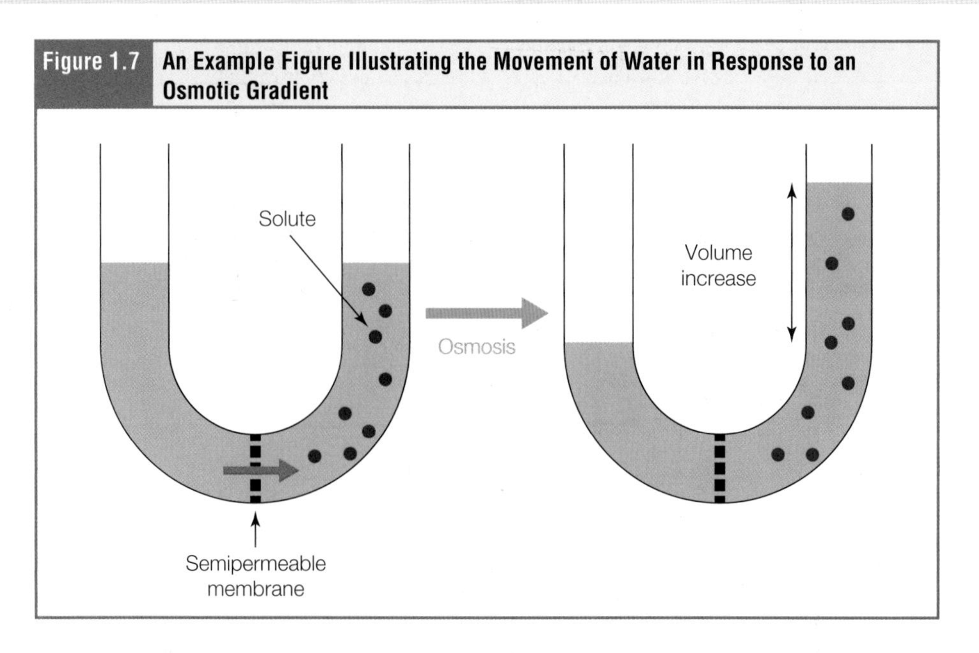

to fortify your knowledge of the key components of the process. You should also take into consideration the figure legend and related text to assist in your interpretation of a diagram.

Cultural Connection

It's All Greek to Me!

The average A&P student will learn more new vocabulary words in each chapter than an introductory language student! Similar to learning any language, A&P students must memorize the words but also master how to use these words in communication with others. Conveniently, there is some redundancy within A&P terminology and so you can often use clues to understand the words. The majority of terms used in A&P are derived from ancient Greek and Latin words. More specifically, a predominant number of anatomical terms come from Latin, and a predominant number of medical terms (e.g., conditions, diseases, professions, and so on) come from ancient Greek. The emphasis on using these two languages likely originated with the very first anatomical study, which occurred in ancient Rome and early medicine and which was a focus in ancient Greece.

Anatomists described structures in the body using plain descriptors, but as language changed over the centuries we were left with very complex terminology. For example, in Chapter 3 you will learn the term *ligand*. The first half of this word, *liga*, is from the Latin *ligare*, which means "to bind." Later, in Chapter 10, you will learn the term *ligament*; a ligament is a structure that connects, or binds, two bones together. Doing the work in Chapter 3 to understand the origin of *liga* makes learning in Chapter 10 easier! Throughout the text we will assist you with the interpretation of these Greek and Latin-derived words.

As a student your primary goal is to master the content of this course. However, this herculean task is much easier if you know about yourself as a learner and have an array of tools to employ in your process. As you go through the course, consider which tools are working for you and which content or skill areas you have the most room to grow in. It is important to make sure you incorporate a variety of diverse learning techniques into your journey to mastery of the material. You will find tips and strategies throughout the book, but your instructors and this chapter are resources that are always available as you consider, observe, and refine your learning techniques.

2

Introduction to the Human Body

Chapter Introduction

An understanding of anatomy and physiology is not only fundamental to any career in the health professions, but it can also benefit you personally. The knowledge and understanding you gain in this course may help you act as an advocate for your own health as well as the health of your family members. Mastery of the language of anatomy and physiology will help you converse with your healthcare providers. Your knowledge and critical thinking skills in this field will help you understand and critique news about nutrition, medications, medical devices, procedures, and both infectious and non infectious diseases.

This chapter begins with an overview of anatomy and physiology and a preview of the body regions and functions. It introduces a set of standard terms for body structures and for planes and positions in the body that will serve as a foundation for more comprehensive information covered later in the text.

2.1 Overview of Anatomy and Physiology

Learning Objectives: By the end of this section, you will be able to:

2.1.1 Define the terms anatomy and physiology.

2.1.2 Give specific examples to show the interrelationship between anatomy and physiology.

Human **anatomy** is the scientific study of the body's structures. Some of these structures are very small and can only be observed and analyzed with the assistance of a microscope (**microscopic anatomy**). Larger structures can readily be seen, manipulated, measured, and weighed (**gross anatomy**). Our visual understanding of the body extends far beyond these two levels of detail. We also have many ways of "seeing" inside the bodies of living individuals with great precision (such as the use of Magnetic Resonance Imaging and X-rays). The "Digging Deeper" feature discusses the physics behind these imaging techniques. Figure 2.1 illustrates their differences by comparing a view of the brain and cranium using a variety of techniques. Figure 2.1A is a gross anatomical view of an intact human brain. Functional Magnetic Resonance Imaging (fMRI, Figure 2.1C) as well as ultrasonography (Figure 2.1D), however, do allow us to see functions in real time, which microscopic anatomy does not. X-ray imaging is best for hard structures such as bones; notice that we cannot visualize the brain through the cranium using an X-ray (Figure 2.1E). CT scans is an imaging technique that uses multiple angles of X-rays at once to produce a more detailed image (Figure 2.1F). Microscopic anatomy (Figure 2.1B) can only be done on cells and tissues removed from the body (which isn't optimal when it comes to brain tissue!). Microscopic anatomy includes **cytology**, the study of cells, and **histology**, the study of tissues. As the technology of microscopes has advanced, anatomists have been able to observe smaller and smaller structures of the body, from slices of large structures like the heart, to the three-dimensional structures of large molecules in the body.

gross anatomy
(macroscopic anatomy)

Anatomists take two general approaches to the study of the body's structures: regional and systemic. **Regional anatomy** is the study of the interrelationships

The Human Anatomy and Physiology Society includes more than 1,700 educators who work together to promote excellence in the teaching of this subject area. The HAPS A&P Learning Outcomes measure student mastery of the content typically covered in a two-semester Human A&P curriculum at the undergraduate level. The full Learning Outcomes are available at https://www.hapsweb.org.

Figure 2.1 **Anatomical Structures in a Variety of Imaging Techniques**

(A) Gross anatomy of the brain considers structures visible with the naked eye. (B) Microscopic anatomy can deal with the same structures, though at a different scale. This is a micrograph of nerve cells from the brain. LM $\times$ 1600. (C) Functional MRI (fMRI) shows regions of the brain active during particular activities. (D) Ultrasound visualization of the fetal brain. (E) X-rays are best for illustrating hard structures such as the skull. (F) CT scans show brain and body structures in a variety of planes.

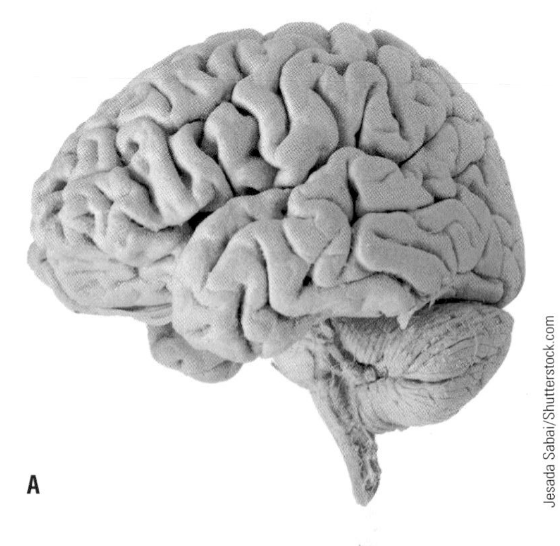
Jesada Sabai/Shutterstock.com

A

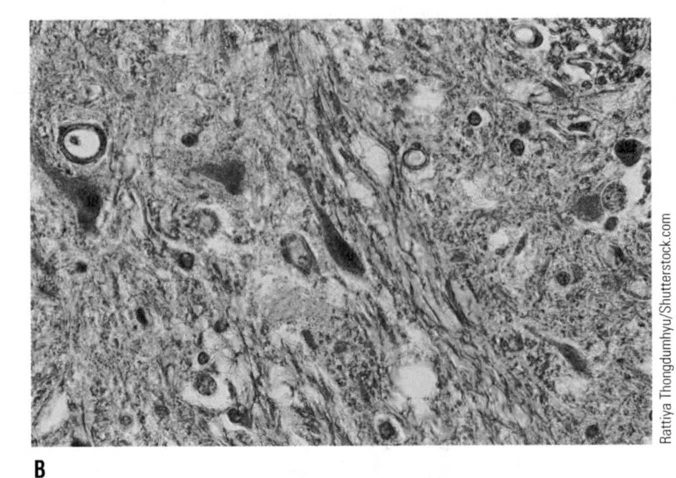
Rattiya Thongdumhyu/Shutterstock.com

B

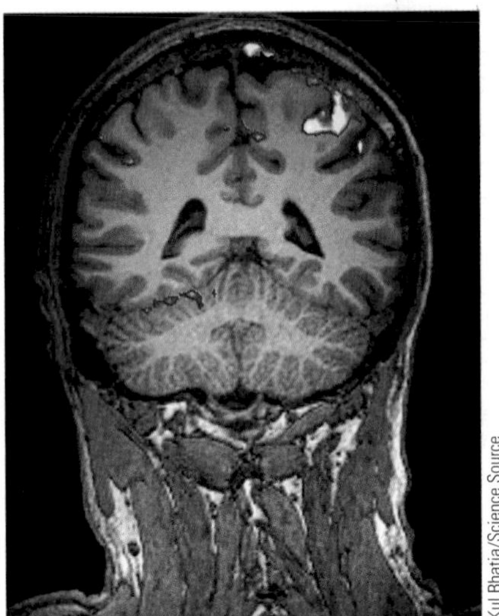
Kul Bhatia/Science Source

C

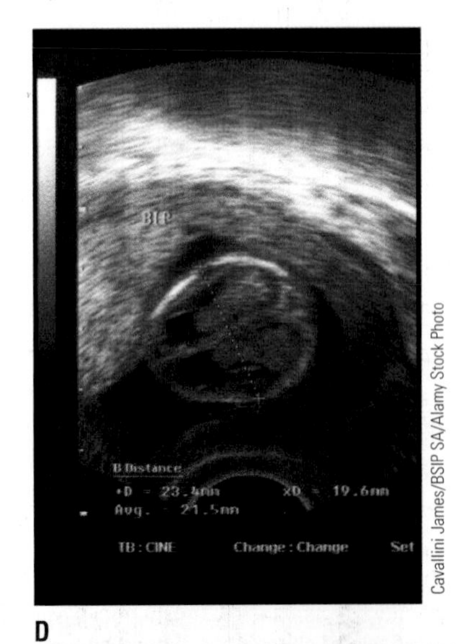

Cavallini James/BSIP SA/Alamy Stock Photo

D

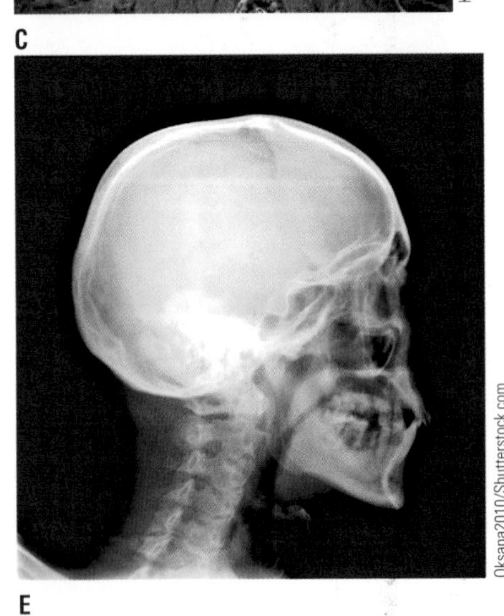
Oksana2010/Shutterstock.com

E

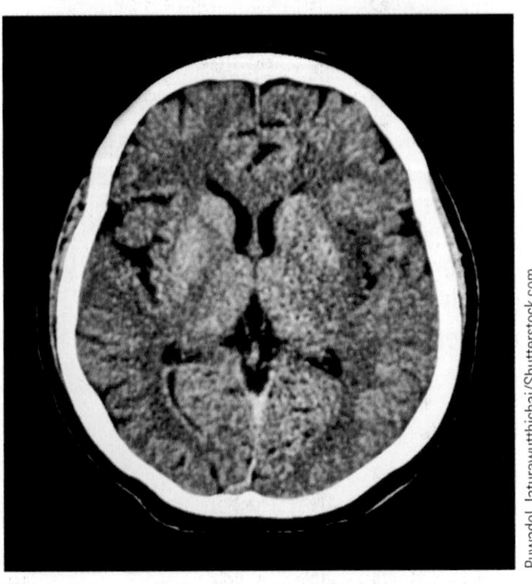
Puwadol Jaturawutthichai/Shutterstock.com

F

of all of the structures in a specific body region, such as the abdomen. Studying regional anatomy helps us appreciate the relatedness of body structures, such as how muscles, nerves, blood vessels, and other structures work together to serve a particular body region. A significant goal of those learning regional anatomy is to build a three-dimensional understanding of the placement and relationships among structures in their minds so that, when presented with images taken from a various perspectives or while performing surgery or dissection, anatomists are able to understand the images. In contrast, **systemic anatomy** is the study of the structures that make up a discrete body system—that is, a group of structures that work together to perform a unique body function. For example, a systemic anatomical study of the muscular system would consider all of the skeletal muscles of the body.

Whereas anatomy is about structure (the *where* of the body's components), physiology is about function (the *how* of the body's components). Human **physiology** is the scientific study of the chemistry and physics of the structures of the body and the ways in which they work together to support the functions of life. For example, if we were examining human sweat, an anatomist would be able to draw the structures of a sweat gland, describe where these glands are found, and compare and contrast different types of sweat glands. A physiologist, on the other hand, would be able to explain how sweat was made, relate sweating to its impacts on the body, and predict what might change about the sweating process in various conditions such as dehydration.

LO 2.1.1

LO 2.1.2

Like anatomists, physiologists typically specialize in a particular branch of physiology. For example, neurophysiology is the study of the brain, spinal cord, and nerves and how these work together to perform functions as complex and diverse as vision, movement, and thinking. Physiologists may work from the organ level (exploring, for example, what different parts of the brain do) to the molecular level (such as exploring how an electrochemical signal travels along neurons).

Learning Connection

Anatomy relies heavily on memorization and understanding skills. Learning physiology relies more heavily on skills such as understanding, analysis, and application. Very little physiology content can be memorized.

Cultural Connection

"We are the ones we have been waiting for."

This line is from Poem for South African Women by June Jordan. From climate change to racial disparities in healthcare, the world is starving for science advocacy. Scientists (and most of us humans!) have a lot of demands on their time, but it is paramount that scientific understanding and data literacy be used to inform public policy and funding. The time for change is now and the wheels of change are pushed by people like you! Student and workforce populations encourage greater creativity, innovation, productivity, and critical thinking in scientific research, healthcare, and public policy. For effective advocacy, consider what it is about healthcare or public policy that makes you passionate. Do you want to see greater gender inclusivity? Do you think our nation should spend more research dollars on a particular heathcare concern? Are you eager to see more attention given to women's healthcare issues? You can bring your scientific knowledge and strong communication skills and get involved at the local, regional, national, and international levels.

To learn more about how to get involved in science and healthcare advocacy, check out: the Annie E. Casey Foundation, the Human Anatomy and Physiology Society, the American Association for the Advancement of Science, and the American Association of Anatomists.

2.1a The Themes of Anatomy and Physiology

There are a few central themes that will appear frequently throughout your study of anatomy and physiology. You can think of these themes as tools that you will pull out and use to build your understanding in different contexts or places within the body. These themes are:

- Structure and function
- Evolution and human variation
- Flow
- Homeostasis

2.2 Structure and Function

Learning Objectives: By the end of this section, you will be able to:

2.2.1* Describe, compare, and contrast various structure–function relationships from molecular to organ level.

2.2.2* Relate the commonly found branching structure to function of an organ.

* Objective is not a HAPS Learning Goal.

LO 2.2.1

Form is closely related to function in all living things. The harmony between form and function can be seen in every aspect of human life from molecular structure to physical traits of the whole organism. As we will examine in Chapter 3, proteins have the most diverse structural variation of all molecules. Protein structure is intimately tied to protein function, and changes in protein shape alter the function of those proteins. A classic example of these changes can be seen when proteins are phosphorylated (a negatively charged phosphate group is added onto a protein). The addition of this negative charge typically changes the shape of the entire protein by drawing positively charged regions of the molecule toward the new addition. Phosphorylation is the most common form of molecular regulation in animal cells. Using phosphorylation, we can turn on or off the activity of enzymes, signaling molecules, or transcription factors (**Figure 2.2**).

We will see many examples of structure and function at every level of organization in this book. One gross anatomical example that is unique to humans is the shape of our pelvises. Humans are the only adult mammals that walk predominately on two feet (**bipedalism**). When early humans evolved to move primarily as bipeds instead of as our tree-dwelling primate ancestors, we evolved a restructured pelvis, one that could support the weight of our abdominal organs as well as accommodate the much larger gluteal (buttock) muscles that were required for stabilization of our torso and efficient forward motion. If we had not needed the function of the pelvis to accommodate

Figure 2.2 Phosphorylation Induces Shape Changes in Proteins

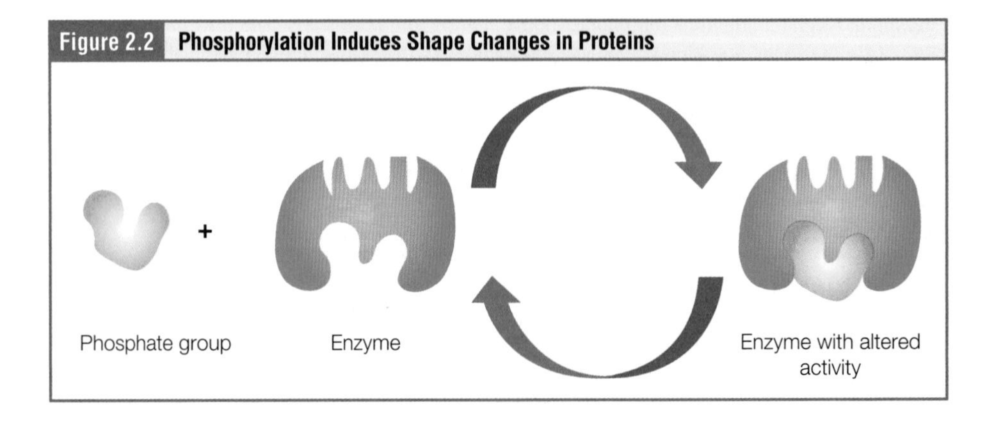

Figure 2.3 Human versus Ape Pelvis

As early humans evolved to become bipedal walkers, the shape of their pelvises had to change to support their new function.

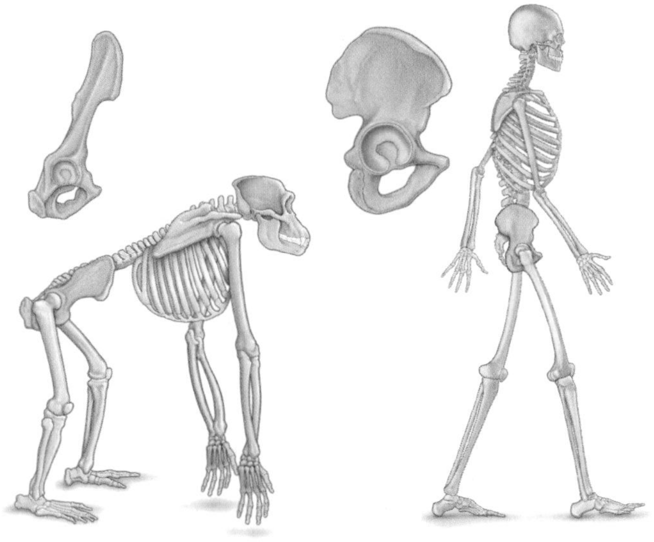

Figure 2.4 The Branching Pattern of the Lungs

Examples of branching structure to maximize surface area can be seen throughout all living organisms.

BSIP/Universal Images Group/Getty Images

bipedalism, we would not have observed the structural change of the pelvis over time (**Figure 2.3**).

LO 2.2.2

A common theme throughout, not just anatomy and physiology but all living things, is that branching structures are efficient for maximizing surface area. From tree branches to vasculature and respiratory conduction tubes, this format is plentiful in anatomy (**Figure 2.4**).

Cultural Connection

Checking Our Lenses

What we know about human evolution is the product of arduous and careful investigative work by archeologists and anthropologists. Occasionally when we look at the past, we see things through a modern lens. Many scientists and science advocates are working to check where assumptions and biases may have worked their way into representations of the past. Lucy, a famous 3.2-million-year-old skeleton, belongs to the species *Australopithecus afarensis*. The discovery of this species was fundamental to our scientific understanding of how bipedalism evolved, *A. afarensis* is hypothesized to be a nonhominid ape that was bipedal—a link, if you will, between tree-dwelling ancestors and bipedal humans.

Lucy and her species have been recently reimagined after a careful look at how this species has been portrayed. In most artistic imaginings of Lucy, she is typically drawn with one or two small children and a male *Australopithecus* nearby. In many representations of this family, the male is hunting or fishing while Lucy is tending to children. However, evolutionary data indicate that small, nuclear families are a very recent feature of human history. Lucy may or may not have had children or lived with male members of their species; we do not even know if *Australopithecus* were monogamous or polygamous, but the frequent representation of them with one male mate and children is decidedly modern and misogynistic. Given the deep roots of racism and sexism in our society, it is important that we reexamine representations of the past to remove modern bias. Recently, an international team of scientists have sought to reimagine Lucy and other historic finds in ways that are free of bias and rooted in the evolutionary data.

2.3 Evolution and Human Variation

Learning Objectives: By the end of this section, you will be able to:

2.3.1* Define the term and explain the concept of *evolution*.

2.3.2* Contrast the impact of selection on traits that affect reproduction and traits that do not; use this to explain examples of anatomical and physiological variation.

* Objective is not a HAPS Learning Goal.

LO 2.3.1

Evolution is a change in gene expression that occurs from generation to generation. Evolution specifically entails genetic changes and does not occur within one's lifetime but across the lifetimes of generations of humans or other organisms. Genetic changes occur randomly and with relative frequency, but they become more common among the individuals of a population when they confer an advantage in a particular environment. The key here to remember is that what is or is not advantageous has to do with the immediate environment.

For example, let's examine our human relationship with UV radiation in sunlight. UV radiation is required for the synthesis of vitamin D, a nutrient that is central to the structure of our bones, as we will learn more about it in Chapter 7. UV radiation is harmful, however, to the structure of a nutrient known as folate. Folate is essential for sperm production and embryonic development. As early humans evolved, they expressed lower and lower amounts of body hair. This likely had to do with these early humans spending less time in trees and more time under the hot sun. Without body hair, these early humans may have been cooler, but they also were at much higher risk of folate damage from UV radiation. Therefore, as successive generations of early humans had less and less body hair, the expression of the skin pigment melanin increased. Producing and expressing the skin pigment **melanin** conferred a tremendous advantage to humans who evolved over generations near the equator or at high altitudes, where UV radiation is the highest. Through genetic changes these populations evolved to express and maintain melanin levels that protected their cells from the strong UV rays of equatorial sunlight, while allowing sufficient UV radiation for the manufacturing of vitamin D. However, as humans migrated to locations with less intense UV radiation, their skin cells developed the adaptation to digest some of the expressed melanin pigment, leading to lighter skin such that vitamin D manufacturing was still possible with lower amounts of UV radiation. We now can see the influence of environment on evolution and genetic change as we look about at the beautiful diversity of skin shades present in our modern human population. The story of melanin is explained in a bit more detail in Chapter 6.

LO 2.3.2

In genetic changes that impact reproduction or the likelihood of living to reproductive age, we tend to see less anatomical or physiological variation among individuals. However, anatomical or physiological variation that does not influence reproduction does not tend to be selected for or against; therefore, a wider degree of variation exists. If we looked inside all of the bodies in a college classroom, for example, we would find many individuals with four pulmonary veins, but some individuals with two and some with three. Most bodies in the classroom would have five lumbar vertebrae, but some may have four, others six. A much wider array of anatomical variation is present in our human population than is represented in most anatomy textbooks.

Physiological variation is much more diverse and widespread. Physiological variables will differ not just from one individual to another, but also between children and adults, women and men, and depending on age. This physiological variation is very important to understand and remember as it can lead to grave consequences in the recognition and treatment of disease. For example, for many years medical students learned that one of the central symptoms of a heart attack is pain in the left shoulder. This was observed over years of studying heart attacks in biological men. As it turns out, this symptom is rarely seen in female heart attack patients; the most common symptom of heart attack in women is nausea. We will never know how many women suffered more significant clinical outcomes because of the delay in or complete failure to recognize their symptoms. Therefore, it is essential to include diverse populations of humans whenever health is being studied. In 2020, when the world raced to develop vaccines to prevent the disease caused by the novel coronavirus, SARS-CoV-2, vaccine developers realized they needed to study the vaccine's efficacy and side effects in many populations of humans, and these clinical trials were among the most diverse ever established.

Learning Check

1. Barbra is learning about the various structures in the area of the body called the pelvic girdle. They are learning about the pelvic joints, the organs within the pelvis such as the urinary bladder, and the muscles of this area. Which type of anatomical study is Barbra using?
   a. Regional anatomy
   b. Systemic anatomy
   c. Cytology
   d. Histology
2. True or False: The shape of a structure will determine the function, but the function will not determine the shape.
   a. True
   b. False
3. Which type of variable would you expect to have the broadest range between adults and infants?
   a. Anatomical variables
   b. Physiological variables

2.4 Flow

Learning Objectives: By the end of this section, you will be able to:

2.4.1* Describe how a gradient determines flow between two regions, and give examples of gradients that exist in different levels of organization in the body.

2.4.2* Predict how changes in a gradient will affect flow rate.

2.4.3* Predict how differences in resistance will affect flow rate.

* Objective is not a HAPS Learning Goal.

LO 2.4.1

In your study of anatomy and physiology you will examine the flow of materials in many different contexts. Some of these include: the flow of ions across the membrane of muscle cells, the flow of air through the lungs, and the flow of blood through the vessels of your body. While the substances and the scale of flow vary widely among these examples, the principles of flow can be applied universally. Substances, whether they are air or sodium ions, flow according to gradients. The three types of gradients that drive flow in physiological contexts are concentration, electrical, and pressure gradients. Pressure gradients drive the flow of fluids and gasses; the pressure generated by the pumping of the heart drives the flow of blood. Concentration gradients drive the flow (here called ***diffusion***) of individual molecules (for example, sodium ions will flow from areas where they are more concentrated to areas where they are less concentrated). Electrical gradients will influence the flow of charged particles, including ions. In all examples of flow, there are also factors

Student Study Tip

Molecules diffusing across a membrane act like a crowd leaving a concert: a larger crowd will rush to exit (larger concentration gradient). If smaller, people will take their time. If the exit door is smaller (membrane surface area), diffusion and exiting will be more difficult. If there is farther to go, as in a large venue, it will be harder to cross the exit/membrane.

LO 2.4.2

Figure 2.5 Flow Rate and Its Determining Factors

Flow is proportional to the size of the gradient and inversely proportional to resistance.

$$\text{Flow} \propto \frac{\text{Gradient}}{\text{Resistance}}$$

LO 2.4.3

that oppose, or resist, flow. We can summarize the rate of flow as being determined by the size of the gradient divided by the resistance (**Figure 2.5**). For example, in the flow of liquids and gasses, the diameter and length of the tubes provide resistance to flow. Different types of flow are illustrated in the "Anatomy of Flow" feature. In Chapter 4 we will look carefully at the flow of molecules across membranes and examine the factors of resistance that oppose these types of flow. In Chapters 20, 22, and 23 we will examine flow and resistance under various physiological and anatomical constraints.

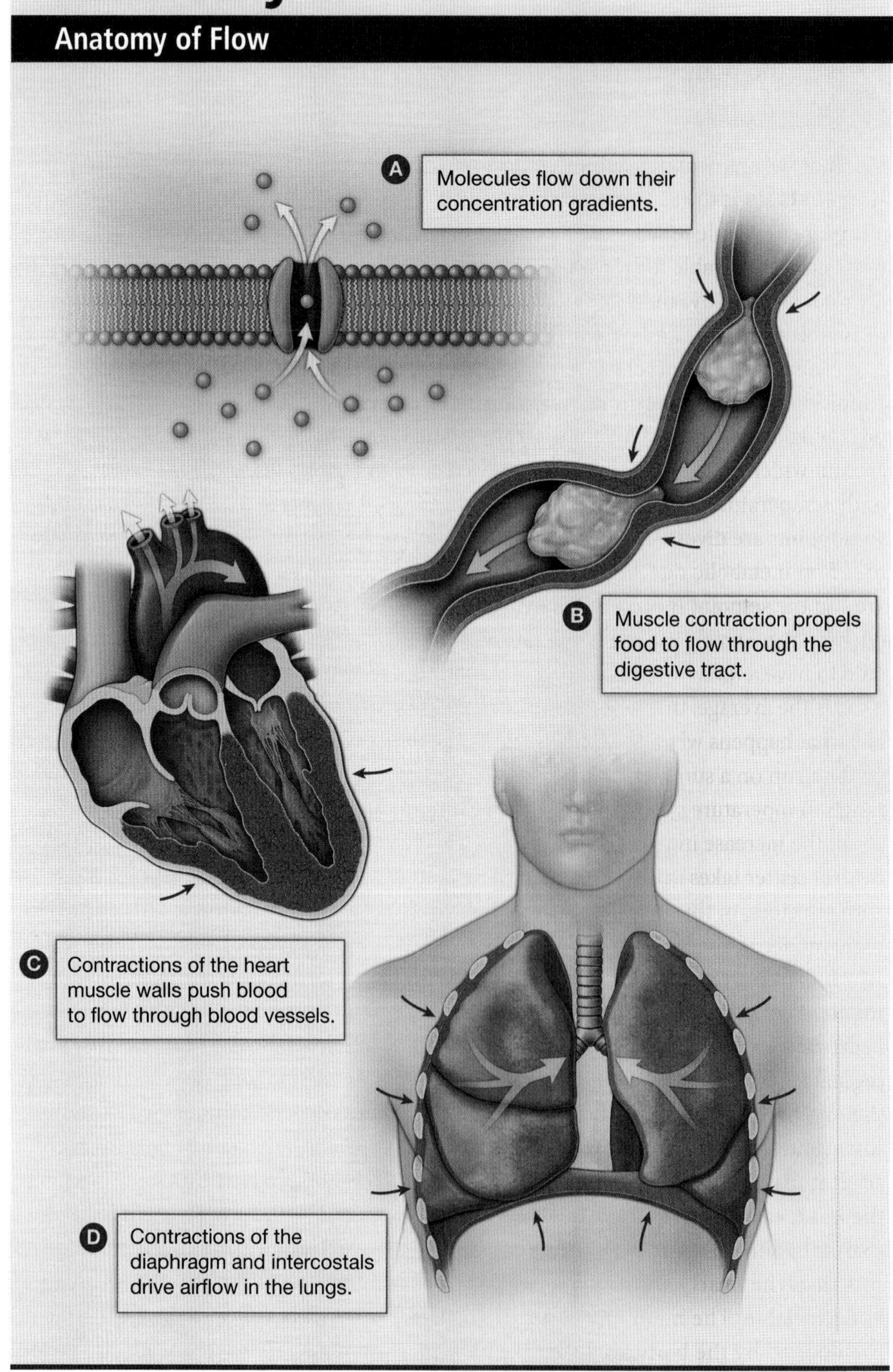

Anatomy of...
Anatomy of Flow
A Molecules flow down their concentration gradients.
B Muscle contraction propels food to flow through the digestive tract.
C Contractions of the heart muscle walls push blood to flow through blood vessels.
D Contractions of the diaphragm and intercostals drive airflow in the lungs.

2.5 Homeostasis

Learning Objectives: By the end of this section, you will be able to:

2.5.1 Define the following terms as they relate to homeostasis: setpoint, variable, receptor (sensor), effector (target), and control (integrating) center.

2.5.2 Explain why negative feedback is the most common mechanism used to maintain homeostasis.

2.5.3 List the main physiological variables for which the body attempts to maintain homeostasis.

2.5.4 List the steps in a feedback mechanism (loop) and explain the function of each step.

2.5.5 Compare and contrast positive and negative feedback in terms of the relationship between stimulus and response, and describe examples of each.

Much of the study of physiology centers on the body's tendency toward homeostasis. **Homeostasis** is the state of dynamic stability of the body's internal conditions. While our bodies are able to withstand a variety of external conditions, many internal conditions must stay stable for the health of our cells. Parameters such as oxygen levels, pH, nutrient availability, and temperature must remain constant for our molecules and cells to be able to survive and perform their functions.

LO 2.5.1

Maintaining homeostasis requires that the body continuously monitor its internal conditions. From body temperature to blood pressure to levels of certain nutrients, each physiological condition has a particular setpoint. A **setpoint** is the physiological value around which the normal range fluctuates. For example, the setpoint for normal human body temperature is approximately 37°C (98.6°F). Of course, humans are diverse and so setpoints are diverse too. Temperature setpoints, for example, appear to vary widely based on metabolic rate, body mass, age, and biological sex. Typically, a range of values around the setpoint is acceptable; we consider a person to be in hypothermia, or dangerously cold, when their body temperature falls below 35°C (95°F) and a person is considered to have a fever when their body temperature registers above 38°C (100.4°F). While 37°C is the average, the range is 35°C to 38°C. Hypothermia and fever are disease states, but what happens when the body warms or cools under healthy conditions? Let's say you go for a run on a summer day. Your body temperature quickly rises above the healthy range. Temperature sensors in your skin detect these changes and provide information about the increase in temperature to a control center, which often is in the brain. The control center takes action to reverse the increase in temperature by communicating with effectors, in this case, sweat glands in the skin, to take action to cool the body. This basic model of homeostasis and homeostatic mechanisms is represented in **Figure 2.6**.

It is important to note that not all physiological variables are homeostatically regulated. Take the following options—blood sugar levels, thyroid hormone level, heart rate, and pH; can you pick out which of the variables is *not* homeostatically regulated? You may not be ready as a physiologist to debate all of these variables, but the one in this list that is not homeostatically regulated is heart rate. Heart rate in a resting adult can be as low as 60 or even 50 beats per minute, and in an exercising adult it may be 195 beats per minute or even higher! This wide variation is because the heart can (within limits) beat as fast or slow as the body needs it to in order to maintain homeostatic levels of blood pressure and oxygenation for healthy cells. In this case, the heart and how frequently it pumps blood is the effector, not the regulated variable. The main physiological variables for which homeostatic ranges are maintained by the body are listed in **Table 2.1**.

Figure 2.6 **A Model of Homeostasis and Homeostatic Mechanisms**

The homeostatic model includes a setpoint for the regulated variable, a sensor, control center, and effector.

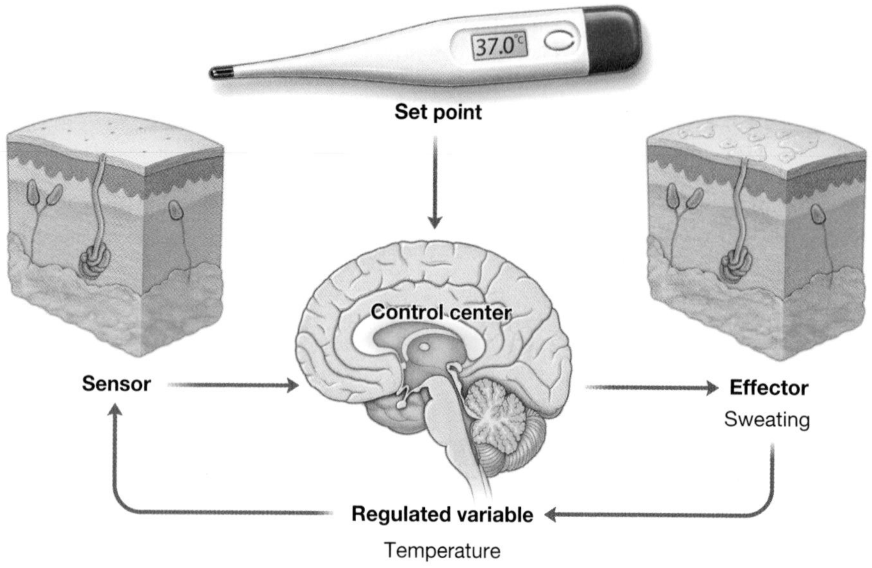

LO 2.5.2

Many of our variables are regulated in a common pattern known as negative feedback. **Negative feedback** is a mechanism that reverses a deviation from the set-point. Therefore, negative feedback maintains body parameters within their normal range. The maintenance of homeostasis by negative feedback goes on throughout the body at all times, and an understanding of negative feedback is thus fundamental to an understanding of human physiology. In a negative feedback pattern, the action of the effectors "turns off" the action of the sensor. An example of negative feedback is illustrated in **Figure 2.7**. In this example, the levels of sugar in the blood fall in an individual who is between meals. This decrease in circulating blood sugar could compromise the function of the body's cells, particularly the brain. The decrease is sensed by the pancreas, which releases a hormone called glucagon to alert the body of this dangerous decrease in nutrients. Upon receiving the glucagon signal, the liver begins to break down glycogen, a storage carbohydrate, and releasing the resulting sugars into the blood. Blood sugar levels rise, and the sensors in the pancreas stop sending signals (**Figure 2.7C**). A parallel system works to control blood glucose levels from going too high. After a meal or glycogen breakdown, blood glucose levels rise. The pancreas releases a hormone known as insulin which

Student Study Tip

Negative feedback is explained by an air conditioner set to a specific temperature: once it has been running for a while, it turns itself off once the setpoint is reachieved!

LO 2.5.3

Table 2.1 Physiological Variables of Homeostatic Ranges Maintained by the Body

Variable	Examples
Blood gas levels	CO_2, O_2
Nutrient levels	Blood glucose
Electrolyte levels	Na^+, Ca^{2+}, K^+
pH	H^+
Blood pressure	
Body temperature	

Figure 2.7 Negative Feedback Loop LO 2.5.4

In a negative feedback loop, a stimulus—a deviation from a setpoint—is resisted through a physiological process that returns the body to homeostasis. (A) A negative feedback loop has four basic parts. (B) Blood sugar levels are regulated by negative feedback. (C) If we graph blood sugar levels over time, we can see how homeostatic mechanisms keep the levels close to the setpoint.

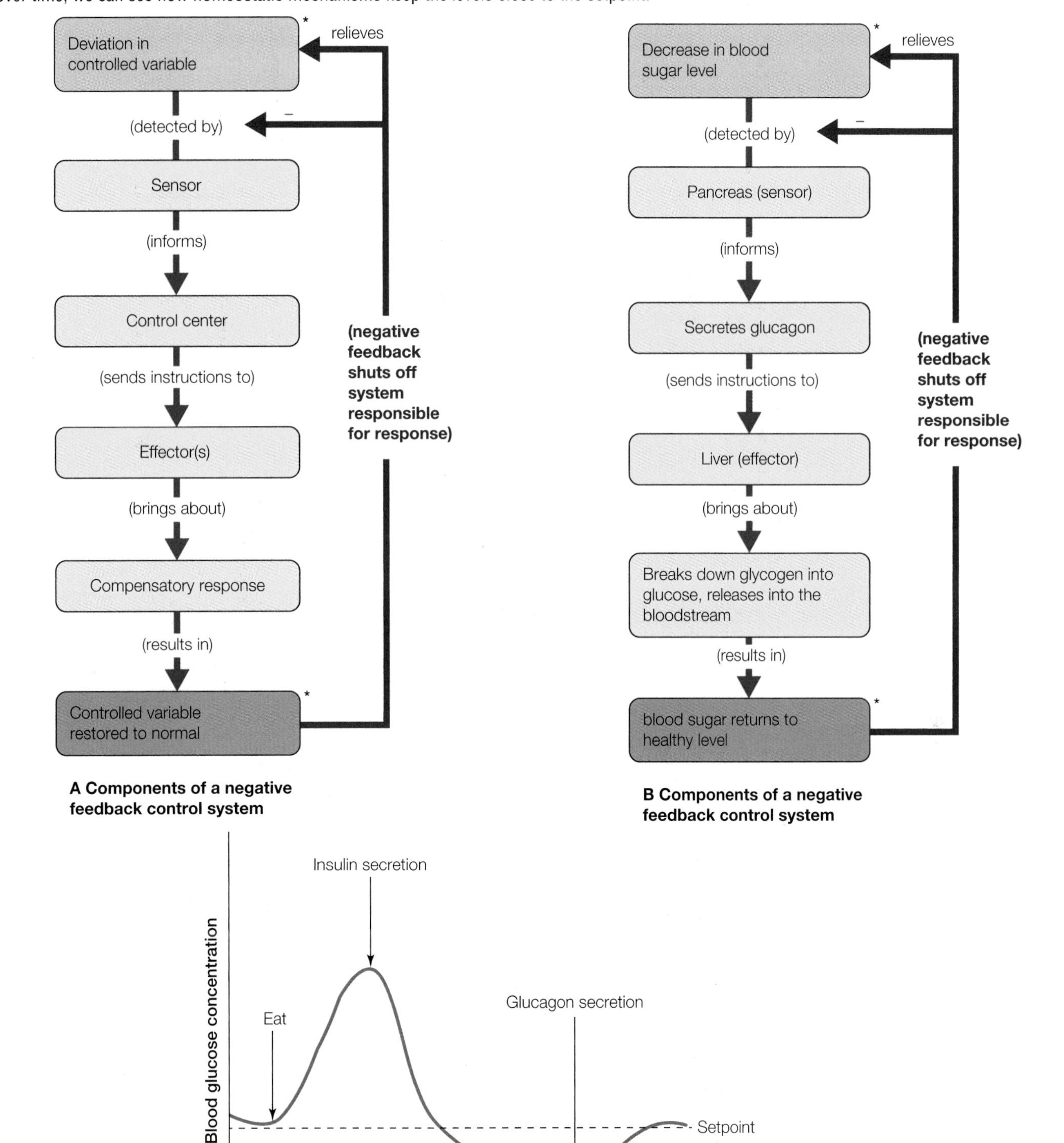

Apply to Pathophysiology

Type I Diabetes

Eliott is nine years old. Lately Eliott has been complaining to their parents of feeling tired. Sometimes lately Eliott gets very thirsty and their parents notice that Eliott can drink an astonishing amount. Orange juice is a particular favorite. Eventually Eliott goes to the doctor and the following is noted on their bloodwork:

Test	Patient value	High/low/within range	Healthy range
Blood glucose	202	H	70–100

1. The doctors suspect that the hormones that control blood glucose are not functioning properly. Which hormone is likely not working?
   A) Glucagon
   B) Insulin
   C) Neither of these
2. The hormone you identified in question 1 can be delivered by an injection under the skin, but delivering the proper dose is critically important. If too much of this hormone were delivered to Eliott, which of the following could occur?
   A) Blood glucose levels would become too high.
   B) Blood glucose levels would sink too low.
   C) The cells of the body would not be able to access enough glucose to live.
3. What organ is not functioning well?
   A) Heart
   B) Stomach
   C) Liver
   D) Pancreas
4. Which type of variable is dysregulated in this scenario?
   A) Anatomical
   B) Physiological

helps cells all over the body to take glucose out of the blood, bringing the blood glucose levels back within the homeostatic range. In either direction, the system effectively turns itself off (**Figures 2.7A** and **2.7B**).

A **positive feedback** system, in contrast, intensifies a change in the body's physiological condition rather than reversing it. A deviation from the normal range results in more change, and the system moves farther away from the normal range. A positive feedback cycle within the body will continue and intensify until there is an interruption. Childbirth is one example of a positive feedback loop that is healthy, but does not work to maintain homeostasis.

LO 2.5.5

Enormous changes in the mother's body are required to expel the fetus at the end of pregnancy. The events of childbirth, once begun, progress rapidly toward a conclusion at which the cycle finally comes to a stop once the fetus and placenta are outside the mother's body. The extreme muscular work of labor and delivery are the result of a positive feedback system (**Figure 2.8**).

The first contractions of labor (the stimulus) push the fetus toward the cervix (the lowest part of the uterus). The cervix contains nerve cells that monitor the degree of stretching (the sensors). These nerve cells send messages to the brain, which in turn causes the pituitary gland to release the hormone oxytocin into the bloodstream. Oxytocin causes stronger contractions of the smooth muscles in the uterus (the effectors), pushing the fetus further down the birth canal. This causes even greater stretching of the cervix. The cycle of stretching, oxytocin release, and increasingly more forceful contractions stops only when the baby is born. At this point, the stretching of the cervix halts, and the cycle comes to a close.

Figure 2.8 Positive Feedback Loop

Childbirth is driven by a positive feedback loop. A positive feedback loop continues on its own until the stimulus is removed or halted. Positive feedback loops do not result in a return to homeostasis.

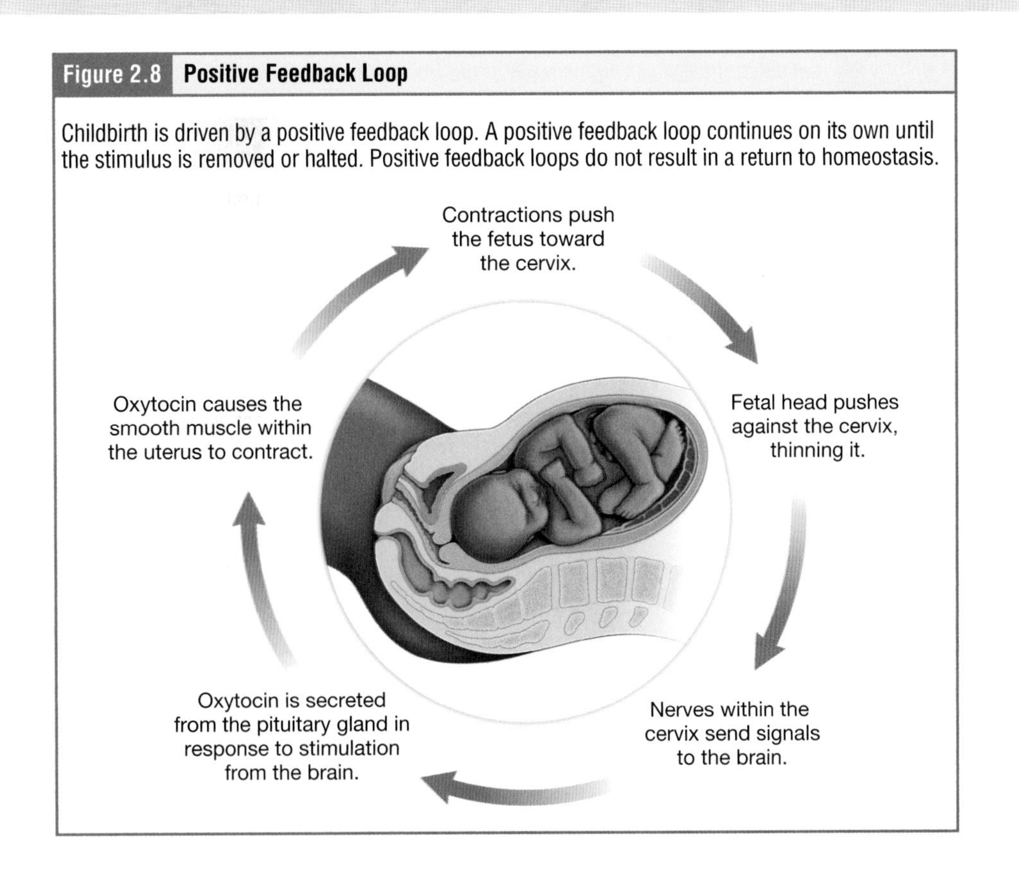

Contractions push the fetus toward the cervix.
Fetal head pushes against the cervix, thinning it.
Nerves within the cervix send signals to the brain.
Oxytocin is secreted from the pituitary gland in response to stimulation from the brain.
Oxytocin causes the smooth muscle within the uterus to contract.

2.6 Structural Organization of the Human Body

Learning Objectives: By the end of this section, you will be able to:

2.6.1 Describe, in order from simplest to most complex, the major levels of organization in the human organism.

2.6.2 Give an example of each level of organization.

2.6.3 List the organ systems of the human body and their major components.

2.6.4 Describe the major functions of each organ system.

In our study of the human body it is helpful to consider its basic architecture; that is, how its smallest parts are assembled into larger structures. We will begin in Chapter 3 with the smallest pieces—subatomic particles, atoms, molecules—and build up to organelles and cells in Chapter 4, tissues in Chapter 5, and the organs in their organ systems in the subsequent chapters. As we go, we will build our understanding of the interconnectedness of the structures within a single human body and the organism with its environment (**Figure 2.9**).

2.6a The Levels of Organization

All matter in the universe is composed of one or more unique pure substances called elements, familiar examples of which are hydrogen, oxygen, carbon, nitrogen, calcium, and iron. The smallest unit of any of these pure substances (elements) is

LO 2.6.1

Figure 2.9 Levels of Structural Organization of the Human Body

The organization of the body often is discussed in terms of six distinct levels of increasing complexity, from the smallest chemical building blocks to a whole human body.

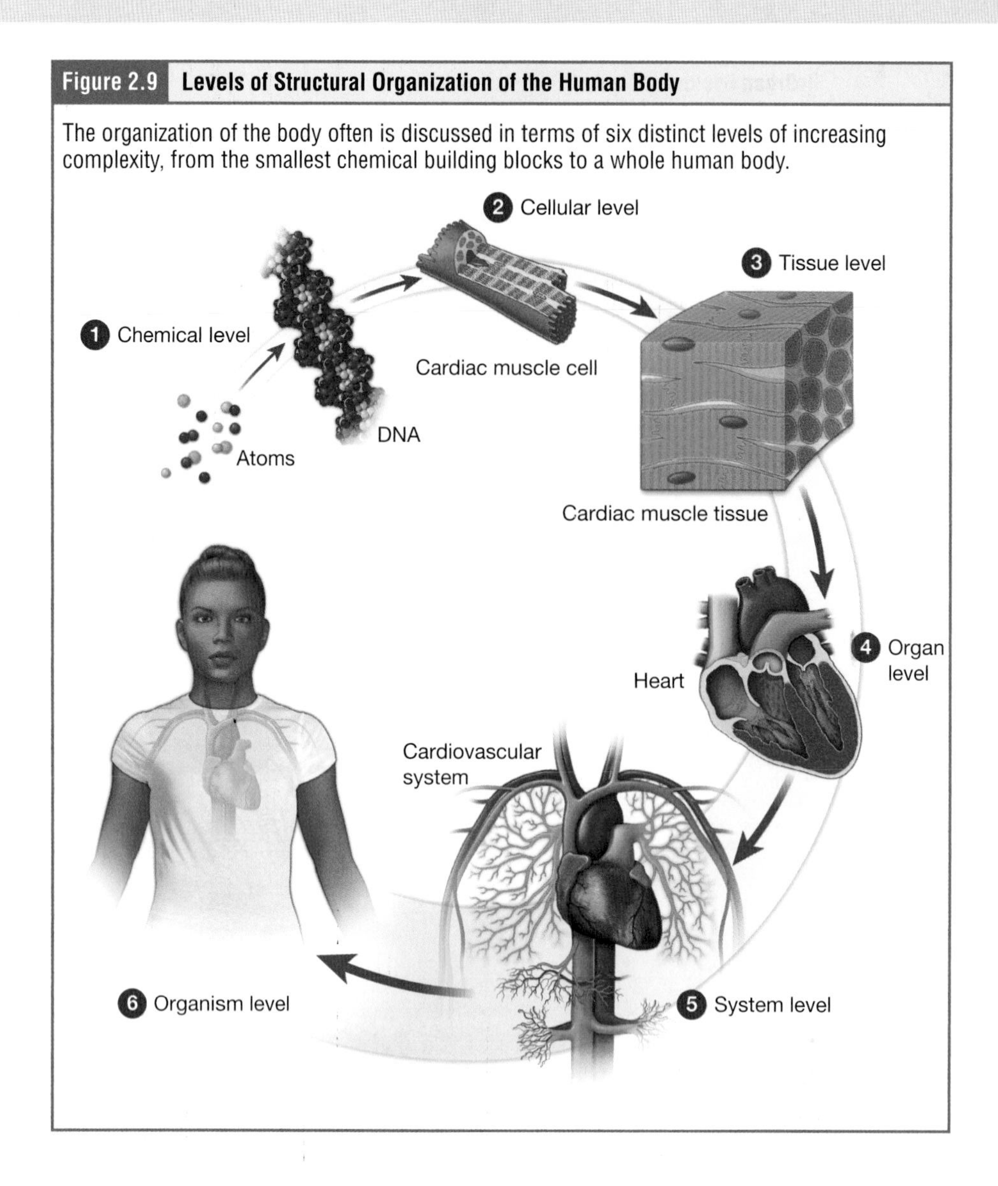

an **atom**. Atoms are made up of subatomic particles such as the proton, electron, and neutron. Two or more atoms combine to form a molecule, such as the water molecules, proteins, and sugars found in living things. Molecules are the chemical building blocks of all body structures. We explore the relationships among molecules in much more detail in Chapter 3.

A **cell** is the smallest independently functioning unit of a living organism. Even bacteria, which are extremely small, unicellular organisms, have a cellular structure. All living structures of human anatomy contain cells, and almost all functions of human physiology are performed in cells or are initiated by cells.

LO 2.6.2

A human cell functions as its own tiny world encased in a protective membrane that encloses a variety of tiny functioning units called **organelles**. In humans, as in all organisms, cells perform all functions of life. A **tissue** is a group of many cells that work together to perform a specific function. An **organ** is a structure of the body that is composed of two or more tissue types; each organ performs one or more specific physiological functions. An **organ system** is a group of organs that work together to perform major functions or meet physiological needs of the body.

This book covers eleven distinct organ systems in the human body (**Figures 2.10A** and **2.10B**). Assigning organs to organ systems can be imprecise since

Figure 2.10A Organ Systems of the Human Body LO 2.6.3

Organs that work together toward one or a set of functions can be grouped into organ systems.

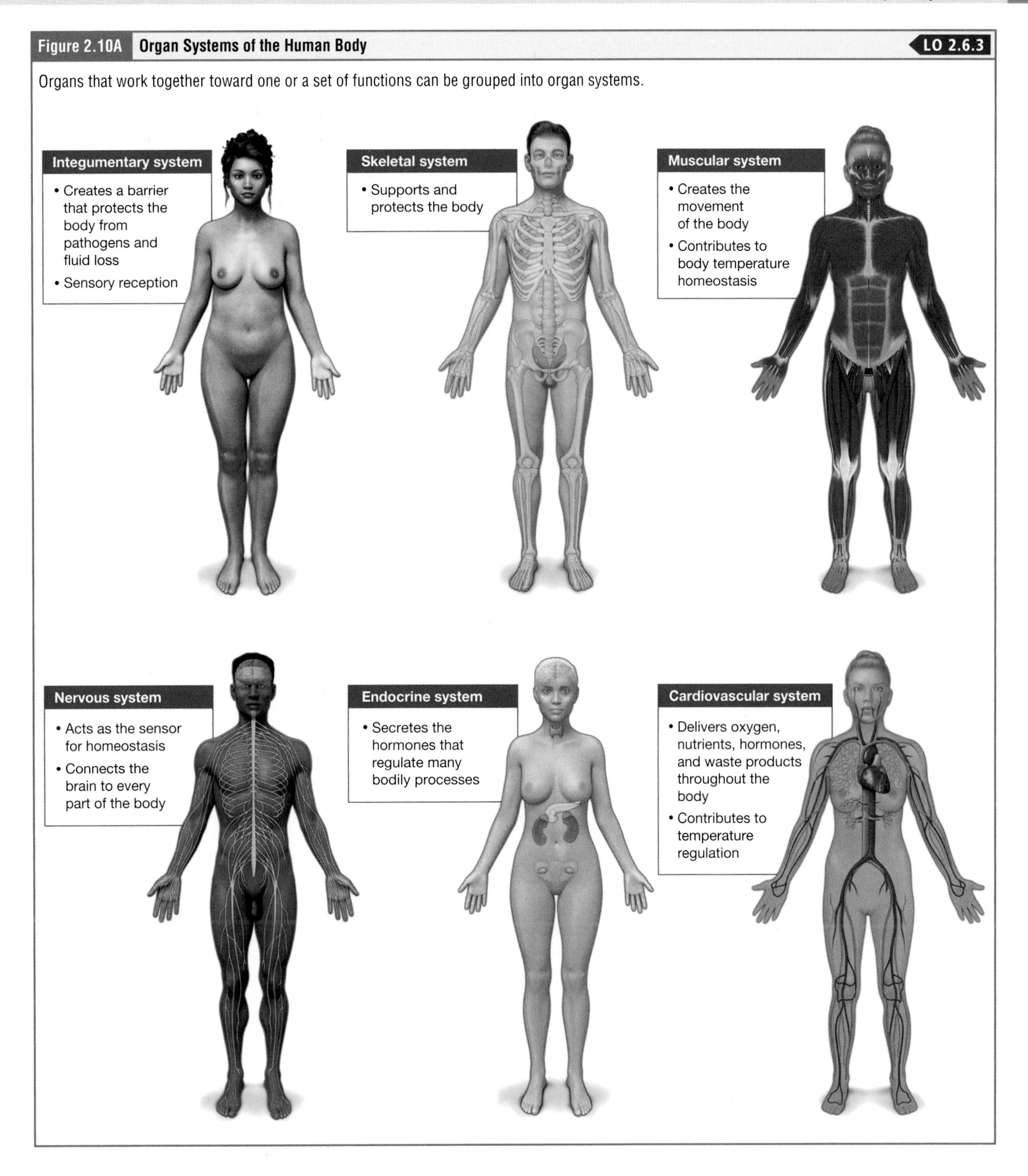

Integumentary system
• Creates a barrier that protects the body from pathogens and fluid loss
• Sensory reception
Skeletal system
• Supports and protects the body
Muscular system
• Creates the movement of the body
• Contributes to body temperature homeostasis
Nervous system
• Acts as the sensor for homeostasis
• Connects the brain to every part of the body
Endocrine system
• Secretes the hormones that regulate many bodily processes
Cardiovascular system
• Delivers oxygen, nutrients, hormones, and waste products throughout the body
• Contributes to temperature regulation

Figure 2.10B Organ Systems of the Human Body (continued) LO 2.6.4

Organs that work together toward one or a set of functions can be grouped into organ systems.

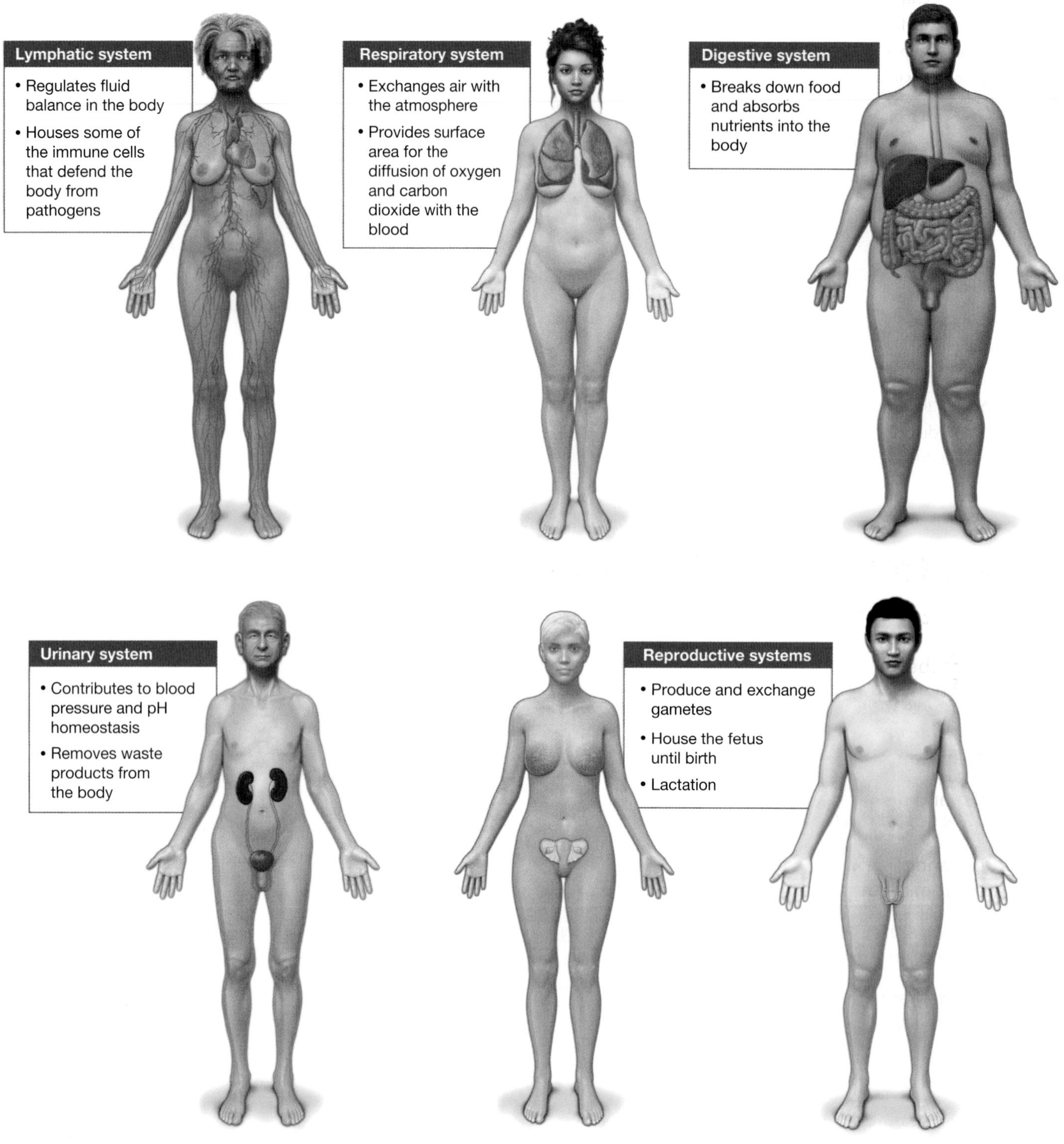

Lymphatic system

- Regulates fluid balance in the body
- Houses some of the immune cells that defend the body from pathogens

Respiratory system

- Exchanges air with the atmosphere
- Provides surface area for the diffusion of oxygen and carbon dioxide with the blood

Digestive system

- Breaks down food and absorbs nutrients into the body

Urinary system

- Contributes to blood pressure and pH homeostasis
- Removes waste products from the body

Reproductive systems

- Produce and exchange gametes
- House the fetus until birth
- Lactation

organs that "belong" to one system can also have functions integral to another system. In fact, most organs contribute to more than one system.

The organism level is the highest level of organization. An **organism** is a living being that has a cellular structure and that can independently perform all physiologic functions necessary for life. In multicellular organisms, including humans, all cells, tissues, organs, and organ systems of the body work together to maintain the life and health of the organism.

Learning Check

1. What is the goal of a positive feedback mechanism?
   a. To restore homeostasis
   b. To return body parameters to the setpoint
   c. To intensify body parameters until there is an interruption
2. Which of the following detects change in a body's internal condition?
   a. Effectors
   b. Sensor
   c. Variable
   d. Setpoint
3. Which of the following organ systems is responsible for removing wastes from blood and excreting them?
   a. Reproductive system
   b. Circulatory system
   c. Urinary system
   d. Digestive system

2.7 Anatomical Terminology

Learning Objectives: By the end of this section, you will be able to:

2.7.1 Describe the human body in anatomical position.

2.7.2 Describe how to use the terms right and left in anatomical reference.

2.7.3 List and define the major directional terms used in anatomy.

2.7.4 List and describe the location of the major anatomical regions of the body.

2.7.5 Describe the location of body structures, using appropriate directional terminology.

2.7.6 Identify and define the anatomic planes in which a body might be viewed.

2.7.7* Classify images based on the section or plane they were taken in.

2.7.8 Identify and describe the location of the body cavities and the major organs found in each cavity.

2.7.9 Identify and describe the location of the four abdominopelvic quadrants and the nine abdominopelvic regions, and the major structures found in each.

2.7.10* Predict, based on location (quadrant, region) of symptoms or pain, what organ may be the issue.

2.7.11* Describe the structure, function, and location of serous membranes.

* Objective is not a HAPS Learning Goal.

Anatomists and healthcare providers use terminology that can be bewildering to the uninitiated. However, the purpose of this language is not to confuse, but rather to increase precision and reduce medical errors. For example, is a scar "above the wrist" located on the forearm two or three inches away from the hand? Or is it at the base of the hand? Is it on the palm side or back side? By using precise anatomical terminology, we eliminate ambiguity. Anatomical terms derive from ancient Greek and Latin words. Building an understanding of anatomically relevant Greek and Latin roots can facilitate your path through learning the language of anatomy and physiology.

For example, in the disorder hypotension, the prefix "hypo-" means "low" or "under"; you will see this root attached to many terms. The hypodermis is the layer of skin under the dermis, hypoglycemia is low blood sugar (gly- is a root for sugar), and hyposecretion refers to an endocrine gland that is secreting less than its typical levels of hormone.

Student Study Tip

Anatomical position is important so no two bones are crossed and you can use anatomical directional terms without confusion!

2.7a Anatomical Position

For precision, anatomists standardize the way in which they view the body. Imagine if you were looking at a body in the position you see in **Figure 2.11**. If you referred to a structure just above the left hand, you would likely be talking about the left wrist. But what if you were referring to a person with their arms crossed? "Above the left" might technically be the right arm. To eliminate confusion, we always discuss a body in a standard position—**anatomical position**—when describing the relative position of one structure to another. Anatomical position, as illustrated in Figure 2.11, is that of the body standing upright, with the feet shoulder width apart and parallel, toes forward. The upper limbs are held out to each side, and the palms of the hands face forward with thumbs out to the sides. The terms *right* and *left* refer to the patient or cadaver's right and left, never to the observer's right and left. Figure 2.11 illustrates the anatomical position with regional terms.

LO 2.7.1

LO 2.7.2

Body position can be described as prone or supine. **Prone** describes a face-down orientation, and **supine** describes a face-up orientation. These terms are sometimes used in describing the position of the body during specific physical examinations or surgical procedures.

2.7b Regional Terms

The human body's numerous regions have specific terms to help increase precision (see Figure 2.11). Notice that the term "brachium" or "arm" is reserved for the upper arm and "antebrachium" or "forearm" is used rather than "lower arm." Similarly, "femur" or "thigh" is correct, and "leg" or "crus" is reserved for the portion of the lower limb between the knee and the ankle. You will be able to describe the body's regions using the terms from the figure.

Anterior (ventral)

Posterior (dorsal)

Superior (cranial)

Inferior (caudal)

2.7c Directional Terms

Certain directional anatomical terms appear throughout this and any other anatomy textbook (**Figure 2.12**). These terms are essential for describing the relative locations of different body structures. For instance, an anatomist might describe one band of tissue as "inferior to" another or a physician might describe a tumor as "superficial to" a deeper body structure. Commit these terms to memory to avoid confusion when you are studying or describing the locations of particular body parts, and remember that we refer to the relative locations of the structures as if the body is in anatomical position.

LO 2.7.3

- **Anterior** Describes the front (belly) of the body. The toes are *anterior* to the foot.
- **Posterior** Describes the back of the body. The spine is *posterior* to the stomach.
- **Superior** Describes a position above or higher than another part of the body proper. The neck is *superior* to the shoulders.
- **Inferior** Describes a position below or lower than another part of the body. The pelvis is *inferior* to the abdomen.
- **Lateral** Describes a structure toward the side of the body. The thumb (pollex) is *lateral* to the digits.
- **Medial** Describes the middle or direction toward the middle of the body. The hallux is the most *medial* toe.

Student Study Tip

Latitude has the Latin root "lat-," which means side. Lateral is farther out to the sides compared to medial, which has the root "medi-," which means middle.

Figure 2.11 Regions of the Human Body LO 2.7.4

The human body is shown in anatomical position in an (A) anterior view and a (B) posterior view.

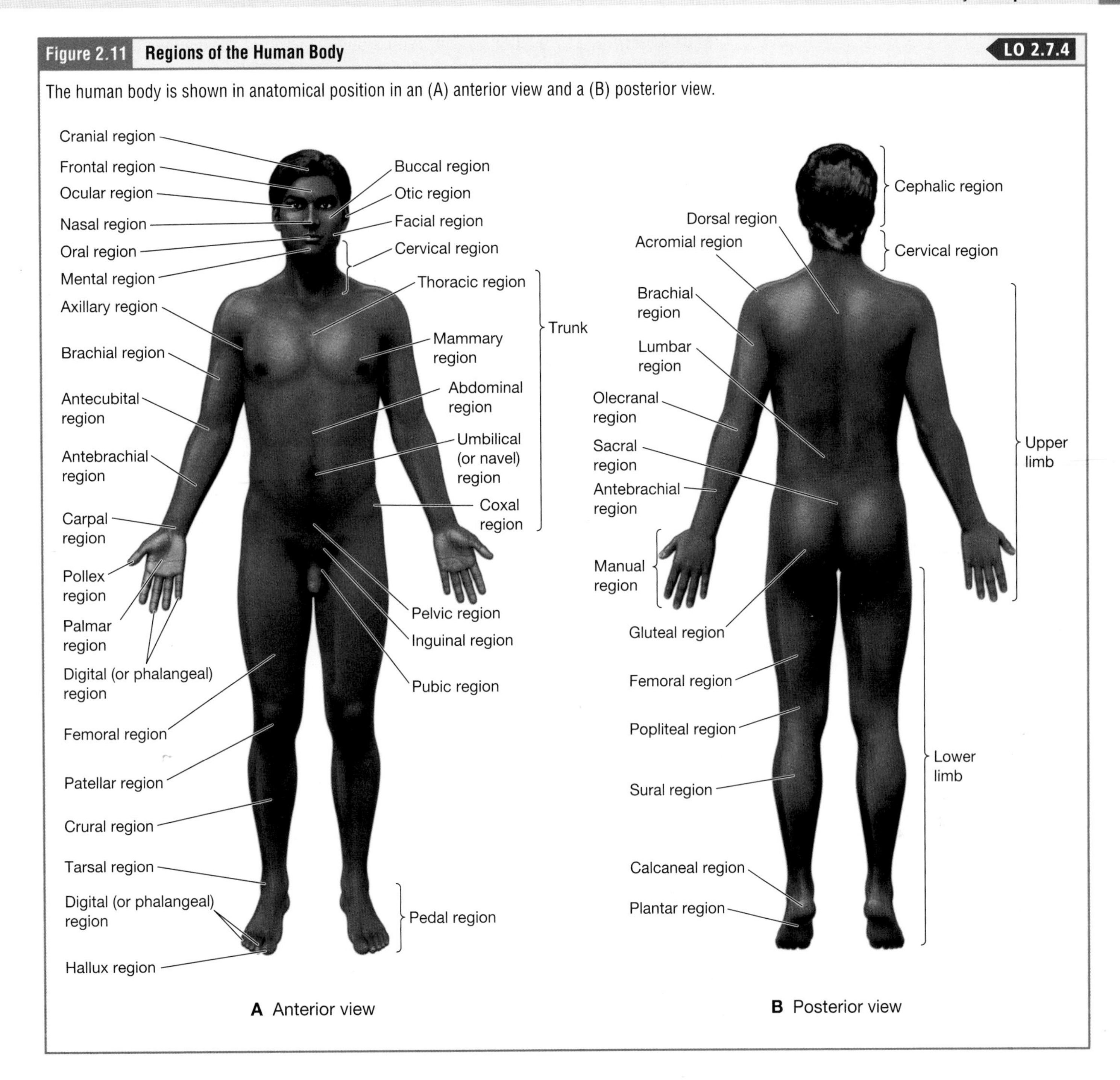

- **Superficial** Describes a position closer to the surface of the body. The skin is *superficial* to the bones.
- **Deep** Describes a position farther from the surface of the body. The brain is *deep* to the skull.
- **Proximal** Describes a position on a limb that is nearer to the point of attachment or the trunk of the body. The brachium is *proximal* to the antebrachium.
- **Distal** Describes a position in a limb that is farther from the point of attachment or the trunk of the body. The crus is *distal* to the femur.

It is worth noting that of these terms, proximal and distal are the ones most often confused by beginning anatomists. Remember that these terms are only used to describe two structures on the same limb. You would not use these terms to describe structures on the trunk or axis of the body.

Figure 2.12 Directional Terms Applied to the Human Body

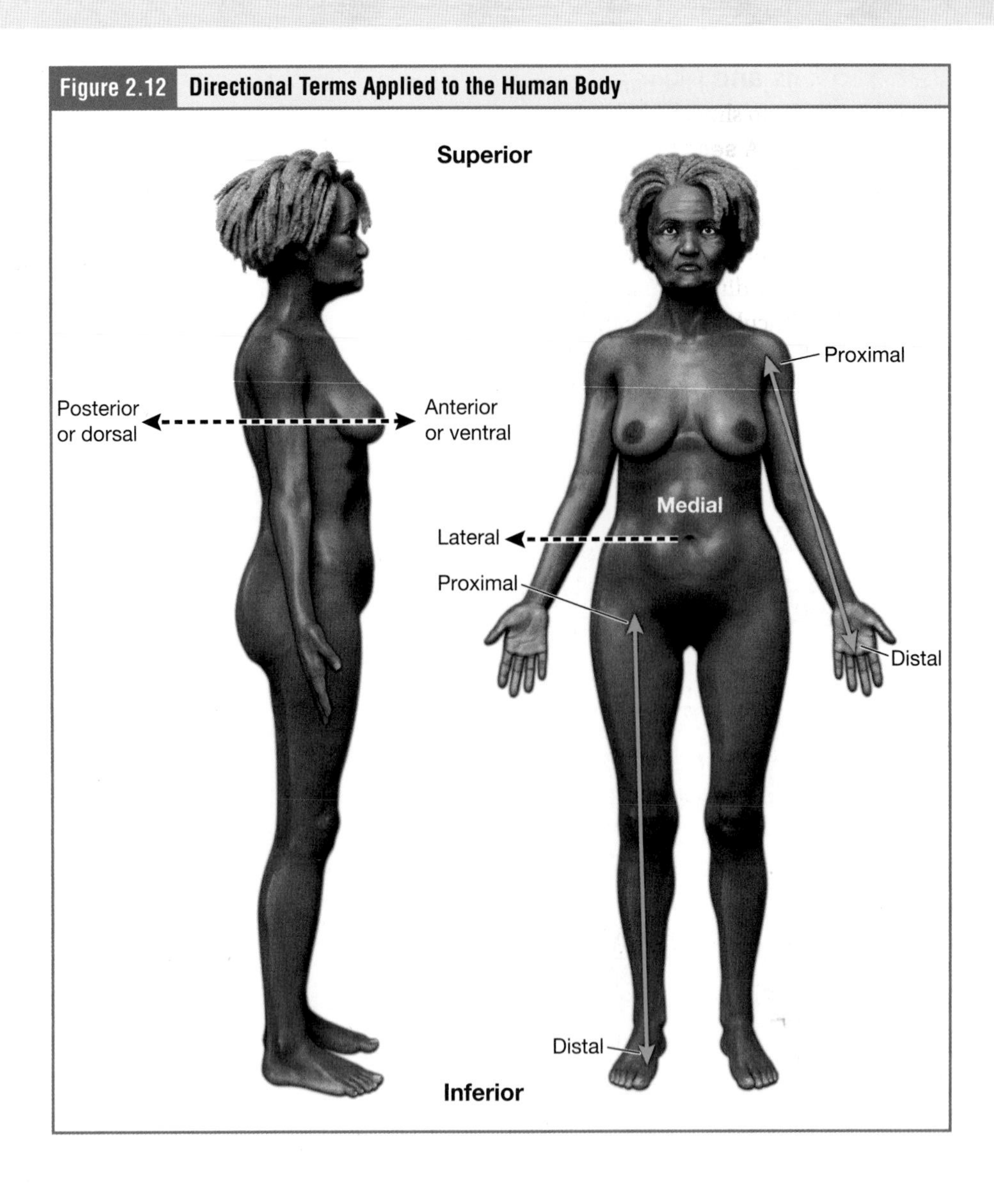

Learning Check

1. What term describes a face-down orientation?
   a. Prone
   b. Inferior
   c. Lateral
   d. Proximal
2. If you have pain in your forearm, which of the following regions appropriately describes the location of your pain?
   a. Brachium
   b. Crus
   c. Femur
   d. Antebrachium
3. In anatomical position, the neck is ________ to the knees.
   a. Superior
   b. Proximal
   c. Inferior
   d. Distal
4. In anatomical position, the wrist is ________ to the elbow. Please select all that apply.
   a. Superior
   b. Inferior
   c. Proximal
   d. Distal

2.7d Sections and Planes

Anatomists refer to slices through structures in the body that enable us to visualize internal anatomy. A **section** is a slice of a three-dimensional structure that has been cut. A **plane** is an imaginary slice through the body used in imaging. Examining a two-dimensional section or plane of a three-dimensional object can be particularly challenging, but this is a skill we must learn and develop in anatomy and physiology. Sections of a three-dimensional organ or structure, particularly microscopic sections, can be difficult to interpret. The following feature, "Anatomy of 2D Sections and Planes," illustrates some of the challenges around visualization of sections. When a three-dimensional structure is folded or coiled and a straight section is cut through

Anatomy of...

Anatomy of 2D Sections and Planes

LO 2.7.6

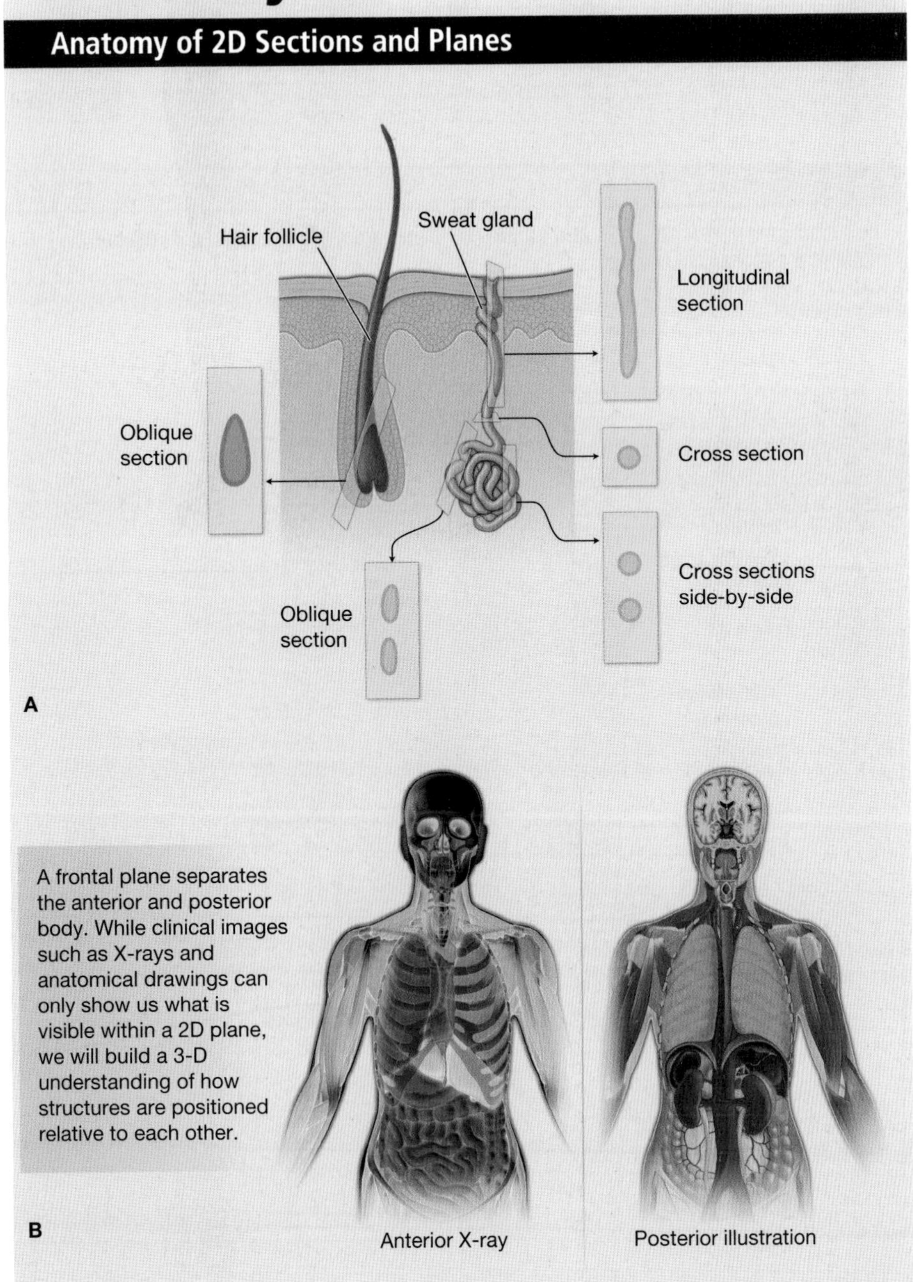
Hair follicle
Sweat gland
Longitudinal section
Oblique section
Cross section
Cross sections side-by-side
Oblique section
A
A frontal plane separates the anterior and posterior body. While clinical images such as X-rays and anatomical drawings can only show us what is visible within a 2D plane, we will build a 3-D understanding of how structures are positioned relative to each other.
B
Anterior X-ray
Posterior illustration

Learning Connection

Image translation is a particularly hard skill to develop within the study of human anatomy. In Chapter 1 we discussed that *skills* require *practice* in order to grow. Try growing your skills through repeated practice of looking at many, many different images of the same structures.

frontal plane (coronal plane)

it, we often end up with what appears to be two side-by-side structures in a section. Anatomy of a 2D anatomical image helps to illustrate some of the ways that planes and sections can be interpreted and misinterpreted.

There are three planes commonly referred to in anatomy and medicine, as illustrated in **Figure 2.13**.

- The **sagittal plane** is the plane that divides the body or an organ vertically into right and left sides. If this vertical plane runs directly down the middle of the body, it is called the *midsagittal* or *median* plane. If it divides the body into unequal right and left sides, it is called a *parasagittal plane* or, less commonly, a *longitudinal section.*
- The **frontal plane** is the plane that divides the body or an organ into an anterior (front) portion and a posterior (rear) portion. The frontal plane is often referred to as a *coronal plane.* ("Corona" is Latin for "crown.")
- The **transverse plane** is the plane that divides the body or organ horizontally into upper and lower portions. Transverse planes produce images referred to as *cross sections.*

Student Study Tip

Find landmarks to orient a section of the body! Transverse cut example: you see a paired structure that looks like the kidneys. Therefore, you are in the abdominal cavity. Identify the vertebrae to find the posterior aspect.

Figure 2.13 Planes of the Body

The sagittal, frontal (or coronal), and transverse planes are the most commonly used in anatomical and medical imaging.

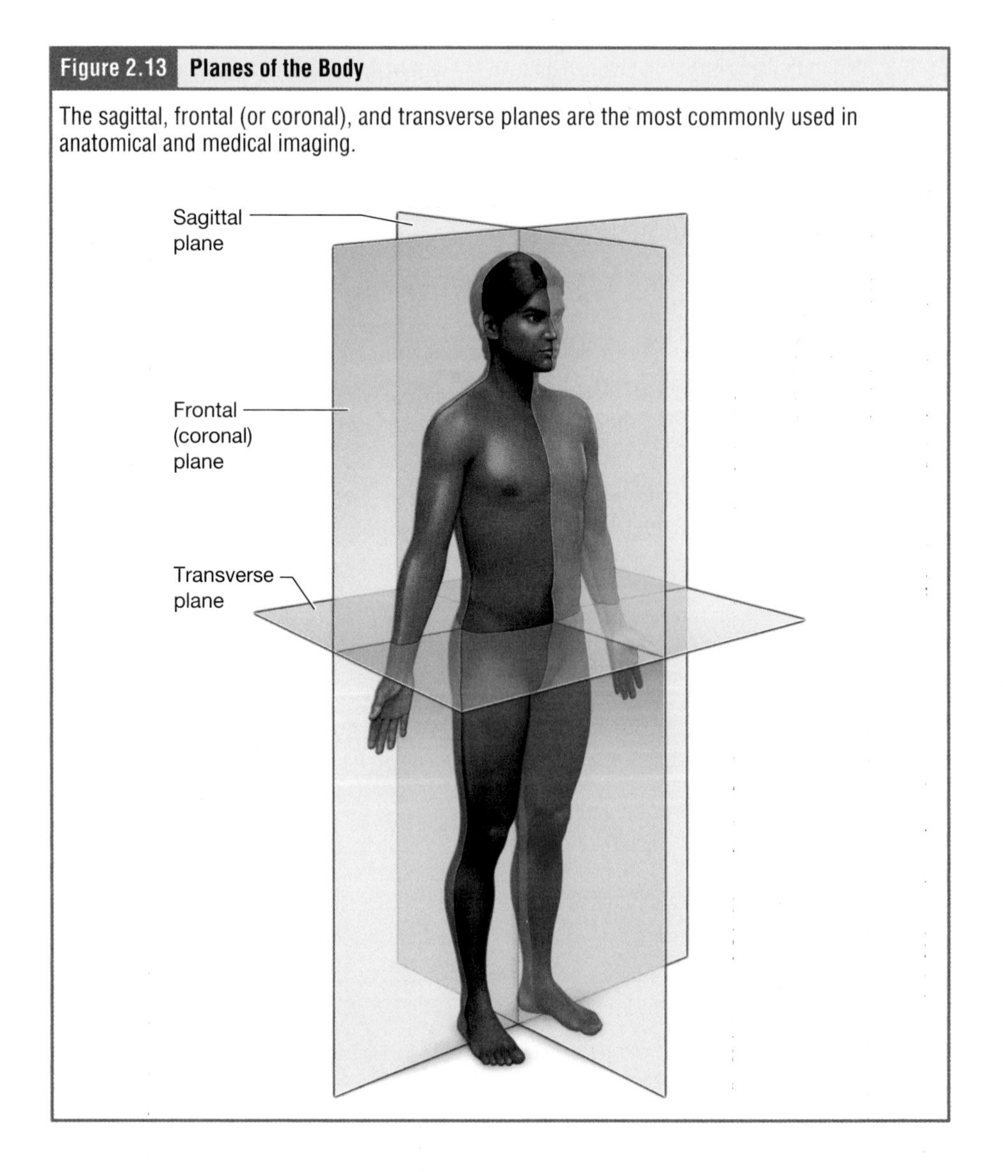

Learning Check

1. Look at the image of the X-ray. It shows an image of the bones of the hand from the front. What anatomical plane best describes the image?
   a. Frontal
   b. Sagittal
   c. Transverse

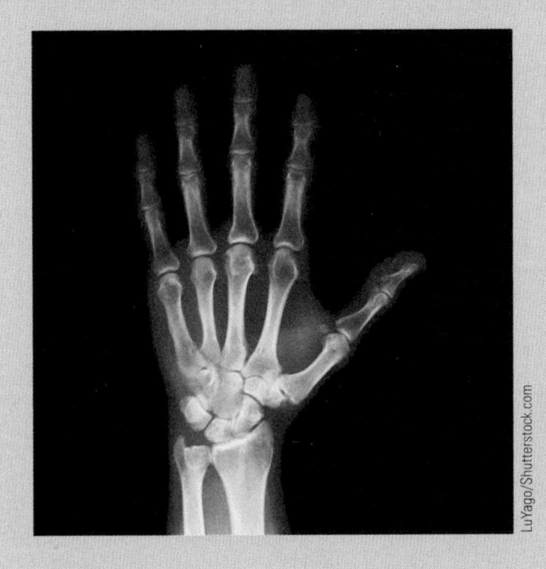

LuYago/Shutterstock.com

2. Look at the CT scan. It shows an image of a brain from the top. What anatomical plane best describes the image?
   a. Sagittal
   b. Transverse
   c. Frontal

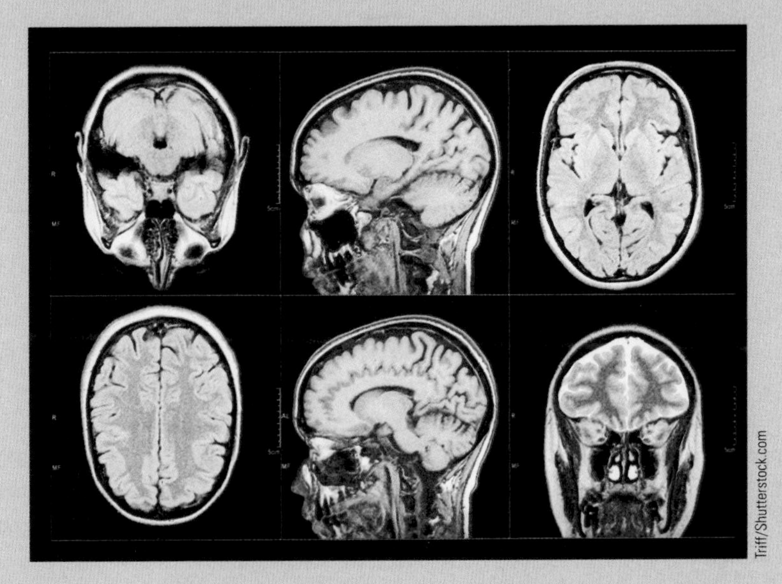

Triff/Shutterstock.com

3. Look at the MRI images. The MRI shows an image of the abdomen in various planes. The *bottom* row in the *second column* is looking at the abdomen from the side. What anatomical plane best describes the image?
   a. Frontal
   b. Sagittal
   c. Inferior

(*Continued*)

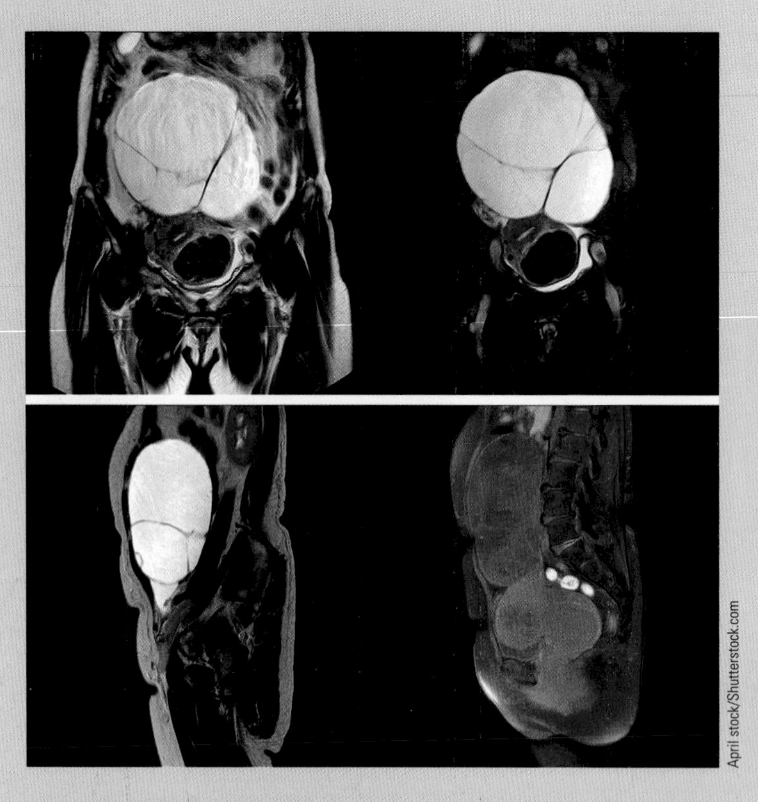

4. Look at the MRI images. The MRI shows an image of the abdomen in various planes. The view shown in the top row, second column is looking at the abdomen from the front. What anatomical plane best describes the image?
   a. Frontal
   b. Transverse
   c. Sagittal
5. Look at the MRI images. The MRI shows an image of the abdomen in various planes. The views in the first column are looking at the abdomen from the top. What anatomical plane best describes the images?
   a. Frontal
   b. Sagittal
   c. Transverse

2.7e Organization and Compartmentalization

A human body consists of approximately 40 trillion human cells. We throw around large numbers all the time, but let's think about that number for a moment. If you had a trillion $1 bills and put them all in a stack, your stack of bills would be over 67,000 miles high. There are almost 8 billion people on the planet; can you imagine trying to get every human to coordinate their activity for one minute? And yet, your body and its 40 trillion cells are working in a coordinated symphony much of the time. If that number seems impressive (and it should!) your body is also home to approximately 100 trillion microbial cells. The body you walk around in is actually more of a floating raft of bacteria than it is human. Your galaxy of human cells is organized in a way that maintains distinct internal compartments. These compartments separate internal body fluids from the countless microorganisms that grow on the surface of the body and the lining of certain tracts within the body.

Compartments are typically divided by sheets that we call membranes. However, in anatomy and physiology we can use the word *membrane* in more than one way. Cells, for example, have a cell membrane (also referred to as the plasma membrane) that keeps the intracellular environment—the fluids and organelles—separate from the extracellular environment. We also use "membrane" to describe a tissue made of many cells that separates, protects, or subdivides body compartments. For example, in Chapter 19 we discuss the membrane, made of many cells, that surrounds and protects the heart, known as the pericardium. As an anatomy and physiology student, it can be helpful to remember the multiplicity of the word *membrane* and clarify for yourself what type of membrane is being referred to in different contexts.

Figure 2.14 illustrates the largest body compartments. The **posterior cavity** and the **anterior cavity** both contain and protect delicate internal organs. The lungs, heart, stomach, and intestines, for example, can expand and contract without distorting other tissues or disrupting the activity of nearby organs.

posterior cavity (dorsal cavity)

anterior cavity (ventral cavity)

spinal cavity (vertebral cavity)

Subdivisions of the Posterior and Anterior Cavities The posterior and anterior cavities are each subdivided into smaller cavities. In the posterior cavity, the **cranial cavity** houses the brain, and the **spinal cavity** (or vertebral cavity) encloses the spinal cord. Just as the brain and spinal cord make up a continuous, uninterrupted structure, the cranial and spinal cavities that house them are also continuous. The brain and spinal cord are protected by the bones of the skull and vertebral column and by cerebrospinal fluid, a colorless fluid produced by the brain, which cushions the brain and spinal cord within the posterior cavity.

The anterior cavity has two main subdivisions: the thoracic cavity and the abdominopelvic cavity (see Figure 2.14). The **thoracic cavity** is the more superior subdivision of the anterior cavity, and it is enclosed by the rib cage. The thoracic cavity contains the lungs and the heart, which are located in a subdivision of the thoracic cavity known as the **mediastinum**. The diaphragm forms the floor of the thoracic cavity and separates it from the more inferior abdominopelvic cavity. The **abdominopelvic cavity** is the largest cavity in the body. Although no membrane physically subdivides the abdominopelvic cavity, it can be useful to distinguish between the abdominal cavity, the division that houses the digestive organs, and the pelvic cavity, the division that houses the organs of reproduction.

abdominopelvic cavity (peritoneal cavity)

Figure 2.14 Posterior and Anterior Body Cavities LO 2.7.8

The anterior cavity includes the thoracic and abdominopelvic cavities and their subdivisions. The posterior cavity includes the cranial and spinal cavities.

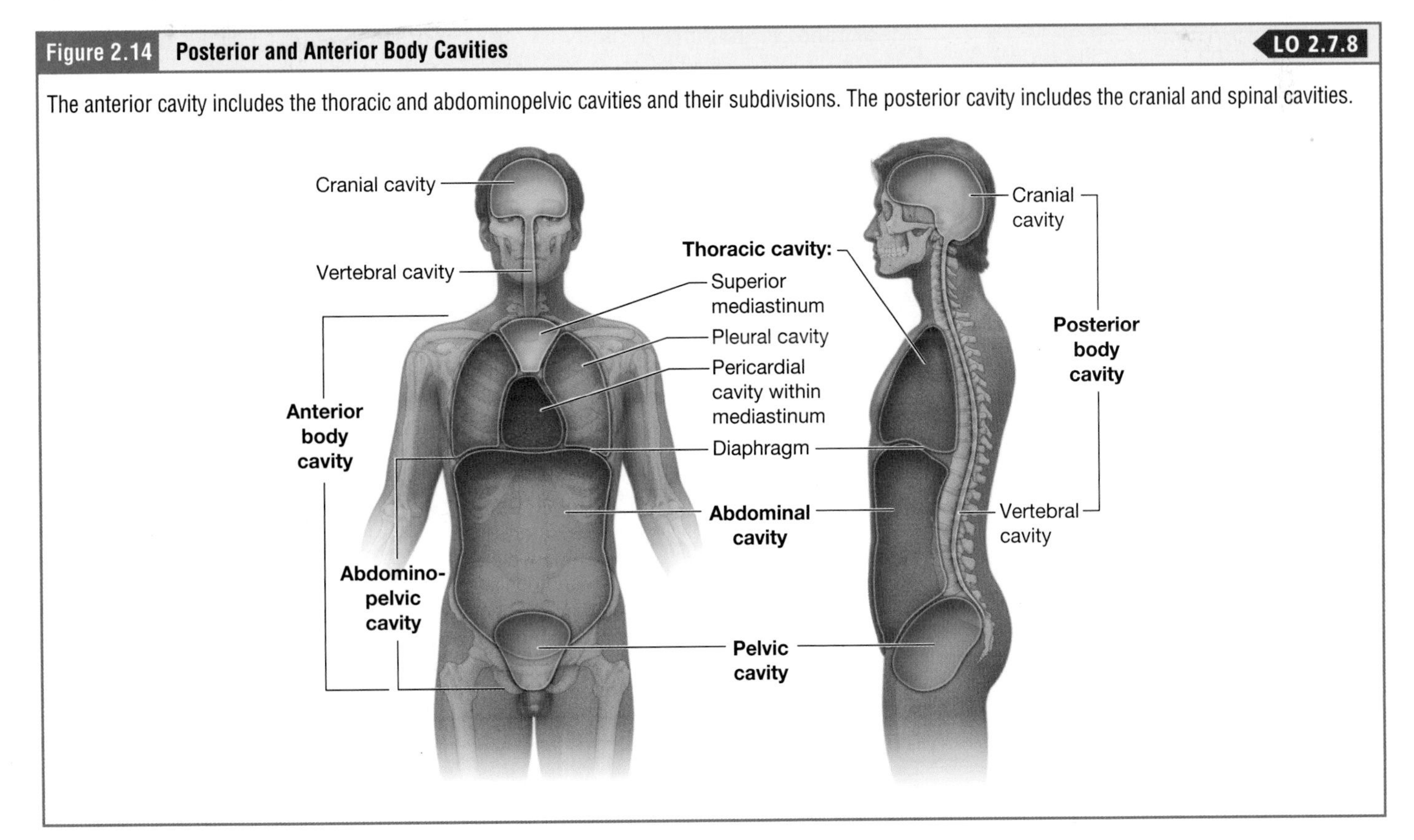

Figure 2.15 Regions and Quadrants

LO 2.7.9

LO 2.7.10

The abdominopelvic cavities can be divided into (A) nine regions or (B) four quadrants.

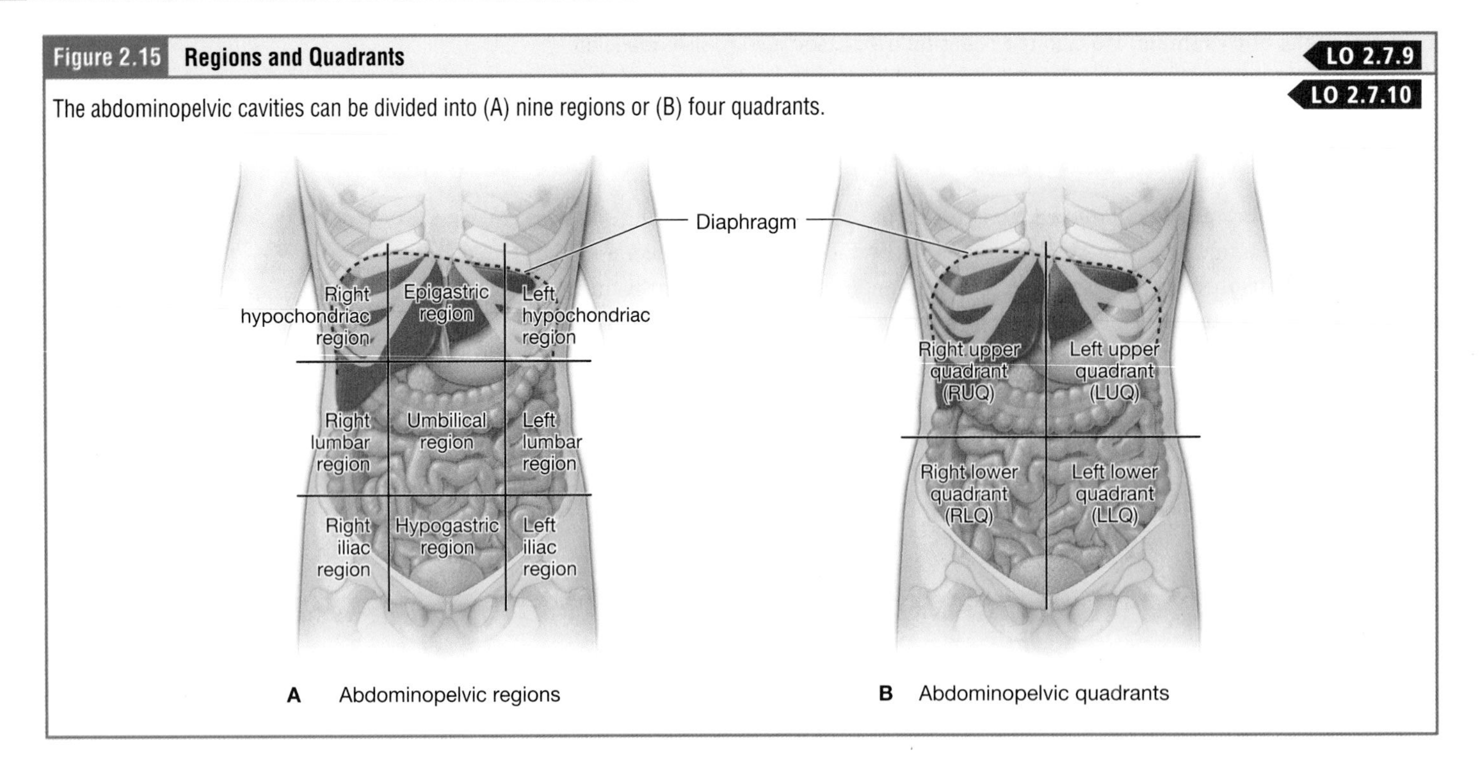

Student Study Tip

To distinguish parietal versus visceral membranes, think about when somebody has a visceral reaction coming from deep within. The visceral layer is deeper than the parietal layer.

LO 2.7.11

serous membrane (serosa)

Learning Connection

Write a list of 10 organs of the body that you already know. Take an opportunity to try chunking them in different ways. Write sentences describing their position in the body relative to each other. Make lists of which ones are found in each cavity, region, and quadrant.

Abdominal Regions and Quadrants Clinically, healthcare providers typically divide up the cavity into either nine regions or four quadrants (**Figure 2.15**). These terms are used most often in describing the location of a patient's abdominal pain or a suspicious mass.

Membranes of the Anterior Body Cavity A **serous membrane** (also referred to as a serosa) is one of the thin membranes that cover the walls and organs in the thoracic and abdominopelvic cavities. The parietal layers of the membranes line the walls of the body cavity (*pariet-* refers to a cavity wall). The visceral layer of the membrane covers the organs (the viscera). Between the parietal and visceral layers is a very thin, fluid-filled serous space, or cavity (**Figure 2.16**).

Figure 2.16 Serous Membranes

A double-layered serous membrane covers the heart. Serous membranes are composed of a visceral and parietal layer with fluid between them. These membranes develop as a fluid-filled sac and the organ develops onto the membrane, causing it to reflect back to cover the heart—much the same way that an underinflated balloon would form two layers surrounding a fist.

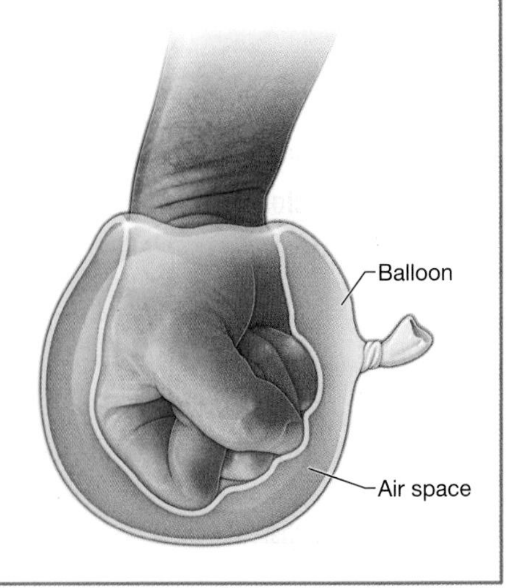

Each of the body's three serous cavities has its own associated membrane. The **pleura** is the serous membrane that surrounds the lungs in the pleural cavity; the **pericardium** is the serous membrane that surrounds the heart in the pericardial cavity; and the **peritoneum** is the serous membrane that surrounds several organs in the abdominopelvic cavity. The serous membranes form fluid-filled sacs, or cavities, that are meant to cushion and reduce friction on internal organs when they move, such as when the lungs inflate or the heart beats. Both the parietal and visceral layers of the serous membrane secrete the thin, slippery serous fluid that fills the cavity and lubricates the membranes. The pericardial and peritoneal cavity fluids reduce friction between the organs and the body or pericardial walls. The pleural cavity fluid provides another function we will discuss more in Chapter 22. Therefore, serous membranes provide additional protection to the viscera they enclose by reducing friction that could lead to inflammation of the organs.

computed tomography (CT)
(CT scans are sometimes called "cat scans")

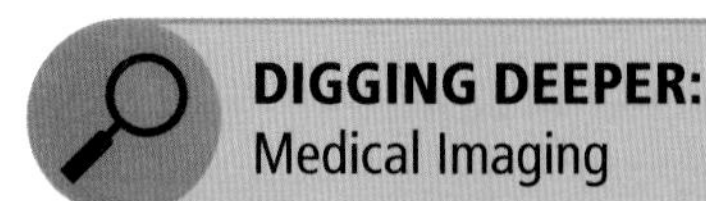

DIGGING DEEPER:
Medical Imaging

X-Rays

The **X-ray** is a form of high-energy electromagnetic radiation capable of penetrating solids. As they are used in medicine, X-rays are emitted from an X-ray machine and directed toward a specially treated metallic plate placed behind the patient's body. The beam of radiation results in darkening of the X-ray plate. Whereas X-rays are slightly impeded by soft tissues, which show up as gray on the X-ray plate, hard tissues such as bone largely block the rays, producing a whiter structure. Thus, X-rays are best used to visualize hard body structures such as teeth and bones.

Computed Tomography

Tomography refers to imaging by planes. **Computed tomography (CT)** is a noninvasive imaging technique that uses computers to analyze several cross-sectional X-rays in order to reveal minute details about structures in the body.

Like many forms of high-energy radiation, X-rays and CT scans are capable of damaging cells and initiating changes that can lead to cancer. These types of imaging must be used sparingly to decrease this risk.

Magnetic Resonance Imaging

Magnetic resonance imaging (MRI) is a noninvasive medical imaging technique based on a phenomenon of nuclear physics in which matter exposed to magnetic fields and radio waves was found to emit radio signals. MRI offers very precise imaging, especially to discover tumors, and also has the major advantage of not exposing patients to radiation.

Drawbacks of MRI scans include their much higher cost, and patient discomfort with the procedure. The MRI scanner subjects the patient to such powerful electromagnets that the scan room must be shielded. The patient must be enclosed in a metal tube-like device for the duration of the scan (sometimes for as long as 30 minutes, which can be uncomfortable and impractical for ill patients). The device is also so noisy that, even with earplugs, patients can become anxious or even fearful. Patients with iron-containing metallic implants (internal sutures, some prosthetic devices, and so on) cannot undergo MRI scanning because it can dislodge these implants.

Functional MRIs (fMRIs), which detect the concentration of blood flow in certain parts of the body, are increasingly being used to study the activity in parts of the brain during various body activities. This has helped scientists learn more about the locations of different brain functions and more about brain abnormalities and diseases.

Positron Emission Tomography

Positron emission tomography (PET) is a medical imaging technique involving the use of so-called radiopharmaceuticals, substances that emit radiation that is short-lived and therefore relatively safe to administer to the body. The main advantage is that PET can illustrate physiologic activity—including nutrient metabolism and blood flow—of the organ or organs being targeted, while CT and MRI scans can only show static images. PET is widely used to diagnose a multitude of conditions, such as heart disease, the spread of cancer, certain forms of infection, brain abnormalities, bone disease, and thyroid disease.

Ultrasonography

Ultrasonography is an imaging technique that uses the transmission of high-frequency sound waves into the body to generate an echo signal that is converted by a computer into a real-time image of anatomy and physiology. Ultrasonography is the least invasive of all imaging techniques, and it is therefore used more freely in sensitive situations such as pregnancy.

Learning Check

1. Which of the following cavities make up the ventral body cavity? Please select all that apply.
   a. Abdominopelvic cavity
   b. Cranial cavity
   c. Thoracic cavity
   d. Vertebral cavity
2. The intestines are organs that help with digestion. In which cavity would you expect to find these organs?
   a. Abdominopelvic cavity
   b. Thoracic cavity
   c. Dorsal body cavity
   d. Cranial cavity
3. Which of the following serous membranes surrounds several organs in the abdominopelvic cavity?
   a. Pleura
   b. Pericardium
   c. Peritoneum

Chapter Review

2.1 Overview of Anatomy and Physiology

The learning objectives for this section are:

2.1.1 Define the terms anatomy and physiology.

2.1.2 Give specific examples to show the interrelationship between anatomy and physiology.

2.1 Questions

1. Which of the following is the definition of anatomy?
   a. Scientific study of the location of the body's structures
   b. Study of cells only
   c. Study of tissues only
   d. Scientific study of how the body's structures function
2. Which of the following correctly describes how anatomy relates to physiology?
   a. Anatomy is concerned with the function of the teeth, while physiology is concerned with tooth structure.
   b. Anatomy describes the double-layered form of serous membranes, while physiology describes how the serous fluid produced by the cells of the membrane provide lubrication.
   c. Anatomy is concerned with the ability of the fingers to grasp objects, while physiology is concerned with the bones and joints that make up the fingers.
   d. Anatomy describes the breathing process, while physiology describes the lobes of the lungs.

2.1 Summary

- Anatomy is the study of body structures, while physiology is the study of how body structures work together to support the functions of life.
- Physiology is determined by anatomy. The function of a specific body part is determined by its shape or structure.
- The study of anatomy consists of the following levels: gross anatomy and microscopic anatomy.
- Approaches to studying anatomy consist of regional anatomy and systemic anatomy.

2.2 Structure and Function

The learning objectives for this section are:

2.2.1* Describe, compare, and contrast various structure–function relationships from molecular to organ level.

2.2.2* Relate the commonly found branching structure to function of an organ.

* Objective is not a HAPS Learning Goal.

2.2 Questions

1. If Micah was working in a loading dock lifting heavy boxes every day, what would you expect to see happen to their muscles over time?

a. Their muscles will get bigger.
b. Their muscles will get longer.
c. Their muscles will get weaker.
d. Their muscles will stay the same.

2. Coronary arteries are branching of blood vessels that provide blood to the heart. This allows the heart to function. Why is it important that the coronary arteries are branching structures?

a. Branching increases the surface area over which the heart can receive blood.
b. Branching structurally supports the heart to prevent it from moving around the body.
c. Branching decreases the volume of blood to the heart to increase blood to the rest of the body.

2.2 Summary

- Form is closely related to function in all living things.
- Function is related to structure at all levels of organization, from the molecular level to the systemic level.
- In all living things, branching patterns provide for maximum surface area.

2.3 Evolution and Human Variation

The learning objectives for this section are:

2.3.1* Define the term and explain the concept of *evolution*.

2.3.2* Contrast the impact of selection on traits that affect reproduction and traits that do not; use this to explain examples of anatomical and physiological variation.

* Objective is not a HAPS Learning Goal.

2.3 Questions

1. What is a change in gene expression that occurs from generation to generation?

a. Melanin
b. Bipedalism
c. Evolution
d. Physiology

2. A group of people discovered a new area of land that has never been inhabited by humans. The climate is hot, sunny, and there is not very much water. Which of the following traits would NOT be advantageous for this group of people to survive?

a. Having more sweat glands to release more body heat
b. Having more melanin to protect the body from UV radiation
c. Having an organ to store water for when the body needs it
d. Having long hair on their head to conserve body heat

2.3 Summary

- Evolution is a change in gene expression that occurs over many generations in a population of living organisms.
- Genetic changes that impact survival or reproduction display less variation than those that do not.
- Anatomical and physiological variations exist among individuals of different biological sex and age, with physiological variation being more diverse than anatomical variation.

2.4 Flow

The learning objectives for this section are:

2.4.1* Describe how a gradient determines flow between two regions, and give examples of gradients that exist in different levels of organization in the body.

2.4.2* Predict how changes in a gradient will affect flow rate.

2.4.3* Predict how differences in resistance will affect flow rate.

* Objective is not a HAPS Learning Goal.

2.4 Question

1. You are observing an experimental setup in a laboratory. The experiment is designed to measure the flow of sodium ions from one side of a cell membrane to the other. The scientist performing the experiment notes that they see an unusually fast rate of flow. Which of the following forces may be at play causing the ions to behave in this manner?
   a. Pressure gradient
   b. Concentration gradient
   c. Resistance
   d. Physiological Variation

Mini Case 1: Flow Rate

In the circulatory system, blood is pumped from the heart around the body through vessels called arteries and veins. Every time the heart contracts, it builds up pressure to pump blood through the arteries.

1. What happens to the flow rate every time the heart contracts?
   a. The flow rate increases.
   b. The flow rate decreases.
   c. The flow rate stays the same.

In some instances, people have plaque buildup in their arteries. This is known as atherosclerosis. When this happens, there is a decrease in the diameter of the arteries.

2. What happens to the flow rate in a person with atherosclerosis?
   a. The flow rate increases.
   b. The flow rate decreases.
   c. The flow rate stays the same.

2.4 Summary

- The flow of substances in the body is driven by three types of gradients: electrical, concentration, and pressure gradients.
- Concentration gradients determine the flow of individual molecules from one location to another.
- Pressure gradients are responsible for the flow of fluids and gasses.
- Electrical gradients cause the movement of charged particles, such as ions, from one location to another.

2.5 Homeostasis

The learning objectives for this section are:

HAPS LEARNING GOALS AND OUTCOMES

2.5.1 Define the following terms as they relate to homeostasis: setpoint, variable, receptor (sensor), effector (target), and control (integrating) center.

2.5.2 Explain why negative feedback is the most common mechanism used to maintain homeostasis.

2.5.3 List the main physiological variables for which the body attempts to maintain homeostasis.

2.5.4 List the steps in a feedback mechanism (loop) and explain the function of each step.

2.5.5 Compare and contrast positive and negative feedback in terms of the relationship between stimulus and response, and describe examples of each.

2.5 Questions

1. Select the most accurate statement about homeostasis:
   a. If the level of a setpoint is too far away from its control center, an effector will communicate with the variable, which will communicate with a sensor to regain homeostasis.
   b. If the level of a variable is too far away from its setpoint, a sensor will communicate with the control center, which will communicate with the effector to regain homeostasis.

c. If the level of an effector is too far away from its variable, a sensor will communicate with the control center, which will communicate with the setpoint to regain homeostasis.
d. If the level of a sensor is too far away from its control center, an effector will communicate with the variable, which will communicate with the setpoint to regain homeostasis.

2. pH levels are homeostatically regulated in the body. When the pH becomes too high or too low, the body systems must return the pH to the setpoint. Which feedback mechanism do you expect to resolve this change and why?
a. Negative feedback mechanism will regain homeostasis to sustain functions of life.
b. Positive feedback mechanism will deviate from homeostasis to sustain functions of life.
c. Negative feedback mechanism will deviate from homeostasis to sustain functions of life.
d. Positive feedback mechanism will regain homeostasis to sustain functions of life.

3. Which of the following is NOT homeostatically regulated?
a. Blood sugar level
b. Thyroid hormone level
c. Heart rate
d. pH

4. In a negative feedback mechanism, the sensor will detect change and send a signal to the effector. What would you expect the effector to do?
a. Send a signal that will intensify the change
b. Send a signal that will reverse a change

5. In a positive feedback mechanism, what is the direct role of the sensor?
a. To reverse a change
b. To intensify a change
c. To maintain homeostasis
d. To achieve the setpoint

2.5 Summary

- Homeostasis is the maintenance of dynamic internal stability in the body.
- Each monitored variable has a set point and normal range. When a variable is outside of its normal range, homeostatic mechanisms will attempt to restore its homeostatic level.
- Negative feedback mechanisms reverse deviations from the set point of a variable and restore the homeostatic level.
- Positive feedback systems temporarily intensify the deviation from the set point of a variable in order to accomplish a specific task. Once the task has been completed, this system will restore homeostasis.

2.6 Structural Organization of the Human Body

The learning objectives for this section are:

2.6.1 Describe, in order from simplest to most complex, the major levels of organization in the human organism.

2.6.2 Give an example of each level of organization.

2.6.3 List the organ systems of the human body and their major components.

2.6.4 Describe the major functions of each organ system.

2.6 Questions

1. List the major levels of organization from simplest to most complex.
a. Organelles < organ system < organ < organism < tissue < cell
b. Cell < organelle < organism < tissue < organ < organ system
c. Organelle < cells < tissue < organ < organ system < organism
d. Organ system < organ < tissue < organism < organelle < cell

2. A skeletal muscle fiber is an example of which level of organization?
a. Atom
b. Cell
c. Tissue
d. Organ

3. Select the pairs that matches the organ system to its major component. Please select all that apply.
a. Skeletal system—peripheral nerves
b. Muscular system—skeletal muscles
c. Endocrine system—adrenal gland
d. Respiratory system—thymus

4. Heloise comes to you with a fever. You run some tests and realize they have pneumonia, which is fluid buildup in the lungs. Which of the following functions are directly impaired?
a. Control of water balance in the body
b. Delivery of oxygen to blood
c. Equalizes temperature in the body
d. Structural support of the body

2.6 Summary

- Organelles are tiny functioning units encased in a membrane to make up a cell. Groups of cells make up a tissue. Two or more tissue types make up an organ. A group of organs make up an organ system. An organism is the highest level of organization that can perform all physiological functions necessary for life.
- The levels of organization in the body are subatomic particles, atoms, molecules, organelles, cells, tissues, organs, organ systems, and the organism. Each level of organization makes up the next level. For example, subatomic particles make up atoms and tissues make up organs.
- The smallest unit of any pure element is an atom.
- An organism is the highest level of organization of the body.
- Each organ system provides different functions in the body. All systems, with the exception of the reproductive system, are necessary for the survival of an individual.

2.7 Anatomical Terminology

The learning objectives for this section are:

HAPS LEARNING GOALS AND OUTCOMES

2.7.1 Describe the human body in anatomical position.

2.7.2 Describe how to use the terms right and left in anatomical reference.

2.7.3 List and define the major directional terms used in anatomy.

2.7.4 List and describe the location of the major anatomical regions of the body.

2.7.5 Describe the location of body structures, using appropriate directional terminology.

2.7.6 Identify and define the anatomic planes in which a body might be viewed.

2.7.7* Classify images based on the section or plane they were taken in.

2.7.8 Identify and describe the location of the body cavities and the major organs found in each cavity.

2.7.9 Identify and describe the location of the four abdominopelvic quadrants and the nine abdominopelvic regions, and the major structures found in each.

2.7.10* Predict, based on location (quadrant, region) of symptoms or pain, what organ may be the issue.

2.7.11* Describe the structure, function, and location of serous membranes.

* Objective is not a HAPS Learning Goal.

2.7 Questions

1. Which of the following describe the body in anatomical position? Please select all that apply.
   a. Thumbs out to the side
   b. Legs crossed over each other
   c. Toes pointing forward
   d. Body lying face down

2. You are doing rounds in a hospital and are about to check on a new patient. You look at the chart and see it says, "removed left kidney." Is this the observer's (your) left, or the patient's left?
   a. Observer's   b. Patient's

3. You are at a gymnastics competition when you see a gymnast do a backflip (see the following figure). In this position, how would you describe the gymnast's knees in relation to their shoulders?
   a. Superior
   b. Inferior
   c. Proximal
   d. Distal

Master1305/Shutterstock.com

4. The posterior surface of the knee is found in the ______ region.

a. Inguinal
b. Patellar
c. Femoral
d. Popliteal

5. You are playing basketball and going for a dunk (see the following figure). How would you describe your right wrist in relation to your right elbow? Please select all that apply.

a. Superior
b. Inferior
c. Proximal
d. Distal

Bbernard/Shutterstock.com

6. A magician that cuts their assistant into anterior and posterior halves would be able to view the assistant in the ______ plane.

a. Sagittal
b. Coronal
c. Transverse
d. Horizontal

7. Rosemarie was in a car accident and hit their head on the dashboard. Below is their X-ray, which looks at the head from the side. What plane is the X-ray taken in?

a. Sagittal
b. Frontal
c. Transverse

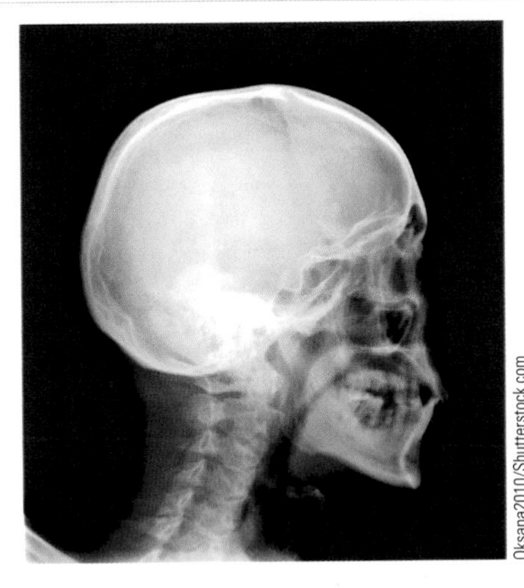

Oksana2010/Shutterstock.com

8. Which of the following organs would you expect to find in thoracic cavity? Please select all that apply.

a. Lungs
b. Heart
c. Mediastinum
d. Stomach
e. Liver

9. Which abdominopelvic region is inferior to the navel (belly button)?

a. Iliac
b. Epigastric
c. Hypogastric
d. Hypochondriac

10. Jim has pain in the left upper quadrant. Which of the following organs could be the source of their pain?

a. Spleen
b. Gallbladder
c. Appendix
d. Liver

11. All of the following are true of serous membranes except:

a. The parietal layer covers the surfaces of organs.
b. The cells of serous membranes secrete a lubricating serous fluid.
c. They are double-layered and contain visceral and parietal layers.
d. They surround the organs of the thoracic and abdominopelvic cavities.

2.7 Summary

- In anatomical position, a person is standing upright, with feet parallel and toes forward, and arms by their side with the thumbs pointing outwards.
- There are three planes that are commonly referred to in anatomy: the sagittal plane divides the body vertically into right and left sides, the frontal plane divides the body into front and back, and the transverse plane divides the body into upper and lower portions.
- The body is divided into compartments known as cavities. The abdominopelvic cavity is also divided into regions and quadrants to describe the location of pain or mass.
- There are three serous membranes: the pleura surrounds the lungs, the pericardium surrounds the heart, and the peritoneum surrounds several organs in the abdominopelvic cavity.

3

The Chemical Level of Organization

Chapter Introduction

The smallest, most fundamental material components of the human body are basic chemical elements. An **element** is a pure substance containing only the same type of atom.

You may have studied chemistry in a dedicated course, but here we will examine human chemistry. This study is somewhat distinct because we will examine organic (carbon-based) molecules, inorganic (non carbon-based) molecules, and biochemicals (those produced by the body) and all of the chemical reactions we examine take place in an aqueous (watery) environment, rather than in air.

This chapter begins by examining elements and how the structures of the smallest units of matter determine the characteristics of the elements.

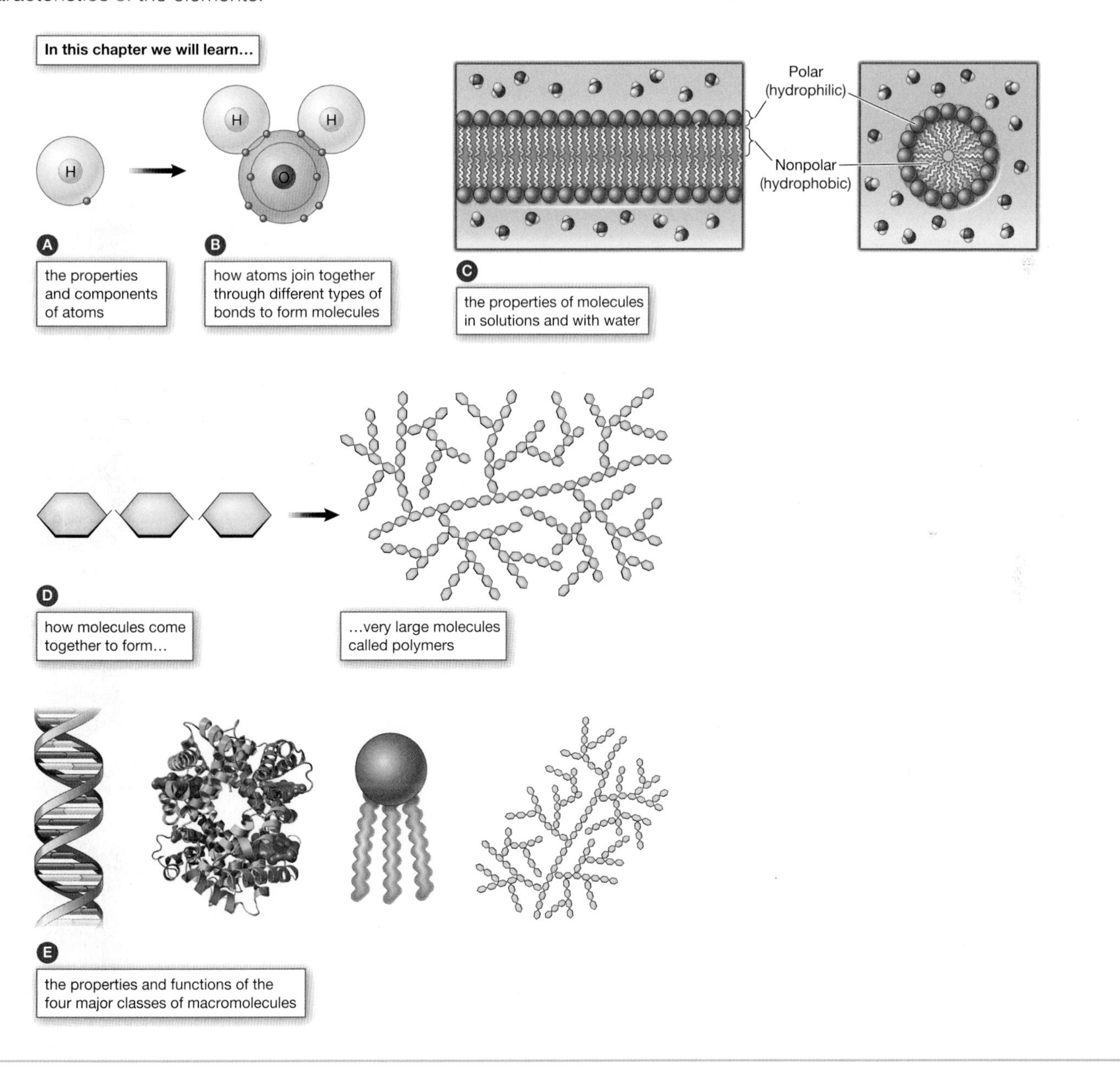

In this chapter we will learn...
H
H
H
O
A
the properties and components of atoms
B
how atoms join together through different types of bonds to form molecules
Polar (hydrophilic)
Nonpolar (hydrophobic)
C
the properties of molecules in solutions and with water
D
how molecules come together to form...
...very large molecules called polymers
E
the properties and functions of the four major classes of macromolecules

3.1 Elements and Atoms: The Building Blocks of Matter

Learning Objectives: By the end of this section, you will be able to:

3.1.1 Compare and contrast the terms atoms, elements, molecules, and compounds.

3.1.2 Describe the charge, mass, and relative location of electrons, protons, and neutrons in an atom.

3.1.3 Distinguish among the terms atomic number, mass number, and atomic weight.

3.1.4 Explain how ions and isotopes are produced by changing the relative number of specific subatomic particles, using one element as an example.

3.1.5 Compare and contrast the terms ion, free radical, isotope, and radioisotope.

3.1.6 Relate the number of electrons in an electron shell to the atom's chemical stability and its ability to form chemical bonds.

3.1.7* Predict, by examining the properties of an atom, the likelihood that it will ionize or form bonds.

* Objective is not a HAPS Learning Goal.

The substance that makes up the universe—whether we are talking about a grain of sand, a star in the sky, or a potato chip—is called **matter**. Scientists define matter as anything that occupies space and has mass. An object's mass and its weight are related concepts, but not quite the same. **Mass** is the amount of matter contained in an object, and the object's mass is the same whether that object is on Earth or in the zero-gravity environment of outer space. An object's weight, on the other hand, is its mass as affected by the pull of gravity. For example, a piece of cheese that weighs a pound on Earth weighs only a few ounces on the moon.

3.1a Elements and Compounds

All matter in the natural world is composed of one or more of the 92 fundamental substances called **elements**. While your body can assemble many of the chemical compounds needed for life from their constituent elements, it cannot make elements. They must come from the environment. A familiar example of an element that you must take in is calcium. Calcium is essential to the human body; it is absorbed and used for a number of processes, including muscle contraction and strengthening bones. Imagine you are snacking on a slice of cheese or a dish of lentils; your digestive system breaks down the food into components small enough to cross into the bloodstream (these processes are discussed in Chapter 25). Among these components, calcium, as an element, cannot be broken down further. The calcium in your food, therefore, is the same as the calcium that forms your bones. Some other elements you might be familiar with are oxygen, sodium, and iron. The elements in the human body are shown in **Figure 3.1**, beginning with the most abundant: oxygen, carbon, hydrogen, and nitrogen. All the elements in your body are derived from the foods you eat and the air you breathe.

LO 3.1.1

In nature, elements rarely occur as individual atoms. Instead, they combine to form either compounds or molecules. A **compound** is a substance composed of two or more elements joined by chemical bonds. For example, the compound glucose is an important body fuel. It is always composed of the same three elements: carbon, hydrogen, and oxygen. The term **molecule** is broader; it can be used to describe any

The Human Anatomy and Physiology Society includes more than 1,700 educators who work together to promote excellence in the teaching of this subject area. The HAPS A&P Learning Outcomes measure student mastery of the content typically covered in a two-semester Human A&P curriculum at the undergraduate level. The full Learning Outcomes are available at https://www.hapsweb.org.

Figure 3.1 Elements of the Human Body

The primary elements that compose the human body are shown from most to least abundant.

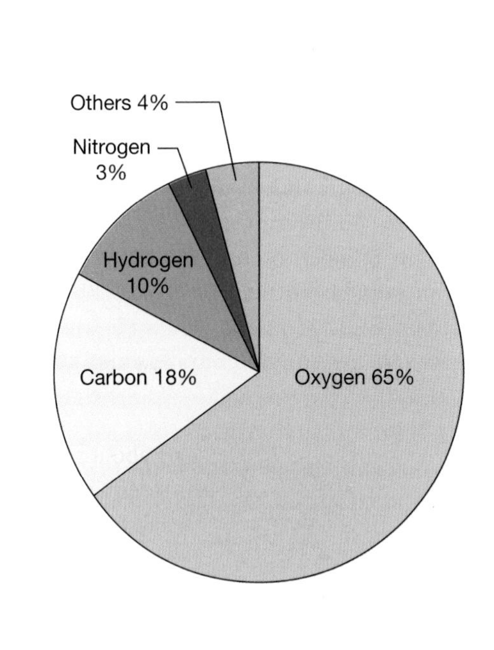

Element and Symbol	Percentage in the Body
Oxygen (O)	65.0
Carbon (C)	18.5
Hydrogen (H)	9.5
Nitrogen (N)	3.2
Calcium (Ca)	1.5
Phosphorus (P)	1.0
Potassium (K)	0.4
Sulfur (S)	0.3
Sodium (Na)	0.2
Chlorine (Cl)	0.2
Magnesium (Mg)	0.1
Trace elements include boron (B), chromium (Cr), cobalt (Co), copper (Cu), fluorine (F), iodine (I), iron (Fe), manganese (Mn), molybdenum (Mo), selenium (Se), silicon (Si), tin (Sn), vanadium (V), and zinc (Zn).	less than 1.0

group of atoms bonded together. Therefore, if we see multiple atoms bonded together, for example, oxygen typically is found in groups of two (O_2) we would describe this as a molecule of oxygen, but the term *compound* would not apply. A single unit of water, H_2O, is both a *molecule* and a *compound*.

3.1b Atoms and Subatomic Particles

An **atom** is the smallest quantity of an element that retains the unique properties of that element. In other words, an atom of hydrogen is a unit of hydrogen—the smallest amount of hydrogen that can exist. As you might guess, atoms are incredibly small. The period at the end of this sentence is larger than millions of atoms together.

LO 3.1.2

Atomic Structure and Energy Atoms are made up of even smaller particles; three types of so-called subatomic particles are important for our understanding of the behavior of atoms and molecules in the human body: the **proton**, **neutron**, and **electron**. As you might guess from their names, protons carry a positive charge, electrons carry a negative charge, and neutrons are neutral, carrying no charge. Protons and neutrons have mass and sit together in the nucleus (center) of the atom, and so the number of each in the nucleus of the atom determines the mass of the element. The electrons of the atom spin around the nucleus quite quickly, actually close to the speed of light; therefore, they do not contribute mass to the atom. Also, an electron has about 1/2000th the mass of a proton or neutron, so even if these orbiters did stay still, they would contribute very little mass.

Figure 3.2 shows two models that can help you imagine the structure of an atom—in this case, helium (He). In the first model (**Figure 3.2A**) helium's two electrons are shown circling the nucleus in a fixed orbit depicted as a ring. Although this model

LO 3.1.3

Figure 3.2 Two Models of Atomic Structure

(A) In this model, the electrons of helium are shown in fixed orbits, depicted as rings, at a precise distance from the nucleus, somewhat like planets orbiting the sun. (B) In this more accurate model, the electrons of carbon are shown in the variety of locations they would have at different distances from the nucleus over time.

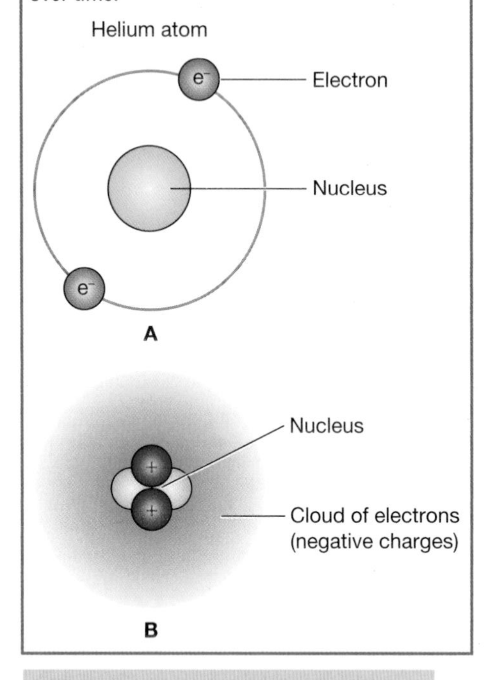

atomic mass units (amu) (daltons)

is helpful in visualizing atomic structure, in reality, electrons do not travel in fixed orbits, but whiz around the nucleus erratically in a so-called electron cloud. **Figure 3.2B** attempts to address this misconception more accurately by drawing in the electron cloud.In this chapter we will use the format of depicting the electrons in a fixed ring.

Atomic Number and Mass Number If atoms are composed of protons, neutrons, and electrons, and one proton is the same as another, whether it is found in an atom of carbon, sodium, or iron, then what gives an element its distinctive properties—what makes carbon so different from sodium or iron? The answer is the *unique quantity* of protons each contains. Carbon, by definition, is an element whose atoms contain six protons. No other element has exactly six protons in its atoms. Moreover, *all* atoms of carbon, whether found in your liver or in a lump of coal, contain six protons. Thus, the **atomic number**, the number of protons in the nucleus of the atom, identifies the element. Because an atom usually has the same number of electrons as protons, the atomic number identifies the usual number of electrons as well. The **atomic weight** of the atom is its weight, measured in **atomic mass units (amu)** Because electrons have so much less mass than protons and neutrons, the atomic weight is roughly equal to the number of protons and neutrons plus a little weight from the electrons. These numbers, as well as the symbol and name of each element can be seen in the periodic table of elements (**Figure 3.3A**). The elements are arranged in order of their atomic number, with hydrogen and helium, the elements with the least mass, at the top of the table.

In their most common form, many elements also contain the same number of neutrons as protons. The most common form of carbon, for example, has 6 neutrons as well as 6 protons, for a total of 12 subatomic particles in its nucleus. Since an atom's atomic number is its number of protons, and its weight is roughly equal to the number of protons and neutrons, you can see in **Figure 3.3B** that atomic weight is approximately twice the value of the atomic number. The number of protons and electrons in an element are equal. The numbers of protons and neutrons may be equal for some elements but are not equal for all.

Learning Check

1. Calcium [Ca] has the atomic number of 20 and atomic mass of 40.078u. Which of the following subatomic particles would you find spinning around the nucleus?
   a. Protons b. Neutrons c. Electrons
2. Calcium [Ca] has the atomic number of 20 and atomic mass of 40.078u. What can you know from the atomic number? Please select all that apply.
   a. The number of protons in a calcium atom
   b. The number of neutrons in a calcium atom
   c. The number of electrons in a calcium atom
3. Calcium [Ca] has the atomic number of 20 and atomic mass of 40.078u. What do atomic mass and atomic number tell you? Please select all that apply.
   a. The number of protons in a calcium atom
   b. The number of neutrons in a calcium atom
   c. The number of electrons in a calcium atom
4. Carbon [C] has the atomic number of 6 and atomic mass of 12.009u. Which of the following best describes the number of protons, neutrons, and electrons in a single carbon atom?
   a. 6 protons, 6 neutrons, 6 electrons
   b. 6 protons, 12 neutrons, 6 electrons
   c. 12 protons, 6 neutrons, 6 electrons
   d. 12 protons, 12 neutrons, 6 electrons

Figure 3.3 The Periodic Table of the Elements

(A) The elements are organized in the periodic table by atomic number. (B) Each element is found within its own box. The atomic number, atomic weight, name, and symbol are the most important pieces of information.

PERIODIC TABLE

Solids
Liquids
Gasses
Artificially Prepared

Period																		
1	1 H Hydrogen 1.008*																	2 He Helium 4.002602
2	3 Li Lithium 6.94*	4 Be Beryllium 9.0121831											5 B Boron 10.81*	6 C Carbon 12.011*	7 N Nitrogen 14.007*	8 O Oxygen 15.999*	9 F Fluorine 18.998403163	10 Ne Neon 20.1797
3	11 Na Sodium 22.98976928	12 Mg Magnesium 24.305*											13 Al Aluminum 26.9815385	14 Si Silicon 28.085*	15 P Phosphorus 30.973761998	16 S Sulfur 32.06*	17 Cl Chlorine 35.45*	18 Ar Argon 39.948
4	19 K Potassium 39.0983	20 Ca Calcium 40.078	21 Sc Scandium 44.955908	22 Ti Titanium 47.867	23 V Vanadium 50.9415	24 Cr Chromium 51.9961	25 Mn Manganese 54.938044	26 Fe Iron 55.845	27 Co Cobalt 58.933194	28 Ni Nickel 58.6934	29 Cu Copper 63.546	30 Zn Zinc 65.38	31 Ga Gallium 69.723	32 Ge Germanium 72.630	33 As Arsenic 74.921595	34 Se Selenium 78.971	35 Br Bromine 79.904*	36 Kr Krypton 83.798
5	37 Rb Rubidium 85.4678	38 Sr Strontium 87.62	39 Y Yttrium 88.90584	40 Zr Zirconium 91.224	41 Nb Niobium 92.90637	42 Mo Molybdenum 95.95	43 Tc Technetium (98)	44 Ru Ruthenium 101.07	45 Rh Rhodium 102.90550	46 Pd Palladium 106.42	47 Ag Silver 107.8682	48 Cd Cadmium 112.414	49 In Indium 114.818	50 Sn Tin 118.710	51 Sb Antimony 121.760	52 Te Tellurium 127.60	53 I Iodine 126.90447	54 Xe Xenon 131.293
6	55 Cs Cesium 132.9054520*	56 Ba Barium 137.327		72 Hf Hafnium 178.49	73 Ta Tantalum 180.94788	74 W Tungsten 183.84	75 Re Rhenium 186.207	76 Os Osmium 190.23	77 Ir Iridium 192.217	78 Pt Platinum 195.084	79 Au Gold 196.966569	80 Hg Mercury 200.592	81 Tl Thallium 204.38*	82 Pb Lead 207.2	83 Bi Bismuth 208.98040	84 Po Polonium (209)	85 At Astatine (210)	86 Rn Radon (222)
7	87 Fr Francium (223)	88 Ra Radium (226)		104 Rf Rutherfordium (267)	105 Db Dubnium (268)	106 Sg Seaborgium (271)	107 Bh Bohrium (270)	108 Hs Hassium (269)	109 Mt Meitnerium (278)	110 Ds Darmstadtium (281)	111 Rg Roentgenium (282)	112 Cn Copernicium (285)	113 Uut Ununtrium (284)	114 Uuq Ununquadium (289)	115 Uup Ununpentium (288)	116 Uuh Ununhexium (293)	117 Uuo Ununoctium (294)	118 Uuo Ununoctium (294)

Lanthanides	57 La Lanthanum 138.90547	58 Ce Cerium 140.116	59 Pr Praseodymium 140.90766	60 Nd Neodymium 144.242	61 Pm Promethium (145)	62 Sm Samarium 150.36	63 Eu Europium 151.964	64 Gd Gadolinium 157.25	65 Tb Terbium 158.92535	66 Dy Dysprosium 162.500	67 Ho Holmium 164.93033	68 Er Erbium 167.259	69 Tm Thulium 168.93422	70 Yb Ytterbium 173.045	71 Lu Lutetium 174.9668
Actinides	89 Ac Actinium (227)	90 Th Thorium 232.0377	91 Pa Protactinium 231.03588	92 U Uranium 238.02891	93 Np Neptunium (237)	94 Pu Plutonium (244)	95 Am Americium (243)	96 Cm Curium (247)	97 Bk Berkelium (247)	98 Cf Californium (251)	99 Es Einsteinium (252)	100 Fm Fermium (257)	101 Md Mendelevium (258)	102 No Nobelium (259)	103 Lr Lawrencium (266)

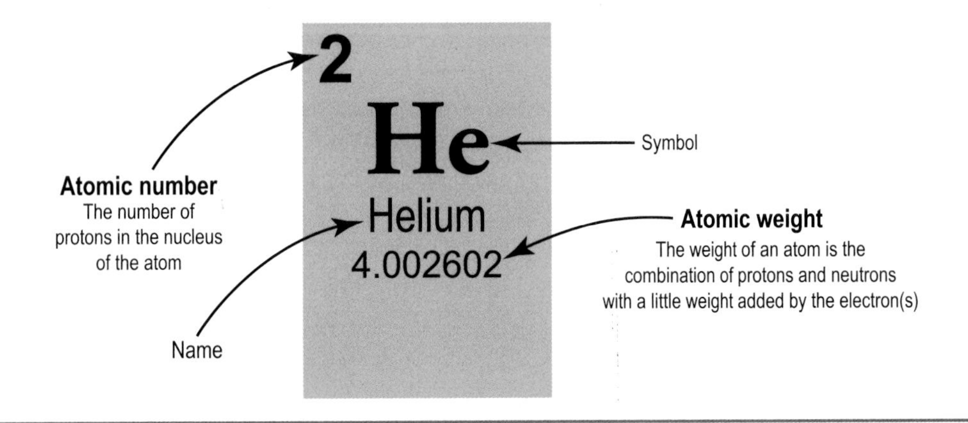

atomic weight (atomic mass)

LO 3.1.4

Figure 3.4 Isotopes of Hydrogen

Protium is by far the most abundant isotope of hydrogen in nature. It has one proton and no neutrons and is designated as ^{1}H. Deuterium has one proton and one neutron and is designated as ^{2}H. Tritium has one proton and two neutrons and is designated as ^{3}H.

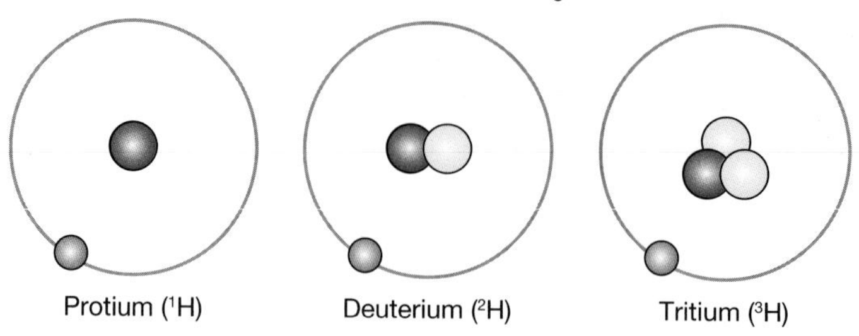

Isotopes Each element has a unique number of protons, and typically the number of protons and neutrons are the same. However, some elements can have unequal numbers of protons and neutrons. These variations are known as *isotopes*. An **isotope** is one of the various forms of an element, distinguished from one another by different numbers of neutrons. Examine the isotopes of hydrogen illustrated in **Figure 3.4**. All three of these atoms have one proton in their nucleus (illustrated in dark pink) and one electron in orbit (illustrated in light purple). Deuterium is the isotope of hydrogen that follows the standard format of an equal number of neutrons (illustrated in light yellow) to the protons and electrons. Protium is the most abundant isotope of hydrogen in nature; it has one proton and one electron but zero neutrons. Another isotope of hydrogen has two neutrons, but still one proton.

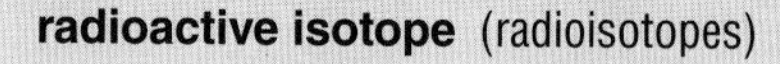

radioactive isotope (radioisotopes)

Some isotopes are unstable, meaning that they may shed their subatomic particles. These unstable forms of elements are known as **radioactive isotopes** or radioisotopes. Of the three forms of hydrogen illustrated in Figure 3.4, tritium is a radioactive isotope. For more information, see the Digging Deeper box on radioactivity.

DIGGING DEEPER:
Radioisotopes

Radioisotopes emit their unstable subatomic particles in predictable ways. As these subatomic particles disseminate, they can destabilize and cause damage to nearby molecules. For these reasons, radioactivity can be very harmful to human cells. This molecular and cellular damage can be harnessed in some cases to treat disease. Interventional radiologists are physicians who treat disease by using minimally invasive techniques involving radiation. Many conditions that could once only be treated with a lengthy and traumatic surgery can now be treated nonsurgically, reducing the cost, pain, length of hospital stay, and recovery time for patients. For example, in the past, the only options for a patient with one or more tumors in their liver were surgery and chemotherapy (the administration of drugs to treat cancer). Some liver tumors, however, are difficult to access surgically, and others could require the surgeon to remove too much of the liver. Moreover, chemotherapy is highly toxic to the liver, and certain tumors do not respond well to it anyway. In some such cases, an interventional radiologist can disrupt the blood supply that feeds the tumors, causing the tumor cells to die. In this procedure, called *radioembolization*, the radiologist inserts tiny radioactive "seeds" into the blood vessels that supply the tumors using a small needle. In the days and weeks following the procedure, the radiation emitted from the seeds destroys the vessels and directly kills the tumor cells in the vicinity of the treatment.

The subatomic particles emitted by radioisotopes can be detected and tracked by imaging technologies. One of the most advanced uses of radioisotopes in medicine is the positron emission tomography (PET) scanner, which detects the activity in the body of a very small injection of radioactive glucose, the simple sugar that cells use for energy. The PET camera reveals to the medical team which of the patient's tissues are taking up the most glucose. Thus, the most metabolically active tissues show up as bright "hot spots" on the images. PET can reveal some cancerous masses because cancer cells consume glucose at a high rate to fuel their rapid reproduction.

Different radioisotopes are used for different clinical purposes. As chemists continue to understand these molecules, the possible clinical applications will grow.

LO 3.1.5

Ions Recall that the number of protons that an atom has defines the element, so this number never changes. We just discussed that the number of neutrons can vary to form isotopes. While isotopes may be more or less stable, isotope formation does not change the charge of the atom because neutrons have no charge. When the number of electrons and the number of protons is the same, the atom is electrically neutral, it has no charge. But when an atom has either more or fewer electrons than it has protons, the electrical balance shifts and the atom has a net charge. This happens frequently to atoms without a full valence shell. An atom may participate in a chemical reaction that results in the donation or acceptance of one or more electrons; the atom then becomes positively or negatively charged. An atom that has an electrical charge—whether positive or negative—is an **ion**.

LO 3.1.6

Potassium [K], for instance, is an important element in all body cells. Its atomic number is 19. It has just one electron in its outermost shell. This characteristic makes potassium highly likely to participate in chemical reactions in which it donates one electron to another atom. After losing the electron, its outermost shell will have eight electrons, a much more stable confirmation. The loss will cause the positive charge of potassium's protons to be more influential than the negative charge of potassium's electrons. In other words, the resulting potassium ion will be slightly positive. A potassium ion is written as K^+, indicating that it has lost a single electron. A positively charged ion is known as a **cation** (**Figure 3.5A**).

Now consider fluorine [F], a component of bones and teeth. Its atomic number is 9, and it has seven electrons in its valence shell. Thus, it is highly likely to bond with other atoms in such a way that fluorine accepts one electron from another atom, filling its outermost shell to eight electrons, a more stable confirmation. When it does, its electrons will outnumber its protons by one, and it will have an overall negative charge. The ionized form of fluorine is called fluoride and is written as F^-. A negatively charged ion is known as an **anion** (**Figure 3.5B**). Note that some ions get a nickname, while others do not. Cations, such as potassium, are simply referred to as *potassium ions*, but anions such as fluorine and chlorine become *fluoride* and *chloride*.

Student Study Tip

Remember that cations are positively charged because the word contains a "t," just like a plus sign, for positive. "Anion" has 2 ns for negative.

Atoms that have more than one electron to donate or accept will end up with stronger positive or negative charges. A cation that has donated two electrons has a net charge of +2. Using magnesium (Mg) as an example, this can be written as Mg^{++} or Mg^{2+}. An anion that has accepted two electrons has a net charge of –2. The ionic form of selenium [Se], for example, is typically written as Se^{2-}.

3.1c The Behavior of Electrons

In the human body, atoms do not hang out independently. Rather, they are constantly reacting with other atoms to form more complex substances. To fully understand anatomy and physiology you must grasp how atoms engage with each other. The key is in understanding the behavior of electrons.

Although electrons do not follow rigid orbits a set distance away from the atom's nucleus, they do tend to stay within certain regions of space called electron shells. An **electron shell** is a layer of electrons that encircle the nucleus at a distinct distance.

The atoms of the elements found in the human body have from one to five electron shells, and all electron shells hold eight electrons except the first shell, which can only hold two. This configuration of electron shells is the same for all atoms. The number of shells of the atoms of an element depends on the total number of electrons it has. Hydrogen and helium have just one and two electrons, respectively, and therefore these atoms have only one electron shell (**Figure 3.6A**). A second shell is necessary to hold the electrons in all elements larger than hydrogen and helium such as carbon (**Figure 3.6B**). Atoms with more than 10 electrons (such as sodium)

Figure 3.5 **Potassium and Fluorine**

(A) Potassium has 19 electrons, with just one lonely electron in its outermost shell. Potassium is likely to engage in reactions in which it donates that electron to another atom. In its new state, potassium has 18 electrons, a more stable 8-electron outer shell, and a net positive charge. (B) Fluorine has nine electrons, with seven electrons in its outermost shell. Fluorine is likely to engage in reactions in which it accepts an electron from another atom to fill up its outermost shell. It now has a net negative charge because it has one more electron than it has protons.

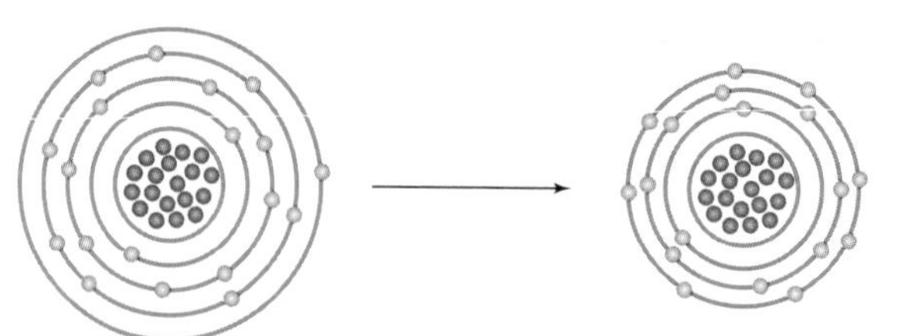

Potassium (K) has 19 protons and 19 electrons, and no charge. The electron in its outermost shell is unstable.

Potassium gives up its outer electron, now it has 19 protons and 18 electrons, so it has positive charge and is written as K^+ or a potassium ion.

A

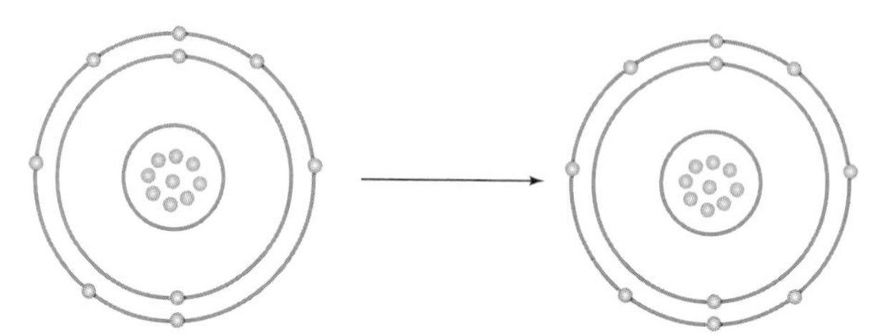

Fluorine (F) has 9 protons and 9 electrons, and no charge. Its outermost shell is unstable because it is almost full. Fluorine gains an electron in its outermost shell.

Now it has 9 protons and 10 electrons, so it has a net negative charge and is written as F^- or a fluoride ion.

B

Learning Check

1. Which of the following is an ion?
   a. Mg^{2+}
   b. O_2
   c. H_2O
   d. CO_2
2. True or False: Mass is gravity dependent, but weight is not.
   a. True
   b. False
3. Which of the following terms refers to variations of an element with more or fewer neutrons than protons?
   a. Ion
   b. Molecule
   c. Isotope
   d. Cation
4. Which of the following terms accurately identifies [Na^{2+}]? Please select all that apply.
   a. Cation
   b. Anion
   c. Ion
   d. Isotope

Figure 3.6 Electron Shells

Electrons orbit the atomic nucleus at distinct levels of energy called electron shells. (A) With one electron, hydrogen only half-fills its electron shell. Helium also has a single shell, but its two electrons completely fill it. (B) The electrons of carbon completely fill its first electron shell, but only half-fills its second. (C) Sodium, with 11 electrons, fills both of its first 2 electron shells and has just 1 electron in its valence shell.

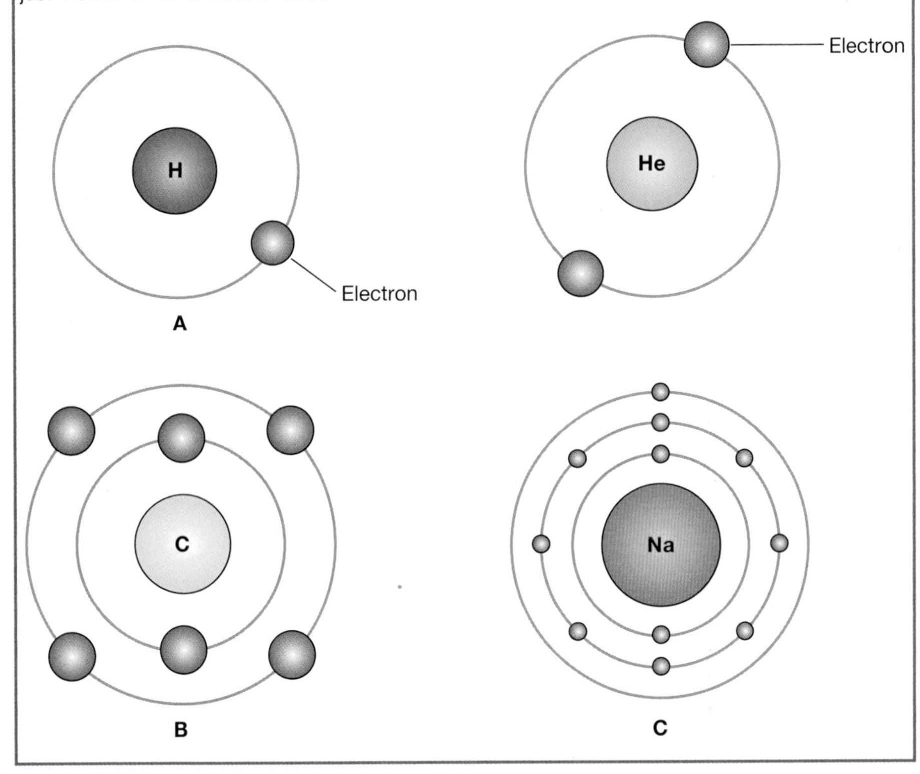

require more than two shells (**Figure 3.6C**). **Valence shell** is the term for an atom's outermost electron shell.

valence shell (outermost electron shell)

LO 3.1.7

The factor that most strongly governs the tendency of an atom to participate in chemical reactions is the number of electrons in its valence shell. All atoms are most stable when their valence shell is happily full. If the valence shell is not full, the atom is reactive, meaning it will tend to react with other atoms in ways that make the valence shell full. Consider hydrogen, with its one electron only half-filling its valence shell. This single electron is likely to be drawn into relationships with the atoms of other elements, so that hydrogen's single valence shell can be stabilized.

In another example, oxygen, with six electrons in its valence shell, is likely to react with other atoms in a way that results in the addition of two electrons to oxygen's valence shell, bringing the number to eight. When two hydrogen atoms each share their single electron with oxygen, so that oxygen fills its valence shell and covalent bonds are formed, resulting in a molecule of water, H_2O.

In nature, atoms of one element tend to join with atoms of other elements in characteristic ways. Learning these common formats allows us to predict the behavior of the elements. **Figure 3.7** reexamines the periodic table. Here we can see that the organization of the table is based on the valence shell of the different elements. Within each vertical column of the table the elements are arranged by the number of electrons their valence shell needs in order to be filled and stabilized. For example, carbon commonly fills its valence shell by linking up with four atoms of hydrogen. In so doing, the

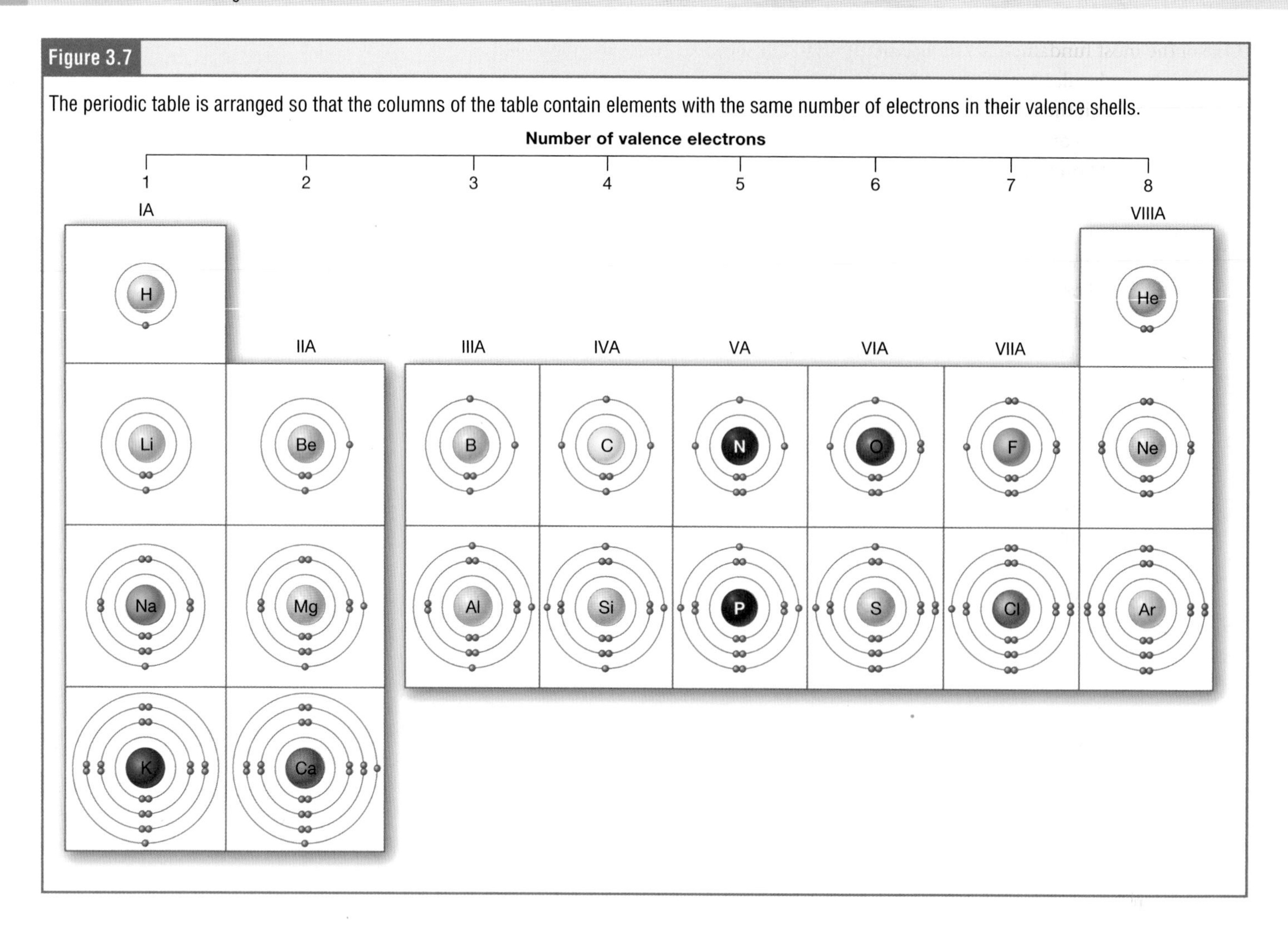

Figure 3.7

The periodic table is arranged so that the columns of the table contain elements with the same number of electrons in their valence shells.

two elements form the simplest of organic molecules, methane, which also is one of the most abundant and stable carbon-containing compounds on Earth. As mentioned previously, another example is water; oxygen needs two electrons to fill its valence shell. It commonly interacts with two atoms of hydrogen, forming H_2O.

3.2 Chemical Bonds

Learning Objectives: By the end of this section, you will be able to:

3.2.1 Explain the mechanism of each type of chemical bond and provide biologically significant examples of each: covalent, ionic, and hydrogen bonds.

3.2.2 List the following types of bonds in order by relative strength: nonpolar covalent, polar covalent, ionic, and hydrogen bonds.

3.2.3 Compare and contrast nonpolar covalent and polar covalent bonds.

One of the most fundamental and useful topics to master in anatomy and physiology may be molecular bonding. A **bond** is an electrical attraction that holds atoms together. Through the formation of bonds, atoms are held together in **molecules** (a grouping of two or more atoms held together by chemical bonds). A molecule may be composed of bonded atoms of the same element, as in the case of O_2 (oxygen). Oxygen as an atom has six electrons in its outer shell, so it is more stable when sharing a pair of electrons with another atom of oxygen or with another element such as H_2O (water). When a molecule is made up of two or more atoms of different elements, it is called a chemical **compound**.

Three types of chemical bonds are important in the human body, because they hold together substances that are used by the body for critical aspects of homeostasis, signaling, and energy production, to name just a few important processes. These three types are ionic bonds, covalent bonds, and hydrogen bonds.

3.2a Ionic Bonds

Have you ever heard the phrase "opposites attract"? Electrical charges certainly do! This is an important concept to keep in mind whenever we are working to understand the behavior of ions.

The opposite charges of cations and anions attract the atoms into such close proximity that they form an ionic bond. An **ionic bond** is an ongoing, close association between ions of opposite charge. The most familiar example of ionic bonding is the table salt you sprinkle on your food. As shown in Figure 3.8, sodium commonly donates an electron to chlorine, becoming the cation Na^+. When chlorine accepts the electron, it becomes the chloride anion, Cl^-. With their opposing charges, these two ions strongly attract each other forming NaCl, or salt.

LO 3.2.1

3.2b Covalent Bonds

Unlike ionic bonds formed by the attraction between a cation's positive charge and an anion's negative charge, molecules formed by a **covalent bond** share electrons in a mutually stabilizing relationship. Like next-door neighbors whose kids hang out first at one home and then at the other, the atoms do not lose or gain electrons permanently. Instead, the electrons move back and forth between the atoms. Because of the close sharing of pairs of electrons (one electron from each of two atoms), covalent bonds are stronger than ionic bonds.

LO 3.2.2

LO 3.2.3

Nonpolar and Polar Covalent Bonds Figure 3.9 shows several common types of covalent bonds. Notice that the two covalently bonded atoms typically share one or two electron pairs. The important concept to take from this is that in covalent bonds, electrons are shared to fill the valence shells of both atoms, ultimately stabilizing both of the atoms involved. In a single covalent bond, a single pair of electrons is shared between two atoms, while in a double covalent bond, two pairs of electrons (two electrons from one atom and two electrons from the other) are shared between two atoms. There even are triple covalent bonds, where three atoms are shared.

You can see that the covalent bonds shown in Figure 3.9 are balanced. The sharing of the negative electrons is relatively equal, as they are anchored between the atoms by the positive pull of the protons on either side. Let's contrast that with the water molecule illustrated in Figure 3.10. The water molecule has three parts: one atom of oxygen, the nucleus of which contains eight protons, and two hydrogen atoms, each hydrogen nucleus containing only one proton. Because every proton exerts an identical positive charge, a nucleus that contains eight protons exerts a charge eight times greater than a nucleus that contains one proton. This means that the negatively charged electrons

Figure 3.8 Ionic Bonding

(A) Sodium readily donates the solitary electron in its valence shell to chlorine, which needs only one electron to have a full valence shell. (B) The opposite electrical charges of the resulting sodium cation and chloride anion result in the formation of a bond of attraction called an ionic bond. (C) The attraction of many sodium and chloride ions results in the formation of large groupings called crystals.

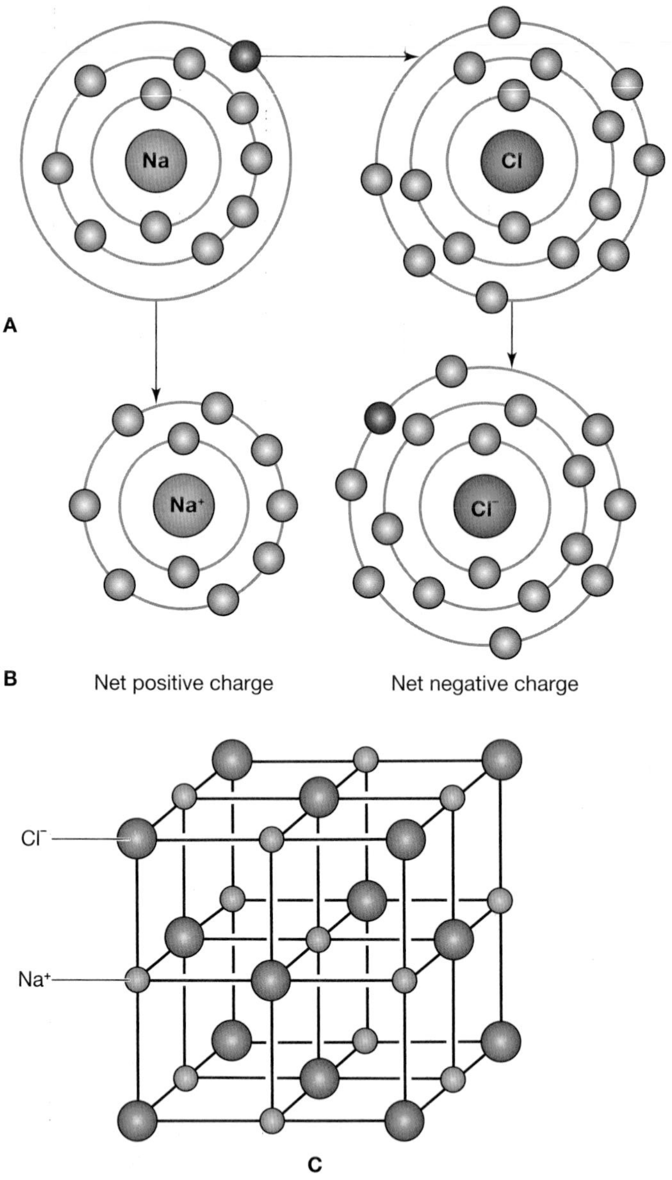

present in the water molecule are more strongly attracted to the oxygen nucleus than to the hydrogen nuclei.

Groups of people with completely opposite views on a particular issue are often described as "polarized." In chemistry, a **polar molecule** is a molecule that has one end that is more positive and one end that is more negative due to the unbalanced sharing of electrons. This is why covalently bonded molecules, such as those in Figure 3.10, that are electrically balanced are referred to as **nonpolar molecules**; that is, no region of the molecule is either more positive or more negative than any other.

Figure 3.9 Covalent Bonding

(A) A single covalent bond joins two atoms of hydrogen. (B) A double covalent bond joins two atoms of oxygen. An atom of oxygen has six electrons in its valence shell; therefore, two more would make it stable. Two atoms of oxygen achieve stability by sharing two pairs of electrons in a double covalent bond. (C) Two double covalent bonds allow an atom of carbon, with four electrons in its valence shell, to achieve stability by sharing two pairs of electrons each with two atoms of oxygen.

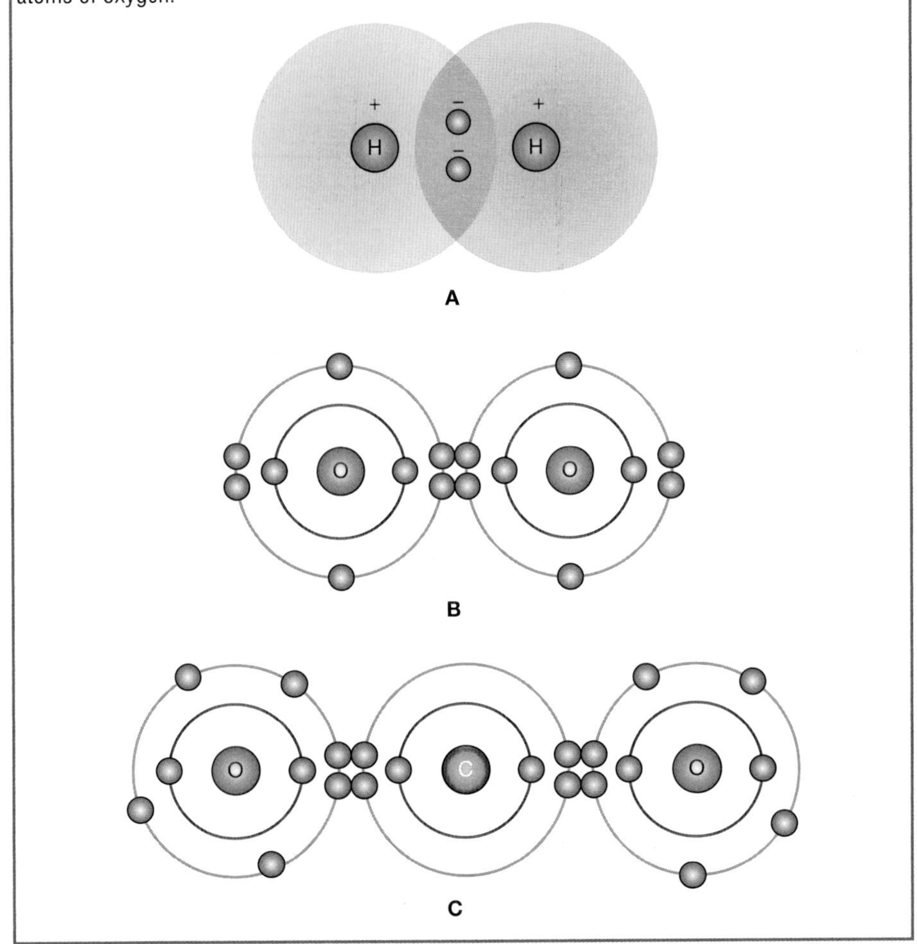

We use the terms *polar* and *nonpolar* to refer both to the bond as well as the molecule as a whole; that is, water (H_2O) is a polar molecule, while oxygen gas (O_2; **Figure 3.10B**) is a nonpolar molecule. The negatively charged regions of polar molecules are referred to as "partial charges" because the strength of the charge is less than one full electron and noted with Greek letter delta (δ^-). The positively charged regions of polar molecules are also indicated with the Greek letter delta (δ^+).

Remember that opposites attract, so charged regions of polar molecules are highly likely to interact with charged regions of other polar molecules. Within the human body, which is largely made up of water molecules, these interactions among polar molecules, called *hydrogen bonds*, are incredibly important.

3.2c Hydrogen Bonds

A **hydrogen bond** is a weak attraction that occurs between the partially positive hydrogen atom (δ^+) within a water molecule and a partially negative region (δ^-) of a neighboring polar molecule (**Figure 3.11A**). In some cases, as we will learn in Section 3.5, hydrogen bonds form within the same molecule.

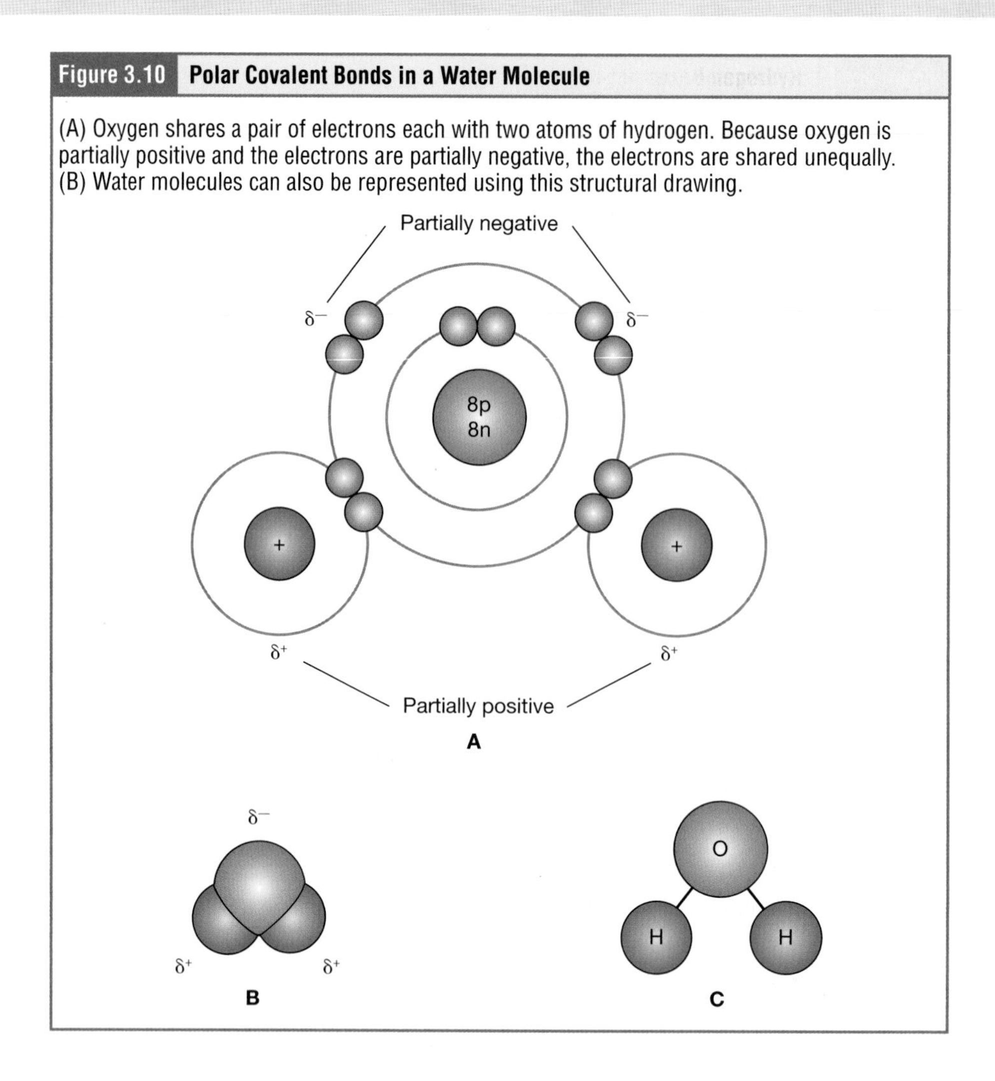

Figure 3.10 Polar Covalent Bonds in a Water Molecule

(A) Oxygen shares a pair of electrons each with two atoms of hydrogen. Because oxygen is partially positive and the electrons are partially negative, the electrons are shared unequally. (B) Water molecules can also be represented using this structural drawing.

The most common example of hydrogen bonding in the natural world occurs between molecules of water. It happens before your eyes whenever two raindrops merge into a larger bead, such as the beads of water seen in **Figure 3.11B**. Hydrogen bonds are individually weak and easily broken. We usually denote that weakness with a dotted line when we illustrate them, as opposed to the solid line used to illustrate the much stronger covalent bond. While they are weak individually, hydrogen bonds are strong collectively. **Figure 3.11C** is a photo of a small lizard that is capable of walking across the surface of water, rather than falling in, held up by the cohesion among the water molecules.

Polar water molecules also strongly attract other polar molecules in the same way that they are attracted to each other (**Figure 3.12A**). Water molecules are also attracted to charged molecules and the charge on ions (**Figure 3.12B**). This explains why NaCl (table salt) dissolves so readily in water. Water readily dissolves all sorts of polar or charged molecules, forming **solutions**. Water cannot, however, form hydrogen bonds with nonpolar molecules. In fact, water molecules repel molecules with nonpolar covalent bonds, like fats, lipids, and oils. You can demonstrate this with a simple kitchen experiment: pour a teaspoon of vegetable oil, a compound formed by nonpolar covalent bonds, into a glass of water. Instead of instantly dissolving in the water, the oil forms a distinct bead because the polar water molecules repel the nonpolar oil.

Figure 3.11 Hydrogen Bonds between Water Molecules

(A) Hydrogen bonds occur between the partially positive charge on the hydrogen atoms and the partially negative charge on the oxygen atoms of neighboring water molecules. Hydrogen bonds are relatively weak, and therefore are indicated with a dotted (rather than a solid) line. (B) Hydrogen bonds hold water molecules together, a property known as water cohesion. (C) While individual hydrogen bonds are weak, hydrogen bonds among many water molecules simultaneously can be quite strong. Here a Green Basilisk or "Jesus Christ" lizard walks quickly across the surface of water in a pond due to the lizard's light weight and the cohesive property of the water.

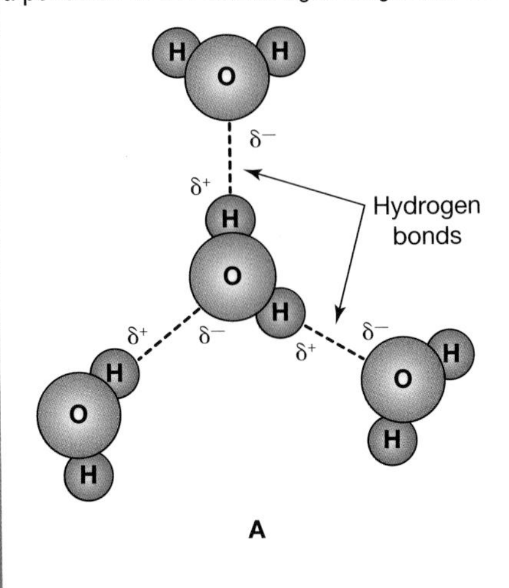

Figure 3.12

(A) Water forms hydrogen bonds with other polar molecules in much the same way that it does with other water molecules. The partially positive charge of a hydrogen atom within water is attracted to the partially negative charge of (in this case) nitrogen within a polar molecule of ammonia. (B) The partially positive and negative ends of water are also attracted to anions and cations, respectively.

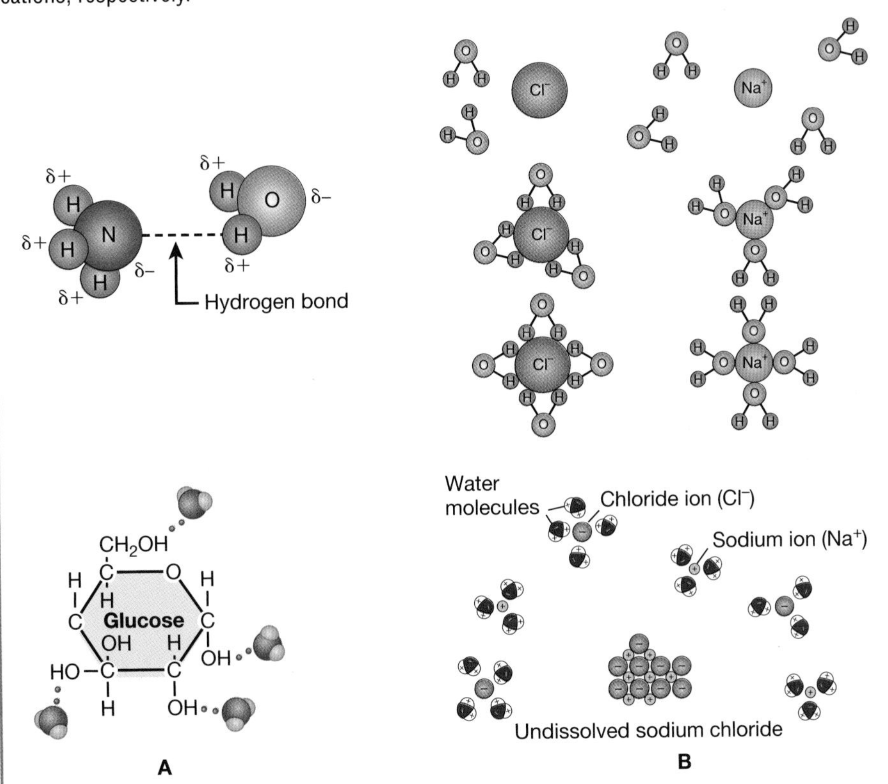

Hydrogen bonds are critical to the functioning of our molecules and cells. However, they are exquisitely vulnerable to environmental changes. Hydrogen bonds are likely to fall apart in the presence of acids or during changes in temperature.

Learning Check

1. Which of the following is true of an ionic bond?
   a. A bond between two ions of opposite charges
   b. Stronger than covalent bonds in water
   c. Must contain a hydrogen atom
2. In a water molecule, what component has a partial negative charge?
   a. Hydrogen
   b. Oxygen
3. Which of the following orders the bonds from weakest to strongest in a watery environment such as the human body?
   a. Hydrogen < ionic < covalent
   b. Covalent < ionic < hydrogen
   c. Ionic < hydrogen < covalent
   d. Ionic < covalent < hydrogen

3.3 Chemical Reactions

Learning Objectives: By the end of this section, you will be able to:

3.3.1* Compare and contrast kinetic and potential energy.

3.3.2* Compare and contrast endergonic and exergonic chemical reactions.

3.3.3* Describe and draw examples of the three basic types of chemical reactions.

3.3.4* List and explain several factors that influence the rate of reactions.

3.3.5 Define enzyme and describe factors that affect enzyme activity.

* Objective is not a HAPS Learning Goal.

One characteristic of a living organism is **metabolism**, which is the sum total of all of the chemical reactions that go on inside the body. The bonding processes you have learned thus far are key components in metabolism. Bonds are formed in order to build new molecules, and bonds are broken in the process of breaking down molecules. **Anabolism** is the process for building new molecules, and **catabolism** is the process for breaking molecules down.

3.3a The Role of Energy in Chemical Reactions

First, when thinking about atoms and molecules interacting, it is useful to define the different types of molecular energy. **Energy** is the capacity to do work, and the work of cells is to maintain order and structure as well as produce molecules such as hormones. The forms of energy relevant to anatomy and physiology are summarized in **Table 3.1**.

LO 3.3.1

- **Potential energy** is stored energy that can be released.
- **Kinetic energy** is the energy of motion. Once potential energy is released, it becomes kinetic energy.
- **Chemical energy** is a form of potential energy in which energy is stored in bonds between atoms and molecules. When those bonds are formed, chemical energy is invested, and when they break, chemical energy is released.

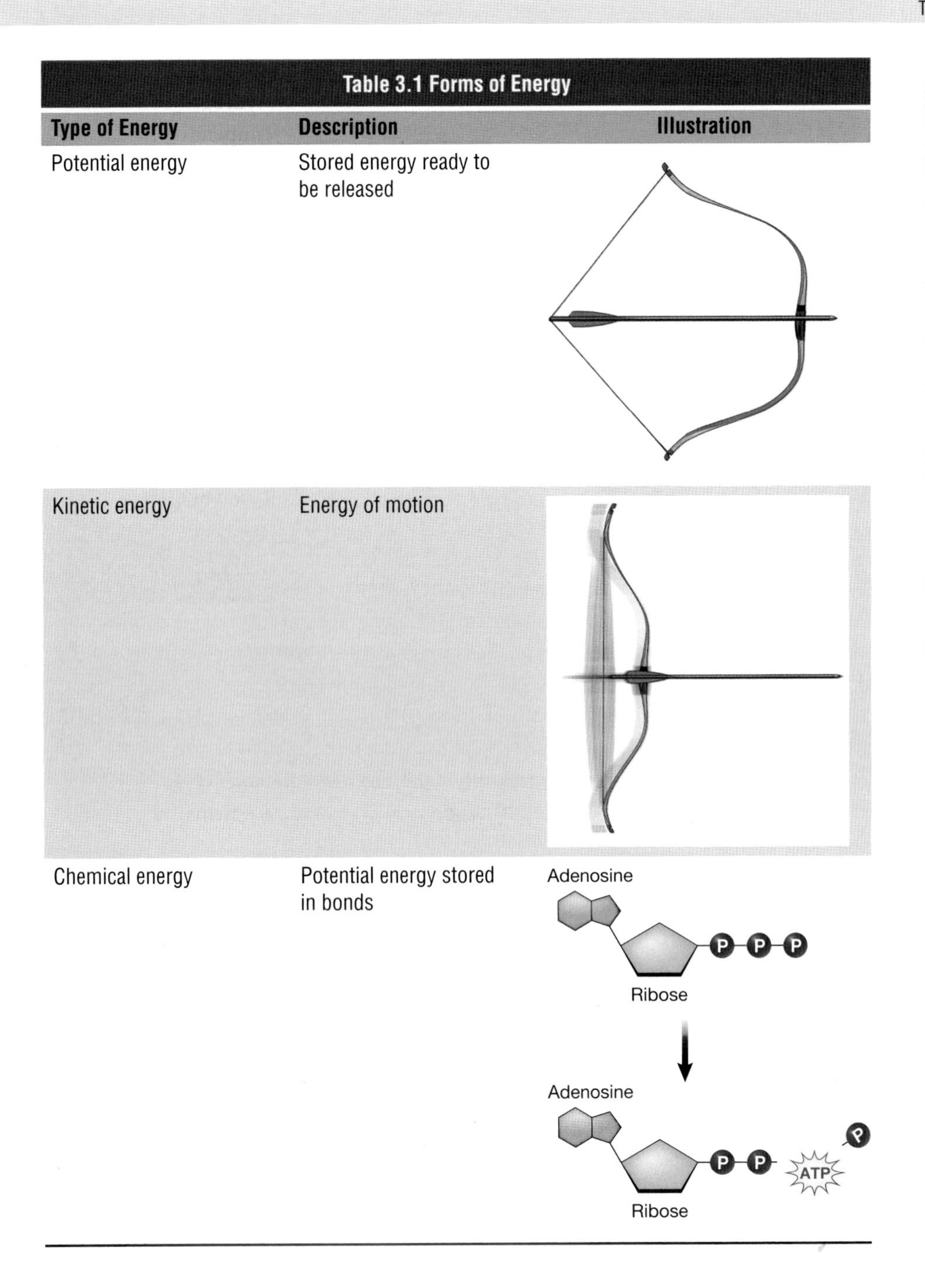

Table 3.1 Forms of Energy

Type of Energy	Description	Illustration
Potential energy	Stored energy ready to be released	
Kinetic energy	Energy of motion	
Chemical energy	Potential energy stored in bonds	

Student Study Tip

To remember that anabolism means "building," try thinking of a builder named Anna. Anna builds = anabolism!

Note that energy, regardless of its form, is never created nor destroyed; rather, it is converted from one form to another. When you eat breakfast before heading out the door for a hike, molecules in your breakfast are catabolized (broken down) and their chemical energy is released. Your muscle cells convert this released chemical energy into the kinetic energy of muscle contraction.

LO 3.3.2

Chemical reactions that release more energy than they absorb are characterized as **exergonic reactions**. The catabolism of the foods in your breakfast is an example. In contrast, chemical reactions that absorb more energy than they release are **endergonic reactions** (**Figure 3.13**). These reactions require energy input, and the resulting molecule stores chemical energy. If energy is neither created nor destroyed, where does the energy needed for endergonic reactions come from? In many cases, it comes from exergonic reactions.

Figure 3.13 Endergonic and Exergonic Reactions

Endergonic reactions require energy; the energy is stored within bonds as potential energy. Exergonic reactions release energy through the breaking of bonds. The most common example of these reactions in A&P is the formation and breakdown of the energy storage molecule, ATP.

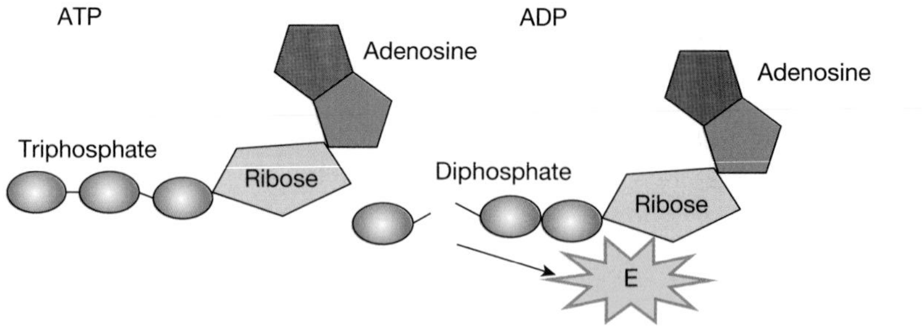

Student Study Tip

As shown here, practice how different types of chemical reactions run using variables, drawings, words, and so on to make sure you understand the mechanisms before adding molecules!

3.3b Characteristics of Chemical Reactions

All chemical reactions begin with a **reactant**, the general term for the one or more substances that enter into the reaction. Sodium and chloride ions, for example, are the reactants in the production of table salt. The one or more substances produced by a chemical reaction are called the **product**. In the figure in the "Anatomy of Chemical Reactions" feature, A and B represent reactants and AB represents the product.

Just as you can express mathematical calculations in equations such as $2 + 7 = 9$, you can use chemical equations to show how reactants become products. As in math, chemical equations proceed from left to right, but instead of an equals sign, they employ an arrow or arrows indicating the direction in which the chemical reaction proceeds. For example, the chemical reaction in which one atom of nitrogen and three atoms of hydrogen produce ammonia would be written as $N + 3H \rightarrow NH_3$. Correspondingly, the breakdown of ammonia into its components would be written as $NH_3 \rightarrow N + 3H$.

LO 3.3.3

Notice that, in the first example, a nitrogen [N] atom and three hydrogen [H] atoms bond to form a compound. This reaction forms bonds, so we can describe it as anabolic and it requires energy, which is then stored within the compound's bonds, so we can describe it as endergonic. Such reactions are referred to as *synthesis reactions.* A **synthesis reaction** is a chemical reaction that results in the synthesis (joining) of components that were formerly separate (part A in the "Anatomy of Chemical Reactions" feature). Again, nitrogen and hydrogen are reactants in a synthesis reaction that yields ammonia as the product. The general equation for a synthesis reaction is $A + B \rightarrow AB$.

In the second example (part B of the "Anatomy of Chemical Reactions" feature), ammonia is catabolized into its smaller components, and the potential energy that had been stored in its bonds is released. Such reactions are referred to as decomposition reactions. A **decomposition reaction** is a chemical reaction that breaks down a molecule into its constituent parts. The general equation for a decomposition reaction is: $AB \rightarrow A + B$.

The third, and the most complex reaction in part C of the "Anatomy of Chemical Reactions" feature is an **exchange reaction**, a chemical reaction in which both synthesis and decomposition occur, chemical bonds are both formed and broken, and chemical energy is absorbed, stored, and released. The exchange reaction shown can be summarized as: $AB + CD \rightarrow AC + BD$. Notice that AB and CD had to be broken down in decomposition reactions to produce these products, whereas A and C and B and D had to bond in synthesis reactions.

Anatomy of...

Chemical Reactions

Chemical reactions in the body can be broken into three types: synthesis, decomposition, and exchange reactions.

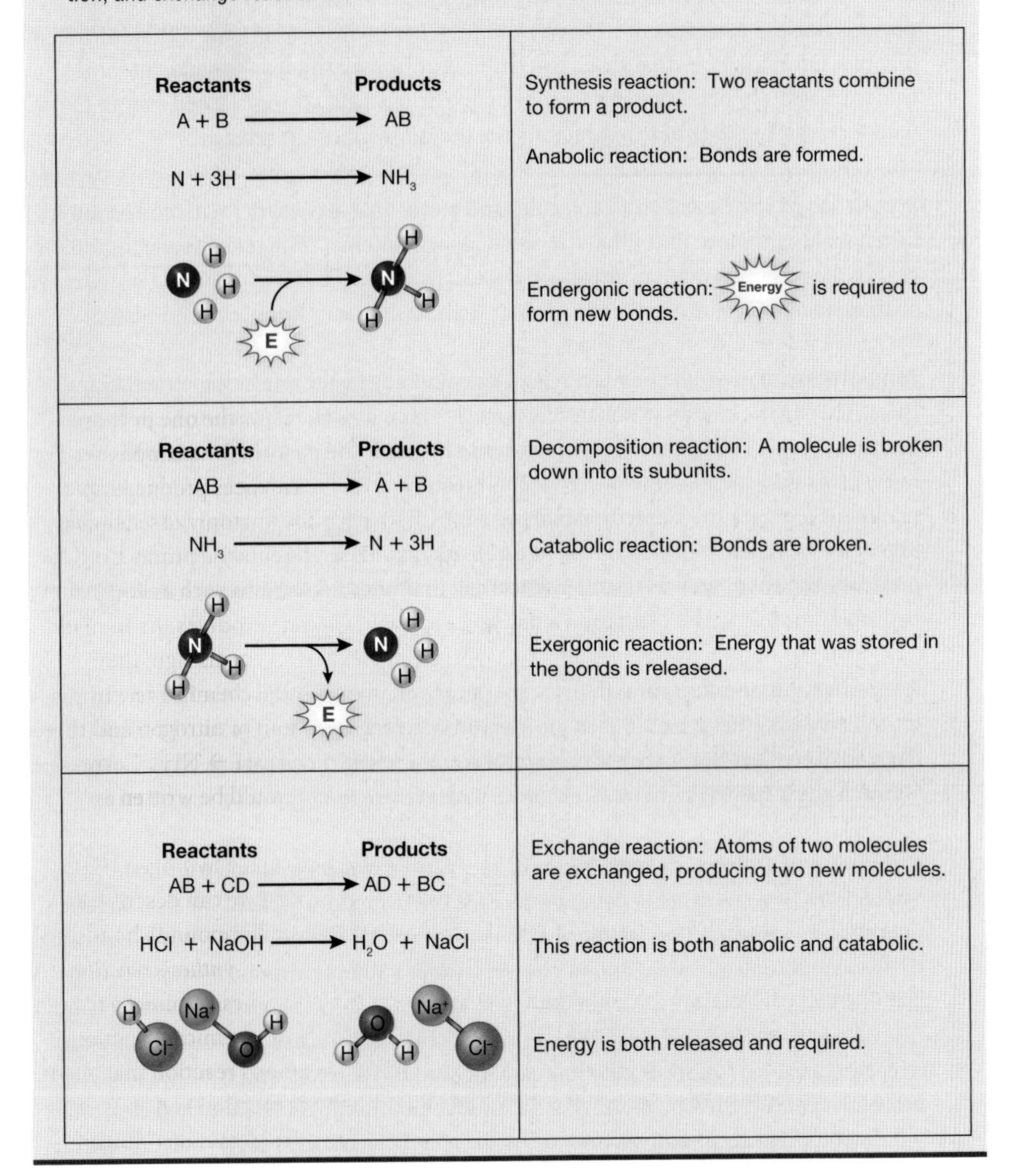

Reactants
Products
A + B → AB
N + 3H → NH3
Synthesis reaction: Two reactants combine to form a product.
Anabolic reaction: Bonds are formed.
Endergonic reaction: Energy is required to form new bonds.
Reactants
Products
AB → A + B
NH3 → N + 3H
Decomposition reaction: A molecule is broken down into its subunits.
Catabolic reaction: Bonds are broken.
Exergonic reaction: Energy that was stored in the bonds is released.
Reactants
Products
AB + CD → AD + BC
HCl + NaOH → H2O + NaCl
Exchange reaction: Atoms of two molecules are exchanged, producing two new molecules.
This reaction is both anabolic and catabolic.
Energy is both released and required.

In theory, any chemical reaction can proceed in either direction under the right conditions. Reactants may synthesize into a product that is later decomposed. This reversibility of a chemical reaction is indicated with a double arrow: $A + BC \rightleftarrows AB + C$. Still, in the human body, many chemical reactions do proceed in a predictable direction, either one way or the other. You can think of this more predictable path as the path of least resistance because, typically, the alternate direction requires more energy. In the watery environment of the human body, we tend to see primarily certain types of reactions that occur in water (dehydration synthesis and hydrolysis reactions).

Student Study Tip

We're warmer when we're running fast than when we are inactive. This is how we can remember that reactions run faster when temperature is higher. We can use this same trick for the diffusion rate, discussed in Chapter 4.

3.3c Factors Influencing the Rate of Chemical Reactions

If you pour vinegar into baking soda, the reaction is instantaneous; the concoction will bubble and fizz. But many chemical reactions take time. A variety of factors influence the rate of chemical reactions. This section, however, will consider only the most important factors in the reactions that take place within human bodies and cells. The most important thing to remember about the likelihood of a synthesis reaction to occur is that the reactants need to be close together in order to interact. That may seem obvious, but remember that these reactants are infinitely small and move randomly. In decomposition reactions, the likelihood of the reaction taking place has more to do with factors that contribute to breaking the bond(s) within the reactant.

LO 3.3.4

Properties of the Reactants Molecules and atoms that are smaller will tend to move faster, making it more likely that the reactants will interact. The phase (solid, liquid, or gas) of the reactants matters as well; molecules within a gas have higher kinetic energy, so these reactions tend to take place faster than reactions in liquid or solid phases.

Temperature Nearly all chemical reactions occur at a faster rate at higher temperatures. The kinetic energy of subatomic particles increases in response to increases in thermal energy. The higher the temperature, the faster the particles move, and the more likely they are to come in contact and react. For this reason, the human body maintains temperature homeostasis around 37°C (98.6°F), allowing for a predictable rate of kinetic energy and therefore rate of reactions. When the body deviates from the homeostatic range, such as during a fever, reaction rates increase but the stability of hydrogen bonds may be compromised.

Concentration and Pressure If just a few people are dancing at a club, they are unlikely to step on each others' toes. But as more and more people get up to dance, collisions are likely to occur. It is the same with chemical reactions: the more particles that are present within a given space, the more likely those particles are to bump into one another.

LO 3.3.5

Enzymes and Other Catalysts If chemical reactions are dependent on reactant interactions and reactants to collide with each other randomly based on their kinetic energy and environmental conditions, the rate of reaction could vary widely depending on those random collisions. But, of course, cellular and organismal function requires stability and predictability. A **catalyst** is a substance that increases the rate of a chemical reaction without itself changing during the reaction. You can think of a catalyst as a reaction enabler. Catalysts help increase the rate and force at which atoms, ions, and molecules collide, thereby increasing the probability that their valence shell electrons will interact. The term **activation energy** describes the barriers to the reaction taking place on its own (**Figure 3.14A**). Activation energy may refer to the reactants being physically close enough together to form bonds, or it may refer to the level of energy needed to break the bonds in the reactants.

The most important catalysts in the human body are enzymes. An **enzyme** is a protein that makes chemical reactions faster and more likely to occur by lowering activation energy. For example, if the activation energy in a synthesis reaction is due to two molecules being close enough together for long enough to form bonds, then an enzyme lowers the activation energy by holding them close together as illustrated in **Figure 3.14B**. Enzymes are critical to the body's healthy functioning. They assist, for example, with the breakdown of food and its conversion to energy and the building of many cellular molecules. In fact, most of the chemical reactions in the body are facilitated by enzymes.

Student Study Tip

Activation energy curves act like a ball being rolled up a hill: once over the hump, the ball will roll down without effort. A smaller hump leads to quicker/easier reactions/rolling; shortening the hump height explains enzyme function.

Learning Connection

Explain a Process

If you were teaching someone else, how would you explain activation energy? Can you think of a metaphor for it?

Figure 3.14 Enzymes and Activation Energy

Enzymes reduce the activation energy required for chemical reactions to occur. (A) Without an enzyme, the energy input needed for a reaction to begin is high and the reaction is unlikely to occur on its own. (B) With the help of an enzyme, less energy is needed for a reaction to begin and the reaction is more likely to occur and will occur faster. (C) Enzymes often reduce activation energy by holding two molecules together close enough and long enough for the reaction to take place.

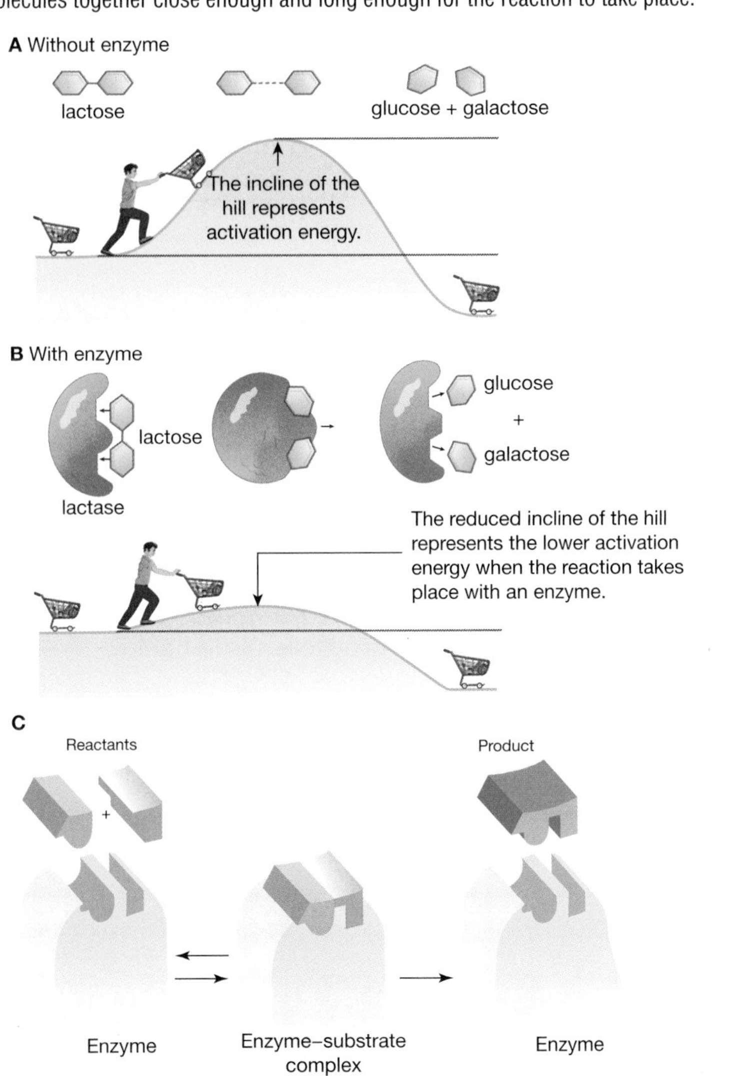
A Without enzyme
lactose
glucose + galactose
The incline of the hill represents activation energy.
B With enzyme
lactose
lactase
glucose
+
galactose
The reduced incline of the hill represents the lower activation energy when the reaction takes place with an enzyme.
C
Reactants
Product
Enzyme
Enzyme–substrate complex
Enzyme

Learning Check

1. The process of breaking down molecules is known as _________. The process of building new molecules is known as _________.
   a. Anabolism; metabolism
   b. Metabolism; catabolism
   c. Catabolism; anabolism
   d. Anabolism; catabolism
2. What kind of energy is chemical energy?
   a. Potential energy
   b. Kinetic energy

(*Continued*)

3. Which of the following does NOT influence the rate of chemical reaction?
   a. Increasing the temperature
   b. Increasing the concentration of products or reactants
   c. The size of the particles
   d. The molecular mass of the reactants
4. Which of the following increases the rate of chemical reactions?
   a. Decreasing the concentration of reactants
   b. Decreasing the concentration of products
   c. Increasing the concentration of catalysts

3.4 Inorganic Compounds Essential to Human Functioning

Learning Objectives: By the end of this section, you will be able to:

3.4.1* Compare and contrast organic and inorganic compounds and give examples of each.

3.4.2 Describe the physiologically important properties of water.

3.4.3 Compare and contrast the terms solution, solute, solvent, colloid suspension, and emulsion.

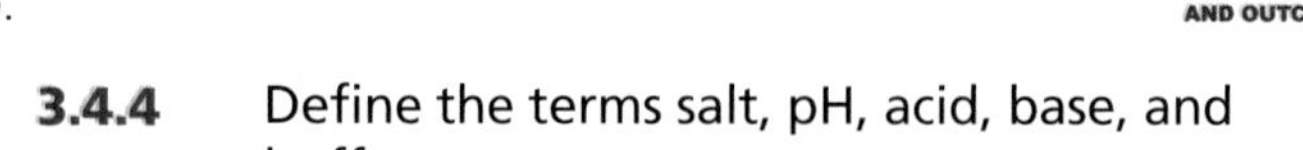

3.4.4 Define the terms salt, pH, acid, base, and buffer.

3.4.5 State the pH values for acidic, neutral, and alkaline (basic) solutions.

* Objective is not a HAPS Learning Goal.

The next two sections of the chapter cover the compounds important for the body's structure and function. In general, these compounds are either inorganic or organic. **Table 3.2** lists some of the common molecules and compounds referenced throughout this book.

LO 3.4.1

- An **inorganic compound** is a substance that does not contain both carbon and hydrogen. A great many inorganic compounds do contain hydrogen atoms, such as

Table 3.2 Commonly Used Chemical Formulas in A&P

Full Name	Chemical Formula
Oxygen	O_2
Carbon dioxide	CO_2
Potassium ions	K^+
Chloride ions	Cl^-
Calcium ions	Ca^{2+} (also expressed Ca^{++})
Glucose	$C_6H_{12}O_6$
Water	H_2O
Hydrochloric acid	HCl
Sodium chloride	NaCl

water (H_2O) and the hydrochloric acid (HCl) produced by your stomach. In contrast, only a handful of inorganic compounds contain carbon atoms. Carbon dioxide (CO_2) is one of the few examples.

- An **organic compound**, then, is a substance that contains both carbon and hydrogen. Organic compounds are synthesized via covalent bonds within living organisms, including the human body.

The following section examines the three groups of inorganic compounds essential to life: water, salts, and acids and bases. Organic compounds are covered later in the chapter.

3.4a Water

LO 3.4.2

Scientists estimate that between 50 and 70 percent of an adult's body is composed of water (this number varies pretty widely and is influenced by age, body composition, and the levels of hormones, particularly estrogen). This water is contained both within the cells and between the cells that make up tissues and organs. Its several roles make water indispensable to human functioning.

Water as a Lubricant and Cushion Water is a major component of many of the body's lubricating fluids. Just as oil lubricates the hinge on a door, water in our joints lubricates the tissues of those joints (explored in Chapter 10), and water inside of cells is critical for their function (explored in Chapter 4). Watery fluids help keep food flowing through the digestive tract and ensure that the movement of adjacent abdominal organs is friction free (Chapter 24).

Water also protects cells and organs from physical trauma, cushioning the brain within the skull, for example, and protecting the delicate nerve tissue of the eyes. Water cushions a developing fetus within the uterus as well. Water is fundamental to human life and plays many different roles in the human body:

- Water is the main component of blood and lymph, and serves a transportation role, moving cells, nutrients, hormones, and wastes around the body.
- Through sweating, water helps to bring elevated body temperatures down.
- A watery fluid acts as a cushion around the brain, protecting it from harm during sudden movements.
- Water is the solvent of the body for hydrophilic molecules. The behavior of hydrophobic molecules is defined by their interactions with the water.
- In our joints and around our organs, water-based fluids lubricate and reduce friction.

Water as a Heat Sink A heat sink is a substance or object that absorbs and dissipates heat but does not experience a corresponding increase in temperature. In the body, water absorbs the heat generated by chemical reactions without greatly increasing in temperature. Moreover, when the environmental temperature soars, the water stored in the body helps keep the body cool. Water brings heat with it as it changes from a liquid to a gas phase. This mechanism, known as *evaporative cooling*, explains how sweat helps the body to cool down once it has warmed past homeostatic range.

LO 3.4.3

Water as a Solvent For cells in the body to survive, they must be kept moist in a water-based liquid called a solution. In chemistry, a liquid **solution** consists of a **solvent** that dissolves a substance called a **solute**. If you were to stir a teaspoon of sugar into a cup of coffee, the coffee would act as the solvent and the sugar in this case would be the solute. As our bodies are largely composed of water, the concentration and characteristics of the solutes is an important focus in physiology.

Examine Figures 3.11 and 3.12. You can see how many different types of molecules can dissolve in water. Water is considered the "universal solvent" because of this. But the ability of molecules to be solvents in water is dependent on their physical properties, especially their charge and polarity since this determines their ability to form hydrogen bonds with water molecules (**Figure 3.15A**). Nonpolar molecules, with no partial or complete charges, are unable to form hydrogen bonds with water molecules and are therefore insoluble (**Figure 3.15B**). Molecules with both polar and nonpolar regions orient themselves so that their hydrophilic regions are oriented toward the water and their hydrophobic regions are protected from the water by their hydrophilic parts (**Figure 3.15C**). It is important to remember that essentially all of the body's chemical reactions occur among compounds dissolved in water.

Concentrations of Solutes There are a number of different, but related terms that we can use to describe solutions:

- The **concentration** of a given solute is the number of particles of that solute in a given space.
    - Example 1: oxygen makes up about 21 percent of atmospheric air.
    - Example 2: In the bloodstream of humans, glucose concentration is usually measured in milligrams (mg) of glucose per deciliter (dL) of blood; a healthy fasting adult average is about 70–100 mg/dL.

Another method of measuring the concentration of a solute is by its **molarity**, which is moles (M) of the molecules per liter (L). A mole of an element is its atomic weight, while a mole of a compound is the sum of the atomic weights of its components,

Figure 3.15 Summary of the Interactions Various Molecules Can Have with Water

(A) Polar or charged substances are hydrophilic; they readily form hydrogen bonds with water, dissolving readily into a watery solution. (B) Nonpolar substances are hydrophobic; they cannot form hydrogen bonds or dissolve in water. (C) Molecules with both hydrophilic and hydrophobic regions cluster within water to display their hydrophilic regions toward the water molecules, hiding their hydrophobic regions.

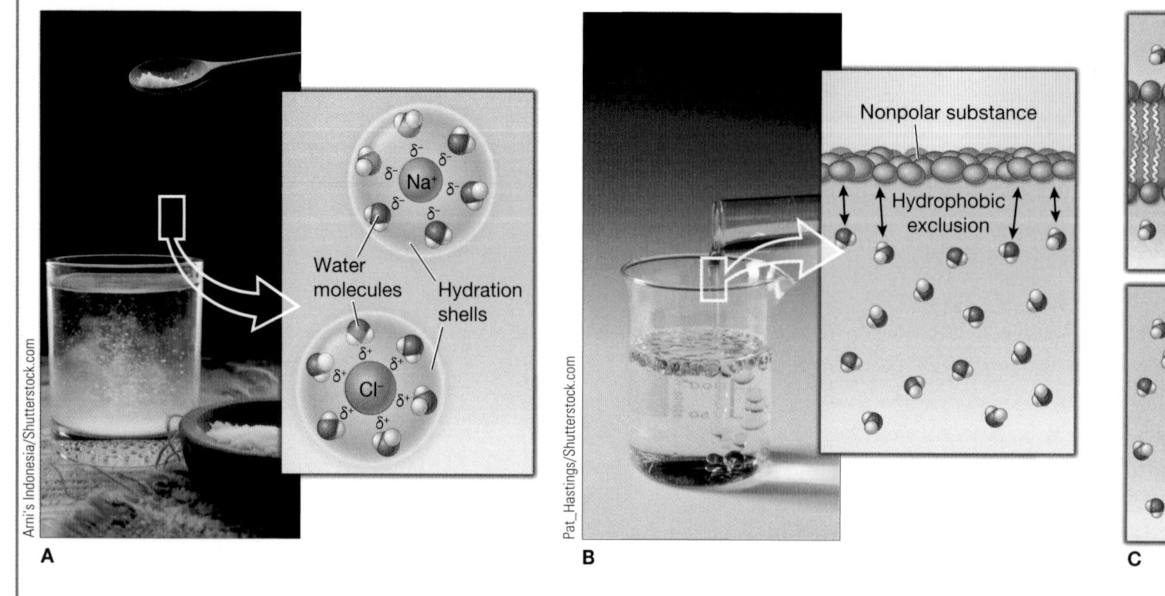

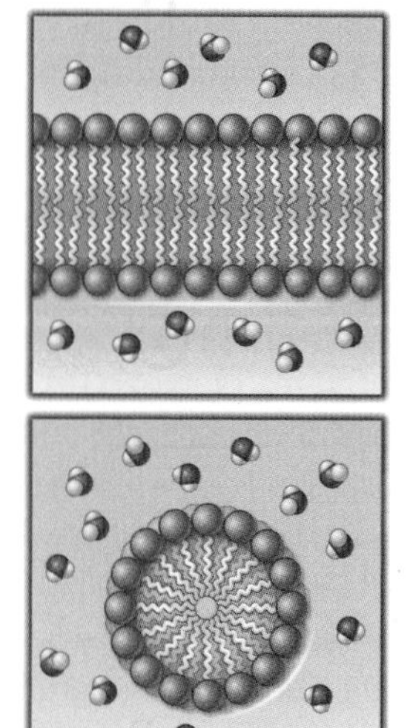

Cultural Connection

Heat Sink to Heat Stroke

Just as water in the body operates as a heat sink, moderating temperature change, water in the environment does the same thing. Air temperatures at the coasts change very little because they are moderated by water, but air temperatures in places with very little water—a desert, for example—can swing wildly without water around to absorb the heat. As Earth's average temperature climbs due to increasing effects of greenhouse gasses, the oceans are holding onto that heat, becoming warmer and warmer over time.

Some of the results of this environmental change include the melting of the polar ice caps and subsequent rising sea levels, the generation of stronger and more devastating storm systems and deadly heat waves, and even longer periods of seasonal allergies. Yes, allergies! Due to increases in global temperature, the number of days of the year with high pollen counts has increased by 21 days over the last 20 years. That's three extra weeks a year of sniffles and sneezes. These longer stretches of allergic symptoms can be very severe for the 10 million Americans with allergy-induced asthma.

Humans aren't the only organisms impacted by the warming oceans. The increased temperature has allowed insects with warm-weather preferences to expand their environments. The United States is now experiencing cases of insect-borne diseases previously unseen in our climate. Zika virus, Dengue fever, and West Nile virus are all examples. Environmental scientists estimate that by the 2080s Earth will be 5–10°F hotter than today's averages, and we can expect these impacts on human health to become more significant over time.

The Oceans Are Getting Warmer

Annual divergence of global ocean temperature from 20th century average (1880-2020)

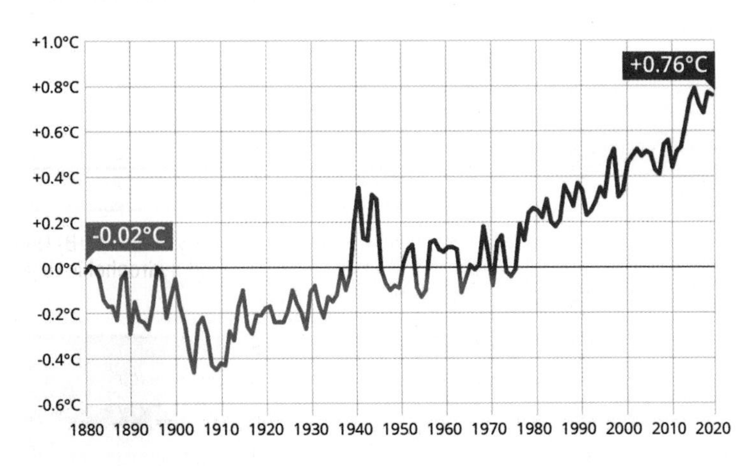

Ocean surface temperatures

Source: NOAA National Centers for Environmental Information (NCEI)

statista

called the *molecular weight*. An often-used example is calculating a mole of glucose, with the chemical formula $C_6H_{12}O_6$. Using the periodic table, the atomic weight of carbon [C] is 12.011 grams (g), and there are six carbons in glucose, for a total atomic weight of 72.066 g. Doing the same calculations for hydrogen [H] and oxygen [O], the molecular weight equals 180.156 g (the "gram molecular weight" of glucose). When water is added to make one liter of solution, you have one mole (1 M) of glucose.

A **colloid** is a mixture that is somewhat like a heavy solution. The solute particles consist of tiny clumps of molecules large enough to make the liquid mixture opaque. Familiar examples of colloids are milk and cream. In the thyroid glands, the thyroid hormone is stored as a thick protein mixture, also called a colloid.

A **suspension** is a liquid mixture in which a heavier substance is suspended temporarily in a liquid but settles out over time. This separation of particles from a suspension is called *sedimentation*. An example of sedimentation occurs in the blood. If left still, the red blood cells in a test tube will settle out of the watery portion of blood (known as *plasma*) over a period of time.

The Role of Water in Chemical Reactions Let's revisit our two main types of chemical reactions, but explore the role of water in them.

Student Study Tip

Water is a third wheel in certain reactions. When it leaves in a dehydration synthesis reaction, the two monomers link and get their alone time! When the third wheel arrives, the monomers are disrupted.

- One form of synthesis reactions, **dehydration synthesis** reactions take place when one reactant gives up an atom of hydrogen and another reactant gives up a hydroxyl group (OH) in the synthesis of a new product. In the formation of their covalent bond, a molecule of water is released as a byproduct (**Figure 3.16**).
- In one form of decomposition reaction, a **hydrolysis reaction**, a molecule of water disrupts a compound, breaking its bonds. The water is itself split into H and OH. One portion of the severed compound then bonds with the hydrogen atom, and the other portion bonds with the hydroxyl group.

These reactions are reversible and play an important role in the chemistry of organic compounds (which will be discussed shortly).

3.4b Salts

LO 3.4.4

Salts are formed when ions combine via their ionic bonds. NaCl, table salt, is the most familiar form of salt, but salts can form through the combination of many different

Figure 3.16 Dehydration Synthesis and Hydrolysis

Monomers, the building blocks of larger molecules, form polymers. (A) In dehydration synthesis reactions, two monomers join together via a covalent bond. In this type of reaction, one molecule gives up a hydroxyl group and the other gives up a hydrogen atom. The lost hydroxyl and hydrogen ions chemically bind to become a molecule of water. (B) In hydrolysis reactions, a covalent bond that joins two monomers is broken. A molecule of water splits and adds a hydrogen atom to one of the monomers and a hydroxyl group to the other.

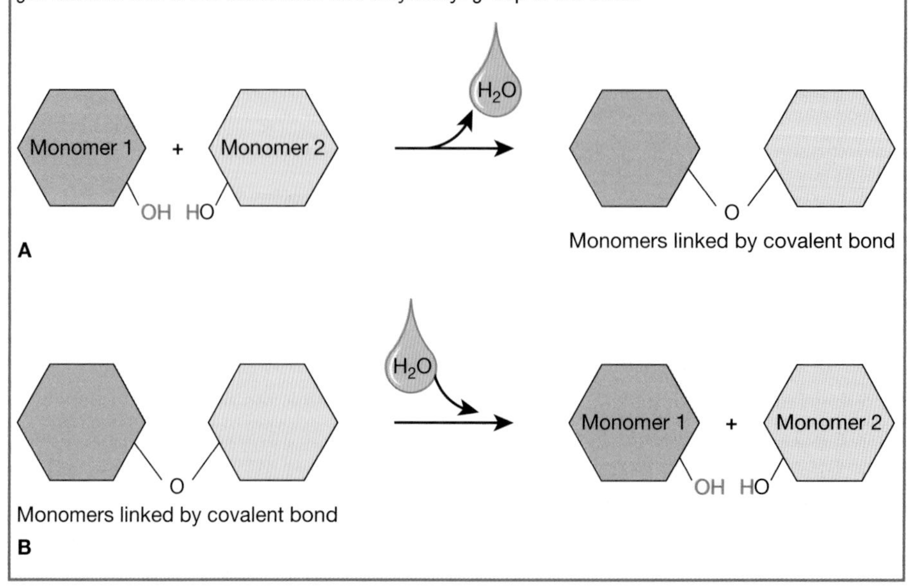

ions. Ions are drawn together by their opposite charges; however, polar water molecules tend to disrupt these ionic bonds. You can now define a **salt** as a substance that, when dissolved in water, dissociates into ions other than H^+ or OH^-. This fact is important in distinguishing salts from acids and bases, discussed next. Salts dissociate completely in water. The positive and negative regions on the water molecule (the hydrogen and oxygen ends, respectively) attract the negative chloride and positive sodium ions, pulling them away from each other. These ions are often referred to as **electrolytes**; particles that are capable of conducting electrical currents in solution. This property is critical to the function of ions in transmitting nerve impulses and prompting muscle contraction. Many other salts are important in the body. For example, bile salts produced by the liver help break apart dietary fats, and calcium phosphate salts form the mineral portion of teeth and bones.

3.4c Acids and Bases

Acids and bases, like salts, dissociate in water into electrolytes. Acids and bases can dramatically change the properties of the solutions in which they are dissolved. Because of this, acid-base homeostasis is critical in the healthy body.

Acids An **acid** is a substance that releases hydrogen ions (H^+) in solution (**Figure 3.17**). Because an atom of hydrogen has just one proton and one electron, a positively charged hydrogen ion is simply a proton. This solitary proton is highly likely to participate in chemical reactions. Strong acids are compounds that release all of their H^+ in solution. Hydrochloric acid (HCl), which is released from cells in the lining of the stomach, is a strong acid because it releases all of its H^+ in the stomach's watery environment. This strong acid aids in digestion and kills ingested microbes.

Weak acids do not ionize completely; that is, some of their hydrogen ions remain bonded within a compound in solution. An example of a weak acid is vinegar, or acetic acid.

Figure 3.17 Acids and Bases

(A) In aqueous solutions, an acid dissociates into hydrogen ions (H^+) and anions. Stronger acids produce solutions with a high concentration of H^+ because nearly every molecule of a strong acid dissociates. (B) In aqueous solutions, a base dissociates into hydroxyl ions (OH^-) and cations. Stronger bases produce solutions with a high concentration of OH^- because nearly every molecule of a strong base dissociates.

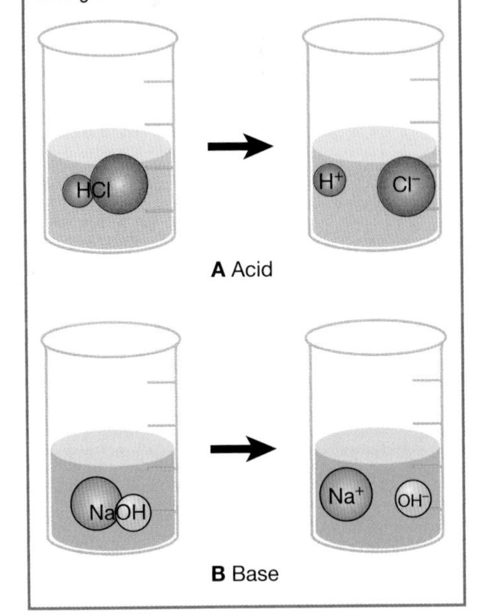

Bases A **base** is a substance that releases hydroxyl ions (OH^-) in solution, or one that accepts H^+ already present in solution (see Figure 3.17). The hydroxyl ions (also known as hydroxide ions) combine with H^+ present to form a water molecule, thereby removing H^+ and reducing the solution's acidity. This absorption of the H^+ ions is known as *neutralization*. Strong bases release most or all of their hydroxyl ions; weak bases release only some hydroxyl ions or absorb only a few H^+. In the body, bases are critical for neutralization and acid–base homeostasis. For example, the partially digested food mixed with hydrochloric acid from the stomach would burn the next portions of the digestive tract if it were not neutralized by bicarbonate (HCO_3^-), a weak base that attracts H^+. Bicarbonate accepts some of the H^+ protons, thereby reducing the acidity of the solution.

The Concept of pH The relative acid or base content of a solution is referred to as acidity or alkalinity and can be measured and given a numeric value on the pH scale (**Figure 3.18**). The **pH scale** stands for the "potential of hydrogen" and measures a solution's hydrogen ion (H^+) concentration. Because this is a logarithmic scale, a solution with a **pH** of 4 has an H^+ concentration that is ten times greater than that of a solution with a pH of 5. Another way to express that is that a solution with a pH of 4 is ten times more acidic than a solution with a pH of 5. A solution with a pH of 7 is considered **neutral**—neither acidic nor basic. Pure water has a pH of 7. The lower the number below 7, the more acidic the solution, or the greater the concentration of H^+. The higher the number above 7, the more basic (alkaline) the solution, or the lower the concentration of H^+.

pH scale (Potential of Hydrogen Scale)

LO 3.4.5

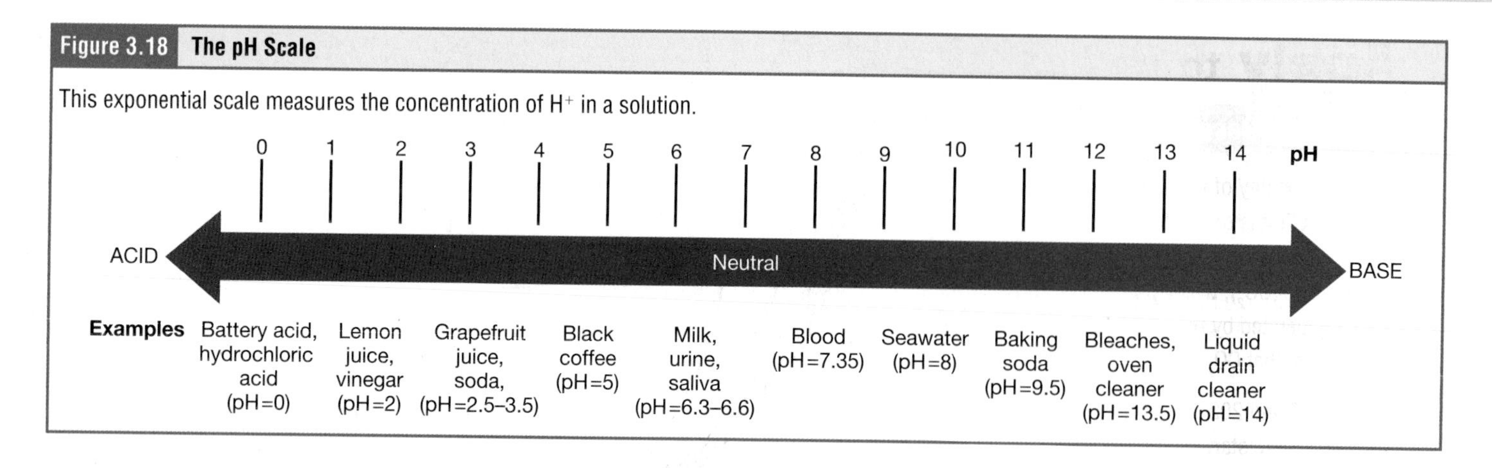

Figure 3.18 The pH Scale

This exponential scale measures the concentration of H^+ in a solution.

Learning Check

1. Solution A has a pH of 4. What is this solvent considered?
   a. Acid
   b. Base
2. Identify the solution you would add to bring the pH of Solution A closer to neutral.
   a. HCl
   b. HNO_3
   c. H_2SO_4
   d. NaOH
3. You find solution C, which has a pH of 6. How much higher is the H^+ concentration in solution A compared to solution C?
   a. 10 times
   b. 20 times
   c. 100 times
   d. 2 times

Buffers The pH of human blood normally ranges from 7.35 to 7.45. At this slightly basic pH, blood can buffer acidic changes that may occur. Homeostatic mechanisms normally keep the pH of blood within this narrow range. This is critical, because fluctuations—either too acidic or too alkaline—can lead to life-threatening disorders. While blood is tightly regulated to stay just slightly alkaline, some body compartments containing other body fluids exhibit different pH properties. The fluid inside the stomach has a pH of 1, for example. Cells consistently exposed to acidic or basic solutions will typically have anatomical protective mechanisms.

All cells of the body depend on homeostatic regulation of acid–base balance at a pH of approximately 7.4. The body therefore has several mechanisms for this regulation, involving breathing, the excretion of chemicals in urine, and the internal release of chemicals collectively called buffers into body fluids. A **buffer** is a solution of a weak acid and its conjugate base. A buffer can neutralize small amounts of acids or bases in body fluids. For example, if there is even a slight decrease below 7.35 in the pH of a bodily fluid, the buffer in the fluid—in this case, acting as a weak base—will bind the excess hydrogen ions. In contrast, if pH rises above 7.45, the buffer will act as a weak acid and contribute hydrogen ions.

Student Study Tip

Why are buffers so important? We move our pH from baseline quite often, for example when we take antacids. These are not lethal because we have buffers in our blood!

Apply to Pathophysiology

Acids and Bases

Excessive acidity of the blood and other body fluids is known as *acidosis*. There are three physiologically significant sources of acid in the body. The first is the stomach, which secretes HCl. The second is exercising muscles, which, in mechanisms we will learn about later in this chapter, sometimes generate lactic acid, which is added to the blood and lowers its pH. The third is related, though a bit more complex. Carbon dioxide (CO_2), when combined with water, can form H_2CO_3, carbonic acid. This acid can dissociate and add H^+ ions to the bloodstream. CO_2 is generated by most cells through their process of utilizing sugar for energy. The main homeostatic mechanism for regulating blood pH is to exhale that CO_2 regularly!

1. In 2020 and 2021, as the novel coronavirus spread around the globe in the worst global pandemic in 100 years, adults and children everywhere started wearing protective masks that covered their nose and mouth. These masks, which proved to be absolutely critical in protecting individuals and curbing the spread of the virus, were occasionally described as uncomfortable, especially while exercising. While this discomfort had a lot of superficial reasons, some poorly designed masks trapped carbon dioxide and caused the exerciser to rebreathe their exhaled carbon dioxide. Imagine a biker who is exercising rigorously wearing one of these poorly designed masks. Do you expect that the biker's blood pH value is
   A) Trending downward (pH below 7.4)?
   B) Rising (pH above 7.45)?
2. The biker stops for a moment when no one else is around to pull their mask down. Do you think they first
   A) Exhale?
   B) Inhale?
3. Now imagine a child who contracts norovirus, commonly known as the "stomach bug." The child has been vomiting for several hours, repeatedly losing their stomach contents and fluid. The child is brought to the ER where their blood is analyzed. Which of the following blood pH values is most likely in this child?
   A) 7.3
   B) 7.5
4. This child, who is exhausted from vomiting, is also afraid of needles. As the nurse approaches to take their blood, they begin to hyperventilate (rapid, short exhales). They are breathing out far more than usual. Take your answer from Question 3. Will this hyperventilation
   A) Nudge the pH values back toward neutral (pH 7.4)?
   B) Make the deviation from neutral even more significant (even more acidic or even more basic)?

Learning Check

1. Identify the organic compound.
   a. NaCl
   b. CO_2
   c. H_2O
   d. $C_6H_{12}O_6$
2. Which of the following is an example of how water acts as a heat sink?
   a. Water lubricates the tissues and joints in our bodies.
   b. Water protects cells and organs from trauma.
   c. Water absorbs heat and changes from liquid to gas.
   d. Water can dissolve many different types of molecules.
3. In a reaction, one reactant gave up a hydrogen atom, and another reactant gave up a hydroxyl group. This reaction produced a water molecule. What is this an example of? Please select all that apply.
   a. Dehydration synthesis
   b. Hydrolysis reaction
   c. Synthesis reaction
   d. Decomposition reaction
4. What ion or molecule does an acid give up in a solution?
   a. OH
   b. H^+
   c. Na^{2+}
   d. Cl^{2-}

3.5 Organic Compounds Essential to Human Functioning

Learning Objectives: By the end of this section, you will be able to:

3.5.1 Define the term organic molecule.

3.5.2 Explain the relationship between monomers and polymers.

3.5.3 Define and provide examples of dehydration synthesis and hydrolysis reactions.

3.5.4 Compare and contrast the general molecular structure of carbohydrates, proteins, lipids, and nucleic acids using chemical formulas.

3.5.5 Describe the building blocks of carbohydrates, proteins, lipids, and nucleic acids, and explain how these building blocks combine with themselves or other molecules to create complex molecules in each class, providing specific examples.

3.5.6 Describe the four levels of protein structure and the importance of protein shape for function.

3.5.7 Define enzyme and describe factors that affect enzyme activity.

LO 3.5.1

Organic compounds consist of carbon atoms covalently bonded to hydrogen. These molecules usually incorporate oxygen, and often other elements as well. Created by living things, they are found throughout the world, in soils and seas and every cell of the human body. The four types most important to human structure and function are carbohydrates, lipids, proteins, and nucleotides. Before exploring these compounds, you need to first understand the chemistry of carbon.

3.5a The Chemistry of Carbon

What makes organic compounds ubiquitous is the chemistry of their carbon core. Carbon atoms have four electrons in their valence shell; therefore, they will tend to react in such a way to fill the four-electron vacancy and complete their valence shell capacity of eight electrons. Carbon atoms do not complete their valence shells by donating or accepting four electrons. Instead, they readily share electrons via covalent bonds.

Carbon atoms commonly share with other carbon atoms, often forming a long carbon chains or carbon rings (**Figure 3.19**). Notice that for the two representative molecules drawn in Figure 3.19 they can be drawn in two ways. In one representation, each carbon atom is illustrated with its symbol of C. In the other representations, the

Figure 3.19

The carbons in organic molecules can form chains (A) or rings (B). Note that both of the molecular drawings in (A) are the same molecule, as are both of the drawings in (B). Chemical structures can be drawn either with the C symbol for carbon or by implying the carbons using lines and angles.

A

B

carbons are implied but not included. Carbon atoms can share electrons with other carbon atoms, or they can share electrons with a variety of other elements, one of which is always hydrogen. Carbon and hydrogen groupings are called *hydrocarbons*. If you study the figures of organic compounds in the remainder of this chapter, you will see several with chains of hydrocarbons in one region of the compound.

Many combinations are possible to fill carbon's four "vacancies." Carbon may share electrons with oxygen or nitrogen or other atoms in a particular region of an organic compound. Moreover, the atoms to which carbon atoms bond may also be part of a functional group. A **functional group** is a group of atoms linked by covalent bonds that tend to function in chemical reactions as a single unit with predictable behaviors. You can think of functional groups as tightly knit "cliques" whose members are unlikely to be parted. Five functional groups are important in human physiology; these are the hydroxyl, carboxyl, amino, and phosphate groups (**Table 3.3**).

LO 3.5.2

Organic biological molecules are often very large. Any large molecule is referred to as a **macromolecule**. These macromolecules are often made by combining several smaller subunits, the way that a brick wall is made by bonding one brick to another until a large structure is formed. The individual units, or bricks in this example, are known as **monomers**; once united, we refer to the larger structure as a **polymer**.

Table 3.3 Functional Groups Important in Human Physiology

Functional Group	Structural Formula	Importance	Image
Hydroxyl	—O—H	• Polar • Involved in dehydration synthesis and hydrolysis reactions • Forms H bonds	**Hydroxyl** (structure: CH_2OH, C, O, H, OH)
Carboxyl	O—C—OH	Carboxyl groups are found within fatty acids, amino acids, and many other acids.	**Carboxyl** (structure: C, O, O^-)
Amino	$—N—H_2$	Amino groups are found within amino acids, the building blocks of proteins.	**Amino** (structure: H, + H, N, H)
Phosphate	$—P—O_4^{2-}$	Phosphate groups are found within phospholipids and nucleotides.	**Phosphate** (structure: O, ^-O—P—O, O^-)
Methyl	$–CH_3$	Methyl groups are essential in DNA regulation.	**Methyl** (structure: H, H—C—, H)

Learning Check

1. In a covalent bond, how many electrons are shared between two atoms?
   a. One
   b. Two
   c. Three
   d. Four
2. How many electrons does carbon have in its valence shell?
   a. One
   b. Two
   c. Three
   d. Four
3. What type of reaction is most likely involved when we combine monomers to form a polymer?
   a. Dehydration synthesis
   b. Hydrolysis

3.5b Carbohydrates

LO 3.5.4

LO 3.5.5

monosaccharide (simple sugar)

A **carbohydrate** is a molecule composed of carbon, hydrogen, and oxygen; in most carbohydrates, hydrogen and oxygen are found in the same two-to-one relative proportions they have in water. In fact, the name "carbohydrate" literally means "carbon + water (hydro)."

Carbohydrates are referred to as saccharides, a word meaning "sugars." The monomer form of carbohydrates is called a **monosaccharide**, and the polymer form is **polysaccharide**. Carbohydrates are commonly found in groups of two monomers joined together with a covalent bond. This form is so common it gets its own name, **disaccharides**.

Monosaccharides There are five monosaccharides that are particularly important in the body. These are glucose, fructose, galactose, ribose, and deoxyribose, shown in Figure 3.20. Notice that all five of these molecules are hydrocarbon rings.

Figure 3.20 Five Important Monosaccharides

These five monosaccharides are combined in different ways to form the polysaccharides of the body and our diet.

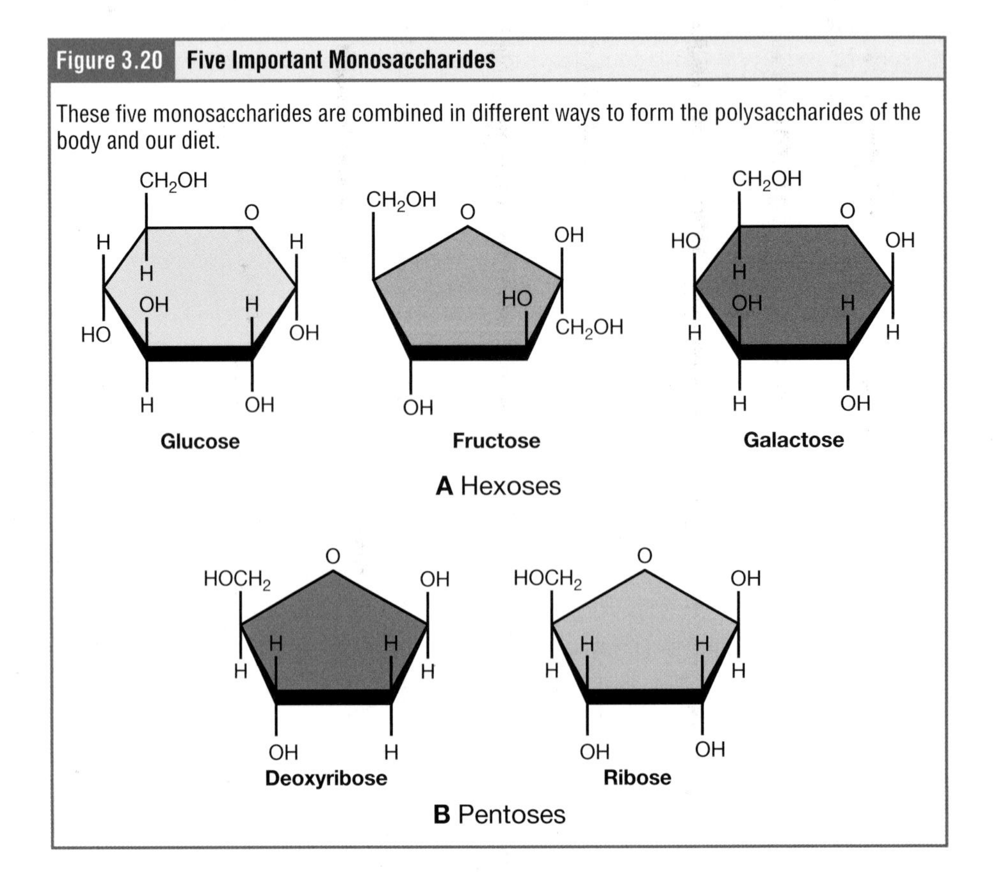

Polysaccharides **Polysaccharides** can contain from a few to a thousand or more monosaccharides. Three of these polysaccharides—starches, glycogen, and cellulose—are important to the body (**Figure 3.21**).

Starches and glycogen are polymers of glucose. These polymers are mechanisms for storing glucose by various organisms. Starches function as glucose storage for plants and are also a glucose-rich and tasty component of the human diet. Glycogen is also a polymer of glucose, but it is stored in the tissues of animals, especially in the muscles and liver. Cellulose, a polysaccharide that is the primary component of the cell wall of green plants, is the component of plant food referred to as "fiber." In humans, cellulose/fiber is not digestible; however, dietary fiber has many health benefits, which we will explore in Chapter 24.

The physiologically important monosaccharides (see Figure 3.21) combine to form three common disaccharides (shown in **Figure 3.22**). These three sugars—sucrose (table sugar), lactose (milk sugar), and maltose (malt sugar)—are common components of human diets. Your cells cannot use these sugars in their disaccharide forms. Instead, in the digestive tract, they are split into their component monosaccharides via hydrolysis. In order to break the covalent bond that holds the sugars together, an enzyme specific for that process must be present.

Functions of Carbohydrates The body obtains carbohydrates from plant-based foods. Grains, fruits, and legumes and other vegetables provide most of the carbohydrate in the human diet, although lactose is found in dairy products. Most of the nutrients taken into the body are used anabolically; that is to say, they are broken into their monomers and then those monomers are used to build new polymers in the cells. Carbohydrates are broken down and can be used to build glycogen as storage, but ultimately most dietary carbohydrate is used for energy.

In the breakdown of glucose, potential energy stored within its bonds is released. The cell can capture that released energy and use it to build a bond within a molecule of adenosine triphosphate, better known as ATP. Glucose can only be completely broken down in the presence of oxygen; when oxygen is absent or low, the glucose molecule is broken into two halves and those halves are converted to lactic acid and only a small amount of ATP can be built. For these reasons, most glucose breakdown occurs in the presence of oxygen. The following formula illustrates the exchange reaction in which glucose is broken down and ATP is built:

$$\underset{\text{glucose}}{C_6H_{12}O_6} + \underset{\text{oxygen}}{6\,O_2} \rightarrow \underset{\text{carbon dioxide}}{6\,CO_2} + 6\,H_2O + ATP$$

Figure 3.21 Three Important Polysaccharides

Three polysaccharides that are particularly important in human health are starches, glycogen, and fiber.

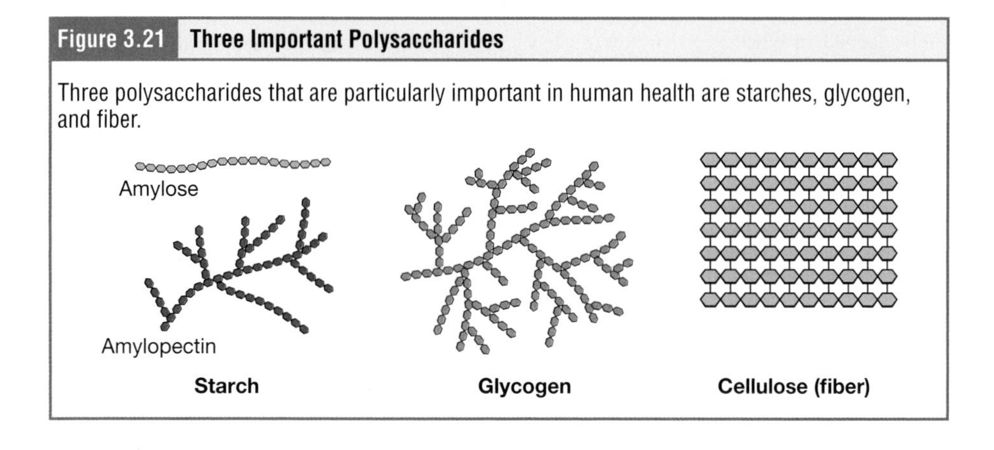

Note that the precursors to ATP are not listed among the reactants in this equation; ATP is considered a byproduct rather than a product. The precursors to ATP formation (ADP and Pi) are always available in cells. While most of the released energy from the glucose molecule is captured in ATP bonds, some of the released energy from the glucose molecule is lost in the form of heat. Note that within this equation and process we can now understand two critical characteristics of humans: One is that our body heat is generated through the process of breaking down glucose and forming ATP; the second is that humans breathe in the oxygen critical for this process and breathe out one of the products (CO_2). Although most body cells can break down other organic compounds for fuel, all body cells can use glucose. Moreover, nerve cells (neurons) in the brain, spinal cord, and throughout the peripheral nervous system, as well as red blood cells, can use only glucose for fuel. For these reasons, it is critical for the body to maintain a constant level of glucose in the blood.

In addition to being a critical fuel source, carbohydrates are present in very small amounts in cells' structure. For instance, some carbohydrate molecules bind with other molecules to produce glycoproteins and glycolipids, both of which function in the cell membrane and contribute to the identity of cells (example: liver cells are "tagged" with different glycolipids and glycoproteins than red blood cells).

Apply to Pathophysiology

Lactase

All mammal species make an enzyme known as lactase. This enzyme's function is to break the covalent bond between the two monomers in the disaccharide lactose (see the following figure). Mammals manufacture this enzyme in early life, typically during the period in which their primary food source is maternal milk.

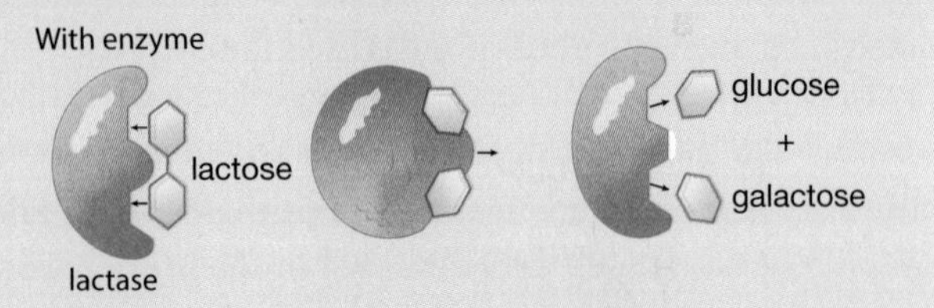

1. When lactase functions, which type of reaction does it enable?
   A) Dehydration synthesis
   B) Hydrolysis
2. When lactase functions, what are the products of the reaction?
   A) Galactose and glucose
   B) Two monomers of glucose
   C) Glucose and fructose

At some point, typically around two years of age in humans, the enzyme lactase stops being produced. At this point lactose cannot be broken down into its monomers and therefore it stays in the inside of the stomach and intestines, it cannot cross over to the inside of the cells and body.

3. Which of the following sentences is the most accurate?
   A) Lactose is a solvent for the contents of the stomach and intestines.
   B) Lactose is a solute in the solution of contents of the stomach and intestines.
   C) Lactose makes the contents of the stomach and intestines more alkaline.

While most humans stop expressing the enzyme lactase, many adults remain capable of breaking down lactose. This is due to the acquisition of lactose-digesting bacteria in the intestines. While the human body itself is incapable of the reaction, the bacteria break down lactose for the individual.

4. Starches are easily digestible by humans, but cellulose is not. Which of the following explanations seems most plausible?
   A) Humans express an enzyme that breaks the bonds between starch monomers, but they do not express an enzyme capable of breaking the bonds between cellulose monomers.
   B) Humans can only absorb the monomer types found in glycogen.
   C) The monomers of cellulose dissociate to become a toxic acid.

Figure 3.22 Three Important Disaccharides

Sucrose, lactose, and maltose are three disaccharides that are particularly relevant to humans. They all are formed by dehydration synthesis reactions that join monosaccharides together.

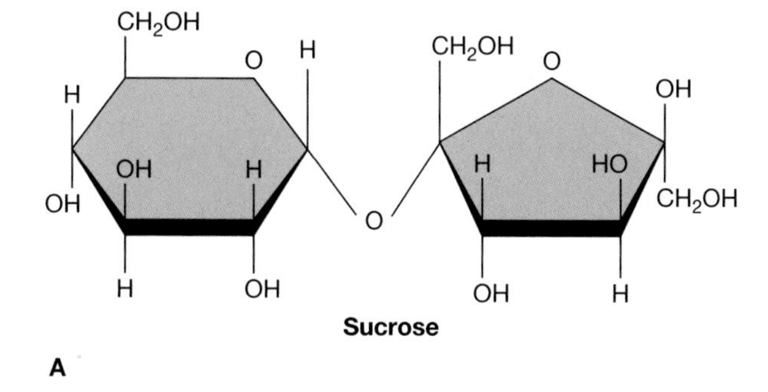

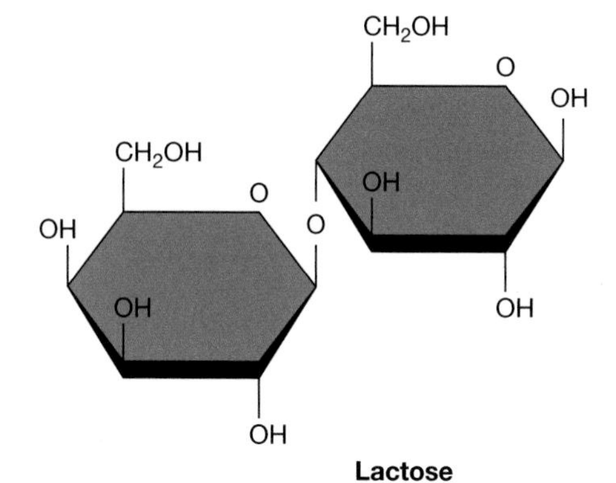

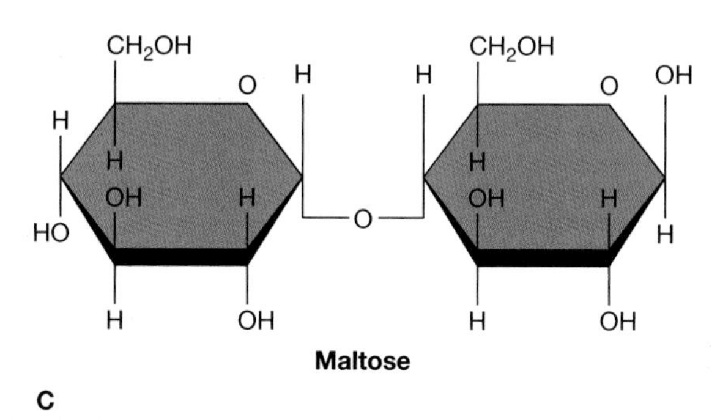

3.5c Lipids

A **lipid** is one of a highly diverse group of compounds made up mostly of hydrocarbons. Their nonpolar hydrocarbons make all lipids hydrophobic. Because of this, lipids do not form a true solution in water.

Triglycerides A **triglyceride** is one of the most common lipids in our diets, and the type found most abundantly in body tissues. This compound, which is commonly referred to as a fat, is formed from the dehydration synthesis of two types of molecules (**Figure 3.23**):

- A glycerol molecule, which consists of three carbon atoms.
- Three fatty acids, which consist of long chains of hydrocarbons that extend from each of the carbon atoms of the glycerol molecule.

Figure 3.23 Triglycerides

Triglycerides, which are dietary lipids, are produced when glycerol is joined to three fatty acid tails via dehydration synthesis.

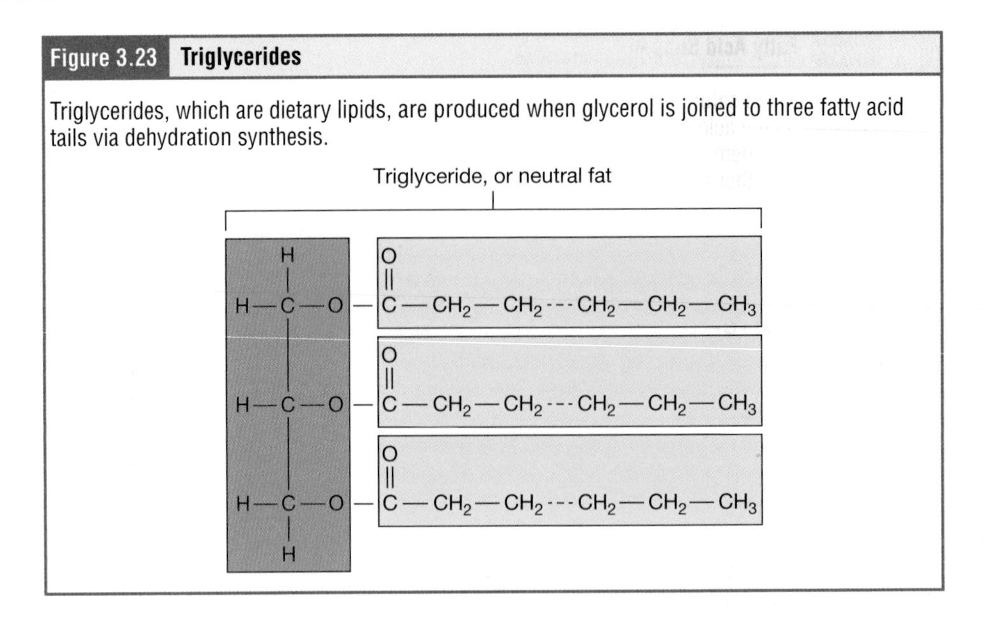

Fatty acid chains are long and consist only of carbons and hydrogens. Remember that carbon can form bonds with up to four other molecules. If you pick one carbon in the middle of the chain in **Figure 3.24**, you see that the carbon is bound to other carbons on each side, as well as two hydrogens. In **Figure 3.24A** you can see that each of the carbons is maximally bonded this way. But notice in **Figure 3.24B** that one of these carbons shared a double bond with its neighboring carbon. In this double bond the carbons are sharing twice as many electrons, so that carbon cannot form as many bonds with hydrogen as its neighbors. This creates a kink in the chain. Lipids, such as the one illustrated in Figure 3.24A, which have their carbons bound to as many hydrogens as possible, are said to be **saturated fats**, that is, saturated with hydrogens. The resulting triglycerides can pack together more tightly (**Figure 3.24C**). Lipids with double bonds, and therefore fatty acid tails that zigzag, are called **unsaturated fats**, and in groups their molecules just cannot get as close together as those of saturated fats (**Figure 3.24D**). The straight, rigid chains of saturated fats pack tightly together and are solid or semisolid at room temperature. Butter and lard are examples, as is the fat found on a steak or in your own body. In contrast, the bent chains of unsaturated fatty acids keep the molecules from being able to pack together tightly, and are therefore liquid at room temperature. Plant oils such as olive oil are one example.

Finally, **trans-fatty acids** found in some processed foods, including some margarines, are thought to be even more harmful to the heart and blood vessels than saturated fatty acids. *Trans* fats are created from unsaturated fatty acids that are chemically treated to add hydrogens and bring these molecules closer to being saturated. It is currently not well understood why these manufactured fats have such an impact on cardiovascular health.

As a group, triglycerides are a major fuel source for the body. When you are resting or asleep, a majority of the energy used to keep you alive is derived from triglycerides stored in your adipose (fat) tissues. Triglycerides also fuel long, slow physical activity such as gardening or hiking, and contribute a modest percentage of energy for vigorous physical activity. Dietary fat also assists the absorption and transport of the nonpolar fat-soluble vitamins A, D, E, and K. Additionally, stored body fat protects and cushions the body's bones and internal organs, and acts as insulation to retain body heat.

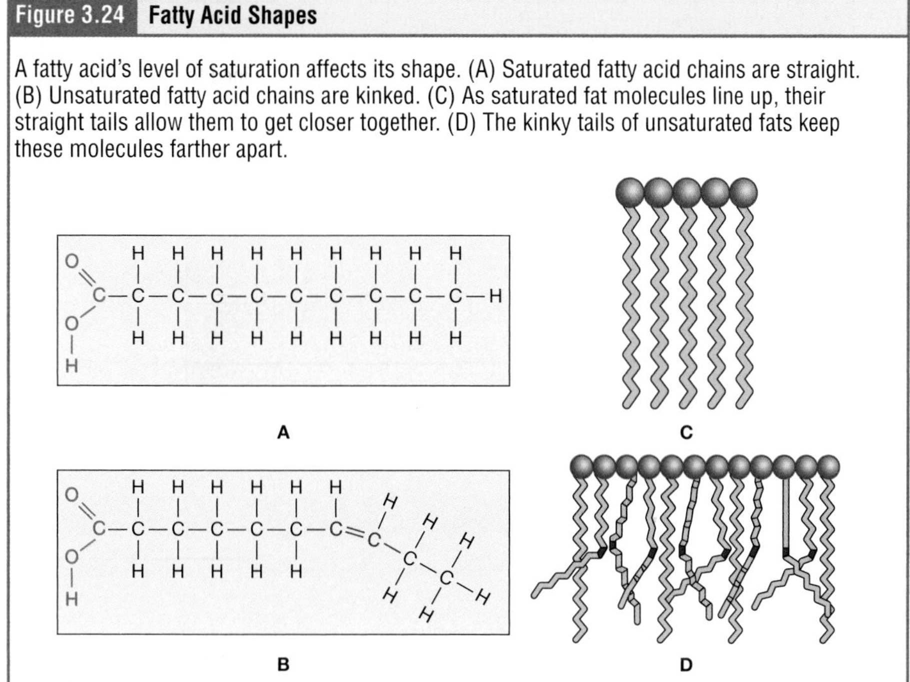

Figure 3.24 Fatty Acid Shapes

A fatty acid's level of saturation affects its shape. (A) Saturated fatty acid chains are straight. (B) Unsaturated fatty acid chains are kinked. (C) As saturated fat molecules line up, their straight tails allow them to get closer together. (D) The kinky tails of unsaturated fats keep these molecules farther apart.

Phospholipids As its name suggests, a **phospholipid** is a lipid molecule that is modified with a phosphate group. Phospholipids share many similarities with triglycerides. Both have fatty acid tails; however, phospholipids have just two tails compared to the three tails of triglycerides (**Figure 3.25**). Where glycerol would form a third bond with a fatty acid tail in triglycerides, in phospholipids that third binding site bonds with a phosphate group. Phosphate groups are highly negatively charged, which makes this head region of the phospholipid polar and hydrophilic. Triglycerides, in contrast, are entirely hydrophobic. Phospholipids are incredibly important in anatomy and physiology. We will discuss them in much greater detail in Chapter 4.

Steroids A **steroid** compound (referred to as a *sterol*) is a nonpolar lipid, but its structure looks much different than the fats we have described so far. Instead of a long tail, its hydrocarbons form rings (**Figure 3.26**). The most significant sterol in humans is cholesterol, which is synthesized by the liver in humans and is also present in most animal-based foods. Like other lipids, cholesterol's hydrocarbons make it hydrophobic; however, it has a polar hydroxyl head that is hydrophilic. Cholesterol is used in a lot of different processes in the body. Cholesterol is also a building block of many hormones, signaling molecules that the body releases to regulate processes at distant sites. Finally, like phospholipids, cholesterol molecules are found in the cell membrane, where their hydrophobic and hydrophilic regions help regulate the flow of substances into and out of the cell.

steroid (sterol)

Prostaglandins Similar to hormones, **prostaglandins** are one of a group of signaling molecules, but prostaglandins are derived from unsaturated fatty acids (see **Figure 3.25C**). Prostaglandins also sensitize nerves to pain. One class of pain-relieving medications called nonsteroidal anti-inflammatory drugs (NSAIDs), such as ibuprofen, works by reducing the effects of prostaglandins.

Figure 3.25 Other Important Lipids

(A) Phospholipids are produced when two fatty acid tails, glycerol, and a phosphate group are connected by covalent bonds. (B) Sterols are a group of lipids, including cholesterol, that are ring-shaped. (C) Prostaglandins are signaling molecules that are produced from fatty acids; shown here is prostaglandin E2 (PGE2).

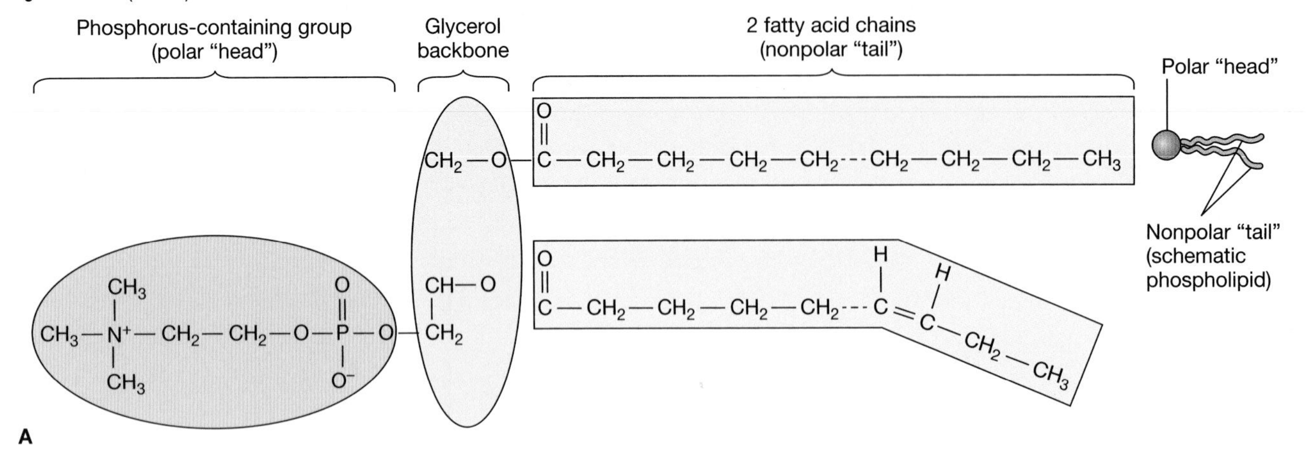

A

B

C

Learning Check

1. What is the difference between a saturated and unsaturated fat? Please select all that apply.
   a. The saturated fat has a double bond.
   b. The saturated fat binds to two fatty acid chains.
   c. The unsaturated fat has a double bond.
   d. The unsaturated fat has a kink in its fatty acid chain.
2. How many fatty acid chains does a phospholipid have?
   a. One
   b. Two
   c. Three
   d. Four
3. Describe the structure of a steroid compound.
   a. Long-tail
   b. Ring
   c. Alpha-helix
   d. Zigzag

3.5d Proteins

A **protein** is an organic molecule composed of amino acids linked by covalent bonds. Proteins include the keratin in your skin, hair, and nails that protects underlying tissues, and the collagen found in your tendons, in your bones, and in the meninges that cover the brain and spinal cord. Proteins comprise many of the body's functional chemicals, including enzymes such as those found in the digestive tract, antibodies help you fight off infections, and signaling molecules, including many of our hormones. While carbohydrates and lipids are composed of hydrocarbons and oxygen, all proteins also contain nitrogen (N), and many contain sulfur (S), in addition to carbon, hydrogen, and oxygen. Most proteins share a characteristic of binding other molecules. Hemoglobin, a protein in your blood, binds oxygen (among other molecules), and protein receptors bind signaling molecules. The molecule that a protein binds is known as its **ligand**.

Microstructure of Proteins The monomers of proteins are **amino acids**. There are 20 amino acids in the human body; however, their structure is almost entirely identical. All amino acids are composed of three functional groups lined to a central carbon atom (Figure 3.26). Two of these three functional groups are the same in all amino acids: an amino group and a carboxyl group (see Table 3.3). Each of the 20 amino acids has a unique third functional group, called its *R group*, that gives the amino acid all of its individual characteristics. For example, the R groups of two amino acids—cysteine and methionine—contain sulfur. Sulfur does not readily participate in hydrogen bonds, but all other amino acids do. This variation influences the way that proteins containing cysteine and methionine are assembled. Some R groups are polar, and some are nonpolar; these properties will determine aspects of protein structure and function within the aqueous solutions of the body.

The carboxyl group and amine group of neighboring amino acids can join via dehydration synthesis to form protein polymers (**Figure 3.27A**). The covalent bond holding amino acids together is called a **peptide bond**. A short sequence of amino acids joined together by peptide bonds is called a peptide. A longer sequence is a protein (see Figure 3.27A).

The body is able to synthesize most of the amino acids from components of other molecules; however, nine cannot be synthesized and have to be consumed in the diet. These are known as the *essential amino acids*.

Figure 3.26

The structure of an amino acid includes an amino group, a carboxyl group, and a third functional group—called an R group—that is unique to each amino acids.

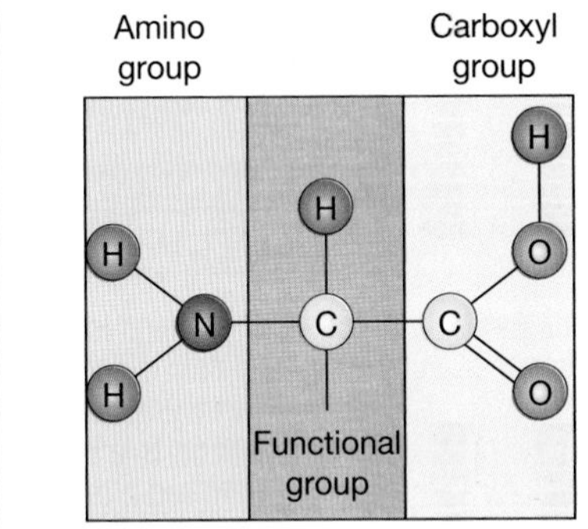

Figure 3.27 Peptide Bond

(A) Different amino acids join together via peptide bonds through dehydration synthesis. (B) Via peptide bonds, amino acids form peptides or proteins.

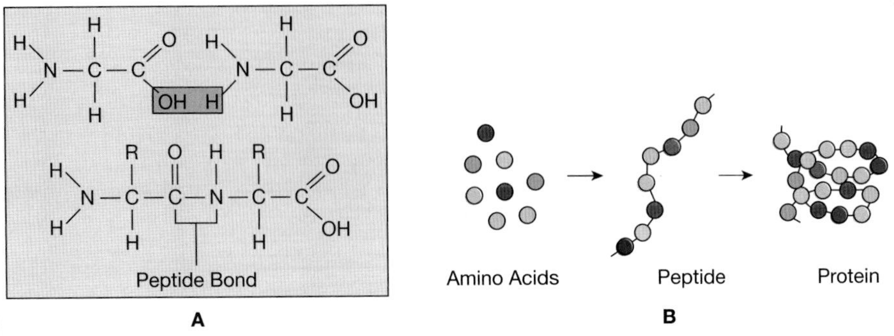

Shape of Proteins Just as a fork cannot be used to eat soup and a spoon cannot be used to spear meat, a protein's shape is essential to its function. A protein's shape is determined, most fundamentally, by the sequence of amino acids of which it is made (**Figure 3.28**). The sequence is called the *primary structure* of the protein.

LO 3.5.6

As amino acids, along with their unique functional groups, are bonded together in peptide bonds, their functional groups may interact with each other (**Figure 3.29**). For example, two amino acids with polar functional groups may be attracted to each other and form a hydrogen bond (**Figure 3.29A**). Amino acids that are hydrophobic will tend to huddle together to exclude water (**Figure 3.29B**), and oppositely charged amino acids may form ionic bonds (**Figure 3.29C**). In this way, proteins quickly become folded structures. The most common initial folding structure is for amino acids to form a coil called an alpha-helix (**Figure 3.28B**). Sometimes instead of a coil, proteins initially fold into a zig-zag structure called a beta-sheet. Both of these are called secondary structures of proteins (the primary structure is just the string of amino acids) and are held together through hydrogen bonding.

The secondary structure of proteins brings new combinations of amino acids into proximity to form further bonds. Additional folds of the proteins are referred to as the protein's tertiary structure (see Figure 3.29C). Often, two or more separate proteins bond to form an even larger protein with a quaternary structure (**Figure 3.28D**). For example, hemoglobin—the protein found in red blood cells—is composed of four tertiary polypeptides.

Learning Connection

Broken Process

Think about a protein in its final, three-dimensional tertiary or quaternary shape. How would changing a functional group, for example swapping a polar for a non-polar functional group, influence protein shape?

Figure 3.28 The Levels of Protein Structure that Determine Their Shape

(A) The primary structure of proteins is a chain of amino acids joined by covalent bonds. (B) The secondary structure, which comes in one of two forms, either alpha-helix or a beta-pleated sheet, is formed when hydrogen bonds between amino acids in different regions of the original chain fold the molecule. (C) The tertiary structure is the result of further folding as more bonds form among amino acids in different regions of the chain. (D) Some proteins additionally form a quaternary structure when interactions between two or more proteins in tertiary structure bond to each other. Hemoglobin, a protein found in red blood cells, is formed by four protein subunits joining together.

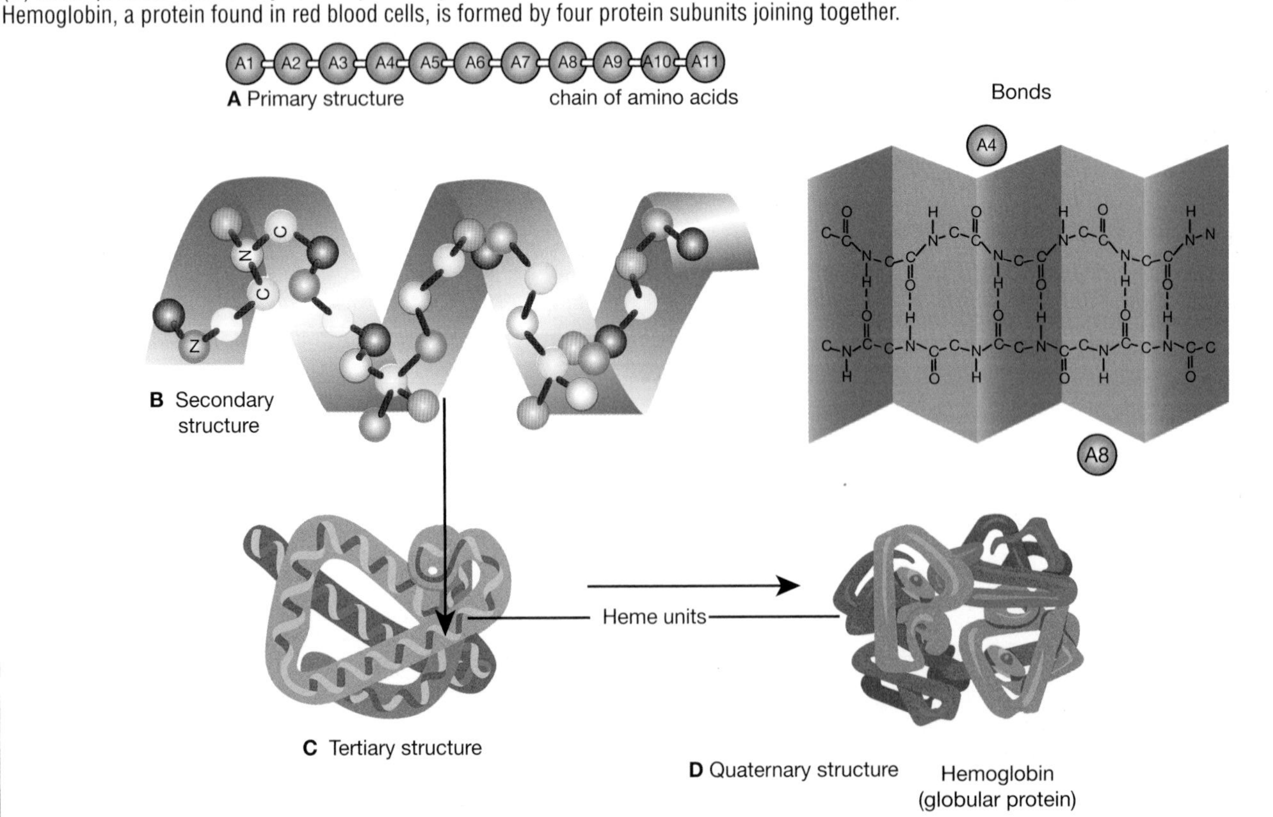

Figure 3.29 Interactions among Amino Acids Determine Protein Shape

As amino acids are added to a growing protein, their functional groups may interact to form bonds or affiliations. In (A), two polar amino acids form a hydrogen bond. In (B), hydrophobic amino acids are huddled together and encircled by polar amino acids to shield them from water. In (C), a positively charged amino acid forms an ionic bond with a negatively charged amino acid.

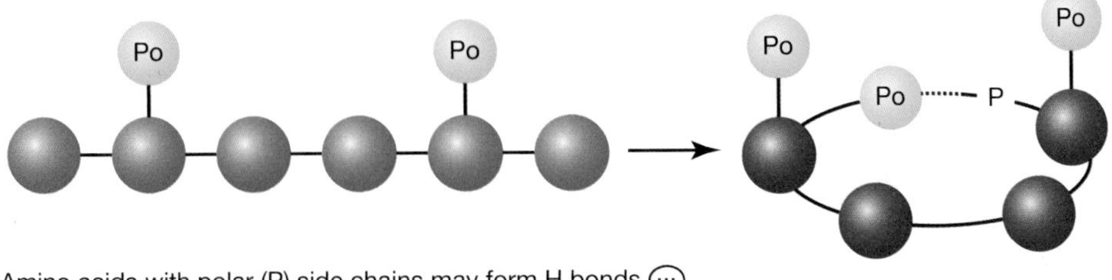

Amino acids with polar (P) side chains may form H bonds
A

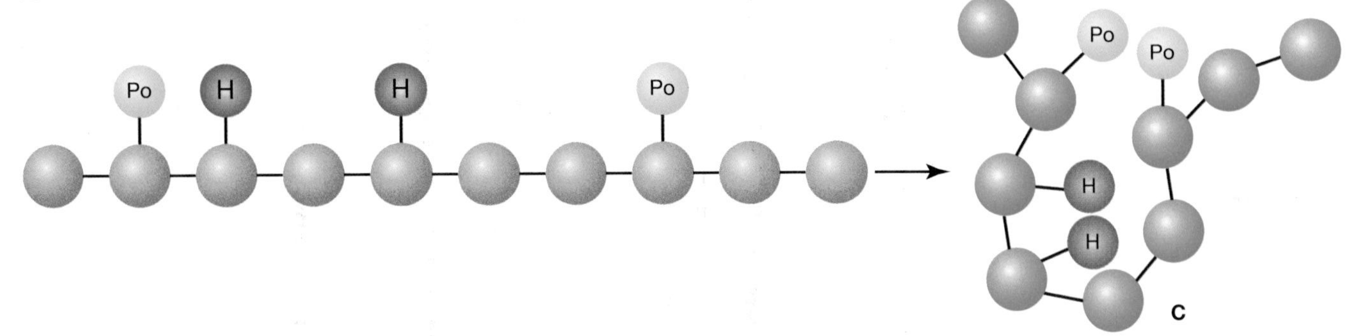

Amino acids that are hydrophobic (H) will huddle together to avoid the hydrophilic area surrounding them.
B

Remember that the glue that holds the secondary and tertiary protein structure together is hydrogen bonds and these bonds are sensitive to extreme heat, acids, bases, and certain other substances. It is critical for the body to maintain homeostasis of these variables for the proteins to avoid the risk of denaturation. **Denaturation** is a change in the structure of a molecule through physical or chemical means. Denatured proteins lose their functional shape and are no longer able to carry out their jobs. An everyday example of protein denaturation is the curdling of milk when acidic lemon juice is added. The contribution of the shape of a protein to its function can hardly be exaggerated. For example, the long slender shape of protein strands that make up muscle tissue is essential to their ability to contract (shorten) and relax (lengthen).

Proteins also change shape in a much more functional way when they form bonds with other molecules. Enzymes, introduced earlier as protein catalysts, are examples of this; their shape changes slightly when they bind to other molecules. The next section takes a closer look at the action of enzymes.

Enzymes Earlier in this chapter we discussed activation energy, which can add barriers or speed bumps to the rate of a reaction. Catalysts, especially enzymes function to lower activation energy.

LO 3.5.7

Enzymatic reactions—chemical reactions catalyzed by enzymes—begin when substrates bind to the enzyme. A **substrate** is a reactant in an enzymatic reaction. This occurs on regions of the enzyme known as *active sites* (**Figure 3.30**). Any given enzyme

Apply to Pathophysiology

Sickle Cell Anemia

Sickle cell anemia is a disease of **hemoglobin**, the protein in red blood cells. Hemoglobin binds oxygen. This specific mutation results in replacement of one amino acid (glutamic acid) with a different one (valine). Glutamic acid is a polar amino acid with a negative charge; valine is a nonpolar, uncharged amino acid.

Glutamic acid → Valine

1. Which of the following is true about protein shape?
   A) Protein shape is inconsequential, as long as the amino acid sequence remains the same.
   B) Physiological conditions (such as pH, temperature, and so on) can induce protein shape changes.
   C) Ligand binding can induce shape changes in proteins.
   D) Both B and C.

2. What might happen if more than one molecule of this mutated protein were found in the cytoplasm of the same cell?
   A) The proteins would repel each other and be found far apart.
   B) The proteins would be drawn toward and/or into the mitochondrion.
   C) The proteins would be drawn together and tend to clump.
   D) The proteins would be found in the same location and orientation as the nonmutated proteins.

3. In cases of protein misfolding, the protein is sometimes no longer able to bind its ligand. Which of the following is/are plausible explanation(s)?
   A) Due to its interaction with other molecules, the binding site is no longer accessible to the ligand.
   B) The binding site now has stronger affinity for a competitor molecule.
   C) The entire protein shape has changed, and the binding site is now on the interior of the protein.
   D) All of the above could be plausible.

catalyzes just one type of chemical reaction. This specificity is due to the fact that the active site is made up of amino acids, and those amino acids determine the shape and electrical charge of the active site. Only the substrate with the particular shape and electrical charge can bind to that active site (**Figure 3.30C**).

Figure 3.30 Steps in an Enzymatic Reaction

When a protein binds a new molecule, often a shape change occurs at the new molecule, and this may alter the relationships among the amino acids that make up the protein. Most enzymes are proteins, so when their substrate binds to the active site of the enzyme, the enzyme changes shape. (A) Enzymes facilitate chemical reactions, often by bringing together substrates so that they can bind. (B) Active sites are specific for their substrates because of shape of the active site as well as the molecular properties, including the charge, of the amino acids that line the active site. (C) In this representation of the protein enzyme, each of the amino acids is represented by a sphere. The amino acids that line the active site are found in various locations along the primary structure of the protein, but when the amino acid chain is folded, amino acids from different regions are brought together to make up the active site.

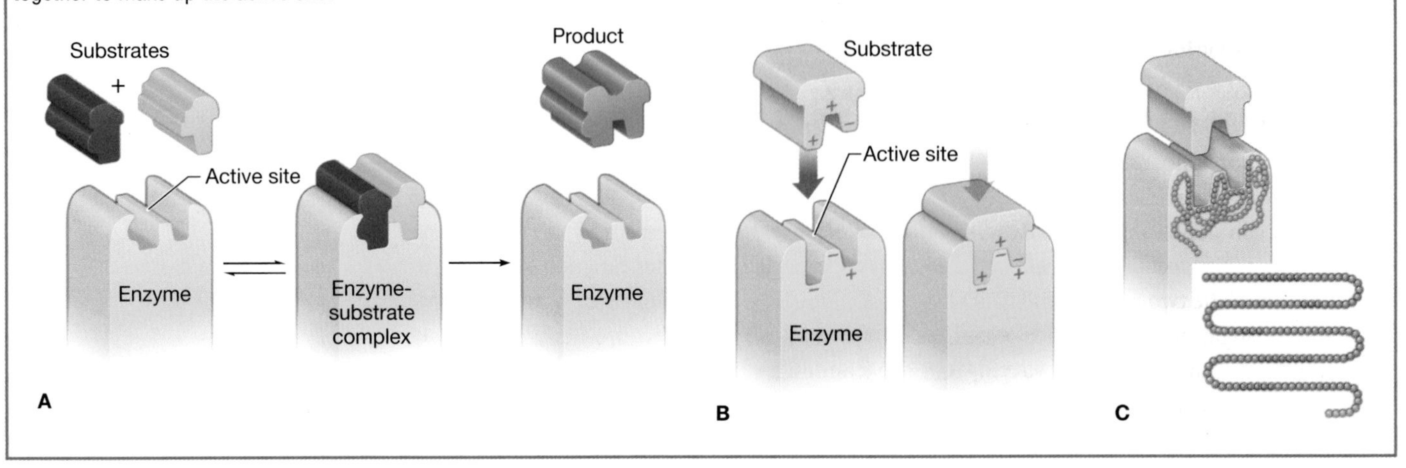

Binding of a substrate produces an enzyme–substrate complex. The enzyme facilitates the interaction of the substrate(s). This promotes increased reaction speed. The enzyme then releases the product(s); the enzyme is unchanged by this process and is free to bind another substrate.

Other Functions of Proteins Proteins play critical roles in the body in addition to their function as enzymes. Proteins act as hormones (chemical messengers that help regulate body functions), provide structural stability, and play protective roles. The body can also use proteins for energy when carbohydrate and fat intake is inadequate, and stores of glycogen and adipose tissue become depleted. However, since there is no storage site for protein except functional tissues, using protein for energy causes tissue breakdown, and results in strength and structural loss.

Learning Check

1. Select all components of an amino acid.
   a. Amino group
   b. Carboxyl group
   c. Central carbon atom
   d. Phosphate group
2. Rank the following from smallest to largest.
   a. Amino acid < peptide < protein
   b. Peptide < amino acid < protein
   c. Protein < peptide < amino acid
   d. Peptide < protein < amino acid
3. How many essential amino acids are there?
   a. 20
   b. 9
   c. 11
   d. 3
4. An increase in body temperature may endanger protein structure. If body temperature increases enough to cause protein denaturation, which of the following may occur? Please select all that apply.
   a. Chemical reactions will slow.
   b. The rate of oxygen transport in the body may decrease.
   c. There will be an increase in free amino acids in the bloodstream due to protein destruction.
   d. There will be an increase in glucose in the bloodstream due to glycogen destruction.

3.5e Nucleic Acids

The fourth type of organic compound important to human structure and function is nucleic acids (**Figure 3.31**). The two groups of nucleic acids in cells are **ribonucleic acid** (**RNA**) and **deoxyribonucleic acid** (**DNA**). Just like proteins and carbohydrates, nucleic acids are composed of monomers linked through bonds. The monomer of nucleic acids is the **nucleotide**, an organic compound composed of three subunits:

- One or more phosphate groups
- A pentose sugar: either deoxyribose or ribose
- A nitrogen-containing base: adenine, cytosine, guanine, thymine, or uracil

Nucleic Acids The two nucleic acids, DNA and RNA, differ in their type of sugar. DNA contains the sugar deoxyribose and is the nucleic acid that stores genetic information. DNA polymers can be built with varying combinations of the monomers adenine, cytosine, guanine, and thymine. Ribonucleic acid (RNA) contains the sugar ribose, and

Figure 3.31 The Nucleotides of DNA and RNA

(A) Nucleotides are composed of a phosphate group (or sometimes more than one phosphate group), a sugar, and a nitrogen-containing base.
(B) There are five nitrogen-containing bases of nucleotides

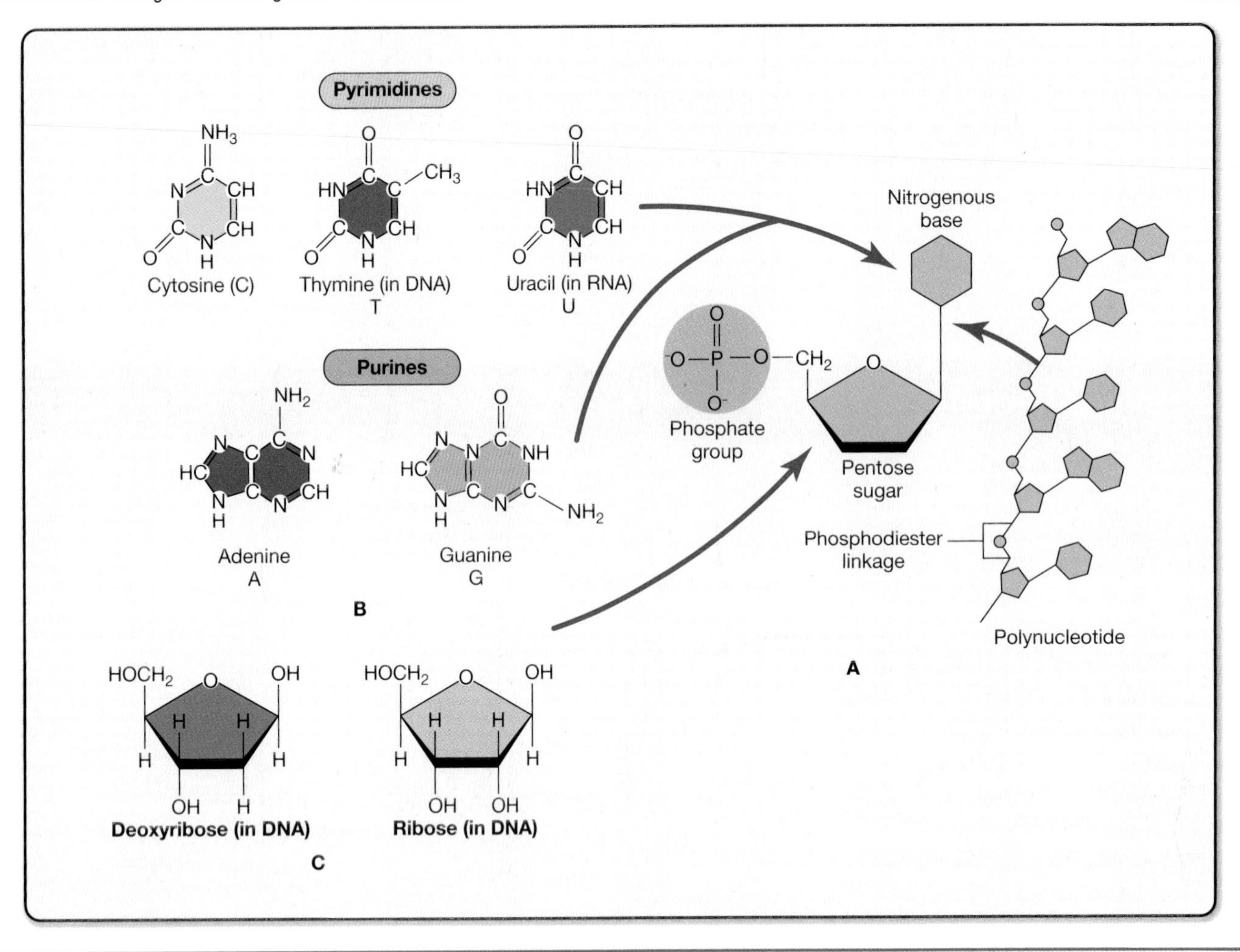

its functions all contribute to the manifestation of the genetic code as proteins. RNA polymers can be built with the monomers adenine, cytosine, guanine, and uracil.

Notice that the monomers adenine and guanine are larger than the others, consisting of double rings; these are classified as **purines**. The monomers cytosine, thymine, and uracil are **pyrimidines** and have a single ring structure.

Let's examine **Figure 3.32** to see how these polymers are built. Just as amino acids are linked at specific molecular locations, DNA and RNA are built via bonds between the pentose sugar of one monomer and the phosphate group of another monomer. In this way the monomers are strung together to form a "strand" or "backbone." The nitrogen-containing bases protrude. In DNA, two such backbones attach at their protruding bases via hydrogen bonds. These twist to form a shape known as a *double helix* (see Figure 3.32).

In contrast, RNA consists of a single strand of sugar-phosphate backbone studded with bases. RNA is used during protein synthesis to carry the genetic instructions from the DNA to the cell's protein manufacturing plants in the cytoplasm, the ribosomes.

Learning Connection

Chunking

Can you use chunking to learn the macromolecules? What are all the monomers? All the polymers? All the types of bonds that join monomers together?

Figure 3.32 Nucleic Acids

RNA is a single-stranded molecule composed of the nucleotides adenine, guanine, cytosine, and uracil. DNA is a double-stranded molecule; the two strands attach via hydrogen bonds between the bases of the component nucleotides. The nucleotides that make up DNA are adenine, guanine, cytosine, and thymine.

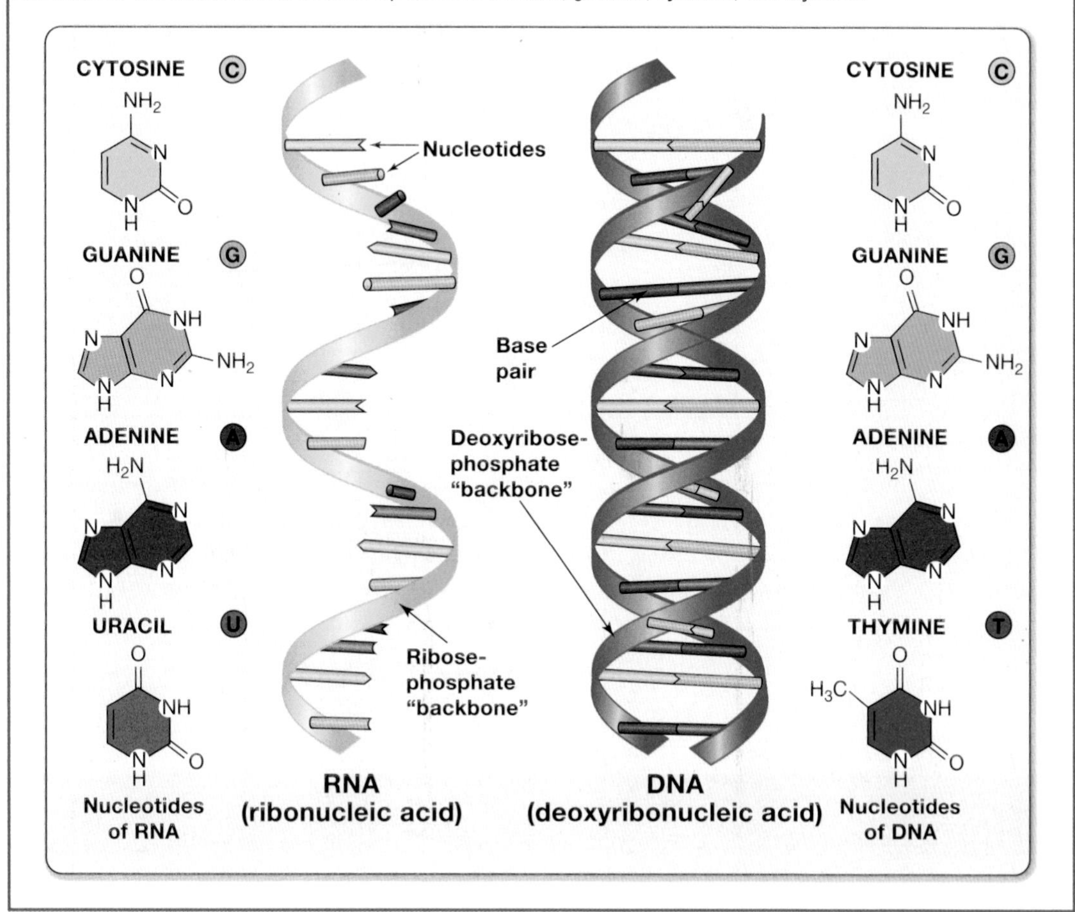

Adenosine Triphosphate The modified nucleotide adenosine triphosphate (ATP) is composed of a ribose sugar, an adenine base, and three phosphate groups (**Figure 3.33**). ATP serves as the cell's energy currency. The covalent bonds among its three phosphates store a significant amount of potential energy. When one of these bonds is

Cultural Connection

Customized Genes for Just the Right Fit

Nucleic acids make up the genetic information of all living things. Until very recently, scientists considered this genetic information to be relatively unchangeable. For example, if you were born with a mutation in the gene for hemoglobin, causing the disease known as sickle cell disease, then you would live with that mutation for the rest of your life. Clinicians can treat the symptoms caused by the mutation, for example by delivering supplemental oxygen, but we have never been able to change the genetic material itself.

That all changed in 2012 when two scientists, Drs Jennifer Doudna and Emmanuelle Charpentier, worked together to discover an enzyme system from bacteria that edits DNA sequences called CRISPR. Doudna and Charpentier won the Nobel Prize in Chemistry in 2020, becoming the first two women to share that prize without a male contributor. In 2021, the US Food and Drug Administration green-lighted the first potential CRISPR-based therapy to be tested in humans. If successful, this drug may be able to cure individuals with sickle cell disease.

Figure 3.33 Structure of Adenosine Triphosphate (ATP)

ATP is a nucleotide found outside of nucleic acids that functions as the cellular currency of energy. It is modified by the addition of three phosphate groups and ribose to its nucleotide structure.

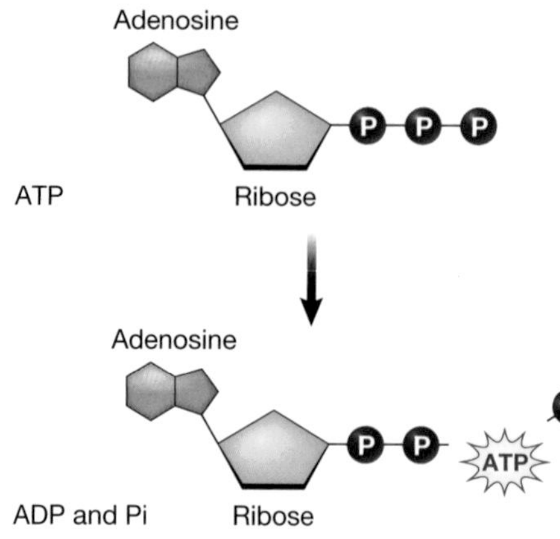

broken, the energy released helps fuel the body's activities, from muscle contraction to the transport of substances in and out of cells to anabolic chemical reactions.

A broken bond results in a phosphate group being cleaved from ATP, resulting in the products adenosine diphosphate (ADP) and a free inorganic phosphate (Pi). This hydrolysis reaction can be written as

$$ATP + H_2O \rightarrow ADP + P_i + energy$$

Learning Check

1. Which component of DNA is highly negatively charged?
   a. Phosphate group
   b. Deoxyribose sugar
   c. Nitrogenous base
2. Which of the following bases are considered purines? Please select all that apply.
   a. Cytosine
   b. Thymine
   c. Uracil
   d. Adenine
3. What type of bond connects the bases of two antiparallel DNA strands?
   a. Hydrogen
   b. Covalent
   c. Ionic

Chapter Review

3.1 Elements and Atoms: The Building Blocks of Matter

The learning objectives for this section are:

3.1.1 Compare and contrast the terms atoms, elements, molecules, and compounds.

3.1.2 Describe the charge, mass, and relative location of electrons, protons, and neutrons in an atom.

3.1.3 Distinguish among the terms atomic number, mass number, and atomic weight.

3.1.4 Explain how ions and isotopes are produced by changing the relative number of specific subatomic particles, using one element as an example.

3.1.5 Compare and contrast the terms ion, free radical, isotope, and radioisotope.

3.1.6 Relate the number of electrons in an electron shell to the atom's chemical stability and its ability to form chemical bonds.

3.1.7* Predict, by examining the properties of an atom, the likelihood that it will ionize or form bonds

* Objective is not a HAPS Learning Goal.

3.1 Questions

1. A single unit of glucose is defined as $C_6H_{12}O_6$. Identify the terms that best describe a single unit of glucose. Please select all that apply.
   a. Molecule
   b. Compound
   c. Element
   d. Atom

2. Which of the following subatomic particles would you find spinning around the nucleus?
   a. Proton
   b. Neutron
   c. Electron

3. The atomic weight of an atom can be approximated by calculating the sum of the:
   a. protons and electrons
   b. neutrons and electrons
   c. protons and neutrons

4. Which of the following shows how sodium ions are produced, and what constitutes sodium isotopes?
   a. Sodium ions are produced when sodium atoms gain an electron during the formation of a chemical bond; sodium isotopes contain different numbers of protons.
   b. Sodium ions are produced when sodium atoms lose an electron during the formation of a chemical bond; sodium isotopes contain different numbers of neutrons.
   c. Sodium ions are produced when sodium atoms lose a neutron during the formation of a chemical bond; sodium isotopes contain different numbers of electrons.
   d. Sodium ions are produced when sodium atoms gain a proton during the formation of a chemical bond; sodium isotopes contain different numbers of neutrons.

Mini Case 1: Exercise Stress Test

You are the first scientist to discover the element sulfur. You have found that its atomic number is 16 and its atomic weight is 32.

1. You studied sulfur further to determine its properties. What would you look for, in order to determine whether isotopes and radioisotopes existed for sulfur?
   a. You would look for atoms with 15 or 17 neutrons, to see if any isotopes existed. You would look for atoms that shed their subatomic particles and became unstable, to determine if any atoms were radioisotopes.
   b. You would look for atoms with 15 or 17 protons, to see if any isotopes existed. You would look for atoms that were stable and did not form chemical bonds, to determine if any atoms were radioisotopes.
   c. You would look for atoms with 15 or 17 electrons, to see if any isotopes existed. You would look for atoms that contained fewer protons than electrons, to determine if any atoms were radioisotopes.

2. Based on the information you have discovered about sulfur, are its atoms stable?
   a. Yes
   b. No

3. How likely is it that sulfur will form chemical bonds?
   a. It is unlikely; sulfur has an even number of electrons in its outermost electron shell, so it is stable.
   b. It is likely; sulfur has 6 electrons in its outermost shell, which can hold 8 electrons. If it forms chemical bonds that result in it containing 8 electrons in its outer shell, it will be more stable.
   c. It is unlikely, since sulfur atoms contain the same number of electrons as protons.
   d It is likely, since sulfur atoms contain an even number of neutrons.

3.1 Summary

- Atoms are made up of protons (positively charged), neutrons (neutrally charged), and electrons. The protons and neutrons make up the nucleus, while the electrons spin around the nucleus in electron shells. The atomic number tells you the number of protons and electrons, and the atomic mass tells you the number of protons and neutrons.
- Usually the numbers of protons, neutrons, and electrons are the same. Protons are the identity of the atom, so protons never leave during reactions. If a neutron is lost or added, the atom is known as an isotope. If an electron is lost or added, the atom is known as an ion.
- The last electron shell is known as the valence shell. During reactions, electrons from the valence shell share bonds with other molecules. If there are fewer than eight electrons in the valence shell, the atom will want to form bonds with other atoms.

3.2 Chemical Bonds

The learning objectives for this section are:

3.2.1 Explain the mechanism of each type of chemical bond and provide biologically significant examples of each: covalent, ionic, and hydrogen bonds.

3.2.2 List the following types of bonds in order by relative strength: nonpolar covalent, polar covalent, ionic, and hydrogen bonds.

3.2.3 Compare and contrast nonpolar covalent and polar covalent bonds.

3.2 Questions

1. What is a unique characteristic of covalent bonds?
   a. They are the weakest bond.
   b. They share electron pairs.
   c. They always involve three atoms.
   d. They involve two opposite charged atoms.

2. Which of the following lists the types of bonds from strongest to weakest?
   a. Polar covalent > nonpolar covalent > ionic > hydrogen
   b. Nonpolar covalent > polar covalent > ionic > hydrogen
   c. Hydrogen > ionic > polar covalent > nonpolar covalent
   d. Hydrogen > ionic > nonpolar covalent > polar covalent

3. Hydrogen fluoride [HF] contains a covalent bond. The fluorine in [HF] has nine protons while the hydrogen only has one. Is this a polar or nonpolar covalent bond?
   a. Polar covalent
   b. Nonpolar covalent

3.2 Summary

- There are three major types of chemical bonds in the body: ionic, covalent, and hydrogen bonds.
- A covalent bond involves the sharing of electrons between atoms in order to fill their valence shells. This is the strongest type of chemical bond.
- An ionic bond is an attraction between oppositely charge atoms. It is formed when one atom transfers one or more electrons to the other atom. Both atoms become ions as the bond is formed and both achieve chemical stability.
- A hydrogen bond is a weak attraction between the slightly positive atom in one polar molecule and the slightly negative atom in a nearby polar molecule.

3.3 Chemical Reactions

The learning objectives for this section are:

3.3.1* Compare and contrast kinetic and potential energy.

3.3.2* Compare and contrast endergonic and exergonic chemical reactions.

3.3.3* Describe and draw examples of the three basic types of chemical reactions.

3.3.4* List and explain several factors that influence the rate of reactions.

3.3.5 Define enzyme and describe factors that affect enzyme activity.

* Objective is not a HAPS Learning Goal.

3.3 Questions

1. You eat a sandwich for lunch. What kind of energy is stored between the atoms and molecules of your sandwich before it is broken down by your body? Please select all that apply.
   a. Potential energy
   b. Chemical energy
   c. Kinetic energy

2. Which of the following describes this reaction: $2Mg + O_2 \rightarrow 2MgO$? Please select all that apply.
   a. Anabolic reaction
   b. Endergonic reaction
   c. Synthesis reaction
   d. Catabolic reaction

3. Which of the following describes this reaction: $H_2O \rightarrow 2H + O$? Please select all that apply.
   a. Endergonic reaction
   b. Exergonic reaction
   c. Catabolic reaction
   d. Decomposition reaction

4. In a fever, the body temperature rises. Identify a possible effect that the increase in temperature has on the human body.
   a. Higher temperatures decrease the number of reactions in the human body.
   b. Higher temperatures increase the potential energy stored in the bonds.
   c. High temperatures increase the speed at which atoms and molecules move, and therefore, the chemical reaction rates.
   d. High temperatures increase the distance between particles.

5. How do enzymes work to increase the rate of chemical reactions?
   a. An enzyme decreases the size of molecules involved in the reaction.
   b. An enzyme decreases the activation energy required in the reaction.
   c. An enzyme increases the speed of molecules involved in the reaction.
   d. An enzyme increases the concentration of molecules in the reaction.

3.3 Summary

- Chemical energy is a type of potential energy, which is energy stored in the bonds of a molecule. When these bonds break, this energy is released and transforms into kinetic energy.
- Chemical reactions occur when reactants form a product. Two reactants can come together to form one product, which is known as a synthesis reaction. One reactant can split up to form two or more products, which is known as a decomposition reaction. Lastly, both of these can occur in a single reaction, and that reaction is known as an exchange reaction.
- The factors that influence the rate of chemical reactions are molecular size, phase of reaction, temperature, concentration, pressure, and presence of a catalyst (or enzyme).

3.4 Inorganic Compounds Essential to Human Functioning

The learning objectives for this section are:

3.4.1* Compare and contrast organic and inorganic compounds and give examples of each.

3.4.2 Describe the physiologically important properties of water.

3.4.3 Compare and contrast the terms solution, solute, solvent, colloid suspension, and emulsion.

3.4.4 Define the terms salt, pH, acid, base, and buffer.

3.4.5 State the pH values for acidic, neutral, and alkaline (basic) solutions.

* Objective is not a HAPS Learning Goal.

3.4 Questions

1. Your classmate asked you whether glucose and sodium chloride were organic or inorganic, and if you would explain why.What would you say?
   a. Both are organic, because they are both found in the human body.
   b. Both are inorganic, because they contain relatively small molecules.
   c. Glucose is organic, since it is composed mainly of carbon.Sodium chloride is inorganic, because it is not composed mainly of carbon.
   d. Glucose is inorganic, because it is nonpolar.Sodium chloride is organic, because it is polar.

2. Why is water called the "universal solvent"?
   a. Water lubricates the tissues and joints in our bodies.
   b. Water protects cells and organs from trauma.
   c. Water absorbs heat and changes from liquid to gas.
   d. Water can dissolve many different types of molecules.

3. Mary Beth is doing a science experiment in class. She leaves her solution overnight. The next morning, they see the heavier particles sitting on the bottom of the tube and the water portion sitting on top. How would you describe the initial solution?
   a. Suspension
   b. Colloid
   c. Concentration
   d. Molarity

4. Which of the following substances can dissociate into ions other than H^+ or OH^-?
   a. Acid
   b. Base
   c. Salt
   d. Buffer

5. Baking soda has a pH of 9.5, vinegar has a pH of 2, and water has a pH of 7. How is each of these substances classified on the pH scale?
   a. Baking soda is a base, vinegar is an acid, and water is neutral.
   b. Baking soda is an acid, vinegar is a base, and water is neutral.
   c. Baking soda is neutral, vinegar is a base, and water is an acid.
   d. Baking soda is a base, vinegar is neutral, and water is an acid.

3.4 Summary

- The inorganic compounds essential to life are water, salts, acids, and bases.
- Water acts as a lubricant, cushion, and heat sink, and is termed the "universal solvent."
- Salts release ions other than H^+ or OH^-. Acids release H^+ ions. Bases release OH^- ions.
- The pH scale is a logarithmic scale with 7 being neutral. Anything below 7 is acidic, while anything above 7 is basic.

3.5 Organic Compounds Essential to Human Functioning

The learning objectives for this section are:

3.5.1 Define the term organic molecule.

3.5.2 Explain the relationship between monomers and polymers.

3.5.3 Define and provide examples of dehydration synthesis and hydrolysis reactions.

3.5.4 Compare and contrast the general molecular structure of carbohydrates, proteins, lipids, and nucleic acids using chemical formulas.

3.5.5 Describe the building blocks of carbohydrates, proteins, lipids, and nucleic acids, and explain how these building blocks combine with themselves or other molecules to create complex molecules in each class, providing specific examples.

3.5.6 Describe the four levels of protein structure and the importance of protein shape for function.

3.5.7 Define enzyme and describe factors that affect enzyme activity.

3.5 Questions

1. Which of the following defines organic molecule?
   a. A substance that contains both carbon and hydrogen.
   b. A substance that contains only carbon.
   c. A substance that contains only hydrogen.
   d. A substance that contains neither hydrogen or carbon.

2. Describe how monomers form polymers.
   a. Hydrolysis reaction
   b. Dehydration synthesis reactions
   c. Ionic bonds

3. Which of the following are produced by a dehydration synthesis of glycerol and three fatty acids?
   a. Carbohydrates
   b. Proteins
   c. Lipids
   d. Amino acids

4. An organic molecule has the following chemical formula: $C_6H_{12}O_6$. Which type of organic compound would you expect this substance to be? Why?
   a. Lipid, because it consists mainly of carbon and hydrogen atoms
   b. Carbohydrate, because it contains a 2:1 ratio of hydrogen to oxygen atoms
   c. Nucleic acid, because it contains equal numbers of carbon and oxygen atoms
   d. Protein, because it contains a 2:1 ration of hydrogen to carbon atoms

5. Explain how phospholipids are different from triglycerides.
   a. In a phospholipid molecule, the head and tail are hydrophobic.
   b. In a phospholipid molecule, the head is polar and the tail is hydrophilic.
   c. In a phospholipid molecule, the head and tail are hydrophilic.
   d. In a triglyceride molecule, the head and tail are hydrophobic.

6. What is unique about proteins? Please select all that apply.
   a. Proteins can fold on top of themselves to form secondary and tertiary structures.
   b. The central carbon of the amino acid monomer is bound to an amino group and a carboxyl group.
   c. The fatty acid chain has a kink in it due to a double covalent bond.
   d. Proteins only bind to carbon, hydrogen, and oxygen to form a ring.

7. What is the function of an enzyme?
   a. Increase the rate of a reaction
   b. Sensitize nerves to pain
   c. Source of fuel for the body

3.5 Summary

- Organic compounds are found in living organisms. They consist mainly of hydrogen atoms covalently bound to carbon atoms.
- There are four major types of organic compounds found in the human body: carbohydrates, lipids, proteins, and nucleic acids.
- Carbohydrates provide energy for body processes and can be stored for later use.
- Lipids provide energy, they provide cushioning and insulation for the body, and they represent an efficient form of energy storage.
- Proteins consist of building blocks called *amino acids*. They function as enzymes, antibodies, plasma proteins, structural proteins, and hormones.
- Nucleic acids (DNA and RNA) are composed of building blocks called *nucleotides*. DNA is the genetic material that stores the instructions for producing proteins for the cells. RNA performs several functions in protein synthesis.

4

The Cellular Level of Organization

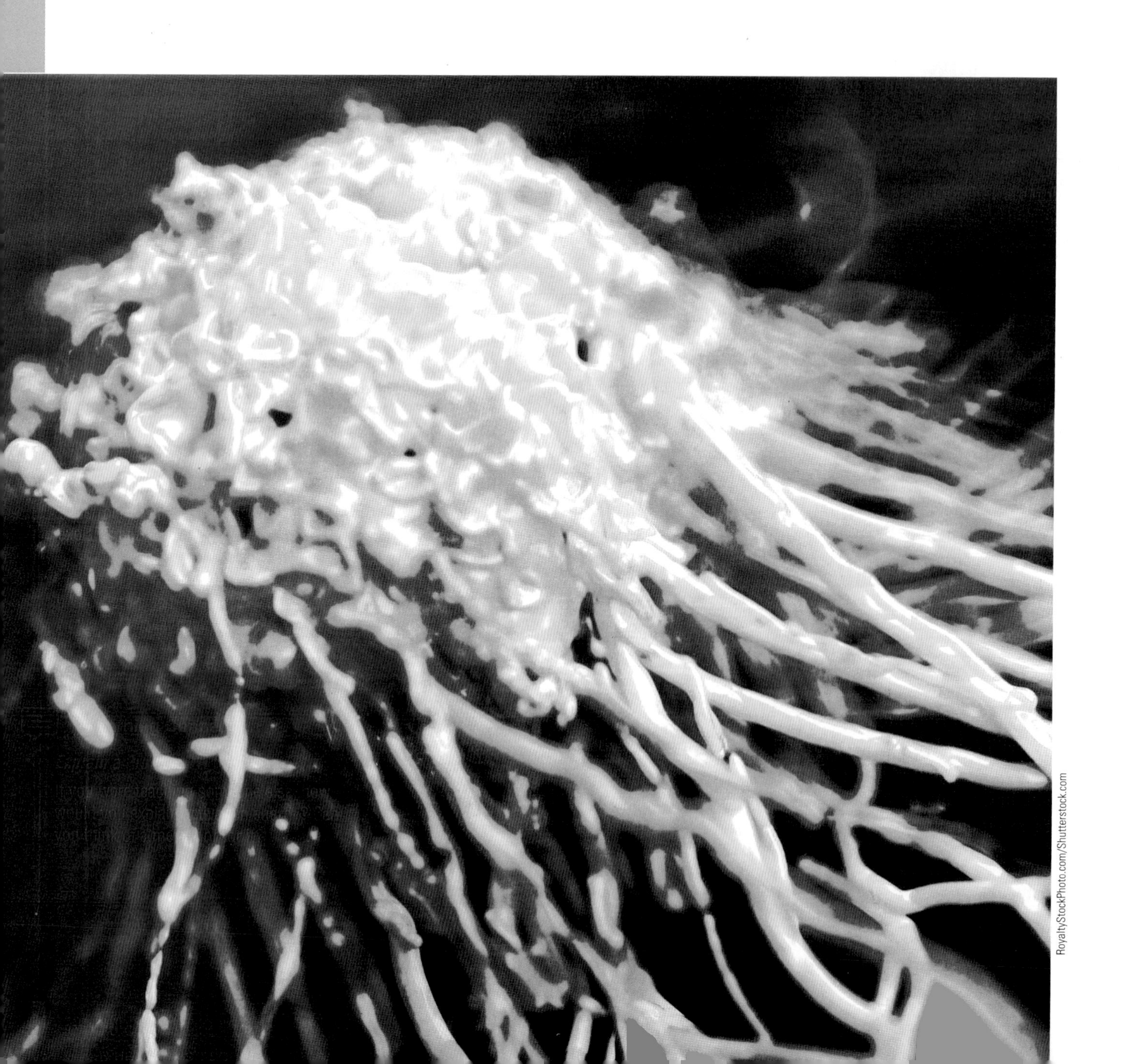

RoyaltyStockPhoto.com/Shutterstock.com

Chapter Introduction

In this chapter we will learn...

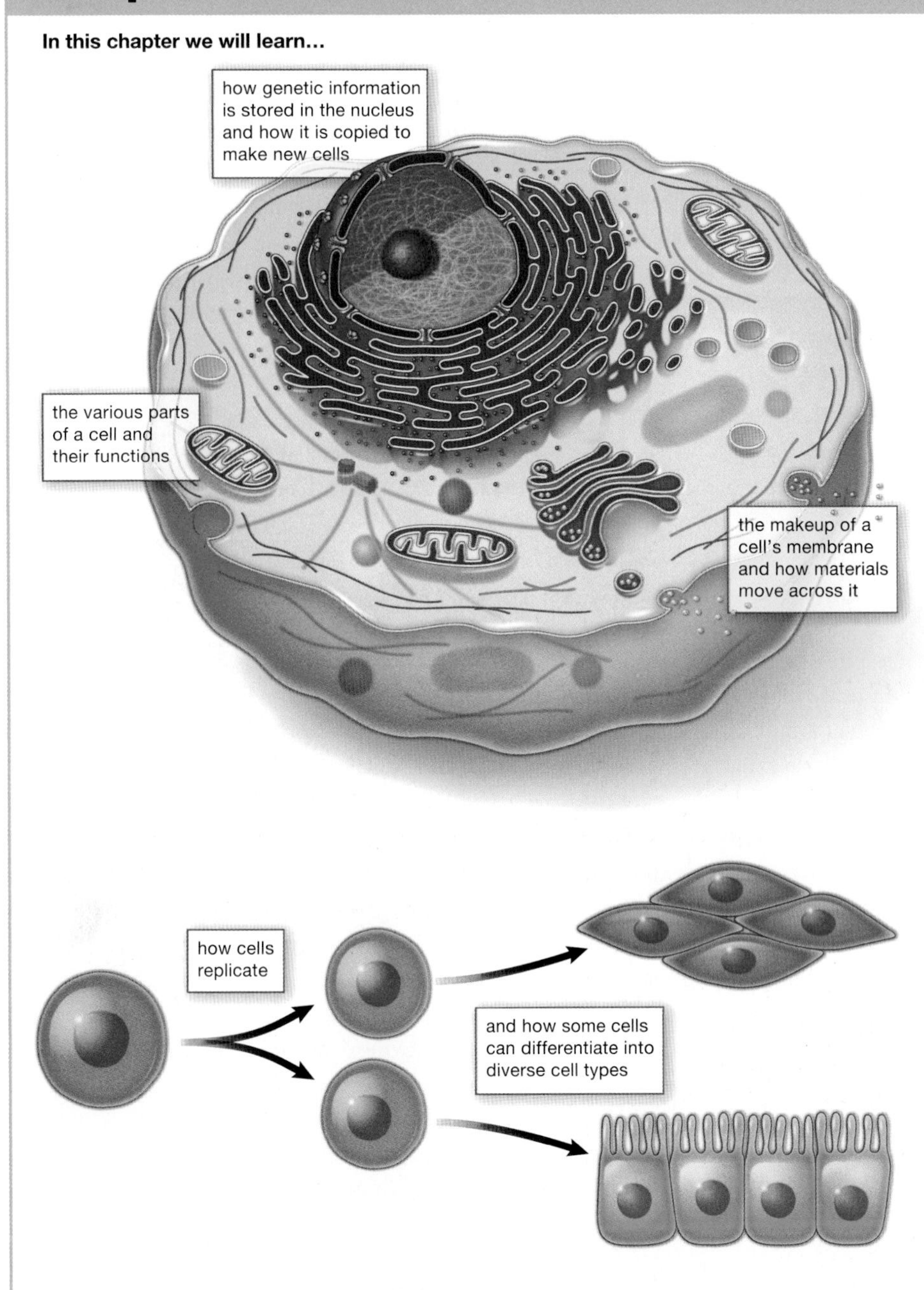
how genetic information is stored in the nucleus and how it is copied to make new cells
the various parts of a cell and their functions
the makeup of a cell's membrane and how materials move across it
how cells replicate
and how some cells can differentiate into diverse cell types

You, like all humans, developed from a single fertilized egg cell into the complex organism containing trillions of specialized cells that you see when you look in a mirror. During this developmental process, early nonspecific cells differentiate and become specialized in their structure and function. These different cell types form specialized tissues that work in concert to perform all of the functions necessary to sustain life.

Consider the difference between a red blood cell and a nerve cell. Red blood cells are among the smallest cells in our body. They are efficient transporters of oxygen and other gasses, and are shaped like a disc that is slightly depressed on each side (they look like a jelly doughnut with all the filling squeezed out). Red blood cells are incapable of duplicating themselves, and so they are disposable, being constantly replaced and removed from circulation. A nerve cell, on the other hand, may be shaped something like a star, sending out long processes up to three feet (one meter) in length! The nerve cells that control the muscles in your toes stretch from mid-spine all the way to those toes. Almost all the nerve cells you have in your brain have been there since you were born. Nerve cells can communicate with one another and with other types of cells using electrical and chemical signals. The differences between nerve and red blood cells illustrate one very important theme that is consistent at all organizational levels of biology: the physical structure of a cell is optimally suited and intimately tied to cellular function. A nerve cell could not perform its function of communication if it were the size of a red blood cell, and a red blood cell would not be able to hold as much oxygen if it had all the intracellular contents of a nerve cell. Keep this theme in mind as you are introduced to the various types of cells in the body throughout the text.

In their own way, each of the body's approximately 40 trillion cells contributes to homeostasis. *Homeostasis* is a term used in biology that refers to a dynamic state of balance within parameters that are compatible with life. For example, living cells require a water-based environment to survive in, and there are various anatomical and physiological mechanisms that maintain water balance in all of the compartments of the body. This is one aspect of homeostasis. When a particular parameter, such as blood pressure or blood oxygen content, moves far enough *out* of homeostasis (too high or too low), illness, disease, or death may result.

4.1 The Cell Membrane and Its Involvement in Transport

Learning Objectives: By the end of this section, you will be able to:

4.1.1 Describe the structure of the cell (plasma) membrane, including its composition and arrangement of lipids, proteins, and carbohydrates.

4.1.2 Compare and contrast intracellular fluid and extracellular fluid with respect to chemical composition and location.

4.1.3 Describe the functions of different plasma membrane proteins (e.g., structural proteins, receptor proteins, channels).

4.1.4 Compare and contrast simple diffusion across membranes and facilitated diffusion in respect to their mechanisms, the type of material being moved, and the energy source for the movement.

4.1.5 Compare and contrast facilitated diffusion, primary active transport, and secondary active transport in respect to their mechanisms, the type of material being moved, and the energy source for the movement.

4.1.6 Define osmosis and explain how it differs from simple diffusion across membranes.

4.1.7 Compare and contrast osmolarity and tonicity of solutions.

4.1.8 Describe the effects of hypertonic, isotonic, and hypotonic solutions on cells.

Despite differences in structure and function, all living cells are surrounded by a membrane. As the outer layer of your skin separates your body from its environment, the cell membrane (also known as the plasma membrane) separates the inner contents of a cell from its external environment and regulates which materials can pass in or out.

4.1a Structure and Composition of the Cell Membrane

LO 4.1.1

cell membrane (plasma membrane)

The **cell membrane** is an elegantly flexible structure composed of phospholipids, cholesterol, and proteins. Phospholipids make up approximately 50 percent of the cell membrane by weight. Let's examine their structure and function first.

A single phospholipid molecule has a phosphate group on one end, called the "head," and two fatty acid tails (**Figure 4.1**). The phosphate group is negatively charged, making the head polar and **hydrophilic** (water-loving); thus, the phospholipid heads are drawn toward the watery environments inside and outside of the cell. The lipid tails, on the other hand, are uncharged, nonpolar, and are **hydrophobic** (repels and is repelled by water). Because cell membranes have both hydrophilic and hydrophobic regions, phospholipids are described as being **amphipathic**. When amphipathic molecules are mixed into water-based solutions, they organize themselves so that their hydrophilic regions are exposed to the water, but their hydrophobic regions are huddled together and protected from the watery environment (**Figure 4.2A**). Human tissues are **aqueous** (watery) environments. Within the aqueous environment, the amphipathic phospholipids arrange themselves into a bilayer so that their hydrophilic heads face the watery inside and watery outsides of the cells, and their lipid tails are hugged into a core of the membrane and protected from the water (**Figure 4.2B**).

LO 4.1.2

The **intracellular fluid (ICF)** is the fluid interior of the cell and the **extracellular fluid (ECF)** is the fluid environment outside of the cell membrane. These two fluids are fairly similar in composition. The term *extracellular fluid* always refers to the fluid

The Human Anatomy and Physiology Society includes more than 1,700 educators who work together to promote excellence in the teaching of this subject area. The HAPS A&P Learning Outcomes measure student mastery of the content typically covered in a two-semester Human A&P curriculum at the undergraduate level. The full Learning Outcomes are available at https://www.hapsweb.org.

Figure 4.1 Phospholipid Structure

A phospholipid molecule consists of a polar phosphate "head," which is hydrophilic, and a nonpolar lipid "tail," which is hydrophobic. Unsaturated fatty acids result in kinks in the hydrophobic tails.

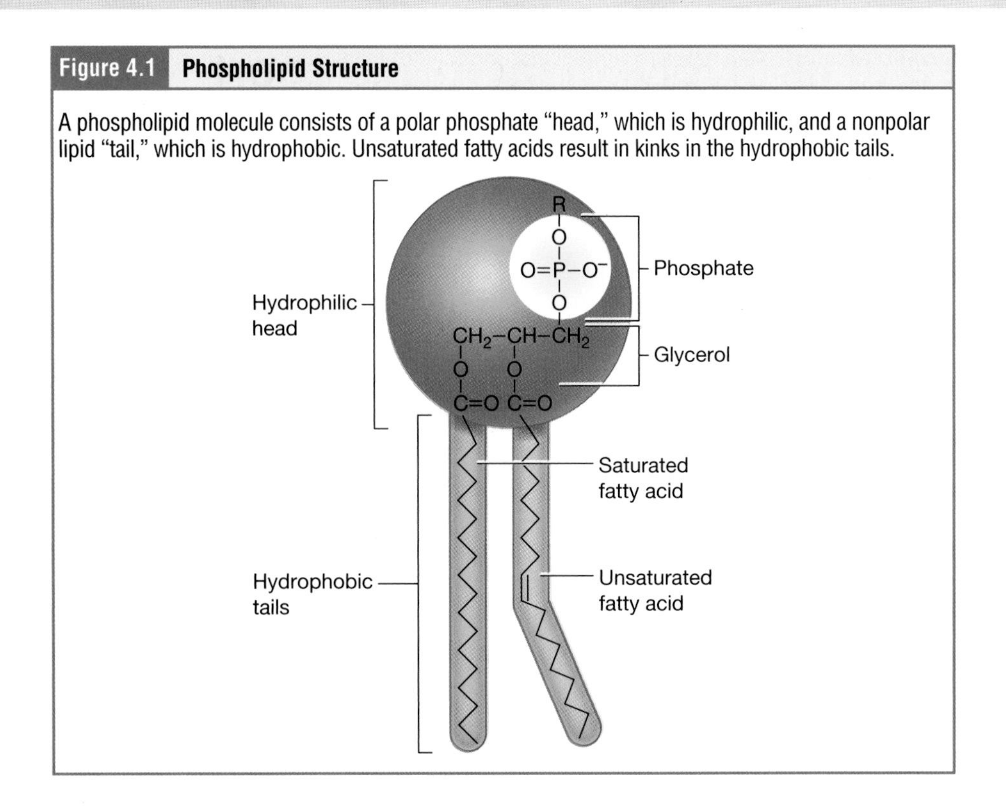

immediately outside of a cell; however, it can be referred to by more regionally specific terms as well: **interstitial fluid (IF)** is the term given to extracellular fluid not contained within blood vessels; plasma is extracellular fluid within blood vessels. The structure of phospholipids is essential to their function. Because the membrane has a hydrophobic core, polar molecules dissolved in the aqueous ICF or ECF cannot readily diffuse across the core of the membrane. Therefore, the phospholipid bilayer makes the cell

Figure 4.2A Arrangements of Amphipathic Molecules in Aqueous Environments

If amphipathic molecules such as phospholipids were mixed into an aqueous solution, there are three forms that they might spontaneously arrange into. In every form, note that the hydrophobic tails are shielded from the water by hydrophilic heads.

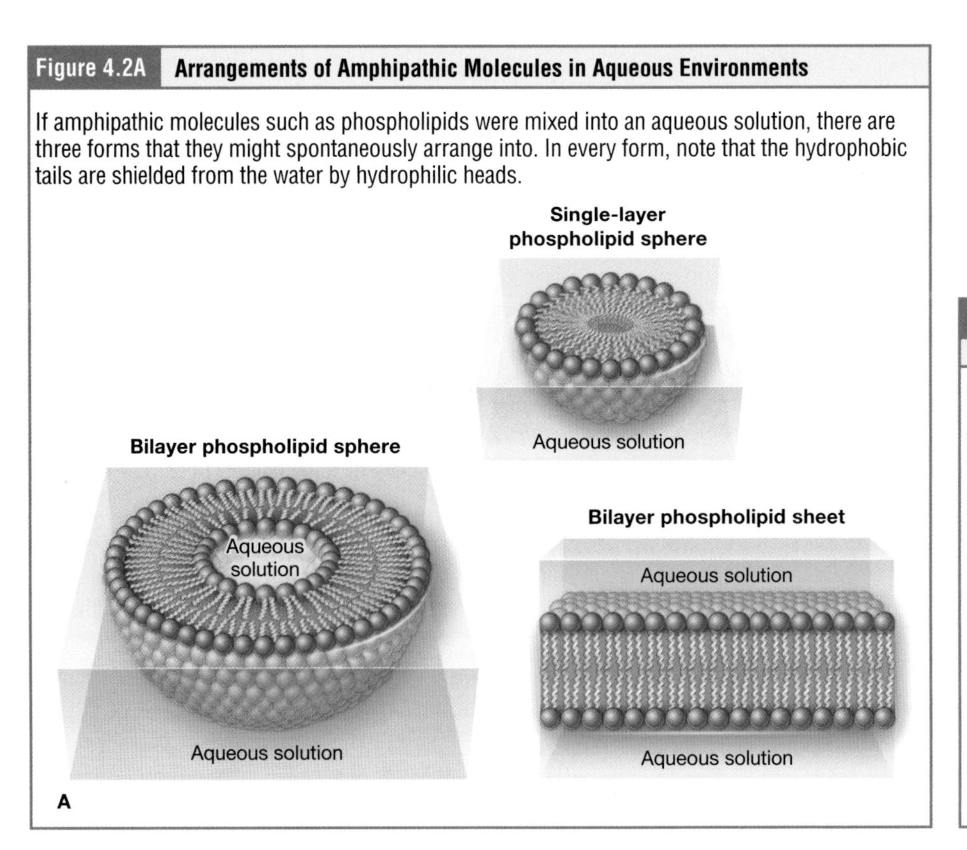

Figure 4.2B Phospholipid Bilayer of Cell Membranes

The phospholipid bilayer consists of two sheets of phospholipids. The hydrophobic tails cluster together and the hydrophilic heads orient themselves toward the watery environments on either side.

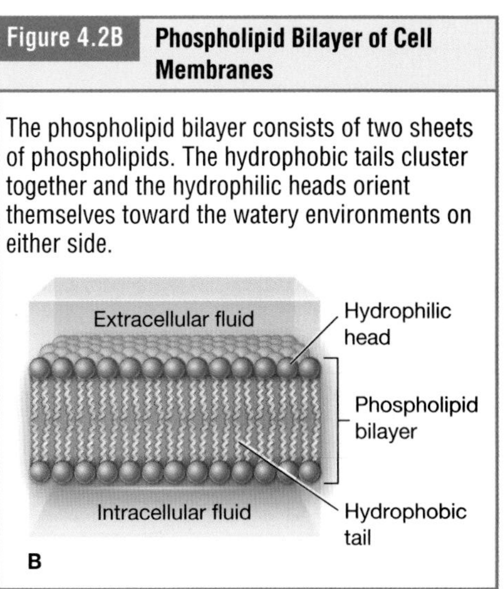

membrane an effective barrier to molecular movement from inside to out or outside to in. The other 50 percent of the cell membrane is made up of proteins and cholesterol. An important feature of the membrane is that it remains fluid and that the lipids and proteins in the cell membrane are not rigidly locked in place. This fluidity is the function of membrane cholesterol.

4.1b Membrane Proteins

LO 4.1.3

The lipid bilayer functions as the barrier between ICF and ECF, but the various membrane proteins add many more functions to the membrane. Membrane proteins come in two flavors: transmembrane and peripheral proteins (**Figure 4.3**). As its name suggests, a **transmembrane protein** is a protein that spans the membrane from its intracellular to its extracellular side. **Channel proteins** are examples of transmembrane proteins that allow select materials, such as certain ions, to pass into or out of the cell.

channel protein (a type of transmembrane protein)

Some transmembrane proteins serve dual roles as both a receptor and an ion channel. **Receptors** are transmembrane proteins found on the extracellular side of the membrane that can selectively bind a specific molecule outside the cell and translate this binding into a chemical reaction inside the cell. Each receptor has a **ligand**, the specific molecule that binds to and engages that receptor.

One example of a receptor–ligand interaction is the receptors on nerve cells that bind neurotransmitters, the chemicals that nerve cells use to communicate. One example of a neurotransmitter is dopamine. When dopamine binds to its receptor protein, a channel within the protein opens to allow certain ions to flow from one side of the membrane to the other.

Some transmembrane proteins are glycoproteins. A **glycoprotein** is a protein that has carbohydrate molecules attached. The attached carbohydrates act as tags on the

Figure 4.3 Cell Membrane

The cell membrane is composed of a bilayer of phospholipids along with many proteins. Like phospholipids, proteins arrange themselves within the membrane according to their affinity for water. Proteins embed in the phospholipid membrane so that the hydrophobic regions of the protein are surrounded by the hydrophobic phospholipid tails of the membrane. Likewise, the hydrophilic regions of the proteins will be aligned with the hydrophilic phospholipid heads. Other molecules within the membrane include cholesterol and carbohydrates.

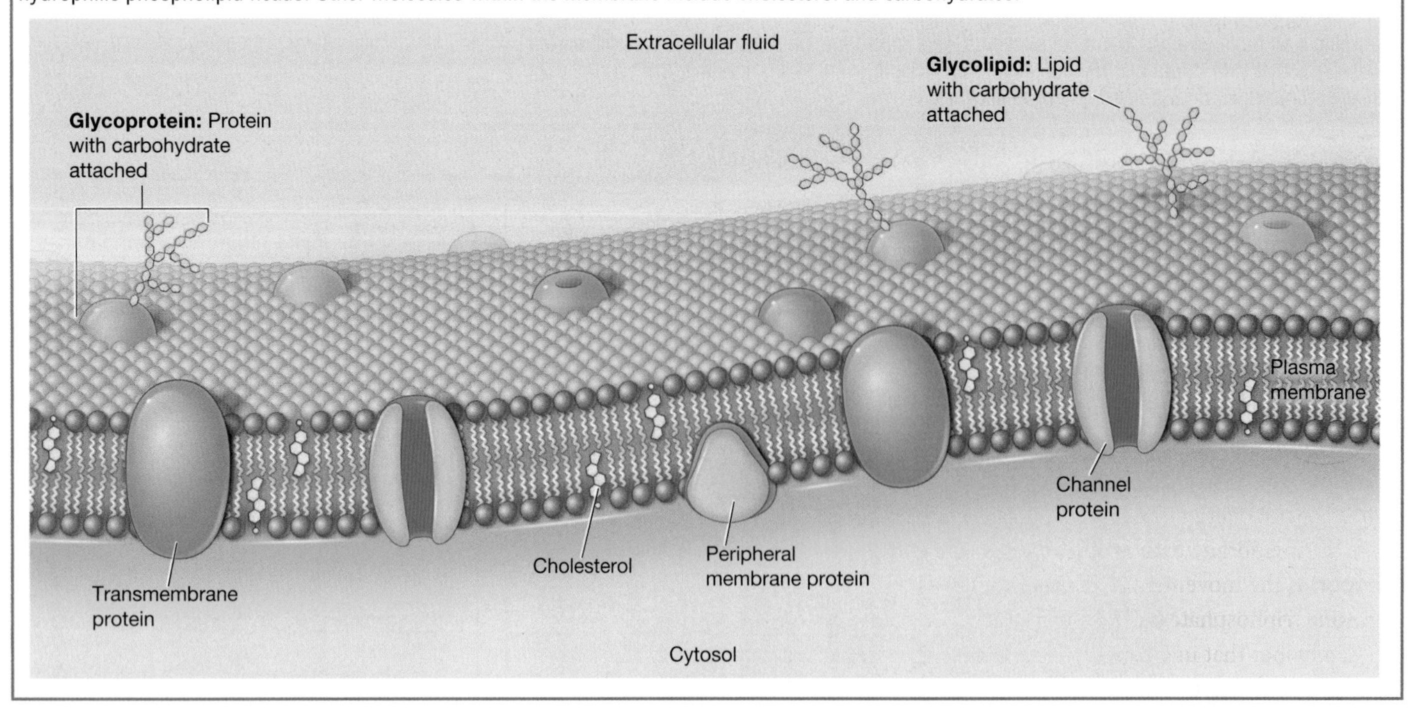

glycoproteins and aid in cell recognition. Some cells carry so many glycoproteins that these form a coat called a **glycocalyx**. A glycocalyx can have various roles. For example, it may have molecules that allow the cell to bind to another cell, it may contain receptors for hormones, or it might have enzymes to break down nutrients. The glycocalyces found in a person's body are products of that person's genetic makeup. They give each of the individual's trillions of cells the "identity" of belonging in the person's body. This identity is the primary way that a person's immune defense cells "know" not to attack the person's own body cells, but it also is the reason organs donated by another person might be rejected.

In contrast to transmembrane proteins, which span the membrane, **peripheral proteins** can be anchored to the interior or exterior membranes of the cell but do not span the membrane (see Figure 4.3). There are several subtypes of each of these proteins.

Learning Check

1. What is the function of the cell membrane? Please select all that apply.
   a. To separate inner contents of the cell from the external environment
   b. To regulate which contents can pass in or out of the cell
   c. To provide rigidity which gives structure to tissues
   d. To move the cell from one location in the body to another
2. How do amphipathic molecules organize when mixed into a water-based solution? Please select all that apply.
   a. Hydrophilic regions are exposed to the watery environment.
   b. Hydrophilic regions are protected from the watery environment.
   c. Hydrophobic regions are exposed to the watery environment.
   d. Hydrophobic regions are protected from the watery environment.
3. What is the function of a channel protein?
   a. To selectively bind to a specific molecule from outside the cell
   b. To translate binding to a specific molecule into chemical energy
   c. To allow select materials to pass into or out of a cell
   d. To bind to and activate a receptor to trigger a chemical reaction inside the cell

4.1c Transport across the Cell Membrane

The membrane's lipid bilayer structure functions to control the movement of all polar, charged, or large molecules across the membrane. The membrane is often referred to as **selectively permeable** because it allows only substances that are relatively small and nonpolar to cross through the lipid bilayer freely.

Some examples of these freely permeable molecules are: lipids, oxygen, carbon dioxide, and alcohol. However, water-soluble (polar) materials—such as glucose, amino acids, and ions—need some assistance to cross the membrane because they are repelled by the hydrophobic tails of the phospholipid bilayer. All substances that move through the membrane do so by one of two general methods, which are categorized based on whether or not energy is required. **Passive transport** is the movement of substances across the membrane without the expenditure of cellular energy. In contrast, **active transport** is the movement of substances across the membrane using energy from adenosine triphosphate (ATP).

Remember that in Chapter 2 we discussed that flow is the product of gradients and resistance (**Figure 4.4**). All molecules will flow down their concentration gradients

Student Study Tip

Active transport moves a substance against its concentration gradient. Diffusion and facilitated diffusion move a molecule down its concentration gradient.

Figure 4.4 Flow Rate and Its Determining Factors

Flow is proportional to the size of the gradient and inversely proportional to resistance.

$$\text{Flow} \propto \frac{\text{Gradient}}{\text{Resistance}}$$

Figure 4.5 Simple Diffusion across the Cell (Plasma) Membrane

The properties of the cell membrane allow small, uncharged molecules such as oxygen and carbon dioxide, and hydrophobic molecules such as lipids, to pass through—down their concentration gradient—by simple diffusion.

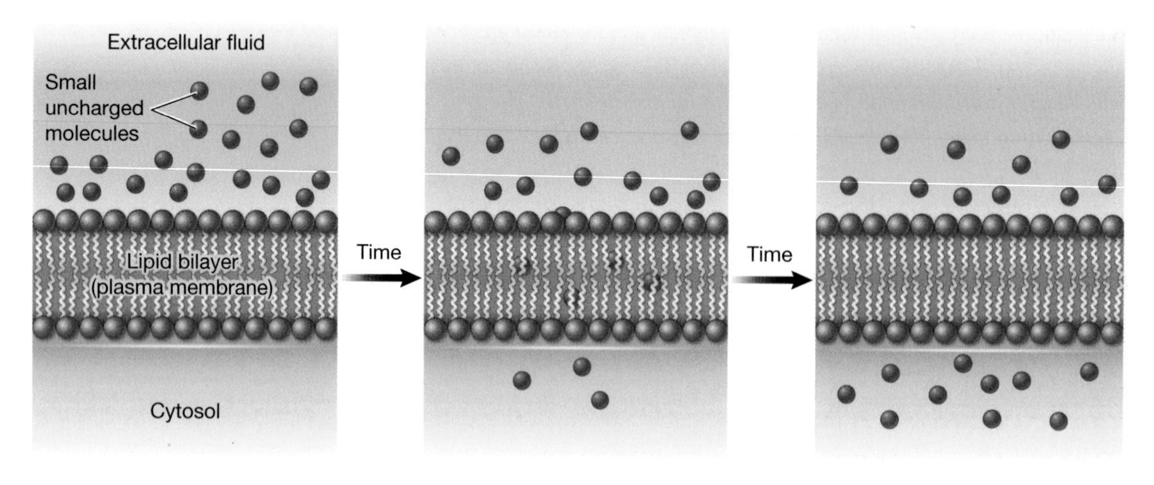

from areas where they are more concentrated to areas where they are less concentrated unless that flow is prevented by sufficient resistance.

Passive Transport In order to understand *how* molecules move across a cell membrane, it is necessary to understand concentration gradients and diffusion. A **concentration gradient** is the difference in concentration of a substance across a space. An example is provided in **Figure 4.5**. Molecules (including ions) will move from where they are more concentrated to where they are less concentrated until they are equally distributed in that space (see Figure 4.5). This movement, from areas where they are more concentrated to areas where they are less concentrated, is referred to as moving *down* the concentration gradient. **Diffusion** is the term for the movement of molecules in this direction (from an area of higher concentration to an area of lower concentration).

Student Study Tip

If you bump into a stranger at a gathering, chances are you'll want to move farther away from them after the collision. The more collisions that happen, the more people/molecules tend to spread apart.

A few common examples help to illustrate this concept:

- Imagine being inside a room where perfume was sprayed (or another scent such as a cup of hot coffee). The scent molecules would naturally diffuse from the spot where they left the bottle to all corners of the room, and this diffusion would continue until no more concentration gradient remained (**Figure 4.6A**).
- Imagine a cube or spoonful of sugar placed in a cup of tea. Eventually the sugar will diffuse throughout the tea until no concentration gradient remains (**Figure 4.6B**).
- Imagine a bit of food coloring or dye dropped into water. The dye molecules slowly spread out until the entire container of water is one uniform color (**Figure 4.6C**).

There are some themes to these examples: diffusion continues until no gradient remains. This is because molecules are always in motion. Their movement never stops but when there are more of them in a closed space, they bounce off each other more, producing more net movement. Once the molecules are relatively equally distributed, this movement is less noticeable until it finally reaches **equilibrium**, the state at which there is no net movement in any direction. Take particular note of

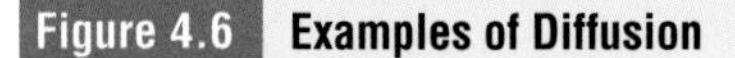

Figure 4.6 Examples of Diffusion

(A) Perfume molecules diffuse from their source, the perfume bottle, throughout the room down their concentration gradient. Eventually they will reach equilibrium and be in an equal concentration in every location within the room. (B) Sugar molecules diffuse throughout a hot cup of coffee. They begin in a group at the bottom of the cup, but slowly diffuse, down their concentration gradient, until they are evenly spread out (equilibrium) throughout the coffee liquid. (C) Molecules of dye spread throughout a container of water, down their concentration gradient.

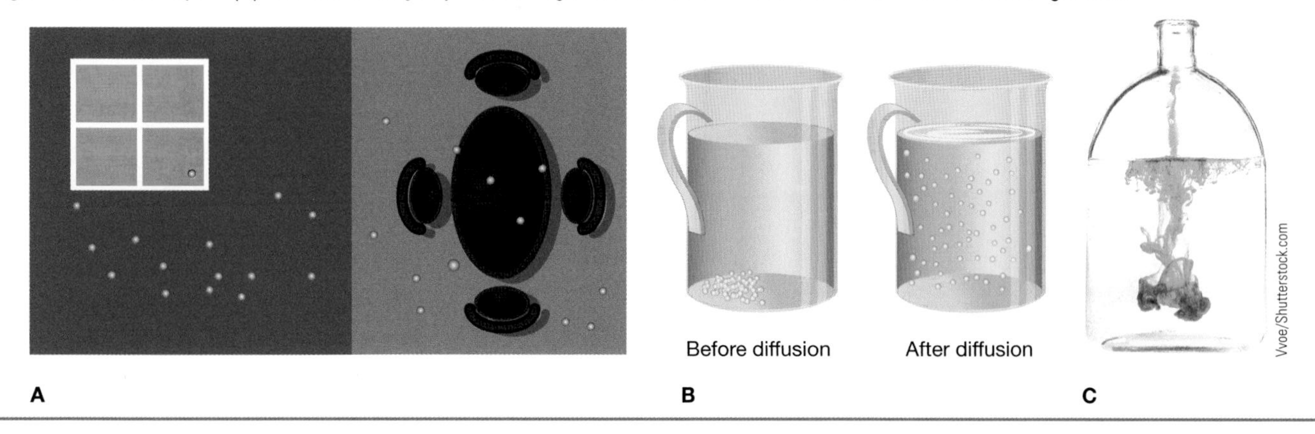

the word *net* in this definition of equilibrium. Molecules still move randomly, but for every molecule that moves in one direction there is an equal amount of movement in another direction. Another theme is temperature. If the room is warmer, the coffee hot instead of iced, or the tea warmer, diffusion occurs even faster. This is because the molecules are bumping into each other and spreading out faster than at cooler temperatures. Let us consider body temperature homeostasis in a new light. If body temperature is steadfastly maintained around 98.6°F, then the rate of molecular motion is stable and predictable. But what happens to the body's molecules during fever or hypothermia?

LO 4.1.4

Different molecules have different relationships with the cell membrane based on their molecular properties. Some substances can easily diffuse through the cell membrane, such as the gasses oxygen (O_2) and carbon dioxide (CO_2). These molecules are small and nonpolar. They dissolve in the ECF, the ICF, and the lipid bilayer. Therefore, they will be able to easily diffuse across the cell membrane according to their concentration gradients.

Because cells rapidly use up oxygen during metabolism, there is typically a lower concentration of O_2 inside the cell than outside. As a result, oxygen will diffuse from the extracellular fluid directly through the lipid bilayer of the membrane and into the cytoplasm. Similarly, because cells produce CO_2 as a byproduct of metabolism, CO_2 concentrations are typically high within the cytoplasm; therefore, CO_2 will move from the cell through the lipid bilayer and into the extracellular fluid, where its concentration is lower.

LO 4.1.5

Polar or ionic molecules, which are hydrophilic, cannot easily cross the phospholipid bilayer. Charged atoms or molecules of any size cannot cross the cell membrane via simple diffusion as the charges are repelled by the hydrophobic tails in the interior of the phospholipid bilayer. Because most substances cannot pass freely through the lipid bilayer of the cell membrane, their movement requires assistance or facilitation. Studded throughout the membrane are transmembrane proteins that act as channels or specialized transport proteins. **Facilitated diffusion** is the diffusion of molecules down their concentration gradients using the assistance of transmembrane proteins. Some molecules cannot cross the membrane on their own (**Figure 4.7**) due to their molecular properties, but proteins can enable their crossing. A common example of

facilitated diffusion (mediated diffusion)

Figure 4.7 Facilitated Diffusion

(A) Facilitated diffusion of ions or molecules crossing the cell membrane takes place with the help of proteins such as channel proteins. Molecules can only move from areas of higher concentration to areas of lower concentration through the channel. (B) Carrier proteins provide another mechanism of facilitated diffusion. An ion or molecule will bind to the carrier protein, triggering a change in the protein's shape such that the ion or molecule can now pass to the other side of the membrane.

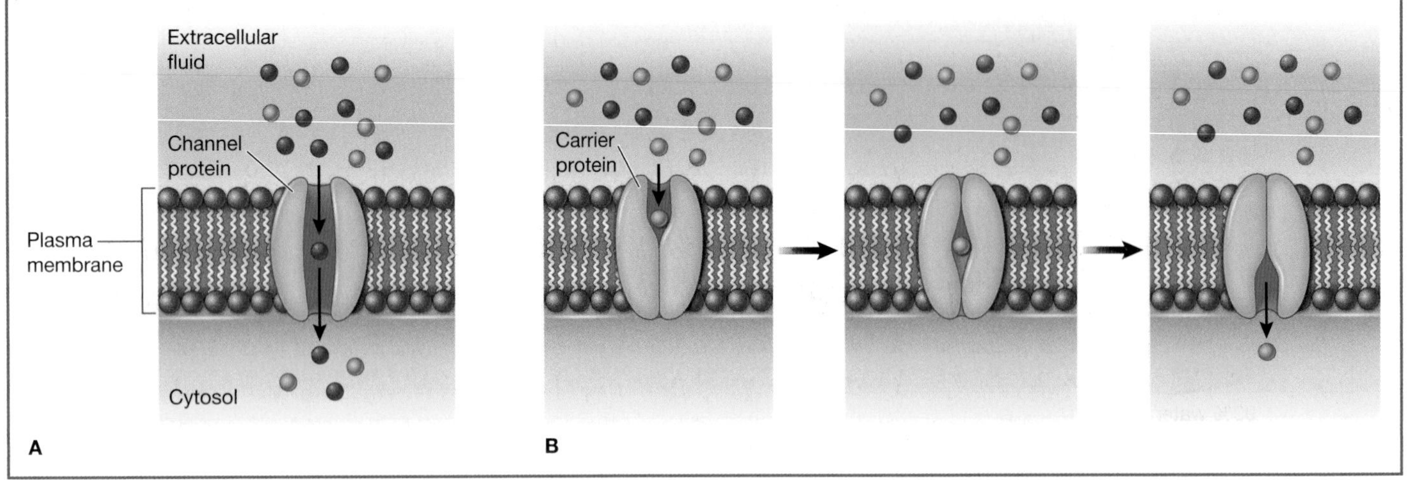

facilitated diffusion is the movement of glucose into the cell, where it is used to make ATP. Although glucose can be more concentrated outside of a cell, it cannot cross the lipid bilayer via simple diffusion because it is both large and polar. To resolve this, a specialized carrier protein called the *glucose transporter* will transfer glucose molecules into the cell to facilitate its inward diffusion.

As an example, even though sodium ions (Na^+) are highly concentrated outside of cells, their charge prevents them from passing through the nonpolar core of the membrane. Their diffusion is facilitated by membrane proteins that form sodium channels, so that Na^+ ions can move down their concentration gradient from outside to inside the cells. There are many other solutes, such as amino acids or glucose, that must undergo facilitated diffusion to move across the cell membrane. Because facilitated diffusion is a passive process, it does not require energy expenditure by the cell.

LO 4.1.6

Water also can move across the cell membrane of all cells, either through protein channels or, to a small degree, by slipping between the lipid tails of the membrane itself. **Osmosis** is the diffusion of water through a semipermeable membrane. Just like the other forms of diffusion we have been discussing, water molecules move randomly but net movement follows water molecule concentration gradients. The fluids of the body are solutions; that is to say, they are mixtures of water and solutes. When we consider a water molecule gradient, we can also look at the distribution of the solute for clues. Examine **Figure 4.8**. Here we are looking at two solutions, each made up of water and solute. The solution on the left is composed almost entirely of water molecules. The solution on the right has a lower concentration of water molecules, but a higher concentration of solute. Now look at **Figure 4.9A**. If these two solutions were adjacent and separated by a membrane that allowed the water molecules to move, but prevented the solute from moving (like the cell membrane), water would move by osmosis to the solution that had a lower water molecule concentration but a higher solute concentration (**Figure 4.9B**). Note that when water molecules move, a volume change occurs; at the end of the osmosis process the two solutions have the same water molecule concentration, but different volumes.

Student Study Tip

Think back to membrane structure: whether osmosis or diffusion occurs between a cell and ECF depends on if the solute can diffuse across the membrane. If the solute trying to cross the membrane is Na^+, for example, will water or the solute cross?

Figure 4.8 Water Molecule Concentration Is Related to Solute Concentration

Think of a solution as being made up of a combination of solute and water molecules; the more solute molecules in the solution, the lower the concentration of water molecules.

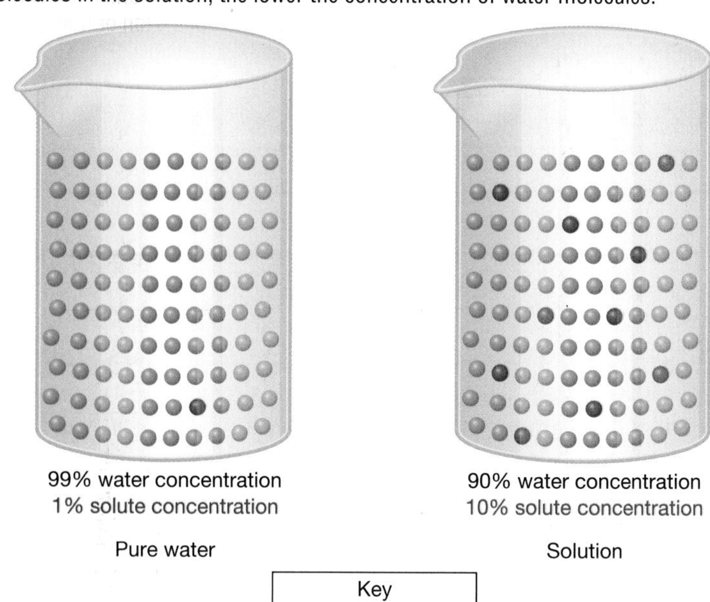

Figure 4.9 Osmosis

Osmosis is the diffusion of water across a membrane down its concentration gradient. The membrane illustrated here is permeable to water, but not to the solute. (A) Depicts the two solutions at the moment they are placed in the beaker. Over time, water molecules will move toward the area where they are in a lower concentration (in other words, where the solute is in a higher concentration) on the right side of the beaker. (B) Illustrates the volumes on each side of the membrane after the water molecules have reached a dynamic equilibrium. Both solutions are now at equal water and solute concentrations, but have different volumes.

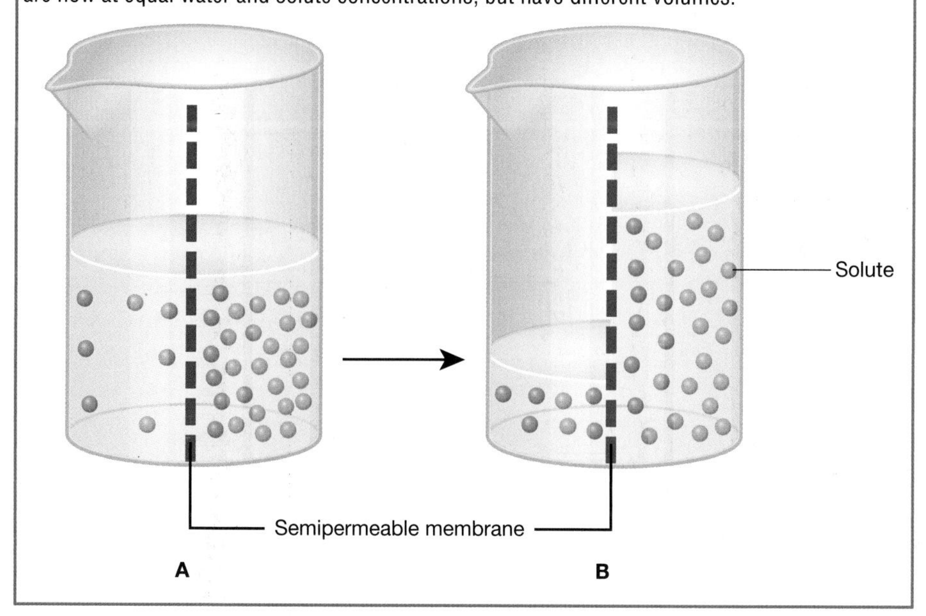

LO 4.1.7

Cells are not able to autonomously regulate the movement of water molecules. Therefore, the water and solute levels of body fluids are under critical homeostatic control. In order for water traffic across the membrane to be regulated, the ICF and ECF must have the same solute concentration. Two solutions that have the same concentration of solutes are said to be **isosmotic.** When comparing two solutions with different solute concentrations, the solution with more solute is called *hyperosmotic* and the solution with less is *hypoosmotic*. These terms (isosmotic, hypoosmotic, hyperosmotic) are used when comparing two solutions (**Figure 4.10**). For example, if we were comparing the blood plasma to the interstitial fluid, we could say that blood plasma is typically hyperosmotic to interstitial fluid.

As important as it is in physiology to consider the water and solute concentrations of the fluids of the body, it is equally important to consider the solute concentration on the inside and outside of the cell. As stated earlier, water can move, to varying degrees, across the cell membrane. Therefore, an imbalance in water concentration can critically alter the volume of cells. To describe the osmolarity of the extracellular fluid compared to the inside of the cell, we use the term **tonicity**.

Figure 4.10 Hyperosmotic, Hypoosmotic, and Isotonic Solutions

The terms hyperosmotic, hypoosmotic, and isosmotic are used to compare two solutions.

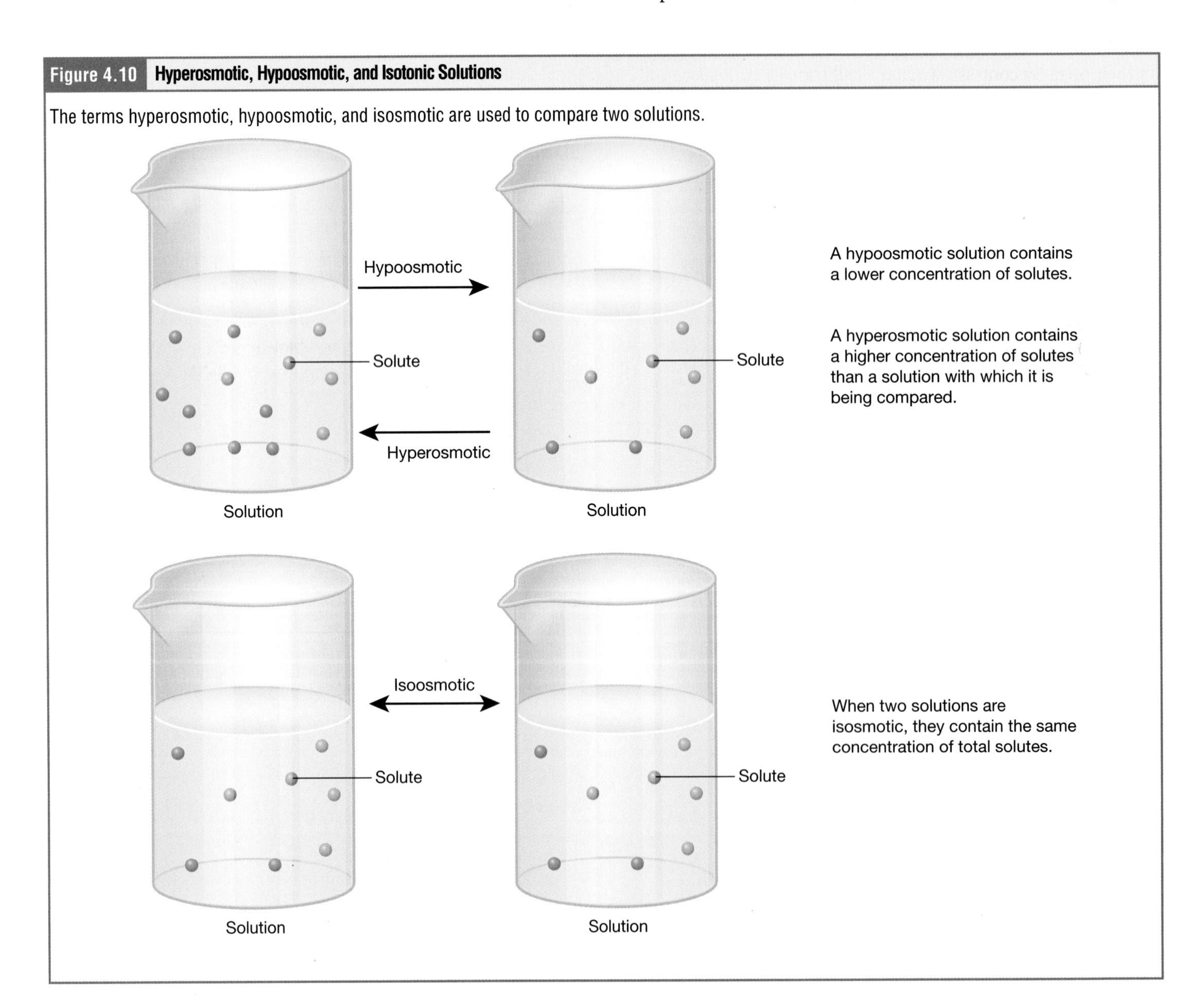
Hypoosmotic
Solute
Hyperosmotic
Solute
Solution
Solution
A hypoosmotic solution contains a lower concentration of solutes.
A hyperosmotic solution contains a higher concentration of solutes than a solution with which it is being compared.
Isoosmotic
Solute
Solute
Solution
Solution
When two solutions are isosmotic, they contain the same concentration of total solutes.

When cells and their extracellular environments are **isotonic**, the concentration of water molecules is the same outside and inside the cells, and the cells maintain their normal function (**Figure 4.11A**). Osmosis will occur when there is an imbalance of solutes on either side of the cellular membrane. When the ECF has a higher concentration of solutes than the ICF, the cell is said to be in a **hypertonic** solution. In this situation, water molecules tend to move into the hypertonic extracellular solution (**Figure 4.11C**); as a result, the cell loses volume and can shrivel. In contrast, a solution that has a lower concentration of solutes than the cell it surrounds is said to be **hypotonic**, and water molecules tend to diffuse into the cell from the ECF. Cells in a hypotonic solution will take on too much water and swell, with the risk of eventually bursting (**Figure 4.11B**). Therefore, it is critical for all the body's approximately 40 trillion cells to maintain solute and water homeostasis such that they are in an isotonic solution. Various organ systems, particularly the kidneys, work to maintain this homeostasis.

LO 4.1.8

Learning Connection

Try Drawing It

Draw a cell within hypertonic, hypotonic, and isotonic solutions. Illustrate the water movement. This is a technique you can use both in studying and in answering exam questions!

Active Transport For all of the transport methods described so far, the cell expends no energy. Membrane proteins that aid in the passive transport of substances do so without the use of ATP because molecules move down their concentration gradients on their own. By contrast, when the cell wants to transport a substance from where it is at a lower concentration to a location where it will be at a higher concentration, ATP is required. Typically, this requires the help of protein carriers, and is referred to as movement *against* its concentration gradient (from an area of lower concentration to an area

Figure 4.11 Cells in Isotonic, Hypotonic, and Hypertonic Solutions

The term *tonicity* is used to compare the concentration of solute in the solution surrounding a cell to the concentration of solute in the intracellular fluid. (A) A red blood cell in an isotonic solution will have no net movement of water across its membrane and no change in size. (B) A red blood cell in a hypotonic solution will have a higher solute concentration, and therefore lower water molecule concentration, on the inside of the cell. Water will travel through aquaporins into the cell, causing the cell to swell in size. (C) A red blood cell in a hypertonic solution will have a lower solute concentration, and therefore a higher water molecule concentration, on the inside of the cell. Water will travel through aquaporins out of the cell, causing the cell to shrink and shrivel.

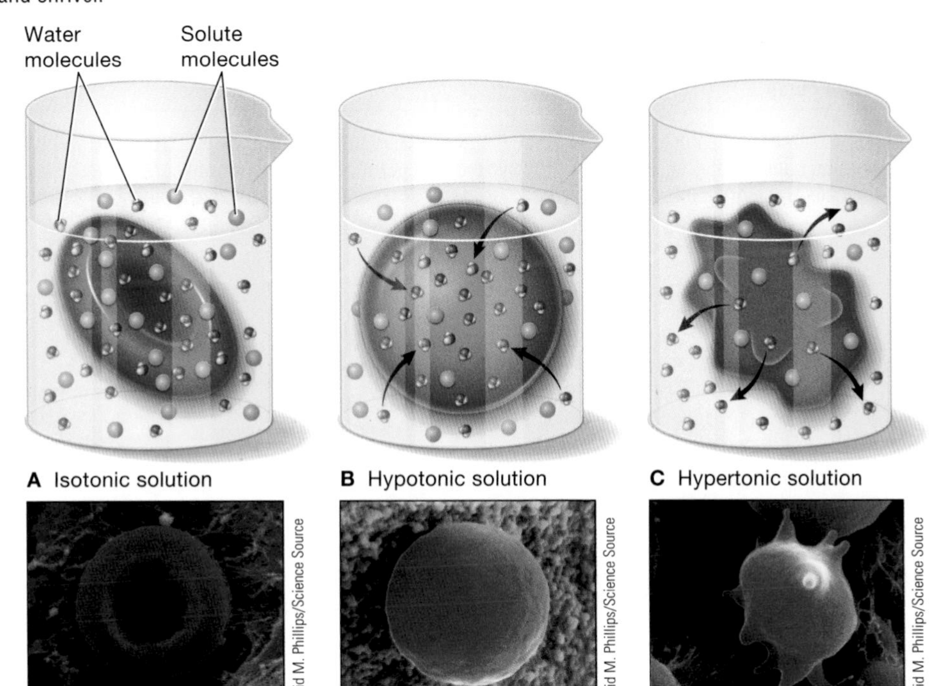

sodium-potassium pump (Na^+/K^+ ATP-ase pump)

of higher concentration). The proteins that are capable of moving molecules against their concentration gradients are typically called pumps.

In physiology, the most common example of a molecular pump is the **sodium-potassium pump**, which transports sodium out of a cell while moving potassium into the cell. The Na^+/K^+ pump is an important active transporter found in the membranes of many cells. Na^+ and K^+ ions both carry a single positive charge. The Na^+/K^+ pump moves three Na^+ ions out of the cell at the same time that it moves two K^+ ions into the cell; therefore, with each turn of the Na^+/K^+ pump, an **electrical gradient**, which is a difference in electrical charge across the membrane, becomes more and more established (**Figure 4.12**). Several cell types, especially nerve and muscle cells, utilize electrical gradients to accomplish cellular work. This process is so important for nerve cells that it accounts for the majority of their ATP usage and a significant proportion of a human's daily calorie use.

As the sodium-potassium pump works, it creates not only an electrical gradient, but two concentration gradients. Sodium is at a greater concentration outside the cell and potassium is at a greater concentration inside the cell. Both of these ions would move down their concentration gradients if the membrane were permeable to them. Electrical gradients contribute to cellular work in nerve cells and muscle cells, as we will discuss in Chapters 13 and 11. Another common use of a gradient is to fuel **secondary active transport**. In this transport mechanism, which is used by many cells the action of an active (ATP-driven) transport pump (the sodium-potassium pump) establishes a concentration gradient that is used in the transport of a second molecule such as glucose.

Symporters are secondary active transporters that move two substances in the same direction. For example, the sodium-glucose symporter allows sodium ions to move down their concentration gradient and into the cell. But this symporter can only complete its shape change once a molecule of glucose binds. In this way, the sodium-glucose symporter "pulls" glucose molecules into the cell against their concentration gradients by using the concentration gradient of sodium.

Antiporters work similarly to symporters, in that they accomplish secondary active transport by utilizing the concentration gradient of one molecule to move a second

Figure 4.12 Sodium-Potassium Pump

The sodium-potassium pump is a membrane feature found in most cells. This active transporter uses ATP to move sodium and potassium ions in opposite directions, each against its concentration gradient. In each cycle of the pump, three sodium ions are moved out of the cell and two potassium ions are imported into the cell. Three gradients are established: a sodium concentration gradient, a potassium concentration gradient, and an electrical gradient.

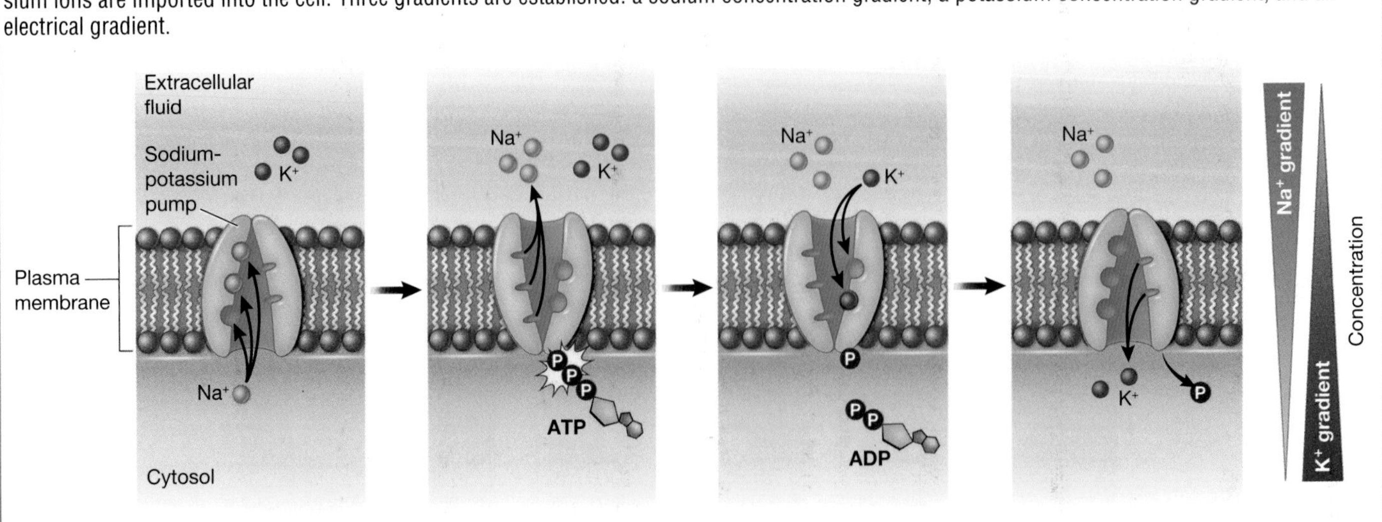

molecule against its concentration gradient. Antiporters differ from symporters, in that they transport the two substances in opposite directions. For example, the sodium-hydrogen ion antiporter uses the energy from the inward flood of sodium ions to move hydrogen ions (H^+) out of the cell. The sodium-hydrogen antiporter is important for maintaining cellular pH homeostasis.

Other forms of active transport do not involve membrane carrier proteins. **Endocytosis** is the process of a cell taking material in by enveloping it in a portion of its cell membrane, and then pinching off that membranous envelope (**Figure 4.13**). Once pinched off, the membrane and its contents become an independent, intracellular vesicle. A **vesicle** is a sac inside the cell, made of the same lipid bilayer as the cellular membrane. Endocytosis often brings materials into the cell so that the material can be broken down or digested. Endocytosis is widely used, but we can fit endocytic processes into three groups. **Phagocytosis** is the endocytosis of large particles (**Figure 4.13A**). Many immune cells engage in phagocytosis of invading pathogens. Like little Pac-men, their job is to patrol body tissues for unwanted matter (such as invading bacterial cells), phagocytize it, and digest it. In contrast to phagocytosis, **pinocytosis** ("cell drinking") brings fluid containing dissolved substances into a cell through membrane-bound vesicles (**Figure 4.13B**).

Phagocytosis and pinocytosis are not highly selective processes. A more selective form of endocytosis is **receptor-mediated endocytosis**, endocytosis that is triggered when membrane-bound receptors bind their ligand (the substance they are selective for). Once ligand binding occurs, the cell will endocytose the part of the cell membrane containing the receptor-ligand complexes (**Figure 4.13C**).

In contrast with endocytosis, **exocytosis** is the process of a cell exporting material by wrapping that material in a vesicle, and then fusing the vesicle with the plasma

Student Study Tip

21^+ tip: pinocytosis functions in cell drinking since "pino-" sounds like "pinot grigio."

Figure 4.13 Three Forms of Endocytosis

Cells can transport extracellular material into the cell through the process of endocytosis. There are three forms of endocytosis, but all require cellular energy (active transport) and each results in a pocket of the cell membrane being formed inside the cell. (A) Cells can engulf large items, such as bacteria or cellular debris, using the nonspecific process of phagocytosis. (B) Cells can take in extracellular fluid, including any small solutes, in pinocytosis. (C) Ligand-binding events of some membrane receptors trigger the cell to invaginate the membrane to take in the ligand in the very selective process of *receptor-mediated* endocytosis.

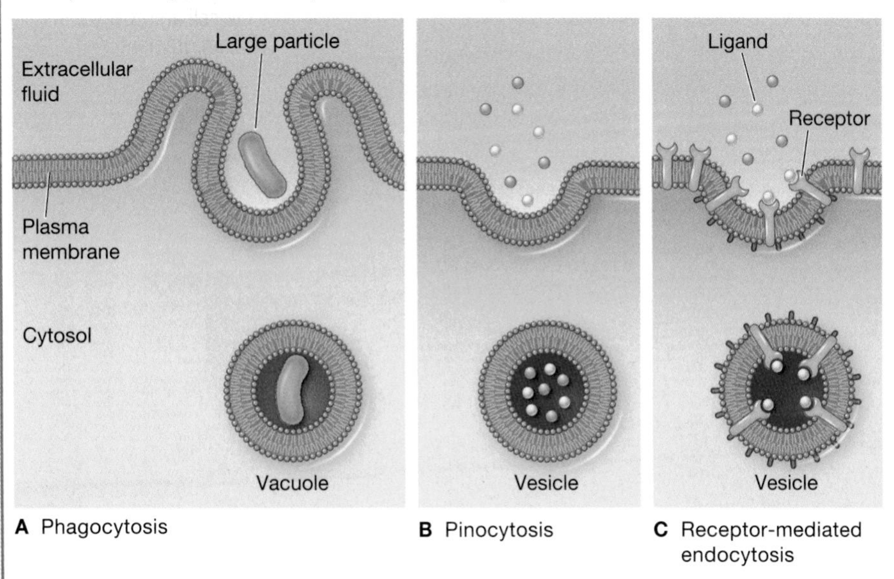

Figure 4.14 Exocytosis

Exocytosis is a process that cells use to export materials. Wastes or products made by the cell, such as signaling molecules and hormones, are packaged into vesicles, the membrane of the vesicle fuses with the cell membrane, and the contents are released into the extracellular space.

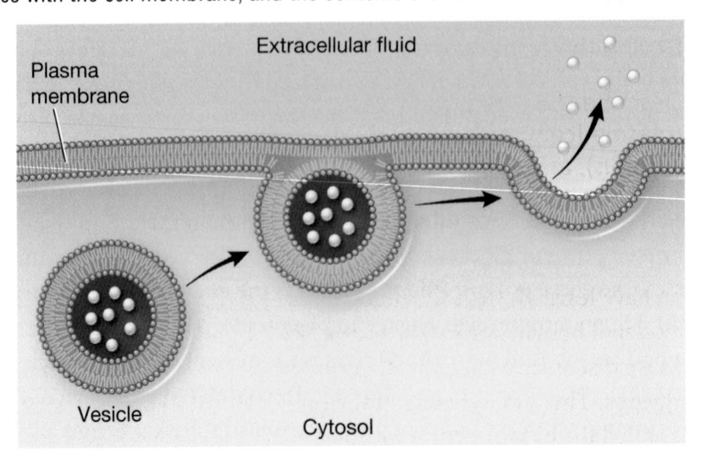

membrane (**Figure 4.14**). When the vesicle membrane fuses with the cell membrane, the vesicle releases its contents into the interstitial fluid. The vesicle membrane then becomes part of the cell membrane. Many cells play manufacturing roles in the body. Examples include cells of the stomach and pancreas that secrete digestive enzymes, endocrine cells that secrete hormones, and some immune cells that secrete large amounts of histamine, a chemical important for immune responses.

Learning Check

1. Which method moves molecules down its concentration gradient through the cell membrane?
   a. Active transport
   b. Passive transport
2. What happens to the flow of molecules once they reach equilibrium?
   a. Intracellular molecules will not exit the cell and extracellular molecules will not enter the cell.
   b. Intracellular molecules will exit the cell at the same rate as the extracellular molecules enter the cell.
   c. Intracellular molecules will exit the cell faster than the extracellular molecules will enter the cell.
   d. Intracellular molecules will exit the cell slower than the extracellular molecules will enter the cell.
3. Solution A has a higher concentration of solutes than Solution B. How would you describe Solution A when compared to Solution B?
   a. Hypotonic
   b. Hypertonic
   c. Hypoosmotic
   d. Hyperosmotic
4. Which of the following is NOT an active transport method?
   a. Sodium-potassium pump
   b. Exocytosis
   c. Endocytosis
   d. Osmosis
5. Which of the following is true about the sodium-potassium pump? Please select all that apply.
   a. It uses ATP and is an example of active transport.
   b. It creates a concentration gradient and an electrical gradient.
   c. It moves two Na^+ ions into the cell and three K^+ ions out of the cell.
   d. Ions move down their concentration gradient.

4.2 The Cytoplasm and Cellular Organelles

HAPS LEARNING GOALS AND OUTCOMES

Learning Objectives: By the end of this section, you will be able to:

4.2.1 Compare and contrast cytoplasm and cytosol.

4.2.2 Define the term organelle.

4.2.3 Describe the three main parts of a cell (plasma [cell] membrane, cytoplasm, and nucleus), and explain the general functions of each part.

4.2.4 Describe the structure and function of the various cellular organelles.

4.2.5 Describe the structure and roles of the cytoskeleton.

LO 4.2.1

LO 4.2.2

LO 4.2.3

Now that you have learned that the cell membrane surrounds all cells, you can dive inside of a model of human cell to learn about its internal components and their functions. As we discussed in Section 4.1 both the external and internal environments are aqueous. The watery inside of the cell along with all its compartments and organelles is called the **cytoplasm**. If we could isolate all of the **organelles**, which are membrane-enclosed bodies within the cell, each of which performs a unique function, we would be left with the **cytosol**, the jellylike substance within the cell that provides the fluid medium necessary for biochemical reactions and functions of the cell. Another way to think of the cell is as a bowl of soup. The bowl—which contains the liquid and keeps the soup from leaking—is the membrane. The broth is the cytosol, and the solids of the soup—whether these are noodles or wontons or carrots—are the organelles. The broth and solids together form the soup, or cytoplasm.

The largest of the cell's organelles is its **nucleus**, which contains the cell's DNA inside of a membrane (Figure 4.15).

4.2a Organelles of the Endomembrane System

LO 4.2.4

In Section 4.1 we discussed that many cells of the body produce and export products that contribute to signaling or homeostasis in the process of exocytosis. But how do those cellular products come to be in the first place? Inevitably, these exports are the product of gene expression and then the cellular processes of production, packaging, and exporting. Just like a factory assembly line, this export work is accomplished through the coordinated work of a set of three major organelles. Together these three are called the *endomembrane system*. The organelles of the endomembrane system include the endoplasmic reticulum, Golgi apparatus, and vesicles.

Endoplasmic Reticulum The **endoplasmic reticulum (ER)** is a system of channels that is continuous with the membrane of the nucleus. It covers the nucleus and is composed of the same lipid bilayer material as the nuclear membrane and the cellular membrane. The ER provides membranous passages throughout much of the cell that function in transporting, synthesizing, and storing materials. The winding structure of the ER results in a large membranous surface area that supports its many functions (Figure 4.16).

Endoplasmic reticulum occurs in two forms: rough ER and smooth ER. Each form performs different functions and can be found in very different amounts depending on the type of cell. **Rough ER (RER)** is so called because of its relationship with ribosomes. Ribosomes are nonmembranous organelles that function to synthesize proteins. When visualizing RER using an electron microscope, scientists observed that RER is associated with so many ribosomes that it has a bumpy, studded appearance

Figure 4.15 A Model Human Cell

While this image is not representative of any one particular human cell (i.e., skin cells and muscle cells do not look like this!), it is a model of a cell containing the primary organelles and internal structures.

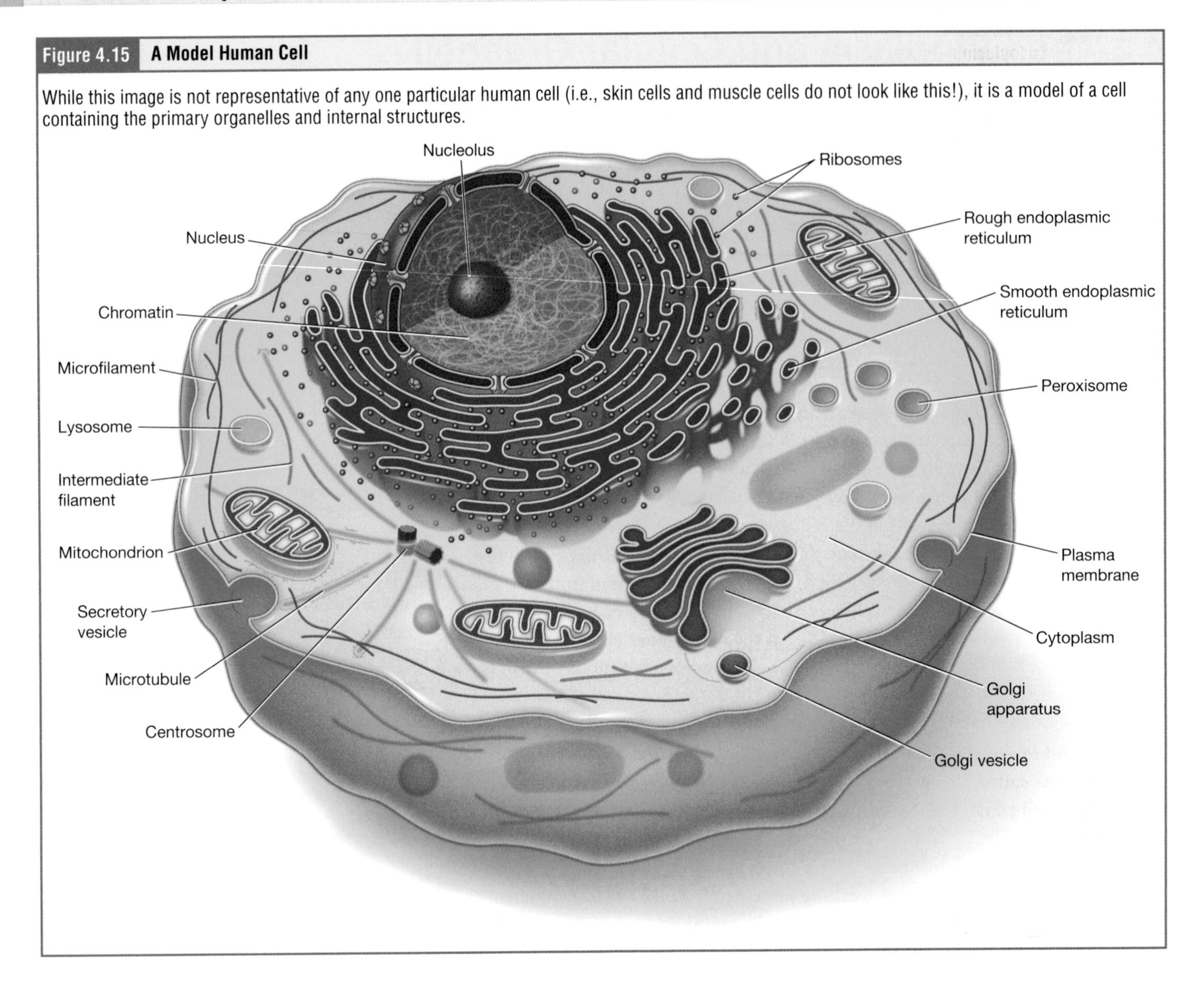

that looked "rough" (see Figure 4.16). **Smooth ER (SER)** lacks these ribosomes, and therefore lacks protein synthesis capacity or a rough appearance.

One of the main functions of the smooth ER is in lipid synthesis. The smooth ER synthesizes phospholipids, the main component of biological membranes, as well as steroid hormones. For this reason, cells that produce large quantities of such hormones, such as those of the ovaries and testes, contain large amounts of smooth ER. In addition to lipid synthesis, the smooth ER also functions as a storage closet in some cells. Skeletal muscle cells, for example, use their smooth ER to store large quantities of Ca^{2+} ions, a function extremely important in their contraction, as we will discuss in Chapter 12. Smooth ER has an additional function of metabolizing some carbohydrates and breaking down certain toxins. In cell types that play bigger detoxification roles (for example, liver cells), there can be an abundance of SER.

In contrast with the many jobs of smooth ER, the primary job of the rough ER and its ribosomes is the synthesis and modification of proteins. Typically, a protein is synthesized (the biological process is known as *translation*) within a

Figure 4.16 Endoplasmic Reticulum (ER)

The ER is a network of thin membrane-bound sacs that surround the nucleus of the cell. The ER can be studded with ribosomes, in which case it is found closer to the nuclear membrane and named *rough ER*, or it can be ribosome-less, in which case it is called *smooth ER*.

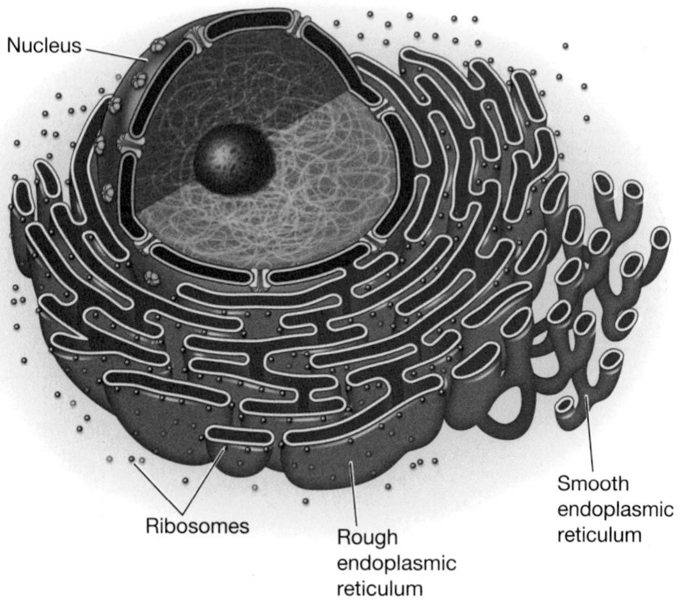

ribosome and released into the channel of the RER, where sugars can be added to it (by a process called *glycosylation*). The proteins made by the RER are packaged into vesicles and transported to the second stop in the endomembrane system: the Golgi apparatus.

The Golgi Apparatus The **Golgi apparatus** is responsible for sorting, modifying, and shipping off the products that come from the rough ER, much like a post office. The Golgi apparatus looks like stacked flattened discs, almost like stacks of thin pancakes. Like the ER, these discs are membranous. The Golgi apparatus has two distinct sides, each with a different role. The side of the Golgi that is closer to the nucleus and ER receives products in vesicles and is often called the *cis*-face. These products are sorted through the apparatus, and then they are released from the opposite side, often referred to as the *trans*-face, after being repackaged into new vesicles. If the product is to be exported from the cell, the vesicle migrates to the cell surface and fuses to the cell membrane, and the cargo is secreted into the extracellular fluid (**Figure 4.17**).

4.2b Organelles for Energy Production and Detoxification

In addition to the jobs performed by the endomembrane system, the cell has many other important functions. Just as you must consume nutrients to provide yourself with energy, so must each of your cells take in nutrients, some of which convert to chemical energy that can be used to power biochemical reactions. Another important function of the cell is detoxification. Humans take in all sorts of toxins from the environment and also produce harmful chemicals as byproducts of cellular processes. Cells called *hepatocytes* in the liver detoxify many of these toxins.

Figure 4.17 Golgi Apparatus

The Golgi apparatus modifies the products made by the rough ER. The modified products are packaged in vesicles and sent to the cell membrane for export (exocytosis) or transported to other areas of the cell for use in cellular processes. The Golgi apparatus also produces new organelles called lysosomes.

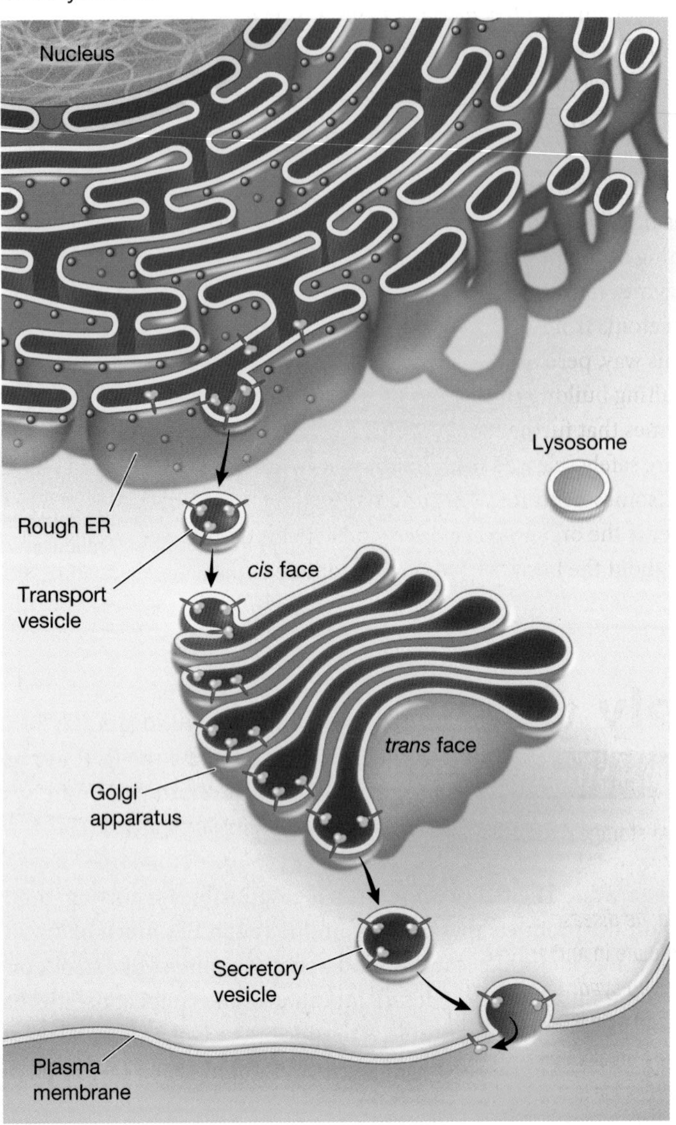

Student Study Tip

Lysosome function can be understood by looking at the root word "*lys-*," meaning to break down. We also see this in the word *analysis*: to break content down and understand it!

Lysosomes As important as it is for a cell to produce new molecules, it is equally important for cells to be able to break down and recycle molecules that are nonfunctional or no longer needed. Some of the proteins made by the RER and packaged by the Golgi are digestive enzymes that function within the cell in digestion and recycling. These enzymes could be dangerous to the cell, and so they are stored within dedicated organelles called lysosomes. **Lysosomes** are organelles that digest unneeded cellular components, such as a damaged organelle. They do so by packing a variety of enzymes that digest different complex molecules into a small pocket of membrane. Lysosomes are also important for breaking down material taken in from the outside of the cell in phagocytosis. For example, when certain immune cells phagocytize bacteria, the vesicle containing the bacterial cell merges with a lysosome and the bacterial cell is digested by the enzymes. As we have seen with other organelles, the number of lysosomes within a cell will vary with the cell's function. Phagocytic immune cells have many more lysosomes than other cells.

Under certain circumstances, lysosomes perform a more grand and dire function. In the case of damaged or unhealthy cells, lysosomes can be triggered to open up and release their digestive enzymes into the cytoplasm of the cell, killing the cell. This "self-destruct" mechanism is called **autolysis**, and is one form of **apoptosis** (purposeful cellular death). This cellular self-destruction is sometimes healthy and necessary during human development when certain cells must be destroyed to accommodate new ones or form the body plan.

apoptosis (programmed cell death)

Peroxisomes Like lysosomes, a **peroxisome** is a membrane-bound cellular organelle that contains mostly enzymes. In function, peroxisomes are most similar to smooth ER, performing both lipid metabolism and chemical detoxification. In contrast to the digestive enzymes found in lysosomes, the enzymes within peroxisomes serve to transfer hydrogen atoms from various molecules to oxygen, producing hydrogen peroxide (H_2O_2). In this way, peroxisomes neutralize poisons such as alcohol.

The resulting buildup of H_2O_2 could also be harmful; however, peroxisomes contain enzymes that further convert H_2O_2 to H_2O and O_2 (water and oxygen)! These byproducts are safely released into the cytoplasm. Like miniature sewage treatment plants, peroxisomes neutralize harmful toxins so that they do not wreak havoc in the cells. The liver is the organ primarily responsible for detoxifying the blood before it travels throughout the body, and liver cells contain an exceptionally high number of peroxisomes.

Apply to Pathophysiology

Lysosomal Storage Diseases

Lysosomal storage diseases are a group of inherited diseases that impact lysosomal function. Three lysosomal diseases are described in the following list:

- *Tay-Sachs disease* is a fatal disorder in which *sphingolipids* (large, complex molecules that combine amino acids and lipid structures) accumulate in and around nerve cells in the brain, disabling nerve cell function and eventually leading to cellular death.
- *Sanfilippo syndrome* is a fatal disorder in which *glycosaminoglycans* (large, complex sugars) accumulate in and around nerve cells in the brain and spinal cord. The sugars interfere with nerve cell function in a variety of ways. Eventually communication between the brain and muscles breaks down and the patients become increasingly immobile.
- *Fabry disease* is a disorder in which *globotriaosylceramides* (large, complex fats) build up in the cells that line blood vessels and other organs. The buildup of these fats can lead to cellular death and blood vessel rupture. Individuals with Fabry disease often experience red spots on their skin where blood vessels have ruptured beneath the surface, kidney issues, and cardiac complications. Patients with Fabry disease have a shorter life expectancy than the general public, but typically live into their fifties.

1. All three of these diseases involve the same organelle, the lysosome, and yet, they have very different symptoms. Which of the following provides an explanation as to why?
   A) The three genetic mutations in these three diseases each impact a different enzyme.
   B) These three diseases each impact different cell types, which have different functions.
   C) Lysosomes are not always critical; therefore, the symptoms vary based on the locations that need lysosomes the most.
2. Considering your answer from Question 1, which of the following is true in these diseases?
   A) In all of these diseases the lysosomes lack function and cannot break down anything.
   B) In each of these diseases the lysosomes are capable of breaking down some materials but not others.
   C) In each of these diseases the cells that are affected are the ones that depend on lysosomal function.

(*Continued*)

3. Lysosomes are expressed in most cells of the body. Why, in these diseases, do the symptoms show up in only particular cells (i.e., nerve cells of the brain versus blood vessel cells)?
   A) The symptoms will only show up in the cells that have the impacted genes.
   B) Differentiation of cells into different cell types with different functions leads to a situation where certain enzymes are particularly critical in some cells and not in others.
   C) The impacted genes are expressed at different phases of the lifespan and therefore affect the cells that are more proliferative and functional in those phases.

Mitochondria A **mitochondrion** (plural = *mitochondria*) is a membranous organelle that is the "energy transformer" of the cell. Mitochondria consist of two lipid bilayers, one outer and one inner (**Figure 4.18**). The inner membrane is highly folded with a great deal of surface area; the folds of the inner membrane are called **cristae**. It is along this inner membrane that a series of proteins, enzymes, and other molecules perform the biochemical reactions of cellular respiration. These reactions convert energy stored in nutrient molecules (such as glucose) into adenosine triphosphate (ATP), which provides usable cellular energy to the cell, discussed further in Chapter 24. You can think of mitochondria similarly to the solar panels on a house. Solar panels do not create energy, but they transform it from energy in sunlight to chemical energy stored in batteries used to, for example, charge your cell phone. Cells use ATP constantly, and so the mitochondria are constantly at work converting the energy in our nutrients into usable ATP. Oxygen molecules are required during cellular respiration, which is why you must constantly breathe it in.

The cells of the body that have the highest demand for energy, such as muscle cells and nerve cells, have the highest concentration of mitochondria. On the other hand, a bone cell, which is not nearly as metabolically active, contains only a fraction of the mitochondria found in a nerve cell.

4.2c The Cytoskeleton

LO 4.2.5

Much like the bony skeleton structurally supports the human body, the cytoskeleton helps the cells to maintain their structure. The **cytoskeleton** is made up of fibrous proteins that provide structural support for cells. These fibers are also critical for cell

Figure 4.18 Mitochondrion

The main function of mitochondria is to produce ATP, the cell's major energy currency, through the process of cellular respiration.

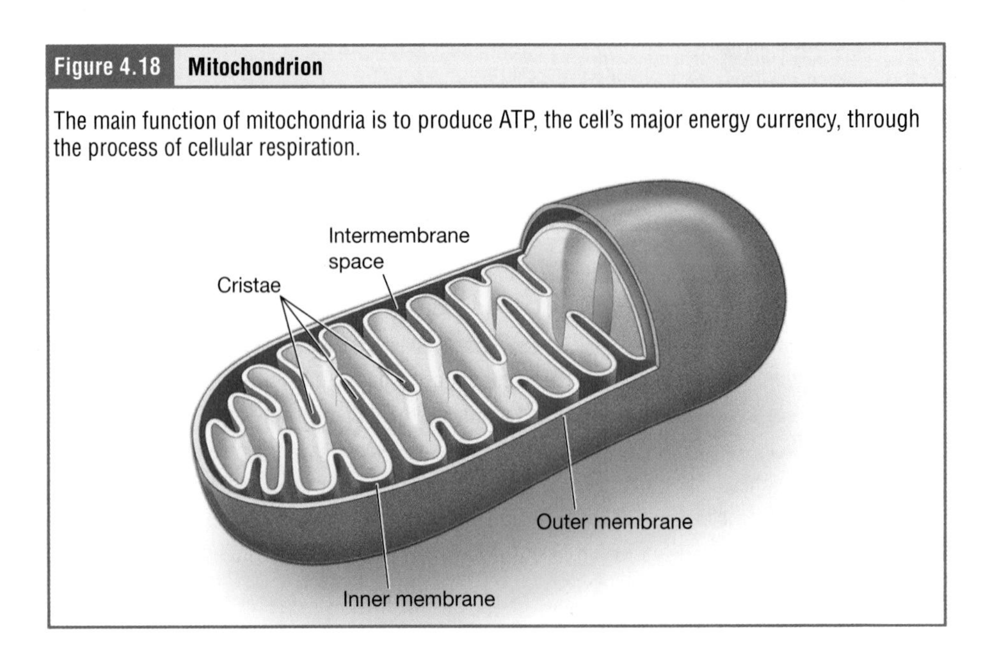

motility, cell reproduction, and transportation of substances and organelles within the cell.

The cytoskeleton forms a complex, threadlike network throughout the cell consisting of three different kinds of protein-based filaments: microfilaments, intermediate filaments, and microtubules (**Figure 4.19**). The thickest of the three is the **microtubule** (**Figure 4.19A**), a structural filament composed of subunits of a protein called *tubulin*. Microtubules maintain cell shape and structure, help resist compression of the cell, and play a role in positioning the organelles within the cell. One very important function of microtubules is to provide the tracks along which vesicles and genetic material can

Figure 4.19 The Three Components of the Cytoskeleton

The cytoskeleton functions to maintain cellular shape and enable movement. Three major classes of cytoskeletal fibers are common in human cells: (A) microtubules, composed primarily of tubulin protein, (B) microfilaments, composed primarily of actin protein, and (C) intermediate filaments, composed primarily of keratin protein.

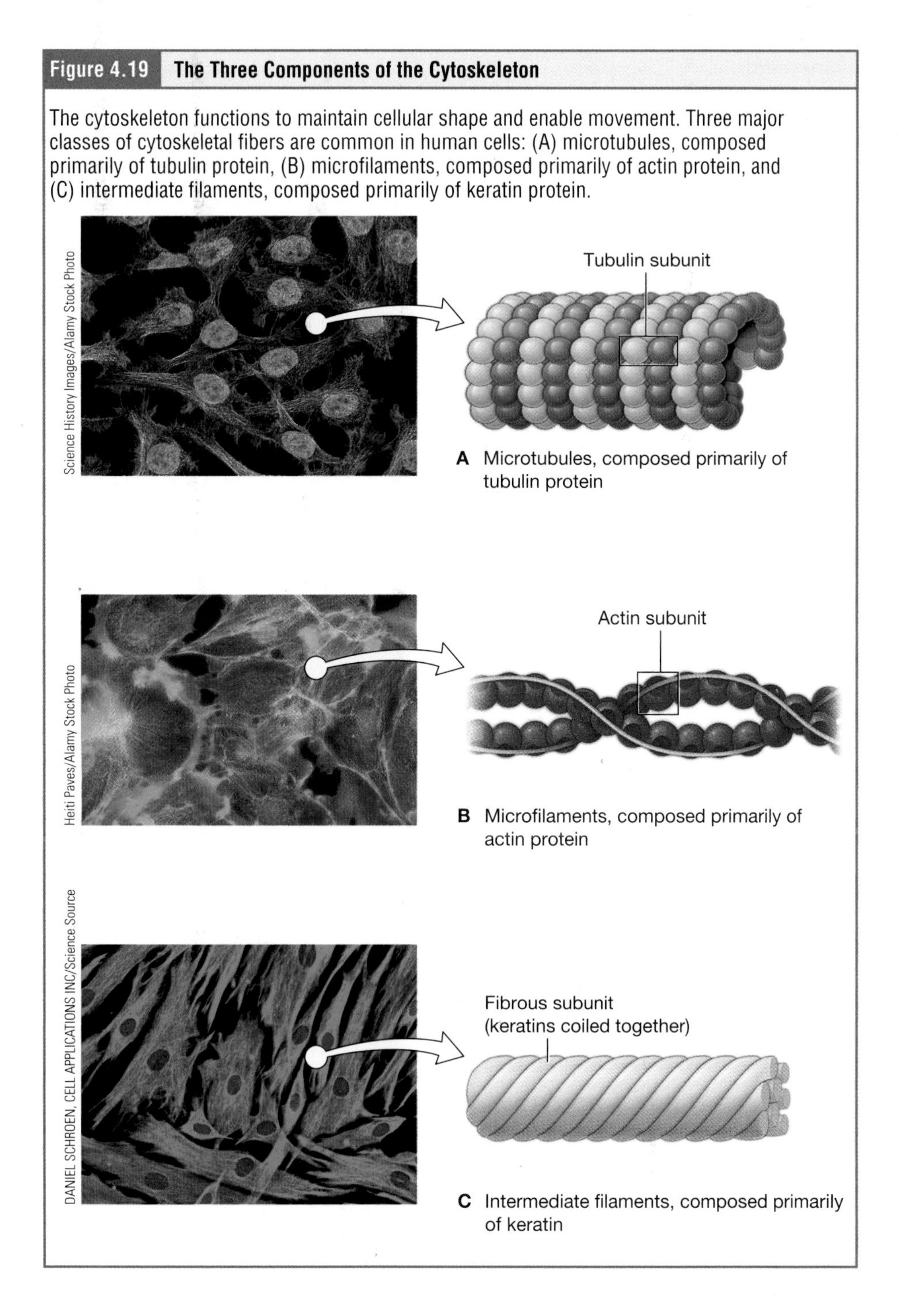

A Microtubules, composed primarily of tubulin protein

B Microfilaments, composed primarily of actin protein

C Intermediate filaments, composed primarily of keratin

be pulled within the cell. Earlier in this section we discussed how the organelles of the endomembrane system work together to manufacture cellular products; these products move within the cell (for example, from the RER to the Golgi) in vesicles that glide along microtubules (**Figure 4.20**).

While the term *cytoskeleton* makes these structures sound like fixed, static scaffolds, microtubules are actually remarkably dynamic. Two short, identical microtubule structures called centrioles are found near the nucleus of cells. A **centriole** can serve as the cellular origination point for microtubules extending outward as cilia or flagella or can assist with the separation of DNA during cell division. Microtubules can elongate out from the centrioles as more tubulin subunits are added, like adding additional links to

Figure 4.20 Vesicles and Intracellular Transport

Vesicles move around the cell, for example from the ER to the Golgi apparatus, along microtubules carried by motor proteins, such as kinesins and dyneins.

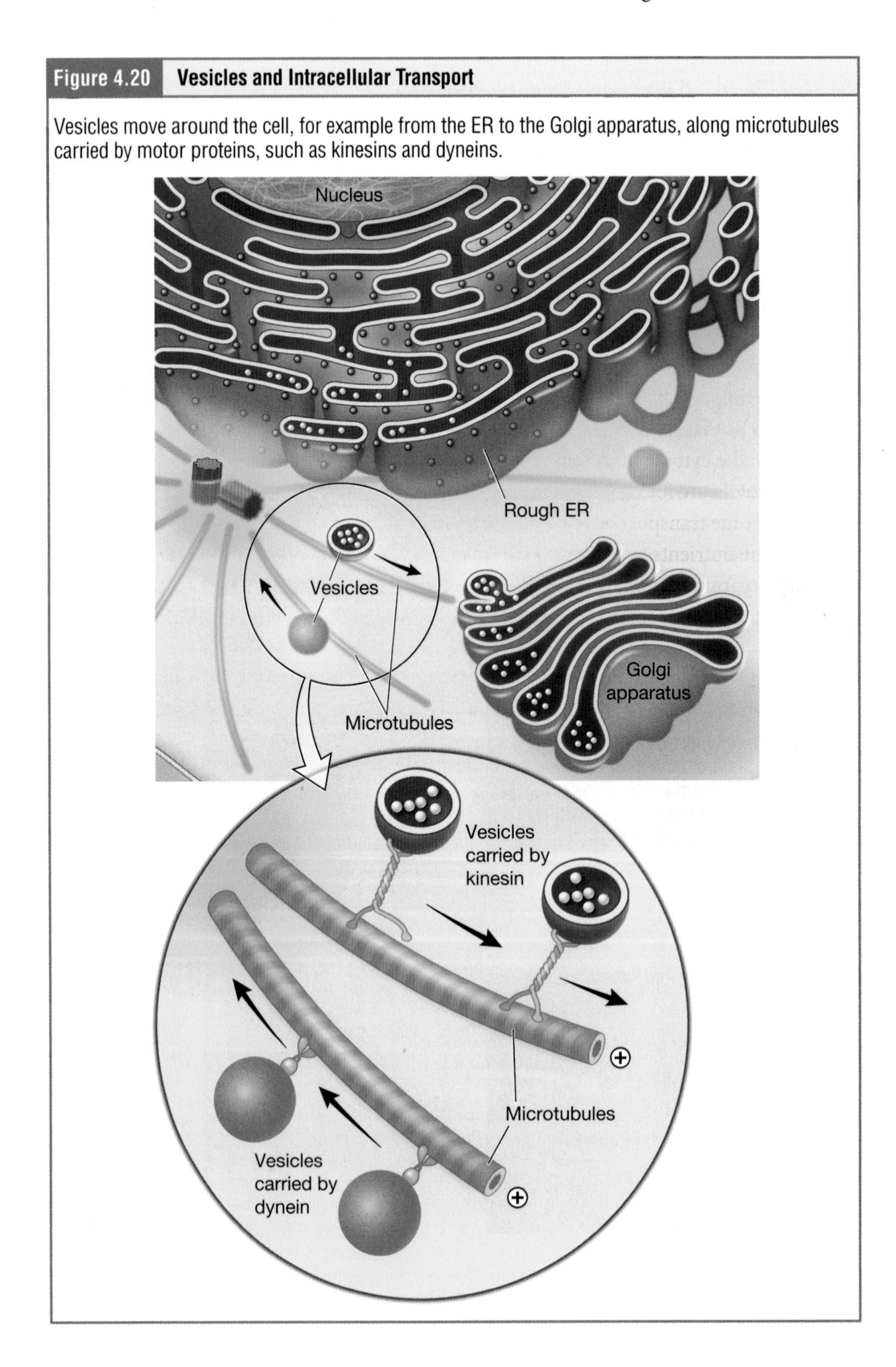

a chain; microtubules can also be quickly shortened as necessary. Through the dynamic changes in microtubule length, objects such as chromosomes can be moved about the cell.

In contrast with microtubules, **microfilaments** are the thinner of cytoskeletal filaments (**Figure 4.19B**) composed primarily of the protein actin. Like microtubules, actin filaments are long chains of single subunits (called actin subunits). Unlike microtubules, which primarily function to move material within the cell, microfilaments primarily function to move the whole cell. Microfilaments also play an important role during cell replication. When a cell undergoing cell division is about to split in half to form two new cells, actin microfilaments work with another protein to create a cleavage furrow that eventually splits the cell down the middle into two new cells.

The final type of cytoskeletal filament is the intermediate filament. As its name would suggest, an **intermediate filament** is a filament intermediate in thickness between the microtubules and microfilaments (**Figure 4.19C**). Intermediate filaments are made up of long fibrous subunits of a protein called keratin that are wound together like the threads that compose a rope. Intermediate filaments, in concert with the microtubules, are important for maintaining cell shape and structure. Intermediate filaments resist any force that could pull cells apart. There are many cases in which cells are prone to tension, such as when epithelial cells of the skin are tugged in different directions. Intermediate filaments help anchor organelles together within a cell and also link cells to other cells by forming special cell-to-cell junctions.

Cell Surface Specializations There are three types of appendages of note on human cells: microvilli, cilia, and flagella. **Microvilli** are tiny and numerous projections on the surface of cells that serve a function of expanding surface area (**Figure 4.21A**). They are anchored to the cytoskeleton through actin filaments, but they are not capable of movement. Microvilli are found wherever a need for a great amount of surface area to accomplish membrane transport is required, such as the cells of the small intestine, which need to absorb the nutrients from our diet across the cells and into the bloodstream.

Cilia are found on many cells of the body, including the epithelial cells that line the airways of the respiratory system. Cilia resemble microvilli, in that there are typically

Figure 4.21 Cellular Appendages

There are three types of human cell surface appendages: (A) Microvilli are associated with microfilaments, but incapable of movement. Their purpose is to expand the surface area of the cell. (B) Cilia are composed of microtubules and are capable of movement. Their function is to sweep material off the surface of the cell. (C) Flagella are very long; the only human cell with a flagellum is a sperm cell. Its function is to propel the sperm toward the egg.

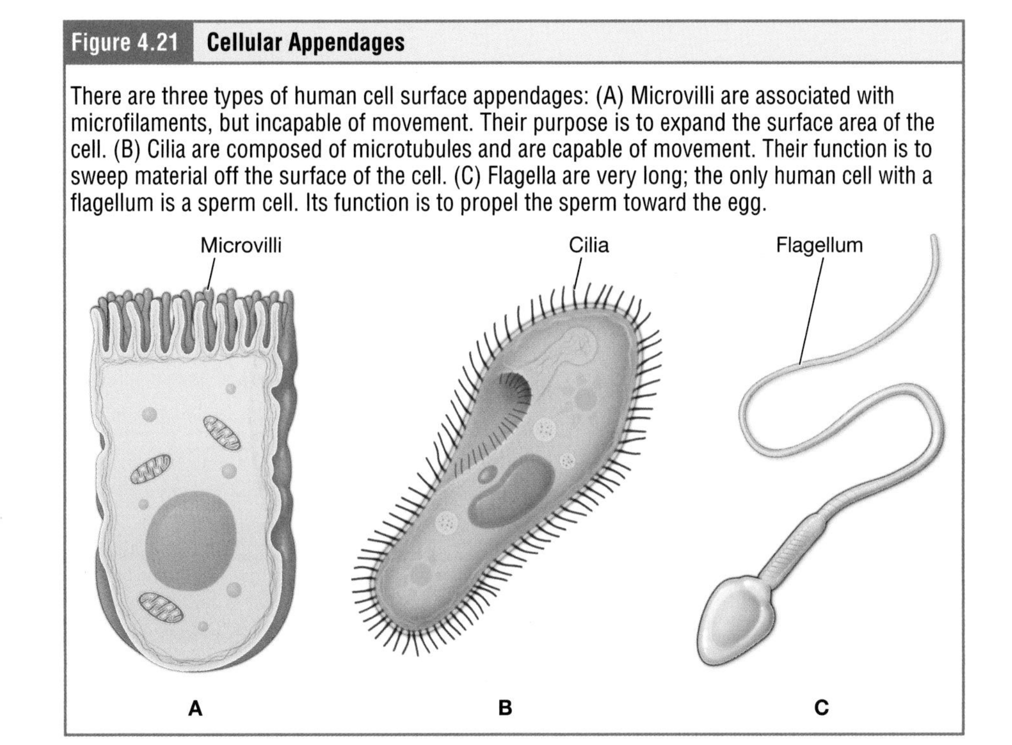

Learning Connection

Chunking

Can you find different ways to group the structures we have talked about? Which of the organelles and cellular structures are visible from the outside versus the inside of the cell? Which organelles have similar functions?

many on the surface of each cell, resembling a hairbrush (**Figure 4.21B**). Unlike microvilli, cilia are capable of movement—they move rhythmically to move materials off the surface of that cell. In the respiratory system, cilia move waste materials such as dust, mucus, and bacteria upward through the airways, away from the lungs and toward the mouth.

A **flagellum** is an appendage larger than a cilium and specialized for cell locomotion. The only flagellated cell in humans is the sperm cell, which must propel itself toward female egg cells (**Figure 4.21C**).

Learning Check

1. Which of the following is a function of rough endoplasmic reticulum but not smooth endoplasmic reticulum?
   a. Synthesizing steroid hormones
   b. Metabolizing carbohydrates
   c. Breaking down certain toxins
   d. Synthesizing proteins
2. Which of the following organelles is responsible for autolysis?
   a. Lysosomes
   b. Smooth ER
   c. Rough ER
   d. Golgi apparatus
3. Which membrane of the mitochondria is the site of ATP synthesis?
   a. Inner membrane
   b. Outer membrane
4. You are designing a synthetic cell to replace cells in the human fallopian tube that have been damaged by infection. These cells live in an area of the reproductive tract where the egg is transported from ovary to uterus. The egg needs to move somewhat slowly and should travel in only one direction. You have to design your cell with a cell surface feature capable of generating a gentle sweeping motion on the cell surface to help the egg move. What cell surface feature would be best for this task?
   a. Cilia
   b. Flagella
   c. Microvilli
   d. None of these

Apply to Pathophysiology

Cystic Fibrosis

Cystic fibrosis (CF) affects approximately 30,000 people in the United States, with about 1,000 new cases reported each year. The genetic disease is most well-known for its damage to the lungs, not only causing breathing difficulties and chronic lung infections, but also affecting the liver, pancreas, and intestines.

The myriad symptoms of CF result from a malfunctioning membrane ion channel called the *cystic fibrosis transmembrane conductance regulator*, or CFTR. In healthy people, the CFTR protein is a transmembrane protein that transports Cl^- ions out of the cell, down their concentration gradients. In a person who has CF, the gene for the CFTR is mutated; thus, the cell manufactures a defective channel protein that typically is not incorporated into the membrane but is instead degraded by the cell.

1. Compare a cell without CFTR expressed (let's call this cell $CFTR^-$) to a cell of a person without cystic fibrosis who expresses typical levels of CFTR (this cell is called $CFTR^+$). Which cell would have a higher intracellular Cl^- concentration?
   A) The $CFTR^+$ cell
   B) The $CFTR^-$ cell
2. Considering where the higher concentration of chloride ions is found, the cytosol of the $CFTR^+$ cell would be ______ compared to the cytosol of the $CFTR^-$ cell.
   A) Hypoosmotic
   B) Hyperosmotic
   C) Isosmotic
3. In the figure, we can visualize the $CFTR^+$ and $CFTR^-$ cells. Imagine that the interstitial fluid surrounding these cells is high in Na^+ ions. Considering the location of the sodium and chloride ions in the interstitial fluid surrounding these cells, toward which of the cells will more cations be drawn?
   A) Toward the $CFTR^+$ cell
   B) Toward the $CFTR^-$ cell

4. Taking your answer to Question 3 into account, compare the extracellular environment of these two cells. The extracellular fluid of which cell could be described as hyperosmotic compared to the extracellular fluid of the other? Assume that water has not yet moved across the plasma membrane of either cell type by osmosis.
   A) The CFTR$^+$ cell
   B) The CFTR$^-$ cell
5. Taking your answer to Question 4 into account, which cell will draw more water molecules to its ECF?
   A) The CFTR$^+$ cell
   B) The CFTR$^-$ cell

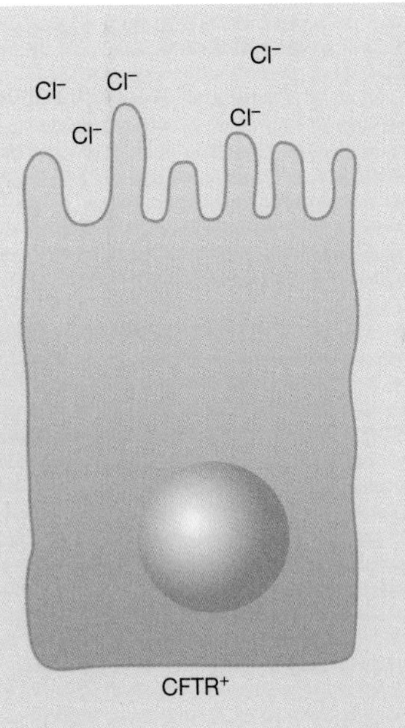

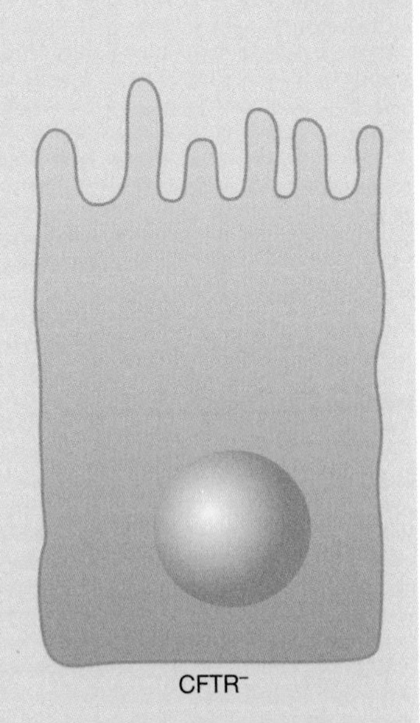

Consider the epithelial lining of the respiratory system. Respiratory epithelial cells secrete mucus, which serves to trap dust, bacteria, and other debris. The cilia of these cells move the mucus and its trapped particles up the airways away from the lungs and toward the outside. In order to be effectively moved upward, the mucus cannot be too viscous; rather it must have a thin, watery consistency. Bacterial infections are also prevented as bacterial cells are carried away from the lung cells by the motion of the cilia and respiratory mucus.

6. Comparing the extracellular fluids of the CFTR$^+$ and CFTR$^-$ cells, consider the functioning of the cilia in the lungs of a person with CF. Which of the following do you imagine is true?
   A) The mucus is thinner and waterier; it does not effectively trap dust and bacteria. The person with CF is at higher risk of infection and has a watery cough.
   B) The mucus is thicker and the cilia have trouble moving. The individual with CF is more prone to infection, and has trouble getting the thick mucus out of their lungs.

4.3 The Nucleus and DNA

Learning Objectives: By the end of this section, you will be able to:

4.3.1* Describe the structure and contents of the nucleus.

4.3.2* Explain the organization of the DNA molecule within the nucleus.

* Objective is not a HAPS Learning Goal.

The nucleus is the largest and most prominent of a cell's organelles (**Figure 4.22**). It stores most of the genetic instructions and is therefore essential in cells that will reproduce themselves. The full set of genetic instructions of an organism is called its **genome**. Interestingly, some cells in the body, such as skeletal muscle cells, are multinucleated, meaning that they contain more than one nucleus. Other cells, such as red blood cells, do not contain a nucleus at all. Red blood cells eject their nuclei as they mature, but without nuclei, the life span of a red blood cell is short, and so the body must produce new ones constantly.

LO 4.3.1

Inside the nucleus lies the blueprint that dictates everything a cell will do and all of the products it will make. This information is stored within DNA. When a cell divides, the DNA must be duplicated so that each new cell receives a full complement of DNA. The following section will explore the structure of the nucleus and its contents, as well as the process of DNA replication.

Figure 4.22 **The Nucleus**

The nucleus is a membranous container for the cell's genome in the form of DNA.

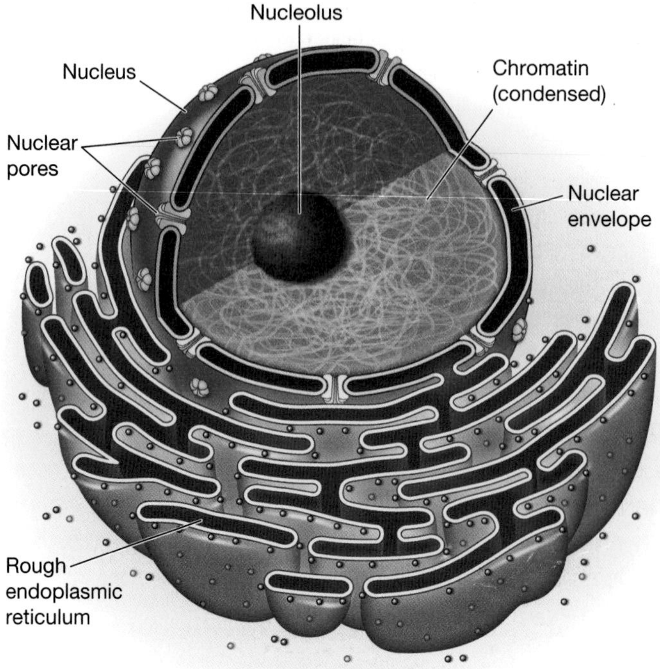

4.3a Organization of the Nucleus and Its DNA

Like most other cellular organelles, the nucleus is surrounded by a membrane. The nuclear membrane is more elaborate than most, consisting of two adjacent lipid bilayers with a thin fluid space in between them. Spanning these two bilayers are nuclear pores. A **nuclear pore** is a tiny passageway for the passage of proteins, RNA, and solutes between the nucleus and the cytoplasm.

The inside of the nucleus is filled with a nucleoplasm that is analogous to the cytoplasm of the cell. Suspended within this liquid are several types of molecules, including the building blocks of nucleic acids. If looking at a nucleus under the microscope, you might notice a dark-staining mass called a **nucleolus**; this is a region of the nucleus that is responsible for manufacturing the RNA necessary for construction of ribosomes. Table 4.1 details the types of nucleic acids in healthy human cells.

LO 4.3.2

With approximately 40 trillion cells, your body is a diverse galaxy. The vast collection of genetic instructions that are used to build and maintain every molecule of every cell is coded in deoxyribonucleic acid (DNA) and stored in the cell's nucleus. These genetic instructions are approximately 3 billion units in length, and so must be kept in an organized fashion so that genes of interest can be found when the cell needs them. When not in the process of replicating, cells store their DNA wrapped around a set of **histone** proteins like thread wrapped around a spool. Scientists use the term **nucleosome** to describe a single, wrapped DNA-histone complex. Multiple nucleosomes along the entire molecule of DNA appear like a beaded necklace, in which the string is the DNA and the beads are the associated histones. Note that in this form the DNA is accessible when it is time to make proteins, but the DNA is also organized. This form of loosely organized DNA is called **chromatin**. When a cell is in the process of division, the chromatin condenses into chromosomes, so that the DNA can be safely transported to the "daughter cells." A **chromosome** is the condensed form of chromatin. It is estimated that humans have almost 22,000 genes distributed on 46 chromosomes.

Table 4.1 The Types of Nucleic Acids in Healthy Human Cells

Nucleic Acid Molecules	Description	Nucleotides	Sugar	Function
DNA	Double-stranded linear Found in chromatin and chromosome forms	ATCG	Deoxyribose	Storage form of the genome
mRNA	Single-stranded linear	UAGC	Ribose	Copy of a single gene that leaves the nucleus to be translated into a protein
(base pairing) tRNA	Single-stranded 3-D, non-linear shape held together by complementary base pairing	AUCG	Ribose	Matches mRNA code with an amino acid
rRNA	Globular with protein interactions	AUCG	Ribose	Provides the structure within which RNA can be translated into a protein

The particular sequence of bases along the DNA molecule determines the genetic code. But a DNA molecule is not made of just one DNA strand, but rather of two strands that "complement" each other in the sense that the molecules that compose the strands fit together and bind to each other, creating a double-stranded molecule that looks much like a long, twisted ladder. Each side rail of the DNA ladder is composed of alternating sugar and phosphate groups (**Figure 4.23A**). The "rungs" of the ladder are made up of pairs of nitrogenous bases. There are four types of nitrogenous bases in DNA: adenine (A), thymine (T), cytosine (C), and guanine (G). Think about a ladder. For stability, each of the rungs must be the same width. Yet, the nitrogenous bases come in two sizes. Adenine (A) and guanine (G) are slightly larger than thymine (T) and cytosine (C) (**Figure 4.23B**). Therefore, there are restrictions on what two bases can form a pair. A and T always pair together, and G and C always pair together. A and G could not pair, because they would form a rung that is wider than the others. Therefore, if the two complementary strands of DNA were pulled apart, you could infer the order of the bases in one strand from the bases in the other, complementary strand. For example, if one strand has a region with the sequence AGTGCCT, then the sequence of the complementary strand would be TCACGGA.

The nucleus and the DNA it contains are a hub of activity throughout cellular life. Throughout each day, DNA's genetic instructions are accessed for making the proteins of the cell, and once in a while the entire DNA collection must be replicated before a cell can reproduce. The macrostructure (large molecular structure) of DNA changes between these two parts of cellular life. When the cell is living, growing, and making proteins, its DNA is loosely wound in the form of chromatin (**Figure 4.24**) for access to the genes needed. When the cell is going to replicate, its DNA needs to be packaged for easier handling and tracking so that it is easier to ensure that each daughter cell receives the correct DNA. During cell replication the DNA is highly condensed in the form of chromatin (see Figure 4.24). Each of these processes are described in this chapter.

Student Study Tip

You can remember that A and T always pair together since As and Tests also do! As and cheating do not go together, as neither do As and guessing.

Figure 4.23 DNA

(A) The DNA molecule is found in a double helix shape that resembles a twisted ladder. (B) DNA is composed of two complementary strands. The strands are bonded together via their nitrogenous base pairs using hydrogen bonds. Each rung of the ladder is a complementary pair of nitrogenous bases. Two pairs of bases are illustrated here. (C) An individual nitrogenous base, guanine.

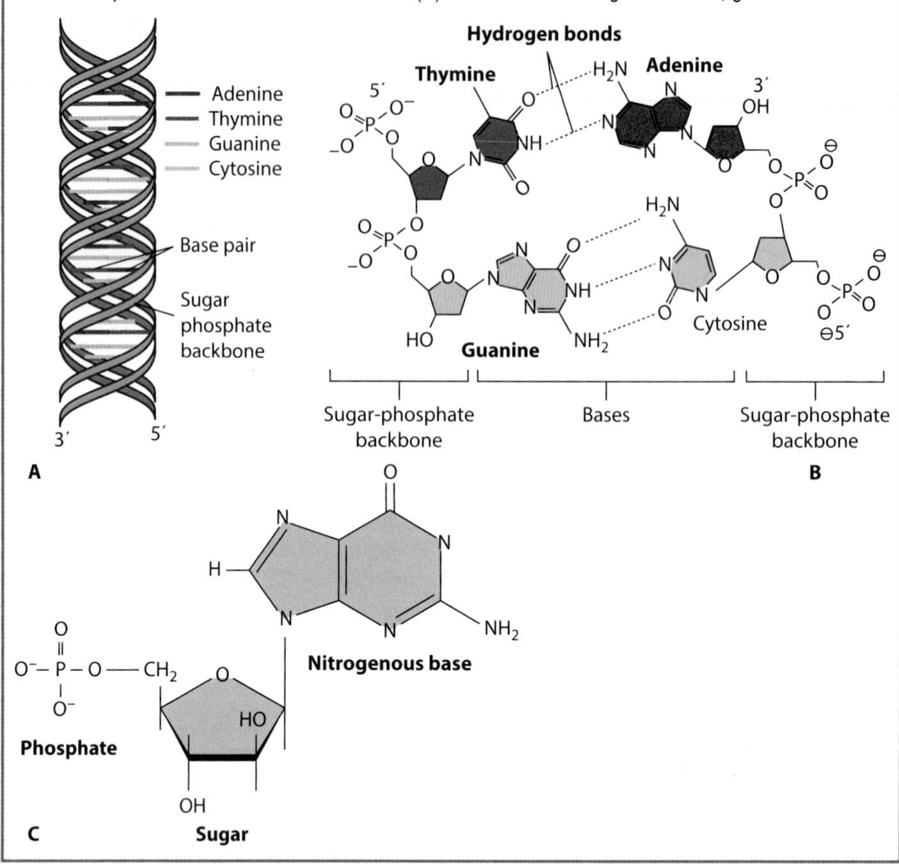

Cultural Connection

The Genome Project of *Some* Humans

From 1990 to 2003, scientists from the United States, the United Kingdom, Japan, France, Germany, Spain, and China worked together in the largest scientific collaboration of all time. The goal of this project was to determine the sequence of all 3 billion nucleotide pairs in the human genome. These scientists created a "reference genome," a mosaic of the genomes of various participants, so that a representative sequence of more than 15,000 human genes would be accessible and searchable. Today this reference genome is used in many scientific projects, including designing drugs, understanding diseases, and diagnosing genetic disease. More recently, a variety of clinical and at-home DNA testing kits have become available for discovering ancestry as well as individual risk of developing certain diseases including cancer, diabetes, and clogged arteries.

Can you think of any potential pitfalls in the design of this project? As it turns out, the majority of participants used to create the mosaic reference genome were of European ancestry. These tests just aren't as accurate in other groups, including humans of Asian, African, and Hispanic origin. Genetic testing on non-European individuals isn't just inaccurate; in some cases it can be practically useless. One recent study describes comparing the DNA of a young Black American patient to the reference genome in hopes of finding a genetic source of a set of mystery symptoms. The comparison turned up many so-called abnormal variants, not because there were many mutations in this patient's genome, but because their genome is not represented well by the European-based reference sequence. In the case of this and many non-European patients, the genetic revolution pioneered by the human genome project doesn't benefit them. Can you think of steps the scientific community could take to make genetic testing more inclusive for all humans?

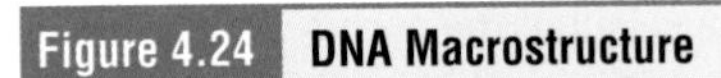

Figure 4.24 DNA Macrostructure

Strands of DNA are organized by wrapping around proteins called histones. During most of cellular life, the proteins and DNA are in a loose formation called chromatin. During cell replication, the chromatin is packed tightly into chromosomes.

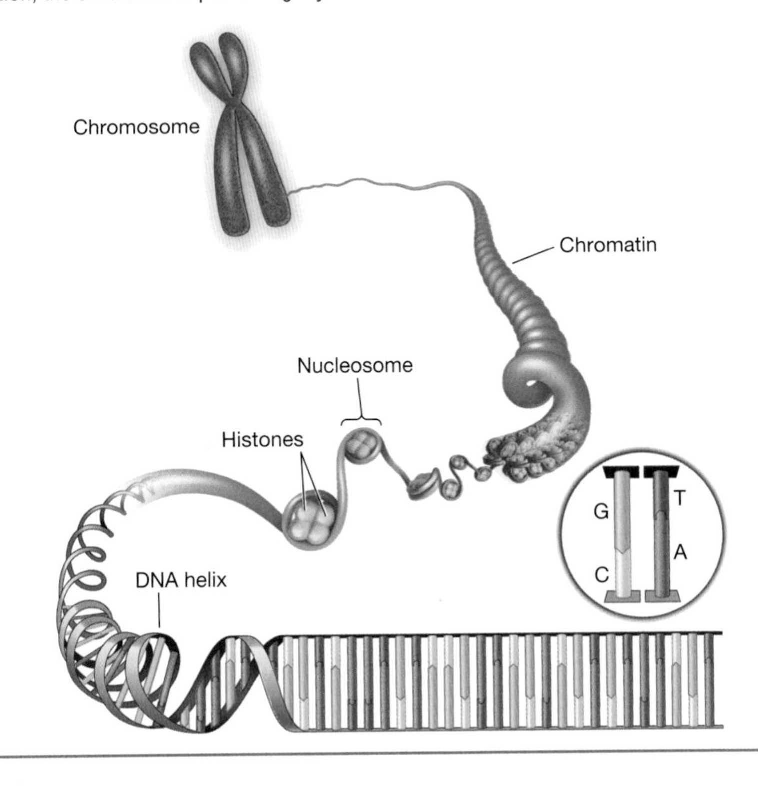

Learning Check

1. How is the nuclear membrane different than the cell membrane?
   a. It does not have any passageways for structures to pass through.
   b. It does not allow any structures through the membrane.
   c. It has two lipid bilayers with thin fluid space between them.
2. Select the complementary strand for AGCCTGATT.
   a. TCGGACTAA
   b. AGCCTGATT
   c. GATTCAGCC
   d. CTAAGTCGG
3. When would you expect the DNA to be highly condensed?
   a. When the cell is reproducing
   b. When the cell is living
   c. When the cell is growing
   d. When the cell is making proteins

4.4 Protein Synthesis

Learning Objectives: By the end of this section, you will be able to:

4.4.1 Define the terms genetic code, transcription and translation.

4.4.2 Explain the process of RNA synthesis.

4.4.3 Explain the roles of tRNA, mRNA, and rRNA in protein synthesis.

DNA contains the information necessary for the cell to build all of the body's components. We say that the DNA is the blueprint for the cell, and yet, only proteins can be made directly from DNA. However, virtually all the functions that a cell carries out are completed with the help of proteins. One of the most important classes of proteins is enzymes, which are necessary for the biochemical reactions inside the cell. Some of these critical biochemical reactions include building larger molecules from smaller components (such as during DNA replication or synthesis of microtubules) and breaking down larger molecules into smaller components (such as when harvesting chemical energy from nutrient molecules). Whatever the cellular process may be, it is almost sure to depend on proteins, and those proteins are made from the cell's DNA. Just as the cell's genome describes its full complement of DNA, a cell's **proteome** is its full complement of proteins. The process of making a protein begins with its gene. A **gene** is a segment of DNA that provides the genetic information necessary to build a single protein.

In Chapter 3 we learned that proteins are polymers of many amino acid building blocks. The sequence of bases in a gene (that is, its sequence of A, T, C, and G nucleotides) translates to an amino acid sequence. Nitrogenous bases are read in groups of three, similar to the way in which the three-letter code d-o-g signals the image of a dog; the three-letter DNA base code, for example CAC, signals the use of a particular amino acid, in this case valine. Therefore, a gene, which is composed of many bases in a unique sequence, provides the code to build an entire protein, with multiple amino acids in the proper sequence (Figure 4.25). All of the DNA of the genome resides in the nucleus. However, unless the cell is replicating, it has only one copy of its genome. If you had a single copy of something very important (for example, your passport), you would want it to keep it in a safe place. The cell does not want the DNA to leave the nucleus, but the machinery for making proteins, the ribosomes, are in the cytoplasm. To bridge these two, a copy of a single gene is made in a messenger molecule, RNA. This messenger RNA (mRNA) copy is made in the nucleus, then exits one of the nuclear pores and heads to the ribosomes of the cytoplasm.

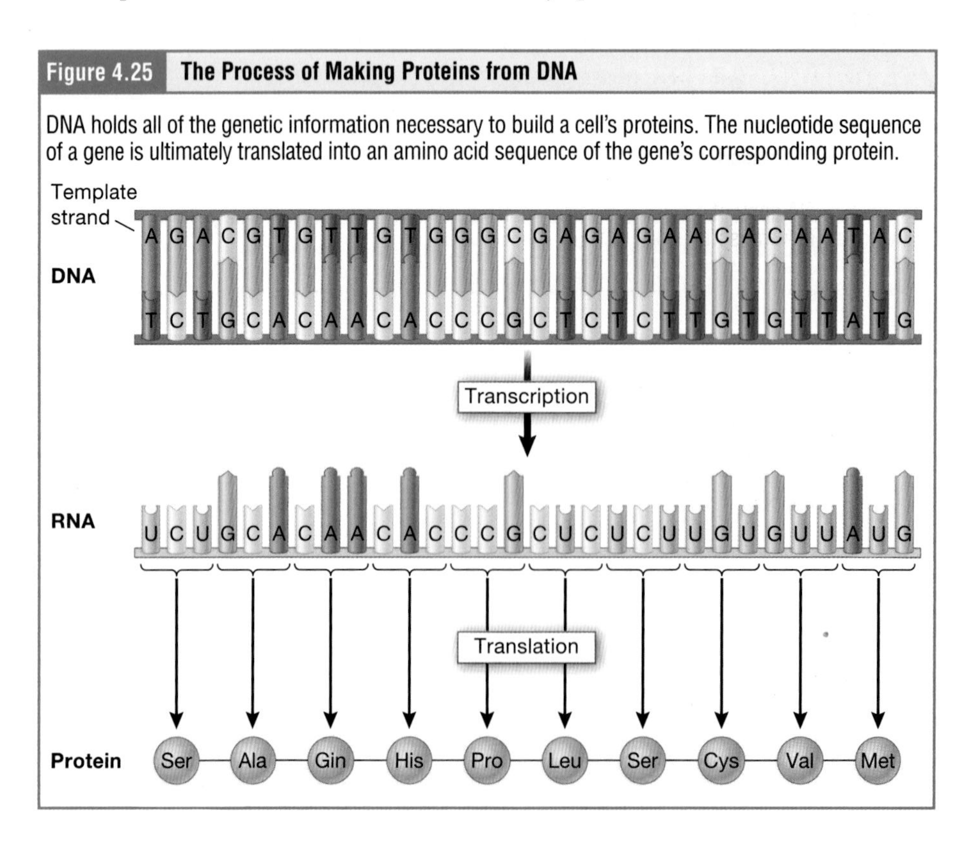

Figure 4.25 The Process of Making Proteins from DNA

DNA holds all of the genetic information necessary to build a cell's proteins. The nucleotide sequence of a gene is ultimately translated into an amino acid sequence of the gene's corresponding protein.

4.4a From DNA to RNA: Transcription

LO 4.4.1

Messenger RNA (mRNA) is a single-stranded nucleic acid that carries a copy of the genetic code for a single gene out of the nucleus and into the cytoplasm where it is used to produce proteins.

There are several different types of RNA, each having different functions in the cell. The structure of RNA is similar to DNA, with a few small exceptions. For one thing, unlike DNA, most types of RNA, including mRNA, are single-stranded and contain no complementary strand. Second, the ribose sugar in RNA contains an additional oxygen atom compared with DNA. Finally, instead of the base thymine, RNA contains the base uracil (U). This means that adenine will always pair up with uracil during the protein synthesis process.

Gene expression—the process of making a molecular product from a gene—begins with the process called **transcription**, in which a messenger RNA strand is made from the gene of interest. In order to make a copy of one DNA strand, the complementary DNA strands need to be separated. Only that small portion of the DNA will be split apart, but the process requires several enzymes to unzip and stabilize the strands. The base code of the gene is used as the template to make the complementary strand of RNA (Figure 4.26). We can break the transcription process into three stages: initiation, elongation, and termination.

LO 4.4.2

1. *Stage 1: Initiation.* The two complementary strands of DNA are separated and **RNA polymerase**, an enzyme that is capable of building RNA strands, begins to synthesize a complementary RNA molecule.
2. *Stage 2: Elongation.* RNA polymerase adds new nucleotides to a growing strand of RNA.
3. *Stage 3: Termination.* RNA polymerase reaches the end of the gene and the polymerase and mRNA transcript are released from the DNA strand. (Summarized in Table 4.2.)

Before the mRNA molecule leaves the nucleus and proceeds to protein synthesis, it is modified in a number of ways. For this reason, it is often called a pre-mRNA at this stage. Your DNA, and thus complementary mRNA, actually contains regions that do

Figure 4.26 **Transcription: Producing an mRNA Transcript from DNA**

A movable RNA copy of a single DNA gene is made through complementary binding. Whereas this copy, the mRNA transcript, can leave the nucleus, DNA is unable to do so.

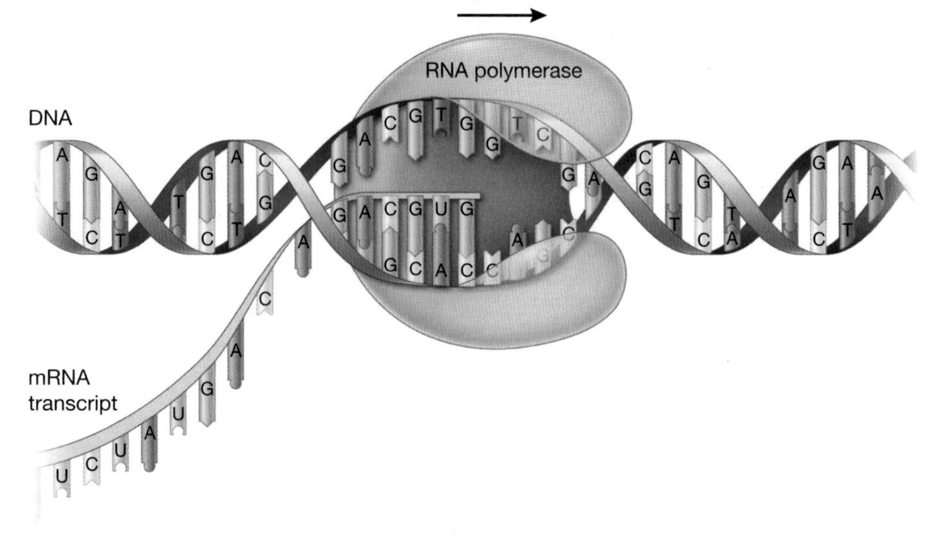

not contribute to the code for amino acids. Their function is still a mystery, but a process called **splicing** removes these noncoding regions from the pre-mRNA transcript to create the mature mRNA molecule (Figure 4.27). Interestingly, during this process the mRNA transcript temporarily ends up in segments and these segments can be joined back together in a variety of ways (Figure 4.28B). When different coding regions of mRNA are spliced out, different variations of the protein will eventually result, with differences in structure and function. This process results in a much larger variety of possible proteins and protein functions. When the mRNA transcript is ready, it travels out of the nucleus and into the cytoplasm.

4.4b From RNA to Protein: Translation

Translation is the process of synthesizing a chain of amino acids called a **polypeptide**. Translation occurs in the ribosome.

LO 4.4.3

Remember that many of a cell's ribosomes are found associated with the rough ER and carry out the synthesis of proteins destined for the Golgi apparatus. Ribosomes are nonmembranous organelles that are composed of **ribosomal RNA (rRNA)** and proteins. When they aren't in the process of translation, ribosomes exist in the cytoplasm in two subunits. When an mRNA molecule is ready to be translated, the two subunits come together and attach to the mRNA. While the ribosome is the organelle where translation takes place, it actually does not take any action in the creation of the new polypeptide. The ribosome simply provides the space for translation, bringing together and aligning the mRNA molecule with another form of RNA, **transfer RNA (tRNA)**. This type of RNA carries amino acids to the ribosome, and attaches each new amino acid to the last, building the polypeptide chain one by one. The tRNA molecules must be able to recognize the codons on mRNA and match them with the correct amino acid. The structure of tRNA is suited for this function. On one end of its structure is a binding site for a specific amino acid. On the other end is a base sequence that complements the mRNA codon specifying its particular amino acid. This sequence of three bases on the tRNA molecule is called an **anticodon**. For example, a tRNA responsible for shuttling the amino acid glycine contains a binding site for glycine on one end. On the other end it contains an anticodon that complements the glycine codon (GGA is a codon for glycine, and so the tRNA's anticodon would read CCU). Equipped with its particular cargo and matching anticodon, a tRNA molecule can read its recognized mRNA codon and bring the corresponding amino acid to the growing chain (Figure 4.28).

Figure 4.27 Creating a Mature mRNA Transcript

The initial mRNA transcript is immature and called a pre-mRNA transcript. It contains all of the code of the gene but also has noncoding regions within the transcript. Before leaving the nucleus, a structure called a *spliceosome* cuts out the noncoding regions (introns) and reconnects the expression regions (exons), resulting in the mature mRNA transcript, ready for translation.

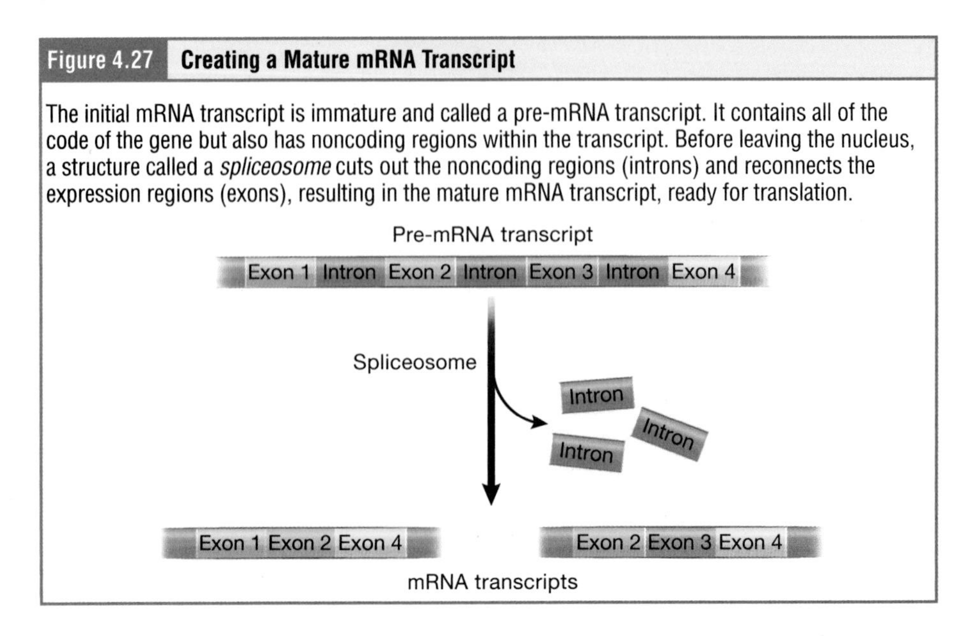

Figure 4.28 Translation from RNA to Amino Acid Sequence

During the process of translation, the mRNA transcript moves through a ribosome in the RER or cytosol. tRNA molecules pair the appropriate amino acids to the mRNA sequence. Each amino acid is joined to the one before it and a polypeptide, or protein, elongates into the cytosol.

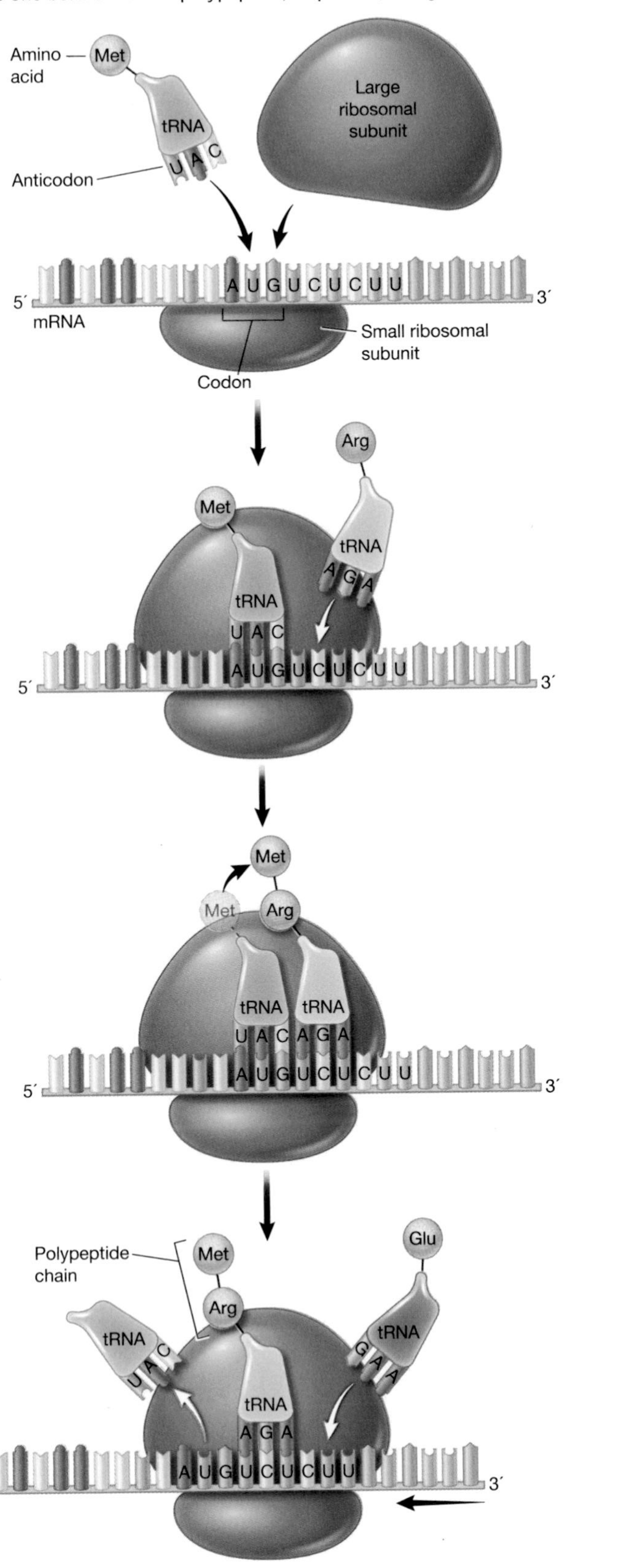

Just like transcription, we can consider translation in three stages: initiation, elongation, and termination (summarized in Table 4.2).

1. *Stage 1: Initiation.* The initial event is the assembly of the ribosome subunits around an mRNA transcript.
2. *Stage 2: Elongation.* In the elongation stage, tRNA molecules are drawn toward the assembled ribosome. If their anticodon complements the next mRNA codon in the sequence, the tRNA attaches its amino acid cargo to the growing polypeptide strand. This attachment takes place with the assistance of various enzymes and requires energy.
3. *Stage 3: Termination.* This process continues until the final codon on the mRNA is reached, which provides a "stop" message that signals termination of translation and triggers the release of the complete, newly synthesized protein. Thus, a gene within the DNA molecule is transcribed into mRNA, which is then translated into a protein product (see Figure 4.28).

Learning Connection

Explain a Process

Try explaining translation from RNA to protein to a nonscientist (a parent, a friend, your dog, and so on). Can you explain why mRNA is required? What is the role of tRNA?

Learning Check

1. Which of the following molecules is made from the genes of a DNA?
   a. Carbohydrates
   b. Proteins
   c. Phospholipids
   d. Steroids
2. Which of the following is unique about RNA? Please select all that apply.
   a. It is single-stranded.
   b. It contains uracil (U).
   c. Adenine pairs with guanine.
   d. Ribose contains less oxygen.
3. What process removes the region of the mRNA that does not contribute to the code for amino acids?
   a. Initiation
   b. Translation
   c. Splicing
   d. Termination
4. What is the anticodon for the mRNA sequence AGCUCAUGU?
   a. UCGAGUACA
   b. AGCUCAUGU
   c. TCGAGTACA
   d. CAUCUGCAC

4.5 Cell Replication

Learning Objectives: By the end of this section, you will be able to:

4.5.1 Compare and contrast somatic cell division (mitosis) and reproductive cell division (meiosis).

4.5.2 Describe the general phases (e.g., G phases, S phase, cellular division) of the cell cycle.

4.5.3 Describe DNA replication.

4.5.4 Compare and contrast chromatin, chromosomes, and chromatids.

4.5.5 Describe the events that take place during mitosis and cytokinesis.

All humans originated from a single fertilized egg. From this one cell, trillions and trillions of rounds of cell replication are needed to occur to result in the body you are sitting in right now. But the work does not end when you reach full height or even adulthood. Inside your body you are constantly making new white blood cells, replacing the lining of your stomach, and producing more skin cells. Cell replication is a constant and lifelong pursuit. Billions of new cells are produced in an adult human every day. There are a few cells in the body that do not undergo cell replication (such as gametes, red blood cells, most neurons, and some muscle cells); most somatic cells replicate often. A **somatic cell** is a general term for a body cell, and all human cells, except for the cells that produce eggs and sperm (which are referred to as **germ cells**), are somatic cells. Somatic cells contain 46 chromosomes. This is the **diploid** number, meaning that within those 46 chromosomes are two copies of each chromosome (one copy received from each parent). A **homologous** pair of chromosomes is the two copies of a single chromosome found in each somatic cell. Human somatic cells contain 23 homologous pairs of chromosomes in each of the somatic cells; 23 is the **haploid** number.

diploid (2n)

Cells in the body replace themselves over the lifetime of a person. For example, the cells lining the gastrointestinal tract must be frequently replaced when constantly "worn off" by the movement of food through the gut. But what triggers a cell to replicate, and how does it prepare for and complete cell replication? The **cell cycle** is the sequence of events in the life of the cell from the moment it is created at the end of a previous cycle of cell division until it then divides itself, generating two new cells.

4.5a The Cell Cycle

One "turn" or cycle of the cell cycle consists of three general phases: interphase, mitosis, and cytokinesis. **Interphase** is the period of the cell cycle during which the cell is not replicating. Most of cellular life is spent in interphase. During interphase, a cell preparing to replicate will copy its genome so that it enters mitosis with one copy for each of the eventual cells. **Mitosis** is the division of genetic material, during which the cell nucleus breaks down and two new, fully functional, nuclei are formed. **Cytokinesis** divides the cytoplasm into two distinctive cells. Mitosis and cytokinesis are two processes within cell replication. There is also a third process worth mentioning, which is meiosis. **Meiosis** is the process in which germ cells, which give rise to the egg and sperm reproductive cells, are made. Meiosis results in cells containing half of the original amount of DNA, in contrast to mitosis, which results in cells that have just as much DNA as the parent cell began with.

LO 4.5.1

Interphase A cell grows and carries out all normal metabolic functions and processes in a period called G_1 (Figure 4.29). **G_1 phase** (gap 1 phase) was described as a "gap" phase by scientists who were seeking cells in mitosis. The majority of cells they observed were in G_1 because this phase is where cells are growing, making proteins, and carrying out the functions of the cell. Some cells stay in G_1 for their entire cellular life. For cells that will divide again, G_1 is followed by replication of the DNA during the S phase. The **S phase** (synthesis phase) is period during which a cell replicates its DNA. After the synthesis phase, the cell proceeds through the G_2 phase. The **G_2 phase** is a second gap phase, during which the cell continues to grow and makes the necessary preparations for mitosis.

LO 4.5.2

G_1 phase (gap one phase)

G_2 phase (gap two phase)

4.5b DNA Replication

DNA replication is the copying of DNA that occurs during S phase (Figure 4.30). Just like transcription and translation, we can describe three phases of DNA replication as initiation, elongation, and termination.

LO 4.5.3

We can describe the DNA replication process in the same terms as transcription and translation (summarized in Table 4.2).

1. *Stage 1: Initiation.* The two complementary DNA strands are separated and stabilized with the help of special enzymes.
2. *Stage 2: Elongation.* Each strand becomes a template along which a new complementary strand is built. **DNA polymerase** brings in the correct bases to complement the template strand, synthesizing a new complementary strand nucleotide

Figure 4.29 The Cell Cycle

(A) The two major phases of the cell cycle include mitosis (cell replication) and nonreplicating cellular life, the phase in which a cell is living and performing its other functions. (B) Some of nonreplicating cellular life is actually spent preparing for cell replication, including replicating organelles or DNA. (C) Cell replication can be divided into mitosis and cytokinesis. Nonreplicating life, or interphase, can be subdivided into G_1 (when the cell grows and performs all of its normal functions), S (in which the cell replicates its DNA), and G_2 phase (in which the cell prepares its intracellular contents for replication).

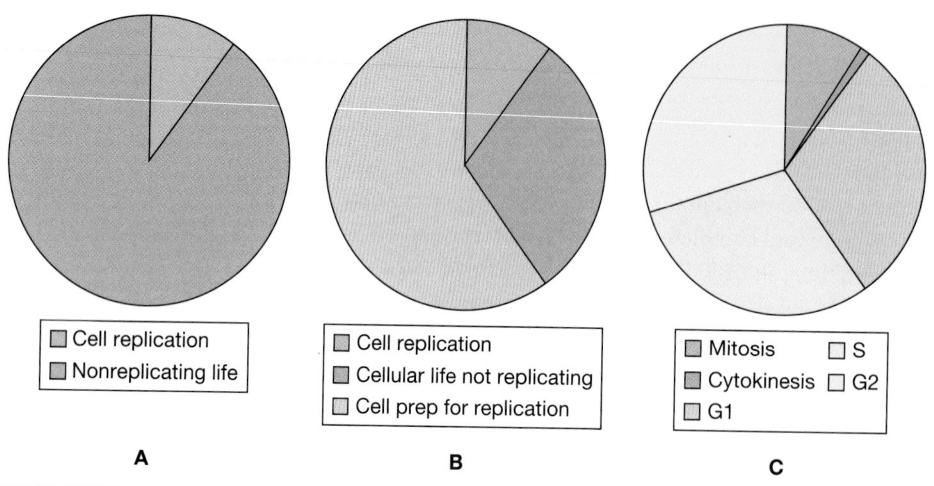

by nucleotide. A DNA polymerase is an enzyme that adds free nucleotides to the end of a chain of DNA, making a new double strand. This growing strand continues to be built until it is a full new complement to the original template strand.

3. *Stage 3: Termination.* Once the two original strands are bound to their own, finished, complementary strands, DNA replication is stopped, and the two new identical DNA molecules are complete.

Each new DNA molecule contains one strand from the original molecule and one newly synthesized strand. This process continues, one chromosome at a time, until the cell's entire genome is replicated. As you might imagine, it is very important that DNA

Figure 4.30 DNA Replication

During the S phase of the cell cycle, a new copy of the DNA genome is made. A number of different enzymes work together. Helicase unwinds the two DNA strands so each strand can be used as a template to synthesize new DNA strands. The result of DNA replication is two copies of each chromosome, one strand of each of which was the original and one newly synthesized strand.

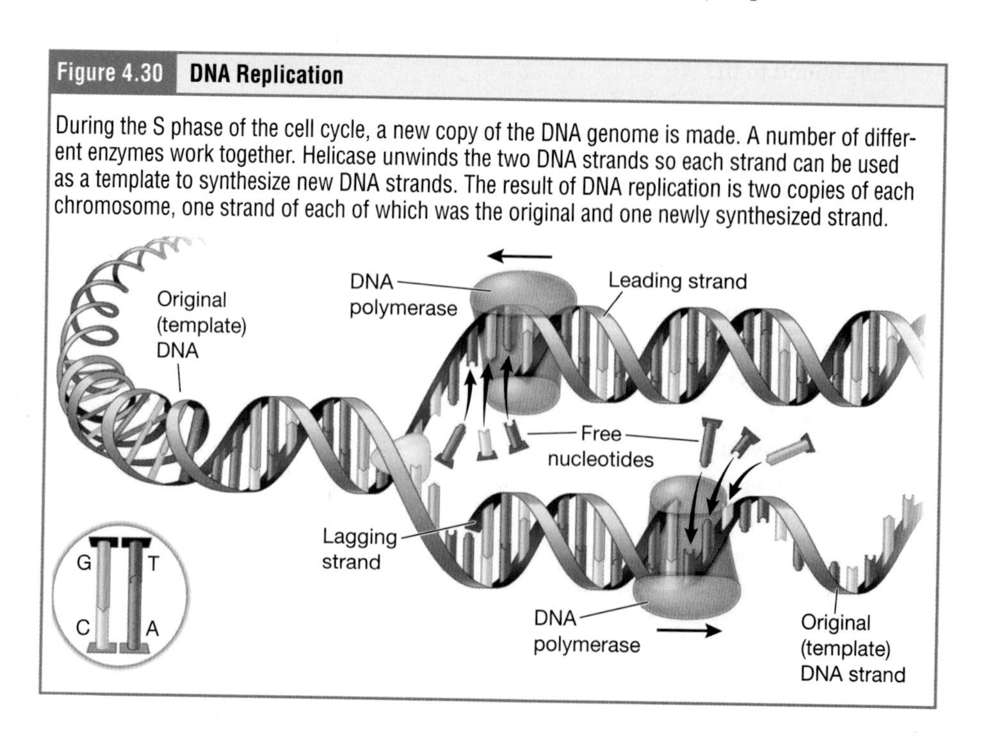

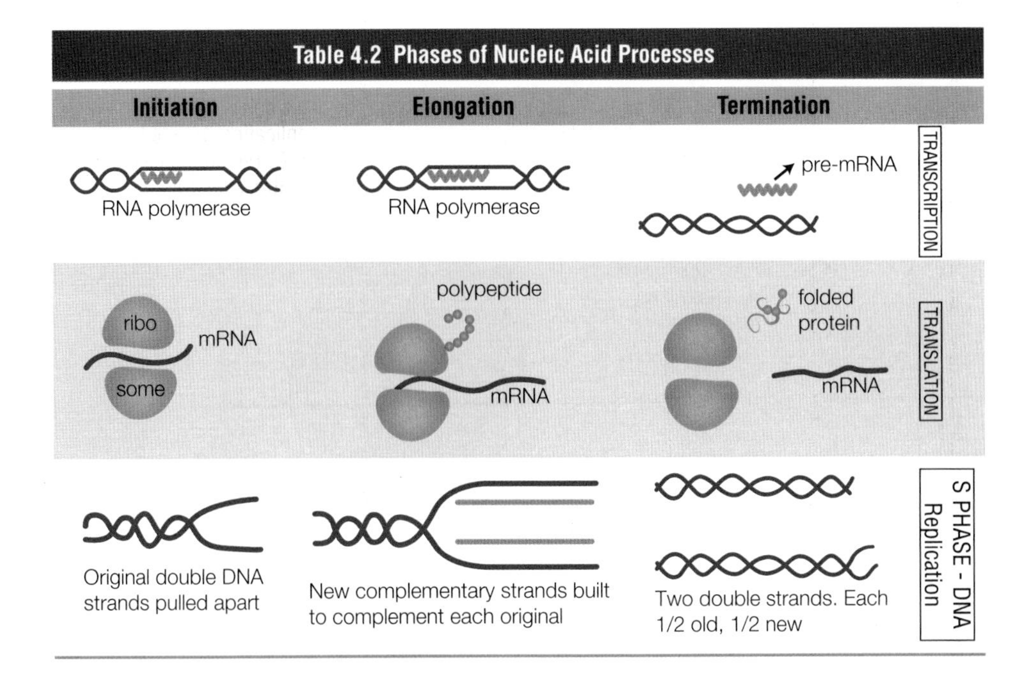

replication take place precisely so that new cells in the body contain the exact same genetic material as their parent cells. Mistakes made during DNA replication, such as the accidental addition of an inappropriate nucleotide, have the potential to render a gene dysfunctional or useless. Fortunately, there are mechanisms in place to minimize such mistakes. A DNA proofreading process enlists the help of special enzymes that scan the newly synthesized molecule for mistakes and corrects them. Once the process of DNA replication is complete, the cell is ready to divide.

Table 4.2 Phases of Nucleic Acid Processes
Initiation
Elongation
Termination
RNA polymerase
RNA polymerase
pre-mRNA
TRANSCRIPTION
ribo
some
mRNA
polypeptide
mRNA
folded
protein
mRNA
TRANSLATION
Original double DNA strands pulled apart
New complementary strands built to complement each original
Two double strands. Each 1/2 old, 1/2 new
S PHASE - DNA Replication

The Structure of Chromosomes During the process of DNA replication, which takes place during S phase of interphase, the amount of DNA within the cell precisely doubled.

In-line Qs: how many homologous pairs? How many chromosomes? If a somatic cell in G_1 is referred to as 2n, what would we call this cell in G_2?

For every chromosome, its replicated copy is called a **sister chromatid** and is physically bound to the other copy at a structure known as a **centromere** (**Figure 4.31**).

LO 4.5.4

Learning Check

1. Which of the following is considered a germ cell?
   a. Cells found in the liver
   b. Cells found in the kidney
   c. Cells found in the heart
   d. Cells that produce sperm
2. Which phase is NOT part of cellular replication?
   a. Interphase
   b. Mitosis
   c. Cytokinesis
3. In what phase of DNA replication does the DNA polymerase add free nucleotides to make a new double strand?
   a. Initiation
   b. Elongation
   c. Termination
   d. G_2 phase
4. How many sister chromatids can be found in a cell during metaphase of mitosis?
   a. 26
   b. 23
   c. 46
   d. 18

Figure 4.31 A Homologous Pair of Chromosomes during Cell Replication

The red and blue colors represent a homologous pair of chromosomes. Remember that human cells have 23 unique chromosomes, but 2 copies (1 that came from sperm and 1 that came from the egg) of each; these are the homologous chromosomes. After DNA replication, each chromosome has been duplicated and has an identical copy called a sister chromatid. Until anaphase of mitosis, these sister chromatids will be bound together at a centromere.

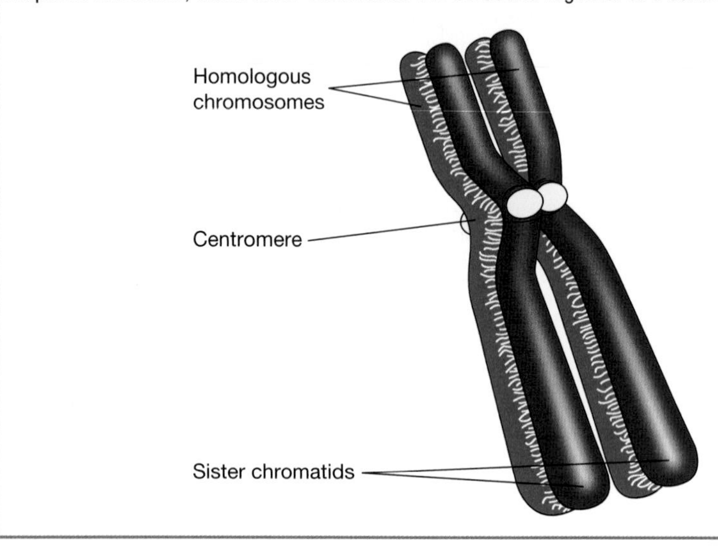

LO 4.5.5

Mitosis and Cytokinesis The **mitotic phase** of the cell includes both mitosis and cytokinesis. In mitosis, the chromosomes are organized and distributed equitably. Cytokinesis then occurs, dividing the cytoplasm into two new cells, each with a full complement of chromosomes. Mitosis is divided into four major stages: prophase, metaphase, anaphase, and telophase (**Figure 4.32**). The process is then followed by cytokinesis.

Figure 4.32 illustrates the phases of mitosis and cytokinesis. Here you can see that in interphase the DNA is in chromatin form. During **prophase**, the first phase of mitosis, the loosely packed chromatin coils and condenses into chromosomes. This transition is noticeable under the microscope as the DNA condenses so completely that the chromosomes are visible as being separate and distinct from each other. Sister chromatids are linked to each other and to the microtubules at their centromere, and now are visible in a long X-shape. In Figure 4.32, there are 6 total chromosomes represented. Remember that in a real human cell, the normal diploid number of chromosomes is 46. The homologous chromosomes are the same size but are drawn in slightly different shades of color to remind us that homologous chromosomes contain the same genes but have different sequences of base pairs as one is inherited from the oocyte and one from the sperm. The nucleolus and the nuclear membrane disappear during this phase and are not visible throughout mitosis.

Throughout the phases of mitosis, the cell needs to organize and distribute its chromosomes. But chromosomes are not capable of moving themselves. To move within the cell, any structure must utilize the cytoskeleton. A major occurrence during prophase is a reorganization of the cell's cytoskeleton itself in order to facilitate the movement of the chromosomes. Recall the cellular structures called centrioles that serve as origin points from which microtubules extend. In prophase, centrioles begin to move apart and head toward the ends, or poles, of the cell. As the centrioles migrate to two different sides of the cell, microtubules begin to extend from each centriole like long fingers from two hands extending toward each other. The **mitotic spindle** is the structure composed of the centrioles and their emerging microtubules.

Learning Connection

Broken Process

What would happen if a pair of sister chromatids did not separate? What if the spindle never formed? What would happen if actin could not form fibers?

Figure 4.32 **Cell Division: Mitosis Followed by Cytokinesis**

The stages of cell division oversee the separation of identical genetic material into two new nuclei, followed by the division of the cytoplasm.

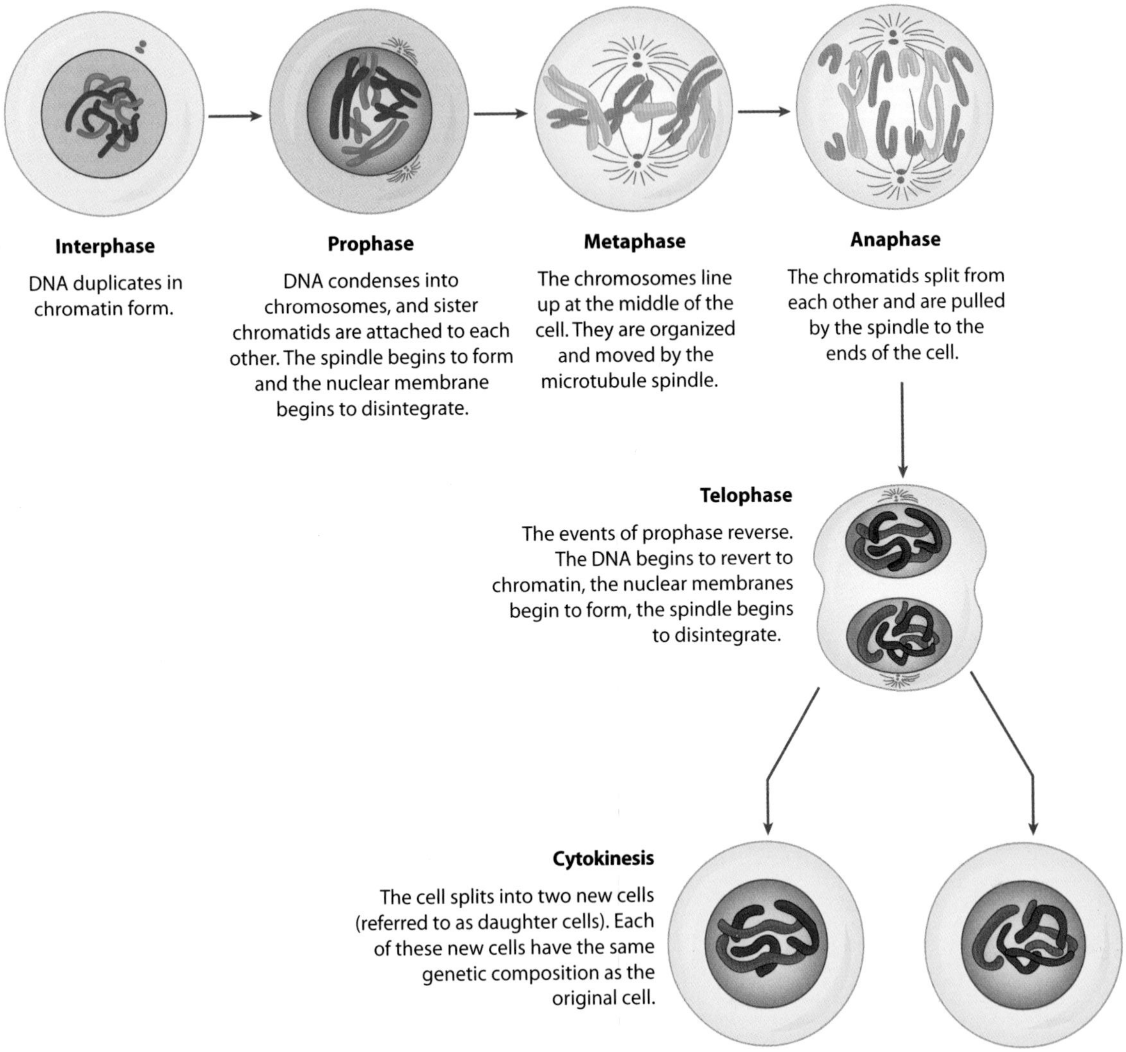

Interphase

DNA duplicates in chromatin form.

Prophase

DNA condenses into chromosomes, and sister chromatids are attached to each other. The spindle begins to form and the nuclear membrane begins to disintegrate.

Metaphase

The chromosomes line up at the middle of the cell. They are organized and moved by the microtubule spindle.

Anaphase

The chromatids split from each other and are pulled by the spindle to the ends of the cell.

Telophase

The events of prophase reverse. The DNA begins to revert to chromatin, the nuclear membranes begin to form, the spindle begins to disintegrate.

Cytokinesis

The cell splits into two new cells (referred to as daughter cells). Each of these new cells have the same genetic composition as the original cell.

Metaphase is the second stage of mitosis. During this stage, the sister chromatids, attached to the microtubules of the mitotic spindle, are pulled into a line along the middle of the cell. A cell in metaphase, viewed through the microscope, appears as if it is wearing a belt, made of tiny X's (see Figure 4.32). The microtubule spindle is now poised to pull apart the sister chromatids and bring one from each pair to each side of the cell.

Anaphase is the third stage of mitosis. In anaphase, the pairs of sister chromatids are separated from one another, forming individual chromosomes once again. As the microtubules shorten, these chromosomes are pulled to opposite ends of the cell. Each end of the cell receives one partner from each pair of sister chromatids, ensuring that the two new daughter cells will contain identical genetic material.

Telophase is the final stage of mitosis. In telophase, many of the events of prophase reverse. Two nuclear membranes begin to form at either end of the dividing cell, enclosing the chromosomes. The DNA uncoils such that the chromosomes return to loosely packed chromatin. Nucleoli also reappear within the new nuclei, and the mitotic spindle breaks apart.

Figure 4.33 Cytokinesis

A deep furrow appears around the cell as rings of actin filaments tighten around the middle of the replicated cell like a belt. Eventually the rings tighten to the extent that the membranes of the daughter cells separate.

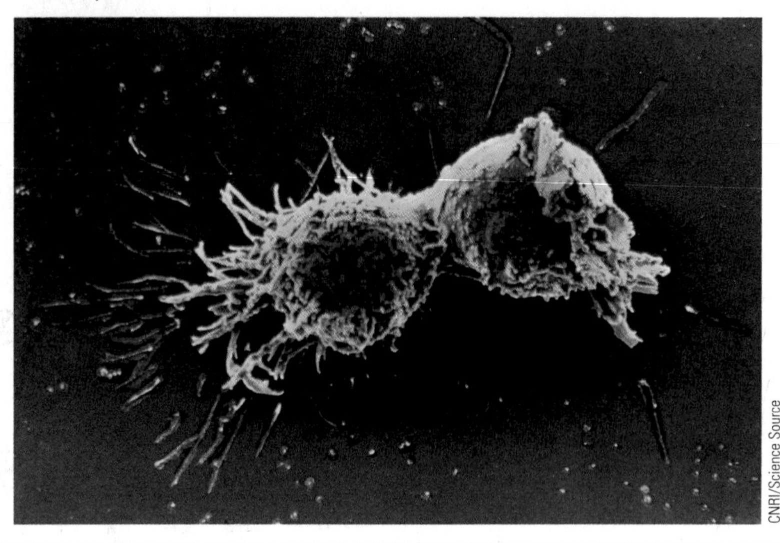

CNRI/Science Source

At the end of telophase, each new cell has received its own complement of DNA, organelles, membranes, and centrioles. A **cleavage furrow** is a contractile band made up of microfilaments (actin fibers) that forms around the midline of the cell during cytokinesis. This contractile band squeezes the two cells apart until they finally separate in the process of cytokinesis (**Figure 4.33**). Two new cells are now formed.

For a human cell, mitosis and cytokinesis takes, on average, one to two hours and occurs in billions of cells each day. If you attended an hour-long lecture, this complex process took place millions of times inside your body while you were focused on learning!

4.5c Cell Cycle Control

The cell cycle is regulated by a number of factors. Precise regulation of the cell cycle is critical for maintaining the health of an organism, and loss of cell cycle control can lead to cancer.

Learning Check

1. Which of the following occurs during prophase? Please select all that apply.
   a. Centrioles begin to move apart and head toward the end.
   b. The chromatin coils condense into chromosomes.
   c. The sister chromatids are aligned in the middle of the cell.
   d. Microtubules shorten to pull the one part of each sister chromatid.
2. What form is the DNA at the end of telophase?
   a. Chromatin
   b. Chromosome
   c. Chromatid
3. What cytoskeletal element creates the cleavage furrow?
   a. Microfilaments
   b. Centrioles
   c. Microtubules
   d. Intermediate filaments

Mechanisms of Cell Cycle Control There are a number of environmental factors that regulate cell replication, as well as some factors internal to the cell (**Figure 4.34**). Environmental factors include:

- Hormones and signals from other cells. Much of the functions of the body are regulated on an inter-system level through communication through hormones. For example, after the birth of a baby, the breasts need to be capable of lactation. Far in advance of the birth, estrogen signaling promotes mitosis in breast tissue so that the breasts can sufficiently remodel to be ready for this task. These signals are typically called **growth factors**.
- Another environmental factor is the number of cells around in the immediate environment. Cell replication can be inhibited through a mechanism called **contact inhibition**; simply put, if a cell is surrounded on all sides by other cells, it won't replicate. Contact inhibition is typically lost in tumor-forming cancer cells.
- Cells may divide simply for efficiency. A large cell, with a large surface area-to-volume ratio, is relatively inefficient because it takes a lot of cellular energy to manage the contents of and transport materials around in a large cell.

Figure 4.34 Factors That Regulate Cell Division

(A) Cells respond to growth factors in their environments. For breast cells, the hormone estrogen serves as a growth factor that promotes mitosis. This is useful in preparing the breasts for lactation during pregnancy. (B) Healthy cells demonstrate contact inhibition. They stop replicating when they are surrounded by other cells on all sides; loss of contact inhibition is one factor in cancer and other tumor development. (C) Surface area-to-volume ratio; cells with large internal volumes are inefficient and may divide simply to increase efficiency in internal processes.

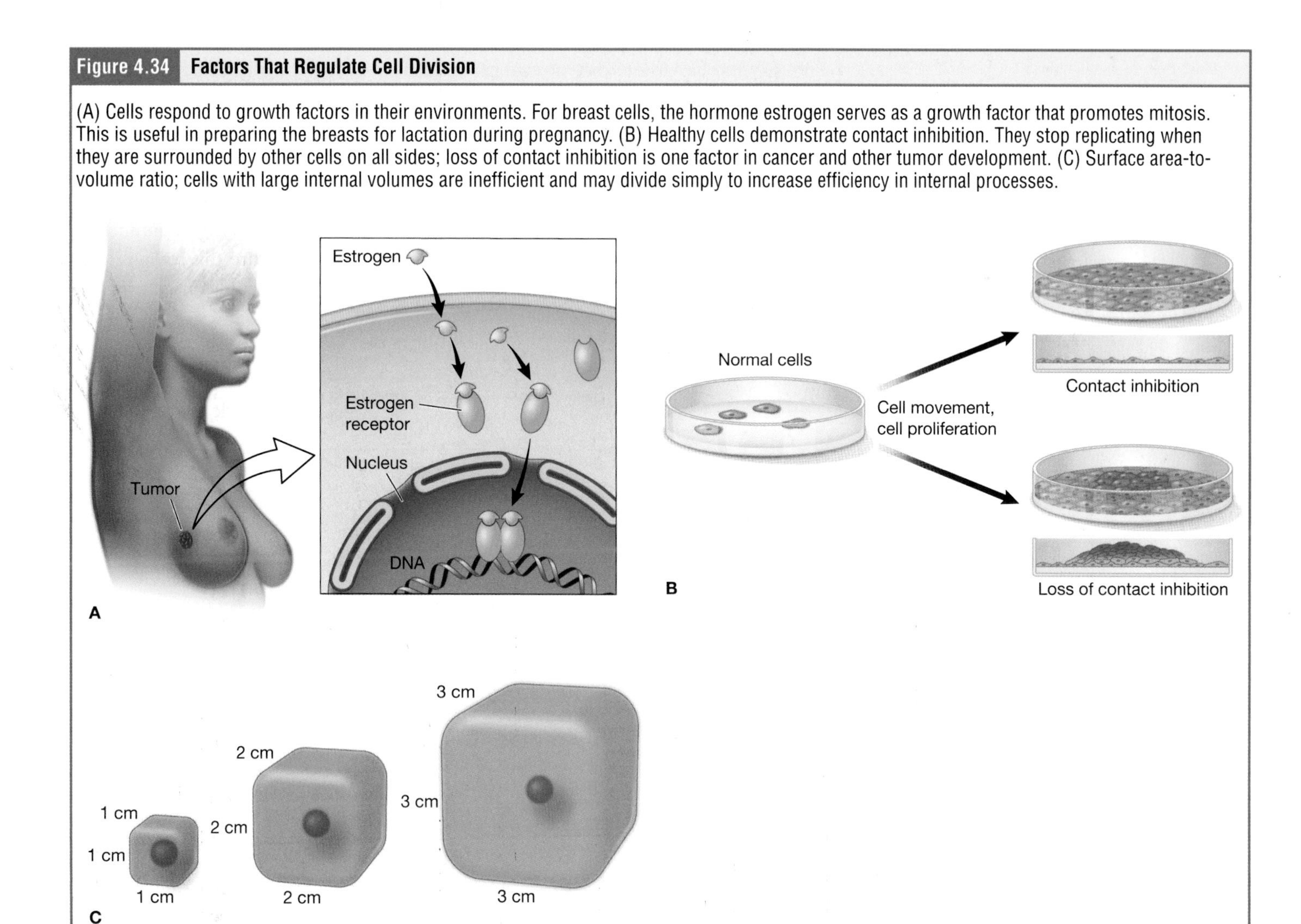

Apply to Pathophysiology

Hormone Replacement Therapy

Hormone replacement therapy (HRT) is a therapeutic approach to alleviate some of the side effects of menopause. In HRT, postmenopausal women are prescribed estrogen, a normally circulating hormone. HRT increases an individual's risk of developing breast cancer by 75 percent; however, some individuals are able to take HRT without developing breast cancer.

1. In the example of HRT and breast cancer development, the estrogen provided to the individual could best be described as
   A) A checkpoint inhibitor.
   B) A transcription factor.
   C) A growth factor.
2. A person would be more likely to develop breast cancer while taking HRT if
   A) A breast cell lost contact inhibition.
   B) A breast cell had a small surface area-to-volume ratio.
   C) A breast cell was stopped at the G_1 checkpoint.
3. If HRT was prescribed along with a drug that inhibited DNA synthesis, the effects on the individual would be
   A) An increased likelihood of developing breast cancer.
   B) A decreased likelihood of developing breast cancer.
4. Whether it is made in the body or prescribed as therapy, estrogen, once bound to its receptor probably serves as a(n)
   A) Transcription factor that increases expression of genes that promote mitosis.
   B) Inhibitor to microtubule formation.
   C) Promotor to alternative splicing of mRNA transcripts.

Cells also have their own internal mechanisms for regulating the cell cycle. As a cell proceeds through the stages of the cell cycle, it reaches particular checkpoints. A **checkpoint** is a point in the cell cycle at which the cycle can be signaled to move forward or stopped. At each of these checkpoints, different varieties of molecules provide the stop or go signals, depending on certain conditions within the cell. At the G_1 checkpoint, for example, the cell must be ready for DNA synthesis to occur. At the G_2 checkpoint the cell must be fully prepared for mitosis. Even during mitosis, a crucial stop and go checkpoint in metaphase ensures that the cell is fully prepared to complete cell division. The metaphase checkpoint ensures that all sister chromatids are properly attached to their respective microtubules and lined up at the metaphase plate before the signal is given to separate them during anaphase.

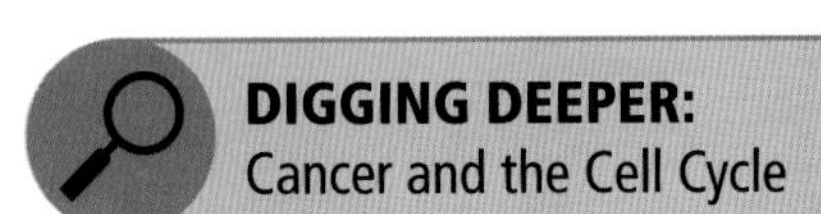

DIGGING DEEPER: Cancer and the Cell Cycle

Cancer is an extremely complex condition, and it may arise from a wide variety of genetic and environmental causes. Even just one form of cancer, such as breast cancer described in the previous "Apply to Pathophysiology" feature, can be caused by many different factors, a combination of factors, or (in many cases) factors that scientists and clinicians have not yet identified. Cancer may be the most well-studied disease that affects human bodies. What scientists have learned so far is that mutations in a cell's DNA that compromise normal cell cycle control systems lead to cancerous tumors. While progressing through the phases of the cell cycle, a large variety of intracellular molecules provide stop and go signals to regulate movement forward to the next phase. These signals are maintained in an intricate balance so that the cell only proceeds to the next phase when it is ready. The cell has both stop and go signals, similar to the throttle and brake systems on a car.

Cyclins are a class of proteins that are encoded by genes, called proto-oncogenes, and provide important signals that regulate the cell cycle and move it forward. Examples of proto-oncogene products include cell-surface receptors for growth factors, or cell-signaling molecules, two classes of molecules that can promote DNA replication and cell division. In contrast, a second class of genes (known as *tumor suppressor genes*) sends stop signals during a cell cycle. For example, certain protein products of tumor suppressor genes signal potential problems with the DNA and thus stop the cell from dividing, while other proteins signal the cell to die if it is damaged beyond repair. Some tumor suppressor proteins function in contact inhibition.

Under normal conditions, these stop and go signals are maintained in a homeostatic balance. Generally speaking, there are two ways that the cell's mitotic engine may lose control: an overactive throttle, or an underactive brake system. When compromised through a mutation, or otherwise altered, proto-oncogenes can be converted to oncogenes, which produce oncoproteins that push a cell forward in its cycle and stimulate cell division even when it is undesirable to do so. For example, a cell that should be programmed to self-destruct (a process called *apoptosis*) due to extensive DNA damage might instead be triggered to proliferate by an oncoprotein. On the other hand, a dysfunctional tumor suppressor gene may fail to provide the cell with a necessary stop signal, also resulting in unwanted cell division and proliferation.

A delicate homeostatic balance between the many proto-oncogenes and tumor suppressor genes controls the cell cycle and ensures that only healthy cells replicate. Therefore, a disruption of this homeostatic balance can cause aberrant cell division and cancerous growths.

Cultural Connection

The Mother of Medicine: Henrietta Lacks

GL Archive/Alamy Stock Photo

Much of what we know about the control of the cell cycle and the initiation of cancer in humans has come from research involving a single kind of cell, the HeLa cell. These cells, the most common human cells in cancer research labs, were isolated from a patient by the name of Henrietta Lacks at Johns Hopkins University in 1951. Ms. Lacks died of cervical cancer when she was just 31 years old, but her cells, which are named with the first two initials of her first and last names, HeLa, continue to replicate in research labs for more than double the length of her lifetime. These cells have contributed a huge legacy to scientific knowledge. HeLa cells are used to study cancer, drugs, toxins, hormones, and even viruses. While her cells are widely famous among scientists, Ms. Lack's name and story was largely unknown for a long time. Ms. Lacks was not asked for her cells, and she was an unknowing contributor to this vast body of scientific knowledge. The social, ethical, scientific, and financial implications of HeLa cells are beautifully and fascinatingly detailed in the book "The Immortal Life of Henrietta Lacks" by Rebecca Skloot (Crown, 2011).

4.6 Cellular Differentiation

Learning Objectives: By the end of this section, you will be able to:

4.6.1* Summarize the properties of stem cells that distinguish them from differentiated cells.

4.6.2* Discuss how the generalized cells of a developing embryo or the stem cells of an adult organism become differentiated into specialized cells.

* Objective is not a HAPS Learning Goal.

How does a complex organism such as a human develop from a single cell—a fertilized egg—into the vast array of cell types such as nerve cells that send signals throughout the body, muscle cells that can contract and produce movement, and white blood cells that can identify and eliminate bacterial invaders? Throughout development and adulthood, the process of cellular differentiation leads cells to assume their final structure and function. **Differentiation** is the process by which unspecialized cells become specialized to carry out distinct functions.

4.6a Stem Cells

A **stem cell** is an unspecialized cell that can replicate as many times as needed and can, under specific conditions, differentiate into specialized cells (**Figure 4.35**). Early embryos are made entirely of stem cells, but as the embryo becomes bigger the cells begin to differentiate to take on the structure necessary for their functions. A skeletal muscle cell needs to produce a lot of the contractile proteins actin and myosin, but it does not need to produce the skin pigment melanin because muscle cells are never exposed to the sun. Therefore, differentiation begins as a genetic process. The cell needs open access to those genes (and thus those proteins) that will be expressed, and not those that will remain silent. The genes that are not required as the cell takes on its differentiated function are turned "off" and their DNA is highly condensed (**Figure 4.35D**). The genes that the cell will need to use more in its new differentiated state are turned "on" through transcription factors (**Figure 4.35C**). A **transcription factor** is one of a class of proteins that bind to specific genes on the DNA molecule and promote its expression. Transcription factors can also inhibit transcription.

LO 4.6.1

LO 4.6.2

While most of the cells in an early embryo are stem cells, an adult body has very few types of stem cells left. Unlike the embryonic stem cells, adult stem cells cannot become any cell in the body, but rather, are limited to cell groups or lineages. A hematopoietic stem cell can become any type of blood cell, but it cannot become a muscle cell. A satellite cell can become new skeletal muscle cells, but it cannot generate heart muscle cells.

Figure 4.35 Cell Differentiation

(A) Stem cells can become any cell in the body. (B) In the process of cell differentiation, stem cells alter their shape, internal organelles, and gene expression in order to take on the specialized function of mature cells. (C) Cells alter their DNA in this process. Genes that are useful in the adult cell fate of the cell are kept accessible. (D) Genes that are not useful in the adult cell fate are packaged so tightly that they are not able to be transcribed.

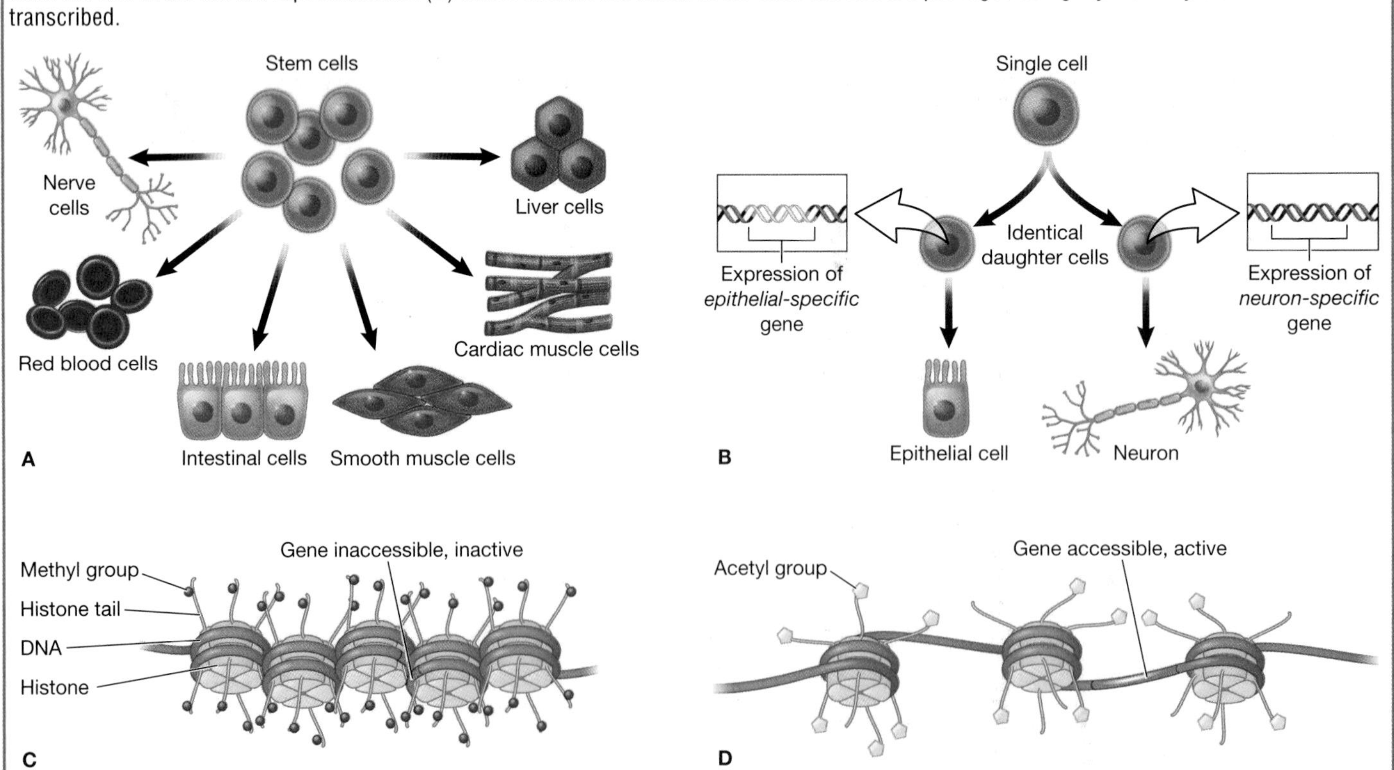

DIGGING DEEPER:
The Promise of Stem Cells

Many cell types in the body are capable of mitosis in order to regenerate and repair damage. For example, if you donated half of your liver to a person with liver disease, your liver cells (hepatocytes) would undergo mitosis and regenerate this organ within you. Over time, most adult cells undergo the wear and tear of aging and lose their ability to divide and repair themselves. Stem cells, on the other hand, have a much higher capacity for mitosis than other cells in your body. Stem cells do not display a particular morphology or function. Adult stem cells, which exist as a small subset of cells in most tissues, keep dividing and can differentiate into a number of specialized cells generally formed by that tissue. These cells enable the body to renew and repair body tissues. For example, in the deeper layer of your skin are a population of stem cells capable of many more rounds of mitosis than other cells. When you damage your skin, mitosis and cell movement help to heal the damage. These stem cells do not look like other skin cells and cannot function as such; however, they are capable of the necessary mitosis when needed.

The mechanisms that induce a nondifferentiated cell to become a specialized cell are poorly understood. In a laboratory setting, it is possible to induce stem cells to differentiate into specialized cells by changing the physical and chemical conditions of growth. Several sources of stem cells are used experimentally and are classified according to their origin and potential for differentiation. Human embryonic stem cells (hESCs) are extracted from embryos and have the ability to differentiate into just about any cell of the body. The adult stem cells that are present in many organs and differentiated tissues, such as bone marrow and skin, are more limited; they can only differentiate to the types of cells found in those specific tissues. An adult muscle stem cell can only differentiate into muscle cells; it cannot give rise to a skin cell.

The 2012 Nobel Prize in Physiology or Medicine was awarded to two researchers who studied stem cells for the discovery that mature cells are capable of *dedifferentiation*, or the reversal of differentiation, and returning to a stem cell state. This process, performed in a laboratory, leads to induced pluripotent stem cells (iPSCs) from mouse and human adult stem cells. These cells are genetically reprogrammed adult cells that function like embryonic stem cells; they are capable of generating differentiated cells.

Because of their capacity to divide and differentiate into specialized cells, iPSCs offer a potential treatment for diseases such as diabetes and heart disease.

An interesting fact: At the time of the award, one of the scientists who won the 2012 Nobel Prize in Physiology or Medicine published a note written by his high school biology teacher that his ambition to become a scientist was "ridiculous." Sir John Gurdon published this note to encourage young scientists everywhere to persevere even when discouraged.

Chapter Review

4.1 The Cell Membrane and Its Involvement in Transport

The learning objectives for this section are:

4.1.1 Describe the structure of the cell (plasma) membrane, including its composition and arrangement of lipids, proteins, and carbohydrates.

4.1.2 Compare and contrast intracellular fluid and extracellular fluid with respect to chemical composition and location.

4.1.3 Describe the functions of different plasma membrane proteins (e.g., structural proteins, receptor proteins, channels).

4.1.4 Compare and contrast simple diffusion across membranes and facilitated diffusion in respect to their mechanisms, the type of material being moved, and the energy source for the movement.

4.1.5 Compare and contrast facilitated diffusion, primary active transport, and secondary active transport in respect to their mechanisms, the type of material being moved, and the energy source for the movement.

4.1.6 Define osmosis and explain how it differs from simple diffusion across membranes.

4.1.7 Compare and contrast osmolarity and tonicity of solutions.

4.1.8 Describe the effects of hypertonic, isotonic, and hypotonic solutions on cells.

4.1 Questions

1. How would you expect phospholipids to arrange themselves within human tissues? Please select all that apply.
   a. The phosphate group will face the inside or outside of the cell.
   b. The phosphate group will hug into the core of the membrane.
   c. The fatty acid tails will face the inside or outside of the cell.
   d. The fatty acid tails will hug into the core of the membrane.

2. All of the following are true of the intracellular and extracellular fluids except:
   a. The intracellular fluid is found within the cells
   b. The extracellular fluid is the fluid outside of the cells.
   c. The interstitial fluid represents a portion of the intracellular fluid.
   d. Blood plasma represents a portion of the extracellular fluid.

3. Plasma membrane proteins that allow specific substances, such as ions, to pass into or out of the cell are called:
   a. Peripheral proteins
   b. Receptor proteins
   c. Glycoproteins
   d. Channel proteins

4. Molecule C cannot freely pass through the cell membrane, so it needs to pass through a channel protein. If there is a higher concentration of Molecule C in the extracellular matrix, what transport method will be used to enter the cell?
   a. Simple Diffusion
   b. Facilitated Diffusion
   c. Primary Active Transport
   d. Secondary Active Transport

Mini Case 1: Peripheral Edema

Blood is composed of cells and a watery fluid known as plasma. In a healthy person, the ratio of water and solutes are the same inside the cell and in the plasma. In some people, the concentration of solutes within the plasma increases. The cell membrane of these cells allows movement of fluid, but not movement of solutes.

1. What method of transport is used to reach an even fluid and solute concentration in the cells and plasma?
   a. Simple Diffusion
   b. Facilitated Diffusion
   c. Osmosis
   d. Active Transport

2. If the sodium chloride (salt) concentration outside a person's cells were higher than that of the intracellular fluid inside the cells, how would the salt concentrations become equilibrated between the intracellular and interstitial fluids?
   a. Water would leave the cells by osmosis and enter the interstitial fluid
   b. Water would leave the cells by diffusion and enter the interstitial fluid
   c. Sodium chloride would enter the cells by diffusion
   d. Sodium chloride would enter the cells by osmosis

3. How would you describe the fluid and solute concentration of plasma in a healthy person?
   a. Isosmotic
   b. Equilibrium
   c. Isotonic
   d. Hypoosmotic

A high-sodium diet often increases the concentration of solutes (sodium) in the plasma.

4. How would you describe the fluid and solute concentration of plasma in people with a high-sodium diet?
   a. Hypertonic
   b. Hypotonic
   c. Isotonic

4.1 Summary

- The cell membrane is created by three structures. The phospholipids form the cell membrane and create the selectively permeable characteristics of the cell membrane. The cholesterol contributes to the fluid component of the cell membrane. The proteins form channels and receptors to allow for movement of structures in and out of the cell.
- Molecules can move down the concentration gradient through passive transport, which requires no energy. Molecules can move up the concentration gradient through active transport, which requires energy.
- Passive transport is done through various forms of diffusion, while active transport is done through the sodium-potassium pump, endocytosis, and exocytosis.

4.2 The Cytoplasm and Cellular Organelles

The learning objectives for this section are:

4.2.1 Compare and contrast cytoplasm and cytosol.

4.2.2 Define the term organelle.

4.2.3 Describe the three main parts of a cell (plasma [cell] membrane, cytoplasm, and nucleus), and explain the general functions of each part.

4.2.4 Describe the structure and function of the various cellular organelles.

4.2.5 Describe the structure and roles of the cytoskeleton.

4.2 Questions

1. True or False: Organelles are inside the cytosol.
   a. True
   b. False

2. Which of the following are membrane-enclosed bodies within a cell that perform a unique function?
   a. Cytoplasm
   b. Cytosol
   c. Cell Membrane
   d. Organelles

3. HIV is a virus that replicates by merging its small viral genome within a human cell's DNA. If you are trying to test to determine if a cell is infected with HIV, where in the cell should you look?
   a. Cell Membrane
   b. Cytoplasm
   c. Nucleus
   d. Cytosol

4. You see a patient that is not feeling well. You look at your patient's blood test and see that there are high levels of alcohol in their blood, though they describe not having much to drink. Which of the following organelles could be dysfunctional?
   a. Peroxisomes
   b. Golgi apparatus
   c. Rough ER
   d. Mitochondria

5. What is the role of cytoskeleton? Please select all that apply.
   a. Structural support for cells
   b. Cell motility
   c. Energy production
   d. Production of H_2O_2

4.2 Summary

- A cell is composed of three portions: the plasma (cell) membrane, the cytoplasm, and the nucleus. The cytoplasm can be further subdivided into the cytosol and the organelles.
- The plasma membrane forms the outer boundary of the cell and regulates the passage of molecules into and out of the cell.
- Some organelles, such as the endoplasmic reticulum, Golgi apparatus, and vesicles, form the endomembrane system, which produces and exports cellular products.
- Other organelles, such as the mitochondria, lysosomes, peroxisomes, function in energy production for the cell and detoxification.
- The cytoskeleton provides structural support and cellular movements; it consists of microtubules, intermediate filaments, and microfilaments.

4.3 The Nucleus and DNA

The learning objectives for this section are:

4.3.1* Describe the structure and contents of the nucleus.

4.3.2* Explain the organization of the DNA molecule within the nucleus.

* Objective is not a HAPS Learning Goal.

4.3 Questions

1. You are looking inside a cell's nucleus under a microscope. You notice a dark-stained mass. What is the function of this structure?
   a. Manufacturing RNA
   b. Storing DNA
   c. Cellular Reproduction
   d. Energy Production

2. When the DNA is not in the process of replicating, what structure does the DNA wrap around?
   a. Nucleosome
   b. Chromatin
   c. Chromosome
   d. Histone

4.3 Summary

- The nucleus consists of two lipid bilayers to form the nuclear membrane. Inside, the nucleus is filled with nucleoplasm. Suspended within the nucleoplasm is the nucleolus.
- DNA consists of Adenine (A), Thymine (T), Cytosine (C), and Guanine (G). A always pairs with T, and C always pairs with G.
- When the cell is living, growing, and producing proteins, the DNA is loosely packed for easy access to the genes. During cellular reproduction, the DNA is highly condensed for easy handling and security of the genes.

4.4 Protein Synthesis

The learning objectives for this section are:

4.4.1 Define the terms genetic code, transcription and translation.

4.4.2 Explain the process of RNA synthesis.

4.4.3 Explain the roles of tRNA, mRNA, and rRNA in protein synthesis.

4.4 Questions

1. Where does translation occur?
   a. Nucleolus
   b. Lysosome
   c. Ribosome
   d. Smooth ER

2. Identify the stage of transcription in which RNA polymerase adds new nucleotides to a growing strand of RNA.
   a. Initiation
   b. Elongation
   c. Termination
   d. Splicing

3. What is the function of tRNA?
   a. Match mRNA with correct amino acid.
   b. Forms the ribosomes with other proteins.
   c. Brings rRNA to the mRNA.
   d. Carries a copy of genetic code for a gene.

4.4 Summary

- DNA stores the genetic code in the nucleus of a cell. The genetic code is a set of instructions for synthesizing proteins.
- Protein synthesis consists of two portions: transcription and translation.
- Transcription consists of copying the genetic code for a protein from DNA onto a new molecule of messenger RNA (mRNA).
- Translation involves the assembly of a protein, using the code stored on mRNA to produce a protein on a ribosome.
- Three types of RNA function in the process of protein synthesis. mRNA transports a copy of the genetic code from DNA to the cytoplasm. Ribosomal RNA (rRNA) makes up a portion of the ribosomes, which read the genetic code from mRNA. Transfer RNA (tRNA) adds amino acids to growing proteins.

4.5 Cell Replication

The learning objectives for this section are:

4.5.1 Compare and contrast somatic cell division (mitosis) and reproductive cell division (meiosis).

4.5.2 Describe the general phases (e.g., G phases, S phase, cellular division) of the cell cycle.

4.5.3 Describe DNA replication.

4.5.4 Compare and contrast chromatin, chromosomes, and chromatids.

4.5.5 Describe the events that take place during mitosis and cytokinesis.

4.5 Questions

1. A hepatic cell can be found in your liver. Will this cell undergo mitosis or meiosis?
   a. Mitosis
   b. Meiosis

2. Which of the following lists the phases of Interphase from beginning to end?
   a. G_1 phase > S phase > G_2 phase
   b. G_1 phase > G_2 phase > S phase
   c. S phase > G_1 phase > G_2 phase
   d. S phase > G_2 phase > G_1 phase

3. What happens in the initiation phase of DNA replication?
   a. Two complementary DNA strands are separated.
   b. DNA polymerase synthesizes a new complementary strand.
   c. There are two identical DNA copies of the original DNA.
   d. The cell grows, makes proteins, and carries out cellular functions.

4. You are looking under the microscope. You see a cell that has a dark line across its middle. The membrane is round and you cannot find the nucleus. What is the most likely structure that makes up that dark line?
   a. Chromatin
   b. Microtubules
   c. Chromatid
   d. Cleavage furrow

5. Identify the phase of cellular replication in which microtubules shorten and separate sister chromatids.
   a. Prophase
   b. Metaphase
   c. Anaphase
   d. Telophase

4.5 Summary

- Cells have 46 total chromosomes, which is 23 pairs of chromatids.
- The cell cycle is split into two phases: Interphase and cell reproduction. The cell spends most of its time in interphase, where it grows and carries out all normal metabolic functions. Cell reproduction is when the cell divides. This is called mitosis for somatic cells and meiosis for reproductive cells.
- Interphase is split into G_1 phase (where the cell grows, makes proteins, and carries out the function of cells), S phase (where the cell replicates its DNA), and G_2 phase (where the cell prepares for mitosis).
- S phase is split into Initiation (where the DNA strands are separated), Elongation (where the DNA polymerase synthesizes a new strand), and Termination (where DNA replication stops).
- Cell replication is split into five phases: Prophase (where the chromatin condenses into chromosomes and the centrioles migrate toward the opposite side), Metaphase (where chromatids align in the middle), Anaphase (where each part of the chromatid is separated toward the opposite side of the cell), Telophase (where the nucleolus starts to form around the chromosomes), and Cytokinesis (where the cleavage furrow separates the cell into two distinct cells).

4.6 Cellular Differentiation

The learning objectives for this section are:

4.6.1* Summarize the properties of stem cells that distinguish them from differentiated cells.

4.6.2* Discuss how the generalized cells of a developing embryo or the stem cells of an adult organism become differentiated into specialized cells.

* Objective is not a HAPS Learning Goal.

4.6 Questions

1. Which of the following are characteristics of stem cells? Check all that apply.
   a. They are unspecialized cells.
   b. They are found in an embryo or fetus, but not in an adult.
   c. They can divide a very limited number of times.
   d. They can differentiate into specialized cells.

2. How does a stem cell undergo differentiation? Please select all that apply.
   a. The parts of the DNA that are not required for adult cell function will be highly condensed.
   b. Transcription factors will bind to the part of the DNA that is frequently needed.
   c. The cells that are not needed will die off and only the required cells will reproduce.
   d. The DNA will alter its genetic makeup to only be composed of the required genes.

4.6 Summary

- Stem cells are unspecialized cells that can replicate as many times as is necessary.
- Stem cells can also produce new cells that can differentiate (specialize) into a variety of cell types.
- Early embryos are mostly composed of stem cells, but as the embryo grows the stem cells will become specialized cells such as cells found in the heart, liver, kidney, and brain.

5

The Tissue Level of Organization

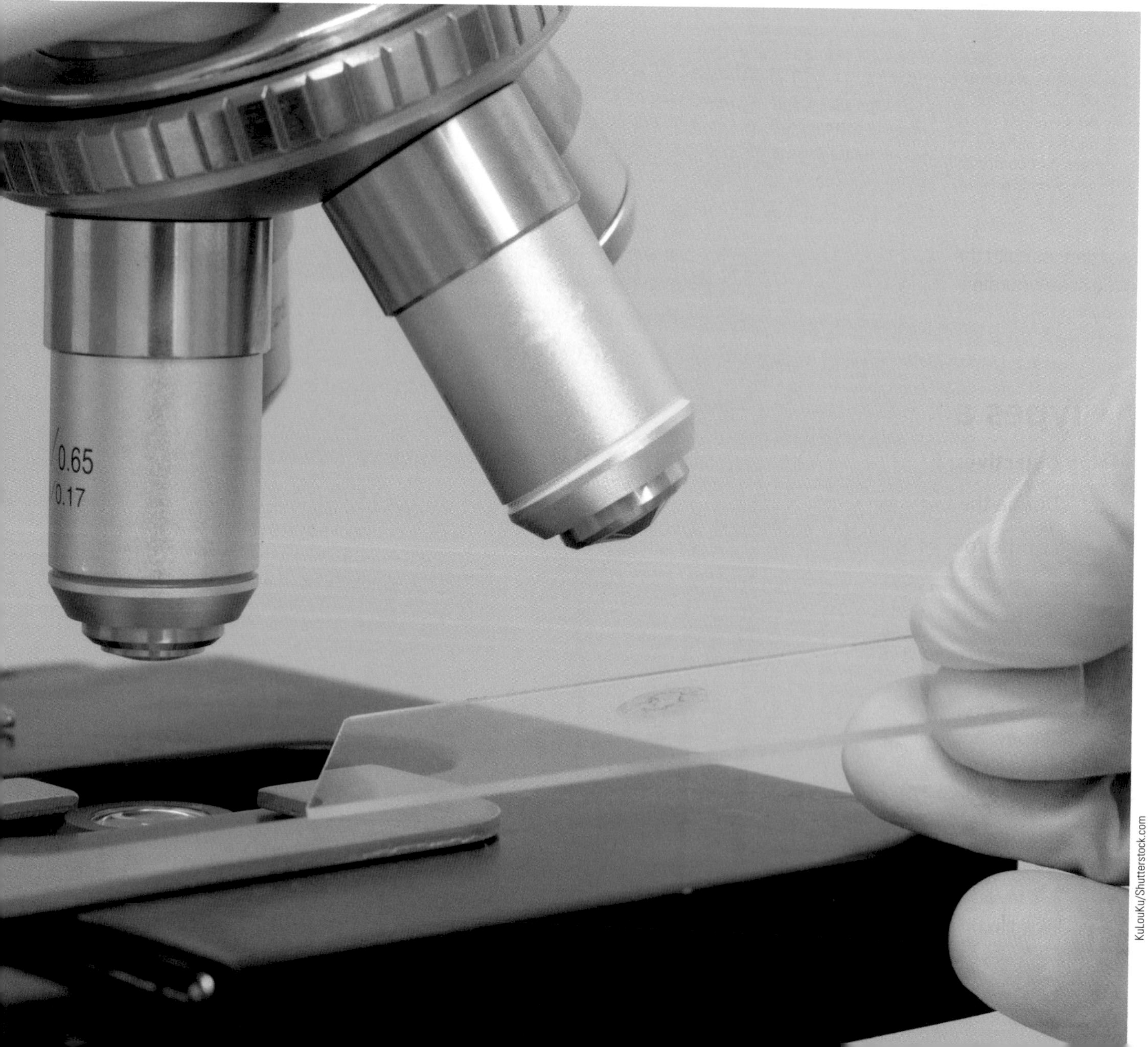

KuLouKu/Shutterstock.com

Chapter Introduction

In this chapter we will learn . . .

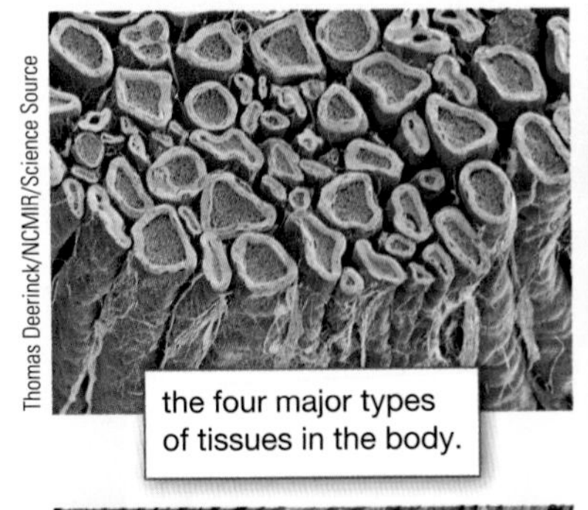

the four major types of tissues in the body.
Thomas Deerinck/NCMIR/Science Source

the various structural properties of each of the four major tissue types.
Steve Gschmeissner/Science Source

the functions of each tissue type and its subtypes, based on the cells that comprise each one.
Callista Images/Image Source/Getty Images

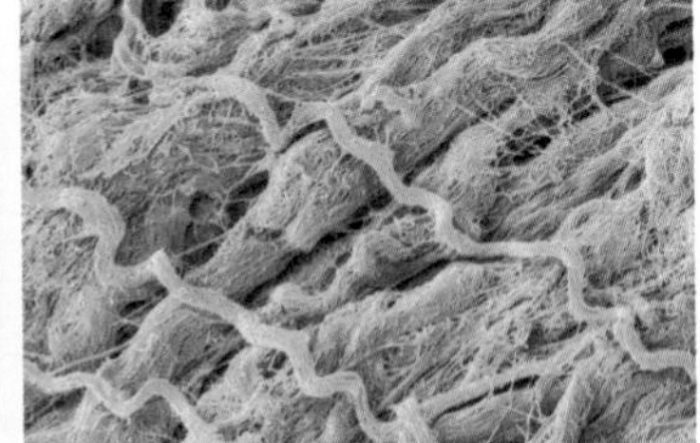

the types of intercellular junctions that allow cells to communicate with each other within a tissue.
Steve Gschmeissner/Science Photo Library/Getty Images

Your body contains at least 200 different types of cells. Most cells contain essentially the same internal structures, yet these vary enormously in shape and function. The different types of cells are not randomly distributed throughout the body; rather they occur in organized units, a level of organization referred to as *tissue*. The next level of organization is the organ, where several types of tissues come together to form a working unit (**Figure 5.1**). Just as understanding the structure and function of cells helps you in your study of tissues, knowledge of tissues will help you understand how organs function.

The human body starts as a single cell at fertilization. As this single cell replicates, it eventually gives rise to trillions of cells, each built from the same blueprint. But as the new organism grows, most cells commit to different developmental pathways to form tissues and eventually organs. The cells within a tissue depend on each other, much the way that humans in a community or society depend on the functions of others. A farmer grows food, but cannot build their own tractor; a nurse knows how to heal the body, but cannot build their own house. Cells within the tissue operate in much the same way; each cell has its own function, and they rely on each other for other functions.

5.1 Types and Components of Tissues

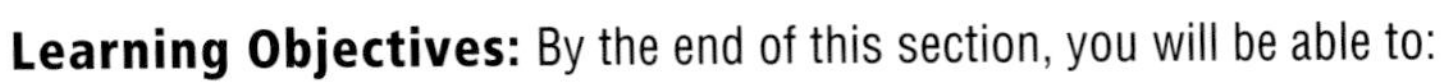

Learning Objectives: By the end of this section, you will be able to:

5.1.1 Define the term histology.

5.1.2 List the four major tissue types.

5.1.3 Compare and contrast the general features of the four major tissue types.

5.1.4* Describe the components and functions of extracellular matrix.

5.1.5 Compare and contrast the types of intercellular connections (cell junctions) with respect to structure and function.

* Objective is not a HAPS Learning Goal.

LO 5.1.1

The term **tissue** is used to describe a group of cells functioning together in the body (**Figure 5.2**). The microscopic study of tissue appearance, organization, and function is called **histology**. A working understanding of histology is critical for mastering the structure and function of the organs; histology is also the foundation for understanding **pathology**, the changes that occur during the course of disease.

The Human Anatomy and Physiology Society includes more than 1,700 educators who work together to promote excellence in the teaching of this subject area. The HAPS A&P Learning Outcomes measure student mastery of the content typically covered in a two-semester Human A&P curriculum at the undergraduate level. The full Learning Outcomes are available at https://www.hapsweb.org.

Figure 5.1 Levels of Structural Organization of the Human Body

The organization of the body often is discussed in terms of six distinct levels of increasing complexity, from the smallest chemical building blocks to a unique human organism.

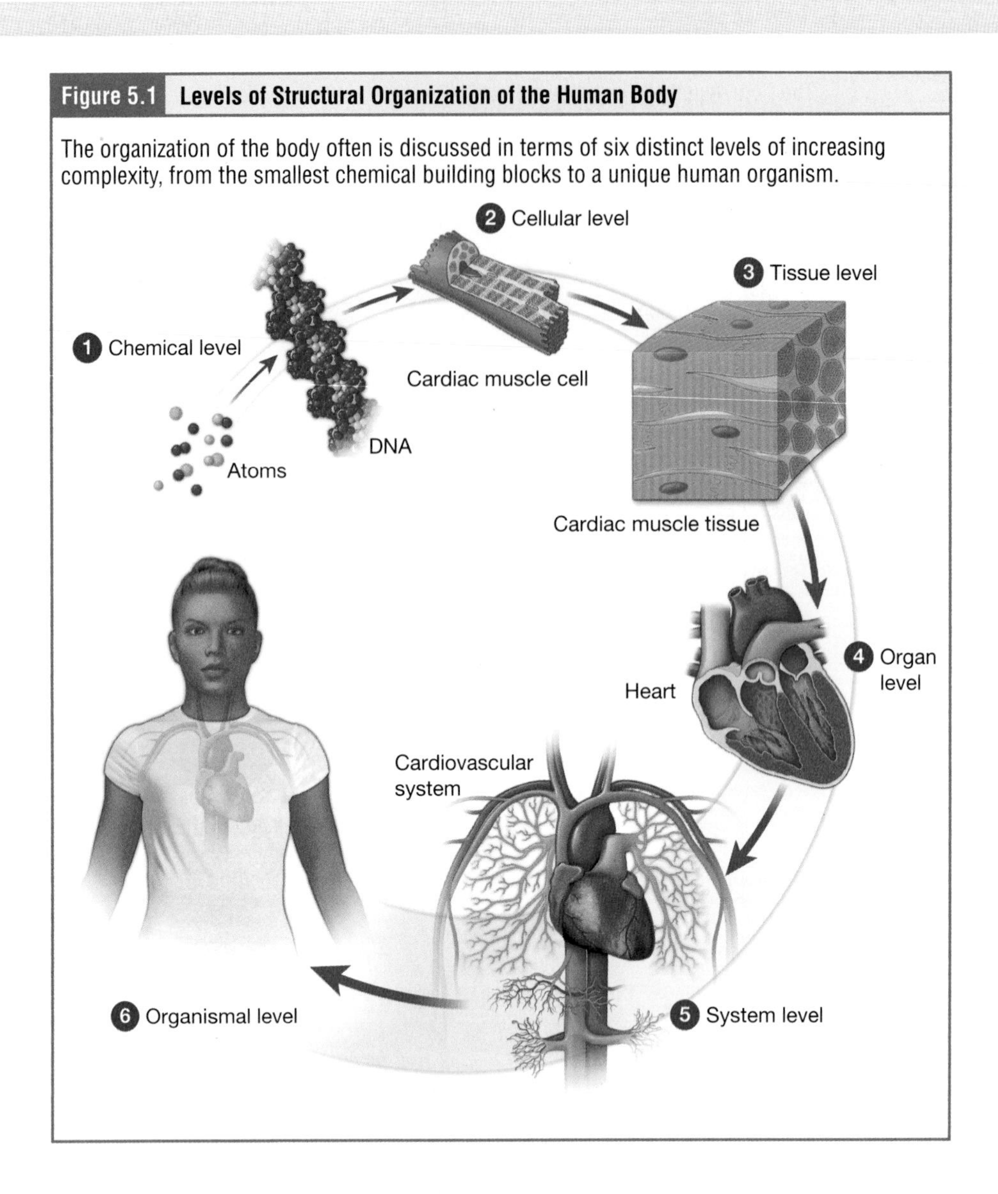

LO 5.1.2

Microscopic observation reveals that the cells in a tissue often look the same and are arranged in an orderly pattern that achieves the tissue's functions. Although there are many types of cells in the human body, they are organized into four categories of tissues: epithelial, connective, muscle, and nervous (see Figure 5.2). Each of these categories is characterized by specific functions that contribute to homeostasis and the multitude of functions of the body. The epithelial and connective tissues are discussed in detail in this chapter. Muscle and nervous tissues will be discussed only briefly in this chapter because they will be discussed more thoroughly in Chapters 11 and 13, respectively. We will close this chapter discussing how tissues work together to form organs (see Figure 5.2).

5.1a The Four Types of Tissues

LO 5.1.3

Epithelial tissue are sheets of cells that cover exterior surfaces of the body, line internal cavities and passageways, and form certain glands. **Connective tissue**, our most diverse category of tissue, binds the cells and organs of the body together and functions in the protection, support, and integration of all parts of the body. **Muscle tissue** is contractile; it provides movement, whether that means moving the skeleton,

Figure 5.2 The Four Types of Tissue in the Body

The four types of tissues in the human body are nervous tissue, epithelial tissue, muscle tissue, and connective tissue.

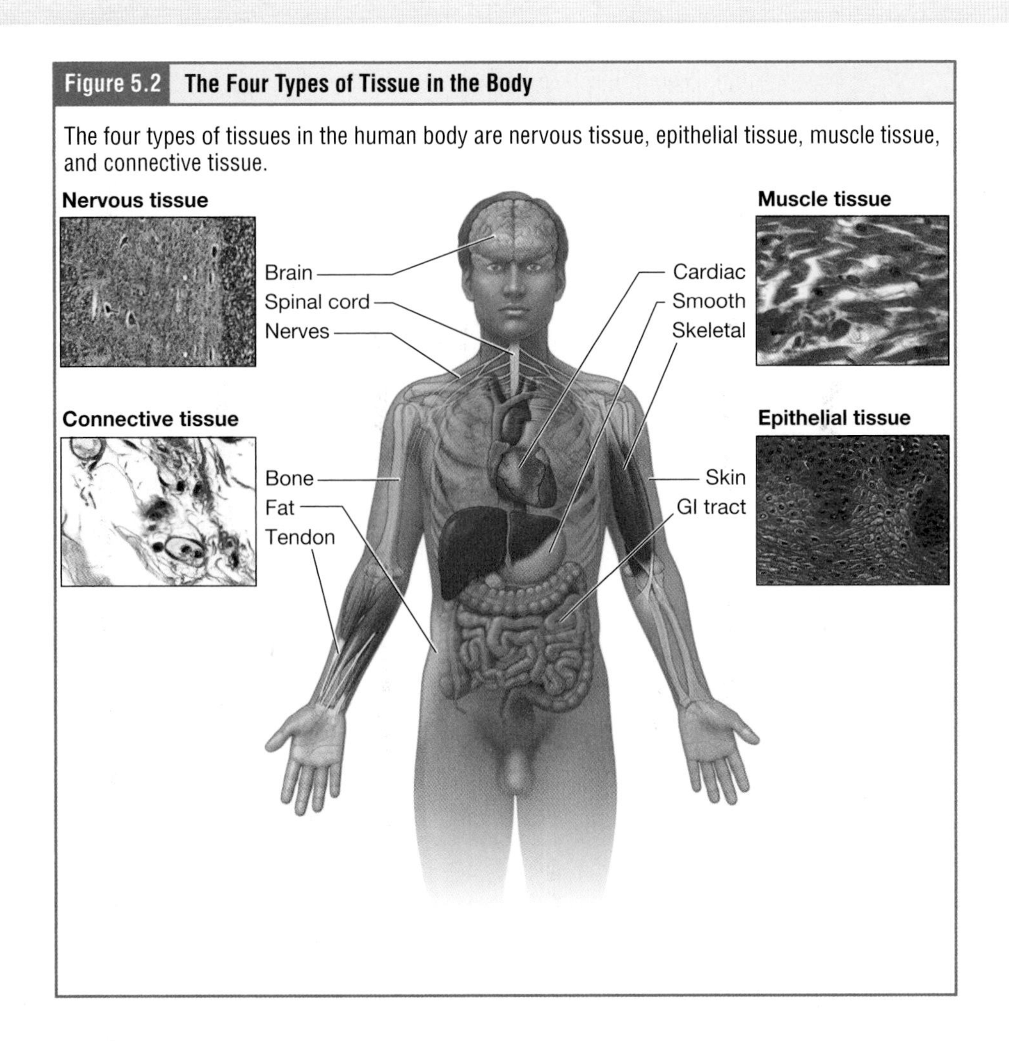

as in walking, or moving the blood through the body as the heart muscle contracts. **Nervous tissue** is capable of short- and long-distance communication throughout the body. In order to wiggle your toes, for example, your brain generates a signal that is sent and propagated down a string of nervous tissue cells that connect to the muscles within your toes.

5.1b Extracellular Matrix

LO 5.1.4

Extracellular matrix (ECM) is a network of substances that surround and support the cells in a tissue. ECM varies widely from one tissue to another in order to best support the functions of the tissue. For example, the ECM of bone tissue is solid while the ECM of the blood is liquid. All ECM has common components, however. Within the ECM, cells are held in place, guided in their movement (for cells that move), and supported. We can consider ECM to be analogous to the walls and floors and air within a building. ECM is made by the cells within the tissue; therefore, it is both unique to each tissue and plastic, or changeable, depending on the activity of the cells that reside there. While there are many molecules that may be within an ECM, the major components of ECM are usually collagen fibers and proteoglycans.

- **Collagen** is the most abundant protein in the human body. Collagen subunits are added together to create long, tough fibers that weave together to form the structure of the ECM. Collagen can be thought of like leather—tough and protective, but flexible and allowing movement.

- **Proteoglycans** are large, negatively-charged molecules that are a combination of carbohydrate and protein. Their negative charge attracts Na^+ and water molecules to the ECM. Once these molecules mix together, the ECM takes on a gel-like consistency. In addition to keeping cells hydrated, the gel also serves to trap and store nutrients and growth factors.

ECM is vital to the health of the cells that reside within it. The growth factors and nutrients trapped within the matrix influence the cells. Cell behavior, gene expression, and even differentiation can be altered by the composition of the ECM.

DIGGING DEEPER: Preparing Tissues for Examination under the Microscope

Our understanding of the structure of tissues comes from studying them under the microscope. But when you first view a magnified tissue, it can be difficult to tell what you are looking at. Examine the micrograph in Figure A. What is this structure? It looks like a wave, and are our insides really bright pink and purple? Let's dig deeper into how these tissues get to the slide under the microscope. A body structure is cut and then embedded in some sort of matrix, often wax. As you can see in Figure B, the *way* the structure is cut will influence the shapes we see under the microscope. The wax/tissue block is then cut into incredibly thin sections using a special blade (Figure C). The sections are placed on slides (Figure D) and then stained (Figure E). There are lots of different stains that can be used and, depending on which ones are chosen, the tissue may take on very different colors and appearances (Figure F). The most common combination of stains is hematoxylin and eosin, or H&E. The hematoxylin binds to DNA so that the nucleus of each cell will stain a dark purple/blue. Eosin stains proteins commonly found in the extracellular matrix and cytosol, turning these pink.

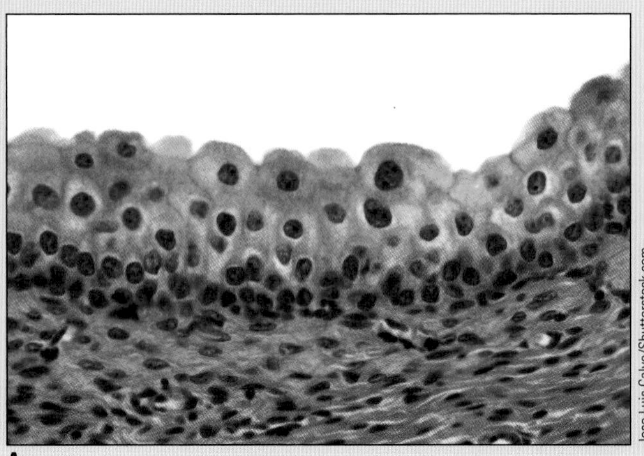

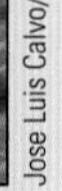

Jose Luis Calvo/Shutterstock.com

A

B

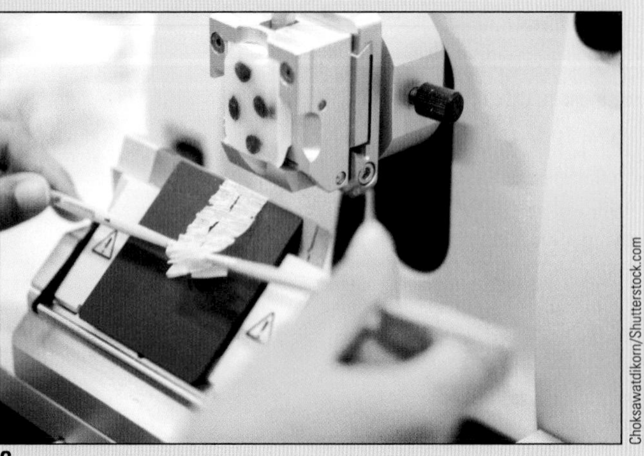

Choksawatdikorn/Shutterstock.com

C

Mauro Fermariello/Science Source

D

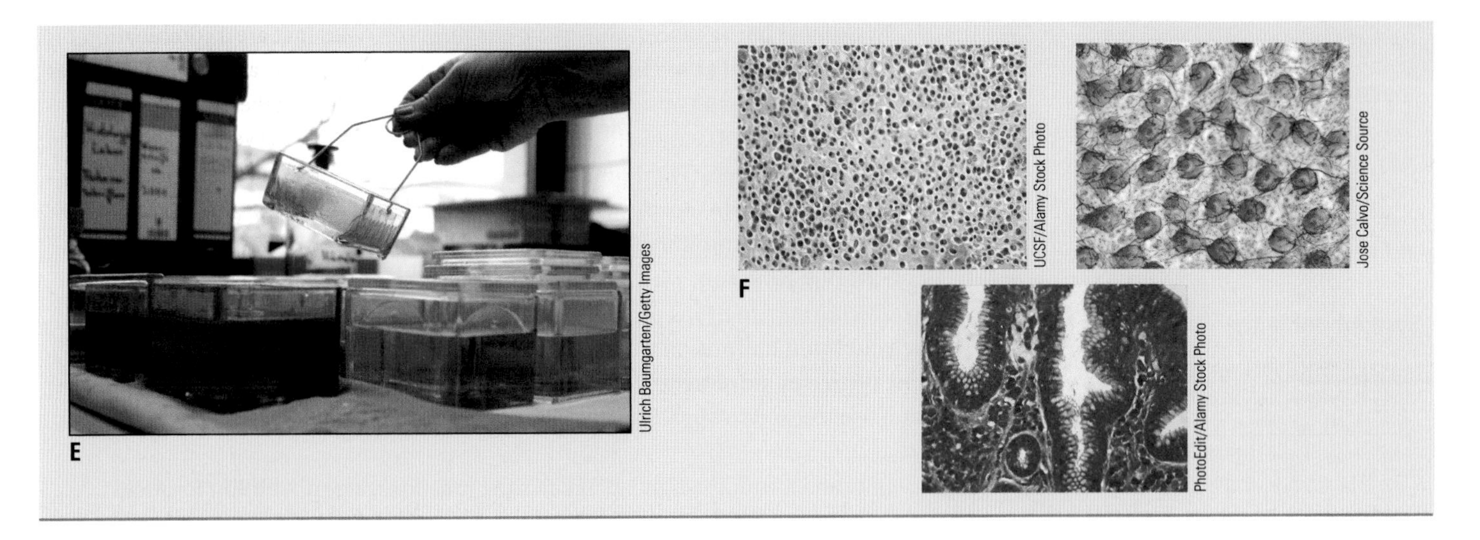

5.1c Cellular Connections

LO 5.1.5

The degree to which cells touch and interact varies from tissue to tissue. In locations where cells are close to each other they can often be connected. The three basic types of connections—tight junctions, desmosomes, and gap junctions (Figure 5.3)—allow varying degrees of interaction between the cells.

Figure 5.3 Types of Cell Junctions

The classes of cell-to-cell junctions are tight junctions, gap junctions, and desmosomes.

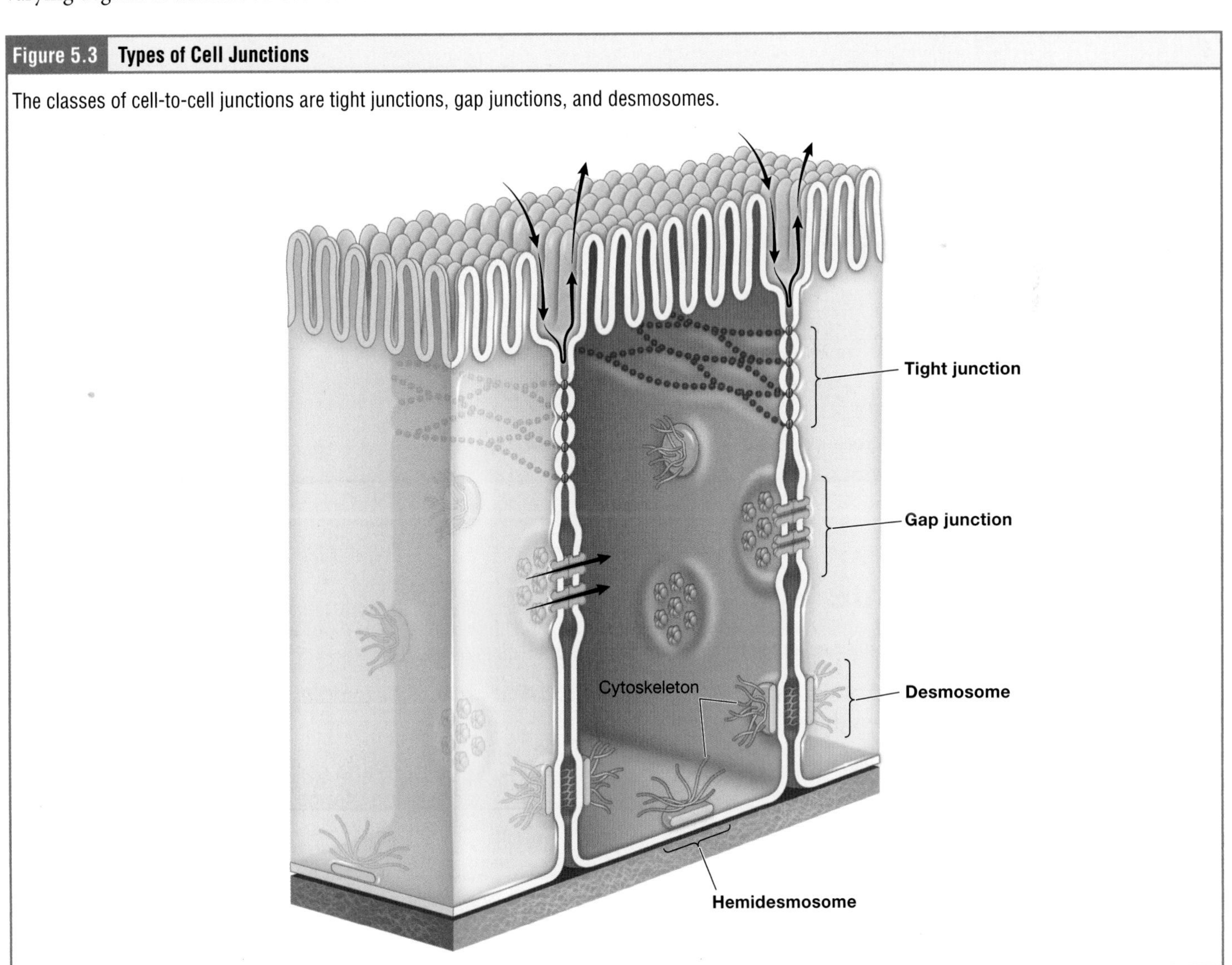

Learning Connection

Try Drawing It

Try drawing cells united by tight junctions, gap junctions, or desmosomes. How can you represent the functions of each structure for yourself so that you remember them?

Student Study Tip

Gap junctions look and act like a tunnel between two cells. You can remember the function this way: a gap junction passes ionic messages between adjacent cells, like a person driving a car through a tunnel carrying a message from a person on one side of the tunnel to a friend waiting on the other side. Gap junctions are found between adjacent cardiac muscle cells; they provide for coordinated heartbeats.

At one end of the spectrum is the **tight junction**, which fuses the membranes of two neighboring cells together tightly so that there is no extracellular space between them, which blocks the movement of substances through the extracellular space between the cells. This prevents material from slipping between the cells to access the underlying tissues. Tight junctions are an important feature of the lining of the intestine, for example, making sure that bacteria cannot slip between the cells and enter the body. **Desmosomes** and their cousins **hemidesmosomes** are two types of **anchoring junction**, which are cell junctions that help stabilize cells structurally. Anchoring junctions provide strong and flexible connections. Think of them like you would buttons on a shirt; the buttons unite the fabric, but still allow for some degree of movement. Desmosomes unite two neighboring cells structurally. Hemidesmosomes, which look like half a desmosome, link cells to the extracellular matrix.

In contrast with the tight and anchoring junctions, a **gap junction** forms an intercellular passageway between the membranes of adjacent cells to facilitate the movement of small molecules and ions among the cytoplasm of adjacent cells. These junctions allow electrical and metabolic coupling of adjacent cells, which coordinates function in large groups of cells.

Learning Check

1. Which of the following is a main category of tissues? Please select all that apply.
   a. Muscle
   b. Nerve
   c. Bone
   d. Skin
2. The extracellular matrix of each tissue has different proportions of collagen and proteoglycans. Which molecule is most abundant in a tissue that is tough and protective, such as a ligament?
   a. Collagen
   b. Proteoglycan
3. What is the difference between anchoring junctions and tight junctions? Please select all that apply.
   a. Anchoring junctions allow electrical and metabolic coupling of adjacent cells.
   b. Anchoring junctions form strong connections to structurally stabilize cells.
   c. Tight junctions tightly fuse cells to the extracellular matrix.
   d. Tight junctions fuse cells to prevent materials from slipping between cells.

5.2 Epithelial Tissue

Learning Objectives: By the end of this section, you will be able to:

5.2.1 Describe the structural characteristics common to all types of epithelia.

5.2.2 Classify different types of epithelial tissues based on structural characteristics.

5.2.3 Describe the microscopic anatomy, location, and function of each epithelial tissue type.

5.2.4 Identify examples of each type of epithelial tissue.

5.2.5 Compare and contrast exocrine and endocrine glands, structurally and functionally.

5.2.6 Compare and contrast the different kinds of exocrine glands based on structure, method of secretion, and locations in the body.

Epithelial tissue is essentially large sheets of cells covering all the surfaces of the body. The main function of epithelial tissues is protection, so they are found in locations that are exposed to the outside world such as your skin and the lining of your mouth. When considering where epithelial tissues are found, one lens with which to view a body structure is whether it is exposed to the "outside." What does it mean to be "outside" the body? One way to consider this question is to remember our human relationship with bacteria. "Inside" spaces are sterile and no bacteria should be found within them. The blood vessels are an example. For these "inside" structures, the epithelial tissue lining them is called **endothelium** (plural = *endothelia*). The airways, the digestive tract, and the urinary and reproductive systems are all connected to the outside world. It is common to find bacteria in these spaces, and so we give the tissues lining these the name **epithelium** (plural = *epithelia*).

LO 5.2.1

All epithelia share some important structural and functional features. This tissue is highly cellular, with little or no extracellular material present between cells. Cells contact other cells in almost every direction, resembling the bricks in a brick wall. Often, adjoining cells are linked by cell junctions. In Chapter 3 we discussed the concept of polarity. In that chapter we discussed that molecules are described as being *polar* when they have two distinct sides, in that case having different charges. Similarly, we describe epithelia as being polar because they have two sides that are distinct. One of their sides, the **apical** side, is exposed to the external environment or an internal space (see the "Anatomy of Epithelia" feature) and the **basal** surface is the only side of the tissue facing the ECM. The specialized ECM that supports and anchors the basal epithelium is known as a **basement membrane**. This supportive membrane consists of two layers: the **lamina lucida**, a mixture of glycoproteins and collagen that provides the immediate attachment site for the epithelium, and a **lamina densa**, which has a denser and more structural weave of tough collagen fibers. The lamina densa connects the epithelium to the underlying connective tissue.

Epithelial tissues are nearly completely **avascular**, meaning that they have no blood vessels. Because no blood vessels cross the basement membrane to enter the tissue, all nutrients and oxygen must travel to the epithelial cells by diffusion from underlying tissues or from the surface. For this reason, epithelial tissues are always found in a relationship with underlying vascularized connective tissue. Because of their free apical edge, many epithelial tissues are exposed to friction. Consider the epithelium that line the gut, for example; the constant traffic of digested food across the apical surface can be damaging to the cells there. For this reason, the sloughing off of damaged or dead cells is a characteristic of surface epithelium and allows our airways, digestive tracts, and skin to rapidly replace damaged cells with new cells. Therefore, epithelial tissues are highly regenerative.

5.2a Generalized Functions of Epithelial Tissue

Epithelial tissues provide the body's first line of protection from physical, chemical, and biological wear and tear. The cells of an epithelium act as gatekeepers of the body, controlling permeability and allowing selective transfer of materials across a physical barrier. Consider what happens when you swim in a pool or the ocean. Your entire body is submerged in a solution, but a very small amount of exchange takes place. For the most part, you emerge from the swim with the same body composition you had when you entered the water. That is because the epithelium of your skin provides an effective barrier, keeping the solutions inside your cells and body in and the solution outside your body out. Now consider another example. Let's say that you eat an apple that has been in your backpack all day. This apple is not sterile, it is likely covered with bacteria and perhaps some fungi. When you eat the microbe-laden apple, your digestive tract

Anatomy of...

Epithelia

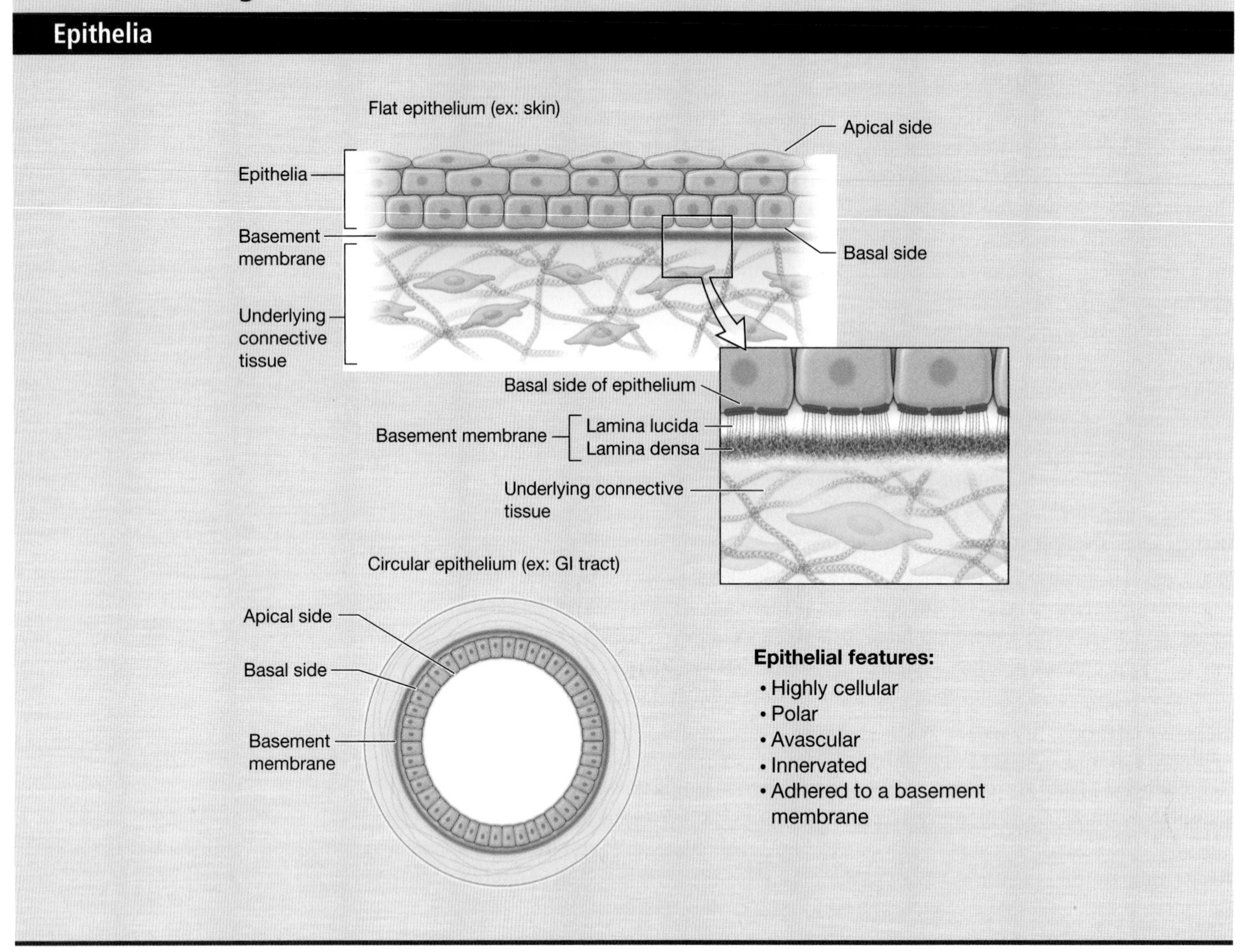

enables the transport of the food molecules into the cells and bloodstream, but keeps the microbes from entering your tissues. In this way, all substances that enter the body must cross an epithelium. Some epithelia often include structural features that allow the selective transport of molecules and ions across their cell membranes. The tight junctions that we read about in the previous section often assist epithelial cells in forming these stringent barriers.

Many epithelial cells are capable of secretion and release mucus and specific chemical compounds onto their apical surfaces. For example, the epithelium of the small intestine releases digestive enzymes. Cells lining the respiratory tract secrete mucus that traps incoming microorganisms and particles. Secretory cells and structures are features of many epithelia. In enclosed spaces that are lined with epithelia, the enclosed space is called a **lumen**. For example, when you eat a snack, the food ends up in the lumen of your stomach.

Learning Check

1. In which of the following areas in your body would you find epithelial tissues?
   a. Outside lining of the stomach
   b. Inside lining of the throat
   c. Inferior lining of the diaphragm
   d. Within the shoulder muscles
2. Identify the layer that is most superficial and prone to friction.
   a. Apical side of epithelium
   b. Basal side of epithelium
   c. Lamina lucida
   d. Lamina densa
3. Identify the layer that is deepest and connects to the underlying connective tissue.
   a. Apical side of epithelium
   b. Basal side of epithelium
   c. Lamina lucida
   d. Lamina densa
4. Explain the advantage of tight junctions that contribute to the function of epithelial cells.
   a. Tight junctions provide stability of the internal organs.
   b. Tight junctions serve as barriers against harmful materials.
   c. Tight junctions communicate with adjacent cells.
   d. Tight junctions contribute to mobility of the internal organs.

5.2b The Epithelial Cell

Epithelial cells are polarized, in that their apical and basal membranes may have different surroundings, but also in the composition of their membranes and the distribution of organelles between their basal and apical surfaces. Whereas certain organelles are closer to the basal sides, other organelles are near the apical surface.

It is a common feature of the apical epithelial surface to have cilia or microvilli (remember these from Chapter 4). Cilia, microscopic extensions of the apical cell membrane, beat in unison and move fluids as well as trapped particles (**Figure 5.4**). Ciliated epithelium lines the ventricles of the brain, where it helps circulate the cerebrospinal fluid. The ciliated epithelium of your airway forms a ciliary escalator that sweeps particles of dust and pathogens trapped in the secreted mucus upward toward

Figure 5.4 The Apical Surface Features of Epithelia

Epithelial cells often have either cilia (A) or microvilli (B) on their apical surface. These two types of surface features have different functions.

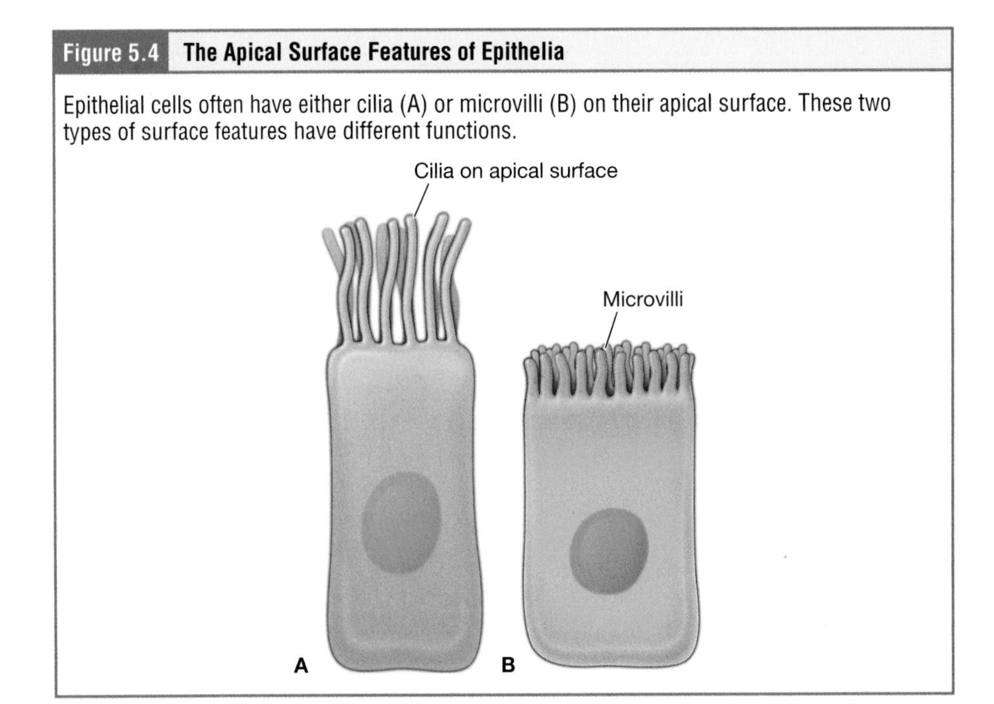

Learning Connection

Micro to Macro

If we consider the function of cilia (to sweep materials off the apical surface) and the function of microvilli (to expand the surface area of the cell), how would you expect the functions of the tissues in which these epithelial accessories are found to differ? What might be the function of an epithelium with microvilli?

the throat. A ciliated epithelium lines the Fallopian tubes; the cilia are responsible for sweeping eggs from the ovaries to the uterus. These very different functions can all be accomplished with the same cellular machinery. Microvilli are similar extensions of the apical cell membrane, but they do not contain cytoskeletal elements and therefore do not have the capacity to move. The function of microvilli is to expand the surface area of the cell to allow more contact with the extracellular fluid on the apical side. Typically, the purpose of this is to allow more membrane transport to occur.

Like all cells, epithelial cell structure is suited to its function. Epithelial cells that are found in locations with a high degree of diffusion—for example, between the air-filled sacs of the lungs and the blood—are thin and flat to enable diffusion to occur rapidly. Epithelial cells in locations of more friction and intracellular transport are considerably larger, often tall and narrow like a column. The three shapes of epithelial cells are: squamous (flat), cuboidal (round), and columnar (tall and narrow).

5.2c Classification of Epithelial Tissues

Epithelial tissues come in various formats according to the functions and environmental factors of their location. There are two characteristics that go into epithelial cell classification: the shape of the cells and the number of the cell layers (**Figure 5.5**). Cell shapes can be squamous (flat and thin), cuboidal (boxy, as wide as it is tall), or columnar (rectangular, taller than it is wide with a nucleus closer to its basal side). The number of cell layers in the tissue can be one (in this type, called a *simple epithelium*, every cell rests on the basal lamina), or more than one (in this type, called a *stratified epithelium*, only the basal layer of cells rests on the basal lamina). Note that in stratified epithelia, the cells in the basal layer often have a cuboidal shape, simply because the cells are stem cells, but the cells in the apical layer take on the differentiated cell shape. Taking these together, most epithelia have a two-part name; the first name describes the number of layers and the second name describes the shape of the cells in their apical layer—for example, *simple squamous* epithelium.

LO 5.2.2

Figure 5.5 Epithelial Tissue Types

Epithelial tissues are named by the number of layers they contain and the shape of their cells. Simple epithelial tissue is the name for tissue with a single layer of cells, and stratified epithelial tissue is the name for tissue consisting of several layers of cells. The second word in the name of each epithelial tissue—squamous, cuboidal, or columnar—is derived from the shape of its cells.

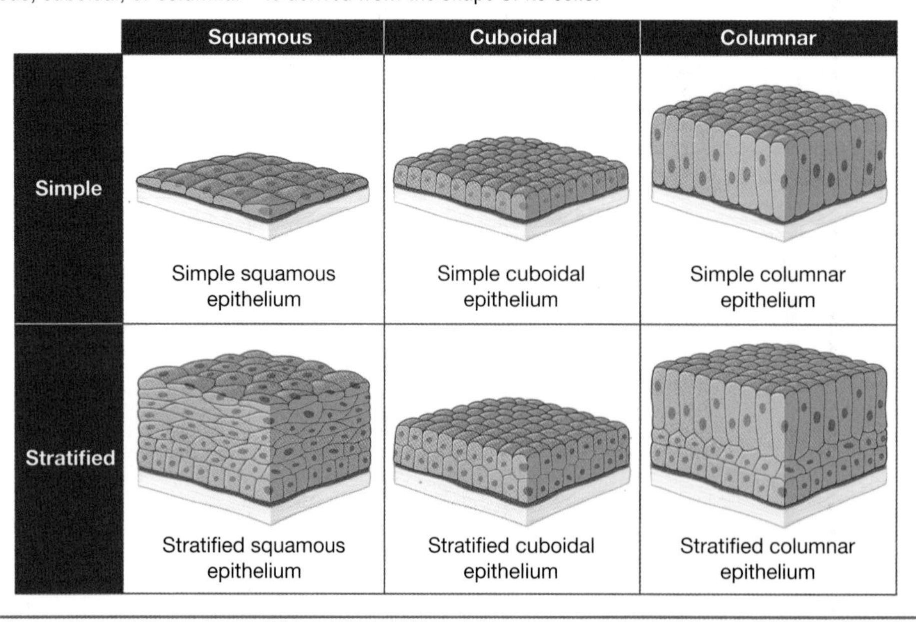

There are two epithelial tissues that do not follow this naming pattern. *Pseudostratified* (*pseudo-* = "false") describes an epithelial tissue with a single layer of irregularly-shaped cells that give the appearance of more than one layer when sectioned for viewing under the microscope. Only columnar epithelia are ever pseudostratified (**Figure 5.6A**). *Transitional* describes a form of specialized stratified epithelium that is found in the urinary bladder. Because the tissue is very stretchy, the shape of the cells can vary depending on how the tissue is stretched. The notable characteristic of transitional epithelium is a scallop pattern of cells at the apical edge (**Figure 5.6B**).

Simple Epithelium The shape of the cells in the single cell layer of simple epithelium reflects the functioning of those cells. The cells in **simple squamous epithelium** have the appearance of a fried egg. Squamous cell nuclei tend to be flat and horizontal, mirroring the form of the cell. Simple squamous epithelium, because of the thinness of the cell, is present where rapid passage of chemical compounds is observed. The alveoli of lungs where gasses diffuse, segments of kidney tubules, and the lining of capillaries are also made of simple squamous epithelial tissue.

LO 5.2.3

LO 5.2.4

In **simple cuboidal epithelium**, the nucleus of the boxlike cells appears round and is generally located near the center of the cell. Simple cuboidal epithelia are typically found in places of transport, such as in the lining of the ducts of glands.

In **simple columnar epithelium**, the nucleus of the tall column-like cells tends to be elongated and located in the basal end of the cells. This epithelium is typically active in the absorption and secretion of molecules. Simple columnar epithelium forms the lining of some sections of the digestive system and parts of the reproductive tract. Columnar epithelia can often have cilia or microvilli on their apical surfaces. For example, ciliated columnar epithelium is found lining parts of the respiratory system, where the beating of the cilia helps remove particulate matter. Microvilli are present on intestinal epithelial cells, where they function to expand the cell's surface area.

Pseudostratified columnar epithelium is a type of epithelium that appears to be stratified, but instead consists of a single layer of irregularly-shaped and differently-sized columnar cells. In pseudostratified epithelium, nuclei of neighboring cells appear at different levels rather than clustered in the basal end. The cells are packed closely and a bit twisted such that when these tissues are sectioned for viewing under the microscope, the arrangement gives the appearance of stratification; but in fact, all the

Figure 5.6 Two Types of Epithelia that Defy the Naming Convention

(A) Pseudostratified columnar epithelium: this epithelium has cells so closely packed together that the tissue occasionally appears stratified when a thin slice is used to create a slide. While every cell contacts the basement membrane, not every cell contacts the apical surface, so it is actually a simple epithelium. (B) Transitional epithelium: this epithelium is stratified, but the cells lack shape because this is a stretchy tissue and the cell shape changes depending on the degree of stretch.

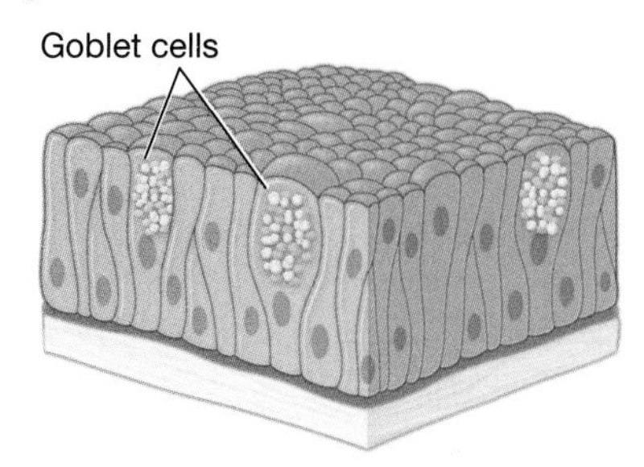

A Pseudostratified columnar epithelium

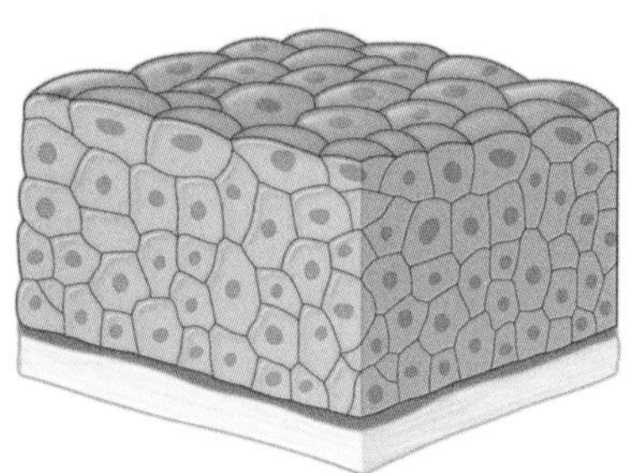

B Transitional epithelium

cells are in contact with the basal lamina. Pseudostratified columnar epithelium is found in the respiratory tract, where some of these cells have cilia.

A common feature of both simple and pseudostratified columnar epithelia is **goblet cells**. These mucus-secreting cells are technically unicellular exocrine glands interspersed between the columnar epithelial cells of mucous membranes (**Figure 5.7**).

> **Student Study Tip**
>
> Try considering an organ's function and matching tissue type(s) to it. Example: The esophagus is a transportation tube between the mouth and stomach in the GI tract. If you swallow a sharp tortilla chip, the lining of the esophagus needs to be protected. What type of epithelium functions in protection?

Stratified Epithelium Stratified epithelia consist of multiple stacked layers of cells. These types of epithelia are found in places where the apical surface is subject to more friction, and the multiple cell layers protect the underlying tissue from wear and tear. Remember that stratified epithelia are named by the shape of the most apical layer of cells. These apical cells are frequently shed due to friction. In order to maintain the number of layers in the tissue, the cells at the basal layer are constantly undergoing mitosis to generate new layers. A newly replicated epithelial cell starts its life at the basal layer and moves upward through the strata as more and more cells are replicated beneath it. Therefore, the cells at the apical layer are the oldest and most differentiated (**Figure 5.8**).

Stratified squamous epithelium is by far the most common type of stratified epithelium in the human body. When stratified squamous epithelia are found in moist places of the body, such as lining the inside of your cheeks, the apical layer of cells consists of living cells that are shed with some frequency. However, when stratified squamous epithelia are found in places of the body that are dry, such as the surface of your skin, the epithelium is modified to help retain moisture. In these tissues, the apical layer of living cells is topped by additional layers of dead cells that are filled with the protein **keratin**. Thus, we may see stratified squamous epithelia described as keratinized (topped with keratin-filled dead cells) or nonkeratinized (top layer is moist living cells). **Stratified cuboidal epithelium** and **stratified columnar epithelium** are uncommon in the human body, appearing only as a lining in a few of the larger glands and ducts.

Transitional epithelium is a specialized stratified epithelium found in the urinary bladder and the tubes that lead to it. The empty bladder is small, about the size of a large strawberry or small plum. The epithelium that lines the empty bladder is crumpled. As the bladder fills with urine, the organ stretches to accommodate up to a liter of liquid and the epithelium becomes taut. As the tissue undergoes these periodic

Figure 5.7 Goblet Cells

(A) Goblet cells, unicellular mucous glands, are common features of columnar epithelium. (B) Goblet cells in a micrograph from the respiratory tract.

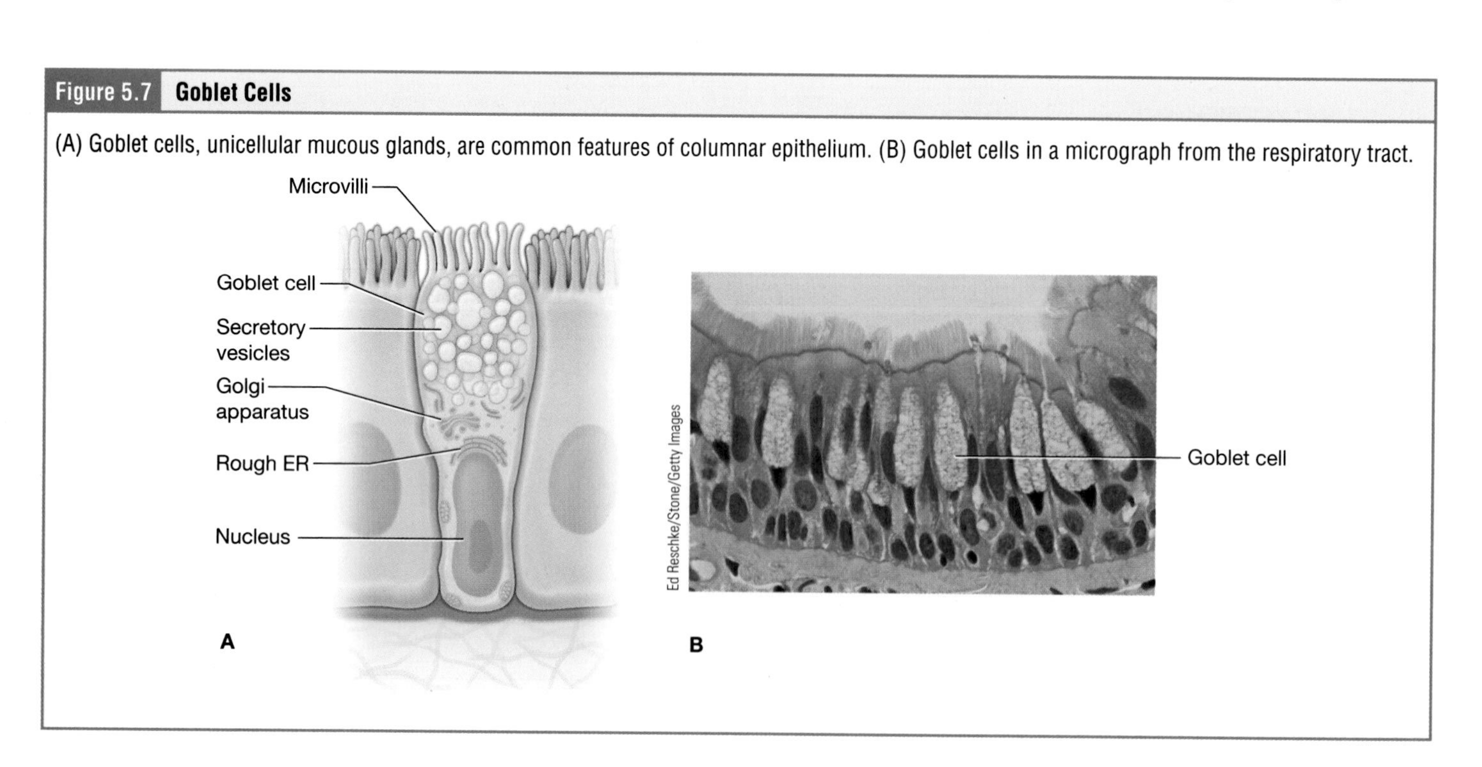

Figure 5.8 Stratified Epithelia

In stratified epithelia, often the basal layer cells are stem cells and do not demonstrate the characteristics of the epithelium.

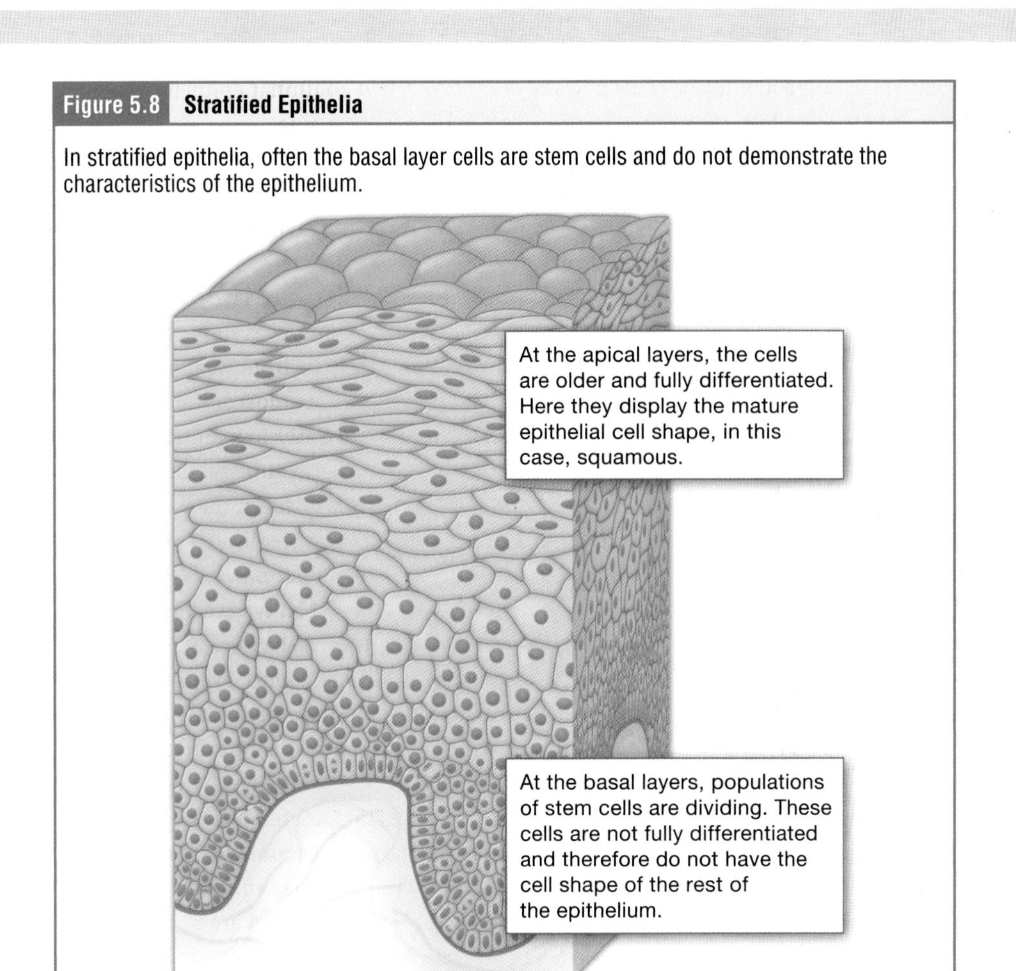

At the apical layers, the cells are older and fully differentiated. Here they display the mature epithelial cell shape, in this case, squamous.
At the basal layers, populations of stem cells are dividing. These cells are not fully differentiated and therefore do not have the cell shape of the rest of the epithelium.

changes, the cells change shape and therefore cannot be classified as squamous or cuboidal. Typically, the apical surface is described as domed, scalloped, or umbrella-shaped when it is viewed on a slide in its unstretched state.

Table 5.1 compares the different types of epithelia.

Table 5.1 Types of Epithelia

Cells	Locations	Functions
Simple squamous epithelium	Air sacs of lungs and the lining of the heart, blood vessels, and lymphatic vessels	Allows materials to pass through by diffusion and filtration, and secretes lubricating substance

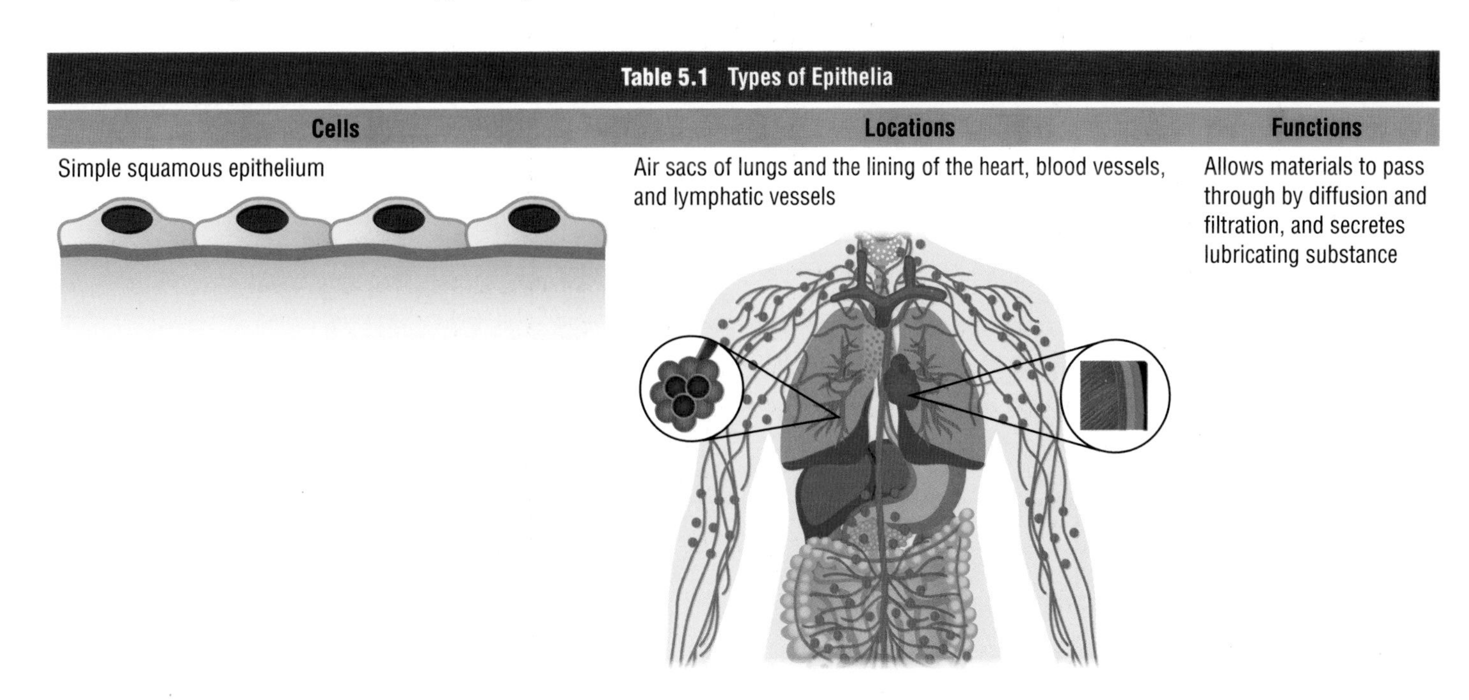

Table 5.1 Types of Epithelia *(Continued)*

Cells	Locations	Functions
Simple cuboidal epithelium	In ducts and secretory portions of small glands and in kidney tubules 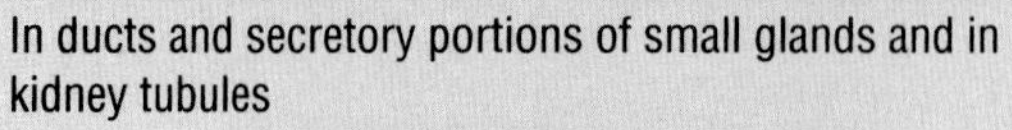 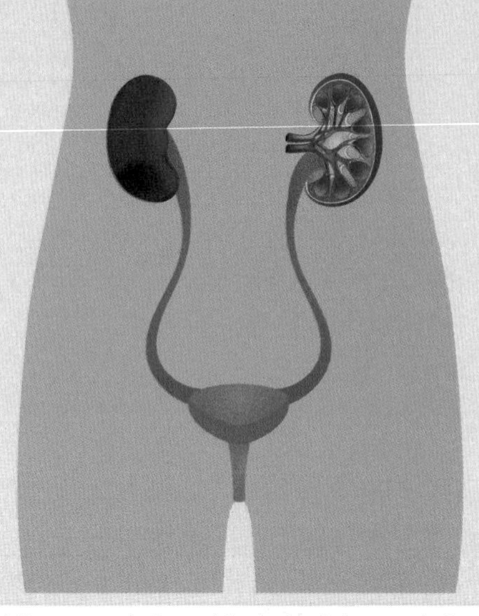	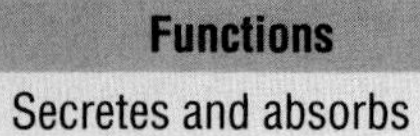 Secretes and absorbs
Simple columnar epithelium 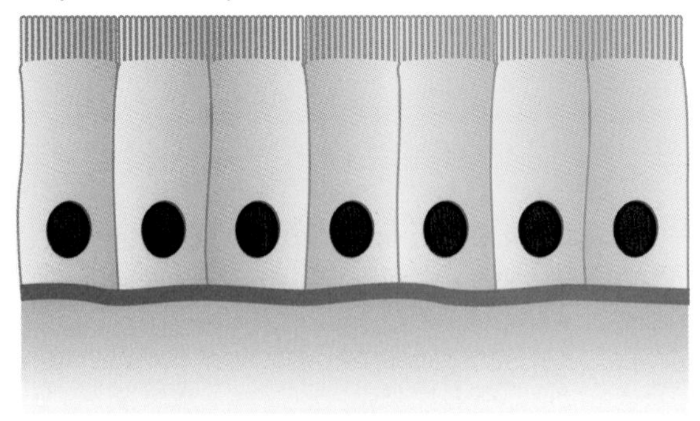	Ciliated tissues are in bronchi, uterine tubes, and uterus; smooth (non-ciliated) tissues are in the digestive tract and bladder 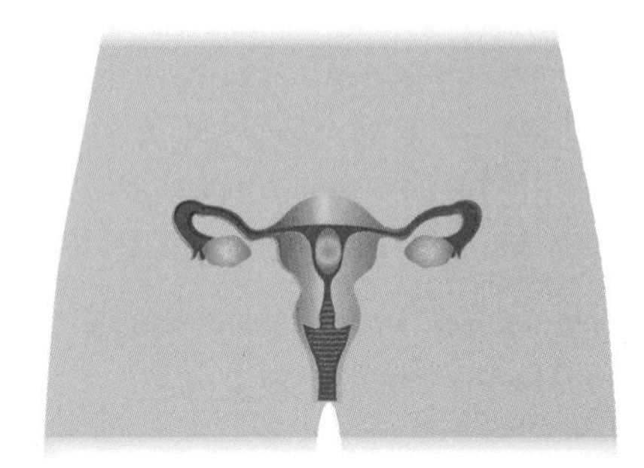	Absorbs; it also secretes mucus and enzymes
Pseudostratified columnar epithelium 	Ciliated tissue lines the trachea and much of the upper respiratory tract 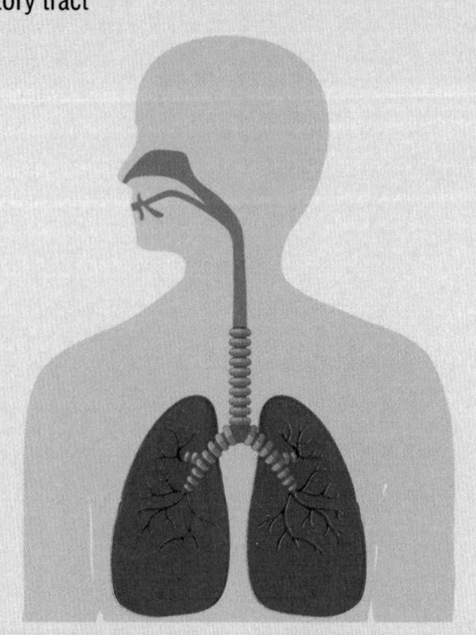	Secretes mucus; ciliated tissue moves mucus

Table 5.1 Types of Epithelia *(Continued)*

Cells	Locations	Functions
Stratified squamous epithelium 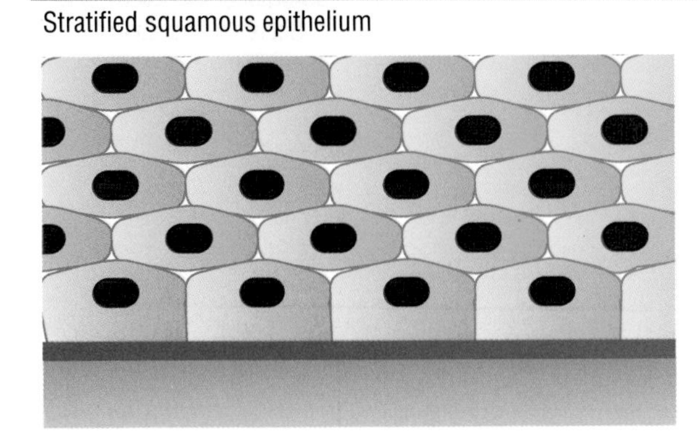	Lines the esophagus, mouth, and vagina 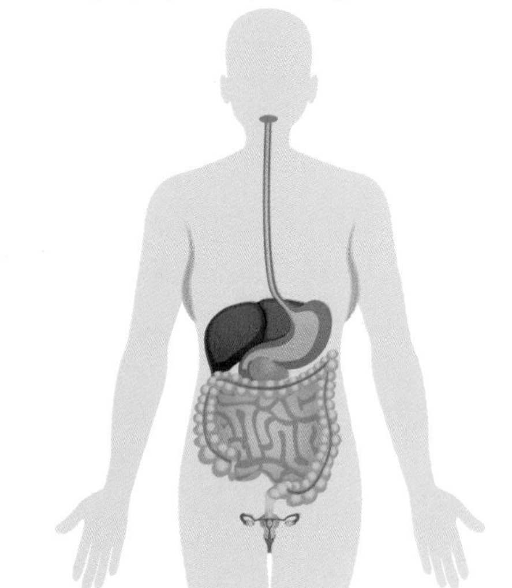	Protects against abrasion
Stratified cuboidal epithelium 	Sweat glands, salivary glands, and the mammary glands 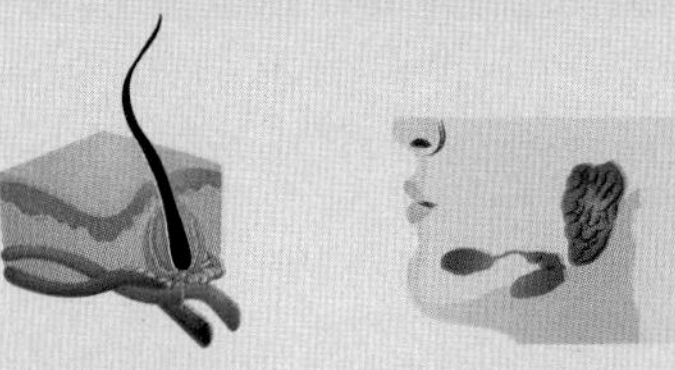	Secretes and protects
Stratified columnar epithelium 	The male urethra and the ducts of some glands 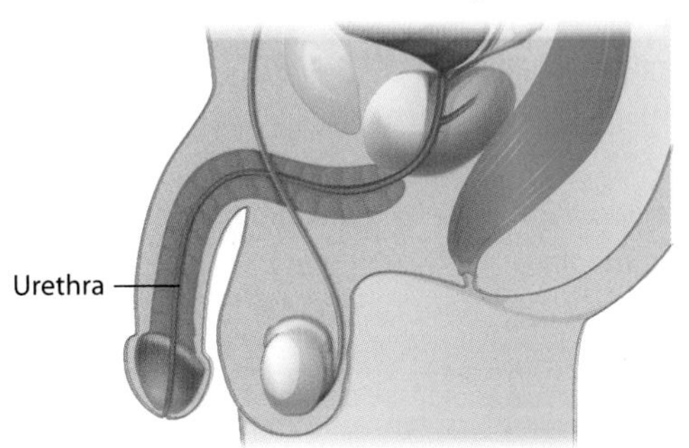	Secretes and protects

Table 5.1 Types of Epithelia *(Continued)*

Cells	Locations	Functions
Transitional epithelium	Lines the bladder, urethra, and ureters	Allows the urinary organs to expand and stretch

Learning Check

1. The small intestine has many folds that increase its surface area. This allows the small intestine to absorb more nutrients. Which of the following would you expect to find in the small intestine to increase its surface area?
   a. Microvilli
   b. Cilia
2. Describe the appearance of stratified squamous epithelium.
   a. One layer of boxy cells
   b. Multiple layers of rectangular cells
   c. One layer of rectangular cells
   d. Multiple layers of flat cells
3. Capillaries are responsible for diffusing nutrients to organs that need them. Describe the appearance of the epithelial tissue you would expect to find in capillaries.
   a. One layer of thin cells
   b. One layer of rectangular cells
   c. Multiple layers of thin cells
   d. Multiple layers of boxy cells
4. Where would you find the oldest cells in a stratified epithelium?
   a. Apical surface
   b. Basal surface
5. Where would you expect to find keratinized stratified squamous epithelium?
   a. On the inside of your mouth
   b. On the inside of your nose
   c. On the inside of your throat
   d. On the surface of your hand

5.2d Glands of Epithelia

LO 5.2.5

A *gland* is a structure that synthesizes and secretes chemicals. Most glands are composed of groups of epithelial cells, but some are unicellular. Keep in mind how we categorize "inside" and "outside" the body; the inside is a sterile place where no bacteria are found, while the outside is those spaces continuous with the outside world, including the lining of the gut. In this way, a gland can be classified as an **endocrine gland** if it secretes within the body and bloodstream, or as an **exocrine gland** if it secretes onto an epithelium to the outside of the body. Sweat glands and the glands of the digestive system are two examples of exocrine glands. A notable anatomical difference between these two categories of glands is that exocrine glands often have a duct through which the gland's secretions leave to reach the epithelium, while endocrine glands are always ductless and release their secretions directly into surrounding tissues and the bloodstream.

Student Study Tip

EXocrine EXiles its secretions to the EXternal of the body, such as the skin or GI tract.

Endocrine Glands The secretions of endocrine glands are called *hormones*. Hormones are released into the interstitial fluid, diffused into the bloodstream, and delivered to targets—in other words, cells that have receptors to bind the hormones. The endocrine system is part of a major regulatory system coordinating the regulation and integration of body responses. A few examples of endocrine glands include the anterior pituitary, thymus, adrenal cortex, and gonads. We will explore this system more in Chapter 17.

Exocrine Glands Exocrine glands release their contents through a duct that leads to the epithelial surface. Mucus, sweat, saliva, and breast milk are all examples of secretions from exocrine glands. Most exocrine secretions are released through tubular ducts. Goblet cells, however, do not have ducts but release their secretion on their apical surface.

Exocrine Gland Structure Exocrine glands are classified by their structure and method of secretion. In terms of structure, exocrine glands are either unicellular or multicellular. The unicellular glands are single cells such as goblet cells scattered among the epithelial cells. The multicellular exocrine glands consist of a single layer of epithelial cells that folds into the deeper tissues below. These glands have deep secretory cells that are connected to the epithelial surface by a tubular duct that can be straight or coiled. Some tubes form pockets called *acini*, while others are simple tubes. The duct can be simple or may be branched (Figure 5.9). More elaborate glands are known as *compound glands* and may even combine the tubular and acinar formats (tubuloacinar compound glands).

LO 5.2.6

Figure 5.9 Structures of Exocrine Glands

Exocrine glands vary by structure and can be broadly classified as simple or compound, and as acinar or tubular.

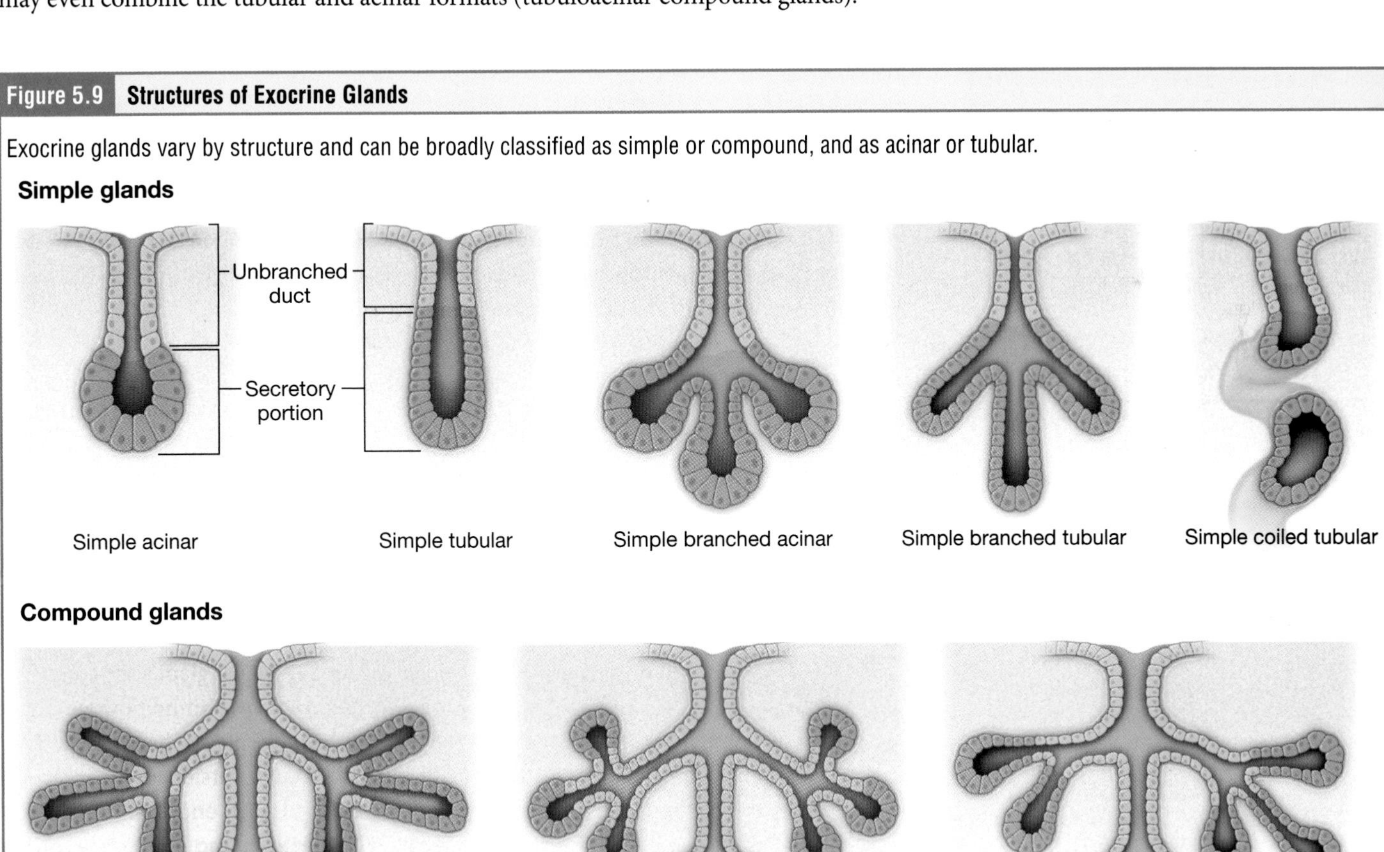

Methods and Types of Secretion Exocrine glands can also be classified by their mode of secretion, which impacts the nature of the substances released (**Figure 5.10**). Remembering that the suffix *-crine* means "to share," we can examine the three modes of exocrine section: merocrine, apocrine, and holocrine. **Merocrine secretion**, the most common type of exocrine secretion, is accomplished through exocytosis. The secretions are enclosed in vesicles that move to the apical surface of the cell. The vesicle merges with the plasma membrane and the contents are released. In merocrine secretion, very little cellular material leaves the cell with the secretion, so materials secreted by this method include sweat and watery mucus.

In **apocrine secretion**, the material to be secreted accumulates near the apical portion of the cell. That portion of the cell and its secretory contents pinch off from the cell and are released. In apocrine secretion, therefore, cellular membrane and cytoplasm are lost along with the secreted material. The sweat glands in the axillary and genital areas release sweat through apocrine secretion. This sweat is therefore more nutrient-rich than sweat produced through merocrine secretion, and bacteria tend to live around these glands to break down these fatty secretions. The bacterial metabolism of this sweat is what causes body odor. The cells of both merocrine and apocrine glands produce and secrete their contents continually; the cells sustain little damage and the nucleus and Golgi regions remain intact after secretion.

In contrast, the process of **holocrine secretion** involves the rupture and destruction of the entire gland cell. The cell accumulates its secretory products and the cell bursts like a piñata, releasing the secretions and killing the cell. New gland cells differentiate from cells in the surrounding tissue to replace those lost by secretion. The secretions of holocrine glands are very rich in lipids and proteins and therefore extremely attractive to many types of bacteria. The glands that produce the oils on the skin and hair are holocrine glands/cells.

Learning Connection

Chunking

When else in A&P have you heard the word *serous* associated with something? What was it? What do you think is common between that structure and a serous exocrine gland? What is different?

Glands are also named after the products they produce. The **serous gland** produces watery secretions derived from blood plasma and typically rich in enzymes, while the **mucous gland** releases watery to viscous products rich in the glycoprotein mucin. In some tissues the glands are strictly one or the other, but occasionally you can find mixed exocrine glands that contain both serous and mucous glands. These mixed glands are common in the salivary glands of the mouth.

Figure 5.10 Mechanisms of Producing Exocrine Secretions

(A) In merocrine secretion, the cells share their contents through exocytosis. (B) In apocrine secretion, the apical half of the cell fills with the secretory product and then separates from the basal half of the cell. (C) In holocrine secretion, the secretory cell is consumed with the secretory product and the whole cell is destroyed as it releases its product.

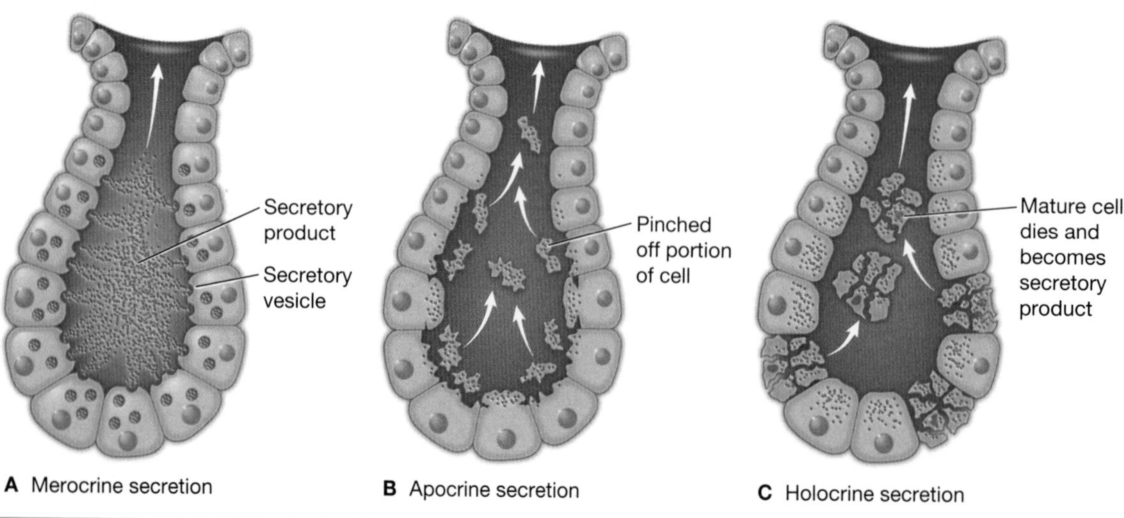

5.3 Connective Tissue

HAPS
LEARNING GOALS AND OUTCOMES

Learning Objectives: By the end of this section, you will be able to:

5.3.1 Describe the structural characteristics common to all types of connective tissue.

5.3.2 Classify different types of connective tissues based on their structural characteristics, functions, and locations in the body.

5.3.3 Identify examples of each type of connective tissue.

LO 5.3.1

Connective tissues are the most structurally diverse category of tissues in the human body. Let's start our examination of connective tissue by learning about the similarities between all types of connective tissues; then we can tackle their differences. Connective tissues come in a vast variety of forms, yet they typically have in common three characteristic components: cells, large amounts of ground substance, and protein fibers (see the "Anatomy of Connective Tissue" feature). The amount and structure of each component correlates with the function of the tissue, as we will illustrate in the coming sections. Unlike epithelial tissue, which is composed of cells that are closely packed with little or no extracellular space in between, connective tissue cells are rarely found touching each other, but are widely dispersed in the extracellular matrix. The matrix plays a major role in the functioning of this tissue. When considering the ECM of connective tissue, we consider the two types of ingredients separately. The major component, the **ground substance**, is the fluid or material between the cells and protein fibers. The density and type(s) of cells, the density and type(s) of protein fibers, and the composition of the ground substance are all characteristics that vary among the types of connective tissues (see the "Anatomy of Connective Tissue" feature).

Anatomy of...

Connective Tissue

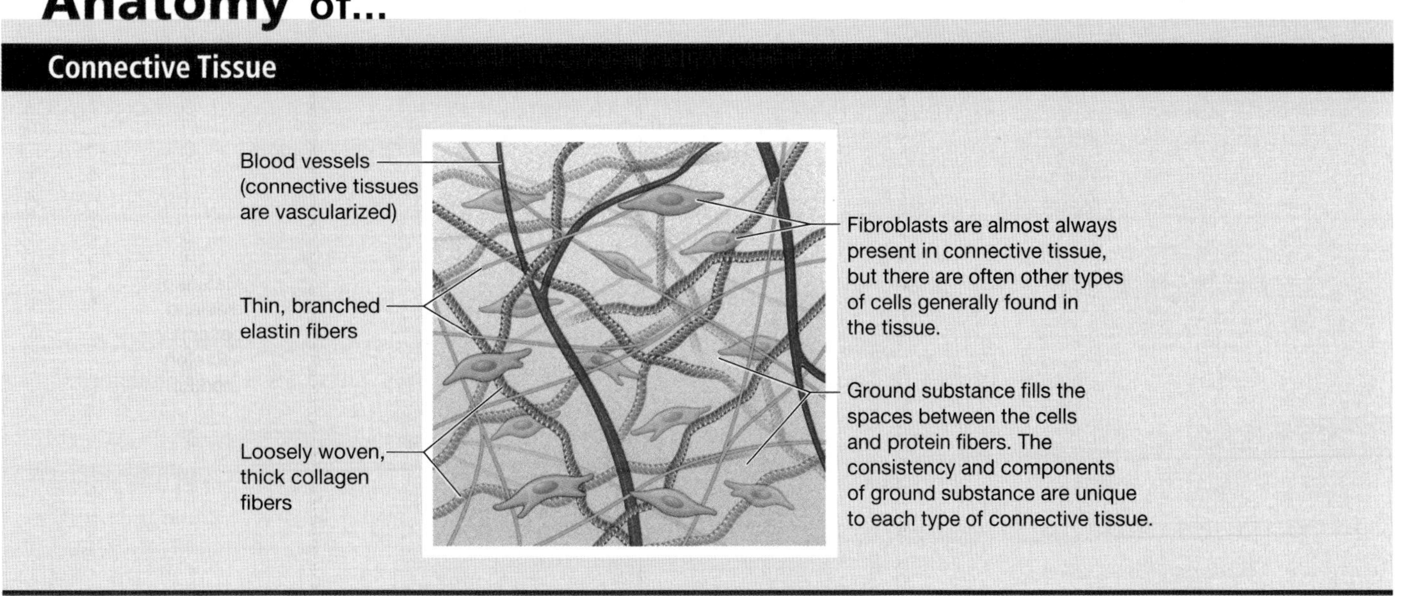
Blood vessels (connective tissues are vascularized)
Thin, branched elastin fibers
Loosely woven, thick collagen fibers
Fibroblasts are almost always present in connective tissue, but there are often other types of cells generally found in the tissue.
Ground substance fills the spaces between the cells and protein fibers. The consistency and components of ground substance are unique to each type of connective tissue.

5.3a Functions of Connective Tissues

As may be obvious from its name, all types of connective tissues function to connect tissues and organs; from the connective tissue sheath that surrounds muscle cells, to the tendons that attach muscles to bones, and to the skeleton that supports the positions of the body. Protection is another major function of connective tissue, in the form of fibrous capsules and bones that protect delicate organs and, of course, the skeletal system. Specialized cells in connective tissue defend the body from microorganisms that enter the body. Transport of fluid, nutrients, waste, and chemical messengers is ensured by specialized fluid connective tissues, such as blood and lymph. Adipose cells store energy in the form of fat and contribute to the thermal insulation of the body.

Student Study Tip

Tendons connect Two different tissue types and Ligaments connect one Lone type of tissue.

5.3b Classification of Connective Tissues

LO 5.3.2

The twelve types of connective tissues can be sorted into broad categories based on general properties. Semisolid, flexible connective tissues are typically grouped together in a category known as **connective tissue proper**. The structural connective tissues—bone and cartilage—are grouped together in a category called **supportive connective tissue**. Liquid connective tissues (blood and lymph) are grouped together in the category called **fluid connective tissue** (Table 5.2). Within each of these categories, connective tissues are further grouped by the density and type of their protein fibers and the types of cells found within the tissue. Remember that these tissues were all studied under the microscope, and so their microscopic appearance played a major role in their categorization. In Table 5.2 the tissues are arranged from loose to dense in terms of how compact their cells, fibers, and minerals, or other solids within their ground substance, appear under the microscope.

Learning Check

1. Identify the component that is most abundant in connective tissues.
   a. Cells
   b. Ground substance
   c. Protein fibers
2. Identify the functions of a connective tissue. Please select all that apply.
   a. Connect tissues and organs
   b. Transport fluids, nutrients, and waste
   c. Communicate with other cells in the body
   d. Control movement of the skeletal system
3. Ligaments and tendons are semi solid, flexible connective tissues. Which category of connective tissues would ligaments and tendons fall under?
   a. Connective tissue proper
   b. Supportive connective tissue
   c. Fluid connective tissue

5.3c Connective Tissue Proper

Tissue types in the *connective tissue proper* category all share a viscous ground substance, abundant protein fibers, and scattered cells. These tissue types often form tendons and ligaments or support epithelial tissues.

Cell Types The types of cells found in any given connective tissue can change periodically. If you take a photo of a connective tissue in a living body at one time and another photo of the same tissue an hour later, you may see an entirely different scene. Therefore, we describe the resident cells—the cells that are always present in connective tissue

Table 5.2 Connective Tissue Types

Connective Tissue Proper	Supportive Connective Tissue	Fluid Connective Tissue
Loose Connective Tissue	**Cartilage**	**Blood**
Areolar Choksawatdikorn/Shutterstock.com	Hyaline Tinydevil/Shutterstock.com	Science Stock Photography/Science Source
Reticular Science Stock Photography/Science Source	Fibrocartilage Tinydevil/Shutterstock.com	
	Elastic Jose Luis Calvo/Science Source	

(Continued)

Table 5.2 Connective Tissue Types

Connective Tissue Proper	Supportive Connective Tissue	Fluid Connective Tissue
Dense Connective Tissue	**Bones**	**Lymph**
Dense regular Ed Reschke/Photolibrary/Getty Images Dense irregular Jose Luis Calvo/Shutterstock.com	Spongy and Compact Bone QA International/Science Source	Lymph Jose Luis Calvo/Science Source
Adipose Tissue		
Biophoto Associates/Science Source		

proper—first, and then describe the cells that may be wandering through at any given moment.

The most abundant cell in connective tissue proper is the **fibroblast.** Fibroblasts are fiber-makers; they are responsible for creating the fibers that weave together to form the structure of connective tissue proper.

Adipocytes (fat cells)

Adipocytes are cells that store lipids as droplets that fill most of the cytoplasm. The number and type of adipocytes depends on the tissue and its location, and these vary among individuals in the population. Adipocytes are found scattered in most forms of connective tissues, but one type of connective tissue—adipose tissue—is composed almost entirely of adipocytes.

Connective tissues have an immune defense role and so immune cells are found patrolling these tissues. They are considered wandering cells because they do not stay put; rather, they may be found in various concentrations. Macrophages and mast cells are the most common types of wandering cells.

Macrophages are large immune cells that wander through tissues searching for waste, debris, or unwanted visitors such as bacteria. Macrophages are an essential component of the immune system. When stimulated, macrophages release **cytokines**, small proteins that act as chemical messengers. Cytokines recruit other cells of the immune system to infected sites and stimulate their activities.

Mast cells are immune cells that secrete chemicals in order to activate immune responses. They contain these chemicals in cytoplasmic vesicles and can release them as necessary. An example of these chemical signals is histamine. When irritated or damaged, mast cells release histamine, an inflammatory mediator, which causes vasodilation and increased blood flow at a site of injury or infection, along with itching, swelling, and redness you recognize as an allergic response. Most allergy medicines are antihistamines; they block this signal from having an effect, thwarting the effect of mast cells on their tissues.

Connective Tissue Fibers and Ground Substance Three main types of fibers are secreted by fibroblasts: collagen fibers, elastic fibers, and reticular fibers. These fibers are woven together like the yarn of a sweater, creating the structure of connective tissue. In some tissues the fiber types are mixed; in other tissues there is only one type of fiber. The fibers can be either crowded together tightly or woven in a loose manner. In broad strokes, we can organize connective tissue proper into the categories of loose and dense depending on how closely packed their fibers are.

Collagen fibers are flexible but also have great tensile strength, resist stretching, and give ligaments and tendons their characteristic resilience and strength. These fibers are able to endure the movement of the body and hold connective tissues together. Collagen, when isolated from animal tissues, continues to provide structure to other substances such as glue and gelatin.

Elastic fibers contain the protein **elastin** along with lesser amounts of other proteins and glycoproteins. The main property of elastin is that after being stretched, compressed, or twisted it will return to its original shape. Elastic fibers are prominent in elastic tissues found in skin and the elastic ligaments of the vertebral column.

Reticular fibers are formed from the same protein subunits as collagen fibers; however, reticular fibers remain narrow and are arrayed in a branching network. They are found throughout the body but are most abundant in the reticular tissue of soft organs, such as liver and spleen, where they anchor and provide structural support to the cells, blood vessels, and nerves of the organ.

Student Study Tip

The root *ret-* means "net" in Latin. The fibers are branched, thin, and netlike. Another use of this root is in the word *retina*, a structure of the eye.

All of these fiber types are embedded in the ground substance. The ground substance varies in stiffness and elasticity depending on the concentrations and proportion of these fiber types and other materials that can be embedded within the ECM. The ground substance is a mixture of the macromolecules secreted by fibroblasts, and the water that is attracted and trapped within the ground substance matrix.

Loose Connective Tissue Loose connective tissue is found between many organs where it acts both to absorb shock and bind tissues together. It allows water, salts, and various nutrients to diffuse through to adjacent or imbedded cells and tissues.

Figure 5.11 Loose Areolar Connective Tissue

Areolar connective tissue is a type of loose connective tissue that consists of a network of loosely arranged collagen and elastin fibers and fibroblasts. (A) The collagen fibers provide strength and the elastin fibers provide flexibility. (B) The dark-staining masses in the micrograph and the nuclei of the fibroblasts.

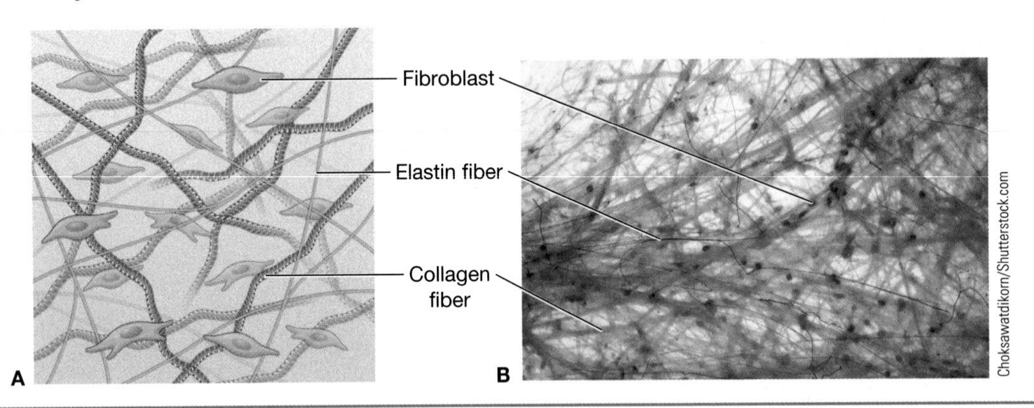

Areolar tissue is a loose connective tissue that resembles a loosely-woven web. It is characterized by predominantly collagen fibers with smaller amount of elastin fibers. Scattered fibroblasts and adipocytes are the main cell types, and these are nourished and supported by the ground substance. Areolar connective tissue is found surrounding blood and lymph vessels, as well as supporting organs in the abdominal cavity. Areolar tissue underlies most epithelia (**Figure 5.11**).

LO 5.3.3

Areolar tissue (loose connective tissue)

Adipose tissue consists mostly of packed cells with few fibers between the cells (**Figure 5.12**). The cells within adipose tissue are known as *adipocytes*; these are specialized and modified for fat storage. The two varieties of adipose tissue are called *white* and *brown*, although this characterization is not entirely accurate; white adipose actually appears more of a dull yellow. White adipose tissue is the most abundant type of adipose in adults, where it contributes mostly to lipid storage and can serve as

Figure 5.12 Adipose Tissue

Adipose tissue is a loose connective tissue that consists of cells with little extracellular matrix. (A) The cells store lipids in a large central vacuole. (B) Brown adipose tissue—common in infants—has more ECM and the lipid is stored in smaller vesicles, with many per cell, rather than a large central vacuole.

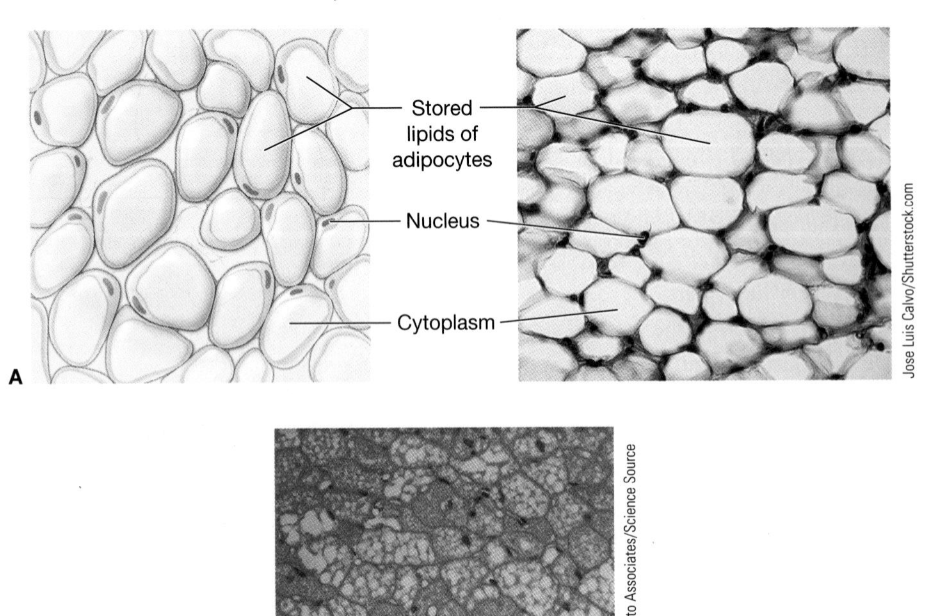

Cultural Connection

The Weight of the World

While adipocytes only store lipids, the most common dietary component that contributes to adipose tissue development is sugar. Sugars can be converted into lipids so that their energy can be stored until needed. The nature of the American diet is that many foods, and the most widely available foods, can be packed with sugars, even if they don't taste sweet. Inequities in food availability can make it complex to find foods low in sugar.

The term *food desert* is used to describe urban or rural areas in which it is difficult to buy affordable high-quality fresh foods. These regions, which are often more affordable to live in, typically have higher rates of obesity. While obesity can be caused by diet, the complex nature of obesity includes many factors that impact body weight. Obesity increases the risk of many other diseases such as heart and cardiovascular disease, stroke, Type 2 diabetes, certain cancers, gastrointestinal problems, gynecological issues, osteoarthritis, and severe COVID-19 symptoms. However, individuals who are overweight or obese often experience bias in the healthcare system that may affect their care or their willingness to seek care. The World Health Organization (WHO) reports that 69 percent of adults with obesity in the United States and Europe report experiences of stigmatization from health care professionals. Weight bias is certainly one factor as it can discourage patients from trusting their primary care providers and prevent them from receiving ideal medical advice and patient-centered care. Thus, weight bias could even cause patients to have even unhealthier increases in weight.

insulation from cold temperatures, as well as shielding some organs from injury. Brown adipose tissue is more common in infants, but very little remains in adult humans. Brown adipose tissue has the unique feature of releasing heat rather than ATP as it breaks down fats—making it particularly helpful in temperature homeostasis, especially when muscle contraction is limited (human babies and hibernating mammals across the animal kingdom). In Figure 5.12, you can compare and contrast white and brown adipose tissue and note the difference in their cellular storage. The stored lipids within the cells repel the stains used in histology preparation (see the "Digging Deeper" feature) and so they appear as holes in the prepared tissue. Whereas brown adipocytes store multiple small lipid droplets in each cell, white adipocytes store lipids in one large droplet, pushing all of the cellular material—including the nucleus—to the periphery.

Reticular tissue is tissue composed of reticular fibers, loosely woven together to form a meshlike supportive framework for the cells of some organs such as lymphatic tissue, the spleen, and the liver (**Figure 5.13**). Most of the cells found in reticular tissue are cells of the immune system.

Figure 5.13 Reticular Tissue

(A) Reticular connective tissue is a loose connective tissue with reticular fibers as the predominant fiber type. These fibers are wavier and more highly branched than elastic fibers, but stain a similar purple/black. (B) The nuclei seen here are of leukocytes because this type of tissue is commonly found in lymphoid organs.

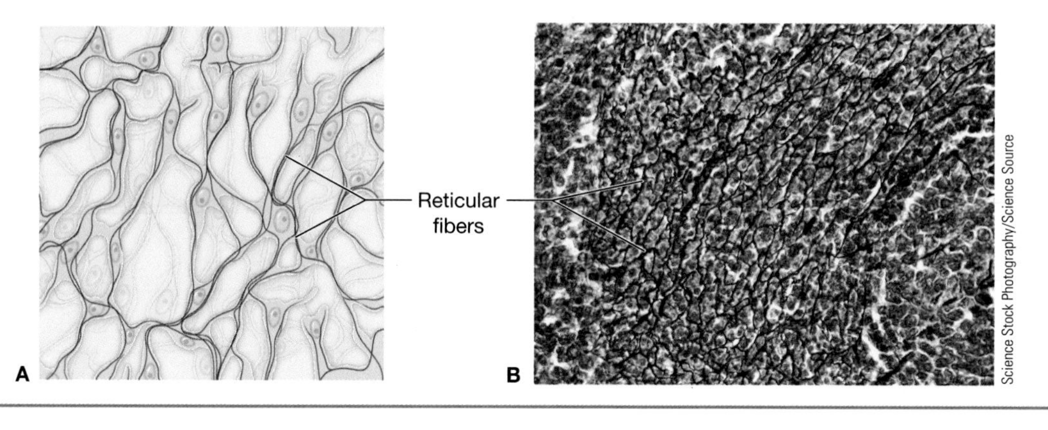

Dense Connective Tissue Dense connective tissue differs from loose, in that it contains far more collagen fibers. An easy visual to imagine the difference between loose and dense is to think of two sweaters, one that is knitted loosely such that there are spaces between the fibers, and another that is knitted so that the strands of yarn are much closer together. The second sweater is stronger when pulled on and provides more of a barrier between one side and another. Dense connective tissue does not allow for as much communication of nutrients and fluids through the tissue, but is more resistant to pulling forces than loose connective tissue. The two types of dense connective tissue contain more collagen fibers than the loose connective tissue types. As a consequence, dense connective tissue is harder to stretch. The two types of dense connective tissue are described by the weave of their collagen fibers. In dense irregular connective tissue, the collagen fibers are woven together in random, net like patters as we saw in areolar tissue (**Figure 5.14**). The fibers, while netlike, are very tightly packed together, and this arrangement gives the tissue tremendous strength. The underlying connective tissue of the skin is an example of dense irregular connective tissue rich in collagen fibers. Skin can be pulled in many directions and therefore needs to be able to withstand forces in any plane.

In dense regular connective tissue (**Figure 5.15**), the collagen fibers are parallel to each other, enhancing tensile strength and resistance to stretching in the direction

Figure 5.14 Dense Irregular Connective Tissue

(A) Dense irregular connective tissue consists of closely-woven collagen fibers and fibroblasts. (B) In dense irregular connective tissue, the fibers are oriented in every direction, like a net, so that force can be withstood in any plane.

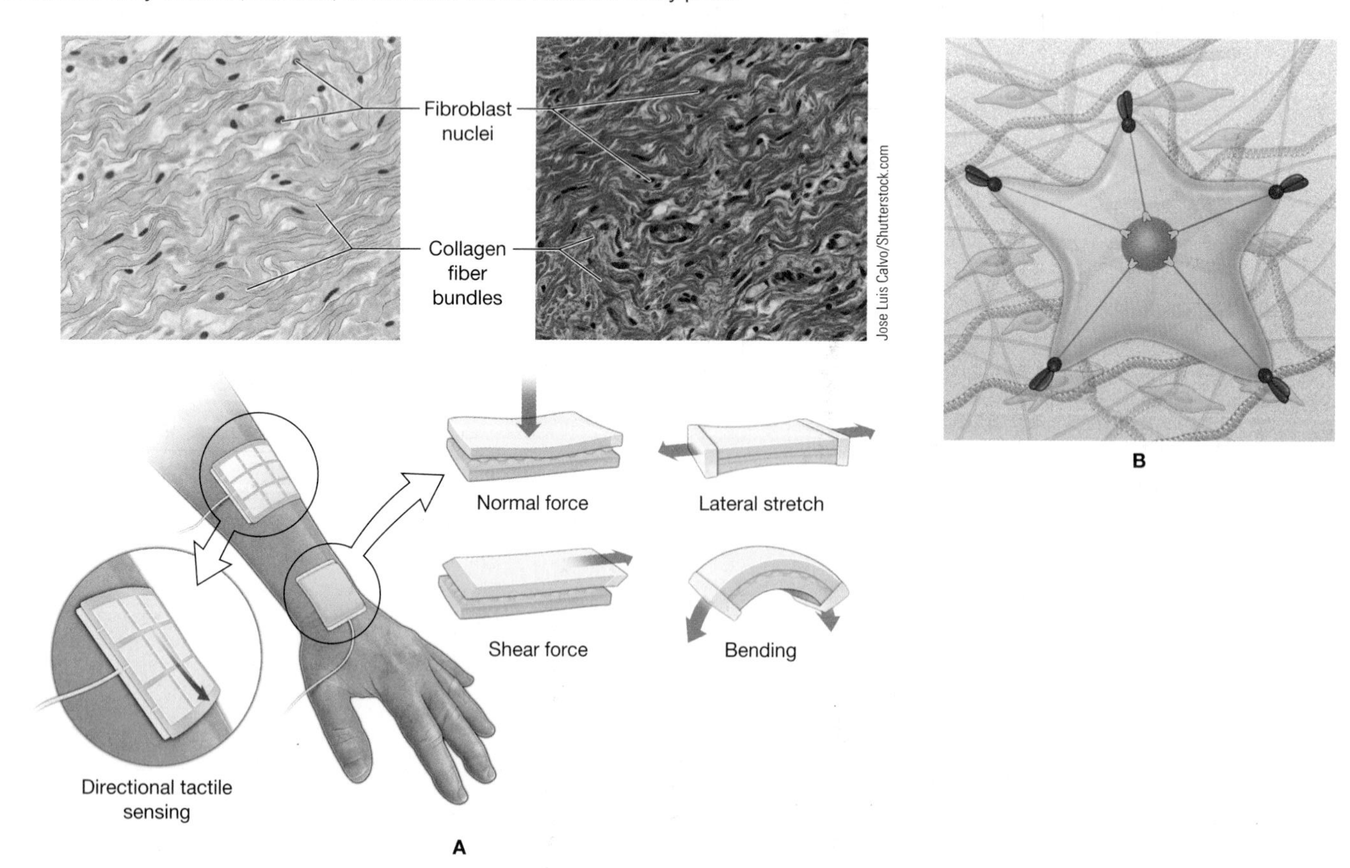

Figure 5.15 Dense Regular Connective Tissue

(A) Dense regular connective tissue consists of closely woven collagen fibers that are oriented in one direction. (B) Because of this, force can be withstood only in one plane.

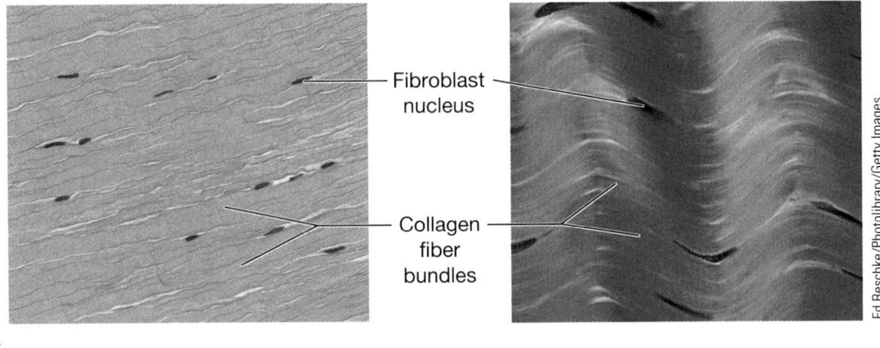

of the fiber orientations. Tendons are made of dense regular connective tissue, and the structure of this tissue is well suited to its function here because tendons are stretched in just one direction (**Figure 5.15B**). Dense regular elastic tissue contains elastin fibers in addition to collagen fibers, which allows the ligament to return to its original length after stretching.

Apply to Pathophysiology

Tendinitis

Arun is a popular and energetic nurse on the cardiology floor of the local hospital. Arun turned 40 this year and decided to take up a hobby of running, with a goal of running a marathon before their half-birthday. Arun is training aggressively on their days off. Today, one of Arun's patients, a retired physical therapist recovering from heart surgery, notices that Arun's footsteps seem to be slower than usual. Arun admits that, following their run yesterday, they have been experiencing a sharp pain in their ankle, just above the heel bone. Arun's patient suggests that Arun has developed tendinitis from increasing the amount and frequency of running too much too fast. The patient recommends making an appointment with a doctor, and until then, icing the tender area, taking nonsteroidal anti-inflammatory medication such as ibuprofen to ease the pain and to reduce swelling, and complete rest for a few weeks. Arun laughs at the suggestion of rest; they can't take time off or pause on their training schedule.

Arun asks, what is tendinitis and how did it happen? Arun remembers that tendons are strips of dense irregular connective tissue that connect muscle to bone.

1. Which type of connective tissue are tendons made of?
   A) Areolar connective tissue
   B) Dense irregular connective tissue
   C) Dense regular connective tissue
   D) Reticular connective tissue
2. The pain associated with tendinitis is due to inflammation as well as the rupture of some of the fibers of the tendon. Which type of fiber is rupturing?
   A) Collagen
   B) Elastic
   C) Reticular

Arun continues to train for the marathon, but during their next long run, they fall suddenly with severe pain in their ankle. Arun's running partner takes them to the ER, where the doctor tells Arun they have ruptured their tendon. The tendon will need to be repaired surgically, and since Arun is young and healthy, the doctor suggests an experimental approach using a synthetic tendon that has been made in a lab.

3. What properties will the synthetic tendon have?
   A) Collagen and elastin fibers that are woven in many different directions
   B) A high proportion of reticular fibers to support more immune cells to stay as residents in the tendon
   C) Primarily collagen fibers that run parallel to each other

5.3d Supportive Connective Tissues

The two major forms of supportive connective tissue—cartilage and bone—allow the body to maintain its posture and protect internal organs. In both of these types of tissues the ground substance takes on a more semisolid or solid consistency due to the molecules that are deposited within the matrix.

Cartilage *Cartilage* is a semisolid material that is able to offer structure and protection to the body while maintaining some flexibility. The ribcage functions to protect the vulnerable heart and lung organs, and therefore benefits from being as strong and rigid as possible. However, the lungs within the ribcage must expand and contract with the breath, so in addition, the ribcage needs a small degree of flexibility. Cartilage offers this flexibility to the ribcage (**Figure 5.16**). The distinctive semisolid property of cartilage is due to the deposition of polysaccharides called *chondroitin sulfates* into the ground substance. Embedded within the cartilage matrix are cartilage cells called **chondrocytes**. These cells are responsible for generating the chondroitin sulfates and other components of the cartilage matrix. They secrete the matrix around themselves and are therefore found encased within it. Microscopic examination of cartilage (**Figure 5.17**) reveals that the chondrocytes appear to sit within small spaces in the otherwise solid matrix. These spaces are called **lacunae** (singular = lacuna).

Just like epithelium, all cartilage tissue is avascular. Thus, all nutrients that the chondrocytes need must diffuse through the matrix to reach them. For this reason, damaged cartilage can be very slow to heal. The nutrients for cartilage come from two main sources. Just like epithelium, most avascular cartilages are supported by vascular connective tissue. In cases in which cartilage is supported by connective tissue, a layer of dense irregular connective tissue called the **perichondrium** encapsulates the cartilage (**Figure 5.18**).

Figure 5.16 The Ribcage is a Merger of Two Types of Supportive Connective Tissues

(A) The majority of the ribcage is made up of bone—the most solid tissue in the body—to protect the delicate lungs. (B) However, the ribcage needs to be slightly flexible to allow for the motion of breathing. To accomplish this, cartilage is employed in the front of the structure.

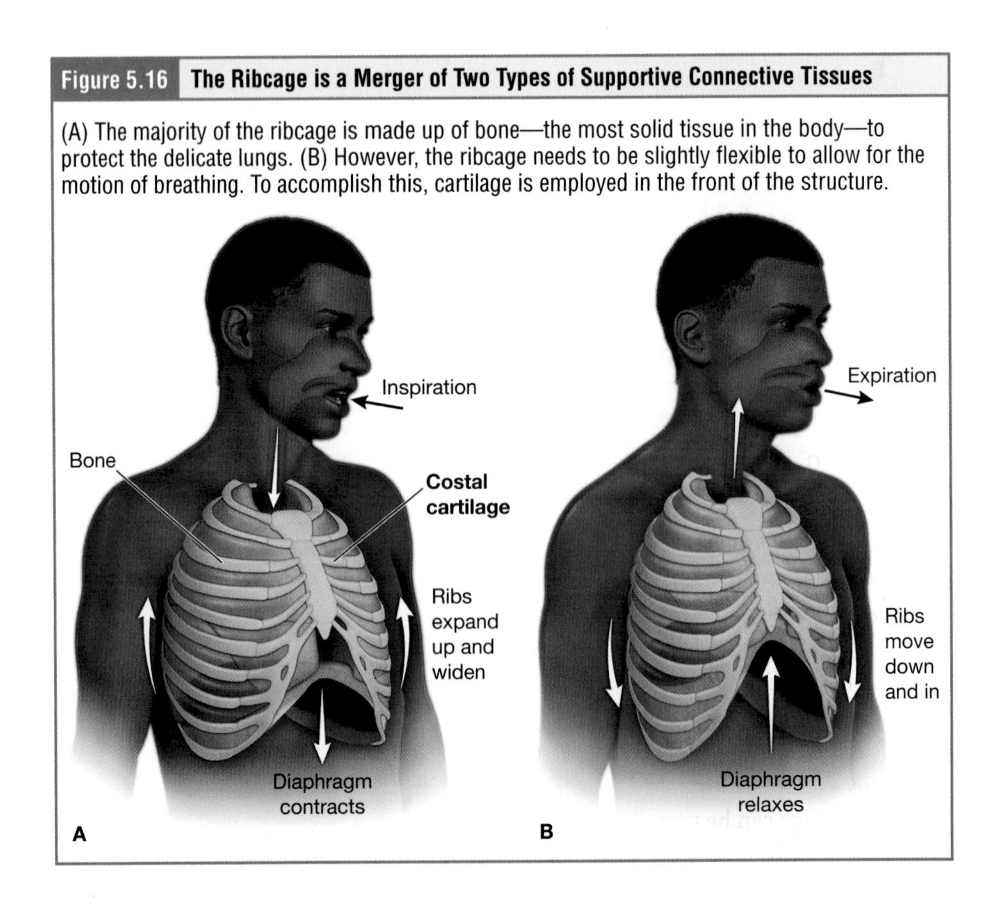

Figure 5.17 Types of Cartilage

Cartilage is a supportive connective tissue consisting of collagenous fibers embedded in a semisolid matrix. (A) Hyaline cartilage has a smooth appearance with no visible fibers; it is slightly flexible and comprises the majority of cartilage in the body. (B) Fibrocartilage is so named because it has many visible collagen fibers. Fibrocartilage is found in weight-bearing locations because of its ability to absorb pressure. (C) Elastic cartilage is notable because of its visible, dark elastic fibers. It is found in places that require the ability to move or change shape. (D) Hyaline cartilage is the most common type of cartilage in the body, and elastic cartilage is the least common.

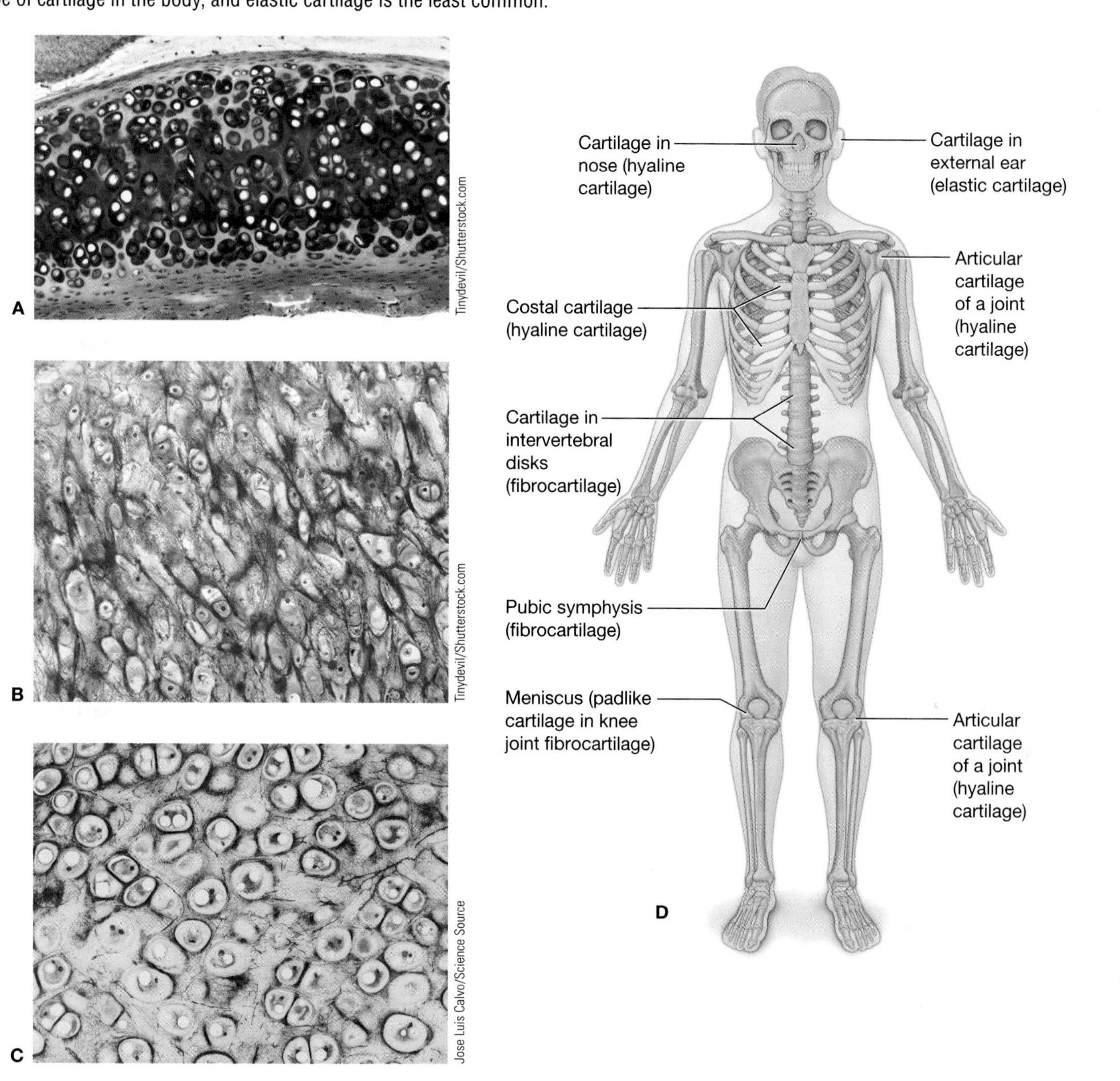

Similar to the organization of connective tissue proper, we describe three different types of cartilage tissue based on the fibers in their matrix (see Figure 5.17). **Hyaline cartilage**, the most common type of cartilage in the body, consists of short and dispersed collagen fibers and contains large amounts of proteoglycans. Due to these proportions, the fibers are not easily visible under the microscope, and the surface of hyaline cartilage is smooth. Both strong and flexible, it is found in the rib cage and covers bones at the joints to provide a smooth, gliding surface. Notably, hyaline cartilage forms a template for bone formation from the fetal period through adolescence. Hyaline cartilage can be remodeled to form bone if the chondrocytes differentiate to become bone cells.

Figure 5.18 Macro and Micro Views of the Perichondrium

The perichondrium is a layer of dense irregular connective tissue that often surrounds a mass of cartilage in the body. This example shows the perichondrium surrounding the articular cartilage that covers the ends of a bone in a joint.

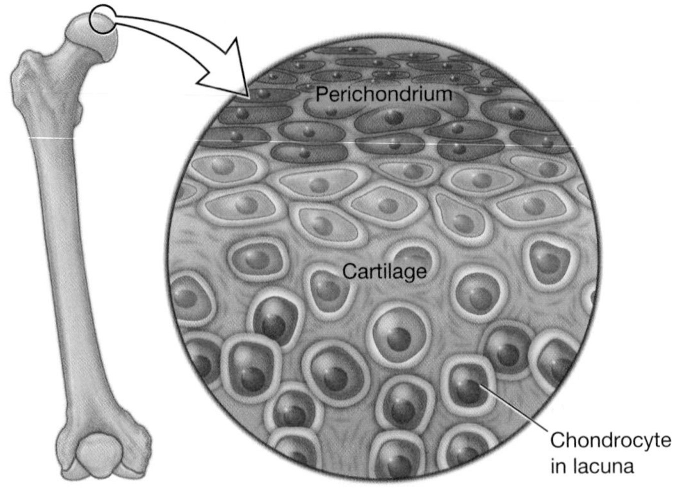

As its name suggests, **fibrocartilage** has thick bundles of collagen fibers dispersed through its matrix. These collagen fibers lend fibrocartilage a toughness and durability to resist impact. Fibrocartilage pads are found in places of intense weight-bearing, such as the knee joint and in between the vertebrae of the spine.

As with the other forms of cartilage, **elastic cartilage** contains collagen and proteoglycans, but this tissue type has far more elastic fibers than the other forms of cartilage. Thus, this tissue gives rigid support as well as elasticity. Tug gently at your ear lobes and notice that the lobes return to their initial shape. The external ear contains elastic cartilage.

Bone *Bone* is the hardest connective tissue. It provides protection to internal organs and supports the body. Bone's rigidity is due to the deposition of calcium and phosphorous onto the collagen fibers of the extracellular matrix. Collagen offers bones both the structure for mineralization and a small degree of flexibility; without collagen, bones would shatter easily. Without mineral crystals, bones would flex and buckle under the weight of the body and provide little support. The two main types of cells—osteocytes and osteoblasts—both play roles in building the matrix of bone tissue. Osteocytes, which play roles similar to chondrocytes, are also located within spaces in the matrix called *lacunae*. Unlike cartilage, bone is a highly vascularized tissue, and so it can recover from injuries in a relatively short time.

Spongy bone earns its name because, rather than a solid tissue, it is a network of beams and bridges surrounded by empty spaces (**Figure 5.19**). It therefore somewhat resembles a kitchen sponge under the microscope with its empty spaces between **trabeculae**, or arches, of bone. Spongy bone is lighter than compact bone; it is found in the interior of some bones and at the ends of long bones. Compact bone is solid and has greater structural strength (see Figure 5.19).

Three of the four forms of supportive connective tissue can be compared with a simple exercise involving your own nose. Take your fingers and palpate (touch) your nose all the way up close to your eyes and forehead. If you wear glasses, you want to feel the place on your nose where your glasses rest. This is bone (specifically, the nasal bones of the skull) and you should be able to feel that it is quite hard and immovable. Next, trace your fingers down about halfway to the end of your nose. Here you can feel that your nose is quite rigid, but with a little more flexibility than the bone portion. Can you guess what

Figure 5.19 Bone: the Most Rigid Type of Connective Tissue

There are two types of bone in the body: compact bone, which forms the periphery of most bones, and spongy bone, which is composed of trabeculae and marrow-filled spaces and forms the deeper portion and the epiphyses of most bones.

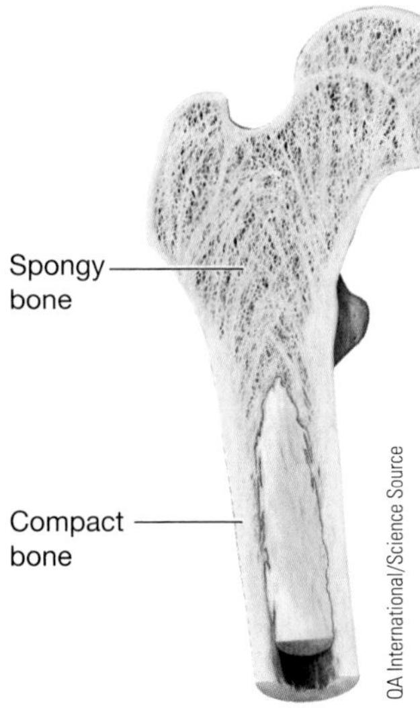

QA International/Science Source

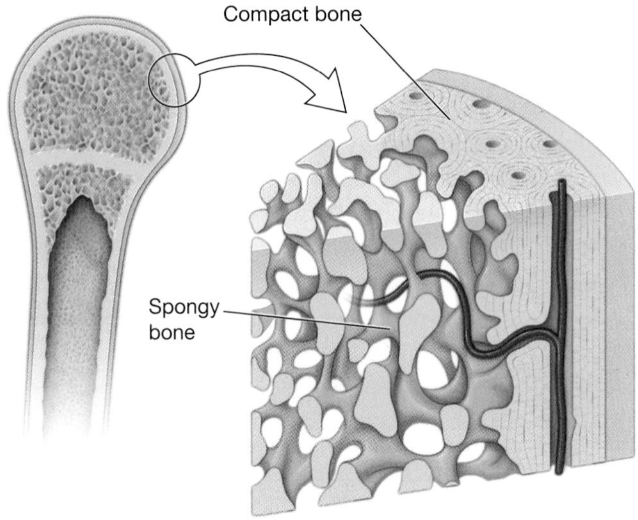

type of connective tissue you are touching now? If you guessed hyaline cartilage, you are absolutely correct. Now take one finger, place it on the tip of your nose, and move it in a circle. Can you feel how bendable the tip of your nose is? It has a distinctive shape that it returns to when you stop touching it, but it is highly flexible. This is elastic cartilage.

5.3e Fluid Connective Tissue

In the fluid connective tissues—blood and lymph—cells circulate in a liquid extracellular matrix called **plasma**. This matrix still has protein in it, but the proteins are mostly in their monomer form rather than strung together in fibers. In the right conditions, they will form polymer fibers similar to those we associate with other connective tissues. This happens during blood clotting, for example. The cell types in the fluid connective tissue category are very different from those in connective tissue proper and supporting connective tissue. While the predominant cells in those tissue types are cells that produce the matrix that surrounds them, in blood—one of the two types of fluid connective tissue—the cells

Student Study Tip

It is easy to forget that blood is a connective tissue, but it connects all body systems together by removing waste and delivering oxygen, nutrients, hormones, and other signaling molecules!

Figure 5.20 Blood: A Fluid Connective Tissue

(A) Blood is a fluid connective tissue containing erythrocytes, five types of leukocytes, and platelets that circulate in a liquid extracellular matrix. (B) Two blood vessels in cross section with blood inside.

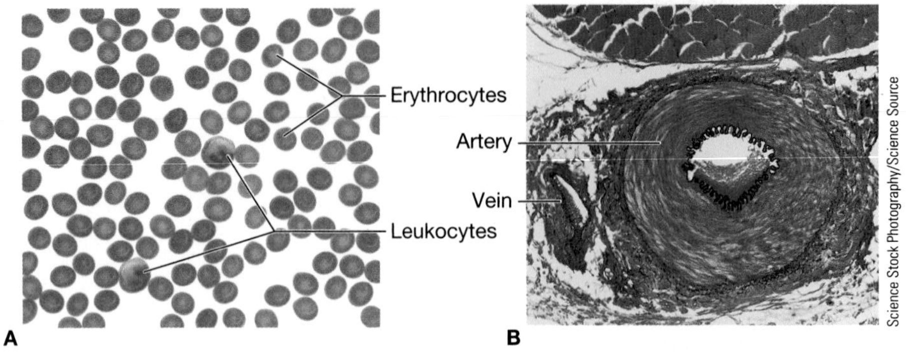

A: Erythrocytes, Leukocytes
B: Artery, Vein

Science Stock Photography/Science Source

do not contribute to the liquid matrix but instead perform very different homeostatic functions. Lymph fluid does not have its own resident cells, though occasionally cells can be found wandering through the lymph. These two tissues—lymph and blood—both serve functions of distributing material (nutrients, salts, gasses, fluid, and wastes) throughout the body. The formed elements circulating in blood are all derived from **hematopoietic stem cells** located in bone marrow (Figure 5.20). **Erythrocytes** (commonly called *red blood cells*) transport oxygen and other gasses, increasing the efficiency of carrying gasses in a liquid (you will read more about this in Chapter 22). *Leukocytes* (commonly called *white blood cells*) are immune cells. They travel within the blood but rarely work there, instead more often leaving the blood to defend against potentially harmful microorganisms in the tissues. *Platelets* are cell fragments involved in blood clotting. Nutrients, salts, and wastes are dissolved in the plasma and transported through the body (see Figure 5.20).

Erythrocytes (red blood cells)

Lymph is a liquid collected from body tissues with excess interstitial fluid. Leukocytes travel within the lymph as well, but the majority of this fluid tissue is acellular (Figure 5.21). Lymphatic capillaries are extremely permeable, allowing larger molecules and excess fluid from interstitial spaces to enter the lymphatic vessels. Lymph drains into blood vessels, delivering molecules to the blood that could not otherwise directly enter the bloodstream. We will discuss lymph and lymphatic vessels in much greater detail in Chapter 21.

Figure 5.21 Lymph within Lymphatic Vessels

Lymph and blood are the two forms of fluid connective tissue. In contrast to blood, lymph is mostly acellular.

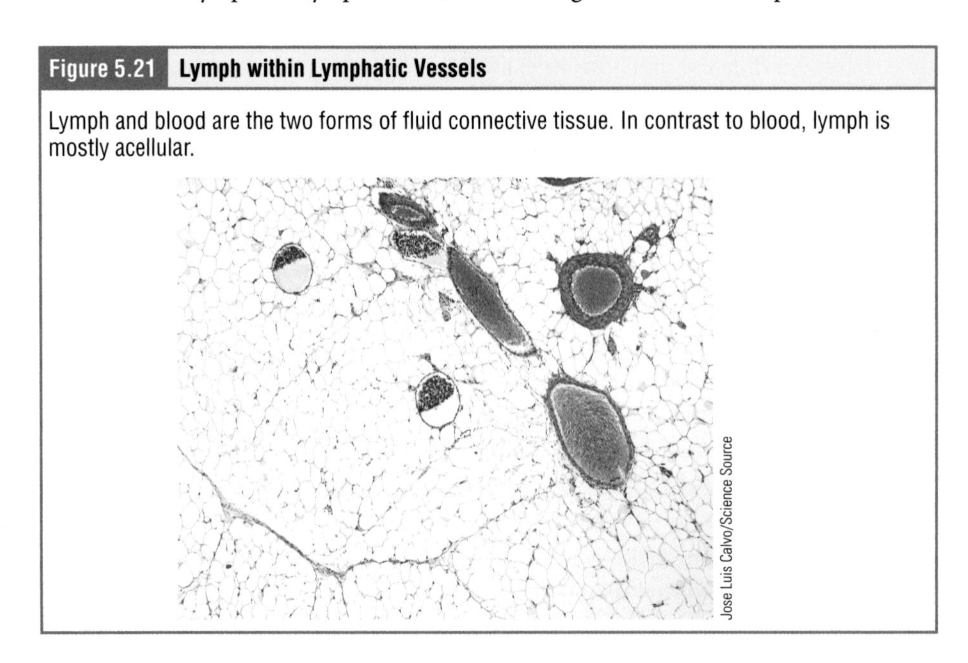

Learning Check

1. Ligaments are stiffer than adipose tissues. Which fiber is abundant in ligaments compared to adipose tissue that contribute to the stiffer characteristic?
   a. Collagen fibers
   b. Elastic fibers
   c. Reticular fibers
2. Like skin, ligaments can withstand forces in all directions. Which category would be most appropriate to classify ligaments?
   a. Dense irregular connective tissue
   b. Dense regular connective tissue
   c. Reticular tissue
   d. Areolar tissue
3. Which of the following is the hardest connective tissue?
   a. Blood
   b. Adipose
   c. Fibrocartilage
   d. Bone
4. What is a similarity between blood and lymph? Please select all that apply.
   a. They both have their own resident cells.
   b. They both distribute materials throughout the body.
   c. They both have liquid extracellular matrix called plasma.
   d. They both have elements derived from hematopoietic stem cells.

5.4 Muscle Tissue

HAPS LEARNING GOALS AND OUTCOMES

Learning Objectives: By the end of this section, you will be able to:

5.4.1 Describe the structural characteristics common to all types of muscle tissue.

5.4.2 Classify different types of muscle tissue based on structural characteristics, functions, and locations in the body.

5.4.3 Identify examples of each type of muscle tissue.

LO 5.4.1

Muscle tissue is all about movement. Muscle cells are often described as excitable, which means that they are at rest until they respond to a stimulus. Their response is to contract, meaning they shorten and generate a pulling force. The three types of muscle tissue have very different functions, but all three accomplish those functions through contraction. The three types of muscle tissue can be differentiated by location, cell shape, and size, and whether their contraction is voluntary (under conscious control; for example, a person decides to open a book to read a chapter on anatomy) or involuntary (not under conscious control, such as the contraction of your pupil in bright light). The three types of muscle tissue are compared in the "Anatomy of Muscle Tissue" feature. Unlike connective tissue, which is made of a mixture of cells, fibers, and ground substance, muscle tissue is packed with cells without room for any other components.

LO 5.4.2

Most **skeletal muscle** is attached to bones, and its contraction makes body movements and posture possible. However, some skeletal muscle is not attached to bones, such as the muscles that attach to the skin to make facial expression possible and the muscles at the top of the throat that allow us to swallow. About 40 percent of your body mass is made up of skeletal muscle. Skeletal muscle uses a tremendous amount of energy, and the increased conversion of nutrients to ATP during skeletal muscle contraction generates a lot of heat. Therefore, skeletal muscles have a role in body temperature homeostasis. When body temperature falls, involuntary contraction of skeletal muscles, shivering, is one mechanism to return body temperature back to the homeostatic range.

Anatomy of...

Muscle Tissue

LO 5.4.3

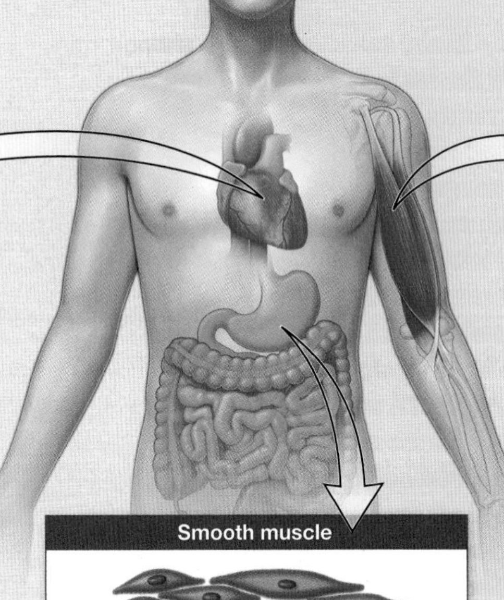

Cardiac muscle

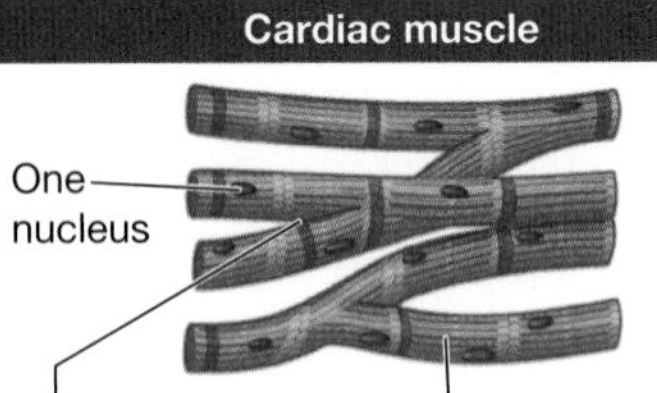

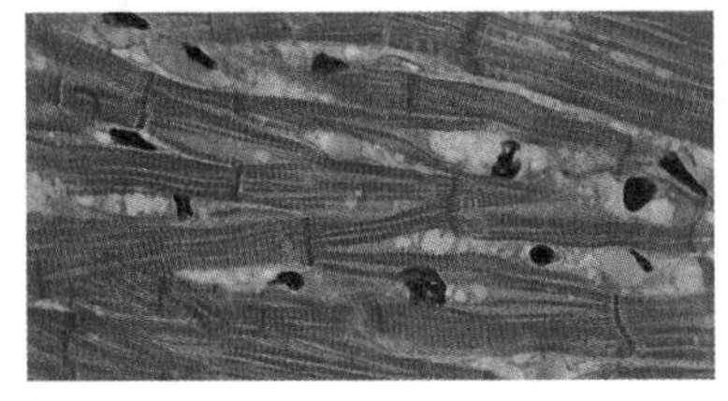

Function
- Involuntary
- Contractions squeeze the heart chambers, forcing blood out of the heart

Location
- Heart

SO. ILL. UNIV./Science Source

Skeletal muscle

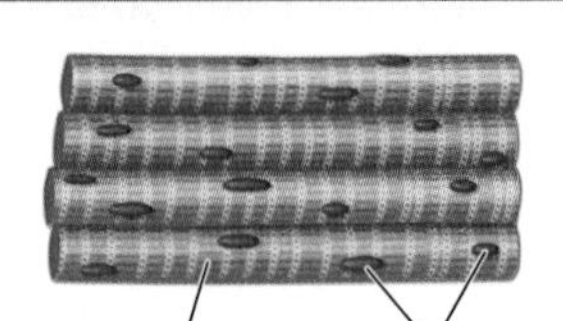

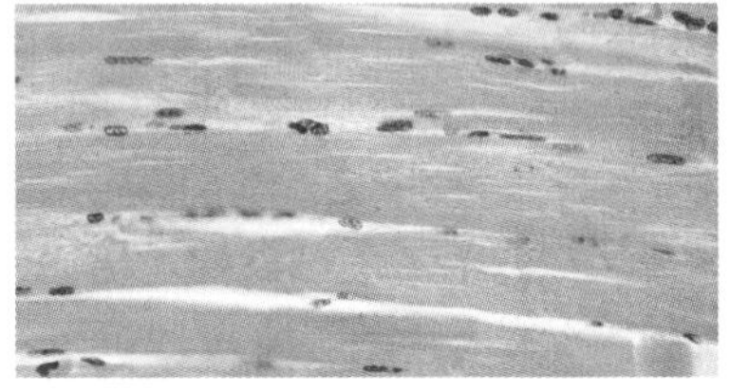

Function
- Voluntary
- Produces movement of the skeleton
- Protects organs
- Generates heat

Location
- Attached to bones
- Openings to body (mouth, anus)

Jose Luis Calvo/Shutterstock.com

Smooth muscle

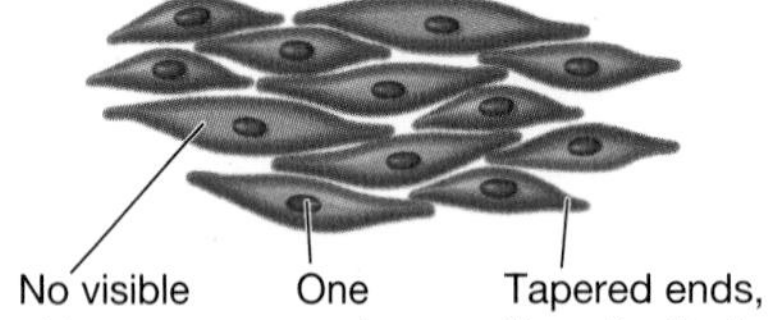

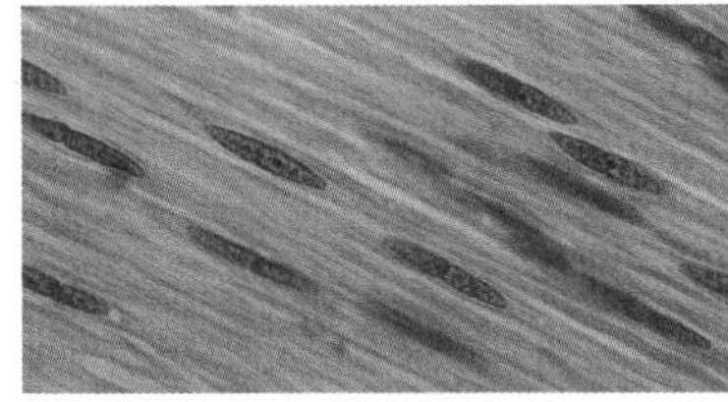

Function
- Involuntary
- Lines organs to move organ contents, such as food, fetus, semen, and uterine lining
- Controls blood flow

Location
- Walls of hollow organs and blood vessels

Ed Reschke/Stone/Getty Images

Skeletal muscle cells, which are also called *skeletal muscle fibers*, remain relatively constant in number throughout life. This fact is surprising to many people, since we can observe changes in skeletal muscle size with increases or decreases in exercise. The reason is that while individual skeletal muscle cells can grow bigger or smaller, skeletal muscle cells are incapable of mitosis due to their shape and multiple nuclei. The cells are multinucleated as a result of the fusion of the many myoblasts that fuse to form each muscle fiber. This fusion results in a very long multinucleated cell; skeletal muscle cells are very long. If you measured your bicep muscle, which runs from your elbow to your shoulder, it might be eight or nine (or more or less depending on your size) inches in length. Your bicep is made up of a bundle of many, many skeletal muscle cells, but each one runs most of the length of the muscle. That is to say, each one is several inches in length.

Skeletal muscle cells are arranged in bundles and surrounded by connective tissue. Under the light microscope, skeletal muscle cells appear striped or striated, with many nuclei that are pushed to the periphery of the cell. This **striation** is due to the consistent arrangement of the contractile proteins, actin and myosin, along with other contractile proteins that compose the cell.

Cardiac muscle forms the contractile walls of the heart. Like skeletal muscle, cells of cardiac muscle, known as **cardiomyocytes**, also appear striated under the microscope. Unlike skeletal muscle fibers, however, cardiomyocytes are smaller and have a single centrally located nucleus. Cardiomyocytes attach to one another with specialized cell junctions called *intercalated discs*. Intercalated discs have both desmosomes and gap junctions. The desmosomes keep the cells physically attached during their unrelenting, forceful contractions. The gap junctions unite the cytoplasm of each cardiomyocyte to its neighbor so that nutrients and electrical signals can pass seamlessly from one to the next. Through the intercalated discs, the attached cells form a mechanical and electrochemical union that allows the cells to synchronize their actions and contract as one. Thus, cardiac muscle effectively squeezes on the blood within the heart and pumps it throughout the body, under involuntary control. The dynamic function of cardiomyocytes is explained in more detail in Chapter 19.

Smooth muscle tissue contraction is responsible for the involuntary movements of internal organs. It forms the contractile component of the digestive, urinary, and reproductive systems as well as the airways and arteries. Each cell is football shaped, with a single nucleus and no visible striations (see the "Anatomy of Muscle Tissue" feature).

✓ Learning Check

1. How is muscle tissue different from the other three types of tissues? Please select all that apply.
   a. Muscle tissues are excitable and respond by generating a contraction.
   b. Muscle tissues are packed with cells without room for other components.
   c. Muscle tissues are rigid and provide protection to internal organs.
   d. Muscle tissues are conducting cells allowing for movement of the body.
2. Alex Morgan is a female soccer player. What muscle does she use to kick a soccer ball into the net?
   a. Skeletal muscle
   b. Cardiac muscle
   c. Smooth muscle
   d. Involuntary muscles
3. Which of the following places would you expect to find smooth muscles?
   a. The muscle that is activated when you want to raise your hand in class
   b. The muscle that allows your heart to beat throughout the day
   c. The muscle that pushes food through your gastrointestinal system
   d. The muscle that allows you to close your eyes when you sleep

5.5 Nervous Tissue

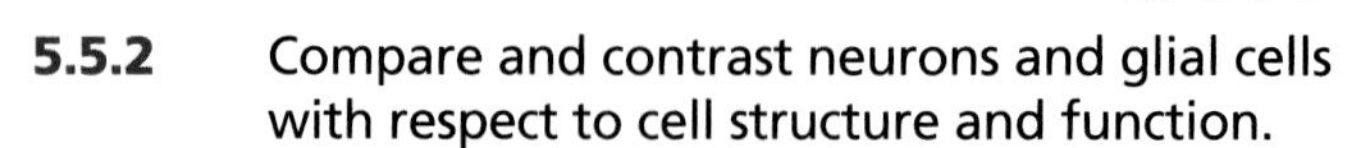

Learning Objectives: By the end of this section, you will be able to:

5.5.1* Describe and draw neurons and glial cells.

5.5.2 Compare and contrast neurons and glial cells with respect to cell structure and function.

* Objective is not a HAPS Learning Goal.

The term *nervous tissue* describes the tissues that make up the brain, spinal cord, and peripheral nerves. Like muscle tissue, nervous tissue is often described as being excitable, meaning it has a resting state and responds to stimulation. Two main classes of cells make up nervous tissue: **neurons** and **glial cells** (see the "Anatomy of Nervous Tissue" feature). Neurons are the cells that respond to stimuli. Neurons send signals to achieve communication throughout the body by way of electrochemical impulses, called **action potentials**. Action potentials result in the release of chemical signaling molecules. Glial cells were long described as playing a supportive role to neurons (the word *glia* is Greek for "glue"); anatomists have long thought that glial cells were simply

LO 5.5.1

Anatomy of... Nervous Tissue

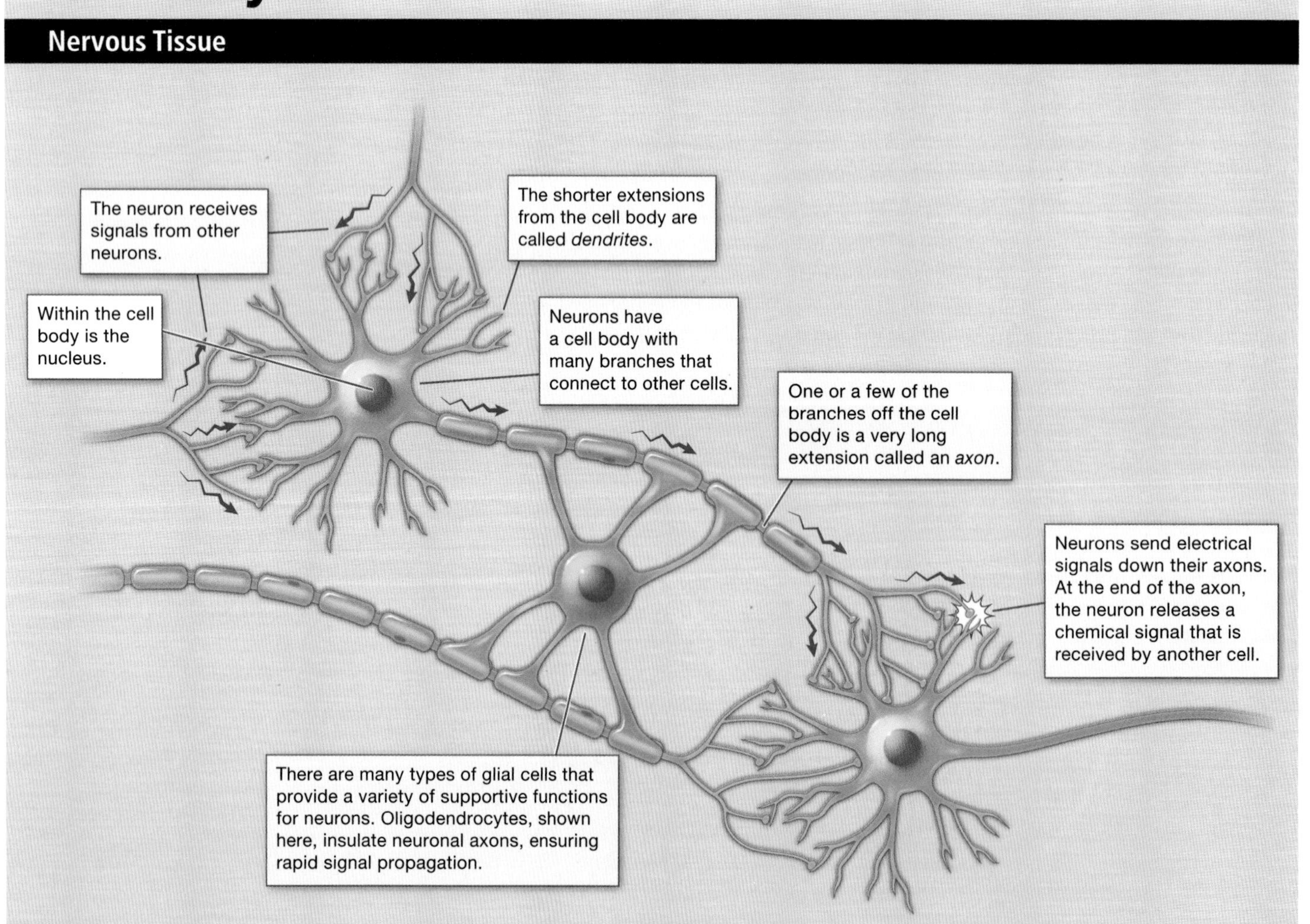

The neuron receives signals from other neurons.
The shorter extensions from the cell body are called *dendrites*.
Within the cell body is the nucleus.
Neurons have a cell body with many branches that connect to other cells.
One or a few of the branches off the cell body is a very long extension called an *axon*.
Neurons send electrical signals down their axons. At the end of the axon, the neuron releases a chemical signal that is received by another cell.
There are many types of glial cells that provide a variety of supportive functions for neurons. Oligodendrocytes, shown here, insulate neuronal axons, ensuring rapid signal propagation.

supporting and holding together the more functional neurons. Recent research has taught us otherwise. When human glial cells are injected into mouse brains, the mice gain ability in terms of problem solving and memory and have reductions in anxious behavior. While glial cells are incapable of sending signals themselves, they enable neurons to function. Glial cells outnumber neurons nine to one in the human nervous system and are absolutely essential to its functioning. You will learn more about how neurons work and the types of glial cells in Chapter 13.

LO 5.5.2

Neurons, the cells of the nervous system that enable communication from one body part to another, have a distinctive morphology, well suited to their role as conducting cells. There are three main parts to a neuron: the **cell body** includes most of the cytoplasm, the organelles, and the nucleus. There are many extensions off of the cell body. **Dendrites** are shorter branches off the cell body that receive signals from neighboring cells. The **axon** is typically much longer, extending from the neuron body and is part of the neuron where signals are sent to other cells (Figure 5.22). Axons can be wrapped in an insulating layer known as **myelin**, which is formed by glial cells. Neurons send electrical signals down their axons; when the electrical impulse reaches the end of the axon, a chemical signaling molecule is released into the **synapse**, the gap between a neuron and its target (for example, another neuron or a muscle cell). Neurons, glial cells, and their functions are discussed in much more depth in Chapter 13.

cell body (soma)

Dendrites (branches)

Figure 5.22 Neurons

Neurons, the conducting cells of nervous tissue, have their own distinctive cellular anatomy. The cell body of a neuron contains the nucleus and other organelles. The dendrites are the recipients of signals or other stimuli. The axon carries the action potential to another cell.

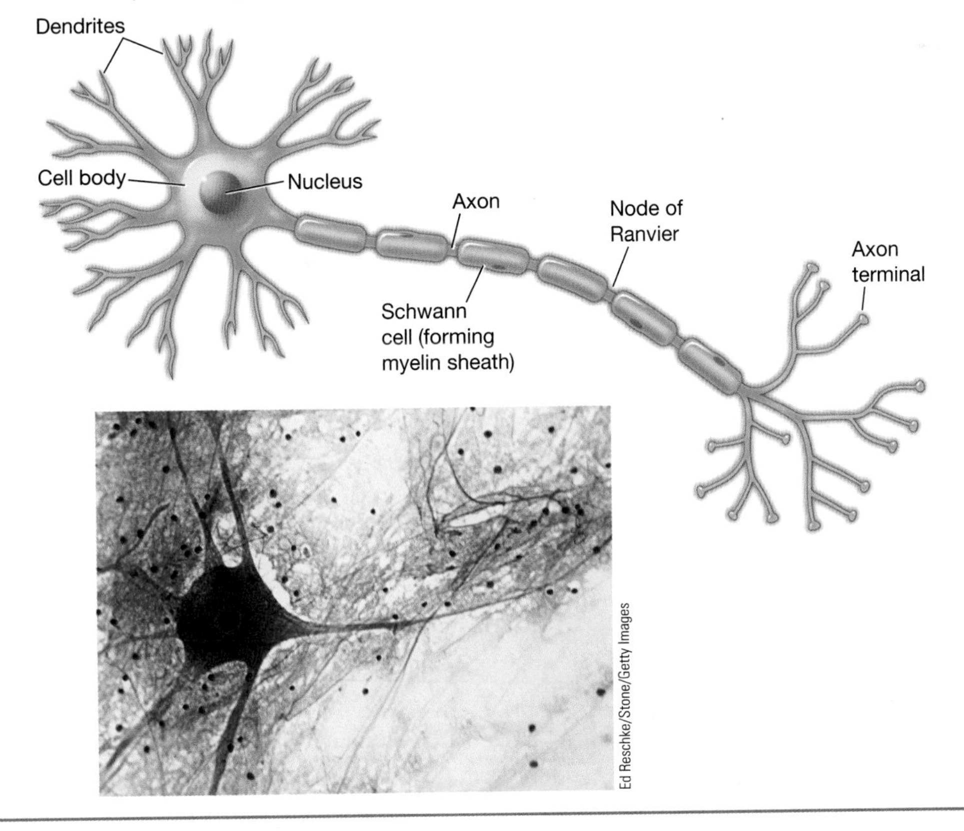

Learning Check

1. Which of the following is not made up of nervous tissue?
   a. Brain
   b. Spinal cord
   c. Peripheral nerves
   d. Vertebral column
2. What is the function of the dendrites?
   a. Receive signals from neighboring cells
   b. Send signal to neighboring cells
   c. Insulate axons of neuron cells
   d. Support neurons and enable them to function
3. You are a neurologist and realize that your patient cannot produce myelin. Which of the following cells could be dysfunctional, leading to decreased myelin production?
   a. Glial cells
   b. Neuron cells

5.6 Membranes

HAPS LEARNING GOALS AND OUTCOMES

Learning Objectives: By the end of this section, you will be able to:

5.6.1 Describe the structure and function of mucous, serous, cutaneous, and synovial membranes.

5.6.2 Describe locations in the body where each type of membrane can be found.

5.6a Tissue Membranes

LO 5.6.1

We most often use the term *membrane* to mean a cell membrane, the phospholipid bilayer that contains a cell. Occasionally, though, we use the word *membrane* to refer to a **tissue membrane**, a thin sheet of cells that covers the outside of the body (for example, skin), the organs (for example, pericardium), internal passageways that lead to the exterior of the body (for example, abdominal mesenteries), or the lining of the moveable joint cavities. There are four types of tissue membranes: mucous, serous, cutaneous, and synovial (**Figure 5.23**).

LO 5.6.2

Mucous membranes are composites of connective and epithelial tissues. These epithelial membranes line the body cavities and hollow passageways that open to the external environment, and include the digestive, respiratory, excretory, and reproductive tracts. The apical surface of mucous membranes is kept moist by exocrine glands, which secrete a covering of mucus.

Serous membranes are epithelial membranes that line the cavities of the body that do not open to the outside. These membranes fold in on themselves and line both the inside wall of the cavity and the surface of the organs located within those cavities. Serous fluid secreted by the cells of the membrane provides lubrication and reduces abrasion and friction between organs. Serous membranes are typically named according to their location (for example, the pericardium lines the heart).

cutaneous membrane (skin)

The skin is an epithelial membrane also called the **cutaneous membrane**. It is a stratified squamous epithelial membrane resting on top of connective tissue. The apical surface of this membrane is exposed to the external environment and is covered with dead, keratinized cells that help protect the body from desiccation and pathogens.

A **synovial membrane** is a type of connective tissue membrane that lines the cavity of a freely movable joint. For example, synovial membranes form a capsule around most of our body's joints, including the shoulder, elbow, and knee. Cells within the synovial membrane release a fluid into the joint called **synovial fluid**, which lubricates the joint and nourishes the cells of the cartilage within joints (because cartilage is avascular and so the cells need nutrients supplied by the fluid).

Figure 5.23 Tissue Membranes

The four categories of tissue membranes in the body are mucous membranes, which line cavities that open to the outside; serous membranes, which are double-layered membranes that surround some organs such as the heart; the cutaneous membrane (the skin); and synovial membranes, which line joints.

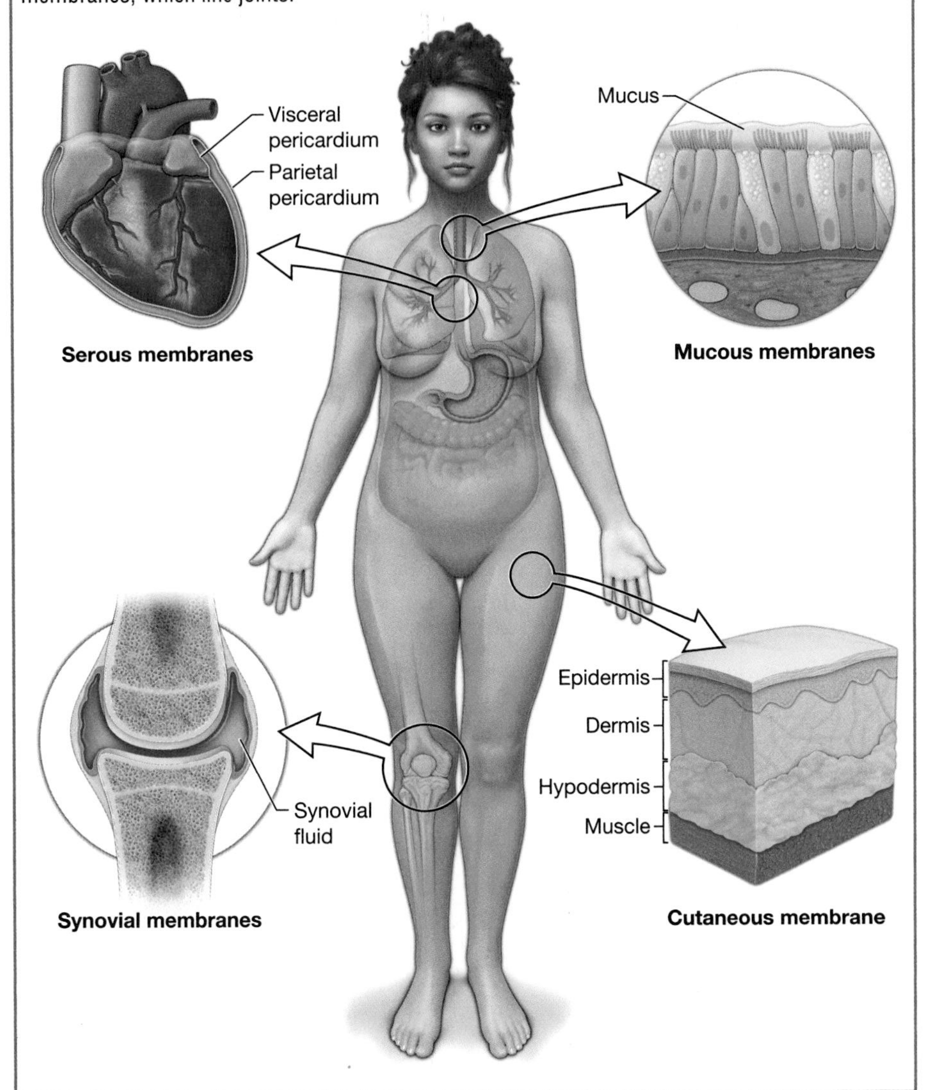

Learning Check

1. The membrane around your lung is known as the pleural membrane. This membrane lines the inside of the wall of the thorax (the wall of the chest) and the surface of the lung. Which of the following membranes is the pleural membrane an example of?
   a. Mucous membrane
   b. Serous membrane
   c. Cutaneous membrane
   d. Synovial membrane
2. Where would you find a synovial membrane?
   a. Around the heart
   b. Inside the throat
   c. Around the knee
   d. On your skin

5.7 Tissue Growth and Healing

Learning Objectives: By the end of this section, you will be able to:

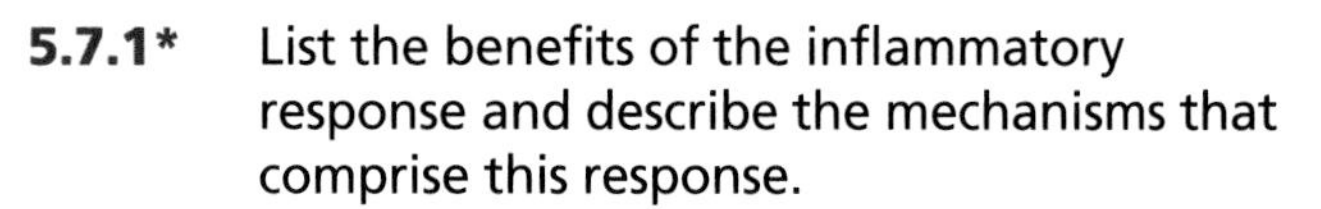

5.7.1* List the benefits of the inflammatory response and describe the mechanisms that comprise this response.

5.7.2 Describe tissue repair following an injury.

5.7.3* Discuss the progressive impact of aging on tissues.

5.7.4* Describe cancerous mutations' effect on tissues.

* Objective is not a HAPS Learning Goal.

Tissues of all types are vulnerable to injury and, inevitably, aging. Understanding both how tissues respond to damage and the impact of aging can help scientists and clinicians search for treatments to various diseases.

5.7a Tissue Injury and Repair

LO 5.7.1

Inflammation is the standard, initial response of the body to injury. Whether the damage is caused by an invading pathogen, a foreign chemical, or physical trauma, all injuries lead to the same sequence of physiological events. Inflammation is an all-purpose tool that limits the extent of injury and initiates repair and regeneration of damaged tissue. **Acute**, or short-term, inflammation resolves over time by the healing of tissue. However, inflammation that goes on too long, becoming **chronic**, can have devastating effects on homeostasis. Arthritis and tuberculosis are two examples of chronic inflammation. The suffix "-itis" denotes inflammation of a specific organ or type; for example, *tendinitis* is the inflammation of a tendon, and *pericarditis* refers to the inflammation of the pericardium, the membrane surrounding the heart.

Upon tissue injury, damaged cells release pro inflammatory chemical signals that begin the process of inflammation (**Figure 5.24A**). Some of these signals cause **vasodilation**, the widening of the blood vessels. Increased blood flow to the area of inflammation allows more leukocytes (white blood cells), the cells of the immune system, to reach the area (**Figure 5.24B**). Because of this vasodilation and resulting increase in blood flow, the area becomes warmer to the touch, and may be noticeably redder, than nearby uninflamed tissues.

The chemical signals released by damaged cells and mast cells—immune cells commonly found in connective tissue proper—recruit white blood cells from the blood toward the site of inflammation. These chemical signals, including the chemical **histamine**, make the lining of the local blood vessel "leaky" to enable white blood cells to move from the blood into the interstitial tissue spaces. Immune cells aren't the only thing from the blood that escapes from the leaky blood vessels; blood plasma also enters the tissues. The excess liquid in tissue causes swelling, more properly called **edema**. Local pain receptors are activated both by signaling molecules, including prostaglandins, and the physical squeeze of the edema. Pain is annoying to the person experiencing it, but it is a useful response that has evolved for protection. Pain helps to prevent further injury by keeping us from using (i.e., walking on) the injured tissue. Pain also keeps us from touching the affected area, preventing the possible introduction of bacteria.

edema (swelling)

LO 5.7.2

These initial responses work to contain the injury and set the stage for healing. The tissue repair phase starts with removal of toxins and waste products (**Figure 5.25**). Waste removal is accomplished by white blood cells called *macrophages*; these cells often live in connective tissue, but also arrive at the injury during the inflammatory stage. If blood vessels are damaged during the injury, then a clotting response works to stop blood loss and prevent infection. The clot will become a scab and then heal. If

Figure 5.24 Inflammation

(A) The inflammatory process begins when damaged cells and/or resident leukocytes of the tissue release signaling molecules, including histamine, to signal danger. (B) These signaling molecules have many effects, including local vasodilation and leakiness of the blood vessel. In response to the signals for help, leukocytes leave the bloodstream to enter the tissue and fight off pathogens.

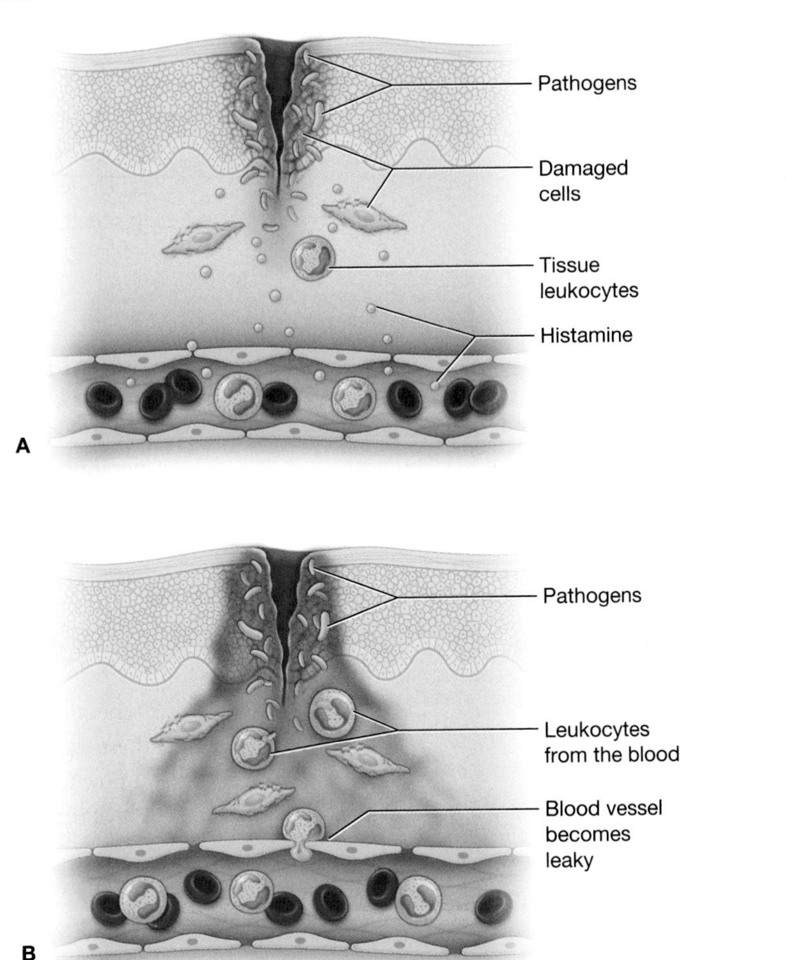

bones are broken or fractured, these tissues must heal as well. These specific types of healing are covered in Chapters 18 and 7, respectively. As healing progresses, fibroblasts from the surrounding connective tissues replace the collagen and extracellular material lost by the injury. **Angiogenesis**, the growth of new blood vessels, results in vascularization of the new tissue. Fibroblasts quickly generate collagen to bridge wounds. If the skin is broken, epithelial cells replicate themselves to cover the break. This healing process occurs as quickly as possible, and the newly generated tissue often has a different composition and texture than the tissue surrounding it.

Learning Connection

Explain a Process

Can you explain the process of inflammation? If you write each step in the process on a sticky note and mix them up, can you put them back in order?

5.7b Tissue and Aging

LO 5.7.3

According to poet Ralph Waldo Emerson, "The surest poison is time." In fact, biology confirms that many body structures change with age, and as a consequence, most functions of the body decline over time. Of course, genetic makeup and lifestyles play a huge role in how aging proceeds. Two major changes at the cellular and tissue level contribute widely to the diverse signs of aging. One change is the rate of mitosis (the

Figure 5.25 Tissue Healing

During wound repair, collagen fibers are laid down randomly by fibroblasts that move in to repair the area.

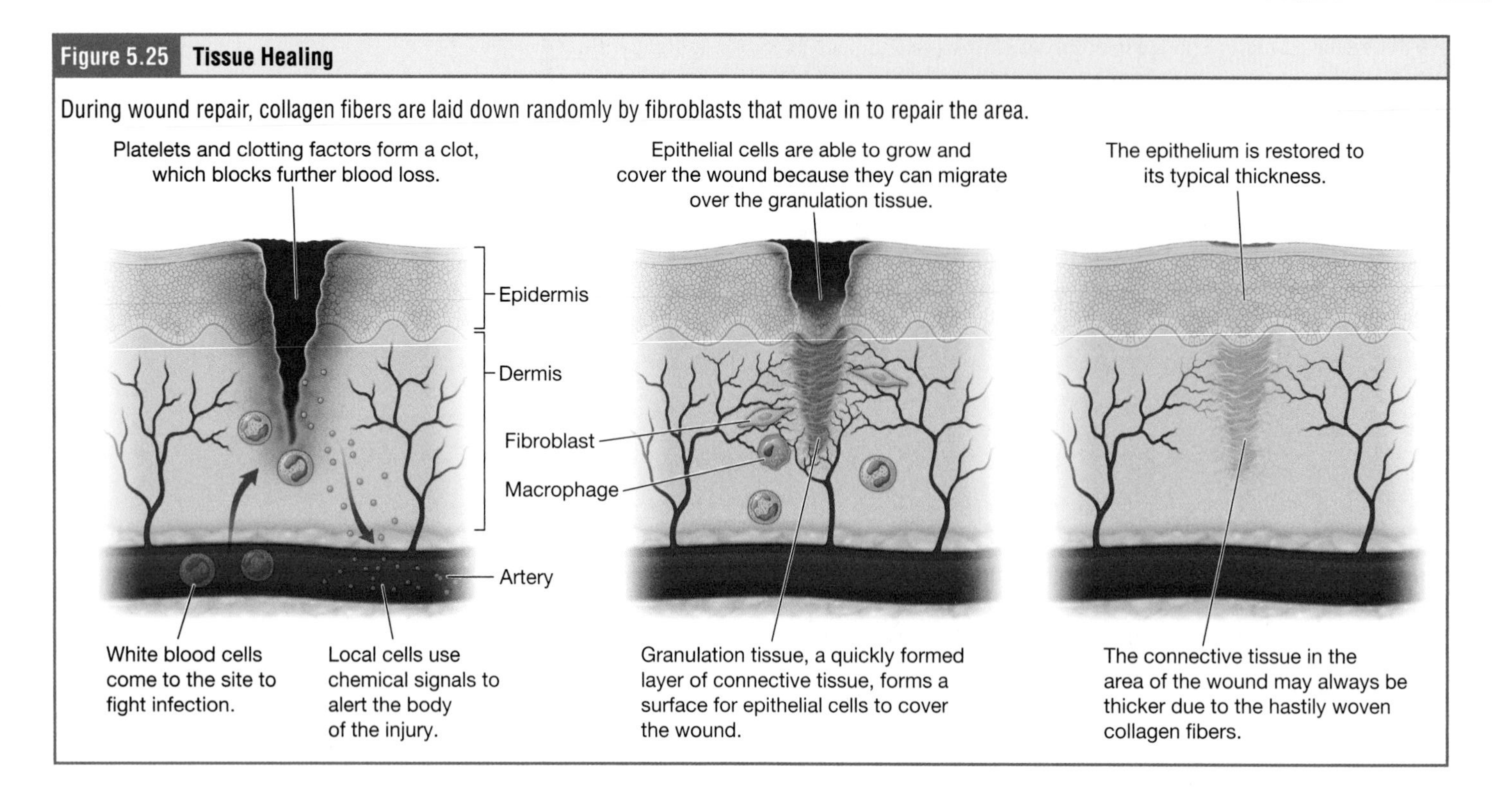

Platelets and clotting factors form a clot, which blocks further blood loss.
Epithelial cells are able to grow and cover the wound because they can migrate over the granulation tissue.
The epithelium is restored to its typical thickness.
Epidermis
Dermis
Fibroblast
Macrophage
Artery
White blood cells come to the site to fight infection.
Local cells use chemical signals to alert the body of the injury.
Granulation tissue, a quickly formed layer of connective tissue, forms a surface for epithelial cells to cover the wound.
The connective tissue in the area of the wound may always be thicker due to the hastily woven collagen fibers.

replication of cells that we learned about in Chapter 4). All cells, including stem cells, lose the ability to replicate as the body ages. The result is that tissue healing, which we learned about in the preceding section, takes much longer. Fewer fibroblasts exist to spin collagen and fewer epithelial cells exist to replicate and cover wounds. If bones are fractured, it takes longer to repair the architecture so that the bones become durable once more. With healing taking longer, the body is affected more by each injury. Bacterial infection becomes more likely, and the muscles are more substantially affected when a structure in the body is forced to remain out of use during healing.

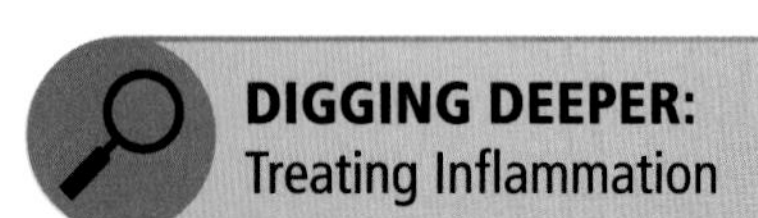

DIGGING DEEPER: Treating Inflammation

Inflammation is an important process for healing, but it is uncomfortable. If it goes on too long, or becomes chronic, it can compromise homeostasis. We recognize inflammation is occurring when four signs are present: heat, pain, redness, and swelling. Of course, redness may be difficult to assess depending on the skin of the patient, so the presence of the other three is sufficient. The pain, as described in the previous text, is due to the effects of prostaglandins on the neurons that are dedicated to communicating pain to the brain. Prostaglandins were first discovered in human semen in 1935 (their name comes from the hypothesis that they are produced by the prostate gland; however, we now know that they are made by cells all over the body), and the 1982 Nobel Prize in Physiology or Medicine went to the scientists who discovered more about these important signaling molecules. The pain of inflammation stems both from tissue damage and the physical constraints of swelling. To help alleviate the pain, a few tactics can be used. Nonsteroidal anti-inflammatory drugs (NSAIDs) like acetaminophen and ibuprofen are drugs that prevent the release of prostaglandins by inhibiting enzymes that produce them. To reduce the swelling and the pain associated with it, the most effective mechanism to induce vasoconstriction—a decrease in blood vessel diameter that results in a reduction in blood flow—is to cool the inflamed area using ice. The body's natural reaction to cold is vasoconstriction, which you will learn more about in Chapter 20. Another mechanism that can be helpful is to compress or elevate the area to assist in fluid drainage through the lymphatic vessels.

Finally, it is very important to not cause further damage to the area by introducing pathogens or physically straining the area. Even though it can be challenging, the patient experiencing inflammation needs rest.

The mnemonic that many people use to remember these treatments is RICE: rest, ice, compression, elevation.

The second major change in the aging body is a decrease in elastic fibers. By the time a human reaches 50, their body contains just 70 percent of the elastic fibers that they had at age 20. By the time they reach 80, they have less than 50 percent. The decrease in elastic fibers has profound effects all over the body. The skin and other tissues become thinner and are less likely to snap back to their original shape after the movement, contributing to wrinkles. Perhaps the most elastic place in the body is our blood vessels, where elasticity allows for the fluctuations in blood pressure that occur with heart contraction and relaxation. The decrease in elasticity leads to higher and higher blood pressure as we age. The lens, a structure in the eye, becomes less flexible, leading to the need for reading glasses. Tendons and ligaments lose the small amount of elasticity they had, leading to joint stiffness and inflexibility. As it becomes harder to move the body, other issues compound, the muscles lose mass, and cartilage at the joints becomes thinner.

LO 5.7.4

There are other, more specific changes with age that we will explore in each chapter. However, it is important to keep in mind that the progressive impact of aging on the body varies considerably among individuals. Studies indicate that exercise and healthy lifestyle choices can slow down the deterioration of the tissues that comes with old age. Choices that we make now affect our future health, flexibility, and cellular function.

5.7c Tissues and Cancer

Cancer is a generic term for many diseases in which cells escape regulatory signals. For scientists who study cancer, it is notable that cancers are not all the same; they develop and progress in ways that are tissue specific and unique. These understandings profoundly impact the ways in which we can prevent and treat cancer. Hallmarks that are true of most cancers are: uncontrolled growth, invasion into adjacent tissues, and, in severe cases, colonization of other organs. The presence of a tumor in and of itself is not always harmful. But tumors may impact the healthy tissues of the body by

Cultural Connection

The Lastin' Elastin

All humans "show their age" at different rates. The age we might guess when we look at someone's face is a combination of factors about their skin and connective tissues, such as the amount of elastin and the quality of the ECM that supports their skin. We have noted that the amount of elastic fibers is a key component. Elastic and collagen fibers are produced by fibroblasts, so the number of fibroblasts that are present in connective tissue later in an individual's lifespan is a key factor in the speed at which age-related changes—such as wrinkles—develop. Elastic fibers are also very susceptible to UV radiation.

In this photograph we can see the effects of UV radiation on elastin. This photo is of a truck driver. The left side of this individual's face was exposed to the sun through their open truck window, while the right side was protected by the truck's roof. Individuals with lighter skin shades are likely to develop age-related changes more quickly.

People of color have a greater deposition of melanin—a UV-blocking pigment—in their skin. One result of this is that fewer UV rays are able to penetrate down to their elastic fibers. Interestingly, it has recently been observed that people of color have, on average, larger and more numerous fibroblasts. If more fibroblasts are present later throughout an individual's lifespan, stronger and more youthful skin is one result. Another result is the speed of collagen production during wound healing. Because of this abundance of fibroblasts, more collagen can be created to repair a wound, leading to a type of scarring called *keloids*.

monopolizing nutrients and blood supply or creating physical constraints that compress or damage nearby structures.

mutation ("gene mutation" = "gene variant")

LO 5.7.4

With 3 billion base pairs to copy each time DNA and the cell replicates, mistakes are bound to happen. An error made while replicating DNA is called a **mutation**. Mutations are relatively common and most have no impact on the cell and body. However, some mutations, especially when made in genes that control the cell cycle, can have profound affects and generate abnormal cells. While mutations happen all the time, they occur more frequently during exposure to environmental agents such as UV radiation from the sun, chemical agents (such as many of the chemicals present in cigarette smoke), or infectious agents (such as some viruses).

malignant (malignant tumor = "metastatic" tumor)

Should a mutation or several mutations impact the cell cycle, a cell may begin to replicate uncontrollably. A mass of cells—a tumor—will result, changing the architecture of the tissue around it. In the most significant instances of tumor formation, groups of cells may break off from the tumor and move around the body. These cancers are known as **malignant**. Tumors that neither metastasize nor cause disease are referred to as **benign**. The specific names of cancers reflect the tissue of origin. Cancers derived from epithelial cells are referred to as *carcinomas*. Cancers in myeloid tissue or blood cells form *myelomas*. *Leukemias* are cancers of white blood cells, while *sarcomas* derive from connective tissue. Cells in tumors differ both in structure and function. Some cells, called *cancer stem cells*, appear to be a subtype of cell responsible for uncontrolled growth. Recent research shows that, contrary to what was previously assumed, tumors are not disorganized masses of cells, but have their own structures.

Cancer treatments vary depending on the disease's type and stage. Traditional approaches, including surgery, radiation, chemotherapy, and hormonal therapy, aim to remove or kill rapidly dividing cancer cells, but these strategies have their limitations. Depending on a tumor's location, for example, cancer surgeons may be unable to remove it. Radiation and chemotherapy are difficult, and it is often impossible to target only the cancer cells. The treatments inevitably destroy healthy tissue as well. To address this, researchers are working on pharmaceuticals that can target specific proteins implicated in cancer-associated molecular pathways.

Learning Check

1. Which of the following makes the lining of the blood vessels "leaky"?
   a. Histamine
   b. Edema
   c. Angiogenesis
   d. Vasodilation
2. You fall and scrape your elbow on the road. You are bleeding, but you do not have any broken bones. Over time, which of the following would you expect to happen as your body tries to heal itself? Please select all that apply.
   a. Angiogenesis to revascularize the new tissue.
   b. Epithelial cells replicate themselves.
   c. Fibroblasts generate collagen to bridge the wound.
   d. Osteoblasts will form new bone cells.
3. Which of the following explains why it takes someone who is 65 years old longer to recover from a wound than someone who is 18? Please select all that apply.
   a. They have fewer fibroblasts to form collagen.
   b. Their cells are less capable of replication.
   c. Their cells no longer undergo mitosis.
   d. Their tissues are less stable due to increased elastic fibers.

Apply to Pathogenesis

Physiological Events in Cancer

As a tumor grows and produces more cells, several physiological constraints will limit its growth. Successful tumors, tumors that are able to continue to grow, are able to trick the tissues into supporting the growth of the tumor.

1. One medicine that can be used to treat cancer is a drug that inhibits VEGF, a signaling molecule that the tumor secretes in order to promote angiogenesis. Why does angiogenesis help tumor formation?
   A) Angiogenesis promotes the breakdown of collagen in the nearby tissue, making more space for the tumor.
   B) Angiogenesis grows new blood vessels, bringing oxygen and nutrients to the growing tumor.
   C) More different types of cells will reach the tumor during angiogenesis, allowing the tumor to become more cellularly diverse.

Growing tumors change the extracellular matrix of the tissue substantially. One change is that they promote collagen fiber building all around the tumor.

2. Many types of immune cells are able to recognize and eliminate cancer cells. Examine Panel B in the figure. What impact might this increase in collagen density have on the local environment?
   A) Immune cells are physically prevented from reaching (and eliminating) the cancer cells.
   B) The collagen provides guidance for the immune cells to reach the tumor.
   C) The collagen helps to replace cells killed by the immune system.
3. An increase in collagen density in breast tissue appears to be a critical factor for the development of this type of cancer. Which of the following roles of collagen might contribute to breast cancer development?
   A) Collagen provides a structural framework for cells to grow and migrate.
   B) Collagen provides tensile strength, allowing tissues to be more flexible and resilient.
   C) Collagen helps tie cells to one another so that they can function together in a community.

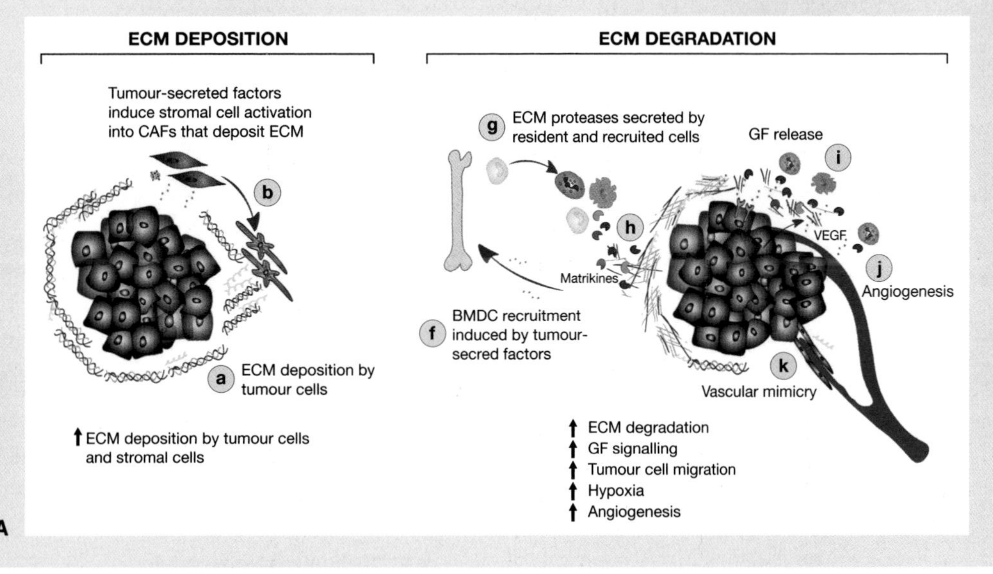

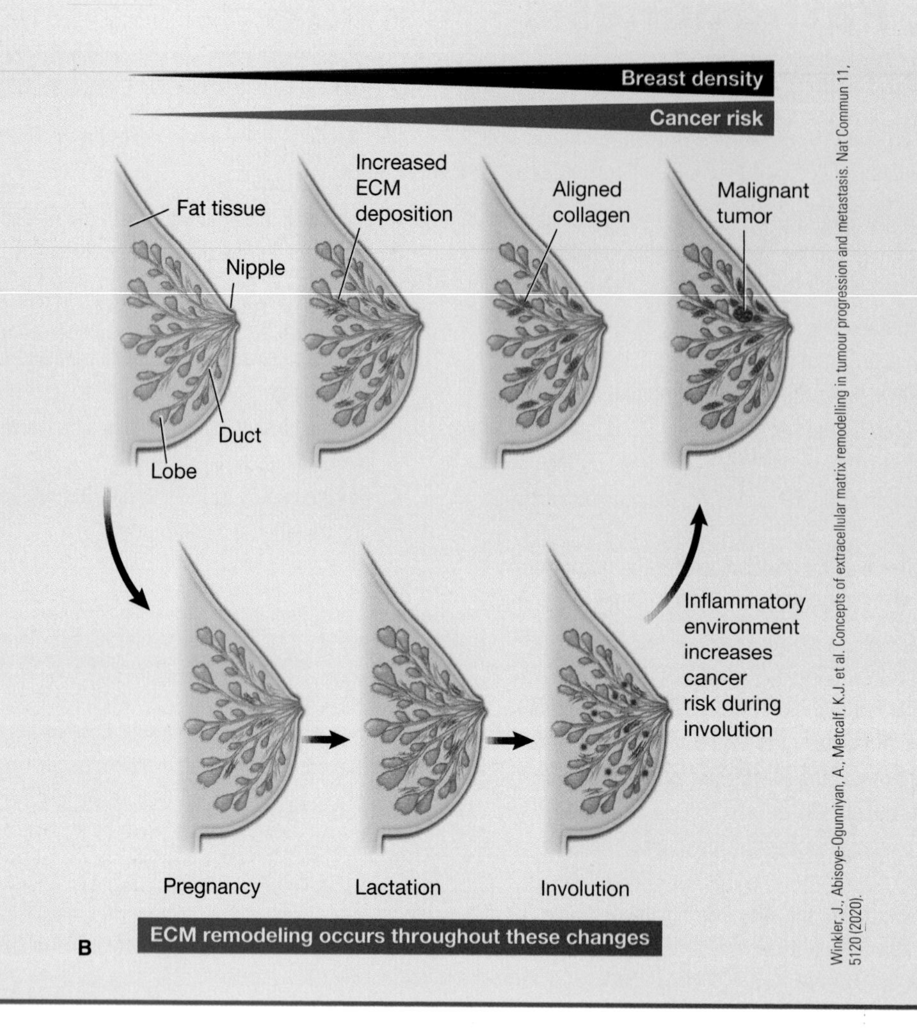

Winkler, J., Abisoye-Ogunniyan, A., Metcalf, K.J. et al. Concepts of extracellular matrix remodelling in tumour progression and metastasis. Nat Commun 11, 5120 (2020).

Chapter Review

5.1 Types and Components of Tissues

The learning objectives for this section are:

5.1.1 Define the term histology.

5.1.2 List the four major tissue types.

5.1.3 Compare and contrast the general features of the four major tissue types.

5.1.4* Describe the components and functions of extracellular matrix.

5.1.5 Compare and contrast the types of intercellular connections (cell junctions) with respect to structure and function.

* Objective is not a HAPS Learning Goal.

5.1 Questions

1. Define the term *histology*.
   a. A group of cells functioning together in the body.
   b. The study of tissue appearance and function.
   c. The study of changes that occur over the lifespan.

2. A soccer player was running when they felt a sharp pain in their knee and fell to the ground. After some tests, the doctors said they had torn a ligament in their knee called the ACL. What type of tissue would the ACL be considered as?
   a. Epithelial
   b. Muscle
   c. Connective
   d. Nervous

3. What is a unique characteristic of muscle tissue?
   a. It provides movement through contraction.
   b. It provides communication throughout the body.
   c. It binds cells and organs together.
   d. It lines internal cavities and passageways.

4. The extracellular matrix of each tissue has different proportions of collagen and proteoglycans. Which molecule is most abundant in a tissue that has a semisolid gel-like consistency?
   a. Collagen
   b. Proteoglycan

5. Your heart is composed of many muscle cells. When stimulated by an electrical signal, all cardiac muscle cells contract together in a coordinated manner. Which intercellular junction would you expect to find in the heart that allows this process to occur?
   a. Desmosomes
   b. Hemidesmosomes
   c. Tight junctions
   d. Gap junctions

5.1 Summary

- The four major types of tissues are: (1) epithelial tissue, which lines internal cavities and exterior surfaces of the body; (2) connective tissue, which binds cells and organs of the body together; (3) muscle tissue, which provides movement through contraction; and (4) nervous tissue, which communicates short and long distances throughout the body.
- Extracellular matrix is composed of: (1) collagen, which is tough and protective but flexible and allows for movement, and (2) proteoglycans, which are negatively-charged molecules that take on a gel-like consistency.
- There are three types of cellular connections: (1) tight junctions, which tightly bind cells together to prevent seepage between them; (2) anchoring junctions, which tightly bind cells together to form structural stability; and (3) gap junctions, which are intercellular passageways that allow large groups of cells to function in a coordinated manner.

5.2 Epithelial Tissue

The learning objectives for this section are:

5.2.1 Describe the structural characteristics common to all types of epithelia.

5.2.2 Classify different types of epithelial tissues based on structural characteristics.

5.2.3 Describe the microscopic anatomy, location, and function of each epithelial tissue type.

5.2.4 Identify examples of each type of epithelial tissue.

5.2.5 Compare and contrast exocrine and endocrine glands, structurally and functionally.

5.2.6 Compare and contrast the different kinds of exocrine glands based on structure, method of secretion, and locations in the body.

5.2 Questions

1. Which of the following are characteristics of epithelial cells? Please select all that apply.
   a. Cells contact each other in almost every direction.
   b. Cells slough off the apical surface.
   c. Cells that are capable of long-distance communication.

2. You are looking in a microscope and see one layer of rectangular cells. What type of tissue are you looking at?
   a. Simple squamous epithelium
   b. Pseudostratified epithelium
   c. Simple columnar epithelium
   d. Stratified cuboidal epithelium

3. The lacrimal gland is a small structure near each of your eyes that is responsible for producing tears. The lacrimal ducts transport tears from the gland to the edge of your eye so you can cry. What type of epithelium would you expect to find in the lacrimal duct?
   a. Simple squamous epithelium
   b. Simple cuboidal epithelium
   c. Simple columnar epithelium
   d. Pseudostratified columnar epithelium

4. The bladder must stretch to accommodate the high volume of urine. What epithelial tissue would you expect to find in the bladder that allows it to stretch?
   a. Stratified columnar epithelium
   b. Simple cuboidal epithelium
   c. Pseudostratified columnar epithelium
   d. Transitional epithelium

5. The anterior pituitary gland is an endocrine gland. How would you describe the structure and function of this gland?
   a. Ductless, secretions released directly into surrounding tissues
   b. Ductless, secretions released through epithelium into extracellular matrix
   c. Duct, secretions released directly into surrounding tissues
   d. Duct, secretions released through epithelium into extracellular matrix

6. Describe merocrine secretion.
   a. Secretions are enclosed in vesicles that move to the apical surface of the cell.
   b. Secretions migrate to apical portion so that cell membrane and cytoplasm are also lost.
   c. Secretions are released by rupturing and destroying the entire cell.

5.2 Summary

- Epithelial tissues cover all surfaces of the body, including the outside and the inside, to protect the body from physical, chemical, and biological wear and tear.
- Epithelial tissues are classified based on the number of rows (simple = 1 row, stratified = multiple rows) and the shape (squamous = flat and thin, cuboidal = boxy, columnar = rectangular). These classifications are unique to their function and therefore are found in specific areas of the body.
- Endocrine glands release hormones into the interstitial fluid, which travels to specific targets that have the receptors for these hormones. Exocrine glands release secretions through a duct that leads to the epithelial surface.

5.3 Connective Tissue

The learning objectives for this section are:

5.3.1 Describe the structural characteristics common to all types of connective tissue.

5.3.2 Classify different types of connective tissues based on their structural characteristics, functions, and locations in the body.

5.3.3 Identify examples of each type of connective tissue.

5.3 Questions

1. Which of the following is NOT a component of connective tissues?
   a. Cells
   b. Ground substance
   c. Protein fibers
   d. Cilia

2. You are looking into a microscope and see circular cells that have a brown tint to them. The slide mostly consists of these cells and very little extracellular matrix. What is a unique feature about this tissue?
   a. Releases heat instead of ATP
   b. Can withstand forces in all directions
   c. Offers structure and protection to the body
   d. Delivers molecules into the blood

3. Your hip has two bones that are connected at the pubic symphysis joint. This connection is very tough, durable, and resists impact. What type of cartilage would you expect to find in this joint?
   a. Hyaline cartilage
   b. Fibrocartilage
   c. Elastic cartilage

5.3 Summary

- Connective tissues have three components: cells, ground substance, and protein fibers. Ground substance is a major component of connective tissues. There are three types of protein fibers: collagen fibers, elastic fibers, and reticular fibers. The proportions of these fibers determine the physical properties of the different types of connective tissue.
- There are three types of connective tissue: connective tissue proper (loose connective tissue and dense connective tissue), supportive connective tissue (cartilage and bones), and fluid connective tissue (blood and lymph).
- The functions of connective tissues include binding structures together, protection of organs, defense of the body against infection by microorganisms, transport of a variety of substances throughout the body, energy storage, and insulation against heat loss.

5.4 Muscle Tissue

The learning objectives for this section are:

5.4.1 Describe the structural characteristics common to all types of muscle tissue.

5.4.2 Classify different types of muscle tissue based on structural characteristics, functions, and locations in the body.

5.4.3 Identify examples of each type of muscle tissue.

5.4 Questions

1. Which of the following characteristics are the same among all types of muscle tissues?
   a. They are all the same size.
   b. They are all voluntary.
   c. There are all involuntary.
   d. They are all excitable.

2. You look through a microscope at a muscle cell. You notice the cell has striations and a single nucleus that is centrally located. From which of the following groups of muscles was this muscle isolated?
   a. Muscles that allow you to kick a ball
   b. Muscles that keeps your heart beating
   c. Muscles that push food through your intestines
   d. Muscle that allows you to control when you urinate

3. In 2016, Lebron James clinched the victory of the NBA Finals when he blocked a shot at the end of Game 7. Which of the following muscles allowed Lebron to jump high enough to block the shot?
   a. Cardiac muscles
   b. Skeletal muscles
   c. Smooth muscles

5.4 Summary

- Muscle tissues are excitable. When excited, the tissue contracts and creates movement.
- There are three types of muscle tissues: (1) skeletal muscle, which controls voluntary movement; (2) cardiac muscle, which involuntarily causes your heart to beat; and (3) smooth muscle, which involuntarily controls movement in your gastrointestinal, urinary, and reproductive systems, as well as in the airways and arteries.
- All three types of muscle tissues consist mainly of densely packed muscle cells (fibers), with very little space between the cells for other components.

5.5 Nervous Tissue

The learning objectives for this section are:

5.5.1* Describe and draw neurons and glial cells.

5.5.2 Compare and contrast neurons and glial cells with respect to cell structure and function.

* Objective is not a HAPS Learning Goal.

5.5 Questions

Mini Case 1: Multiple Sclerosis

Multiple sclerosis (MS) is a degenerative disorder in which your body attacks its own myelin. This decreases the function of nervous tissue.

1. Which of the following cells is wrapped in myelin?

a. Neuron
b. Glial cells

2. What is the role of myelin?

a. Myelin receives signals from neighboring cells.
b. Myelin sends signals to neighboring cells.
c. Myelin insulates the axon of the cell.
d. Myelin plays a supporting role to neurons.

5.5 Summary

- Nervous tissue is made up of neurons and glial cells.
- Nervous tissue makes up the brain, spinal cord, and peripheral nerves.
- Neurons are excitable cells; therefore, they are capable of responding to stimuli.
- Neurons are specialized for communication through action potentials, and glial cells provide structural and nutritional support for neurons.

5.6 Membranes

The learning objectives for this section are:

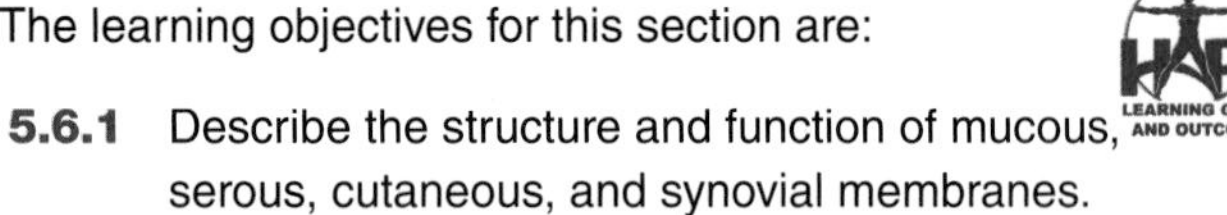

5.6.1 Describe the structure and function of mucous, serous, cutaneous, and synovial membranes.

5.6.2 Describe locations in the body where each type of membrane can be found.

5.6 Questions

1. What is unique about a serous membrane?

a. It lines freely moving joints.
b. It lines passageways that are open to the external environment.
c. It is stratified squamous epithelial cells.
d. It lines the inside wall of the cavity and surface of the organ.

2. You are performing surgery on the knee joint, which is a freely moveable joint. Which of the following membranes will you have to cut through to get inside the knee joint? Please select all that apply.

a. Mucous membrane
b. Serous membrane
c. Cutaneous membrane
d. Synovial membrane

5.6 Summary

- There are four types of tissue membranes in the body: mucous, serous, cutaneous, and synovial.
- Mucous membranes secrete mucus and line body cavities and hollow passageways that open to the external environment.
- Serous membranes secrete watery serous fluid and line body cavities that do not open to the external environment.
- The cutaneous membrane is the skin; it is a dry membrane lined externally with a layer of dead keratinized cells.
- Synovial membranes line the joint cavities of freely movable joints; their cells secrete a viscous synovial fluid.

5.7 Tissue Growth and Healing

The learning objectives for this section are:

HAPS LEARNING GOALS AND OUTCOMES

5.7.1* List the benefits of the inflammatory response and describe the mechanisms that comprise this response.

5.7.2 Describe tissue repair following an injury.

5.7.3* Discuss the progressive impact of aging on tissues.

5.7.4* Describe cancerous mutations' effect on tissues.

* Objective is not a HAPS Learning Goal.

5.7 Questions

1. You are playing hockey when your wrist gets hit by a stick. After a couple of hours you see that your wrist is swollen. After a week, you realize that your wrist is no longer swollen. What is this an example of?
   a. Acute inflammation
   b. Chronic inflammation
2. Which of the following steps of tissue repair occurs immediately after injury?
   a. Chemical signals lead to vasodilation to increase blood flow to the injury site.
   b. Fluid increases in the interstitial space to cause edema.
   c. Macrophages remove waste from the injury site.
   d. Fibroblasts replace the extracellular material lost by injury.
3. Which of the following changes occurs with age? Please select all that apply.
   a. There are fewer elastic fibers in the body as you age.
   b. The number of fibroblasts to replicate collagen decreases.
   c. The epithelial cells replicate faster to increase skin folds.
   d. The arterial wall of vessels stretches excessively to increase blood pressure.
4. Which of the following defines a mutation?
   a. An error made while replicating DNA regardless of the impact on the cell
   b. Anything that impacts the cell cycle to cause cells to replicate uncontrollably
   c. Groups of cells break off and move around the body
   d. Groups of cells that do not metastasize or cause disease

5.7 Summary

- Inflammation is the body's response to injury, which initiates a chain of events to control the damage and heal the tissue.
- Aging leads to a decrease in the rate of mitosis and decrease in the numbers of elastic fibers.
- Cancer occurs when there is a mutation in DNA replication that causes the cell to replicate uncontrollably to form tumors. Those that break off and move around the body are malignant, while those that stay in place and do not cause disease are benign.

6

The Integumentary System

Master1305/Shutterstock.com

Chapter Introduction

In this chapter we will learn . . .

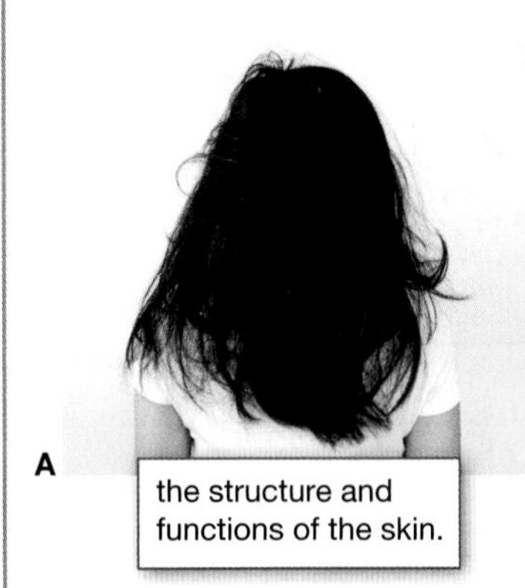

A the structure and functions of the skin.

Sasin Paraksa/Shutterstock.com

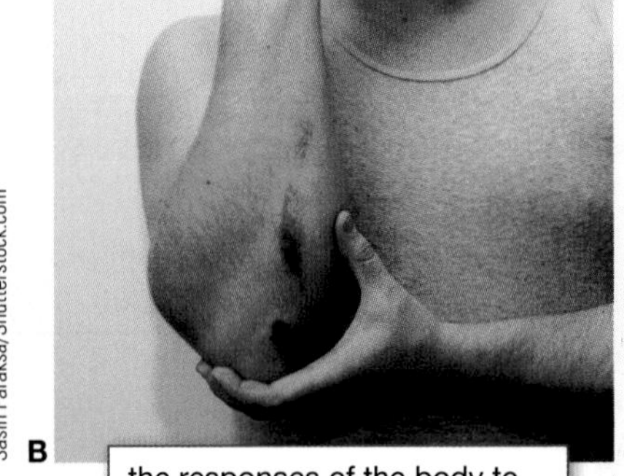

B the responses of the body to injury and disorders of the skin.

Svyatoslav Balan/Shutterstock.com

Our skin ties us to the world. Through sensory reception, our skin lets us know about dangers. Our skin is our most fundamental protection against pathogens and environmental conditions that could harm us. Our skin is a mechanism of self-expression, it is the primary thing that other humans see about us. Unfortunately, in many countries and regions where racism is prevalent and built into our institutions, our skin may even determine much about our experience in the world. Our skin, along with its accessory structures (such as hair), comprises one of the body's most essential and dynamic systems: the integumentary system. **LO 6.1.1**

The **integumentary system** refers to the skin, hair, nails, and exocrine glands, and it is responsible for much more than simply lending to your outward appearance. The skin makes up about 16 percent of your body weight, making the skin and accessory structures the human body's largest organ system. The skin protects the inside of your body, including all your other organs; it is of vital importance to your health. This chapter will introduce the structure and functions of the integumentary system, as well as some of the diseases, disorders, and injuries that can affect this system. **LO 6.1.2**

6.1 Layers of the Skin

HAPS LEARNING GOALS AND OUTCOMES

Learning Objectives: By the end of this section, you will be able to:

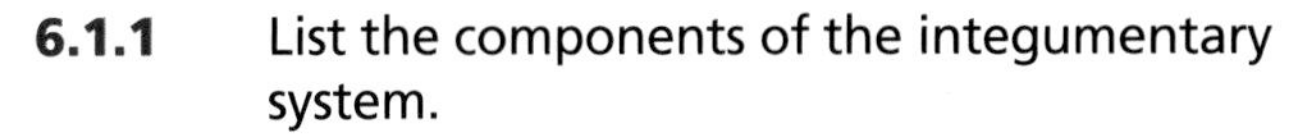

6.1.1 List the components of the integumentary system.

6.1.2 Describe the general functions of the integumentary system and the subcutaneous layer.

6.1.3 Identify and describe the tissue type making up the epidermis.

6.1.4 Identify and describe the layers of the epidermis, indicating which are found in thin skin and which are found in thick skin.

6.1.5 Describe the processes of growth and keratinization of the epidermis.

6.1.6 Compare and contrast thin and thick skin with respect to location and function.

6.1.7 Explain how each of the five layers, as well as each of the following cell types and substances, contributes to the functions of the epidermis: stem cells of stratum basale, keratinocytes, melanocytes, epidermal dendritic (Langerhans) cells, tactile (Merkel) cells and discs, keratin, and extracellular lipids.

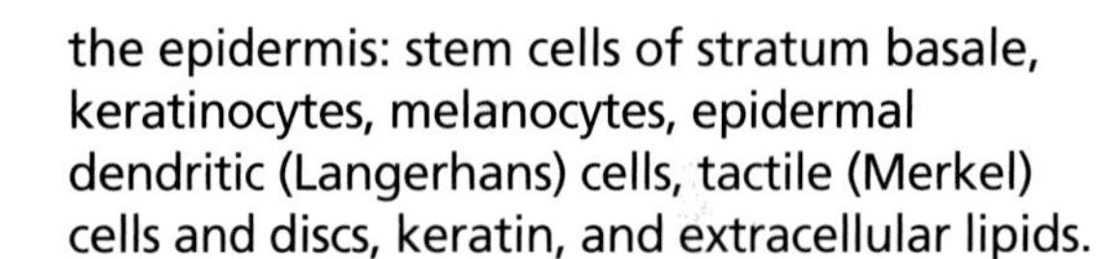

6.1.8 Describe the factors that contribute to skin color.

6.1.9 Describe the functions of the dermis, including the specific function of each dermal layer.

6.1.10 Identify and describe the dermis and its layers, including the tissue types making up each dermal layer.

6.1.11 List and identify the types of tissues found in the hypodermis.

6.1.12 Describe the functions of the subcutaneous layer.

6.1.13 Describe the thermoregulatory role played by adipose tissue in the subcutaneous layer.

The Human Anatomy and Physiology Society includes more than 1,700 educators who work together to promote excellence in the teaching of this subject area. The HAPS A&P Learning Outcomes measure student mastery of the content typically covered in a two-semester Human A&P curriculum at the undergraduate level. The full Learning Outcomes are available at https://www.hapsweb.org.

Although you may not typically think of the skin as an organ, it meets the definition of one, in that it is made of tissues that work together as a single structure to perform unique and critical functions. The skin and its accessory structures make up the integumentary system, which provides the body with overall protection. The skin is made of multiple layers of cells and tissues, which are held to underlying structures by connective tissue (**Figure 6.1**). The deeper layer of skin is well vascularized (has numerous blood vessels). It is also highly innervated; that is to say, it has numerous nerve fibers ensuring communication to and from the brain.

Student Study Tip

Remember that *epi-* means "on" and *hypo-* means "under." Therefore, the three skin layers are literally the dermis, the layer on the dermis, and the layer under the dermis.

You are very familiar with a macro view of the skin. You see skin every day; you may even be looking at skin right now. As we dive into the micro view of the skin, in our broadest view we can divide the skin into three layers: the epidermis, dermis, and hypodermis. Below these layers, a sheet of connective tissue, called *fascia*, ties the skin to the underlying muscles.

6.1a The Epidermis

LO 6.1.3

The **epidermis** is composed of stratified squamous epithelium. It is actually a specialized stratified squamous epithelium called *keratinized stratified squamous*. The lining of structures that are open to the outside is also made of stratified squamous epithelium, but the skin, because it is a dry organ, requires special protection and so is keratinized. Like all epithelia, it is avascular (does not contain blood vessels). This epithelium is made of four, or sometimes five, layers of epithelial cells, depending on body location. From deep to superficial, these layers are the

LO 6.1.4

stratum basale, stratum spinosum, stratum granulosum, and stratum corneum (**Figure 6.2**).

Figure 6.1 Layers of Skin

The skin is composed of many layers. The epidermis, which is the major outermost layer of the skin, is composed of epithelial tissue. The dermis, which lies deep to the epidermis, is composed of connective tissue. Beneath the dermis lies the hypodermis, which has a different composition of connective tissues than the dermis. *Fascia* is the term for the connective tissue that anchors the skin to the muscle beneath.

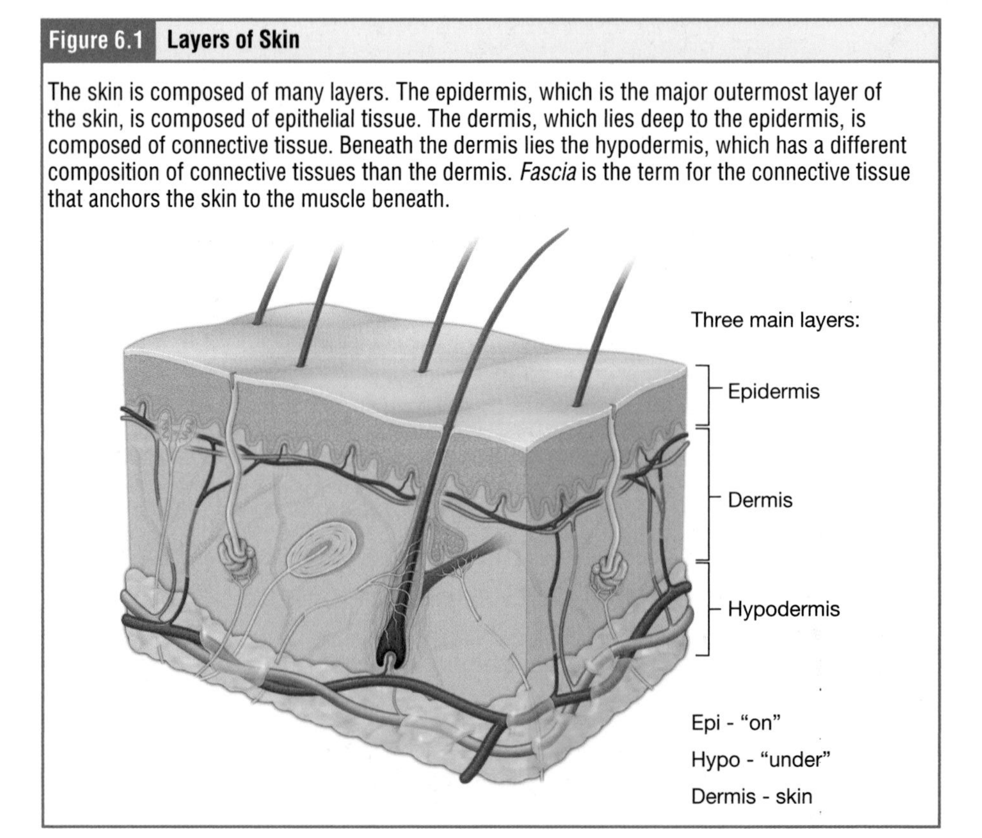

Figure 6.2 Layers of the Epidermis

The epidermis has four layers in most areas of the body: stratum basale, stratum spinosum, stratum granulosum, and stratum corneum.

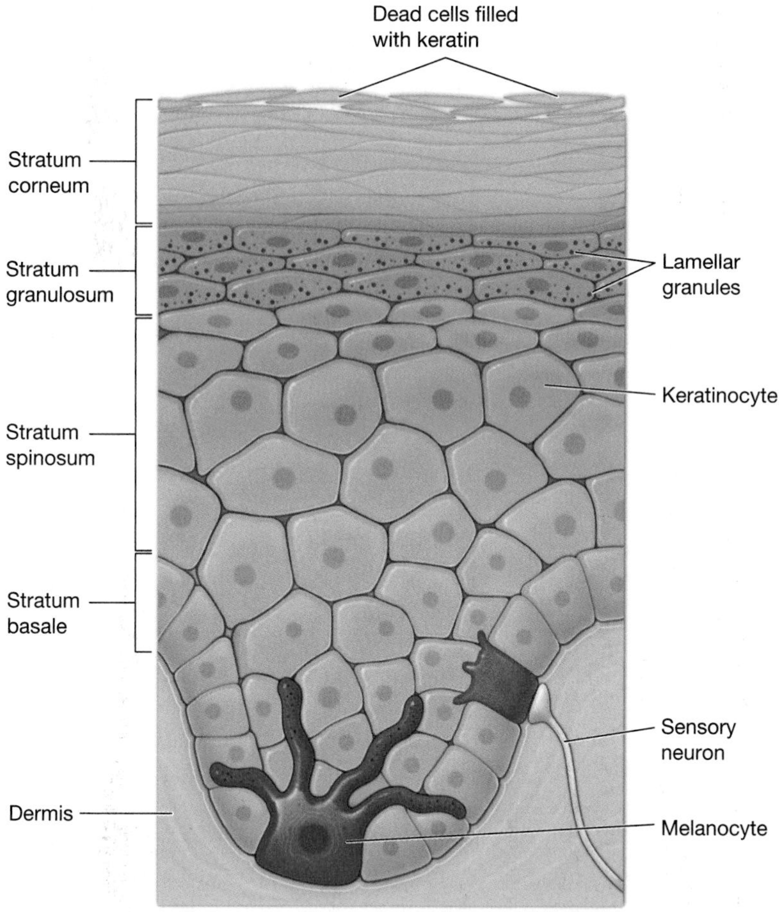

Let's begin by talking about how the epidermal layers are related to each other. It may be tempting to think of these layers as fixed in their relationship to each other—for example you might envision the cells of the stratum granulosum as always being under the cells of the stratum corneum, in the same way that the second floor of a building is always under the third floor. In a way, this is true; however, our skin cell layers are connected to each other in the same way that escalators connect the floors of a building. The cells are "born" through mitosis in the basale, and as they mature, they push up on the epidermis, moving from deep to superficial as the cells at the apical surface are lost to friction and are replaced by the cells beneath. Therefore, the life of a skin epidermal cell begins in the basale, and the cell progresses through all of the layers of the epidermis until it is finally sloughed off of the corneum.

LO 6.1.5

The cells in all of epidermal layers except the stratum basale are called *keratinocytes*. A **keratinocyte** is a cell that manufactures and stores the protein keratin. **Keratin** is an intracellular fibrous protein that gives hair, nails, and skin their hardness and water-resistant properties (**Figure 6.3**).

The skin on the palms of the hands and the soles of the feet has a fifth layer, called the *stratum lucidum*, located between the stratum corneum and the stratum

Figure 6.3 Keratin

(A) Keratin is an intracellular protein that fills keratinocytes. (B) Keratin is a protein composed of long alpha helices. (C) Keratin fibrils are the main component of hair shafts.

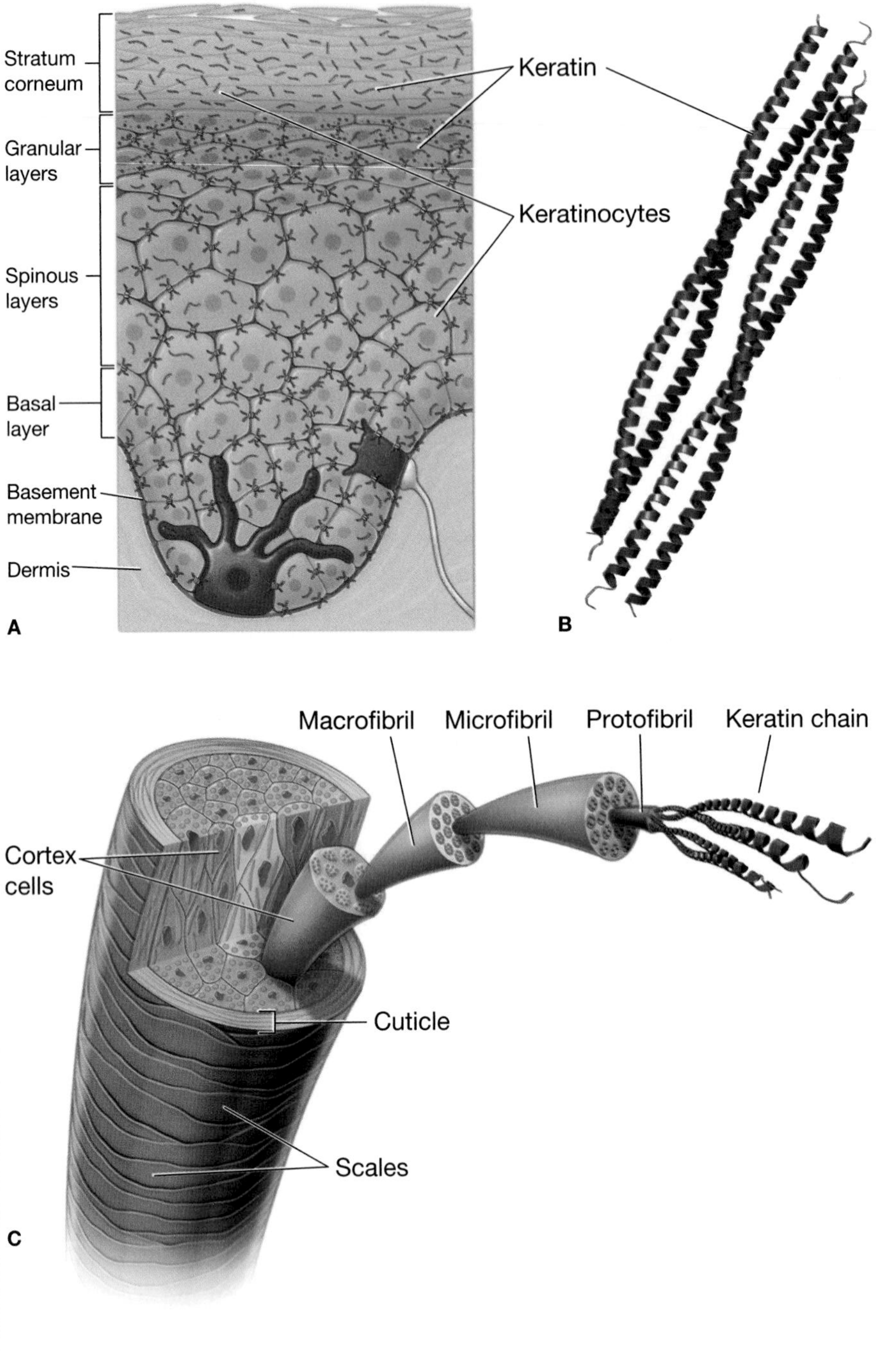

Student Study Tip

Thick skin and stratified squamous epithelium operate under the same principle: More friction means more layers needed for protection!

LO 6.1.6

granulosum (**Figure 6.4**). Skin in these two locations, with its fifth layer of cells, is referred to as "thick" skin, and all other skin is referred to as "thin" skin. That being said, the relative thickness of the skin (for example, the thick skin of your back versus the very thin skin of your eyelids) is contributed to much more by the dermis than the epithelium. Thin skin and thick skin also differ in their glands, as we will discuss later.

Figure 6.4 Thin Skin Versus Thick Skin

(A and B) Thin skin exists all over the body and has four distinguishable cell layers in the epidermis. (C and D) Thick skin, found only on the palms of the hands and soles of the feet, has an additional layer, the stratum lucidum.

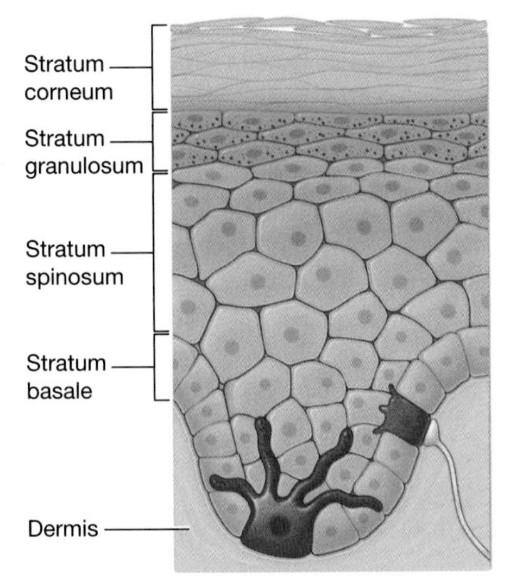

A Thin skin

B

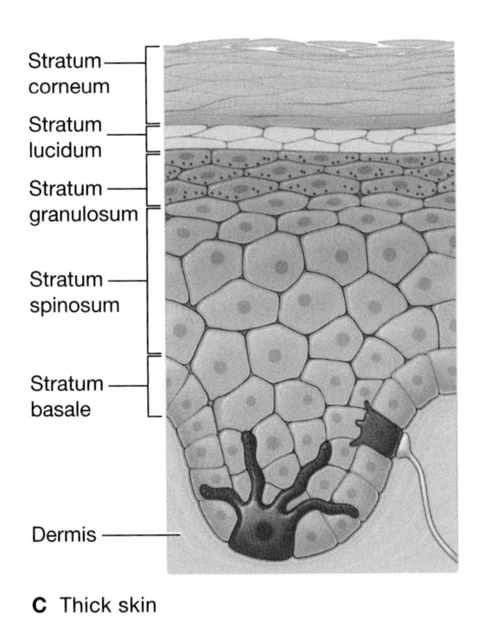

C Thick skin

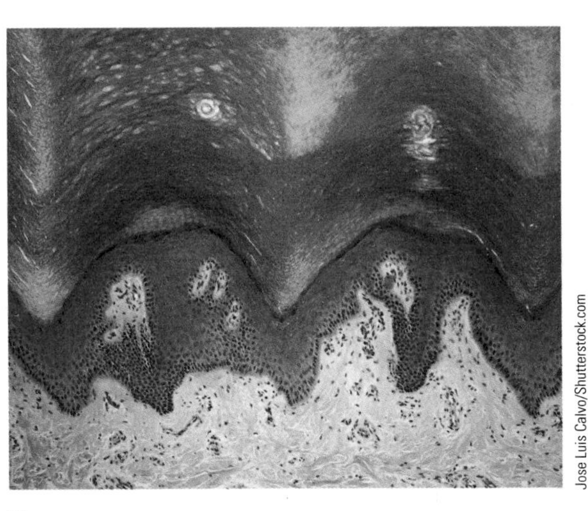

D

Learning Check

1. When you rub your hands together, you may lose some of the epithelial cells. From what layer of the epidermis are cells being lost?
   a. Stratum Corneum
   b. Stratum Basale
   c. Stratum Spinosum
   d. Stratum Granulosum
2. From which layer do keratinocytes mature and become the stratum spinosum?
   a. Stratum granulosum
   b. Stratum basale
   c. Stratum corneum
   d. Stratum lucidum

(Continued)

3. Identify layer D in the figure.
    a. Stratum basale
    b. Stratum spinosum
    c. Stratum granulosum
    d. Stratum lucidum
4. Identify layer C in the figure.
    a. Stratum corneum
    b. Stratum lucidum
    c. Stratum granulosum
    d. Stratum spinosum

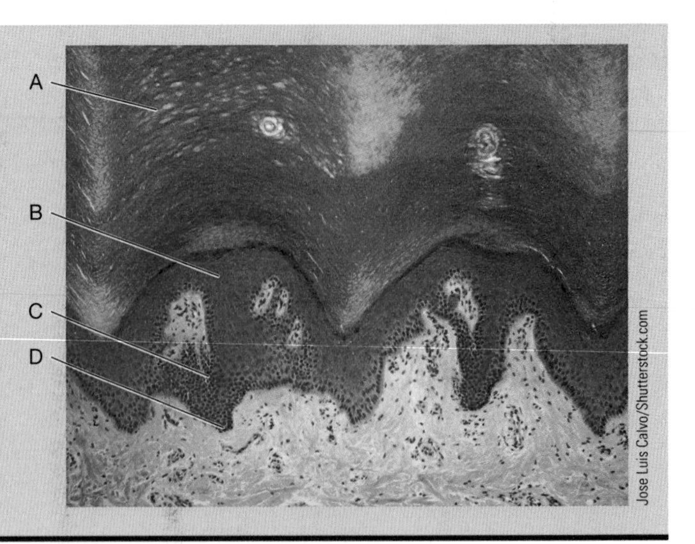

LO 6.1.7

Stratum Basale The **stratum basale** is the deepest epidermal layer and attaches the epidermis to its basement membrane, which ties the epidermis to the dermis below. The stratum basale is a single layer of cells, most of which are stem cells that are a precursor of the keratinocytes of the epidermis. All the keratinocytes are produced from this single layer of cells, which are constantly going through mitosis to produce replacement cells. As new cells are produced, the existing cells of the epidermis are pushed superficially away from the stratum basale. Just like all stem cells, the cells of the stratum basale exhibit decreased function as we age, leading to a thinner epidermis and slower wound healing.

Two other cell types are found dispersed among the stem cells in the stratum basale. The first is a **Merkel cell**, which functions as a sensory receptor and is connected to sensory nerves that send signals about touch to the brain. These cells are especially abundant on the surfaces of the hands and feet. The second is a **melanocyte**, a cell that produces the pigment melanin. **Melanin**, a protein that can be made in two forms, functions to protect cells from ultraviolet (UV) radiation damage.

LO 6.1.8

Melanin occurs in two primary forms. Whereas eumelanin exists as black and brown, pheomelanin provides a more reddish hue. While all humans have a relatively similar concentration of melanocytes, some individuals make less and actively break down the melanin produced by melanocytes, leading to lighter shades of skin. Melanin granules protect the DNA of epidermal cells from UV ray damage; melanin also prevents the breakdown of folic acid, a nutrient necessary for our health and well-being. For these reasons, early humans developed the ability to produce plenty of melanin. However, humans that migrated to environments with less sun exposure suffered from insufficient production of vitamin D, an important nutrient involved in calcium absorption. As evolution of humans at different locations around the globe continued over generations, the amount of melanin produced and expressed in our skin achieved a balance between available sunlight and folic acid destruction, and between protection from UV radiation and vitamin D production.

Learning Connection

Try Drawing It

Draw the cells of the stratum granulosum. They are shaped like an American football, with a central nucleus.Where can you add melanin granules to your drawing so that the nucleus is maximally protected from UV radiation?

Stratum Spinosum In the **stratum spinosum** the young keratinocytes have an American football-shaped appearance, meaning that they are widest in the center and tapered at the ends. The cells are joined to each other through **desmosomes**. The desmosomes interlock the cells with each other such that the skin can undergo a substantial amount of pulling, twisting, and friction and the keratinocytes will remain joined. The stratum spinosum is composed of about eight to ten layers (see Figure 6.4). Interspersed among the keratinocytes of this layer is a type of dendritic cell called the **Langerhans cell**,

Cultural Connection

Fast or Slow, Friend or Foe?

While *Homo sapiens* as a species have been present on Earth for approximately 160,000 years, members of the species likely did not come into contact with other members who looked morphologically different in skin coloration/tone or facial features until about 10,000 years ago. Before this time, members of the species lived in close-knit groups that did not travel as far as humans later would with the first instances of long-distance trade. Imagine if you had only ever seen a very small group of other humans, all of whom looked very similar to you. That existence is a stark contrast to our modern, diverse, global life.

Evolutionary biologists theorize that early humans adapted to having very quick mental processing when encountering organisms that looked different from themselves, as these were likely to be a threat to their safety or food supply. Our quick reactions tend to be rooted in a structure within our brains called the *limbic system*. This system drives responses that are highly emotional and fear-based. Our more rational and logical thinking comes from other regions in the brain, especially the frontal lobe, and this higher-quality thinking takes time—minutes instead of milliseconds. Psychologists theorize that prejudice, stigma, and discriminatory behavior patterns may also have evolutionary roots in these different patterns of thinking. When faced with a new situation or individual, both our limbic system and our frontal lobe get to work; the frontal lobe takes in and categorizes information, while our rapid limbic system may be creating decisions of friend versus foe based on simple visual cues such as expression, features, and skin color. Therefore, fast-brain thinking is linked to human tendencies toward implicit, or unconscious, biases that require awareness and slower-brain thinking (prefrontal cortex) to override. Understanding the evolutionary and neurobiological basis of prejudice will be one important component to combating racism.

which functions similarly to macrophages, engulfing bacteria, foreign particles, and damaged cells that occur in this layer.

The keratinocytes in the stratum spinosum begin the synthesis of keratin and release a water-repelling glycolipid that helps prevent water loss from the body, making the skin relatively waterproof. As new keratinocytes are produced atop the stratum basale, the keratinocytes of the stratum spinosum are pushed into the stratum granulosum.

Stratum Granulosum The **stratum granulosum** is named for its granular appearance. As the keratinocytes are pushed from the stratum spinosum, the cells become flatter, their cell membranes thicken, and they accumulate large amounts of keratin. The cells in the granulosum are characterized by the presence of granules, which are intracellular protein-filled vesicles. Some of the vesicles are filled with a precursor to keratin, **keratohyalin**. Other granules (melanosomes) are filled with melanin. Melanin is made by melanocytes in the stratum basale but transferred to the keratinocytes through long processes of the melanocyte that wind through some tiny channels among the keratinocytes (Figure 6.5). The keratinocytes of the stratum granulosum begin to accumulate so much protein in their cytoplasm that their nuclei and other cell organelles slowly disintegrate. At the most superficial portion of the stratum granulosum the cells begin to die, leaving behind the keratin, keratohyalin, and cell membranes that will form the stratum lucidum, the stratum corneum, and the accessory structures of hair and nails.

Stratum Lucidum The **stratum lucidum**, present in the skin of the palms of the hands and soles of the feet, is a smooth, seemingly translucent layer of the epidermis located just superficial to the stratum granulosum and deep to the stratum corneum. The keratinocytes that compose the stratum lucidum are dead and flattened (see Figure 6.4). In addition to the melanin, keratohyalin, and keratin of the deeper keratinocytes, these cells are densely packed with **eleidin**, a clear protein derived from keratohyalin, which gives these cells their transparent appearance. The high lipid content of this protein provides a barrier to water to these regions.

Figure 6.5 Melanosomes

The relative coloration of the skin depends of the amount of melanin stored in melanosomes inside the keratinocytes.

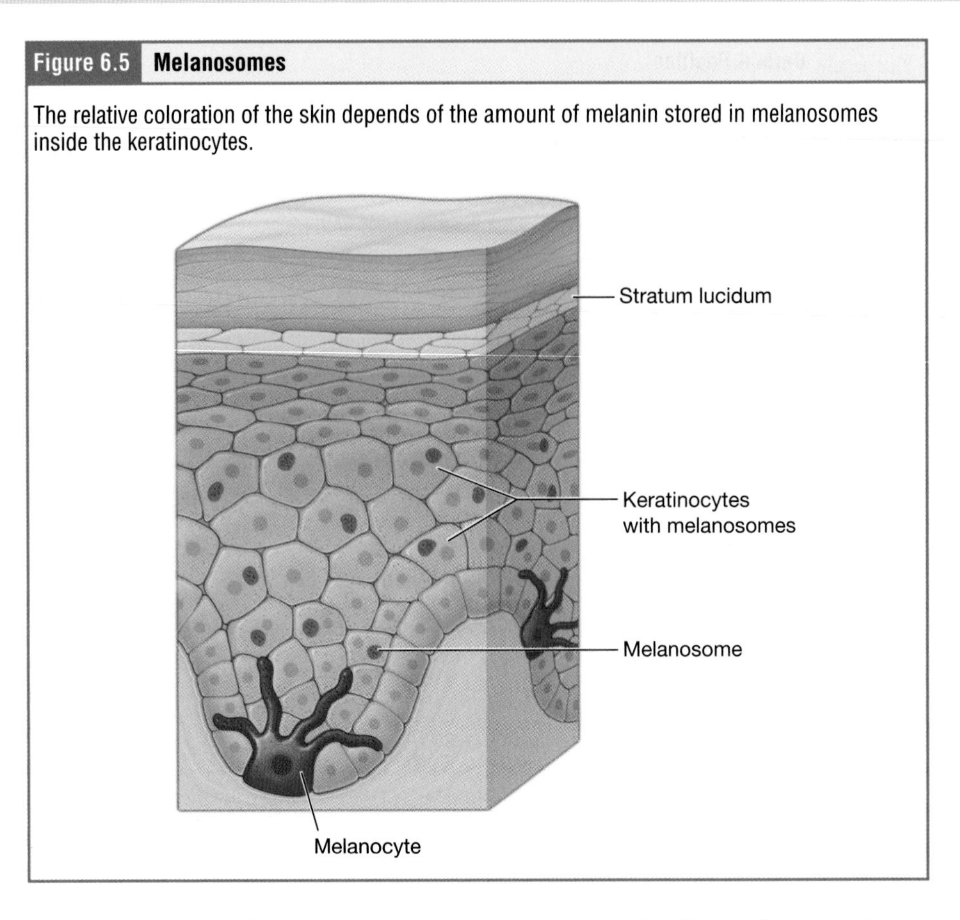

Stratum Corneum The **stratum corneum** is the most superficial layer of every epidermis and is the layer of the skin that is exposed to the outside environment. The stratum corneum is usually composed of about 15 to 30 layers of dead skin cells. This dry, dead layer helps prevent the penetration of microbes and the dehydration of underlying tissues, and provides a mechanical protection against abrasion for the more delicate underlying layers. Cells in this layer are shed periodically and are replaced by cells pushed up from the underlying strata. The entire layer is replaced during a period of about four weeks.

bedsores (decubitis ulcers, pressure ulcers)

Considering Friction Wring your hands, scratch an itch, walk across a sandy beach. Think about how much friction your skin is exposed to daily. The skin has many structural adaptations to meet this functional need. We have discussed already that the keratinocytes of the epidermis are connected by desmosomes to keep them from separating. The stratum corneum, a layer of dead cells, functions by shedding cells in response to contact and friction, preventing the underlying living cells from being damaged. Another structural adaptation to better endure friction is the **dermal papillae**. These fingerlike projections of the dermis interdigitate with downward projections of the epidermis (Figure 6.6A). To imagine how this structure assists with the function of resisting friction, look at your hands. Imagine one hand is the epidermis, the other is the dermis. Put your hands together as if they are high-fiving each other. Now, slide one up or down. See how easy it is for two smooth surfaces to slide past each other? Now, interlace your fingers like the photo in Figure 6.6B. Try sliding your hands now. The interdigitation, or interlocking, of the dermal papillae with the ridges of the stratum basale fortifies the connection between the epidermis and the dermis, so that these two layers do not detach when exposed to friction. These bumps and ridges form during the fetal period and are noticeable externally; you may recognize them as fingerprints. Fingerprints are unique to each individual and are used for forensic analyses because the patterns do not change with the growth and aging processes.

Figure 6.6 Dermal Papillae

(A) The bumps and ridges of the dermis and epidermis interlock to form a combined structure that rarely pulls apart when exposed to friction or shear force. (B) The interdigitation can be likened to the interlocking of fingers.

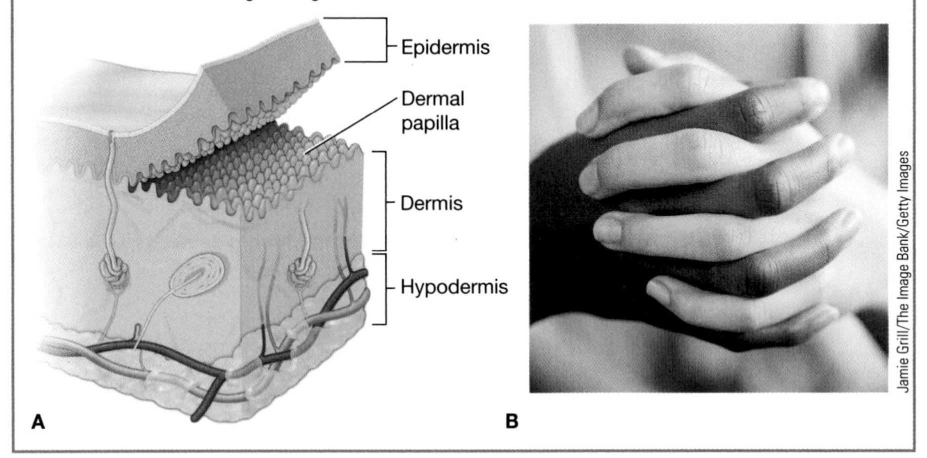

Apply to Pathophysiology

Bedsores

Bedsores, also called *pressure ulcers*, are caused by constant, long-term, pressure necrosis (tissue death) in the epidermis. Bedsores are most common in patients who have conditions that cause them to have reduced mobility.

1. Bedsores occur due to death of epidermal cells when they cannot get adequate nutrients. From where do epidermal (or any epithelial cells) get their nutrients, such as oxygen and glucose?
   A) Blood vessels among the epithelial cells
   B) Blood vessels within the connective tissue underneath the epithelium
   C) Blood vessels in the hypodermis, the adipose-rich connective tissue beneath the skin
2. Bedsores most commonly occur on body parts of the body that endure sustained external pressure and are bony. Why would this combination of external pressure and the presence of bones create an environment where bedsores are more likely to occur?
   A) Because body parts with more adipose tissue have more local nutrients.
   B) Because the squeeze from the external pressure and the bones compresses the blood vessels in between.
   C) Because bones contribute to the starvation of epithelial cells by taking all the nutrients available.
3. The biggest complication with bedsores, besides pain, is the risk of infection. Which type of skin cell functions to watch for microbial invaders?
   A) Merkel
   B) Langerhans
   C) Fibroblast

Learning Check

1. Which of the following statements explains some of the differences in shades of skin color?
   a. Everyone has the same amount of melanocytes, but some melanocytes are nonfunctional and do not produce melanin, leading to lighter shades of skin.
   b. Everyone has the same amount of melanocytes, but some break down more melanin, leading to lighter shades of skin.
   c. Some individuals have more melanocytes, and therefore more melanin than other individuals, leading to darker shades of skin.

(Continued)

2. Imagine you are a scientist working on gene therapy. You have designed a DNA sequence to correct albinisim, a lack of pigment in the skin of some individuals. Your next task is to figure out how to get your gene into the cells that need it. You want the gene to be in the cells that produce the pigment melanin. To which layer of the epidermis do you need to deliver the gene?
   a. Stratum Corneum
   b. Stratum Granulosum
   c. Stratum Spinosum
   d. Stratum Basale
3. Blisters occur when an area of your body undergoes a lot of friction. When this happens, part of your epidermis separates from your dermis and the resulting space fills up with fluid. In a blister, which two structures are now separated that used to be adjacent?
   a. Dermal Papillae and Stratum Corneum
   b. Stratum Corneum and Stratum Granulosum
   c. Derman Papillae and Stratum Basale
   d. Stratum Basale and Stratum Granulosum

6.1b The Dermis

LO 6.1.9

As with all epithelia, the epidermis is avascular and therefore gets its nutrients from an underlying connective tissue layer. This layer, the **dermis**, contains blood and lymph vessels, nerves, and other structures, such as the deeper portions of hair follicles and sweat glands. The dermis itself has two layers, but both are connective tissue proper and composed of an interconnected mesh of elastin and collagenous fibers, produced by fibroblasts (**Figure 6.7**). The thickness of the dermis varies depending on location, with the back of the torso having the most substantial dermal layer and the eyelids having the thinnest. Throughout the dermis, collagen fibers provide structure and tensile strength, with strands of collagen extending throughout the two layers of the dermis and even into the hypodermis. Elastin fibers provide some elasticity to the skin, enabling the skin to stretch and recoil during movement, such as the deformations in the skin when a person's face breaks into a smile. In addition, collagen binds water to keep the skin hydrated. As with all connective tissue, scattered fibroblasts are responsible for the production of these fibers. As we age, the fibroblasts decrease in number and production and the dermis decreases in its density of both collagen and elastin. The

Figure 6.7 The Dermis

The two layers of the dermis—papillary and reticular—are differentiated by the thickness and weave of the collagen fibers.

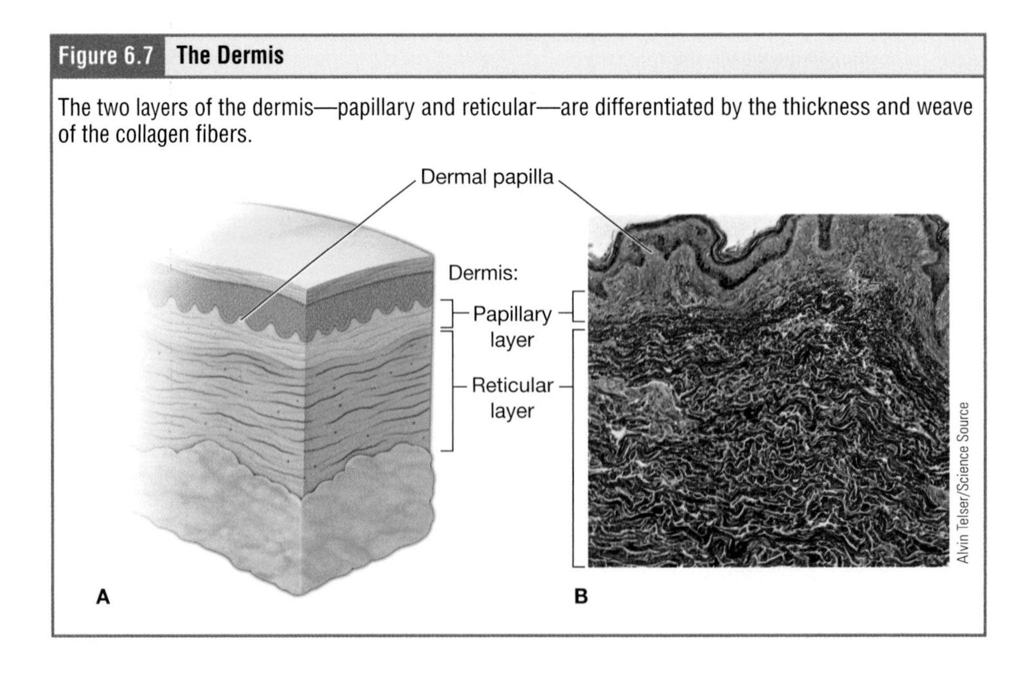

loss of collagen makes the skin thinner and slower to heal, and the decline in elastin leads to a decreased capacity to recoil after stretching, which results in the formation of wrinkles.

The skin can also be affected by pressure associated with rapid growth. A **stretch mark** results when the dermis is stretched beyond its limits of elasticity, as the skin stretches to accommodate the excess pressure and collagen fibers can break. Stretch marks can often accompany rapid weight gain during puberty and pregnancy.

Learning Connection

Quiz Yourself

List all the protein fibers we've learned so far. Which ones can be found in which layers?

LO 6.1.10

Papillary Layer The **papillary layer** is made of loose areolar connective tissue. This superficial layer of the dermis forms the papillae that project into the stratum basale (see Figure 6.7). Within the papillary layer are fibroblasts and an abundance of small blood vessels. In addition, the papillary layer contains nerve fibers, and touch receptors called the **tactile corpuscles** (also known as *Meissner corpuscles*). The collagen fibers of the papillary layer are finer, meaning much smaller in diameter, than the fibers of the deeper reticular layer (see Figure 6.7).

Student Study Tip

The papillary and reticular layers of the dermis can be distinguished since the word *papillary* means "small projections." We see this in other anatomical structures such as bumps on the tongue.

Reticular Layer Underlying the papillary layer is the much more substantial **reticular layer**, composed of dense, irregular connective tissue. This layer is well vascularized and has a rich sensory and sympathetic nerve supply. The reticular layer has a denser weave of collagen fibers, and individual collagen fibers are thicker in diameter than the fibers in the papillary layer.

tactile corpuscles (Meissner corpuscles)

6.1c The Hypodermis

The **hypodermis** (also called the *subcutaneous layer*) is a layer directly below the dermis and serves to connect the skin to the underlying fascia. **Fascia** is the fibrous covering of muscles and organs. You can imagine that some collagen fibers of the hypodermis interweave with the collagen fibers of the fascia, tying these structures together. The hypodermis is not strictly a part of the skin, although the border between the hypodermis and dermis can be difficult to distinguish. The hypodermis consists of well-vascularized, loose areolar connective tissue and adipose tissue, and its functions include fat storage, insulation, and cushioning.

hypodermis (subcutaneous layer)

LO 6.1.11

LO 6.1.12

DIGGING DEEPER:
What Is Fat?

Lipids, adipose tissue, fat, brown fat and white fat, visceral fat. Adipose tissue is found in the hypodermis, in layers around our organs, even inside our bones. What is this substance really about? Hypodermal adipose tissue plays a key role in thermoregulation by preventing the loss of heat from the underlying organs and muscles. But not all fat is the same. For most of us, our adult bodies have plenty of white fat (during cadaver dissection you will discover that this fat is actually more of a yellow color) in the hypodermis, but human infants predominantly have a different form of fat called *brown fat*. Brown fat is also rich in stored lipids, but brown adipose cells have a high proportion of mitochondria, unlike white adipose cells, which have almost none of these organelles. When infant body temperature drops, the brown fat cells break down lipids and use them in cellular respiration, generating heat. In adults experiencing a drop in body temperature, skeletal muscles can generate heat through cellular respiration coupled with shivering. Brown fat is essential for infants because they do not have sufficient skeletal muscle mass to aid in their thermoregulation. LO 6.1.13

Of course, adipose tissue serves a huge role in energy storage. Throughout human evolution, most people have lived in places and times where food is not universally available. The ability to feast during times of plenty (summer, for example) and store excess calories gave a tremendous evolutionary advantage for survival during times of famine or when food was less available, such as winter. Excess calories can be

(Continued)

converted to lipid and stored either in the hypodermis (this is called *hypodermal* or *subcutaneous* fat) or surrounding our organs (*visceral* fat). Visceral fat is very different in makeup than hypodermal fat; visceral fat functions as an endocrine organ, contributing hormone signals to the functioning of the body in addition to energy storage. Where the fat is deposited and accumulates within the hypodermis or hypodermis versus the viscera depends on hormones (testosterone, estrogen, cortisol, insulin, glucagon, leptin, and others), as well as genetic factors and even medications. Fat distribution changes as our bodies mature and age.

The body mass index (BMI) is often used as a measure of fat, although this measure is in fact derived from a mathematical formula that compares body weight (mass) to height. BMI numbers are broadly divided into groups of "underweight," "normal," "overweight," and "obese." This measure, as with a lot of medical reference values, was developed and validated on studies of primarily white individuals who identify as men and therefore may not be accurate when applied to most humans. In fact, only about 25 percent of Americans fall into the BMI range of "normal," not only due to the obesity epidemic in the United States but also because many body types, including individuals who are extremely muscular, will fall outside this range. Epidemiologists do note that BMI can be a broad indicator of mortality risk, with more significant risk of disease among those in the underweight and obese categories, but a person's BMI value tells us little about their individual health.

Apply to Pathophysiology

UV Radiation

The sun emits both visible and ultraviolet light (also called *UV radiation*). Increasing amounts of UV radiation are reaching humans due to the destruction of a protective component of the atmosphere called *ozone*. UV radiation can harm many cellular components.

1. UV radiation damages DNA, causing mutations. Dead cells no longer have DNA that can be damaged, so the area of most concern from UV damage is the most superficial layer of living cells. Which layer is this?
   A) Corneum
   B) Spinosum
   C) Basale
   D) Granulosum
2. UV radiation is especially damaging to elastic fibers and elastin proteins. Which of the following would we expect after years of UV exposure?
   A) Scarring on the skin due to damage to the stratum corneum
   B) Stretch marks or sagging skin due to damage of the structural components
   C) Wrinkles due to an inability of the skin to return to shape after stretching
3. During a sunburn, mast cells will respond to cellular damage and release histamine as a component of an inflammatory response (for more information about inflammatory responses, see Chapter 5). A key component of inflammation is *vasodilation*. Where are the blood vessels located that will dilate in response to histamine?
   A) Stratum Corneum
   B) Stratum Granulosum
   C) Stratum Basale
   D) Papillary Dermis

Learning Check

1. Which layer of the dermis forms the dermal papillae?
   a. Papillary Layer
   b. Reticular Layer
2. Explain why wrinkles form as you age.
   a. The loss of collagen leads to decreased recoil in the skin.
   b. The loss of collagen leads to decreased tensile strength.
   c. The loss of elastin leads to decreased recoil in the skin.
   d. The loss of elastin leads to decreased tensile strength.
3. Discuss the difference between white fat and brown fat. Please select all that apply.
   a. White fat has many mitochondria to generate heat.
   b. Brown fat has many mitochondria to generate heat.
   c. Infants have white fat.
   d. Adults have white fat.

6.2 Accessory Structures of the Skin

Learning Objectives: By the end of this section, you will be able to:

6.2.1 Describe the structure and function of hair.

6.2.2 Describe the growth cycles of hair follicles and the growth of hairs.

6.2.3 Describe the structure and function of nails.

6.2.4 Describe the structure and function of exocrine glands of the integumentary system.

6.2.5 Explain the physiological significance of the presence or absence of sebaceous (oil) glands, sudoriferous (sweat) glands, and hair in the skin of the palms and fingers.

Accessory structures of the skin include hair, nails, sweat glands, and sebaceous glands. These structures span the depth of the epidermis and can extend down through the dermis into the hypodermis. Glands are made of living epithelial cells that are actively producing secretions; in contrast, hair and nails are composed of dead, keratin-filled cells similar to the stratum corneum.

6.2a Hair

LO 6.2.1

You are already very familiar with the part of hair that you can see external to your skin. This externally visible portion, a long filament of keratin, is the **hair shaft**. This shaft of hair is grown within an epidermal structure called the **hair follicle**. The hair follicle is a structure that develops from the epidermis, but as it grows larger it pushes down into the dermis (see the "Anatomy of Hair" feature). It is still an almost entirely epidermal structure, even though a considerable portion of it is found in the dermis. As the hair follicle develops (see "Anatomy of Hair" part (A)), the epidermis invaginates, pushing down into the dermis. It takes with it stem cells of the stratum basale and melanocytes. As these developing epidermal hair cells push in on the basement membrane, a tiny cluster of dermal cells begins to differentiate beneath the developing follicle. This cluster of dermal cells becomes the dermal papilla (see "Anatomy of Hair"). Being connective tissue, the papilla contains blood vessels and nerves, and nutrients that diffuse from the vascularized papilla feed the entire follicle. The deepest portion of the hair follicle is the **hair bulb**. Epithelial cells within the hair bulb form the **hair matrix** and function very similarly to cells of the stratum basale; they divide rather continuously, producing keratinocytes that fill themselves with keratin until they die, leaving membranous sacs of keratin protein behind. These cells become the **hair root**—the portion of the follicle between the bulb and the surface of the skin—and the hair shaft (the hair exposed at the skin's surface).

If we examine the hair itself, the central core of the hair—the **medulla**—is a fragile inner core made up of some living cells and the spaces between them. Thicker hair strands have a medulla, while thinner hair strands do not. Whether (and where) you grow hair with a medulla or not is determined by your genetics. If present, the medulla is surrounded by the **cortex**, a layer of compressed, keratinized cells that is covered by an outer layer of very hard keratinized cells known as the **cuticle**. These layers are depicted in longitudinal and cross sections of the hair (see the "Anatomy of Hair" feature, part (C)). Whether your hair is curly or straight is determined by the structure of the cortex and, to the extent that it is present, the medulla. The shape and structure of these layers are, in turn, determined by the shape of the hair follicle. As new cells are deposited within the hair bulb, the hair shaft is pushed through the follicle toward the surface. Keratinization is completed as the cells are pushed to the skin surface to form the externally visible shaft of hair. The external hair is completely dead and composed entirely of keratin. For this

Anatomy of...

Hair

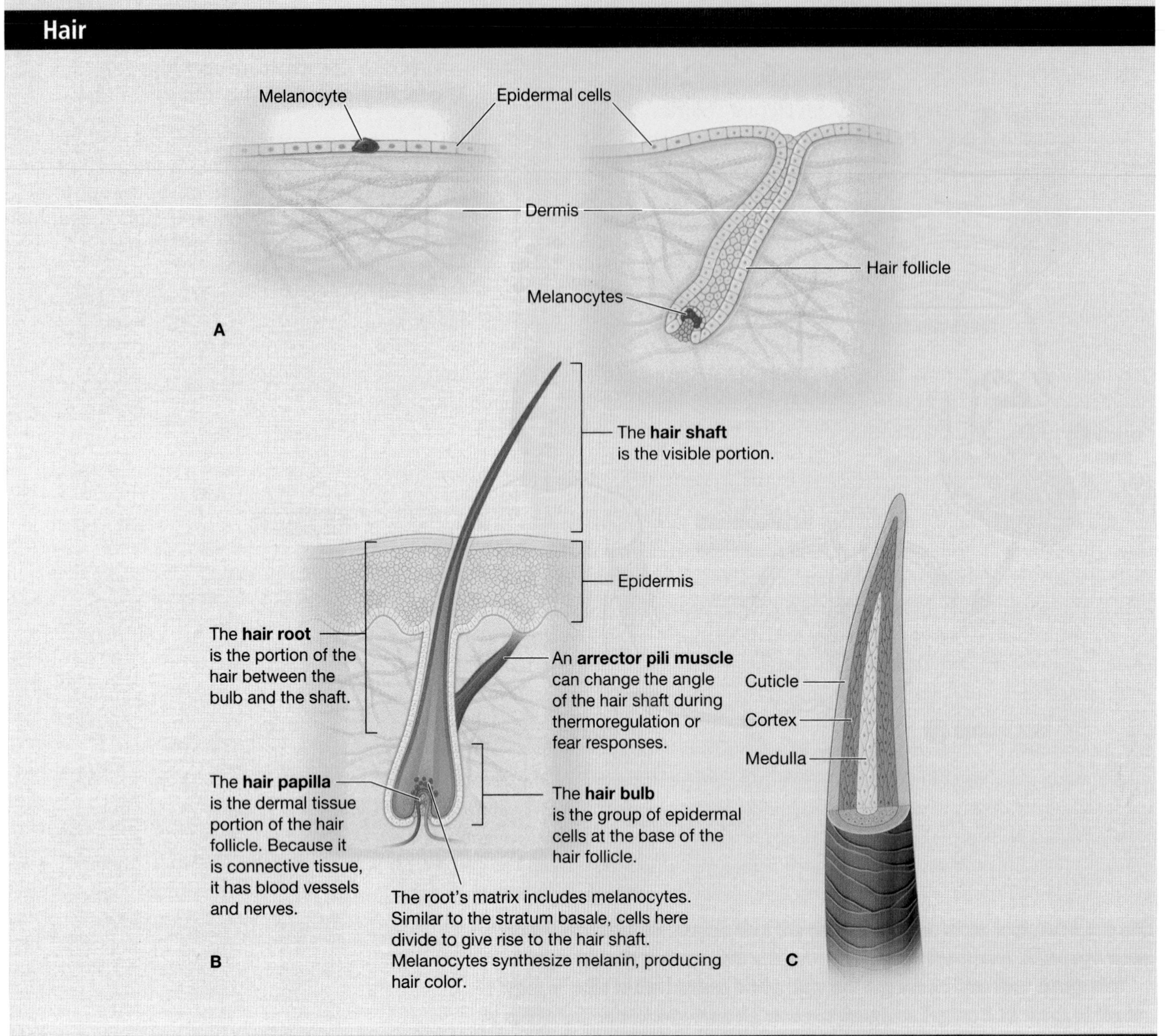

Melanocyte
Epidermal cells
Dermis
Hair follicle
Melanocytes
A
The **hair shaft** is the visible portion.
Epidermis
The **hair root** is the portion of the hair between the bulb and the shaft.
An **arrector pili muscle** can change the angle of the hair shaft during thermoregulation or fear responses.
Cuticle
Cortex
Medulla
The **hair papilla** is the dermal tissue portion of the hair follicle. Because it is connective tissue, it has blood vessels and nerves.
The **hair bulb** is the group of epidermal cells at the base of the hair follicle.
The root's matrix includes melanocytes. Similar to the stratum basale, cells here divide to give rise to the hair shaft. Melanocytes synthesize melanin, producing hair color.
B
C

reason, our hair itself does not have sensation (you know this because it doesn't hurt you to cut or shave your hair). There are nerves within the hair papilla, however, so pulling on hair yields sensation.

Each hair follicle is associated with an oil gland known as a **sebaceous gland**. We will explore the structure and function of these glands in more detail in the next section.

Now that we have learned all about hair's structure, you may be left with some questions about hair function. Why do we have it at all? Why do some people have more of it than others? Why does the hair on our heads grow so long while the hair on our bodies stops at a certain length? Depending on your age, you might be facing a very important question: why does hair go gray? Hair serves a variety of functions, including physical protection, sensory input, thermoregulation, and UV protection. For example, hair around the eyes (eyelashes) and the hair of the eyebrows prevents sweat and other particles from dripping into and bothering the eyes. Hair within the nose defends the body by trapping and excluding dust particles that may contain allergens and microbes. Hair also has a sensory function due to sensory innervation by a hair root plexus surrounding the base of each hair follicle. Hair is extremely sensitive to air movement or other disturbances in the environment, much more so than the skin surface. This feature is also useful for the detection of the presence of insects or other potentially damaging substances on the skin surface. The distribution of hair on the human body changed dramatically during evolution as early humans evolved into a bipedal, walking species, differentiating themselves from their tree-dwelling ancestors. In the trees, primates live in the shade. Their bodies are cooler and sun exposure happens all over the body. Early humans, walking across the open savannah, were exposed to the sun primarily on their heads, as the sun was directly above their upright bodies. They also were likely much hotter than their tree-dwelling ancestors. Therefore, our modern body hair distribution evolved, with longer, denser hair on top of our heads to protect us from the sun and shorter, sparser hair over our bodies. Hair on the human body provides both UV protection and thermoregulation.

Body hair plays a more substantial role in thermoregulation in animals such as dogs and cats, most of which have a much heavier coat than most humans. Each hair root is connected to a small muscle called the **arrector pili** that contracts in response to nerve signals, making the external hair shaft "stand up." These signals are sent for two very different reasons. The upright body hairs trap a layer of air around the body, adding insulation. The skin is pulled up slightly around the hair as well, you may know this thermoregulatory function as "goosebumps." The nervous system may send signals to the arrector pili for a different reason, though, goosebumps can also form during the fear response. Humans still have the arrector pili muscle and associated nerve signals, though it is unlikely that goosebumps contribute much to our thermoregulation. Instead, recent research indicates that the repeated pulling on the hair root by the arrector pili in response to cold temperatures triggers hair growth within the follicle. Researchers found that arrector pili muscles are often lost in the scalps of people with baldness, which prompted them to question whether these muscles play a role in hair growth. The resulting experiments provided data to support this theory.

LO 6.2.2

So why does some hair grow long while other hair stops short? Why can't we braid our leg hair? The answer, like many things, is in your genes. Hair grows in cycles (**Figure 6.8**) with the two main portions of the cycle being the active growth and resting phases. A third phase occurs as a new hair shaft is grown within the bulb, forcing the old hair out (this is why we lose our hairs; on average, we lose about 100 hairs a day). Every one of your hair follicles has this cycle, and the phases last different lengths of time, which are determined genetically. Follicles on the tops of our heads spend a few years in the growth phase before transitioning to the resting phase. For the hair on our bodies, however, the cycle lasts just days or weeks instead of years.

Figure 6.8 Hair Growth Cycle

Hair follicles do not grow continuously. Rather, they alternate between growth and rest cycles. When a new growth phase begins, it pushes the old hair out of the follicle.

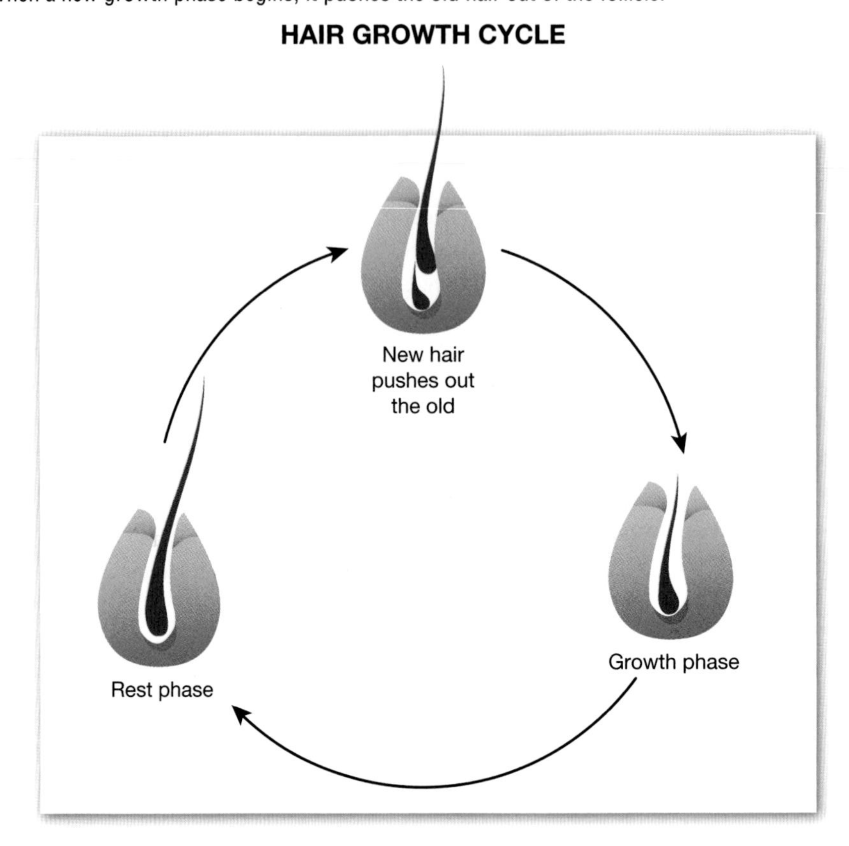

Cultural Connection

The Crown You Never Take Off

All it takes is a look around your classroom or lecture hall to begin to appreciate the abundant diversity in human hair. Curly, wavy, or straight, red, blonde, brown, and black are some basic descriptors, but the complexity is much deeper than those words can give justice to. One of the things that seems to be nearly universally true on every human head is that not all hairs are the same. You may have occasional straight hairs in your otherwise curly mane, or occasional red hairs among your brown ones.

The structure of each follicle and hair determines its unique properties. Hair grows from the bulb within the follicle and emerges out of the skin. The hair that we can see has an outer layer of cuticle, which may be 2–10 layers thick. If we use heat or products, these things affect the cuticle of the hair. Deep to the cuticle, the cortex is built of accumulated keratin and pigment. The relative amount and type of pigment determines your hair color. These pigments are produced within the bulb. Melanin (brown/black) and carotene (orange/red) are the two main pigments of hair. Therefore, if you choose to dye your hair you need the dye to access the cortex, below the cuticle. This is why bleaching and dyeing can be damaging to hair, because these processes affect both the cuticle and the cortex.

Some hairs have a medulla, and some hairs do not. The medulla may continue all the way through the length of the hair, or it may be interrupted. The presence or absence of the medulla determines a lot about your experience with your hair. Some ethnicities, Native Americans and Asians in particular, consistently have medulla in all their hairs. Coarse hair, such as beard hair, has a medulla that is doubly thick. Some hair, such as very fine, straight hair, lacks a medulla altogether.

The last biological and anatomical characteristic of hair is its curliness. Two factors that contribute to hair curl have been identified. The first is anatomical. Within the hair follicle, the growing hair travels through a tunnel to reach the skin surface. The shape of this tunnel may be straight or curly, and this plays a large role in determining the curl of the emerging hair. Another factor is that the keratin proteins themselves are different in curly hair. Within these hairs the keratin molecules form covalent bonds, holding them in a pattern that curls the hair. Chemical straightening products break these bonds, permanently straightening the hair. Many of these anatomical and biological traits are rooted in the genome and, to some degree, ethnicity.

As we age, all stem cells slow their mitotic capability, including those in the hair bulb. Over time, more and more hair will be lost. In addition, eventually the melanocytes in the hair bulb will die, and they will stop lending melanin (and color) to the hair shaft; if the hair continues to grow, it will be gray.

6.2b Nails

LO 6.2.3

The **nail bed** is the living component of nails, a specialized structure of the epidermis, the cells of which produce the **nail body**. The nail body, the hard, bladelike structure you might paint, functions to protect the tips of our fingers and toes, as they are the farthest extremities and the parts of the body that experience the most frequent mechanical stress (Figure 6.9). The nail body also may assist in picking up small objects and, of course, scratching an itch. The nail body is composed of densely packed dead keratinocytes. The nail body forms at the **nail root** (a protected region at the proximal side of the nail bed), which has a matrix of proliferating

Figure 6.9 Nails

Nails are accessory structures of the epithelium made of dead cells packed with keratin.

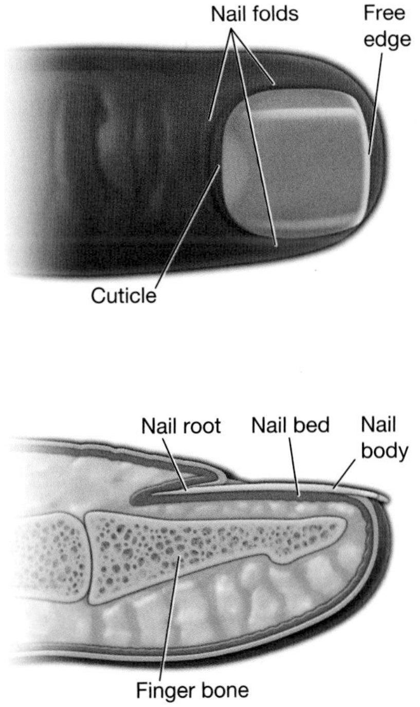

nail cuticle (eponychium)

cells from the stratum basale that enables the nail to grow continuously. All around the perimeter of the nails are flaps of skin called *nail folds* that help to anchor the nail body. At the meeting point between the proximal nail fold and the nail body is a narrow and thin strip of just epidermis, called the **nail cuticle** (also called the *eponychium*). The majority of the nail bed is rich in blood vessels, except at the base, where a thick layer of epithelium overlies the nail bed in a crescent-shaped region called the ***lunula***.

6.2c Sweat Glands

LO 6.2.4

We have talked a lot about human evolutionary adaptations to moving around on two feet. Humans are capable not only of walking bipedally, but also of running. In fact, humans are one of only two mammal species well adapted to long-distance endurance running (horses are the other). Most mammal species do not sweat, but humans, horses, and apes use sweat as a means of thermoregulation both when the body temperature rises due to external temperatures and when it rises due to exercise such as running. Humans sweat all over their bodies, and the vast majority of sweat glands exist for this function of thermoregulation. There is a different type of sweat gland present in the axilla and inguinal regions (armpits and groin) that functions a bit differently. These glands are different in structure, as explored in the "Anatomy of Sweat" feature, but also produce a different kind of sweat that may perform another function altogether. The theory is still under debate, but some scientists hypothesize that the sweat from the glands of the axilla and inguinal region contain chemicals called **pheromones**, a type of chemical signal that organisms can use to communicate with each other.

LO 6.2.5

Eccrine sweat glands are the type of sweat gland found all over the body; they produce sweat for thermoregulation. While these glands are found all over the skin's surface, they are especially abundant on the palms of the hands, the soles of the feet, and the forehead. They are coiled glands lying deep in the dermis, with the duct rising to a pore on the skin surface, where the sweat is released (see the "Anatomy of Sweat" feature). This type of sweat, released by exocytosis, making it merocrine-type secretion, is hypoosmotic compared to interstitial fluid. It is composed mostly of water, with some salt, antibodies, traces of metabolic waste, and **dermcidin**, an antimicrobial chemical. We tend to think of sweat as salty, but it is actually much lower in minerals than our other body fluids. When we release sweat onto the surface of the skin, the temperature of the body causes most of the liquid of sweat to evaporate, which causes a cooling effect. The water turns to vapor, leaving behind the small amounts of salts and other compounds on the surface. The skin surface therefore becomes saltier during the process of sweating. The dermcidin and antibodies left behind help control the growth of bacteria on the skin surface.

Student Study Tip

Apocrine glands develop with Age, After puberty! Eccrine glands are Ever-present at Every stage of life!

Apocrine sweat glands are usually associated with hair follicles and are found in densely hairy areas, such as armpits and genital regions. Apocrine sweat glands are larger than eccrine sweat glands and lie deeper in the dermis, sometimes even reaching the hypodermis, with the duct normally emptying into the hair follicle (see the "Anatomy of Sweat" feature). Apocrine sweat glands are so named because they were originally thought to produce sweat through apocrine-type secretion. Recently, that has been disproven; just like eccrine sweat glands, apocrine sweat glands secrete sweat through exocytosis (merocrine-style secretion). However, the composition of this sweat is different from that of eccrine glands. In addition to water and salts, apocrine sweat includes lipids and proteins. The result is a thicker, more viscous sweat that is also more nutritious for bacteria. Bacteria at the surface of the skin feed off apocrine sweat and produce their own waste products. The result is that sweat from apocrine glands

Anatomy of...

Sweat

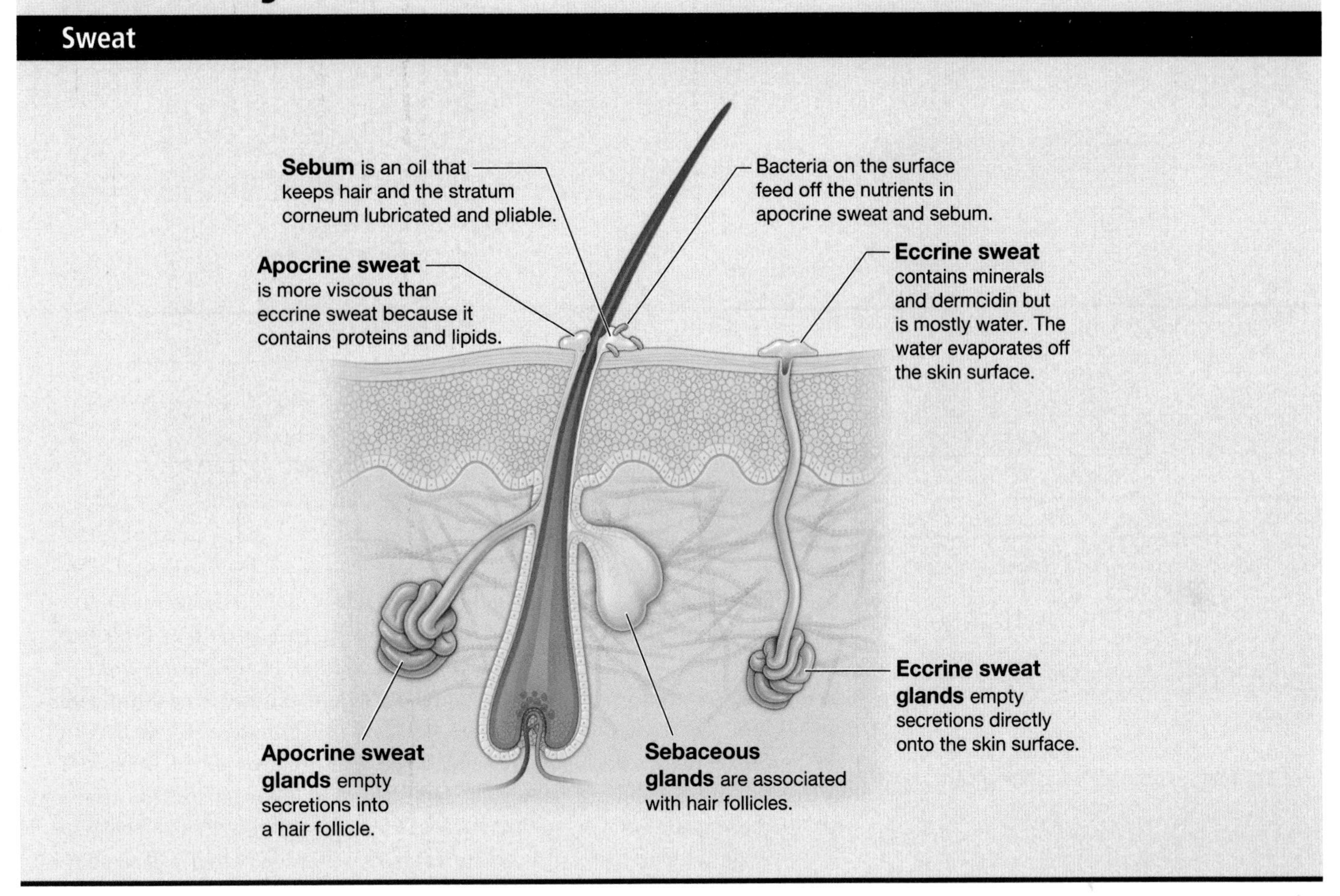

Sebum is an oil that keeps hair and the stratum corneum lubricated and pliable.
Bacteria on the surface feed off the nutrients in apocrine sweat and sebum.
Apocrine sweat is more viscous than eccrine sweat because it contains proteins and lipids.
Eccrine sweat contains minerals and dermcidin but is mostly water. The water evaporates off the skin surface.
Eccrine sweat glands empty secretions directly onto the skin surface.
Apocrine sweat glands empty secretions into a hair follicle.
Sebaceous glands are associated with hair follicles.

is associated with a smell. In addition, as noted previously, there is growing scientific evidence that these apocrine sweat glands may release pheromones, chemical signals that may be involved in sexual attraction. The release of this sweat is under both nervous and hormonal control. Underarm deodorant functions to mask these scents, while antiperspirants use an aluminum-based compound to form a plug and stop sweat from coming out of the pore.

6.2d Sebaceous Glands

A sebaceous gland is a type of oil gland that is found all over the body and helps to lubricate and waterproof the skin and hair. Most sebaceous glands are associated with hair follicles (**Figure 6.10**). They produce an oily mixture called **sebum**, using holocrine secretion. Sebum is a mixture of lipids that lubricates the hair and the dry stratum corneum, keeping it pliable. Sebum also has antibacterial properties and prevents water loss from the skin in low-humidity environments. The secretion of sebum is stimulated by hormones, especially androgens, many of which do not become active until puberty. Thus, sebaceous glands are relatively inactive during childhood.

Figure 6.10 Sebaceous Glands are Associated with Hair Follicles

Sebaceous glands secrete an oily product, sebum, that lubricates the dead cells of the hair shaft and the stratum corneum.

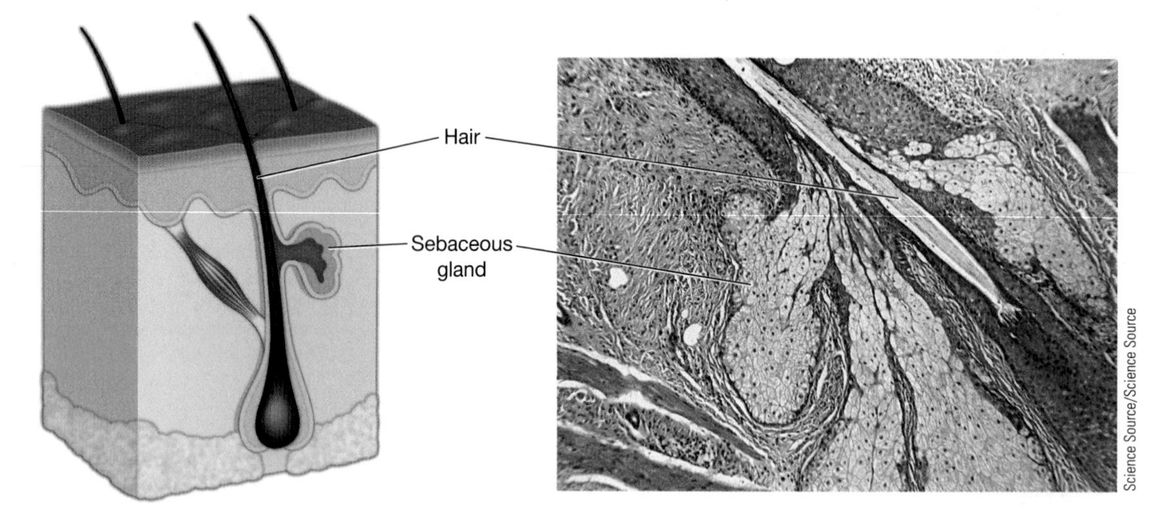

When hormone levels are changing or particularly high, especially in puberty, an overproduction and accumulation of sebum, along with keratin or dead cells, can block hair follicles. The bacteria (*Propionibacterium* and *Staphylococcus* species) that live around the follicle, feeding off sebum and apocrine sweat, will multiply and may trigger an immune reaction and inflammation (**Figure 6.11**). If neutrophils are involved, a white pus may be present (neutrophil reactions are discussed in more detail in Chapter 21), and the pus may turn black over time and with exposure to air.

As we age, the skin's accessory structures also have lowered activity, generating thinner hair and nails and reduced amounts of sebum and sweat. Reduced sweating ability can cause some older adults to be intolerant to extreme heat. Other cells in the skin, such as melanocytes and dendritic cells, also become less active, leading to a paler skin tone and lowered immunity.

Learning Connection

Chunking

Make a drawing of each of the methods of exocrine secretion (merocrine, holocrine, and apocrine). Of the three exocrine secretions we learned about in this chapter (eccrine sweat, apocrine sweat, and sebum), which ones are produced by each type of secretion? Does the method of secretion tell you anything about the content of their products?

Figure 6.11 Acne

Acne is the name for the disorder produced by an immune reaction that can occur when the hair follicle becomes inflamed. The bacteria that feed off sebum can overgrow, triggering inflammation.

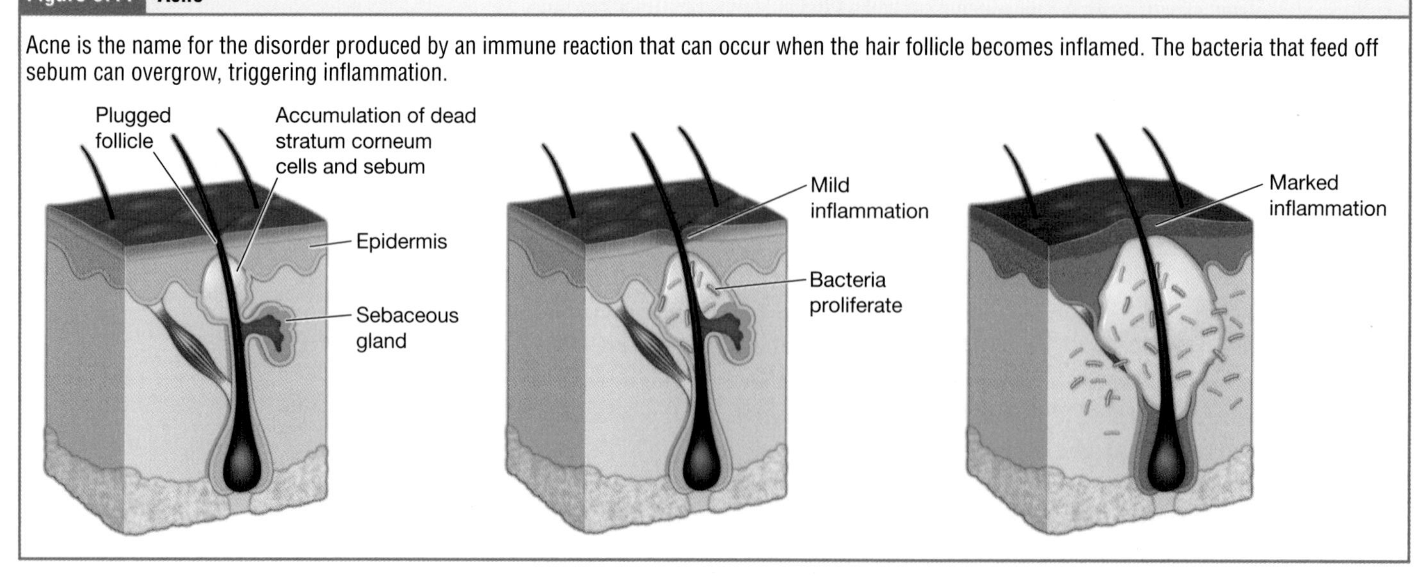

Learning Check

1. Your friend Shreya has curly hair. Which of the following explains why they have curly hair? Please select all that apply.
   a. The shape of the hair follicle causes the hair to grow curled.
   b. The structure of the cortex curls the hair as it grows.
   c. The strength of the arrector pili curls the hair as it grows.
   d. The shape of the dermal papilla curls the hair as it grows.
2. Identify the function of hair on our body.
   a. Keep the body cool when hot.
   b. Protect the body against foreign objects.
   c. Communicate with other organisms.
   d. Improve grip to pick up small objects.
3. Which of the following layers of the epidermis is found at the nail root?
   a. Stratum Corneum
   b. Stratum Granulosum
   c. Stratum Spinosum
   d. Stratum Basale
4. Which gland does deodorant work on to eliminate odors?
   a. Eccrine sweat glands
   b. Apocrine sweat glands
   c. Sebaceous glands
5. Which of the following glands would you expect to find on the back of your hand that would not be present on the palm of your hand? Please select all that apply.
   a. Sebaceous gland
   b. Eccrine sweat gland
   c. Apocrine sweat gland
   d. Hair follicle

6.3 Functions of the Integumentary System

HAPS LEARNING GOALS AND OUTCOMES

Learning Objectives: By the end of this section, you will be able to:

6.3.1 Explain how the integumentary system relates to other body systems to maintain homeostasis.

6.3.2 Explain how the integumentary system maintains homeostasis with respect to thermoregulation and water conservation.

The skin—the largest organ in the body—and its accessory structures are vital to the maintenance of homeostasis. The skin not only serves as a barrier, allowing the world within to maintain homeostatic conditions even when environmental conditions change, but the skin also performs a vital role in connecting other body systems to the outside world so that they can perform their own homeostatic functions.

6.3a Protection

The skin's structure allows it to withstand a variety of environmental conditions, so it can protect the more vulnerable parts of the body from wind, water, and UV sunlight. Due to the presence of layers of keratin and glycolipids in the stratum corneum, the skin also acts as a protective barrier against water loss. It also is the first line of defense against abrasive activity (friction, shear forces) and protects us from contact with grit, microbes, or harmful chemicals. Even sweat has protective functions by deterring microbes with its antibiotic ingredient, dermcidin.

LO 6.3.1

6.3b Sensory Function

An average mosquito weighs 5 mg (about 1/1000 the weight of a piece of paper) and yet, you can feel a mosquito land on your skin, allowing you to flick it off before it bites.

This is because the skin, and especially the hairs projecting from hair follicles in the skin, can sense changes in the environment. Nerve fibers surrounding the hair bulb sense a disturbance and transmit the information to the brain and spinal cord, which can then respond by activating the skeletal muscles of your eyes to see the mosquito and the skeletal muscles of the body to act in response.

The skin acts as a sense organ because the epidermis, dermis, and hypodermis contain specialized sensory nerve structures that detect touch, pressure, surface temperature, and pain. These receptors are more concentrated on the tips of the fingers, which are most sensitive to touch, especially the tactile corpuscles (**Figure 6.12**), which respond to light touch, and the **lamellated corpuscles** (also known as *Pacinian corpuscles*), which respond to pressure and vibration. Merkel cells, seen scattered in the stratum basale, are also touch receptors. In addition to these specialized receptors, there are **nociceptors**, which are nerve fibers specific to communicating pain to the brain and spinal cord, and **thermoreceptors**, which are sensory nerves adapted to detecting changes in temperature. As we learned earlier, there are also nerve fibers connected to each hair follicle. This rich innervation helps us sense our environment and react accordingly.

lamellated corpuscles (Pacinian corpuscle)

6.3c Thermoregulation

LO 6.3.2

The integumentary system helps regulate body temperature through its collaboration with the brain and spinal cord. Body temperature is specifically regulated by a division of the nervous system called the *sympathetic nervous system*; this part of the nervous system is best known for its involvement in our fear or fight-or-flight responses.

Figure 6.12 Sensory Innervation of the Skin

The skin is one of the main structures through which the central nervous system gains information about the environment. Thermoreceptors detect changes in temperature, tactile corpuscles sense touch, nociceptors relay pain information, and deep lamellated corpuscles sense pressure and vibration.

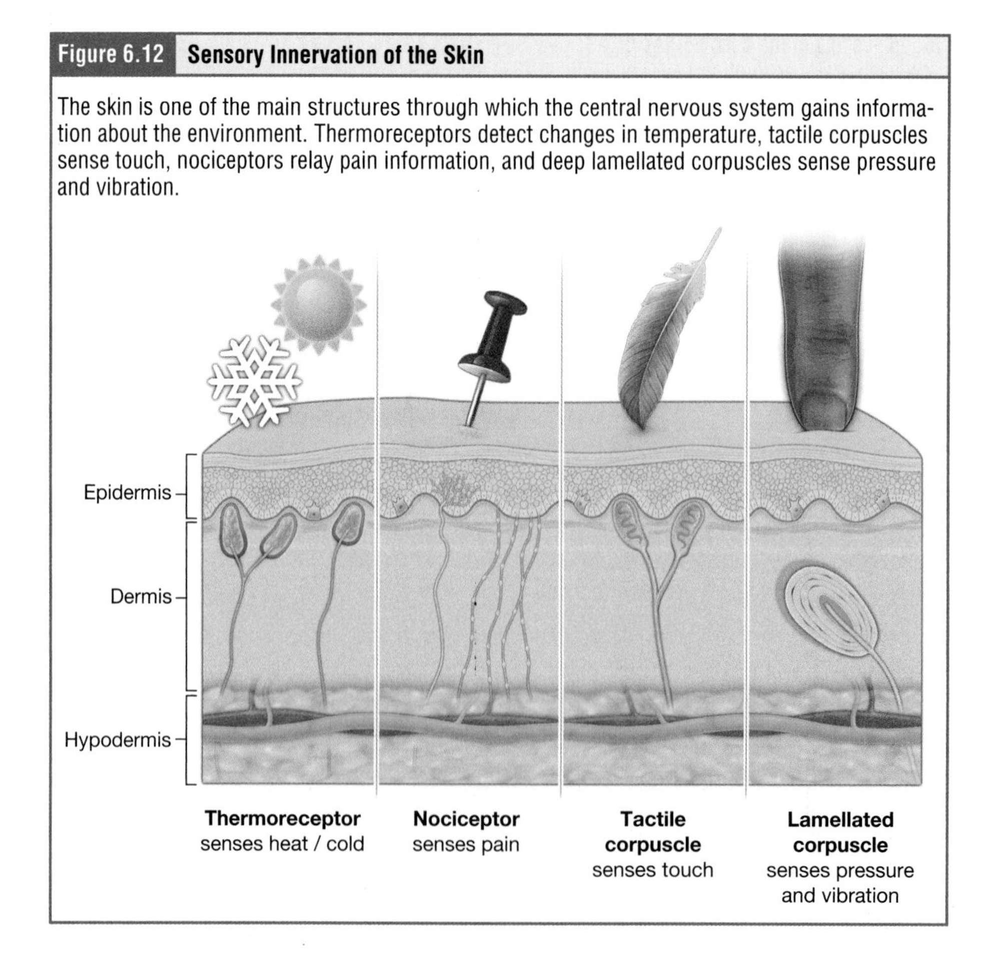

The sympathetic nervous system is continuously monitoring body temperature and initiating appropriate responses. Sweat glands are constantly working to keep the body temperature within the homeostatic range. Even when the body does not appear to be noticeably sweating, approximately 500 mL of sweat is secreted a day. If body temperature rises significantly beyond homeostatic range, either due to a warm environment or exercise, sweat glands will be stimulated by the sympathetic nervous system to produce large amounts of sweat, as much as 0.7–1.5 L per hour! When the sweat evaporates from the skin surface, the body is cooled as body heat is dissipated. In addition to sweating, blood vessels in the dermis dilate (vasodilation) so that excess heat carried in the blood can dissipate through the skin and into the surrounding environment (**Figure 6.13A**). The combination of increased sweat production and a hotter skin surface to promote evaporation is very effective in cooling the body. During times when the body temperature dips below the homeostatic range, these blood vessels in the dermis vasoconstrict, reducing heat loss at the skin's surface (**Figure 6.13B**). If the temperature of the skin drops too much (such as environmental temperatures below freezing), the conservation of body core heat can result in the skin actually freezing, a condition called *frostbite*.

6.3d Vitamin D Synthesis

The epidermal layer of human skin synthesizes **vitamin D** when exposed to UV radiation. In the presence of sunlight, a precursor molecule of vitamin D_3 is synthesized in

Figure 6.13 The Skin Contributes to Temperature Homeostasis

Blood vessels in the dermis vasodilate or vasoconstrict in order to control heat loss at the skin's surface. (A) When a person is hot, dermal blood vessels vasodilate, promoting heat loss. (B) When a person is cold, dermal blood vessels vasoconstrict, retaining heat in the body.

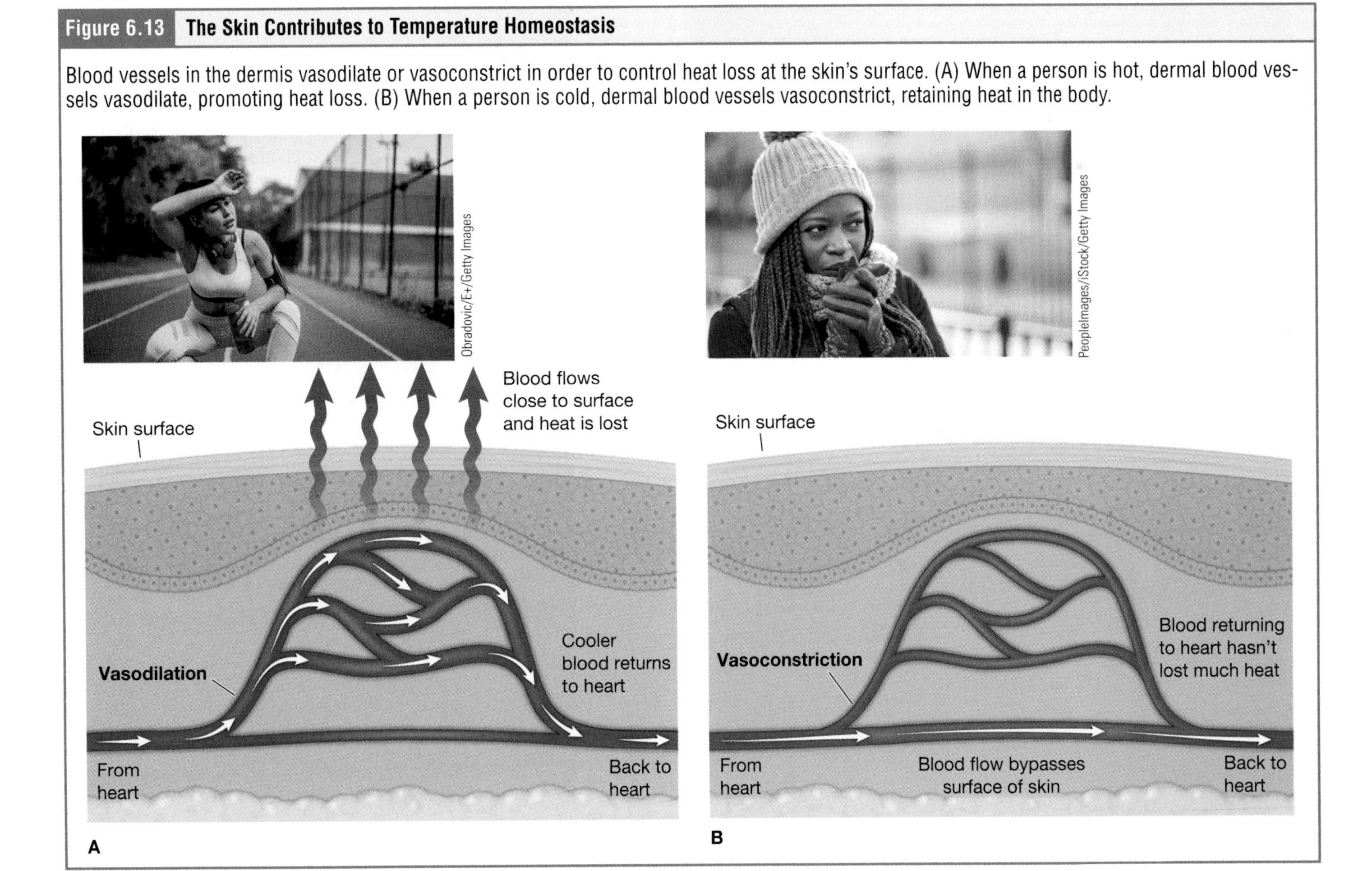

the skin from cholesterol. The liver and kidneys then finish the conversion to vitamin D. This vitamin is essential for normal absorption of calcium and phosphorous, which are required for healthy bones. The absence of sun exposure can lead to a lack of vitamin D in the body. A condition called **rickets**, a painful disorder in children where the bones become misshapen due to a lack of calcium, can occur. Children with rickets may have dramatically weakened bones (**Figure 6.14**). Older adult individuals who suffer from vitamin D deficiency can develop a condition called *osteomalacia*, a softening of the bones. In the United States, vitamin D is added as a supplement to many foods, including milk and orange juice, because many adults do not get enough sun exposure for sufficient vitamin D production.

In addition to its essential role in bone health, vitamin D is essential for general immunity against bacterial, viral, and fungal infections. Vitamin D has been found to be effective in immune responses against tuberculosis and COVID-19, the disease caused by the novel coronavirus. Recent studies are also finding a link between insufficient vitamin D and cancer.

Figure 6.14 Rickets

(A) Rickets is a disease in which insufficient vitamin D is present during bone development. The bones become soft and do not have sufficient strength and rigidity to hold the body's weight. (B) This normal presentation shows an X-ray of the lower limbs of a healthy child.

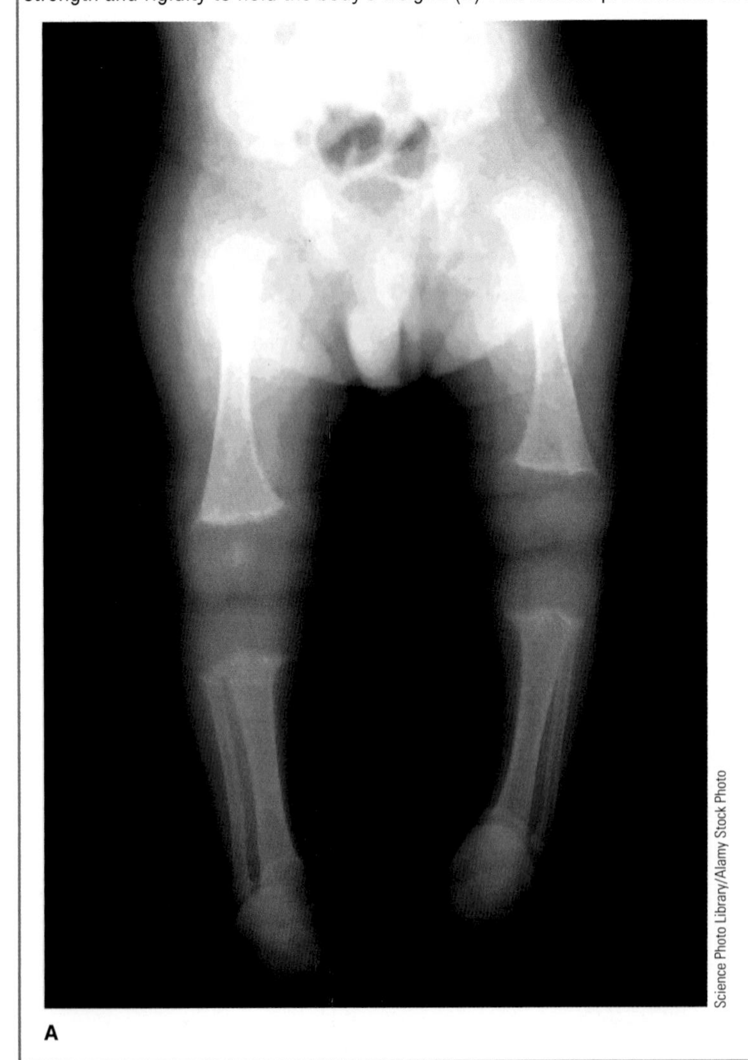

Science Photo Library/Alamy Stock Photo

A

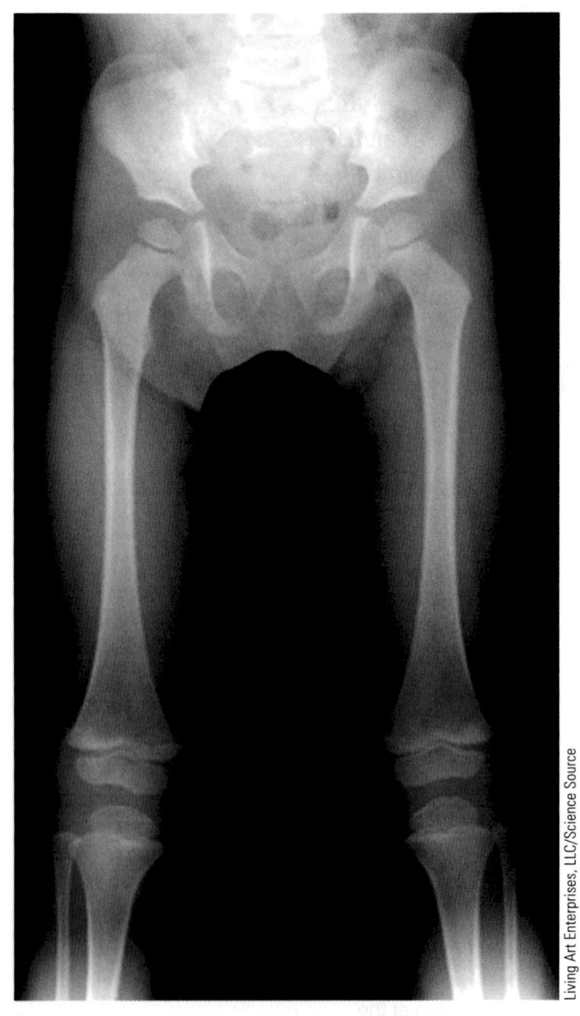

Living Art Enterprises, LLC/Science Source

B

6.4 Healing the Integument

Learning Objectives: By the end of this section, you will be able to:

6.4.1 Given a factor or situation (e.g., second-degree burns [partial-thickness burns]), predict the changes that could occur in the integumentary system and the consequences of those changes (i.e., given a cause, state a possible effect).

6.4.2 *Given a disruption in the structure or function of the integumentary system (e.g., blisters), predict the possible factors or situations that might have caused that disruption (i.e., given an effect, predict possible causes).

6.4a Injuries

Because the skin is the part of our bodies that meets the world most directly, it is especially vulnerable to injury. Injuries include burns and wounds, as well as scars, blisters, and calluses. They can be caused by sharp objects, heat, or excessive pressure or friction to the skin. Skin injuries set off a healing process that occurs in several overlapping stages. The first step to repairing damaged skin is the formation of a blood clot, which helps stop the flow of blood and scabs over with time (Figure 6.15A). Many different types of cells are involved in wound repair, especially if the surface area that needs repair is extensive. Before the basal stem cells of the stratum basale can recreate the epidermis, fibroblasts must mobilize and divide the epidermis, fibroblasts must mobilize, divide, and rapidly deposit collagen to form a bridge across which the epidermis can grow (Figure 6.15B). This quickly-made collagen bridge is called **granulation tissue**. Blood vessels infiltrate the area, bringing nutrients and an oxygen supply to the healing tissue. Immune cells, such as macrophages, roam the area and engulf any debris from the damaged tissue, or pathogens that invaded the wound, to reduce the chance of

Figure 6.15 Wound Healing

(A) The first step in wound healing is clotting of any broken blood vessels. A local inflammatory response includes chemicals that cause vasodilation and the attraction of white blood cells and fibroblasts. (B) Fibroblasts are recruited to the wound, and quickly create a collagen matrix to plug the wound. This collagen-rich tissue is called *granulation tissue*. Once granulation tissue is in place, epithelial cells can begin to migrate over the wound. (C) Epithelial cells fully close the wound; the epithelium and underlying dermis often have a different consistency than the surrounding tissue.

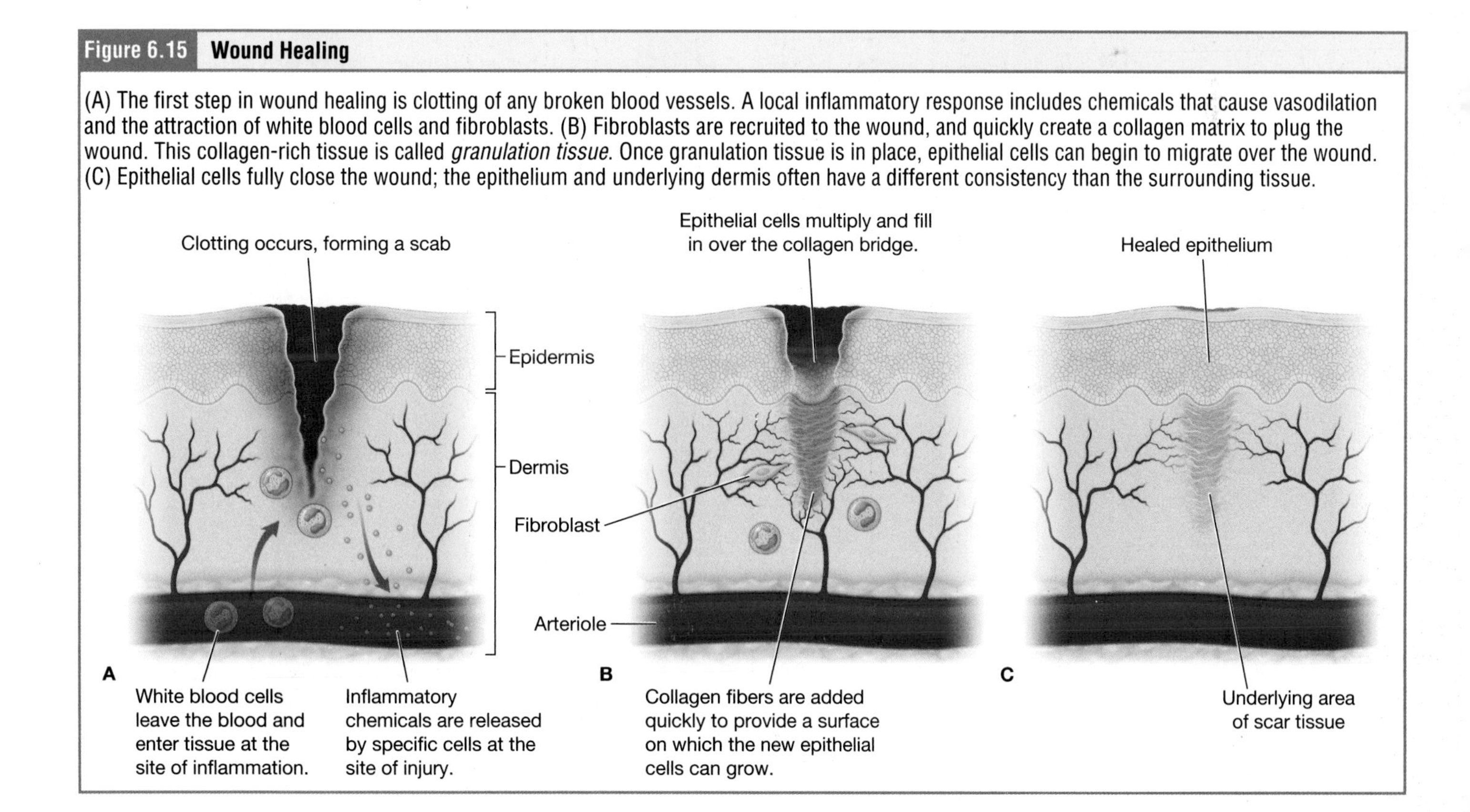

infection. Because the granulation tissue and epithelial covering are created so quickly, this healed area may have a different feel or texture to the surrounding skin. The dermis may be thicker, raised, or more rigid, and the epidermis may be thinner. The intense fibrous nature of the new tissue does not allow for the regeneration of accessory structures, such as hair follicles, sweat glands, or sebaceous glands. The combination of these characteristics is what we know of as a **scar**.

Sometimes the production of scar tissue continues after the wound is healed. The scar becomes raised compared to the tissue around it and is called a **keloid**. Scars can also have a sunken appearance (**atrophic scars**) if insufficient collagen is laid down in the healing process. Atrophic scars are more common following acne and chickenpox.

Burns A **burn** results when the skin is damaged by intense heat, radiation, electricity, or chemicals. The damage results in the death of skin cells. While the skin can typically heal, the danger of burns is that a massive loss of fluid and resulting dehydration, electrolyte imbalance, and infection can occur while the inside tissues of the body are exposed. Renal and circulatory failure can follow, which can be fatal. Burn patients are treated with intravenous fluids to offset dehydration and loss of nutrients.

Burns are sometimes measured in terms of the size of the total surface area affected. This is referred to as the *rule of nines*, which associates specific anatomical areas with a percentage that is a factor of nine (**Figure 6.16**). Burns are also classified by the depth of damage sustained. A **first-degree burn**, such as a mild sunburn, is a superficial burn that affects only the epidermis. Although the skin may be painful and swollen, these burns

Figure 6.16 Estimating the Size of a Burn

The amount of body area affected by a burn is a critical piece of data in the calculation of treatment options.

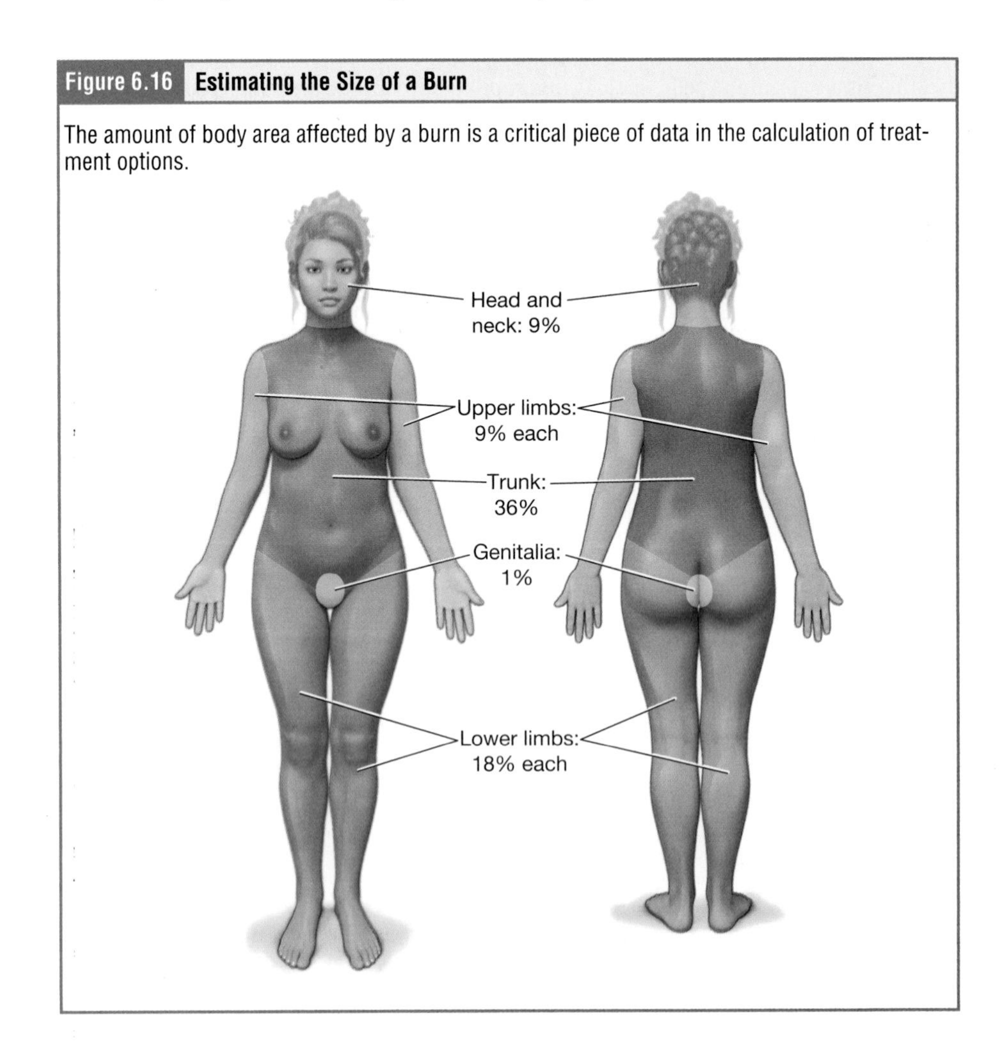

typically heal on their own within a few days. A **second-degree burn** goes deeper and affects both the epidermis and a portion of the dermis. These burns result in swelling and a painful blistering of the skin. It is important to keep the burn site sterile to prevent infection. If this is done, the burn will heal within several weeks. A **third-degree burn** fully extends into the epidermis and dermis, destroying the tissue and affecting the nerve endings and sensory function. These are serious burns that may appear white, red, or black; they require medical attention and will heal slowly without it (Figure 6.17). Counterintuitively, third-degree burns are usually not as painful because the nerve endings themselves are damaged and therefore can no longer communicate pain signals to the brain.

Student Study Tip

1st-degree burns: 1 layer of skin (the epidermis) is affected.

2nd-degree burns: 2 layers of skin (the epidermis and dermis) are affected.

3rd-degree burns: 3 layers of skin (the epidermis, dermis, and hypodermis) are affected.

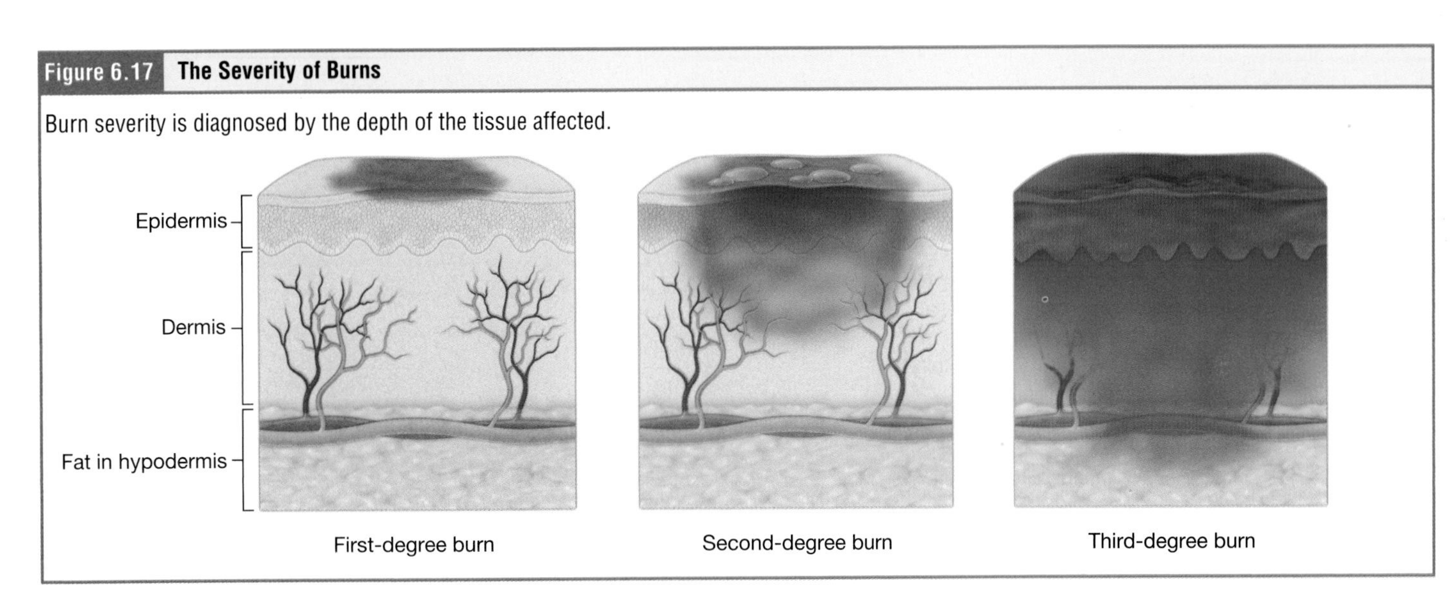

Figure 6.17 The Severity of Burns

Burn severity is diagnosed by the depth of the tissue affected.

DIGGING DEEPER: Skin Cancer

Cancer is a broad term that describes diseases caused by cells in the body that divide uncontrollably. Most cancers are identified by the organ or tissue in which the cancer originates. Skin cancer is extremely common in the United States, with approximately one in five Americans experiencing some type of skin cancer in their lifetime. The degradation of the ozone layer in the atmosphere and the resulting increase in exposure to UV radiation has contributed to its rise. Overexposure to UV radiation damages DNA, which can lead to the cellular changes that make cancer more likely. Although many tumors are benign (harmless), some produce cells that can mobilize and establish tumors in other organs of the body; this process is referred to as **metastasis**. Cancers are characterized by their ability to metastasize. As we have explored in this chapter, not all skin cells are the same and therefore skin cancers vary depending on which type of cell they originate from. Table 6.1 compares the three types of skin cancer.

Table 6.1 Skin Cancer

Cancer Type	Description	Diagram	Photograph
Basal cell carcinoma	**Basal cell carcinoma** is a form of cancer that affects the mitotically active stem cells in the stratum basale of the epidermis. It is the most common of all cancers that occur in the United States and occurs frequently in areas that are most susceptible to long-term sun exposure. Like most cancers, basal cell carcinomas respond best to treatment when caught early.	Basal cell carcinoma	PhotoMix/Alamy Stock Photo

(*Continued*)

Table 6.1 Skin Cancer *(Continued)*

Cancer Type	Description	Diagram	Photograph
Squamous cell carcinoma	**Squamous cell carcinoma** is a cancer that affects the keratinocytes of the stratum spinosum and is the second most common type of skin cancer. While squamous cell carcinomas are less common than basal cell carcinoma, they are more aggressive and can metastasize.	Squamous cell carcinoma	BSIP SA/Alamy Stock Photo
Melanoma	A **melanoma** is a cancer characterized by the uncontrolled growth of melanocytes, the pigment-producing cells in the epidermis. Typically, a melanoma develops from a mole which is a benign overgrowth of melanocytes. A mole becomes melanoma when these melanocytes lose control over the cell cycle and begin to proliferate. Melanomas usually appear as asymmetrical brown and black patches with uneven borders and a raised surface. It is the most fatal of all skin cancers, as it is highly metastatic and can be difficult to detect before it has spread to other organs.	Melanoma	Barry Mason/Alamy Stock Photo

Learning Check

1. Your friend pokes your back to get your attention. Which of the following structures detects this pressure?
   a. Tactile (Meissner's) Corpuscle
   b. Lamellated (Pacinian) Corpuscle
   c. Nociceptors
   d. Thermoreceptors
2. Which of the following could you see in a patient who is otherwise healthy but has had little exposure to sunlight over the past six months?
   a. Dark skin pigmentation
   b. Brittle, weakened bones
   c. Loss of muscle mass and strength
   d. Increased acne
3. You fall and scrape your knee deeply, down to the bone. Over time it heals, but will that area of the skin ever have the same properties as the skin that scraped off?
   a. Yes
   b. No
4. You are babysitting for your neighbor when you hear the child cry from the kitchen. You see the child sitting in front of the stove, holding their hand, which is starting to bleed, and screaming in pain. You look at their hand and see that they burned themself on the stove. What severity of burn could it be?
   a. First-degree
   b. Second-degree
   c. Third-degree

Chapter Review

6.1 Layers of the Skin

The learning objectives for this section are:

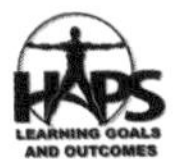

6.1.1 List the components of the integumentary system.

6.1.2 Describe the general functions of the integumentary system and the subcutaneous layer.

6.1.3 Identify and describe the tissue type making up the epidermis.

6.1.4 Identify and describe the layers of the epidermis, indicating which are found in thin skin and which are found in thick skin.

6.1.5 Describe the processes of growth and keratinization of the epidermis.

6.1.6 Compare and contrast thin and thick skin with respect to location and function.

6.1.7 Explain how each of the five layers, as well as each of the following cell types and substances, contributes to the functions of the epidermis: stem cells of stratum basale, keratinocytes, melanocytes, epidermal dendritic (Langerhans) cells, tactile (Merkel) cells and discs, keratin, and extracellular lipids.

6.1.8 Describe the factors that contribute to skin color.

6.1.9 Describe the functions of the dermis, including the specific function of each dermal layer.

6.1.10 Identify and describe the dermis and its layers, including the tissue types making up each dermal layer.

6.1.11 List and identify the types of tissues found in the hypodermis.

6.1.12 Describe the functions of the subcutaneous layer.

6.1.13 Describe the thermoregulatory role played by adipose tissue in the subcutaneous layer.

6.1 Questions

1. Which of the following is NOT a component of the integumentary system?

a. Skin
b. Hair
c. Nails
d. Teeth

2. Zo fell off the playground, scraped their elbow, and was bleeding. You see that only a skin layer has come off, but they do not seem to have broken bones. What larger effects are you concerned about due to the damage to Zo's skin?

a. The nervous tissues will not be able to communicate with Zo's arm.
b. The arm muscles could be weakened due to the damage of muscle tissue.
c. The superficial organs, such as muscles and bones, could be at risk for infection.
d. The bones in Zo's elbow will not be able to support their arm structurally.

3. What type of tissue makes up the epidermis?

a. Keratinized stratified squamous epithelium
b. Keratinized stratified cuboidal epithelium
c. Nonkeratinized stratified squamous epithelium
d. Nonkeratinized stratified columnar epithelium

4. Identify the correct order of epidermal layers from superficial to deep in thin skin.

a. Stratum Corneum, Stratum Granulosum, Stratum Spinosum, Stratum Basale
b. Stratum Basale, Stratum Spinosum, Stratum Lucidum, Stratum Corneum
c. Stratum Lucidum, Stratum Spinosum, Stratum Basale, Stratum Corneum
d. Stratum Corneum, Stratum Spinosum, Stratum Granulosum, Stratum Basale

5. In which layer would you find the oldest keratinocytes?

a. Stratum Basale
b. Stratum Lucidum
c. Stratum Corneum
d. Stratum Granulosum

6. You look at a slide of the epidermis under a microscope. You find that the epidermis does not have a stratum lucidum layer. Where would you expect to find this epithelium? Please select all that apply.

a. Palm of the hand
b. Soles of the feet
c. Surface of the back
d. Forehead

7. You close your eyes and put your hand in a bucket. You feel something wet and gooey. What structures of the epidermis are working to communicate this information to your brain?
   a. Merkel Cells
   b. Stem Cells
   c. Langerhans Cells
   d. Melanocytes

8. Which of the following is true of an individual with lighter skin?
   a. They have fewer melanocytes than do individuals with darker skin.
   b. Their body breaks down more melanin than in an individual with darker skin.
   c. Their melanocytes produce less melanin than those of an individual with darker skin.
   d. Their body breaks down more melanocytes than does that of an individual with darker skin

9. In which layer would you find the tactile corpuscles?
   a. Papillary Layer
   b. Reticular Layer
   c. Epidermis
   d. Hypodermis

10. Which of the following are true of the dermis and its layers? Check all that apply.
    a. The papillary layer is much thicker than the reticular layer.
    b. The reticular layer consists of dense irregular connective tissue.
    c. The dermis is approximately the same thickness as the epidermis.
    d. Fibroblasts produce both elastic and collagen fibers in the dermis.

Mini Case 1: Why is grandma always cold?

As people age, there are several reasons why they feel cold. Some of the causes are a decrease in metabolic rate and a decrease in the speed of circulation of blood through the body. Another cause involves the hypodermis (subcutaneous layer) under the skin.

1. Describe the tissue type(s) found in the hypodermis. Please select all that apply.
   a. Dense irregular connective tissue
   b. Dense regular connective tissue
   c. Areolar connective tissue
   d. Adipose tissue

2. What are the functions of the hypodermis? Check all that apply.
   a. To protect the internal organs from bacteria
   b. To connect the dermis to the underlying fascia
   c. To insulate the organs and skeletal muscles against heat loss
   d. To provide blood flow to the skeletal muscles

3. How does the hypodermis change as we age, that makes grandma always feel cold?
   a. The areolar connective tissue is replaced by dense regular connective tissue, which receives a poor blood supply.
   b. There is a decrease in the thickness of the subcutaneous fat, which reduces the thermoregulatory function of the hypodermis.
   c. The abundant mitochondria in white fat (which makes up the hypodermis) begin to die, which reduces the ability of the hypodermis perform fat breakdown.
   d. The number of sudoriferous glands in the hypodermis decreases as we age.

6.1 Summary

- The skin has three layers: the epidermis, the dermis, and the hypodermis.
- Depending on the location, the epidermis has either four layers or five layers. Every skin has, from superficial to deep, stratum corneum, stratum granulosum, stratum spinosum, and stratum basale. In places that receive a lot of friction, such as the palms of your hands or the soles of your feet, there is also stratum lucidum, which is between stratum corneum and stratum granulosum.
- The dermis is mostly connective tissue, and has two layers: the papillary layer and the reticular layer.
- The hypodermis is the deepest layer of the skin and connects the dermis to the fascia, which surrounds muscles and internal organs.

6.2 Accessory Structures of the Skin

The learning objectives for this section are:

6.2.1 Describe the structure and function of hair.

6.2.2 Describe the growth cycles of hair follicles and the growth of hairs.

6.2.3 Describe the structure and function of nails.

6.2.4 Describe the structure and function of exocrine glands of the integumentary system.

6.2.5 Explain the physiological significance of the presence or absence of sebaceous (oil) glands, sudoriferous (sweat) glands, and hair in the skin of the palms and fingers.

6.2 Questions

1. Which of the following hair structures is found in the deepest location?
   a. Hair shaft
   b. Hair root
   c. Hair bulb
   d. Hair papilla

2. Which of the following is true of hair growth?
   a. Growth cycles for hair all over the body last for several years
   b. When a hair is in its resting phase, it migrates to the skin surface and falls out before a new hair begins to grow in the hair follicle.
   c. Genetics help to determine the length of the phases of the growth cycle of hair.
   d. Arrector pili muscles play no role in hair growth.

3. Which of the following is a function of nails?
   a. Picking something small off the ground
   b. Warming up your body when cold
   c. Cooling your body when hot
   d. Communication with other organisms

4. Which of the following would trigger the sebaceous gland to secrete sebum?
   a. Increased stress, which increases androgenous hormones
   b. Playing outside, which increases your body temperature
   c. Being nervous, which makes your palms sweaty
   d. Going through puberty, which triggers hormonal changes

5. Eccrine sweat glands area abundant in the skin of the palms of the hands. What functions to eccrine sweat glands have in this location? Check all that apply.
   a. They help cool the body by releasing sweat, which evaporates from the skin, transporting the heat into the air.
   b. They produce oil to keep the skin waterproof.
   c. They carry pheromones to other humans that help with sexual attraction.
   d. They transport antimicrobial chemicals to the skin surface, to help prevent infection

6.2 Summary

- From deep to superficial, hair is composed of the hair papilla, the hair bulb, the hair root, and the hair shaft. In the center of the hair shaft is the medulla, which is surrounded by the cortex, which is surrounded by the cuticle. Hair helps maintain thermoregulation, protection, sensory input, and UV protection.
- Nails serve to protect the farthest extremities and parts that experience the maximum mechanical stress.
- There are three types of glands in the skin: (1) the eccrine sweat glands, which produce sweat for thermoregulation; (2) the apocrine sweat glands, which produce sweat and may release pheromones; and (3) sebaceous glands, which release sebum to lubricate hair and the stratum corneum.

6.3 Functions of the Integumentary System

The learning objectives for this section are:

HAPS LEARNING GOALS AND OUTCOMES

6.3.1 Explain how the integumentary system relates to other body systems to maintain homeostasis.

6.3.2 Explain how the integumentary system maintains homeostasis with respect to thermoregulation and water conservation.

6.3 Questions

1. The skin plays an important role in the production of vitamin D from cholesterol. Which of the following could result from a vitamin D deficiency in the body?
   a. A reduction in the production of red blood cells
   b. A decrease in calcium absorption, leading to a weakening of the bones
   c. A decrease in the production of visual pigments in the eye
   d. A reduction in the amount of sweat produced

2. Which of the following explains why you sweat when you have a fever?
   a. The bacteria in your body trigger sweat glands to secrete water vapor.
   b. The increased body heat triggers the nervous system to expel heat through water vapor.
   c. The body expels excess water from the increased intake of fluids through sweating.
   d. The body secretes sweat as a safety mechanism as dermcidin deters bacteria.

6.3 Summary

- The integumentary system functions to protect the body, sense changes in environment, regulate body temperature, and synthesize vitamin D.
- The integumentary system, which includes the skin and its accessory organs, is vital to the homeostasis of the body.
- The skin protects the body against wind, UV radiation, infection, water loss, and injury.
- The skin provides the sensations of touch, pressure, vibration, temperature, and pain.
- The skin helps to regulate body temperature through sweating and changes in blood flow.
- The skin helps in vitamin D synthesis. Vitamin D is important for bone health and immunity.

6.4 Healing the Integument

The learning objectives for this section are:

HAPS LEARNING GOALS AND OUTCOMES

6.4.1 Given a factor or situation (e.g., second-degree burns [partial-thickness burns]), predict the changes that could occur in the integumentary system and the consequences of those changes (i.e., given a cause, state a possible effect).

6.4.2 *Given a disruption in the structure or function of the integumentary system (e.g., blisters), predict the possible factors or situations that might have caused that disruption (i.e., given an effect, predict possible causes).

6.4 Questions

1. Your younger sibling falls at the playground and comes home crying because their knees are bleeding. What is the first step in repairing damaged skin?
   a. Laying granulation tissue
   b. Forming a blood clot
   c. Replicating basal stem cells
   d. Macrophages engulf debris

2. Blisters are caused by fluid buildup between the stratum spinosum and the stratum basale. Which of the following could explain why the blister forms between these two layers?
   a. Increased friction can separate layers of the epidermis.
   b. Increased pressure will separate the epidermis from the dermis.
   c. Infections can lead to increased blood flow between those layers of the epidermis.
   d. Wounds will increase blood flow to that region for healing of the skin.

6.4 Summary

- All wounds heal sequentially. First the blood clots, then there is granulation tissue laid in the region. Next, there is increased blood flow to the region to bring more macrophages to that region. At the same time, there is replication of the stratum basale. Eventually a scar will form.
- There are three types of burns. A first-degree burn affects only the epidermis. A second-degree burn affects the epidermis and part of the dermis. A third-degree burn affects the epidermis and the dermis.
- Three types of cancer can arise in the skin: (1) basal cell carcinoma, (2) squamous cell carcinoma, and (3) melanoma. Squamous cell carcinoma is the most treatable type, while melanoma is most likely to metastasize.

7

Bone Tissue and the Skeletal System

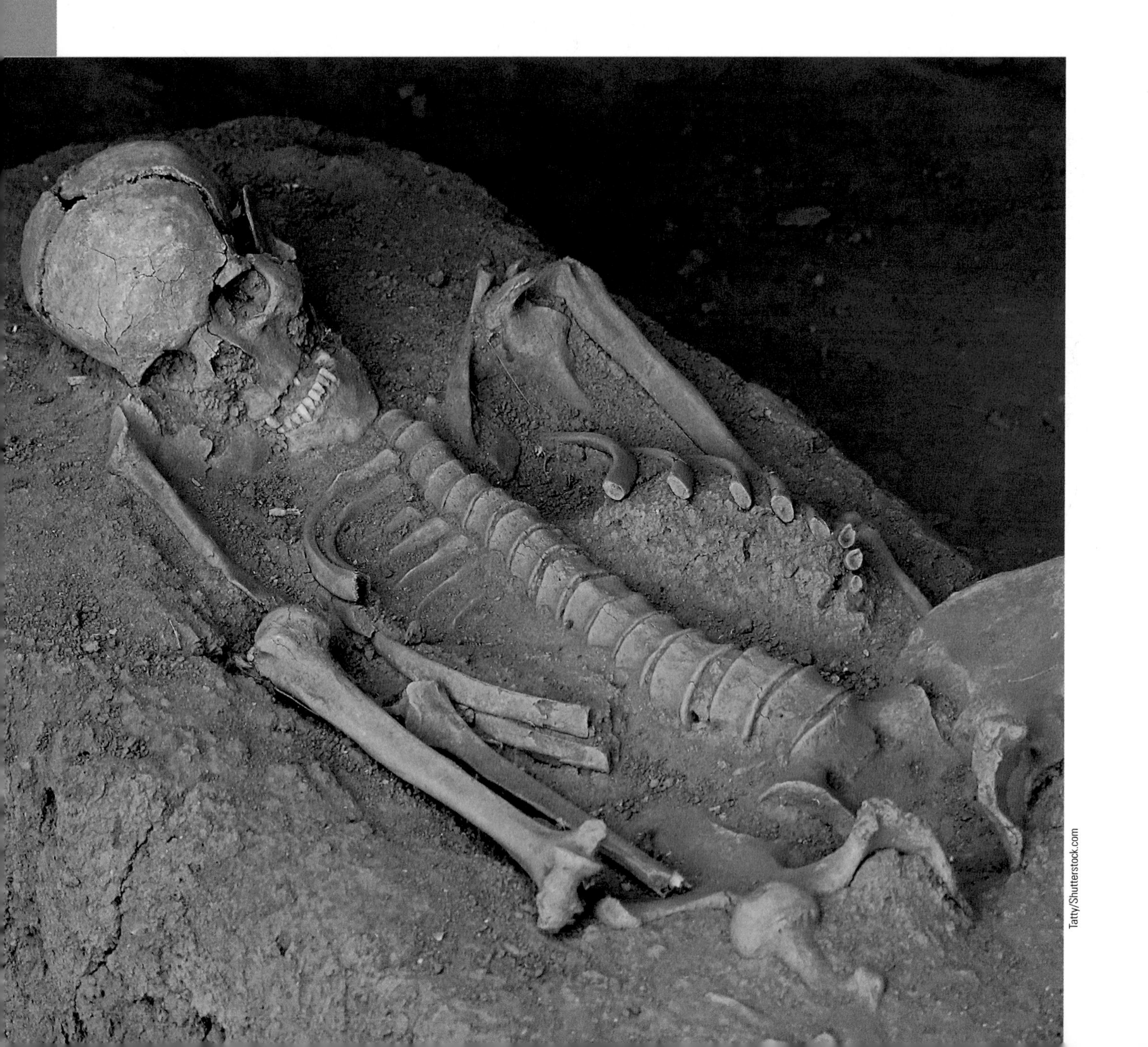

Tatty/Shutterstock.com

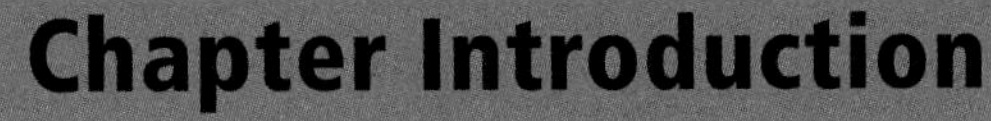

Chapter Introduction

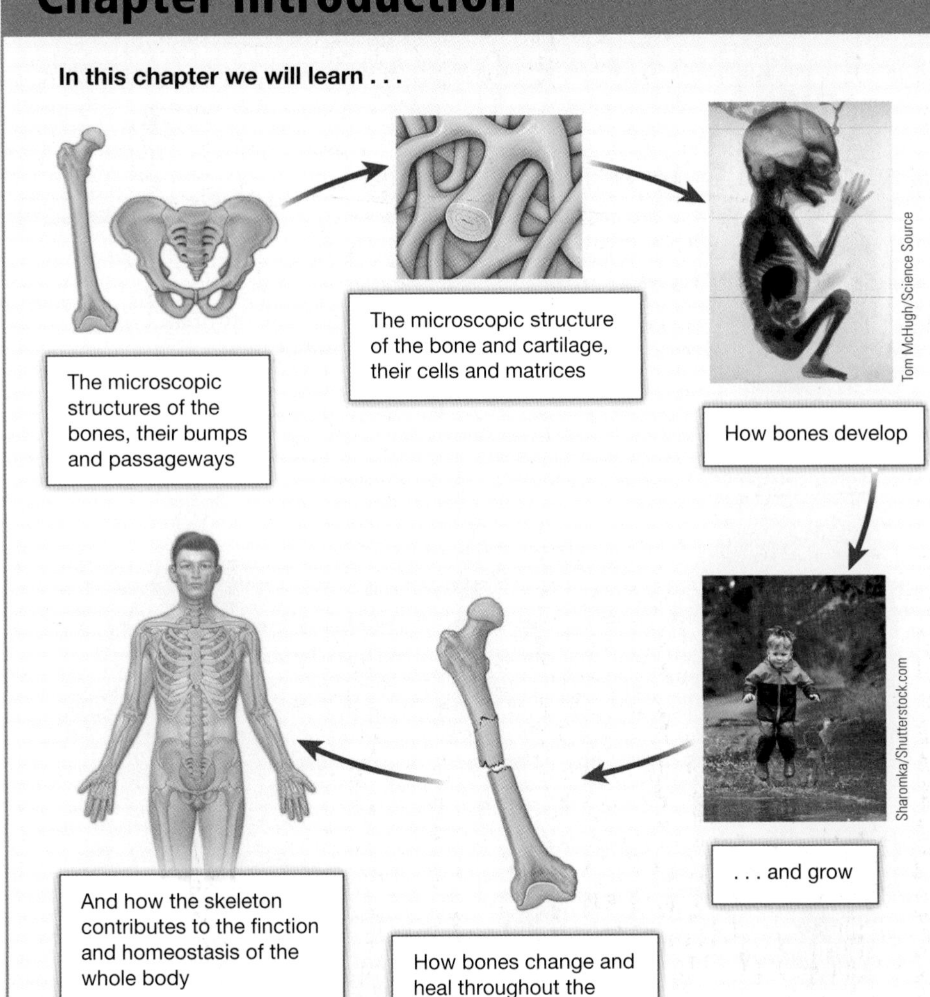

In this chapter we will learn . . .
The microscopic structures of the bones, their bumps and passageways
The microscopic structure of the bone and cartilage, their cells and matrices
How bones develop
Tom McHugh/Science Source
. . . and grow
Sharomka/Shutterstock.com
How bones change and heal throughout the lifespan
And how the skeleton contributes to the finction and homeostasis of the whole body

Skeletons and bones are symbols of death: they last long after the soft tissues of the body have disappeared. We can even visit dinosaur bones in museums millions of years after the animal has passed away! We tend to think of them as sturdy, nonliving parts of the body. The truth is that our bones are perhaps the most dynamic, ever-changing tissue we have. Your skeleton grows, repairs, and renews itself. The bones within it are active and complex organs that serve a number of important functions, including some necessary to maintain homeostasis.

Our skeletons are composed of both bone and cartilage connective tissues. **Bone** is hard and dense, so it is found in the locations where strength and rigidity are required. **Cartilage**, which is less rigid, provides flexibility and smooth surfaces for gliding.

7.1 The Functions of the Skeletal System

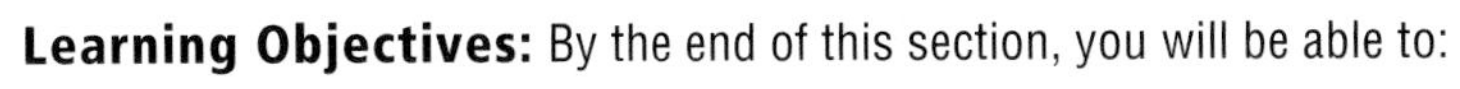

Learning Objectives: By the end of this section, you will be able to:

HAPS LEARNING GOALS AND OUTCOMES

7.1.1 Describe the major functions of the skeletal system.

7.1.2 Describe the locations of the three types of cartilage in the skeletal system.

7.1.3 Describe how the location and distribution of red and yellow bone marrow varies during a lifetime.

7.1.4* Predict disruptions to homeostasis if damage to the skeletal system occurred.

* Objective is not a HAPS Learning Goal.

LO 7.1.1

Just as the steel beams of a building provide a scaffold to support its weight, the bones and cartilage of your skeletal system compose the scaffold that supports the rest of your body. Without the skeletal system, you would be a limp puddle of organs, muscle, and skin! In addition to providing the structure to hold us upright, bones also facilitate

The Human Anatomy and Physiology Society includes more than 1,700 educators who work together to promote excellence in the teaching of this subject area. The HAPS A&P Learning Outcomes measure student mastery of the content typically covered in a two-semester Human A&P curriculum at the undergraduate level. The full Learning Outcomes are available at https://www.hapsweb.org.

movement by providing the rigid attachments for our muscles (**Figure 7.1A**). Bones also protect internal organs from injury by covering or surrounding them. For example, your ribs protect your lungs and heart, the bones of your vertebral column (spine) protect your spinal cord, and the bones of your cranium (skull) protect your brain (**Figure 7.1B**).

7.1a Support, Movement, and Protection

Bones are the organs of the skeletal system. Each bone contributes to body homeostasis in multiple ways. As with any organ, an individual bone is actually a combination of multiple types of tissue. Bone tissue itself can be either compact or spongy, and typically both of these are found in every bone organ (see the "Anatomy of a Typical Bone" feature). While these two types of connective tissues are made of the same components and have the same matrix, their architecture varies.

LO 7.1.2

Cartilage also contributes to the skeletal system. There are three types of cartilage tissue that we introduced in Chapter 5. Elastic cartilage is the rarest of the three

Figure 7.1 Bone Functions

(A) Bones provide muscle attachments that enable movement. (B) Bones provide hard, protective cases for our most vulnerable organs. (C) Bone matrix functions to store calcium and other minerals for the body. Calcium is removed from the bones to serve the other cells of the body as needed, but if too much mineral is removed, the bones may weaken. (D) The hollow interior of bones is filled with red and yellow bone marrow.

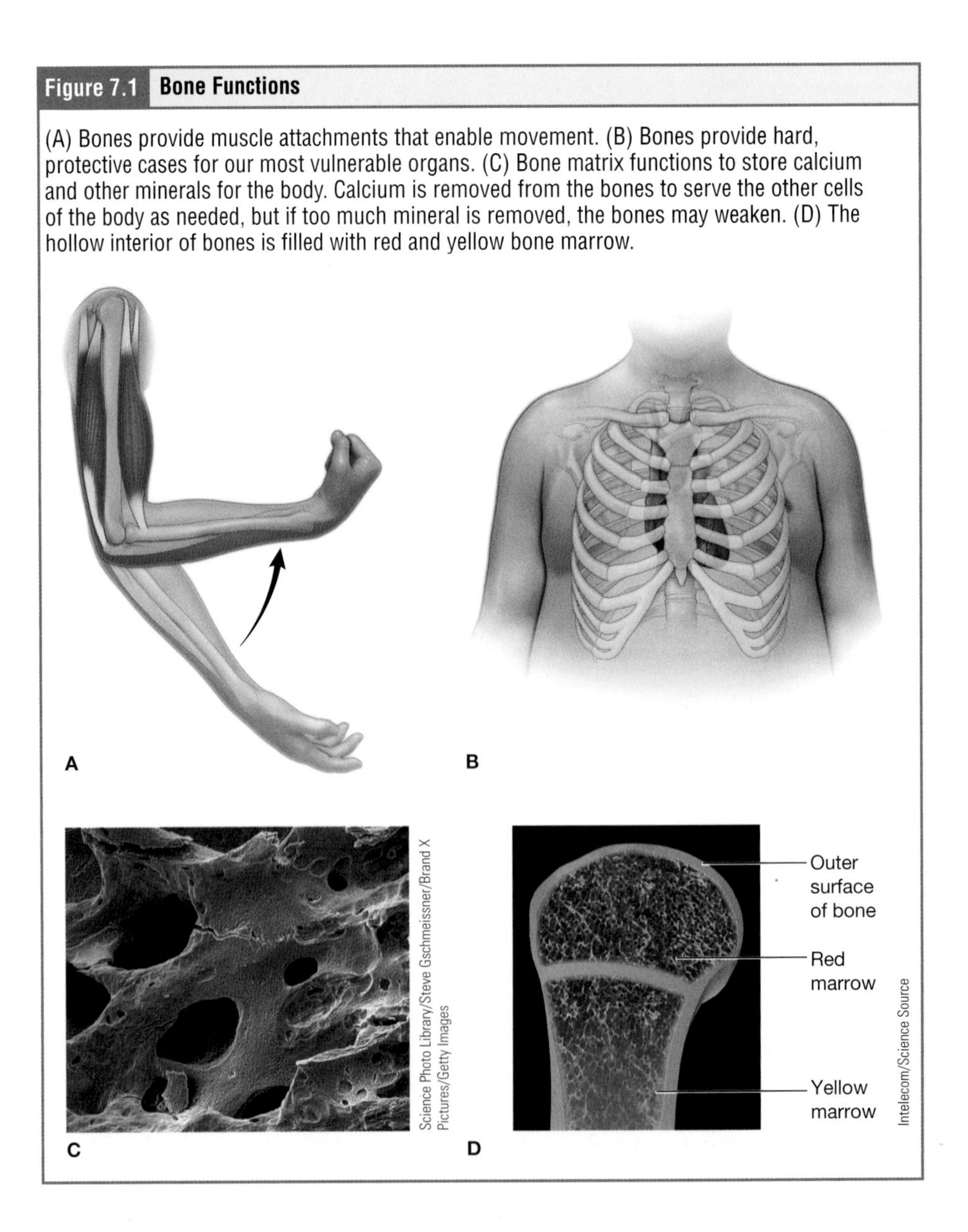

Anatomy of...

A Typical Bone

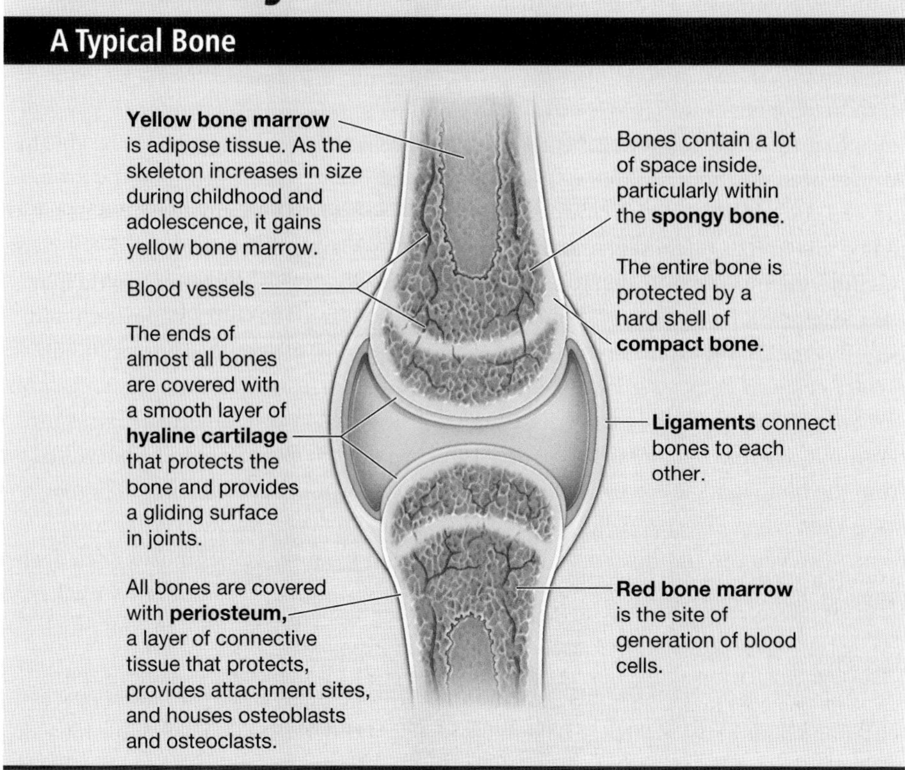
Yellow bone marrow is adipose tissue. As the skeleton increases in size during childhood and adolescence, it gains yellow bone marrow.
Blood vessels
The ends of almost all bones are covered with a smooth layer of **hyaline cartilage** that protects the bone and provides a gliding surface in joints.
All bones are covered with **periosteum,** a layer of connective tissue that protects, provides attachment sites, and houses osteoblasts and osteoclasts.
Bones contain a lot of space inside, particularly within the **spongy bone**.
The entire bone is protected by a hard shell of **compact bone**.
Ligaments connect bones to each other.
Red bone marrow is the site of generation of blood cells.

and does not play many roles in the skeletal system. As with any tissue that contains elastin proteins, **elastic cartilage** is found in places that need to stretch or change shape and then return. The three places where elastic cartilage is found in the human body are: the external ears, the tip of the nose, and the epiglottis, a structure in the airway.

Hyaline cartilage is the most abundant type of cartilage in the skeletal system. Hyaline is a smooth form of cartilage without many fibers. Due to its smooth nature, hyaline cartilage is found at the ends of bones where they form a joint with another bone. These locations, where bones move and glide past each other, depend on the presence of hyaline cartilage to reduce friction. Additionally, since hyaline (like all cartilages) is avascular and aneural—that is, it has no blood vessels or nerves—the cartilage not only physically helps the joints glide but also reduces any pain or damage to the underlying bones. A reduction in the amount of hyaline cartilage at the joints leads to **osteoarthritis**, a very painful disorder that we will learn more about in Chapter 10.

Fibrocartilage, on the other hand, has a dense array of collagen fibers. Because collagen lends strength to tissues, **fibrocartilage** is the form of cartilage that is uniquely suited to withstand pressure or weight-bearing stress. It is found as a component of joints that carry heavy loads, such as those between the vertebrae, within the knees, or at the pubic symphysis (the joint where the two halves of the pelvis meet).

While bone and cartilage compose the structure of the skeleton, there are other tissues that contribute. *Ligaments* are strips of dense regular connective tissue that tie bone to bone at the joints. *Tendons* are similar strips of dense regular connective tissue that attach muscles to the bones that they act on. Adipose tissue and bone marrow are found within our bones, as are blood vessels and nerves (see "Anatomy of a Typical Bone" feature).

7.1b Mineral Storage, Energy Storage, and Hematopoiesis

Bone tissue performs several critical homeostatic functions. Let's consider calcium for a moment. The mineral calcium is critical for body function. As you will learn in Chapter 13, calcium enables neurons to communicate with each other. As you will learn in Chapter 11, calcium is critical for muscle contraction. In Chapter 19 you will learn that the heart cannot beat without calcium. Therefore, you would not be able to survive even one hour without a sufficient amount of calcium in your blood. However, did you consider how much calcium you would need when planning your meals for today? Given that calcium is so critical for you to stay alive, why isn't "ingest sufficient calcium" the number-one item on your to-do list every day? The answer, as you may have suspected, is your bones. Bone extracellular matrix acts as a reservoir for a number of minerals important to the functioning of the body, especially calcium. When dietary calcium intake is greater than what is needed for current homeostasis, calcium can be built into the bone matrix for storage. When dietary calcium intake falls short of what is needed to maintain homeostasis on a given day, calcium can be harvested from the bones and added to the blood. This system works well to maintain homeostasis even when external factors (such as available nutrients) vary. Over the long term, however, the adding and removing of calcium from the bones must be relatively equal. If more calcium is removed from the bone than added, the bones may weaken and be at risk for fracture (**Figure 7.1C**). This process is explored more thoroughly in Section 7.5.

LO 7.1.3

Calcium and other minerals are stored in the bone matrix, but the hollow spaces within the bone serve as storage containers in a very different way. The hollow interior of bone serves as a site for fat storage as well as blood cell production. These two functions are quite different from each other, but because they both occur in tissues found within bones, the tissues are both called *bone marrow* (**Figure 7.1D**). There are two types of bone marrow: yellow marrow and red marrow. **Yellow marrow** is adipose tissue; the triglycerides stored in the adipocytes of the tissue can serve as a source of energy. **Red marrow** is where **hematopoiesis**—the production of blood cells—takes place. Red blood cells, white blood cells, and platelets are all produced in the red marrow. As the child grows and has more and more space in their skeleton, some of the space is filled with yellow bone marrow. By adulthood, red bone marrow is found only within the spaces in the spongy bone, and yellow bone marrow fills the long medullary cavity of long bones.

hematopoiesis (hemopoiesis)

Cultural Connection

Deep in the Marrow

In many cultures, bone marrow is a prized ingredient in cooking. You may have seen ads for bone marrow donation, or even may be a donor or recipient yourself. What is this rare and precious material? Red bone marrow holds the body's ability for hematopoiesis, the generation of blood cells. All the cells circulating in your blood were generated from stem cells that reside in the bone marrow. When you are sick, or even when you are recently vaccinated, the immune cells in the bone marrow reproduce wildly, causing that familiar bone ache sensation that we experience during some illnesses. These stem cells need to not only reproduce to help fight infection and produce immune responses, but they are constantly replenishing the supply of red blood cells in the blood. Thus, these are highly mitotic cells.

Occasionally, cancer can develop in the bone marrow. Multiple myeloma, lymphomas, leukemia, and childhood leukemia are all cancers that occur in the bone marrow. These cancers lead to increased production of blood cells and overcrowding of healthy blood cells in the red bone marrow and blood. Volunteers can offer to donate their bone marrow to patients with these diseases, especially leukemia. To donate, volunteers can sign up for the National Bone Marrow Registry through which they might be matched to a recipient.

Apply to Pathophysiology

The Necessity of Collagen

1. Collagen production accounts for 40 percent of the protein-making activity in the human body. Which of the following are roles of collagen? Please select all that apply.
   A) Provide a waxy barrier in the skin to repel water.
   B) Provide tensile strength to the skin to withstand movement.
   C) Provide a template for bone development and growth.
   D) Allow for the deposition of minerals in bone for rigidity.
2. Osteogenesis imperfecta is a genetic disease in which mutations within the genes for collagen prevent healthy collagen formation. Scurvy is a nutritional disease in which a dietary component (vitamin C) is deficient and healthy collagen cannot form. Which of the following would be a shared characteristic of both diseases?
   A) Bones that break easily **LO 7.1.4**
   B) Digestive issues, including diarrhea
   C) Muscles that cannot relax after contraction (spastic paralysis)
   D) A compromised immune system
3. Osteogenesis imperfecta impacts the formation of stretchy collagen. Given this, what other process would you predict would be impacted by this disease?
   A) Hearing
   B) Vision
   C) Digestion
   D) Excretion
4. In two adult skeletons, one with osteogenesis imperfecta and the other with healthy levels of collagen production, which skeleton will contain more yellow bone marrow?
   A) Osteogenesis imperfecta
   B) Healthy skeleton
   C) They would have the same amount of yellow bone marrow

Learning Check

1. The metacarpal joint is where two bones in your hands form a joint. This is a very mobile joint that allows the bones to move past each other. This joint does not need to withstand weight-bearing stress. What kind of cartilage would you expect to find here?
   a. Hyaline cartilage
   b. Elastic cartilage
   c. Fibrocartilage
2. Which of the following is NOT a function of the skeletal system?
   a. Supporting your body
   b. Protecting your organs
   c. Maintaining homeostasis of minerals
   d. Transporting minerals to your body
3. Explain the role of calcium in increasing bone density.
   a. Calcium ions are compacted into bone.
   b. Calcium ions are stored in the extracellular matrix.
   c. Calcium ions increase production of bone cells.
   d. Calcium ions increase red marrow in the bone.

7.2 Bone Classification

Learning Objectives: By the end of this section, you will be able to:

7.2.1 Classify bones of the skeleton based on their shape.

7.2.2 Identify and describe the structural components of a long bone, and explain their functions.

7.2.3 Define common bone marking terms (e.g., condyle, tubercle, foramen, canal).

The 206 bones in the adult skeleton can be sorted into five categories based on their shape and function (**Figure 7.2**).

LO 7.2.1

Figure 7.2 Classifications of Bones

Bones are classified according to their shape.

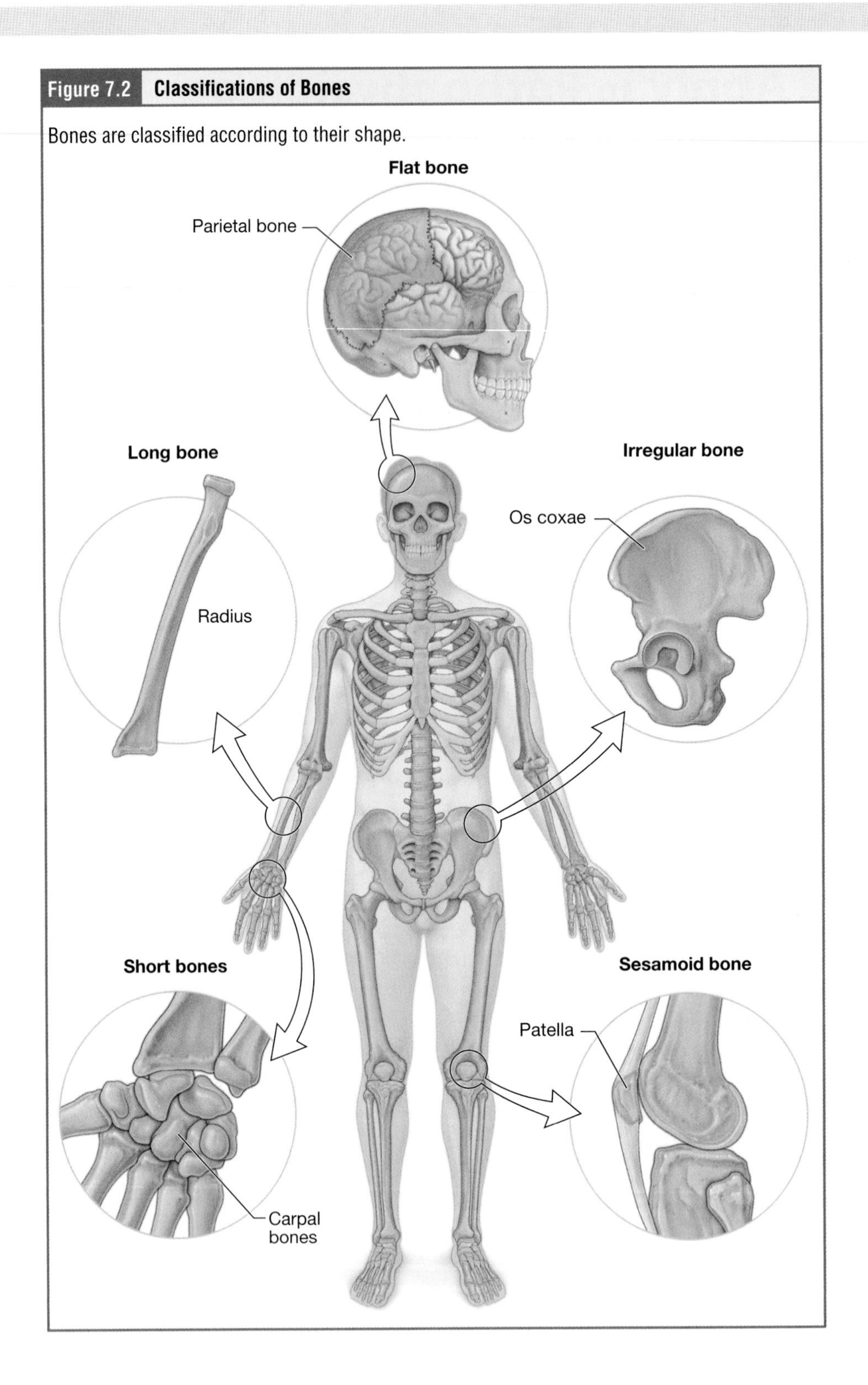

7.2a Long Bones

A **long bone** is one that is cylindrical in shape, longer than it is wide. Keep in mind, however, that the term describes the shape of a bone, not its size. Common examples of long bones are found in the arms (humerus, ulna, radius) and legs (femur, tibia, fibula), but the much smaller bones of the fingers (metacarpals, phalanges) and toes (metatarsals, phalanges) are long bones as well because they are still longer than they are wide. Long bones function as levers; they move when muscles contract.

7.2b Short Bones

A **short bone** is one that is cube-like in shape, being approximately equal in length, width, and thickness. The only short bones in the human skeleton are in the carpals of the wrists and the tarsals of the ankles. Short bones provide stability and support as well as some limited motion.

7.2c Flat Bones

The term ***flat bone*** is somewhat of a misnomer because, although a flat bone is typically thin, it is also often curved. Examples include the cranial (skull) bones, the scapulae (shoulder blades), the sternum (breastbone), and the ribs. Almost all of the flat bones in the body have a primary function of protection of internal organs (skull, sternum, ribs), but they all also serve as points of attachment for muscles.

7.2d Irregular Bones

An **irregular bone** is one that does not have any easily characterized shape and therefore does not fit any other classification. These bones tend to have more complex shapes; one example is the vertebrae, which both protect the spinal cord and attach to many muscles to enable the complex movements of the trunk and torso. Many facial bones, particularly the ones containing sinuses, are classified as irregular bones.

7.2e Sesamoid Bones

A **sesamoid bone** is a small, round bone that is suspended within a tendon or ligament (see Figure 7.2 and **Figure 7.3A**). The unusual name of these bones comes from their resemblance to sesame seeds. These bones form within tendons when a great deal of pressure or friction is generated regularly over time. The sesamoid bones protect tendons by helping them overcome compressive forces. For example, sesamoid bones often develop in the tendons in the soles of the feet—especially in runners (**Figure 7.3B**). Sesamoid bones vary in number and placement from person to

Figure 7.3 Sesamoid Bones

Sesamoid bones develop within tendons over time due to friction.

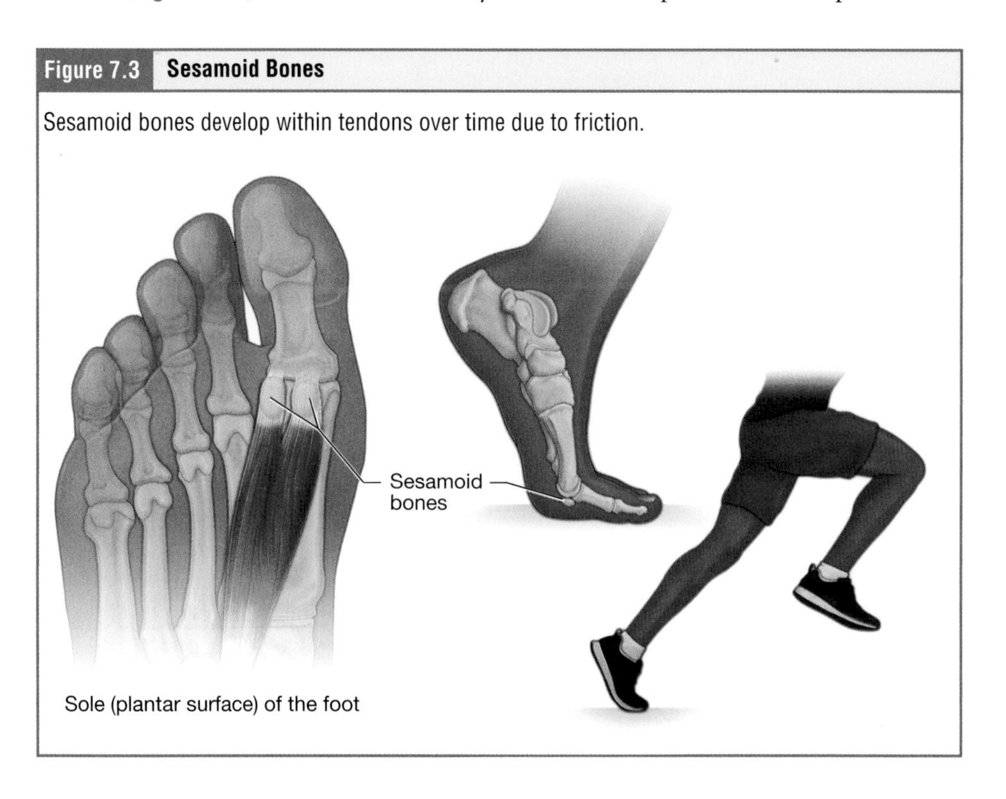

Table 7.1 Bone Classifications

Bone Classification	Features	Function(s)	Examples
Long	Cylinder-like shape, longer than it is wide	Leverage	Femur, tibia, fibula, metatarsals, humerus, ulna, radius, metacarpals, phalanges
Short	Cube-like shape, approximately equal in length, width, and thickness	Provide stability, support, while allowing for some motion	Carpals, tarsals
Flat	Thin and curved	Points of attachment for muscles; protectors of internal organs	Sternum, ribs, scapulae, cranial bones
Irregular	Complex shape	Protect internal organs	Vertebrae, facial bones
Sesamoid	Small and round; embedded in tendons	Protect tendons from compressive forces	Patellae

person but are typically found in tendons associated with the feet, hands, and knees. The patellae (singular = patella) are the only sesamoid bones found in common with every person. **Table 7.1** reviews bone classifications with their associated features, functions, and examples.

Learning Check

1. Classify the humerus (bone in your upper arm) based on its shape.
   a. Long
   b. Short
   c. Flat
   d. Irregular
   e. Sesamoid
2. Classify the lunate (bone in your wrist known as a carpal bone) based on its shape.
   a. Long
   b. Short
   c. Flat
   d. Irregular
   e. Sesamoid
3. Classify the sphenoid bone (bone in your skull that makes up your face) based on its shape.
   a. Long
   b. Short
   c. Flat
   d. Irregular
   e. Sesamoid

LO 7.2.2

Student Study Tip

We see the word *medulla* in many places across the body. It refers to the *internal* region of a structure. Bone marrow resides *inside* the bone, in the *medullary* cavity.

All long bones share a set of characteristics (**Figure 7.4**). A long bone has two ends, which are called the **epiphyses** (singular: epiphysis). Since long bones are found in the limbs, we can use the limb-specific terms to differentiate them. The proximal epiphysis is the end of the bone closest to the trunk, and the distal epiphysis is the end of the bone further from the trunk. The shaft between the two epiphyses is the **diaphysis**. If we cut into the bone, we can see that the walls of the entire bone diaphysis are composed of dense and hard **compact bone**. Bones are not completely solid, however. We can see there is a long, hollow region within the diaphysis called the **medullary cavity**, which is filled with bone marrow and the epiphysis is filled with spongy bone. In a child, all the spaces within bones are filled with red marrow, the type of bone marrow that is capable of generating blood cells.

Each epiphysis meets the diaphysis at the **metaphysis**, the narrow bone region that contains the epiphyseal plate (**Figure 7.5**). The **epiphyseal plate** (growth plate) is a layer of hyaline cartilage within a growing bone. When the bone stops growing in early adulthood (approximately 18–21 years), the cartilage is replaced by osseous tissue and the epiphyseal plate becomes an epiphyseal line. Figure 7.5 compares a long bone from a child with one from an adult. Bone growth is explored more in Section 7.4.

Bone tissue is lined on every surface with a layer of dense irregular connective tissue. The inner surfaces and medullary cavity are lined with **endosteum**, and the outer surface of the bone is covered with the **periosteum** (**Figure 7.6**). These two tissues contain cells capable of creating bone matrix, and so they are the locations where bone growth, repair, and remodeling occur. The periosteum contains blood vessels, nerves, and lymphatic

epiphyseal plate (growth plate)

Learning Connection

Image Translation

Cartilage is much less dense and mineralized than bone, so much so that it is hardly visible in X-rays. What do you think the X-ray of a child's long bone would look like?

Figure 7.4 The Common Structures of Long Bones

A typical long bone with the common features labeled.

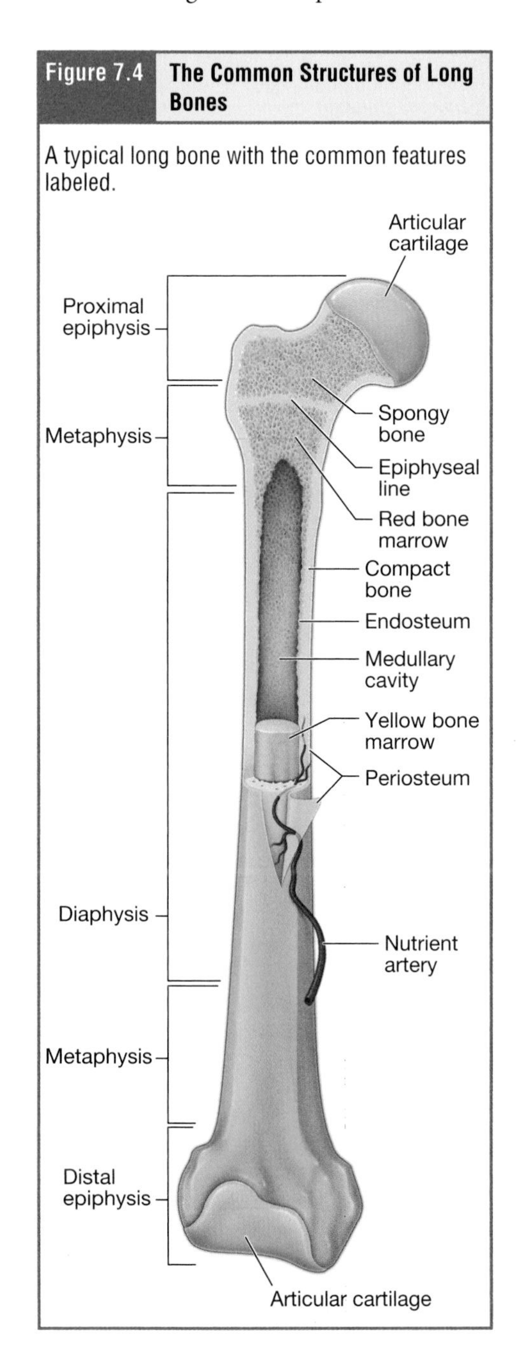

Figure 7.5 Epiphyseal Plates and Epiphyseal Lines

(A) Epiphyseal plates are zones of cartilage in growing bones. (B) At the end of puberty, the cartilage is ossified and becomes bone. An epiphyseal line is still visible.

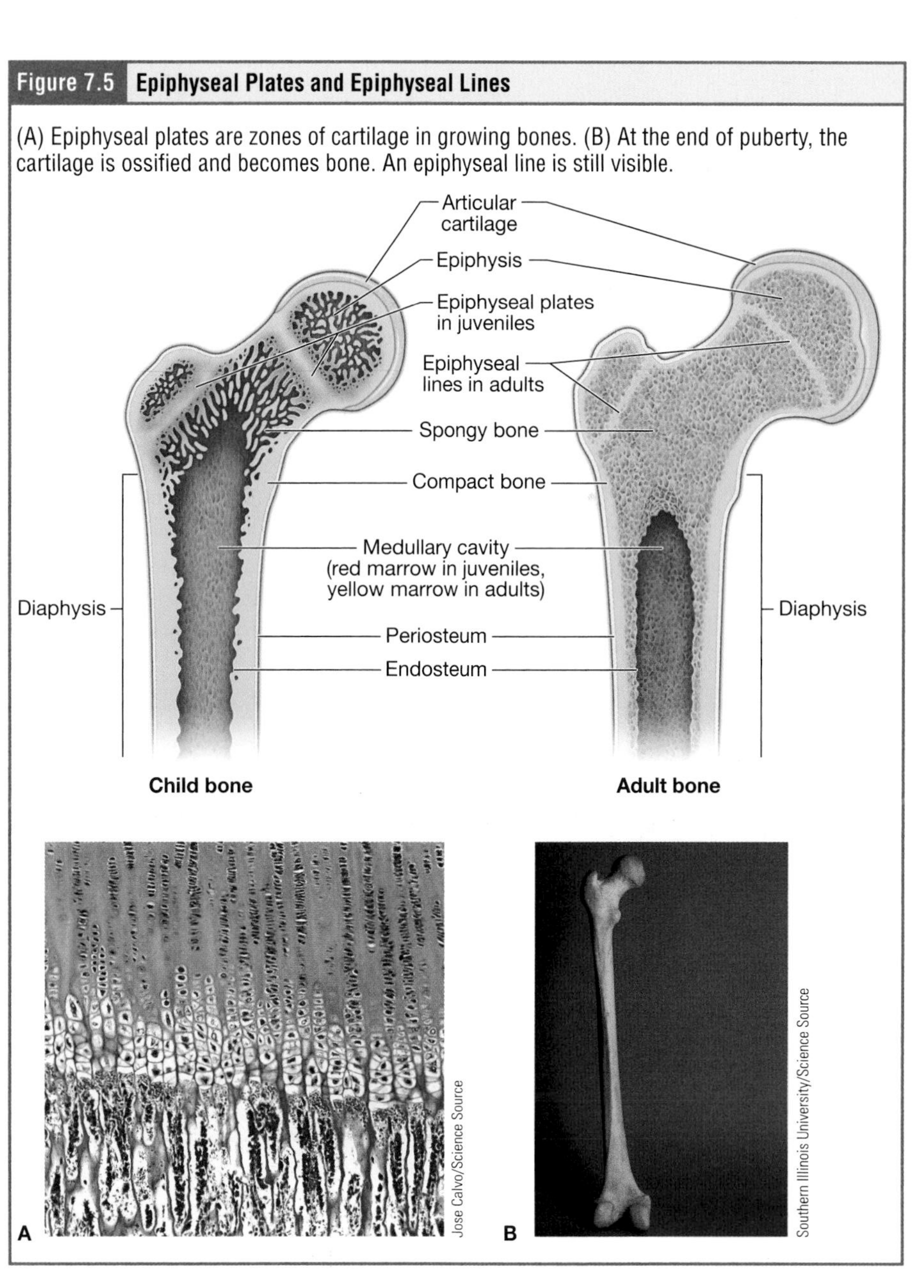

Figure 7.6 The Periosteum and Endosteum

The periosteum forms the outer surface of bone, and the endosteum lines the medullary cavity.

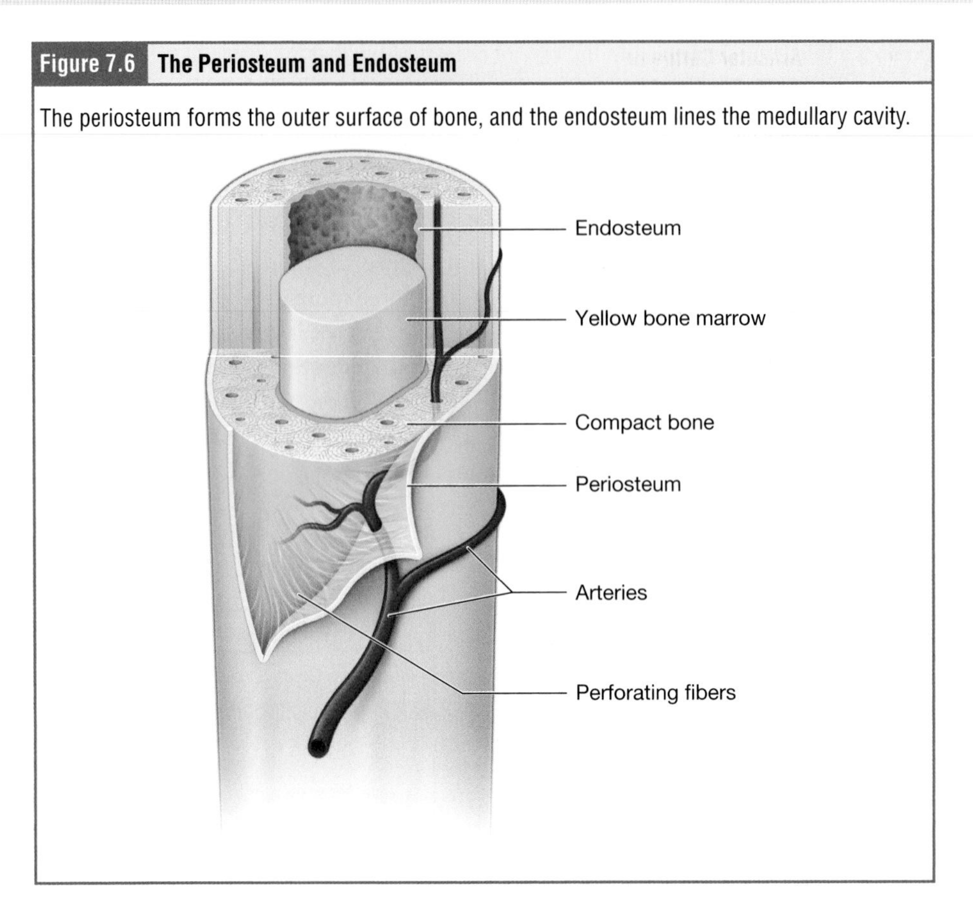

vessels that nourish compact bone. Tendons and ligaments also attach to bones at the periosteum by weaving their collagen fibers into the periosteum matrix (**Figure 7.7**). Thus, the periosteum is subjected to the entire pulling force of muscle contraction. To ensure that the periosteum does not detach from the bone, the collagen fibers of the periosteum weave deep into the compact bone matrix itself. These fibers, called **perforating fibers**, make tiny holes in the outer surface of the bone. The periosteum covers the entire outer surface except where the epiphyses meet other bones to form joints (**Figure 7.8**). In this region, the bone matrix of the epiphyses is covered with **articular cartilage**, a thin

Figure 7.7 The Periosteum

The periosteum provides an attachment site for tendons and ligaments.

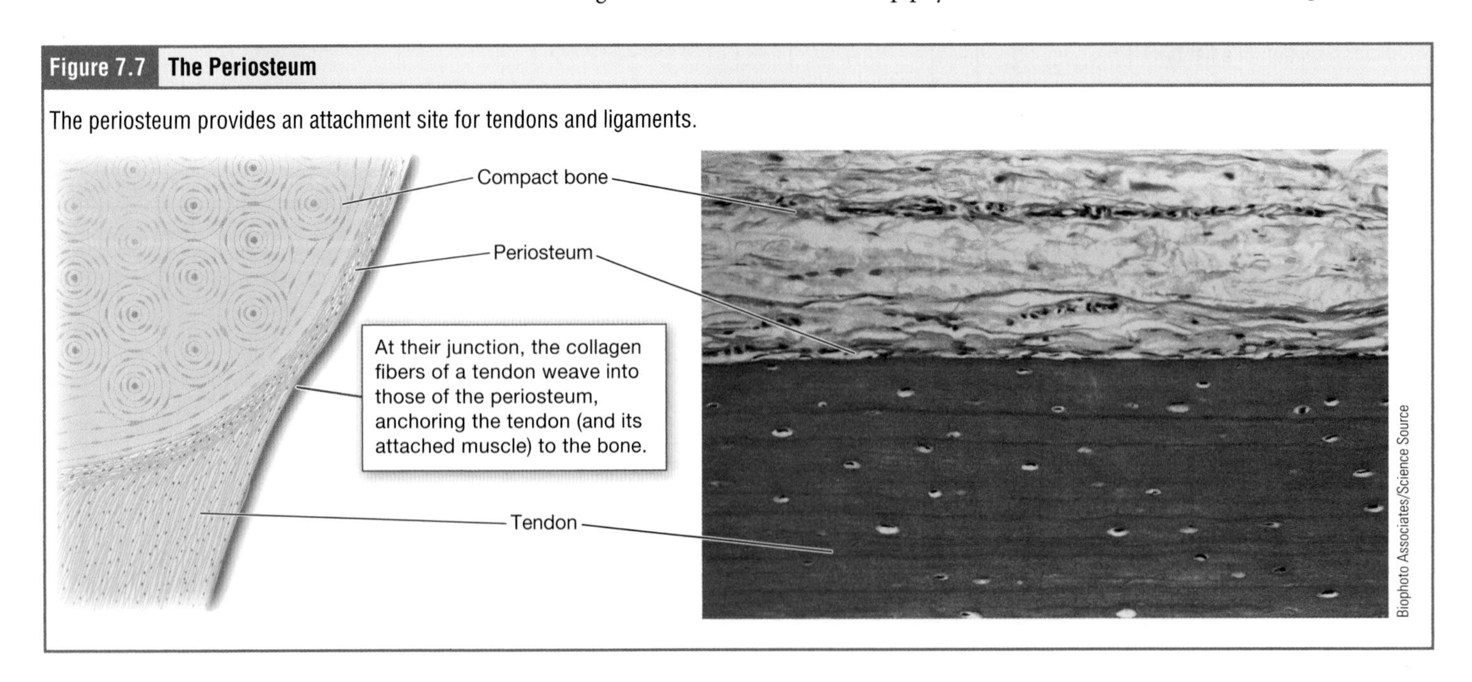

Compact bone
Periosteum
At their junction, the collagen fibers of a tendon weave into those of the periosteum, anchoring the tendon (and its attached muscle) to the bone.
Tendon
Biophoto Associates/Science Source

Figure 7.8 Articular Cartilage

The periosteum does not cover the bones at their joint surfaces. Instead, it forms the fibrous layer of the joint capsule that envelops the joint. The joint-forming surfaces of bones are covered with a layer of hyaline cartilage called *articular cartilage.*

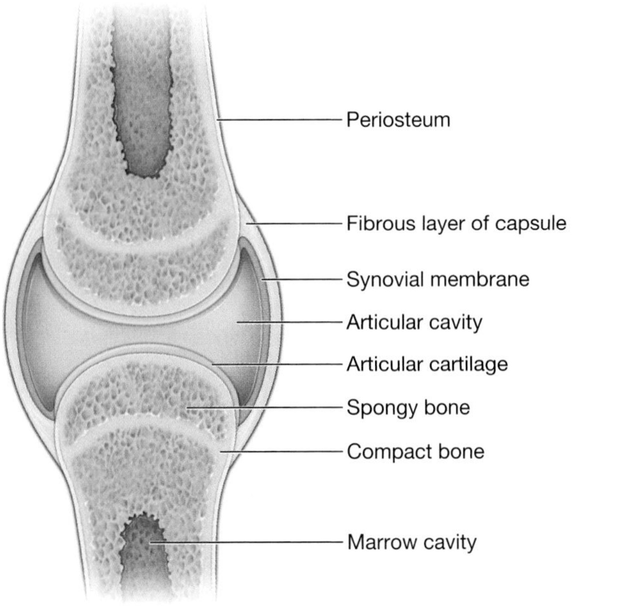

Periosteum
Fibrous layer of capsule
Synovial membrane
Articular cavity
Articular cartilage
Spongy bone
Compact bone
Marrow cavity

layer of hyaline cartilage that reduces friction and acts as a shock absorber. The periosteum separates from the bone at these locations and becomes the **articular capsule**, a fibrous outer covering of each joint.

Flat bones, like those of the cranium, are structures like a sandwich with a core of **spongy bone**, lined on either side by layers of compact bone (**Figure 7.9**). The two layers of compact bone and the interior spongy bone work together to protect the internal organs. If the outer layer of a cranial bone fractures, the brain is still protected by the intact inner layer. Just like in long bones, the spaces within the spongy bone are filled with red bone marrow.

Student Study Tip

Think of bone's layers of dense irregular connective tissue like many layers of a giant spider web. The tissue is incredibly interwoven, which is why it can withstand so much force and movement, like how a spider web is difficult to tear apart.

spongy bone (cancellous bone)

Figure 7.9 The Common Structures of Flat Bones

This cross-section of a flat bone shows spongy bone sandwiched between layers of compact bone.

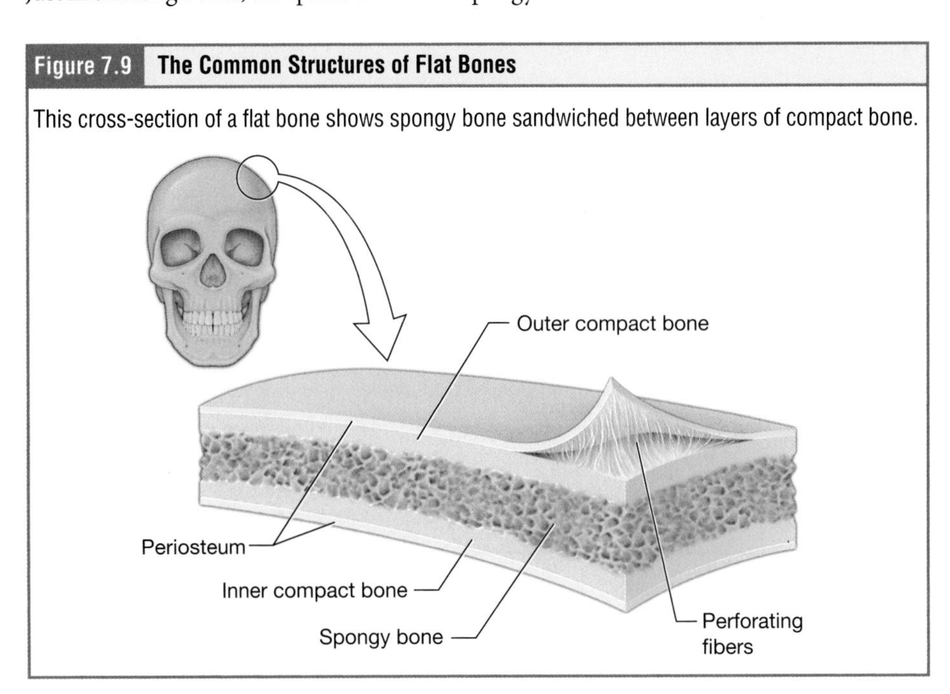

7.2f Bone Markings

LO 7.2.3

The surface features of bones vary considerably, depending on the function and location in the body. **Table 7.2** describes and illustrates the four general classes of bone markings: (1) articulating surfaces, (2) depressions, (3) projections, and

Table 7.2 Bone Markings

Functional Category	Name	Description	Example
Articulating Surfaces	Condyle	Rounded surface	Occipital condyles
	Facet	Flat surface	Vertebrae
	Head	Prominent rounded surface	Head of femur
	Trochlea	Rounded articulating surface	Trochlea of humerus
Depressions	Fossa	Elongated basin	Mandibular fossa
	Sulcus	Groove	Sigmoid sulcus of the temporal bones
Projections	Crest	Ridge	Iliac crest
	Epicondyle	Projection off a condyle	Lateral and medial epicondyles of humerus
	Line	Slight, elongated ridge	Temporal lines of the parietal bones
	Process	Prominent feature	Transverse process of vertebra
	Ramus	Long projection (branch)	Mandibular ramus
	Spine	Sharp process	Ischial spine
	Trochanter	Rough round projection for muscle attachment	Greater trochanter of femur
	Tubercle	Small, rounded process	Tubercle of humerus
	Tuberosity	Rough surface	Deltoid tuberosity
Openings and Spaces	Canal	Passage in bone	Auditory canal
	Fissure	Slit through bone	Auricular fissure
	Foramen	Opening in bone	Foramen magnum in the occipital bone
	Meatus	Opening into canal	External auditory meatus
	Sinus	Air-filled space in bone	Nasal sinus

(4) holes and spaces. An **articulation** is a joint where two bone surfaces come together. These surfaces tend to conform to one another, such as one being rounded and the other cupped, to facilitate the function of the articulation. A depression is typically a sunken-in portion of the bone where a softer structure, such as a blood vessel or organ, rests. A **projection** is an area of a bone that projects above the surface of the bone. These are the attachment points for tendons and ligaments. In general, their size and shape is an indication of the forces exerted through the attachment to the bone. A large muscle, pulling harder on a bone, will have a larger, often rougher projection than a smaller muscle that pulls with less force. A **hole** is an opening or groove in the bone that allows some structure, such as a blood vessel or nerve, to enter or pass through the bone.

✓ Learning Check

1. Which part of the bone would you find yellow marrow? Please select all that apply.
   a. Diaphysis
   b. Medullary cavity
   c. Epiphysis
   d. Growth plate
2. Where would calcium be deposited within the bone? Please select all that apply.
   a. Epiphysis
   b. Compact bone
   c. Medullary cavity
   d. Epiphyseal plate
3. Juanito is a 7-year-old boy who was riding his bike when he fell off onto his hand. The impact fractured the distal radius. Upon X-ray you find that there is a fracture in the metaphysis. Why is a fracture in this place problematic?
   a. He will not be able to move his wrist.
   b. His internal organs will not be well protected.
   c. His bone growth may be stunted.
   d. He will have decreased red blood cell production.
4. The quadriceps are a group of muscles on your thigh that combine to form one tendon, which inserts onto your tibia. These muscles generate large forces when standing still, stepping forward, or jumping up. What type of bone feature would you expect to find on the tibia at the site of this tendon attachment?
   a. Hole
   b. Articulation
   c. Projection
   d. Depression

7.3 The Microscopic Structure of Cartilage and Bone

Learning Objectives: By the end of this section, you will be able to:

7.3.1 Describe the roles of dense regular and dense irregular connective tissue in the skeletal system.

7.3.2 List and describe the cellular and extracellular components of bone tissue.

7.3.3 Using microscopic images, distinguish between the three different types of cartilage.

7.3.4 Identify the microscopic structure of compact bone and spongy bone.

Bone and cartilage tissues differ from other tissues in the body. They also differ significantly from each other! Bone and cartilage share the cellular and fibrous characteristics of many other tissues, but these two tissues offer structure to the body through their unique rigidity. This section will examine the histology, or microscopic anatomy, of these tissues.

7.3a Cartilage Cells and Tissue

The three types of cartilage—hyaline, elastic, and fibrocartilage—share a common set of characteristics. All cartilages are composed of two types of cells: chondrocytes and chondroblasts. **Chondroblasts** are the cells that are capable of generating cartilage matrix. These cells are usually found on the periphery of the tissue, actively secreting matrix. The matrix of cartilage varies a bit from one type to another, but they all share a gel-like quality due to the presence of chondroitin sulfate, a proteoglycan. As the chondroblasts enclose themselves in secrete matrix, their production of matrix slows and the cells differentiate into **chondrocytes**. Chondrocytes occupy small holes in the cartilage called **lacunae**. One thing to note about chondrocytes is that while they are locked in their matrix and cannot move, nutrients can diffuse through the matrix because it is semisolid. Cartilage is avascular but surrounded by a vascularized covering of dense irregular connective tissue called **perichondrium** (**Figure 7.10**). Nutrients leave the blood vessels in the perichondrium and are able to reach the chondrocytes by way of diffusion.

LO 7.3.1

All cartilages contain protein fibers, but the type and concentration of fibers is the factor that varies the most. Remember from Chapter 5 that hyaline cartilage contains collagen, but the fibers are fine enough that they are not easily seen under the microscope. Elastic cartilage contains elastin fibers, and fibrocartilage gets its name from the visible, densely woven collagen fibers that pack its matrix (**Figure 7.11**).

7.3b Bone Cells and Tissue

LO 7.3.2

There are many similarities between bone and cartilage as tissues. Both tissues contain a relatively small number of cells entrenched in a matrix of collagen fibers and chondroitin sulfate. In bone, but not in cartilage, the fibers provide a surface for minerals to adhere. Four types of cells are found within bone tissue: osteoblasts, osteocytes, osteogenic cells, and osteoclasts (see the "Anatomy of Bone Cells" feature).

Figure 7.10 The Perichondrium

Similar to the periosteum, most cartilage is surrounded by layer of dense irregular connective tissue called the *perichondrium*.

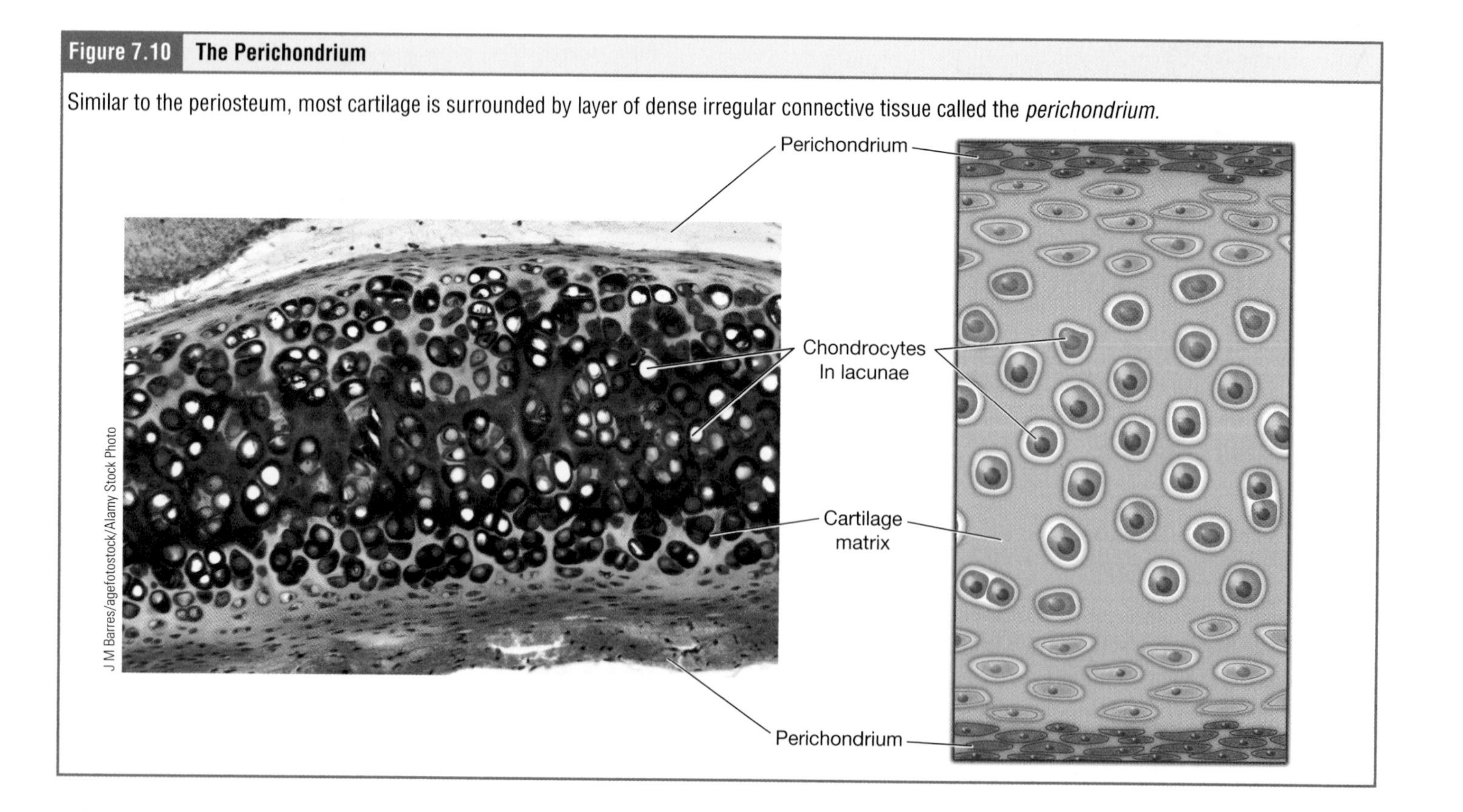

Figure 7.11 The Three Types of Cartilage Found in the Human Body LO 7.3.3

All three types of cartilage, (A) hyaline, (B) elastic, and (C) fibrocartilage have chondrocytes found in lacunae and a semisolid matrix.

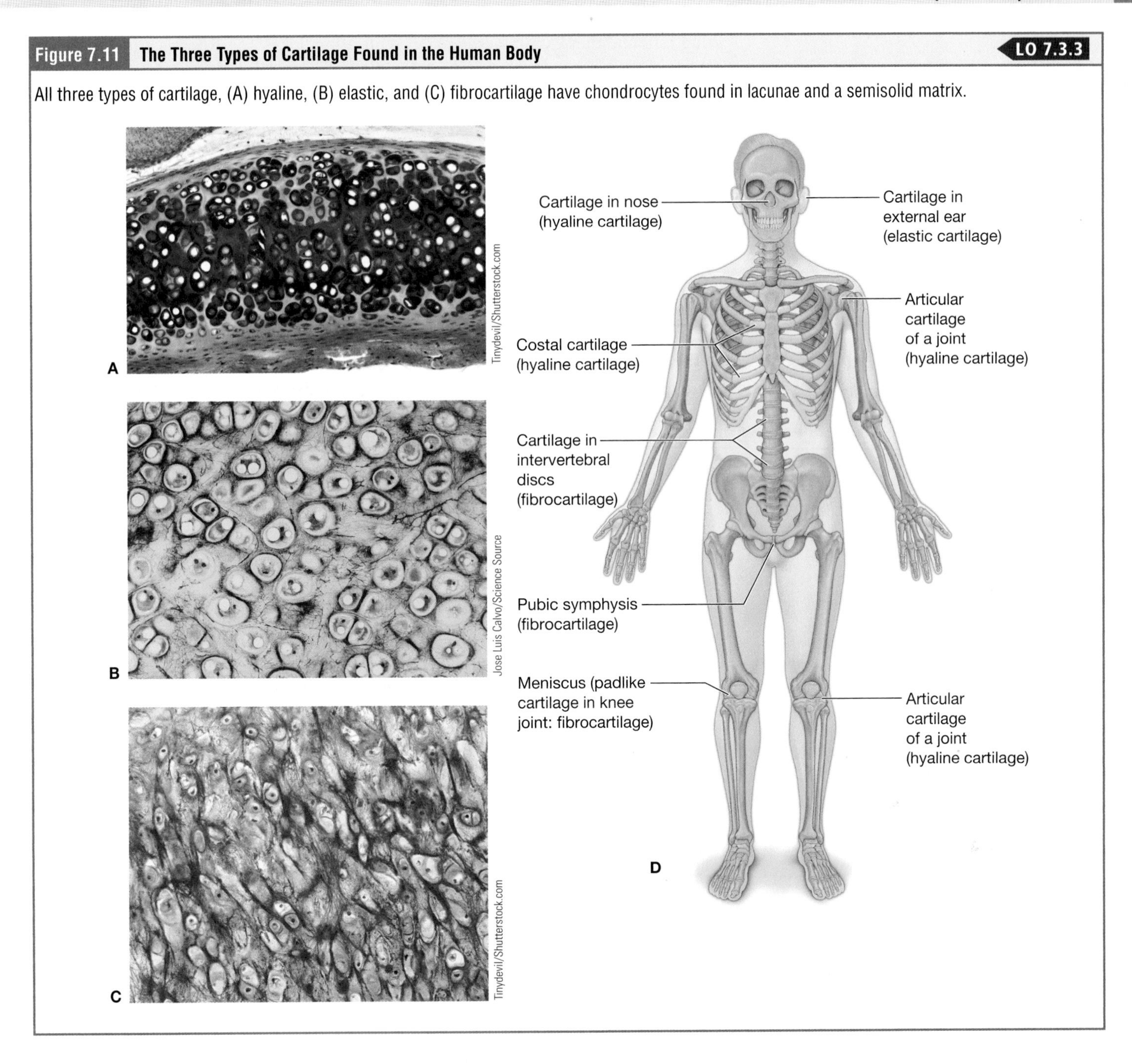

Osteogenic cells are the stem cells of bone. They are undifferentiated with high mitotic activity, and they are the only bone cells that are capable of replication. Osteogenic cells are found in the deep layers of the periosteum and the marrow. Throughout life, osteogenic cells will replicate and one of the daughter cells will differentiate and develop into an osteoblast.

Osteoblasts are the bone cell that form new bone matrix. Osteoblasts, which cannot undergo cell replication, synthesize and secrete the collagen fibers and calcium salts. Osteoblasts are found in the periosteum and endosteum. As the osteoblast calcifies the matrix surrounding it, the osteoblast becomes trapped within. As a result, the osteoblast differentiates and changes in structure, becoming an osteocyte. **Osteocytes** are the most common cells in bone tissue and are fully mature; they cannot change into

Student Study Tip

Use alliteration for the function of osteoblasts: Building Bone is a Blast!

other cells and they cannot undergo cell replication. Like chondrocytes, each osteocyte is located in a space called a *lacuna* and is surrounded by bone tissue. Osteocytes work to maintain the mineral concentration of the matrix via the secretion of enzymes. One crucial difference between osteocytes and chondrocytes is how they access nutrients. Nutrients cannot diffuse through solid bone matrix the way that they diffuse through cartilage matrix. However, osteocytes can communicate with each other and share nutrients and wastes via long cytoplasmic processes that extend through **canaliculi** (singular = canaliculus), channels within the bone matrix.

Student Study Tip

One way to remember who's who in the bone is that *-blasts* build (you've seen other -blast cells, such as fibroblasts, which are the builders of fibers!) and *-clasts* consume.

Bones are incredibly dynamic; they increase and decrease matrix formation in response to changes in friction, weight-bearing or calcium needs of the body. New tissue is constantly formed, and old, injured, or unnecessary bone is dissolved for repair or for calcium release. The cell responsible for bone breakdown is the **osteoclast**. They are found on bone surfaces, are multinucleated, and originate from monocytes and macrophages, two types of white blood cells, not from osteogenic cells. Osteoclasts are continually breaking down old bone while osteoblasts are continually forming new bone. The ongoing balance between osteoblasts and osteoclasts is responsible for the constant but subtle reshaping of bone. The "Anatomy of Bone Cells" feature reviews the bone cells, their functions, and locations.

7.3c Compact and Spongy Bone

Bone tissue comes in two forms, compact and spongy bone. When a building is built, architects combine different structural elements to create a dynamic and strong

Anatomy of...

Bone Cells

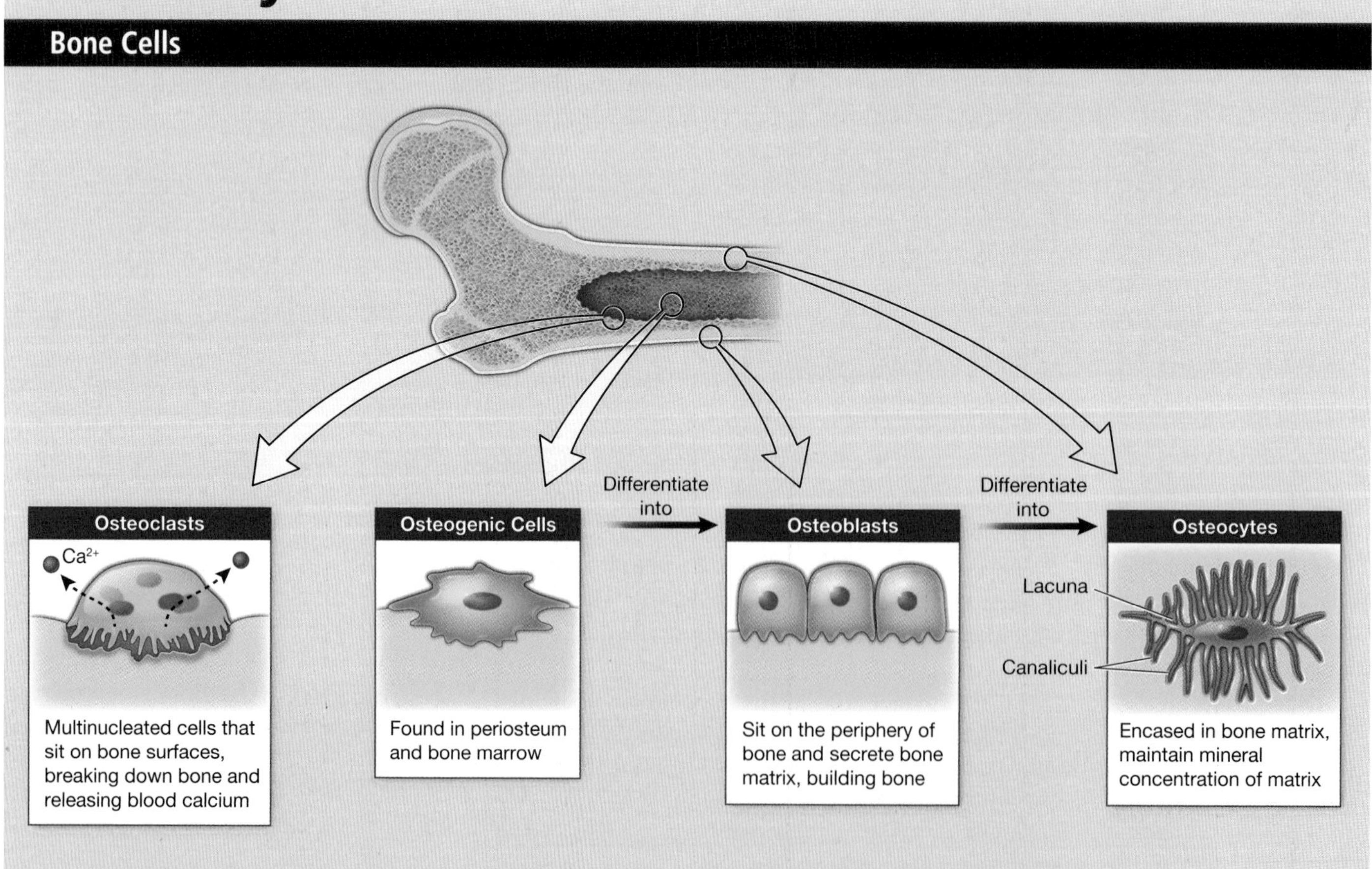
Differentiate into
Differentiate into
Osteoclasts
Ca^{2+}
Multinucleated cells that sit on bone surfaces, breaking down bone and releasing blood calcium
Osteogenic Cells
Found in periosteum and bone marrow
Osteoblasts
Sit on the periphery of bone and secrete bone matrix, building bone
Osteocytes
Lacuna
Canaliculi
Encased in bone matrix, maintain mineral concentration of matrix

structure that is capable of withstanding many different forces. Your bones have evolved two different structural elements for the same reasons. The solid compact bones, like the walls of a building, are solid and capable of holding up a tremendous amount of weight. The beams and girders of a building, however, provide strength in a way that enables the building to withstand weight and pressure in different directions or through shifts. In human bones, spongy bone plays this role. The many intersecting beams, called ***trabeculae***, of spongy bone provide strength and support in different directions, throughout shifts in weight distribution.

Learning Connection

Try Drawing It!

Osteons are highly organized structures, but may not look it at first glance. To master the structure, try making a drawing of an osteon that includes osteocytes in their lacunae, their connections in canaliculi, their arrangement in lamellae and the location and contents of the central canal.

Compact Bone Compact bone is the denser and stronger of the two types of bone tissue (**Figure 7.12**). It can be found under the periosteum, located as an outer shell to each bone, where it provides support and protection.

The microscopic structural unit of compact bone is called an **osteon** (**Figures 7.12A** and **7.12B**). Each osteon acts as a supportive column and runs in the direction of weight bearing (**Figure 7.12C**) and is composed of concentric rings of calcified matrix called

osteon (Haversian system)

Figure 7.12 Diagrams and Photomicrograph of Compact Bone LO 7.3.4

Osteons are the structural unit of compact bone as seen in (A) a cross-section of compact bone, and (B) this photomicrograph of an osteon. (C) Osteons are arranged as columns in the direction of weight-bearing stress.

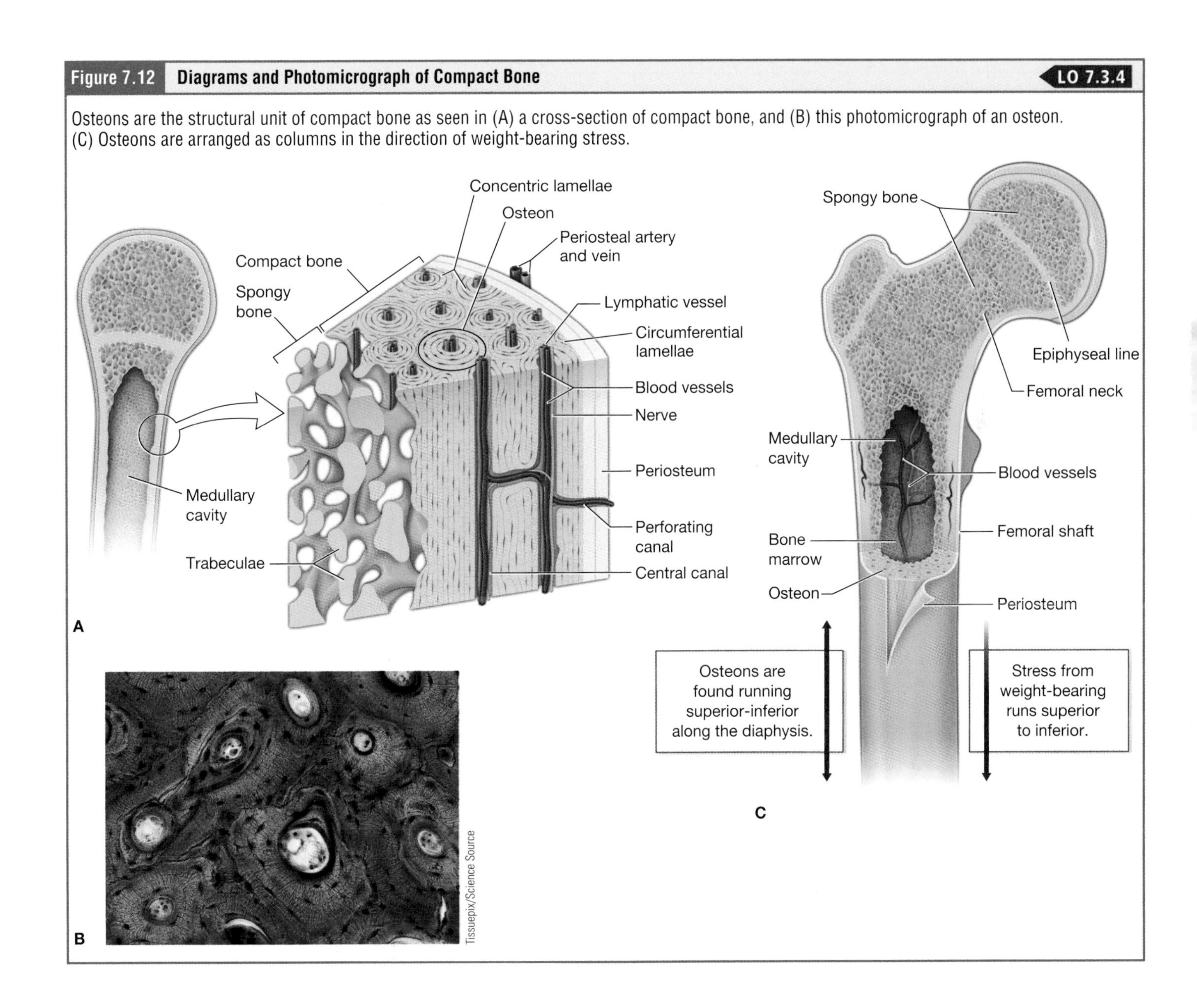

central canal (Haversian canal)

perforating canals (Volkmann's canal)

lamellae. Running down the center of each osteon is the **central canal**, which contains blood vessels, nerves, and lymphatic vessels. These vessels and nerves connect one osteon to another, and the osteons to the endosteum and periosteum, through **perforating canals** (see Figure 7.12A).

Lamellae are sheets of connected osteocytes within their lacunae that wrap around the central canal. Each osteocyte is alone in its lacuna but connected to osteocytes in neighboring lacunae through canaliculi. This system allows nutrients to be transported to the osteocytes and wastes to be removed from them.

Student Study Tip

Although compact bone is rigid, spongy bone resembles an actual sponge. It has a matrix of many small holes, even though it is structurally sound.

Spongy Bone Like compact bone, spongy bone contains osteocytes housed in lacunae, but these are not arranged in concentric circles. Instead, the lacunae and osteocytes are found in a latticelike network of beams called *trabeculae* (**Figure 7.13A**). The trabeculae may appear to be a random network, but each trabecula forms along lines of stress to provide strength to the bone (**Figure 7.13B**). The open spaces within spongy bone serve to make bones lighter so that muscles can move them more easily and provide a space for red marrow, protected by the trabeculae, where hematopoiesis can occur.

Figure 7.13 Spongy Bone

(A) Spongy bone is made up of beams of bone called *trabeculae*. Red marrow fills the spaces in some bones. (B) Trabeculae are constantly remodeled to meet the needs of the bone. Bone matrix is laid down and trabeculae are oriented in the direction(s) of weight-bearing.

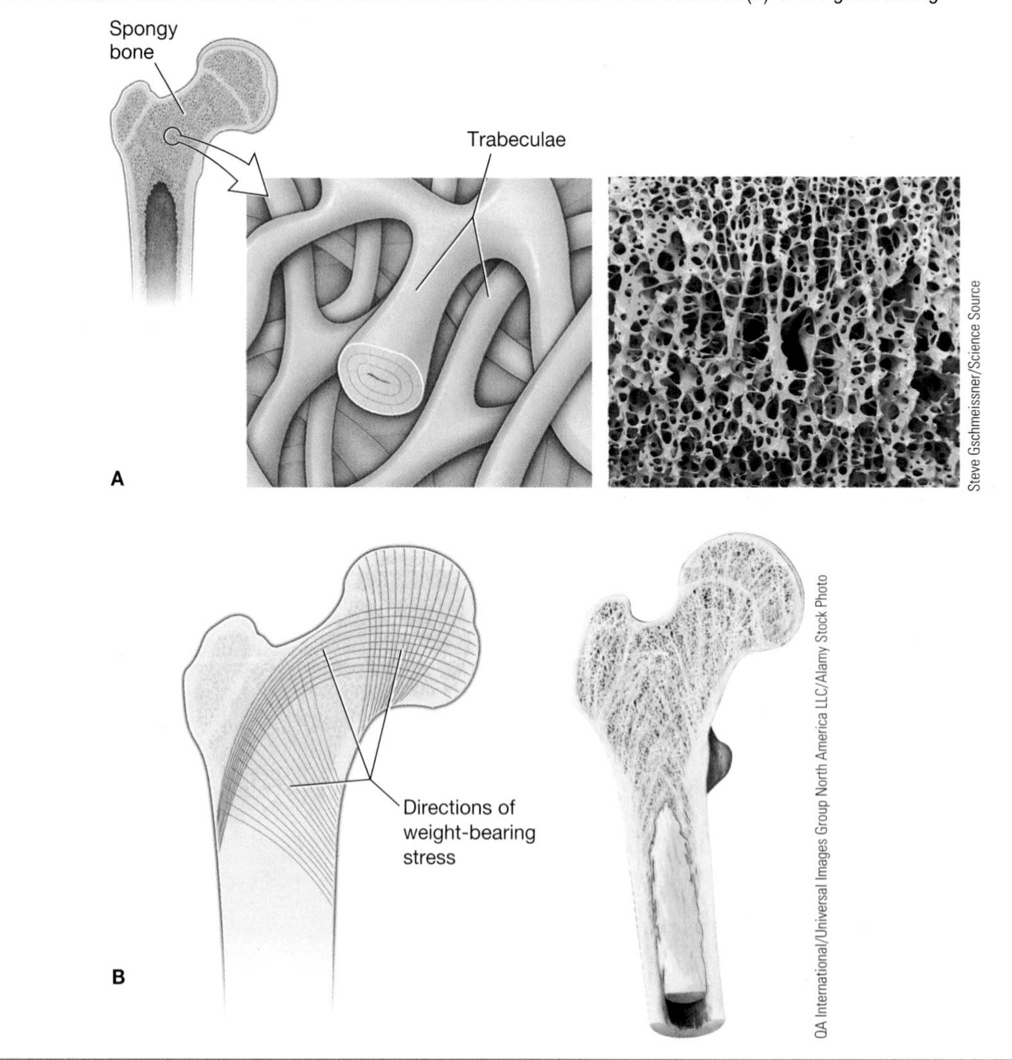

Apply to Pathophysiology

Paget's Disease

Susan just celebrated their sixtieth birthday last week, but they no longer enjoy the active lifestyle they once did. Hikes and short runs leave them feeling a new kind of tired, like their whole skeleton aches. Susan's knees and pelvis especially hurt and they feel like their jeans no longer fit around their thighs. One day Susan takes a minor spill on the stairs and ends up with several fractures.

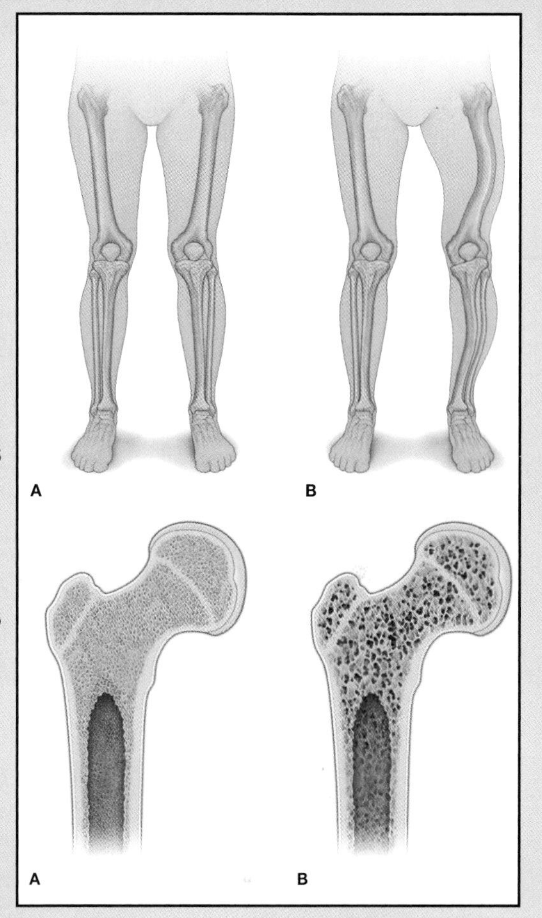

1. At the emergency room a blood test reveals high calcium levels and Susan's X-rays reveal multiple abnormalities. In addition to the multiple fractures, Susan's bones appear less dense overall. The doctors tell Susan that one type of bone cell has become overactive. Which type is it most likely to be?
   A) Osteogenic cell
   B) Osteoblasts
   C) Osteocytes
   D) Osteoclasts
2. The doctors take a large-view X-ray showing Susan that their weakening bones are struggling to support their weight. Which of the images is likely to be Susan's X-ray?
   A) Image A
   B) Image B
3. Susan asks the doctor, "Since this is a problem with my bones getting weaker, can I take calcium dietary supplements to fix it?" Will calcium supplements help? Why or why not?
   A) Yes, Susan's bones need more matrix; raising the blood calcium levels will enable more matrix to be made.
   B) No, Susan's bones are weakening because calcium is being removed from them; adding more calcium to the blood will not help.
   C) No, Susan's bones are weakening because the cells that take calcium from the blood and add it to the bones are dying and need to be replaced.

Learning Check

1. You are looking at cartilage under a microscope and find black dots within circles. What are these black dots called?
   a. Chondroblasts
   b. Lacunae
   c. Chondrocytes
   d. Perichondrium
2. Which of the following highlights the difference between cartilage and bone tissues? Please select all that apply.
   a. Only in bone, nutrients are shared through cytoplasmic processes called canaliculi.
   b. Only in cartilage, there are spaces within the matrix that encompass a cell.
   c. Only in cartilage, there are cells that differentiate into more mature cells.
   d. Only in bone, there are cells that are responsible for breakdown.
3. Complete this analogy: Chondroblasts form new cartilage matrix and differentiate into chondrocytes, while ________ form new bone matrix and differentiate into ________.
   a. Osteogenic cells; osteocytes
   b. Osteoblasts; osteogenic cells
   c. Osteogenic cells; osteoblasts
   d. Osteoblasts; osteocytes
4. Which of the following structures nourishes osteocytes in compact bone?
   a. Canaliculi
   b. Central canal
   c. Lamellae
   d. Trabeculae

7.4 Formation and Growth of Bone and Cartilage

Learning Objectives: By the end of this section, you will be able to:

7.4.1 Explain the roles that specific bone cells play in the formation of bone tissue.

7.4.2 Compare and contrast intramembranous and endochondral (intracartilaginous) bone formation.

In the preceding sections you learned that osteoblasts, the cells capable of building bone, are found on the outer surfaces of the bone beneath the periosteum. The periosteum does not cover the ends of long bones, and so there are no osteoblasts there. This may leave you with the question: how do bones get longer as our bodies grow taller? You may even be wondering how bones develop in the first place. The answers lie within the knowledge we have already built about tissues and fibers. In the early stages of embryonic development, the embryo's skeleton consists of fibrous membranes and hyaline cartilage. By the sixth or seventh week of embryonic development, bone development—**ossification**—begins. Bone can develop from two starter tissues: dense irregular connective tissue or hyaline cartilage. The process of building a bone within a sheet of connective tissue is called **intramembranous ossification** and the process of building a bone from a cartilage template is called ***endochondral ossification***. The end product—bone—is the same, regardless of the pathway that produced it.

ossification (osteogenesis)

7.4a Intramembranous Ossification

LO 7.4.1

During intramembranous ossification, compact and spongy bone develops directly within sheets of connective tissue. As you can imagine, intramembranous ossification typically results in flat bones such as those of the face and cranium.

The process begins when an embryonic cell type, **mesenchymal cells**, gather together and begin to differentiate into specialized cells (**Figure 7.14A**). Some of these cells will differentiate into capillaries, while others will become osteogenic cells and then osteoblasts. Although they will ultimately be isolated in their lacunae and separated from each other by the bone matrix, early osteoblasts appear in a cluster called an **ossification center**.

The osteoblasts secrete **osteoid**, a soft matrix devoid of minerals, which hardens as calcium and other minerals are deposited on it. Osteoblasts become surrounded by matrix and differentiate into osteocytes (**Figure 7.14B**). As osteoblasts transform into osteocytes, osteogenic cells in the surrounding connective tissue differentiate into new osteoblasts.

Osteoid (unmineralized bone matrix) secreted near capillaries will avoid the capillaries and build strips of bone interwoven around the blood vessels, resulting in a network of bony trabeculae that forms spongy bone. Osteoblasts on the surface of the bone contribute to the development of the periosteum (**Figure 7.14C**). The osteoblasts in the periosteum create the protective shell of compact bone last. From the blood vessels within the spongy bone, red bone marrow forms (**Figure 7.14D**).

Intramembranous ossification begins *in utero* during fetal development but continues on into adolescence. At birth, the skull and clavicles are not fully ossified; this is an evolutionary adaptation to the narrow pelvises of bipedal humans. The partially formed fetal skull and shoulders are able to deform during passage through the birth canal. Ossification continues after birth but does not end until the individual is fully grown. The last bones to ossify via intramembranous ossification are

Figure 7.14 Intramembranous Ossification

The formation of bone within a sheet of connective tissue, which is called *intramembranous ossification*, is a process that can be divided into four stages. (A) In the first stage, mesenchymal cells (stem cells) that differentiate into osteoblasts. (B) In the second stage, osteoblasts (bone-forming cells) secrete the unique matrix of bone tissue, which hardens as calcium is added to it. (C) In the third stage, the narrow beams of spongy bone, called *trabeculae*, and the outer periosteum form. Blood vesselsgrow into the developing bone to supply it with oxygen and nutrients. (D) In the fourth stage, compact bone is deposited around the spongy bone.

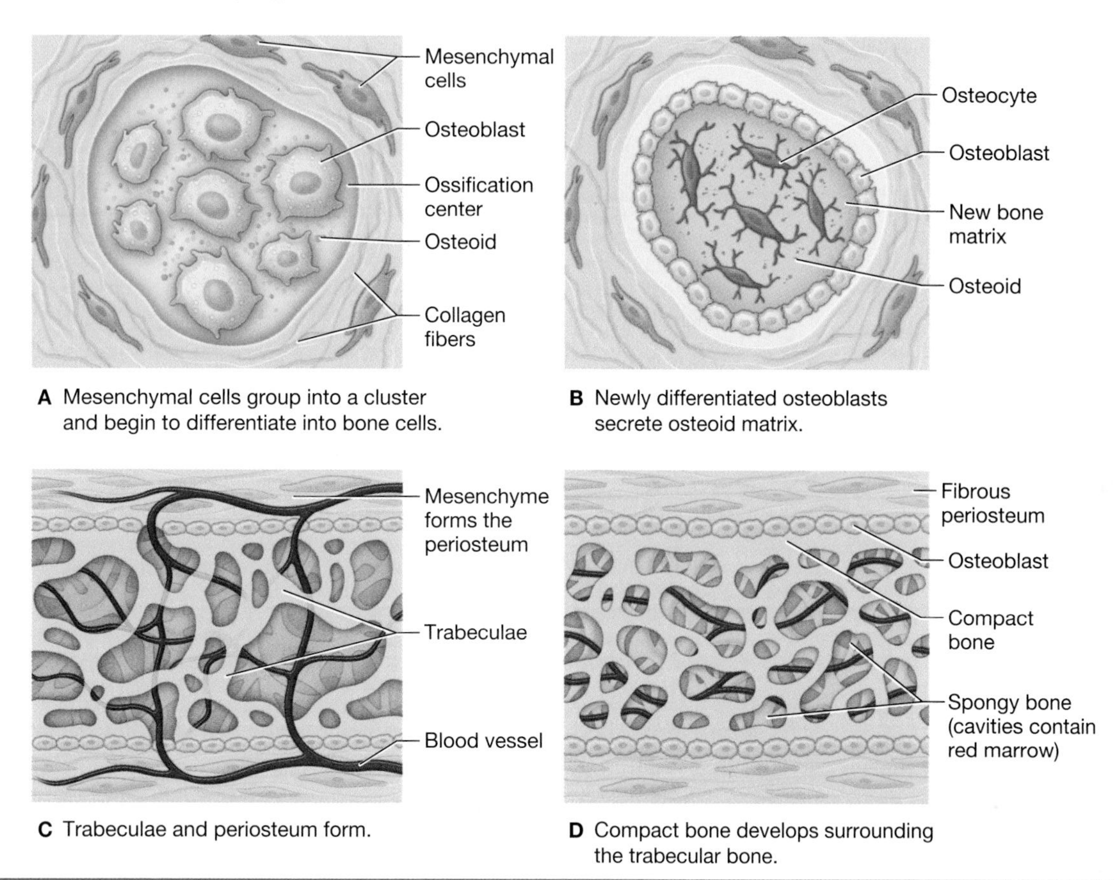

A Mesenchymal cells group into a cluster and begin to differentiate into bone cells.

B Newly differentiated osteoblasts secrete osteoid matrix.

C Trabeculae and periosteum form.

D Compact bone develops surrounding the trabecular bone.

the flat bones of the face, which reach their adult size at the end of the adolescent growth spurt.

7.4b Endochondral Ossification

LO 7.4.2

While intramembranous ossification contributes to the formation of most flat bones, most of our long bones form from a template structure of hyaline cartilage. Throughout fetal development and into childhood growth and development, bone forms on the cartilaginous matrix. The incorporation of hyaline cartilage within the long bones, epiphyseal plates, allows bones to continue to grow throughout the childhood and adolescence.

In endochondral ossification, bone develops by *replacing* hyaline cartilage. Cartilage does not become bone. Instead, cartilage serves as a template to be completely replaced by new bone. Bone formation via endochondral ossification begins with a complete cartilage unit covered by perichondrium (**Figure 7.15A**). Cells within the perichondrium along the diaphysis will differentiate into osteoblasts and begin to form a ring of osteoid matrix (**Figure 7.15B**).

As more matrix is produced and mineralized, nutrients can no longer reach the chondrocytes in the center of the cartilaginous model. These chondrocytes—which function to maintain the cartilage matrix—die, and the matrix disintegrates

Figure 7.15 Endochondral Ossification

Formation of bone from a cartilage template, which is called *endochondral ossification*, begins in the fetal period and continues through adolescence. (A) A hyaline cartilage template forms the majority of bones in the fetal skeleton. (B) As a few cells differentiate into osteoblasts, minerals can be deposited onto the collagen fibers of the cartilage, starting in the diaphysis. Perichondrium forms around the bone. (C) Cartilage is an avascular tissue, but bone is vascularized. Perichondrium transforms into periosteum. Once capillaries penetrate the cartilage, mineralization can increase quickly, forming a densely mineralized area known as the *primary ossification center*. (D) Chondrocytes continue to replicate at the ends of the bone, causing the bone to grow longer. (E) After birth, the epiphyses begin to ossify in areas known as *secondary ossification centers*. (F) Cartilage remains at the epiphyseal plate and as articular cartilage at the ends of the bone.

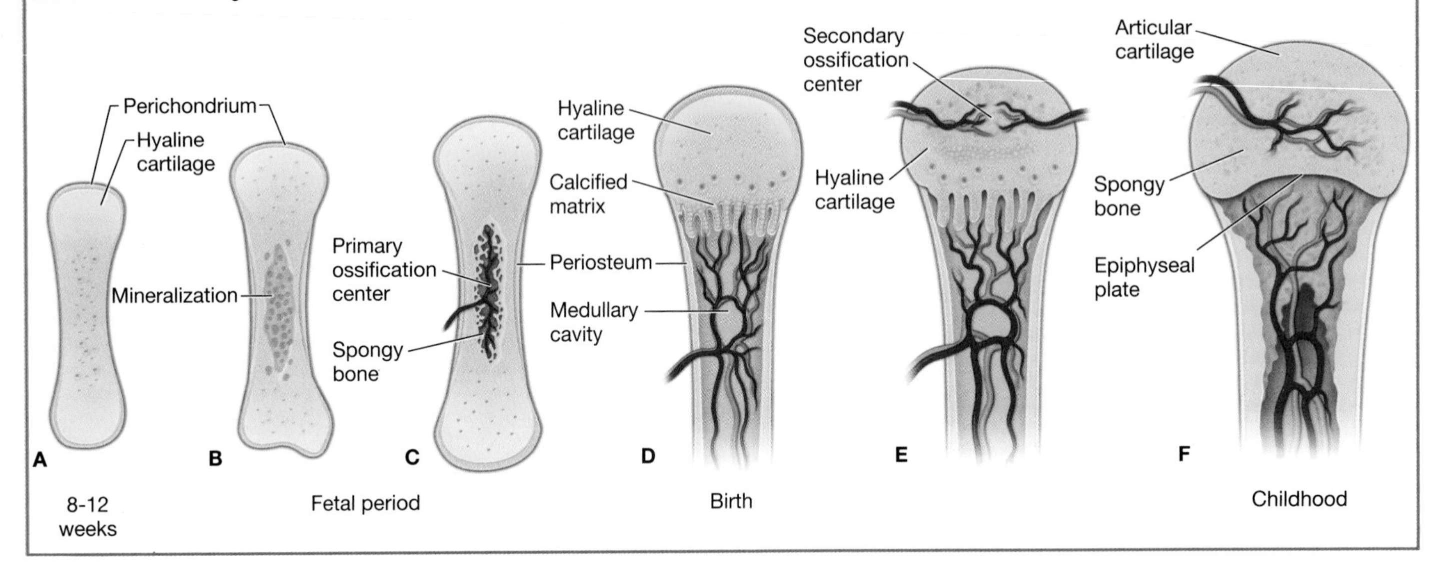

(**Figure 7.15C**). Blood vessels invade the resulting spaces, carrying osteogenic cells with them, many of which will become osteoblasts. These spaces at the core of the forming bone become larger and eventually combine to become the medullary cavity (**Figure 7.15D**).

While these deep changes are occurring, chondrocytes and cartilage continue to grow at the ends of the bone (the future epiphyses), which increases the bone's length at the same time bone is replacing cartilage in the diaphysis. After birth, islands of ossification occur at the epiphyses as well, so that the ends of the bone as well as a plate between the end and the diaphysis are the only regions where cartilage has not yet been replaced (**Figures 7.15E** and **7.15F**). This plate of cartilage between the diaphysis and the end of the bone is the epiphyseal plate and it remains as the active site of growth throughout childhood and adolescence. The difference between early embryonic and fetal skeletons can be seen in **Figure 7.16**.

Learning Check

1. Which of the following ossification processes is from cartilage?
   a. Endochondral ossification
   b. Intramembranous ossification
2. Identify which of the following occurs during endochondral ossification.
   a. Osteoblasts secrete osteoid.
   b. Osteoblasts become surrounded by matrix.
   c. Osteogenic cells differentiate into osteoblasts.
   d. Chondrocytes die and the matrix disintegrates.
3. Identify which of the following occurs during intramembranous ossification.
   a. Chondrocytes become osteocytes.
   b. Osteoblasts are surrounding by osteoid matrix.
   c. Chondrocytes die and the matrix disappears.
   d. Chondrocytes continue to grow at the epiphyses.

Figure 7.16 Ossification in the Embryonic and Fetal Skeletons

Embryonic and fetal skeletons are formed by a combination of intramembranous and endochondral ossification. (A) In this early embryonic skeleton, long bones are formed first via endochondral ossification while flat bones, including the skull, are formed via intramembranous ossification. (B) In this later fetal skeleton, bones of the skull have formed and most long bones have primary ossification centers and appear more densely mineralized.

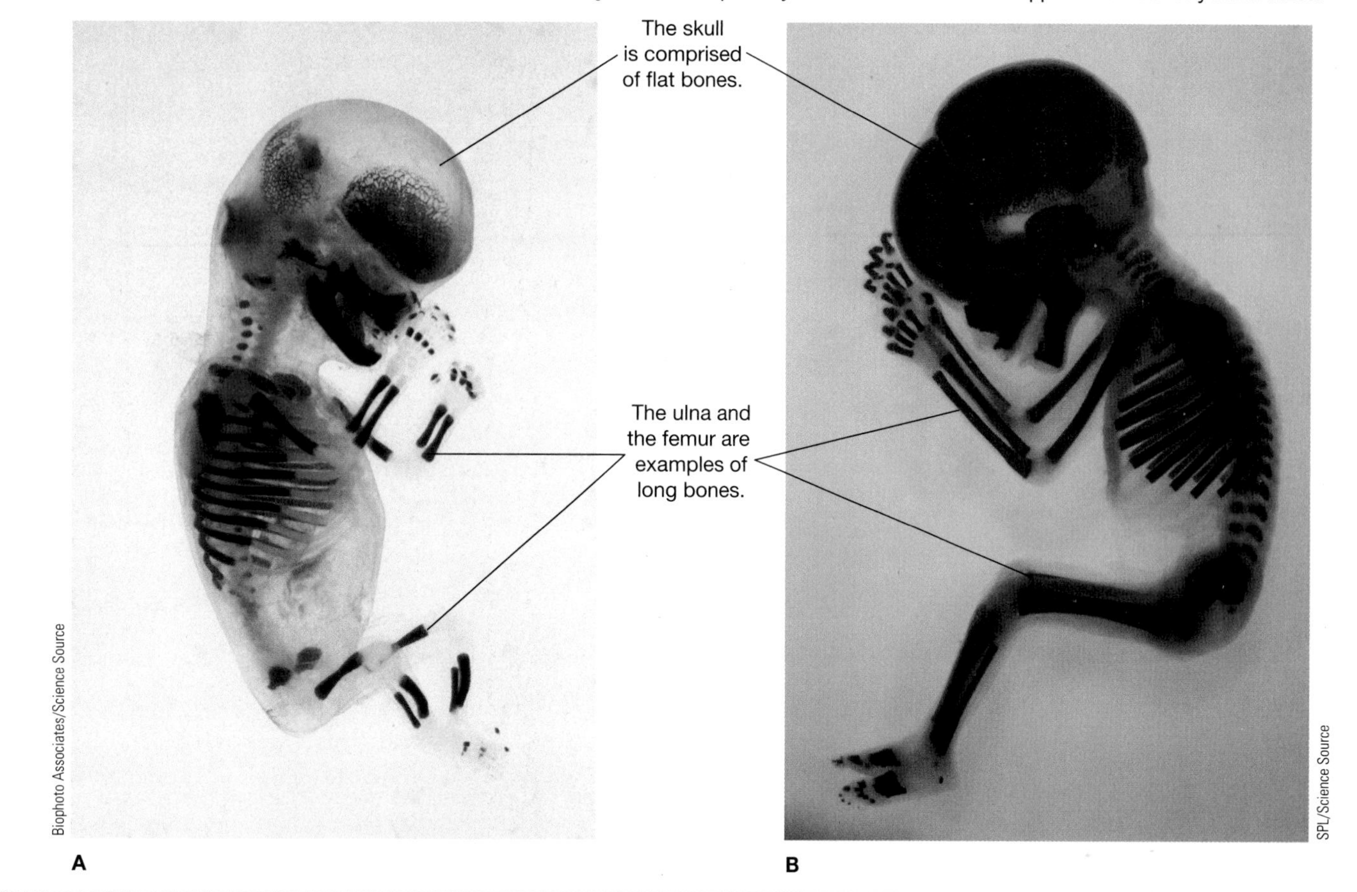
The skull is comprised of flat bones.
The ulna and the femur are examples of long bones.
Biophoto Associates/Science Source
SPL/Science Source
A
B

7.5 Growth, Repair, and Remodeling

HAPS
LEARNING GOALS AND OUTCOMES

Learning Objectives: By the end of this section, you will be able to:

7.5.1 Compare and contrast interstitial (lengthwise) and appositional (width or circumferential) growth in bone tissue.

7.5.2 Compare and contrast the function of osteoblasts and osteoclasts during bone growth, repair, and remodeling.

7.5.3 Describe the bone repair and remodeling process and how it changes as humans age.

7.5.4 Explain the hormonal regulation of skeletal growth.

7.5.5 Explain the roles of parathyroid hormone, calcitriol, and calcitonin in plasma calcium regulation and bone remodeling.

7.5.6 Explain the steps involved in fracture repair.

7.5a Cartilage Growth

LO 7.5.1

Cartilage growth can occur in two different ways. In interstitial cartilage growth, cartilage grows longer due to the mitotic replication of chondrocytes. This process (summarized in **Figure 7.17**) begins with the replication of a chondrocyte within a lacuna. After mitosis, two chondrocytes share a single lacuna, but they soon begin to secrete

Figure 7.17 Interstitial Cartilage Growth

Cartilage grows longer via mitosis of chondrocytes in interstitial cartilage growth.

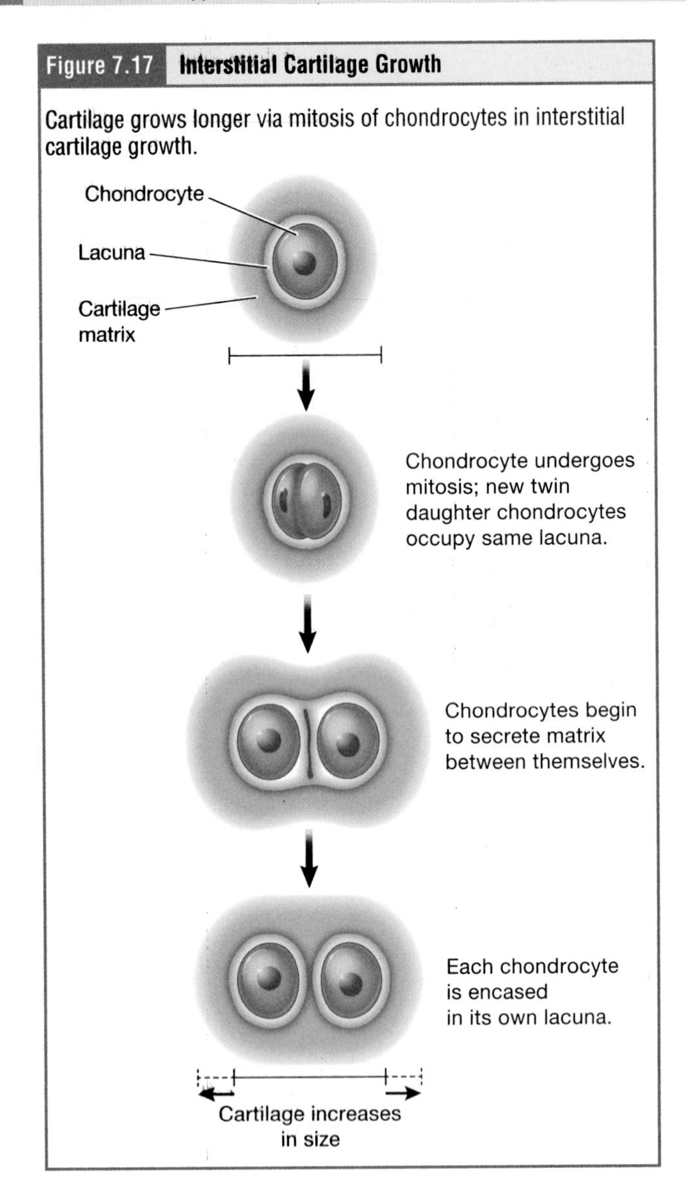

Chondrocyte
Lacuna
Cartilage matrix
Chondrocyte undergoes mitosis; new twin daughter chondrocytes occupy same lacuna.
Chondrocytes begin to secrete matrix between themselves.
Each chondrocyte is encased in its own lacuna.
Cartilage increases in size

Figure 7.18 Appositional Cartilage Growth

Cartilage grows wider in the process of appositional growth. Mesenchymal cells within the perichondrium differentiate into osteoblasts, and then secrete matrix to add to existing cartilage.

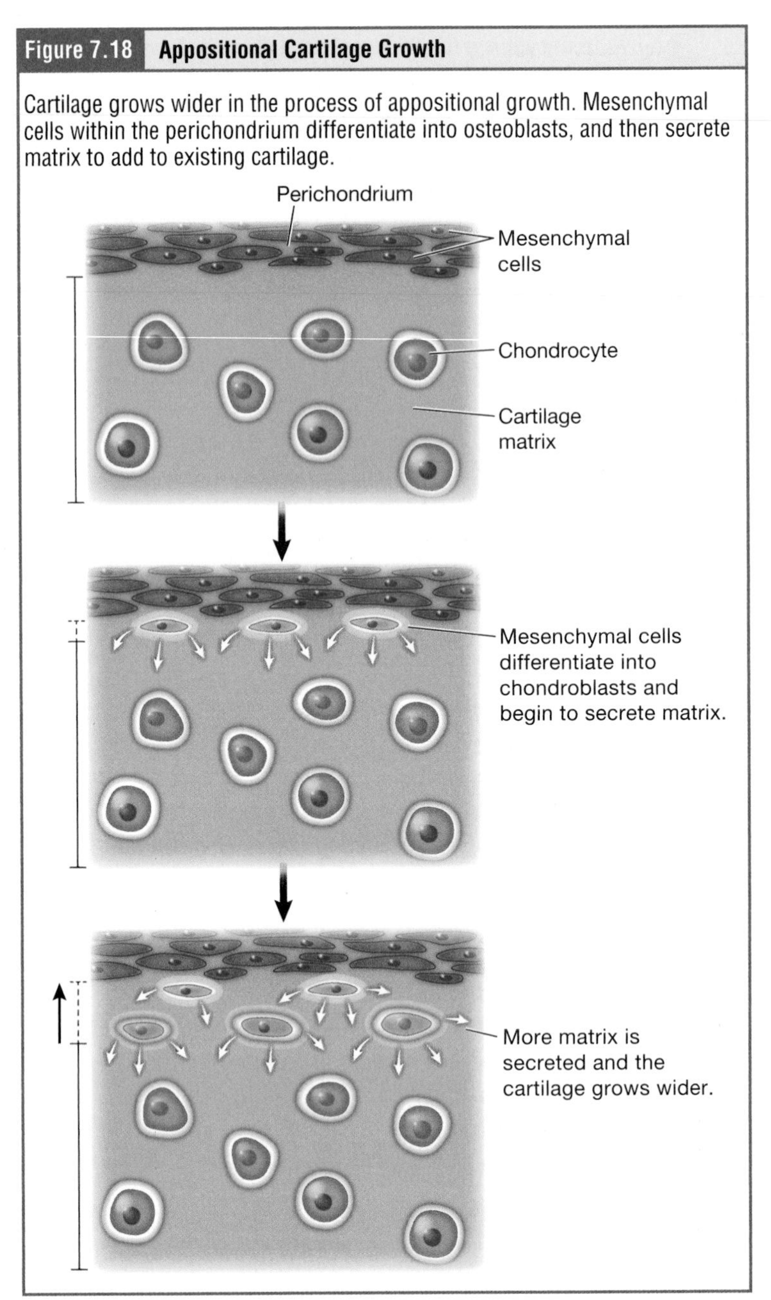

Perichondrium
Mesenchymal cells
Chondrocyte
Cartilage matrix
Mesenchymal cells differentiate into chondroblasts and begin to secrete matrix.
More matrix is secreted and the cartilage grows wider.

matrix between themselves, slowly moving apart. As they push away from each other, the cartilage becomes longer. In **appositional cartilage growth**, cartilage grows wider as mesenchymal cells in the perichondrium differentiate into chondroblasts and then secrete matrix around themselves (**Figure 7.18**).

7.5b How Bones Grow in Length

The epiphyseal plate is a strip of hyaline cartilage at the junction of the diaphysis and epiphysis of a long bone. It is the only site where the bone is capable of growing longer. The process of bone growth in a long bone, which is similar to interstitial growth of cartilage, is summarized in **Figure 7.19**. Chondrocytes within the epiphyseal plate will undergo a period of rapid mitosis, lengthening the epiphyseal plate within the bone and pushing the epiphysis away from the diaphysis (see Figure 7.19, stage 2). The bone is now longer due to the longer epiphyseal plate. The cartilage on the diaphysis side of the epiphyseal plate will then ossify, and the epiphyseal plate returns to its original size, but the bone is now longer (see Figure 7.19, stage 3).

If you photographed an epiphyseal plate in the midst of this growing transition, you would be able to see four zones of cells and activity, summarized in the "Anatomy of an Epiphyseal Plate" feature. The **reserve zone** is the region closest to the epiphyseal end and contains chondrocytes within cartilage matrix. These chondrocytes do not undergo mitosis or participate in bone growth, they function to maintain the cartilage matrix of the epiphyseal plate just like chondrocytes found in other locations.

Moving toward the diaphysis the next zone of cartilage is the **proliferative zone**. Here we find vertical stacks of chondrocytes that are undergoing or have recently undergone mitosis. Chondrocytes in the next layer, the **zone of mature cartilage**, are older and larger than those in the proliferative zone. These cells are beginning to age as the matrix begins to accumulate minerals.

Most of the chondrocytes in the next zone, the **zone of calcified matrix**, are dead because the matrix around them has calcified. The hard calcified matrix prevents nutrients from reaching these chondrocytes and they cannot survive. Capillaries and osteoblasts from the diaphysis penetrate this zone, and the osteoblasts secrete bone tissue on the remaining calcified cartilage. Thus, the zone of calcified matrix connects the epiphyseal plate to the diaphysis.

Bones continue to grow in length, the process is called **interstitial bone growth**, until early adulthood. The rate of growth is controlled by hormones, which will be discussed later. When the chondrocytes in the epiphyseal plate cease their proliferation and bone replaces the cartilage, longitudinal growth stops. All that remains of the epiphyseal plate is the **epiphyseal line**. Once the cartilage of the epiphyseal plate has ossified, longitudinal growth of that bone is impossible. The bones of the human body ossify in a predictable order, but the timing of this within the lifespan varies from one person to another.

7.5c How Bones Grow in Diameter

LO 7.5.2

While bones can increase in length throughout early life, they are capable of increasing in diameter throughout the lifespan. This is called **appositional bone growth** (Figure 7.20). New matrix is laid down on the outer surface of the bone by the osteoblasts in the periosteum. The thickness of compact bone increases and the

Figure 7.19 Interstitial (Lengthwise) Bone Growth

The hyaline cartilage of the epiphyseal plate is the site of bone elongation, shown here in three stages.

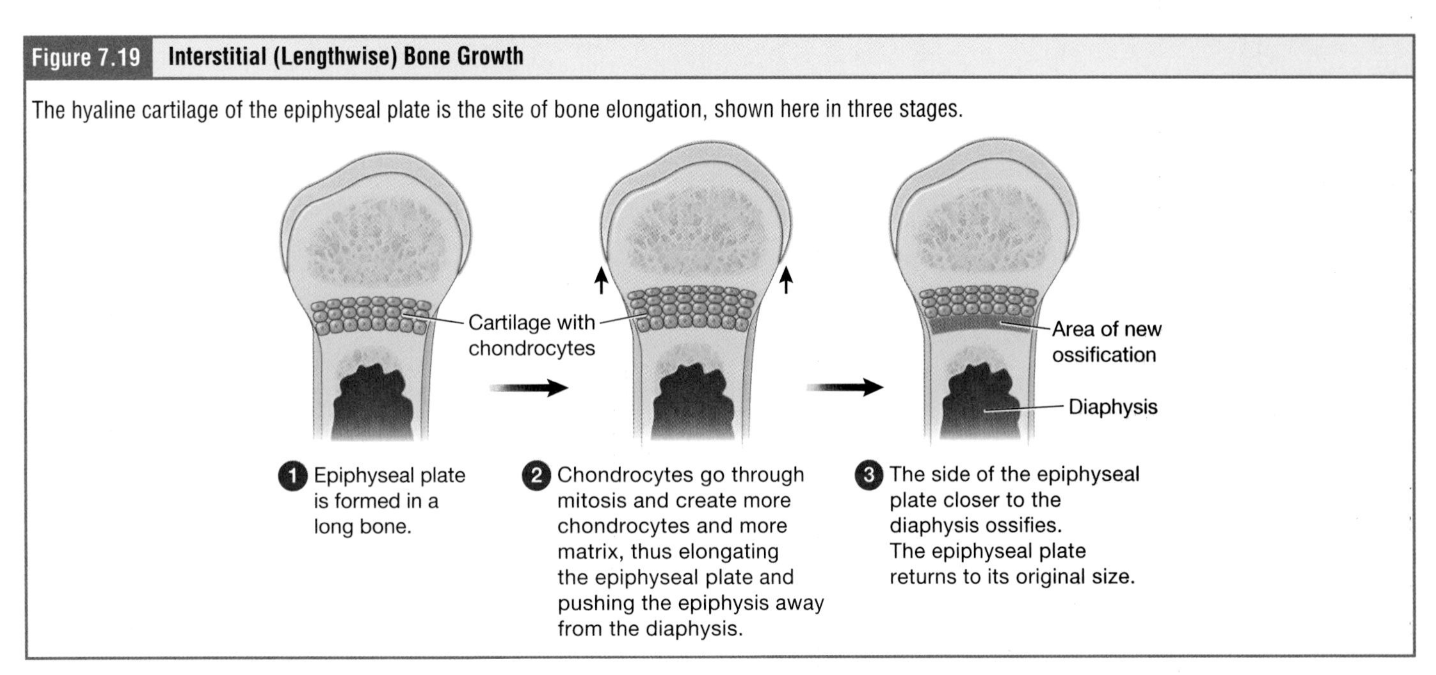
Cartilage with chondrocytes
Area of new ossification
Diaphysis
1 Epiphyseal plate is formed in a long bone.
2 Chondrocytes go through mitosis and create more chondrocytes and more matrix, thus elongating the epiphyseal plate and pushing the epiphysis away from the diaphysis.
3 The side of the epiphyseal plate closer to the diaphysis ossifies. The epiphyseal plate returns to its original size.

Anatomy of...

An Epiphyseal Plate

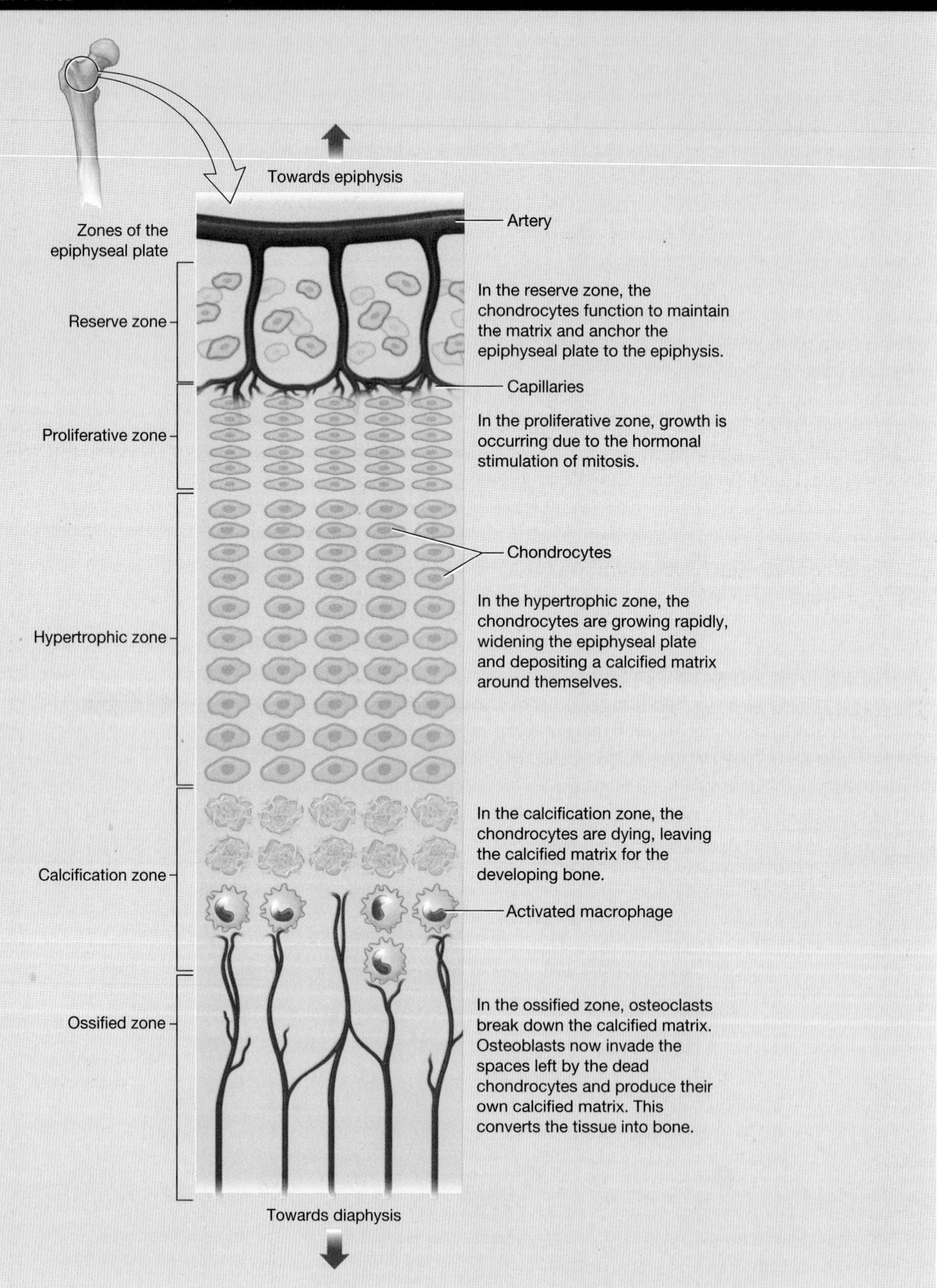

Towards epiphysis
Zones of the epiphyseal plate
Artery
Reserve zone
In the reserve zone, the chondrocytes function to maintain the matrix and anchor the epiphyseal plate to the epiphysis.
Capillaries
Proliferative zone
In the proliferative zone, growth is occurring due to the hormonal stimulation of mitosis.
Chondrocytes
Hypertrophic zone
In the hypertrophic zone, the chondrocytes are growing rapidly, widening the epiphyseal plate and depositing a calcified matrix around themselves.
Calcification zone
In the calcification zone, the chondrocytes are dying, leaving the calcified matrix for the developing bone.
Activated macrophage
Ossified zone
In the ossified zone, osteoclasts break down the calcified matrix. Osteoblasts now invade the spaces left by the dead chondrocytes and produce their own calcified matrix. This converts the tissue into bone.
Towards diaphysis

Figure 7.20 Appositional Bone Growth (Growth in Width)

The osteoblasts within the periosteum respond to many forces, including weight-bearing stress, to lay down bone matrix and grow bones wider. This process occurs throughout the lifespan.

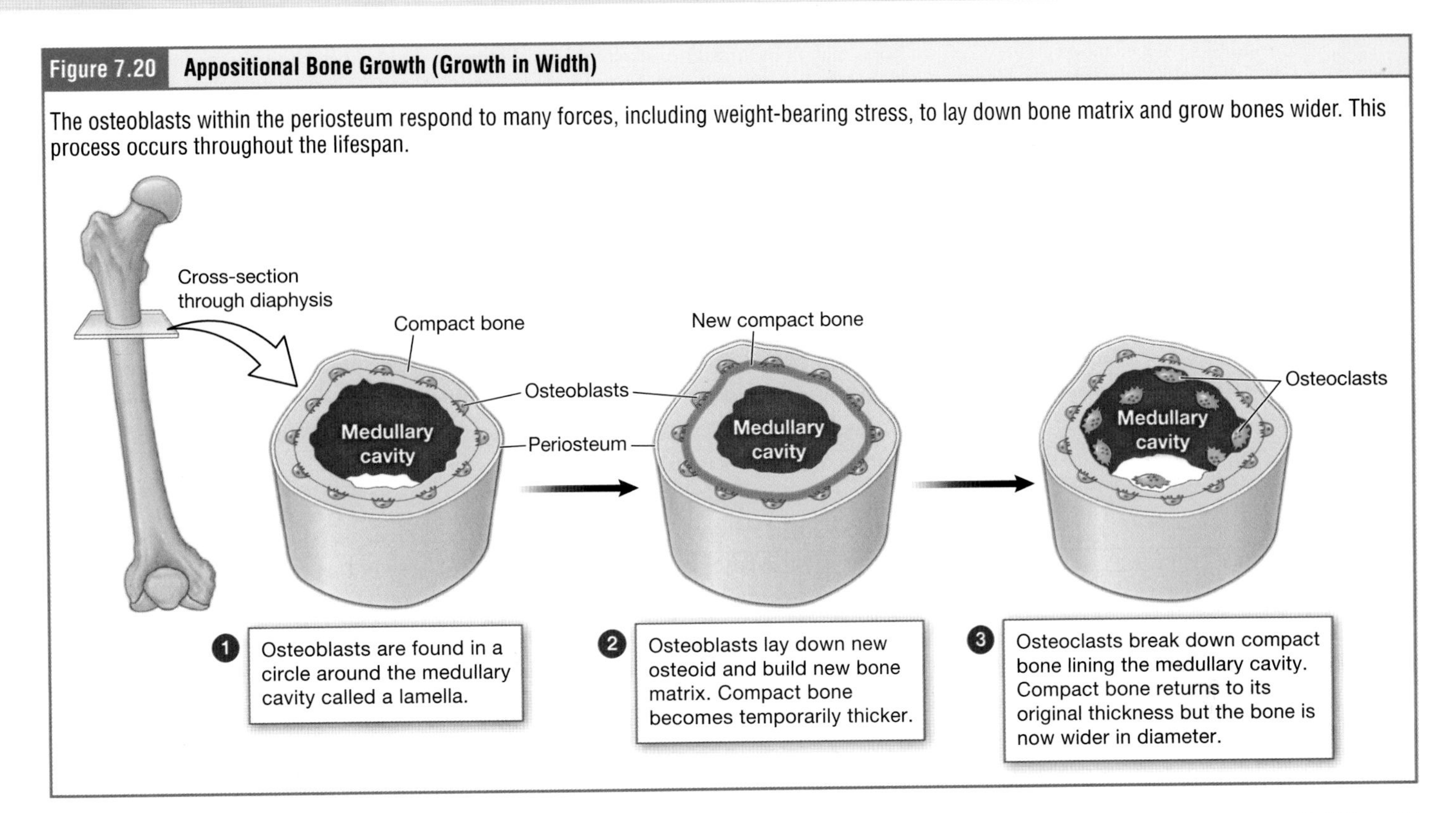

Cross-section through diaphysis
Compact bone
Osteoblasts
Periosteum
Medullary cavity
New compact bone
Medullary cavity
Osteoclasts
Medullary cavity
1 Osteoblasts are found in a circle around the medullary cavity called a lamella.
2 Osteoblasts lay down new osteoid and build new bone matrix. Compact bone becomes temporarily thicker.
3 Osteoclasts break down compact bone lining the medullary cavity. Compact bone returns to its original thickness but the bone is now wider in diameter.

diameter of the diaphysis increases. Next, osteoclasts break down the old bone that lines the medullary cavity, returning the compact bone layer to its original thickness.

Student Study Tip

Appositional growth can be compared to a tree gaining new rings over time as it grows. Bones respond to weight-bearing stress by growing in diameter, similar to a tree's annual growth rings.

7.5d Bone Remodeling

LO 7.5.3

In the two growth processes described so far, interstitial and appositional growth, changes to the bone proceed in a measurable direction, the bone gets longer or wider. However, bones change all the time, every day, without becoming noticeably different. Bones change because they serve as the storage for minerals in the body, and minerals are removed or added depending on the body's homeostatic needs. In adult life, this process of **remodeling** is influenced by the mineral needs of the body and regulated by hormones. About 20 percent of the skeleton is remodeled annually just by destroying old bone and renewing it with fresh bone. Injury, exercise, and other activities can also lead to remodeling.

7.5e Hormones and Bone Tissue

The endocrine system produces and secretes hormones, many of which interact with the skeletal system. These hormones are involved in controlling bone growth, maintaining bone once it is formed, and remodeling it.

LO 7.5.4

Hormones That Influence Osteoblasts and/or Maintain the Matrix Several hormones are necessary for controlling bone growth and maintaining the bone matrix. The pituitary gland secretes growth hormone (GH), which, as its name implies, controls bone growth in several ways. It triggers chondrocyte proliferation. When chondrocytes undergo mitosis in epiphyseal plates, bones get longer. GH also increases the amount of calcium that is absorbed across the intestinal wall, resulting in higher levels of blood and bone calcium, increasing bone density overall.

LO 7.5.5

GH is not alone in stimulating bone growth and maintaining osseous tissue. **Calcitriol**, the active form of vitamin D, is produced by the kidneys and stimulates the absorption of calcium and phosphate from the digestive tract, making more minerals available for bone matrix. Thyroxine (T_4), a hormone secreted by the thyroid gland, promotes the synthesis of bone matrix by stimulating osteoblasts. Estrogen and testosterone also promote osteoblastic activity and production of bone matrix. The increases in estrogen and testosterone during puberty likewise result in increases in osteoblast activity and the growth spurts that often occur during adolescence. The relationship between estrogen and testosterone and bones is not so simple, however. In low and moderate amounts these hormones promote bone growth by promoting osteoblast production of bone matrix. In higher amounts, these hormones stimulate osteoblasts so much that osteoblasts will convert all available cartilage—including the entire epiphyseal plate—into bone. Once the epiphyseal plate is completely converted to bone, it is referred to as the epiphyseal line and bones are incapable of growing any longer. This event, often referred to as *growth plate fusion* or *closure*, occurs at the point at which sex hormones are reaching sufficient levels for reproduction. Evolutionarily, this shift likely occurs so that available energy and nutrients can go to production and nurturing of offspring rather than further growth of the body.

Hormones That Influence Osteoclasts Bone remodeling and blood calcium homeostasis require a balance of activity between osteoclasts and osteoblasts. Two hormones that affect the osteoclasts are parathyroid hormone (PTH) and calcitonin.

PTH stimulates osteoclast proliferation and activity. As a result, calcium is released from the bones into the circulation, thus increasing the calcium ion concentration in the blood and decreasing the density of the bone matrix. PTH prevents loss of calcium in the urine, further increasing blood calcium levels. PTH is released by the parathyroid glands in response to low blood calcium levels (**Figure 7.21**).

Calcitonin, a hormone secreted by the thyroid gland, is released in response to high blood calcium levels (see Figure 7.21). Calcitonin can be thought of as an antagonist to PTH; it has many opposite effects. Calcitonin inhibits osteoclast activity and stimulates calcium uptake by the bones, thus reducing the concentration of calcium ions in the blood. **Table 7.3** summarizes the hormones that influence the skeletal system.

Learning Connection

Broken Process

What would happen to the skeleton if thyroxine (T_4) could not be made? While you're making predictions, think about what might happen if you had too little or too much of each of the hormones in see Table 7.3.

Figure 7.21 Blood Calcium Regulation

Blood calcium homeostasis is controlled through the antagonistic actions of two hormones: calcitonin and parathyroid hormone.

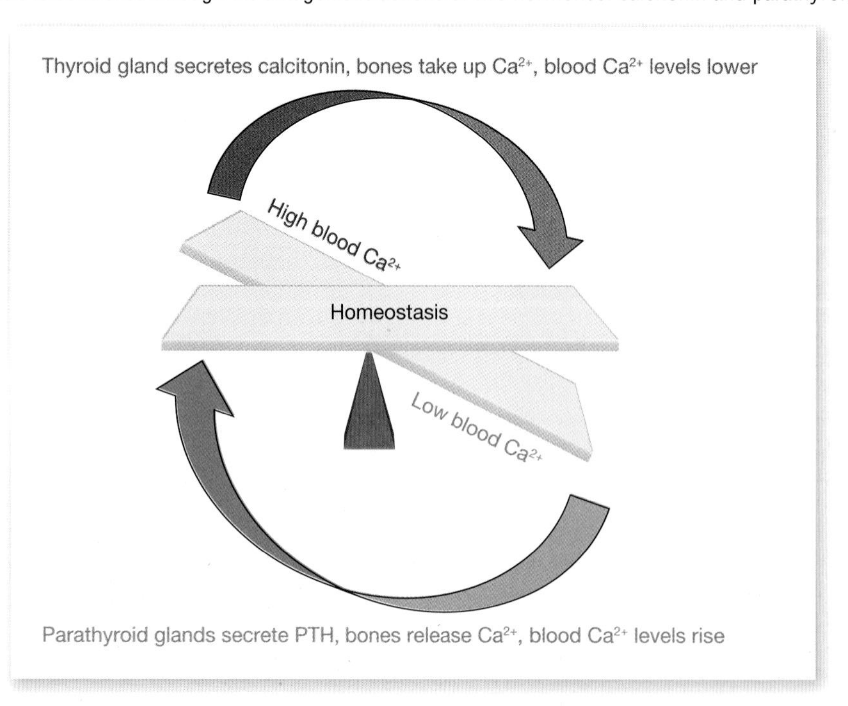
Thyroid gland secretes calcitonin, bones take up Ca^{2+}, blood Ca^{2+} levels lower
High blood Ca^{2+}
Homeostasis
Low blood Ca^{2+}
Parathyroid glands secrete PTH, bones release Ca^{2+}, blood Ca^{2+} levels rise

Table 7.3 Hormones That Influence the Skeletal System

Hormone	Action	Example
GH	Promotes mitosis of chondrocytes, resulting in an increase in the length of bones.	
T_3 and T_4	Increase the activity of osteoblasts, resulting in more bone matrix.	
Estrogen and testosterone	Increase the activity of osteoblasts, resulting in more bone matrix; *at high levels convert epiphyseal plate to epiphyseal line.*	
Calcitriol	Increases the absorption of calcium and phosphate from the digestive tract.	Ca^{2+}
PTH	Increases osteoclast proliferation and activity, resulting in less bone matrix.	
Calcitonin	Increases osteoblast activity, which increases deposition of bone matrix. Decreases osteoclast activity, which prevents resorption of bone matrix.	

Cultural Connection

Really Rickety

Rickets is a disease in which the bones become weaker in children due to insufficient vitamin D. Vitamin D can be acquired through the diet, and this is true for much of the U.S. population for much of the year, but it can also be synthesized in the skin through UV exposure. Vitamin D deficiency in adults causes similar declines in bone strength called *osteomalacia*. Rickets and osteomalacia used to be extremely common around the world and became rarer in the twentieth century as vitamin supplementation became more common. There has been a resurgence of rickets in the past few decades around the world. The populations that are especially vulnerable to modern-day rickets are: children living in near-polar regions such as Siberia and Southern Australia; women and girls living in regions where they wear clothes that cover most of the skin for cultural reasons; and adolescents, particularly those with darker skin, because their need for vitamin D to support their rapidly growing skeletons is especially high. While individuals with lighter skin have historically been at a lower risk for vitamin D deficiencies, more modern practices of sunscreen use and UV-protective clothing can compromise vitamin D levels in even these individuals.

7.5f Bone Repair

LO 7.5.6

All of the cells and processes we have learned about so far can be seen working rapidly during the process of healing from an injury. The complex process of healing is outlined in **Figure 7.22**. Remember that, unlike cartilage, bones are highly vascularized, with blood vessels in the periosteum, osteons, and/or medullary cavity. Therefore, when a bone breaks, bleeding will occur from vessels damaged in the fracture. The first physiological response to the injury is blood clotting. Within hours after the injury the clotting blood will form a **fracture hematoma** (**Figure 7.22A**), which prevents further blood loss. The disruption of blood flow to the bone results in the death of bone cells around the fracture.

In all forms of bone development, a matrix of collagen is required for mineral deposition. Healing from a fracture is no different. Within about 48 hours after the fracture, chondrocytes create a fibrocartilage template for later mineralization called a **callus**. The callus stabilizes the fracture by uniting the two ends of the broken bone (**Figure 7.22B**). Angiogenesis, the growth of new blood vessels, is a key requirement to bring oxygen, nutrients, and minerals to the healing tissue.

Over the next several weeks, osteoclasts break down the dead bone; osteogenic cells become active, replicate, and differentiate into osteoblasts. The osteoblasts will deposit bone matrix, replacing the cartilage callus with a bone callus (**Figure 7.22C**).

Eventually, compact bone replaces spongy bone at the outer margins of the fracture, and healing is complete. A slight swelling may remain on the outer surface of the bone, but quite often, that region undergoes remodeling (**Figure 7.22D**), and no external evidence of the fracture remains.

This healing process will proceed whether the broken bones are in alignment or not (**Figure 7.23A**). Since the healing process begins within hours after the injury, it can be crucial to seek medical treatment if a bone break is suspected. A broken bone can often be manipulated and set into an optimal healing position without surgery, but occasionally the broken bones may need to be held together with surgically placed screws for healing (**Figure 7.23B**). This process of aligning the bones for optimal healing is called **reduction**.

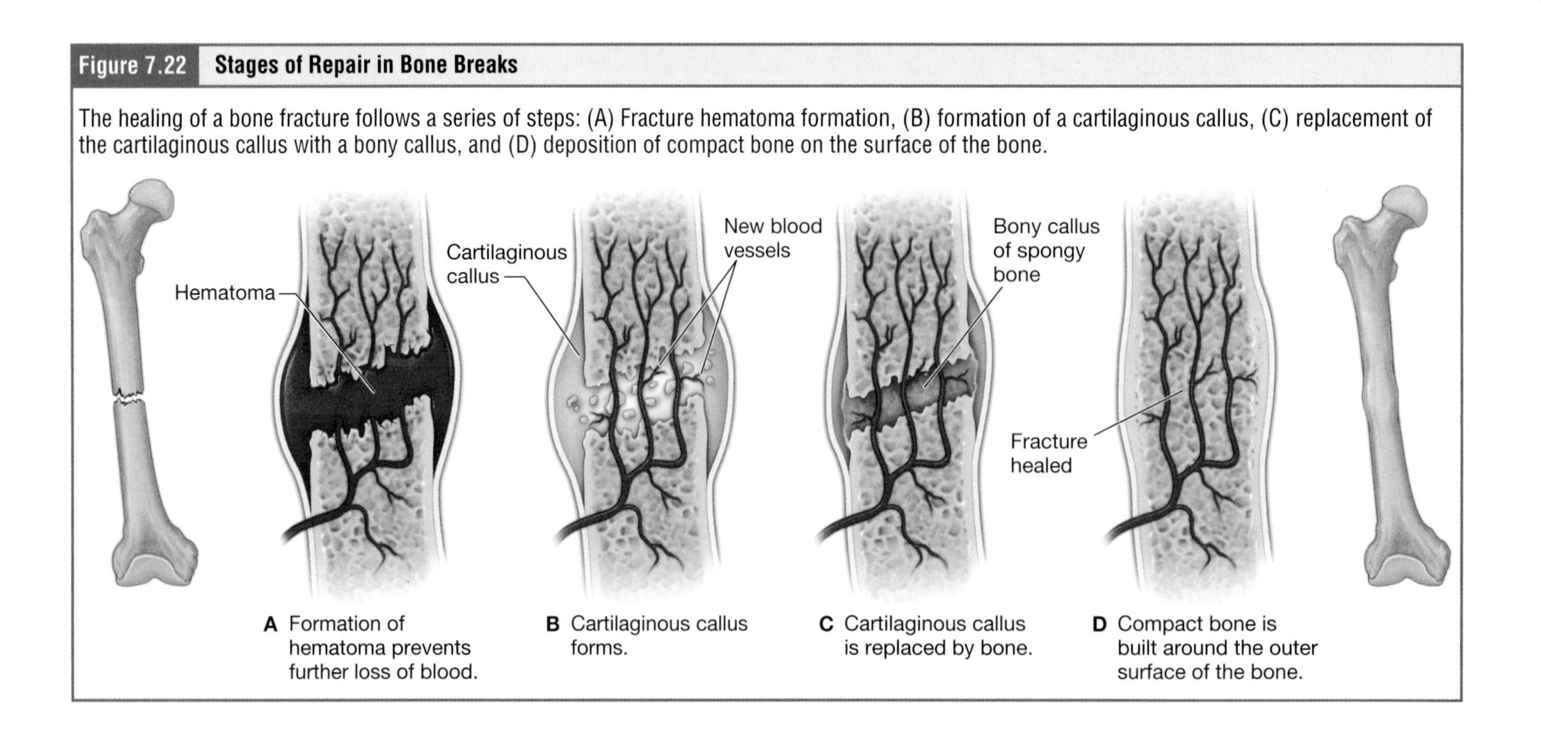

Figure 7.22 Stages of Repair in Bone Breaks

The healing of a bone fracture follows a series of steps: (A) Fracture hematoma formation, (B) formation of a cartilaginous callus, (C) replacement of the cartilaginous callus with a bony callus, and (D) deposition of compact bone on the surface of the bone.

A Formation of hematoma prevents further loss of blood.

B Cartilaginous callus forms.

C Cartilaginous callus is replaced by bone.

D Compact bone is built around the outer surface of the bone.

Figure 7.23 **Stabilization of Bone Fractures with Screws**

(A) Broken bones of the lower leg. (B) To stabilize the bones and make sure they heal in straight cylinders, screws are added surgically.

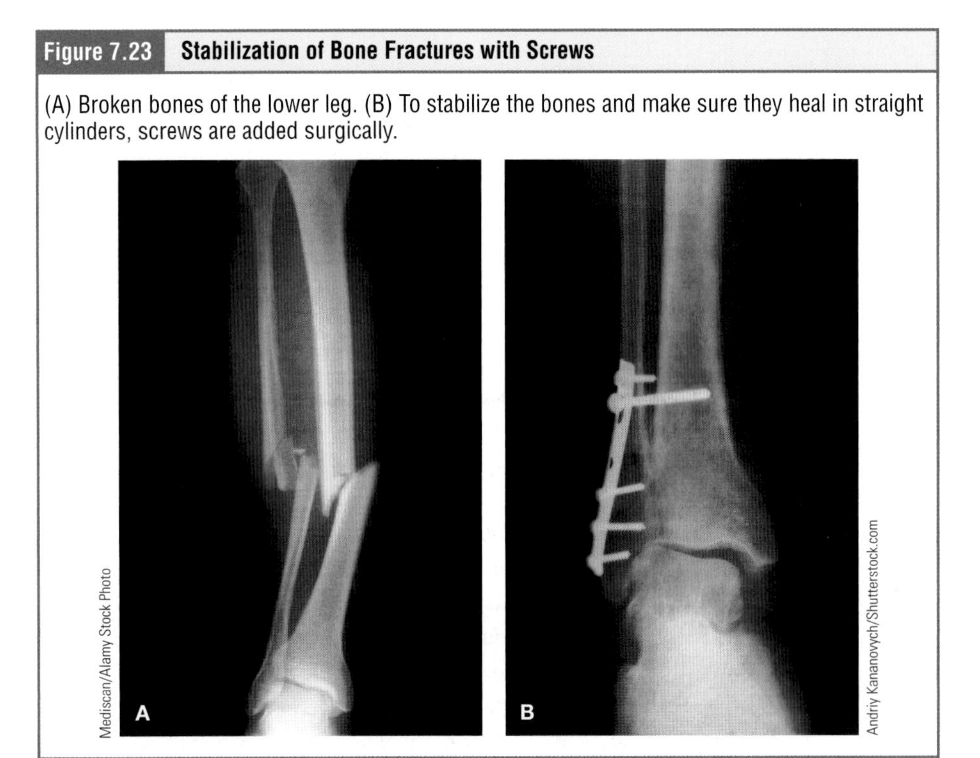

DIGGING DEEPER:
Types of Fractures

Fractures are classified by their complexity, location, and other features. **Table 7.4** outlines common types of fractures. Some fractures may be described using more than one term (e.g., an open transverse fracture) because they may have the features of more than one type.

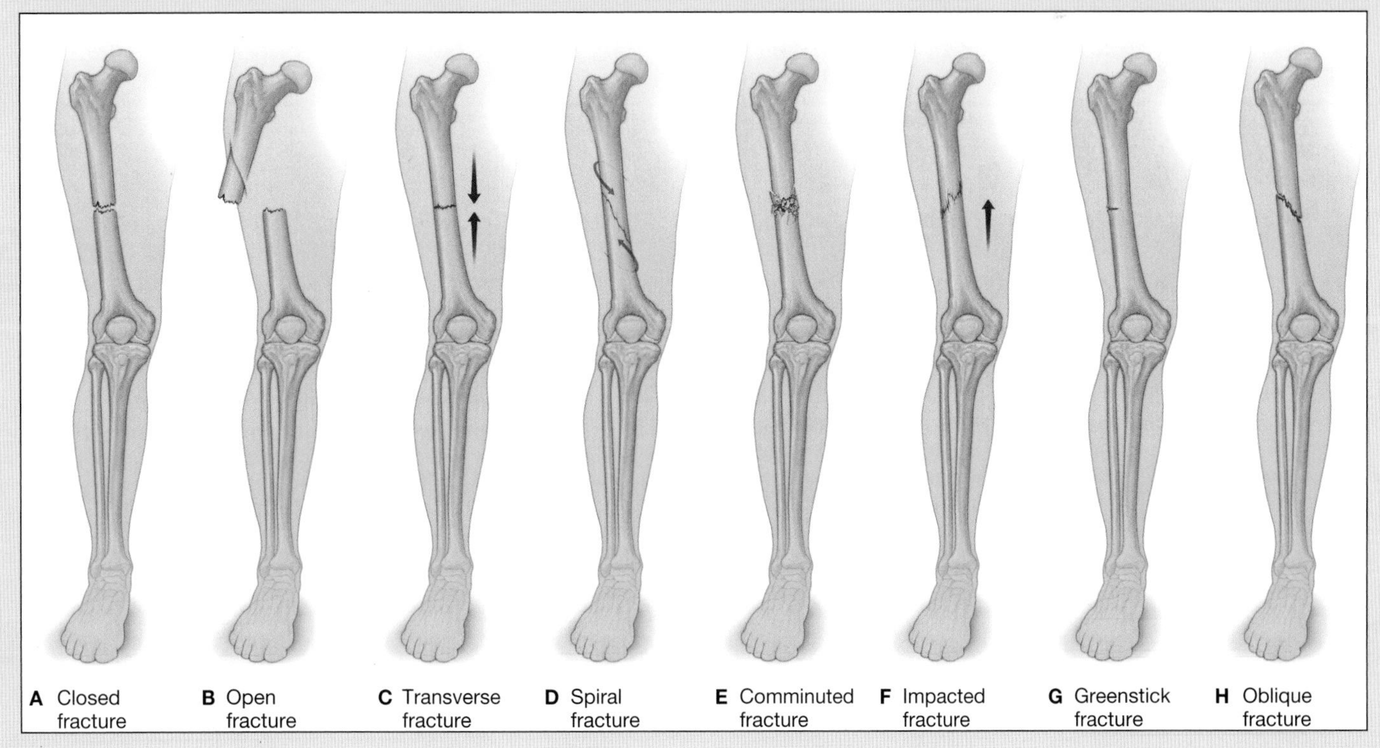

(Continued)

Table 7.4 Common Types of Fractures

Type of Fracture	Description
Transverse	Occurs straight across the long axis of the bone
Oblique	Occurs at an angle that is not 90 degrees
Spiral	Bone segments are pulled apart as a result of a twisting motion
Comminuted	Several breaks result in many small pieces between two large segments
Impacted	One fragment is driven into the other, usually as a result of compression
Greenstick	A partial fracture in which only one side of the bone is broken
Open (or compound)	A fracture in which at least one end of the broken bone tears through the skin; carries a high risk of infection
Closed (or simple)	A fracture in which the skin remains intact

Learning Check

1. Where in the body would you expect interstitial cartilage growth?
   a. Diaphysis of the bone
   b. Growth/epiphyseal plate of the bone
   c. Finger joint
   d. Between vertebrae
2. Which of the following selections lists the zones of the epiphyseal plate in order from the epiphysis to the diaphysis?
   a. Reserve Zone, Proliferative zone, Zone of mature cartilage, Zone of calcified matrix
   b. Reserve Zone, Zone of mature cartilage, Zone of calcified matrix, Proliferative zone
   c. Proliferative zone, Zone of mature cartilage, Reserve zone, Zone of calcified matrix
   d. Zone of calcified matrix, Zone of mature cartilage, Proliferative zone, Reserve zone
3. If a fracture is healed without surgery, why is it important for the bones to be properly aligned throughout the healing process?
   a. To ensure proper angiogenesis formation
   b. To properly align the fibrocartilage network
   c. To decrease osteoclast activity

7.6 Bones and Homeostasis

Learning Objectives: By the end of this section, you will be able to:

7.6.1 Explain how the skeletal system participates in homeostasis of plasma calcium levels.

7.6.2* Explain how exercise impacts bone formation.

7.6.3 Given a factor or situation (e.g., osteoporosis), predict the changes that could occur in the skeletal system and the consequences of those changes (i.e., given a cause, state a possible effect).

7.6.4 *Given a disruption in the structure or function of the skeletal system (e.g., osteoarthritis), predict the possible factors or situations that might have caused that disruption (i.e., given an effect, predict the possible causes).

* Objective is not a HAPS Learning Goal.

Calcium is not only the most abundant mineral in bone; it is also the most abundant mineral in the human body. Calcium ions are needed not only for bone mineralization but also for tooth health, heart contraction, blood clotting, smooth and skeletal muscle cell contraction, and nervous system function. Therefore, skeletal health is a key component to homeostasis. The normal level of calcium in the blood is about 10 mg/dL. Deviations from blood calcium homeostasis are described as *hypo-* or *hypercalcemia*.

Hypocalcemia—when blood levels of calcium are lower than the homeostatic range—can have widespread adverse effects. Without adequate calcium, blood has difficulty clotting, the heart may not be able to maintain its regular rhythm, muscles may have difficulty contracting, nerve cells may not be able to send signals, and bones weaken. Hypocalcemia can result from dietary causes or from changes in hormone concentration. In **hypercalcemia**—the condition of having blood calcium levels higher than the homeostatic range—the nervous system is affected more than muscles or bones.

Obviously, calcium homeostasis is critical. Calcium, being a chemical element, cannot be produced by the body; it must be acquired through diet. This is why the bones act as a storage site for calcium to help the body maintain homeostasis even when blood calcium levels change. The skeletal, endocrine, and digestive and urinary systems all have roles in the maintenance of blood calcium homeostasis (see the "Anatomy of Calcium Homeostasis" feature).

7.6a Nutrition and Bone Tissue

The vitamins and minerals contained in all the food we consume are important for all of our organ systems. However, there are certain nutrients that affect bone health.

Calcium and Vitamin D You already know that calcium is a critical component of bone, and that because the body cannot make its own calcium, it must be obtained

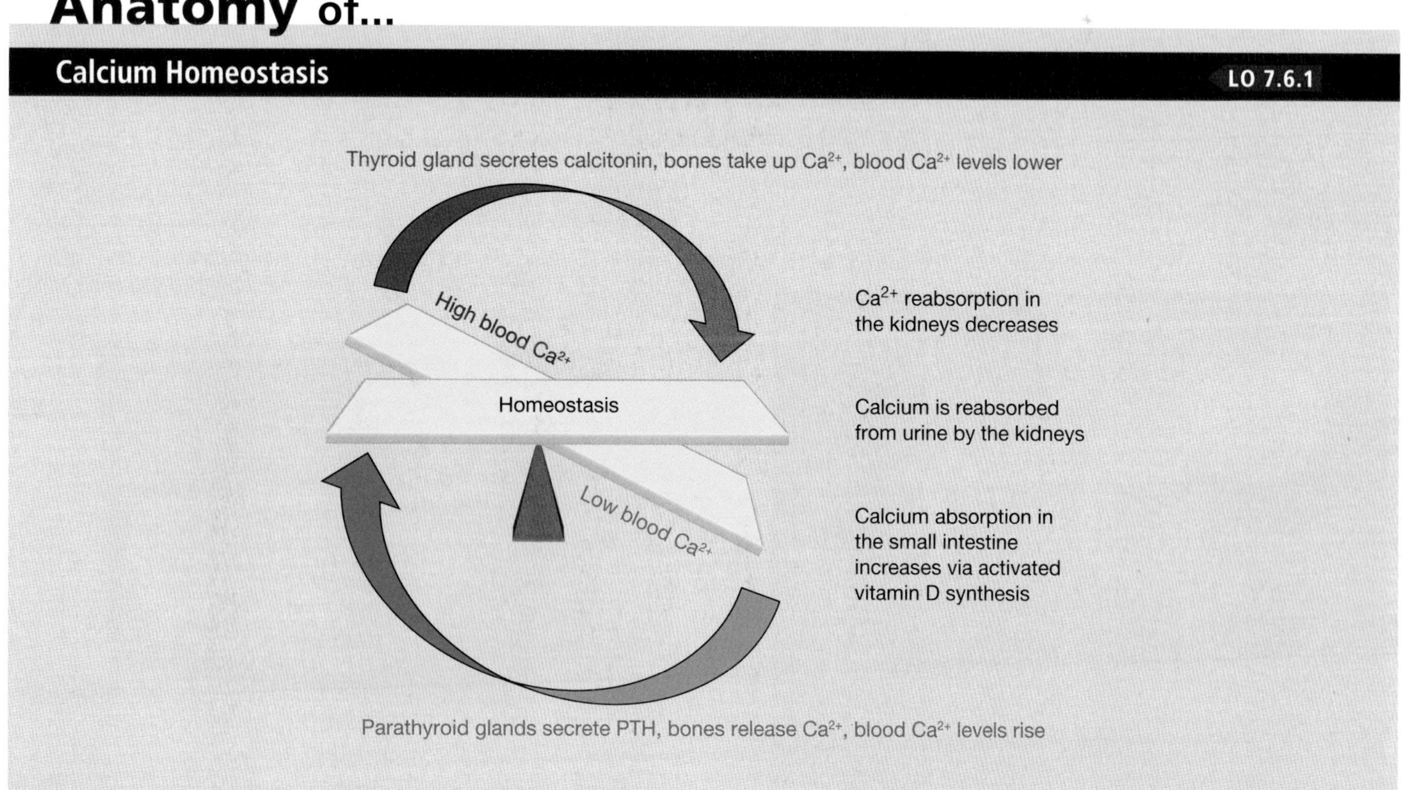
Anatomy of...
Calcium Homeostasis
LO 7.6.1
Thyroid gland secretes calcitonin, bones take up Ca^{2+}, blood Ca^{2+} levels lower
High blood Ca^{2+}
Homeostasis
Low blood Ca^{2+}
Ca^{2+} reabsorption in the kidneys decreases
Calcium is reabsorbed from urine by the kidneys
Calcium absorption in the small intestine increases via activated vitamin D synthesis
Parathyroid glands secrete PTH, bones release Ca^{2+}, blood Ca^{2+} levels rise

from the diet (**Figure 7.24**). However, calcium cannot be absorbed from the small intestine without vitamin D. Therefore, intake of vitamin D is also critical to bone health.

Vitamin D is not found naturally in many foods. Some fish, such as salmon and tuna, have small amounts, but the main dietary source of vitamin D is from foods where vitamin D has been added, such as cow's milk (**Figure 7.25**). Unlike calcium, however, we do not need to obtain Vitamin D from our diets. The action of sunlight on the skin triggers the body to produce its own vitamin D (**Figure 7.26**). The definition of **vitamin** is actually any organic substance essential to human metabolism that is obtained through diet; a hormone, on the other hand, is a chemical substance formed in a tissue that is carried in the blood. Therefore, though we always call it "vitamin D," vitamin D is more often a hormone than a vitamin. In many people who live in climates with warmer temperatures for part of the year and cooler temperatures for other months, vitamin D is sometimes a hormone and sometimes a vitamin. In the United States, 50–60 percent of nursing home and hospitalized patients and 42 percent of other adults are deficient in vitamin D because of inadequate exposure to the sun.

Learning Connection

Broken Process

Imagine that you break one of the bones in your forearm and stabilize the arm in a cast for six weeks. During this time, you do not use the broken arm at all; you perform all tasks with your other arm. At the end of the healing period, imagine that you can take an image of the bone in your upper arm (the humerus), which was not broken, on both sides of the body. Will they be any different?

7.6b Exercise and Bone Tissue

LO 7.6.2

During long space missions, astronauts can lose approximately 1–2 percent of their bone mass per month. This loss of bone mass is thought to be caused by the lack of mechanical stress on astronauts' bones due to the low gravitational forces in space. NASA studies, and tries to prevent, skeletal system changes in astronauts (**Figure 7.27**), and much of what we know about the skeleton and exercise comes from the work of NASA scientists. Lack of weight-bearing stress causes bones to lose both minerals and collagen fibers, becoming weaker. Similarly, weight-bearing stress stimulates the deposition of mineral salts and collagen fibers. The structure of a bone changes dynamically as stress increases or decreases so that the bone is well-suited for the amount of activity it endures. Individuals who exercise regularly will develop thicker, denser bones than individuals with more sedentary lifestyles. The bones undergo remodeling as a result of the forces (or the lack of forces) placed on them.

Figure 7.24 Dietary Calcium

Many different types of foods are good sources of dietary calcium.

Tatjana Baibakova/Shutterstock.com

Figure 7.25 Vitamin D

Vitamin D, which can be synthesized in the body, is not found in many foods. In the United States, vitamin D is added artificially to cow's milk and fortified cereals.

Martin Shields/Science Source

Figure 7.26 Vitamin D Synthesis

Vitamin D can be synthesized in the skin upon exposure to sunlight. Once in the body, it is converted to active forms and facilitates the uptake of calcium from the gut.

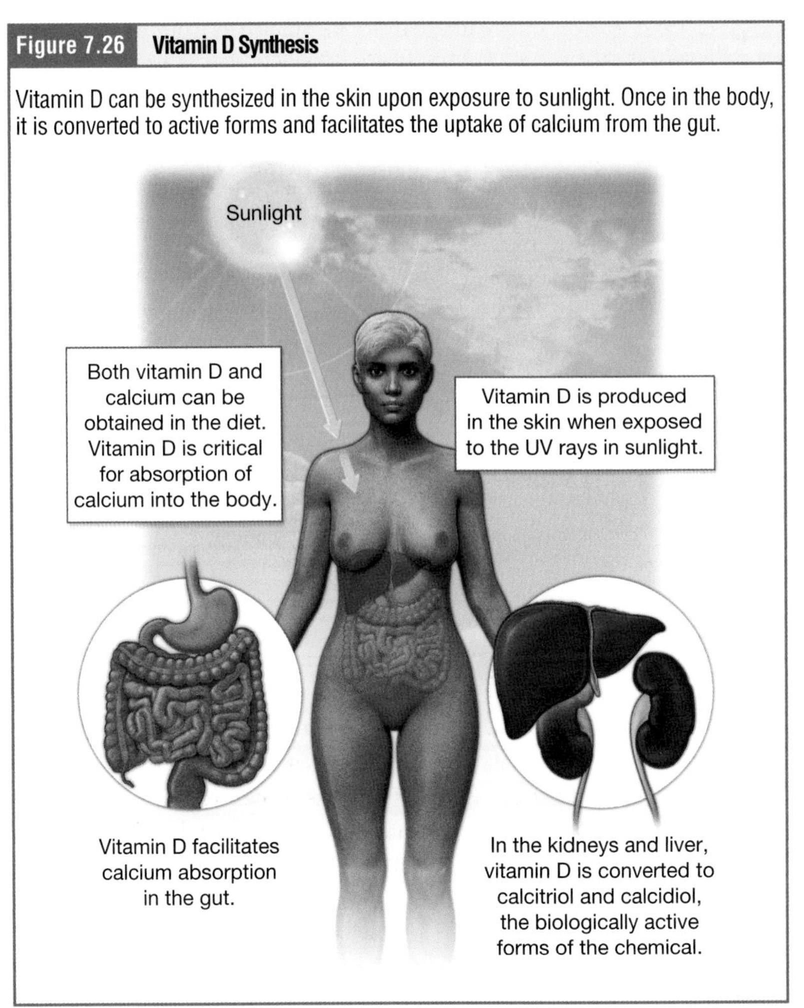

Sunlight
Both vitamin D and calcium can be obtained in the diet. Vitamin D is critical for absorption of calcium into the body.
Vitamin D is produced in the skin when exposed to the UV rays in sunlight.
Vitamin D facilitates calcium absorption in the gut.
In the kidneys and liver, vitamin D is converted to calcitriol and calcidiol, the biologically active forms of the chemical.

Figure 7.27 Exercise Is Essential for Skeletal Maintenance

As observed in both sedentary individuals and astronauts in space for prolonged periods, the absence of weight-bearing exercise causes rapid loss of both collagen and mineral matrix from bone.

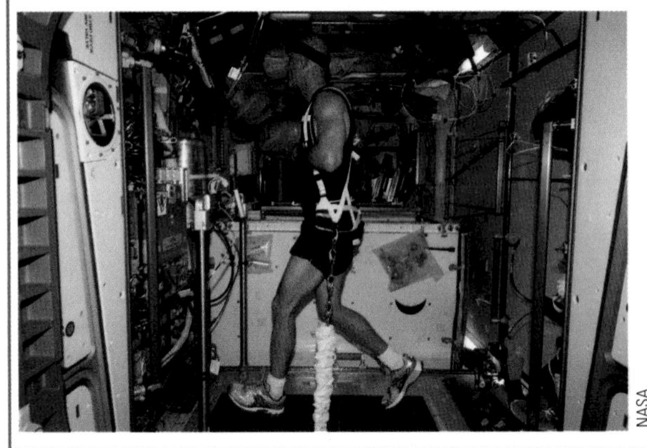

Apply to Pathophysiology

Osteoporosis

LO 7.6.3

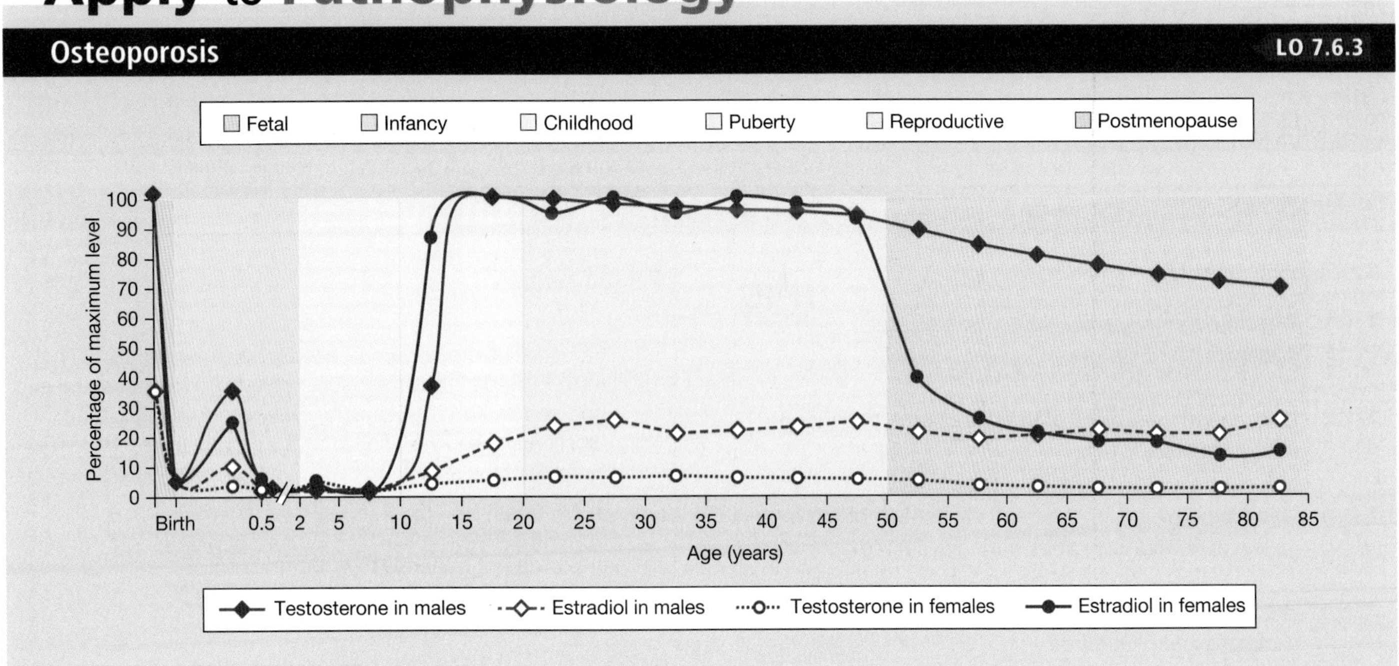

(*Continued*)

Examine the figure. In later life, biological males and females may experience very different physiological consequences to their declining production of sex hormones. Note that in biological females estrogen levels drop off sharply (an event known as menopause) whereas in biological males, the decline in testosterone is gradual occurring over decades.

1. Throughout adulthood, what role do estrogen and testosterone play in bone maintenance?
   A) The sex hormones act to decrease bone matrix, weakening bones.
   B) The sex hormones stimulate bone matrix development, strengthening bones.
   C) The sex hormones have no impact on bone maintenance or development.
2. **Osteoporosis** is a disease characterized by a decrease in bone mass that occurs when the rate of bone resorption exceeds the rate of bone formation, a common occurrence as the body ages. Which cells are more active in osteoporosis?
   A) Osteoclasts are more active than osteoblasts. LO 7.6.4
   B) Osteoblasts are more active than osteoclasts.
   C) Osteocytes are more active than osteogenic cells.
   D) Osteoblasts are more active than osteocytes.
3. Given the data presented in the graph and your answer to Question 1, do you anticipate that there would be a difference between the biological sexes in the risk for osteoporosis?
   A) Biological males would be at a higher risk than biological females.
   B) Biological females would be at a higher risk than biological males.
   C) There would be no significant difference between males and females in the risk for osteoporosis.
4. A significant concern among those with osteoporosis would be:
   A) Digestive issues, including frequent vomiting.
   B) Increased risk of fracture with sudden movements such as sneezing.
   C) Inability to sustain skeletal muscle contraction.

Learning Check

1. Describe the relationship between calcium and vitamin D.
   a. Calcium is critical for vitamin D absorption through the intestines.
   b. Vitamin D is crucial for calcium absorption into the bone.
   c. Vitamin D and calcium must be ingested together to be effective.
   d. Vitamin D breaks down calcium from the bone.
2. Does a person who has a sedentary lifestyle have more or less bone density than someone who has an active lifestyle?
   a. More bone density
   b. Less bone density

Chapter Review

7.1 The Functions of the Skeletal System

The learning objectives for this section are:

7.1.1 Describe the major functions of the skeletal system.

7.1.2 Describe the locations of the three types of cartilage in the skeletal system.

7.1.3 Describe how the location and distribution of red and yellow bone marrow varies during a lifetime.

7.1.4* Predict disruptions to homeostasis if damage to the skeletal system occurred.

* Objective is not a HAPS Learning Goal.

7.1 Questions

1. Faruq fell off the playground and fell on their stomach. They were rushed to the emergency room to find that one of their ribs was broken. The doctors were concerned about the function of their liver and kidney (organs in the body). Which function of the skeletal system is this an example of?
   a. Provide structure to hold the body upright
   b. Protect internal organs from injury
   c. Storage of minerals in bone matrix
   d. Production of red blood cells

2. Identify the locations in the body where fibrocartilage is NOT found.
   a. Vertebrae
   b. Pubic symphysis
   c. Knees
   d. Elbow

3. Josue is 57 years old. Would you expect red or yellow marrow to fill the medullary cavity of their long bones?
   a. Red marrow
   b. Yellow marrow

4. Shekelia is a 65-year-old who does not ingest enough calcium in their diet. Which of the following effects would you expect to see in their body? Please select all that apply.
   a. Decreased density of bone matrix
   b. Decreased red blood cell production
   c. Decreased calcium in the blood
   d. Decreased structural support

7.1 Summary

- The skeletal system is responsible for: (1) Providing structure to hold us upright; (2) Protecting your internal organs from injury; (3) Maintaining homeostasis through mineral deposition and withdrawal; and (4) Production of red blood cells.
- Three types of cartilage exist in the skeletal system: (1) hyaline cartilage, which is rigid and smooth, (2) elastic cartilage, which is flexible, and (3) fibrocartilage, which is tough and durable.
- There are two types of bone marrow: (1) red marrow, which produces blood cells, and (2) yellow marrow, which stores fat.

7.2 Bone Classification

The learning objectives for this section are:

7.2.1 Classify bones of the skeleton based on their shape.

7.2.2 Identify and describe the structural components of a long bone, and explain their functions.

7.2.3 Define common bone marking terms (e.g., condyle, tubercle, foramen, canal).

7.2 Questions

1. Classify the sternum (bone in your chest) based on its shape.
   a. Long
   b. Short
   c. Irregular
   d. Flat
   e. Sesamoid

2. Pooja is going in for a bone marrow transplant. The surgeons will have to go into the bone to take out the infected bone marrow and replace it with healthy bone marrow. Which of the following selections lists the correct order of bone layers through which the surgeon's needle will go?
   a. Endosteum, Compact Bone, Periosteum
   b. Compact Bone, Periosteum, Endosteum
   c. Periosteum, Compact Bone, Endosteum
   d. Periosteum, Endosteum, Compact Bone

3. The rounded end of the femur that articulates with the superior end of the tibia to form the knee joint is called a:
   a. Fossa
   b. Foramen
   c. Tubercle
   d. Condyle
   e. Sulcus

7.2 Summary

- Bones can be classified by shape: (1) Long bones, (2) Short bones, (3) Irregular bones, (4) Flat bones, and (5) Sesamoid bones.
- The epiphysis sits on the proximal and distal ends of the bones. In the center of the bone, you will find the diaphysis. The metaphysis is in the middle of the two, which is where the growth plate is found.
- The most superficial layer is the periosteum, followed by compact bone, followed by the endosteum, until you reach the medullary cavity.

7.3 The Microscopic Structure of Cartilage and Bone

The learning objectives for this section are:

7.3.1 Describe the roles of dense regular and dense irregular connective tissue in the skeletal system.

7.3.2 List and describe the cellular and extracellular components of bone tissue.

7.3.3 Using microscopic images, distinguish between the three different types of cartilage.

7.3.4 Identify the microscopic structure of compact bone and spongy bone.

7.3 Questions

1. Where would you find dense irregular connective tissue in the skeletal system?
a. Cartilage matrix
b. Bone matrix
c. Perichondrium
d. Lamellae

2. Which of the following highlight similarities between cartilage and bone? Please select all that apply.
a. Osteogenic cells are like chondrocytes in that they both form extracellular matrix.
b. Osteocytes are like chondrocytes in that they both are found inside lacunae.
c. The perichondrium is like a canaliculus in that they both are found in the periphery.
d. Osteoblasts are like chondroblasts in that they both differentiate as they mature.

3. Looking under a microscope, you can visibly see thick lines representing bundles of collagen fibers. Identify the cartilage that you are most likely observing.
a. Elastic cartilage
b. Fibrocartilage
c. Hyaline cartilage

4. Which of the following structures would be found in spongy bone that would not be present in compact bone?
a. Trabeculae
b. Central canals
c. Lamellae
d. Osteons

7.3 Summary

- Chondroblasts generate cartilage matrix and differentiate into chondrocytes, which are inside a lacuna. Similarly, osteoblasts generate bone matrix and differentiate into osteocytes.
- Compact bone is found in the periosteum, while spongy bone is found in the epiphysis.
- Cartilage and bone provide the rigidity of the skeletal system. Both tissue types contain a small number of cells in a large amount of extracellular matrix.
- Chondroblasts generate cartilage matrix and differentiate into chondrocytes, which are inside a lacuna. Similarly, osteoblasts generate bone matrix and differentiate into osteocytes.
- The diaphysis of a long bone is composed of compact bone, and the epiphyses are composed of spongy bone.

7.4 Formation and Growth of Bone and Cartilage

The learning objectives for this section are:

7.4.1 Explain the roles that specific bone cells play in the formation of bone tissue.

7.4.2 Compare and contrast intramembranous and endochondral (intracartilaginous) bone formation.

7.4 Questions

1. What is the role of mesenchymal cells during formation of bone tissue?
a. To differentiate into osteogenic cells
b. To secrete matrix devoid of minerals
c. To create protective shell of compact bone
d. To form an ossification center

2. Which of the following is unique to endochondral bone formation?
a. Osteoblasts secrete osteoid matrix.
b. Osteogenic cells differentiate into osteoblasts.
c. Blood vessels invade into disintegrated cartilage matrix.
d. Osteoblasts create protective shell of compact bone.

7.4 Summary

- Bone tissue forms by two methods of ossification, both involving the replacement of existing connective tissue.
- Intramembranous ossification involves the growth of bone tissue within sheets of dense irregular connective tissue. This type of bone formation occurs mainly in flat bones.
- Endochondral ossification involves the replacement of a hyaline cartilage model with bone tissue. This type of bone formation occurs mainly in long bones.

7.5 Growth, Repair, and Remodeling

The learning objectives for this section are:

HAPS LEARNING GOALS AND OUTCOMES

7.5.1 Compare and contrast interstitial (lengthwise) and appositional (width or circumferential) growth in bone tissue.

7.5.2 Compare and contrast the function of osteoblasts and osteoclasts during bone growth, repair, and remodeling.

7.5.3 Describe the bone repair and remodeling process and how it changes as humans age.

7.5.4 Explain the hormonal regulation of skeletal growth.

7.5.5 Explain the roles of parathyroid hormone, calcitriol, and calcitonin in plasma calcium regulation and bone remodeling.

7.5.6 Explain the steps involved in fracture repair.

7.5 Questions

1. Which of the following describes the process of appositional cartilage growth?
   a. Chondroblasts of the perichondrium secrete matrix around themselves.
   b. Chondrocytes secrete matrix between themselves.
   c. Chondroblasts share a single lacuna until the cell matures.
   d. Mitotic replication of chondrocytes occurs.

2. If the calcium level in the blood is too low, which of the following bone cells would remove bone matrix from the bones and add it to the blood for use in nerve impulse conduction and muscle contraction?
   a. Osteocytes
   b. Osteoblasts
   c. Osteogenic cells
   d. Osteoclasts

Mini Case 1: Fractures of the Epiphyseal Plate

Brian, a 12-year-old boy is rushed to the hospital after falling off the jungle gym onto his arm. X-rays indicate that he has a fracture in his forearm going through the epiphyseal plate of the radius.

1. Which of the following is true of the repair and remodeling processes that will occur as Brian's bone is healing? Check all that apply.
   a. The last step of bone repair is the replacement of spongy bone with compact bone.
   b. During bone repair, osteogenic cells will differentiate into osteoblasts, to deposit bone matrix at the injury site.
   c. As Brian's bone is repairing, the bony callus will be produced before the cartilaginous callus.
   d. The healing process in Brian's bone will proceed only if the bone fragments are perfectly aligned.

2. Brian will soon reach the age of puberty, and they will go through a rapid growth phase. What role will growth hormone (GH) play in the growth of their skeleton?
   a. GH will increase osteoblast activity, to increase bone deposition.
   b. GH will inhibit osteoclast activity, to inhibit resorption (removal) of bone matrix.
   c. GH will stimulate the mitosis of chondrocytes, to cause an increase in bone length.
   d. GH will convert the epiphyseal plates to epiphyseal lines, to stop bone growth at the proper time.

3. During recovery, which of the following hormones would ensure that osteoblasts are being stimulated?
   a. Growth hormone
   b. Calcitriol
   c. Thyroxine
   d. Parathyroid hormone

4. Which of the following do you expect to occur first as part of the bone repair process?
   a. Cartilage callus formation
   b. Fracture hematoma formation
   c. Bone callus formation
   d. Fibrocartilage template formation

7.5 Summary

- Cartilage can grow in length (interstitial cartilage growth) or in width (appositional cartilage growth), similar to bone.
- Hormones play a big role in calcium homeostasis to maintain bone density. Growth hormone triggers chondrocyte proliferation and increases calcium absorption across the intestinal wall. Calcitriol also helps absorb calcium. Thyroxine stimulates osteoblasts. Estrogen and testosterone also promote osteoblastic activity and production of bone matrix.
- Parathyroid hormone and calcitonin work together to maintain calcium levels in the blood. Parathyroid hormone stimulates osteoclast proliferation and calcitonin inhibits osteoclast activity and stimulates uptake of calcium by bones.

7.6 Bones and Homeostasis

The learning objectives for this section are:

7.6.1 Explain how the skeletal system participates in homeostasis of plasma calcium levels.

7.6.2* Explain how exercise impacts bone formation.

7.6.3 Given a factor or situation (e.g., osteoporosis), predict the changes that could occur in the skeletal system and the consequences of those changes (i.e., given a cause, state a possible effect).

7.6.4 *Given a disruption in the structure or function of the skeletal system (e.g., osteoarthritis), predict the possible factors or situations that might have caused that disruption (i.e., given an effect, predict the possible causes).

* Objective is not a HAPS Learning Goal.

7.6 Questions

1. Ghazi went to their doctor complaining of tingling, muscle aches, and stiffness. After some tests, the doctor diagnosed them with hypocalcemia. Which of the following steps will the body take to reach homeostasis? Please select all that apply.
   a. Stimulation of osteoblasts
   b. Stimulation of osteoclasts
   c. Decreased reabsorption of calcium by kidneys
   d. Increased absorption of calcium in intestines

2. Explain how exercise increases bone growth.
   a. Increased forces placed on bone by exercise promote bone growth.
   b. Increased activity drives osteoblast activity.
   c. Increased blood flow will increase osteoclast activity.

3. Rheumatoid arthritis is an autoimmune disease, in which antibodies attack a substance in the person's own body. In some cases of rheumatoid arthritis, antibodies attack the hyaline cartilage that covers the ends of long bones in joints. What effect would you expect this to have on the bones? Check all that apply.
   a. Loss of ability to produce new blood cells
   b. Erosion of the ends of the long bones in joints
   c. Loss of ability to store fat
   d. Friction upon movement

4. Osteoarthritis is a condition in which there is thin cartilage between the bones in a joint. Which of the following could explain why a person has this condition?
   a. Increased parathyroid hormone activity
   b. Decreased growth hormone activity
   c. Increased thyroxine
   d. Decreased calcitonin

7.6 Summary

- Calcium is used in the body not only as a major component of the bone matrix, but also in nervous impulse conduction, muscle contraction, and blood clotting.
- Vitamin D, when converted into a hormone, is critical for the absorption of dietary calcium in the small intestine. This is of vital importance in bone health.
- Bones will adapt to the stresses put on them, so activity will promote bone growth and increased bone density.

8

Axial Skeleton

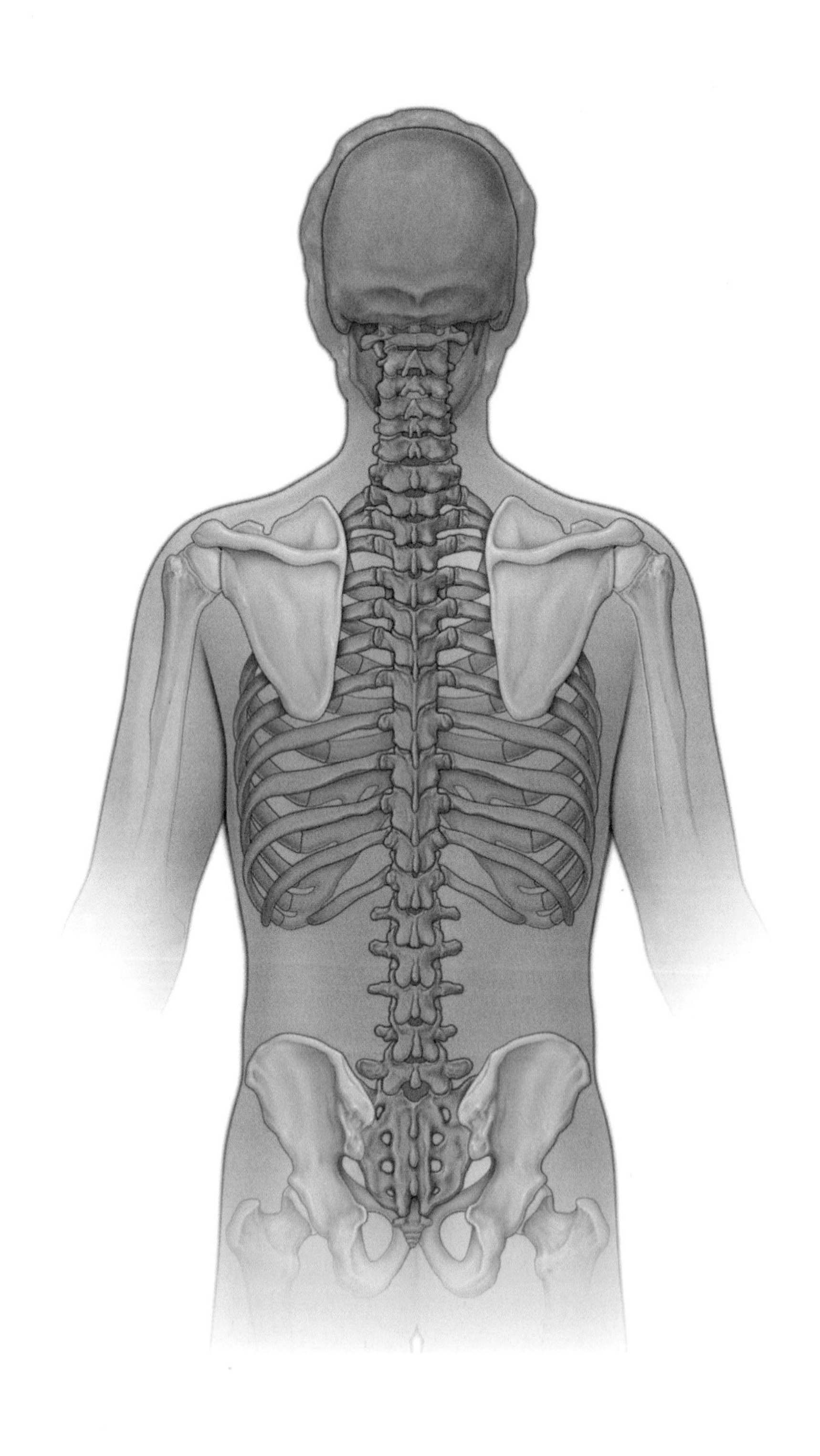

Chapter Introduction

In this chapter we will learn . . .

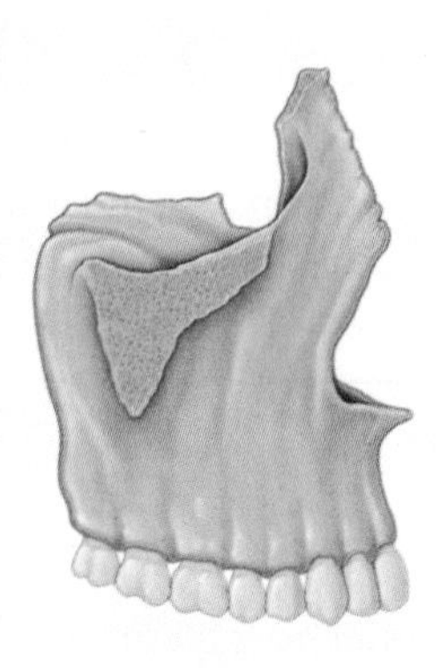

the individual bones of the skull, spine, and rib cage.

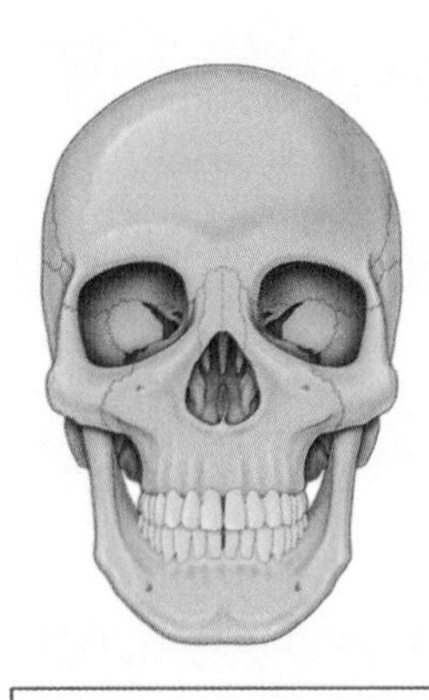

how these bones fit together to form the structures of the axial skeleton.

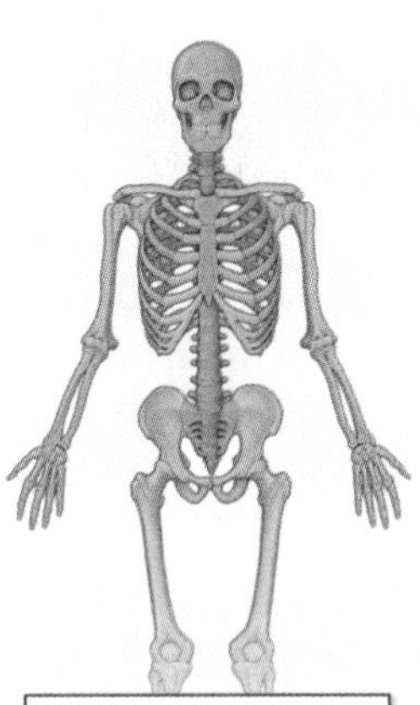

how the axial skeleton contributes to the whole body.

The skeletal system forms the rigid internal framework of the body. It consists of the bones, cartilages, and ligaments. You can learn more about these tissues in Chapters 5 and 7. Bones support the weight of the body, allow for body movements, and protect internal organs. Cartilage provides flexible strength and support for body structures such as the rib cage, the external ear, and the trachea and larynx. At joints of the body, cartilage provides a gliding surface and/or cushioning between bones. Ligaments are the strong connective tissue strips that bind the bones together. Muscles are connected to the bones via tendons. As muscles contract, they pull on the tendons, thereby pulling on the bones to produce movements of the body. Thus, without a skeleton, you would not be able to stand, run, or even feed yourself!

Bones vary in size, shape, and strength based on their functions. For example, examine the bones of the spine; as we descend down the trunk, each vertebra carries more weight than the one superior to it. The vertebrae increase in thickness and strength to support your body weight. Similarly, the size of a bony landmark that serves as a muscle attachment site on an individual bone is related to the strength of this muscle. Muscles can apply very strong pulling forces to the bones of the skeleton. To resist these forces, bones have enlarged bony landmarks at sites where powerful muscles attach. This means that not only the size of a bone, but also its shape, is related to its function. For this reason, the identification of bony landmarks is important during our study of the skeleton.

Bones are dynamic organs that can change in response to changes in muscle strength, muscle use, and body weight. Thus, muscle attachment sites on bones thicken when you begin a workout program that increases muscle strength. Similarly, bones will thicken if you gain body weight or begin pounding the pavement as part of a new running routine. In contrast, a reduction in muscle strength or body weight can lead to thinner bones. This may happen during a prolonged hospital stay, following immobilization in a cast, or going into the weightlessness of outer space.

8.1 Divisions of the Skeletal System

Learning Objectives: By the end of this section, you will be able to:

8.1.1 Distinguish between the axial and appendicular skeletons and list the major bones contained within each.

The Human Anatomy and Physiology Society includes more than 1,700 educators who work together to promote excellence in the teaching of this subject area. The HAPS A&P Learning Outcomes measure student mastery of the content typically covered in a two-semester Human A&P curriculum at the undergraduate level. The full Learning Outcomes are available at https://www.hapsweb.org.

The skeletal system includes all of the bones, cartilages, and ligaments of the body. An adult skeleton has, on average, 206 bones. Children have more bones than adults. Several bones fuse together during childhood and adolescence to form a single adult bone, so while the adult skeleton is larger it contains fewer bones! The primary functions of the skeleton are to provide a rigid structure that can support the body's weight, to provide a structure upon which muscles can act to produce body movements, and to assist the body in calcium homeostasis. The functions and dynamic nature of the skeleton are elaborated on in Chapter 7. Over generations, humans evolved to walk and run in pursuit of food and to carry objects and perform complex tasks. Therefore, the structure of the lower skeleton is specialized for stability during walking or running and the structure of the upper skeleton has greater mobility and ranges of motion, features that allow you to lift and carry objects or turn your head and trunk.

The skeleton is subdivided into two major divisions—the axial and appendicular.

8.1a The Axial Skeleton

LO 8.1.1

The **axial skeleton** forms the vertical, central axis of the body and includes all bones of the head, neck, torso, and back (**Figure 8.1**). The axial skeleton serves to protect the brain, spinal cord, heart, and lungs. It also serves as the attachment site for muscles that move the head, neck, and back, the muscles of respiration, and the abdominals. The easiest way to delineate between bones of the axial and appendicular skeleton is not by the precise location of the bone, but rather the type of movement the bone is involved in. The

Figure 8.1 The Axial and Appendicular Skeletons

The axial skeleton forms the vertical axis of the body. It consists of the skull, vertebral column (including the sacrum and coccyx), and the thoracic cage, formed by the ribs and sternum. The appendicular skeleton is made up of all bones of the upper and lower limbs and the bones that connect the limbs to the axial skeleton.

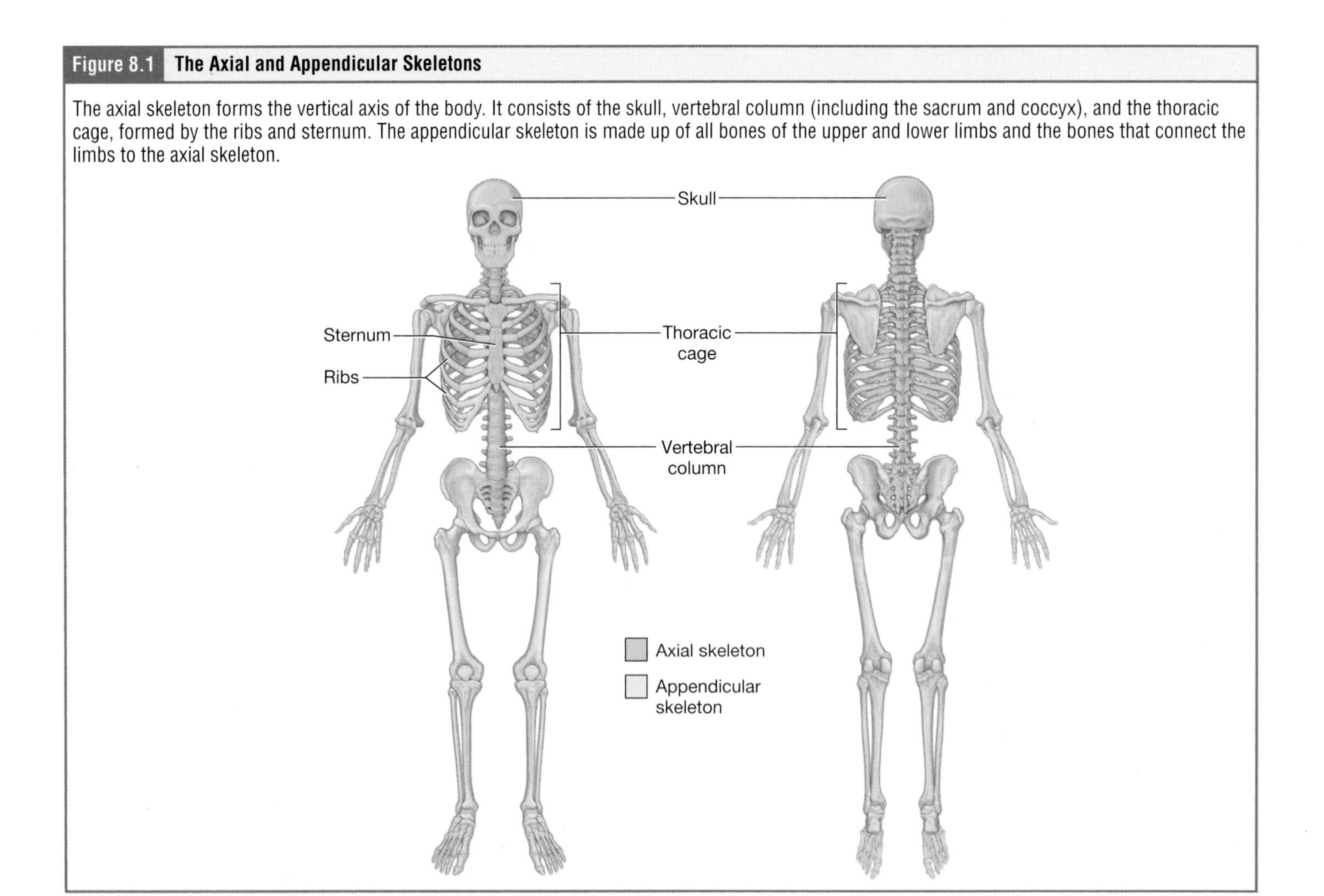

scapulae and pelvis are both centrally located, but they are involved in the movements of the upper and lower limbs and are, therefore, parts of the appendicular skeleton.

The axial skeleton of the adult consists of 80 bones, including the skull, the vertebral column, and the thoracic cage. The skull alone is formed by 22 bones! The adult vertebral column consists of 24 bones, each called a *vertebra*, plus the sacrum and coccyx. The thoracic cage includes the 12 pairs of ribs and the sternum. Additional components of the axial skeleton are the hyoid bone and the ear ossicles (three small bones found in each middle ear).

8.1b The Appendicular Skeleton

The **appendicular skeleton** includes all bones of the upper and lower limbs, plus the bones that attach each limb to the axial skeleton. There are 126 bones in the appendicular skeleton of an adult. The bones of the appendicular skeleton are covered in Chapter 9.

Learning Check

1. Identify the bones that are part of the axial skeleton. Please select all that apply.
   a. Ribs
   b. Skull
   c. Radius
   d. Pelvic girdle
2. Are the ear ossicles part of the axial skeleton or the appendicular skeleton?
   a. Axial skeleton
   b. Appendicular skeleton
3. Jorge is playing baseball when the ball hits their elbow. They go to get it checked and turns out they have fractured their humerus. Is this bone part of the axial or appendicular skeleton?
   a. Axial skeleton
   b. Appendicular skeleton

8.2 The Skull

Learning Objectives: By the end of this section, you will be able to:

8.2.1* Know the bones of the skull and their features.

8.2.2* List and identify the skull sutures.

8.2.3* Identify the skull sinuses and describe their functions.

8.2.4* Describe the changes to the skull throughout development and aging.

* Objective is not a HAPS Learning Goal.

The **skull** is an incredibly intricate and complex anatomical structure. It can be very helpful to learn this material while holding or viewing a skull in person, if you are fortunate enough to have access to one at your school, consider working through the chapter with hands-on learning. We will use colors and a variety of views and planes to illustrate the bones of the skull and their relationships. Remember that with each new image you will want to orient yourself as to the view and perspective.

8.2a Introduction to the Skull

The bones of the skull can be subdivided into the **facial bones** and the **cranium** (**Figure 8.2**). The facial bones form the structures of the face, the nasal cavity, and the mouth, and enclose the eyeballs. The rounded cranium surrounds and protects the brain and houses the middle and inner ear structures.

Figure 8.2 The Bones of the Skull Can Be Divided into Two Groups

The bones that form the rounded brain case that houses the brain are called *cranial bones* and the bones of the upper and lower jaws, nose, orbits, and other facial structures are called *facial bones*.

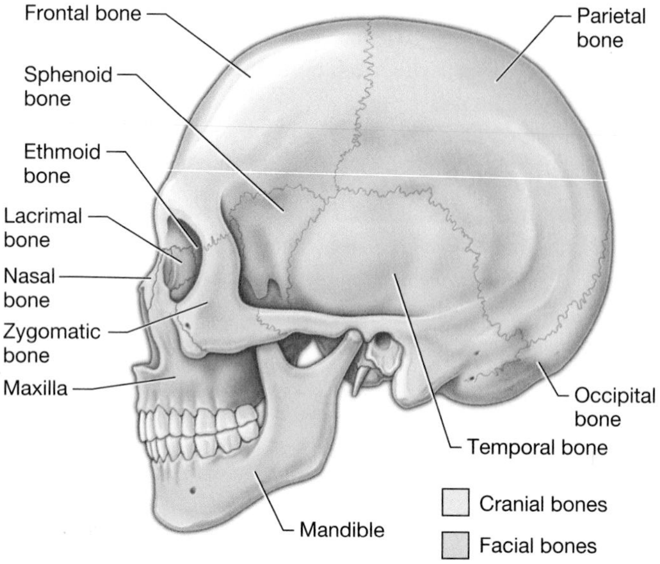

Frontal bone
Parietal bone
Sphenoid bone
Ethmoid bone
Lacrimal bone
Nasal bone
Zygomatic bone
Maxilla
Occipital bone
Temporal bone
Mandible
Cranial bones
Facial bones

In the adult, the skull consists of 22 individual bones, 21 of which are immobile and united into a single unit. The 22nd bone is the **mandible** (lower jaw), which is the only moveable bone of the skull. Some of these bones are *paired*; that is to say, there is one on the right and one on the left (**Figure 8.3**).

The bones of the skull come together to form several cavities that house softer structures. These cavities are the cranial cavity, orbits (eye sockets), nasal cavity, paranasal sinuses, and oral cavity (**Figure 8.4**). To begin our understanding of the skull, we will start by identifying each of the bones. Remember that you learned in Chapter 7 that bones

Figure 8.3 Unpaired and Paired Bones

Some of the bones of the skull are paired, meaning that the same bone can be found on the right and left sides. Some bones, like the frontal bone, are unpaired; there is only one.

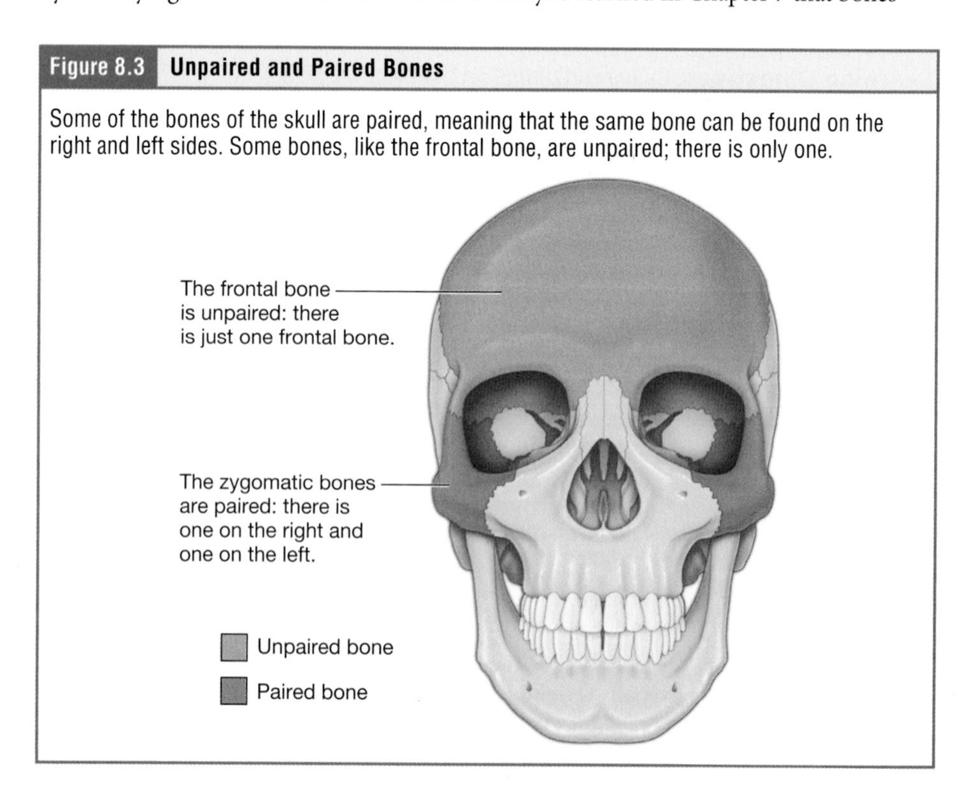
The frontal bone is unpaired: there is just one frontal bone.
The zygomatic bones are paired: there is one on the right and one on the left.
Unpaired bone
Paired bone

Figure 8.4 The Cavities of the Skull

The bones of the skull merge to form several cavities for the soft structures and organs of the head. Smaller, air-filled spaces are called *sinuses*.

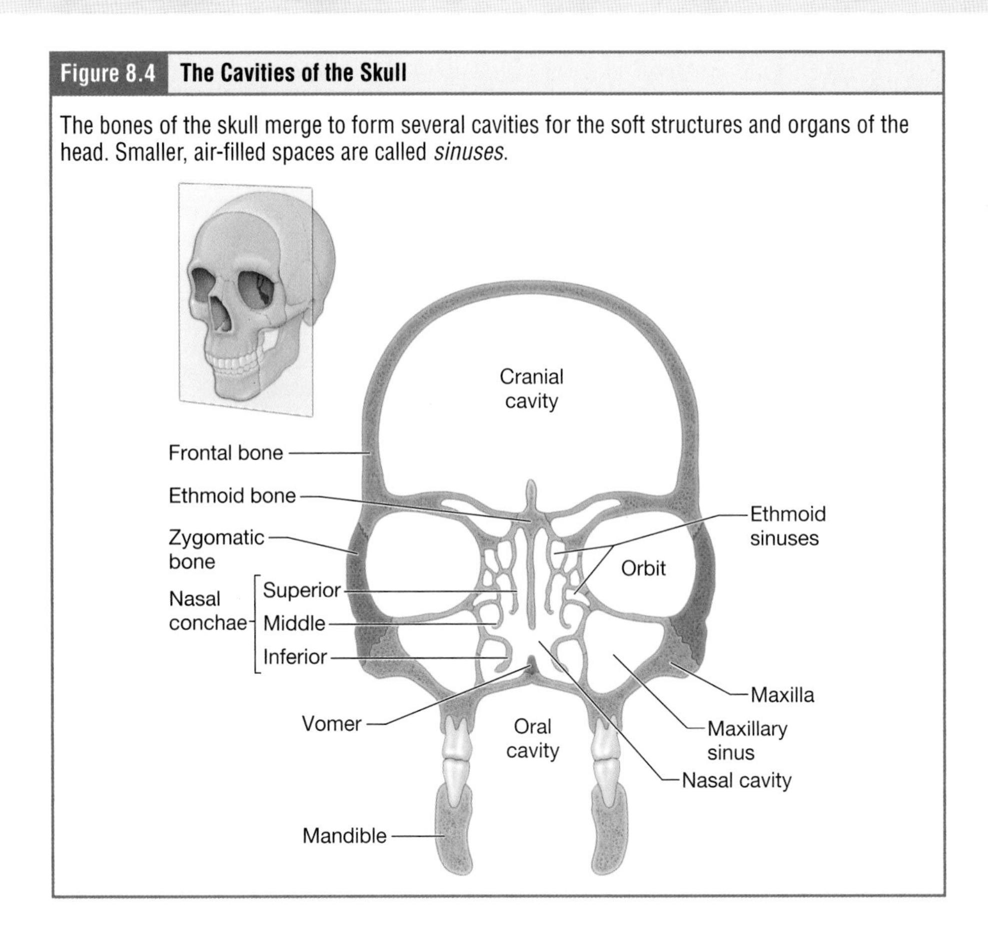

have individual bone markings—the bumps, holes, canals, and depressions of bone at which the bones interact with the other structures (muscles, blood vessels, nerves, and so on) of the body. In this chapter we will describe these features and reference the structures that you will learn about later that interact with the bones at these locations.

Learning Check

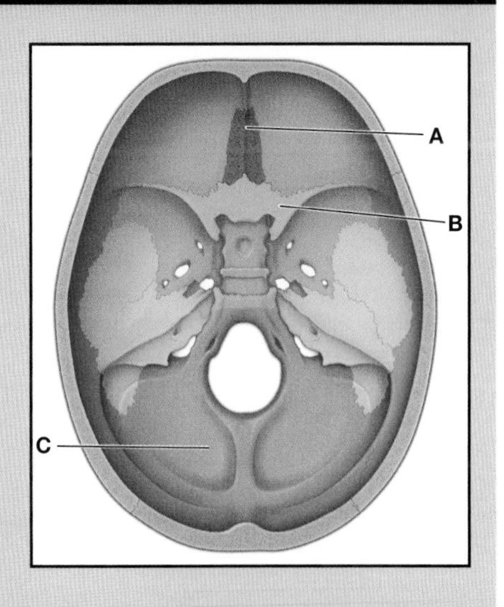

1. Which of the following describes the view of the skull shown in the figure?
   a. Superior
   b. Inferior
   c. Lateral
2. Identify the bone that letter A is pointing to.
   a. Temporal bone
   b. Sphenoid bone
   c. Ethmoid bone
   d. Frontal bone
3. Identify the bone that letter B is pointing to.
   a. Temporal bone
   b. Sphenoid bone
   c. Palatine bone
   d. Occipital bone
4. Identify the bone that letter C is pointing to.
   a. Frontal bone
   b. Temporal bone
   c. Parietal bone
   d. Occipital bone
5. Which of the following bones is not visible in this image?
   a. Frontal bone
   b. Nasal bone
   c. Temporal bone
   d. Occipital bone
   e. Inferior
   f. Lateral

Anatomy of...

The Skull Bones

Frontal Bone

A single frontal bone forms the anterior of the cranium. You probably call it your forehead.

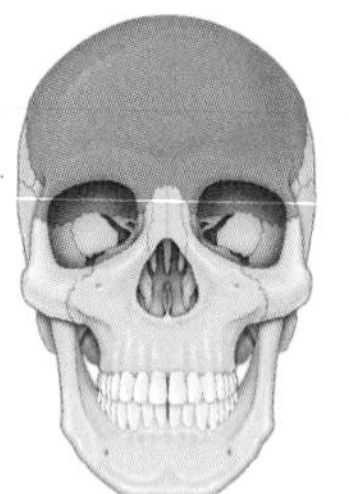
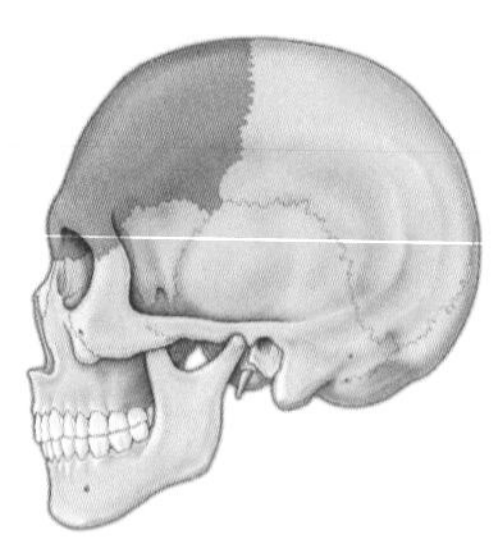

Parietal Bone

Paired parietal bones, joined with a suture, make up the majority of the superior skull.

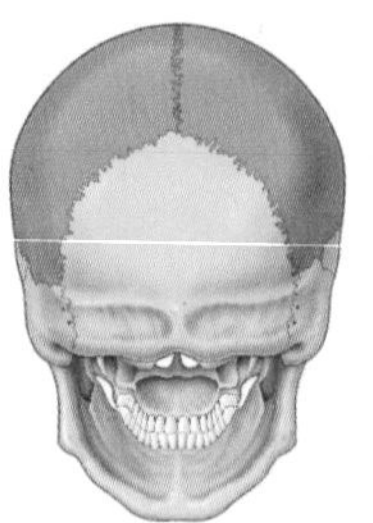
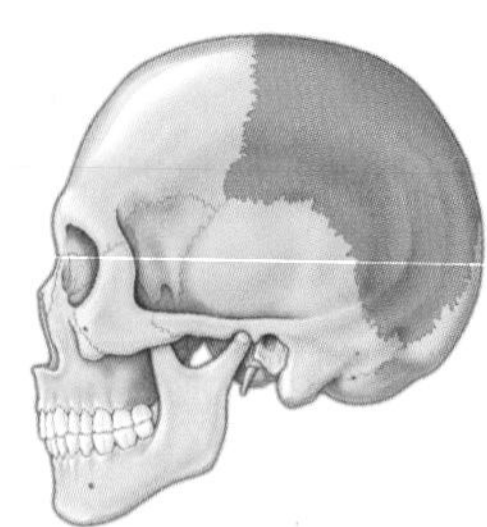

Occipital Bone

The single occipital bone covers the inferior/posterior side of the brain.

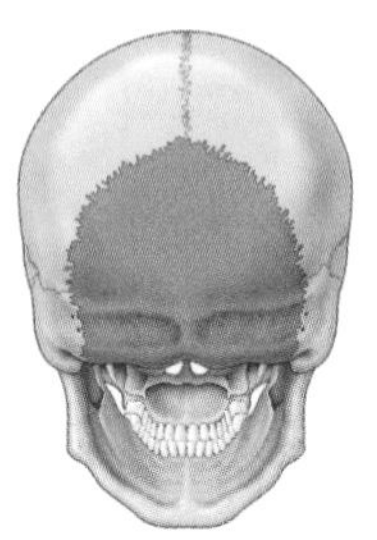
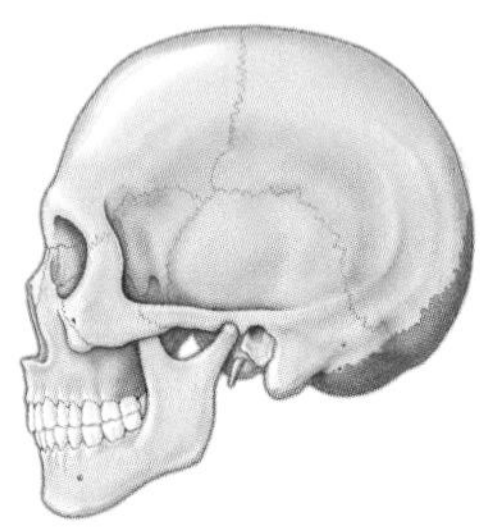

Temporal Bones

The paired temporal bones form the sides of the cranium deep to the external ears.

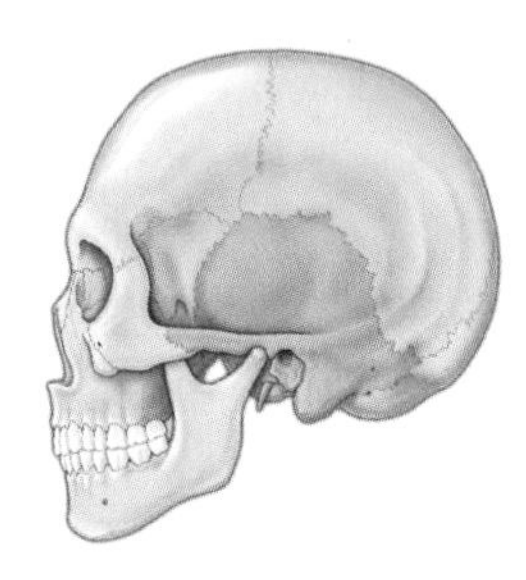

Mandible

In the adult, one mandible forms your lower jaw and has the distinction of being the only movable bone in the skull!

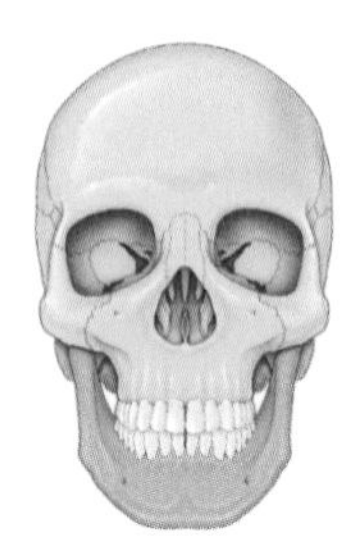
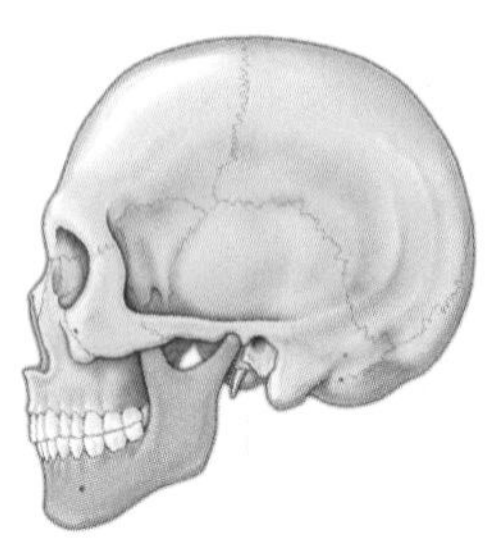

Maxillae

Paired maxillary bones form your upper jaw and the anterior roof of your mouth.

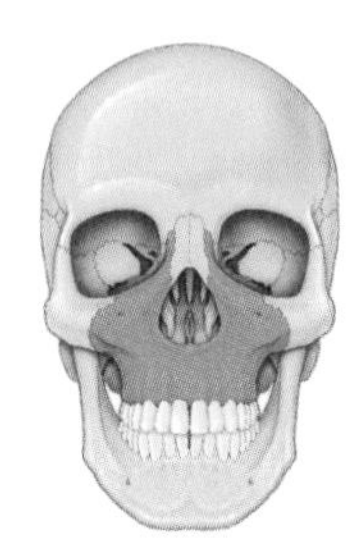
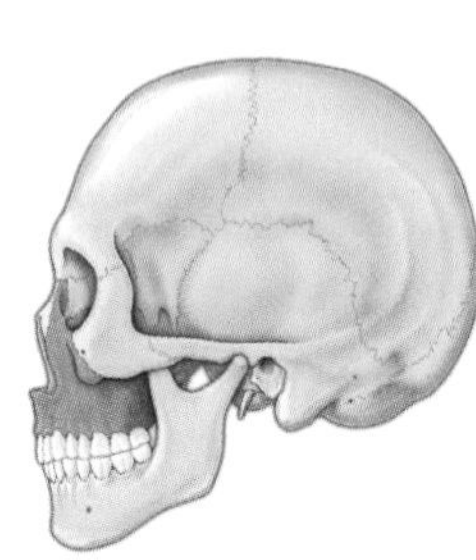

Nasal Bones

The paired nasal bones form the "bridge" of the nose. You might rest your glasses on them if you wear glasses.

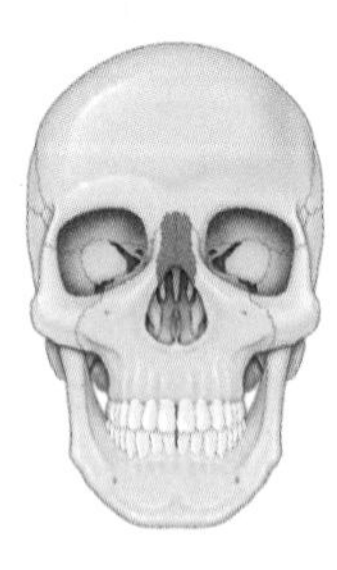
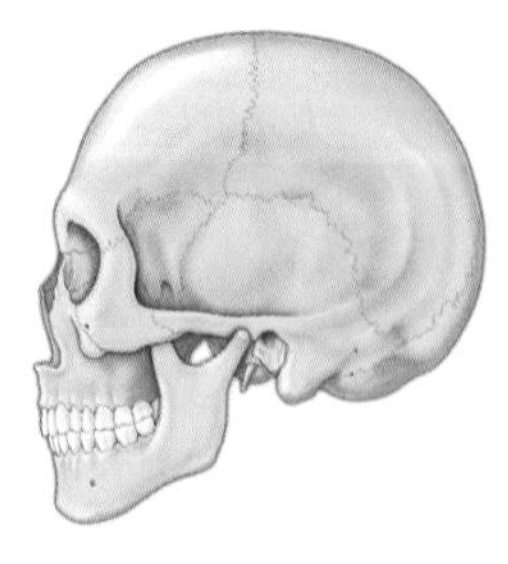

Zygomatic Bones

Paired zygomatic bones provide attachments for many muscles that move the mouth. You probably think of them as your "cheekbones."

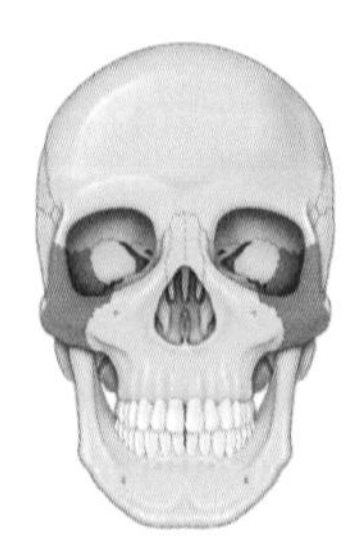
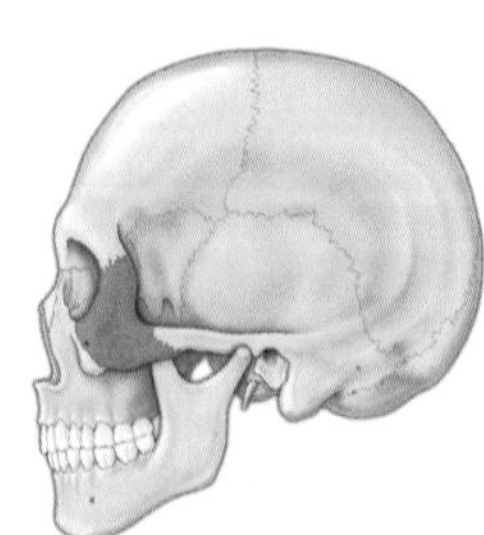

Sphenoid Bone

The sphenoid bone is mostly internal to the skull, making up a portion of the floor of the brain case and the back of the eye orbits. If you put your fingers on your "temples," you are touching part of the sphenoid bone.

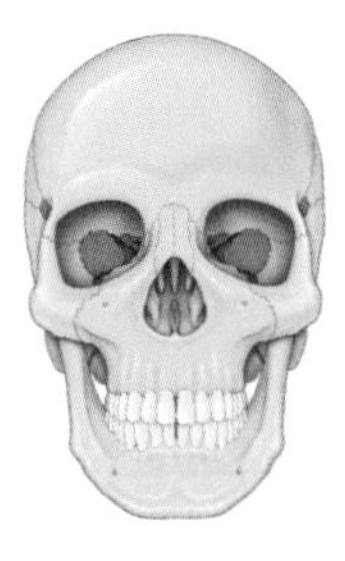
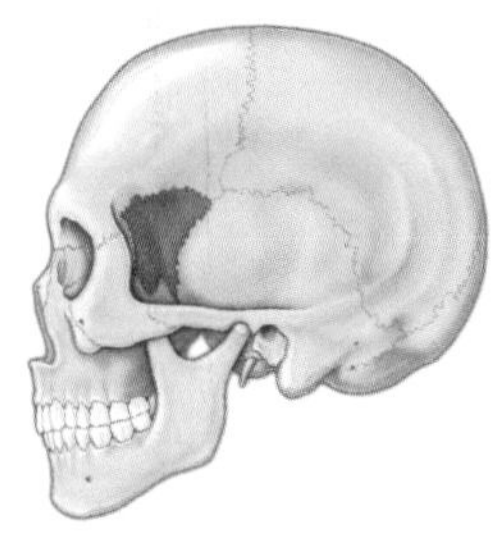
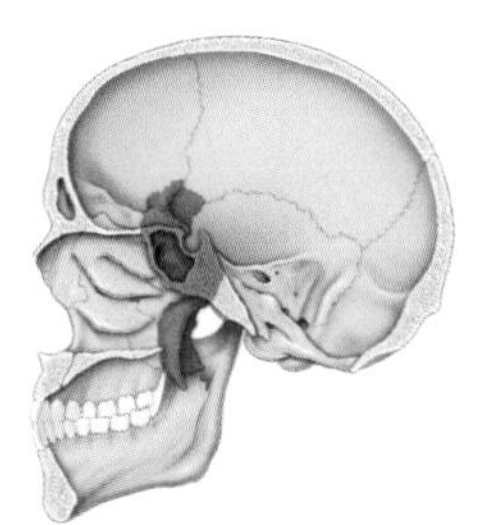
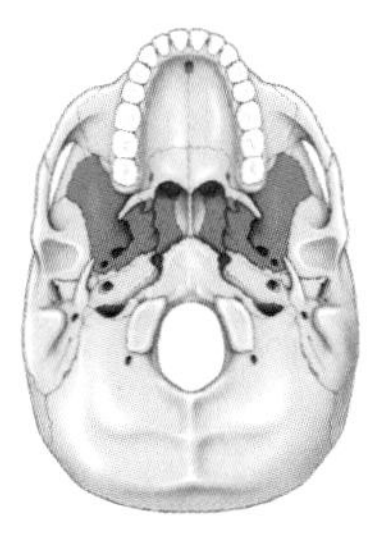
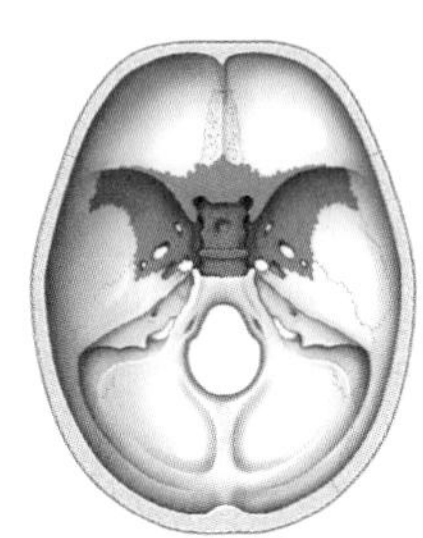

Ethmoid Bone

The ethmoid bone is mostly internal to the skull, and not visible from the outside except through the orbits and nasal cavity. It makes up the barrier between the nasal cavity and the cranial cavity and olfactory (smell) neurons pass through it.

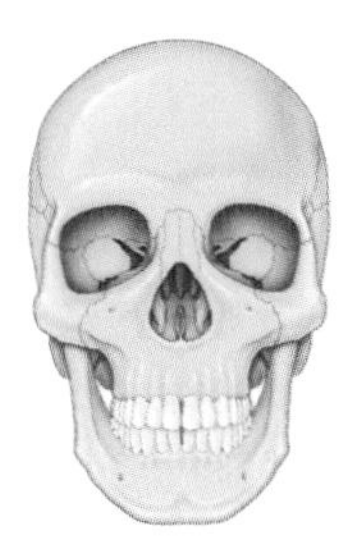
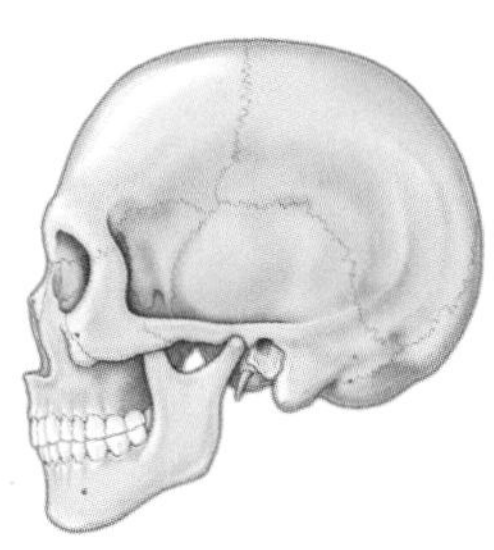
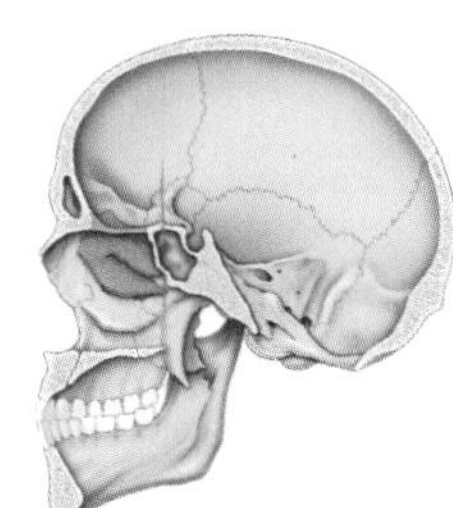
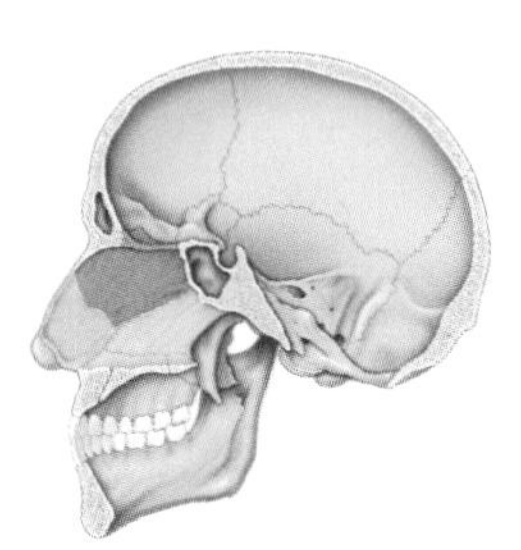

Lacrimal Bones

The tiny, paired lacrimal bones protect our tear ducts.

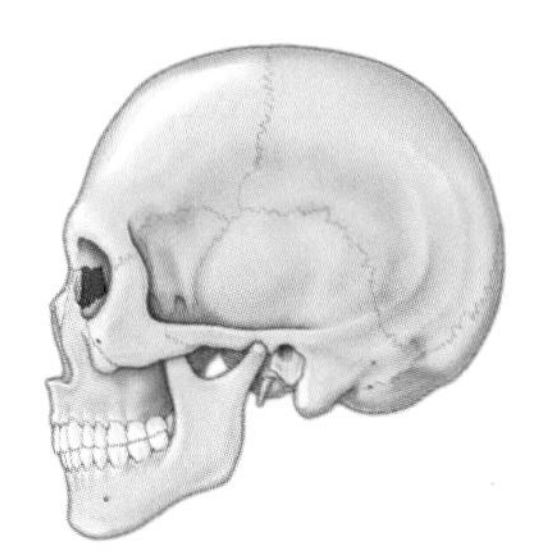

Vomer

The vomer forms the anterior nasal septum (the barrier between your right and left nostrils).

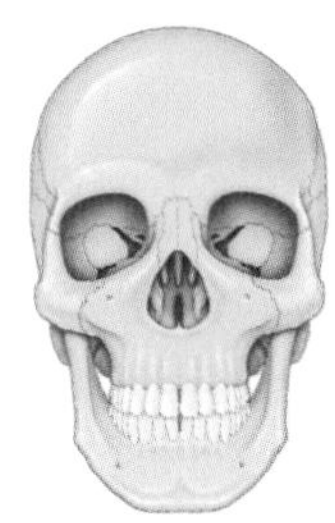
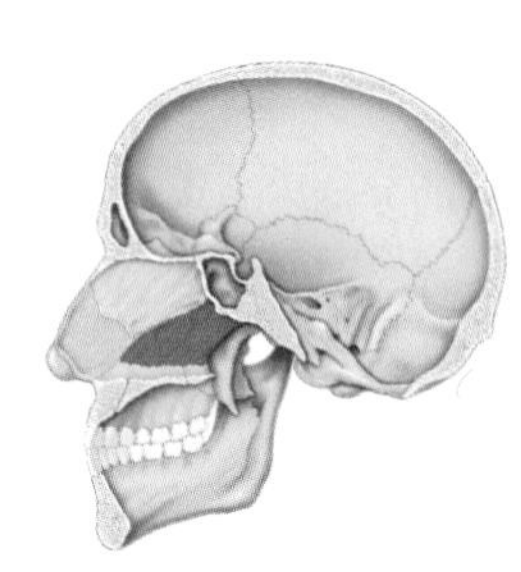

Palatine Bones

The paired palatine bones are completely internal to the skull and cannot be seen from the outside. They contribute to the posterior roof of the mouth.

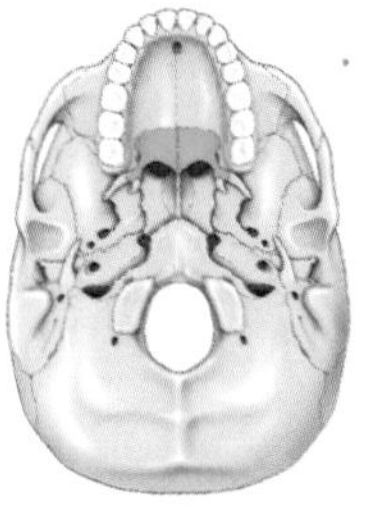
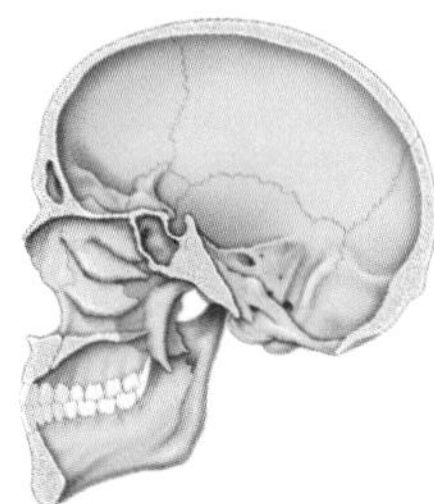

Figure 8.5 The Frontal Bone

The single frontal bone encloses the anterior cranium and forms the upper portions of the eye orbits.

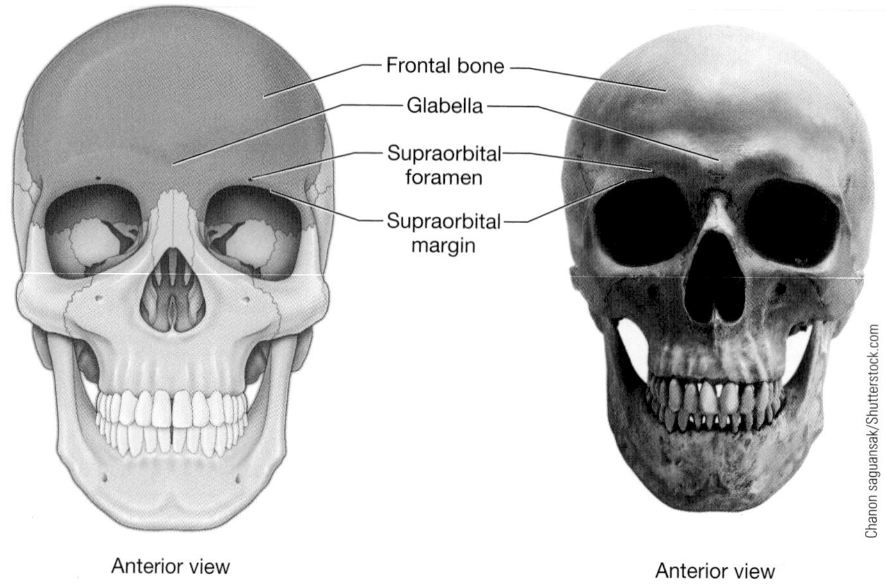

Frontal bone
Glabella
Supraorbital foramen
Supraorbital margin

Anterior view

Anterior view

Chanon saguansak/Shutterstock.com

8.2b Bones of the Skull

LO 8.2.1

Frontal Bone One **frontal bone** forms the forehead. At its anterior midline, between the eyebrows, there is a slight depression called the **glabella** (**Figure 8.5**). The frontal bone also forms the superior rim of the eye orbit; this structure is called the **supraorbital margin** of the orbit. The name *supraorbital* is made by combining the words *superior* and *orbital*; this structure is superior to the orbit. Near the middle of this margin, you can find a small hole, the **supraorbital foramen**, the opening that provides passage for a nerve that provides sensory innervation to the forehead. The frontal bone is thickened just above each supraorbital margin, forming rounded brow ridges. These are located just behind your eyebrows and vary in size among individuals. Inside the cranial cavity, the frontal bone extends posteriorly, forming a portion of the cranium (**Figure 8.6**).

Figure 8.6 The Paired Parietal Bones

The parietal bones make up the middle of the cranium.

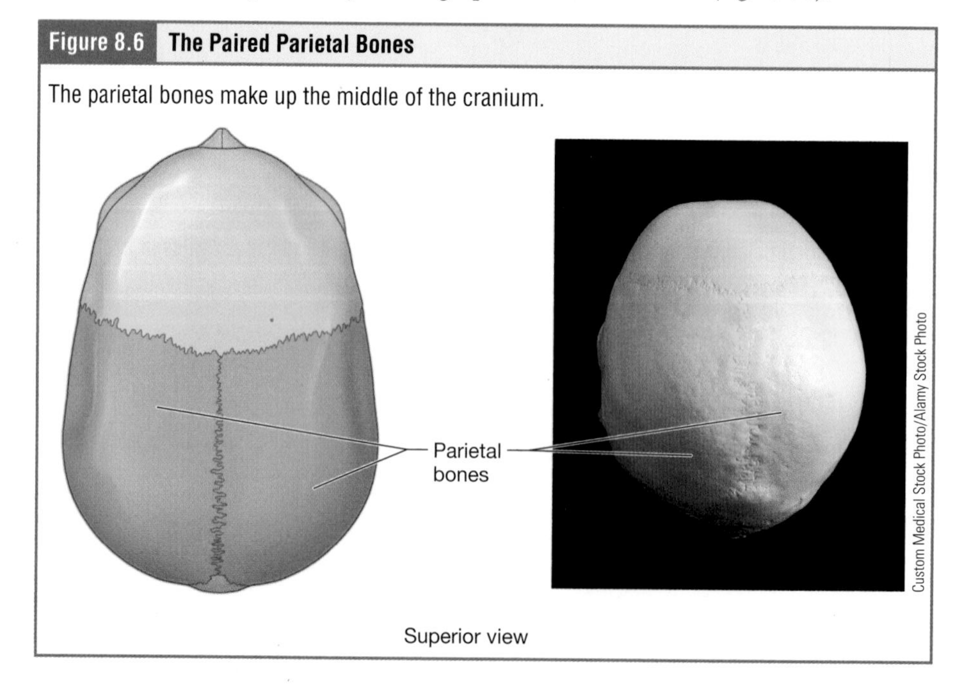

Superior view

Parietal Bone The paired **parietal bones** form most of the superior lateral sides of the skull (see Figure 8.6). These are paired bones, with the right and left parietal bones joining together at the top of the skull. Each parietal bone is also joined to the frontal bone on its anterior edge, the temporal bone on its inferior edge, and the occipital bone on its posterior edge.

Occipital Bone The **occipital bone** is the single bone that forms the posterior skull and posterior base of the cranial cavity (**Figure 8.7A**). On its outside surface, at the posterior midline, is a small protrusion called the **external occipital protuberance**, which serves as an attachment site for a ligament of the posterior neck. On either side of this bump is a **superior nuchal line**. The right and left nuchal lines are important attachment points for muscles of the neck. On the base of the skull (**Figure 8.7B**), the occipital bone contains the large opening of the **foramen magnum**, which allows for passage of the spinal cord as it exits the skull. On either side of the foramen magnum is an oval-shaped **occipital condyle**. These condyles form joints with the first cervical vertebra and thus support the skull on top of the vertebral column (see Figure 8.7A).

Figure 8.7 The Occipital Bone

(A) The single occipital bone forms the back of the cranium. The occipital bone articulates with the first vertebra via the occipital condyles. (B) From the bottom of the skull, we can see that the occipital bone comprises much of the base of the skull, including the large foramen magnum, the opening through which the spinal cord passes.

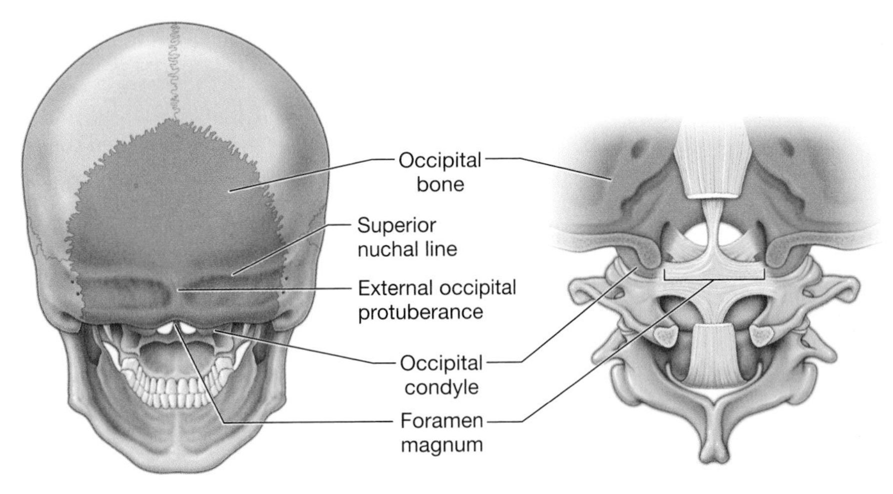

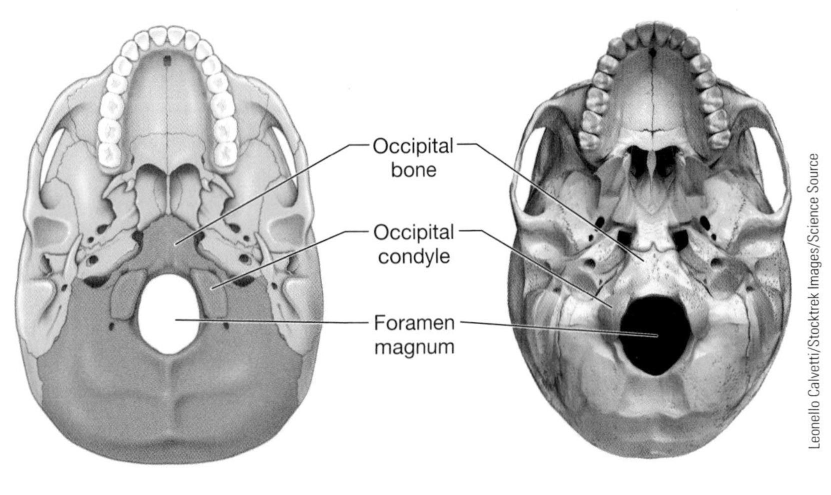

Temporal Bone The **temporal bone** forms the lower lateral side of the skull (see **Figure 8.8**). Folklore about this bone tells us that it is named "temporal" (which means time) because this area of the head (the temple) is where hair typically first turns gray, indicating the passage of time. The temporal bone is subdivided into several regions (**Figure 8.8A**). The flattened, upper portion is the **squamous portion of the temporal bone**. The word *squamous*, just like squamous epithelial cells, means flat. Below this area and projecting anteriorly is the **zygomatic process of the temporal bone**; it earns its name because it forms a joint with the zygomatic bone anterior to it. Posterior to the zygomatic process is the mastoid portion of the temporal bone. Projecting inferiorly from this region is a large prominence, the **mastoid process**, which serves as a muscle attachment site. The mastoid process can easily be felt on the side of the head just behind your earlobe; if you trace your fingers down from the mastoid process, you can feel a large muscle—the *sternocleidomastoid*. Located in the center of the temporal bone is a noticeable canal called the **external acoustic meatus**. This is the tunnel through which sound passes to reach the inner ear.

The **mandibular fossa** is a deep, oval-shaped depression located on the external base of the skull just in front of the external acoustic meatus (**Figure 8.8B**). The mandible joins with the skull at this site as part of the **temporomandibular joint (TMJ)**, which allows for movements of the mandible during opening and closing of the mouth. Two other surface features of the temporal bone are the articular tubercle and the long, thin styloid process. The **articular tubercle** is a smooth ridge located immediately anterior to the mandibular fossa. Both the articular tubercle and mandibular fossa contribute to the temporomandibular joint, the joint that provides for movements between the temporal bone of the skull and the mandible.

Styloid process—Posterior to the mandibular fossa on the external base of the skull is an elongated, downward bony projection called the *styloid process*, so named because of its resemblance to a stylus (a pen or writing tool). This structure serves as an attachment site for several small muscles.

Mandible The mandible forms the lower jaw and is the only moveable bone in the skull (**Figure 8.9A**). At birth, the human mandible consists of paired right and left bones, but these fuse together during the first year to form the single mandible of the adult skull. If viewing the mandible from the lateral side (**Figure 8.9B**), it has an L shape. The horizontal portion of the L-shaped mandible is called the **body of the mandible** and

Figure 8.8 The Temporal Bone

(A) The temporal bone contributes to the lateral walls of the cranium. (B) Zooming in on the isolated temporal bone, we can identify many bone features.

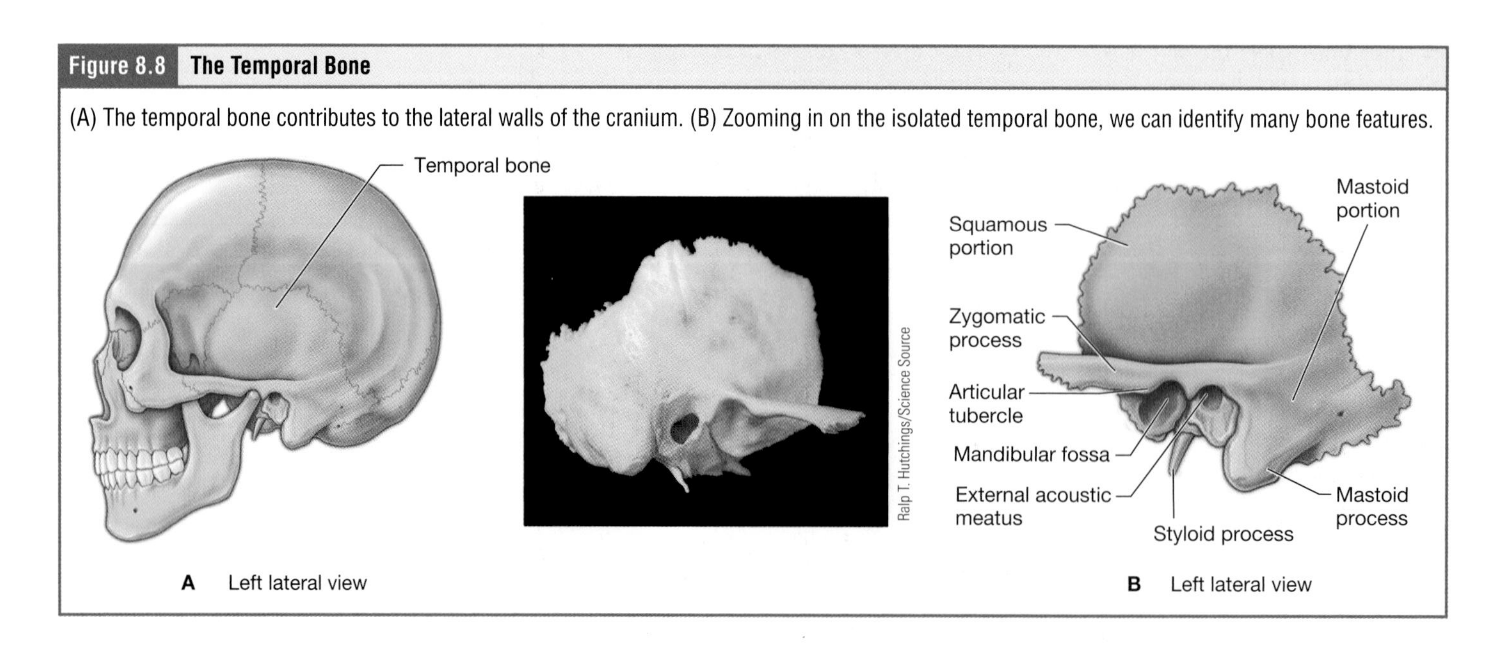

Figure 8.9 The Mandible

(A) The mandible, a single movable bone, forms the lower jaw. (B) In the isolated mandible we can see many of the bone's features, including the portions that form the temporomandibular joint (C).

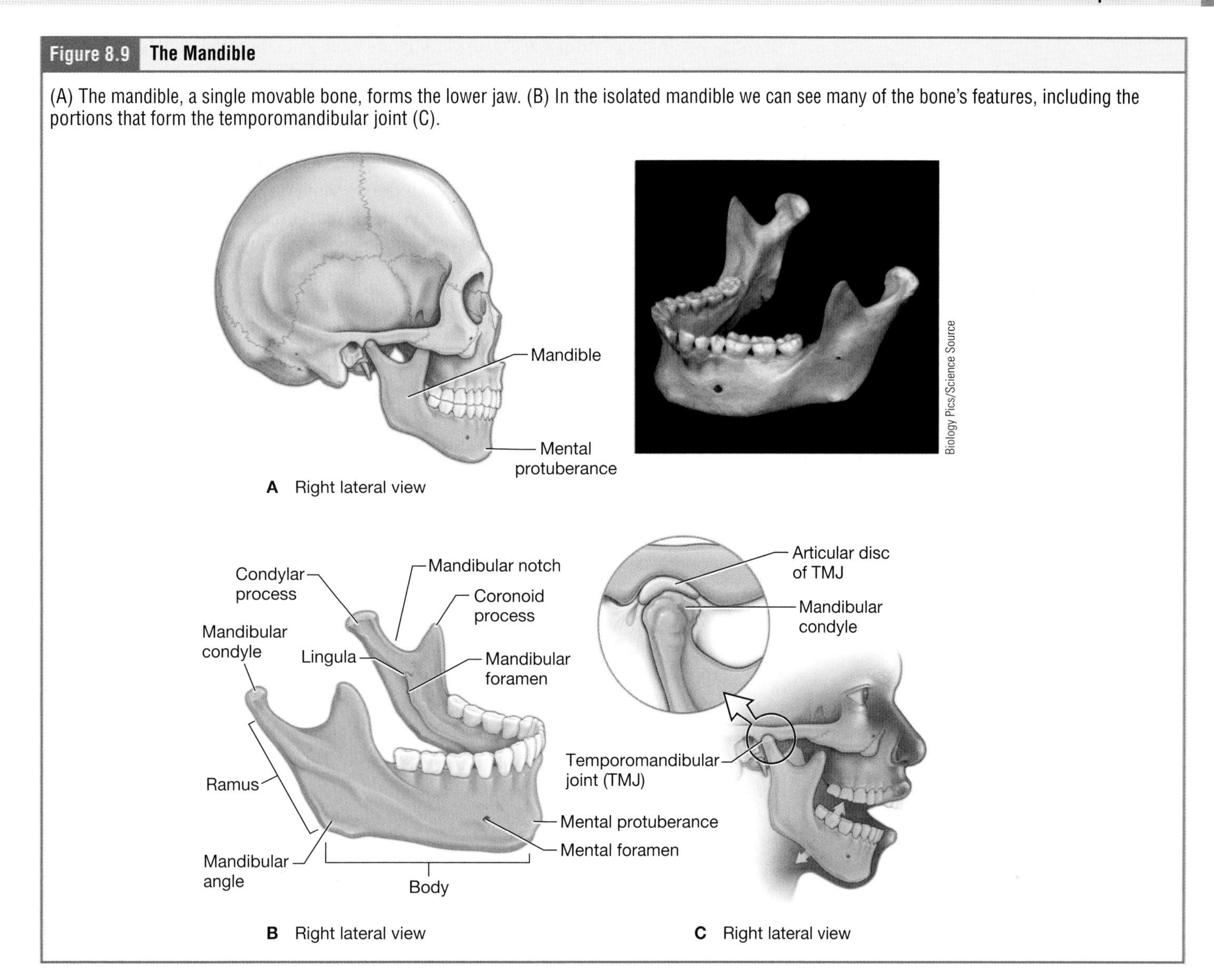

the posterior vertically oriented portion is the **ramus of the mandible**. The corner of the mandible where the body and ramus meet is called the **angle of the mandible**.

Let's examine the ramus in a bit more detail. Notice that it has two upward bony projections. The more anterior projection is the **coronoid process of the mandible**, which provides attachment for one of the biting muscles. The posterior projection is the **condylar process of the mandible**, which is topped by the oval-shaped **condyle**. The condyle of the mandible forms a joint with the temporal bone, specifically the mandibular fossa and articular tubercle of the temporal bone. Together these articulations form the temporomandibular joint, which allows for opening and closing of the mouth (Figure 8.9C). We will discuss this joint in more detail in Chapter 10. Between the coronoid and condylar processes, we see a broad U-shaped curve, the **mandibular notch**.

When we view the mandible from the anterior (Figure 8.10A) we can sometimes see a slight ridge where the two infant mandibles fused in toddlerhood. In some individuals the fusion is incomplete, leaving a small divot at the bottom of this line; this produces a small gathering of the skin of the chin, which is usually called a *cleft chin*. To either side we can observe small holes, the mental foramina. Each mental foramen is an exit site for a sensory nerve that supplies the chin. Along the inferior edge of the anterior mandible is a ridge; in some individuals this can be only slight, while in others it can be very pronounced. This ridge, the *mental protuberance*, forms the chin of an individual.

maxillary bone (maxilla)

Maxillary Bone The **maxillary bone**, often referred to simply as the *maxilla* (plural = maxillae), is one of a pair that together form the upper jaw, most of the roof of the mouth (called the *hard palate*), part of the eye orbit, and the lateral base of the nose (see Figure 8.10). On its lateral sides, the maxilla articulates with the zygomatic bone. As is often the pattern in the facial bones, this projection of the maxilla is named after the bone that it reaches toward, the zygomatic process. The curved, inferior margin of the maxillary bone that forms the upper jaw and contains the upper teeth is the **alveolar process of the maxilla** (Figure 8.10B). Each tooth is anchored into a deep socket called an *alveolus*. The maxilla forms the floor of the eye orbit and on its anterior surface, just below the orbit, is a foramen through which a sensory nerve passes. Since this foramen is inferior to the orbit, it is named the **infraorbital foramen**.

Nasal Bone The paired **nasal bones** are small bones that articulate (join) with each other to form the bony base (bridge) of the nose. They also support the cartilages that form the lateral walls of the nose (see the "Anatomy of the Skull Bones" feature). These are the bones that are damaged when the nose is broken.

Zygomatic Bone The paired **zygomatic bones** are also known as the *cheekbone*. Each zygomatic bone forms much of the lateral wall of the orbit (Figure 8.11). The short temporal process of the zygomatic bone projects posteriorly, where it joins the temporal bone to form the **zygomatic arch**.

Sphenoid Bone The sphenoid, palatine, and ethmoid bones may be the most difficult bones in the skull to learn because they are internal bones and have complex shapes. The **sphenoid bone** is a single bone in the central skull that articulates with almost every other bone of the skull! Portions of the sphenoid—along with the temporal

Figure 8.10 The Maxilla

(A) The two maxillae articulate to form the upper jaw. (B) On the isolated maxilla we can observe many of its features.

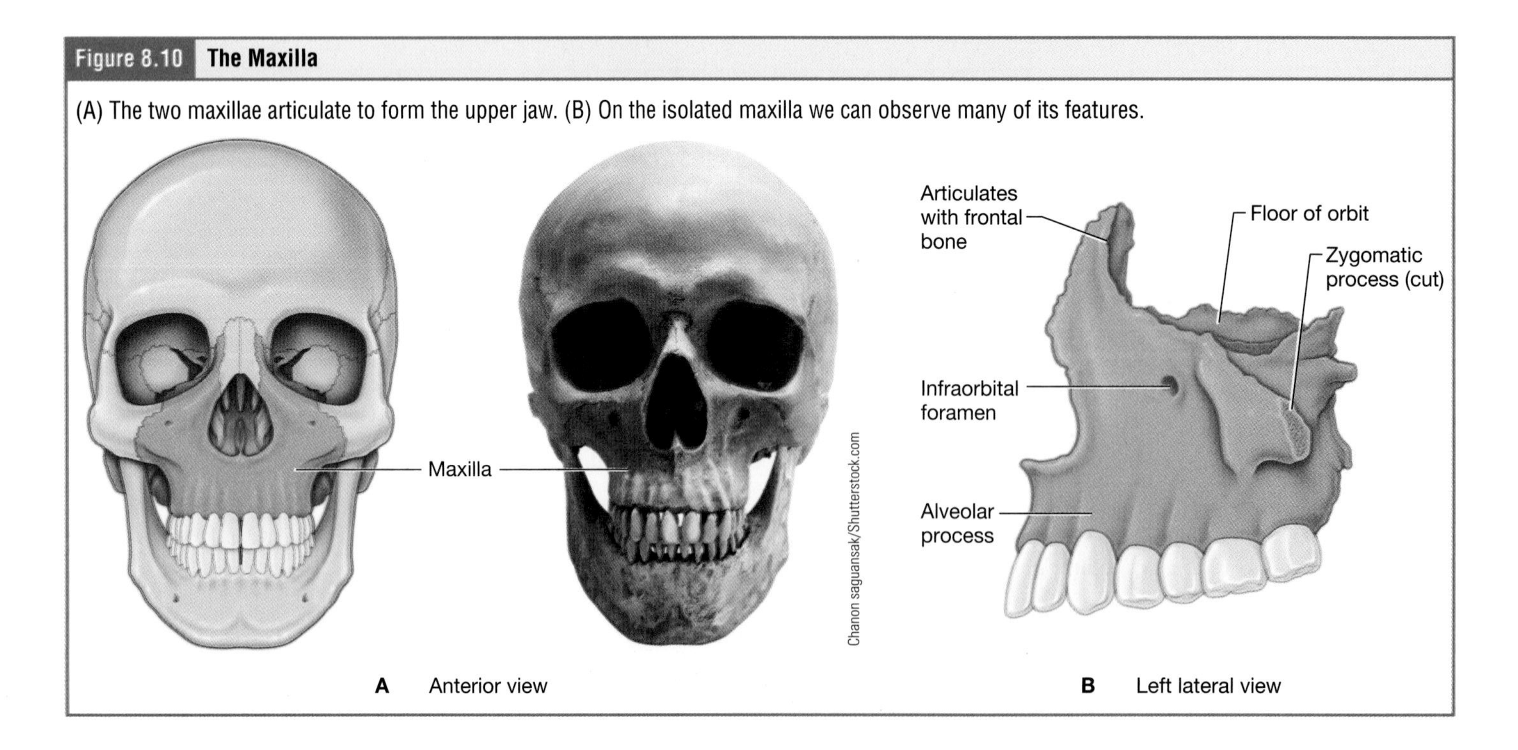

Figure 8.11 The Zygomatic Bone

The zygomatic bone forms the main bony support for the cheek. (A) The temporal process of the zygomatic bone merges with a process from the temporal bone to form the zygomatic arch, a supporting structure for the side of the skull. (B) The zygomatic bone also provides part of the structure of the eye orbit and attachment sites for some of the chewing muscles.

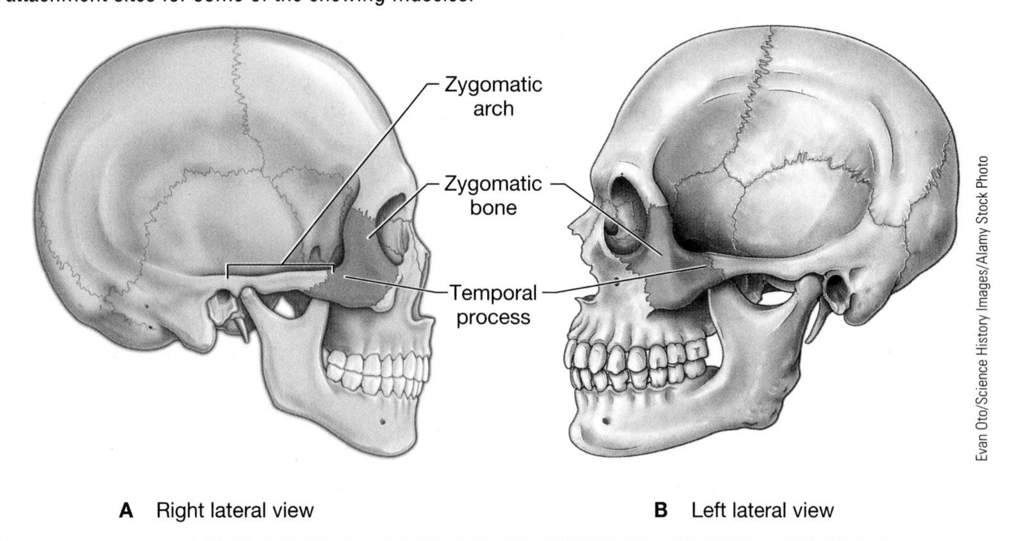

bones—contribute to the temples and can be seen in the lateral view (**Figure 8.12A**). The sphenoid also forms much of the base of the central skull (**Figure 8.12B**). From this view inside the cranial cavity, the right and left **lesser wings of the sphenoid bone** (**Figure 8.12C**), which resemble the wings of a flying bird, form a ridge that marks the boundary between the anterior and middle cranial fossae. The **sella turcica** ("Turkish saddle") is located at the midline of the middle cranial fossa. This bony region of the sphenoid bone is named for its resemblance to the horse saddles used by the Ottoman Turks, with a high back and a tall front. The rounded depression in the floor of the sella turcica is the **pituitary fossa**, which houses and protects the pituitary gland, a key component of the endocrine system. The **greater wings of the sphenoid bone** (see Figure 8.12C) extend laterally to either side away from the sella turcica, where they contribute to the base of the skull.

sella turcica (the name means *Turkish saddle*)

Examining the isolated sphenoid bone (see Figure 8.12C), there are arguably four sets of wings. The most superior are the lesser wings; just below these are the greater wings. Both the lesser and greater wings extend out to the sides of the skull. Projecting downward are two more sets of wings called the medial and lateral *pterygoid processes*. The **medial pterygoid processes** help form the walls of the nasal cavity. The **lateral pterygoid plates** are attachment sites for chewing muscles.

Ethmoid Bone The **ethmoid bone** is a single, centrally located bone that is located between the two eye orbits. It contributes to the walls of the orbits as well as to the nasal cavity (**Figure 8.13A**). On the interior of the skull, the ethmoid also forms a portion of the floor of the cranial cavity (**Figure 8.13B**).

If we examine the isolated ethmoid bone (see Figure 8.13C), we see vertical plates that rise both above and below the bulk of the ethmoid, and elaborate wings out to the sides. The largest of these, the **perpendicular plate of the ethmoid bone**, reaches down from the ethmoid into the nasal cavity, forming the upper portion of the nasal septum. The elaborate side-reaching wings of the ethmoid bone contribute to the lateral walls of the nasal cavity. These elaborate wings, consisting of the **superior nasal concha** and **middle nasal concha**, are thin, curved projections that subdivide the

Figure 8.12 The Sphenoid Bone

The complex, interior sphenoid bone articulates with almost every other bone in the skull. (A) The sphenoid bone is visible on the external aspect of the skull. (B) The sphenoid bone spans the entire width of the skull. and (C) The sphenoid bone has a unique butterfly shape and many intricate features.

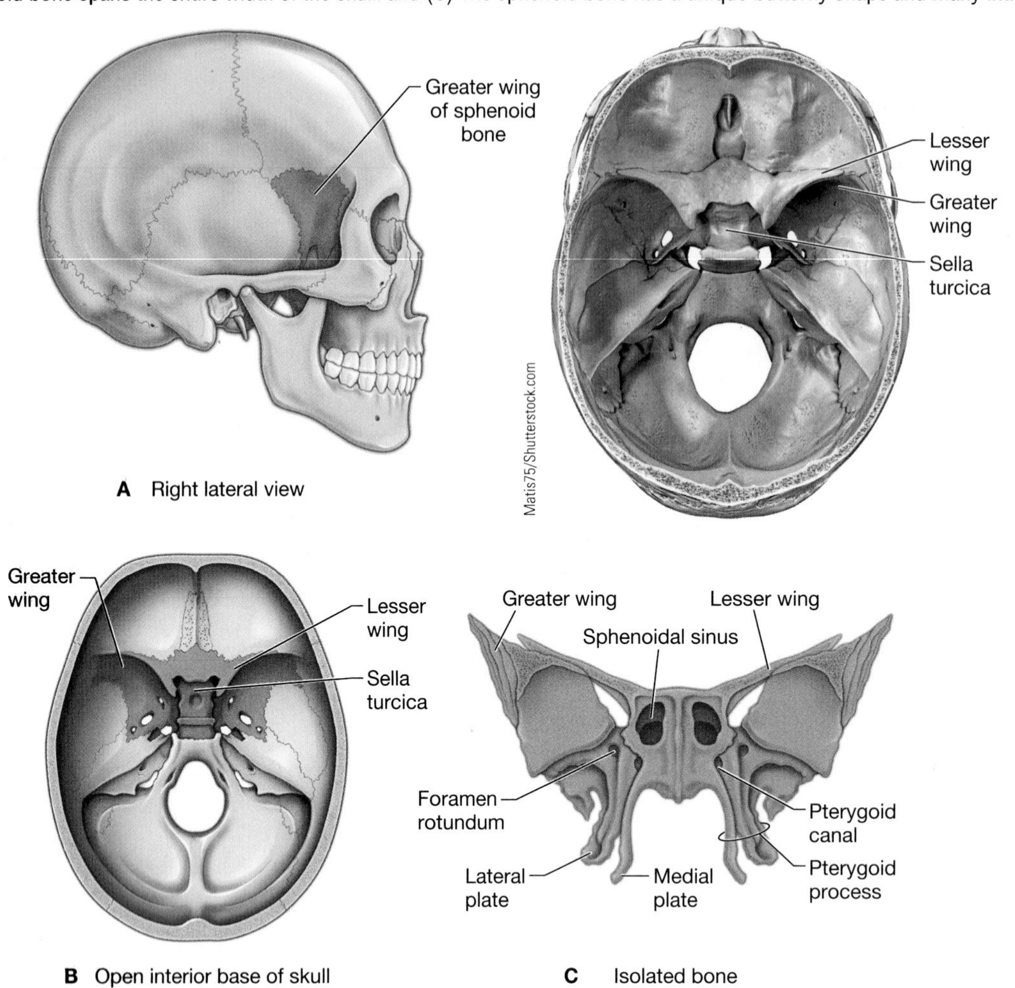

A Right lateral view

B Open interior base of skull

C Isolated bone

Figure 8.13 The Ethmoid Bone

The single, complex interior ethmoid bone separates the nasal cavity from the brain. (A) The ethmoid bone makes up a portion of the eye orbit. (B) The ethmoid bone lies anterior to the sphenoid bone. (C) The perpendicular plate makes up part of the nasal septum.

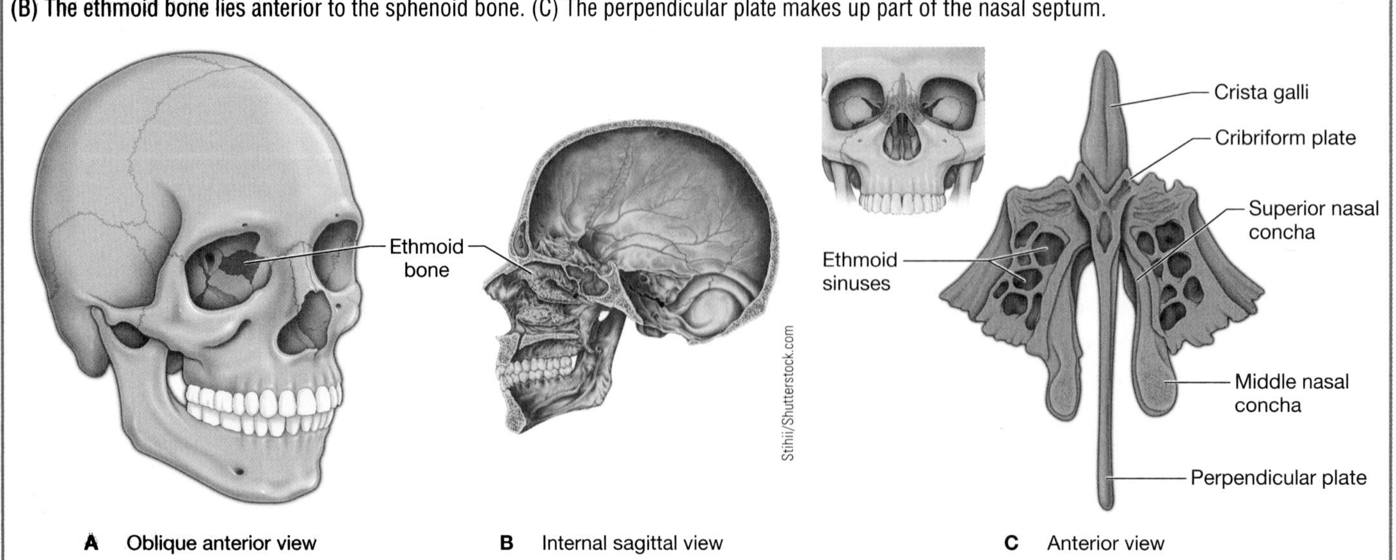

A Oblique anterior view

B Internal sagittal view

C Anterior view

spaces within the nasal cavity. We will discuss the nasal cavity in more detail in later sections.

Viewing the ethmoid bone from within the cranial cavity, we can see it forms a small area at the midline in the cranial floor, forming a barrier between the nose and brain. This part of the ethmoid bone consists of two parts, the *crista galli* and *cribriform plates*. The **crista galli** (named after a rooster's comb) is a small upright bony projection located at the midline. It functions as an anterior attachment point for one of the covering layers of the brain. On the sides of the crista galli are the **cribriform plates**, small, flattened areas riddled with tiny holes, the **olfactory foramina**. In life, olfactory neurons pass through these holes, connecting the brain to the nasal cavity for the purpose of smell.

Lacrimal Bone Each **lacrimal bone** is a small, rectangular bone that is found in the medial eye orbit (see the "Anatomy of the Skull Bones" feature). The lacrimal bones, named for the Latin *lacrima*, which means tear, house a tunnel that connects the inner corner of the eye with the nasal cavity. When a person cries, or when their eyes water, tears collect along the surface of the eye. While some of these tears might spill forward and fall over the cheeks, the majority of the fluid flows through this duct into the nasal cavity, causing a runny nose.

Palatine Bone The **palatine bones**, a pair of irregularly shaped bones internal to the skull (**Figure 8.14A**), contribute small areas of the walls of the nasal cavity and the medial wall of each orbit. Each isolated palatine bone (**Figure 8.14B**) resembles a letter J, with its vertical section contributing to the nasal cavity walls and the **horizontal plate** of each palatine bone joining together at the midline to form the posterior hard palate (the roof of the back of the mouth).

Vomer Bone The single **vomer bone** is a triangular-shaped internal bone that, along with the perpendicular plate of the ethmoid bone, forms part of the nasal septum (see the "Anatomy of the Skull Bones" feature).

Learning Connection

Chunking

Write down all of the bones of the skull that we have just learned. Which bones would be categorized as facial bones and which would be categorized as cranial bones?

Figure 8.14 The Palatine Bone

Two palatine bones form the posterior palate and nasal cavity walls. (A) The palatine bones make up a small posterior portion of the hard (bony) palate and help to form the walls of the nasal cavity. (B) The unusual J shape of the palatine bones allows them to fit among several surrounding bones.

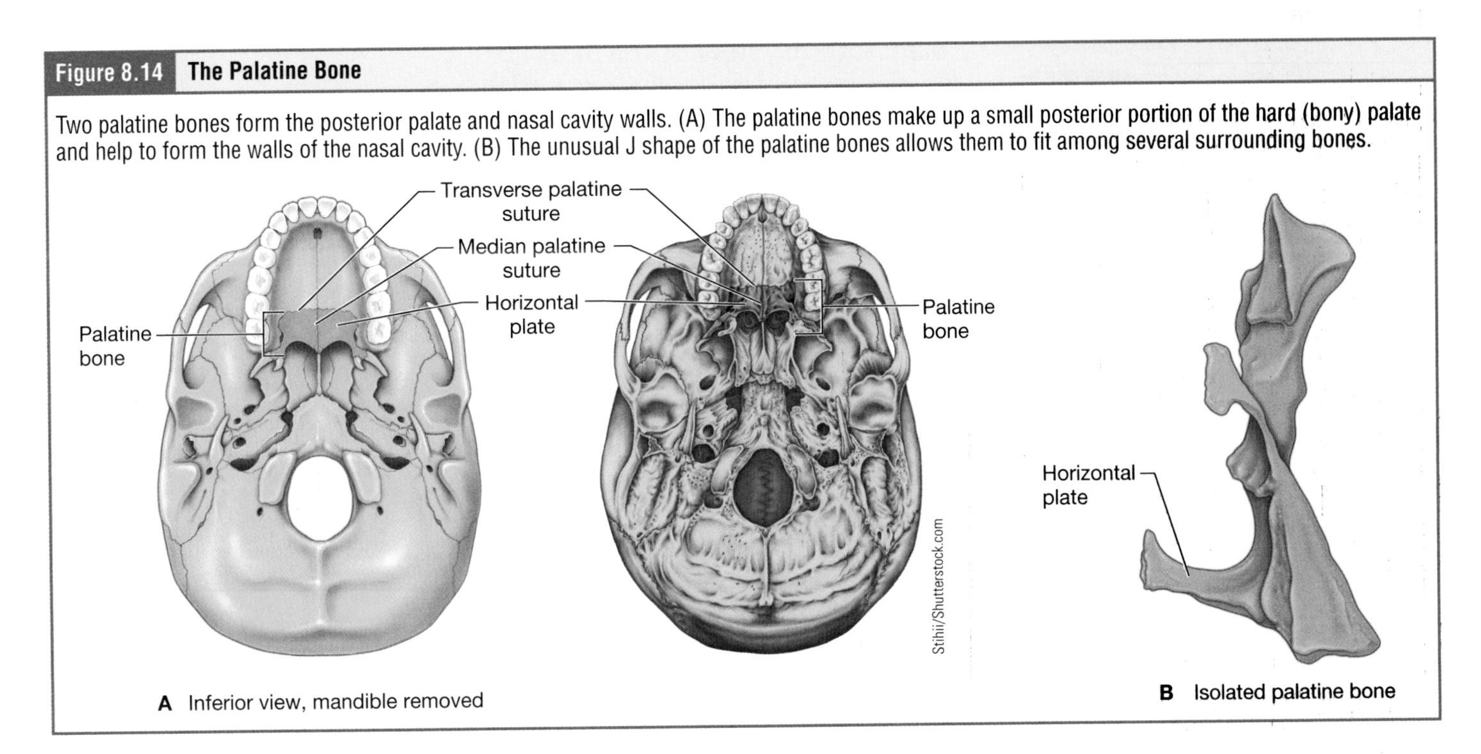

Cultural Connection

Memento Mori

Throughout cultures and religions of the world, perhaps no part of the human body has come to represent death as universally as the skull. Medieval philosophers and theologians were known to have skulls on their desks as a reminder of the short and fleeting nature of life (in modern times it is more likely that your A&P instructor has one on their desk). Skull symbols are featured in both Hindu and Buddhist art of their deities.

The skull and crossbones has come to universally represent something that can cause death, such as poison. In more modern history, many military units—most infamously the Nazi SS—wore insignia featuring a skull, both to strike fear into their opponents and to show their loyalty to their particular ruler "unto death," and in the famous Harry Potter book series, the society of antagonists—the death eaters—used a snake and skull as their symbol.

Perhaps no skull symbolism is as widespread or as joyful as the skull symbology associated with the holiday El Dia de los Muertos, the Day of the Dead. El Dia de los Muertos has its roots in indigenous Aztec culture; the Aztecs were a people who did not believe in mourning their dead, but rather believed in celebrating their lives. On El Dia de los Muertos, celebrated every November 1–2 and coinciding with the Catholic holiday All Souls Day, celebrants honor their ancestors. Cultures who celebrate El Dia de los Muertos believe that death is one part of life and that their deceased loved ones return to their community during this holiday to spend time with their family and friends. As part of this beautiful tradition, skull-shaped candles are often burned, and colorful skull decorations and outfits are worn.

Learning Connection

Quiz Yourself

Try using your web browser's image search function and search for unlabeled images of the intact skull. See if you can label the bones of the skull without looking at your text or notes.

8.2c The Skull as a Whole

Now that we have introduced each of the bones and their features individually, let's put the skull together and examine it and its cavities as a whole. We have examined each bone isolated from the rest, but the skull is a patchwork of these bones, joined by immobile joints called **sutures**. The narrow gap between the bones is filled with dense, fibrous connective tissue that unites the bones. The long sutures located between the bones of the brain case are not straight, but instead follow irregular, tightly twisting paths. These twisting lines serve to tightly interlock the adjacent bones, thus adding strength to the skull for brain protection.

LO 8.2.2

Student Study Tip

The coronal and sagittal sutures both have the same name as the plane along which they run; bringing that information together can help you understand why they are named that way.

8.2d Anterior View of the Skull

The anterior view of the skull is dominated by the large frontal bone, joined to the parietal bones behind it with the **coronal suture**, and the two maxillae. The other dominant features that we can see are the orbits, the nasal cavity, and the mouth (**Figure 8.15**). Let's first define the borders of these spaces. The orbits are rimmed superiorly by the **supraorbital margin**, which are features of the frontal bone. Located near the midpoint of the supraorbital margin is a small opening called the *supraorbital foramen*. This provides for passage of a sensory nerve to the skin of the forehead. The lateral rims of the orbits are formed by the zygomatic bone, and the medial/inferior rims are formed by the maxillae. Within each maxilla we find the infraorbital foramen, which is the point of emergence for a sensory nerve that supplies the anterior face below the orbit.

8.2e The Orbit

The **orbit** provides a protective space for the eyeball and the numerous muscles that move the eyeball or open the upper eyelid. Each orbit is cone-shaped, with a narrow posterior and wider anterior.

The walls of each orbit include contributions from seven skull bones (**Figure 8.16**). The frontal bone, zygomatic bone, and maxilla form the outer edges and the bulk of the walls. The ethmoid bone and lacrimal bone, with a tiny contribution from the palatine bone, make up the medial wall. The sphenoid bone forms the posterior orbit.

Figure 8.15 **The Articulated Skull**

Viewing the skull from the anterior, we can identify many of the bones.

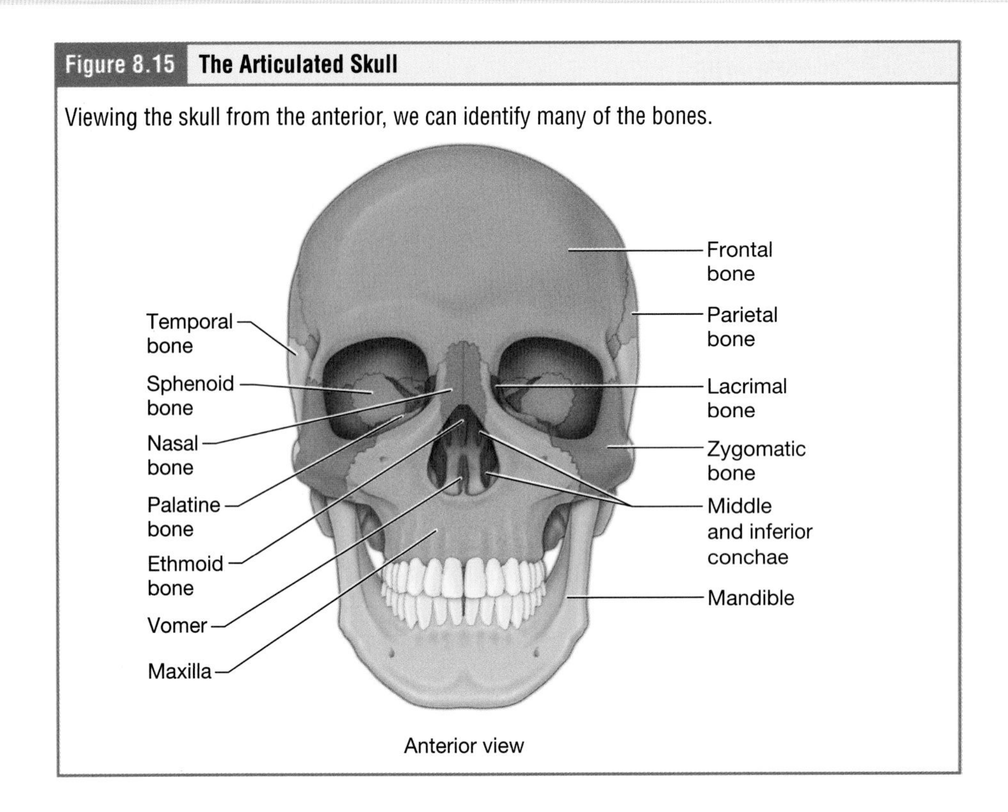

At the posterior apex of the orbit is the opening of the **optic canal**, which allows for passage of the optic nerve from the eye to the brain. Lateral to this is the elongated and irregularly shaped **superior orbital fissure**, which provides passage for the artery that supplies the eyeball, sensory nerves, and the nerves that supply the muscles involved in eye movements.

Moving down in our anterior tour of the skull, we next come to the nasal cavity.

8.2f The Nasal Cavity, Septum, and Conchae

The nasal cavity is bordered by the maxillae and the nasal bones. It is divided into halves by the **nasal septum** (**Figure 8.17A**). The upper portion of the nasal septum is

Figure 8.16 **The Eye Orbit**

Seven skull bones contribute to the walls of the eye orbit.

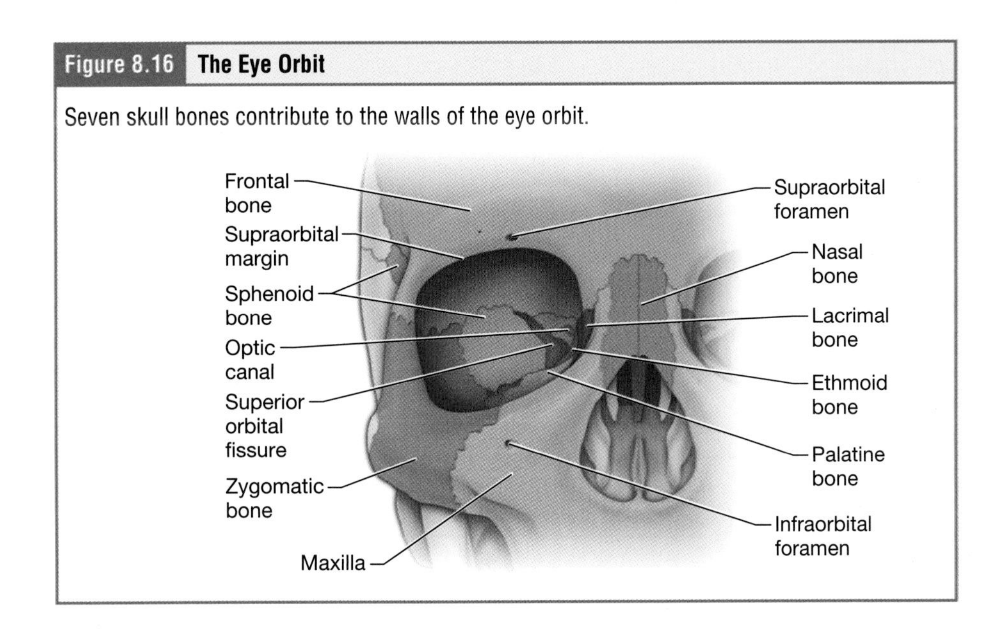

Figure 8.17 The Nasal Cavity

(A) The nasal cavity is bordered by the maxillae and nasal bones. Inside we can see the septum and the conchae. (B) Nasal conchae of animals with more sophisticated senses of smell, such as this lion, are more elaborate.

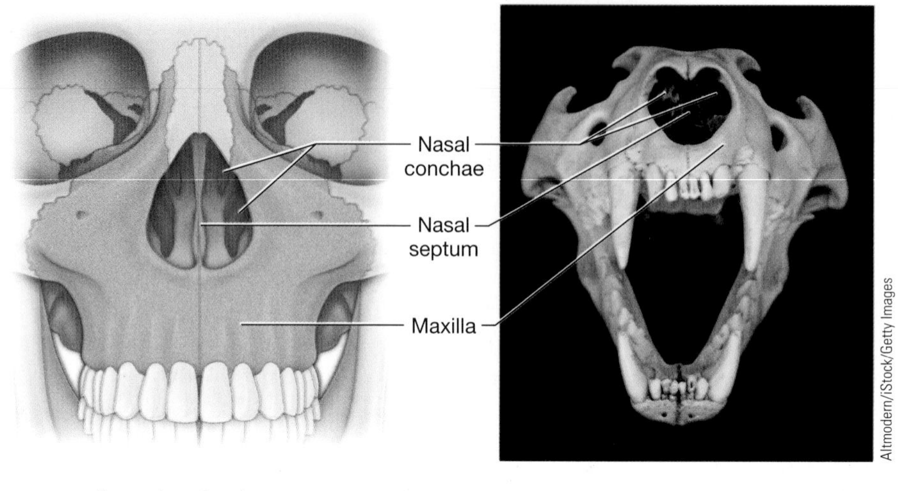

A Anterior view **B** Anterior view

Altmodern/iStock/Getty Images

formed by the perpendicular plate of the ethmoid bone and the lower portion is the vomer bone. When looking into the nasal cavity from the front of the skull, bony plates are seen projecting from each lateral wall, the nasal conchae. In life, the **nasal conchae** are covered with moist mucous membrane and they function to subdivide and add dimension to the nasal cavity. As air is inhaled, it does not travel directly to the back of the nasal cavity and toward the lungs, but rather swirls among the spaces created by the conchae, becoming warmer, moister, and more turbulent as it flows among these spaces. This not only makes the air warmer and moister before it heads to the lungs, it also traps dust, pollen, bacteria, and viruses within the mucus and improves the amount and frequency of air particles that reach the olfactory neurons, improving our sense of smell. Among animals, the more elaborate the nasal conchae system, the more impressive the sense of smell is, as seen in **Figure 8.17B**. The superior and middle nasal conchae are features of the ethmoid bone, but the **inferior nasal conchae** are actually independent bones of the skull.

Student Study Tip

The nasal *conchae* have a very unique shape, much like a *conch* shell inside your skull!

Cultural Connection

The Nose Knows

Differences in nose shape among different ethnicities are due to adaptations to climates and climate change. One of the main functions of the nasal cavity is to condition the air before it travels to the respiratory tract. As the air swirls around the nasal conchae, it is moistened and warmed. This feature would be of utmost importance if you were breathing in the cold dry air of, for example, Northern Europe, but it would be of much less importance if you were breathing in warm tropical air of, for example, the Philippines.

Remember that *homo sapiens* first evolved in the hot climates of Africa, and then slowly migrated around the world, adapting to their environments as they evolved over hundreds of thousands of years. Remains of early humans and related species indicate that humans had shorter, wider nostrils. Groups of humans from hotter, more humid climates kept this nose shape, but some humans who migrated to colder climates slowly evolved longer, narrower nostrils. It is likely that these longer noses provided an advantage in colder climates.

The nasal septum consists of both bone and cartilage components (**Figure 8.18**). The upper portion of the septum is formed by the perpendicular plate of the ethmoid bone. The lower and posterior parts of the septum are formed by the triangular-shaped vomer bone. In an anterior view of the skull (see Figure 8.17A), the perpendicular plate of the ethmoid bone is easily seen inside the nasal opening as the upper nasal septum, but only a small portion of the vomer bone is seen as the inferior septum. A better view of the vomer bone is seen when looking into the posterior nasal cavity with an inferior view of the skull (see Figure 8.18). The anterior nose and nasal septum stick out from the face, where they are at risk of being hit, bent, or deformed. Therefore, instead of bone, the anterior nasal septum is formed by the **septal cartilage**. The septal cartilage is not found in the dry skull.

Learning Connection

Broken Process

In injuries that cause a "broken nose," the conchae often swell to two or three times their original size. What would be the symptoms of a person with swollen conchae?

8.2g Paranasal Sinuses

LO 8.2.3

While technically not visible from the anterior view, we cannot move on from the nasal region without discussing the **paranasal sinuses**, hollow, air-filled spaces located within certain bones of the skull (**Figure 8.19**). All of the sinuses connect to the nasal cavity and are lined with nasal mucosa. They function to lighten the skull; remember that humans have one of the largest head-to-body ratios of any animal that has ever lived (we discuss this in more detail later on in this section), so including air-filled space in the skull helps these already heavy heads to feel lighter. In addition, when humans use their voices, the air travels up from their throats and bounces around the inside of the sinuses, adding a quality called *resonance* to the voice. While this may not be obvious to you every day, it becomes quite apparent once you fill the sinuses with mucus, such as when you have a cold or sinus congestion. The excess mucus production obstructs the narrow passageways between the sinuses and the nasal cavity, causing your voice to sound different to yourself and others. This blockage can also allow the sinuses to

Figure 8.18 Nasal Septum

The nasal septum is a patchwork of three structures: the perpendicular plate of the ethmoid bone, the vomer bone, and the septal cartilage.

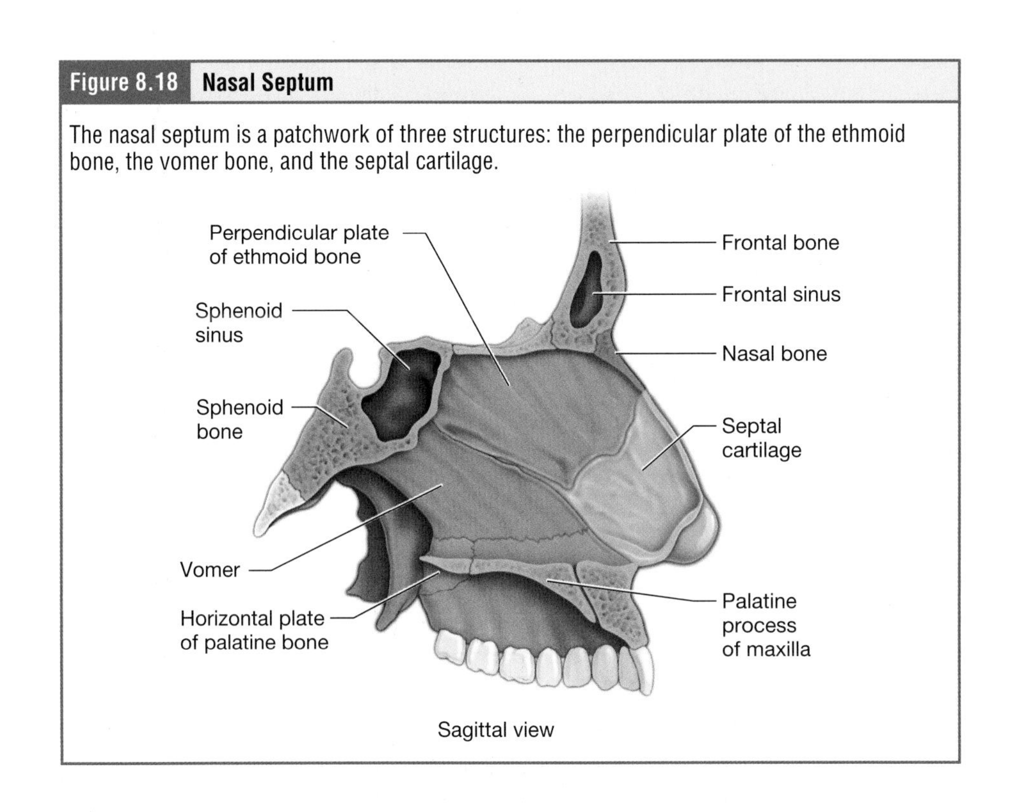

Figure 8.19 The Paranasal Sinuses

Sinuses are hollow, air-filled spaces within bones. They are named for the skull bone that they occupy.

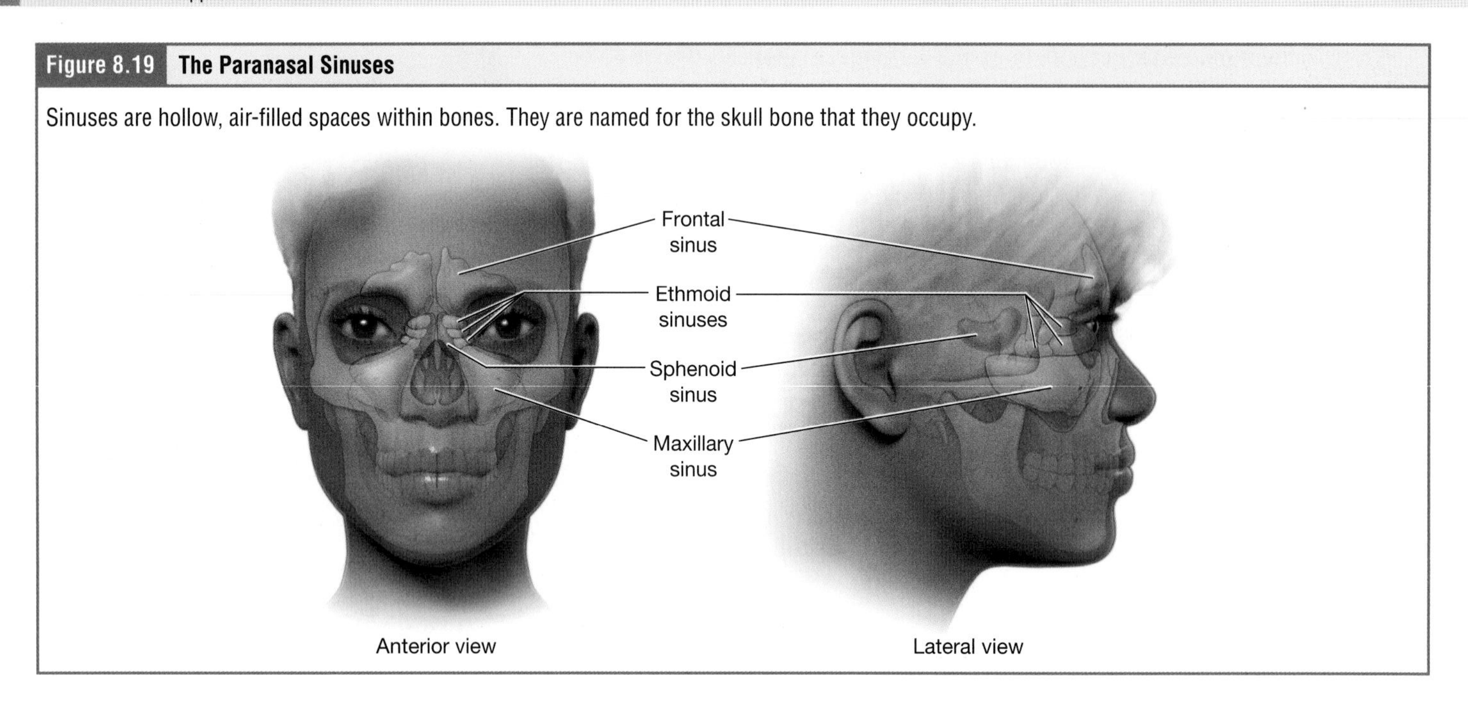

fill with fluid, with the resulting pressure causing pain, discomfort, and a distinctly heavy-feeling head.

Each paranasal sinus is named for the skull bone that it occupies. The **frontal sinus** is located just above the eyebrows, within the frontal bone (see Figure 8.19). Note that this is a source of anatomical diversity; some individuals have two lateral frontal sinuses, while others have a single, fused frontal sinus. The largest sinuses are the paired **maxillary sinuses**. The **sphenoid sinus** is a single, midline sinus. It is located within the body of the sphenoid bone, just anterior and inferior to the sella turcica, thus making it the most posterior of the paranasal sinuses. The lateral aspects of the ethmoid bone contain multiple small spaces separated by very thin bony walls. Each of these spaces is an **ethmoid sinus**. These are located on both sides of the ethmoid bone, between the upper nasal cavity and medial orbit, just behind the superior nasal conchae. The paranasal sinuses can best be seen in a frontal section or frontal CT scan of the skull (**Figure 8.20**).

Marching along on our tour of the anterior view of the skull, we can observe the zygomatic bones forming the cheekbones and the maxillae and mandible forming the oral cavity.

The anterior and lateral walls of the oral cavity are formed by the mandible and two maxillae along with the teeth embedded in their alveolar processes. The floor of the mouth is entirely composed of soft structures (the tongue and associated muscles). The roof of the mouth, often called the *palate*, is formed anteriorly by a joint between the palatine processes of the two maxillae and posteriorly by the joint between the horizontal plates of the two palatine bones (**Figure 8.21A**). In a midsagittal view (**Figure 8.21B**), we can see that the maxillae make up the majority of the roof of the mouth. On the mandible we can see the **mandibular foramen,** an opening located on the medial side of the ramus of the mandible. The opening leads into a tunnel that runs down the length of the mandibular body. The sensory nerve and blood vessels that

Figure 8.20 The Paranasal Sinuses, Posterior View, Coronal Section

The paranasal sinuses can be seen in a frontal section of the skull. Sinuses are found in the frontal, ethmoid, sphenoid, and maxillary bones.

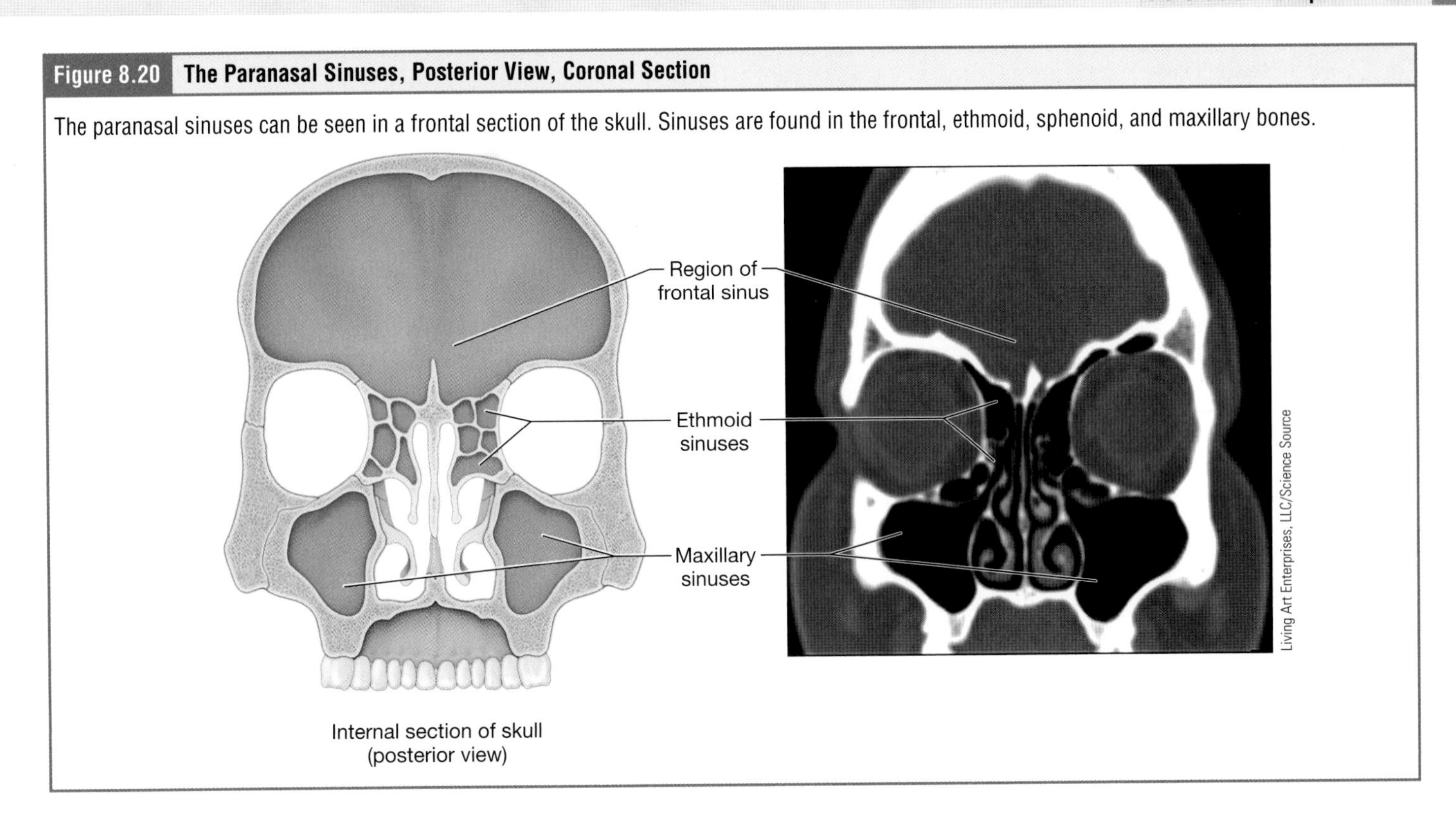

supply the lower teeth enter the mandibular foramen and then follow this tunnel. Thus, to numb the lower teeth prior to dental work, the dentist must inject anesthesia into the lateral wall of the oral cavity at a point prior to where this sensory nerve enters the mandibular foramen.

Figure 8.21 The Oral Cavity

(A) The roof of the mouth, called the *hard palate*, can be seen in an inferior view of the skull. (B) From the In a midsagittal view, the three bones of the oral cavity—the mandible, maxilla, and palatine bone—can be viewed.

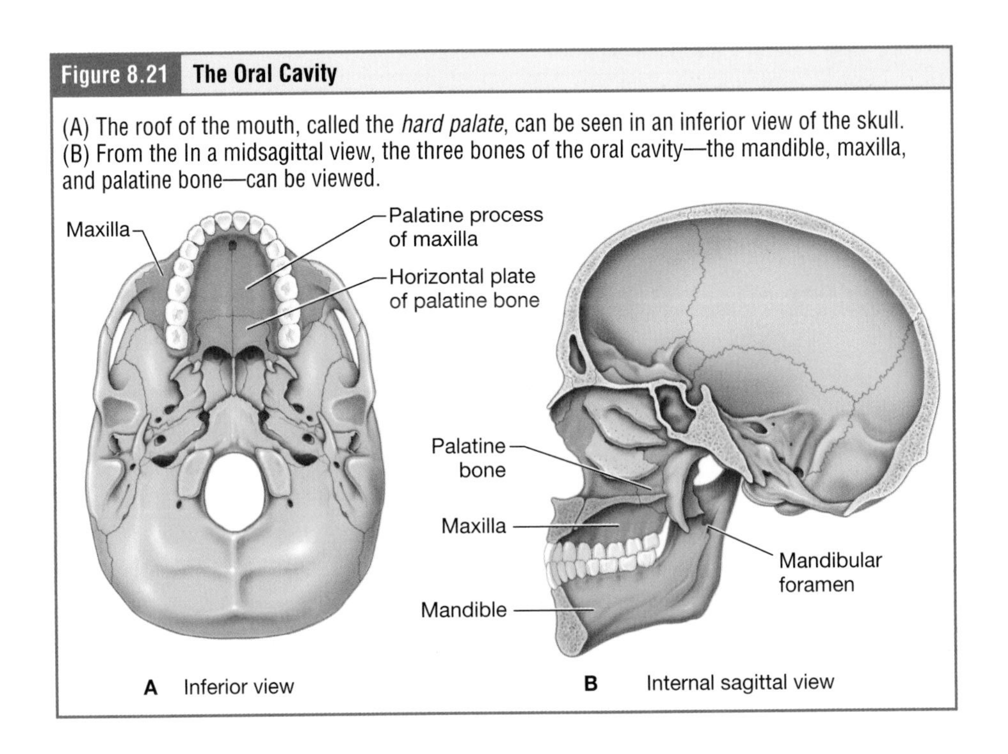

Learning Check

1. Which of the following best describes the view of this bone?
   a. Superior view of the base of the skull
   b. Inferior view of the top of the skull
   c. Lateral view of the left side of the skull
   d. Lateral view of the right side of the skull
2. What is letter A pointing to?
   a. Ramus of the mandible
   b. Maxilla
   c. Zygomatic arch
   d. Foramen magnum
3. What specific parts of the bones make up the structure indicated by letter A?
   a. Zygomatic process of temporal bone and temporal process of zygomatic bone
   b. Temporal process of temporal bone and zygomatic process of zygomatic bone
   c. Ramus of the mandible and alveolar process of the maxilla
   d. Zygomatic process of zygomatic bone and ramus of the mandible
4. Which of the following bones make up the eye orbit? Please select all that apply.
   a. Sphenoid bone
   b. Temporal bone
   c. Maxilla
   d. Mandible

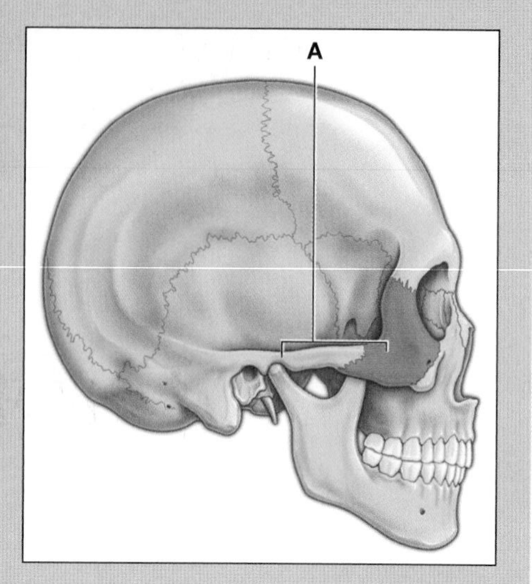

8.2h Lateral View of the Skull

Turning the skull to the side, the lateral view of the skull is dominated by the large, rounded brain case above and the upper and lower jaws with their teeth below (**Figure 8.22**). Separating these areas is the bridge of bone called the *zygomatic arch*. The zygomatic arch, the bony arch on the side of skull that spans from the cheek to just above the ear canal, is formed by the junction of two bony processes: the **temporal process of the zygomatic bone** (the cheekbone) and the longer zygomatic process of the

Figure 8.22 Lateral View of the Skull

The lateral view of the skull shows the cranium, temporal fossa (pink-shaded area), and the upper and lower jaws.

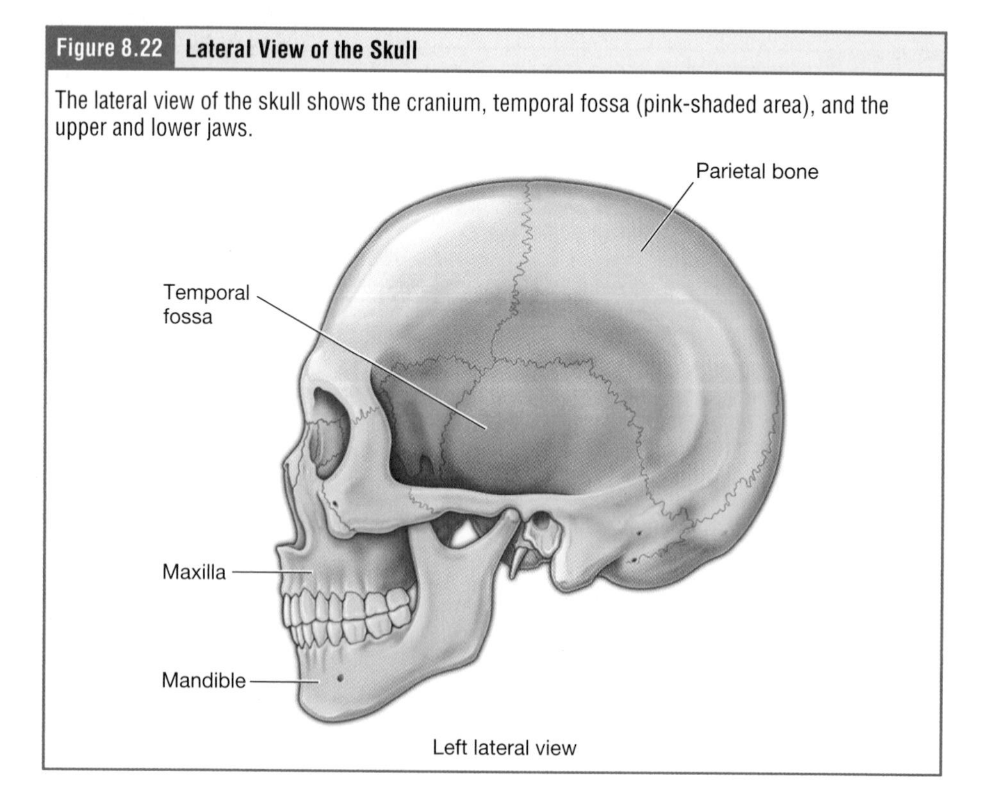

temporal bone. One of the major muscles that pulls the mandible upward during biting and chewing attaches to the zygomatic arch, similarly to the elaboration of the nasal conchae in animal species with greater senses of smell, and animals with powerful bite forces often have thicker, more pronounced zygomatic arches.

On the lateral side of the brain case, above the level of the zygomatic arch, is a shallow space called the **temporal fossa**, which you likely refer to as your temples. The two biggest bones comprising the lateral view are the parietal and temporal bones, which are joined by the **squamous suture**.

8.2i Posterior View of the Skull

From the posterior perspective we can see the two parietal bones, joined at the superior midline by the **sagittal suture**, and the large, single occipital bone. The superior edges of the occipital bone are shaped like an inverted letter V where they articulate with each of the parietal bones. The V-shaped junction is the **lambdoid suture** (Figure 8.23).

Student Study Tip

The lambdoid suture creates the same shape as the Greek letter "Lambda" (Λ).

8.2j The Interior of the Skull—the Brain Case

The brain case, as its name suggests, contains and protects the brain. The interior space, which is almost completely occupied by the brain, is called the *cranial cavity*. Our understanding of the inside of the skull comes from a common transverse cut through the brain case that removes the rounded top of the skull. This "cap," which is called the **calvaria**, is removed in Figure 8.24A. In this view we are looking down on the floor of the brain case, referred to as the **base of the skull**. This is a complex area that varies in depth and has numerous openings for the passage of cranial nerves, blood vessels, and the spinal cord. Inside the skull, the base is subdivided into three large spaces, called the *anterior cranial fossa*, *middle cranial fossa*, and *posterior cranial fossa* (*fossa* = "trench" or "ditch"). From anterior to posterior, the fossae increase in depth. The shape and depth of each fossa corresponds to the shape and size of the brain region that each houses (Figure 8.24B).

calvaria (skullcap)

Figure 8.23 Posterior View of the Skull

This posterior view of the skull is dominated by the parietal and occipital bones.

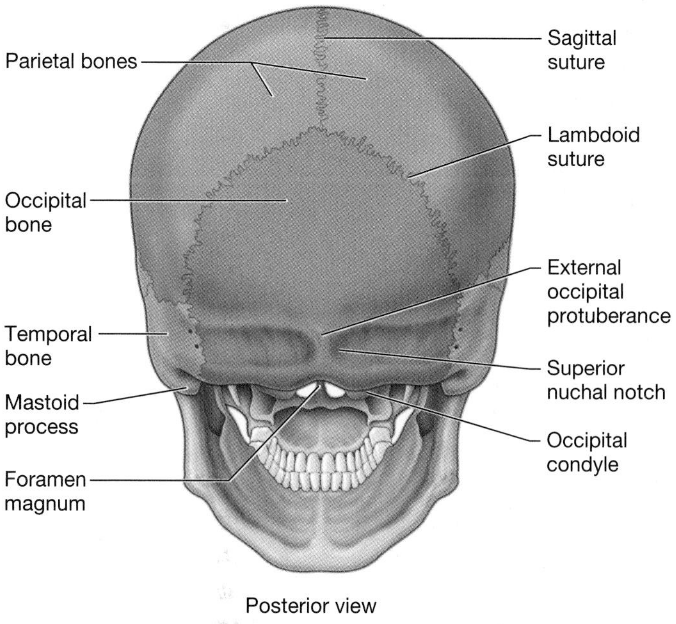

Figure 8.24 The Inside of the Cranial Cavity

(A) The floor of the cranial cavity has three bowl-like depressions. (B) These depressions are the anterior, middle, and posterior cranial fossae. The contributing bones are the frontal, ethmoid, sphenoid, temporal, and occipital bones.

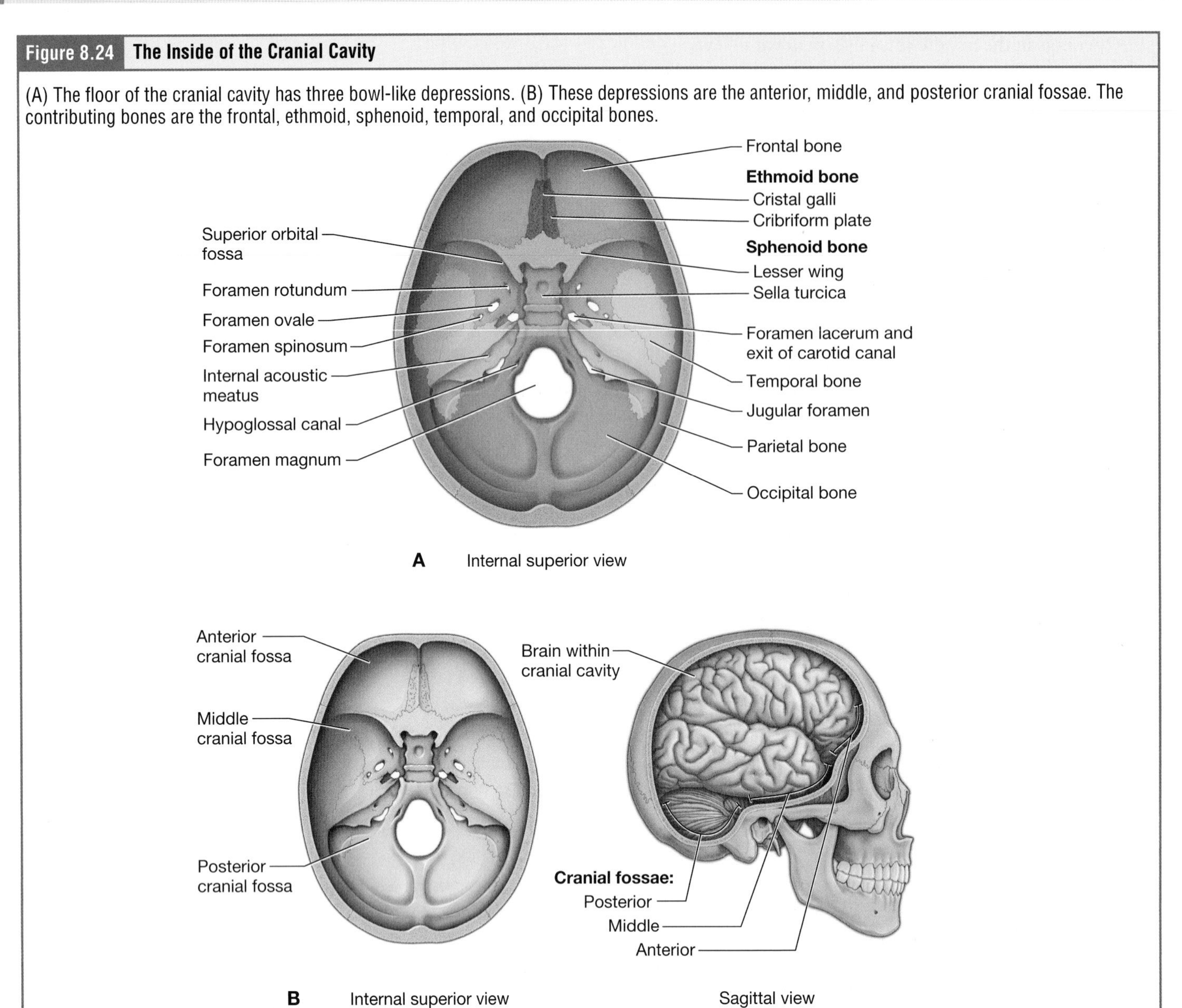

Anterior Cranial Fossa The **anterior cranial fossa** is the most anterior and the shallowest of the three cranial fossae. It overlies the orbits and contains the frontal lobes of the brain. Anteriorly, the anterior fossa is bounded by the frontal bone, which also forms the majority of the floor for this space. The lesser wings of the sphenoid bone form the prominent ledge that marks the boundary between the anterior and middle cranial fossae. Located in the floor of the anterior cranial fossa at the midline is a portion of the ethmoid bone, consisting of the upward-projecting crista galli and, to either side of this, the cribriform plates.

Middle Cranial Fossa The **middle cranial fossa** is deeper and situated posterior to the anterior fossa. It extends from the lesser wings of the sphenoid bone anteriorly, to the bumpy **petrous portion of the temporal bones** posteriorly. The middle cranial fossa is divided at the midline by the upward bony prominence of the sella turcica, a part of the sphenoid bone. The middle cranial fossa has several openings for the passage of blood vessels and cranial nerves (see Figure 8.24A).

Openings in the middle cranial fossa are as follows:

- **Optic canal**—This opening is located at the anterior lateral corner of the sella turcica. It provides for passage of the optic nerve into the orbit.
- **Superior orbital fissure**—This large, irregular opening into the posterior orbit is located on the anterior wall of the middle cranial fossa, lateral to the optic canal and under the projecting margin of the lesser wing of the sphenoid bone. Nerves to the eyeball and associated muscles, as well as sensory nerves to the forehead, pass through this opening.
- **Foramen rotundum**—This rounded opening (*rotundum* = "round") is located in the floor of the middle cranial fossa, just inferior to the superior orbital fissure. It is the exit point for a major sensory nerve that supplies the cheek, nose, and upper teeth.
- **Foramen ovale**—This large, oval-shaped opening in the floor of the middle cranial fossa provides passage for a major sensory nerve to the lateral head, cheek, chin, and lower teeth.
- **Foramen spinosum**—This small opening, located posterior-lateral to the foramen ovale, is the entry point for an important artery that supplies the covering layers surrounding the brain. The branching pattern of this artery forms readily visible grooves on the internal surface of the skull, and these grooves can be traced back to their origin at the foramen spinosum.
- **Stylomastoid foramen**—This small opening is located between the styloid process and mastoid process. This is the point of exit for the cranial nerve that supplies the facial muscles.
- **Carotid canal**—This is the zigzag passageway through which a major artery to the brain enters the skull. The entrance to the carotid canal is located on the inferior aspect of the skull, anteromedial to the styloid process (**Figure 8.25**). From here, the canal runs anteromedially within the bony base of the skull. Just above the foramen lacerum, the carotid canal opens into the middle cranial cavity, near the posterior-lateral base of the sella turcica (see Figure 8.24A).
- **Foramen lacerum**—This irregular opening is located in the base of the skull, immediately inferior to the exit of the carotid canal. This opening is an artifact of the dry skull, because in life it is completely filled with cartilage. All the openings of the skull that provide for passage of nerves or blood vessels have smooth margins; the word *lacerum* ("ragged" or "torn") tells us that this opening has ragged edges and thus nothing passes through it.

Posterior Cranial Fossa The **posterior cranial fossa** is the most posterior and deepest portion of the cranial cavity. The posterior fossa is bounded anteriorly by the petrous ridges, while the occipital bone forms the floor and posterior wall. It is divided at the midline by the large foramen magnum, the opening that provides for passage of the spinal cord.

Located on the medial wall of the petrous ridge in the posterior cranial fossa is the **internal acoustic meatus** (see Figure 8.24A). This opening in the temporal bone provides for passage of the nerve from the hearing and equilibrium organs of the inner ear, and the nerve that supplies the muscles of the face. The foramen magnum is not a flat hole, but a short passageway. Located within its walls along the anterior-lateral edge is the **hypoglossal canal**. These provide passage for an important nerve to the tongue.

Immediately inferior to the internal acoustic meatus is the large, heart-shaped **jugular foramen** (see Figure 8.24A and Figure 8.25). Several nerves from the brain exit the skull via this opening. It is also the exit point through the skull for the jugular vein, which carries all of the venous blood leaving the brain.

For our last view within the interior of the skull, let's examine a skull with the calvarium intact, sectioned down the middle (midsagittally; **Figure 8.26**). In this view we can see the major bones of the brain case—the frontal, parietal, and occipital bones

Figure 8.25 The External View of the Base of the Skull

The floor of the cranial cavity is formed by the frontal, ethmoid, sphenoid, temporal, and occipital bones. Many structures such as blood vessels and nerves enter and exit the brain through openings in these bones.

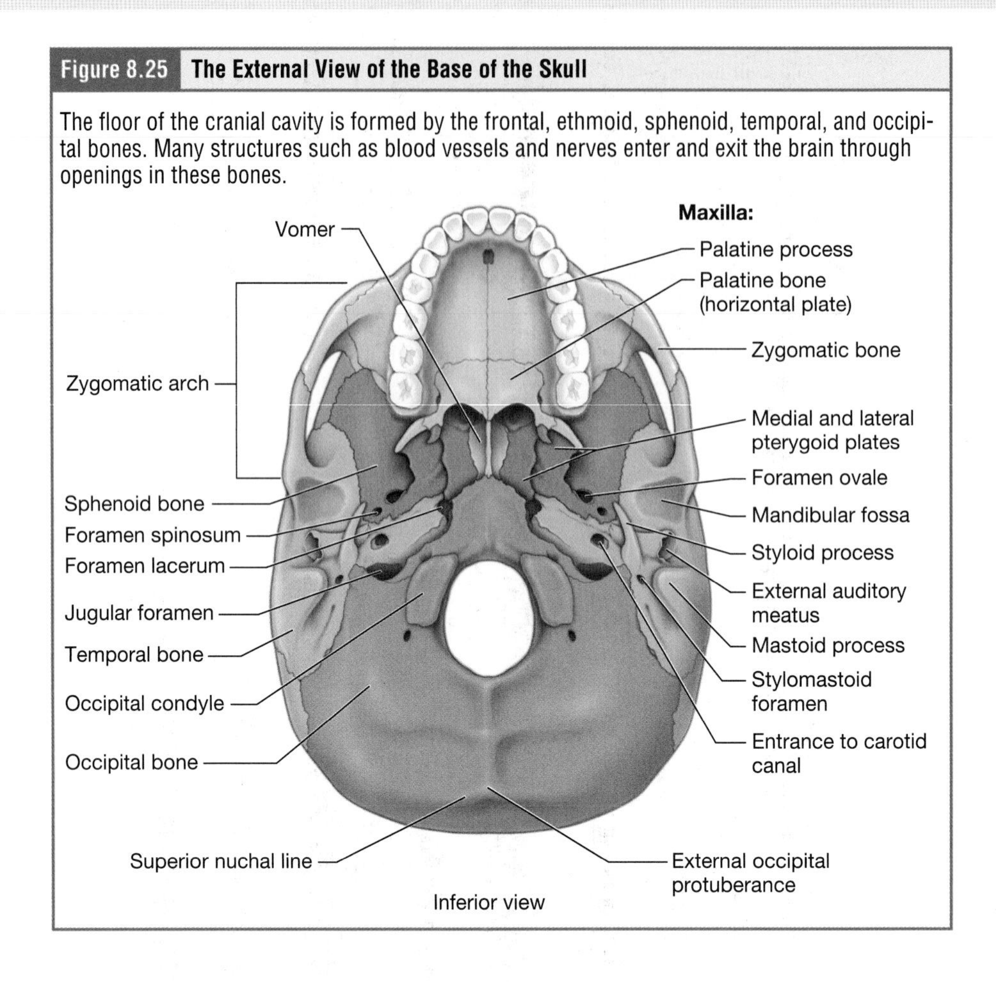

joined by the coronal and lambdoid sutures. Inferiorly, we can see the temporal bone with the internal acoustic meatus and the sphenoid, with its sinus. Anteriorly, we see the perpendicular plate of the ethmoid bone and vomer comprising the nasal septum, and the maxilla and mandible making up the oral cavity.

8.2k Development and Aging of the Skull

LO 8.2.4

Developmentally, the skull may be one of the most fascinating structures in human anatomy. Human births represent a major anatomical constraint. *Homo sapiens*, the modern species of humans, have the largest head-to-body ratio of any species of primate, and nearly the largest of any species that has ever lived (**Figures 8.27A** and **8.27B**). And yet, as bipedal animals, our vertically oriented pelvis is quite compact. Compared to other primates, the head size of human neonates is terribly close to the size of the outlet of the pelvis through which the fetus must pass in birth (**Figure 8.27C**). During embryonic development, the skull bones develop in two different ways. The flat bones of the top and sides of the cranium develop through intramembranous ossification, while some facial bones and some bones at the base of the skull develop through endochondral ossification (see Chapter 7 for a refresher on these processes). Early in embryonic development, the brain is shielded by just a sheet (or membrane) of connective tissue. Throughout the fetal period, islands of bone develop within this sheet (**Figure 8.28A**), eventually forming an incomplete cranial shell at the time of birth. At birth, the bones of the fetal skull remain separated from each other by large areas of dense connective tissue, each of which is called a **fontanelle** (**Figure 8.28B**). The fontanelles are the "soft spots" on an infant's head. While it may seem risky to have a fetus with these vulnerable areas around their brain, fontanelles are important during birth because they allow

Figure 8.26 Midsagittal Section of Skull

This view shows the interior of a skull that has been sectioned sagittally.

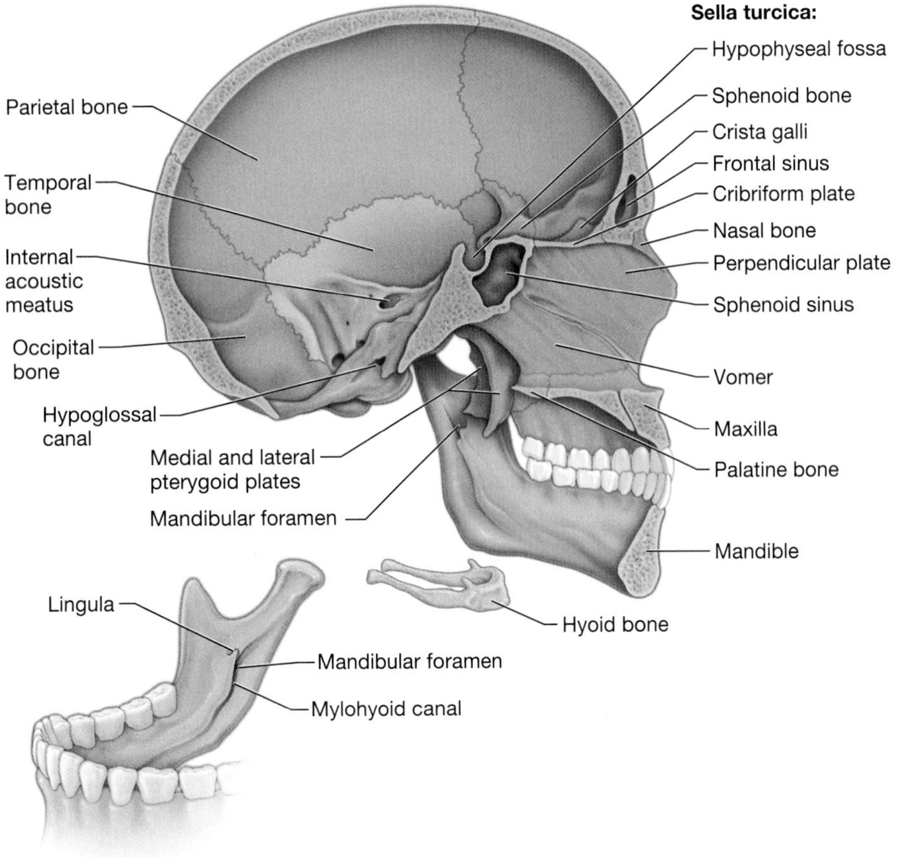

Sella turcica:
Hypophyseal fossa
Sphenoid bone
Crista galli
Frontal sinus
Cribriform plate
Nasal bone
Perpendicular plate
Sphenoid sinus
Vomer
Maxilla
Palatine bone
Mandible
Hyoid bone
Parietal bone
Temporal bone
Internal acoustic meatus
Occipital bone
Hypoglossal canal
Medial and lateral pterygoid plates
Mandibular foramen
Lingula
Mandibular foramen
Mylohyoid canal

the skull to change shape as it squeezes through the birth canal. Babies born vaginally have slightly cone-shaped brains from the overlapping of their skull bones during the squeeze out of the birth canal (**Figure 8.28C**). Babies born via C-section, of course, do not have deformed skulls, as their birth process lacks this squeeze. After birth, the fontanelles allow for continued growth and expansion of the skull as the brain enlarges. The

Figure 8.27 The Brain-to-Body Size Ratios of Humans Compared to Other Animals

(A) Humans have a much greater brain-to-body size ratio than other primates. (B) Among all animals, humans have one of the largest brain-to-body size ratios of any species. (C) These large heads and narrow bipedal hips present a challenge for birth.

1,500
1,250
1,000
750
500
250
0
Brain mass (grams)
Humans
3X
Other primates
0 25 50 75 100
Body mass (kilograms)
A
1,800
1,600
1,400
1,200
1,000
800
600
400
200
0
Brain mass (grams)
Dolphin
Human
Chimpanzee
Gorilla
Octopus
Sheep
Jaguar
Pig
Kangaroo
0 50 100 150 200 250 300
Body mass (kilograms)
B
Orangutan
Chimpanzee
Gorillas
Humans
Head
Hips
C

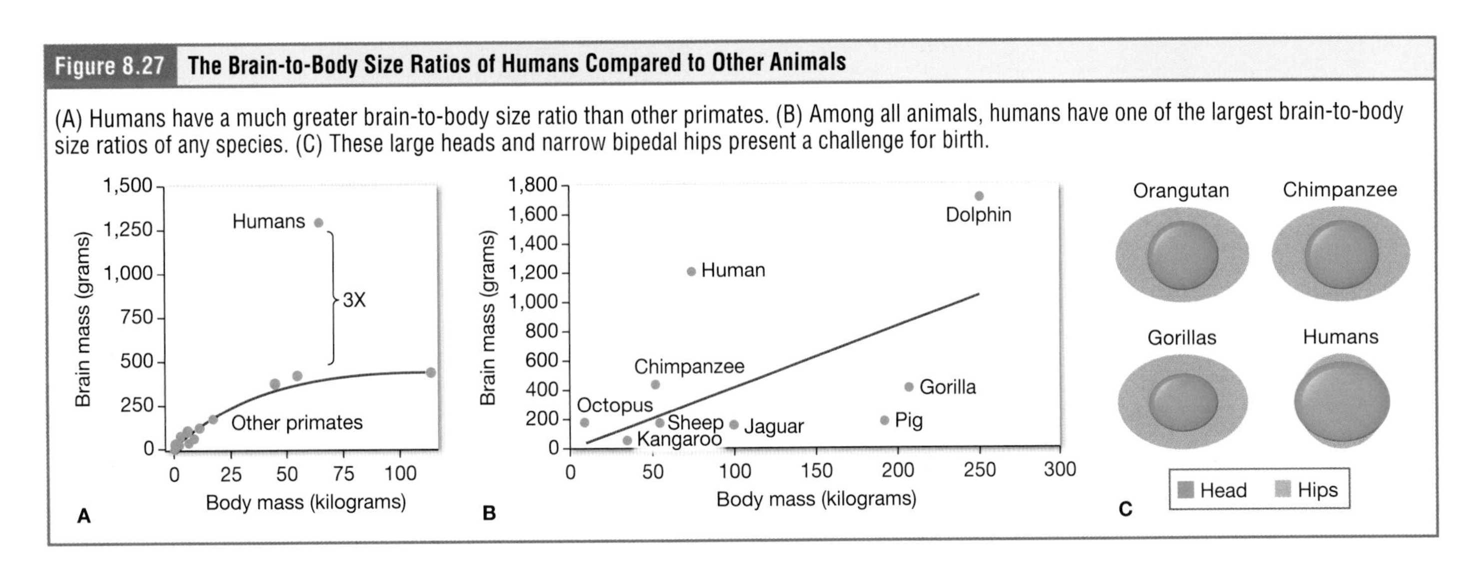

Figure 8.28 The Developing Skull

(A) The early embryonic skull is mostly a sheet of dense irregular connective tissue. Within that sheet, bone begins to form. (B) The fetal skull contains disconnected islands of bone with fontanelles between them. (C) During vaginal birth, the bones of the skull can overlap slightly to fit through the narrow passageway. (D) The sutures of the skull are filled with fibrous connective tissue until later adulthood. Over time, the connective tissue ossifies and the suture closes.

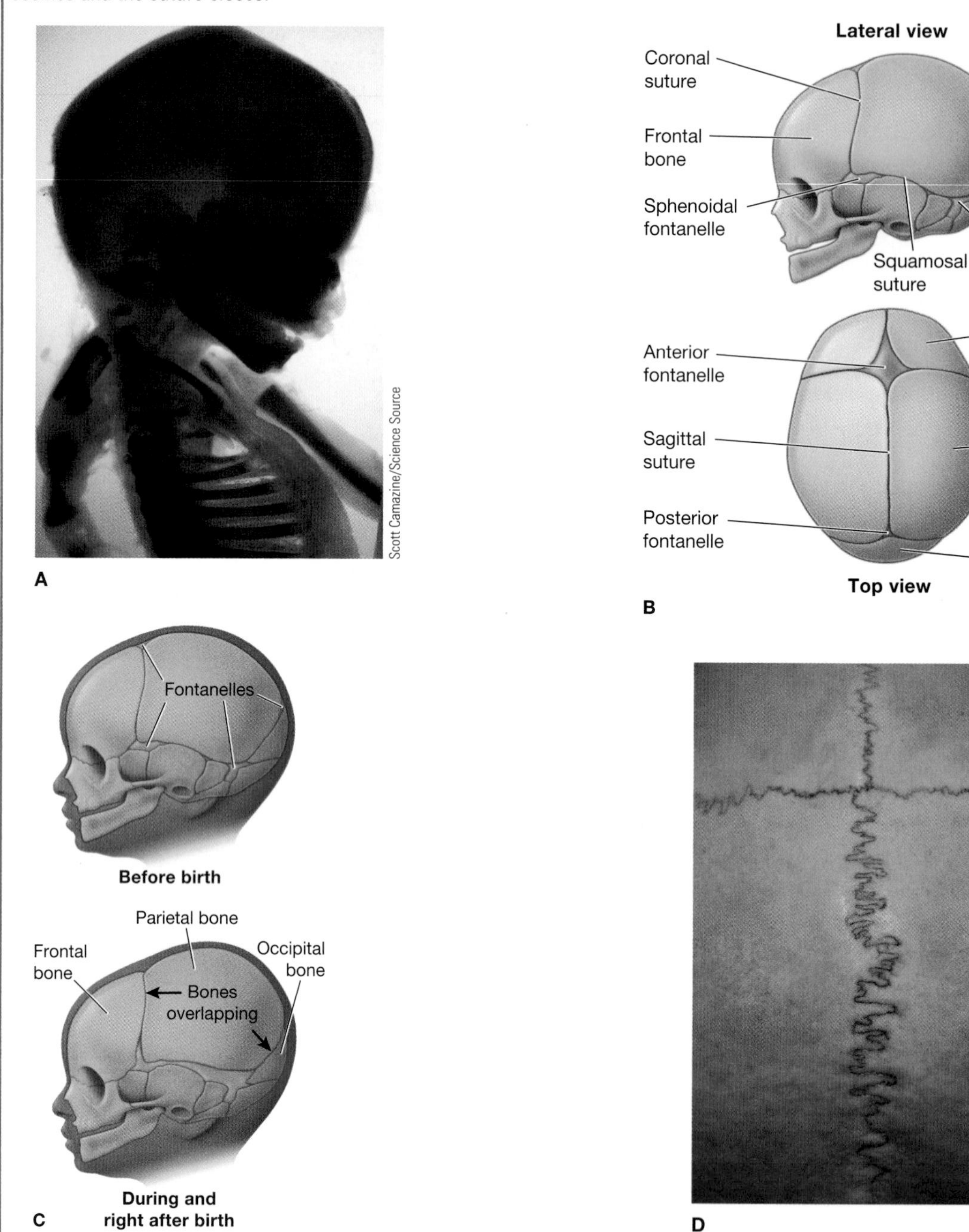

fontanelles decrease in size and disappear by age 2. However, the skull bones remained separated from each other at the sutures, which contain dense fibrous connective tissue that unites the adjacent bones. The connective tissue of the sutures allows for continued growth of the skull bones as the brain enlarges during childhood growth.

In contrast, the facial bones and bones of the floor of the brain case develop through interstitial bone growth using a hyaline cartilage template that slowly ossifies (see Chapter 7 for a refresher on this process). This is a slow process, and the cartilage is not completely converted to bone until the skull achieves its full adult size.

While the fontanelles are closed by approximately age 2, the sutures do not ossify until adulthood. In a predictable pattern, once the connective tissue within each suture ossifies until in the mature adult skeleton, the brain case is entirely fused (**Figure 8.28D**). Cranial suture ossification is used in forensics to estimate the age of skeletal remains.

DIGGING DEEPER: Fusion Disorders

During embryonic development, the right and left maxillae fuse at the midline to form the upper jaw. At the same time, the muscle and skin overlying these bones join together to form the upper lip. Inside the mouth, the palatine processes of the maxillae, along with the horizontal plates of the right and left palatine bones, join together to form the hard palate. If an error occurs in these developmental processes, a birth defect of cleft lip or cleft palate may result.

Cleft lip is a common developmental defect that affects approximately 1 in 1,000 births. This defect involves a partial or complete failure of the right and left portions of the upper lip to fuse together, leaving a cleft.

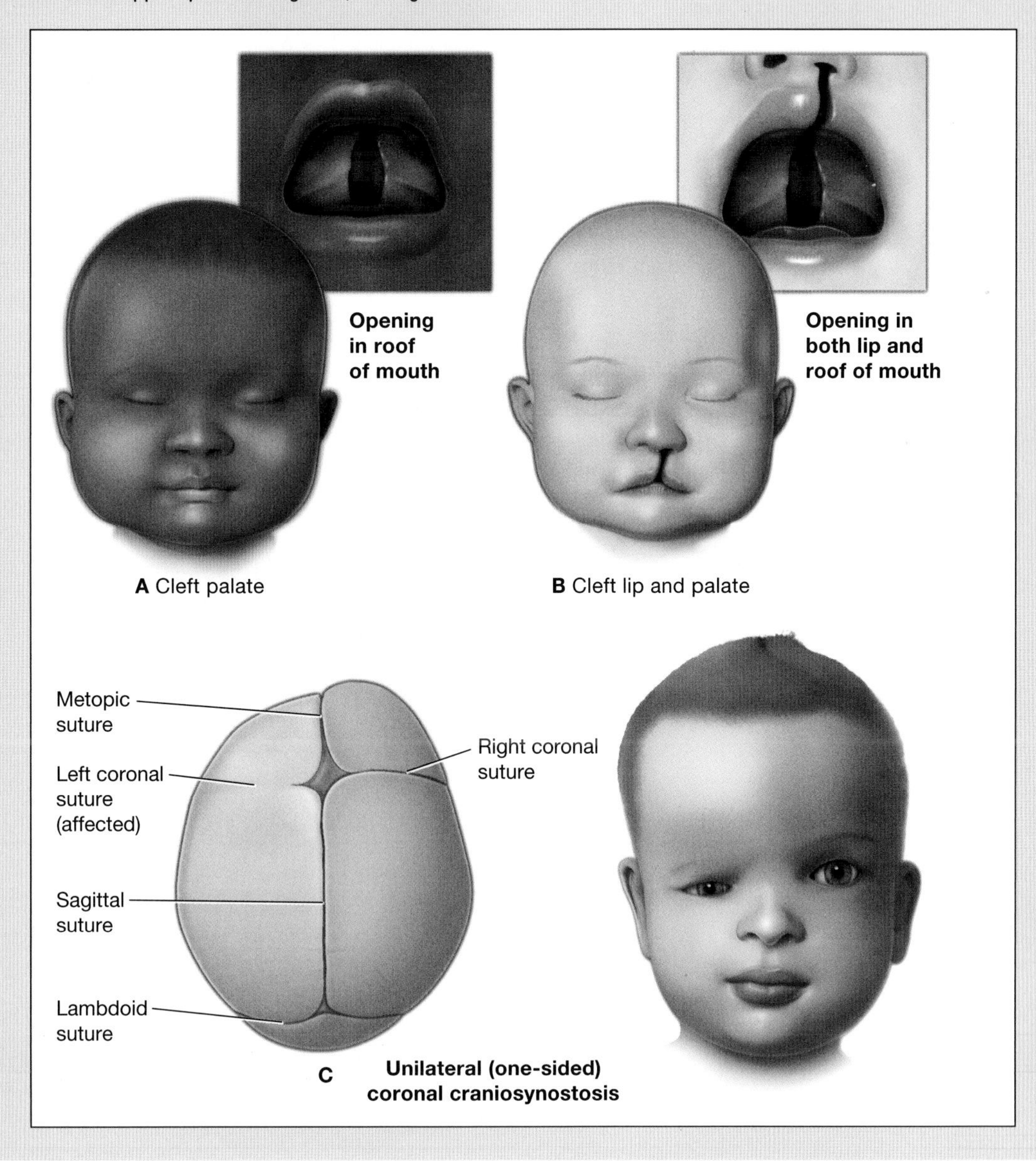

(Continued)

A more severe developmental defect is cleft palate, which affects the hard palate. The *hard palate* is the bony structure that separates the nasal cavity from the oral cavity. It is formed during embryonic development by the midline fusion of the horizontal plates from the right and left palatine bones and the palatine processes of the maxillae. Cleft palate affects approximately 1 in 2,500 births and is more common in females. It results from a failure of the two halves of the hard palate to completely come together and fuse at the midline, thus leaving a gap between them (see Figure A). This gap allows for communication between the nasal and oral cavities. In severe cases, the bony gap continues into the anterior upper jaw where the alveolar processes of the maxilla bones also do not properly join together above the front teeth. If this occurs, a cleft lip will also be seen (see Figure B). Because of the communication between the oral and nasal cavities, a cleft palate makes it extremely difficult for an infant to generate the suckling needed for nursing, thus leaving the infant at risk for malnutrition. Surgical repair is required to correct cleft palate defects.

In contrast, the premature closure (fusion) of a suture line is a condition called *craniosynostosis*. This error in the normal developmental process results in abnormal growth of the skull and deformity of the head. It is produced either by defects in the ossification process of the skull bones or failure of the brain to properly enlarge. Genetic factors are involved, but the underlying cause is unknown. It is a relatively common condition, occurring in approximately 1 in 2,000 births. *Primary craniosynostosis* involves the early fusion of one cranial suture, while *complex craniosynostosis* results from the premature fusion of several sutures.

The early fusion of a suture in primary craniosynostosis prevents any additional enlargement of the cranial bones and skull along this line (see Figure C). Continued growth of the brain and skull is therefore diverted to other areas of the head, causing an abnormal enlargement of these regions. Although the skull is misshapen, the brain still has adequate room to grow and thus there is no accompanying abnormal neurological development.

In cases of complex craniosynostosis, several sutures close prematurely. The amount and degree of skull deformity is determined by the location and extent of the sutures involved. This results in more severe constraints on skull growth, which can alter or impede proper brain growth and development. Cases of craniosynostosis are usually treated with surgery.

8.2l Bones Associated with the Skull: The Ossicles and the Hyoid Bone

The **ossicles** are three tiny bones located within each of the temporal bones. These three bones are part of the ear; we will learn more about them in Chapter 14. Their names are the *malleus*, *incus*, and *stapes*. The **hyoid bone** is an independent bone that does not contact any other bone and thus is not part of the skull (**Figure 8.29**).

Figure 8.29 Hyoid Bone

The hyoid bone is located in the upper anterior neck, tucked under the chin. It does not articulate with any other bone.

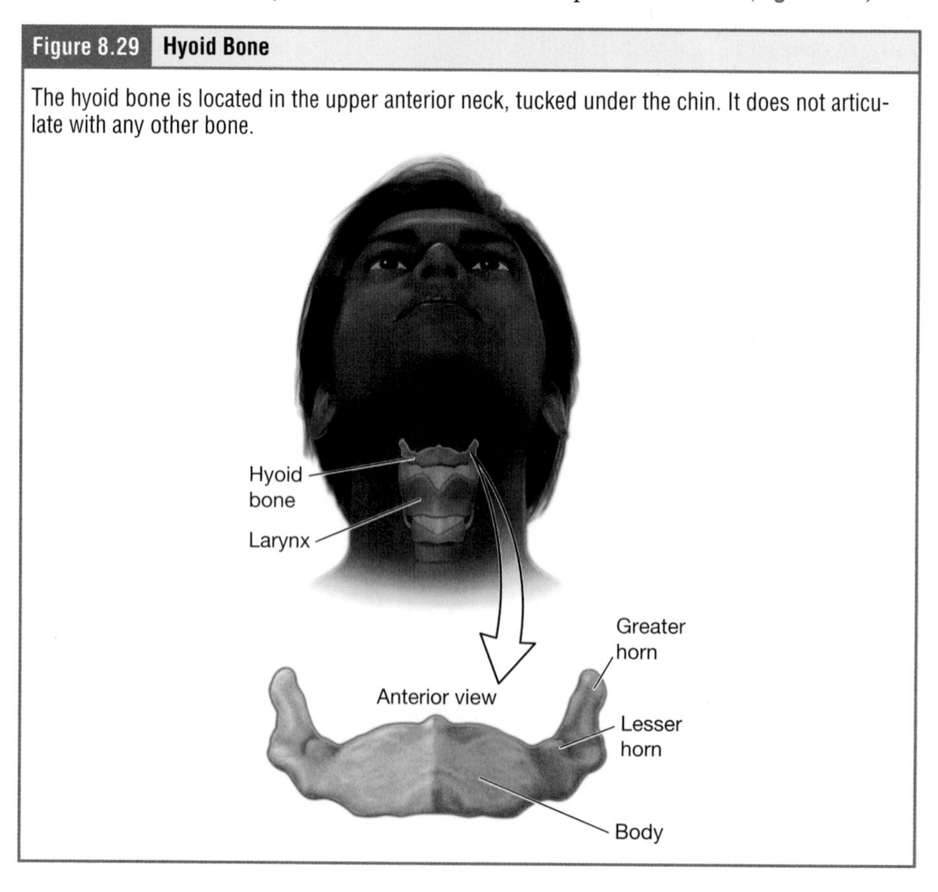

It is a small U-shaped bone located in the upper neck near the level of the inferior mandible, with the tips of the "U" pointing posteriorly. The hyoid serves as an anchor for some muscles of the tongue above. The hyoid is held in position by a series of small muscles that attach to it either from above or below. Movements of the hyoid are coordinated with movements of the tongue, larynx, and pharynx during swallowing and speaking. The hyoid bone is found superficial to the trachea (windpipe) and therefore it occasionally makes an appearance in TV crime dramas as a broken hyoid is the forensic clue used to determine death by strangulation. The last notable fact about the hyoid is that it is the only bone in the human body that does not articulate with another bone!

8.3 The Vertebral Column

HAPS LEARNING GOALS AND OUTCOMES

Learning Objectives: By the end of this section, you will be able to:

8.3.1* Describe the vertebral column as a whole, including the curvatures and regions.

8.3.2 Identify major bone markings (e.g., spines, processes, foramina) on individual vertebrae.

8.3.3* Compare and contrast vertebrae from each region of the spinal column.

8.3.4 Identify and describe the intervertebral joints.

* Objective is not a HAPS Learning Goal.

The **vertebral column** (also known as the *spinal column* or *spine*) is a string of **vertebrae** that forms a flexible column that supports the head, neck, and upper body, allowing for their movements (**Figure 8.30**). Each vertebra is separated and united by an intervertebral disc. Together, the vertebrae and intervertebral discs form the vertebral column. This flexible column also protects the spinal cord, which passes down the back through openings in the vertebrae.

8.3a Regions of the Vertebral Column

The vertebral column has 24 vertebrae, plus the fused vertebrae that make up the sacrum and coccyx. The vertebral column is subdivided into five regions, with the vertebrae in each area named for that region and numbered in descending order. The regions are, from superior to inferior: cervical, thoracic, lumbar, sacral, and coccygeal. In the cervical region (the neck), there are seven cervical vertebrae, each designated with the letter "C" followed by its number. Superiorly, the C1 vertebra articulates with the occipital condyles of the skull. Inferiorly, C1 articulates with the C2 vertebra, and so on. Below these are the 12 thoracic vertebrae, designated T1–T12. The lower back contains the lumbar vertebrae, designated L1–L5. In the sacral region, five sacral vertebrae fuse to form the single **sacrum**, which is also part of the pelvis. Similarly, in the coccygeal region, four small coccygeal vertebrae fuse to form the **coccyx**, or tailbone. However, the sacral and coccygeal fusions do not start until age 20 and are not completed until middle age. Therefore, it is quite possible that you have more vertebrae than your anatomy and physiology instructor!

Student Study Tip

There are 7 cervical, 12 thoracic, and 5 lumbar vertebrae. Traveling down the spine, think of some typical mealtimes. You eat breakfast at 7, lunch at 12, and dinner at 5!

coccyx (tailbone)

An interesting anatomical fact is that almost all mammals—regardless of body size—have seven cervical vertebrae. This means that there are large variations in the size of cervical vertebrae, ranging from the very small cervical vertebrae of a shrew to the greatly elongated vertebrae in the neck of a giraffe. In a full-grown giraffe, each cervical vertebra is 11 inches tall!

Figure 8.30 Vertebral Column

The adult vertebral column consists of 24 individual vertebrae, plus the fused vertebrae of the sacrum and coccyx. The spine is divided into three regions: cervical C1–C7 vertebrae, thoracic T1–T12 vertebrae, and lumbar L1–L5 vertebrae.

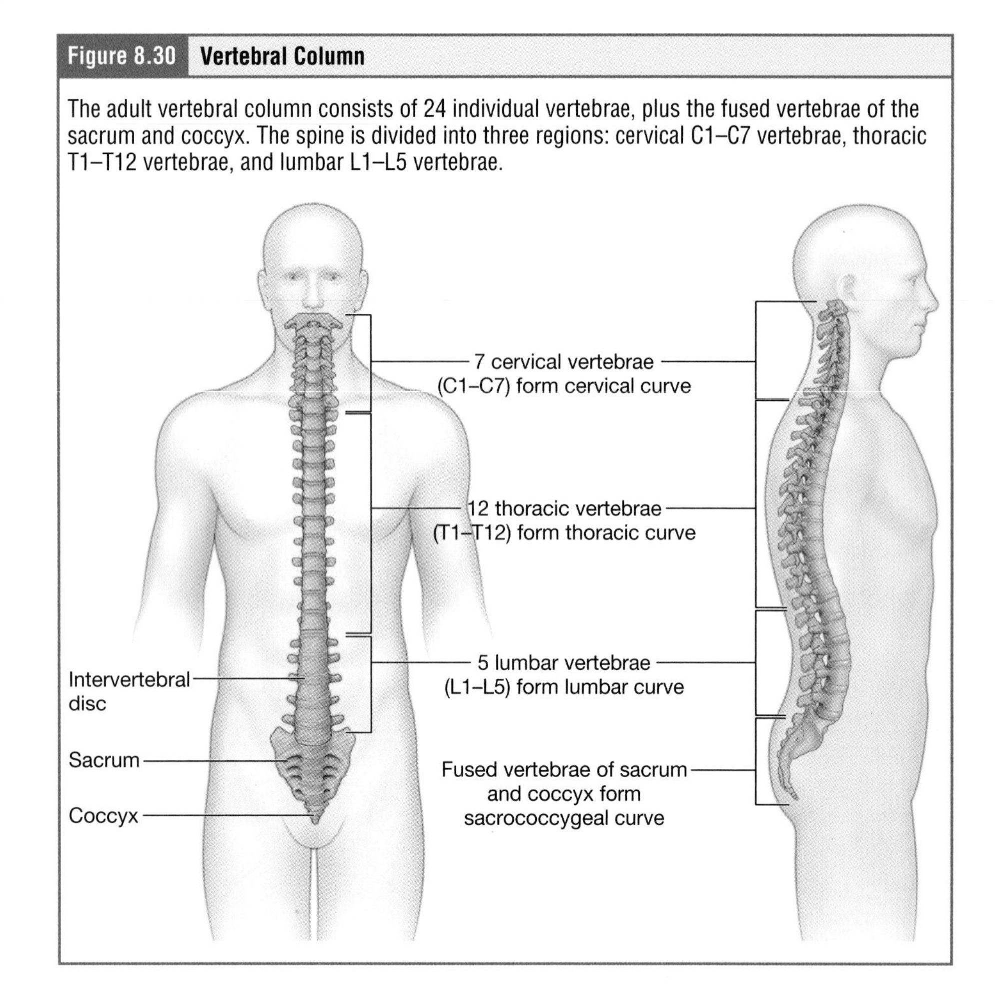

Figure 8.31 The Curvatures of the Vertebral Column

The adult vertebral column is not straight; rather, it has four curvatures to assist with weight bearing and to keep the eyes most available and pointed forward. From superior to inferior, these are the cervical, thoracic, lumbar, and sacrococcygeal curvatures.

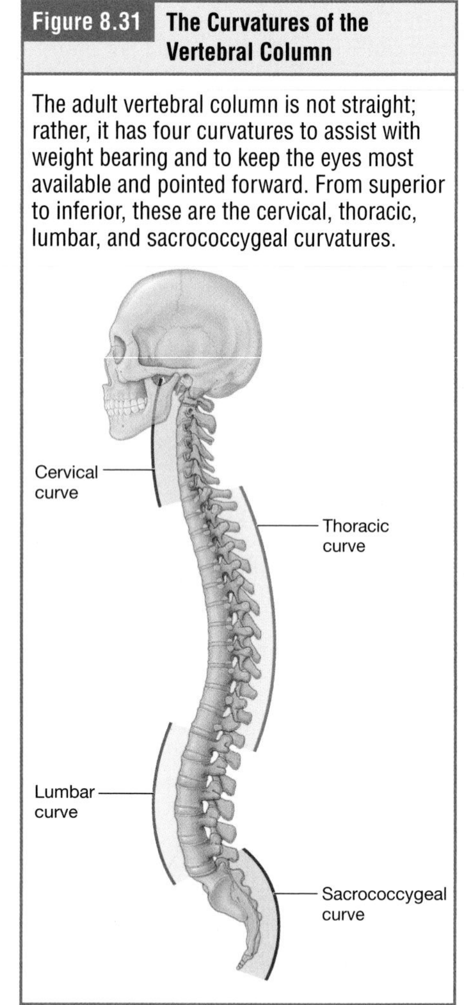

8.3b Curvatures of the Vertebral Column

The adult vertebral column does not form a straight line, but instead has four curvatures along its length (**Figure 8.31**). These curves increase the vertebral column's strength, flexibility, and ability to absorb shock. When the load on the spine is increased (for example, by carrying a heavy backpack), the curvatures increase in depth (become more curved) to accommodate the extra weight. They then return to resting position when the weight is removed.

During fetal development, the body is flexed anteriorly into the "fetal position," giving the entire vertebral column a single curvature that wraps around the abdominal organs. In the adult, this anterior curvature is retained in two regions of the vertebral column as the **thoracic curve**, which involves the thoracic vertebrae, and the **sacrococcygeal curve**, formed by the sacrum and coccyx. As the infant begins to move and support the weight of the body, the curvature of the spine changes in two places. As the infant begins to hold their head upright when sitting, the **cervical curve** of the neck region develops so that the eyes within the face can look up and around at their surroundings. Later, as the child begins to stand and then to walk, the **lumbar curve** of the lower back develops.

Disorders associated with the curvature of the spine include **kyphosis** (an excessive posterior curvature of the thoracic region), **lordosis** (an excessive anterior curvature of the lumbar region), and **scoliosis** (an abnormal lateral curvature, accompanied by twisting of the vertebral column).

kyphosis (humpback or hunchback)

lordosis (swayback)

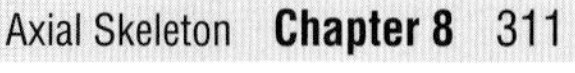

DIGGING DEEPER:
Curvatures of the Vertebral Column

Developmental anomalies, pathological changes, or obesity can enhance the normal vertebral column curves, resulting in the development of abnormal or excessive curvatures (see the figures below). Kyphosis, also referred to as humpback or hunchback, is an excessive posterior curvature of the thoracic region (see Figure A). Kyphosis is explored in more detail in the Apply to Pathophysiology feature. Lordosis, or sway-back, is an excessive anterior curvature of the lumbar region and is most commonly associated with obesity or late pregnancy (see Figure B). The accumulation of body weight in the abdominal region results in an anterior shift in the line of gravity that carries the weight of the body. This causes an anterior tilt of the pelvis and a pronounced enhancement of the lumbar curve.

Scoliosis (see Figure C) is an abnormal lateral curvature, accompanied by twisting of the vertebral column, that typically develops in the growth spurts during late childhood or just before puberty. The cause is usually unknown, but it may result from weakness of the back muscles, defects such as differential growth rates in the right and left sides of the vertebral column, or differences in the length of the lower limbs. When present, scoliosis tends to get worse during adolescent growth spurts. Although most individuals do not require treatment, a back brace may be recommended for growing children. In extreme cases, surgery may be required.

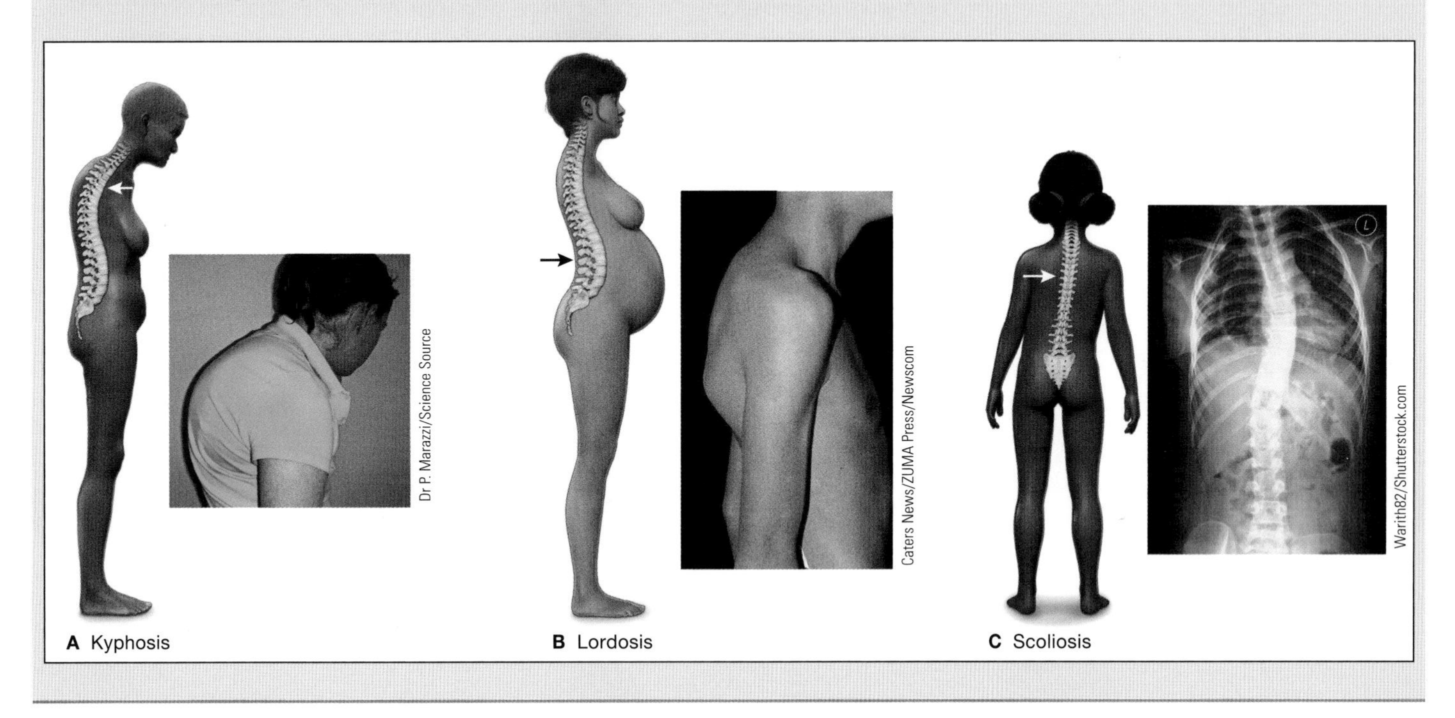

A Kyphosis **B** Lordosis **C** Scoliosis

Vertebrae are beautifully complex and intricate bones. Their technical shape is "irregular" and examination of any of the vertebrae makes it clear that they could not be classified in any of the other shape categories. Not all of our 24 (or perhaps more in your body) vertebrae look the same. To master their structure and function, we will first learn their general structure and the features that are common to all vertebrae, and then we will examine the features unique to the different regions.

8.3c General Structure of a Vertebra

LO 8.3.2

A typical vertebra will consist of a body, a vertebral arch, and seven processes (see the "Anatomy of a Typical Vertebra" feature).

The **body of the rib** is the bulky anterior portion of each vertebra and is the part that supports the body weight. Because of this, the vertebral bodies progressively increase in size and thickness going down the vertebral column. Sandwiched between the bodies of vertebrae are the fibrocartilaginous intervertebral discs. These discs are explored in a bit more detail in the following "Apply to Pathophysiology" feature.

Student Study Tip

The root "ped" means foot. If you flip an image of a vertebra to where the top is anterior, the body appears to be standing on the pedicles, as if they are feet.

The **vertebral arch** forms the posterior portion of each vertebra. It consists of four parts—the right and left pedicles and the right and left laminae. Each **pedicle** forms one of the lateral sides of the vertebral arch. The pedicles are attached to the vertebral body. Together the **laminae** form the roof of the vertebral arch. The arch encloses the **vertebral foramen**, which contains the spinal cord.

Three notable processes project off the vertebral arch. Paired **transverse processes** project laterally at the junction point between the pedicle and lamina. The single **spinous process** projects posteriorly at the midline of the back. The spinous processes can easily be felt as a series of bumps just under the skin down the middle of the back. The transverse and spinous processes serve as important muscle attachment sites.

Now that we have examined one vertebra in isolation, let's begin to put the vertebrae together to see how they connect. Look at a lateral view of vertebrae in **Figure 8.32**. As we have already discussed, between their bodies, a thick pad of fibrocartilage—the intervertebral disc—is found. We can see that the vertebrae also connect posteriorly. Above the spinous process, reaching superiorly is a short, flat process called the **superior articular process**. Reaching down from the anterior base of the spinous process is the **inferior articular process**. These joints, lined with hyaline cartilage, allow for the vertebrae to move and support each other during anterior–posterior motions of the spine, such as arching your back to look up at the ceiling or in a backbend. Notable from the side of the intact vertebral column is the **intervertebral foramen**, the opening through which spinal nerves exit from the vertebral column.

Anatomy of...

A Typical Vertebra

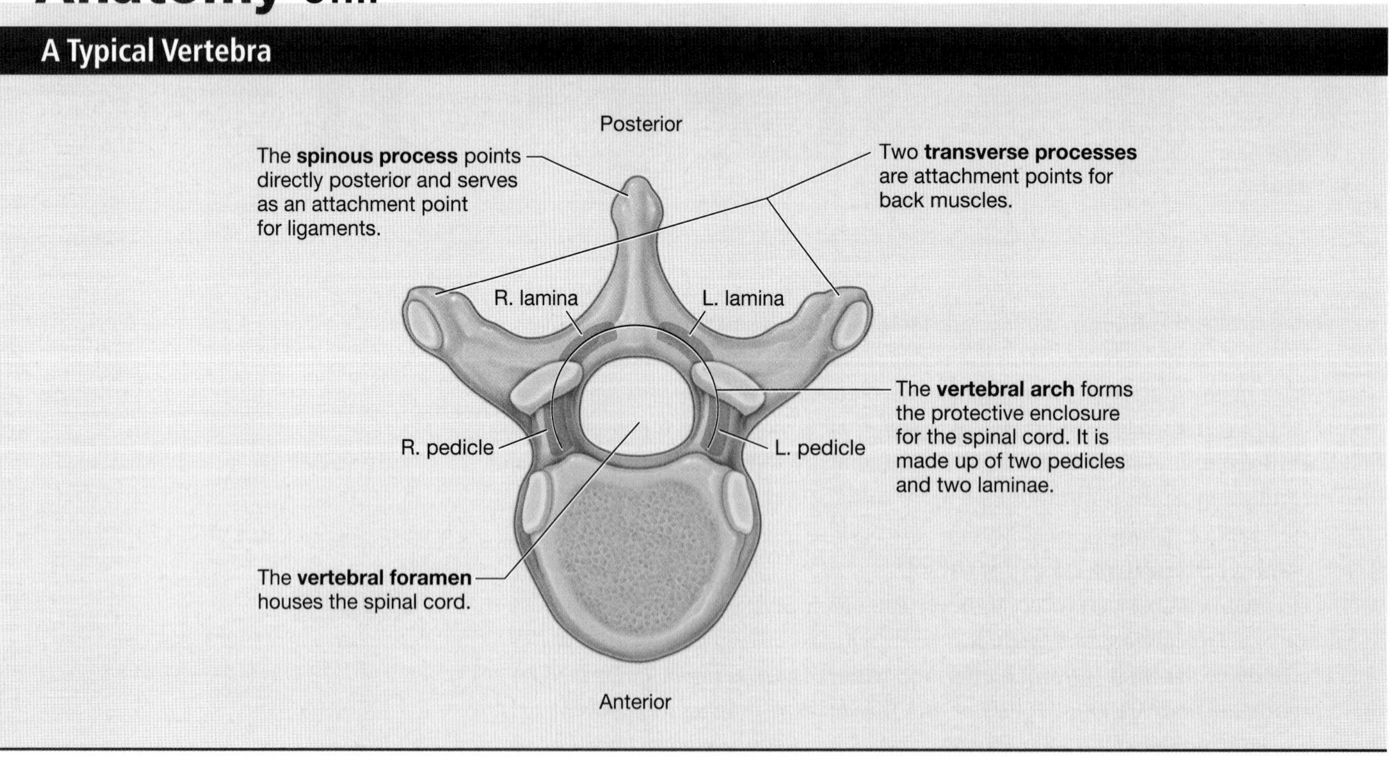
Posterior
The spinous process points directly posterior and serves as an attachment point for ligaments.
Two transverse processes are attachment points for back muscles.
R. lamina
L. lamina
The vertebral arch forms the protective enclosure for the spinal cord. It is made up of two pedicles and two laminae.
R. pedicle
L. pedicle
The vertebral foramen houses the spinal cord.
Anterior

Figure 8.32 Three Articulated Vertebrae

As we stack the vertebrae we can see that they articulate at their bodies—where they sandwich intervertebral discs—and at their articular processes.

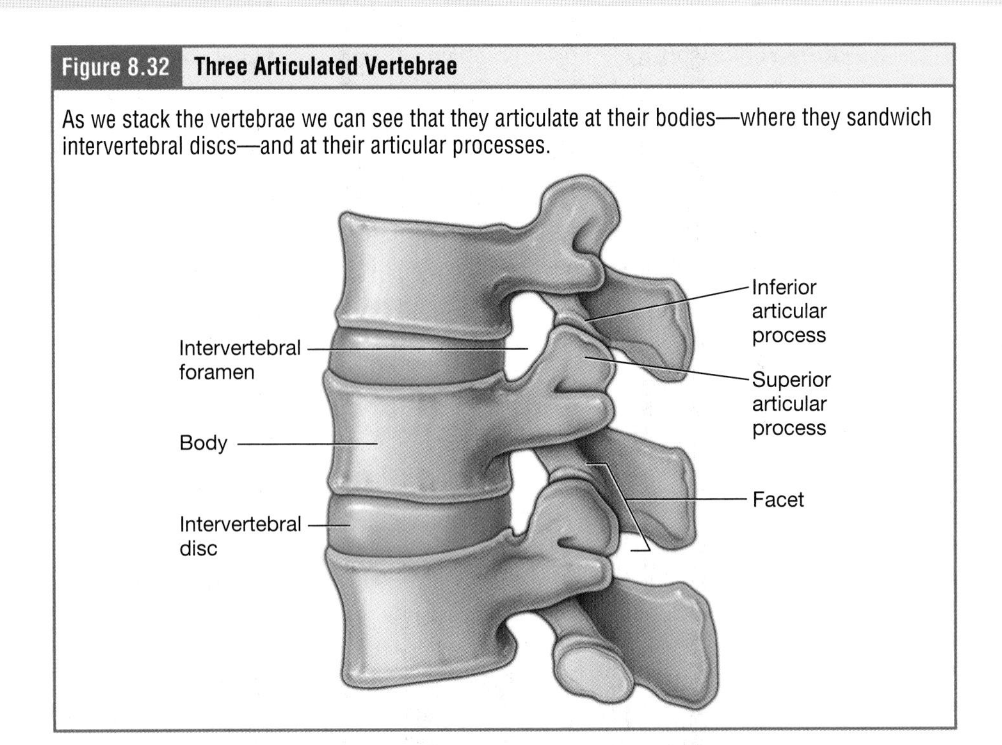

Apply to Pathophysiology

Aging Spines

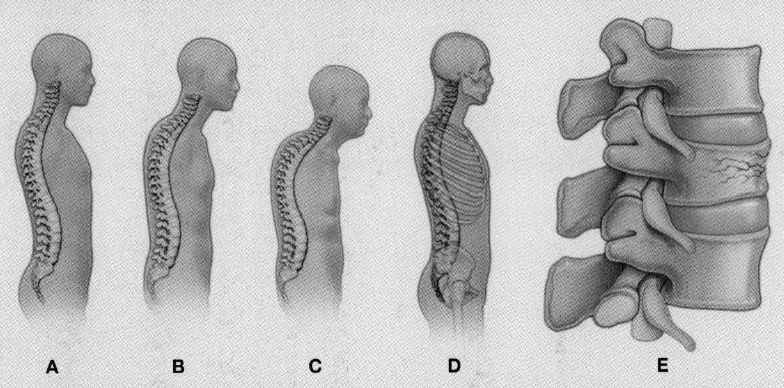

Sally is a sweet, 79-year-old woman who is active in her community, frequently volunteering at the local food bank. Lately, though, she has been in too much pain to engage in the events she used to enjoy. Her children note that her posture is becoming increasingly stooped. They encourage her to go to the doctor, where it is noted in Sally's chart that she has developed kyphosis. Sally's doctor refers her to the hospital for a bone scan.

1. Which of the images is most likely Sally's profile?
   A) Figure (A)
   B) Figure (B)
   C) Figure (C)
   D) Figure (D)
2. The bone scan reveals marked bone loss in the vertebral bodies. Which of the following statements is true?
   A) Sally's vertebrae are collapsing anteriorly, making more space among her spinous processes.
   B) Sally's vertebrae are collapsing posteriorly, pushing the spinous processes together.
   C) Sally's vertebrae are collapsing superior-to-inferiorly, shortening her height and narrowing her intervertebral foramina.
3. It appears that one of Sally's vertebrae may have a compression fracture. This fracture occurs in the weakened vertebral bodies in response to weight bearing (Figure E). In which region of the spine is the compression fraction most likely to occur?
   A) Cervical
   B) Thoracic
   C) Lumbar

8.3d Regional Vertebrae

LO 8.3.3

In addition to the general characteristics of a typical vertebra described in the previous section, vertebrae also display characteristic size and structural features that vary between the different vertebral column regions. For example, the cervical vertebrae, which support the weight of just the head and neck, are smaller than lumbar vertebrae, which support the cumulative weight of the head, neck, thorax, and abdomen. The thoracic vertebrae have sites for rib attachment, and the vertebrae that give rise to the sacrum and coccyx have fused together into single bones.

Cervical Vertebrae Of the seven cervical vertebrae, C1 and C2 have their own unique features and C3–C7 have several characteristic features that differentiate them from thoracic or lumbar vertebrae (**Figure 8.33**). Cervical vertebrae have a small body, reflecting the fact that they carry the least amount of weight. Cervical vertebrae usually have a bifid (Y-shaped) spinous process. The spinous processes of the C3–C6 vertebrae are short, but the spine of C7 is much longer. You can find these vertebrae by running your finger down the midline of the posterior neck until you encounter the prominent C7 vertebra located at the base of the neck. Each transverse process also has a hole within it called the **transverse foramen**. An important artery that supplies the brain ascends up the neck and is protected from damage by passing within these openings.

The first and second cervical vertebrae are unique due to their relationships with the skull. The first cervical (C1) vertebra is often referred to by its unique name—**atlas**—because this vertebra supports the skull on top of the vertebral column (in Greek mythology, Atlas was the god who supported the heavens on their shoulders). The C1 vertebra does not have a body or spinous process. Instead, it is ring-shaped, consisting of an **anterior arch** and a **posterior arch** (**Figure 8.34A**). The transverse

Figure 8.33 Cervical Vertebrae

A typical cervical vertebra has a small body, transverse processes with a transverse foramen, and a bifid spinous process.

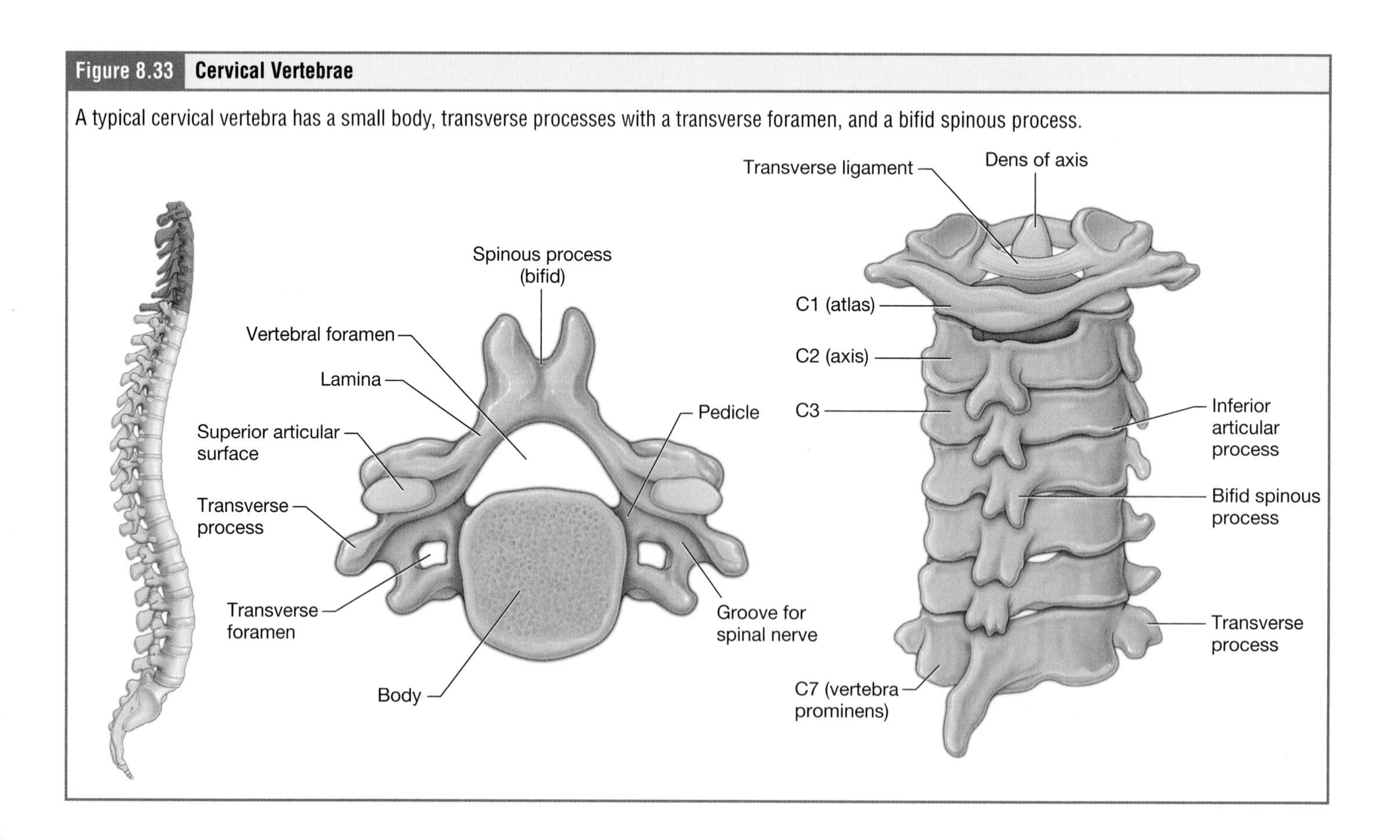

Figure 8.34 Atlas and Axis

(A) The C1 vertebra (atlas) does not have a body or spinous process. It consists of an anterior and a posterior arch and elongated transverse processes. (B) The C2 vertebra (axis) contains the upward-projecting dens, which articulates with the anterior arch of atlas. (C) The anterior view of axis shows the huge upward-projecting dens. (D) The dens of C2 is held in place by the transverse ligament.

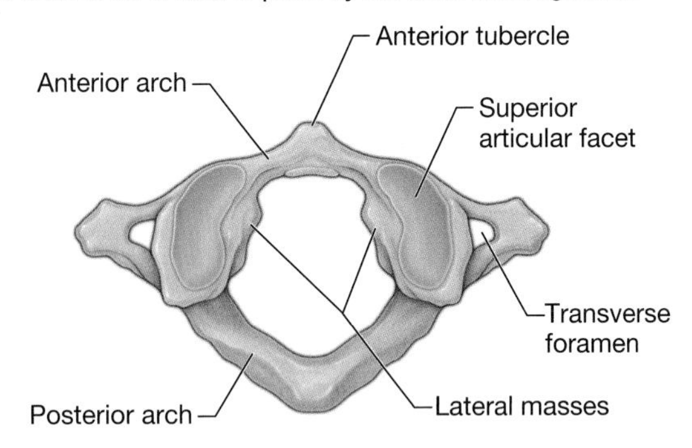

A Superior view of 1st cervical vertebra (atlas)

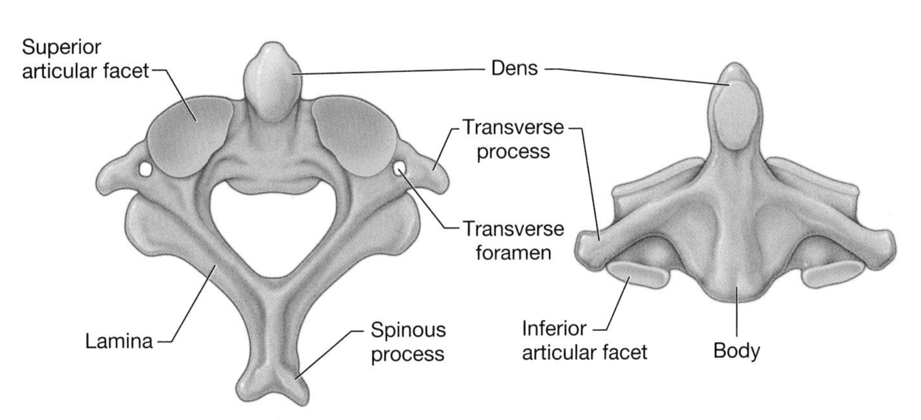

B Superior view of axis

C Anterior view of axis

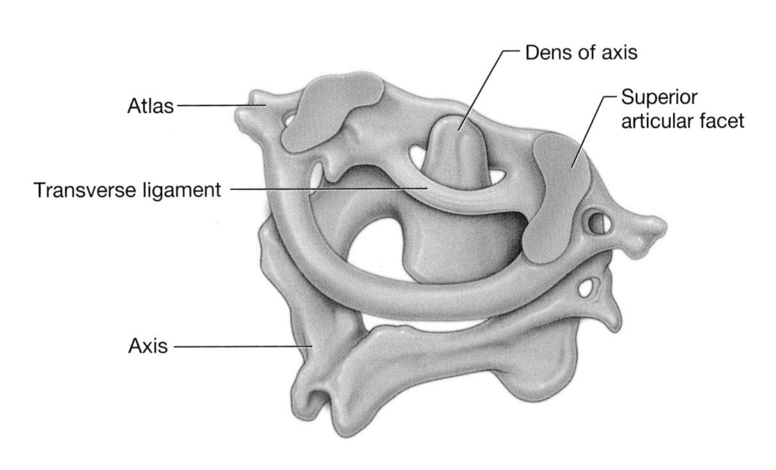

D Oblique view of atlas and axis

processes of the atlas extend more laterally than do the transverse processes of any other cervical vertebrae. The superior articular processes face upward and are deeply curved for articulation with the occipital condyles on the base of the skull. The inferior articular processes are flat and face downward to join with the superior articular processes of the C2 vertebra.

The second cervical (C2) vertebra is called the **axis** because it serves as the axis for rotation when turning the head toward the right or left. The axis resembles typical cervical vertebrae in most respects, but it is easily distinguished by the **dens** (odontoid process), a bony projection that extends upward from the vertebral body on its anterior side (**Figures 8.34B** and **8.34C**). The dens joins with the inner aspect of the anterior arch of the atlas, where it is held in place by transverse ligament (**Figure 8.34D**).

Thoracic Vertebrae The bodies of the **thoracic vertebrae** become larger than those of cervical vertebrae (**Figure 8.35**) as the weight load increases as we descend the spinal column. The characteristic feature for a typical thoracic vertebra is the spinous process, which is long and has a pronounced downward angle that causes it to overlap the next inferior vertebra.

Thoracic vertebrae have additional articulation sites, not present in the vertebrae of other regions, because within the thoracic region each vertebra must articulate with a rib. Each of these articulation sites is called a **facet**. Most thoracic vertebrae have four total facets and articulate with two ribs (**Figure 8.36**). Two facets located on the lateral sides of the vertebral body articulate with the head (end) of a rib. An additional facet is located on each transverse process for articulation with the tubercle of a rib.

Student Study Tip

To make identifying vertebrae more fun, think of the shape of the thoracic vertebrae as a giraffe's head, and the shape of a lumbar vertebra as looking like a moose's head.

Lumbar Vertebrae **Lumbar vertebrae** carry the greatest amount of body weight and are thus characterized by the large size and thickness of the vertebral body (**Figure 8.37**). They have short transverse processes and a short, blunt spinous process that projects posteriorly. The articular processes are large, with the superior process facing backward and the inferior facing forward.

Figure 8.35 Thoracic Vertebrae

A typical thoracic vertebra is distinguished by the downward pointing spinous process.

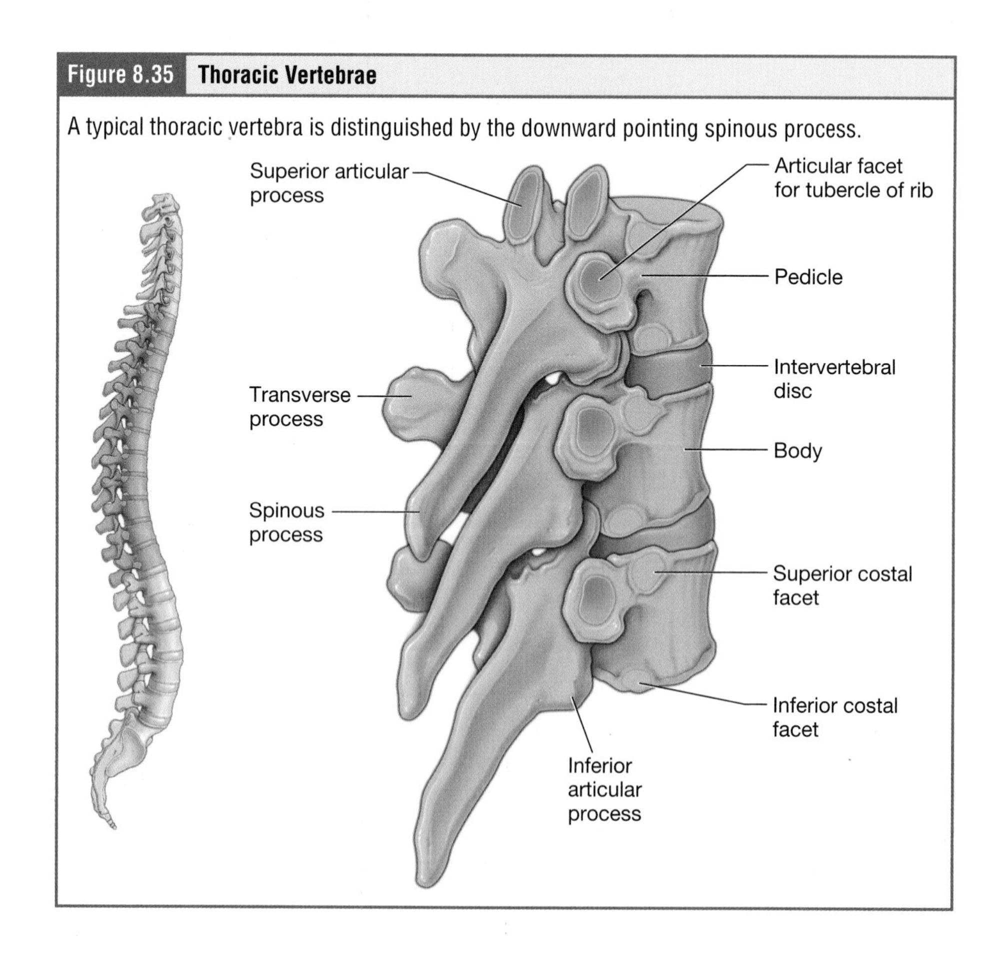

Figure 8.36 Each Rib Articulates with a Thoracic Vertebra

The body of a thoracic vertebra articulates with the head of a rib, and the transverse process articulates with the rib tubercle.

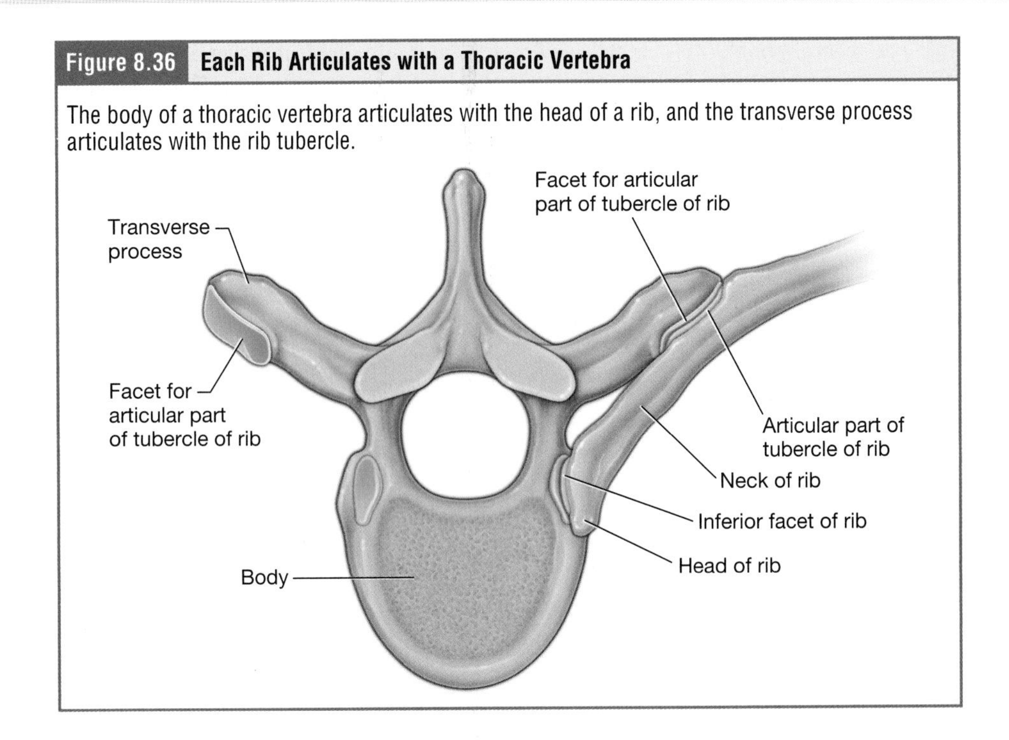

Sacrum and Coccyx The sacrum is a triangular-shaped bone that is thick and wide across its superior base where it is weight bearing and then tapers down to an inferior, non–weight-bearing apex (**Figure 8.38**). It is formed by the fusion of five sacral vertebrae, a process that does not begin until after the age of 20. On the posterior surface,

Figure 8.37 Lumbar Vertebrae

Lumbar vertebrae are characterized by their thick bodies and short, rounded spinous processes.

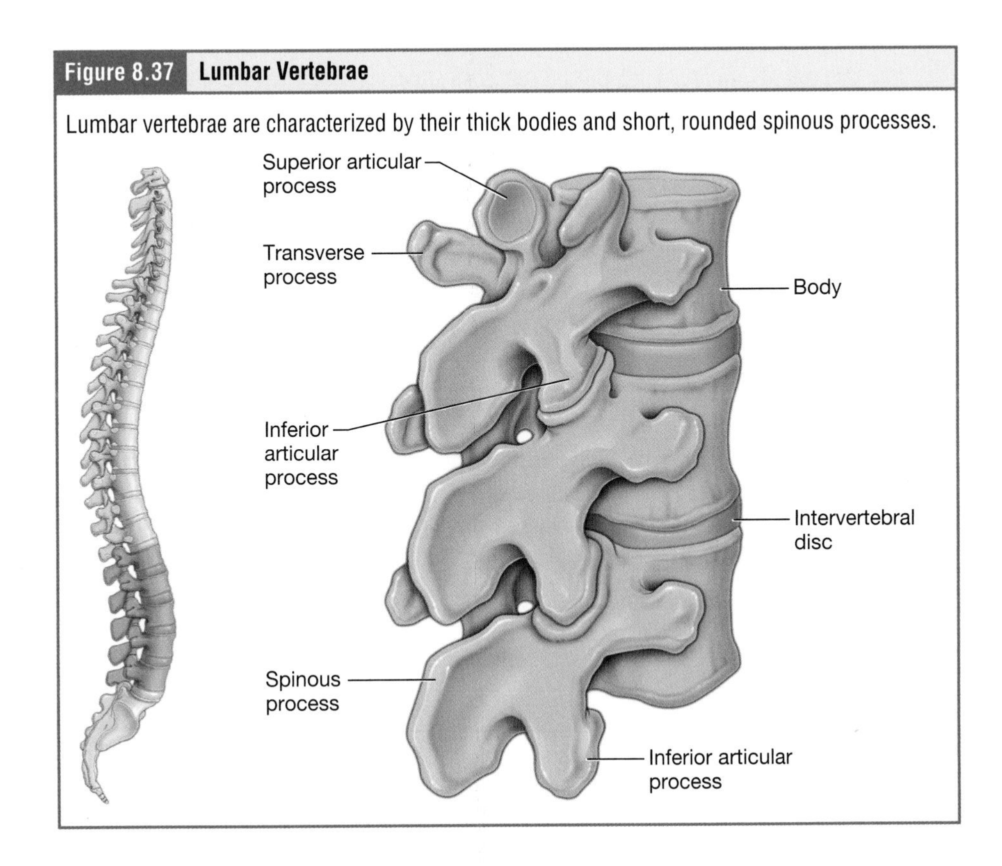

Figure 8.38 Sacrum and Coccyx

The sacrum is formed from the fusion of five sacral vertebrae; the locations of vertebral fusion are indicated by the transverse ridges. The coccyx is formed by the fusion of four small coccygeal vertebrae.

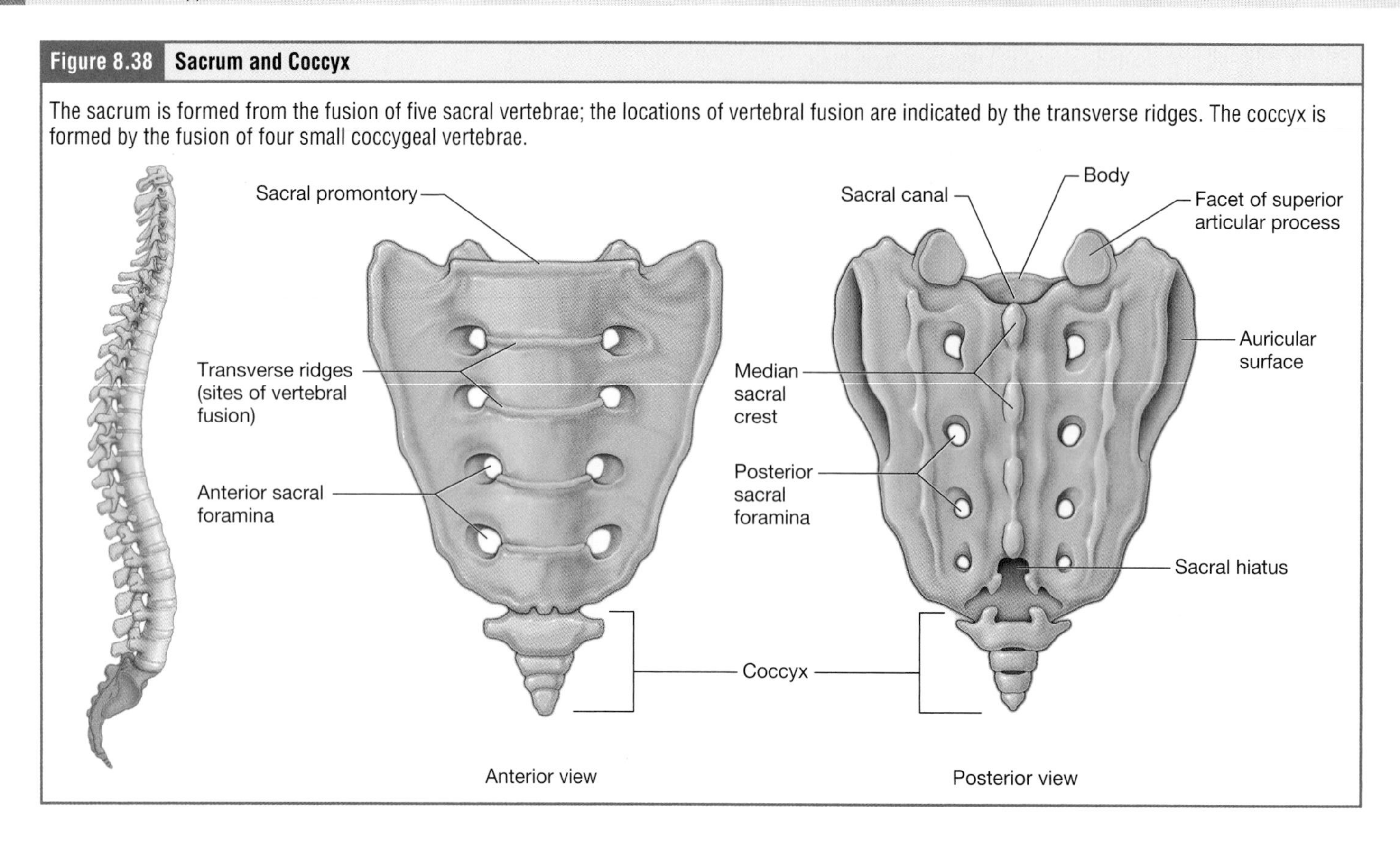

running down the midline, is the **median sacral crest**, a bumpy ridge that is the remnant of the fused spinous processes. Similarly, the fused transverse processes of the sacral vertebrae form the **lateral sacral crest**.

The sacrum articulates with the L5 vertebra superiorly at a region known as the *base of the sacrum*. The term "base" in anatomy can sometimes be confusing when applied because it does not always mean the bottom of a structure but often refers to its flattest surface (that is, the surface it may rest on if you laid the structure on a table). In this case, the **base** of the sacrum is actually its superior surface. The anterior of the base features the **sacral promontory**, which is the anterior lip of the body of the first sacral vertebra. Posteriorly along the base are two **superior articular processes of the sacrum**, which articulate with the inferior articular processes from the L5 vertebra.

Lateral to the base is the roughened auricular surface, which joins with the ilium portion of the hipbone to form the immobile sacroiliac joints of the pelvis. Passing inferiorly through the sacrum is a bony tunnel called the **sacral canal**, the end of which is the **sacral hiatus** near the inferior tip of the sacrum.

The anterior and posterior surfaces of the sacrum have a series of paired openings called **sacral foramina**, which allow nerves to exit the sacral canal.

The coccyx, or tailbone, is derived from the fusion of four very small coccygeal vertebrae (see Figure 8.38). It articulates with the inferior tip of the sacrum. It is not weight bearing in the standing position but may receive some body weight when sitting.

8.3e Intervertebral Discs

LO 8.3.4

The bodies of adjacent vertebrae are strongly anchored to each other by an **intervertebral disc**. This structure provides padding between the bones during weight bearing, and because it can change shape, also allows for movement between the vertebrae. Although the total amount of movement available between any two adjacent vertebrae is small, the sum of these small movements along the entire length of the vertebral column allows large body movements to be produced. Ligaments that extend along the length of the vertebral column also contribute to its overall support and stability.

Intervertebral Disc An *intervertebral disc* is a fibrocartilaginous pad that fills the gap between adjacent vertebral bodies (**Figure 8.39**). The discs provide padding between vertebrae during weight bearing. Because of this, intervertebral discs are thinnest in the cervical region and thickest in the lumbar region, which carries the most body weight. In total, the intervertebral discs account for approximately 25 percent of your body height between the top of the pelvis and the base of the skull. Intervertebral discs are also flexible and can change shape to allow for movements of the vertebral column.

Each intervertebral disc consists of two parts. The **anulus fibrosus** is the tough, fibrous outer layer of the disc. It forms a circle and is firmly anchored to the outer margins of the adjacent vertebral bodies. Inside is the **nucleus pulposus**, consisting of a softer, gel-like material. The nucleus has a high water content that serves to resist compression and thus is important for weight bearing. Over the course of a day (assuming you spend the day sitting upright and walking), the nucleus compresses. If you spend the night resting while lying down, the weight is removed from the intervertebral discs and they return to their original size and shape. Therefore, you are likely to measure slightly taller in the morning than you do in the evening. With increasing age, the water content of the nucleus pulposus gradually declines. This causes the discs to become thinner, decreasing total body height somewhat, and reduces the flexibility and range of motion of the disc, making bending more difficult.

Learning Connection

Macro to Micro

Intervertebral discs are composted of fibrocartilage. What characteristics of this tissue make it well-suited in this location?

Figure 8.39 Intervertebral Discs

Intervertebral discs, made of fibrocartilage, are found between the bodies of adjacent vertebrae. The disc consists of a fibrous outer ring called the *anulus fibrosus* and a gel-like center called the *nucleus pulposus*.

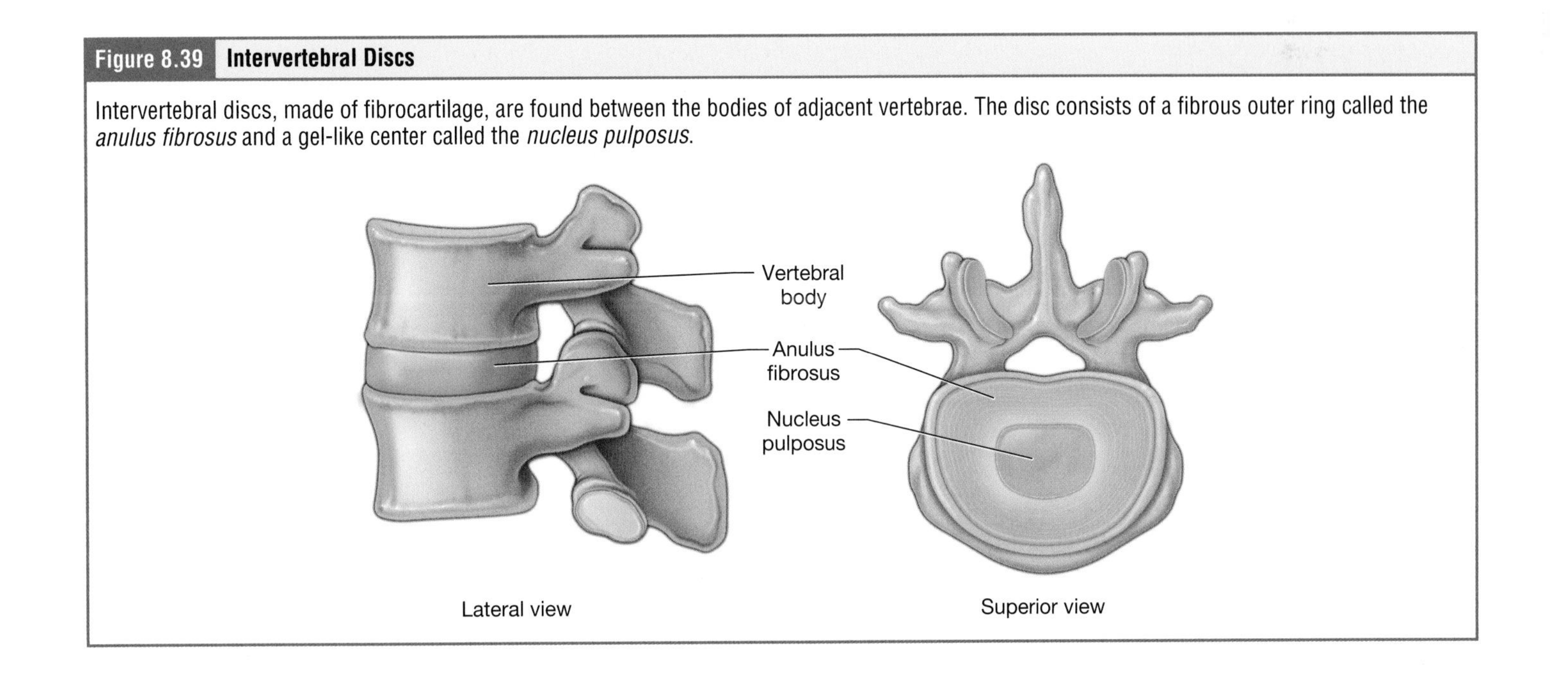

Apply to Pathophysiology

Herniated Disc

Phillip, a 54-year-old male, was working at the factory. It was almost closing time when he just had one more box to move. Since he was in a rush to leave, he quickly bent forward, grabbed the box, and went to pick it up when he felt a sharp pain in his lower back, and tingling and numbness down his right leg. He dropped the box and screamed in pain. As he stood, he found it too painful to rotate his spine. His coworker drove him to the hospital.

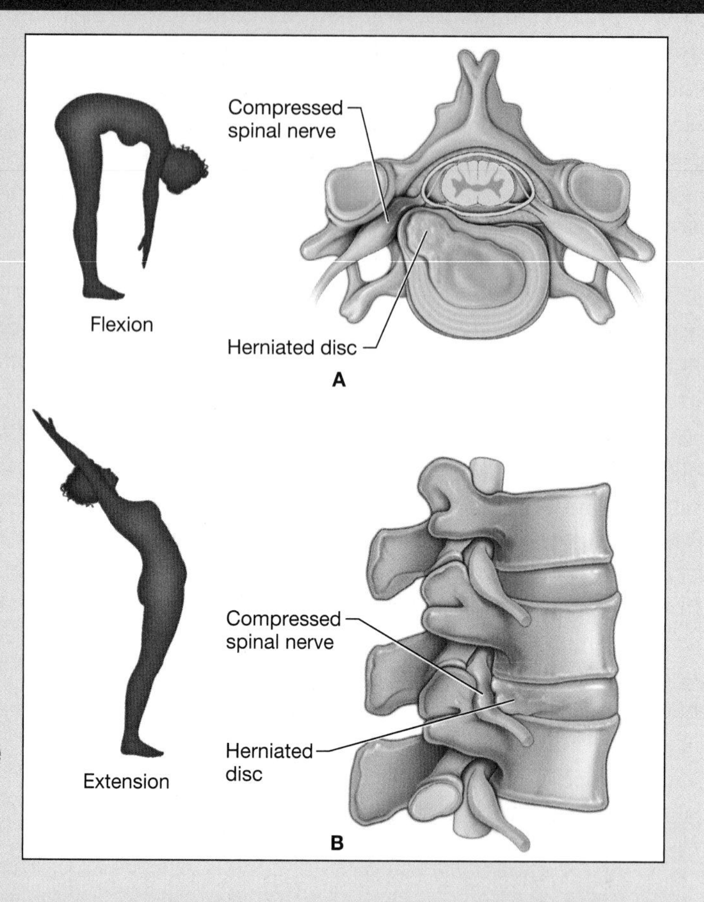

1. Which of the following locations best describes the location of his injury?
   A) Cervical spine
   B) Thoracic spine
   C) Lumbar spine
   D) Sacral spine
2. Using the answer to the previous question, which of the following is the biggest difference of the spinous processes between this region and the others?
   A) Strong, rectangular spinous processes
   B) Small, bifid spinous processes
   C) Thin, long spinous processes
   D) No spinous processes
3. Using the same answer as question 1, which of the following is the biggest difference of the vertebral bodies between this region and the others?
   A) No vertebral bodies
   B) Thick vertebral bodies
   C) Thin vertebral bodies
   D) Fused vertebral bodies

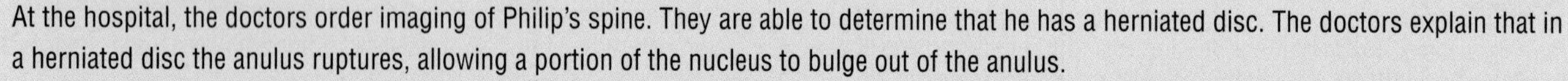

At the hospital, the doctors order imaging of Philip's spine. They are able to determine that he has a herniated disc. The doctors explain that in a herniated disc the anulus ruptures, allowing a portion of the nucleus to bulge out of the anulus.

4. Philip explains to the doctors that he is in much more pain when he extends his back than when he flexes it. These two motions, along with an image of a herniated disc, are shown in Figures (A) and (B). Examine the images. Why do you think Philip is in more pain when he extends his back?
   A) The anulus will protrude into the vertebral foramen during extension.
   B) The nucleus will be able to expand during extension.
   C) The spinous processes compress into the spinal nerves.
   D) The nucleus comes in contact with the spinal nerves and cord.

8.4 The Thoracic Cage

Learning Objectives: By the end of this section, you will be able to:

8.4.1* Identify the three components of the sternum, its position in the thoracic cage, and its features.

8.4.2* Compare and contrast true ribs, false ribs, and floating ribs.

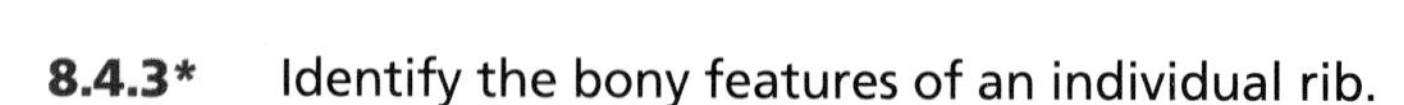

8.4.3* Identify the bony features of an individual rib.

* Objective is not a HAPS Learning Goal.

The **thoracic cage** (rib cage) forms the thorax (chest) portion of the body. It consists of the 12 pairs of ribs with their costal cartilages and the sternum (**Figure 8.40**). The bony thoracic cage protects the heart and lungs.

8.4a Sternum

LO 8.4.1

The **sternum**, sometimes called the *breastbone*, is the elongated bony structure of the anterior thoracic cage. It consists of three parts: the manubrium, body, and xiphoid process (see Figure 8.40). The **manubrium** is the wider, superior portion of the sternum. The top of the manubrium has a shallow, U-shaped border called the **suprasternal notch**. This can be easily felt at the anterior base of the neck, between the medial ends of the clavicles. The **clavicular notch** is the shallow depression located on either side of the manubrium where it articulates with the clavicles (collarbones). Because these bones help the shoulder move, they are considered part of the appendicular skeleton and will be covered in Chapter 9. The first ribs also attach to the manubrium.

The elongated, central portion of the sternum is the body. The manubrium and body join together at the **sternal angle**, so called because the junction between these two components is not flat, but it forms a slight bump (which can be felt on some individuals). The second rib attaches to the sternum at the sternal angle. Since the first rib is hidden behind the clavicle, the second rib is the highest rib that can be felt externally. Thus, the sternal angle and second rib are important landmarks for the identification and counting of the lower ribs. Ribs 3–7 attach to the sternal body.

The inferior tip of the sternum is the **xiphoid process**. This small structure is cartilaginous early in life, but gradually becomes ossified starting during middle age.

Figure 8.40 The Thoracic Cage

The thorax, or thoracic cage, is formed by bone and cartilage to protect the heart and lungs while allowing for the movements of breathing. (A) The sternum has three separate components. (B) Ten of the 12 pairs of ribs are attached to the sternum via costal cartilages. All 12 pairs of ribs are anchored posteriorly to the thoracic vertebrae.

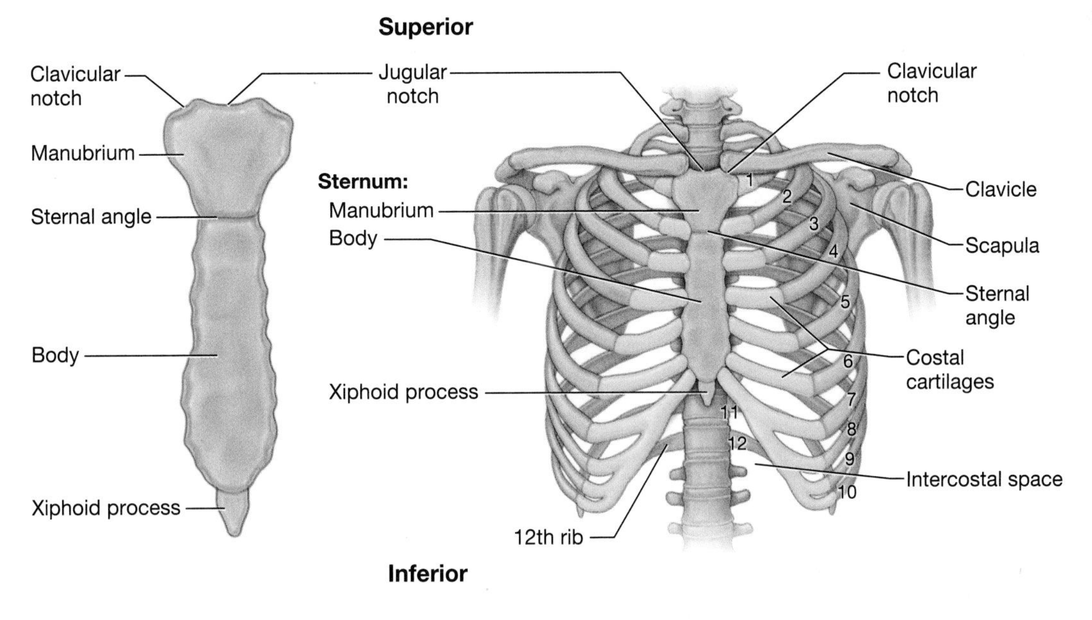

A Anterior view of the sternum

B Anterior view of the thoracic skeleton

LO 8.4.2

Figure 8.41 A Typical Rib

Each rib connects with a thoracic vertebra at the head and tubercle of the rib. Along the rib's inferior edge is a groove in which blood vessels and nerves are nestled and protected. Anteriorly, most ribs connect to costal cartilages.

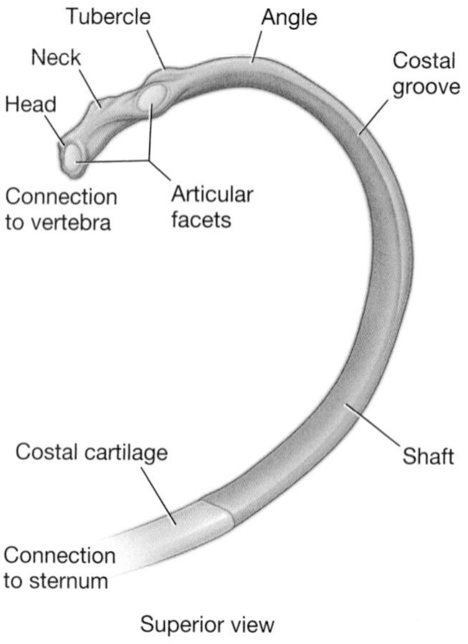

Tubercle
Angle
Neck
Costal groove
Head
Connection to vertebra
Articular facets
Costal cartilage
Shaft
Connection to sternum
Superior view

LO 8.4.3

8.4b Ribs

Each **rib** is a curved, flattened bone. There are 12 pairs of ribs. The ribs articulate posteriorly with the T1–T12 thoracic vertebrae and the ribs are numbered 1–12 to match the thoracic vertebrae they articulate with. None of the ribs articulate directly with the sternum; rather, they attach to the sternum via hyaline cartilage strips called **costal cartilages**. The last two rib pairs, ribs 11 and 12, do not attach to the sternum at all. The ribs are classified into three groups based on their relationship to the sternum. Ribs 1–7 are classified as **true ribs** because their costal cartilages from each of these ribs attaches directly to the sternum. Ribs 8–12 are called **false ribs** because their costal cartilages from these ribs do not attach directly to the sternum. Rather, for ribs 8–10, the costal cartilages are attached to the cartilage of the next higher rib. Thus, the cartilage of rib 10 attaches to the cartilage of rib 9, rib 9 then attaches to rib 8, and rib 8 is attached to rib 7 (see Figure 8.40). The last two false ribs (11–12), which do not articulate with the sternum and have no costal cartilage, are called **floating ribs**.

Parts of a Typical Rib The posterior end of a typical rib is called the **head of the rib** (**Figure 8.41**). This region articulates primarily with the costal facet located on the body of the same numbered thoracic vertebra. Moving laterally along its shaft, the rib narrows slightly after the head. This narrower region is called the **neck of the rib**. On the posterior neck you can find a small bump, the **tubercle of the rib**, which articulates with the facet located on the transverse process of the same numbered vertebra. The remainder of the rib is the **body of the rib**. Progressing laterally along the body, the rib curves dramatically, at the **angle of the rib**, before curving more gradually toward the anterior surface. Along the body, on the inferior edge you find shallow groove, just deep enough to run your fingernail in. Tucked in this **costal groove**, blood vessels and a nerve travel out from the spinal cord and supply the thoracic wall.

Structurally the thorax is a marriage of bone and cartilage which is well suited to its dual function of strength for protection and flexibility for breathing.

Learning Check

1. Which of the following regions has bifid spinous processes?
   a. Cervical
   b. Thoracic
   c. Lumbar
   d. Sacral
2. Is another name for C2 the axis or the atlas?
   a. Axis
   b. Atlas
3. Which region bears the most weight?
   a. Cervical
   b. Thoracic
   c. Lumbar
   d. Sacral
4. How many articulation points does a rib have?
   a. 1
   b. 2
   c. 3
   d. 4
5. Palpate your armpit and walk your hands down six inches until you can palpate your ribs. What structure of the ribs are you palpating?
   a. Head of the ribs
   b. Neck of the ribs
   c. Costal Groove of the ribs
   d. Angle of the rib

Chapter Review

8.1 Divisions of the Skeletal System

The learning objective for this section is:

HAPS LEARNING GOALS AND OUTCOMES

8.1.1 Distinguish between the axial and appendicular skeletons and list the major bones contained within each.

8.1 Question

1. Which of these bones are part of the appendicular skeleton? Please select all that apply.

   a. Ethmoid bone
   b. Ribs
   c. Femur
   d. Scapula

8.1 Summary

- The human skeleton is divided into the axial skeleton and the appendicular skeleton.
- The axial skeleton forms the main axis of the body, and consists of the bones of the skull, vertebral column, thoracic cage, hyoid bone, and auditory ossicles of the ear.
- The appendicular skeleton is composed of the bones of the upper and lower limb, and the bones that attach both limbs to the axial skeleton, such as the scapula and pelvic girdle.

8.2 The Skull

The learning objectives for this section are:

8.2.1* Know the bones of the skull and their features.

8.2.2* List and identify the skull sutures.

8.2.3* Identify the skull sinuses and describe their functions.

8.2.4* Describe the changes to the skull throughout development and aging.

* Objective is not a HAPS Learning Goal.

8.2 Questions

1. Which of the following bones make up the nasal cavity? Please select all that apply.

   a. Vomer
   b. Ethmoid
   c. Frontal
   d. Mandible

2. Identify the two bones which the coronal suture connects. Please select all that apply.

   a. Parietal
   b. Temporal
   c. Occipital
   d. Frontal

3. Explain why our voices change when our sinuses become congested with mucus.

   a. The mucus buildup in the oral cavity decreases the space in the throat for air to exit.
   b. The mucus enters the oral cavity, decreasing the space in the throat for air to exit.
   c. The mucus prevents air from bouncing within the sinus, decreasing the resonance.
   d. The mucus solidifies within the sinus, which reduces the bone's ability to vibrate and produce sound.

4. How is the development of skull bones different than the development of facial bones?

   a. The skull bones are developed through endochondral ossification, while facial bones are developed through intramembranous ossification.
   b. The skull bones are developed through intramembranous ossification, while facial bones are developed through endochondral ossification.
   c. The skull bones are developed into one bone and then broken into several bones after development, while facial bones are developed individually.
   d. The skull bones are primarily cartilage after birth but ossify through development, while facial bones are ossified prior to birth.

8.2 Summary

- The skull is composed of facial bones and cranium. The facial bones are located anteriorly on the skull and form the bony structures of your face. The bones of the cranium surround the brain and protect it from damage.
- The skull contains several sinuses that make the skull lighter in weight. These sinuses are filled with air and are connected to the nasal cavity.
- The skull bones develop through intramembranous ossification, while facial bones develop through endochondral ossification.

8.3 The Vertebral Column

The learning objectives for this section are:

8.3.1* Describe the vertebral column as a whole, including the curvatures and regions.

8.3.2 Identify major bone markings (e.g., spines, processes, foramina) on individual vertebrae.

8.3.3* Compare and contrast vertebrae from each region of the spinal column.

8.3.4 Identify and describe the intervertebral joints.

* Objective is not a HAPS Learning Goal.

8.3 Questions

1. Which of the following are true of the curves and regions of the vertebral column? Select all that apply.
   a. The thoracic curve is the only anterior (kyphotic) curve in the vertebral column.
   b. The cervical and lumbar regions have posterior (lordotic) curves.
   c. The thoracic region contains the largest number of vertebrae in the vertebral column.
   d. The vertebrae in all the spinal regions fuse by the time a person reaches middle age.

Mini Case 1: Low Back Pain

Sheila has been struggling with low back pain for many years. They went to their doctor who, after imaging, said they had spinal stenosis—narrowing of the vertebral foramen.

First, Sheila tried physical therapy for eight weeks, but it did not resolve their symptoms. They went to a neurosurgeon, who suggested a laminectomy to relieve their symptoms. The procedure involves removing the lamina at the site where the spinal canal is narrowed.

1. Why might removing the lamina help relieve Sheila's symptoms?
   a. It would create more space in the vertebral foramen.
   b. It would decrease the vertebral arch to support the spinal cord.
   c. It would remove part of the spinal cord that is the source of pain.
   d. It would increase the space in the intervertebral foramen.

2. Sheila's disorder was found to be in the lumbar region of their vertebral column. How do vertebrae in the lumbar region compare to those of the cervical region? Select all that apply.
   a. Lumbar vertebrae are smaller than cervical vertebrae.
   b. Lumbar vertebrae contain transverse processes, but cervical vertebrae do not.
   c. Cervical vertebrae contain transverse foramina, but lumbar vertebrae do not.
   d. The spinous processes of cervical and lumbar vertebrae are bifid.

3. Which of the following would be considered intervertebral joints? Select all that apply.
   a. Intervertebral discs between adjacent vertebrae
   b. Joints between adjacent transverse foramina
   c. Joints between spinous processes of adjacent vertebrae
   d. Joints formed by the superior and inferior articular processes of adjacent vertebrae

8.3 Summary

- There are five regions of the vertebral column: cervical, thoracic, lumbar, sacral, and coccygeal. The cervical and lumbar regions have lordotic curves, and the thoracic and sacral/coccygeal regions have kyphotic curves.
- A typical vertebra consists of a vertebral body, pedicles, laminae, a spinous process, transverse processes, vertebral foramen, and intervertebral foramen.
- Two typical vertebrae are connected at three joints: two between the superior and inferior articular processes of the adjacent vertebrae, and one intervertebral disc connecting the bodies.

8.4 The Thoracic Cage

The learning objectives for this section are:

8.4.1* Identify the three components of the sternum, its position in the thoracic cage, and its features.

8.4.2* Compare and contrast true ribs, false ribs, and floating ribs.

8.4.3* Identify the bony features of an individual rib.

* Objective is not a HAPS Learning Goal.

8.4 Questions

1. Find the clavicle on your body. Palpate medially along the clavicle all the way until you are in the center of your chest. What is this structure called?

   a. Body of the sternum
   b. Suprasternal notch
   c. Sternal angle
   d. Xiphoid process

2. Maribel attempted to dismount from a piece of gymnastics equipment by doing a backflip. Instead of landing on their feet, however, Maribel landed on their back and immediately felt a sharp pain on the side of their back. After imaging, the doctors said Maribel had broken their 11th rib. Is this considered a true rib, a false rib, or a floating rib?

   a. True rib
   b. False rib
   c. Floating rib

3. Which of the following structures of the rib attaches to the transverse process of the same numbered vertebra?

   a. Head
   b. Neck
   c. Tubercle
   d. Costal groove

8.4 Summary

- The thoracic cage is composed of the sternum (anteriorly) and the ribs (which connect the vertebrae to the sternum).
- The sternum consists of the manubrium, body, and xiphoid process.
- The ribs are classified as true ribs (ribs 1–7) and false ribs (ribs 8–12.) Floating ribs (ribs 11 and 12) are a subset of the false ribs.

9

The Appendicular Skeleton

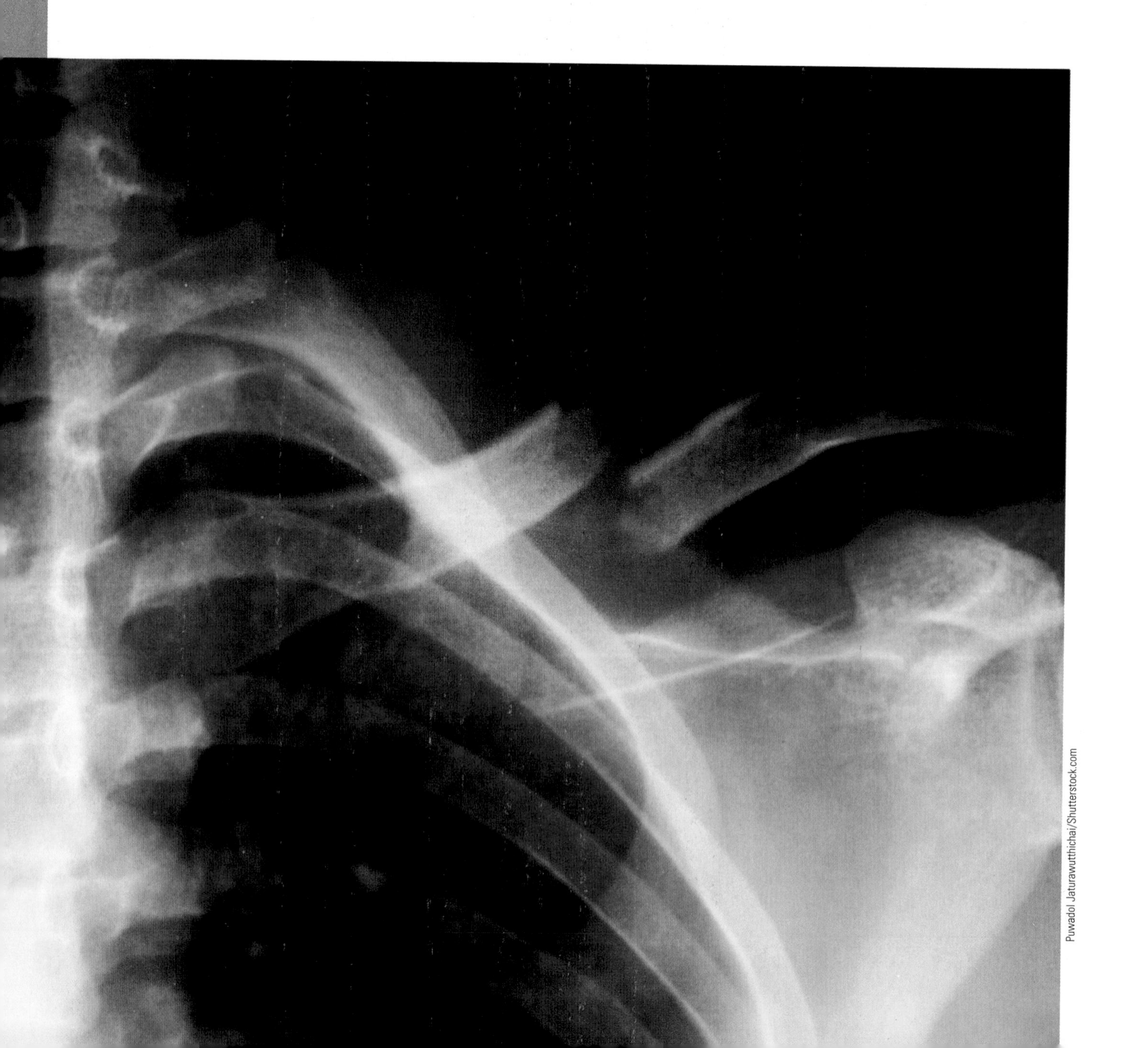

Chapter Introduction

The limbs of the body are the most complex regions of musculoskeletal anatomy; however, just as we discuss in every chapter, they have evolved as a marriage of structure and function. The appendicular skeleton, or the skeleton of the limbs, has 126 adult bones (**Figure 9.1**). In Chapter 8, when we learned the bones of the axial skeleton, we discussed that the division between axial and appendicular is based in function rather than location. For example, the *scapulae*, the shoulder blades, are located on the trunk of the body, lying over the back of the ribcage. They are not, however, part of the axial skeleton; the scapulae are considered part of the appendicular skeleton, and covered in this chapter, because they assist with the movement of the arm. The bones of the pelvis are considered appendicular for the same reason. Because humans stand upright and walk bipedally, the upper limbs and lower limbs have vastly different functional requirements. Early humans were walkers and runners; thus, the bones of the lower limbs are adapted for weight-bearing support, stability, and locomotion. In contrast, our upper limbs were not required for bearing the weight of the body. Instead, our arms evolved for throwing and carrying and are therefore highly mobile and can be utilized for a wide variety of activities. The large range of upper limb movements, coupled with the ability to easily manipulate objects with our hands and opposable thumbs, has allowed humans to construct the modern world in which we live.

The bones of the appendicular skeleton develop via endochondral ossification (a process of developing bone from a cartilage model discussed in more detail in Chapter 7) with one exception, the clavicle, which forms by intramembranous ossification. The development of the

In this chapter we will learn . . .

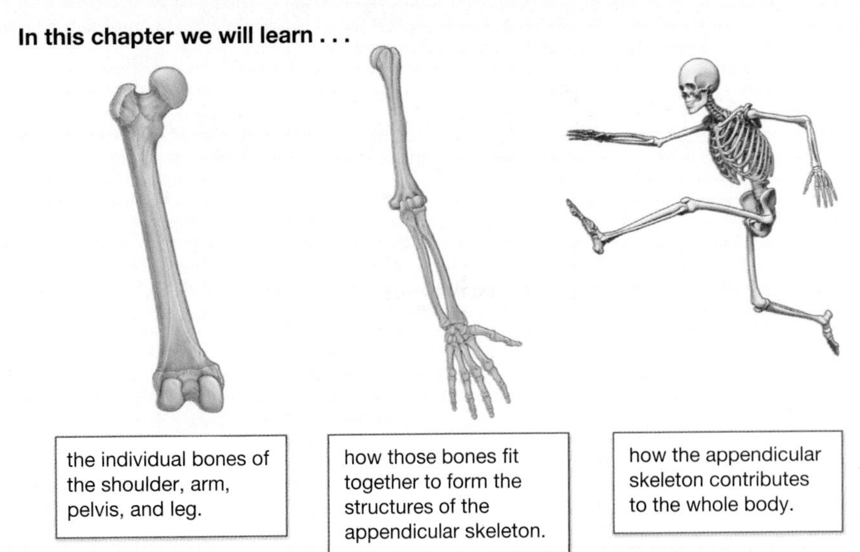

- the individual bones of the shoulder, arm, pelvis, and leg.
- how those bones fit together to form the structures of the appendicular skeleton.
- how the appendicular skeleton contributes to the whole body.

Figure 9.1 The Appendicular Skeleton

The appendicular skeleton consists of the shoulder, pelvis, limb, wrist, ankle, hand, and foot bones.

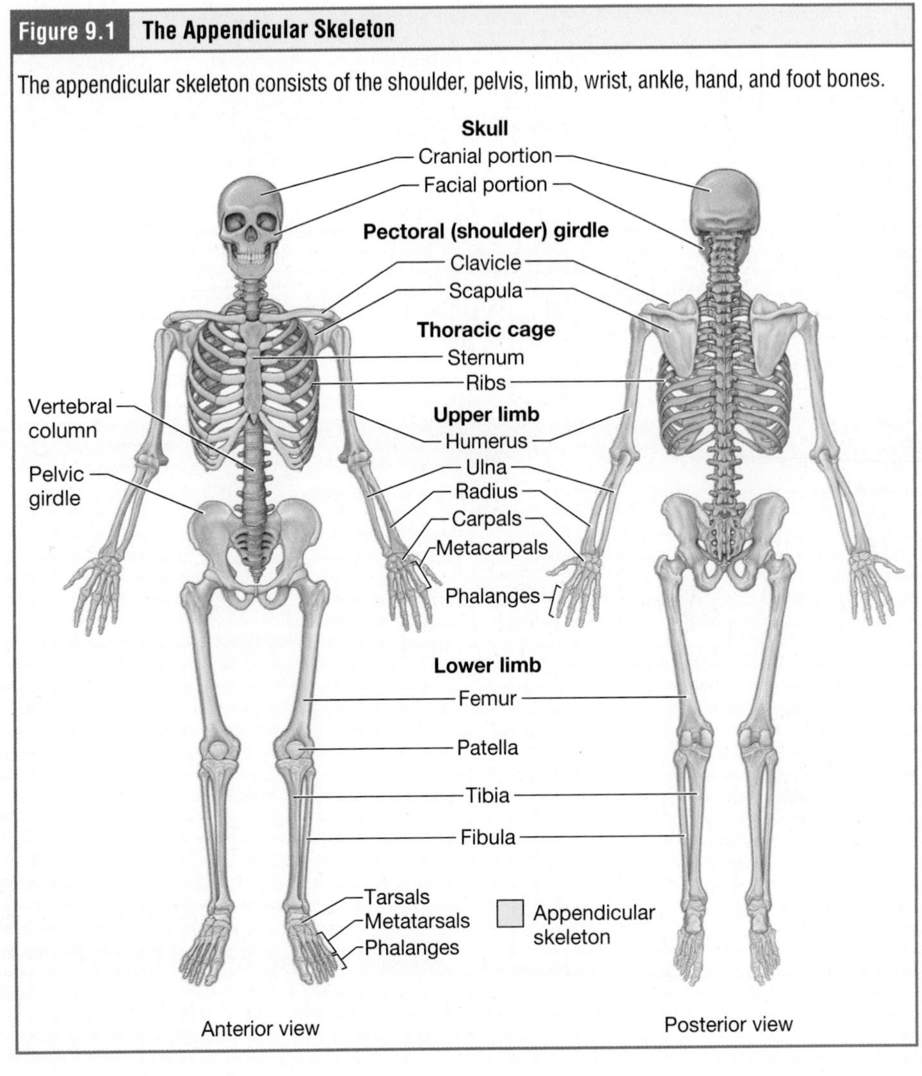

(*Continued*)

appendicular skeleton is one that begins very early in embryonic development but isn't completed until approximately age 25. Therefore, it is likely that some of the readers of this book are still in the process of developing and growing their appendicular skeletons! The major events of skeletal development are outlined in **Figure 9.2**.

This chapter is divided into our exploration of the upper and then lower limbs. Each limb has a "girdle"—the supportive framework that ties it to the axial skeleton—and then the free limb. The limbs follow a pattern (illustrated in the "Anatomy of a Limb" feature); each starts with a single bone in the proximal limb, then a hinge joint (the elbow and knee). The distal limb is composed of two bones, then a more complex joint that contains short bones (wrist and ankle) before the hands and the feet, which share many structural properties. Learning the limbs is a great opportunity to practice chunking and group structures with similar functions. In this chapter we will often use the word "articulate" as a verb; in this case it means "forms a joint."

Figure 9.2 Timeline of Skeletal Development

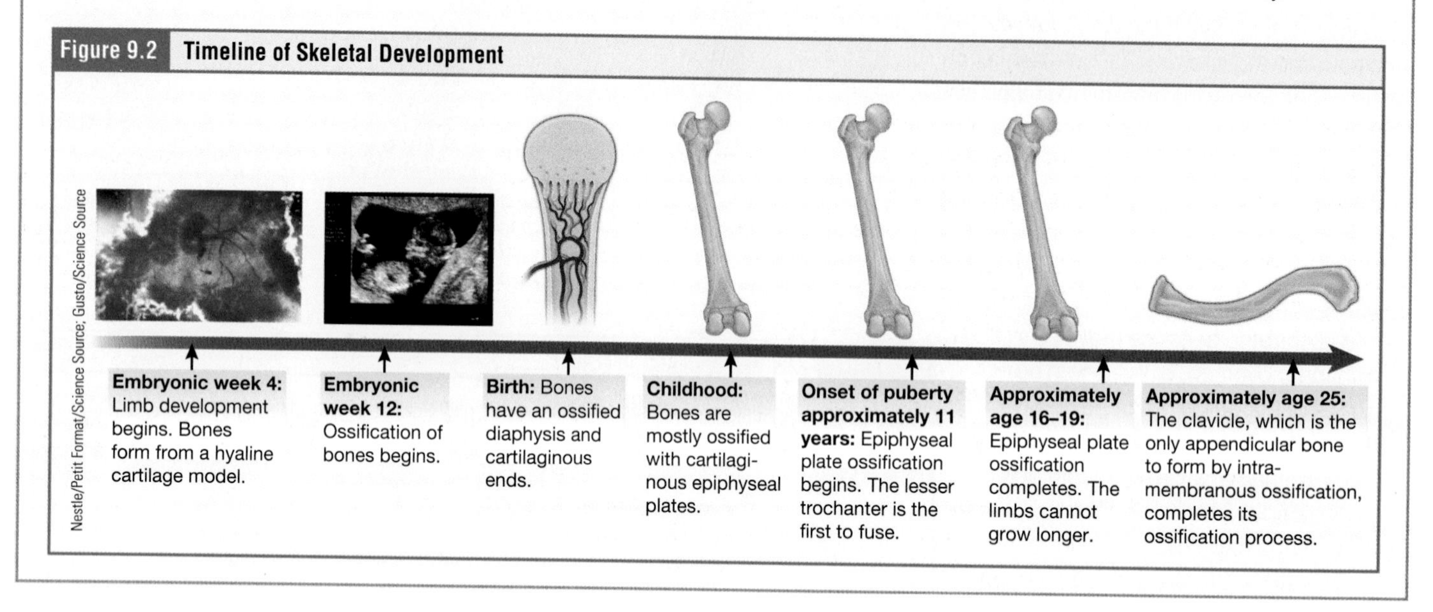

Embryonic week 4: Limb development begins. Bones form from a hyaline cartilage model.
Embryonic week 12: Ossification of bones begins.
Birth: Bones have an ossified diaphysis and cartilaginous ends.
Childhood: Bones are mostly ossified with cartilaginous epiphyseal plates.
Onset of puberty approximately 11 years: Epiphyseal plate ossification begins. The lesser trochanter is the first to fuse.
Approximately age 16–19: Epiphyseal plate ossification completes. The limbs cannot grow longer.
Approximately age 25: The clavicle, which is the only appendicular bone to form by intramembranous ossification, completes its ossification process.
Nestle/Petit Format/Science Source; Gusto/Science Source

Anatomy of...

A Limb

Most mammals have a similar pattern of construction in their limbs, both upper and lower. This pattern is used in both the arm and leg of humans.

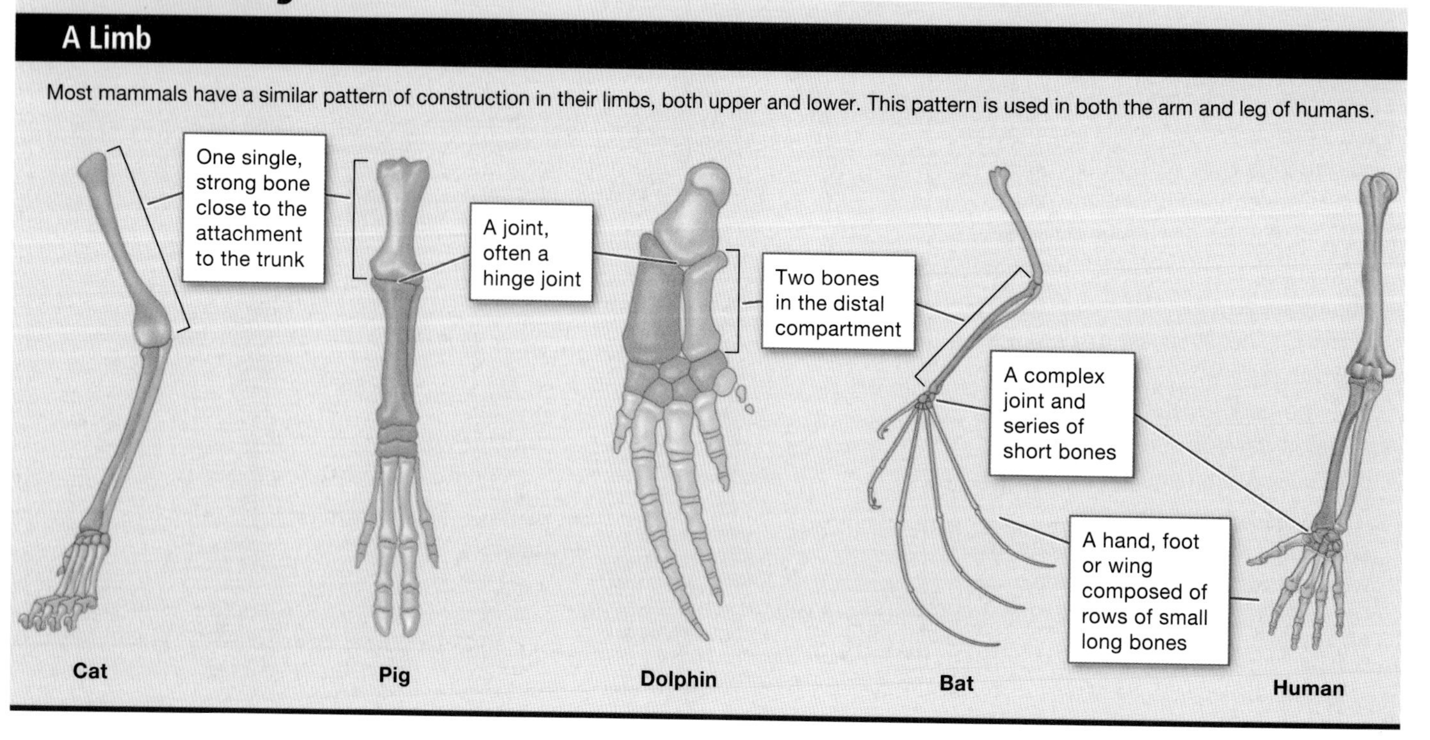

9.1 The Shoulder Girdle

Learning Objectives: By the end of this section, you will be able to:

9.1.1* Identify the bones and bone features of the shoulder girdle.

9.1.2* List the functions of each of the bones of the pectoral girdle.

* Objective is not a HAPS Learning Goal.

The bones that attach each upper limb to the axial skeleton form the **shoulder girdle**. This consists of two bones, the scapula and clavicle (**Figures 9.3A** and **9.3B**). The **clavicle** is an S-shaped bone that connects the sternum to the shoulder. Its medial end articulates with the sternum and its lateral end articulates with the scapula (**Figure 9.3C**). You can easily palpate, or feel with your fingers, the entire length of your clavicle. The clavicle functions to push the shoulder away from the rib cage and prevents anterior collapse even during moments of anterior compression such as doing a push-up or swinging a tennis racket. The **scapula** anchors the shoulder posteriorly. The scapula is a flat, triangular-shaped bone with a prominent ridge running across its posterior surface (see Figure 9.3B). At its lateral edge this ridge forms the bony tip of the shoulder and joins with the lateral end of the clavicle. By following along the clavicle, you can palpate out to the bony tip of the shoulder, and from there, you can move back across your posterior shoulder to follow the ridge of the scapula. Move your shoulder around and feel how the clavicle and scapula move together as a unit. The site where they attach is the **acromioclavicular joint**. These bones serve as important attachment sites for muscles that aid with movements of the shoulder and arm. If you examine their relationship from a superior view (see Figure 9.3C) you can see that together the clavicle and scapula serve to anchor and support the movements of the upper limb from the anterior and posterior.

LO 9.1.1

shoulder girdle (pectoral girdle)

clavicle (collarbone)

scapula (shoulder blade)

acromioclavicular joint (AC joint)

Figure 9.3 The Shoulder Girdle

The shoulder girdle is the anchor for the upper limb, consisting of the clavicle and the scapula. It is shown in three views: (A) Anterior view; (B) Posterior view; and (C) Superior view.

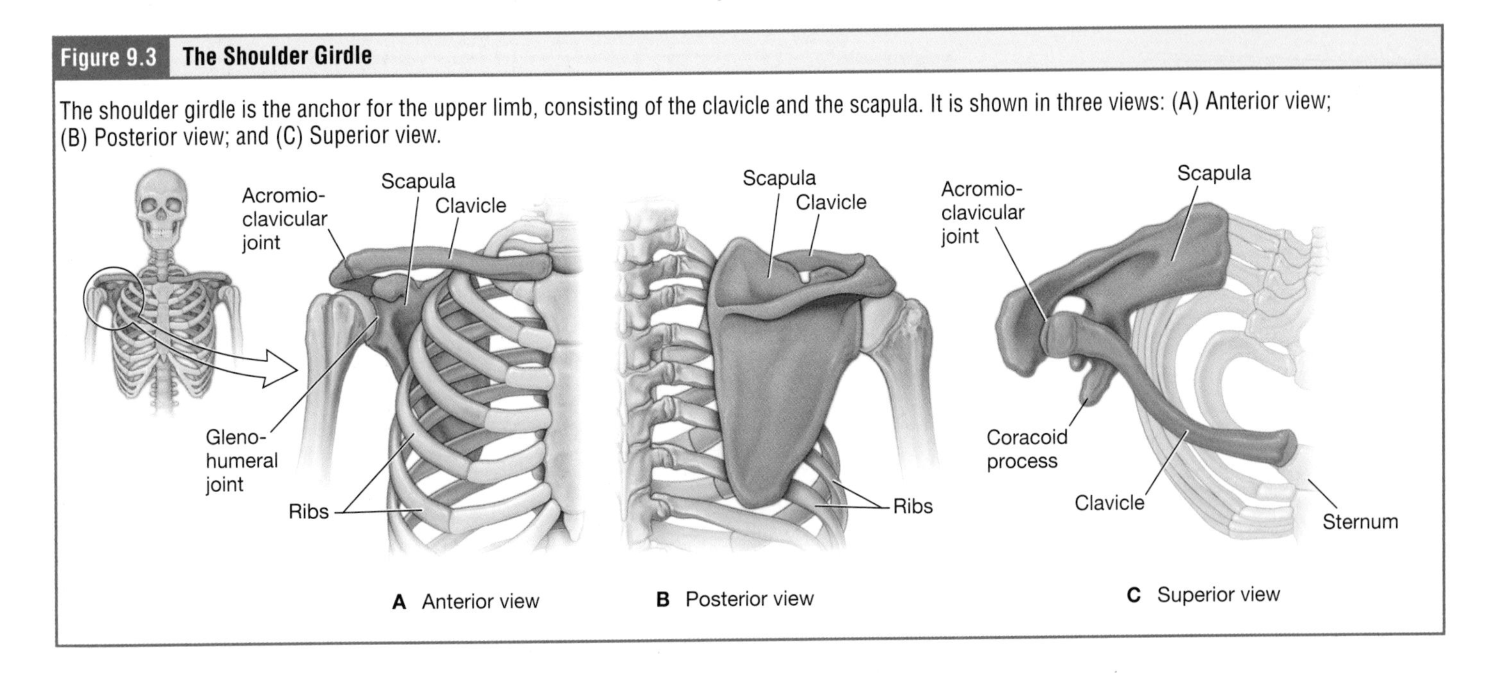

The Human Anatomy and Physiology Society includes more than 1,700 educators who work together to promote excellence in the teaching of this subject area. The HAPS A&P Learning Outcomes measure student mastery of the content typically covered in a two-semester Human A&P curriculum at the undergraduate level. The full Learning Outcomes are available at https://www.hapsweb.org.

9.1a Clavicle

LO 9.1.2

The clavicle has several important functions. The clavicle is technically a long bone, but it shares function with long bones and flat bones. Its function in movement is to support the scapula and to hold the shoulder joint superiorly and laterally away from the ribcage, allowing for maximal freedom of motion for the arm. The clavicle also serves an essential function to protect the underlying nerves and blood vessels as they pass between the trunk of the body and the upper limb.

The **sternal end of the clavicle** has a triangular shape and articulates with the manubrium portion of the sternum (**Figure 9.4**). This juncture between the medial end of the clavicle and the lateral edge of the sternum forms the **sternoclavicular joint.** This joint allows considerable mobility, enabling the clavicle to move in many directions during shoulder movements. The lateral or **acromial end of the clavicle** articulates with the acromion of the scapula, the portion of the scapula that forms the bony tip of the shoulder.

sternal end of the clavicle (medial end of the clavicle)

sternoclavicular joint (SC Joint)

acromial end of the clavicle (lateral end of the clavicle)

The clavicle is one of the bones that can be used in forensics when trying to decide if an adult skeleton is from a biologically male body (that is to say, a body that matured under the influence of higher levels of testosterone) or a biologically female body (that is to say, a body that matured under the influence of higher levels of estrogen). Whereas some clavicles (those from biological females) tend to be shorter, thinner, and less curved, other clavicles (those from the bodies of biological males) tend to be longer and have a larger diameter, greater curvatures, and rougher surfaces where muscles attach.

Learning Connection

Broken Process

Due to a problem during prenatal development, some people are born without clavicles. How would this affect a person's posture?

9.1b Scapula

Along with the clavicle, the scapula forms the other half of the pectoral girdle and thus plays an important role in anchoring the upper limb to the body. The scapula is located on the posterior side of the shoulder. It is surrounded by muscles on both its anterior (deep) and posterior (superficial) sides, and thus does not articulate directly with the ribs.

The scapula has several important landmarks (**Figure 9.5**). The three borders of the scapula, named for their positions within the body, are the **superior border**

Figure 9.4 The Clavicle

The clavicle connects the acromion of the scapula to the sternum of the axial skeleton. Therefore, it has a lateral acromial end and a medial sternal end.

Coracoid process
Clavicle
Acromion
Humerus
Scapula
Acromial end
Superior view
Sternal end
Inferior view

Figure 9.5 The Scapula

The scapula is shown here from its anterior side, which faces the ribcage, and its posterior side, which faces muscles and skin of the back.

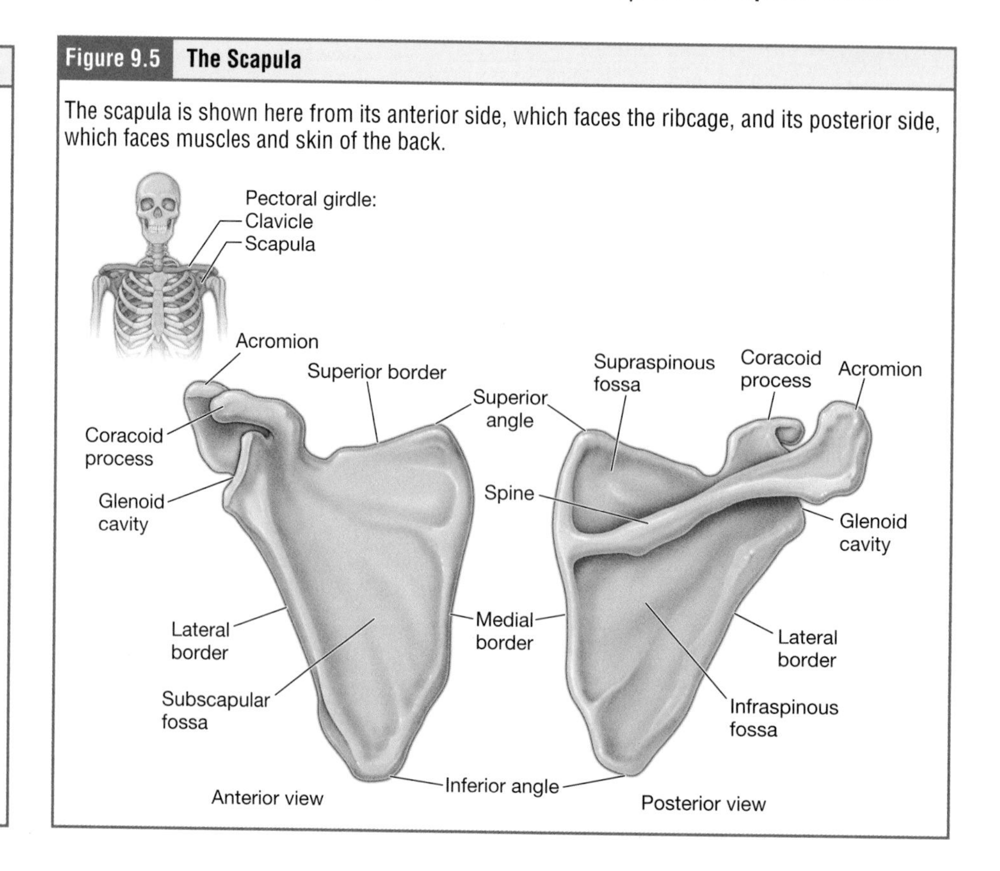

of the scapula, the **medial border of the scapula**, and the **lateral border of the scapula**. The two corners of the triangular scapula are called *angles*, the third corner forms the joint surface. The **superior angle of the scapula** is located between the medial and superior borders, and the **inferior angle of the scapula** is located between the medial and lateral borders. The inferior angle is a particularly significant bone feature, as the most inferior portion of the scapula it serves as the attachment point for several powerful muscles involved in shoulder and upper limb movements. The remaining corner of the scapula, between the superior and lateral borders, is the location of the **glenoid cavity**. This shallow depression cups the head of the humerus, the proximal bone of the arm. Together the humerus and glenoid cavity to form the **glenohumeral joint**.

glenoid cavity (glenoid fossa)

glenohumeral joint (shoulder joint; GH joint)

The scapula also has two prominent projections. Toward the lateral end of the superior border, between the suprascapular notch and glenoid cavity, is the hooklike **coracoid process** (see Figure 9.5, anterior view). This process projects anteriorly and curves laterally. In the superior view (see Figure 9.3C), the coracoid process curves under the lateral end of the clavicle. The coracoid process serves as the attachment site for muscles of the anterior chest and arm. A strong ligament that ties the clavicle to the coracoid process adds stability to the shoulder. On the posterior aspect, the **spine of the scapula** is a long and prominent ridge that runs across its upper portion. Laterally the spine ends in the broad, flat **acromion**. The acromion forms the bony tip of the superior shoulder region and articulates with the lateral end of the clavicle, forming the acromioclavicular joint (see Figure 9.3). Together, the clavicle, acromion, and spine of the scapula form a V-shaped bony line that provides for the attachment of neck and back muscles that act on the shoulder (see Figure 9.3C).

spine of the scapula (scapular spine)

Apply to Pathophysiology

Shoulder Injury

Renee falls while ice skating. As they begin to fall, they stretch out their arms to catch themself. After the fall, their right shoulder is in a lot of pain and they have trouble moving their arm. They hold their arm close to their side as they walk into the ER. Their doctors order X-rays and begin to work through the possibilities with Renee. The doctors tell Renee that the joint between the sternum and the clavicle is very strong, and unlikely to be dislocated, but they are taking images of Renee's clavicle and shoulder girdle to look for damage there.

1. If Renee has a fractured clavicle, which one of the photos would you expect to most closely resemble their posture?
   A) A
   B) B
   C) C

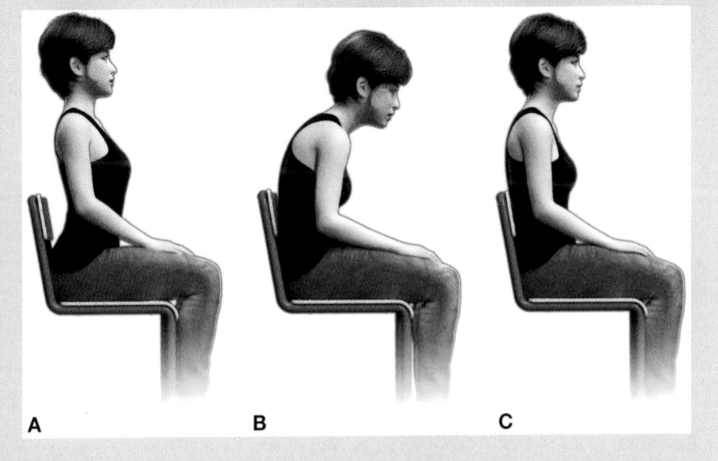

2. Another possibility is that Renee might have a separation of the acromion and clavicle. In this case, would Renee's humerus still be articulated with the glenoid cavity?
   A) Yes
   B) No
3. The X-rays reveal the separation described in Question 2. The doctors tell Renee they will need surgery. In this surgery the doctors will add screws to Renee's distal clavicle to anchor it to the structure immediately below/inferior. Which structure will be used as an anchor?
   A) The acromion
   B) The coracoid process
   C) The subscapular fossa

fossa (cavity)

The scapula has three depressions, each of which is called a **fossa** (plural = fossae). Two of these are found on the posterior surface, one above and one below the scapular spine. They are named for their positions relative to the spine, with the **supraspinous fossa**, named for combining the words *supra* (superior) and *spinous* (spine). The much larger **infraspinous fossa** gets its name by combining *infra* (inferior) with *spinous* (spine). If we turn the scapula over and look at its anterior (deep) surface, as if we were sitting between the scapula and the rib cage, we find the **subscapular fossa**. All of these fossae provide space for the muscles that cross the shoulder joint to act on the humerus.

Learning Check

1. What two bones directly articulate in the glenohumeral joint? Please select all that apply.
   a. Humerus
   b. Scapula
   c. Clavicle
2. If you palpate the spine of your scapula and move your fingers laterally, what structure will you come across?
   a. Coracoid
   b. Clavicle
   c. Acromion
   d. Glenoid Fossa
3. The teres minor is a muscle in the shoulder that is part of the rotator cuff group. It sits inferior to the spine of the scapula. Which of the following depressions would you expect to find the teres minor?
   a. Supraspinous fossa
   b. Infraspinous fossa
   c. Subscapular fossa
   d. Glenoid fossa

9.2 Bones of the Arm

Learning Objectives: By the end of this section, you will be able to:

9.2.1* Identify the bones of the arm and describe the bones features.

9.2.2* List the bones and bony landmarks that articulate at each joint of the upper limb.

* Objective is not a HAPS Learning Goal.

LO 9.2.1

Universally across the animal kingdom, limbs are constructed using a very common layout. Proximally there are fewer, larger bones, and as you move distally down the limb, each region is composed of greater numbers of smaller bones (see the "Anatomy of a Limb" feature). This format allows for stability closer to the trunk and dexterity further away. In humans, the upper limbs and lower limbs are each divided into three regions. For the upper limbs, these regions consist of the **brachium**, located between the shoulder and elbow joints; the **antebrachium**, which is between the elbow and wrist joints; and the **hand**, which is located distal to the wrist (**Figure 9.6**). The brachium contains just a single bone, the **humerus**. The antebrachium contains a pair of bones, the **ulna** and the **radius**. The base of the hand contains eight bones, each called a **carpal bone**, and the palm of the hand is formed by five bones, each called a **metacarpal bone**. The fingers and thumb contain a total of 14 bones, each of which is a **phalanx bone of the hand**.

9.2a Humerus

LO 9.2.2

The humerus is the single bone of the upper arm (**Figure 9.7**). Its proximal end is almost entirely covered by the smooth, rounded **head of the humerus**. This faces medially and articulates with the glenoid cavity of the scapula to form the glenohumeral joint. The head forms a smooth cap to the humerus like an upside-down bowl;

its margins, where the smooth head meets the rougher and narrower rest of the humerus, is known as the **anatomical neck**. In this bone and others, the *neck* is usually a narrower region of the bone, often lateral or distal to the *head* of the bone. If you continue to trace the humerus laterally beyond the anatomical neck, you will find several rough bony ridges. Here and in other bones, rough protrusions are the sites of muscle attachment. The more force that those muscles pull with, the larger and rougher the projections. On the lateral side of the proximal humerus you can find the larger of these projections, the **greater tubercle**. The smaller **lesser tubercle** of the humerus can be seen clearly on the anterior side of the humerus. Both the greater and lesser tubercles serve as attachment sites for muscles that act across the shoulder joint. Between the greater and lesser tubercles is the narrow and deep **bicipital groove**, which provides passage for a tendon of the biceps brachii muscle.

bicipital groove (Intertubercular groove, sulcus intertubercularis)

Proceeding down toward the diaphysis of the humerus, the wide proximal region of the humerus narrows again as the epiphysis becomes the diaphysis or **shaft of the humerus**. This narrowing is a common site of

Figure 9.7 Humerus and Elbow Joint

(A) The humerus is the single bone of the brachium. (B) Its distal end has articulating surfaces where it forms the elbow joint with the radius and ulna.

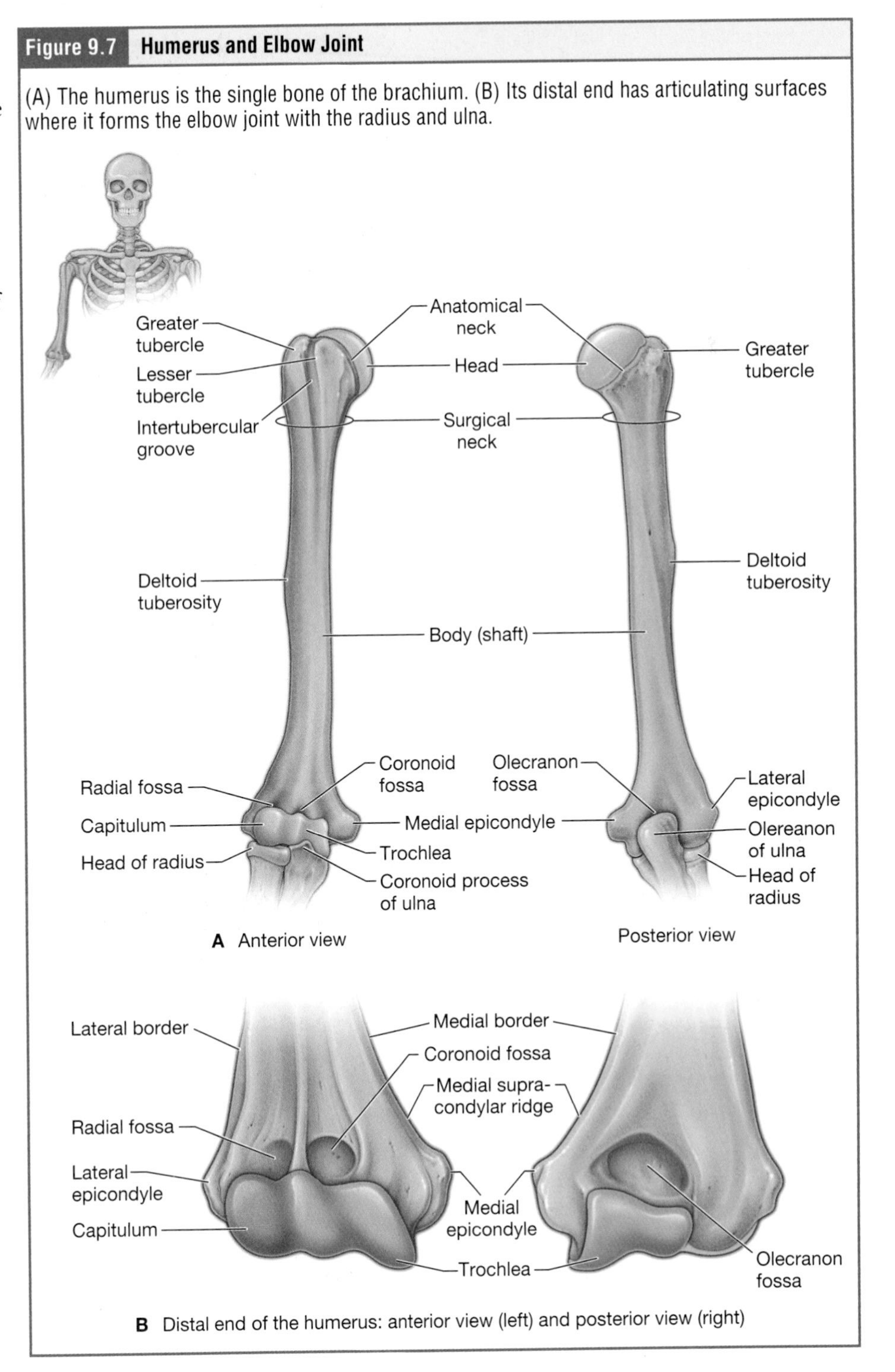

Figure 9.6 The Regions of the Arm

The arm can be divided into three major regions—(A) the brachium, (B) the antebrachium, and (C) the hand—with two major functional joints, the elbow and the wrist. There are also many minor joints between individual bones.

Arm (brachium)
Humerus
Forearm (antebrachium)
Ulna
Radius
Carpals
Metacarpals
Hand
Phalanges

arm fractures and therefore is called the **surgical neck** since it sometimes is a site of fracture-repairing surgeries. The shaft is mostly smooth and unremarkable, except for a roughened protrusion on the lateral middle called the **deltoid tuberosity**. As its name indicates, it is the site of attachment for the deltoid muscle.

Distally, the humerus contributes to the elbow joint and therefore has a number of anatomically-relevant features. Let's start with the anterior view. The medial and lateral sides of the humerus each have a prominent bony projection called an *epicondyle*. Both of these can be palpated (felt) if you wrap your hand around your arm just above the elbow joint. The **medial epicondyle of the humerus** is quite a bit larger than the **lateral epicondyle of the humerus** (**Figure 9.7B**). The roughened ridge of bone above the lateral epicondyle is the **lateral supracondylar ridge**. All of these areas are attachment points for muscles that act on the forearm, wrist, and hand. The powerful grasping muscles of the anterior forearm are attached at the medial epicondyle, which is therefore larger and more robust than the lateral epicondyle that provide attachment for the weaker posterior forearm muscles.

Learning Connection

Micro to Macro

In a given long bone, such as the humerus, where could you find cartilage and where would you find bone? Would your answer be different if you were asked about a 9-year-old humerus and a 90-year-old humerus?

The distal end of the humerus has two smooth areas for articulation with the ulna and radius bones of the forearm to form the **elbow joint**. The more medial of these areas is the **trochlea** which articulates with the ulna bone. Lateral to the trochlea is the **capitulum**, a knoblike structure located on the anterior surface of the distal humerus. The capitulum articulates with the radius bone of the forearm. These two articulation surfaces can be compared in Figure 9.7B. The trochlea, which means "pulley" in Latin, is a gliding surface that spans the anterior and posterior sides of the ulna which, as you will read about in the next section, pivots around this smooth surface. The proximal end of the ulna cups the trochlea and so there are small depressions in the humerus to provide space to accommodate the ulna and radius when the elbow is fully bent (flexed). The processes of the ulna fit into the **coronoid fossa** on the anterior side and the **olecranon fossa** on the posterior. Above the capitulum is the **radial fossa**, which provides space for the head of the radius when the elbow is flexed.

capitulum (lateral condyle of the humerus)

Student Study Tip

The root "capit" means head, and the capitulum's shape resembles a head.

9.2b Radius and Ulna

The forearm fits the limb pattern by having two bones (see the "Anatomy of a Limb" feature). The ulna is the medial bone of the forearm. It runs parallel to the radius, which is the lateral bone of the forearm (**Figures 9.8A** and **9.8B**). The proximal end of the ulna resembles the letter C with its large **trochlear notch** (**Figure 9.8C**). This region fits around the trochlea of the humerus like a baseball glove around a baseball to form the hinge of the elbow joint. The anterior edge of the trochlear notch forms a prominent lip of bone called the **coronoid process of the ulna**. The posterior and superior portions of the proximal ulna make up the **olecranon process**, which forms the bony tip of the elbow. Funnily enough, if you hit your "funny bone," which isn't funny at all but actually quite painful, you're not hitting your humerus, but rather a vulnerable place where a major nerve, the ulnar nerve, passes along your olecranon process.

Student Study Tip

The coronoid process can be easily mixed up with the coracoid process, but you can use their spellings to help. Coro ***n*** oid goes with the ul ***n*** a; both contain the letter N. Cora **c** oid goes with the s **c** apula; both contain the letter C.

The radius and ulna are parallel to each other, and they articulate together at each of their ends. Their shafts bow out to the side slightly and do not touch when the arm is in anatomical position. Between the radius and ulnar shafts, a thin but tough membrane keeps the bones tied together. Because this membrane is between (inter) bones (osseus), it is the **interosseous membrane of the forearm**. Remember that many long bones have a broad or rounded head, and this is true of the radius and ulna as well. The **head of the radius** is proximal and the **head of the ulna** is distal. The head of the radius articulates with the capitulum of the humerus. At their proximal joint, the head

Figure 9.8 The Radius and Ulna

The radius and ulna make up the antebrachium. (A) The medial side of the radius and lateral side of the ulna are attached to each other by an interosseous membrane. (B) This view of the ulna shows the large olecranon process that fits into the olecranon fossa of the humerus when the arm is extended at the elbow. (C) A close-up view of the proximal ulna shows the trochlear notch, which cups the distal end of the humerus in the elbow joint.

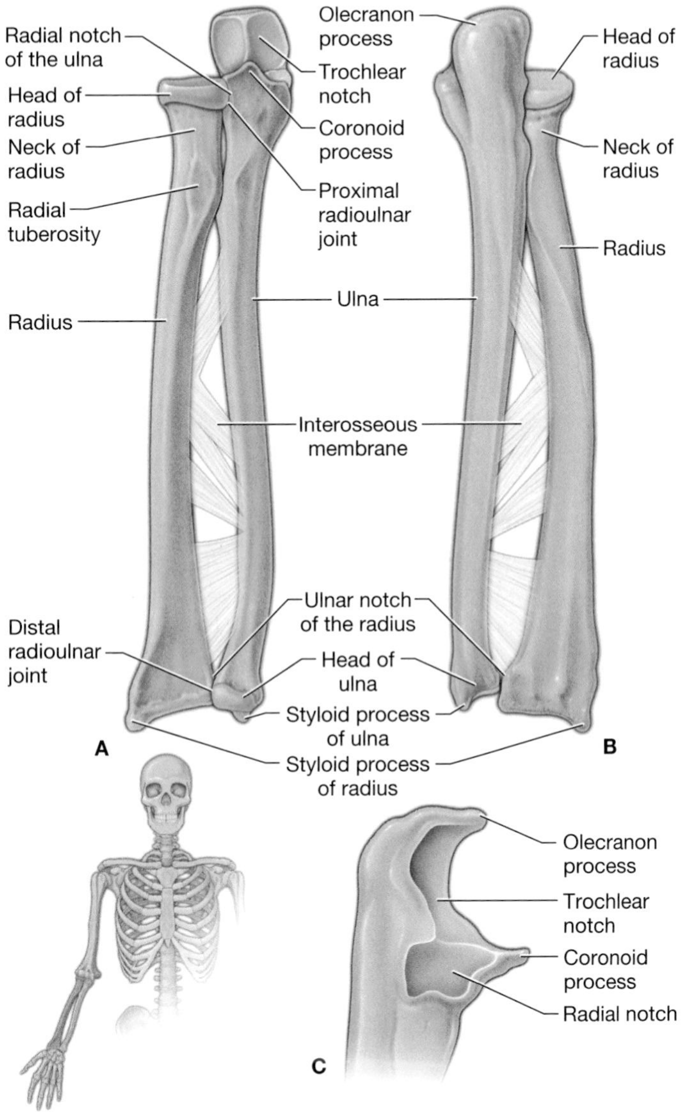

of the radius sits nestled in the **radial notch of the ulna**, forming the **proximal radio-ulnar joint**. This joint can best be understood by examining a skeleton, but let's use our own arms for a moment. Wrap the fingers of one hand around your opposite forearm about halfway between the elbow and the wrist. Start with the inside of your wrist and palm (anterior) facing you, then turn your hand over so that you can see the back of your hand. Did you feel how the radius and ulna twist so that the radius comes up and rolls over the ulna? This movement, called **pronation**, is illustrated in **Figure 9.9**. Now look back at the proximal radio-ulnar joint in Figure 9.8. As the wrist pronates and the distal radius moves, the head of the radius is locked into place at the radial notch by a tough ligament. Within the pocket created by this ligament, the radial head rotates.

Figure 9.9 Supination and Pronation of the Forearm

The forearm is capable of supination and pronation, which allows us to flip the orientation of the hand. This movement is accomplished by the rotation of the radial head, which causes the distal radius to move over and across the distal ulna.

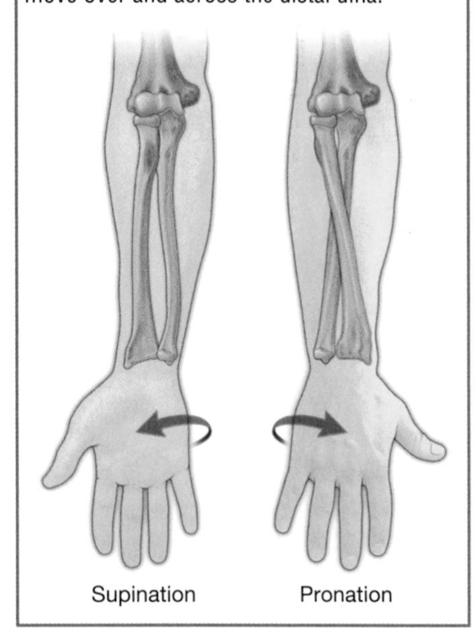

Learning Connection

Quiz Yourself

Both the radius and the ulna contain a styloid process at their distal ends. Which skull bone also contains a styloid process? What structural feature is common to all of these styloid processes?

Student Study Tip

A helpful mnemonic for remembering the arrangement of the carpal bones is "So Long To Pinky, Here Comes The Thumb." This mnemonic starts on the lateral side and names the proximal bones from lateral to medial (scaphoid, lunate, triquetrum, pisiform), then makes a U-turn to name the distal bones from medial to lateral (hamate, capitate, trapezoid, trapezium). Thus, it starts and finishes on the lateral side.

Below the head, the radius narrows slightly. As is the pattern, the narrow region of the bone just after the head is the **neck of the radius**. Inferior to the neck we see a rough protuberance, the **radial tuberosity**, which serves as a muscle attachment point (see Figures 9.8 and 9.9). Now reexamine Figure 9.9. We have mentioned previously that, as the forearm pronates, the radial head rotates in place. In Figure 9.9 the radial tuberosity is marked with an asterisk. Note its location in **supination** (anatomical position) and then in pronation.

On the medial side of the distal radius is the **ulnar notch of the radius**. This shallow depression articulates with the head of the ulna, which together form the **distal radioulnar joint**. At their distal ends, each bone has a delicate inferior projection that reaches toward the wrist. These are the **styloid process of the radius** and the **styloid process of the ulna** (**Figure 9.10A**). The styloid process of the radius provides attachment for ligaments that support the lateral side of the wrist joint. The styloid process of the radius projects quite far distally, and limits lateral the range of motion of the hand. The styloid process of the ulna serves as an attachment point for a connective tissue structure that connects the distal ends of the ulna and radius.

The distal end of the radius has a smooth surface for articulation with two carpal bones to form the **radiocarpal joint** or wrist joint (see Figure 9.10A). The ulna does not technically contribute to the wrist joint; it is slightly shorter and separated from the carpal bones by a fibrocartilage disc (**Figure 9.10B**).

9.2c Bones of the Wrist and the Hand

The wrist and base of the hand are formed by a series of eight small carpal bones (see Figures 9.10A and 9.10B). The carpal bones are arranged in two rows, forming a proximal row of four carpal bones and a distal row of four carpal bones. The bones in the proximal row, running from the lateral (thumb) side to the medial side, are the **scaphoid** ("boat-shaped"), **lunate** ("moon-shaped"), **triquetrum** ("three-cornered"), and **pisiform** ("pea-shaped") bones. The small, rounded pisiform bone articulates with the anterior surface of the triquetrum bone. The pisiform thus projects anteriorly, where it forms the bony bump that can be felt at the medial base of your hand. The distal bones (lateral to medial) are the **trapezium** ("table"), **trapezoid** ("resembles a table"), **capitate** ("head-shaped"), and **hamate** ("hooked bone") bones. The hamate bone is characterized by a prominent bony extension on its anterior side called the **hook of the hamate bone**.

The palm of the hand contains five elongated metacarpal bones. These bones lie between the carpal bones of the wrist and the bones of the fingers and thumb

Cultural Connection

Achondroplasia

Achondroplasia literally translates to "without cartilage growth." This genetic condition results in bodies with typical-length torsos and shorter limbs. Achondroplasia can be inherited through individuals who are affected or are carriers of the genetic mutation, or the genetic mutation can arise spontaneously during gamete formation. The mutation occurs in a single gene, and the result is that growth at the epiphyseal plates of long bones is slowed and then arrested earlier than in individuals without achondroplasia. Because long bones are found primarily in the limbs, this results in individuals with shorter limbs.

Most individuals with achondroplasia live full and long lives, but the onset of motor functions such as crawling and walking may be delayed in children. Individuals with achondroplasia may struggle with physical tasks simply because so many things in the world are designed and built for larger bodies. From cars to toilets, individuals with achondroplasia face daily challenges. There is a lot that the design world can do to make our environment more accessible for individuals of all heights, especially in high-traffic areas such as schools, airports, and grocery stores.

Figure 9.10 Bones of the Wrist and Hand

(A) The wrist is composed of eight carpal bones arranged in two rows. The metacarpal bones form the palm of the hand, and the phalanx bones (phalanges) compose the thumb and fingers. (B) A coronal section through the wrist illustrates the radiocarpal joint. The radius articulates with the scaphoid and lunate. The ulna is shielded from the carpals by the articular disc and does not directly articulate with them.

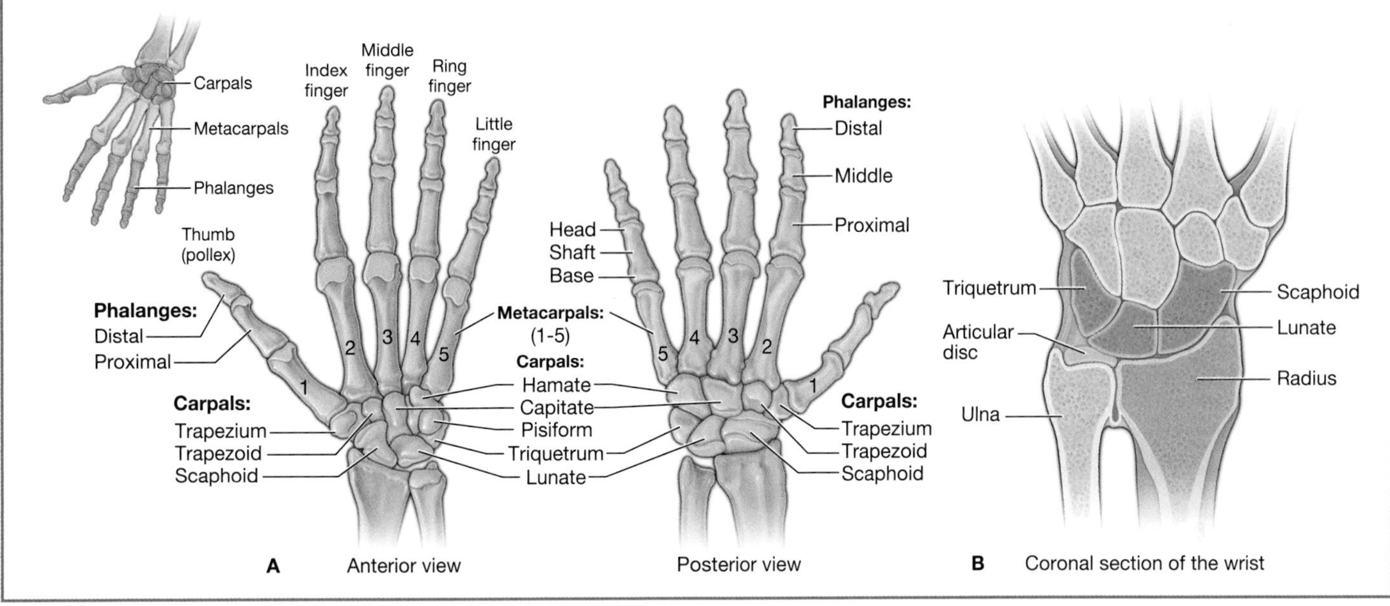

(see Figure 9.10). The metacarpal bones are numbered 1–5, beginning at the thumb. The first metacarpal bone, at the base of the thumb, is separated from the other metacarpal bones. This allows it a freedom of motion that is independent of the other metacarpal bones, which is very important for thumb mobility. The remaining metacarpal bones are united together to form the palm of the hand.

The fingers and thumb contain 14 bones, each of which is called a *phalanx bone* (plural = phalanges). Digit 1 (the thumb, **pollex**) has two phalanges which are called a *proximal phalanx bone* and a *distal phalanx bone*. Digits 2 (index finger) through 5 (little finger) have three phalanges each; between the proximal and distal phalanges is a third bone, the *middle phalanx* (see Figure 9.10).

pollex (thumb)

Learning Connection

Chunking

List all of the bones of the arm. Which ones are classified as long bones? How do we classify the others?

DIGGING DEEPER: Fractures of Upper Limb Bones

When we fall, our first instinct is often to try to "catch" ourselves by stretching out a hand in the direction of the fall. This may be a hard-wired instinct to protect our more vulnerable organs, but often this fall onto an outstretched hand results in fractures of the upper limb bones.

Falls onto the hand or elbow can result in fractures of the humerus (see Figure A) as the force of the impact is transmitted up the limb. The surgical neck of the humerus is a frequent location of fracture, as the distal portion of the humerus is driven into the proximal portion. Falls or blows to the arm can also produce transverse or spiral fractures of the humeral shaft.

In children, a fall onto the tip of the elbow frequently results in a distal humerus fracture. In these, the olecranon of the ulna is driven upward, resulting in a fracture across the distal humerus, a supracondylar fracture.

Another frequent injury following a fall onto an outstretched hand is a Colles fracture ("collees") of the distal radius (see Figure B). This involves a complete transverse fracture across the distal radius that drives the separated distal fragment of the radius posteriorly and superiorly. This is the most frequent forearm fracture and is a common injury in persons over the age of 50, particularly in older individuals with osteoporosis. It also commonly occurs following a high-speed fall onto the hand during activities such as snowboarding or skating.

(*Continued*)

The most commonly fractured carpal bone is the scaphoid, often resulting from a fall onto the hand. Deep pain at the lateral wrist may yield an initial diagnosis of a wrist sprain, but a radiograph taken several weeks after the injury, after tissue swelling has subsided, will reveal the fracture. Due to the poor blood supply to the scaphoid bone, healing will be slow and there is the danger of bone necrosis and subsequent degenerative joint disease of the wrist.

Of particular concern in children is when a fracture occurs at the growth plate (see Figure C). Growth (epiphyseal) plate fractures constitute 20–30 percent of all childhood bone fractures. These fractures can lead to malformations of the bone if growth continues without perfect realignment of the bones. Growth plate fractures therefore more often require surgery than other fractures, which can most often be healed through immobilization in a cast.

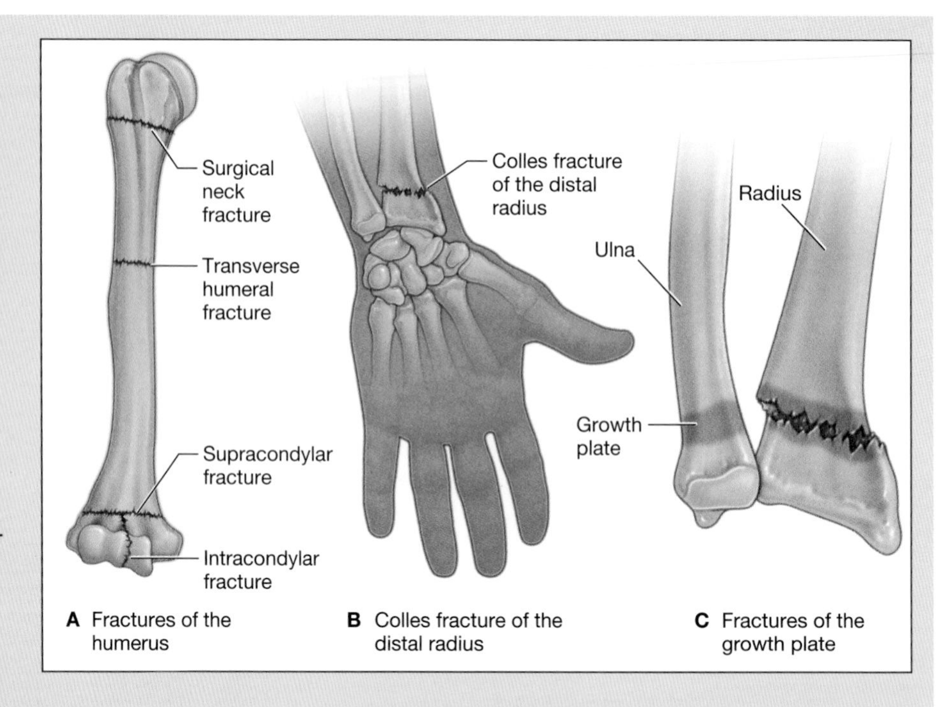

A Fractures of the humerus **B** Colles fracture of the distal radius **C** Fractures of the growth plate

Learning Check

1. Run your fingers distally along the medial side of your upper arm. Proceed in a distal direction until you feel a bump by your elbow. What is this bony landmark called?
   a. Medial epicondyle
   b. Olecranon process
   c. Humeral head
   d. Head of the ulna
2. Julia was playing basketball when the ball hit their finger. They felt a pop and couldn't bend their finger afterward. After getting imaging in the emergency room, the doctors told Julia they had fractured the bones in their finger. What are these bones called?
   a. Carpals
   b. Metacarpals
   c. Phalanges
   d. Clavicles
3. Hold your hand out in front of your and pretend like you are turning a doorknob to open the door. This twisting motion at your forearm can be explained by the ________ rotating within the ________.
   a. Head of the ulna; trochlear notch
   b. Head of the radius; radial notch
   c. Neck of the radius; coronoid process
   d. Head of the radius; ulnar notch

9.3 The Pelvic Girdle and Pelvis

Learning Objectives: By the end of this section, you will be able to:

9.3.1* Describe the three regions of the hip bone and identify their bony landmarks.

9.3.2* Compare and contrast the child and adult pelvis, and the pelvis of biological males and biological females.

9.3.3* Describe the openings of the pelvis and the boundaries of the greater and lesser pelvis.

* Objective is not a HAPS Learning Goal.

os coxae (Coxal bone, hip bone)

The adult **pelvis** is formed by the two hip bones, the sacrum, and, attached inferiorly to the sacrum, the coccyx (**Figure 9.11**). Each right and left hip bone is more properly called an **os coxae**. The os coxae are considered bones of the appendicular skeleton

Figure 9.11 The Adult Pelvis

The adult pelvis is a solid structure made up of four bones—the right and left ossa coxae, the sacrum of the spine, and the coccyx.

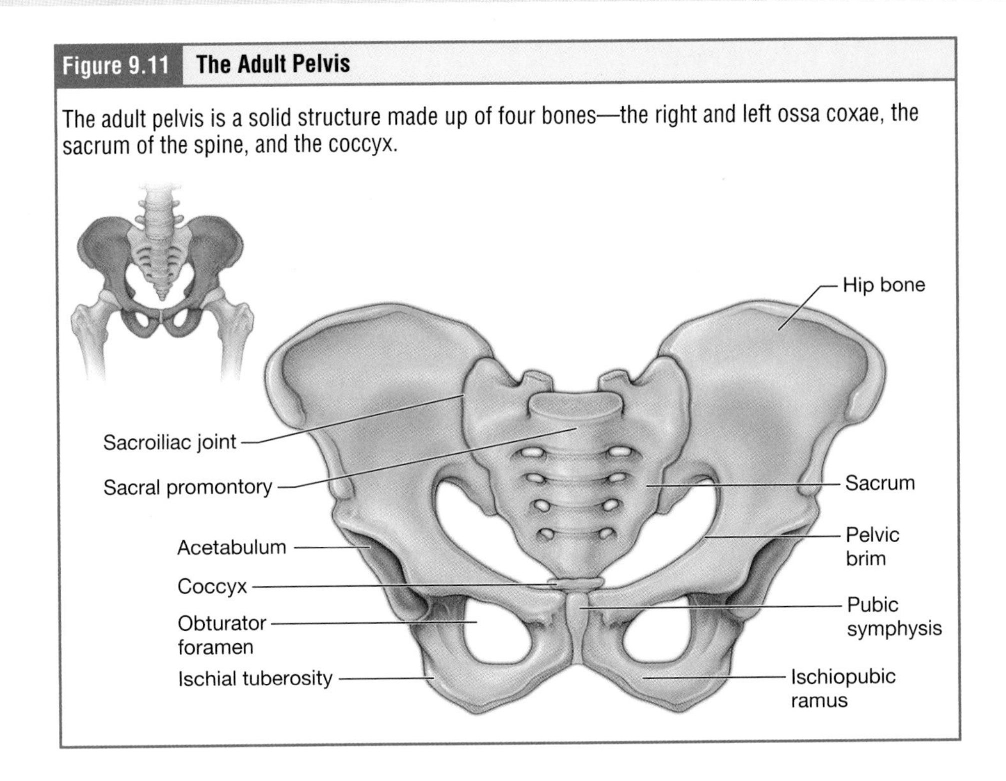

and they are attached to the axial skeleton at their joint with the sacrum of the vertebral column. Anteriorly, the os coxae form a joint with each other called the **pubic symphysis**.

Unlike the bones of the shoulder, which are highly mobile to enable a wide range of arm movements, the bones of the pelvis are strongly united to each other to form a largely immobile, weight-bearing structure. This is important for stability during standing, walking, and running.

9.3a Os Coxae

LO 9.3.1

LO 9.3.2

Each adult os coxae is formed by three separate bones that fuse together during the late teenage years (**Figure 9.12**). The three bones of the childhood pelvis are the *ilium*, the *ischium*, and the *pubis* (**Figure 9.12A**). These names are retained and used to define the three regions of the adult hip bone (**Figure 9.12C**). The superior **ilium** forms the largest part of the hip bone. It is firmly united to the sacrum at the largely immobile **sacroiliac joint** (see Figure 9.11). The **ischium** forms the posteroinferior region of each hip bone; it supports the body when sitting. The **pubis** forms the anterior portion of the hip bone. The pubis curves medially where it joins to the pubis of the opposite hip bone at a specialized joint called the *pubic symphysis*.

sacroiliac joint (SI Joint)

Ilium When you place your hands on your hip, you can feel the arching, superior ridge of the ilium, the **iliac crest** (**Figure 9.13**). The iliac crest ends anteriorly with a rounded rough area, the **anterior superior iliac spine**, which can be felt at your anterolateral hip. Inferior to the anterior superior iliac spine is a rounded protuberance called the **anterior inferior iliac spine**. Both of these iliac spines serve as attachment points for muscles of the thigh. Posteriorly, the iliac crest curves downward to end as the **posterior superior iliac spine**. Muscles and ligaments surround but do not cover this bony landmark, thus sometimes producing a depression seen as a "dimple" located on the lower back. Continuing inferiorly, we find the **posterior inferior iliac spine**. This is located at the inferior end of a large, roughened area called the **auricular surface of the ilium**.

anterior superior iliac spine (ASIS)

posterior superior iliac spine (PSIS)

anterior inferior iliac spine (AIIS)

posterior inferior iliac spine (PIIS)

Figure 9.12 The Child Pelvis

(A) In childhood, there are three individual bones on each side of the pelvis—the ilium, the ischium, and the pubis. (B) An X-ray of child's pelvis. (C) The lateral view and (D) anterior view of the adult os coxae. Each adult os coxae does not form until adolescence. In the adult os coxae, we still use the names *ilium*, *ischium*, and *pubis* to refer to their adult regions after fusion.

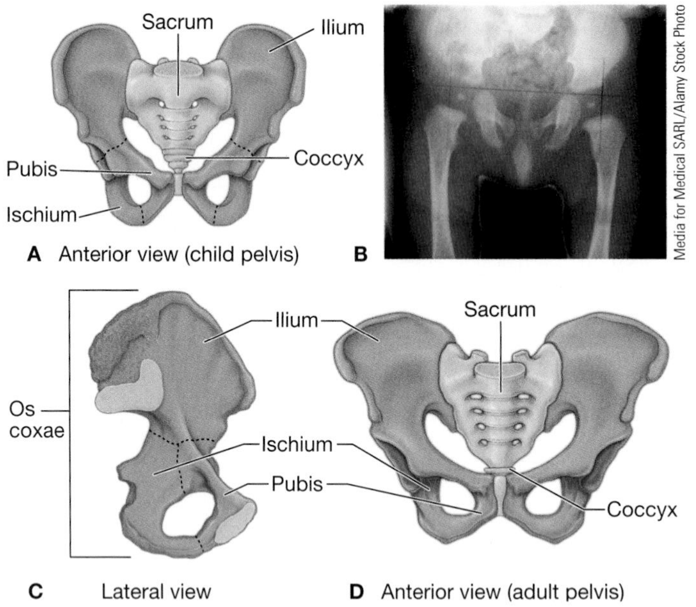

The auricular surface articulates with the auricular surface of the sacrum to form the sacroiliac joint. Both the posterior superior and posterior inferior iliac spines serve as attachment points for the muscles and very strong ligaments that support the sacroiliac joint. The large, inverted U-shaped indentation located on the posterior margin of the lower ilium is called the **greater sciatic notch** because it provides a passageway for the very large sciatic nerve, which carries much of the innervation for the posterior leg.

From the crests, the two ilia have gentle sloping depressions on their anterior and medial surfaces. Each of these is an **iliac fossa** and serves as a bowl to hold the contents of the lower abdominal cavity.

ischial tuberosity ("sit bone")

Ischium The ischium forms the posterior portion of the hip bone (**Figure 9.14**). The large, roughened area of the inferior ischium is the **ischial tuberosity**. This serves as the attachment for the posterior thigh muscles and also carries the weight of the body when sitting. Fitness instructors will often refer to the ischial tuberosities as the *sit bones*. You can feel the ischial tuberosity if you wiggle your pelvis against the seat of a chair. Projecting superiorly and anteriorly from the ischial tuberosity is a narrow segment of bone called the **ischial ramus**. The slightly curved posterior margin of the ischium above the ischial tuberosity is the **lesser sciatic notch**. The bony projection separating the lesser sciatic notch and greater sciatic notch is the **ischial spine**.

Pubis The pubis forms the anterior portion of the hip bone (**Figure 9.15**). The enlarged medial portion of the pubis is the **pubic body**. Located superiorly on the pubic body is a small bump called the **pubic tubercle**. The **superior pubic ramus** is the segment of bone that reaches laterally from the pubic body to join the ilium. The narrow ridge running along the superior margin of the superior pubic ramus is the **pectineal line** of the pubis.

Figure 9.13 The Ilium

The ilium forms the large fan-shaped superior portion of the hip bone. (A) Lateral and medial views. (B) Anterior view.

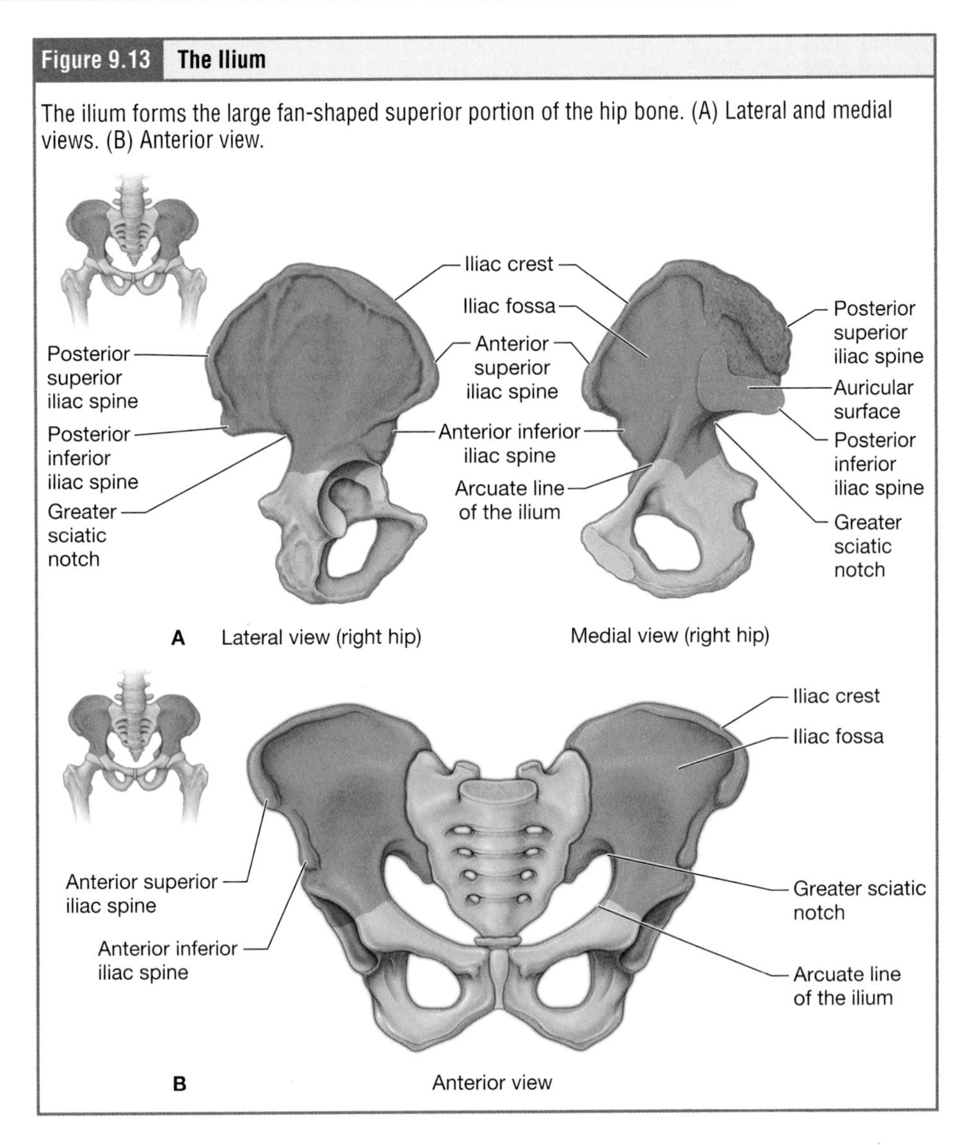

The pubic bodies of each os coxae are joined at the pubic symphysis. Extending downward and laterally from the body is the **inferior pubic ramus**. The inverted V-shape formed at the pubic symphysis is called the **subpubic angle**.

9.3b Features of the Whole Pelvis

The three areas of each hip bone—the ilium, pubis, and ischium—converge centrally to form a deep, cup-shaped cavity called the **acetabulum** (**Figure 9.16**). The acetabulum hosts the rounded head of the femur to form the hip joint. Another notable feature of the pelvis is a large opening formed as the rami of the pubis and ischium meet. The **obturator foramen** is largely filled in by a layer of connective tissue, but, similar to the sciatic notch, a large nerve passes through this space to reach the anterior leg.

The space internal to the bony pelvis houses the organs of the lower abdomen and pelvic cavities. Examine the inside of the pelvis using either the bones in a classroom or Figure 9.16. The internal space can be divided in two by the narrowest internal part of the pelvis, the **pelvic brim**, which is formed by the pubic symphysis anteriorly, and the pectineal line of the pubis, the arcuate line of the ilium, and the sacral promontory (the anterior margin of the superior sacrum) posteriorly. The space superior to the pelvic brim is the **greater pelvis**. This broad area is occupied by portions of the small and large intestines and, because it is more closely associated with the abdominal cavity, it

Learning Connection

Try Drawing It!

Draw a diagram of an os coxae (hip bone). Using different colors for each bone, show how the ilium, ischium, and pubis fuse to form the adult hip bone. What large depression is formed by the fusion of these bones?

pelvic brim (pelvic inlet)

greater pelvis (greater pelvic cavity or false pelvis)

LO 9.3.3

Figure 9.14 The Ischium

The ischium forms the posteroinferior portion of the hip bone. (A) Lateral and medial views. (B) Anterior view.

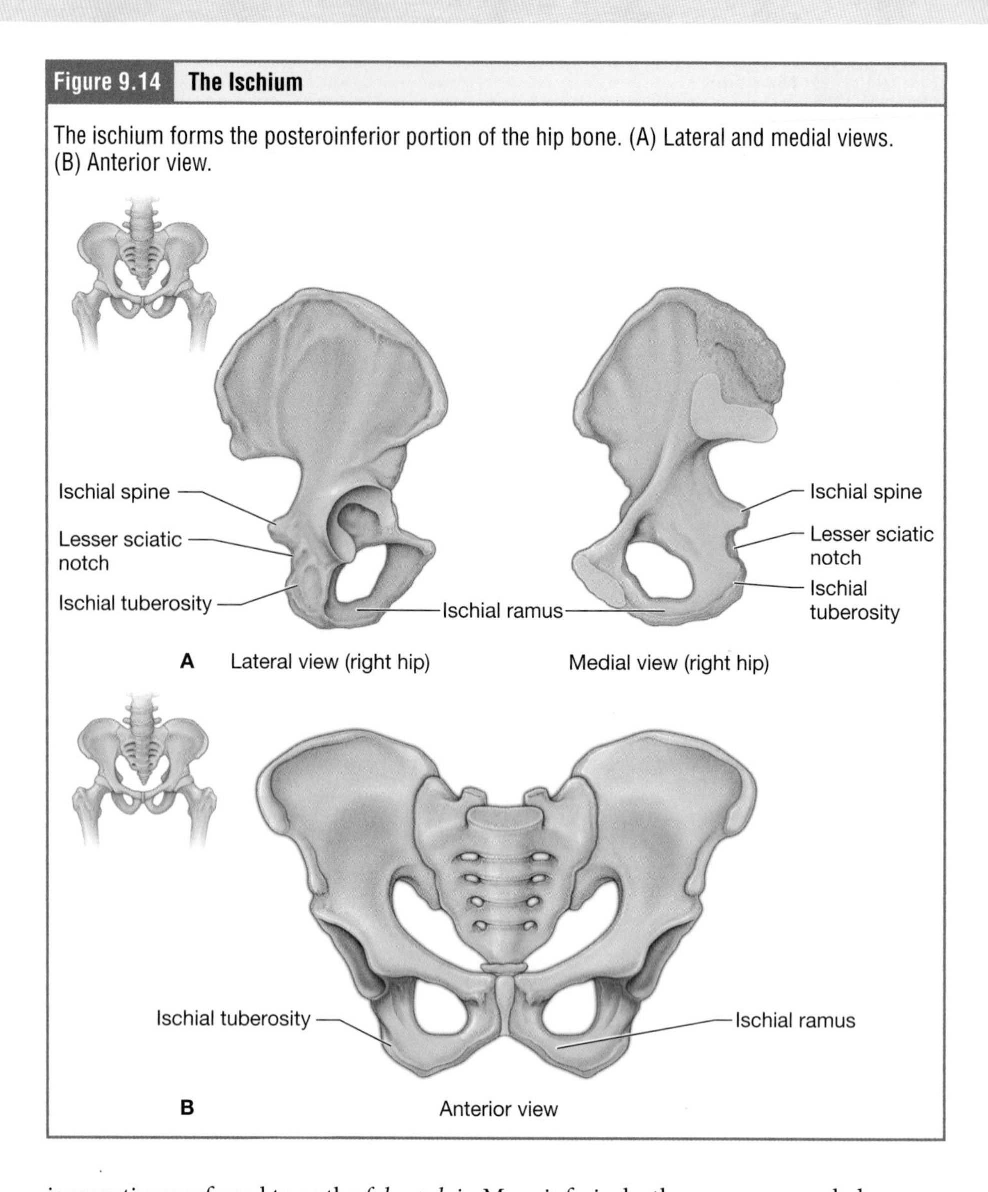

lesser pelvis (lesser pelvic cavity or true pelvis)

is sometimes referred to as the *false pelvis*. More inferiorly, the narrow, rounded space of the **lesser pelvis** contains the bladder, the rectum, and the reproductive organs and thus is also known as the *true pelvis*. The roof of the lesser pelvis is called the **pelvic inlet** (Figure 9.17) because it is the entrance to the pelvis. Similarly, the inferior floor of the lesser pelvis is called the **pelvic outlet**. This opening is defined by the pubic symphysis anteriorly, and the ischiopubic ramus, the ischial tuberosity, and the inferior tip of the coccyx posteriorly. The pelvic inlet and outlet become important landmarks when considering birth as the fetus moves down from the greater pelvis into the lesser pelvis and then passes out of the body through the pelvic outlet. The pelvic inlet and outlet are the narrowest parts of the passage and therefore present anatomical constraints to human birth. This process is considered in more detail in Chapter 27.

Biological Sex and the Pelvis Bones are highly dynamic structures and change over the course of the lifespan in relationship to their use, as well as to other physiological factors such as hormones and diet. Bigger bodies have bigger, wider pelvises and more muscular bodies exert more tension and force on the pelvis, and so these bones—like all bones—will become bigger and rougher in response. A key physiological factor in determining the shape of the pelvis is the hormone estrogen. Estrogen has a profound impact on bone remodeling, as we discussed in Chapter 7, but estrogen also influences

Figure 9.15 The Pubis

The pubis forms the anteromedial portion of the hip bone. (A) Lateral and medial views. (B) Anterior view.

the ligaments that hold the bones of the pelvis together. In childhood and early adolescence, most pelvises look very similar to each other, varying by size but not by shape. In puberty, as estrogen signaling increases dramatically in some individuals, pelvic shape begins to change. Under the influence of high levels of estrogen characteristic of biological females, a few notable changes occur (see the "Anatomy of Hormone-Induced Pelvic Changes" feature). The distance between the anterior superior iliac spines and the ischial tuberosities increases. Together these two changes increase the size of the pelvic outlet, which eases one anatomical constraint of childbirth. Because of this increased pelvic width, the subpubic angle is larger, typically greater than 80 degrees compared to an average of less than 70 degrees in individuals without high estrogen signaling. The female sacrum is wider, shorter, and less curved, and the sacral promontory projects less into the pelvic cavity, thus giving the female pelvic inlet (pelvic brim) a more rounded or oval shape compared to that of males. Lastly, the sciatic notch is narrower and deeper in biological males than the broader notch of biological females. These differences are often so pronounced that the pelvis is often used in sex determination of skeletal remains. Interestingly, once the much higher estrogen levels of the reproductive period decline in middle and late age these characteristic differences also change, and the once wider pelvis of biological females narrows progressively after menopause.

Figure 9.16 The Os Coxae

In adulthood, the three pelvic bones fuse to form each of the os coxae bones. The three bones all contribute to the acetabulum, the bowl-shaped region of articulation of the hip and femur. The ischium and pubis also meet to form an arch around a large opening, the obturator foramen.

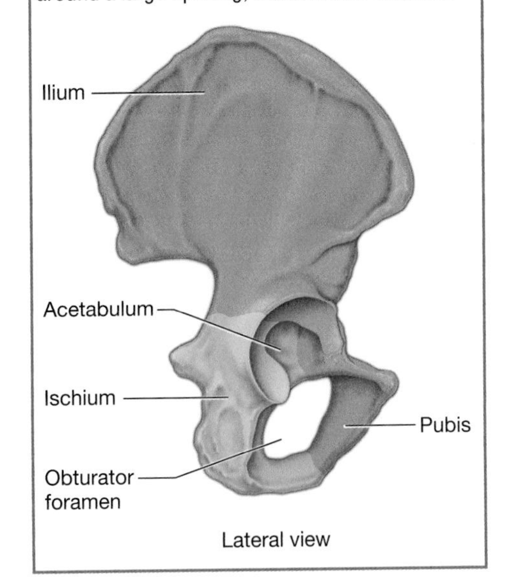

Figure 9.17 The Pelvic Inlet and Outlet

The passageway through the pelvis narrows in two places—the pelvic inlet and the pelvic outlet. These landmarks form the borders of the greater and lesser pelvis. (A) Lateral view. (B) Anterior oblique view. (C) Coronal section.

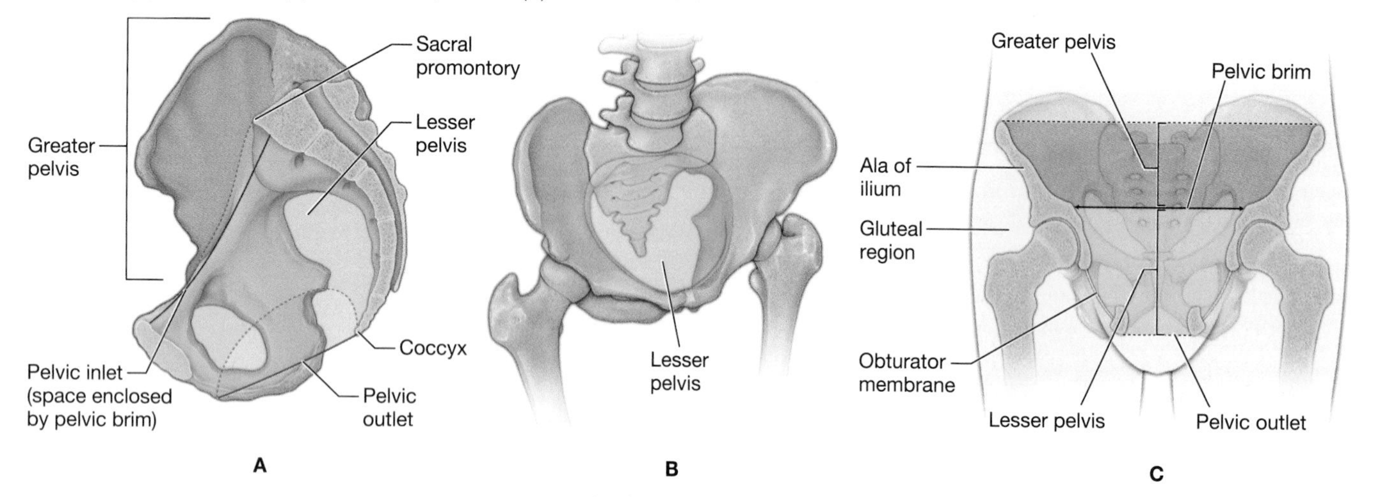

Anatomy of...

Hormone-Induced Pelvic Changes

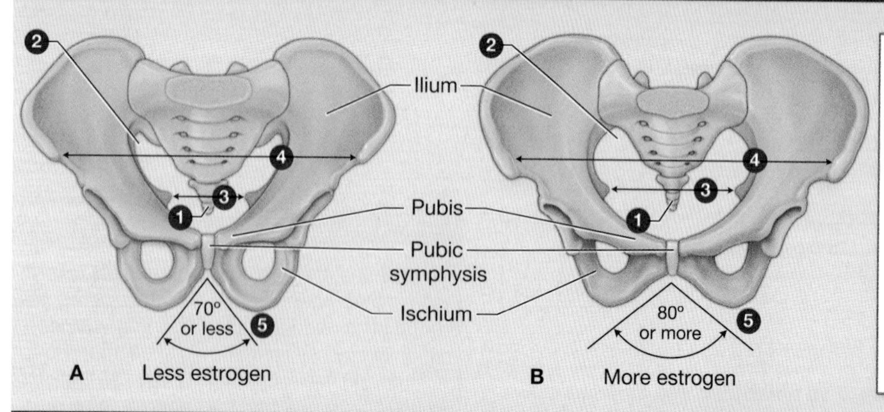

1. A posterior tilting of the coccyx is common in pelvises influenced by estrogen signaling; with less estrogen influence, the coccyx tilts anteriorly.
2. Sciatic notch is narrower under less influence by estrogen.
3. Under the influence of estrogen, there is a greater distance between the ischial tuberosities.
4. Under the influence of estrogen, there is a greater distance between the anterior iliac spines.
5. Subpubic angle widens to greater than 80 degrees under the influence of estrogen.

Learning Check

1. Which of the following changes occurs to the pelvis as a child develops into an adult?
   a. Deeper sciatic notch
   b. Increased bone density
   c. Increased pelvic width
   d. Oval-shaped pelvic inlets
2. Which of the following is most similar to the glenoid fossa of the scapula?
   a. Ilium
   b. Ischium
   c. Pubis
   d. Acetabulum
3. Which bony landmark separates the greater sciatic notch from the lesser sciatic notch?
   a. Iliac fossa
   b. Ischial spine
   c. Anterior superior iliac spine
   d. Anterior inferior iliac spine
4. Which of the following is formed when the two inferior pubic rami are joined at the pubic symphysis?
   a. Pectineal line
   b. Subpubic angle
   c. Pelvic inlet
   d. Pelvic outlet

9.4 Bones of the Leg

Learning Objectives: By the end of this section, you will be able to:

9.4.1* Identify the bones of the leg and describe the bones' features.

9.4.2* List the bones and bony landmarks that articulate at each joint of the lower limb.

* Objective is not a HAPS Learning Goal.

LO 9.4.1

Following the basic limb format (see the "Anatomy of a Limb" feature), the leg has three regions. The **thigh**, stretching between the hip joint and knee joint, has a single long bone, the **femur** (Figure 9.18). The **lower leg** has two parallel long bones that span the distance between the knee joint and the ankle joint. Of these, the **tibia** is the larger, weight-bearing bone located on the medial side of the leg, and the **fibula** is the thinner bone and lateral. The posterior foot and ankle are formed by a group of seven bones, each of which is known as a **tarsal bone**, while the mid-foot contains five long bones, each of which is a **metatarsal bone**. The toes contain 14 small bones, each of which is a **phalanx bone of the foot**.

femur (thigh bone)

tibia (shin bone)

Some key differences between the formats of the arm and leg are found at the knee and the ankle. The leg has a sesamoid bone at the knee; the **patella**, or kneecap, articulates with the distal femur. The ankle is formed by just two bones, rather than the eight of the more complex wrist.

patella (kneecap)

9.4a Femur

The femur is the single bone of the thigh region (Figure 9.19). It is the longest and strongest bone of the body, and accounts for approximately one-quarter of a person's total height. The rounded, proximal end is the **head of the femur**, which articulates with the acetabulum of the hip bone to form the hip joint. If you can look at a femur from its medial side, at the center of the head, normally tucked deep into the acetabulum, is a small indentation called the **fovea capitis**. A ligament attaches here to tie the head of the femur to the acetabulum. An artery also uses the fovea capitis as an entry point into the femur.

LO 9.4.2

Fitting our standard pattern, the head of the femur gives way to a narrower region, the **neck of the femur**. Just like the neck region of other bones, the femoral neck is a common site of fracture. The neck is sandwiched between two much larger regions of bone. To its medial side, of course, is the femoral head. To its lateral side is the **greater trochanter**—a large, upward-pointing, bony projection located above the base of the neck. Multiple muscles that act across the hip joint attach to the greater trochanter, which, because of its projection from the femur, gives additional leverage to these muscles. The greater trochanter can be felt just under the skin on the lateral side of your upper thigh. The bony out pocket continues around the posterior side of the bone (the **intertrochanteric crest**) and gives way to the **lesser trochanter**, a small, bony prominence that lies on the medial aspect of the femur, just below the neck. A single powerful muscle attaches to the lesser trochanter. Running between the greater and lesser trochanters on the anterior side of the femur is the roughened **intertrochanteric line** (Figure 9.20, anterior view).

Running along the posterior **shaft of the femur**, the **gluteal tuberosity** is a roughened area extending inferiorly from the greater trochanter. More inferiorly, the gluteal tuberosity becomes continuous with the **linea aspera** (the name means "rough line"). This is the roughened ridge that passes distally along the posterior side of the mid-femur. Multiple muscles of the hip and thigh regions make long, thin attachments to the femur along the linea aspera (Figure 9.20, posterior view).

Student Study Tip

The gluteal tuberosity is in the gluteal region and is an attachment point for the gluteal muscles. Understanding where a name came from can help you remember it more easily.

Similar to the humerus, the distal end of the femur has rough bony expansions called *epicondyles* on both the medial and lateral sides. The **lateral epicondyle of the femur** provides an attachment point for muscles; it is just lateral and superior to the smooth articulating surface, the **lateral condyle of the femur**. Similarly, the smooth,

Figure 9.18 The Regions of the Leg

The leg can be divided into three major regions—the thigh, the lower leg, and the foot—with two major functional joints, the knee and the ankle. There are also many minor joints between individual bones.

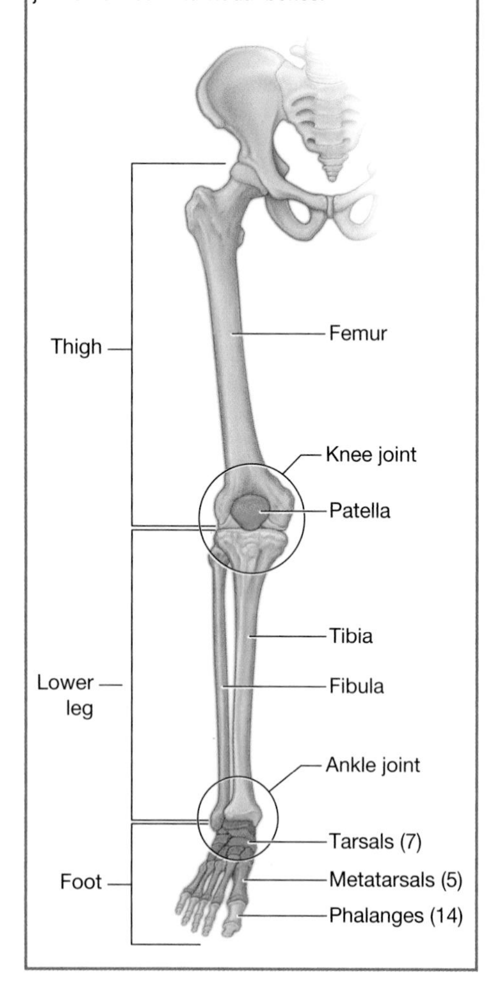

Figure 9.19 The Femur and the Knee Joint

The femur is the single bone of the thigh region. Its head fits into the acetabulum of the os coxae to form the hip joint. Inferiorly, the femur articulates with the tibia at the knee joint. The patella articulates with the distal end of the femur.

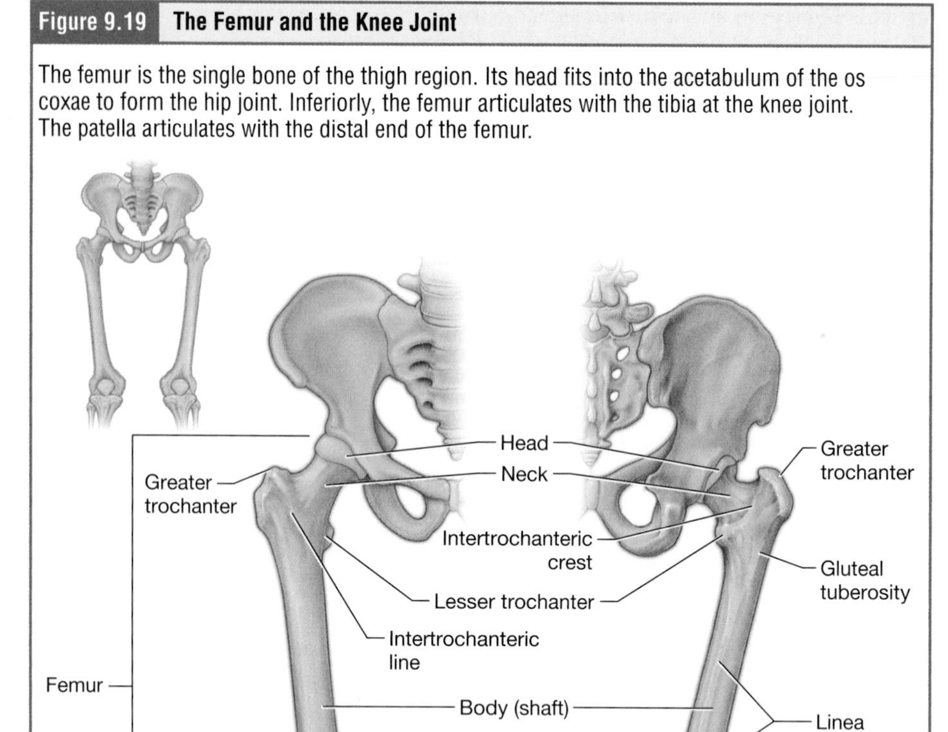

Cultural Connection

Prostheses

One area of medicine that has seen very rapid advancement is the bioengineering of prosthetic limbs. Until relatively recently, amputation of a limb left an individual with a significant disability, and the prosthetics that were generally available attempted to recreate the missing limb's most basic functions; however, modern prostheses have become a mix of mechanics and computers that enable their wearers to resume just about any activity they wish. Prosthetic legs, for example, can have general functionality and allow for walking, jogging, or jumping, but highly specialized prosthetics can also be made that enable their wearers to engage in activities such as sprinting, dancing, and even rock climbing. Aimee Mullins, an actress, model, motivational speaker, and Olympic-level athlete, owns several pairs of prosthetic legs that allow her to pursue many athletic activities, including, perhaps most famously, a pair of legs designed for her to use in sprinting that were modeled after the hind legs of a cheetah. Adrianne Haslet-Davis, a professional dancer who was injured in the Boston Marathon bombings in 2013, was able to return to dancing less than a year later with the assistance of a custom-designed prosthetic leg. The engineers who study and build prostheses are just getting started; future goals in prosthetics include better integration with the nervous system that will enable prostheses to go beyond their current function as helpful tools and provide sensory as well as motor functionality.

articulating region on the medial side, the **medial condyle of the femur**, meets its rough outer side, the **medial epicondyle of the femur**. The lateral and medial condyles articulate with the tibia to form the knee joint. Both epicondyles provide attachment sites for muscles and supporting ligaments of the knee. Posteriorly, the medial and lateral condyles are separated by a deep depression called the **intercondylar fossa**. Anteriorly, the smooth surfaces of the condyles join together to form a wide groove called the **patellar surface**, which provides for articulation with the patella bone. The combination of the medial and lateral condyles with the patellar surface gives the distal end of the femur a horseshoe (U) shape (see Figure 9.20).

Figure 9.20 The Distal Femur

Looking at the femur from an inferior view, we can see the two condyles and the carved-out space—the intercondylar fossa—between them. This space accommodates two major ligaments of the knee.

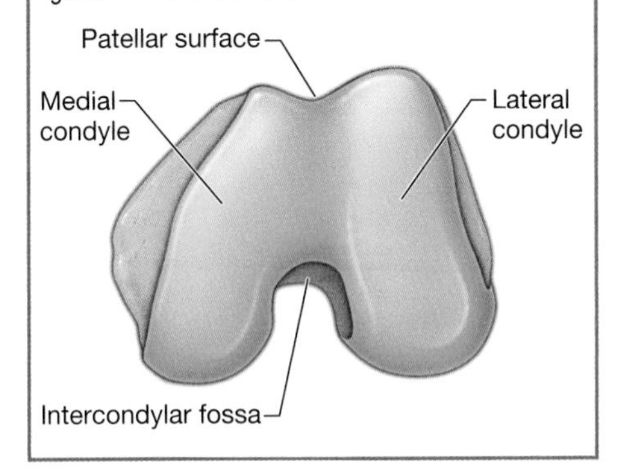

9.4b Patella

The patella (kneecap) is largest sesamoid bone of the body (see Figure 9.20 and **Figure 9.21**). A sesamoid bone is a bone that is incorporated into the tendon of a muscle where that tendon crosses a joint. While sesamoid bone number and placement varies from person to person, the patella is the one sesamoid bone that is universal to all humans (and many other animals). The patella is found embedded in a large tendon that passes across the anterior knee to attach to the tibia. The patella articulates with the patellar surface of the femur and thus prevents rubbing of the muscle tendon against the distal femur. The patella also lifts the tendon away from the knee joint, which increases the leverage power of the quadriceps femoris (thigh) muscle it attaches to. The patella does not articulate with the tibia.

9.4c Tibia and Fibula

The lower leg has two bones—the larger tibia, which is the medial bone, and the smaller fibula (**Figure 9.22**). The tibia is the main weight-bearing bone of the lower leg and the second longest bone of the body, after the femur. The fibula is a slender bone located on the lateral side of the leg. The fibula does not bear weight. It serves primarily for muscle attachments and thus is largely surrounded by muscles. The tibia is located immediately under the skin, allowing it to be easily palpated down the entire length of the lower leg, due to its relationship with the muscles that surround it, the shaft of the fibula cannot be palpated, it can be found below the skin only on its proximal and distal tips.

Student Study Tip

To remember which bone is bigger, think of the Tibia as a Truck and the Fibula as a Fiat. The Fiat is much smaller than a truck!

The proximal end of the tibia is much wider than any other portion of this bone. It has two smooth, flat proximal surfaces, the **medial condyle of the tibia** and the **lateral condyle of the tibia**. In contrast to the femur, there are no muscle attachments to its sides so the tibia does not have epicondyles. The tibial condyles articulate with the medial and lateral condyles of the femur to form the **knee joint**. Between the articulating surfaces of the tibial condyles is the **intercondylar eminence**, an irregular, elevated area that serves as the inferior attachment point for two supporting ligaments of the knee. The fibula does not articulate with the femur; it is lateral and inferior to the knee joint (see Figure 9.22). The **head of the fibula** is the small, knoblike proximal end of the fibula. It articulates with the tibia, nestling into the space just under the lateral tibial condyle, forming the **proximal tibiofibular joint**.

If you palpate (touch) your anterior leg just below the knee, a prominent elevated area can be felt. The **tibial tuberosity**, the final site of attachment for the tendon in which the patella is suspended, attaches here. As we progress down, the **shaft of the tibia** narrows before giving rise to a distal medial expansion, the **medial malleolus**. This forms the large bony bump found on the medial side of the ankle region. The similar bony bump on the lateral side of your ankle is the **lateral malleolus**; it is the distal end of the fibula (see Figure 9.22). Both the smooth surface on the inside of the medial malleolus and the smooth area at the distal end of the fibula articulate with the talus bone of the foot as part of the ankle joint. On the lateral side of the distal tibia is a wide groove called the **fibular notch**. This area articulates with the distal end of the fibula, forming the **distal tibiofibular joint**.

Figure 9.21 The Patella

The patella is a sesamoid bone suspended within a tendon that stretches over the knee. The base is superior and the apex is inferior.

Base

Apex

Figure 9.22 Tibia and Fibula

The tibia is the weight-bearing bone of the lower leg. On its lateral side it articulates with the fibula, which does not bear weight but has many muscle attachments.

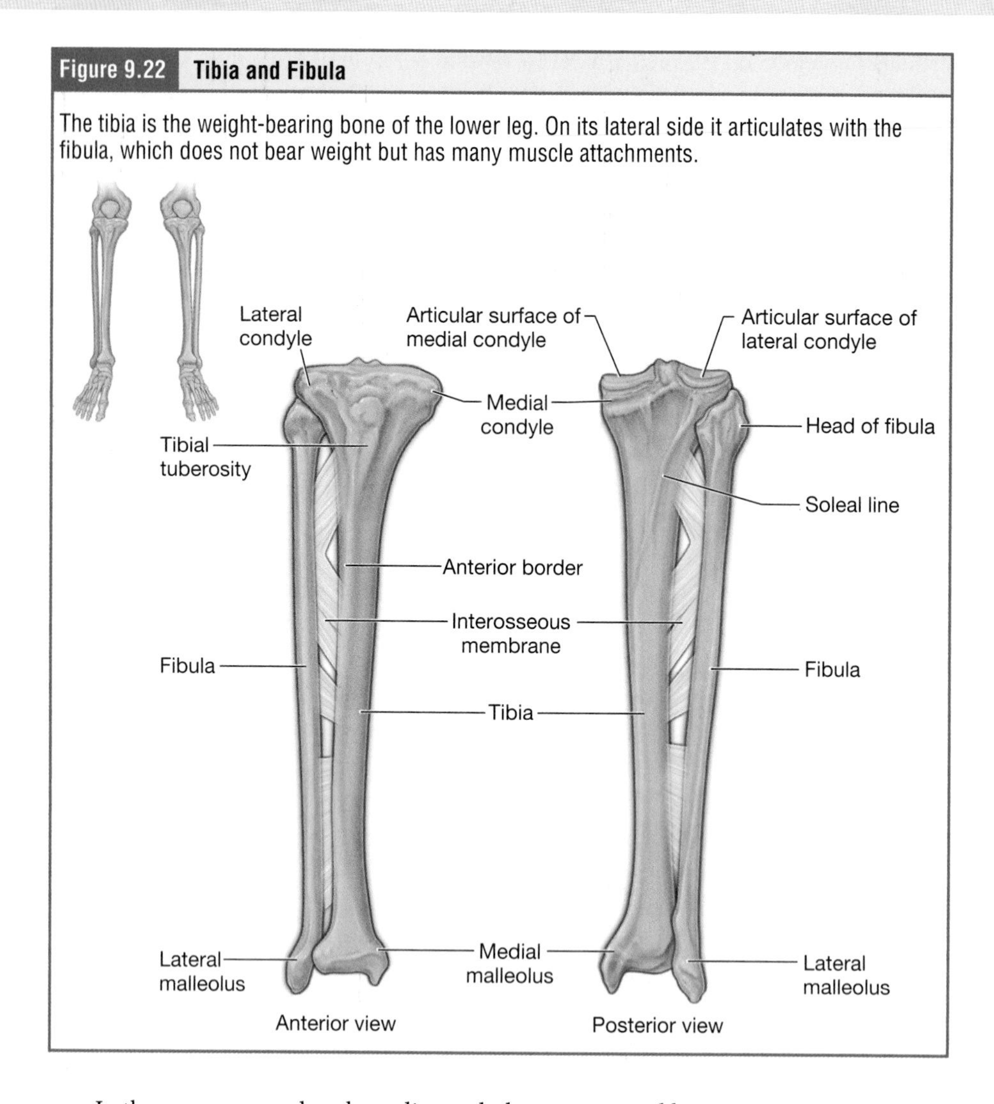

In the arm, we saw that the radius and ulna were united by an interosseous membrane that spans the space between these bones. This format is replayed in the leg, with the tibia and fibula being united by the **interosseous membrane of the leg**, a sheet of dense connective tissue that unites the tibia and fibula bones.

Apply to Pathophysiology

Osgood-Schlatter Disease

Chaviv is a 14-year-old with a love of basketball. Recently they have been complaining of pain in their anterior knee just below the patella and just under the skin.

1. Which bone is most likely the source of the pain?
   A) Femur
   B) Tibia
   C) Fibula

Chaviv goes to the doctor and is diagnosed with Osgood-Schlatter disease, a syndrome in which the ligament that connects the patella to the bone below irritates the epiphyseal plate of that inferior bone.

2. Which feature of the bone is nearest the inflammation?
   A) The shaft of the tibia
   B) The medial malleolus
   C) The lateral malleolus
   D) The tibial tuberosity

The doctor tells Chaviv and their mother that they need to stop playing basketball and rest the knee. The inflammation will go away, but is likely to come back anytime Chaviv overuses their legs until the epiphyseal plate is completely ossified.

3. At which of these ages is Chaviv likely to be able to have a pain-free basketball career?
   A) Age 15
   B) Age 18
   C) Age 25
   D) Age 40

9.4d Bones of the Foot

The ankle joint and posterior half of the foot is formed by seven tarsal bones (**Figure 9.23**). The most superior bone is the **talus**. The talus forms the **ankle joint** by fitting its relatively square-shaped, upper surface into an arch created by the distal ends of the tibia and fibula. Inferiorly, the talus articulates with the **calcaneus**, which is the heel bone and the largest bone of the foot.

The last bone in the proximal row of tarsal bones is the **navicular bone**. The distal row of tarsal bones is defined as the tarsal bones that articulate with the metatarsals.

calcaneus (heel bone)

Student Study Tip

All four bones of the distal row of tarsals are shaped like cubes, with the cuboid as the largest. Going lateral to medial, the three cuneiforms follow the cuboid like three cube-shaped ducklings following the mama cube duck.

Figure 9.23 Bones of the Foot and Ankle

The seven tarsal bones form the inferior ankle and posterior foot, shown in three views—(A) superior view, (B) medial view, and (C) lateral view. The arch of the foot is formed by the metatarsals and the toes are composed of phalanges.

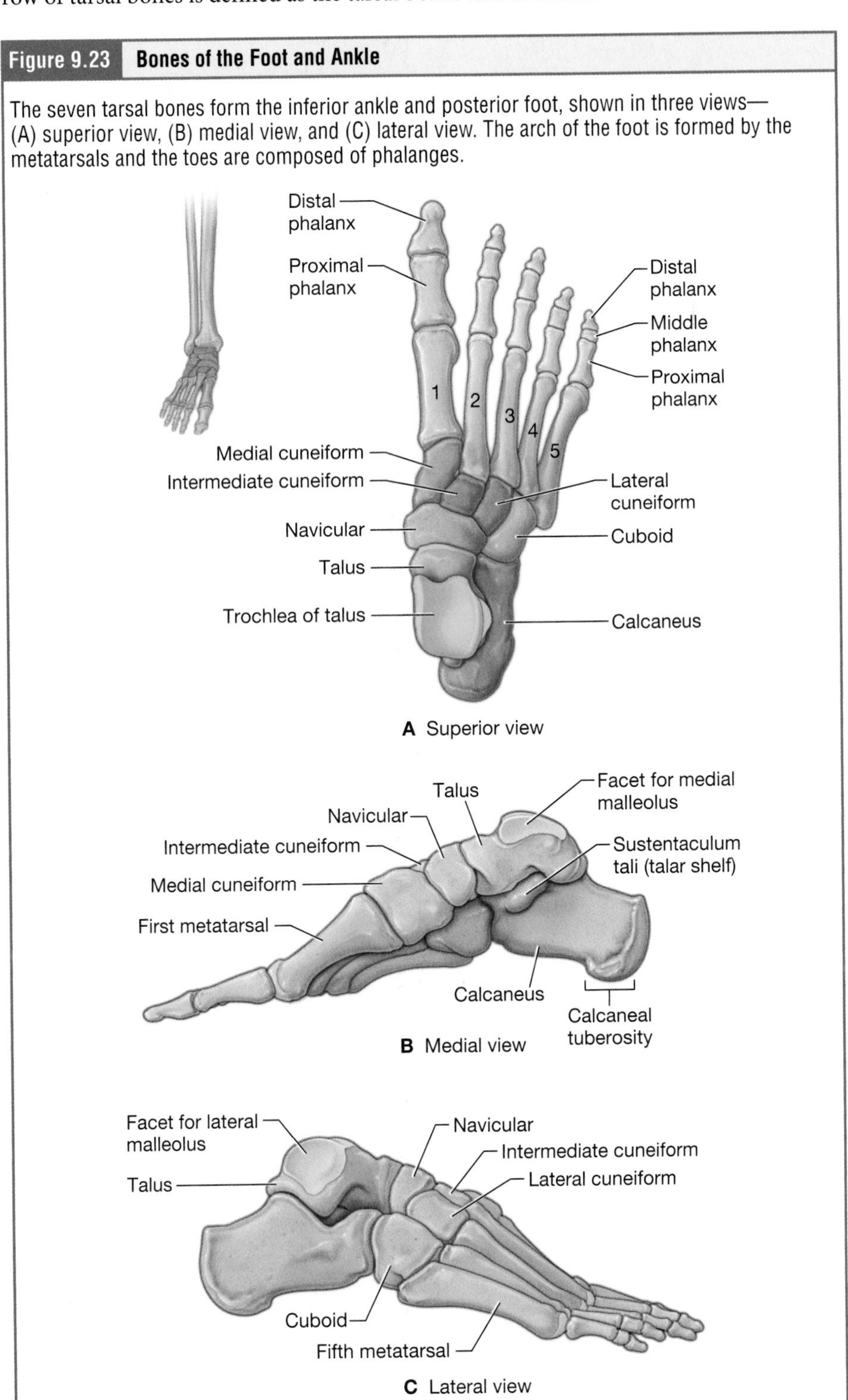

Learning Connection

Image Translation

Look at various online views of the foot (medial, lateral, superior, and inferior). Which tarsal bones are visible in the superior and inferior views, but not visible in the medial and lateral views?

Learning Connection

Chunking

List all of the bones of the leg. Which ones are classified as long bones? How do we classify the others?

hallux (big toe, great toe)

This row includes the **cuboid bone** and the three cuneiform bones: the **medial cuneiform**, the **intermediate cuneiform**, and the **lateral cuneiform.**

The palm of the hand contains five long metacarpal bones that end at the fingers. Similarly, the arch of the foot is formed by five metatarsal bones, which end anteriorly with the phalanges of the toes (see Figure 9.23). These elongated bones are numbered 1–5, starting with the medial side of the foot. The toes contain a total of 14 phalanx bones (phalanges), arranged in a similar manner as the phalanges of the fingers (see Figure 9.23). The toes are numbered 1–5, starting with the big toe (**hallux**). The big toe has two phalanx bones—the proximal and distal phalanges. The remaining toes all have proximal, middle, and distal phalanges. A joint between adjacent phalanx bones is called an *interphalangeal joint.*

DIGGING DEEPER:
Why Can Humans Jump but Elephants Can't?

When the foot comes into contact with the ground during walking, running, or jumping activities, the impact of the body weight puts a tremendous amount of pressure and force on the foot. During running, the force applied to each foot as it contacts the ground can be up to 2.5 times your body weight. The bones, joints, ligaments, and muscles of the foot absorb this force, thus greatly reducing the amount of shock that is passed superiorly into the lower limb and body. The arches of the foot play an important role in this shock-absorbing ability. When weight is applied to the foot, these arches will flatten somewhat, thus absorbing energy. When the weight is removed, the arch rebounds, giving "spring" to the step. The arches also serve to distribute body weight side to side and to either end of the foot. Animals without arches, such as elephants, are not able to jump because the force received by the skeleton without arches would cause bone fractures.

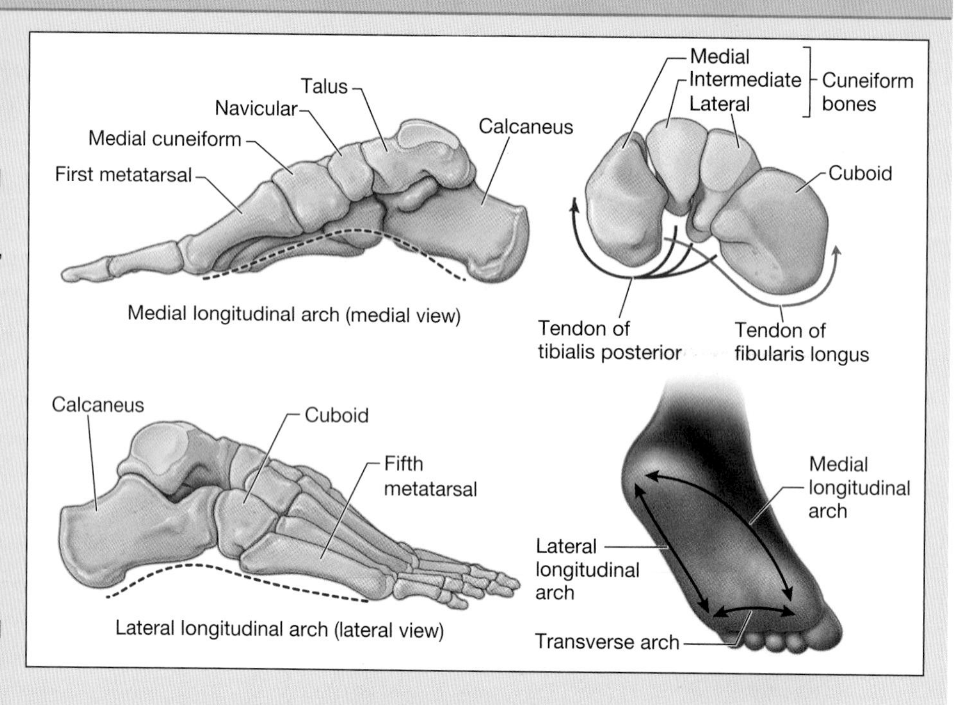

Learning Check

1. The arm and legs are very similar. Which of the following bones does NOT have an equivalent in the arm?
   a. Femur
   b. Fibula
   c. Tarsal
   d. Patella
2. How many tarsal bones are there in the foot?
   a. 5
   b. 7
   c. 9
   d. 3
3. The iliopsoas is a muscle that flexes the hip. This muscle attaches to the small bony prominence that lies on the medial aspect of the femur, just below the neck. Which of the following bony prominences does the iliopsoas attach to?
   a. Greater trochanter
   b. Lesser trochanter
   c. Medial epicondyle of the femur
   d. Lateral epicondyle of the femur

4. The knee joint is similar to the elbow joint in many ways. Identify the similarity between the elbow and the knee. Please select all that apply.
   a. The elbow and knee are hinge joints.
   b. The elbow is formed by three bones like the knee is formed by three bones.
   c. The proximal radioulnar joint is part of the elbow like the proximal tibiofibular joint is part of the knee.
   d. The proximal radioulnar joint and the proximal tibiofibular joint contribute to rotation.

Chapter Review

9.1 The Shoulder Girdle

The learning objectives for this section are:

9.1.1* Identify the bones and bone features of the shoulder girdle.

9.1.2* List the functions of each of the bones of the pectoral girdle.

* Objective is not a HAPS Learning Goal.

9.1 Questions

1. Which of the following does the medial end of the clavicle articulate with?
   a. Manubrium
   b. Glenoid cavity
   c. Acromion
   d. Coracoid process

2. Muhammed is playing on the playground when they fall onto their shoulder. They feel numbness and tingling down their arm. In the emergency room, the doctors think Muhammed fractured a bone in their shoulder girdle that is impinging on the nerves that go from the body to the arm. What bone could be damaged?
   a. Clavicle
   b. Scapula

9.1 Summary

- The appendicular skeleton consists of the bones of the limbs and the girdles that connect the limbs to the axial skeleton.
- The shoulder (pectoral) girdle is composed of the clavicle (collarbone) and the scapula (shoulder blade). It connects the upper limb to the axial skeleton.
- The humerus meets the glenoid cavity of the scapula to form the glenohumeral (shoulder) joint.
- The scapula meets the clavicle to form the acromioclavicular joint.
- The clavicle meets the sternum to form the sternoclavicular joint.

9.2 Bones of the Arm

The learning objectives for this section are:

9.2.1* Identify the bones of the arm and describe the bones' features.

9.2.2* List the bones and bony landmarks that articulate at each joint of the upper limb.

* Objective is not a HAPS Learning Goal.

9.2 Questions

1. Which of the following bones is proximal to the radius?
   a. Humerus
   b. Ulna
   c. Pisiform
   d. Metacarpals

2. Shar was riding their bike in the city, when a car ran a red light and hit them from the side. Shar fell onto their elbow and realized they could not pronate their forearm because it hurt too much. After getting X-rays done in the emergency room, the doctors said Shar fractured their bone, which is making it painful to pronate their forearm and flex their elbow. Which of the following could be fractured?
   a. Head of the ulna
   b. Head of the radius
   c. Head of the humerus
   d. Styloid process of the radius

9.2 Summary

- The upper limb consists of the brachium (upper arm), antebrachium (forearm), and the hand.
- The humerus is the only bone of the brachium.
- The antebrachium contains the radius and ulna.
- The hand consists of the carpals (wrist bones), metacarpals (hand bones), and phalanges (finger bones).

9.3 The Pelvic Girdle and Pelvis

The learning objectives for this section are:

9.3.1* Describe the three regions of the hip bone and identify their bony landmarks.

9.3.2* Compare and contrast the child and adult pelvis, and the pelvis of biological males and biological females.

9.3.3* Describe the openings of the pelvis and the boundaries of the greater and lesser pelvis.

* Objective is not a HAPS Learning Goal.

9.3 Questions

Mini Case 1: Pregnancy

During pregnancy, the fetus develops in a female's reproductive organ known as the uterus.

1. Which region of the os coxae is in contact with the chair when the pregnant female sits down to rest?
   a. Ilium
   b. Ischium
   c. Pubis

2. Which of the following is a border that both regions of the pelvis share?
   a. Pelvic inlet
   b. Pelvic outlet
   c. Pelvic brim
   d. Obturator foramen

During pregnancy, there is an increase in estrogen levels in the body.

3. Which of the following changes would you expect to occur during pregnancy as the estrogen levels increase? Please select all that apply.
   a. Decreased thickness of bone
   b. Increased distance between anterior superior iliac spines
   c. Narrower sciatic notch
   d. Increased size of the pelvis outlet

9.3 Summary

- The hip bone (os coxae) consists of three regions: the ilium (the most superior region), the ischium (the most posterior and inferior region), and the pubis (the most anterior and inferior region).
- The three bones that make up each hip bone converge to form a cup-shaped depression called the acetabulum. The head of the femur articulates with the acetabulum to form the hip joint.
- The pelvis consists of the two hip bones, the sacrum and coccyx. The space inside the pelvis is divided by the pelvic brim into two portions: the greater pelvis (superior portion) and the lesser pelvis (inferior portion).
- The pelvis undergoes many structural changes between childhood and adulthood. The ilium, ischium, and pubis exist as separate bones in a child, but are fused in an adult.
- There are several differences between the pelvis of a biological male and biological female, due to the effects of estrogen. The pelvis of a biological female is wider and more flared than that of a biological male.

9.4 Bones of the Leg

The learning objectives for this section are:

9.4.1* Identify the bones of the leg and describe the bones' features.

9.4.2* List the bones and bony landmarks that articulate at each joint of the lower limb.

* Objective is not a HAPS Learning Goal.

9.4 Questions

1. Which of the following highlights the difference between the humerus and femur?
   a. Only the humerus has a bump distally on the medial and lateral aspects.
   b. Only the femur has one neck that is clearly visible on imaging.
   c. Only the femur has a head that articulates in a concave structure.
   d. Only the humerus articulates with two bones at the distal joint.

2. Identify the bony landmarks that make up the knee joint. Please select all that apply.
   a. Medial epicondyle
   b. Patellar surface
   c. Lateral condyle
   d. Medial malleolus

9.4 Summary

- The lower limb consists of the thigh, the lower leg and the foot.
- The femur is the only bone of the thigh.
- The tibia and fibula are the bones of the lower leg.
- The foot consists of the tarsals (ankle bones), metatarsals (foot bones), and phalanges (toe bones).
- The patella lies against the anterior inferior aspect of the femur. The femur, tibia and patella form the knee joint.

10

Joints

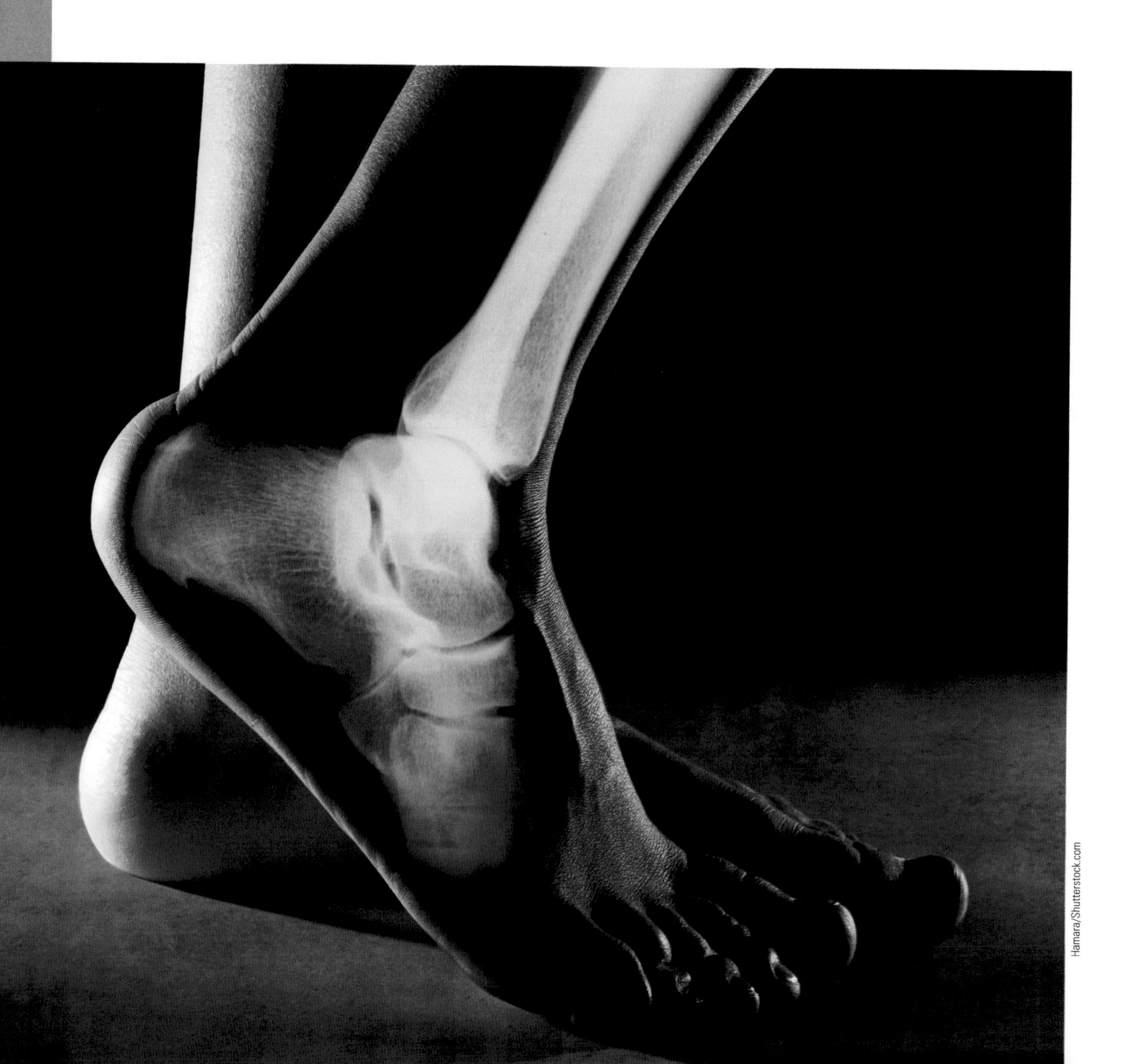

Hamara/Shutterstock.com

Chapter Introduction

In this chapter we will learn...

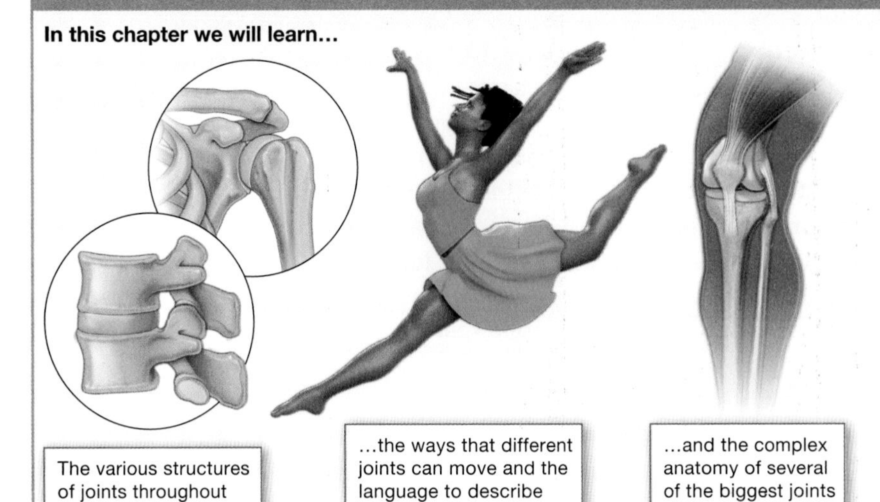

The various structures of joints throughout the human body...
...the ways that different joints can move and the language to describe the movements...
...and the complex anatomy of several of the biggest joints of the body.

With the exception of the hyoid bone in the neck, every bone in the human body is connected to at least one other bone. The locations where bones come together, joints or articulations come in many formats. Some joints, like your knee, allow for movement between the bones. Other joints may bind the bones together so tightly that the bones do not move apart from each other, such as the joints among the bones in your skull. As you can imagine, bones that are tightly bound to each other with connective tissue provide little or no movement are highly stable. Conversely, joints that provide the most movement between bones are the least stable. Understanding the relationship between joint structure and function will help to explain why particular types of joints are found in certain parts of the body.

10.1 Classification of Joints

Learning Objectives: By the end of this section, you will be able to:

10.1.1 Describe the anatomical classification of joints based on structure: fibrous (i.e., gomphosis, suture, syndesmosis), cartilaginous (i.e., symphysis, synchondrosis), and synovial (i.e., planar/gliding, hinge, pivot, condylar, saddle, ball-and-socket), and provide examples of each type.

10.1.2 Explain the relationship between the anatomical classification and the functional classification of joints.

10.1.3 Describe the functional classification of joints (e.g., synarthrosis, diarthrosis) based on amount of movement permitted, and provide examples of each type.

A **joint** is any place where adjacent bones or bone and cartilage come together to form a connection; in these locations we often describe the bones as *articulating* with each other. Joints are classified both structurally and functionally. Structural classifications of joints take into account whether the adjacent bones are directly anchored to each other by fibrous connective tissue or cartilage, or whether the adjacent bones articulate with each other within a fluid-filled space called a **joint cavity**. Functional classifications describe the degree of movement available between the bones, ranging from immobile, to slightly mobile, to freely moveable joints. The amount of movement is directly related to the function of the joint. For example, immobile or slightly moveable joints serve to protect internal organs or give stability to the body. Freely moveable joints, on the other hand allow for much more extensive movements of the body and limbs. As a general rule, the more motion is possible at the joint, the less stable and

joint cavity (synovial cavity)

The Human Anatomy and Physiology Society includes more than 1,700 educators who work together to promote excellence in the teaching of this subject area. The HAPS A&P Learning Outcomes measure student mastery of the content typically covered in a two-semester Human A&P curriculum at the undergraduate level. The full Learning Outcomes are available at https://www.hapsweb.org.

more prone to injury that joint is. For example, the shoulder is the most mobile joint of the human body, it is also the most frequently dislocated.

10.1a Structural Classification of Joints

LO 10.1.1

Learning Connection

Try Drawing It!

Try drawing a cartilaginous joint, a synovial joint, and a fibrous joint. Use the same color each time for the same material (e.g., keep hyaline cartilage blue and fibrocartilage green). Can you now compare and contrast the joints from your drawings?

The structural classification of joints is based on whether the articulating surfaces of the adjacent bones are directly connected by fibrous connective tissue or cartilage, or whether the articulating surfaces are connected via a fluid-filled joint cavity (see the feature "Anatomy of a Joint Structure"). A **fibrous joint** is where the adjacent bones are united by fibrous connective tissue. At a **cartilaginous joint**, the bones are joined by hyaline cartilage or fibrocartilage. At a **synovial joint**, the articulating surfaces of the bones are not directly connected, but instead connect within a joint cavity that is filled with a lubricating fluid. Synovial joints allow for free movement between the bones and are the most common joints of the body.

10.1b Functional Classification of Joints

LO 10.1.2

LO 10.1.3

amphiarthrosis (amphiarthrotic joint)

Whereas the structural classification is primarily concerned with what is between the joints (i.e. fluid, cartilage, connective tissue) the functional classification of joints is all about how the joint moves, specifically the amount of mobility found between the adjacent bones. Joints are thus functionally classified as a **synarthrosis** if no motion is possible at the joint, an **amphiarthrosis** if slight movement is possible at the joint, or as a **diarthrosis**, if a lot of motion is possible at the joint. As always, structure and function are intimately tied. The fibrous type of joint structure may be functionally classified as a synarthrosis (immobile joint) or an amphiarthrosis (slightly mobile joint). Cartilaginous joints are also functionally classified as either a synarthrosis or an amphiarthrosis joint. All synovial joints are functionally classified as a diarthrosis joint.

Synarthrosis *Synarthroses* are immobile or nearly immobile joints. The immobile nature of these joints provides for a strong union between the articulating bones. This is important at locations where the bones provide protection for internal organs.

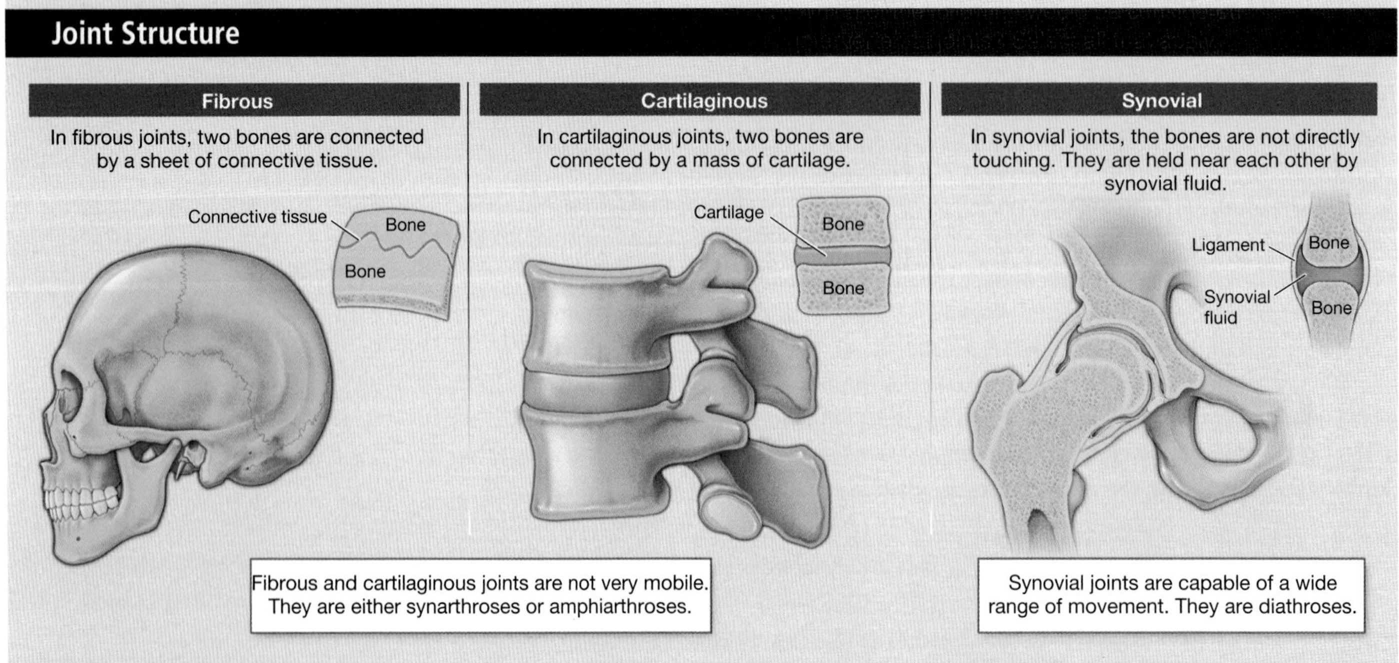

Anatomy of...
Joint Structure
Fibrous
In fibrous joints, two bones are connected by a sheet of connective tissue.
Connective tissue
Bone
Bone
Cartilaginous
In cartilaginous joints, two bones are connected by a mass of cartilage.
Cartilage
Bone
Bone
Synovial
In synovial joints, the bones are not directly touching. They are held near each other by synovial fluid.
Ligament
Bone
Synovial fluid
Bone
Fibrous and cartilaginous joints are not very mobile. They are either synarthroses or amphiarthroses.
Synovial joints are capable of a wide range of movement. They are diathroses.

Examples include the sutures of the skull; the fibrous joints unite the skull bones to protect the brain (**Figure 10.1**).

Figure 10.1 Suture Joints of the Skull

The suture joints of the skull are examples of synarthroses and fibrous joints. They are immobile and in these joints two skull bones are tied together directly with a sheet of connective tissue. Later in life (older adulthood), the connective tissue ossifies and these joints become solid.

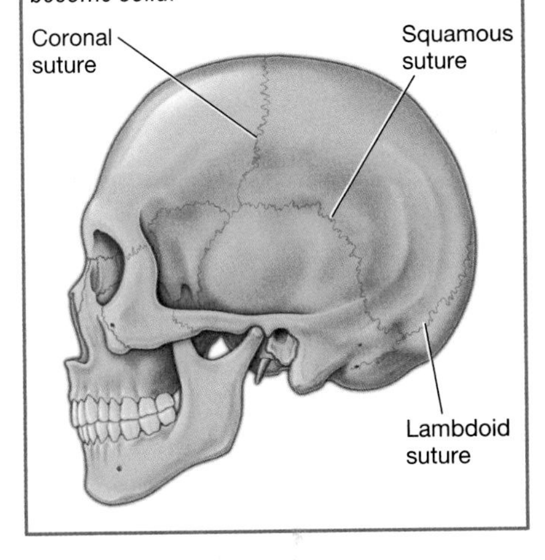

Amphiarthrosis An *amphiarthrosis* is a joint that has limited mobility. An example of this type of joint is the cartilaginous joint that unites the bodies of adjacent vertebrae. Spanning the space between each vertebral bone is a thick pad of fibrocartilage called an *intervertebral disc* (**Figure 10.2A**). The intervertebral disc strongly unites the vertebrae but still allows for a limited amount of movement between them. While each intervertebral joint is capable of only a small degree of motion, the sum of these movements along the length of the vertebral column results in large ranges of body movements, such as when you turn your shoulders to one side while sitting.

Another example of an amphiarthrosis is the pubic symphysis of the pelvis (**Figure 10.2B**). This is a cartilaginous joint in which the pubic regions of the right and left hip bones are strongly anchored to each other by fibrocartilage. This joint normally has very little mobility. The strength of the pubic symphysis is important in conferring weight-bearing stability to the pelvis.

Diarthrosis A *diarthrosis* is a freely mobile joint; the majority of body movements are possible because they occur at diarthroses. Most diarthrotic joints are synovial joints found in the appendicular skeleton. These joints are divided into three categories, based on the number of axes of motion provided by each. An axis in anatomy is described as the movements in reference to the three anatomical planes: transverse, frontal, and sagittal. Thus, diarthroses are classified as uniaxial (for movement in one plane), biaxial (for movement in two planes), or multiaxial joints (for movement in all three anatomical planes).

A **uniaxial joint** only allows for a motion in a single plane (around a single axis). The elbow joint, which only allows for bending or straightening, is an example of a uniaxial joint (**Figure 10.3**). A **biaxial joint** allows for motions within two planes. An example of a biaxial joint is the joint at the base of your finger, where a finger meets the palm of the hand. Examine one of these joints and bend your finger. This is motion in one plane, like the elbow joint. But now spread your fingers away from each other and bring them together. This is motion in a second plane at that same joint. These joints, the *metacarpophalangeal* joints, are biaxial. A joint that allows for the several directions of movement is called a **multiaxial joint**. The shoulder and hip joints are multiaxial joints. They allow the limbs to move in a circle, a much greater range of motion than the uniaxial or biaxial joints.

Learning Connection

Chunking

Functional classifications are all about mobility. Can you arrange functional classes from most to least mobile? Can you arrange structural classifications from most to least mobile?

Figure 10.2 Intervertebral Disc and Pubic Symphysis

Intervertebral discs (A) and the pubic symphysis (B) are examples of amphiarthroses because they are capable of only very small amounts of motion. These two joints are also examples of cartilaginous joints because a pad of cartilage is sandwiched between two bones.

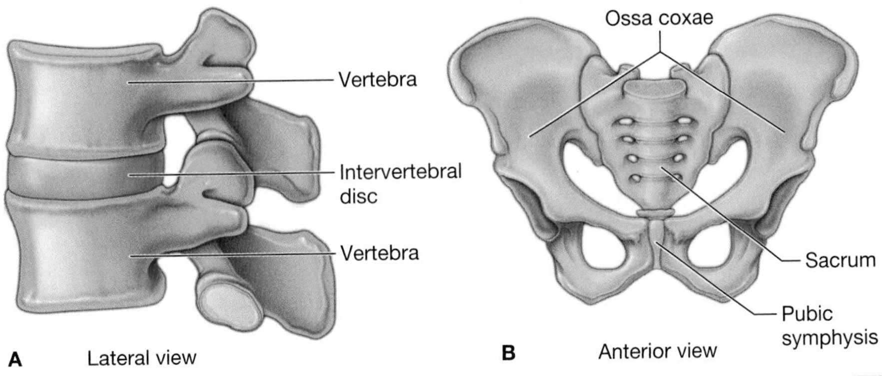

Vertebra
Intervertebral disc
Vertebra
A Lateral view

Ossa coxae
Sacrum
Pubic symphysis
B Anterior view

Cultural Connection

Hip Replacements

Of the approximately 800,000 joints replaced in the United States each year, the majority are hip replacements. Like all joint replacements, they are necessary when the articular cartilage wears thin and exposes the bones to friction. Hip replacements were the first joint replacements ever attempted, with surgeries documented as early as the 1820s. The first materials used in an attempt to replace the cartilage were skin and pig bladders. Knowing what you know now about the properties of cartilage, it's not hard to imagine that these surgeries did not provide long-lasting relief! In addition, these early surgical procedures came with a high risk of infection, since the use of antibiotics in surgery did not become a common practice until the 1940s.

In 1925 the first hip replacement using glass was attempted, but glass, while smooth, was breakable. Surgeons next tried metal, but they found that metal fragments would strip off from the implanted material and become lodged—painfully—in the nearby soft tissues. In the 1960s surgeons finally arrived at the materials that are still used today; the head of the femur is replaced with stainless steel and the acetabulum is replaced with metal. While trial and error was used in medical practice for a long time, we now are able to approach these questions through materials science and biomedical engineering to save lives, money, and pain.

Figure 10.3 Joint Movement Classification

Joints can be classified by the number of anatomical planes or directions in which they allow movement. (A) Uniaxial joints, such as the elbow joint, allow movement in only one plane, a back-and-forth motion. (B) Biaxial joints, such as the metacarpophalangeal joints, allow movement in two directions—in this case forward and backward as well as side to side. (C) The hip joint, which is a multiaxial joint, also allows motion in a circle.

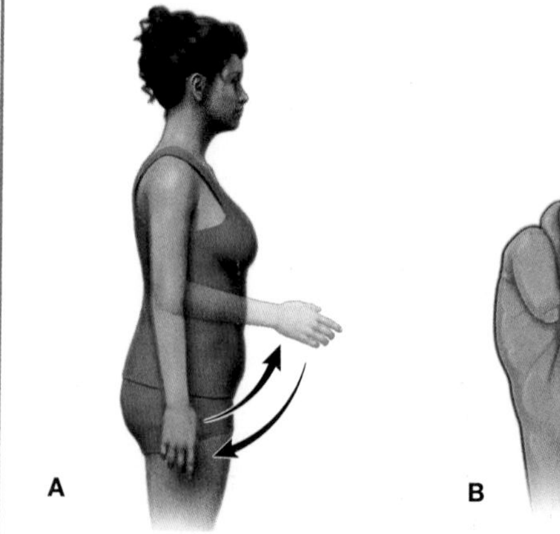

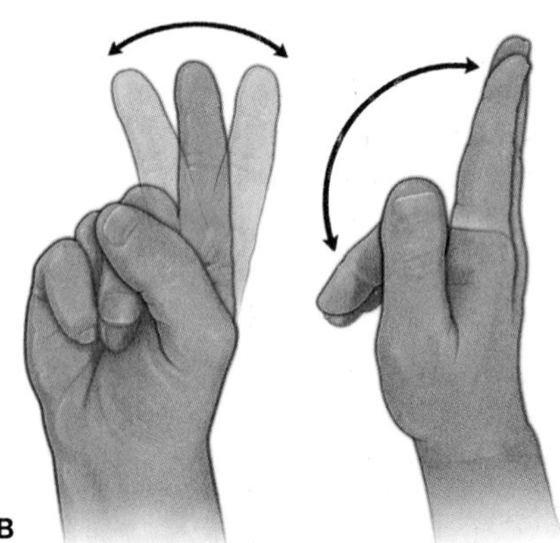

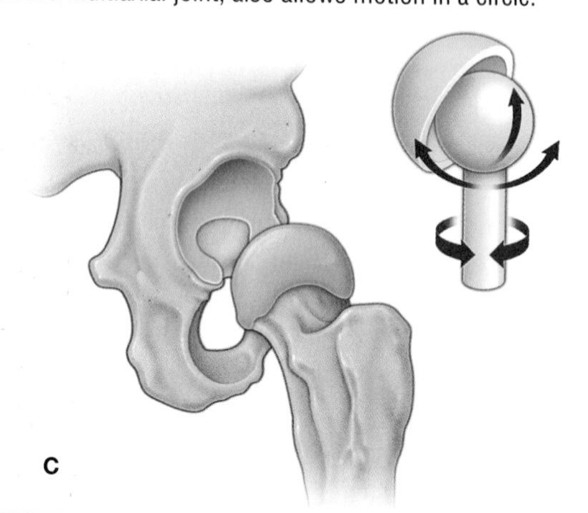

Learning Check

1. Which of the following is a functional classification of a joint? Please select all that apply.
   a. Synarthrosis
   b. Fibrous
   c. Cartilaginous
   d. Diarthrosis
2. The talocalcaneal joint is in the ankle. The articulating surfaces of the two bones (talus and calcaneus) are encased within a joint capsule, which is filled with fluid. This joint is very stable as there is very little movement, if any, at the joint. How would you *functionally* classify the talocalcaneal joint?
   a. Cartilaginous
   b. Synovial
   c. Diarthrosis
   d. Amphiarthrosis
3. The talocalcaneal joint is in the ankle. The articulating surfaces of the two bones (talus and calcaneus) are encased within a joint capsule, which is filled with fluid. This joint is very stable as there is very little movement, if any, at the joint. How would you *structurally* classify the talocalcaneal joint?
   a. Cartilaginous
   b. Synovial
   c. Diarthrosis
   d. Amphiarthrosis

10.2 Fibrous Joints

Learning Objectives: By the end of this section, you will be able to:

10.2.1* Describe the structural features of fibrous joints.

10.2.2* Given the description of a joint, characterize it as a suture, syndesmosis, or gomphosis.

10.2.3* Compare and contrast a suture, syndesmosis, and gomphosis.

* Objective is not a HAPS Learning Goal.

LO 10.2.1

LO 10.2.2

At a fibrous joint, the adjacent bones are directly connected to each other by fibrous connective tissue; thus, the bones do not have a cavity between them, but form a sandwich with a sheet of connective tissue as the filling. There are three types of fibrous joints (**Figure 10.4**). A suture is the narrow fibrous joint found between most bones of the skull. At a syndesmosis joint, the bones are more widely separated but are held together by a sheet of fibrous connective tissue. This type of fibrous joint is found between the shaft regions of the long bones in the forearm and in the leg. Lastly, a gomphosis is the fibrous joint between the roots of a tooth and the bony socket in the jaw into which the tooth fits.

10.2a Suture

LO 10.2.3

All the bones of the skull, except for the mandible, are joined to each other by a fibrous joint called a **suture**. The fibrous connective tissue found at a suture strongly unites the adjacent skull bones and thus helps to protect the brain and form the face. These sutures are not straight lines but are convoluted, like a river, forming a tight union that prevents most movement between the bones (**Figure 10.4A**). Thus, skull sutures are functionally classified as a synarthrosis, although some sutures may allow for slight movements between the cranial bones.

Figure 10.4 Types of Fibrous Joints

Bones joined by a sheet of connective tissue are called fibrous joints. Examples include: (A) sutures between the skull bones, (B) an interosseous membrane, which spans the distance between the shafts of two bones, such as the radius and ulna of the forearm, and (C) a gomphosis, which is a specialized fibrous joint that anchors a tooth in its bony socket.

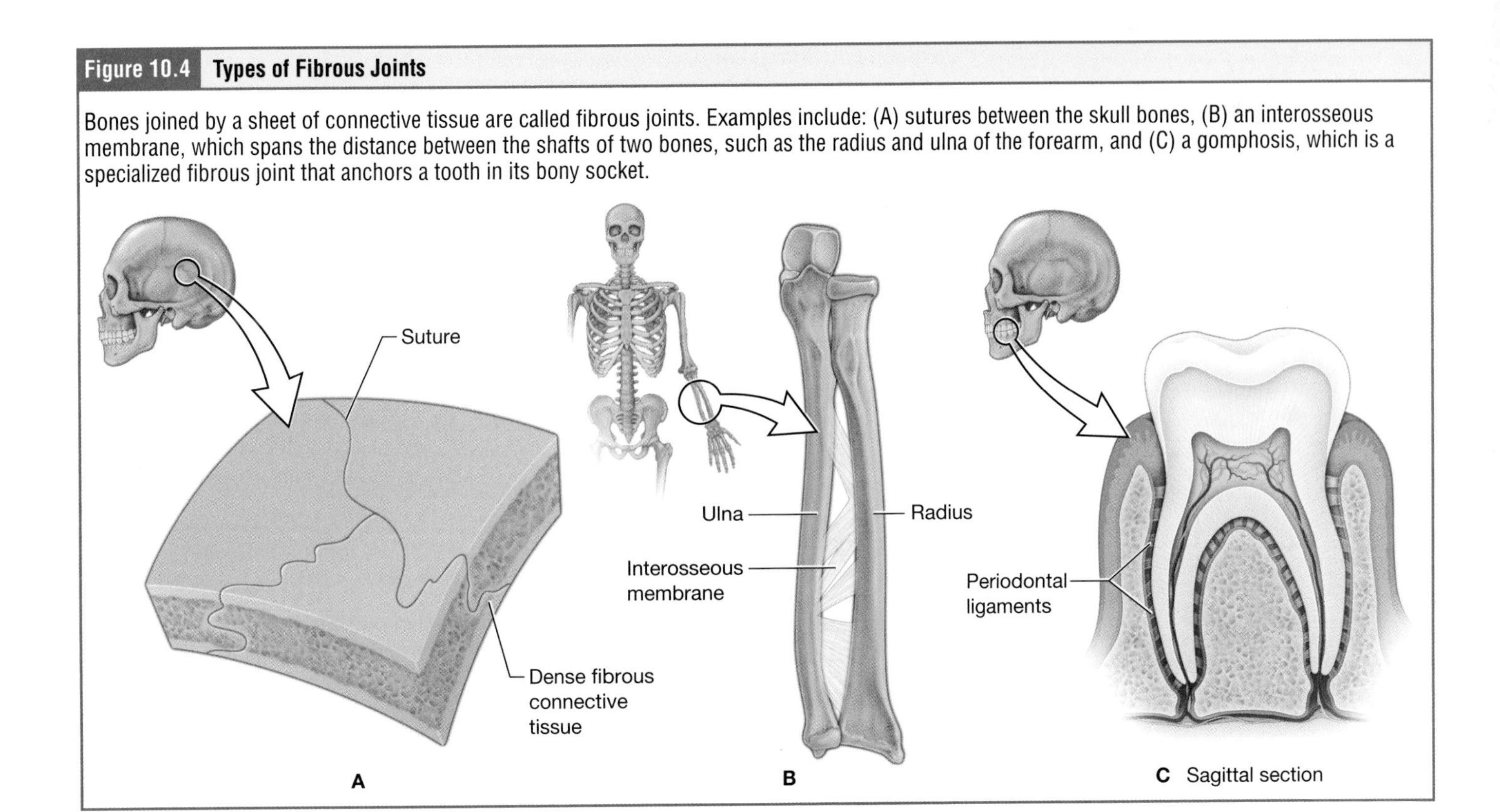

fontanelles ("soft spots")

In newborns and infants, the areas of connective tissue between the bones are much wider, especially in the areas on the top and sides of the skull that will become the sagittal, coronal, squamous, and lambdoid sutures. These broad areas of connective tissue are called **fontanelles** (**Figure 10.5**). During birth, the fontanelles provide flexibility to the skull, allowing the bones to push closer together or to overlap slightly, thus aiding movement of the infant's head through the birth canal. After birth, these expanded regions of connective tissue allow for rapid growth of the skull and enlargement of the brain. The fontanelles greatly decrease in width during the first year after birth as ossification of the skull bones continues. When the connective tissue between the adjacent bones is reduced to a narrow layer, these fibrous joints are now called *sutures* and the fontanelles no longer exist. At some sutures, the connective tissue will ossify and be converted into bone, causing the adjacent bones to fuse to each other later in life.

interosseous membrane (interosseous ligament)

10.2b Syndesmosis

A **syndesmosis** is a type of fibrous joint in which two parallel bones are united to each other by fibrous connective tissue. The gap between the bones may be narrow, with the bones joined by ligaments, or the gap may be wide and filled in by a broad sheet of connective tissue called an **interosseous membrane**.

Student Study Tip

Break down the roots of this word: "inter-" means "between" and "osseous" means "bone," so these joints are sheets of membrane that are found between two parallel bones.

There are two syndesmoses in the human body, each between the parallel bones in the distal limbs. In the forearm, the shaft portions of the radius and ulna bones are strongly united by an interosseous membrane (**Figure 10.4B**). Similarly, in the lower leg the shafts of the tibia and fibula are also united by an interosseous membrane. While these two joints have parallel anatomy, there are some key differences in structure and function. The leg is a site of intense load-bearing both while the body is still during standing and while bearing the impact of the body during walking and running. Thus, stability is a key function for this location. A theme throughout our study of bones is that motion compromises stability, in body locations where stability is required, less motion is desirable. At the distal tibiofibular joint, the articulating surfaces of the bones are bound by fibrous connective tissue and reinforced by ligaments, creating a more stable tibiofibular syndesmosis. In contrast, the arms do not bear the weight of the body, but function in tasks like carrying, grasping, throwing, and even fighting. A more

Figure 10.5 **The Newborn Skull**

Human infants are born with wide soft areas of connective tissue among the bones of the skull. These are called *fontanelles.*

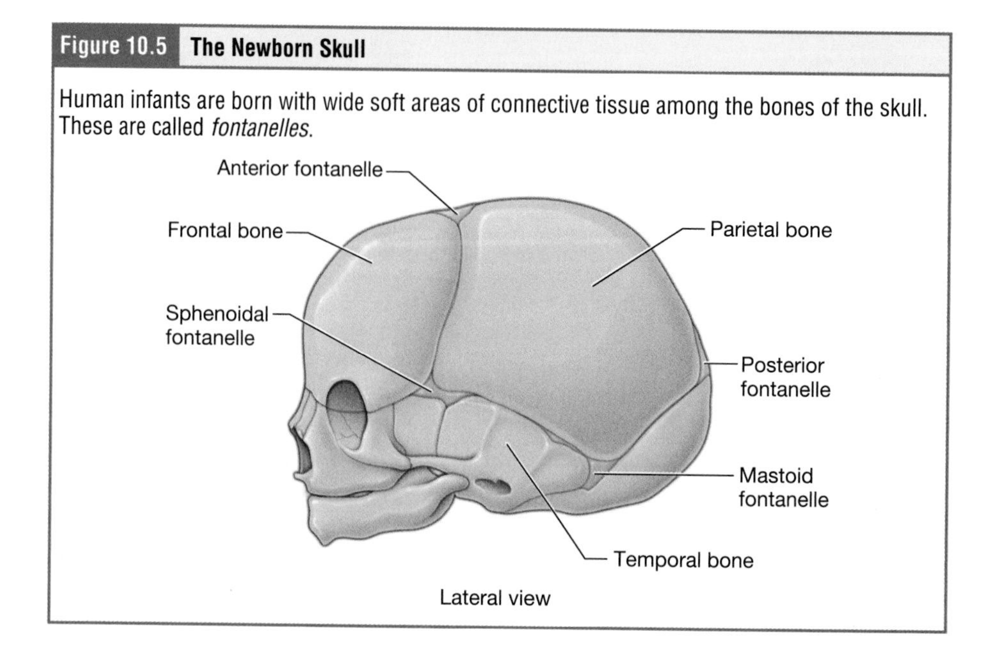

mobile radioulnar syndesmosis has evolved. At this region the proximal radioulnar joint is stabilized by ligaments, but the distal radioulnar joint is much more flexible, allowing the radius to roll over the ulna (this movement is called *pronation*) so that the palm of the hand can face any direction. Because syndesmoses do not prevent all movement between the bones, they are functionally classified as amphiarthroses. The interosseous membranes of the leg and forearm also provide areas for muscle attachment.

10.2c Gomphosis

A **gomphosis** is the specialized fibrous joint that anchors the root of a tooth into its bony socket within the maxillary bone (upper jaw) or mandible bone (lower jaw) of the skull. Spanning between the bony walls of the socket and the root of the tooth are numerous short bands of dense connective tissue, each of which is called a **periodontal ligament** (**Figure 10.4C**). Due to the immobility of a gomphosis, this type of joint is functionally classified as a synarthrosis.

Learning Check

1. Which of the following is NOT a fibrous joint?
   a. Suture
   b. Syndesmosis
   c. Synarthrosis
   d. Gomphosis
2. There is a sheet of connective tissue between the occiput and temporal bone. This prevents the bones from separating. What kind of fibrous joint is found here?
   a. Suture
   b. Syndesmosis
   c. Synarthrosis
   d. Gomphosis

10.3 Cartilaginous Joints

Learning Objectives: By the end of this section, you will be able to:

10.3.1* Describe the structural features of cartilaginous joints.

10.3.2* Given a description of a joint, characterize it as either a synchondrosis or symphysis.

10.3.3* Compare and contrast a synchondrosis and symphysis.

* Objective is not a HAPS Learning Goal.

LO 10.3.1

At a cartilaginous joint, the adjacent bones are united by either hyaline cartilage or fibrocartilage (**Figure 10.6**). Cartilaginous joints are divided into two types depending on if they are joined by hyaline cartilage or fibrocartilage. In a **synchondrosis** the bones are joined by hyaline cartilage, examples include the epiphyseal plates of long bones and places where bone is united to a cartilage structure, such as between the anterior end of a rib and the costal cartilage of the thoracic cage. The second type of cartilaginous joint is a **symphysis**, where the bones are joined by fibrocartilage.

LO 10.3.2

10.3a Synchondrosis

LO 10.3.3

A *synchondrosis* ("joined by cartilage") is a cartilaginous joint where bones are joined together by hyaline cartilage, or where bone is united to hyaline cartilage.

Figure 10.6 The Two Types of Cartilaginous Joints

At cartilaginous joints, bones are united either by hyaline cartilage to form a synchondrosis or by fibrocartilage to form a symphysis. (A) The hyaline cartilage found between the diaphysis and epiphysis of a long bone forms the growth area known as the *epiphyseal plate;* this is an example of a synchondrosis. (B) The pubic portions of the two ossa coxae meet at the midline of the body, but are protected from contacting each other by a pad of fibrocartilage, forming the *pubic symphysis.*

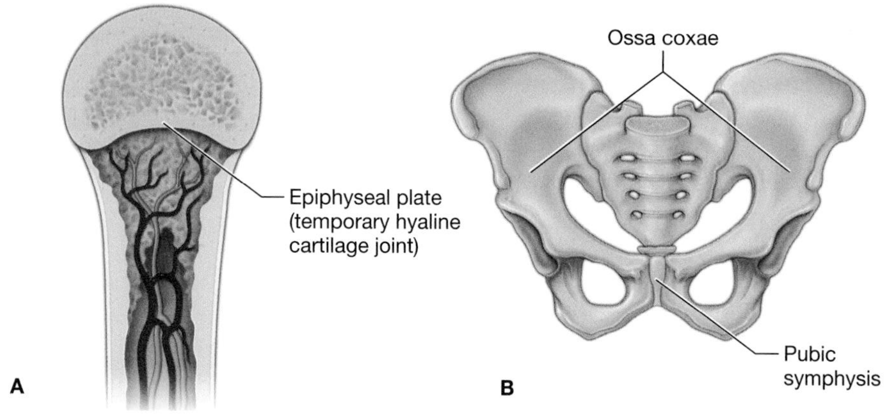

Synchondroses are found in every long bone early in life at the epiphyseal plate (growth plate). The epiphyseal plate is the region of growing hyaline cartilage that unites the diaphysis (shaft) of the bone to the epiphysis (end of the bone) (**Figure 10.6A**). During an individual's late teens and early twenties, the epiphyseal plate in each bone is ossified, the cartilage is completely replaced by bone. Once this occurs, bones cannot grow longer and the synchondrosis is no longer present.

Synchondroses are present in other bones as well. In the young skeleton, synchondroses are present at the sites where the ilium, ischium, and pubic bones meet. When body growth stops, these sites of cartilage disappear and are replaced by bone, forming the single right and left hip bones of the adult.

The joints between the ribs and costal cartilages are also synchondroses and are consistently present throughout life. At these joints, the anterior end of the rib joins to the sternum with a long bar of costal cartilage between. While the sternal end of all but the first ribs is a synovial joint, which you will learn about in Section 9.4, the joint between the rib and its costal cartilage is a synchondrosis. Due to the lack of movement between the bone and cartilage, all synchondroses are functionally classified as synarthroses.

10.3b Symphysis

A joint in which the two bones meet at a fibrocartilage pad is called a *symphysis.* Fibrocartilage is very strong because it contains numerous bundles of thick collagen fibers, thus giving it a much greater ability to resist pulling and bending forces compared with hyaline cartilage. This gives symphyses the ability to strongly unite the adjacent bones, but can still allow for some limited movement to occur. Thus, a symphysis is functionally classified as an amphiarthrosis.

Examples of symphysis joints include the pubic symphysis, the manubriosternal joint, and the intervertebral symphysis. In addition to providing a strong attachment, fibrocartilage pads at all these locations provide cushioning between the bones, which is important when carrying heavy objects or during high-impact activities such as running or jumping.

Learning Check

1. Which of the following could form cartilaginous joints? Please select all that apply.
   a. Hyaline cartilage
   b. Fibrocartilage
   c. Elastic cartilage
   d. Sheet of connective tissue
2. During the development stage, which cartilaginous joint would be more prevalent?
   a. Synchondrosis joint
   b. Symphysis joint

10.4 Synovial Joints

HAPS LEARNING GOALS AND OUTCOMES

Learning Objectives: By the end of this section, you will be able to:

10.4.1 Identify and describe the major structural components of a typical synovial joint.

10.4.2 For each of the six structural types of synovial joints, describe its anatomic features, identify locations in the body, and predict the kinds of movement each joint allows.

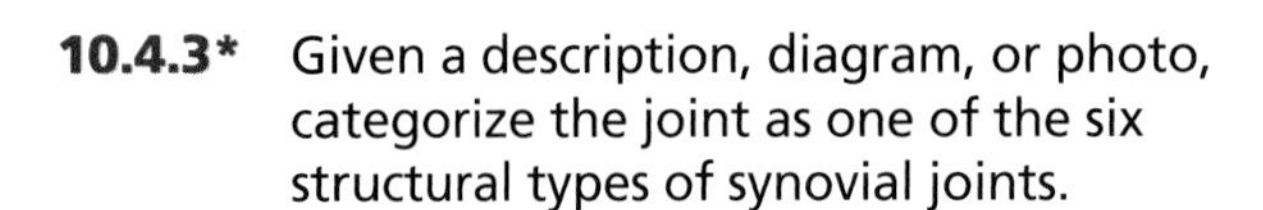

10.4.3* Given a description, diagram, or photo, categorize the joint as one of the six structural types of synovial joints.

* Objective is not a HAPS Learning Goal.

Synovial joints are the most common type of joint in the body (see the "Anatomy of a Typical Synovial Joint" feature). Synovial joints are unique because they feature an enclosed cavity that is filled with fluid. The fluid bathes the ends of the bones, providing both nourishment and lubrication. Also, unlike fibrous or cartilaginous joints, the bone surfaces within a synovial joint do not directly connect to each other, allowing for a greater degree of movement.

Anatomy of...

A Typical Synovial Joint

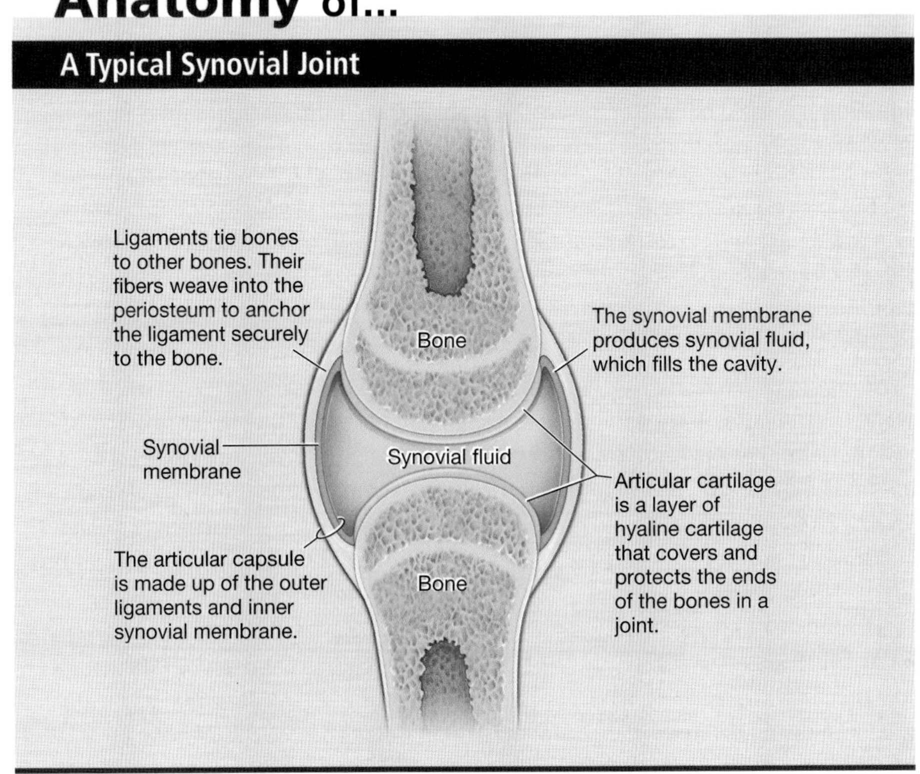

10.4a Structural Features of Synovial Joints

LO 10.4.1

Synovial joints are characterized by the presence of a fluid-filled joint cavity (see the "Anatomy of a Typical Synovial Joint" feature). The walls of this space are formed by the **articular capsule**, a fibrous connective tissue structure that is attached to each bone just outside the area of the bone's articulating surface. The articular capsule has two layers: the ligament that ties the bones together and the inner **synovial membrane**. The cells of this membrane secrete **synovial fluid**, a thick, slimy fluid that provides lubrication to further reduce friction between the bones of the joint. The bones of the joint are tied together by tough ligaments, but they do not touch within the joint cavity. Friction and stress on the bones is eased by the synovial fluid as well as by the presence of the **articular cartilage**, a thin layer of hyaline cartilage that covers the surface of each bone. The articular cartilage allows the bones to move smoothly against each other without damaging the underlying bone tissue. Hyaline cartilage lacks both blood vessels and nerves, so movements of the bones past each other are painless and there is no risk of damage. The synovial fluid provides nourishment to the articular cartilage, which, without its own blood vessels, needs to get nourishment externally. Each time the joint moves, the synovial fluid is compressed and moves through the articular cartilage; think of water moving through a sponge when it is squeezed. Therefore, regular movement at synovial joints is important for their health. An overly sedentary lifestyle compromises joint health because without regular movement the articular cartilage may not get enough nutrients from the synovial fluid and the chondrocytes may suffer. Each synovial joint is functionally classified as a diarthrosis.

Learning Connection

Image Translation

Do an image search in your web browser for an X-ray of synovial joints you know, such as the knee. What does the joint look like in the image? What joint components can you see in the X-ray, and which can you not see?

10.4b Cushioning and Support Structures Associated with Synovial Joints

Ligaments and Tendons Outside of their articulating surfaces, the bones are connected together by ligaments, which are strong bands of fibrous connective tissue. The ligaments form the outer layer of the articular capsule. These strengthen and support the joint by anchoring the bones together and preventing their separation. Ligaments are classified based on their relationship to the fibrous articular capsule. An **extrinsic ligament** is located outside of the articular capsule, an **intrinsic ligament** is fused to or incorporated into the wall of the articular capsule, and an **intracapsular ligament**

Student Study Tip

Use alliteration. "Ligaments" connect "like" structures to one another, so you can remember they connect bone to bone, while tendons connect bone to muscle.

Apply to Pathophysiology

Rheumatoid Arthritis

The suffix *-itis* indicates inflammation at a structure, so *arthro* (joint) with *-itis* indicates inflammation at the joints. Joints can become inflamed for many reasons (some of which are explored in other places in this chapter). In an autoimmune disease called *rheumatoid arthritis*, the immune system mistakenly identifies a protein within the body as a threat and manufactures other proteins called *antibodies* to attack it. Since this protein is found within the synovial membrane, the attack destroys cells within the synovial membrane.

1. What is the product made by the cells of synovial membrane?
   A) Collagen fibers
   B) Elastin fibers
   C) Synovial fluid
2. What is the function of this product?
   A) Adding strength and rigidity to the joint
   B) Adding flexibility to the joint
   C) Providing cushioning and nourishment to the joint structures
   D) Immune defense at the joint
3. The decrease in this product will primarily impact which cells?
   A) Chondrocytes in articular cartilage
   B) Osteocytes in bone matrix
   C) Fibroblasts in the ligaments

is located inside of the articular capsule. At many synovial joints, additional support is provided by the muscles and their tendons that act across the joint. A **tendon** is the dense connective tissue structure that attaches a muscle to bone.

Cushioning Structures A few synovial joints of the body have a pad of fibrocartilage located between the bones within the articular capsule. These might take the form of an **articular disc** or a **meniscus**. These structures can serve several functions, depending on the specific joint. In some places, an articular disc may act to strongly unite the bones of the joint to each other, similar to a cartilaginous joint, but with a synovial capsule. Examples of this structure include the articular discs found at the sternoclavicular joint or between the distal ends of the radius and ulna bones. At other synovial joints, the fibrocartilage provides shock absorption and cushioning between the bones, which is the function of each meniscus within the knee joint. Finally, an articular disc can serve to smooth the movements between the articulating bones, as seen at the temporomandibular joint.

articular disc (meniscus)

Additional structures located outside of a synovial joint serve to prevent friction between the bones of the joint and the overlying muscle tendons or skin (see the "Anatomy of Cushioning at Joints" feature). Bursae and tendon sheaths are both structures that involve additional pockets of synovial fluid. A **bursa** (plural = bursae) is a small sac of synovial fluid. Bursae are located in regions where ligaments, muscles, or tendons might rub against bone (see "Anatomy of Cushioning at Joints" and **Figure 10.7A**). Bursae reduce friction by separating the structures and providing cushioning.

A **tendon sheath** has the same components and structure of a bursa but is elongated and wraps around a muscle tendon at places where the tendon crosses a joint (see "Anatomy of Cushioning at Joints" and **Figure 10.7B**). Tendon sheaths also contain synovial fluid that allows for smooth motions of the tendon during muscle contraction and joint movements.

Fat pads are small structures composed of adipose tissue found at many joints (see "Anatomy of Cushioning at Joints" feature). Their function is not entirely understood and

Figure 10.7 Bursae and Tendon Sheaths

(A) Bursae are fluid-filled sacs that provide cushioning between muscles or tendons and underlying bone. Here, the knee has several bursae, each bursa protecting bone and ligaments/tendons at areas of friction. (B) Tendon sheaths are envelopes of synovial fluid that wrap around tendons in places of friction. Here at the wrist, we see that at the location marked, the tendon sheaths are held close to the bone by the ligament. As the tendons move during thumb movement, they would rub against the bone. To reduce friction at this location, tendon sheaths envelop and protect the tendons.

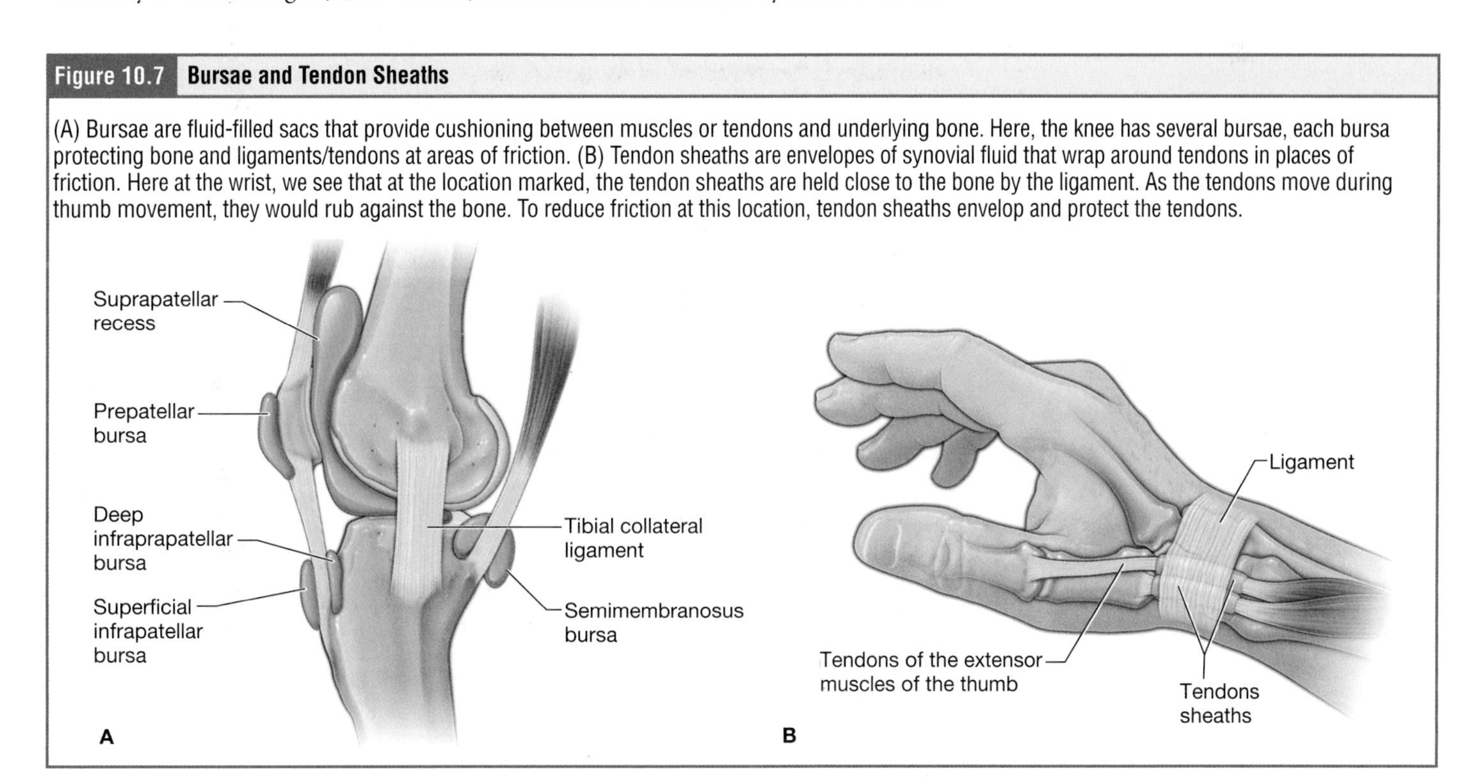

Anatomy of...

Cushioning at Joints

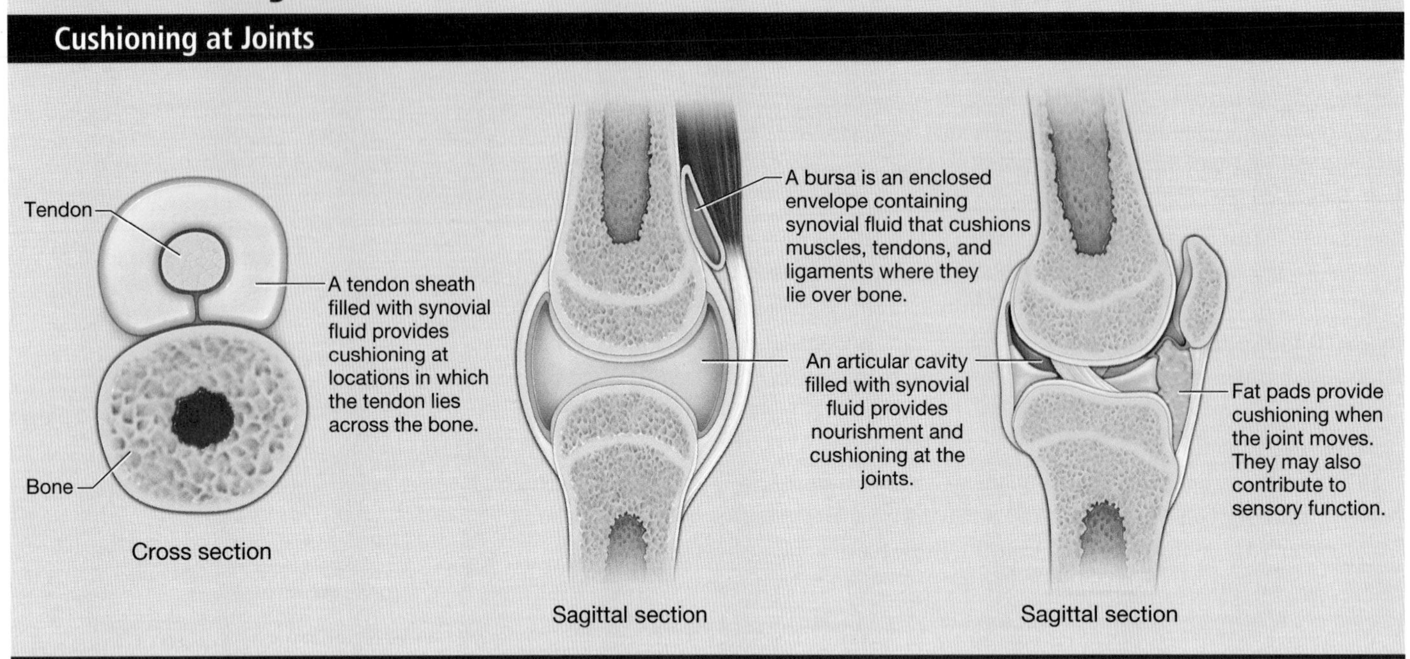

Tendon
A tendon sheath filled with synovial fluid provides cushioning at locations in which the tendon lies across the bone.
Bone
Cross section
A bursa is an enclosed envelope containing synovial fluid that cushions muscles, tendons, and ligaments where they lie over bone.
An articular cavity filled with synovial fluid provides nourishment and cushioning at the joints.
Sagittal section
Fat pads provide cushioning when the joint moves. They may also contribute to sensory function.
Sagittal section

may not be consistent for every fat pad. They provide cushioning, especially when the joint is flexed or in motion, and they may fill newly exposed spaces when a joint moves. Fat pads are innervated and may contribute to sensory function during joint movement.

10.4c Types of Synovial Joints

LO 10.4.2

Synovial joints are subdivided based on the shapes of the articulating surfaces of the bones that form each joint. The six types of synovial joints are pivot, hinge, condyloid, saddle, plane, and ball-and-socket joints (**Figure 10.8**).

pivot joint (trochoid joint)

LO 10.4.3

Pivot Joint At a **pivot joint**, a rounded portion of a bone is enclosed within a ring formed partially by the articulation with another bone and partially by a ligament (**Figure 10.8A**). The bone rotates within this ring. Since the rotation is around a single axis, pivot joints are

DIGGING DEEPER:
Bursitis

The suffix *-itis* means inflammation. Bursae function to help joints move more fluidly by sheltering ligaments and tendons from fiction against the bones of the joint. Bursitis is the inflammation of a bursa. This will cause pain, swelling, or tenderness of the bursa and surrounding area, and may also result in decreased mobility at the joint. Bursitis is most commonly associated with the bursae found at or near the shoulder, hip, knee, or elbow joints. At the knee, inflammation and swelling of the bursa located between the skin and patella bone is prepatellar bursitis ("housemaid's knee"), a condition more commonly seen today in roofers or floor and carpet installers who do not use knee pads. At the elbow, olecranon bursitis is inflammation of the bursa between the skin and olecranon process of the ulna. The olecranon forms the bony tip of the elbow, and bursitis here is also known as "student's elbow."

Bursitis can be either acute (lasting only a few days) or chronic. It can arise from mechanical issues such as muscle overuse, trauma, and excessive or prolonged pressure on the skin. Other issues such as rheumatoid arthritis (which you read about in the "Apply to Pathophysiology: Rheumatoid Arthritis" feature), gout, or infection can all lead to inflammation of the bursae as well. In these cases, only addressing the primary issue (such as the infection if that is the cause) will alleviate the bursitis. Any type of anti-inflammatory agent should help to relieve the inflammation until the source can be found.

Figure 10.8 Types of Synovial Joints

The six types of synovial joints are specialized to accomplish different types of movement. (A) Pivot joints feature rotation around an axis, such as between the first and second cervical vertebrae, which allows for turning of the head side to side (as you might when you say "no"). (B) Hinge joints feature the rounded end of one bone that another bone moves around within one plane. (C) Condyloid joints feature one rounded bone end that is cupped within a bowl-like depression of another bone. (D) Saddle joints feature two curved bones, like saddles, that fit into each other complementarily. (E) Plane joints (such as those between the tarsal bones of the foot) feature flattened, articulating, bony surfaces on both bones, which allow for limited gliding movements. (F) Ball-and-socket joints (such as the hip and shoulder joints) feature a rounded head of a bone moving within a cup-shaped depression in another bone.

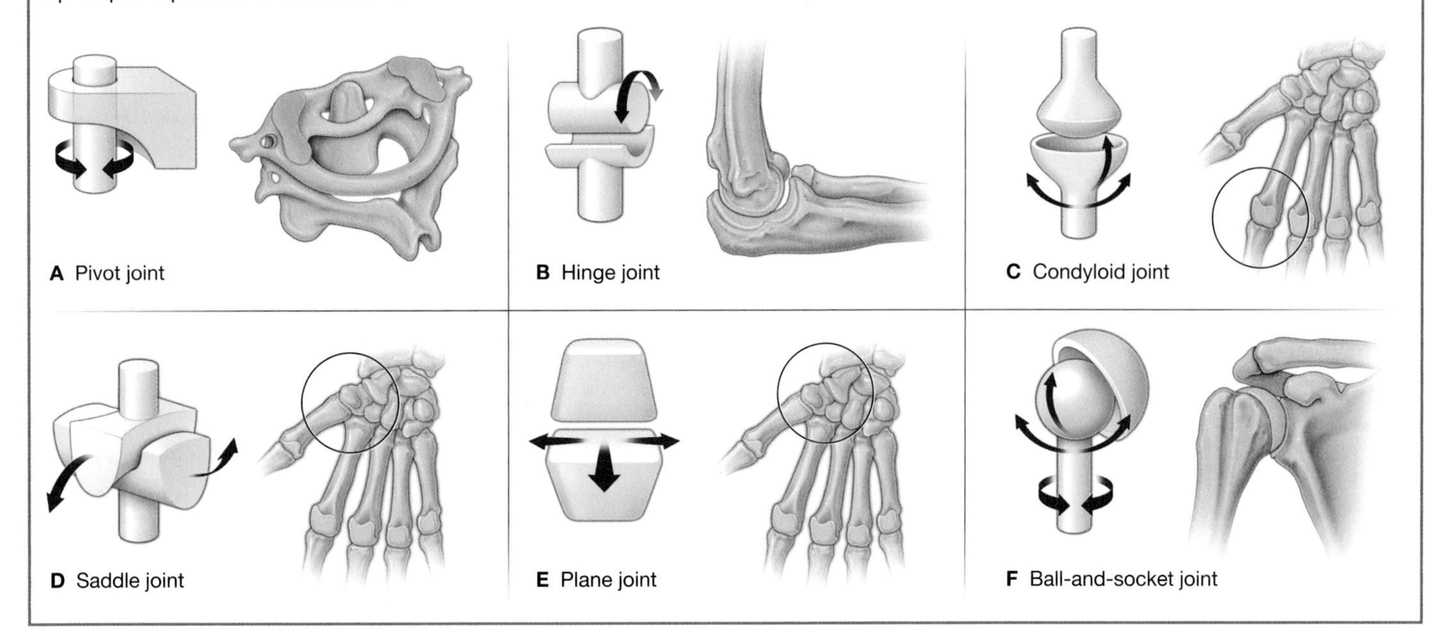

functionally classified as a uniaxial diarthrosis type of joint. An example of a pivot joint is the atlantoaxial joint, found between the C1 (atlas) and C2 (axis) vertebrae. Here, the upward projecting dens of the axis articulates with the inner aspect of the atlas, where it is held in place by a ligament. Rotation at this joint allows you to turn your head from side to side. A second pivot joint is found at the **proximal radioulnar joint**. Here, the head of the radius is largely encircled by a ligament that holds it in place as it articulates with the radial notch of the ulna. Rotation of the radius allows for forearm movements.

Hinge Joint In a **hinge joint**, the convex end of one bone articulates with the concave end of the adjoining bone (**Figure 10.8B**). This type of joint allows only for bending and straightening motions along a single axis, and thus hinge joints are functionally classified as uniaxial joints. A good example is the elbow joint, with the articulation between the trochlea of the humerus and the trochlear notch of the ulna. Other hinge joints of the body include the knee, ankle, and interphalangeal joints between the phalanx bones of the fingers and toes.

Student Study Tip

Hinge joints operate similarly to doors. The only way a door can move while on its hinges is by opening or closing. Place your arm straight out in front of you to flex and extend your elbow, and you can see that the movement resembles a door opening and closing.

Condyloid Joint At a **condyloid joint** (ellipsoid joint), the shallow depression at the end of one bone articulates with a rounded structure from an adjacent bone or bones (**Figure 10.8C**). The knuckle (metacarpophalangeal) joints of the hand between the distal end of a metacarpal bone and the proximal phalanx bone are condyloid joints. Functionally, condyloid joints are biaxial joints that allow for two planes of movement. One movement involves the bending and straightening of the fingers or the anterior-posterior movements of the hand. The second movement is a side-to-side movement, which allows you to spread your fingers apart and bring them together, or to move your hand in a medial/lateral direction.

condyloid joint (ellipsoid joint)

saddle joint (sellar joint)

Saddle Joint At a **saddle joint**, both of the articulating surfaces for the bones have a saddle shape, which is concave in one direction and convex in the other (**Figure 10.8D**). This allows the two bones to fit together like a rider sitting on a saddle. Saddle joints are functionally classified as biaxial joints. The best example in the human body is the first carpometacarpal joint, between the trapezium (a carpal bone) and the first metacarpal bone at the base of the thumb. This joint provides the thumb the ability to move in almost a complete circle. The thumb can move within the same plane as the palm of the hand, which you can see if you put your hand flat on a desk and move your thumb close to and away from the palm. In a second plane of motion, the thumb can move anteriorly, almost laying across the palm. This movement of the first carpometacarpal joint is what gives humans their distinctive "opposable" thumbs. The sternoclavicular joint is also classified as a saddle joint.

plane joint (gliding joint)

Plane Joint At a **plane joint**, the surfaces of the bones are mostly flat and the bones to slide past each other during joint motion (**Figure 10.8E**). The motion at this type of joint is usually small and tightly constrained by surrounding ligaments. While small, the motions available at a plane joint are often in many or even in every direction, thus plane joints can be functionally classified as a multiaxial joint. However, not all of these movements are available to every plane joint due to limitations placed on it by ligaments or neighboring bones. Plane joints are found between the carpal bones (intercarpal joints) of the wrist or tarsal bones (intertarsal joints) of the foot, between the clavicle and acromion of the scapula (acromioclavicular joint), and between the superior and inferior articular processes of adjacent vertebrae (zygapophysial joints).

Ball-and-Socket Joint The joint type with the greatest range of motion is the **ball-and-socket joint**. At these joints, the rounded head of one bone (the ball) fits into the bowl-shaped socket of the adjacent bone (**Figure 10.8F**). The hip joint and the glenohumeral (shoulder) joint are the only ball-and-socket joints of the body. The bone with the "ball" of the ball and socket can move in a circular motion, therefore ball-and-socket joints are classified functionally as multiaxial joints.

DIGGING DEEPER: Joint Damage

Arthritis is a common disorder of synovial joints that involves inflammation of the joint. This often results in significant joint pain, along with swelling, stiffness, and reduced joint mobility. There are more than 100 different forms of arthritis. Arthritis may arise from aging, damage to the articular cartilage, autoimmune diseases, bacterial or viral infections, or unknown (probably genetic) causes.

The most common type of arthritis is osteoarthritis, which is associated with aging and "wear and tear" of the articular cartilage (see Figure A). Osteoarthritis is a disease of the hyaline cartilage at the joint surfaces. Any factors that put stress on the articular cartilage, such as extraordinary weight-bearing, repetitive activities, injuries, or changes to the synovial fluid can cause the chondrocytes to struggle to produce cartilage matrix, causing the cartilage to gradually become thinner. As the articular cartilage layer wears down, more pressure is placed on the bones. The bone tissue underlying the damaged articular cartilage also responds by thickening, producing irregularities and causing the surface of the bone to become rough or bumpy. Joint movement then results in pain and inflammation. In its early stages, symptoms of osteoarthritis may be reduced by mild activity that "warms up" the joint, but the symptoms may worsen following exercise. In individuals with more advanced osteoarthritis, the affected joints can become more painful and therefore are difficult to use effectively, resulting in increased immobility.

For severely arthritic joints, a surgery to replace the hyaline cartilage may be an option. This type of surgery involves replacing the articular surfaces of the bones with artificial materials. For example, in hip replacement, the worn or damaged parts of the hip joint, including the head

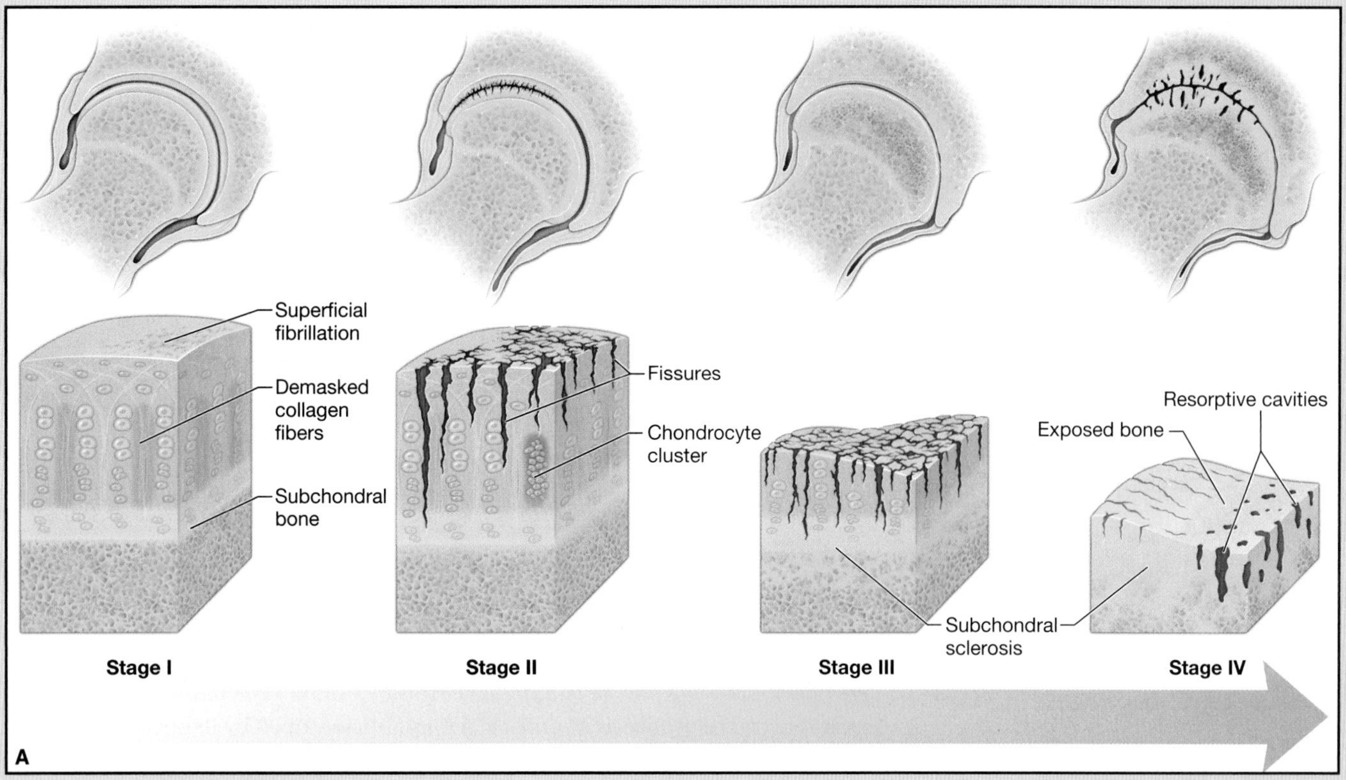

Superficial fibrillation
Demasked collagen fibers
Subchondral bone
Fissures
Chondrocyte cluster
Exposed bone
Resorptive cavities
Subchondral sclerosis
Stage I
Stage II
Stage III
Stage IV
A

and neck of the femur and the acetabulum of the pelvis, are removed and replaced with artificial joint components (see Figure B). The parts, which are always built in advance of the surgery, are sometimes custom-made to produce the best possible fit for a patient.

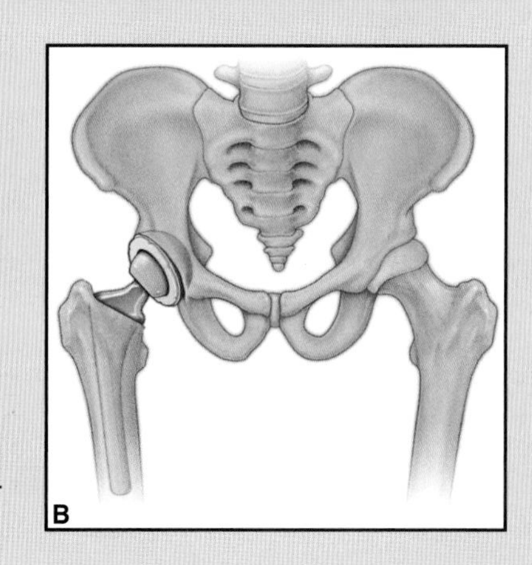

Gout is a form of arthritis that results from the deposition of uric acid crystals within a body joint. Usually only one or a few joints are affected, such as the big toe, knee, or ankle. The attack may only last a few days, but may return to the same or another joint. Gout occurs when the body makes too much uric acid or the kidneys do not properly excrete it. A diet with excessive fructose has been implicated in raising the chances of a susceptible individual developing gout.

Other forms of arthritis are associated with various autoimmune diseases, bacterial infections of the joint, or unknown genetic causes. In all forms of arthritis, there is a decrease in the thickness of the articular cartilage, and the innervated bones begin to respond both with pain as well as inflammation and structural changes.

Learning Check

1. Which of the following would be the most superficial structure in a synovial joint?
   a. Synovial fluid
   b. Ligaments
   c. Synovial membrane
   d. Articular cartilage
2. Which of the following joints has the most range of motion?
   a. Plane
   b. Condyloid
   c. Ball-and-Socket
   d. Pivot
3. Which of the following is an example of a pivot joint?
   a. Atlantoaxial joint
   b. Intervertebral joint
   c. Glenohumeral joint
   d. Intercarpal joint

(*Continued*)

4. The shoulder joint has the greatest number of bursae of any joint in the body: six bursae. The shoulder joint is composed of large number of muscles, ligaments, and bones. It also has the greatest range of motion. Which of the following explains why the shoulder has the most bursae?
   a. The bursae serve to strongly unite bones of the shoulder joint together.
   b. The bursae serve to decrease the friction between the tendons, ligaments, and bones.
   c. The bursae provide cushioning when the shoulder is in motion.
   d. The bursae contribute to sensory functioning during movement.

10.5 Movements at Synovial Joints

Learning Objectives: By the end of this section, you will be able to:

10.5.1 Define the movements that typically occur at a joint (e.g., flexion, extension, abduction, adduction, rotation, circumduction, inversion, eversion, protraction, retraction).

10.5.2* For a given joint, list (if you know) or predict (if a new joint) the motion available based on joint anatomy.

* Objective is not a HAPS Learning Goal.

Learning Connection

Quiz Yourself

Try writing each of these new directional phrases on a flashcard (or your favorite digital study device) and then quiz yourself or a friend. Don't just spit back definitions; move your body to illustrate each motion! Remember that when you take an A&P exam, your body comes to the exam with you. If you learned these words while doing the motion, you can do the motion during the exam to help you remember!

Synovial joints allow the body a tremendous range of movements. The type of movement that can be produced at a synovial joint is determined by its structural type. While the ball-and-socket joint gives the greatest range of movement at an individual joint, in other regions of the body, several joints may work together to produce a particular movement. Remember as we discuss body movements in this section, we are using directional terms and anatomical position of the body. If you need a refresher on these words, reviewing Chapter 2 will help make the following section much easier to follow.

10.5a Flexion and Extension

LO 10.5.1

LO 10.5.2

The simplest way to think of **flexion** and **extension** is that flexion reduces the angle of the joint from its resting position and extension returns the joint to its resting position (**Figure 10.9A** and **10.9B**). For example, start with your body in anatomical position. Now examine your elbow joint. It is a straight-line, or 180 degrees. Now bend your elbow halfway. The angle of the joint is reduced, from 180 degrees to 90 degrees. If you return your elbow back to its straight-line starting position, you are increasing the angle of the joint back to 180 degrees and extending it. For a more complex example, we can examine the vertebral column. Flexion is an anterior (forward) bending of the neck or body, while extension involves a posterior-directed motion, such as straightening from the flexed position. **Hyperextension** is increasing the joint angle beyond 180 degrees (**Figures 10.9C** and **10.9D**). Most joints are not capable of hyperextension, but one example is bending the spine backward as you do if you are seated and you look up toward the ceiling. Since most joints are not capable of hyperextension, this motion sometimes results in injuries. Hyperextension injuries are common at hinge joints such as the knee or elbow. **Lateral flexion** is the bending of the neck or body toward the right or left side.

Student Study Tip

Abduction is moving away from the body, so you can think of when someone has been abducted, they've been taken away.

10.5b Abduction and Adduction

Abduction and **adduction** motions occur on the limbs. Abduction is a motion that pulls a limb, finger, toe, or thumb away from the middle of the body, while adduction is the opposing movement that brings the limb back toward the body or even past the

Figure 10.9 Movements of the Body (Parts A–G)

(A–B) *Flexion* decreases a joint angle and *extension* brings the joint back to its resting position (usually back to a straight line). (C–D) Extension beyond the straight line, increasing the joint angle to greater than 180 degrees, is called *hyperextension*. (E) *Abduction* (moving the structure away from the body, or spreading the fingers and toes) and *adduction* (bringing the structure toward the body or bringing the fingers and toes together) are motions of the limbs, hands, fingers, or toes. *Circumduction* (moving a structure in a circular pattern) combines flexion, adduction, extension, and abduction. (F) Turning of the head or twisting of the trunk is described as *rotation*. The limbs are capable of medial and lateral rotation if they turn the anterior side toward or away from the midline. (G) The radius and ulna are parallel in anatomical position, which is called *supination*. When turning the hand so that the palm faces posteriorly, the radius must rotate over the ulna, a movement called *pronation*.

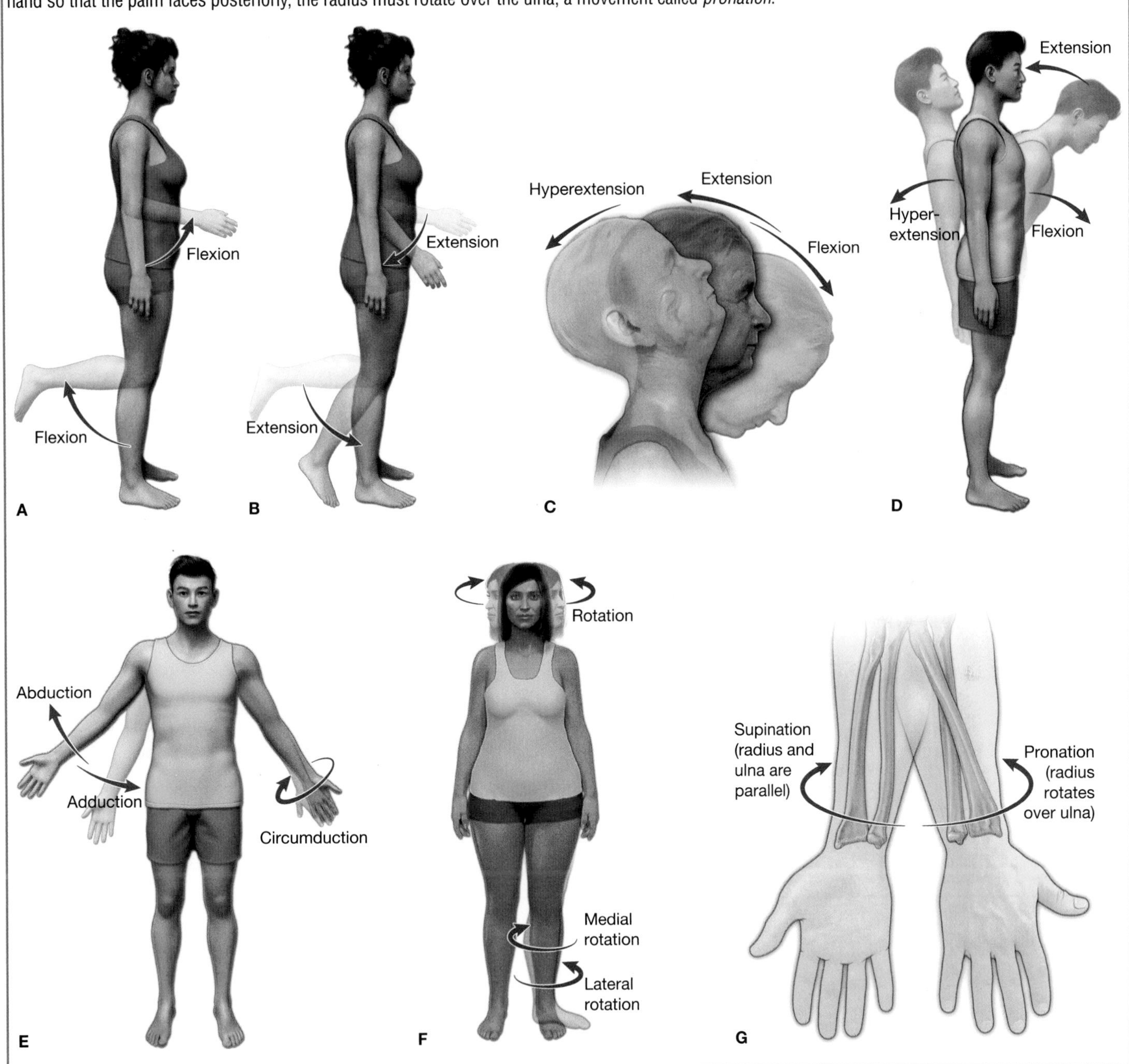

midline of the body (**Figure 10.9E**). An example of abduction is raising the arms straight out to the sides so that your body forms a letter T; adduction brings the arms back down to the side of the body. Similarly, abduction and adduction at the wrist moves the hand away from or toward the midline of the body. Spreading the fingers or toes apart is also abduction, while bringing the fingers or toes together is adduction.

Figure 10.9 Movements of the Body (Parts H–L)

(H) The ankle is a hinge joint capable of two movements: *dorsiflexion*, in which the top of the foot moves closer to the anterior leg, and *plantar flexion*, in which the foot and leg approach a straight line, often referred to as pointing the toes. (I) Eversion and inversion of the foot are accomplished through the plane joints of the tarsal bones. *Eversion* moves the foot so that the sole of the foot is facing laterally, and *inversion* moves the foot so that the sole faces medially. (J) *Protraction* of the mandible pushes the chin forward as the mandible moves forward, and *retraction* returns the mandible to its resting position. (K) The opening and closing of the mouth is accomplished through *depression of the mandible* (opening) and *elevation* (closing). (L) *Opposition* moves the thumb so that it can contact the fingers, and *reposition* restores anatomical position.

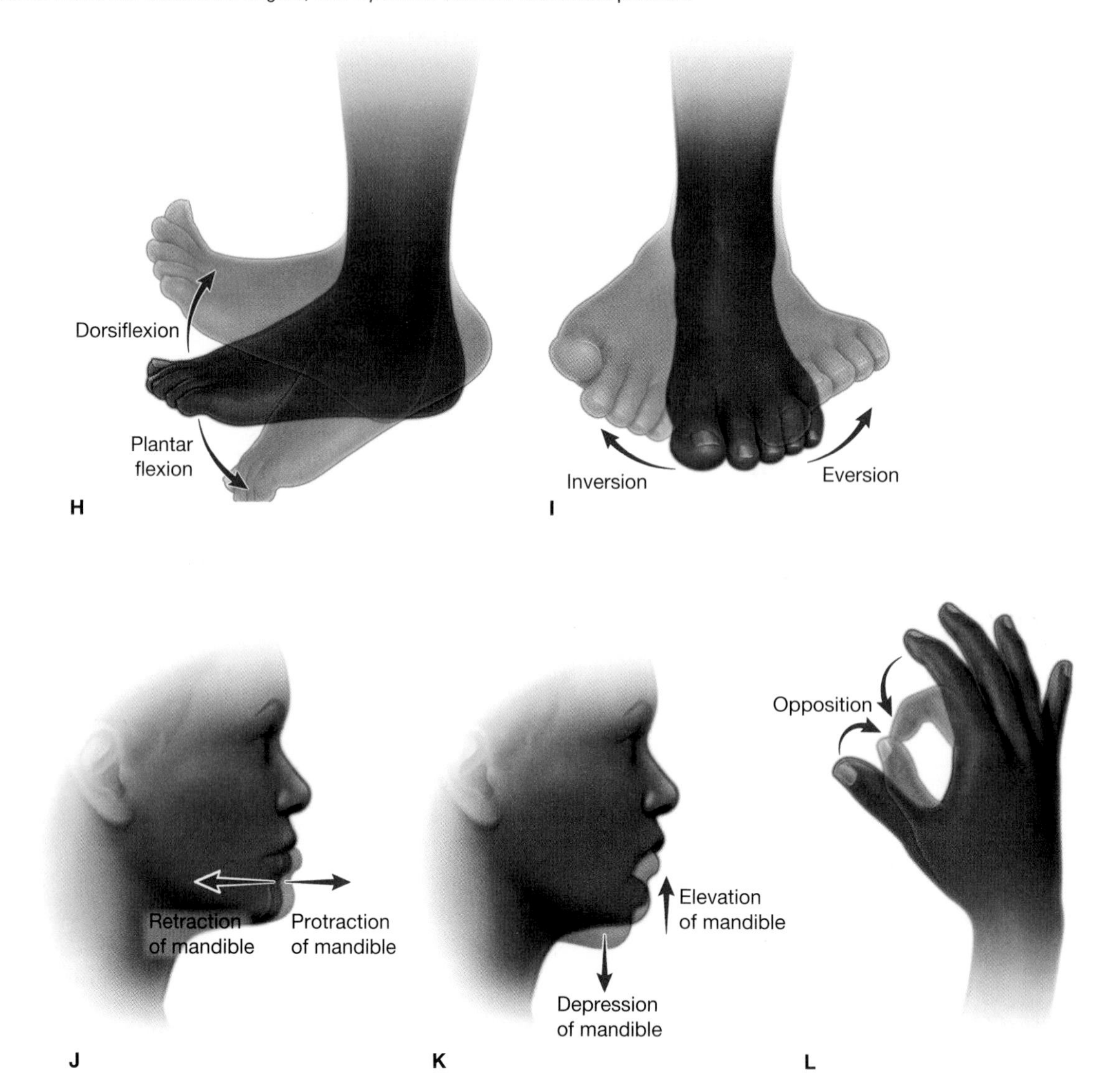

10.5c Circumduction

Circumduction describes movement in a circular manner. It involves the sequential combination of flexion, adduction, extension, and abduction at a joint. This type of motion is found at biaxial condyloid and saddle joints, and at multiaxial ball-and-socket joints (see Figure 10.9E).

10.5d Rotation

Rotation of the neck or body is the twisting movement produced by the summation of the small movements available between adjacent vertebrae (**Figure 10.9F**). Rotation can also occur at the ball-and-socket joints of the shoulder and hip. Here, the humerus and femur rotate around their long axis, which moves the anterior surface of the arm or thigh either toward or away from the midline of the body. Movement that brings the anterior surface of the limb toward the midline of the body is called **medial rotation**. Conversely, rotation of the limb so that the anterior surface moves away from the midline is **lateral rotation** (see Figure 10.9F).

medial rotation (internal rotation)

lateral rotation (external rotation)

10.5e Supination and Pronation

Supination and pronation are movements of the forearm. In the anatomical position, the upper limb is held next to the body with the palm facing forward. In this position, the radius and ulna are parallel to each other and do not cross (you may remember that part of the definition of anatomical position is that no two bones of the body are crossed. This position, with radius and ulna parallel and palms facing forward is the **supinated position** of the forearm. To turn the palm of the hand so that it faces posteriorly the proximal radius rotates within its pivot joint and the distal radius flips over the distal ulna, the two bones are crossed in an X-shape. This is the **pronated position**.

Supination and pronation are the movements of the forearm that go between these two positions. **Pronation** is the motion that moves the forearm from the supinated (anatomical) position to the pronated (palm backward) position. **Supination** is the opposite motion, in which rotation of the radius returns the bones to their parallel positions in anatomical position (**Figure 10.9G**).

Student Study Tip

Supination brings the palm facing upward. Think of how you would hold a bowl of soup, with your palm up, in "soup-ination."

10.5f Dorsiflexion and Plantar Flexion

Dorsiflexion and **plantar flexion** are movements at the ankle joint, which is a hinge joint. Lifting the front of the foot, so that the top of the foot moves toward the anterior leg is dorsiflexion, while lifting the heel of the foot from the ground or pointing the toes downward is plantar flexion. These are the only movements available at the ankle joint (**Figure 10.9H**).

10.5g Inversion and Eversion

Inversion and eversion are complex movements that involve the multiple plane joints among the tarsal bones of the posterior foot. Thus, inversion and eversion are motions of the foot, not the ankle joint. **Inversion** is the turning of the foot to angle the bottom of the foot toward the midline, while **eversion** turns the bottom of the foot away from the midline. These motions are important motions that help to stabilize the foot when walking or running on an uneven surface and aid in the quick side-to-side changes in direction used during sports such as basketball and soccer (**Figure 10.9I**).

10.5h Protraction and Retraction

Protraction and **retraction** are anterior-posterior movements of the mandible or scapula. For the mandible, protraction occurs when the lower jaw is pushed forward, to stick out the chin, while retraction returns the lower jaw to its resting position (**Figure 10.9J**). Protraction of the scapula occurs when the shoulder is moved forward; you can demonstrate this motion by pushing against a wall or pushing a shopping cart forward. Retraction is the opposite motion, with the scapula being pulled posteriorly and medially, toward the vertebral column.

10.5i Depression and Elevation

Depression and **elevation** are downward and upward movements of the scapula or mandible. The upward movement is elevation, while a downward movement is depression. These movements are used to shrug your shoulders. Similarly, elevation of the mandible is the upward movement of the lower jaw used to close the mouth or bite on something, and depression is the downward movement that produces opening of the mouth (**Figure 10.9K**).

10.5j Excursion

Excursion is the side-to-side movement of the mandible. **Lateral excursion** moves the mandible away from the midline, toward either the right or left side. **Medial excursion** returns the mandible to its resting position at the midline.

10.5k Opposition and Reposition

Opposition is the thumb movement that brings the tip of the thumb in contact with the tip of a finger (**Figure 10.9L**). This movement is produced at the first carpometacarpal joint, which is a saddle joint formed between the trapezium carpal bone and the first metacarpal bone. Thumb opposition is produced by a combination of flexion and abduction of the thumb at this joint. Returning the thumb to its anatomical position next to the index finger is called **reposition**.

Cultural Connection

Bipedalism

Of the approximately 250 species of primates, only one species—*Homo sapiens*—spends much time walking around on two feet. In some ways our bodies reflect our tree-climbing ancestry. Our feet, for example, each have 52 bones (or 53, if you have a sesamoid bone) because our ancestors used feet for grasping. Modern human feet are poor graspers; they are used more as springboards to launch our body weight up into the approximately 50 trillion or so steps a human takes in their lifetime. The human pelvis has also undergone substantial changes during bipedal evolution. The human pelvis is more bowl-shaped than other primate pelvises to support the weight of the trunk while walking and to allow more substantial hip extension for springing off into a step. Our hips, while well-suited to walking, are not as durable as our organ systems; more than 300,000 hip fractures are sustained in the United States each year, and about 40 percent of these lead to substantial loss of mobility. While you may not consider it as an important feature in walking, the anatomy of the shoulder also played a role in the evolution of bipedalism. Our extreme range of motion at the shoulder allowed us to throw weapons used in hunting, which ensured that our species was well-fed and successful, and drove the evolution of early humans. Today we know that the wide range of motion in our shoulders comes at the expense of stability; shoulder dislocations are very common among individuals of all ages. The last joint that changed markedly to meet the demands of bipedalism is the joint between the occipital bone and C1 vertebra. The orientation of this joint and the anatomy of the points of articulation evolved as bipedal organisms held their heads on top of the vertebral column rather than somewhat in front of it. In addition, humans are the only primates with a nuchal ligament—the huge band of connective tissue that stabilizes the position of the back of the head with respect to the vertebral column during walking and running.

Learning Check

1. The knee is classified as a hinge joint. It only moves in the sagittal plane. Please predict all the movements available at the knee joint.
   a. Flexion
   b. Extension
   c. Abduction
   d. Adduction

   Use the image provided to answer the following question:

2. Starting from anatomical position, what motions must occur at the leg to end up in the position that is circled in red? Please select all that apply.
   a. Hip extension
   b. Knee flexion
   c. Ankle plantar flexion
   d. Ankle dorsiflexion
3. Which of the following bones can perform protraction and retraction? Please select all that apply.
   a. Mandible
   b. Scapula
   c. Clavicle
   d. Ribs

10.6 Anatomy of Selected Synovial Joints

Learning Objectives: By the end of this section, you will be able to:

10.6.1* Compare and contrast the movements possible and the anatomical constraints of these joints.

10.6.2* Describe the components (bones, bone features, and other structures) that articulate together to form the TMJ, shoulder, elbow, knee, and hip joints.

* Objective is not a HAPS Learning Goal.

This section will examine the anatomy of selected synovial joints of the body. Anatomical names for most joints are derived from the names of the bones that articulate at that joint, such as the radiocarpal joint between the radius and carpal bones. Other joint names that are more familiar to you, such as the elbow, hip, and knee joints, are exceptions to this naming scheme. In this section we will examine the diverse structure function relationships in five joints of the body: the temporomandibular joint, the elbow joint, the hip joint, the knee joint, and the three individual joints that contribute to the shoulder (glenohumeral, sternoclavicular, and acromioclavicular joints).

10.6a Temporomandibular Joint

LO 10.6.1

LO 10.6.2

The **temporomandibular joint (TMJ)** is the joint that allows for depression/elevation, excursion, and protraction/retraction of the mandible. In other words, the TMJ is responsible for opening and closing your mouth, as well as side-to-side and front-back motions during chewing or grinding of your teeth. If we zoom in to the bone features that contribute to the joint, we can examine the articulation formed when the condyle (head) of the mandible fits into the mandibular fossa of the temporal bone. Located between the bones is a flexible articular disc (**Figure 10.10**). This disc serves to smooth the movements between the temporal bone and mandibular condyle. Just anterior to the mandibular fossa, the temporal bone has a smooth flat surface called the *articular tubercle*. This bone feature becomes important in the motion at the TMJ as well.

Movement at the TMJ during opening and closing of the mouth involves both gliding and hinge motions. With the mouth closed, the mandibular condyle and articular disc are located within the mandibular fossa of the temporal bone. During opening of the mouth, the mandible hinges downward and is pulled anteriorly, causing both the condyle and the articular disc to glide forward out of the mandibular fossa and onto

Figure 10.10 Temporomandibular Joint

The temporomandibular joint is the hinge joint of the mouth, where the condyle of the mandible fits within the mandibular fossa of the temporal bone.

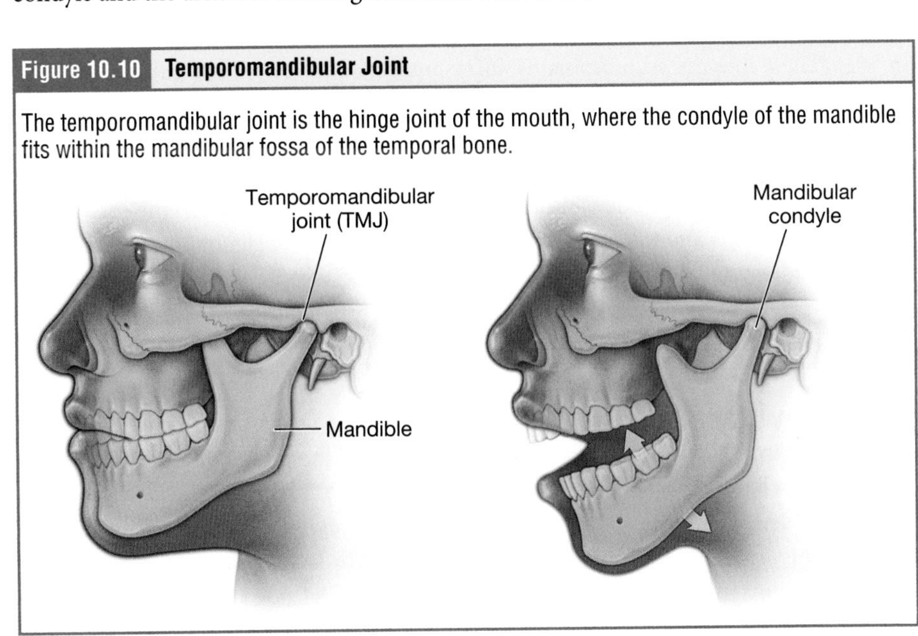

the articular tubercle. The net result is mandibular depression. Elevation moves the mandibular condyle back into the mandibular fossa and the mouth closes.

Dislocation of the TMJ may occur when opening the mouth too wide (such as when taking a large bite) or following a blow to the jaw, resulting in the mandibular condyle moving beyond the articular tubercle. In this case, the individual would not be able to close his or her mouth. Many individuals are prone to wear and tear of the mandible if they overuse the joint by grinding their teeth. Wearing of the articular cartilage covering the bony surfaces of the joint, muscle fatigue from overuse or grinding of the teeth, damage to the articular disc within the joint, or jaw injury may result in headaches, difficulty chewing, or arthritis.

10.6b Shoulder Joint

Movements of the shoulder and upper arm occur because of contributions of three joints: the shoulder joint itself, the joint between the distal end of the clavicle and the scapula, and the joint between the proximal end of the clavicle and the sternum. For our purposes in this text, we will focus on the shoulder joint itself, which is more properly called the **glenohumeral joint**. This is a ball-and-socket joint formed by the articulation between the head of the humerus and the glenoid cavity of the scapula (**Figure 10.11A**). This joint has the largest range of motion of any joint in the body. This freedom of movement is due to a lack of structural support and therefore the glenohumeral joint is both the most mobile and the least stable and most frequently dislocated joint in the human body.

The large range of motion at the shoulder joint is due in part to the size difference between the two bones that contribute to this joint. The head of the humerus is very large and rounded, but the glenoid cavity of the scapula is quite small and shallow, surrounding only about one-third of the surface of the humeral head. The perimeter of the glenoid cavity has a slightly raised ridge of fibrocartilage called the **glenoid labrum**. The articular capsule that surrounds the glenohumeral joint is relatively thin and loose to allow for large motions of the upper limb. Some structural support is provided by ligaments that thicken of the wall of the articular capsule. These include the **coracohumeral ligament**, running from the coracoid process of the scapula to the anterior humerus, and the three **glenohumeral ligaments** that strengthen the anterior and superior sides of the joint.

Student Study Tip

The names of many ligaments are simply a combination of the structures that join together. Use this to help you remember the locations of the ligaments. The glenohumeral ligament is found where the glenoid fossa holds the head of the humerus.

Compared to other joints, the shoulder joint has less surface area between the bones and less support from ligaments. The primary structural support for the shoulder joint is provided by muscles crossing the joint, particularly the four **rotator cuff** muscles. These muscles (supraspinatus, infraspinatus, teres minor, and subscapularis) join the scapula to the humerus and reinforce the joint while maintaining its flexibility. As these muscles cross the shoulder joint, their tendons encircle the head of the humerus and become fused to the anterior, superior, and posterior walls of the articular capsule. As with other joints, bursae help to prevent friction between the rotator cuff muscle tendons and the scapula as these tendons cross the glenohumeral joint. By constantly adjusting their strength of contraction to resist forces acting on the shoulder, these muscles serve as "dynamic ligaments" and thus provide the primary structural support for the glenohumeral joint.

Learning Connection

Image Translation

Use your search engine's image search function to find alternative views of the shoulder such as a posterior view or a superior view. From this new perspective, find a structure, such as the head of the humerus, and from there, try to identify the other components of the shoulder.

10.6c Elbow Joint

The **elbow joint** is a hinge joint formed by the humerus, the radius, and the ulna. A single articular capsule surrounds the ends and articulation of all three bones. **Figure 10.12** depicts a lateral view of the elbow joint through the medial sagittal section.

The articular capsule of the elbow is strengthened by two ligaments that prevent side-to-side movements and hyperextension. On the medial side is the triangular

Figure 10.11 The Glenohumeral Joint

(A) The glenohumeral (shoulder) joint is the most flexible joint in the human body. (B) The wide range of motion is made possible because this ball-and-socket joint has a loose articular capsule and very little bone surface area at the joint. It is supported structurally by ligaments and muscles.

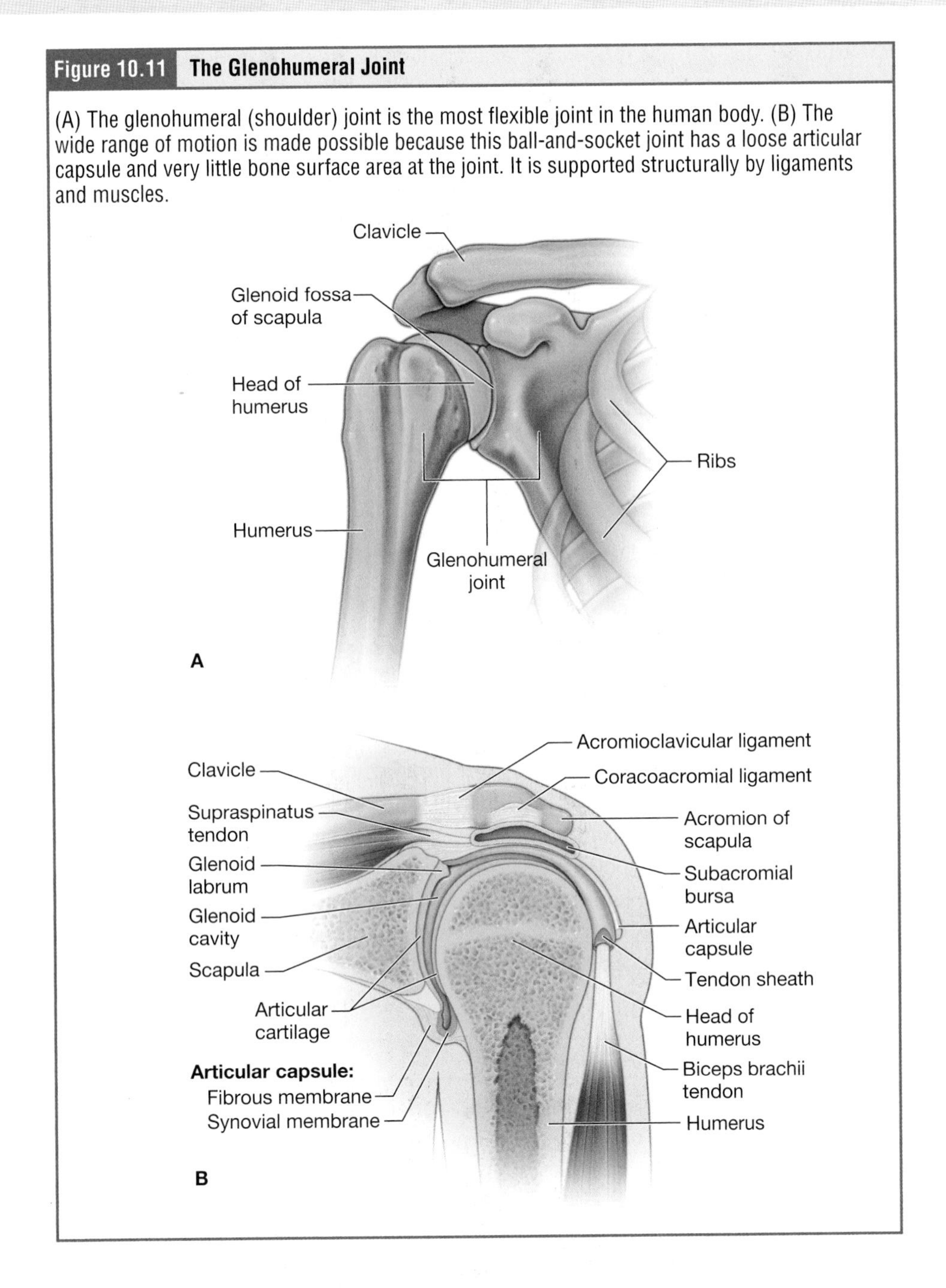

ulnar collateral ligament. The lateral side of the elbow is supported by the **radial collateral ligament**. The **annular ligament** encircles the head of the radius. This ligament surrounds the head of the radius at the proximal radioulnar joint. This is a pivot joint that allows for rotation of the radius during supination and pronation of the forearm.

10.6d Hip Joint

The hip joint is a multiaxial ball-and-socket joint between the head of the femur and the acetabulum of the hip bone (**Figure 10.13**). The weight of the upper body rests on the hips during standing and walking; therefore, strength and stability is favored over mobility and the range of motion at the hip is more limited than at the shoulder joint.

The acetabulum is the socket portion of the hip joint formed at the union of the ilium, the ischium, and the pubis. This space is much deeper than the glenoid fossa and

Figure 10.12 Elbow Joint

The elbow is a hinge joint formed by the trochlea of the humerus and the trochlear notch of the ulna.

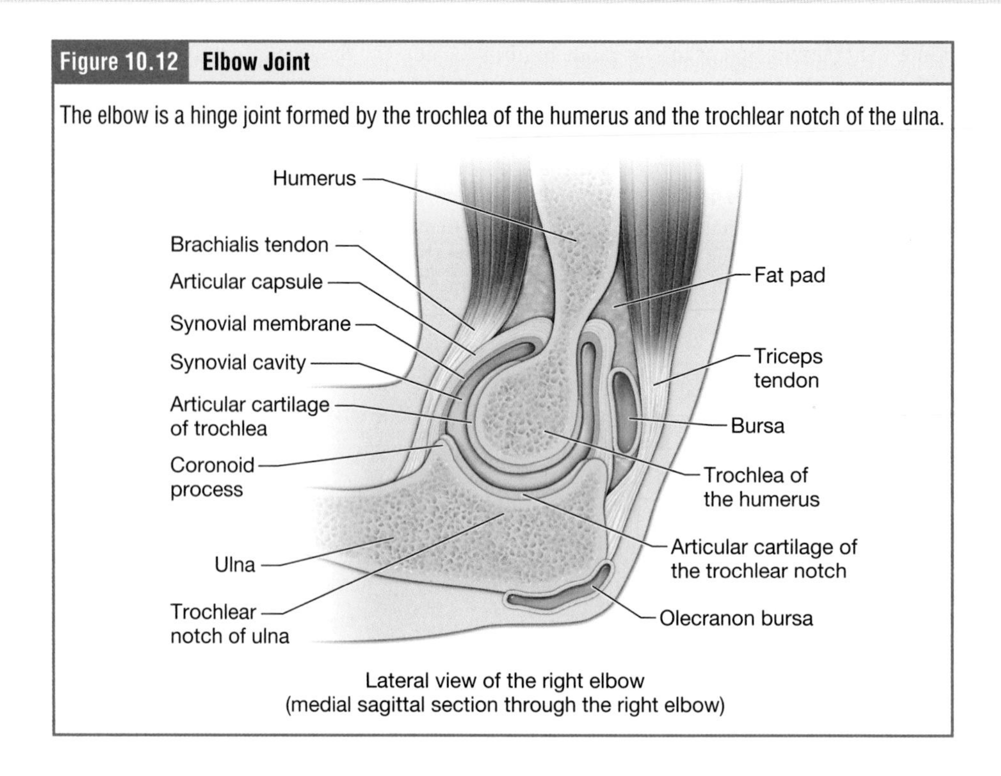

Lateral view of the right elbow
(medial sagittal section through the right elbow)

surrounds more than half of the surface of the femoral head, thus giving stability and weight bearing ability to the joint. The acetabulum holds on even tighter to the femoral head with the **acetabular labrum**, a fibrocartilage lip attached to the outer margin of the acetabulum. Further support for the strength of the joint comes from several thick ligaments that reinforce the joint capsule. These ligaments are the **iliofemoral ligament**, the **pubofemoral ligament**, and the **ischiofemoral ligament**, all of which spiral around the head and neck of the femur (see Figure 10.13). The ligaments are

Figure 10.13 Hip Joint

The hip joint is a ball-and-socket joint that provides greater stability, but a more limited range of motion, than the shoulder.

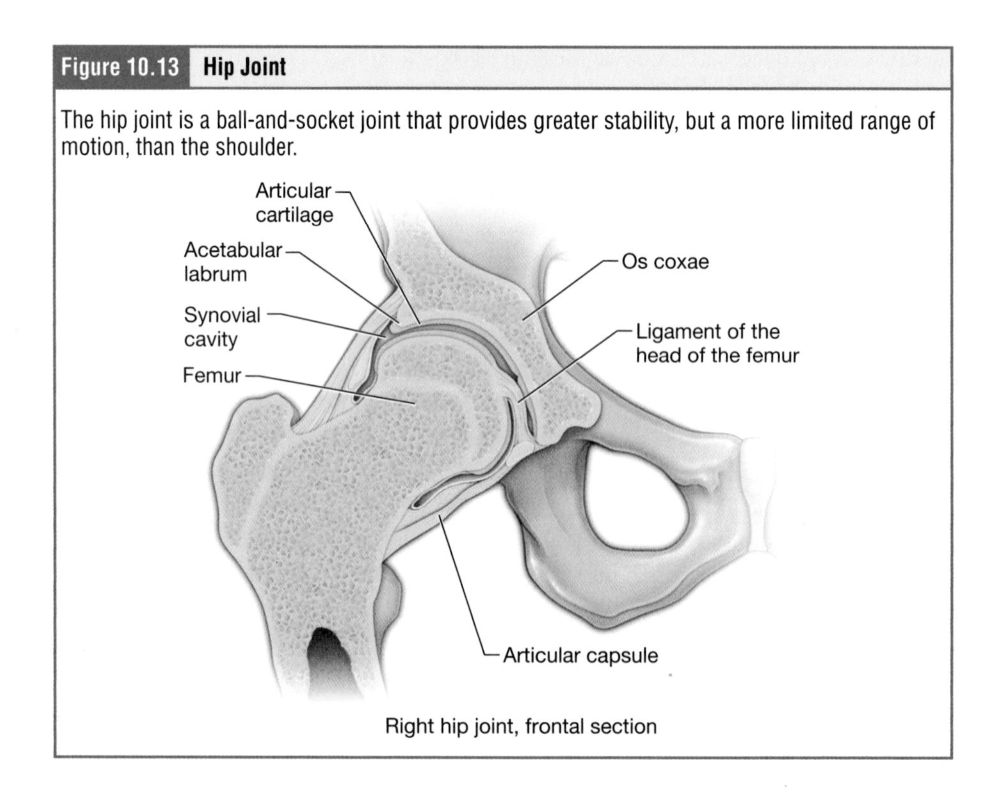

Right hip joint, frontal section

tightened by extension at the hip, thus pulling the head of the femur tightly into the acetabulum when in the upright, standing position. These ligaments thus stabilize the hip joint and allow you to maintain an upright standing position with only minimal muscle contraction. Inside of the articular capsule, the **ligamentum teres** (ligament of the head of the femur) spans between the acetabulum and femoral head. This intracapsular ligament is normally slack and does not provide any significant joint support, but it does provide a pathway for an important artery that supplies the head of the femur.

ligamentum teres (ligament of the head of the femur)

10.6e Knee Joint

The knee joint is the largest joint of the body (**Figure 10.14**). It is an articulation between the femur and the tibia, but it also involves the point of contact between the femur and the patella. The fibula does not articulate with the femur and therefore is not considered part of the knee. While there are three points of contact between bones within the knee joint, there is a single articular capsule. The knee functions as a hinge joint, allowing flexion and extension of the leg. In addition, some rotation of the leg is available when the knee is flexed, but not when extended. The knee is very stable in its extended position, but is vulnerable to injuries associated with hyperextension, twisting, or blows to the medial or lateral side of the joint, particularly while bearing weight.

The patella is a sesamoid bone within the tendon of the quadriceps femoris muscle, the large muscle of the anterior thigh. The patella serves to protect the quadriceps tendon from friction against the distal femur. As you recall, tendons attach muscles to bone while ligaments join one bone to another bone. Therefore, while the patella is suspended within a single sheet of connective tissue, we refer to the quadriceps tendon as the connective tissue superior to the patella, connecting the muscle to the patella, and the **patellar ligament** as the connective tissue that spans the distance between the patella and the tibia just below the knee. Acting via the patella and patellar ligament, the quadriceps femoris is a powerful muscle that acts to extend the leg at the knee.

The rounded condyles of the femur meet the relatively flat condyles of the tibia, forming a hinge joint capable of flexion and extension motions. As the rounded femoral condyles roll across the tibia during flexion or extension, the femur glides forward slightly so that the femoral condyles remain centered over the tibial condyles, thus ensuring weight-bearing support in all knee positions. The lateral condyle of the femur is slightly smaller than the medial condyle. This slight asymmetry has a functional consequence; as the knee comes into full extension, the lateral condyle finishes its rolling motion first and so the femur rotates slightly toward the still-rolling medial condyle. This small rotation serves to "lock" the knee into its fully extended and most stable position. When you go to flex the knee, there is an initial slight rotation to "unlock" from this stable position. This slight rotation is caused by the popliteus muscle. Once unlocked, the hamstring group of muscles can then continue to bring the knee into flexion.

Located between the articulating surfaces of the femur and tibia are two articular discs, the **medial meniscus** and **lateral meniscus** (**Figure 10.14B**). Each is a C-shaped fibrocartilage pad that functions in shock absorption at this weight-bearing joint. The shape of the menisci also helps cup the rounded femoral condyles and keep them in place on the flatter tibial condyles. The menisci are not firmly anchored within the knee and are therefore prone to tearing during twisting motions of the knee. Because cartilage is avascular, the menisci are very limited in their ability to heal if damaged.

The knee joint has multiple ligaments that provide support, particularly in the extended position (**Figure 10.14C**). Outside of the articular capsule, located at the sides of the knee, are two extrinsic ligaments. The **fibular collateral ligament** is on the lateral side and spans from the lateral epicondyle of the femur to the head of the fibula.

fibular collateral ligament (lateral collateral ligament)

Figure 10.14 Knee Joint

(A) The knee joint is the articulation of the femur, the patella, and the tibia. (B) The knee is supported by fibrocartilage discs called *menisci* and held in place by cruciate ligaments and (C) collateral ligaments.

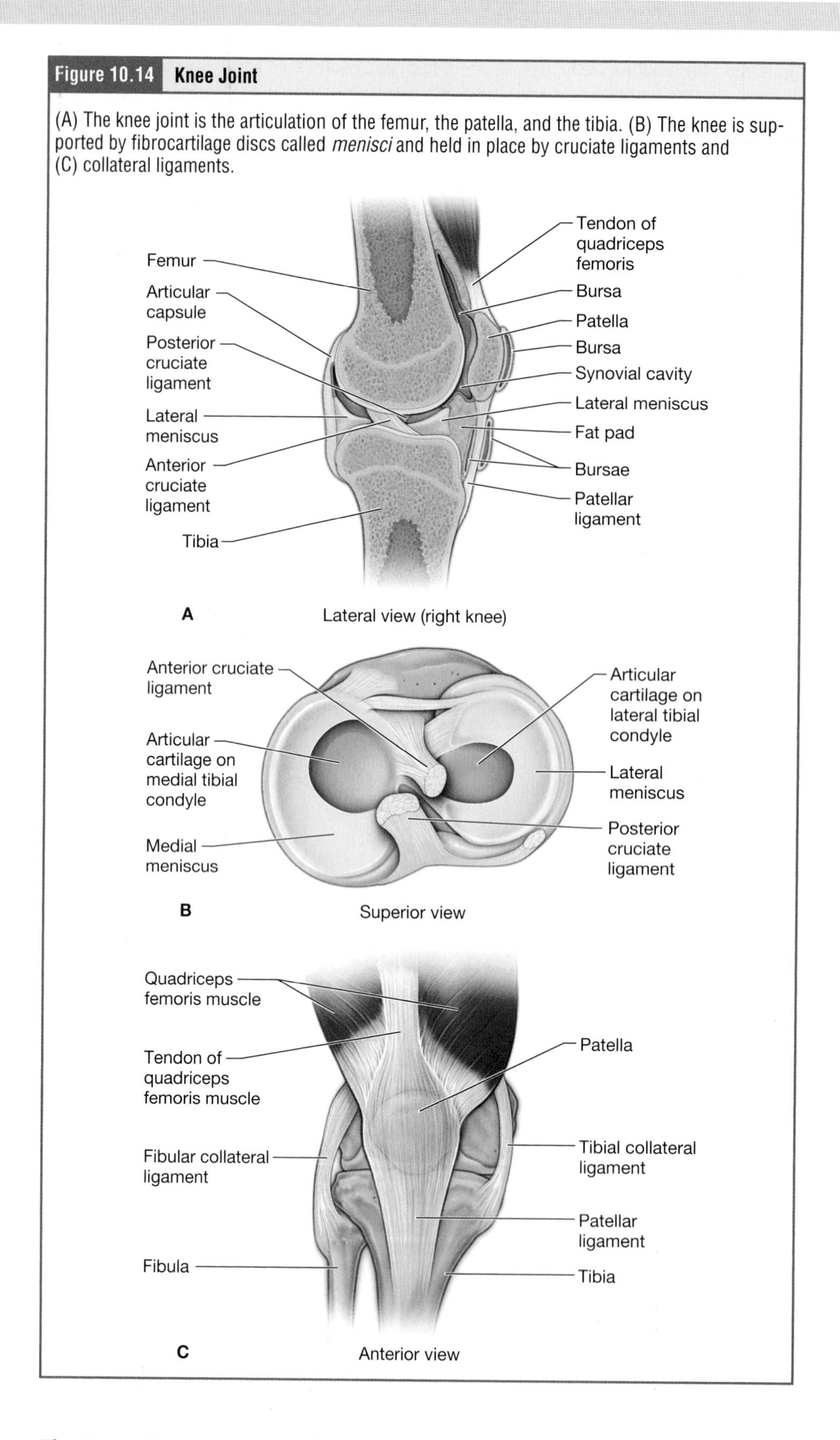

tibial collateral ligament (medial collateral ligament)

The **tibial collateral ligament** of the medial knee runs from the medial epicondyle of the femur to the medial tibia. In the fully extended knee position, both collateral ligaments are taut (tight), thus serving to stabilize the knee and prevent side-to-side motions between the femur and tibia.

Deep within the knee are two ligaments that crisscross and provide stability. The word *cruciate* actually means "cross," and so these two ligaments are named the **anterior cruciate ligament (ACL)** and the **posterior cruciate ligament (PCL)**. The cruciate ligaments help hold the tibia and femur together at all points of flexion. The ACL prevents the femur from sliding backward/posteriorly on the tibia, and the PCL prevents the femur from sliding forward/anteriorly.

Learning Connection

Try Drawing It!

Try drawing a simplified representation of the knee. Be sure to include the two collateral and the two cruciate ligaments. What other structures should you include?

Apply to Pathophysiology

Knee Ligament Tears

Priya is a pedestrian crossing the street. They are receiving an engaging series of text messages from their friend and are watching their phone when they step out into the street. A car is traveling quite fast and tries to brake in order to stop before hitting Priya, but cannot stop in time. Priya hears the squealing brakes and looks up from their phone, turning their body toward the car at the moment of impact. The car's front bumper hits Priya in the anterior knee just below the patella, causing their knee to **hyperextend**.

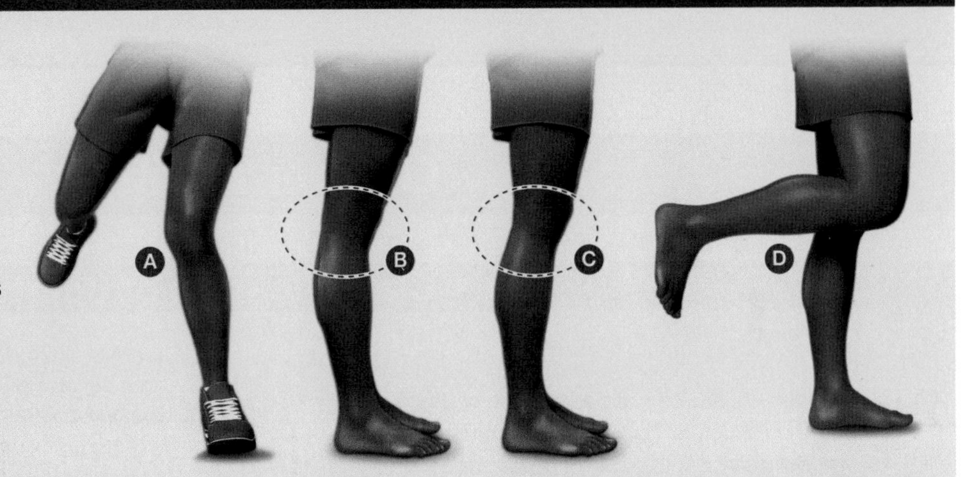

1. Which of these images most closely resembles the position of Priya's knee at the moment of impact?
   A) Image A
   B) Image B
   C) Image C
   D) Image D

As Priya tries to walk home, they find that their knee hurts a lot and feels unstable, and that walking is difficult. In the morning, Priya finds that their knee is markedly swollen and they cannot easily get themself out of bed. As Priya comes to a standing position, their roommate, who is an anatomy student, notices that Priya's lower leg seems to stick out forward in an odd way, as if their tibia now is more anterior than the femur. The roommate brings Priya to the hospital.

2. Of the following, if only one ligament is damaged, which do you think it might be?
   A) The fibular collateral (a.k.a. lateral collateral) ligament is torn.
   B) The tibial collateral (a.k.a. medial collateral) ligament is torn.
   C) The posterior cruciate ligament is torn.
   D) The anterior cruciate ligament is torn.
3. Priya's doctors tell Priya that they will need to perform surgery to repair the torn ligament. Which tissue/material would be a good replacement to maintain the function of the damaged structure?
   A) A rigid but smooth plastic
   B) A strip of tendon from elsewhere in Priya's body
   C) A strip of muscle from a cadaver
   D) A sturdy and strong inflexible metal
4. If, at the moment of impact, Priya hadn't turned toward the car and the car's bumper had struck them from the lateral side of the knee, which injury would have been most likely?
   A) The posterior cruciate ligament would have torn.
   B) The femur would have shattered.
   C) The tibial collateral ligament would have torn.
   D) The fibular collateral ligament would have torn.

Learning Check

1. Describe the motions at the temporomandibular joint during mandibular depression.
   a. The mandible hinges downward, the condyles glide anteriorly, and the disc glides anteriorly.
   b. The mandible hinges upward, the condyles glide posteriorly, and the disc glides posteriorly.
   c. The mandible hinges downward, the condyles glide anteriorly, and the disc glides posteriorly.
   d. The mandible hinges upward, the condyles glide anteriorly, and the disc glides posteriorly.
2. Jackie is a ballet dancer trying to learn a new dance routine. They were standing and started to spin on one foot. Suddenly, they felt a sharp pop as pain shot through the knee of the leg on which they were spinning. Which of the following structures in Jackie's knee could be damaged due to this twisting motion?
   a. Tibial collateral ligament
   b. Anterior cruciate ligament
   c. Medial meniscus
   d. Patellar ligament

Chapter Review

10.1 Classification of Joints

The learning objectives for this section are:

HAPS LEARNING GOALS AND OUTCOMES

10.1.1 Describe the anatomical classification of joints based on structure: fibrous (i.e., gomphosis, suture, syndesmosis), cartilaginous (i.e., symphysis, synchondrosis), and synovial (i.e., planar/gliding, hinge, pivot, condylar, saddle, ball-and-socket), and provide examples of each type.

10.1.2 Explain the relationship between the anatomical classification and the functional classification of joints.

10.1.3 Describe the functional classification of joints (e.g., synarthrosis, diarthrosis) based on amount of movement permitted, and provide examples of each type.

10.1 Questions

1. The metacarpophalangeal joint is a joint in the hand that allows the fingers to move through a large range of motion. The articulating surfaces of the joint are composed of hyaline cartilage. The space between them is filled with fluid, which is contained within a joint cavity. How would you structurally classify the metacarpophalangeal joint?
   a. Fibrous
   b. Cartilaginous
   c. Synovial
   d. Synarthrosis
2. The elbow joint is composed of the articulating surfaces of three bones—the humerus, the ulna, and the radius. The articulating surfaces are composed of hyaline cartilage but are not directly articulating with each other. There is fluid within a joint cavity. The elbow joint moves in one plane—flexion and extension. How would you structurally classify this joint?
   a. Synovial
   b. Diarthrosis
   c. Amphiarthrosis
   d. Fibrous
3. The pubic symphysis joint is a joint that directly connects the two os coxae anteriorly. The joint is composed of fibrocartilage. This is a very stable joint that allows for very little movement at the hip. How would you functionally classify the pubic symphysis joint?
   a. Synarthrosis
   b. Amphiarthrosis
   c. Diarthrosis
   d. Fibrous

10.1 Summary

- Joints are classified either structurally or functionally.
- The structural classification of joints is based on how the articulating surfaces of the bones are connected. In fibrous joints, the bone surfaces are connected by dense connective tissue. In cartilaginous joints, the surfaces are connected by cartilage. Synovial joints have a complex structure; the articulating surfaces of the bones are enclosed in a joint capsule containing synovial fluid.
- Functional classifications are based on the amount of mobility between the adjacent bones. Synarthrosis joints allow for no movement, amphiarthrosis joints allow for little movement, and diarthrosis joints allow for free movement.

10.2 Fibrous Joints

The learning objectives for this section are:

10.2.1* Describe the structural features of fibrous joints.

10.2.2* Given the description of a joint, characterize it as a suture, syndesmosis, or gomphosis.

10.2.3* Compare and contrast a suture, syndesmosis, and gomphosis.

* Objective is not a HAPS Learning Goal.

10.2 Questions

1. Which of the following best describes a fibrous joint?
   a. Bones that are connected to each other by a fibrous connective tissue
   b. Bones that are connected to each other by cartilage
   c. Bones that are connected to each other via a lubricating fluid

2. The ethmoid bone of the skull articulates directly with the sphenoid bone of the skull via strong fibrous connective tissues. The connective tissue that connects the two bones is convoluted to form a tight union and prevent movement. What kind of joint would this be?
   a. Suture
   b. Syndesmosis
   c. Gomphosis

3. Which of the following is true about all fibrous joints?
   a. All fibrous joints are directly connected.
   b. All fibrous joints are synarthrosis joints.
   c. All fibrous joints are found in the skull.
   d. All fibrous joints allow for stability.

10.2 Summary

- In fibrous joints, the bones are connected by fibrous (dense) connective tissue.
- There are three classifications of fibrous joints: (1) suture, (2) syndesmosis, and (3) gomphosis.
- In a suture, the bones are connected by a short layer of dense connective tissue. Many of the joints between adjacent skull bones are sutures.
- In a syndesmosis, bones that lie parallel to each other are connected by fibrous connective tissue. The joint between the tibia and fibula is a syndesmosis.
- In a gomphosis, a cone-shaped root of a tooth is anchored in its socket by a short layer of fibrous connective tissue.

10.3 Cartilaginous Joints

The learning objectives for this section are:

10.3.1* Describe the structural features of cartilaginous joints.

10.3.2* Given a description of a joint, characterize it as either a synchondrosis or symphysis.

10.3.3* Compare and contrast a synchondrosis and symphysis.

* Objective is not a HAPS Learning Goal.

10.3 Questions

1. Which of the following is true about a cartilaginous joint? Please select all that apply.
   a. Two bones are connected by hyaline cartilage.
   b. Two bones are connected by fibrocartilage.
   c. Two bones are connected by synovial fluid.
   d. Two bones are connected by fibrous connective tissue.

2. The sternum is made up of three bones—the manubrium, the body, and the xiphoid process. The manubrium and the body are joined together by a fibrocartilaginous pad. Is this an example of a synchondrosis joint or a symphysis joint?
   a. Synchondrosis
   b. Symphysis

3. Which of the following joints would be classified as a synchondrosis joint? Please select all that apply.
   a. The growth plate of the femur
   b. Ribs articulating with the sternum
   c. Os coxae articulating with each other
   d. The vertebral discs articulating with each other

10.3 Summary

- In a cartilaginous joint, the bones are connected by hyaline cartilage or fibrocartilage.
- There are two classifications of cartilaginous joints: (1) synchondrosis and (2) symphysis.
- In a synchondrosis, the bones are united by hyaline cartilage. An example is an epiphyseal plate of a growing long bone.
- In a symphysis, the bones are united by a pad of fibrocartilage. An example is the pubic symphysis of the pelvis.

10.4 Synovial Joints

The learning objectives for this section are:

10.4.1 Identify and describe the major structural components of a typical synovial joint.

10.4.2 For each of the six structural types of synovial joints, describe its anatomic features, identify locations in the body, and predict the kinds of movement each joint allows.

10.4.3* Given a description, diagram, or photo, categorize the joint as one of the six structural types of synovial joints.

* Objective is not a HAPS Learning Goal.

10.4 Questions

Hailey was playing basketball when they went to block a shot and felt a pop in their shoulder. After hearing the pop, the team athletic trainer took Hailey to get imaging. The imaging showed that the shoulder was dislocated, and there was a fracture on the tip of the humeral head. The shoulder is a type of synovial joint.

1. Which of the following structures are damaged due to the fracture on the tip of the humeral head?
   a. Articular cartilage
   b. Articular capsule
   c. Synovial membrane
   d. Synovial fluid

2. The six types of synovial joints have vastly different structures and ranges of motion. Which of the following is true of the differences between saddle joints and ball-and-socket joints? Select all that apply.
   a. In saddle joints, the bone surfaces are fairly flat, and the bones are able to slide past each other as the joint moves.
   b. Ball-and-socket joints have the widest range of motion of all synovial joints.
   c. In saddle joints, the bones contain both concave and convex surfaces.
   d. The first carpometacarpal joint is an example of a ball-and socket joint.

3. The shoulder joint is formed by the round head of the humerus and a bowl-shaped glenoid fossa of the scapula. What type of synovial joint is the shoulder joint?
   a. Saddle joint
   b. Ball-and-socket joint
   c. Pivot joint
   d. Condyloid joint

10.4 Summary

- Synovial joints are the most common type of joint in the body. They have a more complex structure than fibrous and cartilaginous joints.
- In a synovial joint, the ends of the articulating bones are enclosed in an articular capsule and lubricated by synovial fluid. The bones do not come in direct contact with each other.
- There are six types of synovial joints: pivot, hinge, condyloid, saddle, plane, and ball-and-socket joints.

10.5 Movements at Synovial Joints

The learning objectives for this section are:

HAPS LEARNING GOALS AND OUTCOMES

10.5.1 Define the movements that typically occur at a joint (e.g., flexion, extension, abduction, adduction, rotation, circumduction, inversion, eversion, protraction, retraction).

10.5.2* For a given joint, list (if you know) or predict (if a new joint) the motion available based on joint anatomy.

* Objective is not a HAPS Learning Goal.

10.5 Questions

1. Pretend that you are kicking a soccer ball. What motion at your knee is required for you to kick the ball forward?
   a. Flexion
   b. Extension
   c. Abduction
   d. Inversion

2. You are standing in anatomical position. You want to reach for a cup from the top shelf of your cabinet. What motions at the shoulder are required for you to grab the cup? Please select all that apply.
   a. Extension
   b. Flexion
   c. Pronation
   d. Rotation

10.5 Summary

- These movements include flexion, extension, abduction, adduction, circumduction, elevation, excursion, opposition, and reposition.
- Synovial joints provide a wide range of movements, depending on their structural features.
- When describing the movements of synovial joints, the body is assumed to be in anatomical position.
- The movements of synovial joints include flexion, extension, abduction, adduction, circumduction, rotation, supination, pronation, dorsiflexion, plantar flexion, inversion, eversion, protraction, retraction, depression, elevation, excursion, opposition, and reposition.

10.6 Anatomy of Selected Synovial Joints

The learning objectives for this section are:

10.6.1* Compare and contrast the movements possible and the anatomical constraints of these joints.

10.6.2* Describe the components (bones, bone features, and other structures) that articulate together to form the TMJ, shoulder, elbow, knee, and hip joints.

* Objective is not a HAPS Learning Goal.

10.6 Questions

1. Which of the following motions is found in the elbow joint that is not found in the knee joint?
   a. Flexion
   b. Extension
   c. Pronation
   d. Abduction

2. Which of the following describes how the temporomandibular joint (TMJ) is different than most synovial joints?
   a. The TMJ has hyaline cartilage on the ends of its bones.
   b. The TMJ has an articular disc in the middle of the joint.
   c. The TMJ has an articular capsule composed of fibrous connective tissue.
   d. The TMJ has lubricating fluid within the joint capsule.

10.6 Summary

- The motions available at the temporomandibular joint are depression, elevation, excursion, protraction, and retraction. The joint is formed by the condyle of the mandible articulating with the mandibular fossa of the temporal bone, and a flexible articular disc.
- The shoulder joint is formed by the head of the humerus articulating with the glenoid cavity of the scapula. It is a ball-and-socket joint that allows a wide range of movements; it is the most flexible joint in the body.
- The elbow joint consists of three bones: the humerus, radius, and ulna. It performs flexion, extension, pronation, and supination.
- The hip joint is similar to the shoulder joint in that it is a ball-and-socket joint; in this case, the head of the femur articulates with the acetabulum. The hip joint contains thicker and stronger ligaments and muscles than those of the shoulder. Although the hip joint does not have the same range of motion as the shoulder joint, it is important for bearing the weight of the upper portion of the body.
- The knee joint contains articulations between the condyles of the femur and tibia and between the patella and the patellar surface of the femur. This joint mainly performs flexion and extension, but some rotation occurs under certain circumstances.

11

Muscle Tissue

UfaBizPhoto/Shutterstock.com

Chapter Introduction

In this chapter we will learn...

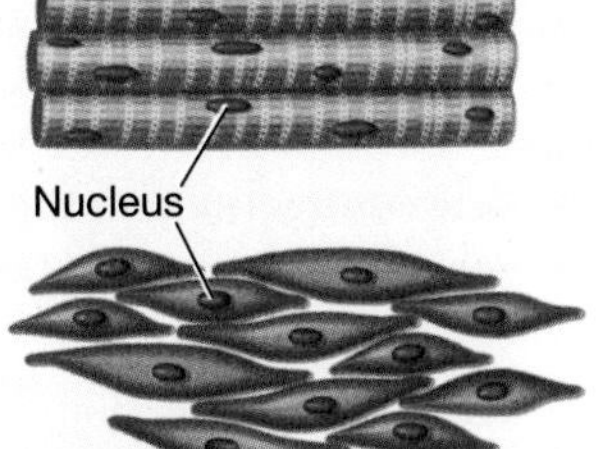

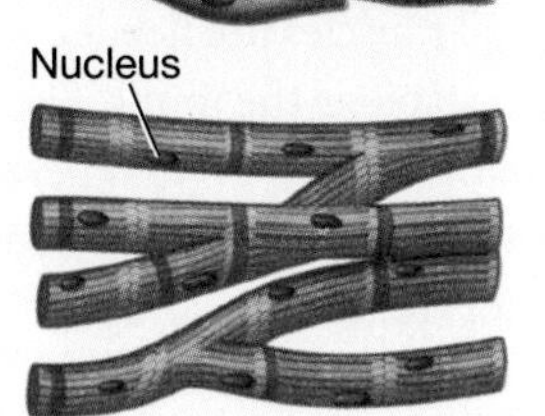

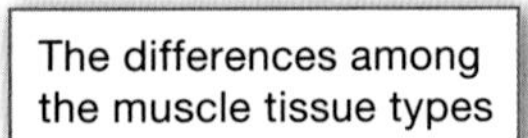

The differences among the muscle tissue types

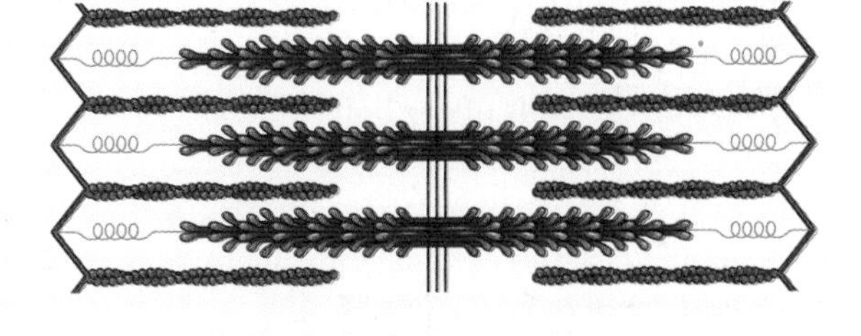

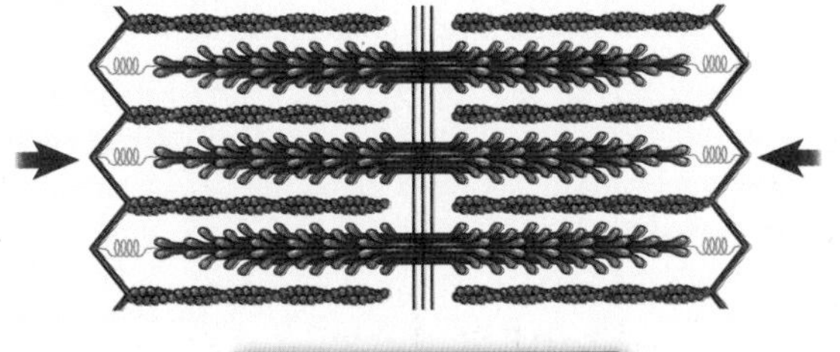

How skeletal muscle contraction is achieved

The word "muscle" in A&P can be used in several ways. We may be talking about an organ (i.e., the biceps brachii muscle in your arm) or a tissue (muscle tissue). Of the tissues, we could be talking about the tissue that comprises a muscle under your skin such as the biceps brachii, or the muscle that lines your stomach or blood vessels, or the tissue that makes up the bulk of the heart. Whether it functions to move food, pump blood, or help you to walk, we will discover all of the muscle tissues of the human body in this chapter.

11.1 Overview of Muscle Tissues

Learning Objectives: By the end of this section, you will be able to:

11.1.1 Describe the major functions of muscle tissue.

11.1.2 Describe the structure, location in the body, and function of skeletal, cardiac, and smooth muscle.

11.1.3 Compare and contrast the general microscopic characteristics of skeletal, cardiac, and smooth muscle.

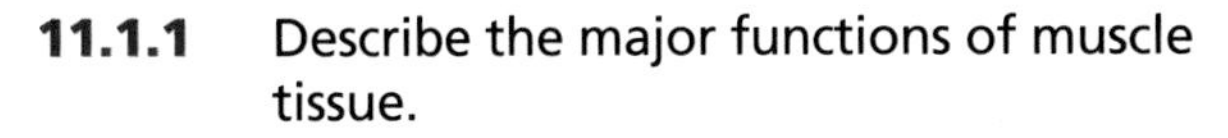

Remember from Chapter 5 that there are four tissue types in the human body. Every organ and organ system is made up of a combination of these four types. One of these—muscle tissue—is the subject of this chapter. One of the challenges learners can sometimes struggle with when discussing muscle tissue is keeping in mind when we are discussing a muscle as an organ (a named muscle such as the abdominals or your deltoid) and when we are considering muscle as a tissue (a group of cells working together to perform one function). In this chapter we discuss the three muscle tissues, while in Chapter 12 we will examine the individual skeletal muscle organs of the body. The three muscle tissues—skeletal, smooth, and cardiac—are very different

The Human Anatomy and Physiology Society includes more than 1,700 educators who work together to promote excellence in the teaching of this subject area. The HAPS A&P Learning Outcomes measure student mastery of the content typically covered in a two-semester Human A&P curriculum at the undergraduate level. The full Learning Outcomes are available at https://www.hapsweb.org.

LO 11.1.1

in structure and appearance. They share some functions and differ in others. Obviously, the best-known feature of all muscle tissue types is their ability to contract and cause movement. When we think of the movement caused by smooth and cardiac muscle contraction, we can think of the movement of internal material. Cardiac muscle makes up the walls of the heart and therefore surrounds the blood-filled chambers. When it contracts it squeezes on that internal material—the blood—and propels it through the circulatory system. Similarly, smooth muscle makes up the walls of our hollow organs such as blood vessels, intestines, the uterus, and so on. When smooth muscle contracts, it propels the internal material, such as food or menstrual blood, in a certain direction. Skeletal muscle contraction, on the other hand, causes external movement. Our limbs move in space, the skin of our face contorts into a smile, and so forth.

Beyond this function of contraction in various forms, skeletal muscle functions to maintain posture. Small, constant adjustments of the skeletal muscles are needed to hold a body upright or balanced in any position. Muscles also stabilize the joints. Skeletal muscles can function to protect internal organs (particularly abdominal and pelvic organs) by acting as an external barrier or shield to external trauma and by supporting the weight of the organs. Think about how you would react, for example, if someone swung a punch toward your belly during a boxing class; you would likely tense your abdominal skeletal muscles to act as a protective shield.

Skeletal muscles contribute to the maintenance of homeostasis in the body by generating heat. Muscle contraction requires energy, and when ATP is broken down, heat is produced. This heat is very noticeable during exercise, when sustained muscle movement causes body temperature to rise, and in cases of extreme cold, when shivering produces random skeletal muscle contractions to generate heat.

LO 11.1.2

You will find that when talking about muscle tissue, skeletal muscle is discussed the most. Due to several physiological and anatomical factors, skeletal muscle is the most well-studied and well-understood. Because skeletal muscle is featured so heavily in our understanding of muscle tissue, let's begin by comparing and contrasting the three tissue types. Skeletal muscle, cardiac muscle, and smooth muscle all exhibit a quality called **excitability**; they can change from relaxed to contracted based on electrical properties at their plasma membranes. The trigger or control over this change is different for the different forms of muscle, however. Smooth and cardiac muscle tissues are considered *involuntary* because the conscious brain cannot control their contraction. Unconscious aspects of the nervous system can influence the excitability of cardiac and smooth muscle to some degree, but you can't stop your heart from beating just by using your willpower. Skeletal muscle, on the other hand, completely depends on signaling from the nervous system to work properly, so we refer to skeletal muscle as *voluntary*. Hormones can also influence muscle contraction, acting primarily on cardiac and smooth muscle. The hormone epinephrine, which is also referred to as *adrenaline*, can increase how hard your heart contracts, leading to the heart-thumping feeling you might experience when startled or on a roller coaster ride.

All muscle tissue can return to its original length after contraction due to a quality of muscle tissue called **elasticity**. As with other tissues, such as the skin, the ability to recoil back to original length is due to elastic fibers. Muscle tissue also has the quality of **extensibility**; it can stretch or extend. **Contractility** allows muscle tissue to pull on its attachment points and shorten with force.

LO 11.1.3

All three types of muscle cells have the same internal components, including contractile proteins, mitochondria, nuclei, and a plasma membrane. However, significant differences exist among the three muscle types in terms of how these components are organized. **Skeletal muscle** cells are long multinucleated structures that compose the skeletal muscle. Skeletal muscle cells are typically as long or almost as long as the muscle organ they are arranged in; for this reason, they are often referred to as *muscle fibers*, which is a synonym for muscle cell. If you glance at your bicep, you can estimate its length to be probably somewhere around nine inches. Within that bicep muscle, many, many skeletal muscle cells are also about the length of that muscle. These skeletal muscle cells result from the fusion of hundreds or thousands of individual cells. Each original cell was much longer and had its own nucleus, and when they fused, they resulted in a longer cell with many nuclei. In contrast, **cardiac muscle** cells each have one to two nuclei and are physically and electrically connected to each other with gap junctions so that a change that occurs in one cell is spread to its neighbors. This structural togetherness allows the entire heart to contract as one unit. Smooth muscle cells are small and shaped like an American football, and each has one single nucleus.

In addition to the differences in cell size, shape, and number of nuclei, there is an additional key microscopic difference among these tissue types. The arrangement of the contractile proteins in skeletal and cardiac muscle follows a precise pattern—so precise, in fact, that these muscle cells look striped under the microscope. This characteristic of being striped or striated is a defining characteristic of skeletal and cardiac muscle cells and tissue. Because the contractile proteins are not arranged in such regular fashion in **smooth muscle**, the cell has a uniform, nonstriated appearance, which early anatomists referred to as "smooth" (**Figure 11.1**).

Figure 11.1 The Three Types of Muscle Tissue

Three types of muscle tissue are found in the body: (A) Skeletal muscle (associated with the bones), (B) Smooth muscle (found in the walls of tubular organs), and (C) Cardiac muscle (found only in the heart).

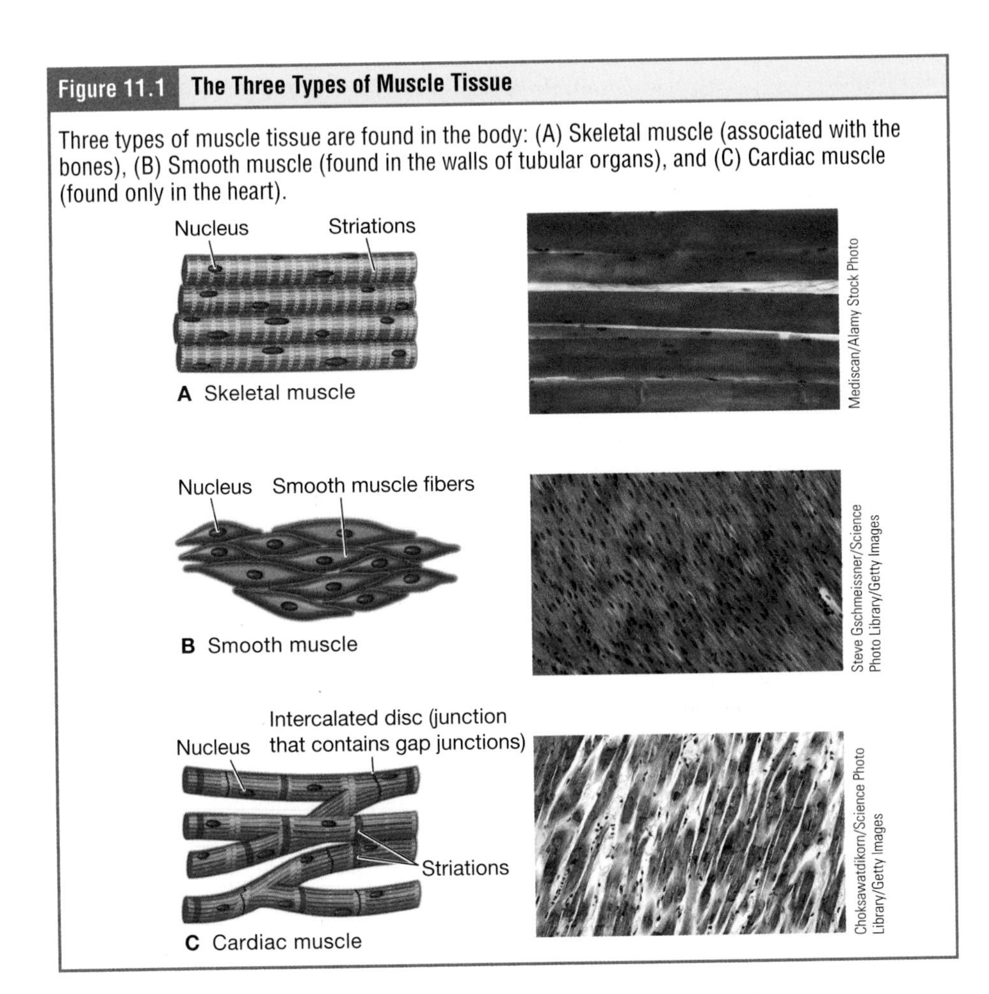

Learning Check

1. Which of the following is not a muscle tissue type?
   a. Cardiac
   b. Smooth
   c. Skeletal
   d. Lung
2. Which of the following do you expect to occur when smooth muscle tissues contract?
   a. Blood will be ejected out of the heart.
   b. The rib cage will expand to allow air in the lungs.
   c. Food will pass through the esophagus.
   d. Flexion will occur at the elbow.
3. Define excitability:
   a. A muscle's ability to change from contracted to relaxed based on electrical properties.
   b. A muscle's ability to return to its original length after contraction.
   c. A muscle's ability to stretch, extend, or recoil back.
   d. A muscle's ability to pull on its attachment point and shorten with force.
4. Describe the microscopic features of smooth muscle tissues. Please select all that apply.
   a. Has striations
   b. Cells (fibers) run the length of the entire muscle
   c. Has one nucleus
   d. American football-shaped

11.2 Skeletal Muscle

HAPS LEARNING GOALS AND OUTCOMES

Learning Objectives: By the end of this section, you will be able to:

11.2.1 Describe the organization of skeletal muscle, from cell (skeletal muscle fiber) to whole muscle.

11.2.2 Name the connective tissue layers that surround each skeletal muscle fiber, fascicle, entire muscle, and group of muscles, and indicate the specific type of connective tissue that composes each of these layers.

11.2.3 Describe the components within a skeletal muscle fiber (e.g., sarcolemma, transverse [T] tubules, sarcoplasmic reticulum, myofibrils, thick [myosin] myofilaments, thin [actin] myofilaments, troponin, tropomyosin).

11.2.4 Define sarcomere.

11.2.5 Describe the arrangement and composition of the following components of a sarcomere: A-band, I-band, H-zone, Z-disc (line), and M-line.

11.2.6 Describe the structure of the neuromuscular junction.

LO 11.2.1

LO 11.2.2

Because skeletal muscle makes up an enormous proportion of the mass of the human body and skeletal muscle cells are the most well studied, we will examine the structure and function of these cells in more detail. Each skeletal muscle is an organ that, like all organs, consists of multiple integrated tissues. These tissues include the skeletal muscle cells, blood vessels, nerve fibers, and wrappings of connective tissue. Each skeletal muscle is wrapped by a sheath of dense, irregular connective tissue called the **epimysium** (see the "Anatomy of a Muscle [Organ]" feature). The epimysium separates muscle from other tissues and organs in the area, allowing the muscle to move independently. An additional layer of dense irregular connective tissue, **fascia**, may be external to the epimysium. Fascia is visible during dissection and also during cooking, if you have ever noticed the thin, iridescent layer of tissue that separates chicken breast from the chicken tender, for example, you're examining the fascia! The fibers of the epimysium are continuous with the fibers of the tendon, which are continuous with the fibers of the periosteum, uniting muscle to bone for effective movements. The tension created by contraction of the muscle fibers is transferred though the epimysium to the tendon, and then to the periosteum to pull on the bone for movement of the skeleton. In other places, the epimysium may fuse with a broad, tendon-like sheet called an **aponeurosis**, or to fascia, the connective tissue between skin and bones.

The inside of each skeletal muscle organ is organized and subdivided. Muscle cells are organized into bundles, each called a **fascicle**. Fascicles are wrapped with a layer of connective tissue called the **perimysium**. The position of the epimysium allows for passage of blood vessels and nerves in the protected spaces between the fascicles. The fascicular organization allows the nervous system to fine-tune the activation of the muscle by triggering only a subset of muscle fibers within a fascicle, or a subset of fascicles within the muscle.

Inside each fascicle, the long muscle cells are each encased in a thin connective tissue layer of collagen and reticular fibers called the **endomysium**. The endomysium is external to the plasma membrane and contains the extracellular fluid and nutrients to support the muscle fiber.

The blood and nervous supply to the muscle is woven throughout these layers of connective tissue. Skeletal muscle requires a tremendous amount of the body's oxygen and calories, which are supplied by blood vessels. Waste generated by muscle cell metabolism is removed through the blood as well. In times of strenuous activity, the blood supply may not be able to provide sufficient nutrients or remove wastes quickly enough. We will discuss adaptations and consequences of this scenario later on in this chapter. Remember that skeletal muscle, unlike cardiac and smooth muscle, only contracts in response to signaling from the nervous system, so muscle organs are highly innervated as well.

Anatomy of...

A Muscle (Organ)

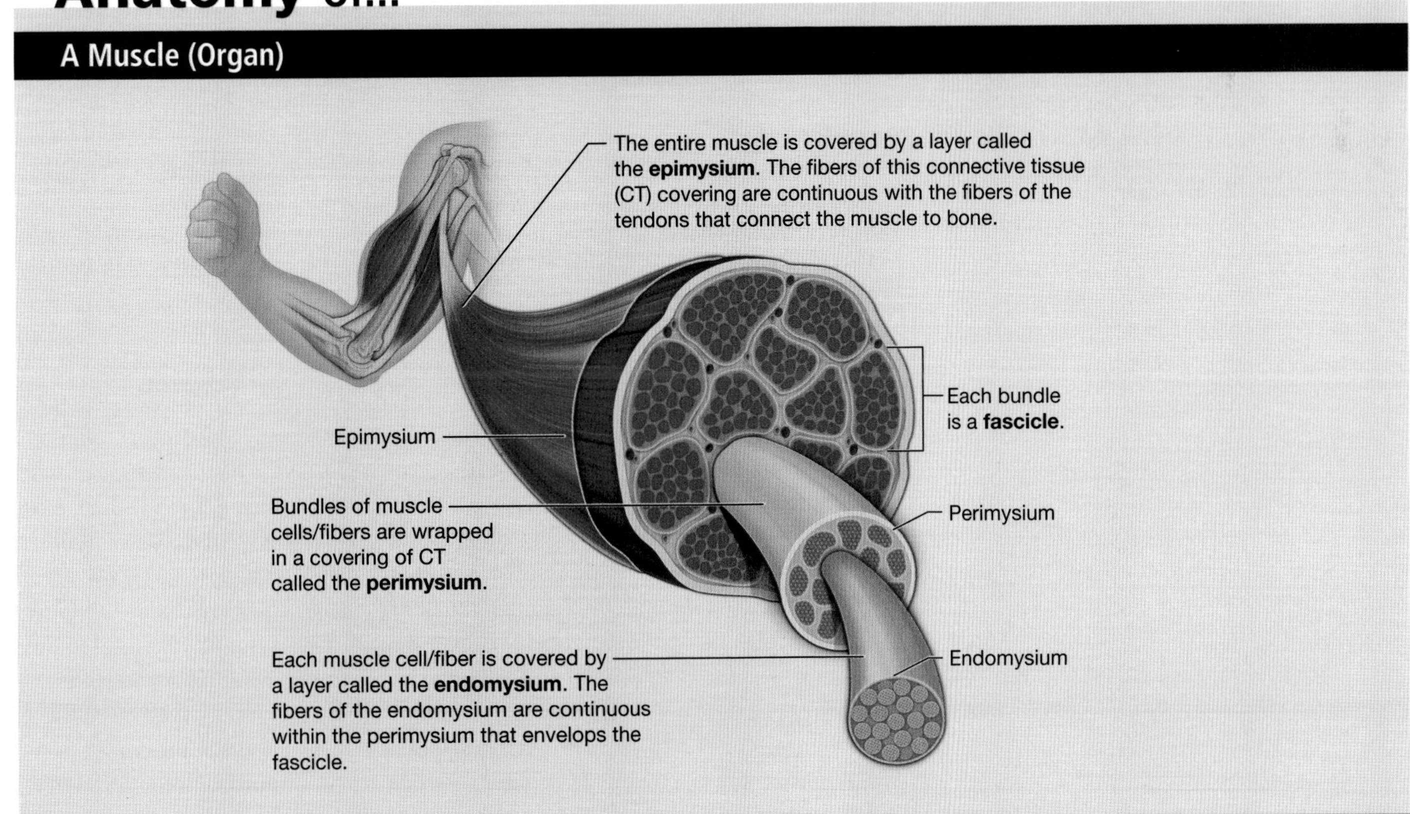

The entire muscle is covered by a layer called the **epimysium**. The fibers of this connective tissue (CT) covering are continuous with the fibers of the tendons that connect the muscle to bone.
Each bundle is a **fascicle**.
Epimysium
Bundles of muscle cells/fibers are wrapped in a covering of CT called the **perimysium**.
Perimysium
Each muscle cell/fiber is covered by a layer called the **endomysium**. The fibers of the endomysium are continuous within the perimysium that envelops the fascicle.
Endomysium

Table 11.1 Muscle Cell Components

Cell Components	Name in a Typical Cell
Sarcolemma	Plasma membrane
Sarcoplasm	Cytoplasm
Sarcoplasmic reticulum (SR)	Endoplasmic reticulum (ER)

LO 11.2.3

11.2a Skeletal Muscle Cells

Much of the terminology associated with muscle cells is rooted in the Greek *sarco*, which means "flesh." The muscle cell has specialized terms for some of the cell components we learned about in Chapter 4, summarized in **Table 11.1**. The plasma membrane is called the **sarcolemma**, the cytoplasm is referred to as **sarcoplasm**, and the specialized smooth endoplasmic reticulum, is called the **sarcoplasmic reticulum (SR)**. These features have counterparts in other cells, but they have a specialized structure to match their function in skeletal muscle cells. The sarcolemma is made of a phospholipid bilayer, just like it is in other cells; however, the sarcolemma of a muscle cell is a bit different than other cells. While the plasma membrane of a typical cell is typically a smooth, rounded surface, the sarcolemma is punctuated by deep invaginations called **T-tubules** ("T" stands for "transverse") (**Figure 11.2**). These invaginations serve a critical function for muscle cells. As we will learn a bit later in this chapter, skeletal muscle cells contract in response to electrical changes on their membrane surfaces. These T-tubules bring the electrical signal deep into the cell so that every myofibril throughout the entire cell can respond to it. The sarcoplasmic reticulum is also a bit different in skeletal muscle cells. Unlike the endoplasmic reticulum, which functions mostly in protein and other molecular manufacturing, the sarcoplasmic reticulum functions primarily in the storage and release of calcium ions. Calcium ions play a critical role in muscle cell contraction (as we will learn a bit later in this chapter) and so the sarcoplasmic reticulum must have a physically intimate relationship with the plasma membrane, where the electrical signal that causes contraction travels. Thus, in Figure 11.2 you can see that the T-tubules dive into the depths of the cell and are surrounded by the sarcoplasmic reticulum. This physical connection allows the translation of the membrane electrical signal to the calcium-storing sarcoplasmic reticulum.

Skeletal muscle cells are multinucleated, as described previously; however, as we examine a muscle cell cross section in the "Anatomy of a Skeletal Muscle Cell" feature, we can see that all of these nuclei are pushed out to the periphery of the cell, almost shoved against the sarcolemma. This is quite different from the typical cell, in which the nucleus is most often seen residing in the middle of

Figure 11.2 The Sarcolemma, the T-Tubules, and the Sarcoplasmic Reticulum

Invaginations of the plasma membrane called T-tubules bring the electrical impulses that travel along the membrane into the depths of the cell. The SR, which stores calcium, abuts the T-tubules. When an electrical impulse travels along a T-tubule, the SR membrane is affected and releases calcium into the sarcoplasm.

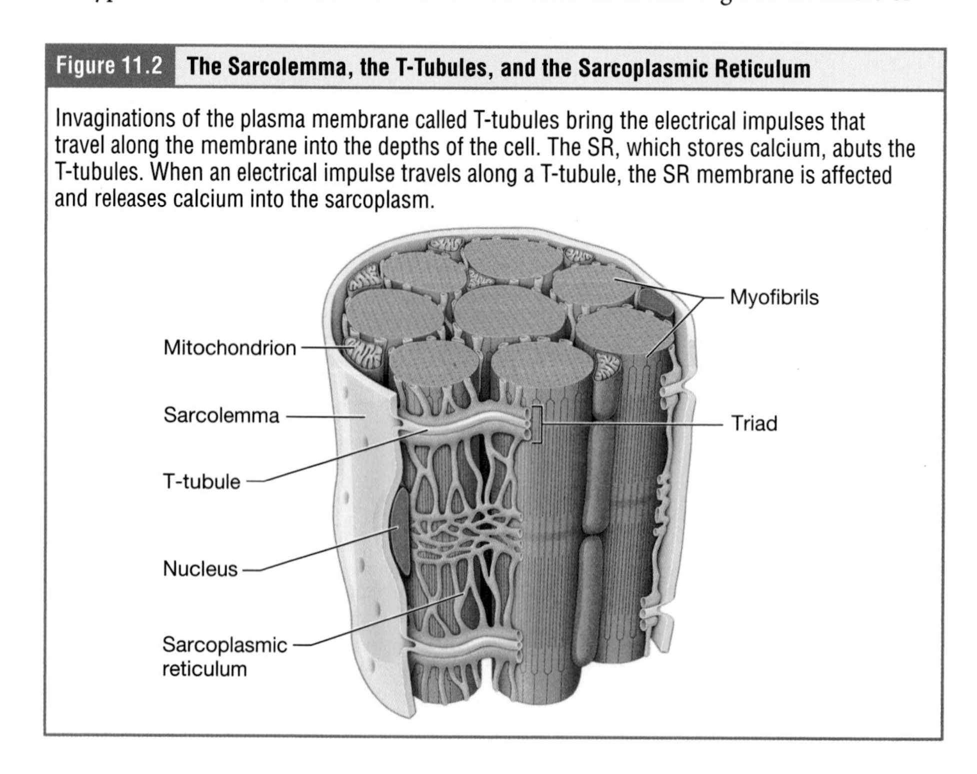

Anatomy of...

A Skeletal Muscle Cell

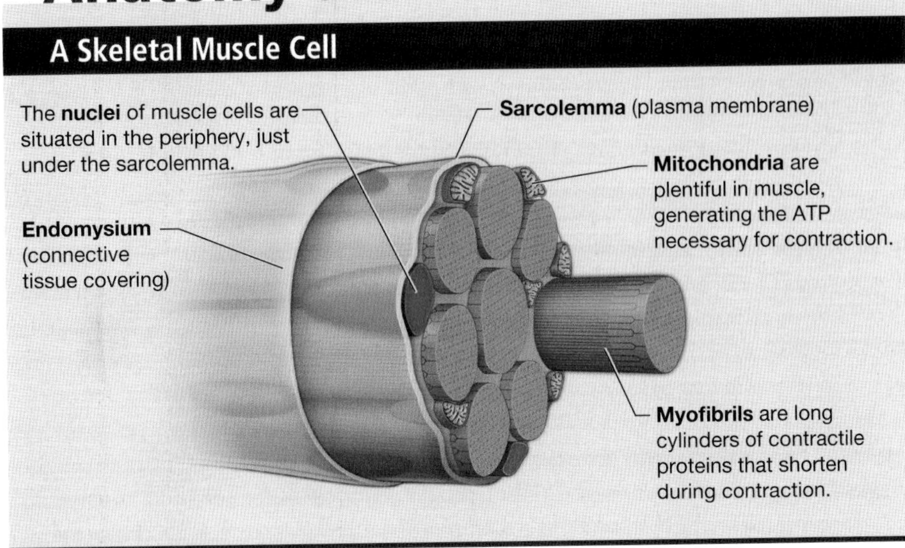
The **nuclei** of muscle cells are situated in the periphery, just under the sarcolemma.
Sarcolemma (plasma membrane)
Mitochondria are plentiful in muscle, generating the ATP necessary for contraction.
Endomysium (connective tissue covering)
Myofibrils are long cylinders of contractile proteins that shorten during contraction.

the cytoplasm; the internal volume of a muscle cell, on the other hand, is almost entirely taken up by myofibrils, which are the contractile machinery. A **myofibril** is a long cylinder of carefully arranged contractile proteins that shorten during contraction. As the myofibrils shorten, they pull on the epimysium at the ends of the muscle, thereby pulling on the tendons of the muscle and moving the bones closer together. Scattered among the myofibrils are mitochondria, organelles that contribute to the conversion of glucose to ATP.

11.2b The Sarcomere

The myofibrils are long cylinders of contractile proteins. The precise arrangement of these proteins within the myofibril leads to the striated appearance of skeletal and cardiac muscle cells. There are several proteins found in sarcomeres, the two main proteins are **myosin** and **actin**. Both myosin and actin are small protein subunits that string together into long fibers called **myofilaments** (**Figure 11.3A**). Along the string of actin subunits, two additional proteins, **troponin** and **tropomyosin** (along with other proteins), are found. Together, the actin, troponin, and tropomyosin filament is called a **thin filament** while the myosin proteins are arranged in a thicker bundled called the **thick filament**. The thin filaments stretch horizontally from a disc (the **Z disc**) that forms the borders of a frame that surrounds the thick filaments; this organized protein unit (called a ***sarcomere***) is the functional unit of the muscle cell (**Figure 11.3B**). The sarcomere is the smallest unit in which contraction occurs; each sarcomere pulls its ends together during contraction. Sarcomeres are incredibly small; you would need to stack 45 sarcomeres end to end to build the thickness of a sheet of printer paper. Imagine how many sarcomeres, laid end to end, it would take to build a muscle cell, which can be inches in length!

Myofibrils, which run the entire length of the muscle fiber, are the thickness of one sarcomere, but are many, many sarcomeres long. The end of the myofibril attaches to the sarcolemma. As sarcomeres contract, the myofibril shortens, pulling on the sarcomere at its ends, thus causing the entire muscle cell to contract.

Examining the sarcomere as a whole, anatomists often use terms to describe the patterns they see under the microscope. The region where the thick filaments are lined

Learning Connection

Macro to Micro

What are the different relationships among the words myofibril, muscle cell, fascicle, muscle, fascia, epimysium, perimysium, and endomysium? Can you group these terms into categories based on whether you would need a microscope to see them or they could be seen with the naked eye?

LO 11.2.4

Z disc (Z line)

Learning Connection

Try Drawing It!

Draw a sarcomere for yourself. Use different colors for each of the proteins.

LO 11.2.5

Figure 11.3 Microscopic Structure of Skeletal Muscle Cells

(A) The sarcomere is the functional and contractile unit of skeletal and cardiac muscle cells. Sarcomeres are bordered by Z lines. Lined up Z line to Z line, long cylinders of sarcomeres called myofibrils (B) comprise the majority of the inside of a muscle cell. Each myofibril is surrounded by a network of sarcoplasmic reticulum.

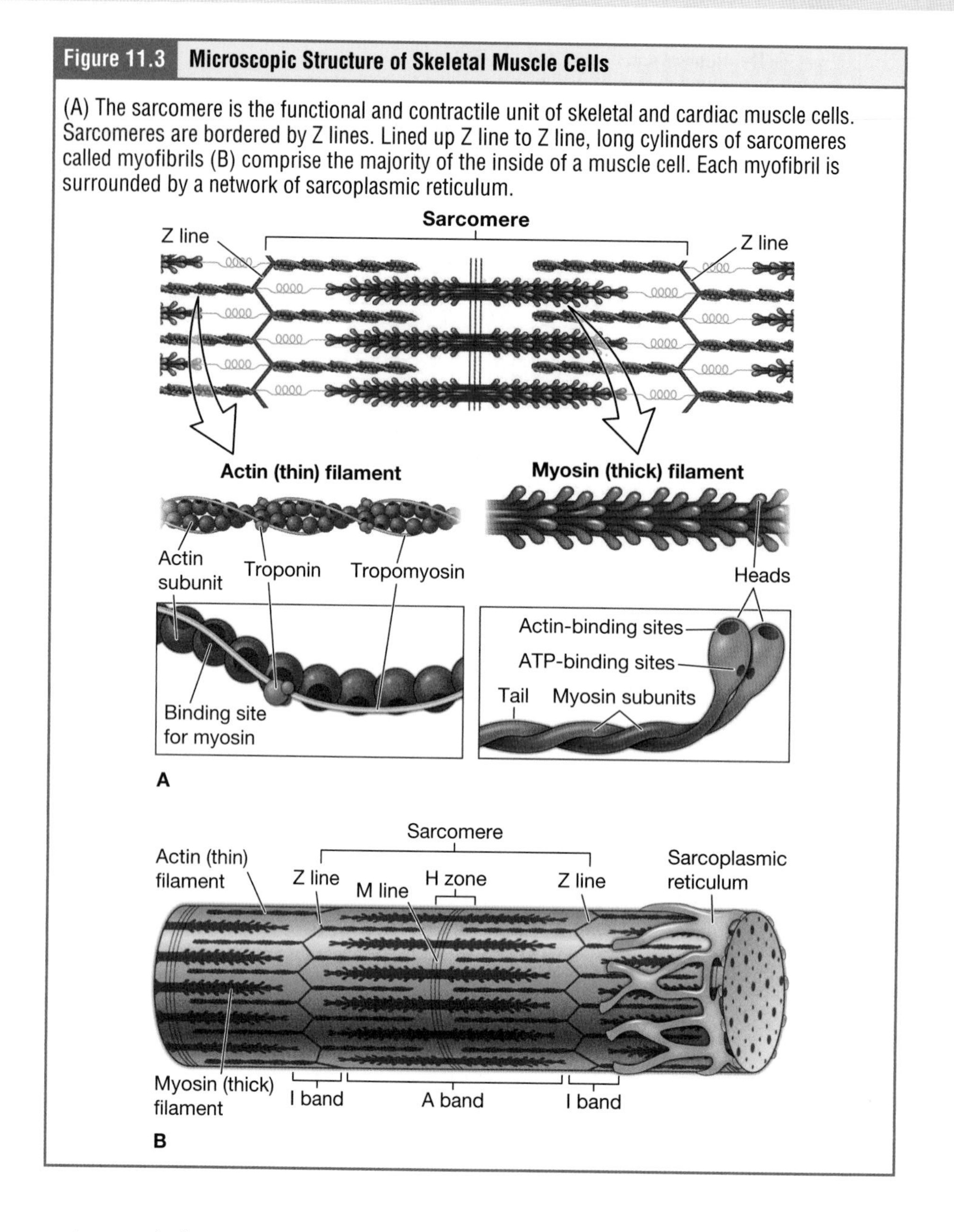

up is quite dark in a microscopic image and is referred to as the **A band**. To the sides of the A band, regions where only thin filaments are found, are the **I bands**. The **M line** is a horizontal line at the exact center of the sarcomere, and the **H zone** extends laterally from the M line; it is the space between the ends of the thin filaments where only thick filaments can be found (see Figure 11.3B).

11.2c The Neuromuscular Junction

LO 11.2.6

Every skeletal muscle cell has one point of contact with the neuron that controls it. Neurons that control skeletal muscle cells are called **motor neurons**. At this site, the **neuromuscular junction (NMJ)** the neuron is able to stimulate an electrical signal that travels along the length of the muscle cell and along its T-tubules (**Figure 11.4**). The region of the sarcolemma at the NMJ is called the **motor end plate**. Excitation signals from the neuron are the only way to cause contraction in a skeletal muscle cell. In contrast, cardiac and smooth muscle cells do not each have an NMJ but have a variety of stimuli that they contract and relax in response to.

Figure 11.4 The Neuromuscular Junction (NMJ)

The neuromuscular junction is a synapse between a motor neuron and a skeletal muscle cell. The surface of the skeletal muscle cell at the synapse is the motor end plate.

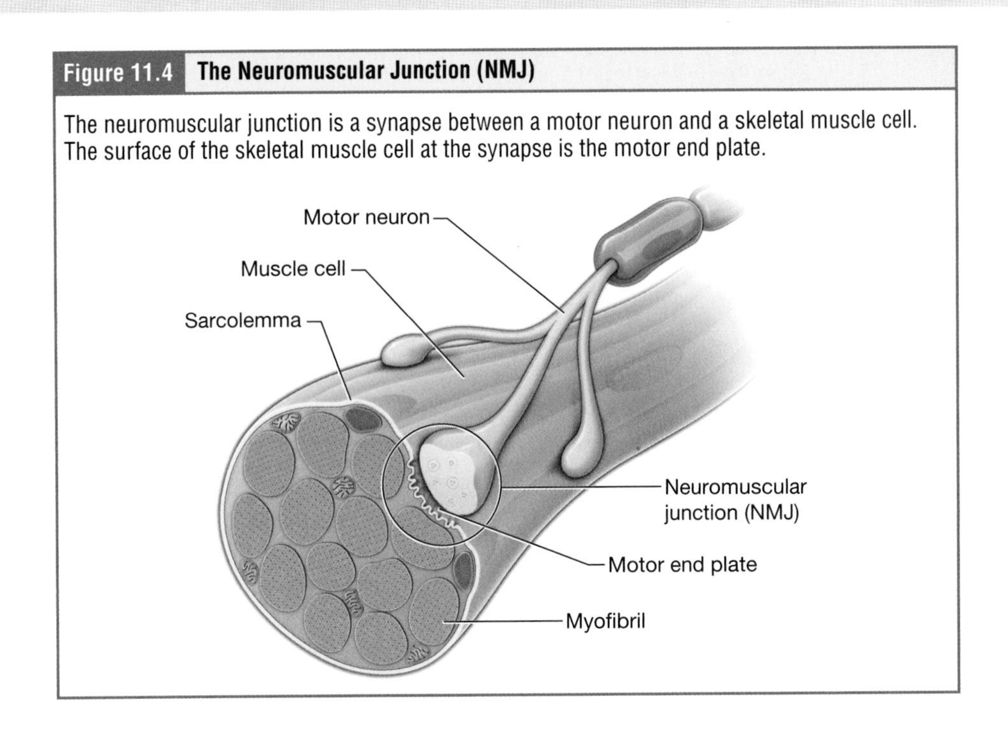

Learning Check

1. Which of the following wraps around a fascicle?
   a. Endomysium
   b. Perimysium
   c. Epimysium
   d. Aponeurosis
2. A calcium ion is inside the T-tubule. When the T-tubule is stimulated, where would the calcium ion go next?
   a. Sarcolemma
   b. Sarcoplasm
   c. Sarcoplasmic reticulum
   d. Myofibrils
3. Which of the following make up a thick filament?
   a. Actin
   b. Myosin
   c. Troponin
   d. Tropomyosin
4. Which of the following would you expect to happen if the motor end plate was desensitized? Please select all that apply.
   a. Decreased calcium release
   b. Weaker muscle contraction
   c. Decreased stimulation of motor neuron
   d. Increased stimulation of sarcolemma

11.3 Skeletal Muscle Cell Contraction and Relaxation

Learning Objectives: By the end of this section, you will be able to:

11.3.1 Define the sliding filament model of skeletal muscle contraction.

11.3.2 Describe the sequence of events involved in the contraction of a skeletal muscle fiber, including events at the neuromuscular junction, excitation-contraction coupling, and cross-bridge cycling.

11.3.3 Describe the sequence of events involved in skeletal muscle relaxation.

Reach down and pick up a pen with your fingers. Seamlessly, in just milliseconds, you were able to coordinate the contraction and relaxation of many muscles to enable this fairly simple movement. You contract and relax your muscles probably hundreds of thousands of times a day without effort or thought, but this series of events, when examined molecularly, is actually quite complex! We discussed earlier that the sarcomere is the functional unit of the muscle cell, that is to say, the smallest single unit of contraction. Let's first examine how the sarcomere contracts and then we will zoom out and look at the series of cellular events that lead to that contraction.

11.3a The Sliding Filament Model of Contraction

LO 11.3.1

For a long time, scientists didn't know exactly how the sarcomere shortened. Did the filaments fold like an accordion? Did the shape of the sarcomere change? After years of careful experimentation, scientists have accepted an explanation of contraction called the *sliding filament model.* The contraction is accomplished with the thin filament framework slides past the static thick filaments, bringing the Z discs closer together (**Figure 11.5**). Since sarcomeres are lined up Z disc to Z disc down the length of the myofibril and cell, the tiny shortening of each sarcomere multiplies and significant muscle shortening occurs. So, how do the filaments slide? The answer lies in the interaction between the myosin protein in the thick filaments and the actin subunits of the thin filaments.

Myosin protein filaments have a head that is held slightly away from the tail portion of the protein (see Figure 11.3A). These heads have a binding site where the myosin can bind to a specific site on the actin subunits. When that site is exposed and there is ATP available in the sarcoplasm, myosin heads will bind to actin; this binding event is called a *cross-bridge.* Like all proteins, when myosin binds to another molecule, it will undergo a shape change. The myosin shape change induced by binding to actin looks almost like a small sit-up of the myosin molecule. The actin thin

LO 11.3.2

Figure 11.5 The Sliding Filament Model of Muscle Contraction

Sarcomeres contract when the myosin heads of the thick filaments pull the actin thin filament framework closer together. The Z lines move closer together and the H zone disappears.

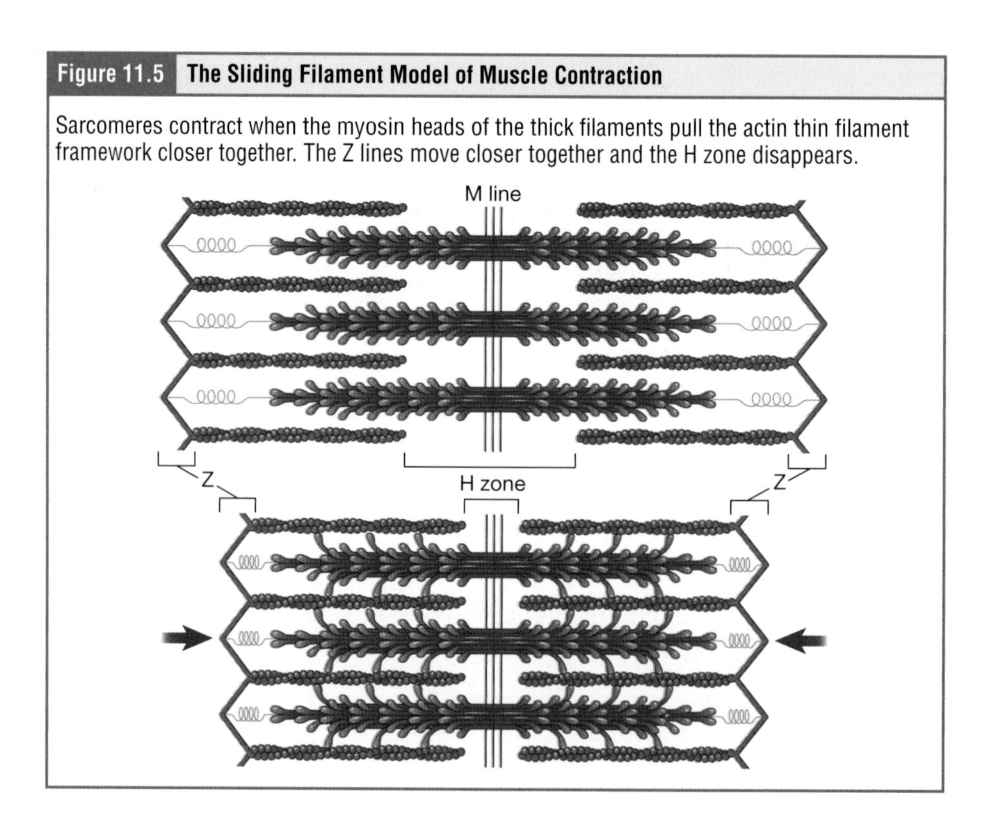

filaments are then pulled by the myosin heads toward the center of the sarcomere, bringing the Z discs closer together. The myosin shape change is small and can only pull a very short distance before it has reached its limit and must be "re-cocked" before it can pull again, a step that requires ATP. If the actin sites for binding myosin are still available and there is still ATP in the sarcoplasm, the myosin heads will reset, bind again, and pull another time, moving the Z discs even closer. This cycle will continue as long as ATP and binding sites remain available, and the Z discs get closer and closer together until maximum contraction is reached. This cycling and progress might be likened to the hand-over-hand motion of pulling an anchor up one hand-grasp and pull at a time.

If cycling and contraction continues whenever ATP and binding sites are available, what keeps us from having fully contracted muscles all the time? The control over muscle contraction comes from the other proteins in the thin filament, the troponin and tropomyosin. Tropomyosin protein winds around the chains of the actin filament like the ribbon on a package and covers the myosin-binding sites, thus preventing myosin from being able to bind actin. Tropomyosin is in this position during muscle relaxation. To enable contraction, something must pull tropomyosin out of the way, allowing cross-bridges to form. Tropomyosin does not sit alone on the actin filament, but rather is part of a structural complex with another protein, troponin. The troponin-tropomyosin complex is responsible for preventing the myosin "heads" from binding to actin, and consequently preventing contraction from occurring. Troponin has a binding site for Ca^{2+} ions.

At the initiation of muscle contraction, Ca^{2+} ions in the sarcoplasm bind to troponin. Like all proteins, when troponin binds to another molecule, it undergoes a shape change. This shape change has the effect of pulling on its partner—tropomyosin—off of its resting position, exposing the binding sites on actin and thus allowing contraction to proceed. In this way, we can consider the troponin-tropomyosin complex to be the brakes on the contraction machinery. So, if contraction is dependent on cross-bridge formation, and cross-bridges can only form when Ca^{2+} ions are present in the sarcoplasm, what regulates the sarcoplasmic concentration of Ca^{2+}?

Learning Connection

Broken Process

What would occur if Ca^{2+} ions were absent in the muscle cell? What would occur if Ca^{2+} ions were always present?

11.3b Excitation-Contraction Coupling

Skeletal muscle contraction is always controlled by the nervous system, so to examine the initiation of a muscle contraction, we must look at the series of events at the NMJ, along the muscle cell sarcolemma, and within the muscle cell. The signals that travel along the motor neuron and the sarcolemma are electrical signals, so let's start by understanding those. All living cells have an electrical gradient across their membranes, meaning that one side of the membrane has more positive and/or fewer negative charges than the other. The difference in electrical charge is called the **membrane potential** and is measured by using a voltmeter that can compare one side to the other (**Figure 11.6**). Recall that neurons and muscle cells are both described as excitable; all excitable cells have a significant membrane potential. The inside of these membranes is usually around −60 to −90 mV, relative to the outside. This means that the inside of the membrane has fewer positively charged molecules, and/or more negatively charged molecules than the outside; the difference between these two sides is between 60 and 90 mV. This difference in charge is established due to the action of the Na^+/K^+ pump. This pump is at work constantly in neurons and muscle cells, and each time it cycles it moves three Na^+ ions to the outside of the cell and two K^+ ions to the inside of the cell. The action of the Na^+/K^+ pump effectively establishes the gradient because more positive charges are found outside than inside. Neurons and muscle cells use their

Student Study Tip

To remember which ion enters the cell and which leaves via the Na^+/K^+ pump, the cell says "nah" to Na^+ and "K" to accepting K^+ into the cell!

Figure 11.6 Measurement of the Membrane Potential

Voltmeters are used to measure the membrane potential, the difference in charge between the inside and outside of the cell.

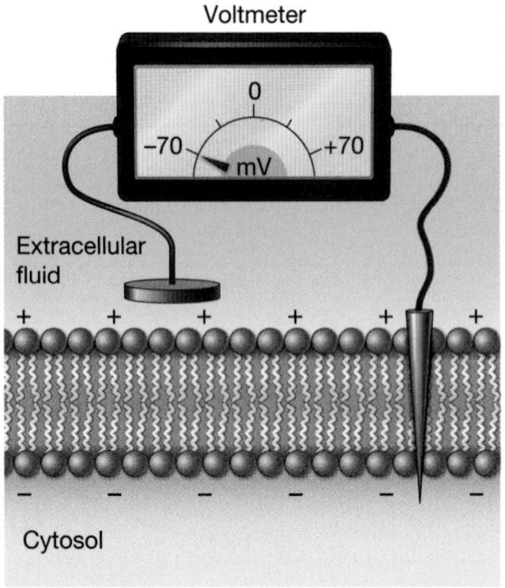

Voltmeter
0
−70
+70
mV
Extracellular fluid
Cytosol

membrane potentials to generate electrical signals. They do this by controlling the movement of charged particles, called *ions*, across their membranes to create electrical currents. The movement of ions is prevented by the membrane, because only small, nonpolar, and uncharged molecules can freely diffuse across the membrane. Therefore, the electrical signal, which occurs due to the movement of these ions, is achieved by opening and closing specialized protein channels in the membrane that allow the ions to move. Although the currents generated by ions moving through these channel proteins are very small, they form the basis of both neural signaling and muscle contraction. When a cell is in a resting state, not contracting or sending a signal, the difference across its membrane is its **resting membrane potential**. When a cell is in an active state, it is contracting or sending a signal, the event, and the changes in charge across the membrane is called an **action potential**.

To examine the process of contraction and relaxation, we will break the events into three parts: the NMJ, the sarcolemma, and the sarcomere.

11.3c Events at the Neuromuscular Junction

The process of contraction begins when a neuronal action potential travels along the motor neuron and reaches the NMJ. At the end of the axon the arrival of the action potential causes the release of a chemical messenger, or **neurotransmitter**, called ***acetylcholine (ACh)***. The ACh molecules diffuse across the small space, the **synaptic cleft**, that spans the distance between the motor neuron and the muscle cell. ACh travels by diffusion and will bind to ACh receptors located within the motor end plate of the sarcolemma on the other side of the synapse. ACh receptors are proteins, and like all proteins, once ACh binds the receptor will change shape. In this case, the binding of ACh opens a channel in the ACh receptor and positively charged ions can pass through the channel. Positively charged ions will be drawn into the muscle fiber because they are attracted to the negative charges there. The resting muscle cell membrane had a significant charge difference across its membrane; it was about 90 mV less positive or more negative on the inside of the membrane compared to the outside (**Figure 11.7**). We can describe the membrane in this resting state as being **polarized** (different on

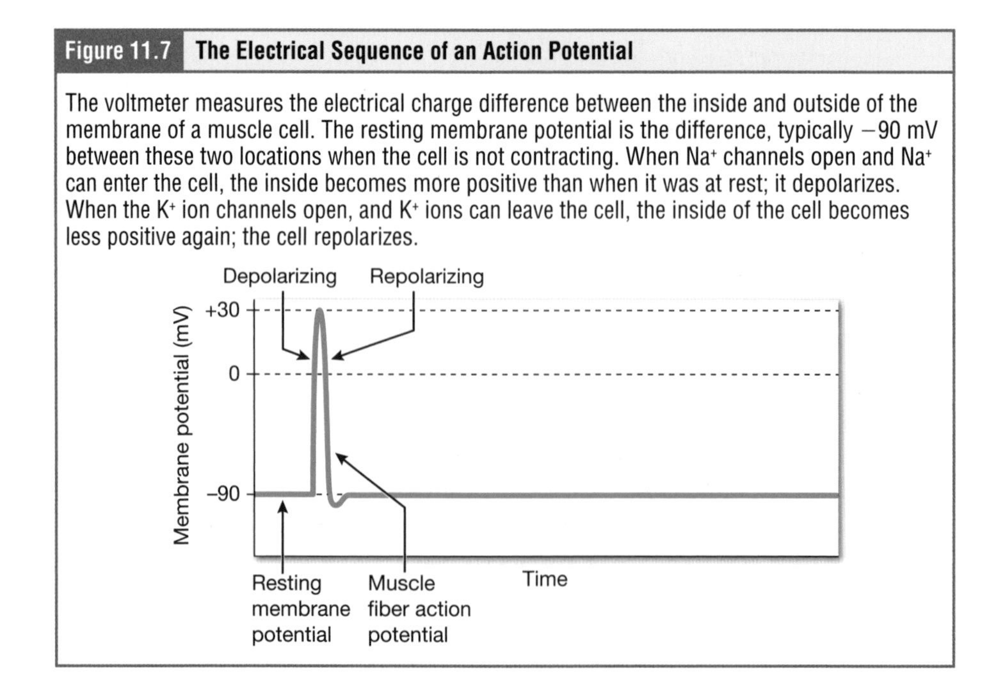

Figure 11.7 The Electrical Sequence of an Action Potential

The voltmeter measures the electrical charge difference between the inside and outside of the membrane of a muscle cell. The resting membrane potential is the difference, typically −90 mV between these two locations when the cell is not contracting. When Na^+ channels open and Na^+ can enter the cell, the inside becomes more positive than when it was at rest; it depolarizes. When the K^+ ion channels open, and K^+ ions can leave the cell, the inside of the cell becomes less positive again; the cell repolarizes.

DIGGING DEEPER:
Interference at the Neuromuscular Junction

A number of toxins (which are naturally occurring) and drugs (made in a laboratory) can affect the neuromuscular junction. Crotoxins, a group of poisons isolated from vipers and other snakes, affect motor neurons and their release of acetylcholine. Botulinum toxin, also called *botox*, similarly prevents the release of acetylcholine at the NMJ, causing paralysis. Curare is a poison isolated from the bark of some South American plants. Curare is a competitive blocker to the acetylcholine receptor, so the acetylcholine released cannot bind to the receptor. The effects of curare look identical to the effects of botox and some of the crotoxins. If acetylcholine cannot bind to its receptor, it will not have an effect on the muscle cell and no muscle contraction will take place. This lack of muscle contraction is called *flaccid paralysis.* Sarin gas also affects the NMJ but produces the opposite effect, resulting in excessive and uncontrolled contraction, called *spastic paralysis.* Sarin gas, a toxin that was originally produced as a pesticide but has been used against humans in warfare and terrorist attacks, prevents acetylcholinesterase from breaking down acetylcholine. This allows acetylcholine to persist in the synapse, causing an extended period of contraction long after the motor neuron has stopped firing signals.

one side compared to the other) and having a membrane potential of −90 mV. Once ACh binds and positive ions rush into the cell, the membrane of the motor end plate will **depolarize**, meaning that the membrane potential of the muscle fiber becomes less negative (closer to zero).

Learning Check

1. Describe the appearance of a contracted muscle cell. Please select all that apply.
   a. Z discs are closer together.
   b. M line is closer together.
   c. I bands are thicker.
   d. A bands are thicker.
2. Which of the following microfilaments is responsible for pulling the other microfilament during a contraction?
   a. Myosin
   b. Actin
3. Where can you find troponin?
   a. Surrounding actin
   b. Surrounding myosin
   c. Embedded within the sarcolemma
   d. Embedded within the sarcoplasmic reticulum
4. The extracellular matrix has a net charge of 250 mV. The cytoplasm in the cell has a net charge of 170 mV. What is the membrane potential?
   a. 250 mV
   b. 170 mV
   c. 420 mV
   d. 80 mV

11.3d Events Along the Sarcolemma

Along the length of the muscle cell the sarcolemma is studded with ion channels. The majority of these channels are described as **voltage-gated channels**. We have described many protein shape changes that occur when a molecule binds to the protein; these voltage-gated channels change shape when there is a change in the membrane potential. The two populations of voltage-gated channels on the muscle cell membrane are both closed (no ions can pass through) when the cell is polarized, and the inside is more negative than the outside. These channels change shape and open (allowing ions to pass through) when the membrane depolarizes. The first to open are **voltage-gated sodium channels**; these channels allow sodium ions (Na^+) to enter the muscle fiber, leading to further depolarization of the cell. The action potential rapidly spreads (or "fires") along the entire membrane, including down each T-tubule as each segment of the membrane depolarizes (**Figure 11.8**). A second population of voltage gated channels opens shortly after the voltage-gated sodium channels. Interspersed along the membrane are **voltage-gated potassium channels**. Once these channels open (they are slower to open than voltage-gated sodium channels potassium ions (K^+) can now traverse the membrane. These ions will flow out of the cell because they are in a greater concentration inside than outside, so they will diffuse down their concentration gradient. The egress of positive charges from the cell **repolarizes** the membrane (brings it back to its polarized state of being more positive outside and more negative inside).

Student Study Tip

Na^+ channels open right away, analogous to a young fit person, but K^+ channels are like creaky grandmas: slow to open and close.

Things happen very quickly in the world of excitable membranes (just think about how quickly you can snap your fingers as soon as you decide to do it). Immediately following depolarization of the membrane, it repolarizes, reestablishing the negative membrane potential. Meanwhile, the ACh in the synaptic cleft is being quickly broken

Figure 11.8 End Plate Potential

The sarcolemma at the NMJ is known as the end plate. The action potential begins at the end plate as acetylcholine receptor channels open and allow positive ions into the cell. This influx of positivity opens voltage-gated channels on the sarcolemma. The action potential proceeds down the membrane.

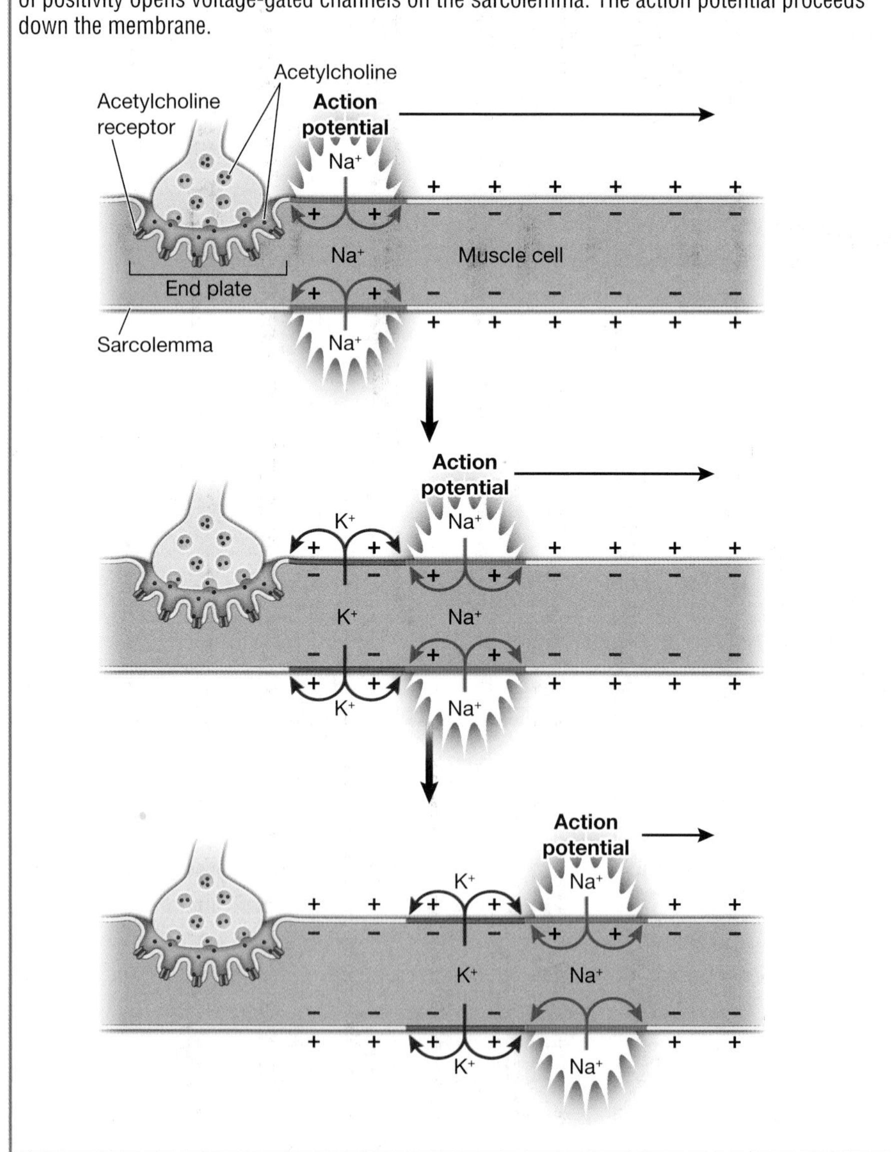

down by an enzyme, **acetylcholinesterase (AChE)**, that is found there. As less and less ACh is in the synapse to bind receptors and the sarcolemma is repolarizing, the action potential ends and the cell goes back to its resting membrane potential.

Learning Connection

Explain a Process

Physiological processes with many steps can be quite difficult to master. Try writing each step of the events at a neuromuscular junction on a separate piece of paper, mixing them up, and then putting them back in order. This activity will not only challenge your memory of the steps, but will help you to understand each function.

11.3e Events at the Sarcomere

Now that we have learned the sequence of electrical, or excitation events in the NMJ and along the sarcomere, we can relate them to the mechanical contraction that occurs at the sarcomere. The union of these physiological events is often called **excitation-contraction coupling**. The muscle cell action potential, which sweeps along the sarcolemma as a wave, is "coupled" to the actual contraction through the release of calcium ions (Ca^{2+}) from its storage in the cell's SR. For the action potential to reach the membrane of the SR, there are periodic invaginations in the sarcolemma, called *T-tubules*. These T-tubules ensure that the membrane can get close to the SR in the sarcoplasm. The arrangement of a T-tubule with the membranes of SR on either side is illustrated in **Figure 11.9**. Once released, the Ca^{2+} interacts with the troponin-tropomyosin protein complex, moving the tropomyosin aside so that the actin-binding sites are exposed for attachment by myosin heads. Cross-bridge cycling can then occur; each sarcomere shortens, and the muscle itself contracts. Cross-bridge cycling is sustained by ATP, so as long as both ATP and Ca^{2+} are available in the cytoplasm, contraction will continue. The full sequence of events, from motor neuron to sarcomere, is illustrated in Figure 11.9.

Student Study Tip

T-tubules look like a capital "T" and transmit signals deep into the cell.

Figure 11.9 The Steps in Muscle Cell Contraction

Muscle cell contraction begins when acetylcholine, released by the motor neuron, depolarizes the motor end plate. As the sarcolemma and T-tubules depolarize, calcium is released into the cell, triggering the movement of troponin and tropomyosin on the thin filaments away from the myosin binding sites. Myosin heads, now able to form cross-bridges with actin, contract the sarcomere, the entire muscle cell and the entire muscle.

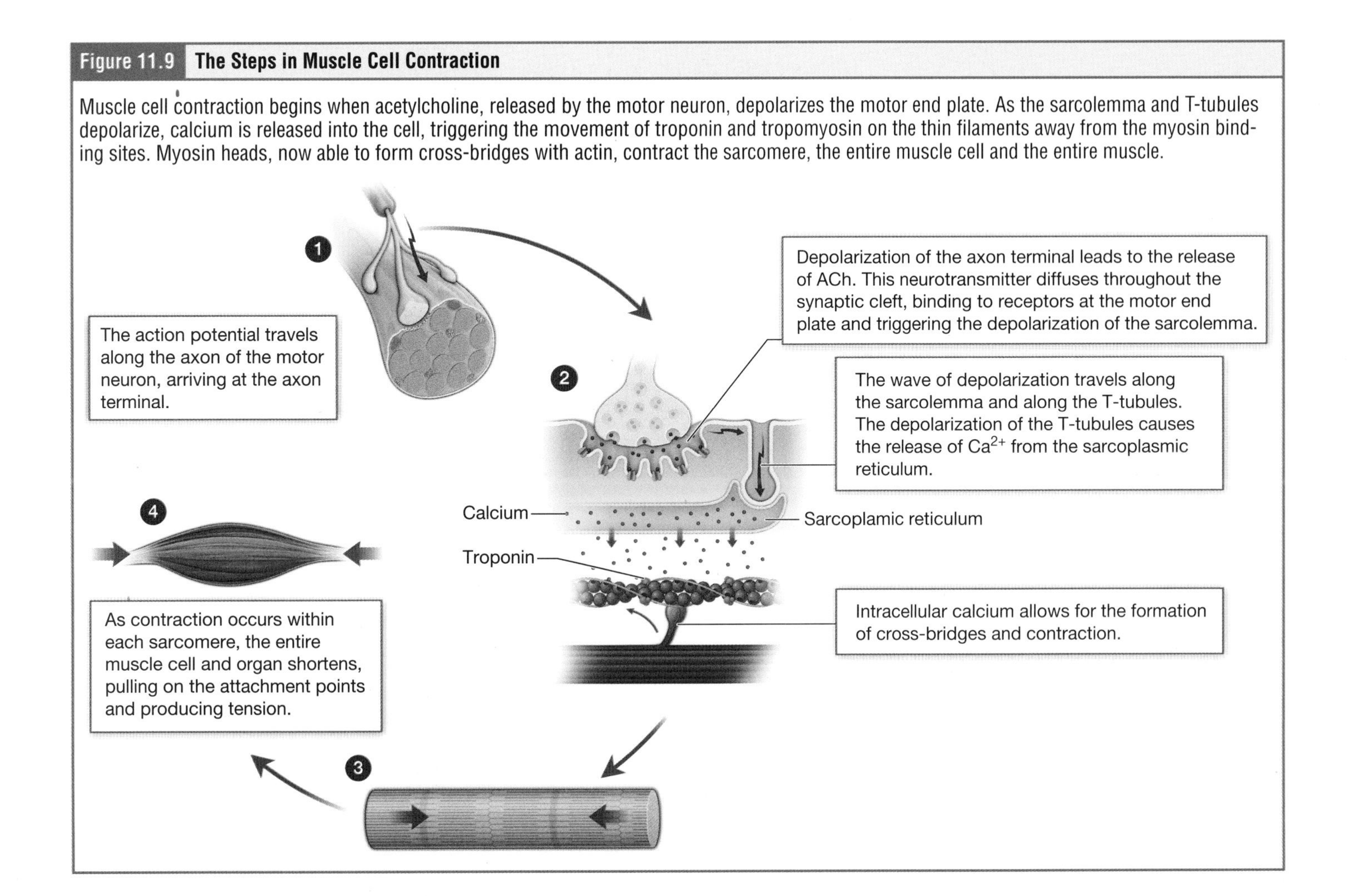

1
The action potential travels along the axon of the motor neuron, arriving at the axon terminal.
Depolarization of the axon terminal leads to the release of ACh. This neurotransmitter diffuses throughout the synaptic cleft, binding to receptors at the motor end plate and triggering the depolarization of the sarcolemma.
2
The wave of depolarization travels along the sarcolemma and along the T-tubules. The depolarization of the T-tubules causes the release of Ca2+ from the sarcoplasmic reticulum.
Calcium
Sarcoplamic reticulum
Troponin
Intracellular calcium allows for the formation of cross-bridges and contraction.
3
4
As contraction occurs within each sarcomere, the entire muscle cell and organ shortens, pulling on the attachment points and producing tension.

11.3f ATP and Muscle Contraction

Our muscle cells use a majority of the calories we need each day. It is clear to see why when we examine the few different roles ATP plays in the events of contraction and relaxation and multiply that out by the number of sarcomeres in one myofibril, the number of myofibrils in one muscle cell, the number of muscle cells in a muscle, and the number of muscles in the human body (truly an impossible calculation, but you can understand that the number would be mind-boggling!). For thin filaments to continue to slide past thick filaments during muscle contraction, myosin heads must pull the actin at the binding sites, detach, return to their original shape, attach to more binding sites, pull, detach, return to their original shape, and so on. This motion of the myosin heads is similar to that of the oars when an individual rows a boat: The blades of the oars (the myosin heads) pull, are lifted from the water (detach), repositioned (re-cocked), and then immersed again to pull (**Figure 11.10**). Each cycle requires energy, and the action of the myosin heads in the sarcomeres repetitively pulling on the thin filaments also requires energy, which is provided by ATP.

Cross-bridge formation occurs when the myosin head attaches to the actin. At this point in the cycle, adenosine diphosphate (ADP) and inorganic phosphate (P_i) are still bound to myosin (see "Anatomy of Cross-Bridge Cycling," Step 1) from the previous cycle. The binding of myosin to actin triggers the shape change in myosin. This movement, called the **power stroke**, resembles a sit-up as the angle between myosin head and the thick filament decreases. As the myosin power stroke occurs the actin, bound to the myosin head, is pulled toward the center of the sarcomere and the ADP and P_i are released from myosin (see "Anatomy of Cross-Bridge Cycling," Step 2). The ATP binding of myosin is now empty, and if more ATP is available in the sarcoplasm a new ATP molecule will bind. This binding causes the myosin head to detach from the actin (see "Anatomy of Cross-Bridge Cycling," Step 3). After this occurs, ATP is converted to ADP and Pi by the enzyme **ATPase** on myosin. The energy released during the breakdown of ATP allows the myosin head to detach from the actin thin filament and return to its original shape (see "Anatomy of Cross-Bridge Cycling," Step 4). When the myosin head is in its original, cocked shape, myosin is in a high-energy configuration. You can think of this myosin head as resembling a compressed spring; it is loaded with energy and ready

Figure 11.10 Sarcomere Shortening

In the fully contracted state, the thin filament network of a sarcomere is brought so close together that the thin filaments actually overlap. Certain regions present in the relaxed sarcomere (such as the I band and H zone) are practically nonexistent in the contracted sarcomere.

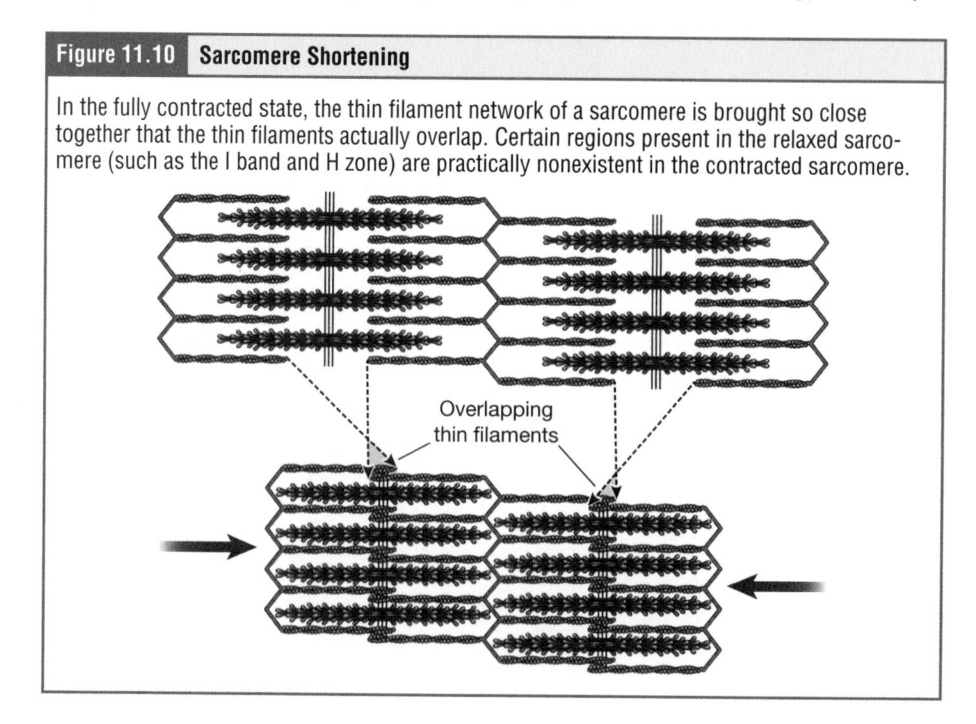

to move. This energy is expended as the myosin head moves through the power stroke, and at the end of the power stroke, the myosin head is in a low-energy position. After the power stroke, ADP is released; however, the formed cross-bridge is still in place, and actin and myosin are bound together. As long as ATP is available, it readily attaches to myosin; the cross-bridge cycle can recur and muscle contraction can continue.

When insufficient ATP is present in the sarcoplasm, myosin heads will remain attached to the actin filaments, unable to be released, and prolonged contraction will occur. This can happen during fatigue and lead to cramping (there are other causes of cramping as well), but it also occurs in death, when new ATP molecules can no longer

Anatomy of...

Cross-Bridge Cycling

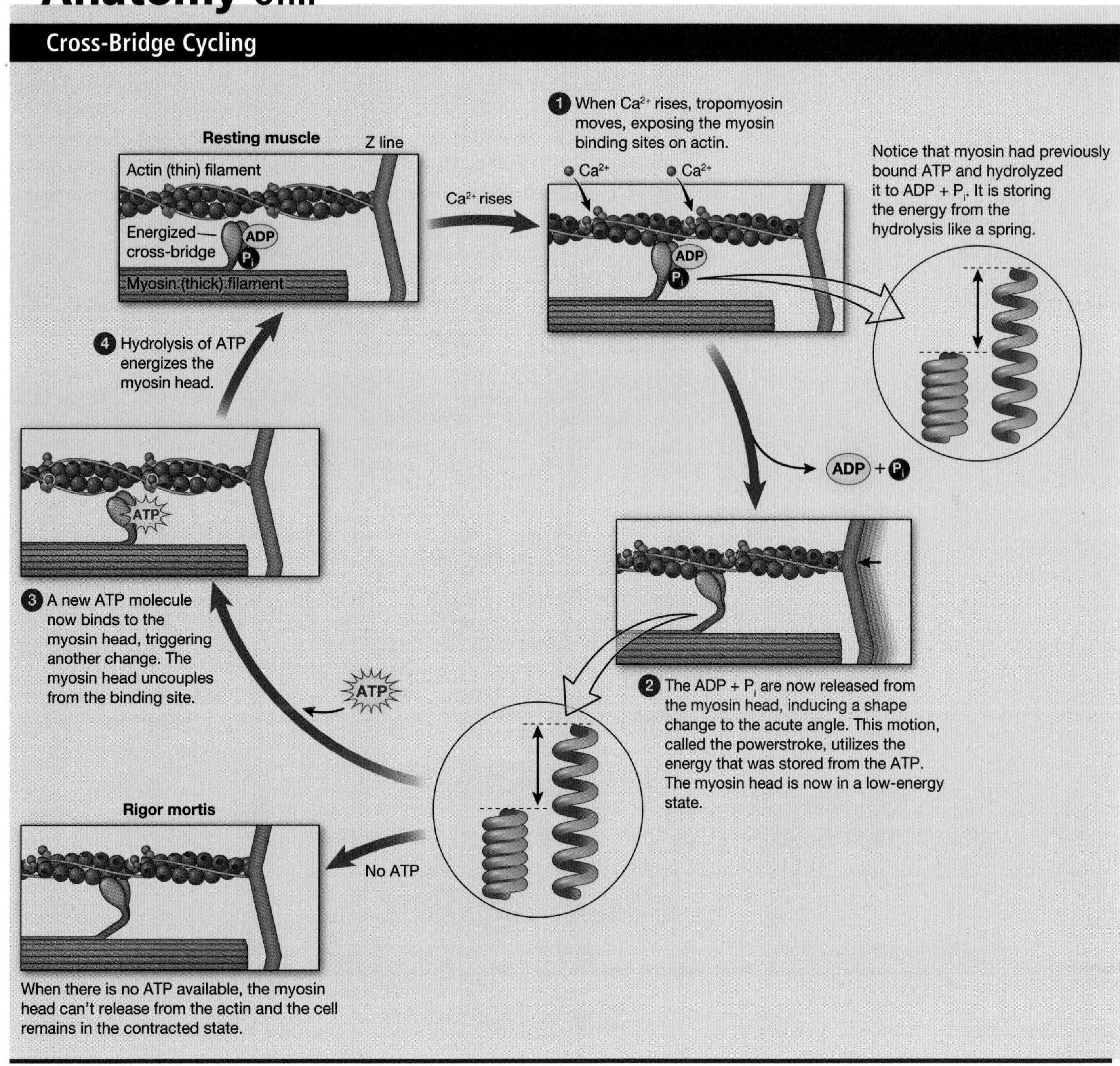

Resting muscle
Z line
Actin (thin) filament
Energized cross-bridge
ADP
P_i
Myosin (thick) filament
Ca^{2+} rises
1 When Ca^{2+} rises, tropomyosin moves, exposing the myosin binding sites on actin.
Ca^{2+}
Ca^{2+}
Notice that myosin had previously bound ATP and hydrolyzed it to ADP + P_i. It is storing the energy from the hydrolysis like a spring.
4 Hydrolysis of ATP energizes the myosin head.
ADP + P_i
ATP
3 A new ATP molecule now binds to the myosin head, triggering another change. The myosin head uncouples from the binding site.
ATP
2 The ADP + P_i are now released from the myosin head, inducing a shape change to the acute angle. This motion, called the powerstroke, utilizes the energy that was stored from the ATP. The myosin head is now in a low-energy state.
Rigor mortis
No ATP
When there is no ATP available, the myosin head can't release from the actin and the cell remains in the contracted state.

be made. Muscles all over the body will enter a prolonged state of rigid contraction known as *rigor mortis*, which eventually ends as the contractile proteins deteriorate.

ATP therefore must be available for crossbridge cycling, but ATP plays another critical role in muscle relaxation as well. The two substances—ATP and Ca^{2+}—must be present in the sarcoplasm in order for contraction to occur. A critical step in relaxation is removing the Ca^{2+} ions from the cytosol. Ca^{2+} ions are released from the SR during the excitation of the muscle cell, and they are resequestered in the SR by ATP-powered protein pumps. Without sufficient energy in the form of ATP, both muscle contraction and muscle relaxation are impossible.

11.3g Skeletal Muscle Cell Relaxation

LO 11.3.3

The contraction and relaxation of skeletal muscle cells is controlled by the motor neuron. To initiate relaxation, the neuron stops releasing its chemical signal (ACh) into the synapse at the NMJ. The ACh that has already been released is broken down by acetylcholinesterase, and the channels within the ACh receptors close. The muscle fiber repolarizes, which closes the gates in the SR where Ca^{2+} was being released. ATP-driven pumps move Ca^{2+} out of the sarcoplasm back into the SR. Without Ca^{2+} bound, troponin and tropomyosin slide back into place, covering the actin-binding sites on the thin filaments. The formation of cross-bridges between the thin and thick filaments is prevented, and the muscle fiber loses its tension and relaxes (**Figure 11.11**).

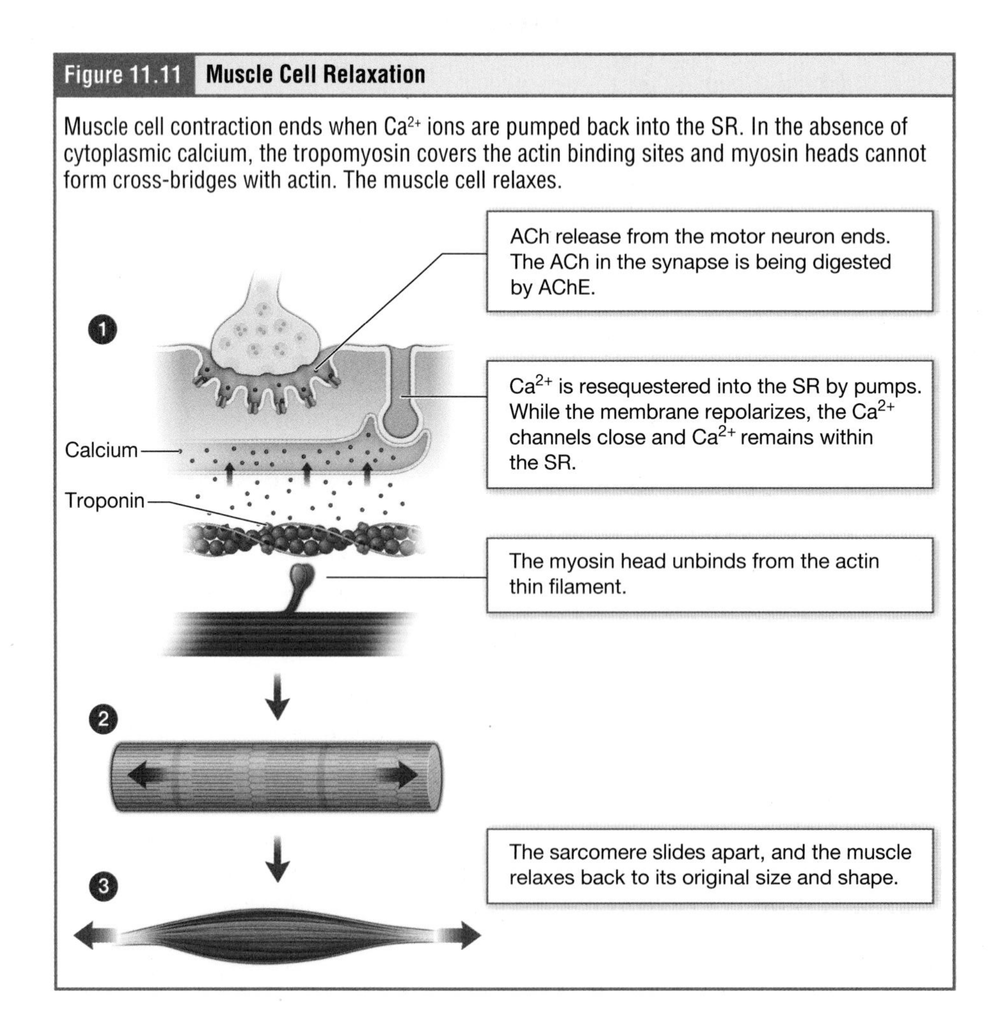

1
2
3
Calcium
Troponin
ACh release from the motor neuron ends. The ACh in the synapse is being digested by AChE.
Ca2+ is resequestered into the SR by pumps. While the membrane repolarizes, the Ca2+ channels close and Ca2+ remains within the SR.
The myosin head unbinds from the actin thin filament.
The sarcomere slides apart, and the muscle relaxes back to its original size and shape.

Figure 11.11 Muscle Cell Relaxation

Muscle cell contraction ends when Ca^{2+} ions are pumped back into the SR. In the absence of cytoplasmic calcium, the tropomyosin covers the actin binding sites and myosin heads cannot form cross-bridges with actin. The muscle cell relaxes.

Cultural Connection

Botox on the Face and the Palms, Just Not in the Mouth

Botulinum toxin, or botox, is a toxin secreted by the bacteria *Clostridium botulinum* that prevents the release of acetylcholine into the neuromuscular junction. Because of acetylcholine's role in causing contraction, when its release is prevented muscles cannot contract and a condition called *flaccid paralysis* occurs. Botulism, the illness that occurs when botox affects the whole body, can be lethal. Botulism can occur when botox is ingested. Because the *Clostridium botulinum* bacteria are resistant to moderate heat, botulism is associated with food poisoning when food has been inadequately cooked. Botulism is especially common in foods that have been canned improperly, and fear of this illness was the major driver for the United States to determine food safety standards. These days botulism is fairly rare, but botox has been found to have a number of more helpful uses. Quite famously, botox injections to the facial muscles are used to treat and prevent wrinkles. Wrinkles form when muscles hold the skin in a folded position repeatedly over time; botox prevents muscles in the area of the injection from contracting. This not only prevents wrinkle formation but also changes the individual's facial expression. Other uses for botox are in the prevention of headaches, muscle spasms, and excessive sweating.

11.4 Skeletal Muscle Metabolism

11.4.1 Describe the sources of ATP (e.g., glycolysis, oxidative phosphorylation, creatine phosphate) that muscle fibers use for skeletal muscle contraction.

11.4.2 Describe the events that occur during the recovery period from skeletal muscle activity.

11.4.3 Explain the factors that are believed to contribute to skeletal muscle fatigue.

11.4.4 Compare and contrast the anatomical and metabolic characteristics of slow oxidative (Type I), fast oxidative (Type IIa, intermediate, or fast twitch oxidative glycolytic), and fast glycolytic (Type IIb/IIx or fast twitch anaerobic) skeletal muscle fibers.

11.4a Sources of ATP

LO 11.4.1

Muscle cells use three mechanisms to generate ATP. Immediately upon contraction, muscle cells will generate ATP using creatine phosphate. After the first few seconds, muscle cells will need to begin either glycolysis or aerobic cellular respiration (**Figure 11.12**).

Creatine phosphate is a molecule that, similarly to ATP, can store energy in its phosphate bonds. A resting muscle builds up a store of creatine phosphate. This acts as an energy reserve that can be used to quickly create more ATP. When the muscle starts to contract and needs energy, creatine phosphate transfers its phosphate to ADP to form ATP, leaving creatine as a byproduct. This reaction occurs very quickly; thus, creatine phosphate–derived ATP powers the first few seconds of muscle contraction. However, creatine phosphate can only provide approximately 15 seconds' worth of energy, at which point another energy source has to be used (**Figure 11.12A**).

As the ATP produced by creatine phosphate is depleted, muscles turn to glycolysis as an ATP source. **Glycolysis** is an anaerobic (oxygen-independent) process that breaks down glucose to produce ATP; however, glycolysis cannot generate ATP as quickly as creatine phosphate does. Thus, the switch to glycolysis results in a slower rate of ATP availability to the muscle. During glycolysis, the cell breaks each glucose molecule in half, producing two molecules of pyruvate. The energy that was stored in the chemical bonds is released and is used to attach P_i to two ADP molecules, producing two ATP molecules. This new ATP can now be used for muscle contraction or relaxation (**Figure 11.12B**).

glycolysis (anaerobic respiration)

Figure 11.12 Generation of ATP in Muscle Fibers

(A) Resting muscles store a small amount of ATP, but it is depleted within seconds once contraction begins. Creatine phosphate generates enough ATP to last for about another 15 seconds. (B) The anaerobic breakdown of a glucose molecule (glycolysis) generates two ATP and two pyruvic acid molecules. If oxygen is available, pyruvic acid will enter the aerobic respiration pathways. In the absence of oxygen, pyruvic acid is converted into lactic acid; this is thought to be one of the causes of muscle fatigue. Lactic acid formation is common during strenuous exercise, when oxygen availability is insufficient to meet the demands of the exercising muscles. (C) Aerobic respiration, which requires oxygen, involves the complete breakdown of glucose into water, carbon dioxide, and ATP. This process occurs in the mitochondria, and produces about 95 percent of the ATP used by resting or moderately exercising muscles.

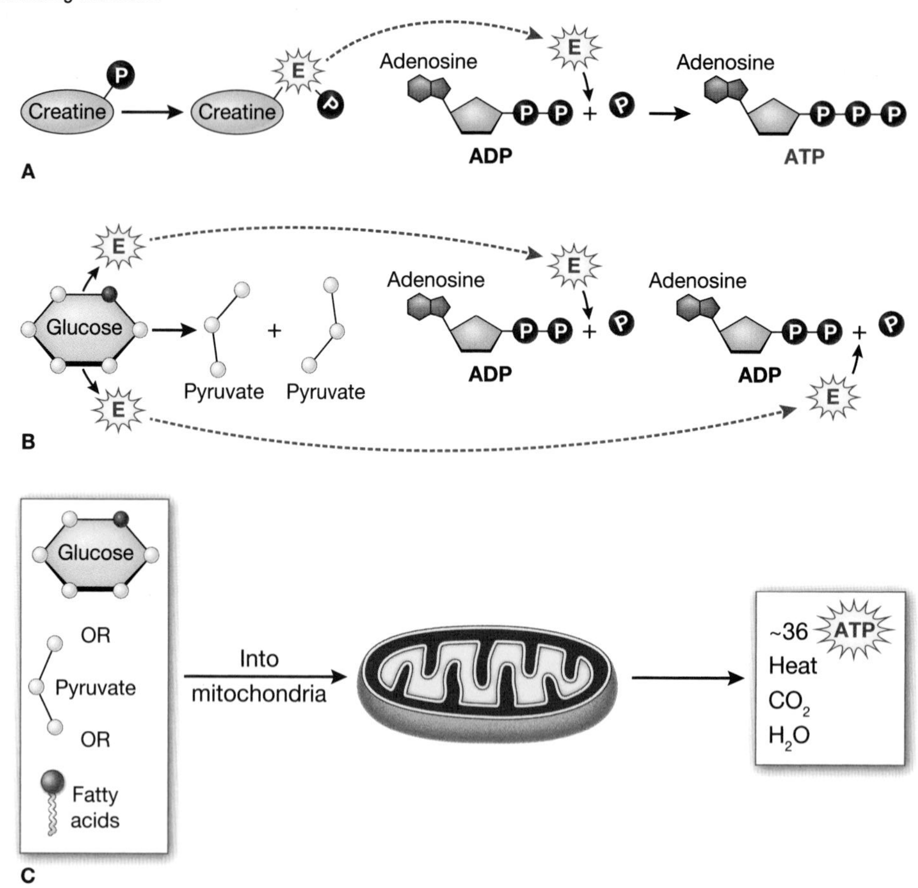

If oxygen is available, the pyruvate from glycolysis is further broken down in aerobic respiration. However, if oxygen is not available, pyruvate is converted to **lactic acid**. It is not common for oxygen to be unavailable, but this can occur during strenuous exercise when high amounts of energy are needed but sufficient oxygen cannot be delivered to muscle. Glycolysis itself cannot be sustained for very long (approximately one minute of muscle activity), but it is useful in facilitating short bursts of high-intensity output. Lactic acid can be recycled by the liver but does contribute to a lowering of the cellular and blood pH, which can have physiological consequences.

aerobic respiration (oxidative phosphorylation)

More commonly, sufficient oxygen is available, and the pyruvate is further broken down in **aerobic respiration**, a process that occurs in the mitochondria and is capable of producing high amounts of ATP. Pyruvate, as well as other nutrients, can be broken down in the presence of oxygen (O_2). The energy in the bonds of these molecules is released when the bonds are broken and used to convert ADP to ATP. Some energy is lost in the form of heat. The atoms of the broken-down molecules remain as carbon dioxide (CO_2) and water (H_2O). Aerobic respiration is much more efficient than anaerobic glycolysis, producing approximately 36 ATPs per molecule of glucose versus 4 ATPs from glycolysis. However, aerobic respiration cannot be sustained without a steady supply of O_2 to the skeletal muscle and is much slower (**Figure 11.12C**). Under most circumstances, approximately 95 percent of the ATP required for resting or moderately active muscles is provided by aerobic respiration.

The aerobic respiration of muscle fibers contributes a tremendous amount of heat to the body. Our body temperature homeostasis is dependent on a certain amount of muscle cell contraction. When we have a significant decrease in muscle contraction and therefore aerobic respiration, such as occurs during sleep, our body temperature drops. Conversely, when we have an increase in muscle contraction, such as during exercise, our body temperature rises. When body temperature homeostasis is in peril, such as on a very cold day, the nervous system can activate muscle cell contraction in the form of shivering and the required aerobic respiration will increase the temperature of the body.

Muscle cells therefore have a high demand for oxygen and rely primarily on oxygen delivered by the bloodstream. Muscle cells can also store small amounts of oxygen on a protein called *myoglobin* in their cytoplasm. In fact, one way that muscle cells adapt to the demands of exercise is by increasing their cellular stores of myoglobin and therefore their oxygen-storage capacity. When exercise demands exceed the ability of the cardiovascular system to deliver oxygen, skeletal muscle cells utilize the oxygen stored on their myoglobin. After exercise is over, the breath rate may continue to be high for a short period of time to replenish the myoglobin stores in advance of the next burst of activity. Regular exercise also catalyzes changes in the circulatory system so that O_2 can be supplied to the muscles more efficiently and for longer periods of time.

LO 11.4.2

Muscle cells are also able to store their own glucose in the form of glycogen. Some muscle fibers (fast-glycolytic) store more glycogen than others. Glycogen storage can increase with muscle use.

LO 11.4.3

Muscle fatigue occurs when a muscle can no longer contract in response to signals from the nervous system. Scientists understand many causes of muscle fatigue, but still do not have a comprehensive understanding. It is clear that one cause of muscle fatigue is a depletion of ATP. ATP deficiency is more likely to play a role in fatigue following brief, intense muscle output rather than sustained, lower intensity efforts. Scientists also know that lactic acid buildup as a result of dependance on glycolysis for ATP generation lowers the intracellular pH, affecting enzyme and protein activity and therefore contributing to fatigue. A significant factor is ion flow. The concentrations of Na^+ and K^+ are important for muscle cell action potentials. After repeated excitation of a muscle cell, imbalances in Na^+ and K^+ may occur, disrupting Ca^{2+} flow out of the SR. Long periods of sustained exercise may damage the SR and the sarcolemma, resulting in impaired Ca^{2+} regulation.

One anatomical adaptation that contributes to muscle organ homeostasis is that all muscles are a blend of different types of muscle cells. The cells, also called *muscle fibers*, come in three varieties. The varieties of muscle fibers vary by two criteria: how fast the fibers are able to contract, and how the fibers produce their ATP. Using these criteria, we can sort muscle cells into three types: slow oxidative, fast oxidative, and fast glycolytic. Most skeletal muscles in a human contain all three types, although in varying proportions (**Figure 11.13A**). The speed of contraction is dependent on how quickly myosin's ATPase hydrolyzes ATP to enable cross-bridge action. Fast fibers hydrolyze ATP approximately twice as quickly as slow fibers, resulting in much quicker cross-bridge cycling (which pulls the thin filaments toward the center of the sarcomeres at a faster rate). Muscles that function in endurance activities, such as the muscles of the back that function in maintaining our posture all day, have a higher proportion of slow oxidative fibers. Muscles that need to react quickly, such as the muscles of facial expression, as well as those of the fingers, have a higher proportion of fast glycolytic fibers. You can demonstrate their quick fatigue when you try to hold a smile for a photograph for a few minutes.

Figure 11.13 Three Types of Muscle Fibers

(A) Muscles are a blend of all three fiber types. Slow-oxidative fibers are dark red due to the presence of myoglobin. Fast oxidative fibers have some myoglobin so they are a lighter shade of red. Fast glycolitic fibers are pale as they lack myoglobin. (B) Slow-oxidative fibers make ATP efficiently and use it slowly, so they are able to continue contracting without fatiguing. Fast oxidative fibers make ATP efficiently but use it quickly, so they fatigue after some time. Fast glycolitic fibers make ATP inefficiently and use it quickly, so they fatigue rapidly. (The forward slashes in the X axis of these graphs indicate elapsed time.)

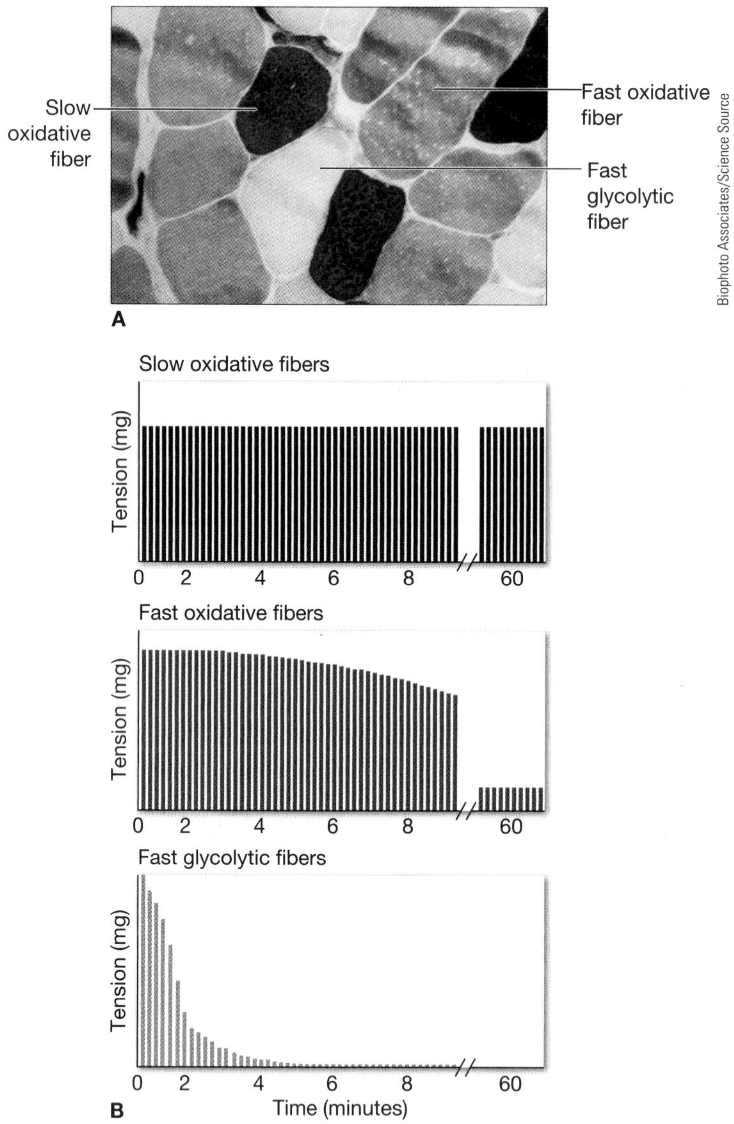

Biophoto Associates/Science Source

LO 11.4.4

slow oxidative (SO) (Slow twitch muscle fibers, Type I muscle fibers, fatigue-resistant muscle fibers)

Slow oxidative (SO) fibers contract relatively slowly and use aerobic respiration to produce ATP. Because oxidative fibers depend on aerobic respiration, which can only occur in the mitochondria, they contain many more mitochondria than the glycolytic fibers. The SO fibers also possess myoglobin, an internal O_2-storing molecule. Both myoglobin in muscle fibers and a similar protein, hemoglobin, in red blood cells contain the element iron, so their presence gives both of these cells a distinctive red color (see Figure 11.13A). Together these features allow SO fibers to produce large quantities of ATP, which can sustain muscle activity without fatiguing for long periods of time (**Figure 11.13B**). The fact that SO fibers can function for long periods without fatiguing makes them useful in maintaining posture and stabilizing bones and joints. Because they do not produce high tension, they are not used for powerful, fast movements that require high amounts of energy and rapid cross-bridge cycling.

Fast glycolytic (FG) fibers have fast contractions and primarily use glycolysis. The FG fibers fatigue more quickly than the others (see Figure 11.13B). FG fibers primarily use glycolysis as their ATP source. Remember that glycolysis does not require oxygen but is also less efficient than aerobic respiration. Therefore, these cells do not have substantial numbers of mitochondria or store oxygen on myoglobin, and they lack the red color of oxidative fibers (see Figure 11.13A). However, because they utilize more glucose than SO fibers, they store large amounts of glycogen, a starch that liberates glucose molecules when it breaks down. FG fibers are used to produce rapid, forceful contractions to make quick, powerful movements. These fibers fatigue quickly, permitting them to only be used for short periods (see Figure 11.13B).

Fast oxidative (FO) fibers are able to contract quickly and rely on a mix of aerobic respiration and glycolysis. Because of this incorporation of glycolysis and the speed at which these fibers contract and use ATP, FO fibers will fatigue more quickly than SO fibers. FO fibers are sometimes called *intermediate fibers* because they possess characteristics that are intermediate between FG and SO fibers. They produce ATP relatively quickly—more quickly than SO fibers—and thus can produce relatively high amounts of tension. They are oxidative because they produce ATP aerobically, possess high amounts of mitochondria, and do not fatigue quickly (see Figure 11.13B). However, FO fibers do not possess significant myoglobin, giving them a lighter color than the red SO fibers (see Figure 11.13A). FO fibers are used primarily for movements, such as walking, that require more energy than postural control but less energy than an explosive movement, such as sprinting. FO fibers are useful for this type of movement because they produce more tension than SO fibers, but they are more fatigue-resistant than FG fibers.

Fast glycolytic (FG) (Fast-twitch muscle fibers, Type IIb muscle fibers; Type Iix fibers)

Student Study Tip

Think of dark and white meat placement on a chicken. Dark meat is associated with slow oxidative fibers and white with fast glycolytic fibers. Whereas chicken breasts, used for flying, are white and fatigue easily, the thighs, used for running, are darker and are therefore more endurance-based.

fast oxidative (FO) (Type IIa muscle fibers, intermediate fibers)

Apply to Pathophysiology

Muscular Dystrophy

Duchenne muscular dystrophy (DMD) is genetic disease in which a structural protein, dystrophin, is not produced. Dystrophin helps the thin filaments of the myofibrils bind to the sarcolemma, and without it muscle contractions can cause the sarcolemma to tear.

1. DMD is a genetic disease but the symptoms do not appear until muscle contractions with significant force are attempted. At what point in the lifespan would you guess that the first symptoms of DMD would appear?
   A) At birth, when the baby is born.
   B) At toddlerhood, when the child begins walking and running.
   C) At adolescence, when the child engages in competitive sports.
   D) At old age, as the muscle cells begin to die.
2. Why does contraction cause the issues in DMD pathogenesis?
   A) During contraction force is applied to the thin filament framework by the myosin filaments.
   B) During contraction, more blood flows to the muscle causing inflammation.
   C) During contraction, the thick filaments move around the cell, pulling on the sarcolemma.
3. When the sarcolemma tears, what happens to the muscle cell?
   A) The cell would be flooded with waste products from outside the cell.
   B) The stored Ca^{2+} would leak out of the cell.
   C) The cell will no longer be able to maintain a resting membrane potential or fire action potentials.

Learning Check

1. Explain the role of ATP during muscle contractions.
   a. The ATP binds to actin to form cross-bridges with myosin.
   b. The ATP binds to myosin to change shape of the head.
   c. The ATP triggers an action potential through the sarcolemma.
   d. The ATP is stored in T-tubules until they are triggered to be released.
2. When you participate in explosive sports that require you to expel a lot of energy in short amounts of time, you are not getting enough oxygen in your system. These sports include sprinting, football, weightlifting, and many more. If you are participating in these sports, which of the following sources of ATP would you primarily be relying on? Please select all that apply.
   a. Creatine phosphate
   b. Glycolysis
   c. Aerobic respiration
3. Which of the following ions contributes to action potentials in skeletal muscles? Please select all that apply.
   a. Na^+
   b. H^+
   c. K^+
   d. Ca^{2+}
4. You are entering off-season training for hockey. During your off-season training, you are performing exercises for your quadriceps muscle that force the muscle to contract quickly, for a short period. This way of training will fatigue the muscle quickly but will give you the most power. What type of muscle fiber are you training your quadriceps muscle to be?
   a. Slow oxidative
   b. Fast glycolytic
   c. Fast oxidative

11.5 Whole Muscle Contraction

Learning Objectives: By the end of this section, you will be able to:

11.5.1 Define the following terms: tension, contraction, twitch, motor unit, and myogram.

11.5.2 Compare and contrast isotonic and isometric contraction.

11.5.3 Compare and contrast concentric and eccentric contraction.

11.5.4 Interpret a graph of the length-tension relationship and describe the anatomical basis for that relationship.

11.5.5 Interpret a myogram of a twitch contraction with respect to the duration of the latent, contraction, and relaxation periods and describe the events that occur in each period.

11.5.6* Use a graph of tension and muscle stimulus to explain the physiology of summation and tetanus.

11.5.7 Interpret a myogram or graph of tension versus stimulus intensity and explain the physiological basis for the phenomenon of recruitment.

* Objective is not a HAPS Learning Goal.

LO 11.5.1

To move an object, referred to as *load*, the sarcomeres in the muscle fibers of the skeletal muscle must shorten. The force generated by the contraction of the muscle is called **muscle tension**. Muscle tension must match the load. If, for example, you lifted a glass of water with the same muscle tension generated to move a couch, you would end up with water on your face very quickly! Not all muscle contraction is used to move loads, though. The description for a contraction that does not move a load, such as holding a plank position in exercise or carrying a suitcase with a straight, unmoving arm, is an

isometric contraction. Muscle tension that does move an object, such as the contraction that raises that water glass to your mouth, is an **isotonic contraction**.

LO 11.5.2

In isotonic contractions, a load is moved as muscle shortens. Two terms are used to describe the types of isotonic contractions: *concentric* and *eccentric*. These two types of contractions, which are opposite movements, will often occur together in sequence. A **concentric contraction** involves the muscle shortening to move a load. An example of this is the biceps brachii muscle contracting when a hand weight is brought upward with increasing muscle tension (**Figure 11.14A**). As the biceps brachii contracts, the angle of the elbow joint decreases as the forearm is brought toward the body. An **eccentric contraction** occurs as the muscle tension decreases and the muscle lengthens. In this case, the weight is lowered in a slow and controlled manner (**Figure 11.14B**) as the amount of cross-bridge formation decreases, and the angle of the elbow joint increases.

LO 11.5.3

An isometric contraction occurs as the muscle produces tension without changing the angle of a skeletal joint (**Figure 11.14C**). Isometric contractions involve sarcomere

Figure 11.14 Types of Muscle Contractions

Isotonic contractions are contractions in which a load moves and the muscle changes length. (A) In concentric isotonic contractions, the muscle shortens as it works. (B) During eccentric isotonic contractions, the muscle lengthens as it works. (C) Isometric contractions are contractions in which muscle length does not change because the load exceeds the tension the muscle can generate.

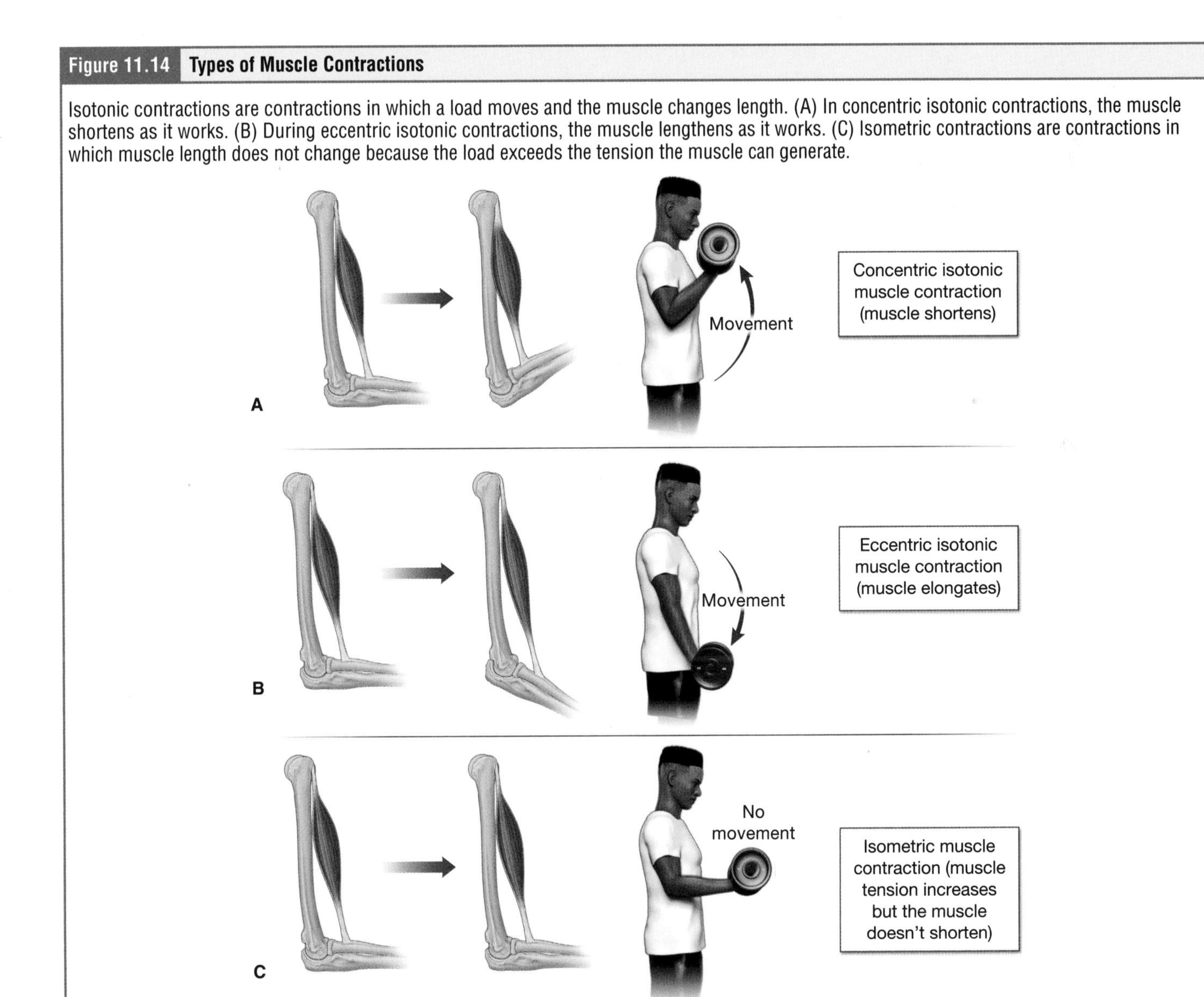

shortening and increasing muscle tension, but do not move a load, as the force produced cannot overcome the resistance provided by the load. For example, if you attempt to lift a weight that is too heavy, sarcomeres shorten to a point, and muscle tension builds, but no change in the angle of the elbow joint occurs. In everyday living, isometric contractions are active in maintaining posture and maintaining bone and joint stability. However, holding your head in an upright position occurs not because the muscles cannot move the head, but because the goal is to remain stationary and not produce movement. Most actions of the body are the result of a combination of isotonic and isometric contractions working together to produce a wide range of outcomes.

11.5a The Length-Tension Range of a Sarcomere

LO 11.5.4

Student Study Tip

Myosin in the cross-bridge cycle can be compared to somebody pulling a rope (actin) hand over hand. If the rope (actin) is too far away, myosin can't grab it.

For skeletal muscle to contract, myosin heads attach to actin to form the cross-bridges that result in sarcomere shortening (muscle tension). Structurally, cross-bridges can only form at locations where thin and thick filaments already overlap. In **Figure 11.15** you can see that in a stretched sarcomere there may not be sufficient overlap of the thin and thick filaments for many cross-bridges to form, so not much tension can be produced. The length of the sarcomere has a direct influence on the force generated in muscle contraction. This is called the *length-tension relationship*. When the sarcomere is already contracted, there is not much space for the thin filaments to come closer together and not much tension can be generated by further contraction. There is an optimal length of overlap between these states that maximizes the overlap of actin-binding sites and myosin heads.

11.5b Sustained Muscle Contraction

A single action potential from a motor neuron that produces a single contraction in the muscle cells it is attached to will produce an isolated contraction called a **twitch**.

Figure 11.15 **Sarcomere Length**

Sarcomeres produce maximal tension when there is optimal overlap between thick and thin filaments.

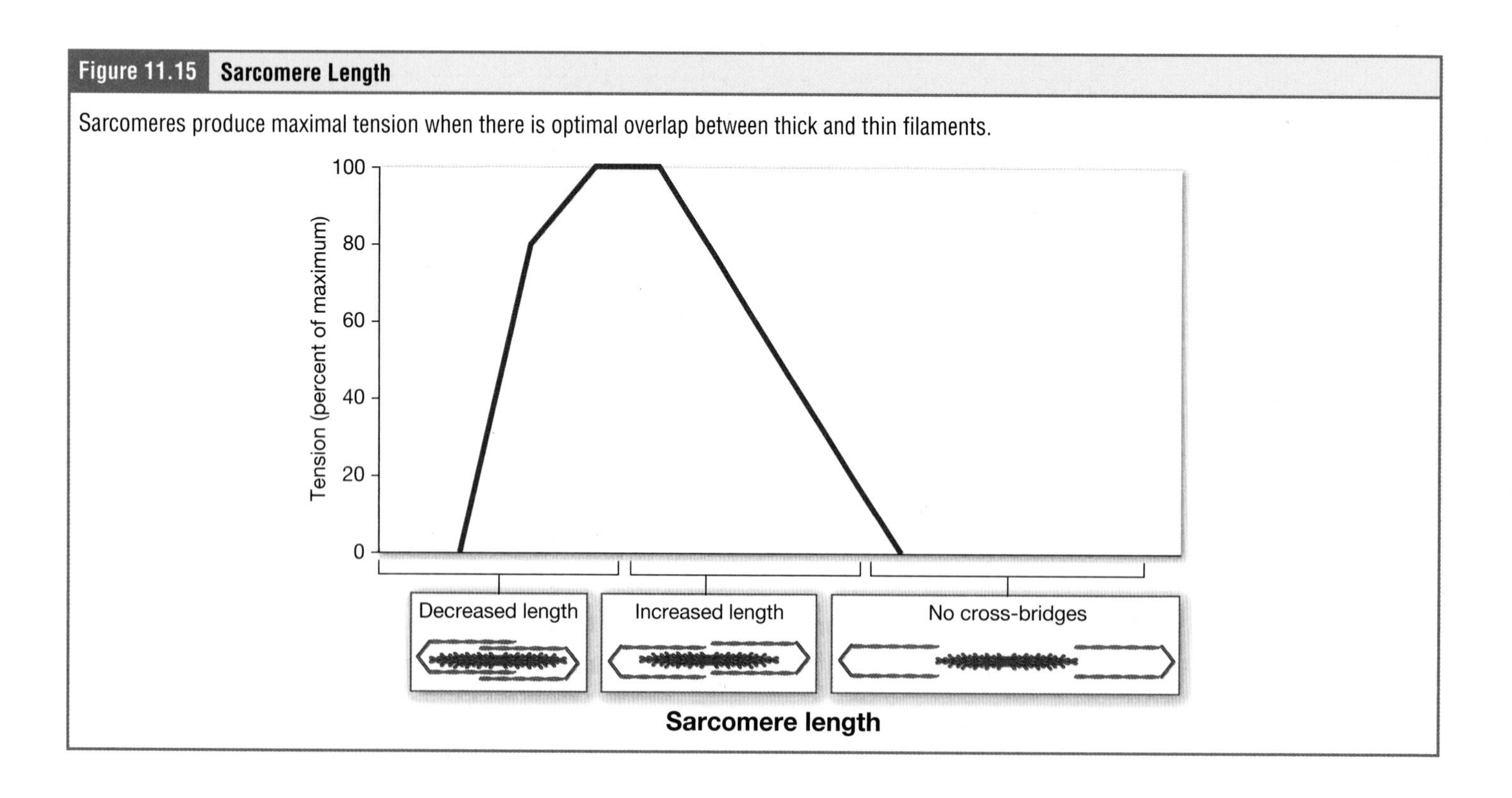

A twitch can last from a few milliseconds to 100 milliseconds, depending on the muscle type. The tension produced by a single twitch can be measured by a **myogram**, a representation of the amount of tension produced over time (**Figure 11.16**). Each contraction undergoes three phases. The first phase is the **latent period**, during which the action potential is being propagated along the sarcolemma and Ca^{2+} ions are released from the SR. This is the phase during which excitation and contraction are being coupled but contraction has yet to occur. The **contraction phase** occurs next. The Ca^{2+} ions in the sarcoplasm have bound to troponin, tropomyosin has shifted away from actin-binding sites, cross-bridges have formed, and sarcomeres are actively shortening to the point of peak tension. The last phase is the **relaxation phase**, when tension decreases as contraction stops. Ca^{2+} ions are pumped out of the sarcoplasm into the SR, and cross-bridge cycling stops, returning the muscle fibers to their resting state.

LO 11.5.5

Although a person can experience a muscle "twitch," a single twitch does not produce any significant muscle activity in a living body. Notice in Figure 11.16 that the action potential begins and ends before the mechanical contraction. What do you think would happen if another action potential followed? Another action potential would release more Ca^{2+} ions into the sarcoplasm, allowing for the continued exposure of actin's binding sites and the continuation of cross-bridge cycling. A series of action potentials would produce a sustained muscle contraction known as **summation** (**Figure 11.17**). Think about how you use muscles in life, muscles sometimes sustain contraction for a short period of time (such as lifting your leg to take a step) or a long time (carrying an object across the room), but in all cases of work the muscle needs to contract for some period of time. The nervous system fine-tunes the duration of muscle contraction by continuing to fire action potentials until it is time for relaxation. In addition to determining the duration of contraction through continued action potential delivery, the nervous system can also adjust the frequency of action potentials and the number of motor neurons transmitting action potentials to adjust the tension produced in skeletal muscle. At high, continued frequency of action potential delivery,

LO 11.5.6

Figure 11.16 A Myogram of a Muscle Twitch

The contraction produced from a single action potential (top panel) is a muscle twitch (bottom panel). Twitches have a latent period, a contraction phase of increasing tension, and a relaxation phase of decreasing tension. The action potential is completed during the latent period.

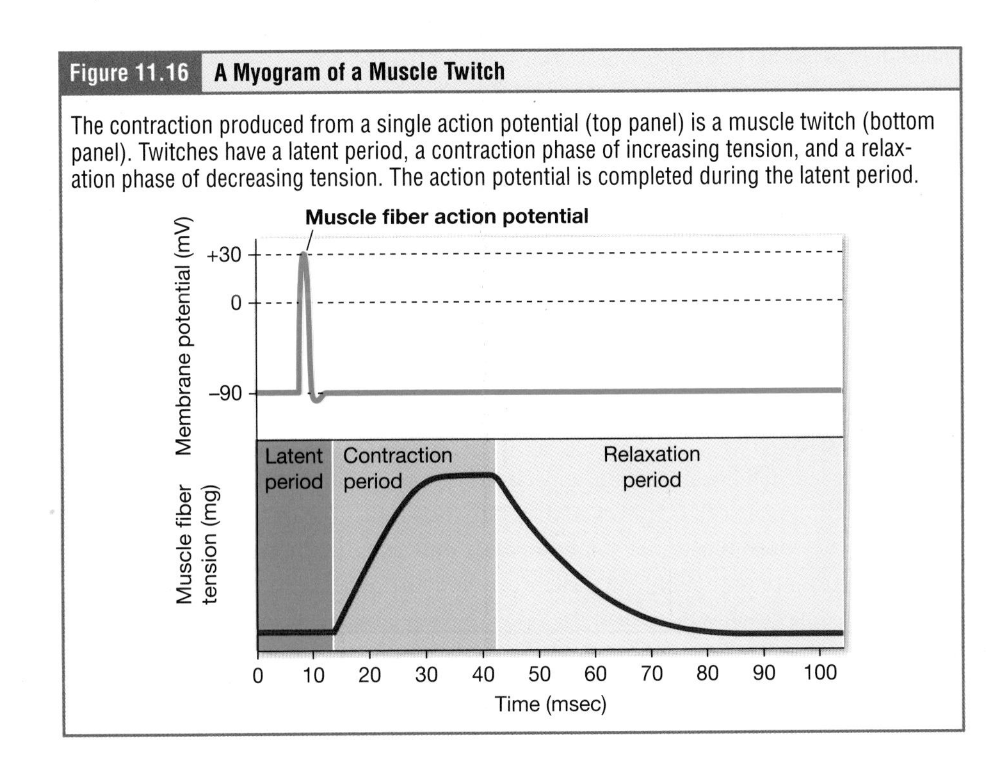

Figure 11.17 Muscle Twitch Summation and Tetanus

A single action potential triggers a short contraction called a muscle twitch. If action potentials are separated by enough time, they will elicit individual twitches. If the action potentials are sufficiently close together, they will produce a second contraction before the first has relaxed; this is called *summation*. If enough action potentials occur in a short time, they build on each other until the muscle is maximally contracted, a state known as *tetanus*.

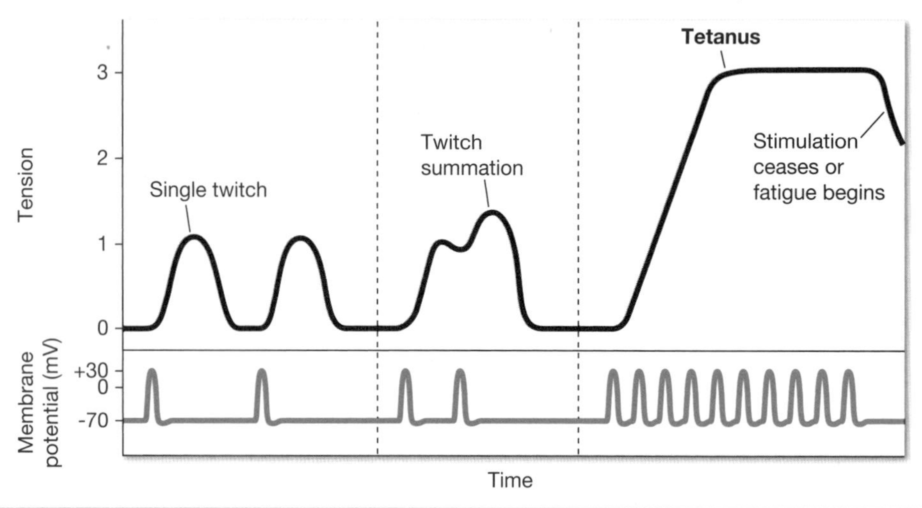

the tension in the muscle continues to rise until it reaches a peak point of continuous contraction known as **tetanus** (see Figure 11.17).

During tetanus, the concentration of Ca^{2+} ions in the sarcoplasm allows virtually all of the sarcomeres to form cross-bridges and shorten, so that a contraction can continue uninterrupted (until the muscle fatigues and can no longer produce tension).

11.5c Muscle Tone

Skeletal muscles are rarely completely relaxed, or flaccid. Even if a muscle is not producing movement, it is usually contracted a small amount, a baseline of contraction called **muscle tone**. The tension of muscle tone contributes to the stabilization of the joints and maintenance of posture.

Muscle tone is accomplished by a complex interaction between the nervous system and skeletal muscles that results in the activation of a few groups of muscle at a time, with a rotation so that no single muscle cell reaches fatigue and some muscle cell groups can recover while others are active.

All of these muscle activities are under the exquisite control of the nervous system. Neural control regulates concentric, eccentric and isometric contractions, muscle fiber recruitment, and muscle tone. A crucial aspect of nervous system control of skeletal muscles is the role of motor units.

11.5d Motor Units

As you have learned, every skeletal muscle cell is controlled by a single motor neuron. Each muscle cell is innervated by only one motor neuron and has only one NMJ. However, each motor neuron innervates multiple muscle cells (**Figure 11.18A**). Motor neurons are incredibly long, their cell body is found in the spinal cord, but their long axons reach out to form the NMJ in muscles around the body. Along the way, the axon may split many times, with each branch forming an NMJ with a different muscle cell. The group of muscle fibers in a muscle innervated by a single motor neuron is called a **motor unit**. They are described this way, as a unit, because they all contract together. A neuron firing an action potential cannot send the action potential to just some of its branches. Rather, the end of each of the axon branches will release ACh into the NMJ synapse with its muscle cell. So, each muscle cell of the motor unit contracts and relaxes at the same time. The size of a motor unit is variable depending on the nature of the muscle. All muscles in the body are composed of more than one motor unit (**Figures 11.18B** and **11.18C**). When a muscle contracts, the nervous system may activate all of the motor units in the muscle or just a few; this gives the nervous system control over the degree of contraction. A muscle that is composed of many small motor units would be able to be fine-tuned to an exquisite level. The smallest motor units exist in muscle capable of very fine motor movements, such as your fingers or face. In contrast, a muscle that is composed of fewer larger motor units would be capable of only gross levels of control. The larger muscle groups of the body, such as the thigh muscles or back muscles, are composed of fewer, larger motor units.

Imagine that you attempt to lift a surprisingly heavy object. At first you produce isometric contraction, and the object doesn't move, but you increase your attempt to lift it. As more strength is needed, more motor units are employed, or recruited to the task. The increase in muscle contraction, known as **recruitment**, occurs as the nervous system is adjusting to a task. As more motor units are recruited, the muscle contraction grows progressively stronger. This allows a feather to be picked up using the biceps brachii arm muscle with minimal force, and a heavy weight to be lifted by the same muscle by recruiting more and larger motor units.

Each motor unit is composed of a single muscle fiber type (see Figure 11.18C). The nervous system can control not only how many motor units are recruited to a task, but also which kinds.

LO 11.5.7

Figure 11.18 Motor Units

(A) All of the muscle cells innervated by a single motor neuron contract in concert and are therefore called a motor unit. (B) A muscle is a blend of motor units. (C) Each motor unit involves only a single type of muscle fiber. Muscle fiber diversity exists in all muscles.

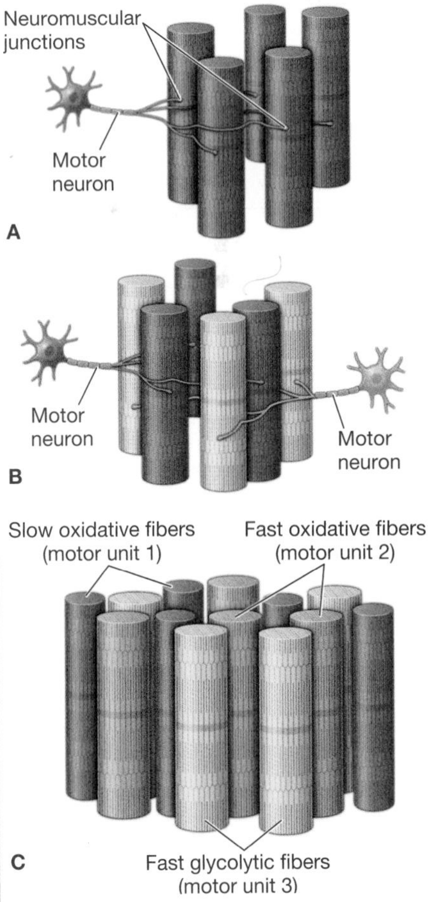

Cultural Connection

Athletic Bodies

Individual differences in muscle mass and performance can be observed, but researchers are still unraveling the physiological and anatomical roots of these differences. Testosterone plays a huge role in promoting the increases in the anatomical muscle mass that produce strength. While elite biological male athletes average about 10 percent faster than elite biological female athletes in speed sports such as running and swimming, biological male athletes can average from 29 percent to more than 50 percent higher when it comes to sports involving explosive strength (especially upper body strength)—such as golf, cricket, baseball, and weightlifting. It seems that testosterone may not only make a short-term difference in skeletal muscle mass but may also influence changes at puberty that have long-lasting effects. The International Olympic Committee (IOC) recently changed their rules to allow transgender women athletes to compete in the women's division of sports if they had their testosterone levels pharmacologically suppressed to within a particular range for at least 12 months beforehand. The reduction in muscle mass and strength during this timeframe was about 5 percent, suggesting that the effects of testosterone during puberty may make long-term anatomical changes. Earlier use of testosterone blockers may be welcome in trans individuals. Another factor to consider is ethnicity. Scientists in the United States have found that ethnicity may play a role in muscle loss during aging, with Asian Americans losing muscle mass faster during aging than other groups, and African Americans and Hispanic Americans retaining muscle strength and performance longer into their later years. While these observations have been reliably noted across large populations of study participants, researchers are still working to understand the physiology behind the differences.

DIGGING DEEPER: Exercise and Muscle Performance

Physical training alters the appearance of skeletal muscles and can produce changes in muscle performance. Conversely, a lack of use can result in decreased performance and muscle appearance. Externally, we often notice that muscles appear to get larger with regular exercise. This is not because new muscle cells are formed—skeletal muscle is nonmitotic—but because each individual skeletal muscle cell is adding myofibrils and may therefore become larger. The reverse occurs when muscles are not used; structural proteins and myofibrils are lost and muscle mass decreases. Cellular components of muscles can also undergo changes in response to changes in the type and amount of muscle use.

Slow fibers are predominantly used in endurance exercises that require little force but involve numerous repetitions. The aerobic metabolism used by SO fibers allows them to maintain contractions over long periods. Endurance training modifies these slow fibers to make them even more efficient by producing more mitochondria to enable more aerobic respiration and more ATP production. Endurance exercise can also increase the amount of myoglobin in a cell, as increased aerobic respiration increases the need for oxygen.

Exercise training can trigger the formation of more extensive blood vessel supply to a muscle to supply oxygen and glucose, and to remove metabolic waste.

Resistance exercises, as opposed to endurance exercise, require large amounts of FG fibers to produce short, powerful movements that are not repeated over long periods. The high rates of ATP hydrolysis and cross-bridge formation in FG fibers result in powerful muscle contractions. Muscles used for power have a higher ratio of FG to SO/FO fibers. Resistance exercise, such as weightlifting, affects muscles by increasing the formation of myofibrils, thereby increasing the thickness of muscle fibers. This added structure causes hypertrophy, or the enlargement of muscles, exemplified by the large skeletal muscles seen in bodybuilders and other athletes.

Except for the hypertrophy that follows an increase in the number of sarcomeres and myofibrils in a skeletal muscle, the cellular changes observed during endurance training do not usually occur with resistance training. There is usually no significant increase in mitochondria or capillary density. However, resistance training does increase the development of connective tissue, which adds to the overall mass of the muscle and helps to contain muscles as they produce increasingly powerful contractions. Tendons also become stronger to prevent tendon damage, as the force produced by muscles is transferred to tendons that attach the muscle to bone.

Apply to Pathophysiology

Aging

Skeletal muscle cells cannot replicate themselves through mitosis; therefore when skeletal muscle cells die they are typically replaced by connective tissue and adipose tissue.

1. Which of the following are true of aging muscles that have some replacement of muscle cells with connective tissues? Please select all that apply.
   A) The muscle will be less stretchy.
   B) The muscle will have less strength during contraction.
   C) The muscle may bulge or appear swollen.
   D) The muscle may become avascular.
2. For reasons we do not completely understand, FG fibers appear to decline faster in an aging body than SO fibers. Which of the following is likely true?
   A) The movements and functions of the legs and back may reach fatigue more quickly in old age because these locations have a predominance of FG fibers.
   B) All muscles will be affected as the number of FG fibers decreases, but some locations will be affected more than others.
   C) A significant increase in muscle fatigue will occur as the aging body depends more and more on SO fibers.
3. Betty and Stu, an older couple, are describing their observations about aging to their grandchild, who is in medical school. Both Betty and Stu were runners. Betty, a former sprinter, is noticing a lot of changes as they age. They no longer feel that they can sprint on the track and are upset by the loss of this pastime. But Stu, a long-distance marathon runner, says that they are slower, but can still run long distances and do not describe a marked decline in their fitness. The grandchild proposes a few guesses as to why Betty and Stu are experiencing different age-related changes. Which of their proposals is most likely correct?
   A) The loss of FG fibers in both aging bodies is affecting Betty's sprinting more than Stu's endurance running.
   B) Because the fibers that maintain joint stability are aging, Betty experiences more impact while sprinting than Stu does while running.
   C) As an endurance runner, Stu has built more muscle over time and so can weather the changes of aging for longer.

Learning Check

1. You are playing volleyball and you jump up as high as you can to make a block. In order to do so, your gastrocnemius has to shorten so it can propel the load of your body upward against gravity. What kind of contractions does your gastrocnemius undergo? Please select all that apply.
   a. Isometric contraction
   b. Isotonic contraction
   c. Concentric contraction
   d. Eccentric contraction
2. You are trying to lift a heavy box onto the top shelf of your closet. In order to do so, you contract your deltoid muscles. When you have your arms by your side, the deltoid muscles are the longest. When you have your arms straight over your head, the deltoid muscles are the shortest. Which position will generate the most force so you can lift the box onto the top shelf of your closet? Please select all that apply.
   a. Arms down by your side
   b. Arms straight overhead
   c. Arms straight out in front of you
   d. Arms straight out to the side of you
3. You are studying a myogram when you notice that there is no relaxation phase after the contraction phase. Which of the following are you observing?
   a. Summation
   b. Tetanus
4. Which of the following motor units are recruited first during a contraction?
   a. Small motor units
   b. Medium motor units
   c. Large motor units

11.6 Cardiac Muscle Tissue

Learning Objectives: By the end of this section, you will be able to:

11.6.1* Compare and contrast the function and structure of cardiac and skeletal muscle cells.

11.6.2* Describe the cellular features of a cardiac muscle cell.

* Objective is not a HAPS Learning Goal.

There are three types of muscle tissue, and while we know and understand the most about skeletal muscle tissue, cardiac and smooth muscle tissues have many of the same features. In this section we will compare and contrast the features of cardiac muscle with skeletal muscle, but a deeper examination of cardiac muscle is done in Chapter 19. Cardiac muscle is found only in the heart. One distinctive function unique to cardiac muscle is the highly coordinated contractions required of this tissue. Similar to skeletal muscle, cardiac muscle is striated and organized into sarcomeres, possessing the same banding organization as skeletal muscle (**Figure 11.19**). However, cardiac muscle fibers are shorter than skeletal muscle fibers and usually contain only one nucleus, which is located in the central region of the cell. Cardiac muscle fibers also possess many mitochondria and a large number of myoglobin molecules, as ATP is produced primarily through aerobic metabolism. Unlike skeletal muscle, where separate motor units contract asynchronously, every cardiac muscle cell of the heart chambers must contract at once in order to effectively pump blood into the vessels of the circulatory system. Several of the structural features of cardiac muscle cells have evolved to enable this coordination. The cells are extensively branched and are connected to one another at their ends by intercalated discs. An **intercalated disc** contains desmosomes to hold the cells

LO 11.6.1

LO 11.6.2

Figure 11.19 Cardiac Muscle Tissue

Cardiac muscle tissue is found only in the walls of the heart. The cells are branched, contain one or two nuclei, and are united by intercalated discs.

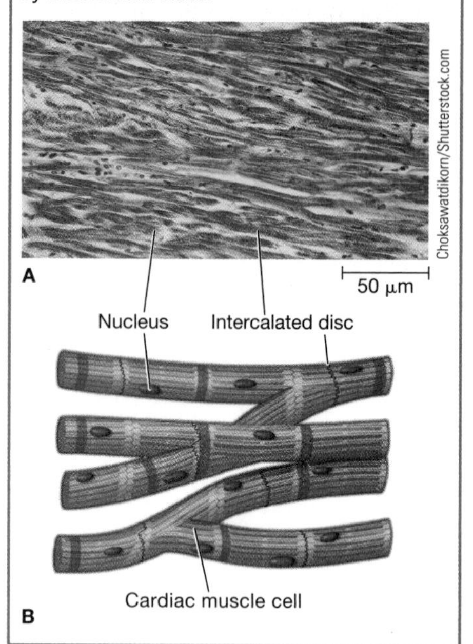

Figure 11.20 Cardiac Muscle Cells

Intercalated discs unite cardiac muscle cells structurally and functionally. Their desmosomes hold the cells together and their gap junctions allow the action potentials to spread from one cell to neighboring cells.

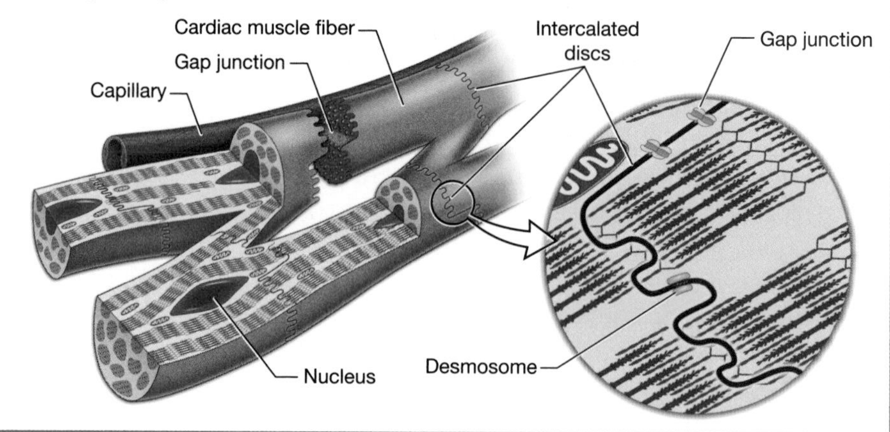

together structurally and gap junctions that allow the action potential to pass from the inside of one cell to the inside of the cells that it is connected to (**Figure 11.20**). The gap junctions of the intercalated disc eliminate the need for an NMJ for each cardiac cell, as many cells are activated by one action potential.

Not only is it unnecessary for each cell to have an NMJ, but the nervous system does not control cardiac muscle contractions in the same way that it does skeletal muscle contractions. Cardiac muscle tissue contractions (heartbeats) are controlled by specialized cardiac muscle cells called *pacemaker cells* that initiate the action potential independent of the nervous system. One way independence from the nervous system can best be illustrated is that when cardiac muscle cells are grown in dishes in a laboratory, the cardiac muscle cells can be seen contracting under the microscope in the right environmental conditions. Cardiac muscle cannot be consciously controlled, but the pacemaker cells respond to signals from the autonomic nervous system (the fight-or-flight system) to speed up or slow down the heart rate. The pacemaker cells can also respond to various hormones that influence heart rate. Cardiac muscle cells are largely nonmitotic, which limits resiliency from tissue damage resulting from heart attack or COVID-19 illness.

Learning Connection

Broken Process

When cardiac muscle cells die, they are usually replaced by fibrous connective tissue that does not have the same properties. What do you think happens to the spread of the electrical signal when there are patches of dead tissue within the wall of the heart?

Learning Check

1. Describe the appearance of a cardiac muscle fiber. Please select all that apply.
   a. Striations present
   b. Few mitochondria
   c. Intercalated discs
   d. Long muscle fibers
2. Which of the following controls cardiac muscle contractions? Please select all that apply.
   a. Motor neurons
   b. Autonomic nervous system
   c. Sympathetic nervous system
   d. Pacemaker cells

11.7 Smooth Muscle

Learning Objectives: By the end of this section, you will be able to:

11.7.1 Describe the sources of calcium in smooth muscle contraction and explain how an increase in cytoplasmic calcium initiates contraction.

11.7.2 Compare the signals that initiate smooth muscle contraction to the signal that initiates skeletal muscle contraction.

Smooth muscle was named "smooth" because of the absence of notable striations, which are present when viewing cardiac and skeletal muscle under the microscope. Smooth muscle still has thin and thick filaments, but they are not arranged in precise linear columns of sarcomeres and so the proteins do not create the same banding pattern when viewed under the microscope (Figure 11.21). Smooth muscle is present in the walls of hollow organs like the urinary bladder, uterus, stomach, intestines, and in the walls of passageways such as the arteries and veins of the circulatory system and the tracts of the respiratory, urinary, and reproductive systems.

Smooth muscle fibers are shaped like an American football (wide in the middle and tapered at both ends) and have a single nucleus; they are thousands of times shorter than skeletal muscle cells. Although they do not have striations and sarcomeres, smooth muscle thin filaments are anchored by dense bodies. A **dense body** is analogous to the Z discs of skeletal and cardiac muscle fibers and is fastened to the sarcolemma (Figure 11.22). Like skeletal muscle, calcium ions are stored in the SR, however Ca^{2+} also enters the cell from the extracellular fluid.

LO 11.7.1

Smooth muscle cells lack T-tubules, but their surface is pockmarked with small pockets called **caveolae**. Because smooth muscle cells do not contain troponin, cross-bridge formation is not regulated by the troponin-tropomyosin complex but instead by the regulatory protein **calmodulin**. In a smooth muscle fiber, external Ca^{2+} ions passing through opened calcium channels in the sarcolemma or SR, bind to calmodulin. The Ca^{2+}-calmodulin complex then activates an enzyme called *myosin kinase*, which, in turn, activates the myosin heads. The heads can then attach to actin-binding sites and pull on the thin filaments. The thin filaments also are anchored to the dense bodies, which are tied together through a network of thin filaments (see Figure 11.22). This arrangement causes the entire muscle fiber to contract in a manner whereby the ends are pulled toward the center, causing the midsection to bulge in a corkscrew motion.

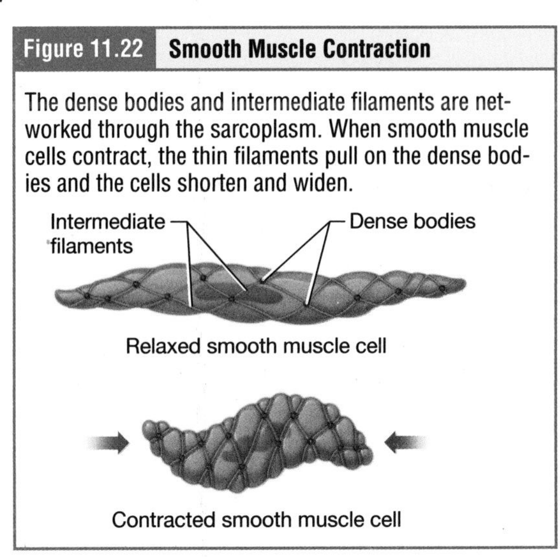

Figure 11.22 Smooth Muscle Contraction

The dense bodies and intermediate filaments are networked through the sarcoplasm. When smooth muscle cells contract, the thin filaments pull on the dense bodies and the cells shorten and widen.

Figure 11.21 Smooth Muscle Tissue

(A) Smooth muscle is the only muscle tissue type that is unstriated. These small, American football-shaped cells are found in the walls of hollow organs, tracts, and blood vessels throughout the body. (B) Histological image of smooth muscle.

Nucleus Smooth muscle fibers

A

B

Steve Gschmeissner/Science Photo Library/Getty Images

Muscle contraction continues until ATP-dependent calcium pumps actively transport Ca^{2+} ions back into the SR and out of the cell. However, a low concentration of calcium remains in the sarcoplasm to maintain muscle tone. This remaining calcium keeps the muscle slightly contracted, which is important in certain locations such as blood vessels. A decrease in smooth muscle tone can cause a precipitous drop in blood pressure and subsequent fainting.

Because most smooth muscles must function for long periods without rest, their power output is relatively low, but contractions can continue without using large amounts of energy. Smooth muscle is not under voluntary control; thus, it is called *involuntary muscle*. The triggers for smooth muscle contraction include hormones, stimulation from the nervous system, and local factors. In certain locations, such as the walls of visceral organs, stretching the muscle can trigger its contraction (the stress-relaxation response).

LO 11.7.2

Neurons do not form the highly organized NMJs with smooth muscle that we examined between motor neurons and skeletal muscle fibers. Instead, one neuron will wind around smooth muscle cells with a series of neurotransmitter-filled bulges called *varicosities* along the axon (**Figure 11.23**). When an action potential travels down the neuron, each **varicosity** releases neurotransmitters into its synapse, triggering contraction. Some smooth muscle—especially the smooth muscle found in the walls of the hollow organs—contains pacemaker cells similar to those described in cardiac muscle.

Smooth muscle is organized in two ways: as single-unit smooth muscle (which is much more common) and as multiunit smooth muscle. The two types have different locations in the body and have different characteristics. Single-unit smooth muscle has its muscle fibers joined by gap junctions so that the muscle contracts as a single-unit in the same way as cardiac muscle. This type of smooth muscle is found in the walls of all visceral organs except the heart (which has cardiac muscle in its walls). Single-unit smooth muscle has a **stress-relaxation response**. This means that as the muscle of a hollow organ is stretched when the organ fills, the mechanical stress of the stretching

Figure 11.23 Smooth Muscle Innervation

The neurons that innervate smooth muscle cells have long axons with swellings called varicosities along their length. Each varicosity is capable of releasing neurotransmitters and causing contraction in each of the smooth muscle cells it forms a synapse with.

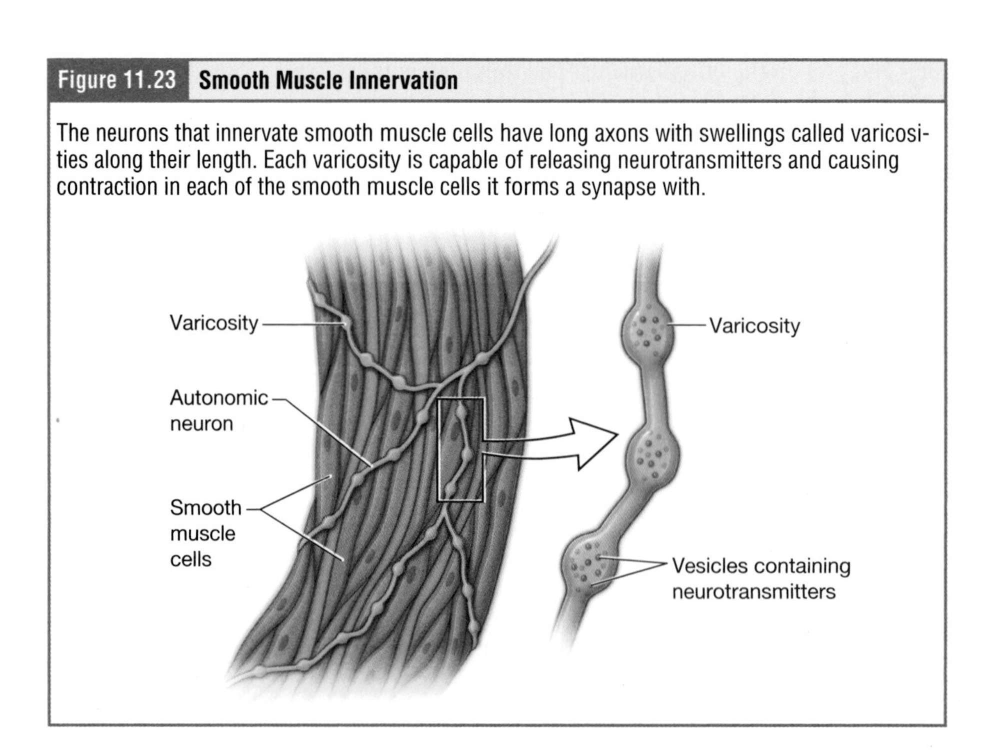

will trigger contraction, but this is immediately followed by relaxation so that the organ does not empty its contents prematurely. This is important for hollow organs, such as the stomach or urinary bladder, which continuously expand as they fill. In general, smooth muscle produces slow, steady contractions that allow substances, such as food in the digestive tract, to move through the body.

Multiunit smooth muscle cells rarely possess gap junctions, and thus the cells are not electrically coupled. As a result, contraction does not spread from one cell to the next but is instead confined to the cell that was originally stimulated. This type of tissue is found around large blood vessels, in the respiratory airways, and in the eyes.

Learning Check

1. Identify the location where you would find smooth muscle fibers. Please select all that apply.
   a. Heart
   b. Biceps muscle
   c. Artery
   d. Stomach
2. Which of the following is unique to smooth muscles compared to cardiac and skeletal muscles? Please select all that apply.
   a. Presence of dense bodies
   b. American football-shaped appearance
   c. Contractile capabilities
   d. Presence of microfilaments
3. Which of the following can occur due to decreased calcium in smooth muscle cells?
   a. Widening of the urinary tract
   b. Narrowing of blood vessels
   c. Increased tone of organs
   d. Drop in blood pressure

Chapter Review

11.1 Overview of Muscle Tissues

The learning objectives for this section are:

11.1.1 Describe the major functions of muscle tissue.

11.1.2 Describe the structure, location in the body, and function of skeletal, cardiac, and smooth muscle.

11.1.3 Compare and contrast the general microscopic characteristics of skeletal, cardiac, and smooth muscle.

11.1 Questions

1. Which of the following properties is shared among all muscle tissues?
   a. Squeeze blood through the body
   b. Contracts to cause movement
   c. Propels food through the intestines
   d. Allows the body to move in space
2. Where would you expect to find smooth muscles?
   a. In the heart chambers
   b. Around the femur
   c. Between the ribs
   d. Inside the stomach
3. Which of the following characteristics are found in the cells of all muscle tissue types?
   a. They have striations.
   b. They have gap junctions.
   c. They are long.
   d. They have nuclei.

11.1 Summary

- There are three types of muscle tissue in the body: skeletal, cardiac and smooth.
- All three types of muscle tissues exhibit the qualities of excitability, contractility, extensibility, and elasticity.
- Skeletal muscle is attached to the bones of the skeleton, and is the most extensive of the three types of muscle tissue. It is responsible for voluntary movements of the body.
- Cardiac muscle is only found in the heart. It is involuntary and is responsible for the beating of the heart.
- Smooth muscle is found in the tubular organs of the body; it is responsible for many types of involuntary movements in the internal organs.
- The cells of each type of muscle tissue has unique microscopic properties. A skeletal muscle cell is cylindrical, striated, and multinucleated. A cardiac muscle cell is a striated, branching cylinder with a single nucleus. A smooth muscle cell is football-shaped and non-striated, and contains one nucleus.

11.2 Skeletal Muscle

The learning objectives for this section are:

HAPS LEARNING GOALS AND OUTCOMES

11.2.1 Describe the organization of skeletal muscle, from cell (skeletal muscle fiber) to whole muscle.

11.2.2 Name the connective tissue layers that surround each skeletal muscle fiber, fascicle, entire muscle, and group of muscles, and indicate the specific type of connective tissue that composes each of these layers.

11.2.3 Describe the components within a skeletal muscle fiber (e.g., sarcolemma, transverse [T] tubules, sarcoplasmic reticulum, myofibrils, thick [myosin] myofilaments, thin [actin] myofilaments, troponin, tropomyosin).

11.2.4 Define sarcomere.

11.2.5 Describe the arrangement and composition of the following components of a sarcomere: A-band, I-band, H-zone, Z-disc (line), and M-line.

11.2.6 Describe the structure of the neuromuscular junction.

11.2 Questions

1. Define a fascicle.
   a. Connective tissue between skin and bones
   b. Bundles of muscle cells wrapped together
   c. Broad, tendon-like sheet
   d. Smallest functional unit where contraction occurs

2. Select the order of connective tissue layers from most superficial to deep.
   a. Endomysium, perimysium, epimysium
   b. Perimysium, epimysium, endomysium
   c. Epimysium, perimysium, endomysium
   d. Epimysium, endomysium, perimysium

3. Which of the following organelles in a skeletal muscle fiber stores calcium ions for use in skeletal muscle contraction?
   a. Transverse tubules
   b. Troponin
   c. Sarcoplasmic reticulum
   d. Sarcolemma

4. Define a sarcomere.
   a. Z disc to Z disc
   b. H zone plus I band
   c. H zone plus A band
   d. A band plus I band

5. Which of the following regions shortens when a sarcomere contracts? Please select all that apply.
   a. M line
   b. H zone
   c. A band
   d. I band

6. What is the pathway of an electrical impulse to contract muscle tissues?
   a. Motor neuron → Neuromuscular Junction → Motor End Plate → T-tubules
   b. T-tubules → Motor Neuron → Neuromotor Junction → Motor End Plate
   c. Neuromotor Junction → Motor Neuron → Motor End Plate → T-tubules
   d. Neuromotor Junction → Motor End Plate → Motor Neuron → T-tubules

11.2 Summary

- A skeletal muscle is composed of groups of skeletal muscle cells (fibers) called fascicles.
- An entire skeletal muscle is surrounded by a layer of connective tissue called the epimysium. Each fascicle is surrounded by perimysium, and each skeletal muscle cell is surrounded by endomysium.
- Each skeletal muscle cell consists of typical organelles, and specialized organelles such as the sarcoplasmic reticulum and the transverse (T) tubules.
- A skeletal muscle fiber consists of cylindrical units called myofibrils. Each myofibril is divided into contractile units called sarcomeres.
- Sarcomeres consist of protein filaments composed of actin and myosin.
- Neuromuscular junctions are points of contact between motor neurons and muscle fibers, which allow motor neurons to stimulate contraction in the muscle fibers.

11.3 Skeletal Muscle Cell Contraction and Relaxation

The learning objectives for this section are:

11.3.1 Define the sliding filament model of skeletal muscle contraction.

11.3.2 Describe the sequence of events involved in the contraction of a skeletal muscle fiber, including events at the neuromuscular junction, excitation-contraction coupling, and cross-bridge cycling.

11.3.3 Describe the sequence of events involved in skeletal muscle relaxation.

11.3 Questions

1. The influx of __ ion into the cell leads to an action potential.
   a. Na^+
   b. K^+
   c. Ca^{2+}
   d. H^+

Mini Case 1: Myasthenia Gravis

Myasthenia Gravis is a rare autoimmune disorder that results in skeletal muscle weakness and rapid muscle fatigue. It does this by targeting and attacking the acetylcholine receptors.

1. Which of the following is true of the sliding filament model of skeletal muscle contraction? Select all that apply.
   a. The only neurotransmitter used in neuromuscular junctions is acetylcholine.
   b. As skeletal muscle fibers contract, the thin and thick filaments shorten.
   c. Each myosin cross-bridge binds to only one actin binding site during the process of muscle contraction.
   d. As skeletal muscle fibers contract, the overlap of the thin and thick filaments increases.

2. Predict what would happen due to decreased or damaged acetylcholine receptors. Please select all that apply.
   a. Increased influx of calcium into the cytosol
   b. Increased buildup of acetylcholine in the synaptic cleft
   c. Decreased influx of potassium ions during an action potential
   d. Decreased formation of cross-bridges

3. After a skeletal muscle fiber has finished contracting, several mechanisms cause it to relax. Treatment for a patient with myasthenia gravis involves increasing muscle contraction and decreasing the normal relaxation process, so that the patient has less muscle weakness. Which oft the following types of treatments could potentially help the patient? Please select all that apply.
   a. Medications that allow the motor neurons that control skeletal muscle fibers to continue releasing acetylcholine
   b. Acetylcholinesterase inhibitors
   c. Medications that decrease the action of troponin
   d. Medications that prevent reuptake of calcium ions by the sarcoplasmic reticulum

11.3 Summary

- The sliding filament model is an explanation of the mechanism by which muscle fibers shorten during muscle contraction. The thin and thick filaments in the sarcomeres of muscle fibers do not shorten during muscle contraction; instead, they increase the extent of their overlap.
- When stimulated by a motor neuron, an action potential is initiated in the motor end plate of a muscle fiber. This is coupled to the contraction of the muscle fiber.
- Stimulation of a muscle fiber results in the release of calcium ions from the sarcoplasmic reticulum (SR). This leads to cross-bridge cycling, in which myosin heads on the thick filaments bind to binding sites on the actin of the thin filaments. The myosin heads then flip over, pulling the thin filaments toward the center of the sarcomere. This results in the shortening of the sarcomere and the muscle fiber.
- Muscle fiber relaxation occurs when the motor neuron stops stimulating the muscle fiber. The calcium ions are pumped back into the SR, stopping the cross-bridge cycling. The thin and thick filaments then slide back to their resting length.

11.4 Skeletal Muscle Metabolism

The learning objectives for this section are:

HAPS LEARNING GOALS AND OUTCOMES

11.4.1 Describe the sources of ATP (e.g., glycolysis, oxidative phosphorylation, creatine phosphate) that muscle fibers use for skeletal muscle contraction.

11.4.2 Describe the events that occur during the recovery period from skeletal muscle activity.

11.4.3 Explain the factors that are believed to contribute to skeletal muscle fatigue.

11.4.4 Compare and contrast the anatomical and metabolic characteristics of slow oxidative (Type I), fast oxidative (Type IIa, intermediate, or fast twitch oxidative glycolytic), and fast glycolytic (Type IIb/IIx or fast twitch anaerobic) skeletal muscle fibers.

11.4 Questions

1. Athletes that require energy for a long period of time, like soccer players or marathon runners, will require large amounts of ATP. Which of the following systems would they rely on the most?
   a. Glycolysis
   b. Oxidative phosphorylation
   c. Creatine phosphate

2. When a person finishes an exercise session, the breathing rate remains elevated for a few minutes. How does this help the skeletal muscles recover from the exercise and prepare for the next exercise session?
   a. The increased rate of breathing causes fatty acids to be broken down for energy instead of glucose.
   b. The increased breathing rate replenishes the oxygen reserves in the muscles, which are bound to myoglobin.
   c. The increased rate of breathing clears nitrogen out of the lungs.
   d. The increased breathing rate prevents excess ATP production.

3. Explain how ATP deficiency can contribute to muscle fatigue.
   a. Decreased ATP will contribute to excessive cross-bridging formation.
   b. Decreased ATP will contribute to prolonged contraction.
   c. Decreased ATP will contribute to decreased contraction frequency.
   d. Decreased ATP will contribute to decreased contraction power.

4. The gastrocnemius is a muscle in your calf. You use this muscle to stand and walk. An average person stands 2–4 hours per day and takes 4,000 steps per day. Describe the anatomical features would you expect to find in this muscle fiber.
   a. Many mitochondria
   b. Many glycogen molecules
   c. Decreased myoglobin molecules
   d. Decreased mitochondria

11.4 Summary

- ATP is generated by one of three mechanisms. If oxygen is available, then oxidative phosphorylation takes place. If oxygen is not available, ATP is generated through creatine phosphate and glycolysis.
- Muscle fatigue is believed to be caused by several mechanisms, including ATP depletion, lactic acid buildup, and imbalances in the levels of sodium, potassium, and calcium ions.
- Muscle fibers are categorized based on how quickly they contract and how they generate ATP. Slow oxidative fibers contract slowly and use aerobic respiration to produce ATP. Fast glycolytic fibers contract quickly and generate ATP through glycolysis. Lastly, fast oxidative fibers contract quickly and rely on mix of glycolysis and aerobic respiration and possess characteristics of both slow oxidative and fast glycolytic fibers.

11.5 Whole Muscle Contraction

The learning objectives for this section are:

HAPS LEARNING GOALS AND OUTCOMES

11.5.1 Define the following terms: tension, contraction, twitch, motor unit, and myogram.

11.5.2 Compare and contrast isotonic and isometric contraction.

11.5.3 Compare and contrast concentric and eccentric contraction.

11.5.4 Interpret a graph of the length-tension relationship and describe the anatomical basis for that relationship.

11.5.5 Interpret a myogram of a twitch contraction with respect to the duration of the latent, contraction, and relaxation periods and describe the events that occur in each period.

11.5.6* Use a graph of tension and muscle stimulus to explain the physiology of summation and tetanus.

11.5.7 Interpret a myogram or graph of tension versus stimulus intensity and explain the physiological basis for the phenomenon of recruitment.

* Objective is not a HAPS Learning Goal.

11.5 Questions

1. What is a motor unit?
   a. The response of a single muscle fiber to neural stimulation
   b. A graph of the tension produced over time during a single twitch
   c. The force generated by muscle contraction
   d. A motor neuron and all of the muscle fibers that it controls

2. If you want to move a box from the ground to the top shelf, which of the following types of contraction is required?
   a. Isotonic contraction
   b. Isometric contraction

3. When you go to sit down in the chair, the quadriceps muscles are controlling how slowly your knees are bending. In other words, they generate more force as the knee bends. Which of the following describes the type of contraction of the quadriceps?
   a. Concentric contraction
   b. Eccentric contraction

4. You are trying to do a biceps curl. You have your arm straight, so the biceps muscle is stretched. You realize lifting the weight is too hard. Which of the following can you do to make it easier to lift the weight?
   a. Straighten your elbow more to stretch your biceps muscle.
   b. Bend your elbow to shorten your biceps muscle.

5. Ca^{2+} ions are bound to troponin, tropomyosin shifts away from actin-binding sites, and cross-bridges form. Which of the following phases in a myogram does this describe?
   a. Latent period
   b. Contraction phase
   c. Shortening phase
   d. Relaxation phase

6. "Lockjaw" is a disease that occurs when a bacterium from metal gets into your bloodstream. The bacterium affects your nerves so that the electrical signal from the nerve is constantly triggering the sarcolemma. Which of the following symptoms could result from this bacterial infection?
   a. Weakness
   b. Tetanus
   c. Low tone
   d. Paralysis

7. When trying to lift the following objects, which object would require your muscles to recruit the greatest number of motor units?
   a. A pencil
   b. A textbook
   c. A desk
   d. A car

11.5 Summary

- Muscle tension is a force generated by the contraction of the muscle. Muscle tension that does not move a load is called an isometric contraction, while muscle tension that moves a load is called an isotonic contraction. There are two types of isotonic contractions—a concentric contraction involves the muscle shortening to move a load, while an eccentric contraction occurs as the muscle tension decreases and the muscle lengthens.
- There is an optimal length of overlap between microfilaments. If the muscle is too stretched, there is little overlap between microfilaments, resulting in less force generation. If the muscle is too short, there is not much space for the thin filaments to come close together, resulting in less force generation.
- Myograms can help identify whether there is summation or whether there is tetanus. The myogram has three phases—the latent period, the contraction phase, and the relaxation phase.
- A myogram displays the amount of tension produced in a twitch of a single muscle fiber over time. The phases of muscle fiber contraction are the latent period, the contraction phase, and the relaxation phase.
- If a second stimulus is applied to a muscle fiber before tension has subsided from the previous stimulus, summation will occur; this results in increased tension generation.
- Continuous frequent stimulation of a muscle fiber results in a sustained contraction of maximal tension called tetanus.
- Lifting objects of different weights (loads) requires the recruitment of different numbers of motor units. Lightweight objects require muscles to recruit only a small number of motor units. As the load increases, the muscle must recruit a progressively larger number of motor units.

11.6 Cardiac Muscle Tissue

The learning objectives for this section are:

11.6.1* Compare and contrast the function and structure of cardiac and skeletal muscle cells.

11.6.2* Describe the cellular features of a cardiac muscle cell.

* Objective is not a HAPS Learning Goal.

11.6 Questions

1. Which of the following is a similarity between cardiac and skeletal muscles?
   a. Both fibers contain one nucleus.
   b. Both fibers are striated.
   c. Both fibers are found in the heart.
   d. Both fibers are contracted together.

2. Which of the following cellular features can be found in cardiac muscle cells? Please select all that apply.
   a. The cells are parallel to each other.
   b. The cells are connected by intercalated discs.
   c. The cells have very few mitochondria.
   d. The cells have a large number of myoglobin molecules.

11.6 Summary

- Cardiac muscle cells are only found in the heart. They are similar to skeletal muscle cells in that they are striated. They differ from skeletal muscle cells in that they are branched cells that contain only one nucleus, and they are involuntary.
- Cardiac muscle cells differ from skeletal muscle cells in that they are not controlled by motor neurons; instead, they are controlled by pacemaker cells within the heart.
- Cardiac muscle cells must contract as a unit in order to produce coordinated heartbeats. Once some of the muscle cells are stimulated by the pacemaker cells of the heart, they distribute the contraction command to nearby cells through intercalated discs, gap junctions that pass ionic messages from one cell to the next.

11.7 Smooth Muscle

The learning objectives for this section are:

HAPS LEARNING GOALS AND OUTCOMES

11.7.1 Describe the sources of calcium in smooth muscle contraction and explain how an increase in cytoplasmic calcium initiates contraction.

11.7.2 Compare the signals that initiate smooth muscle contraction to the signal that initiates skeletal muscle contraction.

11.7 Questions

1. Which of the following store calcium for smooth muscle contractions? Please select all that apply.
   a. T-tubules
   b. Sarcoplasmic reticulum
   c. Extracellular fluid
   d. Sarcoplasm
2. Define a varicosity.
   a. A swelling at the synapse that is filled with neurotransmitters
   b. A regulatory protein to initiate contractions in muscle
   c. A small pocket that lines the cell membrane of muscle cells

11.7 Summary

- Smooth muscles are organized in two ways. A single-unit smooth muscle is a group of smooth muscle cells that contract together. A multiunit smooth muscle is one in which the electrical impulse is restricted to one cell instead of triggering a group of cells.
- Smooth muscle is involuntary, and is found in the hollow organs of the body.
- Smooth muscle is the only type of muscle tissue that is not striated. It contains thin and thick filaments, but the filaments are not organized into sarcomeres.
- Smooth muscle cells are football-shaped and contain a single nucleus.
- Smooth muscle is activated by a variety of stimuli, such as stretch, hormones, and autonomic neurotransmitters.

12

The Muscular System

Avpics/Alamy Stock Photo

Chapter Introduction

In this chapter we will learn . . .

The different formats, structures and ways of describing muscles

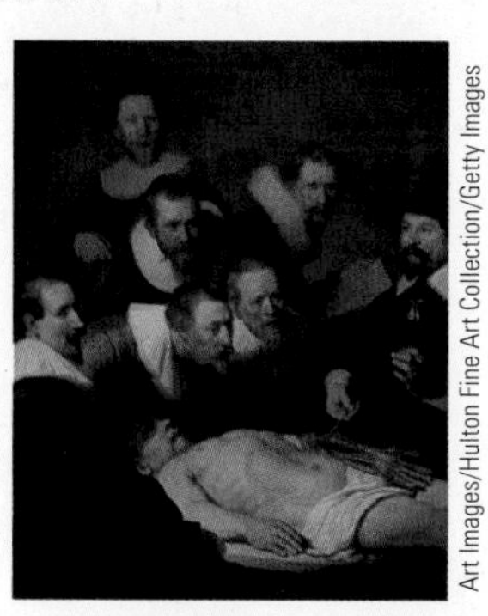

How early anatomists named muscles and how their descriptions can help us learn the muscles today

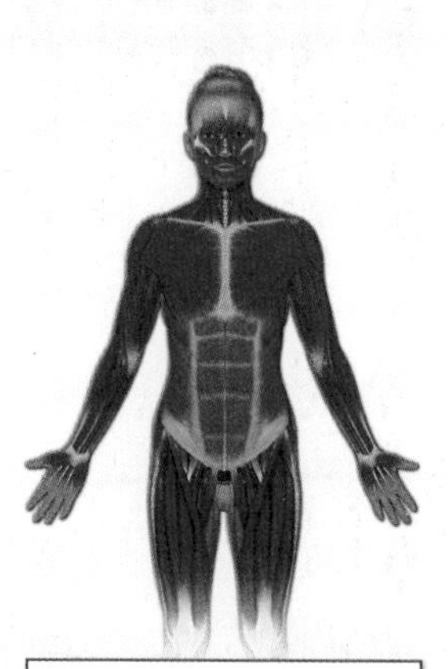

The individual muscles of the human body

Think about the things that you do each day—talking, walking, sitting, standing, and running—all of these activities require movement of particular skeletal muscles. Skeletal muscles are used even during sleep. The precise movements of the body, throwing a ball, for example, require the coordinated contraction and relaxation of a small symphony of muscles working together. In this chapter we will learn the organization of the skeletal muscle system, and more precisely, the names of the skeletal muscles of the human body. As you can imagine, there are a lot of names! To master them all, it helps to learn the methods by which skeletal muscles are named; in some cases, the muscle is named by its shape, and in other cases it is named by its location or attachments to the skeleton. In this chapter we will explore how skeletal muscles are arranged to accomplish movement, and how other muscles may assist, or be arranged on the skeleton to resist or carry out the opposite movement. The actions of the skeletal muscles will be covered in a regional manner, working from the head down to the toes.

12.1 Interactions of Skeletal Muscles, Their Fascicle Arrangement, and Their Lever Systems

Learning Objectives: By the end of this section, you will be able to:

12.1.1 For a given movement, differentiate specific muscles that function as prime mover, antagonist, synergist, or fixator.

12.1.2 Define the terms prime mover (agonist), antagonist, synergist, and fixator.

12.1a Interactions of Skeletal Muscles in the Body

To pull on a bone and change the angle of its joint, a skeletal muscle contracts and pulls on the fibers of its tendon. The tendon is attached to the bone's periosteum. Each muscle is attached in two locations, but generally only produces movement at one of them. For example, let's consider the biceps brachii muscle located in your upper arm (if you already know this muscle on your own body, you can use your own arm; otherwise, the muscle is labeled in **Figure 12.1**). The biceps brachii, like all muscles,

The Human Anatomy and Physiology Society includes more than 1,700 educators who work together to promote excellence in the teaching of this subject area. The HAPS A&P Learning Outcomes measure student mastery of the content typically covered in a two-semester Human A&P curriculum at the undergraduate level. The full Learning Outcomes are available at https://www.hapsweb.org.

LO 12.1.1

Figure 12.1 **Prime Movers and Synergists**

The action shown in this illustration is flexion of the elbow. (A) The biceps brachii is the prime mover, the main muscle responsible for this action. (B) The brachioradialis and brachialis are both synergists that aid in this motion.

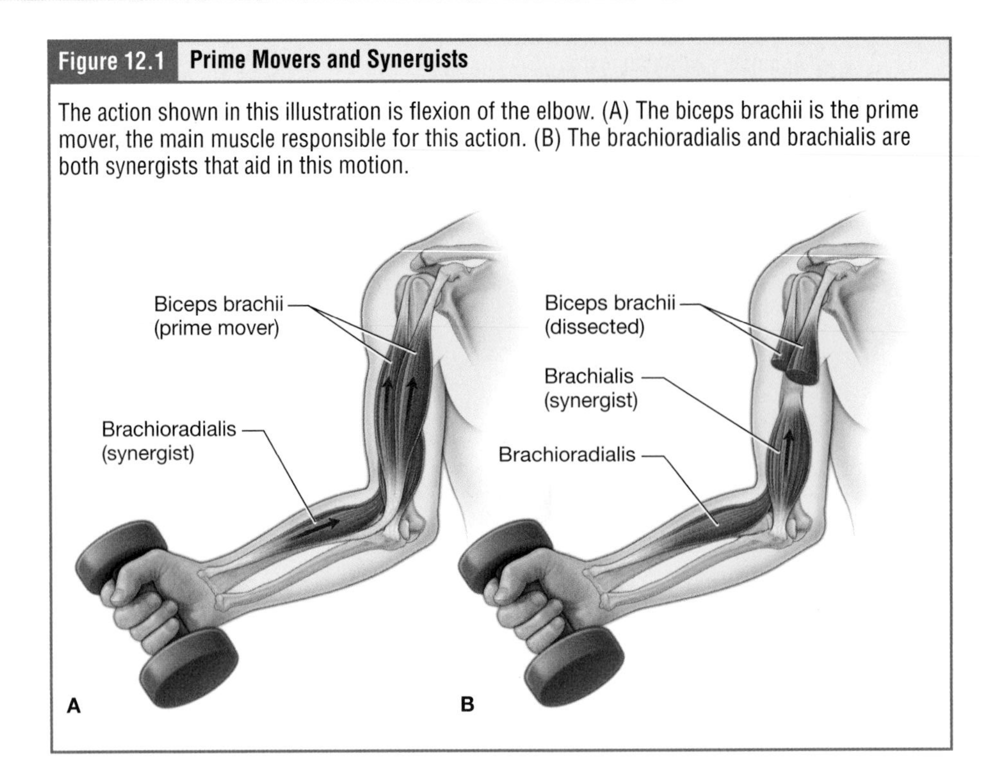

is attached to multiple bones. Its proximal attachments are at the shoulder, but try contracting your bicep, does any movement occur at your shoulder? No, because its attachments at the shoulder are its **origin**, its unmovable end. The moveable end of the muscle that attaches to the bone being pulled is called the muscle's **insertion**; when you contract your bicep, you'll notice that your elbow flexes. So, its distal end is where it inserts.

prime mover (agonist)

LO 12.1.2

Although a number of muscles may be involved in an action, the principal muscle involved is called the **prime mover**. In the act of flexing your elbow, the biceps brachii is the prime mover, but other muscles—brachioradialis and brachialis—also contribute to the motion either by flexing the joint or by stabilizing the insertion. These muscles are called **synergists** in this action (see Figure 12.1), if the muscle acts by stabilizing the insertion it can be described both as a synergist and a **fixator**. Every action in the body has an opposite, a motion required to undo the action. Now that you have flexed your elbow, eventually you will need to straighten it again. Another word that can be used to describe the prime mover of an action is **agonist**. In this case, the biceps brachii is the agonist of elbow flexion. The opposite action, elbow straightening or elbow extension, is accomplished by the triceps brachii (**Figure 12.2**). Since the actions of biceps and triceps brachii oppose each other, the triceps is the **antagonist** of elbow flexion. The antagonist is the muscle that produces the opposite action. Antagonists play two important roles in muscle function: (1) they maintain body or limb position, such as holding the arm out or standing erect and (2) they help to provide resistance to some movements to help with movement fluidity. These terms are reversed for the opposing action. For elbow extension, the triceps brachii is the agonist and the biceps brachii is the antagonist (see Figure 12.2).

There are also skeletal muscles—such as the muscles that produce facial expressions—that do not pull against the skeleton for movement. The insertions and origins of facial muscles are in the skin, so that certain individual muscles contract to form a smile or frown, form sounds or words, and raise the eyebrows. There also are skeletal muscles in the tongue, and the external urinary and anal sphincters that allow for voluntary

Figure 12.2 Agonists and Antagonists

Each action has an opposite. Flexing the elbow has the opposite action of extending the elbow. (A) In the action of flexing the elbow, the biceps brachii is the agonist, the muscle accomplishing the action. (B) In the action of extending the elbow, the biceps brachii is the antagonist that causes the opposite action of the triceps brachii.

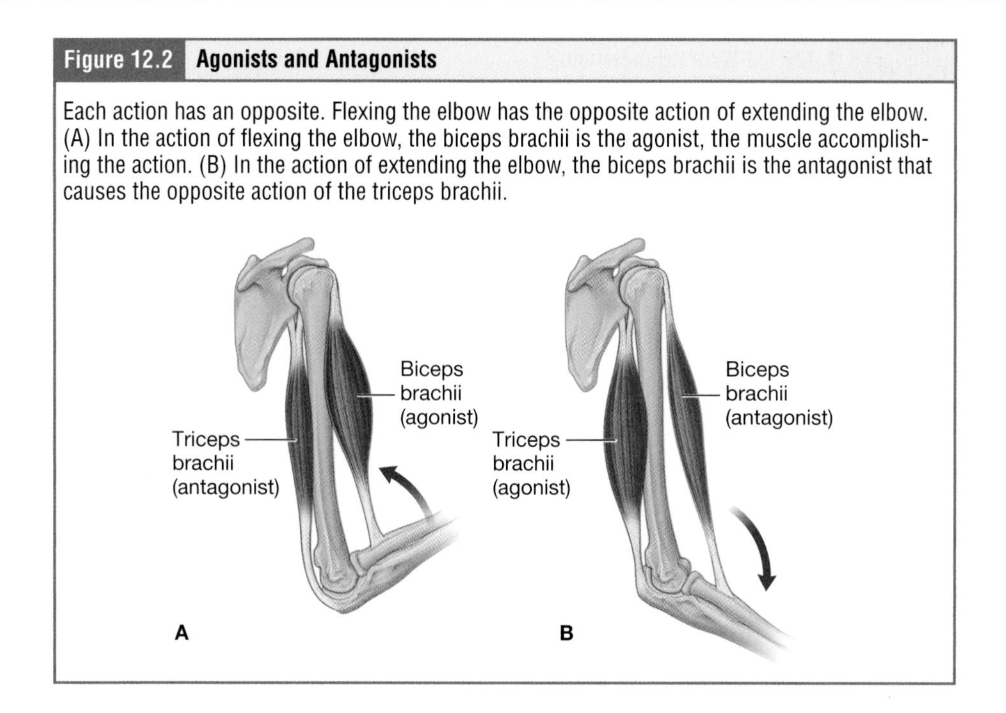

regulation of urination and defecation, respectively. In addition, the diaphragm contracts and relaxes to change the volume of the pleural cavities, but it does not move the skeleton to do this.

12.1b Patterns of Fascicle Organization

Recall from Chapter 11 that skeletal muscle is enclosed in connective tissue scaffolding at three levels. Each muscle fiber (cell) is covered by endomysium and the entire muscle is covered by epimysium. Within the muscle, groups of muscle fibers are "bundled" as a unit and wrapped with a covering of a connective tissue called *perimysium*; this bundled group of muscle fibers is called a **fascicle**. Fascicle arrangement is related to the force generated by a muscle and affects the muscle's range of motion. Skeletal muscles can be classified in several ways, one of which is their fascicle arrangement. What follows are the most common fascicle arrangements.

Parallel muscles have fascicles that are arranged in the same direction as the long axis of the muscle (**Figure 12.3**). The majority of skeletal muscles in the body have this type of organization. Some parallel muscles resemble straps, while others are plumper in the center and taper out toward the tendons. The plump center of this muscle is its **belly**. When a muscle contracts, the contractile fibers shorten it to an even larger bulge (**Figure 12.4**). These parallel muscles with large bellies are called **fusiform**.

Circular muscles have fibers that wrap in a circle (see Figure 12.3). When they relax, the size of the opening increases, and when they contract, the size of the opening shrinks to the point of closure. Several circular muscles get the name *orbicularis*; one example is the orbicularis oris muscle, which surrounds the mouth. When it contracts, the oral opening becomes smaller, as when puckering the lips for whistling. Many circular muscles are in places you cannot see externally. A muscle in one of these locations is called a **sphincter**; its function is to open and close passageways within the body. The external urethral sphincter, for example, is a voluntary skeletal muscle at the end of your urethra (the passageway for urine from your bladder to the outside

Figure 12.3 Muscle Fiber Arrangements

The skeletal muscles of the body consist of fascicles, arranged in different patterns: (A) fusiform, (B) parallel, (C) circular, (D) convergent, (E) unipennate, (F) bipennate, and (G) multipennate.

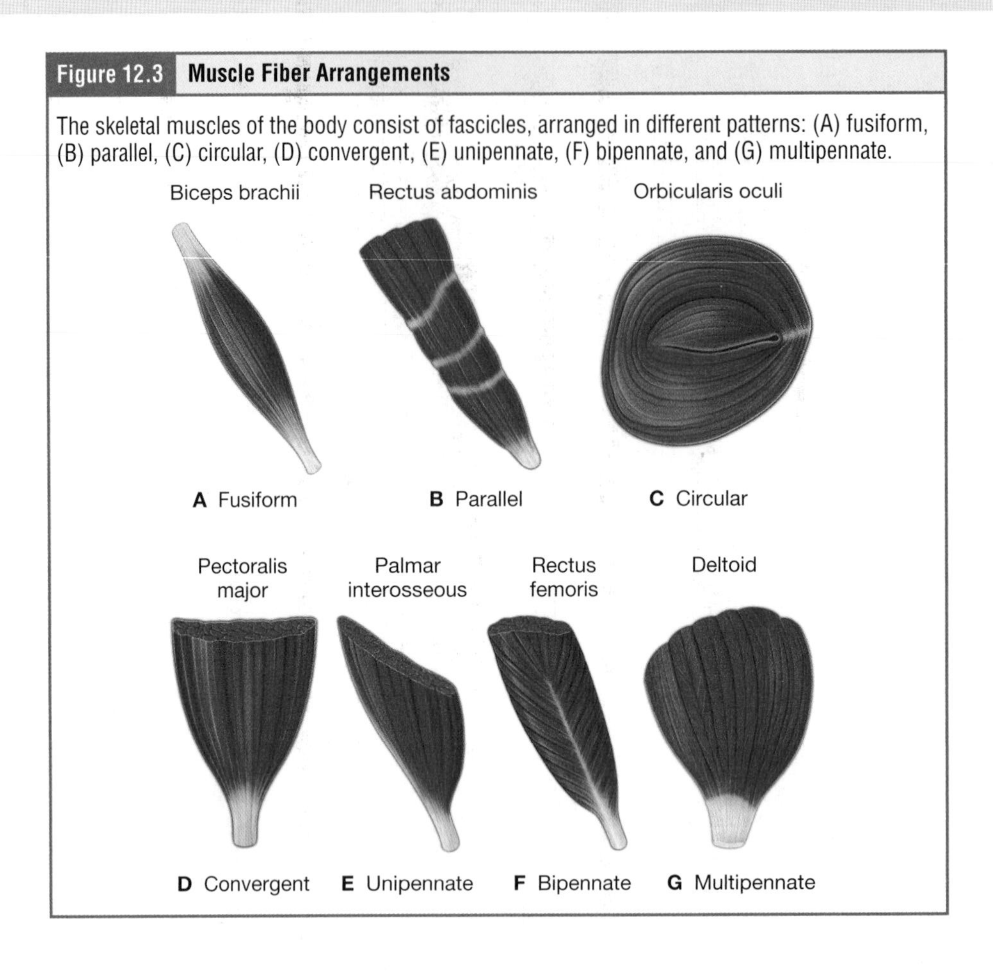

Figure 12.4 Muscle Bellies

(A) The large meaty center of a muscle is called the *belly*. (B) During contraction, the contracting sarcomeres shift the mass of the muscle and the belly becomes larger and more noticeable, and the tendons on either end pull their points of attachment closer together.

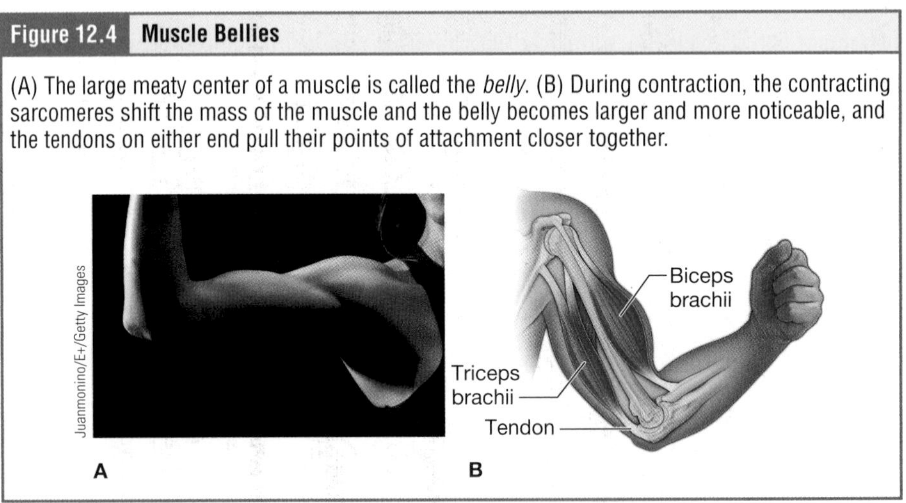

world). Relaxing your external urethral sphincter is a voluntary action that allows you to urinate.

When a muscle has a widespread origin over a sizable area, but then the fascicles come to a single, narrow insertion point, the muscle is called **convergent**. The largest muscle on the chest, the pectoralis major, is an example of a convergent muscle because it converges on the greater tubercle of the humerus via a tendon.

Pennate muscles (penna = "feathers") blend into a tendon that runs through the central region of the muscle for its whole length, somewhat like the quill of

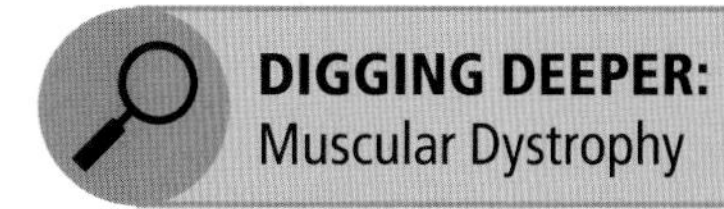

DIGGING DEEPER:
Muscular Dystrophy

Muscular dystrophies are a set of related disorders, all of which feature muscles that progressively weaken over time. In Duchenne muscular dystrophy (DMD), the muscles of the pelvis, thighs, and calves weaken first. As they lose strength, the muscle cells and organs shorten. The hamstrings and gastrocnemius shorten first, leading to difficulty walking and a frequent early observation of a child with DMD walking on their toes. As the gluteal muscles shorten, the pelvis tilts anteriorly and an arch in the lower back can develop, which is often quite painful. With weakened and shortened posterior compartment leg muscles, the affected individual has trouble with leg extension at the hip, as the push off from an extended leg is a key component of fluid walking and running.

DMD has no cure, but treatments include the use of steroids to increase muscle strength and decrease inflammation. DMD and other muscular dystrophies have been the target of many drug design and trial attempts, including recent gene therapy trials.

a feather with the muscle arranged similar to the feathers. Pennate muscles as a group tend to not produce a lot of motion but are very strong and produce a lot of force. Pennate muscles are described by the arrangement of fascicles relative to the tendon.

In a **unipennate** muscle, the fascicles are located on one side of the tendon. The extensor digitorum of the forearm is an example of a unipennate muscle. A **bipennate** muscle has fascicles on both sides of the tendon. The tendon of **multipennate** muscles branches within the muscle, producing a muscle that resembles many feathers arranged together. A common example of this type of muscle is the deltoid muscle of the shoulder.

Learning Connection

Try Drawing It

Can you draw an example of each type of fascicle arrangement? If you are given a drawing of muscles, can you identify an example of each kind?

Learning Check

1. The flexor digitorum superficialis muscle is responsible for the flexing the fingers and wrist. Its attachments sites are the radial head, proximal ulna, and middle phalanges of the fingers. Which of the following accurately describes the origin and insertion of the flexor digitorum superficialis muscles? Please select all that apply.
   a. The origin is the radial head and proximal ulna.
   b. The origin is the middle phalanges of the fingers.
   c. The insertion is the radial head and proximal ulna.
   d. The insertion is the middle phalanges of the fingers.
2. Which of the following correctly defines *synergist*?
   a. Muscles involved in the principal movement
   b. All muscles that contribute to the same action
   c. Muscles that oppose each other
   d. Muscles that act to stabilize the insertion
3. What pattern of muscle has fascicles that blend into a tendon that runs through the center of the muscle for its whole length?
   a. Parallel
   b. Circular
   c. Fusiform
   d. Pennate
4. The rectus abdominis is an abdominal muscle, or your "six-pack abs." The muscle fibers are all in line with each other running from the top to the bottom. Which of the following best describes the rectus abdominis muscle?
   a. Parallel
   b. Circular
   c. Fusiform
   d. Pennate

12.2 Naming Skeletal Muscles

Learning Objectives: By the end of this section, you will be able to:

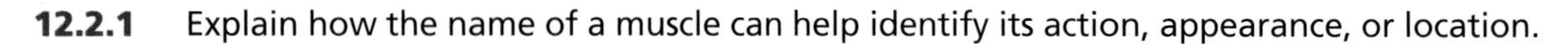

12.2.1 Explain how the name of a muscle can help identify its action, appearance, or location.

Like so many of the anatomy and physiology terms, much of the naming of skeletal muscles is based on Latin and Greek root words. Understanding these terms may help you to not just memorize the names of the body's muscles, but also to understand them so that, should you ever be presented with an unfamiliar muscle name (or one you have forgotten), you may be able to describe or find the muscle based on the name alone (see the "Anatomy of a Muscle Name" feature).

Early anatomists named the skeletal muscles as they were observing them through dissection and in living bodies. Therefore, the names—mostly rooted in Greek and Latin—are mostly derived from easily observed characteristics of the muscle. Most muscles are named after their shape, their size compared to other muscles in the area, their location, the orientation of their fibers, how many origins they have, or their action (see the "Anatomy of a Muscle Name" feature).

Table 12.1 Greek and Latin Words That Pertain to Muscle Size

Term	Meaning
Maximus	The largest of a group
Medius	Medium-sized in a group
Minimus	The smallest of a group
Brevis	Short
Longus	Long
Major	The larger of two
Minor	The smaller of two
Longissimus	The longest

- ***Muscle shape:*** some muscles are named based on their resemblance to a shape, as in the case of the rhomboids or the deltoid muscles (**Figure 12.5**).
- ***Muscle size:*** in several locations in the body, individual muscles in a group are named for their size relative to one another. For example, in the gluteal region—the buttocks—there are three muscles; the largest of these is the gluteus maximus, followed by the gluteus medius (medium) and the gluteus minimus (smallest). Greek and Latin words that describe size are listed in **Table 12.1**.

LO 12.2.1

- ***Location:*** muscles are often named for the region they are located in. For example, both the biceps brachii and the triceps brachii are located in the brachial region. Sometimes muscles in a group are named for their position relative to the midline: **lateralis** can be used to describe a muscle located laterally in relation to the

Figure 12.5 Some Muscles Are Named for Their Shape

(A) The rhomboid muscles of the upper back are named for their shape, which is similar to that of a rhombus. (B) The deltoid muscle of the shoulder is so named because it resembles an upside-down Greek letter delta.

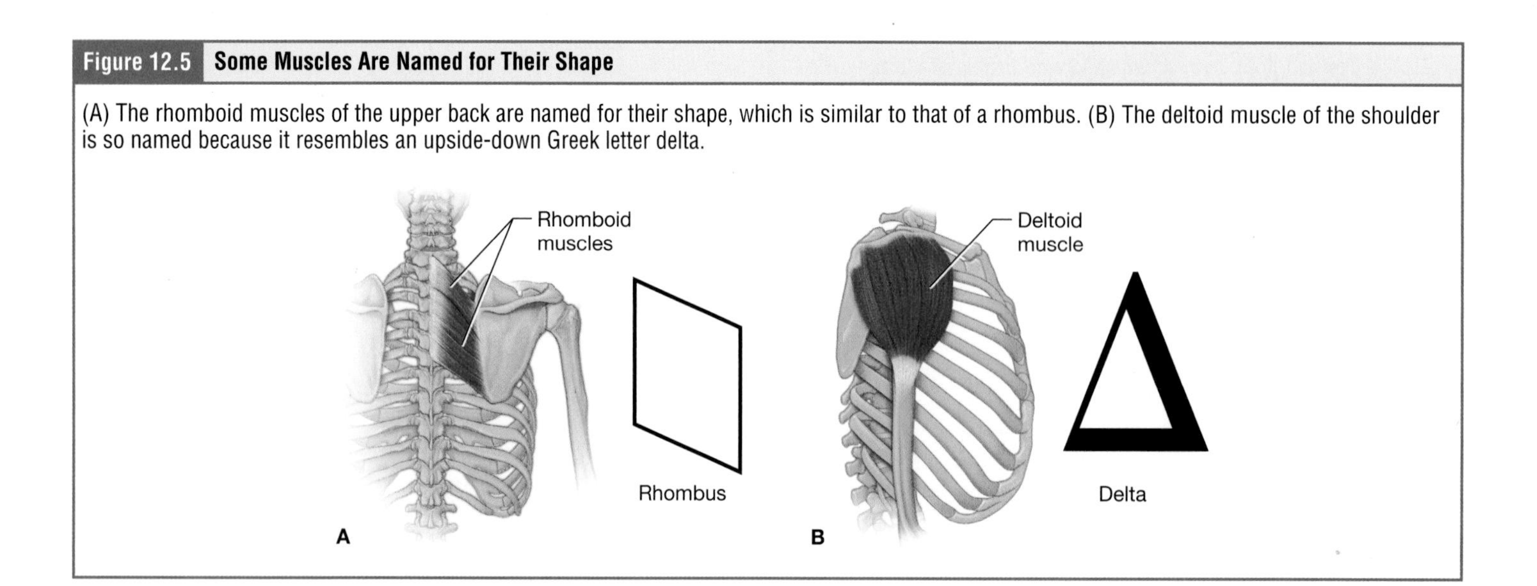

midline of the body or another structure, and **medialis** located medially in relation to the midline of the body or another structure.

- ***Orientation of fibers:*** the orientation of the muscle fibers and fascicles is used to describe some muscles, such as the **rectus** (straight) abdominis, or the **oblique** (at an angle) muscles of the abdomen.
- ***Number of origins:*** the number of origins a muscle has can differentiate it from other nearby muscles. Biceps and triceps brachii are differentiated by the number of "heads" and therefore attachment sites. Greek and Latin prefixes that indicate number are summarized in Table 12.2.
- ***Action:*** many muscles are named by their action. For example, the adductor muscles of the thigh are responsible for bringing the leg back to its medial position after abduction. Remember that muscles act on the joint they cross in predictable ways. A muscle that crosses the anterior side of a joint produces flexion, for example. Muscle actions are summarized in Figure 12.6.
- ***Attachment location:*** attachment location can also appear in a muscle name. When the name of a muscle is based on its attachments, the origin is always named first. For instance, the sternocleidomastoid muscle of the neck has a dual origin on the sternum (sterno-) and clavicle (cleido), and it inserts on the mastoid process of the temporal bone.
- ***Grouping:*** some muscles exist in groups such as the quadriceps, a group of four muscles located on the anterior thigh, and the hamstrings, which are located on the posterior thigh.

Table 12.2 Greek and Latin Prefixes That Indicate Number

Prefix	Meaning
Uni	One
Bi / Di	Two
Tri	Three
Quad	Four
Multi	Many

Learning Connection

Chunking

Find similar names among muscles in different parts of the body. What are all the muscles you can find that include the term *biceps*? What about all the flexor muscles?

Anatomy of...

A Muscle Name

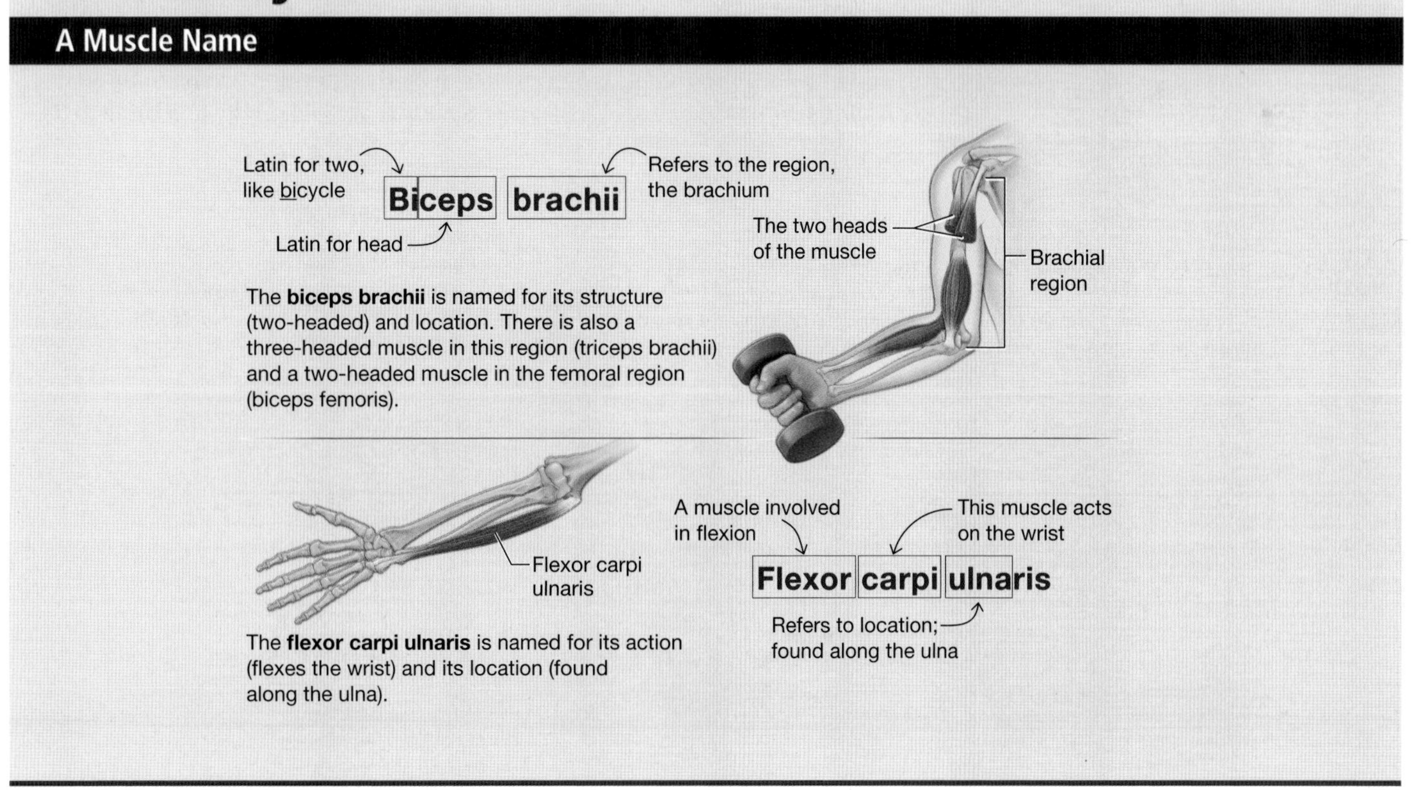
Latin for two, like bicycle
Biceps brachii
Refers to the region, the brachium
Latin for head
The two heads of the muscle
Brachial region
The **biceps brachii** is named for its structure (two-headed) and location. There is also a three-headed muscle in this region (triceps brachii) and a two-headed muscle in the femoral region (biceps femoris).
Flexor carpi ulnaris
A muscle involved in flexion
This muscle acts on the wrist
Flexor carpi ulnaris
Refers to location; found along the ulna
The **flexor carpi ulnaris** is named for its action (flexes the wrist) and its location (found along the ulna).

Figure 12.6 Muscle Actions

Muscles cross the joint on which they act; therefore, their actions follow a predictable pattern. (A and B) Muscles that are on the lateral side of the joint produce abduction of the limbs or lateral flexion of the trunk and neck. Muscles on the medial side of the joint produce adduction. (C and D) With the exception of the knee, muscles that cross the anterior side of a joint produce flexion and muscles that cross the posterior side cause extension.

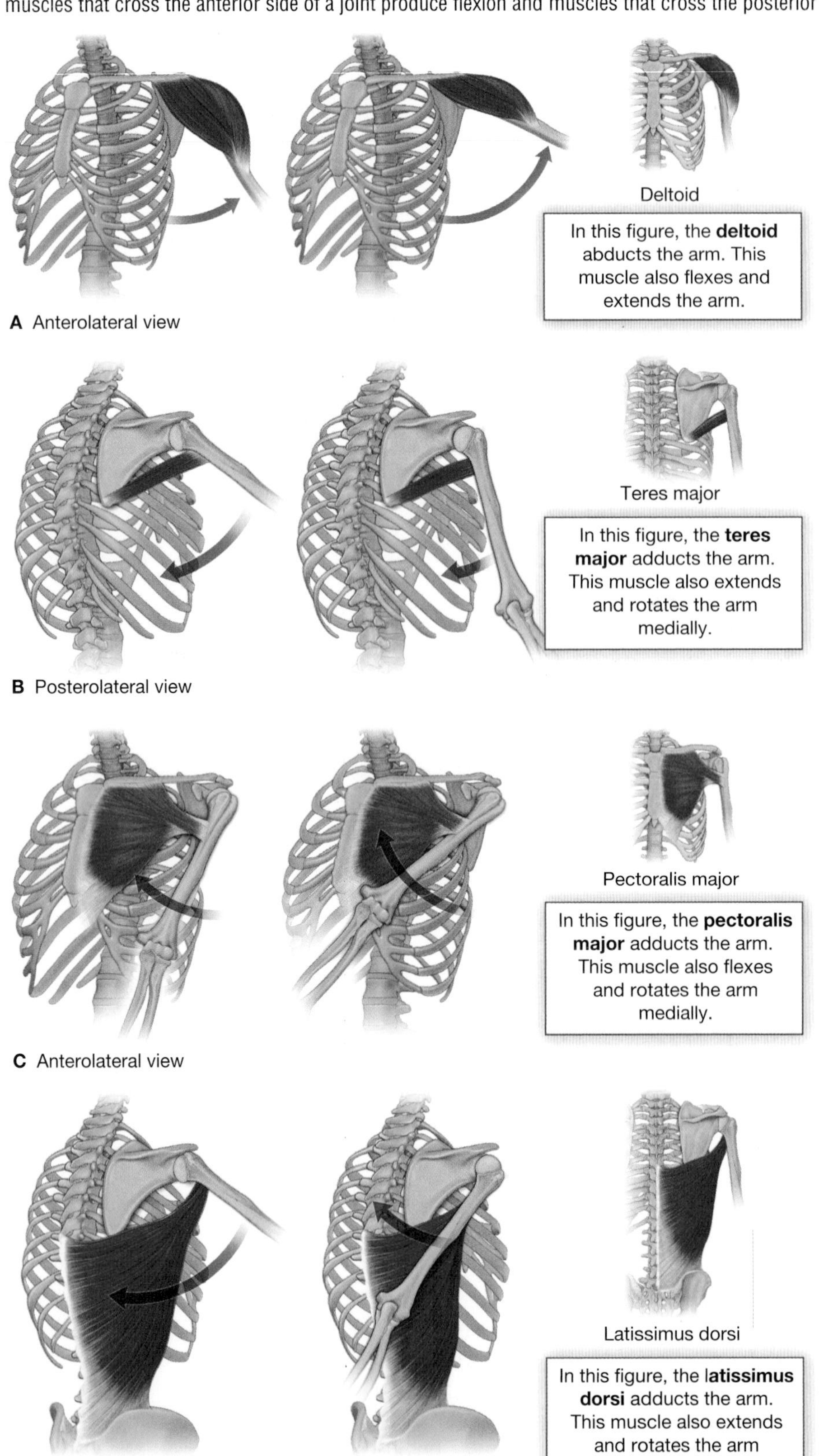

Deltoid
In this figure, the **deltoid** abducts the arm. This muscle also flexes and extends the arm.
A Anterolateral view
Teres major
In this figure, the **teres major** adducts the arm. This muscle also extends and rotates the arm medially.
B Posterolateral view
Pectoralis major
In this figure, the **pectoralis major** adducts the arm. This muscle also flexes and rotates the arm medially.
C Anterolateral view
Latissimus dorsi
In this figure, the **latissimus dorsi** adducts the arm. This muscle also extends and rotates the arm medially.
D Posterolateral view

Learning Check

1. The rhomboid minor is a muscle that is posterior of the shoulder. It contributes to movements of the shoulder blades. Looking at the name, which of the following is this muscle most likely named after?
   a. Shape
   b. Location
   c. Direction of muscle fibers
   d. Action
2. The vastus medialis is a muscle that is part of the quadriceps muscle group. This muscle sits on the anterior of the thigh and toward the medial aspect of the femur. Looking at the name, which of the following is this muscle most likely named after?
   a. Shape
   b. Location
   c. Number of origins
   d. Action
3. The external oblique is a muscle that makes up the sides of your trunk. The fibers are oriented at an angle. There are several layers, but this muscle is the most superficial layer. Looking at the name, which of the following is this muscle most likely named after? Please select all that apply.
   a. Direction of muscle fibers
   b. Location
   c. Size
   d. Action

12.3 Axial Muscles

Learning Objectives: By the end of this section, you will be able to:

12.3.1 Identify the location, general attachments, and actions of the major skeletal muscles.

12.3.2 Describe similar actions (functional groupings) of muscles in a particular compartment (e.g., anterior arm) or region (e.g., deep back).

Like the bones of the skeleton (Chapters 8 and 9), the skeletal muscles are divided into **axial** (muscles of the trunk and head) and **appendicular** (muscles of the arms and legs) categories. Some of the axial muscles may seem to blur the boundaries because they cross over to the appendicular skeleton.

12.3a Muscles of Facial Expression

Most of the muscles in the body originate and attach to bones. The muscles of facial expression are unusual in that they originate on the bones of the skull but insert in the skin of the face. Because the muscles insert in the skin rather than on bone, when they contract, the skin moves to create facial expression. These muscles are illustrated in **Figure 12.7**.

LO 12.3.1

Two circular muscles are found in the face, both named *orbicularis* due to their shape. The **orbicularis oris** surrounds the mouth, and the **orbicularis oculi** is a circular muscle that closes the eye. The **occipitofrontalis**, named for the two skull bones it lies over, moves the scalp and eyebrows (if you raise your eyebrows in surprise, you are contracting your occipitofrontalis). The muscle has a frontal belly and an occipital (near the occipital bone on the posterior part of the skull) belly. The **frontalis** belly and the **occipitalis** belly are connected by a broad tendon called the **epicranial aponeurosis**, but there is no muscle across the top of the head.

epicranial aponeurosis (galea aponeurosis)

A large portion of the face is composed of the **buccinator** muscle, which compresses the cheek. This muscle allows you to whistle, blow, and suck; and it contributes to the action of chewing. Humans rely heavily on social cues from facial expression, and one of the most important of these is smiling. Two muscles are responsible for raising the corners of the mouth: **zygomaticus major** and **zygomaticus minor**. We can fake a smile for a photo by contracting these two muscles, but true smiles generally involve the orbicularis oculi as well, which crinkles the sides of the eyes during a warm smile.

Figure 12.7 Muscles of Facial Expression

This illustration shows (A) anterior and (B) lateral views of the muscles of facial expression, with the largest ones being labeled. These muscles insert into the skin of the face.

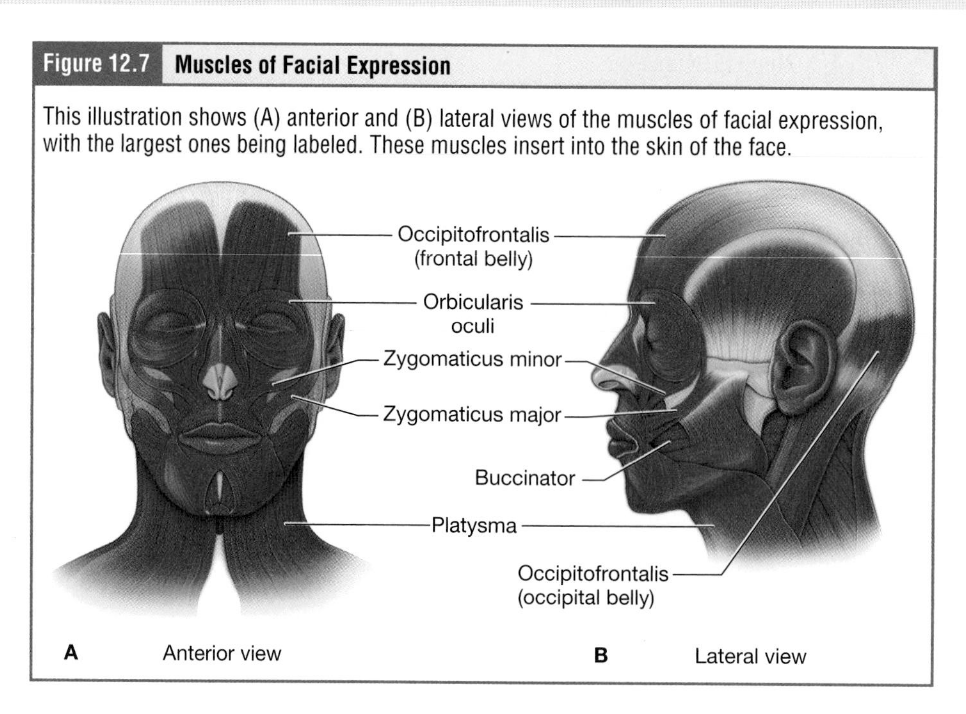

Learning Connection

Quiz Yourself

Try writing down each muscle on a piece of paper or flashcard (or digital device) and bringing them with you to a mirror to see if you can contract each of the muscles on cue.

Last, the skin of the neck can be tensed and the corners of the mouth pulled down (try saying "Ewww" as you imagine something unpleasant) by contracting the **platysma** of the neck.

The muscles of facial expression are organized in **Table 12.3**. You will learn the nerves of the body in Chapter 15, but we have included the innervation of each muscle here for your reference.

Table 12.3 Muscles of Facial Expression

Muscle	Description	Origin (O) and Insertion (I)	Action	Nerve Supply	You May Know This Muscle from...
Orbicularis oris	Circular muscle surrounding the mouth	(O) Medial aspects of the maxilla and mandible (I) Skin and mucous membrane of the lips	Puckers the lips	Cranial nerve VII	Playing a wind instrument such as the trumpet
Orbicularis oculi	Circular muscle surrounding the eye	(O) Nasal part of frontal bone, maxilla, and lacrimal (I) Skin of orbital region	Opens and closes the eye	Cranial nerve VII	Winking at someone or shutting your eyes forcefully
Occipitofrontalis	Muscle with two bellies spanning the top and back of the head	(O) Skin of the eyebrow, muscles of the forehead, and the superior nuchal line (I) Epicranial aponeurosis	Moves the scalp and the eyebrows	Cranial nerve VII	Raising your eyebrows in surprise
Buccinator	Makes up the cheeks and contributes to facial expression	(O) Maxilla, mandible, and temporomandibular joint (I) Fibers of the orbicularis oris	Compresses the cheek	Cranial nerve VII	Whistling or sucking in your cheeks to make a "fish face"
Zygomaticus major	One of the main muscles involved in facial expression, attaching at the corners of the mouth	(O) Anterior lateral surface of the zygomatic bone (I) Muscles of the upper lip	Raising the corners of the mouth	Cranial nerve VII	Smiling for a photograph
Zygomaticus minor	One of the main muscles involved in facial expression, attaching at the corners of the mouth	(O) Posterior lateral surface of the zygomatic bone (I) Muscles of the upper lip	Raising the corners of the mouth	Cranial nerve VII	Smiling for a photograph
Platysma	Very superficial muscle of the neck	(O) Fascia over the upper, lateral chest (I) Skin, fascia, and muscles of the mandible	Tenses the skin of the neck	Cranial nerve VII	

Figure 12.8 Muscles of the Eyes

(A) The extrinsic eye muscles originate on the skull and insert onto the surface of the eye.
(B) Muscles surround the eye to enable a wide range of motion.

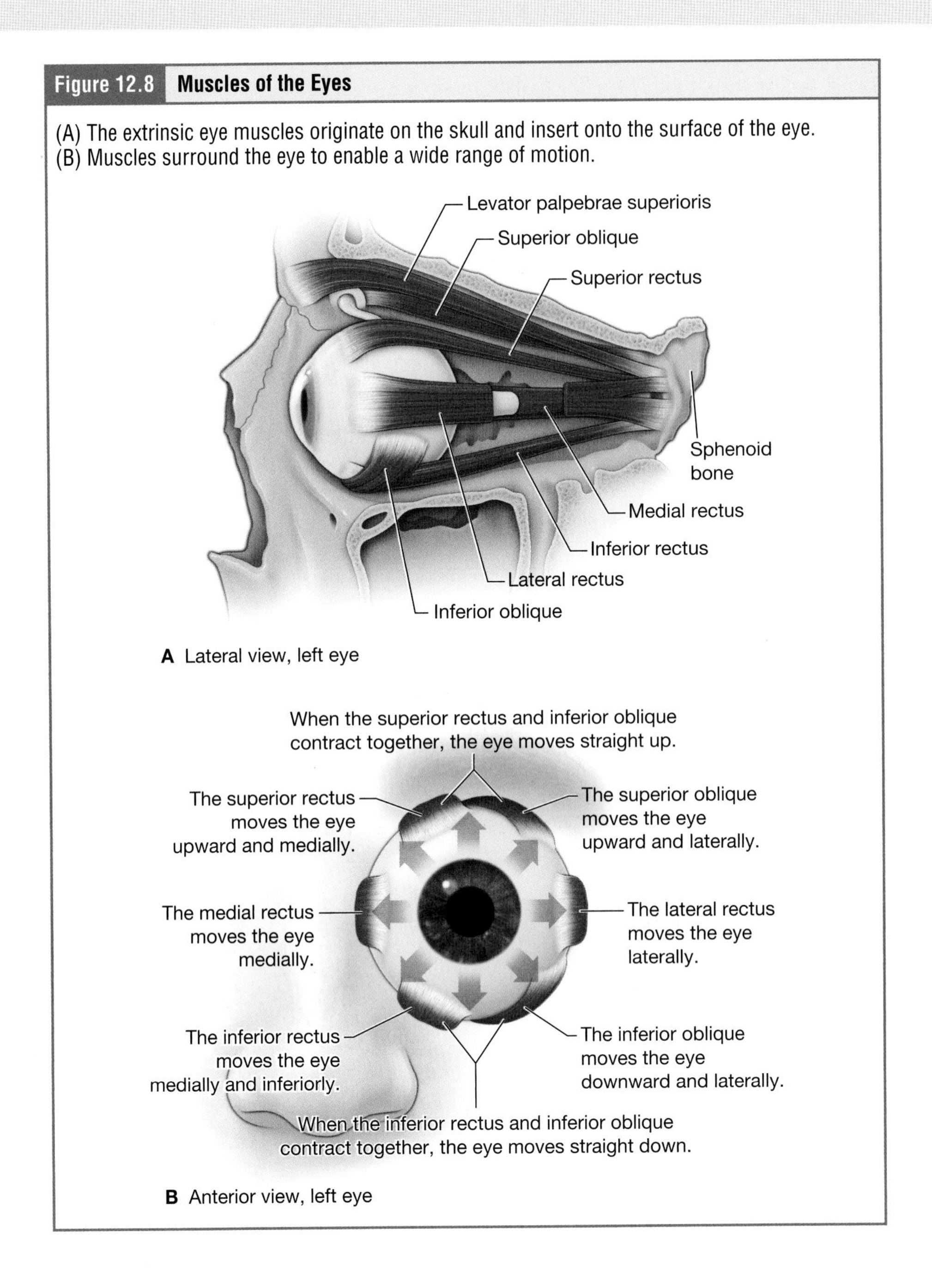

12.3b Muscles That Move the Eyes

The movement of the eyeball is under the control of the **extrinsic eye muscles**, which originate outside the eye and insert onto the outer surface of the white of the eye. These muscles are located inside the eye socket and cannot be seen on any part of the visible eyeball (Figure 12.8 and Table 12.4). We can roll our eyes in a full circle due to the coordinated action of these muscles. Nearly 40 percent of the human brain mass is dedicated to vision, so it makes some sense that we have exquisite control over eye motion.

12.3c Muscles That Move the Lower Jaw

In anatomical terminology, chewing is called **mastication**. Muscles involved in chewing must be able to exert enough pressure to bite through and then chew food before it is swallowed (Figure 12.9 and Table 12.5). The **masseter** muscle is the main muscle used for chewing because it elevates the mandible (lower jaw) to close the mouth, and it is assisted by the **temporalis** muscle, which retracts the mandible. You

Table 12.4 Muscles of Eye Movement

Muscle	Description	Origin (O) and Insertion (I)	Action	Nerve Supply	You May Know This Muscle from...
Medial rectus	Muscle of the eye involved in moving eye within orbit	(O) Common tendinous ring (I) Surface of eye	Moves the eye medially	Cranial nerve III	Looking around
Lateral rectus	Muscle of the eye involved in moving eye within orbit	(O) Common tendinous ring (I) Surface of eye	Moves the eye straight to the (lateral) side	Cranial nerve VI	Looking around
Inferior rectus	Muscle of the eye involved in moving eye within orbit	(O) Common tendinous ring (I) Surface of eye	Moves the eye medial inferior (toward the tip of your nose)	Cranial nerve III	Looking around
Superior rectus	Muscle of the eye involved in moving eye within orbit	(O) Common tendinous ring (I) Surface of eye	Moves the eye up and medially	Cranial nerve III	Looking around
Inferior oblique	Muscle of the eye involved in moving eye within orbit	(O) Maxilla bone (I) Surface of eye	Moves the eye down and lateral	Cranial nerve III	Looking around
Superior oblique	Muscle of the eye involved in moving eye within orbit	(O) Sphenoid bone (I) Surface of eye	Moves the eye up and laterally	Cranial nerve VI	Looking around

Figure 12.9 Muscles of Mastication

The muscles that move the lower jaw enable chewing (mastication) as well as language: (A) Superficial chewing muscles and (B) deep chewing muscles.

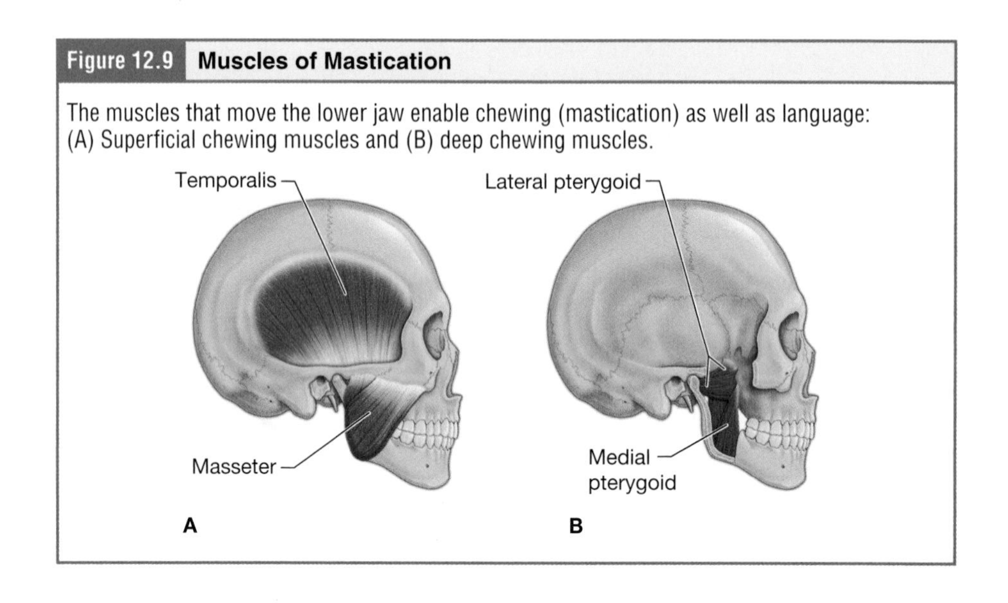

can feel the temporalis move by putting your fingers to your temple as you chew. Deep in the skull and not visible externally, the two sets of **pterygoid muscles** are synergists for opening and closing the mouth, and they are the prime side-to-side movers of the jaw.

12.3d Muscles That Move the Tongue

Although the tongue is obviously important for tasting food, it is also necessary for mastication, swallowing, and speech (**Figure 12.10** and **Table 12.6**). Because it is so moveable, the tongue facilitates complex speech patterns and sounds.

Tongue muscles can be extrinsic or intrinsic. Extrinsic tongue muscles insert in the tongue but have outside origins, and the intrinsic tongue muscles insert and originate within the tongue. Whereas the extrinsic muscles move the whole tongue in different directions, the intrinsic muscles allow the tongue to change its shape (such as curling the tongue in a loop or flattening it).

Table 12.5 Muscles of Mastication

Muscle	Description	Origin (O) and Insertion (I)	Action	Nerve Supply	You May Know This Muscle from...
Masseter	Lateral muscle acting on the jaw and mainly responsible for chewing	(O) Zygomatic arch (I) Angle and ramus of the mandible	Elevation of the mandible	Cranial nerve V	Chewing something sticky like a piece of candy or toffee and you can feel the muscles in your cheek get sore
Temporalis	Muscle just above the jaw and spanning over the lateral bones of the skull	(O) Frontal, parietal, and temporal bones (I) Coronoid process of the mandible	Retraction of the mandible	Cranial nerve V	Headache (some people feel that they have a headache in their "temple" and it is often this muscle)
Medial pterygoid	Deep facial muscle assisting in chewing	(O) Pterygoid plate of the sphenoid bone (I) Medial surface of the ramus and angle of the mandible	Prime mover: Side-to-side movement of the jaw, opening and closing the mouth	Cranial nerve V	
Lateral pterygoid	Deep facial muscle assisting in chewing	(O) Pterygoid plate of the sphenoid bone (I) Joint capsule of the temporomandibular joint and the condyloid process of the mandible	Prime mover: Side-to-side movement of the jaw, opening and closing the mouth	Cranial nerve V	

Figure 12.10 Muscles that Move the Tongue

The names of muscles that act on the tongue end with the suffix *-glossus*.

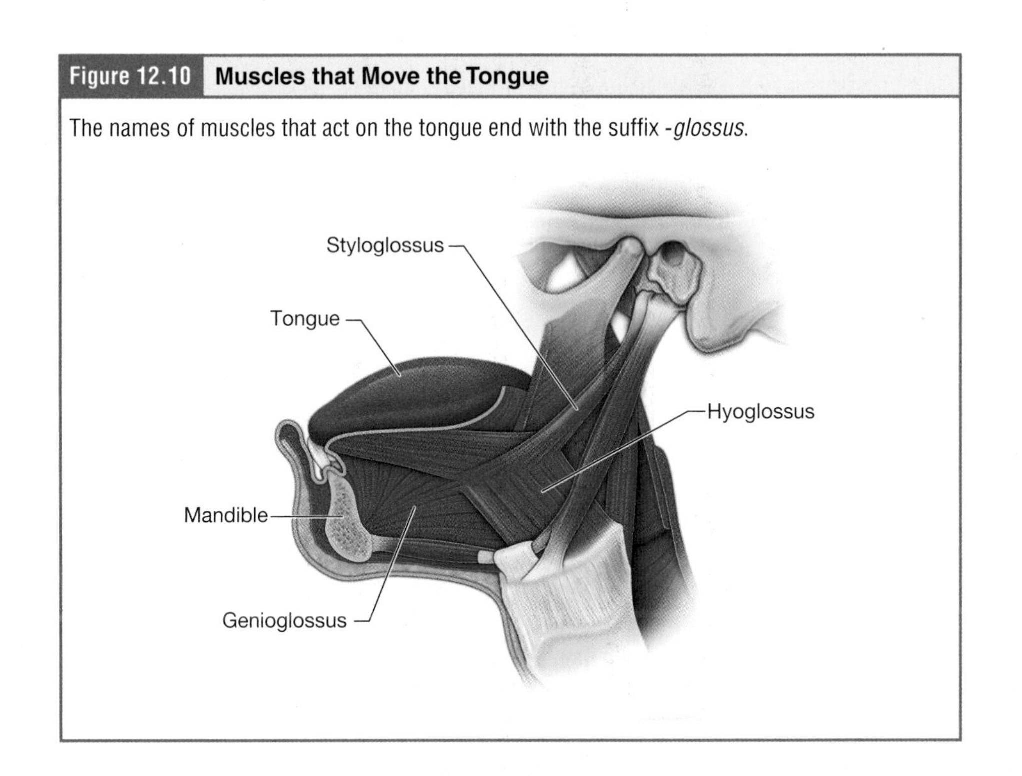

The extrinsic muscles all include the word root *glossus* (glossus = "tongue"), and the muscle names are derived from where the muscle originates. The **genioglossus** (genio = "chin") originates on the mandible and allows the tongue to move downward and forward. The **styloglossus** originates on the styloid bone and allows upward and backward motion. The **palatoglossus** originates on the soft palate to elevate the back of the tongue, and the **hyoglossus** originates on the hyoid bone to move the tongue downward and flatten it.

Table 12.6 Muscles That Move the Tongue

Muscle	Description	Origin (O) and Insertion (I)	Action	Nerve Supply	You May Know This Muscle from...
Genioglossus	Fan-shaped extrinsic muscle of the tongue	(O) Mandible bone (I) Tongue and body of the hyoid	Depression and protraction of the tongue	Hypoglossal nerve	Chewing, word and sound formation in singing and talking
Styloglossus	Smallest of the extrinsic tongue muscles	(O) Styloid bone (I) Merges with the hyoglossus and inferior longitudinal muscle of the tongue	Elevation and retraction of the tongue	Hypoglossal nerve	Chewing, word and sound formation in singing and talking
Palatoglossus	Extrinsic muscle of the tongue in the lateral pharyngeal wall	(O) Soft palate (I) Posterolateral tongue	Elevation of the back of the tongue	Cranial nerve X	Chewing, word and sound formation in singing and talking
Hyoglossus	Thin muscle in the upper neck and floor of the mouth	(O) Hyoid bone (I) Lateral tongue	Depression and flattening of the tongue	Hypoglossal nerve	Chewing, word and sound formation in singing and talking

The muscles of the anterior neck assist in deglutition (swallowing) and speech. The muscles of the neck are categorized according to their position relative to the hyoid bone (**Figure 12.11**). **Suprahyoid muscles** are superior to it, and the **infrahyoid muscles** are located inferiorly.

The suprahyoid muscles raise the hyoid bone, the floor of the mouth, and the larynx during swallowing. These include the **digastric**, which has anterior and posterior bellies

Figure 12.11 Muscles of the Anterior Neck

The anterior muscles of the neck facilitate swallowing and speech. The suprahyoid muscles originate above the hyoid bone in the chin region. The infrahyoid muscles originate below the hyoid bone in the lower neck.

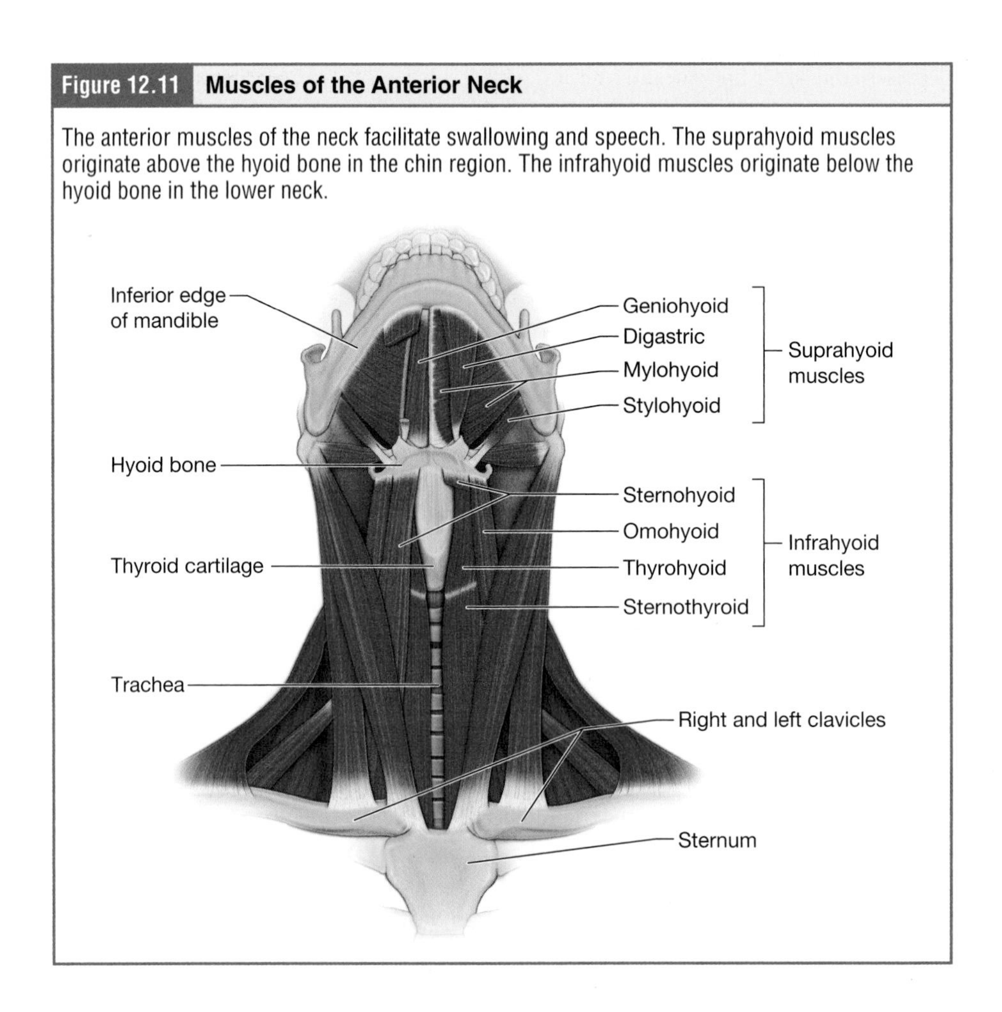

Figure 12.12 Lateral and Posterior Neck Muscles

The muscles of the neck move the head, cervical vertebrae, and scapulae.

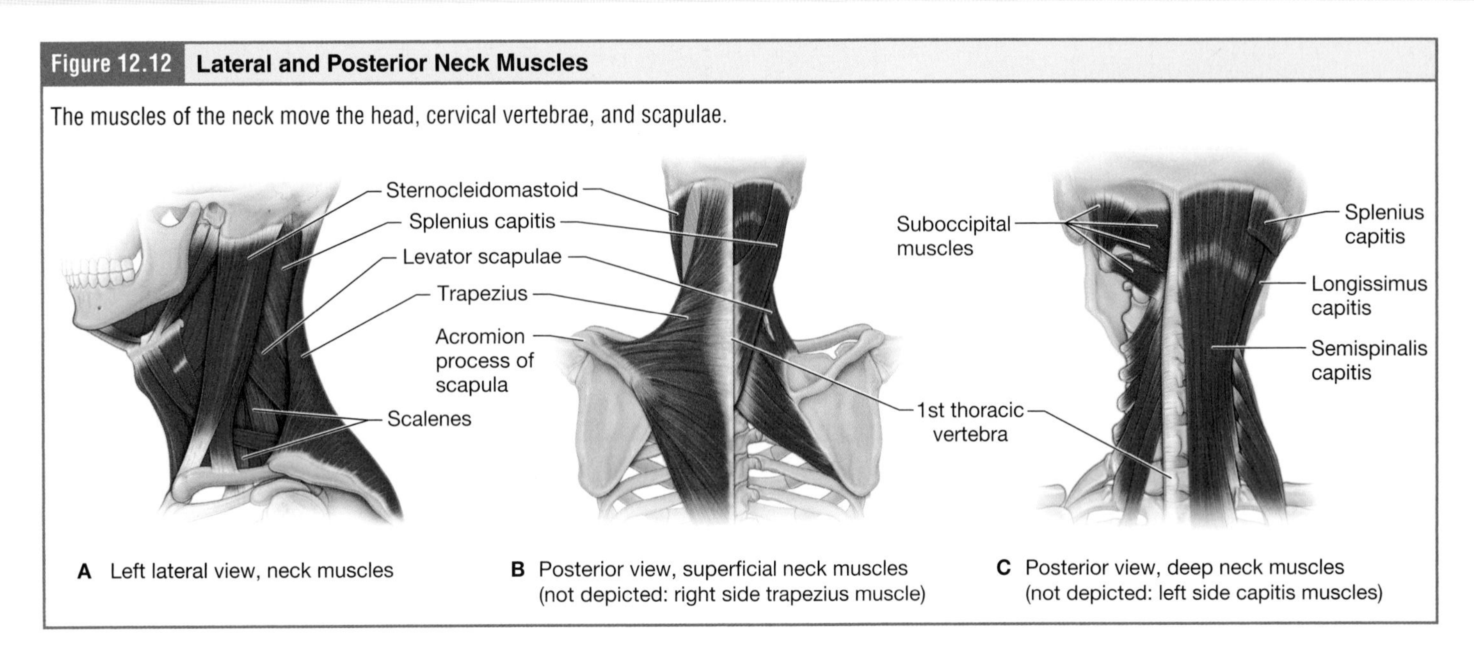

A Left lateral view, neck muscles

B Posterior view, superficial neck muscles (not depicted: right side trapezius muscle)

C Posterior view, deep neck muscles (not depicted: left side capitis muscles)

that work to elevate the hyoid bone and larynx during swallowing. The **stylohyoid** moves the hyoid bone posteriorly, elevating the larynx, and the **mylohyoid** lifts it and helps press the tongue to the top of the mouth. The **geniohyoid** depresses the mandible in addition to raising and pulling the hyoid bone anteriorly.

The straplike infrahyoid muscles generally depress the hyoid bone and control the position of the larynx. The **omohyoid**, which has superior and inferior bellies, depresses the hyoid bone in conjunction with the **sternohyoid** and **thyrohyoid**. The thyrohyoid muscle also elevates the larynx's thyroid cartilage, while the **sternothyroid** depresses it to create different tones of voice.

12.3e Muscles That Move the Head

The head is balanced, moved, and rotated by the neck muscles (**Figure 12.12** and **Table 12.7**). The head can move in all directions, and most of the muscles described here are paired, with one on each side. When just one of the pair contracts, the head moves to that side (lateral flexion). When they contract together on both sides, the head flexes anteriorly or extends posteriorly. The major muscle (prime mover) that laterally flexes the head and allows the head to rotate on the spine is the **sternocleidomastoid**. In addition, both sternocleidomastoids working together flex the head. Place your fingers on both sides of the neck and turn your head to the left and to the right.

Table 12.7 Muscles That Move the Head

Muscle	Description	Origin (O) and Insertion (I)	Action	Nerve Supply	You May Know This Muscle from...
Sternocleidomastoid	Thick muscle attaching the sternum, clavicle, and mastoid together	(O) Manubrium of the sternum and medial clavicle (I) Mastoid process of the temporal bone	Prime mover: flexion of the head and neck	Cranial nerve XI	Turning your head to one side or flexing your head, such as resting your chin on your chest
Scalenes	Muscles of the neck that assist the sternocleidomastoid	(O) Transverse processes of cervical vertebrae (I) First and second ribs	Flexion of the neck	Spinal nerves C3–C8	Neck pain (when these muscles become stiff, they pull on their insertion points and can cause pain and limited range of motion)

Table 12.8 Muscles of the Posterior Neck and Back

Muscle	Description	Origin (O) and Insertion (I)	Action	Nerve Supply	You May Know This Muscle from...
Splenius capitus	Intrinsic straplike muscle of the posterior neck working to keep the head upright	(O) Ligamentum nuchae, lower cervical and upper thoracic spinous processes (I) Occipital bone and mastoid process	Extension of the head and neck	Posterior rami of spinal nerves C2–C3	Keeping your head up (but overexertion of this muscle can cause headaches)
Splenius cervicis	Intrinsic muscle of the neck superficial to splenius capitis	(O) Thoracic spinous processes (I) Atlas and axis	Bilateral extension of the head and neck, and unilateral rotation	Posterior rami of spinal nerves C3–C4	Keeping your head up (but overexertion of this muscle can cause headaches)
Iliocostalis	Deep, erector spinae group	(O) Posterior ribs, thoracolumbar aponeurosis (I) Cervical transverse processes and posterior ribs	Vertebral column extension and control of flexion, unilateral flexion (side-bending), and rotation	Posterior rami of spinal nerves	Standing up all day long
Longissimus	Deep, erector spinae group	(O) Lower cervical and thoracic transverse processes, thoracolumbar aponeurosis (I) Mastoid process and cervical and thoracic transverse processes	Vertebral column extension and control of flexion, unilateral flexion (side-bending), and rotation	Posterior rami of spinal nerves	Standing up all day long
Spinalis	Deep, erector spinae group	(O) Ligamentum nuchae and spinous processes (I) Thoracic spinous processes	Vertebral column extension and control of flexion, unilateral flexion (side-bending), and rotation	Posterior rami of spinal nerves	Standing up all day long
Transversopinalis (group)	Small muscles that connect the vertebrae to one another	(O) Transverse processes of each vertebra (I) Spinous processes of each vertebra	Stabilizing the vertebral column contributes to rotation	Posterior rami of spinal nerves	Rotating to the side, as if swinging a baseball bat or a golf club

You will feel the movement originate there (see Figure 12.12). The **scalenes** are the synergists to the sternocleidomastoids in neck flexion and rotation.

12.3f Muscles of the Posterior Neck and the Back

LO 12.3.2

The posterior muscles of the neck are primarily concerned with head movements, like extension. The back muscles stabilize and move the vertebral column and are grouped according to the lengths and direction of the fascicles (**Table 12.8**).

The **splenius** muscle originates at the midline and runs laterally and superiorly to its insertion. From the sides and the back of the neck, the **splenius capitis** inserts onto the head region, and the **splenius cervicis** extends onto the cervical region. These muscles can extend the head, laterally flex it, and rotate it (**Figure 12.13**).

The **erector spinae group** forms the majority of the muscle mass of the back and it is the primary extensor of the vertebral column. It controls flexion, lateral flexion, and rotation of the vertebral column, and maintains the lumbar curve. The erector spinae contains three groups of muscles, from lateral to medial they are: the **iliocostalis**, the **longissimus**, and the **spinalis**.

Student Study Tip

To remember the erector spinae muscles in order, just use the mnemonic I Love Spines: iliocostalis, longissimus, and spinalis.

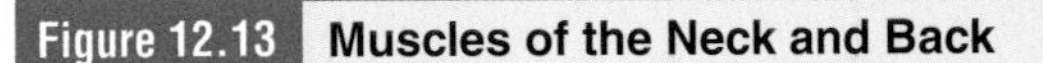

Figure 12.13 Muscles of the Neck and Back

Keeping the spinal column erect or extending it is the work of the long muscles to either side of the spine. Smaller muscles act as synergists and also contribute to lateral flexion.

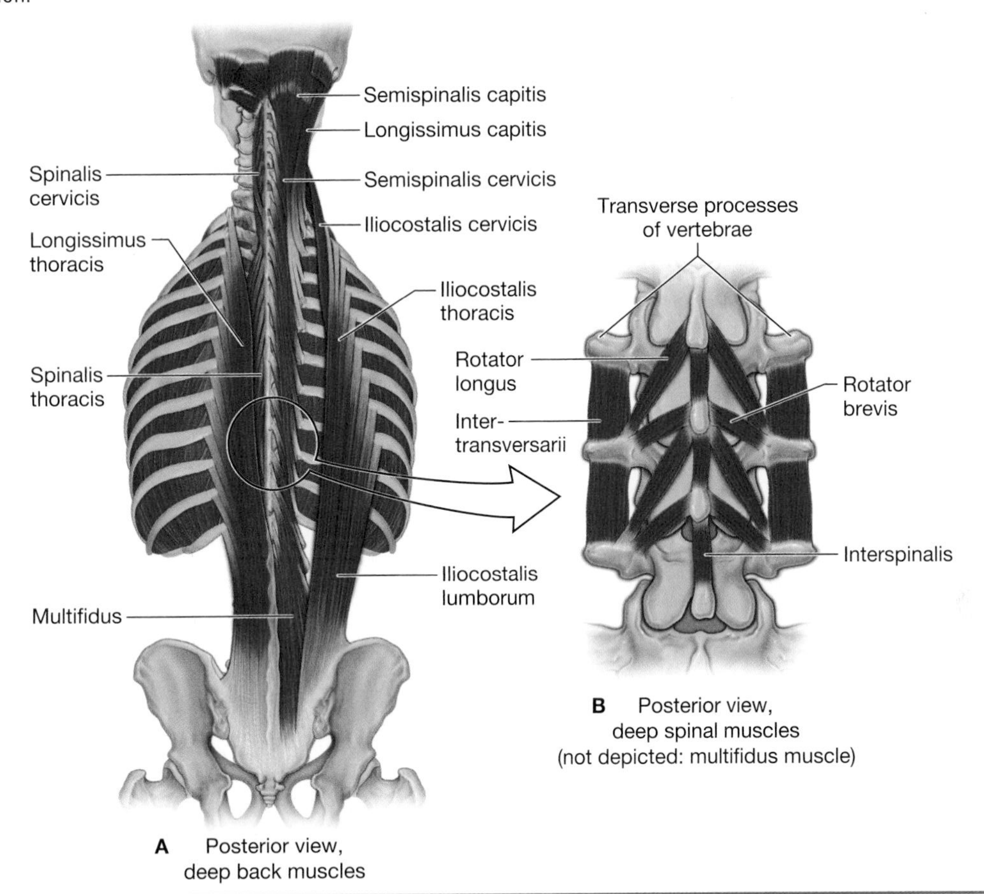

A Posterior view, deep back muscles

B Posterior view, deep spinal muscles (not depicted: multifidus muscle)

The **transversospinalis muscles** run from the transverse processes to the spinous processes of the vertebrae. The **quadratus lumborum** muscles are found within the lumbar region. When one side contracts, they contribute to lateral flexion; when both sides contract, they extend the spine along with the erector spinae.

Learning Check

1. Which of the following muscles is involved in a true smile? Please select all that apply.
   a. Orbicularis oculi
   b. Zygomaticus major
   c. Buccinator
   d. Orbicularis oris
2. You are blowing out your candles for your birthday. What muscles are you using?
   a. Buccinator
   b. Orbicularis oculi
   c. Platysma
   d. Masseter
3. Which of the following muscles moves the tongue downward?
   a. Genioglossus
   b. Styloglossus
   c. Palatoglossus
   d. Hyoglossus
4. You are driving and turning your head to check the side-view mirrors. What muscles are you using to allow you to rotate your head to the side?
   a. Splenius capitis
   b. Sternocleidomastoid
   c. Digastric
   d. Styloglossus

12.3g Muscles of the Abdomen

The muscles of the anterolateral abdominal wall can be divided into four groups: the external obliques, the internal obliques, the transversus abdominis, and the rectus abdominis (**Figures 12.14** and **12.15** and **Table 12.9**). Note that there are four groups of muscles, but there are only three layers (**Figure 12.14D**); the rectus abdominis muscles share the same layer as the internal obliques.

Our examination of the abdomen starts at the sides of the body, the anterolateral abdominal wall. The **external oblique**, closest to the surface, is composed of fibers in a diagonal (oblique) orientation running inferiorly and medially. If you were to slide your hand into the front pocket of your pants, you would be sliding in the same orientation as the external oblique. The next deepest layer, with fibers that run perpendicular, is the **internal oblique**; its fibers run superiorly and medially. The deepest muscle, the **transversus abdominis**, is arranged transversely around the abdomen, similar to wearing a belt. This arrangement of three bands of muscles in different orientations allows various movements and rotations of the trunk. The three layers of muscle also help to protect the internal abdominal organs in an area where there is no bone.

The **linea alba** is a white, fibrous band that is made of the bilateral **rectus sheaths** that join at the anterior midline of the body. These enclose the **rectus abdominis** muscles (a pair of long, linear muscles, commonly called the *"sit-up" muscles*) that originate at the

Figure 12.14 Muscles of the Abdomen

(A and B) The anterior view of the abdomen dominated by the rectus abdominis, which is covered by a sheet of connective tissue called the *rectus sheath*. (C) On the sides of the abdomen, the external oblique muscles form the superficial layer, while the internal oblique muscles form the middle layer and the transversus abdominis forms the deepest layer.

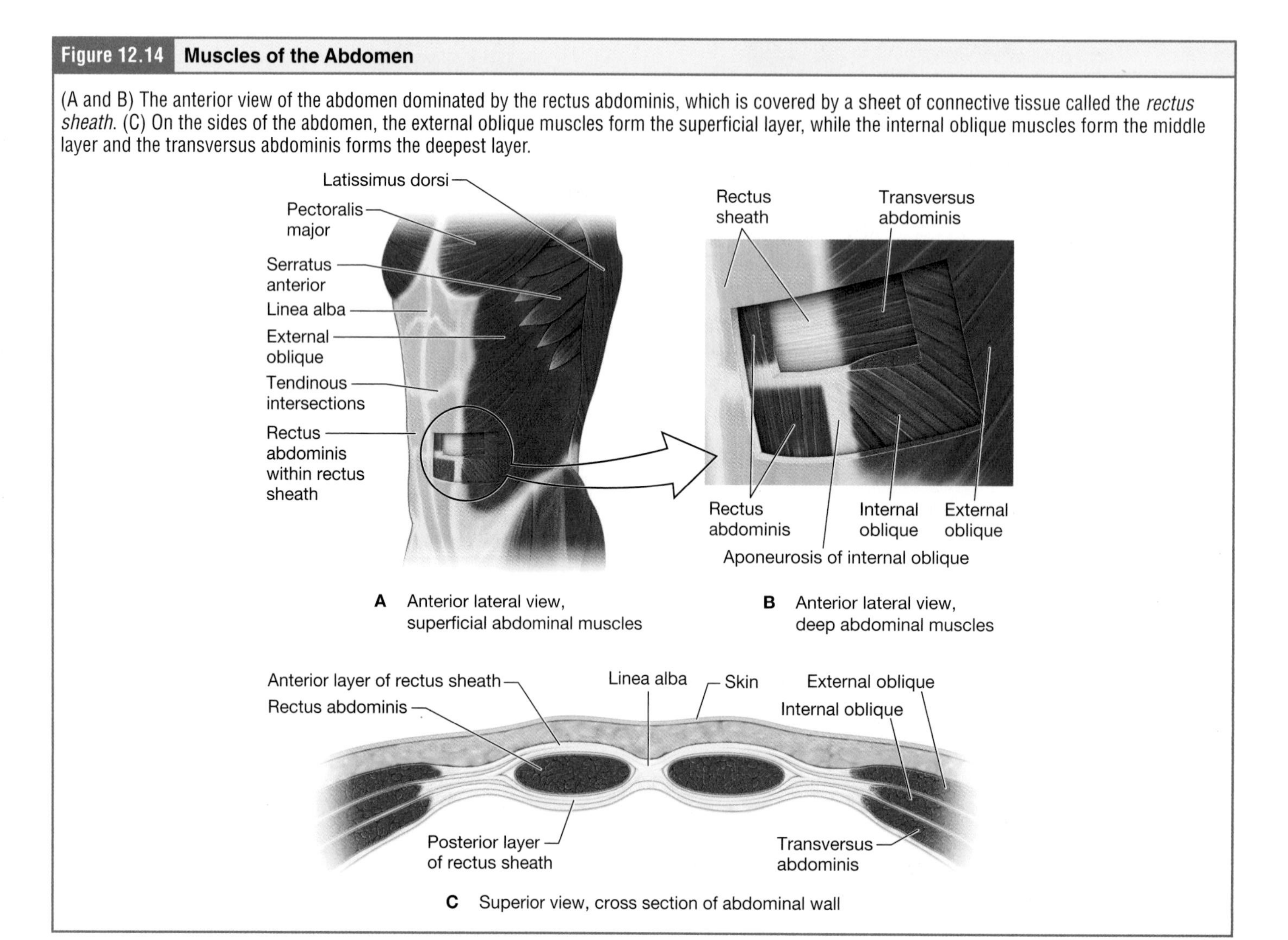

Table 12.9 Muscles of the Anterolateral Abdominal Wall

Muscle	Description	Origin (O) and Insertion (I)	Action	Nerve Supply	You May Know This Muscle from...
External abdominal oblique	Outermost layer of abdominal muscles with fibers running inferomedially	(O) Lower ribs (I) Pubis and abdominal aponeurosis	Flexion, lateral flexion, and rotation of the trunk	Intercostal, iliohypogastric, and ilioinguinal nerves	Doing twisting exercises as you rotate your spine
Internal abdominal oblique	Intermediate layer of abdominal muscles with superomedial fibers	(O) Iliac crest and thoracolumbar aponeurosis (I) Costal cartilages and abdominal aponeurosis	Flexion, lateral flexion, and rotation of the trunk	Intercostal, iliohypogastric, and ilioinguinal nerves	Doing twisting exercises as you rotate your spine
Transverse abdominis	Deepest layer of abdominal muscles, fibers run transversely	(O) Iliac crest and thoracolumbar aponeurosis (I) Pubis and abdominal aponeurosis	Compression of abdomen	Intercostal, iliohypogastric, and ilioinguinal nerves	If someone were to punch you in the stomach, think of this muscle as what would contract immediately to protect your internal organs
Rectus abdominis	Superficial abdominal muscle that runs along the middle of the abdomen	(O) Pubic symphysis (I) Costal cartilages	Flexion of the trunk	Intercostal nerves	"Six-pack abs" or the movement involved in sit-ups and crunches
Psoas major	Posterior abdominal muscle, comes together with iliacus to become a major hip flexor—iliopsoas	(O) Lumbar vertebrae (I) Lesser trochanter of the femur	Flexion of the hip and trunk	Femoral nerve and lumbar ventral rami	
Iliacus	Comes together with psoas major to become a major hip flexor—iliopsoas	(O) Anterior ilium (I) Lesser trochanter of the femur	Flexion of the hip and trunk	Femoral nerve	
Quadratus lumborum	Deep layer of muscle forming the posterior wall of the abdomen	(O) Iliac crest (I) Transverse processes of lumbar vertebrae	Lateral flexion of the trunk	Lumbar anterior rami	Lower back pain is often related to this muscle

pubic crest and symphysis, and extend the length of the body's trunk. Each muscle is segmented by transverse bands of collagen fibers. This results in the look of "six-pack abs," as each segment hypertrophies on individuals at the gym who do many sit-ups. The rectus abdominis sits within the middle layer of the abdomen, along with the internal obliques.

The posterior abdominal wall is formed by the lumbar vertebrae, parts of the ilia of the hip bones, psoas major and iliacus muscles, and quadratus lumborum muscles. This part of the core plays a key role in stabilizing the rest of the body and maintaining posture (see Figure 12.15).

12.3h Muscles of the Thorax

The muscles of the chest serve to facilitate breathing by changing the size of the thoracic cavity (**Figure 12.16**; see also **Figure 12.17** and **Table 12.10**). First you can understand the motions on your own body, then learn the muscles responsible. Put your hands on the sides of your chest and inhale. You can observe that your ribs move up and out to the sides, like the curtains on a stage. Now, keeping your hands in place, exhale; you can observe that the thoracic cavity decreases in size. Next, place your hands on your belly and repeat the inhale. You may notice that your belly protrudes during the inhale. When you exhale, try forcing the exhale to squeeze all the air out; do you feel your abdominal muscles contracting? The abdomen and thorax are separated by the diaphragm, but they borrow space from each other in various physiological circumstances.

Figure 12.15 Muscles of the Abdomen

The muscles in the posterior wall of the abdomen contribute to both lumbar spine and hip movements.

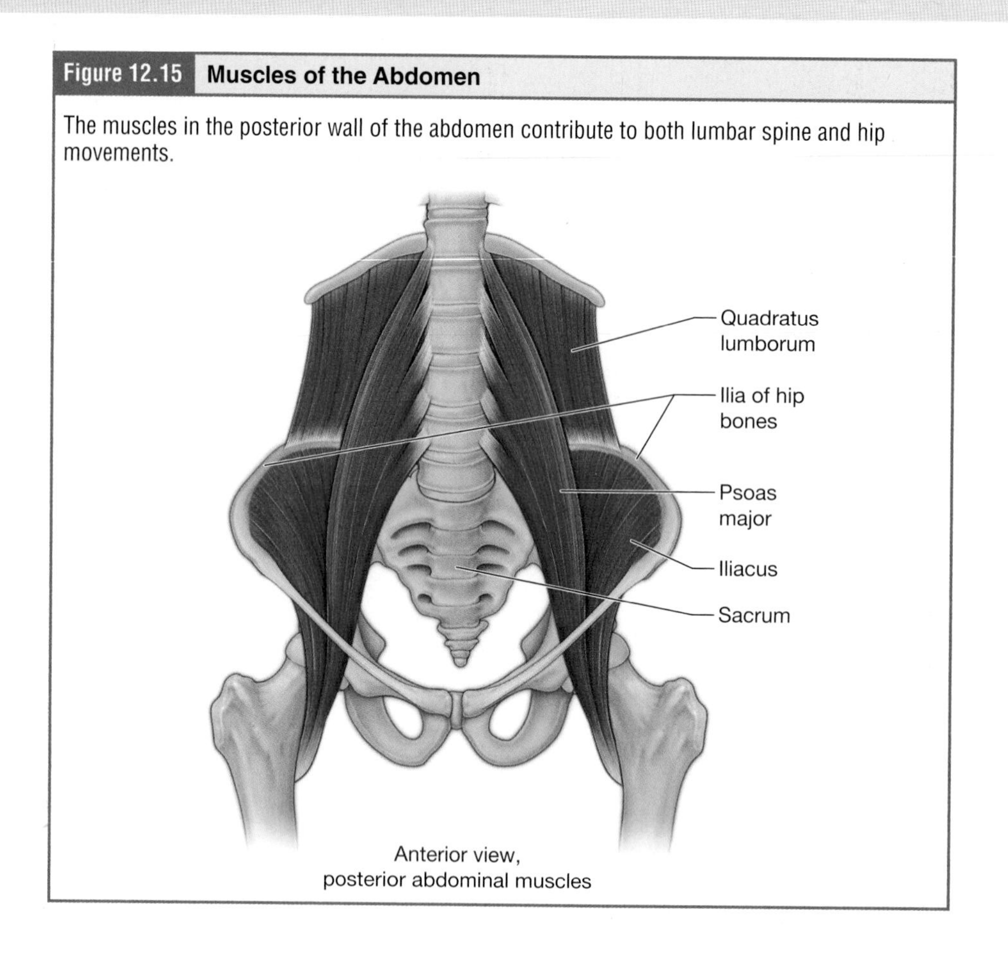

Anterior view, posterior abdominal muscles

Figure 12.16 The Muscles of the Diaphragm

The diaphragm divides the thoracic and abdominal cavities and functions in respiration. (A) In the anterior view, the diaphragm is a parachute- or bell-shaped muscle. (B) The inferior view shows the large central tendon of the diaphragm.

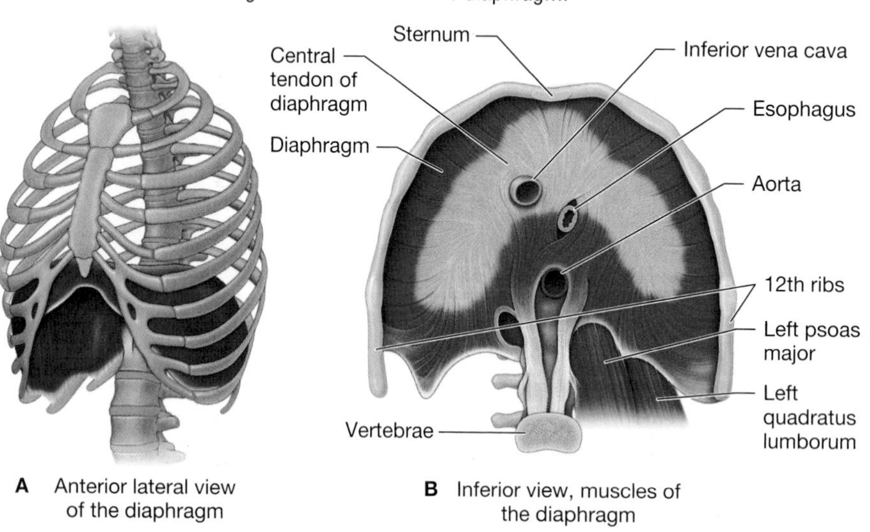

A Anterior lateral view of the diaphragm

B Inferior view, muscles of the diaphragm

Student Study Tip

The movement of the diaphragm is like a bouncing trampoline—the muscle is anchored laterally, and the movement (and insertion) occurs in the central tendon.

The Diaphragm The **diaphragm** is a bell- or parachute-shaped muscle that divides the thoracic cavity from the abdominal cavity (see Figure 12.16). When relaxed, the diaphragm is curved in a dome, but when it contracts during inhalation, it flattens, pushing down on the abdominal contents. This is why you observed your belly protruding slightly during the inhalation; the flattening of the contracted diaphragm makes

Table 12.10 Muscles of the Thorax

Muscle	Description	Origin (O) and Insertion (I)	Action	Nerve Supply	You May Know This Muscle from...
Diaphragm	Most important breathing muscle, separates the thorax from the abdomen	(O) Costal cartilages and lumbar vertebrae (I) Central tendon of the diaphragm	Depresses the central tendon to create more space in the thorax during inspiration	Phrenic nerve	Hiccups are spasms of the diaphragm
External intercostals	Most external layer of muscles between the ribs, fibers are oriented inferomedially	(O) Inferior margins of ribs (I) Superior margins of ribs	Elevates ribs during inspiration	Intercostal nerves	When you breathe in, you can feel your rib cage expand, and this is assisted by the intercostal muscles
Internal intercostals	Middle layer of muscles between the ribs, fibers are oriented superomedially	(O) Inferior margins of ribs (I) Superior margin of ribs	Depress and draw the ribs together during expiration	Intercostal nerves	
Innermost intercostals	Deepest layer of muscles between the ribs, the fibers are horizontal	(O) Costal grooves of ribs (I) Superior border of rib below	Assist the internal intercostals during expiration	Intercostal nerves	

Figure 12.17 Intercostal Muscles

The intercostal muscles span the spaces between adjacent ribs. There are three layers of intercostal muscles—external, internal, and innermost.

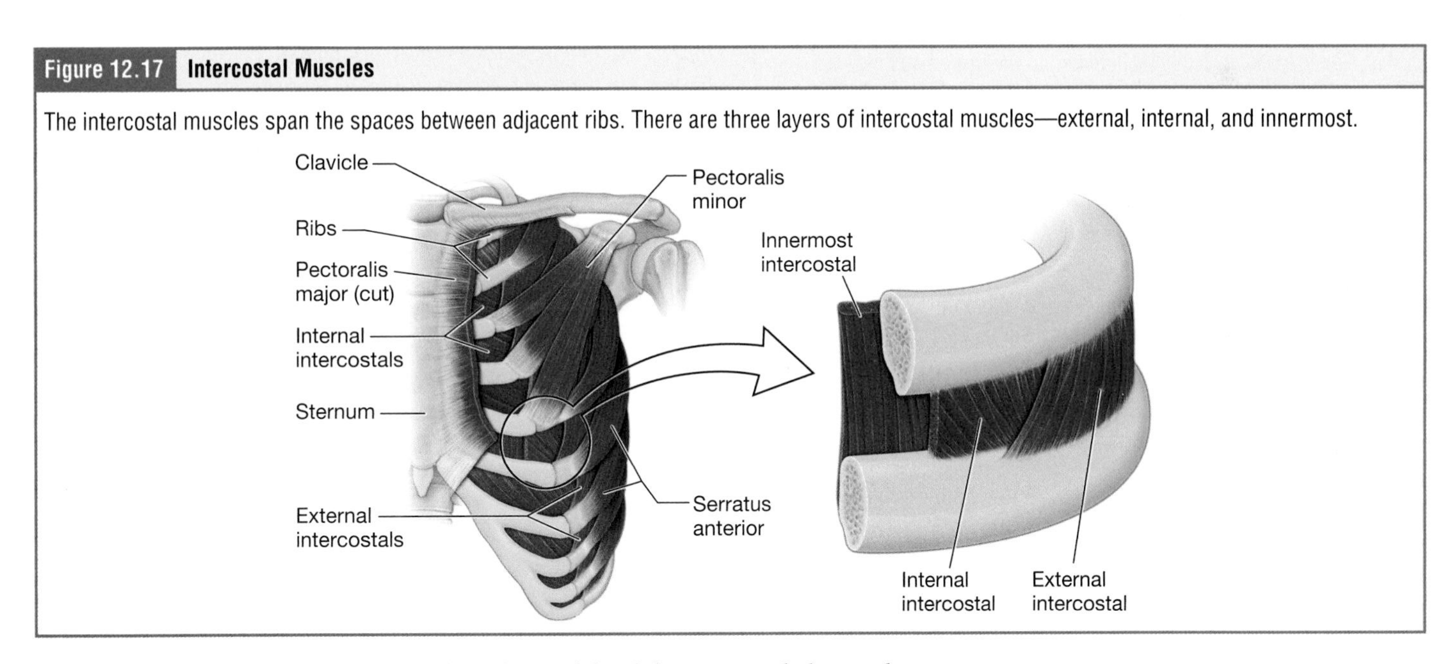

more space in the thorax but decreases the volume of the abdomen. In exhalation, the diaphragm simply relaxes back to its dome shape. To force breath out in an exhalation, the abdominal muscles contract, pushing the abdominal contents up against the inferior side of the diaphragm and further decreasing the space in the thorax.

The diaphragm is a round muscle with a central tendon. The inferior surface of the pericardium pleural membranes (parietal pleura) fuse onto the central tendon of the diaphragm.

The diaphragm also includes three openings for the passage of structures between the thorax and the abdomen. The inferior vena cava passes through the **caval opening**, and the esophagus and attached nerves pass through the **esophageal hiatus**. The aorta, thoracic duct, and azygous vein pass through the **aortic hiatus** of the posterior diaphragm.

The Intercostal Muscles There are three sets of muscles, called **intercostal muscles**, which span each of the spaces between the ribs. The principal role of the intercostal muscles is to assist in breathing by changing the dimensions of the rib cage (see Figure 12.17).

Apply to Pathophysiology

Fractured Rib

Imani is an 18-year-old female who plays basketball for her school team. She was going up for a layup, when a defensive player caught her arm, causing Imani to fall to the ground. Immediately, she got up with severe pain in her right side, grabbing her right side and hunched over to that side. She was rushed to the hospital.

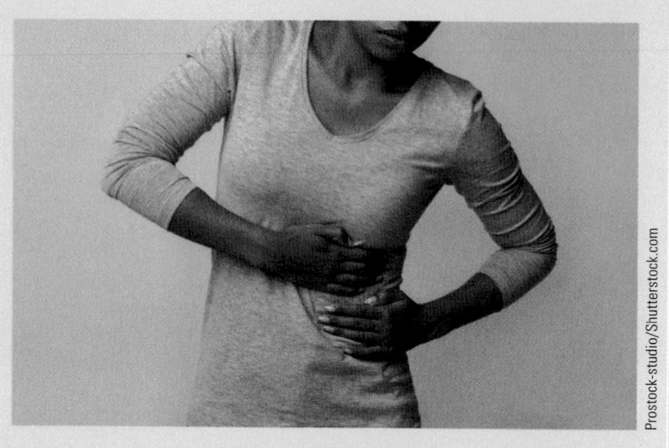

Prostock-studio/Shutterstock.com

1. Imani walks into the ER hunched to the right with holding her side protectively. Which of these muscles is the most contracted?
   A) Illiacus
   B) Psoas major
   C) Internal obliques
   D) Gluteus medius
2. You notice that Imani has shortness of breath. You note that her superior thorax is expanding a lot, but that her inferior thorax does not seem to be moving much; you also note that her belly does not seem to move with her breath at all. Imani's breathing is notably shallow but her respiratory rate has increased. Which of the following muscles are not working as effectively after the injury?
   A) Serratus anterior
   B) Intercostals
   C) Pectoralis minor
   D) B and C only
3. The muscle(s) you identified in Question 2 are not working effectively. Which muscles might be compensating?
   A) The diaphragm is working harder than the pectoralis minor.
   B) The intercostals are working harder than the sternocleido-mastoid (if you don't know this muscle, look it up!).
   C) The serratus anterior is working harder than the diaphragm.
   D) The pectoralis minor is working harder than the serratus anterior.
4. When focusing on restoring normal breathing, what muscles need to be strengthened?
   A) Diaphragm
   B) Pectoralis minor
   C) Sternocleidomastoid
   D) Serratus anterior

The 11 pairs of superficial **external intercostal** muscles aid in inspiration of air during breathing because when they contract, they raise the ribs both up and out to the side, which expands the thorax. The 11 pairs of **internal intercostal** muscles, just under the externals, are used for expiration because they draw the ribs together to constrict the rib cage. The **innermost intercostal** muscles are the deepest, and they act as synergists for the action of the internal intercostals.

12.3i Muscles of the Pelvic Floor and Perineum

ischiococcygeus (coccygeus)

The pelvic floor is a muscular sheet that provides a base for the pelvic cavity. The **pelvic diaphragm**, spanning anteriorly to posteriorly from the pubis to the coccyx, is comprised of the **levator ani** and the **ischiococcygeus**. Its openings include the anal canal and urethra, and the vagina in women.

The large levator ani consists of two skeletal muscles, the **pubococcygeus** and the **iliococcygeus** (Figure 12.18). The levator ani is considered the most important muscle of the pelvic floor because it supports the pelvic viscera. It resists the pressure produced by contraction of the abdominal muscles when that pressure is applied to the colon to aid in defecation and to the uterus to aid in childbirth (assisted by the ischiococcygeus, which pulls the coccyx anteriorly). This muscle also creates skeletal muscle sphincters at the urethra and anus.

Figure 12.18 Muscles of the Pelvic Floor

The pelvic floor muscles form the floor of the pelvic and abdominal cavities, supporting and regulating the functions of the organs within.

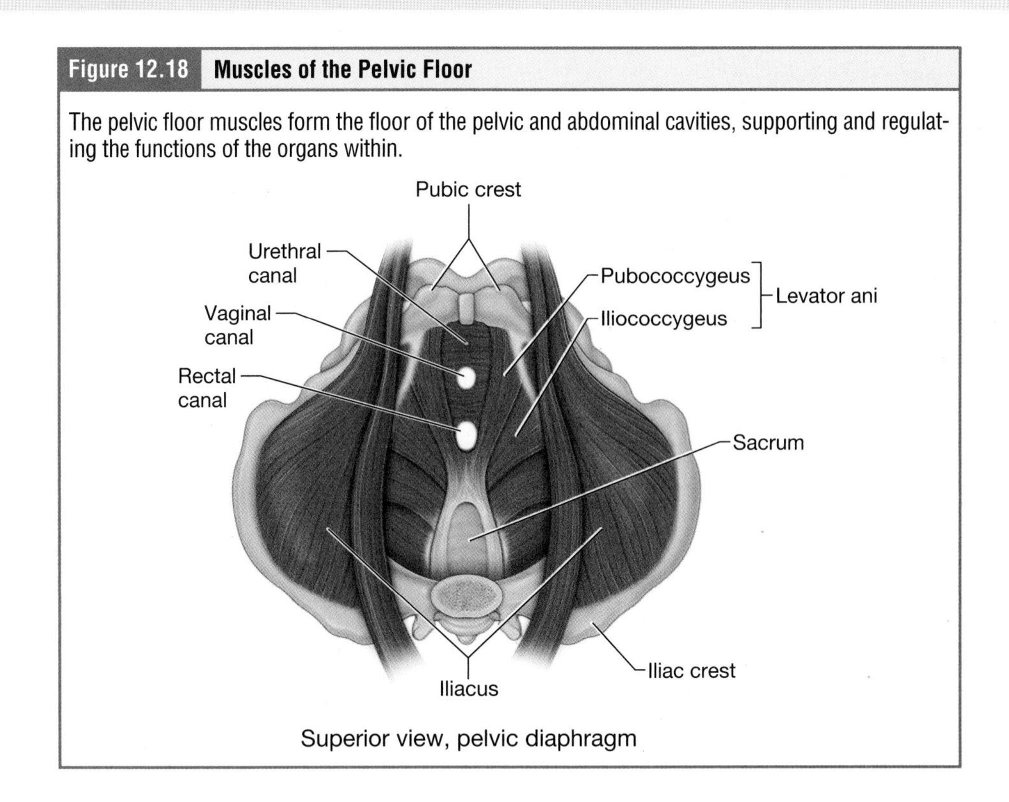

The **perineum** is the diamond-shaped space between the pubic symphysis (anteriorly), the coccyx (posteriorly), and the ischial tuberosities (laterally), lying just inferior to the pelvic diaphragm (levator ani and ischiococcygeus). Divided transversely into triangles, the anterior is the **urogenital triangle**, which includes the external genitals. The posterior is the **anal triangle**, which contains the anus (**Figure 12.19**). The perineum is also divided into superficial and deep layers by muscles common to individuals of both genders.

Figure 12.19 Muscles of the Perineum

The same perineal muscles are present in the pelvises of all individuals, but are arranged in slightly different patterns around the genitals.

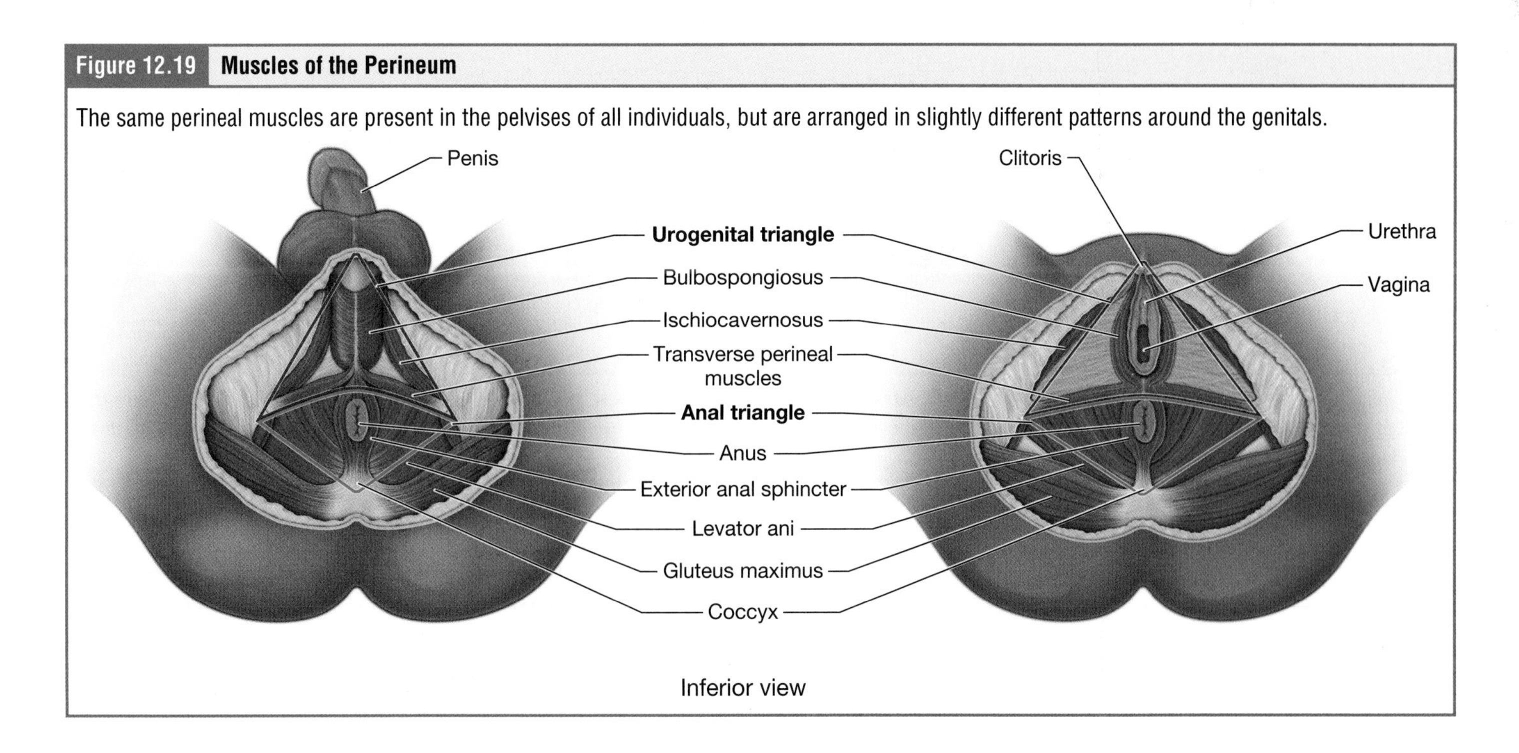

Cultural Connection

Pelvic Floor Dysfunction and Injury

Dysfunction of the pelvic floor is remarkably common, though rarely talked about in U.S. culture. While pelvic floor dysfunction can happen to bodies of any age or sex, for obvious reasons pelvic floor damage is very common after pregnancy and birth. Approximately 36 percent of vaginal births and 25 percent of postpartum individuals overall (averaging those who gave birth vaginally or by C-section) are left with pelvic floor injury. If left untreated, these injuries can cause issues with future pregnancies as well as with walking, sexual satisfaction, and urinary and bowel function throughout the lifespan. Pelvic floor injury or dysfunction is highly treatable through physical therapy. In France, 100 percent of postpartum patients receive pelvic floor therapy as part of the standard care following the birth of a baby. In the United States, by comparison, only a fraction of the 25 percent of postpartum individuals with pelvic floor injury are able to receive therapy. There are a number of barriers to therapy, including cost, insurance coverage, available childcare during the visits, the lack of pelvic floor PTs in the country—especially in rural areas—and recognition of the need and referrals by a patient's primary care doctor or obstetrician.

Learning Check

1. Which of the following abdominal muscles have muscle fibers that are oriented superiorly and medially?
   a. External oblique
   b. Internal oblique
   c. Transversus abdominis
   d. Rectus abdominis
2. Take a deep breath in and blow out all the air from your lungs. When you blow out the air, are you contracting or relaxing the diaphragm?
   a. Contracting
   b. Relaxing
3. Which of the following muscles supports the pelvic viscera? Please select all that apply.
   a. Pubococcygeus
   b. Iliococcygeus
   c. Ischiococcygeus
   d. Perineum

12.4 Appendicular Muscles

Learning Objectives: By the end of this section, you will be able to:

12.4.1 Identify the location, general attachments, and actions of the major skeletal muscles.

12.4.2 Describe similar actions (functional groupings) of muscles in a particular compartment (e.g., anterior arm) or region (e.g., deep back).

Muscles of the shoulder and upper limb can be divided into four groups: muscles that stabilize and position the shoulder, muscles that move the arm, muscles that move the forearm, and muscles that move the wrists, hands, and fingers. These classifications are about where the action of the muscle is, but muscles may not be in the same region as their action. The term **intrinsic** describes muscles that are located in the same region as their action. If you pinch the skin between your thumb and hand and wiggle your thumb, you can feel intrinsic muscles of the hand at work. Now wrap your hand around your opposite forearm just distal to the elbow and wiggle your fingers. The muscles at work are the extrinsic muscles of the hand. The term **extrinsic** means that muscles are located in a region some distance from where they act.

12.4a Shoulder Muscles

LO 12.4.1

Muscles that stabilize and move the shoulder are located either on the anterior thorax or on the posterior thorax (**Figure 12.20** and **Table 12.11**). The anterior muscles include the **subclavius**, **pectoralis minor**, and **serratus anterior**. The posterior

Student Study Tip

The serratus anterior muscle is shaped like a serrated knife. It has sharp ridges that span over the ribs.

Figure 12.20 Shoulder Muscles

Many muscles stabilize the shoulder so that it can serve as a steady base for the movements of the arm. (A) Left anterior lateral view and (B) posterior view of the shoulder muscles.

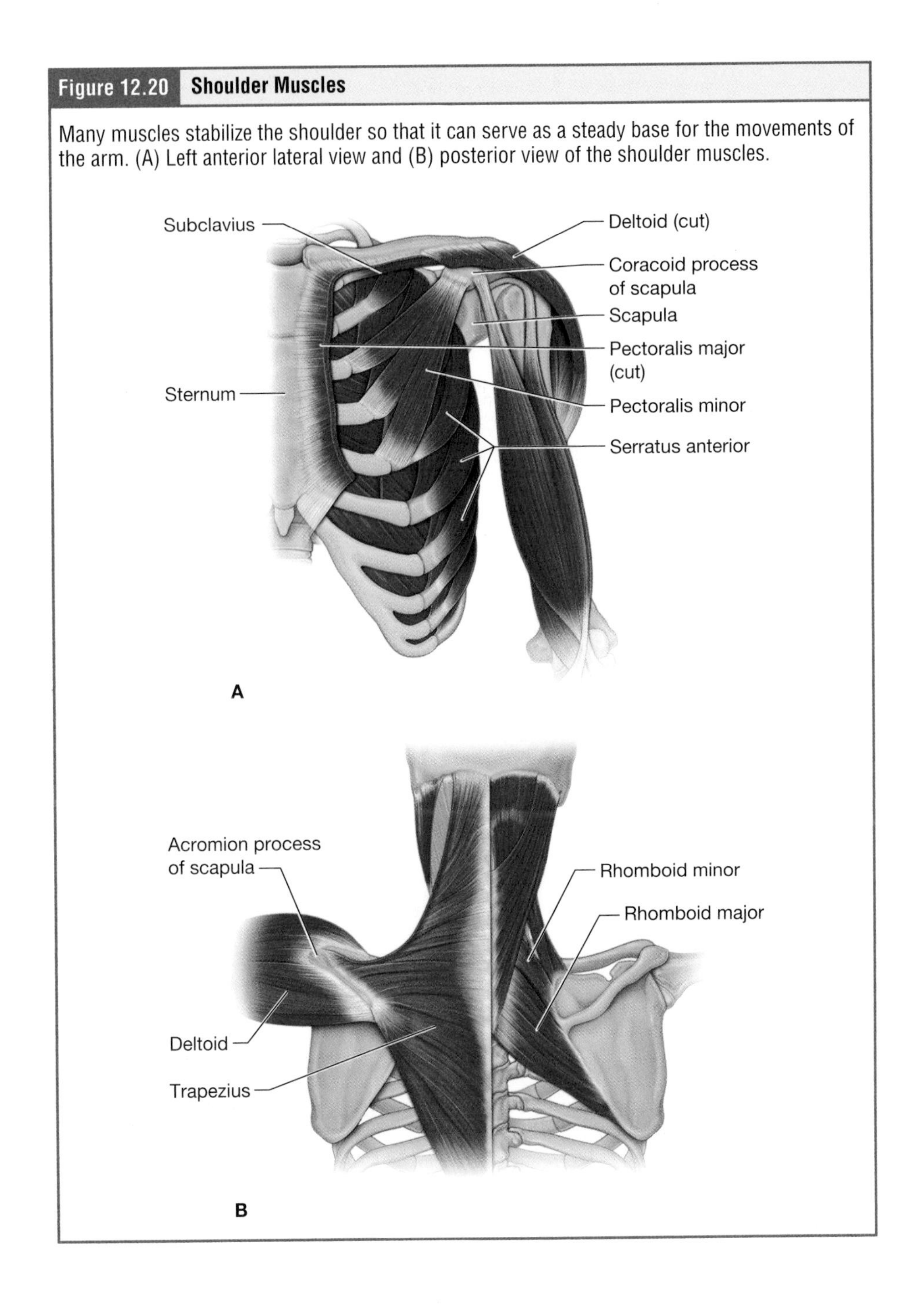

Table 12.11 Shoulder and Brachial Muscles

Muscle	Description	Origin (O) and Insertion (I)	Action	Nerve Supply	You May Know This Muscle from...
Subclavius	Muscle sitting just underneath the clavicle	(O) Costal cartilage of first rib (I) Inferior clavicle	Stabilizes the clavicle	Nerve to subclavius	
Pectoralis minor	Smaller anterior chest muscle, deep to pectoralis major	(O) Anterior surfaces of upper ribs (I) Coracoid process of the scapula	Protraction and depression of the scapula	Medial pectoral nerve	"Drop and give me twenty push-ups"
Serratus anterior	Thin muscle laying right on top of the anterior ribs	(O) Outer surface of upper ribs (I) Anterior surface of the scapula	Protraction and stabilization of the scapula	Long thoracic nerve	
Trapezius	Superficial muscle of the back and shoulder	(O) Occipital bone, ligamentum nuchae, spinous processes of cervical and thoracic vertebrae (I) Lateral clavicle and spine of scapula	Elevation, retraction, and depression of the scapula	Spinal accessory nerve (cranial nerve XI)	People have sore "traps" from wearing a heavy backpack, lifting heavy objects, or engaging in repetitive activities such as swimming
Rhomboid major	Superficial muscle just deep to traps	(O) Spinous processes of upper thoracic vertebrae (I) Vertebral border of the scapula	Retraction of the scapula	Dorsal scapular nerve	When you think of squeezing your shoulders back, the rhomboids are what you are using
Rhomboid minor	Superficial muscle just deep to the traps and above rhomboid major	(O) Spinous processes of cervical and thoracic vertebrae (I) Root of spine of scapula	Retraction of the scapula	Dorsal scapular nerve	When you think of squeezing your shoulders back, the rhomboids are the muscles you are using
Pectoralis major	Anterior superficial muscle of the chest	(O) Medial clavicle, sternum, and costal cartilages (I) Proximal humerus	Prime mover: flexion, medial rotation, and adduction of the shoulder	Medial and lateral pectoral nerves	Prime muscle in push-ups
Latissimus dorsi	Convergent, covers much of the superficial lower back	(O) Thoracolumbar aponeurosis and spinous processes of thoracic vertebrae (I) Proximal humerus	Prime mover: extension, medial rotation, and adduction of the shoulder	Thoracodorsal nerve	Wood choppers (pull downs)
Deltoid	Thick muscle covering the majority of the shoulder superficially and laterally	(O) Clavicle and posterior scapula (I) Lateral and proximal humerus	Prime mover for abduction of the arm, flexion and medial rotation, extension and lateral rotation	Axillary nerve	The muscle you probably think of when you hear "shoulder muscle"
Subscapularis	Part of the rotator cuff, the only muscle of this group that is anterior to the scapula	(O) Anterior surface of the scapula (I) Lesser tubercle of the humerus	Medial rotation and stabilization of the shoulder and arm	Upper and lower subscapular nerves	Shoulder tendinitis is often caused by an injury to a rotator cuff muscle, due to repetitive overhead movements such as playing tennis or pitching a baseball
Supraspinatus	Part of the rotator cuff, located on the posterior scapula	(O) Surface of the scapula above the spine (I) Greater tubercle of the humerus	Abduction and stabilization of the shoulder and arm	Suprascapular nerve	Shoulder tendinitis is often caused by an injury to a rotator cuff muscle, due to repetitive overhead movements, such as playing tennis or pitching a baseball

continued

Table 12.11 (*continued*)

Muscle	Description	Origin (O) and Insertion (I)	Action	Nerve Supply	You May Know This Muscle from...
Infraspinatus	Part of the rotator cuff, located on the posterior scapula	(O) Surface of the scapula below the spine (I) Greater tubercle of the humerus	Lateral rotation and stabilization of the shoulder and arm	Suprascapular nerve	Shoulder tendinitis is often caused by an injury to a rotator cuff muscle, due to repetitive overhead movements, such as playing tennis or pitching a baseball
Teres major	Short but powerful muscle, inferior to teres minor	(O) Lateral border of the posterior and inferior scapula (I) Proximal humerus	Extension, adduction, and medial rotation of the arm	Lower subscapular nerve	
Teres minor	Part of the rotator cuff, located on the posterior scapula	(O) Lateral border of the scapula (I) Greater tubercle of the humerus	Lateral rotation, extension, and stability of the shoulder and arm	Axillary nerve	Shoulder tendinitis is often caused by an injury to a rotator cuff muscle, due to repetitive overhead movements, such as playing tennis or pitching a baseball
Coracobrachialis	Muscle extending from the scapula to the middle of the brachium anteriorly	(O) Coracoid process of the scapula (I) Medial surface of the humerus	Flexion and adduction of the arm	Musculocutaneous nerve	

muscles include the **trapezius**, **rhomboid major**, and **rhomboid minor**. In general, the anterior muscles serve to pull the scapula forward—a motion called **protraction**—and the posterior muscles serve both to **retract** (pull back) the scapula and to pull the scapula toward the midline. Both sets of muscles provide stability when the humerus moves.

LO 12.4.2

12.4b Muscles That Move the Humerus

The glenohumeral joint, where the humerus meets the scapula, has the widest range of motion of any joint in the human body (**Figure 12.21**). As a consequence, there is an unusually large array of muscles that cross the shoulder joint and move the humerus (**Figure 12.22** and see Table 12.11). The **pectoralis major** is a large, fan-shaped muscle that covers much of the superior portion of the anterior thorax. The broad, triangular **latissimus dorsi** is located on the inferior part of the back. Both of these muscles are prime movers of the humerus, and both are convergent muscles, lending a tremendous amount of strength to arm movements.

The rest of the muscles that move the humerus originate on the scapula. The **deltoid**, the thick muscle that caps the lateral shoulder, is the major abductor of the arm, but it also facilitates flexing and medial rotation as well as extension and lateral rotation. The **subscapularis** originates on the anterior scapula and medially rotates the arm. Named for their locations, the **supraspinatus** (superior to the spine of the scapula) and the **infraspinatus** (inferior to the spine of the scapula) abduct the arm and laterally rotate the arm, respectively. The thick and flat **teres major** is inferior to the teres minor and extends the arm, as well as assisting in adduction and medial rotation of it. The long **teres minor** laterally rotates and extends the arm. Finally, the **coracobrachialis** flexes and adducts the arm.

Figure 12.21 Shoulder Movements

The movements available at the shoulder joint include (A) retraction and protraction, (B) flexion and extension, (C) abduction and adduction, and (D) internal and external rotation.

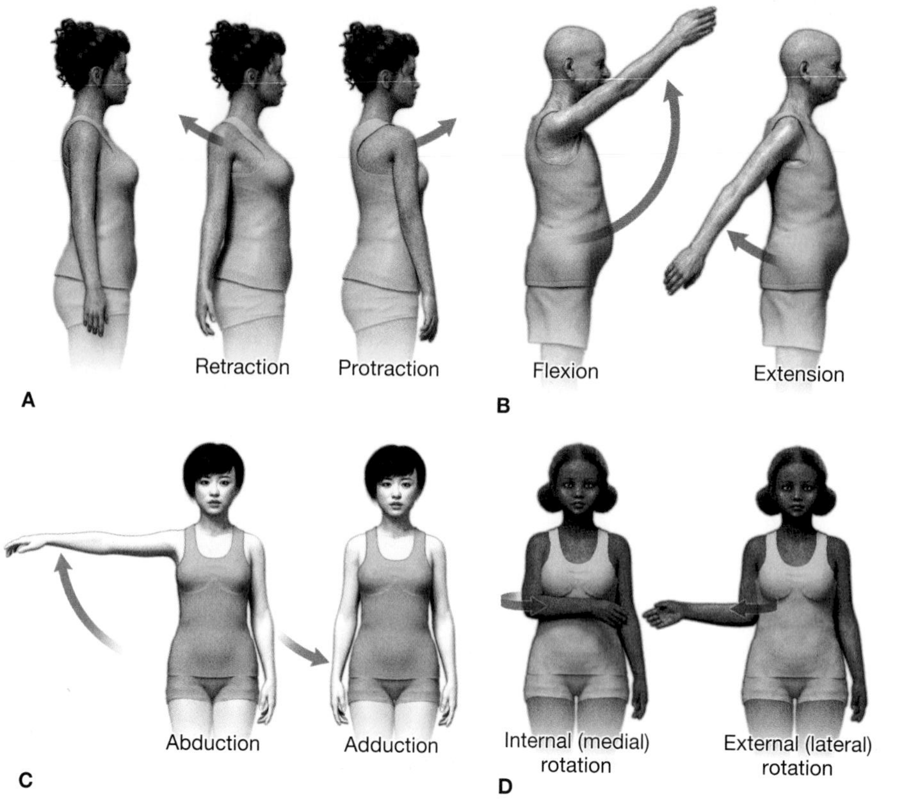

Retraction
Protraction
A
Flexion
Extension
B
Abduction
Adduction
C
Internal (medial) rotation
External (lateral) rotation
D

rotator cuff (musculotendinous cuff)

The tendons of the deep subscapularis, supraspinatus, infraspinatus, and teres minor connect the scapula to the humerus, forming the **rotator cuff**, the circle of tendons around the shoulder joint. The rotator cuff lends structure and stability to the shoulder joint.

12.4c Muscles That Move the Forearm

The forearm, made of the radius and ulna bones, has four main types of action at the hinge of the elbow joint: flexion and extension of the elbow joint, and pronation and supination of the forearm (**Figure 12.23**). The elbow flexors include the biceps brachii, brachialis, and brachioradialis (see Figure 12.23 and **Table 12.12**). The two-headed **biceps brachii** crosses the shoulder and elbow joints to flex the forearm, also taking part in supinating the forearm at the radioulnar joints and flexing the arm at the shoulder joint. Deep to the biceps brachii, the **brachialis** provides additional power in flexing the forearm. Finally, the **brachioradialis** can flex the forearm quickly or help lift a load slowly. The extensors are the **triceps brachii** and **anconeus**. The pronators are the **pronator teres** and the **pronator quadratus**, and the only prime supinator is **supinator**.

12.4d Muscles That Move the Wrist, Hand, and Fingers

The extrinsic muscles of the hand and wrist originate on the humerus and are connected to the hand by long tendons. Most of the tendons pass through an arch provided

Figure 12.22 Muscles That Move the Humerus

(A and C) The muscles that flex or move the humerus anteriorly are generally located on the anterior side of the joint. (B) The muscles that abduct the humerus are found on the lateral side of the joint. The muscles that adduct the humerus generally originate from the middle or lower back. (D) The muscles that extend or move the humerus posteriorly are located on the posterior side of the joint.

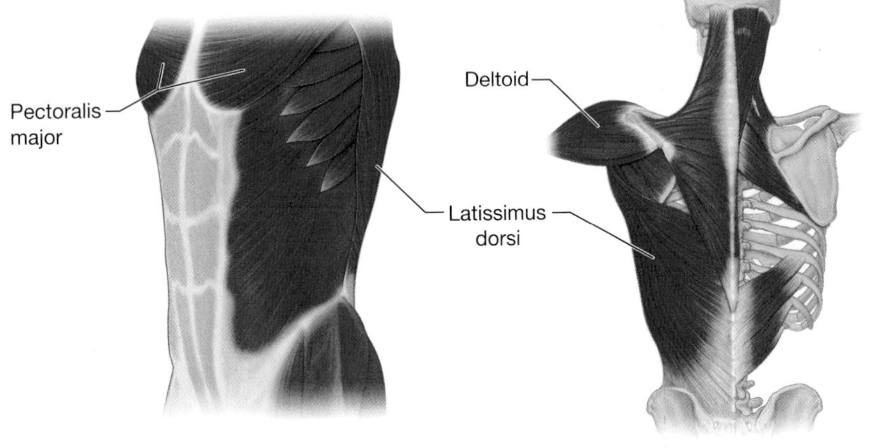

A Left anterior lateral view, pectoralis major and latissimus dorsi

B Left anterior lateral view, left deltoid and left latissimus dorsi

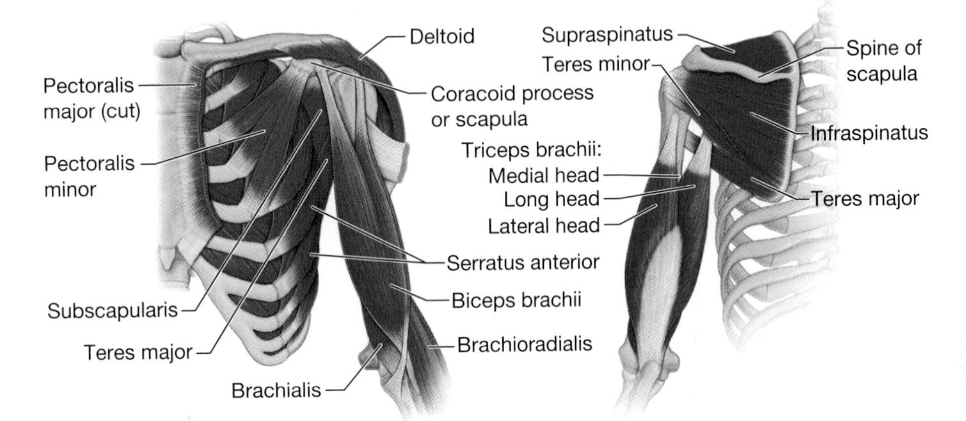

C Anterior view, deep muscles of the left shoulder and left upper arm

D Posterior view, deep muscles of the left shoulder and left upper arm

by the carpal bones of the wrist known as the carpal tunnel (**Figure 12.24**). Fibrous bands called **retinacula** sheathe the tendons at the wrist. The **flexor retinaculum** extends over the palmar surface of the hand while the **extensor retinaculum** extends over the dorsal surface of the hand. **Figure 12.25** depicts movements of the forearm, wrist, and fingers.

flexor retinaculum (transverse carpal ligament)

The anterior forearm muscles produce flexion of the wrist and the hand, and the posterior forearm muscles produce extension. Most of the muscles of the forearm have three-part names. The first part of their name describes their action, and the second part of their name describes the location of their action, either at the wrist (carpi) or on the fingers (digitorum). The third part of their name may describe either their location—either on the radial side of the forearm, the ulnar side of the

Figure 12.23 Muscles That Move the Forearm

The muscles that flex and extend the elbow as well as pronate and supinate the forearm originate in the brachial region. The muscles in the forearm move the wrists, hands, and fingers. (A and B) Anterior and posterior views of the left upper arm, (C and D) palmar and dorsal views of the superficial muscles of the left forearm, and (E and F) palmar and dorsal views of the deep muscles of the left forearm.

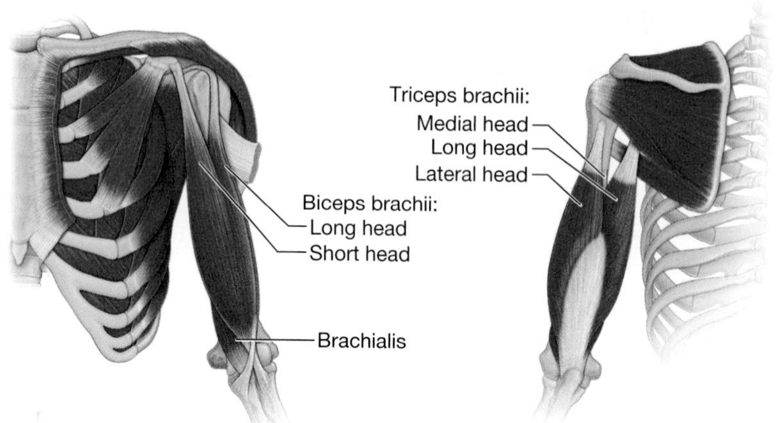

A Anterior view, left upper arm

B Posterior view, left upper arm

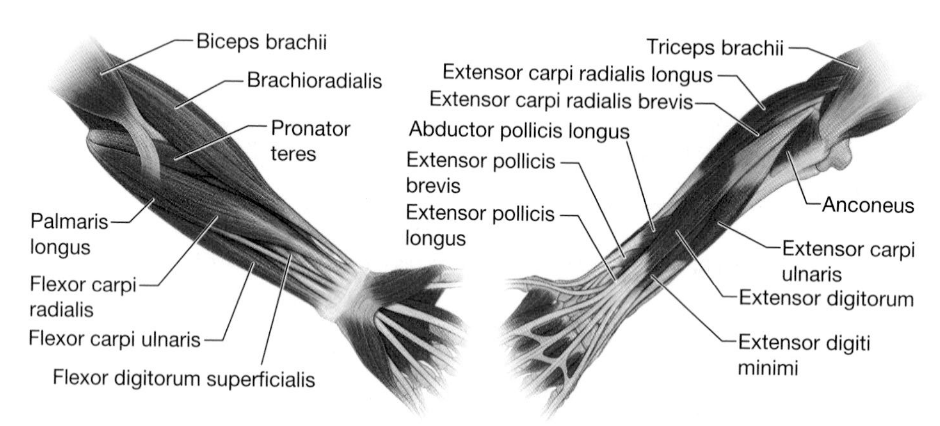

C Palmar view, superficial muscles of the left forearm

D Dorsal view, superficial muscles of the left forearm

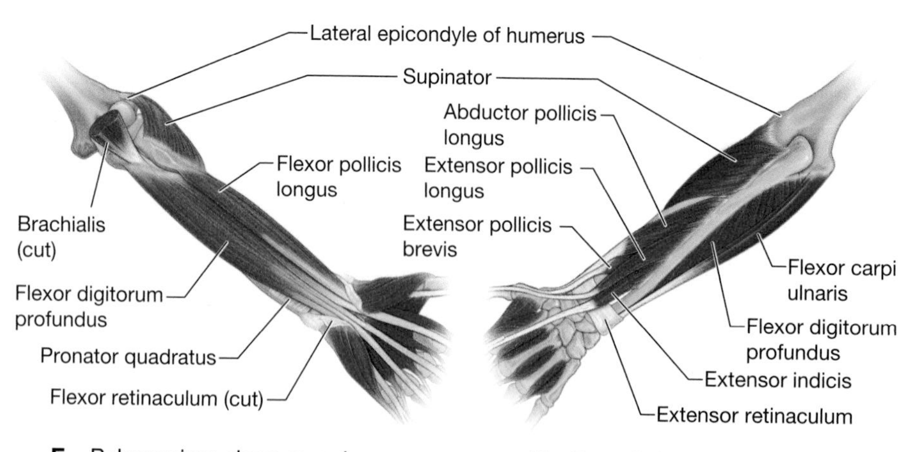

E Palmar view, deep muscles of the left forearm

F Dorsal view, deep muscles of the left forearm

Table 12.12 Forearm, Wrist, and Hand Muscles

Muscle	Description	Origin (O) and Insertion (I)	Action	Nerve Supply	You May Know This Muscle from...
Biceps brachii	Superficial, two-headed muscle that crosses two joints and sits on the anterior brachium	(O) Distal scapula (I) Radial tuberosity of the radius	Prime mover in forearm flexion, supination of forearm	Musculocutaneous nerve	Bicep curls, body builders flexing their biceps
Brachialis	Deep brachial muscle crossing the elbow joint	(O) Lower anterior humerus (I) Coronoid process of the ulna	Flexion of the forearm	Musculocutaneous nerve	
Brachioradialis	More distal muscle separating the anterior and posterior compartments of the forearm	(O) Distal and lateral humerus (I) Distal radius	Flexion of the forearm with the arm in neutral position	Radial nerve	This muscle is referred to as "the drinking muscle" because of the position of the arm where the muscle is activated
Triceps brachii	Superficial muscle spanning the posterior brachium with three heads	(O) Distal scapula and posterior, proximal humerus (I) Olecranon process of the ulna	Extension of the shoulder and elbow	Radial nerve	Tricep pull-downs
Anconeus	Small muscle crossing the elbow joint	(O) Lateral epicondyle of the humerus (I) Lateral surface of the olecranon	Extension and stabilization of the forearm	Radial nerve	
Pronator teres	Superficial and proximal muscle of the anterior antebrachium	(O) Medial epicondyle of the humerus and proximal ulna (I) Middle of lateral radius	Pronation of the forearm	Median nerve	
Pronator quadratus	Deepest layer of the anterior antebrachial muscles, this muscle is shaped like a quadrangle	(O) Distal anteromedial ulna (I) Distal anterolateral radius	Pronation of the forearm	Median nerve	
Supinator	Deep posterior muscle of the antebrachium	(O) Lateral epicondyle of the humerus (I) Lateral surface of the proximal radius	Prime mover for supination	Radial nerve	
Flexor carpi radialis	Lateral and superficial antebrachial muscle in the anterior compartment and acting on the wrist	(O) Medial epicondyle of the humerus (I) Bases of second and third metacarpals	Flexion and radial deviation of the wrist	Median nerve	
Palmaris longus	Very thin and long superficial muscle, not everyone has this muscle	(O) Medial epicondyle of the humerus (I) Flexor retinaculum and proximal phalanges	Weak wrist flexion	Median nerve	If you make a claw with your hand, the tendon right in the center of your wrist is the palmaris longus
Flexor carpi ulnaris	Medial, anterior, and superficial antebrachial muscle acting on the wrist	(O) Medial epicondyle of the humerus and proximal posterior ulna (I) Carpal bones and fifth metacarpal	Flexion and ulnar deviation of the wrist	Ulnar nerve	

continued

Table 12.12 (*continued*)

Muscle	Description	Origin (O) and Insertion (I)	Action	Nerve Supply	You May Know This Muscle from...
Flexor digitorum superficialis	Superficial muscle in the anterior compartment and acting on the fingers	(O) Medial epicondyle of the radius and proximal ulna (I) Middle phalanges of digits 2–5	Flexion of digits 2–5 and flexion of the wrist	Median nerve	Grip strength exercises
Flexor pollicis longus	Deep and extrinsic anterior antebrachial muscle	(O) Anterior surface of the radius (I) Distal phalanx of the thumb	Flexion of the thumb	Median nerve	
Flexor digitorum profundus	Deep anterior antebrachial muscle acting on the fingers	(O) Proximal anterior ulna (I) Distal phalanges of digits 2–5	Flexion of digits 2–5 and flexion of the wrist	Median and ulnar nerves	Grip strength exercises
Extensor carpi radialis longus	Superficial posterior antebrachial muscle	(O) Lateral epicondyle of the humerus (I) Posterior surface of the second metacarpal	Extension and radial deviation of the wrist	Radial nerve	
Extensor carpi radialis brevis	Superficial posterior antebrachial muscle running just beneath ECRL	(O) Lateral epicondyle of the humerus (I) Posterior surface of the third metacarpal	Extension of the wrist	Radial nerve	
Extensor digitorum	Superficial posterior antebrachial muscle	(O) Lateral epicondyle of the humerus (I) Digits 2–5	Extension of the digits and wrist	Radial nerve	Tendons on the back of your hand that you can see going to the fingers, especially while you are doing something like typing or playing the piano
Extensor digiti minimi	Superficial posterior antebrachial muscle	(O) Lateral epicondyle of the humerus (I) Digit 5	Extension of digit 5 and wrist	Radial nerve	
Extensor carpi ulnaris	Superficial posterior antebrachial muscle	(O) Lateral epicondyle of the humerus and proximal posterior ulna (I) Fifth metacarpal	Extension and ulnar deviation of the wrist	Radial nerve	
Abductor pollicis longus	Deep extrinsic muscle running laterally along the antebrachium to the thumb	(O) Posterior surfaces of the radius and ulna (I) First metacarpal	Abduction of the thumb and radial deviation of the wrist	Radial nerve	If you make a thumbs-up, you can see this tendon pop out along the lateral side of your wrist and thumb
Extensor pollicis brevis	Deep posterior muscle of the antebrachium	(O) Posterior radius (I) Proximal phalanx of the thumb	Extension of the thumb and radial deviation of the wrist	Radial nerve	
Extensor pollicis longus	Deep posterior muscle of the antebrachium	(O) Posterior ulna (I) Distal phalanx of the thumb	Extension and radial deviation of the wrist	Radial nerve	If you make a thumbs-up, you can see this tendon closer to the middle of the back of your hand, still closer to the thumb side

continued

Table 12.12 *(continued)*

Muscle	Description	Origin (O) and Insertion (I)	Action	Nerve Supply	You May Know This Muscle from...
Extensor indicis	Deep posterior antebrachial muscle	(O) Posterior ulna (I) Digit 2	Extension of digit 2 and wrist	Radial nerve	Using your pointer finger (index)
Abductor pollicis brevis	Intrinsic thenar muscle, most lateral	(O) Flexor retinaculum and lateral carpal bones (I) Proximal phalanx of the thumb	Abduction of the thumb	Median nerve	
Opponens pollicis	Intrinsic thenar muscle	(O) Flexor retinaculum and the trapezium (I) first metacarpal	Opposition of thumb	Median nerve	If you make the number 4 sign with your hand, you are using your opponens pollicis to bring your thumb across your hand
Flexor pollicis brevis	Intrinsic thenar muscle, deep to opponens pollicis and abductor pollicis brevis	(O) Flexor retinaculum and lateral carpal bones (I) Proximal phalanx of the thumb	Flexion of the thumb	Median and ulnar nerves	
Adductor pollicis brevis	Intrinsic thenar muscle, fibers run between digits 1 and 2	(O) Second and third metacarpals, lateral carpal bones (I) Proximal phalanx of the thumb	Adduction of the thumb	Ulnar nerve	
Abductor digiti minimi	Intrinsic hypothenar muscle	(O) Pisiform (I) Proximal phalanx of digit 5	Abduction of digit 5	Ulnar nerve	
Flexor digiti minimi	Intrinsic hypothenar muscle	(O) Flexor retinaculum and the hamate (I) Proximal phalanx of digit 5	Flexion of digit 5	Ulnar nerve	
Opponens digiti minimi	Intrinsic hypothenar muscle	(O) Flexor retinaculum and the hamate (I) Fifth metacarpal	Opposition of digit 5	Ulnar nerve	
Lumbricals	Intermediate intrinsic muscles, the only group in the body with no bony attachments!	(O) Tendons of flexor digitorum profundus (I) Extensor expansions of digits 2–5	Flexion and extension of the digits	Median and ulnar nerves	
Palmar interossei	Small intrinsic intermediate muscles, between the bones of the metacarpals on the palmar side	(O) Palmar surfaces of metacarpals (I) Proximal phalanges	Adduction, flexion, and extension of the digits	Ulnar nerve	
Dorsal interossei	Small intrinsic intermediate muscles, between the bones of the metacarpals on the dorsal side	(O) Posterior surfaces of metacarpals (I) Proximal phalanges	Abduction, flexion, and extension of the digits	Ulnar nerve	

Figure 12.24 The Carpal Tunnel

The carpal tunnel is a passageway for the tendons of the extrinsic muscles of the hand, which are protected by the carpal bones posteriorly and wrapped anteriorly by the flexor retinaculum.

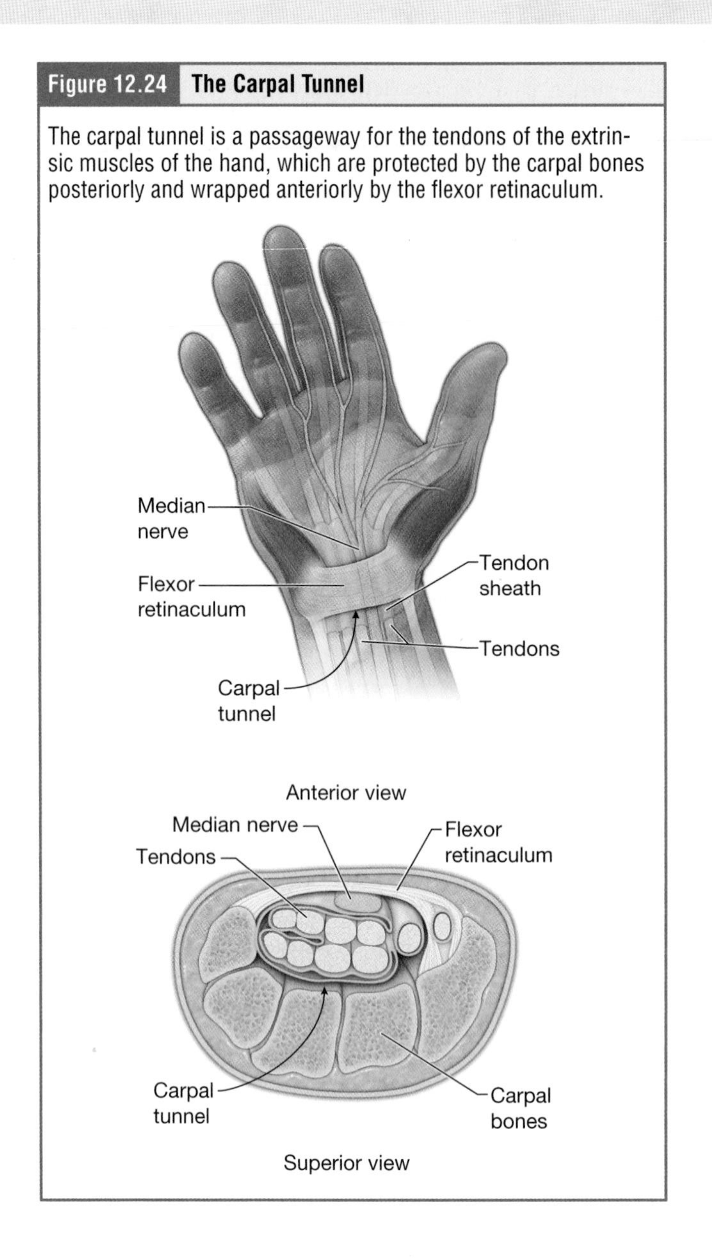

forearm, superficial (superficialis) or deep (profundus)—or their size, such as the longer muscle (longus) or the shorter muscle (brevis) in a group. These small muscles are arranged in two layers, one superficial and one deep. On the anterior side from lateral to medial, the superficial muscles are the **flexor carpi radialis, palmaris longus**, **flexor carpi ulnaris**, and **flexor digitorum superficialis** (see Table 12.12). The flexor digitorum superficialis flexes the hand as well as the digits at the knuckles, which allows for rapid finger movements, as in typing or playing a musical instrument. The deep anterior muscles are the **flexor pollicis longus** and the **flexor digitorum profundus**.

The superficial posterior muscles include the **extensor radialis longus**, **extensor carpi radialis brevis**, **extensor digitorum**, **extensor digiti minimi**, and the **extensor carpi ulnaris**. The deep posterior muscles include the **abductor pollicis longus**, **extensor pollicis brevis**, **extensor pollicis longus**, and **extensor indicis**.

Figure 12.25 Movements of the Forearm, Wrist, and Fingers

(A) The elbow is capable of flexion and extension. Muscles on the anterior arm are flexors, and muscles on the posterior are extensors. (B) The radius and ulna are capable of pronation and supination. The movements of the wrist are (C) ulnar and radial deviation, (D) flexion and extension, and (E) pronation and supination. The movements of the fingers are (F) flexion, extension, and hyperextension, (G) abduction and adduction, and (H) circumduction.

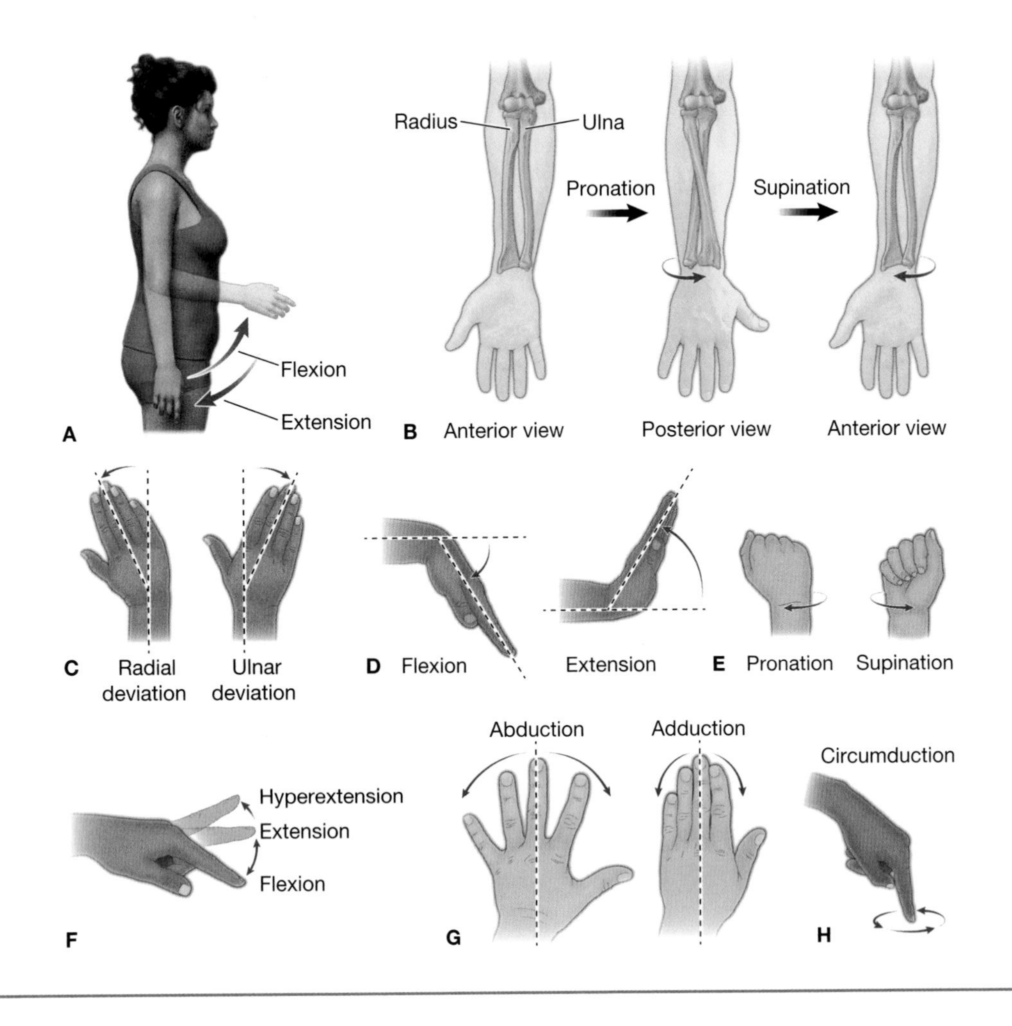

Intrinsic Muscles of the Hand The **intrinsic muscles of the hand** both originate and insert within it (**Figure 12.26** and see Table 12.12). These muscles allow your fingers to make precise movements, such as typing or writing. These muscles are divided into three groups. The **thenar** muscles are on the radial/thumb side of the palm, the **hypothenar** muscles are on the medial aspect of the palm, and the **intermediate** muscles are between them.

The thenar muscles include the **abductor pollicis brevis**, **opponens pollicis**, **flexor pollicis brevis**, and **adductor pollicis**. These muscles all act on the thumb. The movements of the thumb play an integral role in most precise movements of the hand.

The hypothenar muscles include the **abductor digiti minimi**, **flexor digiti minimi brevis**, and **opponens digiti minimi**. These muscles all act on the little finger. Finally, the intermediate muscles act on all the fingers and include the **lumbrical**, the **palmar interossei**, and the **dorsal interossei**.

Student Study Tip

To remember the term *lumbrical*, think of a lumberjack chopping wood and getting splinters in their hands and lumbricals.

Figure 12.26 **Intrinsic Muscles of the Hand**

The intrinsic muscles of the hand have origins and insertions within the hand. These muscles provide fine motor control of the fingers. (A and B) Palmar and dorsal views of the superficial muscles of the right hand. (C and D) Palmar and dorsal views of the interossei (interosseous muscles) of the right hand.

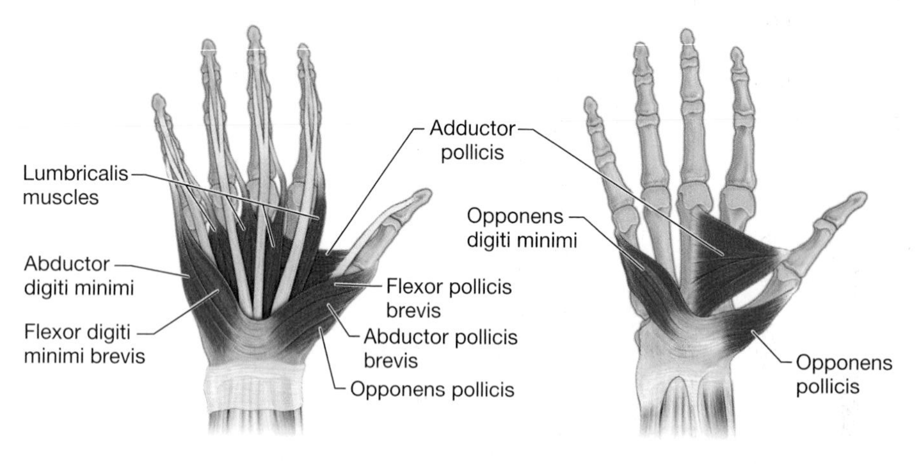

A Palmar view, superficial muscles of the right hand

B Palmar view, second superficial muscles of the right hand

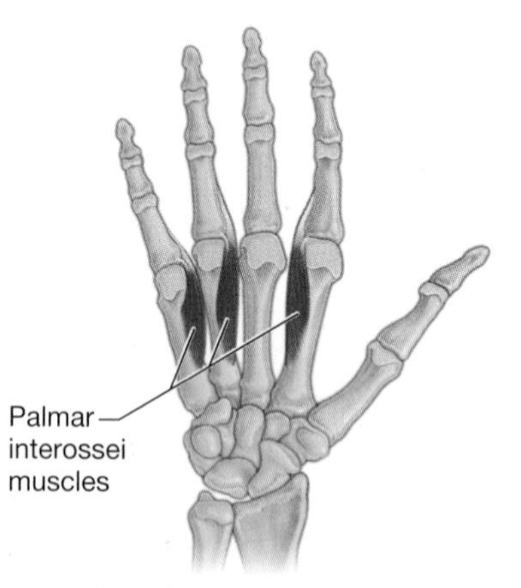

C Palmar view, interossei muscles of the right hand

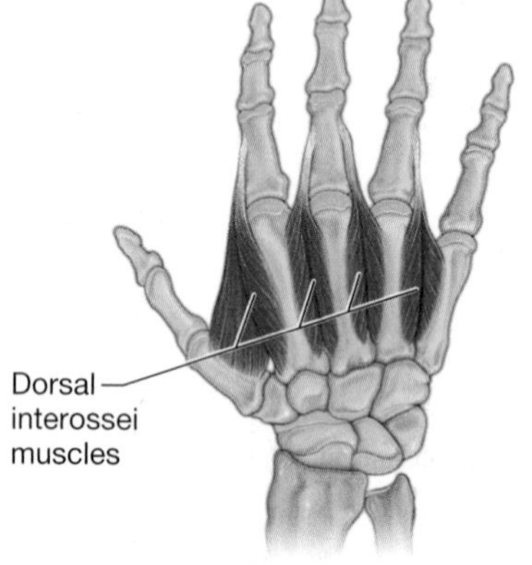

D Dorsal view, interossei muscles of the right hand

Comparatively, there is much more movement at the shoulder than at the hips. There is very little movement of the pelvis because of its connection with the sacrum at the base of the axial skeleton. The pelvis evolved a more restricted range of motion in humans because our bipedal nature requires stability and support of body weight during walking and running. Apes, which have a greater range of motion at their hip joints, use more energy to stabilize their torsos during bipedial walking; they are also less stable than humans when moving bipedally. What the leg muscles lack in range of motion and versatility, they make up for in size and power, facilitating the body's stabilization, posture, and movement. The motions of the hip are illustrated in **Figure 12.27**.

Learning Connection

Image Translation

Use your web browser to look up at least three common physical exercises (or use another source, like a friend who is an athlete). For each exercise image, name the muscles at work.

Gluteal Region Muscles That Move the Femur Most muscles that insert on and move the femur originate on the pelvis. The **psoas major** and **iliacus** together make up the **iliopsoas group**. Some of the largest and most powerful muscles in the body are the gluteal muscles. The **gluteus maximus** is the largest; deep to the gluteus maximus is the **gluteus medius**, and deep to the gluteus medius is the **gluteus minimus**, the smallest of the trio (**Figure 12.28** and **Table 12.13**).

Apply to Pathophysiology

Carpal Tunnel Syndrome

The carpal tunnel is a passageway for the tendons of the extrinsic muscles of the hand. They are protected posteriorly by the carpal bones and held in place anteriorly by a broad sheet of connective tissue, the flexor retinaculum. Almost all of the extrinsic muscle tendons pass through the carpal tunnel along with one nerve, the median nerve. If nerves are compressed, they can have trouble firing signals, leading to pain or numbness.

George is a 54-year-old administrative assistant. They spend much of their day typing emails for their boss. George has recently been diagnosed with Type II Diabetes. A few months ago, they began to feel periodic pain and tingling (pins and needles) in their right hand.

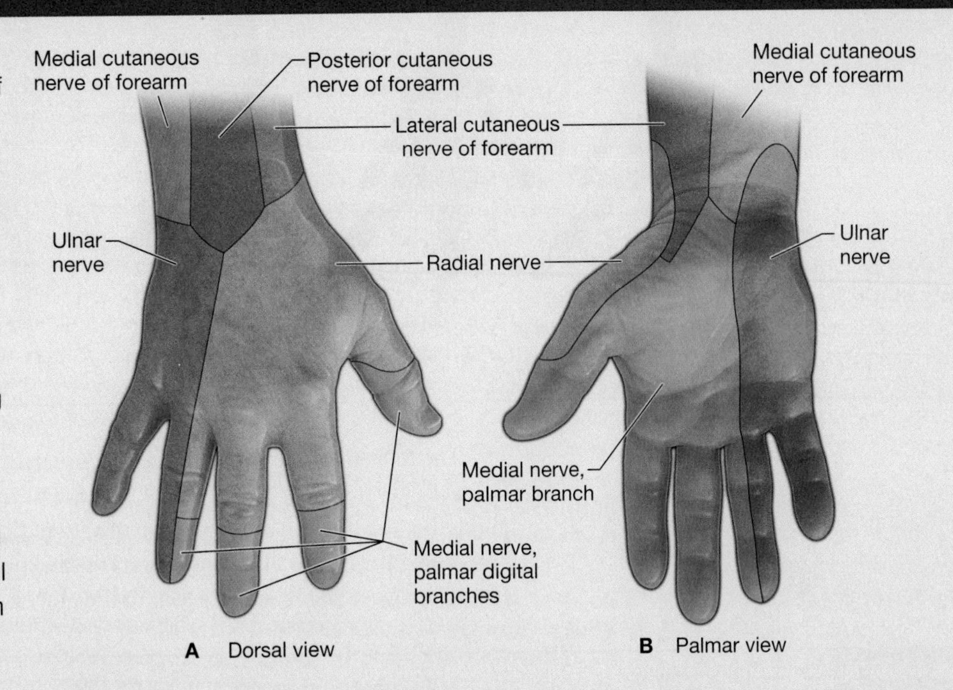

1. The image provided is a map of the sensory innervation of the hands. All of the sensation in each shaded region sends its sensory information to the brain through each of the nerves. If George really does have carpal tunnel syndrome, which of the regions shown would be the location of his pain and tingling?
   A) The thumb and index finger
   B) The back of the hand (dorsal) across the metacarpals
   C) The pinky finger
2. George goes to see their doctor, who indicates that a lot of computer users complain of carpal tunnel syndrome symptoms. What is the link between keyboard use and these symptoms?
   A) Holding the wrist in an extended position on the keyboard compresses the median nerve within the tunnel.
   B) Extension of the wrist during typing inflames the tendons of the wrist extensor muscles, putting pressure on the nearby median nerve.
   C) Extension of the wrist during typing inflames the tendons of the wrist extensor muscles, putting pressure on the nearby ulnar nerve.
3. George's doctor refers them to an orthopedist who immediately takes note of George's recent diabetes diagnosis. Diabetes can contribute to inflammation all over the body. In this case, what may be inflamed that would cause George's carpal tunnel syndrome?
   A) The extensor retinaculum
   B) The tendon sheaths of the extrinsic muscles of the hand
   C) The tendon sheaths of the intrinsic muscles of the hand
4. George is diagnosed with carpal tunnel syndrome. The doctor suggests a surgery called *carpal tunnel release*. This surgery cuts a ligament and has a high rate of relief of symptoms. Cutting which ligament would provide more space for the inflamed structure?
   A) Extensor retinaculum
   B) Interosseus membrane
   C) Flexor retinaculum
5. Over the months following the surgery, the ligament will heal as the cut ends reattach, usually allowing more space within the carpal tunnel. During the recovery period, the tendons within the tunnel are inflamed and the actions of their muscles become weaker. Which actions can George expect to be weaker?
   A) Moving the wrist laterally
   B) Extending the wrist
   C) Flexing the fingers, as in making a fist

Figure 12.27 Movements of the Hip

The hip joint has a limited range of motion compared to the shoulder. The leg can (A) abduct, (B) adduct, (C) flex, and (D) extend at the hip.

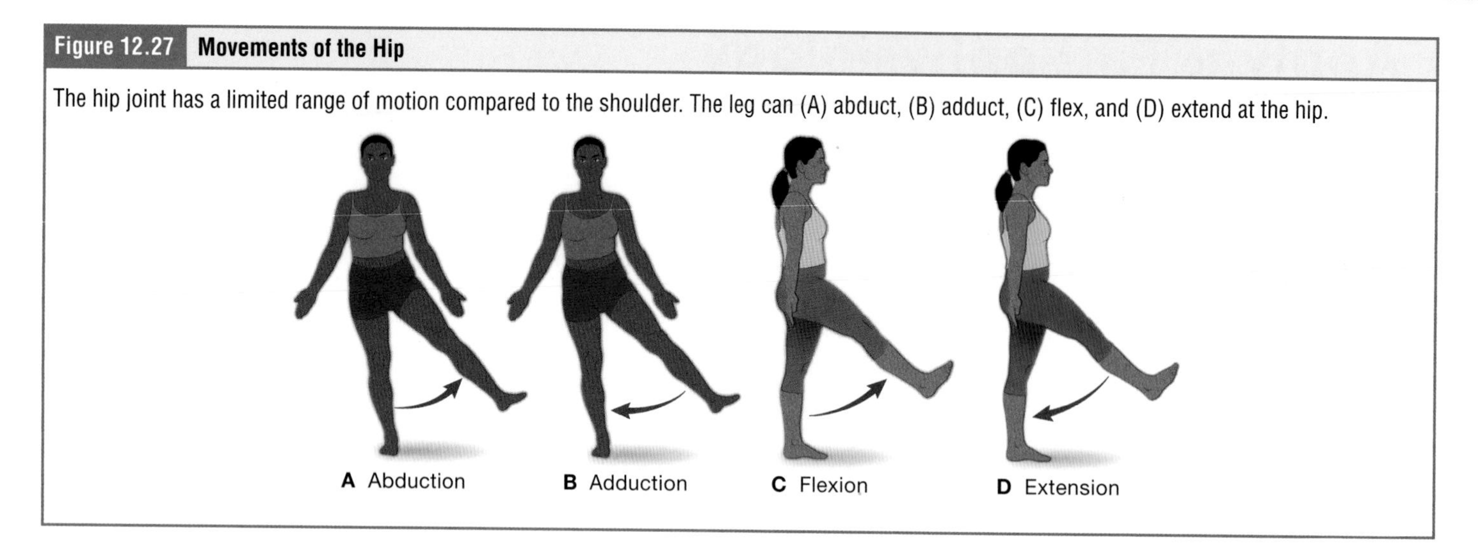

iliotibial tract (Iliotibial band, IT band)

The **tensor fasciae latae** is a thick, squarish muscle in the superior aspect of the lateral thigh. It acts as a synergist of the gluteus medius and iliopsoas in flexing and abducting the thigh. It sits at the top of a long, flat tendon called the **iliotibial tract** that runs along the lateral thigh and inserts at the knee. Pulling on the iliotibial tract contributes to knee stabilization. Deep to the gluteus maximus, the **piriformis**, **obturator**

Figure 12.28 Hip and Thigh Muscles

The powerful muscles of the hip originate on the pelvis, cross the hip joint, and insert on the femur. The muscles that flex and extend the knee originate on the femur and insert into the bones of the knee joint or lower leg. Anterior hip muscles provide flexion; posterior muscles provide extension; lateral muscles provide abduction, and medial muscles provide adduction. Knee flexors are on the posterior side, while extensors are on the anterior side. Pelvic and thigh muscles, right leg: (A) anterior view, deep muscles; (B) anterior view of the muscles, and (C) posterior view, superficial muscles.

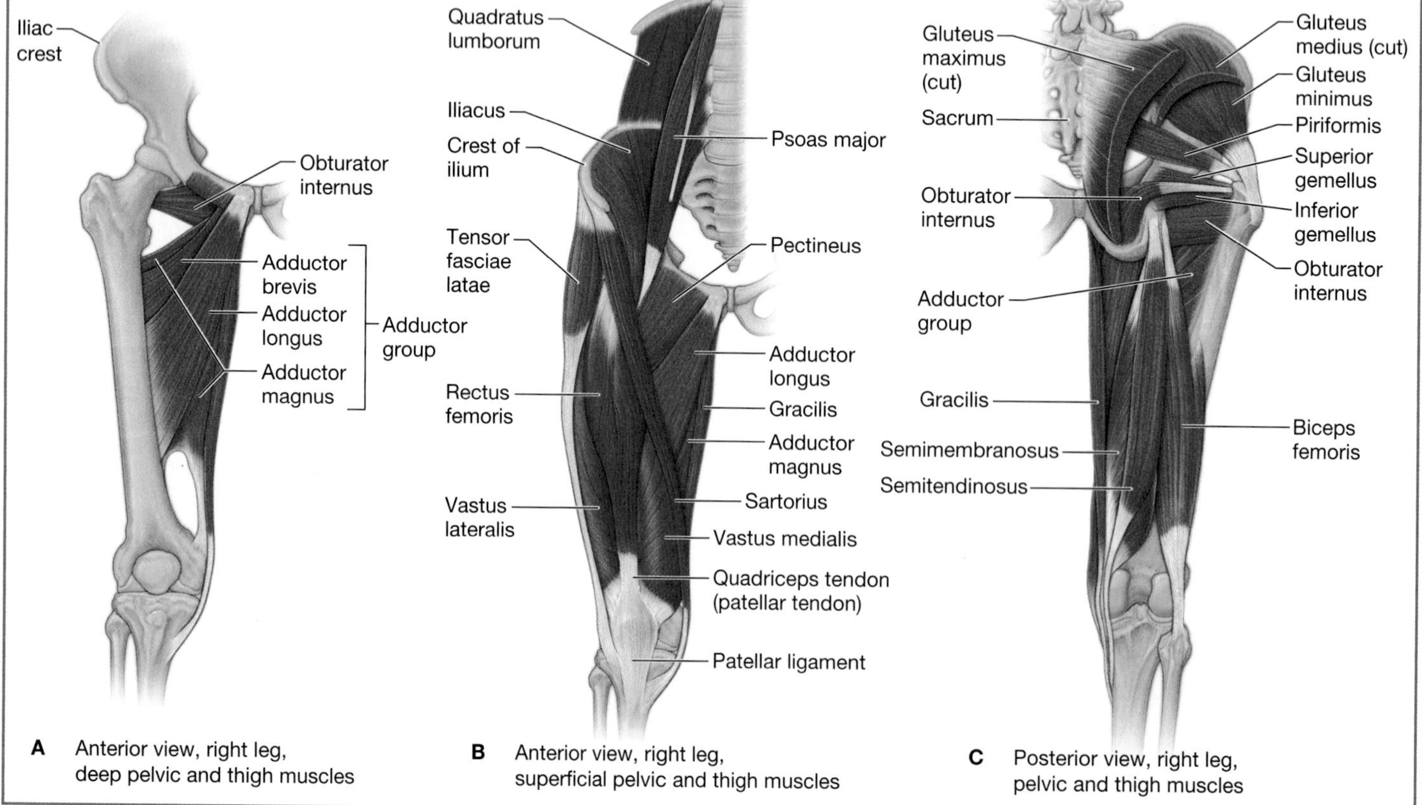

A Anterior view, right leg, deep pelvic and thigh muscles

B Anterior view, right leg, superficial pelvic and thigh muscles

C Posterior view, right leg, pelvic and thigh muscles

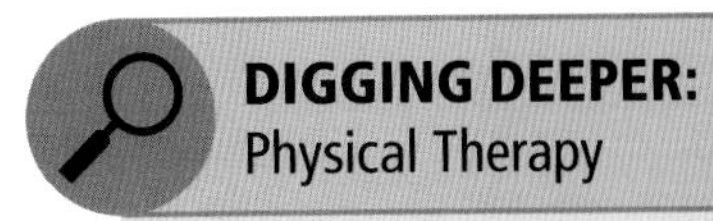

DIGGING DEEPER:
Physical Therapy

Physical therapists help their patients heal from injury or disease. They focus on movement-based therapeutic approaches. There are many types of physical therapy, including cardiopulmonary physical therapy for heart and lung disorders, orthopedic physical therapy for skeletal muscle and bone injury, pelvic floor physical therapy for problems of the pelvic floor or disease, and neurological physical therapy that focuses on recovery or improvement of symptoms during neurological diseases such as stroke or multiple sclerosis. Some physical therapists focus on one part of the lifespan, such as geriatric or pediatric (elderly patients and children).

Physical therapists often report enjoying their jobs because they develop relationships with their patients and see them improve on a day-to-day basis and over time. Since each person is different, physical therapists need to constantly use their problem-solving skills to figure out how to treat and where to progress the patient. Physical therapy can be absolutely essential for healing correctly from injury, surgery, or disease.

Physical therapy is more than just muscles; it is about movement as a whole and the complex anatomical magic that happens inside the body with movement. Movement is the difference between whether someone develops pneumonia in the hospital or not. It is one of the first lines of defense against hospital-induced delirium. And of course, it is the path any person takes to overcome musculoskeletal pathologies. All in all, physical therapy is a popular and rapidly expanding profession.

Learning Check

1. The abductor pollicis brevis is a muscle that originates from the carpal bones and inserts on the phalanges of the thumb. It is responsible for abducting the thumb, or moving it away from the rest of the hand. Is the abductor pollicis brevis muscle an intrinsic or extrinsic hand muscle?
   a. Intrinsic
   b. Extrinsic
2. Identify the actions of the deltoid. Please select all that apply.
   a. Abduction
   b. Adduction
   c. Flexion
   d. Lateral rotation
3. Which of the following rotator cuff muscles work in synergy to laterally rotate the shoulder? Please select all that apply.
   a. Supraspinatus
   b. Infraspinatus
   c. Teres minor
   d. Subscapularis
4. Describe the location of the flexor carpi radialis.
   a. A muscle that sits anteriorly and spans the wrist on the radial side.
   b. A muscle that sits anteriorly and spans the fingers on the radial side.
   c. A muscle that sits posteriorly and spans the wrist on the radial side.
   d. A muscle that sits posteriorly and spans the fingers on the ulnar side.

DIGGING DEEPER:
Muscle Changes

Like bones, muscles are incredibly dynamic. They can increase or decrease in size, length, and strength. Imagine, for example, a cast for a broken ulna that covers the elbow, holding it in a flexed position. If the cast stays on for a few months, the flexor compartment muscle cells and their sarcomeres will shorten and tighten, and consequently will have a hard time stretching out when the cast is removed. The extensor compartment, on the other hand, will be stretched throughout the time the cast is on. These sarcomeres and their muscle cells will lengthen over this time. These same principles apply to any positions held over time. Many Americans, for example, spend much of their days seated, with their legs flexed at the hip. Over time, the iliopsoas shortens and tightens. This can lead to a tilt in the pelvis, which then strains the lower back. Nearly 65 million Americans report back pain every year, many of them because of overly tight iliopsoas muscles.

Table 12.13 Muscles of the Hip and Thigh

Muscle	Description	Origin (O) and Insertion (I)	Action	Nerve Supply	You May Know This Muscle from...
Gluteus maximus	Largest and most superficial gluteal muscle	(O) Posterior sacrum (I) Posterior proximal femur, iliotibial tract	Prime mover of hip extension, medial rotation of the hip	Inferior gluteal nerve	Walking (you use this muscle every time you take a step), glute bridges
Gluteus medius	Intermediate layer of gluteal muscles	(O) Posterior ilium (I) Greater trochanter of the femur	Abduction and medial rotation of the hip	Superior gluteal nerve	Glute bridges
Gluteus minimus	Smallest and deepest gluteal muscle	(O) Posterior ilium (I) Greater trochanter of the femur	Abduction and medial rotation of the hip	Superior gluteal nerves	Glute bridges
Tensor fascia latae	Muscle belly that continues to the iliotibial tract, on the superior lateral aspect of the thigh	(O) Anterior ilium (I) Iliotibial tract	Stabilization of the knee, hip flexion, abduction, and medial rotation	Superior gluteal nerve	Many people get tight iliotibial bands and have to roll them out with a foam roller
Piriformis	Deep muscle and a short lateral rotator of the hip	(O) Anterior sacrum (I) Greater trochanter of the femur	Lateral rotation of the hip	Anterior rami S1–S2	
Obturator internus	Deep posterior muscle that is a part of the short lateral rotator group	(O) Posterior rim of the obturator foramen (I) Greater trochanter of the femur	Lateral rotation of the hip	Nerve to obturator internus	
Obturator externus	Deep anterior muscle that is a part of the short lateral rotator group	(O) Anterior rim of the obturator foramen (I) Greater trochanter of the femur	Lateral rotation of the hip	Obturator nerve	
Superior gemellus	A small and deep gluteal muscle assisting the hip external rotators	(O) Ischial spine (I) Greater trochanter of the femur	Lateral rotation and abduction of the hip	Nerve to obturator internus	
Inferior gemellus	A small and deep gluteal muscle assisting the hip external rotators	(O) Ischial tuberosity (I) Greater trochanter of the femur	Lateral rotation and abduction of the hip	Nerve to quadratus femoris	
Quadratus femoris	Deep posterior muscle part of the short lateral rotator group	(O) Ischium (I) Posterior proximal femur	Lateral rotation of the hip	Nerve to quadratus femoris	
Adductor longus	Anterior and medial muscle of the thigh, part of the adductor group	(O) Anterior pubis (I) Posterior and medial femur	Adduction and flexion of the hip	Obturator nerve	Lateral squats (when you feel the burn on the inside of your thigh, that's your adductors firing)
Adductor brevis	Anterior and medial muscle of the thigh, part of the adductor group	(O) Inferior pubis (I) Posterior and medial femur	Adduction and flexion of the hip	Obturator nerve	Lateral squats (when you feel the burn on the inside of your thigh, that's your adductors firing)
Adductor magnus	Deepest layer of the medial thigh, this muscle makes up the floor of the posterior thigh and has two heads	(O) Ischium and pubis (I) Posterior medial femur and distal medial femur	Adduction and extension of the hip	Obturator and sciatic nerves	
Pectineus	Short and deep adductor muscle in the anterior thigh	(O) Superior pubis (I) Proximal and medial femur	Adduction and flexion of the hip	Femoral nerve	

continued

Cultural Connection

Table 12.13 (*continued*)

Muscle	Description	Origin (O) and Insertion (I)	Action	Nerve Supply	You May Know This Muscle from...
Gracilis	Most medial muscle of the thigh, straplike, and attaching at the knee joint	(O) Pubis (I) Proximal tibia	Adduction of the hip and flexion of the knee	Obturator nerve	
Rectus femoris	Anterior compartment of the thigh, part of the quadriceps group (this is the only muscle of the group that also acts on the hip)	(O) Anterior ilium (I) Tibia via quadriceps tendon	Knee extension and hip flexion	Femoral nerve	Squats
Vastus lateralis	Anterior compartment of the thigh, part of the quadriceps group	(O) Proximal posterior femur and greater trochanter (I) Tibia via quadriceps tendon	Knee extension	Femoral nerve	Squats
Vastus medialis	Anterior compartment of the thigh, part of the quadriceps group	(O) Proximal posterior femur (I) Tibia via quadriceps tendon	Knee extension	Femoral nerve	Squats
Vastus intermedius	Anterior compartment of the thigh, part of the quadriceps group. This muscle is deep to rectus femoris	(O) Anterior and lateral femur (I) Tibia via quadriceps tendon	Knee extension	Femoral nerve	Squats
Sartorius	Anterior and superficial thigh muscle performing similar actions on both the hip and knee	(O) Anterior ilium (I) Proximal medial tibia	Flexion, abduction, and lateral rotation of the hip, flexion and medial rotation of the knee	Femoral nerve	Sitting cross-legged
Biceps femoris	Posterior thigh muscle with two heads, part of the hamstring group	(O) Ischium and posterior femur (I) Lateral surfaces of tibia and fibula	Knee flexion and hip extension	Sciatic nerve	Hamstring curls, feeling a stretch while touching your toes
Semitendinosus	Posterior superficial thigh muscle, part of the hamstring group	(O) Ischium (I) Proximal medial tibia	Knee flexion and hip extension	Sciatic nerve	Hamstring curls, feeling a stretch while touching your toes
Semimembranosus	Posterior thigh muscle, part of the hamstring group and deep to semitendinosus	(O) Ischium (I) Posterior and medial tibia	Knee flexion and hip extension	Sciatic nerve	Hamstring curls, feeling a stretch while touching your toes

internus, **obturator externus**, **superior gemellus**, **inferior gemellus**, and **quadratus femoris** laterally rotate the femur at the hip.

The **adductor longus**, **adductor brevis**, and **adductor magnus** can both medially and laterally rotate the thigh depending on the placement of the foot. Whereas the adductor longus flexes the thigh, the adductor magnus extends it. The **pectineus** adducts and flexes the femur at the hip as well. The pectineus is located in the **femoral triangle**, which is formed at the junction between the hip and the leg and also includes the femoral nerve, the femoral artery, the femoral vein, and the deep inguinal lymph nodes.

Student Study Tip

Since the tensor fasciae latae attaches to the iliotibial tract, remember its name by thinking of how it pulls the iliotibial tract and "tenses the fascia a lot during contraction."

Thigh Muscles That Move the Femur, Tibia, and Fibula Deep fascia in the thigh separates it into medial, anterior, and posterior compartments (see Figure 12.28). The muscles in the **medial compartment of the thigh** are responsible for adducting the femur at the hip. Along with the adductor longus, adductor brevis, adductor magnus, and pectineus, the straplike **gracilis** adducts the thigh in addition to flexing the knee.

Cultural Connection
Sartorius Muscle
The name of the sartorius muscle comes from the Latin word *sartor*, which means "tailor." Historically, before the invention of sewing machines, tailors used to sit in a cross-legged position atop tables—sometimes in groups—to do their work, as shown in Figure 12.28B. The sartorius muscle crosses both the hip and the knee, flexing both joints. It is rare for muscles in the body to act on two joints at the same time, so its action—to produce the dual flexion required to sit cross-legged—earned the sartorius muscle its name.

The muscles of the **anterior compartment of the thigh** flex the thigh and extend the knee. This compartment contains the **quadriceps femoris group**, which actually consists of four muscles that extend and stabilize the knee. The **rectus femoris** is on the anterior aspect of the thigh, the **vastus lateralis** is on the lateral aspect of the thigh, the **vastus medialis** is on the medial aspect of the thigh, and the **vastus intermedius** is between the vastus lateralis and vastus medialis and deep to the rectus femoris. All four muscles converge onto a common tendon, the **quadriceps tendon**, which blankets the anterior knee and inserts on the tibial tuberosity. Suspended within the quadriceps tendon is the patella, a sesamoid bone that divides the tendon in half. The inferior half, between the patella and the tibia, is the **patellar ligament**. In addition to the quadriceps femoris, the **sartorius** is a bandlike muscle that extends from the anterior superior iliac spine to the medial side of the proximal tibia. This versatile muscle flexes the knee and flexes, abducts, and laterally rotates the thigh at the hip. The sartorius crosses and acts at both the hip and the knee; it allows us to sit cross-legged.

sartorius ("Tailor's" muscle)

The **posterior compartment of the thigh** includes muscles that flex the knee and pull the thigh posteriorly, extending the hip. The three long muscles on the back of the knee are the **hamstring group**, which flexes the knee. These are the **biceps femoris**,

Apply to Pathophysiology

Knee Injury

Mimi is a 16-year-old female soccer player. She was dribbling the ball toward the goal when a player on the opposite team came up on her left side and kicked at the ball. The player accidentally struck Mimi's lower leg from the side and Mimi collapsed as pain shot through her knee. The team trainer helped her off the field and took her to the hospital for X-rays and an MRI. The doctors came back with some tough news: Mimi had torn her ACL, a critical ligament within her knee. Her doctors will replace the torn ligament surgically and use a strip from the medial side of Mimi's own quadriceps tendon as the replacement tissue.

1. Borrowing tissue from the quadriceps tendon will weaken the muscles' action. Which of these actions will be weakened?
   A) Extending the knee
   B) Flexing the knee
   C) Flexing the hip
   D) All of these
2. A year after the surgery, Mimi goes back to the surgeon for a follow-up. She has been experiencing a lot of pain right at her patella, her kneecap. When the doctor examines her, they are looking for signs that Mimi's patella is beginning to shift to one side, away from the middle of the knee; this can be a side effect to the surgery. Which side is the patella more likely to shift toward?
   A) Lateral
   B) Medial
3. In order to restore Mimi's patella to the center of the knee, the strength of which two muscles should be in balance?
   A) Rectus femoris and vastus medialis
   B) Rectus femoris and vastus intermedius
   C) Vastus lateralis and vastus medialis

semitendinosus, and **semimembranosus**. The tendons of these muscles form the **popliteal fossa**, the diamond-shaped space at the back of the knee.

> **Student Study Tip**
>
> The tendon for semitendinosus is the most prominent hamstring tendon and can be easily palpated. Use this to help remember the name *semi"tendin"osus.*

12.4e Muscles That Move the Feet and Toes

Similar to the thigh muscles, the muscles of the lower leg are divided by deep fascia into compartments, although the leg has three: anterior, lateral, and posterior (**Figure 12.29**).

The muscles in the anterior compartment of the lower leg are the **tibialis anterior** (a long and thick muscle on the lateral surface of the tibia), the **extensor hallucis longus**, deep under it, and the **extensor digitorum longus**, lateral to it; all of these contribute to raising the front of the foot when they contract. Also found in this compartment is the **fibularis tertius**, a small muscle that originates on the anterior surface of the fibula; this is associated with the extensor digitorum longus and sometimes fused to it, but is not present in all people. The **superior extensor retinaculum** and the **inferior extensor retinaculum**, thick bands of connective tissue, hold the tendons of these muscles in place during dorsiflexion.

> **fibularis tertius** (peroneus tertius)

Figure 12.29 Muscles of the Lower Leg

The muscles of the lower leg act on the ankle and foot. (A) The anterior compartment muscles are dorsiflexors, and (B) the muscles of the posterior compartment are plantar flexors. (C) The lateral and medial muscles invert, evert, and rotate the foot.

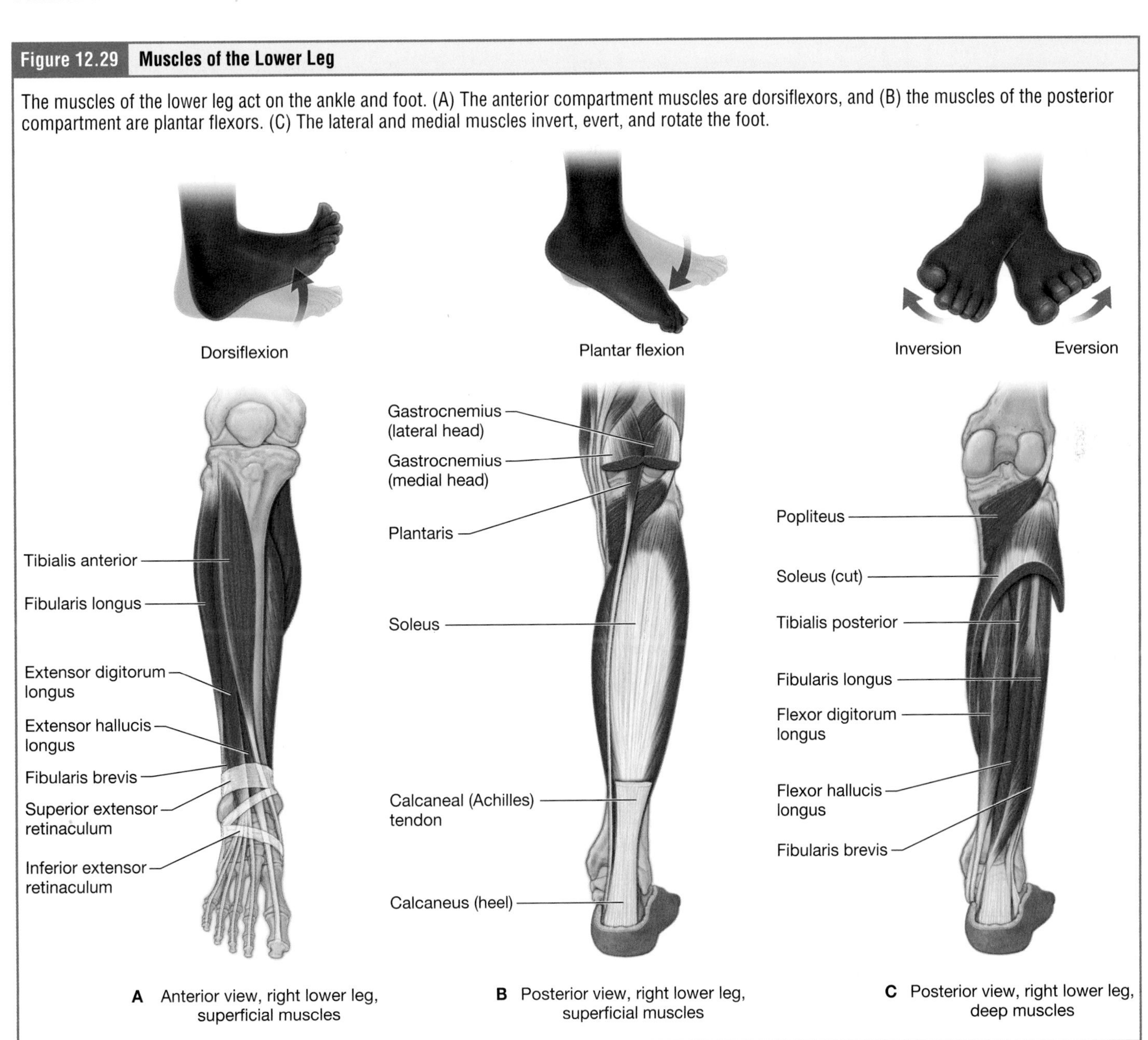

fibularis longus (peroneus longus)

fibularis brevis (peroneus brevis)

calcaneal tendon (Achilles tendon)

tibialis posterior (posterior tibialis)

The lateral compartment of the lower leg includes two muscles, the **fibularis longus** and the **fibularis brevis**; these muscles function in eversion and plantar flexion of the foot. The superficial muscles in the posterior compartment of the lower leg all insert onto the **calcaneal tendon**, a strong tendon that inserts into the calcaneal bone of the ankle. The muscles in this compartment are large and strong and keep humans upright. The most superficial and visible muscle of the calf is the **gastrocnemius**. Deep to the gastrocnemius is the wide, flat **soleus**. The **plantaris** runs diagonally between these two. The plantaris is a site of anatomical variation; some people have no plantaris, most people have one plantaris, and some individuals have two. In addition, there are four other deep muscles in the posterior compartment of the leg: the **popliteus**, the **flexor digitorum longus**, the **flexor hallucis longus**, and the **tibialis posterior**.

Like the hand, the foot also has intrinsic muscles, which originate and insert within. These muscles primarily provide support for the foot and its arch and contribute to movements of the toes (**Figure 12.30** and **Table 12.14**). The sole of the foot is

Figure 12.30 Intrinsic Muscles of the Foot

The muscles of the sole flex the toes. There are three layers of plantar muscles: (A) superficial, (B) intermediate, and (C) deep. (D) The muscles along the dorsal side of the foot extend the toes.

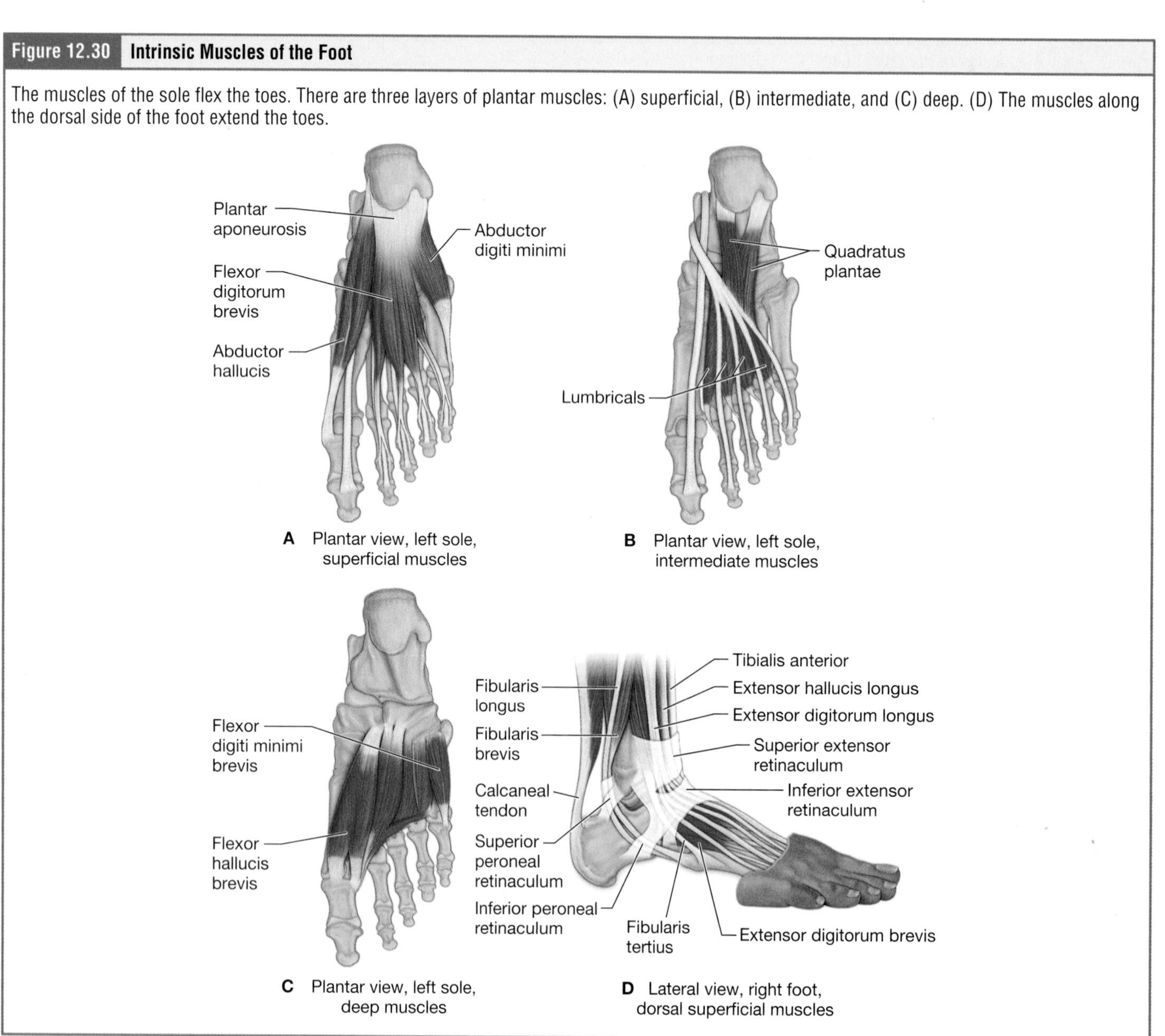

Table 12.14 Muscles of the Leg and Foot

Muscle	Description	Origin (O) and Insertion (I)	Action	Nerve Supply	You May Know This Muscle from...
Tibialis anterior	Anterior lower leg, overlaying the lateral portion of the tibia	(O) Lateral condyle and upper portion of the shaft of the tibia (I) Tendon inserts on the inferior surface of medial cuneiform and first metatarsal	Prime mover for dorsiflexion, inversion of the foot	Deep fibular nerve	Shin splints (a pain along the shin associated with running) are inflammation of the tibialis anterior
Extensor hallucis longus	Extrinsic muscle in the anterior compartment of the lower leg	(O) Anterior shaft of the fibula (I) Distal phalanx of the great toe	Extension of the great toe	Deep peroneal nerve	
Extensor digitorum longus	Extrinsic muscle in the anterior compartment of the lateral lower leg	(O) Proximal anterior shaft of the fibula (I) Extensor expansion of toes 2–5	Extension of toes 2–5 and dorsiflexion of the ankle	Deep peroneal nerve	Seeing visible tendons on the top of your foot when lifting your toes off the ground
Fibularis tertius	Small and deep muscle on the lateral lower leg whose tendon is visible but whose muscle belly is not in a superficial view	(O) Distal anterior fibula (I) Base of fifth metatarsal	Eversion of the foot and dorsiflexion of the ankle	Deep peroneal nerve	
Fibularis longus	Superficial muscle of the lateral compartment of the lower leg	(O) Proximal and lateral shaft of the fibula (I) First metatarsal and medial cuneiform	Eversion of the foot and plantarflexion of the ankle	Superficial peroneal nerve	
Fibularis brevis	Muscle of the lateral compartment of the lower leg, deep to fibularis longus	(O) Distal and lateral shaft of the fibula (I) Lateral surface of the fifth metatarsal	Eversion of the foot and plantarflexion of the ankle	Superficial peroneal nerve	
Soleus	Wide and flat intermediate muscle of the posterior lower leg	(O) Posterior tibia and proximal posterior fibula (I) Calcaneus via the calcaneal tendon	Plantarflexion of the ankle	Tibial nerve	When standing on your feet, this is the muscle that contracts to keep postural stability in your lower leg
Gastrocnemius	Large and powerful superficial muscle with two heads, found in the posterior compartment of the lower leg	(O) Medial and lateral epicondyles of the femur (I) Calcaneus via the calcaneal tendon	Prime mover for plantarflexion of the ankle and flexion of the knee	Tibial nerve	Calf raises
Plantaris	Very thick, tendinous muscle running between the soleus and gastrocnemius	(O) Lateral supracondylar ridge of the femur (I) Posterior calcaneus	Plantarflexion of the ankle and flexion of the knee	Tibial nerve	
Popliteus	Small muscle behind the knee joint in the deep layer of the posterior lower leg	(O) Lateral condyle of the femur (I) Proximal posterior tibia	Medial rotation of the tibia and lateral rotation of the femur to assist in knee flexion	Tibial nerve	
Flexor digitorum longus	Deep and extrinsic muscle of the lower leg, tendons run medially	(O) Posterior tibia (I) Distal phalanges of toes 2–5	Flexion of toes 2–5 and plantarflexion of the ankle	Tibial nerve	

continued

Table 12.14 (*continued*)

Muscle	Description	Origin (O) and Insertion (I)	Action	Nerve Supply	You May Know This Muscle from...
Flexor hallucis longus	Deep and extrinsic muscle of the lower leg, tendons run medially	(O) Distal posterior fibula (I) Distal phalanx of great toe	Flexion of the great toe	Tibial nerve	
Tibialis posterior	Deep muscle in the posterior leg running along the surface of the tibia	(O) Posterior tibia and proximal posterior fibula (I) Plantar surfaces of tarsals and metatarsals	Inversion of the foot and plantarflexion of the ankle	Tibial nerve	
Extensor digitorum brevis	Intrinsic muscle of the foot on the dorsal surface	(O) Anterior and lateral calcaneus (I) Extensor expansion of toes 1–4	Extension of toes 1–4	Deep peroneal nerve	

Student Study Tip

You may have heard of plantar fasciitis, which is inflammation of the plantar fascia or plantar aponeurosis.

supported by a thick sheet of connective tissue called the **plantar aponeurosis**, which runs from the calcaneus bone to the toes. The intrinsic muscles of the foot consist of two groups: the **dorsal group** includes only one muscle, the **extensor digitorum brevis**. The second group is the **plantar group**, which consists of three layers: superficial, intermediate, and deep.

Learning Check

1. Which of the following describes the location of the gluteal region muscles from superficial to deep?
   a. Gluteus maximus, gluteus medius, gluteus minimus
   b. Gluteus medius, gluteus minimus, gluteus maximus
   c. Gluteus minimus, gluteus medius, gluteus, maximus
   d. Gluteus maximus, gluteus minimus, gluteus medius
2. Which of the following contribute to medial rotation of the femur at the hip? Please select all that apply.
   a. Piriformis
   b. Quadratus femoris
   c. Adductor magnus
   d. Iliacus
3. Which of the following actions can the sartorius produce? Please select all that apply.
   a. Hip extension
   b. Knee flexion
   c. Knee adduction
   d. Lateral rotation of the femur
4. You are sitting on the bus listening to music. You hear your favorite song and can't help but tap your feet to the beat. Which of the following muscles allow you to lift your foot off the ground so you can tap your feet?
   a. Tibialis anterior
   b. Popliteus
   c. Gastrocnemius
   d. Posterior tibialis

Chapter Review

12.1 Interactions of Skeletal Muscles, Their Fascicle Arrangement, and Their Lever Systems

The learning objectives for this section are:

12.1.1 For a given movement, differentiate specific muscles that function as prime mover, antagonist, synergist, or fixator.

12.1.2 Define the terms prime mover (agonist), antagonist, synergist, and fixator.

12.1 Questions

1. The flexor carpi radialis and extensor carpi radialis are muscles that produce radial deviation at the wrist. Which of the following terms best describes these muscles as they contribute to radial deviation? Please select all that apply.
   a. Synergist
   b. Agonists
   c. Antagonists
   d. Prime mover

2. A muscle that contributes to an action, but is not the main muscle for that action, is called a(n):
   a. Fixator
   b. Antagonist
   c. Agonist
   d. Synergist

12.1 Summary

- A muscle runs from the origin to the insertion.
- Muscles that produce the main action are prime movers or agonists. The muscles that produce the opposite action of the main action are antagonists. Muscles that produce the same action are called *synergists*.
- Muscle fascicles are organized in many different ways that explain the amount of force generated by the muscle and the range of motion available at the joint. These arrangements are parallel, circular, and pennate muscle fibers.

12.2 Naming Skeletal Muscles

The learning objective for this section is:

12.2.1 Explain how the name of a muscle can help identify its action, appearance, or location.

12.2 Question

1. Identify the following muscles that are named based on their location. Please select all that apply.
   a. Trapezius
   b. Intercostals
   c. Posterior tibialis
   d. Quadriceps

12.2 Summary

- The names of muscles can provide important information about their location, action, or appearance.
- Muscles can be named for their size, shape, fiber orientation, action, location, or grouping.
- Some muscles are named using two or more naming conventions.

12.3 Axial Muscles

The learning objectives for this section are:

12.3.1 Identify the location, general attachments, and actions of the major skeletal muscles.

12.3.2 Describe similar actions (functional groupings) of muscles in a particular compartment (e.g., anterior arm) or region (e.g., deep back).

12.3 Questions

1. What is the action of the masseter muscle?
   a. Elevates the mandible
   b. Depresses the mandible
   c. Retracts the mandible
   d. Protracts the mandible

2. Which of the following muscles work in synergy to flex the neck? Please select all that apply.
   a. Sternocleidomastoid
   b. Scalenes
   c. Sternothyroid
   d. Splenius

12.3 Summary

- Muscles of the axial region are found in the head and trunk of the body.
- Axial muscles contribute to facial expression, movement of the head, neck and abdomen, breathing, and support of the pelvic floor.
- Some axial muscle cross over to the appendicular skeleton.

12.4 Appendicular Muscles

The learning objectives for this section are:

12.4.1 Identify the location, general attachments, and actions of the major skeletal muscles.

12.4.2 Describe similar actions (functional groupings) of muscles in a particular compartment (e.g., anterior arm) or region (e.g., deep back).

12.4 Questions

Mini Case 1: Spinal Cord Injury

Emory was involved in a traumatic car accident in which they injured their spinal cord. This resulted in their losing function in their legs and most of their abdomen. Because of this, Emory has to work with Physical Therapists for functional training.

The first task Emory is working on is transferring from a wheelchair to their bed. To do so, they slide a sliding board under their hips, and push themself up and to the side with their arm.

1. Which of the following muscles does Emory use to push themself up and to the side?
   a. Biceps brachii
   b. Triceps
   c. Pectoralis major
   d. Quadriceps

The next task is bed mobility, where Emory has to learn to roll in bed. Since Emory does not have trunk control, their rolling has to be initiated by their arms. In order to do so, Emory has to swing their arms over their body, and use that momentum to grab onto the side railings of the bed, then pull the rest of the body so they are lying on their side.

2. Which of the following muscles would allow Emory to grip the side rails? Please select all that apply.
   a. Flexor pollicis longus
   b. Extensor pollicis brevis
   c. Brachialis
   d. Fibularis longus

12.4 Summary

- Appendicular muscles are found in the upper and lower limbs and around the joints that connect the limbs to the axial skeleton.
- Appendicular muscles are responsible for movements of the arm, forearm, hands, fingers, legs, feet and toes.
- Some appendicular muscles stabilize and move the shoulder and hip.

13

The Nervous System and Nervous Tissue

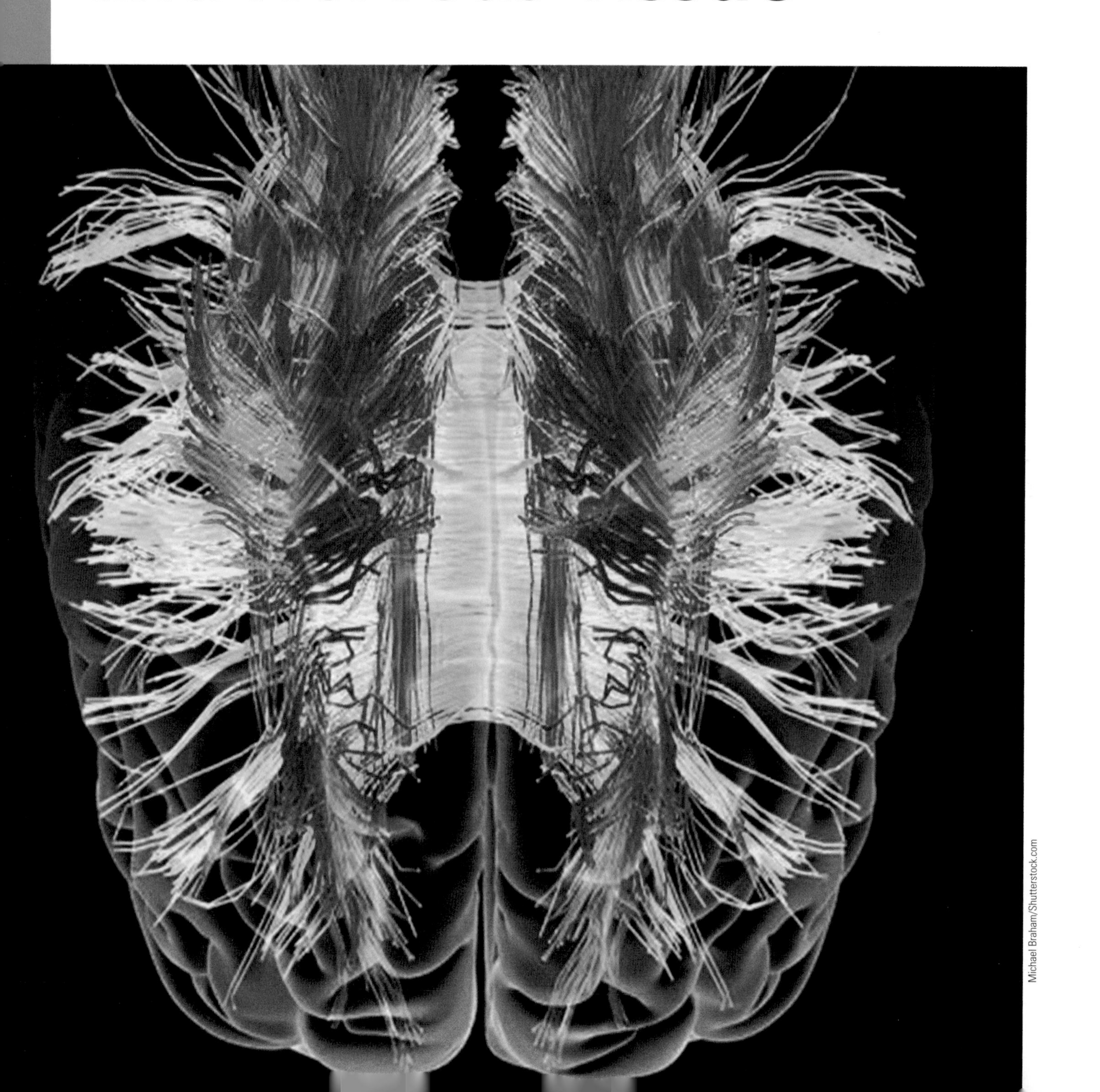

Chapter Introduction

In this chapter we will learn...

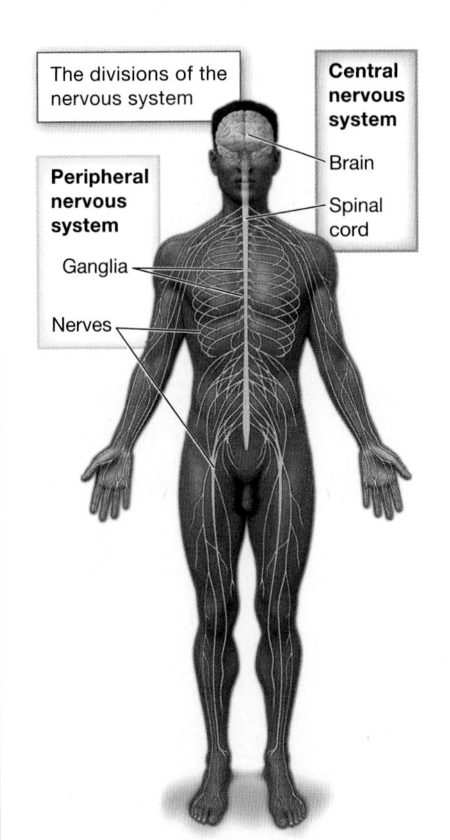

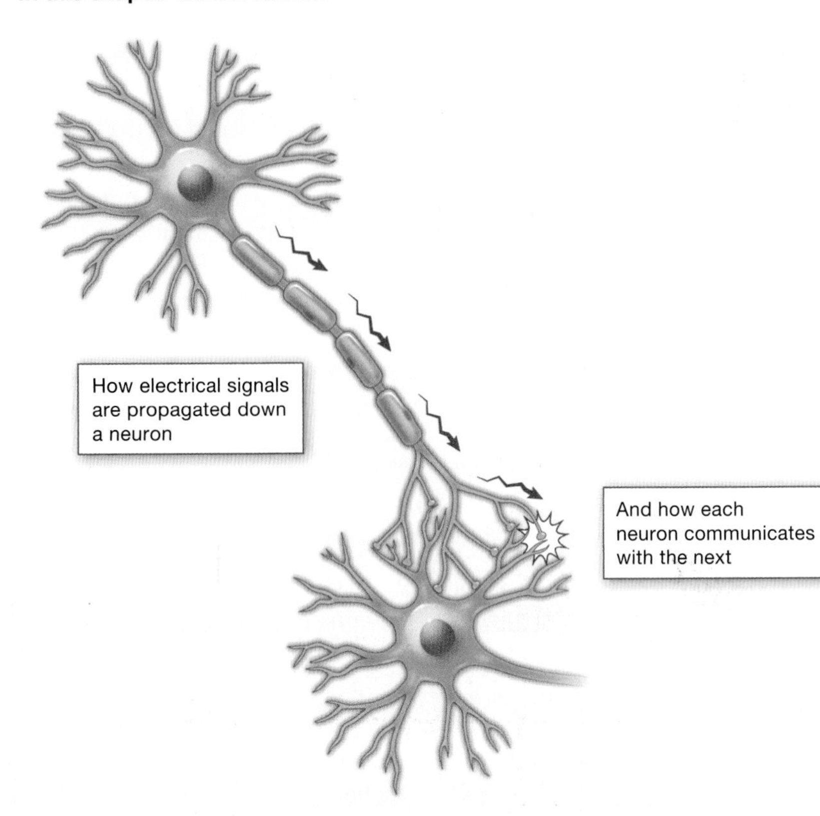

The nervous system is vast, complex, and largely still a mystery to us. There are so many human functions we still do not understand. Because the brain only functions while intact and within the body, the study of the nervous system has been challenging for most of scientific history. It is only recently that scientists have begun to make significant strides in advancing our understanding. The nervous system can be considered from a zoomed-out view, as a functioning organ system, but much of our coverage will be on the main organ—the brain—as well as on the micro view of a single nervous system cell, the neuron. One of our early goals in understanding this system is to gain versatility in our language and conceptual understanding to be able to see how these views fit together.

13.1 Organization and Functions of the Nervous System

Learning Objectives: By the end of this section, you will be able to:

13.1.1 Describe the general functions of the nervous system.

13.1.2 Compare and contrast the central nervous system (CNS) and the peripheral nervous system (PNS) with respect to structure and function.

13.1.3 Differentiate between the motor (efferent) and sensory (afferent) components of the nervous system.

The Human Anatomy and Physiology Society includes more than 1,700 educators who work together to promote excellence in the teaching of this subject area. The HAPS A&P Learning Outcomes measure student mastery of the content typically covered in a two-semester Human A&P curriculum at the undergraduate level. The full Learning Outcomes are available at https://www.hapsweb.org.

13.1.4 Describe the nervous system as a control system with the following components: sensory receptors, afferent pathways, control (integrating) center, efferent pathways, and effector (target) organs.

13.1.5 Compare and contrast the somatic motor and autonomic motor divisions of the nervous system.

13.1.6 Compare and contrast the somatic sensory and visceral sensory divisions of the nervous system.

Consider the perspective of early anatomists. The **brain**, the nervous tissue contained within the skull, appears as a dull gray, solid mass in the skull, attached to the long, solid **spinal cord**, the extension of nervous tissue within the vertebral column. These organs, and the branching nerves they attach to, appear simple. Over time, scientists have discovered that the nervous system actually has a very complex structure. Within the brain, many different and separate regions are responsible for many different and separate functions. It is as if the nervous system is composed of many organs that all look similar and can only be differentiated using modern scientific tools. In comparison, it is easy to see that the stomach is different than the esophagus or the liver, so you can imagine the digestive system as a collection of specific organs.

13.1a The Functions of the Nervous System

LO 13.1.1

The nervous system is involved in receiving information about the environment around us (sensation) and generating responses to that information (motor responses). When we consider this vast and complex organ system, we can view it functionally or structurally. Functionally, the nervous system can be divided into regions that are responsible for **sensation** (sensory functions) and for the **response** (motor functions). But there is a third function that needs to be included. Sensory input needs to be integrated with other sensations, as well as with memories, emotional state, or learning (cognition). Some regions of the nervous system are termed **association areas**. The process of integration combines sensory perceptions and higher cognitive functions such as memories, learning, and emotion to produce a response. For example, consider that a chocolate chip cookie (or another favorite treat) is in front of you. As you smell and see the cookie (sensation), your association areas are comparing the incoming sensory information to a vast library of past experiences. Your association areas allow for the specific recognition of this item as a cookie, and that it is something that you enjoy. You then access other areas of your brain involved in decision-making. Is this a good time to eat a cookie? Is it in your best interest to eat the cookie? Are you hungry? Should you save half of the cookie for a friend? Eventually the response regions of your brain coordinate the motor function to reach out, grab the cookie, take a bite, chew, and swallow.

We most often think of the sensation-integration-response cycle through the lens of our conscious perception and voluntary responses as described in the preceding cookie example. However, a good amount of the sensory information received by the brain never results in conscious awareness. Can you, for example, sense the shift in blood pressure that occurs when you move from a seated to a standing position? No? But your brain received information about that change in blood pressure, integrated it and sent responses to change the beating of the heart and the contraction of smooth muscle in the blood vessel walls. The conscious versus unconscious sensation and voluntary versus involuntary responses are controlled by different branches of the nervous system; these are discussed in the next section.

13.1b The Central and Peripheral Nervous Systems

LO 13.1.2

One way to divide the structures of the nervous system is anatomically. The two major anatomical regions are the central and peripheral nervous systems. The **central nervous system (CNS)** consists of the brain and spinal cord. From these two structures, many nerves branch off, connecting the brain and spinal cord to the far reaches of the body. All the nerves make up the **peripheral nervous system (PNS)**. The PNS is everything else besides the brain and spinal cord (**Figure 13.1**). The brain is contained within the cranial cavity of the skull, and the spinal cord is protected within vertebral cavity of the vertebral column. The peripheral nerves, for the most part, are found outside of bony protection, but instead traverse through muscles and near or within the walls of organs.

Both the CNS and PNS are composed almost entirely of nervous tissue, which contains two basic types of cells: neurons and glial cells. **Neurons** are the cells capable of generating communication between different locations. For example, if you feel a bug land on your skin, the conscious awareness in the brain is possible because communication about the sensation was sent along a chain of neurons beginning at the skin under the bug and ending in a region of the brain dedicated to interpreting sensory information for that location. Early anatomists and physiologists discovered the functionality of neurons, but they did not understand what glial cells contributed to the function of the nervous system. They used the word *glia*, from the Greek term for glue, to describe these cells which do not perform communication tasks within the nervous system. As we now understand, there are several varieties of glial cells, but as a group each **glial cell** provides structure and support to the neurons and their activities.

glial cell (neuroglial cells)

LO 13.1.3

While the central and peripheral nervous systems' divisions are strictly anatomical, the nervous system can also be divided by its functions. There are two ways to consider

Figure 13.1 The Divisions of the Nervous System

The vast human nervous system can be subdivided anatomically or functionally.

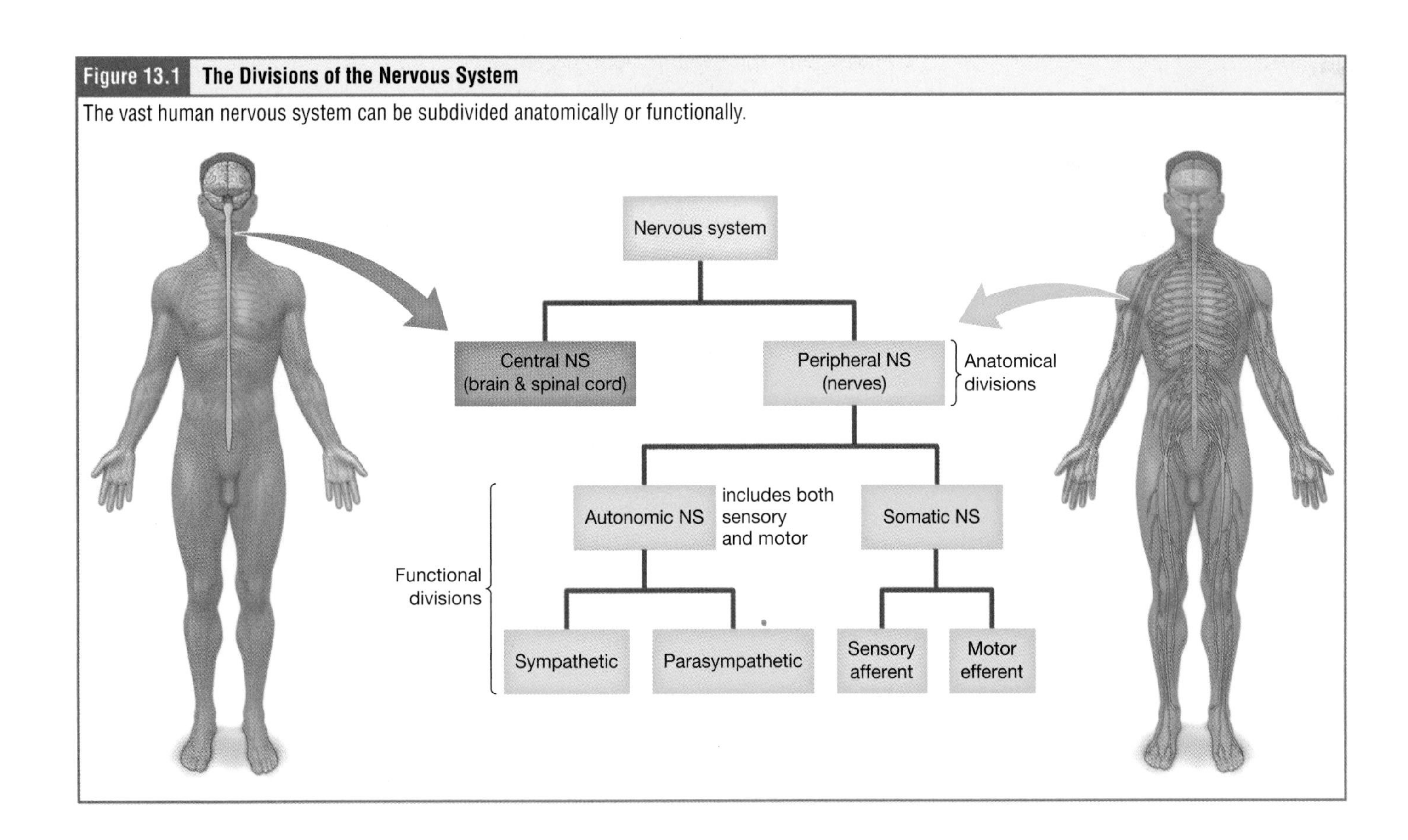

Student Study Tip

Efferent and exiting both start with an "E," and toward has an "A" in it.

how the nervous system is divided functionally. First, the basic functions of the nervous system are sensation, integration, and response (**Figure 13.2**). The sensory division sends information toward the integration location, and responses are sent out. Integration occurs in the brain and spinal cord, and the neurons that bring sensory information here are sensory or afferent neurons. The responses are carried along motor or efferent neurons (see Figure 13.1). The terms *afferent* and *efferent* are anatomical directional terms that are used throughout the body. **Afferent** means "to" or "toward" and **efferent** means "away" or "exiting." The efferent signal is sent to an **effector**, an organ or muscle that responds.

LO 13.1.4

LO 13.1.5

The last system for dividing the nervous system is based on what type of location the nerve innervates, or communicates with. The **somatic nervous system (SNS)** is responsible for conscious perception and voluntary motor responses. Voluntary motor response means the contraction of skeletal muscle, but those contractions are not always "voluntary" (i.e., the result of a conscious decision). Some somatic motor responses are reflexes, and often happen without a conscious decision. Other motor responses become automatic as a person learns motor skills. For example, you once needed to use conscious effort to walk, but in time this became a motor pattern that could proceed automatically.

Figure 13.2 The Nervous System Divisions of Sensation, Integration, and Response

The nervous system has sensory components that receive and deliver sensory information to the brain and spinal cord. The central nervous system is the location where information is integrated and processed. A response is sent to an effector along a motor pathway.

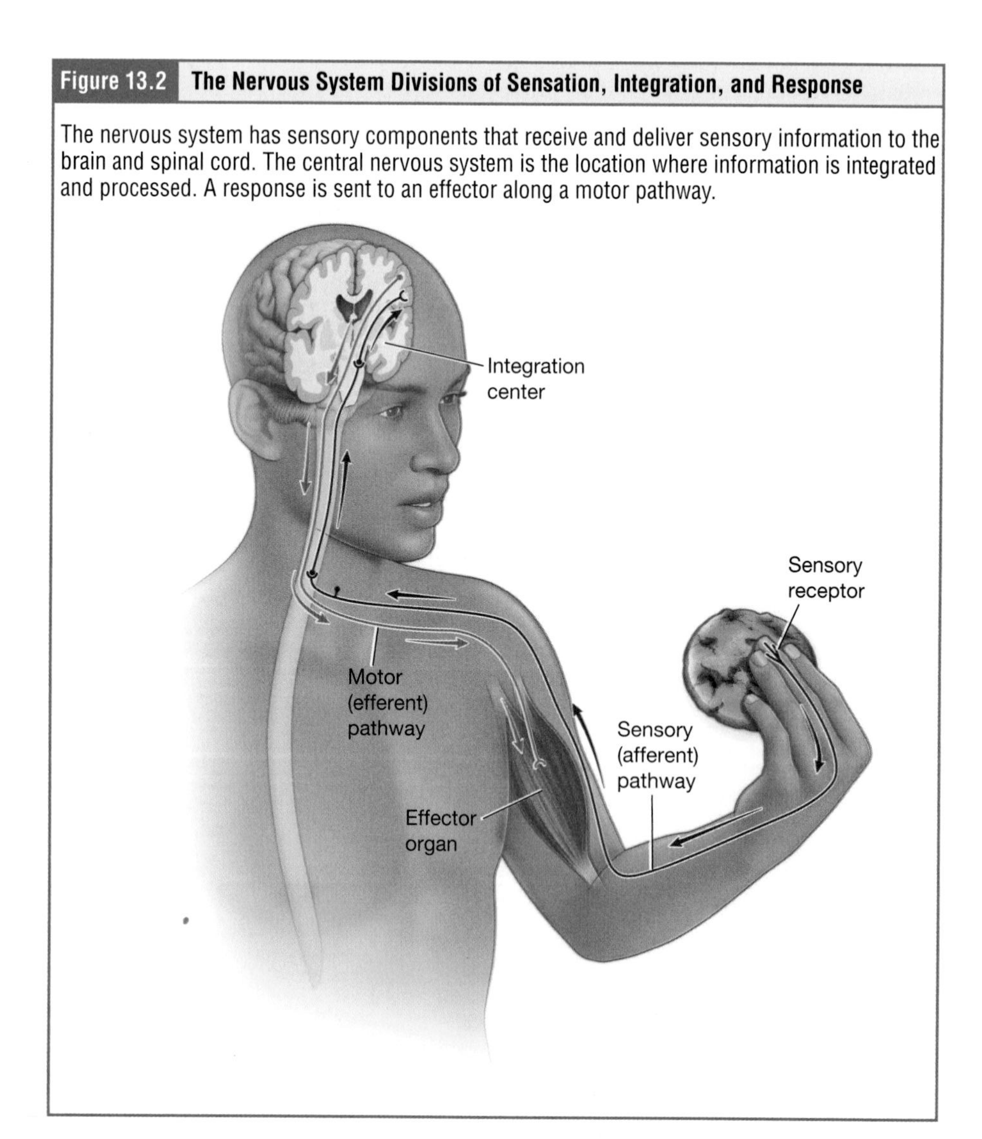

The **autonomic nervous system (ANS)** is responsible for involuntary control of the body, usually for the sake of homeostasis (regulation of the internal environment). The sensory input to the autonomic nervous system can involve both external and internal stimuli. Examples of sensory input to the ANS include things such as stretch in the wall of the stomach that indicates relative hunger or fullness, concentrations of different molecules in the blood, or the smell or sight of nearby food. Autonomic motor output extends to smooth and cardiac muscle as well as glands. Examples include the secretion of saliva, contraction of the smooth muscle lining the intestines, and changes in respiration rate. The role of the autonomic system is to control the actions of the organ systems of the body to preserve homeostasis. Note that both the autonomic and somatic divisions of the nervous system have sensory input and motor output.

Student Study Tip

The word *soma* means "body" in Greek, and "autonomic" sounds like "automatic," for involuntary movement.

LO 13.1.6

DIGGING DEEPER: How fMRI Has Changed Our Understanding of the Brain

The brain and nervous system have long been a bit of a mystery to scientists. The brain simply does not function outside of the body, and since neurons are largely nonmitotic, it is impossible to remove any portion of the brain for study without harming the subject. By studying individual neurons in the lab, scientists have come to understand that neurons are highly metabolic, requiring a great deal of oxygen and glucose to function. Unlike muscle cells, neurons do not store their own myoglobin or glycogen, so their continued function relies entirely on delivered nutrients from the blood. This necessity impacts blood glucose and oxygen homeostasis in a myriad of ways that we will discuss throughout the coming chapters of this text, but scientists also figured out that in regions of the brain that were sending more signals at any given time, there will be greater blood flow to support those neurons. Not only did this turn out to be true, but it provided the basis for unraveling many of the mysteries of the brain! Functional magnetic resonance imaging (fMRI) is a technique in which changes in blood flow in the brain can be measured. Scientists have used this technique to discover the functions of various regions of the brain. A 2012 study used fMRI imaging while asking patients to look at famous works of art or listen to samples of music. For each experience, the subject was asked to rate the art or music on a scale of what they considered to be "beautiful." The fMRI results showed that the same region of the brain had increased blood flow when the subjects were experiencing beauty, regardless of whether that beauty was auditory or visual or whether it was heavy metal or a classical symphony. This region of the brain has subsequently been found to be active in attraction between human individuals as well.

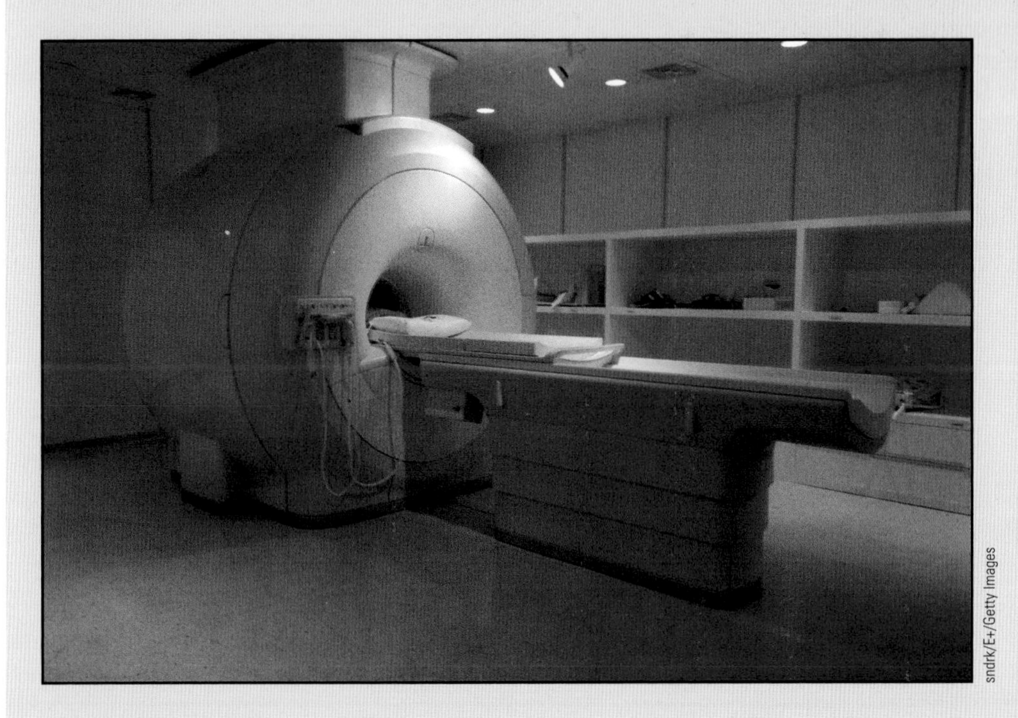

sndrk/E+/Getty Images

Learning Check

1. You are walking your dog in the park. You see a pothole in front of you. Which of the following functions of the nervous system is responsible for developing a motor plan to avoid the pothole?
   a. Sensory input
   b. Sensory integration
   c. Motor response
2. Which of the following systems are encased in bony protection?
   a. Central nervous system
   b. Peripheral nervous system
3. There is a receptor in the carotid artery that detects the amount of oxygen circulating in the blood. If there is less oxygen, a neuron sends a signal from the receptor to the CNS. From the CNS, another neuron sends a signal to the heart to increase the heart rate. This allows more oxygen to circulate in the body. Which of the following functionally describes the neuron that sends a signal from the receptor to the CNS?
   a. Afferent neuron
   b. Efferent neuron
   c. Somatic neuron
   d. Autonomic neuron
4. In the digestive system, the intestines are surrounded by muscles that push the food through subconsciously. Which of the following anatomically describes the system of nerves that targets those muscles?
   a. Somatic nervous system
   b. Autonomic nervous system
   c. Afferent nervous system
   d. Efferent nervous system

13.2 Nervous Tissue and Cells

HAPS LEARNING GOALS AND OUTCOMES

Learning Objectives: By the end of this section, you will be able to:

13.2.1 Identify and describe the major components of a typical neuron (e.g., cell body, nucleus, nucleolus, chromatophilic substance [Nissl bodies], axon hillock, dendrites, and axon) and indicate which parts receive input signals and which parts transmit output signals.

13.2.2 Compare and contrast the three structural types of neurons (i.e., unipolar [pseudounipolar], bipolar, and multipolar) with respect to their structure, location, and function.

13.2.3 Compare and contrast the three functional types of neurons (i.e., sensory [afferent] neurons, interneurons [association neurons], and motor [efferent] neurons) with respect to their structure, location, and function.

13.2.4 Describe the structure, location, and function of each of the six types of neuroglial (glial) cells.

13.2.5 Define myelination and describe its function, including comparing and contrasting how myelination occurs in the CNS and PNS.

Nervous tissue is composed of two types of cells: neurons and glial cells. Neurons, or neural cells, generate and propagate electrical signals into, within, and out of the nervous system. These cells are responsible for communication. Glial cells are various cell types that are responsible for maintaining and supporting neurons. They make up neural tissue, excluding neurons.

13.2a Neurons

Neurons are responsible for the communication that the nervous system provides. Your body is home to approximately 40 trillion cells, getting even a fraction of those cells to coordinate their functions is an enormous task. Imagine the distance between the brain and the muscles in your toes; there are more cells between these locations then there are humans on the planet. Neurons accomplish the task of communication over these long distances through a combination of electrical and chemical signals. As with all cells of the body, their structure is uniquely suited to their function. Neurons have long processes that traverse these distances. Between the brain and the toe muscles are just two neurons carrying the message all this way! Glial cells do not directly send communication messages but play instrumental and vital roles in the communication process.

LO 13.2.1

Parts of a Neuron Neurons are cells and contain many of the typical organelles such as nucleus, endoplasmic reticulum (ER) and ribosomes. The main part of a neuron, where the majority of these organelles reside, is the cell body (see the "Anatomy of a Neuron" feature). The cell bodies of neurons are notable because they often have dark spots or splotches. These dark regions, called *chromatophilic substance*, are areas of rough ER. Neurons have a great deal of protein production and therefore have a higher proportion of rough ER than most cells. Neurons have many extensions of their cell membranes. All these extensions, or processes, enable the neuron to connect with other cells. One of the extensions is an axon—a process that extends from the cell body that allows the neuron to send signals toward other cells. That single axon can branch repeatedly to communicate with many target cells. The other processes of the neuron are **dendrites**, which receive information from other neurons. Dendrites are usually highly branched processes, providing locations for other neurons to communicate with the neuron. Neurons in the brain may receive signals from thousands or tens of thousands of other neurons, all through these highly branched dendrites. The junction where neurons connect with other cells is called a **synapse**. Synapses are not physical connections between the cells but are locations where two cells are so close to each other they can efficiently communicate through secreted chemical signals. Information, in the form of an electrical signal, flows through a neuron from the dendrites, across the cell body, and down the axon.

dendrites (dendrons)

Anatomy of...

A Neuron

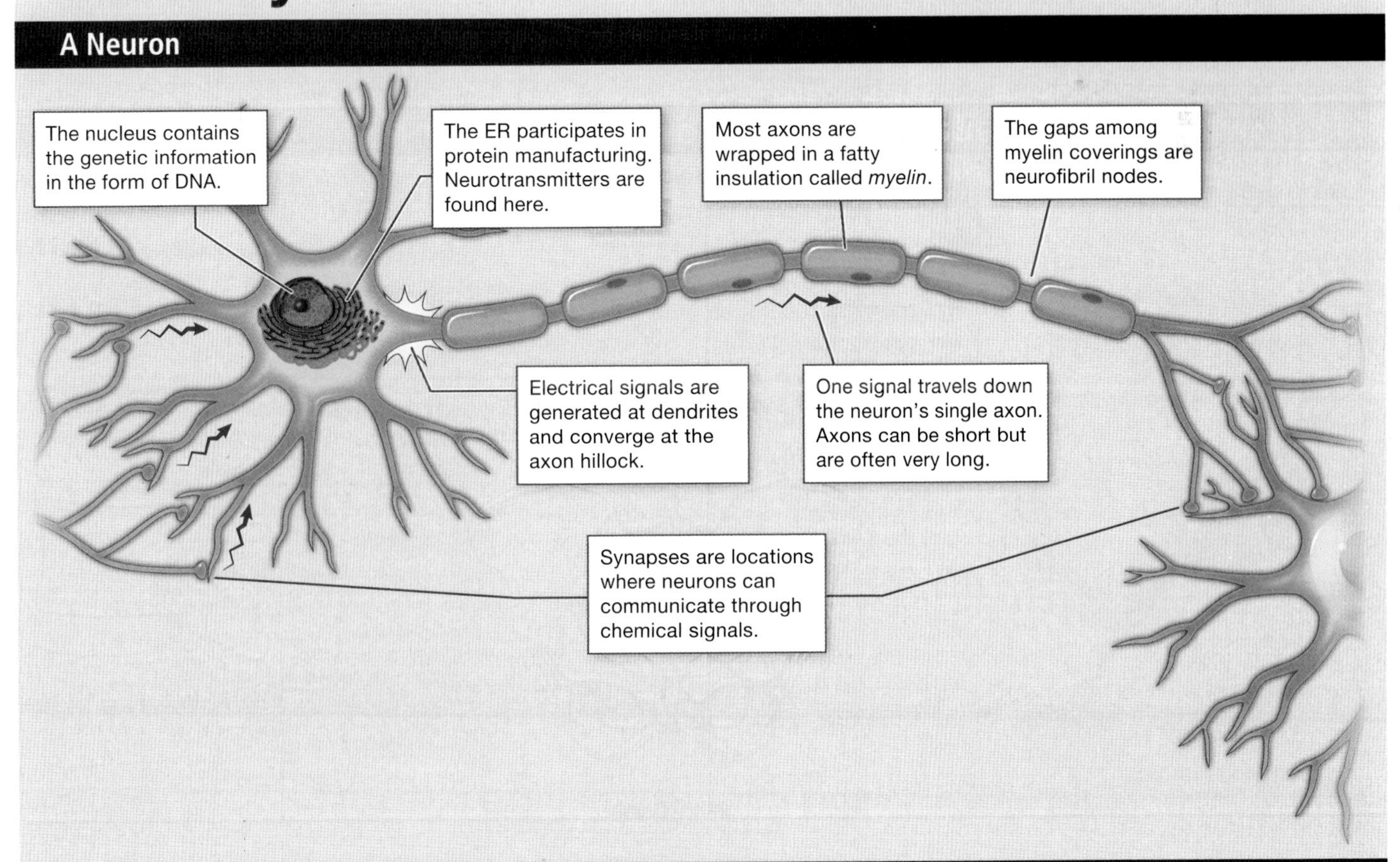

The nucleus contains the genetic information in the form of DNA.
The ER participates in protein manufacturing. Neurotransmitters are found here.
Most axons are wrapped in a fatty insulation called *myelin*.
The gaps among myelin coverings are neurofibril nodes.
Electrical signals are generated at dendrites and converge at the axon hillock.
One signal travels down the neuron's single axon. Axons can be short but are often very long.
Synapses are locations where neurons can communicate through chemical signals.

Where the axon emerges from the cell body, there is a special region referred to as the **axon hillock**. This tapered region of the cell body is notable because it is the location where signals that have begun at different dendrites converge to become a single signal that travels down the axon.

Many axons are wrapped by a fatty insulating substance called **myelin**. When this is present the neuron is said to be *myelinated*. Myelin is made from glial cells. Myelin acts as insulation, much like the plastic or rubber that is used to insulate electrical wires. A key difference between myelin and the insulation on a wire is that there are gaps in the myelin covering of an axon. Each gap is called a **neurofibril node**, and, as we will explore later on in this chapter, these locations are important to the way that electrical signals travel down the axon. Most axons branch, ending in several **axon terminals** and several synapses, so that the neuron can communicate with many other cells.

neurofibril node (node of Ranvier)

axon terminals (synaptic boutons, terminal boutons, end feet)

LO 13.2.2

Types of Neurons Anatomists often describe neurons by the number of processes attached to the cell body (**Table 13.1**).

Unipolar cells have only one process emerging from the cell body that then splits into an axon and dendrites. Unipolar cells are always sensory neurons that send information about the environment toward the brain and spinal cord. Their dendrites receive sensory information, for example the dendrites may be embedded in the skin

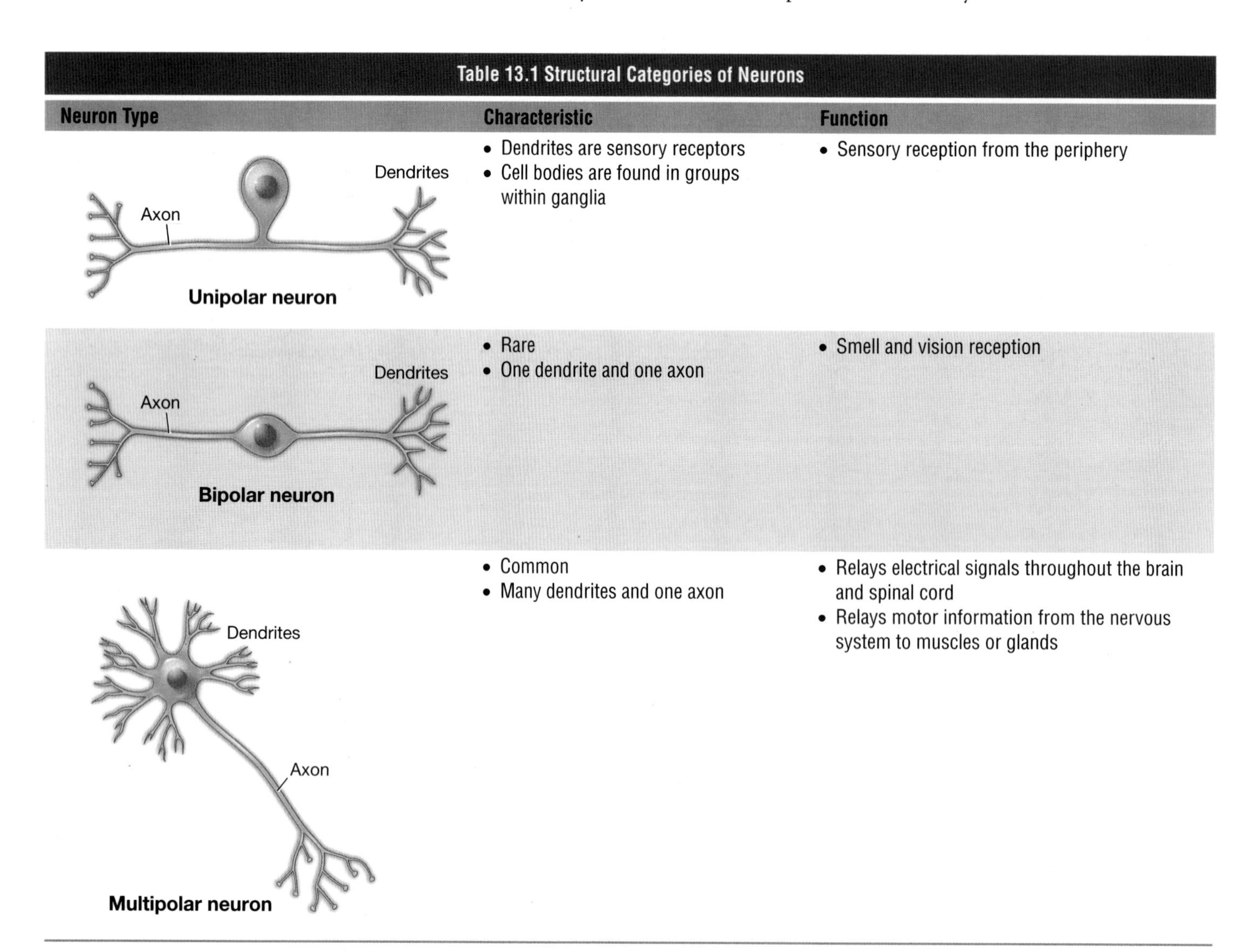

Table 13.1 Structural Categories of Neurons

Neuron Type	Characteristic	Function
Unipolar neuron (Axon, Dendrites)	• Dendrites are sensory receptors • Cell bodies are found in groups within ganglia	• Sensory reception from the periphery
Bipolar neuron (Axon, Dendrites)	• Rare • One dendrite and one axon	• Smell and vision reception
Multipolar neuron (Dendrites, Axon)	• Common • Many dendrites and one axon	• Relays electrical signals throughout the brain and spinal cord • Relays motor information from the nervous system to muscles or glands

to communication touch or temperature information to the brain. The cell bodies of unipolar neurons are always found grouped together in structures known as ganglia. Their axons project into the CNS.

Bipolar cells have two processes—an axon and a dendrite—which extend from each end of the cell body. Bipolar cells are not very common; in humans they are found only in the nose and eyes.

Multipolar neurons are the most common type of neuron. They have one axon and many dendrites.

sensory neurons (afferent neurons)

interneurons (association neurons)

motor neurons (efferent neurons)

Remembering that the three basic functions of the nervous system are sensation, integration, and response, we can also classify neurons for how they contribute to these functions. **Sensory neurons** collect information about the internal and external environment and send it toward the CNS for integration and processing on **interneurons**. Responses to that sensory information (for example, sweating when it is hot outside) are completed through **motor neurons** that emanate from the CNS and connect to effectors. Thus, we can classify neurons as sensory neurons, interneurons, and motor neurons in addition to their anatomical classifications of unipolar, bipolar, and multipolar.

LO 13.2.3

13.2b Glial Cells

Glial cells outnumber neurons 9 to 1; they were long considered to be supporting cells and were not the focus of much research until recently. Today scientists know much more about the roles that these cells play, but there is likely more to learn.

LO 13.2.4

There are six types of glial cells. Four of them are found in the CNS and two are found in the PNS. Table 13.2 outlines some common characteristics and functions.

Learning Connection

Try Drawing It!

Draw a multipolar neuron. Label the cell body, nucleus, chromatophilic substance, dendrites, and axon. Then add arrows to show information traveling from the dendrites to the cell body and from the cell body to the axon.

Glial Cells of the CNS **Astrocytes**—so named because they appear star-shaped under the microscope—have many processes extending from their main cell body (Figure 13.3A). Remember that glial cells cannot send signals, so their processes, rather than being axons and dendrites as in the case of neurons, are cell extensions resembling the arms of an octopus. Those processes extend to interact with neurons and blood vessels. They support the neurons by maintaining the concentration of chemicals in the extracellular space, removing excess signaling molecules, and reacting to tissue damage. Astrocytes also function as gatekeepers to the brain by wrapping their processes around the brain's blood vessels and regulating the flow of materials from the blood and into the nervous system environment. This physiological barrier, the **blood-brain barrier (BBB)**, keeps many molecules that circulate in the rest of the body from getting into the CNS. Most neurons in the CNS are not capable of mitosis, and therefore neurons that die are not replaced. Humans have the highest number of neurons in their lives on the day they are born; that number decreases throughout the lifespan. Therefore, the BBB evolved to help protect the CNS environment from any damage. Nutrients, such as glucose or amino acids, can pass through the highly selective BBB,

Table 13.2 Glial Cell Types by Location and Basic Function

CNS Glia	PNS Glia	Basic Function
Astrocyte	Satellite cell	Regulate extracellular environment
Oligodendrocyte	Neurilemma cell	Myelination
Microglia	-	Immune defense and waste removal
Ependymal cell	-	CSF production

Figure 13.3 Glial Cells of the CNS

(A) Astrocytes make up the blood-brain barrier and regulate the environment around neurons. (B) Microglia serve as resident immune cells in the CNS. (C) Oligodendrocytes myelinate CNS axons. (D) Ependymal cells create the cerebrospinal fluid.

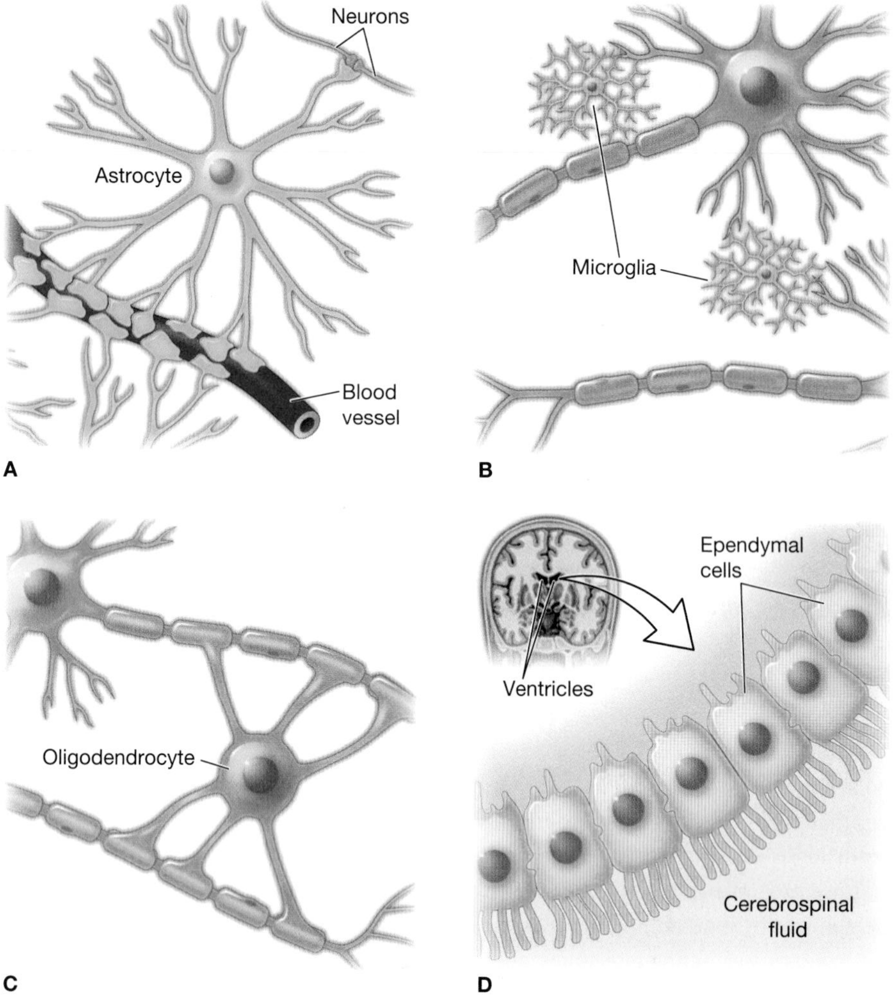

but most other molecules cannot. A side effect of the BBB's protective structure is that it interferes with drug delivery to the CNS. Pharmaceutical companies are challenged to design drugs that can cross the BBB as well as have an effect on the nervous system.

Another consequence of the BBB is that white blood cells—the body's main line of defense against infection—cannot cross from the blood, where these cells travel, into the brain environment. While this barrier protects the CNS from exposure to toxic or pathogenic substances, it also keeps out the cells that could protect the brain and spinal cord from disease and damage. Instead, **microglia** act as resident immune cells. Microglia originate as white blood cells, called *macrophages*, which become part of the CNS during early development. When microglia encounter diseased or damaged cells, they ingest and digest those cells or the pathogens that cause disease. Microglia function in the healthy brain as well; they contribute to neurological development by cleaning up wastes and cells that are no longer needed.

Also found in CNS tissue is the **oligodendrocyte**, which is the glial cell type that insulates axons in the CNS. Similar to many of the cell types we have explored in this chapter, oligodendrocytes have processes reaching out from a main cellular structure (**Figure 13.3C**) Each process reaches out and surrounds an axon to insulate it in myelin. One oligodendrocyte will provide the myelin for multiple axon segments, either for the same axon or for separate axons. The function of myelin will be discussed later in this chapter.

Ependymal cells are glial cells that filter blood to make **cerebrospinal fluid (CSF)**, the fluid that circulates through the CNS. CSF is the extracellular fluid of the neurons, and its specific composition enables optimal neuronal function. CSF is produced in spaces within the brain called **ventricles**, where ependymal cells line the walls (**Figure 13.3D**). These glial cells appear similar to epithelial cells, making a single layer of cells with little intracellular space and tight connections between adjacent cells. They also have cilia on their apical surface to help move the CSF through the ventricle. Ependymal cells are highly selective about which materials are included in CSF and which are kept out of the nervous system.

Glial Cells of the PNS One of the two types of glial cells found in the PNS is the **satellite cell**. Satellite cells surround the cell bodies of sensory neurons (**Figure 13.4A**). This accounts for the name, based on their appearance under the microscope. They provide support, performing functions in the periphery similar to those of astrocytes in the CNS—except, of course, there is no BBB in the PNS.

The second type of glial cell is the **neurilemma cell**, which insulates axons with myelin in the PNS. Neurilemma cells differ from oligodendrocytes in that neurilemma cells do not have processes (**Figure 13.4B**). Rather, the entire cell wraps around a single portion of one axon like a tortilla wrapping to form a burrito. The nucleus and cytoplasm of the neurilemma cell are flattened and found within the myelin sheath.

Neurilemma cell (Schwann cell)

Figure 13.4 Glial Cells of the PNS

(A) Satellite cells surround the cell bodies of neurons when they are clustered together in ganglia. (B) Schwann cells serve a function similar to that of oligodendrocytes, myelinating the long axons of peripheral neurons.

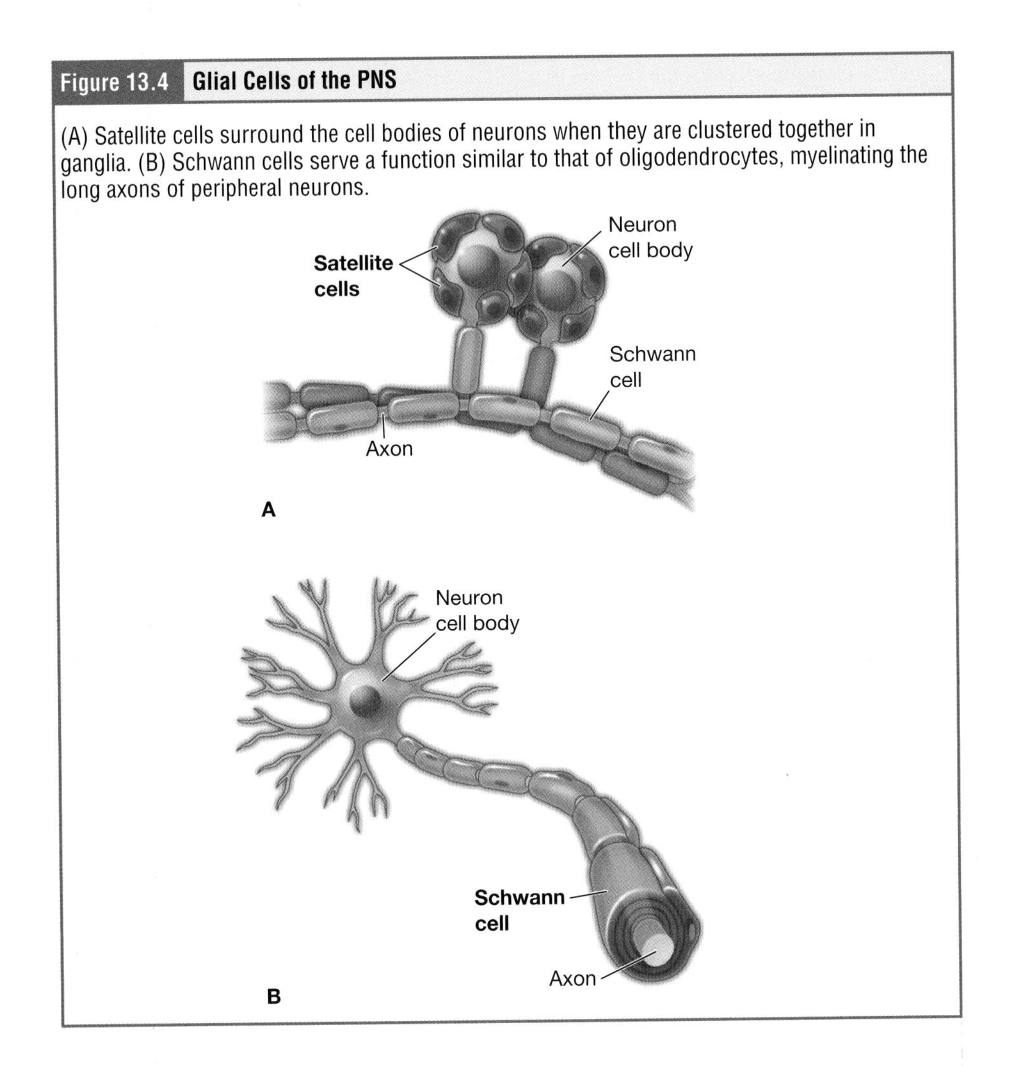

Figure 13.5 The Process of Myelination

In both the peripheral (A) and central (B) nervous system, glial cells wrap their membranes many times around the axons of neurons, providing electrical insulation to prevent ion leakage.

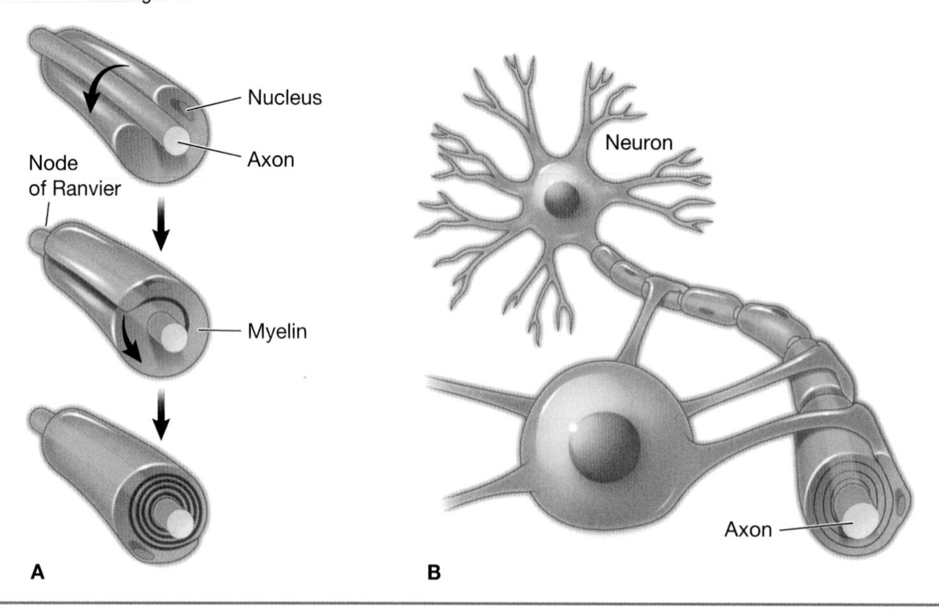

LO 13.2.5

Learning Connection

Quiz Yourself

Quiz yourself on each of the glial cell names, the locations where they are found, and what their functions are.

Myelin Both oligodendrocytes and neurilemma cells form myelin, the insulation for axons in the nervous system. The structure of these two cells differs slightly, but the function is the same. In all cells, the plasma membrane forms a selective barrier that prevents the random crossing of most molecules. Axons have a plasma membrane just like every cell, but in myelination the plasma membrane of the oligodendrocyte or neurilemma cell provides an additional barrier. Not just one additional barrier, actually, but each of these cell types wrap around the axon many times, further and further restricting the movement of molecules between the inside of the neuron and its external environment. In this way, the **myelin sheath** facilitates the transmission of electrical signals down the axon by preventing the leak of electrical charge out of the neuron. Myelin forms during development (**Figure 13.5A**), as the glial cell wraps progressively around the axon until it forms many layers (**Figure 13.5B**). Myelin sheaths can be 100–1,000 times the diameter of the axon.

Apply to Pathophysiology

Demyelination

Rachel is a 37-year-old female experiencing back pain as well as numbness and tingling in her legs. A sample of her cerebrospinal fluid obtained by lumbar puncture shows the presence of an abnormally high concentration of the disease-fighting proteins called *antibodies*, which suggests excess immune system activity within her brain and spinal cord. Magnetic resonance imaging (MRI) is used to visualize her nervous system tissues, and several abnormal spots are noted in her spinal cord, brainstem, and brain. Her condition is diagnosed as **Multiple Sclerosis (MS)**.

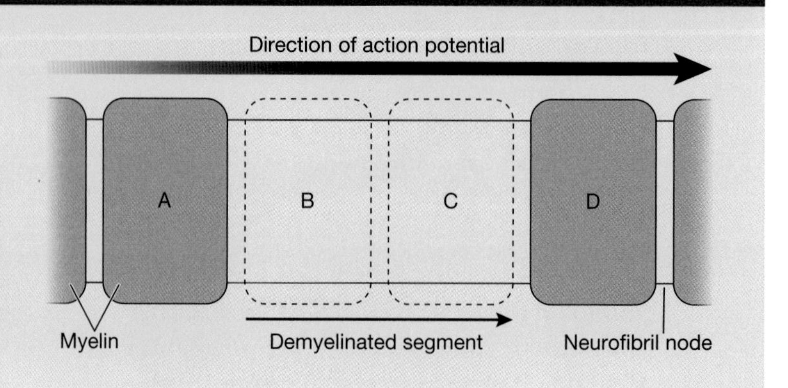

(*Continued*)

MS is a disease in which a loss of myelin occurs at one or several places in the nervous system. It is an autoimmune condition in which the myelin sheaths are attacked and destroyed by antibodies from the patient's own immune system.

1. Which point on the membrane will have the highest permeability to Na^+ at rest?
   A) A
   B) B
   C) D
   D) All points of the axon will have the same permeability to Na^+ at rest.
2. If a scientist were able to isolate the antibodies responsible, what cell would they react against?
   A) Oligodendrocytes
   B) Astrocytes
   C) Satellite cells
3. In which location on the membrane do you expect to have the highest concentration of Na^+/K^+ pumps?
   A) A
   B) B
   C) C

Learning Check

1. What part of the neuron is responsible for converging multiple signals into a single signal that travels down the axon?
   a. Axon terminal
   b. Axon hillock
   c. Dendrites
   d. Nucleus
2. Which of the following best describes the neuron in the figure at right?
   a. Sensory
   b. Motor
   c. Interneuron
   d. Multipolar
3. Which of the following are functions of astrocytes? Please select all that apply.
   a. Maintaining concentration of chemicals in the extracellular space
   b. Surrounding an axon of the PNS to insulate it with myelin
   c. Filtering blood within ventricles to produce cerebrospinal fluid
   d. Regulating the flow of materials from the blood into the nervous system
4. Which of the following would occur if the myelin sheath were damaged? Please select all that apply.
   a. There would be a decreased amount of CSF.
   b. The axon would have excessive leakage of electrical charge.
   c. The nervous system would be unable to clean up waste.
   d. The nerve conduction would slow down.

13.3 Neurophysiology

Learning Objectives: By the end of this section, you will be able to:

13.3.1 Define and describe depolarization, repolarization, hyperpolarization, and threshold.

13.3.2 Describe the role of the sodium-potassium ATPase pump in maintaining the resting membrane potential.

13.3.3 List the major ion channels of neurons and describe them as leak (leakage or passive) or voltage-gated channels, mechanically gated channels, or ligand-gated (chemically-gated) channels, and identify where they typically are located on a neuron.

13.3.4 Describe the physiological basis of the resting membrane potential (RMP) in a neuron including the ion channels involved, the relative ion concentrations, and the electrochemical gradient.

13.3.5 Compare and contrast graded potentials and action potentials, with particular attention to their locations in the neuron and the ions and ion channels involved in each.

13.3.6* Compare and contrast IPSPs and EPSPs with respect to the ions involved and effects on the postsynaptic cell.

13.3.7 Explain temporal and spatial summation of postsynaptic potentials.

13.3.8 Distinguish between absolute and relative refractory periods and compare the physiological basis of each.

13.3.9 Explain the impact of absolute and relative refractory periods on the activity of a neuron.

13.3.10 Describe the physiological process involved in the conduction (propagation) of an action potential, including the types and locations of the ion channels involved.

13.3.11 Compare action potential conduction (propagation) in an unmyelinated versus a myelinated axon.

13.3.12 Explain how axon diameter and myelination affect conduction velocity.

* Objective is not a HAPS Learning Goal.

Now that we have met all the cell types that make up nervous tissue, we can develop our understanding of how nervous tissue is capable of communication. Before getting to the nuts and bolts of these communication signals, an illustration of how the components come together will be helpful (see the "Anatomy of Neural Communication" feature). In this example a sensory neuron, in this case one that responds to temperature changes, responds to an increase in temperature by generating an electrical signal called an **action potential**. This signal travels within the sensory neuron and causes the release of a chemical signal, a **neurotransmitter**, at the end of the neuron. This neurotransmitter causes the next neuron in the chain to fire its own action potential, and the set of events repeats until the signal reaches a dedicated area of the brain for reception of sensory information from this region of the body.

The idea of electrical signaling, which we are more likely to associate with light bulbs and cell phones, occurring within the body can seem quite strange. However, we can remember that electricity—whether we are talking about a battery or a neuron—is simply the movement of ions. Therefore, if positive and negative charges can be separated, electrical signals can be generated. The membrane is an excellent barrier to the movement of ions because its nonpolar core is impermeable to charged particles. Another core idea when considering electrical particles is to remember that opposites attract. Positive charges are drawn toward negative charges and vice versa (**Figure 13.6A**). Since the charges cannot readily cross the membrane, the result is that oppositely charged molecules gather on either side of the membrane. Transmembrane proteins—specifically proteins that are channels—make it possible for charged particles to cross the membrane. Several transport channels, as well as active transport pumps, are necessary for neurons to generate electrical signals. Neuron action potentials are very similar to the muscle cell potentials covered in Chapter 11. Of special interest is the carrier protein referred to as the **sodium/potassium (Na^+/K^+) pump** that moves sodium ions (Na^+) out of a cell and potassium ions (K^+) into a cell, thus regulating ion concentration on both sides of the cell membrane.

sodium/potassium (Na^+/K^+) pump (sodium potassium ATPase)

Student Study Tip

The action of the Na^+/K^+ pump is very important to remember for charge and concentration differences across the membrane. Na^+ has three characters while K^+ has only two, so three Na^+ ions exit while only two K^+ ions enter.

As the Na^+/K^+ pump works, it creates two chemical gradients. The pump moves more and more Na^+ ions to the outside of the cell, creating an area with a high concentration of Na^+ ions in contrast to the interior of the cell, which has a far lower concentration. The Na^+/K^+ pump is likewise creating a high concentration of K^+ ions in the interior of the cell, with a low concentration of K^+ ions on the outside. The Na^+/K^+ pump also is creating an electrical gradient. Each time it turns it moves three Na^+ ions to the outside of the cell but only two K^+ ions to the inside of the cell. Therefore, the charges are distributed across the membrane unequally, with the outside of the

Anatomy of...

Neural Communication

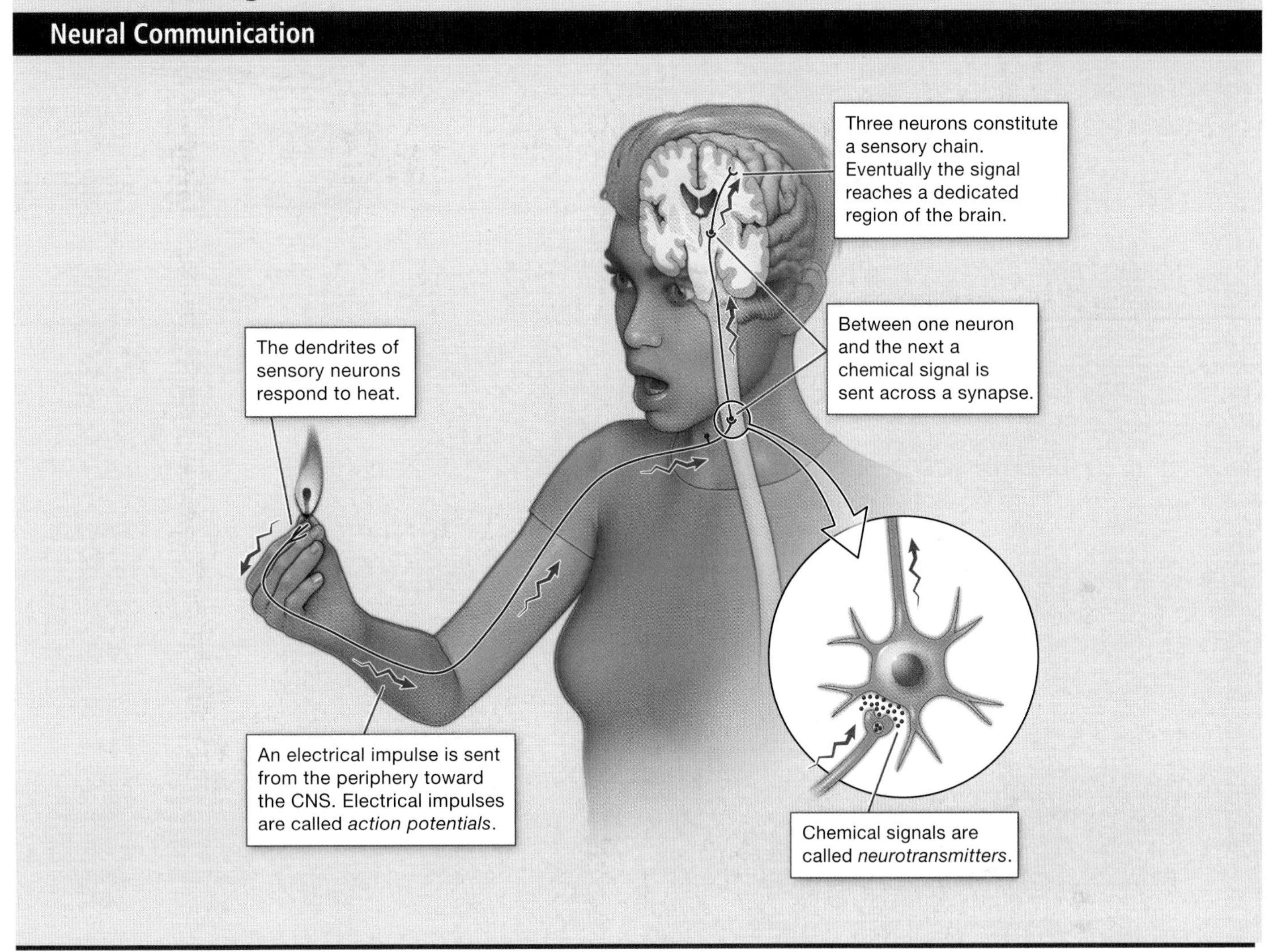

Three neurons constitute a sensory chain. Eventually the signal reaches a dedicated region of the brain.
The dendrites of sensory neurons respond to heat.
Between one neuron and the next a chemical signal is sent across a synapse.
An electrical impulse is sent from the periphery toward the CNS. Electrical impulses are called *action potentials*.
Chemical signals are called *neurotransmitters*.

cell being more positive than the inside. The inside of the cell also has more negative charges; the intracellular proteins often contribute negative charges through their amino acid side groups and added phosphate groups. The inside of the cell is therefore more negative in addition to being less positive. The difference in charge can be measured in volts using a voltmeter (**Figure 13.6B**).

LO 13.3.1

With each side of the membrane having such different environments, we can say that the membrane is polarized, a term we also used to describe molecules with different sides and epithelia, which have different environments on their two sides. The neuron stays in this resting polarized state whenever it is not firing a signal. If the amount of charge difference between the two sides of the membrane decreases and the membrane is less polarized than it was, we say that it is *depolarizing* (**Figure 13.6C**). If the cell becomes even more polarized than it usually would be, we say it is *hyperpolarizing*. If the cell becomes depolarized and then returns to its polarized state, we say it is *repolarizing*.

The sodium/potassium pump requires energy in the form of adenosine triphosphate (ATP), so it is also referred to as an ATPase pump. Throughout the body

Figure 13.6 Membrane Potential LO 13.3.2

(A) The Na^{+}/K^{+} pump distributes charge unevenly across the membrane so that the neuron cell membrane has a more positive environment on its external surface and a more negative environment on its internal surface. (B) A voltmeter uses electrodes placed inside and outside the cell to measure the difference in charge. (C) That difference in charge and changes can be plotted on a graph of membrane potential over time. The membrane is typically in a polarized state, but that can change depending on cellular events.

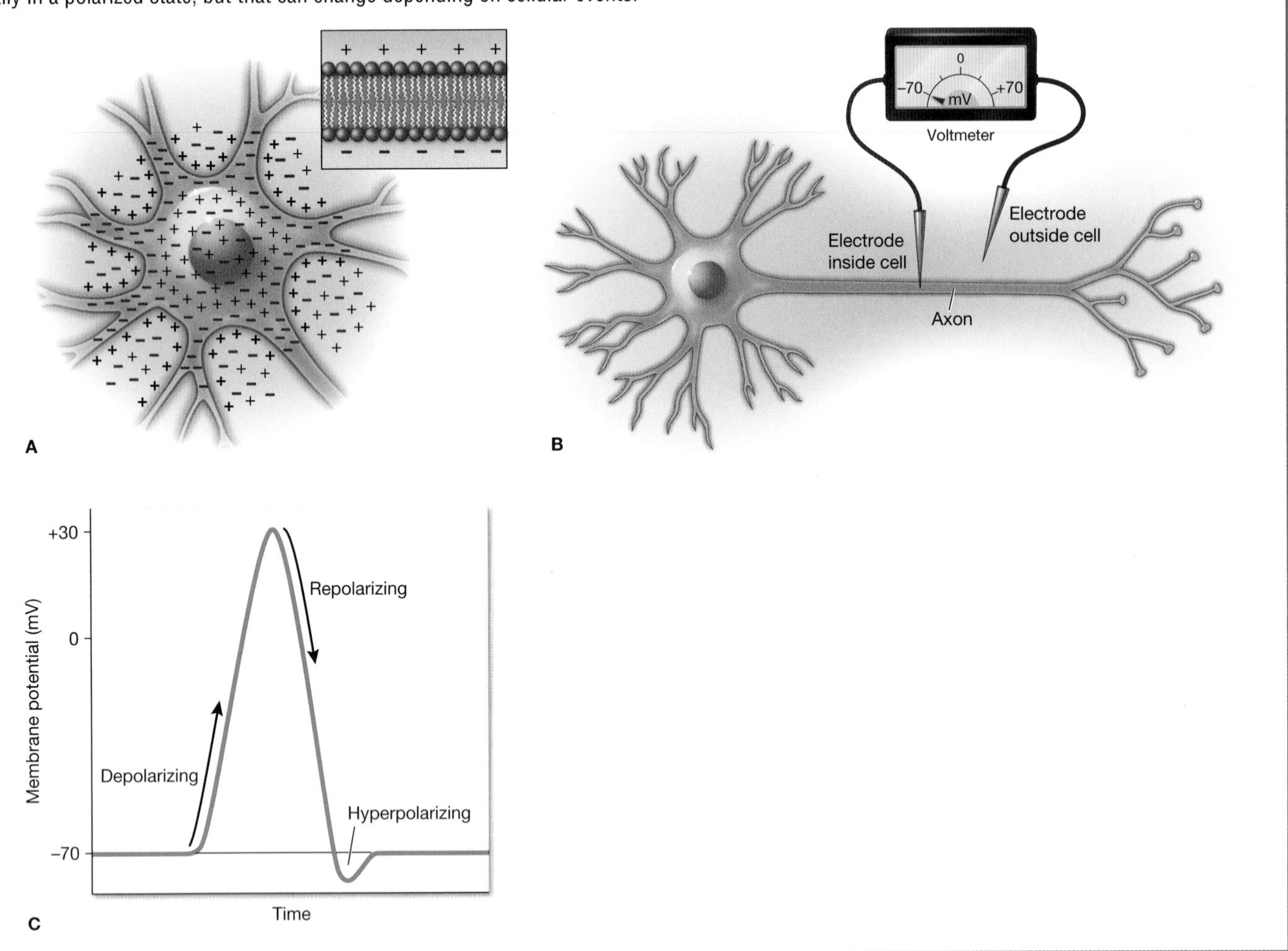

the extracellular concentration of Na^+ is higher than the concentration inside of cells, and the concentration of K^+ is higher inside the cell than outside (**Figure 13.7**). These gradients are more exaggerated in muscle cells and neurons because these cells express higher levels of the Na^+/K^+ pump. As the pump continues to work, it pushes more and more Na^+ ions out of the cell, where they are already at a higher concentration. This is why the pump requires energy, it is moving the ions against their concentration gradients.

Some of the transmembrane proteins within these cell membranes are ion channels. Ion channels function as tunnels that allow specific ions to cross the membrane in response to an existing concentration gradient. Ion channels have specificity for the ions they permit to pass through them. Some channels will allow any cations (positive ions) or any anions (negative ions), while others allow only a select few. Still others will allow just one specific ion to pass.

Remember that all molecules will move based on their own chemical gradients as well; for example, sodium ions will diffuse away from areas where there are many

Figure 13.7 The Na^+/K^+ Pump

The Na^+/K^+ pump is an active transporter that uses ATP to move three Na^+ ions out of the cell and two K^+ ions into the cell.

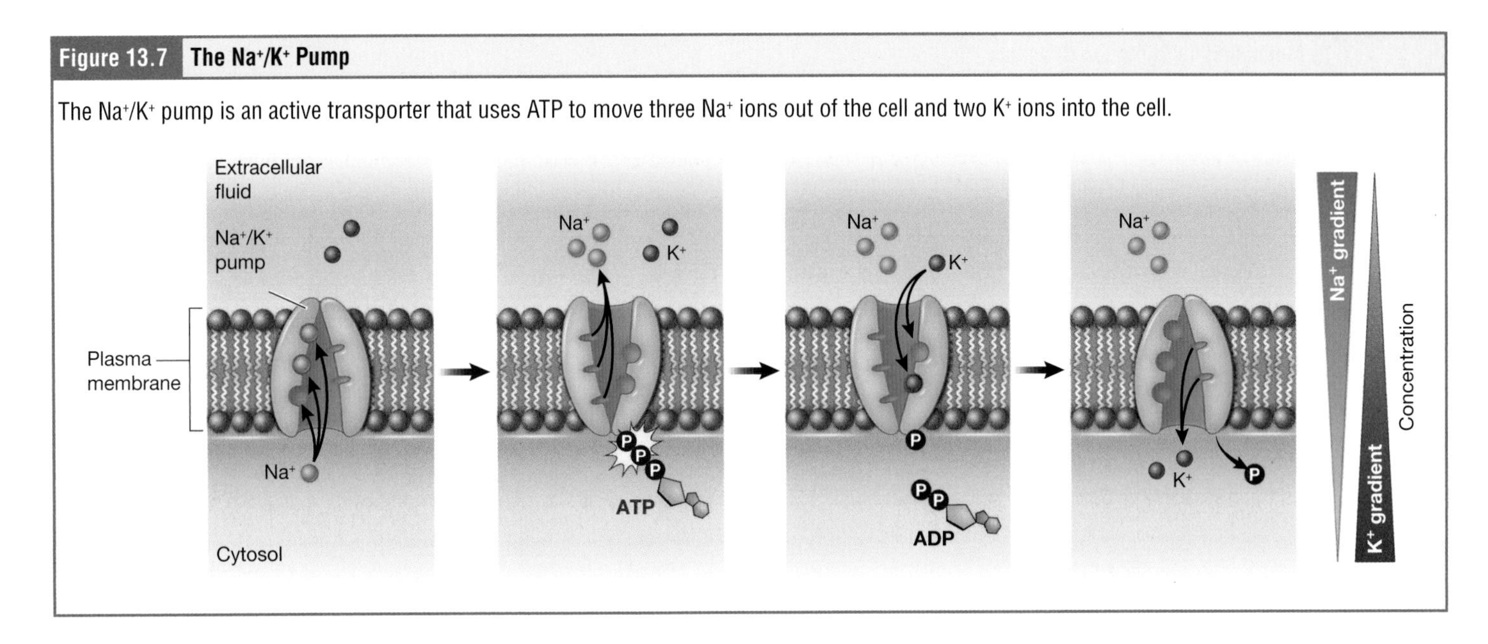

other sodium ions. Ion channels do not always freely allow ions to diffuse across the membrane. Some are opened or closed by certain events; physiologists refer to these channels as *gated*. So, we can categorize channels on the basis of what ions can travel through them—for example, a sodium channel—or how they are gated. These classes of ion channels are found primarily in the cells of nervous or muscular tissues.

ligand-gated channel (ionotropic receptors)

A **ligand-gated channel** opens because a molecule—a ligand—binds to the extracellular region of the channel (**Figure 13.8**). An example of a ligand-gated channel is the acetylcholine receptor we met when learning about muscles in Chapter 11.

LO 13.3.3

A **mechanically gated channel** opens when the cell membrane physically distorts, usually under external pressure (**Figure 13.9**). Many channels associated with the sense of touch are mechanically gated. For example, as pressure is applied to the skin,

Figure 13.8 Ligand-Gated Channels

Ligand-gated channels open when the ligand—in this case the neurotransmitter acetylcholine—binds. In its open shape, the channel allows specific ions, in this case Na^+, to pass through.

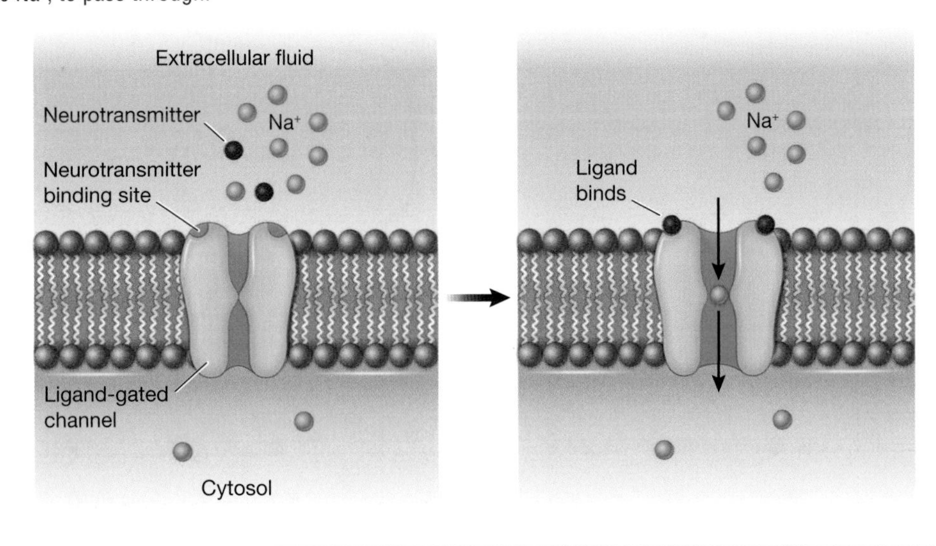

Figure 13.9 Mechanically Gated Ion Channels

Mechanically gated ion channels open or close in response to a change in the shape of the channel. Mechanically gated channels will open when the membrane is deformed by touch stimuli, such as when pressure is applied to the surface of the skin.

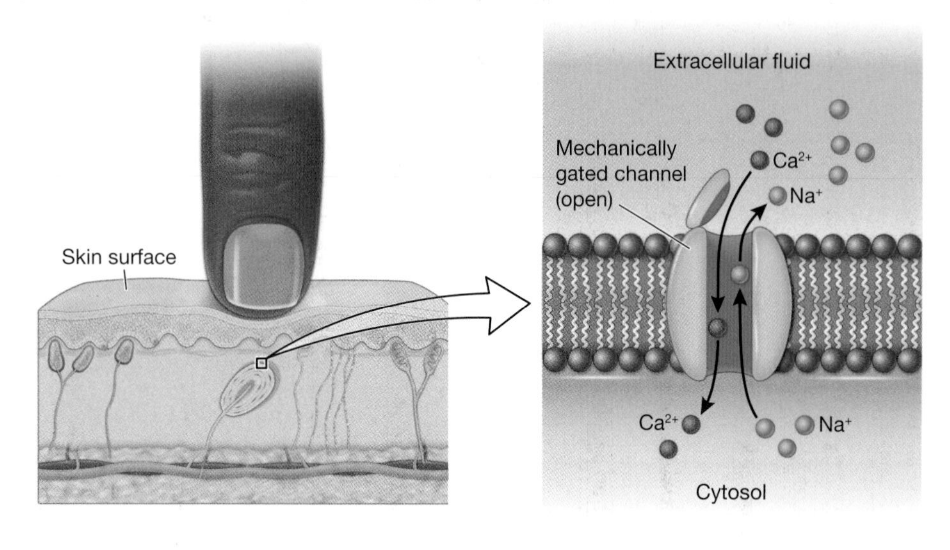

these channels open and allow ions to enter the cell, initiating the electrical signal that will travel toward the CNS.

A **voltage-gated channel** is a channel that responds to changes in the electrical properties of the membrane in which it is embedded. As we have previously discussed in this chapter, the properties of the membrane and actions of the Na^+/K^+ pump establish a charge difference across the membrane. If this charge difference changes, the channel changes shape, moving from open to closed or closed to open (**Figure 13.10**).

Figure 13.10 Voltage-Gated Channels

Voltage-gated channels open and close based on the polarization of the membrane.

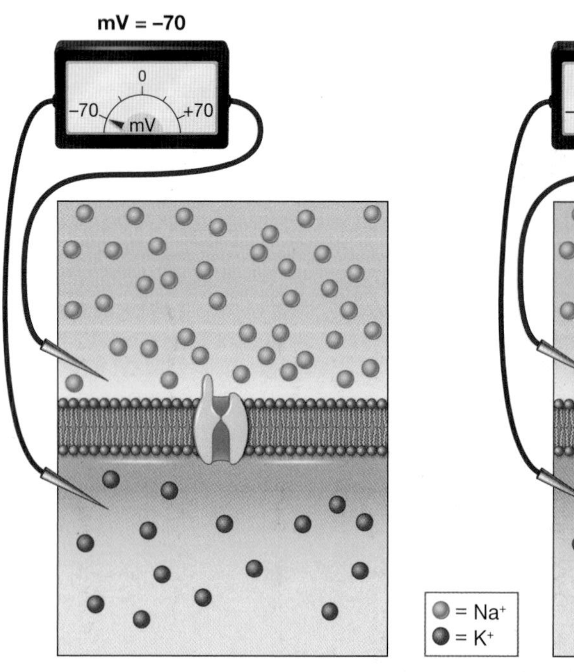

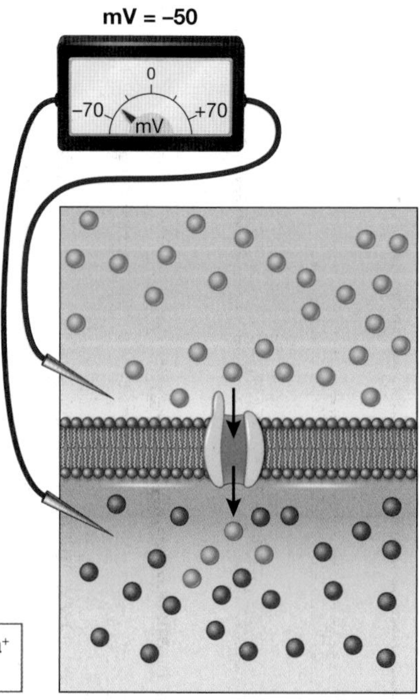

A **leak channel** is either always open or opening and closing at random. Because no single cellular event controls this opening and closing, it is referred to as "leaking." Leak channels are present in different proportions in different types of cells, leading to slight variations in their physiology.

leak channel (non-gated channel)

13.3a Membrane Potential

LO 13.3.4

When the ion channels are closed (except for leak channels) and the Na^+/K^+ pump is at work, the charge difference between the inside and outside of the membrane reaches a steady state. For neurons, this charge difference is typically about 70 mV of difference, with the inside being more negative and less positive than the outside of the cell. Since the neuron in this state, with all of its channels closed, is not firing a signal, we describe the cell as being *at rest*. The membrane's voltage difference in this state, typically −70 mV, is its **resting potential**. In contrast, the electrical signal that enables communication down the length of the cell is an action potential.

resting potential (resting membrane potential)

There are several factors that contribute to the resting potential:

- *Ion concentration*. When the cell is at rest, the concentration of Na^+ outside the cell is 10 times greater than the concentration inside. Also, the concentration of K^+ inside the cell is greater than outside. The cytosol contains a high concentration of negative charges in the form of phosphate ions and negatively charged proteins. This unequal distribution of ions is due to the action of the Na^+/K^+ pump.
- *Intracellular proteins*. The proteins inside the cell are largely negatively charged, especially when considering that many of them carry phosphate groups due to regulation through phosphorylation. Since there are far more proteins inside the cell than outside, these molecules contribute to the overall charge difference between the inside and outside of the cell.
- *Ion leak*. The membrane resists the movement of ions from one side to the other because the charged ions are relatively impermeable to the phospholipid bilayer. However, the membrane contains some leak channels that allow ions to move randomly. There are far more K^+ leak channels than Na^+ leak channels. Because K^+ is at a higher concentration inside the cell, it tends to leak out of the cell through these channels. This means more positive charges leaving the cell, and an even greater charge difference between the positive exterior of the membrane and the more negative interior.

The Action Potential *Resting membrane potential* describes the steady state of the cell when it is not firing a signal. This is a dynamic process that is balanced by ion leakage and ion pumping. The voltage difference between the outside and inside of the cell is about 70 mV, and this will not change without any outside influence (see the "Anatomy of an Action Potential" feature, step 1). The signal—the action potential—starts when an external event causes the membrane potential to change.

LO 13.3.5

The action potential is a rapid depolarization of the membrane that travels from the axon hillock to the axon terminal. It is initiated in all neurons with an external depolarization event called a **graded potential**, which occurs at the dendrites or cell body. For the membrane to depolarize, it must become more positive on the inside of the cell. This happens when a channel opens that permits Na^+ to cross the membrane. Because the concentration of Na^+ is higher outside the cell than inside the cell by a factor of 10, ions will rush into the cell. Because sodium is a positively charged ion, it will change the voltage differences across the membrane. The resting potential is typically around -70 mV, so the Na^+ ions entering the cell will cause the interior of the cell to become less negative and the difference between the outside and inside of the cell to become smaller, depolarization (see the "Anatomy of an Action Potential" feature, step 2).

The initial depolarization of the membrane will lead to the opening of voltage-gated ion channels. Remember that voltage-gated channels exist in one shape (either closed or open) at rest and then change their shape when the charge difference across the membrane changes. In the case of the voltage-gated ion channels of the neuron, the two types—Na^+ voltage-gated channels and K^+ voltage-gated channels—are both closed when the cell is at its resting membrane potential, but then open when the cell depolarizes to -55 mV or above. These two types of channels are found all along the length of the axon. They open when the cell is depolarized past their **threshold voltage** of -55 mV, and they close when the cell repolarizes to greater than that threshold voltage. In other words, the channels are closed at -70 mV; they are triggered to open at -55 mV, and then they close again if the cell repolarizes to -60 mV or -70 mV. There are three differences between these types of voltage-gated channels. The first difference, as you can guess from their names, has to do with which ions are permitted to pass through them. Na^+ voltage-gated channels permit only Na^+ ions to pass through; K^+ voltage-gated channels permit only K^+ ions to pass through. The second difference is the speed at which the channel can pivot from open to closed. Na^+ voltage-gated channels open and close quickly because the change in the shape of the protein channels required for channel opening and closing occurs quickly. K^+ voltage-gated channels, on the other hand, open and close slowly; this protein takes time to complete its change. So, while both channels are *triggered* to open at -55 mV, the Na^+ voltage-gated channels open right away, but a few milliseconds (ms) pass before the K^+ voltage-gated channels can open. The same difference happens when the cell repolarizes; both channels are triggered to close, but the Na^+ voltage-gated channel closes quickly and the K^+ voltage-gated channel closes slowly. The third difference is that the Na^+ voltage-gated channels have an extra gate called an *inactivation gate*, which acts like a protein arm that can swing into place covering their opening. K^+ voltage-gated channels do not have inactivation gates.

Let's examine what events take place if the graded potential depolarizes the cell past the threshold voltage of -55 mV. Both Na^+ and K^+ voltage-gated channels are triggered to open, but Na^+ voltage-gated channels open first. The concentration gradient for Na^+ is so strong that Na^+ ions will enter the cell, rapidly depolarizing the axon (see the "Anatomy of an Action Potential" feature, step 3). In fact, the concentration gradient is great enough that Na^+ ions will enter the cell even after the membrane potential has become zero, so that the voltage across the membrane begins to become positive. The electrical gradient also plays a role, as negative proteins below the membrane attract the sodium ion. Around the time that the membrane potential reaches $+30$ mV, the K^+ channels finally reach their open confirmation (see the "Anatomy of an Action Potential" feature, step 4). Another event occurs at this time as well; the inactivation gate of the Na^+ voltage-gated channel swings into place over the channel's opening, blocking the passage of any further Na^+ ions into the cell. The combined effect of these two events is that the membrane stops depolarizing and starts to repolarize as positive K^+ ions leave the cell through their open channel following their concentration gradient (see the "Anatomy of an Action Potential" feature, step 5). This is the **repolarization phase**, meaning that the membrane voltage moves back toward the -70 mV value of the resting membrane potential.

Learning Connection

Explain a Process

Write out each step of the action potential on an individual piece of paper or post-it note. Mix them up. Can you put them back in order?

As the membrane potential repolarizes past the threshold value of -55 mV, both sets of voltage-gated channels are triggered to close. As with opening, the Na^+ channel moves first, returning to its closed conformation from its inactivated conformation

Anatomy of...

An Action Potential

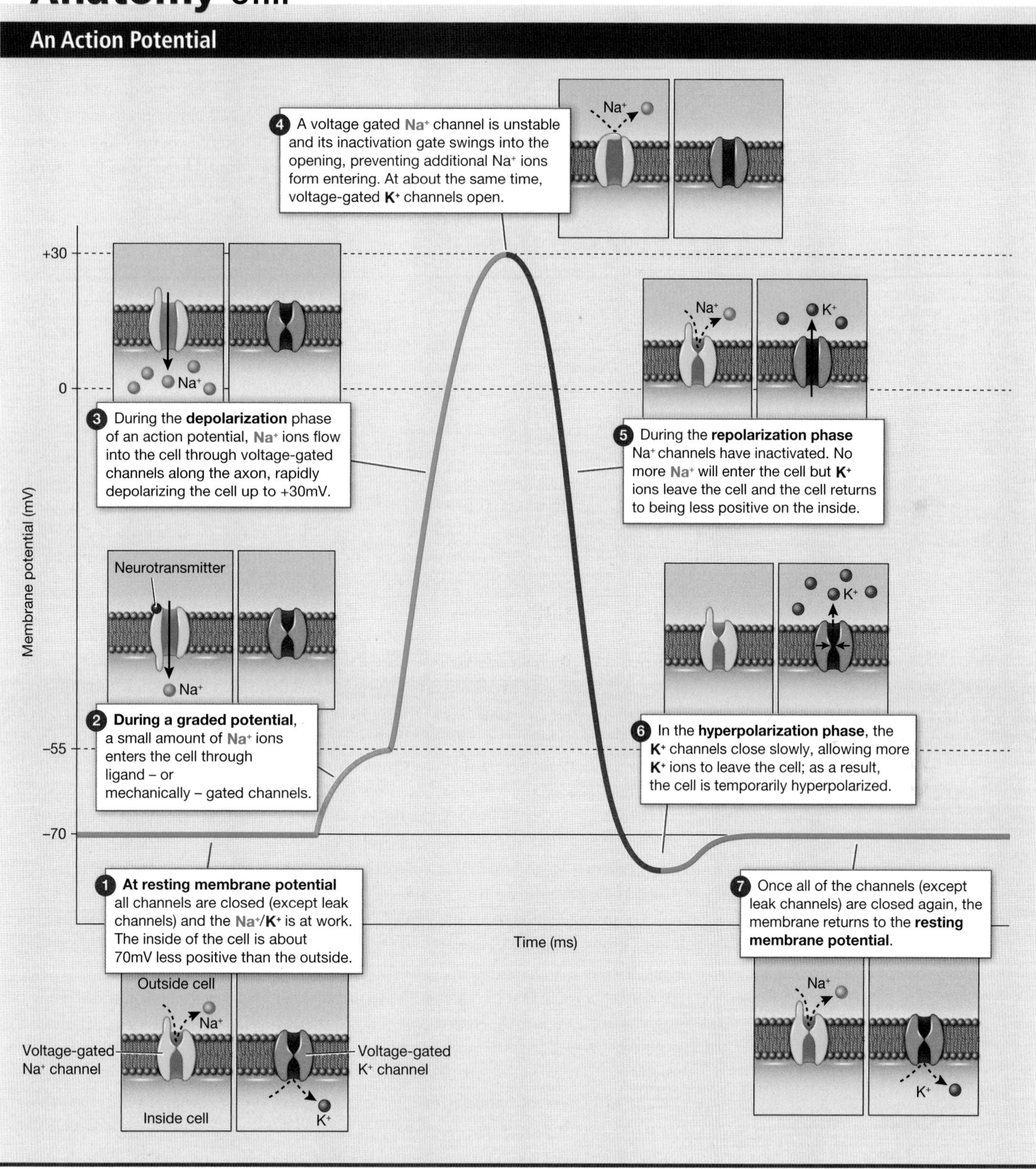

(**Figure 13.11**). The K^+ channel closes shortly after this, but not before allowing so many K^+ ions to leave that the cell briefly hyperpolarizes (see the "Anatomy of an Action Potential" feature, step 6). Once the K^+ channels have closed, the membrane returns to its resting potential due to the action of the Na^+/K^+ pump. The action potential has resolved, and the cell will soon be ready to fire another signal (see the "Anatomy of an Action Potential" feature, step 7).

Figure 13.11 The Three Conformations of the Na^+ Voltage-Gated Channels

Sodium voltage-gated channels can be closed, open, or open but inactivated. Once inactivated, the channel must return to a closed conformation before being able to open again.

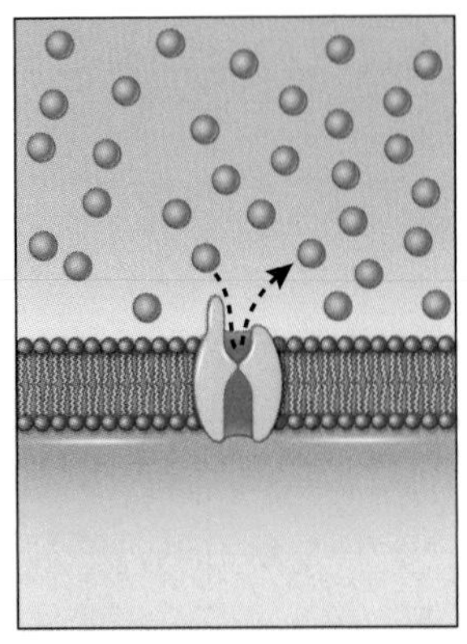

Closed. Na^+ ions cannot pass.

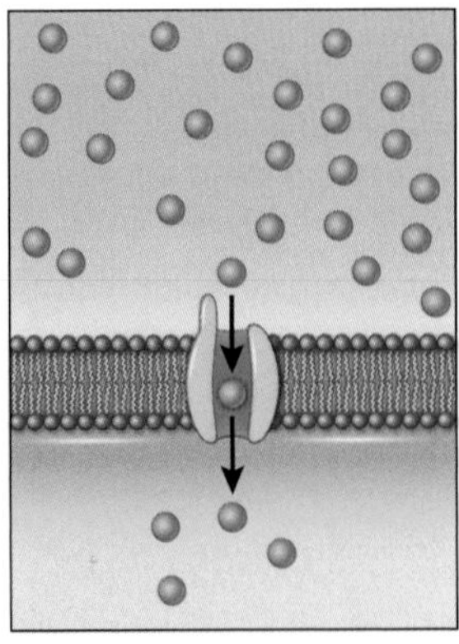

Open. Na^+ ions can pass.

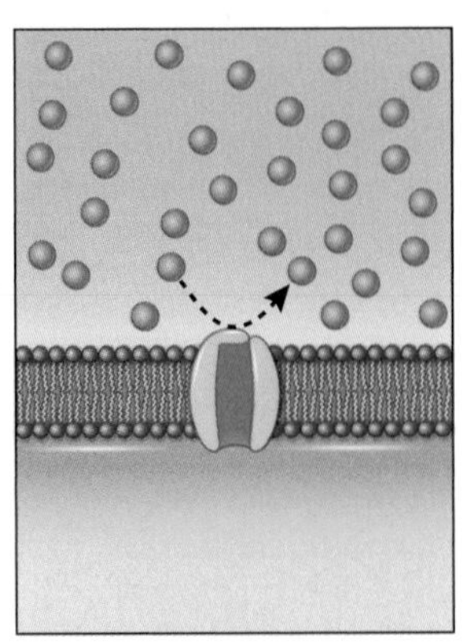

Inactivated. Channel is open but blocked by inactivation gate. Na^+ ions cannot pass.

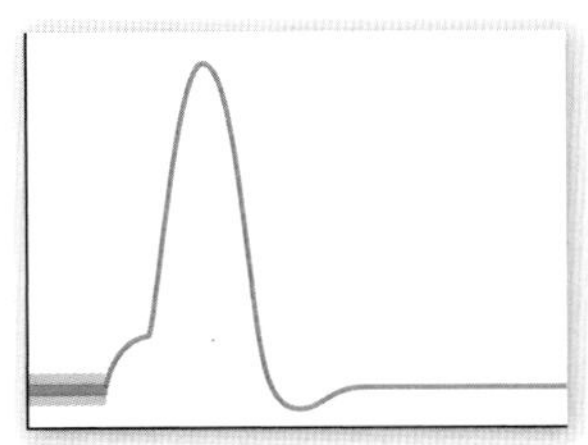

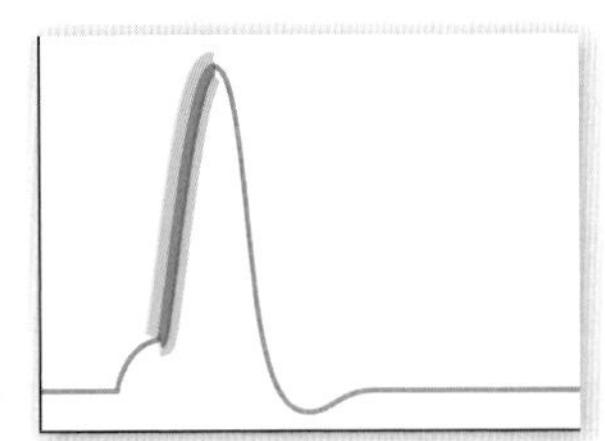

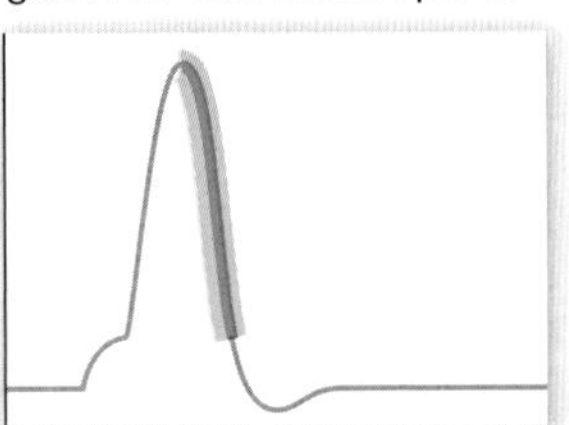

Learning Check

1. The resting membrane potential of a neuron is −70 mV. Currently, the membrane potential is −55 mV. Which of the following best describes the state of the neuron?
   a. Depolarized
   b. Hyperpolarized
   c. Polarized
   d. Repolarized
2. Which of the following contribute to a neuron having a resting membrane potential of −70 mV? Please select all that apply.
   a. The higher concentration of potassium inside the cell than outside the cell
   b. The higher concentration of proteins outside of the cell than inside the cell
   c. The high number of potassium leak channels throughout the axon
   d. The high number of sodium leak channels throughout the axons
3. The threshold voltage for a neuron is −30 mV. What is the first ion channel to open to initiate an action potential when the electrical charge reaches that threshold?
   a. Sodium
   b. Potassium
   c. Calcium
   d. Phosphate
4. The threshold voltage for a neuron is −30 mV. What type of channel is the first to open to initiate an action potential when the electrical charge reaches that threshold?
   a. Ligand-gated channel
   b. Mechanically gated channel
   c. Voltage-gated channel
   d. Leak channel

The Graded Potential All action potentials are initiated by a graded potential, but not all graded potentials lead to an action potential. Graded potentials are named "graded" because they can be any size (just like you can earn different levels of grades in your classes). The graded potential will cause an action potential only if it is large enough to depolarize the membrane to the threshold voltage, or if smaller graded potentials can add together to depolarize the membrane up to threshold voltage. Once the beginning of the axon depolarizes to threshold, an action potential is guaranteed because the voltage-gated channels will open. Graded potentials usually occur on the dendrites or cell body of a neuron, and the amount of change in the membrane potential is determined by the size of the stimulus that causes it.

Graded potentials can be of two sorts: either they are depolarizing or hyperpolarizing (see the "Anatomy of a Graded Potential" feature). For a membrane at the resting potential, a graded potential represents a change in that voltage either above or below −70 mV due to an external event. Depolarizing graded potentials are often the result of Na^+ or Ca^{2+} entering the cell. Both of these ions have higher concentrations outside the cell than inside; because they have a positive charge, they will move into the cell, causing it to become less negative relative to the outside. Hyperpolarizing graded potentials can be caused by K^+ leaving the cell or Cl^- entering the cell. If a positive charge moves out of a cell, the cell becomes more negative; if a negative charge enters the cell, the same thing happens. All graded potentials are due to the opening of either mechanically gated or ligand-gated channels, which differentiates them from action potentials, which involve the opening of voltage-gated channels. Many graded potentials occur because of events at the synapse.

Anatomy of...

A Graded Potential

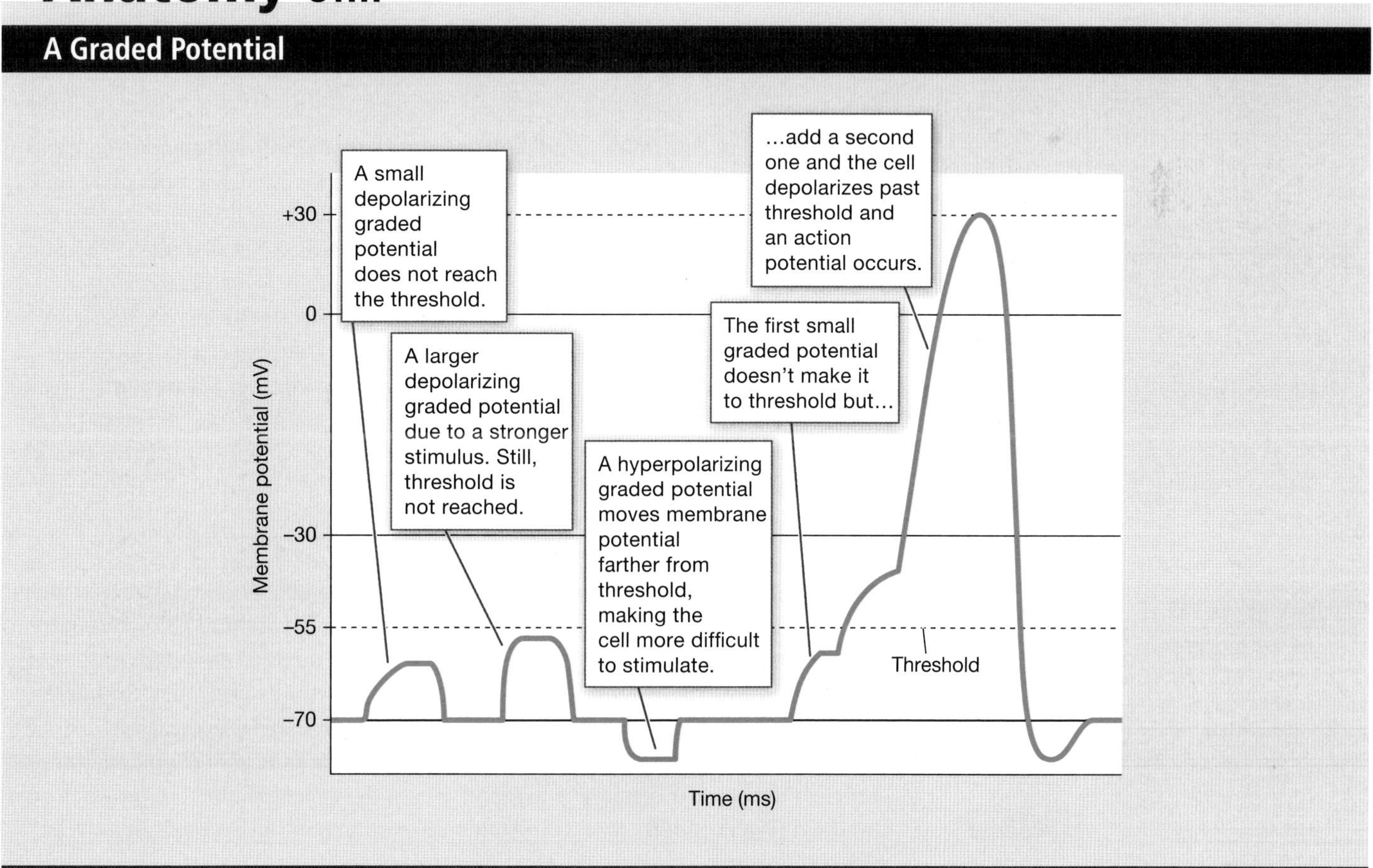

Apply to Pathophysiology

Septocaine

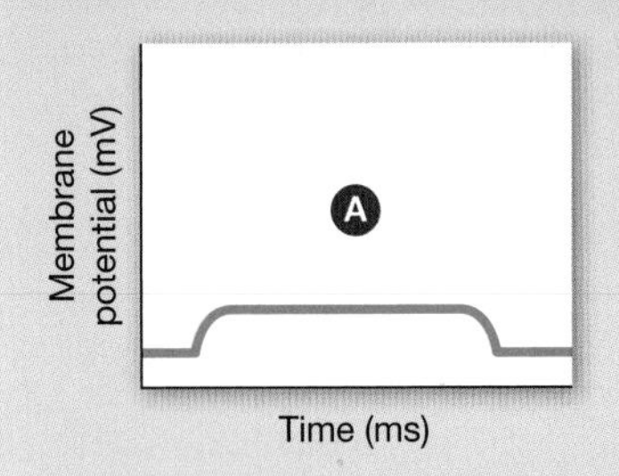

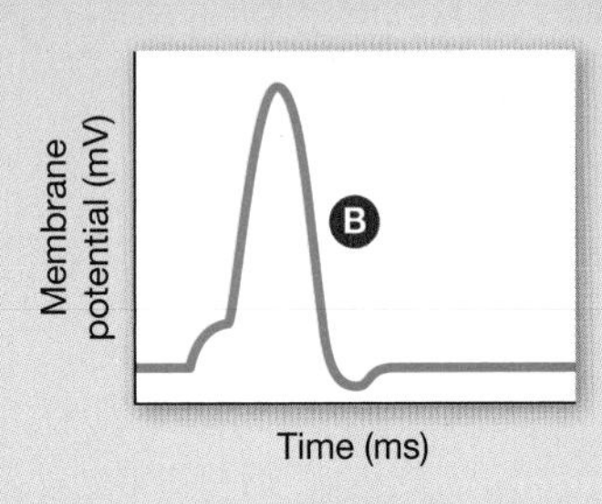

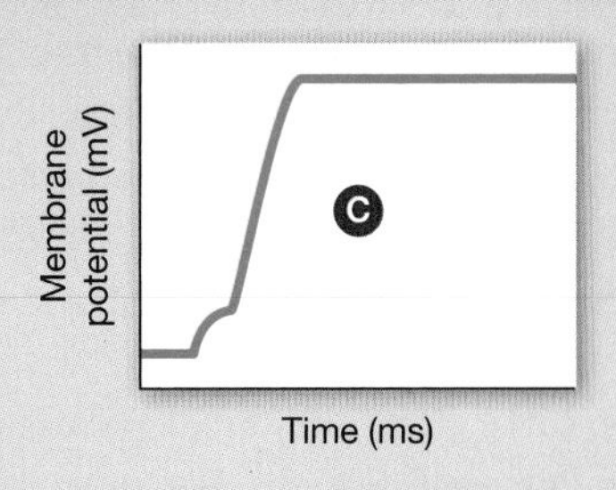

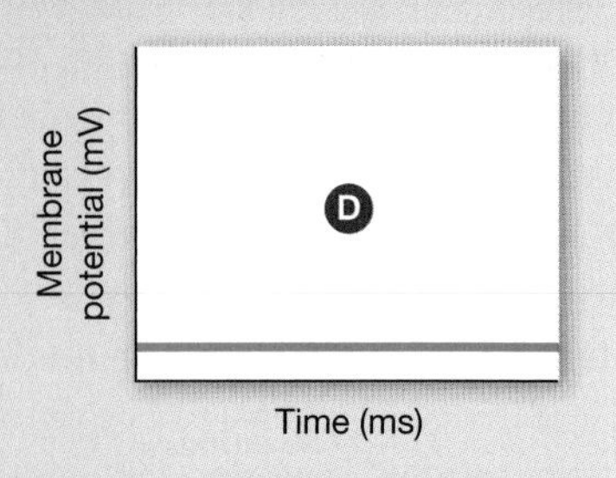

Novocaine is a drug that was used to numb, or block sensation; it was most commonly used during dental procedures such as root canals, filling cavities, tooth extractions, and so on. Due to the prevalence of allergic reactions to novocaine, however, between the 1980s and early 2000s it was gradually replaced as a dental anesthetic by septocaine and lidocaine.

Septocaine (articaine HCl and epinephrine) is injected at precise locations in order to affect the neurons closest to where the dental work will take place. The mechanism of Septocaine is that it prevents voltage-gated Na^+ channels from opening.

1. In an individual undergoing what would be painful dental work, but who is numb to the pain through the use of Septocaine, which of the following will still occur?
   A) Stimulus will occur but no graded potential and no action potential.
   B) Stimulus and graded potential will occur but no action potential.
   C) Stimulus, graded potential, and action potential will occur but the CNS will not integrate the signal.
2. Which type of neuron will be directly affected by the Septocaine?
   A) Afferent neuron
   B) Interneuron
   C) Efferent neuron
3. Which graph best represents the membrane potential of the affected neuron during the use of Septocaine?
   A) A
   B) B
   C) C
   D) D

Student Study Tip

Being excited is a positive feeling, making the membrane potential more positive.

summate (sum)

Types of Graded Potentials A **postsynaptic potential (PSP)** is the graded potential in the dendrites of a neuron that is receiving signals across a synapse from other cells. Postsynaptic potentials can be depolarizing or hyperpolarizing. Depolarization in a postsynaptic potential is called an **excitatory postsynaptic potential (EPSP)** because it causes the membrane potential to move toward threshold. Hyperpolarization in a postsynaptic potential is an **inhibitory postsynaptic potential (IPSP)** because it causes the membrane potential to move away from threshold. These graded potentials can lead to the neuron reaching threshold if the changes add together or **summate**. The combined effects of different types of graded potentials are illustrated in **Figure 13.12**. If the total change in voltage in the membrane of the axon hillock is a positive 15 mV, meaning that the membrane depolarizes from -70 mV to -55 mV, then the graded potentials will result in the membrane reaching threshold and an action potential will occur.

LO 13.3.6

Student Study Tip

Spatial starts with an "S" for multiple Sources of GPs; Temporal starts with "T" for one source firing many APs in a very short Time!

Since summation is the adding together of graded potentials, it can occur in many ways. In one example, EPSPs might occur in several synapses within a close period of time (**Figure 13.12A**). Because most neurons have hundreds or thousands of synaptic inputs, this type of summation—**spatial summation**—is likely to occur often. **Temporal summation**, on the other hand, is when a single synapse causes graded potentials within a short period of time. The time between the graded potentials is not sufficient to allow the membrane to repolarize, so the depolarization progresses toward threshold (**Figure 13.12B**). Spatial and temporal summation can act together as well.

LO 13.3.7

Figure 13.12 Summation of Graded Potentials

Multiple graded potentials can contribute to the membrane potential. (A) In spatial summation, multiple presynaptic neurons cause EPSPs within a short period of time. Their effects add together to bring the cell to threshold. (B) In temporal summation, a single presynaptic neuron causes EPSPs within a short enough period of time that the postsynaptic cell reaches threshold. (C) In this case of spatial summation, two synapses cause different types of graded potentials. The effect of IPSPs can cancel out the effects of EPSPs, preventing the postsynaptic cell from reaching threshold.

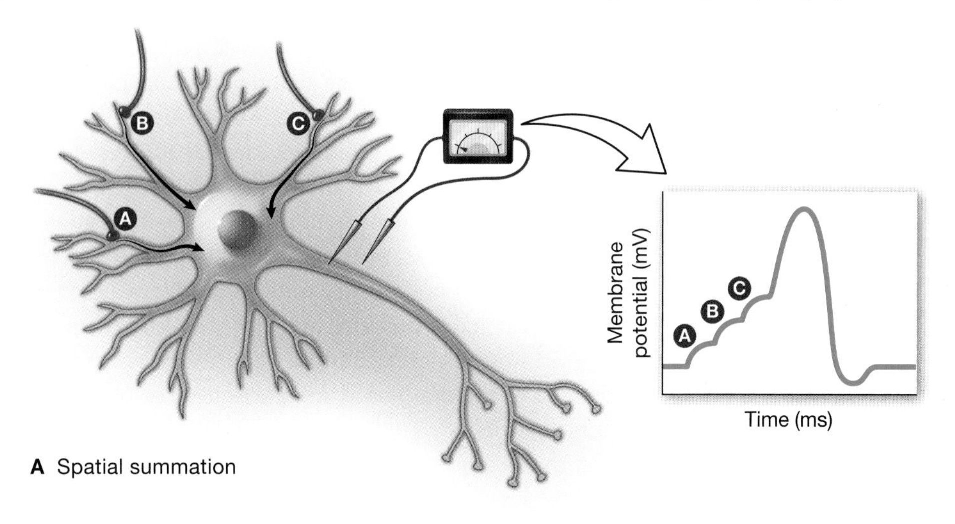

A Spatial summation

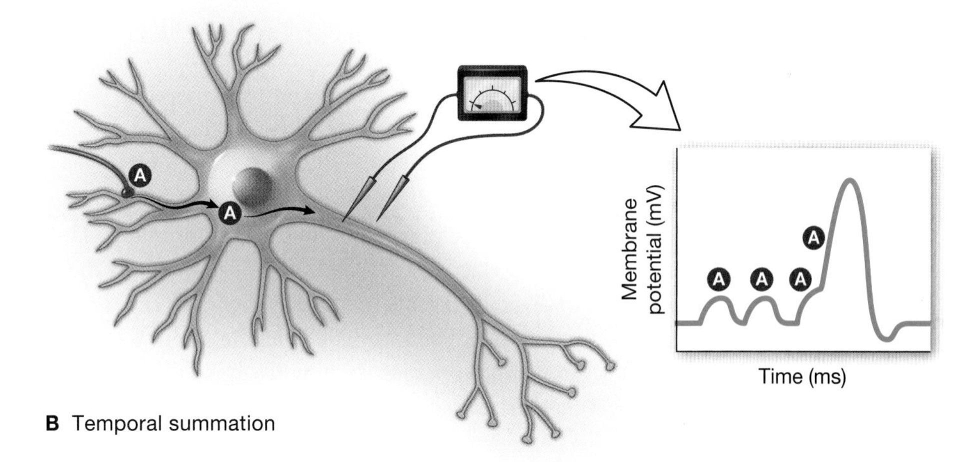

B Temporal summation

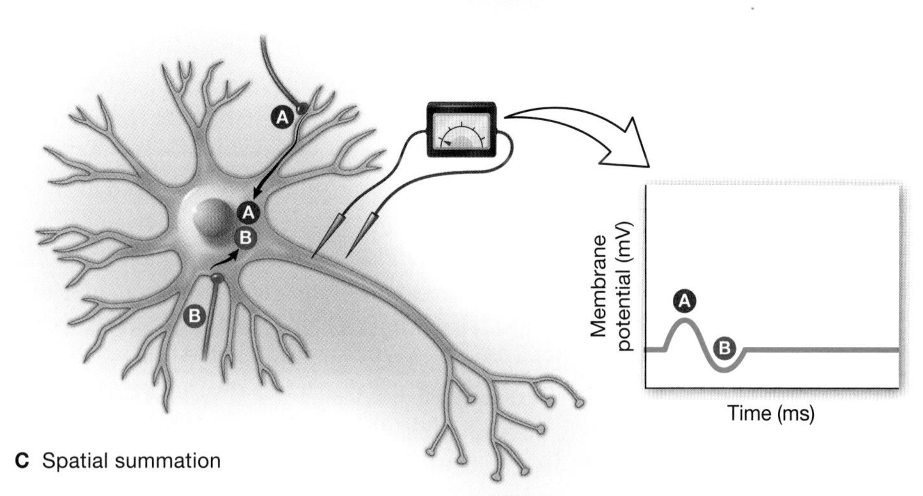

C Spatial summation

Learning Connection

Chunking

List the differences between graded and action potentials on separate index cards. Here are some ideas to get you started:

- One occurs first and one occurs as a result of the other.
- One has greater magnitude (size) and the other has lower magnitude.
- One involves a subthreshold potential change and the other occurs from a threshold potential.
- One remains in one local area and the other propagates down the axon.
- One can be a depolarization or a hyperpolarization, and the other can only begin as a depolarization.

After listing all of these differences, mix up your index cards, and then sort them into two piles, one for graded potentials and one for action potentials.

Student Study Tip

APs are "all or nothing" events, like flushing a toilet: you can't do it halfway. When you press the lever you get the same amount of power every flush, just like you get the same AP amplitude every time threshold is reached. If you don't press the lever hard enough, you don't get any flush at all.

Summation can occur between an excitatory and inhibitory synapse. In these cases, the two events may cancel each other out, making it unlikely that the excitatory event will lead the cell toward threshold.

LO 13.3.8

Refractory Periods Action potentials are often said to be "all-or-none," meaning that they either occur or they do not. This property is due to the threshold voltage. If the threshold for opening the voltage-gated channels is reached, the action potential will occur; if the threshold is not reached, the action potential will not occur. Once threshold is reached, the set of events is almost exactly the same for every action potential. However, once an action potential is initiated, the next action potential cannot begin. For example, once a Na^+ voltage-gated channel is already open, it cannot be stimulated to stay open or open further; it must reset back to its closed position before opening again. This property distinguishes neurons from muscle cells, which can continue to stay in a depolarized and contracted state until fatigue is reached. The period of time after the start of the first action potential and before a second action potential can begin is referred to as the **refractory period**. There are two phases of the refractory period: the **absolute refractory period** and the **relative refractory period**. During the absolute refractory period, another action potential cannot start, regardless of the strength of the stimulus. This is because the Na^+ voltage-gated channels are either open or inactivated. Once some of the channels are back to their resting conformation (less than -55 mV), a new action potential could be started, but only by a stronger stimulus than the one that initiated the current action potential; this phase is the relative refractory period (**Figure 13.13**).

A significant factor in the refractory period, and therefore the delay between one action potential and the ability to fire another, is the inactivation gates of Na^+ voltage-gated channels. You may be asking yourself why these channels have inactivation gates, and why K^+ voltage gated channels do not have them. Both types of voltage-gated channels open and close in response to the same stimulus. They open when the membrane depolarizes above -55 mV and they close when the membrane repolarizes below -55 mV. However, the events that occur after each channel opens are markedly different. Let's examine the K^+ voltage-gated channels first. K^+ voltage-gated channels open at depolarization; once open, they allow K^+ ions to leave the cell, repolarizing the cell. The cell will repolarize due to the outflow of K^+ ions through the channel until the cell repolarizes back to below the threshold voltage. The K^+ channel will then be triggered to close. Effectively, K^+ voltage-gated channels follow a negative feedback loop; the events that follow their opening lead to their closure (**Figure 13.14B**). Na^+ voltage-gated channels, on the other hand, follow a positive feedback loop pattern. These channels open at depolarization and then they allow Na^+ ions to flow into the cell, further depolarizing the membrane. In order to close again, the channel must wait for the membrane to repolarize, but the membrane will proceed further and further from repolarization as long as the channel is open. Thus, the inactivation gate is essential to stop the positive feedback loop (**Figure 13.14A**).

LO 13.3.9

Refractory periods play an important role in the functioning of the nervous system. Notice that neurons have only one signal—the action potential—and these signals are all-or-none, they are always the same. However, the nervous system must be capable of nuance. For example, we need to be able to feel the difference between something that is pleasantly warm and something that is dangerously hot. The neurons cannot fire louder or more intense action potentials; their only way of communicating intensity is through the frequency of action potentials. The refractory period prevents action potentials from being fired too frequently, but because very intense stimuli can cause action potentials during the relative refractory period, the neurons are able to communicate very intense stimuli by firing rare relative refractory period action potentials.

Figure 13.13 Refractory Periods

(A) A small amount of stimulus is able to depolarize the neuron past threshold and cause an action potential. (B) While in the absolute refractory period, no stimulus, even a tremendously strong one, can trigger a second action potential because all of the Na^+ voltage-gated channels are either open or inactivated. (C) During the relative refractory period, some Na^+ voltage-gated channels have closed again; a very strong stimulus could trigger a second action potential, though the action potential might not depolarize the cell to the typical +30 mV value.

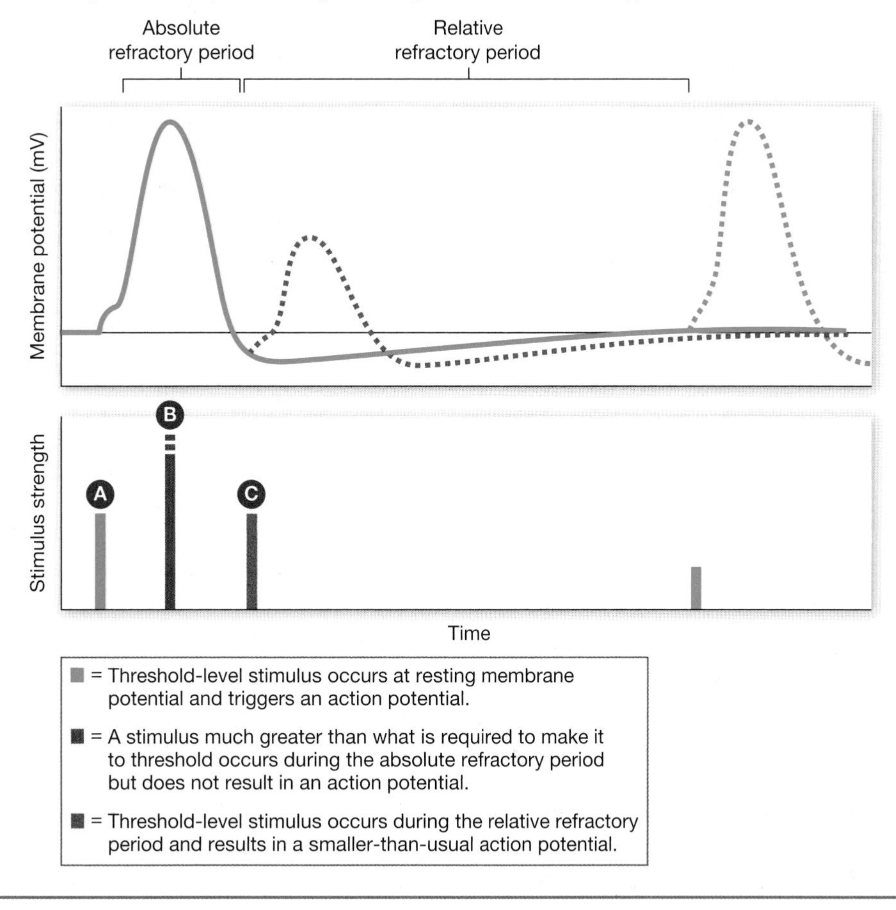

Figure 13.14 The Function of Inactivation Gates of Na^+ Voltage-Gated Channels

(A) Na^+ voltage-gated channels require inactivation gates because the opening of the channels sets off a series of events that follow a positive feedback loop pattern. (B) K^+ voltage-gated channels do not have inactivation gates because opening of these channels sets off a series of events that follow a negative feedback loop pattern.

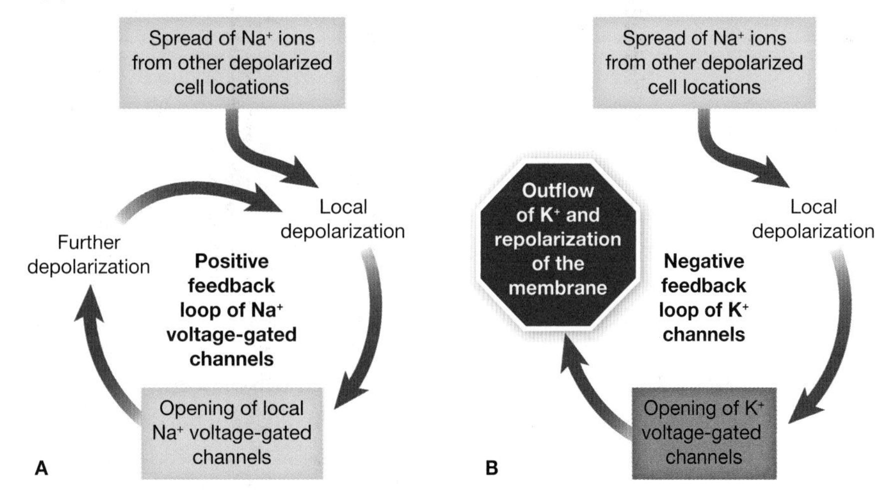

Propagation of the Action Potential So far, we have discussed the events of the action potential at a single, isolated location along the axon. In reality, though, we know that axons can be very long—sometimes inches or even feet in length—and the action potential must spread along the axon in order to enable communication from cell body to axon terminal. The action potential is initiated at the beginning of the axon, at the axon hillock. There is a high density of Na^+ voltage-gated channels here, and this is the location where voltage changes from the dendrites or cell body are initiated. If the axon hillock depolarizes to threshold, these Na^+ voltage-gated channels open, allowing a deluge of Na^+ ions to enter the beginning of the axon. Stop for a moment and remember that ions (or any molecule) will always move randomly but according to their concentration gradients. Thus, the inflowing Na^+ ions are less likely to flow backward toward the cell body, which has already depolarized, and are more likely to flow toward the axon terminal, a region of the cell that is still less positive and has fewer Na^+ ions. As these Na^+ ions diffuse through the axon, they depolarize the next segment of the axon, triggering the opening of voltage-gated channels in this location. The panel of drawings in **Figure 13.14** illustrates this series of events.

LO 13.3.10

Because Na^+ voltage-gated channels are inactivated at the peak of the depolarization, they cannot be opened again for a brief time—the absolute refractory period. Because of this, any depolarization spreading back toward previously opened channels has no effect. At the same time as the Na^+ channels inactivate, the K^+ channels finally open, bringing this segment of the membrane back to a depolarized state. As the membrane repolarizes to become more negative than -55 mV, the voltage-gated channels reset back to their closed state and this segment of the axon enters its relative refractory period. The next segment of the axon is currently depolarizing and the segment beyond that is about to begin depolarizing. Thus, like dominoes, the axon depolarizes and recovers in segments until the wave of depolarization and repolarization reaches the axon terminals.

The action potential propagation, as described so far and depicted in **Figure 13.15**, describes the events that occur in unmyelinated axons. Remember that myelination, a wrapping of plasma membranes of glial cells, acts as an insulator to the electrical current down the axon. In these myelinated neurons, the action potential propagates differently. Ions that enter at the axon hillock start to spread along the length of the axon segment, but there are no Na^+ voltage-gated channels embedded in the membrane except at the neurofibril nodes. In the segments of the axon that are covered with glial cells, there are no voltage-gated channels. The myelination prevents any leak of the ions across the membrane, so once they have entered the cell, the Na^+ ions flow toward the next node and depolarize the node, opening the voltage-gated channels there (**Figure 13.16**). Because there is not constant opening of these channels along the axon segment, the depolarization spreads at a much faster speed. Therefore, myelinated axons send action potentials faster than unmyelinated axons.

Propagation along an unmyelinated axon is referred to as **continuous conduction**; along the length of a myelinated axon, it is **saltatory conduction**. Continuous conduction is slow because each channel shape change takes a bit of time, and the axon is covered continuously with voltage-gated channels, each of which must change shape to propagate the action potential. Saltatory conduction is faster because there are far fewer channel shape changes required. The action potential is often described as jumping from one node to the next because all of the events of the action potential occur only at the nodes. In fact, *saltare* means "to jump" in Latin.

LO 13.3.11

Along with the myelination of the axon, the diameter of the axon can influence the speed of conduction. You will remember from earlier chapters that the rate of any kind of

Figure 13.15 **Action Potential Propagation down an Unmyelinated Axon**

An action potential is propagated down an unmyelinated axon more slowly than a myelinated axon until it reaches the axon terminal. At each successive portion of the axon, depolarization is quickly followed by repolarization.

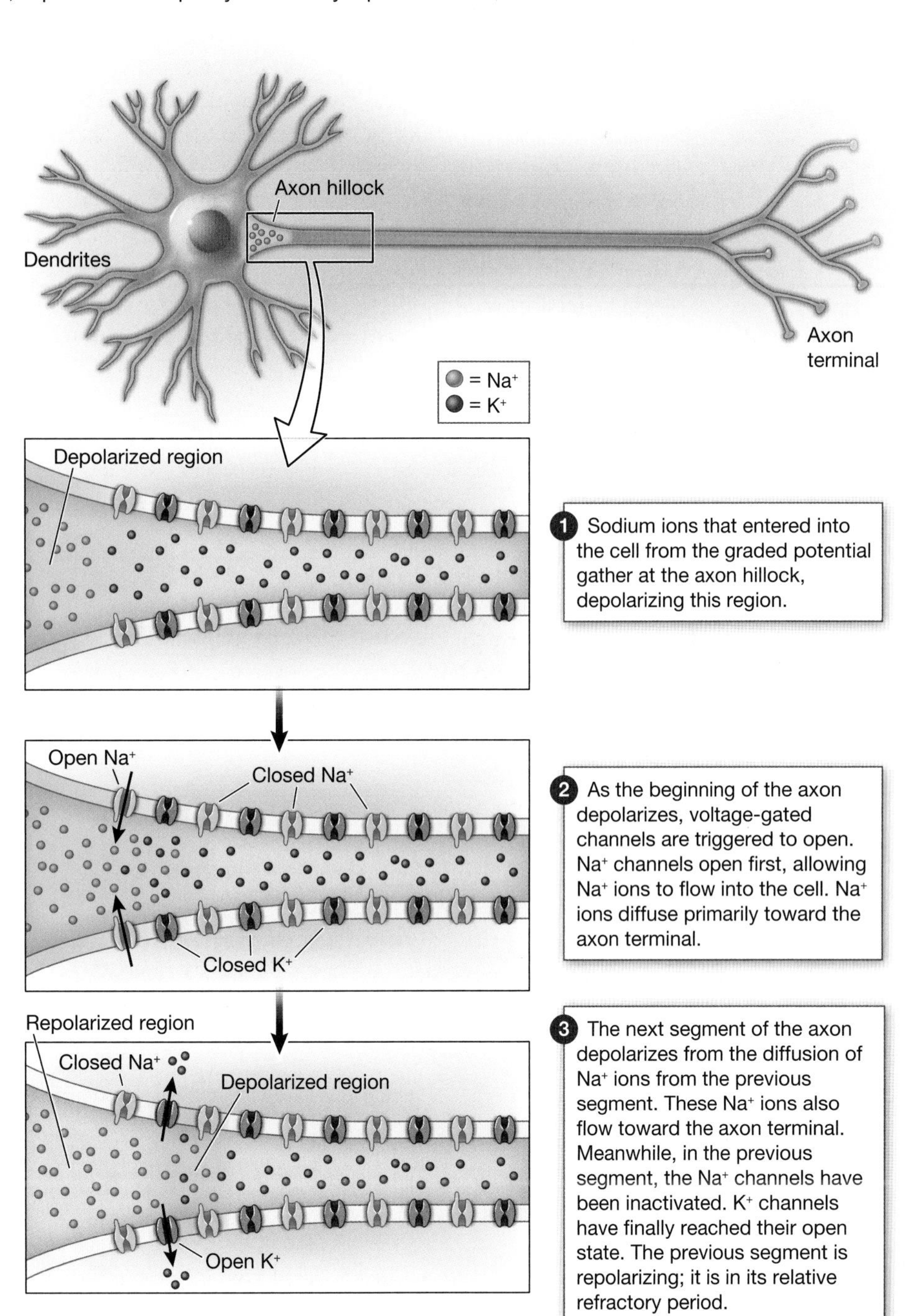

1 Sodium ions that entered into the cell from the graded potential gather at the axon hillock, depolarizing this region.

2 As the beginning of the axon depolarizes, voltage-gated channels are triggered to open. Na^+ channels open first, allowing Na^+ ions to flow into the cell. Na^+ ions diffuse primarily toward the axon terminal.

3 The next segment of the axon depolarizes from the diffusion of Na^+ ions from the previous segment. These Na^+ ions also flow toward the axon terminal. Meanwhile, in the previous segment, the Na^+ channels have been inactivated. K^+ channels have finally reached their open state. The previous segment is repolarizing; it is in its relative refractory period.

Figure 13.16 Action Potential Propagation down a Myelinated Axon

An action potential is propagated down a myelinated axon very quickly, proceeding from one node to the next, until it reaches the axon terminal. At each node, depolarization is quickly followed by repolarization.

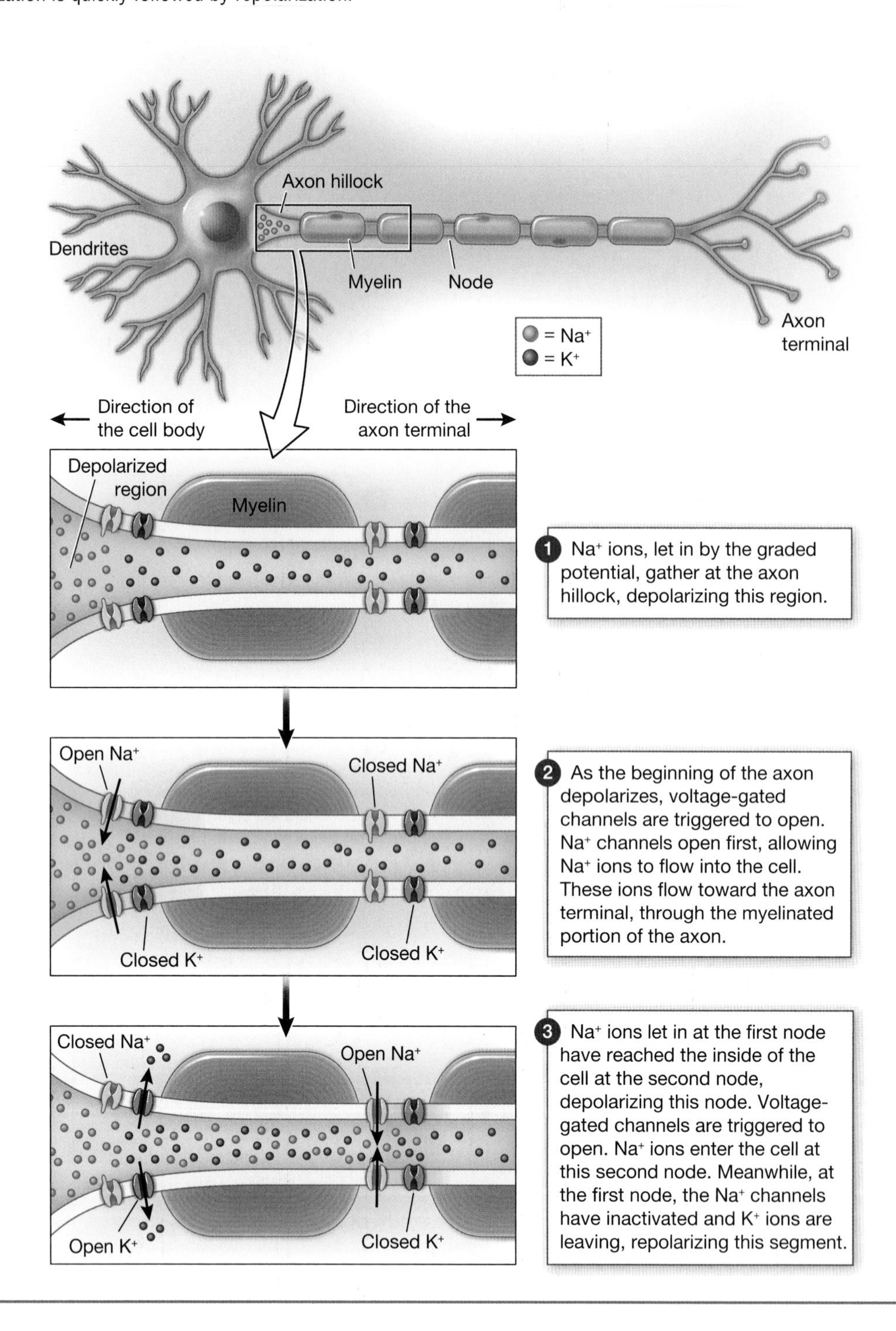

Axon hillock
Dendrites
Myelin
Node
Axon terminal
= Na+
= K+
Direction of the cell body
Direction of the axon terminal
Depolarized region
Myelin
1 Na+ ions, let in by the graded potential, gather at the axon hillock, depolarizing this region.
Open Na+
Closed Na+
Closed K+
Closed K+
2 As the beginning of the axon depolarizes, voltage-gated channels are triggered to open. Na+ channels open first, allowing Na+ ions to flow into the cell. These ions flow toward the axon terminal, through the myelinated portion of the axon.
Closed Na+
Open Na+
Open K+
Closed K+
3 Na+ ions let in at the first node have reached the inside of the cell at the second node, depolarizing this node. Voltage-gated channels are triggered to open. Na+ ions enter the cell at this second node. Meanwhile, at the first node, the Na+ channels have inactivated and K+ ions are leaving, repolarizing this segment.

flow in physiology is a product of its gradient and resistance to its flow (**Figure 13.17**). In this case, the rate of flow of Na^+ ions down the axon from axon hillock to axon terminal is influenced by two factors. The first is the concentration gradient that exists between one segment of the axon and the next. Because of the consistent action of the Na^+/K^+

Figure 13.17 Flow Rate Equation as It Applies to the Flow of Cations inside of an Axon

The flow rate equation is used to help us understand many different types of physiological flow. Here it is used to understand that the gradient contributes to driving the flow of cations from the cell body to the axon terminal. One of the forces that slows this flow is the diameter of the axon.

$$\text{Flow} \propto \frac{\Delta P}{R}$$

Flow of Na^+ ions down the axon

Gradient of $[Na^+]$ between segments of the axon

Resistance: diameter of axon

More Na^+

Less Na^+

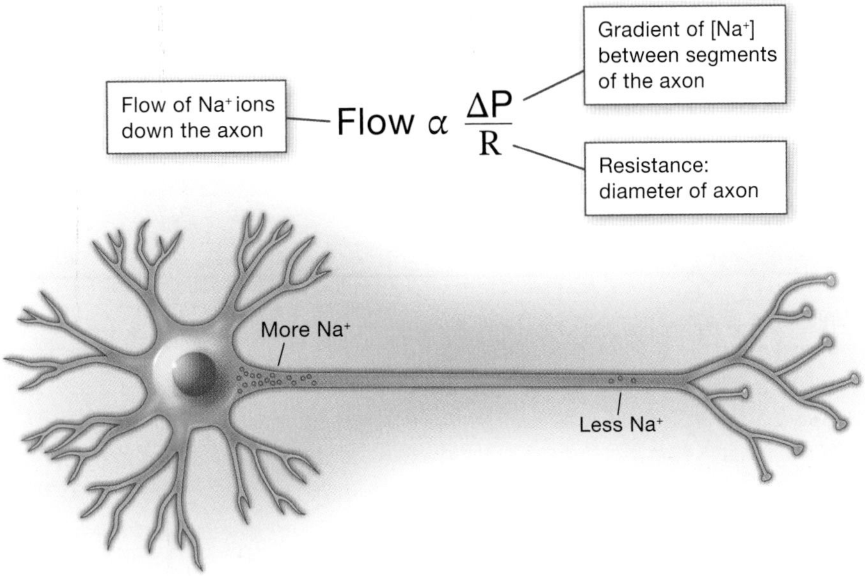

pump, this gradient is likely to be relatively similar from one axon to the next. The other factor, resistance to flow, is influenced by the diameter of the axon. Much as water runs faster in a wide river than in a narrow creek, the wave of depolarization spreads faster down a wide axon than down a narrow one. This resistance factor is generally true for electrical wires or plumbing, just as it is true for axons and blood vessels, although the specific conditions are different at the scales of ions or cells versus water in a river.

LO 13.3.12

Learning Connection

Micro to Macro

Myelination has an important effect on the conduction velocity of pain fibers (axons). Fast pain (type A delta, myelinated) fibers are stimulated when you experience acute pain, such as that of a cut; their conduction velocity is 5–40 m/sec. In contrast, slow pain (type C, unmyelinated) fibers, which respond mainly to visceral pain, such as that of a stomach ache, conduct at speeds of only 0.5–2 m/sec. The events occurring in these microscopic neurons influence how quickly you feel different types of pain.

Cultural Connection

Rattled Brains

Chronic traumatic encephalopathy (CTE) is a brain condition that occurs after repeated head trauma. Most of what we know about CTE comes from studies done on both mice and former athletes. During traumatic brain injury, the soft tissue of the brain can be damaged as it collides with the hard skull or even the tough dura mater within the cranium. Symptoms of CTE include personality changes, confusion, erratic behavior, and memory and thinking problems.

CTE is similar to Alzheimer's in many ways in that both conditions are progressive, meaning that they both get worse over time and they both cause a loss of brain function. One hallmark of CTE is that it often disproportionately affects the frontal lobe, leading to personality and behavioral changes. In both diseases, proteins that are typically found inside neurons build up in the extracellular space. These proteins interfere with neuronal communication across the synapses. Failure to receive signals may cause neurons to die. Over time this leads to a loss of brain mass as well as brain function.

CTE is especially common in boxers and players of American-style football, both of whom incur frequent and repeated blows to the head. There has been a significant amount of scientific research focused on football players. In one study that included the brains of 110 former NFL (National Football League) football players, 109 of the brains exhibited evidence of CTE. The force and frequency of repeated blows to the head of these athletes has led to progressive changes in their brains. The NFL has invested over $100 million in CTE research, and hopefully changes can be made to the helmets or style of play in American-style football so that we can protect the brains of those who enjoy participating in this sport.

Learning Check

1. What is the difference between an action potential and a graded potential?
   a. Action potentials are in the dendrites while graded potentials are in the axon.
   b. Action potentials are smaller in voltage compared to graded potentials.
   c. Action potentials are initiated once graded potentials summate to the threshold.
   d. Action potentials are hyperpolarizing while graded potentials are depolarizing.
2. Gamma-aminobutyric acid (GABA) is a neurotransmitter in the body. When GABA binds to its receptor, the neuron increases the influx of Cl^- and outflow of K^+. Which of the following describes the graded potential that is formed by GABA?
   a. Excitatory postsynaptic potential
   b. Inhibitory postsynaptic potential
3. Which of the following describes the period of an action potential in which another action potential cannot start, regardless of the strength of the stimulus?
   a. Absolute refractory period
   b. Relative refractory period
4. There are three zones within an axon. Zone A is at the end of the action potential; the sodium ion channels have just closed and potassium ion channels are just opening. Zone B is just distal to Zone A, where an action potential is just starting where the sodium ion channels just opened. Zone C is just distal to Zone B, where an action potential has not been reached yet and the resting membrane potential is still below the threshold. Which of the following zones has the most sodium ions present?
   a. Zone A
   b. Zone B
   c. Zone C

13.4 Communication between Neurons

Learning Objectives: By the end of this section, you will be able to:

13.4.1 Define a synapse, and explain the difference between an electrical synapse and a chemical synapse.

13.4.2 Describe the structures involved in a typical chemical synapse (e.g., axon terminal [synaptic knob], voltage-gated calcium channels, synaptic vesicles of presynaptic cell, synaptic cleft, neurotransmitter receptors of the postsynaptic cell).

13.4.3 Describe the events of synaptic transmission in proper chronological order from the release of neurotransmitter by synaptic vesicles to the effect of the neurotransmitter on the postsynaptic cell.

13.4.4 Describe the different mechanisms (e.g., reuptake, enzymatic breakdown, diffusion) by which neurotransmitter activity at a synapse can be terminated.

13.4.5 List the most common excitatory and inhibitory neurotransmitter(s) used in the nervous system.

13.4.6 Explain how a single neurotransmitter can elicit different responses at different postsynaptic cells.

So far, we have described the events, within one neuron, from graded potential to the action potential reaching the axon terminal. Remember that communication from periphery to the brain, for example, to communicate a touch sensation or signal pain, occurs along a chain of neurons. In this section we will tackle the events that occur between neurons; that is, the end result of an action potential in one neuron and how it initiates an action potential in its neighbor.

13.4a Synapses

LO 13.4.1

There are two types of connections between excitable cells—chemical synapses and electrical synapses. In a **chemical synapse**, a chemical signal—namely, a neurotransmitter—is released from one cell and diffuses across a small extracellular region to bind to receptors on the other cell. The neuromuscular junction that we described in Chapter 11 is an example of a chemical synapse. In an **electrical synapse**, there is a direct physical connection between the two cells so that ions of the electrical signal can pass directly from within one cell to within its neighbor (**Figure 13.18**). Electrical synapses are much less common in the human nervous system. This section will concentrate on the chemical type of synapse.

Figure 13.18 Chemical and Electrical Synapses

(A) Chemical synapses represent most of the synapses in the adult nervous system. In a chemical synapse, the cells do not physically touch, but communicate via secreted chemicals—neurotransmitters—across a small space, the synaptic cleft. (B) Electrical synapses occur in developing human nervous systems, between cardiac muscle cells, and in other organisms. In an electrical synapse, the two cells are joined by gap junctions so that the action potential of one cell passes into the neighboring cell.

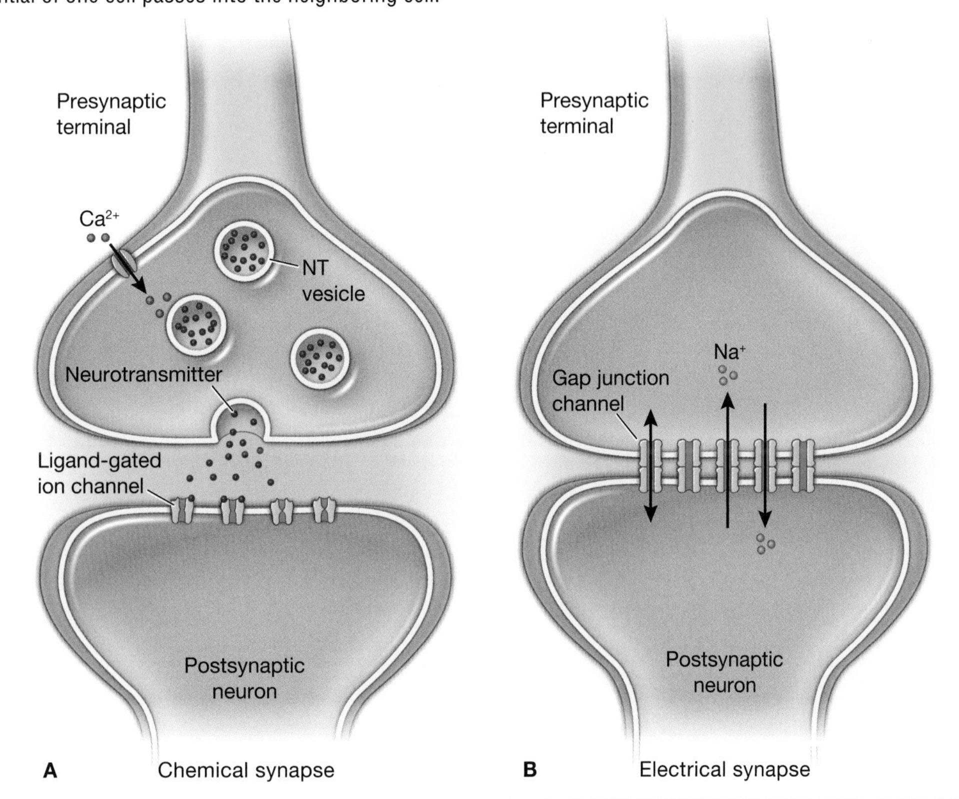

All chemical synapses have the following common components (see the "Anatomy of a Synapse" feature):

- Presynaptic cell
- One type of neurotransmitter
- Synaptic cleft
- Receptors for the neurotransmitter
- Postsynaptic cell
- A system for clearing the neurotransmitter from the synapse

LO 13.4.2

The synaptic events begin when the action potential reaches the axon terminal of the presynaptic neuron (**Figure 13.19**). Along the axon of each neuron, we have two types—Na^+ and K^+—of voltage-gated channels. Here at the axon terminal, we meet a third type of voltage-gated channel, the Ca^{2+} voltage-gated channel. When an action potential reaches the axon terminals, the Ca^{2+} voltage-gated channels in the membrane of the synaptic end bulb open. Ca^{2+} ions will enter the cell, flowing down their concentration gradients. The Ca^{2+} ions bind to proteins on the neurotransmitter vesicles. The Ca^{2+} ions facilitate the merging of the vesicle with the plasma membrane so that the neurotransmitter is released through exocytosis into the small gap between the cells, known as the **synaptic cleft**.

LO 13.4.3

Figure 13.19 The Structure of a Chemical Synapse

A synapse between two neurons includes the axon terminal of the presynaptic neuron and the surface of the postsynaptic cell. The entrance of extracellular Ca^{2+} into the axon terminal causes vesicle fusion and neurotransmitter release. The released neurotransmitter diffuses across the synaptic cleft and binds to receptors on the postsynaptic cell.

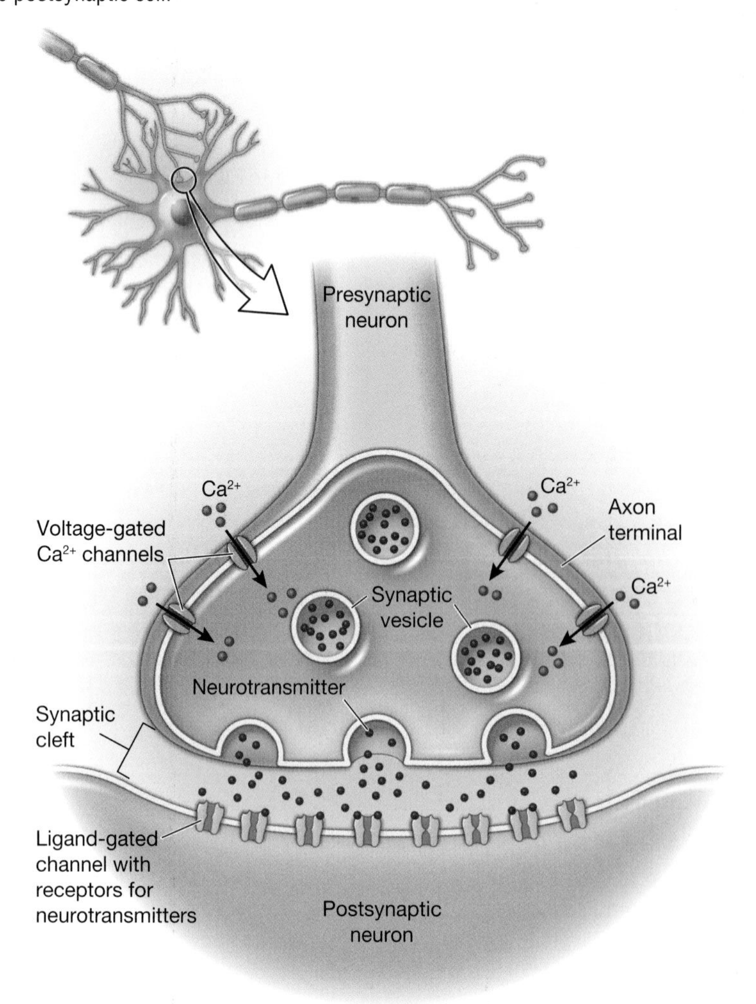

The neurotransmitter molecules diffuse within the synaptic cleft. This movement is random, but some of the neurotransmitter molecules will interact with neurotransmitter receptors expressed on the surface of the postsynaptic cell. These receptors are ligand-gated ion channels that are specific for the neurotransmitter. The binding of the neurotransmitter to its receptor, similar to a key turning in a lock, opens the channel, allowing ions to flow. This is the graded potential. If the receptor is a Na^+ ligand-gated channel, as shown in the "Anatomy of a Synapse" feature, then Na^+ ions will flow into the postsynaptic cell, depolarizing it. This is an EPSP, and the membrane potential of the postsynaptic cell moves closer to threshold. If the receptor is a Cl^- ligand-gated ion channel, Cl^- ions may flow into the cell—if it is in a greater concentration outside the cell—and will hyperpolarize the cell; this is an IPSP. Finally, if the receptor is a K^+ ligand-gated ion channel, K^+ ions will flow out of the cell, hyperpolarizing it.

The effect on the postsynaptic cell will continue until the neurotransmitter is cleared from the synapse. All chemical synapses are open spaces and so many of the neurotransmitter molecules will diffuse away from the area over time. Outside of the synapse, these molecules are likely cleaned up and destroyed or recycled by glial cells. However, most synapses have additional mechanisms for clearing neurotransmitter faster than simply waiting for it to diffuse away. In addition to diffusion, there are two methods that may shorten the time that neurotransmitter is present in the synapse. One is that some presynaptic cells have the ability to take the released neurotransmitter back into the presynaptic cell. In this process, called **reuptake**, the endocytosed neurotransmitter may be stored for the next use, or destroyed and recycled. In the second option, *breakdown*, the postsynaptic cell expresses enzymes on its membrane that are able to break down neurotransmitter in the synapse. This mechanism is depicted in the "Anatomy of a Synapse" feature.

LO 13.4.4

Neurotransmitters There are several classes of neurotransmitters that are found at various synapses in the nervous system. Neurons are often named by the type of neurotransmitter they release. Most neurons produce and release just one neurotransmitter, even if they have many axon terminals; all axon terminals release the same neurotransmitter. There are often many different receptors for one neurotransmitter; the neurotransmitter serotonin, for example, has at least 15 different known receptors (this number might grow as science advances).

Cholinergic is the term for neurons that release the neurotransmitter acetylcholine. This includes all of the neuromuscular junctions, as well as many other cholinergic synapses that are found in the nervous system.

There are two types of cholinergic receptors—the **nicotinic receptor**, which is found in the NMJ as well as the adrenal medulla and some autonomic synapses, and the **muscarinic receptor**, which is found in many organs that receive autonomic input. Both of these receptors are named for drugs that interact with the receptor in addition to acetylcholine. The drug nicotine, which is found in all tobacco products and some vaping devices (vape pens), is an agonist for nicotinic receptors—meaning that it will bind to the nicotinic receptor and activate it in the same way as acetylcholine. Muscarine, which is a product of certain mushrooms, is an agonist for the muscarinic receptor. However, nicotine will not bind to the muscarinic receptor and muscarine will not bind to the nicotinic receptor. Acetylcholine is cleared from the synapse through breakdown by the enzyme acetylcholinesterase.

LO 13.4.5

Another group of neurotransmitters is amino acids. These include GABA (gamma-aminobutyric acid, a derivative of glutamate), and glycine. The neurons that produce and release these neurotransmitters are glutamatergic, GABAergic, and glycinergic, respectively. They each have their own receptors. Amino acid neurotransmitters are eliminated

Anatomy of...

A Synapse

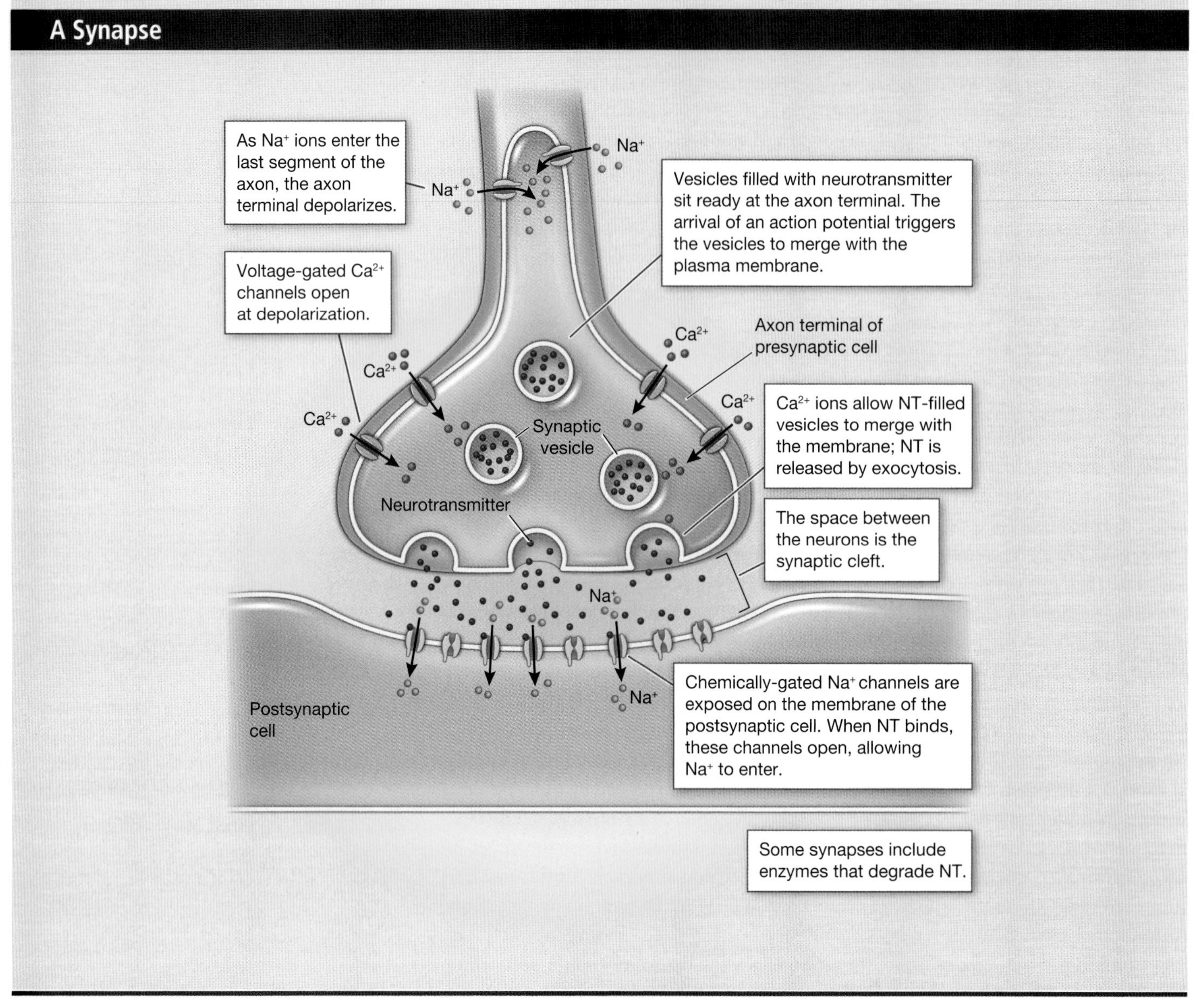

from the synapse by reuptake. A pump in the cell membrane of the presynaptic neuron will clear the amino acid from the synaptic cleft so that it can be recycled, repackaged in vesicles, and released again. GABA is a notable neurotransmitter because it is the primary mechanism for inhibiting neurons via IPSPs. GABA receptors are Cl^- channels, and GABA binding hyperpolarizes cells and makes them less likely to fire action potentials. Substances that block GABA release can induce convulsions.

As Na^+ ions enter the last segment of the axon, the axon terminal depolarizes.
Na^+
Na^+
Vesicles filled with neurotransmitter sit ready at the axon terminal. The arrival of an action potential triggers the vesicles to merge with the plasma membrane.
Voltage-gated Ca^{2+} channels open at depolarization.
Ca^{2+}
Axon terminal of presynaptic cell
Ca^{2+}
Ca^{2+}
Ca^{2+}
Ca^{2+} ions allow NT-filled vesicles to merge with the membrane; NT is released by exocytosis.
Synaptic vesicle
Neurotransmitter
The space between the neurons is the synaptic cleft.
Na^+
Chemically-gated Na^+ channels are exposed on the membrane of the postsynaptic cell. When NT binds, these channels open, allowing Na^+ to enter.
Na^+
Postsynaptic cell
Some synapses include enzymes that degrade NT.

Another class of neurotransmitter is the **biogenic amine**, a group of neurotransmitters that are made from amino acids. Serotonin is one example. Serotonin is cleared from the synapse through reuptake. An important class of drugs, selective serotonin reuptake inhibitors (SSRIs) are used to treat mood disorders, including depression. Depression is more common than you might expect. While there are several potential causes of depression, drugs that increase the availability of serotonin are effective for many people.

Other biogenic amines include dopamine, norepinephrine, and epinephrine. Neurons that release dopamine are dopaminergic. Dopamine is removed from the synapse primarily by reuptake, and there are five known types of dopamine receptors. Norepinephrine and epinephrine are two very similar neurotransmitters that actually share receptors. Neurons that release norepinephrine and epinephrine are adrenergic neurons and the receptors and their four receptors are alpha and beta adrenergic receptors. Norepinephrine and epinephrine are cleared from the synapse through either reuptake or through breakdown by an enzyme called *monoamine oxidase*. Norepinephrine and epinephrine can also be secreted into the bloodstream, where they act as hormones. The common neurotransmitters and their characteristics are summarized in **Table 13.3**.

The important thing to remember about neurotransmitters, and signaling chemicals in general, is that the effect is entirely dependent on the receptor. The release of a neurotransmitter may have different, even completely opposite, effects on different cells, depending on which receptor the postsynaptic cell expresses. Many of the neurotransmitter receptors are, as we have previously discussed, ligand-gated ion channels such as the nicotinic receptor for acetylcholine. Other receptors, instead of functioning as ion channels, function as ligand-binding sites; when ligands bind to these receptors, it induces a variety of metabolic changes in the cell. These receptors may also cause changes in the cell, such as the activation of genes in the nucleus, and therefore the increased synthesis of proteins. In neurons, these kinds of changes are often the basis of stronger connections between cells at the synapse and may be the basis of learning and memory.

Learning Connection

Broken Process

What do you think would happen in the nervous system if we altered the clearance of a neurotransmitter, leaving the NT in the synapse for a longer period of time?

LO 13.4.6

Table 13.3 Characteristics of Neurotransmitter Systems

Neurotransmitter Family	Specific Neurotransmitters	Receptor Names	Clearance from Synapse	Roles
Cholinergic	Acetylcholine (Ach)	Nicotinic Muscarinic	Diffusion Degradation by acetylcholinesterase (AChE)	The NT of all NMJs as well as widespread CNS function. A decrease in ACh in certain brain areas is found in Alzheimer's disease.
Biogenic Amines	Norepinephrine/ Epinephrine	Alpha (α) and beta (β) adrenergic receptors	Diffusion Reuptake	Roles in mood and regulation of organ function
	Dopamine	Dopamine R (5 known)	Diffusion Reuptake	Reward system neurotransmitter; encourages our brain to do things again, such as succeed on a test, help a friend, eat, and so on. Deficient signaling involved in Parkinson's disease.
	Serotonin	5HT receptors (15 known)	Diffusion Reuptake	Widespread uses throughout the NS; involved in appetite and digestion regulation, mood, and others. SSRIs are drugs commonly used to treat depression and anxiety.
Amino Acids	GABA		Diffusion Reuptake	Principal inhibitory neurotransmitter of the NS
	Glutamate		Diffusion Reuptake	Important in learning and memory
	Glycine		Diffusion Reuptake	Principal inhibitory neurotransmitter of the spinal cord
Peptides	Endorphins	Opioid receptors		Involved in "runner's high"; inhibit pain-sensing neurons
	Substance P			Principal pain NT; secreted by pain-sensing neurons to signal to the CNS

DIGGING DEEPER: Proteopathies

The Na^+/K^+ pump is a protein. The ion channels are proteins. The neurotransmitter receptors are proteins. Many of the neurotransmitters are amino acids—building blocks of proteins. The nervous system runs on and relies on protein function. The function of a protein is intrinsically tied to its shape. The underlying cause of many neurodegenerative diseases, such as Alzheimer's and Parkinson's, appears to be related to proteins—specifically, to protein shape. One of the strongest theories of what causes Alzheimer's disease is based on the accumulation of dense conglomerations of a particular protein; this protein has a function in all brains, but in brains with Alzheimer's it has changed shape and is not functioning correctly. Similarly, Parkinson's disease is linked to an increase in a particular protein that is toxic to the cells of one particular area of the brain.

For proteins to function correctly, they are dependent on their three-dimensional shape. The linear sequence of amino acids folds into a three-dimensional shape that is based on the interactions between and among those amino acids. When this folding is disturbed, causing proteins to take on a different shape, they stop functioning correctly. Disease may result from loss of function of these proteins, or these altered proteins start to accumulate and may become toxic to the cells around them. For example, in Alzheimer's, the hallmark of the disease is the accumulation of protein plaques in the cerebral cortex. The term coined to describe this sort of disease is "proteopathy" and it includes other diseases. Creutzfeldt-Jacob disease, the human variant of the prion disease known as Mad Cow disease in cattle, also involves the accumulation of protein plaques, similar to that seen in Alzheimer's. Diseases of other organ systems can fall into this group as well, such as cystic fibrosis or type 2 diabetes. Recognizing the relationship between these diseases has suggested new therapeutic possibilities. Interfering with the accumulation of the proteins—possibly as early as their original production within the cell—may unlock new ways to alleviate these devastating diseases.

Cultural Connection

Individual Brains

Just like the rest of our bodies, but to a greater extent, no two brains are anatomically structured in exactly the same way. From one individual to another, our brains function dramatically differently. We can consider differences such as variations in the way that we focus, how creative we are, or our degrees of sensory awareness. In some cases, these differences can be categorized. Some categories within neurodiversity include autism spectrum disorders, attention deficit hyperactivity disorder, dyslexia, dyscalculia, or dyspraxia. Even the brains of biological males and biological females differ somewhat.

Often when we see anatomical or physiological differences between biological males and biological females, those differences have arisen in response to circulating hormones. In the case of brains, however, there may also be some genetic factors that differentiate brain function in biological males from that in biological females. Interestingly, there is a theory called *sexual diergism*, which indicates that even though biological male and biological female brains function differently, they have the same outcomes. An example of sexual diergism is the hypothalamus, which—although it has some different functions in male and female brains—allows biologically male and biologically female bodies to display remarkably similar behaviors and maintain similar physiological functions, such as maintaining body temperature within a set range, though the ranges typically differ.

While all brains function differently by recognizing and using the talents and strengths of different types of brains, our society cannot function better as a whole. For example, individuals whose brains operate at a consistently higher level of anxiety are often very clear-headed in emergencies because they are well practiced at emergency thinking. Individuals on the autism spectrum can often see entire structures or situations in a unique way. Scientist, animal behaviorist, and author Temple Grandin, for example, identifies as an individual on the autism spectrum. The unique ways that her brain operates has enabled her to be able to understand and reimagine systems of living and slaughtering livestock to increase welfare, comfort, and ethical treatment of farm animals. She has written more than 100 books and more than 60 scientific articles.

Learning Check

1. Which of the following components of a synapse is responsible for the release of a neurotransmitter?
   a. Presynaptic cell
   b. Postsynaptic cell
2. Glutamate is a neurotransmitter that is responsible for stimulating neural activity. Which of the following ions allows glutamate to be released into the synaptic cleft?
   a. Sodium
   b. Potassium
   c. Calcium
   d. Chlorine
3. Glutamate is a neurotransmitter that is responsible for stimulating neural activity. It is part of the amino acid group of neurotransmitters. How is glutamate removed from the synaptic cleft?
   a. The presynaptic cell reuptakes the excess neurotransmitters.
   b. The excess glutamate dissolves in the extracellular matrix over time.
   c. The glutamate stays in the synapse until all of it is used.
   d. The vesicle only releases the exact amount needed.

Chapter Review

13.1 Organization and Functions of the Nervous System

The learning objectives for this section are:

13.1.1 Describe the general functions of the nervous system.

13.1.2 Compare and contrast the central nervous system (CNS) and the peripheral nervous system (PNS) with respect to structure and function.

13.1.3 Differentiate between the motor (efferent) and sensory (afferent) components of the nervous system.

13.1.4 Describe the nervous system as a control system with the following components: sensory receptors, afferent pathways, control (integrating) center, efferent pathways, and effector (target) organs.

13.1.5 Compare and contrast the somatic motor and autonomic motor divisions of the nervous system.

13.1.6 Compare and contrast the somatic sensory and visceral sensory divisions of the nervous system.

13.1 Questions

1. Which of the following is/are a general function of the nervous system? Please select all that apply.
   a. You feel someone tickling your feet.
   b. You move your arm to slap a mosquito that is biting your leg.
   c. You smell flowers and it reminds you of spring.
   d. Your skin turns pink on a hot day.

2. Which of the following is/are found in the peripheral nervous system? Please select all that apply.
   a. Brain
   b. Spinal cord
   c. Nerves
   d. Neurons

3. You get hit in the nose during a wrestling match. Your nose starts to swell and you start to feel tears roll down your cheeks. This happens when the lacrimal gland releases salty water—known as tears—through the "tear duct" and into your eyes. In this example, which of the following describes the afferent pathway?
   a. The nose swelling
   b. Tears rolling down cheek
   c. Lacrimal gland producing tears
   d. Feeling pain after getting hit

4. After falling off your bike, you are rushed to the hospital. There, the doctor performs a series of tests to make sure there is not significant damage in your body. Your doctor hits your patellar tendon and you notice your leg kick up. There is a reflex in your body that allows this to happen. The reflex takes the input from the stretch of the patellar tendon and responds by contracting your quadriceps muscle. Which of the following is the effector?

a. Patellar tendon
b. Spinal cord
c. Quadriceps
d. Lower leg

5. A female gymnast is about to perform a routine. She starts by raising her hands up and proceeds to do flips, turns, and jumps. Which of the following nervous systems is responsible for her voluntarily performing these stunts for her routine?

a. Somatic nervous system
b. Autonomic nervous system

6. Kobe has been playing basketball for 40 minutes, but their team is down by 1 point with 2 seconds left in the game. In Kobe's body, a receptor detects that there is less oxygen in their blood stream. This triggers a cascade of reactions that results in the heart beating faster. At this point, Kobe's heart is racing. Which of the following best describes the receptor that detects the decreased oxygen in Kobe's blood?

a. Somatic sensory receptor
b. Visceral sensory receptor
c. Effector organ
d. Afferent pathway

13.1 Summary

- The nervous system is responsible for sensory input, motor response, and the integration of sensory input with other sensations such as memory, emotional state, or learning.
- The nervous system is composed of the central nervous system (brain and spinal cord), and the peripheral nervous system (nerves that connect the periphery of the body to the central nervous system).
- The nervous system can be classified by function into afferent, integrating, and efferent components. Afferent (sensory) pathways transmit sensory information from receptors to the central nervous system. Integrating centers in the central nervous system coordinate sensory information an formulate motor commands. Efferent (motor) pathways transmit motor commands from the central nervous system to muscles and glands.
- The nervous system is also categorized based on the location of the organ that the nerve innervates. The somatic nervous system is responsible for conscious perception and voluntary motor responses, while the autonomic nervous system is responsible for the involuntary control of the body, usually for the sake of homeostasis.

13.2 Nervous Tissue and Cells

The learning objectives for this section are:

HAPS LEARNING GOALS AND OUTCOMES

13.2.1 Identify and describe the major components of a typical neuron (e.g., cell body, nucleus, nucleolus, chromatophilic substance [Nissl bodies], axon hillock, dendrites, and axon) and indicate which parts receive input signals and which parts transmit output signals.

13.2.2 Compare and contrast the three structural types of neurons (i.e., unipolar [pseudounipolar], bipolar, and multipolar) with respect to their structure, location, and function.

13.2.3 Compare and contrast the three functional types of neurons (i.e., sensory [afferent] neurons, interneurons [association neurons], and motor [efferent] neurons) with respect to their structure, location, and function.

13.2.4 Describe the structure, location, and function of each of the six types of neuroglial (glial) cells.

13.2.5 Define myelination and describe its function, including comparing and contrasting how myelination occurs in the CNS and PNS.

13.2 Questions

1. Which of the following parts of the neuron receives signals?

a. Cell body
b. Axon terminal
c. Dendrites
d. Nucleus

2. The foot has many sensory receptors that converge into one dendrite, which merges with the axon to attach to the cell body. Therefore, the cell body has only one projection. What type of neuron does this describe at the foot?

a. Unipolar neurons
b. Bipolar neurons
c. Multipolar neurons

3. The back is less sensitive than the hand in detecting touch, vibration, and pain. Which of the following transmits these signals from the periphery to the brain?

a. Sensory neurons
b. Interneurons
c. Motor neurons

4. Which of the following glial cells are responsible for producing myelin in the central nervous system?

a. Astrocytes
b. Oligodendrocytes
c. Schwann cells
d. Satellite cells

5. What are the functions of myelin? Please select all that apply.

a. Relays electrical signaling
b. Insulates axons of neurons
c. Produces input for neurons
d. Prevents leakage of electrical charge

13.2 Summary

- The neuron is the main cell of the nervous system. It consists of many typical organelles such as the nucleus, ER, and ribosomes, which are found in the cell body. The dendrites receive signals and transmit them to the cell body, which integrates that information and sends an output through the axon, which synapses to another neuron's dendrites.
- Neurons are classified based on their function or anatomical features. Based on the function, the cells can be sensory neurons, interneurons, or motor neurons. Based on their anatomical features, the cells can be unipolar, bipolar, or multipolar.
- Glial cells are supporting cells of the nervous system, with very important roles. There are four types of glial cells that support central nervous system neurons: astrocytes, oligodendrocytes, ependymal cells and microglia. The two types of glial cells that support the neurons of the peripheral nervous system are satellite cells and neurilemma (Schwann) cells.
- Myelin insulates the axons to allow for electrical signals to transmit through the axon quickly. It is produced by the oligodendrocytes in the central nervous system and by neurilemma cells in the peripheral nervous system.

13.3 Neurophysiology

The learning objectives for this section are:

HAPS LEARNING GOALS AND OUTCOMES

13.3.1 Define and describe depolarization, repolarization, hyperpolarization, and threshold.

13.3.2 Describe the role of the sodium-potassium ATPase pump in maintaining the resting membrane potential.

13.3.3 List the major ion channels of neurons and describe them as leak (leakage or passive) or voltage-gated channels, mechanically gated channels, or ligand-gated (chemically-gated) channels, and identify where they typically are located on a neuron.

13.3.4 Describe the physiological basis of the resting membrane potential (RMP) in a neuron including the ion channels involved, the relative ion concentrations, and the electrochemical gradient.

13.3.5 Compare and contrast graded potentials and action potentials, with particular attention to their locations in the neuron and the ions and ion channels involved in each.

13.3.6* Compare and contrast IPSPs and EPSPs with respect to the ions involved and effects on the postsynaptic cell.

13.3.7 Explain temporal and spatial summation of postsynaptic potentials.

13.3.8 Distinguish between absolute and relative refractory periods and compare the physiological basis of each.

13.3.9 Explain the impact of absolute and relative refractory periods on the activity of a neuron.

13.3.10 Describe the physiological process involved in the conduction (propagation) of an action potential, including the types and locations of the ion channels involved.

13.3.11 Compare action potential conduction (propagation) in an unmyelinated versus a myelinated axon.

13.3.12 Explain how axon diameter and myelination affect conduction velocity.

* Objective is not a HAPS Learning Goal.

13.3 Questions

1. Which of the following describes the phase of an action potential in which the charge inside the neuron changes from −70 mV to perhaps −10 mV?
   a. Depolarization
   b. Repolarization
   c. Hyperpolarization
   d. Polarization

2. What is the role of the sodium/potassium pump?
   a. To initiate an action potential
   b. To repolarize the neuron
   c. To maintain the resting membrane potential
   d. To depolarize the neuron past the threshold

3. Roberto is playing a game where they have to close their eyes, put their hands in a box, and guess what item is in the box by feeling the object. What kind of ion channel is responsible for sensory input of the object touching Roberto's hand?
   a. Ligand-gated channel
   b. Mechanically gated channel
   c. Voltage-gated channel
   d. Leak channel

4. Which of the following contribute to a resting membrane potential in a neuron? Please select all that apply.
   a. The concentrations of K^+ inside the cell and Na^+ outside the cell being the same
   b. The higher concentration of proteins inside the cell than in the extracellular matrix
   c. The membrane allowing sodium and potassium to freely pass through
   d. The greater number of K^+ leaky channels than Na^+ leaky channels

5. Where can a graded potential occur within a neuron? Please select all that apply.
   a. Dendrite
   b. Cell body
   c. Axon hillock
   d. Axon terminal

6. Which of the following would bring the electrical charge of a neuron closer to threshold?
   a. Excitatory postsynaptic potential
   b. Inhibitory postsynaptic potential

7. In neurological pathologies that have increased spasticity, it is common for a single neuron to have several graded potentials from many nearby neurons at the same time. These inputs summate in the cell body and lead to many action potentials exiting through the axon to muscles. The muscles stay contracted due to constant action potentials going through the axon. Which of the following best describe the graded potentials that occur in spastic muscles? Please select all that apply.
   a. Excitatory postsynaptic potential
   b. Inhibitory postsynaptic potential
   c. Spatial summation
   d. Temporal summation

8. Which of the following best represents a relative refractory period?
   a. An action potential cannot start regardless of the magnitude of the stimulus.
   b. An action potential can start with the same magnitude of the stimulus.
   c. An action potential can only start with a larger magnitude of the stimulus.
   d. An action potential can only start with a smaller magnitude of the stimulus.

9. Which of the following explains why a neuron cannot produce another action potential during the absolute refractory period?
   a. The membrane potential is below the resting membrane potential.
   b. The K^+ ion channels are already open or inactivated.
   c. The Na^+ ion channels are already open or inactivated.
   d. There are no more ions in the cell to produce another action potential.

10. Which of the following allows the action potential to propagate forward through the axon?
    a. The influx of sodium into the cell body
    b. The influx of potassium into the cell body
    c. The myelin preventing backflow of ions
    d. The axon terminal being negatively charged

11. In Multiple Sclerosis (MS), the myelin sheath around axons is damaged. Which of the following would you expect to see in someone with MS?
    a. Fast conduction through the axon
    b. Slow conduction through the axon
    c. Excessive leaking of sodium ions into the extracellular matrix
    d. Excessive buildup of potassium ions in the cell

12. The femoral nerve is wider in diameter and has more myelin than the axillary nerve. Which of the following is/are true about the femoral nerve? Please select all that apply.
    a. It will propagate faster due to the wider diameter.
    b. It will propagate slower due to the wider diameter.
    c. It will propagate faster due to the increased myelin.
    d. It will propagate slower due to the increased myelin.

13.3 Summary

- Neurons communicate with other cells through changes in charge across plasma membranes.
- Several types of membrane channels allow ions to enter or exit a neuron in order to change the charge difference (membrane potential) across the plasma membrane: ligand-gated channels, mechanically-gated channels, voltage-gated channels, and leak channels.
- Ions are not uniformly distributed across the plasma membrane of a neuron. Sodium ions are in higher concentration outside the cell, and potassium ions and negatively charged ions and proteins are in higher concentration inside the cell. This causes the resting potential (the charge difference across the membrane) of most neurons to be about −70 mV.
- Changes in the resting potential occur in response to various stimuli. Depolarization is a decrease in the negativity inside a neuron, due to an influx of sodium ions. Repolarization often follows depolarization; it is an increase in negativity inside a neuron, due to an efflux of potassium ions. Hyperpolarization occurs when the inside of a neuron becomes more negative than the resting potential.
- Changes in the resting potential of a neuron can be classified as graded potentials or action potentials. A graded potential is a slight change in the resting potential, which is not sufficient to reach the threshold potential, so it does not generate an action potential. In an action potential, a threshold depolarization causes propagation of a wave of depolarization down the axon to the axon terminal. Once at the axon terminal, the neuron passes an electrical message to another neuron of other cell through a neurotransmitter.
- An action potential proceeds in a specific sequence. Depolarization is followed by repolarization, and then a slight hyperpolarization. Finally, the neurons returns to its resting potential due to the actions of the sodium-potassium ATPase pump.
- After an action potential has occurred in a neuron, it goes through two periods of recovery before it can generate another action potential, called the absolute and relative refractory periods.

13.4 Communication between Neurons

The learning objectives for this section are:

13.4.1 Define a synapse, and explain the difference between an electrical synapse and a chemical synapse.

13.4.2 Describe the structures involved in a typical chemical synapse (e.g., axon terminal [synaptic knob], voltage-gated calcium channels, synaptic vesicles of presynaptic cell, synaptic cleft, neurotransmitter receptors of the postsynaptic cell).

13.4.3 Describe the events of synaptic transmission in proper chronological order from the release of neurotransmitter by synaptic vesicles to the effect of the neurotransmitter on the postsynaptic cell.

13.4.4 Describe the different mechanisms (e.g., reuptake, enzymatic breakdown, diffusion) by which neurotransmitter activity at a synapse can be terminated.

13.4.5 List the most common excitatory and inhibitory neurotransmitter(s) used in the nervous system.

13.4.6 Explain how a single neurotransmitter can elicit different responses at different postsynaptic cells.

13.4 Questions

1. What synapse is most common in the human body and sends electrical signaling through neurotransmitters?
   a. Electrical synapse
   b. Chemical synapse

2. Which of the following is/are true of the structures that make up a chemical synapse? Please select all that apply.
   a. Gap junctions allow ions to pass from the presynaptic neuron to the postsynaptic neuron.
   b. The axon terminal releases neurotransmitters from the presynaptic neuron into the synaptic cleft.
   c. K^+ ions entering the axon terminal through voltage-gated K^+ channels help with the release of neurotransmitters.
   d. The receptors on the postsynaptic neuron consist of ligand-gated Na^+ channels in excitatory synapses.

Mini Case 1: Parkinson's Disease

Each year, many people develop Parkinson's Disease, which is a neurological disease that results in resting tremors, rigidity, decreased balance, gait deficits, and cognitive impairments. This occurs when there is less dopamine produced in and released from the brain. The lack of dopamine plays a major role in the progression of the disease.

1. What is the first step in synaptic transmission?
   a. Calcium ions enter the axon terminal.
   b. The vesicle merges with the plasma membrane.
   c. Neurotransmitters are released into the synaptic cleft.
   d. Ions will flow into the postsynaptic neuron.

While there is less dopamine produced in the synapse, the same rate of dopaminergic cleaning up occurs.

2. Which of the following is a possible way that dopamine can get cleaned up within the synaptic cleft? Please select all that apply.
   a. Dopamine can diffuse throughout the synaptic cleft.
   b. The presynaptic neuron can reuptake neurotransmitters.
   c. The postsynaptic neuron can break down neurotransmitters.
   d. The neurotransmitters break down independently.

3. What type of neurotransmitter is dopamine?
   a. Cholinergic
   b. Nicotinic
   c. Amino acid
   d. Biogenic amine

4. Dopamine affects several processes in the body, including attention, control of impulses, movement, sleep, cognition, memory, and decision-making. Each of these processes is controlled by neurons in different areas of the brain. How is it possible that dopamine could affect a diverse group of neurons in so many different ways?
   a. Dopamine is used in chemical synapses in some neurons and in electrical synapses in other neurons.
   b. Different types of dopamine receptors are found in different types of neurons.
   c. Voltage-gated dopamine channels are found in the membranes of some types of postsynaptic neurons, while ligand-gated dopamine channels are found in other types.
   d. The effects of dopamine vary depending on whether it is released from the presynaptic neuron or the postsynaptic neuron.

13.4 Summary

- There are two types of synapses: electrical synapses, which pass electrical signals directly between cells, and chemical synapses, which pass electrical signals through neurotransmitters.
- A chemical synapse consists of a presynaptic cell, a neurotransmitter, a synaptic cleft, a postsynaptic cell, receptors for neurotransmitters, and a system for clearing neurotransmitters from the synaptic cell.
- Typically, as an action potential reaches the axon terminal, calcium ion channels bring calcium into the cell, which bind to the presynaptic vesicle. This allows the presynaptic vesicle to bind to the plasma membrane and release the contents—usually neurotransmitters—into the synaptic cleft. The neurotransmitters bind to the receptors of the postsynaptic neuron, and this initiates a reaction—usually an action potential—throughout the postsynaptic neuron.
- There are three types of neurotransmitters: acetylcholine, amino acids (such as GABA and glycine), and biogenic amines (such as epinephrine, norepinephrine, dopamine, and serotonin.)

14

Anatomy of the Nervous System

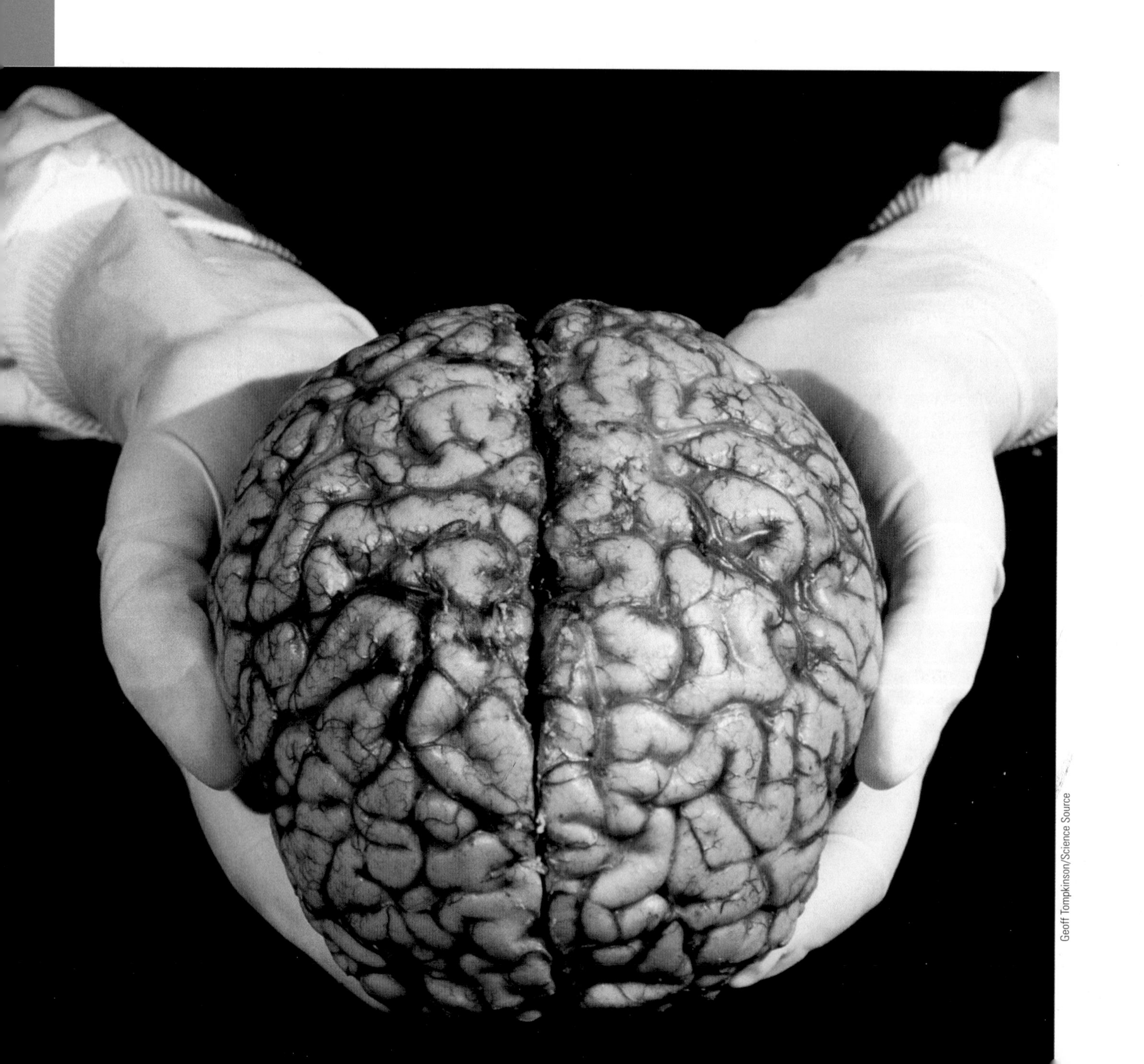

Geoff Tompkinson/Science Source

Chapter Introduction

In this chapter we will learn...

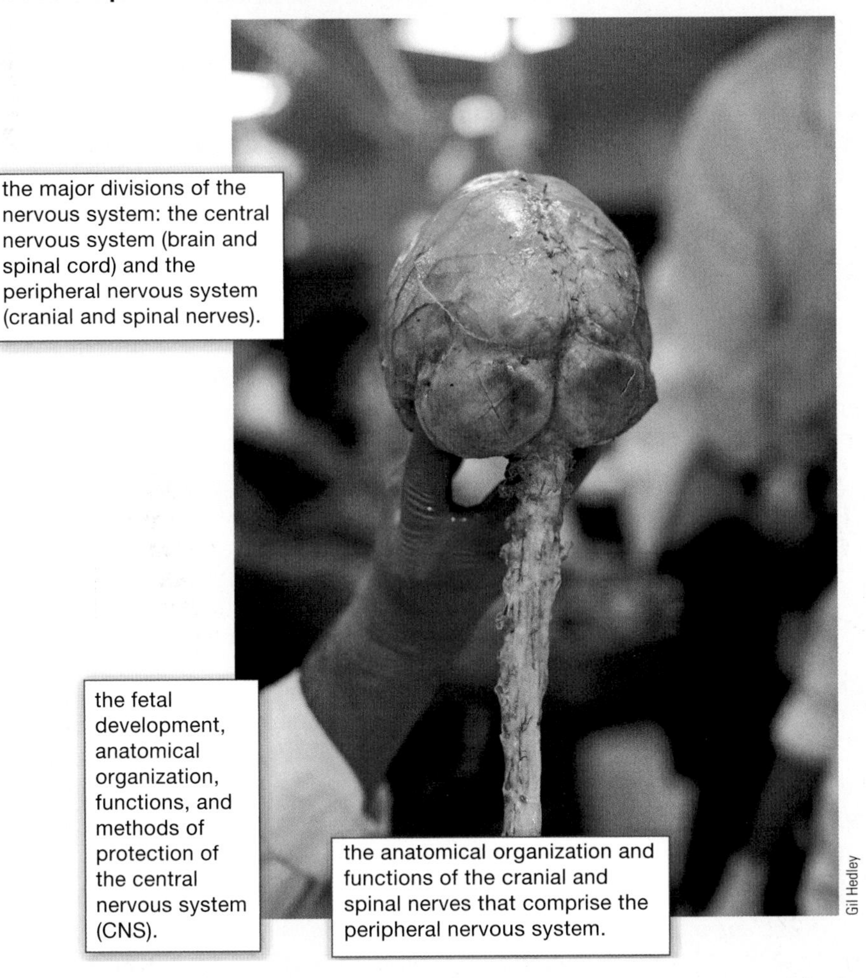

the major divisions of the nervous system: the central nervous system (brain and spinal cord) and the peripheral nervous system (cranial and spinal nerves).
the fetal development, anatomical organization, functions, and methods of protection of the central nervous system (CNS).
the anatomical organization and functions of the cranial and spinal nerves that comprise the peripheral nervous system.
Gil Hedley

The nervous system is responsible for controlling much of the body, both through somatic (voluntary) and autonomic (involuntary) functions. The structures of the nervous system must be described in detail to understand how many of these functions are possible. There is a physiological concept known as *localization of function* that states that certain structures are specifically responsible for prescribed functions. It is an underlying concept in all of anatomy and physiology, but the nervous system illustrates the concept very well.

Fresh, unstained nervous tissue can be described as gray or white matter, and within those two types of tissue it can be very hard to see any detail. However, as our knowledge of the brain has grown, specific regions and structures have been described and related to specific functions. Understanding these structures and the functions they perform requires a detailed description of the anatomy of the nervous system, delving deep into what the central and peripheral structures are.

The place to start this study of the nervous system is the beginning of the individual human life, within the womb. The embryonic development of the nervous system allows for a simple framework on which progressively more complicated structures can be built. With this framework in place, a thorough investigation of the nervous system is possible.

14.1 General Anatomy of the Nervous System

Learning Objectives: By the end of this section, you will be able to:

14.1.1 Describe the composition and arrangement of the gray and white matter in the CNS.

14.1.2 Compare and contrast the structure and location of a tract and nerve.

The Human Anatomy and Physiology Society includes more than 1,700 educators who work together to promote excellence in the teaching of this subject area. The HAPS A&P Learning Outcomes measure student mastery of the content typically covered in a two-semester Human A&P curriculum at the undergraduate level. The full Learning Outcomes are available at https://www.hapsweb.org.

14.1.3 Compare and contrast the structure and location of a nucleus and ganglion.

14.1.4 Identify the layers of the meninges and describe their anatomical and functional relationships to the CNS (brain and spinal cord).

14.1.5 Identify and describe the epidural space, subdural space and subarachnoid space associated with the brain and the spinal cord, and identify which space contains cerebrospinal fluid.

14.1.6 Describe the structure and location of the dural venous sinuses, and explain their role in drainage of blood from the brain.

14.1.7 Identify and describe the structure and function of the cranial dural septa.

14.1.8 Describe how the cranial bones and the vertebral column protect the CNS.

14.1.9* Draw or describe the structure of a typical nerve, including the connective tissue wrappings and any associated cells.

14.1.10 Identify and describe the ventricular system components.

14.1.11 Describe the production, flow, and reabsorption of cerebrospinal fluid (CSF), from its origin in the ventricles to its eventual reabsorption into the dural venous sinuses.

14.1.12 Describe the general functions of cerebrospinal fluid (CSF).

14.1.13 Compare and contrast the structure of the dura mater surrounding the brain and the spinal cord.

* Objective is not a HAPS Learning Goal.

Learning Connection

Chunking

Many of the new anatomical structures you will learn in this chapter share parts of their names with the cranial bones you may have learned in Chapter 8. Consider listing all of the new terms that share those names (for example, everything with the word *occipital* in its name).

14.1a Anatomical Patterns of Nervous Tissue

The nervous system can be subdivided and its components can be categorized in a number of ways based on both their anatomy and their function. The broadest of these divisions is the central nervous system (CNS) and the peripheral nervous system (PNS), which are discussed in more detail in Chapter 13. In this chapter we are primarily discussing the anatomy of the nervous system (though we cannot depart far from discussing the function, as these two are inextricably linked!).

One anatomical pattern that we can easily see when examining the nervous system is that neuron cell bodies tend to be gathered together, and neuron axons tend to be bundled together. In the brain and the spinal cord, there are regions that predominantly contain cell bodies and regions that are largely composed of just axons. These two regions within nervous system structures are often referred to as **gray matter** (the regions with many cell bodies and dendrites) or **white matter** (the regions with many axons). The color distinction between gray and white matter is observable in gross dissection of unstained, nervous tissue (**Figure 14.1**). The appearance of the white matter is due to the myelination around the axons. The myelin lipids can appear as white ("fatty") material, much like the fat on a raw piece of chicken or beef. The gray matter, with its cell bodies, dendrites, and some short unmyelinated axons, lacks this fat component and therefore appears a dull gray or tan.

LO 14.1.1

The distinction between gray matter and white matter is used to describe regions within the CNS because the axons and cell bodies of neurons are gathered in large regions that can be seen with the unaided eye. However, clusters of cell bodies and axons occur in the PNS as well. Whereas a bundle of axons found in the CNS that appears white and can be described as white matter is also called a **tract**, a bundle of axons in the PNS is called a **nerve** (**Figure 14.2**). There is an important point to make about these terms, which is that they can both be used to refer to the same bundle of axons. When those axons are in the PNS, the term is *nerve*, but if they are CNS, the

LO 14.1.2

Figure 14.1 **Gray Matter and White Matter of the Brain**

In human brains the gray matter forms a shell—the cortex—around the white matter.

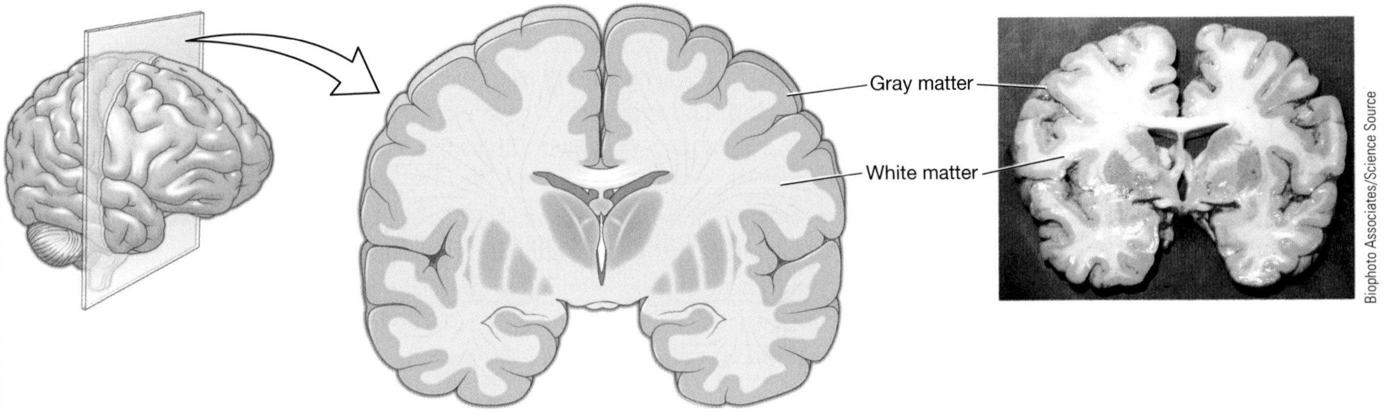

Gray matter

White matter

Biophoto Associates/Science Source

Figure 14.2 **Nervous System Structures**

Axons are gathered into structures called *nerves* in the PNS and *tracts* in the CNS. Clusters of cell bodies are called *nuclei* in the CNS and *ganglia* in the PNS.

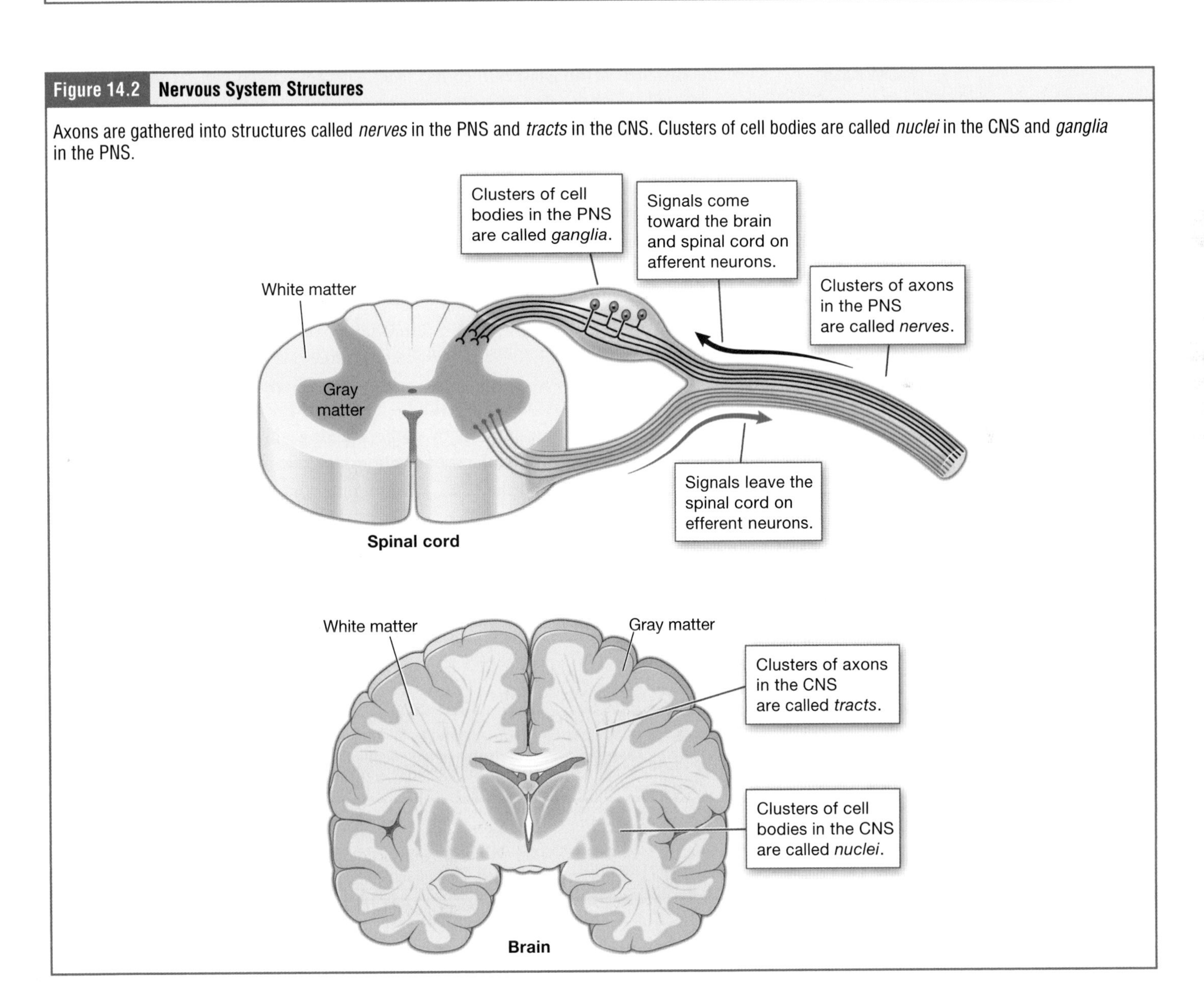
Clusters of cell bodies in the PNS are called *ganglia*.
Signals come toward the brain and spinal cord on afferent neurons.
White matter
Clusters of axons in the PNS are called *nerves*.
Gray matter
Signals leave the spinal cord on efferent neurons.
Spinal cord
White matter
Gray matter
Clusters of axons in the CNS are called *tracts*.
Clusters of cell bodies in the CNS are called *nuclei*.
Brain

term is *tract*. The most obvious example of this is the axons that project from the retina into the brain. Those axons are called the *optic nerve* as they leave the eye, but when they are inside the cranium, they are referred to as the *optic tract*. There is a specific place where the name changes, which is the optic chiasm, but they are still the same axons. Similarly, a collection of neuron cell bodies in the CNS is referred to as a **nucleus**. In the PNS, a cluster of neuron cell bodies is referred to as a **ganglion** (see Figure 14.2). The terms *nucleus* and *membrane* are used in various ways within anatomy and physiology. A nucleus can be the center of an atom, where protons and neutrons are found; it is the center of a cell, where the DNA is found; and it is a center of some function in the CNS (**Figure 14.3**).

LO 14.1.3

You will notice that in the brain, the gray matter is superficial, forming a shell around underlying white matter. There are deeper gray matter nuclei, often called *basal nuclei*. The opposite pattern holds true in the spinal cord, here the white matter is superficial and gray matter is deep (see Figure 14.2).

Connective Tissue Protection Remember that neurons are largely nonmitotic; they cannot reproduce themselves. Therefore, they are surrounded by connective tissue and (in some cases) bone to protect them from damage. In the CNS, the brain and spinal cord are shrouded in three layers of connective tissue, collectively called **meninges**. Each of these layers has a first name that describes its appearance, and its second name is *mater*, which is Latin for "mother." Early anatomists described these layers as protective mothers, hovering over the brain and spinal cord to shield them from damage. The three maters are, from superficial to deep: the *dura mater*, the *arachnoid mater*, and the *pia mater*.

LO 14.1.4

The **dura mater** is a thick, collagen-rich fibrous layer that acts as a strong protective sheath over the entire brain and spinal cord. Its name, dura, comes from the same root as "durable"; it is named for its thick, tough nature. It is anchored to the inner surface of the cranium and vertebral cavity, and in some locations it separates into two layers, one just inside the bony structures of the skull and the other overlaying the surface of the brain. When the dura forms two layers, the more superficial layer—associated with the skull—is called the *periosteal dura mater* and the deeper layer—which sits atop the two other meningeal layers—is the *meningeal dura mater*. Beneath the dura is the **arachnoid mater**, a thinner fibrous membrane that forms a loose sac around the CNS. The arachnoid membrane sits on top of a

Figure 14.3 The Term "Nucleus" in A&P

The term *nucleus* can mean: (A) The nucleus of an atom; (B) The nucleus of a cell, the organelle that contains DNA; (C) A cluster of unmyelinated material, usually cell bodies and/or short unmyelinated axons in the brain.

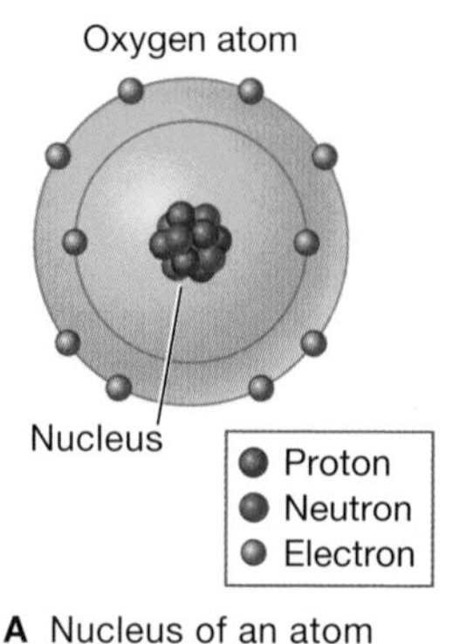

A Nucleus of an atom

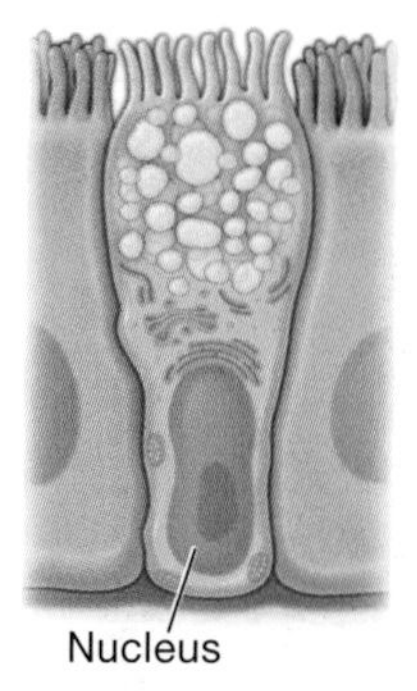

B Nucleus of a cell

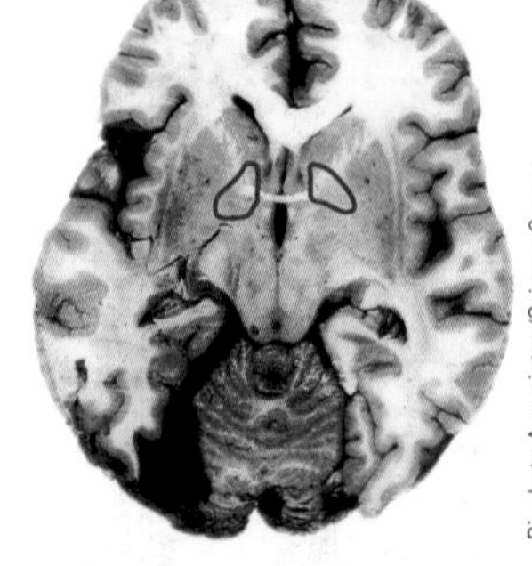

C Nucleus in the brain

thin, filamentous mesh that looks like a spider web, giving this layer its name. Under the arachnoid, and sitting directly on top of the surface of the CNS, is the **pia mater**, the thinnest membrane, which hugs all of the bumps, grooves, and indentations of the brain like a thin layer of paint (**Figure 14.4**). The name *pia mater* comes from the

Figure 14.4 The Meninges of the Brain

There are three layers of connective tissue membranes that surround the central nervous system; these are called the *meninges*. (A) The dura, arachnoid, and pia maters are continuous between the brain and spinal cord. (B) At some points within the cranium, the dura splits into two layers, the periosteal and meningeal layers. The space between them accommodates venous blood as it drains from the brain and returns to the right side of the heart.

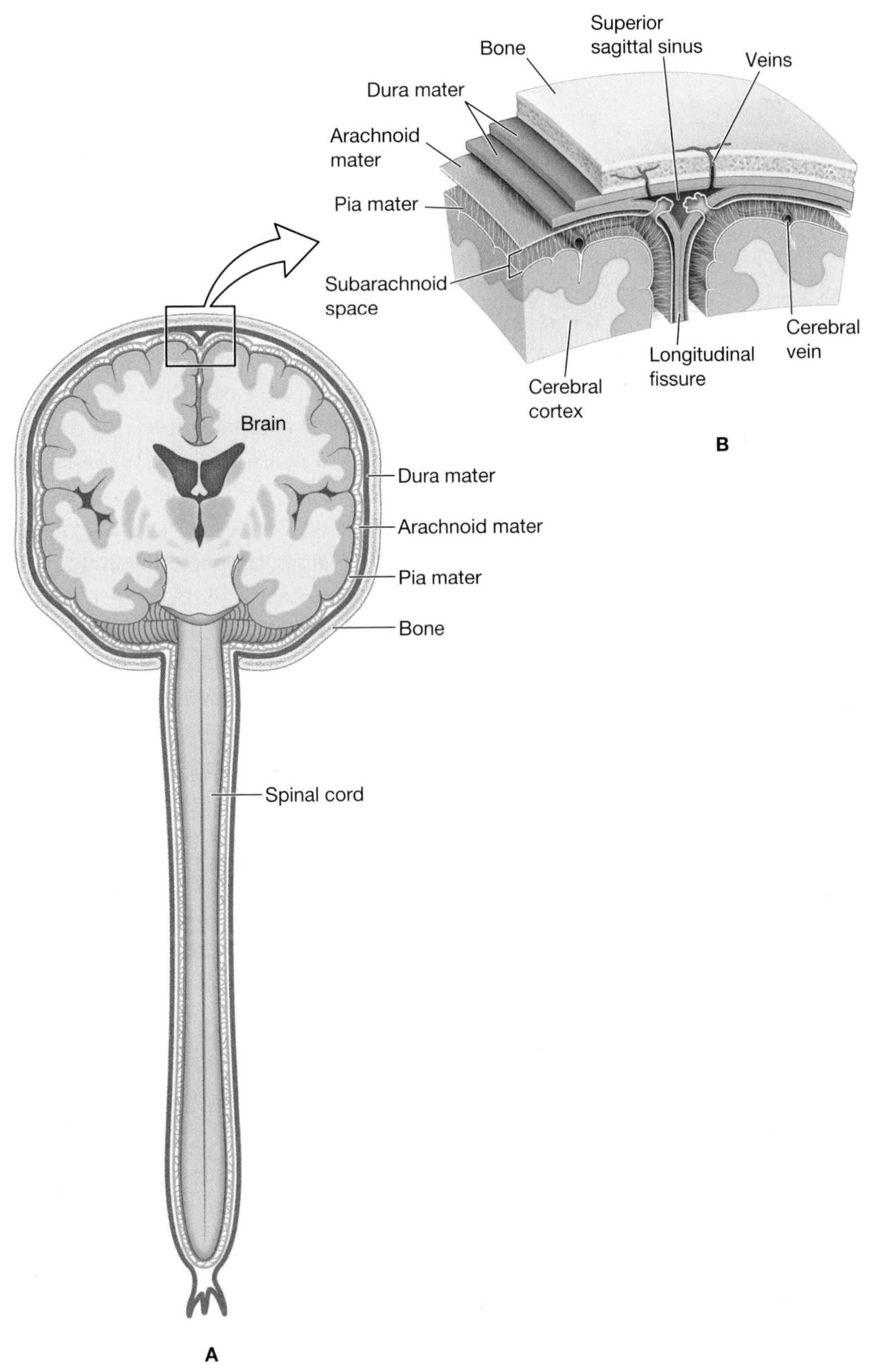

Latin for "tender mother," describing this thin membrane as a gentle covering for the brain.

The three meningeal layers cover both the brain and the spinal cord continuously (**Figure 14.5**). The arachnoid is held away from the pia mater by a network of filaments that creates a space called the **subarachnoid space** between these two meningeal layers. Within this space flows **cerebrospinal fluid (CSF)**, a fluid that provides a liquid cushion to the brain and spinal cord. Similar to clinical blood work, a sample of CSF can be withdrawn to find chemical evidence of infection or metabolic changes in the functions of nervous tissue (CSF will be discussed in more detail in the next section). The pia has a continuous layer of cells providing a fluid-impermeable membrane, which prevents leakage of the CSF into the neurons below.

LO 14.1.5

The subarachnoid space is one of the fluid-filled spaces formed by the meninges, but in the locations where the dura splits into two layers, there are fluid-filled cisterns between the layers of dura mater. These locations, illustrated in **Figure 14.6**, are called **dural venous sinuses** and are collecting spaces for the venous blood leaving the brain and returning to the heart. The dura also folds in to provide separating structures to stabilize the brain within the cranium. There are four of these **cranial dural septa**, which subdivide the cranial space. These four structures are: the *falx cerebri, tentorium cerebelli, falx cerebelli*, and *diaphragma sellae* (**Figure 14.7**). The **falx cerebri**, the largest of these dural septa, is located midsagittally, hanging down directly from the center of the skull and separating the right and left sides of the brain. Running within the falx cerebri are two of the dural venous sinuses. The **superior sagittal sinus** runs along the superior edge of the falx cerebri, just inside of the cranial bones. On its deep edge, the **inferior sagittal sinus** is sandwiched between the halves of the brain as the deepest or most inferior portion of the falx cerebri (see Figure 14.7).

LO 14.1.6

LO 14.1.7

A similar, but much smaller, vertical midsagittal dural septum—the **falx cerebelli**—separates the two halves of the cerebellum, a smaller structure in the cranium. This falx cerebelli also holds a small sinus, the **occipital sinus** (note that these structures are directly beneath the occipital bone).

There are two horizontal dural septa as well—the tentorium cerebelli and the diaphragma sella. The **diaphragma sella**, an anterior structure, provides a roof over the sella turcica, a depression in the sphenoid bone that houses the pituitary gland. The **tentorium cerebelli** provides a roof over the cerebellum, separating it from the rest of the brain. The folds of the tentorium cerebelli contain two venous sinuses; the **straight**

Learning Connection

Micro to Macro

Imagine the falx cerebri or examine it in Figure 14.7. Which skull bones does it come into contact with?

Figure 14.5 The Meninges of the CNS

The dura mater (A) is thick, durable, and opaque due to the density of collagen fibers within. The arachnoid mater (B) sits above the gyri and fissures encasing both the brain and CSF. The pia mater (C) closely surrounds every gyrus and fissure, and is tightly adhered to the surface of the brain.

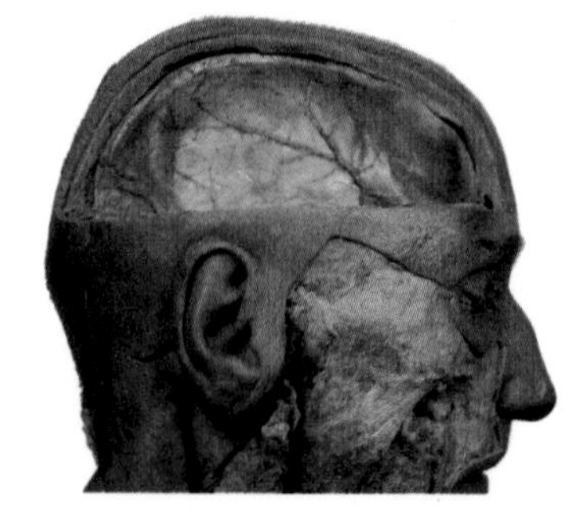

A

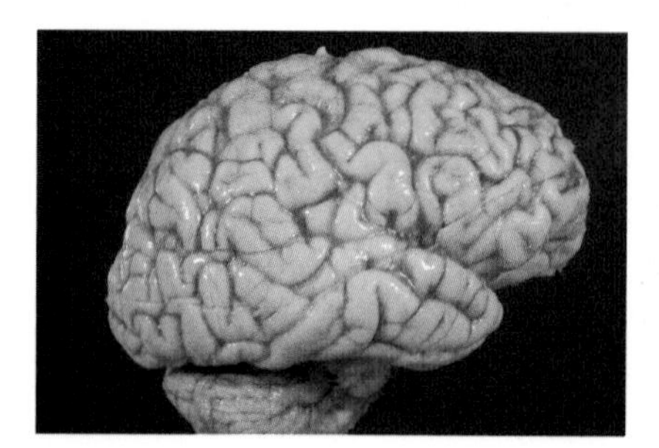

B

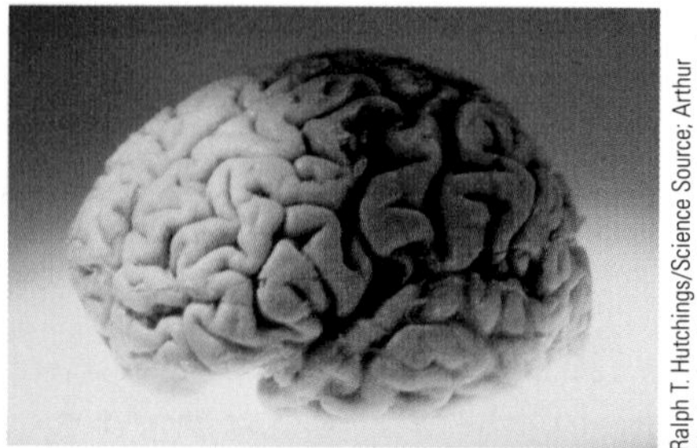

C

Figure 14.6 Dural Venous Sinuses

Blood that is leaving the brain drains first through spaces within the dura mater called *dural venous sinuses* before joining the veins that lead to the heart.

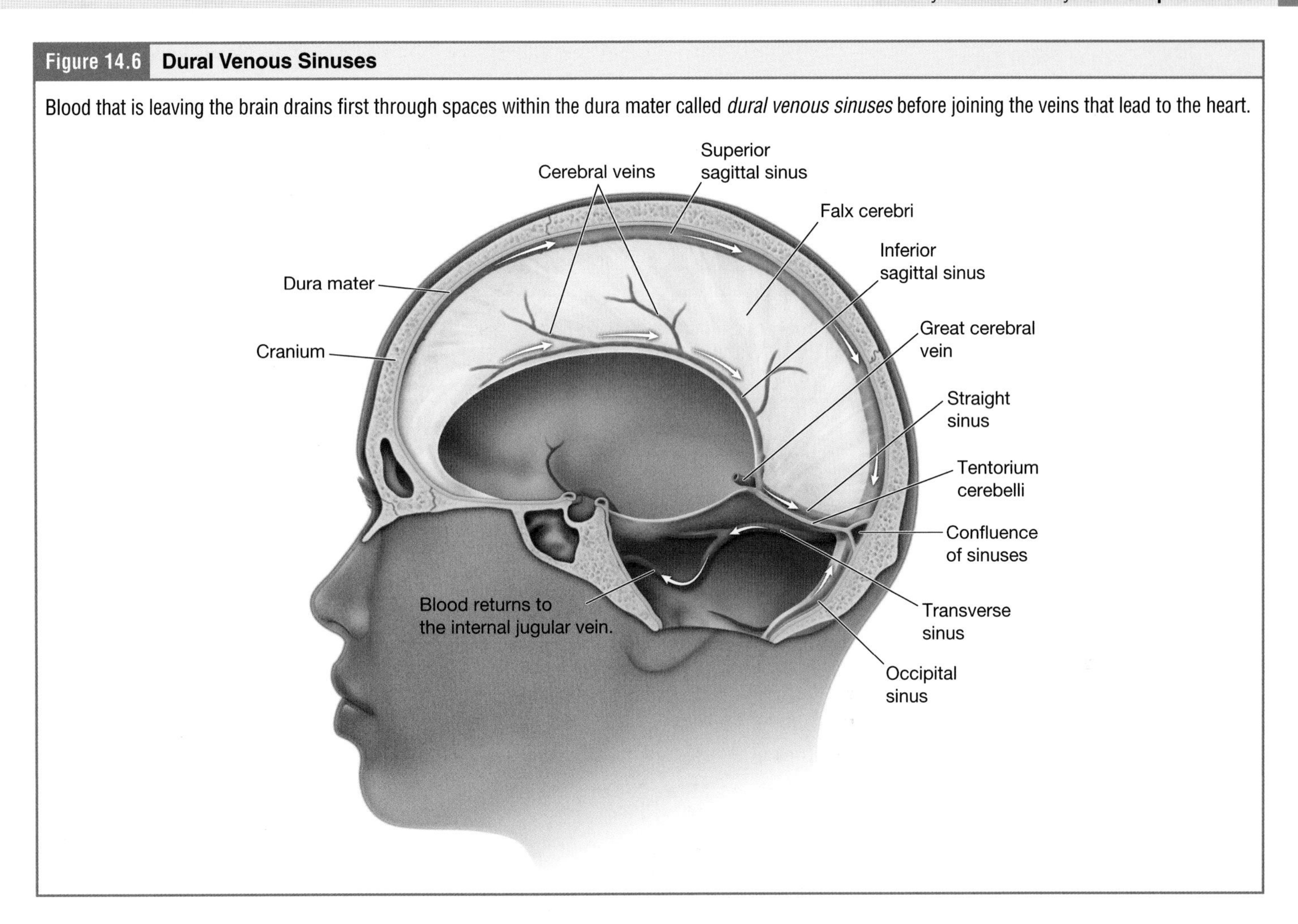

sinus runs vertically along its deepest border and the superficial **transverse sinuses** run in right and left transverse planes out from the midline.

transverse sinuses (lateral sinuses)

In a living, uninjured body, the meninges are directly in contact with each other except for the fluid-filled subarachnoid space and the dural venous sinuses. When an injury occurs, however, and there is bleeding within the cranium, blood may force the meninges apart. If blood accumulates between the dura and the arachnoid, then these two layers separate, forming a **subdural space**. Because the subdural space is not always present, but can be in certain circumstances, it is called a *potential* space. Likewise, a potential **epidural space** can form between the dura and the cranial bones, if blood becomes trapped here.

Student Study Tip

Remember your roots: The subdural space is *under* the dura mater, so it is found between the dura mater and the arachnoid space. The epidural space is *above* the dura mater, so it is between the dura mater and the cranial bones.

LO 14.1.8

In addition to its protection by the meninges, the CNS is shielded from damage by hard bone tissue. The bones of the cranium are knitted together by immobile joints (sutures) to form an impenetrable case for the brain. The vertebrae protect the spinal cord while providing space for the spinal nerves to exit and a small amount of mobility at each joint to allow for body movements.

LO 14.1.9

Like the CNS, neuronal structures in the PNS are also protected by connective tissue wrappings. Then nerves, bundles of axons in the PNS, are composed of more than just nervous tissue. They have connective tissues mixed into their structure, as well as blood vessels supplying the cells with nourishment. The outer surface of a nerve is clothed in a layer of fibrous connective tissue called the **epineurium**. Within the nerve, axons are further bundled into **fascicles**, which are each surrounded by their own layer of fibrous connective tissue, called **perineurium**. Finally, individual axons are surrounded by loose connective tissue called the **endoneurium** (Figure 14.8). This

Figure 14.7 Cranial Dural Septa

At some locations, the dura mater invaginates deeply into the cranium, separating and stabilizing the parts of the brain.

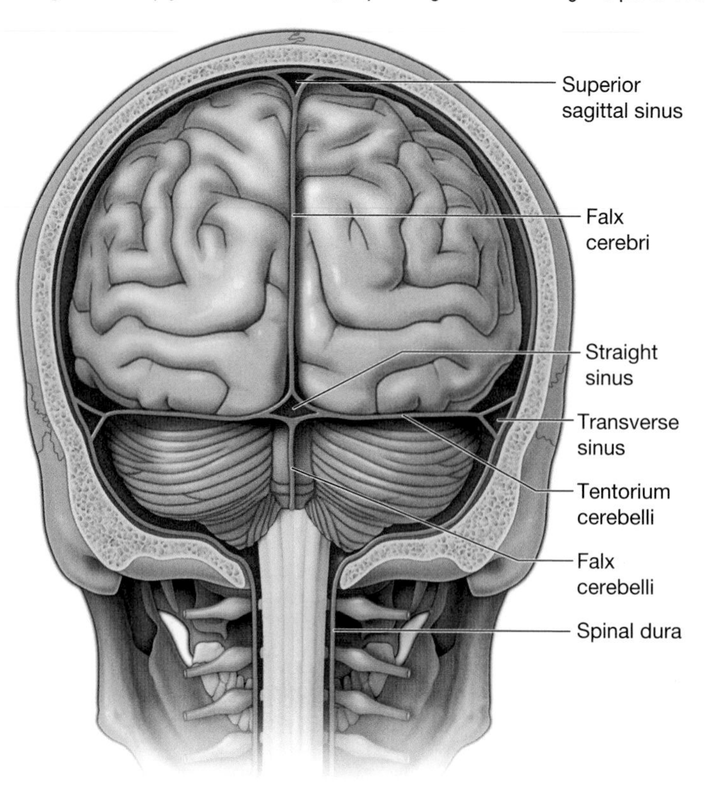

Figure 14.8 Nerve Structure

Nerves are bundles of axons found outside the spinal cord or brain. The axons within nerves are organized into fascicles, each of which is wrapped with a sheet of connective tissue called a *perineurium*. Individual axons are protected by a thin connective tissue layer called the *endoneurium*, and the entire nerve, which may contain small blood vessels, is wrapped in a connective tissue layer called the *epineurium*.

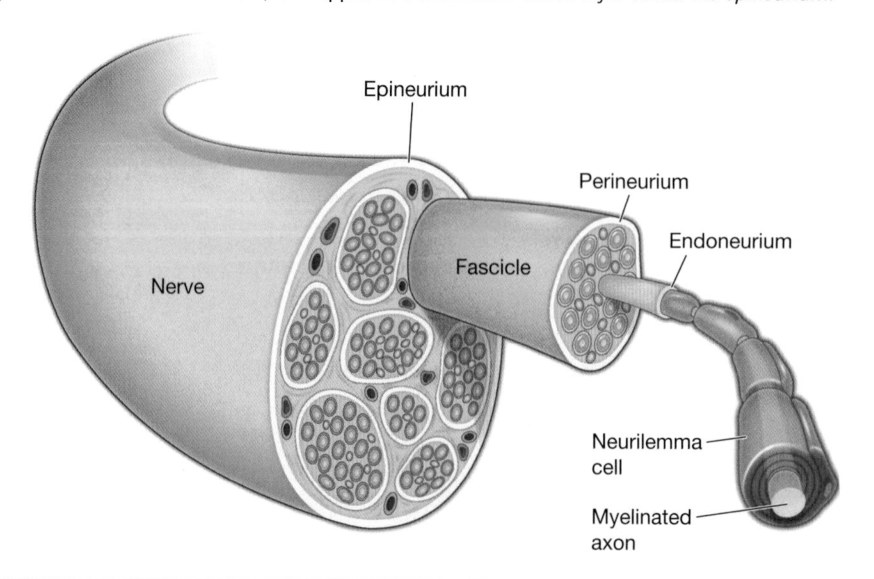

Apply to Pathophysiology

Meningitis

The suffix *-itis* indicates inflammation; for example, bronchitis is inflammation of the bronchi. Meningitis is a condition that occurs when a viral or bacterial infection makes its way into the brain or spinal cord. Shane is an 18-year-old college student who has been complaining of a sore throat, but assumed it was a mild cold or allergies. Yesterday, they experienced nausea and vomited once. Today, they were feeling better and didn't want to miss any more class time. Shane attends classes for the day but feels increasingly tired. By the end of the day Shane has a headache and a fever and heads to the ER. The ER doctors suspect that Shane may have **meningitis** and want to sample their cerebrospinal fluid.

1. The ER nurse prepares Shane for "lumbar puncture" to obtain cerebrospinal fluid. Where exactly will they insert the needle?
   A) Under the epithelium
   B) Into the subarachnoid space
   C) Into the epidural space
   D) Beneath the connective tissue covering any nerve

   The CSF has leukocytes (white blood cells) as well as *Neisseria meningitidis* bacteria, a species that can cause meningitis. The doctors understand that it is likely that all three meningeal layers are inflamed.

2. What will happen as the meninges swell?
   A) They will push the cranial bones outward, expanding the space within the cranial sutures.
   B) They will push the facial bones forward, distorting Shane's features and creating unbearable sinus pressure.
   C) They will push the cerebrum inward, damaging axonal tracts.
   D) They will push the cerebrum inward, damaging neuronal cell bodies.

3. While in the ER, Shane's neck begins to ache. This is likely because:
   A) Shane has been sitting stiffly in the ER waiting room and in classes all day.
   B) The spinal cord meninges are continuous with the brain meninges and both are inflamed.
   C) Shane's head is heavier than usual due to the infection, so the neck muscles are overtired.
   D) The infection has spread from the brain's meninges to the anatomically separated spinal cord meninges through the bloodstream.

organization is similar to the connective tissue sheaths for muscles. Nerves are associated with the region of the CNS to which they are connected, either as cranial nerves connected to the brain or spinal nerves connected to the spinal cord.

14.1b Blood and Cerebrospinal Fluid

Cerebrospinal fluid is a clear fluid derived from blood that circulates within the tissues of the central nervous system. The peripheral nervous system does not have CSF or an equivalent. CSF is produced by ependymal cells (one of the types of glial cells discussed in Chapter 13) within the four spaces of the brain called ventricles (**Figure 14.9**). Ependymal cells surround capillaries within the pia mater and filter blood plasma to generate CSF. In dissection, the ventricles are empty spaces with tangles of capillaries shrouded with ependymal cells. These capillary-membrane tangles are known as the **choroid plexus**.

At the choroid plexus, the ependymal cells extract from the blood a fluid (CSF) that is similar in composition to blood plasma by selectively moving solutes out of the blood and into the ventricles and allowing water to move in that direction by osmosis. CSF is produced in the ventricles, which are anatomically connected to each other. The **lateral ventricles** are paired (there is one right and one left) superior ventricles that are housed within the cerebral hemispheres. On their posterior ends they each connect to a small drain called the **interventricular foramen**. CSF produced within the lateral ventricles flows through the interventricular foramina (plural) to the **third ventricle**. The third ventricle is connected to the **fourth ventricle** through the **cerebral aqueduct**. All four ventricles have a choroid plexus and produce CSF. Within the wall

interventricular foramen (foramen of Monro)

LO 14.1.10

cerebral aqueduct (Aqueduct of Sylvius)

Figure 14.9 The Ventricles of the Brain

There are four ventricle structures deep within the brain. Each is a fluid-filled space with a choroid plexus. (A) Lateral and posterior views and (B) frontal section. The choroid plexus generates cerebrospinal fluid from blood plasma. (C) In the choroid plexus, ependymal cells are glial cells that generate the CSF.

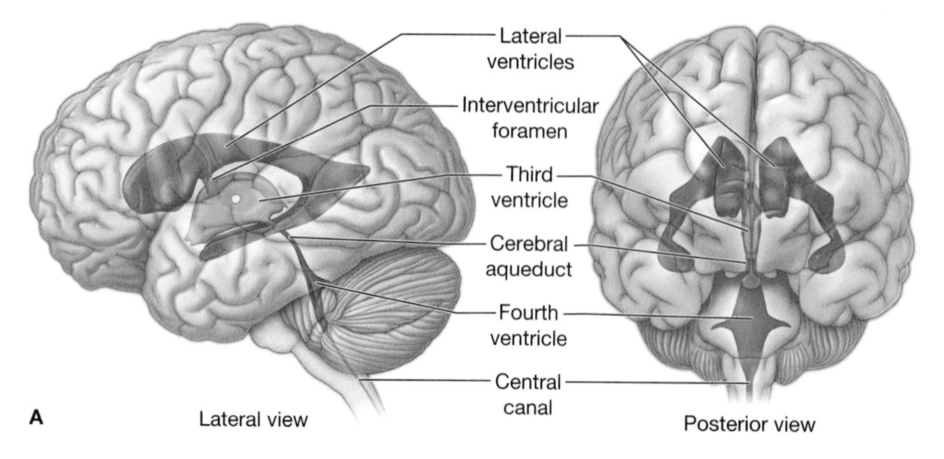

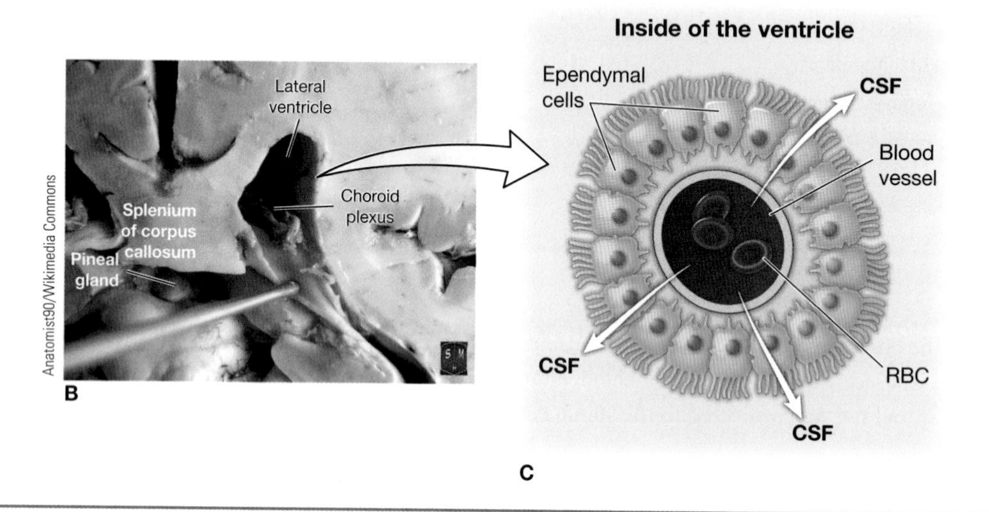

of the fourth ventricle are small holes that allow any CSF collected there to exit and fill the subarachnoid space. Inferiorly, the fourth ventricle sits on top of, and connects to, a hollow column in the center of the spinal cord called the **central canal**. This canal does not produce CSF but is constantly filled with it. CSF flows from the bottom of the central canal and fills the subarachnoid space of the spinal cord (**Figure 14.10**). CSF volume is tightly regulated; as the pressure increases, the CSF is absorbed by the arachnoid membrane and added back to the venous blood. In this way, CSF can be considered somewhat analogous to blood in that it has a role in delivering nutrients and carrying away wastes. CSF does not carry all the nutrients to the CNS; the brain and the spinal cord are both highly vascularized. CSF has several additional roles. Importantly, CSF provides physical protection to the brain, allowing it to float within the cranium rather than sink into the foramen magnum. CSF cushions the brain during sudden movements such as turning your head rapidly to the side or jolting forward in a car that is braking quickly. If you enjoy roller coasters or contact sports, you can thank your CSF for protecting your soft brain from being damaged by hitting against your hard skull.

LO 14.1.11

LO 14.1.12

Figure 14.10 **Cerebrospinal Fluid Circulation**

CSF is produced in the ventricles and flows from the lateral ventricles to the third and fourth ventricles. The fourth ventricle empties inferiorly into the central canal of the spinal cord. At the bottom of the spinal cord it flows upward through the CNS in the subarachnoid space.

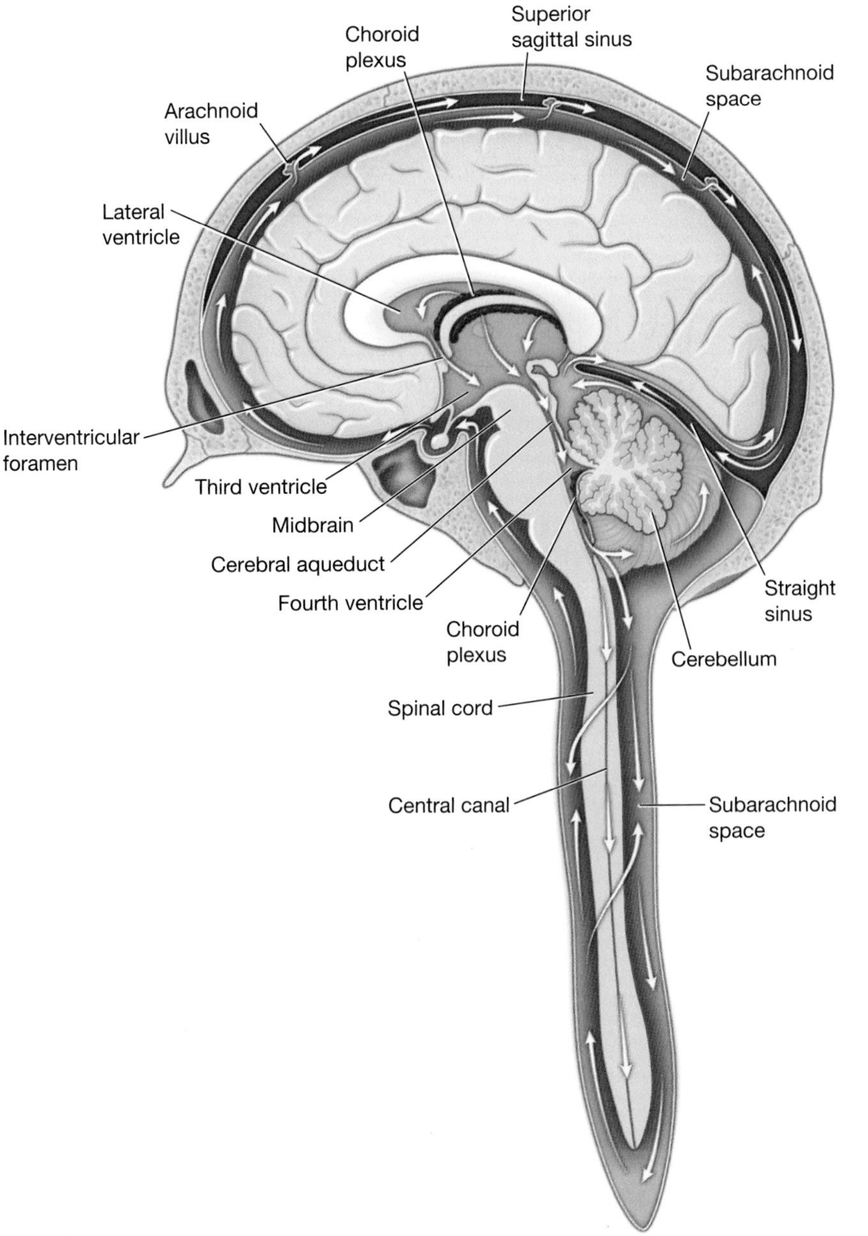

Blood serves the primary role of nutrient delivery to both the CNS and the PNS. Blood vessels that provide oxygen and nutrients and carry away wastes are typically found sandwiched among the fascicles in a peripheral nerve. Nerves, however, are bundles of axons; the cell bodies of neurons are more vulnerable. Remember that most neurons are nonmitotic; they cannot be replaced if they die, so protecting these cells from harm is critical. Therefore, both the PNS and the CNS have anatomical structures to protect the vulnerable neuronal cell bodies. In the PNS, as you may remember from Chapter 13, cell bodies are surrounded by satellite cells, one type of glial cell. The satellite cells help regulate the materials that come and go from the cell bodies. In the CNS, glial cells also serve a protective role, forming the blood-brain barrier. In this case the glial cells are astrocytes, which surround the brain's blood vessels and join together with tight junctions, forming a tight seal. This anatomical feature makes the brain's capillaries the most impermeable capillaries in the body. However, all of the principles of membrane transport still apply in this critical location. Glucose can cross the blood-brain barrier via transporters; gasses such as oxygen and carbon dioxide, as well as any lipid-soluble molecule, including the drugs caffeine and nicotine, can cross by simple diffusion. Many molecules that are not lipid soluble cannot cross the blood-brain-barrier; this is both protective and detrimental, as it causes significant challenges to the treatment of disease.

Learning Check

1. In the brain, gray matter is found in the ——— while in the spinal cord, the gray matter is found in the ———.
   a. Periphery; center
   b. Center; periphery
   c. Center; center
   d. Periphery; periphery
2. Which of the following is the deepest meningeal structure?
   a. Dura mater
   b. Arachnoid mater
   c. Subarachnoid space
   d. Pia mater
3. When a nerve is stretched, the inner connective tissue of the nerve can be damaged, but the most superficial connective tissue is not damaged. As long as the most superficial layer is intact, the nerve can heal. What layer must remain intact?
   a. Epineurium
   b. Perineurium
   c. Endoneurium
   d. Interneurium
4. Which of the following structures connects the third and fourth ventricles?
   a. Cerebral aqueduct
   b. Choroid plexus
   c. Interventricular foramen
   d. Lateral ventricles

   LO 14.1.13
5. How does the structure of the spinal cord's dura mater compare with that of the brain's dura mater?
   a. The dura mater of both the brain and spinal cord contains a periosteal and meningeal layer.
   b. The dura mater around some areas of the brain contains a periosteal and meningeal layer, while that of the spinal cord consists of only one layer.
   c. The dura mater of the spinal cord contains a periosteal and meningeal layer, while that of the brain consists of only one layer.
   d. The structure is identical; the dura mater of both the brain and spinal cord consists of one layer.

14.2 The Central Nervous System

Learning Objectives: By the end of this section, you will be able to:

14.2.1 Identify and describe the 3 primary brain vesicles formed from the neural tube.

14.2.2 Identify and describe the 5 secondary brain vesicles formed from the neural tube and name the parts of the adult brain arising from each.

14.2.3 Identify and describe the four major parts of the adult brain (i.e., cerebrum, diencephalon, brainstem, cerebellum).

14.2.4 Identify and describe the three major cerebral regions (i.e., cortex, white matter, cerebral nuclei [basal nuclei]).

14.2.5 Identify and define the general terms gyrus, sulcus, and fissure.

14.2.6 Identify and describe the major landmarks of the cerebrum (e.g., longitudinal fissure, lateral sulcus [fissure], central sulcus, transverse fissure, precentral gyrus, postcentral gyrus).

14.2.7 Identify and describe the cerebral hemispheres and the five lobes of each (i.e., frontal, parietal, temporal, occipital, insula).

14.2.8 Identify and describe the primary functional cortical areas of the cerebrum (e.g., primary motor cortex, primary somatosensory cortex, primary auditory cortex, primary visual cortex, primary olfactory cortex, primary gustatory cortex).

14.2.9 Compare and contrast the cerebral location and function of the motor speech area (Broca area) and Wernicke area.

14.2.10 Compare and contrast the three cerebral white matter tracts (i.e., association, commissural, projection).

14.2.11* Describe the location and functions of two of the basal ganglia.

14.2.12 Name the major components of the diencephalon.

14.2.13 Describe the structure, location, and major functions of the thalamus.

14.2.14 Describe the structure, location, and major functions of the hypothalamus, including its relationship to the autonomic nervous system and the endocrine system.

14.2.15 Describe and identify the epithalamus, including the pineal gland and its function.

14.2.16 Name the three subdivisions of the brainstem.

14.2.17 Describe the structure, location, and major functions of the midbrain (mesencephalon), including the cerebral peduncles, superior colliculi, and inferior colliculi.

14.2.18 Describe the structure, location, and major functions of the pons.

14.2.19 Describe the structure, location, and major functions of the medulla oblongata (medulla), including the pyramids and decussation of the pyramids.

14.2.20 Describe the structure, location, and major functions of the cerebellum.

14.2.21 Identify and describe the cerebellar hemispheres, vermis, arbor vitae (cerebellar white matter), cerebellar peduncles, and cerebellar cortex (folia, cerebellar gray matter).

14.2.22 Describe the major components and functions of the reticular activating system (RAS).

14.2.23 Describe the major components and functions of the limbic system.

14.2.24 Identify and describe the gross anatomy of the spinal cord, including its enlargements (i.e., cervical and lumbar), conus medullaris, cauda equina, and filum terminale.

14.2.25 Identify and describe the anatomical features seen in a cross-sectional view of the spinal cord (e.g., anterior horn, lateral horn, posterior horn, gray commissure, central canal, anterior funiculus [column], lateral funiculus [column], posterior funiculus [column]).

14.2.26 Compare and contrast the location, composition, and function of the anterior (ventral) roots, posterior (dorsal) roots, and posterior (dorsal) root ganglion with respect to the spinal cord.

14.2.27 Describe the structure, location, and function of ascending and descending spinal cord tracts.

* Objective is not a HAPS Learning Goal.

The brain is a complex organ composed of gray parts and white matter, which can be hard to distinguish. Starting from an embryologic perspective allows you to understand more easily how the parts relate to each other. The embryonic nervous system begins as a very simple structure—essentially just a straight line, which then gets increasingly complex. Looking at the development of the nervous system with a couple of early snapshots makes it easier to understand the whole complex system.

14.2a Embryonic Development of the Nervous System

Early in embryonic development a groove forms along one side of the embryo, stretching from one end to the other. Eventually this groove invaginates and pinches off, forming a hollow tube—the **neural tube**—along the posterior of the embryo. A collection of scattered cells—**neural crest cells**—surround the neural tube. Local signals and hormones drive the differentiation and development of the neural crest cells and neural tube cells into neurons. The neural tube becomes the brain, eyes, and spinal cord, and the neural crest cells become the peripheral nerves (**Figure 14.11**).

LO 14.2.1

forebrain (prosencephalon)

midbrain (mesencephalon)

hindbrain (rhombencephalon)

As the anterior end of the neural tube starts to develop into the brain, it forms a couple of enlargements. Similar to a balloon animal, the long, straight neural tube begins to take on a new shape. In the first stage, three vesicles form, which are called **primary vesicles** (**Figure 14.12A**). These are the **forebrain**, the **midbrain**, and the **hindbrain**. The brain continues to develop, and the vesicles differentiate further (see **Figure 14.12B**). The three primary vesicles become five **secondary vesicles**. The forebrain becomes two new vesicles called the **telencephalon** and the **diencephalon**. The telecephalon will become the cerebrum. The diencephalon gives rise to several adult structures, including the retinas of the eyes, the thalamus, and the hypothalamus. The midbrain does not differentiate into any finer divisions. The rest of the brain develops around it and constitutes a large percentage of the mass of the brain. The hindbrain develops into the **metencephalon** and the **myelencephalon**. The metencephalon becomes the adult structures known as the *pons* and the *cerebellum*. The myelencephalon becomes the adult structure known as the *medulla oblongata* (**Figure 14.12C**).

LO 14.2.2

Figure 14.11 Development of the Nervous System

The outer layer of the embryo folds in to form a pouch and then pinches off to form a tube. The new outer layer becomes the epidermis and the pinched-off tube becomes the neural tube, which will later become the spinal cord. Small clusters of neural cells—the neural crest cells—become the peripheral nerves. A rigid cylinder of tissue internal to the neural tube—the notochord—later segments to become the bodies of the vertebrae.

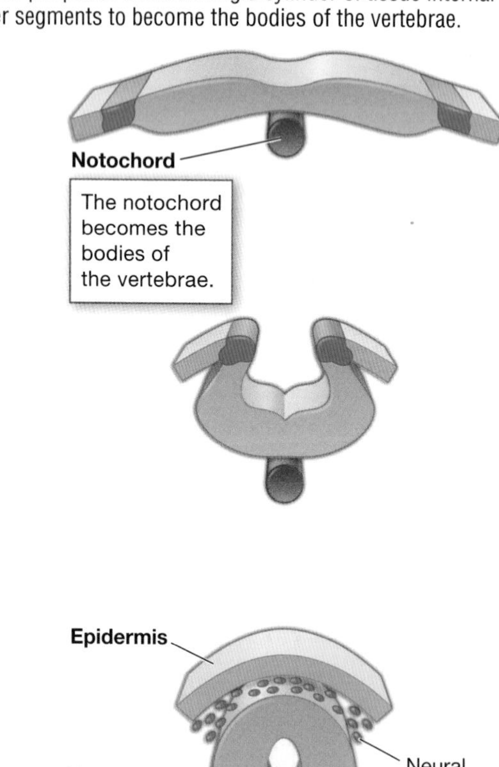

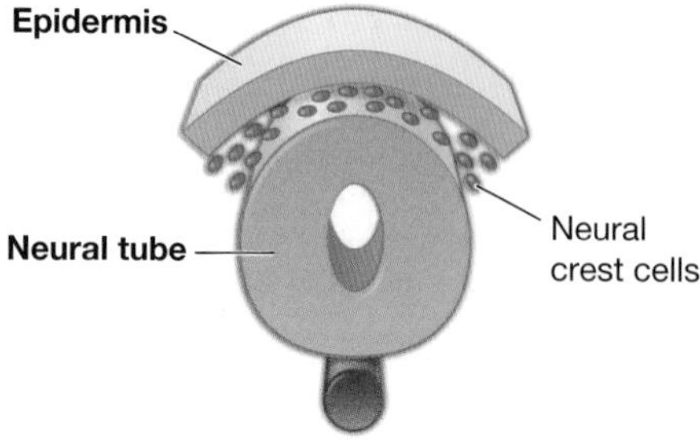

Figure 14.12 The Development of the Central Nervous System

(A) The neural tube forms three swellings—the forebrain, the midbrain, and the hindbrain. (B) The forebrain and the hindbrain each split into two additional swellings. The forebrain splits into the telencephalon and the diencephalon, and the hindbrain splits into the metencephalon and the myelencephalon. (C) These structures develop into adult brain structures. The forebrain becomes the cerebrum, the thalamus, the hypothalamus, and the eyes. The midbrain remains as the midbrain. The hindbrain develops into the pons, the cerebellum, and the medulla oblongata.

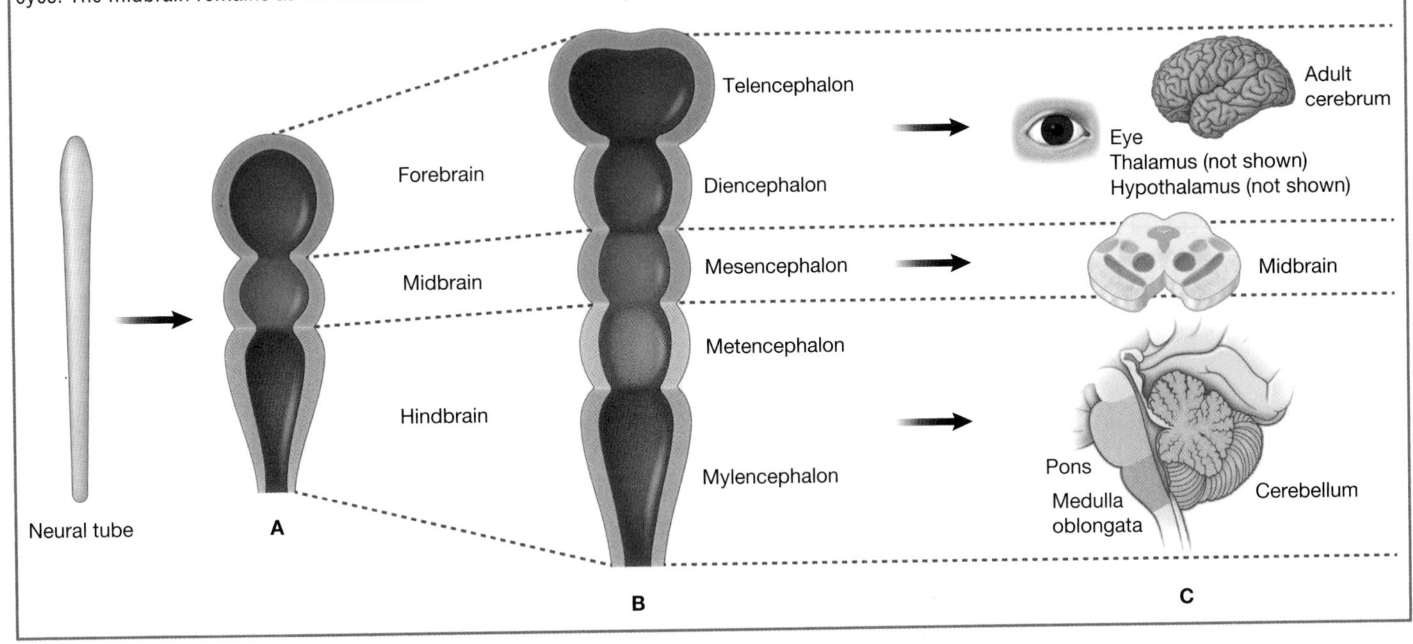

14.2b The Brain

LO 14.2.3

The adult brain is described in terms of four major regions: the cerebrum, the diencephalon, the brainstem, and the cerebellum. A person's conscious experiences are based on neural activity in the brain. The regulation of homeostasis is governed by a specialized region in the brain. The brain receives sensory input from the body and controls muscle movements through chains of motor neurons.

LO 14.2.4

LO 14.2.5

LO 14.2.6

The Cerebrum The iconic wrinkled, gray globe of the human brain, which appears to make up most of the mass of the brain, is the **cerebrum** (Figure 14.13). The wrinkled, superficial portion is the **cerebral cortex**, and the rest of the structure is deep to the cortex. As the brain develops within the cranium, it runs out of space for all of the cell bodies of the cerebral cortex; therefore, the developing brain folds in on itself repeatedly. Each fold is a **gyrus** (plural: gyri) and the grooves between the folds are **sulci** (singular: sulcus). Larger grooves are fissures; the largest of these is the **longitudinal fissure**, the separation between the two sides of the cerebrum. It separates the cerebrum into two distinct halves, a right and left **cerebral hemisphere**. Deep within the cerebrum, the white matter of the **corpus callosum** is the major tract of communication between the two hemispheres of the cerebral cortex (Figure 14.14).

Many of the higher neurological functions—such as memory, emotion, language, and consciousness—are the result of cerebral function. The two cerebral hemispheres are largely mirror images of each other; for example, each hemisphere contains a region for receiving sensory information and controlling motor output; these regions are all approximately the same size. Interestingly, each hemisphere receives somatosensory information and controls motor movements in the opposite side of the body, so the left side receives input from, and sends commands to, the right arm and right leg and vice versa. There are very few functional differences between the hemispheres. Language functions, however, are

Student Study Tip

There's a *colossal* amount of communication via the corpus *callosum*.

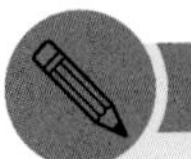

Learning Connection

Try Drawing It

On an unlabeled drawing of a brain, or on your own drawing, label the right and left hemispheres, the longitudinal fissure, a gyrus, and a sulcus.

Figure 14.13 The Cerebrum

The cerebrum is the largest brain structure. Its surface is characterized by many gyri and sulci. It is divided into right and left hemispheres by the longitudinal fissure.

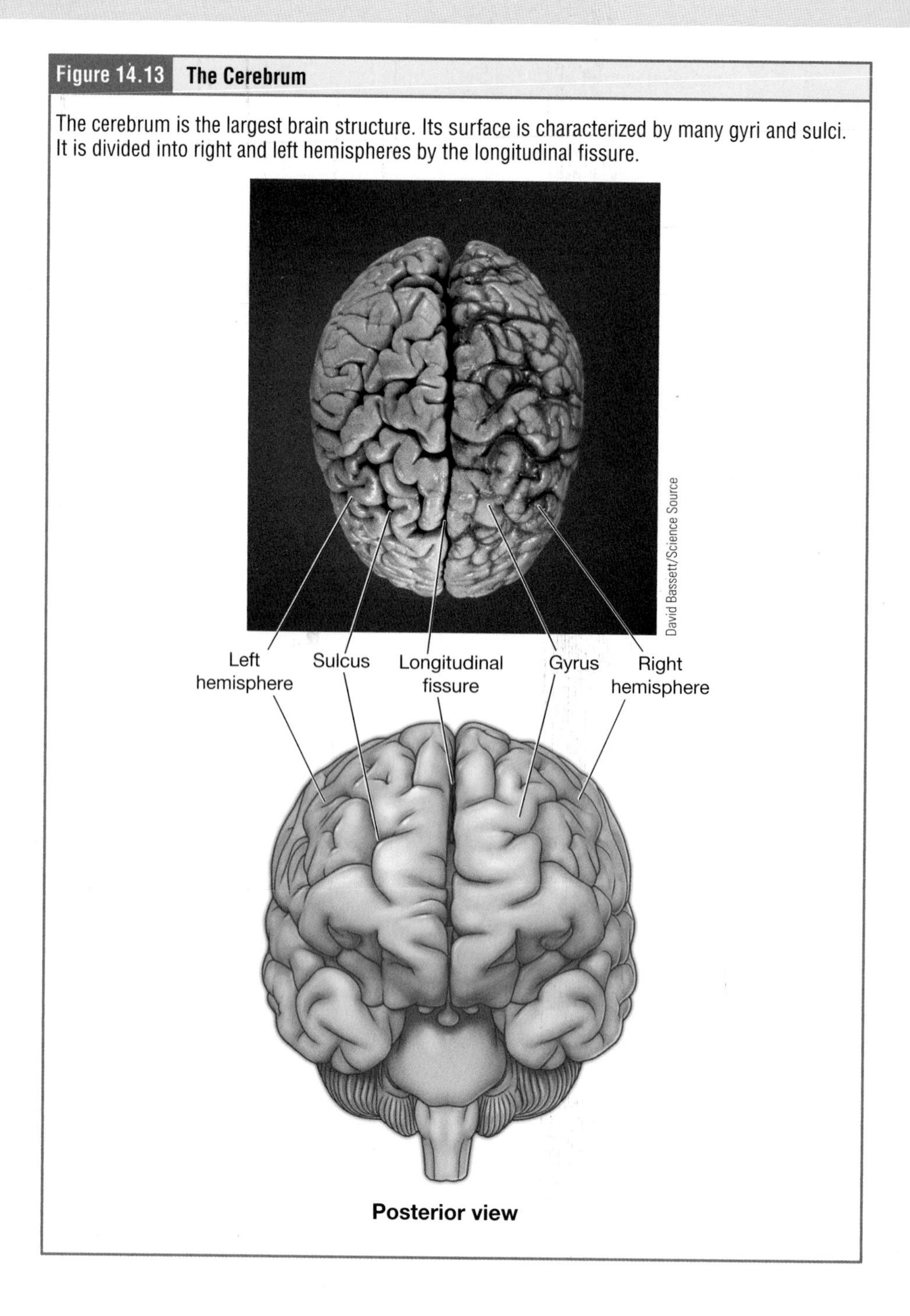

the exception. Language functions—both the understanding of language and the muscle contractions required to form words—appear to be left-sided functions in most individual brains.

The cortex of each cerebral hemisphere can be divided into five anatomical and functional areas called *lobes*. Deep to the cortex, the basal nuclei are responsible for cognitive processing, their most important function being that associated with planning movements. The **basal forebrain** contains nuclei that are important in learning and memory. The **limbic system** is a collection of cerebral structures involved in emotion, memory, and behavior. These structures are collectively called *subcortical structures* and are discussed in more detail following our discussion of the cortex.

Figure 14.14 The Corpus Callosum

The right and left cerebral hemispheres are joined by a huge bundle of axons called the corpus callosum. The corpus callosum unites the two hemispheres functionally.

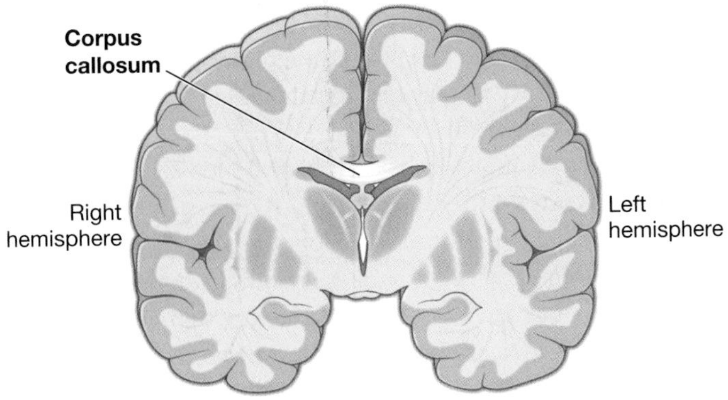

Cerebral Cortex The outer shell of the cerebrum is a continuous layer of gray matter that contains the cell bodies of the neurons responsible for the higher functions of the nervous system. Using the largest sulci as landmarks, the cortex can be divided into four major lobes: the frontal, parietal, temporal, and occipital lobes (**Figure 14.15**). Notably, there is one frontal lobe across the anterior brain, one occipital lobe across the posterior brain, and then paired temporal and occipital lobes (a right and a left of each). The **lateral sulcus**, which separates the **temporal lobe** from the other regions, is one such landmark. Superior to the lateral sulcus are the **parietal lobe** and the **frontal lobe**, which are separated from each other by the **central sulcus**. The **occipital lobe** has no obvious anatomical border between it and the surrounding lobes—most likely because it is intensely functionally connected to other portions of the brain.

LO 14.2.7

central sulcus (Sulcus of Rolando)

Figure 14.15 Lobes of the Cerebrum

The cerebral cortex is divided into five functional regions. The frontal lobe, parietal lobes, occipital lobes, and temporal lobes are visible here; the insula is the fifth region, which lies deep to the temporal lobe; part of the insula is visible in this figure.

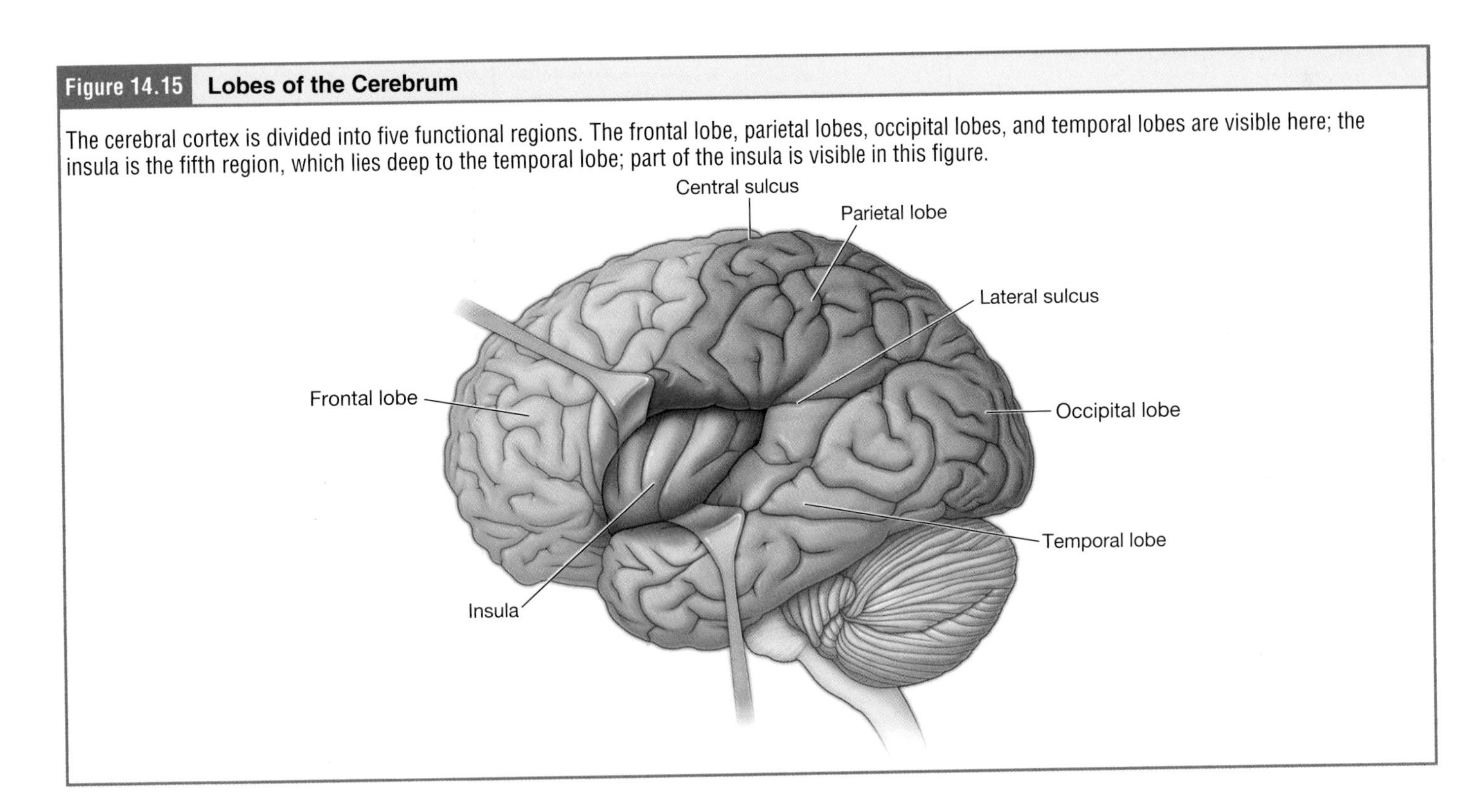

Learning Connection

Try Drawing It

Now that you can see the anatomy of the lobes of the cerebrum from the lateral perspective in Figure 14.13, try drawing or labeling the lobes on an unlabeled diagram from the posterior, superior or anterior views.

Except for the gyri and fissures, and the gray and white matter, the regions of the brain are relatively indistinguishable from each other. For a long time, anatomists did not have much insight into how functions were localized within the brain until doctors began cataloging case studies of brain injuries. As it turns out, time and time again, injuries in the same location would produce similar functional losses in different patients. From careful observation, and now through more advanced techniques like functional MRI imaging, we have come to understand that the cerebral cortex exhibits exquisite anatomical specificity.

The single (unpaired) frontal lobe is found just under the frontal bone. It occupies the entire anterior cortex and ends inferiorly at the *central sulcus*, a deep sulcus that divides the frontal lobe from the parietal lobes. Inferiorly, the frontal lobe is bordered by the *lateral sulcus*, which separates it from the temporal lobe. The frontal lobe contains areas that plan and initiate muscle contraction, including the contractions that allow for speech formation and eye movements. The frontal lobe is also involved in a lot of decision-making and other higher-order cognitive behaviors. The gyrus just anterior to the central sulcus is the **precentral gyrus** (see Figure 14.15). This region is the **primary motor cortex**. Cells from this region of the cerebral cortex are the upper motor neurons that instruct cells in the spinal cord to move skeletal muscles. Interestingly, this small slice of brain controls the motor function of the entire body. Its mass is proportional to the complexity of movement available, so the region of the brain that controls the face and the hands—small regions with complex movements—is larger than the region of the brain that controls the trunk—a large region with limited movement (**Figure 14.16**). The reason for this proportionality has to do with the size of the motor units serving these locations. Anterior to this region are a few areas that are associated with planned movements. Anterior to these regions is the **prefrontal lobe**, which serves cognitive functions that can be the basis of personality, short-term memory, and consciousness. Until relatively recently, personality disorders were sometimes treated using a prefrontal lobotomy, a controversial and outdated mode of treatment that profoundly affected the patient's personality.

LO 14.2.8

The main function associated with the parietal lobe is **somatosensation**, meaning the general sensations associated with the body. The anterior border of the parietal lobe is the central sulcus. The first gyrus posterior to the central sulcus is the

Cultural Connection

Right-Brained and Left-Brained People

Pop media articles are full of references to being "right-brained" or "left-brained." Right-brained individuals are described as being more creative and emotional, while left-brained people are described as being more analytical and organized. If you search the Internet, you can find cute infographics and even dating advice based on how you identify in these two categories. Is being "left-brained" an anatomical distinction? Largely, no. There are a few functional differences between the two hemispheres of the brain (language being the most important of these one-sided functions). However, in most respects scientists currently believe that the hemispheres are mirror images of each other. Some neurobiologists think that the hemispheres may work on the same tasks but function slightly differently. It may be more interesting to consider the anatomical variation of the corpus callosum, the wide tract that connects the right and left hemispheres. This connection does demonstrate anatomical variation with function. Individuals who are left-handed or ambidextrous (use both hands) have larger corpora callosa. As a group, musicians have the largest corpora callosa. Much, much anatomical study has gone into validating or refuting the observation that biological women have larger corpora callosa than biological males. While this is still hotly debated in the field to this day, and likely due to more heterogeneity within these classifications than biologists account for, most studies concur that they do.

postcentral gyrus, and this region is known as the **primary somatosensory cortex**. All of the senses from the skin—including touch, pressure, tickle, pain, itch, and vibration—are processed in this area. Similar to the primary motor cortex, this region is proportioned to receive sensory information from the areas that send the most signals. The hands, face, mouth, and throat send the most signals, and these small areas take up almost half of the primary somatosensory cortex (see Figure 14.16). **Proprioception**, the sense of body position, is also processed in the primary somatosensory cortex.

proprioception (kinesthesia)

The paired temporal lobes are found beneath the temporal bone on the lateral/inferior regions of the brain (see Figure 14.15). Neurons within the temporal lobes are dedicated to the tasks of hearing and smelling.

Figure 14.16 The Primary Motor Cortex and Primary Somatosensory Cortex

The two gyri to either side of the central sulcus are drawn here. The precentral gyrus, anterior to the central sulcus, contains the primary motor cortex. Here the cell bodies for the neurons that influence the motor neurons reside. They are organized anatomically so that cell bodies of all of the neurons that influence a particular region of the body are clustered together. The postcentral gyrus, which contains the primary somatosensory cortex, is the final anatomical destination for the chains of neurons bringing somatosensory information into the CNS for processing. Here, again, the gyrus is organized anatomically so that all axons from a particular body region end in the same location. This organization is the basis for CNS localization of incoming information.

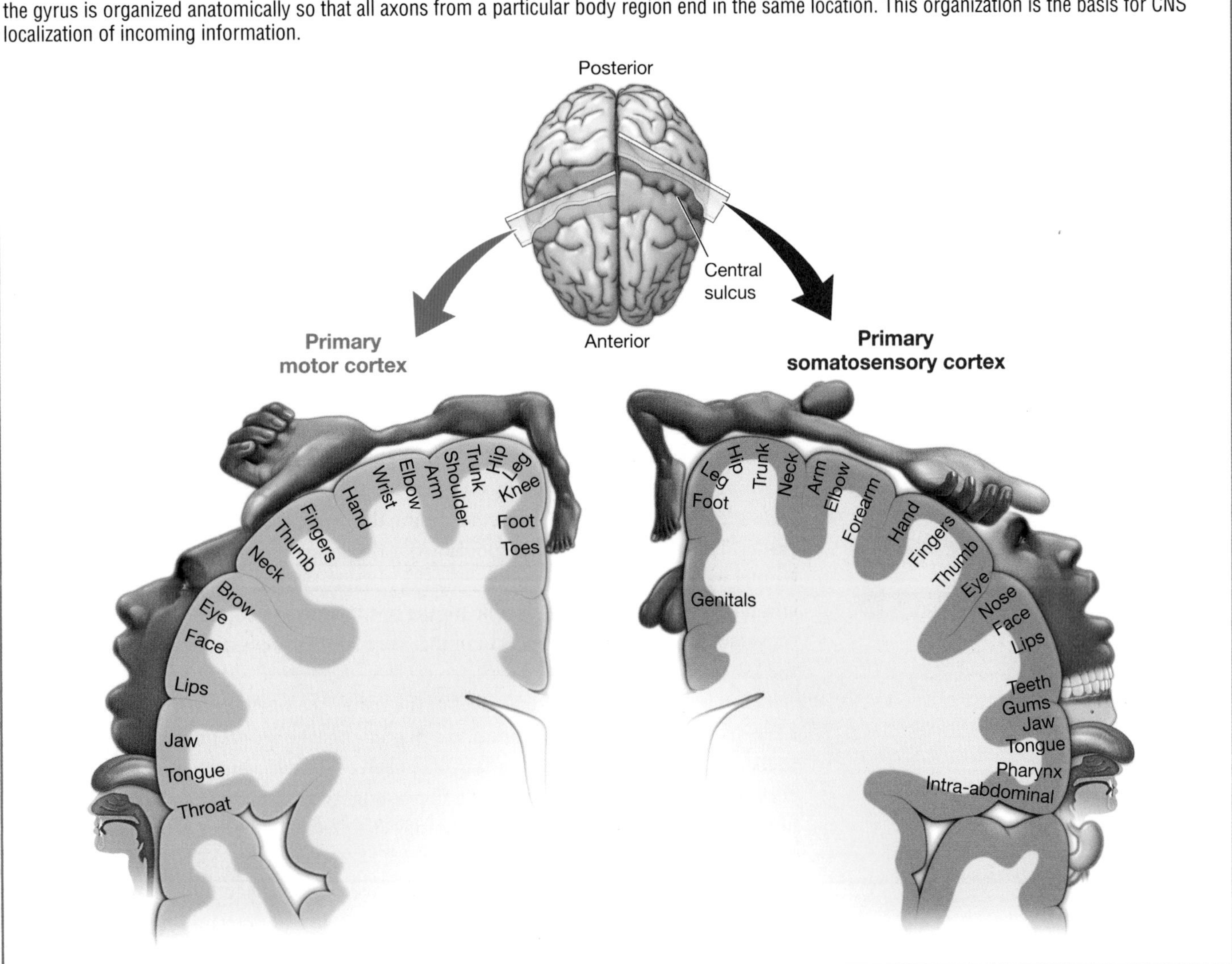

The paired insula lobes are not visible from the external surface, but are folded regions of cortex deep to the temporal lobes. The insula is involved in taste associations, meaning that it functions to help you connect a new taste to taste memories.

The single occipital lobe spans the posterior of the brain, a mirror of the frontal lobe. This large region is dedicated entirely to processing visual information and visual memories.

Brain Functions In general, the brain regions are described using the words *primary* or *association*. A primary region is the region that directly receives sensory input or controls motor output. When you see the words of this text, your eyes take in the shapes of the letters and send signals directly to a primary visual area in the occipital lobe for interpretation. The direct translation of that image (or the direct translation of other sensory information) occurs here; you can, for example, interpret that the letter *o* is a round shape that is empty in the center. These primary centers connect to association areas which compares this information to your library of past experiences. That round, hollow letter *o* that you've just seen is compared to your memories, where you can interpret that it is an *o*, makes a particular sound, and is found starting words like *octopus* and *octagon*. The collaboration between primary and association areas of the brain allows you to recognize smells, read language, stop at a stop sign, and interpret the feel of your favorite sweater. Note that the primary and association regions for sensory information coming from the skin—the primary somatosensory cortex—is in the parietal lobe, but the special senses—the senses that occur at the head such as sight and smell—have primary and association areas in the other lobes.

Visual stimuli are processed by primary and association areas found within the occipital lobe. As a species, humans rely on their visual information more than their other senses; in fact, nearly 40 percent of the cerebral cortex is dedicated to vision.

The auditory primary and association areas are located within the temporal lobe. As with the other senses, the auditory association area helps to compare a sound that you are hearing at the current time to sounds from your past experiences. When you hear a voice, for example, you may be able to recognize it as one you have heard before or even specifically match it to a friend or an instructor that you know. Likewise, you have primary and association areas of your brain for smell and taste, which are located in the temporal and insula lobes, respectively.

Motor areas also have a primary area—the primary motor cortex—as well as secondary or association areas that assist in planning coordinated motor activities such as typing on a keyboard, walking, and forming the sounds for speech. All of the motor areas are found in the frontal lobe, including the primary motor cortex discussed earlier. The **premotor area** is responsible for thinking of a movement to be made. The **frontal eye fields** are important in eliciting eye movements and in attending to visual stimuli. **Broca's area** is responsible for the production of language or controlling movements responsible for speech; in the vast majority of people, it is located only on the left side.

Broca's area (motor speech area)

LO 14.2.9

Two other major functional areas of the cortex are worth noting. Wernicke's area, a region on the left side of the brain, is a major language area. This region contributes to the recognition and understanding of spoken or written language. The prefrontal cortex is involved in our intellectual and personality-based functions such as humor, judgement, decision-making, and planning. The prefrontal cortex is one of the last regions of the brain to mature; myelination of axons and elimination of unnecessary synapses continues into the twenties, leading neuroscientists to understand that teenagers may struggle with complex planning, impulsivity, or determining the possible consequences of their actions.

stroke (cerebrovascular accident, CVA)

Apply to Pathophysiology

Strokes

The supply of blood to the brain is crucial to its ability to perform its many functions. Without a steady supply of oxygen and glucose, the nervous tissue in the brain cannot keep up its extensive electrical activity. These nutrients enter the brain through the blood, and if blood flow is interrupted, neurological function is compromised.

The common name for a disruption of blood supply to the brain is a **stroke**. It is caused by a blockage to an artery or arteriole in the brain. Many of the arteries that serve the brain are protected within the meninges. The blockage is typically derived from some type of embolism: a blood clot, a fat embolus, or an air bubble. When the blood cannot travel through the artery, the surrounding tissue that is deprived starves and dies. Strokes often result in the loss of very specific functions.

1. Which of the following structures is most likely to be damaged by a stroke?
   A) A tract
   B) The gray matter of a nucleus
   C) The gray matter of the cortex
2. If a stroke occurred in the occipital lobe, what kinds of functions may be affected?
   A) Decision-making
   B) Understanding spoken language
   C) Controlling muscle movement
   D) Coordinating visual information
3. Knowing what you've learned about mitosis and neurons, how likely do you think full recovery is after a stroke?
   A) Full recovery is always possible
   B) Partial recovery is possible, especially if new ways of completing tasks are learned

DIGGING DEEPER: Famous Brains

The technique of functional magnetic resonance imaging (fMRI) was discovered in 1990 by Seiji Ogawa, a Japanese neuroscientist. Before the emergence of fMRI machines, our understanding of brain functionality was extremely limited. The majority of our collective scientific understanding came from case studies of injuries. Doctors and scientists would document the location of the injury and the functional changes that occurred in the individual post-injury. Because it would be completely unethical to damage someone's brain on purpose, these observations were always on individual humans, and we had little understanding of the scope of anatomical variation among individuals. Of these case studies, perhaps the most famous was the case of Phineas Gage. He was a railroad worker who had a metal spike driven through his prefrontal cortex in 1848 (see Figure). He survived the accident, but according to secondhand accounts, his personality changed drastically. Friends described him as no longer acting like himself. Whereas he was a hardworking, amiable person

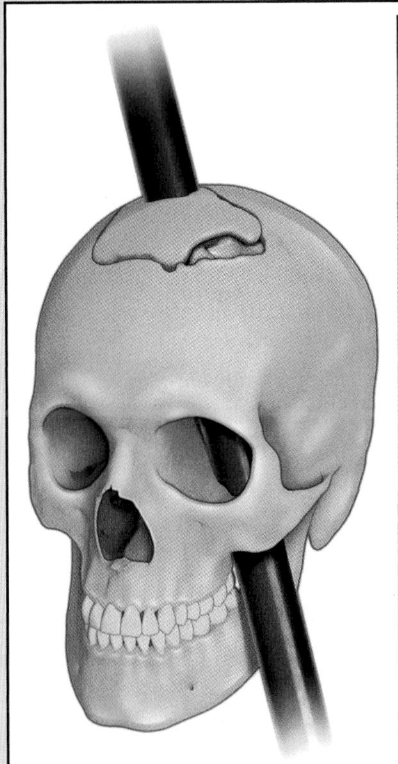

GL Archive/Alamy Stock Photo

(*Continued*)

before the accident, he became irritable, temperamental, and lazy after the accident. Many of the accounts of his change may have been inflated in the retelling, and some behavior was likely attributable to alcohol used as a pain medication. However, this seminal case changed the field of neurobiology forever, suggesting that personality was linked to one location in the brain. Indeed, now we know this to be true. A second famous case is the stroke of Jill Bolte Taylor, a Harvard neuroscientist who documented her stroke and recovery in the *New York Times* bestselling book *My Stroke of Insight.* Dr. Taylor suffered a stroke that impacted her language centers in her left hemisphere. While recovering, Dr. Taylor found that not only was she unable to communicate with others through language, but that her "internal monologue"—the thoughts inside her head—were devoid of language. Her recounting of the stroke as it progressed, as well as her recuperation and how the event changed her life even after her language function returned, are fascinating.

Subcortical Structures Immediately deep to the gray matter of the cortex are central white matter tracts. These tracts are connecting regions of the brain to each other—for example, the occipital lobe to Wernicke's area—to process and understand the written language you are reading right now! Some of the tracts are composed of sensory and motor axons traveling up to the brain from the spinal cord or vice versa (**Figure 14.17**). **Association tracts** (**Figure 14.17A**) are small white matter tracts that connect association areas in the same hemisphere. The shortest association tracts, **arcuate fibers**, connect areas within a single lobe. Arcuate fibers would, for example, connect the premotor area of the frontal lobe to the primary motor cortex. Longer tracts, **longitudinal fasciculi**, connect functional areas between one lobe and another within the same hemisphere. Cerebral hemispheres are connected through **commissural tracts**, the largest of which is the corpus callosum. Finally, **projection tracts** are the large bundles of motor and sensory axons coming into the brain from the spinal cord (sensory) or leaving the brain and descending into the spinal cord (motor).

commissural tracts (commissural fibers)

LO 14.2.10

projection tracts (projection fibers)

Beneath the cerebral cortex and among the white matter tracts are sets of nuclei known as **basal nuclei** that contribute to cortical processes (note: in some older books these structures may be called *basal ganglia*; the field of anatomy now agrees that the term *ganglia* should only apply to the peripheral nervous system). These nuclei contribute to forebrain cortical function, particularly filtering and metering responses. They may, for example, control the intensity of a motion such as how hard to grip during a handshake. They may also play a role in cognition and emotion, filtering the responses that we make to stimuli. The major structures of the basal nuclei are the **caudate**, the **putamen**, and the **globus pallidus**, which are located deep in the cerebrum. The caudate is a long nucleus that follows the basic C-shape of the cerebrum from the frontal lobe, through the parietal and occipital lobes, into the temporal lobe. The putamen is mostly deep in the anterior regions of the frontal and parietal lobes. Together, the caudate and putamen are called the **striatum**. The globus pallidus is a layered nucleus that lies just medial to the putamen; they are called the *lenticular nuclei* because they look like curved pieces fitting together like lenses. The globus pallidus has two subdivisions—the external and internal segments—which are lateral and medial, respectively. These nuclei are depicted in a frontal section of the brain in **Figure 14.18**. The basal nuclei in the cerebrum are connected with a few more nuclei in the brainstem that together act as a functional group that forms a motor pathway, including the subthalamic nuclei and the substantia nigra of the midbrain.

basal nuclei (basal ganglia)

LO 14.2.11

caudate (caudate nucleus)

Alzheimer's disease is associated with a loss of neurons in the basal nuclei of the forebrain, the symptoms of which give scientists a hint at their functions. Two midbrain nuclei, the **hippocampus** and the **amygdala**, are involved in long-term memory formation and emotional responses. The basal nuclei are a set of regions in the cerebrum that compare cortical processing with the general state of activity in the nervous system to influence the likelihood of movement taking place. For example, while a student is sitting in a classroom listening to a lecture, the basal nuclei will keep the urge to jump up and scream from actually happening. (The basal nuclei are also referred to as the *basal ganglia*, although that is potentially confusing because the term *ganglia* is typically used for peripheral structures.)

Figure 14.17 **White Matter Tracts of the Brain**

Different functional areas of the brain are connected through tracts of axons that make a notable pattern in the brain. (A) Frontal sections demonstrate the long tracts of descending motor and ascending sensory axons that are continuous from the brain to the spinal cord. (B) Midsagittal sections show the connections from the cortex to underlying structures throughout the brain.

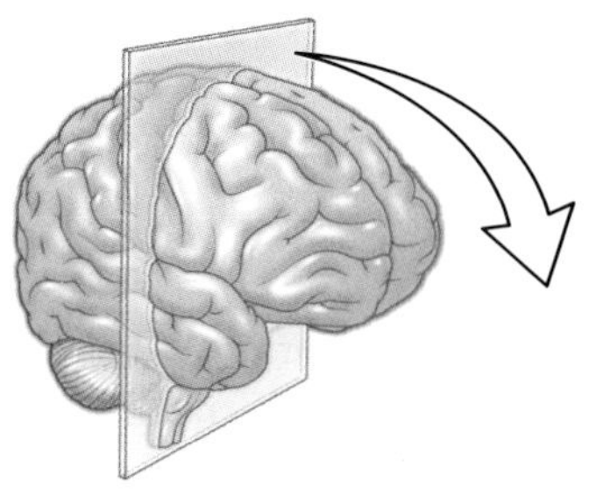

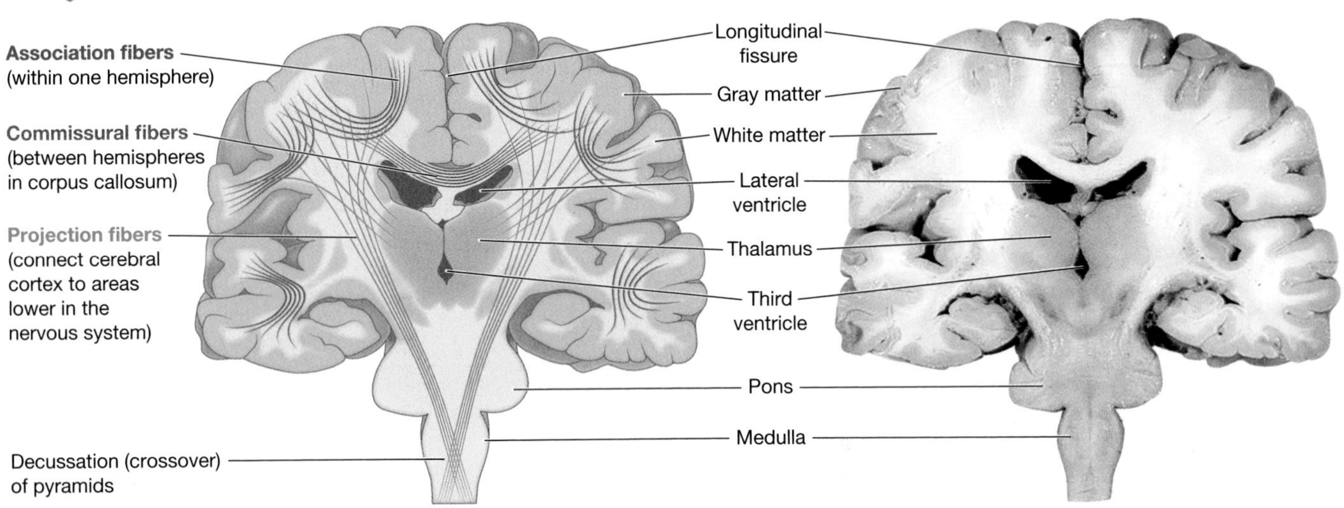

A Frontal section

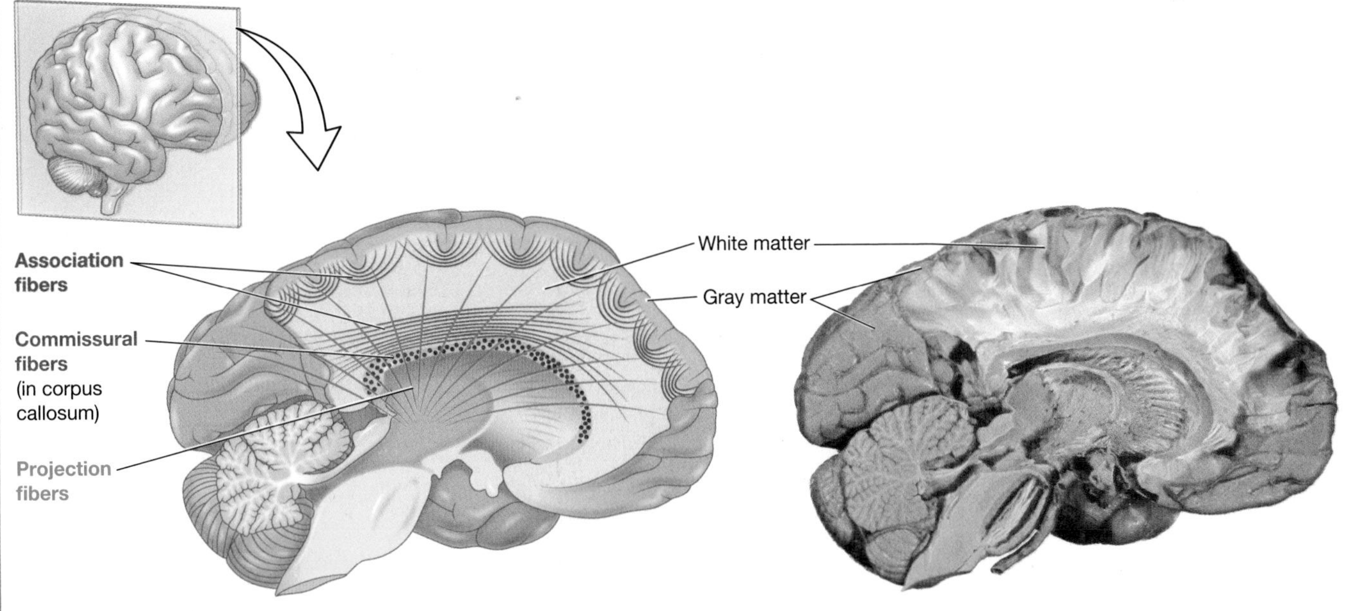

B Midsagittal section

Figure 14.18 Nuclei of the Brain

Clusters of cell bodies appear as underlying gray matter deep within the brain. Each nucleus has one or a few functions.

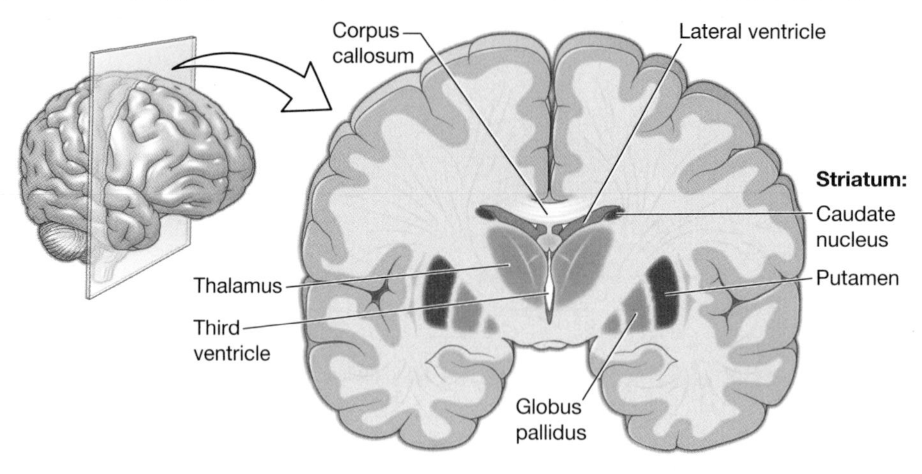

DIGGING DEEPER:
Types of Headaches

The brain itself has no pain receptors; it is impossible for you to perceive any pain from your brain itself. Because of this, brain surgery can be done while the patient is awake, with local anesthetic to the superficial structures of the scalp, skull, and meninges (see photo). This is tremendously helpful to brain surgeons who are searching for the exact location of a problem, such as a tumor, in the brain. But if the brain can't hurt, how do we develop headaches? The short answer is that typically headaches are pain from the dura mater. But there are actually over 150 different recognized types of headaches. Headaches can occur due to changes in blood flow to the brain. These may occur if blood pressure changes throughout the body, or if there is not enough glucose to feed the brain. Dehydration, especially dehydration following alcohol consumption, can also cause headaches in a slightly different way. When the blood pressure falls too low due to a lack of hydration, insufficient CSF production can be a consequence. As the pressure within the cranium falls, the dura mater sags slightly, causing a pain sensation. Likewise, pulling on the dura when the neck muscles are overly contracted, such as during a tension headache, also triggers pain. When the headache is dural in origin, the pain is felt bilaterally on both sides of the head because the dura is not separated by the hemispheres of the brain. Migraine headaches, however, are typically felt only on one side of the head, a hint that migraines may be neurological in nature, rather than originating from the dura mater. Other clues as to the neurological origin of migraines is the incorporation of different senses. Rather than just pain, migraines may come with alterations in the sense of smell (often a heightened sense of smell) and alterations in vision (typically spots, flashes, or a heightened sensitivity to light). Other senses may also be involved, such as a ringing in the ears and numb or tingling skin. Additionally, about half of migraines involve nausea or vomiting. While we don't know exactly what causes migraines, and not all migraines are the same, scientists have developed several effective treatments for migraines and those who suffer from migraines have more control over their symptoms than ever before.

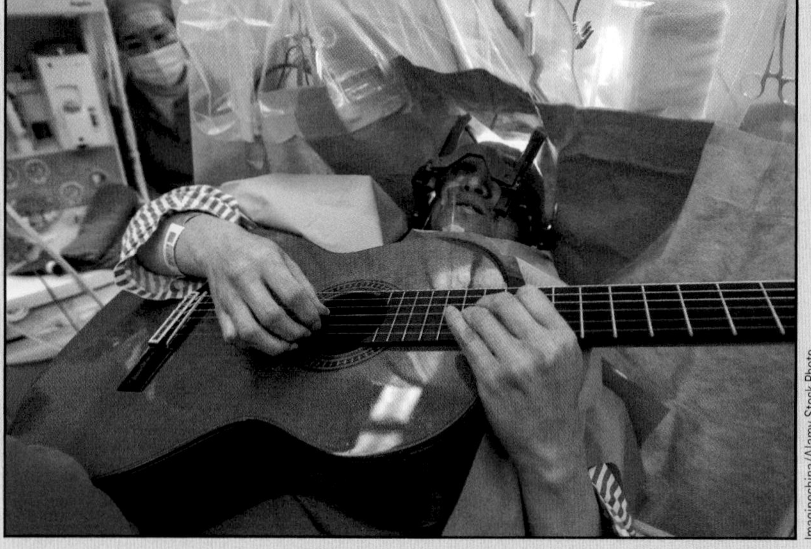

Imaginechina/Alamy Stock Photo

Learning Check

1. Which of the following vesicles were part of the primary vesicles at the first stage of development? Please select all that apply.
   a. Forebrain
   b. Diencephalon
   c. Metencephalon
   d. Hindbrain
2. Which of the following divides the two cerebral hemispheres?
   a. Corpus callosum
   b. Longitudinal fissure
   c. Central sulcus
   d. Precentral gyrus
3. On January 20 every four years, the newly sworn-in U.S. president has to give a speech. What area of the brain is responsible for production of language for the speech?
   a. Broca's area
   b. Wernicke's area
4. Which of the following components is/are part of the basal nuclei? Please select all that apply.
   a. Caudate
   b. Commissural tract
   c. Substantia nigra
   d. Globus pallidus

14.2c The Diencephalon

LO 14.2.12

The diencephalon is the one region of the adult brain that retains its name from embryologic development. Structurally, it is the connection between the cerebrum and the rest of the nervous system, with one exception. The rest of the brain, the spinal cord, and the PNS all send information to the cerebrum through the diencephalon. Output from the cerebrum passes through the diencephalon. The single exception is the system associated with **olfaction**, the sense of smell, which connects directly with the cerebrum. Olfaction, which we will discuss in more detail in Chapter 15, is the exception to many patterns of nervous system anatomy, likely due to its evolutionary history in other mammal and vertebrate species.

The diencephalon is deep beneath the cerebrum; its hollow core is the third ventricle and its solid structures surround this ventricle. The two major regions of the diencephalon are the thalamus and the hypothalamus (**Figure 14.19**). There smaller

Student Study Tip

Any region of the brain with "thalamus" in its name is part of the diencephalon.

Figure 14.19 The Diencephalon

The diencephalon includes the hypothalamus, the thalamus, and the pineal gland.

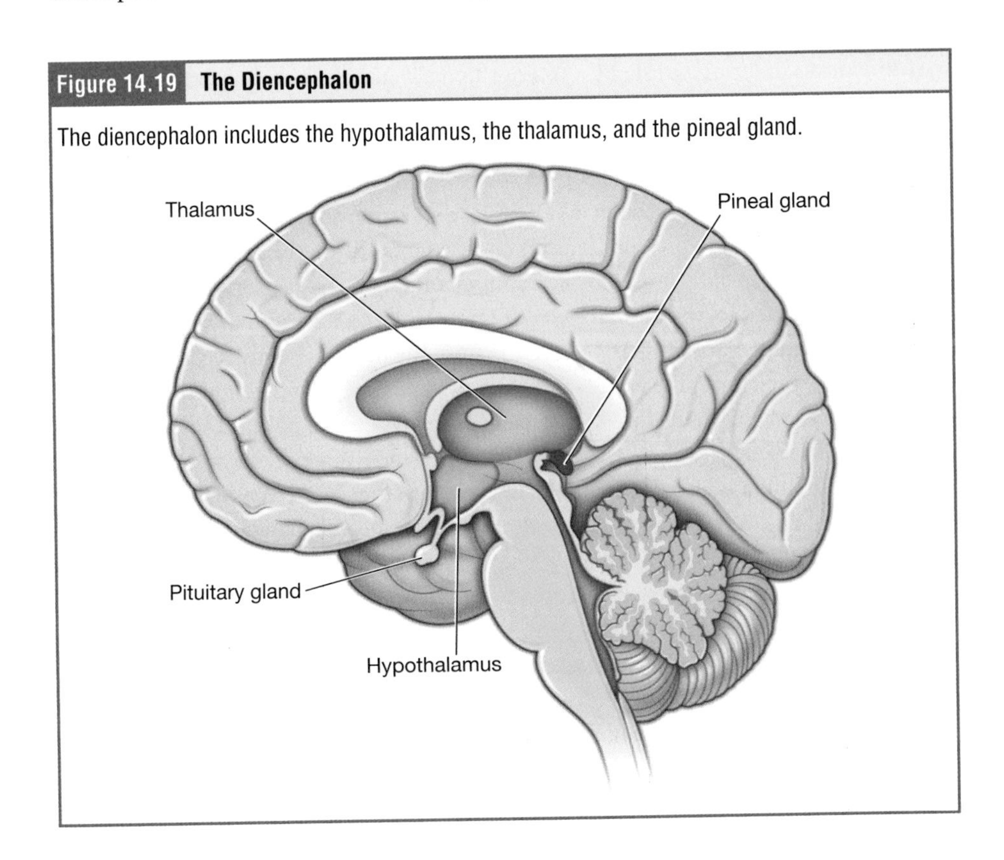

structures of the diencephalon are the **epithalamus**, which contains the pineal gland, and the **subthalamus**, which includes the subthalamic nuclei that collaborate with basal nuclei discussed earlier.

LO 14.2.13

interthalamic adhesion (intermediate mass)

Thalamus The **thalamus** is referred to as one structure, but it is a collection of nuclei that relay information between the cerebral cortex and the periphery, spinal cord, or brainstem. There are two paired egg-shaped masses that form the thalamus, and in most (but not all) brains they are connected by an **interthalamic adhesion**. All sensory information, except for the sense of smell, passes through the thalamus before being processed by the cortex. The thalamus is able to edit, amplify, or diminish sensory information before sending it along to the primary regions in the cerebrum. Axons from the peripheral sensory organs, including those that function in somatosensation, have a synapse in the thalamus and thalamic neurons project directly to the cerebrum. It is a requisite synapse in any sensory pathway, except for olfaction. The "Anatomy of Sensory Information" feature information illustrates the basic pathway of sensory information to the brain. The thalamus does not just pass the information on; it also processes that information. For example, the portion of the thalamus that receives auditory information will influence what auditory stimuli are important, or what receives attention. For example, if you walk into a café with your friend, you are surrounded by other conversations, the music playing in the background, the sounds of coffee being made, and so on. Your thalamus helps to edit the extraneous auditory information, so that you can focus on the words being said by your friend.

The cerebrum also sends information down to the thalamus, which usually communicates motor commands. This involves interactions with the cerebellum and other nuclei in the brainstem. The cerebrum interacts with the basal nuclei, which involves connections within the thalamus. The primary output of the basal nuclei is to the thalamus, which relays that output to the cerebral cortex. The cortex also sends information to the thalamus that will then influence the effects of the basal nuclei.

LO 14.2.14

Hypothalamus Inferior and slightly anterior to the thalamus is the **hypothalamus**, the other major region of the diencephalon (see Figure 14.19). The hypothalamus is a collection of nuclei that are largely involved in regulating homeostasis. The hypothalamus is the chief regulator of the autonomic nervous system (ANS); it controls the activity of ANS centers in the brainstem and therefore is responsible for the regulation of heart contractions, blood pressure, muscle contractions within the digestive tract, and many other activities of the organs. The hypothalamus sits directly superior to, and is physically connected to, the pituitary gland. Through its own hypothalamic hormones, as well as its regulation of pituitary gland function, the hypothalamus is considered to be the central regulator of the entire endocrine system. The hypothalamus has other regions that interact with the limbic system and are involved in physical responses to emotion such as skeletal muscle tension when experiencing anger.

The hypothalamus also directly regulates several other homeostatic functions. The body thermostat is in the hypothalamus; it functions to detect changes in temperature and respond, such as initiating sweating when body temperature rises. The hypothalamus directly and indirectly regulates appetite and food intake by monitoring blood levels of nutrients and controlling the brain's feelings of hunger or satiation. Similarly, the hypothalamus regulates our feelings of thirst by monitoring the ratio of fluid to ions in the blood. When ion concentrations increase, for example, when you eat a salty snack, the hypothalamus sends signals to the cerebrum for you to feel thirsty and seek out water or another beverage.

Anatomy of...

Sensory Information

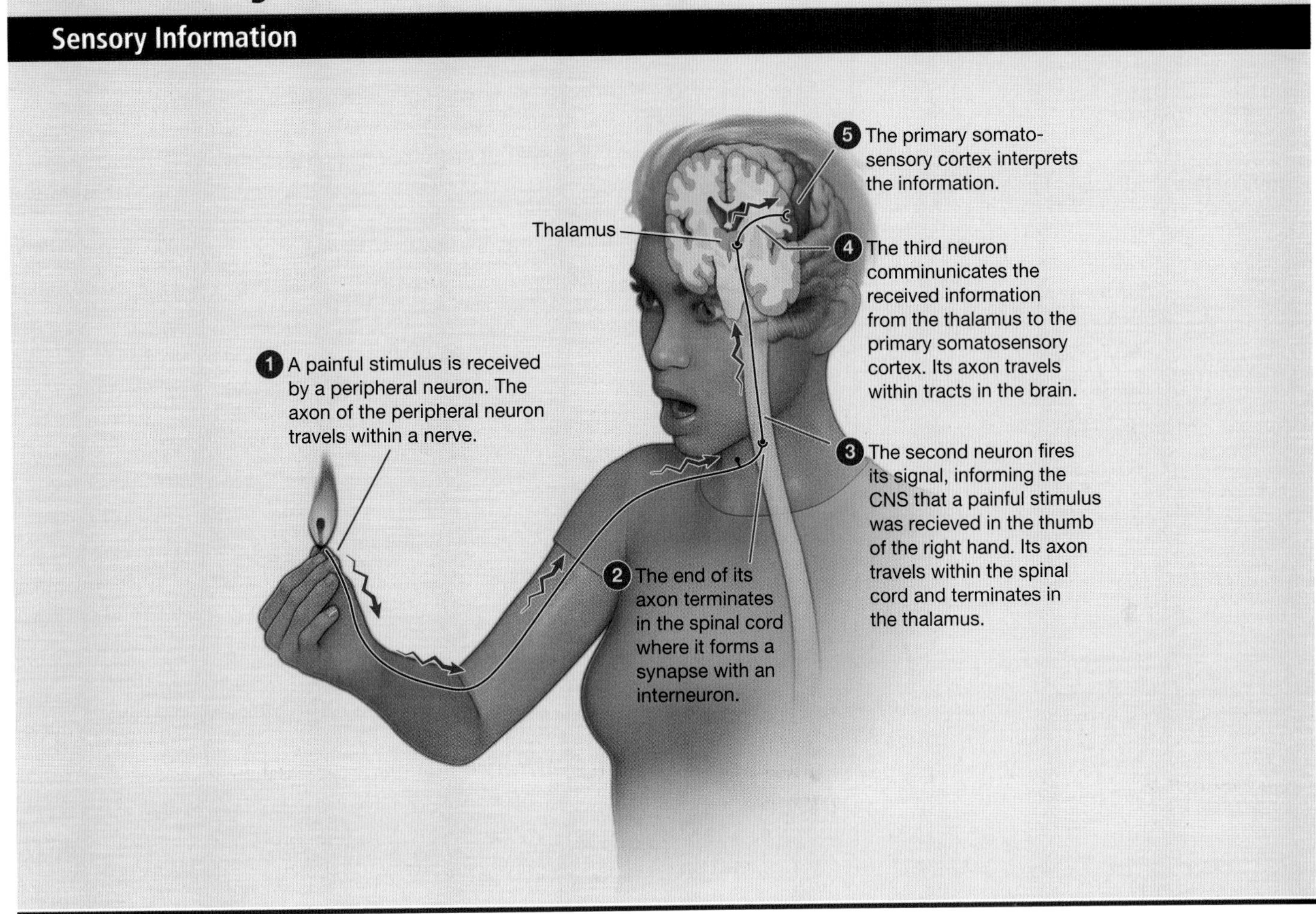

Thalamus
1 A painful stimulus is received by a peripheral neuron. The axon of the peripheral neuron travels within a nerve.
2 The end of its axon terminates in the spinal cord where it forms a synapse with an interneuron.
3 The second neuron fires its signal, informing the CNS that a painful stimulus was recieved in the thumb of the right hand. Its axon travels within the spinal cord and terminates in the thalamus.
4 The third neuron comminunicates the received information from the thalamus to the primary somatosensory cortex. Its axon travels within tracts in the brain.
5 The primary somato-sensory cortex interprets the information.

LO 14.2.15

The hypothalamus works with the **pineal gland**, a structure within the epithalamus, to regulate sleep-wake cycles. The hypothalamus incorporates light/dark information from the visual system to set the biological clock of sleeping and wakefulness. The pineal gland secretes melatonin, a hormone that causes us to feel sleepy, at times dictated by the hypothalamus.

14.2d Brainstem

LO 14.2.16

The brainstem consists of the midbrain, the pons, and the medulla (Figures 14.20). Inferiorly, the brainstem is continuous with the spinal cord. Attached to the brainstem, but considered a separate region of the adult brain, is the cerebellum, which will be discussed later.

LO 14.2.17

Midbrain The midbrain is a small region between the thalamus and the pons. The cerebral aqueduct, which connects the third and fourth ventricles, passes through the center of the midbrain. On the anterior surface of the midbrain are large bundles of ascending and descending tracts, the **cerebral peduncles** (Figure 14.21).

Figure 14.20 Brainstem

The three regions of the brainstem are the midbrain, the pons, and the medulla.

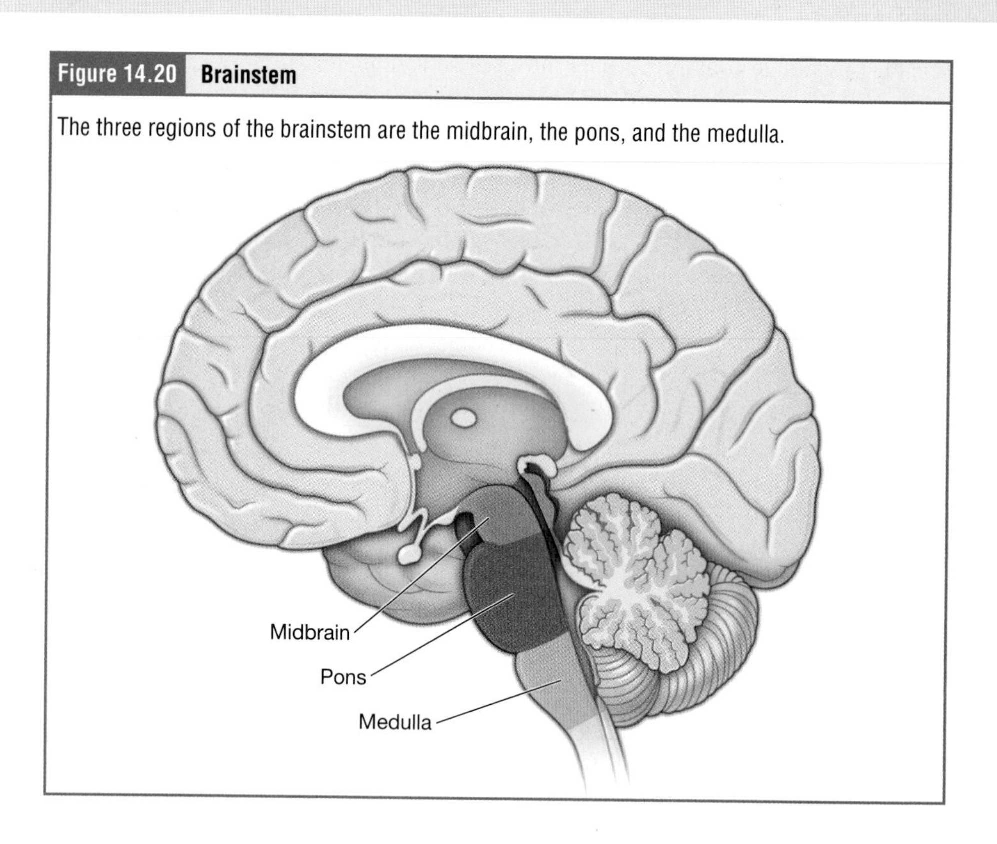

Along the anterior surface of the midbrain are four noticeable bumps, the *corpora quadrigemina*. Each of these four bumps is known as a *colliculus*. The two most inferior of these are the **inferior colliculi** and are part of the auditory brainstem pathway. Neurons of each inferior colliculus project to the thalamus, which then sends auditory information to the cerebrum for the conscious perception of sound. The **superior**

Figure 14.21 Cross Section of the Midbrain

The midbrain includes several key nuclei, each with its own function, as well as the cerebral peduncles.

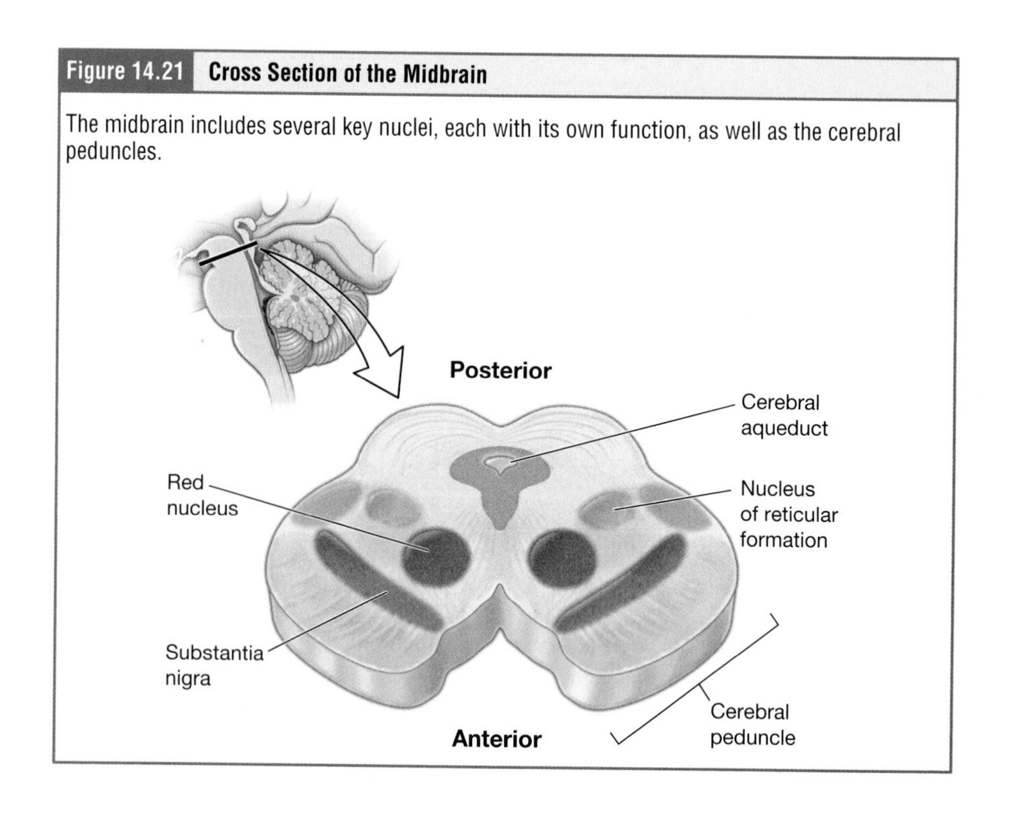

colliculi function to combine sensory information with motor output, such as moving your head to the source of a touch stimulus.

Two other noteworthy nuclei of the midbrain are the substantia nigra and the red nucleus. The **substantia nigra** is a nucleus deep within the midbrain that functions to inhibit motor neurons and contributes to the fluidity of motor movements. In Parkinson's disease the neurons of the substantia nigra decrease their production of dopamine, their neurotransmitter, which results in tremors and other uncontrolled muscle movements. The substantia nigra gets its name because of its dark appearance. Melanin, the skin pigment molecule, is a precursor to dopamine and it is found here in large amounts, lending a dark appearance to the tissue. The red nuclei are paired, small nuclei located slightly posterior to the substantia nigra. These nuclei have a rich blood supply and the neurons themselves store iron, which gives these regions their red hue. The red nuclei contribute to the reticular formation, discussed later, which helps to regulate attention and wakefulness.

LO 14.2.18

Pons The **pons** is a visible bulge on the anterior surface of the brainstem. It is the main connection between the cerebellum and the brainstem. The bridgelike white matter is only the anterior surface of the pons; the gray matter beneath this is a continuation of the nuclei of the midbrain.

LO 14.2.19

Medulla The medulla oblongata, or simply the medulla, is the most inferiorly located structure in the brain. It houses the fourth ventricle and its choroid plexus along its posterior (**Figure 14.22**). Its anterior surface is taken up by tracts of ascending and descending neurons. These tracts, called **pyramidal tracts**, cross over on the medulla's surface. The crossing over of the tracts is what leads to the right cerebral hemisphere controlling and receiving sensory information from the left side of the body and the left cerebral hemisphere controlling and receiving sensory information from the right

pyramidal tracts (corticospinal tracts)

Figure 14.22 Cross Section of the Medulla

The medulla oblongata is the most inferior structure of the brainstem. It includes many nuclei with individual functions, the pyramidal tracts, and the fourth ventricle, which is filled with CSF.

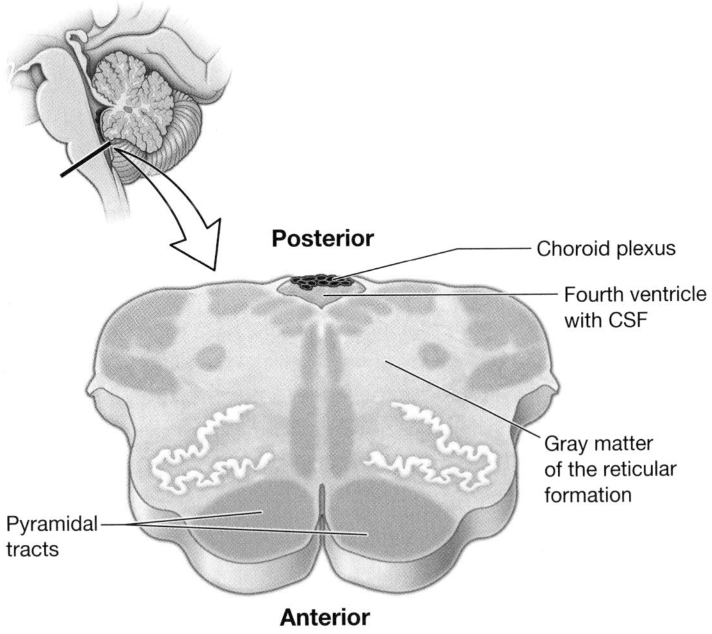

side of the body. The interior of the medulla contains nuclei responsible for the regulation of our most basic homeostatic functions. There are three functional clusters:

- The cardiovascular center—this includes several nuclei that function to regulate the force and rate of heart contraction as well as the diameter of blood vessels to adjust blood pressure.
- The respiratory centers—these function to adjust the rate and depth of breathing.
- The abdominal and thoracic motor centers—these nuclei are responsible for various coordinated muscle activities of the thorax and abdomen, including swallowing, vomiting, sneezing, and coughing.

Learning Check

1. Neural pathways for of the following does not go through the thalamus?
   a. Vision
   b. Hearing
   c. Smell
   d. Sensation
2. Which of the following would you expect to find in someone with hypothalamic dysfunction?
   a. Difficulty understanding language
   b. Inability to control involuntary movement
   c. Difficulty staying asleep at night
   d. Inability to hear
3. Which of the following structures are found in the midbrain? Please select all that apply.
   a. Cerebral peduncles
   b. Basal nuclei
   c. Pyramidal tracts
   d. Substantia nigra
4. What would you expect to see if someone had a stroke in the medulla oblongata? Please select all that apply.
   a. Difficulty maintaining a steady heart rate
   b. Weakness in the opposite side of the body
   c. Inability to hear in busy environments
   d. Visual deficits that result in unsteady gaze

14.2e The Cerebellum

LO 14.2.20

LO 14.2.21

The **cerebellum**, as the name suggests, is the "little brain." It is covered in gyri and sulci like the cerebrum, and looks like a miniature version of that part of the brain, in fact, it accounts for approximately 10 percent of the mass of the brain (**Figure 14.23**). Like the cerebrum, the cerebellum also has an outer cortex of gray matter and inner white matter tracts. These tracts form a treelike shape, so early anatomists named them the **arbor vitae** (tree of life). The cerebellum contributes to motor activities, but unlike the motor activities initiated by the cerebrum we have no conscious control over the cerebellum. The cerebellum has a role in strategizing and fine-tuning motor movements initiated by the cerebrum. A key part of the contributions of the cerebellum is that it incorporates sensory information, particularly proprioception, the understanding of where the body is in space, in its blueprint for motor movements. For example, let's say that the cerebrum wants to initiate a familiar activity, such as writing; however, instead of writing with a pen on a piece of paper, you're writing with a whiteboard marker on a wall-mounted whiteboard. You need to adjust the way you hold the marker, the angle of your hands, and the size and shape of the letters. Your initiation of the movements comes from the cerebrum, but the cerebellum fine-tunes the blueprint for this activity to account for the sensory information it has. It then sends the blueprint back to the primary motor cortex for implementation.

14.2f Brain Systems and Functions that Bridge Brain Regions

The Reticular Formation A diffuse region of connected gray matter throughout the brainstem, diencephalon, and spinal cord, known as the **reticular formation**, functions

Figure 14.23 **The Cerebellum**

The cerebellum consists of a small cortex of gray matter and underlying white matter tracts.

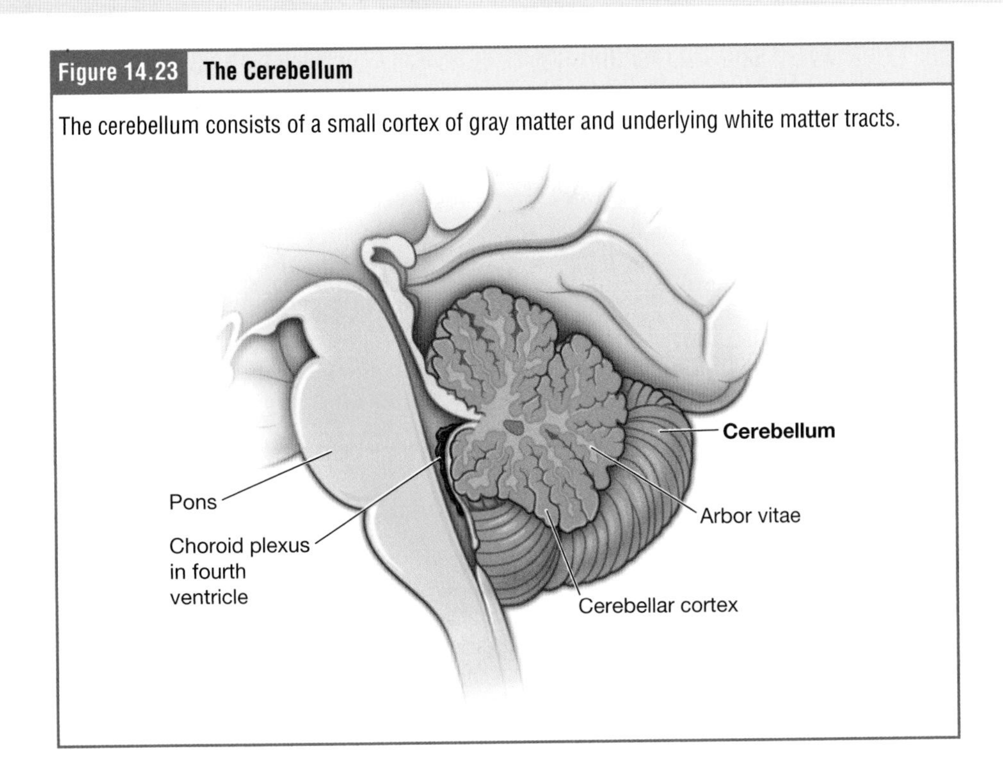

in the regulation of the various states of sleep and wakefulness, including general brain activity and attention. This network of interconnected structures incorporates both sensory and motor components. The sensory portion, called the *reticular activating system (RAS)*, processes the visual, auditory, and touch stimuli that influence alertness. Falling asleep to a gentle sound, waking at the sound of your alarm, and not being able to sleep on a plane when someone is kicking the back of your seat are all examples of the reticular activating system influencing your level of alertness. The motor components control our muscle tone when sleeping and the relaxation of our skeletal muscles, as well as the changes in breath rate and heart rate between sleeping and wakefulness. The activity of the reticular formation operates on a spectrum; rather than having just two modes of awake and asleep, there are many modes between, such as drowsy.

LO 14.2.22

Cultural Connection

Sleep Hygiene

Sleep plays an essential role in restoration of the nervous system. Sleep is an important time for memories to be consolidated, which helps the memory to be retrievable later. Sleep is also a time when cerebrospinal fluid flow increases, carrying away more built-up toxins or waste products than during waking hours. Sleep benefits are not accumulated in a single night of sleep; rather, it seems that a month or more of regular sleep is required for improvements to brain function to set in. A recent study, published in the journal *Nature*, found that college students scored higher on exams when they had slept more.[1] However, it wasn't just the night before the exam that mattered; the study found that students who had slept better during the month preceding the exam had higher average scores (the study was performed on students taking introductory chemistry classes). The term *sleep hygiene* refers to sleep habits such as regulating the amount of screen time before bed, going to bed at a consistent time each night, and committing to sleeping for a regular number of hours each night.

[1]Okano, K., Kaczmarzyk, J. R., Dave, N., et al. Sleep quality, duration, and consistency are associated with better academic performance in college students. *NPJ Sci. Learn.* **4**, 16 (2019).

While asleep, cortical function slows but does not completely stop. The brainstem, particularly the reticular formation, is still very active. Sleep can be divided into two main classes—REM and non-REM sleep—based on the type of brain activity occurring at the time. **REM (rapid eye movement) sleep** is the type of sleep in which dreaming occurs. REM sleep is believed to be a time when the brain is organizing memories, which may be why we often dream about things that are ongoing in our daily lives. Non-REM sleep is believed to be more restorative to the brain. Recently, it has been discovered that CSF flows differently in the brain during sleep—in slow pulses that act to clear CSF and any accumulated waste or toxins from the brain. This finding may be a key to understanding why sleep is so essential for brain function.

Cultural Connection

Beauty Is in the Brain of the Beholder

Functional magnetic resonance imaging (fMRI) is a brain imaging technique that maps blood flow during certain tasks. Using fMRI, doctors and scientists can observe which brain regions are highly metabolic, or in other words, functioning more than usual, during particular tasks. fMRI data has dramatically expanded our understanding of the anatomical specificity of the brain. One study asked participants to experience works of art and rate them on a scale of beautiful to unpleasant. The works of art included both visual art, such as famous paintings, and auditory art, or music. Regardless of which type of stimulus and which particular painting or song, a single small frontal cortex region lit up with activity when the participant reported that they found what they were experiencing beautiful. This "beauty center" does not function in all perceptions of beauty. As it turns out, in later studies scientists found that three different frontal lobe regions became activated when the stimulus was a beautiful face, rather than a beautiful work of art.

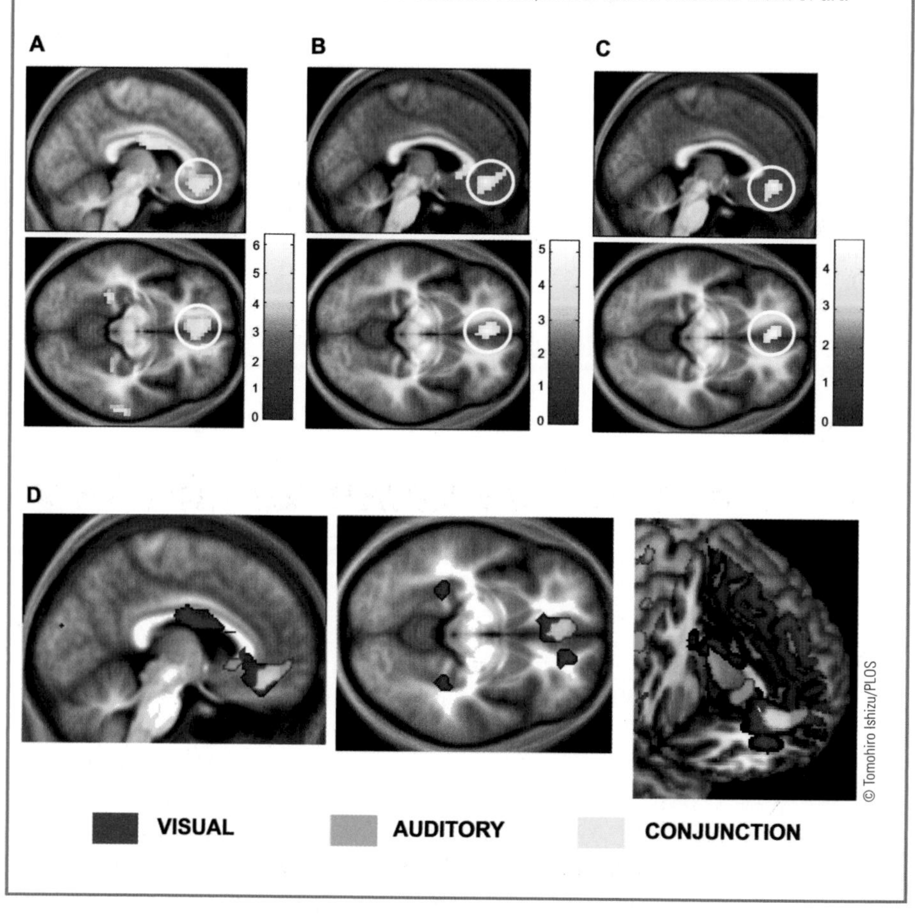

14.2g The Limbic System

LO 14.2.23

The limbic system is a network of structures in the diencephalon and cerebrum that work together to process emotion. The structures in the brain that have been identified to contribute to limbic function are: the hippocampus, the amygdala, the **olfactory bulbs**, tracts, and cortex, the **cingulate gyrus**, the **parahippocampal gyrus**, the **fornix**, the **anterior thalamic nuclei**, the **habenular nuclei**, the **septal nuclei**, and the **mammillary bodies**. Of these, the hippocampus and amygdala are of particular interest, as we understand their functions more completely.

The paired hippocampi are found deep in the temporal lobe (**Figure 14.24**). These structures are one of only two places in the nervous system where neurogenesis—the production of new neurons—has been observed in humans. The hippocampus is involved in memory formation and navigation (creating mental maps). Before smartphones, scientists noted that London taxi drivers, who have memorized the complex streets of London, averaged larger hippocampi than regular London citizens, establishing the first suggestion that the hippocampus was capable of growing. Later, scientists established that neurons in the hippocampus are capable of mitosis.

The two *amygdalae* (singular: amygdala) are paired structures that sit at the end of each hippocampus. These structures are specialized to process our fear responses. Individuals with damage to their amygdalae have altered or reduced fear responses and cannot recognize menacing facial expressions in others. The amygdalae seem to store fear-based memories and may play a role in post-traumatic stress disorder.

Student Study Tip

Remember that "olfactory" refers to smell by thinking about how it would smell if you were in an "old factory." (Probably sort of musty!)

Student Study Tip

Think of the memory center like a college "campus" for hippos. They need to study so that they don't forget what they've learned.

14.2h The Spinal Cord

LO 14.2.24

The spinal cord is the other organ of the central nervous system. Whereas the brain develops out of expansions of the neural tube, the spinal cord maintains the tube structure and is relatively consistent in its anatomy throughout its regions. The length of the spinal cord is divided into regions that correspond to the regions of the vertebral column. The name of each spinal cord region corresponds to the level at which spinal nerves that exit it pass through the intervertebral foramina. The most superior is the

Figure 14.24 The Limbic System

The limbic system is a system of connected structures deep under the cerebral cortex. These structures function in emotion and memory.

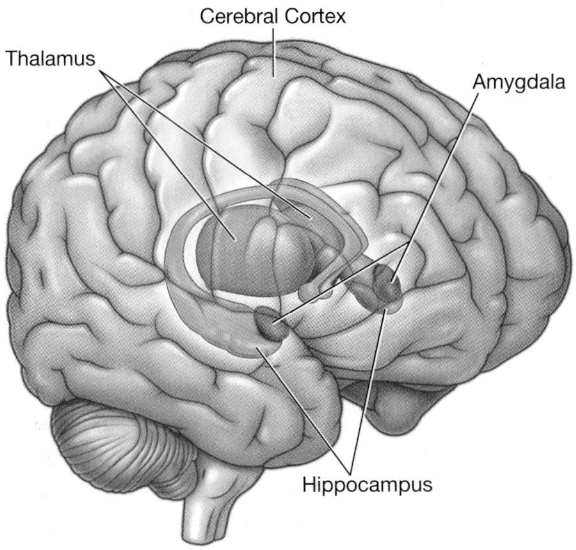

Cerebral Cortex
Thalamus
Amygdala
Hippocampus

cervical region, followed by the thoracic, lumbar, and finally the sacral region. The spinal cord is shorter than the full length of the vertebral column because the spinal cord stops growing longer after the second year of life, but the skeleton continues to grow through adolescence. While the solid column of the spinal cord ends early within the lumbar region, the rest of the spinal column is not empty; rather, a long bundle of nerves hangs down encased within the inferior spine. Early anatomists thought that this bundle of nerves resembled a horse's tail and named it the **cauda equina**.

The External Surface of the Spinal Cord The anterior surface of the spinal cord is marked by the **anterior median fissure**, and the posterior midline is marked by the **posterior median sulcus** (Figure 14.25). These two grooves run the entire length of the spinal cord. The spinal cord consistently is shaped like a column, but it is wider at some locations than others. In general, you can think of the spinal cord as being similar to a highway. Almost all of its structure is about connecting the PNS to the brain with axons coming in from and leaving for structures outside the CNS. Just as a highway might have more lanes to accommodate more traffic close to a major city, the spinal cord is larger at its connection to the brainstem and narrowest inferiorly. Likewise, the spinal cord is widest at the points of more "traffic" where more axons are coming and going from the cord. There are two such enlargements in the spinal cord; the cervical and lumbar enlargements occur in the regions that contain an increased number of axons entering and exiting the cord from the upper and lower limbs, respectively.

LO 14.2.25

Figure 14.25 Cross Section of the Spinal Cord

Cross sections of the spinal cord reveal the central gray matter to have a butterfly shape. (A) The anterior horns contain the cell bodies of motor neurons, and the posterior horns contain the axons of afferent sensory neurons. In the thoracic region, the lateral horns contain the cell bodies of autonomic neurons. (B) A stained cross section of the spinal cord reveals the gray and white matter when viewed under the microscope.

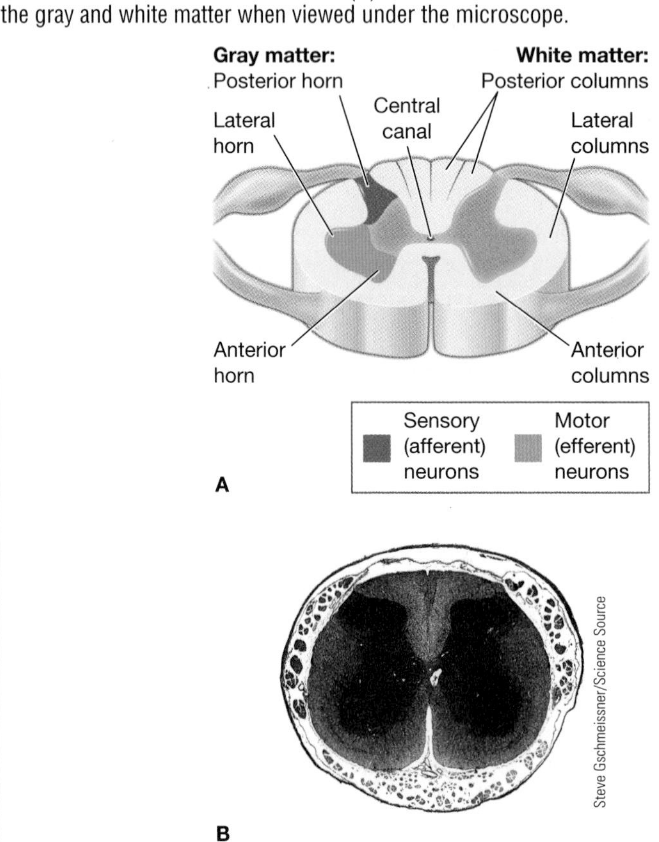

The Internal Anatomy of the Spinal Cord Axons enter the posterior side through the **dorsal (posterior) nerve root**, which marks the **posterolateral sulcus** on either side. The axons emerging from the anterior side do so through the **ventral (anterior) nerve root**. Note that it is common to see the terms *dorsal* (dorsal = "back") and *ventral* (ventral = "belly") used interchangeably with posterior and anterior, particularly in reference to nerves and the structures of the spinal cord. You should learn to be comfortable with both.

LO 14.2.26

Gray Horns In cross-section, the gray matter of the spinal cord has the appearance of a capital letter H or a butterfly (Figure 14.25A). Each of the four projections of gray matter, what we would consider to be the wings if we were comparing the gray matter to a butterfly, are referred to as *horns*. The **posterior horn** contains the axons of sensory neurons. Their short axons enter the spinal cord through a bundle called the *posterior root*. Along this root, the cell bodies of these neurons are bundled in the **posterior root ganglion**. The **anterior horn** contains the cell bodies of motor neurons that are sending out motor signals to the skeletal muscles. In the thoracic and upper lumbar regions of the spinal cord, another set of horns are visible; these **lateral horns** contain neurons of the autonomic nervous system.

posterior root ganglion (dorsal root ganglion)

anterior horn (ventral horn)

Some of the largest neurons of the spinal cord are the multipolar motor neurons in the anterior horn. Their axons exit the spinal cord in the **anterior root** and stretch out toward the skeletal muscles, where they terminate at the neuromuscular junction and cause skeletal muscle contractions. The motor neuron that causes contraction of the big toe, for example, is located in the sacral spinal cord. Depending on your height, the axon that has to reach all the way to that muscle may therefore be two to three feet in length. The neuronal cell body that maintains that long axon must be quite large, making it one of the largest cells in the body.

anterior root (ventral root)

LO 14.2.27

White Columns Just as the gray matter is separated into horns, the white matter of the spinal cord is separated into columns. Whereas **ascending tracts** of axons in these columns carry sensory information up to the brain, **descending tracts** carry motor commands from the brain to the muscles. Looking at the spinal cord longitudinally, the columns extend along its length as continuous bands of white matter. Between the two posterior horns of gray matter are the **posterior columns**. Between the two anterior horns, and bounded by the axons of motor neurons emerging from that gray matter area, are the **anterior columns** (Figure 14.26). The white matter on either side of the spinal cord, between the posterior horn and the axons of the anterior horn neurons, is known as the **lateral columns**. The posterior columns are composed of axons of ascending tracts. The anterior and lateral columns are composed of many different groups of axons of both ascending and descending tracts—the latter carrying motor commands down from the brain to the spinal cord to control output to the periphery.

posterior columns (posterior funiculi)

anterior columns (ventral columns, anterior funiculi)

lateral columns (lateral funiculi)

Learning Check

1. What would you expect to see if someone had a stroke in which the cerebellum was impaired?
   a. Inability to control voluntary movements
   b. Excessive uncontrolled movement
   c. Inability to maintain a steady sleep-wake cycle
   d. Difficulty processing emotion
2. The dorsal nerve root enters the spinal cord ______ while the ventral nerve root exits ______.
   a. Anteriorly; posteriorly
   b. Posteriorly; anteriorly
   c. Anteriorly; anteriorly
   d. Posteriorly; posteriorly
3. The anterior horn of the spinal cord is responsible for ______ function, while the posterior horn is responsible for ______ function.
   a. Motor; sensory
   b. Sensory; ANS
   c. ANS; motor
   d. Motor; ANS

Figure 14.26 White Matter of the Spinal Cord

The white matter tracts are arranged in columns. The posterior columns carry ascending sensory tracts; the anterior and lateral columns carry a mix of tracts.

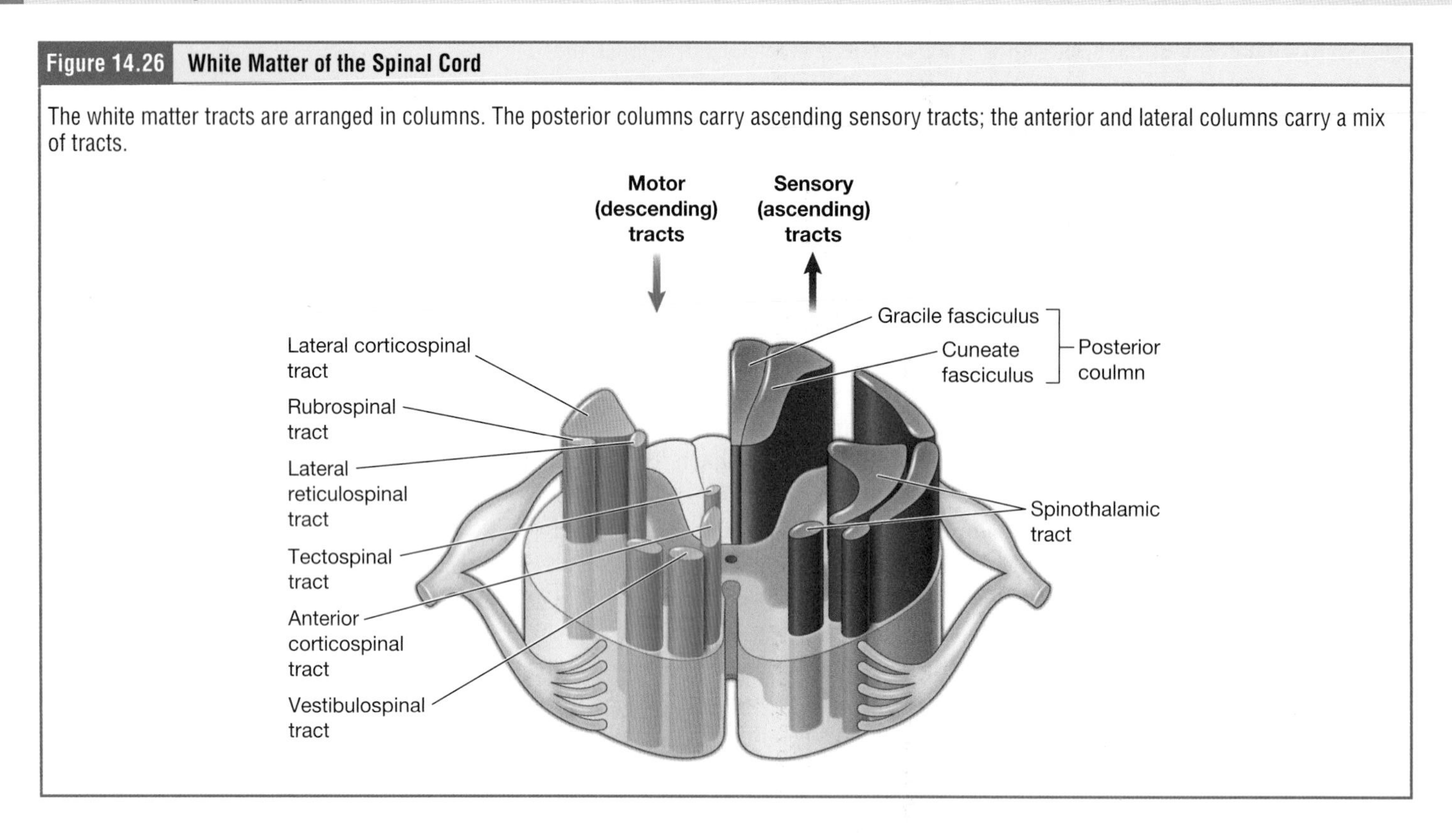

14.3 The Peripheral Nervous System

Learning Objectives: By the end of this section, you will be able to:

14.3.1 List and identify the cranial nerves by name and number.

14.3.2 Describe the major functions of each cranial nerve and identify each cranial nerve as predominantly sensory, motor, or mixed (i.e., sensory and motor).

14.3.3 List the cranial nerves that have parasympathetic (ANS) components.

14.3.4 Identify and describe the formation, structure, and branches of a typical spinal nerve, including the roots and the rami (e.g., anterior [ventral], posterior [dorsal]).

14.3.5 List the number of spinal nerve pairs emerging from each spinal cord region (i.e., cervical, thoracic, lumbar, sacral, coccygeal).

14.3.6 Define a spinal nerve plexus.

14.3.7 For the cervical, brachial, lumbar, and sacral nerve plexuses, list the spinal nerves that form each plexus, describe the plexus' major motor and sensory distributions, and list the major named nerves that originate from each plexus.

The PNS is organized into nerves and ganglia; however, it contains a multitude of functional elements. Some peripheral structures are incorporated into organs, including the skin, as well as visceral organs such as the stomach and intestines. In describing the anatomy of the PNS, most of the structures we discuss are connected to the CNS, motor, and sensory elements. However, a special subset of the PNS is the **enteric nervous system**—the system that controls and regulates the gastrointestinal tract and operates somewhat independently of the CNS.

14.3a Ganglia

A ganglion is a group of neuron cell bodies in the periphery. Ganglia can be categorized, for the most part, as either sensory ganglia or autonomic ganglia; there are no motor ganglia. We are already familiar with the posterior root ganglion. These ganglia hold the cell bodies of sensory neurons that extend into the CNS through the posterior root. From the outside, the ganglion looks like an enlarged bump on the posterior root. Beneath its connective tissue covering, it includes the cell bodies of these afferent neurons (**Figure 14.27**). Within the ganglion, the small round nuclei of satellite cells can be seen surrounding—as if they were orbiting—the neuron cell bodies.

There are many more ganglia; most of these are associated with the autonomic nervous system, which is discussed in more detail in Chapter 16.

14.3b Nerves

There are two main classes of nerves in the body—cranial nerves and spinal nerves. These nerves are categorized based on where they attach to the CNS. **Cranial nerves** attach directly to the brain. They are mostly, but not entirely, found within the cranium. **Spinal nerves** attach to the spinal cord.

Figure 14.27 Posterior Root Ganglion

(A) The posterior root ganglion is a cluster of cell bodies of the sensory axons that travel within the posterior root. (B) Under the microscope, we can observe that each cell body is surrounded by a ring of satellite cells, a type of glial cell found in peripheral ganglia.

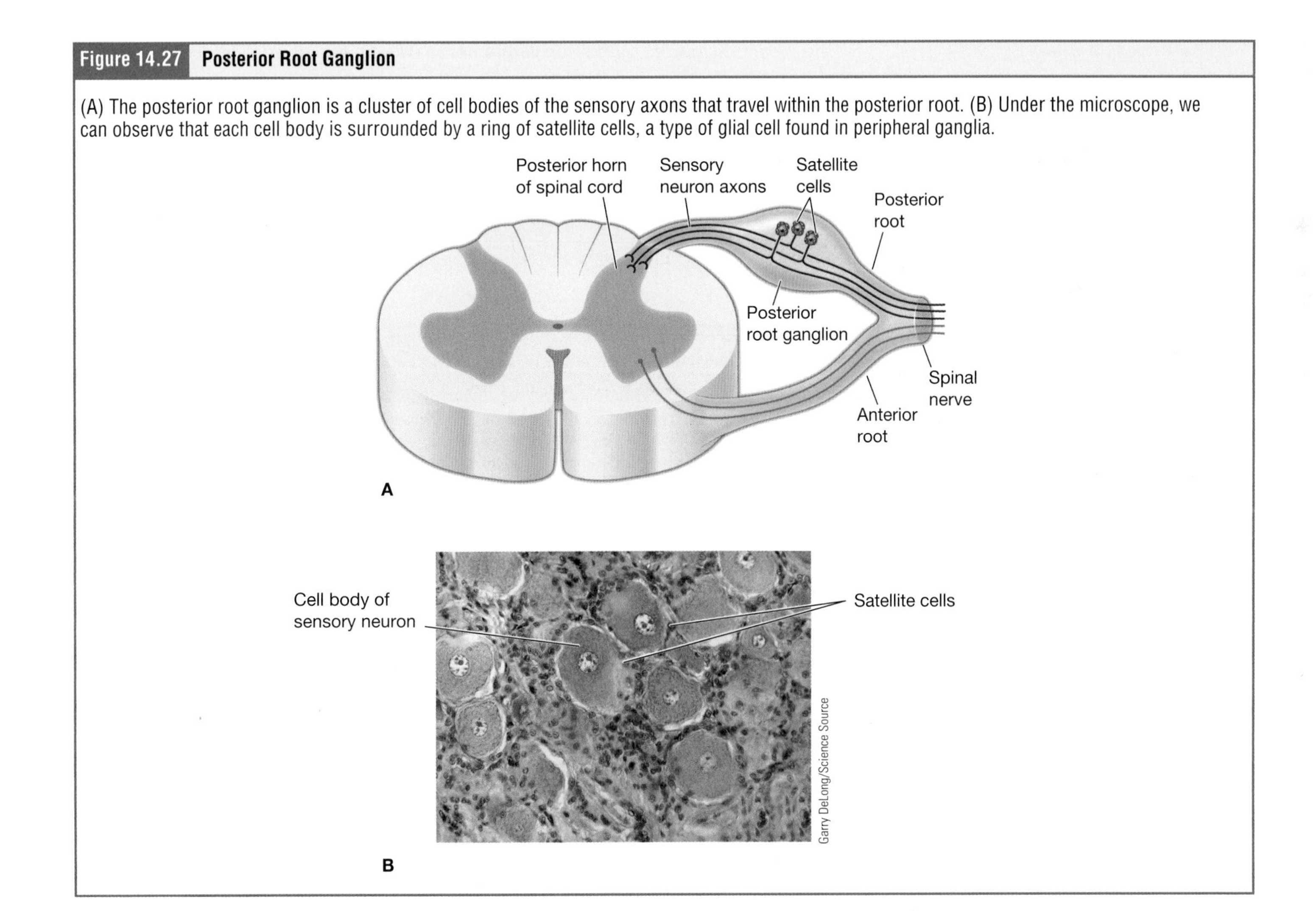

Student Study Tip

A mnemonic for remembering the cranial nerves in order: *On Occasion Our Trusty Truck Acts Funny, Very Good Vehicle Any How.*

Cranial Nerves The cranial nerves are primarily responsible for the sensory and motor functions of the head and neck (one of these nerves targets organs in the thoracic and abdominal cavities as part of the parasympathetic nervous system). There are 12 cranial nerves; these are designated CN I through CN XII for "Cranial Nerve," using Roman numerals for 1 through 12. Depending on the axons that are found within, they can be classified as sensory nerves, motor nerves, or a combination of both.

The names of the cranial nerves are listed in Table 14.1 along with a brief description of their function, their source, and their target.

LO 14.3.1

The cranial nerves are numbered for the order in which they exit the brain from anterior to posterior (Figure 14.28). The **olfactory nerve (CN I)** and **optic nerve (CN II)** are responsible for the senses of smell and vision, respectively. The **oculomotor nerve (CN III)** is responsible for eye movements by controlling four of the **extraocular muscles.** It is also responsible for lifting the upper eyelid, and for constriction of the pupil. The **trochlear nerve (CN IV)** and the **abducens nerve (CN VI)** both contribute to eye movement through their control of different extraocular muscles. The **trigeminal nerve (CN V)** is responsible for cutaneous sensations of the face and controlling some of the mouth muscles. The **facial nerve (CN VII)** is responsible for the muscles involved in facial expressions, as well as for part of the sense of taste and the production of saliva. The **vestibulocochlear nerve (CN VIII)** is responsible for the senses of hearing and balance. The **glossopharyngeal nerve (CN IX)** is responsible for controlling muscles in the oral cavity and upper throat, as well as for part of the sense of taste and the production of saliva. The **vagus nerve (CN X)** is responsible for contributing to homeostatic control of the organs of the thoracic and upper abdominal cavities. The **spinal accessory nerve (CN XI)** is responsible for controlling certain muscles of the neck and back; it arises from the rootlets of several of the cervical spinal nerves. The **hypoglossal nerve (CN XII)** is responsible for controlling the muscles of the lower throat and tongue.

Student Study Tip

Here's another mnemonic for Sensory, Motor, and Both: *Some Say Marry Money, But My Brother Says Big Brains Matter More.*

vestibulocochlear nerve (CN VIII) (auditory nerve, acoustic nerve)

spinal accessory nerve (CN XI) (accessory nerve)

LO 14.3.2

Three of the cranial nerves also contain autonomic fibers, and a fourth is almost purely a component of the autonomic system. The oculomotor, facial, and

Table 14.1 Cranial Nerves

Nerve	Function	Source	Target
Olfactory I	Sensory	Cerebrum	Nose
Optic II	Sensory	Cerebrum	Eyes
Oculomotor III	Motor	Brainstem: Midbrain-pontine junction	Muscles of the eyes
Trochlear IV	Motor	Brainstem: Midbrain	Superior oblique muscle of the eye
Trigeminal V	Sensory and Motor	Brainstem: Pons	Face, sinuses, teeth (sensory), and muscles of mastication (motor)
Abducens VI	Motor	Brainstem: Pontine-medulla junction	Lateral rectus muscle of the eye
Facial VII	Sensory and Motor	Brainstem: Pons	Muscles of the face and sensory for a portion of the tongue.
Vestibulocochlear VIII	Sensory	Brainstem: Pontine-medulla junction	Inner ear
Glossopharyngeal IX	Sensory and Motor	Brainstem: Medulla oblongata	Posterior part of the tongue, tonsils, pharynx (sensory), and pharyngeal musculature (motor)
Vagus X	Sensory and Motor	Brainstem: Medulla oblongata	Heart, lungs, bronchi, gastrointestinal tract
Accessory XI	Motor	Brainstem: Medulla oblongata	Sternocleidomastoid and trapezius muscles
Hypoglossal XII	Motor	Brainstem: Medulla oblongata	Muscles of the tongue

Figure 14.28 The Cranial Nerves

The cranial nerves are numbered for the order in which they attach to the brain.

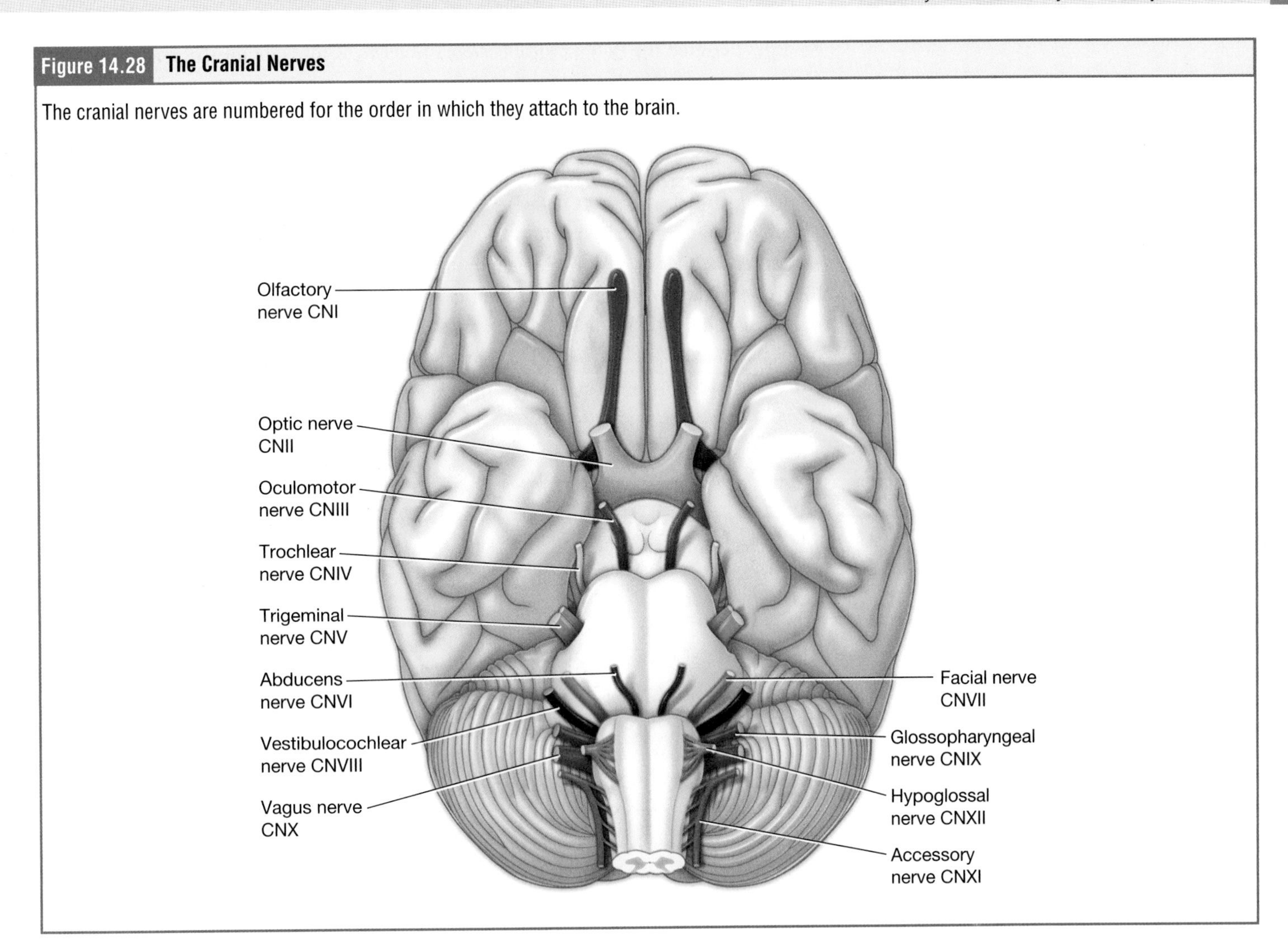

LO 14.3.3

glossopharyngeal nerves contain fibers that contact autonomic ganglia. Whereas the oculomotor fibers initiate pupillary constriction, the facial and glossopharyngeal fibers both initiate salivation. The vagus nerve primarily targets autonomic ganglia in the thoracic and upper abdominal cavities.

Another important aspect of the cranial nerves that lends itself to a mnemonic is the functional role each nerve plays. The nerves fall into one of three basic groups. They may contain only sensory axons, only motor axons, or a mixture of both.

LO 14.3.4

Spinal Nerves The nerves connected to the spinal cord are the spinal nerves. The arrangement of these nerves is much more regular than that of the cranial nerves. All of the spinal nerves are combined sensory and motor axons that form from the convergences of the two nerve roots. The sensory axons enter the spinal cord as the posterior nerve root. The motor fibers, both somatic and autonomic, emerge as the anterior nerve root.

LO 14.3.5

There are 31 spinal nerves, named for the level of the spinal cord at which each one emerges. There are eight pairs of cervical nerves designated C1 to C8, twelve thoracic nerves designated T1 to T12, five pairs of lumbar nerves designated L1 to L5, five pairs of sacral nerves designated S1 to S5, and one pair of coccygeal nerves. The nerves are numbered from the superior to inferior positions, and each emerges from the vertebral column through the intervertebral foramen at its level. The first nerve, C1, emerges between the first cervical vertebra and the occipital bone. The second nerve, C2, emerges between the first and second cervical vertebrae. The same occurs for C3 to C7, but C8 emerges between the seventh cervical vertebra and the first thoracic vertebra.

Just after the spinal nerves form from the convergence of the anterior and posterior roots, the spinal nerves begin to branch once again.

Spinal nerves extend outward from the vertebral column to innervate the organs and muscles. If we are examining a region in the body, for example the forearm region, the nerves found within are named, for example, the radial and ulnar nerves. These nerves are not straight continuations of the spinal nerves, but rather the reorganization of the axons in the spinal nerves, which occurs in a plexus (**Figure 14.29**). There are four

LO 14.3.6

nerve plexuses, which are formed when spinal nerves weave together. Arising from

Figure 14.29 Spinal Nerve Plexuses

Spinal nerves emerge individually from the spinal cord, traveling between the vertebrae. The axons within the spinal nerve weave together in a braidlike structure called a plexus. The nerves that emerge have new bundles of axons and are named for the regions they innervate.

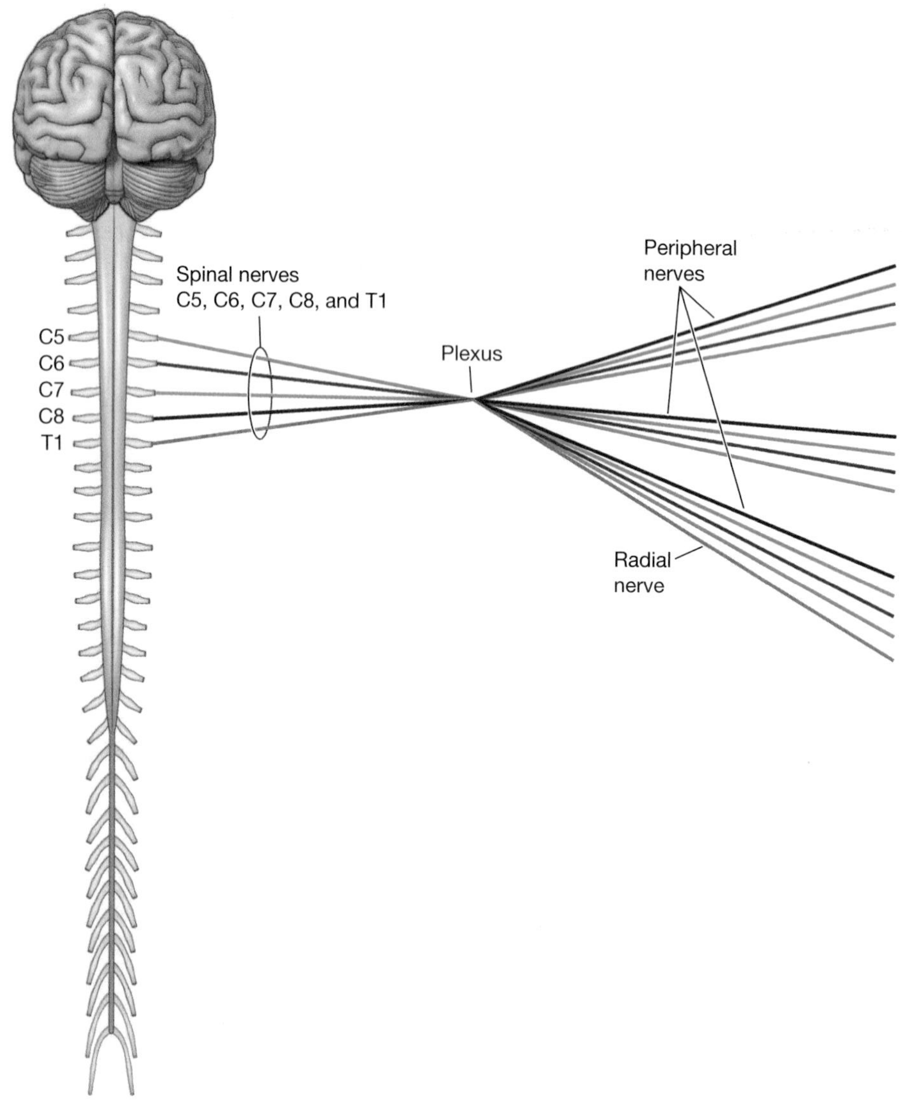

each plexus are nerves with names that innervate specific structures, usually within on region of the body.

Of the four nerve plexuses, two are found at the cervical level, one at the lumbar level, and one at the sacral level (**Figure 14.30**). The **cervical plexus** is composed of axons from spinal nerves C1 through C5 and branches into nerves in the posterior neck and head, as well as the **phrenic nerve**, which connects to the diaphragm at the base of the thoracic cavity. The other plexus from the cervical level is the **brachial plexus**. Spinal nerves C4 through T1 reorganize through this plexus to give rise to the nerves of the arms, as the name "brachial" suggests. A large nerve from this plexus is

Figure 14.30 The Four Spinal Nerve Plexuses

The spinal nerves that exit throughout the thoracic region travel as individual nerves. Everywhere else, the spinal nerves braid into plexuses. There are four plexuses in the human body: the cervical, the brachial, the lumbar, and the sacral plexuses.

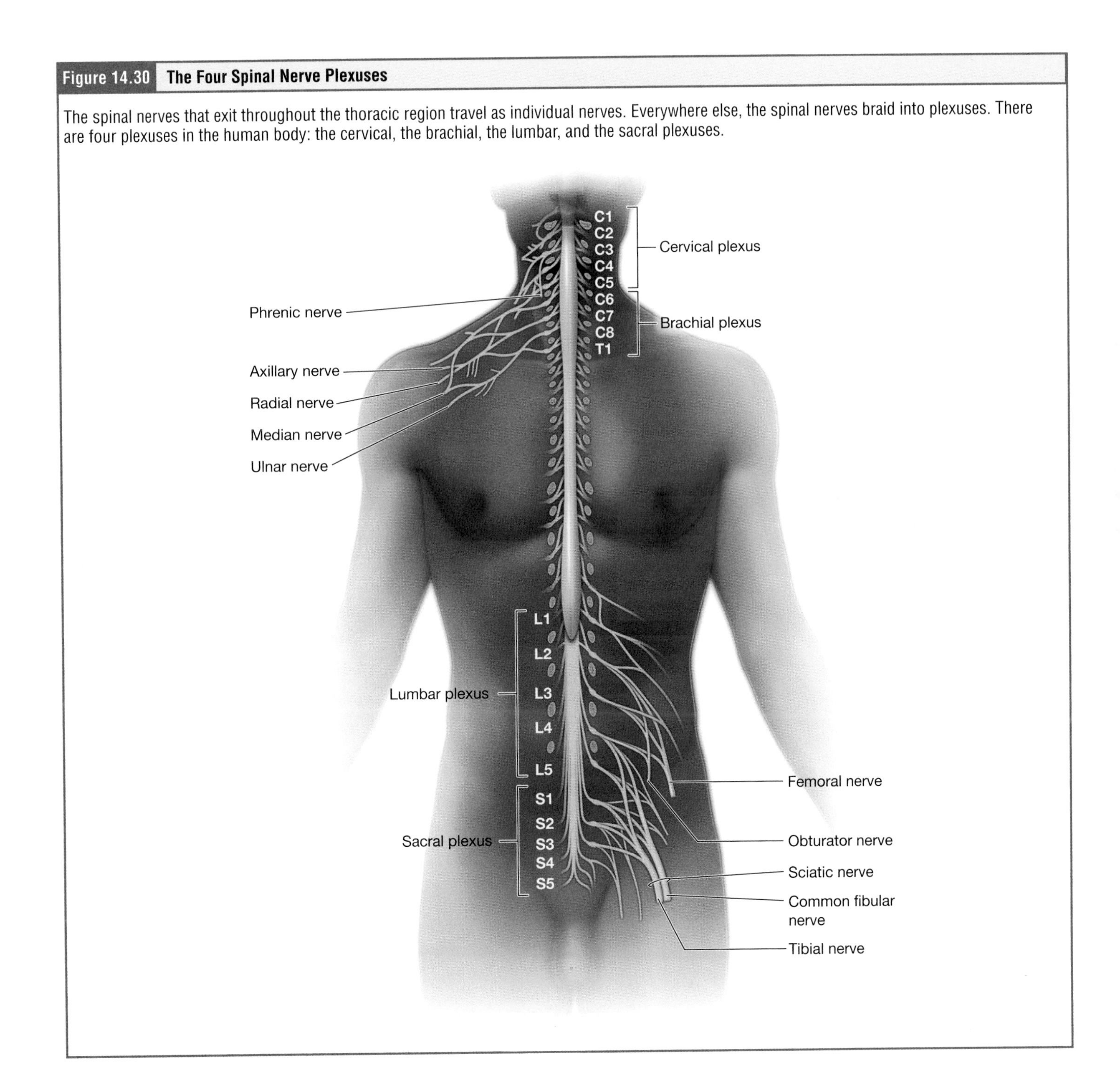

the **radial nerve** from which the **axillary nerve** branches to go to the armpit region. The radial nerve continues through the arm and is paralleled by the **ulnar nerve** and the **median nerve**. The **lumbar plexus** arises from all the lumbar spinal nerves and gives rise to nerves enervating the pelvic region and the anterior leg. The **femoral nerve** is one of the major nerves from this plexus, which gives rise to the **saphenous nerve** as a branch that extends through the anterior lower leg. The **sacral plexus** comes from the lower lumbar nerves L4 and L5 and the sacral nerves S1 to S4. The most significant systemic nerve to come from this plexus is the **sciatic nerve**, which is a combination of the **tibial nerve** and the **fibular nerve**. The sciatic nerve extends across the hip joint and is most commonly associated with the condition **sciatica**, which is the result of compression or irritation of the nerve or any of the spinal nerves giving rise to it. The named nerves that arise from each of the four plexuses are detailed in Table 14.2.

fibular nerve (common fibular nerve, common peroneal nerve, external popliteal nerve, lateral popliteal nerve)

LO 14.3.7

These plexuses are described as arising from spinal nerves and giving rise to certain systemic nerves, but they contain fibers that serve sensory functions or fibers that serve motor functions. This means that some fibers extend from cutaneous or other peripheral sensory surfaces and send action potentials into the CNS. Those are axons of sensory neurons in the posterior root ganglia that enter the spinal cord through the dorsal nerve root. Other fibers are the axons of motor neurons of the anterior horn of the spinal cord, which emerge in the ventral nerve root and send action potentials to cause skeletal muscles to contract in their target regions. For example, the radial nerve contains fibers of cutaneous sensation in the arm, as well as motor fibers that move muscles in the arm.

Spinal nerves of the thoracic region, T2 through T11, are not part of the plexuses but rather emerge and give rise to the **intercostal nerves** found between the ribs, which articulate with the vertebrae surrounding the spinal nerve.

Table 14.2 The Four Plexuses

Plexus	Nerve	Rami	Motor Innervation	Cutaneous Innervation
Cervical Plexus: C1–C5	Phrenic Nerve	C3–C5	Diaphragm	Sensation to the central tendon of the diaphragm
Brachial Plexus: C5–T1	Radial Nerve	C5–T1	Triceps brachii, posterior compartment of the antebrachium	Posterior brachium and dorsal aspect of the hand
	Axillary Nerve	C5–C6	Teres minor and deltoids	Lateral shoulder
	Median Nerve	C5–T1	Anterior compartment of the antebrachium and thenar musculature	Anterior and lateral aspect of the hand, dorsal and distal digits
	Ulnar Nerve	C8–T1	Flexor carpi ulnaris, flexor digitorum profundus, and intrinsic muscles of the hand	Medial aspect of the hand, digits 4 and 5
Lumbar Plexus: L1–L4	Femoral Nerve	L2–L4	Anterior compartment of the thigh	Branches into the saphenous nerve
	Saphenous Nerve	L3–L4	No motor innervation	Anterior surface of the lower leg
Sacral Plexus: L4–S4	Sciatic Nerve	L4–S3	Biceps femoris, semimembranosus, semitendinosus	No sensory innervation
	Tibial Nerve	L4–S3	Muscles of the posterior lower leg and intrinsic foot muscles	Skin of the posterolateral side of the leg and the lateral aspect of the foot
	Fibular Nerve	L4–S2	Muscles of the lateral compartment of the lower leg	Skin of the posterolateral side of the leg and the lateral aspect of the foot

Learning Check

1. For which of the following cranial nerves are the name and number correctly matched?
   a. Optic Nerve (CN III)
   b. Facial nerve (CN V)
   c. Vestibulocochlear Nerve (CN VIII)
   d. Spinal accessory Nerve (CN XII)
2. Raphael comes to the emergency room "feeling funny." You do a thorough neurological evaluation and notice that Raphael has sensation impairments on their face. Raphael also has difficulty smiling or making a funny face. Which of the following cranial nerves do you think is impaired?
   a. Vagus nerve
   b. Trigeminal nerve
   c. Facial nerve
   d. Olfactory nerve
3. How many pairs of spinal nerves are in the lumbar region?
   a. 8
   b. 12
   c. 5
   d. 6
4. Which of the following nerves is part of the lumbar plexus?
   a. Femoral nerve
   b. Phrenic nerve
   c. Tibial nerve
   d. Median nerve

Chapter Review

14.1 General Anatomy of the Nervous System

The learning objectives for this section are:

HAPS LEARNING GOALS AND OUTCOMES

14.1.1 Describe the composition and arrangement of the gray and white matter in the CNS.

14.1.2 Compare and contrast the structure and location of a tract and nerve.

14.1.3 Compare and contrast the structure and location of a nucleus and ganglion.

14.1.4 Identify the layers of the meninges and describe their anatomical and functional relationships to the CNS (brain and spinal cord).

14.1.5 Identify and describe the epidural space, subdural space and subarachnoid space associated with the brain and the spinal cord, and identify which space contains cerebrospinal fluid.

14.1.6 Describe the structure and location of the dural venous sinuses, and explain their role in drainage of blood from the brain.

14.1.7 Identify and describe the structure and function of the cranial dural septa.

14.1.8 Describe how the cranial bones and the vertebral column protect the CNS.

14.1.9* Draw or describe the structure of a typical nerve, including the connective tissue wrappings and any associated cells.

14.1.10 Identify and describe the ventricular system components.

14.1.11 Describe the production, flow, and reabsorption of cerebrospinal fluid (CSF), from its origin in the ventricles to its eventual reabsorption into the dural venous sinuses.

14.1.12 Describe the general functions of cerebrospinal fluid (CSF).

14.1.13 Compare and contrast the structure of the dura mater surrounding the brain and the spinal cord.

* Objective is not a HAPS Learning Goal.

14.1 Questions

1. Which of the following is/are true about white matter in the CNS? Please select all that apply.
   a. Region with many cell bodies
   b. Superficial location in the brain
   c. Superficial location in the spinal cord
   d. Consists of myelinated axons

2. Where would you expect to find the corticospinal tract, a neuronal pathway that is responsible for voluntary movement?
   a. Central nervous system
   b. Peripheral nervous system

3. The thalamus is a structure in the central nervous system that contains many cell bodies. Which of the following accurately describes the group of cell bodies within the thalamus?
   a. Nucleus
   b. Ganglion
   c. Tract
   d. Nerve

4. Which of the following layers is the thinnest membrane of the meninges that tightly surrounds the CNS?
   a. Dura mater
   b. Arachnoid mater
   c. Pia mater
   d. Falx cerebri

5. Darnell walks into the hospital with a high fever, stiff neck, headache, and sensitivity to light. You suspect meningitis. In order to confirm your hypothesis, you order a cerebrospinal fluid (CSF) test. Which layers will the needle have to pass in order to drain CSF? Please select all that apply.
   a. Dura mater
   b. Arachnoid mater
   c. Pia mater
   d. Cranial dural septa

6. Which of the following sinuses can you find directly superior to the cerebellum?
   a. Straight sinus
   b. Occipital sinus
   c. Superior sagittal sinus
   d. Inferior sagittal sinus

7. What is the role of the cranial dural septa?
   a. They subdivide the cranial space.
   b. They return venous blood from the brain to the heart.
   c. They protect the entire brain and spinal cord.
   d. They form when blood accumulates.

8. Which of the following pairs of words correctly completes the statement "The brain is protected by three meningeal layers and the bones of the cranium, while the spinal cord is protected by ______ and ______."?
   a. Two meningeal layers; the bones of the vertebrae
   b. Three meningeal layers; no bony protection
   c. Zero meningeal layers; the bones of the vertebrae
   d. Three meningeal layers; the bones of the vertebrae

9. Which of the following structures would you expect to find within fascicles?
   a. Epineurium
   b. Perineurium
   c. Endoneurium
   d. Exoneurium

10. Which of the following is/are true of the ventricular system of the brain? Please select all that apply.
    a. The only paired ventricles in the brain are the lateral ventricles.
    b. The third ventricle is connected to the fourth ventricle by the interventricular foramen.
    c. All four of the ventricles contain a choroid plexus and produce cerebrospinal fluid.
    d. After the cerebrospinal fluid has circulated around the brain and spinal cord, it is transported to the kidneys for excretion in the urine.

11. Which of the following structures would the CSF travel to next if it is currently in the fourth ventricle?
    a. Third ventricle
    b. Cerebral aqueduct
    c. Subarachnoid space
    d. Interventricular foramen

12. What is the function of cerebrospinal fluid (CSF)?
    a. Provide physical protection to the CNS
    b. Transport nutrients to the CNS
    c. Transport solutes to the PNS
    d. Connect the four ventricles

13. Within the cranium, the dura matter occasionally splits into two layers. Which of the following statements is true of the dura mater in the vertebral column?
    a. The dura mater in the vertebral column, like the dura mater in the cranium, is often found in two layers. These layers share the same names, periosteal and meningeal, as the cranial dural layers.
    b. The dura mater in the vertebral column, like the dura mater in the cranium, is often found in two layers, but the layers have different names in the vertebral column.
    c. Unlike the dura mater in the cranium, the dura mater within the vertebral column has only one layer.

14.1 Summary

- In the brain, the white matter is deep and the gray matter is superficial. In the spinal cord, the reverse is true: the white matter is superficial and the gray matter is deep. Grey matter consists of neuron cell bodies, dendrites, and unmyelinated axons. White matter consists of myelinated axons.
- In the peripheral nervous system, a group of cell bodies is called a *ganglion* and a bundle of axons is called a *nerve*. In the central nervous system, a group of cell bodies is called a *nucleus* while a bundle of axons is called a *tract*.
- There are three layers of connective tissues (called *meninges*), which protect the brain and the spinal cord. The dura mater is the most superficial, followed by the arachnoid mater, and the deepest is the pia mater. The subarachnoid space (between the arachnoid and the pia mater) houses the cerebrospinal fluid to provide nutrition and a liquid cushion to the brain and spinal cord.
- The dural venous sinuses are found throughout the brain to transport venous blood from the brain back to the heart. The cranial dural septa subdivide the cranial space and can be found throughout the brain as well.
- The cerebrospinal fluid (CSF) circulates throughout the brain but is produced in the four ventricles, which are anatomically connected to each other.

14.2 The Central Nervous System

The learning objectives for this section are:

HAPS LEARNING GOALS AND OUTCOMES

14.2.1 Identify and describe the 3 primary brain vesicles formed from the neural tube.

14.2.2 Identify and describe the 5 secondary brain vesicles formed from the neural tube and name the parts of the adult brain arising from each.

14.2.3 Identify and describe the four major parts of the adult brain (i.e., cerebrum, diencephalon, brainstem, cerebellum).

14.2.4 Identify and describe the three major cerebral regions (i.e., cortex, white matter, cerebral nuclei [basal nuclei]).

14.2.5 Identify and define the general terms gyrus, sulcus, and fissure.

14.2.6 Identify and describe the major landmarks of the cerebrum (e.g., longitudinal fissure, lateral sulcus [fissure], central sulcus, transverse fissure, precentral gyrus, postcentral gyrus).

14.2.7 Identify and describe the cerebral hemispheres and the five lobes of each (i.e., frontal, parietal, temporal, occipital, insula).

14.2.8 Identify and describe the primary functional cortical areas of the cerebrum (e.g., primary motor cortex, primary somatosensory cortex, primary auditory cortex, primary visual cortex, primary olfactory cortex, primary gustatory cortex).

14.2.9 Compare and contrast the cerebral location and function of the motor speech area (Broca area) and Wernicke area.

14.2.10 Compare and contrast the three cerebral white matter tracts (i.e., association, commissural, projection).

14.2.11* Describe the location and functions of two of the basal ganglia.

14.2.12 Name the major components of the diencephalon.

14.2.13 Describe the structure, location, and major functions of the thalamus.

14.2.14 Describe the structure, location, and major functions of the hypothalamus, including its relationship to the autonomic nervous system and the endocrine system.

14.2.15 Describe and identify the epithalamus, including the pineal gland and its function.

14.2.16 Name the three subdivisions of the brainstem.

14.2.17 Describe the structure, location, and major functions of the midbrain (mesencephalon), including the cerebral peduncles, superior colliculi, and inferior colliculi.

14.2.18 Describe the structure, location, and major functions of the pons.

14.2.19 Describe the structure, location, and major functions of the medulla oblongata (medulla), including the pyramids and decussation of the pyramids.

14.2.20 Describe the structure, location, and major functions of the cerebellum.

14.2.21 Identify and describe the cerebellar hemispheres, vermis, arbor vitae (cerebellar white matter), cerebellar peduncles, and cerebellar cortex (folia, cerebellar gray matter).

14.2.22 Describe the major components and functions of the reticular activating system (RAS).

14.2.23 Describe the major components and functions of the limbic system.

14.2.24 Identify and describe the gross anatomy of the spinal cord, including its enlargements (i.e., cervical and lumbar), conus medullaris, cauda equina, and filum terminale.

14.2.25 Identify and describe the anatomical features seen in a cross-sectional view of the spinal cord (e.g., anterior horn, lateral horn, posterior horn, gray commissure, central canal, anterior funiculus [column], lateral funiculus [column], posterior funiculus [column]).

14.2.26 Compare and contrast the location, composition, and function of the anterior (ventral) roots, posterior (dorsal) roots, and posterior (dorsal) root ganglion with respect to the spinal cord.

14.2.27 Describe the structure, location, and function of ascending and descending spinal cord tracts.

* Objective is not a HAPS Learning Goal.

14.2 Questions

1. Identify the primary vesicles that are formed from the neural tube. Please select all that apply.
   a. Forebrain
   b. Diencephalon
   c. Hindbrain
   d. Myelencephalon

2. Which of the following secondary brain vesicles develop from the forebrain, and which parts of the adult brain do the secondary brain vesicles become?
   a. Metencephalon and myelencephalon; pons, medulla oblongata, and cerebellum
   b. Midbrain; midbrain
   c. Telencephalon and diencephalon; cerebrum, retinas of the eye, thalamus, and hypothalamus
   d. Midbrain and metencephalon; midbrain and cerebellum

3. Which of the following matches a major part of the adult brain with its description?
   a. The cerebrum exerts some control over motor function, coordinating sensory input from the eyes, inner ears, and proprioceptors in the joints.
   b. The brainstem consists of the midbrain, pons and medulla oblongata.
   c. The cerebellum consists of the mainly of the thalamus and hypothalamus.
   d. The diencephalon is responsible for many of the higher mental functions, such as learning, memory, emotions, and states of consciousness.

4. Which of the following structures is the deepest of the cerebral regions?
   a. Cortex
   b. White matter
   c. Cerebral nuclei

5. What is a “fissure,” as the term applies to brain anatomy?
   a. A deep groove on the surface of the brain
   b. A shallow groove on the surface of the brain
   c. A tubular passageway in the brain
   d. A raised ridge on the surface of the brain

6. Filipe comes into the hospital exhibiting difficulty moving their right arm, numbness, and speech deficits. Upon imaging, you see that there is a stroke of the left middle cerebral artery, which supplies the left side of the brain and goes up toward the top. You are trying to find the impaired region on an image of the left side of the brain. Which of the following landmarks will help you identify the impaired area? Please select all that apply.
   a. Central sulcus
   b. Longitudinal fissure
   c. Lateral fissure
   d. Precentral gyrus

7. Which of the following are true of the cerebral hemispheres and their lobes? Please select all that apply.
   a. The corpus callosum separates the left and right cerebral hemispheres.
   b. Each parietal lobe contains a somatosensory cortex and is responsible for the coordination of sensory information received from the body.

c. All of the lobes of the cerebrum are named for the bones that they underlie, except the insula.
d. The frontal lobe is responsible for higher order cognitive functions, behaviors, and decision-making.

8. If a person is having difficulty moving their left arm, which of the following functional cortical areas would you suspect is impaired?

a. Primary motor cortex
b. Primary sensory cortex
c. Primary gustatory cortex
d. Primary olfactory cortex

9. Which of the following functional cortical areas do you suspect is impaired in a person who is able to understand words but is having difficulty producing words?

a. Broca's area
b. Wernicke's area
c. Primary motor cortex
d. Primary sensory cortex

10. Which of the following tracts connects the two cerebral hemispheres?

a. Commissural tract
b. Projection tracts
c. Longitudinal fasciculi
d. Association tracts

11. All of the following are true of the basal nuclei except:

a. They are sometimes referred to as the basal ganglia.
b. They consist of the caudate, putamen, and globus pallidus.
c. They are the main structures of the brain that control states of consciousness.
d. They function in the regulation of various aspects of motion.

12. Which of the following are components of the diencephalon? Please select all that apply.

a. Thalamus
b. Pineal gland
c. Hippocampus
d. Amygdala

13. You wake up on a Sunday morning to the smell of pancakes. You follow your nose to the kitchen. Which of the following neural information is NOT processed through the thalamus?

a. Walking to the kitchen
b. Seeing around your home
c. Getting out of bed
d. Smelling the pancakes

14. When exercising, the muscles and brain demand more oxygen. The body accommodates by decreasing the diameter of your blood vessels, increasing the blood pressure, and increasing your heart rate. Which of the following structures primarily regulates this to maintain homeostasis in the body?

a. Hypothalamus
b. Thalamus
c. Subthalamus
d. Epithalamus

15. Which of the following is true about the pineal gland?

a. It secretes melatonin into the bloodstream.
b. It incorporates light/dark information from the visual system.
c. It regulates appetite and food intake.
d. It detects temperature changes within the body.

16. Which of the following structures is/are found within the brainstem? Please select all that apply.

a. Midbrain
b. Medulla oblongata
c. Thalamus
d. Cerebellum

17. What is the function of the superior colliculi?

a. Connects the third and fourth ventricles
b. Sends auditory information to the thalamus
c. Controls fluidity of motor movements
d. Combines sensory with motor output

18. Where is the pons located?

a. Within the brainstem
b. Inferior to the medulla
c. Superior to the cerebrum
d. Posterior to the midbrain

19. The neurons that control your left arm cross over before reaching the right half of the cerebral hemisphere. Where do these neurons cross over?

a. Pons
b. Medulla oblongata
c. Midbrain
d. Thalamus

20. What is/are the function of the cerebellum? Please select all that apply.

a. Voluntary movement
b. Involuntary movement
c. Proprioception integration
d. Nociceptive integration

21. During a dissection lab, you are looking at the cerebellum when you see a treelike structure. Is this white matter or gray matter?

a. White matter
b. Gray matter

22. Which of the following contributes to the sensory component of alertness?

a. Rapid eye movement
b. Reticular activating system
c. Fornix
d. Cingulate gyrus

23. Which of the following structures are part of the limbic system? Please select all that apply.

a. Mammillary bodies
b. Amygdala
c. Arbor vitae
d. Pyramidal tracts

Mini Case 1: Spinal Nerve Compression

1. Juanita comes to a clinic complaining of sensory deficits between their legs and difficulty walking. After a thorough neurological screening, the doctor thinks that nerves in a structure on the inferior-most aspect of the spinal cord are being compressed. Which of the following spinal cord structures is most likely to be compressed?

a. Cauda equina
b. Posterolateral sulcus
c. Cerebral peduncles
d. Substantia nigra

2. Compression of nerves in which of the following spinal cord regions is most likely causing Juanita's sensory deficits?

a. Anterior horn
b. Posterior horn
c. Lateral horn
d. Posterolateral sulcus

3. Since Juanita is also having difficulty walking, the doctor explains that Juanita's motor neurons are also being compressed. Which structure emerges from the spinal cord and transports the axons of motor neurons into spinal nerves?

a. Posterior (dorsal) roots
b. Posterior (dorsal) root ganglia
c. Lateral root ganglia
d. Anterior (ventral) roots

4. Since Juanita can feel their sensory and motor issues, nerve impulses must be reaching areas of their brain that regulate conscious awareness. Through which of the following structures do these nerve impulses travel from the spinal cord to the brain?

a. Descending spinal cord tracts
b. Anterior spinal roots
c. Ascending spinal cord tracts
d. Central canal of the spinal cord

14.2 Summary

- The brain is composed of the cerebrum, the diencephalon, the brainstem, and the cerebellum.
- The cerebrum is the largest portion of the brain; it is responsible for learning, reasoning, language skills, emotion, and states of consciousness. Each cerebral hemisphere consists of lobes, each of which is responsible for a specific function. The primary sensory and motor cortices provide sensory integration and motor output, respectively.
- The corpus callosum connects the two cerebral hemispheres and allows for neural communication between them. The basal nuclei are subcortical nuclei that function in the planning and initiation of movement.
- The diencephalon contains the thalamus and the hypothalamus. The thalamus is responsible for all sensory and motor integration, except for olfaction. The hypothalamus is responsible for regulating the autonomic nervous system and maintaining homeostasis for many variables.
- The brainstem contains the midbrain, the pons, and the medulla. It contains the control centers for several vital processes, such as respiratory and cardiovascular function.
- The cerebellum fine-tunes motor control.
- The reticular formation is related to sleep and wakefulness.
- The limbic system is made up of several structures that work together to process emotion.
- The spinal cord runs inferiorly from the brain through the vertebral canal. It relays sensory and motor information between the brain and the peripheral nerves, and is a center for spinal reflexes.

14.3 The Peripheral Nervous System

The learning objectives for this section are:

HAPS LEARNING GOALS AND OUTCOMES

14.3.1 List and identify the cranial nerves by name and number.

14.3.2 Describe the major functions of each cranial nerve and identify each cranial nerve as predominantly sensory, motor, or mixed (i.e., sensory and motor).

14.3.3 List the cranial nerves that have parasympathetic (ANS) components.

14.3.4 Identify and describe the formation, structure, and branches of a typical spinal nerve, including the roots and the rami (e.g., anterior [ventral], posterior [dorsal]).

14.3.5 List the number of spinal nerve pairs emerging from each spinal cord region (i.e., cervical, thoracic, lumbar, sacral, coccygeal).

14.3.6 Define a spinal nerve plexus.

14.3.7 For the cervical, brachial, lumbar, and sacral nerve plexuses, list the spinal nerves that form each plexus, describe the plexus' major motor and sensory distributions, and list the major named nerves that originate from each plexus.

14.3 Questions

1. What is cranial nerve III?

a. Optic nerve
b. Olfactory nerve
c. Oculomotor nerve
d. Trochlear nerve

2. Which cranial nerve is responsible for senses of hearing and balance?

a. Trochlear nerve
b. Glossopharyngeal nerve
c. Vestibulocochlear nerve
d. Spinal accessory nerve

3. Which of the cranial nerves have autonomic nervous system components? Please select all that apply.

a. Glossopharyngeal nerves
b. Vagus nerve
c. Optic nerve
d. Trigeminal nerve

4. Once the anterior and posterior spinal roots converge, outside the spinal cord what structure do they form?

a. Cervical plexus
b. Thoracic plexus
c. Spinal nerve
d. Sciatic nerve

5. How many spinal nerve pairs emerge from the cervical region?

a. 8
b. 12
c. 5
d. 6

6. How is a spinal nerve plexus formed?

a. From a group of spinal roots that converges
b. From a group of spinal nerves that weaves together
c. From a group of spinal nerves that runs parallel
d. From a group of spinal roots that runs parallel

7. Jacquise fell on their shoulder during gymnastics practice. Jacquise goes to the emergency room because they cannot bend their fingers and they feel sharp pain throughout their arm. After thorough testing, you find that Jacquise has thoracic outlet syndrome where there is compression of the brachial plexus. Which of the following nerves could be affected? Please select all that apply.

a. Tibial nerve
b. Phrenic nerve
c. Axillary nerve
d. Radial nerve

14.3 Summary

- The peripheral nervous system consists of cranial nerves, spinal nerves, ganglia, and the enteric nervous system.
- There are 12 pairs of cranial nerves arising from the brain; some are sensory, some are motor, and some are mixed. They provide a variety of functions, such as the senses of vision, hearing, taste, and the motor control of muscles of the eyes, face, neck, back and digestive organs.
- There are 31 pairs of spinal nerves arising from the spinal cord. They are all mixed (both sensory and motor) nerves that provide sensation to and motor control of various organs and muscles.
- In four regions of the spinal cord, the nerves weave together to form nerve networks called spinal nerve plexuses. These are the cervical, brachial, lumbar, and sacral plexuses. Each plexus innervates a different region of the body.

15

The Somatic Nervous System

Cipolina/Shutterstock.com

Chapter Introduction

In this chapter you will learn...

How the sensory receptors in our skin communicate to the brain about our environment . . .

. . . along with the anatomy and physiology of vision, taste, smell, hearing, and balance.

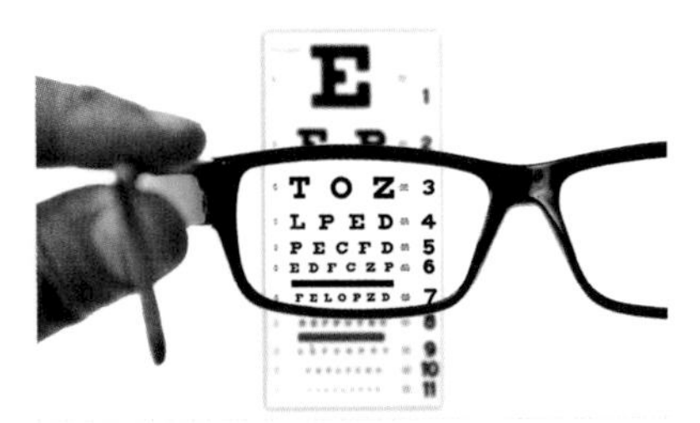

Rungnapa Tarasiri/Shutterstock.com; DimaBerlin/Shutterstock.com; Lapina/Shutterstock.com; Castaldostudio.com/Shutterstock.com; Jose Luis Pelaez Inc/DigitalVision/Getty Images; TinnaPong/Shutterstock.com

The peripheral nervous system can be divided into two functional divisions—the somatic and autonomic divisions. The most obvious distinction between these two divisions is that the somatic nervous system yields conscious perception of the environment and voluntary responses by means of skeletal muscles. In other words, the somatic nervous system involves the sensory and motor activities that we are aware of. For example, your somatic nervous system includes the visual system you are using to read this book, and the muscles you contract to turn its pages (or scroll down if you are reading digitally). In contrast, the visceral sensory system operates almost entirely without conscious perception. You cannot, for example, request that your digestive system slow its motility so that you don't experience diarrhea. In this chapter we will explore the peripheral sensory neurons and how they send their signals into the central nervous system (CNS). We will discuss the motor responses that originate in the CNS and traverse the somatic peripheral nervous system to reach their target muscles. We will also illustrate how these two sides of the nervous system interact fluidly (for example, when you touch a hot stove, you pull your hand away).

When we discuss the senses, we often use one chief distinction, general versus special senses. *General senses* are the senses that occur all over your body such as sensing touch or a change in temperature. *Special senses* are limited to your head and have unique anatomical components as well. The special senses include vision, taste, smell, hearing, and balance.

The Human Anatomy and Physiology Society includes more than 1,700 educators who work together to promote excellence in the teaching of this subject area. The HAPS A&P Learning Outcomes measure student mastery of the content typically covered in a two-semester Human A&P curriculum at the undergraduate level. The full Learning Outcomes are available at https://www.hapsweb.org.

15.1 Structure and Function of Sensory and Motor Pathways

Learning Objectives: By the end of this section, you will be able to:

15.1.1 Describe the locations and functions of the first-, second- and third-order neurons in a sensory pathway.

15.1.2 Describe the locations and functions of the upper and lower motor neurons in a motor pathway.

15.1.3 Describe the concept of decussation and its functional implications.

15.1.4 Define the term reflex.

15.1.5 Describe reflex responses in terms of the major structural and functional components of a reflex arc.

15.1.6 Distinguish between each of the following pairs of reflexes: intrinsic versus learned, somatic versus visceral, monosynaptic versus polysynaptic, and ipsilateral versus contralateral.

15.1.7 Describe the following reflexes and name all components of each reflex arc: stretch reflex, (Golgi) tendon reflex, flexor (withdrawal) reflex, and crossed-extensor reflex.

Sensory information comes into the central nervous system (CNS) and is delivered to the brain region dedicated to decoding it through a chain of neurons. The first neuron in that chain is the **sensory receptor**; these neurons help us learn about the environment around us, or about the environment inside the body. Sensory stimuli come from a variety of sources, and in a variety of modalities. Sound stimuli are vibrations, hot and cold stimuli are temperature, taste stimuli are chemicals. All sensory stimuli are received and changed into the electrochemical signals of the nervous system. In other words, the stimulus changes the cell membrane potential of a sensory neuron. The stimulus causes a graded potential for the sensory neuron, and if the stimulus is sufficient to bring the membrane potential past threshold, the sensory neuron will fire an action potential that is relayed into the CNS, where it is integrated with other sensory information—or sometimes higher cognitive functions—to become a perception of that stimulus (see the "Anatomy of a Sensory Pathway" feature). Some sensory integration may lead to conscious awareness of the stimuli (and some does not); some sensory integration may lead to a motor response, while some does not.

LO 15.1.1

LO 15.1.2

Motor responses are also enacted through a chain of neurons. However, there are typically just two neurons in a motor pathway. The first, or primary, neuron in a motor pathway has a cell body in the primary motor cortex. Its axon descends through the white matter tracts in the brain, crossing over to the opposite side of the body at the decussation of the pyramids in the medulla oblongata. Its axon descends all the way down the spinal cord to somewhere approximately on the transverse plane of the muscle it acts on. There, within the anterior horn, it forms a synapse—the only synapse in a motor pathway—with the cell body of the motor neuron. The axon of the motor neuron exits the spinal cord in the anterior root; it joins axons of primary sensory neurons in the spinal nerve. It may weave with other axons in a plexus, but eventually the motor neuron ends in axon terminals that form the neuromuscular junction (NMJ) with muscle cells (see the "Anatomy of a Motor Pathway" feature).

LO 15.1.3

One thing you will notice as you examine both the motor and sensory pathways is that the pathway crosses sides of the body. If the somatic sensory stimulus is being received on the left side, it crosses over or **decussates** to the right. Decussation occurs

Anatomy of...

A Sensory Pathway

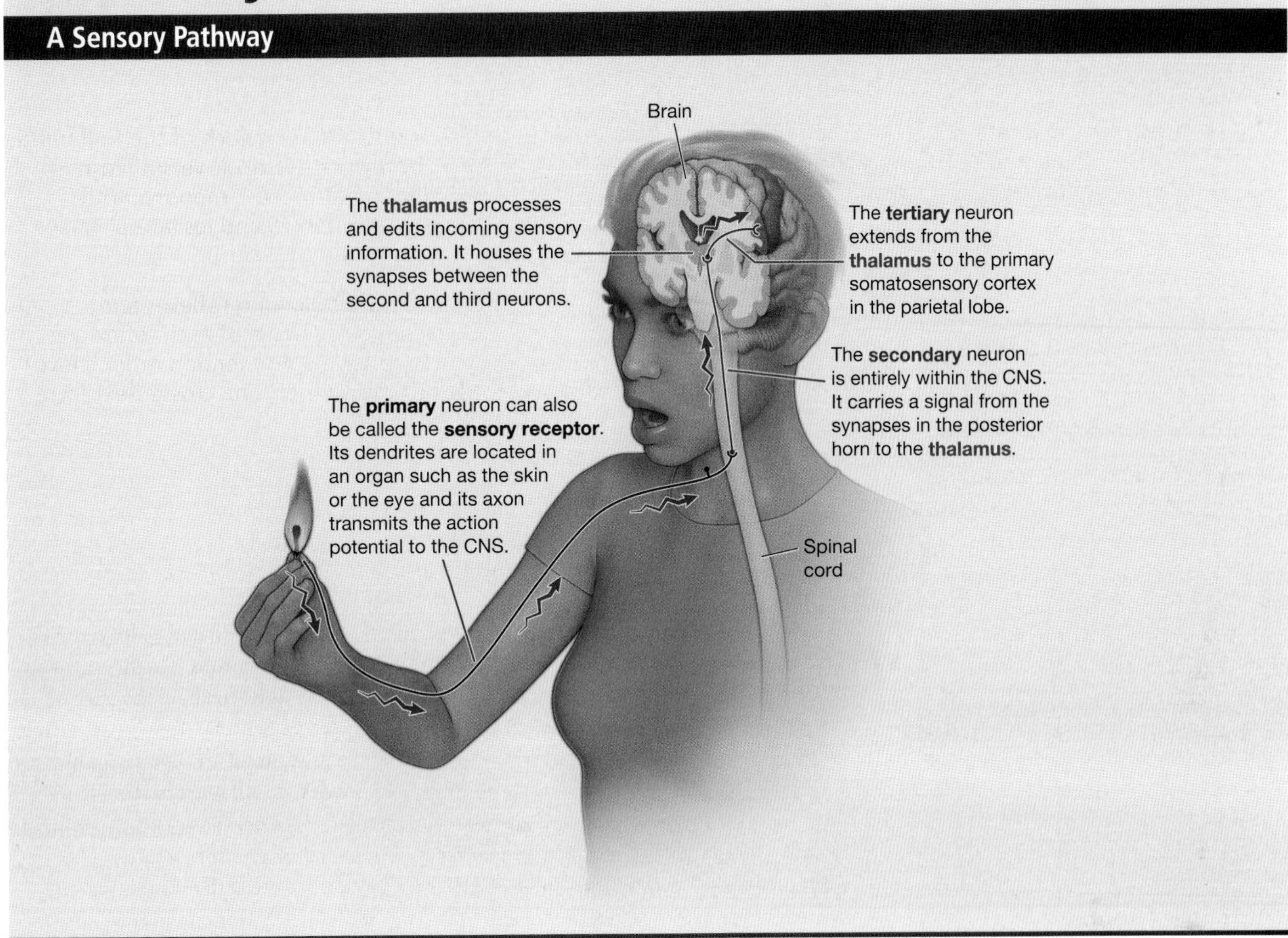

Brain
The thalamus processes and edits incoming sensory information. It houses the synapses between the second and third neurons.
The tertiary neuron extends from the thalamus to the primary somatosensory cortex in the parietal lobe.
The secondary neuron is entirely within the CNS. It carries a signal from the synapses in the posterior horn to the thalamus.
The primary neuron can also be called the sensory receptor. Its dendrites are located in an organ such as the skin or the eye and its axon transmits the action potential to the CNS.
Spinal cord

at the medulla oblongata of the brainstem. This crossing over has functional implications in brain injury. Right-brain injury may impact motor or sensory function on the left side of the body, and vice versa. Some special sense information does not cross over.

15.1a Reflexes

LO 15.1.4

We use the word *reflex* in everyday language to mean something that is instinctual or automatic. True to their name, anatomical **reflexes** are connections between the sensory and motor neurons that do not include the higher brain centers or include conscious or voluntary aspects of movement. **Figure 15.1** illustrates a simple reflex.

The process of a reflex begins at the sensory receptor. The sensory receptor receives a stimulus, which causes a graded potential. If it is sufficient to bring the sensory receptor membrane to threshold, the sensory receptor fires an action potential, which travels along the length of its fiber from dendrite to axon terminal within the spinal cord. Here, the sensory neuron releases its neurotransmitter, causing a graded potential in its postsynaptic neuron, which is, in this simple example, a motor neuron. The motor neuron's axon stretches out from the anterior horn, possibly traveling within the same spinal nerve, and terminates at the NMJ, causing a muscle contraction.

LO 15.1.5

Anatomy of...

A Motor Pathway

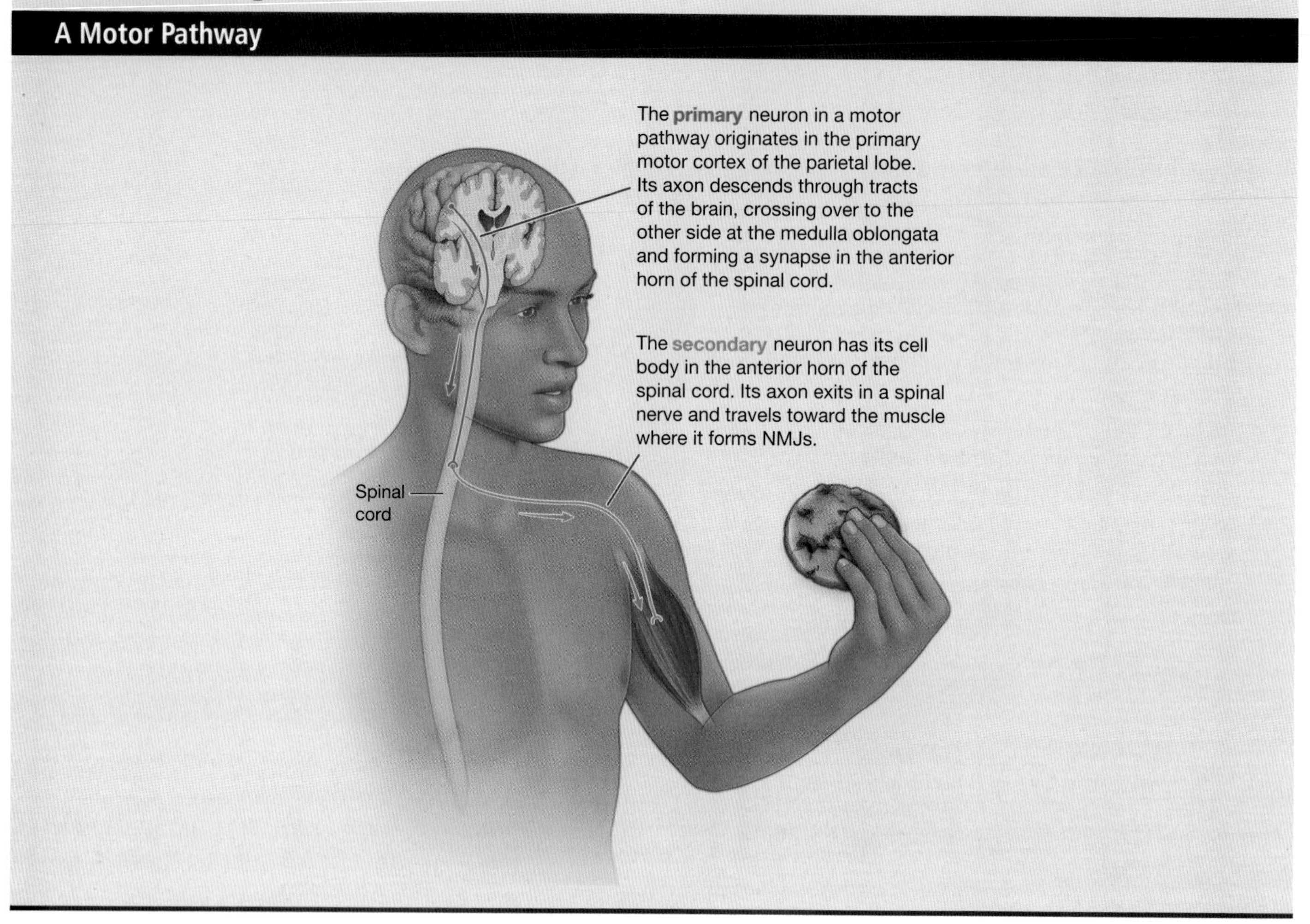

The primary neuron in a motor pathway originates in the primary motor cortex of the parietal lobe. Its axon descends through tracts of the brain, crossing over to the other side at the medulla oblongata and forming a synapse in the anterior horn of the spinal cord.
The secondary neuron has its cell body in the anterior horn of the spinal cord. Its axon exits in a spinal nerve and travels toward the muscle where it forms NMJs.
Spinal cord

The reflex arc illustrated in Figure 15.1 involves the spinal cord and is therefore considered a **spinal reflex**. Some reflexes involve the brain (these are known as **cranial reflexes**). Either way, the synapse or synapses that connect the sensory neuron to the motor neuron will always occur in the CNS.

Figure 15.1 A Simple Reflex Arc

A simple reflex arc contains one sensory neuron and one motor neuron. The synapse that joins them is located in the spinal cord.

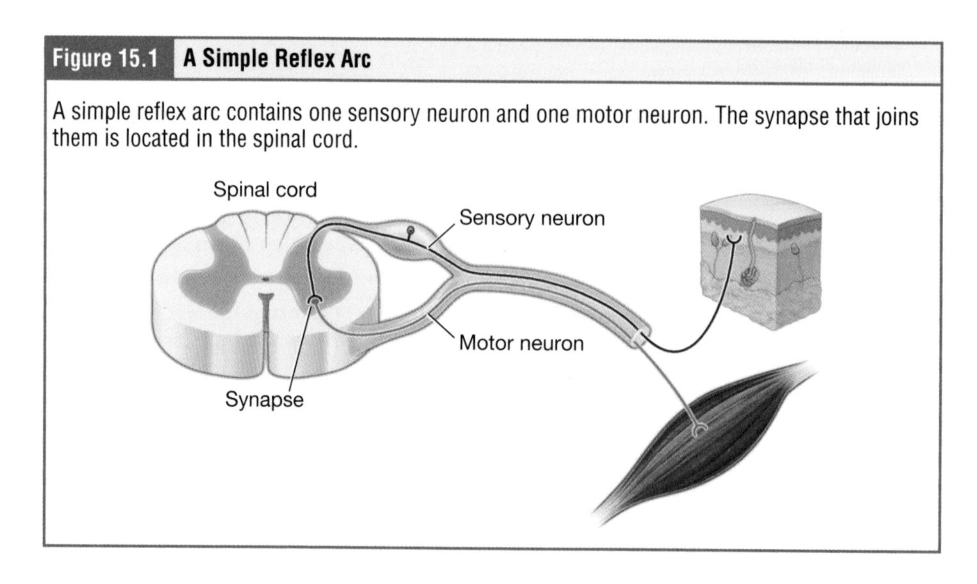

LO 15.1.6

Reflexes vary in complexity and function. There are four general ways of categorizing reflexes:

- Intrinsic versus learned: this category of reflex describes whether a reflex developed in the fetal period in which case at birth it is already present and considered an **intrinsic reflex**, or if it developed after birth, in which case it is a **learned reflex**.
- Somatic versus visceral: reflexes are considered to be a **somatic reflex** if the motor response is carried out by a skeletal muscle, or a **visceral reflex** if the motor response involves smooth or cardiac muscle or a gland.
- Monosynaptic versus polysynaptic: this category describes reflexes simply by how many synapses are in the reflex arc. Figure 15.2 illustrates the difference between a **monosynaptic reflex**, which has just one synapse between the sensory and motor neurons, or a **polysynaptic reflex**, which has more than one synapse and therefore involves interneurons in addition to the sensory and motor neurons.
- Ipsilateral versus contralateral: an **ipsilateral reflex** begins and ends on the same side of the body—for example, tapping on the patellar tendon of the right knee causes a movement in the right lower leg. A **contralateral reflex** begins and ends on opposite sides of the body—for example, stepping on a sharp object with your right foot causes you to shift your weight to your left foot (Figure 15.3).

Student Study Tip

Another time the term *intrinsic* appears is relating to muscles, meaning a muscle found in the area in which its action occurs. In both cases, intrinsic means "within."

Figure 15.2 Monosynaptic versus Polysynaptic Reflex Arcs

(A) The simplest reflex arcs (monosynaptic reflex arcs) contain just one synapse between the sensory and motor neurons. (B) More complex reflex arcs (polysynaptic reflex arcs) contain two or more synapses and one or more interneurons.

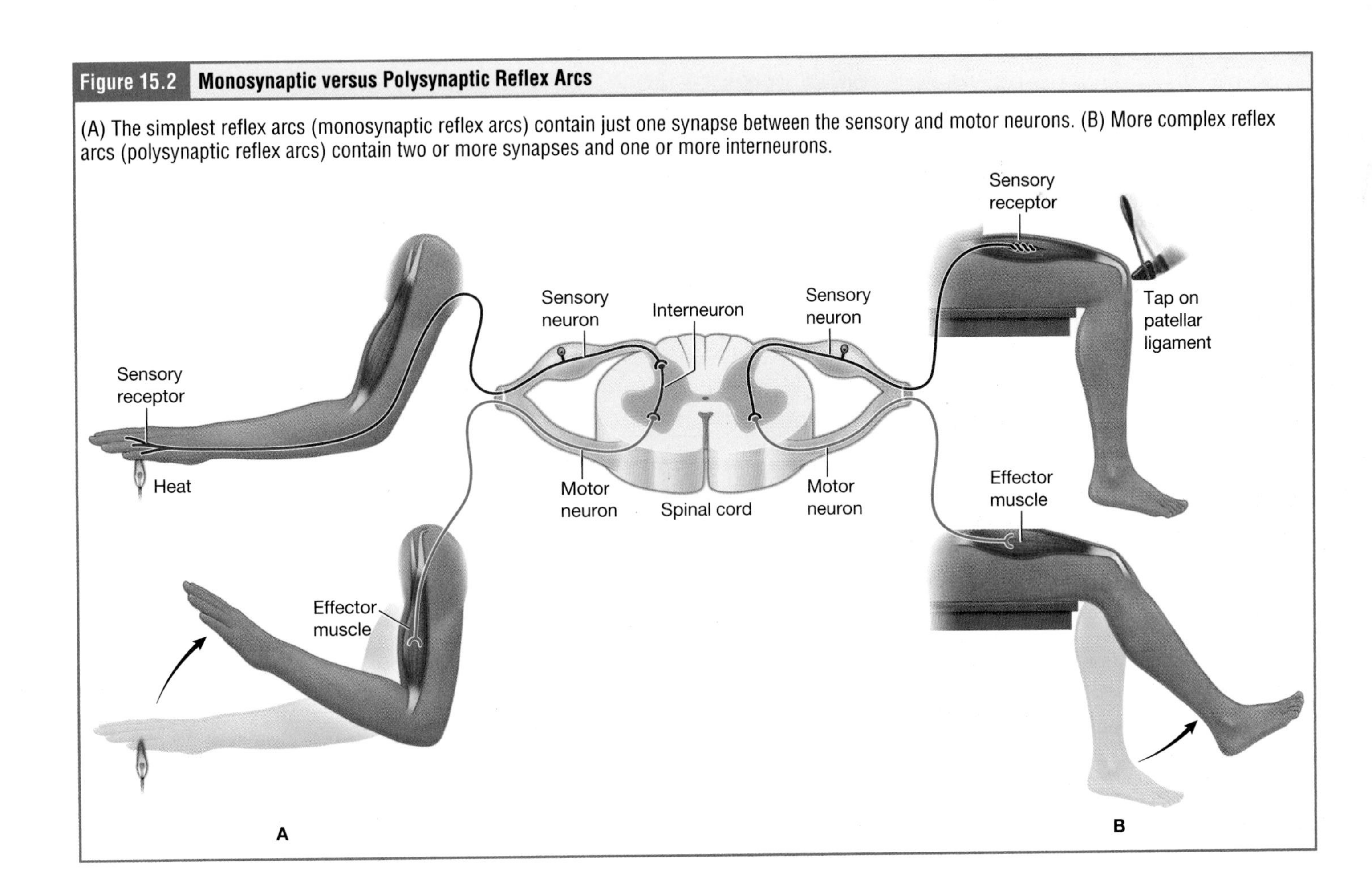

Figure 15.3 Ipsilateral versus Contralateral Reflexes

(A) In ipsilateral reflexes, the sensory stimulation and motor output occur on the same side of the body. (B) In contralateral reflexes, the sensory stimulation and motor output occur on opposite sides of the body.

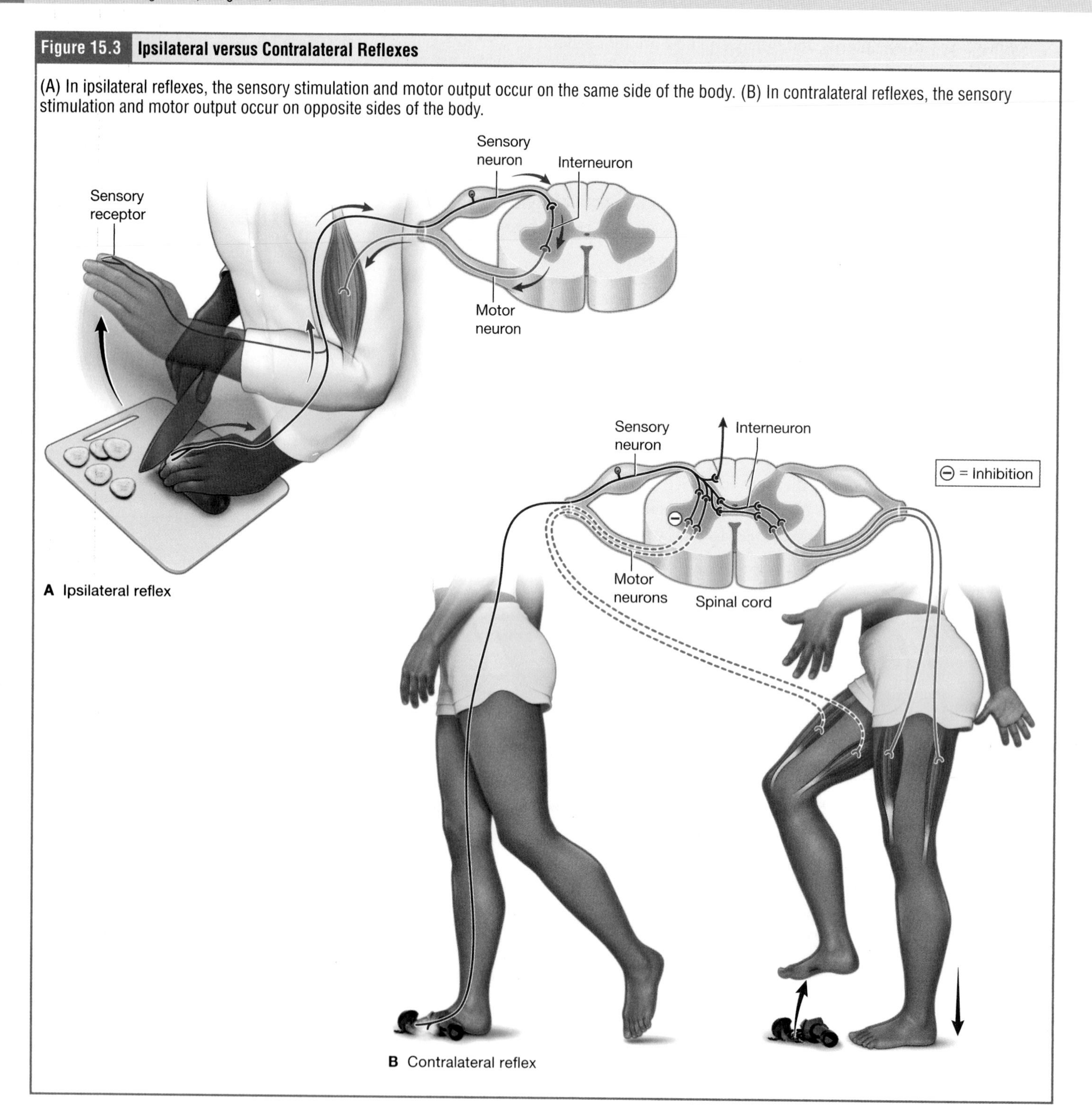

We can also describe reflexes functionally. These categories are: withdrawal, stretch, tendon, and crossed-extensor reflexes. An example of a **withdrawal reflex** occurs when you step on a painful stimulus, such as a tack or a sharp rock. The nociceptors that are activated by the painful stimulus activate the motor neurons responsible for contraction of the tibialis anterior muscle, causing dorsiflexion of the foot or withdrawal of the foot from the painful stimulus (**Figure 15.4**).

LO 15.1.7

Another type of reflex is a **stretch reflex**. In this reflex, when a skeletal muscle is stretched, a structure called a **muscle spindle** is activated. Muscle spindles are bundles of muscle cells that are wrapped with sensory neuron dendrites (**Figure 15.5**)

Figure 15.4 Withdrawal Reflex

The withdrawal reflex helps to protect the body from damage by quickly activating the muscles to pull away from a harmful stimulus.

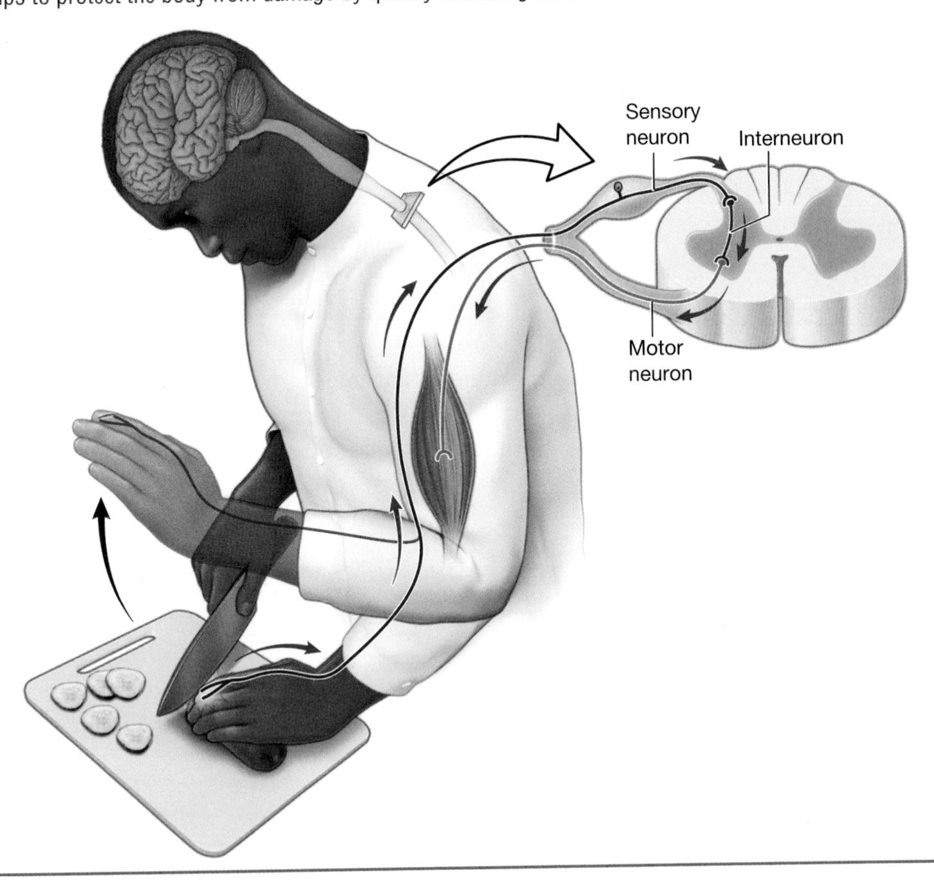

Figure 15.5 Stretch Reflexes

In stretch reflexes, sensory neurons detect stretch in a muscle (Step 1) and transmit a signal into the CNS (Step 2). They activate a motor neuron (Step 3), which causes muscle contraction to reverse the stretch (Step 4).

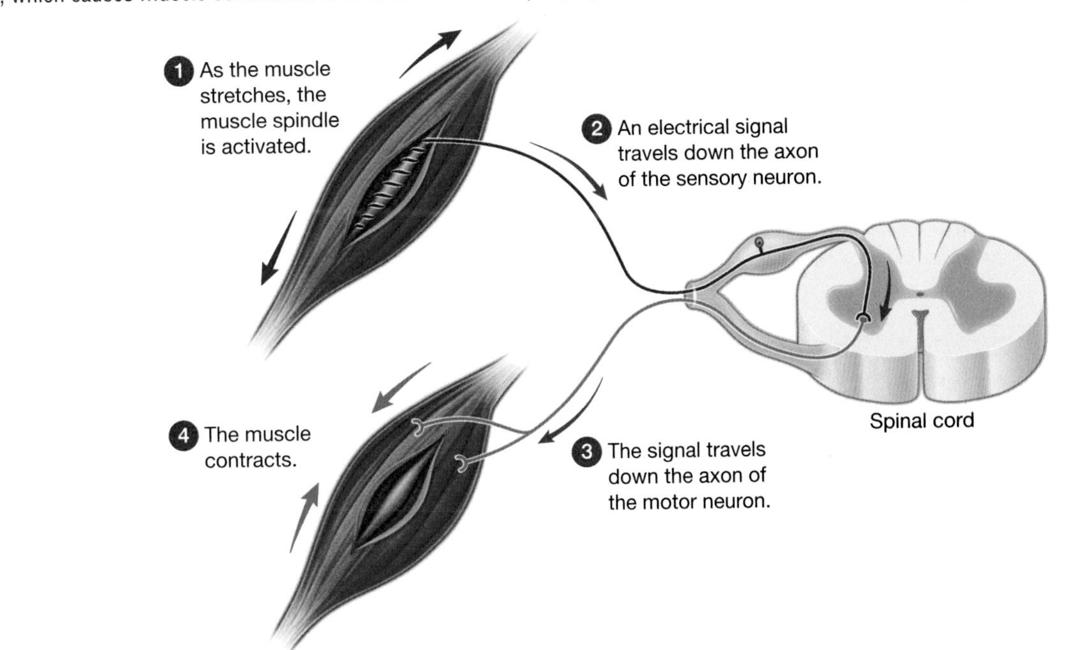

1 As the muscle stretches, the muscle spindle is activated.
2 An electrical signal travels down the axon of the sensory neuron.
Spinal cord
3 The signal travels down the axon of the motor neuron.
4 The muscle contracts.

This receptor sends a signal into the CNS, and the motor neuron of this reflex arc will cause contraction of the muscle. The reflex helps to maintain muscles at a constant length. A common example of this reflex is the knee jerk that is elicited by a rubber hammer struck against the patellar ligament in a physical exam.

Tendon reflexes provide the opposite protection to stretch reflexes. Stretch reflexes work to protect the muscles during stretch, but **tendon reflexes** work to protect the muscles during contraction. During a tendon reflex, the tendon will become stretched during muscle contraction. Sensory receptors within the tendon signal into the CNS and synapse on an interneuron. The interneuron forms a synapse with the motor neuron, but this is an inhibitory synapse. Instead of causing an action potential within the motor neuron, the interneuron inhibits it through hyperpolarization, preventing action potentials and causing muscle relaxation (**Figure 15.6**).

Crossed-extensor reflexes are contralateral reflexes as described earlier and illustrated in Figure 15.3B. In these reflexes, activation of the opposite side of the body occurs, usually due to a painful stimulus that needs to be avoided.

Learning Connection

Chunking

For each of the reflexes described (for example, a tendon reflex), describe it with as many reflex terms as you can. Monosynaptic or polysynaptic? Ipsilateral or contralateral?

Figure 15.6 Tendon Reflexes

Tendon reflexes respond when the tendon that connects a muscle to bone is overly stretched due to contraction of the muscle (Step 1). The sensory neuron transmits electrical signals into the CNS (Step 2), which causes an inhibitory interneuron to inhibit the motor neuron (Step 3). As the motor neuron stops firing signals, the muscle relaxes (Step 4).

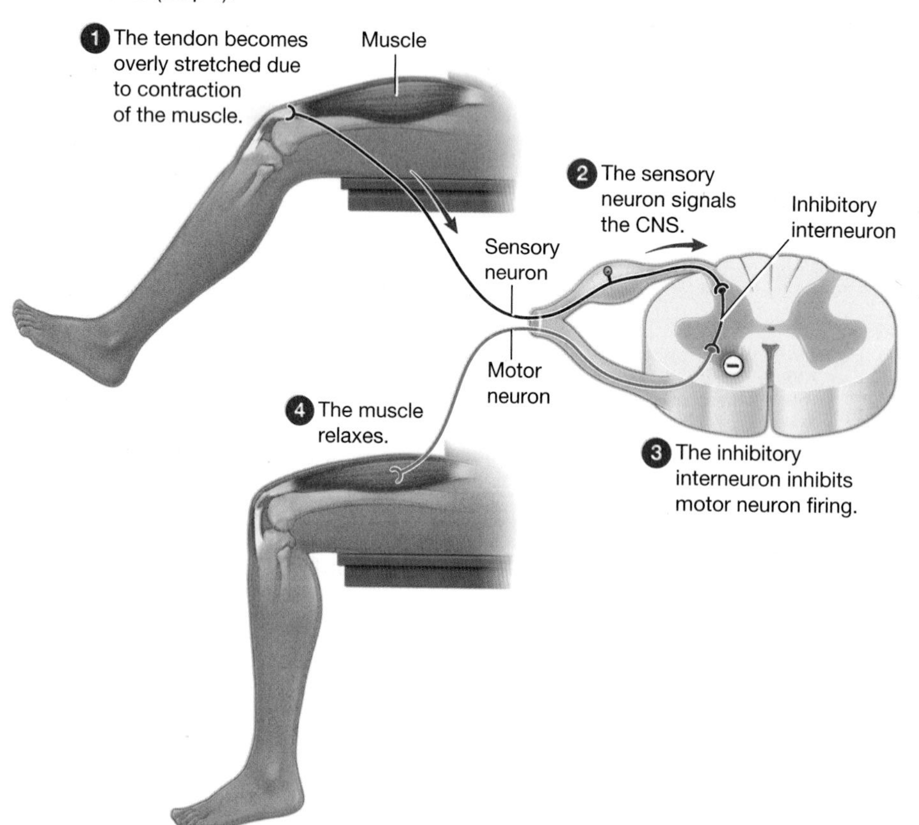
1 The tendon becomes overly stretched due to contraction of the muscle.
Muscle
2 The sensory neuron signals the CNS.
Inhibitory interneuron
Sensory neuron
Motor neuron
4 The muscle relaxes.
3 The inhibitory interneuron inhibits motor neuron firing.

Learning Check

1. When something is hot, the body has a defensive reflex that pulls your body part away from the hot source. If you touch a hot pan with your finger, you will pull away immediately. Which of the following is responsible for detecting heat?
   a. Sensory receptor
   b. Sensory neuron
   c. Motor neuron
   d. Effector
2. When something is hot, the body has a defensive reflex that pulls your body part away from the hot source. If you touch a hot pan with your finger, you will pull away immediately. Which of the following structures sends a signal to the arm muscles to pull away from the heat source?
   a. Sensory receptor
   b. Sensory neuron
   c. Motor neuron
   d. Effector
3. When something is hot, the body has a defensive reflex that pulls your body part away from the hot source. If you touch a hot pan with your finger, you will pull away immediately. Which of the following accurately describe this reflex? Please select all that apply.
   a. Somatic
   b. Visceral
   c. Withdrawal
   d. Stretch
4. Which of the following describe a reflex where the sensory neuron synapses with the motor neuron directly in the spinal cord? Please select all that apply.
   a. Monosynaptic
   b. Polysynaptic
   c. Spinal
   d. Cranial

15.2 Sensory Receptors

Learning Objectives: By the end of this section, you will be able to:

15.2.1 Define sensory receptor.

15.2.2 Define transduction, perception, sensation, and adaptation.

15.2.3 Distinguish between tonic and phasic receptors.

15.2.4 Compare and contrast the types of sensory receptors based on the type of stimulus (i.e., thermoreceptor, photoreceptor, chemoreceptor, baroreceptor, nociceptor [pain receptor], mechanoreceptor).

15.2.5 Compare and contrast the three types of sensory receptors, based on their stimulus origin (i.e., exteroceptors, interoceptors [visceroceptors], proprioceptors).

LO 15.2.1

The sensory receptor is the cell or structures that detect sensations. A receptor cell is changed directly by a stimulus; more specifically, its membrane potential is changed by the stimulus. In Chapter 13 we discussed that neurons generally receive input at their dendrites or cell bodies, and then send information along their axons. The sending of the signal—an action potential—occurs due to the opening and closing of voltage-gated ion channels that allow ions to move across the membrane. The movement of charged particles is electricity, and so these action potential signals are electrical signals. They begin when the cell body or dendrites receive a stimulus or multiple stimuli. The stimulus causes the opening of either mechanically gated or chemically gated ion channels. The opening of these channels initiates the electrical signal that eventually becomes the action potential.

With this in mind, we can now learn about sensory reception. If sound stimuli are vibrations, we can understand that those vibrations cause mechanically gated ion channels to open. If taste stimuli are chemicals, we can now understand that those chemicals cause chemically gated ion channels to open. These stimuli generate the graded potential that, if large enough, leads to an action potential and a signal sent from into the CNS about the stimulus. Because receptor cells are able to convert one type of stimulus—for example, vibration—into an electrical signal, they are often referred to as **transducers**. *Transduction* describes the conversion of one form of energy into another.

LO 15.2.2

Sensory receptors are named *receptors* because they receive sensory stimuli; they are called *transducers* because they transduce the information into an electrical signal. However, they do not automatically provide sensation. **Sensation**, the conscious awareness of stimuli, requires that the information be received and interpreted by the brain. The brain is able to piece together three key facets of information about the stimulus. The brain is able to understand the type of stimulus—for example, touch and not pain—based on the type of neuron that is sending signals. Likewise, the brain can combine information from more than one type of receptor—for example, painful touch. The brain is able to understand the location that the stimulus is coming from due to the anatomical specificity of the somatosensory cortex. Sensory information from touch receptors in the arm and leg is transmitted through different chains of neurons that end in different locations in the somatosensory cortex. Lastly, the brain is able to understand the intensity of the stimulus. This communication into the CNS is particularly fascinating given the simplicity of neuronal function. Action potentials are often referred to as "all-or-none" because, after depolarizing to threshold, they nearly always reach the same amplitude and follow the same series of events. How, then, can a sensory neuron impart the intensity of the stimulus to the nervous system if it cannot fire larger action potentials? The answer is that intensity is communicated by the frequency of action potentials. More frequent firing indicates a stronger stimulus (see the "Anatomy of a Sensation" feature).

Student Study Tip

Remember graded potentials adding up in summation when we summed muscle twitches? That same principle is what we are looking at again.

Have you ever noticed that you stop sensing something? For example, when you walk into a home for the first time you may be keenly aware of the smells of that home (especially if it is not your home). Perhaps you smell the onions that are cooking on the stove, or you can smell that the rooms were recently painted, or the odor of the dog that lives in the home. But when you check in with yourself 10 minutes later, it is likely that not only have you stopped paying attention to the smells; you might not even be able to smell them again even if you tried. The phenomenon of **adaptation**—becoming less sensitive to a constant stimulus—allows for the nervous system to become aware of a change in stimulus (such as stepping into a new place) and then move on to prioritizing other information. This decrease in sensitivity is useful when it comes to some sensory information; it allows the brain to fine-tune its focus. Your brain has already taken in an interpreted the smells of the house; it does not need to keep that in the forefront of your consciousness. The sensory receptors that adapt quickly are called **phasic receptors**. There is some sensory information, however, that your brain must constantly monitor. For example, proprioception—your sense of where your body is in space—must be constantly monitored and adjusted. Imagine you raised your arm in class to answer a question; it would not be good if your brain adapted to the sensation of your arm being above your head and then forgot to lower it. Proprioceptors, as well as some other sensory receptors, are **tonic receptors**; they do not adapt (or adapt very slowly) even when the stimulus remains constant. **Figure 15.7** illustrates the pattern of action potentials fired by phasic and tonic receptors during equal and constant stimuli.

LO 15.2.3

Learning Connection

Micro to Macro

Adaptation involves a sensory neuron that stops or slows its rate of neurotransmitter release even though it is still receiving graded potentials. What are the steps between graded potential generation and neurotransmitter release? Which steps might *not* be occurring during adaptation to lead to the reduced neurotransmitter release?

Anatomy of...

A Sensation

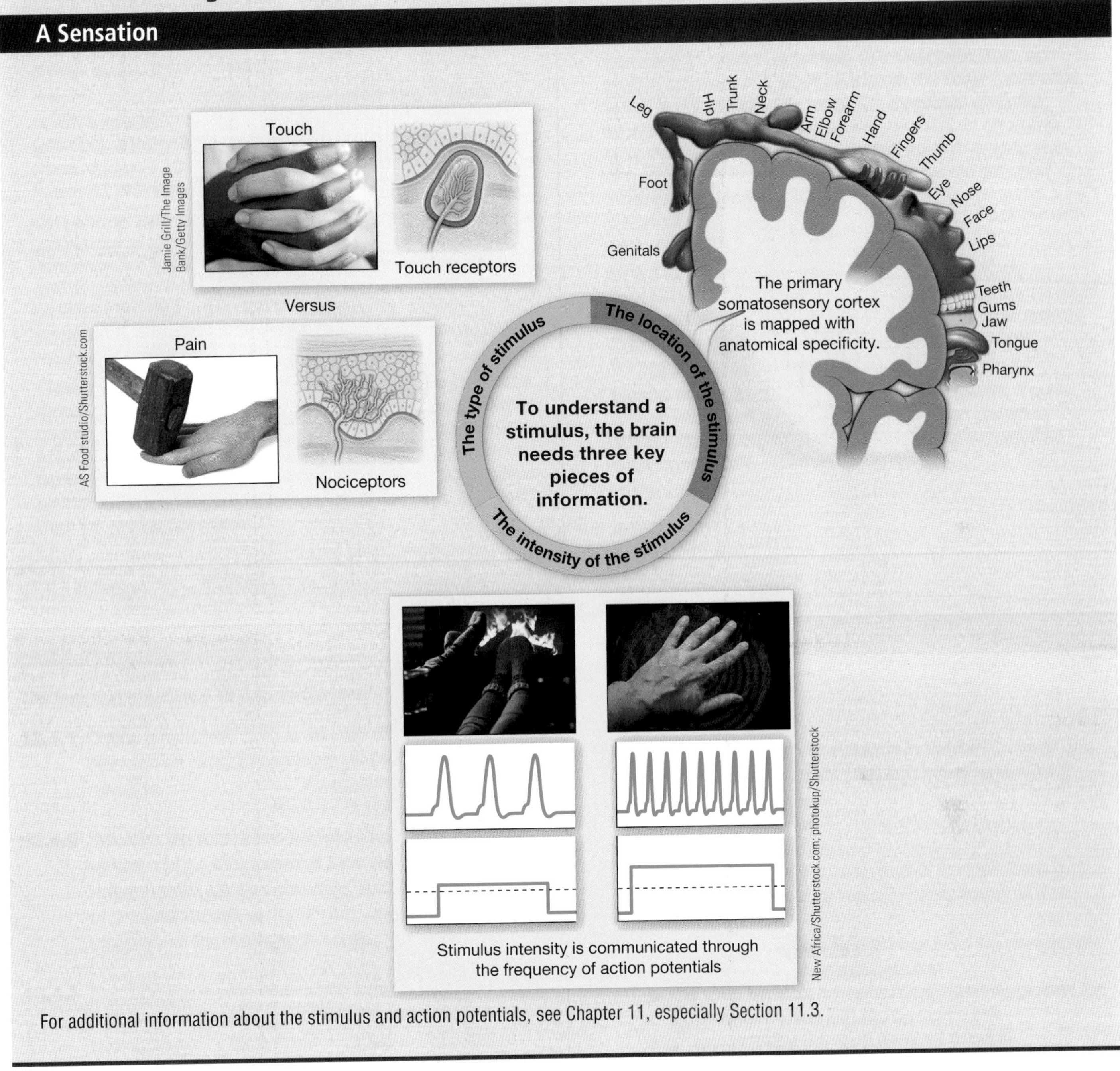

For additional information about the stimulus and action potentials, see Chapter 11, especially Section 11.3.

LO 15.2.4

The structure of each sensory receptor cell is specialized for its function; different types of stimuli are sensed by different types of receptor cells. Chemical stimuli, such as ions and macromolecules, are received by **chemoreceptors**. As you will see, some other types of receptors can respond to chemicals, but chemoreceptors are sensory receptors that send the CNS information about chemicals. For example, taste and smell are both senses that respond to chemical stimuli. **Osmoreceptors** are one type of chemoreceptor; these sensory receptors respond to solute concentrations of body fluids.

Figure 15.7 Adaptation

Peripheral sensory neurons may adapt—that is, stop sending signals about sensory stimuli—rapidly or slowly. Neurons that adapt rapidly are called *phasic receptors*; neurons that adapt slowly are called *tonic receptors*.

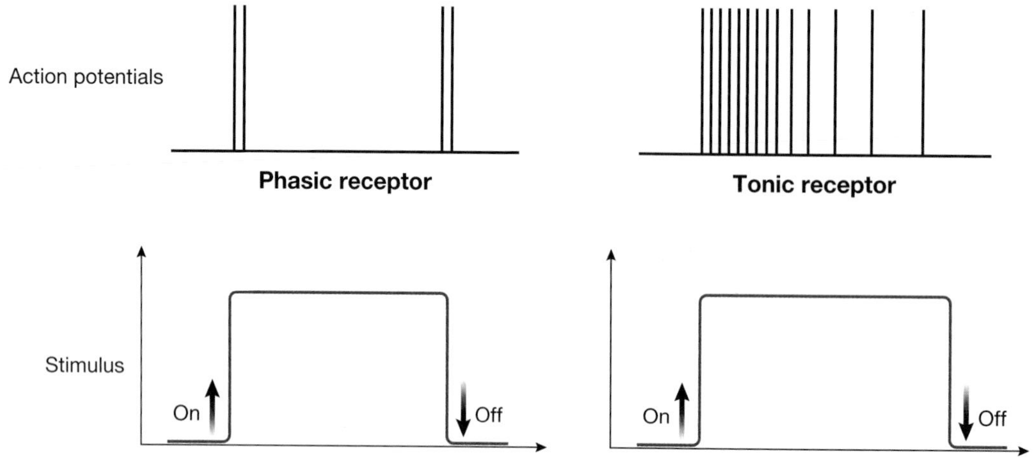

Action potentials

Phasic receptor

Tonic receptor

Stimulus

On

Off

On

Off

Thermoreceptors respond to changes in temperature; their ion channels open or close within certain temperature ranges. Physical stimuli, such as pressure and vibration, as well as the sensation of sound and body position (balance), are interpreted through **mechanoreceptors**.

Baroreceptors are a type of mechanoreceptor with a specialized structure that allows them to respond to pressure. Baroreceptors are important in monitoring external as well as internal pressure.

Photoreceptors are a unique population of receptor that is found only in the eye. These receptors respond to photons (particles) of light.

The last form of sensory receptor, **nociceptors**, are specialized receptors for pain. Somatic pain—the pain felt, for example, when you scrape your knee—is primarily a chemical sense that interprets chemicals released when cells and tissues are damaged. However, visceral pain—for example, damage to an internal organ—is primarily a mechanical sense. Stretching, tearing, or damage to internal tissues and cells will trigger these nociceptors to fire signals.

LO 15.2.5

Sensory receptors can also be classified by the location—internal or external—from which they receive their stimuli. For example, a chemoreceptor such as a taste receptor can be classified as an exteroceptor because it receives stimuli from outside the body. A chemoreceptor can also be an osmoreceptor; it is receiving stimuli from inside of the blood vessels. An **exteroceptor** is a receptor reacts to stimuli in the external environment, such as the somatosensory receptors that are located in the skin. An **interoceptor** is one that interprets stimuli from internal organs and tissues. Finally, a **proprioceptor** is a receptor that receives stimuli from a moving part of the body, such as a muscle or joint. Proprioceptors sense limb position and movement so that the brain can accurately move our body parts without looking at them. For example, when a baby crawls, it needs to move its arms and legs while looking forward (or perhaps for you, it's more that you are able to walk while looking at your phone); proprioception provides the brain with the information about where the limbs are at all times.

exteroceptor (outer ear)

The last way we can classify receptors is based on their distribution. Receptors found all over the body are called *general sense receptors*; receptors for the senses limited to the head are *special sense receptors*.

Learning Check

1. Which of the following informs the brain of the intensity of a stimulus?
   a. Magnitude of action potentials
   b. Frequency of action potentials
   c. Duration of action potentials
   d. Length of the action potentials
2. Which of the following types of sensory receptors allows you to detect pain?
   a. Nociceptor
   b. Chemoreceptor
   c. Thermoreceptor
   d. Mechanoreceptor
3. Which of the following types of receptors do not adapt to continuous stimulation?
   a. Phasic receptors
   b. Tonic receptors
   c. Chemoreceptors
   d. Baroreceptors
4. The kidneys have receptors that detect the concentration of ions within the blood. If the concentration of ions is too high, it will filter those ions out and excrete them in the urine. Which of the following best describes the sensory receptors found in the kidney?
   a. Osmoreceptors
   b. Mechanoreceptors
   c. Baroreceptors
   d. Photoreceptors

15.3 General Senses

Learning Objectives: By the end of this section, you will be able to:

15.3.1 Compare and contrast the location, structure, and function of the different types of tactile receptors (e.g., tactile [Merkel] corpuscle, lamellated [Pacinian] corpuscle).

15.3.2* Define and connect the concepts of receptive field and acuity.

15.3.3* Explain how the anatomical pathway of pain sensation can confuse pain information in referred pain and phantom pain.

* Objective is not a HAPS Learning Goal.

15.3a Touch

The general senses, which can be sensed everywhere from your fingers to your eyeballs, include pressure, vibration, light touch, tickle, itch, temperature, pain, and proprioception. The receptors of these general senses are not associated with a specialized organ but are spread throughout the body in a variety of organs. Many of the receptors are located in the skin, but receptors are also found in muscles, tendons, joint capsules, ligaments, and in the walls of visceral organs.

LO 15.3.1

In broad terms, we can divide general sense receptors into those that are anatomically simple and those that are slightly more complex. **Unencapsulated receptors** are anatomically simple, just the dendrites of the sensory neuron enmeshed in the surrounding tissue. **Encapsulated receptors** are a bit more anatomically complex; the dendrites of the sensory neuron are wrapped in such a way that enables their function (**Table 15.1**). Free nerve endings are the simplest of unencapsulated receptors. Both pain and temperature signals are transduced by free nerve endings. Temperature receptors are called *thermoreceptors*; their ion channels open at particular temperature ranges, so one thermoreceptor may only respond to hot temperatures and its neighbor may only respond to cold temperatures. The brain is able to assess overall temperatures by which combination of thermoreceptors are firing signals. Fascinatingly, thermoreceptors can also respond to certain chemicals. For example,

unencapsulated receptors (unencapsulated nerve endings)

encapsulated receptors (encapsulated nerve endings)

Table 15.1 The Distribution of Touch Receptors in the Skin

Encapsulated versus Unencapsulated Receptors	Name	Example	Also known as	Location(s)	Stimuli
Unencapsulated Receptors	Merkel cells		-----	Stratum basale	Low-frequency vibration
	Free nerve endings		-----	Papillary layer of dermis and/or epidermis	Pain, temperature, mechanical deformation
	Hair follicle receptors		Root hair plexus	Wrapped around hair follicles in the dermis	Movement of hair
Encapsulated Receptors	Lamellated corpuscle		Pacinian corpuscle	Deep dermis and hypodermis	Deep pressure
	Tactile corpuscle		Tactile epithelial cells, Meissner's corpuscles	Dermal papillae. Highly represented in sensitive locations such as lips, hands, genitals, nipples, eyeballs	Light touch
	Bulbous corpuscle		Ruffini's corpuscle	Dermis and hypodermis	Stretch

the sensation of heat associated with spicy foods involves **capsaicin**, the active molecule in hot peppers such as jalapenos. Capsaicin molecules bind to an ion channel in heat-sensing nociceptors, bringing them to threshold and causing an action potential. Remember that the brain interprets the type of stimulus based on the type of sensory receptor sending signals. Therefore, the brain understands that a warm temperature receptor is firing a signal and interprets this as an increase in temperature, even though no temperature change has occurred.

If you drag your finger across a textured surface, the skin of your finger will vibrate. Such low-frequency vibrations are sensed by mechanoreceptors called **Merkel cells**, another type of unencapsulated receptor. Merkel cells are located in the stratum basale of the epidermis. Hair follicle receptors are the third type of unencapsulated receptor. These nerve endings detect the movement of hair at the surface of the skin, such as when an insect walks along your skin.

Merkel cells (Merkel-Ranvier cells, tactile epithelial cells)

Deep pressure and vibration are transduced by **lamellated corpuscles**, which are encapsulated receptors found deep in the dermis, or subcutaneous tissue. Light touch is transduced by the encapsulated endings known as **tactile corpuscles**, which are found in the dermal papillae. Stretching of the skin is transduced by stretch receptors known as **bulbous corpuscles**. Bulbous corpuscles are also present in joint capsules, where they measure stretch in the components of the skeletal system within the joint. Both the structure of the receptor—especially if it is encapsulated or unencapsulated—and its location within the skin contribute to its sensory function. Whereas bulbous corpuscles and lamellated corpuscles function in coarser sensations such as pressure, finer sensations such as discriminating touch, light touch, and temperature are detected closer to the surface and with less elaborately wrapped receptors such as free nerve endings and tactile corpuscles (see **Figure 15.8**).

lamellated corpuscles (Pacinian corpuscle)

bulbous corpuscles (Ruffini ending, Ruffini corpuscle)

tactile corpuscles (Meissner's corpuscles)

LO 15.3.2

Each touch receptor has a certain amount of tissue space—its **receptive field** (**Figure 15.9A**)—within which it receives information. Any stimulus within that receptive field can cause a sufficient depolarization of the receptor so that the receptor signals to the brain. Remember, however, that the brain can only understand that this touch receptor has fired a signal; it cannot determine from where within the receptor's receptive field the stimulus came. Even if there are multiple stimuli within a receptive field, the brain cannot perceive any differences; all it can perceive is that this neuron

receptive field (sensory space)

Figure 15.8 The Distribution of Touch Receptors in the Skin

Touch receptors for acuity are found more superficially; pressure and vibration are sensed deeper in the skin.

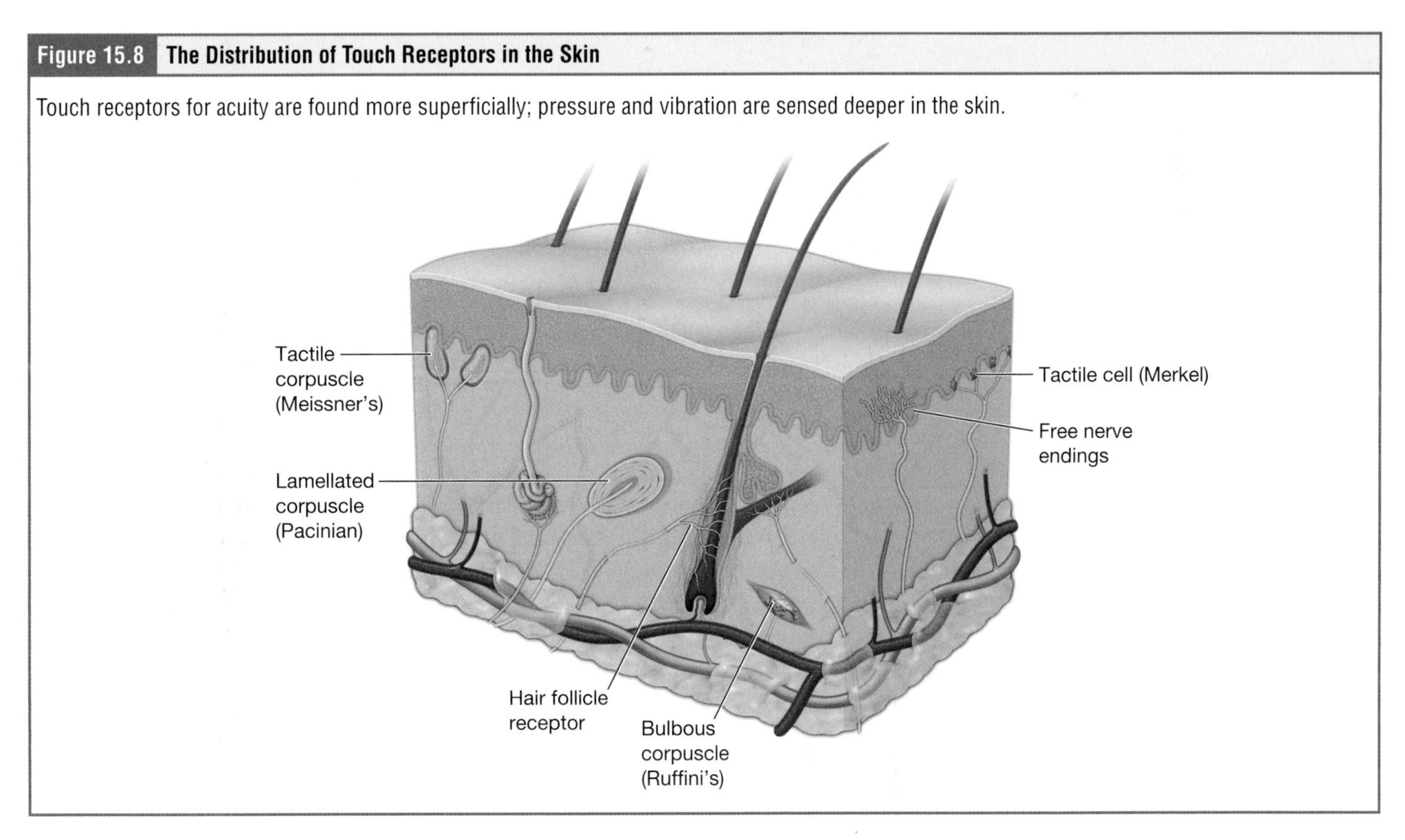

Figure 15.9 Receptive Fields

(A) The receptive field of a sensory neuron is the area from which its receptors can detect stimuli. (B) When a touch stimulus is applied anywhere within the receptive field of a sensory neuron, the neuron sends an electrical stimulus to the CNS. (C) The number of electrical signals sent to the CNS from multiple touch stimuli is determined by the size and density of the receptive fields and sensory neurons present. Touch stimuli activating small, densely-packed receptive fields from 3 adjacent sensory neurons cause 3 electrical signals to be sent to the CNS. Touch stimuli activating 3 regions within the same large receptive field of a sensory neuron cause 1 electrical signal to be sent to the CNS.

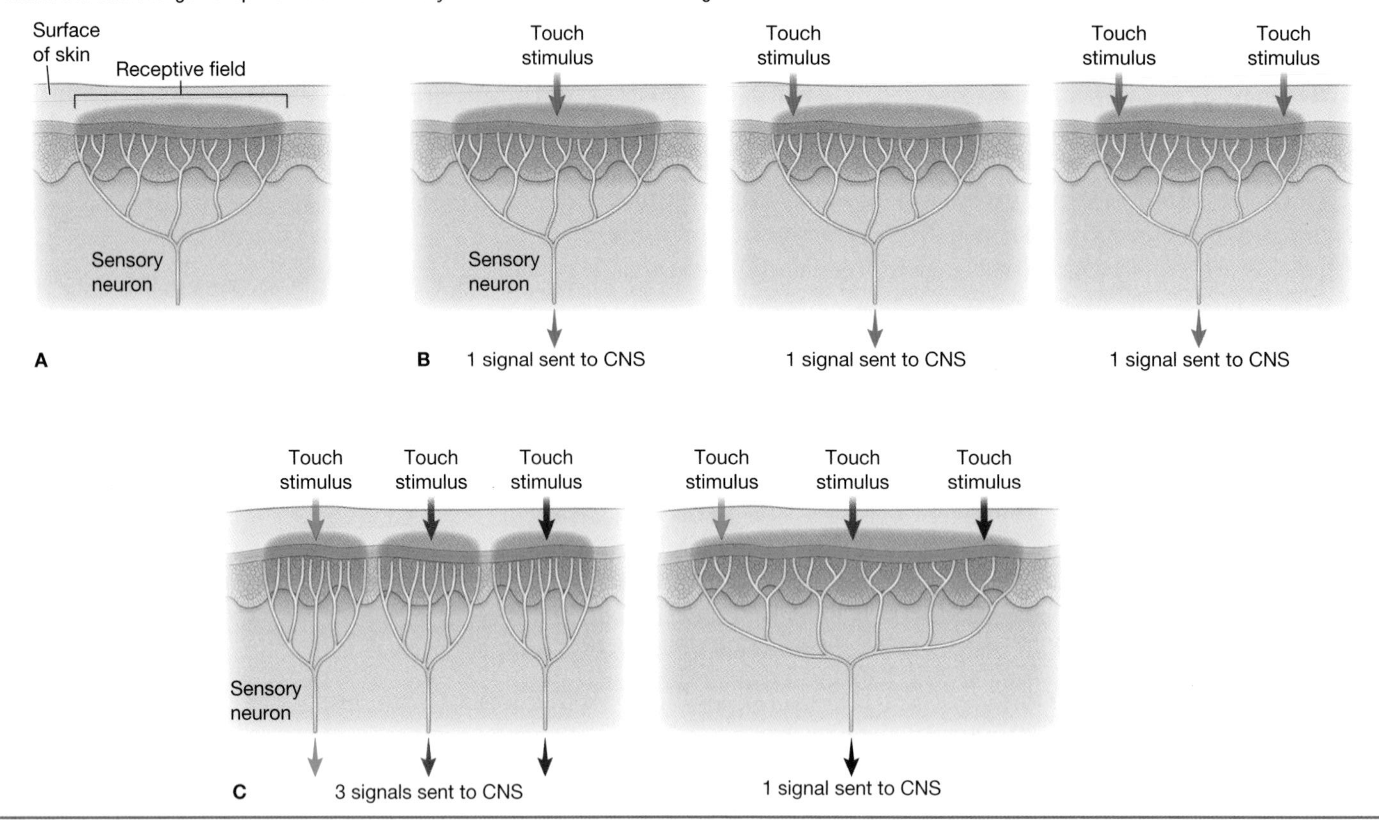

is receiving stimulation (**Figure 15.9B**). An area of the skin that has a higher density of small receptive fields would therefore be able to communicate more nuanced information to the brain than an area of the skin that had fewer, larger receptive fields (**Figure 15.9C**). The term **acuity** is used to describe how accurately our brains can understand a stimulus. Whereas touch receptors for acuity are found more superficially, those for pressure and vibration are sensed deeper in the skin. A higher density of smaller receptive fields lends more acuity to our sense of touch. Whereas this pattern is more common in places like the face and hands, places such as the back more commonly have fewer, larger receptive fields and poor touch acuity.

Student Study Tip

This follows the same principle as motor neurons and muscle acuity. Drawing this connection can help you understand the significance of large and small receptive fields in touch.

Pain While unpleasant, our ability to feel pain is incredibly important to our survival. Pain is a signal to stop before further damage to the cells and tissues of the body can occur. The stimulus for pain is damage to tissues or cells. Nociceptors are the receptors for this stimulus, and they transmit pain information to the CNS and up to the brain for conscious processing in chains of three neurons like all other sensory stimuli (**Figure 15.10**). However, somatic pain and visceral pain can sometimes share the same ascending pathway, leading to a lack of acuity in the brain as to where the pain is coming from. Because somatic pain is so much more common than pain from within the organs, the brain may assume that visceral pain is actually somatic in origin. This idea, that somatic pain is perceived when the stimulus is visceral, is called **referred pain**. Referred pain is important for clinicians to learn about so that they can investigate all

LO 15.3.3

Figure 15.10 Referred Pain

Visceral and somatic pain pathways often share components, leading to some confusion over the source of pain stimuli when pain is visceral in origin. In this case, pain originating in the heart is perceived as being derived from the left arm.

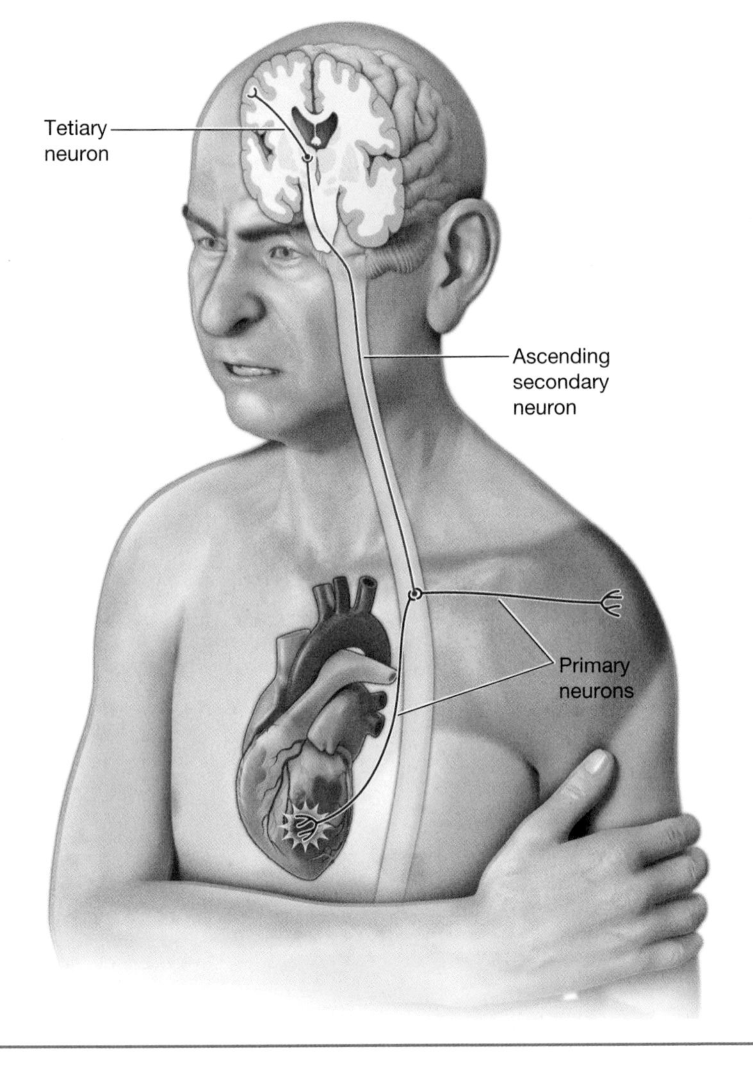

of the potential sources of tissue damage. For example, among biological males, it is common for tissue damage of the heart to be perceived as shoulder pain, so if a patient in the emergency room complains of shoulder pain and also has shortness of breath or other symptoms of cardiac issues, testing for damage to the heart may be a good idea. Biological females, however, are much less likely to experience referred pain with a heart attack but are more likely to experience nausea and vomiting, so even if a patient arrives at the ER without shoulder pain, cardiac damage is still a possibility.

Examine the pain pathway illustrated in Figure 15.10. Now imagine that this patient lost their arm in a car accident. The majority of the primary neuron from their arm would be lost as well, but the rest of the anatomical pathway would still exist in the CNS. This anatomical setup can lead to the continued perception of sensation from the missing limb after it is gone, which is called **phantom pain**. While pain is one possible sensation, other patients with missing limbs may perceive itchy sensations or even touch sensation. Sometimes these can be very specific. It can be psychologically taxing or even frustrating to the patient to continue to experience sensation long after the limb is no longer present.

substance P (SP)

DIGGING DEEPER: Endorphins and Opioids

Most nociceptors use a neurotransmitter called **substance P** to communicate to the secondary neuron in the pain pathway (see figure, lower right portion). The amount of substance P within the synapse will determine how intensely, and for how long, the pain is experienced. Most of these synapses have an additional neuron—an inhibitory neuron—that can inhibit the nociceptor and prevent or diminish its release of substance P. The neurotransmitter released by the inhibitory neuron is a family of molecules called **endogenous opioids**. Endogenous opioids or *endorphins*, as they are sometimes called, are released in a variety of physiological circumstances, including during intense exercise. Their release sometimes prevents pain from being fully felt at times; for example, a player might sprain their ankle during a soccer game, but not really feel the pain from the injury until they are resting after the end of the game. Other molecules can also bind to the endogenous opioid receptors. These molecules, collectively called *opioids*, have been used both medicinally and recreationally for millennia. Examining the illustration, it is easy to understand how opioids help relieve pain. Opioid receptors are expressed not only on nociceptors but also in regions of the brain associated with addiction and respiration. Opioids are among the most addictive substances in the world. It is unfortunately very common to develop an addiction after using opioids prescribed by a doctor to relieve pain. Opioid addiction is a severe public health crisis in the United States; it is imperative to treat those currently battling addiction and to find alternatives to opioid use in pain management. One hope as an opioid alternative is acupuncture, a form of medicine that involves inserting thin needles into specific locations on the patient's body. Acupuncture is thought to work by causing the body to release endorphins, utilizing the body's own mechanism for pain relief.

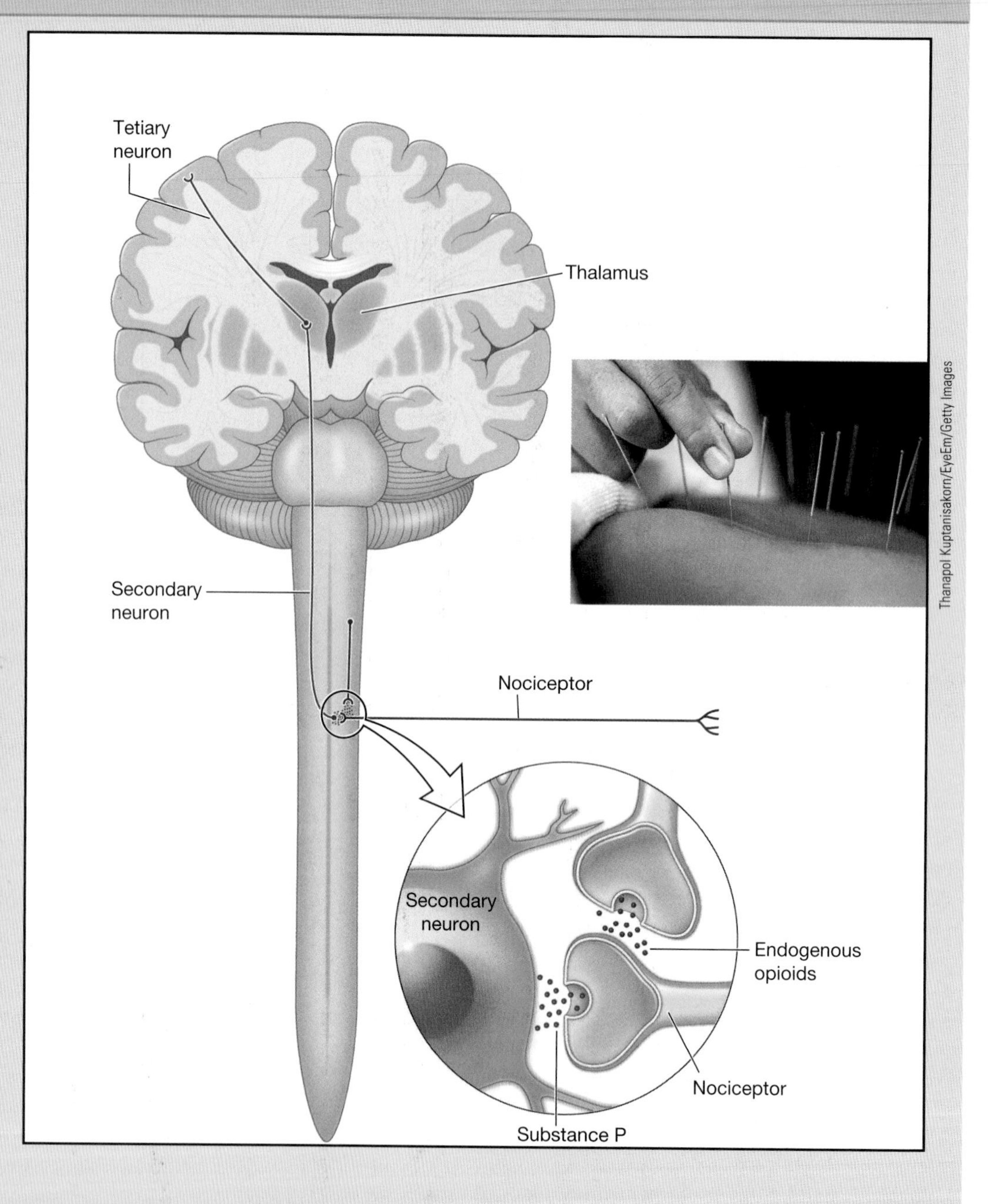

endogenous opioids (endorphins)

Learning Check

1. Which of the following receptors are unencapsulated receptors? Please select all that apply.
   a. Merkel cells
   b. Thermoreceptors
   c. Lamellated corpuscles
   d. Bulbous corpuscles
2. Tactile corpuscles detect ______ while lamellated corpuscles detect ______.
   a. Vibration; light touch
   b. Stretch; pressure
   c. Light touch; pressure
   d. Pain; stretch
3. The thigh is an area that has large receptive fields. What does this suggest about the acuity in the thigh?
   a. High acuity
   b. Low acuity

15.4 Special Senses

Learning Objectives: By the end of this section, you will be able to:

15.4.1 Identify and describe the accessory eye structures (e.g., conjunctiva and lacrimal apparatus).

15.4.2 Identify the tunics of the eye and their major components (e.g., cornea, sclera, iris, ciliary body), and describe the structure and function of each.

15.4.3 Describe the lens and its role in vision.

15.4.4 Identify and describe the anterior and posterior cavities of the eye and their associated humors.

15.4.5 Describe phototransduction (i.e., how light activates photoreceptors) and explain the process of light and dark adaptation.

15.4.6 Trace the path of light as it passes through the eye to the retina, and describe which structures are responsible for refracting the light rays.

15.4.7 Compare and contrast the functions and locations of rods and cones.

15.4.8 Relate changes in the anatomy of the eye to changes in vision.

15.4.9 Given a factor or situation (e.g., macular degeneration), predict the changes that could occur in the affected sense and the consequences of those changes (i.e., given a cause, state a possible effect).

15.4.10 Trace the signal pathway from the retina through the optic nerve, optic chiasm, optic tract, and to the various parts of the brain.

15.4.11 Explain how the sense organs relate to other body organs and systems to maintain homeostasis.

15.4.12 Classify gustatory receptor cells based on the type of stimulus (i.e., modality).

15.4.13 Describe the primary taste sensations.

15.4.14 Identify and describe the location and structure of taste buds.

15.4.15 Explain the process by which tastants activate gustatory receptors.

15.4.16 Trace the path of gustation from gustatory receptors through specific cranial nerves to various parts of the brain.

15.4.17 Identify and describe the composition and location of the olfactory epithelium.

15.4.18 Explain the process by which odorants activate olfactory receptors.

15.4.19 Trace the path of olfaction, from the olfactory receptors to the initiation of an action potential in the olfactory nerves, through the olfactory bulb, the olfactory tract, and to the various parts of the brain.

15.4.20 Classify the receptor cells for hearing based on the type of stimulus (i.e., modality).

15.4.21 Identify the macroscopic structures of the outer (external), middle, and inner (internal) ear and their major components (e.g., auditory ossicles, auditory [pharyngotympanic] tube), and describe the structure and function of each.

15.4.22 Identify and describe the microscopic structures within the inner (internal) ear (e.g., maculae, cristae ampullares, spiral organ [of Corti]).

15.4.23 Explain the process by which an action potential is generated at the spiral organ (of Corti).

15.4.24 Trace the path of sound from the external ear to the inner ear, including where sound is amplified.

15.4.25 Explain how the structures of the ear enable differentiation of pitch, intensity (loudness), and localization of sounds.

15.4.26 Trace the signal path from the spiral organ (of Corti) to the cochlear branch of the vestibulocochlear nerve (CN VIII) and to the various parts of the brain.

15.4.27 Compare and contrast static and dynamic equilibrium.

15.4.28 Describe the structure of a macula and its function in static equilibrium.

15.4.29 Describe the structure of an ampulla and its function in dynamic equilibrium.

15.4.30 *Given a disruption in the structure or function of one of the senses, (e.g., vertigo), predict the possible factors or situations that might have caused that disruption (i.e., given an effect, predict the possible causes).

15.4.31 Trace the signal path from the maculae and cristae ampullares to the vestibular branch of the vestibulocochlear nerve (CN VIII) and to the various parts of the brain.

15.4a Vision

Vision is the special sense of sight that is based on the transduction of light by the photoreceptors of the eyes. In order to understand sight fully, or at least to the extent that scientists currently understand it, we will discuss the anatomy of the eyes, the physiology of photoreception, and the interpretation of visual information in the brain. Let's begin with eye anatomy.

Eye Anatomy Every structure within and around the eye either serves a function that contributes to physical protection or to vision. The bony orbits surround the eyeballs, protecting them and anchoring the soft tissues of the eye (**Figure 15.11**). The eyelids and eyelashes help to protect the eye from abrasions by blocking particles from landing on the surface of the eye. The eyebrows help to prevent sweat from running into the eyes from the forehead. The **conjunctiva** is a layer of stratified columnar epithelium rich with goblet cells, blood vessels, and sensory neurons. The conjunctiva extends over the white areas of the eye and folds in on itself to line the inside of the eyelids, serving a sensitive and protective role for these structures. The mucus of the conjunctiva helps to keep the surface of the eyes moist. Contributing to the moisture of the eyes, tears are produced by the **lacrimal glands**, located just above the lateral side of the eye (**Figure 15.12**). Tears produced by this gland flow across the eye and drain into the **lacrimal duct**, which sits in the medial corner of the eye. As the tears flow over the conjunctiva, they wash away

LO 15.4.1

Figure 15.11 The External Features of the Eye

The eye has hard and soft layers of protection. It sits within the hard, bony shell of the orbit, and is protected by the eyelids, eyebrows, and conjunctiva.

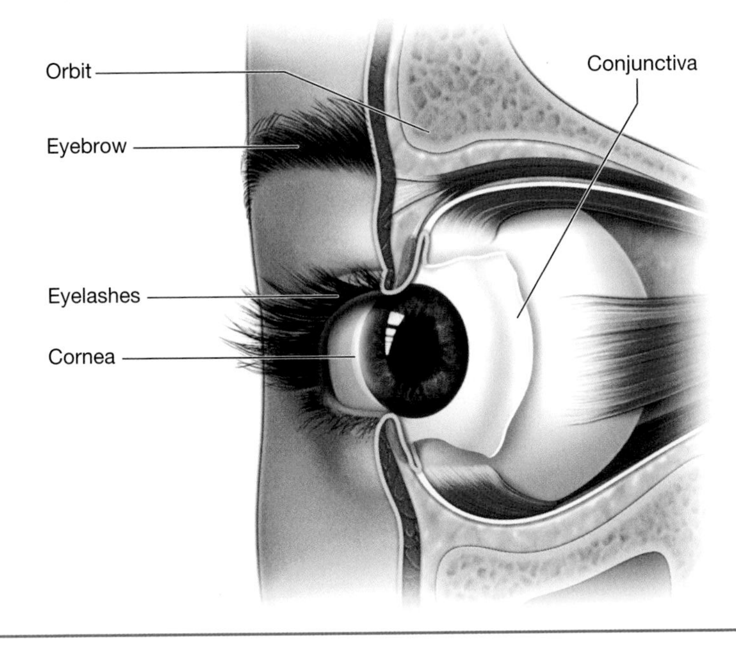

Figure 15.12 The Nasolacrimal Apparatus

The lacrimal gland produces tears on the lateral side of the eye. The tears sweep across the eye from the lateral to the medial side and are drained into the nose by the nasolacrimal duct.

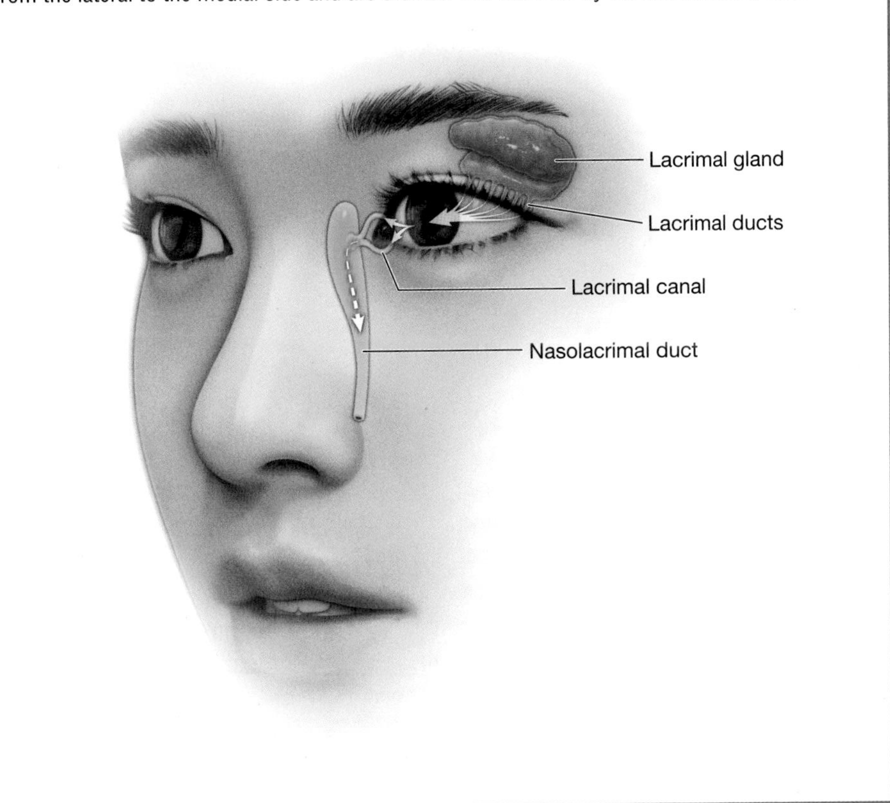

Student Study Tip

Your nose runs when you cry because the nasolacrimal duct empties tears into the nasal cavity after they have washed the surface of the eye. Thinking about your own bodily functions can help you remember this anatomy.

foreign particles. The lacrimal duct ends within the nasal cavity, so this production of tears keeps the mucous membranes of the nose moist as well. Tears are produced at a steady level throughout every day, but tear production can increase in certain environmental conditions, such as on a windy day, or during emotional responses including feeling sad or laughing. When the pace of tear production increases beyond the pace at which the tears can be drained by the lacrimal duct, they may spill over the bottom eyelid—we typically call this *crying*. Physiologists are not sure why we increase tear production in response to emotions, but one theory is that crying provides a silent cue to the people around you that you are in distress or need help. For example, imagine a group of early humans that are trying to evade a predator. If one of the humans twists their ankle and is having trouble walking, crying would be a mechanism for the other people around to understand that they are in pain and come to help without necessarily alerting the predator that one person is more vulnerable than the others.

The eye itself is a fluid-filled ball composed of three layers of tissue. The layers (in some sources they are called *tunics*), from outside to inside, are the fibrous layer, the vascular layer, and the neural layer.

LO 15.4.2

The outermost **fibrous layer** is made of dense connective tissue and provides protection and scaffolding to the parts of the eye within. The fibrous layer includes the white **sclera** and the clear **cornea** (see the "Anatomy of the Eye" feature). The sclera—the white of the eye—is continuous with the transparent cornea that covers the anterior

Anatomy of...

The Eye

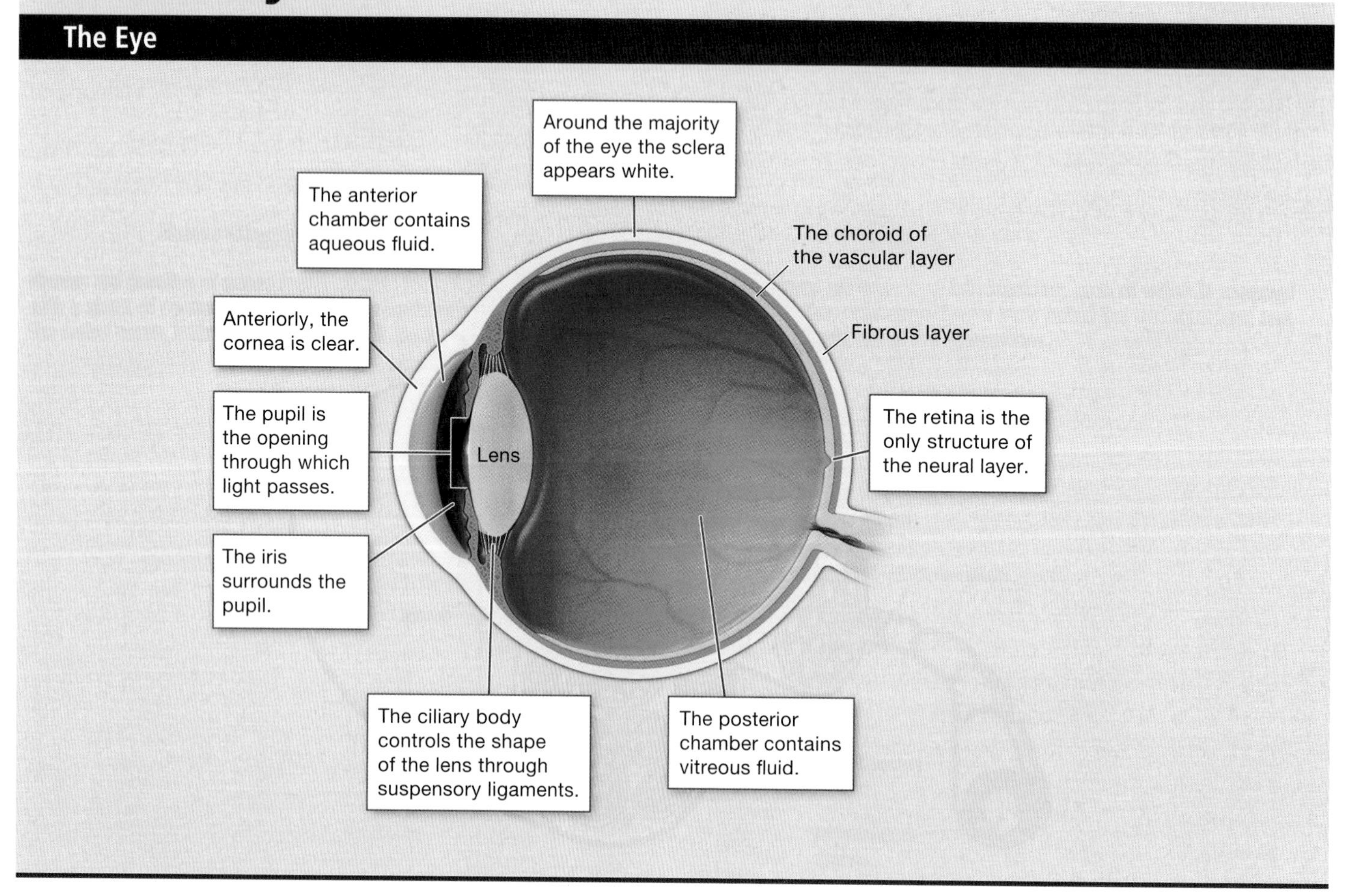

tip of the eye and allows light to enter the eye. The sclera and the cornea are both made of dense connective tissue, but the difference in collagen fiber orientation makes the sclera opaque and the cornea clear.

The middle **vascular layer** is also primarily connective tissue, but is rich with blood vessels and intrinsic muscles. There are three individual structures of the vascular layer: the choroid, the ciliary body, and the iris. The **choroid** is highly vascularized loose connective tissue that provides a blood supply to the eyeball. The choroid lines the majority of the eyeball, except anteriorly where the ciliary body and iris are found. The **ciliary body** is a muscle that is attached to the **lens** by **suspensory ligaments**. The lens is a clear disc through which light enters the eye. When the muscles within the ciliary body contract, they pull on the lens, distorting its shape and changing the angle at which light enters the eye. The adjustments of the lens are used to direct light to where it can be received with the most acuity. Overlaying the ciliary body, and visible when viewing the anterior eye, is the **iris**. When you identify the color of someone's eyes, you are speaking specifically of the iris. The iris is actually a smooth muscle that widens or narrows around the **pupil**, the hole at the center of the eye that allows light to enter. The iris constricts the pupil in response to bright light and dilates the pupil in response to dim light. The pupil is an opening in the center of the iris which, because it leads to the dark enclosed space posterior to the iris, appears from the outside as a black circle—much the same way that a window to a dark room appears black.

choroid (choroida, choroid coat)

LO 15.4.3

The innermost layer of the eye is the **neural layer**, or **retina**, which contains the nervous tissue responsible for photoreception.

neural layer (retina)

LO 15.4.4

The eye is also divided into two cavities: the anterior cavity and the posterior cavity. The *anterior cavity* is the space between the cornea and lens, including the iris and ciliary body. It is filled with a watery fluid called the **aqueous humor**. The *posterior cavity* is the space behind the lens that extends to the posterior side of the interior eyeball, where the retina is located. The posterior chamber is filled with a jellylike fluid called the **vitreous humor**.

LO 15.4.5

The retina is composed of several layers and contains specialized cells for transduction of light. The photoreceptor cells are the most posterior layer of the retina (**Figure 15.13**). They experience a change in their membrane potential when stimulated by light energy. This change in membrane potential alters the amount of neurotransmitter that the photoreceptor cells release onto **bipolar cells**. The bipolar cells are arranged in a layer just anterior to the photoreceptor cells. The bipolar cells in the retina have synaptic connections with **ganglion cells** of the most anterior retinal layer. This arrangement is counterintuitive; we might expect that the photoreceptor cells are located in the most anterior layer of the retina. However, the particles of light are able to traverse through the ganglion and biopolar cell layers to be received by the photoreceptor cells posteriorly. Some of the light is absorbed by these structures before the light reaches the photoreceptor cells. The axons of the ganglion cells, which lie in the innermost layer of the retina, converge along the anterior lining of the posterior chamber, and leave the eye as the **optic nerve** (see Figure 15.13).

Learning Connection

Chunking

Write every bolded eye structure from this section on a separate sticky note or scrap of paper. Mix them up. Now separate them into two piles—one for structures that light passes through and the other for structures that light does not pass through. With the pile that permits light, order them from anterior to posterior. Now reorder them for the order that light passes through them, starting with the most external structure and ending with the photoreceptor.

LO 15.4.6

LO 15.4.7

There are two types of photoreceptors—rods and cones—which differ slightly in the shape and significantly in function. The **rod photoreceptors** contain the pigment **rhodopsin**. The **cone photoreceptors** contain a variety of pigments called **opsins**, each of which is sensitive to a particular wavelength of light. *Wavelength* is the pattern in which a particle of light travels, and its wavelength determines its color. The pigments in human eyes are specialized in perceiving three different primary colors: red, green, and blue (**Figure 15.14**). Because we can only see the light that we have pigments for, we cannot see other forms of light—for example, ultraviolet light—because we do not have pigments that are sensitive to them.

Figure 15.13 Photoreceptor

The tissue of the retina has consists of three layers of cells. The most posterior layer consists of the photoreceptor cells themselves. Anterior to this layer are bipolar cells. The most anterior (bordering on the posterior cavity of the eye) are the ganglion cells. Light enters anteriorly through the pupil, and the photons of that light pass between the ganglion cells, then between the cells of the bipolar layer, and then finally to the photoreceptor cells.

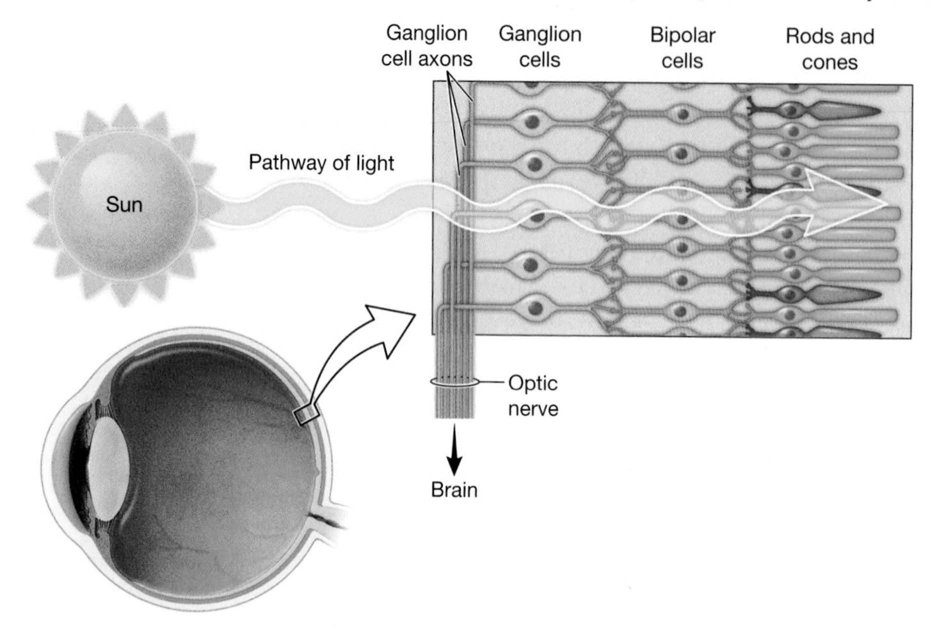

At the molecular level, the arrival of particles of light will cause changes in the photopigment molecules that lead to changes in membrane potential of the photoreceptor cells. A single unit of light is called a **photon**. In Figure 15.14 you can see that each type of cone accepts photons traveling within a very narrow range of wavelengths. Rods, however, accept photons traveling within a wider range of wavelengths. This means that rods cannot discriminate among different wavelengths of light and therefore can only tell the brain about the presence (white) or absence (black) of light.

Our bodies are amazingly versatile. We can survive in warm and cold temperatures, eat a feast, or fast for a day; we can also see in both very dim places—such as a movie

Figure 15.14 Comparison of Color Sensitivity of Cones

This is a comparison of the wavelength sensitivity of the four photopigments.

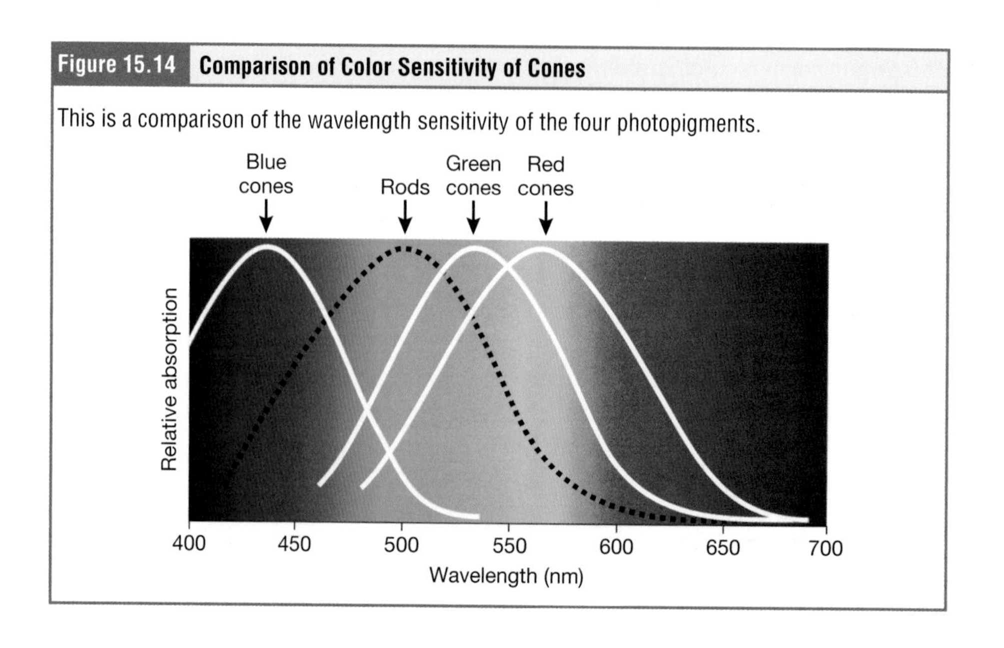

theater—and outside on a sunny day. But while we are able to see in both environments, we see differently in each. Outside on a sunny day, our cones are doing the work, taking in the multiple wavelengths of light and allowing us to see in vivid color and with strong acuity. In the bright light our rods bleach, meaning that they store rhodopsin for later use and cease responding to photons of light. Therefore, when we step into a dark space from a brightly lit one the rods are not immediately available for vision and initially all we see is darkness. The rods come back into function quickly, and within the first few minutes or so, we begin to see in the dark space. This process, called **dark adaptation**, can take up to 20 or even 30 minutes to complete, but eventually we come to see in the dark space. Our dark vision will always lack color vision and have less acuity than our vision in lightness. The opposite event, **light adaptation**, occurs as we step out from the dark space and into a brightly lit one. Immediately our irises constrict to reduce the amount of incoming light, and rods and cones both send signals initially, creating a confusing visual signal that we might call a glare. The rods turn off, or bleach, quickly and the cones adjust their signaling within the first few minutes outdoors.

Student Study Tip

Cones do the work when we Come outside, and Rods do the work when we Return.

DIGGING DEEPER: How Pigments Lead to Color Vision

Opsin pigments are actually transmembrane proteins that contain a cofactor known as **retinal**. Retinal is a hydrocarbon molecule related to vitamin A. When a photon hits retinal, the long hydrocarbon chain of the molecule is biochemically altered. Specifically, photons cause some of the double-bonded carbons within the chain to switch from a *cis* to a *trans* conformation in a process called **photoisomerization**. Before interacting with a photon, retinal's flexible double-bonded carbons are in the *cis* conformation. This molecule is referred to as 11-*cis*-retinal. A photon interacting with the molecule causes the flexible double-bonded carbons to change to the *trans* conformation, forming all-*trans*-retinal, which has a straight hydrocarbon chain (see figure).

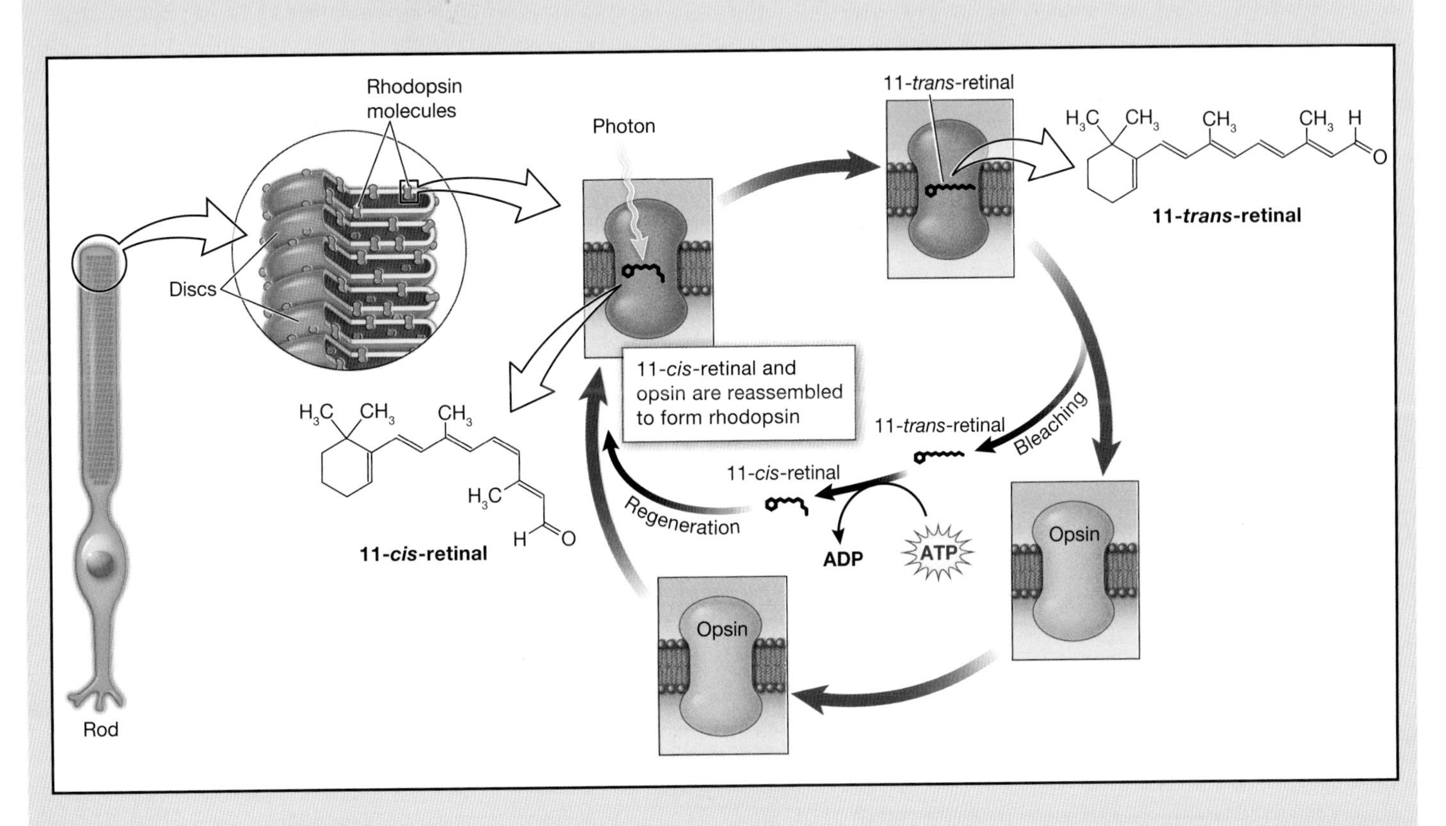

The shape change of retinal in the photoreceptors initiates visual transduction in the retina. Activation of retinal and the opsin proteins result in activation of a G protein. The G protein changes the membrane potential of the photoreceptor cell, which then releases less neurotransmitter into the outer synaptic layer of the retina. Until the retinal molecule is changed back to the 11-*cis*-retinal shape, the opsin cannot respond again to light energy, a phenomenon which is called *bleaching*. When bleaching occurs in a large group of photopigments, it causes the retina to effectively reverse its perceptions of light and dark in the affected areas. After a bright flash of light, afterimages are usually seen in negative. The photoisomerization is reversed by a series of enzymatic changes so that the retinal responds to more light energy.

The opsins are sensitive to limited wavelengths of light. Rhodopsin, the photopigment in rods, is most sensitive to light at a wavelength of 498 nanometers (nm). The three color opsins have peak sensitivities of 564 nm, 534 nm, and 420 nm, corresponding roughly to the primary colors of red, green, and blue. Rods are so sensitive to light that a single photon can result in an action potential from a rod's corresponding retinal ganglion cell.

The three types of cone opsins, being sensitive to different wavelengths of light, provide us with color vision. By comparing the activity of the three different cones, the brain can extract color information from visual stimuli. For example, a bright blue light that has a wavelength of approximately 450 nm would activate the "red" cones minimally, the "green" cones marginally, and the "blue" cones predominantly. The relative activation of the three different cones is calculated by the brain, which perceives the color as blue. However, cones cannot react to low-intensity light, and rods do not sense the color of light. Therefore, our low-light vision is—in essence—in grayscale. In other words, in a dark room, everything appears as a shade of gray. If you think that you can see colors in the dark, it is most likely because your brain knows what color something is and is relying on that memory.

fovea centralis (fovea)

The retina coats the entire inner surface of the eye, but it is not uniform in its composition. The periphery of the retina is composed almost entirely of rods. Posteriorly, the density of cones increases steadily. There is a broad region of the posterior retina with a high proportion of cones called the **macula lutea** (**Figure 15.15**). The center of the macula lutea is the **fovea centralis**, the region with the highest density of cones and almost no rods. The fovea centralis also lacks blood vessels, so there are no obstructions to photons reaching the cone photoreceptors. This is the singular spot where visual acuity is the highest.

Light enters the eye through the pupil and passes through the lens. The function of the lens is to focus the light at the fovea centralis. As a person moves in either direction, and the light does not focus directly on this central point of the retina, visual acuity and color perception drop significantly.

Apply to Pathophysiology

Aging Eyes

LO 15.4.8

1. As we age, the lens becomes less flexible and the ciliary muscles have trouble bending it. Think about the functions of the lens; which of the following would occur if the lens became less flexible?
   A) The eyes would become dry and particles would stick to the surface.
   B) Acuity may be compromised, particularly when shifting focus from far away to near.
   C) Dark spots may appear in the visual field.
2. Macular degeneration is a common disease of aging. In it, blood vessels grow across the retina, on the anterior surface of the ganglion cell axons at the macula lutea. Draw the arrangement of cells in the retina and add blood vessels as described here. The red blood cells can absorb light due to their pigmentation. Which of the following may occur in macular degeneration?
   A) Acuity may be compromised, particularly when shifting focus from far away to near.
   B) Dark spots may appear in the visual field.
   C) Light will reflect from the blood vessels, making the visual field brighter.
   D) Peripheral vision will be compromised.

LO 15.4.9

Figure 15.15 Pathway of Visual Information

Each eye senses visual information from a majority of the visual field. The axons of the retinal ganglion cells travel to the brain in the optic nerves, which cross over at the optic chiasm. At this location, visual information segregates by location, so that axons from retinal cells that are stimulated by the right side of the visual field synapse in the left side of the occipital lobe of the cerebrum and vice versa.

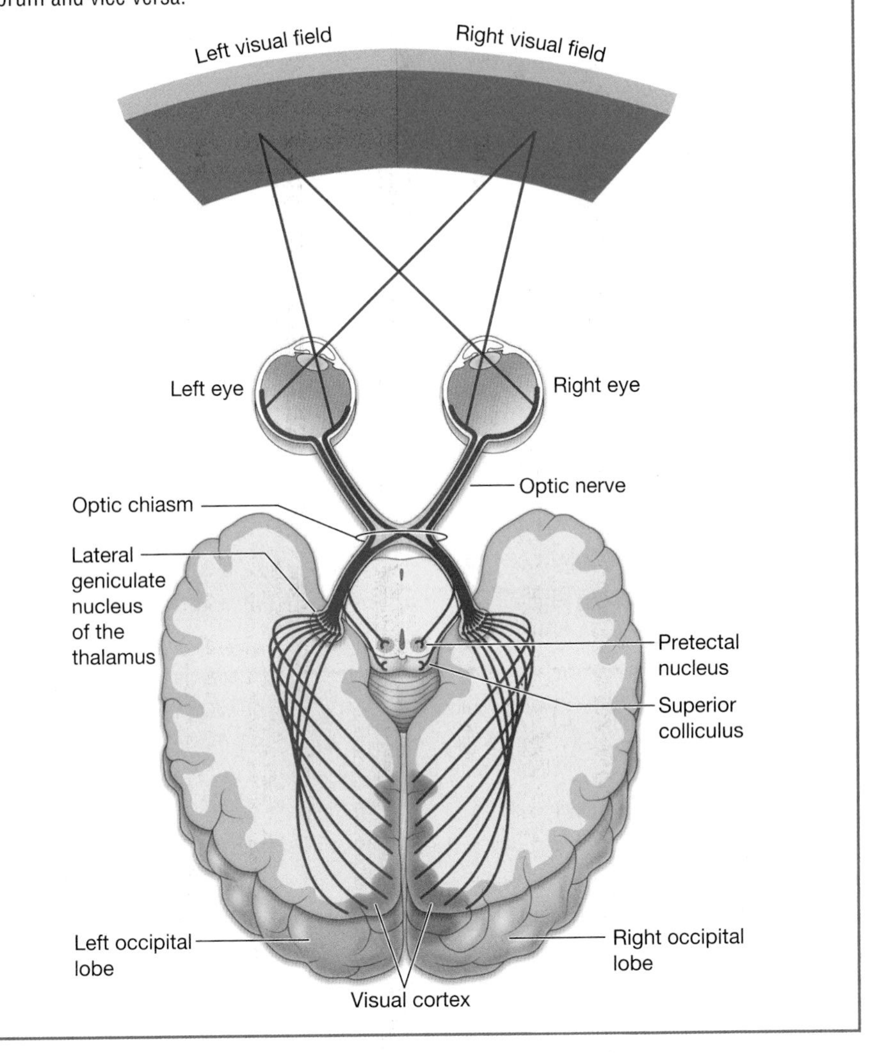

Within each eye, the axons of the ganglion cells gather to form the optic nerve; this passes through bony tunnels in the skull, the optic canals. Like somatosensory tracts, the optic nerves cross over. This physical crossing point is the **optic chiasm** (see Figure 15.15). This crossing over is more complex than the crossing over of the somatosensory tracts. The axons from the medial retina of the left eye cross over to the right side of the brain at the optic chiasm. However, within each eye, the axons projecting from the lateral side of the retina do not cross. Therefore, the left half of the field of view of each eye is processed on the right side of the brain, whereas the right half of each field of view of each eye is processed on the left side of the brain.

LO 15.4.10

Extending from the optic chiasm, the majority of the axons of the visual system project to the thalamus—specifically, the **lateral geniculate nucleus**. As is its role with sensory information, the thalamus edits, amplifies, or diminishes the incoming information. Axons from this nucleus then project to the primary visual cortex, located in the occipital lobe of the cerebrum. A small number of the axons do not pass through the thalamus, but instead travel to a reflex center, the superior colliculus that controls the eye muscles in a reflexive manner (for example, closing your eyes tightly when an object approaches them). In addition, a very small number of the axons project from the optic chiasm to the **suprachiasmatic nucleus** of the hypothalamus. The suprachiasmatic nucleus uses information about sunlight and darkness to establish the **circadian rhythm** of our bodies, allowing certain physiological events to occur at approximately the same time every day.

LO 15.4.11

Learning Check

1. Which of the following structures does light hit first to enter the eye?
   a. Sclera
   b. Cornea
   c. Retina
   d. Pupil
2. Which of the following structures is mainly responsible for the vasculature in your eyes?
   a. Ciliary body
   b. Iris
   c. Retina
   d. Choroid
3. Which of the following would you expect if some of a person's cones were genetically absent or mutated?
   a. Color blindness
   b. Blurred vision
   c. Partial blindness
   d. Eye pain
4. Which of the following structures contains the highest density of cones?
   a. Fovea centralis
   b. iris
   c. Suprachiasmatic nucleus
   d. Aqueous humor

15.4b Taste (Gustation)

LO 15.4.12

LO 15.4.13

Humans vary significantly from one person to another in terms of how much acuity they have in their sense of taste. Taste is a chemical sense. Only a few individual tastes have been recognized by scientists. These are: sweet, salty, sour, bitter, umami, and fat (*umami*, a word derived from the Japanese for "essence of deliciousness," is a savory taste that is based on amino acids). What does it take to be a recognized taste? It means that scientists have discovered molecular receptors on the tongue that respond to specifically to those molecules. Therefore, when you taste a delicious hamburger with ketchup and pickles, what is really occurring physiologically is a symphony of signaling from these six types of receptors.

LO 15.4.14

papillae (lingual papillae)

taste receptor cells (gustatory cells)

basal epithelial cells (basal cells)

The surface of the tongue, along with the rest of the oral cavity, is lined by a stratified squamous epithelium. Raised bumps on the tongue are not tastebuds but are **papillae** (singular = papilla) and contain our taste buds. **Taste buds** are composed of specialized epithelial cells called **taste receptor cells** that are the receptors and transducers of taste stimuli (**Figure 15.16**). These receptor cells are sensitive to the chemicals contained within foods that are ingested, and they release neurotransmitters based on the amount of the chemical in the food. The neurotransmitters from the taste receptor cells activate underlying taste neurons, the axons of which carry taste information toward the CNS. Also within the taste bud are a few **basal epithelial cells**. These are stem cells that can reproduce to replace the taste receptor cells if they age or are damaged. For example, if you burn your tongue on a hot food or beverage, you may experience a few days of reduced taste sensation. In time, the basal epithelial cells will replace the lost taste cells. As we age, our capacity for

Figure 15.16 Taste Buds

Taste molecules enter a taste bud through a taste pore. The molecular receptors that can bind taste molecules are expressed on hairs atop the apical surface of taste cells. Taste cells transmit information to the brain along sensory neurons.

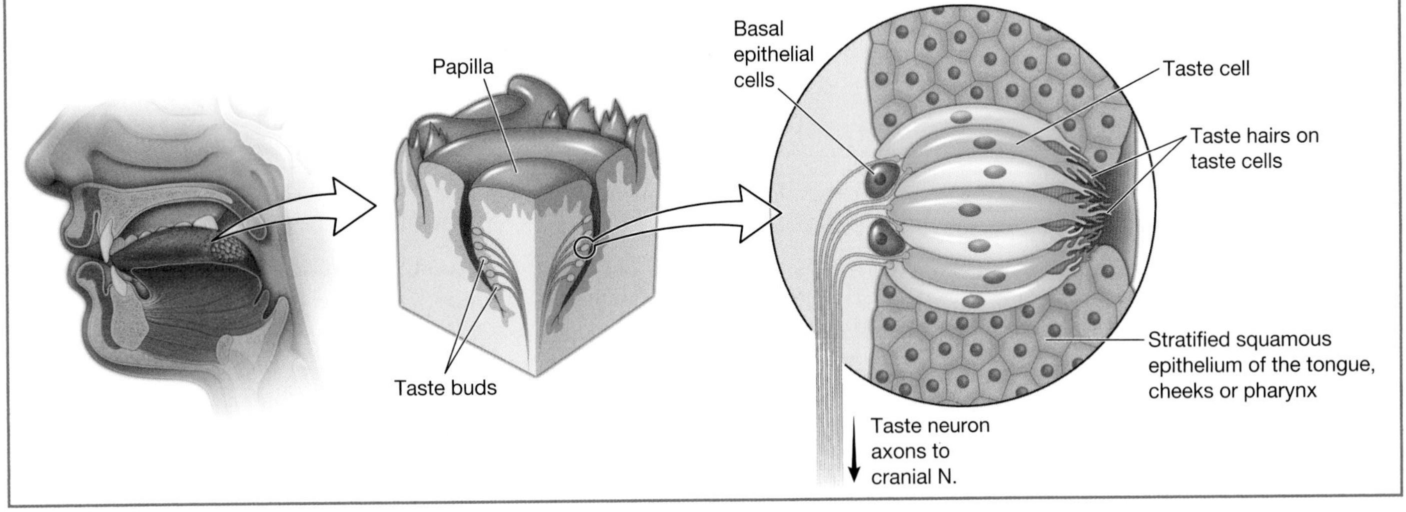

mitosis reduces all over the body. We heal from injury more slowly and we produce fewer and fewer taste receptor cells. Because of this, children have a stronger sense of taste on average than adults and young adults have a stronger sense of taste on average than elderly adults.

LO 15.4.15

The structure of the taste bud includes a relatively small pore through which the taste molecules must pass in order for reception to occur. One of the many functions of saliva is to solubilize taste molecules so that more of them pass into the pore to reach the taste receptor cells. When sodium (Na^+) ions solubilize in the saliva and reach the taste receptor cells, they can pass through Na^+ channels and enter the receptor. The entry of Na^+ into these cells results in the depolarization of the cell membrane and the generation of a receptor potential, which leads to your perception of a salty taste.

Sour taste is the perception of H^+ concentration. Just as with sodium ions in salty flavors, these hydrogen ions enter the cell and trigger depolarization. Sour flavors are, essentially, the perception of acids in our food. Increasing hydrogen ion concentrations in the saliva (lowering saliva pH) triggers progressively stronger graded potentials in the gustatory cells.

The first two tastes (salty and sour) are triggered by the cations Na^+ and H^+. The other tastes result from food molecules binding to specific receptor. The activation of those receptors ultimately leads to depolarization of the gustatory cell. Sweet tastes are due to gustatory cells responding to the presence of glucose dissolved in the saliva. Other monosaccharides such as fructose, or artificial sweeteners such as aspartame (NutraSweet™), saccharine, or sucralose (Splenda™) also activate the sweet receptors. The affinity for each of these molecules varies, and some will taste sweeter than glucose because they bind to the receptor differently.

Bitter taste is similar to sweet in that food molecules bind to specific receptors. However, there are a number of different ways in which this can happen because there is a large diversity of bitter-tasting molecules. There are at least 23 different identified bitter receptors, but not all humans express all of them. This leads to major individual differences and may explain some of our diversity in taste preferences. For example, some individuals love coffee, a bitter-flavored beverage that is often repulsive to others. Our bitter reception is overexpressed in children; evolutionary biologists think that this is a protective mechanism to protect children from poisons. The consequence is that children are very sensitive to bitter flavors and may be unlikely to tolerate bitter foods such as broccoli.

Like sweet and bitter, umami is based on the activation of a specific receptor. The molecule that activates this receptor is the amino acid L-glutamate. Therefore, the umami flavor is often perceived while eating protein-rich foods. The food additive monosodium glutamate (MSG) is a strong activator of these receptors.

Our sense of taste for fats is the most recently discovered taste stimulus.

Evolutionarily taste is thought to be important for helping humans to identify foods that are safe to eat and not spoiled. Taste also prepares the body for digestion; the signaling from taste receptors in the tongue helps to influence the secretion of digestive enzymes in the gut. Because humans universally enjoy fatty foods, and because fat digestion is complex, scientists long hypothesized that fat was its own taste. Recently this hypothesis was confirmed with the identification of receptors that bind specifically to fatty acids and signal into the brain.

LO 15.4.16

Once the gustatory cells are activated by the taste molecules, they release neurotransmitters onto the dendrites of taste neurons. These neurons are part of the facial and glossopharyngeal cranial nerves, as well as a component within the vagus nerve dedicated to the gag reflex. The facial nerve collects signals from taste buds in the anterior third of the tongue. The glossopharyngeal nerve connects to taste buds in the posterior two-thirds of the tongue. The vagus nerve connects to taste buds in the extreme posterior of the tongue—verging on the pharynx—which are more sensitive to unpleasant stimuli such as bitterness. All three pathways converge on the solitary nucleus in the medulla oblongata. From there, a secondary neuron brings signals to the thalamus and then to the taste cortex of the insula (**Figure 15.17**).

Cultural Connection

Supertasting

Humans vary from one individual to another in many, many anatomical ways. But perhaps no single facet of our anatomy demonstrates as much interhuman variation as our sense of taste. Two individuals may express vastly different numbers of tastebuds. To put it a different way, the person sitting next to you may express up to 10,000 more tastebuds than you do! Even identical twins express different numbers of taste buds. These anatomical differences can lead to extraordinary differences in experience. In simple terms, the more receptors you express on your tongue, the more information the brain can receive about flavor. Therefore, if two people are sharing a meal and one person expresses thousands more taste buds, that person will be much more sensitive to the flavors in the food. People who express very high numbers of taste buds are referred to as *supertasters*. Not only sheer numbers vary among individuals; expression patterns do as well. People descended from ancient humans who inhabited colder climates tend to express very different bitter receptor profiles than do modern humans descended from ancient tropical populations. Among these colder-climate populations, bitter foods such as coffee, tea, and beer tend to be more tolerated or even enjoyed. Among tropical-region descendants, spicier foods tend to be enjoyed. It all has to do with receptor expression! Receptor expression is highest in children, making these individuals very sensitive to almost all flavors. This is likely why adults find many children to be "picky eaters." The truth is that they are not picky, just very sensitive.

Figure 15.17 The Pathway of Taste Information

Taste information travels to the CNS via three of the cranial nerves. The first synapse occurs in the medulla oblongata; the second synapse is in the thalamus. Taste information is processed in the insula.

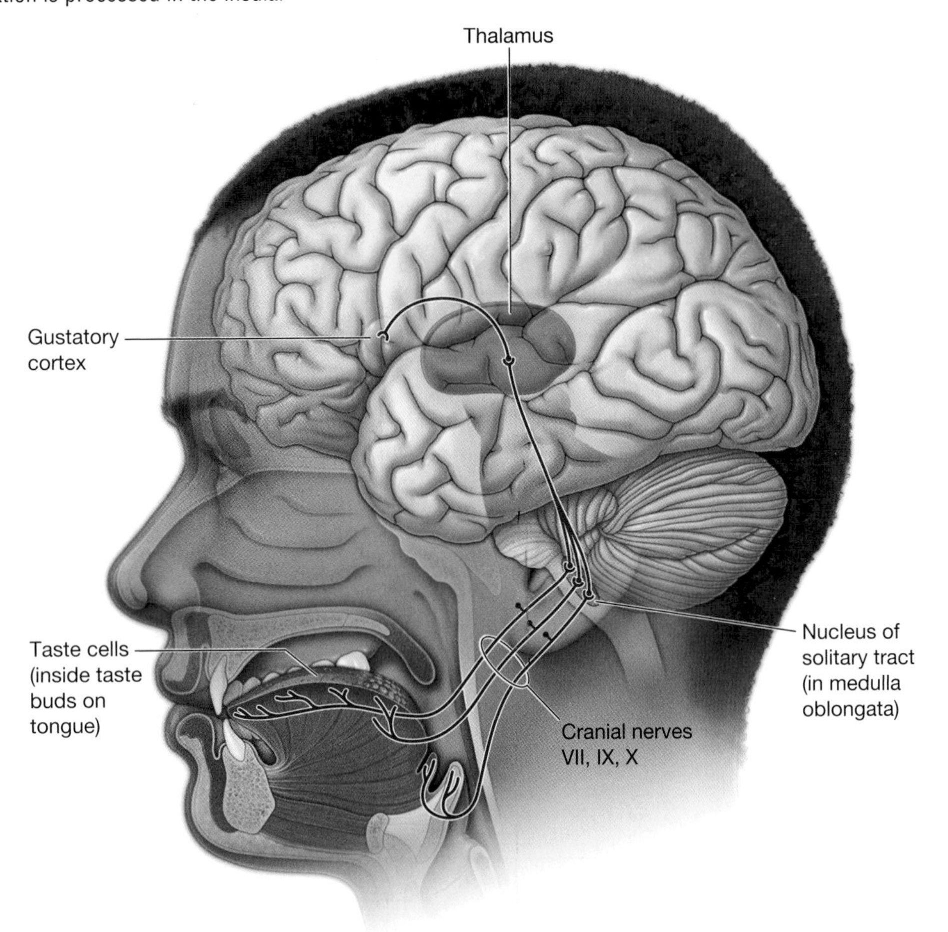

Learning Check

1. Which of the following are true about taste buds? Please select all that apply.
   a. Taste buds contain papillae.
   b. Taste buds contain taste receptor cells.
   c. Basal epithelial cells produce lost taste receptor cells.
   d. All taste buds detect different tastes.
2. If someone has cancer in the mouth or throat, the typical treatments are chemotherapy and radiation therapy. This is known to kill the cells of the salivary glands, leading to decreased production of saliva. Which of the following symptoms would you expect due to the lack of saliva production?
   a. Increased sensitivity to sour tastes
   b. Decreased sensitivity to all tastes
   c. Inability to regenerate taste buds
   d. Impaired gag reflex
3. Which of the following symptoms would you expect if the vagus nerve were impaired? Please select all that apply.
   a. Impaired taste in the anterior third of the tongue
   b. Impaired taste in the posterior two-thirds of the tongue
   c. Impaired taste in the extreme posterior of the tongue
   d. Impaired ability to gag

15.4c Smell (Olfaction)

LO 15.4.17

LO 15.4.18

LO 15.4.19

Like taste, the sense of smell, or **olfaction**, is also responsive to chemical stimuli. The olfactory receptor neurons are located in a small region within the superior nasal cavity (**Figure 15.18**). This region is referred to as the **olfactory epithelium** and contains bipolar sensory neurons interspersed among supporting cells. Each **olfactory sensory neuron** has dendrites that extend down into the mucous lining of the nasal cavity. As airborne molecules are inhaled through the nose, they pass over the olfactory epithelial region and dissolve into the mucus. These **odorant molecules** may bind to receptors on the olfactory dendrites. The binding of the receptors will produce a graded membrane potential in the olfactory neurons. While scientists have only identified six distinct types of taste receptors, they have identified thousands of different odorant receptors.

The axon of the olfactory neurons extends from the basal surface of the epithelium, through the foramina in the cribriform plate of the ethmoid bone, and into the **olfactory bulb** on the inferior surface of the frontal lobe. From there, the axons gather into the olfactory tract and then split to travel to several brain regions. Some travel to the cerebrum—specifically to the primary olfactory cortex, which is located in the temporal lobe.

Others project to structures within the limbic system and hypothalamus, where smells become associated with long-term memory and emotional responses. This is how certain smells trigger emotional memories, such as the smell of food associated with one's birthplace. Smell is the one sensory modality that does not travel through the thalamus before connecting to the cerebral cortex. This intimate connection between the olfactory system and the cerebral cortex is one reason why smell can be a potent trigger of memories and emotion.

Figure 15.18 The Olfactory Epithelium

The olfactory receptor cells, bipolar neurons, dangle into the olfactory epithelium. Their axons travel within the foramina of the cribriform plate of the ethmoid bone, and the first synapse is within the olfactory bulb.

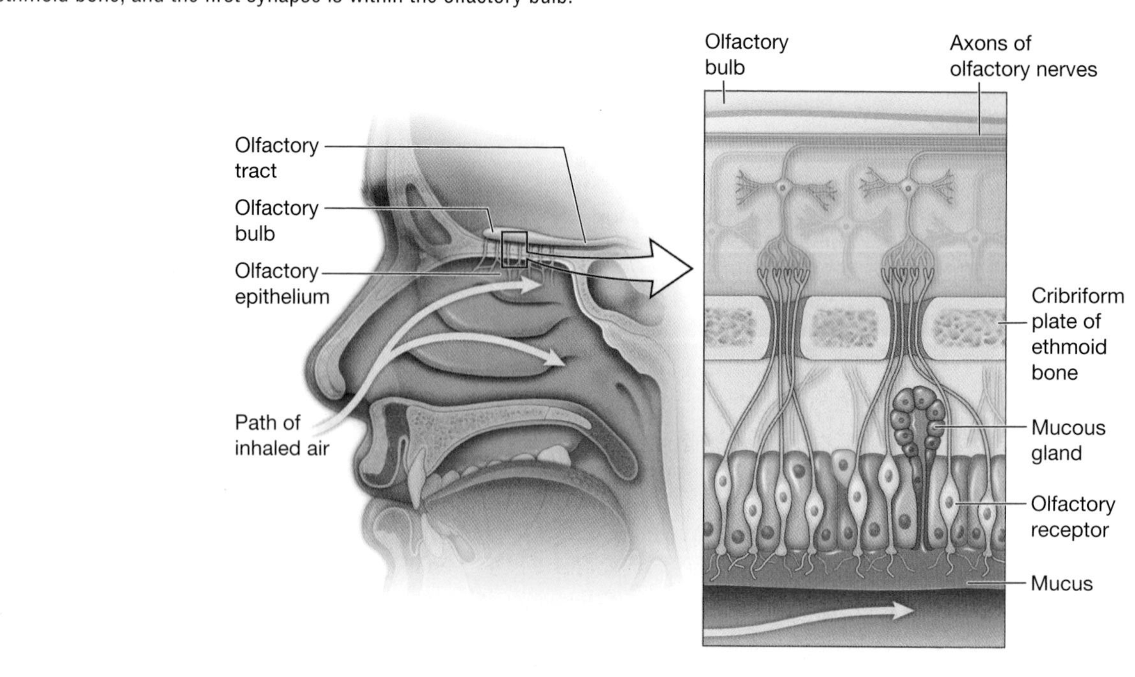

Apply to Pathophysiology

Smell and COVID-19

Loss of smell, or *anosmia*, is a common symptom of the disease COVID-19, which is caused by the virus SARS-CoV-2. Since the virus started circulating in 2019, scientists have been trying to unravel how this virus damages the sense of smell without necessarily causing inflammation of the nasal epithelium.

1. The SARS-CoV-2 virus may infect or cause damage to either the supporting cells or the olfactory neurons. Given this information, if you were a scientist looking for the virus, in which of these locations would you be most likely to find it?
   A) Inside the gray matter of the brain
   B) Within the olfactory cranial nerve
   C) Within the nasal cavity
2. The infectious period typically lasts for about 10 days, but many of the symptoms may linger. Knowing what you know about olfactory neurons, if some of these cells are damaged, will the sense of smell be able to be recovered?
   A) Yes, because other cells can perform this function too.
   B) Yes, because these neurons are capable of replicating themselves.
   C) No, damage to neurons always leads to permanent loss of function.
3. After loss of sensory information, for example during blindness, the regions of the cerebral cortex responsible for processing that type of sensory information may undergo structural changes. A group of scientists wants to investigate if there are changes to the cerebral cortex following COVID anosmia; which lobe should they study?
   A) The insula
   B) The occipital lobe
   C) The temporal lobe

The olfactory sensory neurons are one of two neuron types that are known to undergo mitosis. Therefore, our sense of smell can change throughout our lifespan. One identified trigger for olfactory neurogenesis is the hormone estrogen. Estrogen levels rise dramatically during pregnancy, and pregnant individuals report a heightened sense of smell. The new sensitivity to smells may be one component to the so-called morning sickness of pregnancy.

15.4d Hearing (Audition)

LO 15.4.20

LO 15.4.21

Hearing is the transduction of sound waves (vibrations) into electrical signals interpretable by the brain. As with our other senses, there is ultimately one type of receptor cell that is responsible for this transduction; however, the process is made possible through the assistance of the other structures of the ear. The large, fleshy structure on the lateral aspect of the head is known as the **auricle** (**Figure 15.19**). The curves of the auricle funnel and direct sound waves toward the **auditory canal**. The canal enters the skull through the external auditory meatus of the temporal bone. At the internal end of the auditory canal is the **tympanic membrane**, or eardrum, which vibrates as it is struck by sound waves. The auricle, ear canal, and tympanic membrane are often referred to as the **external ear**. The **middle ear** consists of a space spanned by three small bones called the **ossicles**. The three ossicles are the **malleus**, the **incus**, and the **stapes**. The malleus is attached to the tympanic membrane and articulates with the incus. The incus, in turn, articulates with the stapes. The stapes rests against the oval window, the door to the **inner ear**. We can think of the two drums—the tympanic membrane and oval window—as having roles in vibration according to sound waves. The ossicles are responsible for amplifying the sound between them so that the oval window receives stronger

auricle (auricula)

auditory canal (external auditory canal, external auditory meatus, external acoustic meatus)

tympanic membrane (eardrum)

malleus (hammer)

incus (anvil)

stapes (stirrup)

Figure 15.19 Structure of the Ear

The auricle, ear canal, and tympanic membrane are all part of the external ear. The middle ear is an air-filled space that houses the ossicles. The cochlea and the vestibule are responsible for audition and equilibrium, respectively.

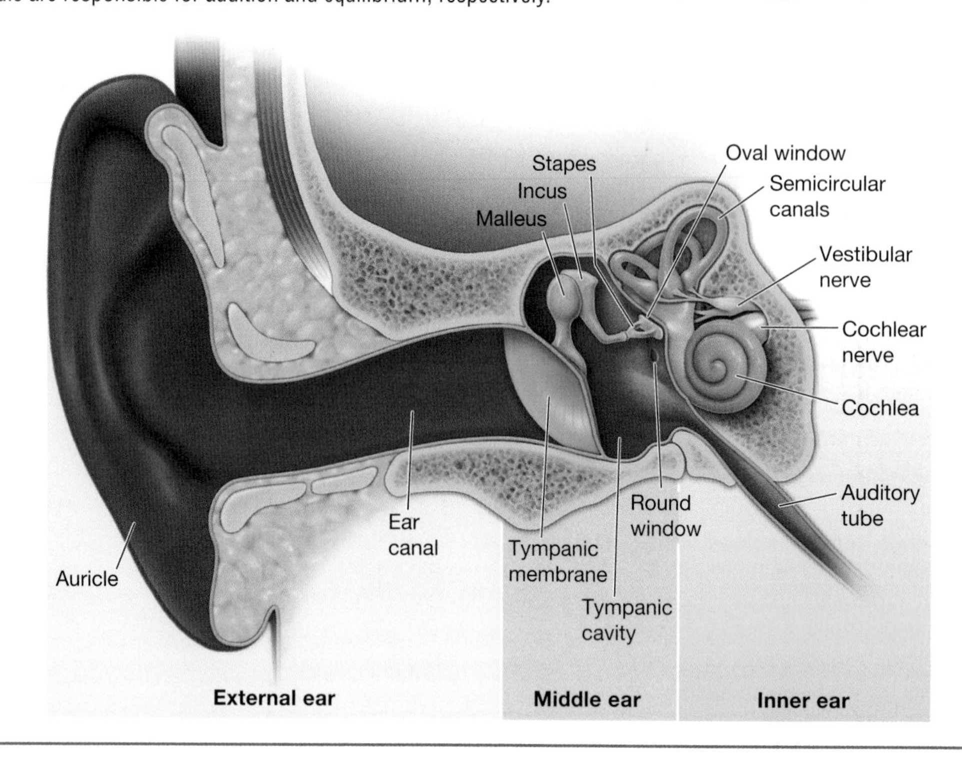

vibrations than the tympanic membrane (**Figure 15.20**). The oval window requires amplification because on its internal side it faces a fluid-filled space. The middle ear is an air-filled space connected to the pharynx through the auditory tube. This connection to the throat helps regulate air pressure within the middle ear. During a respiratory infection, however, it can become filled with mucus, which dampens the ability of the bones to amplify the vibrations. The tube is lined by the muscles of the pharynx, which contract during swallowing or yawning.

Student Study Tip

You can remember the order of the malleus, incus, and stapes bones (superficial to deep) because if you don't listen you will MIS something.

Figure 15.20 The Function of the Ossicles

The ossicles amplify the vibrations received by the tympanic membrane. They occupy the air-filled middle ear, but the vibrations must be amplified in order to impact the fluid-filled inner ear.

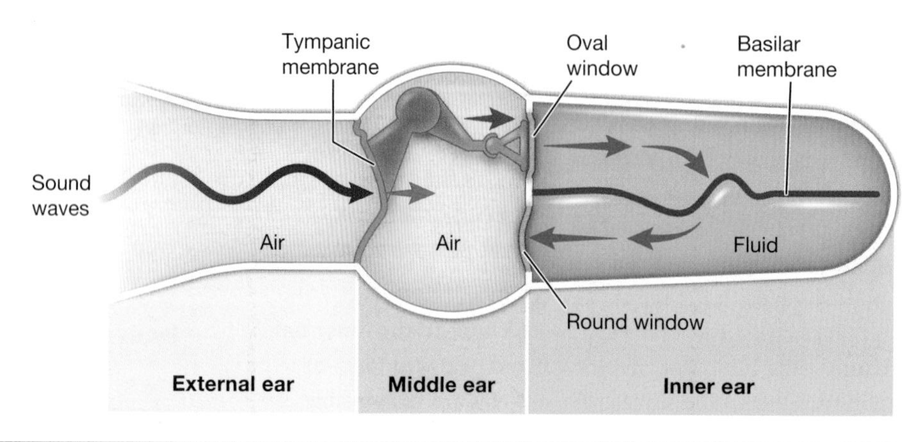

The inner ear occupies the bony labyrinth, a set of tunnels within the temporal bone. It is functionally divided into three regions, the **cochlea**, the **vestibule**, and the semicircular canals (**Figure 15.21**). The cochlea is responsible for hearing; the vestibule and the semicircular canals are responsible for our sense balance.

LO 15.4.22

Extending out from the oval window is the long fluid-filled tube of the cochlea. The tube, if cut in cross section, is really three tubes—the **scala vestibuli**, the **scala tympani**, and the **cochlear duct** (**Figure 15.22A**). The cochlear duct sits between the scala vestibuli and the scala tympani. The cochlear duct contains several **spiral organs**, which transduce the wave motions of the two scala into neural signals (**Figure 15.22B**). The organs of Corti lie on top of the **basilar membrane**, which is the floor of the cochlear duct and the roof of the scala tympani. As the fluid waves move through the scala vestibuli and scala tympani, the basilar membrane moves up and down. Sitting atop the basilar membrane are **hair cells**, the receptors and transducers of sound stimuli. Hair cells are so named because they have groups of immobile cilia, here called *stereocilia*, that project from their apical surfaces like mohawk hairdos. The tips of the stereocilia brush against the tectorial membrane suspended above them within the cochlear duct. As sound waves traverse through the scala tympani and the scala vestibuli, the basilar membrane and its hair cells bounce. Louder sounds produce more movement than quiet sounds. The stereocilia of the hair cells crash against the tectorial membrane. This will depolarize the hair cell membrane, triggering nerve impulses that travel down the afferent nerve fibers attached to the hair cells. Through this process, sound waves are transduced to electrical impulses that can be received and interpreted by the brain.

scala vestibuli (vestibular duct, vestibular ramp)

cochlear duct (scala media)

LO 15.4.23

scala tympani (tympanic duct, tympanic ramp)

LO 15.4.24

Figure 15.21 The Structures of the Inner Ear

The inner ear contains three separate anatomical and functional regions: the vestibule, the semicircular canals, and the cochlea.

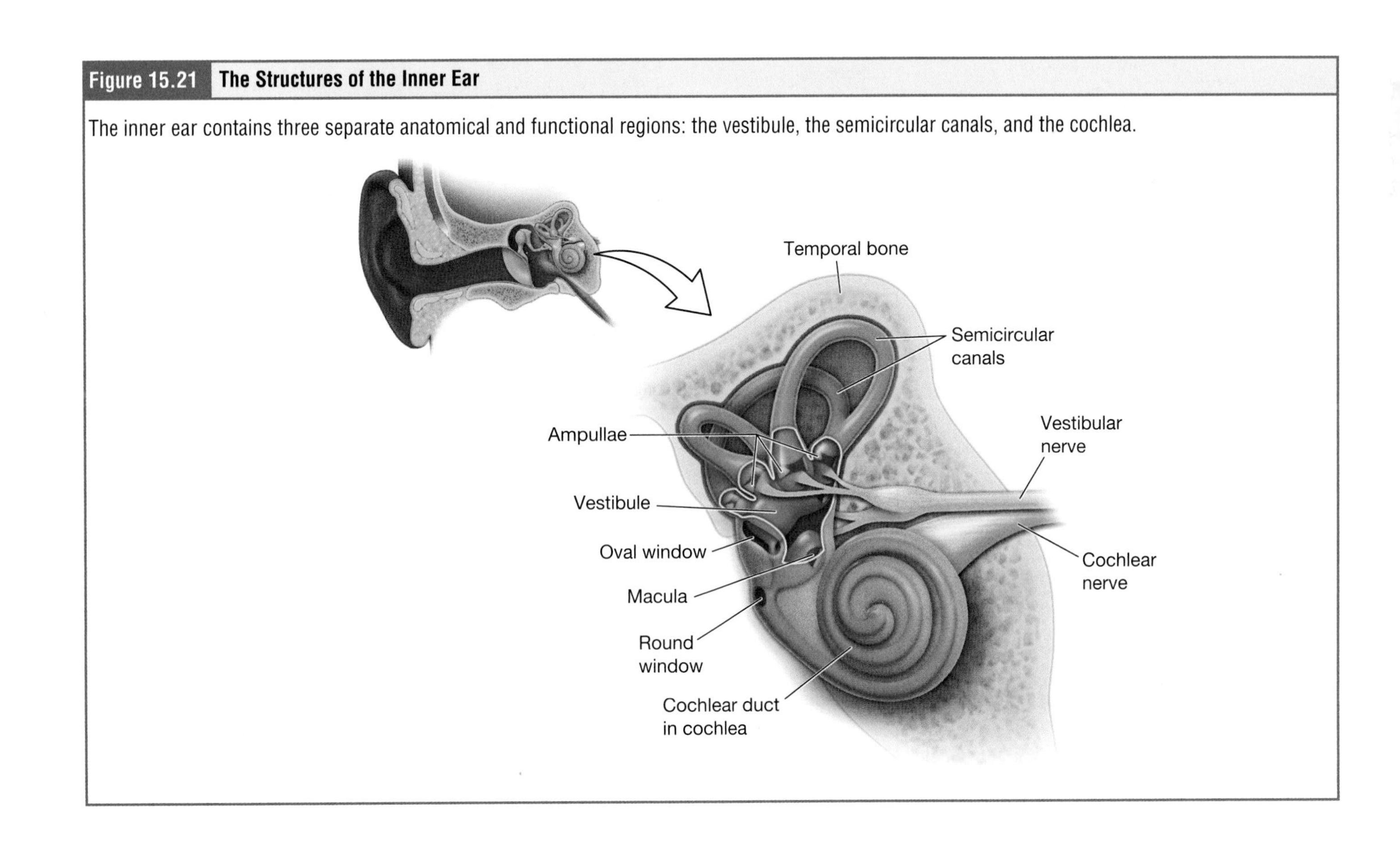

Figure 15.22 The Cochlea

(A) The cochlea is a long tube curled like a snail. Unspooled and cross-sectioned, we can see that it actually consists of three anatomically separate tubes rolled together. (B) Within the cochlear duct, hair cells sit on top of the basilar membrane. Their stereocilia project upward toward the unmoving tectorial membrane.

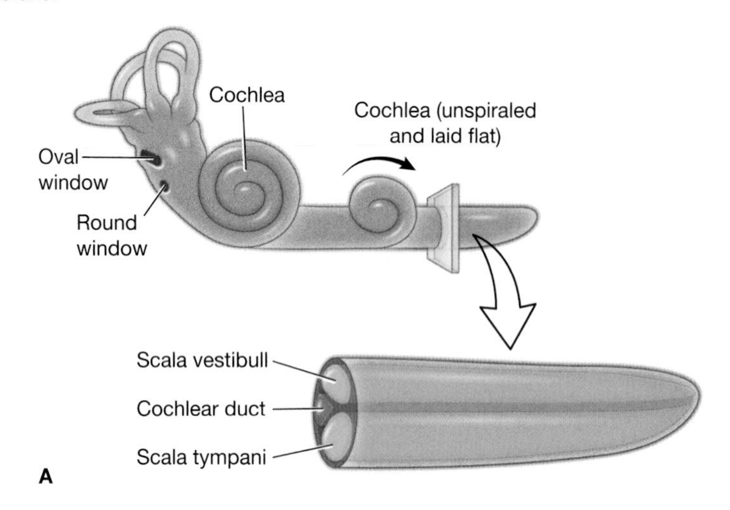

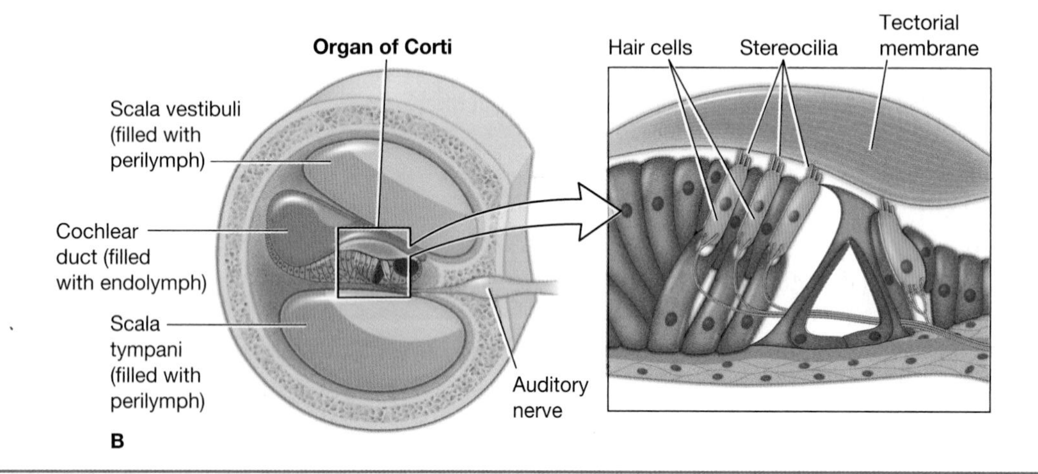

Cultural Connection

Deaf Culture

Historically, and to this day in many sectors of society, deafness, or hearing loss, has been viewed as a disability. However, the Deaf community rejects this notion, instead viewing deafness as a difference in the human way of life, compared to that of hearing people. There is a rich community and culture around deafness that embraces visual communication through American Sign Language; this is called *Deaf Culture.* In addition to the use of sign language, Deaf Culture consists of several behaviors, values, customs, and even the subject matter used in artwork, which are considered unique to the Deaf community.

Behaviors typical in Deaf Culture include eye contact during signing, touching a person or tapping on a table to gain a person's attention, and sitting or standing farther away from another deaf person while signing than hearing individuals would do while speaking. The values of the Deaf community include showing respect for the use of sign language and acceptance of deafness as a normal state for people with impaired hearing. In Deaf Culture, it is customary to ask for many personal details when meeting others of the Deaf community, and to set up the next meeting of the community before leaving a social gathering. Deaf Culture artwork is typically centered around drawings of the faces and hands of people, and often uses bold colors to attract attention to subjects such as signing hands.

Learning Connection

Explain a Process

From the moment a friend speaks your name to when your brain processes the sound, which structures do the sound waves hit or pass through? What is the function of each?

Deafness can arise in a variety of ways and exists in many levels of severity. Let's examine the two major types of deafness. One type of hearing loss occurs if sound entering the outer ear cannot reach the inner ear; this is called *conductive hearing loss*. It is sometimes temporary (as in the case of a large accumulation of ear wax in the auditory canal) and sometimes permanent (as in the case of otosclerosis, in which bone tissue is abnormally deposited at the base of the stapes). *Sensorineural hearing loss*, on the other hand, results from damage to the cochlea or the nerve pathways that transmit sensory information from the cochlea to the brain; this is more commonly permanent.

So far, we have discussed how transduction happens, as well as how we can interpret volume. But how do we differentiate between one sound and another? How can you recognize a friend's voice compared to a stranger's, or the sound difference between a violin and a piano? The answer is in the anatomy of the basilar membrane. The basilar membrane is not the same width all along its length; instead, it is wider at the beginning—near the oval window—and steadily narrows along its length (**Figure 15.23A**). The different widths of the basilar membrane mean that different spots along its length bounce in response to different frequency of sound waves. Higher frequency waves move the region of the basilar membrane that is close to the base of the cochlea; lower-frequency waves move the region of the basilar membrane that is near the tip of the cochlea. Sound wave frequency has to do with the pitch of the sound, with higher-pitched sounds (a squeak, a soprano singing, a scream) having higher frequencies and lower-pitched sounds (drumbeats, deep voices, thunder, bass guitars) having lower frequencies (**Figure 15.23B**).

LO 15.4.25

The basal side of the hair cell forms a synapse with auditory neurons. Their axons gather together and form the cochlear branch of the vestibulocochlear nerve. Their first synapse with the second nerve in the auditory pathway is within the **cochlear nucleus** in the medulla oblongata (**Figure 15.24**). From here, secondary neurons travel to the **superior olivary nucleus** of the pons. Tertiary neurons carry the signals to the midbrain, and the fourth neurons in the chain carry signals to the

LO 15.4.26

Figure 15.23 The Basilar Membrane

(A) The basilar membrane, on which the organ of Corti sits, separates the cochlear duct from the scala tympani. It narrows along its length within the cochlea. (B) Sound pitch is related to the frequency of the sound waves and different widths; therefore, different locations along the basilar membrane length will vibrate in response to different pitches/wavelengths. In this way, the ear can differentiate among different pitches.

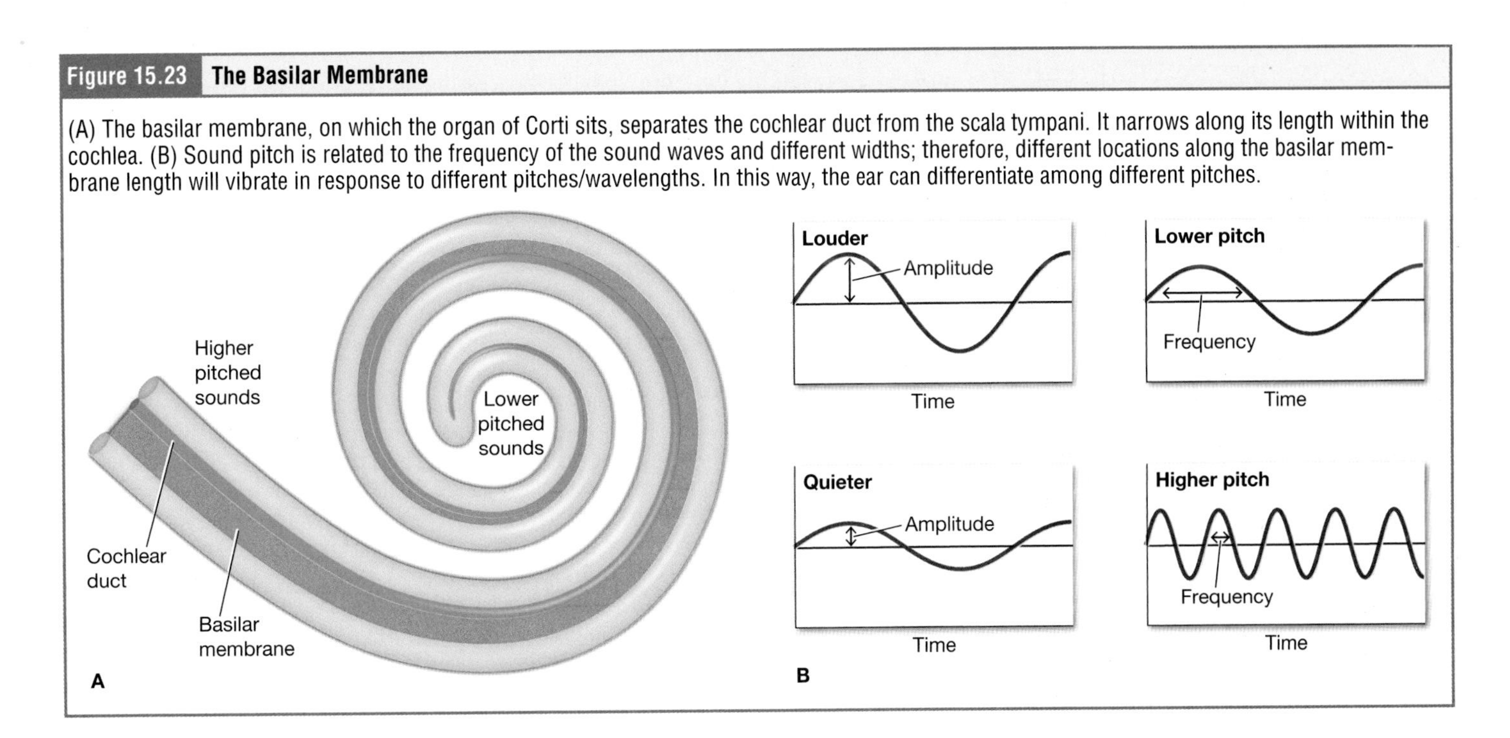

Figure 15.24 The Pathway for Auditory Information

The cochlear branch of the vestibulocochlear nerve carries the axons from the cochlea. With a total of five nerves before the information reaches the primary auditory cortex in the temporal lobe, the auditory pathway is more complex than other sensory pathways.

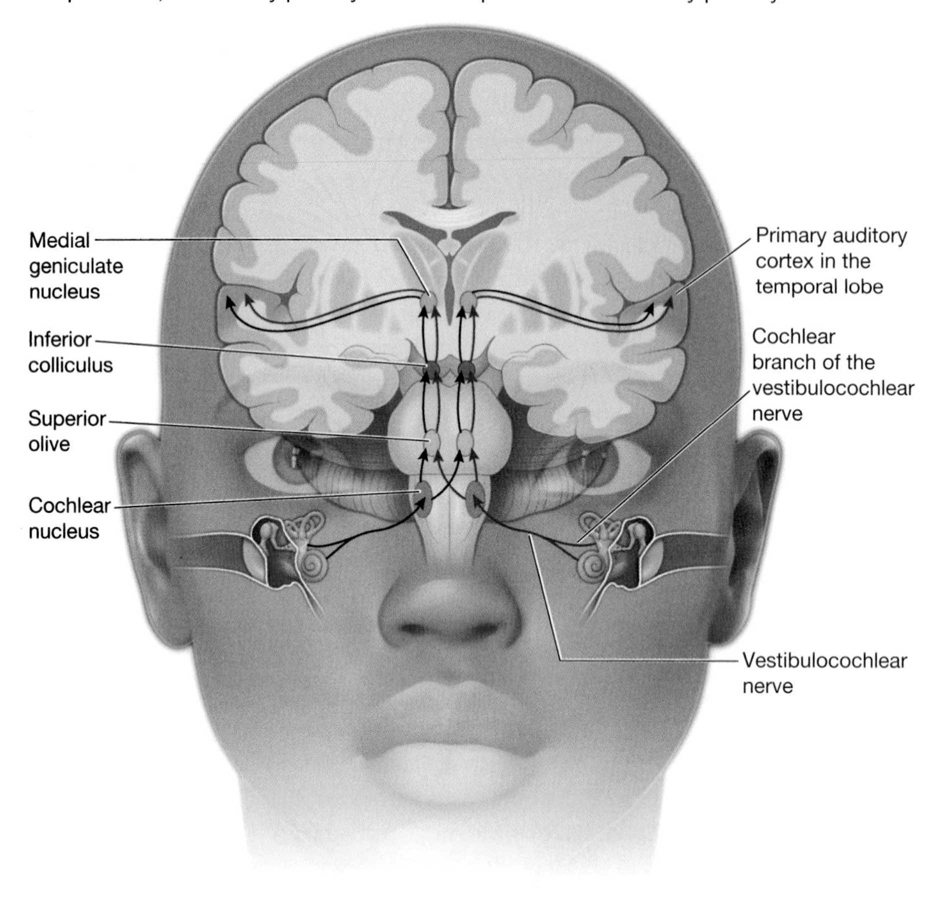

thalamus. The fifth and final neuron in the chain carries information to the primary auditory cortex in the temporal lobe for processing. Further connections to auditory association areas of the brain help us to connect sounds to our library of past experiences and recognize them.

Cultural Connection

Timbre

Most humans enjoy music very much, and it has a tremendous ability to stir emotions within us. We listen to our favorite song when we need a pick-me-up; many baseball players have a walk-up song playing in the stadium to build their confidence as they approach the plate; the dark dramatic music playing in the background contributes to our sadness while watching a movie. In this chapter we have explained how pitch and volume are transduced. But it is likely that you are able to interpret much more from sound than just these two modalities. Listen to a single note played on a piano versus a violin, or one sung by a human versus plucked on a guitar. You can tell which sound came from which instrument, but how? The qualities beyond pitch and volume are referred to as a sound's *timbre*. Timbre is often referred to as a sound's color or quality, but there are many tiny sounds within that provide clues that our cochlea and our auditory processing centers can pick up on. Musicians can vary timbre by the techniques they use to produce sound. (You can use your web search tool for examples, or use your own voice or instruments around your living space if you can.) Our appreciation for sounds and timbres may vary by culture and past experiences. Timbre themes are common in different types of music from around the globe.

15.4e Equilibrium (Balance)

Along with hearing, the inner ear is responsible for encoding information about **equilibrium**, the sense of balance. The hair cell is also the receptor and transducer, but the anatomical structure surrounding it lets it respond to different stimuli. The hair cells of the vestibule and semicircular canals sense head position, head movement, and whether our bodies are in motion. Our sense of head position and acceleration, referred to as **static equilibrium**, is determined by the vestibule. The rotation of the head or body, **dynamic equilibrium**, is determined by the semicircular canals.

LO 15.4.27

The hair cells of the vestibule sit, surrounded by supporting cells, on a flat membrane called a **macula** (Figure 15.25). Similar to the microanatomy of the spiral organ in the cochlea, the hair cells have upward projections of their stereocilia. Here, though, the

LO 15.4.28

Figure 15.25 The Otolithic Membrane

The maculae with their otolithic membranes function to inform the brain about head tilt. Like all parts of the inner ear, the hair cell is the receptor in the maculae.

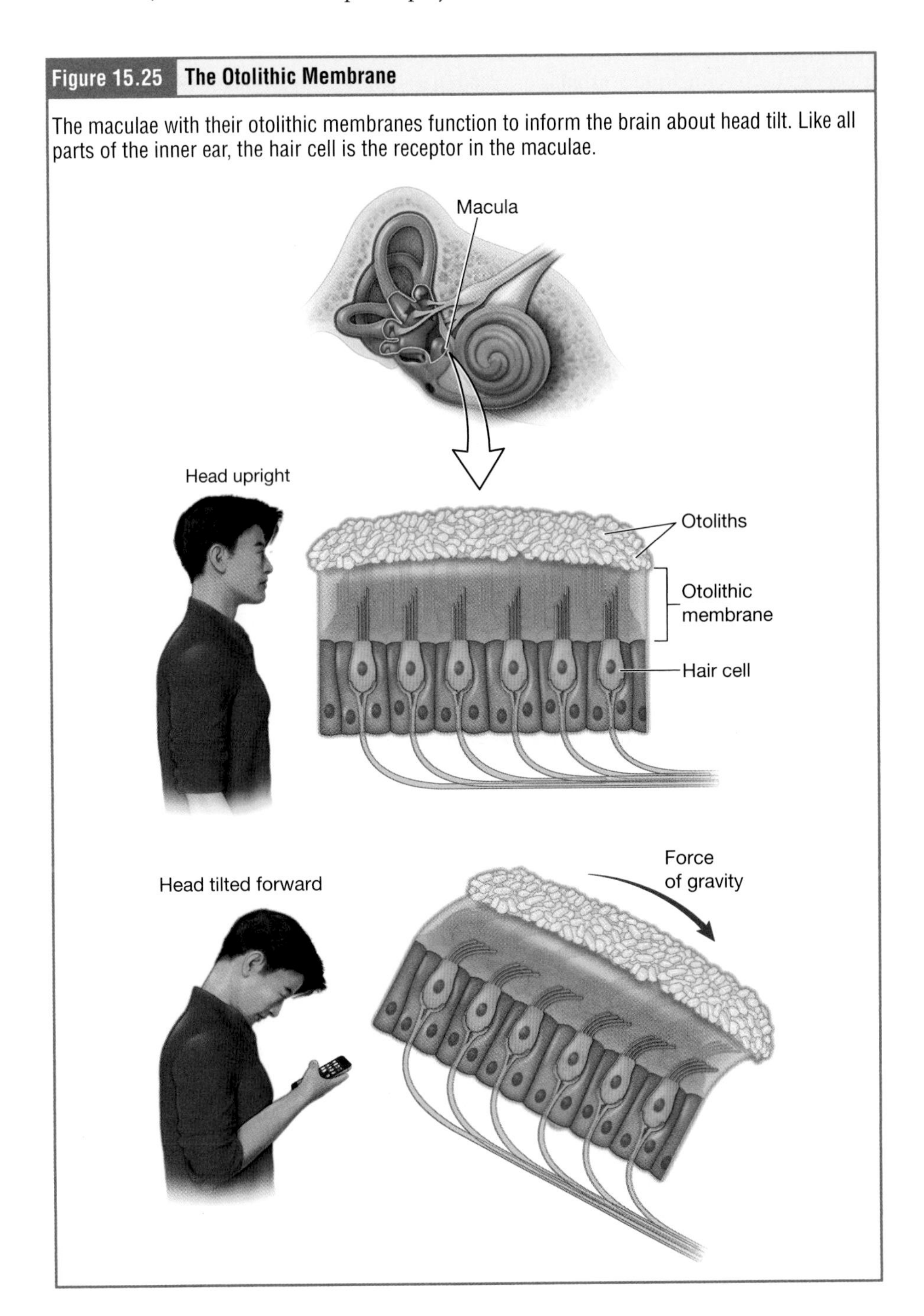

stereocilia are enmeshed in a gelatinous layer called the *otolithic membrane*. Resting on the otolithic gel are calcium carbonate crystals, called *otoliths*. The otoliths essentially make the otolithic membrane top-heavy. So, imagine that you tilt your head forward, as if you are nodding yes to a question. The cell membrane is relatively stable, but the gel with its otoliths slides forward, bending the sterocilia of the hair cells. Just as we saw in the cochlear hair cells, this movement changes the membrane potential of the hair cells, allowing them to signal to underlying neurons. Bending the stereocilia in the forward tilting of the head depolarizes the hair cell. Bending the stereocilia in the opposite direction during backward tilting of the head hyperpolarizes these cells. The exact position of the head is interpreted by the brain based on the pattern of hair-cell depolarization. It is not only helpful to keep track of your head position while standing still, but the vestibule also helps us interpret speed, for example, when traveling in a car or riding a roller coaster.

As their name implies, the semicircular canals are three ring-like extensions of the inner ear. Each semicircular canal is oriented in a slightly different plane (**Figure 15.26**). At the base of each semicircular canal, is a region known as the **ampulla**. The ampulla contains the hair cells that respond to rotational movement, such as turning the head from side to side while saying "no." The stereocilia of these hair cells extend into a jelly-like cap, the **cupula**, which sits on top of the ampulla. As the head rotates, the fluid lags and the cupula bends in the direction opposite to the head movement. The semicircular canals contain several ampullae, with some oriented horizontally and others oriented vertically. By combining the information generated from all of these ampullae, the nervous system can detect the direction of most head movements within three-dimensional (3-D) space.

LO 15.4.29

endolymph (Scarpa fluid)

The cupula sits within a watery environment; the specific fluid that that surrounds each cupula is **endolymph**, a fluid derived from blood plasma but with a slightly different composition. The endolymph flowing past the cupula as the head rotates influences the degree to which the cupula (and therefore the stereocilia) bend. Changes in the volume or viscosity of the endolymph may create a false sense of movement during alcohol use, vertigo, or dizziness. When incoming information to the brain about equilibrium does not agree with incoming information about vision or the incoming information from proprioceptors, it can lead to nausea. This is the mechanism behind motion sickness.

LO 15.4.30

Learning Check

1. Which of the following structures is found in the outer ear?
   a. Incus
   b. Cochlea
   c. Malleus
   d. Auricle
2. Which of the following would you expect if there were a lesion in the left superior olivary nucleus?
   a. Loss of hearing in the left ear
   b. Loss of hearing in the right ear
   c. Inability to hear higher frequencies
   d. Inability to hear lower frequencies
3. In which of the following scenarios would you lose balance if you had a lesion in the vestibule? Please select all that apply.
   a. Standing on one leg
   b. Looking side to side
   c. Standing with your eyes closed
   d. Throwing a baseball

Figure 15.26 The Ampulla

(A) The ampulla is the functional unit of the semicircular canals. An ampulla sits at the base of each fluid-filled semicircular canal. (B) As the head rotates, the fluid flows down from the canal and around the ampulla, swirling the gelatinous cupula and bending the hair cell stereocilia.

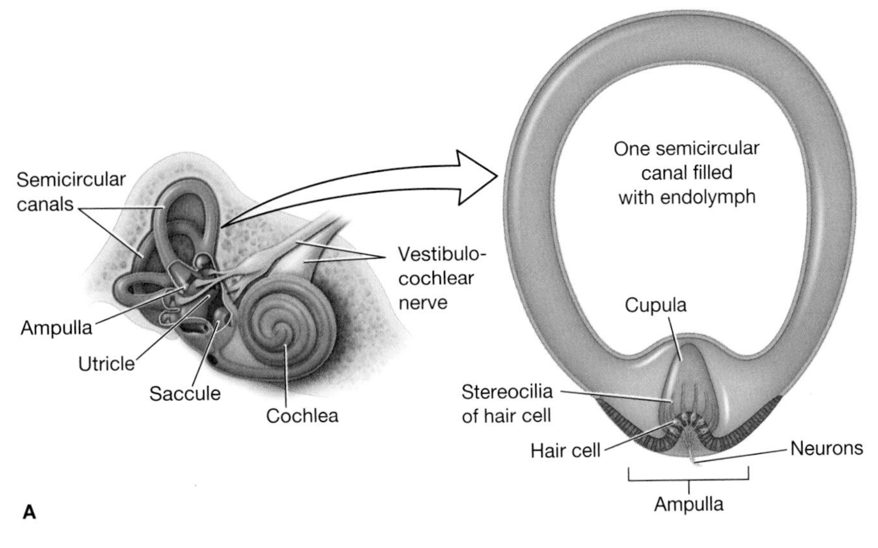

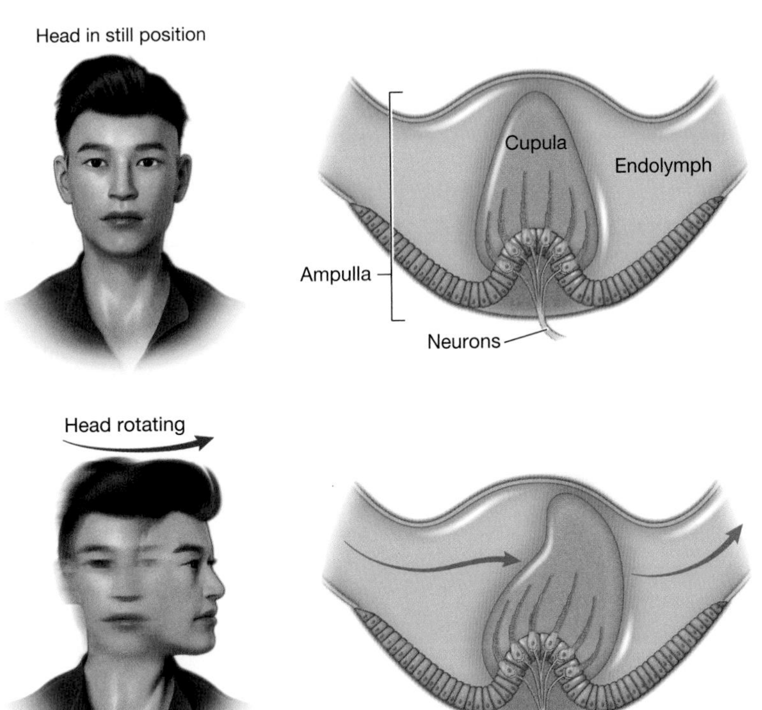

At the bases of the ampulla and macula, neurons synapse with the hair cells. The axons of these neurons gather together to form the vestibular branch of the vestibulocochlear nerve. These axons terminate in a variety of nuclei in the pons and medulla. These nuclei process and relay information to a variety of locations in the brain, including the cerebellum, the thalamus, the reticular formation, and nuclei of a variety of other functions. Through these connections, the incoming information from the vestibule and semicircular canals influences eye movement, body position, posture, alertness, and conscious awareness.

LO 15.4.31

Learning Connection

Chunking

Compare and contrast how hair cells function in different parts of the ear.

Chapter Review

15.1 Structure and Function of Sensory and Motor Pathways

The learning objectives for this section are:

HAPS LEARNING GOALS AND OUTCOMES

15.1.1 Describe the locations and functions of the first-, second- and third-order neurons in a sensory pathway.

15.1.2 Describe the locations and functions of the upper and lower motor neurons in a motor pathway.

15.1.3 Describe the concept of decussation and its functional implications.

15.1.4 Define the term reflex.

15.1.5 Describe reflex responses in terms of the major structural and functional components of a reflex arc.

15.1.6 Distinguish between each of the following pairs of reflexes: intrinsic versus learned, somatic versus visceral, monosynaptic versus polysynaptic, and ipsilateral versus contralateral.

15.1.7 Describe the following reflexes and name all components of each reflex arc: stretch reflex, (Golgi) tendon reflex, flexor (withdrawal) reflex, and crossed-extensor reflex.

15.1 Questions

1. You are walking on the beach. To feel the sand on the bottom of your foot, Neuron A will transmit sensory information from the receptor to Neuron B in the spinal cord. Neuron B will transmit the sensory information to Neuron C in the brain. Neuron C will transmit the sensory information from one part of the brain to the part that integrates the sensory information. Which of the following describes Neuron B?
   a. First-order neuron
   b. Second-order neuron
   c. Third-order neuron
   d. Fourth-order neuron

2. Your baby cousin is crying, so to cheer them up, you make funny faces. To make funny faces, a neuron (Neuron X) travels from one part of your brain to another. There, neuron X synapses with Neuron Y, which travels from your brain to your facial muscles. Which of the following best describes Neuron Y?
   a. Upper motor neuron
   b. Lower motor neuron
   c. First-order neuron
   d. Third-order neuron

3. Janice presents to the emergency room with a stroke in the right motor cortex, which is a region in the brain that is responsible for motor response. Which of the following deficits would you expect to find?
   a. Janice would be unable to move their right arm.
   b. Janice would be unable to feel on their left leg.
   c. Janice would be unable to move their left elbow.
   d. Janice would be unable to feel on their right shoulder.

4. Which of the following describe a spinal reflex? Please select all that apply.
   a. Connections between motor neurons that do not involve sensory input
   b. Connections between neurons that do not involve higher brain centers
   c. Connections between sensory and motor neurons
   d. Connections between neurons that involve voluntary movement

5. From which of the following parts of the spinal cord does motor response exit in a reflex?
   a. Dorsal horn
   b. Anterior horn
   c. Lateral horn

6. When someone has a traumatic brain injury, a neurologist checks to see whether or not the patient is responsive. In order to do so, the neurologist screams their name. If they do not respond, the neurologist inflicts some sort of pain to elicit any reaction. Which of the following best describe the reflex that occurs in response to pain? Please select all that apply.
   a. Intrinsic reflex
   b. Learned reflex
   c. Cranial reflex
   d. Somatic reflex

7. Which of the following describes a reflex in which a tendon is stretched to elicit a motor response?
   a. Tendon reflex
   b. Cross-extensor reflex
   c. Stretch reflex
   d. Withdrawal reflex

15.1 Summary

- Sensory receptors detect sensory input, and send a neurological signal up a cascade of neurons to the brain for sensory integration.
- Reflexes occur when a sensory neuron is connected to a motor neuron without the involvement of higher brain functioning or voluntary control.
- Many sensory and motor pathways decussate, or cross to the other side of the body, in the medulla oblongata of the brainstem. Therefore, in many cases, sensory information from one side of the body is sent to the opposite side of the brain. Also, motor information originating on one side of the brain controls muscle contraction on the opposite side of the body.
- The main types of reflexes in the body are: (1) stretch reflexes, in which stretched muscles respond by contracting, (2) withdrawal reflexes, which remove a limb from a painful stimulus, (3) tendon reflexes, which protect stretched tendons from tearing, and (4) crossed-extensor reflexes, which maintain posture during a withdrawal reflex in the opposite leg.
- Stretch reflexes are monosynaptic, because they contain only one synapse. Their reflex arcs contain only two neurons (a sensory and a motor neuron). Withdrawal reflexes are polysynaptic, because they contain two or more synapses. Their reflex arcs contain at least one interneuron.

15.2 Sensory Receptors

The learning objectives for this section are:

15.2.1 Define sensory receptor.

15.2.2 Define transduction, perception, sensation, and adaptation.

15.2.3 Distinguish between tonic and phasic receptors.

15.2.4 Compare and contrast the types of sensory receptors based on the type of stimulus (i.e., thermoreceptor, photoreceptor, chemoreceptor, baroreceptor, nociceptor [pain receptor], mechanoreceptor).

15.2.5 Compare and contrast the three types of sensory receptors, based on their stimulus origin (i.e., exteroceptors, interoceptors [visceroceptors], proprioceptors).

15.2 Questions

1. Which of the following defines a sensory receptor?
   a. A neuron that transmits sensory information to the spinal cord
   b. A structure that detects sensory input from the environment
   c. A structure that receives sensory input from a sensory neuron
   d. A neuron that sends motor information to an effector

2. Which of the following phenomena is defined by a conscious awareness of stimuli that are received and interpreted by the brain?
   a. Transducers
   b. Sensations
   c. Adaptation
   d. Receptors

3. You jump into the pool on a hot summer day. Your body is met with a cold, refreshing sensation from the pool. After two minutes, your body tunes out the sensation and all you focus on is the fun you are having with your friends. Which type of receptors are the sensory receptors on your skin that tune out the cold sensation after two minutes?
   a. Phasic receptors
   b. Tonic receptors

4. Which of the following types of receptors detects changes in solute concentrations in body fluids? Please select all that apply.
   a. Osmoreceptors
   b. Chemoreceptors
   c. Baroreceptors
   d. Photoreceptors

5. Robert went to a neurologist because they found themself falling a lot. The neurologist performed some tests to assess Robert's sensation to touch, ability to detect joint position, and reflexes of their internal organs. Robert struggled with detecting their limb position in space. Which of the following receptors is most likely impaired?
   a. Exteroceptors
   b. Interoceptors
   c. Proprioceptors

15.2 Summary

- Sensory receptors transduce stimuli into an electrical charge that is interpreted by your brain as sensation.
- Sensory receptors can adapt quickly (phasic receptors) or slowly/not at all (tonic receptors).
- Sensory receptors can be classified on the basis of the origin of the stimulus, as follows (1) exteroceptors respond to stimuli outside of the body, (2) interoceptors respond to stimuli in internal organs, and (3) proprioceptors respond to stimuli in muscles and joints.
- Receptors can also be classified based on the type of sensory information to which they respond: (1) chemoreceptors (chemical concentration), (2) thermoreceptors (temperature), (3) photoreceptors (light), (4) baroreceptors (blood pressure), (5) nociceptors (pain), and (6) mechanoreceptors (touch and pressure).

15.3 General Senses

The learning objectives for this section are:

HAPS LEARNING GOALS AND OUTCOMES

15.3.1 Compare and contrast the location, structure, and function of the different types of tactile receptors (e.g., tactile [Merkel] corpuscle, lamellated [Pacinian] corpuscle).

15.3.2* Define and connect the concepts of receptive field and acuity.

15.3.3* Explain how the anatomical pathway of pain sensation can confuse pain information in referred pain and phantom pain.

* Objective is not a HAPS Learning Goal.

15.3 Questions

Mini Case 1: Diabetes

Simone is a 65-year-old female who is diagnosed with diabetes. She does not take good care of her diabetes. She does not exercise, does not follow a low-sugar diet, and does not take her medication. One day she takes her shoes off after a long day of work and finds her left foot bleeding. She had no idea! She rushes to the hospital where they treat her for her foot wounds. A neurologist performs a test in which he lightly touches a cotton ball to the bottom of her foot.

1. Which of the following types of receptors are responsible for feeling the light touch of the cotton ball at the bottom of the foot?

a. Lamellated corpuscles
b. Tactile corpuscles
c. Bulbous corpuscles
d. Unencapsulated receptors

2. The bottom of the foot is highly sensitive. It is also easy to tell exactly where something is on the bottom of the foot without having to look. Which of the following explains why the bottom of the foot is highly sensitive?

a. It has a high density of larger receptive fields.
b. It has a high density of smaller receptive fields.
c. It has a low density of larger acuity fields.
d. It has a low density of smaller acuity fields.

Over time, Simone's foot wounds spread, which lead to her foot becoming infected. Unfortunately, the doctors decided that Simone's foot had to be amputated to save her life. The surgeons amputated her left foot from below her knee.

3. During the days immediately after the amputation, she is awakened by excruciating pain in her left ankle, or what was her left ankle, which was amputated. This is very common in people who have had an amputation—they perceive pain from the region of the amputated limb. Which of the following terms describes this phenomenon?

a. Referred Pain
b. Phantom Pain

15.3 Summary

- Tactile receptors can be classified as either unencapsulated or encapsulated. Unencapsulated receptors are free nerve endings or Merkel cells. Encapsulated receptors are lamellated corpuscles, tactile corpuscles, or bulbous corpuscles.
- Each sensory receptor or sensory neuron has a specific receptive field, which is the area from which it can detect sensory information. The acuity of a sensory receptor or neuron is the precision with which it can locate a stimulus.
- Nociceptors detect pain, and relay that information to the brain for sensory integration.
- Pain pathways do not always tramsmit sensory information as expected. Referred pain is pain that feels as if it originates in a site other than its actual site of origin; it is due to common interneurons for somatic and visceral pathways. Phantom pain is perceived pain originating in a missing limb; it occurs because part of the pain pathway is still intact after the loss of the limb.

15.4 Special Senses

The learning objectives for this section are:

HAPS LEARNING GOALS AND OUTCOMES

15.4.1 Identify and describe the accessory eye structures (e.g., conjunctiva and lacrimal apparatus).

15.4.2 Identify the tunics of the eye and their major components (e.g., cornea, sclera, iris, ciliary body), and describe the structure and function of each.

15.4.3 Describe the lens and its role in vision.

15.4.4 Identify and describe the anterior and posterior cavities of the eye and their associated humors.

15.4.5 Describe phototransduction (i.e., how light activates photoreceptors) and explain the process of light and dark adaptation.

15.4.6 Trace the path of light as it passes through the eye to the retina, and describe which structures are responsible for refracting the light rays.

15.4.7 Compare and contrast the functions and locations of rods and cones.

15.4.8 Relate changes in the anatomy of the eye to changes in vision.

15.4.9 Given a factor or situation (e.g., macular degeneration), predict the changes that could occur in the affected sense and the consequences of those changes (i.e., given a cause, state a possible effect).

15.4.10 Trace the signal pathway from the retina through the optic nerve, optic chiasm, optic tract, and to the various parts of the brain.

15.4.11 Explain how the sense organs relate to other body organs and systems to maintain homeostasis.

15.4.12 Classify gustatory receptor cells based on the type of stimulus (i.e., modality).

15.4.13 Describe the primary taste sensations.

15.4.14 Identify and describe the location and structure of taste buds.

15.4.15 Explain the process by which tastants activate gustatory receptors.

15.4.16 Trace the path of gustation from gustatory receptors through specific cranial nerves to various parts of the brain.

15.4.17 Identify and describe the composition and location of the olfactory epithelium.

15.4.18 Explain the process by which odorants activate olfactory receptors.

15.4.19 Trace the path of olfaction, from the olfactory receptors to the initiation of an action potential in the olfactory nerves, through the olfactory bulb, the olfactory tract, and to the various parts of the brain.

15.4.20 Classify the receptor cells for hearing based on the type of stimulus (i.e., modality).

15.4.21 Identify the macroscopic structures of the outer (external), middle, and inner (internal) ear and their major components (e.g., auditory ossicles, auditory [pharyngotympanic] tube), and describe the structure and function of each.

15.4.22 Identify and describe the microscopic structures within the inner (internal) ear (e.g., maculae, cristae ampullares, spiral organ [of Corti]).

15.4.23 Explain the process by which an action potential is generated at the spiral organ (of Corti).

15.4.24 Trace the path of sound from the external ear to the inner ear, including where sound is amplified.

15.4.25 Explain how the structures of the ear enable differentiation of pitch, intensity (loudness), and localization of sounds.

15.4.26 Trace the signal path from the spiral organ (of Corti) to the cochlear branch of the vestibulocochlear nerve (CN VIII) and to the various parts of the brain.

15.4.27 Compare and contrast static and dynamic equilibrium.

15.4.28 Describe the structure of a macula and its function in static equilibrium.

15.4.29 Describe the structure of an ampulla and its function in dynamic equilibrium.

15.4.30 *Given a disruption in the structure or function of one of the senses, (e.g., vertigo), predict the possible factors or situations that might have caused that disruption (i.e., given an effect, predict the possible causes).

15.4.31 Trace the signal path from the maculae and cristae ampullares to the vestibular branch of the vestibulocochlear nerve (CN VIII) and to the various parts of the brain.

15.4 Questions

1. Which of the following defines the conjunctiva?
   a. The layer of the eye that folds on itself to line the inside of the eyelids
   b. A gland that produces tears and is located above the lateral side of the eyes
   c. Highly vascularized connective tissue that provides the blood supply to the eyes
   d. A jellylike fluid in the posterior cavity between the lens and retina

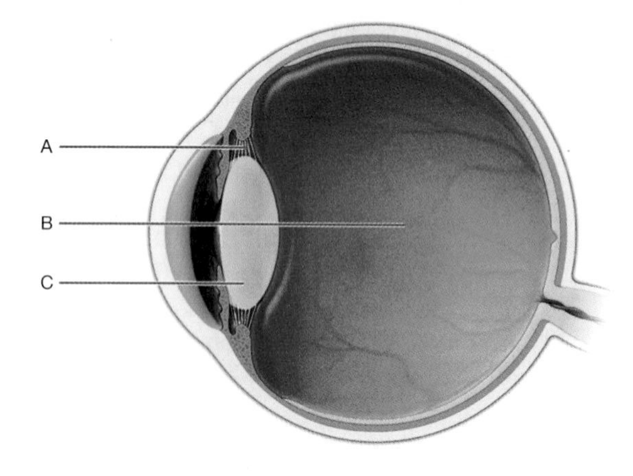

2. Identify structure A in the figure.
   a. Suspensory ligaments
   b. Sclera
   c. Iris
   d. Ciliary body

3. You are driving through California, with sunglasses on to avoid the bright sun. You admire the different scenery: the beaches on one side, the mountains on the other. Explain the role structure C plays in allowing you to see.
   a. Refracts (bends) the light as it enters the eyes
   b. Focuses light directly on the fovea centralis
   c. Constricts or dilates in response to bright or dim light
   d. Contains the nervous tissue responsible for photoreception

4. Which of the following fluids fills structure B?
   a. Aqueous humor
   b. Vitreous humor

5. Helen Frankenthaler is a female artist who was extremely innovative in her use of color, which evolved throughout her career. In the 1970s, she began to use thicker paints so she could employ even brighter colors. Which of the following allow you to see and appreciate the different colors? Please select all that apply.
   a. Rods
   b. Cones
   c. Rhodopsin
   d. Opsin

6. Which of the following describes the correct sequence of structures through which light passes in the eye?
   a. Cornea → Lens → Retina → Optic Nerve
   b. Lens → Optic Nerve → Retina → Sclera
   c. Pupil → Cornea → Neural Layer → Ganglion Cells
   d. Conjunctiva → Sclera → Choroid → Rhodopsin

7. Which of the following describe the location of rods and cones? Please select all that apply.
   a. The rods are found in the peripheral regions of the retina.
   b. The cones are found in the peripheral regions of the retina.
   c. The fovea centralis has the highest concentration of rods.
   d. The fovea centralis has the highest concentration of cones.

8. Which of the following allows you to see objects at far distances?
   a. The macula lutea integrates more light photons.
   b. The vitreous humor will decrease in density.
   c. The ciliary body contracts to tighten flatten the lens, making it less convex.
   d. The iris dilates to increase the field of view.

9. The stapes is a bone that lies against the oval window of the inner ear. Otosclerosis is a condition that involves the deposition of extra bone tissue on the stapes of the middle ear. How would you expect this condition to affect a person's hearing or equilibrium?
   a. It would cause a loss of balance while the person was standing still, due to damage to the otolithic membrane.
   b. It would cause impaired hearing, because of a reduction in vibrations transmitted to the cochlea.

c. It would cause a loss of balance while the person was moving, due to damage to the hair cells in the semicircular canals.
d. It would cause impaired hearing, because of a stiffening of the tympanic membrane (eardrum).

10. Which of the following traces the pathway from the retina to the brain?

a. Optic nerve → Optic Chiasm → Optic Tract → Lateral Geniculate Nucleus
b. Optic Tract → Optic Chiasm → Optic Nerve → Lateral Geniculate Nucleus
c. Optic Chiasm → Optic Tract → Optic Nerve → Suprachiasmatic Nucleus
d. Optic Chiasm → Optic Nerve → Suprachiasmatic Nucleus → Lateral Geniculate Nucleus

11. To which of the following structures do the balance organs in the vestibule and semicircular canals relay information in order to maintain homeostasis? Please select all that apply.

a. Cerebellum
b. Auditory cortex
c. Reticular formation
d. Spinal cord

12. Classify gustatory receptor cells based on the type of stimuli to which they respond.

a. Chemoreceptors
b. Thermoreceptors
c. Mechanoreceptors
d. Photoreceptors

13. Which of the following are primary taste sensations? Please select all that apply.

a. Sweet
b. Spicy
c. Sour
d. Bland

14. Which of the following are true about the papillae on the tongue? Please select all that apply.

a. They have pores where molecules can enter to perceive taste.
b. They are adjacent to other papillae to prevent excessive molecules from entering.
c. They contain taste buds, which are composed of taste receptor cells that transduce taste stimuli.
d. They cannot regenerate, causing the sense of taste to weaken with age.

15. You are enjoying ice cream on a hot summer day. Explain how you can perceive the sweet taste of the ice cream.

a. Gustatory receptors respond to glucose to perceive sweetness.
b. Na^+ ions depolarize the cell membrane to generate an action potential.
c. Thermoreceptors send action potentials to perceive the cold object.
d. Chemoreceptors depolarize when any food touches them.

16. Identify the correct pathways of gustation from the gustatory receptors to the brain. Please select all that apply.

a. Facial nerve → Vagus nerve → Glossopharyngeal nerve → Taste cortex
b. Glossopharyngeal nerve → Solitary nucleus → Thalamus → Insula
c. Vagus nerve → Medulla oblongata → Thalamus → Taste cortex
d. Solitary nucleus → Vagus nerve → Taste cortex → Thalamus

17. Where are the olfactory receptor neurons located?

a. Superior portion of the nasal cavity
b. Inferior portion of the nasal cavity
c. Surface of the tongue
d. Auditory canal

18. It is Sunday morning, and you wake up to the wonderful smell of pancakes cooking downstairs. Which of the following explains how your nose processes the sweet smell of pancakes?

a. The odorant molecules travel into the brain via olfactory neurons.
b. The odorant molecules bind to olfactory neuron receptors.
c. The odorant molecules bind to the optic neuron dendrites.
d. The odorant molecules travel into the cribriform plate.

19. To which of the following structures is an electrical impulse transmitted, immediately after leaving the olfactory bulb on its way to the brain?

a. Olfactory neuron
b. Olfactory tract
c. Cribriform plate
d. Olfactory cortex

20. Which of the following types of receptors transduce sound in the ear?

a. Chemoreceptors
b. Mechanoreceptors
c. Photoreceptors
d. Thermoreceptors

21. The auditory (eustachian) tube connects the middle ear to the pharynx; it allows excess fluid to drain out of the middle ear. In children, this tube is shorter and has less of a slope than in adults. This impedes the tube's ability to drain fluid properly. Which of the following explains the hearing loss that could be associated with an ear infection?

a. The fluid buildup prevents sound waves from traveling efficiently.
b. The excess fluid inhibits the tympanic membrane from vibrating.
c. The excess fluid overflows into the ossicles, inhibiting sensory transmission.

22. Which of the following components of the inner ear contains the hearing receptor organ?

a. Cochlea
b. Vestibule
c. Semicircular canals
d. Tympanic membrane

23. Habil is a music lover. Just last weekend, they went to a concert of their favorite band. They had the opportunity to stand in the front row in front of the loudspeakers, singers, and musicians. Which of the following explains what happens in Habil's ear to perceive the loud music?

a. The sound waves travel faster, hitting the tectorial membrane.
b. The sound waves move the hair cells faster, sending more frequent impulses.
c. More sound waves hit the tectorial membrane, sending larger impulses.
d. The sound waves move the hair cells at a higher magnitude.

24. If a sound wave is currently in the malleus, which of the following structures would the sound wave immediately pass through next?

a. Auricle
b. Tympanic membrane
c. Incus
d. Cochlea

25. The piccolo is one of the highest-pitched instruments in an orchestra. Explain how the ear perceives the sound of the piccolo.

a. The high pitch will move the hair cells at a faster frequency.
b. The high pitch will move the hair cells that are closest to the base of the cochlea.
c. The high pitch will move the hair cells that are at the tip of the cochlea.
d. The high pitch will move more hair cells, sending more impulses.

26. Gene is rushed to the hospital, where they are diagnosed as having suffered a basilar stroke. The stroke is specific to the pons, causing this region to become impaired. Which of the following symptoms would you expect to see in Gene?

a. Gene will have difficulty interpreting visual input.
b. Gene will have difficulty hearing from at least one ear.
c. Gene will have difficulty distinguishing different tastes.
d. Gene will have difficulty interpreting pain.

27. Shyna is a female figure-skater who is very good at spinning in circles on one leg. Which of the following structures allows Shyna to maintain her balance while she is spinning?

a. Semicircular canals
b. Vestibule
c. Maculae
d. Cochlea

28. You are walking when you notice your shoe is untied, so you look down. Explain what happens in the macula of your inner ear when you bend your head forward to look at your shoes.

a. The otoliths move the stereocilia forward, depolarizing the hair cell.
b. The otoliths move the stereocilia backward, depolarizing the hair cell.
c. The otoliths move the stereocilia forward, hyperpolarizing the hair cell.
d. The otoliths move the stereocilia backward, hyperpolarizing the hair cell.

29. If the cupula in the ampulla of a semicircular canal is bending to the right, toward what side is the head rotating?

a. Left
b. Right
c. Up
d. Down

30. Your 60-year-old grandparent is complaining about blurry vision in one eye. This issue seems to have developed over time but is now causing your grandparent to have difficulty reading and seeing objects clearly. Which of the following could be the cause of the cloudy vision in the affected eye?

a. The sclera has become slightly hardened over time, due to scar tissue formation that occurs with aging.
b. The pupil has become slightly constricted, due to increased sensitivity of the affected eye to light.
c. The lens has become cloudy, instead of clear, due to the breakdown of proteins inside the lens.
d. The ciliary body has become attached to the suspensory ligaments over time; this is preventing some light from entering the eye.

31. To what structure do the axons of the vestibular branch of the vestibulocochlear nerve relay information for sensory integration? Please select all that apply.

a. Pons
b. Medulla
c. Primary auditory cortex
d. Thalamus

15.4 Summary

- The special senses include vision, taste, smell, hearing, and balance.
- The eye contains structures that allow a certain amount of light to pass through the eyeball. This light is perceived by photoreceptors in the back of the eye, which then send sensory signals to the brain to perceive vision.
- The tongue has chemoreceptors within taste buds, which react to certain molecules or ions binding to specific receptors. These send action potentials through cranial nerves into the brain so it can perceive different tastes.
- The superior portion of the nasal cavity contains olfactory sensory neurons, which have dendrites that extend down into the mucous lining. Once odorant molecules bind to the receptors on those dendrites, a graded potential is sent through the olfactory nerves, olfactory bulb, and olfactory tract, toward various brain regions, one of which is the primary olfactory cortex.
- The ear consists of several parts that work together to transduce sound waves into mechanical stimuli so the brain can perceive sound. Vibration in the ear creates sound waves, which move hair cells in the cochlea of the inner ear, causing them to contact the tectorial membrane; this results in the generation of a nerve impulse. This impulse is transmitted through several regions before it reaches the primary auditory cortex in the temporal lobe of the cerebrum.
- Balance is also detected within the ear in the semicircular canals and vestibule. The cilia of hair cells, which are embedded in a gelatinous substance, bend in response to changes in body position. This results in sending sensory information to various nuclei in the pons and medulla Next, the information is relayed to various other structures in the brain and spinal cord to maintain upright posture and equilibrium.

16

The Autonomic Nervous System

Rawpixel.com/Shutterstock.com

Chapter Introduction

In this chapter we will learn . . .

how the nervous system balances emergency and exercise with rest and homeostasis.

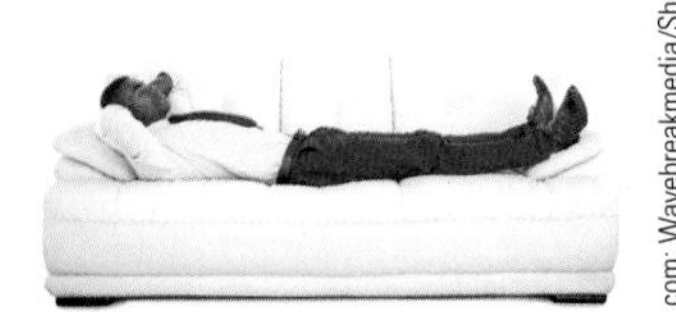

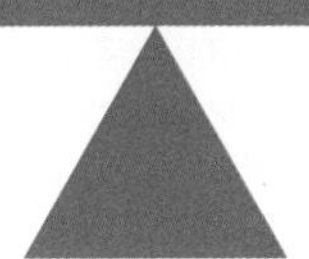

Romariolen/Shutterstock.com; Wavebreakmedia/Shutterstock.com

The autonomic nervous system (ANS) is the arm of the nervous system that provides the main innervation for the organs. There are two branches of the ANS that work together to influence the physiology of our organ systems. One branch of the ANS is often associated with the "fight-or-flight response," which refers to the preparation of the body to either run away from a threat or to stand and fight in the face of that threat. Our understanding of both the physiology of this system and its evolutionary origins has advanced significantly in the last 30 years. Early research was done primarily on rats, but humans, and their responses, are much more complex! In basic terms, you can think of this as your emergency system. To illustrate, consider an imaginary circumstance where you are hiking with a group of friends and suddenly encounter a wild bear. Survival may require things such as: staying alert, having oxygen and nutrients for muscle contraction, staying in motion, being aware of your surroundings. While this is not a very common threat that humans deal with in the modern world, it represents the type of environment in which the human species evolved and adapted. This branch of the ANS in modern humans can become activated through other threats to homeostasis, regardless of whether those threats are real or perceived. Fasting to observe a religious holiday, for example, requires the activation of the ANS to keep nutrient levels high. Facing a difficult exam may activate the ANS even though the "threat" is perceived and not an actual threat to homeostasis. Exercise is another example of a moment when activation of the emergency system occurs in a non-emergency.

The signs that this branch of your autonomic system is activated are things we normally associated with stress or nervousness such as heart rate increases, sweating, and dilated pupils. You may not see it, but on the inside of your body your bronchi increase in diameter, your blood pressure increases, and blood vessels dilate in skeletal muscles.

The emergency response system does not look the same for every human; the hormone profiles of stress can be markedly different from one individual to another, as we will discuss later.

The ANS is not just about responding to threats. Besides the branch responsible for responding to emergencies, there is a branch that is often referred to as "rest and digest" responses. Without pending emergency or stress situations, the ANS works to maintain homeostasis by regulating blood pressure and digestion, among other functions.

The Human Anatomy and Physiology Society includes more than 1,700 educators who work together to promote excellence in the teaching of this subject area. The HAPS A&P Learning Outcomes measure student mastery of the content typically covered in a two-semester Human A&P curriculum at the undergraduate level. The full Learning Outcomes are available at https://www.hapsweb.org.

16.1 Overview of the Autonomic Nervous System

HAPS LEARNING GOALS AND OUTCOMES

Learning Objectives: By the end of this section, you will be able to:

16.1.1 Compare and contrast the autonomic nervous system (ANS) to the somatic nervous system (SNS) with respect to site of origination, number of neurons involved in the pathway, effectors, receptors, and neurotransmitters.

16.1.2 Name the two main divisions of the ANS and compare and contrast the major functions of each division, their neurotransmitters, the origination of the division in the CNS, the location of their preganglionic and postganglionic (ganglionic) cell bodies, and the length of the preganglionic versus postganglionic axons.

16.1.3 Describe the major components of the sympathetic and parasympathetic divisions (e.g., sympathetic trunk [chain], white and gray rami communicantes, splanchnic nerves, pelvic splanchnic nerves, CN III, CN VII, CN IX, CN X) and the major ganglia of each division (e.g., terminal ganglia, intramural ganglia, sympathetic trunk [chain] ganglia, prevertebral [collateral] ganglia).

16.1.4 Describe the different anatomical pathways through which sympathetic and parasympathetic neurons reach target effectors.

Student Study Tip

Intramural sports are played within a school; intramural ganglia are found within the walls of a target organ.

LO 16.1.1

The nervous system can be divided into two functional parts: the somatic nervous system (SNS) and the autonomic nervous system (ANS). The major differences between the two systems are evident in the responses that each produces. The SNS causes contraction of skeletal muscles. The ANS controls cardiac and smooth muscle, as well as glandular tissue. The SNS is associated with voluntary responses (though many, such as breathing, can happen without conscious awareness), and the ANS is associated with involuntary responses related to homeostasis such as blood pressure regulation.

LO 16.1.2

LO 16.1.3

In addition to the endocrine system, the ANS is instrumental in homeostatic mechanisms in the body. The ANS regulates many of the internal organs through a balance of two aspects, or divisions. The two divisions of the ANS are the **sympathetic division** and the **parasympathetic division** (Figure 16.1). The sympathetic system is associated with the preservation of life in emergency situations. The parasympathetic activity is dominant during times between emergencies. Homeostasis is maintained by the balance between the two systems. Almost all organs (effectors) receive input by both the sympathetic and parasympathetic branches, with a few exceptions. **Dual innervation**, the input by both the sympathetic and parasympathetic divisions, determines the level of activity or output of an organ or tissue. For example, the heart receives connections from both the sympathetic and parasympathetic divisions. Whereas one causes the heart rate to increase, the other causes the heart rate to decrease.

16.1a Sympathetic Division of the Autonomic Nervous System

To respond to a threat or enable survival, the sympathetic system causes divergent effects because many different effector organs are activated together for a common purpose. For the purposes of understanding the physiological response to stressors, we need to define stress in both evolutionary and modern terms. **Stress** is a real or perceived threat to our homeostasis. As we discussed in the introduction to this chapter, an evolutionarily relevant example of such a threat would be the need to escape from a predator. A more modern example might be getting through a week of

Figure 16.1 The Divisions of the Nervous System

The sympathetic and parasympathetic nervous systems are divisions of the autonomic nervous system, the branch of the peripheral nervous system that innervates our organs.

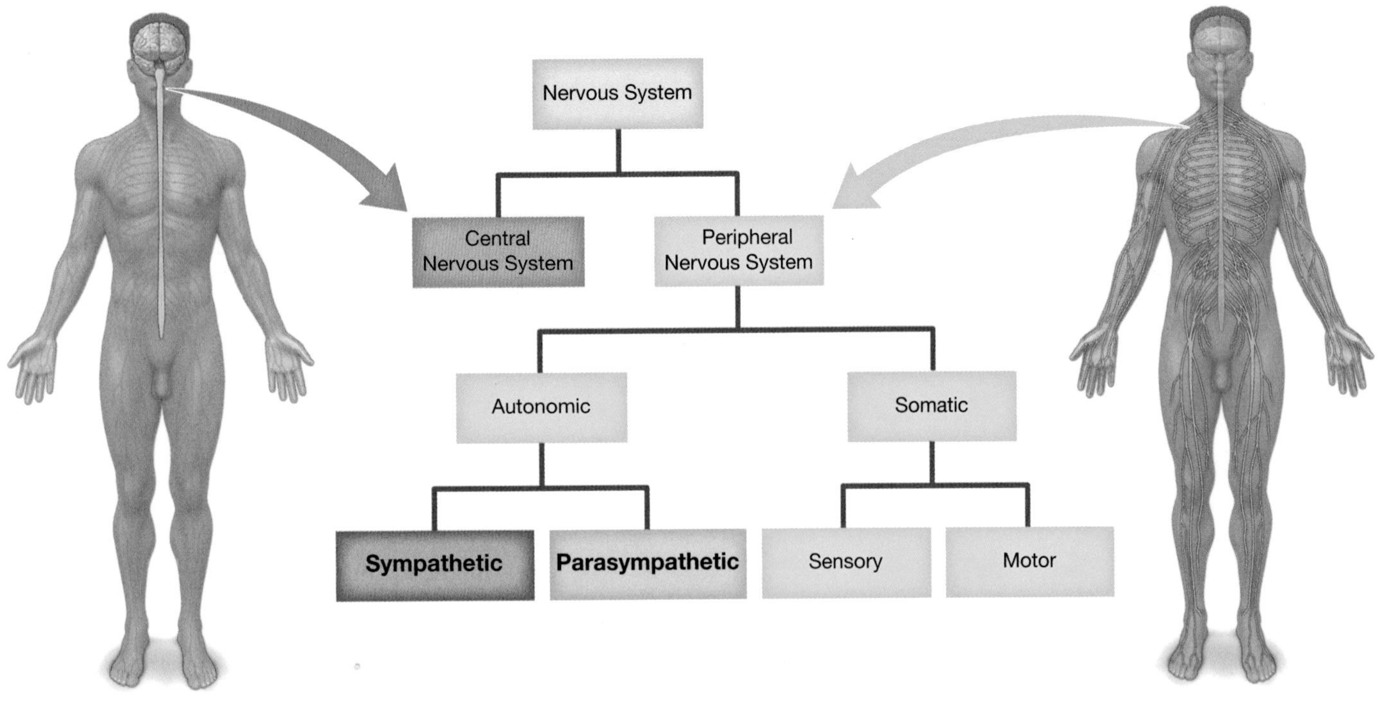

exams, struggling to find money for rent, or getting through a period with less food than needed. While the first two, the exams and rent, may not seem directly linked to homeostasis, our concern during those stressors activates the same components of the nervous system as escaping from a predator. While stress gets a bad reputation (as it should, for stress that goes on too long is very damaging to your body), the sympathetic nervous system responses are absolutely essential in a day-to-day way. The sympathetic nervous system helps you to maintain your blood sugar levels in between meals and your blood pressure throughout a variety of circumstances.

To easily remember the different effects of sympathetic nervous system activation, you can imagine the body's response as it quickly prepares to run from a predator:

- More oxygen needs to be inhaled and delivered to skeletal muscle.
- Sweating keeps the excess heat that comes from muscle contraction from causing the body to overheat.
- Blood is shifted away from the digestive system, and its activity is therefore slowed. Blood flow is shifted toward the task of delivering oxygen to skeletal muscles; this is not the time to stop and eat a snack.
- Pupils dilate and the brain becomes alert to take in more information about potential dangers.

LO 16.1.4

The sympathetic nervous system directives must be disseminated quickly and to an array of effectors in the body. To coordinate all these responses simultaneously, the connections in the sympathetic system diverge from a limited region of the central nervous system (CNS). The connections for the sympathetic nervous system emerge from the thoracic and upper lumbar spinal cord (**Figure 16.2**). It is referred to as the **thoracolumbar system** to reflect this anatomical basis. The format for connections to the effectors is summarized in the "Anatomy of the Sympathetic Nervous System Pathway" feature

thoracolumbar system
(sympathetic system)

Figure 16.2 Pathways of the Sympathetic and Parasympathetic Divisions

The sympathetic division neurons exit from the thoracic and lumbar regions of the spinal cord; the parasympathetic neurons exit from the brainstem and sacral spinal cord. While both pathways diverge widely, the extensively branching neurons are the preganglionic neurons of the parasympathetic division and the postganglionic neurons of the sympathetic division.

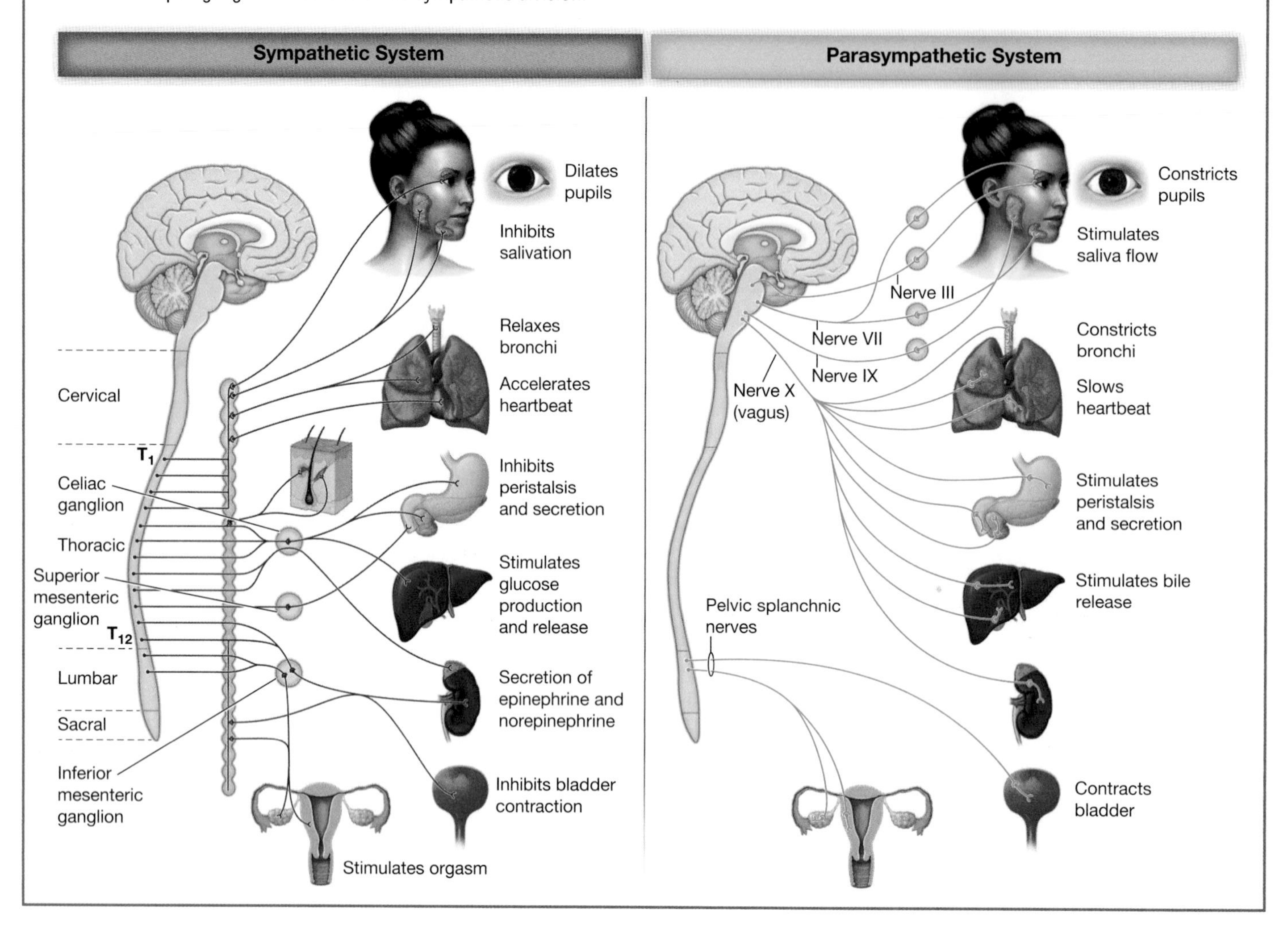

sympathetic chain ganglia (paravertebral ganglia)

and is as follows: the first neuron has a cell body in the lateral horn of the thoracic or upper lumbar spinal cord. Its axon projects through the anterior root and reaches a ganglion. Because all ANS pathways involve two neurons with a synapse in a ganglion, the first neuron in the system is referred to as the **preganglionic neuron** and the second neuron in the system is the **postganglionic neuron**. The majority of ganglia of the sympathetic system are in a long string—the **sympathetic chain ganglia**—that runs alongside the vertebral column. There are typically 23 ganglia in the chain on either side of the spinal column. Three correspond to the cervical region, twelve are in the thoracic region, four are in the lumbar region, and four correspond to the sacral region. The cervical and sacral levels are not connected to the spinal cord directly through the spinal roots, but through ascending or descending connections through the bridges within the chain. Some primary sympathetic neurons form their synapse in ganglia outside the sympathetic chain.

There are three different routes that the axon of a preganglionic neuron in the sympathetic system might take to reach its synapse (**Figure 16.3**). The first type is most direct: the preganglionic axon projects to the chain ganglion at the same level as its target effector (the organ, tissue, or gland it innervates). In this example, **Figure 16.3A**,

Anatomy of...

A Sympathetic Nervous System Pathway

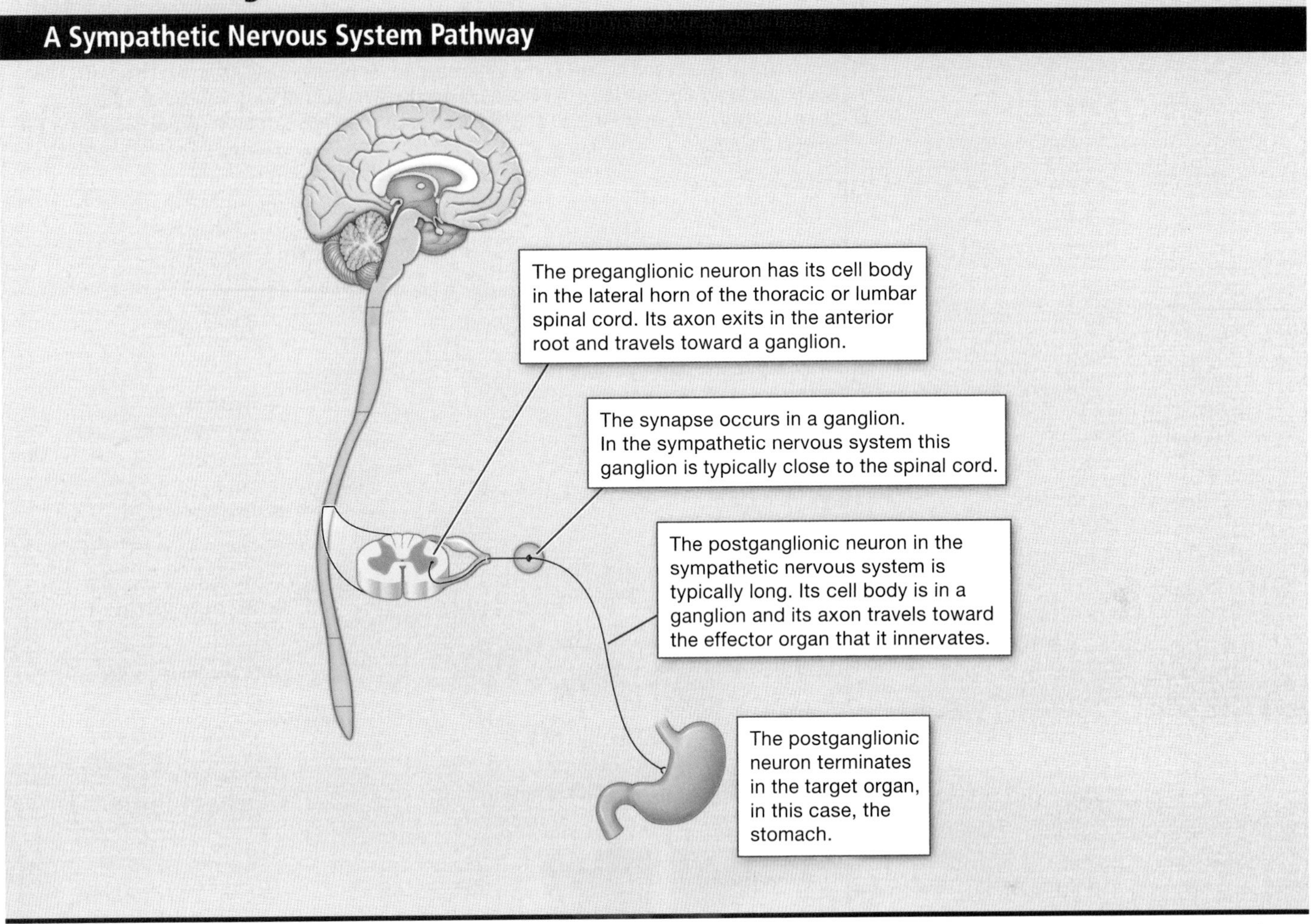

the preganglionic axon synapses within the chain ganglion at the same level as its target effector. All of these structures are within a single transverse plane in the body. The axons of the preganglionic neurons travel within the **white rami communicantes** (singular = *ramus communicans*); which are myelinated and therefore referred to as "white." The preganglionic neuron synapses with the postganglionic neuron within the T1 sympathetic chain ganglion. This neuron then projects to a target effector—in this case, the trachea—via **gray rami communicantes**, which are unmyelinated axons.

The preganglionic neuron has its cell body in the lateral horn of the thoracic or lumbar spinal cord. Its axon exits in the anterior root and travels toward a ganglion.
The synapse occurs in a ganglion. In the sympathetic nervous system this ganglion is typically close to the spinal cord.
The postganglionic neuron in the sympathetic nervous system is typically long. Its cell body is in a ganglion and its axon travels toward the effector organ that it innervates.
The postganglionic neuron terminates in the target organ, in this case, the stomach.

In some cases, the target effectors are located superior or inferior to the spinal segment at which the preganglionic fiber emerges. In these cases, the synapse with the postganglionic neuron occurs at chain ganglia superior or inferior to the location of the preganglionic neuron (**Figure 16.3B**). An example of this are the preganglionic sympathetic neurons that innervate the eye. The spinal nerve tracks up through the chain until it reaches the **superior cervical ganglion**, where it synapses with the postganglionic neuron.

Not all axons from the preganglionic neurons terminate in the chain ganglia. Additional axons from the anterior nerve root continue through the chain and on to one of the prevertebral ganglia as the **greater splanchnic nerve** or **lesser splanchnic nerve**. For example, the greater splanchnic nerve at the level of T5 synapses with a collateral ganglion outside the chain before making the connection to the postganglionic nerves that innervate the stomach (**Figure 16.3C**).

white rami communicantes (white communicating branch, white communcating ramus)

gray rami communicantes (gray communicating branch, gray communicating ramus)

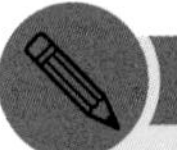

Learning Connection

Try Drawing It

In Chapter 14 you may have gained practice drawing a cross section of the spinal cord. Try reviewing those drawings and add the rami communicantes to your drawing. Remember to include where the axons begin and terminate.

Figure 16.3 Sympathetic Pathways

The preganglionic axon can follow a few pathways. (A) The preganglionic neuron can project out to the ganglion at the same level and synapse on a postganglionic neuron. (B) A preganglionic axon can reach and synapse on a more superior or inferior ganglion in the chain. (C) The axon can project through the white ramus communicans, but not form a synapse within the chain. Instead, it projects through one of the splanchnic nerves to a prevertebral ganglion or the adrenal medulla.

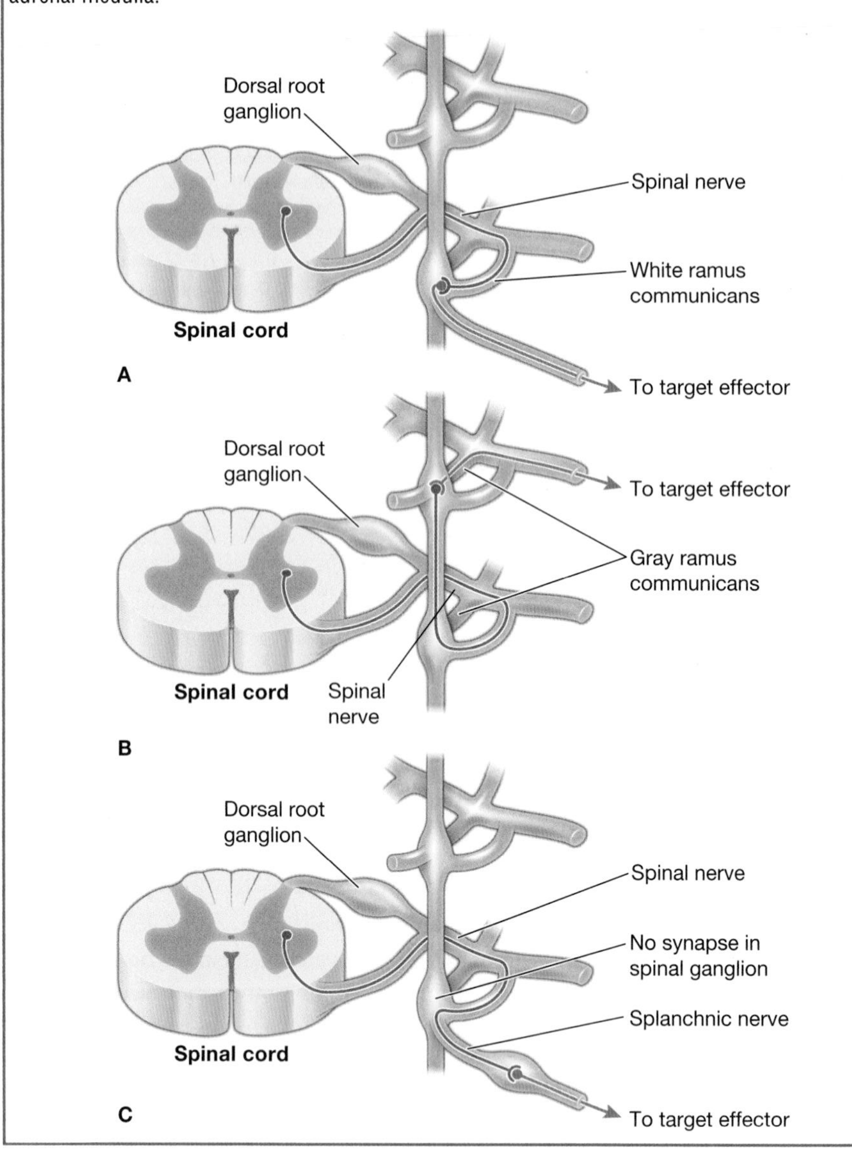

collateral ganglia (prevertebral ganglia)

Collateral ganglia are situated anterior to the vertebral column and receive inputs from splanchnic nerves as well as from central sympathetic neurons. Due to their location in front of the vertebral column, they are also called *prevertebral ganglia*. They help regulate the activities of organs in the abdominal cavity and are also considered part of the enteric nervous system. The three collateral ganglia are the **celiac ganglion**, the **superior mesenteric ganglion**, and the **inferior mesenteric ganglion** (see Figure 16.2).

Regardless of where the ganglion is located, the synapse between the preganglionic and postganglionic sympathetic neurons always involves the same neurotransmitter, acetylcholine (ACh). The preganglionic neuron releases ACh, and the postganglionic fiber has ACh receptors on its dendrites or cell body surface. Most postganglionic neurons in the sympathetic division release norepinephrine onto their target cells of the effector organ.

Compared with the preganglionic fibers, postganglionic fibers of the sympathetic nervous system are long because of the relatively greater distance from the ganglion to the target effector. These fibers are unmyelinated.

The coordination of responses during sympathetic nervous system activation is supplemented by a hormonal response. Many cells throughout the body express receptors for two related hormones that are produced and released from an endocrine gland, the **adrenal medulla**, the interior portion of a structure called the *adrenal gland*. This response is detailed further in this chapter. These hormones are released in response to action potentials traveling down a preganglionic axon that terminated within the adrenal gland. There is no ganglion in the chain (**Figure 16.4**).

adrenal medulla (suprarenal medulla)

Figure 16.4 A Comparison of Synapse Locations and Neurotransmitter Use in the Different Branches of the Nervous System

Somatic motor pathways have only one peripheral neuron that extends from the anterior horn of the spinal cord to the muscle. The synapse—the neuromuscular junction—releases acetylcholine (ACh). In the sympathetic nervous system, there are two neurons in a chain; the synapse is close to the spinal cord. The preganglionic axon releases ACh. The postganglionic axon releases norepinephrine. In the parasympathetic nervous system, there are also two neurons in a chain, but the synapse is typically much farther from the spinal cord. The preganglionic and postganglionic axons both release ACh. In the adrenal pathway, a synapse occurs within the adrenal gland. When stimulated, the postganglionic neuron releases epinephrine and norepinephrine into the bloodstream.

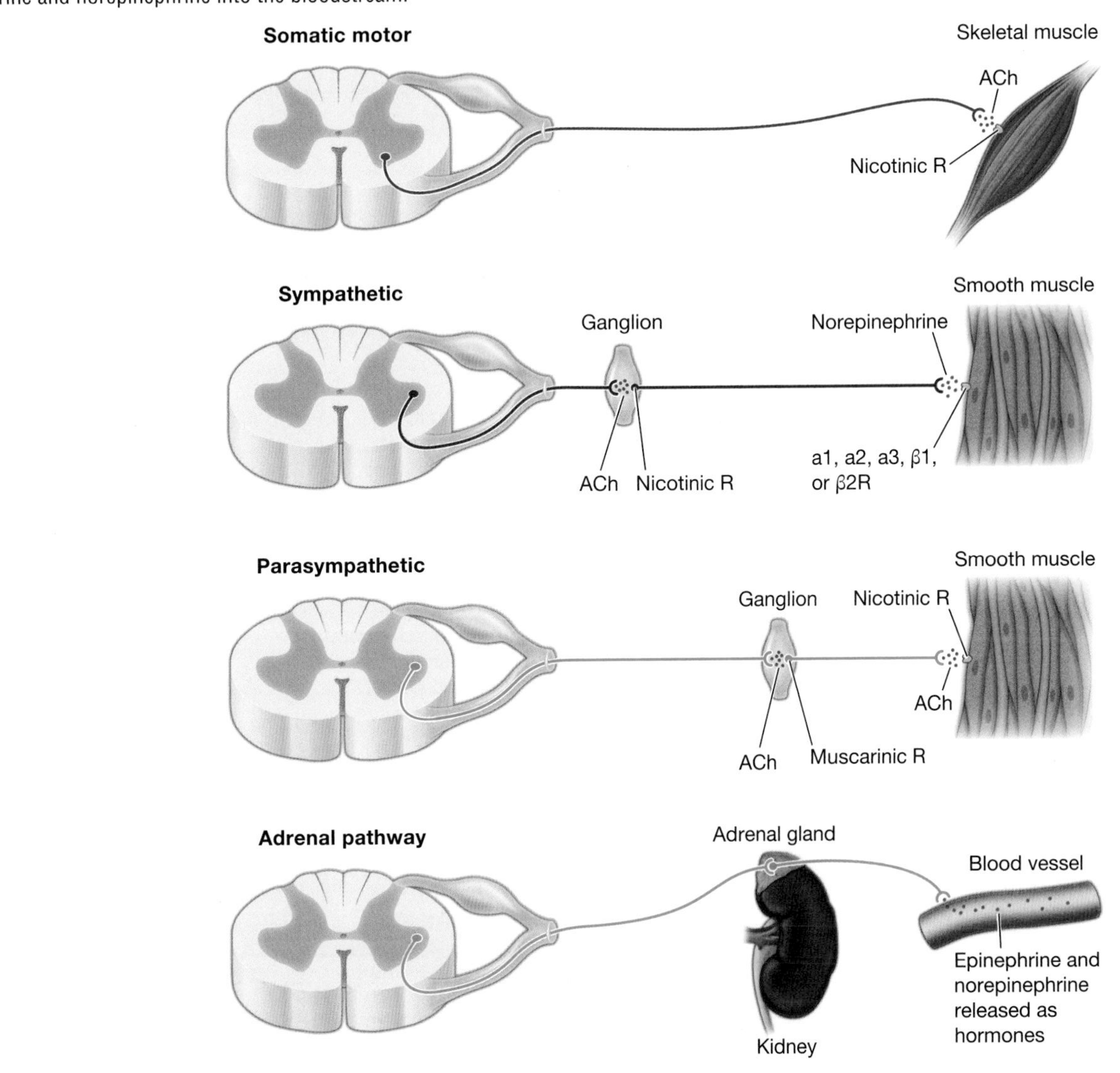

The projections of the sympathetic division of the ANS branch widely, resulting in a broad influence of the system throughout the body. This divergence is seen in the branching patterns of postganglionic sympathetic neurons—a single postganglionic sympathetic neuron may have 10–20 targets. This anatomical divergence results in simultaneous functional changes. As a response to a threat to homeostasis, the sympathetic nervous system will increase heart rate and breathing rate as well as shift blood flow toward the skeletal muscles and away from the digestive system. Sweat gland secretion will also increase as part of the integrated response. The sympathetic nervous system also works to liberate stored nutrients so that the muscles have plenty of energy even without eating. All of those physiological changes are going to be required to occur simultaneously to either run away from or fight the threat. In modern stress responses these physiological effects may be less helpful.

16.1b Parasympathetic Division of the Autonomic Nervous System

The parasympathetic division of the ANS is named because its preganglionic neurons are located on either side of the thoracolumbar region of the spinal cord (para- = "beside" or "near"). The parasympathetic system can also be referred to as the **craniosacral system** because the cell bodies of the preganglionic neurons are located in the brain stem and the lateral horn of the sacral spinal cord.

craniosacral system (parasympathetic system)

The pathway format of the parasympathetic division is detailed in the "Anatomy of the Parasympathetic Nervous System Pathway" feature. Whereas the long preganglionic fibers from the cranial region travel in cranial nerves, preganglionic fibers from the sacral

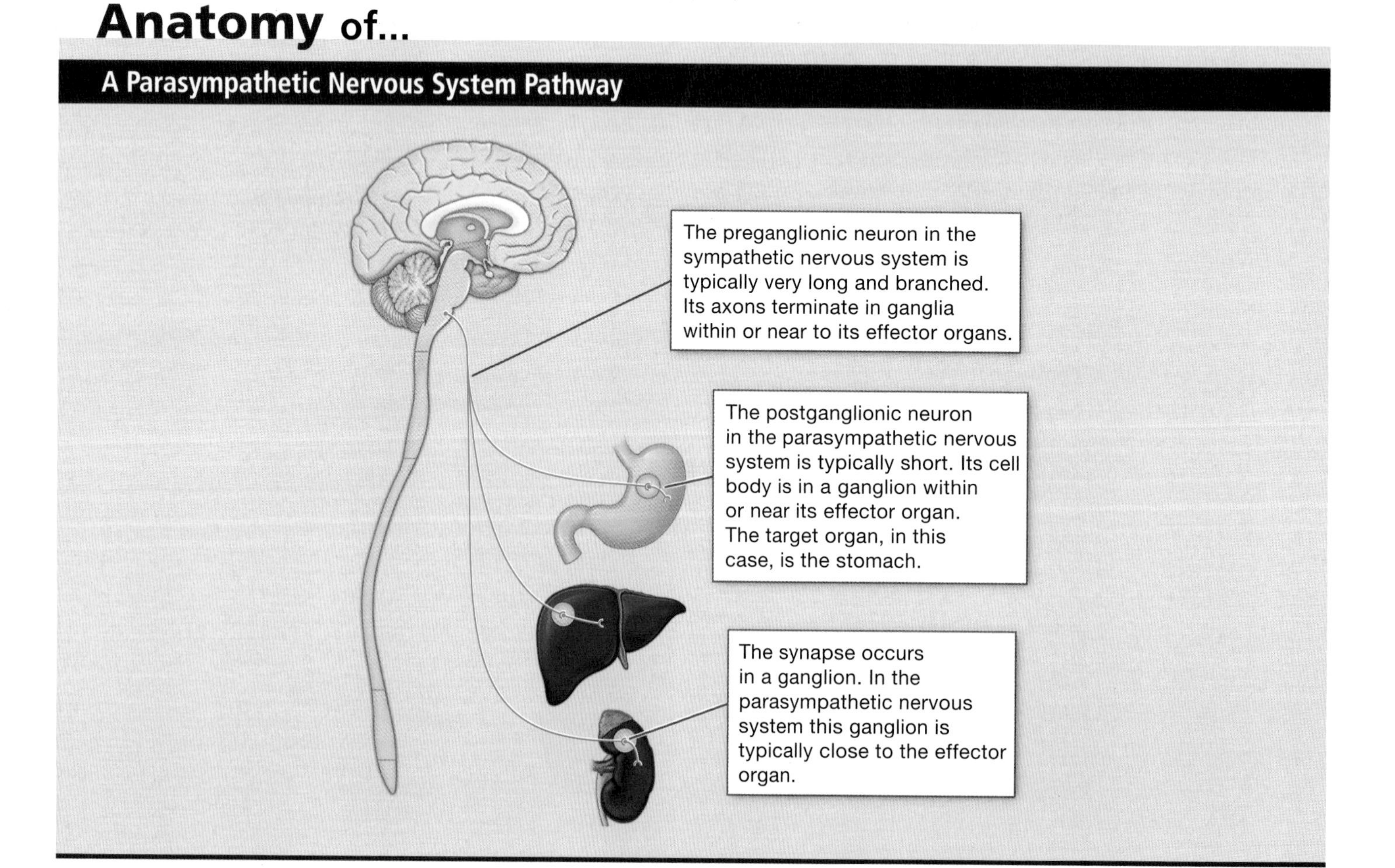
Anatomy of...
A Parasympathetic Nervous System Pathway
The preganglionic neuron in the sympathetic nervous system is typically very long and branched. Its axons terminate in ganglia within or near to its effector organs.
The postganglionic neuron in the parasympathetic nervous system is typically short. Its cell body is in a ganglion within or near its effector organ. The target organ, in this case, is the stomach.
The synapse occurs in a ganglion. In the parasympathetic nervous system this ganglion is typically close to the effector organ.

region travel in peripheral nerves. The targets of these fibers are **terminal ganglia**, which are located near the target effector, or **intramural ganglia**, which are located within the walls of the target organ. The short postganglionic fiber projects from the ganglion to the specific target tissue within the organ. Comparing the relative lengths of axons in the parasympathetic system, the preganglionic fibers are long and the postganglionic fibers are short because the ganglia are close to—and sometimes within—the target effectors.

There are five main paths of the parasympathetic system:

- The oculomotor nerve (CN III) innervates muscles that move the eyeball, allowing you to look in different directions. The preganglionic axons within CN III originate within nuclei in the brain and terminate in the **ciliary ganglion**, which is located in the orbit. The postganglionic neurons then project to the smooth muscle of the iris to control pupillary size.
- The facial nerve (CN VII) contains axons of preganglionic neurons that control salivary, mucous, and tear glands. These preganglionic axons diverge and travel through different branches of CN VII. Some axons terminate in the **pterygopalatine ganglion** and other axons follow a different branch and terminate in the **submandibular ganglion**.
- The glossopharyngeal nerve (CN IX) also contributes to the production of saliva. Preganglionic neurons from CN IX terminate in the **otic nucleus**, where they synapse with the postganglionic neurons. There are three pairs of salivary glands, and each gland produces a slightly different saliva mixture. You will learn more about these glands in Chapter 23; for now, however, you can understand that two of the pairs are innervated by the CN VII and one pair is innervated by CN IX.
- The vagus nerve (CN X) innervates organs in the thoracic, abdominal, and pelvic cavities. CN X gets its name, *vagus*, which means "wanderer" in Latin, because of the meandering and widespread path of its neurons. Many of the preganglionic neurons weave together with sympathetic postganglionic neurons in plexuses throughout the thoracic and abdominal cavities. The names and locations of these plexuses are outlined in Figure 16.4.
- The **pelvic splanchnic nerves** are formed by the preganglionic neurons of the sacral portion of the parasympathetic division. These axons reach through the hypogastric plexus and toward terminal or intermural ganglia throughout the pelvic cavity.

pterygopalatine ganglion (sphenopalatine ganglion, nasal ganglion)

otic nucleus (otic ganglion)

Learning Connection

Quiz Yourself

List the major ganglia of the sympathetic and parasympathetic divisions of the ANS. After studying their locations and pathways in the text and art figures, stand in front of a mirror and try pointing to their general locations on yourself as you say their names.

Learning Check

1. Which of the following is/are associated with the somatic nervous system? Please select all that apply.
   a. Causes contraction of skeletal muscle
   b. Controls cardiac and smooth muscle
   c. Associated with voluntary movement
   d. Has two divisions that regulate organs
2. Which division of the autonomic nervous system do you expect to become activated right before taking a stressful test?
   a. Somatic nervous system
   b. Visceral nervous system
   c. Sympathetic nervous system
   d. Parasympathetic nervous system
3. Which of the following is/are a component of the parasympathetic nervous system? Please select all that apply.
   a. Prevertebral ganglia
   b. Celiac ganglion
   c. Terminal ganglia
   d. Intramural ganglia
4. Where would a nervous impulse be transmitted next as it emerges from the ciliary ganglion?
   a. Oculomotor nerve (CN III)
   b. Vagus nerve (CN X)
   c. Otic nucleus
   d. Smooth muscles of the iris

16.2 Chemical Components of the Autonomic Responses

HAPS LEARNING GOALS AND OUTCOMES

Learning Objectives: By the end of this section, you will be able to:

16.2.1 Compare and contrast cholinergic and adrenergic receptors with respect to neurotransmitters that bind to them, receptor subtypes, receptor locations, target cell response (i.e., excitatory or inhibitory), and examples of drugs, hormones, and other substances that interact with these receptors.

16.2.2 Explain the relationship between chromaffin cells in the adrenal medulla and the sympathetic division of the nervous system.

Where an autonomic neuron connects with a target, there is a synapse. The electrical signal of the action potential causes the release of a signaling molecule, which will bind to receptor proteins on the target cell. Synapses of the autonomic system are classified as either **cholinergic**, meaning that **acetylcholine (ACh)** is released, or **adrenergic**, meaning that **norepinephrine** is released. The term *adrenergic* comes from the same root as the common term for norepinephrine and epinephrine (adrenaline). The terms *cholinergic* and *adrenergic* refer not only to the signaling molecule that is released, but also to the receptors they bind to.

LO 16.2.1

The cholinergic (ACh) system includes two classes of receptors: **nicotinic receptors** and **muscarinic receptors**. The nicotinic receptor is a **chemically gated ion channel**, while muscarinic receptors are **receptors** that trigger changes within the cell without allowing ions to pass through the membrane. The receptors are named for their exogenous ligands, the molecules from outside the body that bind them. The endogenous ligand for both nicotinic and muscarinic receptors is ACh.

chemically gated ion channel (ligand-gated ion channel)

Student Study Tip

You might have heard of beta blockers, which dampen the effects of neurotransmitters that bind to beta receptors. They have various effects, one being slowing heart rate.

The adrenergic system also has two broad classes of receptors, named the **alpha (α)-adrenergic receptor** and the **beta (β)-adrenergic receptor**. There are two types of α-adrenergic receptors, termed α1 and α2, and three types of β-adrenergic receptors, termed β1, β2, and β3. The adrenergic receptors can bind both norepinephrine and a related molecule, **epinephrine** (receptors of the autonomic Nervous system are summarized in **Figure 16.5**).

epinephrine (adrenaline)

All preganglionic neurons, both sympathetic and parasympathetic, release ACh. All postganglionic neurons—the targets of these preganglionic axons—have nicotinic receptors in their cell membranes. The nicotinic receptor is a ligand-gated cation

Figure 16.5 The Cholinergic and Adrenergic Synapses of the Autonomic Nervous System

The autonomic nervous system has two types of synapses, cholinergic and adrenergic. ACh is the neurotransmitter used in cholinergic synapses. Epinephrine and norepinephrine are used in adrenergic synapses.

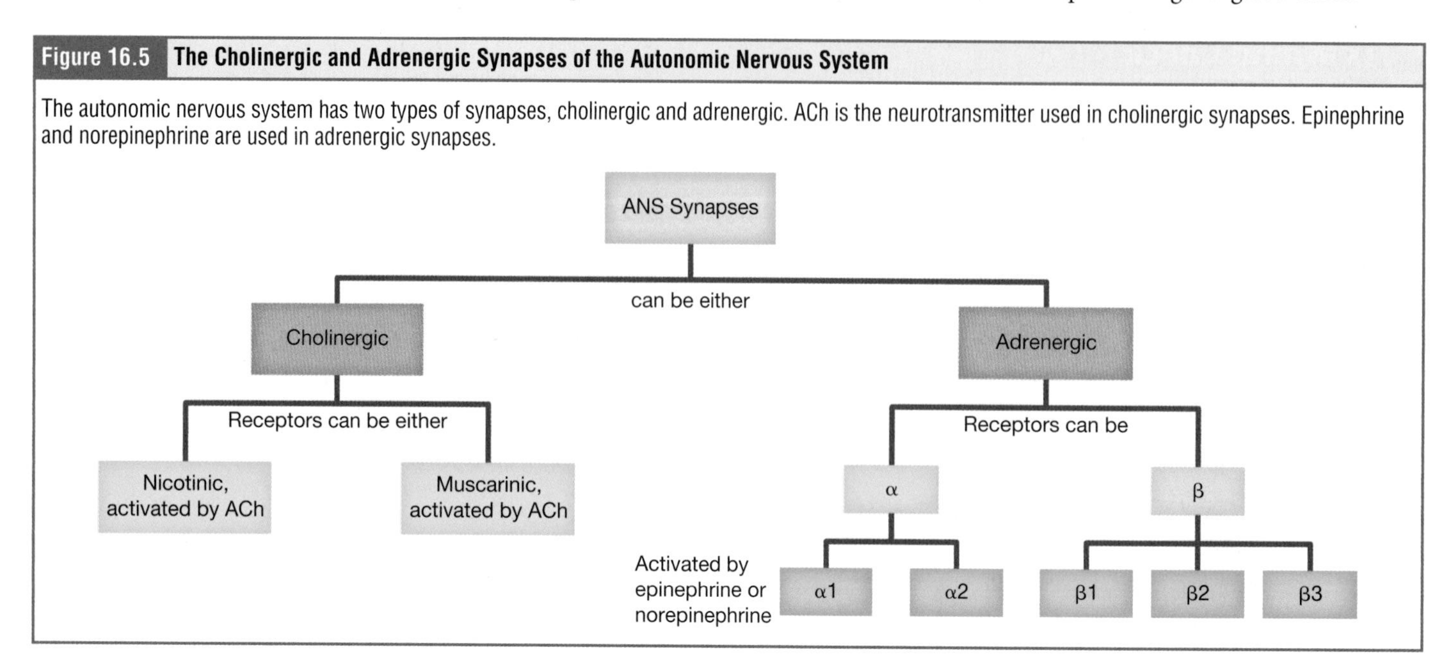

DIGGING DEEPER:
Ephedra as a Weight Loss Supplement

Weight loss can be very difficult. Historically, some individuals have tried taking diet pills to help speed up the process. Ephedra, an herb that is derived from the plant *Ephedra sinica*, which grows around the world, including in China (where it is known as *Ma Huang*) and India, has been used historically as a dietary supplement to promote moderate weight loss and increase energy. The active ingredient in ephedra is called *ephedrine*, which has been shown in several studies to increase resting metabolic rate and the rate at which your body uses fat metabolically. Both of these effects can promote weight loss.

Ephedrine belongs to a group of chemical agents called *sympathomimetics*, whose name is derived from the fact that their effects mimic those of sympathetic nervous system activation. Ephedrine promotes norepinephrine release from sympathetic postganglionic neurons, increases norepinephrine activity on α- and β-adrenergic receptors, and prevents reuptake of norepinephrine by the sympathetic neurons that secrete it. These actions increase the availability and effects of norepinephrine in the body. Systemically, this activates your stress response. Effects include increased heart rate, breathing rate, and blood pressure, dilation of the pupils of the eyes, bronchodilation of the major airways, and inhibition of digestive processes. All of these actions result in increased metabolic rate in the body, which causes a person to burn more Calories per hour and, assuming the amount of food intake has not increased, weight loss is likely to occur. The structure of ephedrine is similar to that of amphetamines (such as methamphetamine), so it is not surprising that they produce similar effects.

However, due to reports of many adverse and dangerous side effects, the use of ephedra in diet pills has been banned in some countries. The United States banned its use for this purpose in 2004. Some of the reported side effects include nausea and vomiting, anxiety, seizures, hypertension, strokes, cardiac arrhythmias, heart attacks, and even death. Some of these effects were intensified when ephedra was combined with caffeine in diet pills. However, it is important to realize that the ephedra ban in the United States applied to its use in unregulated, over-the-counter dietary supplements. Its active ingredient, ephedrine, is still used in some FDA-regulated medications that treat hypotension (low blood pressure), asthma, and nasal congestion (for example, Sudafed).

channel that results in depolarization of the postsynaptic membrane. The postganglionic parasympathetic fibers also release ACh, but the receptors on their targets are muscarinic receptors, which are receptors but not ion channels and do not cause depolarization of the postsynaptic membrane. Postganglionic sympathetic axons release norepinephrine, except for axons that project to sweat glands and to blood vessels associated with skeletal muscles, which release ACh (summarized in Figure 16.4).

Apply to Pathophysiology

Treating Asthma (Atropine and Albuterol)

Asthma is a disease in which the tubes of the lungs—the bronchi—can narrow unexpectedly. There are a number of diverse triggers for asthma, from allergies to exercise, but once it begins, the set of events is always the same. The smooth muscle that wraps the bronchi begins to constrict, reducing the available space for air to flow from inside to out and outside to in. In order to restore airflow, a doctor needs to relax the smooth muscle of the bronchi.

1. Which division of the ANS is responsible for narrowing the airways?
   A) Parasympathetic
   B) Sympathetic
2. Eliot is a 7-year-old experiencing an asthma attack. Eliot's mother delivers an inhaled dose of albuterol, which is a beta-2 adrenergic receptor agonist. An agonist is a drug that activates that receptor. What other molecule binds to beta-adrenergic receptors and activates them?
   A) Muscarine
   B) Acetylcholine
   C) Norepinephrine
3. Even after the albuterol, Eliot is struggling with their asthma. Eliot's mother drives Eliot to the emergency room, where they are given another inhaled medicine: a mixture of albuterol and atropine. Atropine is a muscarinic receptor antagonist. An antagonist is a molecule that binds to a receptor and blocks it, preventing the body's endogenous molecules from binding their receptor. What molecule is being blocked now that atropine is in Eliot's system?
   A) Muscarine
   B) Acetylcholine
   C) Norepinephrine

We use the term *synapse* throughout this chapter, but autonomic postganglionic neurons in a synapse are unlike those that we have studied at the neuromuscular junction or between neurons. Rather than the typical axon terminals we are used to, the structures of these synapses are chains of swellings along the length of a postganglionic axon, each of which is called a **varicosity** (**Figure 16.6**).

LO 16.2.2

adrenal gland (suprarenal gland)

The term *adrenergic* may remind you of the word *adrenaline*. Adrenaline and epinephrine are two names for the same molecule. The **adrenal gland** secretes epinephrine and norepinephrine as hormones in response to stimulation from a preganglionic sympathetic neuron (see Figure 16.4). Norepinephrine also functions as a neurotransmitter. Remember that whereas a hormone is a signaling molecule that travels in the blood, a neurotransmitter is a local signaling molecules used at a synapse. The same molecule can be a neurotransmitter or a hormone, as we see here; it just depends on where they are released and therefore how widespread their effects are. The adrenal gland secretes a mix of epinephrine and norepinephrine that is about 80 percent epinephrine and 20 percent norepinephrine. These hormones spread rapidly throughout the body and operate on adrenergic receptors systemically. The release of these molecules into the bloodstream allows the sympathetic nervous system response to go on longer than it would if it were based on neurotransmitters—which are quickly

Figure 16.6 Varicosities

The release of neurotransmitter from an autonomic postganglionic neuron occurs at many places along the axon rather than from an axon terminal.

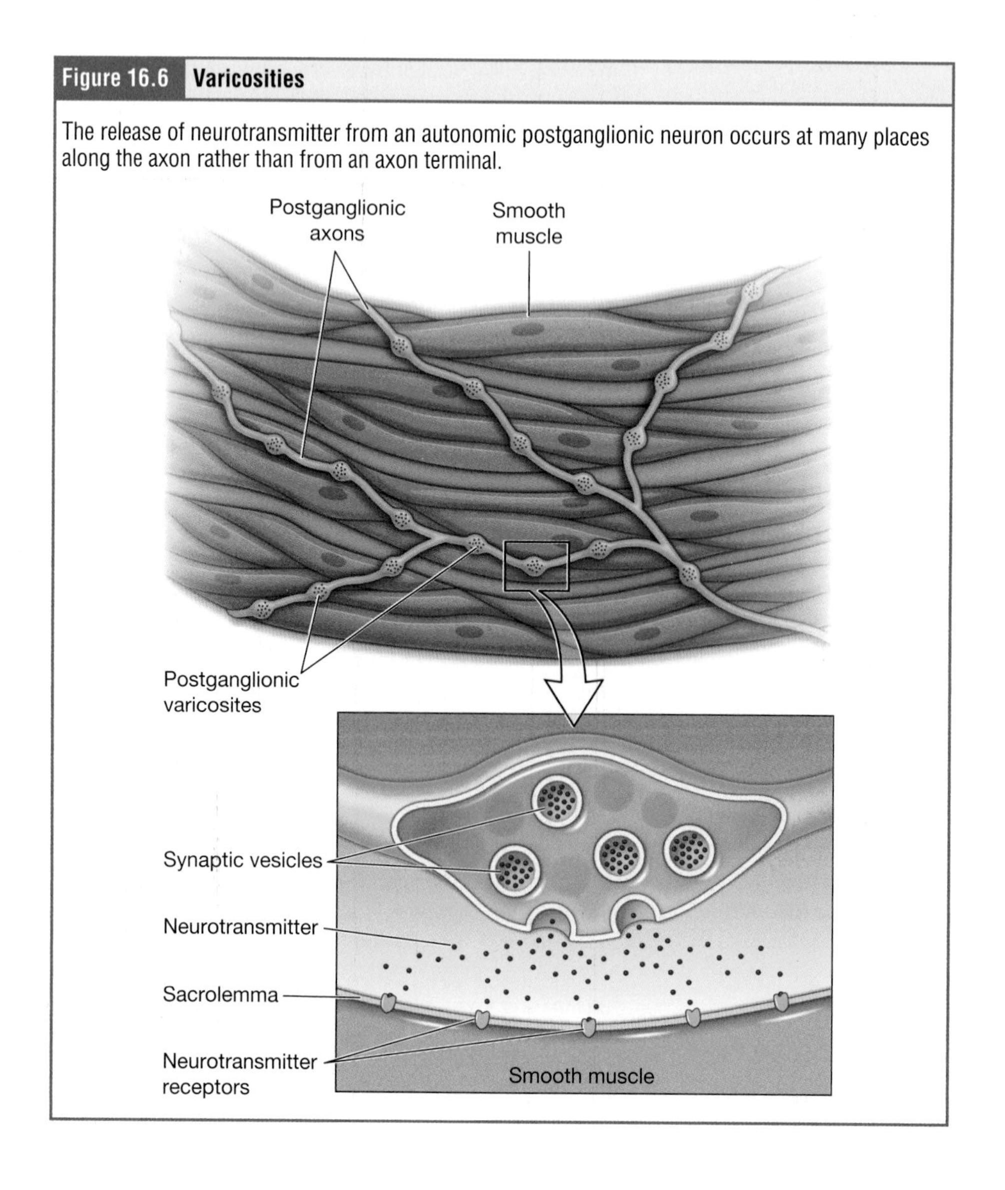

cleared from the synapse—alone. You have undoubtedly experienced the prolonging of a sympathetic nervous system response before. Think about how you feel when someone jumps out from behind a corner and surprises you. Even when the person is a friend or a sibling, it takes a minute or two for your heart and breathing rate to return to normal. Evolution may have driven the incorporation of a hormone into the sympathetic nervous system response due to the advantage of having prolonged alertness and readiness after an initial activation.

You may remember from Chapter 4 that receptors bind molecules based on their size and molecular properties. For example, epinephrine and norepinephrine have very similar chemical structures, varying only slightly. As a consequence, they both bind the same receptors. Therefore, exogenous molecules—those that originate from outside the body—can bind to our internal receptors if they share shape and molecular similarities with our own signaling molecules. Some naturally occurring molecules bind our receptors by coincidence. Examples of this include opium (derived from the *Papaver somniferum* variety of the poppy plant), which binds to endogenous opioid receptors, and muscarine (made by certain mushrooms), which binds to muscarinic receptors. Other molecules are made in a lab and purposely designed to bind to our receptors; one example of this is albuterol, a drug that binds to β_2 adrenergic receptors and is used to treat asthma. **Table 16.1** details the adrenergic receptors, their roles, and the molecules—both endogenous and exogenous—that bind to them. **Table 16.2** details the cholinergic receptors, and their roles and ligands.

Learning Connection

Explain a Process

Examine Table 16.1 and imagine that a blood vessel that leads to skeletal muscle expresses β_2 adrenergic receptors and a blood vessel that leads to the intestines expresses α_1 adrenergic receptors. In a sympathetic nervous system response in which epinephrine is released into the bloodstream, what will happen to blood flow to each of these locations? How might this be advantageous in the sympathetic nervous system response?

Table 16.1 Adrenergic Receptors

Receptor Class	Receptor Sub-class	Site(s) in the Body	Roles	Endogenous Ligand(s)
Alpha receptors	Alpha-1	Skin, GI and pelvic organs, blood vessels	Causes smooth muscle contraction in most organs Causes arrector pili muscles to tense in skin Causes vasoconstriction of blood vessels	Both epinephrine and norepinephrine
	Alpha-2	Pancreas, platelets, CNS	Inhibits insulin release, leading to higher levels of sugar in the blood Promotes clotting Reduces secretion of norepinephrine in the brain and spinal cord	
Beta receptors	Beta-1	Heart and kidney	Increase heart rate and force of contraction Increases renal secretion of renin, which leads to higher blood pressure	Both epinephrine and norepinephrine
	Beta-2	Blood vessels (select locations), lungs, uterus, stomach, and intestines	Causes relaxation and dilation of blood vessels that lead to the heart, skeletal muscles, and liver Causes relaxation of the bronchioles in the lungs Causes stomach and intestines to relax and expand	
	Beta-3	Adipose tissue	Stimulates the breakdown of stored fat, increasing the amount of lipids in the bloodstream	

Table 16.2 Cholinergic Receptors

Receptor Type	Location(s)	Roles	Endogenous Ligand(s)	Example of an Exogenous Ligand
Nicotinic Receptors	Adrenal medulla, all skeletal neuromuscular junctions, postganglionic neurons throughout the ANS, some CNS synapses	Depolarizes the postsynaptic cell, causing contraction or a new action potential	Acetylcholine	Nicotine—A drug in cigarettes and some vaping devices, causes skeletal muscle activation, but the CNS locations lead to addiction
Muscarinic Receptors	All target tissues of the parasympathetic nervous system	Can have excitatory effects or inhibitory effects, but is not an ion channel so does not directly affect membrane potential	Acetylcholine	Muscarine—A type of mushroom poison; causes salivation, intestinal cramping, slowing of the heart rate

Learning Check

1. Which of the following receptors do norepinephrine bind to? Please select all that apply.
   a. Nicotinic receptor
   b. Muscarinic receptors
   c. Alpha adrenergic receptors
   d. Beta adrenergic receptors
2. Which neurotransmitter is released when the autonomic nervous system wants to increase heart rate?
   a. Acetylcholine
   b. Norepinephrine
   c. Epinephrine
   d. Dopamine
3. Which of the following hormones does the adrenal gland release in the largest amount?
   a. Acetylcholine
   b. Norepinephrine
   c. Epinephrine

16.3 Autonomic Reflexes and Homeostasis

Learning Objectives: By the end of this section, you will be able to:

16.3.1 Describe visceral reflex arcs, including structural and functional details of sensory and motor (autonomic) components.

The ANS regulates organ systems through reflexes similar to the somatic reflexes described in Chapter 15. Remember that a reflex arc integrates incoming sensory information (afferent information) with an automatic outgoing motor response (efferent response). The main difference between somatic and autonomic reflexes is in what target tissues are effectors. Somatic responses are based solely on skeletal muscle contraction. The effectors of the autonomic system, however, are cardiac muscle, smooth muscle, and glands.

16.3a The Structure of Reflexes

In comparing somatic and visceral reflexes, we can compare the afferent and efferent branches separately. The **afferent branch**, carrying sensory information for integration,

Figure 16.7 Comparison of Somatic and Visceral Reflexes

The outgoing motor pathway differs between autonomic and somatic reflexes. In autonomic reflexes there are two neurons in a chain; their synapse is in a ganglion. The synapse between the postganglionic neuron and the target cells occurs at varicosities. In somatic reflexes the motor pathway consists of a single motor neuron, which ends at a neuromuscular junction with a skeletal muscle cell.

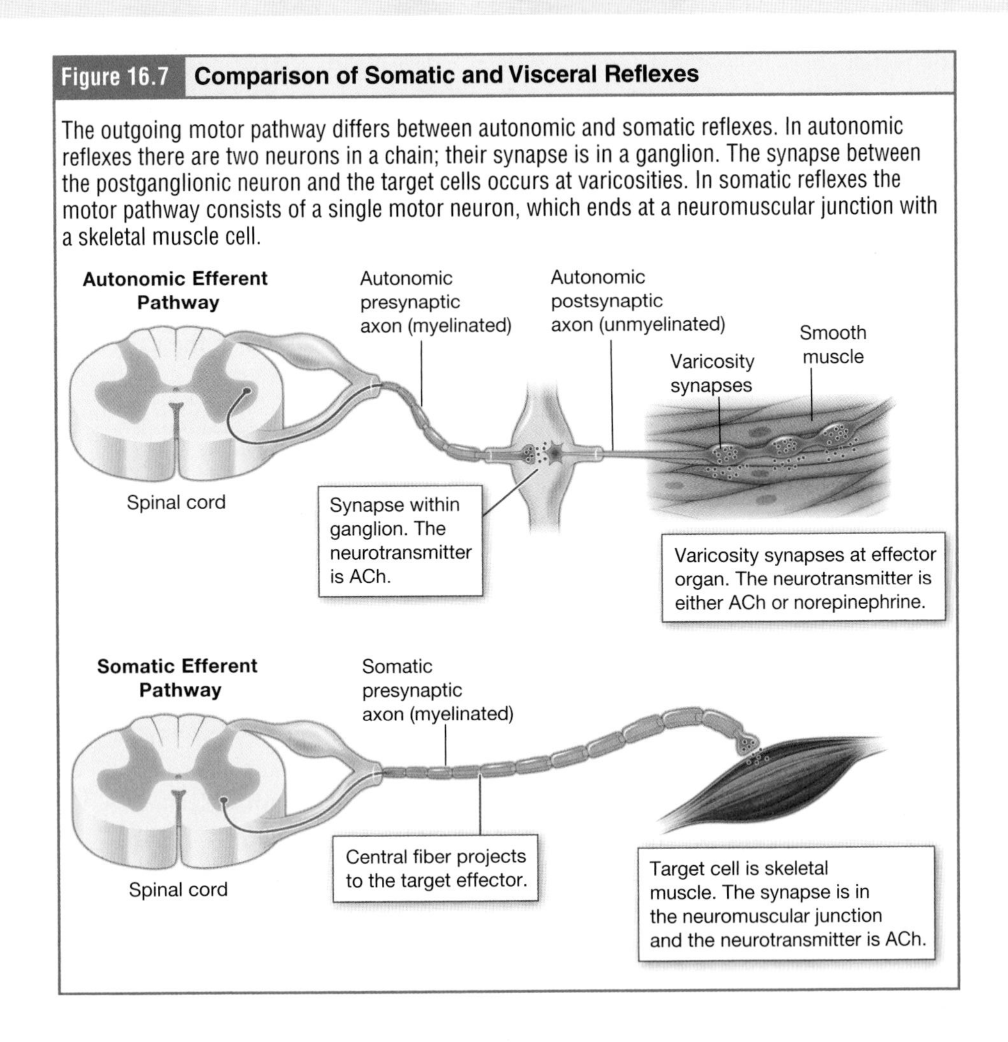

looks similar in both somatic and visceral reflexes. A single sensory neuron receives input from the periphery and projects into the CNS to initiate the reflex (**Figure 16.7**).

The afferent branch of a reflex arc does differ between somatic and visceral reflexes in some instances. The inputs to visceral reflexes can come from special or somatic senses, similar to the inputs for somatic reflexes. But some visceral reflex arcs receive sensory input from the viscera; these visceral sensations never result in conscious perception. For example, **baroreceptors**, special mechanoreceptors in the walls of the aorta and carotid sinuses that sense stretch in the walls of those vessels when blood pressure increases. This information feeds into a reflex that helps regulate blood pressure and keep it within homeostatic range. You do not have a conscious perception of having high blood pressure, but this information is critical for your survival.

Though visceral senses are not primarily a part of conscious perception, but some of those sensations may make it to conscious awareness. If you swallow a large chunk of a bagel or other solid food, for instance, you will probably feel the lump of food as it is pushed through your esophagus. If you inhale especially cold air, you can feel it as it enters your trachea. You do not feel smaller chunks of food or air at typical temperatures, so these visceral sensations only reach conscious awareness when they are especially stimulating. Other visceral sensations will never make it to conscious perception; you will never be able to feel changes in blood pressure or blood sugar levels.

The **efferent branch** of most somatic reflexes is a single motor neuron that exits the anterior horn of the spinal cord and reaches toward a skeletal muscle, causing

Student Study Tip

Afferent signals are **A**rriving at the CNS.

LO 16.3.1

baroreceptors (pressoreceptors)

Student Study Tip

Efferent signals are **E**xiting the CNS.

contraction. In contrast, the output of a visceral reflex consists of a two-neuron pathway of the ANS, which we have just described in this chapter. This two-neuron pathway consists of a preganglionic neuron and a postganglionic neuron. The preganglionic neuron emerges from the lateral horn of the spinal cord and ends in a ganglion, where it synapses with a postganglionic neuron. The postganglionic neuron projects to its target (or effector) tissue.

The effector organs that are the targets of the autonomic system range from the iris and ciliary body of the eye to the urinary bladder and reproductive organs. The thoracolumbar output, through the various sympathetic ganglia, reaches all of these organs. The cranial component of the parasympathetic system projects from the eye to part of the intestines. The sacral component innervates the majority of the large intestine and the pelvic organs of the urinary and reproductive systems.

Autonomic Plexuses In several locations within the thoracic, abdominal, and pelvic cavities, axons of both sympathetic and parasympathetic pathways weave together in networks called **autonomic plexuses** (**Figure 16.8**).

- The **cardiac plexus**, located in the thoracic cavity, contains both sympathetic postganglionic axons and parasympathetic preganglionic axons. Activation of the sympathetic neuron pathways that pass through this plexus will lead to increases in both heart rate and the force of cardiac contractions. If you've ever felt your heart thumping in your chest after a fright, these sympathetic neurons were at work. The parasympathetic pathways that pass through the cardiac plexus slow heart rate but have no impact on the force of cardiac contractions.
- The **pulmonary plexus** is also located in the thoracic cavity and contains both sympathetic postganglionic axons and parasympathetic preganglionic axons. Activation of the sympathetic neuron pathways that pass through this plexus will lead to bronchodilation and allow more air to travel through the tubes of the lungs to reach the depths of the lungs. Parasympathetic pathway activation has the opposite effect, reducing the diameter of the bronchi.
- The **esophageal plexus** is also in the thoracic cavity. Activation of the sympathetic neuron pathways that pass through this plexus will inhibit muscle contraction and parasympathetic pathway activation will increase muscle contraction in

Apply to Pathophysiology

Orthostatic Hypotension

Have you ever stood up quickly and felt dizzy for a moment? This is because, for one reason or another, blood is not getting to your brain and your neurons are briefly deprived of oxygen. When you change position from sitting or lying down to standing, your cardiovascular system has to adjust for a new challenge, keeping blood pumping up into the head while gravity is pulling more and more blood down into the legs.

1. In this acute moment, as you change position and there is less and less blood traveling to your brain, which division of the ANS will become dominant in its effects on the blood vessels?
   A) Parasympathetic
   B) Sympathetic
2. What type of receptor is likely signaling to the CNS about the change in body position as you stand up?
   A) Proprioceptors
   B) Nociceptors
   C) Chemoreceptors
3. The heart may temporarily speed up in order to pump more blood toward the brain. Signaling through which of the following receptors will result in an increase in heart rate?
   A) Alpha-1 adrenergic
   B) Beta-1 adrenergic
   C) Muscarinic

Figure 16.8 The Autonomic Plexuses

In several locations, axons within both sympathetic and parasympathetic pathways weave together to form a nerve plexus. These plexuses contain axons that are headed to the same effector or region.

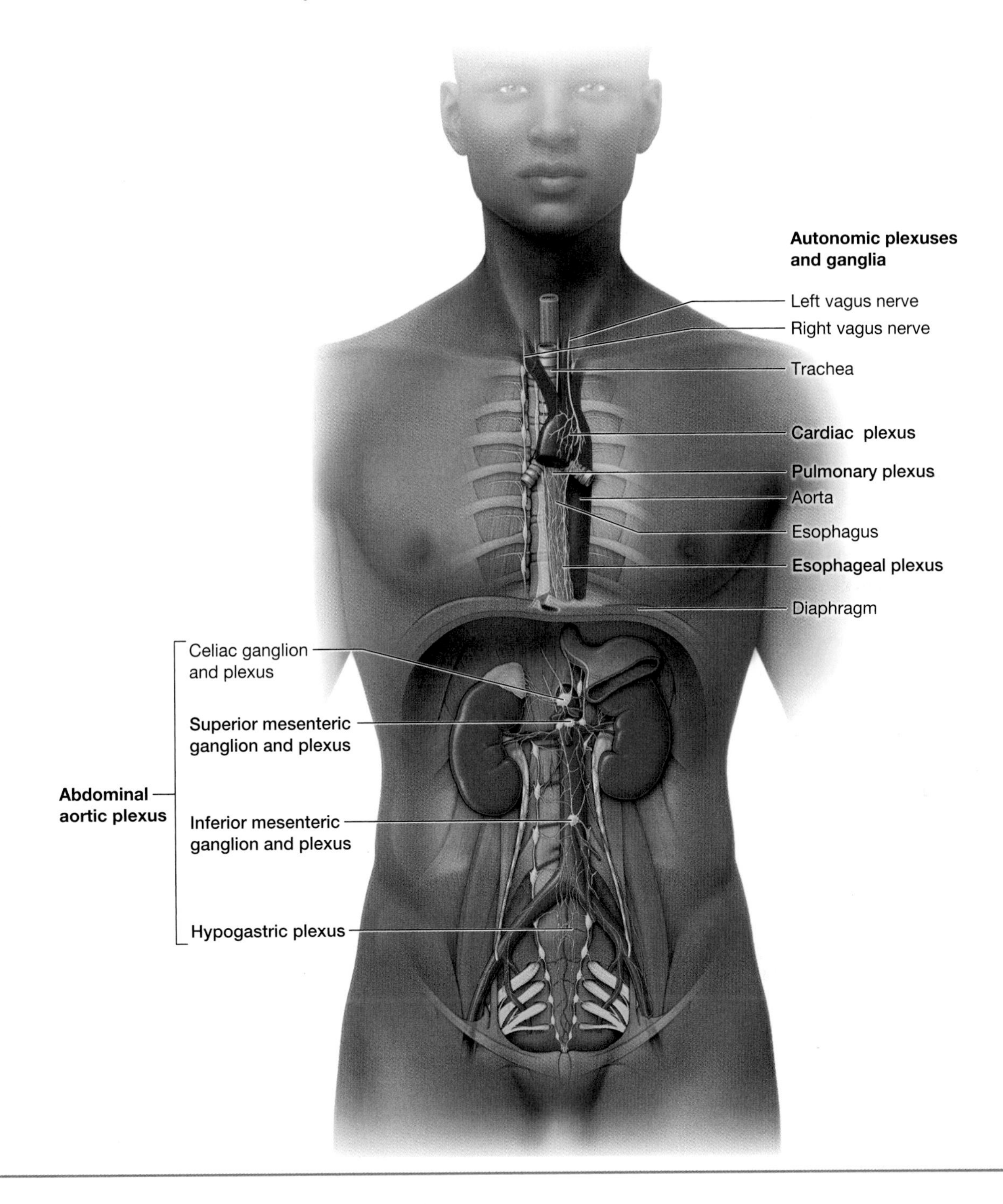

the wall of the esophagus. The esophagus is a long tube through which food and drink passes to reach the stomach. The food and drink that we swallow can only reach the stomach through waves of muscle contraction called *peristalsis*. Remember that the parasympathetic division is often referred to as the "rest and digest" system; here we see one example of how parasympathetic activation promotes digestion. If you have ever been nervous or upset and have felt like it was difficult

solar plexus (celiac plexus)

Cultural Connection

The Solar Plexus

The celiac plexus is commonly nicknamed the **solar plexus** because the nerves emanating from it resemble the rays of the sun. A fall or a strike to the abdomen can sometimes make it temporarily difficult to breathe and is commonly referred to as "getting the wind knocked out of you" or a "blow to the solar plexus." The "wind knocked out of you" sensation is due to reflexive spasms of the diaphragm that prevent it from assisting with breath and not due to damage to the plexus itself. In some cultures the solar plexus is referred to as the third chakra (named manipura in the South Asian language, Sanskrit). It is thought to be the center of digestion and metabolism, but is also sometimes thought to influence self-esteem, willpower, and other personality traits. Other cultures also attribute this region to intuition or emotion. The phrase "gut feeling," referring to the idea that the viscera have emotions, can be traced back in history for thousands of years.

to swallow a bite of food, you have experienced the reduction in activity of the parasympathetic innervation that travels through this plexus.

celiac plexus (solar plexus)

- The **abdominal aortic plexus** is a compound plexus that contains three interconnected plexuses. The **celiac plexus**, the **superior mesenteric plexus**, and the **inferior mesenteric plexus** together innervate all of the abdominal organs along with a few pelvic target tissues. The remaining pelvic innervation comes from the **hypogastric plexus**.

Learning Check

1. Mechanoreceptors in the esophagus release neurological signals whenever food is present. Which of the following best describe the neuron bringing signals from the mechanoreceptors to the spinal cord? Please select all that apply.
   a. Afferent branch
   b. Sensory neuron
   c. Efferent branch
   d. Motor neuron
2. Which of the following exit the spinal cord to activate an organ during a reflex response? Please select all that apply.
   a. Afferent branch
   b. Sensory neuron
   c. Efferent branch
   d. Motor neuron
3. Which anatomical plexus describes sympathetic and parasympathetic pathways weaving together to innervate the pelvic target tissues?
   a. Cardiac plexus
   b. Pulmonary plexus
   c. Esophageal plexus
   d. Hypogastric plexus

16.4 Broad Impacts of Autonomic Responses

Learning Objectives: By the end of this section, you will be able to:

16.4.1 Explain the role of the nervous system in the maintenance of homeostasis and give examples of how the nervous system interacts with other body systems to accomplish this.

16.4.2 Compare and contrast the effects (or lack thereof) of sympathetic and parasympathetic innervation on various effectors (e.g., heart, airways, gastrointestinal tract, iris of the eye, blood vessels, sweat glands, arrector pili muscles).

16.4.3* Define autonomic tone and explain its relevance in homeostasis.

16.4.4* Connect the functions of the ANS to the stress response and explain how chronic modern stress can impact long-term health.

*Objective is not a HAPS Learning Goal.

Organ function is held in the balance between the input from the sympathetic and parasympathetic divisions. For example, parasympathetic activation will stimulate smooth muscle to narrow the airways; sympathetic activation will inhibit those muscles, causing airways to dilate. Therefore, the diameter of the airways at any given moment depends on the types of inputs they are receiving from each arm of the ANS. Almost all organs receive input from both the sympathetic and parasympathetic divisions. This is called *dual innervation*. In general, the sympathetic and parasympathetic divisions of the ANS have opposite effects on any one organ. Increasing and decreasing heart rate, widening or narrowing blood vessels, and stimulating or inhibiting smooth muscle contractions are all examples. At any given moment, in order to accomplish a particular function, one or the other divisions of the ANS may be sending more signals to a given organ. For example, to enable digestion, smooth muscle contraction is required in the intestines. The parasympathetic division promotes digestion by sending more signals to the intestinal smooth muscle in order to enable those contractions and therefore digestion. At this moment we would say that the intestines are in **parasympathetic dominance**. Later, to accomplish another function, such as running to catch a bus, these contractions are not useful and waste precious energy, so the intestines go into **sympathetic dominance** and smooth muscle activity is inhibited. The reproductive organs, uniquely, experience cooperative effects from the two arms of the ANS. Some events that occur in these organs do not have typical homeostatic ranges; for example, orgasm and childbirth both involve deviations from homeostasis in which both divisions of the ANS contribute to the body's return to homeostasis. We will explore these topics further in Chapter 27.

Student Study Tip

Think of parasympathetic dominance versus sympathetic dominance as a sort of competition between the two systems. Whichever one is winning has more influence.

LO 16.4.1

Learning Connection

Chunking

Make a table to compare the opposing effects of the sympathetic and parasympathetic divisions of the autonomic nervous system on the following structures or variables: heart rate (increase or decrease), pupil of the eye (dilation or constriction), breathing rate (decrease or increase), trachea and bronchi (constriction or dilation), and digestive activities (increase or decrease). Then you will have a small study guide of autonomic nervous system effects.

There are a few targets that receive innervation from the sympathetic nervous system only. Most blood vessels, all sweat glands, and all arrector pili muscles receive sympathetic innervation without antagonistic parasympathetic innervation. For these organs, their responses are based solely on the level of activation they receive. For example, the more signals being sent by sympathetic neurons, the more sweat will be produced. Fewer signals will result in less sweating.

LO 16.4.2

Student Study Tip

The parasympathetic system *slows* the heart rate, while the sympathetic system *raises* blood pressure. Use these examples to help solidify the role of each system: one ramps things up, the other slows things down.

For each organ system, there may be a bigger role for sympathetic or parasympathetic dominance in the resting state, which is known as **autonomic tone**. One easy-to-consider example of this is heart rate. The frequency of heart contractions, or heart rate, for a resting adult human is around 70 beats per minute (bpm). Heart rate is influenced by both the sympathetic and parasympathetic divisions, with the sympathetic system causing increases in rate and the parasympathetic system causing decreases in rate. (The heart also controls its own rate, as we will learn in Chapter 19.) The resting heart rate is the result of the parasympathetic system's slowing the heart down from its intrinsic rate; therefore, in a state of rest the heart can be said to be in *parasympathetic tone*.

LO 16.4.3

In a similar fashion, another aspect of the cardiovascular system is primarily under sympathetic control. Blood pressure is influenced by the level of contraction of smooth muscle in the walls of blood vessels. The release of norepinephrine from postganglionic sympathetic causes the constriction of these smooth muscle cells, thereby increasing blood pressure. The hormones released from the adrenal medulla—epinephrine and norepinephrine—will also bind to the adrenergic receptors on these muscle cells. Those hormones travel through the bloodstream where they can easily interact with the receptors in the vessel walls. The parasympathetic system has no significant input to the systemic blood vessels, so the sympathetic system determines their tone.

16.4a Stress

LO 16.4.4

We associate the sympathetic nervous system with emergency control; its nickname of "fight-or-flight" is well accepted and engrained in us. However, organ function and maintaining homeostasis require the actions of the sympathetic nervous system every hour of every day. Further, modern humans have varying levels of psychological stress that may be activating their sympathetic nervous system to varying degrees. Acute stress events, such as taking an exam, interviewing for a job, surviving a minor car accident, or even watching a scary movie may activate this so-called emergency system. Physiological stress events, such as going through a period of fasting, adjusting to blood loss after donation, or running a marathon, also send our organ systems into sympathetic dominance.

Learning Connection

Micro to Macro

When a person is under stress, the sympathetic nervous system is activated. Dilation of the large airways and constriction of certain blood vessels both occur due to sympathetic action on smooth muscle cells in the walls of these organs. What is the advantage of having smooth muscle in these organs (as opposed to skeletal muscle) during a stress response?

The ANS is not the only thing controlling the human stress response; there are several hormones that contribute. Notably, a hormone called *cortisol* increases in our blood during and after stressful events. Remembering that stress is a real or perceived threat to our homeostasis, cortisol, together with the sympathetic nervous system, enacts a series of physiological changes to help us survive when homeostasis is threatened. The themes are: more nutrients and oxygen into the blood, and more blood to the head and the skeletal muscles.

- Nutrients: The sympathetic nervous system contributes to releasing stored lipids and increasing the blood sugar levels through its action on beta-3 and alpha-2 receptors, respectively. Cortisol contributes to blood nutrient levels by promoting the breakdown of proteins.
- Oxygen: The sympathetic nervous system causes the dilation of the airways to facilitate more oxygen getting into the body, and the release of more carbon dioxide waste.
- Blood: Blood flow changes in several ways. The action of the sympathetic nervous system shifts blood flow toward the muscles and away from the viscera. The sympathetic nervous system also increases heart rate, which delivers the nutrient- and oxygen-rich blood throughout the body and increases blood pressure. Cortisol contributes to blood pressure by constricting all the blood vessels throughout the body. High blood pressure is advantageous in short-term stress because it helps make sure that adequate blood will reach the brain, even if blood volume falls due to dehydration or blood loss.

These shifts away from homeostasis are all advantageous while escaping a predator or even succeeding during finals week. However, if stress goes on for too long, these changes can have adverse effects on the body. Chronic high blood pressure can damage blood vessels and affect the heart, and high blood nutrient levels can become difficult to control, permanently altering our homeostatic mechanisms. Chronic stress may have implications in the risk of many long-term diseases, including type 2 diabetes and cardiovascular disease.

Learning Connection

Broken Process

The adrenal gland secretes hormones that are vital to the stress response. If a person had to have both adrenal glands removed due to the presence of noncancerous tumors, which hormones would the person have to take regularly in order to be able to respond normally to stressful situations?

In addition to its association with the fight-or-flight response and rest-and-digest functions, the ANS is responsible for certain everyday functions. For example, it comes into play when homeostatic mechanisms dynamically change, such as the physiological changes that accompany exercise. Getting on the treadmill or heading out for a bike ride will cause the heart rate to increase, breathing to be stronger and deeper, sweat glands to activate, and the digestive system to suspend activity. These are the same physiological changes associated with the fight-or-flight response, but there is nothing chasing you on that treadmill.

This is an acceptable deviation from homeostasis. Normally, homeostatic mechanisms work because "maintaining the internal environment" would mean getting all those changes back to their set points. Instead, the sympathetic system has become active during exercise so that your body can engage in the exercise. At the end of the exercise, the body puts homeostatic mechanisms into action and returns to healthy homeostatic ranges in terms of variables such as heart rate and sweating. Exercise can be a great way to "turn off" sympathetic nervous system responses that are due to stress because at the end of the exercise, parasympathetic dominance will kick in, even if the stressor has not changed.

Learning Check

1. Which of the following describe a heart rate that is constantly high? Please select all that apply.
   a. Parasympathetic dominance
   b. Sympathetic dominance
   c. Parasympathetic tone
   d. Sympathetic tone
2. Which of the following divisions responds to stress?
   a. Parasympathetic division
   b. Sympathetic division
3. Which hormones increase with increased stress? Please select all that apply.
   a. Cortisol
   b. Epinephrine
   c. Norepinephrine
   d. Acetylcholine
4. Which of the following physiological differences would you expect to see in someone who worked a stressful job versus someone who worked a less stressful job? Please select all that apply.
   a. Increased blood sugar
   b. Breakdown of protein
   c. Constricted airways
   d. Decreased blood pressure

Cultural Connection

Fight or Flight? Fright and Freeze? Tend and Befriend?

The original usage of the epithet "fight or flight" comes from a scientist named Walter Cannon, who published his scientific theories and observations in 1932. Both scientists and nonscientists were interested in learning more about human stress responses, and research continues on understanding our physiological responses to stress to this day. Over time, the common phrase of "fight or flight" has been expanded to "fight, flight, or fright," or even "fight, flight, fright, or freeze." Cannon's original contribution was a catchy phrase to express some of what the nervous system does in response to a threat, but it is incomplete. Freeze responses are very common during acute stress.

Until the mid-1990s, much of what we learned about the hormone and nervous system responses had come from studies involving only rats. Of the studies that incorporated humans, 83 percent involved exclusively male-identifying individuals. Over the last 30 years, studies involving humans have gradually begun to include individuals regardless of their gender identity. The largest steps in understanding the sympathetic nervous system came in the 1990s and early 2000s as researchers began studies of more diverse human subjects. Scientists noted that throughout human evolution, and before the widespread availability of birth control in the 1960s, biological women spent much of their lifespan pregnant, nursing, or parenting. The idea that half of the human population would be unable to fight or flee gave researchers the idea to look for complexity in the response to stress. Indeed, they found another hormone that is secreted in high levels in some humans during stress responses. This hormone, oxytocin, is also secreted at high levels during childbirth, lactation, and social bonding behaviors such as spending time with friends or having sexual intercourse. Researchers found that in some individuals, primarily biological women, high levels of oxytocin are secreted during stress, driving both the promotion of social bonds as well as bond-seeking behavior. In other words, these individuals tend to reach out to their friends and loved ones for support rather than exhibit fleeing or fighting behaviors. Researchers referred to this slightly different stress response as "tend and befriend." Looking across the animal kingdom, researchers found that in species such as horses, where newborns can run soon after birth, all animals of that species exhibit the traditional profile of sympathetic nervous system activation along with cortisol. But in species such as humans and dogs, where the young must be cared for during an emergency, both oxytocin and cortisol are secreted in high amounts. Furthermore, oxytocin helps to activate the parasympathetic nervous system, bringing the physiological state back toward homeostasis more rapidly than in individuals for whom oxytocin secretion does not increase during the stress response. The name "sympathetic" (sym- = "together"; pathos = "suffering,") can be translated to "suffering together."

DIGGING DEEPER: Corticosteroids

Cortisol is a hormone that is featured in the stress response; we will discuss the wide-ranging effects of this hormone in more detail in Chapter 17. Cortisol is one of a group of hormones called *corticosteroids*, which are steroid hormones produced in the adrenal cortex that help regulate glucose balance, especially during the stress response. One of its effects during the stress response is on the immune system. The *immune system* is a network of cells and tissues that keeps us healthy by fighting off pathogens. It also helps our tissues heal from damage, such as cuts or sprained ligaments, by initiating inflammation. The process of inflammation helps prevent the spread of infection by walling off the infection site and bringing white blood cells and chemical mediators to the area to speed the healing process. Healing from injury, whether it is a sprain or an infection, has two requirements in addition to the response of the immune system: time and energy. In fact, the actions of the immune system require so much energy that we usually must rest, either resting the affected area or resting our entire bodies. A side effect of resting is that we are unlikely to further damage or infect ourselves. If you sprain your ankle, it will be vulnerable to further damage if you continue to use it. Instead, we generally recommend resting the ankle and using crutches or a wheelchair until it has healed enough to support your weight. As we have discussed in this chapter, the human stress response evolved to support escape and survival. Running from a predator is not the moment to rest a sprain. And so, to aid in surviving the acute stress event, cortisol inhibits the immune system, preventing inflammation or energy-requiring immune activation until the stress has passed. This, of course, can have profound effects on human health. For example, it can be difficult to recover from surgery during a time of intense stress. Stress is linked to shingles outbreaks, cold sores, and other infections. We have borrowed this function of cortisol pharmacologically in order to treat unwanted inflammation. The cream you might use to treat an immune reaction to poison ivy or insect bites is hydrocortisone, a topical application of cortisol. For a systemic immune response, your doctor might prescribe prednisone, a tablet of cortisol. Patients who are treated systemically with cortisol often feel stress-like symptoms.

Chapter Review

16.1 Overview of the Autonomic Nervous System

The learning objectives for this section are:

HAPS LEARNING GOALS AND OUTCOMES

16.1.1 Compare and contrast the autonomic nervous system (ANS) to the somatic nervous system (SNS) with respect to site of origination, number of neurons involved in the pathway, effectors, receptors, and neurotransmitters.

16.1.2 Name the two main divisions of the ANS and compare and contrast the major functions of each division, their neurotransmitters, the origination of the division in the CNS, the location of their preganglionic and postganglionic (ganglionic) cell bodies, and the length of the preganglionic versus postganglionic axons.

16.1.3 Describe the major components of the sympathetic and parasympathetic divisions (e.g., sympathetic trunk [chain], white and gray rami communicantes, splanchnic nerves, pelvic splanchnic nerves, CN III, CN VII, CN IX, CN X) and the major ganglia of each division (e.g., terminal ganglia, intramural ganglia, sympathetic trunk [chain] ganglia, prevertebral [collateral] ganglia).

16.1.4 Describe the different anatomical pathways through which sympathetic and parasympathetic neurons reach target effectors.

16.1 Questions

Mini Case 1: Spinal Cord Injury

Joshua was surfing in Hawaii. They were trying to ride a big wave toward the shore, but they ventured to an area that was shallower than it appeared; the surfboard ran aground and stopped suddenly, causing Joshua to fall forward and slam their back onto the rocks. Joshua's friends called the lifeguard, who stabilized Joshua onto a board and jet-skied them back to shore and rushed them to the hospital. A thorough examination revealed Joshua had suffered a complete T2 spinal cord injury that impacted the autonomic nervous system's control over the heart.

1. How do the autonomic nervous system differ from the somatic nervous system? Please select all that apply.
   a. The autonomic nervous system exerts control over smooth and cardiac muscle, while the somatic nervous system exerts control over the skeletal muscles.
   b. The autonomic nervous system is involved in voluntary responses, while the somatic nervous system is involved in involuntary responses.
   c. Autonomic pathways consist of two neurons, while somatic motor pathways consist of one neuron.
   d. Autonomic synapses use acetylcholine as a neurotransmitter, while somatic motor synapses use norepinephrine.

2. Upon examination and testing, Joshua was found to be unable to increase their heart rate. Which of the following divisions of the autonomic nervous system would you expect to have been impacted by the injury?
   a. Sympathetic division
   b. Parasympathetic division
   c. Somatic division

3. Which of the following ganglia are part of the sympathetic division of Joshua's autonomic nervous system? Please select all that apply.
   a. Terminal ganglia
   b. Celiac ganglion
   c. Paravertebral ganglia
   d. Intramural ganglia

4. By which of the following pathways can Joshua's sympathetic neurons reach their effector organs? Please select all that apply.
   a. The preganglionic axon can synapse in a sympathetic ganglion, and the postganglionic axon can reach the effector organ.
   b. The preganglionic axon can synapse in a ganglion within its effector organ, and the postganglionic axon can exert control over the effector organ.
   c. The preganglionic axon can travel to superior or inferior ganglia of the sympathetic chain.
   d. The preganglionic axon can travel through the white ramus communicans without synapsing and proceed to the adrenal medulla.

16.1 Summary

- The nervous system can be classified functionally into two portions: (1) the somatic nervous system, which controls the voluntary responses of the skeletal muscles, and (2) the autonomic nervous system, which controls the involuntary responses of the cardiac and smooth muscles, glands and organs.
- The autonomic nervous system can be divided into two divisions: (1) the sympathetic division, or "fight-or-flight" branch, and (2) the parasympathetic division, or "rest and digest" branch.
- Most organs have dual innervation; they are regulated by both the sympathetic and parasympathetic divisions. Most often, the two divisions produce opposite effects.
- The sympathetic nervous system is activated by physical or emotional stress. Sympathetic stimulation causes increased heart and breathing rates, increased blood pressure, dilation of the pupils, dilation of the major airways, and decreased digestive activities and immune responses.
- The parasympathetic nervous system is activated when a person is in a relaxed state, such as after eating a meal. Parasympathetic stimulation causes decreased heart and breathing rates, decreased blood pressure, constriction of the pupils, constriction of the major airways, and increased digestive activities and immune responses.
- In the somatic nervous system, the pathway between the central nervous system and an effector organ consists of one neuron. In the autonomic nervous system, the pathway consists of a chain of two neurons, the preganglionic neuron and the postganglionic neuron. The two neurons in the pathway often synapse in a ganglion.

16.2 Chemical Components of the Autonomic Responses

The learning objectives for this section are:

HAPS LEARNING GOALS AND OUTCOMES

16.2.1 Compare and contrast cholinergic and adrenergic receptors with respect to neurotransmitters that bind to them, receptor subtypes, receptor locations, target cell response (i.e., excitatory or inhibitory), and examples of drugs, hormones, and other substances that interact with these receptors.

16.2.2 Explain the relationship between chromaffin cells in the adrenal medulla and the sympathetic division of the nervous system.

16.2 Questions

1. Which neurotransmitter is released by neurons that project to sweat glands or skeletal muscles?
   a. Acetylcholine
   b. Norepinephrine
   c. Epinephrine
   d. Nicotine

2. Which of the following is true about chromaffin cells?
   a. They transport epinephrine from the adrenal medulla.
   b. They produce norepinephrine in the adrenal medulla.
   c. They stimulate preganglionic sympathetic neurons.
   d. They emit neurotransmitters into the bloodstream.

16.2 Summary

- The autonomic nervous system (ANS)uses the neurotransmitters acetylcholine (ACh) and norepinephrine (NE) in its synapses.
- ACh is released by all preganglionic fibers of the sympathetic and parasympathetic divisions of the ANS and by all postganglionic fibers of the parasympathetic division.
- NE is released by most postganglionic sympathetic fibers. However, ACh is released by the axons that project to sweat glands and blood vessels associated with skeletal muscles.
- The adrenal glands release epinephrine and norepinephrine as sympathetic hormones, in response to stimulation by preganglionic sympathetic neurons.
- Acetylcholine receptors are classified as either nicotonic or muscarinic receptors, depending on whether the receptors respond to nicotine or muscarine.

16.3 Autonomic Reflexes and Homeostasis

The learning objective for this section is:

HAPS LEARNING GOALS AND OUTCOMES

16.3.1 Describe visceral reflex arcs, including structural and functional details of sensory and motor (autonomic) components.

16.3 Question

1. Which parts of an autonomic reflex arc would be impaired after a T2 spinal cord injury that affected the beating action of the heart? Please select all that apply.
   a. Afferent
   b. Efferent
   c. Sensory
   d. Motor

16.3 Summary

- Reflex arcs are pathways through the nervous system that coordinate sensory (afferent) information with motor (efferent) information.
- In contrast to somatic reflexes, whose target organs are skeletal muscles, autonomic reflexes target smooth and cardiac muscle and glands.
- Somatic efferent pathways consist of a single motor neuron that synapses on its target organ (a skeletal muscle). Autonomic efferent pathways consist of two neurons in a chain, which synapse in a ganglion and release autonomic neurotransmitters to their target organs through varicosities.
- In several locations within the major body cavities, the axons of sympathetic and parasympathetic pathways weave together into networks called autonomic plexuses; these are the cardiac, pulmonary, esophageal and abdominal aortic plexuses.

16.4 Broad Impacts of Autonomic Responses

The learning objectives for this section are:

HAPS LEARNING GOALS AND OUTCOMES

16.4.1 Explain the role of the nervous system in the maintenance of homeostasis and give examples of how the nervous system interacts with other body systems to accomplish this.

16.4.2 Compare and contrast the effects (or lack thereof) of sympathetic and parasympathetic innervation on various effectors (e.g., heart, airways, gastrointestinal tract, iris of the eye, blood vessels, sweat glands, arrector pili muscles).

16.4.3* Define autonomic tone and explain its relevance in homeostasis.

16.4.4* Connect the functions of the ANS to the stress response and explain how chronic modern stress can impact long-term health.

* Objective is not a HAPS Learning Goal.

16.4 Questions

1. The lungs are innervated by both the sympathetic and parasympathetic division. What would you expect to happen to maintain homeostasis if the respiration rate increased?
   a. Increased sympathetic drive
   b. Increased parasympathetic drive

2. The digestive system is innervated by the sympathetic and parasympathetic division. What would you expect to see if there were parasympathetic dominance?
   a. Increased motility of the food through the digestive tract
   b. Decreased motility of the food through the digestive tract

3. Which of the following defines autonomic tone?
   a. An organ's resting state
   b. An organ's active state
   c. An organ's resistance to change
   d. An organ's ability to adapt

4. Which of the following effects are caused by chronic stress?
   a. High blood pressure
   b. Decreased heart rate
   c. Increased protein formation
   d. Decreased respiration rate

16.4 Summary

- The autonomic nervous system (ANS)maintains homeostasis by balancing the effects of the sympathetic and parasympathetic branches under changing circumstances.
- Dual innervation of most organs by both branches of the ANS allows the body to respond to sympathetic stimulation when under stressful conditions and parasympathetic stimulation when under relaxed conditions.
- The stress response is regulated by the sympathetic nervous system and hormones such as cortisol. When under stress, the body responds in certain ways that provide more blood flow, nutrients and oxygen to active organs, such as the skeletal muscles, heart and brain.
- During resting or relaxed conditions, parasympathetic stimulation provides more blood flow, oxygen and nutrients to such organs as those of the digestive system, while maintaining the homeostatic needs for oxygen and nutrients in the other organs.

17

The Endocrine System

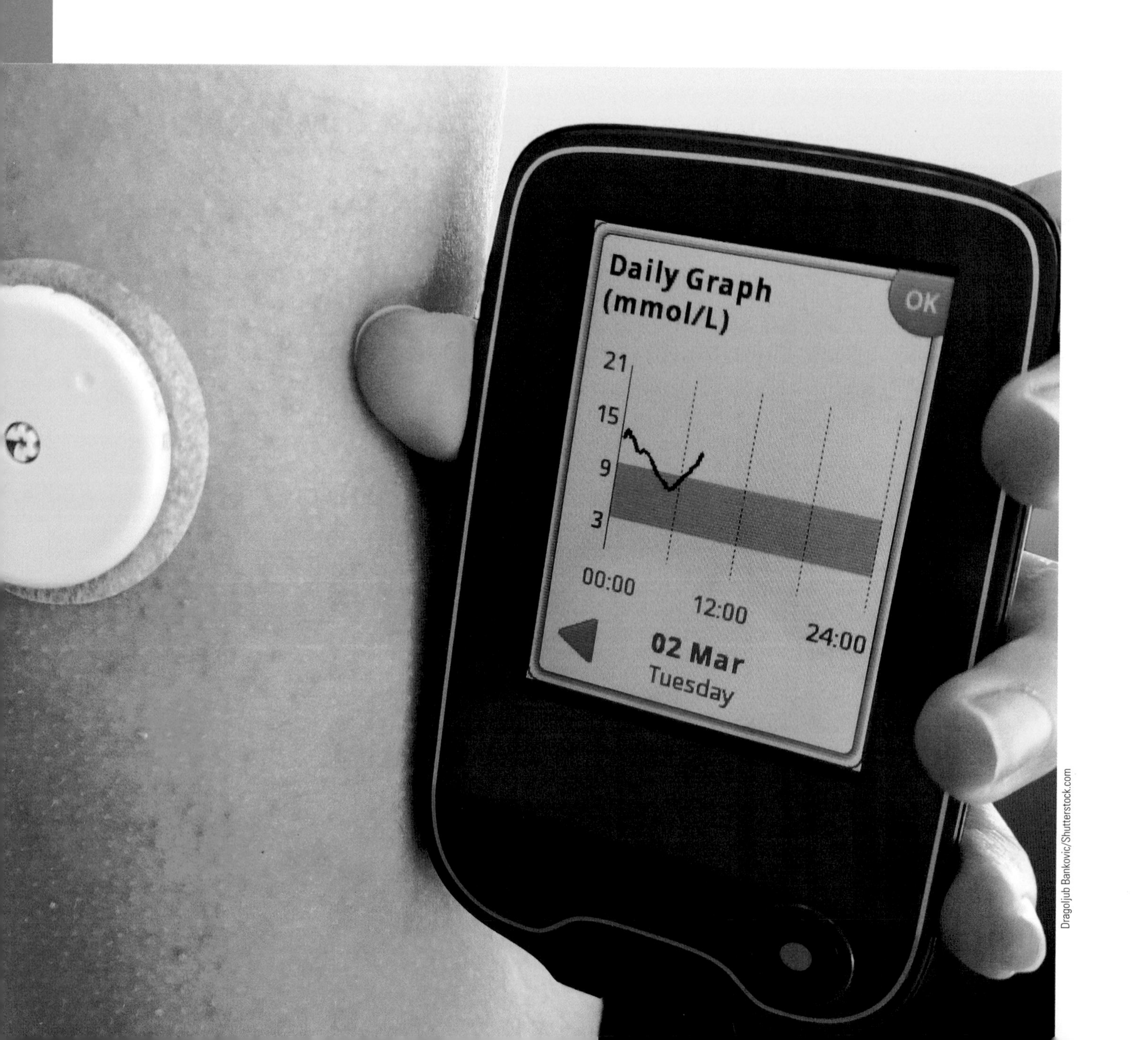

Dragoljub Bankovic/Shutterstock.com

Chapter Introduction

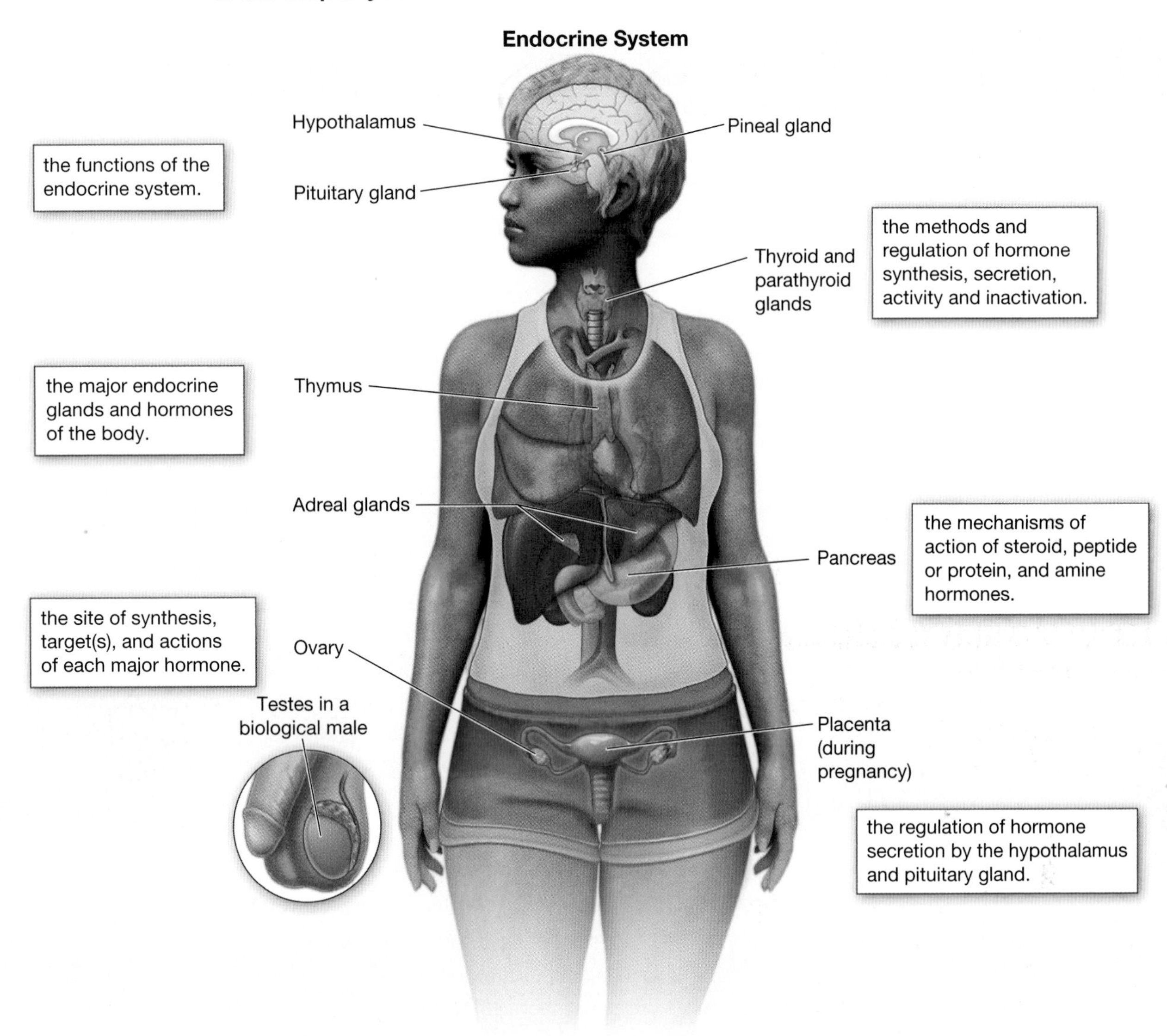
In this chapter you will learn...
Endocrine System
the functions of the endocrine system.
the major endocrine glands and hormones of the body.
the site of synthesis, target(s), and actions of each major hormone.
the methods and regulation of hormone synthesis, secretion, activity and inactivation.
the mechanisms of action of steroid, peptide or protein, and amine hormones.
the regulation of hormone secretion by the hypothalamus and pituitary gland.
Hypothalamus
Pineal gland
Pituitary gland
Thyroid and parathyroid glands
Thymus
Adreal glands
Pancreas
Ovary
Testes in a biological male
Placenta (during pregnancy)

Your body is made up of around 40 trillion cells. That's a much higher number than the number of people that live on this planet. Imagine trying to create a signal that reaches a majority of the planet's people and coordinates their behavior within a few seconds. It may seem impossible and yet, within your own body, widespread communication and coordination is made possible all the time. Efficient, widespread, and rapid systemic communication is made possible by the endocrine system.

17.1 An Overview of the Endocrine System

Learning Objectives: By the end of this section, you will be able to:

17.1.1 Describe the major functions of the endocrine system.

17.1.2 Define the terms hormone, endocrine gland, endocrine tissue (organ), and target cell.

17.1.3 Compare and contrast how the nervous and endocrine systems control body functions, the anatomical pathways by which the signals reach their targets, what determines the target of the pathway, the speed of the target response(s), the duration of the response, and how signal intensity is coded.

LO 17.1.1

Communication is a process in which a sender transmits signals to one or more receivers to control and coordinate actions. In the human body, two major organ systems participate in relatively "long-distance" communication: the nervous system and the endocrine system. The endocrine system uses secreted chemical signals, called *hormones*, as its means of communication. **Hormones** are signaling molecules released into the bloodstream; they can reach most cells of the body and have widespread effects. The nervous system uses electrical signals that traverse down the long axons of neurons. Chemical signals are utilized by the nervous system just for communication between one cell and another. The communication and coordination these two systems achieve together maintains homeostasis in the body. The endocrine system is responsible for regulating the body's use of calories and nutrients, the secretion of wastes, the maintenance of blood pressure and osmolarity, growth, fertility, sex drive, lactation, and sleep. The hormones of the endocrine system also impact mood and emotion. As discussed in Chapter 16, our ability to respond to stress—a real or perceived threat to our homeostasis—involves both the nervous and endocrine systems.

Student Study Tip

You can remember the function of the endocrine system by its name: the prefix *endo* means "in" and *crine* means "secretion," which tells us how hormones are secreted into the bloodstream!

17.1a Chemical Signaling

Universal chemical signaling involves exocytosis, which releases molecules that diffuse throughout body fluids and are bound by a molecular receptor on or in a target cell. There are four types of chemical signaling in the body:

- Local intercellular communication, in which a chemical is released by one cell and induces a response in neighboring cells, is **paracrine signaling**.
- Local communication can also affect the cell that releases the signal; this is known as **autocrine signaling**.
- Chemicals released by neurons to communicate with a local cell or cells are **neurotransmitters**.
- Releasing chemical signals into the bloodstream that induce responses in specific target cells throughout the body is **endocrine signaling**.

In endocrine signaling, hormones secreted into the extracellular fluid diffuse into the blood or lymph, and can then travel great distances throughout the body. In contrast, autocrine signaling affects the same cell that released the signal. Paracrine signaling is limited to the local environment. Although paracrine signals may enter the bloodstream, their concentration is generally too low to elicit a response from distant

The Human Anatomy and Physiology Society includes more than 1,700 educators who work together to promote excellence in the teaching of this subject area. The HAPS A&P Learning Outcomes measure student mastery of the content typically covered in a two-semester Human A&P curriculum at the undergraduate level. The full Learning Outcomes are available at https://www.hapsweb.org.

Figure 17.1 A Comparison of Endocrine and Exocrine Glands

Secretions from both endocrine and exocrine glands are released through exocytosis. In endocrine secretion, exocytosis releases the secretion into the bloodstream or surrounding extracellular fluid. In exocrine secretion exocytosis releases the secretion into a duct that carries the secretions to a surface.

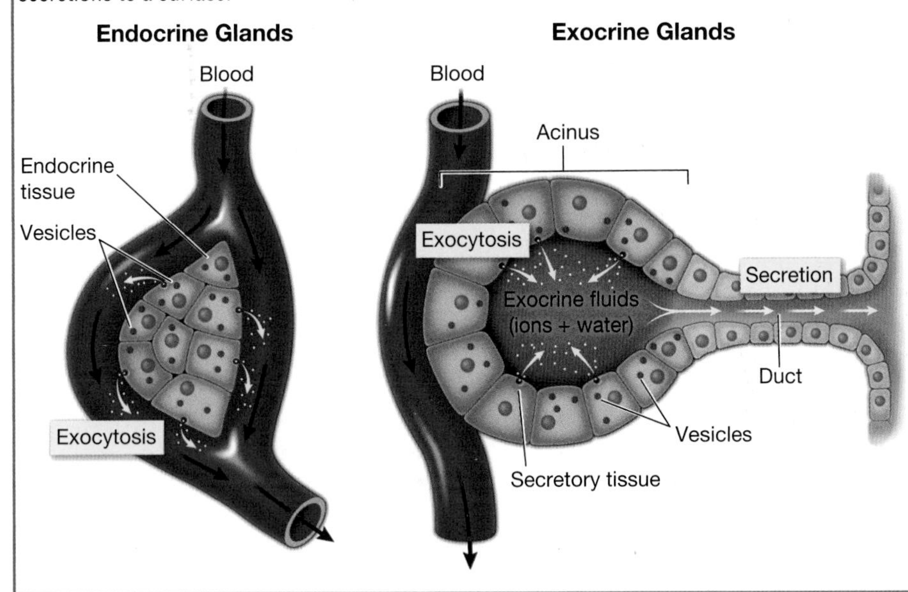

tissues. The neurotransmitters of the nervous system are a specific example of a paracrine signal that acts only locally within the synaptic cleft.

The endocrine system consists of cells, tissues, and organs that secrete hormones. **Endocrine glands** are the major players in this system. The primary function of endocrine glands is to secrete their hormones directly into the surrounding fluid; unlike exocrine glands, they do not have ducts that carry their secretions away (**Figure 17.1**). The hormones are carried throughout the body in the blood. The endocrine system includes the pituitary, thyroid, parathyroid, adrenal, and pineal glands (**Figure 17.2**). Some of these glands have both endocrine and non-endocrine functions. For example, the pancreas contains cells that function in digestion as well as cells that secrete hormones. The hypothalamus, thymus, heart, kidneys, stomach, small intestine, liver, adipose tissue, ovaries, and testes are other organs that contain cells with endocrine function (**Figure 17.3**).

Learning Connection

Chunking

Which of these systems are local? Which achieve long-distance communication? Which system(s) depend on receptors? Which system(s) use electrical signals?

LO 17.1.2

Let's track a hormone as it travels throughout the body in the bloodstream. Depending on the rate that the heart is beating, it will take somewhere around a minute for a hormone molecule to make a loop throughout the circulatory system. Given enough loops, it has the opportunity to make its way to just about any of the 40 trillion cells of the body. Which cell (or cells) will it affect? A given hormone only impacts the activity only of its **target cells**—that is, cells with receptors for that particular hormone. Once the hormone binds to the receptor, a chain of events is initiated that leads to the target cell's response. Hormones play a critical role in the regulation of physiological processes because of the target cell responses they regulate. These responses contribute to human reproduction, growth and development of body tissues, metabolism, fluid, and electrolyte balance, sleep, and many other body functions.

Student Study Tip

A hormone reaches everywhere blood flows in the body, but only where there is a lock (target cell's receptor) that fits the hormone key will the hormone effect change. This is like how your house key doesn't open every door in your neighborhood.

17.1b Neural and Endocrine Long-Distance Signaling

LO 17.1.3

The body is a vast galaxy of cells, and two different systems—the nervous system and the endocrine system—are capable of long-distance communication. The nervous system uses two forms of communication—electrical and chemical signaling. Neurotransmitters,

Figure 17.2 The Primary Endocrine Glands of the Human Body

The primary function of the pineal, pituitary, thyroid, and adrenal glands is to produce hormones. The pancreas is an organ that functions equally in endocrine and exocrine secretions.

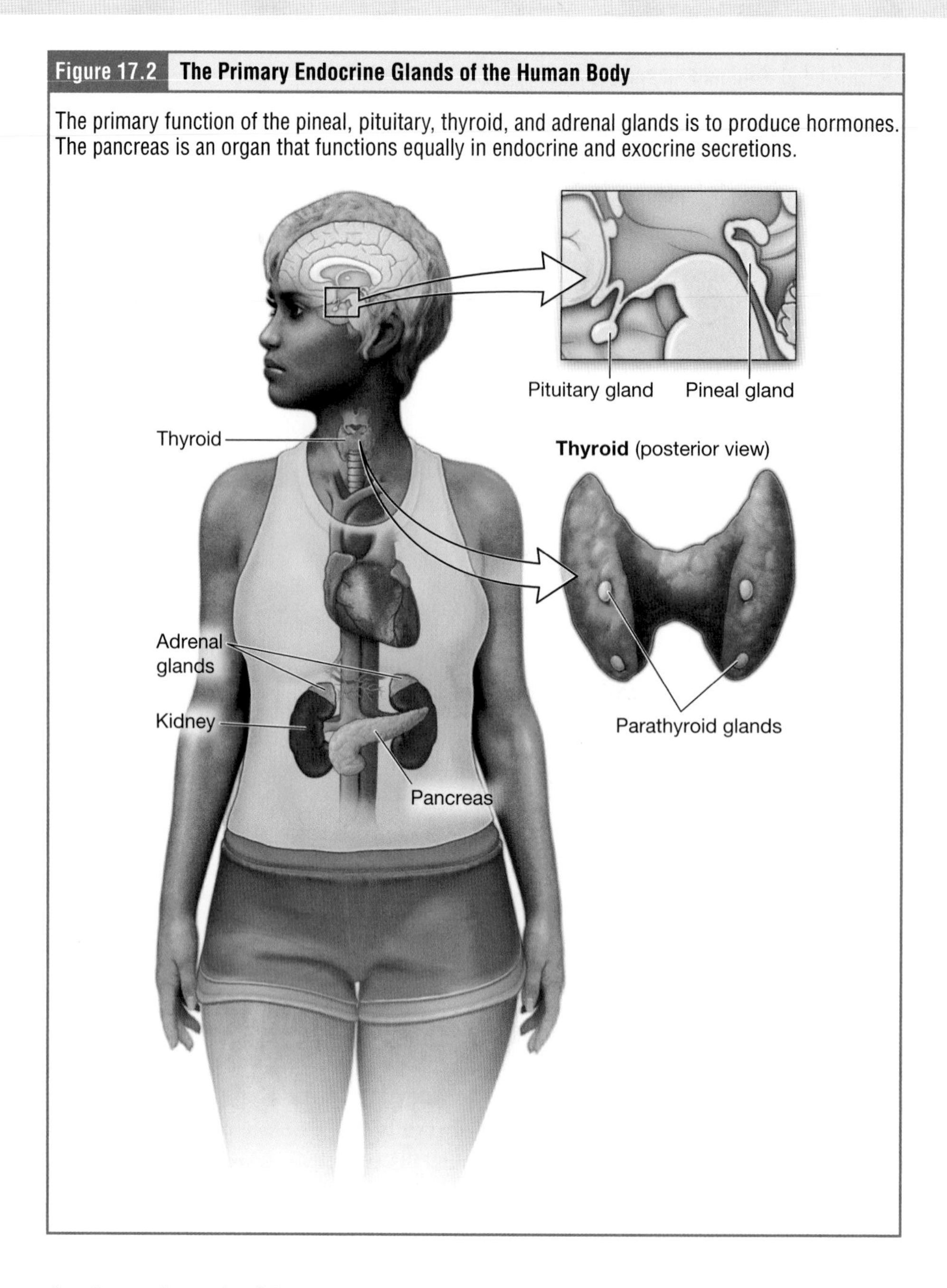

the chemical signals of the nervous system, act locally and rapidly. Once the neurotransmitters bind to receptors on the postsynaptic cell, the receptor stimulation triggers a cellular response. The target cell responds within milliseconds of receiving the chemical stimulation and the response then ends very quickly. In this way, neural communication enables body functions that involve quick, brief actions, such as movement, sensation, and cognition. In contrast, the **endocrine system** uses just one method of communication: chemical signaling. These signals are sent by the endocrine organs, which secrete chemicals—the **hormones**—into the extracellular fluid. Hormones are transported throughout the body by the blood and circulatory system, where they bind to receptors on target cells and induce a response. There are two consequences to these differences in structure. One is that endocrine signaling can be more widespread, affecting many target cells at once. Another is that endocrine signaling requires more time than neural signaling to prompt a response. Some target cell responses are rapid, while other responses can take longer. For example, the emergency stress hormones released when you are

Figure 17.3 Organs That Have Endocrine Roles

The hypothalamus, thymus, heart, kidneys, stomach, small intestine, liver, adipose tissue, ovaries, and testes all have other functions in the body, but each of these organs secretes one or more hormones and is therefore also considered an endocrine gland.

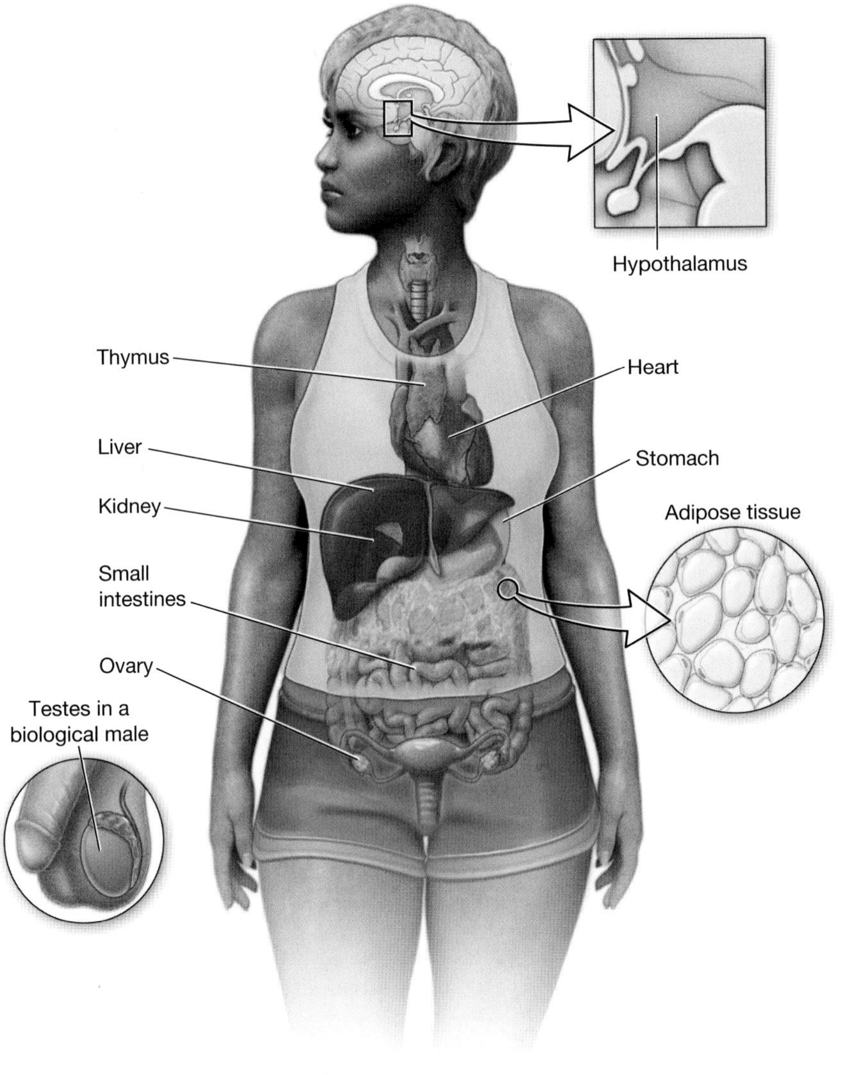

confronted with a dangerous or frightening situation—epinephrine and norepinephrine—will trigger responses in their target tissues within seconds. In contrast, it may take up to 48 hours for target cells to respond to certain reproductive hormones.

One note about the efficiency of chemical signals in the body: The body produces only a few dozen individual signaling molecules; however, these molecules can have different effects in different tissues. One hormone may have multiple receptors and therefore cause different types of changes in its different target cells. A single signaling molecule may be a paracrine signal in some instances or a hormone in others, depending on where it is released. Because of this redundancy in the body, drugs that target the production, release, or reception of different signaling molecules can have widespread effects.

In broad strokes we can generalize that the nervous system involves quick responses to changes in the internal or external environment, and the endocrine system is usually slower acting—making changes to maintain homeostasis and working toward longer-term goals such as nutrient storage, growth, and reproduction.

Learning Check

1. Which of the following is the chemical signal used in the endocrine system?
   a. Hormones
   b. Neurotransmitters
   c. Sodium
   d. Calcium
2. The femoral nerve synapses on the quadriceps muscles. In order to make the muscles contract, the neurotransmitter is released from the axon terminal into the synapse, which binds to the receptors of the neighboring muscle cell. What type of chemical signaling is this an example of?
   a. Paracrine signaling
   b. Autocrine signaling
   c. Synaptic signaling
   d. Endocrine signaling
3. The adrenal gland releases a group of hormones called *corticosteroids*. These hormones act as anti-inflammatory agents, maintain blood sugar levels, maintain blood pressure, and regulate water balance. Which of the following explains how these hormones can affect many organs in the body?
   a. These hormones bind to all the organs in the body.
   b. These hormones can bind to several different receptors.
   c. The body has one receptor throughout all organs.
   d. Hormones travel in the blood and do not target specific organs.

17.2 Hormones

Learning Objectives: By the end of this section, you will be able to:

17.2.1 List the three major chemical classes of hormones (i.e., steroid, peptide, amino acid-derived [amine]) found in the human body.

17.2.2 Compare and contrast how steroid and peptide hormones are produced and stored in the endocrine cell, released from the endocrine cell, and transported in the blood.

17.2.3 Compare and contrast the locations of target cell receptors for steroid and peptide hormones.

17.2.4 Compare and contrast the mechanisms of action of plasma membrane hormone receptors and intracellular hormone receptors, including the speed of the response.

17.2.5 Describe the various signals that initiate hormone production and secretion (e.g., monitored variables, direct innervation, neurohormones, other hormones).

17.2a Types of Hormones

LO 17.2.1

The hormones of the human body can be divided into three main groups on the basis of their chemical structure. The chemical structure will impact the distribution and location of reception for the hormones. Some hormones are derived from lipids. These so-called steroid hormones are lipid-based and lipophilic (literally, "fat-loving"), so they cross membranes easily. This means they cannot be stored or excluded from a cell (see the "Anatomy of a Steroid Hormone" feature). The two other groups of hormones are both based on amino acids, the monomers of proteins (see the "Anatomy of a Protein Hormone" feature). Amine hormones are modified versions of single amino acids. As such they are very small and transport easily. Their precise characteristics are based on the chemical makeup of their functional groups. Peptide hormones are produced when multiple amino acids are joined together.

Anatomy of...

A Steroid Hormone

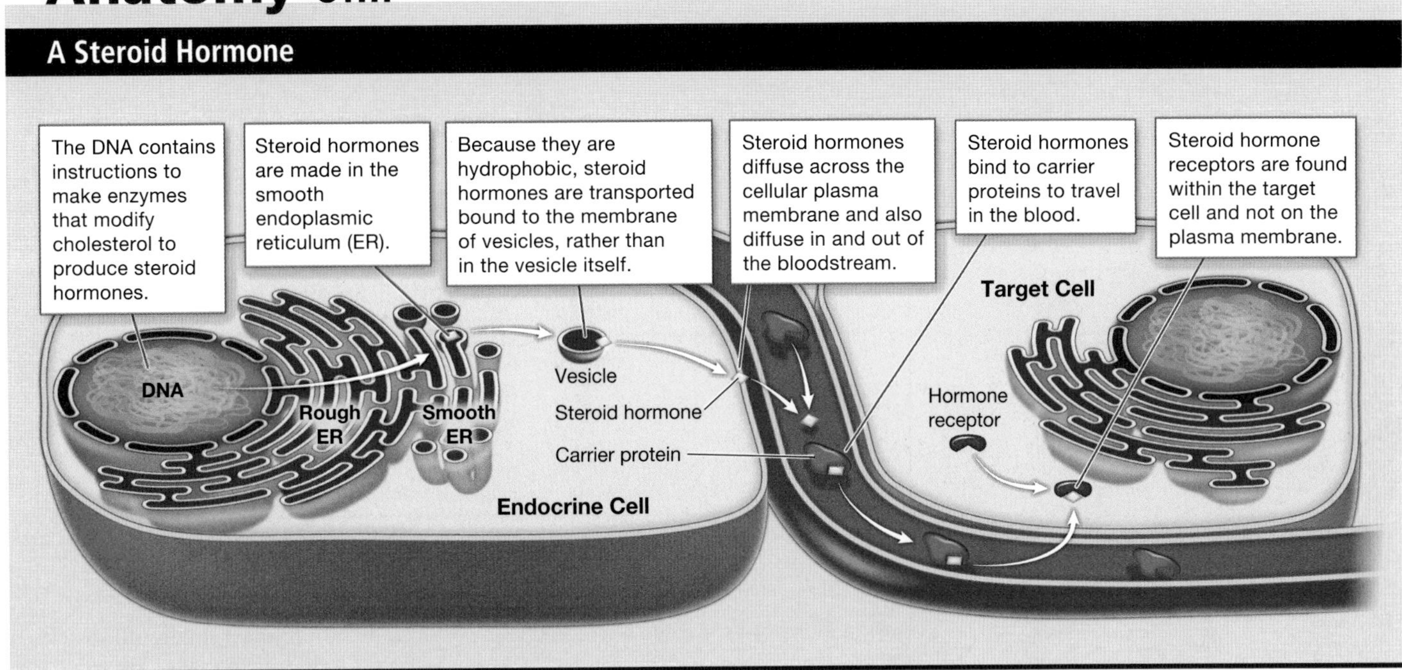

The DNA contains instructions to make enzymes that modify cholesterol to produce steroid hormones.
Steroid hormones are made in the smooth endoplasmic reticulum (ER).
Because they are hydrophobic, steroid hormones are transported bound to the membrane of vesicles, rather than in the vesicle itself.
Steroid hormones diffuse across the cellular plasma membrane and also diffuse in and out of the bloodstream.
Steroid hormones bind to carrier proteins to travel in the blood.
Steroid hormone receptors are found within the target cell and not on the plasma membrane.
Target Cell
DNA
Rough ER
Smooth ER
Vesicle
Steroid hormone
Carrier protein
Hormone receptor
Endocrine Cell

Anatomy of...

A Protein Hormone

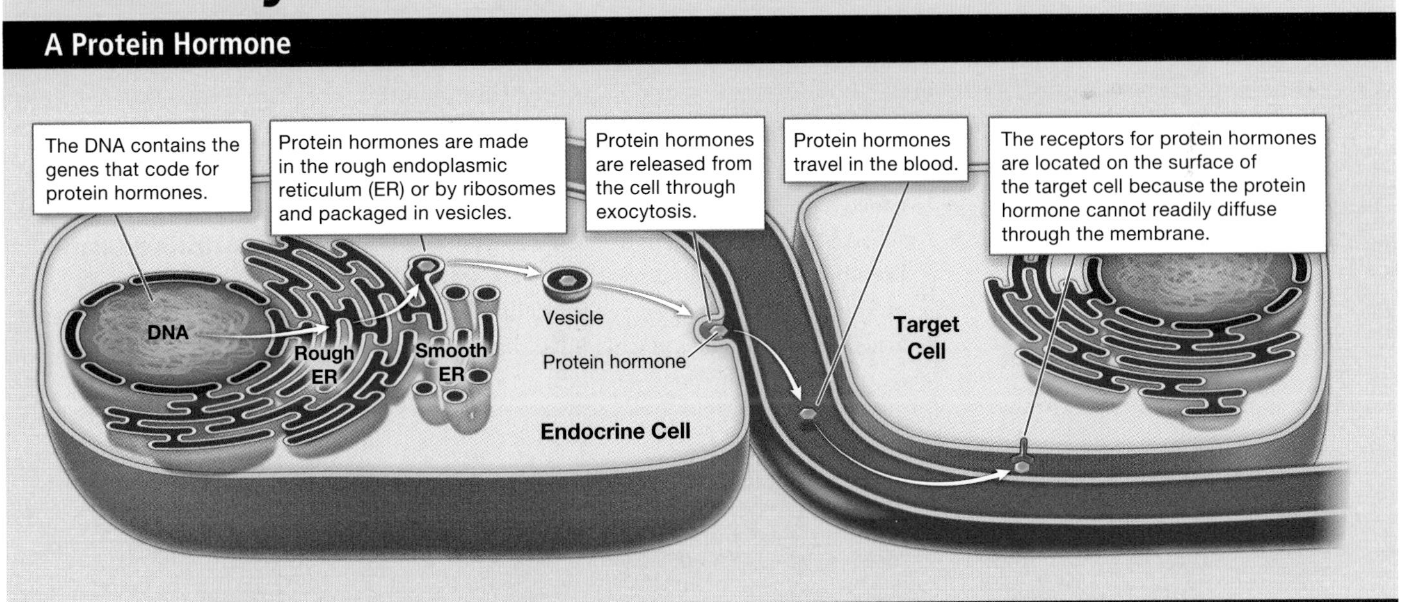

The DNA contains the genes that code for protein hormones.
Protein hormones are made in the rough endoplasmic reticulum (ER) or by ribosomes and packaged in vesicles.
Protein hormones are released from the cell through exocytosis.
Protein hormones travel in the blood.
The receptors for protein hormones are located on the surface of the target cell because the protein hormone cannot readily diffuse through the membrane.
DNA
Rough ER
Smooth ER
Vesicle
Protein hormone
Target Cell
Endocrine Cell

Steroid Hormones The steroids, or steroid hormones, are produced by modifying cholesterol (**Figure 17.4**). Like cholesterol, steroid hormones are not soluble in water (they are hydrophobic). Two famous steroid hormones, testosterone and the estrogens, are produced by the gonads.

Figure 17.4 Steroid Hormones

Steroid hormones are derived from cholesterol and maintain the characteristic ring shape. Because they are lipids, they are lipophilic and diffuse through membranes easily.

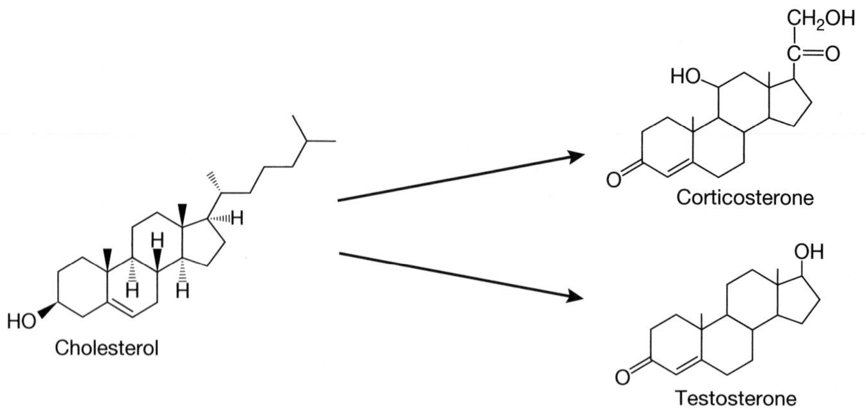

Cholesterol

Corticosterone

Testosterone

Amine Hormones Amine hormones are produced by modifying a single amino acid (**Figure 17.5**). There are amine hormones produced from the modification of either tryptophan or tyrosine. An example of a hormone derived from tryptophan is melatonin, which is secreted by the pineal gland and helps regulate circadian rhythm. Tyrosine derivatives include epinephrine and norepinephrine. Epinephrine and norepinephrine are secreted by the adrenal medulla and play a role in the fight-or-flight response.

Peptide and Protein Hormones Whereas the amine hormones are derived from a single amino acid, peptide and protein hormones consist of multiple amino acids that link to form an amino acid chain (**Figure 17.6**). By convention, a chain of up to 50 amino acids is called a *peptide*, while a chain of greater than 50 amino acids is called a *protein*.

An example of a peptide hormone is antidiuretic hormone (ADH), a pituitary hormone consisting of nine amino acids, which is important in fluid balance. An example of a protein hormone is insulin, a pancreatic hormone consisting of 51 amino acids, which promotes glucose uptake into many body cells.

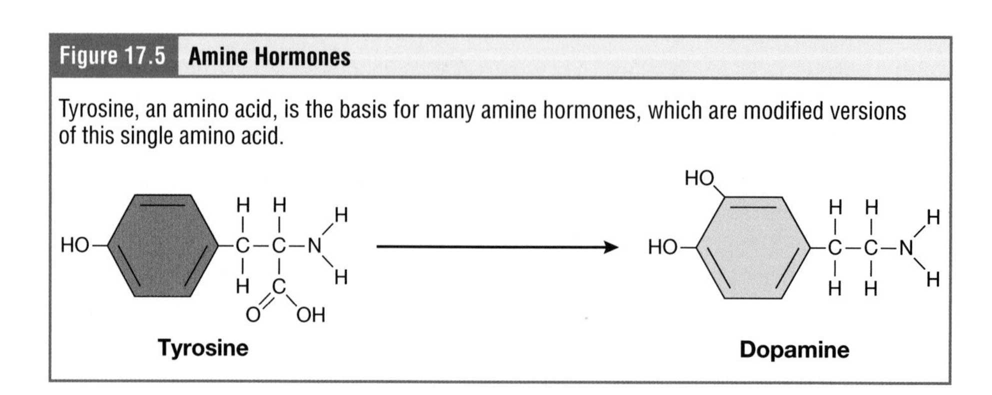

Figure 17.5 Amine Hormones

Tyrosine, an amino acid, is the basis for many amine hormones, which are modified versions of this single amino acid.

17.2b Pathways of Hormone Action

Production All hormones are produced within the cells of endocrine glands. Humans do not carry genes in their DNA that directly code for steroid hormones. Rather, our genomes contain the instructions to produce enzymes that are capable of modifying cholesterol into a new structure as a steroid hormone. Whereas peptide hormones consist of short chains of amino acids, protein hormones are longer polypeptides. Both types are synthesized like other proteins: DNA is transcribed into mRNA, which is translated into amino acids at a ribosome or rough endoplasmic reticulum.

LO 17.2.2

Once produced, like most proteins, protein hormones may be modified by the golgi apparatus and then packaged into vesicles. Once the cell is triggered to release the hormone, the vesicles merge with the plasma membrane and the protein hormones are released through the process of exocytosis. Not steroid hormones, though; steroid hormones are unable to be stored in vesicles. Because they are lipophilic, they will readily diffuse out of the vesicle. Steroid hormones, therefore, cannot be kept in storage waiting to be released, but must be made on demand when the endocrine cell is triggered to begin production. As soon as the steroid hormone is made, it will diffuse out of the cell and into the extracellular fluid.

Figure 17.6 Peptide Hormones

Peptide hormones are composed of up to 50 amino acids joined by peptide bonds.

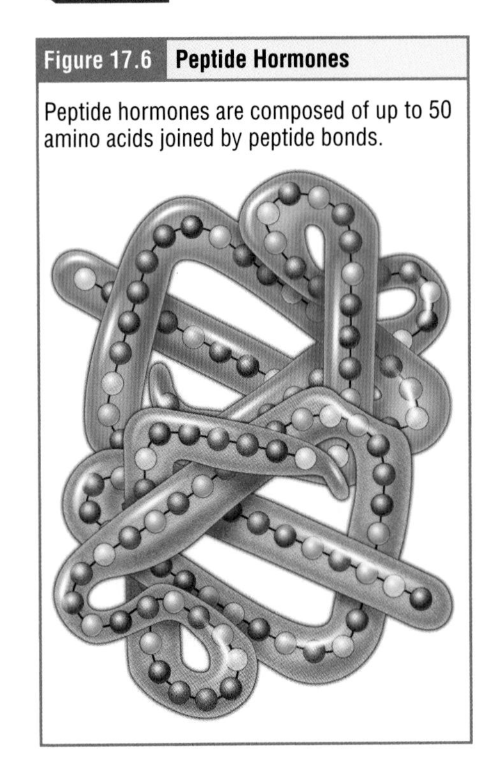

Travel Similarly, because of the difference in their hydrophobicity, protein and steroid hormones travel differently in the blood. Because blood is water-based, lipid-derived hormones must travel to their target cell bound to a transport protein. With their structure and traveling mode of being bound to proteins, steroid hormones tend to last longer in the bloodstream; they are likely to persist for 30–90 minutes after release, compared to only a few minutes for a protein-based hormone.

Reception Once released, every hormone has the ability to reach just about any cell in a body. Which cells are its targets is determined by their expression of a **hormone receptor**, a protein located either inside the cell or within the cell membrane. The receptor is specific to the hormone. Figure 17.7 shows a target cell that expresses several receptor molecules. This target cell contains hormone receptors specific for steroid hormone A and protein hormone B. The cell is not the target of, and will not respond to, protein hormone A. You may notice that steroid hormone A is found both inside and outside the cell; because the steroid is lipophilic, it diffuses easily through the cell membrane and can be found inside of cells regardless of whether they have receptors for the steroid. Steroid hormone receptors are usually found inside the cell. The same type of receptor may be located on cells in different body tissues, and trigger somewhat different responses. And there is often more than one type of receptor for a single hormone. Thus, the response triggered by a hormone depends not only on the hormone, but also on the presence or absence of specific hormone receptors in or on the target cell.

LO 17.2.3

LO 17.2.4

Once the target cell receives the hormone signal, it can respond in a variety of ways. The response may include the stimulation of protein synthesis, activation or deactivation of enzymes, alteration in the permeability of the cell membrane, altered rates of mitosis and cell growth, or stimulation of the secretion of products. Moreover, a single hormone may be capable of inducing different responses in a given cell.

Figure 17.7 Hormone Receptors

Proteins are too large to diffuse through the membrane; therefore, their receptors are expressed on the cell surface. Steroid hormones and thyroid hormones are lipid soluble, so they diffuse into and out of the cell randomly. Cells express receptors for these hormones internally rather than on the surface. The hormone receptor may be in the cytosol or even within the nucleus. Once bound, the hormone-receptor complex acts as a transcription factor. The cell shown in this figure is a target cell of steroid hormone A and protein hormone B, but it is not a target cell of protein hormone A.

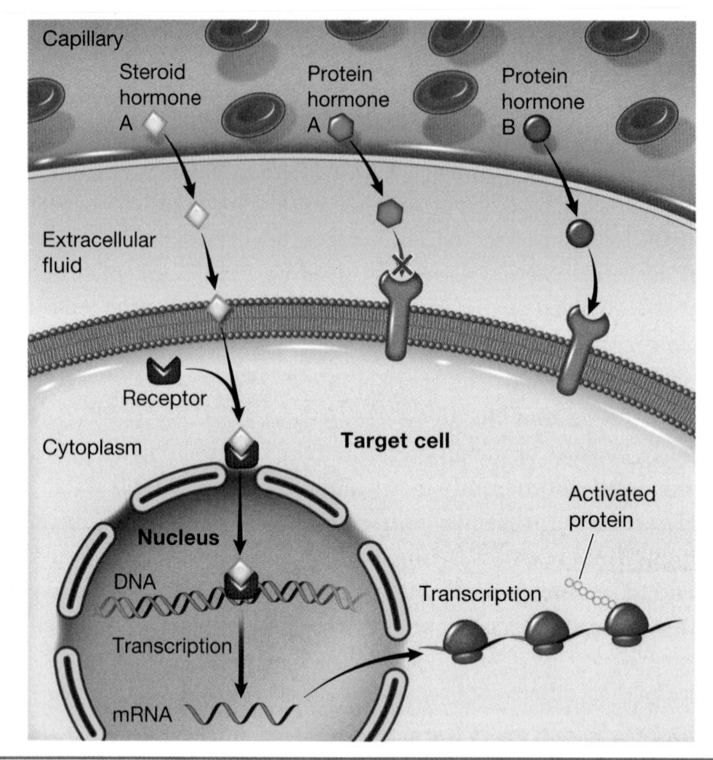

Pathways Involving Intracellular Hormone Receptors Intracellular hormone receptors are located inside the cell. Hormones that bind to this type of receptor must be able to cross the cell membrane. Steroid hormones are derived from cholesterol and therefore can readily diffuse through the lipid bilayer of the cell membrane to reach the intracellular receptor. Thyroid hormones are also lipid-soluble due to their unique structure and can enter the cell.

An internal hormone receptor may reside within the cytosol or within the nucleus. In either case, when it binds its hormone, the hormone and receptor move together within the cell nucleus and bind to a particular segment of the cell's DNA. The binding of the hormone-receptor complex to DNA triggers transcription of a target gene, generating an mRNA transcript, which moves to the cytosol and directs protein synthesis by ribosomes (see Figure 17.7).

Pathways Involving Membrane-bound Hormone Receptors Hydrophilic, or water-soluble, hormones are unable to diffuse through the lipid bilayer of the cell membrane. Cells express the receptor for these hormones on their cell surface. Except for thyroid hormones, which are lipid-soluble, all amino acid–derived hormones bind to cell membrane receptors that are located on the extracellular-facing side of the cell membrane. In order to effect a change in the cell, these receptors are connected to a cascade of signaling molecules within the cell. Like a domino effect or a relay race, each signaling molecule effects a change in the next molecule in the cascade. In many cases we call the internal signaling molecules *messengers* with the internal signaling molecule being called a **second messenger** and the hormone itself called a **first messenger**.

The second messenger used most commonly is **cyclic adenosine monophosphate (cAMP)** (Figure 17.8).

1. The cAMP second messenger pathway begins when a hormone binds to its receptor in the cell membrane.
2. This receptor is associated with an intracellular component called a **G protein**, and binding of the hormone activates the G-protein component.
3. The activated G protein in turn activates an enzyme called **adenylyl cyclase**, also known as *adenylate cyclase*.
4. Adenylyl cyclase converts adenosine triphosphate (ATP) to cAMP.
5. As the second messenger, cAMP activates a type of enzyme called a **protein kinase** that is present in the cytosol.
6. Activated protein kinases initiate a **phosphorylation cascade**, in which multiple protein kinases phosphorylate (add a phosphate group to) numerous and various cellular proteins, including other enzymes.

G protein (guanine nucleotide-binding protein)

The phosphorylation of cellular proteins can trigger a wide variety of effects, including cellular movement, the rate of cellular respiration, or triggering the synthesis

Figure 17.8 cAMP as a Second Messenger

Water-soluble hormones cannot diffuse through the plasma membrane and so their receptors are expressed on the cell surface. Many of these receptors are associated with a G-protein-coupled receptor, which triggers an intracellular signaling cascade that leads to a cellular response.

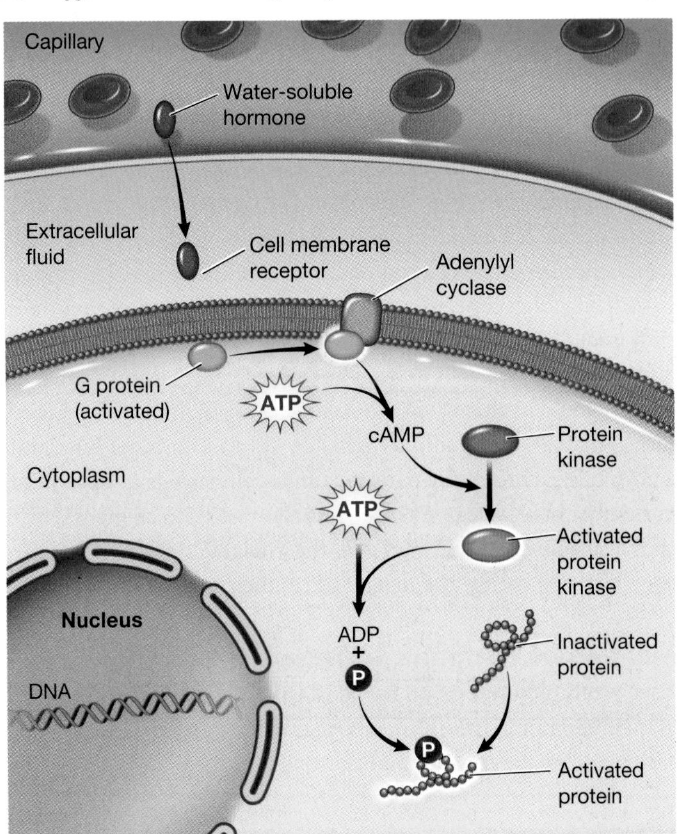

of new molecules. The effects vary according to the type of target cell, the G proteins and kinases involved, and the phosphorylation of proteins. Examples of hormones that use cAMP as a second messenger include calcitonin, which is important for bone construction and regulating blood calcium levels; glucagon, which plays a role in blood glucose levels; and thyroid-stimulating hormone, which causes the release of T_3 and T_4 from the thyroid gland.

Overall, the signaling cascade functions to amplify the efficacy of a hormone. Since each step within the cascade can affect multiple downstream molecules, a significant cellular response can emerge when triggered by even a very low concentration of hormone in the bloodstream. This amplification also makes for a very rapid response to hormone reception. However, the duration of the cellular response to the hormone is short, as cAMP is quickly deactivated by the enzyme **phosphodiesterase (PDE)**, which is located in the cytosol. The action of PDE helps to ensure that a target cell's response ends quickly. If new hormone molecules arrive at the cell membrane, then the cellular response will continue.

Not all water-soluble hormones initiate the cAMP second messenger system. One common alternative system uses calcium ions as a second messenger. In this system, G proteins activate the enzyme phospholipase C (PLC), which functions similarly to adenylyl cyclase. Once activated, PLC cleaves a membrane-bound phospholipid into two molecules: **diacylglycerol (DAG)** and **inositol triphosphate (IP_3)**. Like cAMP, DAG activates protein kinases that initiate a phosphorylation cascade. At the same time, IP_3 causes calcium ions to be released from storage sites within the cytosol, such as from within the smooth endoplasmic reticulum. The calcium ions then act as second messengers in two ways: they can influence enzymatic and other cellular activities directly, or they can bind to calcium-binding proteins, the most common of which is calmodulin. Upon binding calcium, calmodulin is able to modulate protein kinase within the cell. Examples of hormones that use calcium ions as a second messenger system include angiotensin II, which helps regulate blood pressure through vasoconstriction, and growth hormone–releasing hormone (GHRH), which causes the pituitary gland to release growth hormones.

17.2c Factors Affecting Target Cell Response

You will recall that the only cells that will respond to a given hormone are those that express receptors for that hormone. However, in addition to receptor expression, several other factors influence the target cell response. For example, when a hormone circulates in the bloodstream at a high concentration for a persistent length of time, target cells may decrease their number of expressed receptors for that hormone. This process is called **downregulation**, and it allows cells to become less reactive to the excessive hormone levels. An analogy would be that when the volume is too high on your music, you might cover your ears so that less of the sound makes its way to your inner ear. Similarly, when the level of a hormone is chronically reduced, target cells engage in **upregulation** to increase their number of receptors. This process allows cells to be more sensitive to the hormone that is available. Cells may also alter the availability of their downstream signaling molecules, which will change the magnitude of their response to bound hormone.

17.2d Regulation of Hormone Secretion

LO 17.2.5

Hormones have profound impacts on homeostasis and therefore must be tightly controlled to maintain homeostasis. Ideally, to maximize control, we would want to be able to carefully regulate the release of a hormone into the bloodstream. We would also want the hormone to only act for a short time so its effects would not go on too long; more hormone could always be released as needed. Thus, the body maintains control over hormone action by regulating hormone production and degradation. Feedback loops are one central tool in hormone regulation.

Role of Feedback Loops Most hormones are regulated through negative feedback loops. Some hormones, oxytocin being the most notable example, follow positive feedback loop patterns. Negative feedback is characterized by the inhibition of further secretion of a hormone in response to adequate levels of that hormone. In other words, negative feedback loops turn themselves off through their own actions. As a negative feedback loop begins, the events leading to its end are set in motion. Figure 17.9 illustrates a negative and positive feedback loop. In the negative feedback loop, the production and release of the hormone calcitonin is regulated by the levels of calcium ions in the blood. A rise in blood calcium ions following a meal triggers the release of calcitonin. The action of calcitonin is that calcium is removed from the blood and added to

Student Study Tip

One real-life example of a negative feedback loop is an air conditioner. When it runs, the room gets cold, so you turn the AC off. The action of the cycle leads to it turning off.

Figure 17.9 Negative and Positive Feedback Loops

In a negative feedback loop, the product of the cycle resolves the stimulus, ending the cycle. In a positive feedback loop, the cycle will not stop on its own, but the products of the loop perpetuate the cycle until an external interruption occurs.

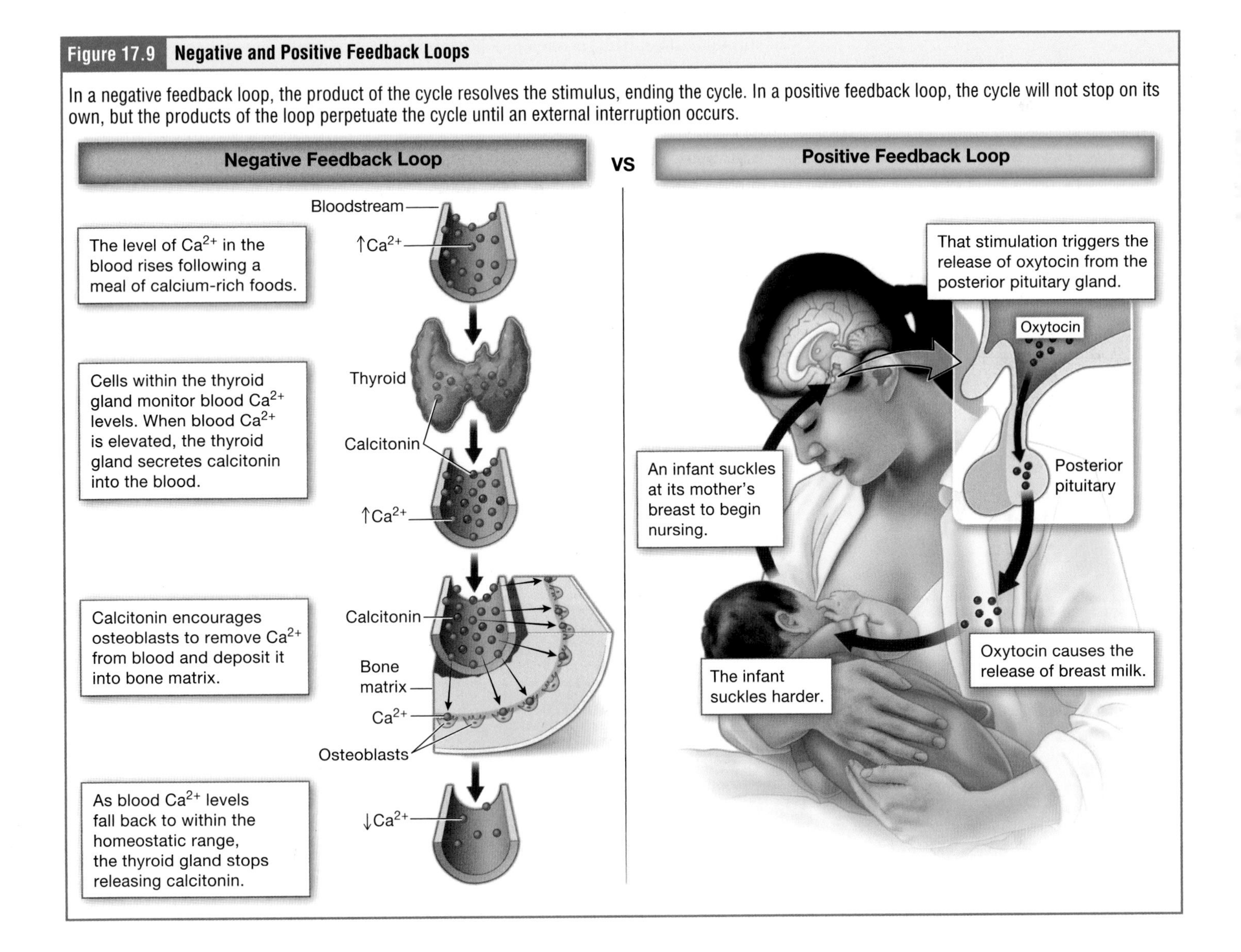
Negative Feedback Loop
VS
Positive Feedback Loop
Bloodstream
↑Ca2+
The level of Ca2+ in the blood rises following a meal of calcium-rich foods.
Thyroid
Calcitonin
Cells within the thyroid gland monitor blood Ca2+ levels. When blood Ca2+ is elevated, the thyroid gland secretes calcitonin into the blood.
↑Ca2+
Calcitonin
Bone matrix
Ca2+
Osteoblasts
Calcitonin encourages osteoblasts to remove Ca2+ from blood and deposit it into bone matrix.
↓Ca2+
As blood Ca2+ levels fall back to within the homeostatic range, the thyroid gland stops releasing calcitonin.
That stimulation triggers the release of oxytocin from the posterior pituitary gland.
Oxytocin
Posterior pituitary
An infant suckles at its mother's breast to begin nursing.
Oxytocin causes the release of breast milk.
The infant suckles harder.

the bone matrix. As this action takes place in the bone tissue, it causes the concentration of calcium ions in the blood to decrease, thereby turning off the stimulus for further calcitonin release. Hormones regulated by negative feedback allow blood levels of the hormone to be regulated within a narrow range.

Positive feedback loops, on the other hand, perpetuate themselves. They will not turn off until an external event stops them. Oxytocin is a hormone whose release follows a positive feedback loop pattern. One trigger for oxytocin release is sensory stimulation of the nipples. In the case of a lactating person, an infant suckling at the nipples will cause the release of oxytocin. One of oxytocin's actions is to cause the release of breast milk. If milk is not released, the infant may stop suckling, but with milk to drink, the infant likely continues suckling, perhaps more vigorously. With continued stimulation of the nipples, oxytocin continues to be released, milk continues to flow, and the feeding continues. Unlike a negative feedback loop, nothing intrinsic will turn this cycle off. It is only by an external event—when the infant is full and stops sucking at the nipple—that oxytocin secretion wanes and the cycle resolves (Figure 17.9).

Learning Connection

Explain a Process

Try explaining negative and positive feedback to a person who has never studied these processes. For example, negative feedback can be compared to the regulation of temperature in your house by a thermostat. Can you think of other examples?

Role of Endocrine Gland Stimuli In general, we can say there are three different types of stimuli that control the release of hormones (**Figure 17.10**).

Figure 17.10 Three Triggers of Hormone Release

Three types of stimuli control the release of hormones. These include non-hormone chemicals such as calcium ions (A), tropic hormones like TRH (B), and stimulation from the nervous system (C).

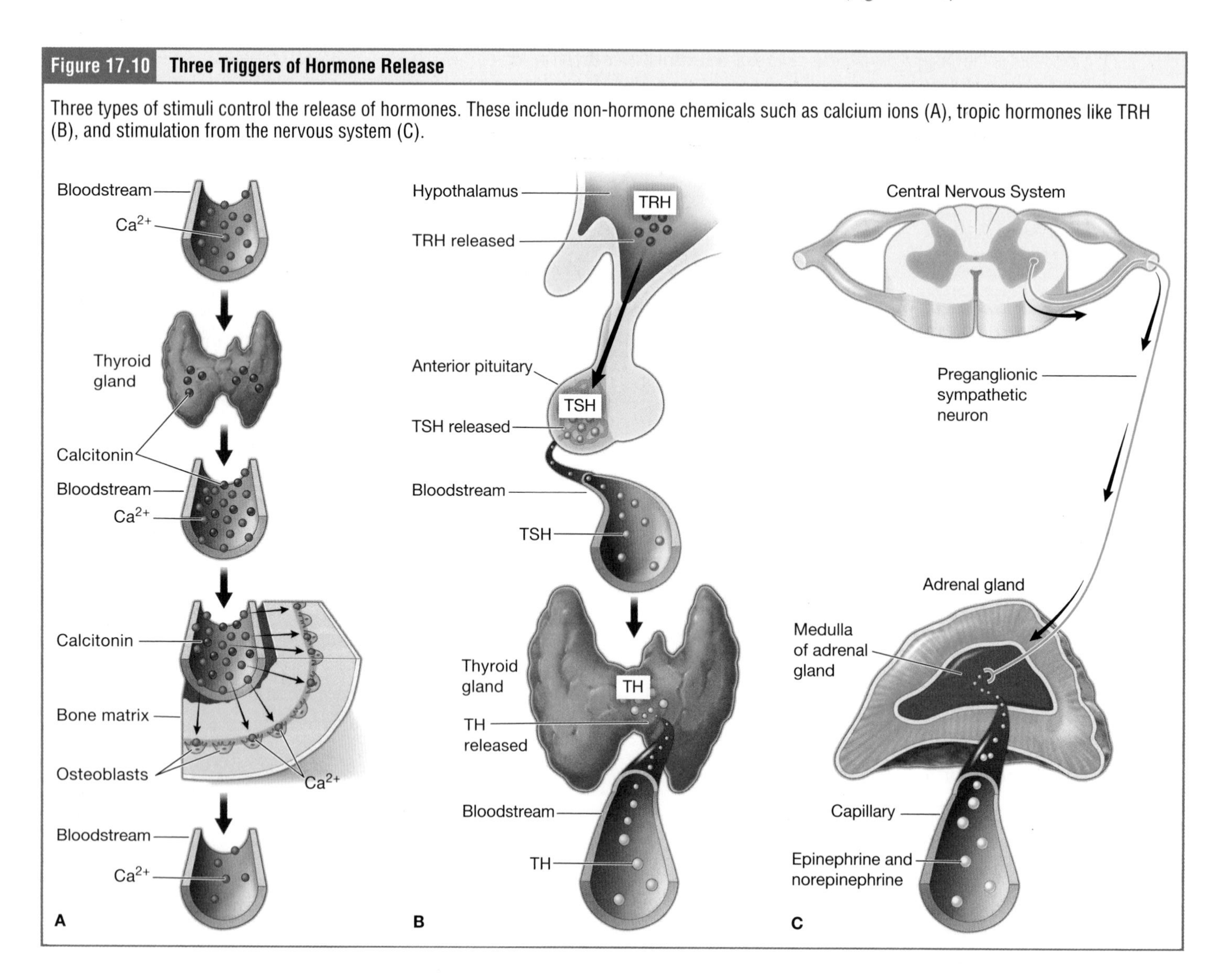

One trigger can be when blood levels of non-hormone chemicals, such as nutrients or ions, increase or decrease outside of the homeostatic range. In this example in **Figure 17.10A**, as well as the example of a negative feedback loop in Figure 17.9, the endocrine gland directly monitors the blood levels of a nutrient.

An endocrine gland may also secrete its hormone in response to the presence of another hormone produced by a different endocrine gland. These hormones that act on endocrine glands to trigger the release of other hormones are called **tropic hormones**. Such hormonal stimuli often involve the hypothalamus (**Figure 17.10B**).

In addition to these chemical signals, hormones can also be released in response to stimulation from the nervous system. A common example is the adrenal medulla, which secretes both the hormones norepinephrine and epinephrine in response to stimulation by the sympathetic nervous system (**Figure 17.10C**).

Learning Check

1. How are steroid hormones transported across a cell membrane?
   a. Steroid hormones are actively transported through protein channels.
   b. Steroid hormones are passively transported through protein channels.
   c. Steroid hormones are actively transported through the cell membrane.
   d. Steroid hormones are passively transported through the cell membrane.
2. Someone who has Type 1 diabetes will have less insulin in their body. People with Type 1 diabetes often take an oral form of insulin as medication. This is an agonist and binds to the same receptors as insulin. When it binds to the hormone receptor, it activates a G protein. What is the next step after the G protein is activated?
   a. Activation of adenylyl cyclase
   b. Conversion of ATP to cAMP
   c. Activation of protein kinase
   d. Initiation of phosphorylation cascade
3. Which of the following are true about the production of protein hormones? Please select all that apply.
   a. Cholesterol is modified to produce the hormone.
   b. DNA and mRNA synthesize the hormone.
   c. The hormone will be made upon demand.
   d. Vesicles store the hormones until needed.
4. What is the role of a negative feedback loop?
   a. To promote further secretion of hormones
   b. To inhibit further secretion of hormones
   c. To initiate hormone breakdown

17.3 Endocrine Control by the Hypothalamus and Pituitary Gland

Learning Objectives: By the end of this section, you will be able to:

17.3.1 Describe the locations and the anatomical relationships of the hypothalamus, anterior pituitary, and posterior pituitary, including the hypothalamic-hypophyseal portal system.

17.3.2 Explain the role of the hypothalamus in the release of hormones from the posterior pituitary.

17.3.3 Name the two hormones produced by the hypothalamus that are stored in the posterior pituitary, and the hormones' primary targets and effects.

17.3.4 Explain the role of hypothalamic neurohormones (regulatory hormones) in the release of anterior pituitary hormones.

17.3.5 Describe major hormones secreted by the anterior pituitary, their control pathways, and their primary target(s) and effects.

The hypothalamus–pituitary complex is often thought of as the "command center" of the endocrine system because it plays a central role in the release of many different hormones. The hypothalamus and the pituitary glands secrete several hormones that directly produce responses in target tissues, as well as many tropic hormones that regulate the synthesis and secretion of hormones of other glands.

LO 17.3.1

The **hypothalamus** is a structure of the diencephalon of the brain located anterior and inferior to the thalamus (**Figure 17.11**). It has both neural and endocrine functions, produces and secretes many hormones, and integrates the nervous system and

Figure 17.11 The Hypothalamus and the Pituitary Gland

The hypothalamus is part of the diencephalon. The pituitary gland, which is suspended beneath it, really consists of two glands. The anterior pituitary is made of glandular tissue and secretes six different hormones. The posterior pituitary is an extension of the hypothalamus and contains neurons that originate in the hypothalamus.

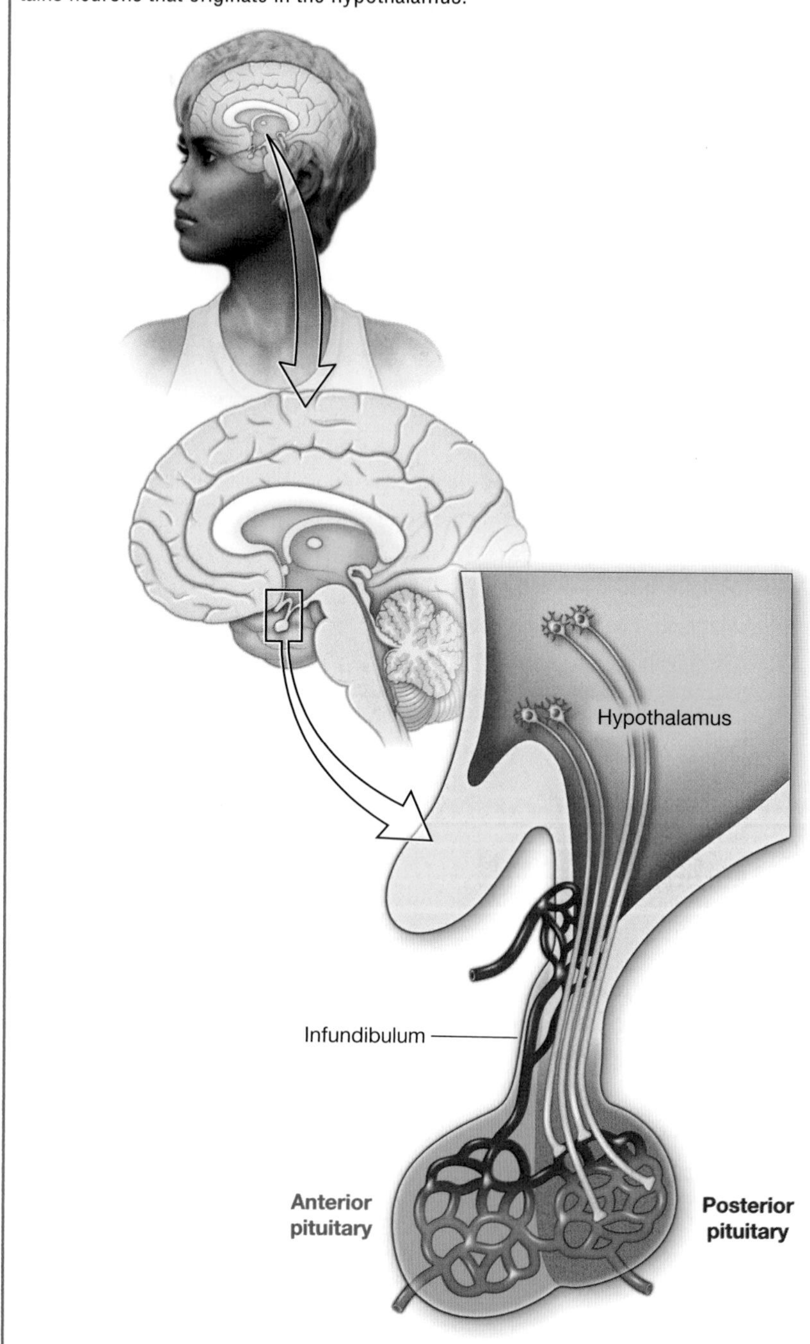

the endocrine system. In addition, the hypothalamus is anatomically and functionally linked to the **pituitary glands**, which can be seen directly inferior to the hypothalamus. While the pituitary has long been described as a single gland, it is in fact two separate structures, the anterior and posterior pituitary glands. The anterior pituitary (**Figure 17.12A**) is an anatomically distinct gland that is physically attached to the posterior pituitary. The posterior pituitary is connected to the hypothalamus by a stem called the **infundibulum**. The infundibulum is similar to a tract in that it is filled with neuron axons. These neurons have their cell bodies in the hypothalamus and their axon terminals in the posterior pituitary (**Figure 17.12B**). Two hormones are made in these neurons; their production occurs in the cell bodies in the hypothalamus. The hormones are transported down the length of the axons and released from the axon terminals and into the blood of the posterior pituitary.

infundibulum (pituitary stalk)

The anterior pituitary is made of glandular tissue that has a different (non-neural) embryonic origin. It hangs onto the infundibulum like a child getting a piggyback ride, but it is not an extension of the hypothalamus the way that the posterior pituitary is. The anterior pituitary is intimately connected to the hypothalamus in a different way though. A portal system is a system of blood vessels that connects two locations. In a **portal system**, one tissue or region has a capillary bed (a system of small blood vessels) through which it receives nutrients and releases any products or wastes; a second tissue or region has another capillary bed that serves the same function. These two tissues are connected because their capillary beds are connected to each other and blood flows directly from the first tissue to the second without circulating elsewhere or returning to

Figure 17.12 The Anterior and Posterior Pituitary

(A) The anterior pituitary is anatomically and functionally separate from the hypothalamus. They are connected by the hypothalamic-hypophyseal portal system. (B) The posterior pituitary contains the axons and axon terminals of neurons that originate in the hypothalamus. Hormones are produced by these neurons and released from the posterior pituitary into the bloodstream.

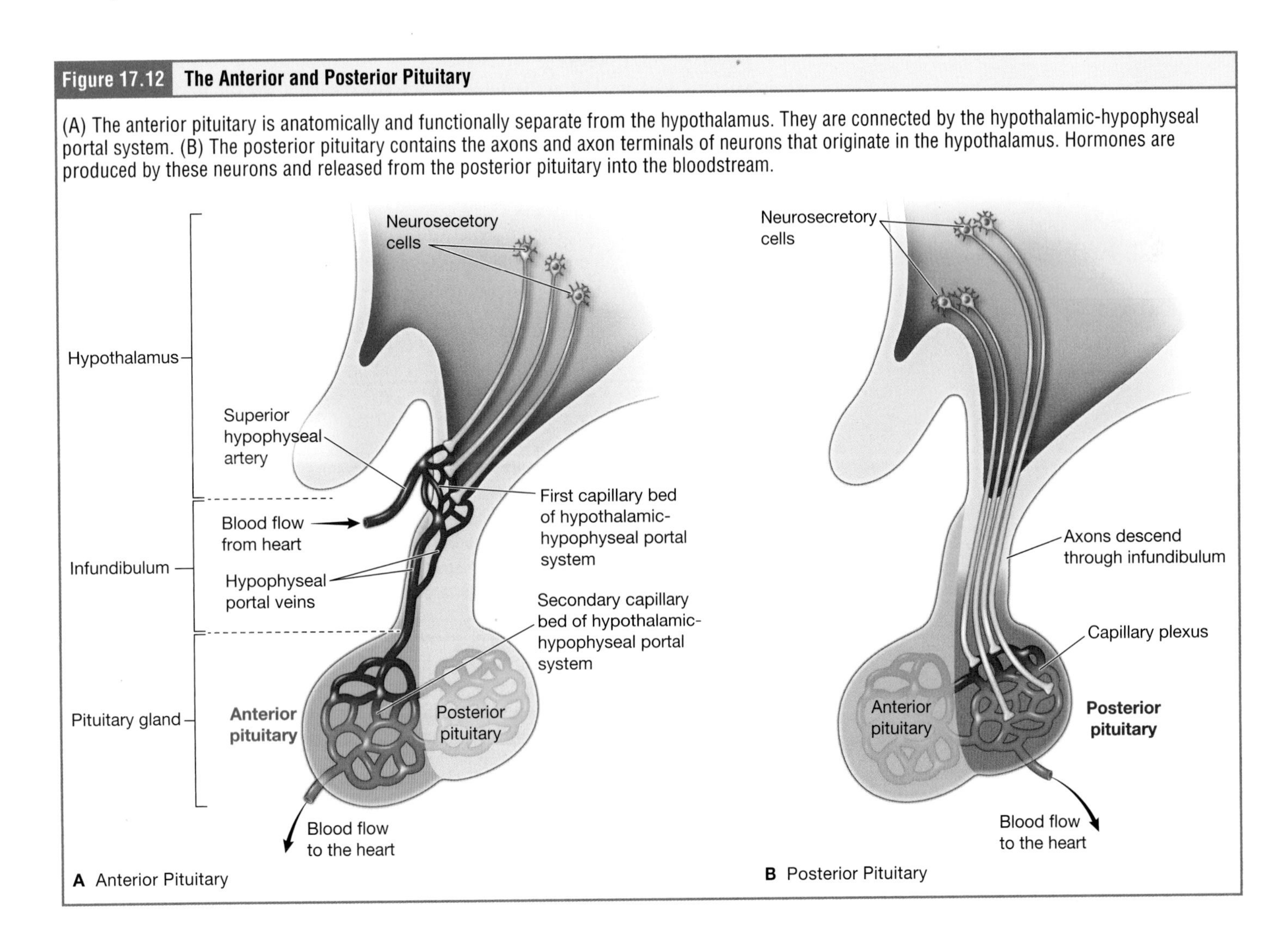

hypothalamic-hypophyseal portal system (hypophyseal portal system)

the heart. The hypothalamus and anterior pituitary gland are connected by the **hypothalamic-hypophyseal portal system.**

17.3a Posterior Pituitary

LO 17.3.2

The posterior pituitary gland does not produce hormones, but the gland is composed of the axons and axon terminals of the neurons of the hypothalamus and therefore stores and secretes hormones produced by the hypothalamus. The cell bodies of these neurons reside in the paraventricular nucleus and the supraoptic nucleus, two regions of the hypothalamus.

LO 17.3.3

Whereas the paraventricular nuclei produce the hormone oxytocin, the supraoptic nuclei produce ADH. These hormones travel along the axons into storage sites in the axon terminals of the posterior pituitary. In response to signals from the same hypothalamic neurons, the hormones are released from the axon terminals into the bloodstream.

Oxytocin Historically, the hormone **oxytocin** was known for its role in childbirth and lactation. Throughout most of pregnancy, oxytocin hormone receptors are not expressed at high levels in the uterus. Toward the end of pregnancy, uterine smooth muscle upregulates expression of oxytocin receptors in the uterus. Oxytocin, a peptide hormone, stimulates uterine contractions. The contraction of the uterus will push the fetal body, usually the head, against the cervix, the doughnut-shaped structure at the base of the uterus. Over several hours (or sometimes days), the cervix stretches and thins in response to the contractions, and once it is sufficiently thinned and dilated, the fetus may be able to pass through. Thus, oxytocin, which is released in a positive feedback loop pattern, is essential to the process of vaginal childbirth.

Oxytocin is also necessary for the milk ejection reflex (commonly referred to as "let-down") in breastfeeding individuals as described earlier. As the newborn begins suckling, sensory receptors in the nipples transmit signals to the hypothalamus. In response, oxytocin is secreted and released into the bloodstream. Within seconds, cells in the mother's milk ducts contract, ejecting milk into the infant's mouth.

Oxytocin receptors are also expressed throughout the brain, including in olfactory processing regions and limbic system structures. The level of oxytocin receptor expression in the brain is similar in all individuals, regardless of biological sex or gender identity. This led researchers to question if oxytocin may play other roles beyond childbirth and lactation. We now understand that oxytocin contributes to social bonding behavior. Bonding

Cultural Connection

Oxytocin as a Social Hormone

Oxytocin levels vary among individuals but, given the hormone's role in reproduction and parenting behavior, the simplest observations come from studies in children. Researchers have found that oxytocin levels vary among children and correlate to social skills and impairments. Children with lower oxytocin levels have more significant observed social impairments than children with higher oxytocin levels. Autism spectrum disorders (ASDs) are a related group of developmental disorders that are characterized by social, behavioral, and communication-related challenges. The inclusion of the word *spectrum* here conveys a wide diversity of symptoms, challenges, and strengths of those diagnosed with ASDs. Researchers and clinicians have studied whether treating individuals on the autism spectrum with oxytocin produces changes in their social skills. Among adults with ASDs, the results have not been persuasive that there is a benefit to oxytocin treatment. But among children, a significant benefit has been observed with oxytocin treatment. Researchers found that the most social skill improvement with oxytocin treatment occurred among children with low oxytocin levels. Children with already higher oxytocin levels did not show significant improvement, indicating that a threshold level of oxytocin may be useful in social skill development. Oxytocin is delivered by a nasal spray as it is thought to best reach the central nervous system through this route.

is an essential part of parent–child relationships and our emotional attachment to romantic partners and friends. Oxytocin is also thought to be involved in feelings of love and closeness, as well as in the sexual response. Oxytocin can be prescribed to initiate labor in pregnant individuals and is being investigated as a treatment for individuals who may not form social bonds as readily as others, such as some individuals on the autism spectrum.

Antidiuretic Hormone (ADH) The solute concentration of the blood, or *blood osmolarity*, may change in response to the consumption of certain foods and fluids, as well as in response to disease, injury, medications, or other factors. Blood osmolarity is constantly monitored by **osmoreceptors**—specialized cells within the hypothalamus that are particularly sensitive to the concentration of ions and solutes.

In response to high blood osmolarity, which can occur during dehydration or following a very salty meal, the osmoreceptors signal the posterior pituitary to release **antidiuretic hormone (ADH)** into the blood. The target cells of ADH are located in the kidneys. The action of ADH is to decrease the amount of body water that becomes urine; in other words, to conserve water and keep it in the body rather than losing it through urination. You may notice that your urine is sometimes very pale yellow and sometimes a very dark yellow; you are observing the action of ADH in real time. If you are urinating small volumes of dark urine, ADH is at work to retain water in the body. If you are urinating large volumes of pale urine, ADH is probably not being released in very significant amounts at this moment. When released in very high concentrations ADH can also cause constriction of blood vessels.

antidiuretic hormone (ADH) (vasopressin)

Interestingly, drugs can affect the secretion of ADH. For example, alcohol consumption inhibits the release of ADH, resulting in increased urine production that can eventually lead to dehydration. Extreme post-alcohol dehydration is often called a *hangover*. A disease called *diabetes insipidus* is characterized by chronic underproduction of ADH that causes chronic dehydration. Because little ADH is produced and secreted, not enough water is reabsorbed by the kidneys. Although patients feel thirsty, and increase their fluid consumption, this doesn't effectively decrease the solute concentration in their blood because ADH levels are not high enough to trigger water reabsorption in the kidneys. Ion imbalances can occur in severe cases of diabetes insipidus.

Learning Connection

Broken Process

What would you expect to happen if the pituitary stalk (infundibulum) were accidentally cut during brain surgery? Which hormones would be affected? What effects would this have on the body?

17.3b Anterior Pituitary

LO 17.3.4

Unlike the posterior pituitary, which does not synthesize hormones but merely stores them, the anterior pituitary does produce hormones. The six anterior pituitary hormones are regulated via tropic hormones from the hypothalamus. The hypothalamic tropic hormones are secreted by neurons, but enter the anterior pituitary through the blood vessels of the hypothalamic-hypophyseal portal system (see Figure 17.12). Within this network, hypothalamic hormones are transported to the anterior pituitary without first entering the systemic circulation. This anatomical feature prevents the hypothalamic hormones from being diluted in the circulatory system and speeds communication between the two locations. Hormones produced by the anterior pituitary (in response to releasing hormones) enter a secondary capillary bed, and from there drain into thecirculation. The anterior pituitary hormones are summarized in **Figure 17.13** and **Table 17.1**.

Student Study Tip

Anterior pituitary involves Another hormone, Posterior involves one Primary hormone.

Growth Hormone The endocrine system regulates the growth of the human body, protein synthesis, and cellular replication. A major hormone involved in this process is **growth hormone (GH)**, a protein hormone produced and secreted by the anterior pituitary gland. Its primary function is anabolic; it promotes protein synthesis and tissue building through direct and indirect mechanisms (**Figure 17.14**). GH levels are controlled by the release of GHRH and GHIH (growth hormone-inhibiting hormone) (also known as *somatostatin*) from the hypothalamus.

Figure 17.13 Summary of the Hormones Released by the Pituitary Gland

The anterior pituitary gland secretes six hormones and the posterior pituitary gland secretes two. The hormones have target tissues throughout the body.

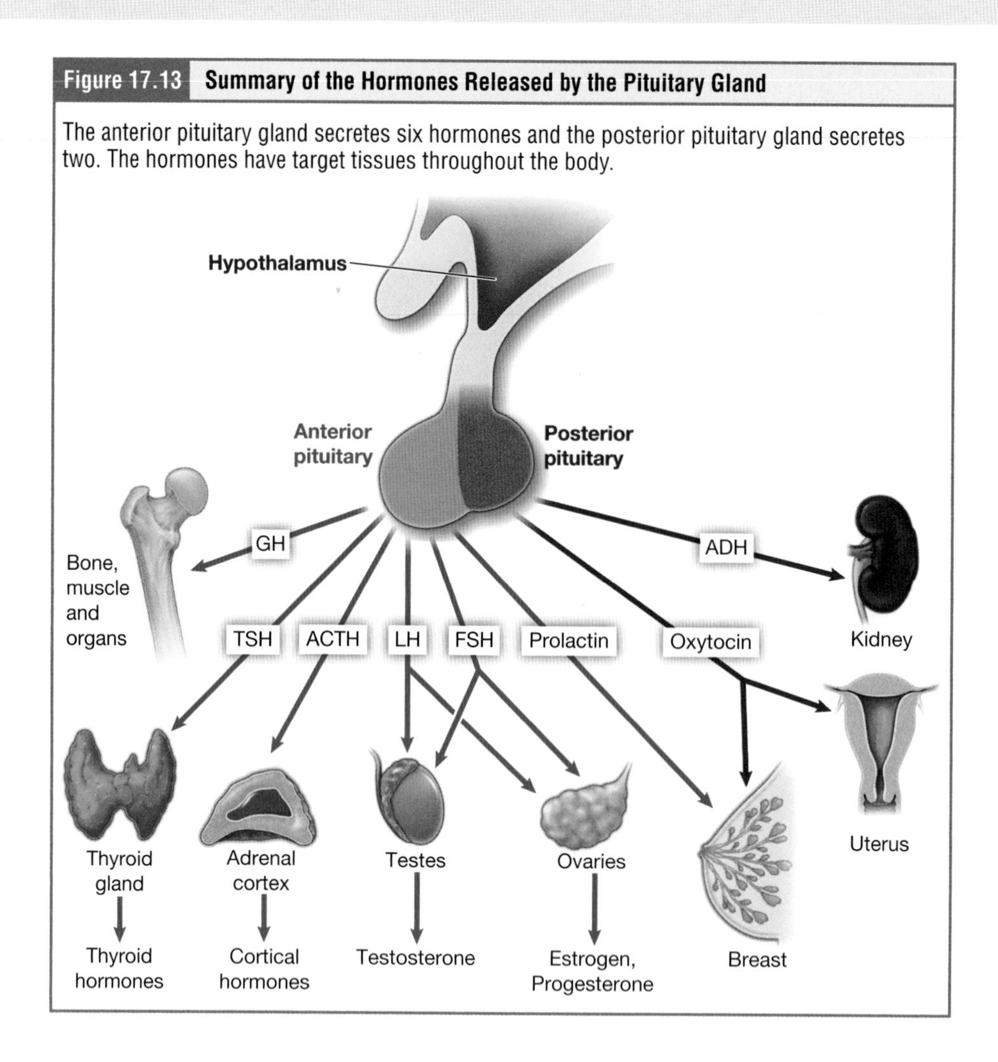

Table 17.1 Anterior Pituitary Hormones LO 17.3.5

Name	Type	Target tissue(s)	Effects	Tropic activity	Related hypothalamic hormone
Growth Hormone (GH)	Protein	Chondrocytes in cartilage, muscle cells	Promotes growth and mitosis	No	Stimulated by growth hormone–Releasing hormone (GHRH) inhibited by somatostatin (GHIH)
Prolactin (PRL)	Peptide	Mammary glands	Promotes milk production	No	Inhibited by dopamine (PIH, or prolactin-inhibiting hormone)
Thyroid Stimulating Hormone	Protein	Thyroid gland	Stimulates release of thyroid hormone	Yes	Stimulated by thyroid-releasing hormone; inhibited by negative feedback
Adrenocorticotropic Hormone (ACTH)	Peptide	Adrenal cortex	Stimulates the release of corticotropic hormones	Yes	Stimulated by corticotropin-releasing hormone; inhibited by negative feedback
Luteinizing Hormone (LH)	Protein	Gonads	Stimulates androgen production, controls ovulation	Yes	Stimulated by gonadotropin-releasing hormone; inhibited by negative feedback
Follicle Stimulating Hormone (FSH)	Protein	Gonads	Stimulates gamete (sperm and egg) production	No	Stimulated by gonadotropin-releasing hormone; inhibited by negative feedback

Figure 17.14 **Growth Hormone**

Growth hormone (GH) is an important anterior pituitary hormone that regulates the growth process during fetal life, childhood and adolescence. However, it also has several vital functions all throughout our lifespan, such as adding energy sources (nutrients) to the blood when necessary and sparing glucose for use by the central nervous system. GH secretion is regulated by a releasing and an inhibiting hormone from the hypothalamus. GH exerts some of its growth-promoting effects through insulin-like growth factors (IGFs).

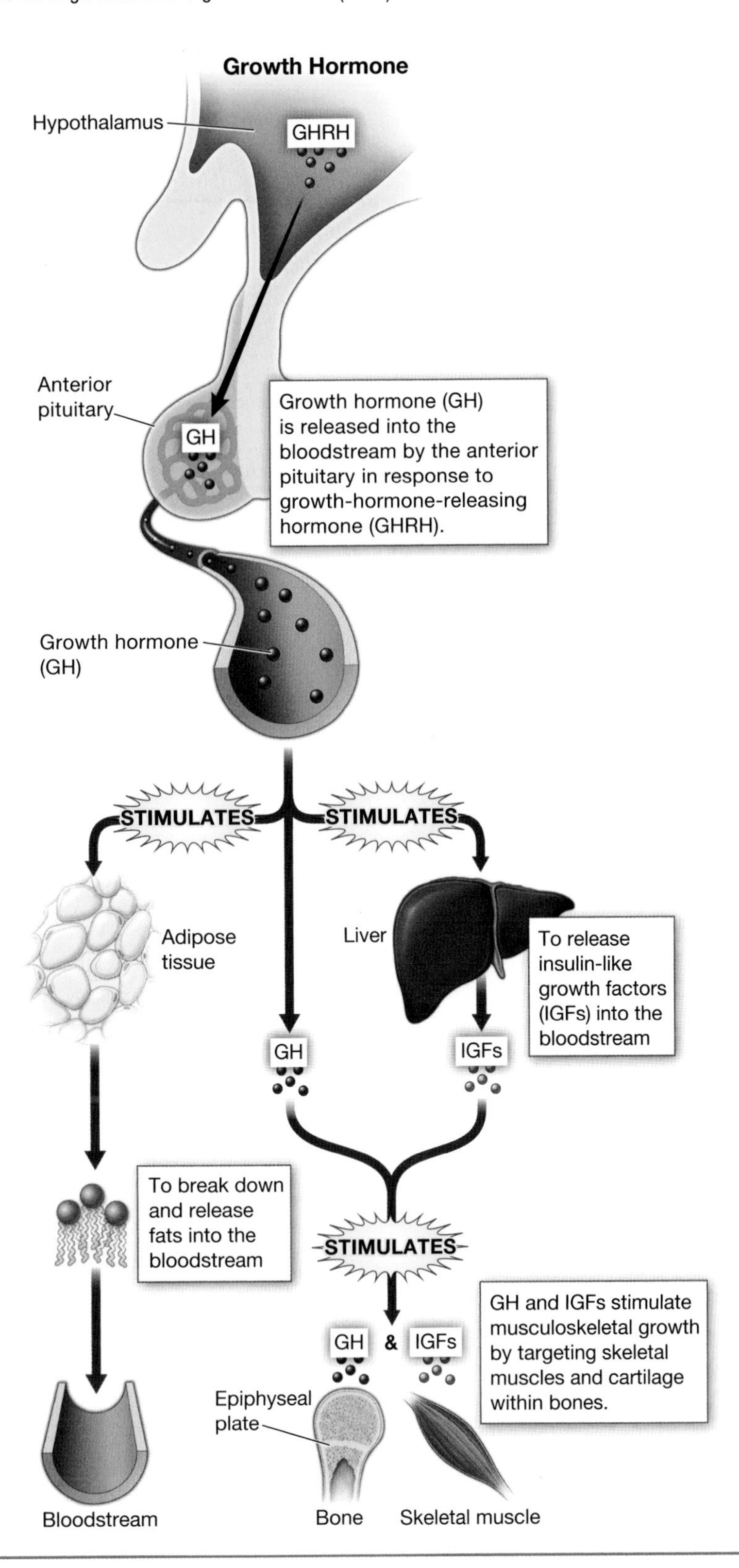
Growth Hormone
Hypothalamus
GHRH
Anterior pituitary
GH
Growth hormone (GH) is released into the bloodstream by the anterior pituitary in response to growth-hormone-releasing hormone (GHRH).
Growth hormone (GH)
STIMULATES
STIMULATES
Adipose tissue
Liver
To release insulin-like growth factors (IGFs) into the bloodstream
GH
IGFs
To break down and release fats into the bloodstream
STIMULATES
GH and IGFs stimulate musculoskeletal growth by targeting skeletal muscles and cartilage within bones.
GH
&
IGFs
Epiphyseal plate
Bloodstream
Bone
Skeletal muscle

DIGGING DEEPER: Growth Hormone Diversity and Disorders

Growth hormone (GH) is secreted by the anterior pituitary and has effects throughout the body. It is best known for its promotion of mitosis in cartilage and its hypertrophy of skeletal muscle cells. Notably, GH does not act directly on bone cells, but instead promotes mitosis in cartilage cells; as a result, the epiphyseal plates of growing bones elongate, and therefore the bones themselves become longer. Decreased levels of GH during childhood or adolescence cause decreased overall stature. Significant decreases in GH signaling can cause some of the types of dwarfism. The opposite—excessive secretion of GH—can also occur. When excess GH is secreted prior to the end of puberty, the affected individual experiences substantial gains in height and the condition is known as **gigantism**. Individuals with gigantism typically grow to between seven and nine feet in height, creating substantial issues for these individuals in terms of their interaction with a world in which everything is built for a smaller population. When GH secretion occurs or is continued past epiphyseal (growth) plate fusion, bone length is no longer impacted but the cartilage at the joints grows in response to the excess GH. This condition, called **acromegaly**, is characterized by larger-than-average hands, feet, and ears. Because GH can affect other tissues as well, individuals with any of these disorders can have cardiovascular or other organ system complications. Gigantism and acromegaly are typically caused by tumors in the anterior pituitary; if these tumors are significant enough in size, they can compress the optic nerves, which cross over in the brain just in front of the anterior pituitary. Thus, vision problems are a common complaint of those with gigantism or acromegaly.

Cultural Connection

Little People/Dwarfism

There are approximately 300 different physiological forms of dwarfism. Many of the forms develop due to issues with growth hormone, which is secreted from the anterior pituitary gland, although other forms of short stature are due to mutations or aberrations in other signaling pathways. A person with **dwarfism**, or short stature, is defined as having a height of less than 58 inches (4’10”), or a height in the lowest 2.3 percent of the human population. There are communities and cultures around short stature and dwarfism. One of the largest organizations in the United States is the Little People of America. Organizations like these connect and empower individuals with short stature though conferences, online forums, and regional groups. Within these communities, individuals with short stature network and engage with each other and with their own and other families.

Growth requires energy, so one of the effects of GH is to stimulate **lipolysis**, the breakdown of adipose tissue, releasing fatty acids into the blood. These fatty acids can fuel the other growth-related effects of GH. GH also stimulates the liver to break down glycogen to glucose, further increasing energy levels available in the blood.

GH indirectly mediates growth by triggering the liver and other tissues to produce a group of proteins called **insulin-like growth factors (IGFs)**. These proteins enhance cellular proliferation and inhibit apoptosis, or programmed cell death. Skeletal muscle and cartilage cells are particularly sensitive to stimulation from IGFs. Dysfunction of the endocrine system’s control of growth can result in several disorders.

thyroid-stimulating hormone (TSH) (thyrotropin)

Thyroid-Stimulating Hormone The activity of the thyroid gland is regulated by **thyroid-stimulating hormone (TSH)**. TSH is released from the anterior pituitary in response to thyrotropin-releasing hormone (TRH) from the hypothalamus. As discussed shortly, it triggers the secretion of thyroid hormones by the thyroid gland. In a classic negative feedback loop, elevated levels of thyroid hormones in the bloodstream then trigger a drop in production of TRH and subsequently TSH.

Adrenocorticotropic Hormone The **adrenocorticotropic hormone (ACTH)** stimulates the adrenal cortex to secrete corticosteroid hormones such as cortisol. The release of ACTH is regulated by the corticotropin-releasing hormone (CRH) from the hypothalamus in response to normal physiologic rhythms. A variety of stressors can also influence its release, and the role of ACTH in the stress response is discussed later in this chapter.

adrenocorticotropic hormone (ACTH) (corticotropin)

Follicle-Stimulating Hormone and Luteinizing Hormone The endocrine glands secrete a variety of hormones that control the development and regulation of the reproductive system. Much of the development of the reproductive system occurs during puberty, and the puberty phase is initiated by gonadotropin-releasing hormone (GnRH), a hormone produced and secreted by the hypothalamus. GnRH stimulates the anterior pituitary to secrete follicle-stimulating hormone (FSH) and luteinizing hormone (LH). The levels of GnRH are regulated through a negative feedback loop; high levels of reproductive hormones inhibit the release of GnRH.

Follicle-stimulating hormone (FSH) stimulates the production and maturation of sex cells, or gametes, including oocytes in ovaries and sperm in testes. **Luteinizing hormone (LH)** triggers the release of oocytes in a process called *ovulation*, as well as the production of estrogens and progesterone by the ovaries. LH stimulates production of testosterone in the testes. We will take a much deeper dive into the functions of these hormones in Chapter 27.

Prolactin **Prolactin (PRL)** promotes the production of milk in individuals who are lactating. During pregnancy, it contributes to development of the mammary glands, and after birth, it stimulates the mammary glands to produce breast milk.

Prolactin secretion is inhibited by dopamine, which typically acts as a neurotransmitter but in this case is released from neurons in the hypothalamus into the blood. Prolactin levels only rise in response to prolactin-releasing hormone (PRH) from the hypothalamus during pregnancy and lactation.

DIGGING DEEPER:
Breast Milk and Breastfeeding

Breastfeeding is controlled through the collaboration of two pituitary hormones: prolactin and oxytocin. Prolactin, together with estrogen, promotes mitosis in breast tissue, enlarging the breasts during pregnancy in preparation for lactation. Once the infant begins to suckle at the breast, prolactin promotes the production of breast milk. Breast milk is a secretion that has a high lipid content, but also contains some surprising ingredients. The second largest ingredient in breast milk is an oligosaccharide that is indigestible by the infant. This starchy molecule serves to feed the infant's gut microbiome—the garden of bacteria that occupies human intestines. Since infant formula does not contain this ingredient, the intestines of breast-fed and formula-fed infants differ. Breast milk also contains a number of immune ingredients. Lactoferrin, the second most abundant protein in breast milk, helps the infant immune system fight off bacterial infections. Antibodies—immune-fighting proteins that are specific to particular pathogens—are passed from the mother's immune system to the infant through breast milk. Lastly, lysozyme—another bacteria-fighting immune compound—is also found in breast milk. All of these ingredients help the infant immune system and body develop. Prolactin does not work alone in breastfeeding; oxytocin plays a critical role. In addition to its role in social bonding, oxytocin is responsible for controlling the ejection of milk from the breast, marrying these two functions physiologically. Milk is ejected because of an oxytocin surge when the infant suckles at the breast, but this in turn promotes bonding behavior. Some breastfeeding individuals find that they eject milk in response to other social cues, especially when seeing someone cry. Oxytocin promotes the activity of the parasympathetic nervous system, decreasing heart rate and blood pressure, so some individuals experience a relaxing sensation when breastfeeding.

Learning Check

1. Which endocrine gland produces oxytocin?
   a. Hypothalamus
   b. Anterior pituitary
   c. Posterior pituitary
   d. Thalamus
2. Which endocrine gland produces adrenocorticotropic hormone?
   a. Hypothalamus
   b. Anterior pituitary
   c. Posterior pituitary
   d. Thalamus
3. When someone has low blood pressure, there is a decreased volume of blood in the arteries. Low blood pressure can be treated with a medication called vasopressin, which acts in the same manner as antidiuretic hormone. Which of the following is an effect of vasopressin?
   a. Increased water reabsorption
   b. Increased protein synthesis
   c. Increased secretion of cortisol
   d. Increased sodium reabsorption
4. Which of the following would result in a reduction of thyroid hormone released into the bloodstream?
   a. A decrease in circulating thyroid hormone levels
   b. A decrease in circulating thyroid-stimulating hormone levels
   c. An increase in circulating thyroid-stimulating hormone levels

17.4 The Major Hormones of the Body

HAPS LEARNING GOALS AND OUTCOMES

Learning Objectives: By the end of this section, you will be able to:

17.4.1 Describe the anatomy of the thyroid gland, its location, the major hormones secreted, the control pathway(s) for hormone secretion, and the hormones' primary targets and effects.

17.4.2 Describe the anatomy of the parathyroid glands, their location, the major hormone secreted, the control pathway(s) for hormone secretion, and the hormone's primary targets and effects.

17.4.3 Describe a simple endocrine pathway in which the response is the negative feedback signal (e.g., parathyroid hormone, insulin).

17.4.4 Compare and contrast negative feedback for hypothalamic-anterior pituitary-peripheral endocrine gland pathways to negative feedback for most simple endocrine pathways.

17.4.5 Describe the anatomy of the adrenal cortex, its location, the major hormones secreted, the control pathway(s) for hormone secretion, and the hormones' primary targets and effects.

17.4.6 Describe the anatomy of the pancreas, its location, the major hormones secreted, the control pathway(s) for hormone secretion, and the hormones' primary targets and effects.

17.4.7 Describe the anatomy of the thymus gland, its location, the major hormones secreted, the control pathway(s) for hormone secretion, and the hormones' primary targets and effects.

17.4.8 Provide some examples of hormones that are secreted from diffuse endocrine tissues or single endocrine cells.

LO 17.4.1

17.4a The Thyroid Gland

A butterfly-shaped organ, the **thyroid gland** lies over the anterior surface of the trachea, just inferior to the larynx (**Figure 17.15**). The medial region, called the *isthmus*, is flanked by wing-shaped left and right lobes. Under the microscope we can see that the tissue of the thyroid gland is composed mostly of thyroid follicles. The **thyroid follicles** are made up of a ring of thyroid epithelial cells surrounding a central cavity filled with a sticky fluid called **colloid**, which is the precursor to thyroid hormone.

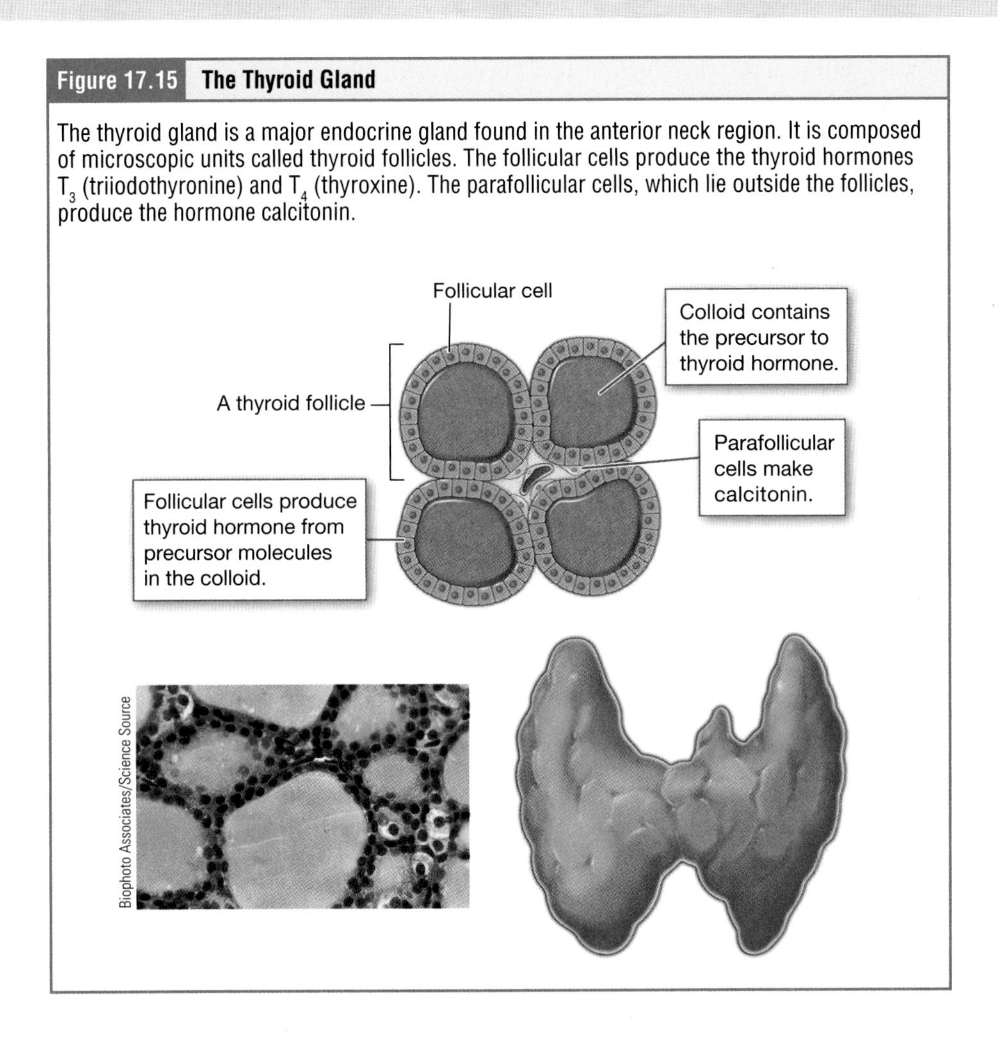
Follicular cell
Colloid contains the precursor to thyroid hormone.
A thyroid follicle
Parafollicular cells make calcitonin.
Follicular cells produce thyroid hormone from precursor molecules in the colloid.
Biophoto Associates/Science Source

Figure 17.15 The Thyroid Gland

The thyroid gland is a major endocrine gland found in the anterior neck region. It is composed of microscopic units called thyroid follicles. The follicular cells produce the thyroid hormones T_3 (triiodothyronine) and T_4 (thyroxine). The parafollicular cells, which lie outside the follicles, produce the hormone calcitonin.

Synthesis and Release of Thyroid Hormones The thyroid follicles produce two hormones, which are really two versions of the same hormone—thyroid hormone (TH). Between thyroid follicles are clusters of cells that are not part of the thyroid follicle. These *parafollicular cells* produce a third hormone, calcitonin.

Iodine is a key ingredient in the production of TH, so TH synthesis can only begin when iodine is bound by the follicular cells. TSH from the anterior pituitary acts as a tropic hormone that kickstarts the production of TH. TH synthesis then proceeds as follows:

1. Binding of TSH to its receptors in the follicle cells of the thyroid gland causes the cells to actively transport iodide ions (I^-) across their cell membrane, from the bloodstream into the cytosol. As a result, the concentration of iodide ions "trapped" in the follicular cells is many times higher than the concentration in the bloodstream.
2. Iodide ions then move to the lumen of the follicle cells that border the colloid. There, the ions undergo oxidation (their negatively charged electrons are removed). The oxidation of two iodide ions ($2\ I^-$) results in iodine (I_2), which passes through the follicle cell membrane into the colloid.
3. In the colloid, peroxidase enzymes link the iodine to amino acids in colloid to produce two intermediaries—one with one iodine and one attached to two iodines. When one of each of these intermediaries is linked by covalent bonds, the resulting compound is **triiodothyronine** (T_3), a TH with three iodines. Much more commonly, two copies of the second intermediary bond to form tetraiodothyronine, also known as **thyroxine** (T_4), a TH with four iodines.

These hormones remain in the colloid center of the thyroid follicles until TSH stimulates endocytosis of colloid back into the follicle cells. There, lysosomal enzymes break apart the thyroglobulin colloid, releasing free T_3 and T_4. Both T_3 and T_4 are able

triiodothyronine (T_3)

thyroxine (T_4 hormone, tetraiodothyronine)

to freely diffuse across membranes, which makes them difficult to store and control once they are in these forms. T_3 and T_4 diffuse across the follicle cell membrane and enter the bloodstream. In the bloodstream, T_3 and T_4 circulate bound to transport proteins. This "packaging" prevents their free diffusion into body cells. When blood levels of T_3 and T_4 begin to decline, bound T_3 and T_4 are released from these plasma proteins and readily cross the membrane of target cells. T_3 is more potent than T_4, and many cells convert T_4 to T_3 through the removal of an iodine atom.

Regulation of TH Synthesis Blood levels of T_3 and T_4 are regulated in a classic negative feedback pattern. As shown in Figure 17.14, low blood levels of T_3 and T_4 stimulate the release of TRH from the hypothalamus, which triggers secretion of TSH from the anterior pituitary. In turn, TSH stimulates the thyroid gland to secrete T_3 and T_4. The levels of TRH, TSH, T_3, and T_4 are regulated by a negative feedback system in which increasing levels of T_3 and T_4 decrease the production and secretion of TSH (**Figure 17.16**).

Figure 17.16 The Regulation of Thyroid Hormone Production

Thyroid hormone is produced in thyroid follicles; its production is regulated in a classic negative feedback pattern.

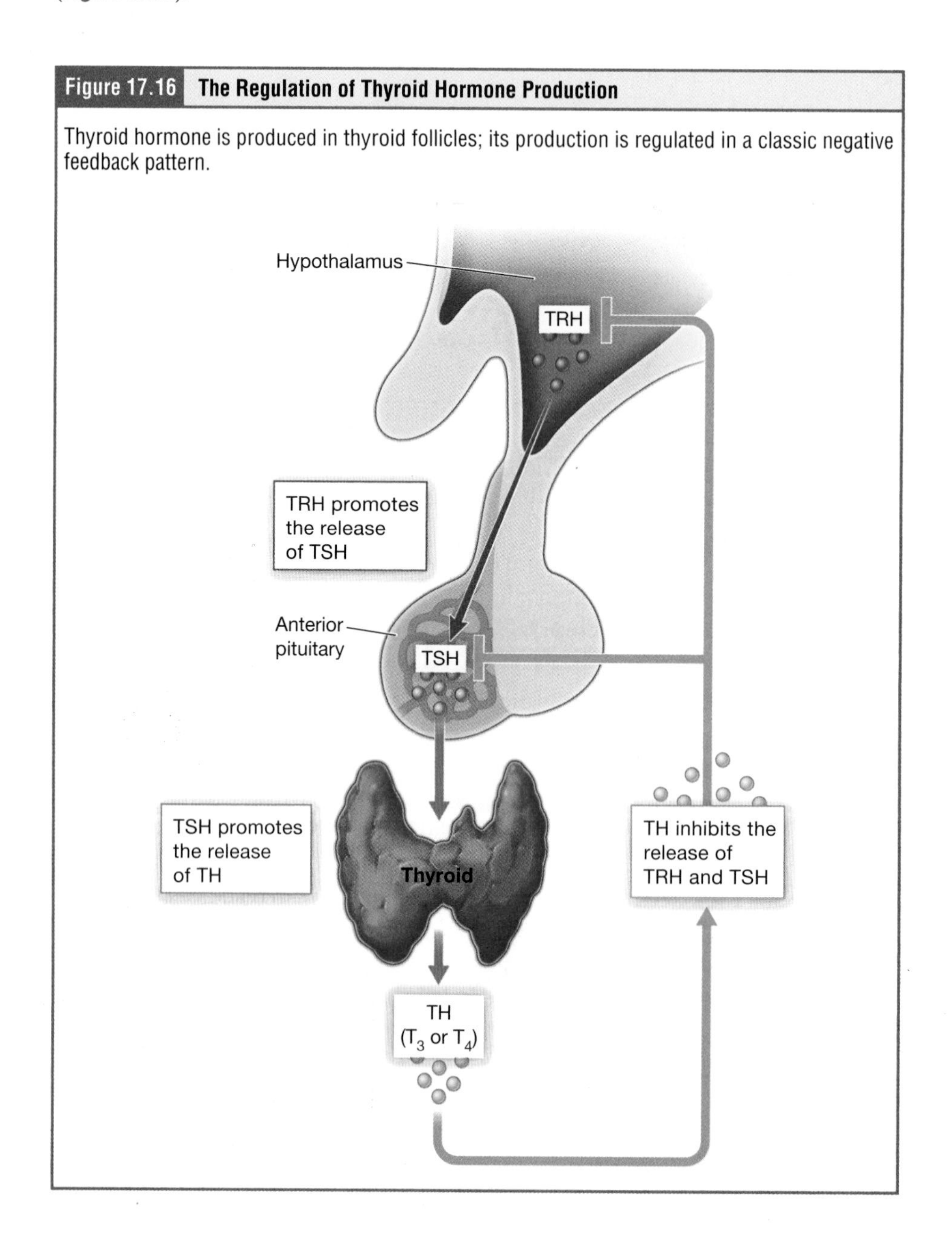

Functions of Thyroid Hormones The THs, T_3 and T_4, are often referred to as *metabolic hormones* because their levels influence the body's **basal metabolic rate**—the amount of energy used by the body at rest. THs directly impact the Na^+/K^+ pump within every cell, causing these active transporters to work faster. Na^+/K^+ pumps are dependent on ATP, so as they start to work harder, each cell must increase its production of ATP. As you may recall, heat is a byproduct of the cellular respiration process through which ATP is generated; therefore, as the cells' demand for ATP increases, so does body temperature.

Adequate levels of THs are also required for protein synthesis, especially in nervous tissue. Adequate TH is therefore essential for fetal and childhood development and growth. These THs have a complex interrelationship with reproductive hormones, and deficiencies can influence libido, fertility, and other aspects of reproductive function. Notably, THs increase the body's sensitivity to epinephrine and norepinephrine. THs do not directly impact the heart or blood vessels, but when levels of T_3 and T_4 hormones are high, they potentiate the effects of epinephrine and norepinephrine to allow them to further accelerate the heart rate, strengthen the heartbeat, and increase blood pressure more than they would in the absence of TH. Because THs regulate metabolism, heat production, protein synthesis, and many other body functions, thyroid disorders can have severe and widespread consequences.

Dietary iodine deficiency can result in the impaired ability to synthesize T_3 and T_4. When T_3 and T_4 cannot be produced, TSH is secreted in increasing amounts. As a result of this accelerating stimulation, more and more colloid accumulates in the thyroid follicles. The accumulation of colloid increases the overall size of the thyroid gland, a condition called a **goiter** (**Figure 17.17**). A goiter is a visible indication that either insufficient TH is being produced, too much TSH is being released, or the TSH receptors are being stimulated by something else. **Hypothyroidism** is the condition that results when too little TH

Figure 17.17 Goiter

A goiter is an enlarged thyroid gland, which occurs as a result of overproduction of TSH. This can occur in cases of both hyperthyroidism and hypothyroidism.

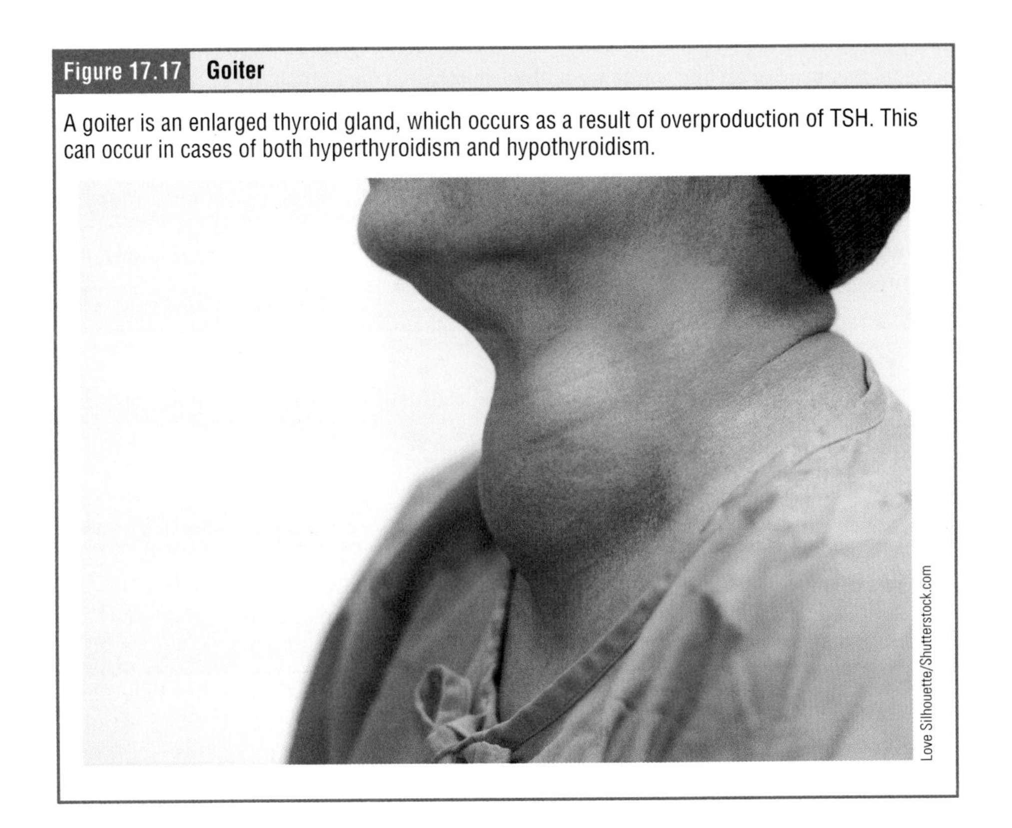

Apply to Pathophysiology

Graves Disease

Swapna, a 23-year-old woman, visits her physician with complaints of feeling nervous and irritable for the past four months. She feels very warm in rooms where everyone else feels comfortable. Her skin is unusually warm and moist to the touch. She has lost 30 pounds of body weight over this period despite having a voracious appetite and increased food intake. Her doctor notes that she has slightly higher blood pressure and heart rate values than in other visits, her skin is warm and moist, and her eyes appear to be bulging out slightly (exopthalmos). The lower portion of her neck has a noticeable bulge. Her hands exhibit a slight shakiness and her reflexes are hyperreactive.

1. The doctor also notes a higher breath rate than is typical. The increase in respiratory rate compared to healthy values could be explained by:
   A) A change in neurological activity in the cerebrum
   B) A change in the amount of acetylcholine being produced
   C) A change in the amount of cellular respiration occurring
   D) A change in the degree of consciousness
2. The physician decides to do some bloodwork. Blood tests reveal the presence of antibodies that bind to the TSH receptor. These antibodies might bind to the receptor site and act as an agonist or an antagonist, or they may block the receptor's binding site altogether. Which of these explanations may help explain this patient's symptoms?
   A) The antibodies bind to another site, but not the receptor's active site.
   B) The antibodies act as an antagonist.
   C) The antibodies act as an agonist.
3. Which of the following would NOT be a successful strategy for treating this patient?
   A) Removal or damage to some of the thyroid gland tissue
   B) Prednisone—a systemic immune system suppressor
   C) Supplemental iodine

is being produced, possibly because of an iodine deficiency. In hypothyroidism, less ATP is being generated; therefore, weight gain and feeling cold in rooms where others do not are both symptoms, as are fatigue and lethargy. **Hyperthyroidism** is when too much TH is being generated. Na^+/K^+ pumps go into overdrive, creating a desperate need for ATP. Energy is used up rapidly and nutrient stores are depleted. Weight loss and an increase in body temperature or feeling warm are both symptoms, as are excitability or irritability, as the effects of even low levels of epinephrine and norepinephrine are potentiated.

Calcitonin The parafollicular cells of the thyroid gland secrete a hormone called **calcitonin** that contributes to the regulation of blood calcium. Calcitonin is released in response to a rise in blood calcium levels (see **Figure 17.19** in the following section). It functions to decrease blood calcium concentrations by:

- Inhibiting the activity of osteoclasts (bone cells that release calcium into the circulation by degrading bone matrix)
- Increasing the activity of osteoblasts (bone cells that deposit bone matrix containing calcium into the bones)
- Decreasing calcium absorption in the intestines
- Increasing calcium loss in the urine

Calcium is critical for many biological processes. It is a second messenger in many signaling pathways, and is essential for muscle contraction, nerve impulse transmission, and blood clotting. Given the critical importance of calcium in cells throughout the body, blood calcium levels are tightly regulated by the endocrine system (**Figure 17.18**). If the body did not have sufficient calcium for even a day, vital body functions might shut down. It would be too risky to rely on a steady intake of calcium from the diet; instead, the body stores any excess calcium in the bones. When blood calcium levels fall, those

Figure 17.18 Calcium Regulation

Calcitonin and parathyroid hormone (PTH) are antagonistic hormones that regulate calcium balance in the blood. Calcitonin is secreted from the thyroid gland when the blood calcium level is too high, and PTH is secreted from the parathyroid glands when it is too low.

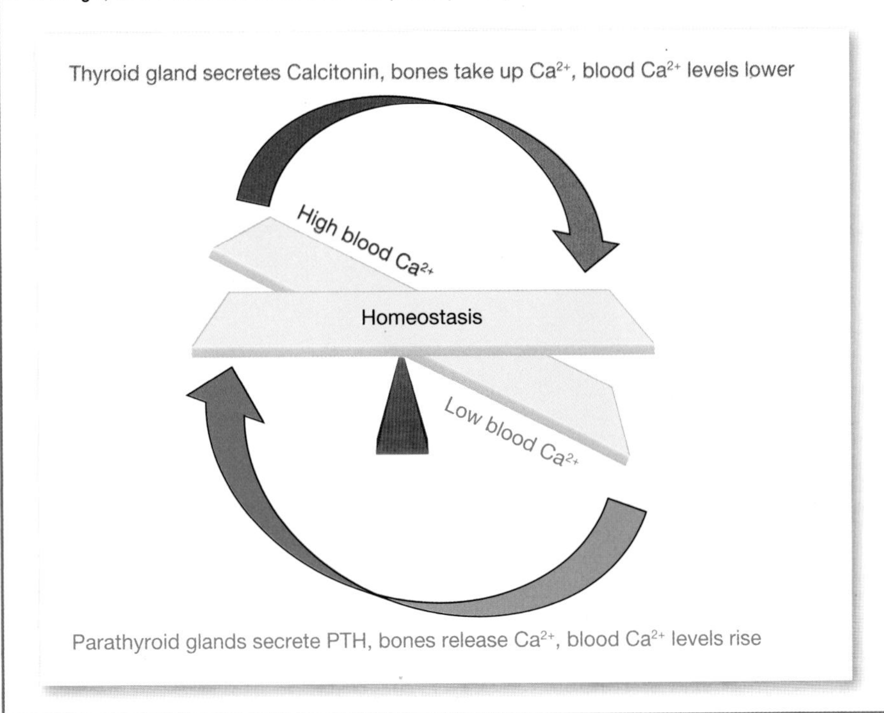
Thyroid gland secretes Calcitonin, bones take up Ca^{2+}, blood Ca^{2+} levels lower
High blood Ca^{2+}
Homeostasis
Low blood Ca^{2+}
Parathyroid glands secrete PTH, bones release Ca^{2+}, blood Ca^{2+} levels rise

stores can be tapped to bring blood calcium levels back up to within the homeostatic range. Calcitonin functions when blood calcium levels are high and excess calcium can be stored. An antagonist hormone, parathyroid hormone (discussed in the following section) releases calcium from the bones when blood calcium levels are too low.

17.4b The Parathyroid Glands

LO 17.4.2

The **parathyroid glands** are tiny, round structures found embedded in the posterior surface of the thyroid gland (see Figure 17.19). The exact number of parathyroid glands varies from one person to another, but most people have between two and six. The parathyroid glands of most people are limited to the posterior thymus, but occasionally there are more in tissues of the neck or chest. There are two types of cells that have been identified in the parathyroid glands. The function of one type of parathyroid cells, the **oxyphil cells**, is not yet known. **Chief cells**, the other cell type, are known to produce and secrete the **parathyroid hormone (PTH)**, a regulator of blood calcium levels and the antagonist hormone to calcitonin (see Figure 17.18).

Student Study Tip

PTH and calcitonin exist in a seesaw-like balance relationship. When one level goes up, the other goes down!

PTH, a peptide hormone, is secreted in response to low blood calcium levels. PTH secretion causes the release of calcium from the bones by stimulating osteoclasts, which secrete enzymes that degrade bone matrix and release calcium into the blood. PTH also inhibits osteoblasts, preventing calcium from being deposited back into the bone matrix as blood calcium levels rise. In addition, PTH initiates the production of the steroid hormone calcitriol (also known as *active vitamin D or 1,25-dihydroxyvitamin D*) in the kidneys. Calcitriol functions to increase absorption of dietary calcium by the intestines. A negative feedback loop regulates the levels of PTH, with rising blood calcium levels inhibiting further release of PTH.

LO 17.4.3

Figure 17.19 The Parathyroid Glands

The parathyroid glands are most often found on the posterior surface of the thyroid gland. Most individuals have four parathyroid glands, but it is possible to have as few as two or as many as six.

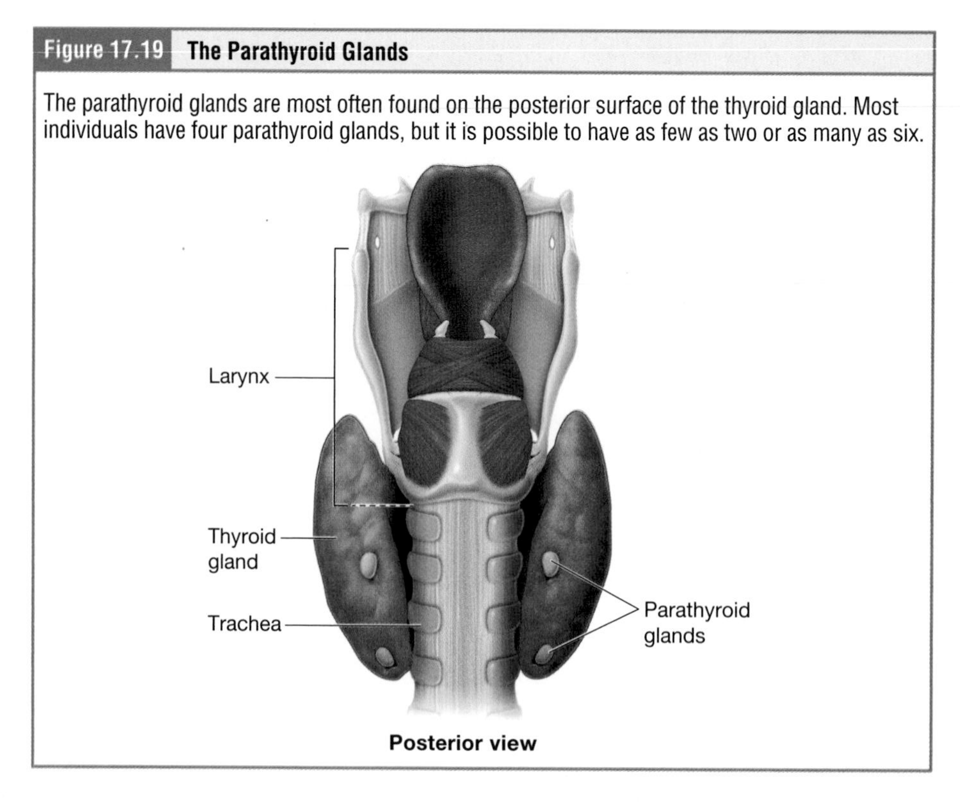

Abnormally high activity of the parathyroid gland can cause **hyperparathyroidism**, a disorder caused by an overproduction of PTH that results in excessive calcium resorption from bone. Hyperparathyroidism can significantly decrease bone density, leading to spontaneous fractures or deformities. As blood calcium levels rise, cell membrane permeability to sodium is decreased, and the responsiveness of the nervous system is reduced. At the same time, calcium deposits may collect in the body's tissues and organs, impairing their functioning.

In contrast, abnormally low blood calcium levels may be caused by PTH deficiency, called **hypoparathyroidism**, which may develop following injury or surgery involving the thyroid gland. Low blood calcium increases membrane permeability to sodium, resulting in muscle twitching, cramping, spasms, or convulsions. Severe deficits can paralyze muscles, including those involved in breathing, and can be fatal.

Learning Check

LO 17.4.4

1. Which of the following occurs once thyroid-stimulating hormone binds to its receptors in the follicle cells? Please select all that apply.
   a. Iodine ions are actively transported into the cell
   b. Iodine ions are passively transported into the cell
   c. Sodium ions are actively transported into the cell
   d. Synthesis of triiodothyronine and thyroxine
2. Which of the following are effects of thyroid hormone? Please select all that apply.
   a. Increased body temperature
   b. An increase in metabolic rate
   c. Production of insulin
   d. Regulation of appetite
3. Which of the following produces and secretes parathyroid hormone?
   a. Oxyphil cells
   b. Chief cells
   c. Thyroid follicles
   d. Colloid
4. Which of the following effects would you expect to see in someone who has hyperparathyroidism?
   a. Decreased bone density
   b. Decreased calcium resorption
   c. Muscle twitching
   d. Muscle cramping

(Continued)

5. How does the negative feedback mechanism for regulating PTH secretion differ from that of the thyroid hormones, T_3 and T_4?
   a. PTH secretion is regulated simply by the blood level of calcium, while T_3 and T_4 secretion is regulated through a feedback loop involving hormones of the hypothalamus and anterior pituitary gland.
   b. PTH secretion is regulated through a feedback loop involving hormones of the hypothalamus and anterior pituitary gland, while T_3 and T_4 secretion is regulated simply by the blood level of sodium.
   c. PTH secretion is regulated by the blood level of calcium, while T_3 and T_4 secretion is regulated by the blood level of sodium.
   d. The secretion of PTH and the thyroid hormones T_3 and T_4 are both regulated by negative feedback loops involving hormones of the hypothalamus and anterior pituitary gland.

17.4c The Adrenal Glands

The **adrenal glands** are wedges of tissue that sit on the tops of the kidneys in a fibrous capsule (**Figure 17.20**). The adrenal glands have a rich blood supply and experience one of the highest rates of blood flow in the body. Similar to the pituitary glands, which seem like one gland when examined at the gross or macroscopic level, the adrenal glands are more accurately two glands, with completely separate functions. To get an idea of what the adrenal gland looks like in cross section, try cutting an orange in half. The inner pulp, the **adrenal medulla**, is connected to the central nervous system and functions as a component of the sympathetic nervous response. When stimulated by neuronal signals, the adrenal medulla secretes epinephrine and norepinephrine into the

adrenal glands (suprarenal gland)

adrenal medulla (suprarenal medulla)

Figure 17.20 The Adrenal Glands

The adrenal glands are wedges of tissue that sit atop each kidney. Anatomically, the cortex completely surrounds the medulla in each adrenal gland. Physiologically, the cortex and medulla function as separate endocrine glands—the cortex secreting steroid hormones and the medulla secreting amine hormones.

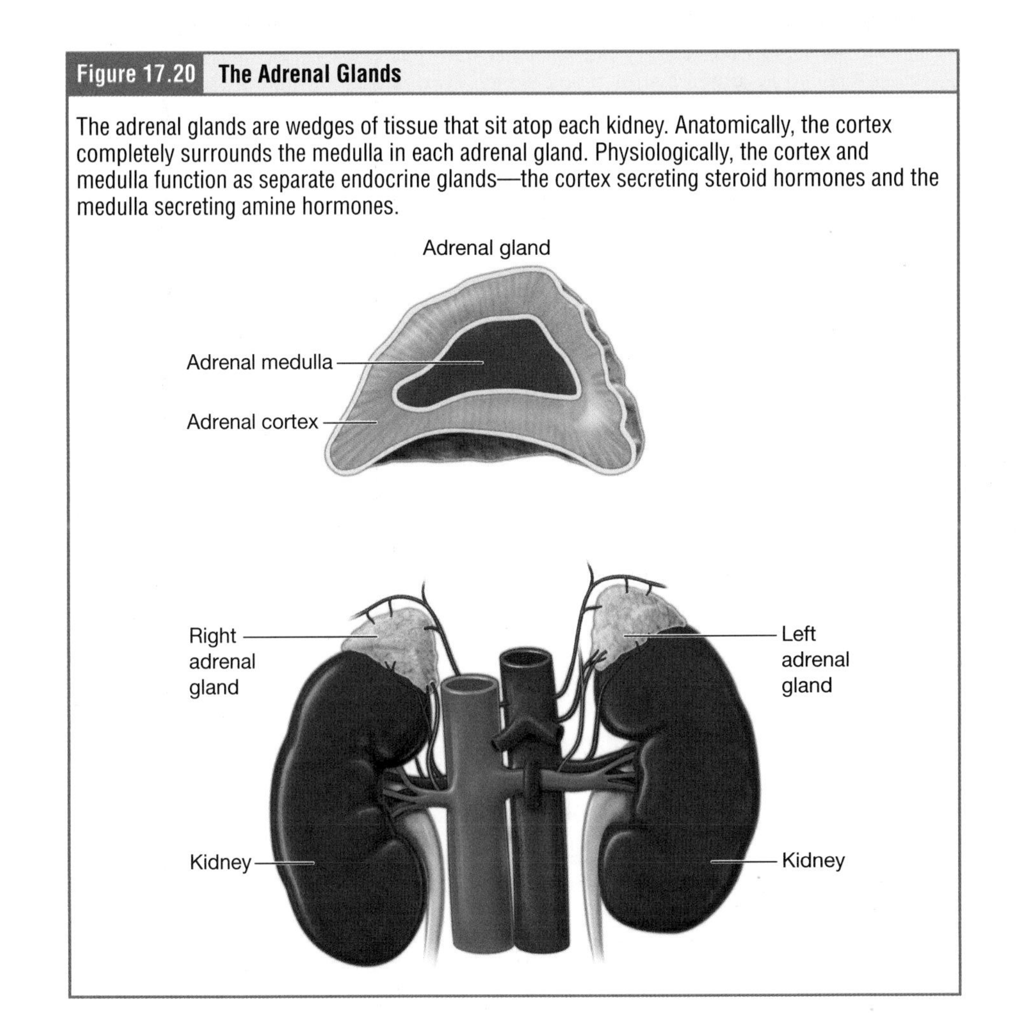

adrenal cortex (suprarenal cortex)

LO 17.4.5

zona reticularis (reticulate zone)

bloodstream. The **adrenal cortex**, analogous to the pithy orange peel, is anatomically and functionally distinct glandular tissue that secretes a group of its own steroid hormones, collectively called **corticoids**. Collectively, the corticoids are important for the regulation of the long-term stress response, blood pressure and blood volume, nutrient uptake and storage, fluid and electrolyte balance, and inflammation. The adrenal cortex consists of three zones: the **zona glomerulosa**, the **zona fasciculata**, and the **zona reticularis**. Each region secretes its own set of hormones.

Hormones of the Zona Glomerulosa The most superficial region of the adrenal cortex is the zona glomerulosa, which produces a group of hormones collectively referred to as **mineralocorticoids** because of their effect on body minerals, especially sodium and potassium. These hormones are essential for fluid and electrolyte balance.

Aldosterone is the major mineralocorticoid. It is important in regulating the levels of sodium and potassium in the blood. For example, it is released in response to elevated blood K^+, low blood Na^+, low blood pressure, or low blood volume. In response, aldosterone increases the excretion of K^+ and the retention of Na^+, which in turn increases blood volume and blood pressure. Its secretion is prompted when CRH from the hypothalamus triggers ACTH release from the anterior pituitary.

Aldosterone is also a key component of the renin-angiotensin-aldosterone system (RAAS), which is a hormone system that carefully regulates blood pressure. This system will be covered in more detail in Chapter 25.

Hormones of the Zona Fasciculata The intermediate region of the adrenal cortex is the zona fasciculata, named as such because the cells form columns that resemble bundles of sticks separated by tiny blood vessels (you may recall that the term *fascicle* is used to describe bundles of axons in nerves and bundles of muscle cells in skeletal muscles). The cells of the zona fasciculata produce hormones called **glucocorticoids** because of their role in glucose metabolism. The most important of these is **cortisol**. In response to long-term stress, the hypothalamus secretes CRH, which in turn triggers the release of ACTH by the anterior pituitary. ACTH triggers the release of the glucocorticoids. Their overall effect is to inhibit tissue building while stimulating the breakdown of stored nutrients to maintain adequate nutrient supplies. Remember from Chapter 16 that the body's response to stress is to run, fight, or survive. It makes sense that in these real or perceived threats to our homeostasis, we would liberate stored nutrients to enable these survival activities. In conditions of long-term stress, for example, cortisol promotes the breakdown of glycogen to glucose, the breakdown of stored triglycerides into fatty acids and glycerol, and the breakdown of muscle proteins into amino acids. These raw materials can then be used to fuel cellular work. The adrenal gland's roles in the stress response are illustrated in **Figure 17.21**.

You are probably familiar with prescription and over-the-counter medications containing glucocorticoids, such as cortisone injections into inflamed joints, prednisone tablets and steroid-based inhalers used to manage severe asthma, and hydrocortisone creams applied to relieve itchy skin rashes. These drugs reflect another role of cortisol—the downregulation of the immune system, which inhibits the inflammatory response.

Hormones of the Zona Reticularis The deepest region of the adrenal cortex is the zona reticularis, which produces small amounts of a class of steroid sex hormones called *androgens*. During puberty and most of adulthood, androgens are produced in the gonads. The androgens produced in the zona reticularis supplement the gonadal androgens. They are produced in response to ACTH from the anterior pituitary and are converted to testosterone or estrogens in the tissues.

Figure 17.21 The Adrenal Gland and the Stress Response

The adrenal medulla secretes epinephrine and norepinephrine into the bloodstream in response to neural signaling activated by stress stimuli. The adrenal cortex secretes cortisol into the bloodstream in response to a tropic hormone cascade initiated by the hypothalamus in response to stress stimuli.

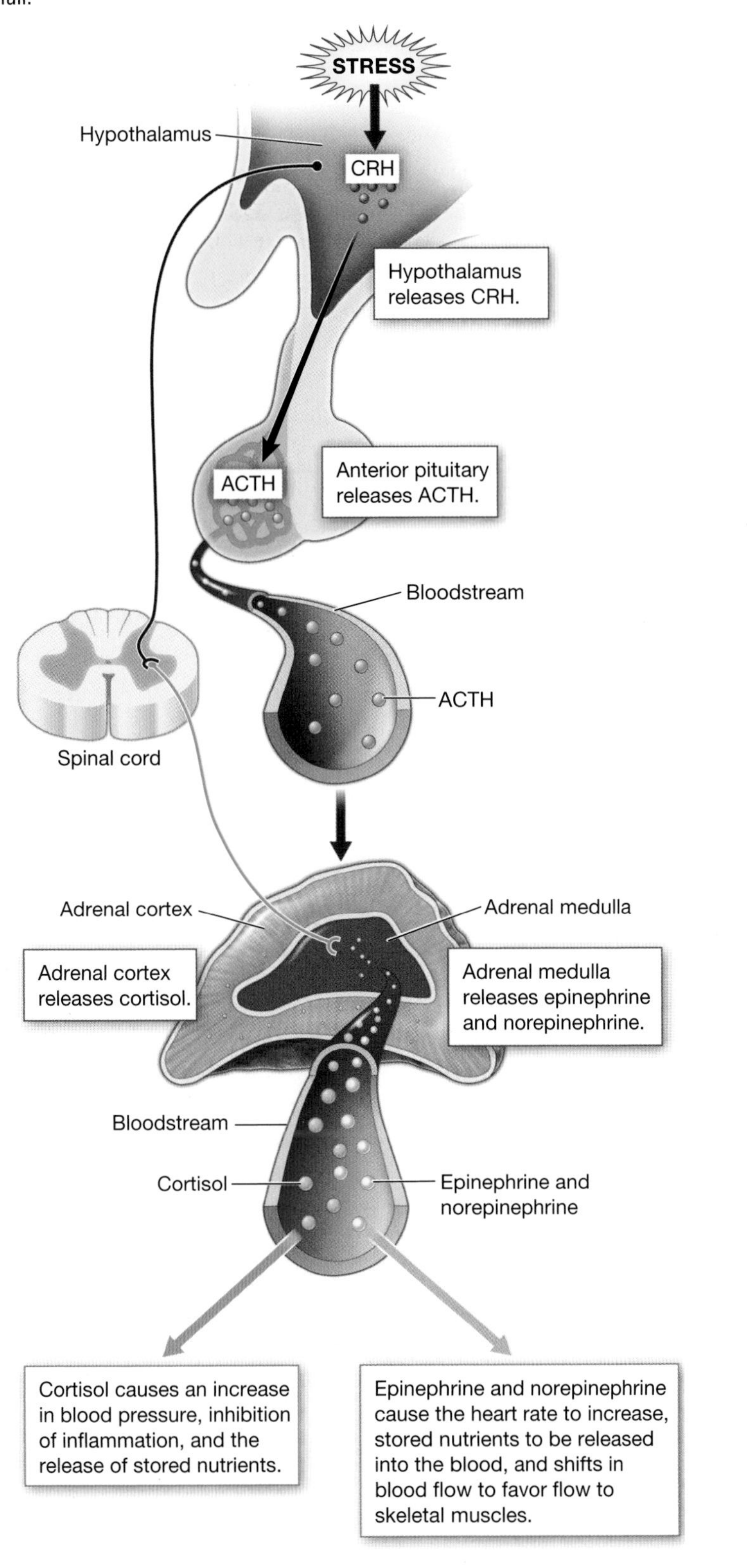
STRESS
Hypothalamus
CRH
Hypothalamus releases CRH.
ACTH
Anterior pituitary releases ACTH.
Bloodstream
ACTH
Spinal cord
Adrenal cortex
Adrenal medulla
Adrenal cortex releases cortisol.
Adrenal medulla releases epinephrine and norepinephrine.
Bloodstream
Cortisol
Epinephrine and norepinephrine
Cortisol causes an increase in blood pressure, inhibition of inflammation, and the release of stored nutrients.
Epinephrine and norepinephrine cause the heart rate to increase, stored nutrients to be released into the blood, and shifts in blood flow to favor flow to skeletal muscles.

Apply to Pathophysiology

Adrenal Insufficiency

Most hormones and secreted products follow a negative feedback loop pattern of production. In a negative feedback loop, when blood or body fluid levels of a particular hormone rise, production of that hormone is inhibited. Many different disease symptoms are controlled through the use of corticosteroid drugs. Allergies, asthma, arthritis, and many skin conditions are treated with corticosteroids.

1. As exogenous corticosteroid levels increase, what will happen to the production of endogenous corticosteroids such as cortisol?
   A) Production will increase.
   B) Production will decrease.
   C) Production will be unchanged.
2. Which of the following are roles of endogenous corticosteroids?
   A) Controlling heart rate
   B) Controlling blood pressure
   C) Controlling lactation
   D) Controlling body temperature
3. Individuals, particularly small children, who have been treated with corticosteroids for long periods of time may develop a condition called *adrenal insufficiency*, where the adrenal glands no longer produce adequate amounts of their hormones. In a child who has been diagnosed with adrenal insufficiency, which of the following physiological processes may not proceed in a typical way or at a typical rate?
   A) Puberty
   B) Musculoskeletal growth
   C) Cognitive gains such as learning to read

17.4d The Adrenal Medulla

As noted in Chapter 16, the adrenal medulla releases epinephrine and norepinephrine in response to acute, short-term stress mediated by the sympathetic nervous system.

epinephrine (adrenaline)

The medullary tissue is composed of **chromaffin** cells, which produce the chemicals **epinephrine** (also called *adrenaline*) and **norepinephrine** (or *noradrenaline*). Epinephrine is produced in greater quantities—approximately a 4-to-1 ratio with norepinephrine—and is the more powerful hormone. Because the chromaffin cells release epinephrine and norepinephrine into the systemic circulation, where they travel widely and exert effects on distant cells, they are considered hormones. These two compounds are also released from neurons elsewhere in the body. In those locations they are considered neurotransmitters because they are released into synapses.

Learning Connection

Try Drawing It

Certain endocrine glands consist of two portions that secrete different types of hormones. Try drawing a picture of the two portions of the adrenal gland and the pituitary gland. In each portion of your drawings, list the hormones that are secreted there.

17.4e Disorders Involving the Adrenal Glands

Several disorders are caused by the dysregulation of the hormones produced by the adrenal glands. For example, Cushing's disease is a disorder characterized by high blood glucose levels and the accumulation of lipid deposits on the face and neck. It is caused by hypersecretion of cortisol. The most common source of Cushing's disease is a pituitary tumor that secretes cortisol or ACTH in abnormally high amounts. Other common signs of Cushing's disease include the development of a moon-shaped face, a buffalo hump on the back of the neck, rapid weight gain, and hair loss. Chronically elevated glucose levels are also associated with an elevated risk of developing type 2 diabetes. In addition to hyperglycemia, chronically elevated glucocorticoids compromise immunity, resistance to infection, and memory, and can result in rapid weight gain and hair loss.

In contrast, the hyposecretion of corticosteroids can result in Addison's disease, a rare disorder that causes low blood glucose levels and low blood sodium levels. The signs and symptoms of Addison's disease are vague and are typical of other disorders as well, making diagnosis difficult. They may include general weakness, abdominal pain, weight loss, nausea, vomiting, sweating, and cravings for salty food.

17.4f The Pancreas

LO 17.4.6

The **pancreas** is a long, slender organ, located posterior to the stomach (**Figure 17.22**). The pancreas functions entirely in secretion, but uniquely it is both an exocrine and an endocrine organ. We will discuss its exocrine functions in Chapter 23. Its endocrine function is completed by groups of cells called **pancreatic islets** that secrete the hormones glucagon, insulin, somatostatin, and pancreatic polypeptide (PP). Each of the hormones is produced by a dedicated cell type. The islets, therefore, are composed of four types of cells:

- The **alpha cell** produces the hormone glucagon. Glucagon plays an important role in blood glucose regulation; low blood glucose levels stimulate its release.
- The **beta cell** produces the hormone insulin. Elevated blood glucose levels stimulate the release of insulin.
- The **delta cell** secretes the peptide hormone somatostatin. Recall that somatostatin is an inhibiting hormone; it can inhibit the release of GH, glucagon, and insulin.
- The **PP cell** secretes the pancreatic polypeptide hormone. It is thought to play a role in appetite.

Student Study Tip

The name *glucagon* is intuitive because it sounds like "glucose gone," which is when it is secreted!

PP cell (pancreatic polypeptide cells, gamma cells, F cells)

Regulation of Blood Glucose Levels by Insulin and Glucagon Glucose is required for cellular respiration and is the preferred fuel for all body cells. The body takes in glucose from food, but any glucose that is not immediately taken up by cells for fuel

Figure 17.22 The Pancreas

The majority of the tissue in the pancreas is exocrine tissue. However, the pancreas plays a significant endocrine role as well, producing and releasing, among others, the hormones glucagon and insulin. The endocrine cells of the pancreas are clustered in pancreatic islets.

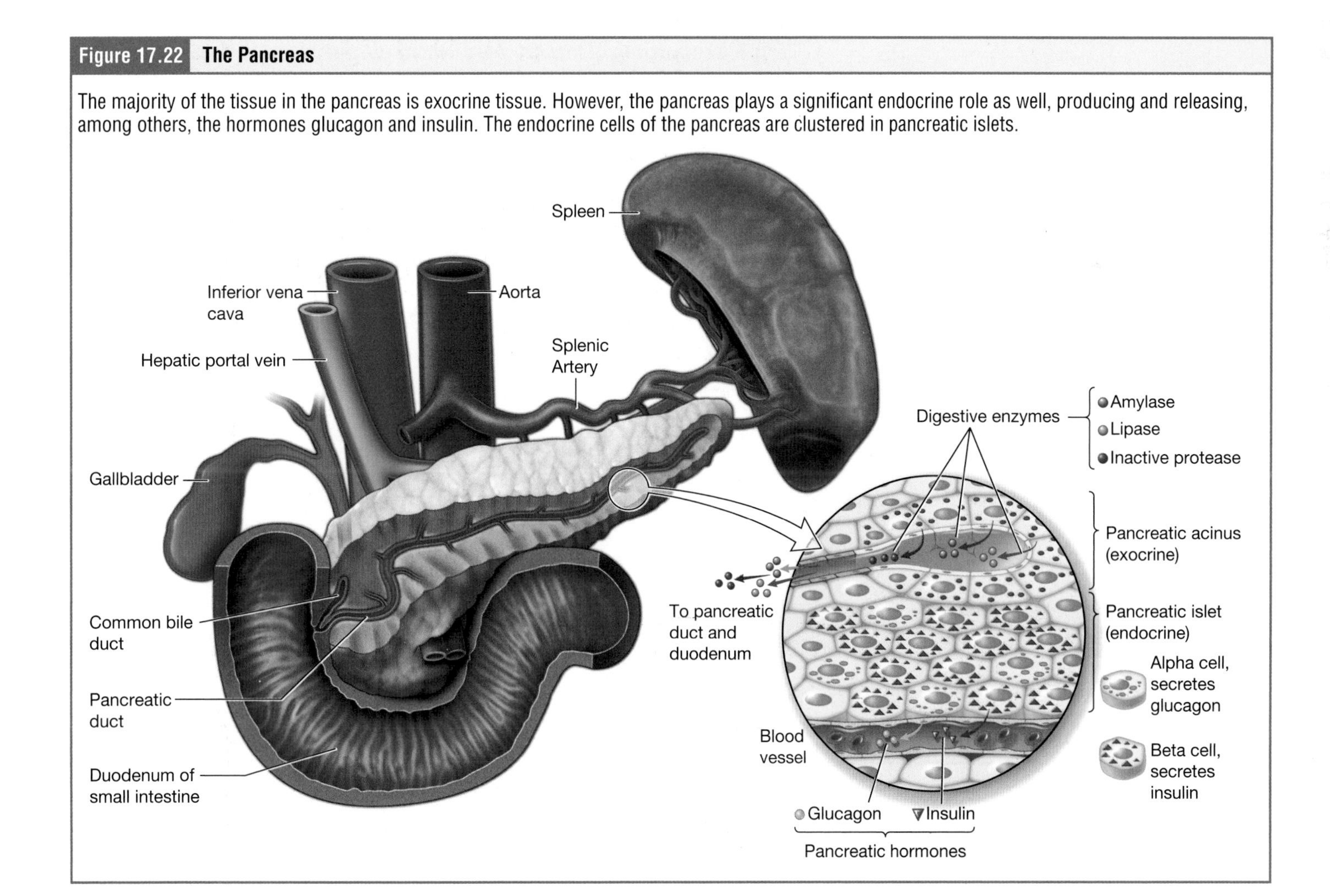

Figure 17.23 Blood Glucose Regulation

The level of glucose is kept within the homeostatic range by the actions of insulin, which promotes cellular uptake of glucose, and glucagon, which liberates stored glucose from the liver.

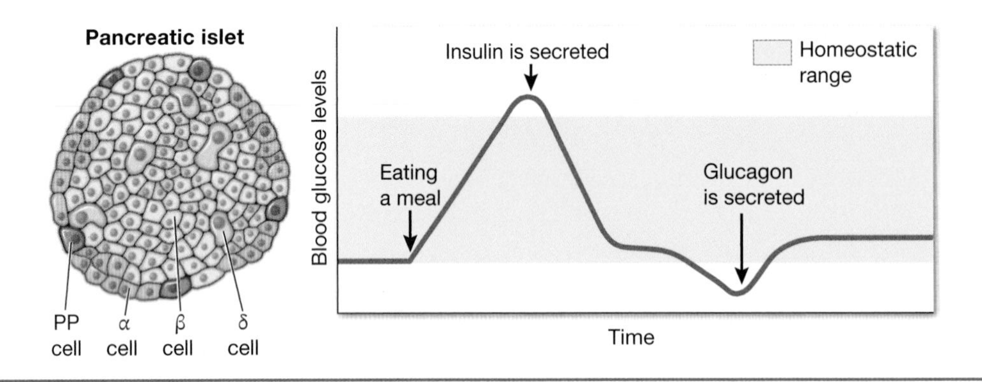

can be stored by the liver and muscles as glycogen, or converted to triglycerides and stored in the adipose tissue. Hormones regulate both the storage and the utilization of glucose. Receptors located in the pancreas sense blood glucose levels, and subsequently the pancreatic cells secrete glucagon or insulin to maintain normal levels. When blood glucose levels rise outside of the homeostatic range, for example, after eating a meal, insulin is secreted by the beta cells of the pancreas (**Figure 17.23**).

The primary function of **insulin** is to facilitate the uptake of glucose into body cells. Red blood cells, as well as cells of the brain, do not have insulin receptors on their cell membranes and do not require insulin for glucose uptake. All other body cells, including skeletal muscle cells and adipose cells, require insulin if they are to take glucose from the bloodstream.

These insulin target cells, such as skeletal muscle cells, keep their glucose transporters in vesicles on the inside of the cell. Once insulin binds the insulin receptor, a second messenger system is activated, and the result is the exocytosis of the glucose transporter vesicle and the expression of glucose transporters on the cell surface. The target cell can now transport glucose from the blood to the cytosol for use in cellular respiration (**Figure 17.24**). In the absence of insulin, or if the insulin reception

Figure 17.24 The Mechanism of Insulin's Action

When insulin binds to its receptor (R), it triggers a cellular response that upregulates the surface expression of glucose transporters, allowing the cell to take glucose in from the surrounding environment.

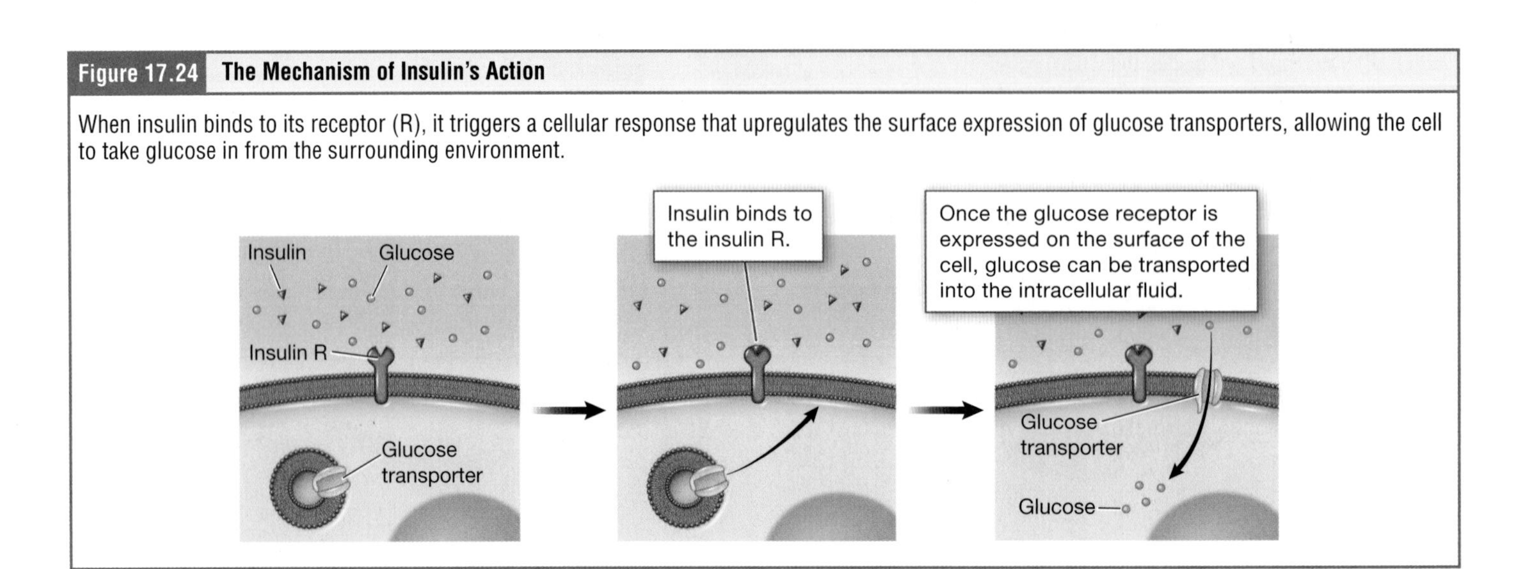

pathway desensitizes, these target cells are left without a mechanism to access glucose from the bloodstream.

Insulin also stimulates the liver to convert excess glucose into glycogen for storage, and it inhibits enzymes involved in the breakdown of nutrient stores. Finally, insulin promotes triglyceride synthesis in adipose tissue. The secretion of insulin is regulated through a negative feedback mechanism. As blood glucose levels decrease, further insulin release is inhibited.

Receptors in the pancreas can also sense any decline in blood glucose levels, such as during periods of fasting or during exercise. In response, the alpha cells of the pancreas secrete the hormone **glucagon**, which has several effects:

- It stimulates the liver to convert its stores of glycogen back into glucose. This response is known as **glycogenolysis**. The glucose is then released into the circulation for use by body cells.
- It stimulates the liver to take up amino acids from the blood and convert them into glucose. This response is known as **gluconeogenesis**.
- It stimulates lipolysis, the breakdown of stored triglycerides into free fatty acids and glycerol. Some of the free glycerol released into the bloodstream travels to the liver, which converts it into glucose. This is also a form of gluconeogenesis.

Taken together, these actions increase blood glucose levels. The activity of glucagon is regulated through a negative feedback mechanism; rising blood glucose levels inhibit further glucagon production and secretion.

Apply to Pathophysiology

T2DM

Dysfunction of insulin production and secretion, as well as the target cells' responsiveness to insulin, can lead to a condition called **diabetes mellitus**. An increasingly common disease, the incidence of diabetes has risen steadily in the United States for the last 50 years.

There are two main forms of diabetes mellitus. Type 1 diabetes (T1DM) is an autoimmune disease affecting the beta cells of the pancreas. With the immune attack on beta cells in type 1 diabetics, insulin production declines and synthetic insulin must be administered by injection or infusion. This form of diabetes accounts for less than 5 percent of all diabetes cases in the United States.

Type 2 diabetes (T2DM) is acquired, meaning that it develops during the lifespan and lifestyle factors such as diet and exercise seem to play a role.

1. In T2DM, target cells do not respond to insulin, which of the following may occur?
   A) Downregulation (decreased expression) of the insulin receptor
   B) Decreased production of insulin
   C) Decrease in glucose intake from the diet
2. As a result, sugars such as glucose circulate in the blood longer. Which of the following would be true of the blood of someone with T2DM compared to someone who did not have any form of diabetes?
   A) On average, the blood of the individual with T2DM will have higher osmolarity than the blood of the individual without diabetes.
   B) On average, the blood of the individual with T2DM will have lower osmolarity than the blood of the individual without diabetes.
   C) On average, the blood of the two individuals will be isosmotic to each other.
3. One common complication of T2DM over time is damage to small blood vessels. Using your answer to Question 2, what may be occurring that causes damage to the blood vessels?
   A) The blood vessels shrivel and become stiff.
   B) The blood vessels stretch and break.
   C) The above average cellular content of the blood creates tears in the vessel walls.
4. Individuals with diabetes often measure the amount of sugar in the blood. They are monitoring to make sure they do not have too much sugar in the blood. The relative amount (compared to the healthy range) can be described by the word part *-glycemia*. Which of the following words would mean above average amount of blood sugar?
   A) Hypoglycemia
   B) Isoglycemia
   C) Hyperglycemia

LO 17.4.7

17.4g The Thymus

The **thymus** is an organ of the immune system that is larger and more active during infancy and early childhood and begins to atrophy as we age. Its endocrine function is the production of a group of hormones called **thymosins** that contribute to the development and differentiation of T lymphocytes, which are immune cells. Although the role of thymosins is not yet well understood, it is clear that they contribute to the immune response. Thymosins have been found in tissues other than the thymus and have a wide variety of functions, so the thymosins cannot be strictly categorized as thymic hormones.

Throughout the body, other organs whose primary function is not the production of hormones nonetheless have secondary functions in the production and release of hormones. Some of these are discussed elsewhere in this text, but are briefly illustrated here.

LO 17.4.8

17.4h The Heart

When the body experiences an increase in blood volume or pressure, the cells of the heart's atrial walls stretch. In response, specialized cells in the wall of the atria produce and secrete the peptide hormone **atrial natriuretic peptide (ANP)**. ANP signals the kidneys to allow more sodium into the urine. Sodium will bring water along by osmosis and the volume of water in the body and therefore in the blood will decrease.

17.4i The Gastrointestinal Tract

The endocrine cells of the gastrointestinal tract are located in the wall of the stomach and small intestine. Some of these hormones are secreted in response to eating a meal, and aid in digestion. An example of a hormone secreted by the stomach cells is **gastrin**, a peptide hormone secreted in response to stomach distention that stimulates the release of hydrochloric acid, which aids in digestion. Other hormones produced by the intestinal cells aid in glucose metabolism, such as by stimulating the pancreatic beta cells to secrete insulin, reducing glucagon secretion from the alpha cells, or enhancing cellular sensitivity to insulin.

17.4j The Kidneys

The kidneys participate in several complex endocrine pathways and produce certain hormones. The kidneys participate in the RAAS, our body's major method for controlling blood pressure. The kidneys also play a role in regulating blood calcium levels through the production of calcitriol from vitamin D_3, which is released in response to the secretion of PTH. In addition, the kidneys produce the hormone **erythropoietin (EPO)** in response to low oxygen levels. EPO stimulates the production of red blood cells (erythrocytes) in the bone marrow, thereby increasing oxygen delivery to tissues. You may have heard of EPO as a performance-enhancing drug (in a synthetic form).

Learning Connection

Chunking

In the endocrine system, there are several examples of antagonistic hormones, which produce opposite effects in the body. Prepare a table of antagonistic hormones, stating the factor that they control, and the effect of each hormone on that factor. For example, which hormones have opposite effects on the level of blood glucose? See how many sets of antagonistic hormones you can come up with!

17.4k Adipose Tissue

Adipose tissue produces and secretes several hormones, collectively called **adipokines**. Most adipokines are involved in metabolism and nutrient storage, but others have cardiovascular roles and influence inflammation. One important example is **leptin**, a protein manufactured by adipose cells that circulates in amounts directly proportional to levels of body fat. Leptin is released in response to food consumption and acts by binding to brain neurons involved in energy intake and expenditure. Binding of leptin produces a feeling of satiety after a meal, thereby reducing appetite. It also appears that the binding of leptin to brain receptors triggers the sympathetic nervous system to regulate bone metabolism, increasing deposition of cortical bone. Adiponectin—another hormone synthesized by adipose cells—appears to reduce cellular insulin resistance and to protect blood vessels from inflammation and atherosclerosis. Its levels are lower in people who are obese, and they rise following weight loss.

17.4l The Skin

The skin functions as an endocrine organ in the production of vitamin D_3. When cholesterol present in the epidermis is exposed to ultraviolet radiation, it is converted to inactive vitamin D, which then enters the blood. The liver then can convert this molecule to an intermediate that travels to the kidneys and is then converted into the active form of vitamin D_3. Vitamin D is important in a variety of physiological processes, including intestinal calcium absorption and immune system function. Vitamin D seems to play a protective role in the pathophysiology of some lung infections, including tuberculosis.

17.4m The Liver

The liver secretes a few hormones or hormone precursors. As discussed earlier in this chapter, the liver releases IGF in response to GH. The liver also produces angiotensinogen, the precursor to angiotensin, which increases blood pressure. Thrombopoietin is a hormone that stimulates the production of the blood's platelets. Hepcidins regulate iron homeostasis in our body fluids.

Learning Connection

Quiz Yourself

After studying the locations of the major endocrine glands and the organs that contain endocrine cells, write the names of the glands and organs on a piece of paper. While standing in front of a mirror, go down your list and try to point out the endocrine glands and organs on yourself. Say their names as you point them out; this will help you remember them.

Learning Check

1. Prednisone is a synthetic form of a glucocorticoid used as a medication. Which zone of the adrenal gland produces natural glucocorticoids?
   a. Zona glomerulosa
   b. Zone fasciculata
   c. Zona reticularis
   d. Zona medullaris
2. Identify the effects of glucagon. Please select all that apply.
   a. Storage of excess glucose as glycogen
   b. Breakdown of glycogen into glucose
   c. Formation of triglycerides from fatty acids
   d. Absorption of amino acids from the blood into the liver for gluconeogenesis
3. During surgery, patients sometimes lose a lot of blood. Therefore, it is common for a patient's red blood cell count to be low after surgery. Which organ is responsible for stimulating the production of red blood cells?
   a. Heart
   b. Kidneys
   c. Liver
   d. Skin
4. What is the function of thrombopoetin?
   a. To stimulate production of platelets
   b. To stimulate production of red blood cells
   c. To stimulate production of white blood cells
   d. To stimulate production of Vitamin D_3

Chapter Review

17.1 An Overview of the Endocrine System

The learning objectives for this section are:

17.1.1 Describe the major functions of the endocrine system.

17.1.2 Define the terms hormone, endocrine gland, endocrine tissue (organ), and target cell.

17.1.3 Compare and contrast how the nervous and endocrine systems control body functions, the anatomical pathways by which the signals reach their targets, what determines the target of the pathway, the speed of the target response(s), the duration of the response, and how signal intensity is coded.

17.1 Questions

1. Identify the functions of the endocrine system. Please select all that apply.
   a. Maintenance of blood pressure
   b. Secretion of wastes
   c. Transportation of nutrients
   d. Detection of changes in homeostatic levels of substances

2. Which of the following secretes hormones into the bloodstream?
   a. Endocrine tissue
   b. Target cells
   c. Cell receptors
   d. Goblet cells

3. What type of communication is used by both the endocrine and nervous systems?
   a. Hormones
   b. Chemical signaling
   c. Electrical signaling
   d. Neurotransmitters

17.1 Summary

- The endocrine and nervous systems regulate organ and system activities through long-distance communication via signaling molecules.
- Both the endocrine and nervous systems use chemical messengers to communicate with their target cells and organs. The endocrine system secretes hormones into the blood, while the nervous system secretes neurotransmitters into synapses.
- Hormones secreted by endocrine glands travel through the blood to their specific target cells and organs. Target cells contain receptors that bind only to the specific hormones that affect them.

17.2 Hormones

The learning objectives for this section are:

17.2.1 List the three major chemical classes of hormones (i.e., steroid, peptide, amino acid-derived [amine]) found in the human body.

17.2.2 Compare and contrast how steroid and peptide hormones are produced and stored in the endocrine cell, released from the endocrine cell, and transported in the blood.

17.2.3 Compare and contrast the locations of target cell receptors for steroid and peptide hormones.

17.2.4 Compare and contrast the mechanisms of action of plasma membrane hormone receptors and intracellular hormone receptors, including the speed of the response.

17.2.5 Describe the various signals that initiate hormone production and secretion (e.g., monitored variables, direct innervation, neurohormones, other hormones).

17.2 Questions

1. Which class of hormone best describes antidiuretic hormone (ADH), a hormone that consists of nine amino acids?
   a. Steroid b. Peptide
   c. Amine d. Protein

Mini Case 1: Running Away from a Thunderstorm—the Endocrine Response to Stress

Alka was out for a walk in the woods when a thunderstorm suddenly blew into the area. Alka ran to the car as quickly as possible, to get out of the wind, rain, and lightning. Alka's endocrine system helped their body respond to the new level of activity necessary to reach the safety of the car quickly.

1. One of the hormones that helped Alka increase their metabolic rate in order to run quickly to the car is epinephrine. Which of the following are true of the hormone epinephrine? Please select all that apply.
   a. Epinephrine is produced by modifying cholesterol.
   b. Epinephrine is secreted by the adrenal medulla.
   c. Epinephrine can pass through the plasma membrane of its target cells.
   d. Epinephrine can travel through the blood without binding to a transport protein.

2. Another group of two hormones that will help Alka run quickly to the car is the thyroid hormones (triiodothyronine and thyroxine). Thyroid hormones have a particular structure that causes them to be lipophilic, even though they are not produced from cholesterol. Where are the receptors for thyroid hormones located?
   a. In the blood
   b. On the plasma membrane of the target cells
   c. Inside the target cells
   d. Thyroid hormones do not require receptors to act on their target cells

3. Epinephrine and thyroid hormones interact with their target cells by different mechanisms, due to their different solubilities in the plasma membrane of their target cells. Which of the following is true of the mechanism of action of these hormones with their target cells?
   a. Epinephrine binds to hormone receptors on the plasma membrane of its target cells.
   b. Thyroid hormones require the use of a second messenger to exert their effects on their target cells.
   c. Due to its structure, epinephrine would be expected to stimulate DNA transcription in its target cells.
   d. Thyroid hormones must initiate a cascade of signaling molecules and reactions before they can exert their effects on their target cells.

4. Which of the following factors would initiate Alka's epinephrine secretion while they are running to the car to escape the storm? Please select all that apply.
   a. ACTH secretion from the pituitary gland
   b. Stimulation by the sympathetic nervous system
   c. A feeling of stress or fear
   d. CRH secretion from the hypothalamus

17.2 Summary

- There are three types of hormones: steroid hormones, amine hormones, and peptide, and protein hormones.
- Steroid hormones are synthesized from cholesterol. They enter their target cells by diffusing through the phospholipid bilayer of the cell membrane. They react with intracellular receptors, and act as transcription factors.
- Amine hormones are synthesized by modifying single amino acids. They cannot cross the phospholipid bilayer of the cell membrane, so they bind to receptors bound to the cell membrane. They bring about their cellular changes via second messengers.
- Peptide and proteins hormones consist of chains of amino acids. They cannot cross the cell membrane, so they bind to receptors bound to the cell membrane. They bring about their cellular changes via second messengers.
- The response of a target cell to a hormone can be increased or decreased by upregulation or downregulation of their receptors.
- Hormone secretion can be regulated through negative and positive feedback, tropic hormones, ion or nutrient levels, or nervous system stimulation.

17.3 Endocrine Control by the Hypothalamus and Pituitary Gland

The learning objectives for this section are:

17.3.1 Describe the locations and the anatomical relationships of the hypothalamus, anterior pituitary, and posterior pituitary, including the hypothalamic-hypophyseal portal system.

17.3.2 Explain the role of the hypothalamus in the release of hormones from the posterior pituitary.

17.3.3 Name the two hormones produced by the hypothalamus that are stored in the posterior pituitary, and the hormones' primary targets and effects.

17.3.4 Explain the role of hypothalamic neurohormones (regulatory hormones) in the release of anterior pituitary hormones.

17.3.5 Describe major hormones secreted by the anterior pituitary, their control pathways, and their primary target(s) and effects.

17.3 Questions

1. Which of the following structures is most superior?
   a. Hypothalamus
   b. Anterior pituitary
   c. Posterior pituitary
   d. Infundibulum

2. Which of the following organs stores hormones produced by the hypothalamus?
   a. Anterior pituitary
   b. Posterior pituitary
   c. Adrenal gland
   d. Thalamus

3. After a long soccer game outdoors, Elizabeth was dehydrated. When dehydrated, your body works hard to retain water. Which of the following hormones is responsible for retaining water in the body?
   a. Oxytocin
   b. Antidiuretic hormone
   c. Adrenocorticotropic hormone
   d. Prolactin

4. Which of the following are a benefits of the hypothalamic-hypophyseal portal system? Please select all that apply.
   a. The effects are long-lasting.
   b. The reaction happens quickly.
   c. The effects will reach distant organs.
   d. The hormones will not be diluted.

5. Acromegaly is a condition in which there is an excessive production of growth hormone. Which of the following structures may be impaired, causing acromegaly?
   a. Thymus
   b. Thalamus
   c. Anterior pituitary
   d. Posterior pituitary

17.3 Summary

- The hypothalamus regulates hormone secretion from the anterior pituitary gland by secreting tropic hormones into the hypothalamic-hypophyseal portal system.
- The anterior pituitary gland produces six major hormones: growth hormone, thyroid-stimulating hormone, adrenocorticotropic hormone, follicle-stimulating hormone, luteinizing hormon, and prolactin.
- The hypothalamus is connected to the posterior pituitary gland by the infundibulum. The hypothalamus synthesizes oxytocin and antidiuretic hormone. These hormones are stored in the posterior pituitary gland, and secreted in response to nerve impulses from the hypothalamus.

17.4 The Major Hormones of the Body

The learning objectives for this section are:

HAPS LEARNING GOALS AND OUTCOMES

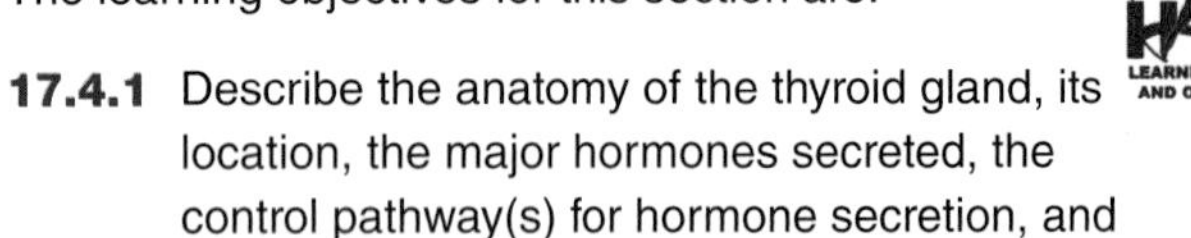

17.4.1 Describe the anatomy of the thyroid gland, its location, the major hormones secreted, the control pathway(s) for hormone secretion, and the hormones' primary targets and effects.

17.4.2 Describe the anatomy of the parathyroid glands, their location, the major hormone secreted, the control pathway(s) for hormone secretion, and the hormone's primary targets and effects.

17.4.3 Describe a simple endocrine pathway in which the response is the negative feedback signal (e.g., parathyroid hormone, insulin).

17.4.4 Compare and contrast negative feedback for hypothalamic-anterior pituitary-peripheral endocrine gland pathways to negative feedback for most simple endocrine pathways.

17.4.5 Describe the anatomy of the adrenal cortex, its location, the major hormones secreted, the control pathway(s) for hormone secretion, and the hormones' primary targets and effects.

17.4.6 Describe the anatomy of the pancreas, its location, the major hormones secreted, the control pathway(s) for hormone secretion, and the hormones' primary targets and effects.

17.4.7 Describe the anatomy of the thymus gland, its location, the major hormones secreted, the control pathway(s) for hormone secretion, and the hormones' primary targets and effects.

17.4.8 Provide some examples of hormones that are secreted from diffuse endocrine tissues or single endocrine cells.

17.4 Questions

1. After your blood test, the doctor diagnoses you with hypothyroidism, a condition where your body produces less than the typical amounts of thyroid hormones. Which of the following symptoms would you expect to see? Please select all that apply.
   a. Weight loss
   b. Lethargy
   c. Increase in body temperature
   d. Fatigue

2. Which of the following is an effect of the hormone released by the parathyroid gland?
   a. An increase in blood calcium levels
   b. A decrease in blood pressure
   c. An increase in basal metabolic rate
   d. A decrease in blood glucose levels

3. Which of the following would you expect to happen in a negative feedback loop if the hormone levels were decreasing?
   a. The body will continue to produce that hormone.
   b. The body will stop producing that hormone.

4. Which of the following are true of the negative feedback mechanisms for insulin and cortisol, two hormones that affect the concentration of glucose in the blood? Please select all that apply.
   a. Insulin and cortisol are both regulated directly by the blood level of glucose.
   b. Insulin is regulated by the blood level of glucose.
   c. Cortisol is regulated by a negative feedback loop involving hormones of the hypothalamus and anterior pituitary gland.
   d. Insulin and cortisol are each regulated by a negative feedback loop involving hormones of the hypothalamus and anterior pituitary.

5. Which of the following lists the order of the three zones of the adrenal cortex in order from superficial to deep?
   a. Zone reticularis → Zona fasciculata → Zone glomerulosa
   b. Zona reticularis → Zona glomerulosa → Zona fasciculata
   c. Zona fasciculata → Zona glomerulosa → Zona reticularis
   d. Zona glomerulosa → Zona fasciculata → Zona reticularis

6. Georgia was putting up holiday lights outside their home when they slipped and fell off the ladder. Georgia fell on their back and was rushed to the hospital. Through imaging, the doctors found left rib fractures and stomach lacerations. Which of the following glands are you most concerned about?
   a. Thyroid gland
   b. Parathyroid gland
   c. Adrenal cortex
   d. Pancreas

7. What is the function of thymosins?
   a. Development of T lymphocytes
   b. Regulate blood calcium levels
   c. Produce parathyroid hormone
   d. Differentiate into red blood cells

8. Excessive alcohol consumption can lead to liver failure, which occurs when many liver cells start to die. Which of the following hormones would be affected due to liver failure? Please select all that apply.
   a. Insulin-like growth factor
   b. Angiotensin
   c. Vitamin D
   d. Adipokines

17.4 Summary

- The thyroid gland is found on the anterior and lateral surfaces of the trachea. It synthesizes thyroid hormone, which influences the body's metabolic rate, and calcitonin, which helps to regulate the blood calclium level.
- The parathyroid glands are found on the surface of the thyroid gland and release parathyroid hormone, which regulates blood calcium levels and is an antagonist to calcitonin.
- The adrenal glands are found on top of the kidneys. The adrenal cortex consists of three distinct layers, each synthesizing their own hormones with unique functions. The adrenal medulla synthesizes epinephrine and norepinephrine.
- The pancreas is found posterior to the stomach. It produces and secretes several hormones, such as insulin and glucagon, which are responsible for regulating the glucose level in the blood.
- The thymus is mostly found in infants and is responsible for secreting thymosins, which develop immature T cells into mature, functional immune cells.
- Organs such as the heart, GI tract, kidneys, adipose tissue, skin, and liver produce their own hormones.
- The secretion of many hormones is regulated by negative feedback. Some hormones respond to the blood level of a particular ion or nutrient. Other hormones are regulated through more complex pathways, involving hormones of the hypothalamus and anterior pituitary gland.

18

The Cardiovascular System: Blood

Elke Meitzel/Image Source/Getty Images

Chapter Introduction

In this chapter we will learn...

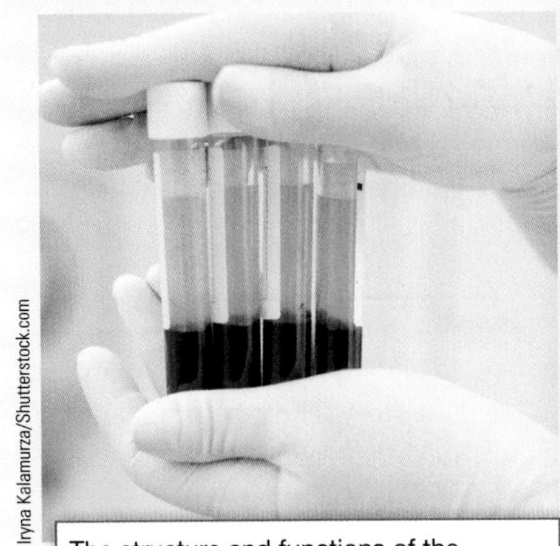

The structure and functions of the different components of human blood. . .

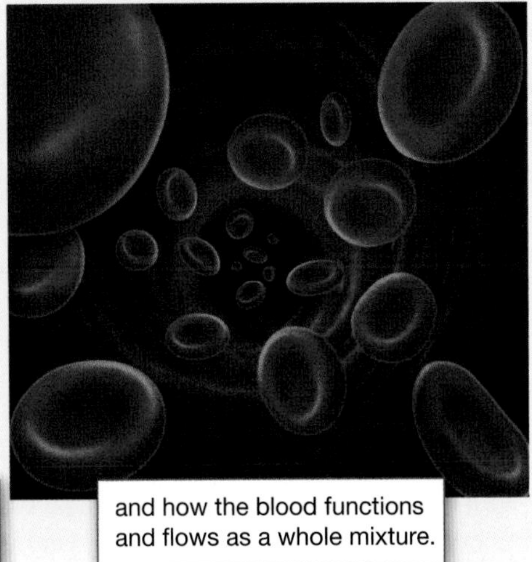

and how the blood functions and flows as a whole mixture.

Single-celled organisms do not need blood. They obtain nutrients directly from, and excrete wastes directly into, their environment. The human organism cannot do that. Our large, complex bodies need blood to deliver nutrients to and remove wastes from our trillions of cells. The heart pumps blood throughout the body in a network of blood vessels. Together, these three components—blood, heart, and blood vessels—make up the cardiovascular system. This chapter focuses on the medium of transport: blood.

18.1 The Composition of Blood

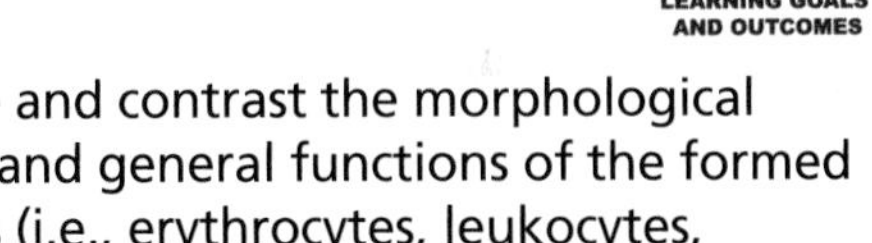

Learning Objectives: By the end of this section, you will be able to:

18.1.1 Describe the major functions of each component of the cardiovascular system (i.e., blood, heart, blood vessels).

18.1.2 Describe the general composition of blood (e.g., plasma, formed elements).

18.1.3* Describe the functions of blood and connect those functions to the maintenance of homeostasis.

18.1.4* Compare and contrast bulk flow and diffusion.

18.1.5 Describe the composition of blood plasma.

18.1.6 List the major types of plasma proteins, their functions, and sites of production.

18.1.7 Compare and contrast the morphological features and general functions of the formed elements (i.e., erythrocytes, leukocytes, platelets).

18.1.8 Describe the structure and function of hemoglobin, including its breakdown products.

18.1.9* Define hematocrit and discuss healthy ranges for adults and common reasons for deviation from homeostasis.

18.1.10 List the five types of leukocytes in order of their relative prevalence in normal blood, and describe their major functions.

* Objective is not a HAPS Learning Goal.

The Human Anatomy and Physiology Society includes more than 1,700 educators who work together to promote excellence in the teaching of this subject area. The HAPS A&P Learning Outcomes measure student mastery of the content typically covered in a two-semester Human A&P curriculum at the undergraduate level. The full Learning Outcomes are available at https://www.hapsweb.org.

LO 18.1.1

LO 18.1.2

leukocytes (white blood cells or WBCs)

Recall that **blood** is a connective tissue. Like all connective tissues, it is made up of cells, protein fibers, and an extracellular matrix. The extracellular matrix of blood is a liquid, and the protein fibers are dissolved within. The cells, or in this case cellular elements include *erythrocytes* (red blood cells or RBCs), *leukocytes* (white blood cells or WBCs), and cell fragments called *platelets*. Because the blood includes both cells and cell fragments, we use the term **formed elements** to describe these three components together. The extracellular matrix, called **plasma**, suspends the formed elements and enables them to circulate throughout the body within the vessels of the cardiovascular system. The heart functions to create a pressure gradient that moves the blood through the system.

18.1a The Functions of Blood

LO 18.1.3

The primary function of blood is to deliver oxygen and nutrients to and remove wastes from body cells, but that is only the beginning of the story. The specific functions of blood also include defense, distribution of heat, and maintenance of homeostasis.

Transportation Nutrients from the foods you eat are absorbed in the digestive tract. Yet, every one of the 40 trillion cells in your body—a veritable galaxy of cells—needs access to those nutrients. The bloodstream enables the delivery of nutrients to body cells. We will learn more about nutrient absorption in Chapter 23. Oxygen from the air you breathe diffuses into the blood at the lungs, which we will discuss in Chapter 22. That oxygen-rich blood then travels from the lungs to the heart, which then pumps it out to the rest of the body. As we discussed in Chapter 17, endocrine glands scattered throughout the body release their products—called *hormones*—into the bloodstream, which carries them to distant target cells. Blood also picks up cellular wastes and byproducts from each and every cell and transports them to various organs for removal.

Defense Many types of WBCs protect the body from infection, from SARS CoV-2—the virus that causes COVID-19—to bacteria that are found on the surfaces of our environment. Other WBCs seek out and destroy internal threats, such as cancerous cells.

When damage to the body results in bleeding, blood platelets and fibers dissolved in the plasma interact to seal the ruptured blood vessels, protecting the body from further blood loss.

Maintenance of Homeostasis Recall that body temperature is regulated via a classic negative-feedback loop. If you exercise on a hot day, your rising core body temperature will trigger homeostatic mechanisms, including increased transport of blood from your core to your skin, to allow for heat loss off the surface. In contrast, on a cold day, blood is diverted away from the skin to prevent heat loss and maintain your core body temperature.

Blood also helps to maintain the chemical balance of the body. Proteins and other compounds in blood act as buffers, which thereby help to regulate the pH of body tissues. Blood also helps to regulate the water content of body cells.

18.1b Whole Blood

Because of the central role in uniting and serving the different systems, organs, and cells of the body, sampling the blood of an individual can provide a lot of information about their overall health at any moment. Whole blood travels within and comes out of the body as the opaque dark red liquid you have undoubtedly seen (**Figure 18.1A**), but

Anatomy of...

Leukocytes

Granules filled with parasite-fighting chemicals

Multilobed nucleus

Eosinophils are characterized by their bright pink/red granules and their multilobed nucleus. Eosinophils evolved to function in anti-parasite responses but in modern humans often contribute to allergic reactions.

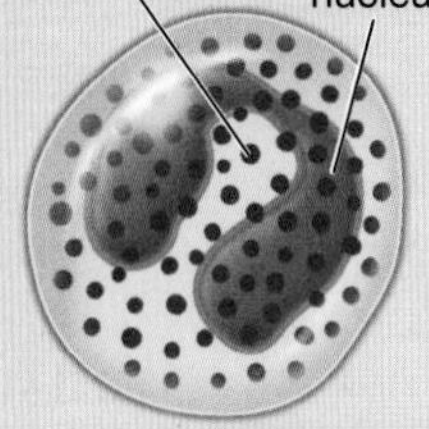

Basophils are characterized by dark blue/purple granules and a bi- or multilobed nucleus. Like eosinophils, basophils evolved to fight parasites but are more often associated with allergic responses in modern humans.

No granules

One large nucleus

Monocytes have a large, often kidney-bean-shaped nucleus and no granules. Monocytes do not participate directly in immune responses but when they leave the blood and move into tissues, they differentiate into macrophages.

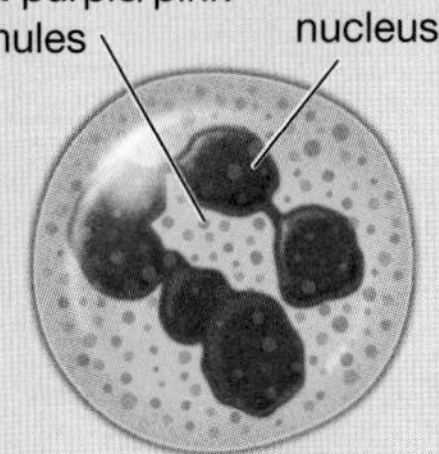

Neutrophils are the most abundant type of leukocyte in the blood. These cells can fight infection in a number of ways but commonly they will engulf a bacterium and then undergo their own cell death, killing the bacterium in the process. During some bacterial infections, dead neutrophils pile up and create a substance called *pus*.

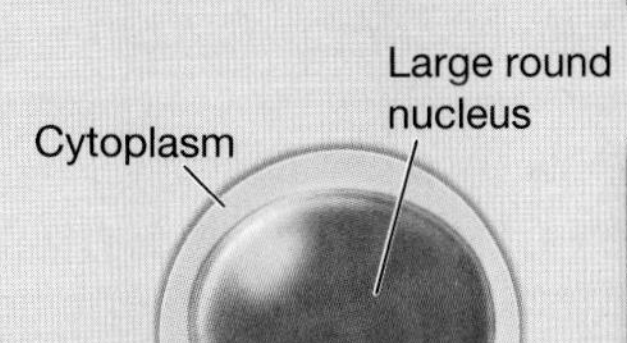

Lymphocytes are smaller than other leukocytes. Almost the entire cell is taken up by their round nucleus. Lymphocytes are strategic immune fighters and are capable of developing immunological memory, remembering pathogens that have caused previous infections.

Figure 18.6 Diapedesis

Leukocytes use the blood as a transportation system when they are called to the site of infection. When they arrive in the blood vessel closest to the infection site, they leave the blood in a process called *diapedesis*, and head to the tissues.

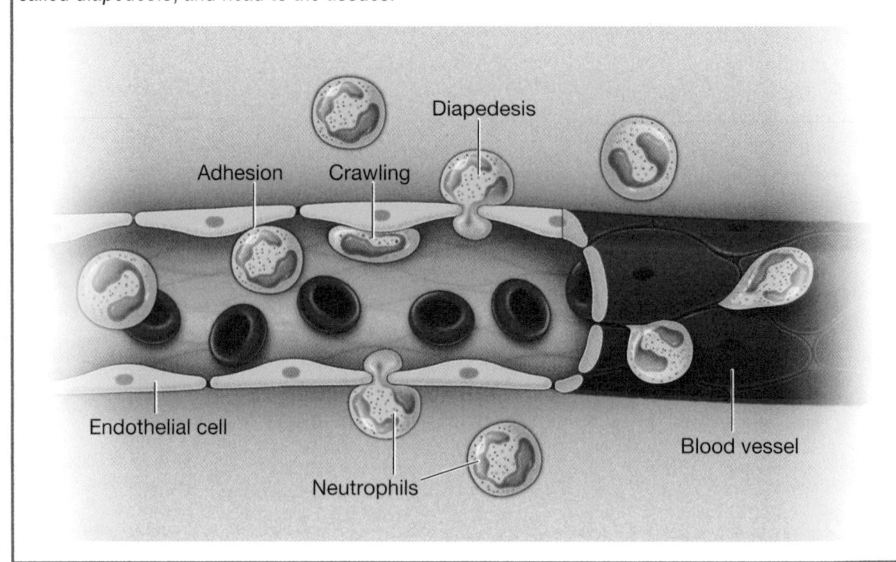

infection by secreting chemical signals. This attracting of leukocytes occurs because of **positive chemotaxis**. Because each type of leukocyte specializes in a particular type of immune response, taking a count of the types and percentages of leukocytes present in a blood sample can provide evidence as to the type of infection and help lead to a diagnosis and treatment.

18.1g Classification of Leukocytes

LO 18.1.10

When scientists first began to observe stained blood slides, it quickly became evident some leukocytes were marked with dots all over their cytoplasm. These dots are **granules**, specialized vesicles that contain chemicals used in the cell's response to a **pathogen**, a disease-causing agent. Leukocytes can be divided into two groups, according to whether they contain granules:

Student Study Tip

The prefix *path* refers to a disease, like in the word *pathology*, the study of disease.

- **Granular leukocytes** contain abundant granules within the cytoplasm. They include neutrophils, eosinophils, and basophils. See the "Anatomy of Leukocytes" feature and **Figures 18.7A–C**.
- **Agranular leukocytes** have far fewer and less obvious granules. Agranular leukocytes include monocytes and lymphocytes (see **Figures 18.7D** and **18.7E**).

Granular Leukocytes We will consider the granular leukocytes in order from most common to least common. All of these are produced in the red bone marrow and have a short lifespan of hours to days. They typically have a lobed nucleus and are classified according to which type of stain best highlights their granules.

The most common of all the leukocytes, **neutrophils** will normally comprise 40–60 percent of the total leukocytes in the blood. Their granules are numerous but appear a faint, light purple color. The nucleus has a distinct lobed appearance and may have two to five lobes; the number increases as the cell ages.

Figure 18.7 Leukocytes

(A) Human neutrophils are granulocytes, but their granules are much fainter than those of basophils or eosinophils. The most notable feature of neutrophils is their unusually shaped nuclei, which are multi-lobed. (B) Human eosinophils are characterized by many dark red or pink granules that fill their cytoplasm and sometimes obscure the view of their nucleus. During parasitic infections or allergic reactions eosinophils swell larger and move their granules toward the membrane to release the chemicals contained within them. (C) Human basophils are characterized by many dark blue or purple granules that fill their cytoplasm and sometimes obscure the view of their nucleus. Basophils are found during parasitic infections and severe allergic responses. (D) Human monocytes are agranular (have no granules) and as a result, their cytoplasm looks smoother than that of granulocytes. Their nuclei are well defined and usually in a kidney bean or horseshoe shape. (E) Human lymphocytes are much smaller than any of the other leukocytes (they are about twice the size of a red blood cell, rather than three to five times the size). They have no granules, and their nuclei are perfectly round and take up most of the cell. There is often very little cytoplasm showing around the nucleus, so these cells often appear as small, dark dots among red blood cells.

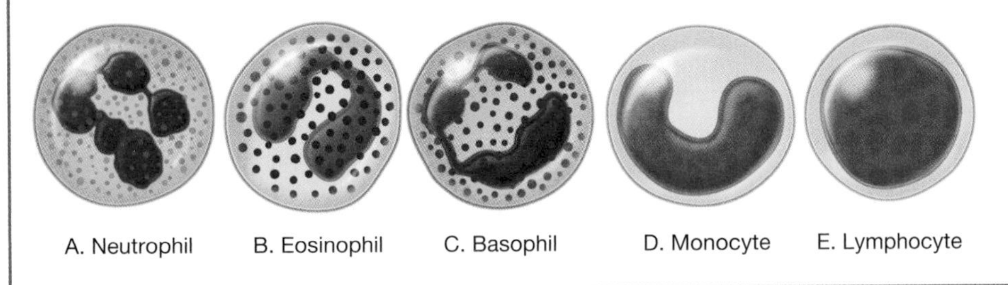

Neutrophils are rapid responders to the site of infection and will quickly phagocytize, or engulf, bacteria. Their granules include enzymes that can break down bacterial cell walls. High counts of neutrophils indicate infection, most likely bacterial infection.

Eosinophils typically represent 2–4 percent of total leukocyte count. The granules of eosinophils stain best with an acidic stain known as *eosin*. The nucleus of the eosinophil will typically have two to three lobes and the granules will have a bright pink or red color.

Some eosinophil granules contain molecules toxic to parasitic worms, which can enter the body through the integument, or when an individual consumes raw or undercooked fish or meat. High counts of eosinophils are typical of patients experiencing allergies, parasitic worm infestations, and some autoimmune diseases. Low counts may be due to drug toxicity and stress.

Basophils are the least common leukocytes, typically comprising less than 1 percent of the total leukocyte count. The granules of basophils stain best with basic (alkaline) stains. Basophils contain large granules that pick up a dark blue stain and are so common they may make it difficult to see the two-lobed nucleus.

Student Study Tip

Basophils are bilobed.

The granules of basophils release **histamine**, an inflammatory chemical. High counts of basophils are associated with allergies, parasitic infections, and hypothyroidism. Low counts are associated with pregnancy, stress, and hyperthyroidism.

Agranular Leukocytes Agranular leukocytes contain smaller, less visible granules in their cytoplasm than do granular leukocytes. The nucleus is simple in shape, sometimes with an indentation but without distinct lobes. There are two major types of agranulocytes: lymphocytes and monocytes.

Lymphocytes form initially in the bone marrow; much of their subsequent development and reproduction occurs in the lymphatic tissues. Lymphocytes are the second most common type of leukocyte, accounting for about 20–30 percent of all leukocytes,

B lymphocytes (B cells)

T lymphocytes (T cells)

and are essential for the immune response. Lymphocytes are typically smaller than the other leukocytes with a large nucleus that takes up almost the entire cell.

The three major groups of lymphocytes include **natural killer (NK) cells**, **B lymphocytes**, and **T lymphocytes**, all of which play prominent roles in defending the body against specific pathogens. While the three different types of leukocytes have very different roles, we cannot distinguish among them under the light microscopes. We will discuss the functions of these subclasses in more detail in Chapter 21.

Abnormally high lymphocyte counts are characteristic of viral infections as well as some types of cancer. Abnormally low lymphocyte counts are characteristic of prolonged (chronic) illness or immunosuppression, including that caused by HIV infection and drug therapies that often involve steroids.

Monocytes are easily recognized by their large size and horseshoe-shaped nuclei. Monocytes are cells in transit, but once they leave the blood and reside in the tissues they differentiate into **macrophages**, cells capable of phagocytizing debris, foreign pathogens, worn-out erythrocytes, and many other damaged cells. Macrophages also release antimicrobial chemicals that harm pathogens and attract other leukocytes to the site of an infection. Whereas some macrophages occupy fixed locations, others wander through the tissue fluid.

Abnormally high counts of monocytes are associated with viral or fungal infections, tuberculosis, and some forms of leukemia and other chronic diseases. Abnormally low counts are typically caused by suppression of the bone marrow.

When examining leukocytes under the microscope, a few questions can be used to differentiate among the leukocyte types and determine what kind of cell is being examined (**Figure 18.8**).

Learning Connection

Quiz Yourself

Write down the names and functions of all of the formed elements on separate pieces of paper or index cards. Then mix them up and see if you can match the names of the formed elements with their functions.

Figure 18.8 Leukocyte Decision Tree

When examining leukocytes and trying to determine their type, a decision tree can be a useful tool.

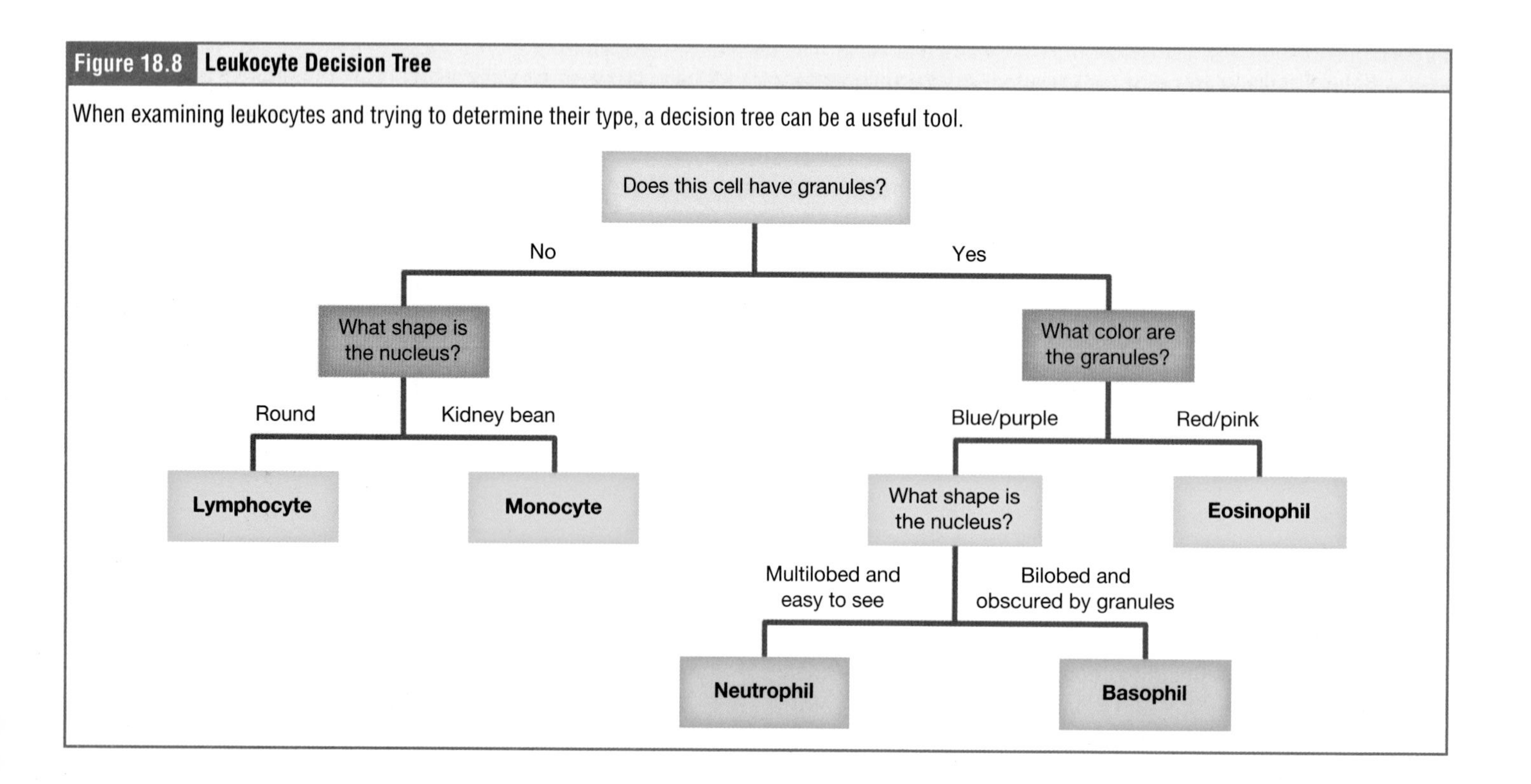

Apply to Pathophysiology

Leukocyte Responses to Disease

A complete blood count with differential (or CBCdiff) is a common blood analysis ordered by doctors who suspect infection or particular diseases in their patient. Examine the cases below and answer the questions about the most likely diagnosis.

LaToya is a 32-year-old mother of 2 children. It is September 2021 and LaToya's two children are returning to school. September is "ragweed season" in the area where LaToya lives, and ragweed can prompt allergic responses. There is also a high circulating level of coronavirus infections in the community. One of LaToya's children tests positive for COVID and the next day, LaToya wakes up with a stuffy nose and sore throat. Her doctor orders a COVID test and also a CBCdiff. The CBC comes back first.

Test	Normal Range	Value
White Blood Cell	4.0–11.0 Th/μL	15.7
Neutrophils	40–60%	50%
Lymphocytes	20–40%	20%
Monocytes	2–8%	4%
Eosinophils	1–4%	11%
Basophils	0.5–1%	5%
Red Blood Cell Count	3.90–5.20 M/μL	4.12
Hematocrit	34.1–44.9%	39.0

1. Which of the following is most likely causing LaToya's symptoms?
   A) A bacterial infection
   B) A viral infection of SARS CoV-2, the virus that causes COVID
   C) Allergies
2. Which of the following cell type values helped you arrive at your conclusion?
   A) Neutrophils
   B) Eosinophils
   C) Lymphocytes

Manolo works for a renovation company as a tiler. Last week they cut themself while using their tile saw. The wound has mostly healed, so Manolo is back at work; they are just careful of the wound, which is a bit swollen and painful, so they keep it tightly bandaged. A week after the initial injury, Manolo develops a fever. Their partner was sick with the flu the week before so Manolo assumes they must be fighting that same viral infection. Their doctor is not so sure, so orders a CBCdiff. Below are Manolo's values.

Test	Normal Range	Value
White Blood Cell	4.0–11.0 Th/μL	25.7
Neutrophils	40–60%	84%
Lymphocytes	20–40%	11%
Monocytes	2–8%	2%
Eosinophils	1–4%	2%
Basophils	0.5–1%	1%
Red Blood Cell Count	3.90–5.20 M/μL	5.02
Hematocrit	34.1–44.9%	44.0

3. Which of the following is most likely causing Manolo's symptoms?
   A) A bacterial infection of the wound
   B) A viral infection of influenza
   C) Allergies
4. Which of the following cell type values helped you arrive at your conclusion?
   A) Neutrophils
   B) Eosinophils
   C) Lymphocytes

platelets (thrombocytes)

Learning Connection

Try Drawing It

After studying pictures of the blood cells and fragments in your textbook or on the Internet, draw pictures of erythrocytes, the five types of leukocytes, and platelets. Draw them to scale to show which ones are largest and smallest. Which of the formed elements lack(s) a nucleus?

18.1h Platelets

Platelets are not cells but rather fragments of cytoplasm surrounded by a plasma membrane. Platelets are produced by **megakaryocytes**, a type of cell found in the bone marrow. Megakaryocytes release thousands of cytoplasmic fragments, each enclosed by a bit of plasma membrane. These enclosed fragments are platelets. Each megakaryocyte releases 2000–3000 platelets during its lifespan (Figure 18.9).

Platelets are small, but numerous. After entering the circulation, approximately one-third migrate to the spleen for storage for later release in response to any rupture in a blood vessel. They are critical to hemostasis, the stoppage of blood flow following damage to a vessel. They also secrete a variety of growth factors essential for growth and repair of tissue, particularly connective tissue.

Figure 18.9 Megakaryocytes

Platelets are cell fragments formed in the bone marrow by megakaryocytes. These large cells pinch off small chunks of cytoplasm and cell membrane to form platelets.

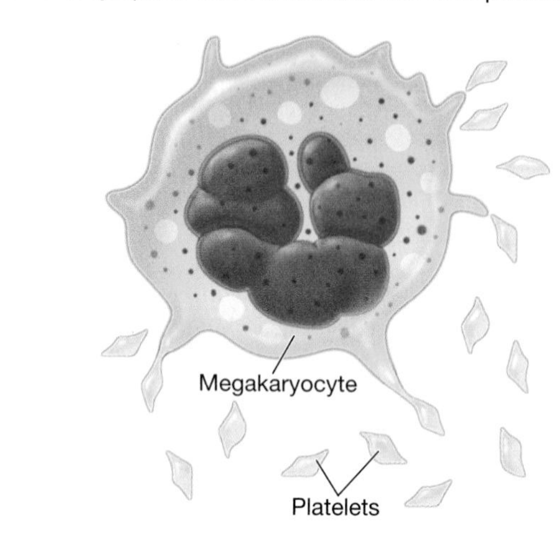

Cultural Connection

Types of Blood Donation

Whole blood donation is usually used for incidents of trauma to replace lost blood, as well as for people undergoing surgery. Red cell donation is a type of blood donation in which the donor has more blood taken from them, but the blood is separated, and the plasma, platelets, and white blood cells are returned to the donor. The isolated red blood cells are used for patients who have lost large amounts of blood during trauma or birth, as well as for patients with sickle cell anemia. Platelet donation involves whole blood donation followed by return of the red and white blood cells to the donor. Because there is no loss of oxygen-carrying capacity, those that donate platelets can do so quite frequently. Donated platelets are most often used to treat patients undergoing cancer treatment. Similarly, plasma donation takes just the donor's plasma, returning the formed elements to the donor. This process is relatively easy, though the donor will be encouraged to rehydrate after. Plasma is often used to treat individuals who need assistance in blood clotting, since the clotting factors that are dissolved in the plasma will function in the recipient of the donation. All blood donation is restricted by blood type in one way or another, and some blood types are able to be used in wider sets of circumstances, making these blood types particularly sought after by blood donation agencies.

Learning Check

1. You are in oxygen deficit as you start to exercise. Will hemoglobin have a higher or lower affinity for oxygen compared to resting conditions?
   a. Higher
   b. Lower
   c. No change
   d. Impossible to tell with the given information
2. Whole blood moving from an area of higher to lower concentration is an example of ______. Solutes crossing a membrane to equalize concentration is an example of ______.
   a. Bulk flow; bulk flow
   b. Diffusion; bulk flow
   c. Bulk flow; diffusion
   d. Bulk flow; osmosis
3. Fick's Law states that the diffusion rate of a gas will increase with greater surface area for diffusion, greater concentration gradient of the ions, and greater membrane permeability. It will decrease with increasing membrane thickness. The biconcave shape of an erythrocyte increases which factor of Fick's Law?
   a. Surface area
   b. Concentration gradient
   c. Membrane permeability
   d. Membrane thickness
4. You have a case of strep throat. Which of the following extremely common granular leukocytes will have elevated levels in your blood?
   a. Basophils
   b. Eosinophils
   c. Lymphocytes
   d. Neutrophils

18.2 Production of the Formed Elements

HAPS LEARNING GOALS AND OUTCOMES

Learning Objectives: By the end of this section, you will be able to:

18.2.1 Describe the locations of hematopoiesis (hemopoiesis) and the significance of the hematopoietic stem cell (HSC or hemocytoblast).

18.2.2 Explain the basic process of erythropoiesis, the significance of the reticulocyte, and regulation through erythropoietin (EPO).

18.2.3 Explain the basic process of thrombopoiesis.

18.2.4 Explain the basic process of leukopoiesis.

The lifespan of the formed elements is very brief. Although one type of leukocyte called *memory cells* can survive for years, most erythrocytes, leukocytes, and platelets normally live only a few hours to a few weeks. Thus, the body must form new blood cells and platelets quickly and continuously. When you donate a unit of blood, your body typically replaces the donated plasma within 24 hours, but it takes about 4–6 weeks to replace the blood cells. This restricts the frequency with which donors can safely contribute their blood. The process by which this replacement occurs is called **hematopoiesis**.

18.2a Sites of Hematopoiesis

LO 18.2.1

Most hematopoiesis occurs in the red bone marrow, a connective tissue within the spaces of spongy (cancellous) bone tissue. In children, red bone marrow occupies all

Learning Connection

Broken Process

As a person ages, much of the red bone marrow is converted into yellow bone marrow, which contains stem cells for the production of fats, cartilage and bone, and stores fats. What would happen if all of a person's red bone marrow was converted into yellow bone marrow? Production of which types of blood cells would be affected? What would the end result be?

of the hollow spaces within bones but in adults, the process is largely restricted to the cranial and pelvic bones, the vertebrae, the sternum, and the proximal epiphyses of the femur and humerus.

18.2b Differentiation of Formed Elements from Stem Cells

Hematopoietic stem cells reside in the bone marrow and give rise to all of the formed elements of blood.

Hematopoiesis begins when the hematopoietic stem cell divides. One of the daughter cells remains a stem cell and the other differentiates into one of two types of stem cells (**Figure 18.10**).

- **Lymphoid stem cells** give rise to lymphocytes, which include the T cells, B cells, and natural killer (NK) cells, all of which function in immunity.
- **Myeloid stem cells** give rise to all the other formed elements, including the erythrocytes; and the other leukocytes: neutrophils, eosinophils, and basophils and monocytes.

Lymphoid and myeloid stem cells do not immediately divide and differentiate into mature formed elements. As you can see in Figure 18.10, there are intermediate stages of precursor cells.

Figure 18.10 Hematopoiesis

The formation of all of the formed elements in blood occurs from a single hematopoietic stem cell in the bone marrow.

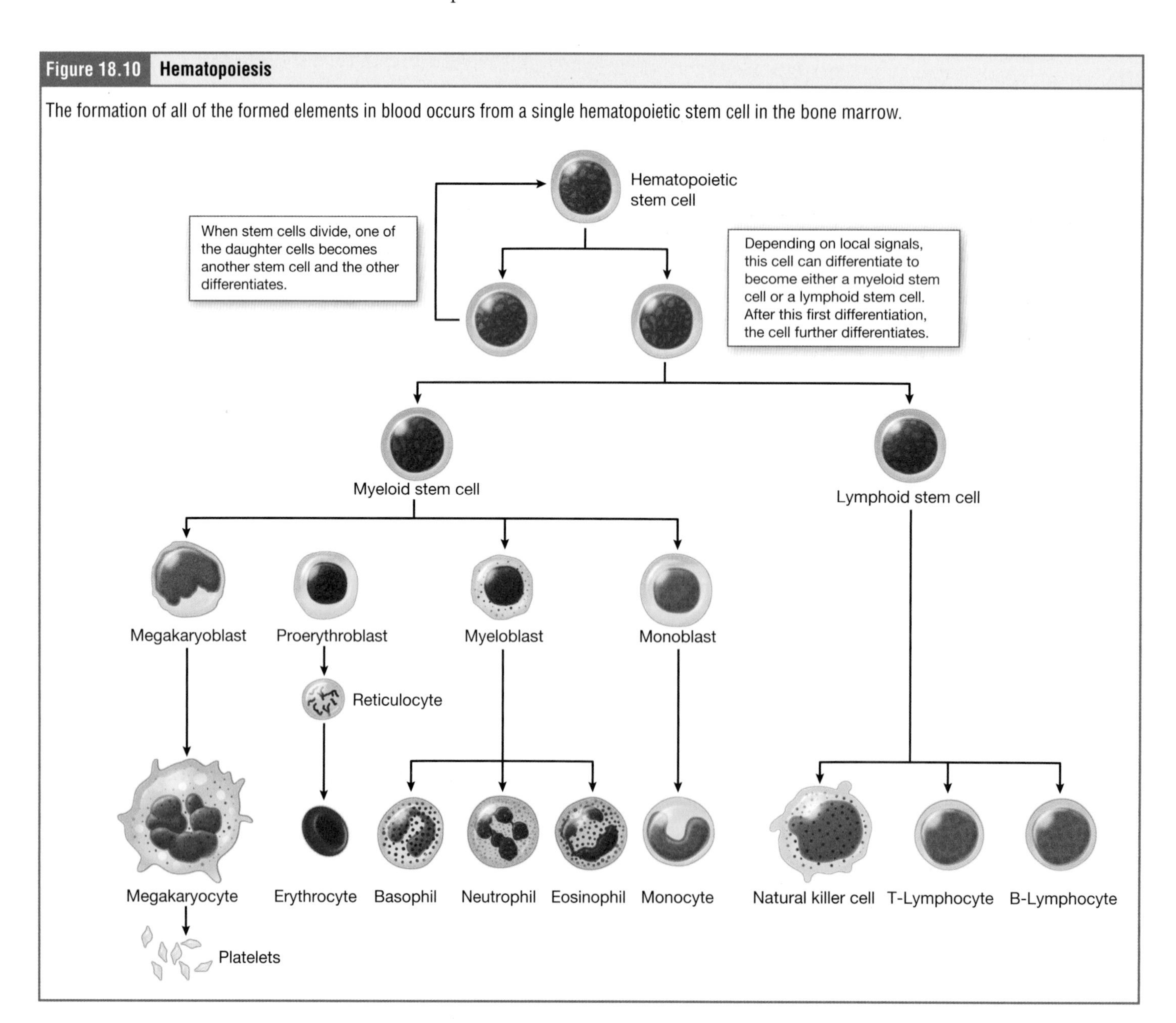

18.2c Hemopoietic Growth Factors

Development from stem cells to precursor cells to mature cells is guided by secreted chemicals called *growth factors*. These include the following:

- **Erythropoietin (EPO)** is a hormone secreted by the kidneys in response to low oxygen levels. It prompts the production of erythrocytes, which is called *erythropoiesis*. LO 18.2.2 LO 18.2.3
- **Thrombopoietin**, another hormone, is produced by the liver and kidneys. It triggers the development of megakaryocytes into platelets. LO 18.1.4
- **Cytokines** are chemical signals secreted by a wide variety of cells, including red bone marrow, leukocytes, macrophages, fibroblasts, and endothelial cells. They act locally as autocrine or paracrine factors, stimulating the proliferation of progenitor cells and helping to stimulate both nonspecific and specific resistance to disease. Some of these cytokines stimulate leukopoiesis, the production of leukocytes from hematopoietic stem and progenitor cells.

Learning Check

1. With consistent treadmill running, our body's need for oxygen increases. Therefore, we need more oxygen carried to our tissues, which is accomplished by a/an ______ production of erythropoietin by the liver. This helps diminish the oxygen deficit, so we call this homeostatic mechanism ______.
   a. Increased; negative feedback
   b. Decreased; negative feedback
   c. Increased; positive feedback
   d. Decreased; positive feedback
2. In which of the following situations would a cytokine be released to affect change in the body? Please select all that apply.
   a. A cell targeting itself
   b. A cell targeting a neighboring cell
   c. Synaptic signaling
   d. A cell targeting a distant cell through the bloodstream
3. Myeloid stem cells give rise to what type of leukocytes?
   a. Granular
   b. Agranular
   c. Both
   d. Neither

18.3 Hemostasis

Learning Objectives: By the end of this section, you will be able to:

18.3.1 Describe the vascular phase of hemostasis, including the role of endothelial cells.

18.3.2 Describe the role of platelets in hemostasis and the steps involved in the formation of the platelet plug.

18.3.3 Explain how the positive feedback loops in the platelet and coagulation phases promote hemostasis.

18.3.4 Describe the basic steps of coagulation resulting in the formation of the insoluble fibrin clot.

18.3.5 Differentiate among the intrinsic (contact activation), extrinsic (cell injury), and common pathways of the coagulation cascade.

18.3.6 Explain the role of vitamin K in blood clotting.

18.3.7 Describe the process of fibrinolysis, including the roles of plasminogen, tissue plasminogen activator, and plasmin.

Student Study Tip

The suffix *stasis* means *static*, which is how we can remember that hemostasis means that blood is no longer flowing out of the body! We also see this in homeostasis (when our body is in equilibrium).

Platelets are key players in **hemostasis**, the process by which the body plugs a ruptured blood vessel to prevent blood loss (**Figure 18.11**, Step 1). Although rupture of larger vessels usually requires medical intervention, hemostasis is quite effective in dealing with small, simple wounds. There are three steps to the process: vascular spasm, the formation of a platelet plug, and coagulation (blood clotting). Failure of any of these steps will result in **hemorrhage**—excessive bleeding.

Figure 18.11 Clot Formation

Step 1: The formation of a blood clot is essential when a wound occurs that impacts a blood vessel. Step 2: The immediate reaction is vascular spasm, a narrowing of the vessel. Step 3: Soon a platelet plug forms. Step 4: Fibrin, a fibrous protein formed from fibrinogen, stabilizes the clot as coagulation occurs.

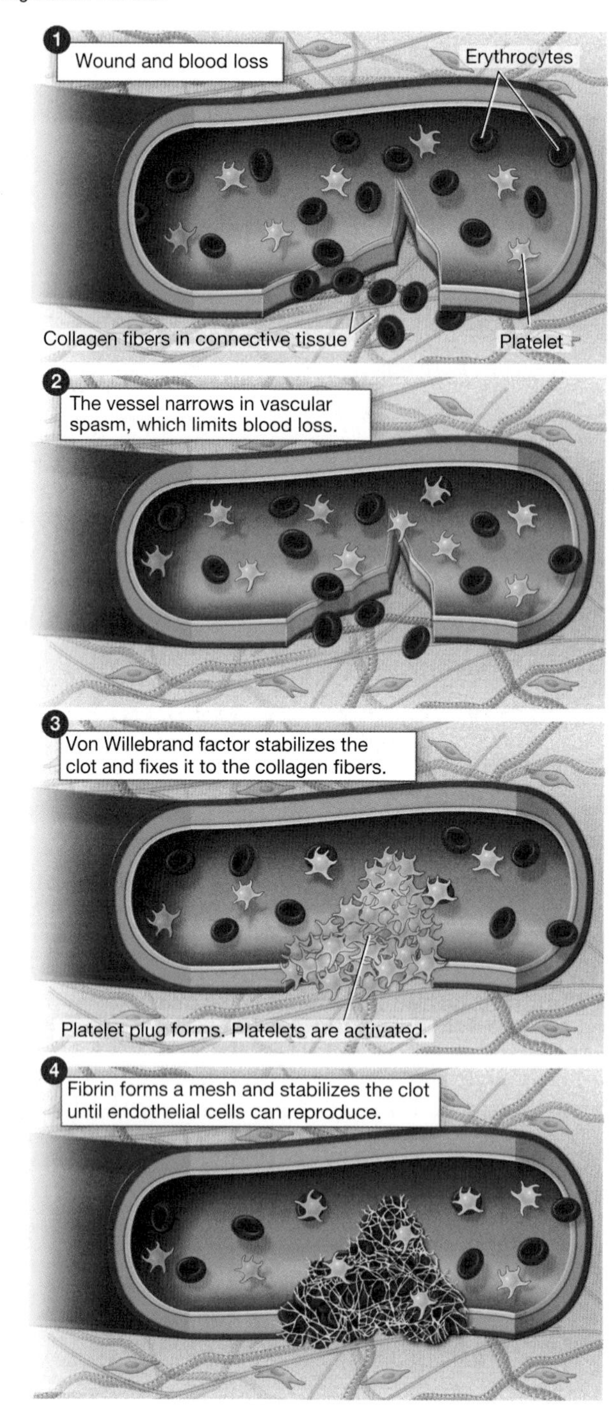

1
Wound and blood loss
Erythrocytes
Collagen fibers in connective tissue
Platelet
2
The vessel narrows in vascular spasm, which limits blood loss.
3
Von Willebrand factor stabilizes the clot and fixes it to the collagen fibers.
Platelet plug forms. Platelets are activated.
4
Fibrin forms a mesh and stabilizes the clot until endothelial cells can reproduce.

18.3a Vascular Spasm

LO 18.3.1

When a vessel is severed or punctured, or when the wall of a vessel is damaged, vascular spasm occurs. In **vascular spasm**, the smooth muscle in the walls of the vessel contracts dramatically (Figure 18.11, Step 2).

The vascular spasm response is believed to be triggered by chemicals that are released by **endothelial cells**, the cells that line blood vessels. This phenomenon typically lasts for up to 30 minutes, which allows the other steps to take place and contribute to hemostasis.

18.3b Formation of the Platelet Plug

LO 18.3.2

In the second step, platelets, which normally float free in the plasma, arrive at the area of vessel rupture. At this location, the underlying connective tissue, rich with collagenous fibers, is exposed. Platelets are not normally exposed to collagen, as there are no collagen fibers in the blood, so the presence of collagen provides an important signal that activates the platelets. They begin to clump together, become spiked and sticky, and bind to the exposed collagen and endothelial lining (Figure 18.11, Step 3). This process is assisted by a plasma protein called *von Willebrand factor*, which will help stabilize the growing **platelet plug** and adhere it to the collagen of the connective tissue. As platelets collect, they simultaneously release chemicals from their granules into the plasma that cause further platelet activation and adhesion. This is a positive feedback loop that contributes further to hemostasis. Among the substances released by the platelets are:

LO 18.3.3

- Adenosine diphosphate (ADP), which helps additional platelets to adhere to the injury site, reinforcing and expanding the platelet plug.
- Serotonin, which maintains vasoconstriction.
- Prostaglandins and phospholipids, which also maintain vasoconstriction and help to activate further clotting chemicals.

Student Study Tip

The more platelets that come to the injury, the more that can release chemicals that call for even more platelets. This is a positive feedback loop!

The platelet plug temporarily seals the opening in a blood vessel. Plug formation, in essence, buys the body time while more sophisticated and durable repairs are being made.

18.3c Coagulation

LO 18.3.4

LO 18.3.5

Those more sophisticated and more durable repairs are collectively called **coagulation**, the formation of a blood clot. The process is sometimes characterized as a cascade, because one event prompts the next as in a multilevel waterfall. The result is the production of a gelatinous but robust clot made up of a mesh of **fibrin** in which platelets and blood cells are trapped (Figure 18.11, Step 4). Fibrin forms from the polymerization (making of a polymer) of fibrinogen, one of the plasma proteins. You can think of this process sort of like building a brick wall. The fibrinogen molecules are floating individually, but when you link them together they form a large netlike protein that is capable of stabilizing the clot (**Figure 18.12**).

clotting factors (coagulation factors)

Figure 18.12 Fibrin

Netlike fibrin forms during blood clotting to stabilize the clot and prevent blood loss while the vessel heals.

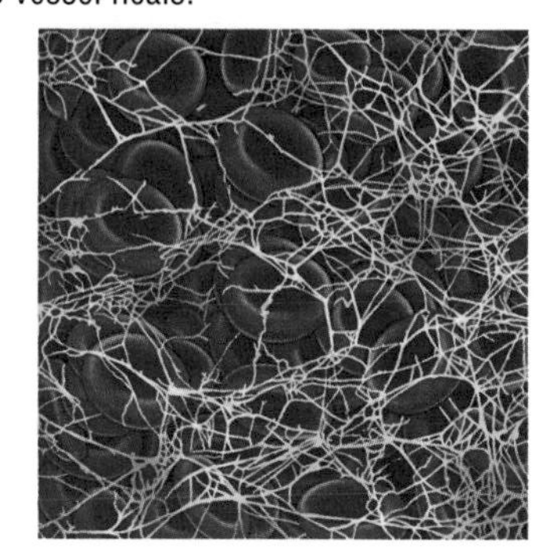

Clotting Factors Involved in Coagulation In the coagulation cascade, chemicals called **clotting factors** (or *coagulation factors*) prompt reactions that activate still more coagulation factors. The process is complex, but is initiated along two basic pathways:

- The extrinsic pathway, which normally is triggered by trauma that breaks the wall of the blood vessel.
- The intrinsic pathway, which begins in the bloodstream and is triggered by internal damage to the wall of the vessel.

Both of these merge into a third pathway, referred to as the *common pathway*. All three pathways are dependent upon the 12 known clotting factors, including Ca^{2+} and vitamin K (**Table 18.2**). Clotting factors are secreted primarily by the liver and the platelets.

Table 18.2 Clotting Factors

Factor Number	Name	Type of Molecule	Source	Pathway(s)
I	Fibrinogen	Plasma protein	Liver	Common; converted into fibrin
II	Prothrombin	Plasma protein	Liver*	Common; converted into thrombin
III	Tissue thromboplastin or tissue factor	Lipoprotein mixture	Damaged cells and platelets	Extrinsic
IV	Calcium ions	Inorganic ions in plasma	Diet, platelets, bone matrix	Entire process
V	Proaccelerin	Plasma protein	Liver, platelets	Extrinsic and intrinsic
VI	Not used	Not used	Not used	Not used
VII	Proconvertin	Plasma protein	Liver*	Extrinsic
VIII	Antihemolytic factor A	Plasma protein factor	Platelets and endothelial cells	Intrinsic; deficiency results in hemophilia A
IX	Antihemolytic factor B (plasma thromboplastin component)	Plasma protein	Liver*	Intrinsic; deficiency results in hemophilia B
X	Stuart–Prower factor (thrombokinase)	Protein	Liver*	Extrinsic and intrinsic
XI	Antihemolytic factor C (plasma thromboplastin antecedent)	Plasma protein	Liver	Intrinsic; deficiency results in hemophilia C
XII	Hageman factor	Plasma protein	Liver	Intrinsic; initiates clotting in vitro also activates plasmin
XIII	Fibrin-stabilizing factor	Plasma protein	Liver, platelets	Stabilizes fibrin; slows fibrinolysis

*Vitamin K required

LO 18.3.6

The liver requires the fat-soluble vitamin K to produce many of them. Vitamin K (along with biotin and folate) is somewhat unusual among vitamins in that it is not only consumed in the diet but is also synthesized by bacteria residing in the large intestine. The calcium ion, considered factor IV, is derived from the diet and from the breakdown of bone. Some recent evidence indicates that activation of various clotting factors occurs on specific receptor sites on the surfaces of platelets.

The 12 clotting factors are numbered I through XIII according to the order in which they were discovered. Factor VI was once believed to be a distinct clotting factor, but is now thought to be identical to factor V. Rather than renumber the other factors, factor VI was allowed to remain as a placeholder and also a reminder that knowledge changes over time.

Extrinsic Pathway The quicker responding and more direct **extrinsic pathway** begins when damage occurs to the surrounding tissues, such as in a traumatic injury or the wound scenario described earlier. Upon contact with blood plasma, damaged cells outside (or extrinsic to) the blood vessel, release factor III. Sequentially, Ca^{2+} and then factor VII, which is activated by factor III, are added, forming an enzyme complex. This enzyme complex leads to activation of factor X, which activates the common pathway (discussed in the following "Common Pathway" section). The events in the extrinsic pathway are completed in a matter of seconds.

Intrinsic Pathway The **intrinsic pathway** is longer and more complex. In this case, the factors involved are inside, or intrinsic to, the blood vessel. The pathway can be prompted events such as arterial disease; however, it is most often initiated when factor XII, which is always circulating in the bloodstream, comes into contact with foreign materials, such as when a blood sample is put into a glass test tube. Within the body, factor XII is typically activated when it encounters negatively charged molecules, such as inorganic polymers and phosphate produced earlier in the series of intrinsic pathway reactions. Factor XII sets off a series of reactions that in turn activate factor XI and then factor IX. In the meantime, chemicals released by the platelets increase the rate of these activation reactions. Finally, factor VIII from the platelets and endothelial cells combines with factor IX to form an enzyme complex that activates factor X, leading to the common pathway. The events in the intrinsic pathway are completed in a few minutes.

Common Pathway Both the intrinsic and extrinsic pathways lead to the **common pathway**, in which fibrin is produced to seal off the vessel. Once factor X has been activated by either the intrinsic or extrinsic pathway, the enzyme prothrombinase converts factor II, the inactive enzyme prothrombin, into the active enzyme **thrombin**. Note that if the enzyme thrombin were not normally in an inactive form, clots would form spontaneously, and blood would not flow! Then, thrombin converts factor I, the soluble fibrinogen, into the insoluble fibrin protein strands. Factor XIII then stabilizes the fibrin clot.

Learning Connection

Explain a Process

If you were trying to help a classmate understand the three processes involved in hemostasis, would you explain them in simple terms? Which process occurs first? Which one is the most effective? How are platelets involved?

18.3d Fibrinolysis

The stabilized clot is acted upon by contractile proteins within the platelets. As these proteins contract, they pull on the fibrin threads, bringing the edges of the clot more tightly together, somewhat like tightening loose shoelaces.

LO 18.3.7

To restore normal blood flow as the vessel heals, the clot must eventually be removed. **Fibrinolysis** is the gradual degradation of the clot. Again, there is a fairly complicated series of reactions that involves factor XII and protein-catabolizing enzymes. During this process, the inactive protein plasminogen is converted into the active **plasmin**, which gradually breaks down the fibrin of the clot. Additionally, bradykinin—a vasodilator—is released, reversing the effects of the serotonin and prostaglandins from the platelets. This allows the smooth muscle in the walls of the vessels to relax and helps to restore the circulation.

18.3e Plasma Anticoagulants

An **anticoagulant** is any substance that opposes coagulation. Several circulating plasma anticoagulants play a role in limiting the coagulation process to the region of injury and restoring a normal, clot-free condition of blood.

- **Antithrombin** inactivates factor X and opposes the conversion of prothrombin to thrombin in the common pathway.
- **Heparin** also opposes prothrombin. Heparin is also found on the surfaces of cells lining the blood vessels. Because heparin is present, an intact, healthy endothelium will prevent clotting. A pharmaceutical form of heparin is often administered therapeutically, for example, in surgical patients at risk for blood clots.

Blood transfusions in humans were risky procedures until the discovery of the major human blood types in 1900. Until that point, physicians did not understand why some blood transfusions resulted in death, while in other cases blood transfusions helped patients heal. Blood type refers to presence or absence of specific marker molecules on the plasma membranes of erythrocytes. With their discovery, it became possible for the first time to match patient–donor blood types and prevent transfusion reactions and deaths.

Learning Check

1. In vascular spasm, blood vessels (vasoconstrict/vasodilate). This allows (more/less) blood to reach the site of blood vessel rupture.
   a. Vasoconstrict; less
   b. Vasoconstrict; more
   c. Vasodilate; more
   d. Vasodilate; less
2. Which of the following triggers the common pathway to be activated at the end of both the extrinsic and intrinsic pathways of blood coagulation?
   a. Release of factor X
   b. Release of factor III
   c. Ca^{2+} release
   d. Release of factor IX
3. Clotting factors activate more clotting factors. This is an example of a (positive/negative) feedback loop.
   a. Positive
   b. Negative
4. Vitamin K is used in our liver to produce many clotting factors. In which ways do we maintain healthy vitamin K levels in our bodies? Please select all that apply.
   a. Diet
   b. Sunlight
   c. Produced by bacteria
   d. Our bodies convert foods to vitamin K

18.4 Blood Typing

Learning Objectives: By the end of this section, you will be able to:

18.4.1 Explain the role of surface antigens on erythrocytes in determining blood groups.

18.4.2 Describe how the presence or absence of Rh antigen results in blood being classified as positive or negative.

18.4.3 List the type of antigen and the type of antibodies present in each ABO blood type.

18.4.4 Describe the development and clinical significance of anti-Rh antibodies.

18.4.5 Predict which blood types are compatible and what happens when the incorrect ABO or Rh blood type is transfused.

18.4a Antigens, Antibodies, and Transfusion Reactions

The term **antigen** refers to a molecule or group of molecules that the body does not recognize as belonging to the "self" and that therefore trigger a defensive response from the leukocytes of the immune system. In Chapter 21 we will discuss the antigens of pathogens in detail, and you may have heard of "antigen testing" as a way to check for the presence of certain viruses or other pathogens. Here, scientists named the molecules on the surface of erythrocytes as antigens because these molecules triggered the immune reactions, called *transfusion reactions*, that made some forms of blood incompatible with others. One of our immune tools is a small protein called an *antibody* that is made in response to an antigen (**Figure 18.13**).

More than 50 antigens have been identified on the surface of erythrocytes, but only three are strongly antigenic, meaning that they are the most likely to provoke an immune response in certain blood transfusion recipients. These antigens are known as antigen A, antigen B, and antigen D, which is also called the *Rh factor* (**Figure 18.14**).

LO 18.4.1

Although the **ABO blood group** name consists of three letters, ABO blood typing designates the presence or absence of just two antigens, A and B. Both are glycoproteins. People whose erythrocytes have A antigens on their erythrocyte membrane surfaces are designated blood type A, and those whose erythrocytes have B antigens are blood type B. People can also have both A and B antigens on their erythrocytes, in

Figure 18.13 Pathogens, Antigens, and Antibodies

The term *pathogen* refers to an entire disease-causing entity such as a bacterium or a virus. The term *antigen* is a feature of a pathogen, or of a foreign material that the immune system reacts against. Antigens can be parts of viruses, of cells, or even food or pollen components that cause allergies. Antibodies are proteins made by the immune system and are important tools in the immune response.

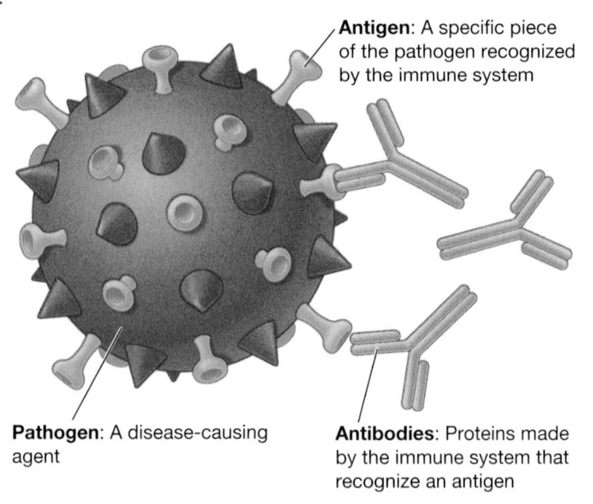

Figure 18.14 All Erythrocytes (RBCs) Express Antigens

There are three antigens that are particularly inflammatory to the immune system. Individuals who express antigen A are considered blood type A. Individuals who express antigen B are considered blood type B. Individuals who express both are type AB and individuals who express neither of these antigens are considered blood type O. In addition, individuals are positive or negative for the Rh factor (green diamonds). So an individual who expresses all three antigens is considered type AB positive and an individual who expresses none of them is considered type O negative.

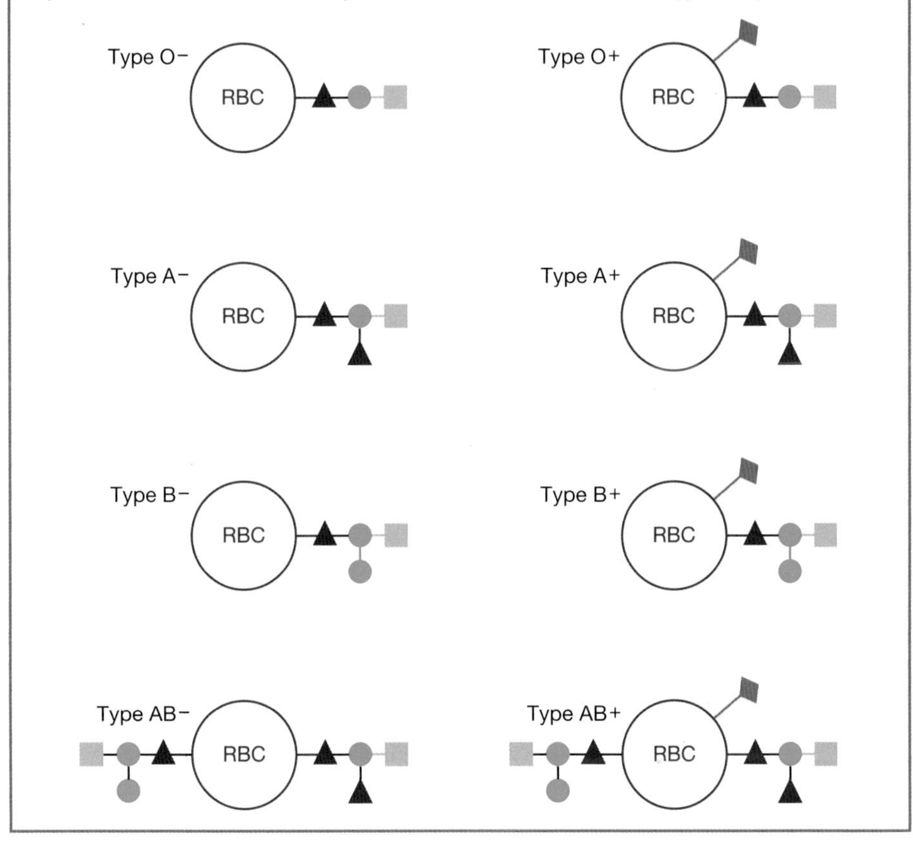

which case they are blood type AB. People with neither A nor B antigens are designated blood type O. ABO blood types are genetically determined.

Normally the body must be exposed to a foreign antigen before an antibody can be produced. This is not the case for the ABO blood group. Individuals with type A blood—without any prior exposure to incompatible blood—have preformed antibodies to the B antigen circulating in their blood plasma. These antibodies, referred to as anti-B antibodies, will cause agglutination and hemolysis if they ever encounter erythrocytes with B antigens. Similarly, an individual with type B blood has pre-formed anti-A antibodies. Individuals with type AB blood, which has both antigens, do not have preformed antibodies to either of these. People with type O blood lack antigens A and B on their erythrocytes, but both anti-A and anti-B antibodies circulate in their blood plasma. **Table 18.3** summarizes the characteristics of the blood types in the ABO blood group.

LO 18.4.2

The **Rh blood group** is classified according to the presence or absence of a second erythrocyte antigen identified as Rh. (The Rh is short for "rhesus"; it was first discovered in a rhesus macaque, a type of primate often used in research because its blood is similar to that of humans.) Although dozens of Rh antigens have been identified, only one (designated D) is clinically important. Those who have the Rh D antigen present on their erythrocytes—about 85 percent of Americans—are described as Rh positive (Rh+) and those who lack it are Rh negative (Rh–). Note that the Rh group is distinct

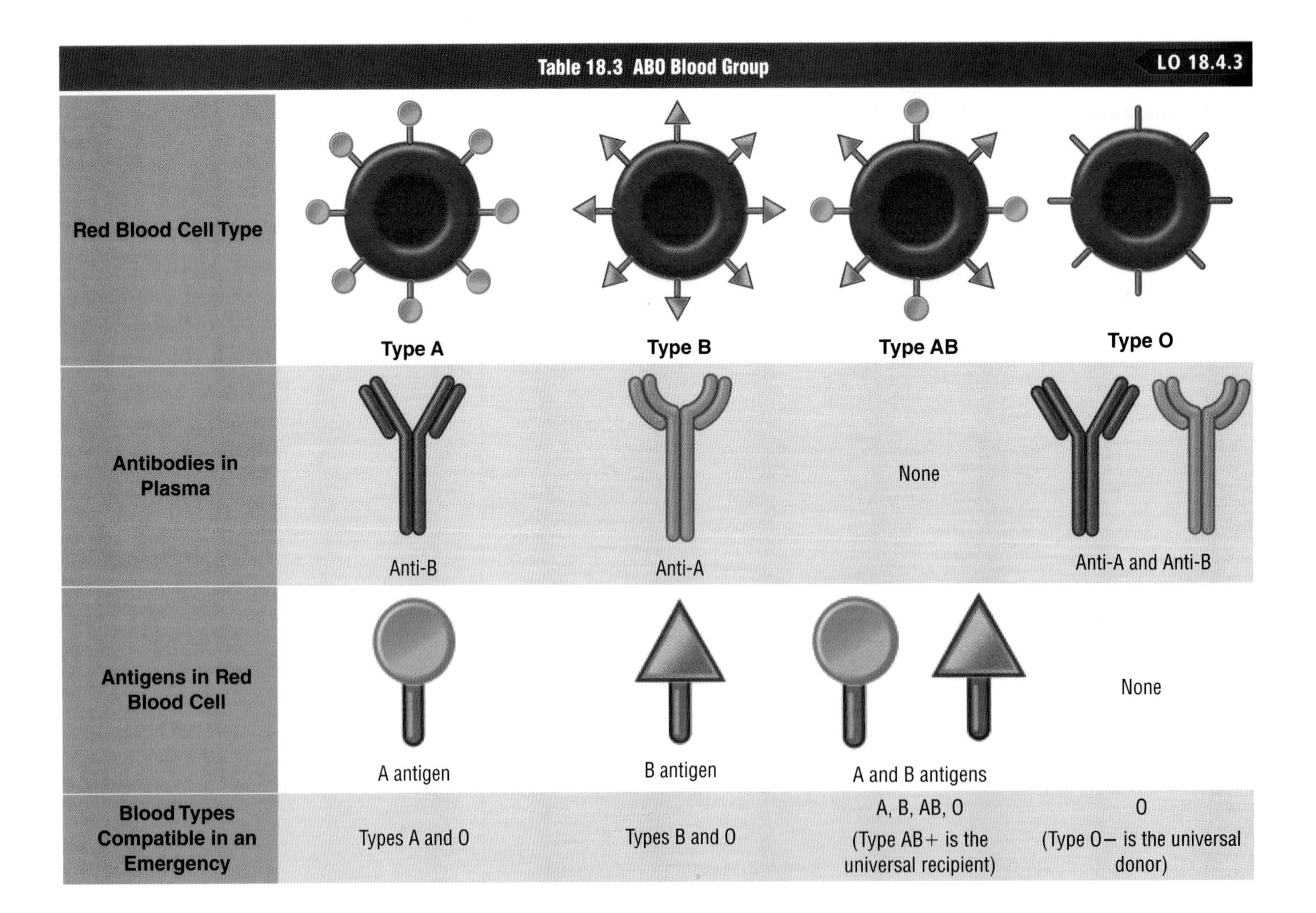

Table 18.3 ABO Blood Group — LO 18.4.3

Red Blood Cell Type	Type A	Type B	Type AB	Type O
Antibodies in Plasma	Anti-B	Anti-A	None	Anti-A and Anti-B
Antigens in Red Blood Cell	A antigen	B antigen	A and B antigens	None
Blood Types Compatible in an Emergency	Types A and O	Types B and O	A, B, AB, O (Type AB+ is the universal recipient)	O (Type O– is the universal donor)

from the ABO group, so any individual, no matter their ABO blood type, may have or lack this Rh antigen (**Figure 18.15**). When identifying a patient's blood type, the Rh group is designated by adding the word positive or negative to the ABO type. For example, A positive (A+) means ABO group A blood with the Rh antigen present, and AB negative (AB–) means ABO group AB blood without the Rh antigen.

In contrast to the ABO group antibodies, which are preformed, antibodies to the Rh antigen are produced only in Rh negative individuals after exposure to the antigen. This process, called *sensitization*, occurs following a transfusion with Rh negative incompatible blood or, more commonly, with the birth of an Rh positive baby to an Rh negative mother. Problems are rare in a first pregnancy, since the fetus's Rh+ cells rarely cross the placenta (the organ of gas and nutrient exchange between the baby and the mother). However, during the process of birth, the Rh negative mother can be exposed to the baby's Rh positive cells. After exposure, the mother's immune system begins to generate anti-Rh antibodies. If the mother should then conceive another Rh positive fetus, the Rh antibodies she has produced can cross the placenta into the fetal bloodstream and destroy the fetal RBCs. This condition, known as **hemolytic disease of the newborn (HDN)** or *erythroblastosis fetalis*, may cause anemia in mild cases, but the agglutination and hemolysis can be so severe that without treatment the fetus may die in the womb or shortly after birth.

LO 18.4.4

hemolytic disease of the newborn (HDN) (erythroblastosis fetalis)

Transfusion Reactions Antibodies are Y-shaped and have two antigen-binding sites. During an antibody response, clumps of antigen-bearing erythrocytes can form. This process is called **agglutination** (see Figure 18.15). The clumps of erythrocytes block small blood vessels throughout the body, depriving tissues of oxygen and nutrients. As the erythrocyte clumps are degraded, in a process called **hemolysis**, their hemoglobin

Figure 18.15 Agglutination

(A) Antibodies against antigens expressed on red blood cells will bind to red blood cells at each of their two binding sites. (B) This can cause clumping of the red blood cells, which can clog small vessels.

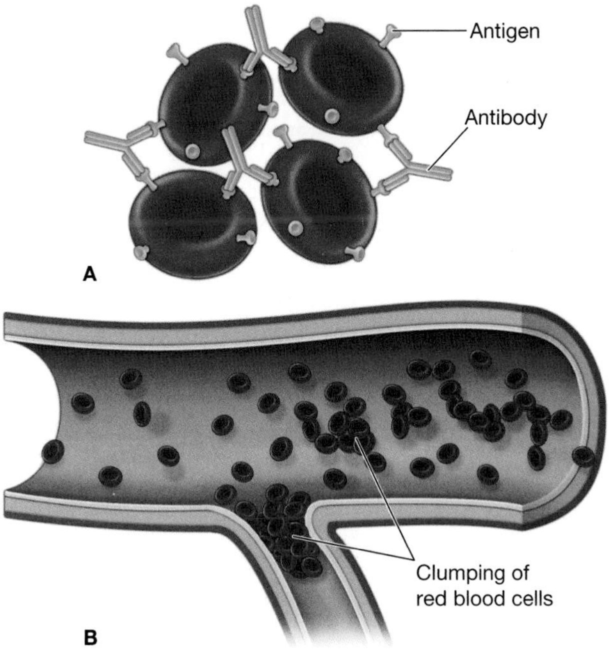

Learning Connection

Micro to Macro

A patient with type AB+ blood needed a large volume transfusion. Type O+ donor blood was available. The patient was given only the erythrocytes from the donor blood, in order to prevent an adverse reaction. What is present in the donor plasma that could cause a harmful reaction in the patient?

is released into the bloodstream. This hemoglobin travels to the kidneys, which are responsible for filtration of the blood. However, the load of hemoglobin released can easily overwhelm the kidney's capacity to clear it, and the patient can quickly develop kidney failure. Therefore an individual can only receive blood from blood types that they do not have antibodies against. If, for example, a person is A+ (A positive) they can receive blood in which the RBCs contain A and/or Rh antigens, but they would react against B antigens and therefore should not receive any blood type that includes B antigens (type B or AB).

LO 18.4.5

Learning Check

1. A patient with B– blood, who has never been exposed to the Rh antigen, receives a blood transfusion from someone B+. Will an agglutination reaction occur?
   a. Yes
   b. No
2. You have blood with antibodies against both A and B antigens. What blood type are you?
   a. A
   b. B
   c. AB
   d. O
3. Say you have type B blood. Antibodies against which antigens are present?
   a. B
   b. A
   c. Antigens A and B
   d. Neither antigens A nor B
4. There is a patient in the emergency room that needs a blood transfusion. The patient has type AB blood. Which of the following people would they be able to accept the transfusion from? Please select all that apply.
   a. Person with type A blood
   b. Person with type B blood
   c. Person with type O blood
   d. Person with type AB blood

Chapter Review

18.1 The Composition of Blood

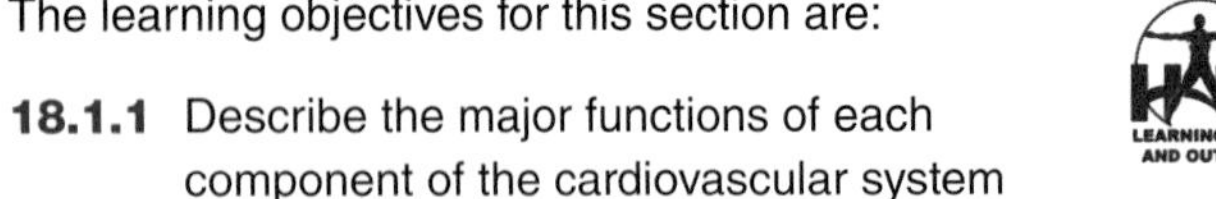

The learning objectives for this section are:

18.1.1 Describe the major functions of each component of the cardiovascular system (i.e., blood, heart, blood vessels).

18.1.2 Describe the general composition of blood (e.g., plasma, formed elements).

18.1.3* Describe the functions of blood and connect those functions to the maintenance of homeostasis.

18.1.4* Compare and contrast bulk flow and diffusion.

18.1.5 Describe the composition of blood plasma.

18.1.6 List the major types of plasma proteins, their functions, and sites of production.

18.1.7 Compare and contrast the morphological features and general functions of the formed elements (i.e., erythrocytes, leukocytes, platelets).

18.1.8 Describe the structure and function of hemoglobin, including its breakdown products.

18.1.9* Define hematocrit and discuss healthy ranges for adults and common reasons for deviation from homeostasis.

18.1.10 List the five types of leukocytes in order of their relative prevalence in normal blood, and describe their major functions.

* Objective is not a HAPS Learning Goal.

18.1 Questions

1. All of the following are functions of the blood except:
   a. Transportation of oxygen, carbon dioxide, nutrients, wastes, and electrolytes throughout the body
   b. Defense against pathogens that enter the body
   c. Production of hormones and plasma proteins
   d. Regulation of body temperature

2. Which blood component is the least dense, therefore rises to the top of the tube after being spun in the centrifuge?
   a. White blood cells
   b. Plasma
   c. Erythrocytes
   d. Platelets

3. You are lying in bed scrolling on your phone when you realize you lost track of time and are now late to class. You jump out of bed and start running. The blood vessels supplying your digestive system will ______ and those that supply skeletal muscle will ______.
   a. Vasoconstrict; vasoconstrict
   b. Vasodilate; vasodilate
   c. Vasoconstrict; vasodilate
   d. Vasodilate; vasoconstrict

4. Electrolytes like Na^+ and K^+ cross membranes due to ____ gradients, and blood flows due to ____ gradients.
   a. Pressure; concentration
   b. Resistance; pressure
   c. Pressure; resistance
   d. Concentration; pressure

5. Sally is donating plasma tomorrow and the blood drive issues a recommendation of how to prepare for it. Which of the following recommendations will they issue?
   a. Drink a lot of water on the day prior to and the day of your appointment.
   b. Eat fatty foods on the day of your appointment.
   c. Eat a lot of carbohydrates on the day of your appointment.

6. What is the most abundant plasma protein? Where is it produced?
   a. Globulins; The kidneys
   b. Fibrinogen; The liver
   c. Albumin; The kidneys
   d. Albumin; The liver

7. Which of the following features of erythrocytes contribute to their oxygen-carrying capacity? Please select all that apply.
   a. Smooth surface
   b. Biconcave shape
   c. Lack of organelles
   d. Abundance of organelles

8. Which of the following are true of hemoglobin? Please select all that apply.
   a. Deoxygenated hemoglobin contains no bound oxygen molecules.
   b. Hemoglobin consists of four subunits, each containing an iron ion.
   c. Hemoglobin's breakdown products, such as biliverdin and bilirubin, are added to the bile and excreted in the feces.
   d. In oxygen-poor conditions, the affinity of hemoglobin for oxygen increases.

9. Which of the following individuals would be expected to have the lowest hematocrit?
   a. A menstruating person with high muscle mass
   b. A nonmenstruating person with low muscle mass
   c. A nonmenstruating person with high muscle mass
   d. A menstruating person with low muscle mass

10. Which of the following is the least common type of leukocyte? What is a feature of their function?
   a. Eosinophils; destroy bacteria
   b. Neutrophils; destroy bacteria
   c. Basophils; release histamine
   d. Basophils; poison parasites that enter the body

18.1 Summary

- Blood is a connective tissue that consists of formed elements and blood plasma. Blood accounts for about 8% of the body weight.
- Blood functions in the transportation of respiratory gasses, nutrients, and wastes, as well as participating in defense against infection and the maintenance of homeostasis.
- Blood plasma is the liquid matrix of the blood; it contains dissolved substances and plasma proteins. The three types of plasma proteins are albumins, globulins, and fibrinogen; they function in osmotic balance, lipid transport, the immune response, and blood coagulation.
- The three types of formed elements of the blood are (1) erythrocytes, or red blood cells, which transport oxygen and carbon dioxide between the lungs and body cells, (2) leukocytes, or white blood cells, which defend the body against infection and disease, and (3) platelets, which participate in the stoppage of bleeding.
- Blood plasma is the liquid matrix of the blood. It transports red and white blood cells, platelets, nutrients, wastes, oxygen, carbon dioxide, electrolytes, enzymes and plasma proteins around the body.

18.2 Production of the Formed Elements

The learning objectives for this section are:

18.2.1 Describe the locations of hematopoiesis (hemopoiesis) and the significance of the hematopoietic stem cell (HSC or hemocytoblast).

18.2.2 Explain the basic process of erythropoiesis, the significance of the reticulocyte, and regulation through erythropoietin (EPO).

18.2.3 Explain the basic process of thrombopoiesis.

18.2.4 Explain the basic process of leukopoiesis.

18.2 Questions

1. Where do hematopoietic stem cells reside?
   a. The blood plasma
   b. The kidneys
   c. The adrenal cortex
   d. The bone marrow

2. Erythropoiesis is a homeostatic regulator of oxygen. Does the process follow a positive or negative feedback loop?
   a. Positive feedback
   b. Negative feedback
   c. Impossible to tell with the given information

3. Suppose you have a condition in which your body does not release sufficient amounts of thrombopoietin. Which of the following homeostatic mechanisms would be impaired?
   a. Nutrient regulation
   b. Waste products filtering out
   c. Blood clotting
   d. Body temperature regulation

4. Which of the following cell types develop from lymphoid stem cells through leukopoiesis? Please select all that apply.
   a. Platelets
   b. T cells
   c. Natural killer cells
   d. B cells

18.2 Summary

- Hematopoiesis, the production of formed elements, occurs in the red bone marrow.
- Hematopoietic stem cells give rise to two other types of stem cells: (1) lymphoid stem cells, which give rise to lymphocytes, and (2) myeloid stem cells, which give rise to erythrocytes, neutrophils, eosinophils, basophils, monocytes and platelets.
- Erythropoiesis, or red blood cell formation, is stimulated by the hormone erythropoietin, which is produced by the kidneys in response to a low blood level of oxygen.
- Leukopoiesis, or white blood cell formation, is stimulated by various cytokines, in order to defend the body against infection and disease.
- Thrombopoiesis, or platelet formation, is activated by the hormone thrombopoietin; this keeps the platelet level sufficient to participate in hemostasis.

18.3 Hemostasis

The learning objectives for this section are:

18.3.1 Describe the vascular phase of hemostasis, including the role of endothelial cells.

18.3.2 Describe the role of platelets in hemostasis and the steps involved in the formation of the platelet plug.

18.3.3 Explain how the positive feedback loops in the platelet and coagulation phases promote hemostasis.

18.3.4 Describe the basic steps of coagulation resulting in the formation of the insoluble fibrin clot.

18.3.5 Differentiate among the intrinsic (contact activation), extrinsic (cell injury), and common pathways of the coagulation cascade.

18.3.6 Explain the role of vitamin K in blood clotting.

18.3.7 Describe the process of fibrinolysis, including the roles of plasminogen, tissue plasminogen activator, and plasmin.

18.3 Questions

1. In vascular spasm, the damaged blood vessel ______ triggered by chemicals released by ______.
   a. Vasodilates; platelets
   b. Vasodilates; epithelial cells
   c. Vasoconstricts; endothelial cells
   d. Vasoconstricts; white blood cells

2. Which of the following substances released by platelets function in activating further clotting chemicals? Please select all that apply.
   a. Adenosine diphosphate
   b. Serotonin
   c. Prostaglandins
   d. Phospholipids

3. The platelet plug phase of blood clotting contributes to homeostasis. Does it do so by using a positive or a negative feedback loop?
   a. Positive
   b. Negative
   c. Impossible to tell with the given information

4. Which clotting factor stabilizes the insoluble fibrin clot?
   a. Clotting factor I
   b. Clotting factor III
   c. Clotting factor X
   d. Clotting factor XIII

5. Which of the following are true of the intrinsic pathway of blood coagulation? Please select all that apply.
   a. It begins in the bloodstream.
   b. It is initiated by trauma that breaks the wall of a blood vessel.
   c. It merges with the extrinsic pathway into a common pathway that completes the blood coagulation.
   d. It has a faster response time than the extrinsic pathway.

6. Vitamin K is required to produce which of the following clotting factors? Please select all that apply.
   a. Prothrombin
   b. Antihemolytic factor A
   c. Hageman factor
   d. Proconvertin

7. In the absence of bradykinin, what step(s) of fibrinolysis could not happen? Please select all that apply.
   a. Breaking down the fibrin of the clot
   b. Vasodilation of the previously constricted vessel
   c. Stabilization of contractile proteins in the platelets
   d. Restoration of normal circulation

18.3 Summary

- Hemostasis is the stoppage of bleeding. It consists of three processes: vascular spasm, platelet plug formation, and coagulation.
- Hemostasis begins with vascular spasm, in which blood vessels vasoconstrict to restrict blood loss.
- Then, the exposed collagen from the wall of the broken blood vessel signals for platelet activation and a platelet plug is formed. This limits further blood loss.
- The final and most effective phase of hemostasis is coagulation, in which liquid plasma is converted into a solid gel, and fibrin threads provide a network to stabilize the clot. Coagulation occurs as a stepwise cascade of chemical reactions.
- Fibrinolysis occurs after healing has occurred, to restore normal blood flow to the previously affected area.

18.4 Blood Typing

The learning objectives for this section are:

18.4.1 Explain the role of surface antigens on erythrocytes in determining blood groups.

18.4.2 Describe how the presence or absence of Rh antigen results in blood being classified as positive or negative.

18.4.3 List the type of antigen and the type of antibodies present in each ABO blood type.

18.4.4 Describe the development and clinical significance of anti-Rh antibodies.

18.4.5 Predict which blood types are compatible and what happens when the incorrect ABO or Rh blood type is transfused.

18.4 Questions

1. A person with type A blood expresses which of the following antigens on their red blood cells?
   a. A
   b. B
   c. Neither A nor B
   d. Both A and B

Mini Case 1: Blood Typing and Transfusions

Jane was brought to the hospital after a car accident. They had a serious cut on their left leg, and had lost a significant amount of blood. The doctor determined that Jane needed a blood transfusion. After determining Jane's blood type, compatible donor blood had to be found.

1. When Jane's blood was mixed on separate glass slides with anti-A, anti-B, and anti-Rh antibodies, their blood was agglutinated by the anti-B, but not by the anti-A or anti-Rh antibodies. What is Jane's blood type?
   a. AB+
   b. O−
   c. B−
   d. A+

2. Which type(s) of blood antibodies, if any, would you expect to be found in Jane's blood plasma?
   a. Anti-B and anti-Rh
   b. Anti-A and anti-B
   c. None
   d. anti-A

3. Which of the following blood types could be used as donor blood, to eliminate the possibility of a transfusion reaction? Please select all that apply.
   a. O−
   b. A+
   c. B−
   d. AB+

4. The blood bank reports that they have no blood that matches Jane's exact blood type. Which blood type is the second choice and why?
   a. AB+, because it is the universal donor
   b. O−, because its red blood cells do not contain antigens A, B, or Rh
   c. B+, because its red blood cells contain the Rh factor
   d. A−, because its red blood cells do not contain the Rh factor

18.4 Summary

- The ABO blood group system is based on the presence or absence of A and B antigens on the cell membranes of a person's red blood cells. The four blood types are A (antigen A only), B (antigen B only), AB (antigens A and B), and O (neither antigen A nor B).
- In each blood type, the body produces antibodies against antigens that are absent from the red blood cell membranes. The exception is that people with type AB blood do not produce blood antibodies, because their red blood cells contain both antigens A and B.
- Part of a person's blood type is determined by the presence or absence of the Rh antigen (or Rh factor) on the red blood cell membranes. A person whose red blood cells have the Rh antigen is said to be Rh positive; those whose red blood cells lack it are said to be Rh negative.
- To determine a successful donor for a blood transfusion, the recipient's antibodies and the donor's antigens must be taken into account. A person cannot receive donor blood from anyone who has red blood cell antigens that could be attacked by the recipient's antibodies.
- In an unsuccessful blood transfusion, agglutination results, in which clumps of erythrocytes block small blood throughout the body, depleting tissues of oxygen and nutrients. The donor red blood cells are destroyed, which is called hemolysis. This renders the transfusion useless and causes a severe transfusion reaction in the recipient's body. The overload of hemoglobin can also overwhelm the kidney's filtration system and cause kidney failure.

19

The Cardiovascular System: The Heart

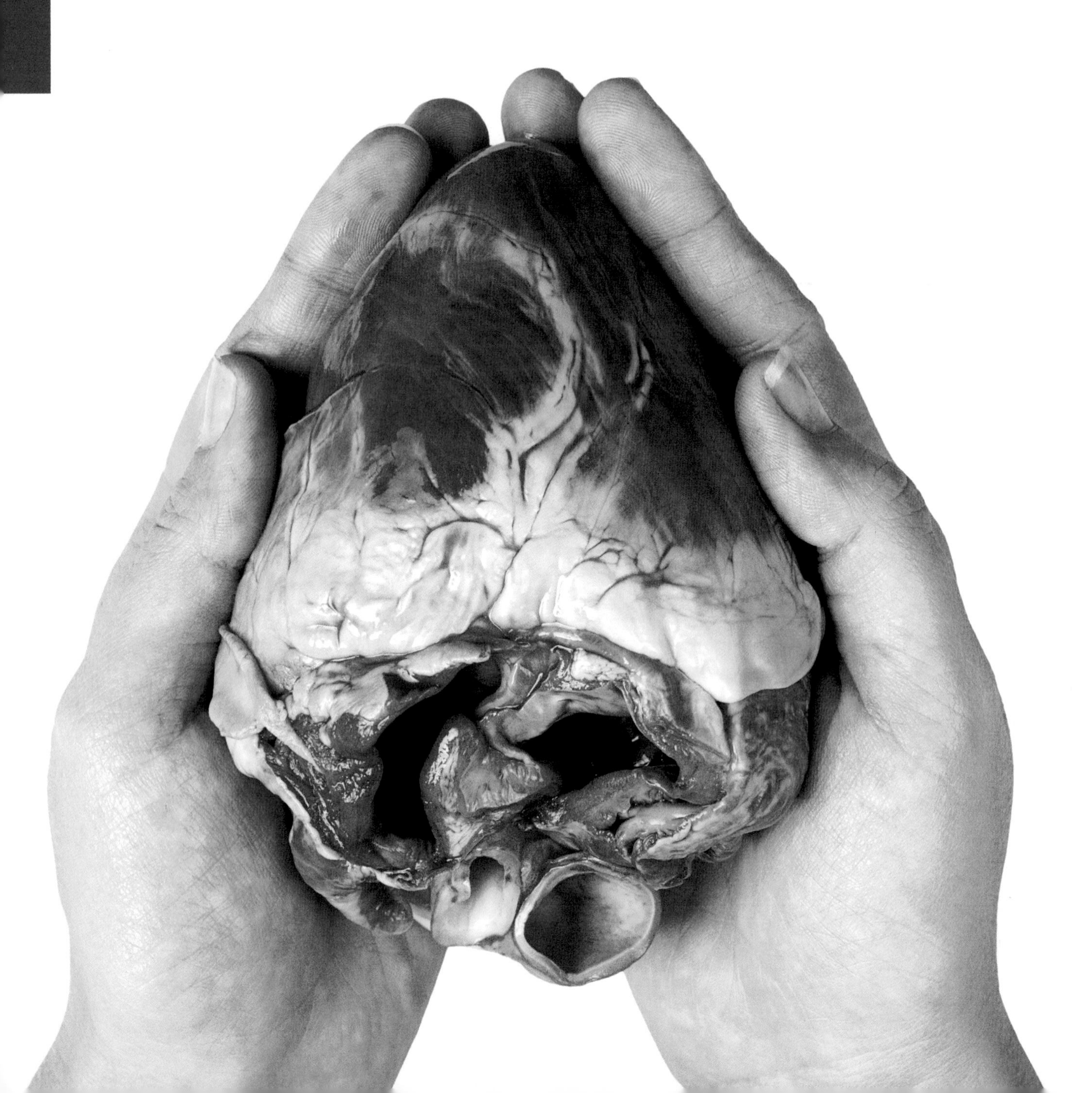

Balkonsky/Shutterstock.com

Chapter Introduction

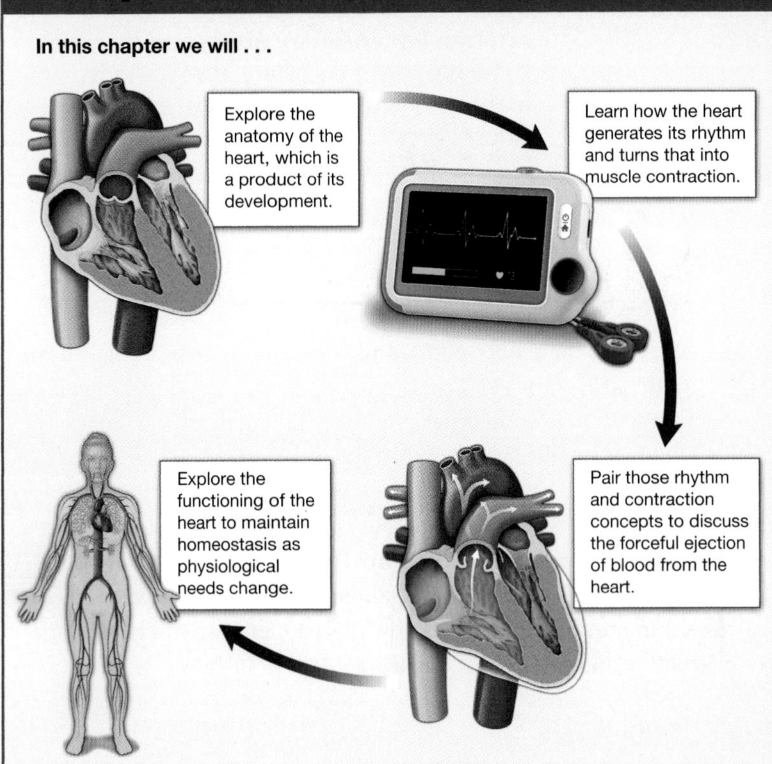

In this chapter we will . . .
Explore the anatomy of the heart, which is a product of its development.
Learn how the heart generates its rhythm and turns that into muscle contraction.
Pair those rhythm and contraction concepts to discuss the forceful ejection of blood from the heart.
Explore the functioning of the heart to maintain homeostasis as physiological needs change.

In this chapter, you will explore the remarkable pump that propels the blood into the vessels. There is no single better word to describe the function of the heart other than "pump," since it develops the pressure that ejects blood into the blood vessels throughout the body.

Although the term "heart" is an English word, cardiac (heart-related) terminology can be traced back to the Latin term, "kardia." Cardiology is the study of the heart, and cardiologists are the physicians who deal primarily with the heart.

In this chapter you will occasionally read "average" values, particularly in reference to volumes of blood moving through the heart. ***Remember*** that these numbers, as with so much of our scientific and medical information, has resulted from decades of scientific study. Historically, the individuals in these studies have most often been White men and so our "average values" are skewed toward a minority of individuals. As a scientific society, we have so much work to do to understand anatomical and physiological diversity. Commonly used physiological averages are usually referring to an adult who weighs approximately 150 pounds and has a moderately active lifestyle.

19.1 Heart Anatomy

Learning Objectives: By the end of this section, you will be able to:

19.1.1 Describe the position of the heart in the thoracic cavity.

19.1.2 Describe the changes in major fetal cardiovascular structures (i.e., umbilical vessels, ductus venosus, ductus arteriosus, foramen ovale) that typically occur beginning at birth, and the ultimate postnatal remnants (fates) of these structures.

19.1.3 Explain the structural and functional differences between atria and ventricles.

19.1.4 Identify and describe the location, structure, and function of the fibrous pericardium, parietal and visceral layers of the serous pericardium, serous fluid, and the pericardial cavity.

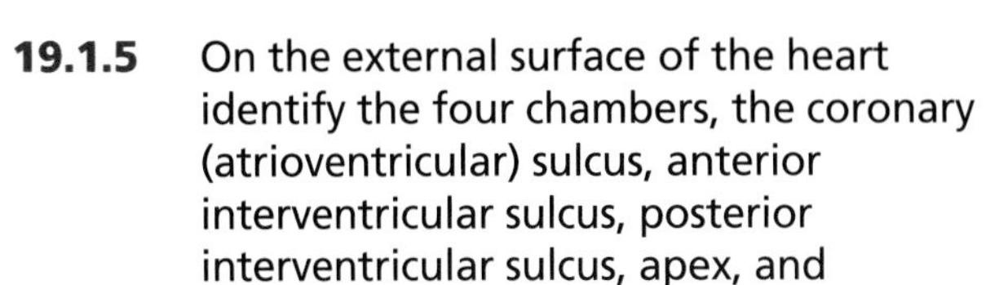

19.1.5 On the external surface of the heart identify the four chambers, the coronary (atrioventricular) sulcus, anterior interventricular sulcus, posterior interventricular sulcus, apex, and base.

19.1.6 *Given a disruption in the structure or function of the cardiovascular system (e.g., pulmonary edema), predict the possible factors or situations that might have created that disruption (i.e., given an effect, predict possible causes).

19.1.7 Describe the structure and functions of each layer of the heart wall (i.e., epicardium, myocardium, endocardium).

The Human Anatomy and Physiology Society includes more than 1,700 educators who work together to promote excellence in the teaching of this subject area. The HAPS A&P Learning Outcomes measure student mastery of the content typically covered in a two-semester Human A&P curriculum at the undergraduate level. The full Learning Outcomes are available at https://www.hapsweb.org.

19.1.8 Identify and describe the structure and function of the primary internal structures of the heart, including chambers, septa, valves, papillary muscles, chordae tendineae, fibrous skeleton, and venous and arterial openings.

19.1.9 Trace the path of blood through the right and left sides of the heart, including its passage through the heart valves, and indicate whether the blood is oxygen-rich or oxygen-poor.

19.1.10 Describe the blood flow to and from the heart wall, including the location of the openings for the left and right coronary arteries, left coronary artery and its major branches, right coronary artery and its major branches, cardiac veins, and coronary sinus.

In this chapter, we will discuss the events, effort, and factors involved in a single heart contraction; however, the scale of the work performed by the heart is staggering. If one assumes an average rate of 75 contractions per minute, a human heart will contract approximately 108,000 times in one day, more than 39 million times in one year, and nearly 3 billion times during a 75-year lifespan. Each of the major pumping chambers of the heart ejects approximately 70 mL of blood per contraction in a resting adult. This is equal to 5.25 L of fluid per minute and approximately 14,000 L per day. Over one year, that equals 10,000,000 L or 2.6 million gallons of blood sent through roughly 60,000 miles of vessels. In order to understand how this happens, we will explore the micro and macro structure and function of the heart.

LO 19.1.1

19.1a Location of the Heart

The human heart is located medially within the thoracic cavity, in a space between the lungs known as the **mediastinum** (Figure 19.1). Within the mediastinum, the heart is encased in a tough membrane known as the pericardium. This sac envelops the heart.

Figure 19.1 Position of the Heart in the Thorax

The heart is located within the thoracic cavity within the mediastinum. It is wide at the top and tapered inferiorly.

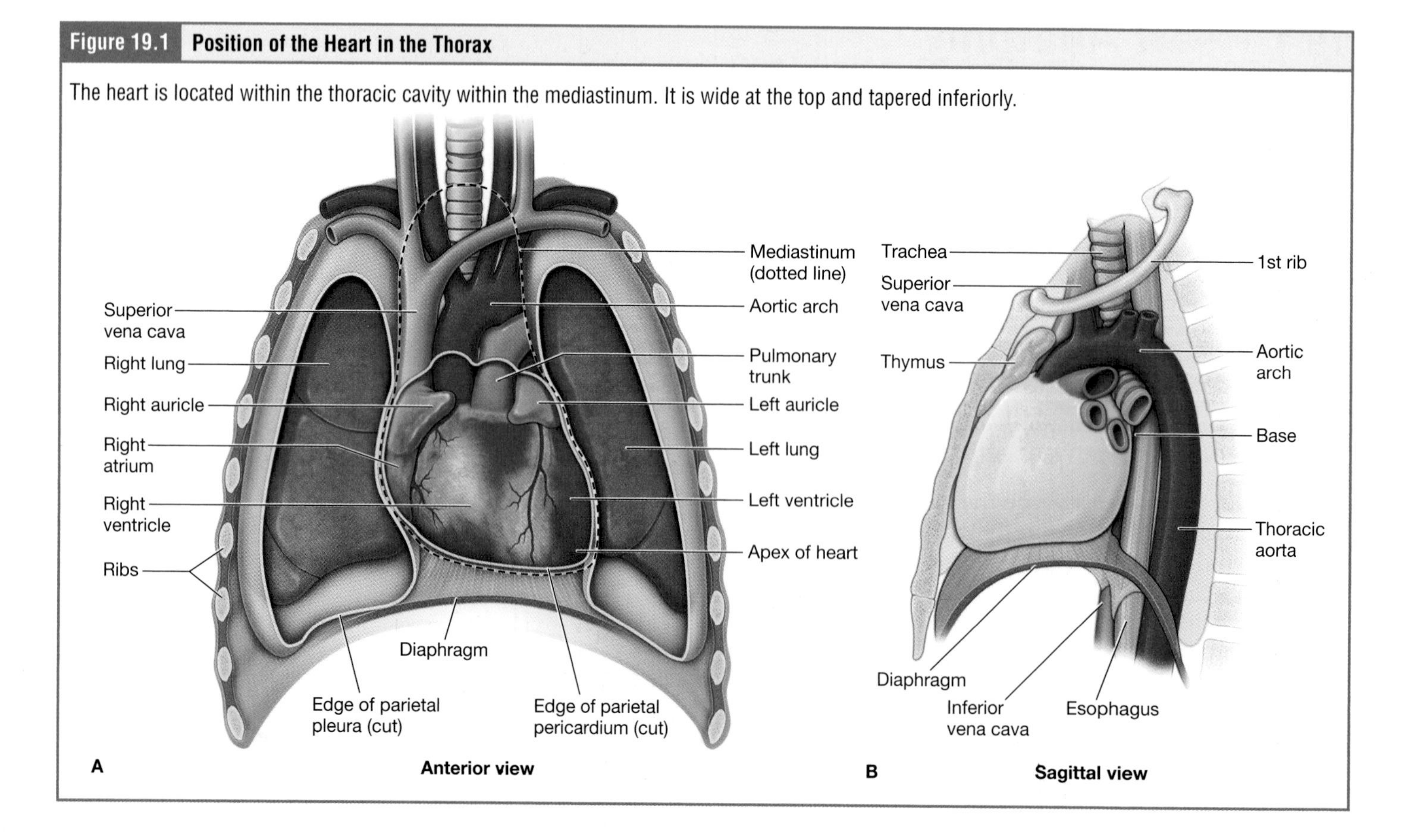

The heart takes up the depth of the thorax; its posterior surface is close to the bodies of the vertebrae, and its anterior surface nearly reaches the sternum and costal cartilages. The body's largest veins, the superior and inferior venae cavae, and the great arteries, the aorta and pulmonary trunk, emerge from the superior surface of the heart. These vessels are often referred to as the *great vessels* due to their size. The superior edge of the heart is located approximately at the level of the third costal cartilage, as seen in (**Figure 19.1B**). The inferior tip of the heart, the **apex**, can be found just to the left of the sternum between the junction of the fourth and fifth ribs near their articulation with the costal cartilages. It is notable that while the majority of the heart's mass sits midsagittally, directly behind the sternum, the inferior heart does deviate slightly to the left. This leftward shift impacts the lungs, the left lung accommodates the heart with a depression in its medial surface, called the **cardiac notch**. The right lung is slightly bigger as it has more thoracic space than the left (see Chapter 22).

This deviation from midline occurs during embryonic development of the heart. The embryonic heart folds on itself twice, once at around 24 and again at 35 days of gestation. It is the first of these folds that pushes the majority of the embryonic ventricle toward the left (**Figure 19.2**). You can see in **Figure 19.2A** that before these folds, at around day 23, the early embryonic heart is a tube with five distinct zones. The most superior, the **truncus arteriosus**, becomes the two great arteries. The next two zones, the **bulbus cordis** and the ventricle, will eventually be the most inferior parts of the heart, the left and right ventricles. The most inferior zones are the embryonic atrium and the **sinus venosus**; these areas, after the folding of the organ, will become the atria, which are superior in the adult heart (see Figure 19.2A). Then, walls between the chambers must form (**Figure 19.2B**). **Congenital heart defects** (CHDs) are anatomical heart issues present from birth and can result if these embryonic steps are incomplete or occur in atypical ways. CHDs affect about 1 percent of births in the United States each year.

Figure 19.2 The Development of the Heart During the Embryonic Period

LO 19.1.2

(A) The heart initially develops as a tube, then folds on itself twice once at approximately 24 days of gestation and again at 35 days of gestation. (B) As the external structure is forming, the separations among the heart chambers begin to form. The atria are separated from each other incompletely, a large hole called the foramen ovale remains until around the time of birth. The ventricles are separated by a thick septum. Each atrium is separated from its ventricle by a valve.

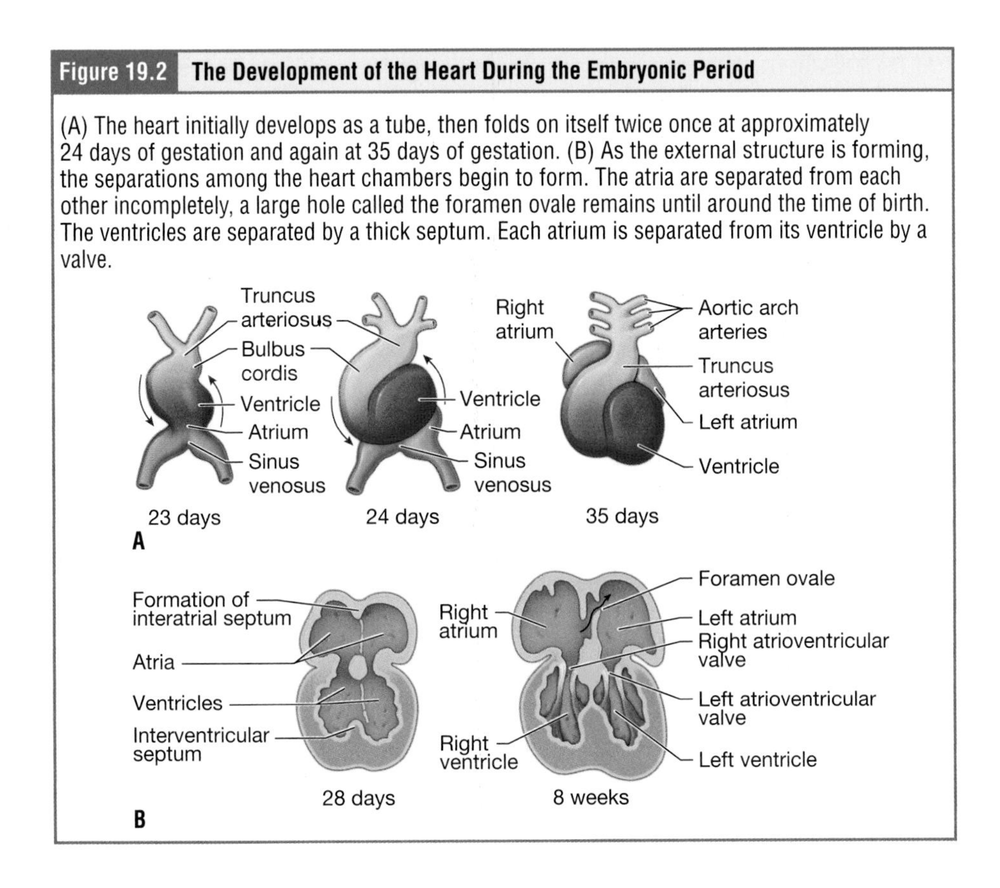

Student Study Tip

The entrance of a fancy building is often referred to as the atrium. You can apply this to the heart: Blood enters the atrium before it enters the ventricle.

LO 19.1.3

atrium (plural = atria)

19.1b Chambers and Circulation through the Heart

The human heart consists of four chambers: The left side and the right side each have one superior **atrium (plural: atria)** above an inferior **ventricle** (Figure 19.3A). The atria act as receiving chambers; they receive venous blood and contract to push the blood into the ventricles. The ventricles are the primary pumping chambers of the heart, propelling blood out the great arteries toward the lungs or the rest of the body.

There are two distinct systems of blood vessels in human circulation, the pulmonary and systemic circuits (Figure 19.3B). Within each of these circuits we have different types of vessels. As a rule, **arteries** carry blood from the ventricles away from the heart, and **veins** return blood to the atria. Although both circuits transport blood and everything it carries, the blood in the two circuits differs in the composition of its gasses. The **pulmonary circuit** transports blood to and from the lungs. Whereas blood leaving the heart in the pulmonary circuit is low in oxygen, or **deoxygenated**, blood returning to the heart from the lungs is **oxygenated**. The **systemic circuit** transports oxygenated blood to virtually all of the tissues of the body and returns relatively deoxygenated blood and carbon dioxide to the heart to be sent back to the pulmonary circulation.

The right ventricle pumps deoxygenated blood into the **pulmonary trunk**, which leads toward the lungs and bifurcates into the left and right **pulmonary arteries** carrying blood to each of the lungs. These vessels then branch many times, and eventually give way to tiny **pulmonary capillaries**, where gas exchange occurs: carbon dioxide exits the blood and oxygen enters. The pulmonary trunk arteries and their branches are the only arteries in the postnatal body that carry deoxygenated blood. Highly oxygenated blood returning from the pulmonary capillaries in the lungs passes through a series of vessels that join together to form the **pulmonary veins**—the only postnatal veins in the body that carry highly oxygenated blood. The pulmonary veins bring oxygenated blood from the lungs into the left atrium, which pumps the blood into the left ventricle.

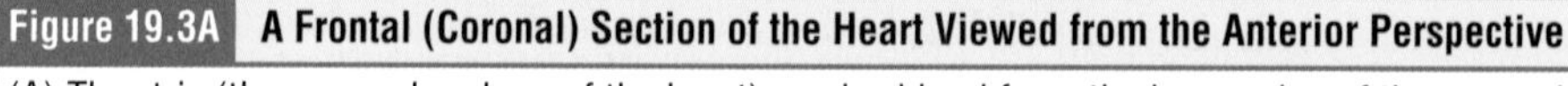

Figure 19.3A **A Frontal (Coronal) Section of the Heart Viewed from the Anterior Perspective**

(A) The atria (the upper chambers of the heart) receive blood from the large veins of the systemic and pulmonary circuits, and transport it into the ventricles (the lower chambers). The ventricles then pump the blood into the systemic and pulmonary circuits through the aorta and pulmonary trunk respectively.

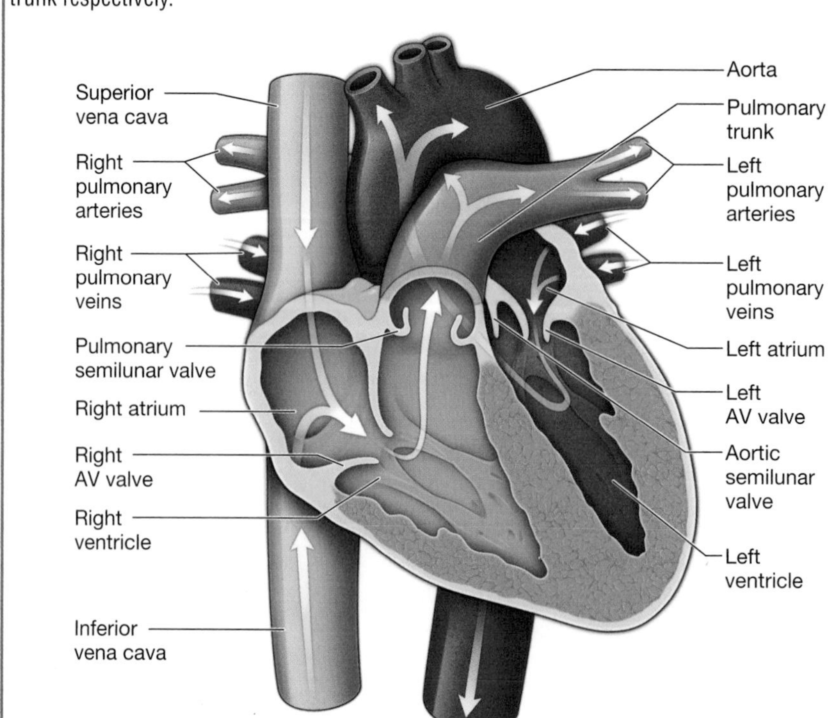

Figure 19.3B The Intertwined Pulmonary and Systemic Circuits of Blood Flow

(B) The systemic circuit transports oxygen-rich blood from the left ventricle to the body cells through the aorta; oxygen-poor blood returns to the right atrium of the heart through the venae cavae. The pulmonary circuit transports oxygen-poor blood from the right ventricle to the lungs through the pulmonary trunk; oxygen-rich blood returns to the left atrium through the pulmonary veins.

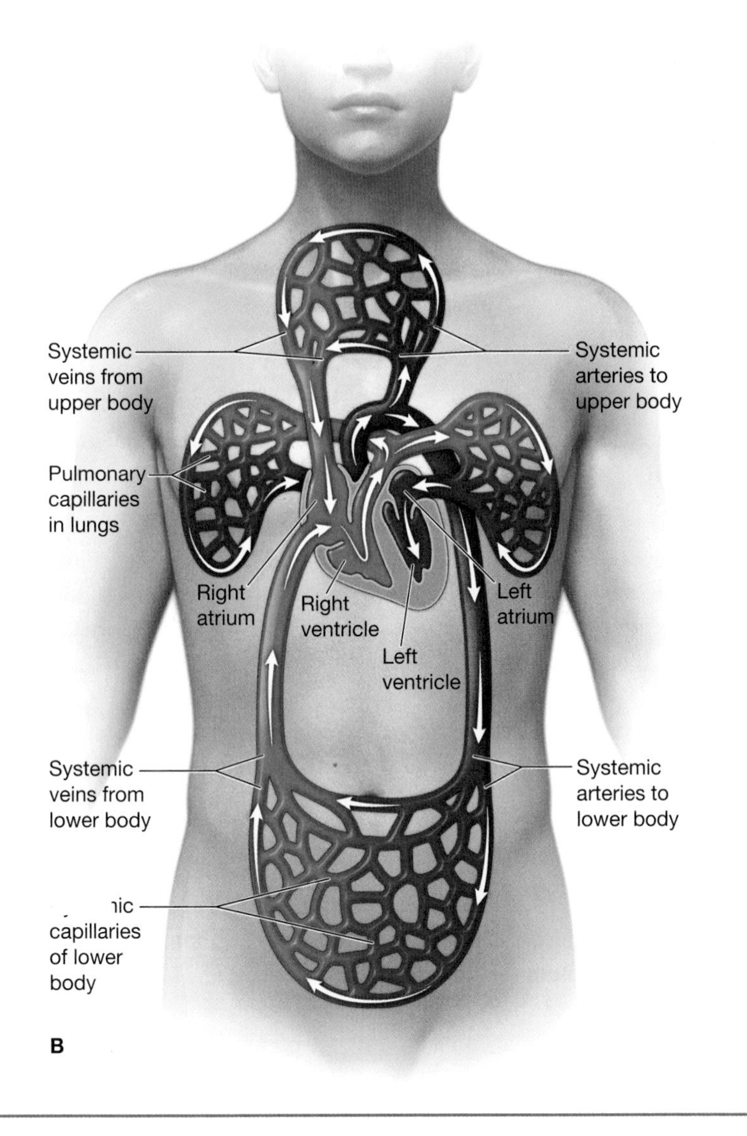

When the left ventricle contracts, it pumps oxygenated blood into the aorta and on to the many branches of the systemic circuit. Eventually, these vessels lead to the systemic capillaries, where exchange with the tissues and cells of the body occurs. In this case, oxygen and nutrients exit the blood to be used by the cells in their metabolic processes, and carbon dioxide and waste products enter the blood.

The systemic capillaries merge to form venules, joining to form ever-larger veins, eventually merging to form the two major systemic veins, the **superior vena cava** and the **inferior vena cava** (these two veins together are referred to by the plural: **venae cavae**), which return blood to the right atrium. This deoxygenated blood in the right atrium then flows into the right ventricle. Thus, we can consider the right side of the heart as the location that receives deoxygenated blood from the systemic circuit and sends it to the pulmonary circuit. Congruently, the left side receives oxygenated blood from the pulmonary system and sends it out to the systemic circuit.

Cultural Connection

How Do You Heal a Broken Heart?

You can eat heartily, act heartless, or feel lighthearted. We might hear a story that makes us feel heartsick or even heartbroken. The heart functions as a mechanical pump and yet we have placed so much importance on it. Why? The heart of this matter is almost universally historical. Chinese culture has a very intriguing concept of the heart (心; pronounced: "sheen") as the seat of human thought and emotion. Aristotle, an ancient Greek philosopher, hypothesized that the heart controls sensation, thought, and movement, functions we now know belong to the brain. In Ancient Egypt the heart was believed to be the seat of the soul. The heart determined the fate of Egyptians after they died. It was believed that the god Anubis would weigh the heart of a dead person. If the heart were heavy it would be considered guilty, but if it were light the owner of the heart would be allowed to go to Heaven. For this reason the heart was left in the body even after most organs were removed in the process of mummification.

Today we still hold many beliefs about the heart in cultures around the world. But it is not only privately that we may view the heart this way. Takotsubo cardiomyopathy, also known as stress cardiomyopathy or broken heart syndrome, can occur when the level of epinephrine, a stress hormone, overwhelms the heart. The left ventricle actually changes shape during this event, ballooning outwards. Takotsubo cardiomyopathy typically occurs during an acutely stressful event such as the death of a loved one, a traumatic event such as an earthquake, or in response to the administration of epinephrine. It predominantly affects biological women, for reasons that have yet to be explained.

Learning Connection

Micro to Macro

Here we are moving from a macro examination to a micro examination as we introduce more structures. Try putting them in order from superficial to deep, in the order in which you would encounter them if you were dissecting a cadaver.

19.1c Membranes, Surface Features, and Layers

Our exploration of heart structures begins by examining the membranes that surround the heart, the prominent surface features of the heart, and the layers that form the wall of the heart. Each of these components plays its own unique role in terms of function.

LO 19.1.4

pericardium (pericardial sac)

Membranes The **pericardium** consists of two distinct sublayers: the sturdy outer fibrous pericardium and the inner serous pericardium (**Figure 19.4**). The fibrous pericardium is made of tough, dense, irregular connective tissue that protects the heart and maintains its position in the thorax. The fibrous pericardium is attached inferiorly to the diaphragm and superiorly to the great vessels. The more delicate serous pericardium consists of two layers: the parietal pericardium, a simple squamous epithelial layer which is fused to the inner side of the fibrous pericardium, and a deeper visceral pericardium, or **epicardium**, which is bound to the heart and is part of the heart wall (see Figure 19.4); the epicardium is composed of a simple squamous epithelium and a thin layer of areolar connective tissue. The serous pericardium, like other serous membranes, forms when a fluid-filled sac folds in on itself (**Figure 19.5**). Therefore, the two serous pericardial membranes are separated by a fluid-filled **pericardial cavity**. The two serous membranes themselves produce about 50 mL (3.5 tablespoons) of serous fluid, which serves to reduce friction when the heart contracts. As blood returns to the heart from the veins, the heart muscle relaxes against the pericardial layers; in this way the pericardium is important in preventing the heart from overfilling.

LO 19.1.5

sulcus (plural = sulci)

Surface Features of the Heart If the pericardium is removed, the features of the surface of the heart are visible. There is a superficial pouchlike extension of each atrium called an **auricle**—a name that means "earlike"—because its shape resembles a folded-over dog ear (**Figure 19.6A**). Auricles are relatively thin-walled structures connected to each atrium that can lend each atrium extra capacity, filling with blood when the amount of blood returning to the heart from the veins is excessive. (We will explore the factors that determine venous return in Chapter 20.) Also prominent along the surfaces of the heart is a series of grooves, each of which is known as a **sulcus** (plural = sulci).

Figure 18.1 Whole Blood

(A) Blood is a mixture of plasma, erythrocytes, leukocytes, platelets, and dissolved substances. (B) The components can be separated by their densities. This process will occur slowly if blood is left in a tube or bag, but it can be sped up by centrifugation. (C) After centrifugation or settling, the blood separates into three components: plasma (containing the dissolved substances) in the top of the tube or bag, leukocytes (white blood cells) and platelets in the middle, and red blood cells at the bottom. The small middle layer is often called the *buffy coat.*

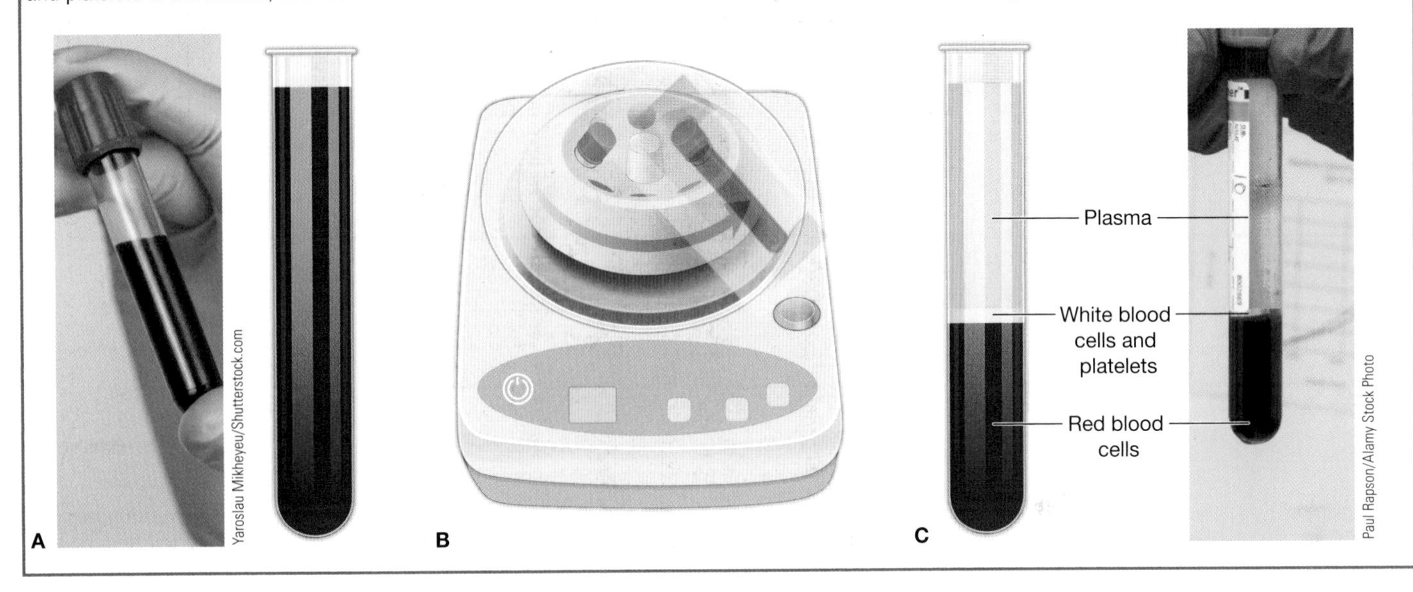

if left sitting over the course of a day, or spun quickly in a machine called a *centrifuge* (**Figure 18.1B**), the components of blood, plasma, WBC and platelets, and RBCs will separate because they have different densities (**Figure 18.1C**). Plasma, being the least dense, rises to the top. RBCs are the densest and settle at the bottom of the tube. A thin layer of WBCs and platelets separates the plasma from the RBCs.

LO 18.1.4

Throughout the study of anatomy and physiology, we spend a lot of time discussing the movement of individual materials through the processes of diffusion, osmosis, endocytosis, and exocytosis. We often refer to a concentration gradient as being a driving force for diffusion and osmosis. It is important to note the difference here when we discuss whole blood. The movement of blood through the circulatory system is based on a **pressure gradient**; it will move from where there is more pressure (the contracting heart) to where there is less pressure. Since every component of blood is equally subject to that pressure, the fluid moves all together, in bulk. The term **bulk flow** refers to the movement of a mixture according to its pressure gradient (see the "Anatomy of Flow" feature).

Student Study Tip

Bulk flow is like pouring a bowl of soup straight into the sink: all of the chicken, noodles, and so on from the soup all move together.

Whole blood is viscous and somewhat sticky to the touch. Viscosity, or thickness, is one factor in its resistance to flow. You can illustrate this for yourself by sampling two fluids through a straw. If you use the same straw to drink water and then place it in a milkshake or smoothie, you can notice how much more difficult it is to move the more viscous fluid through the straw. Blood travels through strawlike tubes—the blood vessels—and therefore its viscosity is one determining factor in its flow. The viscosity of blood is influenced by the presence of the plasma proteins and formed elements within the blood.

18.1c Plasma

Do you know your body weight? Approximately 8 percent of adult body weight is the blood volume. As you can imagine, smaller individuals have a lower blood volume and larger individuals have a higher blood volume. Other changes to the blood occur due

Anatomy of...

Flow

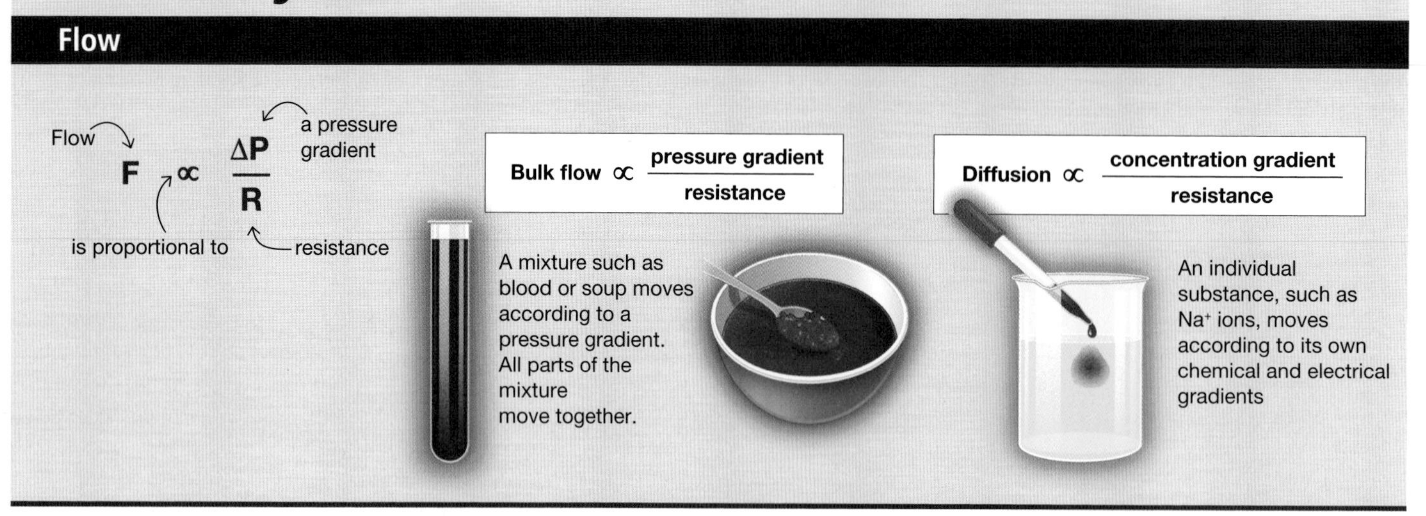

to anatomical and physiological differences as we will discuss throughout the chapter. Individuals who menstruate regularly may have slightly lower blood volumes than those who do not menstruate. Adult blood volume ranges from four to six liters; for consistency, we will use five liters as our reference value throughout the text. Of that five liters of whole blood volume, approximately three liters is blood plasma.

LO 18.1.5

Like other fluids in the body, plasma is composed primarily of water; in fact, it is about 92 percent water. Dissolved or suspended within this water is a mixture of substances, including many proteins. There are literally hundreds of substances dissolved or suspended in the plasma, although many of them are found only in very small quantities. There could not be one consistent and true list of the substances within plasma, because the plasma is a reflection of the current homeostatic state of the blood. If you have just eaten lunch, for example, your plasma may contain different substances at this moment than it will in the middle of the night while you are sleeping (and therefore not eating or digesting).

Plasma Proteins About 7 percent of the volume of plasma—nearly all that is not water—is made of proteins. These include several plasma proteins (proteins that are unique to the plasma), plus a much smaller number of other proteins, including enzymes and hormones. The major components of plasma are summarized in **Table 18.1**. The three major groups of plasma proteins are as follows:

LO 18.1.6

- **Albumins** are the most abundant of the plasma proteins. Made by the liver, albumins serve as transport vehicles for fatty acids and steroid hormones. Recall that lipids are hydrophobic and therefore would not dissolve readily in the watery plasma, however, their binding to albumin enables them to travel in the blood. Albumin is also a regulator of osmotic pressure of blood; that is, its presence helps to retain water inside the blood vessels and prevents too much water from being drawn toward the tissues. This in turn helps to maintain both blood volume and blood pressure. Albumin is not just an abundant protein in human bodies, but it is abundant throughout the animal kingdom. Egg whites, for example, are almost entirely composed of albumin.
- The second most common plasma proteins are the **globulins**. A heterogeneous group, there are three main subgroups known as *alpha*, *beta*, and *gamma* globulins. The alpha and beta globulins are transport proteins; they help shuttle iron, lipids, and the fat-soluble vitamins to the cells. Like albumin, they also contribute to osmotic pressure. The gamma globulins are proteins involved in immunity and are better known as **antibodies** or **immunoglobulins**. Although other plasma proteins

antibodies (immunoglobulins, gamma globulins)

Table 18.1 Major Blood Components

Component and percentage of blood	Subcomponent and percentage of component	Type and percentage (where appropriate)	Site of production	Major function(s)
Plasma: 46–63 percent	Water: 92 percent	Fluid	Absorbed by intestinal tract or produced by metabolism	Transport medium
	Plasma proteins: 7 percent	Albumin: 54–60 percent	Liver	Maintain osmotic concentration, transport lipid molecules
		Globulins: 35–38 percent	Alpha globulins: liver	Transport, maintain osmotic concentration
			Beta globulins: liver	Transport, maintain osmotic concentration
			Gamma globulins (immunoglobulins): plasma cells	Immune responses
		Fibrinogen: 4–7 percent	Liver	Blood clotting in hemostasis
	Regulatory proteins: <1 percent	Hormones and enzymes	Various sources	Regulate various body functions
	Other solutes: 1 percent	Nutrients, gasses, and wastes	Absorbed by intestinal tract, exchanged in respiratory system, or produced by cells	Numerous and varied
Formed elements: 37–54 percent	Erythrocytes: 99 percent	Erythrocytes	Red bone marrow	Transport gasses, primarily oxygen and some carbon dioxide
	Leukocytes: <1 percent Platelets: <1 percent	Granular leukocytes: neutrophils Eosinophils basophils	Red bone marrow	Nonspecific immunity
		Agranular leukocytes: lymphocytes monocytes	Lymphocytes: bone marrow and lymphatic tissue Monocytes: red bone marrow	Lymphocytes: specific immunity Monocytes: nonspecific immunity
	Platelets: <1 percent		Megakaryocytes: red bone marrow	Hemostasis

are produced by the liver, immunoglobulins are produced by leukocytes. Antibodies and how they are made will be discussed thoroughly in Chapter 21.

- The least abundant plasma protein is **fibrinogen**. Like albumin and the alpha and beta globulins, fibrinogen is produced by the liver. It is essential for blood clotting, a process described in Section 18.3.

Other Plasma Solutes In addition to proteins, plasma contains a wide variety of other substances. These include various ions, such as sodium, potassium, and calcium; dissolved gasses, oxygen, carbon dioxide, and nitrogen; various organic nutrients, such as vitamins, lipids, glucose, and amino acids; and metabolic wastes. Combined, all of these nonprotein solutes contribute approximately 1 percent to the total volume of plasma.

Learning Connection

Chunking

List all of the plasma proteins. Then group them in different ways. Which plasma proteins function in maintaining osmotic pressure? Which of them play transport roles? Which plasma proteins are important during bleeding and clotting? Which plasma proteins are associated with immune defenses?

erythrocytes (red blood cells or RBCs)

18.1d Erythrocytes

Erythrocytes, commonly known as *red blood cells (or RBCs)*, are by far the most common formed element. A single drop of blood contains millions of erythrocytes and just thousands of leukocytes. In fact, erythrocytes are estimated to make up about 25 percent of the total cells in the body. They are one of the smallest cell types in the human body. The function of erythrocytes is to transport gasses. Erythrocytes can carry several gasses—including carbon monoxide and carbon dioxide, as well as hydrogen ions—but their primary function is to carry oxygen.

LO 18.1.7

Erythrocytes are sleek and efficient carriers of blood gasses. They have few organelles and no nucleus. Without much in the way of internal contents, the shape of an erythrocytes can be described as a biconcave disc; that is, they are plump at their periphery and very thin in the center (**Figure 18.2**). Since they lack most organelles, there is more interior space for the presence of the hemoglobin molecules that transport gasses. The biconcave shape also provides a greater surface area across which gas exchange can occur, relative to its volume; a sphere of a similar diameter would have a lower surface area-to-volume ratio. Capillaries, the smallest blood vessels, are extremely narrow, slowing the passage of the erythrocytes and providing an extended opportunity for gas exchange to occur. However, the space within capillaries can be so minute that, despite their own small size, erythrocytes may have to fold themselves like a taco in order to make their way through.

LO 18.1.8

Hemoglobin **Hemoglobin** is a large molecule made up of proteins and iron. It consists of four folded chains of a protein called **globin**, designated alpha 1 and 2, and beta 1 and 2 (**Figure 18.3**). Each of these globin molecules is bound to a red pigment molecule called **heme**, which contains an ion of iron (Fe^{2+}).

To give you an idea of the vastness of the human blood supply, each iron ion in the heme can bind to one oxygen molecule; therefore, each hemoglobin molecule can transport four oxygen molecules. An individual erythrocyte may contain about 300 million hemoglobin molecules, and therefore can bind to and transport up to 1.2 billion oxygen molecules.

When hemoglobin molecules are in oxygen-rich environments, they will bind oxygen molecules. However, in oxygen-poor environments, hemoglobin's affinity for oxygen decreases and oxygen will readily fall off the hemoglobin molecule. Therefore,

Figure 18.2 Red Blood Cell Structure

Red blood cells can be described as bioconcave discs, meaning that they are relatively flat with a curve inward on each side. This shape allows them a maximum surface area-to-volume ratio; it also allows them to bend as they travel through blood vessels.

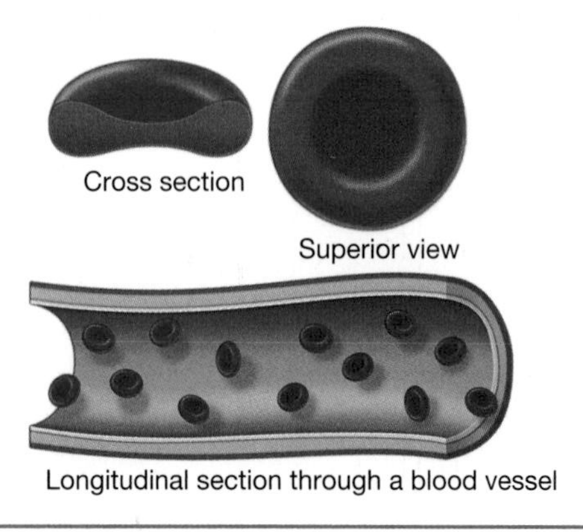

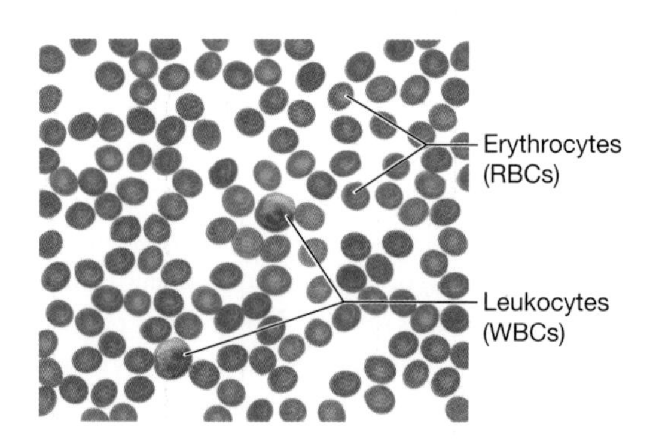

Figure 18.3 **Hemoglobin Structure**

Hemoglobin is the gas-carrying protein inside red blood cells. It has four subunits, each containing an iron ion, which lends hemoglobin a high affinity for oxygen.

β chain 1
Fe^{2+}
Heme
β chain 2
α chain 1
α chain 2

when the hemoglobin—and therefore the blood environment—is carrying more oxygen, we can describe the blood as being **oxygenated**. Blood is the most oxygenated as it leaves the lungs, where hemoglobin picks up oxygen. The oxygenated blood then travels to the body tissues, where it releases some of the oxygen molecules, becoming **deoxygenated** (deoxygenated hemoglobin is sometimes referred to as *reduced hemoglobin*). Oxygen release depends on the need for oxygen in the surrounding tissues, so hemoglobin rarely, if ever, leaves all of its oxygen behind; in fact, "deoxygenated" blood is usually still 70–80 percent oxygenated.

oxygenated (oxygen-rich)

deoxygenated (oxygen-poor)

DIGGING DEEPER:
Pulse Oximetry

The saturation of the blood with oxygen can be measured in an easy and noninvasive fashion by a device known as a *pulse oximeter*, which is applied to a thin part of the body, typically the tip of the patient's finger (see figures at right). The device works by sending two different wavelengths of light (one red, the other infrared) through the finger and measuring the light with a photodetector as it exits. Depending on its oxygen saturation, hemoglobin absorbs light differentially. The machine can read how much light was absorbed and how much is reflected back. Saturation levels of 95–100 percent are considered healthy pulse oximeter readings. Lower percentages reflect **hypoxemia**, or low blood oxygen. Monitoring oxygen levels in the blood is critical in the care of individuals with lung infections or diseases such as asthma or COVID. Pulse oximeter readings are often used to determine when a patient needs supplemental oxygen or to be put on a ventilator. Like many medical studies and devices, human diversity may not have played a big enough role in the development of these critical devices. The U.S. Food and Drug Administration, as well as several medical journals—including the *New England Journal of Medicine*—have recently warned that pulse oximeters may not be accurate enough in Black patients or patients with darker skin. Better technology must be developed in order to ensure that all patients are monitored correctly and given accurate life-saving care.

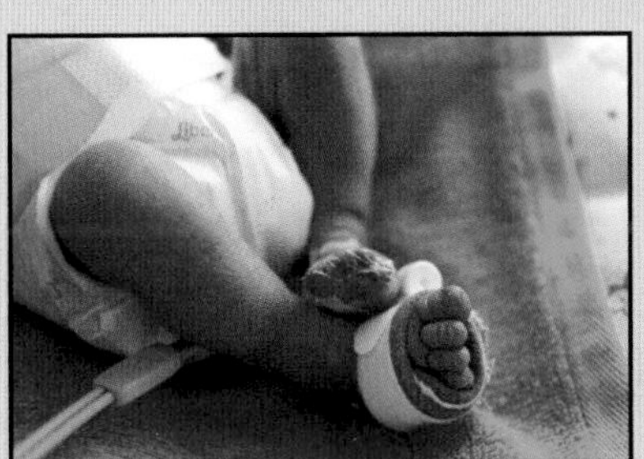

Kristina Bessolova/Shutterstock.com

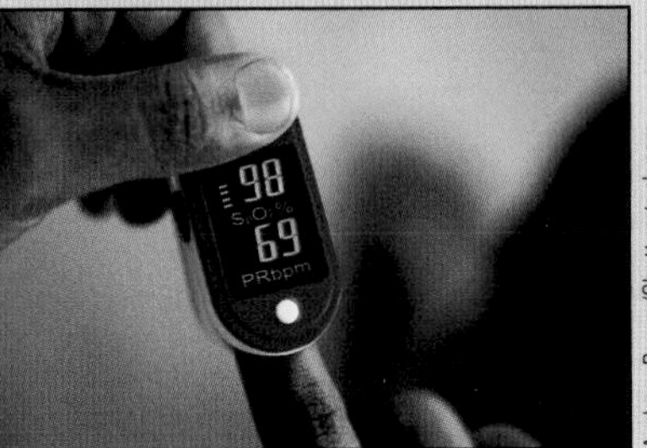

Andrey_Popov/Shutterstock.com

Changes in the levels of RBCs or the number of hemoglobin molecules they carry can have significant effects on the body's ability to effectively deliver oxygen to the tissues. Any condition that results in insufficient numbers of RBCs or insufficient ability to carry oxygen can be called **anemia**. There are many different physiological issues that can lead to insufficient RBC numbers, detailed in the "Digging Deeper: Anemias" feature in this chapter. An overproduction of RBCs produces a condition called **polycythemia**. The primary drawback with polycythemia is not a failure to directly deliver enough oxygen to the tissues, but rather the increased viscosity of the blood. Since blood flow is opposed by resistance, and one factor of resistance is viscosity, polycythemia makes it more difficult for the heart to circulate the blood throughout the circulatory system.

The kidneys are responsible for filtering some wastes out of the blood and producing urine. So much blood flows through the kidneys at any time that by then end of the day they have filtered about 180 liters (~380 pints) of blood if we are considering an adult with 5 liters of total blood volume. Thus, the kidneys serve as ideal sites for receptors to monitor if the blood is sufficiently saturated with oxygen. In other words, if, over the course of a day, the kidneys were receiving blood that was consistently low in oxygen, then it stands to reason that the blood is not carrying enough oxygen for the body's daily needs. In response to consistently low oxygen saturation in the blood, the kidneys will secrete **erythropoietin (EPO)**, a hormone that promotes the generation of new erythrocytes in the bone marrow. As long as the other necessary ingredients for **erythropoiesis**, the making of new RBCs, are available in the bone marrow, new RBCs will be produced, thereby increasing the oxygen-carrying capacity of the blood. Iron and vitamin B_{12} are two important components to the erythropoiesis process. In a classic negative feedback loop, as oxygen saturation rises, EPO secretion falls, and vice versa, thereby maintaining homeostasis (**Figure 18.4**).

Figure 18.4

The kidneys receive a tremendous amount of blood flow each day. When they detect a reduction in oxygen-carrying capacity, they secrete erythropoietin, a hormone that induces erythropoiesis—the synthesis of new red blood cells—in the red bone marrow.

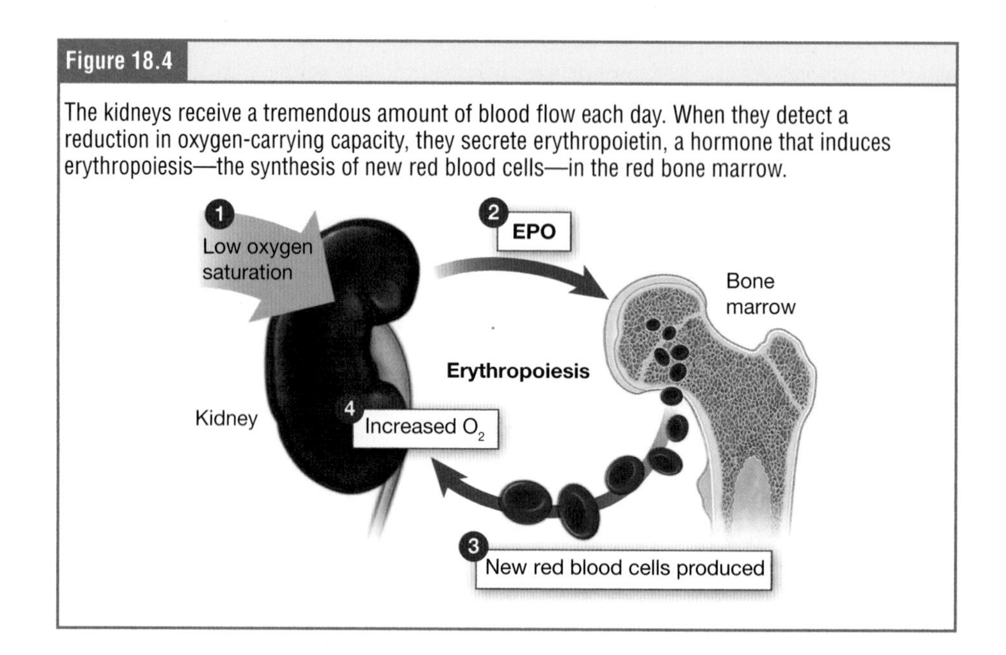

Cultural Connection

Blood Doping

There is a relationship between altitude or elevation and the amount of oxygen available in atmospheric air. We will explore this relationship in greater detail in Chapter 22. People traveling to high elevations may experience symptoms of hypoxemia, such as fatigue, headache, and shortness of breath, for a few days after their arrival. In response to the hypoxemia, the kidneys secrete erythropoietin (EPO) to step up the production of erythrocytes until homeostasis is achieved once again. To avoid the symptoms of hypoxemia, or altitude sickness, mountain climbers typically rest for a period ranging from several days to a week or more at a series of camps situated at increasing elevations to allow EPO levels and, consequently, erythrocyte counts, to rise. Some athletes will train at higher altitudes to increase their hematocrits before their races or athletic performances. With higher hematocrits, the athletes' bodies will be able to deliver more oxygen to their muscles at the peak intensity of their athletic performance. There are other, illicit methods of increasing one's hematocrit. Some athletes have been known to inject more RBCs into their bodies prior to a race or competition. Others may abuse exogenous EPO to enhance their own erythropoietic capacity. Lance Armstrong, one of the most famous athletes in cycling history, was discovered to have abused EPO; consequently, he was stripped of many of his achievements. The International Olympic Committee has long screened athletes for EPO and many athletes have either been denied the opportunity to compete or stripped of their Olympic medals after testing positive for EPO doping.

LO 18.1.9

hematocrit (packed cell volume)

While the kidneys are constantly monitoring the oxygen-carrying capacity of the blood, clinically we estimate this capacity using the **hematocrit**, which measures the percentage of RBCs in a blood sample. When blood is separated into its components as depicted in Figure 18.1, the percentage of total blood volume that is composed of erythrocytes can be measured. Hematocrit can vary widely among individuals; however, a range of about 36–50 percent is considered healthy. On average, individuals with more muscle mass tend to have higher hematocrits because their bodies have a higher demand for oxygen. Menstruation is another factor that can impact hematocrit; whereas individuals who menstruate have lower average values—often between 37 and 47—individuals who do not menstruate range more typically between 42 and 52. Another factor is athleticism; runners and bikers typically have higher hematocrits than do those who live at higher altitudes. The percentage of other formed elements, the WBCs and platelets, is extremely small, so it is not normally considered with the hematocrit.

18.1e The Life Cycle of Erythrocytes

Production of erythrocytes occurs in the bone marrow. Although adults only have red bone marrow—the type of bone marrow capable of erythropoiesis—at the ends of long bones and in some of their irregular and flat bones, the average adult produces somewhere around 2 million new RBCs per second. That means you probably made about 20 million RBCs while you were reading the last sentence of this text! For this production to occur, a number of raw materials must be present in adequate amounts. These include the same nutrients that are essential to the production and maintenance of any cell, such as glucose, lipids, and amino acids. However, erythrocyte production also requires several trace elements:

- Iron. We have said that each heme group in a hemoglobin molecule contains an ion of iron. Iron is relatively uncommon in foods and humans have limited ability to absorb the iron in our diets. In fact, less than 20 percent of the iron we consume is absorbed, and we absorb even less if our iron is in plant sources rather than meat. Therefore, humans spend energy recycling the iron that is already in the body. The bone marrow, liver, and spleen can store iron in the protein compounds **ferritin** and **hemosiderin**. When EPO stimulates the production of erythrocytes, iron is released from storage, bound to the blood protein **transferrin**, and carried to the red marrow, where it can be added to hemoglobin.

- B vitamins. The B vitamins folate and vitamin B_{12} facilitate DNA synthesis. Thus, both are critical for the synthesis of new cells, including erythrocytes.

Erythrocytes live up to 120 days in the circulation, after which the worn-out cells are removed by the liver and spleen. Typically, the rate of breakdown of old erythrocytes balances the production of erythrocytes in order to maintain a constant hematocrit. The following text and **Figure 18.5** summarize the further breakdown of the components of the degraded erythrocytes' hemoglobin.

- Globin—the protein portion of hemoglobin—is broken down into amino acids, which can be sent back to the bone marrow to be used in the production of new erythrocytes. Hemoglobin that is not phagocytized is broken down in the circulation, releasing alpha and beta chains that are removed from circulation by the kidneys.
- The iron contained in the heme portion of hemoglobin may be stored in the liver or spleen, primarily in the form of ferritin or hemosiderin, or carried through the bloodstream by transferrin to the red bone marrow for recycling into new erythrocytes.
- The non-iron portion of heme is degraded into the waste product **biliverdin**, a green pigment, and then into another waste product, **bilirubin**, a yellow pigment. Bilirubin is removed from the blood by the liver, which uses it in the manufacture of bile, a compound released into the intestines to help emulsify dietary fats. It is then eliminated from the body in the feces. The brown color of feces is due to the breakdown products of bilirubin within.

The breakdown pigments formed from the destruction of hemoglobin can be seen under the skin during bruising in many individuals. At the beginning of the bruise,

Figure 18.5 Red Blood Cell Recycling

Step1: New erythrocytes are made every second in the red bone marrow. Step 2: Erythrocytes circulate for approximately four months. Step 3: The liver and spleen are responsible for removing old red blood cells from circulation and recycling their components. Step 4: Heme, in particular, is conserved through careful recycling. The iron ions are cleaved off and transported by transferrin to storage sites; the remaining portions of the heme groups are converted first to biliverdin and then to bilirubin.

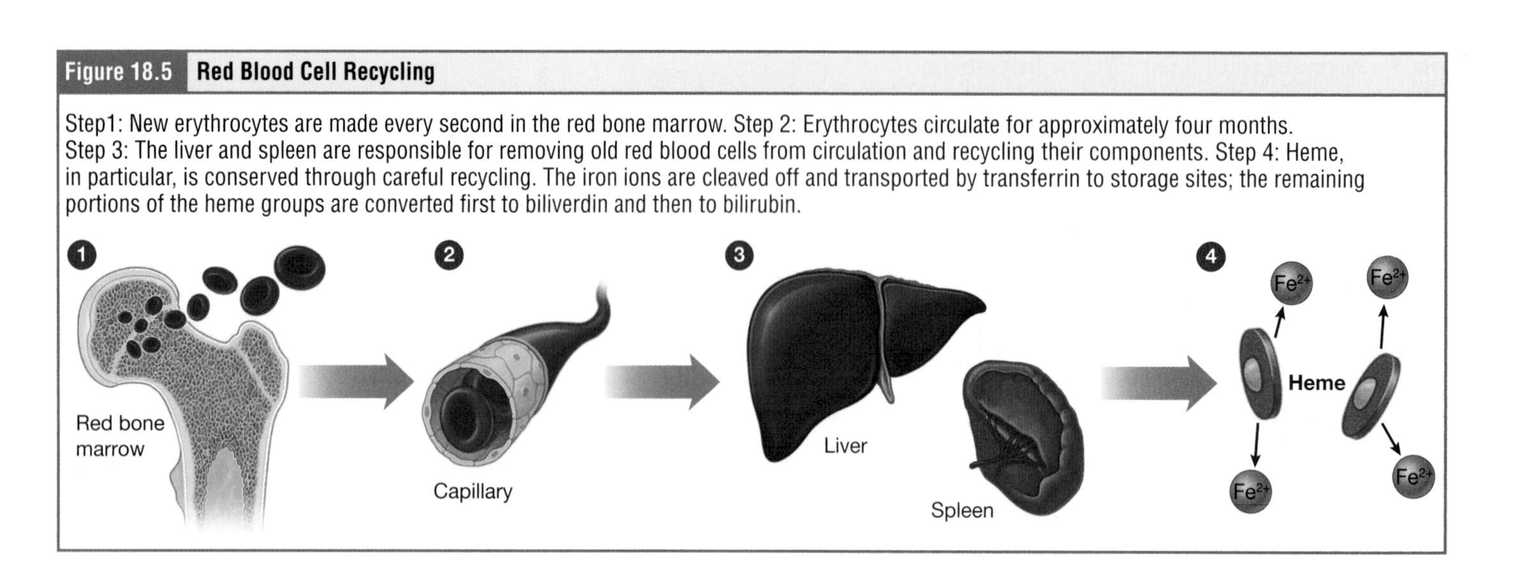

blood collects under the skin due to broken blood vessels. Over time, local cells break down and recycle the hemoglobin. As the hemoglobin is converted to biliverdin, the bruise may turn various shades of green. As it is further converted to bilirubin, the bruise may appear yellow.

sickle cell disease (sickle cell anemia)

DIGGING DEEPER: Anemias

The size, shape, and number of erythrocytes, and the number of hemoglobin molecules, can all have a major impact on a person's health. When the number of RBCs or hemoglobin is deficient, the general condition is called *anemia*. There are more than 400 types of anemia, and more than 3.5 million Americans suffer from this condition. Anemia can be broken down into three major groups: those caused by blood loss, those caused by faulty or decreased RBC production, and those caused by excessive destruction of RBCs. Normal-sized cells are referred to as *normocytic*, smaller-than-normal cells are referred to as *microcytic*, and larger-than-normal cells are referred to as *macrocytic*. The effects of the various anemias are widespread because reduced numbers of RBCs or hemoglobin will result in lower levels of oxygen being delivered to body tissues. Since oxygen is required for tissue functioning, anemia produces fatigue, lethargy, and an increased risk for infection. An oxygen deficit in the brain impairs the ability to think clearly and may prompt headaches and irritability. Lack of oxygen leaves the patient short of breath, even as the heart and lungs work harder in response to the deficit.

Blood loss anemias may occur due to bleeding from wounds or ulcers, hemorrhoids, inflammation of the stomach, and some cancers of the gastrointestinal tract.

Anemias caused by faulty or decreased RBC production include **sickle cell disease** (discussed in the following "Apply to Pathophysiology" feature), iron deficiency anemia, vitamin deficiency anemia, and diseases of the bone marrow and stem cells.

- Iron deficiency anemia is the most common type and results when the amount of available iron is insufficient to allow production of sufficient heme. This condition can occur in individuals with a deficiency of iron in the diet and is especially common in teens and children as well as in vegans and vegetarians. Additionally, iron deficiency anemia may be caused by an inability to either absorb or transport iron.
- Vitamin-deficient anemias generally involve insufficient vitamin B_{12} and folate.
  - Megaloblastic anemia involves a deficiency of vitamin B_{12} and/or folate, and often involves diets deficient in these essential nutrients. Lack of meat or a viable alternate source, and overcooking or eating insufficient amounts of vegetables, may lead to a lack of folate.
  - Pernicious anemia is caused by poor absorption of vitamin B_{12} and is often seen in patients with Crohn's disease (a severe intestinal disorder often treated by surgery), intestinal parasites, or AIDS, as well as in those who have undergone surgical removal of the intestines or stomach (common in some weight loss surgeries).
  - Pregnancies, some medications, excessive alcohol consumption, and some diseases such as celiac disease are also associated with vitamin deficiencies. It is essential to provide sufficient folic acid during the early stages of pregnancy to reduce the risk of neurological defects, including spina bifida, a failure of the neural tube to close.
- Assorted disease processes can also interfere with the production and formation of RBCs and hemoglobin. If myeloid stem cells are defective or replaced by cancer cells, there will be insufficient quantities of RBCs produced.
- Aplastic anemia is the condition in which there are deficient numbers of RBC stem cells. Aplastic anemia is often inherited, or it may be triggered by radiation, medication, chemotherapy, or infection.
- **Thalassemia** is an inherited condition typically occurring in individuals from the Middle East, the Mediterranean, African, and Southeast Asia, in which maturation of the RBCs does not proceed normally. The most severe form is called *Cooley's anemia*.
- Lead exposure, whether from industrial sources, dust from paint chips of lead-containing paints, or even pottery that has not been properly glazed, may also lead to destruction of the red marrow.
- Various disease processes also can lead to anemias. These include chronic kidney diseases often associated with a decreased production of EPO, hypothyroidism, some forms of cancer, lupus, and rheumatoid arthritis.

Apply to Pathophysiology

Sickle Cell Disease

A characteristic change in the shape of erythrocytes is seen in sickle cell disease (also referred to as *sickle cell anemia*). A genetic disorder, it is caused by production of an abnormal type of hemoglobin, called *hemoglobin S*, which delivers less oxygen to tissues and causes erythrocytes to assume a sickle (or crescent) shape, especially at low oxygen concentrations. These abnormally shaped cells can then become lodged in narrow capillaries because they are unable to fold in on themselves to squeeze through, blocking blood flow to tissues and causing a variety of serious problems, ranging from painful joints to delayed growth and even blindness and cerebrovascular accidents (strokes). Sickle cell anemia is a genetic condition; in particular, it is found in individuals of African descent. Sickle cell traits likely persisted in African populations because those carrying one or both copies of the sickle cell trait are protected from malaria. Individuals with just one copy, heterozygotes, are protected from malaria but do not have significant sickle cell disease symptoms. Individuals with two copies, homozygotes, are also protected from malaria, but have particularly serious sickle cell disease symptoms.

18.1f Leukocytes and Platelets

Leukocytes, commonly known as *white blood cells (WBCs)*, are the soldiers of the immune response. Leukocytes protect the body against invading microorganisms, eliminate body cells with mutated DNA, and clean up debris. See the "Anatomy of Leukocytes" feature for a summary of leukocyte types. It is helpful to note that only a tiny fraction of the body's leukocyte population can be found in the blood. These cells spend most of their time in other structures and tissues of the body. We will explore leukocytes in much more detail in Chapter 21. Platelets are essential for the repair of blood vessels when damage to them has occurred; they also provide growth factors for healing and repair.

hematopoiesis (hemopoiesis)

Characteristics of Leukocytes Leukocytes and erythrocytes both originate from stem cells in the bone marrow; the term for the production of these blood cells is *hematopoiesis*. Leukocytes and erythrocytes are very different from each other in many significant ways. For instance, leukocytes are far less numerous than erythrocytes: Recall that all of the leukocytes together represent just a thin line of the hematocrit. Leukocytes are much larger than erythrocytes and are the only formed elements of the blood that are complete cells, possessing a nucleus and organelles. And although there is just one type of erythrocyte, there are many types of leukocytes. While erythrocytes are regularly replaced after 120 days on average, leukocytes have variable lifespans depending on their type and the current immune responses ongoing in the body. Some leukocyte types live only few hours (or even a few minutes in the case of acute infection), while others can circulate for years.

One of the most distinctive characteristics of leukocytes is their movement. Whereas erythrocytes spend their days circulating within the blood vessels, leukocytes routinely leave the bloodstream to perform their defensive functions in the body's tissues. For leukocytes, the vascular network is simply a highway they travel and soon exit to reach their true destination. As shown in **Figure 18.6**, they leave the capillaries—the smallest blood vessels—through a process known as **diapedesis**, in which they squeeze between the cells that make up the blood vessel wall.

diapedesis (emigration)

Once they have exited the capillaries, some leukocytes will take up residence in lymphatic tissue, bone marrow, the spleen, the thymus, or other organs. Others will move about through the tissue spaces wandering freely or moving in the direction of chemical signals. Leukocytes may call to each other to recruit help when fighting an

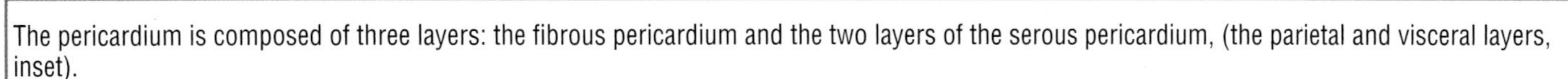

Figure 19.4 **Frontal Section of the Heart Within the Pericardium**

The pericardium is composed of three layers: the fibrous pericardium and the two layers of the serous pericardium, (the parietal and visceral layers, inset).

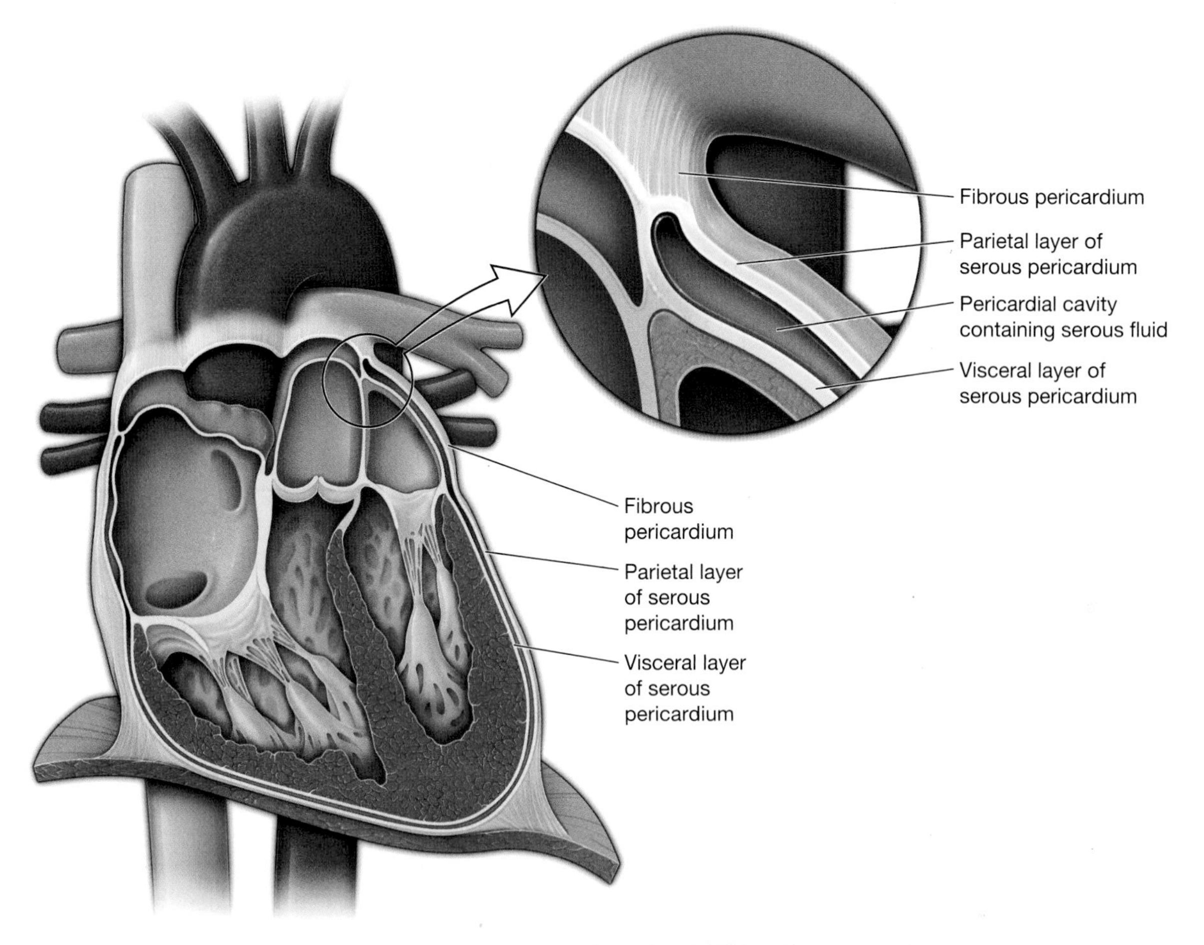

Figure 19.5 **Serous Membranes Develop Around Many Organs**

Serous membranes develop as a fluid-filled sac. The organ develops, enlarges, and pushes into the serous sac, which folds in on itself, like a fist pushing into a balloon filled with air. At the end of the developmental period, the mature organ is surrounded by a sac composed of a double membrane. The membrane closest to the organ is named the visceral layer and the membrane farther away is the parietal layer. Between them is a fluid-filled cavity.

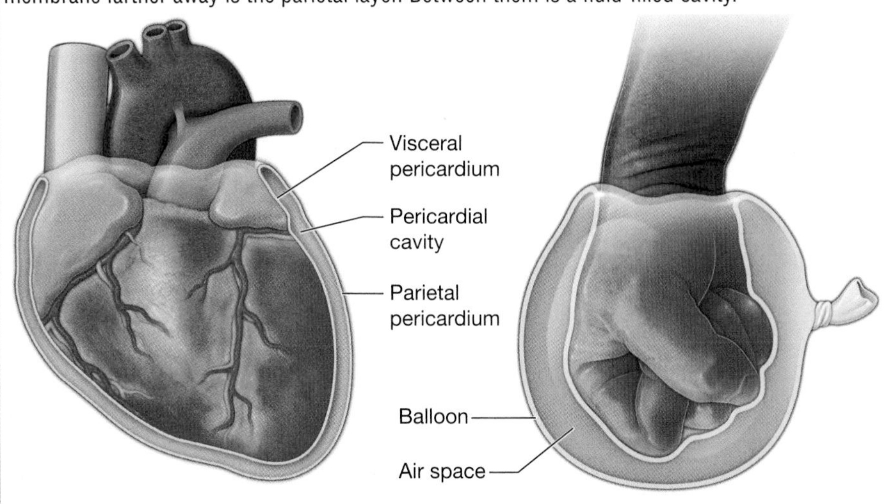

Apply to Pathophysiology

LO 19.1.6

Pericarditis

A patient is brought into the Emergency Department with a gunshot wound to the chest. X-rays reveal that the bullet missed the heart but did puncture the fibrous and parietal pericardia. Blood from the surrounding tissue damage is filling the pericardial cavity. Based on the radiologist's estimates from the X-ray image, it appears as if the pericardial cavity contains three times the normal amount of fluid.

1. What are the functions of the pericardium?
2. Which of these functions will be impacted by an accumulation of fluid in the pericardial cavity?
3. Do you think, in this patient, the heart is filling with and ejecting the same amount of blood into the systemic circuit as before the trauma?
   A) Same amount
   B) More
   C) Less
4. Removal of this excess fluid requires insertion of drainage tubes into the pericardial cavity. A medical student prepares to insert a long needle to remove the accumulated fluid. Which is the correct order of membranes that she will puncture with her needle?
   A) Visceral, parietal, fibrous
   B) Fibrous, parietal, visceral
   C) Fibrous, parietal
   D) Parietal, visceral

Learning Connection

Quiz Yourself

How does the structure and location of the auricle contribute to its function?

Blood vessels that transport blood to and from the wall of the heart (**coronary vessels**) are nestled within these sulci (Figure 19.6A–B) shows this relationship on a cadaveric heart). The **coronary sulcus** is located between the atria and ventricles. Located between the left and right ventricles are two additional sulci: the **anterior interventricular sulcus** on the anterior surface of the heart and the **posterior interventricular sulcus** on the posterior surface. By identifying these landmarks you can estimate the location of the chambers beneath. On the posterior surface of the heart, you can note that the pulmonary veins converge to make a broad, flat superior surface known as the **base** (Figure 19.6B).

19.1d Layers

LO 19.1.7

The wall of the heart is composed of three layers. From superficial to deep, these are the epicardium, the myocardium, and the endocardium (see Figure 19.4). The outermost layer of the wall of the heart, the epicardium, is also the innermost or visceral layer of the pericardium, discussed earlier.

The middle and thickest layer is the **myocardium**, composed almost entirely of cardiac muscle cells. Its other components are: a framework of collagenous fibers that provide structure, the blood vessels that supply the myocardium, and the nerve fibers that help regulate the heart. It is the contraction of the myocardium that pumps blood through the heart and into the arteries. The muscle pattern is elegant and complex to create the most effective contraction. As depicted in Figure 19.7, the heart muscle wraps each atrium; as this muscle contracts it squeezes the blood down toward the ventricles. The ventricular muscle arrangement is swirled from the apex up toward the arteries, which exit the heart superiorly. As these muscles contract, blood is squeezed up from the apex and out, like squeezing a tube of toothpaste from the bottom to the nozzle.

Although the ventricles on the right and left sides pump the same amount of blood per contraction, the muscle wall of the left ventricle is much thicker and more developed than that of the right ventricle. Like skeletal muscle, cardiac muscle cells increase in size when they need to produce more force. If we compare the systemic and pulmonary circuits, the systemic circuit is much longer and the blood within the systemic circuit is at a higher pressure. Both of these factors (length and pressure) offer resistance to the ejection of blood into the circuit. In order to overcome the high resistance required to pump blood into the systemic circuit, the left ventricle must generate

Figure 19.6A–B Anterior and Posterior Views of the Heart

(A) On the anterior view, the following structures are prominent: the large ventricles, the smaller auricles (which form the outer walls of the atria), the aorta, which emerges from the left ventricle, and the pulmonary trunk, which emerges from the right ventricle. (B) In the posterior view, the prominent structures include the pulmonary veins, which enter the left atrium, and the superior and inferior venae cavae, which enter the right atrium. The coronary blood vessels are visible in both views.

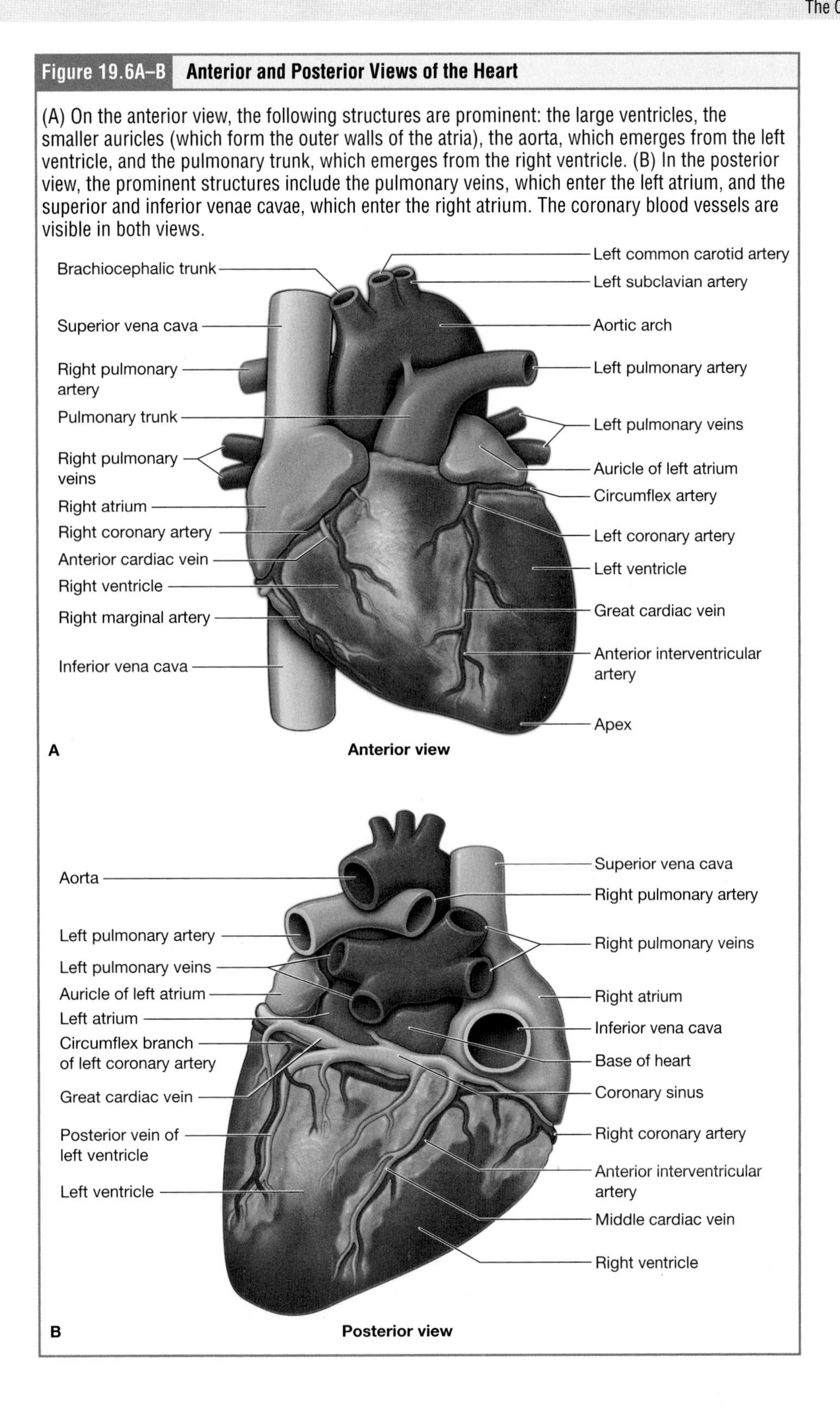

a greater amount of pressure than the right ventricle. **Figure 19.8** illustrates the differences in muscular thickness needed for each of the ventricles. The fetal heart walls have no difference in thickness (see Figure 19.2). However, as these resistance factors become more established in early life, the left ventricle **hypertrophies**, or thickens, in response to the need for more force.

Student Study Tip

Use alliteration for the sides of the heart: Left is Larger. Pair that with your understanding of hypertrophy to help you remember that the left side of the heart supplies the entire body.

Figure 19.6C Anterior Surface of the Heart, Cadaver Dissection

(C) In this anterior surface view of the heart, the left and right ventricles and the aorta are the predominant features. The fat deposits protect the coronary blood vessels that run through the anterior interventricular sulcus.

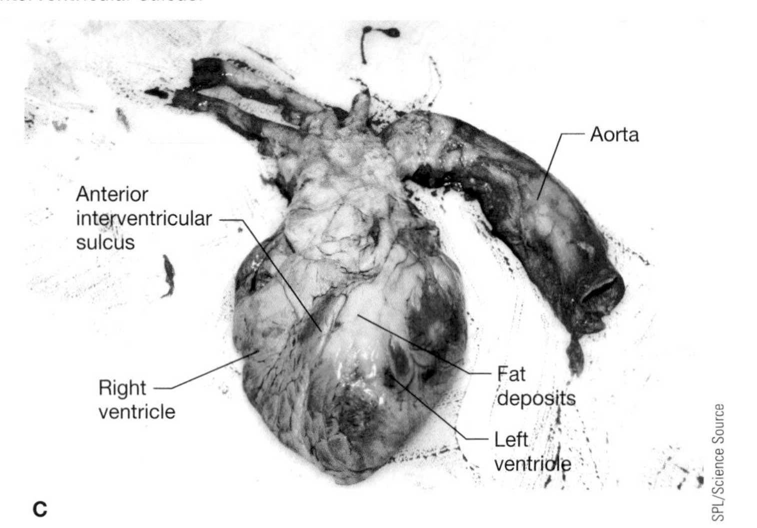

Figure 19.7 The Myocardial Fibers Are Oriented in a Spiral Pattern for Efficient Muscle Contraction

The muscles wrap around each ventricle (A) and then wrap in a spiral around the ventricles (A and B). When these spirally-oriented muscles contract they squeeze the blood inside the heart from bottom to top, forcing it toward the great vessels.

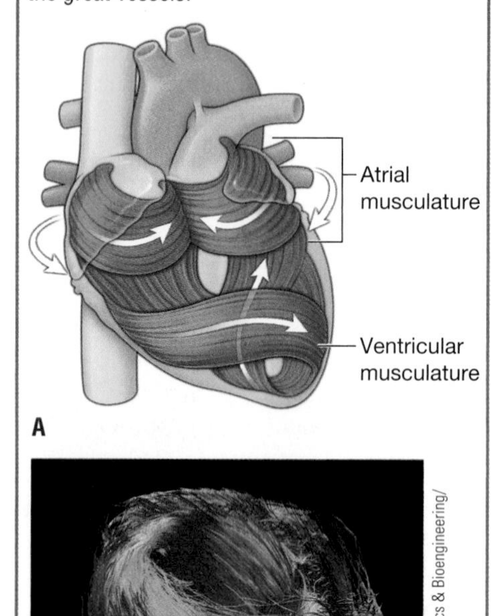

The innermost layer of the heart wall, the **endocardium**, is joined to the myocardium with a thin layer of connective tissue. The endocardium lines the chambers where the blood circulates and covers the heart valves. It is made of simple squamous epithelium called **endothelium**, which is continuous with the endothelial lining of the blood vessels (see Figure 19.4).

Figure 19.8 Transverse and Coronal Sections of the Right and Left Ventricles

(A) A transverse section and (B) a coronal section through the ventricles reveal a substantial thickness in the left ventricular walls. The right ventricle shares the thick inner wall (the interventricular septum) with the left ventricle, but its outer wall is significantly thinner.

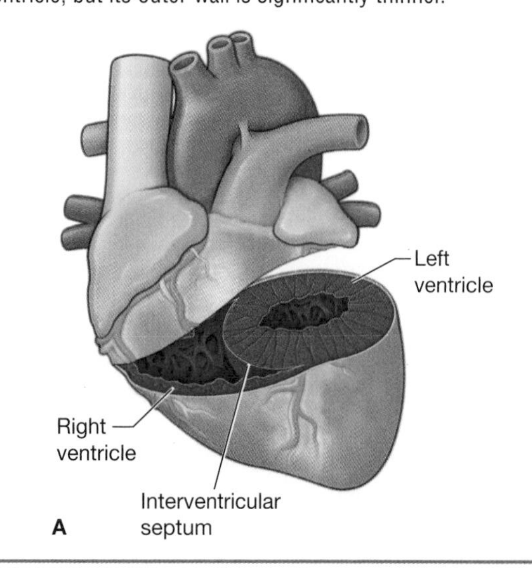

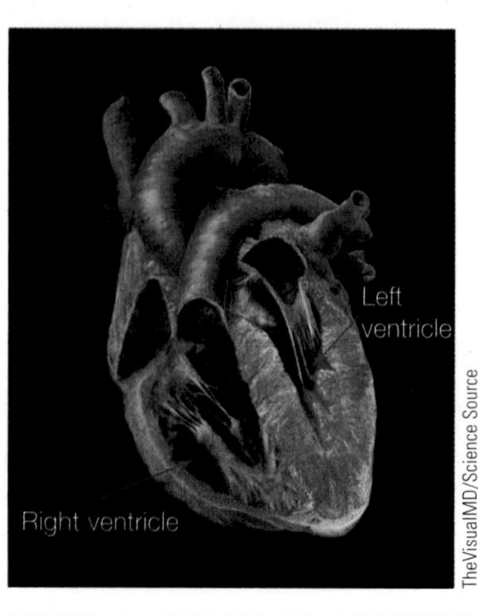

Learning Check

1. Which zone of the developing heart contributes to the formation of the ventricles?
   a. Bulbus cordis
   b. Truncus arteriosus
   c. Sinus venosus
   d. Atrium
2. Fill in the blanks: The pulmonary artery carries ______ blood while the pulmonary vein carries ______ blood.
   a. Oxygenated; oxygenated
   b. Oxygenated; deoxygenated
   c. Deoxygenated; deoxygenated
   d. Deoxygenated; oxygenated
3. Which of the following options correctly describes the path of blood as it travels from the lungs, through the body, back to the lungs?
   a. Right atrium → Right ventricle → Body → Left atrium → Left ventricle → Lungs
   b. Right ventricle → Right atrium → Body → Left ventricle → Left atrium → Lungs
   c. Left atrium → Left ventricle → Body → Right atrium → Right ventricle → Lungs
   d. Left ventricle → Left atrium → Body → Right ventricle → Right atrium → Lungs
4. Explain why the pericardium is important in preventing the heart from overfilling.
   a. The heart relaxes against the fluid-filled pericardial cavity, which prevents the cardiac muscles from overstretching.
   b. The fibrous pericardium is composed of a thick, fibrous connective tissue, which forms an inflexible cover that prevents the heart from expanding.
   c. The pericardium is a single-layered structure that allows the heart to expand freely during filling of the chambers.
5. Which layer of the heart wall is made almost entirely of cardiac muscle cells?
   a. Epicardium
   b. Pericardium
   c. Myocardium
   d. Endocardium

19.1e Internal Structure of the Heart

Recall that the heart's contraction cycle creates a dual pattern of circulation—the pulmonary and systemic circuits—in order to develop a more precise understanding of cardiac function, it is first necessary to explore the internal anatomical structures in more detail.

LO 19.1.8

septum (plural = septa)

Septa of the Heart The heart is divided into four distinct chambers by walls called septa (singular = **septum**). Located between the two atria is the **interatrial septum**. Typically, in an adult heart, the interatrial septum bears an oval-shaped depression known as the **fossa ovalis**, a remnant of an opening in the fetal heart known as the **foramen ovale**. The foramen ovale allowed blood in the fetal heart to pass directly from the right atrium to the left atrium, allowing some blood to bypass the pulmonary circuit. In the fetus, blood becomes oxygenated from the placenta, not the lungs, so the pulmonary circuit does not have the same critical function that it will have beginning at the first breath. Within seconds after birth, a flap of tissue known as the **septum primum** closes the foramen ovale and establishes the separation of the deoxygenated blood in the right side of the heart from the oxygenated blood in the left side of the heart.

Between the two ventricles is the **interventricular septum**. The interventricular septum is substantially thicker than the interatrial septum and contributes to the much greater pressure generated when the ventricles contract. The interventricular septum is visible in **Figure 19.9**, but note that the interatrial septum is not visible, since its location is covered by the aorta and pulmonary trunk in this view.

The separation between the atria and ventricles is structurally different than the septa that divide the left and right sides of the heart. Whereas the interatrial and

Figure 19.9 **A Frontal Section of the Heart Viewed From the Anterior Perspective**

The left and right ventricles are separated by the thick, muscular interventricular septum, and the left and right atria are separated by the thinner interatrial septum. The left and right atrioventicular valves are much larger than the pulmonary and aortic valves.

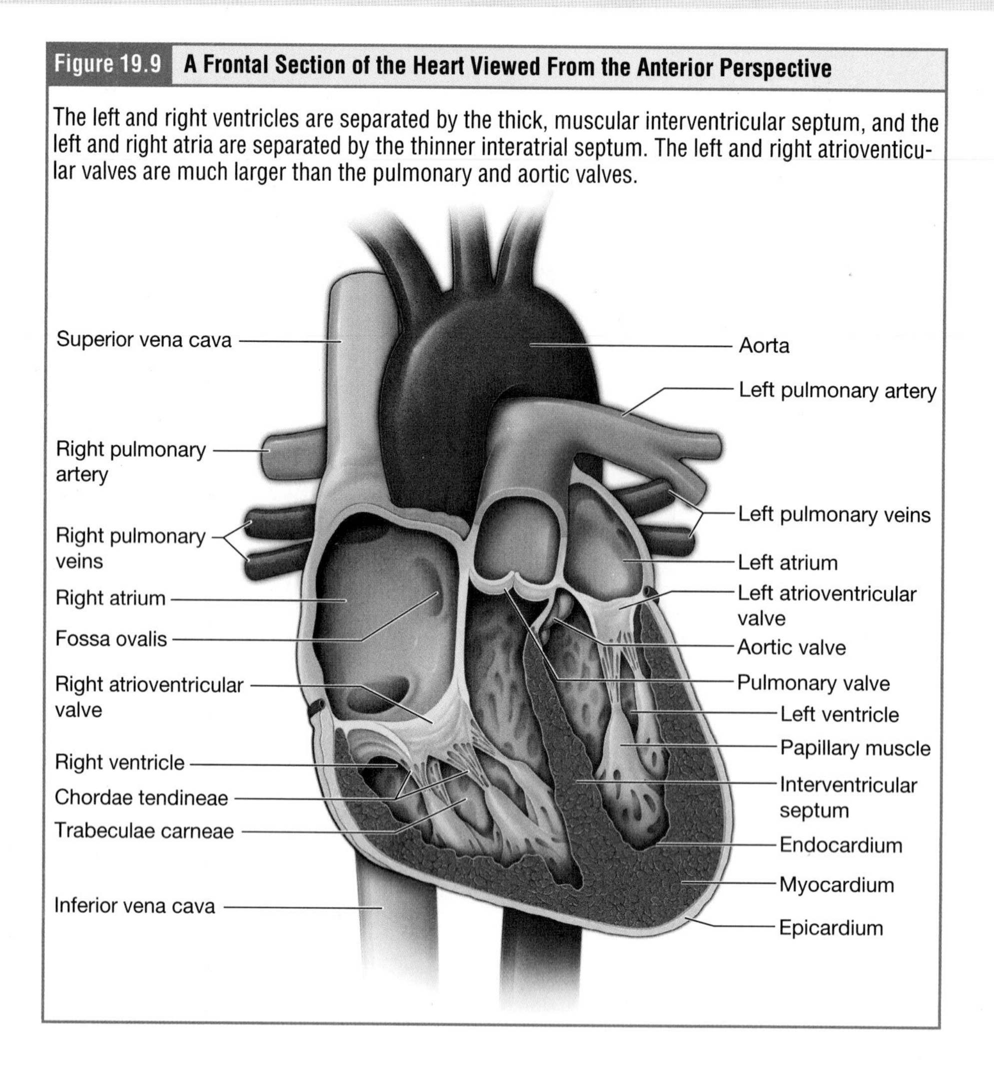

interventricular septa are composed of a mixture of muscle and fibrous tissue, the atria are separated from the ventricles by purely fibrous tissue coated with the endothelium. This structure, known as the **fibrous skeleton of the heart**, contains four openings that allow blood to move from the atria into the ventricles and from the ventricles into the pulmonary trunk and aorta (**Figure 19.10**). Each of these openings is guarded by a **valve**, a specialized structure that ensures one-way flow of blood. The valves between the atria and ventricles are known as **atrioventricular valves**. The valves at the openings that lead to the pulmonary trunk and aorta are known as **semilunar valves**. Each valve is reinforced with a rigid ring of fibrous tissue. Importantly, the fibrous skeleton is electrically nonconductive. The presence of the fibrous skeleton, in addition to its structural role, separates the electrical signals that control the contraction of the atrial and the ventricles.

Right Atrium The right atrium receives deoxygenated blood returning to the heart from the systemic circulation through three vessels. The two major systemic veins, the superior and inferior venae cavae, and the **coronary sinus**, the large coronary vein that drains the heart myocardium, all empty into the right atrium. The superior vena cava drains blood from regions superior to the diaphragm: the head, neck, upper limbs, and the thoracic region. The inferior vena cava drains blood from areas inferior to the diaphragm: the lower limbs and abdominal and pelvic cavities (see Figure 19.9).

There are a few notable exceptions to the otherwise smooth internal surface of the right atrium. If you were running your gloved finger along the medial wall of the right atrium (along the interatrial septum) you would find a depression where the septum feels incredibly thin. This is the fossa ovalis. As your gloved finger made its way around the atrial walls, the

Figure 19.10 The Valves of the Heart

The valves are composed of fibrous connective tissue covered with endothelium. Because they are not made out of cardiac muscle tissue, they are nonconductive (electrical signals stop and do not pass through). The fibrous connective tissue that surrounds the valves and holds them together is sometimes referred to as the fibrous skeleton of the heart.

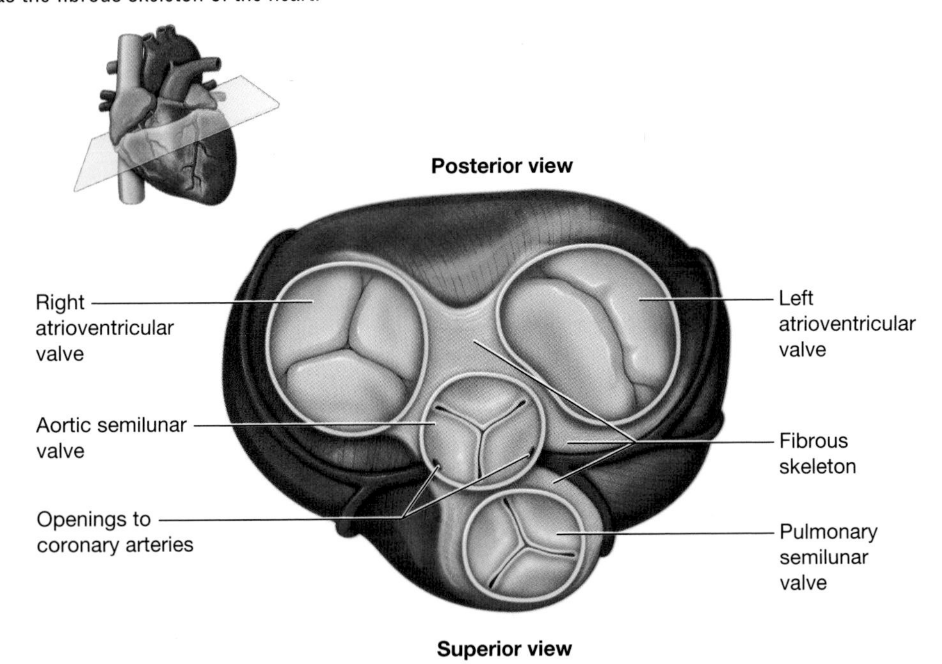

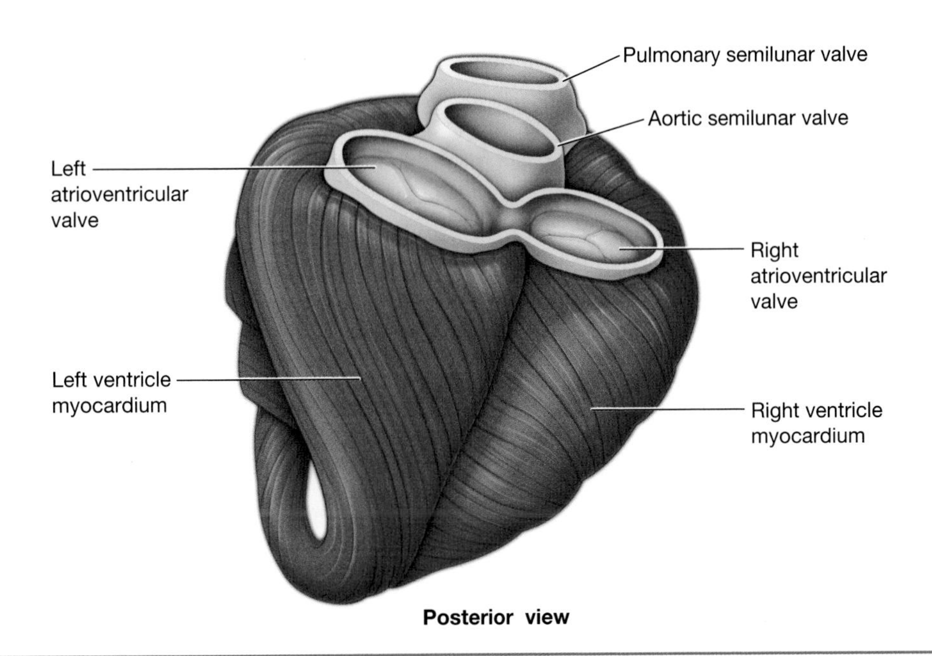

smooth surface would suddenly give way to rough, prominent ridges of muscle called the **pectinate muscles** along the anterior wall and through the auricle. The right auricle also has pectinate muscles. This presence of pectinate muscles is a notable difference between the left and right atria, the left atrium does not have pectinate muscles except inside the auricle.

LO 19.1.9

The atria receive venous blood on a nearly continuous basis. Most (about 80 percent) of the blood that returns to the atria (often referred to as **venous return**) occurs while the atria are relaxed, passing through the atrioventricular valves and filling the ventricles mostly by gravity. When the atria contract, they pump the remaining blood into the ventricles just prior to ventricular contraction.

Apply to Pathophysiology

Patent Foramen Ovale

One very common form of interatrial septum pathology is patent foramen ovale, which occurs when the fossa ovalis is unable to fully fuse at birth. The word *patent* is from the Latin root patens for "open." It may be benign or asymptomatic, perhaps never diagnosed, or in extreme cases, it may require surgical repair to close the opening permanently. As much as 20–25 percent of the general population may have a patent foramen ovale, but fortunately, most have the benign, asymptomatic version. Despite its prevalence in the general population, the causes of patent foramen ovale are unknown, and there are no known risk factors.

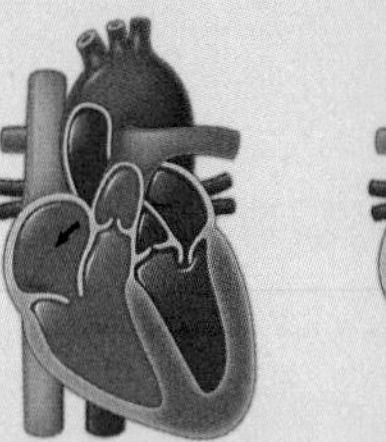

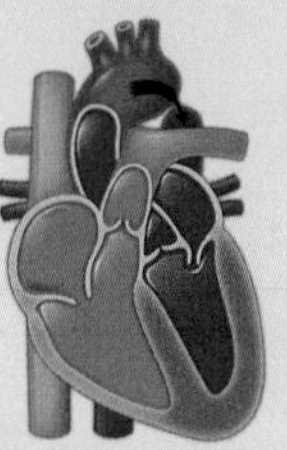

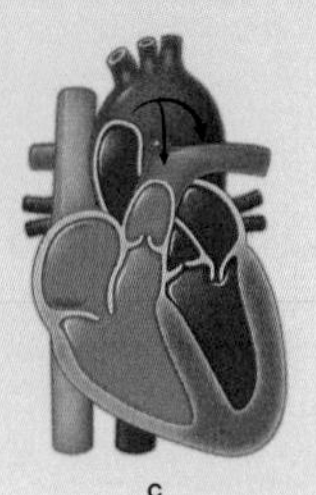

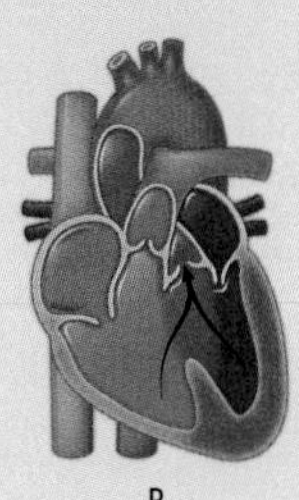

1. Which of the images in the figure represents patent foramen ovale?
2. While most cases are benign, what do you think some of the complications might be?

Coarctation of the aorta is a congenital abnormal narrowing of the aorta that occurs when the aorta and pulmonary trunk divide. In this heart defect, the pulmonary trunk and aorta are physically divided; however, the aorta develops a distinctive narrow band just after the aortic arch.

3. Which of the images in the figure represents coarctation of the aorta?
4. Do you think that this CHD is often benign, as is the case with patent foramen ovale? Or do you think it represents a severe physiological issue for the infant born with this CHD?
5. What do you think would be the symptoms of an infant with coarctation of the aorta?

A patent ductus arteriosus is a congenital condition in which the pulmonary trunk and aorta do not fully divide and separate.

6. Which of the images in the figure represents patent ductus arteriosus?

Patent ductus arteriosus is a CHD that ranges from severe to benign depending on the size of the opening.

7. What are the consequences for blood flow in patent ductus arteriosus? Can you draw it or describe it?
8. What are the consequences for blood pressure in patent ductus arteriosus?

Tetralogy of Fallot is a CHD that may occur from exposure to unknown environmental factors; it is a complex disease with four different defects. The first is an opening in the ventricular septum that allows blood to flow freely between the left and right ventricles. The second has to do with the positioning of the aortic opening, which is directly superior to this septal defect. The result is that blood from both ventricles enters the aorta. The third defect in tetralogy of Fallot is that the pulmonary semilunar valve is stenotic (thicker and more difficult to open). The fourth defect is that the right ventricular wall is thicker; it is equally thick as the left ventricular wall.

9. Which of the images in figure represents tetralogy of Fallot?
10. Considering the other three defects, why is the right ventricular wall thicker? How did that develop?

right atrioventricular valve (tricuspid valve)

chordae tendineae (tendinous cords)

Right Ventricle Blood passes through the three flaps or cusps of the **right atrioventricular valve** to reach the right ventricle. The valve flaps are made of fibrous connective tissue covered by endocardium (see Figure 19.10). Each flap of the valve is attached to strong **chordae tendineae**, cords composed largely of collagen with a small amount of elastin, making them incredibly strong and a bit stretchy. They connect each of the flaps to a **papillary muscle** that extends from the floor of the ventricle (**Figure 19.11**).

When the myocardium of the ventricle contracts, pressure within the ventricular chamber rises. Blood, like any fluid, flows from higher pressure to lower pressure areas, in this case, from the rising pressure of the ventricle toward the pulmonary trunk and the atrium. Examine the orientation of the valves in **Figure 19.12**. As blood is pushed

Figure 19.11 A Frontal Section of a Cadaveric Heart

A frontal section of a cadaveric heart reveals the chordae tendineae and papillary muscles.

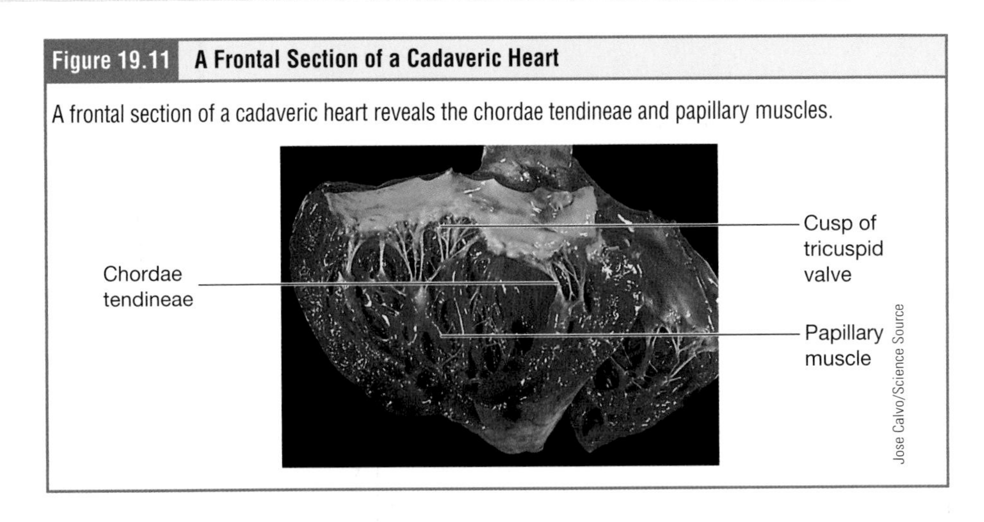

Figure 19.12 The Structure of the Heart Valves Is Simple, As Is Their Function

As blood pushes downward on the atrioventricular valve flaps, it opens them, allowing blood to flow from the atria to the ventricles. As the ventricles contract, blood is pushed against the bottom of the atrioventricular flaps, forcing them closed. Similarly, as blood pushes against the bottom of the semilunar valves, the flaps are forced open and blood can flow into the leaving vessels. As the blood flows backwards, it catches on the upside-down cusps of the semilunar valves and forces them closed.

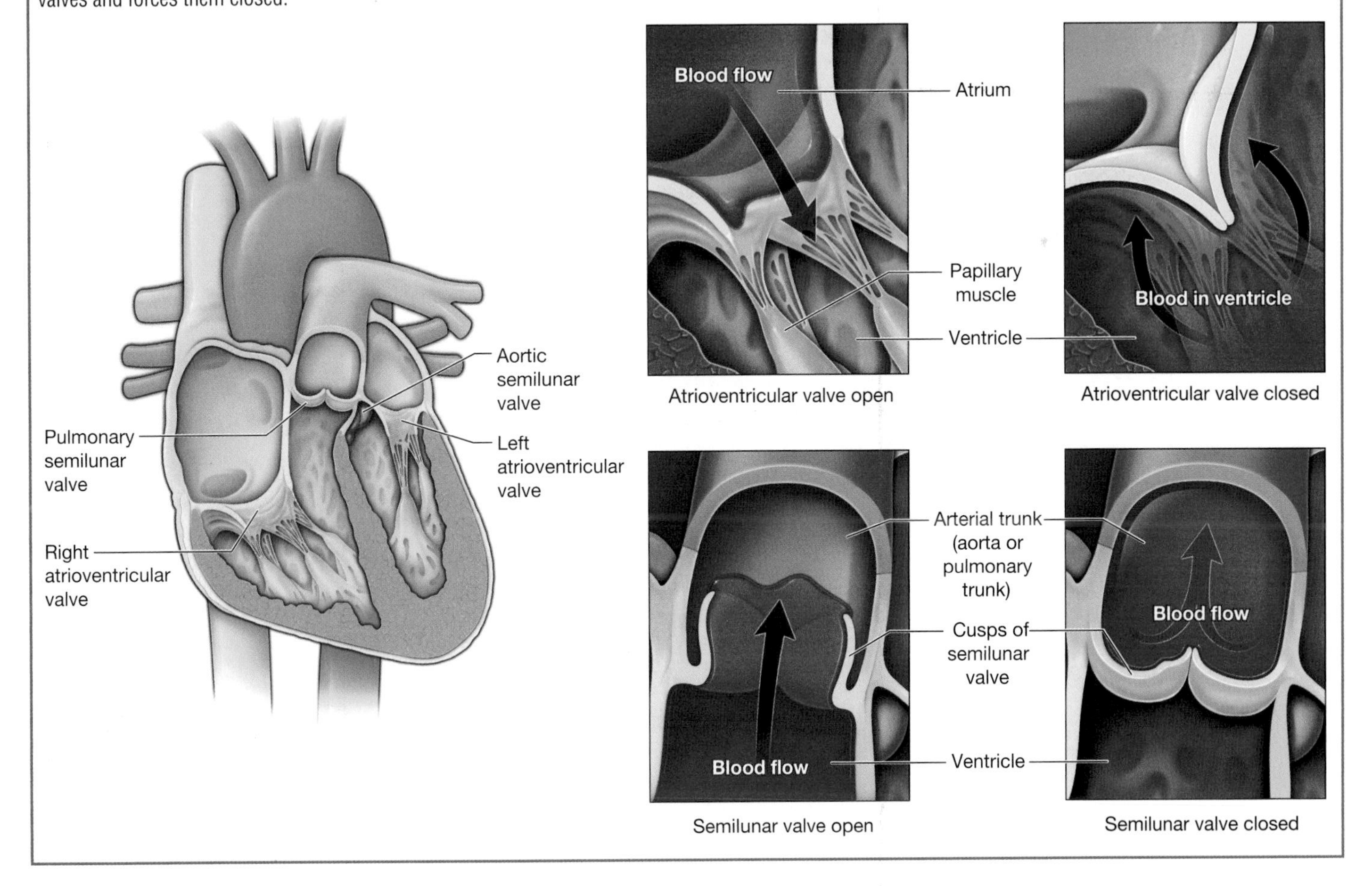

against the underside of the flaps with ventricular contraction, it forces the valve flaps together. The force of this closure is so dramatic that it produces a sound that can be heard outside the heart. The first heart sound or the "lub" of the "lub-dub." As the ventricular muscle cells contract, the papillary muscles also contract, generating tension on the chordae tendineae. This prevents the flaps of the valves from being flipped into the atria and prevents any flow of blood back into the atria during ventricular contraction. Figure 19.11 shows papillary muscles and chordae tendineae attached to the right atrioventricular valve.

The walls of the ventricle are rough with ridges of cardiac muscle called **trabeculae carneae.** In addition to these muscular ridges, a bridge of cardiac muscle, also covered by endocardium, known as the **moderator band** (see Figure 19.9) arises from the inferior interventricular septum and crosses the interior space of the right ventricle to connect with a papillary muscle. The moderator band provides reinforcement of the thin walls of the right ventricle; it also plays a crucial role in cardiac conduction.

When the right ventricle contracts, it forces blood up against the right atrioventricular valve, closing it, then ejects the blood into the pulmonary trunk, which branches into the left and right pulmonary arteries that carry blood to each lung. At the base of the pulmonary trunk is the **pulmonary semilunar valve**, which prevents backflow from the pulmonary trunk. After blood passes through the semilunar valves, these valves close to prevent backflow into the ventricle. The closure of the semilunar valves produces the second audible heart sound, the "dub" of the "lub-dub." Notice in Figure 19.12 that the flap orientation of the semilunar valves is opposite to the orientation of the atrioventricular valves. This is because whereas the atrioventricular valves are preventing backflow of blood from inferior (ventricles) to superior (atria), the semilunar valves are preventing backflow of blood from superior (aorta and pulmonary trunk) to inferior (ventricles). Both sets of valves share the same basic structure; the flaps are concave in the direction that backflow of blood could come from.

pulmonary semilunar valve (pulmonary valve, the pulmonic valve, or the right semilunar valve)

Left Atrium After exchange of gasses in the pulmonary capillaries, oxygenated blood returns to the left atrium via one of the pulmonary veins. (Anatomical note: most humans have four pulmonary veins, but some degree of anatomical variation exists. Each person has between two and four pulmonary veins.) While the left atrium does not contain pectinate muscles, it does have an auricle that includes these pectinate ridges. Like the right atrium, blood flows nearly continuously from the pulmonary veins to fill the atrium and from here through the open **left atrioventricular valve** into the left ventricle. Most of this blood flow is passive during a time when both the atria and ventricles are relaxed; when the left atrium contracts, it pumps any remaining atrial blood into the left ventricle. On both sides of the heart, this atrial contraction accounts for the final 20 percent of ventricular filling, with 80 percent of the filling occurring passively while the atrium is relaxed and filling.

left atrioventricular valve (mitral valve; bicuspid valve)

Left Ventricle Recall that, although both sides of the heart will pump the same amount of blood, the muscular layer is much thicker in the left ventricle compared to the right (see Figure 19.8). Like the right ventricle, the left also has trabeculae carneae, but there is no moderator band. Like its partner on the right, the left atrioventricular valve is connected to papillary muscles via chordae tendineae. The left ventricle is the major pumping chamber for the systemic circuit; it ejects blood into the aorta through the **aortic semilunar valve.**

aortic semilunar valve (aortic valve)

Apply to Pathophysiology

Valvular Heart Disease: Valve Regurgitation and Stenosis

When heart valves do not function properly, the result is valvular heart disease, which can range from benign to lethal. Whereas some of these conditions are congenital, others may be attributed to trauma or develop over time. Some malfunctions are treated with medications, others require surgery, and still others may be mild enough that the condition is merely monitored since treatment might trigger more serious consequences.

Valvular disorders are often caused by inflammation of the heart. One common trigger for this inflammation is infection with the bacterium *Streptococcus pyogenes*, which causes a variety of infections, including strep throat. A rare but serious complication of untreated or incompletely treated strep throat can be rheumatic fever, in which the bacterial infection spreads and triggers endocarditis (inflammation of the endocardium).

We have just read that complete and forceful closure of the valves produces the audible heart sounds. With this in mind, consider an 8-year-old patient that comes in for a pediatric well-check appointment. During auscultation, or listening to a patient's heart sounds, instead of the typical "lub-dub" a whooshing sound is heard in place of the first heart sound. Examine **Figure 19.32** (page 778); the doctor hears "whoosh-dub" clearly when listening at the location in purple. The "whoosh" is not heard as clearly in other locations and is faintest when listening at the location indicated in green. At the green location the doctor hears the more typical "lub-dub" sound with a faint muffling in the background.

1. The doctor suspects valve regurgitation. Which valve is most likely allowing backflow of blood in this case?
   A) Right atrioventricular valve
   B) Pulmonary semilunar valve
   C) Left atrioventricular valve
   D) Aortic semilunar valve
2. The doctor sends the patient for an echocardiogram, an ultrasound of the heart that allows blood flow to be seen. Backflow of blood is verified. Which of the following is likely to also be true?
   A) The foramen ovale is still open, allowing some blood flow from the right atrium to the left atrium.
   B) One of the chordae tendineae is ripped and no longer tethered to its papillary muscle.
   C) Blood can be seen in the pericardial cavity.

Stenosis is a condition in which the heart valves become rigid and may calcify over time. The loss of flexibility of the valve interferes with normal function and may cause the heart to work harder to propel blood through the valve. Aortic stenosis affects approximately 2 percent of the population over 65 years of age, and the percentage increases to approximately 4 percent in individuals over 85 years. Examine the stenosis of the aortic valve seen in the following figure.

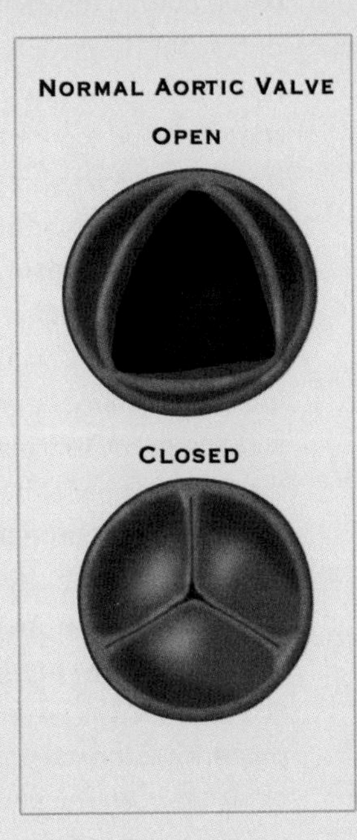

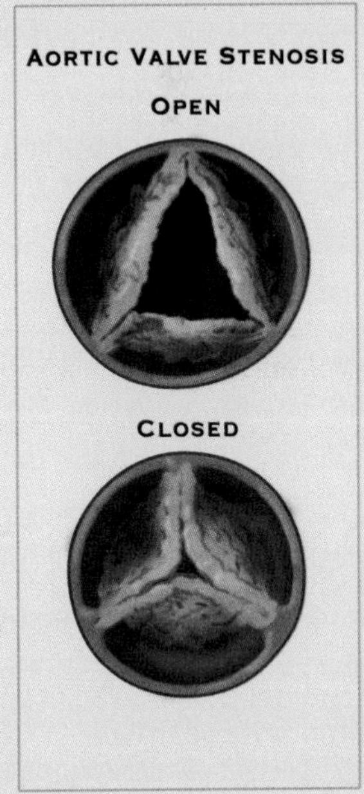

Monica Schroeder/Science History Images/Alamy Stock Photo

3. In a patient with aortic valve stenosis, which type of blood flow would be directly reduced?
   A) Blood flow from the right atrium to right ventricle
   B) Blood flow from the right ventricle to the pulmonary circuit
   C) Blood flow from the left atrium to left ventricle
   D) Blood flow from the left ventricle to the systemic circuit
4. Valvular stenosis is a condition that develops slowly over years. Which of the following pathologies is most likely to have developed along with the aortic valve stenosis seen in the figure?
   A) Left ventricle hypertrophy
   B) Right atrial dilation (getting bigger)
   C) Coarctation (narrowing) of the aorta

19.1f Coronary Circulation

LO 19.1.10

Recall that the heart is a remarkable pump composed largely of cardiomyocytes, cardiac muscle cells, that are incredibly active. Like all other cells, a **cardiomyocyte** requires a reliable supply of oxygen and nutrients and a way to remove wastes, so it needs a dedicated and extensive circulation. While other cells typically have periods of rest and periods of activity, a cardiomyocyte performs critical and nearly ceaseless activity of the heart throughout life; therefore, its need for a blood supply is even greater than for a typical cell. The network of blood vessels that serve the myocardium is called the coronary circulation. Despite the constant need of the cardiomyocytes for oxygen and nutrients, blood flow through the coronary circulation is not continuous; rather, it cycles, reaching a peak when the heart muscle is relaxed and nearly ceasing while it is contracting. The reason for this is anatomical.

Blood vessels in the coronary circulation are considered part of the systemic circuit; therefore, coronary arteries carry oxygenated blood away from the aorta. They branch into smaller and smaller vessels, eventually splitting into the tiny, thin-walled capillaries that allow oxygen to diffuse from the blood to the tissues, in this case the myocardium. Small veins will drain these capillaries, and merge together to become larger and larger vessels. Therefore, the terms *supply*, *branch*, and *arise* are often used to describe arteries, and the terms *drain* and *merge* are often used to describe veins. The term *companion vessel* refers to an artery and vein that are found together in the body. The artery in a companion vessel pair supplies, or brings oxygenated blood to, an area or areas; the companion vein drains, or takes the deoxygenated blood away from, the area.

Coronary Arteries **Coronary arteries** supply blood to the myocardium and other components of the heart. **Figure 19.13** illustrates the close anatomical relationship between the openings for the coronary arteries and the flaps of the aortic semilunar valve. When the semilunar valves open, they almost completely obstruct the entrance to the coronary arteries; it is when they close as the ventricle relaxes that blood can flow into the coronary circulation. Consider the impact of this anatomical feature on heart function. The more times per minute the heart contracts, the less time there is for blood to flow into the coronary circulation and feed the heart muscle. This is one reason that makes rapid heart rate a potentially complex issue. Figure 19.13 illustrates the coronary circulation. Note the right coronary artery and left coronary artery. The **left coronary artery** distributes blood to the left side of the heart (the left atrium and ventricle), and the interventricular septum. The left coronary artery gives rise to two branches: the circumflex artery and the anterior interventricular artery. The **circumflex artery** supplies the left atrium and posterior walls of the left ventricle. Eventually, it fuses with the small branches of the right coronary artery. The larger **anterior interventricular artery**, also known as the left anterior descending artery (LAD), trails the anterior interventricular sulcus and supplies the interventricular septum and the majority of the anterior-inferior heart wall.

anterior interventricular artery (left anterior descending artery or LAD)

The **right coronary artery** proceeds along the coronary sulcus and distributes blood to the right atrium, portions of both ventricles, and the heart conduction system. Normally, one or more marginal arteries arise from the right coronary artery inferior to the right atrium. The **marginal arteries** supply blood to the superficial portions of the right ventricle. On the posterior surface of the heart, the right coronary artery gives rise to the **posterior interventricular artery**, also known as the posterior descending artery. It runs along the posterior portion of the interventricular sulcus toward the apex of the heart, giving rise to branches that supply the interventricular septum and portions of both ventricles.

posterior interventricular artery (posterior descending artery)

Coronary Veins **Coronary veins** drain the heart wall and generally parallel the coronary arteries in location (see **Figure 19.13C**). The **great cardiac vein** can be seen

Figure 19.13 Coronary Artery and Coronary Vein Blood Flow

(A) The coronary arteries; (B) the coronary veins; (C) the coronary arteries branch off of the aorta immediately following the aortic semilunar valve. When this valve is open (ventricular contraction), it almost completely obstructs the entrance to the coronary vessels. When the valve is closed (ventricular relaxation), blood can flow into the vessels and throughout the coronary circulation.

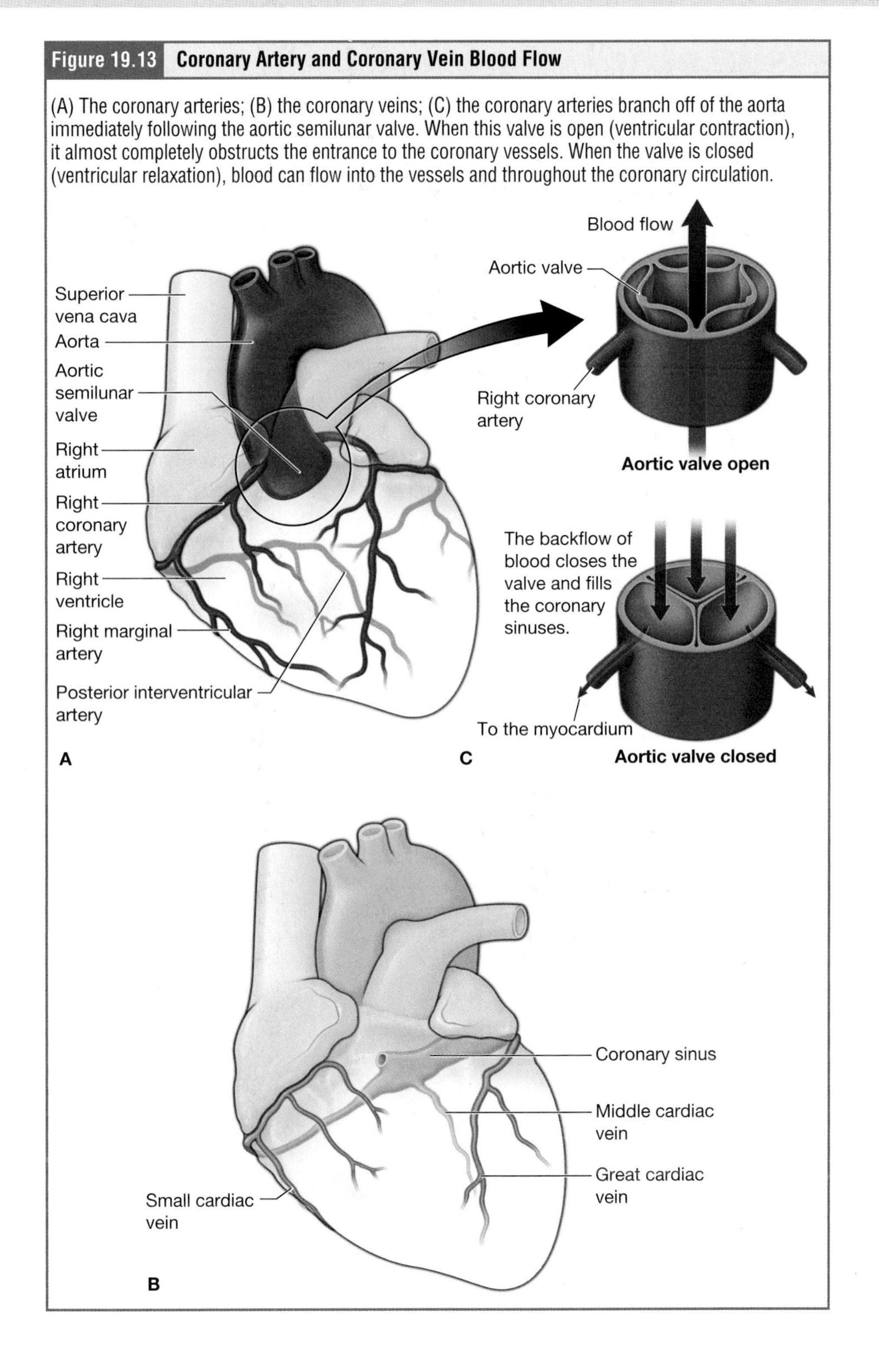

on the anterior surface of the heart following the interventricular sulcus; it eventually flows along the coronary sulcus, where it merges with the coronary sinus on the posterior surface. The great cardiac vein can be considered the companion vessel to the anterior interventricular artery and drains the same areas this artery supplies (interventricular septum and the majority of the anterior-inferior heart wall). It receives blood from several merging branches, including the posterior cardiac vein, the middle cardiac vein, and the small cardiac vein. The **posterior cardiac vein** is the companion

Apply to Pathophysiology

Coronary Artery Disease

Coronary artery disease is the leading cause of death worldwide. It occurs when the buildup of plaque—a fatty material including cholesterol, connective tissue, white blood cells, and some smooth muscle cells—within the walls of the arteries obstructs the flow of blood and decreases the flexibility of the vessels. An additional complication of these plaques is that they can occasionally become unstable and break off, traveling elsewhere in the circulation and potentially becoming lodged in smaller vessels. The figure at right shows the blockage of coronary arteries highlighted by the injection of dye.

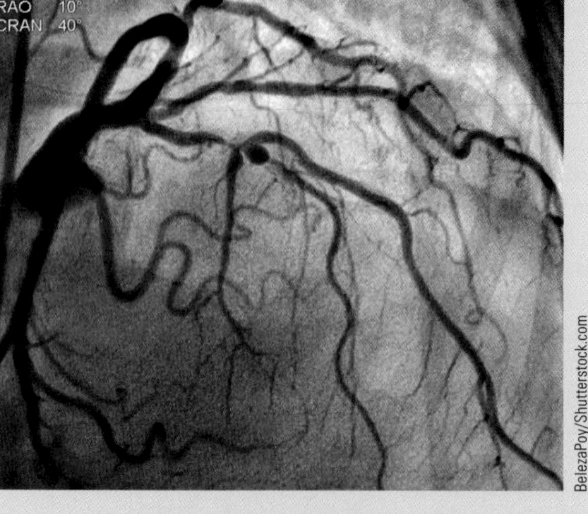
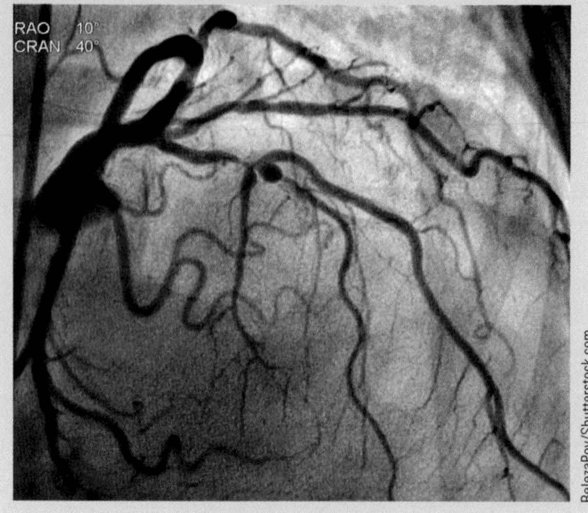

This disease often begins in childhood and progresses slowly throughout the lifespan. Autopsies of children who have died for other reasons reveal fatty "streaks" in the vessels; these fatty streaks become more rigid and established over time. Coronary artery disease typically reaches a sufficient severity to cause symptoms much later in life. The following figure shows the progression of atherosclerosis from fatty streak to mature plaque.

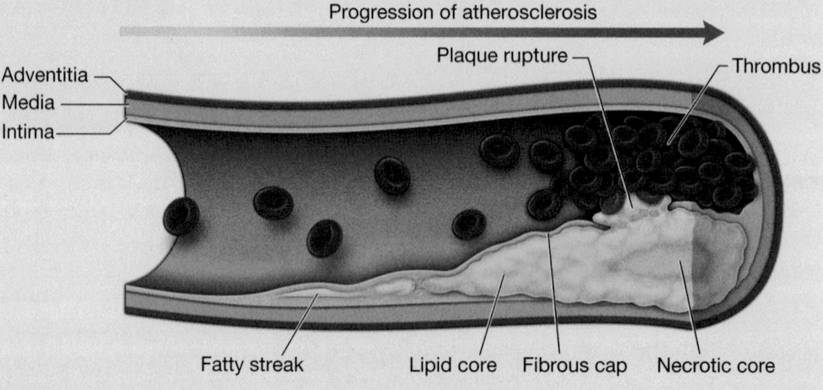

When blood flow is interrupted to any cells, two concerns are **hypoxia** (a lack of oxygen) and **ischemia** (death of cells from lack of blood flow).

1. If the patient whose heart is imaged in the first figure arrived in the ER, what is the primary concern for her clinically?
   A) Ischemia of neurons in the brain
   B) Ischemia of smooth muscle lining the aorta
   C) Ischemia of the interventricular septum
   D) Ischemia of the left atrium
2. Why is plaque buildup of such concern in the coronary arteries? Would clinicians be equally concerned with similar plaques in the coronary veins? Why or why not?
3. A risk factor for coronary artery disease is excess cholesterol and saturated fat in the bloodstream. Which of the following is a plausible explanation for why?
   A) Hydrophobic interactions.
   B) Cholesterol is foreign and causes vascular spasm.
   C) Cholesterol is foreign and the immune system attacks it.

vessel of the circumflex artery. The **middle cardiac vein** is the companion vessel of the posterior interventricular artery. The **small cardiac vein** is the companion vessel of the right coronary artery and drains the blood from the posterior surfaces of the right atrium and ventricle. The coronary sinus is a large, thin-walled vein on the posterior surface of the heart, lying within the atrioventricular sulcus and emptying directly into the right atrium. The **anterior cardiac veins** parallel the small cardiac arteries and drain the anterior surface of the right ventricle. Unlike the other cardiac veins, the anterior cardiac veins bypass the coronary sinus and drain directly into the right atrium.

Learning Check

1. What is the name of the structure that divides the two ventricles within the heart?
   a. Interventricular Septum
   b. Interatrial septum
   c. Semilunar valve
   d. Papillary muscle
2. Which of the following structures prevents the flaps of the atrioventricular valves from flipping into the atria?
   a. Trabeculae carneae
   b. Moderator band
   c. Chordae tendineae
   d. Pectinate muscles
3. Which heart chamber pumps blood through the aorta into the systemic circulation?
   a. Left atrium
   b. Left ventricle
   c. Right atrium
   d. Right ventricle
4. When does the blood enter the coronary circulation?
   a. Ventricular systole (contraction)
   b. Ventricular diastole (relaxation)
5. Which of the following coronary veins drain directly into the right atrium? Choose all that apply.
   a. Anterior cardiac veins
   b. Coronary sinus
   c. Great cardiac vein
   d. Middle cardiac vein

19.2 Cardiac Muscle and Electrical Activity

Learning Objectives: By the end of this section, you will be able to:

19.2.1 Describe the microscopic anatomy of the myocardium, including the location and function of the intercalated discs.

19.2.2 List the parts of the electrical conduction system of the heart in the correct sequence for one contraction and explain how the electrical conduction system functions.

19.2.3 Explain why the SA node normally paces the heart.

19.2.4 Explain how the cardiac conduction system produces coordinated heart chamber contractions.

19.2.5 Contrast the initiation of action potentials in cardiac autorhythmic cells, in cardiac contractile cells, and in skeletal muscle cells.

19.2.6 List the phases of contractile and autorhythmic cardiac muscle action potentials and explain the ion movements that occur in each phase.

19.2.7 Explain the significance of the plateau phase in the action potential of a cardiac contractile cell.

19.2.8 Compare the refractory periods of cardiac contractile muscle and skeletal muscles.

19.2.9 Compare and contrast the molecular events of cardiac muscle contraction/relaxation and skeletal muscle contraction/relaxation.

19.2.10 Explain the role of calcium in determining the force of myocardial contraction (contractility).

19.2.11 Name the waveforms in a normal electrocardiogram (ECG or EKG) and explain the electrical events represented by each waveform.

19.2.12* Predict changes to EKGs in various disease states (e.g., an increase in extracellular K^+ concentration).

19.2.13 Compare and contrast the role of autonomic innervation in the depolarization of cardiac pacemaker cells, ventricular contractile cells, and skeletal muscle cells.

19.2.14* Predict changes to heart function if the permeabilities of ions were altered in electrical conduction system cells or cardiomyocytes.

* Objective is not a HAPS Learning Goal.

Recall that cardiac muscle shares a few characteristics with both skeletal muscle and smooth muscle, but it has some unique properties of its own. Perhaps most exceptional is its ability to initiate an electrical potential at a fixed rate that spreads rapidly from cell to cell to trigger contraction. This property is known as **autorhythmicity**. This unique feature means that if you severed all connections between the heart and the brain, the heart could still beat. In fact, heart transplantation has been made more successful by the invention of organ care systems (a.k.a. heart in a box) that circulate oxygenated blood through a heart after it has been removed from the donor and is en route to the transplant recipient (**Figure 19.14**). These hearts, remarkably, continue to beat, outside the body, through the property of autorhymicity. In the body, **heart rate (HR)** is initiated by the heart's autorhythmicity center, but the rate is modulated by the endocrine and nervous systems, as we will read about later in this chapter.

There are two major types of cardiac muscle cells: myocardial contractile cells and myocardial conducting cells. The **myocardial contractile cells** constitute the bulk (99 percent) of the cells in the myocardium. Contractile cells are able to pass the impulses that are responsible for contractions from one cell to another, but they cannot generate these impulses on their own; in other words, contractile cells do not have autorhythmicity. The **myocardial conducting cells** (1 percent of the cells) form the conduction system of the heart. These cells are responsible for generating and conducting the action potential (the electrical impulse) that drives heart contractions. These cells are generally much smaller than the contractile cells and have few of the myofibrils or filaments needed for contraction.

19.2a Structure of Contractile Cardiac Muscle

Compared to the giant cylinders of skeletal muscle, cardiac muscle cells (or contractile cardiomyocytes) are considerably shorter and smaller in diameter. Similar to skeletal muscle tissue, cardiac muscle also demonstrates striations, the alternating pattern of dark A bands and light I bands due to the precise arrangement of the myofilaments and sarcomeres in these cells (**Figure 19.15C**). Though there are fewer of them in cardiac

Figure 19.14 The "Heart-in-a-Box" Transplant Device

The so-called "heart in a box" is a transplant technology that allows a donor heart to continue to beat while it is transported to the transplant recipient.

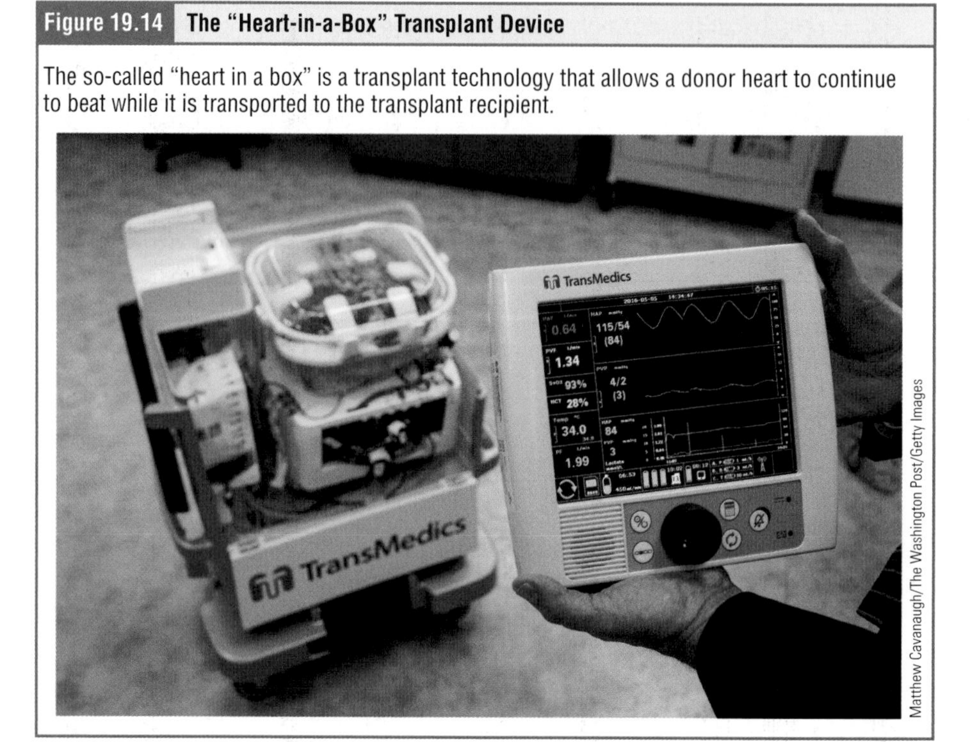

Matthew Cavanaugh/The Washington Post/Getty Images

muscle cells, T (transverse) tubules penetrate from the surface plasma membrane (the sarcolemma) to the interior of the cell, allowing the electrical impulse to reach the interior just as they do in skeletal muscle cells (**Figure 19.15A**). Both skeletal and contractile cardiac muscle cells have sarcoplasmic reticulum. In cardiac muscle cells, however, the sarcoplasmic reticulum stores few calcium ions, so most of the calcium ion influx must come from outside the cells. Mitochondria are plentiful, providing energy for the contractions of the heart. Typically, cardiomyocytes have a single central nucleus, but two or more nuclei may be found in some cells.

Cardiac muscle cells are branched or forked, typically connecting to more than one cardiomyocyte neighbor on at least one of its ends (Figure 19.15A). The junctions between adjoining cells, **intercalated discs**, are unique in structure (**Figure 19.15B**). At these sites the sarcolemmas from adjacent cells are bound together with three kinds of cellular junctions: desmosomes, tight junctions, and large numbers of gap junctions (see Chapter 5). Together the desmosomes and tight junctions unite the cells structurally, tying them together throughout contraction and relaxation. The gap junctions allow the passage of ions between the cells, uniting the cells functionally by allowing the electrical signal to pass from one cell to its neighbors and helping to synchronize the contraction (**Figure 19.15C**).

LO 19.2.1

Figure 19.15 The Structure of Intercalated Discs

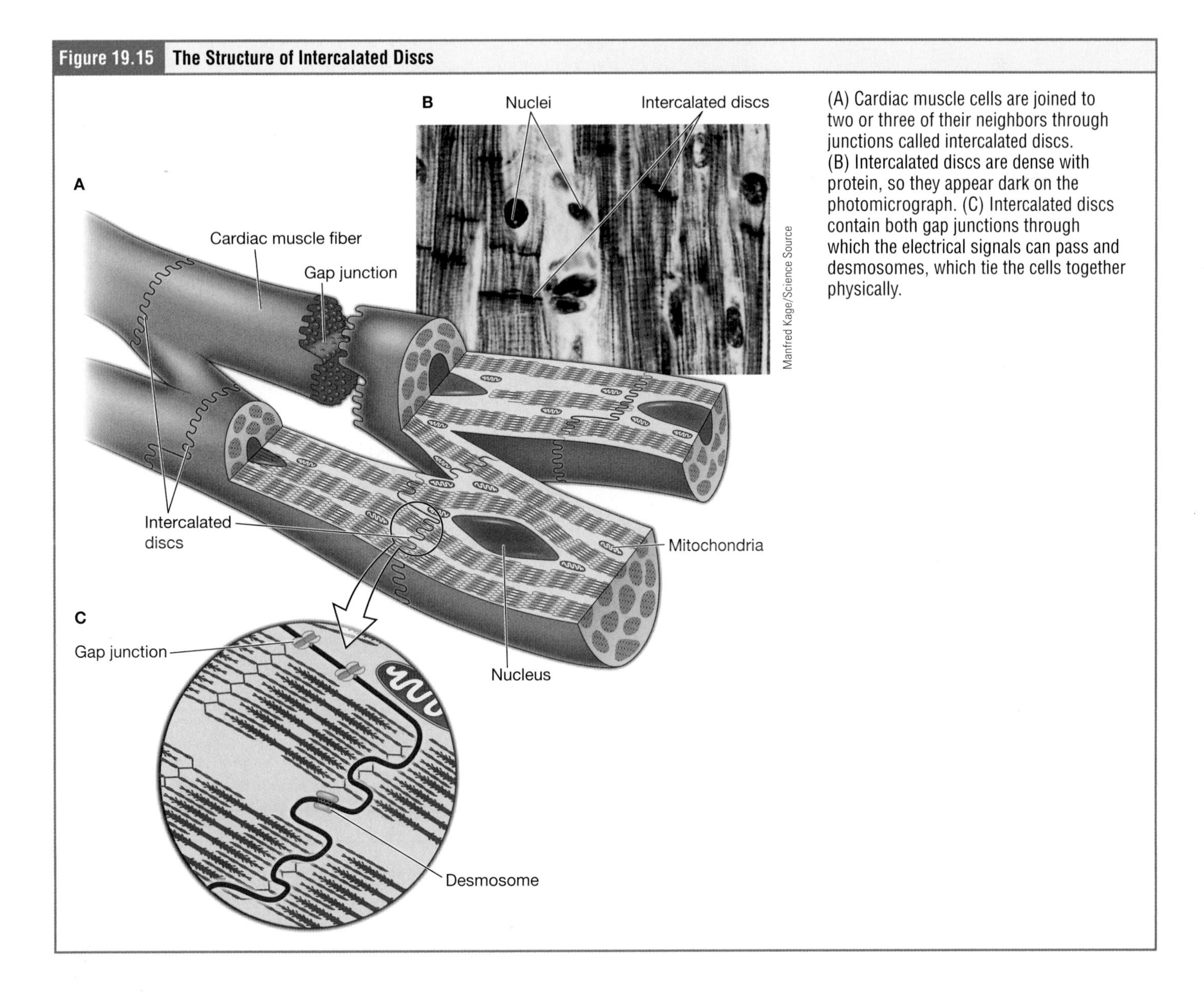

(A) Cardiac muscle cells are joined to two or three of their neighbors through junctions called intercalated discs. (B) Intercalated discs are dense with protein, so they appear dark on the photomicrograph. (C) Intercalated discs contain both gap junctions through which the electrical signals can pass and desmosomes, which tie the cells together physically.

DIGGING DEEPER:
Heart Disease: Damage to Cardiac Muscle Tissue

Heart disease causes one in every four deaths in the United States and about 7 percent of American adults are living with diagnosed heart or coronary artery disease. The significant issue with heart attacks and heart disease is that damaged cardiac muscle cells appear to be unable to replace themselves by mitosis. Unlike smooth muscle cells, which have an adult stem cell population that can aid in restoration of the tissue after cells have been lost, cardiac muscle tissue has a very limited capacity for regeneration. This means that in the event of cardiac muscle cell death, only rarely can the lost cells be replaced with new contractile cardiac muscle cells.

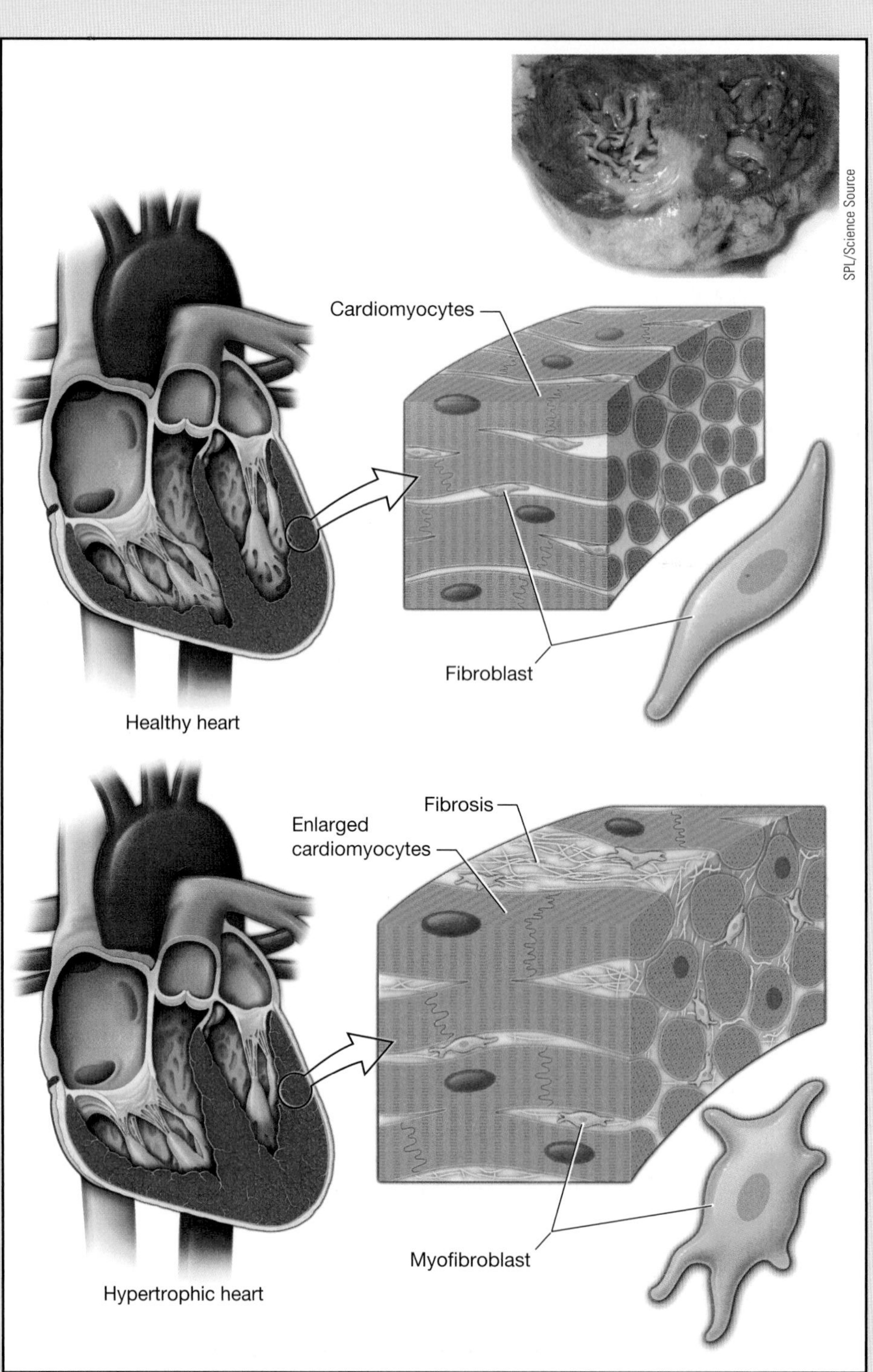

What causes the death of cardiac muscle cells? A few different factors. The largest contributing factor is ischemia, loss of oxygen, due to blockages or narrowing in the coronary arteries. A blockage in the coronary arteries, which starves cardiac muscle cells of much-needed oxygen, is called a heart attack or myocardial infarction. But other issues can cause damage of the heart too. SARS-CoV-2, the virus that causes the disease COVID-19, can directly infect and kill cardiac muscle tissue, leading to tissue damage as well. One of the key clinical tests for diagnosing damage to heart muscle cells is to test the blood for the presence of troponin, a protein that should only be found inside muscle cells but is released when they die. In one study, 78 percent of COVID infections, including infections that produced no other symptoms, resulted in cardiac muscle cell death. This means that millions of Americans (and people all across the world) are likely to suffer long-lasting cardiac complications following this viral illness. The areas where cardiac muscle cells die are generally replaced with fibrous scar tissue (see the lower right figure). Since this scar tissue cannot contribute to contraction, the remaining cardiac muscle cells must work harder to maintain blood flow. In addition, the interruptions in passage of the action potential from cell to cell can have implications for cardiac rhythm, leading to arrhythmias.

Research and therapeutic sciences have been searching for a long time for ways to improve cardiac function after heart damage. While cardiac muscle tissue has lower numbers of stem cells than smooth and skeletal muscle cells, there are a few adult stem cells in this tissue. Researchers have found that they can find and extract these cells, replicate them in the laboratory, and then reinject them into patients following heart attacks. This experimental treatment reduces the size of the damaged area and improves cardiac function.

Another experimental approach is to create stem cells through a mechanism established in 2012 called *induced pluripotency*. Induced pluripotent stem cells (iPSCs) can be made from any adult cell and then differentiated into cardiac muscle cells and implanted. This technique has been successful in helping mice heal from cardiac damage in the lab, but is still experimental in humans, having only been attempted in a handful of individuals.

A growing field of therapeutic science is cardiac rehabilitation. Cardiac rehabilitation is a careful and personalized approach to improving cardiac function through exercise and nutrition counseling. This form of rehabilitation was found to be effective in a dose-dependent fashion, that is to say, for every session attended, there were more and more improvements in the patient. As we discussed previously, once cardiac muscle cells have been replaced by scar tissue, the burden on each of the remaining cardiac muscle cells is increased. Cardiac rehabilitation focuses on methods for increasing the strength and durability of the remaining cardiac muscle cells as well as improving cardiovascular health throughout the entire body.

Currently, cardiac rehab presents an important path forward for individuals recovering from cardiac damage. Hopefully someday we will have more cellular therapeutics to replace damaged cardiac muscle cells.

19.2b Cardiac Muscle Metabolism

Different types of metabolism are available to most cells. Cardiac muscle metabolism, however, is typically limited to aerobic respiration. The required oxygen from the lungs is brought to the heart, and every other organ, attached to the hemoglobin molecules within the erythrocytes. Heart cells also store appreciable amounts of oxygen on the protein myoglobin. Normally, these two mechanisms, circulating oxygen and oxygen attached to myoglobin, can supply sufficient oxygen to the heart, even during peak performance. Like all muscles, cardiac muscles change with use. As athletes train, they add more glycogen and myoglobin to their cardiomyocytes, enabling these cells to be able to function with less support from the blood supply.

Skeletal muscle and cardiac muscle cells are both flexible in what nutrients they are able to use to generate ATP. In fact, cardiomyocytes are even more flexible than skeletal muscle cells in that they can use lactic acid as a fuel in addition to fatty acids and glucose. But, in contrast to skeletal muscle cells, cardiac muscle cells are much more dependent on the presence of oxygen to convert the energy within nutrients to ATP. Therefore, one of the most dangerous scenarios for cardiomyocytes is disruption in the flow of oxygenated blood (for example, if coronary artery disease creates a blockage in the blood supply). Additionally, new blood vessels may grow, a process called angiogenesis, to feed the increasingly active cells.

19.2c Conduction System of the Heart

If embryonic heart cells are grown in a Petri dish, each cell is capable of generating its own electrical impulse and contraction. When two independently beating embryonic cardiac muscle cells are placed together, the cell with the higher inherent rate sets the pace, and the impulse spreads from the faster to the slower cell to trigger a contraction. As more cells are joined together, the fastest cell continues to assume control of the rate. A fully developed adult heart maintains the capability of generating its own electrical impulse, triggered by the fastest cells, as part of the cardiac conduction system. The components of the cardiac conduction system include the sinoatrial node, the atrioventricular node, the atrioventricular bundle, the atrioventricular bundle branches, and the Purkinje fibers (**Figure 19.16**).

LO 19.2.2

LO 19.2.3

Sinoatrial (SA) Node Normal cardiac rhythm is established by the **sinoatrial (SA) node**, a specialized clump of myocardial conducting cells located in the superior/posterior wall of the right atrium. The SA node has the highest inherent rate of depolarization and is known as the pacemaker of the heart. It initiates the **sinus rhythm**, or the normal electrical pattern followed by contraction of the heart.

Figure 19.16 The Conduction System of the Heart

The conduction system of the heart is a collection of cardiac muscle cells that generates and distributes the electrical impulses that cause the heart to beat.

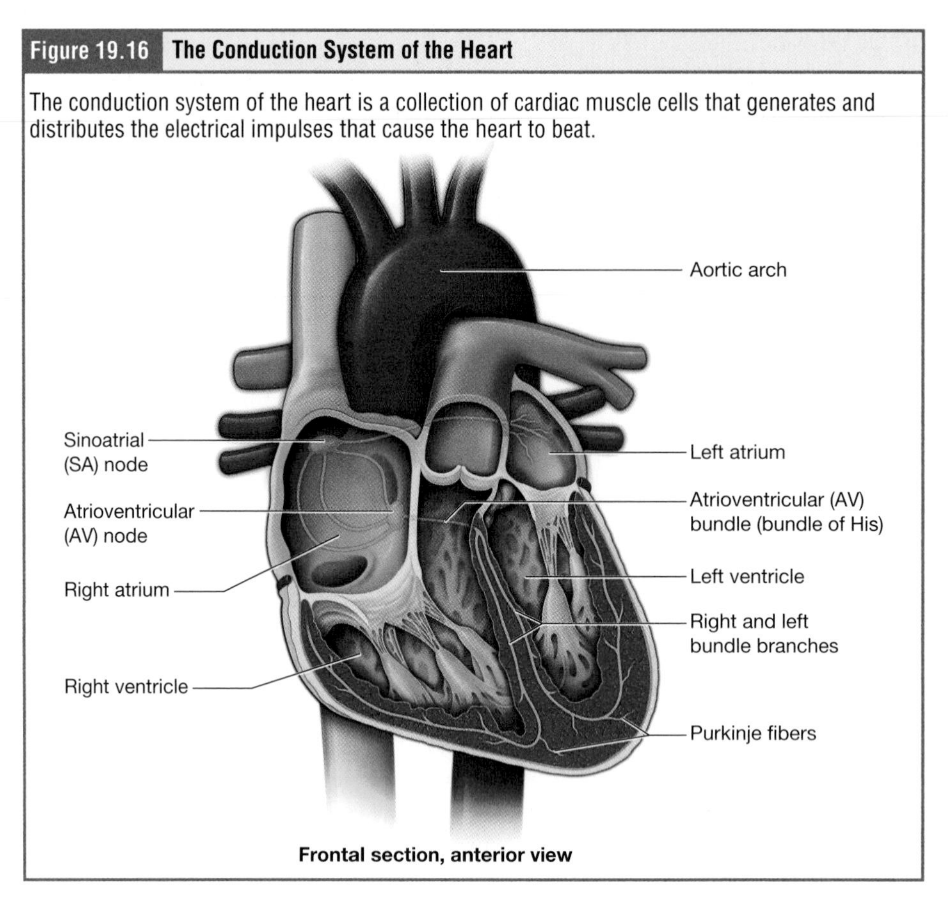

Frontal section, anterior view

LO 19.2.4

This impulse spreads from its initiation in the SA node somewhat slowly throughout the atrial cardiomyocytes through their gap junctions. The signal also passes more rapidly along conduction cells of the **internodal pathway** toward the atrioventricular node (see Figure 19.16). The signals traveling among the atrial cardiomyocytes do not continue to the ventricular cardiomyocytes because these groups of contractile cells are anatomically separated by the nonconductive tissue of the fibrous skeleton of the heart. Therefore, the only impulse to reach the ventricles is the impulse that traveled the internodal pathway to the atrioventricular node.

Figure 19.17 illustrates the initiation of the impulse in the SA node that then spreads throughout the atria toward the atrioventricular node (step 2). The electrical event, the wave of depolarization, is the trigger for muscular contraction. The wave of depolarization begins in the right atrium, and the impulse spreads across the superior portions of both atria and then down through the contractile cells (shown in yellow in Figure 19.17). The electrical impulse triggers a contraction in contractile cells. This wave of contraction (shown in purple in Figure 19.17) moves from superior to the inferior portions of the atria, efficiently pumping blood into the ventricles (step 3).

Atrioventricular (AV) Node The **atrioventricular (AV) node** is a second clump of specialized myocardial conductive cells, located in the inferior floor of the right atrium. The signal slows down slightly when it reaches the AV node. This delay in transmission is attributable to two anatomical factors: the AV nodal cells are slightly smaller in diameter and they have fewer gap junctions connecting them. This pause is critical to heart function, as it allows the atria to complete their contraction (moving all of the atrial blood into the ventricles) before the impulse is transmitted to the cells of the ventricle itself. With extreme stimulation by the SA node, the AV node can transmit impulses maximally around 220 per minute. This establishes the typical maximum heart rate in a healthy young individual. Damaged hearts or those stimulated by drugs can contract at higher rates, but at these rates, the heart can no longer effectively pump blood.

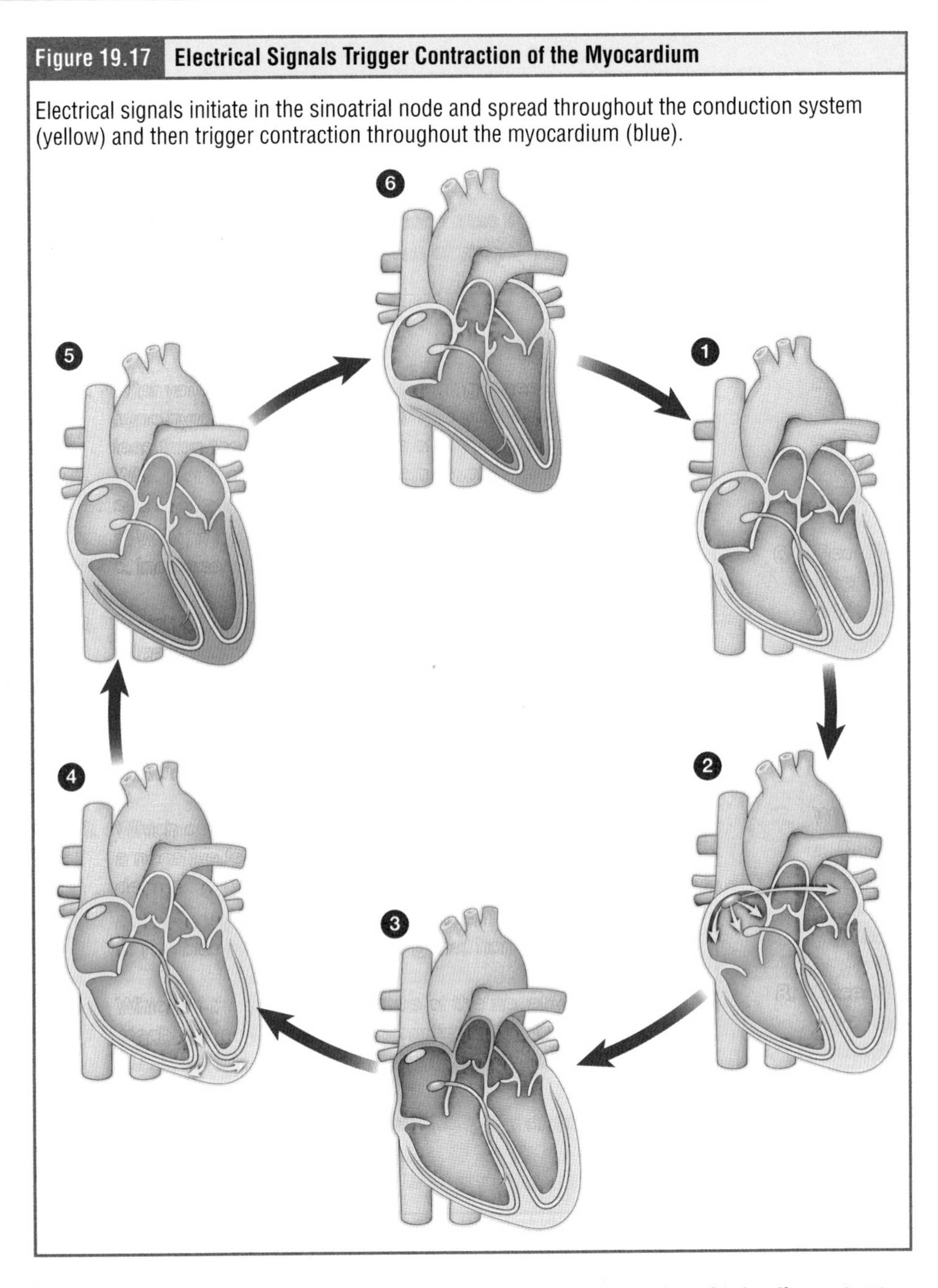

Figure 19.17 Electrical Signals Trigger Contraction of the Myocardium

Electrical signals initiate in the sinoatrial node and spread throughout the conduction system (yellow) and then trigger contraction throughout the myocardium (blue).

Atrioventricular Bundle (Bundle of His), Bundle Branches, and Purkinje Fibers Arising from the AV node, the **atrioventricular bundle**, or bundle of His, proceeds through the interventricular septum before dividing into two **atrioventricular bundle branches**, commonly called the left and right bundle branches. The left bundle branch supplies the left ventricle, and the right bundle branch the right ventricle. Since the left ventricle is much larger than the right, the left bundle branch is also considerably larger than the right. Portions of the right bundle branch are found in the moderator band and supply the right papillary muscles. Because of this connection, each papillary muscle receives the impulse at approximately the same time, so they begin to contract simultaneously just prior to the remainder of the myocardial contractile cells of the ventricles. This is believed to allow tension to develop on the chordae tendineae prior to right ventricular contraction. There is no corresponding moderator band on the left. Both bundle branches descend and reach the apex of the heart where they connect with the Purkinje fibers (see Figure 19.17, step 4).

atrioventricular bundle (bundle of His)

atrioventricular bundle branches (bundle branches)

The **Purkinje fibers** are conductive fibers that spread the impulse throughout the ventricle walls. They branch through the myocardium from the apex of the heart superiorly toward the atria. The Purkinje fibers transmit the impulse quickly (see Figure 19.17, step 5). Since the electrical stimulus begins to spread from cardiomyocyte to cardiomyocyte at the apex, the contraction also begins at the apex and travels toward the tops of the ventricles. This allows the blood to be pumped out of the ventricles superiorly aorta and pulmonary trunk.

Learning Check

1. Identify the type of junction that is responsible for uniting the adjoining cells functionally by synchronizing cardiac contractions.
   a. Desmosomes
   b. Tight junctions
   c. Gap junctions
   d. Sarcolemmas
2. What is the function of the sinoatrial (SA) node?
   a. Carries impulses within the interventricular septum
   b. Initiates the sinus rhythm of the heart
   c. Responsible for atrial contraction
   d. Spread impulses throughout the ventricular wall
3. What is the function of the atrioventricular (AV) node?
   a. Delays the impulse initiated by the SA node, allowing the atria to finish contracting before the ventricles contract
   b. Initiates the sinus rhythm
   c. Responsible for atrial contraction
   d. Spread impulses throughout the ventricular wall
4. Which of the following represents the conduction pathway of an electrical impulse through the heart?
   a. SA Node → AV node → Atrioventricular Bundle → Left and Right Branches → Purkinje Fibers
   b. AV node → SA node → Purkinje Fibers → Atrioventricular Bundle → Left and Right Branches
   c. SA Node → Purkinje Fibers → AV node → Atrioventricular Bundle → Left and Right Branches
   d. AV Node → Atrioventricular Bundle → SA Node → Purkinje Fibers → Left and Right Branches

LO 19.2.5

Membrane Potentials and Ion Movement in Nodal Cells The impulses, or action potentials, we have been discussing are considerably different between cardiac conduction cells and cardiac contractile cells. While sodium (Na^+) and potassium (K^+) play essential roles, calcium (Ca^{2+}) is also critical for both cell types. The skeletal muscle and neuronal cells we explored in Chapters 10 and 12 have stable resting membrane potentials and populations of ion channels that are triggered to open when the membrane potential becomes more positive than the resting membrane potential. Cardiac conduction cells do not have a stable resting potential due to the fact that they have distinct populations of ion channels, some of which open when the membrane potential drifts more positive, and a unique population of sodium ion channels that open when the cells return to a more negative membrane potential. That is to say, while skeletal muscle cells and neurons return to a resting membrane potential after a depolarization and remain at rest until stimulated again, cardiac conduction cells begin to depolarize as soon as they return to a negative membrane potential, they never stay at rest. This **spontaneous depolarization** is what gives the heart its autorhythmicity.

spontaneous depolarization (prepotential depolarization)

LO 19.2.6

Examine **Figure 19.18**. In phase one, voltage-gated calcium ion channels open when the membrane depolarizes past threshold for those channels. Ca^{2+} enters the cell, further depolarizing it at a more rapid rate until it reaches a value of approximately +5 mV. At this point, the calcium ion channels close and voltage-gated K^+ channels open, allowing outflux of K^+ and the cell repolarizes (phase two). When the membrane potential reaches approximately −60 mV, the K^+ channels close and the unique populations of Na^+ channels open, depolarizing the cell slowly, bringing it toward the threshold voltage at which Ca^{2+} channels will open (phase 3).

Figure 19.18 Spontaneous Depolarization of a Cardiac Nodal Cell

(1) As a cardiac nodal cell repolarizes following a depolarization, its membrane potential falls to around –60 mV. At this low membrane potential, a population of Na^+ channels opens, allowing Na^+ to enter the cell and depolarizing the cell steadily. (2) When the cell depolarizes past its threshold, a population of voltage-gated Ca^{2+} channels open, allowing Ca^{2+} to flow into the cell and depolarizing the cell rapidly. (3) As these channels close, and K^+ channels open, the cell repolarizes.

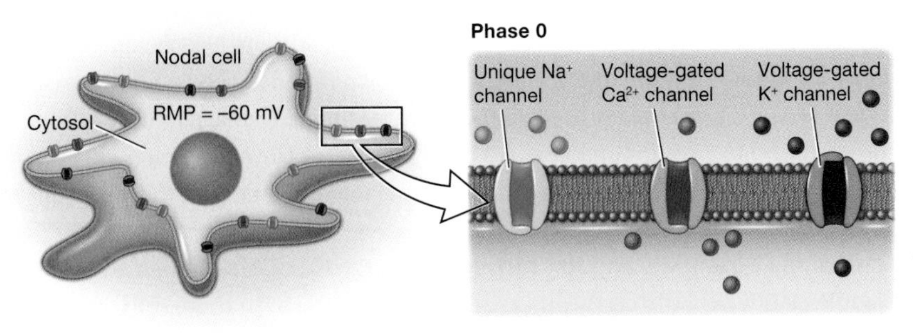

Phase 1

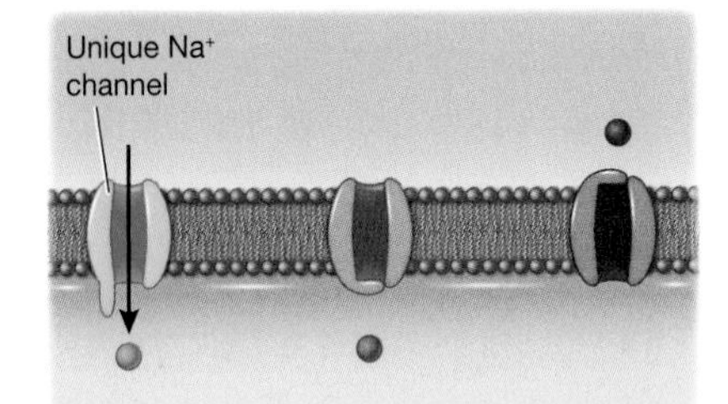

❶ Na^+ flows into the cell, depolarizing the membrane.

Phase 2

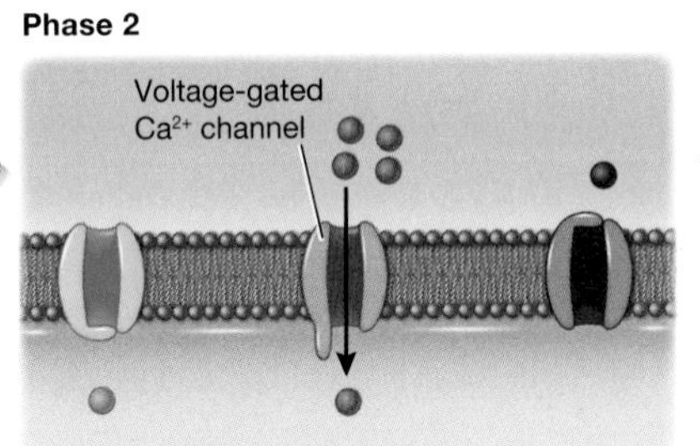

❷ Voltage-gated Ca^{2+} channels open when the membrane depolarizes to its threshold. Ca^{2+} ions enter the cell, rapidly depolarizing it.

Phase 3

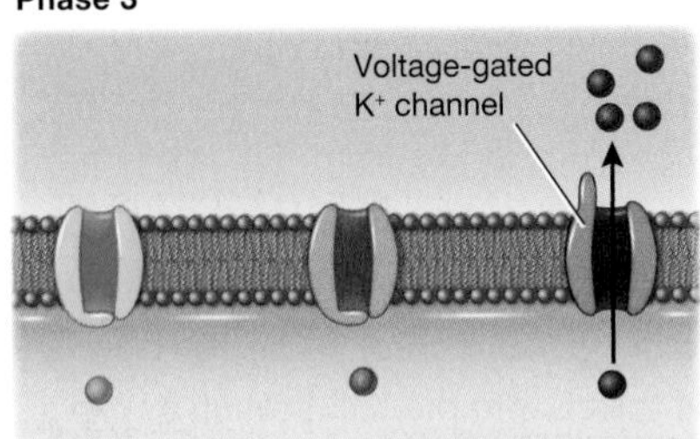

❸ At approximately the same time, voltage-gated K^+ channels open as voltage-gated Ca^{2+} channels close. The membrane voltage quickly repolarizes as cations are now flowing out in much greater numbers than they are flowing in. Membrane potential returns toward –60mV at which point K^+ channels will close and the unique population of Na^+ channels will open once again.

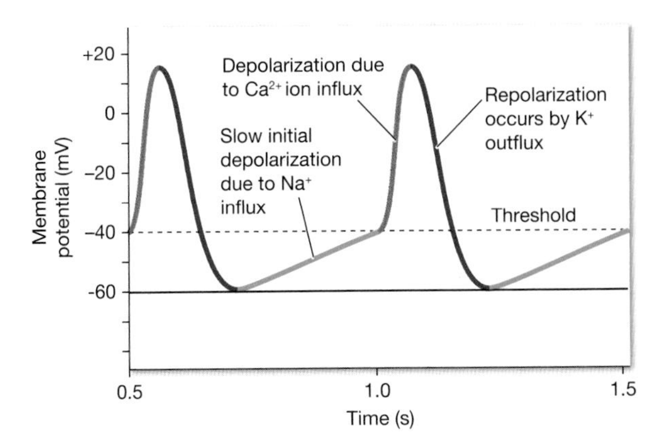

Comparative Rates of Conduction System Firing The pattern of spontaneous depolarization, followed by rapid depolarization and repolarization just described, can be seen not just in the SA node but in the other conductive cells of the heart as well. Since the SA node expresses a higher density of specialized Na^+ channels, it reaches threshold faster than any other component of the conduction system and therefore it is the pacemaker. The SA node, without nervous or endocrine control, would initiate a heart impulse approximately 80–100 times per minute (**Table 19.1**). Although each component of the conduction system is capable of generating its own impulse, the rate slows progressively as you proceed from the SA node to the Purkinje fibers. Damage to the SA node, therefore, can lead to a condition known as **bradycardia**, or slow heart rate. This is one clinical reason that an artificial pacemaker may be implanted to take over the job of the SA node.

Membrane Potentials and Ion Movement in Cardiac Contractile Cells The contractile cells have a distinctly different electrical pattern. In these cells, there is a rapid depolarization, followed by a plateau phase and then repolarization. This phenomenon accounts for the long **refractory periods** required for the cardiac muscle cells to pump blood effectively before they are capable of firing for a second time.

LO 19.2.7

LO 19.2.8

Contractile cells demonstrate a stable resting membrane potential, similar to skeletal muscle cells and neurons, but in contrast to conduction cells. They lack the unique population of sodium channels that open at a negative repolarization, and so they do not demonstrate autorhythmicity. Their resting membrane potential is approximately −80 mV for cells in the atria and −90 mV for cells in the ventricles. For all cardiac contractile cells, when stimulated by an action potential that travels into the cell through its gap junctions, voltage-gated sodium channels rapidly open, beginning the positive-feedback mechanism of depolarization (see phase 1 in **Figure 19.19**). The rapid influx of Na^+ raises the membrane potential to approximately +30 mV, at which point the sodium channels close (see phase 2 in Figure 19.19). At this point, voltage-gated K^+ channels begin to open and K^+ ions leave

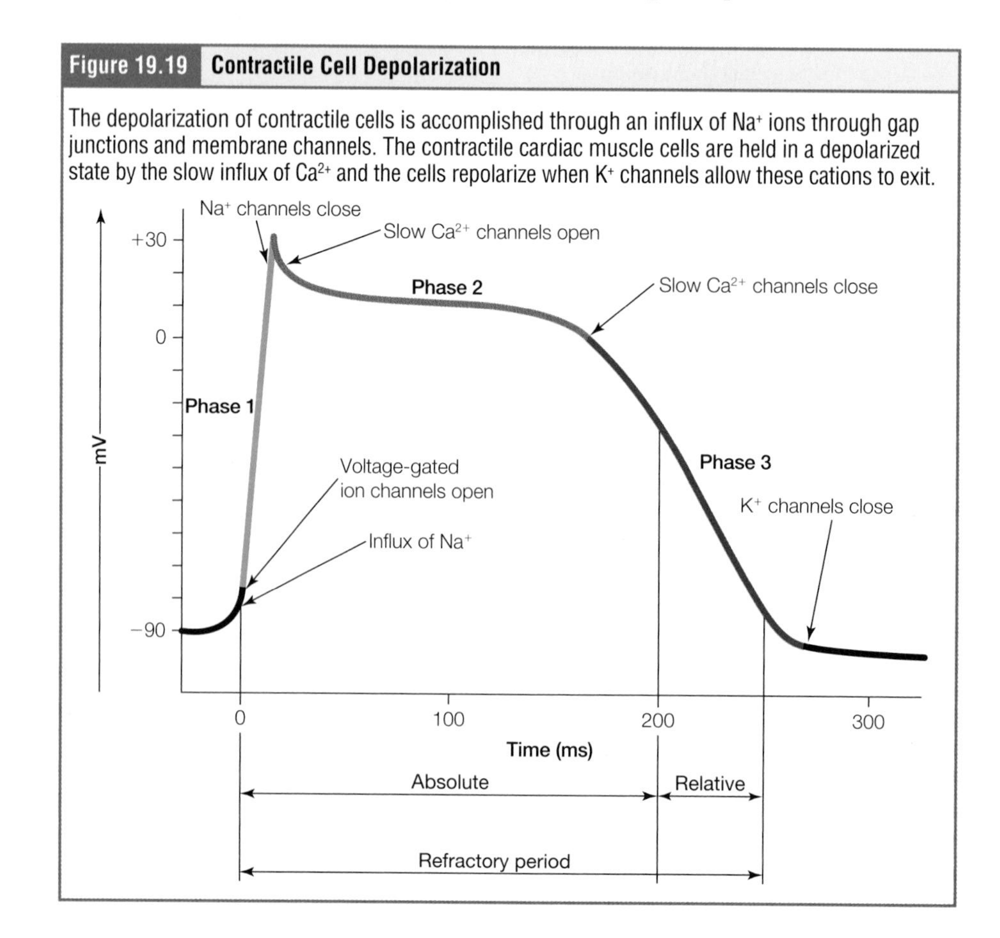

Figure 19.19 Contractile Cell Depolarization

The depolarization of contractile cells is accomplished through an influx of Na^+ ions through gap junctions and membrane channels. The contractile cardiac muscle cells are held in a depolarized state by the slow influx of Ca^{2+} and the cells repolarize when K^+ channels allow these cations to exit.

Table 19.1 Pacemaker Potential by Heart Location

Area of Conduction System	Number of Pacemaker Signals It Can Fire in 10 Seconds
Sinoatrial node 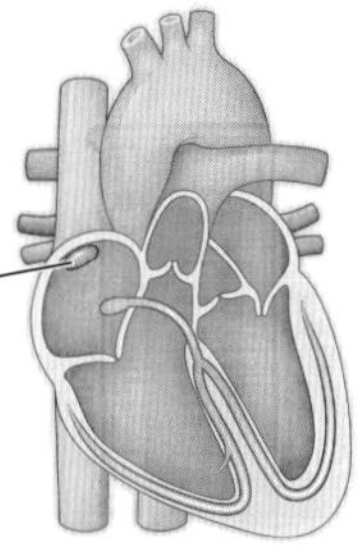	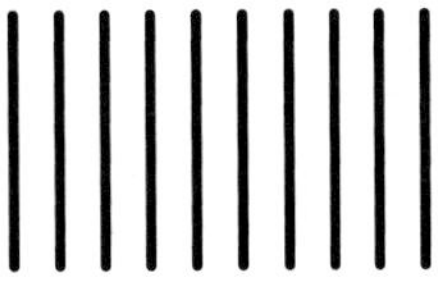
Atrioventricular node 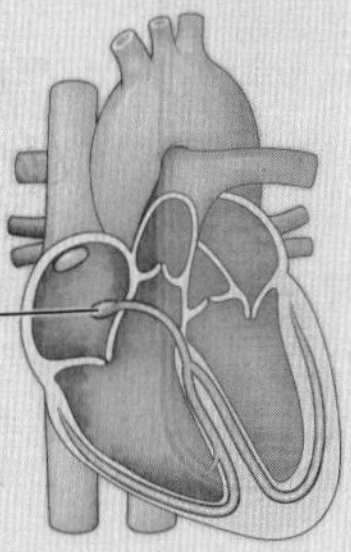	
AV bundle 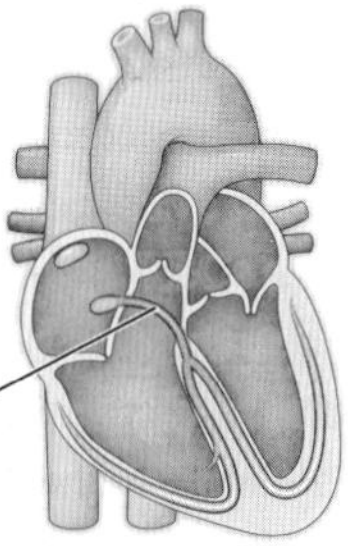	
Bundle branches 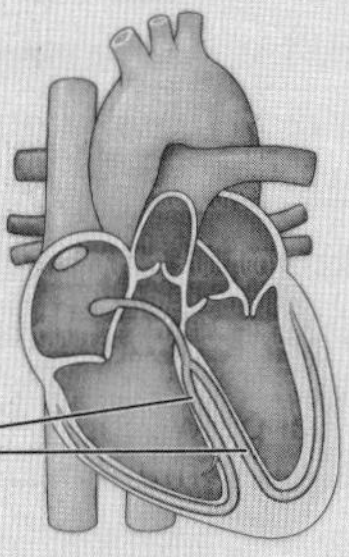	
Purkinje fibers 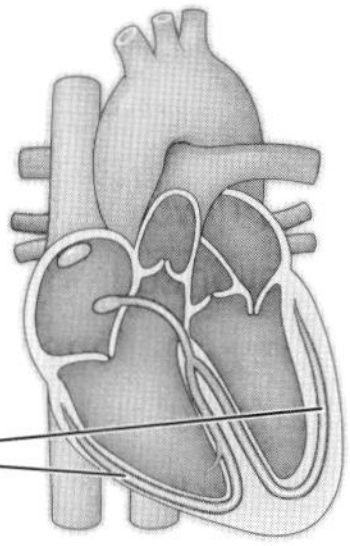	

the cell. The cell begins to repolarize, but a population of slow Ca^{2+} channels also opens around this time, allowing small amounts of Ca^{2+} to enter the cell at the same time as K^+ channels are just beginning to open and allow for K^+ to exit. With positively charged ions moving in both directions across the cell membrane, the cell is held in a steady state called the plateau phase (see phase 3 in Figure 19.19). Once the membrane potential reaches approximately 0 mV, the Ca^{2+} channels close and K^+ channels remain open, now that more positively charged ions are exiting the cell than entering, the cell repolarizes more rapidly.

LO 19.2.9

Student Study Tip

Think of cardiac muscle contraction as pumping a bicycle tire with air: One continuous push on the pump will not fill the tire. It needs to be released and reset in order to create enough pressure to push in more air.

Figure 19.20 compares the action potentials and mechanical contractions of skeletal and cardiac muscle cells. Both of these types of muscle cells have sodium and potassium voltage-gated channels, and yet the shape of their action potential graphs looks very different. This is due to the presence of the slow Ca^{2+} channels in cardiomyocytes. This extended plateau period is critical to the overall function of the heart. Examine in **Figure 19.20A** that the action potential of the skeletal muscle cell ends as the contraction begins. Therefore, the cell can fire another action potential while still contracting, permitting summation (**Figure 19.20B**). In contrast, the cardiomyocyte action potential is held in the plateau phase for almost the duration of the contraction, ensuring that summation cannot occur, and the cell will relax before another action potential and contraction can occur. This is critical because the filling of the ventricles can only happen when the cardiomyocytes are relaxed.

LO 19.2.10

Calcium Ions Calcium ions play two critical roles in the physiology of contractile cardiac muscle. Their influx through slow calcium channels accounts for the prolonged plateau phase that prevents summation in cardiac muscle cells and ensures cardiac muscle relaxation between contractions. Calcium ions also bind to troponin, shifting the troponin-tropomyosin complex and removing the inhibition that prevents the heads of the myosin molecules from forming cross bridges with actin. Essentially, just like in skeletal muscle cells, the presence of calcium in the cytosol releases the brakes on the contractile machinery and allows contraction to proceed. In skeletal muscle, calcium comes from only one source, the sarcoplasmic reticulum. In cardiac muscle cells, however, calcium ions come from both the sarcoplasmic reticulum as well as entering the cell from the extracellular environment through the slow Ca^{2+} channels during the plateau phase.

19.2d Electrocardiogram

electrocardiogram (ECG) (EKG)

By careful placement of surface electrodes (leads) on the body, it is possible to record the movement of charged ions through the heart as the myocardium depolarizes and repolarizes. This tracing of electrical signals is the **electrocardiogram (ECG)**. Careful analysis of the ECG reveals a detailed picture of heart function and is an indispensable and noninvasive clinical diagnostic tool. The ECGs can be performed using three, five, or twelve leads. The greater the number of leads an electrocardiograph uses, the more information the ECG provides; therefore, twelve-lead ECGs are most common. **Figure 19.21A** illustrates the placement of the leads. Note that only ten electrodes are placed on the skin; the other two "leads" are calculated from these. In an ECG we are taking a global snapshot of ion movement; this is different from the graphs of membrane potential we have been examining up until now. One way to consider an ECG is the ripples in a puddle caused by dropping a pebble. **Figures 19.21B** and **19.21C** illustrate the difference in scale of these two modes of examining electrical activity.

LO 19.2.11

Figure 19.22 shows a healthy ECG tracing. There are five prominent points on the ECG: The P wave, the three points of the QRS complex, and the T wave. The small **P wave** represents the ion concentration shifts caused by the depolarization of the atria. The atria begin contracting a short period of time after the start of the P wave. The large **QRS complex** represents the repolarization of the atria and the depolarization of the ventricles. These two events occur so closely in time that you cannot distinguish them on the ECG. The ventricles begin to contract as the QRS reaches the peak of the R wave. Lastly, the **T wave** represents the repolarization of the ventricles.

Figure 19.20 **The Relationship Between Action Potentials and Contraction**

In skeletal muscle (A) the action potential (blue) precedes and is completed before the mechanical contraction (purple) occurs. This allows for multiple electrical events to take place in rapid succession and results in a contraction that builds and does not relax (C). In contrast, in cardiac muscle (B) the action potential is held in a plateau while the contraction occurs and resolves. The cell does not repolarize and therefore cannot fire another action potential until relaxation has occurred. This prevents the contractions from building upon each other.

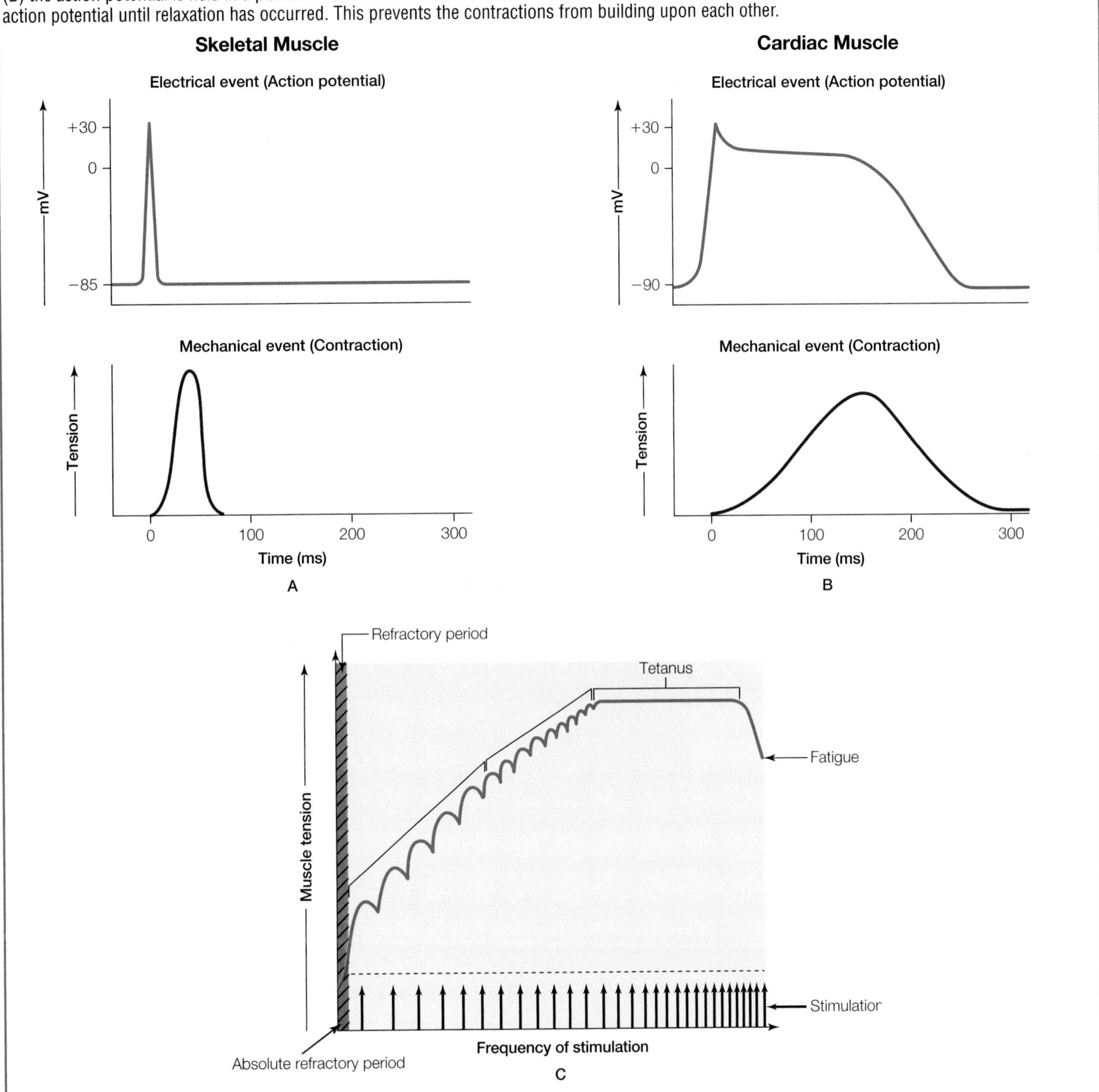

The repolarization of the atria occurs during the QRS complex, which masks it on an ECG. The ECG events, ventricular muscle cell depolarization, and contraction are all compared in **Figure 19.23**.

As a clinician examines an ECG, they are typically looking less at the waves themselves than at the intervals of time between them. The segments and intervals of an ECG tracing are indicated in **Figure 19.24**. Segments are defined as the regions between two waves. Intervals include one segment plus one or more waves. For example, the PR

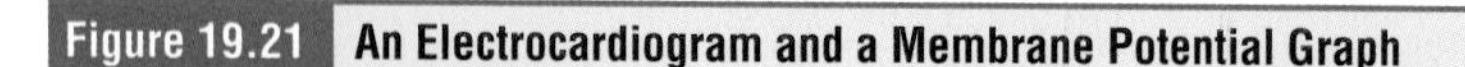

Figure 19.21 An Electrocardiogram and a Membrane Potential Graph

(A) The position of the ten leads in an EKG. (B) An EKG is a sampling of the ion movement all over the body. (C) In contrast, a membrane potential graph is examining ion movement on a much smaller scale, just from one side of a cell membrane to the other.

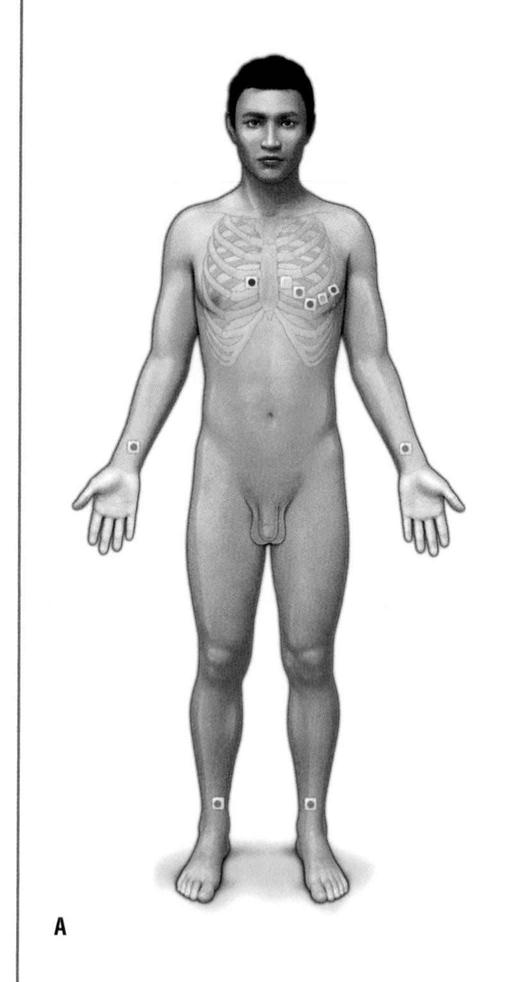

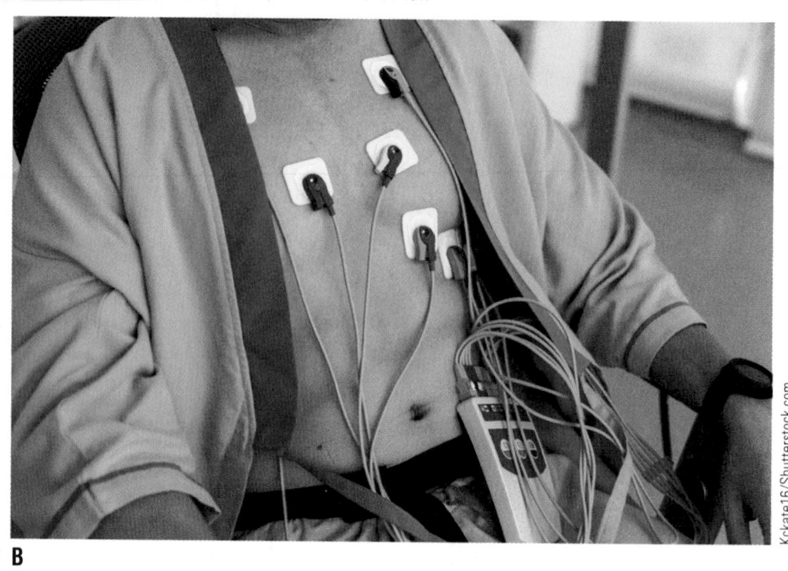

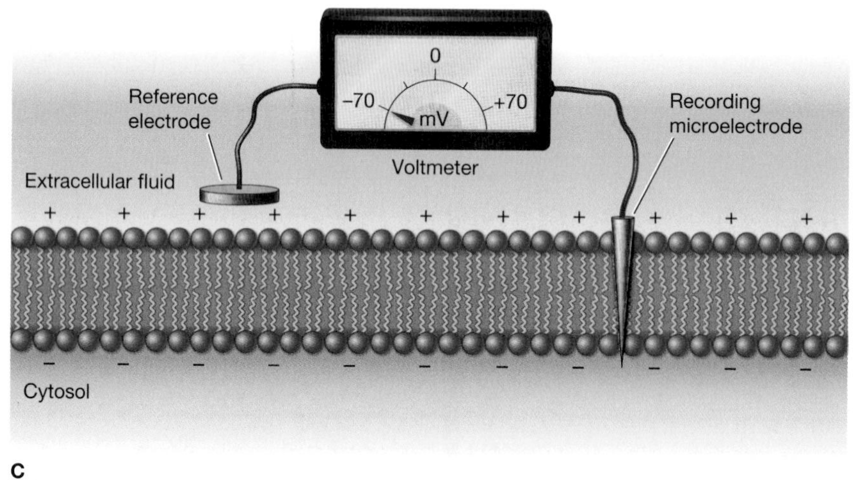

segment begins at the end of the P wave and ends at the beginning of the QRS complex. The PR interval starts at the beginning of the P wave and ends with the beginning of the QRS complex. The PR interval is more clinically relevant; should there be a delay in passage of the impulse from the SA node to the AV node, it would be visible in the PR interval. Figure 19.24 correlates events of heart contraction to the corresponding segments and intervals of an ECG.

LO 19.2.12

Coordinated heart contraction is dependent upon the fluid current of the action potential through the heart's conduction system as well as the seamless flow of the action potential from one contractile cardiomyocyte to the next. But what would happen if there were patches of dead or nonconductive cardiomyocytes? In cases where the action potential cannot pass seamlessly throughout the contractile cardiomyocytes of the atria or the ventricles, a condition known as **fibrillation** can result. In fibrillation, the heart beats in a wild, uncoordinated manner, which prevents it from being able to pump effectively. Earlier in this chapter we compared the coordinated inferior-to-superior squeeze of the ventricles to squeezing a tube of toothpaste. If we can extend this analogy

Figure 19.22 The Waveforms, Segments, and Intervals of an ECG

An ECG represents the events that occur during a heartbeat. It consists of waveforms, segments and intervals, each of which represents a specific event in the cardiac cycle.

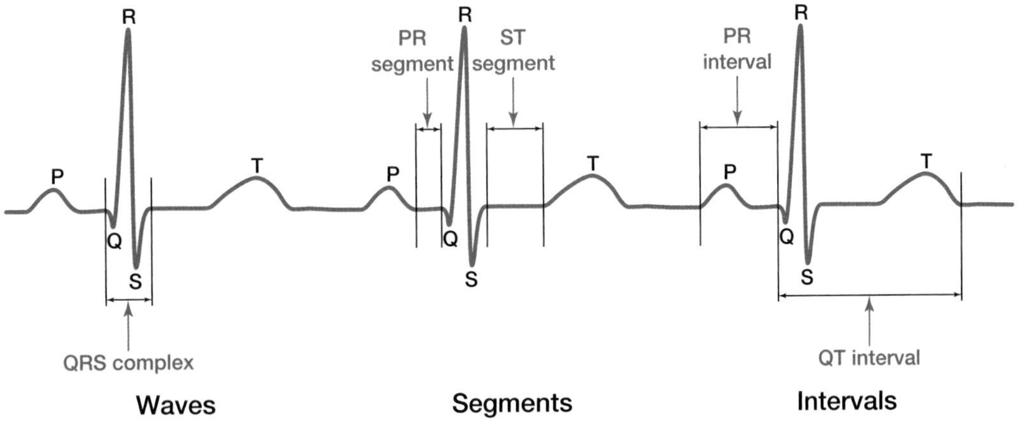

Figure 19.23 The Relationship Between the Membrane Potential Graph and the QRS Complex

The membrane potential graph of individual contractile cardiomyocytes corresponds with the QRS complex of the ECG.

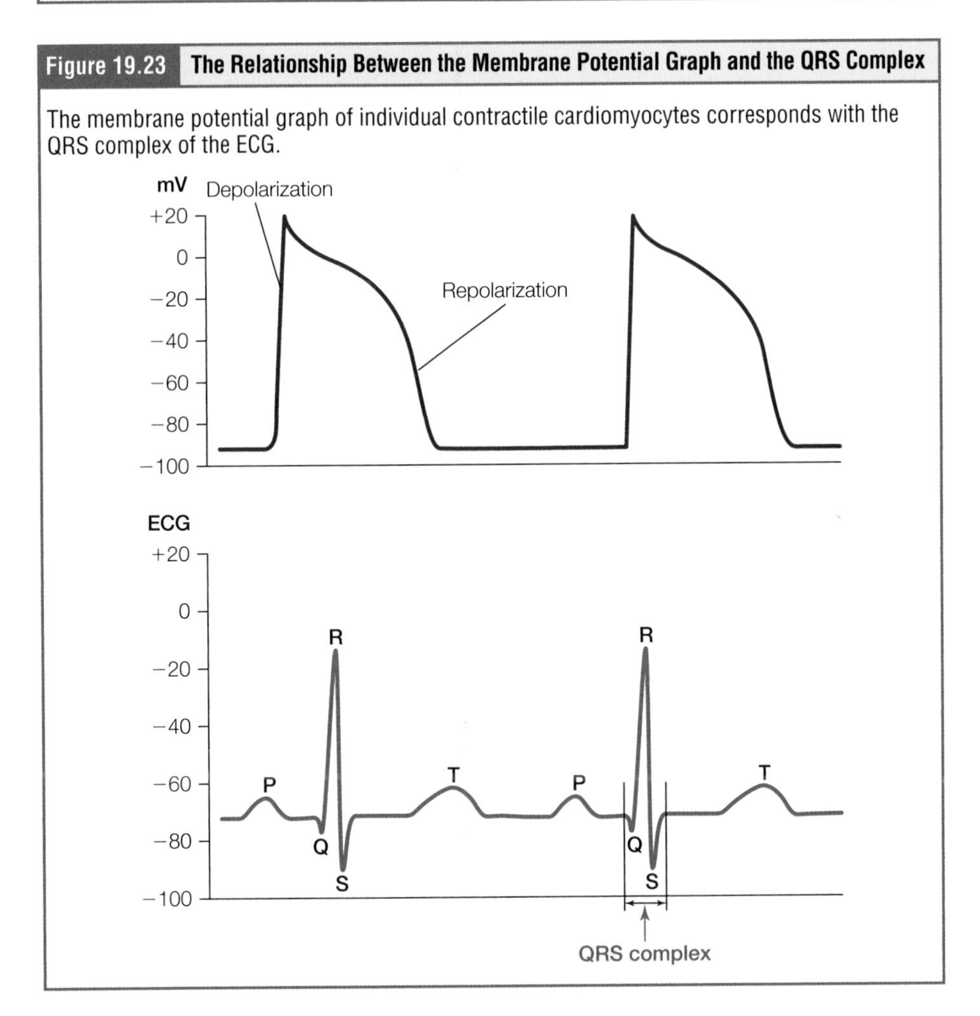

to fibrillation, the uncoordinated contractions of fibrillation would be akin to poking the tube of toothpaste here and there without organizing a squeeze all the way around its diameter. Atrial fibrillation is a serious condition, but as long as the ventricles continue to pump blood, the patient's life may not be in immediate danger. This is because about 80 percent of ventricular filling occurs by gravity; atrial contraction accounts for just 20 percent of ventricular filling. Ventricular fibrillation, on the other hand, is a medical emergency that requires life support, because the ventricles are not effectively pumping

Figure 19.24 The Relationship Between an ECG Tracing and the Electrical Events of the Cardiac Cycle

The segments and waves of the ECG correspond to the electrical events in the conduction system of the heart.

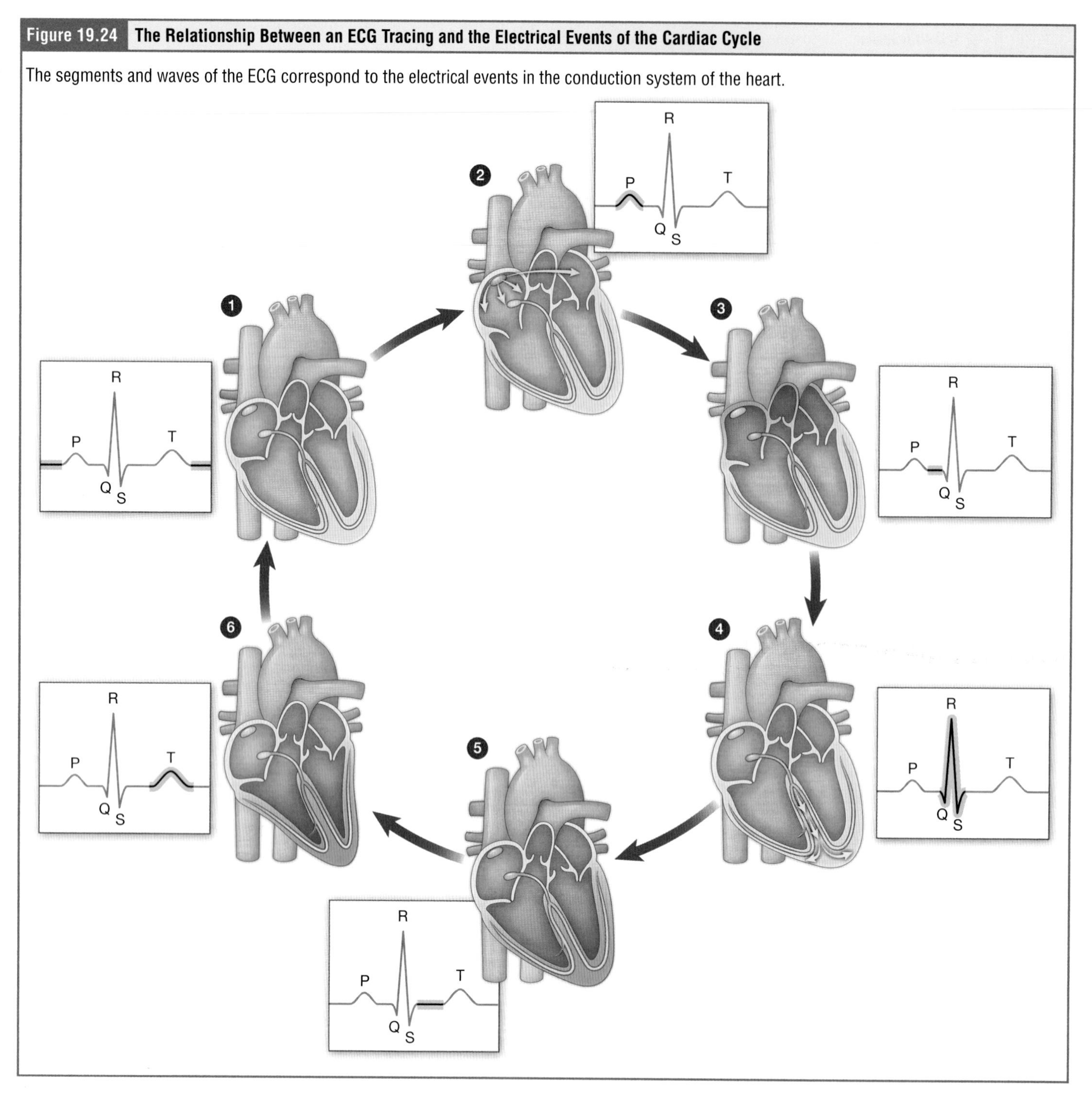

blood. In a hospital setting, it is often described as "code blue." If untreated for as little as a few minutes, ventricular fibrillation may lead to brain death. The most common treatment is defibrillation, in which special paddles are used to apply an electrical charge to the heart in an attempt to establish a normal sinus rhythm. A defibrillator (**Figure 19.25**) effectively stops the heart. If the SA node is able to trigger a normal conduction cycle, the heart will begin to beat again. Because of their effectiveness in reestablishing a normal sinus rhythm, automated external defibrillators (AEDs) are being placed in areas frequented by large numbers of people, such as schools, restaurants, and airports. These devices contain simple and direct verbal instructions that can be followed by nonmedical personnel in an attempt to save a life.

Figure 19.25 **A Defibrillator**

A defibrillator is a device that is used to treat abnormal heart rhythms. It essentially stops the heart in an attempt to allow the SA node to reestablish a normal heart rhythm.

Dario Lo Presti/Shutterstock.com

Apply to Pathophysiology

Heart Block

A **heart block** refers to an impairment in the conduction pathway. A block can occur at any point in the conduction pathway, but we will consider blocks between the SA node and the AV node.

AV blocks are often described by degrees. A first-degree or partial block indicates a delay in conduction between the SA and AV nodes. Every action potential still reaches the AV node, but it takes longer to get there than is typical. This condition is typically asymptomatic.

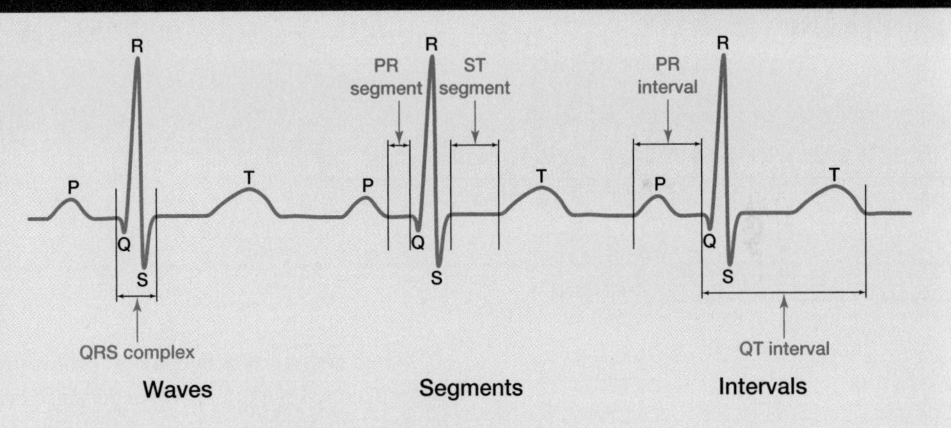

1. In a case of first-degree AV block, would the atria still contract normally?

The figure at right relates to questions 2 through 5.

2. The figure shows a healthy ECG as an example. Which of the other ECGs likely was taken from an individual with first-degree AV block?

A second-degree or incomplete block occurs when some impulses from the SA node reach the AV node and continue, while others do not. What do you expect to see on the ECG of an individual with second-degree AV block? Second-degree AV block results in symptoms of inadequate blood flow such as dizziness, lightheadedness, and fatigue.

3. The figure shows a healthy ECG as an example. Which of the other ECGs likely was taken from an individual with second-degree AV block?

(*Continued*)

In third-degree or total AV block, none of the impulses from the SA node reach the AV node. In these cases, the ventricles will not contract and a medical emergency ensues.

4. The figure shows a healthy ECG as an example. Which of the other ECGs likely was taken from an individual with third-degree AV block?
5. Compare and contrast B and C. These ECGs were both taken from patients in fibrillation (one atrial, the other ventricular). Which tracing represents atrial fibrillation and which one represents ventricular fibrillation?

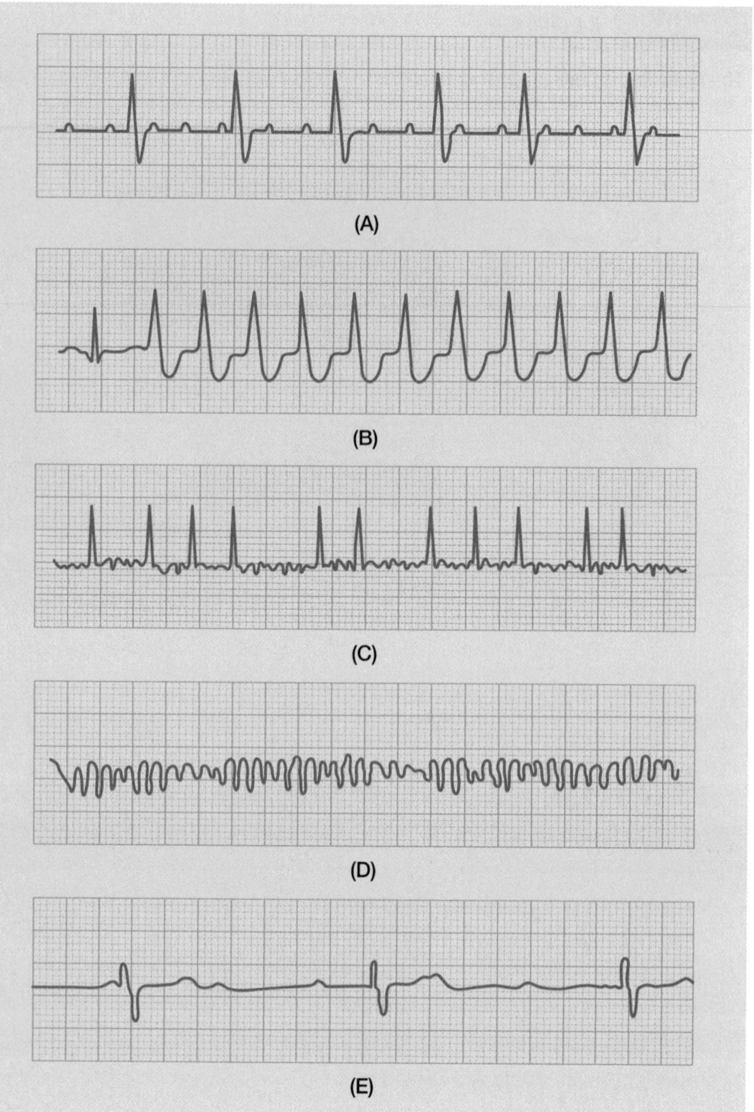

When arrhythmias become a chronic problem, the AV node may take over the role of pacemaker. The patient will experience bradycardia and may struggle with adequate blood flow. In order to speed up the heart rate and restore full sinus rhythm, a cardiologist can implant an **artificial pacemaker**, which delivers electrical impulses to the heart muscle to ensure that the heart continues to contract and pump blood effectively. These artificial pacemakers are programmable by the cardiologists and can either provide stimulation temporarily upon demand or on a continuous basis. Some devices also contain built-in defibrillators.

19.2e Nervous System Influence on Heart Activity

LO 19.2.13

While the SA node is responsible for initiating the electrical signals that initiate contraction, the autonomic nervous system influences both heart rate and the strength of ventricular contraction based on physiological need. Nervous system control over HR is centralized within the two cardiovascular centers (cardioaccelerator and cardioinhibitory centers) of the medulla oblongata (**Figure 19.26**). The inputs to these centers come from baroreceptors and chemoreceptors found in the aortic arch and carotid arteries that fire signals toward the cardiovascular centers based on blood pressure (baroreceptors) and oxygen and acid content

Figure 19.26 Cardiac Autorhythmicity

The heart has its own autorhythmicity but is influenced by the autonomic nervous system.

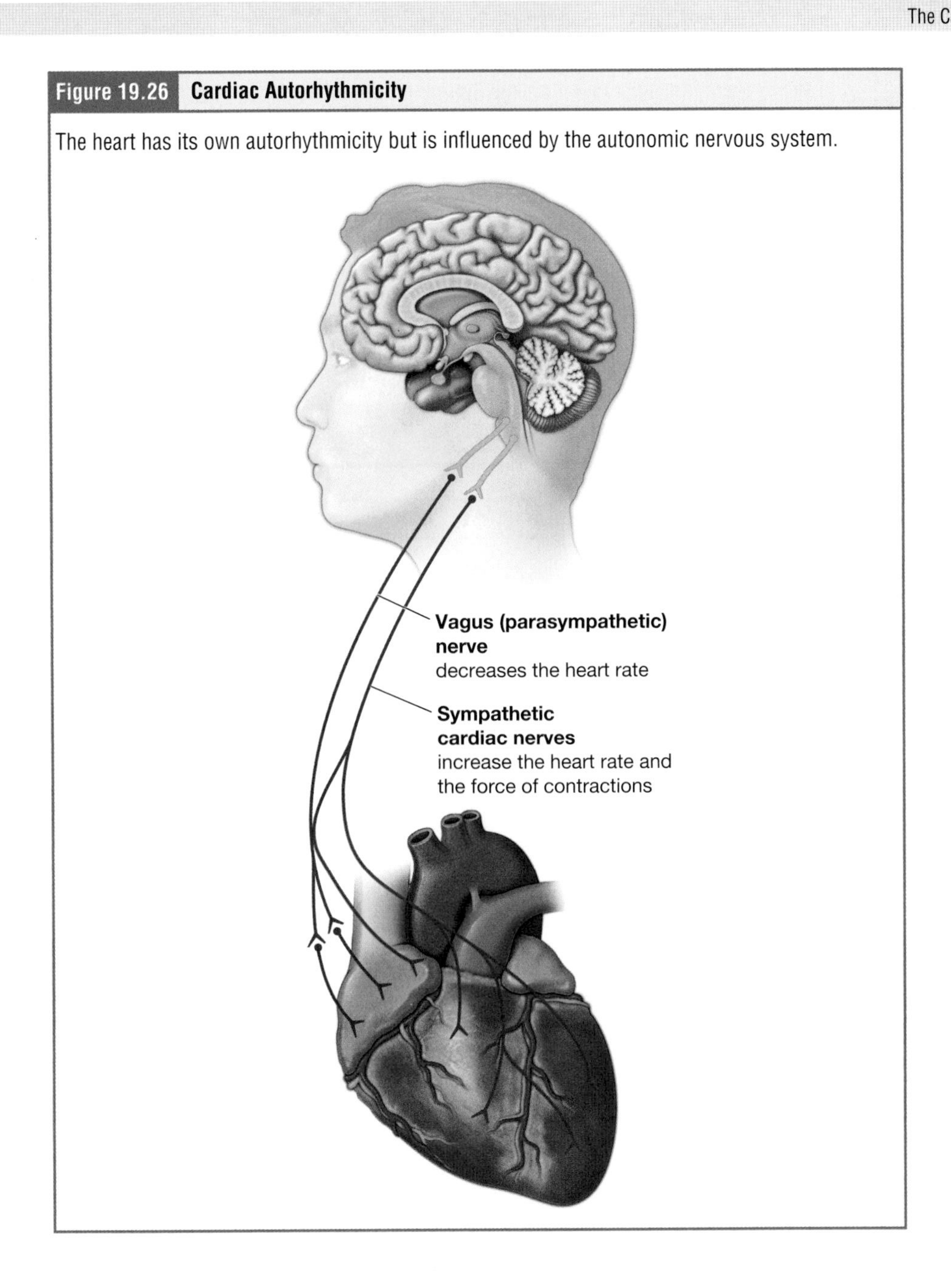
Vagus (parasympathetic) nerve
decreases the heart rate
Sympathetic cardiac nerves
increase the heart rate and the force of contractions

(chemoreceptors). The cardioaccelerator regions send signals to the heart via sympathetic stimulation of the cardioaccelerator nerves, and the cardioinhibitory centers send signals to the heart via parasympathetic stimulation as one component of the vagus nerve, cranial nerve X.

The vagus nerve innervates many thoracic and abdominal organs (see Chapter 15). Branches of the vagus nerve innervate the SA and AV nodes, increase departure of K^+ ions from nodal cells, and decrease their rate of firing. Thus, stimulation from the cardioinhibitory center, via the parasympathetic vagus nerve, decreases heart rate.

LO 19.2.14

The cardioaccelerator region sends signals to the SA node, the AV node, and also the ventricular myocardium via sympathetic nerve fibers altering their permeability to Ca^{2+} ions. Ca^{2+} ions play a role both in the depolarization of nodal cells and in the contraction of cardiac contractile cells. Therefore, the cardioaccelerator region is capable of both speeding up the heart rate and increasing the force of ventricular contraction (causing the ventricles to contract faster and harder). It is likely you have felt the impact of the sympathetic nervous system on your heart at some point, perhaps

Figure 19.27 The Effects of the Autonomic Nervous System on Heart Rate

The autonomic nervous system influences heart rate.

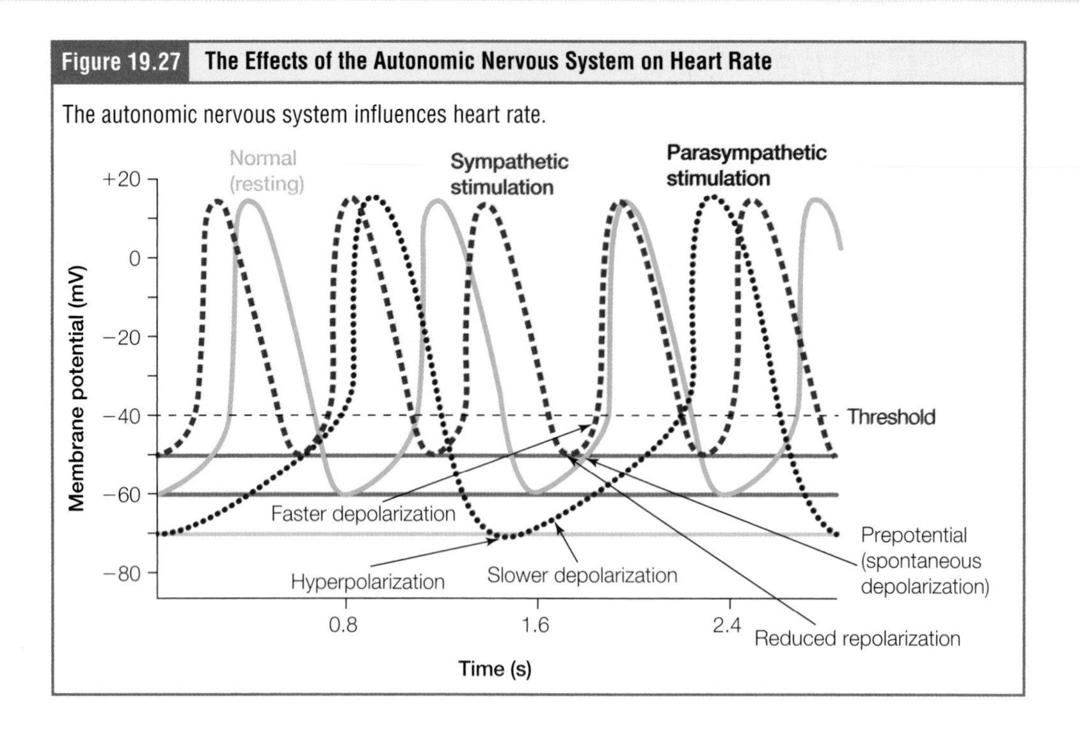

when watching a scary movie or riding a roller coaster. During these moments you can feel your heart beating harder than it typically does; this is due to the sympathetic innervation to the ventricle walls. By contrast, the parasympathetic nervous system can slow heart rate, but it cannot directly affect the force of contraction. **Figure 19.27** illustrates the effects of parasympathetic and sympathetic stimulation on the normal sinus rhythm.

Learning Check

1. At which point during an action potential do the calcium ion channels open in a cardiac *conducting* cell?
   a. When the membrane depolarizes past the threshold
   b. When the membrane reaches the peak of depolarization
   c. When the membrane is repolarized
2. At which point during an action potential do the calcium ion channels open in a cardiac *contractile* cell?
   a. When the membrane depolarizes past the threshold
   b. Just after the membrane reaches the peak of depolarization
   c. When the membrane is repolarized
3. Which of the following defines the QRS complex? Please check all that apply.
   a. Depolarization of the atria
   b. Repolarization of the atria
   c. Depolarization of the ventricle
   d. Repolarization of the ventricle
4. Which of the following events is caused by the cardioinhibitory center, in order to decrease heart rate?
   a. Increasing permeability of potassium ions
   b. Increasing permeability of sodium ions
   c. Decreasing permeability of potassium ions
   d. Decreasing permeability of sodium ions

19.3 Cardiac Cycle

Learning Objectives: By the end of this section, you will be able to:

19.3.1 Define cardiac cycle, systole, and diastole.

19.3.2 Describe the phases of the cardiac cycle, including ventricular filling, isovolumic (isovolumetric) contraction, ventricular ejection, and isovolumic (isovolumetric) relaxation.

19.3.3 Relate the electrical events represented on an electrocardiogram (ECG or EKG) to the normal mechanical events of the cardiac cycle.

19.3.4 Explain how atrial systole is related to ventricular filling.

19.3.5 Define end diastolic volume (EDV) and end systolic volume (ESV), and calculate stroke volume (SV) given values for EDV and ESV.

19.3.6 Relate the opening and closing of specific heart valves in each phase of the cardiac cycle to pressure changes in the heart chambers and the great vessels (i.e., blood vessels entering and leaving the heart).

19.3.7 Compare and contrast pressure and volume changes of the left and right ventricles during one cardiac cycle.

19.3.8 Define systolic and diastolic blood pressure and interpret a graph of aortic pressure versus time during the cardiac cycle.

19.3.9 Relate the heart sounds to the events of the cardiac cycle.

19.3.10* Draw and explain the waves of Wiggers Diagram.

* Objective is not a HAPS Learning Goal.

LO 19.3.1

The period of time and set of events that begins with contraction of the atria and ends with the relaxation of the ventricles is known as the **cardiac cycle** (**Figure 19.28**). The terms **systole** and **diastole** are used to discuss the contraction and relaxation, respectively, of the chambers. For example, *atrial diastole* refers to when the atria are relaxed, and *ventricular systole* refers to ventricular contraction. There is a period of time when both the atria and the ventricles are relaxed and we can refer to this as *total diastole*; the atria and ventricles are never in systole at the same time.

19.3a Pressures and Flow

When we consider the movement of individual molecules, we discuss their movement in response to concentration and sometimes electrical gradients. But when we consider the movement of a lot of molecules together in a solution (bulk flow), another force dictates the movement of the solution. Both air and blood are examples of fluids, and all fluids flow according to pressure gradients—that is, they move from regions that are higher in pressure to regions that are lower in pressure. The heart is a muscle pump; the contraction of the chambers increases the pressure on the fluid within. Accordingly, when the heart chambers are relaxed (diastole), the internal pressure is low, and so blood will flow into the atria from the veins, which are higher in pressure. As blood flows into the atria, the interatrial pressure will rise, so the blood will initially move passively from the atria into the ventricles (gravity is also a factor in this flow). When the action potential triggers the atrial myocardium to contract (atrial systole), the pressure within the atria rises further, pumping blood into the ventricles. During ventricular systole, pressure rises in the ventricles, pumping blood into the lower-pressure pulmonary trunk

Figure 19.28 The Stages of the Cardiac Cycle

(1) In total diastole, the heart is completely relaxed. It is the point of lowest pressure in the system and so blood flows here filling up the atria and the ventricles passively. (2) In atrial systole, the atrial walls contract, squeezing the blood within the atria through the atrioventricular valves and into the ventricles. (3) In early ventricular systole, the ventricular walls begin to contract, squeezing on the blood within the ventricles. The blood is pushed up against the bottom of the atrioventricular valves, closing them. (4) In late ventricular systole, the ventricles squeeze hard, forcing the blood up against the bottom of the semilunar valves, opening the valves and pushing blood through and into the arteries. (5) In early ventricular diastole, the ventricular walls begin to relax. They become the point of lowest pressure and the blood in the arteries sloshes backwards toward them. The blood catches on the underside of the semilunar valve cusps, forcing the valves closed and preventing backflow.

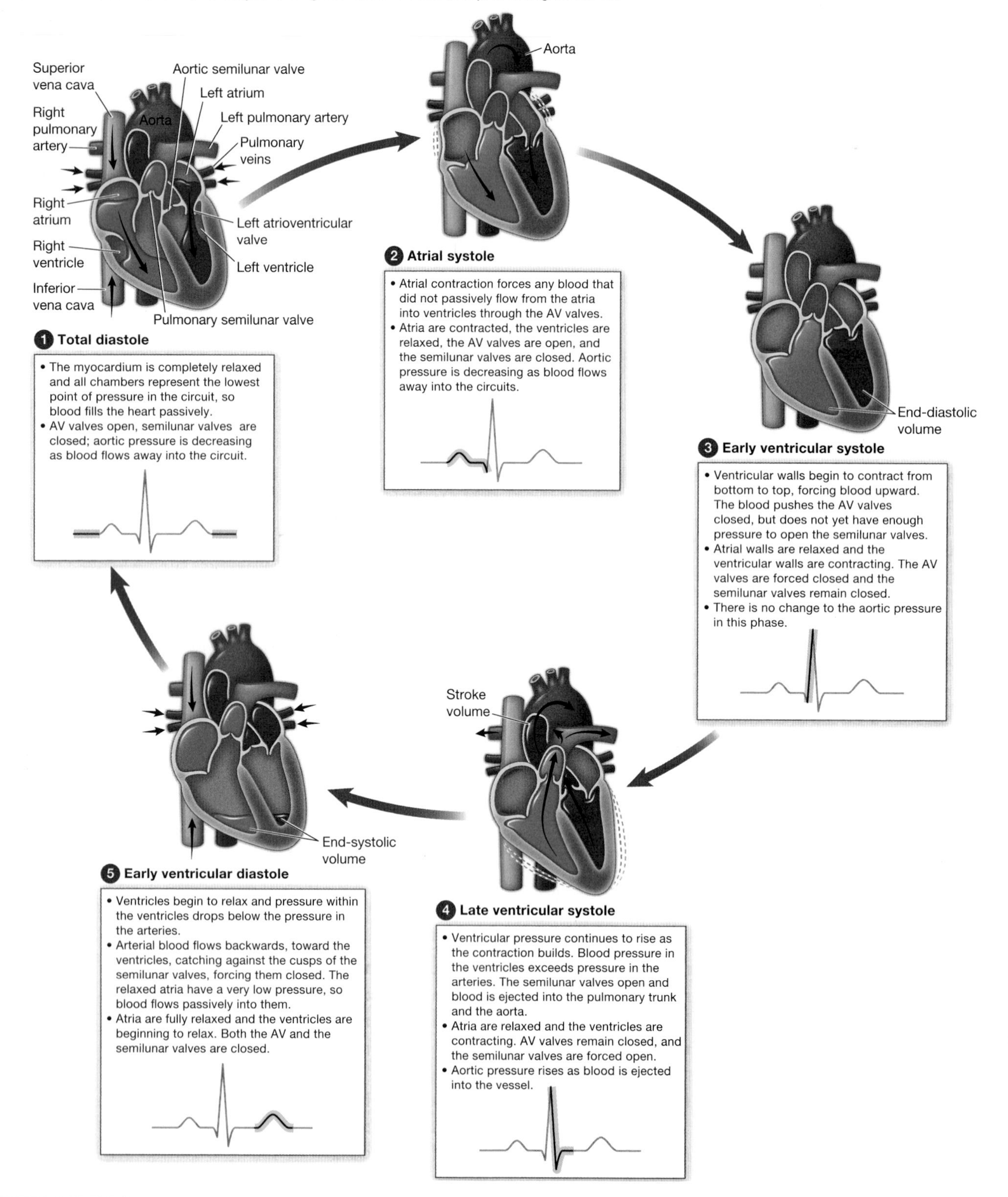

Superior vena cava
Aortic semilunar valve
Left atrium
Right pulmonary artery
Aorta
Left pulmonary artery
Pulmonary veins
Right atrium
Left atrioventricular valve
Right ventricle
Left ventricle
Inferior vena cava
Pulmonary semilunar valve
1 Total diastole
• The myocardium is completely relaxed and all chambers represent the lowest point of pressure in the circuit, so blood fills the heart passively.
• AV valves open, semilunar valves are closed; aortic pressure is decreasing as blood flows away into the circuit.
Aorta
2 Atrial systole
• Atrial contraction forces any blood that did not passively flow from the atria into ventricles through the AV valves.
• Atria are contracted, the ventricles are relaxed, the AV valves are open, and the semilunar valves are closed. Aortic pressure is decreasing as blood flows away into the circuits.
End-diastolic volume
3 Early ventricular systole
• Ventricular walls begin to contract from bottom to top, forcing blood upward. The blood pushes the AV valves closed, but does not yet have enough pressure to open the semilunar valves.
• Atrial walls are relaxed and the ventricular walls are contracting. The AV valves are forced closed and the semilunar valves remain closed.
• There is no change to the aortic pressure in this phase.
Stroke volume
4 Late ventricular systole
• Ventricular pressure continues to rise as the contraction builds. Blood pressure in the ventricles exceeds pressure in the arteries. The semilunar valves open and blood is ejected into the pulmonary trunk and the aorta.
• Atria are relaxed and the ventricles are contracting. AV valves remain closed, and the semilunar valves are forced open.
• Aortic pressure rises as blood is ejected into the vessel.
End-systolic volume
5 Early ventricular diastole
• Ventricles begin to relax and pressure within the ventricles drops below the pressure in the arteries.
• Arterial blood flows backwards, toward the ventricles, catching against the cusps of the semilunar valves, forcing them closed. The relaxed atria have a very low pressure, so blood flows passively into them.
• Atria are fully relaxed and the ventricles are beginning to relax. Both the AV and the semilunar valves are closed.

and aorta (remember that blood cannot flow backwards into the atria because the AV valves closed in early (isovolumetric) ventricular systole). When we consider pressure within the cardiovascular system, we can consider two sources of pressure—one being caused by the contraction of muscular walls and the other being caused by the addition of more fluid to a chamber, increasing the pressure within. To visualize this, it can be helpful to imagine a water balloon. Simply adding more water to the balloon increases the pressure within as is evident by the stretching of the balloon walls. Squeezing on the balloon from the outside further drives up the pressure (**Figure 19.29**).

19.3b Phases of the Cardiac Cycle

LO 19.3.2

At the beginning of the cardiac cycle, (see #1 in Figure 19.28) both the atria and ventricles are relaxed (total diastole). At this point, with no contraction from the cardiac muscle and very low blood volume within the chambers, pressure within the heart is at its lowest point. Since blood flows from areas of greater pressure to areas of lesser pressure, blood is flowing into the right atrium from the superior and inferior venae cavae and the coronary sinus. Blood is flowing into the left atrium from the pulmonary veins. The two atrioventricular valves are both open, so blood flows unimpeded from the atria and into the ventricles. Approximately 80 percent of ventricular filling occurs passively during total diastole because the ventricles are at a lower pressure than the rapidly filling atria. The two semilunar valves are closed, preventing any flow of blood between the ventricles and the pulmonary trunk and aorta.

Figure 19.29 Hydrostatic Pressure

Hydrostatic pressure can be illustrated using water balloons (DO try this at home!).

Exopixel/Shutterstock.com

Desislava Panteva/EyeEm/Getty Images

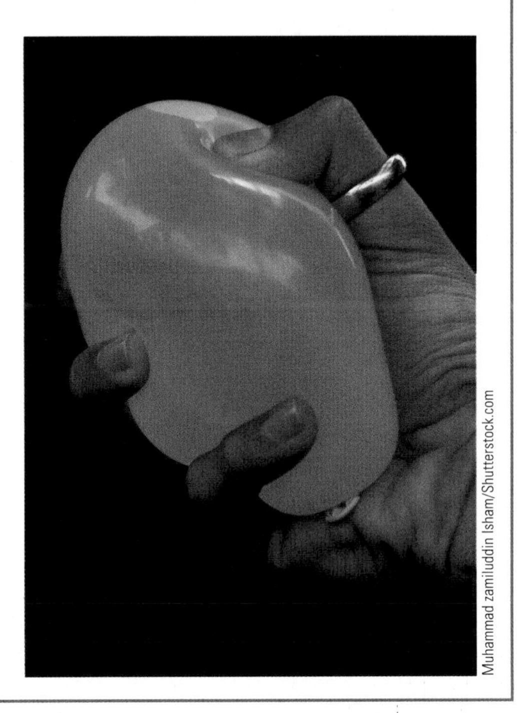

Muhammad zamiluddin Isham/Shutterstock.com

Atrial Systole and Diastole The terms systole and diastole describe both phases of the cardiac cycle (see Figure 19.28) and the contraction and relaxation of the chambers themselves. As the action potential passes from the SA node throughout the atrial cardiomyocytes, the atria depolarize and the P wave should be evident on the ECG (see #2 in Figure 19.28). As the atrial muscles depolarize and contract, pressure rises within and blood is pumped into the ventricles through the open atrioventricular valves. At the start of atrial systole, the ventricles are already filled with approximately 80 percent of their capacity due to inflow during diastole (see #1 in Figure 19.28). Atrial contraction contributes the remaining 20 percent of filling. As the impulse reaches the AV node and proceeds down the bundle branches, the atria repolarize and relax (repolarization of the atria is part of the QRS complex on the ECG). At the end of atrial systole and just prior to ventricular contraction, the ventricles contain approximately 130 mL blood in a resting adult in a standing position. This volume is known as the **end-diastolic volume (EDV)** or preload. After systole, the atria begin to relax once again. Once the walls have relaxed sufficiently so that the pressure within the atria is lower than the pressure within the venae cavae and pulmonary veins, the atria begin to fill once again.

LO 19.3.3

LO 19.3.4

LO 19.3.5

end-diastolic volume (EDV) (preload)

Ventricular Systole The action potential leaves the AV node and travels down the bundle branches. As the conduction system branches into the Purkinje fibers at the heart's apex, the action potential begins to travel throughout the ventricular wall from cardiomyocyte to neighboring cardiomyocytes through their gap junctions. Both because of this anatomical direction of electrical signal spread as well as because of arrangement of cardiomyocytes (the myocardial vortex), the ventricle walls contract from inferior apex toward the atrial. This ventricular systole (see #3 in Figure 19.28) follows the depolarization of the ventricles represented by the QRS complex in the ECG. As the muscles in the ventricle contract, the pressure of the blood within the chamber rises quickly (refer to Figure 19.31). Even in the initial phases of ventricular contraction, blood pressure in the ventricles quickly rises above that of the atria that are now relaxed and in diastole. This combination of pressure differences and inferior-to-superior direction of contraction causes blood to flow toward the atria, pushing against the underside of the atrioventricular valves, forcing them closed. This early phase of ventricular contraction is called **isovolumic contraction**, also called isovolumetric contraction (see #2 in Figure 19.28) because in these initial phases the ventricular blood pressure is not yet great enough to force the semilunar valves open; therefore, though there is a significant rise in blood pressure, there is no change to blood volume. Compare and contrast the tracings of ventricular pressure and ventricular volume in Figure 19.31. This phase of the cardiac cycle is somewhat akin to squeezing on a tied-off water balloon; the pressure is increasing dramatically because of the contraction of the ventricular walls although the blood has nowhere to go.

LO 19.3.6

isovolumic contraction (isovolumetric contraction)

LO 19.3.7

In the second phase of ventricular systole, the **ventricular ejection phase**, the continued contraction of the ventricular muscle has raised the pressure within the ventricle to the point that it is greater than the pressures in the pulmonary trunk and the aorta (the pressures within the aorta and pulmonary trunk that the ventricles must overcome is known as afterload). The semilunar valves are forced open and blood flows from the greater pressure within the ventricles to the lower pressure in the leaving vessels. Pressure generated by the left ventricle needs to be appreciably greater than the pressure generated by the right ventricle because the existing pressure in the aorta will be so much higher than the existing pressure in the pulmonary trunk and arteries.

Nevertheless, both ventricles pump the same amount of blood. The quantity of blood ejected from the ventricles in one systole is referred to as stroke volume. Stroke volume will normally be in the range of 70–80 mL. Since ventricular systole began with an EDV of approximately 130 mL of blood, this means that there is still 50–60 mL of blood remaining in the ventricle following contraction. This volume of blood is known as the **end-systolic volume (ESV)**.

One thing you may have noticed is that in all of these measurements we refer to the left ventricle. Do you think the same things are happening in the right ventricle? We know that the action potential signal travels down both bundle branches and throughout the Purkinje fibers at the same time; therefore, both ventricles contract together. They do not, however, exert the same pressure. Remember that the left ventricle has a much thicker myocardium than the right; it is therefore capable of exerting much more force on the blood within the chamber. This is important because blood exiting the left side of the heart has a much larger circuit to travel in than the smaller pulmonary circuit. What about volume? Do you think both ventricles eject the same volume? Let's consider this. The ventricles are intimately connected both at their shared septum but also through the pulmonary circuit. It is imperative that the right and left ventricles eject the same amount of blood, so that blood, ejected from the right and traveling through the pulmonary circuit, eventually fills the left ventricle to be ejected to the body.

Learning Connection

Broken Process

What happens if the ventricles eject different volumes of blood? Imagine the scenario that would result if the right ventricle ejected more blood than the left? What about if the left ejected a higher volume than the right?

Ventricular Diastole Ventricular relaxation, or diastole, follows repolarization of the ventricles. Ventricular repolarization is represented by the T wave of the ECG. Like ventricular systole, it also has two distinct phases.

During the early phase of ventricular diastole, as the ventricular muscle relaxes, pressure on the remaining blood within the ventricle begins to fall. When pressure within the ventricles drops below pressure in both the pulmonary trunk and aorta, blood flows back toward the lower pressure ventricles. As this backward-flowing blood approaches the semilunar valves, it catches on the underside of the valve cusps, closing the semilunar valves and producing the second heart sound, the "dub" of the "lub-dub." Since the atrioventricular valves remain closed at this point, there is no change in the volume of blood in the ventricle, so the early phase of ventricular diastole is called the **isovolumic ventricular relaxation phase**, also called isovolumetric ventricular relaxation phase (see #2 in Figure 19.28). If we examine the tracing of the blood pressure in the aorta, we can see the pressure was rising steadily throughout ventricular contraction as blood was pushed into the aorta (see Figure 19.31), then during isovolumic ventricular relaxation, as the blood in the aorta sloshes backward toward the low-pressure ventricles the aortic pressure takes a small dip. As the blood catches on the semilunar valve cusps and turns around, aortic pressure rises once again. This momentary dip in aortic pressure is known as the **dicrotic notch** (see Figure 19.31).

Student Study Tip

"Iso-" means equal. So "isovolumetric" tells you the volume of blood doesn't change. Thus, no valves are open during these phases because blood is not moving chambers.

In the second phase of ventricular diastole, called **late ventricular relaxation**, the ventricular muscle continues to relax and pressure on the blood within the ventricles drops even further. Eventually, it drops below the pressure in the atria. When this occurs, blood flows from the now higher-pressure atria into the now lower-pressure ventricles, pushing open the atrioventricular valves. As pressure drops within the ventricles, blood flows from the major veins into the relaxed atria and from there into the ventricles. Both chambers are in diastole, the atrioventricular valves are open, and the semilunar valves remain closed (see #4 in Figure 19.28). The cardiac cycle is complete.

LO 19.3.8

Are the words *systole* and *diastole* familiar to you? If you have had your blood pressure taken before, you might remember that the values of blood pressure are recorded as systolic and diastolic, for example, 120 systolic/80 diastolic. The connection between these concepts is that blood pressure is measured, usually from the brachial artery in the arm, at the point of highest pressure in that artery, which corresponds to ventricular systole and ejection of blood into the artery, and the point of lowest pressure in the artery, which corresponds to ventricular diastole. In **Figure 19.30**, you can see how the pressure within the aorta varies during the cardiac cycle. Parallel (though lower pressure) variations occur in all of the arteries of the body.

Student Study Tip

Apply the cardiac cycle to an ECG tracing. The atria do not have as strong of a contraction or as much of an electrical impulse as the ventricles.

19.3c Heart Sounds

LO 19.3.9

As noted in other portions of this chapter, the heart sounds are produced by the closing of the valves of the heart. In a healthy heart, there are only two audible **heart sounds** referred to as S_1 and S_2. S_1 is the sound created by the closing of the atrioventricular valves during ventricular contraction and is normally described as a "lub," or first heart sound. The second heart sound, S_2, is the sound of the closing of the semilunar valves during early ventricular diastole and is described as a "dub" (see Figure 19.31).

The term **murmur** is used to describe an unusual sound coming from the heart that is caused by the turbulent flow of blood. During auscultation, it is common practice for the clinician to ask the patient to breathe deeply. This procedure not only allows for listening to airflow, but it may also amplify heart murmurs. Inhalation increases blood flow into the right side of the heart and may increase the amplitude of right-sided heart murmurs. Expiration partially restricts blood flow into the left side of the heart and may amplify left-sided heart murmurs. **Figure 19.32** indicates proper placement of the stethoscope bell to facilitate auscultation.

Learning Connection

Try Drawing It

Universally, Figure 19.31 is hard for students to interpret. Try this: On a blank piece of paper, draw each line for yourself while thinking through or narrating the events. Understand when pressures and volumes decrease and when they increase.

Figure 19.30 ECG

The waves and segments of an ECG correspond to the parts of a cardiac cycle.

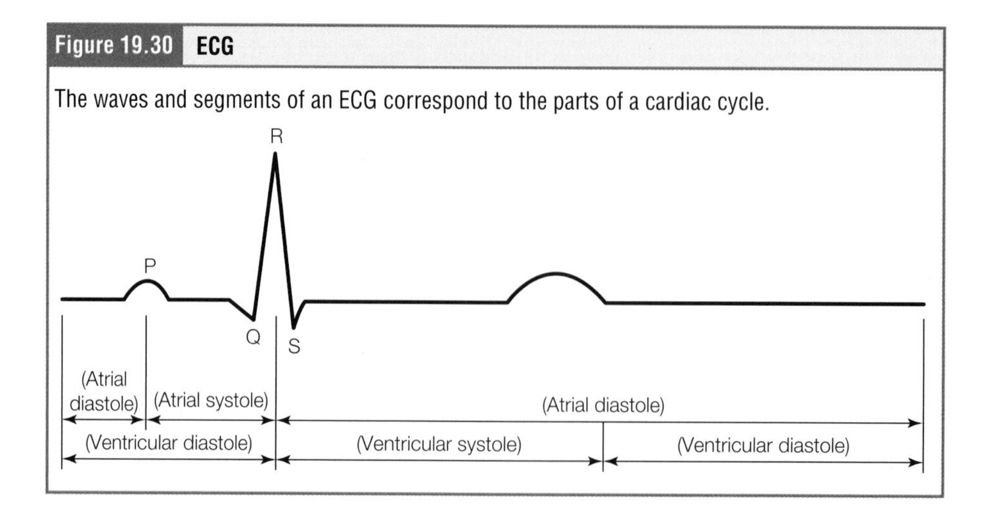

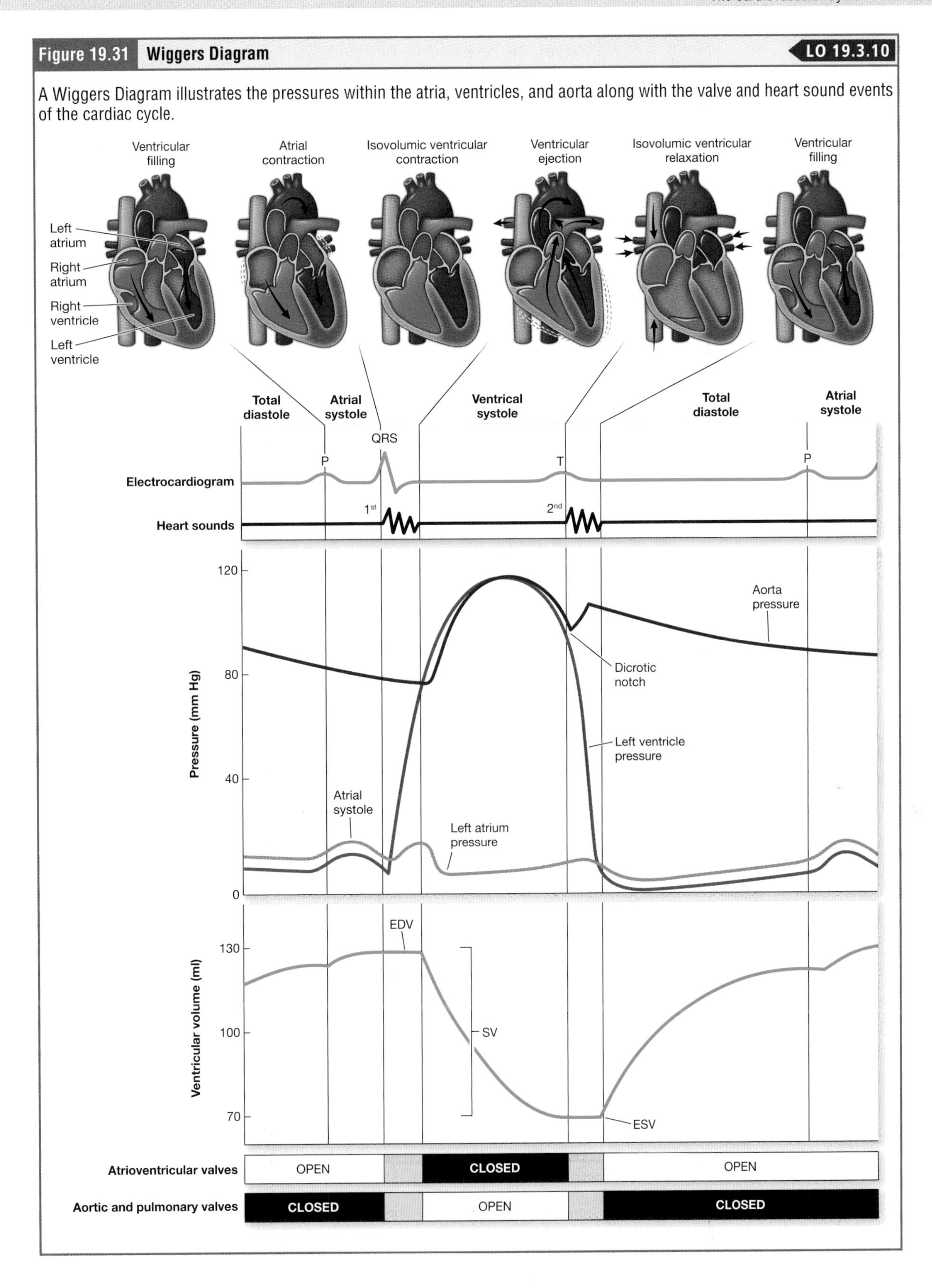

Figure 19.31 Wiggers Diagram

A Wiggers Diagram illustrates the pressures within the atria, ventricles, and aorta along with the valve and heart sound events of the cardiac cycle.

Figure 19.32 Stethoscope Placement for Heart Auscultation

A stethoscope must be placed in a different area in order to hear the heart sounds (sounds of auscultation) derived from each heart valve.

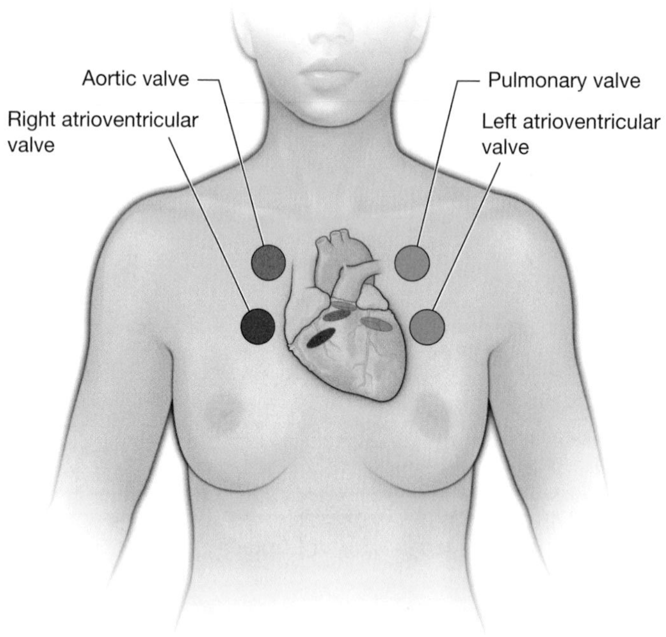

Learning Check

1. Ventricular systole is defined as:
   a. Contraction of the ventricles
   b. Relaxation of the ventricles
   c. Contraction of the atria
   d. Relaxation of the atria
2. During the T wave in an EKG, which of the following do you expect to happen in the heart?
   a. The atria will depolarize.
   b. The ventricles will repolarize.
   c. The atria will repolarize and the ventricle will depolarize.
   d. The atria will depolarize and the ventricles will repolarize.
3. What will happen when the pressure in the ventricles is higher than the pressure in the atria?
   a. Blood will flow backward and close the atrioventricular valves.
   b. Blood will flow backward and close the semilunar valves.
   c. Blood will flow from the atrium to the ventricles during atrial diastole.
   d. Blood will stay in the ventricles and eject into the aorta.

19.4 Cardiac Physiology

Learning Objectives: By the end of this section, you will be able to:

19.4.1 Define cardiac output (CO) and state its units of measurement.

19.4.2 Calculate the cardiac output given the stroke volume and the heart rate.

19.4.3 Predict how changes in heart rate (HR) and/or stroke volume (SV) will affect cardiac output (CO).

19.4.4 Describe the concepts of ejection fraction and cardiac reserve.

19.4.5 Describe the role of the autonomic nervous system in the regulation of cardiac output.

19.4.6 Describe the influence of positive and negative chronotropic agents on HR.

19.4.7 Explain the influence of positive and negative inotropic agents on stroke volume (SV).

19.4.8 Define venous return, preload, and afterload, and explain the factors that affect them.

19.4.9 Explain how venous return, preload, and afterload each affect end diastolic volume (EDV), end systolic volume (ESV), and stroke volume (SV).

19.4.10 State the Frank-Starling Law of the heart and explain its significance.

19.4.11 Explain the relationship between changes in HR and changes in filling time and EDV.

19.4.12* Predict changes to cardiac output in various conditions (e.g., increase in HR after caffeine intake).

* Objective is not a HAPS Learning Goal.

The autorhythmicity inherent in cardiac cells keeps the heart beating at a regular pace; however, heart rate is not homeostatically regulated. Heart rate climbs dramatically to meet the physiological demands of exercise, for example. Therefore, the heart rate needs to be influenced by and respond to outside factors. Neural and endocrine controls are vital to the regulation of cardiac function. In addition, the heart is sensitive to several environmental factors, including local ion (electrolyte) concentration.

19.4a Resting Cardiac Output

In the following paragraphs we discuss the regulation of cardiac function in various circumstances. First, we examine an individual at rest. These variables may look similar to you right now, assuming you are reading this while sitting relatively still or lying down.

Heart rate (HR) is not homeostatically regulated. Rather, it can increase or decrease dramatically (within a range of approximately 60–180 bpm) in order to meet the demands for how much blood the tissues need in a given moment. **Cardiac output (CO)** is a measurement of the amount of blood pumped by each ventricle in *one minute*. It is a factor of HR and **stroke volume (SV)**, the amount of blood pumped by each ventricle. It can be represented mathematically by the following equation:

LO 19.4.1

$$CO = HR \times SV$$

LO 19.4.2

These two variables can be regulated independently, but they influence each other. For example, a child has a smaller heart than an adult; the smaller heart has a lower capacity and therefore a lower stroke volume, and so the smaller heart needs to contract more times in a minute in order to maintain resting cardiac output. An adult athlete, by contrast, is likely to have a larger, stronger heart, so stroke volume is higher and heart rate can be lower at rest. In all individuals HR and SV are regulated independently and cardiac output is the value of physiologic importance. It is essential to maintain cardiac

LO 19.4.3

output so that there is always sufficient blood flow to the brain and other organs. The two components of CO (stroke volume and heart rate) can vary as needed. In the event of a deviation from homeostasis, the relationship between these variables is even more obvious. Take an individual who just donated blood. With reduced blood volume, SV will decrease. To maintain CO, HR can increase. SV is normally measured using an **echocardiogram** to record EDV and ESV, and calculating the difference: SV = EDV − ESV. SV can also be measured using a specialized catheter, but this is an invasive procedure and far more dangerous to the patient. Average resting SV ranges from 50 to 100 mL. There are several important variables, including size of the heart, cardiovascular condition of the individual, contractility, duration of contraction, preload or EDV, and afterload or resistance. Average resting HR varies widely from individual to individual based on their age, cardiovascular condition, and level of fitness, as well as other factors such as smoking and caffeine intake. In general, the resting heart rate for most individuals will be 60–100 bpm, though athletes often have resting heart rates between 50 and 60 bpm.

Let's take an individual with a heart rate of 75 bpm and a stroke volume of 70 mL. Using these numbers, the CO is 5.25 L/min; these are pretty typical resting values. What is most striking about the cardiac output is that the average blood volume for a 150-lb man is 5 L. This means that for most people at rest, the entire volume of blood in the body passes through the heart each minute.

Factors influencing CO are summarized in **Figure 19.33**.

LO 19.4.4

Another useful calculation is **ejection fraction**, which is the *portion* of the blood that is pumped or ejected from the heart with each contraction. In comparison, SV is the amount of blood ejected, rather than the proportion. A strong, efficient heart will eject more of its EDV because the muscle is capable of greater contraction. SV will change both based on the strength of the heart, but also based solely on the EDV. To calculate ejection fraction, SV is divided by EDV. Despite the name, the ejection fraction is normally expressed as a percentage. Ejection fractions range from approximately 55–70 percent.

Student Study Tip

To understand influences on cardiac output, walk through scenarios that could alter different steps to measuring CO: What happens to CO if EDV is lowered? How would increased afterload impact SV?

LO 19.4.5

Figure 19.33 Several Factors That Affect Cardiac Output

There are many factors that influence cardiac output (CO). Since CO is the product of heart rate (HR) and stroke volume (SV), any factor that changes either of these variables also changes the CO.

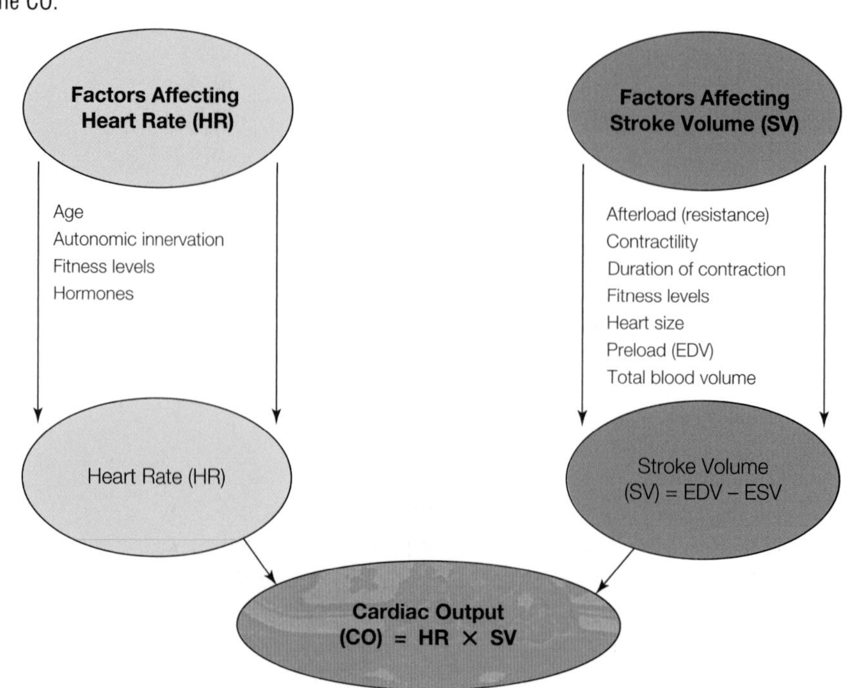

19.4b Exercise and Cardiac Function

In healthy young individuals, HR may increase to 190 bpm or higher during exercise (in general, max heart rate is calculated as 220–your age, so if you are 20 years old, max HR is around 200). HRs vary considerably, not only with age but also with exercise and fitness levels.

SV can also increase during exercise from approximately 70–130 mL due to increased strength of contraction, Frank-Starling mechanism, and EDV. This would increase CO to approximately 19.5 L/min, 4–5 times the resting rate. Even higher levels can be observed in athletes. At their peak performance, they may increase resting CO by 7–8 times in order to deliver the increased amounts of oxygen required and remove the increased wastes generated during this time. Think about that for a moment. If, at rest an individual typically moves their entire blood volume through their heart in a minute, that means that an athlete, while exercising, moves their entire blood volume through their heart up to eight times in a single minute!

Since the heart is a muscle, exercising it increases its efficiency. Athletes have stronger muscles and therefore more efficient heart function. Think back to our water balloon example; the harder the squeeze, the more fluid is ejected. Athletes tend to have lower resting HRs because they have higher resting SVs. Each systole is more effective, so they need fewer contractions within a minute in order to have the same cardiac output. The difference between maximum and resting CO is known as the **cardiac reserve**. This number is smaller in hearts that have lower resting SV, and higher in hearts that have higher resting SV. This is because the factor that is most important for maintaining homeostasis is the cardiac output, the amount of blood ejected *over time*.

19.4c Other Factors Influencing Heart Rate and Contractility

With the combination of autorhythmicity and innervation, the cardiovascular center is able to provide relatively precise control over HR and stroke volume. However, there are a number of other factors both endogenous and exogenous that have an impact on heart function as well. Factors that influence HR are referred to as chronotropic factors; factors that influence contractility are inotropic factors. These can be sorted into positive and negative factors. For example, a factor that speeds heart rate is called a **positive chronotropic factor**, while a factor that slows the heart rate is called a **negative chronotropic factor**. We will consider, in depth, the major physiologic chrono- and inotropic factors. A wider list is summarized in **Table 19.2**.

LO 19.4.6

LO 19.4.7

Epinephrine and Norepinephrine The catecholamines (epinephrine and norepinephrine) are secreted by the adrenal medulla and are important components of the fight-or-flight mechanism. Epinephrine and norepinephrine have similar effects: binding to the beta-1 adrenergic receptors expressed on the SA and AV nodes as well as the ventricular myocardium. In these locations, beta-adrenergic stimulation leads to increased permeability to sodium and calcium ions. This change in permeability increases the rate of depolarization so that threshold is reached more quickly and the period of repolarization is shortened. Thus, heart rate and contractility both increase.

Thyroid Hormones One of the effects of thyroid hormone (TH) is that it potentiates the effects of epinephrine, that is to say, it makes the effects of epinephrine stronger. Therefore, TH causes an increase in HR and contractility. The physiologically active form of thyroid hormone (T_3 or triiodothyronine) has been shown to directly enter cardiomyocytes and alter cellular activity. T_3 appears to increase contractility directly by altering the availability of calcium ions.

Cultural Connection

On the Basis of Sex

Recent research has indicated that biological women are more than twice as likely to die from a heart attack as biological men. In investigating the cause of this statistic, researchers have uncovered a tangled web of reasons. For starters, biological women are 50 percent more likely to be misdiagnosed when they first arrive at the hospital with heart attack symptoms. Why? Because heart attacks look different for different bodies. Most medical students are taught to recognize chest pain and left shoulder pain as symptoms of heart attack and to confirm their hypothesis with a blood test for troponin (troponin, a sarcomere protein, is released into the bloodstream when cardiomyocytes are damaged). The symptoms of heart attack in biological women may not include either of these symptoms; more often they include fatigue, shortness of breath, pain in the back or neck, nausea, and/or vomiting. Further, the diagnostic threshold for "normal" values of troponin may be too high for biological women. So, more frequently than with biological men, the heart attacks of biological women go on, untreated, for longer.

The majority of the research on cardiovascular disease involved studies on predominantly biological male populations. Once scientists began more equitable research, they have discovered that not only are heart attacks perceived differently when they occur in biological male and biologically female patients, but the healthy hearts of these individuals may function slightly differently as well. On average, biological females have smaller hearts with thinner myocardial layers. Can you think about what this difference might mean physiologically? If you are thinking that biological females may have, on average, a slightly lower stroke volume, you would be correct! However, there is no average difference in the cardiac output of biological males and females. Why? Because on average, biological females have higher heart rates. Another key difference is in the way our bodies respond to stress. In response to sympathetic nervous system activation, biological females experience an increase in heart rate and therefore cardiac output, but biological males also experience a systemic increase in the constriction of their arteries.

Over time, these differences may have long-lasting implications for cardiovascular health. At this point so much of what we know and teach about the cardiovascular system has come from studies done predominantly on males. Many of these studies were of rather homogenous populations and skewed toward younger, more athletic, and often White males. In order to be able to address disparities in health outcomes, we need to start by understanding the diversity of our hearts and bodies.

The function of both the pacemaker cells and the contractile cells is dependent on the movement of ions. Ions, as we know, move according to their concentration gradients. Therefore, heart function can be widely altered by changes in the extracellular concentration of the ions that impact these cells (sodium, potassium, calcium).

19.4d Stroke Volume

Many of the same factors that regulate HR also impact cardiac function by altering SV. While a number of variables are involved, SV is ultimately dependent upon the difference between EDV and ESV. The two primary factors to consider are the preload, or the stretch on the ventricles prior to contraction, and the contractility, the force or strength of the contraction itself.

LO 19.4.8

Preload Preload is another way of describing EDV. Preload/EDV has two effects on stroke volume. One is that the higher the volume of the blood at the start of contraction (the EDV) the more blood it is possible to eject; the other is that the blood in the ventricle causes the wall to stretch, and more blood leads to more stretch.

LO 19.4.9

Therefore, the greater the EDV is, the greater the preload is, and SV is likely to increase. With increasing ventricular filling, both EDV or preload increase, and the cardiac muscle itself is stretched to a greater degree. At rest, there is little stretch of

Apply to Pathophysiology

Patient Presenting with Bradycardia

Imagine you are an intern in a doctor's office. Your job is to enter patient information into their digital medical records. The patient whose information you are entering now came in with a heart rate of 50 bpm, technically bradycardia.

1. Which of the following could be reasons for this patient's low heart rate? (Choose all that apply)
   A) The patient has high blood pressure and high afterload.
   B) The patient is an athlete.
   C) The patient is a child.
   D) The patient has damage to their SA node.

You look back at the previous visits in the medical record and find that until about a year ago the patient had average resting heart rates in the 70–80 bpm range. So, you hypothesize that the bradycardia is new and caused by some recent event.

2. Which of the following symptoms is the patient most likely experiencing?
   A) A feeling of heart pounding in their chest
   B) Dizziness
   C) Tingling in fingers and toes
   D) Elevated body temperature

The office has a portable ECG and was able to capture a few minutes of ECG tracing.

3. What is the most likely abnormality?
   A) Missing P waves
   B) Long T-Q segments
   C) Short PR interval
   D) Long ST segment

the ventricular muscle, and the sarcomeres remain short. With increased ventricular filling, the ventricular muscle is increasingly stretched, and the sarcomere length increases. As the sarcomeres reach their optimal lengths, they will contract more powerfully, because more of the myosin heads can bind to the actin on the thin filaments, forming cross bridges and increasing the strength of contraction and SV. This relationship between ventricular stretch and contraction has been stated in the well-known **Frank-Starling mechanism**. If the stretch on the cardiac walls were to continue and the sarcomeres were stretched beyond their optimal lengths, the force of contraction would decrease. The presence of the tough, fibrous pericardium physically prevents the heart from overfilling, preventing any overstretch of the cardiomyocyte sarcomeres.

LO 19.4.10

When considering preload, one of the primary factors is filling time, or the duration of ventricular diastole during which filling occurs. The more rapidly the heart contracts, the shorter the filling time becomes, and the lower the EDV and preload are. This effect can be partially overcome by increasing the second variable (contractility) and raising SV, but over time the heart is unable to compensate for decreased filling time, and preload also decreases. Herein lies one of the dangers of tachycardia.

LO 19.4.11

A second factor affecting preload is the physiological conditions impacting the venous system. As discussed in Chapter 20, increases in both breath rate and skeletal muscle activity contribute pressure to the veins and increase venous return to the heart, which contributes to ventricular filling, EDV, and preload.

Table 19.2 Factors That Influence Heart Rate

Factor	Action	Effect on the Heart
Cardioaccelerator (sympathetic) nerves	Stimulate SA and AV nodes	Increases HR
	Stimulate ventricular myocardium	Increases force of contraction
Cardioinhibitory (parasympathetic) nerves	Inhibit SA and AV nodes	Decreases HR
Chemoreceptors (aorta and carotid body)	Influence cardioaccelerator and cardioinhibitory centers in medulla	Can affect sympathetic or parasympathetic stimulation to the heart
Baroreceptors (aorta and carotid body)	Influence cardioaccelerator and cardioinhibitory centers in medulla	
Limbic system	Anticipation of exercise or strong emotions	Increase in HR and force of contraction through sympathetic NS stimulation
Catecholamines	Epinephrine and Norepinephrine, released by the adrenal gland bind to beta-adrenergic receptors on the SA node and ventricular myocardium	Increased HR and force of contraction
Thyroid hormones	Increases expression of beta receptors, making heart more sensitive to circulating epinephrine	Increased HR and force of contraction
Calcium	Increase in calcium in the extracellular fluid around the heart increases the length of the plateau as well as the amount of calcium that floods the sarcoplasm of cardiomyocytes	Decreased HR with increased length and force of contraction
Potassium	Increase in potassium in the extracellular fluid around the heart changes the concentration gradient, making it less likely that K^+ will leave the cardiomyocytes, freezing them in contraction.	Myocardium contracts, and cannot relax. Cardiac arrest.
Body temperature	Body temperature sets the tempo for chemical reactions; when temperature increases, enzymes work faster and more chemical reactions take place each minute, increasing the need for oxygen and glucose.	Increased temperature increases heart rate
Nicotine	Stimulates sympathetic nervous system	Increased HR and contractility
Opiates	Depresses function in cardioaccelerator region	Decreases HR
Caffeine	Stimulates SA node	Increases HR

You can begin to get a sense of the impacts of cardiovascular exercise on cardiac function. Cardiovascular exercises such as running or biking will increase skeletal muscle activity and respiration rate, thus dramatically increasing venous return to the heart. This increased preload will drive up SV by providing both increased EDV and increased myocardial stretch. In addition, exercise increases the activity of the sympathetic nervous system, which affects contractility of the ventricles as well as heart rate (see Table 19.2).

Contractility It is virtually impossible to consider preload or ESV without including an early mention of the concept of contractility. Indeed, the two parameters are intimately linked by the Frank-Starling mechanism. Additional factors that increase contractility are described as **positive inotropic factors**, and those that decrease contractility are described as **negative inotropic factors**.

Not surprisingly, whereas sympathetic stimulation is a positive inotrope, parasympathetic stimulation is a negative inotrope. In addition to their stimulatory effects on HR, they also bind to both alpha and beta receptors on the cardiac muscle cell membrane to increase metabolic rate and the force of contraction. This combination of actions has the net effect of increasing SV and leaving a smaller residual ESV in the ventricles. In comparison, parasympathetic stimulation increases the action of the vagus nerve, decreasing heart rate but having minimal impacts on SV.

Some inotropic agents are summarized in Table 19.2. Of course, many synthetic drugs, including dopamine, digoxin, and beta-blockers, have been developed to influence the force of heart contraction exogenously.

Afterload Stroke volume (the amount of blood pushed out of the ventricle in a single contraction) is impacted both by factors within the ventricle itself as well as factors from outside of it. As we know, blood (and all fluids) flow from areas of greater pressure to areas of lesser pressure. Therefore the ventricles must always be able to generate *more* pressure than that which exists in the locations they are ejecting blood into (the aorta and the pulmonary trunk). The term **afterload** refers to the tension that the ventricles must develop to pump blood effectively against any resistance in the vascular system. Any condition that increases resistance requires a greater afterload to force open the semilunar valves and pump the blood. One way to think about afterload is by using a weight-lifting analogy. Imagine an athlete lifting a dumbbell. The afterload is the amount of force the muscle needs to generate in order to move the dumbbell. The heavier the dumbbell, the harder the muscles need to work to move it. Damage to the valves, such as stenosis, which makes them harder to open, will also increase afterload. In contrast, any decrease in resistance decreases the afterload. **Table 19.3** summarizes the major factors influencing SV, and **Figure 19.34** summarizes the major factors influencing CO. Assess your understanding of these complex dynamics by attempting to complete the tables in the Chapter Review.

Table 19.3 Factors That Affect Stroke Volume (SV)

	Preload	Contractility	Afterload
Raised due to:	• fast filling time • increased venous return **Increases end-diastolic volume, Increases stroke volume**	• sympathetic stimulation • epinephrine and norepinephrine • high intracellular calcium ions • high blood calcium level • thyroid hormones • glucagon **Decreases end-systolic volume, Increases stroke volume**	• increased vascular restistance • semilunar valve damage **Increases end-systolic volume, Decreases stroke volume**
Lowered due to:	• decreased thyroid hormones • decreased calcium ions • changes in extracellular potassium concentration • changes in extracellular sodium concentration • low body temperature • hypoxia • abnormal pH balance • drugs (i.e., calcium channel blockers) **Decreases end-diastolic volume, Decreases stroke volume**	• parasympathetic stimulation • acetylcholine • hypoxia • hyperkalemia **Increases end-systolic volume, Decreases stroke volume**	• decreased vascular resistance **Decreases end-systolic volume, Increases stroke volume**

Figure 19.34 Flow Chart of Factors That Affect Cardiac Output LO 19.4.12

Cardiac output (CO) is the product of heart rate (HR) and stroke volume (SV), so the factors that influence HR or SV also affect the CO. A variety of factors increase or decrease HR or SV.

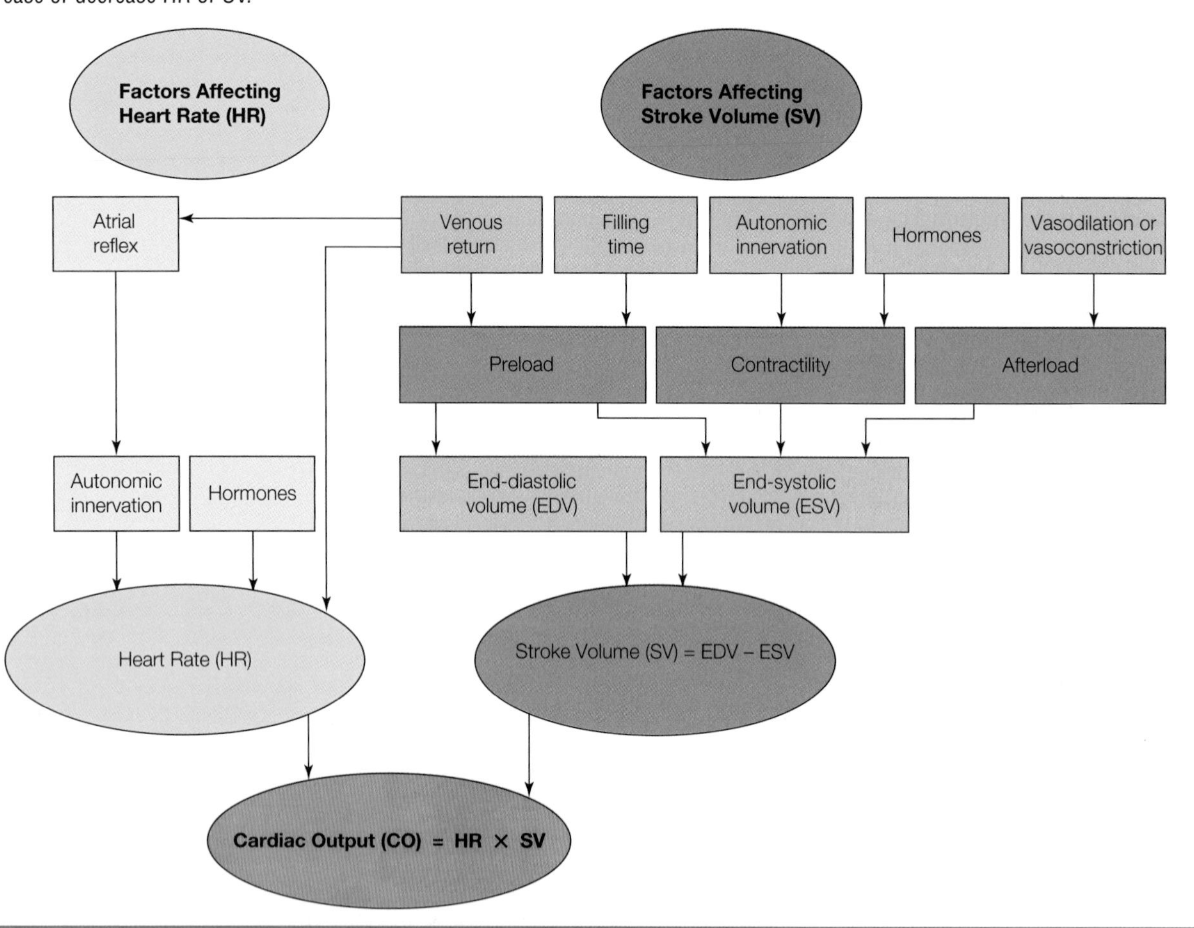

Learning Check

1. What is cardiac output?
   a. The amount of blood pumped by each ventricle in one minute
   b. The amount of blood pumped by each ventricle each time the heart beats
   c. The portion of the blood that is pumped or ejected from each ventricle with each contraction
   d. The difference between maximum and resting stroke volume
2. All of the following explain why cardiac output increases during exercise EXCEPT:
   a. HR increases
   b. Increased stroke volume
   c. Increased strength of contraction
   d. Decreased cardiac reserve
3. Which of the following do NOT affect stroke volume?
   a. Preload
   b. Contractility
   c. Atrial reflex
   d. Afterload
4. Alisha has a presentation in class today and she is very nervous. Wanting to be alert and ready, she is now drinking her third cup of coffee. Her heart rate, normally around 72 bpm, has increased to 105. What change, if any, has likely occurred in terms of cardiac output?
   a. Increased above her typical value range
   b. Decreased below her typical value range
   c. Has not changed

Chapter Review

As a reminder, learning objectives (LOs) clarify what learners should be able to do after reading each section of the text. One approach to your studying is to try to write or talk out a response for each LO.

In addition, try testing your mastery of the material with the questions provided, which target some of the LOs. If you need a refresher, a synopsis of each chapter section has been provided.

19.1 Heart Anatomy

The learning objectives for this section are:

19.1.1 Describe the position of the heart in the thoracic cavity.

19.1.2 Describe the changes in major fetal cardiovascular structures (i.e., umbilical vessels, ductus venosus, ductus arteriosus, foramen ovale) that typically occur beginning at birth, and the ultimate postnatal remnants (fates) of these structures.

19.1.3 Explain the structural and functional differences between atria and ventricles.

19.1.4 Identify and describe the location, structure, and function of the fibrous pericardium, parietal and visceral layers of the serous pericardium, serous fluid, and the pericardial cavity.

19.1.5 On the external surface of the heart identify the four chambers, the coronary (atrioventricular) sulcus, anterior interventricular sulcus, posterior interventricular sulcus, apex, and base.

19.1.6 *Given a disruption in the structure or function of the cardiovascular system (e.g., pulmonary edema), predict the possible factors or situations that might have created that disruption (i.e., given an effect, predict possible causes).

19.1.7 Describe the structure and functions of each layer of the heart wall (i.e., epicardium, myocardium, endocardium).

19.1.8 Identify and describe the structure and function of the primary internal structures of the heart, including chambers, septa, valves, papillary muscles, chordae tendineae, fibrous skeleton, and venous and arterial openings.

19.1.9 Trace the path of blood through the right and left sides of the heart, including its passage through the heart valves, and indicate whether the blood is oxygen-rich or oxygen-poor.

19.1.10 Describe the blood flow to and from the heart wall, including the location of the openings for the left and right coronary arteries, left coronary artery and its major branches, right coronary artery and its major branches, cardiac veins, and coronary sinus.

19.1 Questions

1. From what perspective could you best view the base of the heart?
a. Anterior
b. Superior
c. Posterior
d. Inferior

2. Which of the following are true of the foramen ovale? Please select all that apply.
a. It allows fetal blood to flow directly from the right ventricle to the left ventricle.
b. It allows fetal blood to flow directly from the right atrium to the left atrium.
c. It is closed after birth to keep blood from the pulmonary and systemic circuits separate.
d. It remains open after birth to provide extra blood flow to the lungs.

3. Which of the following are true of the differences between the atria and the ventricles? Please select all that apply.
a. The atria are the upper chambers of the heart and the ventricles are the lower chambers.
b. The atria transport blood directly into the pulmonary trunk and aorta.
c. The atria have thicker muscular walls than the ventricles.
d. The atria are the receiving chambers and the ventricles are the pumping chambers.

4. A patient has pericarditis (inflammation of the pericardium). In this patient, excess serous fluid has built up in the pericardial cavity. A surgeon would like to use a needle to drain the accumulated fluid. Which of the following layers will the needle pierce to reach the fluid? Choose all that apply.

a. Myocardium
b. Epicardium
c. Parietal layer of serous pericardium
d. Fibrous pericardium

5. If you were looking at an anterior view of the heart, which of the following structures would divide the left ventricle from the right ventricle?

a. Posterior interventricular sulcus
b. Coronary sulcus
c. Anterior interventricular sulcus
d. Interatrial septum

6. Ventricular septal defect is a congenital heart condition in which the wall between the two ventricles fails to form properly. In infants with this condition, blood mixes between the left and right atria through a large hole. Which of these drawings represents an embryonic heart in which the septum between the ventricles should be forming?

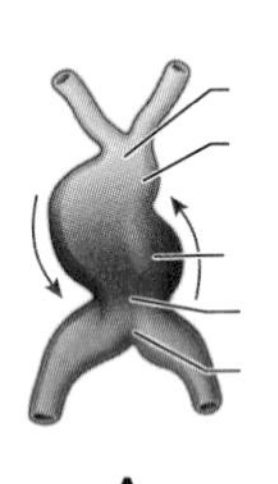

A

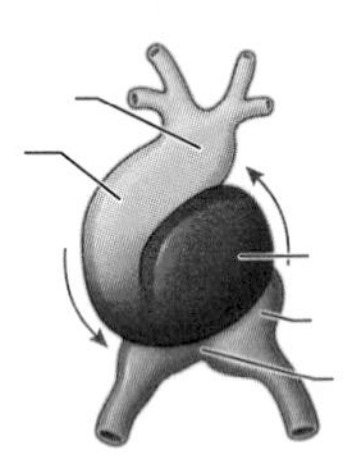

B

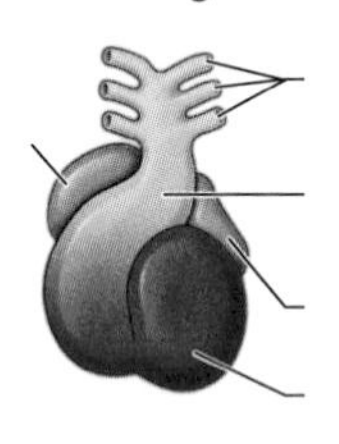

C

a. A
b. B
c. C

7. Which layer of the wall of the heart is composed of simple squamous epithelium, and is continuous with the inner lining of blood vessels?

a. Myocardium
b. Epicardium
c. Visceral pericardium
d. Endocardium

8. Which of the following is **not** important in preventing backflow of blood from the ventricles into the atria?

a. Chordae tendineae
b. Papillary muscles
c. AV valves
d. Endocardium

9. Which of the following lists the valves in the order through which the blood flows from the superior and inferior venae cavae through the heart?

a. Tricuspid, pulmonary semilunar, bicuspid, aortic semilunar
b. Mitral, pulmonary semilunar, bicuspid, aortic semilunar
c. Aortic semilunar, pulmonary semilunar, tricuspid, bicuspid
d. Bicuspid, aortic semilunar, tricuspid, pulmonary semilunar

10. Which of the following are true of the coronary circulation? Please select all that apply.

a. The great cardiac vein is the largest coronary vein.
b. The coronoary sinus is the largest cornrary artery.
c. The left and right coronary arteries branch from the aorta close to the aortic semilunar valve.
d. The major branches of the left coronary artery are the circumflex artery and the anterior interventricular artery.

19.1 Summary

- The heart resides within the pericardium and is located in the mediastinal space within the thoracic cavity.
- The pericardium consists of two fused layers: an outer fibrous pericardium and an inner double-layered serous pericardium that consists of a parietal and visceral layer.
- The walls of the heart are composed of an outer epicardium, a thick myocardium, and an inner endocardium.
- The human heart consists of a pair of atria, which receive blood and pump it into a pair of ventricles, which pump blood into the great vessels.
- The right atrium receives systemic blood relatively low in oxygen and pumps it into the right ventricle, which pumps it into the pulmonary circuit.
- Exchange of oxygen and carbon dioxide occurs in the lungs, and blood high in oxygen returns to the left atrium, which pumps blood into the left ventricle, which in turn pumps blood into the aorta and the remainder of the systemic circuit.
- The septa are the partitions that separate the chambers of the heart. They include the interatrial septum, the interventricular septum, and the fibrous skeleton of the heart.
- The atria and ventricles are separated by the atrioventricular valves, which prevent the backflow of blood when ventricular pressure rises. Each is attached to chordae tendineae that extend to the papillary muscles, which are extensions of the myocardium, to prevent the valves from being blown back into the atria.
- The pulmonary semilunar valve is located at the base of the pulmonary trunk, and the aortic semilunar valve is located at the base of the aorta.
- The right and left coronary arteries are the first to branch off the aorta and arise from two of the three sinuses located near the base of the aorta.
- Cardiac veins parallel the small cardiac arteries and generally drain into the coronary sinus.

19.2 Cardiac Muscle and Electrical Activity

The learning objectives for this section are:

HAPS LEARNING GOALS AND OUTCOMES

19.2.1 Describe the microscopic anatomy of the myocardium, including the location and function of the intercalated discs.

19.2.2 List the parts of the electrical conduction system of the heart in the correct sequence for one contraction and explain how the electrical conduction system functions.

19.2.3 Explain why the SA node normally paces the heart.

19.2.4 Explain how the cardiac conduction system produces coordinated heart chamber contractions.

19.2.5 Contrast the initiation of action potentials in cardiac autorhythmic cells, in cardiac contractile cells, and in skeletal muscle cells.

19.2.6 List the phases of contractile and autorhythmic cardiac muscle action potentials and explain the ion movements that occur in each phase.

19.2.7 Explain the significance of the plateau phase in the action potential of a cardiac contractile cell.

19.2.8 Compare the refractory periods of cardiac contractile muscle and skeletal muscles.

19.2.9 Compare and contrast the molecular events of cardiac muscle contraction/relaxation and skeletal muscle contraction/relaxation.

19.2.10 Explain the role of calcium in determining the force of myocardial contraction (contractility).

19.2.11 Name the waveforms in a normal electrocardiogram (ECG or EKG) and explain the electrical events represented by each waveform.

19.2.12* Predict changes to EKGs in various disease states (e.g., an increase in extracellular K^+ concentration).

19.2.13 Compare and contrast the role of autonomic innervation in the depolarization of cardiac pacemaker cells, ventricular contractile cells, and skeletal muscle cells.

19.2.14* Predict changes to heart function if the permeabilities of ions were altered in electrical conduction system cells or cardiomyocytes.

* Objective is not a HAPS Learning Goal.

19.2 Questions

1. Which of the following is unique to cardiac muscle cells compared to skeletal and smooth muscle cells?

a. Only cardiac muscle contains a sarcoplasmic reticulum.
b. Only cardiac muscle has gap junctions.
c. Only cardiac muscle is capable of autorhythmicity.
d. Only cardiac muscle has branching, cylindrical cells.

2. Which of the following true of the cardiac conduction system? Please select all that apply.

a. The AV node is the pacemaker of the heart.
b. The Purkinje fibers distribute electrical impulses to the contractile cells of the ventricles.
c. The AV bundle sends impulses directly to the SA node.
d. In the cardiac conduction system, depolarization leads to contraction of a heart chamber.

3. Which component of the heart conduction system would have the fastest rate of firing?

a. Atrioventricular node
b. Atrioventricular bundle
c. Sinoatrial node
d. Purkinje fibers

4. Which of the following allows the cardiac conduction system to produce coordinated contractions of the heart chambers? Please select all that apply.

a. The AV bundle transmits impulses to the left and right bundle branches simultaneously.
b. The Purkinje fibers on both sides of the heart transmit impulses to the SA node at the same time.
c. The left and right bundle branches transmit impulses to the Purkinje fibers on both sides of the heart simultaneously.
d. The SA node transmits impulses directly to the Purkinje fibers on both sides of the heart simultaneously, bypassing the AV node.

5. Which of the following ions is necessary to initiate action potentials in cardiac nodal cells and cardiac contractile cells, but not in skeletal muscle cells?

a. Ca^{2+}
b. K^+
c. Na^+
d. Cl^-

6. Which of the following are true of the action potentials of cardic nodal (autorhythmic) cells? Please select all that apply.

a. They have a long plateau phase.
b. They have no stable resting potential.
c. They exhibit spontaneous depolarization.
d. They have an exceptionally long refractory period.

7. What is the importance of the plateau phase in the action potentials in cardiac contractile cells?
   a. It prevents a toxic accumulation of potassium ions in the cells.
   b. It allows the heart to beat faster during exercise.
   c. It prevents summation in the contractions of cardiac contractile cells.
   d. It prevents hyperpolarization in cardiac contractile cells.

8. All of the following are true of the refractory period in cardiac contractile muscle cells except:
   a. The refractory period is longer than it is in skeletal muscle cells.
   b. The refractory period is important for increasing the strength of muscle contraction in cardiac contractile cells.
   c. The refractory period lasts until the muscle contraction is completed.
   d. The refractory period is due to the action of slow sodium channels.

9. All of the following are true of the contractions of both skeletal muscle and cardiac muscle cells except:
   a. Both types of muscle cells depolarize via an influx of sodium ions.
   b. Both types of muscle cells repolarize via an efflux of potassium ions.
   c. Both types of muscle cells contain voltage-gated channels for sodium, potassium, and calcium ions.
   d. Both types of muscle cells respond to action potentials by contracting.

10. All of the following are true of the roles of calcium ions in the contraction of cardiac contractile cells except:
    a. Calcium ions increase contractility of cardiac contractile cells.
    b. Calcium ions enhance summation in cardiac contractile cells.
    c. Calcium binds to troponin to initiate crossbridging.
    d. Calcium ions are responsible for the plateau phase of the action potentials.

Mini Case 1: Atrial Flutter

David had coronary artery disease and hypertension (high blood pressure), and went to see the cardiologist because of palpitations, dizziness, fatigue and shortness of breath. The cardiologist ordered an electrocardiogram (ECG of EKG).

1. Which of the following electrical events is represented by the P wave of an ECG?
   a. Ventricular depolarization
   b. Atrial contraction
   c. Atrial depolarization
   d. Atrial repolarization

2. David was diagnosed with atrial flutter, a condition in which the atria beat too rapidly. How would you expect this to appear on an ECG?
   a. Extra QRS complexes
   b. Extra P waves
   c. Extra T waves
   d. Extra P waves and QRS complexes

3. The cardiologist prescribed a beta adrenergic blocker (beta blocker) to help with the atrial flutter. This medication blocks the effects of a certain branch of the nervous system in order to lower the rapid beating of the atria. Which branch of the nervous system would you expect a beta blocker to affect?
   a. Sympathetic
   b. Somatic
   c. Parasympathetic
   d. Enteric

4. Another medication used to treat both atrial flutter and hypertension is a calcium channel blocker. By what mechanism would you expect this medication to help with David's atrial flutter?
   a. Blocking the calcium channels would hyperpolarize the atrial muscle cells, preventing action potentials.
   b. Blocking the calcium channels would prevent repolarization of the atria.
   c. Blocking the calcium channels would prevent some of the crossbridge cycling in the atrial contractile cells.

19.2 Summary

- The heart is regulated by both neural and endocrine factors, yet it is capable of initiating its own action potential followed by muscular contraction.
- The conductive cells within the heart establish the heart rate and transmit it through the myocardium. The contractile cells contract and propel the blood through the circuits.
- The normal path of transmission for the conducting cells is the sinoatrial (SA) node, internodal pathways, atrioventricular (AV) node, atrioventricular (AV) bundle (of His), bundle branches, and Purkinje fibers.
- An action potential for the conducting cells consists of a prepotential phase with a slow influx of Na^+ followed by a rapid influx of Ca^{2+} and outflux of K^+.
- Contractile cells have an action potential with an extended plateau phase that results in an extended refractory period to allow complete contraction atria before the ventricles begin contracting.
- Recognizable points on the ECG include the P wave, which corresponds to atrial depolarization, the QRS complex, which corresponds to ventricular depolarization, and the T wave, which corresponds to ventricular repolarization.

19.3 Cardiac Cycle

The learning objectives for this section are:

19.3.1 Define cardiac cycle, systole, and diastole.

19.3.2 Describe the phases of the cardiac cycle, including ventricular filling, isovolumic (isovolumetric) contraction, ventricular ejection, and isovolumic (isovolumetric) relaxation.

19.3.3 Relate the electrical events represented on an electrocardiogram (ECG or EKG) to the normal mechanical events of the cardiac cycle.

19.3.4 Explain how atrial systole is related to ventricular filling.

19.3.5 Define end diastolic volume (EDV) and end systolic volume (ESV), and calculate stroke volume (SV) given values for EDV and ESV.

19.3.6 Relate the opening and closing of specific heart valves in each phase of the cardiac cycle to pressure changes in the heart chambers and the great vessels (i.e., blood vessels entering and leaving the heart).

19.3.7 Compare and contrast pressure and volume changes of the left and right ventricles during one cardiac cycle.

19.3.8 Define systolic and diastolic blood pressure and interpret a graph of aortic pressure versus time during the cardiac cycle.

19.3.9 Relate the heart sounds to the events of the cardiac cycle.

19.3.10* Draw and explain the waves of Wiggers Diagram.

* Objective is not a HAPS Learning Goal.

19.3 Questions

1. Most blood enters the ventricles during ________.
a. Atrial systole
b. Atrial diastole
c. Ventricular systole
d. Isovolumic contraction

2. Which of the following are true of the isovolumic contraction phase of the cardiac cycle? Please select all that apply.
a. The blood pressure is decreasing in the ventricles.
b. All four heart valves are closed.
c. The blood volume is not changing within the ventricles.
d. The phase that immediately follows isovolumic contraction is the ventricular filling phase.

3. Ventricular relaxation immediately follows ________.
a. Atrial depolarization
b. Ventricular repolarization
c. Ventricular depolarization
d. Atrial repolarization

4. Which of the following are true of the process of ventricular filling? Please select all that apply.
a. Atrial systole causes about 20% of ventricular filling.
b. Atrial systole is responsible for 100% of ventricular filling.
c. A large percentage of ventricular filling occurs as blood passes through open atrioventricular valves.
d. Atrial diastole is responsible for 100% of ventricular filling.

5. If the end-diastolic volume (EDV) is 125 mL and the end-systolic volume is 50 mL, what is the stroke volume?
a. 175 mL
b. 50 mL
c. 75 mL
d. 2.5 mL

6. When the pressure in the ventricles exceeds the pressure in the atria and the great vessels that leave the heart, what effect does this have on the heart valves? Please select all that apply.
a. The tricuspid valve closes.
b. The mitral (bicuspid) valve opens.
c. The pulmonary semilunar valve closes.
d. The aortic semilunar valve opens.

7. All of the following are true of the pressure and volume changes in the ventricles during the cardiac cycle except:
a. Throughout the cardiac cycle, increases or decreases in pressure and volume that occur in the left ventricle also occur in the right ventricle.
b. During the isovolumic phases, the ventricular volume remains constant.
c. During the isovolumic phases, the ventricular pressure remains constant.
d. During the late ventricular relaxation (ventricular filling) phase, the pressure in the ventricles eventually decreases to the point at which it is lower than atrial pressure.

8. What is the best definition of the systolic pressure?
a. The average pressure that drives blood to the tissues
b. The difference between the maximum and minimum pressures in the arteries
c. The minimum pressure remaining in the arteries before the next ventricular contraction
d. The maximum pressure occurring in the arteries during ventricular contraction

9. Which of the following events is occurring during the first heart sound of the cardiac cycle?
   a. Atrial systole
   b. Ventricular diastole
   c. Closing of the atrioventricular valves
   d. Closing of the semilunar valves

10. According to the Wiggers Diagram, when ventricular pressure is at its highest level, which of the following is true?
    a. The atria are in systole.
    b. Aortic pressure is also at its highest level.
    c. The second heart sound is occurring.
    d. The QRS complex of the ECG is occurring.

19.3 Summary

- The cardiac cycle comprises a complete relaxation and contraction of both the atria and ventricles (total diastole).
- Beginning with all chambers in diastole, blood flows passively from the great veins into the atria and through the atrioventricular valves into the ventricles.
- The atria begin to contract (atrial systole), following depolarization of the atria, and pump blood into the ventricles.
- The ventricles begin to contract (ventricular systole), raising pressure within the ventricles.
- When ventricular pressure rises above the pressure in the atria, blood flows toward the atria, closing the AV valves and producing the first heart sound, S_1 or "lub."
- As pressure in the ventricles rises above that in the two major arteries, blood pushes the two semilunar valves open and moves into the pulmonary trunk and aorta in the ventricular ejection phase.
- Following ventricular repolarization, the ventricles begin to relax (ventricular diastole), and pressure within the ventricles drops.
- As ventricular pressure drops, there is a tendency for blood to flow back into the venticles from the major arteries, producing the dicrotic notch in the ECG and closing the two semilunar valves.
- The second heart sound, S_2 or "dub," occurs when the semilunar valves close.
- When the ventricular pressure falls below that of the atria, blood moves from the atria into the ventricles, opening the atrioventricular valves and marking one complete heart cycle.
- The valves prevent backflow of blood. Failure of the valves to operate properly produces turbulent blood flow within the heart; the resulting heart murmur can often be heard with a stethoscope.

19.4 Cardiac Physiology

The learning objectives for this section are:

19.4.1 Define cardiac output (CO) and state its units of measurement.

19.4.2 Calculate the cardiac output given the stroke volume and the heart rate.

19.4.3 Predict how changes in heart rate (HR) and/or stroke volume (SV) will affect cardiac output (CO).

19.4.4 Describe the concepts of ejection fraction and cardiac reserve.

19.4.5 Describe the role of the autonomic nervous system in the regulation of cardiac output.

19.4.6 Describe the influence of positive and negative chronotropic agents on HR.

19.4.7 Explain the influence of positive and negative inotropic agents on stroke volume (SV).

19.4.8 Define venous return, preload, and afterload, and explain the factors that affect them.

19.4.9 Explain how venous return, preload, and afterload each affect end diastolic volume (EDV), end systolic volume (ESV), and stroke volume (SV).

19.4.10 State the Frank-Starling Law of the heart and explain its significance.

19.4.11 Explain the relationship between changes in HR and changes in filling time and EDV.

19.4.12* Predict changes to cardiac output in various conditions (e.g., increase in HR after caffeine intake).

* Objective is not a HAPS Learning Goal.

19.4 Questions

1. What is meant by the cardiac output, and in what units is it expressed?
   a. The volume of blood leaving each ventricle with each heartbeat; L/min

b. The number of heartbeats that occur in one minute; beats/min or bpm
c. The volume of blood leaving each ventricle in one minute; L/min
d. The volumne of blood leaving both ventricles in one minute; mL/min

2. If the stroke volume is 65 mL/beat, and the heart rate is 60 beats/min, what is the cardiac output?
a. 3.9 L/min
b. 5 mL/beat
c. 135 mL/min
d. 3900 mL/beat

3. In a healthy young adult, what happens to the cardiac output when the heart rate increases?
a. It increases.
b. It decreases.
c. It remains constant.
d. There is no way to predict.

4. Which of the following are true of the ejection fraction? Please select all that apply.
a. It is the same as the stroke volume.
b. It is usually about 60% of the EDV.
c. It is the percentage of blood pumped out of the ventricles during each contraction.
d. It is calculated by dividing the EDV by the SV.

5. You are about to present an oral report in your class and you are suddenly very nervous. How would this affect your cardiac output (CO) and why?
a. Your CO would increase due to sympathetic stimulation.
b. Your CO would decrease due to parasympathetic stimulation.
c. Your CO would decrease due to sympathetic stimulation.
d. Your CO would increase due to parasympathetic stimulation.

6. Which of the following are negative chronotropic factors that affect the heart? Please select all that apply.
a. Cardioaccelerator nerve stimulation of the heart
b. Ingestion of caffeine
c. An increase in calcium in the extracellular fluid
d. Parasympathetic stimulation of the heart

7. Which of the following are true of inotropic factors that affect the heart? Please select all that apply.
a. They are factors that affect the contractility of the heart.
b. They are factors that affect the heart rate.
c. A positive inotropic factor is nicotine.
d. A negative inotropic factor is epinephrine.

8. The force in the vascular system that the heart must overcome to pump blood out of the ventricles is known as ________.
a. preload
b. afterload
c. cardiac output
d. stroke volume

9. Which of the following factors would increase the stroke volume? Please select all that apply.
a. An increase in venous return to the heart
b. An increase in vascular resistance (afterload)
c. An increase in end-diastolic volume (EDV)
d. Parasympathetic stimulation

10. The Frank-Starling mechanism is a law that states that:
a. The higher the blood volume, the higher the blood pressure.
b. The higher the vascular resistance, the lower the end-systolic volume.
c. The greater the stretching of the ventricular muscle cells during ventricular filling, the greater the force of the contraction that will follow.
d. The greater the venous return, the greater the heart rate will be.

11. What effects would an increase in heart rate have on the ventricular filling time and the end-diastolic volume (EDV)?
a. It would lengthen the filling time and increase the EDV.
b. It would lengthen the filling time and decrease the EDV.
c. It would shorten the filling time and increase the EDV.
d. It would shorten the filling time and decrease the EDV.

12. Which of the following factors would increase the cardiac output? Please select all that apply.
a. An increase in end-diastolic volume
b. Ingestion of caffeine
c. A decrease in body temperature
d. Taking an opiate medication

19.4 Summary

- Many factors affect HR and SV, and together, they contribute to cardiac output. HR is regulated by autonomic stimulation and hormones.
- There are several feedback loops that contribute to maintaining homeostasis dependent upon activity levels, such as the atrial reflex, which is determined by venous return.
- SV is regulated by autonomic innervation and hormones, and also by filling time and venous return.
- Venous return is determined by activity of the skeletal muscles, blood volume, and changes in peripheral circulation. Venous return determines preload and the atrial reflex.
- Filling time directly related to HR also determines preload. Preload then impacts both EDV and ESV.
- Autonomic innervation and hormones largely regulate contractility. Contractility impacts EDV as does afterload.
- CO is the product of HR multiplied by SV. SV is the difference between EDV and ESV.

20

The Cardiovascular System: Blood Vessels and Circulation

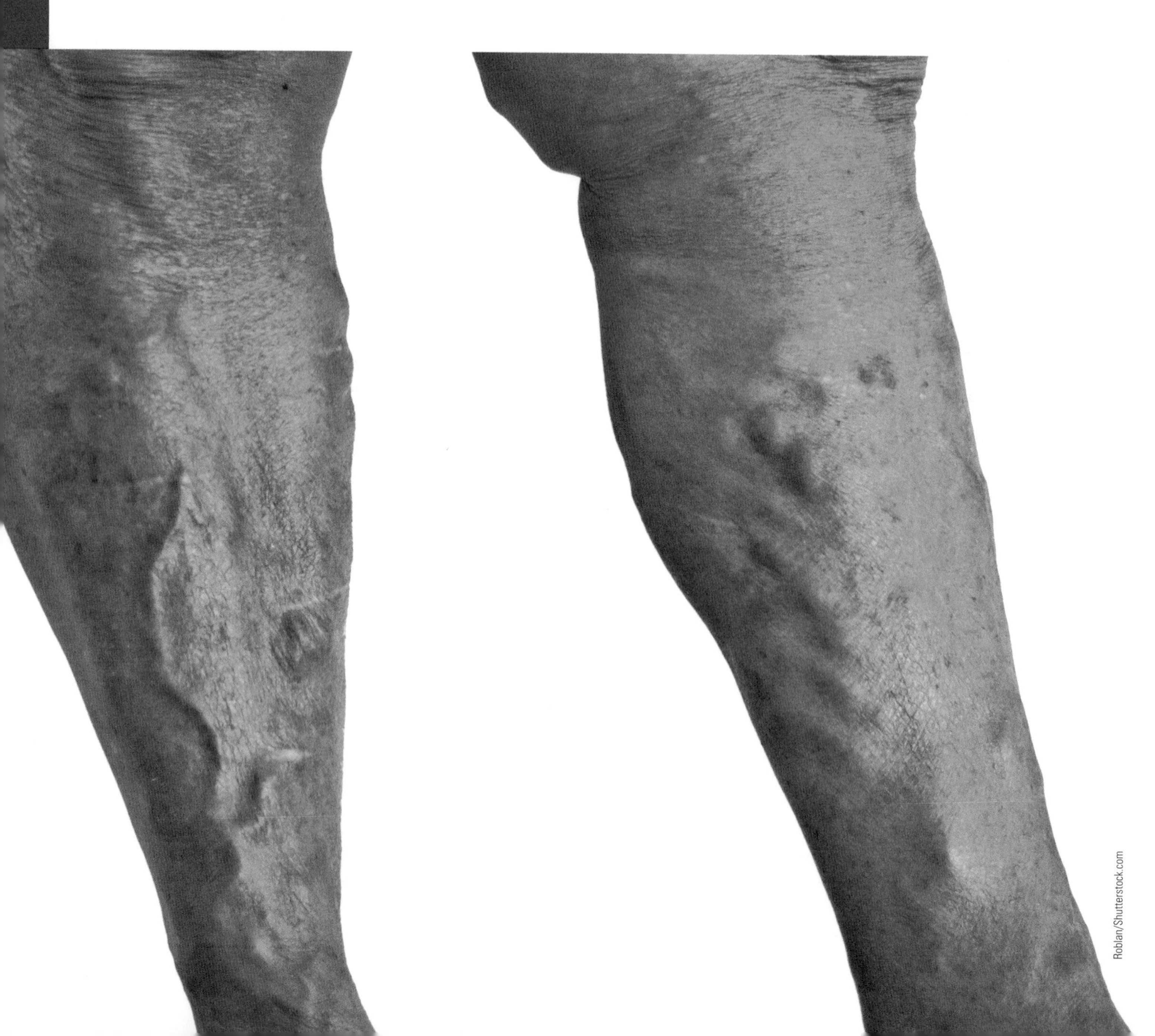

Roblan/Shutterstock.com

Chapter Introduction

In this chapter we will learn . . .

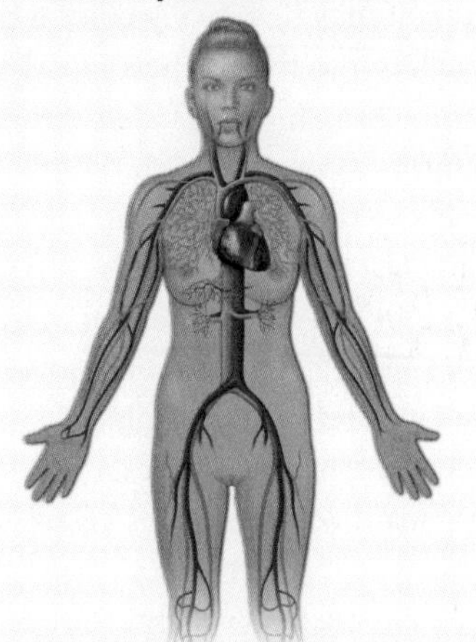

The network of blood vessels that keeps all the tissues of the body supplied with nutrients and oxygen . . .

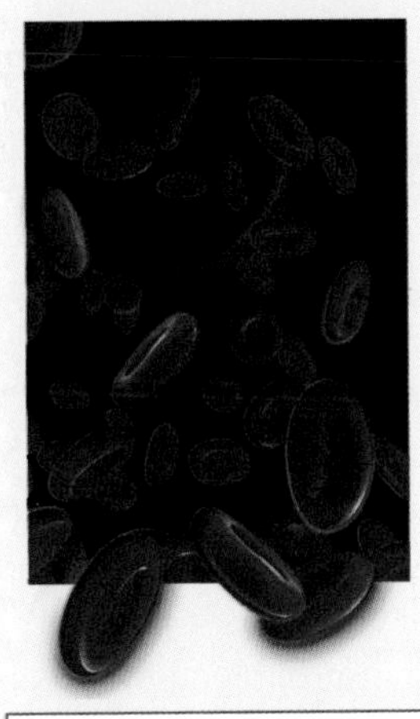

and the physiological factors that keep blood flowing or help to change blood flow when needed.

In the previous two chapters we discussed the composition of blood and how the heart functions to create the pressure gradient that drives its flow. We will now discuss the structure of the vessels in which blood flows so that we can understand the dynamics that control its flow and the exchange with the tissues that enables its function.

20.1 Anatomy of Blood Vessels

Learning Objectives: By the end of this section, you will be able to:

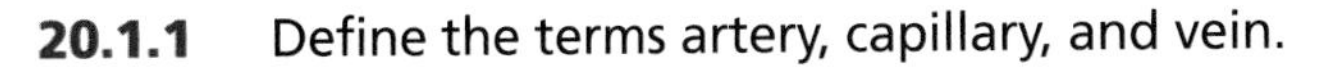

20.1.1 Define the terms artery, capillary, and vein.

20.1.2 List the three tunics associated with most blood vessels and describe the composition of each tunic.

20.1.3 Compare and contrast tunic thickness, composition, and lumen diameter among arteries, capillaries, and veins.

20.1.4 Define vasoconstriction and vasodilation.

20.1.5 Identify and describe the structure and function of specific types of blood vessels (i.e., elastic [conducting] arteries, muscular [distributing] arteries, arterioles, capillaries, venules, veins).

20.1.6 List types of capillaries, state where in the body each type is located, and correlate their anatomical structures with their functions.

20.1.7 Describe the functional significance of the venous reservoir.

20.1.8 Define anastomosis and explain its functional significance (e.g., cerebral arterial circle [Circle of Willis]).

The Human Anatomy and Physiology Society includes more than 1,700 educators who work together to promote excellence in the teaching of this subject area. The HAPS A&P Learning Outcomes measure student mastery of the content typically covered in a two-semester Human A&P curriculum at the undergraduate level. The full Learning Outcomes are available at https://www.hapsweb.org.

For mastery of the topic of blood flow, it is helpful to review some concepts introduced in other chapters (**Figure 20.1**). Blood travels throughout the body within blood vessels, and whole blood never leaves the vessel system except in the case of injury. Arteries carry blood away from the heart and branch into ever-smaller vessels. This branching format is repeated throughout nature, it may help to visualize a tree with a single trunk that divides repeatedly and branches into the smallest of twigs. In that tree, as with in our circulatory system, the branching format is an effective structure for maximizing surface area. In our circulatory system, as with the tree, the larger vessels require more support and have walls so thick that no nutrients or gasses can make their way across. Eventually, the smallest arteries—vessels called *arterioles*—further branch into tiny capillaries, where the vessel walls are so thin that nutrients and wastes can be easily exchanged. These small capillaries merge together to form venules, small blood vessels that carry blood to a vein, a larger blood vessel that returns blood to the heart. Throughout the body the word **artery** describes blood vessels that carry blood away from the heart and the word **vein** describes vessels that bring blood back toward the heart.

Student Study Tip

Use alliteration to remember that Arteries carry blood Away from the heart.

LO 20.1.1

Arteries and veins transport blood in two distinct circuits: the systemic circuit and the pulmonary circuit (**Figure 20.2**). The *systemic circuit* carries blood throughout the body, with the function of providing oxygen and nutrients to the skeletal muscles

Figure 20.1 Review of Important Concepts

Several concepts that we have discussed in other places are instrumental in understanding the physiology of the circulatory system. Bulk flow is how mixed gasses and fluids move. Flow is directly proportional to a pressure gradient and inversely proportional to resistance. In many places in biology, we see branching structures as ways to maximize surface area and volume relationships. The circulatory system has maximized surface area through extensive branching of blood vessels. The greater the ratio of surface area to volume, the more diffusion is possible; in this case the extensive branching of blood vessels allows for maximum diffusion of gasses and nutrients.

Review of important concepts

In mixtures like blood and soup, all components of the mixture move together in bulk flow, which is determined by the pressure gradients.

$$F \propto \frac{\textbf{gradient}}{\textbf{resistance}}$$

The flow equation can help us understand all kinds of flow, including bulk flow and diffusion.

Bulk flow differs from diffusion, which is determined by concentration and electrical gradients.

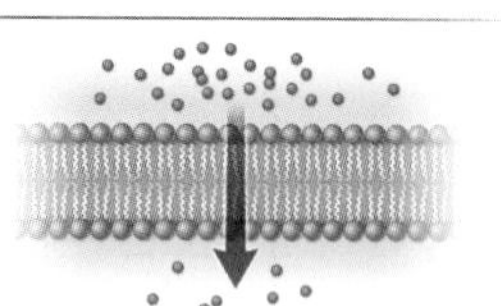

Throughout biology, branching structures are used to maximize surface-area-to-volume relationships.

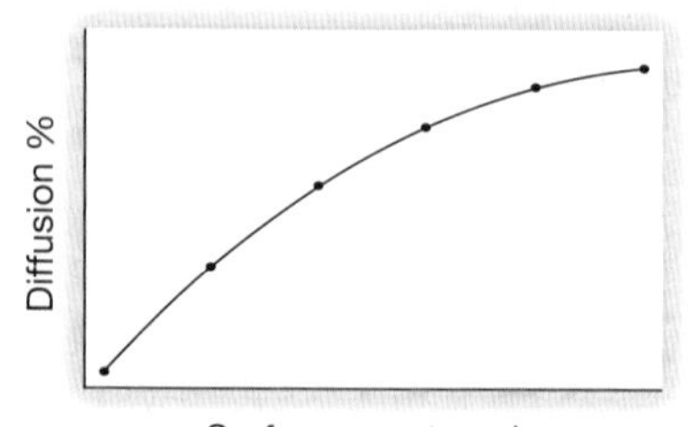

There are many factors that influence the rate of diffusion, which can by summarized by Fick's Law. In particular, surface area proportionally increases the diffusion rate.

Figure 20.2 Cardiovascular Circulation Pathways

The body's vast system of blood vessels is grossly divided into two main circuits. The pulmonary circulation transports blood traveling from the right side of the heart toward the lungs for oxygenation; the systemic circuit transports blood from the left side of the heart toward the tissues of the body.

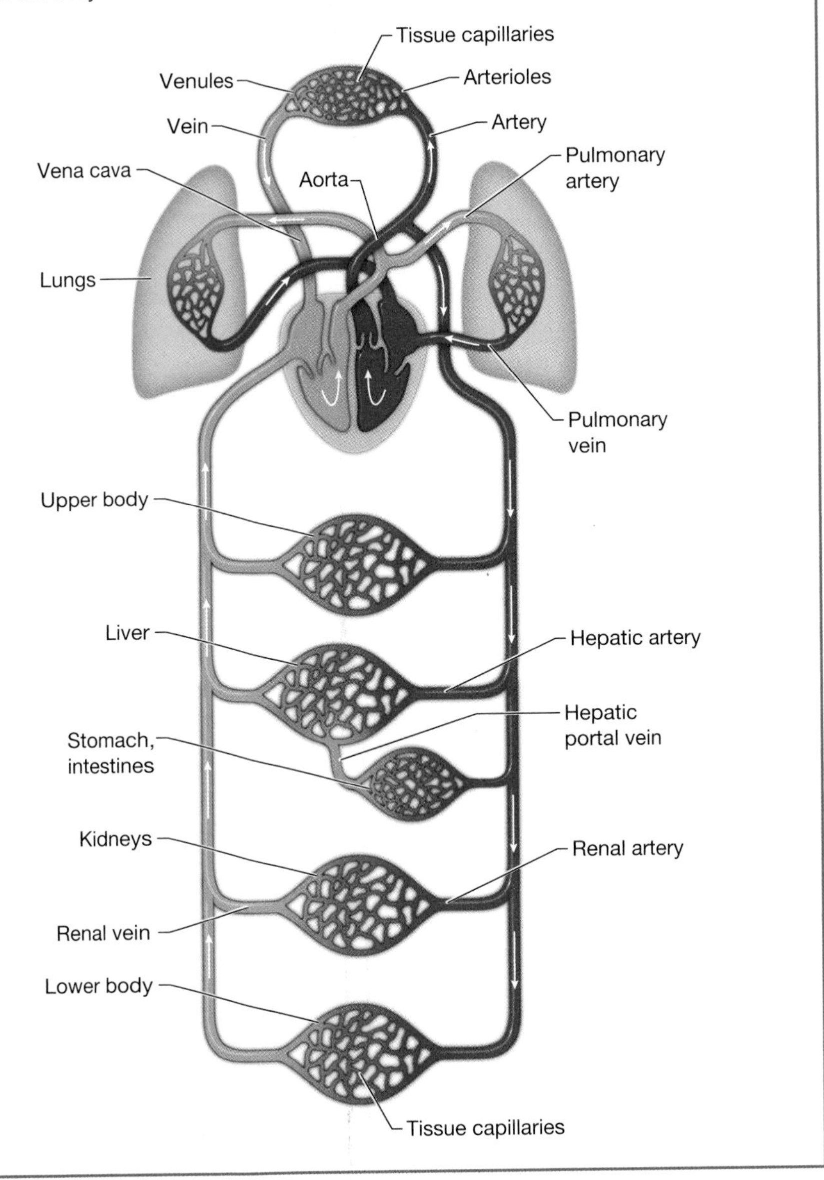

and organ systems. The *pulmonary circuit* carries blood to the lungs, with the goal of taking on more oxygen and getting rid of carbon dioxide. Systemic arteries carry high-pressure blood that has just been expelled from the heart. This blood is rich in oxygen. The blood returned to the heart through systemic veins has less oxygen, since much of the oxygen carried by the blood in the arteries has been delivered to the cells. This blood is also some of the lowest-pressure blood in the system because it has traveled a long way from the heart, where most pressure is generated. Pulmonary arteries carry blood away from the heart and toward the lungs for gas exchange. Pulmonary veins then return freshly oxygenated blood from the lungs to the heart to be pumped back out into systemic circulation.

20.1a Structure-Function Relationships

The types of blood vessels vary slightly in their structures, but they share the same general features. Every vessel has a **lumen**—a hollow passageway through which blood flows—and a wall that surrounds the lumen. Arteries and arterioles have thicker walls than do veins and venules because they are closer to the heart and receive blood that is surging at a far greater pressure (**Figure 20.3**). Their walls have greater elasticity to cope with the differences in pressure between systole (heart contraction) and diastole (heart relaxation) and more smooth muscle to direct the flow of blood. Together these two characteristics give the walls of arteries and arterioles a flexible yet somewhat rigid rounded shape. When dissecting, you might squeeze an artery between your gloved fingers. The artery will compress, and then spring back to its rigidly open shape, like a macaroni noodle that has been cooked *al dente*. Veins, on the other hand, experience much less pressure from the blood traveling within them, so their walls lack elasticity. Similarly, there is very little smooth muscle in the wall of a vein. During dissection, squeezing a vein between your gloved fingers might feel more like compressing a folded piece of cloth, veins lack the rigid springiness of their artery counterparts. Blood travels faster and spends less time overall in arteries compared to veins, which explains why veins have a larger lumen-to-wall ratio compared with arteries, which have a smaller lumen and thicker, more substantial walls. The larger diameter allows more blood to flow with less vessel resistance.

Figure 20.3 **A Comparison of Arteries and Veins under the Microscope**

Arteries carry blood away from the heart, and therefore are exposed to blood traveling at a higher velocity and pressure than veins. Their walls must be able to accommodate that pressure and their lumens are smaller as the blood spends less time there. Veins, by comparison, do not have wall structure that tolerates high pressure, and their walls are thinner, their structure baggier and less rigid. Because blood moves more slowly through veins, they have larger lumens.

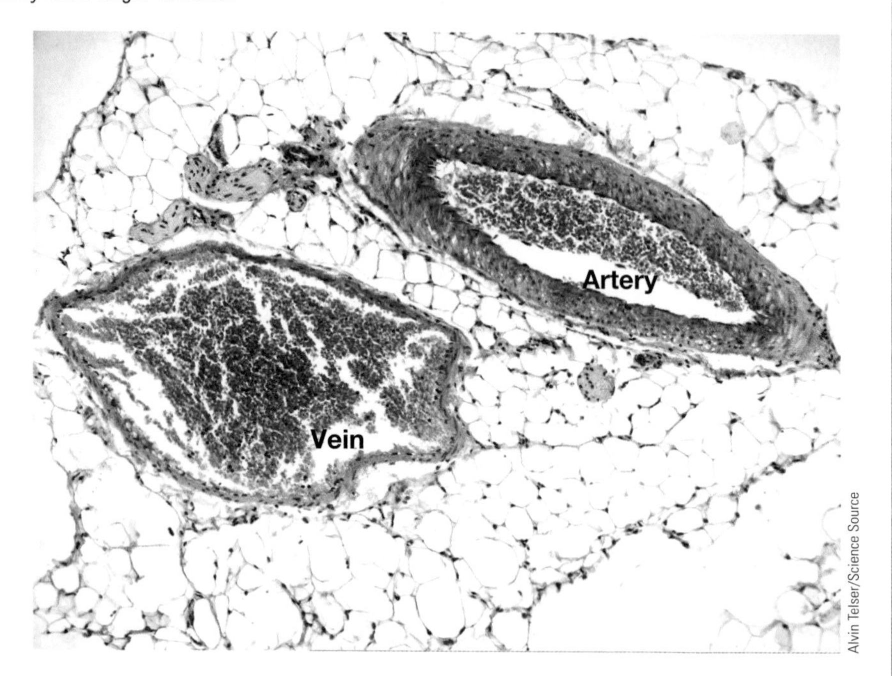

As blood travels away from the heart, it steadily loses pressure. By the time blood has passed through the capillaries and entered the venules, the pressure supplied by heart contractions is almost entirely gone. In other words, in comparison to arteries, venules and veins withstand a much lower pressure from the blood that flows through them. For this reason, their walls are considerably thinner and the diameter of their lumens is larger. In addition, many veins of the body, particularly those of the limbs, contain valves. The valves assure one-way flow toward the heart; once blood has moved past a valve, the valve closes and prevents its backflow. This is critical because blood flow becomes sluggish in the extremities as a result of the lower pressure and the effects of gravity.

LO 20.1.2

Both arteries and veins have the same three distinct layers, called *tunics* (from the Latin term *tunica*). From the most interior layer to the outer, these tunics are the *tunica intima*, the *tunica media*, and the *tunica externa* (**Figure 20.4**). While both arteries and veins have these layers, the layers are structured differently in the two different types of vessels.

tunica intima (tunica interna)

The **tunica intima**, the innermost layer of blood vessels, is composed of endothelium, the simple squamous epithelium and its basement membrane that lines the entire inner surface of the cardiovascular system and is continuous with the endocardium of the heart. Damage to this endothelial lining and exposure of blood to the collagenous fibers beneath is one of the primary causes of blood clot formation. The endothelium plays an active role in the regulation of blood flow. This layer can release local chemicals called **endothelins** that can constrict the smooth muscle within the walls of the vessel.

Beneath the endothelium is the basement membrane that anchors the endothelium to the connective tissue. The basement membrane provides strength while maintaining flexibility and is permeable, allowing materials to diffuse through it. The base of the tunica intima contains a small amount of areolar connective tissue that consists

Figure 20.4 Structure of Blood Vessels

Both arteries and veins have three layers, called *tunics*. However, the microanatomy and relationship among the three layers is different in arteries and veins.

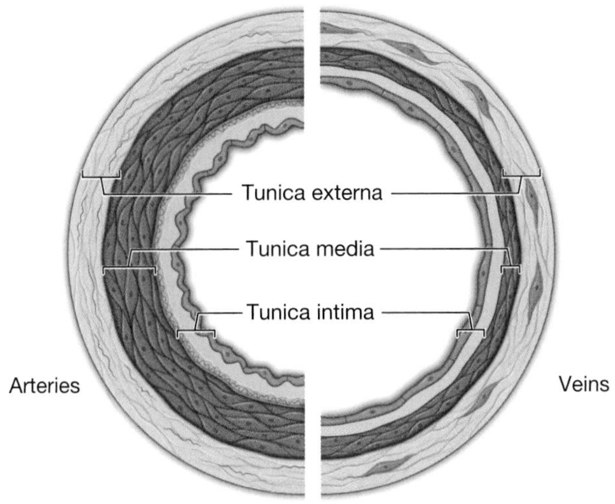

primarily of elastic fibers to provide the vessel with additional flexibility; it also contains some collagenous fibers to provide additional strength.

LO 20.1.3

internal elastic membrane (internal elastic lamina)

In larger arteries, there is also a thick, distinct layer of elastic fibers known as the **internal elastic membrane** (also called the *internal elastic lamina*) at the boundary with the tunica media. Like the other components of the tunica intima, the internal elastic membrane provides structure while allowing the vessel to stretch. It is permeated with small openings that allow exchange of materials between the tunics. The internal elastic membrane is not apparent in veins. In contrast, many veins, particularly in the lower limbs, contain valves of thickened endothelium that are reinforced with connective tissue, extending into the lumen. These valves function to prevent backflow in the low-pressure veins. Arteries do not have valves.

Under the microscope, the lumen and the entire tunica intima of a vein will appear smooth, while those of an artery will normally appear wavy because of the partial constriction of the smooth muscle in the tunica media—the next layer of blood vessel walls.

LO 20.1.4

Tunica Media The **tunica media**, the middle layer of blood vessel walls, is composed largely of smooth muscle (see Figure 20.4). The contraction of this smooth muscle enables **vasoconstriction**, a narrowing of the vessel that restricts flow within. When the smooth muscle of the tunica media relaxes, the vessel widens in a process called **vasodilation**. Both vasoconstriction and vasodilation are regulated in part by small vascular nerves, known as **nervi vasorum** or "nerves of the vessel," that run within the walls of blood vessels. These are generally all sympathetic fibers with a few anatomical exceptions. Notably, parasympathetic innervation is found within the walls of the blood vessels of the genitalia, and stimulation of these parasympathetic fibers causes vasodilation in the external genitalia. This vasodilation is a key factor in arousal and function of the erectile tissue in all genitalia, which is covered in more detail in Chapter 27. For the other vessels, however, the contraction or relaxation of the smooth muscle within the tunica media is controlled by sympathetic innervation. Local factors, discussed later in the chapter, can also influence this muscle tissue.

The smooth muscle layers of the tunica media are supported by a framework of collagen fibers that binds the tunica media to the inner and outer tunics. Along with the collagen fibers are large numbers of elastic fibers that appear as wavy lines in prepared slides. In larger arteries there is a thick **external elastic membrane**, a wavy elastic fiber layer that provides stretchiness to the vessel. This structure is not usually seen in smaller arteries, nor is it seen in veins.

The tunica media is generally the thickest layer in arteries, and it is much thicker in arteries than it is in veins.

tunica externa (tunica adventitia)

Tunica Externa The outer tunic, the **tunica externa**, is a substantial sheath of connective tissue composed primarily of collagen fibers. Some bands of elastic fibers and groups of smooth muscle fibers can be found in the tunica externa of some vessels. This layer is normally the thickest tunic in veins (but usually thinner than the tunica media in arteries). The outer layers of the tunica externa are not distinct but rather blend with the surrounding connective tissue outside the vessel, helping to hold the vessel in place. For example, if you can locate a superficial vein on one of your upper limbs and try to move it, you will find that the tunica externa prevents any repositioning of the vein.

These layers of the walls of arteries and veins are built of living cells as well as the collagen and elastic fibers the cells produce; therefore, their structure and integrity require nourishment. These cells also produce waste. Since blood passes through the larger vessels relatively quickly, and their walls are very thick, there is insufficient time for enough diffusion of nutrients and wastes with the blood passing through. For this reason, larger arteries and veins contain small blood vessels within their walls known as the **vasa vasorum**—which in Latin means "vessels of the vessel"—to provide them with exchange of nutrients and wastes. The vasa vasorum must function in the outer layers of the vessel (see Figure 20.4) or the pressure exerted by the blood passing through the vessel would collapse it, preventing any exchange from occurring. The lower pressure within veins allows the vasa vasorum to be located closer to the lumen. The restriction of the vasa vasorum to the outer layers of arteries is thought to be one reason that arterial diseases are more common than venous diseases; its location makes it more difficult to nourish the cells of the arteries and remove waste products.

20.1b Arteries

LO 20.1.5

The structure of arteries enables these vessels to withstand the high pressure of blood ejected from the heart. Those closest to the heart have the thickest walls, and the highest percentage of elastic fibers in all three of their tunics. This type of artery is known as an **elastic artery** (Figure 20.5). Their abundant elastic fibers allow them to expand, to accommodate blood pumped from the ventricles as it passes through them, and then to recoil after the surge has passed. If artery walls were rigid and unable to expand and recoil, their resistance to blood flow would greatly increase. The elastic recoil of the vascular wall helps to propel the blood through the arterial system.

elastic artery (conducting artery)

As the blood moves farther from the heart the pressure generated from the heart decreases. For the vessels, this means that the walls are subjected to reduced swings in pressure, and do not have to expand and recoil to the same extent. Therefore, the percentage of elastic fibers in an artery's tunica intima decreases and the amount of smooth muscle in its tunica media increases. The artery at this point is described as a **muscular artery**. Their smooth muscle layer allows muscular arteries to play a leading role in vasoconstriction. There is not one particular anatomical point where elastic arteries become muscular arteries, rather, there is a gradual transition as the vascular tree repeatedly branches.

muscular artery (distributing artery)

Figure 20.5 Types of Arteries and Arterioles

The arteries closest to the heart have the highest proportion of elastic fibers to help them stretch. They are the largest arteries and are called *elastic arteries*. Elastic arteries branch into smaller arteries and eventually the proportion of muscle in the wall of the vessel becomes substantial. These arteries, muscular arteries, contribute to blood flow, pressure, and thermoregulation through vasoconstriction and vasodilation. Muscular arteries branch into smaller arteries, which branch into arterioles.

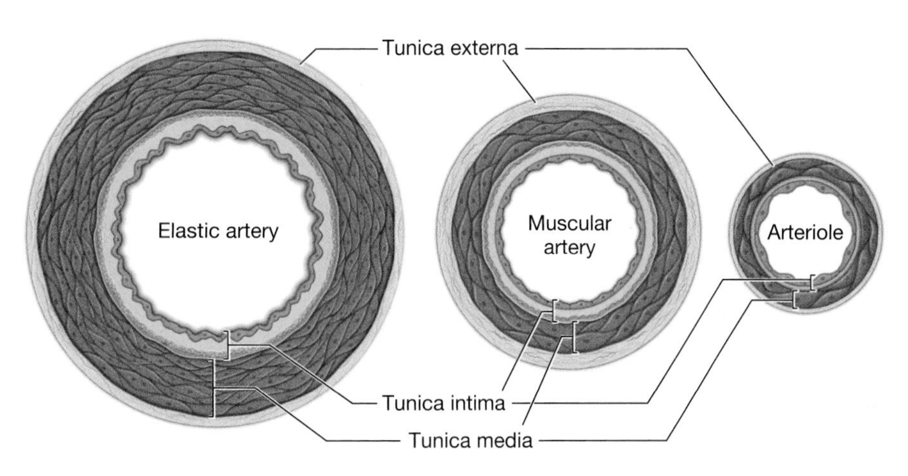

arteriole (resistance vessel)

20.1c Arterioles

An **arteriole** is a very small artery that leads to a capillary. Arterioles have the same three tunics as the larger vessels, but the thickness of each is greatly diminished. The tunica media of an arteriole is typically only one or two smooth muscle cell layers in thickness. The tunica externa in arterioles is very thin (see Figure 20.5).

The lumen of arterioles is much smaller in diameter than that of muscular arteries. This reduction in lumen space through which to pass slows down the flow of blood, which also causes a substantial drop in blood pressure. Arterioles are the site of most resistance in the cardiovascular system. The muscle fibers in arterioles are normally slightly contracted, causing arterioles to maintain a consistent muscle tone—in this case referred to as **vascular tone**—in a manner similar to the muscular tone of skeletal muscle. Vascular tone is maintained through innervation from the sympathetic nervous system, but the smooth muscle contraction of arterioles is also influenced by local chemical signaling. The importance of the arterioles is that they will be the primary site of both resistance and regulation of blood pressure, and vasoconstriction and vasodilation in the arterioles are the primary mechanisms for distribution of blood flow.

20.1d Capillaries

A **capillary** is a thin-walled microscopic channel that supplies blood to the tissues themselves (a process called **perfusion**). Exchange of gasses and other substances between the blood and the surrounding cells and their tissue fluid (interstitial fluid) occurs across the walls of the capillaries. Capillaries are quite small; some are just barely wide enough for an erythrocyte to squeeze through, and their walls consist of a single endothelial cell surrounded by a basement membrane with occasional smooth muscle fibers. For capillaries to function, their walls must be leaky, allowing substances to pass through. There are three major types of capillaries—continuous, fenestrated, and sinusoid capillaries (**Figure 20.6**)—which differ according to their degree of "leakiness."

LO 20.1.6

Continuous Capillaries The most common type of capillary, found in almost all tissues, is the continuous capillary. **Continuous capillaries** are characterized by a complete endothelial lining, meaning that the endothelial cells all touch each other and do not contain large holes. Tight junctions knit endothelial cells together. Despite the presence of scattered tight junctions, there are some spaces between the cells that allow for exchange of water and other very small molecules between the blood plasma and the interstitial fluid. Substances that can pass between cells include glucose, water, ions, and leukocytes. In addition, small hydrophobic molecules like gasses and hormones can cross the endothelial cells by diffusing across their membranes. Other substances can cross capillary endothelial cells through endo- and exocytosis.

Fenestrated Capillaries A **fenestrated capillary** is one that has pores (or fenestrations) through the endothelial cells to allow larger molecules to cross the capillary wall easily. The number of fenestrations and their degree of permeability vary, however, according to their location. Fenestrated capillaries are common in the small intestine, which is the primary site of nutrient absorption, as well as in the kidneys, which filter the blood. They are also found in the choroid plexus of the brain and many endocrine structures, where they allow hormones to pass into the blood from the cells that make them.

sinusoid capillaries (sinusoidal capillaries, sinusoids)

Sinusoid Capillaries The least common capillary type, **sinusoid capillaries** are capillaries with extensive intercellular gaps and incomplete basement membranes.

Figure 20.6 **Types of Capillaries**

(A) The vast majority of capillaries in the body are continuous, meaning that each endothelial cell connects to adjacent endothelial cells without major openings or gaps. (B) In places where more exchange between the tissues and the blood is required, fenestrated capillaries, which contain small pores, can be found. (C) Sinusoidal capillaries contain even larger gaps than fenestrated capillaries.

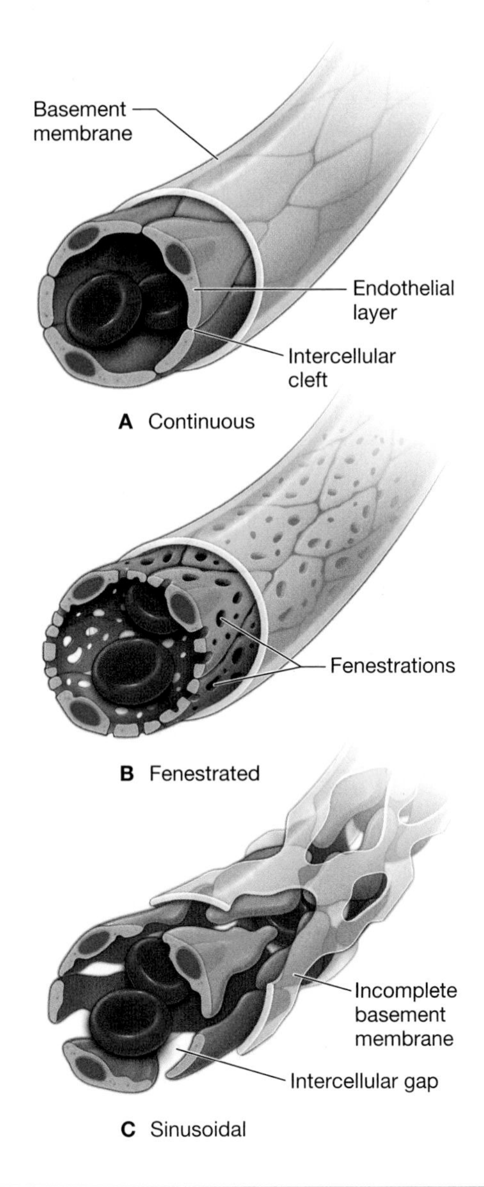

These very large openings allow for the passage of the largest molecules, including plasma proteins and even cells. Blood flow through sinusoids is very slow, allowing more time for exchange of gasses, nutrients, and wastes. Sinusoids are found in the liver and spleen, bone marrow, and many endocrine glands including the pituitary and adrenal glands. Sinusoid capillaries are only found in organs that rely on the passage of cells or large proteins for their functions. For example, when bone marrow forms new blood cells, the cells must enter the blood supply and can only do so through the large openings of a sinusoid capillary; they cannot pass through the small openings of continuous or fenestrated capillaries. The liver utilizes sinusoid capillaries in order to remove nutrients and wastes brought to it by the hepatic portal vein from both the digestive tract and spleen, to recycle red blood cells and to release plasma proteins into the blood.

20.1e Metarterioles and Capillary Beds

A **metarteriole** is a type of vessel that has structural characteristics of both an arteriole and a capillary. A metarteriole is slightly larger than the typical capillary and contains rings of smooth muscle. A metarteriole brings blood toward a **capillary bed** that may consist of 10–100 capillaries (**Figure 20.7**).

The **precapillary sphincters**, circular smooth muscle cells that surround the capillary at its origin with the metarteriole, tightly regulate the flow of blood from a metarteriole to the capillaries it supplies. Their function is critical: If all of the capillary beds in the body were to open simultaneously, there would be too many vessels to fill, the total blood volume would fill the capillaries alone and there would be none in the arteries, arterioles, venules, veins, or the heart itself. Much of the time, the precapillary sphincters are closed. When the surrounding tissues need oxygen or have excess waste products, the precapillary sphincters open, allowing blood to flow through and exchange to occur (**Figures 20.7A** and **20.7B**). If all of the precapillary sphincters in a

Figure 20.7 Capillary Beds

Capillary beds are small networks of connected capillary vessels. A metarteriole brings blood to the capillary bed and a thoroughfare channel carries the oxygenated blood away and into a venule. (A) When the precapillary sphincter is closed, the blood flows directly through the thoroughfare channel. (B) When precapillary sphincters are open, some of the blood flows through the capillary bed before draining into the venule.

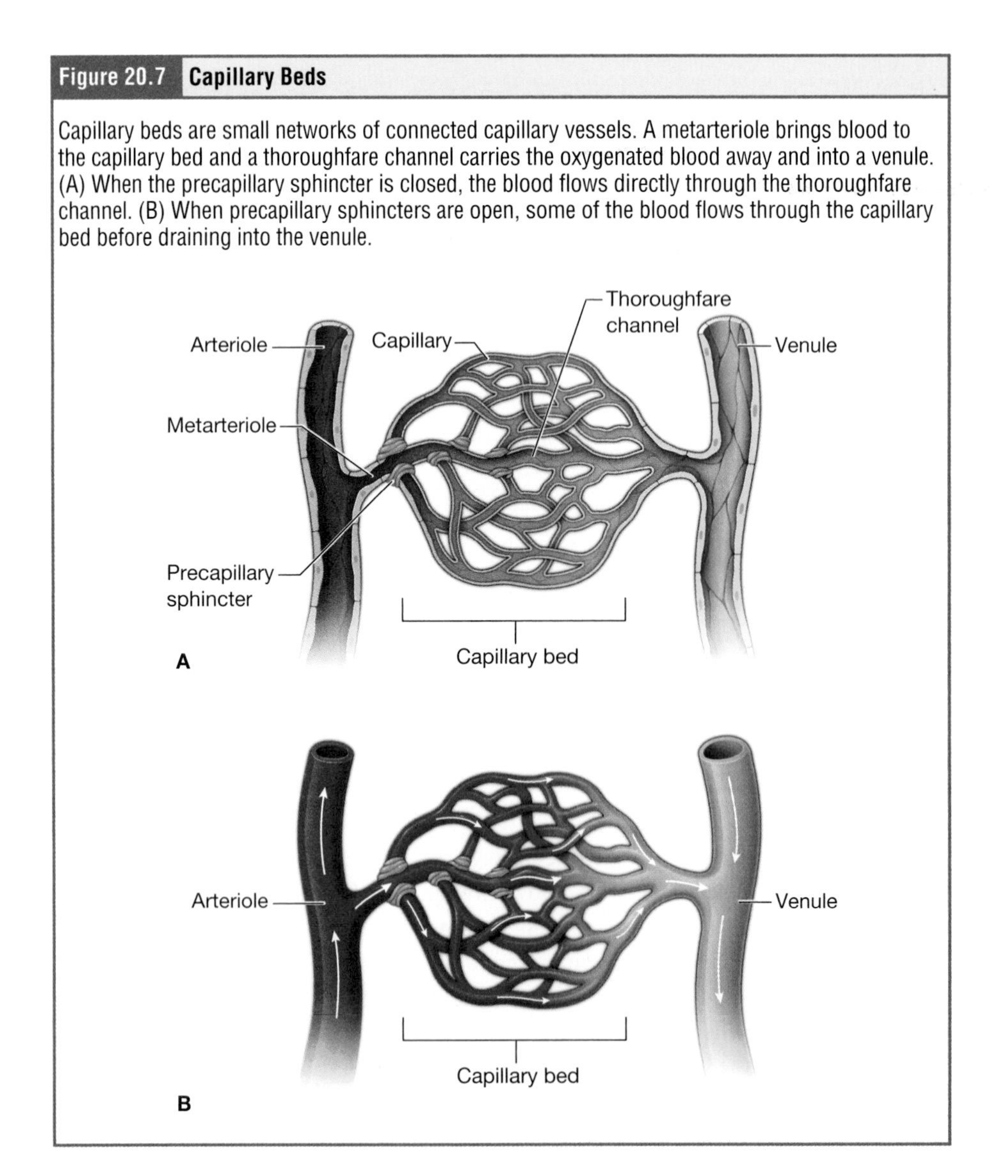

capillary bed are closed, blood will flow from the metarteriole directly into a **thoroughfare channel** and then into the venous circulation, bypassing the capillary bed entirely.

The smooth muscle cells of a precapillary sphincter are regulated by changes in internal conditions, such as oxygen, carbon dioxide, hydrogen ion, and lactic acid levels. For example, during strenuous exercise when oxygen levels decrease and carbon dioxide, hydrogen ion, and lactic acid levels all increase, the precapillary sphincters relax and blood perfuses the capillary beds within the active skeletal muscle. Similarly, the presence of nutrients in the digestive tract causes these sphincters to open and allow the nutrients to be absorbed into the blood. During sleep or rest periods, capillary beds in both areas are largely closed; they open only occasionally to allow oxygen and nutrient supplies to travel to the tissues to maintain cellular life.

Learning Connection

Micro to Macro

Smooth muscle and elastic connective tissue are important components in the walls of blood vessels. In which types of blood vessels (arteries, arterioles, capillaries, venules, veins) are each of these tissues found? What roles do these tissues play in each of the vessels that contain them?

20.1f Venules

A **venule** is an extremely small vein. Multiple venules join to form veins. The walls of venules consist of an inner layer (the endothelium), a thin middle layer containing a few muscle cells and elastic fibers, and a very thin outer layer of connective tissue fibers that constitutes the tunica externa (**Figure 20.8**).

20.1g Veins

The term *vein* describes any blood vessel that carries blood toward the heart. Compared to arteries, veins are thin-walled vessels with large lumens (see Figure 20.8). Because they are low-pressure vessels, larger veins are commonly equipped with valves that promote the unidirectional flow of blood toward the heart and prevent backflow toward the capillaries. **Table 20.1** compares the features of arteries and veins.

20.1h Veins as Blood Reservoirs

LO 20.1.7

Because blood moves more slowly and at a lower pressure when traveling through the veins, at any given moment a majority—about 64 percent—of the blood volume will

Figure 20.8 Venules and Veins

Veins have the same three tunics as arteries; their tunica externa is the most significant of their layers. Veins do not have much in the way of a tunica media and their tunica intima has folds that form valves. These valves ensure one-way flow. Both medium and large veins have valves but venules do not.

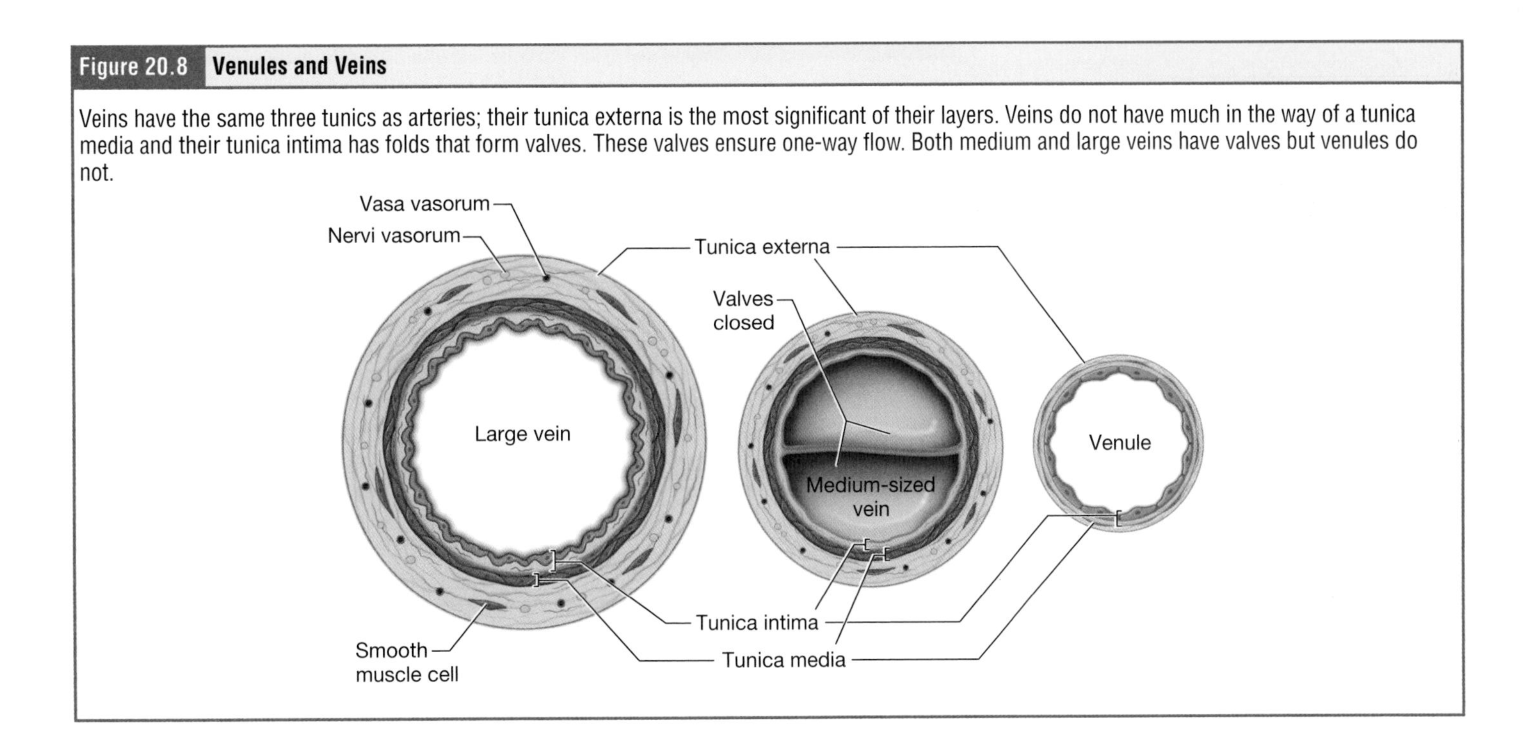

Table 20.1 Comparison of Arteries and Veins

Features	Arteries	Veins
Direction of blood flow	Conducts blood away from the heart	Conducts blood toward the heart
General appearance	Rounded	Irregular, often collapsed
Pressure	High	Low
Wall thickness	Thick	Thin
Relative oxygen concentration	Higher in systemic arteries Lower in pulmonary arteries	Lower in systemic veins Higher in pulmonary veins
Valves	Not present	Present most commonly in limbs and in veins inferior to the heart

Cultural Connection

Safe Injection Sites

AP Images/Paul Chiasson

The opioid crisis in the United States is undoubtedly the worst drug crisis in the country's history. Approximately 50,000 Americans die each year of opioid overdoses, and an estimated 10 million Americans are currently suffering from opioid addiction. To put that number in perspective, that is approximately the population of the states of North Carolina or Ohio. Opioid use comes with many risks, including death by accidental overdose, but the act of injection itself comes with a significant number of risks. Approximately 1 million people, or 10 percent of opioid users, inject drugs into their veins. Injection can come with risks of increasing scar tissue at the injection site, as well as a risk of infection. The infectious risk increases dramatically when the supplies or environment where the drug use occurs are not clean or stable. Depending on the conditions of the needle, the skin, and the environment, bacterial, fungal, or viral infections are all possible. When the tunics of the veins themselves are damaged, scar tissue, track marks, emboli or collapse of a vein can occur. When the needle misses the vein, abscesses—localized infections with pus-filled cores—can form. When the needle carries a pathogen and is injected correctly into a vein, systemic bacterial or fungal infections may occur; if these spread to the heart, endocarditis—infection of the inner lining of the heart—may develop. Endocarditis represents an acute health crisis and is likely to permanently damage the heart or cause death. If the bacteria or fungi spread to the brain, meningitis can occur. HIV and Hepatitis C virus infections are common among intravenous (IV) drug users. IV drug users typically began their relationship with opioids through prescription. Opioids have been overprescribed and prescribed for the wrong reasons by doctors in the United

States for decades. Recently the makers of opioid drugs have admitted to persuading doctors to prescribe these dangerous drugs to drive up their profits. Now, many IV drug users are struggling to eliminate drug use from their lives. In many U.S. cities, safe injection sites are being established by local governments. These locations are staffed with medical professionals who can treat an overdose if one should occur. The sites are stocked with clean, sterile injection supplies as well as instructions for their safe use. Most importantly, the sites offer a variety of support to drug users seeking treatment for their addiction. These safe injection sites reduce the harm and stigma that comes with opioid use and pave the way for treatment of the underlying addiction. As we work to reverse the course of the U.S. opioid crisis, which was fueled by pharmaceutical companies and exacerbated by the economic and mental health aspects of the coronavirus pandemic, safe injection sites will provide a safe way to reduce the anatomical and immune implications of this disease.

be found in the veins. For this reason, veins are often referred to as *blood reservoirs*, or *venous reservoirs*. Given their thin walls, which lack the muscle and elastic fibers of arteries, veins are able to expand as needed to contain the slow-moving blood.

When blood flow needs to be redistributed to other portions of the body, the small amount of smooth muscle in venous walls will constrict. Less dramatic than the vasoconstriction seen in arteries and arterioles, venoconstriction may be likened to a "stiffening" of the vessel wall. This increases pressure on the blood within the veins, speeding its return to the heart.

20.1i Alternative Blood Flow Pathways

In most cases the pattern of blood flow away from and toward the heart follows a predictable pattern:

Artery → Arteriole → Capillary bed → Venule → Vein

But there are a few anatomical variations to this pattern (see the "Anatomy of Blood Flow Pathways" feature). An **arterial anastomosis** is an anatomical pattern where multiple arteries converge to supply their blood to a common capillary bed. A **venous anastomosis** is an anatomical pattern where venules split and contribute blood to multiple veins. Veinous anastomoses are more common than arterial anastomoses, but arterial anastomoses can be anatomically critical. For example, the capillaries of the brain are served by several arteries. If blood is not able to travel through one of these arteries due to injury, the blood supplied by the other arteries will still fill the brain's capillaries and ensure neuronal function. An **arteriovenous anastomosis** provides a detour for the blood around the route that leads to a capillary bed. In these blood vessels, blood never travels to the capillary bed and never exchanges gasses or nutrients with the tissues. Arteriovenous anastomoses are found in the hands and feet of humans, and they are used during dips in body temperature. When hypothermia (becoming too cold to maintain core body temperature) is a concern, blood within these regions will travel through the shunt rather than perfuse the capillaries. This alternative pathway helps to prevent heat loss, but it also compromises the cellular function of the tissues in these areas.

LO 20.1.8

A **portal system** is a blood flow pattern in which two capillary beds are linked between the artery and vein:

Artery → Arteriole → Capillary bed → Connecting vessel → Venule → Vein

Portal systems function to connect and provide exchange between two tissues that are not located close together. In a typical portal system, materials such as nutrients or hormones are added to the blood in one capillary bed and removed in the second capillary bed.

Anatomy of...

Blood Flow Pathways

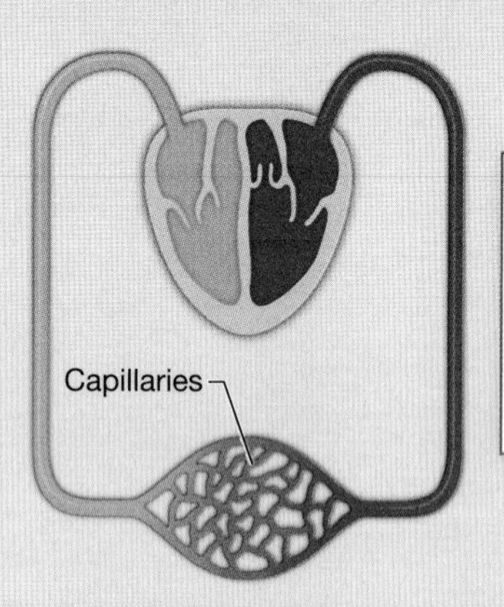

In the standard pathway, blood flows away from the heart in an artery that branches into arterioles, and then gas and material exchange takes place in the capillaries. Deoxygenated blood is drained from the capillaries into a venule and then flows through veins to return to the heart.

In an anastomosis, multiple arteries, all of which are carrying oxygenated blood, merge together and combine their blood to feed an area. Anastomoses have evolved to guarantee blood flow to critical areas like the brain.

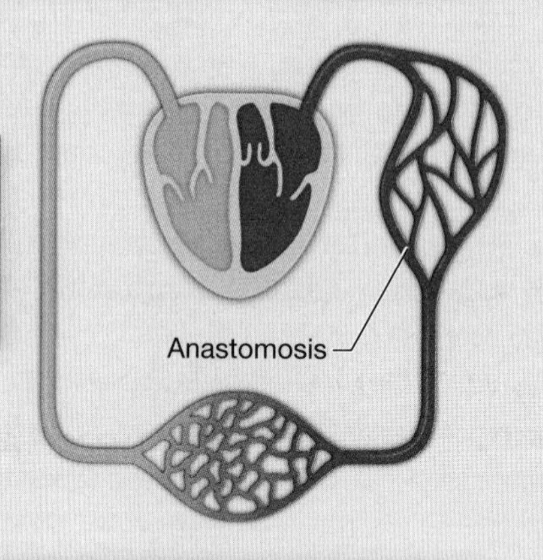

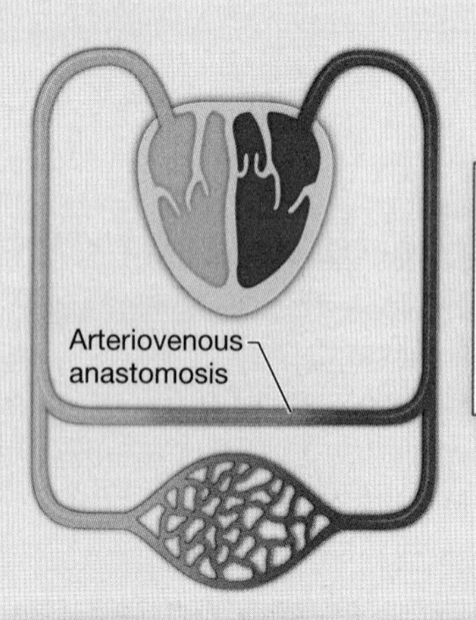

An arteriovenous anastomosis provides a stent or bypass that avoids the capillary bed. These anatomical blood flow patterns are present in the hands and feet. During cold temperatures, blood flows through the shunt and avoids the capillary bed to prevent heat loss.

In a portal system, two capillary beds are linked between the artery and vein. Typically material such as nutrients or hormones are added to the blood at the first capillary bed and then removed at the second capillary bed. Portal systems enable exchange between two tissues of the body.

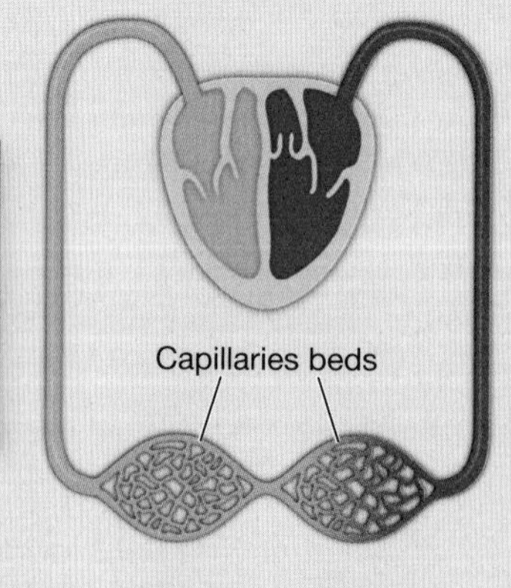

Learning Check

1. Which circuit supplies blood to the muscles of the heart?
   a. Systemic circuit
   b. Pulmonary circuit
   c. Both
2. Which layer of the blood vessel narrows the vessel to restrict blood blow?
   a. Tunica intima
   b. Tunica media
   c. Tunica externa
   d. Tunica interna
3. Where are the vasa vasorum located?
   a. Tunica interna
   b. Tunica externa
   c. Tunica intima
   d. Tunica media
4. What is the difference between fenestrated capillaries and sinusoid capillaries?
   a. The endothelial cells of fenestrated capillaries all touch each other.
   b. The endothelial cells of sinusoid capillaries have small pores.
   c. The capillary walls of sinusoid capillaries have the largest openings.
   d. Fenestrated capillaries are the most common type of capillary.
5. The spleen is a lymphatic organ where debris and waste can be removed from the blood. Many white blood cells also enter the blood from the spleen to fight infection in the blood. What type of capillary would you expect to find in the spleen?
   a. Fenestrated capillaries
   b. Sinusoidal capillaries
   c. Continuous capillaries

20.2 Blood Flow, Blood Pressure, and Resistance

Learning Objectives: By the end of this section, you will be able to:

20.2.1 Define blood flow, blood pressure, and peripheral resistance.

20.2.2 State and interpret the equation that relates fluid flow to pressure and resistance.

20.2.3* Explain the clinical and physiological significance of pulse pressure and mean arterial pressure values.

20.2.4* List and explain the contribution of the components that contribute to vascular resistance and mean arterial pressure.

20.2.5 Describe the role of arterioles in regulating tissue blood flow and systemic arterial blood pressure.

20.2.6 Interpret relevant graphs to explain the relationships between vessel diameter, cross-sectional area, blood pressure, and blood velocity.

20.2.7* Describe the maintenance of driving pressure in veins including the skeletal and pulmonary pumps.

20.2.8 Given a factor or situation (e.g., left ventricular failure), predict the changes that could occur in the cardiovascular system and the consequences of those changes (i.e., given a cause, state a possible effect).

*Objective is not a HAPS Learning Goal.

LO 20.2.1

Blood flow refers to the movement of blood through a vessel, tissue, or organ, and is usually expressed in terms of volume of blood per unit of time. We know that blood flow is an example of bulk flow, that all components of blood flow together. Flow rates are influenced by the pressure gradient generated by the heart and opposed by the factors of resistance, which we will discuss in more detail in this chapter. As a reminder, the following equation can be used to compare the factors that influence blood flow:

LO 20.2.2

Flow α pressure gradient/resistance

As noted earlier, hydrostatic pressure is the force exerted by a fluid on the walls of its container. One form of hydrostatic pressure is **blood pressure**, the force exerted by blood upon the walls of the blood vessels or the chambers of the heart. Blood exerts pressure on all of the vessel walls, but it is generally measured in the brachial artery just above the elbow. The device used to measure blood pressure, a **sphygmomanometer**, measures blood pressure in millimeters of mercury (mm Hg).

20.2a Arterial Blood Pressure

When systemic arterial blood pressure is measured, it is recorded as a ratio of two numbers (e.g., 120/80 mm Hg is a typical blood pressure value for a healthy adult). The top number represents the systolic pressure, and the bottom number represents the diastolic pressure. The **systolic pressure** reflects the arterial pressure in the brachial artery during ejection of blood when the heart's ventricles contract. Recall from Chapter 19 that heart contraction is called *systole*. The **diastolic pressure** is the arterial pressure of blood during ventricular relaxation, or *diastole*.

Student Study Tip

Remember what happens during systole, your blood vessels squish. Think of it as "squishtole!"

LO 20.2.3

Pulse Pressure The difference between the systolic pressure and the diastolic pressure is the **pulse pressure**. For example, an individual with a systolic pressure of 120 mm Hg and a diastolic pressure of 80 mm Hg would have a pulse pressure of 40 mm Hg. As shown in **Figure 20.9**, the pulse pressure is the greatest in the arteries closest to the heart. These arteries have the most elastic fibers in their walls and so they can stretch and recoil dramatically between systole and diastole. Pulse pressure decreases in individuals who experience a loss or deterioration of elastic fibers in their artery walls due to age or chronic high blood pressure. With less elastic fibers, the heart has to work harder to eject blood.

Figure 20.9 Blood Pressure Decreases as Blood Flows Farther from the Heart

Blood pressure fluctuates wildly in the vessels close to the heart where ejected blood comes in waves during every heartbeat. During systole the pressure increases significantly, and during diastole the pressure drops dramatically. As the blood travels farther from the heart it encounters more friction and its pressure decreases between the arterioles and the capillaries. The blood in the venules and veins is under very low pressure.

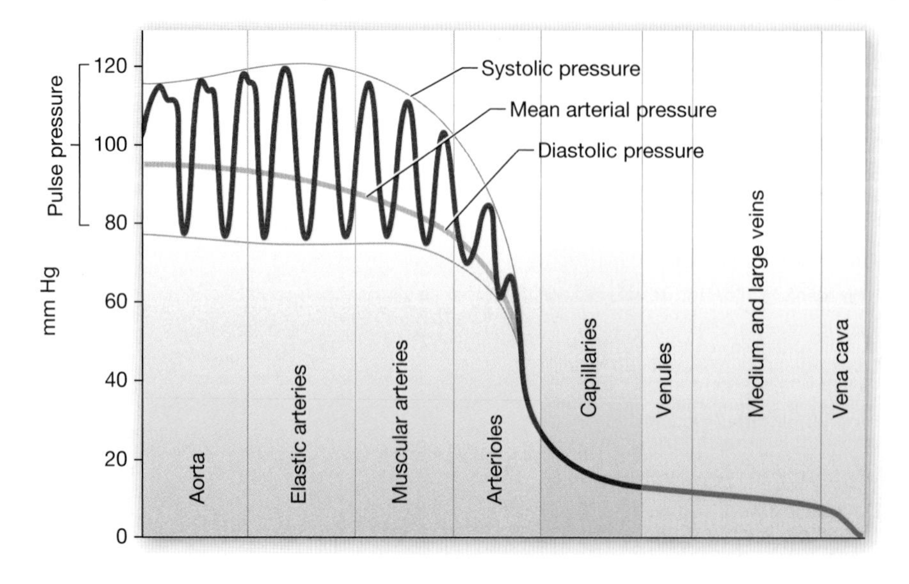

LO 20.2.4

Mean Arterial Pressure **Mean arterial pressure (MAP)** represents the "average" pressure that the arteries experience over time. Although complicated to measure directly and complicated to calculate, MAP can be approximated by adding the diastolic pressure to one-third of the pulse pressure or systolic pressure minus the diastolic pressure:

$$MAP = diastolic\ BP + \frac{(systolic - diastolic\ BP)}{3}$$

For an individual with a measured blood pressure of 120/80 mm Hg, MAP is approximately 80 + (120 − 80) / 3, or 93.33. Normally, the MAP falls within the range of 70–110 mm Hg. MAP is an important value to consider physiologically. If the MAP value falls for an extended time, blood pressure will not be high enough to ensure circulation to and through the tissues, which results in **ischemia**, or insufficient blood flow. A condition called **hypoxia**—inadequate oxygenation of tissues—commonly accompanies ischemia. The term *hypoxemia* refers to low levels of oxygen in systemic arterial blood. Neurons are especially sensitive to hypoxia and may die or be damaged if blood flow and oxygen supplies are not quickly restored. MAP also helps us consider the conditions that the artery walls are exposed to over time. A high average MAP indicates a level of pressure and stress on the artery walls.

Student Study Tip

The suffix *-emia* means "condition of the blood." Use this to differentiate between hypoxia and hypoxemia, and other conditions that will be covered in this chapter.

20.2b Pulse

After blood is ejected from the heart, elastic fibers in the arteries help maintain a high pressure gradient as they expand to accommodate the blood and then recoil, pushing back on the contents of the artery. This expansion and recoiling effect, known as the **pulse**, can be felt through the skin. Although the effect diminishes as the distance from the heart increases, elements of the systolic and diastolic components of the pulse are still evident down to the level of the arterioles. Because pulse indicates heart rate, it is measured clinically to provide clues to a patient's state of health. It is recorded as beats per minute.

Pulse can be palpated manually by placing the tips of the fingers across an artery that runs close to the body surface and pressing lightly. While this procedure is normally performed using the radial artery in the wrist or the common carotid artery in the neck, any superficial artery that can be palpated may be used (**Figure 20.10**). Measuring the pulse allows us to get an indirect measurement of heart rate, as well as to evaluate blood force to different locations in the body. For example, if you suspected that a wound in the thigh was severe, you might take the pulse more distally on that leg to see if adequate perfusion was still occurring.

Student Study Tip

There is a pulse in your thumb as well, and if you are taking your own pulse, using your thumb will help magnify it.

Learning Check

1. Orthostatic hypotension occurs when the systolic pressure drops greater than 20 mm Hg and the diastolic pressure drops greater than 10 mm Hg when a person who is lying down suddenly stands up. Sharron's blood pressure was 120/80 when lying down, but 95/65 when standing up. Mike's blood pressure was 125/80 when lying down, but 115/70 when standing up. Who has orthostatic hypotension?
   a. Sharron
   b. Mike
2. Upon standing, Sharron's blood pressure becomes 95/65 mm Hg. What is Sharron's pulse pressure?
   a. 95 mm Hg
   b. 65 mm Hg
   c. 160 mm Hg
   d. 30 mm Hg
3. Lying down, Mike has a blood pressure of 125/80 mm Hg. What is Mike's approximate mean arterial pressure?
   a. 95 mm Hg
   b. 140 mm Hg
   c. 170 mm Hg
   d. 40 mm Hg

Figure 20.10 Pulse Points

There are several locations on the human body where pulse pressure can be felt on the surface of the skin.

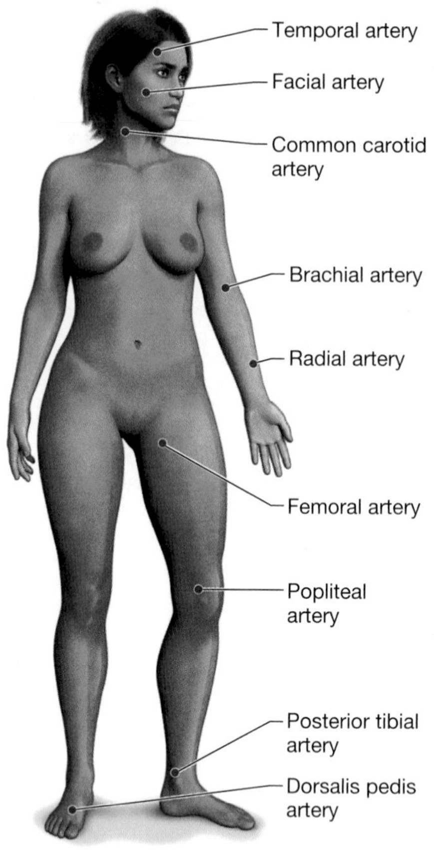

Temporal artery
Facial artery
Common carotid artery
Brachial artery
Radial artery
Femoral artery
Popliteal artery
Posterior tibial artery
Dorsalis pedis artery

20.2c Variables Affecting Blood Flow and Blood Pressure

Five variables influence blood flow and blood pressure:

- Cardiac output
- Volume of the blood
- Vessel compliance
- Viscosity of the blood
- Blood vessel length and diameter

Recall that blood moves from higher pressure to lower pressure, a pressure gradient is required for blood to flow (**Figure 20.11**). Therefore, in order to drive blood out of the heart the ventricles need to generate more pressure than exists in the arteries. In order for blood to return to the heart the pressure in the atria during diastole must be lower than that in the veins. Resistance opposes blood flow. Any factor that increases resistance will impede blood flow and increase the work of the heart to create a pressure gradient that can drive flow. In this section we will explore these five variables of pressure gradients and resistance.

Figure 20.11 Factors That Contribute to Flow Rate

The flow rate is proportional to a pressure gradient. Both cardiac output and blood volume contribute to that pressure gradient. Flow is opposed by resistance. The three components of resistance are vessel compliance, blood viscosity, and vessel length and diameter.

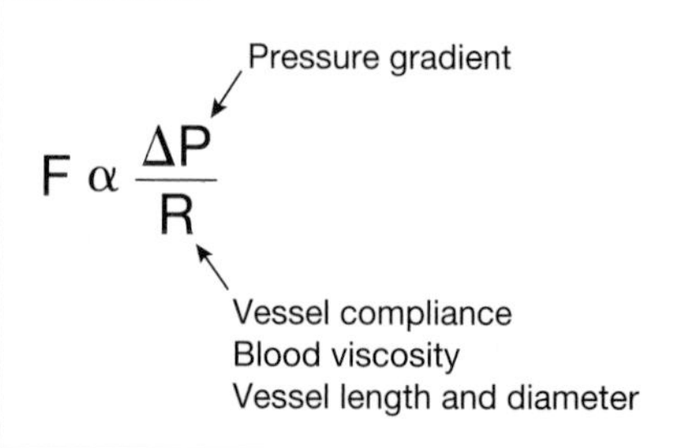

20.2d Factors That Contribute to the Pressure Gradient

Two different factors contribute to the pressure gradient established by the left ventricle: cardiac output and blood volume.

Cardiac Output Cardiac output (CO) is the measurement of the amount of blood that leaves the ventricles per minute. Blood is ejected during every systole, so one factor that contributes to cardiac output is how many systolic events occur per minute—in other words, the heart rate (HR). The other factor in cardiac output, of course, how much blood the ventricles eject each time they contract. This measurement, known as *stroke volume* (SV), is determined by a number of factors discussed in Chapter 19. If we multiply the amount of blood ejected from the heart per contraction with the number of contractions that occur in one minute, we end up with the amount of blood ejected from the heart within a minute—the cardiac output:

$$\mathrm{CO} = \mathrm{HR} \times \mathrm{SV}$$

Blood Volume The relationship between blood volume, blood pressure, and blood flow is obvious. Water may merely trickle along a creek bed in a dry season, but rush quickly and under great pressure after a heavy rain. Similarly, as blood volume decreases, pressure and flow rate decrease. As blood volume increases, pressure and flow rate increase.

Under normal circumstances, blood volume is regulated through a variety of homeostatic mechanisms and therefore does not vary substantially. In certain disease states such as bleeding, dehydration, or severe burns, low blood volume, called **hypovolemia**, may occur.

Hypervolemia, or excessive fluid volume, may be caused by retention of water and sodium, as seen in patients with heart failure, liver cirrhosis, and some forms of kidney or endocrine diseases. Restoring homeostasis in these patients depends upon reversing the condition that triggered the hypervolemia.

In case of a change in either blood volume or cardiac output, the gradient established by the heart's contractions would vary, impacting the rate of blood flow.

20.2e Factors That Contribute to Resistance

Three factors influence the resistance to fluid flow. These factors are the same for any fluid.

A Mathematical Approach to Resistance Factors Jean-Louis-Marie Poiseuille was a French physician and physiologist who devised a mathematical equation describing the resistance to blood flow called *Poiseuille's equation*. The same equation also applies to engineering studies of the flow of fluids. Although understanding the math behind the relationships among the factors affecting blood flow is not necessary and you would not be asked to calculate resistance using Poiseuille's equation, the representation of these parameters in equation format can help solidify an understanding of their relationships. A variation of Poiseuille's equation examines the factors that contribute to resistance (**Figure 20.12**).

Examining this equation, we can see that there are only three variables: viscosity, vessel length, and radius, since 8 and π are both constants. The important thing to remember is this: Two of these variables—viscosity and vessel length—will change slowly in the body. Only one of these factors—the radius—can be changed rapidly by vasoconstriction and vasodilation, thus dramatically impacting resistance and flow. Further, small changes in the radius will greatly affect flow, since it is raised to the fourth power in the equation.

Let's consider the three factors of resistance (vessel length, blood viscosity, and vessel radius) in more detail.

Figure 20.12 Resistance to Flow

Resistance to flow can be summarized in a mathematical equation. In this equation, the length of the vessel and the viscosity of the blood are both proportional to resistance. This means that as these numbers increase, the vessel is longer, or the blood is thicker, there is more resistance to flow. The radius of the vessel is inversely proportional to resistance, and the radius is raised to the fourth power. This means that as the radius increases, there is less resistance to flow. The radius of the vessel has a much greater impact on resistance than the other two factors.

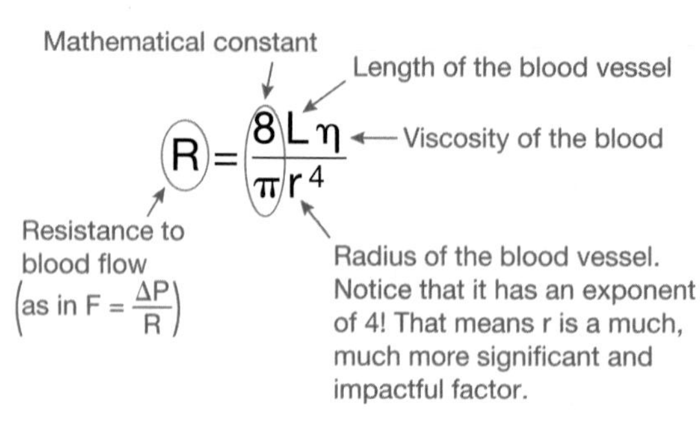

Vessel Length As we can see in the equation in Figure 20.12, the length of a vessel is directly proportional to its resistance: the longer the vessel, the greater the resistance and the lower the flow. The reason for this has to do with friction. As blood flows within a vessel, or any fluid flows within any tube, the blood encounters friction as it contacts the walls of the tube. The longer this goes on, the more friction is encountered and the greater the resistance to flow (**Figure 20.13**).

The length of our blood vessels increases throughout childhood as we grow, of course, but in adults this remains relatively unchanged outside of instances of weight gain. Even then, because adipose tissue does not have an extensive vascular supply, it changes very little. Overall, vessels decrease in length only during weight loss or amputation. An individual weighing 150 pounds has approximately 60,000 miles of vessels in their body. Gaining about 10 pounds adds from 2,000 to 4,000 miles of vessels, depending upon the nature of the gained tissue. One of the great benefits of weight reduction is the reduced stress to the heart, which does not have to overcome the resistance of as many miles of vessels.

Blood Viscosity Viscosity is the thickness of fluids, which affects their ability to flow. Drinking water through a straw, for example, is easier than drinking a milkshake, which has a much higher viscosity. Blood viscosity is a factor of resistance because of

Figure 20.13 The Impact of Friction on Blood Flow

Blood encounters friction at the walls of the blood vessel.

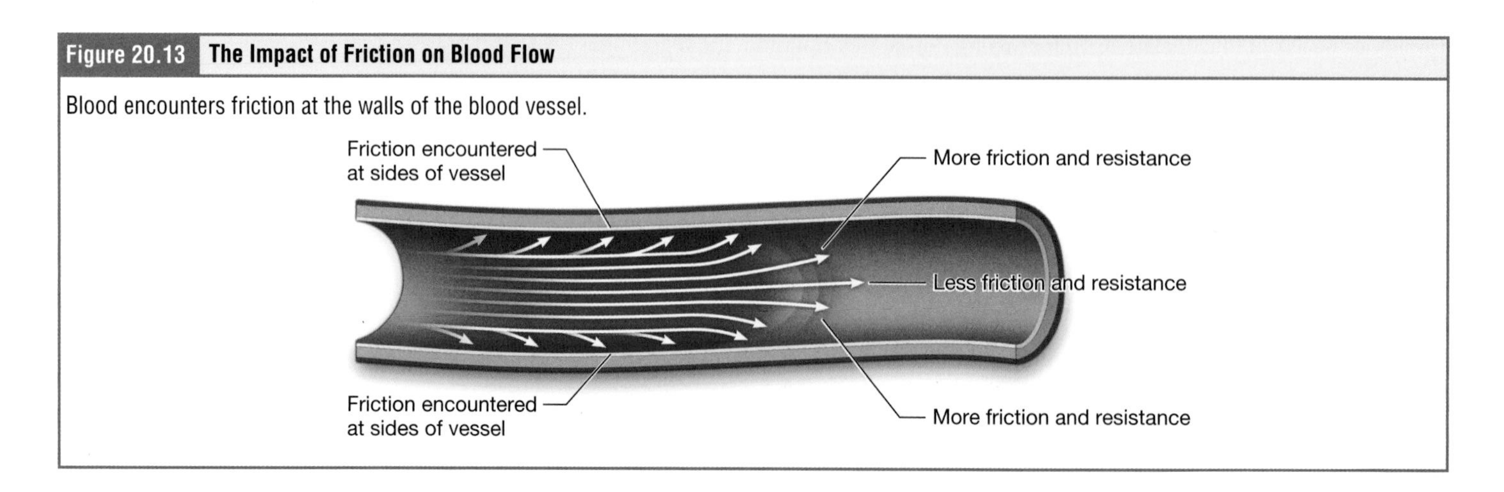

friction. The thicker the fluid, the more friction it encounters along the blood vessel walls, creating more resistance to flow.

The two primary determinants of blood viscosity are the formed elements and plasma proteins. Since the vast majority of formed elements are erythrocytes, any condition affecting the making or recycling of erythrocytes can alter viscosity. Since most plasma proteins are produced by the liver, any condition affecting liver function can also change the viscosity slightly and therefore alter blood flow. While leukocytes and platelets are normally a small component of the formed elements, there are some rare conditions in which severe overproduction can impact viscosity as well.

Blood Vessel Radius In contrast to length, the radius of blood vessels changes anatomically throughout the body according to the type of vessel, as we discussed earlier. The radius of any given vessel may also change frequently throughout the day in response to various physiological circumstances that trigger vasodilation and vasoconstriction. The *vascular tone* of the vessel is the contractile state of the smooth muscle and the primary determinant of radius, and thus of resistance and flow. The effect of vessel radius on resistance is inverse: an increased radius means there is less blood contacting the vessel wall, thus lower friction and lower resistance, subsequently increasing flow (**Figure 20.14**). A decreased diameter means more of the blood contacts the vessel wall, and resistance increases, subsequently decreasing flow.

The influence of lumen diameter on resistance is dramatic: A slight increase or decrease in diameter causes a huge decrease or increase in resistance. This is because resistance is inversely proportional to the radius of the blood vessel (one-half of the vessel's diameter) raised to the fourth power ($R = 1/r^4$). This means, for example, that if an artery or arteriole constricts to one-half of its original radius, the resistance to flow will increase sixteen times. And if an artery or arteriole dilates to twice its initial radius, then resistance in the vessel will decrease to one-sixteenth of its original value and flow will increase sixteen times.

Compliance is the ability of any compartment to expand to accommodate increased content. A metal pipe, for example, is not compliant, while a balloon is highly compliant. Sweatpants are more compliant than jeans, and so on. The greater

Learning Connection

Explain a Process

If you were trying to explain the effects of blood vessel radius on blood flow and blood pressure to a friend, what would you tell your friend?

Figure 20.14 Friction and Vessel Size

Since the blood encounters friction only at the blood vessel walls, vessels with larger diameters have more space in the middle where blood can flow without friction. Therefore, the diameter or radius of a blood vessel is inversely proportional to the resistance. Blood flows with less resistance in larger vessels.

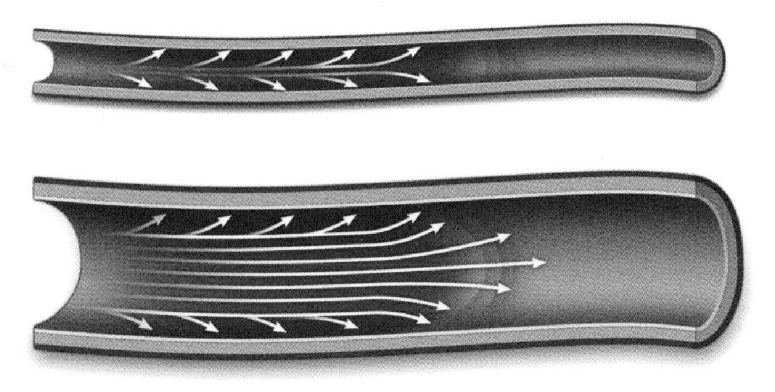

the compliance of an artery, the more effectively it is able to expand to accommodate surges in blood flow generated by the ejection of blood during systole. Therefore, in the moment of blood ejection from the ventricles, arteries decrease their resistance by temporarily increasing their radius. Due to their thinner walls, veins are more compliant than arteries and can expand to hold more blood. When the arteries are exposed to high pressure over time, their elastic fibers can wear down, leading to arterial stiffening. In the case of stiff arteries, compliance is reduced and resistance to blood flow is increased. The result is more turbulence, higher pressure within the vessel, and reduced blood flow. This increases the workload on the heart.

LO 20.2.5

Arterioles are the site of greatest resistance in the entire vascular network. This may seem surprising, given that capillaries have a smaller size. How can this phenomenon be explained?

LO 20.2.6

Figure 20.15 compares vessel diameter and total cross-sectional area through the systemic vessels. While the diameter of an individual capillary (**Figure 20.15A**) is significantly smaller than the diameter of an arteriole, there are vastly more capillaries in the body than there are other types of blood vessels (**Figure 20.15B**). **Figure 20.15C** shows that blood pressure drops as blood travels from arteries to arterioles, capillaries, venules,

Figure 20.15 Changes in Blood Vessel Diameter, Total Cross-Sectional Area, Blood Pressure, and Blood Flow Velocity that Occur as Blood Flows through the Systemic Circuit

(A) Blood vessel diameter decreases as arteries split into smaller and smaller arteries and then arterioles and capillary beds. The vessels increase in diameter once again as venules merge together to form veins and veins merge to form larger veins. (B) The total cross-sectional area of vessels helps us to understand how much space these vessels take up in the body. By far, capillaries and venules take up the most space. (C) Blood pressure decreases steadily as we move away from the heart; the right side of the heart has the lowest pressure in the entire system. (D) Blood flow is fastest close to the heart; the velocity or speed of flow decreases as blood moves farther away from the heart, and only picks back up in the venae cavae.

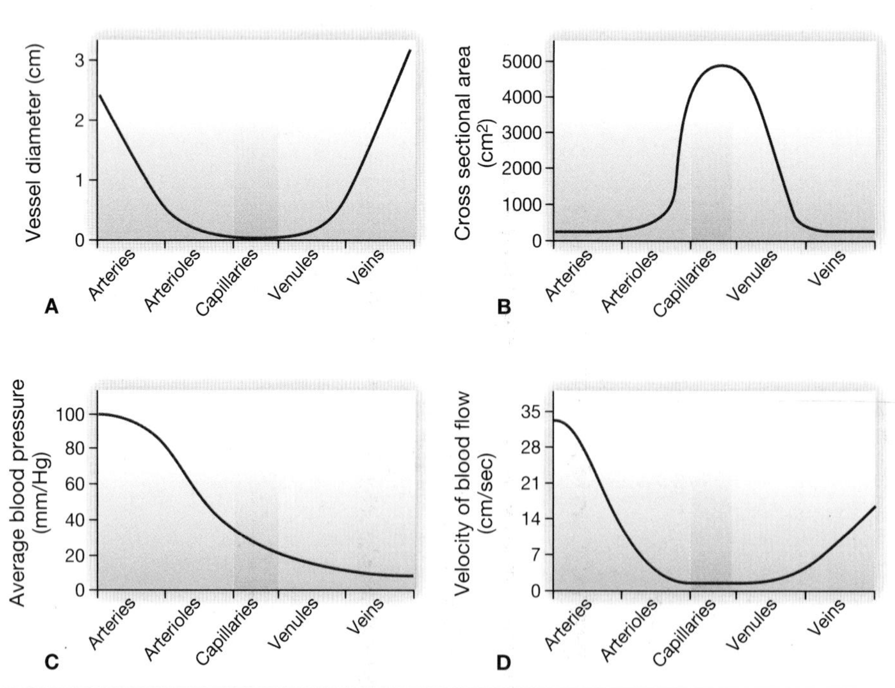

Apply to Pathophysiology

Atherosclerosis

Atherosclerosis is a disease that begins with injury to the endothelium of an artery, which may be caused by irritation from high blood glucose, infection, tobacco use, excessive blood lipids, or high blood pressure. Tissue injury can cause inflammation. As inflammation spreads into the artery wall, it weakens and scars it, leaving it less elastic.

1. During this inflammatory process, which of the following traits of an artery would be reduced?
   A) Muscularity
   B) Compliance
   C) Surface area

This process occurs often and throughout the lifespan. However, when circulating levels of lipids are high, lipids and cholesterol can seep between the damaged endothelial cells and become trapped within the artery wall, exacerbating the inflammation. Eventually, this buildup is called a *plaque*.

2. In the equation that determines resistance, shown on page 814, which of the factors will decrease?
   A) η
   B) L
   C) r^4
3. Decreasing this value will have what effect on overall resistance?
   A) Resistance will increase.
   B) Resistance will decrease.
   C) Resistance will be unchanged.

and veins, and encounters greater resistance. However, the site of the most precipitous drop, and the site of greatest resistance, is the arterioles. Furthermore, arterioles play a significant role in regulating blood pressure due to vasodilation and vasoconstriction.

Figure 20.15D shows that the velocity (speed) of blood flow decreases as the blood moves from arteries to arterioles to capillaries. This slow flow rate allows more time for exchange processes to occur. As blood flows through the veins, the rate of velocity increases, as blood is returned to the heart.

hypertension (high blood pressure)

DIGGING DEEPER: Hypertension

Hypertension, or high blood pressure, is defined as having persistent arterial blood pressure readings of greater than 140/90 mm Hg. According to this definition, an estimated 34 percent of the U.S. adult population and 26 percent of the world adult population have hypertension. It is more common in African American adults than in other U.S. demographic populations, and slightly more common in biological males than in biological females. While the causes of these differences are unclear, it may have to do with genetics, protective effects of estrogen signaling, activity level, or diet. Because high blood pressure has no visible symptoms in its early stages—nearly half of all individuals who have hypertension are unaware that they have it—it tends not to be noticed until after it has caused noticeable damage to the organ systems, which can take years in many cases. Blood pressure screening is one of the many reasons that routine medical care is so important. If left untreated, damage to the heart, blood vessels, and organ systems will occur. When symptoms do occur, they include headaches, dizziness, nosebleeds, vision changes, buzzing in the ears, nausea and vomiting, confusion, fatigue, and anxiety.

Although high blood pressure can result from a variety of causes, about 90 percent of cases are of unknown origin; this is called *primary hypertension*. The other 10 percent of cases have a known cause (secondary hypertension), such as kidney disease, hyperthyroidism, renal artery obstruction, narrowing (stenosis) of the renal arteries, and Cushing's syndrome (in which excess cortisol causes sodium and water retention, resulting in increased blood volume and pressure).

The known contributing factors to primary hypertension are heredity, obesity, age, smoking, alcohol use, stress, a diet high in Na^+ or cholesterol, and low in K^+, Ca^{2+}, Mg^{2+}, and diabetes. Since blood pressure is the product of cardiac output and peripheral resistance, an increase in either of these factors would increase the blood pressure. However, it seems that the major direct cause of hypertension is increased peripheral resistance due to the vasoconstriction of arterioles, and the resulting increase in friction between the blood and the walls of the blood vessels.

Vasoconstriction can occur in many physiological circumstances, but systemic vasoconstriction is one side effect of the stress hormone cortisol, and stress is likely a major contributing factor for the high rates of hypertension among U.S. adults.

There is no cure for hypertension. As this disease progresses, it puts a strain on the heart, making the left ventricle work harder to pump blood into the systemic circuit against higher blood pressure; this leads to a thickening of the myocardium of the left ventricle and enlargement of the heart (left ventricular hypertrophy). When coronary blood flow cannot keep up with the increased demands of the extra muscle tissue, cardiac muscle cells die and are replaced by dense connective tissue. If left untreated, this can lead to complications such as chest pain, coronary artery disease, atherosclerosis, heart failure, myocardial infarction (heart attack), stroke, and kidney failure.

The good news is that lifestyle changes and medication can keep hypertension under control in most people. Losing weight, adopting a diet low in salt and cholesterol and high in fruits and vegetables, quitting smoking, limiting alcohol intake, getting regular exercise, and controlling stress contribute greatly to lowering blood pressure. Medications exert their effects on the factors that contribute to blood pressure regulation, such as blood volume, cardiac output, or peripheral resistance; they include diuretics, angiotensin converting enzyme (ACE) inhibitors, beta blockers, and calcium channel blockers.

20.2f Venous System

The pumping of the heart propels the blood into the arteries, from an area of higher pressure toward an area of lower pressure. As we see in Figure 20.15C, pressure gradients exist not just between the heart and the arteries, but among all the vessels; for example, the pressure in the veins is slightly lower than that in the capillaries. If blood is to flow from the veins back into the heart, the pressure in the veins must be greater than the pressure in the atria of the heart. However, an additional factor complicates venous return: For most of the body, when upright in a sitting or standing position the heart is above the majority of the veins, and blood returning to the heart must fight gravity. Two factors help drive blood flow back toward the heart. First, the pressure in the atria during diastole is very low, often approaching zero when the atria are relaxed (atrial diastole). Second, two physiologic "pumps" increase pressure in the venous system.

LO 20.2.7

Skeletal Muscle Pump In the limbs, the pressure within the veins is increased during skeletal muscle contraction due to the anatomical relationship between the veins and the skeletal muscles. The term **skeletal muscle pump** (**Figure 20.16**) describes veins that are sandwiched among muscles in the limbs. As muscles contract (for example, during walking or running) they exert pressure on the veins, squeezing blood up toward the heart. Veins have numerous one-way valves, so when the increased pressure causes blood to flow upward, the blood flow opens valves superior to the contracting muscles, and blood flows through. As the muscle relaxes, blood flows downward, but catches on the valves, closing them. With the next skeletal muscle contraction, the blood will be propelled further upward, toward the heart.

Respiratory Pump Similar to the muscle pump, the **respiratory pump** aids blood flow through the veins of the thorax and abdomen. During inhalation, the volume of the thorax increases; this volume increase causes pressure within the thorax to decrease. The decreased thoracic pressure is related to the flow of air into the lungs, down its pressure gradient, but it also means that the pressure decreases within the veins of the thorax, driving blood flow to these newly lower-pressure vessels. During exhalation, air pressure increases within the thoracic cavity and the thoracic veins, driving blood flow toward the heart. As in the skeletal muscle veins, valves prevent backflow.

Figure 20.16 The Skeletal Muscle Pump

(A) Large veins are sandwiched among the fibers of skeletal muscles. (B) During muscle contraction, the vessel is squeezed and blood is propelled through the vein. This mechanism helps to push blood back toward the heart even in low-pressure veins.

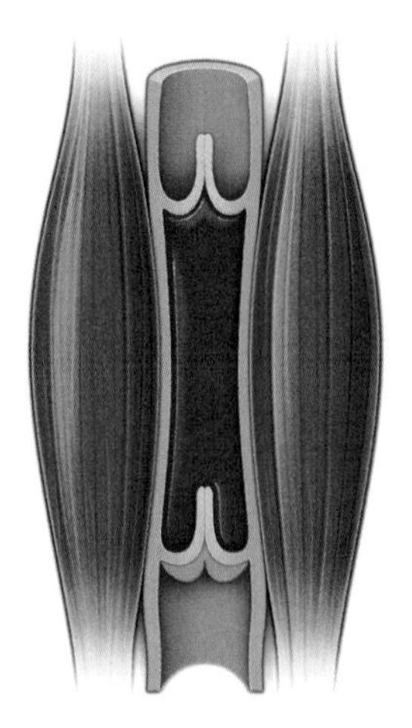

A The valves are closed and the muscles are relaxed.

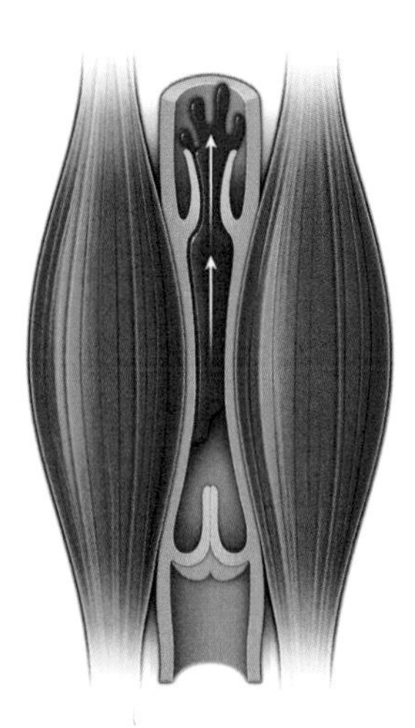

B The valve above the muscle opens and the muscles are contracted.

DIGGING DEEPER: Varicose Veins

Varicose veins are enlarged, bulging, or twisted veins that develop most commonly in the superficial veins of the lower legs. They often cause pain, aching, or itching around the affected veins, a feeling of heaviness in the legs, and swelling in the ankles and feet. The pain typically becomes more severe after long periods of standing or sitting. In spite of the discomfort, however, varicose veins are not usually of serious medical concern. Occasional complications include venous ulcers on the skin around the veins, rupture and bleeding from affected veins, blood clots, and pulmonary embolism.

Superficial veins are particularly susceptible to becoming varicose veins because they are not surrounded by skeletal muscles, so they cannot utilize the skeletal muscle pump to help return blood to the heart. Due to their distance from the heart, lower-leg veins have a greater risk than those of other areas of the body. Although all venous blood is under relatively low blood pressure, that of the legs must also overcome the force of gravity on its long journey back to the heart. Over time, this can lead to a weakening and stretching of one-way venous valves, and an accumulation of blood below the valves. This can result in stretching and twisting of the veins in the area, and therefore, the formation of varicose veins.

It is estimated that approximately 25 percent of adults in the United States have varicose veins. They occur more frequently in people that regularly stand or sit for long periods of time, those over the age of 50, and individuals with increased abdominal weight such as during pregnancy and abdominal obesity. In these cases, the weight of a fetus in the uterus or the presence of adipose tissue in the abdominal area exerts pressure on the pelvic veins, which hinders the upward flow of venous blood from the legs toward the heart. Increased levels of estrogen and progesterone during pregnancy also cause vasodilation and a thinning of the walls of the veins, which can lead to the weakening and excessive dilation that create varicose veins.

Treatments include regular exercise, elevating the legs above the waist, and wearing compression stockings to apply pressure to the swollen veins. If the pain becomes severe, varicose veins can be surgically removed, tied off, or blocked by chemical injections or laser therapy. While varicose veins do seem to have a genetic component, prevention includes exercise, a frequent change of position often throughout the day, and maintaining blood pressure through a diet high in fiber and low in sodium and refined sugar.

Apply to Pathophysiology

Exercise and Venous Return

During exercise, many changes occur within the body. At the most microscopic level, within the skeletal muscle cells, more sarcomere contractions are occurring. As you remember from Chapter 11, crossbridge cycling requires ATP. This increased demand for ATP drives up the rate of cellular respiration. Cellular respiration can be summarized by the equation:

$$C_6H_{12}O_6 \text{ (glucose)} + O_2 \text{ (oxygen)} \rightarrow CO_2 \text{ (carbon dioxide)} + H_2O \text{ (water)} + \text{ATP} + \text{heat}$$

What does this have to do with the circulatory system? The blood ferries that glucose and oxygen to the muscle cells and carries away the carbon dioxide before it builds up and harms the muscle cells. During exercise, the skeletal muscle cells that are hard at work require more blood flow than usual.

1. Which of the following processes can help get that increased blood flow toward the muscles that need it most?
   A) Vasoconstriction
   B) Vasodilation
   C) Angiogenesis
2. In order to meet the demand for both oxygen and the increased carbon dioxide generation, the breath rate typically increases during exercise. What effect, if any, will this have on the blood flow through the veins?
   A) No effect
   B) Increased blood flow through the veins of the thorax and abdomen due to the work of the pulmonary pump
   C) Decreased blood flow through the veins of the thorax and abdomen because more blood is in the arteries at this time
3. During exercise the skeletal muscles of the limbs are typically working hard, with alternating periods of contraction and relaxation. What effect, if any, will this have on the blood flow through the veins?
   A) No effect
   B) Increased blood flow through the veins of the limbs due to the work of the skeletal muscle pump
   A) Decreased blood flow through the veins of the limbs because more blood is in the arteries at this time
4. Cardiac output (the amount of blood ejected from the heart in one minute) is determined by both heart rate and stroke volume (the amount of blood ejected in each ventricular contraction). Given your answers to Questions 2 and 3, if heart rate stayed the same during the period of exercise, would cardiac output change?
   A) Yes, cardiac output would decrease because much of the blood is traveling slowly in the veins.
   B) Yes, cardiac output would increase because of increased volume of blood returning to the heart from the veins.
   C) No, if heart rate does not increase, cardiac output will not change.

Cultural Connection

Birth Control and Deep Vein Thrombosis

The family of reproductive hormones called *estrogens* increases the risk of arterial and venous thrombosis (clot formation). Clots are most likely to form in any individual within the deep veins of the legs. Movement of blood within these vessels is dependent on skeletal muscle contraction, so during times of rest or standing still, it is possible for the blood to stagnate, pool, and form clots. While the mechanism isn't completely understood, scientists do know that estrogen promotes several aspects of clot formation. Promotion of clot formation is correlated to very high levels of estrogen, outside the range that is typically found during the menstrual cycle. However, these very high levels of estrogen are often found at the end of pregnancy or in individuals taking exogenous estrogen, such as trans women, postmenopausal women on hormone replacement therapy, or women taking some forms of birth control pills. Earlier forms of birth control contained higher levels of estrogen, and deep vein thrombosis (DVT) occurred occasionally in individuals taking birth control. However, in recent years doctors have found that birth control medications can be just as effective at preventing ovulation with lower doses of estrogen, which reduces the risk of DVT as well as other unwanted side effects such as weight gain.

Learning Check

1. Sodium is an ion that causes water retention. If you had a high-sodium diet, how would you expect your blood pressure to change?
   a. Increase
   b. Decrease
2. In people with hyperlipidemia, there is plaque and cholesterol floating around in the arteries. This can stick to the walls of the arteries. What would you expect to happen to the blood pressure as more plaque becomes lodged in the walls of the arteries?
   a. Increase
   b. Decrease

LO 20.2.8

3. How can leg exercise benefit someone who has leg swelling?
   a. The muscles will utilize the blood pooling in the legs.
   b. The muscles will pump the blood toward the heart.
   c. The muscles will drain blood from the vessels.
   d. The muscle will push the blood toward the extremities.

20.3 Capillary Exchange

HAPS LEARNING GOALS AND OUTCOMES

Learning Objectives: By the end of this section, you will be able to:

20.3.1 Explain the mechanisms of capillary exchange of gases, nutrients, and wastes.

20.3.2* Describe the forces that regulate movement of fluid and substances between the blood and the interstitial fluid.

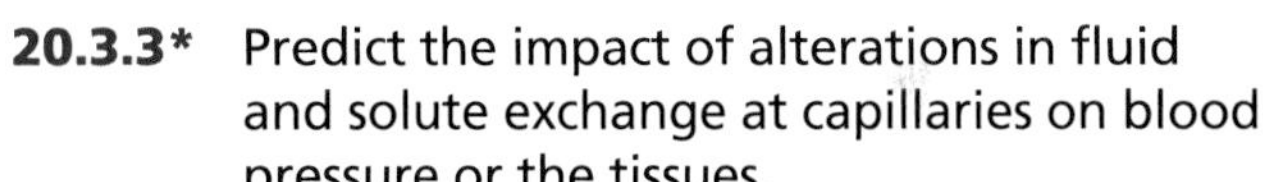

20.3.3* Predict the impact of alterations in fluid and solute exchange at capillaries on blood pressure or the tissues.

* Objective is not a HAPS Learning Goal

The primary purpose of the cardiovascular system is to circulate gasses, nutrients, wastes, and other substances to and from the cells of the body. Much of the circulatory system functions in transportation of blood only; no exchange or change of blood composition occurs in arteries, arterioles, venules, and veins. However, at the capillaries, small molecules, such as gasses, lipids, and lipid-soluble molecules, can diffuse directly through the membranes of the endothelial cells and move between the blood and the surrounding tissues. Glucose, amino acids, and ions—including sodium, potassium, calcium, and chloride—use transporters to move through specific channels in the membrane by facilitated diffusion. Glucose, ions, water, and larger molecules may also leave the blood via bulk flow through intercellular clefts. Larger molecules can pass through the pores of fenestrated or sinusoidal capillaries, and large plasma proteins can only pass through the gaps in the sinusoid capillaries. Some large proteins in blood plasma can move into and out of the endothelial cells packaged within vesicles by endocytosis and exocytosis. Diffusion and facilitated diffusion occur due to concentration gradients, and bulk flow and osmosis occur due to pressure gradients; the two types of pressure gradients are known as *hydrostatic pressure* and *osmotic pressure*. The pressures dictating movement in and out of capillaries are summarized in the "Anatomy of Capillary Transport" feature.

LO 20.3.1

Hydrostatic Pressure The primary force driving bulk flow is hydrostatic pressure, which can be defined as the pressure of any fluid enclosed in a space. **Blood hydrostatic pressure** is the force exerted by the blood onto the walls of the blood vessels or heart chambers. Blood hydrostatic pressure is the force that drives fluid by bulk flow out of capillaries and into the tissues; specifically, fluid leaves by bulk flow through the spaces between endothelial cells.

Anatomy of...

Capillary Transport

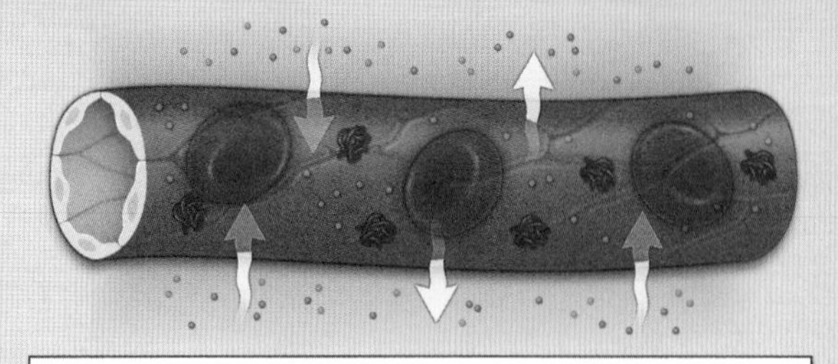

Diffusion occurs across endothelial cells according to concentration gradients.

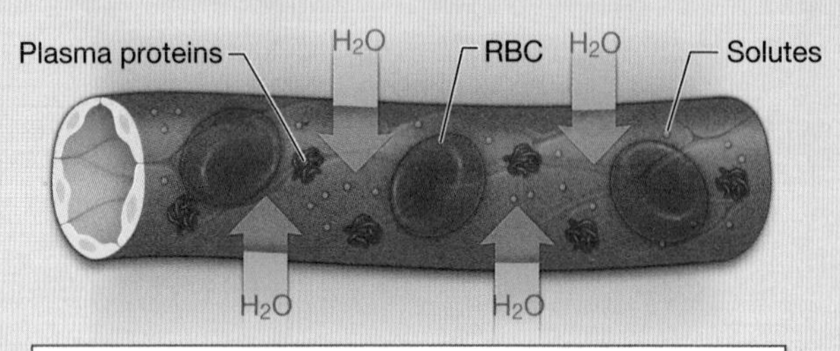

Osmotic pressure is the likelihood that H_2O will move due to an osmotic gradient. The osmolarity of the blood is always higher than that of the tissues because of the presence of plasma proteins. Therefore, osmotic pressure favors fluid flow into the blood.

Hydrostatic pressure is the force exerted by a fluid onto the walls of its container.

Bulk flow driven by hydrostatic pressure occurs through small spaces between the cells.

LO 20.3.2

Osmotic Pressure Osmotic pressure, the force that draws water toward an area with higher solute concentration, works in opposition to hydrostatic pressure. A region higher in solute concentration (and lower in water concentration) draws water across a semipermeable membrane from a region higher in water concentration (and lower in solute concentration). The concentration of most solutes will more or less equalize across the capillary membrane if given enough time. However, plasma proteins are too large to cross the wall of the capillary and so they do not leave the blood under most physiological circumstances. Plasma proteins represent the main osmotic regulators of the blood, their presence pulls water back into the bloodstream, preventing too much water from leaving and becoming interstitial fluid.

The **net filtration pressure (NFP)** represents the balance of the hydrostatic and osmotic pressures. It is equal to the difference between the hydrostatic and the osmotic pressure. Filtration is the amount of fluid leaving the blood and entering the interstitial fluid, and the net pressure is the hydrostatic pressure driving fluid flow out, minus the osmotic pressure that drives it back in (**Figure 20.17**).

Student Study Tip

Think of hydrostatic and osmotic pressures as though they are in a fight, and the bigger fighter wins. The greatest pressure is always the pressure that drives the fluid movement.

Figure 20.17 Capillary Exchange Factors

The volume of plasma that leaves the capillary at the tissues is a balance between the opposing forces of osmotic pressure and hydrostatic pressure.

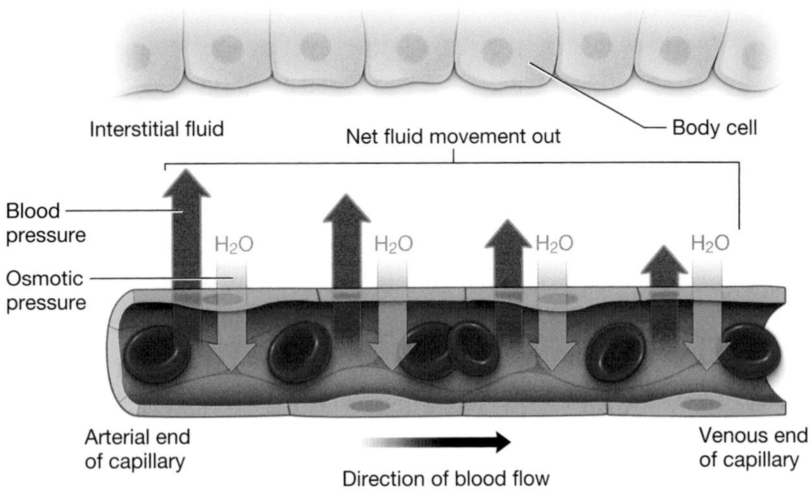

20.3a The Role of Lymphatic Capillaries

Since overall hydrostatic pressure is higher than osmotic pressure, it is inevitable that more net fluid will exit the capillary than is reabsorbed. Considering all capillaries over the course of a day, this can be quite a substantial amount of fluid—approximately three to four liters of lost fluid per day. Since an average adult human has two to three total liters of plasma, it is essential to drain this excess fluid and return it to the blood. As fluid leaves the capillaries and becomes interstitial fluid, the interstitial fluid pressure rises, and fluid enters the lymphatic system. These extremely thin-walled vessels have copious numbers of valves that ensure unidirectional flow through ever-larger lymphatic vessels that eventually drain into the bloodstream close to the heart. In this way, lymph may be thought of as recycled blood plasma. We will discuss this process and its benefits in more detail in Chapter 21.

Learning Check

1. Define hydrostatic pressure of the blood.
   a. The force exerted by blood onto the walls of the blood vessels
   b. The force that draws water toward an area with high solute concentration
   c. Force on the brachial vein during ejection of blood during contraction
2. Which of the following explains why more fluid exits the capillary than is reabsorbed?
   a. The fluid pressure is greater than the hydrostatic pressure
   b. The hydrostatic pressure is greater than the osmotic pressure
   c. The osmotic pressure is greater than the hydrostatic pressure
   d. The systolic pressure is greater than diastolic pressure

LO 20.3.3

20.4 Homeostatic Regulation of the Vascular System

Learning Objectives: By the end of this section, you will be able to:

20.4.1 Explain the role of the autonomic nervous system in regulation of blood pressure and volume.

20.4.2 Explain the steps of the baroreceptor reflex and describe how this reflex maintains blood pressure homeostasis when blood pressure changes.

20.4.3 Provide specific examples to demonstrate how the cardiovascular system maintains blood pressure homeostasis in the body.

20.4.4 Explain the role of the precapillary sphincter in autoregulation.

20.4.5 Explain how local control mechanisms and myogenic autoregulation influences blood flow to tissues.

20.4.6 List some chemicals that cause either vasodilation or vasoconstriction and explain the circumstances in which they are likely to be active.

20.4.7 List the local, hormonal and neural factors that affect peripheral resistance and explain the importance of each.

Blood provides essential gasses and nutrients to the cells of the body. But within a day, cells vary in their physiological needs. At one moment you may be exercising, and the skeletal muscle cells will have a tremendous physiological need; an hour later, those skeletal muscle cells are at rest, and you are refueling with a meal, the nutrients of which must be absorbed into the bloodstream for transport around the body. In order to maintain homeostasis, the cardiovascular system continually redirects the flow of blood to favor the tissues that need it most. In a very real sense, the cardiovascular system engages in resource allocation, because there is not enough blood flow to distribute blood equally to all tissues simultaneously.

Three homeostatic mechanisms—neural, endocrine, and autoregulatory—ensure adequate blood flow, blood pressure, and blood distribution.

20.4a Neural Regulation

The nervous system plays a critical role in the regulation of vascular homeostasis. The primary regulatory sites include the cardiovascular centers in the brain that control both cardiac and vascular functions. In addition, more generalized neural responses from the limbic system and the autonomic nervous system are factors.

The Cardiovascular Centers in the Brain Neurological regulation of blood pressure and flow depends on the cardiovascular centers located in the medulla oblongata.

These nuclei respond to changes in blood pressure as well as blood concentrations of oxygen, carbon dioxide, and hydrogen ions. There are three components:

LO 20.4.1

- The cardioaccelerator centers stimulate cardiac function by regulating heart rate and stroke volume via sympathetic stimulation from the (sympathetic) cardiac accelerator nerves.
- The cardioinhibitory centers slow cardiac function by decreasing heart rate and stroke volume via parasympathetic stimulation from the vagus nerve.
- The vasomotor centers control vessel tone or contraction of the smooth muscle in the tunica media. Changes in diameter affect peripheral resistance, pressure, and flow, which affect cardiac output. The majority of these neurons act via the release of the neurotransmitter norepinephrine from sympathetic neurons.

LO 20.4.2

Baroreceptor Reflexes Baroreceptors are specialized stretch receptors located within thin areas of blood vessels and heart chambers that respond to the degree of stretch caused by the hydrostatic pressure of the blood. They send impulses to the cardiovascular center to regulate blood pressure. Baroreceptors are found in two locations: the aorta and carotid arteries. The **aortic baroreceptors** are found in the walls of the ascending aorta just superior to the aortic valve; the **carotid baroreceptors** are found at the carotid bodies, the point of divergence of the internal and external carotid arteries.

When blood pressure increases or decreases, the baroreceptors communicate with the cardiovascular center in the medulla oblongata and trigger a reflex that maintains homeostasis (**Figure 20.18**):

LO 20.4.3

- When blood pressure rises too high, the baroreceptors trigger parasympathetic stimulation of the heart. As a result, heart rate, and therefore cardiac output, decreases. Sympathetic stimulation of the peripheral arterioles will also decrease, resulting in vasodilation. Combined, these activities cause blood pressure to decrease.
- When blood pressure drops too low, the rate of baroreceptor firing decreases. This will trigger an increase in sympathetic stimulation of the heart, causing both heart rate and the force of ventricular contraction to increase. As a result, cardiac output increases. It will also trigger sympathetic stimulation of the peripheral vessels, resulting in vasoconstriction. Combined, these activities cause blood pressure to rise.

Chemoreceptor Reflexes Located alongside the baroreceptors are chemoreceptors that monitor levels of oxygen, carbon dioxide, and hydrogen ions (pH), and thereby contribute to vascular homeostasis. They signal the cardiovascular center as well as the respiratory centers in the medulla oblongata.

Since tissues consume oxygen and produce carbon dioxide and acids as waste products, when the body is more active, oxygen levels fall and carbon dioxide levels rise as cells undergo cellular respiration to meet the energy needs of activities. Often due to carbon dioxide buildup or lactic acid generation, the blood pH also drops. When the body is resting, oxygen levels are higher, carbon dioxide levels are lower, more hydrogen is bound, and pH rises. The chemoreceptors respond to increasing changing levels of carbon dioxide, pH, or oxygen and signal the cardioaccelerator, cardioinhibitor, and vasomotor centers to facilitate the cardiovascular system to transport these gasses to the lungs for exchange. This interrelationship of cardiovascular and respiratory control cannot be overemphasized.

Other neural mechanisms can also have a significant impact on cardiovascular function. These include the limbic system that links physiological responses to psychological stimuli, as well as generalized sympathetic and parasympathetic stimulation.

Figure 20.18 Blood Pressure Control

Blood pressure is monitored through baroreceptors at two locations: the aortic bodies and the carotid bodies.

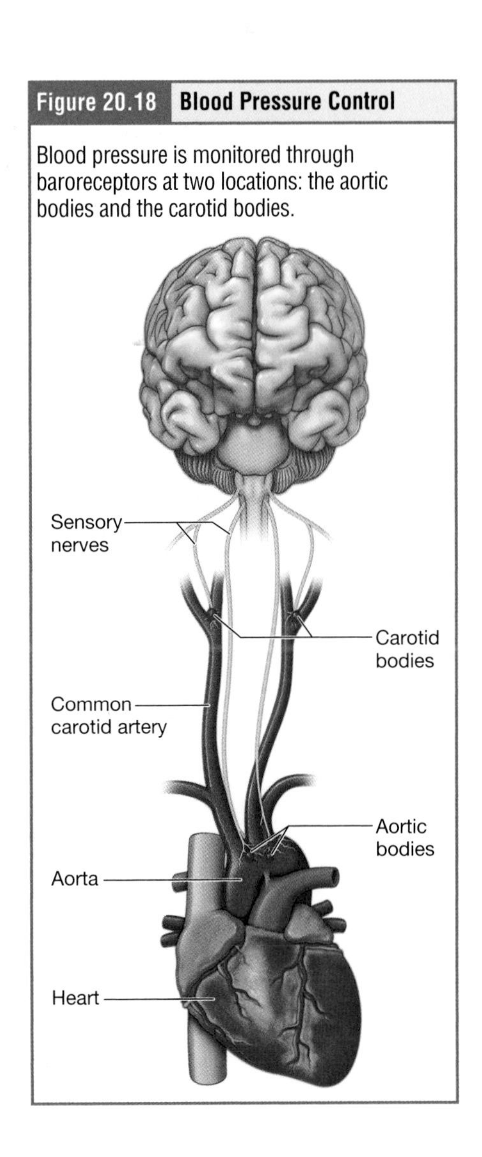

20.4b Endocrine Regulation

Endocrine control over the cardiovascular system involves the catecholamines (epinephrine and norepinephrine), as well as several hormones that interact with the kidneys in the regulation of blood volume.

Epinephrine and Norepinephrine The catecholamines epinephrine and norepinephrine are released by the adrenal medulla, and enhance and extend the body's sympathetic or "fight-or-flight" response. They increase heart rate and force of contraction, encourage vasoconstriction of blood vessels to organs not essential for flight-or-fight responses (such as the intestines), and vasodilate the blood vessels that serve the skeletal muscles and the heart.

Hormones That Connect the Renal and Cardiovascular Systems One of the primary homeostatic mechanisms for regulating blood volume is through the creation of urine. When blood volume rises, some of the plasma is filtered by the kidney and becomes urine, lowering blood volume and blood pressure. When the blood volume and pressure are low, the kidneys produce less urine volume, conserving fluid in the bloodstream. Antidiuretic hormone, renin, angiotensin, aldosterone, and atrial natriuretic hormone are hormones that connect the functions of these systems, and contribute to the regulation of blood volume through urine production. We will discuss them in more detail in Chapter 25.

Erythropoietin Erythropoietin (EPO) is released by the kidneys when blood flow and/or oxygen levels of the blood decrease. EPO stimulates the production of erythrocytes within the bone marrow. Erythrocytes make up the majority of the formed elements of the blood and are a significant factor of viscosity, and therefore resistance. In addition, EPO is a vasoconstrictor, further increasing resistance.

20.4c Autoregulation of Blood Flow

As the name would suggest, autoregulation mechanisms require neither specialized nervous stimulation nor endocrine control. Rather, these are local, self-regulatory mechanisms that allow each region of tissue to adjust local blood flow.

LO 20.4.4

Chemical Signals Involved in Autoregulation Chemical signals work at the level of the precapillary sphincters to trigger either constriction or relaxation. As you read earlier in this chapter, opening a precapillary sphincter allows blood to flow into that particular capillary, while constricting a precapillary sphincter temporarily shuts off blood flow to that region.

- ***Relaxation and opening of the precapillary sphincter:*** Symptoms of increased metabolism all cause the opening of precapillary sphincters. These include decreased oxygen concentrations, increased carbon dioxide concentrations, increasing levels of lactic acid concentration, and increasing potassium ion concentrations. Many of these same triggers will also stimulate the release of nitric oxide (NO), a powerful vasodilator.
- ***Contraction and closing of the precapillary sphincter:*** This is triggered by the opposite levels of these same regulators. Increased oxygen concentration, decreased carbon dioxide, or other symptoms of decreasing metabolic activity will also prompt the release of endothelins, powerful vasoconstricting peptides secreted by endothelial cells.

LO 20.4.5

The Myogenic Response The **myogenic response** is a reaction to the stretching of the smooth muscle in the walls of arterioles as changes in blood flow occur through the vessel. When blood flow is low, the vessel's smooth muscle will be only minimally stretched. In response, it relaxes, allowing the vessel to dilate and thereby increasing the movement of blood into the tissue. When blood flow is too high, the smooth muscle will contract in response to the increased stretch, prompting vasoconstriction that reduces blood flow. It often seems counterintuitive that when a vessel is stretched, and we presume hydrostatic pressure within to be higher, it constricts in response. The myogenic response is a protective function. Excessive blood flow could damage smaller and more fragile vessels, so the myogenic response is a localized reaction that stabilizes blood flow in the capillary network that follows that arteriole.

These blood flow regulators, and a few others, are summarized in **Table 20.2**.

Table 20.2 Factors That Control Peripheral Resistance LO 20.4.6 LO 20.4.7

Control	Factor	Vasoconstriction	Vasodilation
Neural	Sympathetic stimulation	Arterioles within skin mostly	Arterioles within heart and some skeletal muscles
	Parasympathetic	No innervation	Arterioles in external genitalia
Endocrine	Epinephrine	Arterioles within skin	Arterioles within heart and skeletal muscles
	Norepinephrine	Similar to epinephrine	Similar to epinephrine
	Angiotensin II	Powerful vasoconstrictor	n/a
	Atrial natriuretic hormone (ANH) (peptide)	n/a	Powerful vasodilator
	Antidiuretic hormone (ADH)	Moderately strong vasoconstrictor	n/a
Other factors	Decreasing levels of oxygen	n/a	Opens precapillary sphincters
	Decreasing pH	n/a	Opens precapillary sphincters
	Increasing levels of carbon dioxide	n/a	Opens precapillary sphincters
	Increasing levels of potassium ion	n/a	Opens precapillary sphincters
	Increasing levels of prostaglandins	Closes precapillary sphincters for many	Opens precapillary sphincters for many
	Increasing levels of adenosine	n/a	Vasodilation
	Increasing levels of NO	n/a	Vasodilation and opens precapillary sphincters
	Increasing levels of lactic acid and other metabolites	n/a	Vasodilation and opens precapillary sphincters
	Increasing levels of endothelins	Vasoconstriction	n/a
	Increasing levels of platelet secretions	Vasoconstriction	n/a
	Increasing hyperthermia	n/a	Vasodilation
	Stretching of vascular wall (myogenic)	Vasoconstriction	n/a
	Increasing levels of histamines from basophils and mast cells	n/a	Vasodilation

Apply to Pathophysiology

Edema

Edema is the clinical term for tissue swelling. Edema occurs when the interstitial fluid level is too high.

1. Which of the following would cause the interstitial fluid volume to rise to an abnormally high level? Please select all that apply.
   A) An increase in blood osmotic pressure
   B) An increase in blood hydrostatic pressure
   C) A decrease in blood osmotic pressure
   D) A decrease in blood hydrostatic pressure
2. The level of which of the following substances is the major regulator of the osmolarity of the blood?
   A) Sodium
   B) Potassium
   C) Erythrocyte
   D) Plasma proteins

Learning Check

1. What is the role of the cardioaccelerator center?
   a. Stimulates cardiac function and increases heart rate
   b. Inhibits cardiac function and decreases heart rate
   c. Increases the diameter of arterioles
   d. Decreases the peripheral resistance of arterioles
2. In acute stress situations, your blood pressure spikes. Which of the following receptors will signal the heart to decrease the heart rate?
   a. Chemoreceptors
   b. Baroreceptors
   c. Sympathetic receptors
   d. Tactile receptors
3. Which of the following would you expect to occur with an increase in erythropoietin? Please select all that apply.
   a. Increase in blood flow
   b. Increase in red blood cells
   c. Increase in resistance
   d. Increase in bone deposition
4. Which type of receptor would you find in the smooth muscle of arterioles that regulates blood flow during a myogenic response?
   a. Mechanoreceptor—ability to detect stretch
   b. Baroreceptor—ability to detect pressure
   c. Chemoreceptor—ability to detect pH levels
   d. Thermoreceptor—ability to detect temperature differences

20.5 Circulatory Pathways

HAPS LEARNING GOALS AND OUTCOMES

Learning Objectives: By the end of this section, you will be able to:

20.5.1 Describe the systemic and pulmonary circuits (circulations) and explain the functional significance of each.

20.5.2 Identify the major arteries and veins of the pulmonary circuit.

20.5.3 Identify the major arteries and veins of the systemic circuit.

20.5.4 Describe the structure and functional significance of the hepatic portal system.

LO 20.5.1

The vast network of blood vessels, analogous to the myriad subway tunnels of New York City or the intertwining freeways of Los Angeles, can be grossly separated into two circuits—the pulmonary and systemic circuits (**Figure 20.19**). The **pulmonary circuit** brings blood from the right side of the heart to the lungs for gas exchange. It returns oxygenated blood to the left side of the heart. The **systemic circuit** brings the blood that has been ejected from the left ventricle to all of the tissues of the body, returning the oxygen-depleted blood back to the right side of the heart. As you learn about the vessels of the systemic and pulmonary circuits, notice that many arteries and veins share the same names, parallel one another throughout the body, and are very similar on the right and left sides of the body, meaning that there is, for example, both a left femoral artery and a right femoral artery. Arteries and veins in the same location often, but not always, share a name. There is a femoral artery and a femoral vein, but the carotid artery has a nearby vein named the jugular. There are more veins in the body than there are arteries. In the limbs there are superficial veins, such as the great saphenous vein in the femoral region, that have no arterial counterpart. A single vessel may change names as it traverses through different locations. For example, the left subclavian artery becomes the axillary artery as it passes through the body wall and into the axillary region, and then becomes the brachial artery as it flows from the axillary

Figure 20.19 The Intertwining Circuits of Blood Flow

The cells of the body require oxygen and nutrients and take these items out of the blood when if flows through the systemic circuit. The blood, then poorer in oxygen content, returns to the right side of the heart, where it is kept separate from oxygenated blood as it is pumped to the lungs for gas exchange.

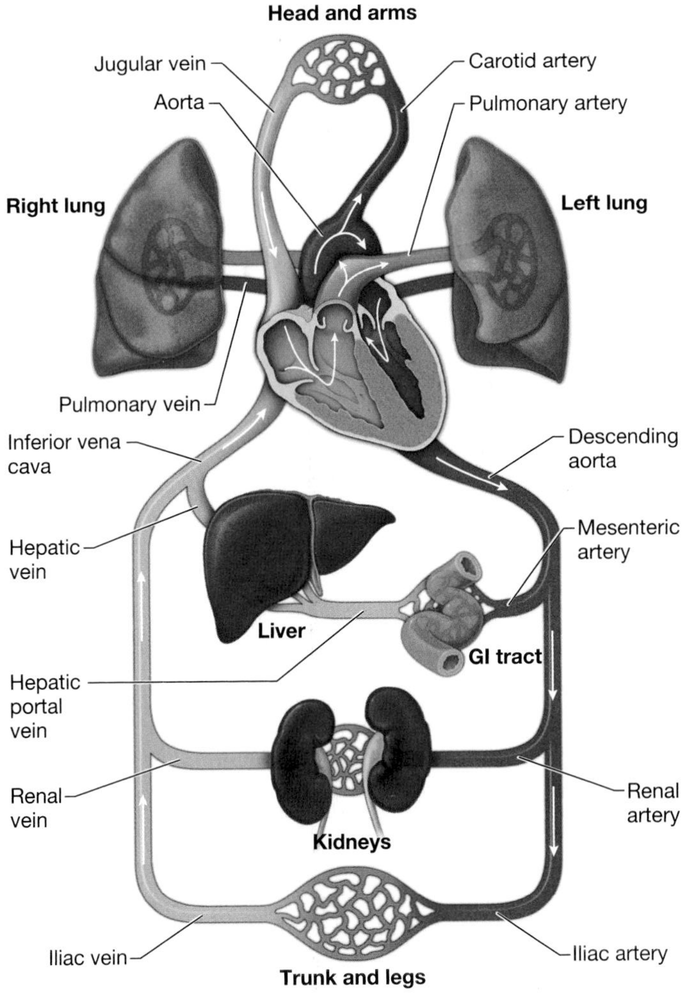

region into the upper arm (or brachium). You will also find examples of anastomoses, where two blood vessels that previously branched reconnect. Anastomoses are especially common in veins, where they help maintain blood flow even when one vessel is blocked or narrowed, although there are some important ones in the arteries supplying the brain.

A very large artery that is short and branches quickly into two arteries is typically referred to as a **trunk**, a term indicating that the vessel gives rise to several smaller arteries. A short, wide vein that collects blood from many smaller veins is often termed a *sinus*, such as the dural venous sinus in the brain.

Student Study Tip

This is where the tree analogy really comes in. Remember trunks as where the smaller vessels "branch" off.

20.5a Pulmonary Circulation

Recall that blood returning from the systemic circuit enters the right atrium (**Figure 20.20**). This blood is relatively low in oxygen and relatively high in carbon dioxide, since it is returning to the heart from the tissues. From the right atrium, blood

Figure 20.20 The Pulmonary Circuit

The pulmonary arteries carry blood away from the right side of the heart and toward the lungs, where the blood gives off accumulated carbon dioxide and picks up oxygen before returning, via the pulmonary veins, to the left side of the heart.

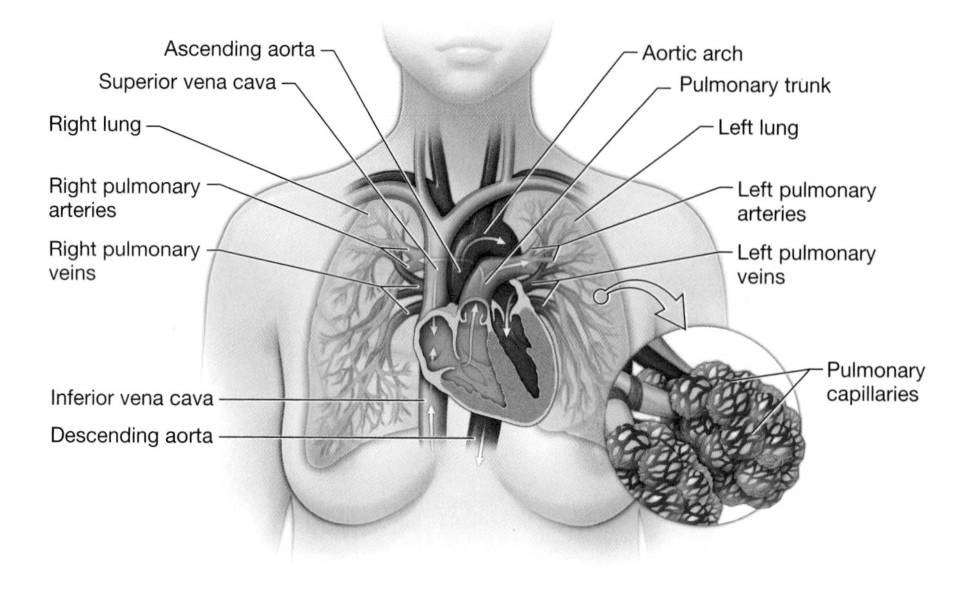

moves into the right ventricle, which pumps it to the lungs for gas exchange. This system of vessels, which carries blood from the right side of the heart to the lungs and then to the left side of the heart, is the pulmonary circuit.

LO 20.5.2

The single vessel exiting the right ventricle is the **pulmonary trunk**. At the base of the pulmonary trunk is the pulmonary semilunar valve, which prevents backflow of blood into the right ventricle during ventricular diastole. The pulmonary trunk splits into two branches, the left and right **pulmonary arteries**. Each pulmonary artery branches many times within the lung, forming a series of smaller arteries and arterioles that eventually lead to the pulmonary capillaries. The pulmonary capillaries surround lung structures known as *alveoli* that are the sites of oxygen and carbon dioxide exchange.

Once gas exchange is completed, oxygenated blood flows from the pulmonary capillaries into a series of pulmonary venules that eventually lead to a series of larger **pulmonary veins**. In most humans four pulmonary veins, two on the left and two on the right, return blood to the left atrium. This is a site of anatomical variation; the two left or right pulmonary veins sometimes merge, so any given individual may have two, three, or four pulmonary veins. Regardless of their number, the pulmonary veins return blood to the left atrium, completing the pulmonary circuit.

20.5b Systemic Arteries

LO 20.5.3

From the left atrium, blood moves into the left ventricle, which pumps blood into the aorta. The aorta and its branches—the systemic arteries—send blood to virtually every organ of the body (Figure 20.21).

20.5c The Aorta

The **aorta** is the largest artery in the body (Figure 20.22). It carries blood ejected from the left ventricle; it arches around the top of the heart and descends through the thorax

Figure 20.21 **The Major Arteries of the Systemic Circuit**

These arteries carry oxygenated blood. Oxygenated blood is a brighter red than venous blood, and so these vessels are often colored red in anatomical illustrations.

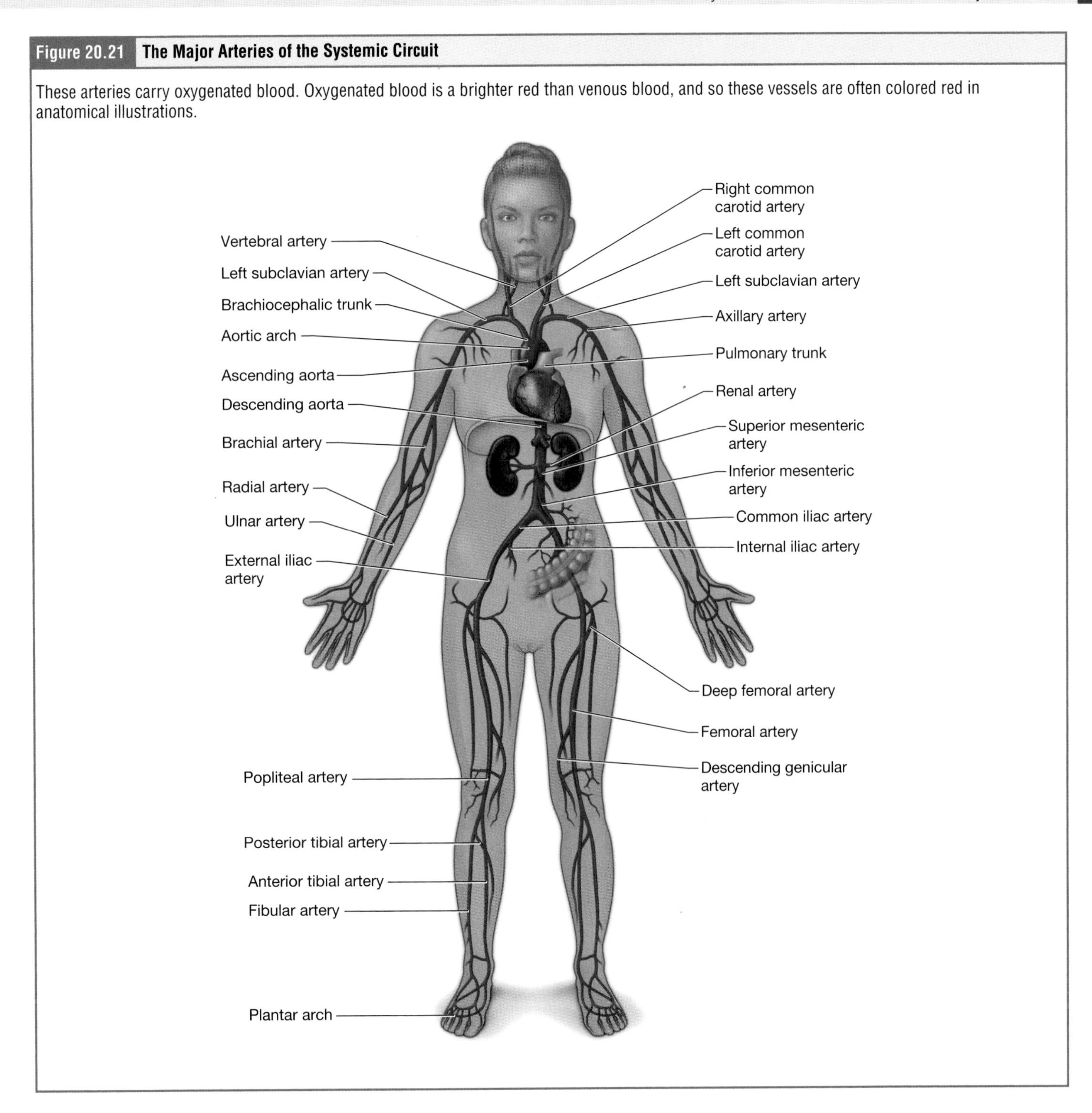

and to the abdominal region, where it splits into the two common iliac arteries. The aorta has three regions: the ascending aorta, the aortic arch, and the descending aorta. Arteries originating from the aorta ultimately distribute blood to virtually all tissues of the body. The aorta begins at the **aortic semilunar valve**, which prevents backflow of blood into the left ventricle while the heart is relaxing. The **ascending aorta** carries blood superiorly for a short distance before forming a graceful arch, called the **aortic arch**, which crests over the heart and then dips posteriorly, forming the **descending aorta** and then the **thoracic aorta** as the vessel continues down through the thorax

Figure 20.22 The Aorta

The ascending aorta emerges from the middle of the heart, beginning at the aortic semilunar valve, which separates the aorta from the left ventricle. The aorta arches, curving around the top of the heart, before descending through the thorax posterior to the heart.

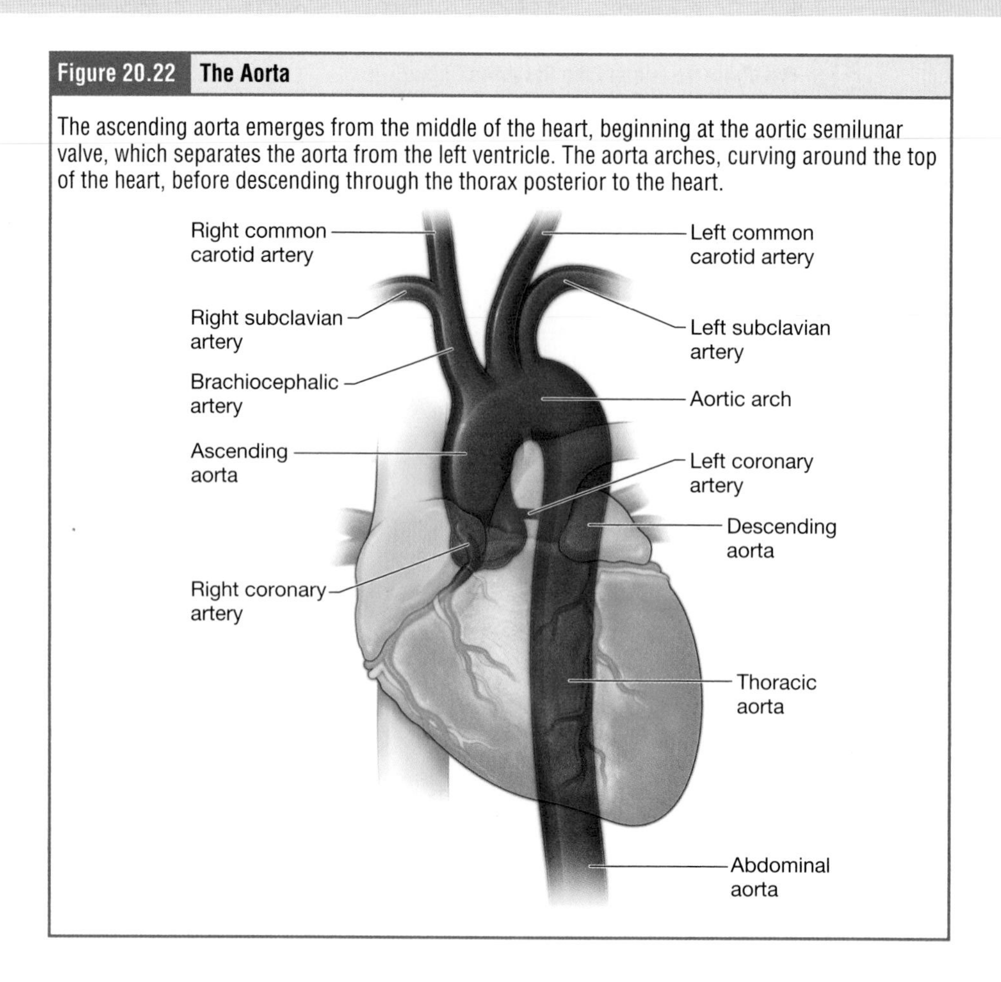

(see Figure 20.22). The thoracic aorta has paired branches at the level of each vertebra; these branches are the *intercostal arteries*. The aorta passes through an opening in the diaphragm known as the **aortic hiatus**; inferior to the diaphragm it is called the **abdominal aorta**. The abdominal aorta then splits into the two common iliac arteries at the level of the fourth lumbar vertebra (**Figure 20.23**).

Coronary Circulation The first vessels that branch from the ascending aorta are the paired coronary arteries (**Figure 20.24**), which arise from the ascending aorta just superior to the aortic semilunar valve. The coronary arteries encircle the heart, forming a ringlike structure that divides into the next level of branches that supplies blood to the heart tissues. These vessels are discussed in more detail in Chapter 19.

Aortic Arch Branches There are three major branches of the aortic arch—the brachiocephalic trunk, the left common carotid artery, and the left subclavian artery. As you would expect based upon proximity to the heart, each of these vessels is classified as an elastic artery.

The brachiocephalic trunk is the first branch off the aortic arch. As a trunk, it is a short vessel that splits into two arteries, the right subclavian artery and the right common carotid artery. The left subclavian and left common carotid arteries arise independently from the aortic arch but otherwise follow a similar pattern and distribution to the corresponding arteries on the right side (see Figure 20.22).

Each **subclavian artery** ascends to the shoulder, passing beneath the clavicle. The subclavian arteries supply blood to the arms, chest, shoulders, back, and central

Figure 20.23 The Thoracic and Abdominal Aorta

As the aorta descends through the thorax, it carries oxygenated blood to the lower body. The descending thoracic aorta is renamed the abdominal aorta when it crosses the diaphragm.

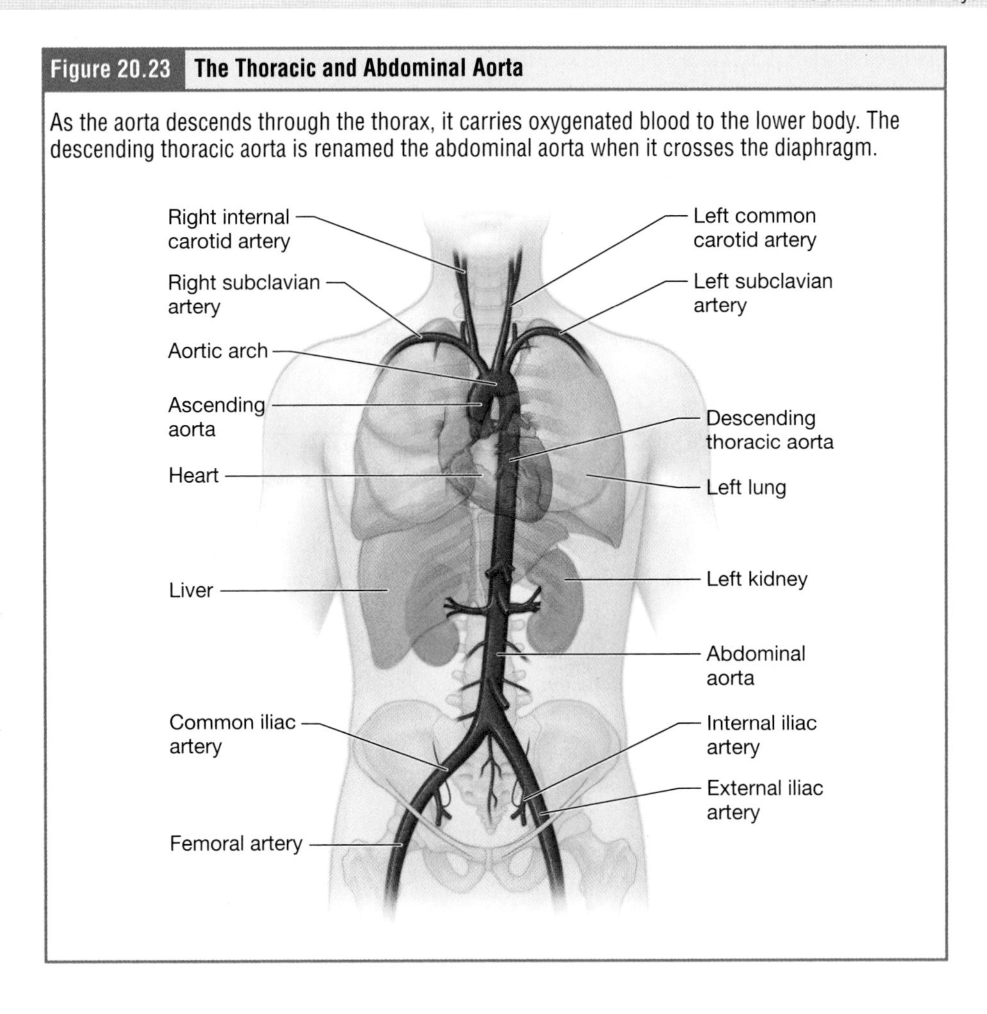

Figure 20.24 The Coronary Vessels

The coronary vessels are the first branches off the aorta. They branch throughout the wall of the heart to supply the cardiac muscle cells with oxygen and nutrients.

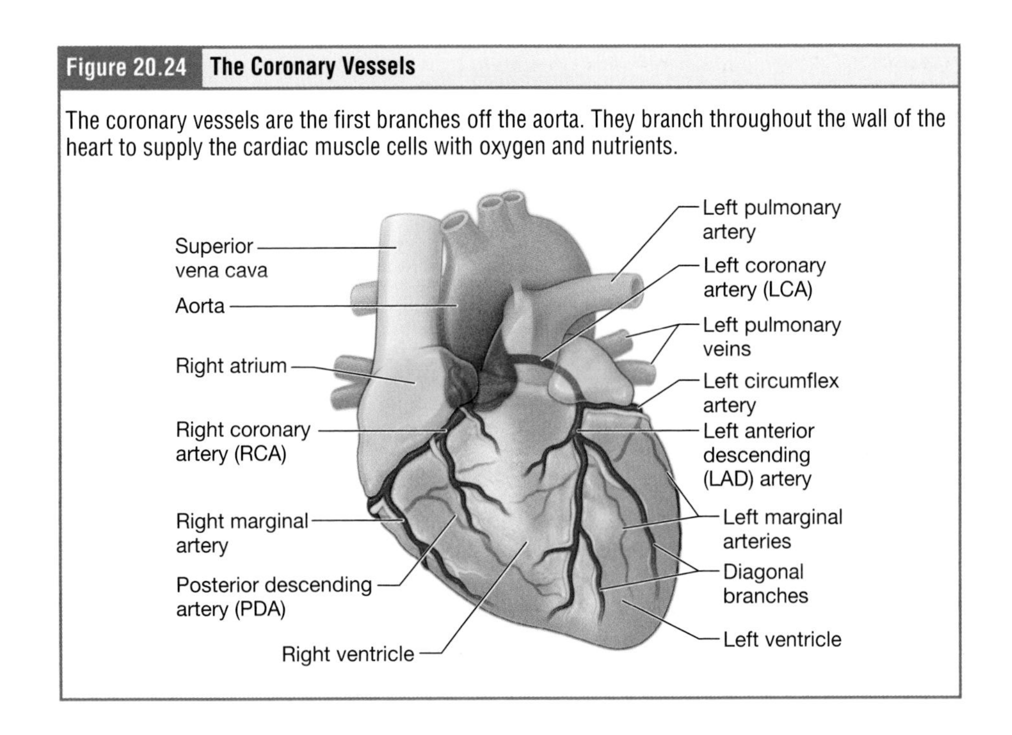

internal thoracic artery (mammary artery)

nervous system. Each subclavian artery then gives rise to three major branches: the internal thoracic artery, the vertebral artery, and the thyrocervical artery (**Figure 20.25**). The **internal thoracic artery**, or mammary artery, supplies blood to the thymus, the pericardium of the heart, and the anterior chest wall. The **vertebral artery** passes through the vertebral foramen of the cervical vertebrae and then through the foramen magnum into the cranial cavity to supply blood to the brain and spinal cord. The paired vertebral arteries join together to form the large *basilar artery* at the base of the medulla oblongata (**Figure 20.26**). This is an example of an anastomosis. The subclavian artery also gives rise to the **thyrocervical artery**, which provides blood to the thyroid, the cervical region of the neck, and the upper back and shoulder.

The left **common carotid artery** arises from the aortic arch and ascends toward the neck, where it divides into internal and external carotid arteries. The right common carotid artery arises first from the brachiocephalic trunk before splitting into the internal and external carotid arteries. The **external carotid artery** supplies blood to the face, lower jaw, neck, esophagus, and larynx. These branches include the lingual, facial, occipital, maxillary, and superficial temporal arteries. The **internal carotid artery** initially forms an expansion known as the **carotid body**, containing the carotid baroreceptors and chemoreceptors. Like their counterparts in the aorta, the information provided by these receptors is critical to maintaining cardiovascular homeostasis (see Figure 20.25).

The internal carotid arteries and the vertebral arteries are the two primary suppliers of blood to the human brain. Given the central role and vital importance of the brain to life, it is critical that blood supply to this organ remains uninterrupted. Recall that blood flow to the brain is remarkably constant, with approximately 20 percent of blood flow directed to this organ at any given time. When blood flow is interrupted, even for just a few seconds, loss of consciousness or temporary loss of neurological function can result.

Figure 20.25 Arteries Supplying the Head and Neck

The common carotid arteries branch off the brachiocephalic trunk or aorta and carry oxygenated blood to the neck, scalp, face, and brain.

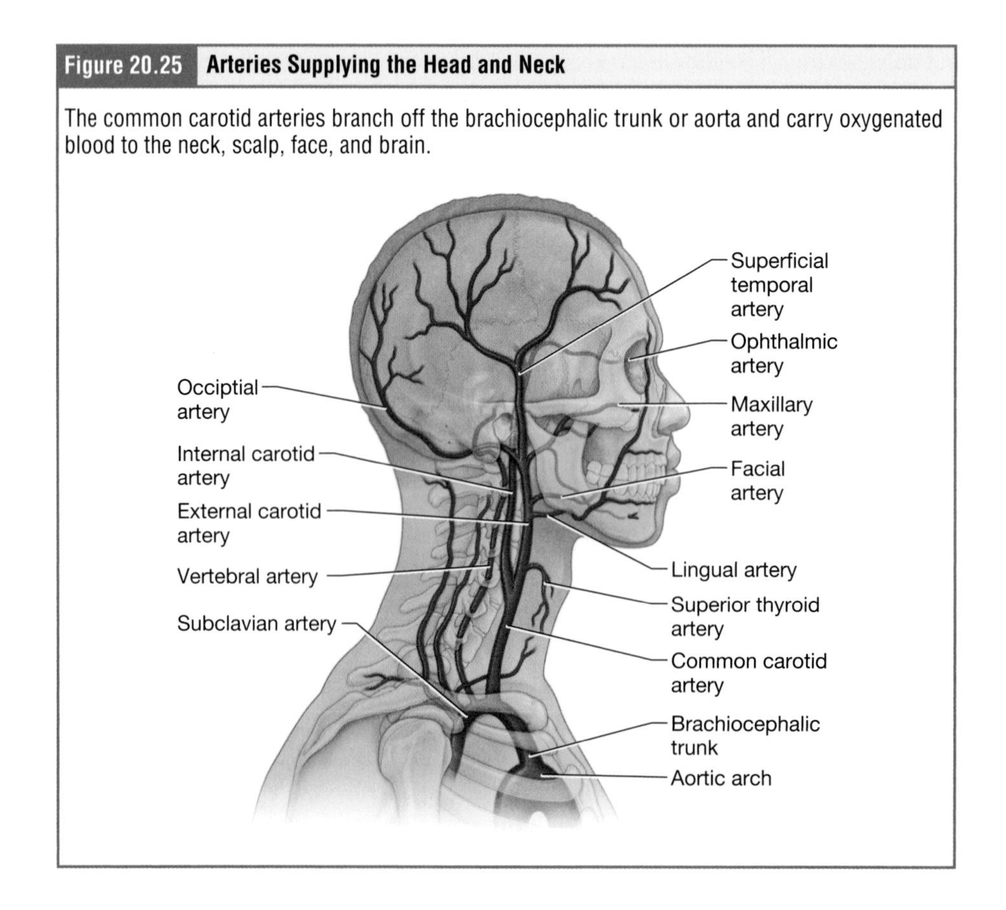

Figure 20.26 Arteries of the Brain

The internal carotid arteries and the vertebral arteries carry blood upward toward the brain. The vertebral arteries join at the brainstem to form the basilar artery. The two internal carotid (and one basilar) arteries contribute to an arterial circle—the cerebral arterial circle—at the base of the brain.

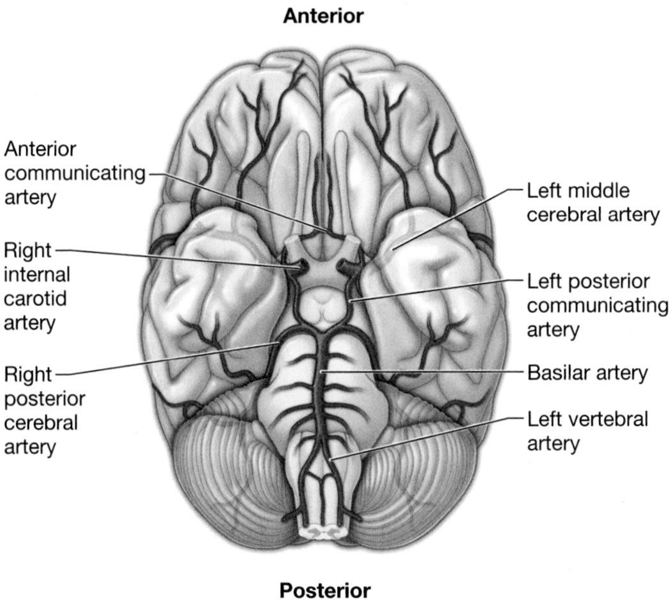

In some cases, the damage may be permanent. The locations of the arteries in the brain not only provide blood flow to the brain tissue but also prevent interruption in the flow of blood. Both the carotid and vertebral arteries branch once they enter the cranial cavity, and some of these branches form a structure known as the **cerebral arterial circle**, an anastomosis that is remarkably like a traffic circle that sends off branches (in this case, arterial branches to the brain) illustrated in Figure 20.26 and **Figure 20.27**.

cerebral arterial circle (circle of Willis)

The internal carotid artery continues through the carotid canal of the temporal bone and enters the base of the brain through the carotid foramen, where it gives rise to several branches (see Figures 20.25 and 20.26). One of these branches is the **anterior cerebral artery**, which supplies blood to the frontal lobe of the cerebrum. Another branch, the **middle cerebral artery**, supplies blood to the temporal and parietal lobes, which are the most common sites of CVAs. The **ophthalmic artery**, the third major branch, provides blood to the eyes.

The right and left anterior cerebral arteries join together to form an anastomosis called the **anterior communicating artery**. The initial segments of the anterior cerebral arteries and the anterior communicating artery form the anterior portion of the arterial circle. The posterior portion of the arterial circle is formed by a left and a right **posterior communicating artery** that branches from the **posterior cerebral artery**, which arises from the basilar artery. It provides blood to the posterior portion of the cerebrum and brain stem. The **basilar artery** is an anastomosis that begins at the junction of the two vertebral arteries and sends branches to the cerebellum and brainstem. It flows into the posterior cerebral arteries.

Thoracic Aorta and Major Branches The thoracic aorta begins at the level of vertebra T5 and continues through to the diaphragm at the level of T12, initially traveling within

Figure 20.27 Graphical Representation of the Cerebral Arterial Circle

The cerebral arterial circle ensures blood flow to the brain even if blood flow is interrupted in one of the supplying vessels.

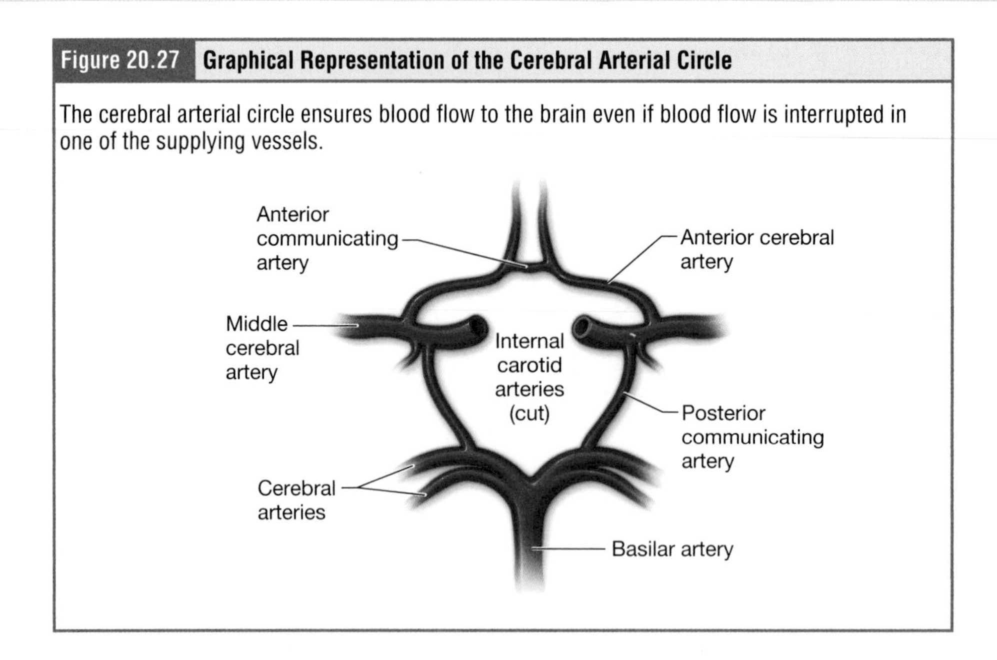

the mediastinum. As it passes through the thoracic region, the thoracic aorta gives rise to several branches (**Figure 20.28**). Each **bronchial artery** (typically two on the left and one on the right) supplies systemic blood to the lungs and visceral pleura. Note that the lungs have the pulmonary circuit, but this blood is less oxygen-rich and does not feed the cells of the lungs with oxygen and nutrients; the bronchial arteries do that. Each **pericardial artery** supplies blood to the pericardium, the **esophageal artery** provides blood to the esophagus, and the **mediastinal artery** provides blood to the mediastinum. The remaining thoracic aorta branches include the intercostal and superior phrenic arteries. Each **intercostal artery** provides blood to the muscles of the thoracic cavity and vertebral column. The **superior phrenic artery** provides blood to the superior surface of the diaphragm.

celiac trunk (celiac artery)

Abdominal Aorta and Major Branches After crossing the diaphragm through the aortic hiatus, the thoracic aorta is called the *abdominal aorta*. This vessel descends through the abdominal cavity, and its branches supply oxygenated blood to the organs. A single **celiac trunk** emerges from the aorta and divides into three arteries: the **left gastric artery**, which supplies blood to the stomach and esophagus; the **splenic artery**, which supplies blood to the spleen; and the **common hepatic artery**, which supplies blood to the liver. Branches off the common hepatic artery include the **hepatic artery proper** to supply blood to the liver, the **right gastric artery** to supply blood to the stomach, the **cystic artery** to supply blood to the gallbladder, and several other branches that supply blood to the duodenum and the pancreas. The next two vessels that arise from the abdominal aorta are the superior and inferior mesenteric arteries. The **superior mesenteric artery** branches into several major vessels that supply blood to the small intestine (duodenum, jejunum, and ileum), the pancreas, and a majority of the large intestine. The **inferior mesenteric artery** supplies blood to the distal segment of the large intestine, including the rectum.

In addition to these single branches, the abdominal aorta gives rise to several significant paired arteries along the way. These include the inferior phrenic arteries, the adrenal arteries, the renal arteries, the gonadal arteries, and the lumbar arteries. Each **inferior phrenic artery** is a counterpart of a superior phrenic artery and supplies blood to the inferior surface of the diaphragm. The **adrenal artery** supplies blood to

Figure 20.28 The Arterial Supply to the Thorax and Abdomen

(A) All of the arteries in the thorax and abdomen can be traced back to branches off the thoracic and abdominal aortas. (B) The branches of the celiac trunk.

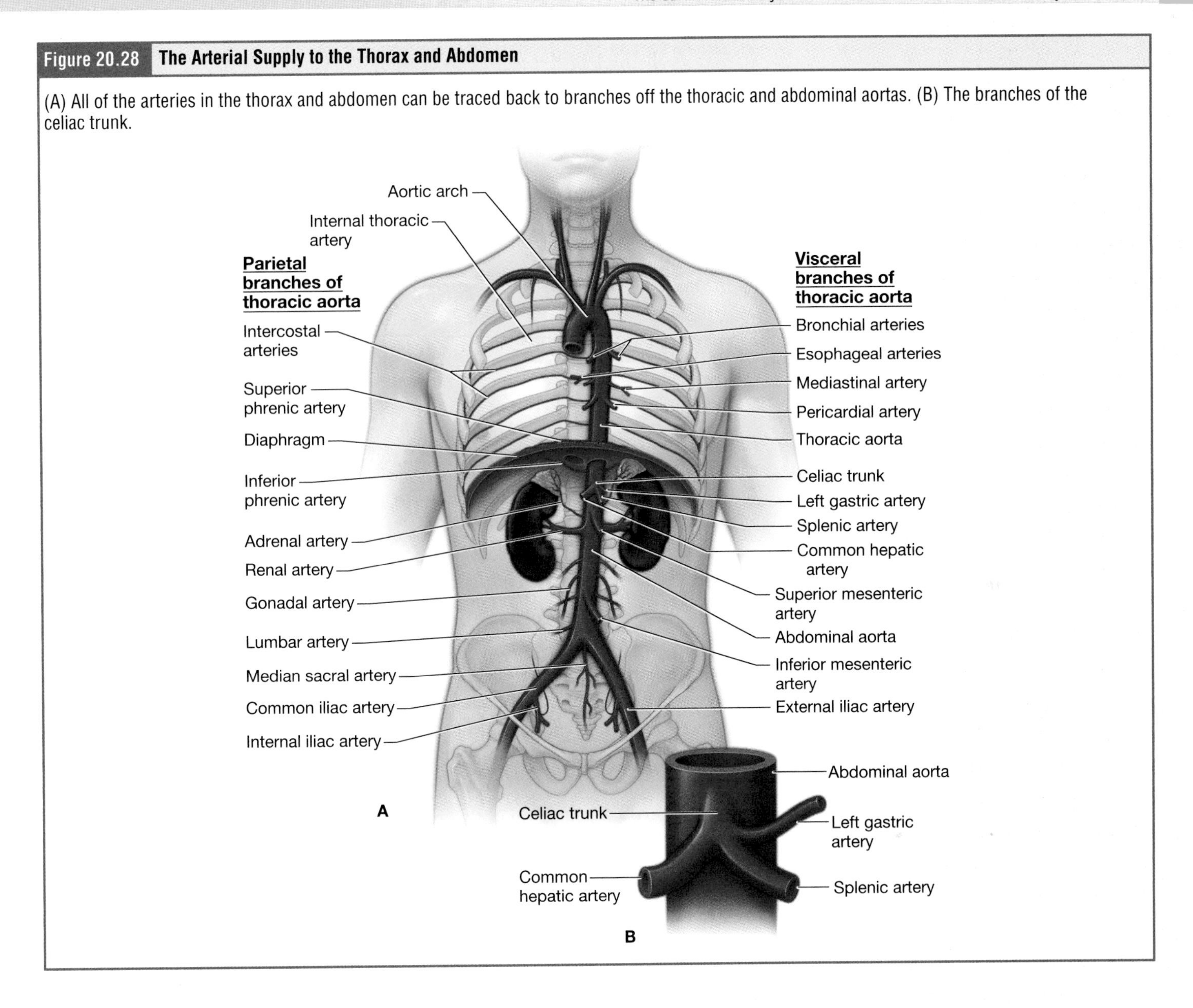

the adrenal glands. Each **renal artery** supplies a kidney. Each **gonadal artery** supplies blood to the gonads, or reproductive organs. The four paired **lumbar arteries** are the counterparts of the intercostal arteries and supply blood to the lumbar region, the abdominal wall, and the spinal cord. In some instances, a fifth pair of lumbar arteries emerges from the median sacral artery.

The aorta divides at approximately the level of vertebra L4 into a left and a right **common iliac artery** but continues as a small vessel—the **median sacral artery**—into the sacrum. The common iliac arteries provide blood to the pelvic region and ultimately to the lower limbs. They split into external and internal iliac arteries approximately at the level of the lumbar-sacral articulation. Each **internal iliac artery** sends branches to the urinary bladder, the walls of the pelvis, the external genitalia, and the medial portion of the femoral region. When a uterus and vagina are present, these organs are supplied by the internal iliac artery as well. The much larger **external iliac artery** supplies blood to each of the lower limbs. **Figure 20.28A** shows the distribution of the major branches of the aorta into the thoracic and abdominal regions.

Figure 20.28B shows a graphical representation of the branches off the celiac trunk.

20.5d Arteries Supplying the Upper Limbs

As the subclavian artery exits the thorax into the axillary region, it is renamed the **axillary artery**. Although it does branch and supply blood to the region near the head of the humerus (via the humeral circumflex arteries), the majority of the vessel continues into the upper arm, or brachium, and becomes the brachial artery (Figure 20.29). The **brachial artery** supplies blood to much of the brachial region and divides at the elbow into several smaller branches, including the **deep brachial arteries**, which provide blood to the posterior surface of the arm, and the **ulnar collateral arteries**, which supply blood to the region of the elbow. As the brachial artery approaches the coronoid fossa, it splits into the radial and ulnar arteries, which continue into the forearm, or antebrachium. The **radial artery** and **ulnar artery** parallel their namesake bones, giving off smaller branches until they reach the wrist, or carpal region. At this level, they fuse to form the superficial and deep **palmar arches** that supply blood to the hand, as well as the **digital arteries** that supply blood to the digits.

20.5e Arteries Serving the Lower Limbs

deep femoral artery (deep artery of the thigh)

The external iliac artery exits the body cavity and enters the femoral region of the lower leg (Figure 20.30). As it passes through the abdominal wall and into the leg, it is renamed the **femoral artery**. It gives off several smaller branches as well as the lateral **deep femoral artery**, which in turn gives rise to a **lateral circumflex artery**. These

Figure 20.29 Arterial Supply to the Upper Limb

The subclavian artery, which branches off the aorta on the left and the brachiocephalic trunk on the right, carries oxygenated blood to the upper limbs.

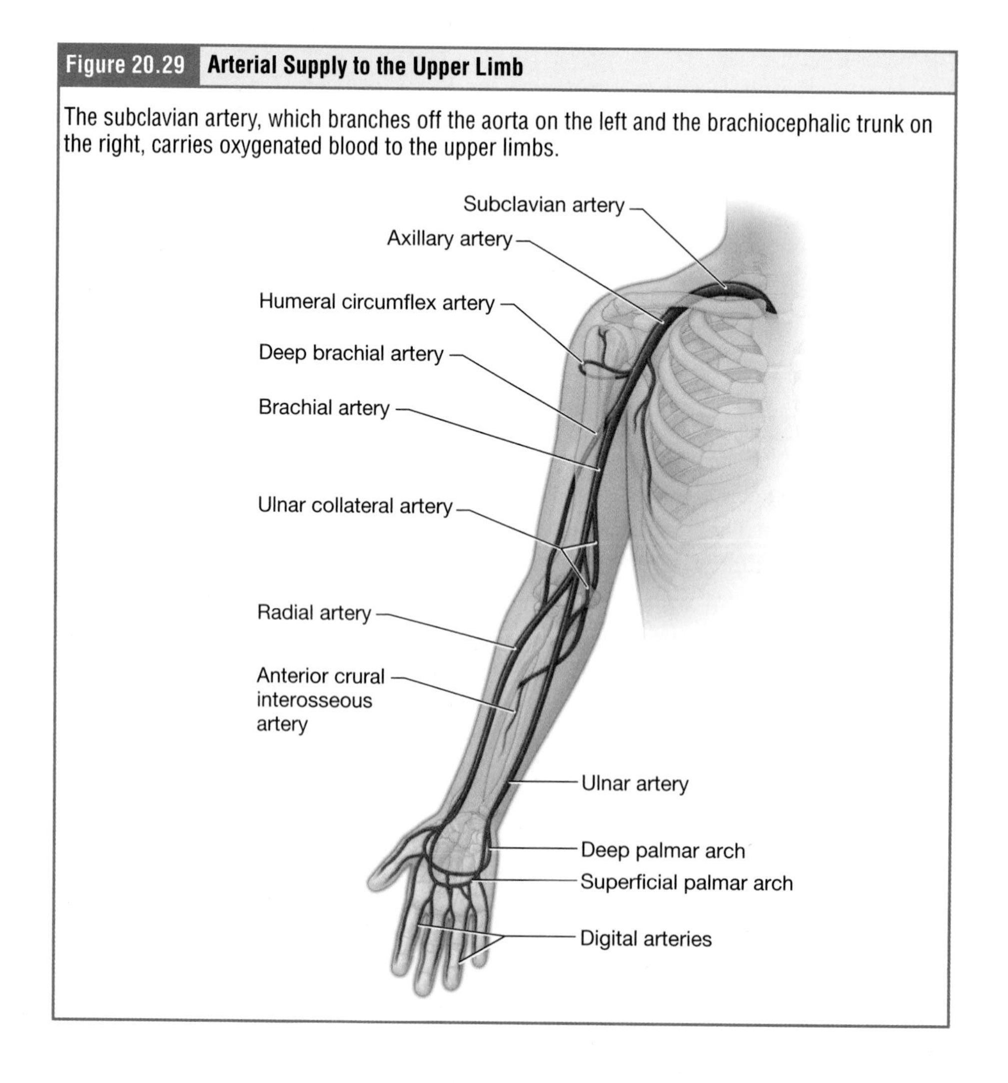

Figure 20.30 Arteries in the Lower Limb

The common iliac arteries branch from the abdominal aorta and supply the pelvic organs, genitals and lower limb with oxygenated blood.

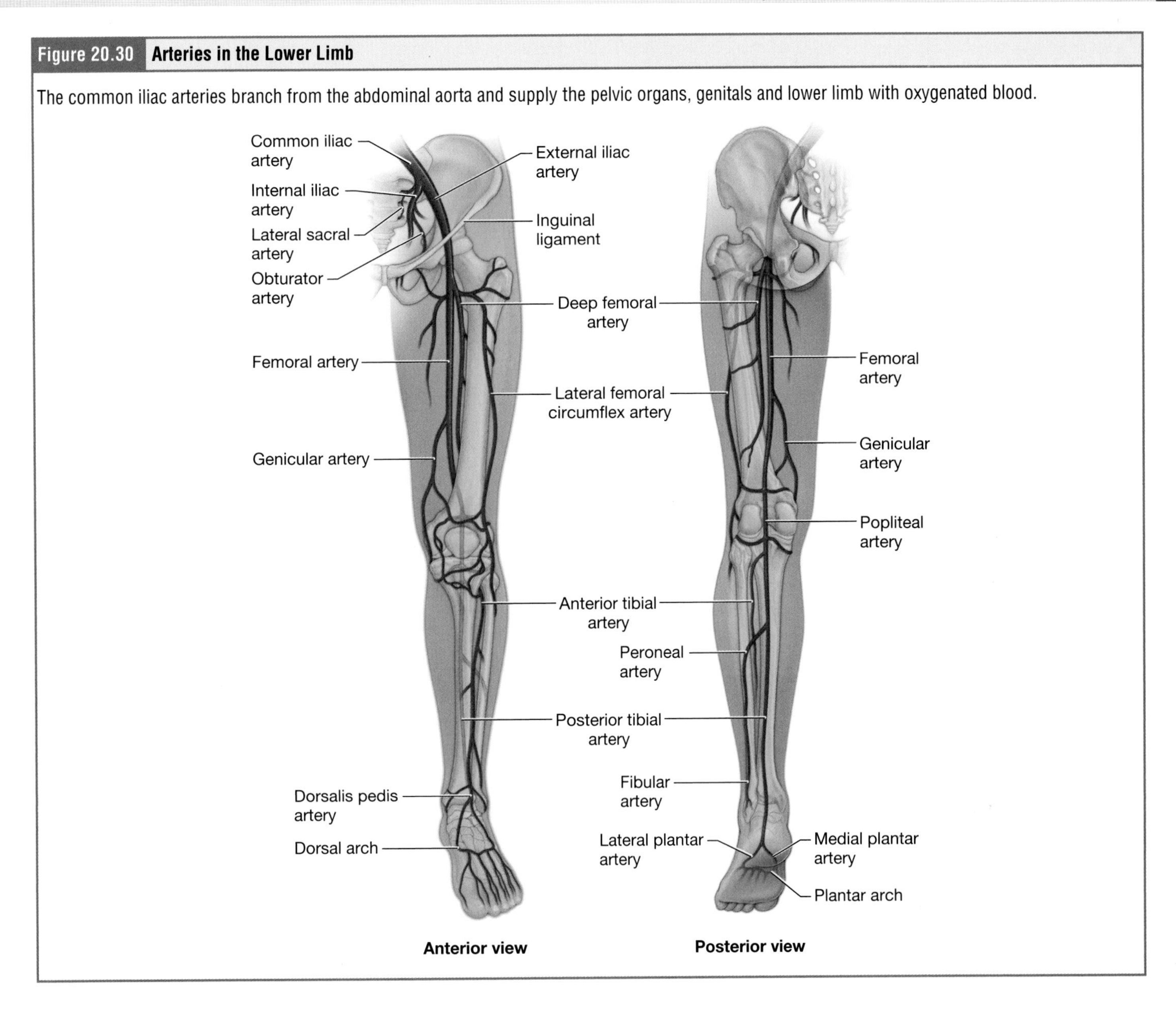

arteries supply blood to the deep muscles of the thigh as well as ventral and lateral regions of the skin. The femoral artery also gives rise to the **genicular artery**, which provides blood to the region of the knee. As the femoral artery passes posterior to the knee, it is called the *popliteal artery*. The **popliteal artery** branches into the anterior and posterior tibial arteries.

The **anterior tibial artery** is located between the tibia and fibula, and supplies blood to the muscles and integument of the anterior tibial region. Upon reaching the tarsal region, it becomes the **dorsalis pedis artery**, which branches repeatedly and provides blood to the tarsal and dorsal regions of the foot. The **posterior tibial artery** provides blood to the muscles and integument on the posterior surface of the tibial region. The fibular or peroneal artery branches from the posterior tibial artery. It bifurcates and becomes the **medial plantar artery** and **lateral plantar artery**, providing blood to the plantar surfaces. There is an anastomosis with the dorsalis pedis artery, and the medial and lateral plantar arteries form two arches called the **dorsal arch** (also called the *arcuate arch*) and the **plantar arch**, which provide blood to the remainder of the foot and toes.

Learning Connection

Quiz Yourself

After studying the major systemic arteries, trace the path of a drop of blood from the aorta to the tarsal region of the foot.

dorsal arch (arcuate arch, arcuate artery)

Learning Check

1. Which circuit transports blood to the lungs?
   a. Pulmonary
   b. Systemic
2. Which of the following is the first vessel that branches from the ascending aorta?
   a. Subclavian artery
   b. Carotid artery
   c. Coronary artery
   d. Thoracic artery
3. Someone that has a lot of cholesterol floating in their blood can have a thrombosis, in which fat can form a clot in a smaller artery and reduces blood flow. What organ would be affected if the person had a hepatic artery thrombosis?
   a. Liver
   b. Spleen
   c. Kidney
   d. Stomach
4. Peripheral vascular disease is a condition in which there is reduced blood flow distally due to increased plaque buildup within an arterial wall. A common artery that becomes occluded is the popliteal artery. Which of the following arteries would have reduced blood flow in this case?
   a. Femoral artery
   b. Brachial artery
   c. Axillary artery
   d. Dorsalis pedis artery

20.5e Overview of Systemic Veins

Systemic veins return blood to the right atrium. Since the blood has already passed through the systemic capillaries, it will be relatively low in oxygen concentration, pressure, and velocity. In many cases, veins run parallel to arteries and share the same name (**Figure 20.31**).

In both the neck and limb regions, there are often both superficial and deeper levels of veins. The deeper veins generally correspond to the complementary arteries. The superficial veins do not normally have corresponding arteries, but in addition to returning blood, they also make contributions to the maintenance of body temperature. When the ambient temperature is warm, more blood is diverted to the superficial veins, from which it can give off heat to the environment. In colder temperatures, there is more constriction of the superficial veins and blood is diverted deeper to where the body can retain more of the heat.

The right atrium receives all of the systemic venous return. Most of the blood flows into either the superior vena cava or the inferior vena cava. Most of the venous blood from above the diaphragm flows into the superior vena cava; this includes blood from the head, neck, chest, shoulders, and upper limbs. The exception to this is that most venous blood flow from the coronary veins flows directly into the coronary sinus and from there directly into the right atrium. Beneath the diaphragm, systemic venous blood enters the inferior vena cava.

brachiocephalic vein (innominate vein)

internal thoracic vein (internal mammary vein)

The Superior Vena Cava The **superior vena cava** drains most of the body superior to the diaphragm (**Figure 20.32**). On both the left and right sides, the **subclavian vein** is the extension of the axillary vein; it runs in parallel to the subclavian artery. It fuses with the external and internal jugular veins from the head and neck to form the **brachiocephalic vein**. Each **vertebral vein** also flows into the brachiocephalic vein close to this fusion. These veins arise from the base of the brain and the cervical region of the spinal cord, and flow largely through the transverse foramina in the cervical vertebrae. They are the counterparts of the vertebral arteries. Each **internal thoracic vein**, also

Figure 20.31 The Major Systemic Veins

The veins of the systemic circuit drain the capillary beds and return blood to the right side of the heart.

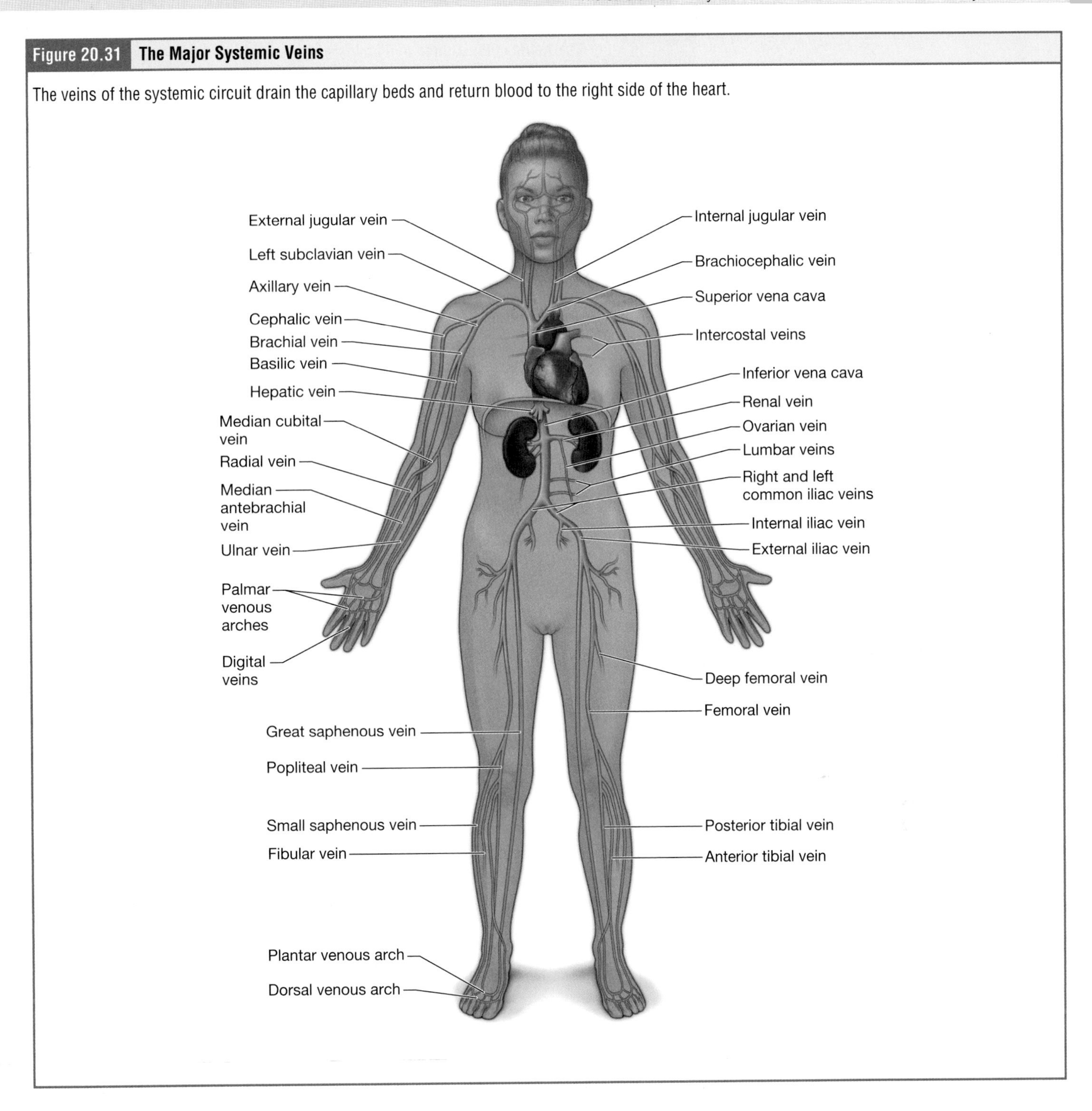

known as an *internal mammary vein*, drains the anterior surface of the chest wall and flows into the brachiocephalic vein.

The remainder of the blood supply from the thorax drains into the azygos vein. Each **intercostal vein** drains muscles of the thoracic wall, each **esophageal vein** delivers blood from the inferior portions of the esophagus, each **bronchial vein** drains the systemic circulation from the lungs, and several smaller veins drain the mediastinal region. Bronchial veins carry approximately 13 percent of the blood that flows into the bronchial arteries; the remainder intermingles with the pulmonary circulation and

Figure 20.32 The Veins of the Thorax and Abdomen

The veins of the thorax and abdomen drain into the inferior vena cava.

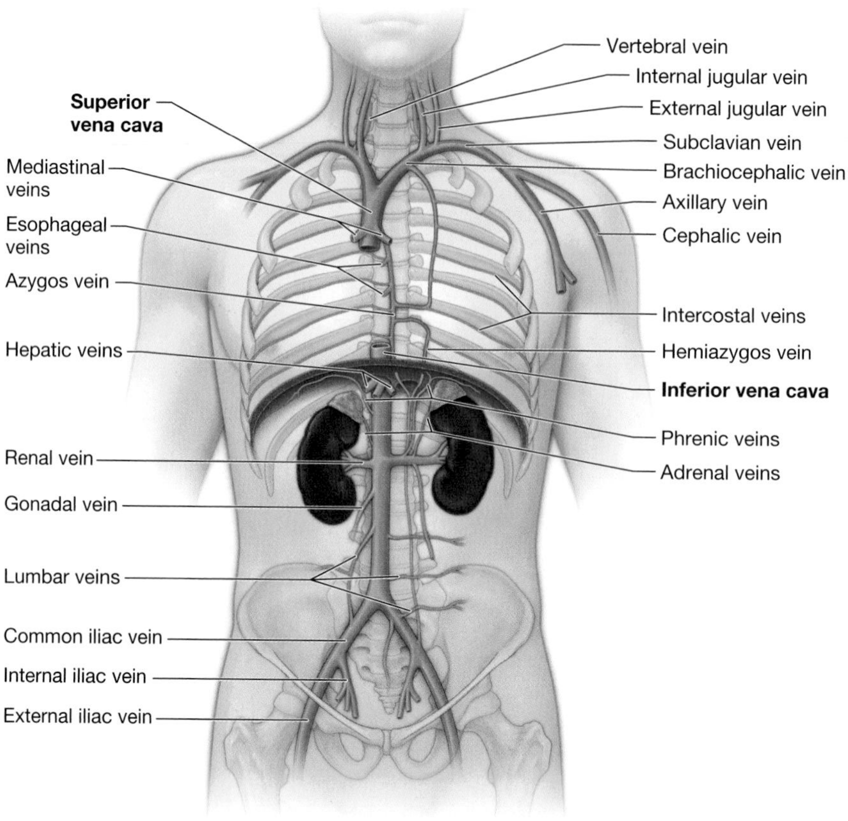

hemiazygos vein (hemi- = "half")

returns to the heart via the pulmonary veins. These veins flow into the **azygos vein**, and with the smaller **hemiazygos vein** (hemi- = "half") on the left of the vertebral column, drain blood from the thoracic region. The hemiazygos vein does not drain directly into the superior vena cava, but enters the brachiocephalic vein via the superior intercostal vein.

The azygos vein passes through the diaphragm from the thoracic cavity on the right side of the vertebral column and begins in the lumbar region of the thoracic cavity. It flows into the superior vena cava at approximately the level of T2, making a significant contribution to the flow of blood. It combines with the two large left and right brachiocephalic veins to form the superior vena cava.

temporal vein (superficial temporal vein)

Veins of the Head and Neck Blood from the brain and the superficial facial vein flow into each **internal jugular vein** (**Figure 20.33**). Blood from the more superficial portions of the head, scalp, and cranial regions, including each **temporal vein** and **maxillary vein**, flows into each **external jugular vein**. Although the external and internal jugular veins are separate vessels, there are anastomoses between them close to the thoracic region. Blood from the external jugular vein empties into the subclavian vein.

Venous Drainage of the Brain Circulation to the brain is both critical and complex (**Figure 20.34**). Many smaller veins of the brainstem, and the superficial veins of the cerebrum, lead to pockets within the dura mater called **dural venous sinuses**.

Figure 20.33 **The Venous Drainage of the Head and Neck**

Within the cranium, venous blood collects in dural venous sinuses and then drains into the internal jugular vein. All blood returns to the heart through the internal jugular, external jugular, or vertebral veins.

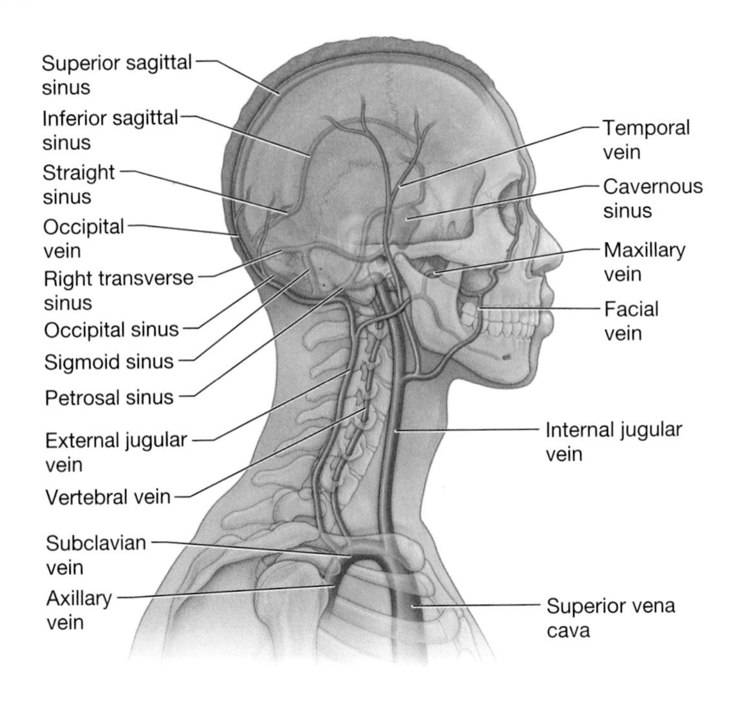

Figure 20.34 **The Dural Venous Sinuses**

Venous blood and absorbed cerebrospinal fluid are collected within the cranium through pockets in the dura mater called *dural venous sinuses*. These regions drain into the internal jugular veins.

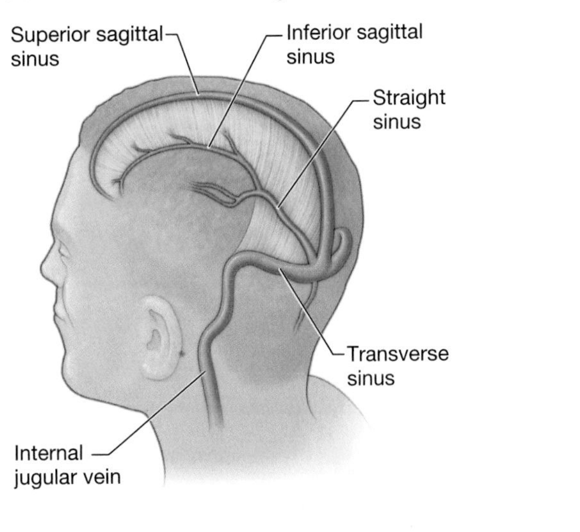

These sinuses include the superior and inferior sagittal sinuses, and the straight sinus. Ultimately, sinuses will lead back to either the inferior jugular vein or the vertebral vein.

Most of the veins on the superior surface of the cerebrum flow into the largest of the sinuses, the **superior sagittal sinus**. It is located midsagittally between the meningeal and periosteal layers of the dura mater within the falx cerebri. In addition to venous blood, most excess cerebrospinal fluid is absorbed and added into the superior sagittal sinus. Blood from most of the smaller vessels in the inferior cerebrum flows into the **straight sinus**. All of the dural venous sinuses drain into the internal jugular vein. The internal jugular vein flows parallel to the common carotid artery and is more or less its counterpart. It empties into the brachiocephalic vein. The veins draining the cervical vertebrae and the posterior surface of the skull, including some blood from the occipital lobe, flow into the vertebral veins. These parallel the vertebral arteries and travel through the transverse foramina of the cervical vertebrae. The vertebral veins also flow into the brachiocephalic veins.

Veins Draining the Upper Limbs The **digital veins** in the fingers come together in the hand to form the **palmar venous arches** (**Figure 20.35**). From here, the veins come together to form the radial vein, the ulnar vein, and the median antebrachial vein. The **radial vein** and the **ulnar vein** parallel the bones of the forearm and join together at the

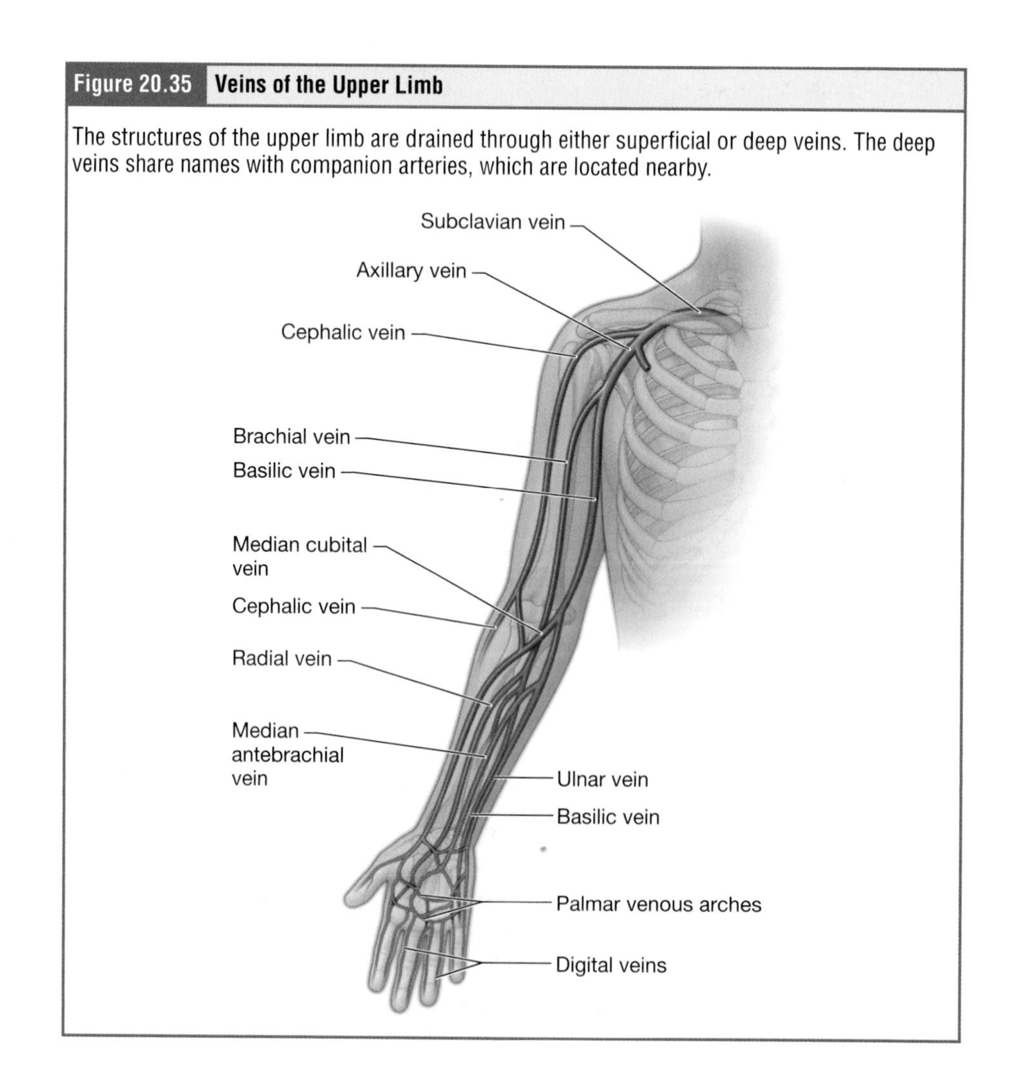

Figure 20.35 Veins of the Upper Limb

The structures of the upper limb are drained through either superficial or deep veins. The deep veins share names with companion arteries, which are located nearby.

antebrachium to form the **brachial vein**, a deep vein that flows into the axillary vein in the brachium.

The **median antebrachial vein** parallels the ulnar vein, is more medial in location, and joins the **basilic vein** in the forearm. As the basilic vein reaches the antecubital region, it gives off a branch called the **median cubital vein** that crosses at an angle to join the cephalic vein. The median cubital vein is the most common site for drawing blood in clinical settings. The basilic vein continues through the arm medially and superficially to the axillary vein.

The **cephalic vein** begins in the antebrachium and drains blood from the superficial surface of the arm into the axillary vein. It is extremely superficial and easily seen along the surface of the biceps brachii muscle in individuals with good muscle tone and in those without excessive subcutaneous adipose tissue in the arms.

The **subscapular vein** drains blood from the subscapular region and joins the cephalic vein to form the **axillary vein**. As it passes through the body wall and enters the thorax, the axillary vein becomes the subclavian vein.

Learning Connection

Quiz Yourself

After studying the systemic veins, trace the path of a drop of blood from the digital veins of the hand to the superior vena cava.

The Inferior Vena Cava Other than the small amount of blood drained by the azygos and hemiazygos veins, most of the blood inferior to the diaphragm drains into the inferior vena cava before it is returned to the heart. Lying just beneath the parietal peritoneum in the abdominal cavity, the **inferior vena cava** parallels the abdominal aorta, where it can receive blood from abdominal veins. The lumbar portions of the abdominal wall and spinal cord are drained by a series of **lumbar veins**, usually four on each side. The superior lumbar veins drain into either the azygos vein on the right or the hemiazygos vein on the left (**Figure 20.36A**) and

Cultural Connection

Why Are Some People's Veins Harder to Find Than Others?

When you've needed to have blood drawn at the doctor's office or hospital, has the nurse or medical technician been able to "find" your veins easily? The median cubical vein is always pretty easy to find, of course; that is why it is the most common vein used for blood draws. But depending on a few factors, the veins may be harder or easier to access with the needle. Significant factors that make it more likely to be able to have a quick and painless blood draw include hydration level, percent body fat, and athleticism. The more hydrated you are, the higher your blood volume. Since veins hold 60 percent of the body's blood at any given time, this number will play a significant role in the volume of the veins and therefore, their plu
ness. Large veins travel within the hypodermis, or subcutaneous layer of the skin. T
can have more or less fat embedded in it, depending on the individual. The s
surrounding the vein can shroud it, making it harder to see and access
the level of athleticism plays a key role in how accessible the vei
that this connection—athleticism and vein prominence—so
tend to be much more visible than the veins of nonathletes.
skeletal muscles, the demand for oxygenated blood in those ar
and arterioles are capable of vasodilation during times of increa
volume through the veins near skeletal muscles increases as well,
veins. Over time, the veins increase in diameter and are more visib
athletes enjoy this visible change, while other athletes may not like th
prominent veins. These visible veins are a symbol of the physiological
work. Athletes are actually much more likely to develop varicose veins (
veins). Another factor that may influence vein visibility is whether or not
previous chest or breast surgery. Because the lymphatic system returns its
the blood within the thorax, these surgeries may alter the lymphatic drainag
these cases, fluid backup in the tissues can result and place more of the work
back to the heart on the veins, causing them to swell. Genetics may also play a

Figure 20.36 The Hepatic Portal System and Inferior Vena Cava

(A) The inferior vena cava drains the abdominal cavity. Several pairs of lumbar veins contribute blood from the body wall. (B) The venous blood from the spleen, stomach, and intestines drains into the hepatic portal vein, which brings blood to the liver. The renal veins that drain the kidneys connect directly to the inferior vena cava. (C) The drainage of the abdominal organs into the hepatic portal system is represented graphically.

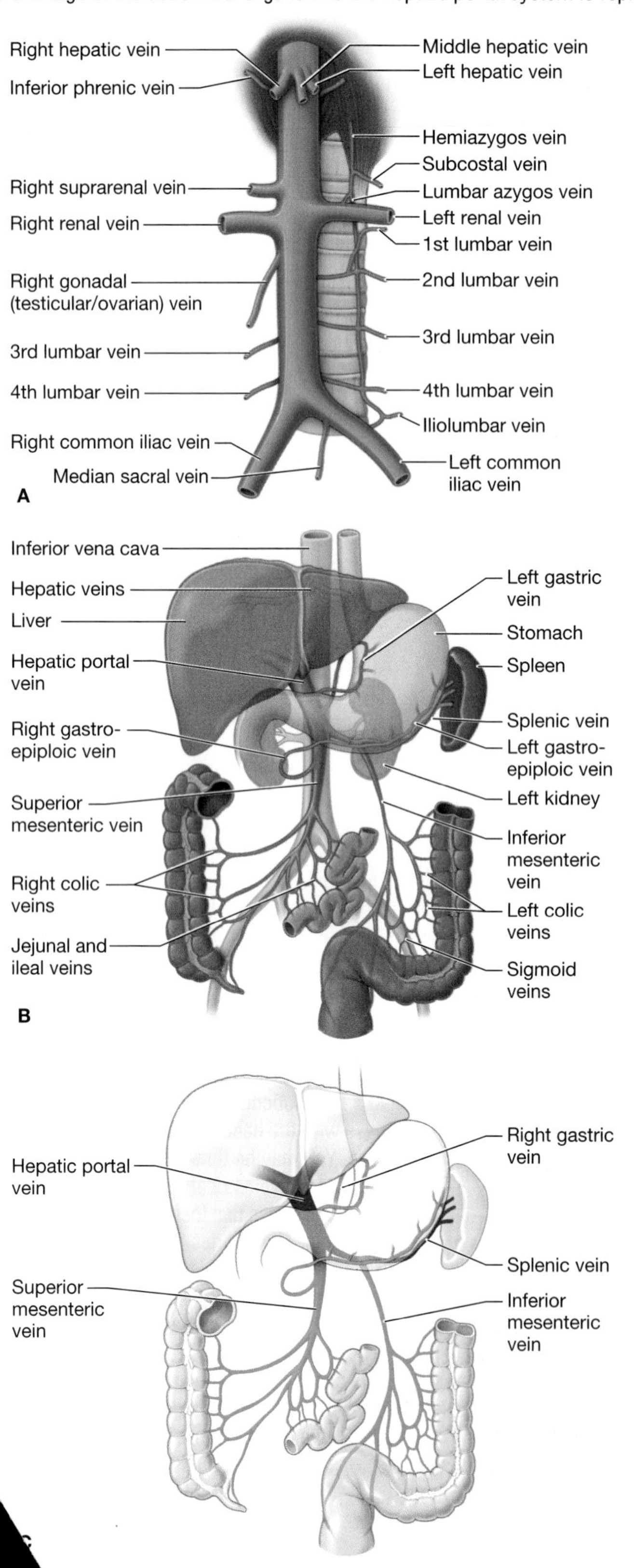

return to the superior vena cava. The remaining lumbar veins drain directly into the inferior vena cava.

Blood supply from the kidneys flows into each **renal vein**, normally the largest veins entering the inferior vena cava (**Figure 20.36B**). A number of other, smaller veins empty into the left renal vein. Each **adrenal vein** drains the adrenal or suprarenal glands located immediately superior to the kidneys. Whereas the right adrenal vein enters the inferior vena cava directly, the left adrenal vein enters the left renal vein.

The gonads drain into **gonadal veins**. The right gonadal vein empties directly into the inferior vena cava, and the left gonadal vein empties into the left renal vein.

Each side of the diaphragm drains into a **phrenic vein**; whereas the right phrenic vein empties directly into the inferior vena cava, the left phrenic vein empties into the left renal vein. Blood supply from the liver drains into each **hepatic vein** and directly into the inferior vena cava. Since the inferior vena cava lies primarily to the right of the vertebral column and aorta, the left renal vein is longer, as are the left phrenic, adrenal, and gonadal veins.

20.5f The Hepatic Portal System

LO 20.5.4

The liver is a complex biochemical processing plant. It packages nutrients absorbed by the digestive system; produces plasma proteins, clotting factors, and bile; and disposes of worn-out cell components and waste products. The liver also plays a critical function in filtering materials absorbed from the intestines. Instead of entering the circulation directly absorbed nutrients (or toxins) travel to the liver via the **hepatic portal system** for filtering (**Figure 20.36**). Portal systems include two capillary beds (see Figure 20.7). In this case, the initial capillaries from the stomach, small intestine, large intestine, and spleen lead to the hepatic portal vein and end in specialized capillaries—the hepatic sinusoids—within the liver.

The hepatic portal vein also receives blood from the inferior mesenteric vein, plus the splenic veins. The superior mesenteric vein receives blood from the small intestine, two-thirds of the large intestine, and the stomach. The inferior mesenteric vein drains the distal third of the large intestine, including the descending colon, the sigmoid colon, and the rectum. The splenic vein is formed from branches from the spleen, pancreas, and portions of the stomach, and the inferior mesenteric vein. The hepatic portal vein also receives branches from the gastric veins of the stomach and cystic veins from the gallbladder. The hepatic portal vein delivers materials from these digestive and circulatory organs directly to the liver for processing.

Because of the hepatic portal system, the liver receives its blood supply from two different sources: normal systemic circulation via the hepatic artery and the hepatic portal vein. The liver processes the blood from the portal system to remove certain wastes and excess nutrients, which are stored for later use. This processed blood, as well as the systemic blood that came from the hepatic artery, exits the liver via the right, left, and middle hepatic veins, and flows into the inferior vena cava.

Learning Connection

Try Drawing It

Draw a representation of the hepatic portal system. Draw the organs that drain into the hepatic portal vein instead of the inferior vena cava and show how the veins from these organs converge to form the hepatic portal vein.

Veins Draining the Lower Limbs The superior surface of the foot drains into the digital veins, and the inferior surface drains into the **plantar veins**, which flow into a complex series of anastomoses in the feet and ankles, including the **dorsal venous arch** and the **plantar venous arch** (**Figure 20.37**). From the dorsal venous arch, blood supply drains into the anterior and posterior tibial veins. The **anterior tibial vein** drains the area near the tibialis anterior muscle and combines with the posterior tibial vein and the fibular vein to form the popliteal vein. The **posterior tibial vein** drains the posterior surface of the tibia and joins the popliteal vein. The **fibular vein** drains the muscles and integument in proximity to the fibula and also joins the popliteal vein.

Figure 20.37 Venous Drainage of the Lower Limb

The structures of the lower limb are drained by superficial and deep veins. The deep veins share names with companion arteries, which are located nearby.

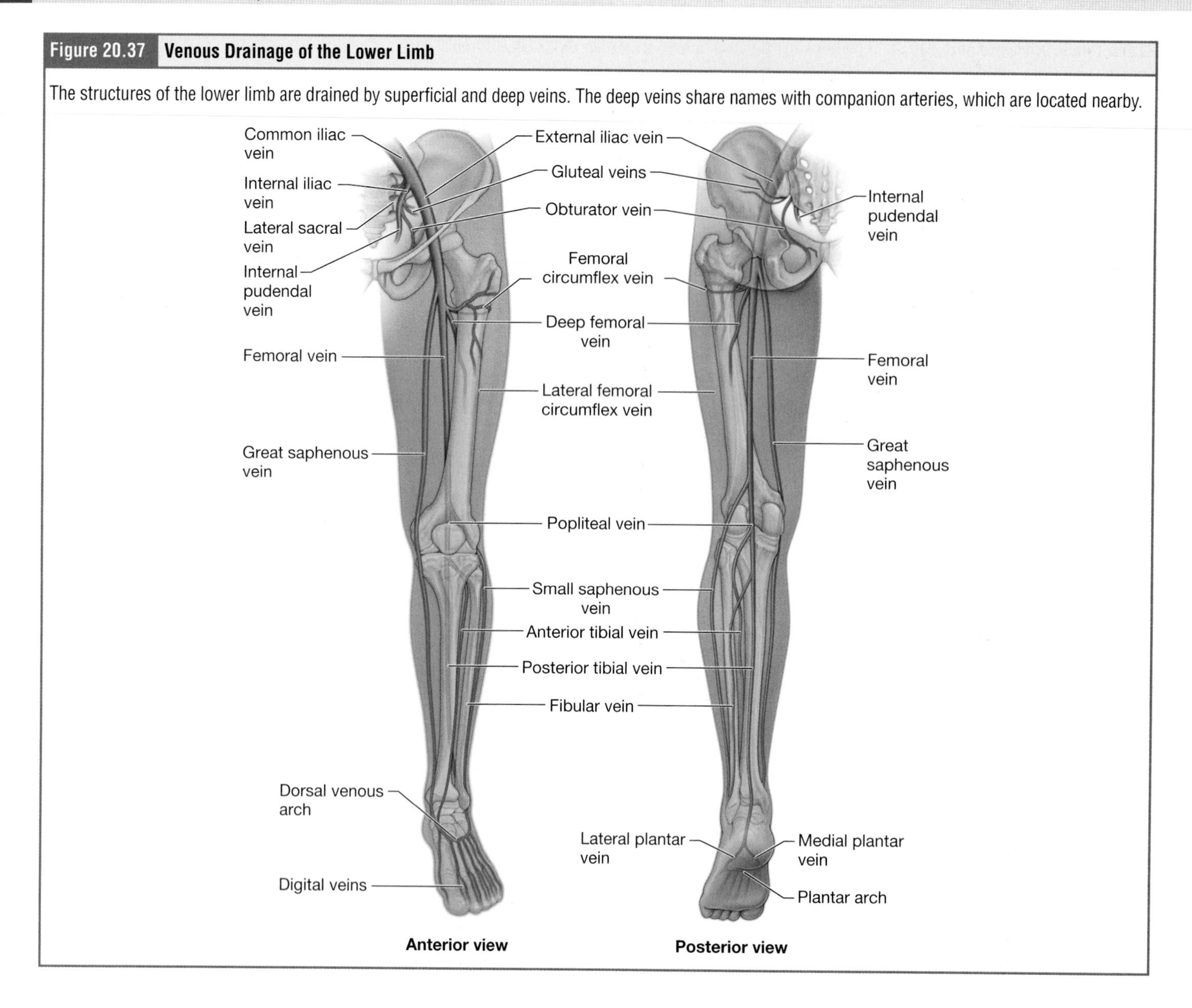

deep femoral vein (profunda femoris vein)

middle sacral vein (median sacral vein)

Learning Connection

Broken Process

The great saphenous vein is often removed from a person's leg to be used as graft material in coronary bypass surgery. How would this affect venous drainage from the leg?

The **small saphenous vein**, located on the lateral surface of the leg, drains blood from the superficial regions of the lower leg and foot, and flows into to the **popliteal vein**. As the popliteal vein passes behind the knee in the popliteal region, it becomes the femoral vein. It is palpable in patients without excessive adipose tissue.

Close to the body wall, the great saphenous vein, the deep femoral vein, and the femoral circumflex vein drain into the femoral vein. The **great saphenous vein** is a prominent surface vessel located on the medial surface of the leg and thigh that collects blood from the superficial portions of these areas. The **deep femoral vein**, as the name suggests, drains blood from the deeper portions of the thigh. The **femoral circumflex vein** forms a loop around the femur just inferior to the trochanters and drains blood from the areas in proximity to the head and neck of the femur.

As the **femoral vein** penetrates the body wall from the femoral portion of the upper limb, it becomes the **external iliac vein**, a large vein that drains blood from the leg to the common iliac vein. The pelvic organs and integument drain into the **internal iliac vein**, which forms from several smaller veins in the region, including the umbilical veins that run on either side of the bladder. The external and internal iliac veins combine near the inferior portion of the sacroiliac joint to form the common iliac vein. In addition to blood supply from the external and internal iliac veins, the **middle sacral vein** drains the sacral region into the **common iliac vein**. Similar to the common iliac arteries, the common iliac veins come together at the level of L5 to form the inferior vena cava.

Apply to Pathophysiology

Pulmonary Embolism

An **embolus** is a type of traveling blood clot. A blood clot is called a **thrombus** when it develops, and thrombi can develop within blood vessels anywhere in the body, sometimes in response to vessel damage and sometimes spontaneously. If the thrombus becomes dislodged and starts to travel through the circulatory system, it is now called an **embolus**. If it becomes lodged in an artery and reduces or blocks blood flow, it is called an **embolism**. The most common sites of thrombus development are in the deep veins of the lower limb.

1. Which of the following is a site where thrombi are likely to develop?
   A) The great saphenous vein
   B) The popliteal vein
   C) The cephalic vein
2. Assuming a thrombus developed in a deep vein of the lower limb and then became an embolus and began to travel, which of the following represents its path?
   A) External iliac vein ➔ hepatic portal vein ➔ inferior vena cava ➔ right atrium
   B) Internal iliac vein ➔ inferior vena cava ➔ azygos vein ➔ right atrium
   C) External iliac vein ➔ inferior vena cava ➔ right atrium
3. Since all of the veins listed in Question 2 increase in size as they approach the heart, the embolus will keep flowing within the venous blood without getting stuck. At what anatomical location do the blood vessels begin to branch and their diameters narrow, leading to a possible point when the embolus could become lodged and block flow of blood?
   A) Pulmonary arteries
   B) Pulmonary veins
   C) Left semilunar valve
4. To avoid further emboli, the doctors implant a filter into the bloodstream to catch any emboli moving from the limbs or pelvis *before reaching the heart*. Would the doctors be able to place *one* filter that would catch emboli originating in both the upper and lower extremities?
   A) Yes, one filter will catch emboli from both locations
   B) No, more than one filter would be needed
5. To catch any emboli from just lower limb or pelvis, which vein would be the best location for the filter?
   A) Superior vena cava
   B) Inferior vena cava
   C) Hepatic portal vein

Learning Check

1. Which of the following would return blood to the heart via the superior vena cava? Please select all that apply.
   a. Right arm
   b. Stomach
   c. Left thigh
   d. Head
2. Where would the superior sagittal sinus drain into?
   a. Straight sinus
   b. Inferior vena cava
   c. Internal jugular vein
   d. Temporal vein
3. Which of the following veins drains blood from the diaphragm?
   a. Phrenic vein
   b. Gonadal vein
   c. Hepatic vein
   d. Adrenal vein
4. Why is it important for the liver to have a portal system?
   a. The liver is a big organ, so it needs more blood.
   b. The liver filters wastes and adjusts the concentration of absorbed nutrients.
   c. The liver cannot drain blood through veins.

20.6 Development of Blood Vessels and Fetal Circulation

HAPS LEARNING GOALS AND OUTCOMES

Learning Objectives: By the end of this section, you will be able to:

20.6.1 Describe the role of the placenta, umbilical vessels, ductus venosus, foramen ovale, and ductus arteriosus in fetal circulation.

20.6.2 Trace the pathway of blood flow from the placenta, through the fetal heart and body, and back to the placenta.

20.6.3 Describe the changes in major fetal cardiovascular structures (i.e., umbilical vessels, ductus venosus, ductus arteriosus, foramen ovale) that typically occur beginning at birth, and the ultimate postnatal remnants (fates) of these structures.

Circulation patterns established around the fourth week of embryonic development. As in the adult, the cells and tissues of the embryo rely on the circulatory system for nutrients and gasses, and to remove waste products. Throughout embryonic and fetal development, **angiogenesis**—the creation of new blood vessels—is ongoing as growth proceeds. Angiogenesis continues throughout childhood and even in adulthood as needed during healing or weight gain.

As the embryo grows within the uterus, its requirements for nutrients and gas exchange also grow. The placenta—a circulatory organ unique to pregnancy—develops from embryonic tissue and interfaces with the uterine wall to provide a site of exchange of nutrients, gasses, and waste products (**Figure 20.38**). The placenta is a large, flat structure (about two pounds of tissue and approximately the size of a dinner plate at full term) that is a large nest of fetal capillaries. Blood from the uterine arteries bathes the surface of the placenta and, while maternal and fetal blood never mix, nutrients, gasses, and wastes can diffuse between the two circulations. Emerging from the placenta is the **umbilical vein**, which carries oxygen-rich blood from the placenta to the fetal inferior vena cava via the ductus venosus. Two **umbilical arteries** carry oxygen-depleted fetal blood, including wastes and carbon dioxide, to the placenta from the fetal systemic circulation.

LO 20.6.1

The fetus does not rely on its lungs for oxygenating the blood; therefore, the pulmonary circuit is largely bypassed until birth. Two shunts—alternate paths for blood flow—divert blood from the pulmonary to the systemic circuit. These shunts close shortly after birth, however, when the newborn begins to breathe (**Figure 20.39**):

LO 20.6.2

LO 20.6.3

- The **foramen ovale** is an opening in the interatrial septum that allows blood to flow from the right atrium to the left atrium. A valve associated with this opening prevents backflow of blood during the fetal period. As the newborn begins to breathe (after birth), this shunt closes. After it closes it remains a very thin portion of tissue, much thinner than the surrounding interatrial septum. In the adult, this thinly closed foramen is the fossa ovalis.
- The **ductus arteriosus** is a short, muscular vessel that connects the pulmonary trunk to the aorta. Most of the blood pumped from the right ventricle into the pulmonary trunk is thereby diverted into the aorta. Only enough blood reaches the fetal lungs to maintain the developing lung tissue. When the newborn takes its first breath, pressure within the lungs drops dramatically, and both the lungs and the pulmonary vessels expand. As the amount of oxygen increases, the

Figure 20.38 The Placenta

(A) A full-term placenta is slightly larger than a dinner plate and weighs about two pounds.
(B) The structure of the placenta keeps fetal blood within capillary beds while maternal blood bathes the outer surface of the capillaries. Thus, fetal and maternal blood never mix, but gas and nutrient exchange is possible.

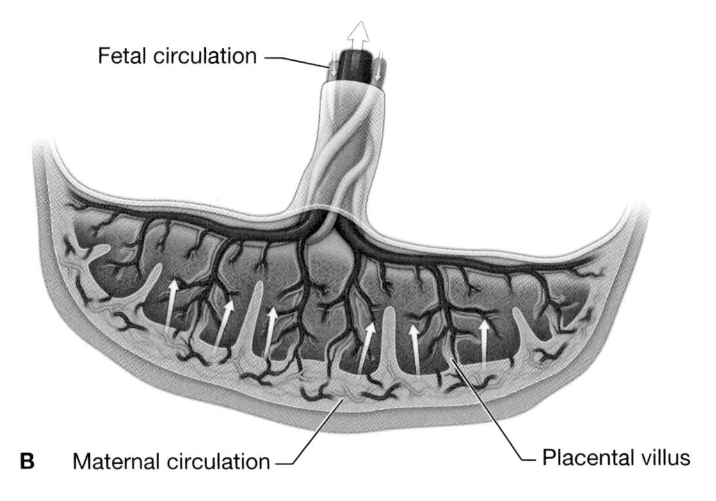

smooth muscles in the wall of the ductus arteriosus constrict, sealing off the passage. Eventually, the muscular and endothelial components of the ductus arteriosus degenerate, leaving only the connective tissue component of the ligamentum arteriosum.

- The **ductus venosus** is a third shunt, which acts as a temporary blood vessel that branches from the umbilical vein, allowing much of the freshly oxygenated blood from the placenta—the organ of gas exchange between the mother and fetus—to bypass the fetal liver and go directly to the fetal heart. The ductus venosus closes slowly during the first weeks of infancy and degenerates to become the ligamentum venosum.

Figure 20.39 Fetal Circulation

Because the fetus obtains oxygenated blood from the placenta rather than from the pulmonary circuit, there are three major alterations to the circulatory system that must occur after birth.

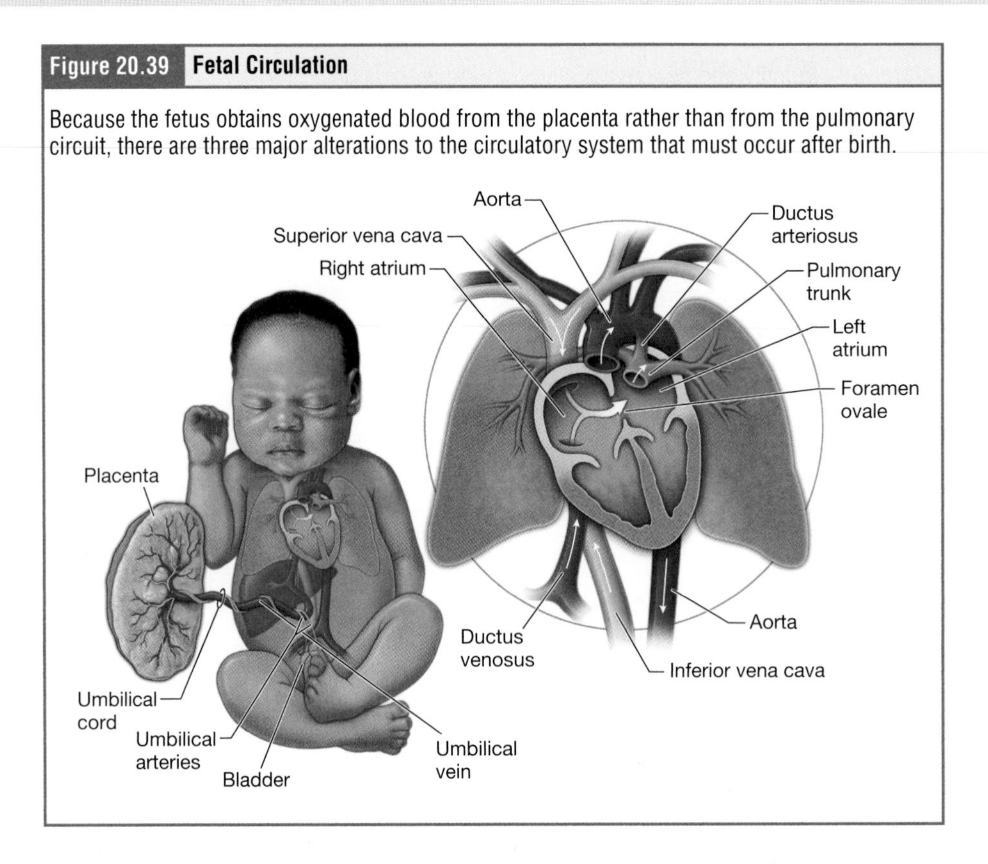

Learning Check

1. Which of the following blood vessels carry oxygen-rich blood toward the fetal heart? Please select all that apply.
   a. Umbilical vein
   b. Umbilical artery
   c. Ductus arteriosus
   d. Ductus venosus
2. Which of the following is an opening in the interatrial septum that allows blood to flow from the right atrium to the left atrium?
   a. Foramen ovale
   b. Ductus arteriosus
   c. Ductus venosus

Chapter Review

20.1 Anatomy of Blood Vessels

The learning objectives for this section are:

20.1.1 Define the terms artery, capillary, and vein.

20.1.2 List the three tunics associated with most blood vessels and describe the composition of each tunic.

20.1.3 Compare and contrast tunic thickness, composition, and lumen diameter among arteries, capillaries, and veins.

20.1.4 Define vasoconstriction and vasodilation.

20.1.5 Identify and describe the structure and function of specific types of blood vessels (i.e., elastic [conducting] arteries, muscular [distributing] arteries, arterioles, capillaries, venules, veins).

20.1.6 List types of capillaries, state where in the body each type is located, and correlate their anatomical structures with their functions.

20.1.7 Describe the functional significance of the venous reservoir.

20.1.8 Define anastomosis and explain its functional significance (e.g., cerebral arterial circle [Circle of Willis]).

20.1 Questions

1. Define the term *capillary*.
   a. Blood vessels that carry blood away from the heart
   b. Blood vessels that carry blood toward the heart
   c. Blood vessels that supply blood directly to tissues

2. Joshua is running outside. To supply muscles with more oxygen, the body decreases blood supply to organs such as the intestines. What layer of the vessel releases local chemicals to decrease blood supply to the intestines?
   a. Tunica intima
   b. Tunica adventitia
   c. Tunica media
   d. Tunica externa

3. The lumen of which of the following blood vessels has the greatest diameter?
   a. Artery
   b. Venules
   c. Capillary
   d. Metarteriole

4. The superior mesenteric artery supplies oxygen to the small intestine. When you eat a large meal, the small intestine needs to work hard to digest the food, which requires increased blood flow. What happens to the superior mesenteric artery after you eat a big meal?
   a. Vasodilation
   b. Vasoconstriction

5. The popliteal artery is found running through the back of the knee, far away from the heart. Which type of artery would you expect the popliteal artery to be?
   a. Elastic Artery
   b. Muscular Artery

6. After eating a meal, nutrients are absorbed into the bloodstream in the small intestine. What type of capillaries would you expect to find at the small intestine?
   a. Continuous capillaries
   b. Fenestrated capillaries
   c. Sinusoid capillaries
   d. Porous capillaries

7. What is the functional significance of a venous reservoir?
   a. To reserve blood until needed by organs
   b. To contain slow-moving blood
   c. To promote unidirectional flow toward the heart
   d. To prevent backflow toward the capillaries

8. What is the functional significance of an arterial anastomosis?
   a. To supply blood to a widespread area
   b. To provide alternative blood supplies to an area
   c. To link two or more capillary beds
   d. To return blood from capillaries toward the heart

20.1 Summary

- Blood vessels are classified as one of the following: (1) arteries transport blood away from the heart, (2) veins transport blood toward the heart, or (3) capillaries exchange gasses, nutrients and wastes between the blood and the body cells or lungs.
- The cardiovascular system consists of two pathways. The systemic circuit transports blood from the heart to the body cells and then back to the heart. In this circuit, blood drops off oxygen and nutrients to the body cells, and picks up carbon dioxide and wastes.
- The pulmonary circuit transports blood from the heart to the lungs and then back to the heart; while in the lungs, the blood picks up oxygen and drops off carbon dioxide.
- All arteries and veins consist of three layers. From superficial to deep, the layers are the tunica externa, tunica media, and tunica intima. Capillaries consist of only one layer; they are composed of simple squamous epithelium, and are simple a continuation of the tunica intima of arterioles.
- The walls of arteries are much thicker than those of veins, especially in the tunica media, which is mainly composed of

smooth muscle tissue. This imparts strength to the walls of arteries, so they can withstand the fluctuations in diameter that result from the pumping action of the heart.

- Blood vessels can change diameter under certain conditions. Vasoconstriction is a decrease in the diameter of a blood vessel; this decreases the blood flow through the vessel. Vasodilation is an increase in the diameter of a blood vessel; this increases the blood flow through the vessel.
- There are three types of capillaries. The most common type is continuous capillaries, which are found in almost all tissues. Fenestrated capillaries have small pores to allow for larger molecules to cross the capillary wall. Lastly, sinusoid capillaries have extensive intercellular gaps that allow the passage of large molecules or waste.

20.2 Blood Flow, Blood Pressure, and Resistance

The learning objectives for this section are:

HAPS LEARNING GOALS AND OUTCOMES

20.2.1 Define blood flow, blood pressure, and peripheral resistance.

20.2.2 State and interpret the equation that relates fluid flow to pressure and resistance.

20.2.3* Explain the clinical and physiological significance of pulse pressure and mean arterial pressure values.

20.2.4* List and explain the contribution of the components that contribute to vascular resistance and mean arterial pressure.

20.2.5 Describe the role of arterioles in regulating tissue blood flow and systemic arterial blood pressure.

20.2.6 Interpret relevant graphs to explain the relationships between vessel diameter, cross-sectional area, blood pressure, and blood velocity.

20.2.7* Describe the maintenance of driving pressure in veins including the skeletal and pulmonary pumps.

20.2.8 Given a factor or situation (e.g., left ventricular failure), predict the changes that could occur in the cardiovascular system and the consequences of those changes (i.e., given a cause, state a possible effect).

* Objective is not a HAPS Learning Goal.

20.2 Questions

1. Which of the following would you expect to occur if someone has a high density of proteins in their blood, making their blood highly viscous? Please select all that apply.
   a. Increase in blood pressure
   b. Decrease in blood pressure
   c. Increase in blood flow
   d. Decrease in blood flow

2. When arteries vasoconstrict, the diameter of the vessel decreases. What immediately happens to the fluid flow?
   a. It increases.
   b. It decreases.

3. Which of the following are true of the physiological significance of the mean arterial pressure (MAP)? Please select all that apply.
   a. The MAP represents the minimum pressure of the blood flowing through the major arteries during ventricular diastole.
   b. If the MAP is too low, blood flow will not be sufficient to deliver oxygen and nutrients to the body cells, and the cells will become hypoxic.
   c. If the MAP is too low, it signifies a degeneration in the elastic fibers in the walls of arteries, which means that the heart must work harder to pump blood into the arteries.
   d. The MAP represents the average pressure driving blood through the arteries to the body cells.

4. Which of the following factors of blood pressure are affected in arteries that contain a significant amount of cholesterol in the blood and plaque accumulation? Please select all that apply.
   a. Cardiac output
   b. Blood volume
   c. Blood vessel diameter
   d. Blood viscosity

5. Which of the following plays a significant role in regulating blood flow through capillaries?
   a. Arteries
   b. Arterioles
   c. Venules
   d. Veins

Mini Case 1: Peripheral Vascular Disease

George arrives at the emergency room complaining of a cramping and aching pain in their right lower extremity. George complains that the pain is worse with walking and only improves with rest. After doing a blood test, the medical professionals find high quantities of lipids and cholesterol circulating within the blood.

1. As the doctor is conducting tests to check George's blood flow and pressure, the doctor explains that blood flow and blood pressure change throughout the systemic circuit. Using the following figure, explain why George's capillaries can have the smallest diameter but not the lowest blood pressure.

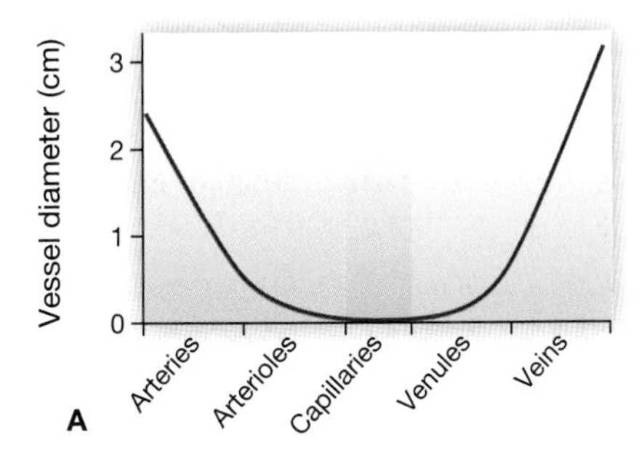

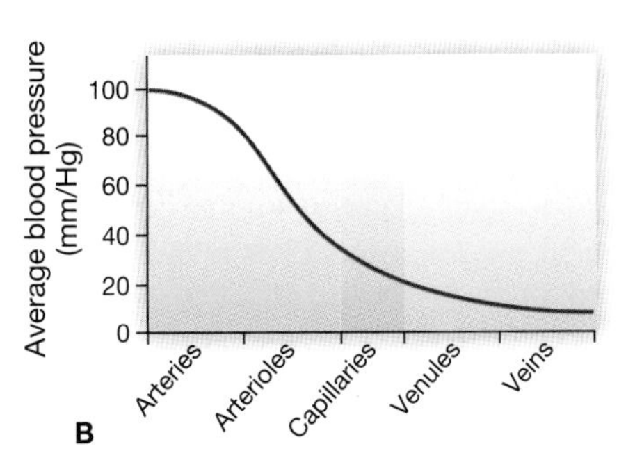

a. A decrease in diameter leads to a decrease in blood pressure.
b. Vessels farther away from the heart have lower blood pressures.
c. There are many capillaries which cover a large surface area.
d. Capillaries have less compliance than veins.

2. The doctor explains that a change of diet and adding daily exercise would be good for George's condition. Why would jogging every day improve the circulation in George's legs? Please select all that apply.

a. Contraction of George's leg muscles during jogging would squeeze the blood vessels in their legs, which would help the blood return to the heart.
b. Increasing the rate and depth of breathing during exercise would help venous blood return to the heart.
c. Exercise would constrict the arterioles in George's legs, which would increase blood flow to their leg muscles.
d. Exercise would cause a decrease in blood viscosity, which would increase blood flow.

The doctors diagnose George with peripheral vascular disease. This occurs when there is an occlusion in an artery that decreases blood flow to the parts of the body distal to the occlusion. The occlusion is so severe that the body cannot compensate enough to maintain regular blood flow.

3. Which of the following would you expect to occur at the ankle with an unresolvable occlusion at the knee?

a. Increased blood pressure
b. Decreased blood pressure

20.2 Summary

- Arterial blood pressure is stated as the systolic blood pressure over the diastolic blood pressure. It is also important to know the pulse pressure, which indicates the condition of someone's arteries, and the mean arterial pressure, which represents the average pressure driving blood to the body cells.
- Blood flow is the volume of blood moved through a blood vessel or organ per unit of time (such as Liters/minute). Blood flow is directly proportional to the pressure gradient between two points and inversely proportional to the resistance.
- Blood flow and blood pressure are affected by five factors: (1) cardiac output, (2) blood volume, (3) vessel compliance, (4) blood viscosity, and (5) blood vessel length and diameter.
- The pressure gradient is determined by the cardiac output and blood volume. The resistance is affected by the blood viscosity, blood vessel length, and blood vessel diameter. The diameter of blood vessels can change the resistance quickly via vasodilation and vasoconstriction.
- The venous system relies on pressure gradients as blood will travel from high to low pressure areas. Two pumps, the skeletal muscle pump and the respiratory pump, facilitate changes in pressure to guide venous blood flow back to the heart.

20.3 Capillary Exchange

The learning objectives for this section are:

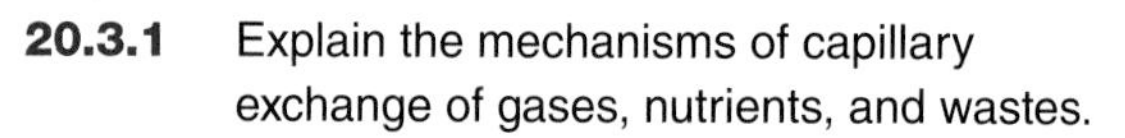

20.3.1 Explain the mechanisms of capillary exchange of gases, nutrients, and wastes.

20.3.2* Describe the forces that regulate movement of fluid and substances between the blood and the interstitial fluid.

20.3.3* Predict the impact of alterations in fluid and solute exchange at capillaries on blood pressure or the tissues.

* Objective is not a HAPS Learning Goal.

20.3 Questions

1. Which feature of capillaries allow for rapid exchange of small molecules?
   a. The small vessel diameter
   b. The thin wall of the endothelial cells
   c. The overlapping endothelial cells
   d. The shorter vessel length

2. Which of the following pressures drives blood flow from the interstitial fluid into vessels?
   a. Hydrostatic pressure
   b. Osmotic pressure
   c. Systolic pressure
   d. Diastolic pressure

3. Which of the following pressures would you expect to increase in someone who has a high-sodium diet?
   a. Hydrostatic pressure
   b. Osmotic pressure
   c. Systolic pressure
   d. Diastolic pressure

20.3 Summary

- Substances are transported between capillaries and the interstitial fluid by diffusion across the cell membrane, diffusion through ion channels, bulk flow through intercellular clefts, or passage through the pores of certain types of capillaries. The properties of each substance, such as its size or solubility in lipids, determines the method by which it crosses the cell membrane.
- There are two major pressures that dictate movement in and out of capillaries. Hydrostatic pressure is the pressure exerted by fluids against the inner walls of the capillaries; this pressure causes water to move from the capillaries into the interstitial fluid. Osmotic pressure is the pressure exerted by solutes, which draws water toward regions of higher solute concentration; this force draws water from the interstitial fluid into the capillaries.
- Because more fluid leaves the capillaries than returns to them each day, the excess fluid is picked up by lymphatic capillaries, transported through the lymphatic system, and returned to the systemic circulation via the subclavian veins.

20.4 Homeostatic Regulation of the Vascular System

The learning objectives for this section are:

20.4.1 Explain the role of the autonomic nervous system in regulation of blood pressure and volume.

20.4.2 Explain the steps of the baroreceptor reflex and describe how this reflex maintains blood pressure homeostasis when blood pressure changes.

20.4.3 Provide specific examples to demonstrate how the cardiovascular system maintains blood pressure homeostasis in the body.

20.4.4 Explain the role of the precapillary sphincter in autoregulation.

20.4.5 Explain how local control mechanisms and myogenic autoregulation influences blood flow to tissues.

20.4.6 List some chemicals that cause either vasodilation or vasoconstriction and explain the circumstances in which they are likely to be active.

20.4.7 List the local, hormonal and neural factors that affect peripheral resistance and explain the importance of each.

20.4 Questions

1. Which of the following are true of the role of the autonomic nervous system on the regulation of blood pressure? Please select all that apply.
   a. The cardioinhibitory center consists of sympathetic neurons, which increase blood pressure.
   b. The vagus nerve decreases heart rate and stroke volume, which results in a decrease in blood pressure.
   c. The vasomotor center exerts only parasympathetic control over blood vessel diameter.
   d. The cardioaccelerator center, the cardioinhibitory center, and the vasomotor center are all located in the medulla oblongata of the brainstem.

2. When blood pressure drops too low, which of the following happens first?
   a. Increased parasympathetic stimulation
   b. Increased sympathetic stimulation
   c. Increased heart rate
   d. Increased vasodilation

3. The body compensates for an occlusion in blood flow to maintain homeostasis and regular blood flow. Which of the following are compensations to maintain blood flow homeostasis? Please select all that apply.
   a. Increase in heart rate
   b. Decrease in heart rate
   c. Increase in blood pressure
   d. Decrease in blood pressure

4. Which of the following structures will direct blood flow into a specific capillary?
   a. Venule
   b. Arteriole
   c. Capillary
   d. Sphincter

5. Which of the following responses would increase blood flow only to a local area?
   a. Myogenic response
   b. Baroreceptor reflex
   c. Chemoreceptor reflex
   d. Epinephrine release

6. You have just awakened in the morning and stood up immediately. Suddenly, you feel a "head rush," but then you regain typical blood flow to your brain. Which of the following chemicals would you expect to cause increased blood flow to the brain?
   a. Epinephrine
   b. Acetylcholine
   c. ADH
   d. Endothelin

7. An increase in sympathetic nervous system activity would result in an (increase/decrease) of the systemic blood pressure through its effects on (cardiac output/peripheral resistance/both of these factors).
   a. Increase; cardiac output
   b. Decrease; peripheral resistance
   c. Increase; both of these factors
   d. Decrease; both of these factors

20.4 Summary

- Blood pressure is regulated by changing its determining factors, such as blood volume, heart rate, stroke volume, and peripheral resistance.
- The cardiovascular centers in the medulla oblongata regulate heart function and blood pressure. The sympathetic cardioaccelerator center increases heart rate and stroke volume, while the parasympathetic cardioinhibitory center decreases heart rate and stroke volume. The vasomotor center adjusts peripheral resistance by changing the diameter of arterioles.
- Baroreceptor reflexes monitor blood pressure and relay information to the cardiovascular centers, to increase or decrease the blood pressure when it is out of balance.
- Chemoreceptors monitor oxygen, carbon dioxide, and pH levels. This information is sent to the respiratory and cardiovascular centers in order to adjust blood flow and pressure. This ensures proper delivery of oxygen and nutrients to the body cells.
- The endocrine system also plays a role in the homeostasis of blood flow, blood volume, and blood pressure through the secretion of hormones such as epinephrine, norepinephrine, erythropoietin, renin, aldosterone, antidiuretic hormone and atrial natriuretic hormone.
- Precapillary sphincters adjust blood flow to specific organs and tissues through vasoconstriction and vasodilation.

20.5 Circulatory Pathways

The learning objectives for this section are:

HAPS LEARNING GOALS AND OUTCOMES

20.5.1 Describe the systemic and pulmonary circuits (circulations) and explain the functional significance of each.

20.5.2 Identify the major arteries and veins of the pulmonary circuit.

20.5.3 Identify the major arteries and veins of the systemic circuit.

20.5.4 Describe the structure and functional significance of the hepatic portal system.

20.5 Questions

1. People with high cholesterol in their arteries often develop plaque buildup in the arterial walls. If a piece of this plaque breaks away and becomes lodged in a smaller artery, the condition is known as an *embolism*. What would you expect to occur immediately if someone had a pulmonary embolism?
   a. Death of muscle tissue
   b. Decreased blood oxygen levels
   c. Weakness of lower extremities
   d. Difficulty with digestion

2. Which of the following are true of the pulmonary trunk? Please select all that apply.
   a. It transports blood toward the heart.
   b. It transports oxygen-poor blood.
   c. It is classified as an artery.
   d. It arises from the pulmonary veins.

3. Which of the following arteries supplies blood to the head?
   a. Abdominal aorta
   b. Common carotid artery
   c. Mediastinal artery
   d. Common iliac artery

4. How are dietary nutrients absorbed into the liver?
   a. Hepatic vein
   b. Hepatic portal vein
   c. Hepatic artery

20.5 Summary

- Blood flows through the body in two pathways: the systemic circuit and the pulmonary circuit.
- The pulmonary circuit transports blood containing carbon dioxide from the heart to the lungs; the blood transfers the carbon dioxide to the alveoli of the lungs and returns to the heart carrying oxygen from the lungs.
- The systemic circuit transports blood containing oxygen and nutrients from the heart to the body cells; blood then returns to the heart carrying carbon dioxide and wastes from the cells.
- In the systemic circuit, blood is transported from the left ventricle to the aorta, and then into systemic arteries, arterioles, capillaries, venules and veins. Blood returns to the right atrium of the heart through the superior or inferior vena cava.
- In the pulmonary circuit, blood is transported from the right ventricle to the pulmonary trunk, and then into pulmonary arteries, arterioles, capillaries, venules and veins. Blood returns to the left atrium of the heart through the pulmonary veins.
- The hepatic portal system transports venous blood from the absorbing digestive organs into the liver, before it drains into the inferior vena cava. While in the liver, nutrient concentrations are adjusted and toxins are removed. Then the blood leaves the liver through the hepatic veins, and drains into the inferior vena cava to join the general circulation.

20.6 Development of Blood Vessels and Fetal Circulation

The learning objectives for this section are:

20.6.1 Describe the role of the placenta, umbilical vessels, ductus venosus, foramen ovale, and ductus arteriosus in fetal circulation.

20.6.2 Trace the pathway of blood flow from the placenta, through the fetal heart and body, and back to the placenta.

20.6.3 Describe the changes in major fetal cardiovascular structures (i.e., umbilical vessels, ductus venosus, ductus arteriosus, foramen ovale) that typically occur beginning at birth, and the ultimate postnatal remnants (fates) of these structures.

20.6 Questions

1. Which of the following fetal structures connects the pulmonary trunk to the aorta?

a. Foramen ovale
b. Ductus arteriosus
c. Ductus venosus
d. Umbilical vein

2. Describe the blood flow of a fetus from the placenta to the arm.

a. Umbilical vein > ductus venosus > inferior vena cava > aorta > brachial artery > arm
b. Umbilical artery > ductus arteriosus > pulmonary vein > lungs > aorta > brachial artery > arm
c. Umbilical artery > pulmonary artery > aorta > brachial artery > arm
d. Umbilical vein > ductus arteriosus > pulmonary vein > aorta > brachial artery > arm

3. Patent ductus arteriosus is a very rare condition in which the ductus arteriosus does not seal off. Which of the following would you expect to find in patients with ductus arteriosus?

a. Deoxygenated blood in the systemic circulation
b. Oxygenated blood in the pulmonary circulation
c. Decreased muscle mass of cardiac muscle
d. Increased muscle mass of cardiac muscle

20.6 Summary

- Oxygen and nutrients are supplied to a fetus by the placenta. Oxygen-rich blood travels through the umbilical vein to the fetus via the ductus venosus. Carbon dioxide and fetal wastes are transported to the placenta via two umbilical arteries.
- Since the fetal lungs do not function, there are two shunts that bypass the pulmonary circuit and provide oxygen-rich blood to the fetus. The foramen ovale is an opening between the atria, through which blood flows from the right atrium into the left atrium. The ductus arteriosus transports blood from the pulmonary trunk to the aorta.
- The foramen ovale, ductus arteriosus, and ductus venosus close shortly after birth.

21

The Lymphatic and Immune System

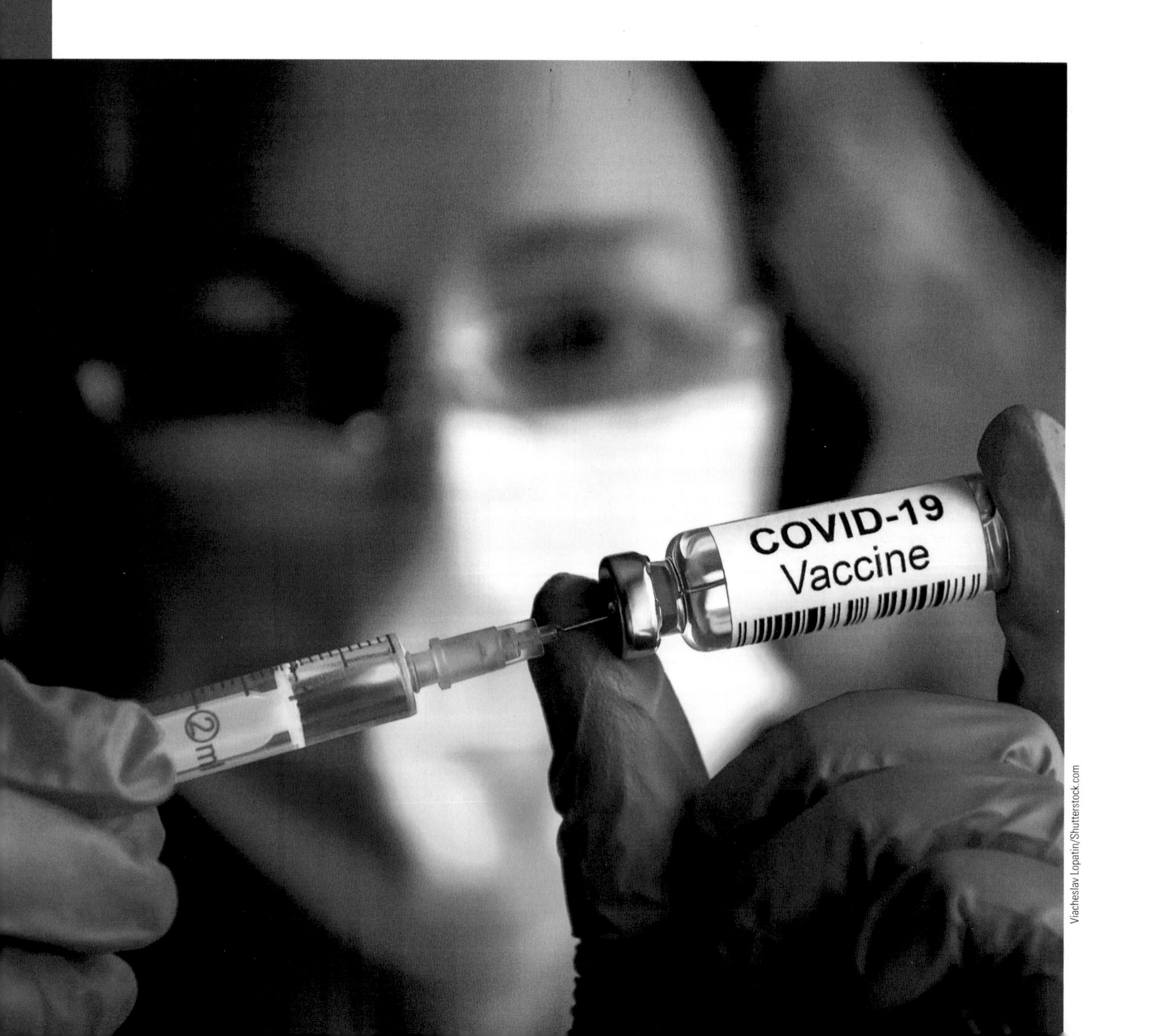

Chapter Introduction

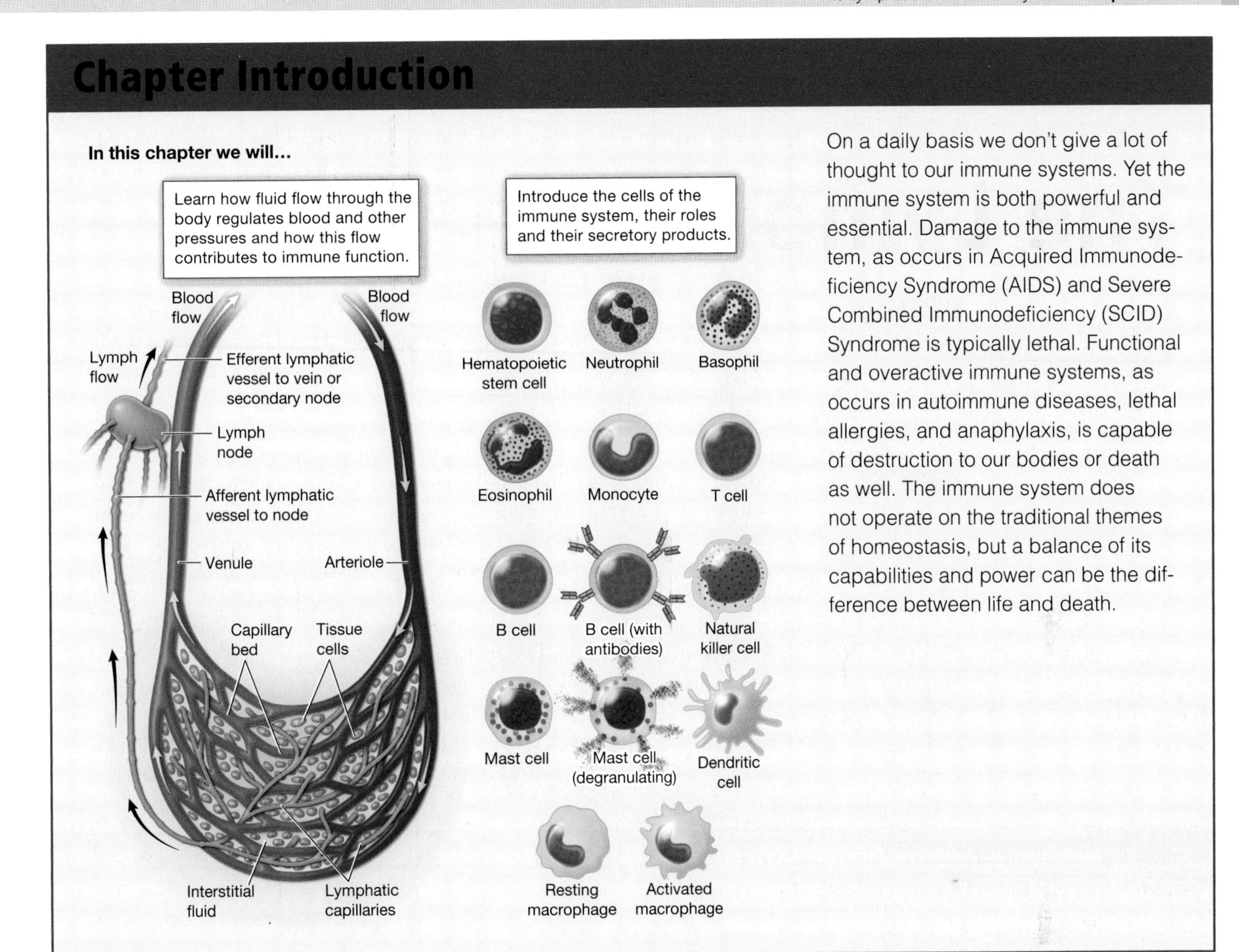

On a daily basis we don't give a lot of thought to our immune systems. Yet the immune system is both powerful and essential. Damage to the immune system, as occurs in Acquired Immunodeficiency Syndrome (AIDS) and Severe Combined Immunodeficiency (SCID) Syndrome is typically lethal. Functional and overactive immune systems, as occurs in autoimmune diseases, lethal allergies, and anaphylaxis, is capable of destruction to our bodies or death as well. The immune system does not operate on the traditional themes of homeostasis, but a balance of its capabilities and power can be the difference between life and death.

21.1 Anatomy of the Lymphatic and Immune Systems

Learning Objectives: By the end of this section, you will be able to:

21.1.1 Describe the major functions of the lymphatic system.

21.1.2 Compare and contrast whole blood, plasma, interstitial fluid, and lymph.

21.1.3 Compare and contrast lymphatic vessels and blood vessels in terms of structure and function.

21.1.4 Describe the path of lymph circulation.

21.1.5 Describe the structure, functions, and major locations of the following lymphatic organs: lymph nodes, thymus, and spleen.

21.1.6 Describe the structure, function, and major locations of lymphatic nodules (e.g., mucosa-associated lymphoid tissue [MALT], tonsils).

The Human Anatomy and Physiology Society includes more than 1,700 educators who work together to promote excellence in the teaching of this subject area. The HAPS A&P Learning Outcomes measure student mastery of the content typically covered in a two-semester Human A&P curriculum at the undergraduate level. The full Learning Outcomes are available at https://www.hapsweb.org.

The **immune system** is the complex collection of cells and organs that destroys invaders that could cause disease or death. The immune system depends on the lymphatic system for its function and so we typically discuss these two systems together. The **lymphatic system** is the system of vessels, cells, and organs that carries excess fluids to the bloodstream and filters pathogens from the blood. Its primary function is to regulate fluid levels throughout the body. It acts as a partner to the blood vessels, collecting interstitial fluid from the tissues and returning it to the blood.

21.1a Functions of the Lymphatic System

LO 21.1.1

A major function of the lymphatic system is to drain body fluids and return them to the bloodstream. Blood pressure causes leakage of fluid from the capillaries, resulting in the accumulation of fluid in the interstitial space—that is, spaces between individual cells in the tissues. Each day an amount of plasma equal to approximately three liters (the body's entire plasma volume) leaks out of the capillaries and into the tissues—not just once, but *five to seven times*, meaning that if this volume of fluid were not collected and returned to the blood, the body's blood volume would be depleted in only a few hours on an average day. Plasma leaks out of the capillaries into the interstitial spaces surrounding the cells, where it becomes interstitial fluid. Interstitial fluid is the fluid that bathes the cells in the tissues. It creates an important component of the homeostasis for each cell. Nutrients and dissolved gasses arrive at each cell through the interstitial fluid, and wastes leave the cells by dissolving in the interstitial fluid. This fluid is essential to the health of the cells both in its composition as well as its volume. The lymphatic system drains excess interstitial fluid, preventing levels from getting too high. Fluid drained from the tissues is channeled back into the bloodstream via a series of vessels, trunks, and ducts. **Lymph** is the term used to describe interstitial fluid once it has entered the lymphatic system (**Figure 21.1**). The lymph is carried in a series of

LO 21.1.2

Figure 21.1 Fluid Recycling in the Body

The body is more than 50 percent water; this water is held in various compartments. The fluid of the blood is known as *plasma*. When blood plasma leaves the blood vessels due to various forces, becomes interstitial fluid. When interstitial fluid volume or pressure becomes high enough, some of this fluid is collected into the lymphatic vessels, where it is known as *lymph*.

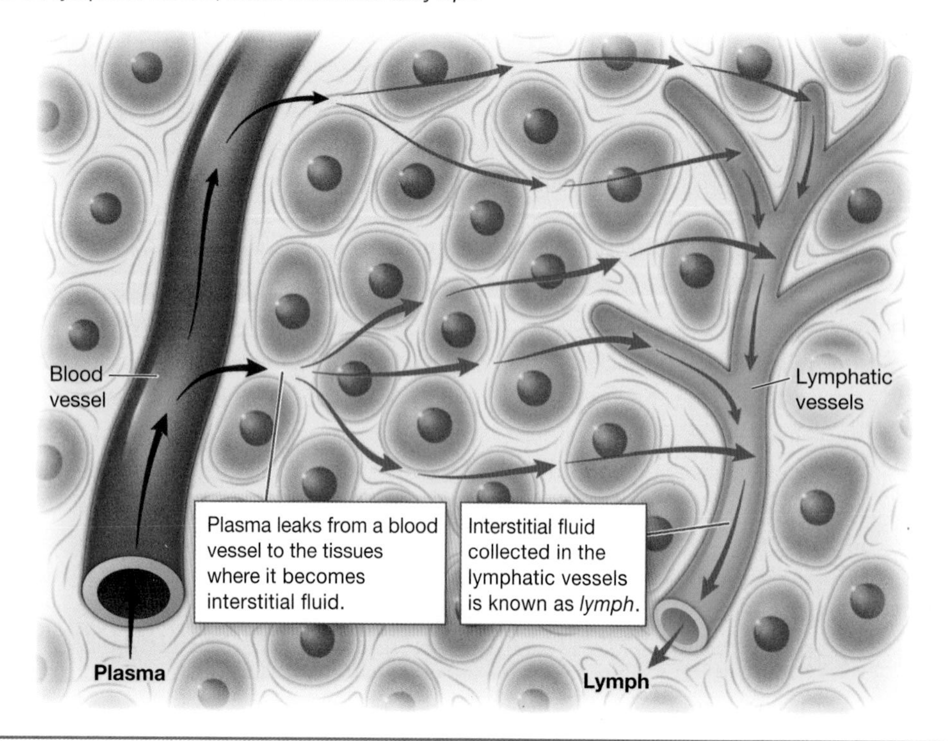
Blood vessel
Lymphatic vessels
Plasma leaks from a blood vessel to the tissues where it becomes interstitial fluid.
Interstitial fluid collected in the lymphatic vessels is known as *lymph*.
Plasma
Lymph

vessels that merge together, similar to veins. Close to the heart, the lymph vessels meet the venous system, returning lymph to the blood. When the lymphatic system is damaged in some way, such as a blockage or injury, fluid backs up in the lymph vessels as well as in the tissue spaces. This accumulation of fluid is referred to as **edema**; the area may appear swollen and, if left unresolved, may lead to serious medical consequences (**Figure 21.2**).

Student Study Tip

Think of the lymphatic system as recycling services for the blood. Vessels drop off their recycling (interstitial fluid) and the lymphatic system collects it, brings it back to the blood where it is needed, and if this does not happen, it is bad for the "environment."

As we discuss throughout this text, the body, with its 40 trillion cells, is a vast landscape. The immune system has just a small network of cells that survey this entire body. The lymph fluid provides an extraordinarily efficient mechanism for monitoring the tissues of the body. Fluid is collected from all the tissues and then travels through lymph nodes, where immune system cells can check the drained fluid for evidence of infection or tissue damage. The network of lymphatic vessels also acts as avenues for transporting the cells of the immune system. In this way, the lymphatic system's homeostasis-maintaining function in fluid regulation of the body dovetails with the functions of the immune system, and so we often consider these two systems together in anatomy and physiology.

LO 21.1.3

Cells of the immune system not only use lymphatic vessels to make their way from interstitial spaces back into the circulation, but they also use lymph nodes as major staging areas for the development of critical immune responses. A *lymph node* is one of the several small, bean-shaped organs located throughout the lymphatic system. Lymph passes through lymph nodes as it travels to rejoin the blood. When there is evidence of infection in the lymph, the leukocytes in the lymph nodes become activated and begin to proliferate. This can cause the node to swell and become tender. If you have ever noticed swollen lymph nodes in your neck or armpit during an infection, you now understand why.

Learning Connection

Micro to Macro

You have just learned that edema can occur when lymphatic drainage is blocked in some way. Look up the terms *mastectomy* and a disease called *elephantiasis*. How do these cause edema? What is the relationship between each and the lymphatic system?

Figure 21.2 Edema

Under most circumstances, the amount of interstitial fluid is kept relatively constant due to the balance between capillary flow and lymph collection. When fluid flow out of the capillary exceeds the rate of lymph collection, swelling, known as *edema*, can result.

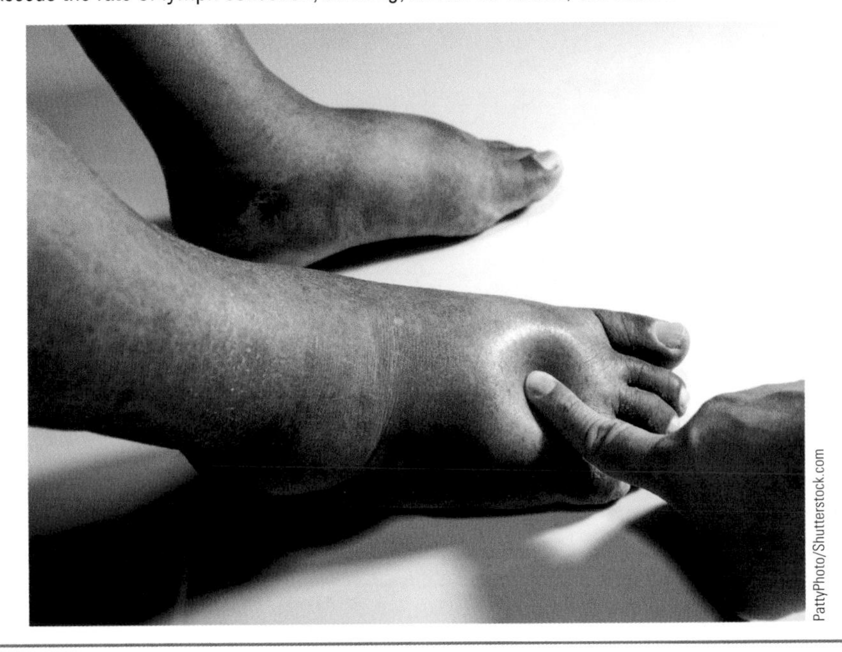

21.1b Structure of the Lymphatic System

The smallest lymphatic vessels are *lymphatic capillaries*, which feed into larger and larger lymphatic vessels, and eventually empty into the bloodstream via a series of ducts. Along the way, the lymph travels through the lymph nodes, which are commonly found near the groin, armpits, neck, chest, and abdomen. Humans have about 500–600 lymph nodes throughout the body (**Figure 21.3**).

A major distinction between the lymphatic and cardiovascular systems in humans is that lymph is not actively pumped by the heart, but the pressure that drives lymphatic flow occurs due to the movements of the body, the contraction of skeletal muscles during body movements, and breathing. Lymphatic vessels are studded with one-way valves that prevent backflow while the lymph moves toward the heart. Lymph flows from the lymphatic capillaries, through lymphatic vessels, and then is dumped

Figure 21.3 Lymph Nodes

The lymphatic vessels bring their lymph through organs called *lymph nodes* before returning the fluid to the circulation. There are hundreds of lymph nodes in the body, but they are clustered in the neck, armpits, abdomen, and groin.

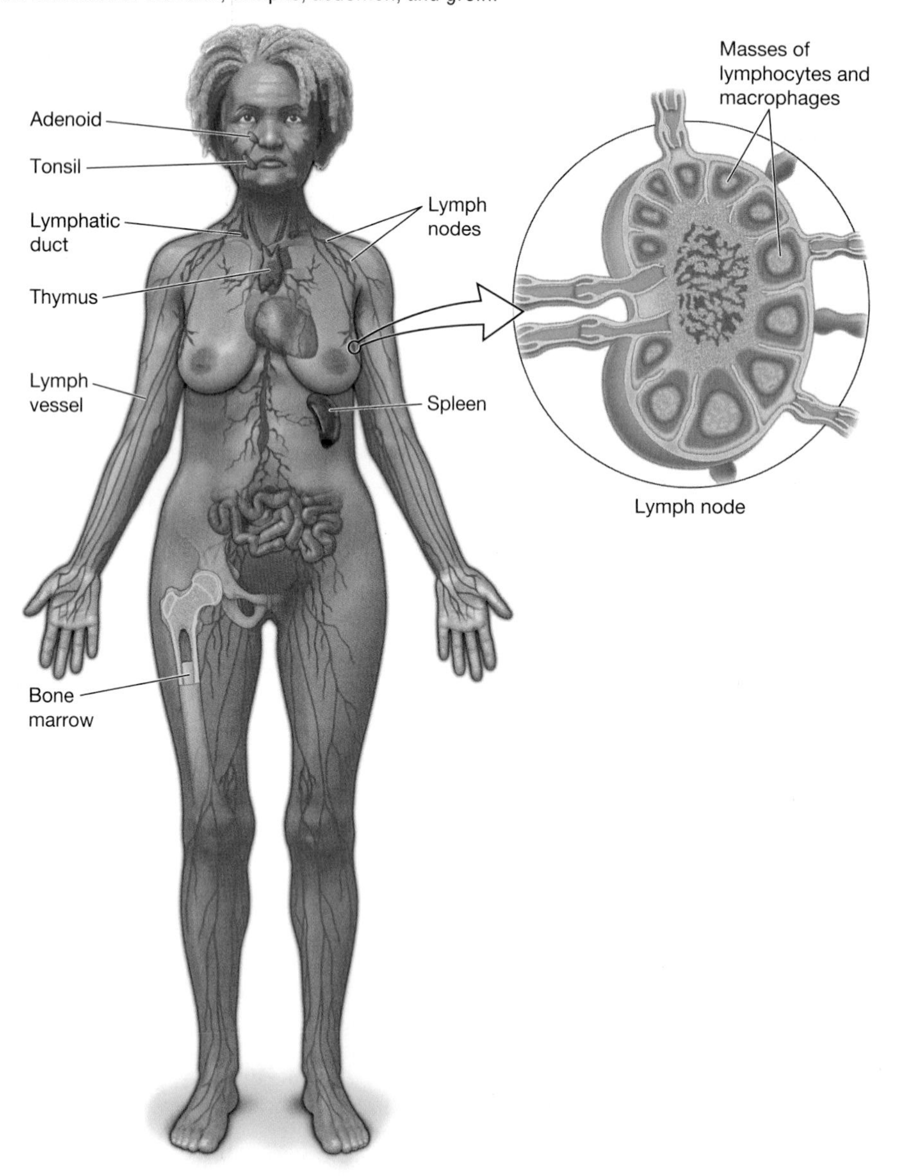

into the circulatory system via the lymphatic ducts located at the junction of the jugular and subclavian veins in the neck.

Lymphatic Capillaries **Lymphatic capillaries** are vessels where interstitial fluid enters the lymphatic system. Once it is inside a lymphatic capillary, we now refer to the fluid as *lymph fluid*. Located in almost every tissue in the body, these vessels are interlaced among the arterioles and venules of the circulatory system in the soft connective tissues of the body (**Figure 21.4A**). The only parts of the body that do not contain lymph

Figure 21.4 Blood and Lymphatic Vessels

(A) The smallest blood vessels—capillaries—are intertwined with lymphatic capillaries. (B) The structure of lymphatic capillaries includes overlapping endothelial cells. When interstitial fluid pressure rises, the flaps open, allowing interstitial fluid to flow into the lymphatic vessel, where it is now called *lymph*. (C) Fluid flows according to pressure gradients. When interstitial fluid pressure is greater than the pressure within the lymphatic vessels, fluid flows into the lymphatic vessel. But when pressure inside the lymphatic vessel exceeds pressure outside, the lymph flows toward the vessel walls and presses on the endothelial flaps, forcing them closed and preventing fluid egress.

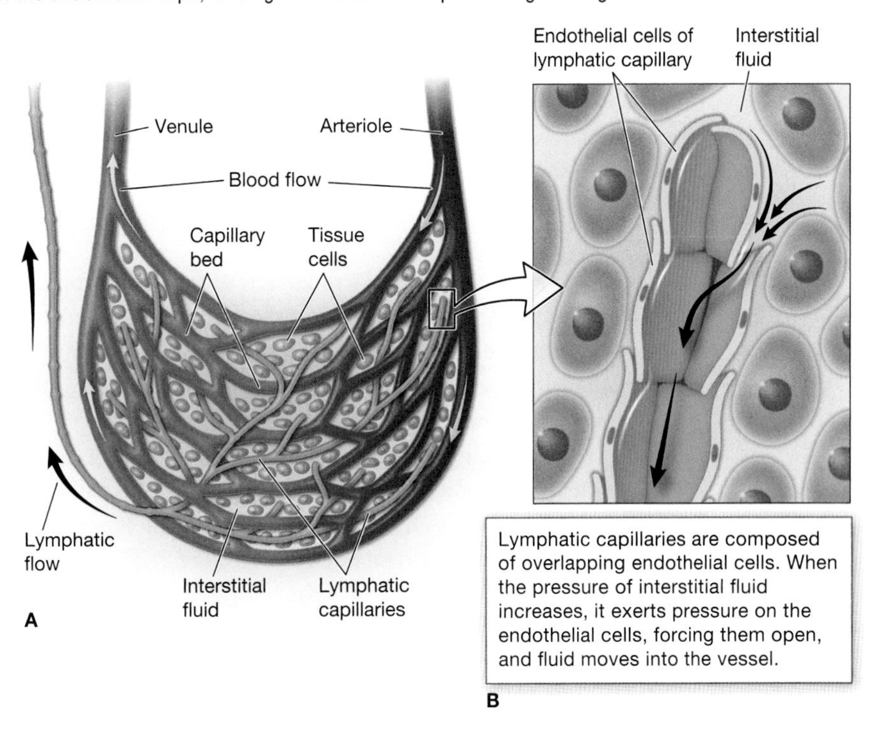

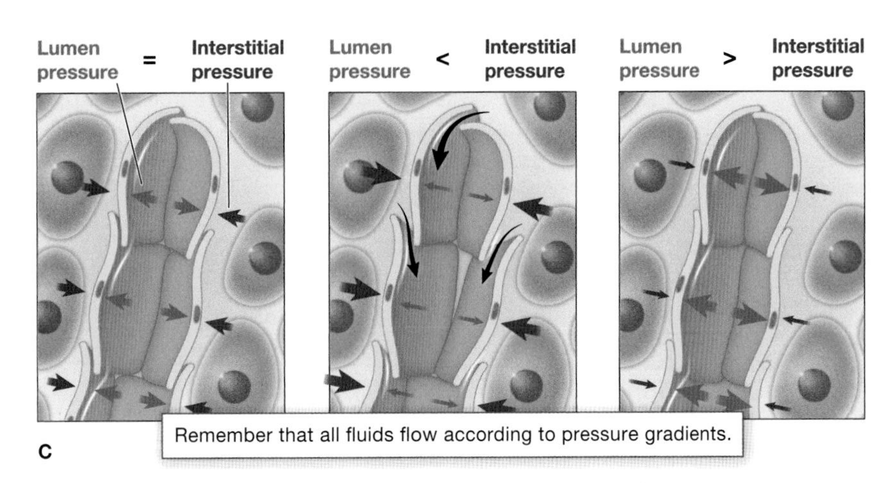

Student Study Tip

The lymphatic capillaries are similar to one-way cat flaps on doors, only in reverse; whereas a cat flap allows the cat to exit the house, but requires the door to be opened to let the cat back in, lymphatic capillaries allow fluid to enter, but their structure prevents fluid from leaving.

vessels are the central nervous system (CNS), bone marrow, bones, teeth, and the cornea of the eye.

The walls of lymphatic capillaries are only one cell thick, similar to blood vessel capillaries. However, their structure has a very significant difference from their blood-carrying counterparts. The lymphatic endothelial cells overlap with each other slightly on their ends (see **Figure 21.4B**). As fluid pressure leaks from the blood capillaries and fluid pressure builds up on the interstitial side of the vessel wall, that pressure pushes the endothelial overlap open, like a door. Fluid can now flow inside. Should the fluid pressure build up on the lumen side of the lymphatic vessel, that pressure pushes back on the endothelial cell, closing it, and preventing the influx of more fluid or the leakage of any of the lymph into the interstitial fluid (**Figure 21.4C**).

The structure of lymphatic capillaries elegantly permits flow into the lymphatic system when a tissue needs draining and prevents backflow into the tissue of the lymph. These capillary walls also serve another handy purpose. In the small intestine, lymphatic capillaries called **lacteals** are critical for the transport of dietary lipids and lipid-soluble vitamins to the bloodstream. These lipids are too large to make their way easily into the blood vessel capillaries, but the overlapping endothelial cell flaps of the lymphatic vessels easily accommodate their entry into the lacteal. These dietary lipids travel through the lymphatic system and then rejoin the blood near the heart. The consequence to this modality of lipid transport is that these lipids are not filtered by the liver in the same way that the bloodborne nutrients are (we will return to this topic in Chapter 23).

LO 21.1.4

Larger Lymphatic Vessels, Trunks, and Ducts The lymphatic capillaries empty into larger lymphatic vessels, which are similar to veins in terms of their three-tunic structure and the presence of valves. These one-way valves are located fairly close to one another, and each one causes a bulge in the otherwise flaccid vessel wall, giving the larger vessels a beaded appearance (see Figure 21.3).

The smaller lymphatic vessels eventually merge to form larger lymphatic vessels known as **lymphatic trunks**. On the right side of the body, the right sides of the head and thorax, as well as the right upper limb, drain lymph fluid into the right subclavian vein via the right lymphatic duct (**Figure 21.5**). On the left side of the body, the remaining lymphatic vessels of the body drain into the larger thoracic duct, which drains into the left subclavian vein. The thoracic duct itself begins just beneath the diaphragm in the **cisterna chyli**, a saclike chamber that receives lymph from the lower abdomen, pelvis, and lower limbs by way of the left and right lumbar trunks and the intestinal trunk.

cisterna chyli (receptaculum chyli)

The overall drainage system of the body is asymmetrical (see Figure 21.5). The **right lymphatic duct** receives lymph from only the upper right side of the body. The lymph from the rest of the body enters the bloodstream through the **thoracic duct** via all the remaining lymphatic trunks. In general, lymphatic vessels of the subcutaneous tissues of the skin—that is, the superficial lymphatics—follow the same routes as veins, while the deep lymphatic vessels of the viscera generally follow the paths of arteries.

LO 21.1.5

Lymph Nodes As the lymph passes through the vessel system on route to the ducts, it will inevitably pass through lymph nodes or nodules. **Lymph nodes** function to remove debris and pathogens from the lymph, and are thus sometimes referred to as the "filters of the lymph" (**Figure 21.6**). Any bacteria that infect the tissues will be found in the interstitial fluid of tissues and will be taken up by the lymphatic capillaries along with the fluid and transported to a regional lymph node. Cells

Figure 21.5 Lymphatic Drainage

Lymphatic capillaries merge together to form larger and larger vessels until they meet with, and drain their contents into, veins close to the heart. The two sides of the body do not drain symmetrically.

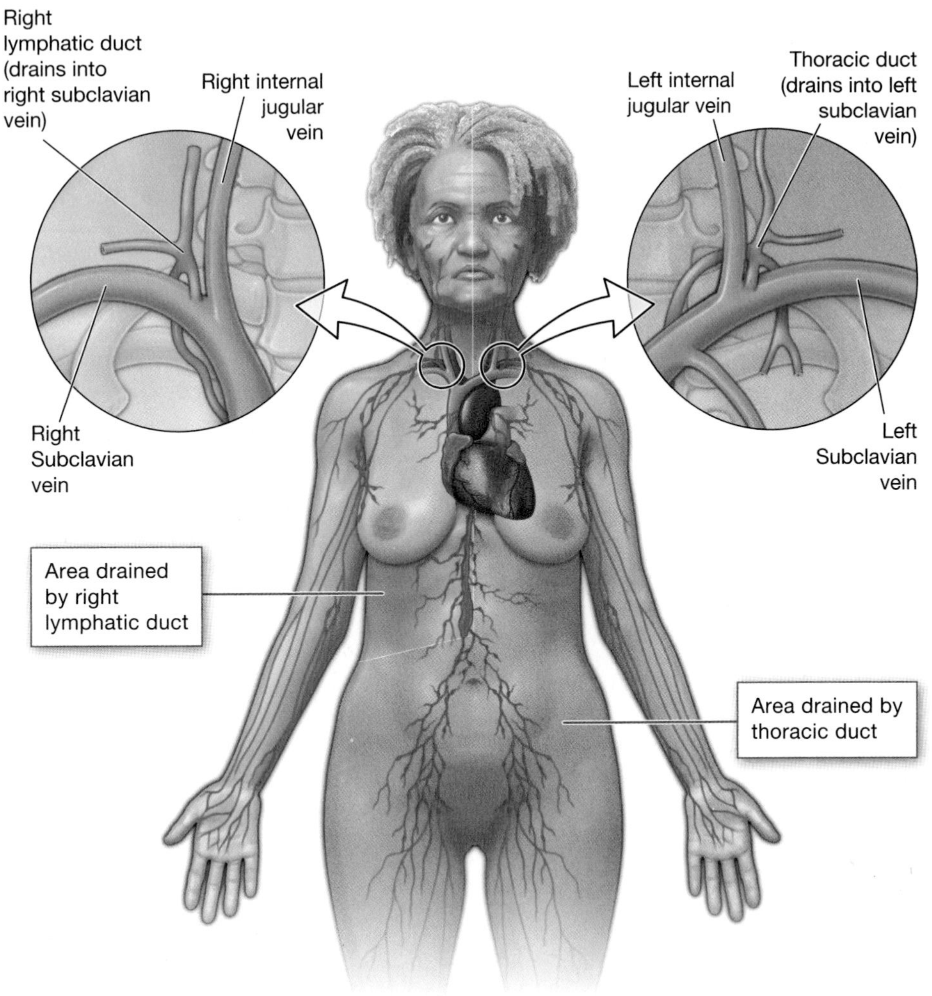

within this organ can engulf and kill many of the pathogens that pass through, thereby removing them from the body. Other nodal cells may become activated by signaling molecules or other evidence of infection in the lymph. Structurally, lymph nodes are round or bean-shaped lymphatic organs and are surrounded by a tough capsule of connective tissue and are separated into compartments by trabeculae, the extensions of the capsule. In addition to the structure provided by the capsule and trabeculae, the structural support of the lymph node is provided by a series of reticular fibers.

The major routes into the lymph node are via **afferent lymphatic vessels** (see Figure 21.6). Cells and lymph fluid that leave the lymph node may do so by another set of vessels known as the **efferent lymphatic vessels**. Lymph enters the lymph node and first flows through the filtration zone, anatomically known as the **subcapsular sinus**, which is occupied by cells that are capable of engulfing and destroying bacteria and debris. Lymph then flows through the dividing zone, or **cortex**, which consists of follicles (gatherings of rapidly dividing lymphocytes). If

Figure 21.6 Lymph Nodes

Lymph nodes are dotted along the path of lymphatic vessels returning lymph to the blood. Within these nodes are areas of lymphocytes that monitor the lymph for signs of infection.

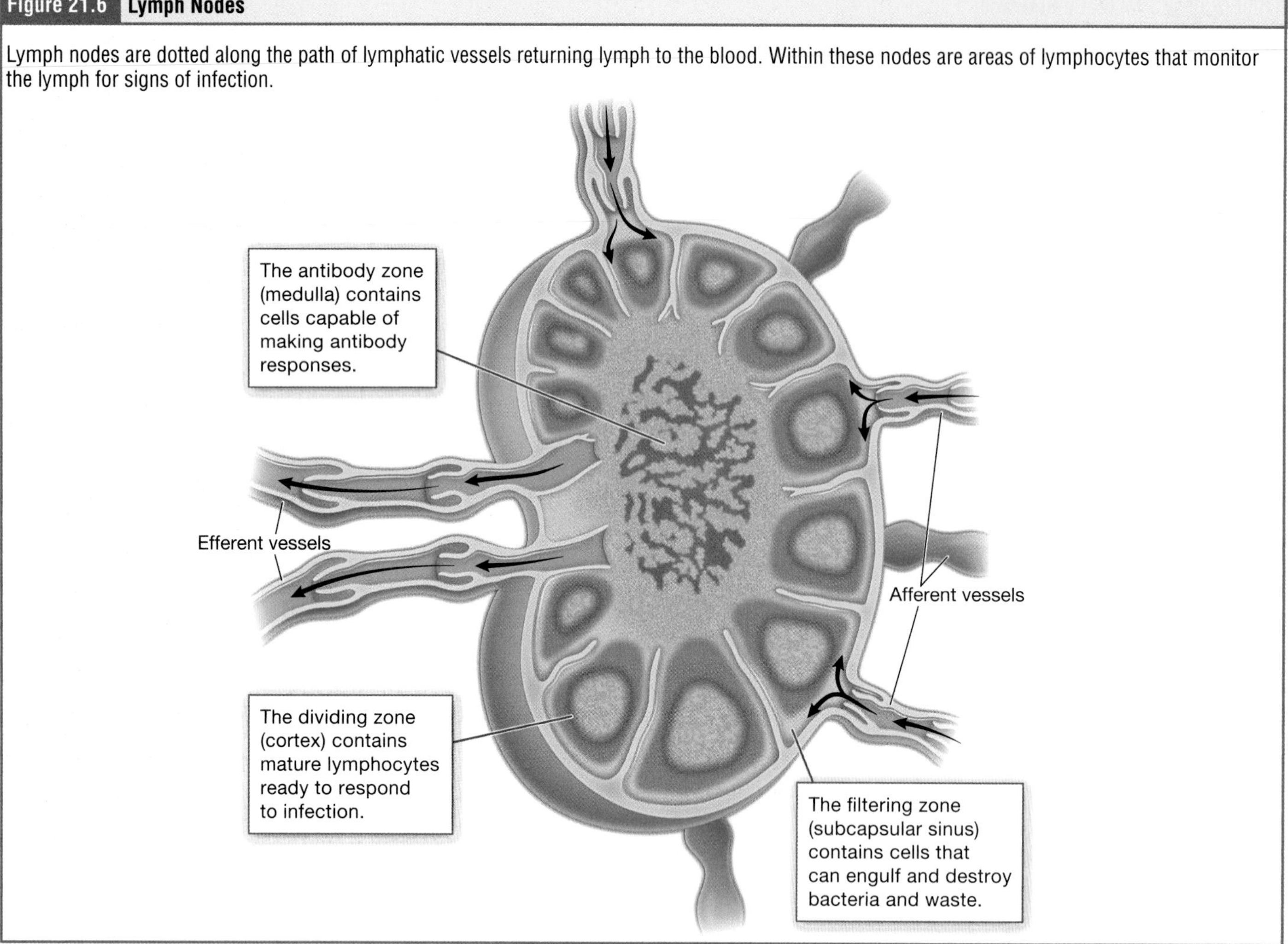

The antibody zone (medulla) contains cells capable of making antibody responses.
Efferent vessels
Afferent vessels
The dividing zone (cortex) contains mature lymphocytes ready to respond to infection.
The filtering zone (subcapsular sinus) contains cells that can engulf and destroy bacteria and waste.

these cells are activated by materials within the lymph, they replicate faster. As the lymph continues to flow through the node, it enters the medulla, which consists of cells that specialize in antibody responses, before leaving the node via the efferent lymphatic vessels.

LO 21.1.6

Lymphoid Nodules The other lymphoid tissues, the **lymphoid nodules**, have a simpler architecture than the lymph nodes. They do not have a capsule, but instead are a dense cluster of lymphocytes lodged within the walls of other tissues. Lymphoid nodules are abundant in the respiratory and digestive tracts, areas routinely exposed to environmental pathogens.

Tonsils are specific lymphoid nodules located along the inner surface of the throat and are important in developing immunity to oral pathogens (**Figure 21.7**). These tonsils are sometimes referred to as *adenoids* when swollen. Such swelling is an indication of an active immune response to infection. Histologically, tonsils do not contain a complete capsule, and the epithelial layer invaginates deeply into the interior of the tonsil to form **tonsillar crypts**. These crevices accumulate all sorts of materials taken into the body through eating and breathing. Bits of food, mucus, and microbes from our food may collect here and interact with the numerous lymphoid follicles within the tonsils. This interaction may be important both for serving as vigilant sentinels for potential invaders and for learning the difference between a friend and a

Figure 21.7 Tonsils

(A) Tonsils are specialized lymphatic tissue in the pharynx. (B) The surface of tonsils is deeply cracked or invaginated to capture food particles and expose the lymphocytes to them. These tissues are thought to be particularly important for the building and maintenance of immune tolerance.

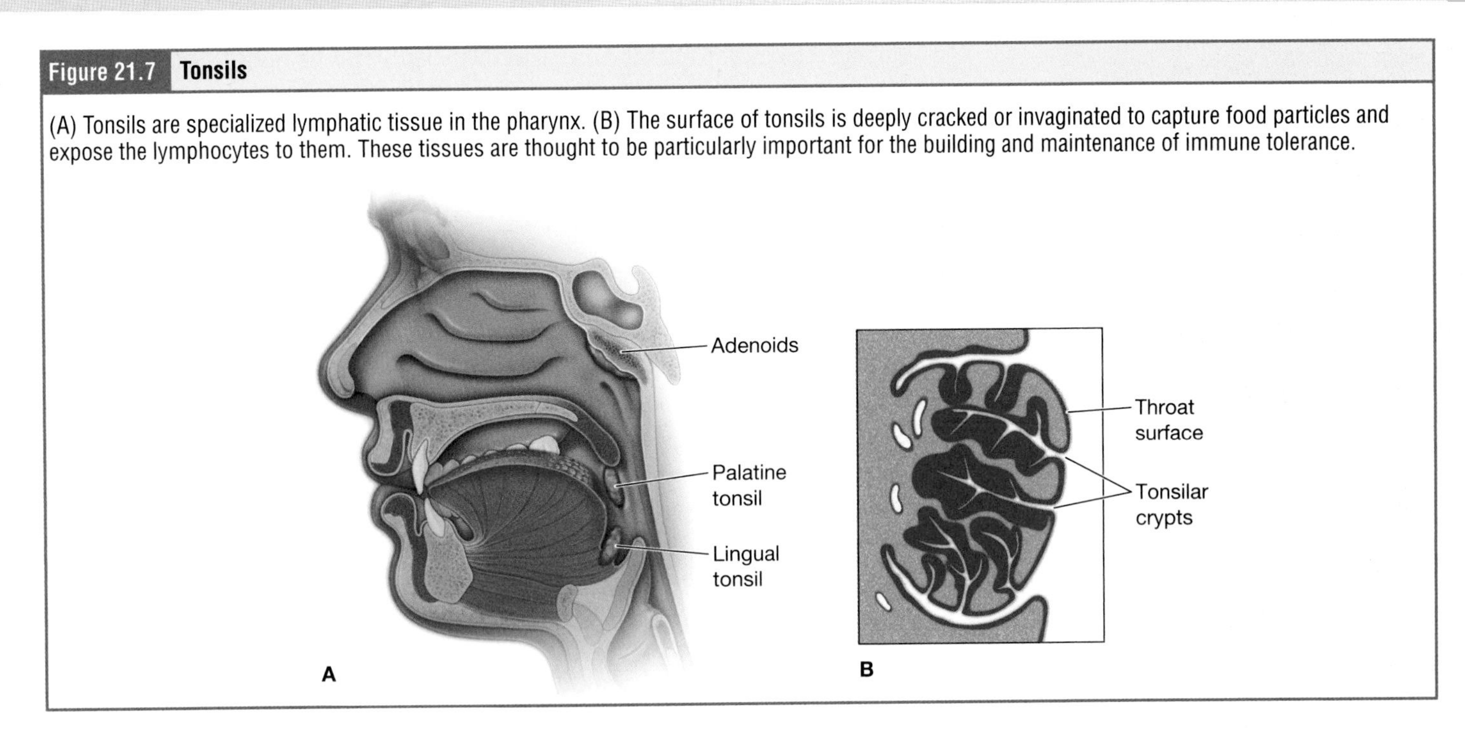

foe. The lymphoid tissue of tonsils may help children's bodies recognize, destroy, and develop immunity to common environmental pathogens so that they will be protected in their later lives. Tonsils are often removed in those children who have recurring throat infections.

The term *mucosa* refers to a specific layer of tissue found within many organs. **Mucosa-associated lymphoid tissue (MALT)** is a widely distributed group of lymphoid nodules found within this layer in the gastrointestinal tract, breast tissue, lungs, and eyes (**Figure 21.8**). Like tonsils, MALT nodules contain specialized cells that sample material from the intestinal lumen and transport it to nearby follicles so that adaptive immune responses to potential pathogens can be mounted. MALT is important both in mounting immune responses to gastrointestinal pathogens (they are responsible for the diarrhea response to pathogenic bacteria, for example) and for

Figure 21.8 Mucosa-Associated Lymphoid Tissue (MALT)

MALT nodules are found within the walls of mucosal tissues, including the intestines, shown here.

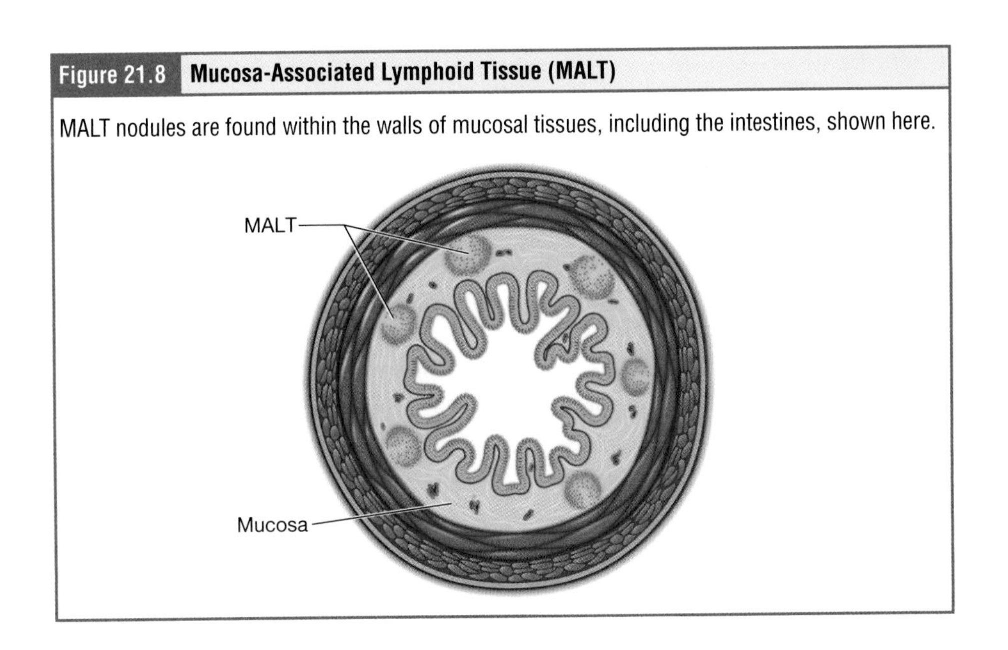

developing tolerance to our foods, a subject we will discuss in more depth later in this chapter.

Bronchus-associated lymphoid tissue (BALT) consists of lymphoid follicular structures found along the walls of the bronchi in the lungs. These tissues, in addition to the tonsils, are effective against inhaled pathogens.

Lymph Organs The **thymus** is a lymphoid organ which is larger and has a more complex structure than a lymph node. The thymus is found in the space between the sternum and the aorta of the heart (Figure 21.9). Connective tissue forms a capsule and leaves of the capsule divide the inner thymus tissue into lobules via extensions. The outer region of the organ is known as the *cortex* and the inner region is the *medulla.* The thymus is large in the fetal and infant stages of life and becomes steadily smaller throughout the lifespan, to the extent that it is difficult to isolate in a cadaver of advanced age. The primary function of the thymus is the maturation of one type of lymphocyte that happens during infancy and childhood. We will discuss this function in more detail later in this chapter.

In addition to the thymus, the **spleen** is a major lymphoid organ. It is found in the posterior abdominal cavity, posterior to the stomach (Figure 21.10). The spleen is wrapped in a thin capsule but is somewhat fragile due to its extensive vascularization. The spleen is unique among the lymph organs because it filters blood rather than lymph. It clears the blood of both microbes and dying red blood cells. The spleen also functions as the location of immune responses to bloodborne pathogens.

The spleen is divided into nodules by strips of connective tissue, and within each nodule is an area reddish in color and one that stains purplish in color when using traditional staining methods. The red area is referred to as **red pulp** and consists mostly of red blood cells. The purplish area is called **white pulp** and is filled with leukocytes. Upon entering the spleen, blood from the capillaries passes through both "pulp" areas. The red pulp monitors for the age and quality of red blood cells, taking old or damaged cells out of circulation. The white pulp monitors the blood for the presence of bloodborne pathogens.

Figure 21.9 The Thymus

The thymus is the site of T call development. (A) It is large in the fetus, covering the entire superficial half of the heart. (B) It atrophies over time, becoming quite small in the adult.

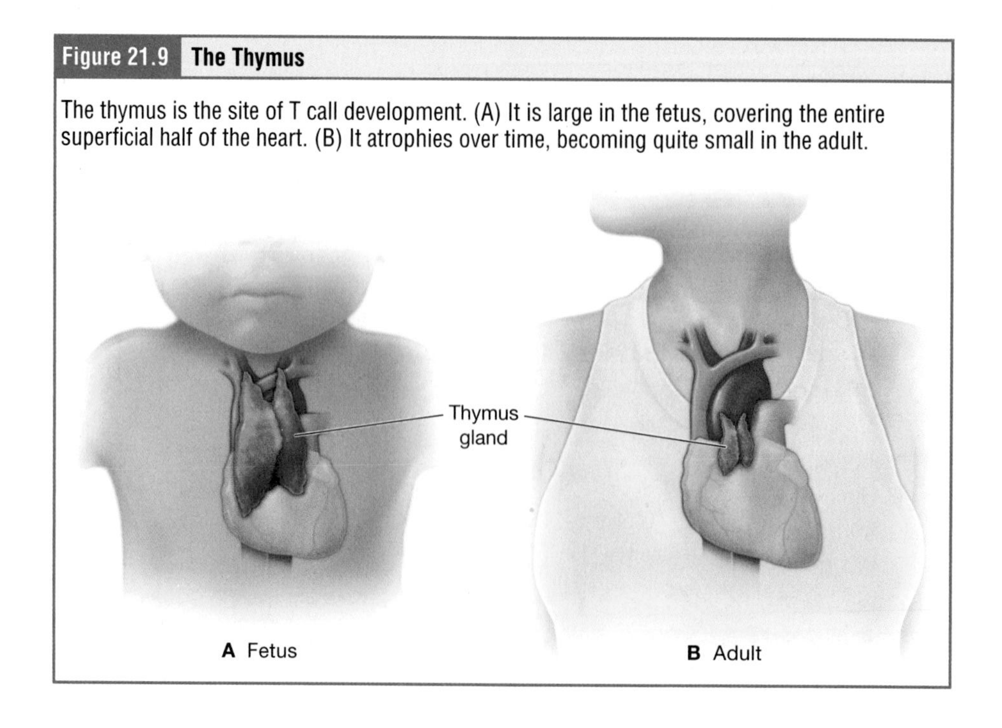

Figure 21.10 The Spleen

The spleen is located in the superior abdomen posterior to the stomach. It functions to filter blood; the white pulp examines the blood for pathogens and the red pulp recycles red blood cells.

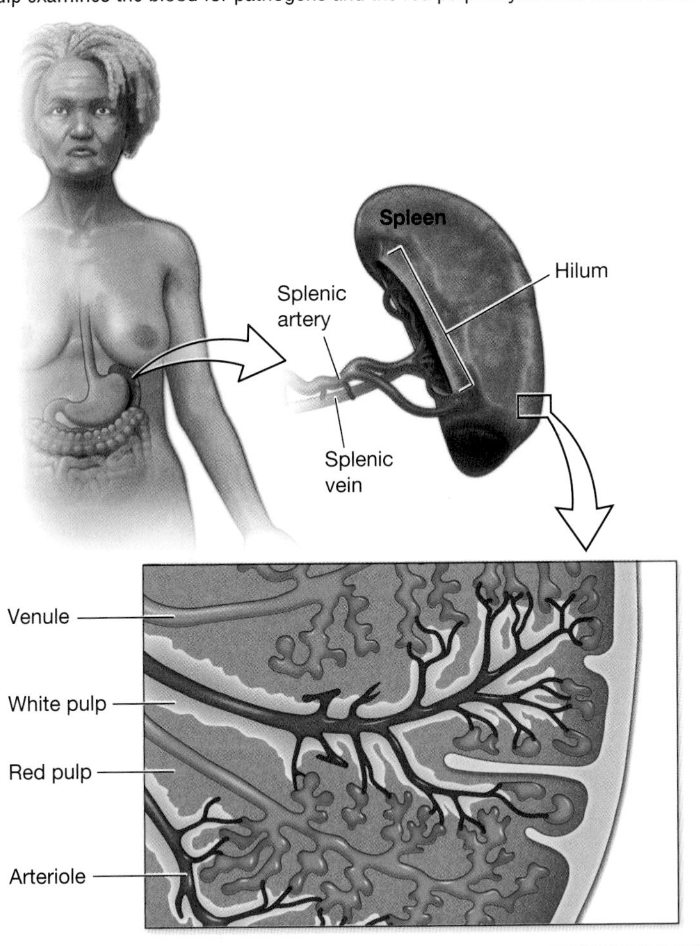

Learning Check

1. Rahim is trail running when they lose their footing and sprain their ankle. Almost immediately, there is an accumulation of fluid around Rahim's ankle. What is this called?
   a. Lymph
   b. Edema
   c. Plasma
   d. Lymph node
2. Which of these is a difference between lymphatic capillaries and blood capillaries?
   a. Fluid enters in blood capillaries while fluid exits in lymphatic capillaries.
   b. Cells in lymph capillaries overlap, while cells in blood capillaries are tightly bound.
   c. Lymph capillary walls consist of one layer of cells, while blood capillary walls consist of several layers.
   d. Blood capillaries are larger in diameter than lymph capillaries.
3. Which of the following structures does the right lymphatic duct drain into?
   a. Right subclavian vein
   b. Thoracic duct
   c. Cisterna chyli
   d. Lymphatic trunks
4. Streptococcus, or "strep throat," is a common oral infection that usually occurs from sharing drinks or breathing in droplets from someone else who has the infection. Which of the following lymph nodules are responsible from protecting you against this oral infection?
   a. Tonsils
   b. Mucosa-associated lymphoid tissue
   c. Bronchus-associated lymphoid tissue
   d. Spleen

21.2 Overview of the Immune Response

Learning Objectives: By the end of this section, you will be able to:

21.2.1* Describe the three levels of defense in the immune system.

21.2.2 Compare and contrast innate (nonspecific) with adaptive (specific) defenses.

21.2.3 Describe the roles of various types of leukocytes in innate (nonspecific) and adaptive immune responses.

21.2.4 Explain ways in which the innate (nonspecific) and adaptive (specific) immune responses cooperate to enhance the overall resistance to disease.

* Objective is not a HAPS Learning Goal.

When you consider the immune system, you can make a number of metaphors for any sort of security system. For example, consider a bank that is concerned about theft. The bank would want to have as many barriers as possible to prevent a thief from entering the parts of the bank where the money is kept, as a start. Should the barrier be breached, the bank would want to have an emergency response system, for example a panic button that immediately dialed 911. If that emergency response system were to be triggered, help would be on the way immediately. But the bank might also have something savvier, something to prevent future vulnerabilities if an initial attack occurred. The bank might, for example, have cameras to learn about the thieves and how they were able to breach the system. The bank might request fingerprinting to learn more about the thieves and identify them. Based on what is learned from an initial attack, the bank might make some changes to make it more formidable in the future.

LO 21.2.1

The immune system works much the same way. We can consider the features of the immune system to be able to be grouped three ways:

- Barriers such as the skin and mucous membranes, which act instantaneously to prevent pathogenic invasion into the body tissues.
- The rapid "emergency responders" of the **innate immune response**. A variety of cells and soluble factors that activate quickly but have limited efficacy.
- The slower but more specific and effective **adaptive immune response**, which involves many cell types and soluble factors and is notable for its long-lasting memory capability.

LO 21.2.2

white blood cells (leukocytes; leucocytes)

phagocytic cells (phagocytes)

The cells of the blood, including all those involved in the immune response, arise in the bone marrow via various differentiation pathways from hematopoietic stem cells as discussed in Chapter 18. Red blood cells and platelets are also made in the bone marrow, but here we will discuss **white blood cells** or **leukocytes**. Functionally, white blood cells are divided into three broad classes:

LO 21.2.3

LO 21.2.4

- **Phagocytic cells.** *Phagocytosis* is a form of endocytosis in which the engulfed material is destroyed inside the phagocytic cell. *Phagocytic cells* can ingest pathogens or other cellular debris for destruction. Phagocytic cells are mostly involved in the innate immune response, but many of them are capable of kickstarting the adaptive immune response.
- **Lymphocytes.** A specialized class of leukocytes, which coordinate the activities of adaptive immunity.
- **Granular cells.** Cells that containing granules, which are packets of noxious materials that can be released in an attack against pathogens. Granular cells are important components of the innate immune response.

These broad groupings of immune defenses are summarized in the "Anatomy of Immune Defenses" feature.

Anatomy of...

Immune Defenses

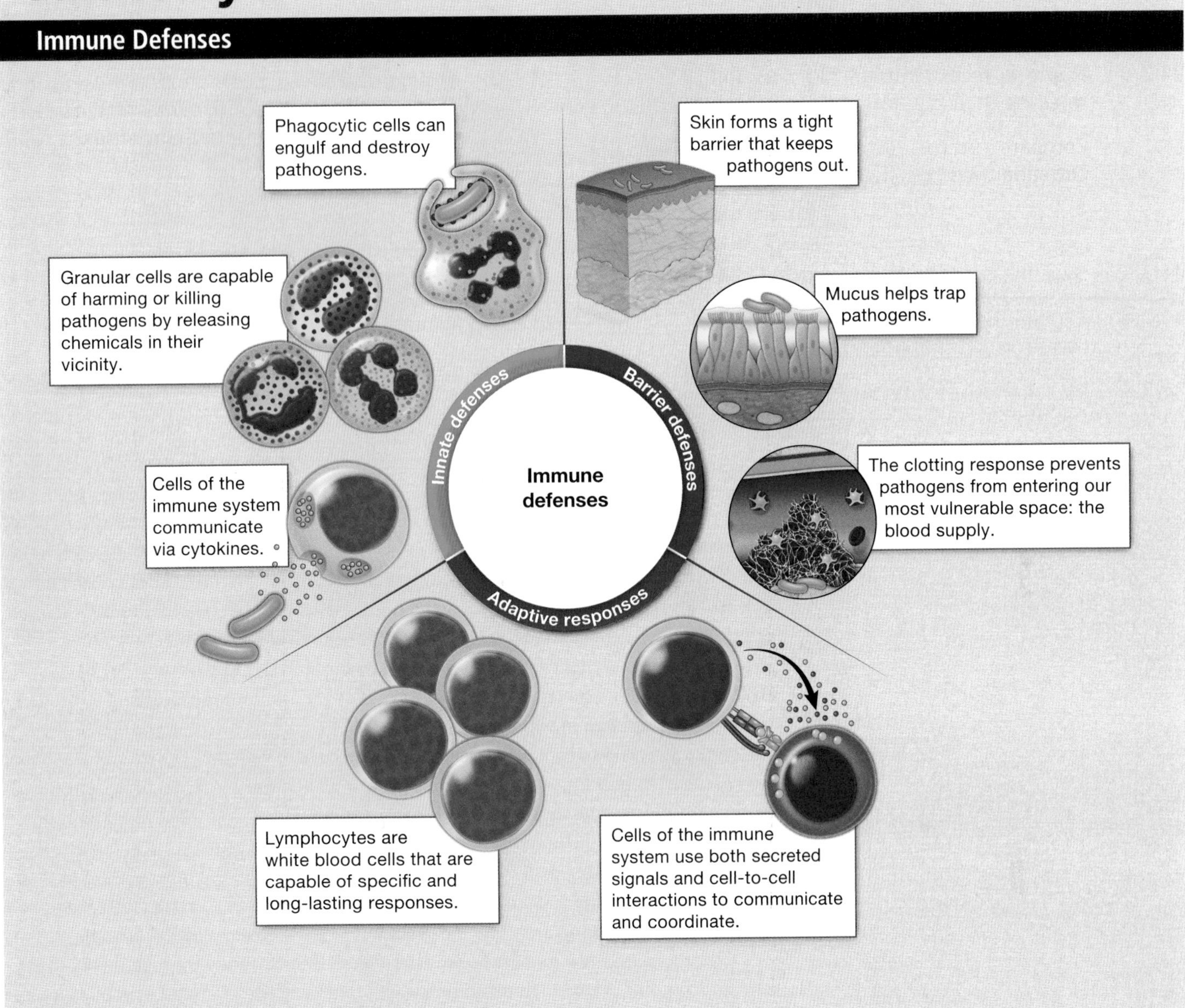

Phagocytic cells can engulf and destroy pathogens.
Skin forms a tight barrier that keeps pathogens out.
Granular cells are capable of harming or killing pathogens by releasing chemicals in their vicinity.
Mucus helps trap pathogens.
Innate defenses
Barrier defenses
Immune defenses
Cells of the immune system communicate via cytokines.
The clotting response prevents pathogens from entering our most vulnerable space: the blood supply.
Adaptive responses
Lymphocytes are white blood cells that are capable of specific and long-lasting responses.
Cells of the immune system use both secreted signals and cell-to-cell interactions to communicate and coordinate.

Learning Check

1. In the presence of an infection, eosinophils release chemicals that are harmful to the infected cell. The chemicals quickly recruit other leukocytes to attack the cell. This is quick but has a limited efficacy. What is this an example of?
   a. Innate immune response
   b. Adaptive immune response
2. Rami had the flu earlier in the year. Unfortunately, the same pathogens have entered their body again. Which of the following immune responses would you expect to occur?
   a. Innate immune response
   b. Adaptive immune response

21.3 Barrier Defenses and the Innate Immune Response

HAPS LEARNING GOALS AND OUTCOMES

Learning Objectives: By the end of this section, you will be able to:

21.3.1 Name surface membrane barriers and describe their physical, chemical, and microbiological mechanisms of defense.

21.3.2 Describe the steps involved in phagocytosis and provide examples of important phagocytic cells in the body.

21.3.3 Describe the functions of natural killer cells.

21.3.4 Explain the role of pattern-recognition receptors in innate defenses.

21.3.5 Explain how complement and interferon function as antimicrobial chemicals.

21.3.6 Define diapedesis, chemotaxis, opsonization, and membrane attack complex, and explain their importance for innate defenses.

21.3.7 Describe the mechanisms that initiate inflammation.

21.3.8 Summarize the cells and chemicals involved in the inflammatory process.

21.3.9 Explain the benefits of inflammation.

21.3.10 Describe the mechanism of fever, including the role of pyrogens.

21.3.11 Explain the benefits of fever.

Any discussion of the innate immune response usually begins with the physical barriers that prevent pathogens from entering the body, flush them out before they can establish infection, or create a sufficiently inhospitable environment that pathogens cannot survive. Barrier defenses are the body's most basic defense mechanisms. The barrier defenses are not a response to infections, but they are continuously working to protect against a broad range of pathogens.

LO 21.3.1

The different modes of barrier defenses are associated with the external surfaces of the body, where pathogens may try to enter (**Table 21.1**). The primary barrier to the entrance of microorganisms into the body is the skin. Not only is the skin covered with a layer of dead, keratinized epithelium that is too dry to enable bacterial growth, but as these cells are continuously sloughed off from the skin, they carry bacteria and other pathogens with them. Additionally, sweat and other skin secretions may lower pH, contain toxic lipids, and physically wash microbes away (see the "Anatomy of the Skin as a Barrier" feature).

Our foods may bring microbes and pathogens into the body through the mouth and GI tract. For example, think about an apple that you carry around in your school bag all day before you take it out with your unwashed hands, maybe rub it on your shirt, and begin to eat it. There are likely to be loads of microbes on its surface. But the apple will not make you sick. For one thing, it is unlikely that any of the microbes are actually harmful to you. But of course, the GI system is a veritable fortress of barriers to invasion. The mouth is lined with a stratified epithelium, as one barrier. But it is also

Table 21.1 Barrier Defenses

Site	Specific Defense	Protective Aspect
Skin	Epidermal surface	Keratinized cells of surface, Langerhans cells
Skin (Sweat/Secretions)	Sweat glands, sebaceous glands	Low pH, washing action
Oral Cavity	Salivary glands	Lysozyme
Stomach	Gastrointestinal tract	Low pH
Mucosal Surfaces	Mucosal epithelium	Nonkeratinized epithelial cells
Normal Flora (Nonpathogenic Bacteria)	Mucosal tissues	Prevent pathogens from growing on mucosal surfaces

Anatomy of... The Skin as a Barrier

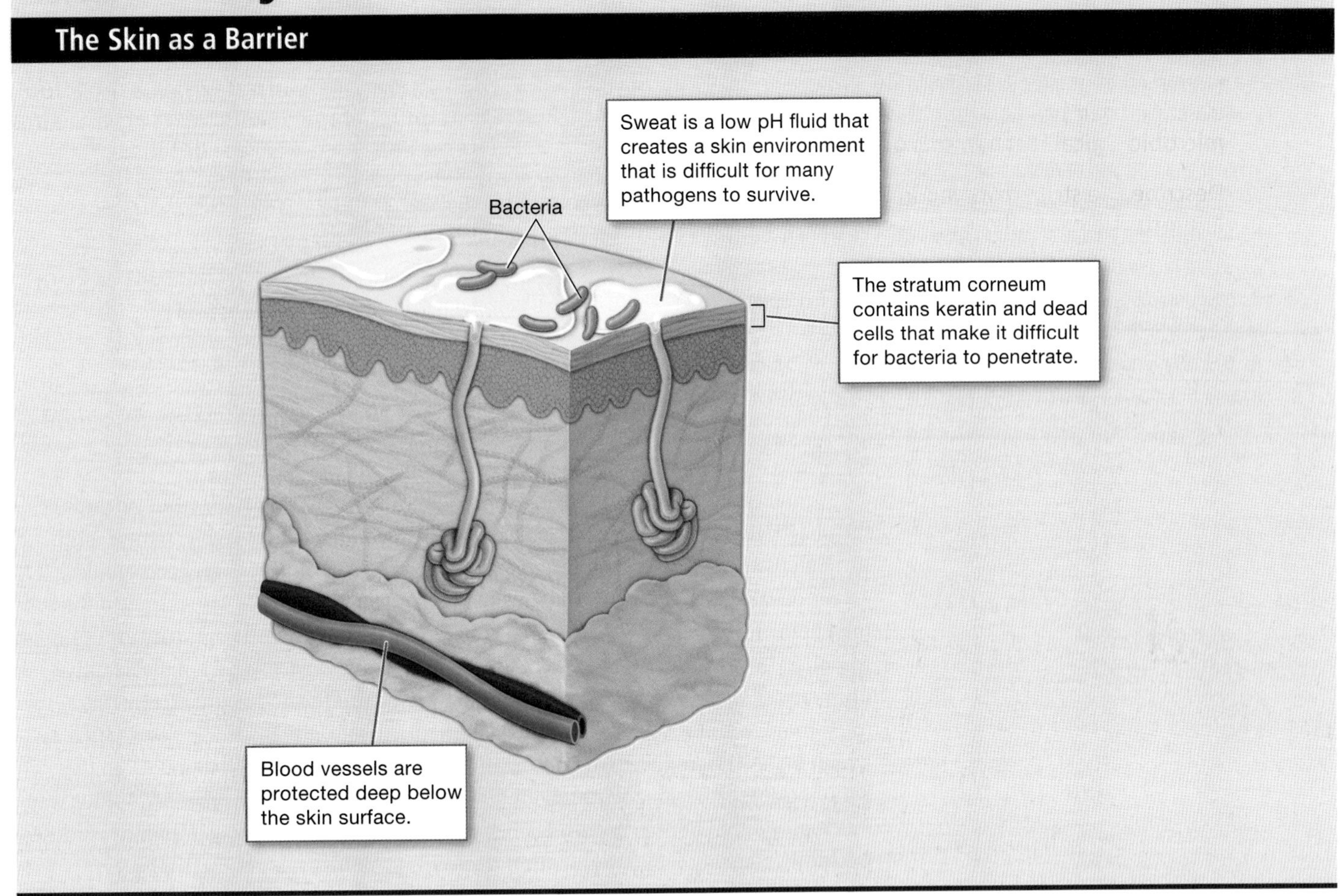

The Skin as a Barrier
Sweat is a low pH fluid that creates a skin environment that is difficult for many pathogens to survive.
Bacteria
The stratum corneum contains keratin and dead cells that make it difficult for bacteria to penetrate.
Blood vessels are protected deep below the skin surface.

coated with saliva, which is rich in **lysozyme**—an enzyme that destroys bacteria by digesting their cell walls. The acidic environment of the stomach, which is fatal to many pathogens, is also a barrier. In many regions of the GI tract, a thick layer of mucus sits atop the epithelium. This mucus, which is also found in the respiratory tract, the reproductive tract, the eyes, the ears, and the nose, traps both microbes and debris and helps with their removal. In the case of the respiratory tract, ciliated epithelial cells move mucus and anything trapped inside it upward to the mouth, where it is then swallowed into the digestive tract, ending up in the harsh acidic environment of the stomach.

Student Study Tip

The respiratory tract has a "mucociliary escalator" in which you can imagine the pathogens traveling upward in an escalator to enter the GI tract via the mouth.

21.3a Cells of the Innate Immune Response

Phagocytosis, the destruction of debris or pathogens through engulfment, is a central tool in the immune system fight. The phagocytes are the body's fast-acting first line of immunological defense against organisms that have breached barrier defenses and have entered the vulnerable tissues of the body. The phagocyte takes the organism inside itself as a phagosome, which subsequently fuses with a lysosome and its digestive enzymes, effectively killing many pathogens.

Student Study Tip

Phagocytes are like Pac-Man; they eat everything in front of them.

Phagocytes: Macrophages and Neutrophils Many of the cells of the immune system have a phagocytic ability, at least at some point during their life cycles. Macrophages, neutrophils, and dendritic cells are the major phagocytes of the immune system.

LO 21.3.2

Apply to Pathophysiology

Smoking and the Mucociliary Escalator

One of the effects of smoking any substance that sheds particulate matter into the lungs (i.e., tobacco or marijuana) is that the ciliated epithelial cells of the lungs can lose their cilia.

1. What is the function of these cilia?
   A) To sweep material from the apical surface of the cells
   B) To enable the epithelial cells to move within the lungs
   C) To interact with the cells of the immune system
2. As cilia number and density declines, which of the following is likely to occur?
   A) Adaptive immunity activation is delayed.
   B) Communication among the cells of the lungs is impaired.
   C) Mucus and foreign matter get stuck in the lungs.
3. Considering your response to Question #2, which of the following is likely TRUE?
   A) More robust antibody responses
   B) More frequent, longer lasting respiratory infections
   C) Damage to the epithelial blood vessels within the lungs

A **macrophage** is an irregularly shaped phagocytic cell that is the most versatile of the phagocytes in the body. Macrophages are highly mobile and are able to move both through tissues and squeeze through capillary walls. Macrophages have long cellular extensions that can reach out like arms to grab and pull bacteria or debris toward the cell membrane for engulfment (**Figure 21.11**). They not only participate in innate immune responses but have also evolved to cooperate with lymphocytes as part of the adaptive immune response. Macrophages exist in many tissues of the body, either freely roaming through connective tissues or fixed to reticular fibers within specific tissues such as lymph nodes. Macrophages often migrate to the lymph nodes to collaborate with the lymphocytes that reside there and jump-start the adaptive immune response. In this way, macrophages can be seen as a bridge between the innate and adaptive immune responses. **Dendritic cells**, though fewer in number, play very similar roles to macrophages.

Neutrophils are a type of white blood cell found in tremendous numbers in the bloodstream. Neutrophils are both granular cells and phagocytic cells. Their

Figure 21.11 Macrophages are Phagocytic Cells That Live in the Tissues

This macrophage is capturing bacteria with its long, slender extensions called *pseudopodia* (false feet). The macrophage will then surround and engulf the bacteria and digest them with its lysosomal enzymes.

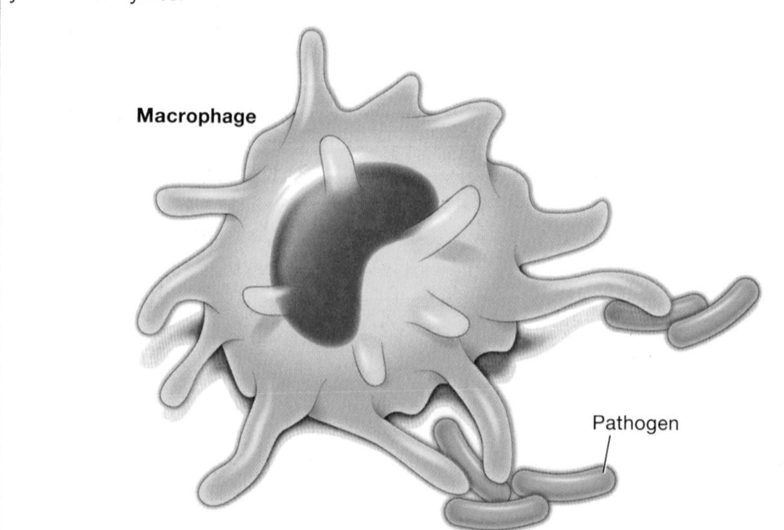

cytoplasmic granules contain a variety of vasoactive mediators such as histamine. Neutrophils are attracted to an area of infection via chemotaxis; in response to chemical signals, neutrophils leave the bloodstream to enter the infected tissues.

Whereas macrophages act like sentries, always on guard against infection, neutrophils can be thought of as military reinforcements that are called into a battle to hasten the destruction of the enemy once it has been detected. Neutrophils phagocytose bacteria primarily, and then they begin a process of autolysis (self-destruction). Both the neutrophil and the engulfed bacteria are killed in the process. This is another way that neutrophils differ from macrophages: whereas macrophages engulf bacteria and present evidence of that enemy to other members of the immune response, neutrophils engulf bacteria and self-destruct. In sites of bacterial infection, so many neutrophils may die so quickly that the cell debris cannot be cleared fast enough. In these instances, pus—a white viscous material comprised of dead neutrophil debris—collects in the area.

Student Study Tip

Remember that neutrophils are granular AND phagocytic because they "love" to be "neutral."

LO 21.3.3

Natural Killer (NK) Cells NK cells are a type of lymphocyte that have the ability to induce apoptosis, or programmed cell death, in cells infected with intracellular pathogens such as viruses. When a cell becomes infected with a pathogen, it changes some of its expression of membrane proteins. NK cells are able to recognize these changes and then induce apoptosis, in which a cascade of events inside the cell causes its own death. One of the things that is fascinating about apoptosis is that it is an organized process. Instead of an explosion and resulting debris, the cell eliminates its internal contents using its lysosomes before death. The result is the elimination of an infected cell without releasing the pathogen or leaving behind much cell debris.

Granulocytes Basophils and eosinophils are granulocytes that circulate in the blood. Their granules contain chemicals that are particularly harmful to parasitic infection. Another form of granulocyte, the **mast cell**, lives only in the tissues. The granules of mast cells primarily contain **histamine**, an important cytokine in the inflammation pathway, and **prostaglandins**, signaling molecules with a wide range of effects, including stimulating nociceptors to send pain signals to the CNS.

mast cell (mastocyte; labrocyte)

Apply to Pathophysiology

Macrophages and Tuberculosis

The bacterial species *Mycobacterium tuberculosis* (Mtb) causes the disease tuberculosis (TB). TB is the world's oldest known infectious disease; scientists have observed evidence of TB in mummies estimated to be about 6,000 years old! TB infection can become chronic and last for years, in part because of the bacteria's ability to evade macrophages.

1. Which of these are functions of macrophages? Please select all that apply.
   A) Kill infected cells
   B) Engulf/phagocytose bacteria
   C) Activate the adaptive immune response
   D) Secrete interferon

One of the unique traits of Mtb is that the bacteria can prevent the macrophage phagosome from merging with a lysosome.

2. What would this mean for the bacteria?
   A) The bacteria would not be killed, and would persist inside the macrophage.
   B) The bacteria would die inside of the phagosome, which has a very low pH.
   C) The bacteria would die inside of the phagosome, which has bacteria-killing enzymes.
3. Taking your answer to Question #2, which of the following components of the immune response may be delayed?
   A) Complement cascade
   B) Pyrogen release
   C) Adaptive immune activation

21.3b Recognition of Pathogens

The two big functional differences between the innate and adaptive responses can be summarized with the words *time* and *specificity*.

- ***Time***: Whereas the innate response can kick in immediately or within hours of infection, the adaptive immune response takes several days to be effective. The adaptive immune response, in fact, typically requires the innate immune response to activate it, so it always is effective later in infections with novel pathogens.
- ***Specificity***: The adaptive immune response is exquisitely specific in its ability to recognize details of pathogens. For example, the adaptive immune response is tailored not just to one pathogen compared to another—such as influenza virus instead of Ebola virus—but can distinguish even between different versions of the same pathogen, such as the alpha and delta variants of SARS-COV-2. The innate immune response, on the other hand, recognizes patterns of pathogen-related molecules, such as bacterial cell wall components. In this way, the innate immune system responds to all bacteria in the same way. Innate immune cells use pattern recognition receptors to recognize common features of pathogens. LO 21.3.4 A **pattern recognition receptor (PRR)** is a membrane-bound receptor that recognizes characteristic features of a pathogen or molecules released by stressed or damaged cells. PRRs, which are thought to have evolved prior to the adaptive immune response, are present on the cell surface whether they are needed or not. The innate immune system expresses only a limited number of receptors that are active against as wide a variety of pathogens as possible. This strategy is in stark contrast to the approach used by the adaptive immune system, which uses large numbers of different receptors, each highly specific to a particular pathogen.

Should the cells of the innate immune system come into contact with a species of pathogen they recognize, the cell will bind to the pathogen and initiate phagocytosis (or cellular apoptosis in the case of an intracellular pathogen) in an effort to destroy the offending microbe. Receptors vary somewhat according to cell type, but they usually include receptors for bacterial components and for complement, discussed in Section 21.3c.

Learning Check

1. Which cells of the innate immune response are considered granular cells?
   a. Macrophages
   b. Dendritic cells
   c. Neutrophils
   d. Natural killer cells
2. Which of the following chemokines are released by mast cells? Please select all that apply.
   a. Histamine
   b. Prostaglandins
   c. Endorphins
3. Which of the following are true about the innate immune response? Please select all that apply.
   a. The innate immune response takes several days.
   b. The innate immune response begins immediately.
   c. The innate immune cells respond to all bacteria the same way.
   d. The innate immune cells distinguish different pathogens.

21.3c Soluble Mediators of the Innate Immune Response

The previous discussions have alluded to chemical signals that can induce cells to change various physiological characteristics, such as the expression of a particular receptor. These soluble factors are secreted during innate or early induced responses, and later during adaptive immune responses.

Cytokines and Chemokines A **cytokine** is a signaling molecule that allows cells to communicate with each other over short distances. Cytokines are secreted into the intercellular space, and the action of the cytokine induces the receiving cell to change its physiology. A **chemokine** is a soluble chemical mediator similar to cytokines except that its function is to attract cells (chemotaxis) from longer distances.

LO 21.3.5

Interferons are one group of cytokines that are used by cells infected with viruses. An infected cell will secrete interferons into the local extracellular fluid to trigger local cells to make antiviral proteins. The infected cell is essentially warning its neighbors that the virus may come for them too, and as a result the neighboring cells are able to prepare their defenses.

Other compounds of the innate immune response also help activate or alert other cells. For example, the liver makes a few proteins that recognize and bind to bacterial cell walls. Phagocytes such as macrophages have receptors for these proteins, and they are thus able to recognize them as they are bound to the bacteria. This brings the phagocyte and bacterium into close proximity and enhances the phagocytosis of the bacterium by the process known as *opsonization*. **Opsonization** is the tagging of a pathogen for phagocytosis by the binding of an antibody or an antimicrobial protein.

LO 21.3.6

complement system (complement cascade)

Complement System The **complement system** is a series of proteins that are always found in the blood plasma. While the proteins are always present, they are awaiting activation to serve their pathogen-fighting functions. Made in the liver, they have a variety of functions in the innate and adaptive immune responses. The complement system works as a cascade, like a domino effect in which each protein sequentially activates the next protein in the system. Once activated, the series of reactions is irreversible. The complement system is capable of the following functions (summarized in **Figure 12.12**):

- Opsonization
- Acting as chemotactic agents to attract phagocytic cells to the site of inflammation
- Poking holes in the plasma membrane of the pathogen, killing it

21.3d Inflammatory Response

The hallmark of the innate immune response is **inflammation**. Inflammation is something everyone has experienced. Stub a toe, cut a finger, or do any activity that causes tissue damage and inflammation will result, with its four characteristics: heat, redness, pain, and swelling. It is important to note that inflammation does not have to be initiated by an infection, but can also be caused by tissue injuries. The initiation of inflammation is usually either the presence of a pathogen or release of internal cellular contents into the extracellular fluid due to damage from an injury. The major goal of the inflammatory reaction is to bring cells to the damaged area to clear cellular debris, respond to any pathogens that have entered, and set the stage for wound repair (**Figure 21.13**).

LO 21.3.7

There are four parts to the inflammatory response:

LO 21.3.8

- ***Tissue Injury.*** The released contents of injured cells stimulate the release of *mast cell* granules and their potent inflammatory mediators. *Histamine* increases the

Figure 21.12 Complement

Complement is an innate immune system group of factors that can (A) opsonize a pathogen, (B) recruit phagocytic cells to engulf the pathogen, and (C) attack the membrane of the pathogen, poking an opening in its cell membrane or cell wall and causing the cell to die.

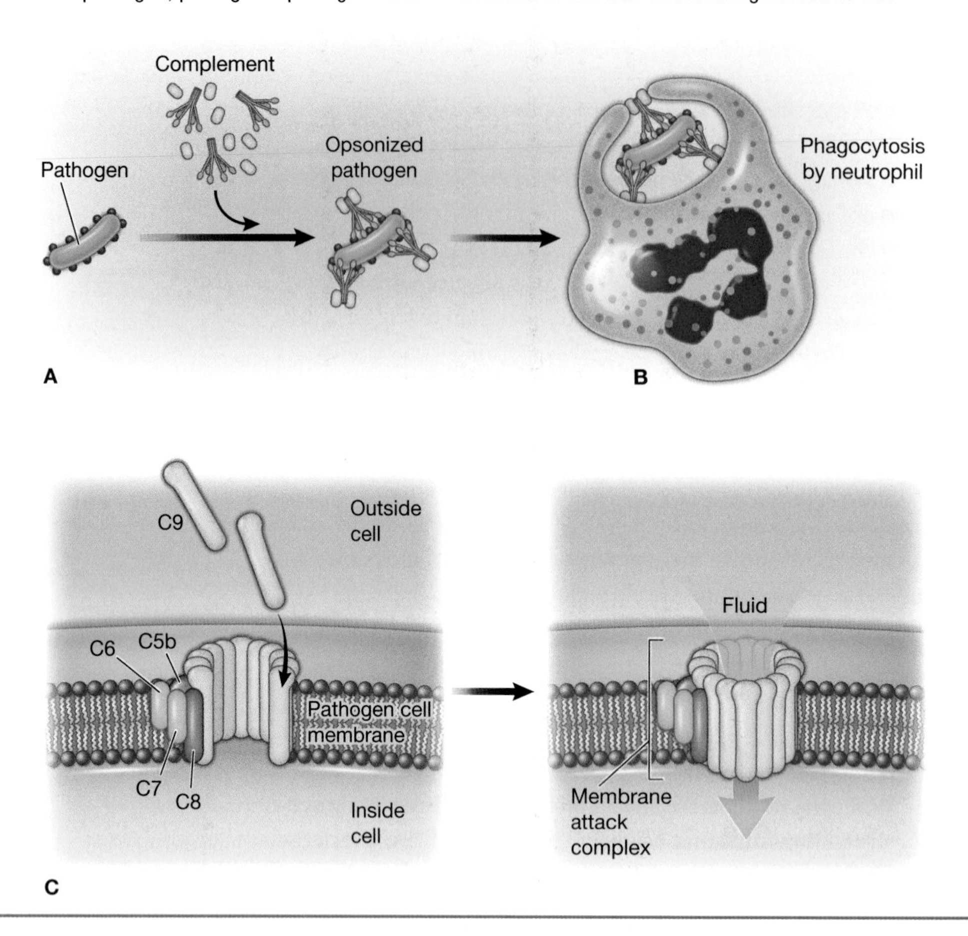

diameter of local blood vessels (vasodilation), causing an increase in blood flow to the injured site. Histamine also increases the permeability of local capillaries, causing more plasma than usual to leak out and form interstitial fluid. This causes the swelling associated with inflammation.

Prostaglandins cause vasodilation by relaxing vascular smooth muscle and are a major cause of the pain associated with inflammation. Nonsteroidal anti-inflammatory drugs (NSAIDs) such as aspirin and ibuprofen relieve pain by inhibiting prostaglandin production.

- ***Vasodilation.*** Prostaglandins and histamine are vasodilators that increase the diameters of local capillaries. This causes increased blood flow and is responsible for the heat and redness of inflamed tissue. It allows greater access of the blood and the cells that travel in blood to the site of inflammation.
- ***Increased Vascular Permeability.*** These and other cytokines increase the permeability of the capillaries, which allows the cells to exit the blood more easily. A side effect of this increased permeability is the leakage of fluid into the interstitial space, resulting in the swelling, or edema, associated with inflammation.

Figure 21.13 Inflammation

The inflammatory response can be triggered by either (A) infection by a pathogen or (B) injury to the tissue. Although started by different stimuli, the inflammatory response proceeds similarly.

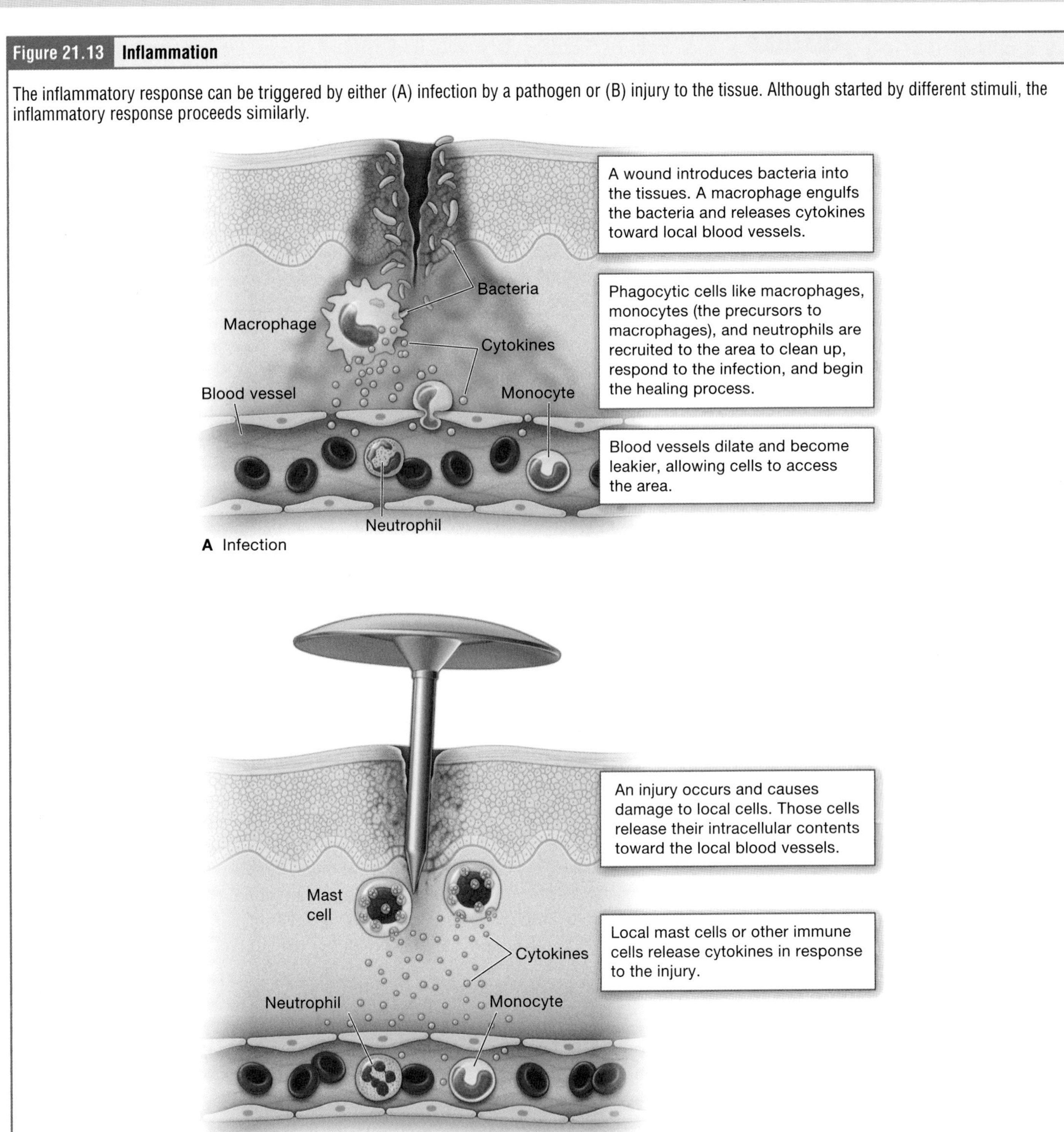

A wound introduces bacteria into the tissues. A macrophage engulfs the bacteria and releases cytokines toward local blood vessels.
Phagocytic cells like macrophages, monocytes (the precursors to macrophages), and neutrophils are recruited to the area to clean up, respond to the infection, and begin the healing process.
Blood vessels dilate and become leakier, allowing cells to access the area.
Bacteria
Macrophage
Cytokines
Blood vessel
Monocyte
Neutrophil
A Infection
An injury occurs and causes damage to local cells. Those cells release their intracellular contents toward the local blood vessels.
Local mast cells or other immune cells release cytokines in response to the injury.
Mast cell
Cytokines
Neutrophil
Monocyte
B Injury

- ***Recruitment of Phagocytes.*** Neutrophils flock in incredible numbers to the site of injury, where they can phagocytose any infiltrated pathogens and kill the invader along with themselves. Monocytes, the precursors to macrophages, also are recruited by the inflammatory cytokines. Once they exit the blood vessel, they mature into macrophages, enabling them to participate in both the cleanup of debris from the injury and the immune response.

LO 21.3.9

Overall, inflammation is valuable for many reasons. Not only are the pathogens killed and debris removed, but the increase in vascular permeability encourages the entry of clotting factors, the first step toward wound repair. Inflammation also facilitates the transport of antigens to lymph nodes by dendritic cells for the development of the adaptive immune response. Gene expression studies have shown that African ancestry is associated with a more robust inflammatory response, which may lead to more resiliency to certain infections. If the cause of the inflammation is not resolved, however, it can lead to chronic inflammation, which is associated with major tissue destruction and fibrosis. Chronic inflammation has a role in increasing the risk of or exacerbating several diseases such as atherosclerosis, arthritis, diabetes mellitus, and some forms of cancer.

21.3e Fever

LO 21.3.10

Fever, an increase of more than 1°C (1.8°F) above the set point for body temperature, is a very useful tool of the innate immune system. It occurs when any of a group of cytokines, collectively called *pyrogens*, is released. These molecules impact the hypothalamus and cause the release of one prostaglandin molecule in particular—prostaglandin E2. With this molecule acting as the go-ahead signal, the hypothalamus works to increase the body temperature by changing blood flow dynamics and generating shivering, which work in concert to drive the body temperature up.

LO 21.3.11

Fever has several benefits. Generally, the increased temperature can interfere with replication of bacteria and viruses as well as increase the actions of interferons, which help cells protect themselves against viral infection. Because all chemical reactions take place faster at higher temperatures, the enzymes required to produce and release cytokines can work faster. Fever also impacts the binding of iron to its transport and storage proteins, making this precious resource less available in the fluids of the body, which inhibits bacterial replication and lifespan. If it is safe and can be tolerated, allowing a patient to endure a fever may limit the duration of the infection.

Prostaglandins affect nociceptors, making them more likely to fire pain signals into the CNS. Prostaglandins increase vasodilation and vascular permeability, contributing to swelling; one prostaglandin, E2, can trigger the hypothalamus to allow a fever to develop. NSAIDs are a group of related drugs that inhibit prostaglandin synthesis. This group includes ibuprofen, acetaminophen, and aspirin. By inhibiting prostaglandin synthesis, these drugs are capable of reducing pain, inflammation, and fever.

Cultural Connection

Sex Hormones and the Immune Response

In general, estrogen has a stimulating effect on the immune system while testosterone has an immunosuppressive effect. Scientists don't know exactly why this is, but there are several evolutionary theories. One is that estrogen is secreted in very high amounts during pregnancy (about ten times higher at the end of pregnancy than before or after pregnancy) and that this stimulation of the immune system may help to protect the growing fetus from pathogens. Another theory is that testosterone, which is secreted during the stress response as well as impacting reproductive function, may help dampen the immune response during fight-or-flight scenarios (for more on this topic, see the "Stress and the Immune Response" Cultural Connection feature). Regardless of how this difference evolved, there are several impacts on

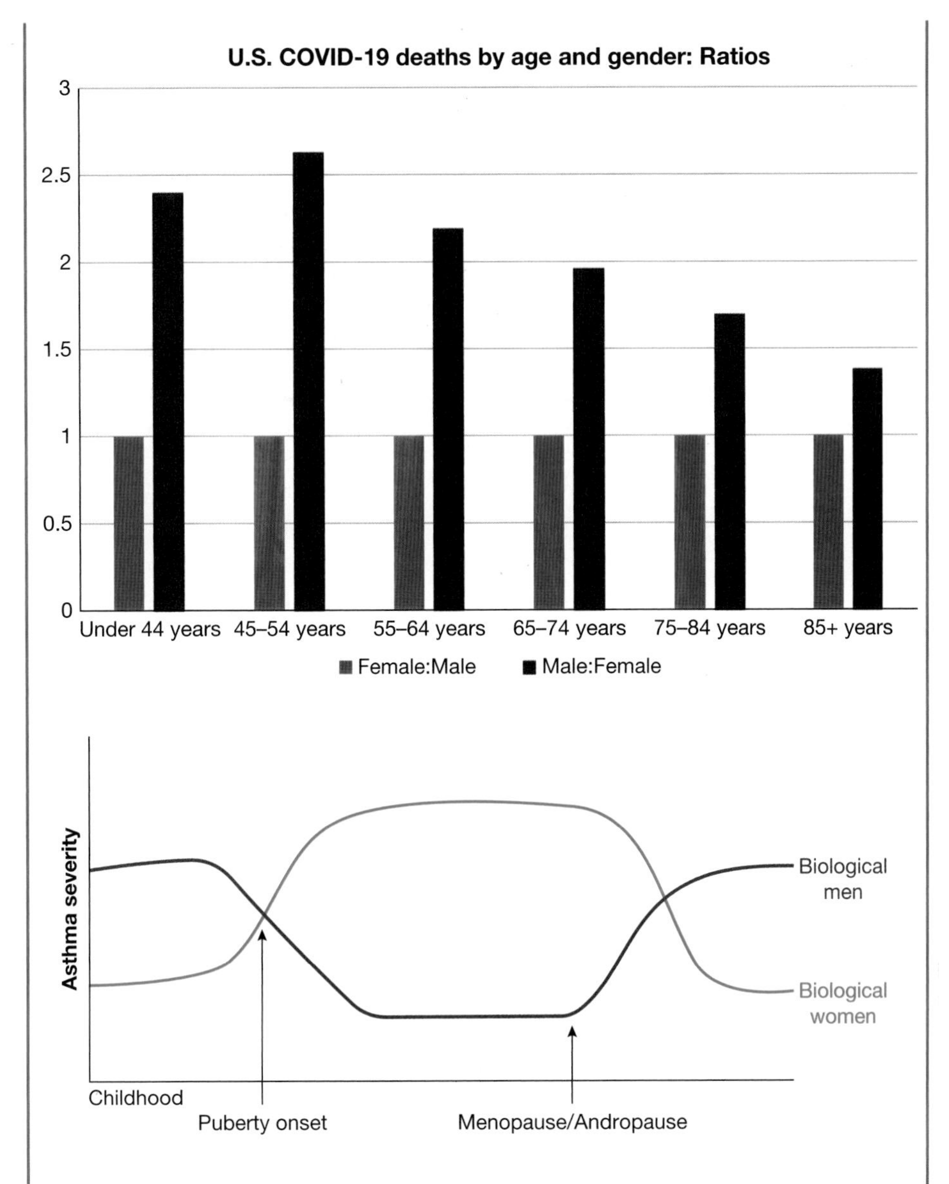

the lifespan that are notable. One is in infection. Throughout the ongoing COVID-19 pandemic, officials around the world and in the United States have noted that biological males are much more likely than age-matched biological females to die from COVID-19. There are several biological factors that play into this, but it is suspected that testosterone may have played a role in dampening the immune response against the virus. Similarly, a study of immune responses following the seasonal influenza (flu) vaccine found that biological females needed half as much vaccine as biological males to produce the same level of anti-flu antibody responses. Biological females are more than twice as likely as biological males to develop autoimmune diseases. Adding to this evidence are interesting data collected from individuals with asthma. Asthma is an immune-mediated overreaction that takes place in the lungs, causing narrowing of the bronchioles, which restricts air flow in and out of the body. Asthma symptoms are generally more prevalent and more severe in young biological males and less severe in young biological females. At the onset of puberty, however, these profiles change. With higher circulating testosterone, biological males have a reduction in—sometimes a complete break from—their symptoms, only to find that they return in mid- or later life when testosterone levels decline. Biological females, on the other hand, often experience an increase in their asthma symptoms, with severity peaking in young adulthood and then declining around menopause.

Learning Check

1. An infected cell in the body releases a particular group of cytokines into the extracellular fluid. This triggers neighboring cells to make antiviral proteins to protect them from the infection. What is the name of this group of cytokines?
   a. Histamine
   b. Prostaglandin
   c. Interferon
   d. Macrophage
2. What is the term that describes the process in which pathogens are tagged by antibodies to bring phagocytes to eliminate the pathogens?
   a. Inflammation
   b. Opsonization
   c. Fever
   d. Phagocytosis
3. What are the roles of prostaglandins in inflammation? Please select all that apply.
   a. Prostaglandins release histamine.
   b. Prostaglandins vasodilate blood vessels.
   c. Prostaglandins increase vascular permeability.
   d. Prostaglandins phagocytize pathogens.
4. Sharpe wakes up shivering, sweating, and sneezing. After taking their body temperature, they realize they have a fever. Which of the following describes what is happening at the cellular level during Sharpe's fever?
   a. The hypothalamus is decreasing blood flow to lower body temperature.
   b. The higher temperature speeds up enzyme production to accelerate immune responses.
   c. The iron readily binds to its transport protein, inhibiting bacterial replication.

21.4 The Adaptive Immune Response

Learning Objectives: By the end of this section, you will be able to:

21.4.1 Define antigen, self-antigen, and antigen receptor.

21.4.2 Compare and contrast antibody-mediated (humoral) and cell-mediated (cellular) immunity.

21.4.3 Describe the immunological memory (anamnestic) response.

21.4.4 Explain the roles of antigen-presenting cells (APCs) and provide examples of cells that function as APCs.

21.4.5 Define major histocompatibility complex (MHC).

21.4.6 Explain the functions of class I and class II major histocompatibility complex (MHC) proteins in adaptive (specific) immunity.

21.4.7 Describe where class I and class II major histocompatibility complex (MHC) proteins are found.

21.4.8 Describe where B and T cells originate, and contrast where they attain their immunocompetence.

21.4.9 Compare and contrast the mechanisms of antigen challenge and the clonal selection processes of B and T cells, including effector cells, helper cells, memory cells, and important cytokines.

21.4.10 Describe the general structure and functions of the various types of lymphocytes (e.g., helper T cells, cytotoxic T cells, regulatory [suppressor] T cells, B cells, plasma cells, memory cells).

21.4.11 Compare and contrast the defense mechanisms and functions of B and T cells.

21.4.12 Interpret a graph of the primary and secondary immune response, in terms of the relative concentrations of different classes of antibodies produced over time.

21.4.13 Describe antibody structure.

21.4.14 Compare and contrast the structure and functions of the classes of antibodies.

21.4.15 Describe mechanisms of antibody action and correlate mechanisms with effector functions.

21.4.16* Compare and contrast the anatomical locations and immune responses against different types of pathogens.

* Objective is not a HAPS Learning Goal.

Innate immune responses are not always sufficient to eliminate a pathogen. In many cases, the innate immune system works with the adaptive immune system, guiding specialized immune system cells in how to attack the pathogen. Thus, these are the two important arms of the immune response.

Learning Connection

Quiz Yourself

After studying the innate and adaptive immune responses, write the characteristics of each on separate pieces of paper or index cards. After mixing up your papers, try to sort them into two piles—one for innate responses and the other for adaptive responses. Start with characteristics such as speed of onset (slow or fast), types of cells involved, types of responses involved, and whether they are specific for one type of pathogen or nonspecific.

21.4a The Benefits of the Adaptive Immune Response

LO 21.4.1

The specificity of the adaptive immune response—its ability to specifically recognize and make an individual, tailored response against different of pathogens—is its great strength. **Antigens**, the molecules or groups of molecules that the body does not recognize as belonging to the "self," are recognized by receptors on the surface of B and T lymphocytes. Antigens elicit responses by the immune system, even though they may not be harmful or even foreign. The immune system might react against an antigen on the surface of a bacterial cell, or a protein found in peanuts, or even a self molecule. The adaptive immune response to these antigens is incredibly versatile. We can divide the actions of the adaptive immune response into cellular responses and antibody-mediated responses. **Cellular responses** are those enacted by the cells themselves. T lymphocytes, often simply called *T cells*, are the central actors in cellular responses. **Antibody-mediated responses** are the set of actions made possible by **antibodies**, secreted proteins of the adaptive immune response.

LO 21.4.2

cellular responses (cell-mediated immunity)

Primary Disease and Immunological Memory The immune system's first exposure to a pathogen elicits a **primary response**. Many of the symptoms of an infection are simply our observation of the immune response in action. The production of mucus is an attempt to clear invaders from our mucous membranes; coughing, sneezing, vomiting, and diarrhea are all attempts to eject pathogens that have invaded past the body's barriers. Pain, inflammation, and fever are all symptoms of an innate response. Every time we are infected with a pathogen for the first time—called *primary disease*—the impact is always relatively severe because it takes time for an initial adaptive immune response to a pathogen to become effective.

primary response (primary immune response)

Upon reexposure to the same pathogen, a secondary adaptive immune response begins, which is stronger and faster that the primary response. The **secondary adaptive response** often eliminates a pathogen before it can cause significant tissue damage or any symptoms. Without symptoms, there is no disease, and the individual is not even aware of the infection. This secondary response is the basis of **immunological memory**, which protects us from getting diseases repeatedly from the same pathogen. This helps us to understand why children suffer more frequent infections than adults. Adults often have such sophisticated adaptive defenses that they do not exhibit symptomatic responses to the same pathogens. To some degree, childhood illness is useful because an individual's exposure to pathogens early in life spares the person from these diseases later in life.

LO 21.4.3

Self Recognition A third important feature of the adaptive immune response is its ability to distinguish between self-antigens—those that are normally present in the body—and foreign antigens (those that might be on a potential pathogen). As T and B cells mature, there are mechanisms in place that prevent them from recognizing self-antigens, preventing a damaging immune response against the body. These mechanisms are not 100 percent effective, however, and when cells persist that are reactive against self-antigens, immune system responses to self-cells or tissues can result. When consistent or severe, these diseases are called **autoimmune diseases** (see the "Digging Deeper" feature later in this chapter).

Figure 21.14 T Cell Receptors

T cells express T cell receptors on their surface. T cell receptors are composed of two chains of proteins that contain a binding site for antigens.

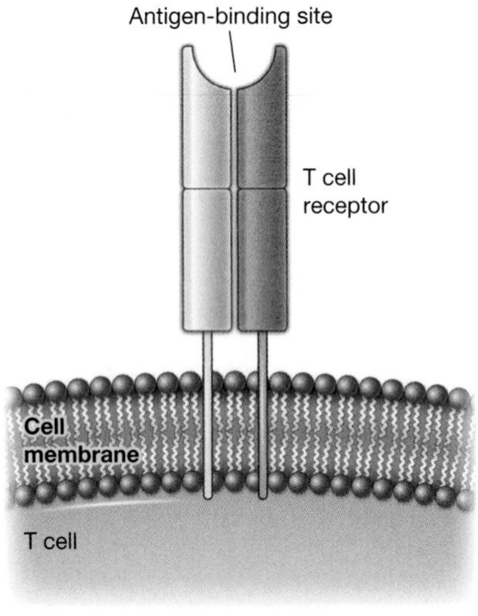

21.4b T Cell-Mediated Immune Responses

The primary cells that control the adaptive immune response are the lymphocytes, the T and B cells. T cells are particularly important, as they not only control a multitude of immune responses directly but also influence B cell responses. Thus, many of the decisions about how to attack a pathogen are made at the T cell level.

T lymphocytes recognize antigens based on a cell surface antigen receptor (**Figure 21.14**).

The T cell receptor is a cell surface protein. As such, its amino acid sequence is determined by the genome. The end of the receptors furthest from the cell membrane features the **antigen-binding site**, which is formed by both of the two chains of proteins that form the receptor. The amino acid sequences of those two areas combine to determine its antigenic specificity. You may remember from Chapter 3 that the individual properties of amino acids contribute to how they can bind other molecules—for example, a positively charged amino acid will be attracted to a negatively charged region on an antigen. Each T cell produces only one type of receptor, and thus each cell is specific for a single antigen.

21.4c Antigens

Antigens on pathogens are usually large and complex, and consist of many individual characteristics, called *antigenic determinants.* An **antigenic determinant** is one of the small regions within an antigen to which a receptor can bind, and antigenic determinants are limited by the size of the receptor itself. It is the interaction of the shape and characteristics of the antigen and the complementary shape and characteristics of the amino acids of the antigen-binding site that accounts for the chemical basis of specificity.

antigenic determinant (epitope)

antigen-presenting cells (APCs) (accessory cells)

LO 21.4.4 **LO 21.4.5** **LO 21.4.6**

Antigen Processing and Presentation The interaction between T cells and their antigens is quite complex. T cells do not recognize free-floating or cell-bound antigens as they appear on the surface of the pathogen. They only recognize antigens on the surface of specialized cells called **antigen-presenting cells (APCs)**. Antigens are internalized by these cells, and they break the antigen into smaller pieces through enzymatic reactions. The antigen fragments are then brought to the cell's surface by an antigen-presenting protein known as a **major histocompatibility complex (MHC)** molecule. MHC proteins operate almost like an ice cream scoop, sampling the antigenic material on the inside of the cell and then holding it on the surface of the cell to show it to a T cell. MHC proteins, like all proteins, are encoded in the genome. The coupling antigen and MHC occurs inside the cell, and then the complex is brought to the surface for recognition by the T cell receptor (**Figure 21.15**).

LO 21.4.7

Two distinct types of MHC molecules, **MHC class I** and **MHC class II**, play different roles in antigen presentation. They each bring antigens from inside the cell to the surface and present the antigens to the T cell and its receptor. However, MHC class I and class II molecules are expressed on different types of cells. Almost all cells of the body express MHC class I. They are continually sampling their internal environment and presenting those molecules on MHC class I. T cells will continually monitor the material presented on MHC class I in case there is evidence of an intracellular pathogen such as a virus, in which case the immune system would work to eliminate the infected cell and contain the infection. MHC class II molecules, on the other hand, are expressed only by phagocytic APCs. These cells are sampling from their lysosomes and showing the T cells the materials they have been taking in through phagocytosis.

Thus we can define antigen presentation in two ways, since many cells of the body present potential antigens via MHC class I and phagocytic cells present ingested material that is much more likely to be harmful to the body through MHC class II. For this reason, phagocytic cells that can present antigen on MHC class II are often referred to as *professional APCs.* Macrophages, dendritic cells, and B cells are all professional APCs.

Student Study Tip

An antigen presented in MHC class I protein can be thought of as "look what I made," while an antigen presented in MHC class II protein can be thought of as "look what I ate."

Figure 21.15 Antigen Presentation by Cells Containing MHC Class II Proteins

After an antigen-presenting cell (APC), such as a macrophage, engulfs a pathogen, the phagosome merges with a lysosome. Digestion of the pathogen by lysosomal enzymes results in the formation of antigenic fragments, which then bind to MHC class II proteins. The antigenic fragment-MHC-class II protein complex is then displayed on the cell membrane of the APC, where a helper T cell is finally able to recognize and bind to its antigen.

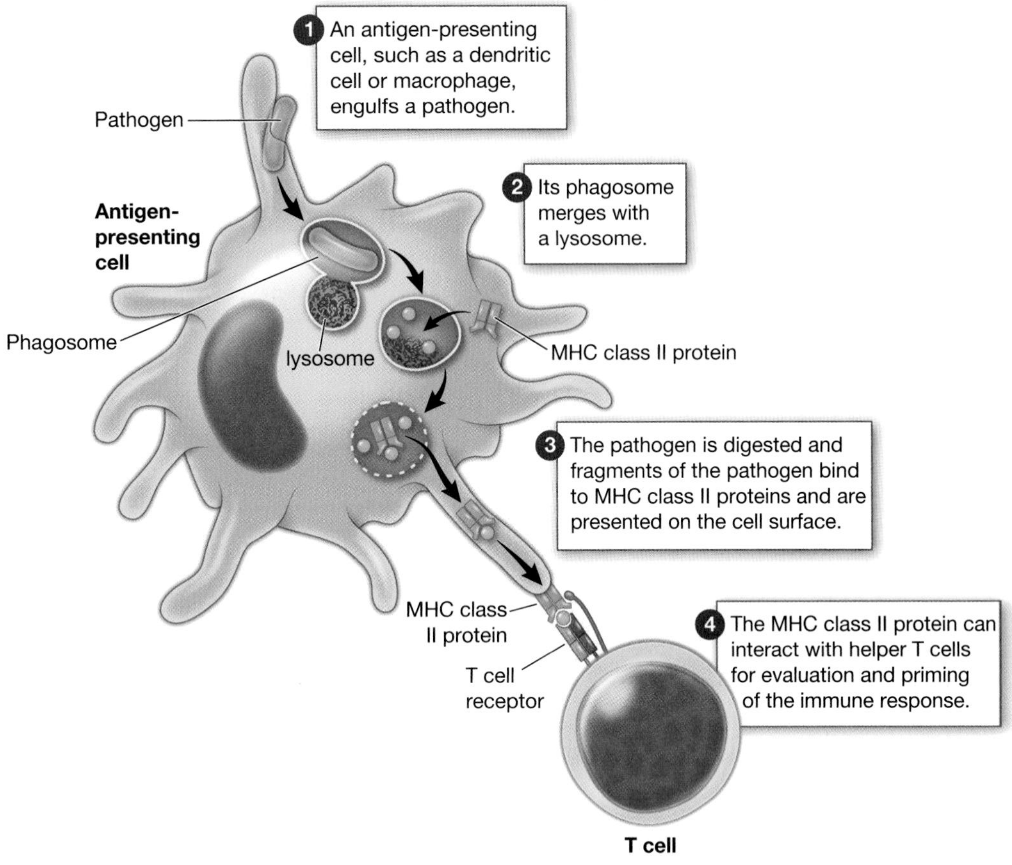

1 An antigen-presenting cell, such as a dendritic cell or macrophage, engulfs a pathogen.

Pathogen

2 Its phagosome merges with a lysosome.

Antigen-presenting cell

Phagosome

lysosome

MHC class II protein

3 The pathogen is digested and fragments of the pathogen bind to MHC class II proteins and are presented on the cell surface.

MHC class II protein

4 The MHC class II protein can interact with helper T cells for evaluation and priming of the immune response.

T cell receptor

T cell

21.4d T Cell Development and Differentiation

The exact amino acids that make up the binding site of an antigen receptor are encoded by the genome. Our genome possesses the ability, through complex mechanisms, to create an army of receptors capable of recognizing a dizzying array of antigens. Yet, we rarely have lymphocytes that react against our own cells or tissues. It is not that we never produce self-reactive lymphocytes, but that we have a process of eliminating T cells that might attack our own antigens, referred to as **T cell tolerance**. T cells are produced in the bone marrow, but while they are still immature and inactive, they travel to the thymus. While in the thymus, T cells are exposed to specialized epithelial cells that express every possible self-antigen. Any T cells that react against these self-antigens will be eliminated. In fact, only about 2 percent of the T cells that enter the thymus leave it as mature, functional T cells. The T cells that leave are both self-tolerant and immunocompetent, meaning that they are tolerant—or will not react against—self-antigens and they are competent to react against foreign antigens.

LO 21.4.8

21.4e Mechanisms of T Cell-Mediated Immune Responses

Mature T cells become activated by recognizing processed foreign antigen in association with an MHC molecule; this recognition launches them into mitosis, creating many new, identical T cell clones in a short period of time. This proliferation of

T cells is called **clonal expansion** and is necessary to make the immune response strong enough to effectively control a pathogen. How does the body select only those T cells that are needed against a specific pathogen? Again, the specificity of a T cell is based on the amino acid sequence and the three-dimensional shape of the antigen-binding site formed by the variable regions of the two chains of the T cell receptor (**Figure 21.16**). **Clonal selection** is the process of antigen binding only to those T cells that have receptors specific to that antigen. Each T cell that is activated has a specific receptor "hard-wired" into its DNA, and all of its progeny will have identical DNA and T cell receptors, forming clones of the original T cell. By the time this process is complete, the body will have large numbers of specific lymphocytes available to fight the infection.

LO 21.4.9

21.4f The Cellular Basis of Immunological Memory

As already discussed, one of the major features of an adaptive immune response is the development of immunological memory.

During clonal expansion, some of the T cells generated become memory T cells. **Memory T cells** are long-lived and can even persist for decades. Memory cells are primed to act rapidly. Thus, any subsequent exposure to the pathogen will elicit a very

Figure 21.16 Clonal Expansion of T Cells

Once T cells recognize an antigen bound to an MHC molecule, they will begin to reproduce themselves. Some of the daughter cells will become clones that can go out into the tissues and fight the infection; others will become a memory T cell population.

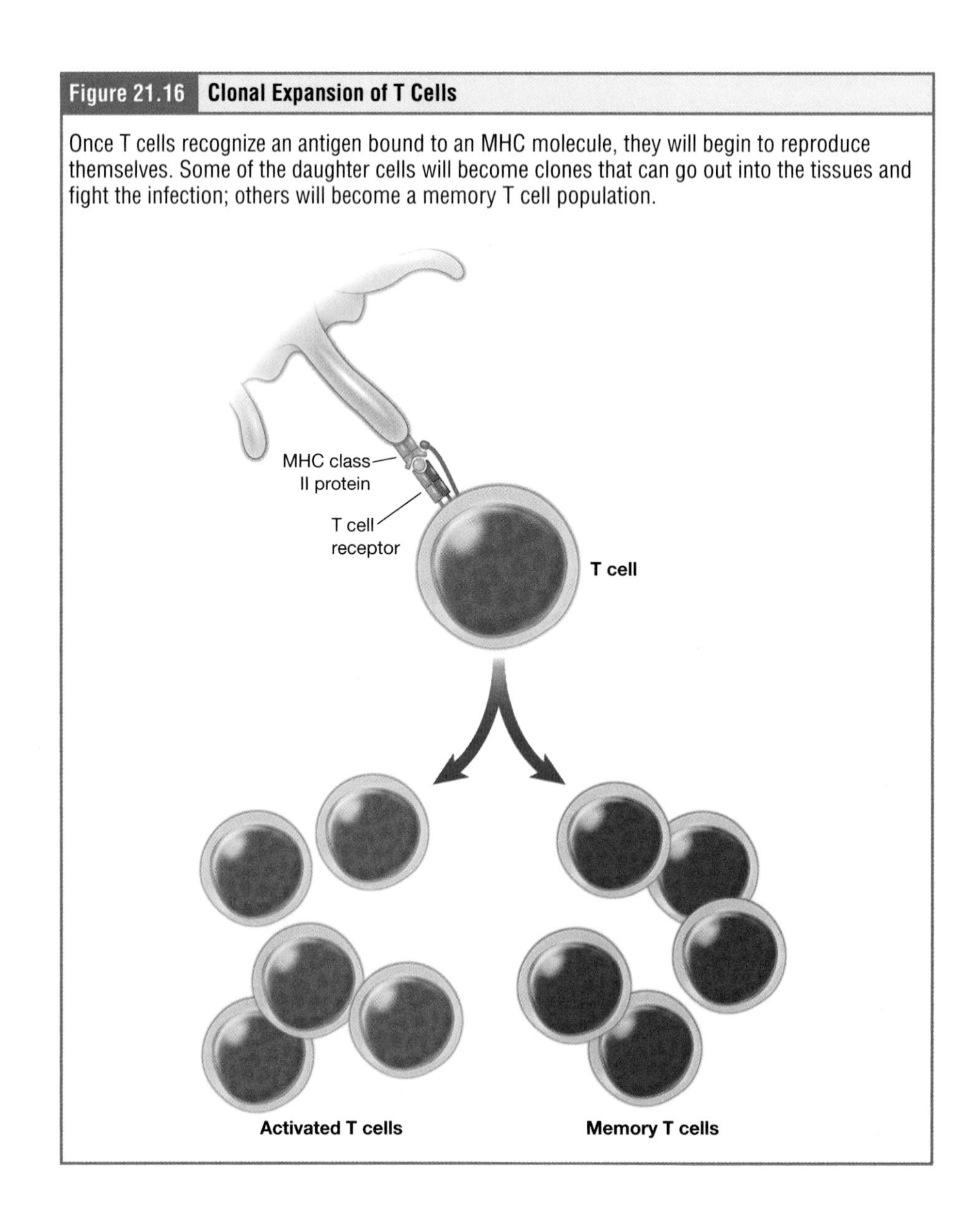

rapid T cell response. This rapid, secondary adaptive response generates large numbers of effector T cells so fast that the pathogen is often overwhelmed before it can cause any symptoms of disease. This is what is meant by immunity to a disease. The same pattern of primary and secondary immune responses occurs in B cells and the antibody response, as will be discussed in Section 21.4i.

21.4g T Cell Types and their Functions

LO 21.4.10

There are several types of T cells. While these cells all look the same under the light microscope, more sensitive scientific methods have determined that we can differentiate among different functional types of T cells based on the proteins they express on their cell surface. Of particular significance are two cell surface markers called *CD4* and *CD8*. These proteins not only help us determine the functional role of the T cell, but the proteins themselves are cell adhesion molecules that keep the T cell in close contact with the APC by directly binding to the MHC molecule. Thus, T cells and APC are held together in two ways: by CD4 or CD8 attaching to MHC and by the T cell receptor binding to antigen (**Figure 21.17**).

Figure 21.17 CD4 and CD8 Proteins

(A) Helper T cells express CD4 proteins next to their T cell receptors. Together the two molecules interact with MHC class II proteins on the surface of antigen-presenting cells. (B) Cytotoxic T cells express CD8 proteins next to their T cell receptors. Together the two molecules interact with MHC class I on the surface of cells throughout the body.

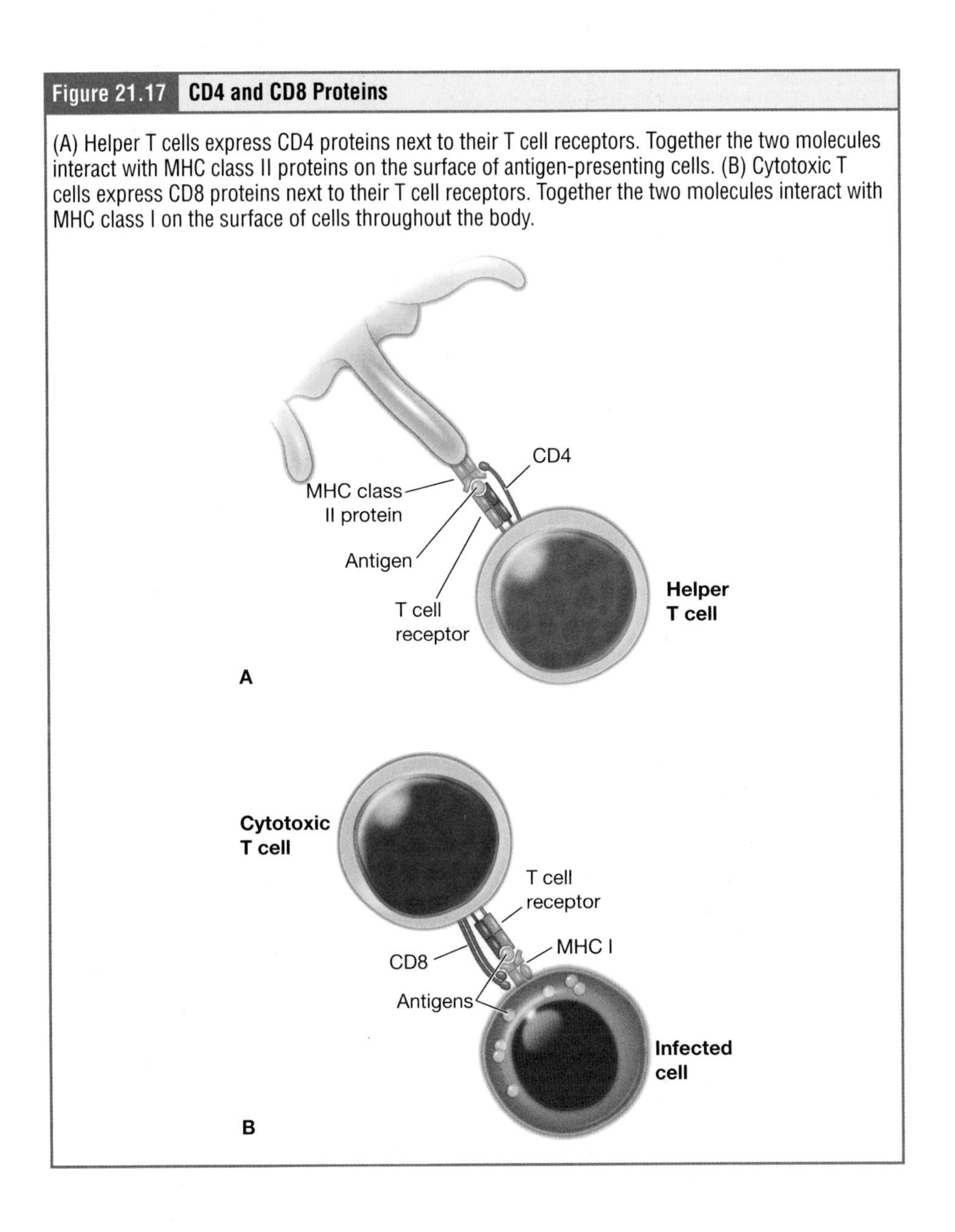

CD4-bearing T cells are called *helper T cells* and CD8-bearing T cells are called *cytotoxic T cells* (**Figure 21.18**).

LO 21.4.11

helper T cells (CD4+ T cells) (T_h cells)

Helper T Cells **Helper T cells (CD4+ T cells)** function by secreting cytokines that act to enhance other immune responses. The cytokines secreted by helper T cells help regulate the immunological activity and development of a variety of cells, including macrophages and other types of T cells. Cytokines secreted by T cells also drive B cell differentiation into antibody-generating cells. In fact, T cell help is required for antibody responses. Helper T cells recognize and respond to MHC class II proteins and antigens presented here.

cytotoxic T cells (CD8 T cells) (T_c cells)

Cytotoxic T Cells **Cytotoxic T cells (CD 8 T cells)** are T cells that kill target cells by inducing apoptosis, using a similar mechanism as NK cells. As was discussed earlier with NK cells, killing a virally infected cell before the virus can complete its replication cycle results in the production of no infectious particles. As more cytotoxic cells are

Figure 21.18 **The Functions of Helper and Cytotoxic T Cells**

Helper T cells secrete cytokines, which activate other cells of the immune system, so they can participate in the immune response against pathogens. Cytotoxic T cells destroy pathogens by secreting enzymes or substances that puncture the cell membranes or cell walls of the pathogens, or by promoting apoptosis in body cells that are infected with viruses or have become cancerous.

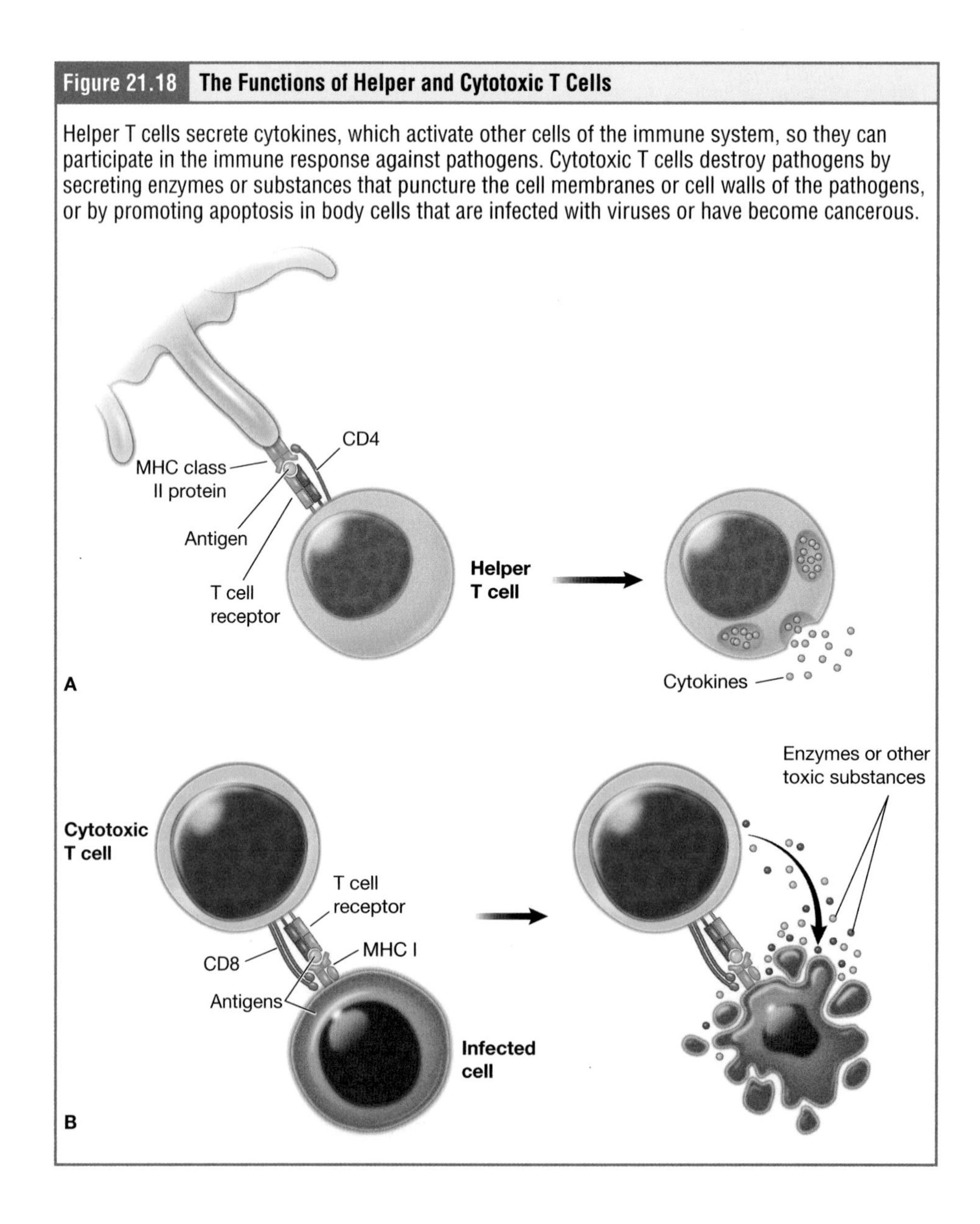

activated during an immune response, they overwhelm the ability of the virus to cause disease. In addition, each cytotoxic cell can kill more than one target cell, making them especially effective. Cytotoxic T cells recognize and respond to antigens presented in MHC class I proteins.

Regulatory T Cells **Regulatory T cells** are a specialized type of helper T cell. Regulatory T cells help to turn off the immune response once the threat of infection has passed. Specifically, they suppress other T cell immune responses. Overactive or inappropriate immune activity can lead to autoimmune disease, allergies, or other chronic inflammatory conditions, so regulatory T cells play an important role in the resolution and quieting of an immune response.

regulatory T cells (suppressor T cells, T_{reg} cells)

Learning Check

1. Carl had the flu three months ago. They were coughing, had a fever, and felt tired. Unfortunately, Carl is currently infected with the same flu pathogen; however, they have no symptoms. Which of the following best describes the current adaptive immunological response?
   a. Primary response
   b. Secondary response
2. Which of the following MHC molecules is expressed on most types of cells?
   a. MHC Class I
   b. MHC Class II
   c. MHC Class III
   d. MHC Class IV
3. Which of the following types of cells will kill a virally infected cell?
   a. Helper T cells
   b. CD4+ T cells
   c. Cytotoxic T cells
   d. Regulatory T cells

21.4h B Cells and Their Functions

Antibodies were the first component of the adaptive immune response to be characterized by scientists working on the immune system. It was already known that individuals who survived a bacterial infection were immune to reinfection with the same pathogen. Early microbiologists took serum from an immune patient and mixed it with a fresh culture of the same type of bacteria, then observed the bacteria under a microscope. The bacteria became clumped in a process called *agglutination*. When a different bacterial species was used, the agglutination did not happen. Thus, there was something in the serum of immune individuals that could specifically bind to and agglutinate bacteria (**Figure 21.19**).

Scientists now know the cause of the agglutination is antibodies, proteins by secreted B cells. B cells express receptors on their surface, **B cell receptors**, that are specific for antigen, just like T cells. After activation, B cells are able to secrete these antibodies in a process detailed below.

There are five different classes of antibody found in humans: IgM, IgD, IgG, IgA, and IgE. Each of these has specific functions in the immune response, so by learning about them, researchers can learn about the great variety of antibody functions critical to many adaptive immune responses.

T cells require antigen to be processed by other cells before they can recognize it. B cell function is simpler in that B cells can recognize native, unprocessed antigen and do not require the participation of MHC molecules and APCs.

Learning Connection

Broken Process

Considering what you have learned about the functions of the thymus, what would happen if a young child's thymus were destroyed in an accident? Which types of cells would be missing? How would this affect antibody production and the immune response? Would this affect the innate immune response, the adaptive immune response, or both?

B cell receptors (BCR)

Figure 21.19 **Agglutination**

The shape of an antibody and its two binding sites allows it to cross-link antigens. This results in large clusters of antigens or pathogens, a process called *agglutination*.

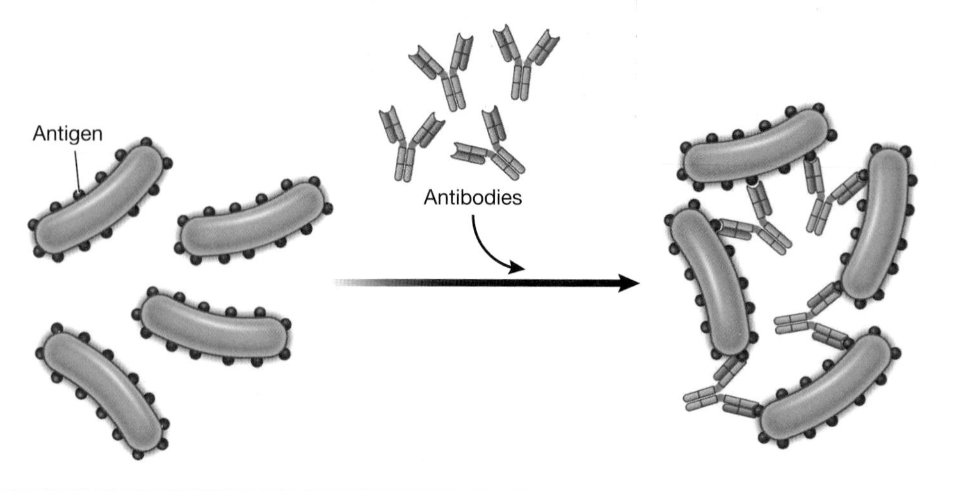

Learning Connection

Explain a Process

Explain to a classmate the difference between B cell and T cell activation. Which one requires an APC? Which one produces cytokines? Which one requires the help of cytokines from the other?

21.4i B Cell Differentiation and Activation

B cells are both produced and differentiate in the bone marrow. During the process of maturation, up to 100 trillion different B cells are generated, each specific to a different antigen, similar to the diversity of antigen receptors seen in T cells.

B cell differentiation and the development of tolerance are not quite as well understood as they are in T cells, but we do know that B cells that recognize self-antigens are destroyed or inactivated in the bone marrow. Some B cell tolerance occurs in the periphery as well because B cells require T cell signals in order to activate. Therefore if B cells react against self-antigen, but no helper T cells are specific for the same antigen and therefore do not provide cooperative signals, the B cells will not produce a response and in some cases the self-reactive B cell may be destroyed.

B cell activation occurs with the support of helper T cells in lymph nodes. After B cells are activated by binding to their antigen, they differentiate into **plasma cells**. Plasma cells often leave the lymph system and migrate back to the bone marrow, where the whole differentiation process started. After secreting antibodies for a specific period, they die, as most of their energy is devoted to making antibodies and not to maintaining themselves. These antibodies are essential in the fight against a pathogen. After an infection is over, antibodies can be measured circulating in the blood for a long time.

LO 21.4.12

Like T cells, B cells are also capable of differentiating into memory B cells. Memory B cells function in a way similar to memory T cells. They lead to a stronger and faster secondary response when compared to the primary response (**Figure 21.20**). In addition to the secondary response taking far less time after the initial exposure, the antibodies made in secondary and subsequent exposures are much higher quality.

21.4j Antibody Structure

Antibodies are proteins consisting of four chains of amino acids and decorated with carbohydrates. The two **heavy chains** and the two **light chains** form the antibody. The antigen-binding sites are composed of antibodies from both the heavy chain and the light chain (**Figure 21.21**). There are five different styles of antibodies that are slightly different in structure and function but mainly different in the time and location in which they are found.

Figure 21.20 Memory Responses

The first time the body is exposed to a pathogen it takes time (1) for the adaptive immune response to become activated and effective. When exposed a second or subsequent time, the already primed memory cells react swiftly (2).

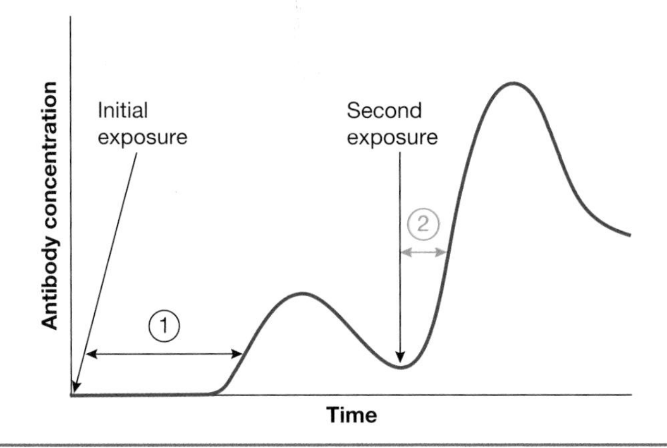

Figure 21.21 Antibody Structure

Antibodies are composed of four chains of proteins, two heavy and two light chains. The heavy chains contribute to both the antigen-binding site and the Fc region. The light chains contribute only to the antigen-binding sites.

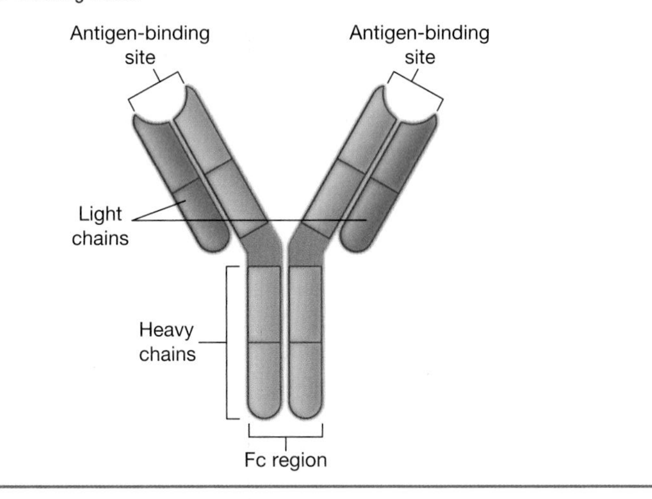

DIGGING DEEPER:
HIV/AIDS

The critical nature of helper T cells was fully understood when a novel virus began to infect humans and circulate around the globe in the late 1970s. Human immunodeficiency virus (HIV) specifically infects helper T cells and, while there are very few of these cells in the body, they are critical to all facets of the adaptive immune response. When a person becomes infected with HIV, the viral load swells initially but then subsides once the adaptive immune response begins. As long as there are significant numbers of helper T cells present, the viral load remains low. But once the number of helper T cells in the body begins to fall as a result of the HIV infection, the viral levels swell and a devastating positive feedback loop ensues. Acquired Immunodeficiency Syndrome (AIDS) is diagnosed when the helper T cell count falls below a critical threshold. The vast majority of HIV positive individuals, if untreated, will lose their entire adaptive immune response, including the ability to make antibodies, during the final stages of AIDS. There are a handful of individuals who have not yet progressed to AIDS even though they have been infected with HIV for decades. Current antiviral treatments are very effective and, when treated early and consistently, most individuals can live long and productive lives after HIV diagnosis. In fact, thorough research has found that when individuals take anti-retroviral therapy, their viral levels are often undetectable for many years. A U.S. public health campaign aims to spread the word that "U=U," which means that undetectable viral loads mean that the infection is untransmittable, and these individuals can lead active and safe sex lives without risk of spreading HIV to their partner(s).

All antibody molecules have a Y-shape; the base of the Y is known as the **Fc region**. The Fc region of the antibody is formed by the two heavy chains coming together (see Figure 21.21). The Fc portion of the antibody is important in that many cells of the immune system have Fc receptors. Cells having these receptors can then bind to antibody-coated pathogens and phagocytose or destroy them. At the other end of the molecule are two identical antigen-binding sites.

LO 21.4.14

Five Classes of Antibodies and Their Functions In general, antibodies have two basic functions. They can act as the B cell antigen receptor or they can be secreted, circulate, and bind to a pathogen, often labeling it for identification by other forms of the immune response. All of the antibody types are named with "Ig" for immunoglobulin, followed by a letter. The five classes are IgM, IgG, IgA, IgE, and IgD (**Figure 21.22**).

IgM is described as a pentamer, a molecule with five distinct subunits. It resembles a star with five individual antibody structures that meet at their Fc regions. IgM is usually the first antibody made during a primary response. Its 10 antigen-binding sites and large shape allow it to bind well to many bacterial surfaces. It activates the complement cascade, and promotes chemotaxis, opsonization, and cell lysis. Thus, it is a very effective antibody against bacteria at early stages of a primary antibody response. As

Figure 21.22 The Five Classes of Antibodies

IgM antibodies are joined in a pentameter, IgA antibodies are joined in a dimer, and IgE, IgG, and IgD antibodies are all monomers.

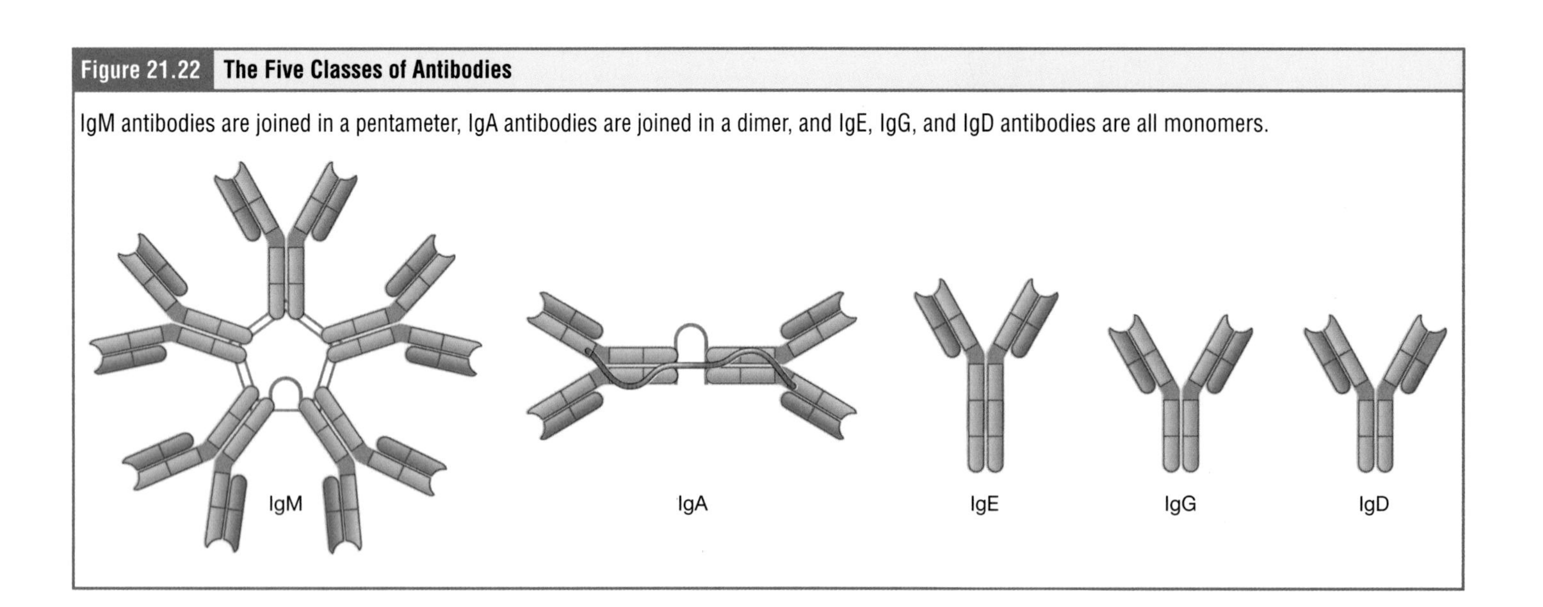

the primary response proceeds, the B cell that has been producing IgM switches to producing IgG, IgA, or IgE by the process known as *class switching*. **Class switching** is the change of one antibody class to another. While the class of antibody changes, the specificity and the antigen-binding sites do not. Thus, the antibodies made are still specific to the pathogen that stimulated the initial IgM response.

LO 21.4.15

IgG is a major antibody of late primary responses and the main antibody of secondary responses in the blood. This is because class switching occurs during primary responses. IgG is a monomeric antibody that clears pathogens from the blood and can activate complement proteins (although not as well as IgM), taking advantage of its antibacterial activities. Furthermore, this class of antibody is the one that crosses the placenta to protect the developing fetus from disease; it exits the blood to the interstitial fluid to fight extracellular pathogens.

Figure 21.23 compares IgM and IgG responses in infection.

IgA exists in two forms, but is most often found as a dimer, a structure consisting of two antibodies joined at their Fc regions. The cells that make IgA antibodies take up residence in the exocrine glands and their antibodies are included in the secretions of these glands, which include mucus, saliva, and tears. Thus, IgA is the only antibody to leave the interior of the body to protect body surfaces. IgA is also of importance to newborns because this antibody is present in mother's breast milk. During late pregnancy and lactation, IgA-secreting B cells will leave other mucous membranes and migrate to the breasts to enrich breast milk with antibodies, which serves to protect the infant from disease. This is particularly important because the immune systems of newborns are so immature; these infants are extremely vulnerable to disease.

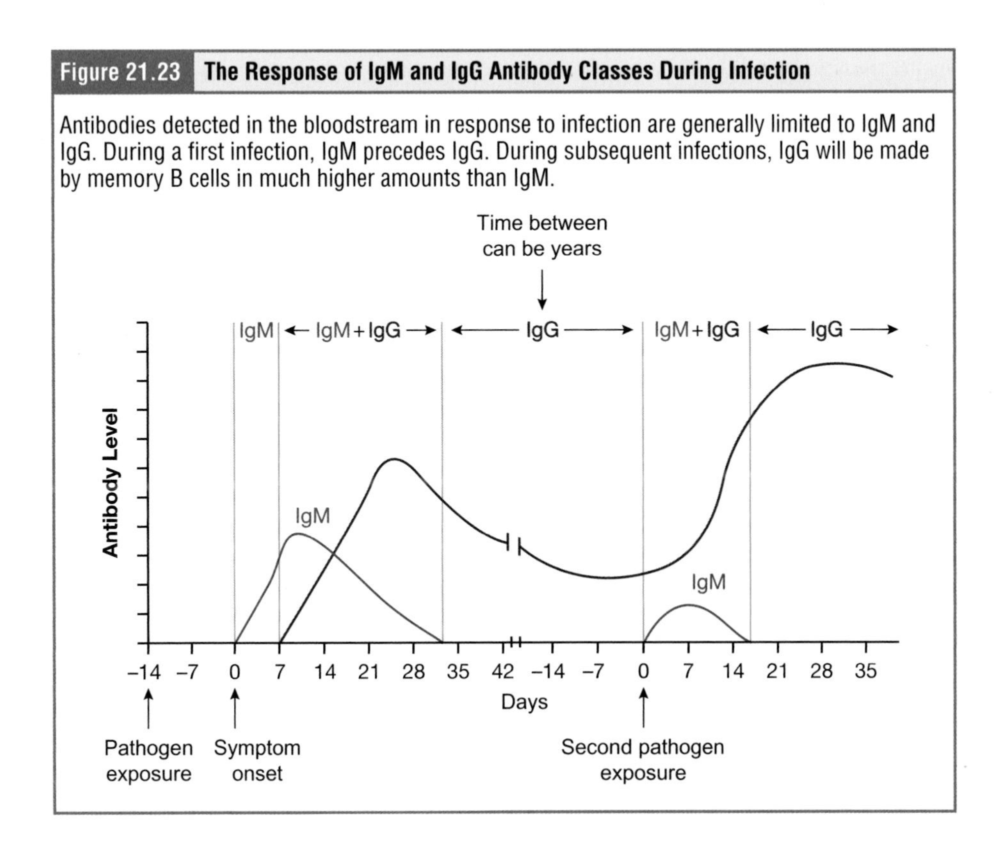

Figure 21.23 **The Response of IgM and IgG Antibody Classes During Infection**

Antibodies detected in the bloodstream in response to infection are generally limited to IgM and IgG. During a first infection, IgM precedes IgG. During subsequent infections, IgG will be made by memory B cells in much higher amounts than IgM.

IgE is usually associated with allergies and anaphylaxis. It is present in the lowest concentration in the blood, but is found in much higher concentrations in interstitial fluid. Its Fc region binds strongly to an IgE-specific Fc receptor on the surfaces of mast cells. IgE makes mast cell degranulation very specific and is known to recruit mast cells to participate in allergy responses.

IgD is mostly found expressed on B cell plasma membranes, but is secreted in very small amounts into the bloodstream.

Mucosal tissues are major barriers to the entry of pathogens into the body. The IgA (and sometimes IgM) antibodies in mucus and other secretions can bind to the pathogen, and in the cases of many viruses and bacteria, neutralize them. **Neutralization** is the process of coating a pathogen with antibodies, making it physically impossible for the pathogen to bind to receptors. Neutralization, which occurs in the blood, lymph, and other body fluids and secretions, protects the body constantly. Neutralizing antibodies are the basis for the disease protection offered by vaccines. Vaccinations for diseases that commonly enter the body via mucous membranes, such as influenza, are usually formulated to enhance IgA production.

Clonal Selection of B Cells Clonal selection and expansion in B cells works much the same way as in T cells. Only B cells with appropriate antigen specificity are selected for and expanded (**Figure 21.24**). Eventually, the plasma cells secrete antibodies with antigenic specificity identical to those that were on the surfaces of the selected B cells. Notice in the figure that both plasma cells and memory B cells are generated simultaneously.

21.4k Active versus Passive Immunity

Immunity to pathogens, and the ability to control pathogen growth so that damage to the tissues of the body is limited, can be acquired by (1) the active development of an immune response in the infected individual or (2) the passive transfer of immune components from an immune individual to a nonimmune one. Both active and passive immunity have examples in the natural world and as part of medicine.

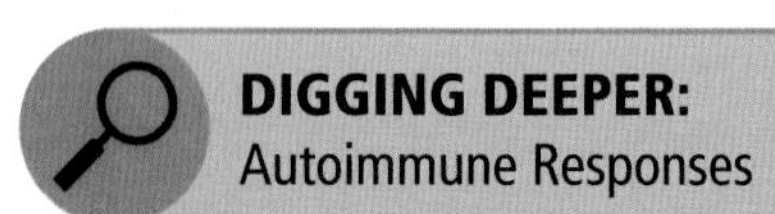

DIGGING DEEPER:
Autoimmune Responses

The immune system generally works on a principle of "self" versus "non-self" with B and T cell tolerance being built on elimination of self-reactive cells. But this is clearly not the whole story. We put "non-self" materials in our bodies all the time (you did the last time you ate a meal) and we have immune tolerance to those materials—except when we don't; rates of food allergies are increasing all the time. Likewise, we have "non-self" materials such as pollen that may or may not trigger an immune reaction. Keep in mind that these materials are limited to our mucous membranes. Another aberration to the "self" versus "non-self" rule is when the immune system reacts against "self" antigens. The trigger for these diseases is, more often than not, unknown, and the treatments are usually based on suppressing the actions of the immune system, which of course has side effects when it comes to pathogens. Environmental triggers seem to play large roles in autoimmune responses. One explanation for the breakdown of tolerance is that, after certain bacterial infections, an immune response to a component of the bacterium cross-reacts with a self-antigen. This mechanism is seen in rheumatic fever, a result of infection with *Streptococcus* bacteria, which causes strep throat. The antibodies to one of this pathogen's proteins cross-react with a component of heart muscle cells. The antibody binds to these molecules and activates complement proteins, causing damage to the heart, especially to the heart valves. There are genetic factors in autoimmune diseases as well. Some diseases are associated with the MHC genes that an individual expresses. The reason for this association is likely because if one's MHC molecules are not able to present a certain self-antigen, then that particular autoimmune disease cannot occur. Overall, there are more than 80 different autoimmune diseases.

Figure 21.24 **Clonal Expansion of B Cells**

(A) Once a B cell binds to the antigen for which it is specific, it will undergo clonal expansion. Some of the new cells will become plasma cells, which are capable of producing antibodies. Others will become memory B cells, which will not respond during the first exposure to the pathogen, but will circulate in the blood for years. (B) During a secondary response, the memory B cells reproduce quickly to yield an army of antibody-producing cells.

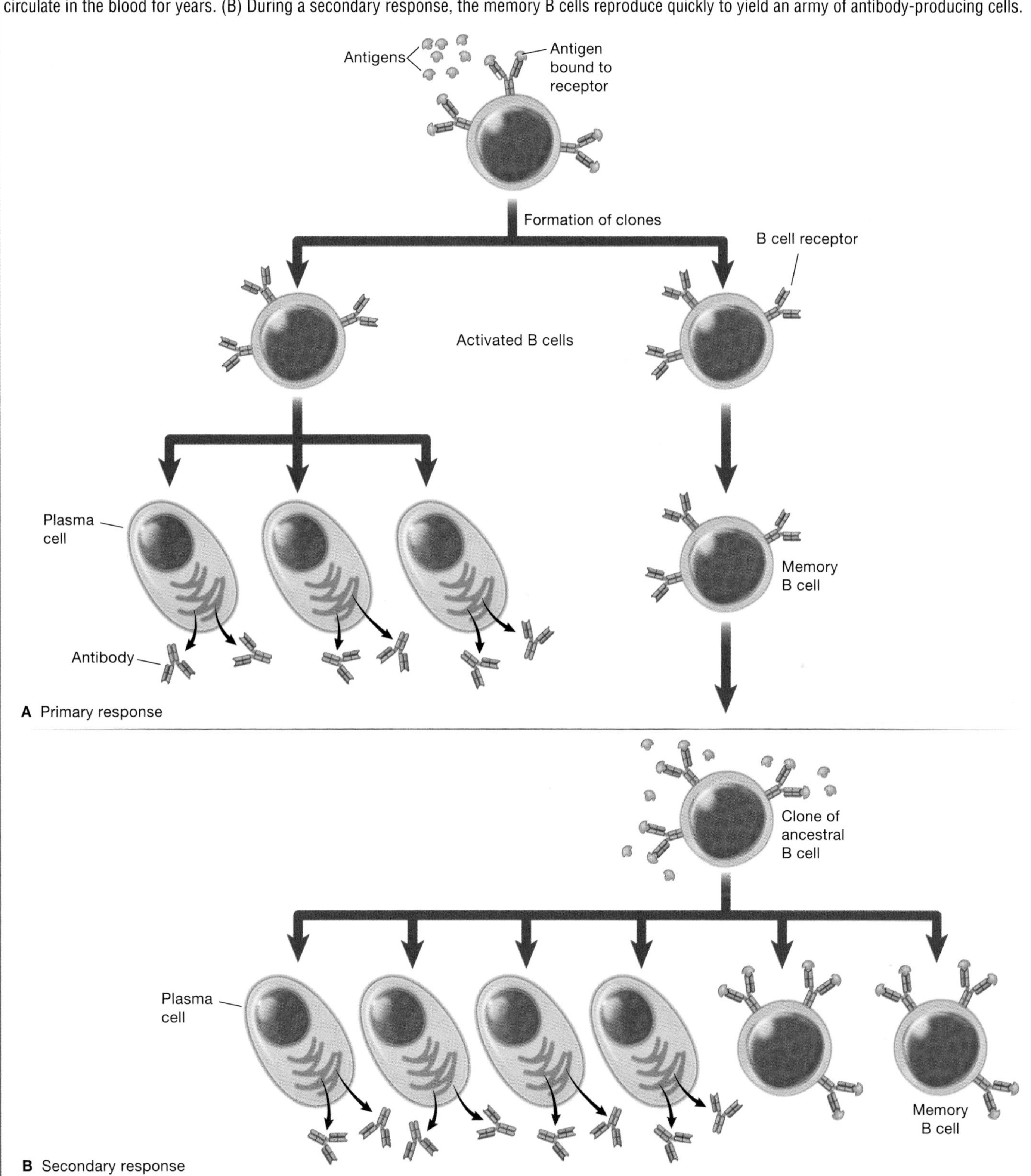

Learning Connection

Chunking

Now that you have studied the various types of immune cells and responses, make a list of all of the different types of immune cells (such as macrophages, the different types of leukocytes, and the various types of lymphocytes). Categorize each cell in your list by function. Which ones engage in phagocytosis and which ones help destroy pathogens using the chemicals that they secrete? Which cells are able to directly kill another cell?

Active immunity is the resistance to pathogens acquired during an adaptive immune response within an individual (**Table 21.2**). Naturally acquired active immunity, the response to a pathogen, is the focus of this chapter. Artificially acquired active immunity involves the use of vaccines. A vaccine is a killed or weakened pathogen or its components that, when administered to a healthy individual, leads to the development of immunological memory (a weakened primary immune response) without causing much in the way of symptoms. Thus, with the use of vaccines, one can avoid the damage from disease that results from the first exposure to the pathogen, yet reap the benefits of protection from immunological memory. The advent of vaccines was one of the major medical advances of the twentieth century and led to the eradication of smallpox and the control of many infectious diseases, including polio, measles, and whooping cough.

Passive immunity arises from the transfer of antibodies to an individual without requiring them to mount their own active immune response. Naturally acquired passive immunity is seen during fetal development. IgG is transferred from the maternal circulation to the fetus via the placenta, protecting the fetus from infection and protecting the newborn for the first few months of its life. As already stated, a newborn benefits from the IgA antibodies it obtains from milk during breastfeeding. The fetus and newborn thus benefit from the immunological memory of the mother to the pathogens to which she has been exposed. In medicine, artificially acquired passive immunity usually involves injections of immunoglobulins, taken from animals previously exposed to a specific pathogen. This treatment is a fast-acting method of temporarily protecting an individual who was possibly exposed to a pathogen. The downside to both types of passive immunity is the lack of the development of immunological memory. Once the antibodies are transferred, they are effective for only a limited time before they degrade.

Table 21.2 Active versus Passive Immunity

	Natural	Artificial
Active	Adaptive immune response	Vaccine response
Passive	Transplacental antibodies/breastfeeding	Immune globulin injections

Cultural Connection

Stress and the Immune Response

The two hormones common to all stress responses in all humans are cortisol and epinephrine. Epinephrine circulates in high amounts during short-term acute stress responses, driving up breath and heart rates during an emergency. Cortisol plays a role in long-term stress, such as a tough final exam period or a stressful period at work. Both of these stress hormones alter the actions of the immune system. At low levels, both stress hormones act to encourage immune cells to leave the blood and enter the tissues; at high levels, however, epinephrine is thought to "freeze" immune cells, hindering them from making their way to sites of infection. Cortisol's effects are longer lasting and more nuanced. Cortisol inhibits both immune cell replication and the release of certain cytokines, blocking the cells from being able to communicate and inhibiting adaptive cells from clonal expansion. The results of stress can be devastating. Often physiological or psychological stress can inhibit the immune system to the degree that pathogens that have been hiding in our bodies in low numbers are able to replicate and cause an infection, as happens in yeast infections, cold sores or genital herpes outbreaks, and shingles. Another common effect is that infections are acquired during a period of stress, but the immune system fails to respond to them until the stress has subsided. The immune system is only able to respond to the infection once the stressful period is over, as often happens with college students who make it through the end of the semester only to find themselves sick once they get home. Both epinephrine and cortisol are used exogenously to suppress the immune system clinically. An EpiPen is an injectable form of epinephrine that is used to treat anaphylaxis (a lethal form of allergic response). Cortisol is used in oral medications such as prednisone, topical medications such as hydrocortisone cream, and inhaled medications such as Flovent or Pulmicort, to treat a variety of unwanted immune reactions from poison ivy responses to asthma.

21.4l Responses to Different Pathogens

LO 21.4.16

The human immune response is very complex, as you now can appreciate. We can be infected with viruses, which are a thousandth the size of the smallest human cell, and we can also be infected with multicellular parasites. Tapeworms, for example, can occupy the human intestine and grow to be dozens of feet in length! The immune system can recognize and respond to an incredible diversity of pathogens (**Figure 21.25**). Worm parasites such as helminths are seen as the primary reason why the mucosal immune response, IgE-mediated allergy and asthma, and eosinophils evolved. These parasites were at one time very common in human society. When infecting a human, often via contaminated food, some worms take up residence in the gastrointestinal tract. Eosinophils are attracted to the site by T cell cytokines, which release their granule contents upon their arrival. Mast cell degranulation also occurs, and the fluid leakage caused by the increase in local vascular permeability is thought to have a flushing action on the parasite, expelling its larvae from the body. Furthermore, if IgE labels the parasite, the eosinophils can bind to it by its Fc receptor. The primary mechanisms against viruses are NK cells, interferons, and cytotoxic T cells. Antibodies are effective against viruses mostly during protection, where an immune individual can neutralize them based on a previous exposure. Antibodies have no effect on viruses or other intracellular pathogens once they enter the cell, since antibodies are not able to penetrate the plasma membrane of the cell. Many cells respond to viral infections by downregulating their expression of MHC class I molecules. This is to the advantage of the virus, because without class I expression, cytotoxic T cells have no activity. NK cells, however, can recognize virally infected class I-negative cells and destroy them. Thus, NK and cytotoxic T cells have complementary activities against virally infected cells.

Interferons have activity in slowing viral replication and are used in the treatment of certain viral diseases, such as hepatitis B and C, but their ability to eliminate the

Figure 21.25 Types of Pathogens

Viruses are very small and cannot reproduce themselves, so they must infect host cells and exist inside them. Fungal, bacterial, and parasitic pathogens most often are extracellular pathogens, although some of these species can infect cells.

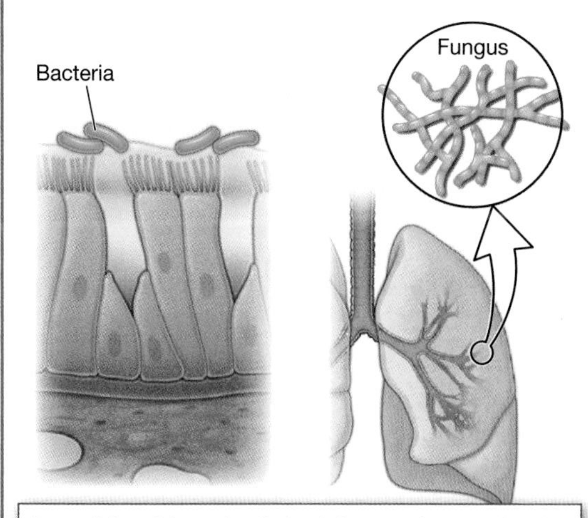

Bacterial and fungal cells both have unique molecules on their cell walls that can trigger immune responses. Both are typically extracellular pathogens, though some small bacterial species can live inside human cells. Antibodies, neutrophils, and macrophages are particularly essential components of the immune responses against bacterial and fungal pathogens.

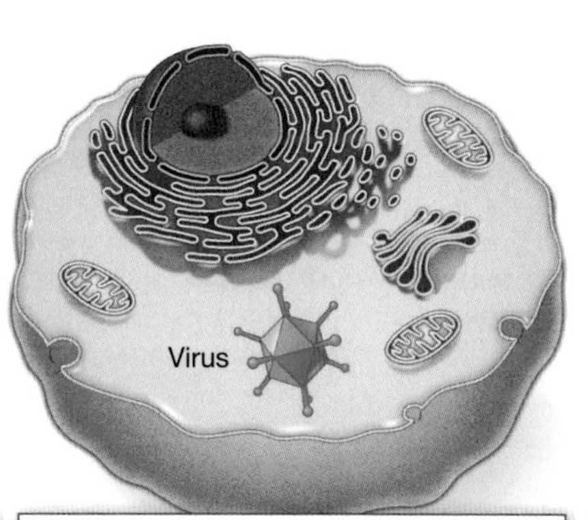

Viruses are intracellular pathogens. They are unable to replicate themselves; therefore, they must live inside a host cell and borrow their ribosomes. Immune system components that can recognize infected cells, such as T cells and NK cells, are particularly useful.

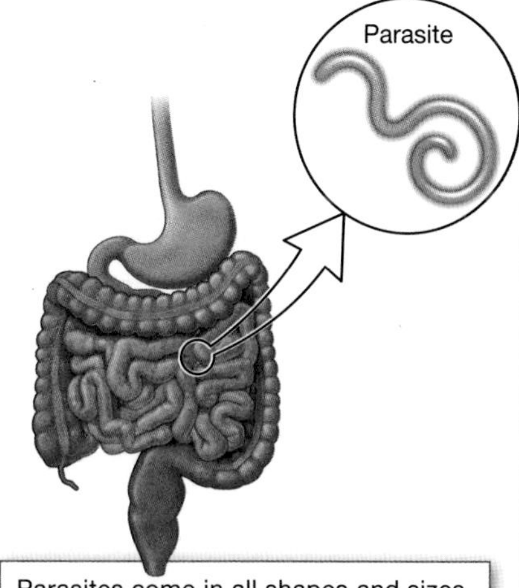

Parasites come in all shapes and sizes. Some, such as malarial parasites, can be intracellular, but most are multicellular extracellular pathogens. IgE and granulocytes such as mast cells, eosinophils, and basophils, are specialized parasite fighters.

virus completely is limited. The cytotoxic T cell response, though, is key, as it eventually overwhelms the virus and kills infected cells before the virus can complete its replicative cycle. Clonal expansion and the ability of cytotoxic T cells to kill more than one target cell make these cells especially effective against viruses. In fact, without cytotoxic T cells, it is likely that humans would all die at some point from a viral infection (if no vaccine were available).

Learning Check

1. Within which of the following structures are B cells that react to self-antigens destroyed?
   a. Thymus
   b. Hypothalamus
   c. Bone marrow
   d. Spleen
2. Kwami has an ear infection for the first time. The infection began a few days ago. If Kwami had a blood test for antibodies, which of the following classes of antibodies would you expect to have increased to the greatest extent?
   a. IgM
   b. IgG
   c. IgA
   d. IgE
3. Chicken pox (varicella) is a viral disease that causes rashes and is highly contagious. Children are recommended to be fully vaccinated by the time they are 6 years old. Anyone that has had the vaccine is expected to be protected against the virus. What type of immunity does the chicken pox vaccine provide?
   a. Active immunity
   b. Passive immunity

Chapter Review

21.1 Anatomy of the Lymphatic and Immune Systems

The learning objectives for this section are:

21.1.1 Describe the major functions of the lymphatic system.

21.1.2 Compare and contrast whole blood, plasma, interstitial fluid, and lymph.

21.1.3 Compare and contrast lymphatic vessels and blood vessels in terms of structure and function.

21.1.4 Describe the path of lymph circulation.

21.1.5 Describe the structure, functions, and major locations of the following lymphatic organs: lymph nodes, thymus, and spleen.

21.1.6 Describe the structure, function, and major locations of lymphatic nodules (e.g., mucosa-associated lymphoid tissue [MALT], tonsils).

21.1 Questions

1. All of the following are functions of the lymphatic system except:
   a. Draining excess interstitial fluid from the tissues
   b. Transporting oxygen through the bloodstream
   c. Regulating the fluid levels around the body
   d. Filtering pathogens out of the blood

2. What is the difference between interstitial fluid and lymph, if any?
   a. Plasma proteins are present in interstitial fluid, but not in lymph.
   b. Red blood cells are present in lymph, but not in interstitial fluid.
   c. Interstitial fluid is found in the spaces surrounding the cells, but lymph is found in the vessels of the lymphatic system.
   d. There is no difference between interstitial fluid and lymph; they are chemically identical and both are found in the spaces surrounding the cells and the vessels of the lymphatic system.

Mini Case 1: Lymphedema

Sharron is a 54-year-old female who has a history of right breast cancer. During surgery to remove a breast tumor, the doctors also removed several axillary lymph nodes.

1. When Sharron's lymph nodes were removed, some lymphatic vessels were also removed. How do lymphatic vessels compare to blood vessels? Please select all that apply.
   a. Blood capillaries can absorb much larger molecules than lymphatic capillaries.
   b. Lymphatic capillaries receive fluid from arterioles, just like blood capillaries.
   c. Lymphatic capillaries consist of cells that overlap, unlike blood capillaries.
   d. Lymphatic vessels contain one-way valves similar to those of veins.

2. By which of the following pathways is lymph flowing out of the right axillary lymph nodes returned to the bloodstream?
   a. It drains into the cisterna chyli, the thoracic duct, and then the left subclavian vein.
   b. It drains into the right lymphatic duct and then into the right subclavian vein.
   c. It drains directly into the left subclavian vein.
   d. It drains directly into the right subclavian vein.

3. Without axillary lymph nodes, which of the following symptoms could be present?
   a. Excess swelling in the arm
   b. Increased pain in the shoulder
   c. Decreased sensation in the arm
   d. Bluish color in the fingers

4. The removal of axillary lymph nodes is a common procedure in in breast cancer surgery. Other lymphatic tissue, called MALT, is also found in the breast. Which of the following are true of MALT? Please select all that apply.
   a. MALT is found in the submucosa of several organs.
   b. MALT is also found in the gastrointestinal tract, lungs and eyes.
   c. MALT consists of lymphoid nodules.
   d. MALT removes damaged red blood cells from the bloodstream and breaks them down.

21.1 Summary

- The lymphatic system drains interstitial fluid from the tissues and returns it to the blood. The plasma of the blood seeps into the extracellular space, becoming interstitial fluid. The endothelial cells of lymphatic capillaries overlap, so once the interstitial pressure builds up, the fluid pushes the endothelial cells apart, flowing into the lymphatic capillaries. Once in the lymphatic capillaries, the interstitial fluid is called *lymph*, which travels to one of the subclavian veins and back into the bloodstream.
- The major organs of the lymphatic system are the lymph nodes, thymus and spleen.
- Lymph nodes remove debris and pathogens from the lymph. The thymus is responsible for the maturation of T cells (T lymphocytes) that originate in the red bone marrow. The spleen filters the blood, removes pathogens and damaged red blood cells from the blood, and initiates the immune response in the body.
- Lymphatic capillaries are specialized to easily absorb interstitial fluid, because their cells overlap and act like valves.
- Lymphatic vessels are similar to veins in that they are thin-walled and contain one-way valves.
- The flow of lymph through the lymphatic system proceeds from lymphatic capillaries into lymphatic vessels. After being filtered by lymph nodes, the lymph continues through larger lymphatic vessels, and then into even larger lymphatic trunks. Next, the lymph drains into either the thoracic duct or the right lymphatic duct; it is finally emptied into the bloodstream through one of the subclavian veins.

21.2 Overview of the Immune Response

The learning objectives for this section are:

21.2.1* Describe the three levels of defense in the immune system.

21.2.2 Compare and contrast innate (nonspecific) with adaptive (specific) defenses.

21.2.3 Describe the roles of various types of leukocytes in innate (nonspecific) and adaptive immune responses.

21.2.4 Explain ways in which the innate (nonspecific) and adaptive (specific) immune responses cooperate to enhance the overall resistance to disease.

* Objective is not a HAPS Learning Goal.

21.2 Questions

1. When someone sneezes in your direction, what are the first lines of defense to prevent pathogenic invasion into the body tissues? Please select all that apply.
   a. Skin
   b. Mucous membrane
   c. White blood cells
   d. Phagocytic cells

2. Which of the following immune responses are slower, but more specific and effective than the other?
   a. Innate immune response
   b. Adaptive immune response

3. Which of the following leukocytes coordinates the activities of adaptive immunity?
   a. Lymphocytes
   b. Granular cells
   c. Phagocytic cells
   d. Red blood cells

4. Describe the interplay between the leukocytes of the innate and adaptive immune responses.
   a. Granular cells and lymphocytes work together to engulf pathogens.
   b. Granular cells and phagocytes are active in innate immune response, while lymphocytes are active in the adaptive immune response that follows. Phagocytes can present antigens to some types of lymphocytes.
   c. Lymphocytes detect pathogens while granular cells release noxious materials.
   d. Phagocytic cells and lymphocytes embed pathogens into long-term memory.

21.2 Summary

- The immune system consists of three levels of defense: (1) barrier defenses, such as the skin and mucous membranes; (2) innate immune response; and (3) adaptive immune response.
- Barrier defenses, such as the skin, act to prevent pathogens from entering the internal tissues of the body.
- Innate (nonspecific) defenses act quickly and respond to the presence of many types of pathogens. These defense mechanisms are performed by granular white blood cells and phagocytes.
- Adaptive (specific) defenses act more slowly, but are more effective than innate responses. Adaptive defenses are performed by lymphocytes (T cells and B cells), each of which responds to the presence of only one type of pathogen.
- The three levels of defense cooperate in act in a specific order. The barrier defenses act first, by attempting to keep pathogens from entering the body. Once pathogens do enter the body, the innate defenses act quickly to attempt to prevent infection. Finally, the adaptive defenses act to prevent an infection by a specific type of pathogen.

21.3 Barrier Defenses and the Innate Immune Response

The learning objectives for this section are:

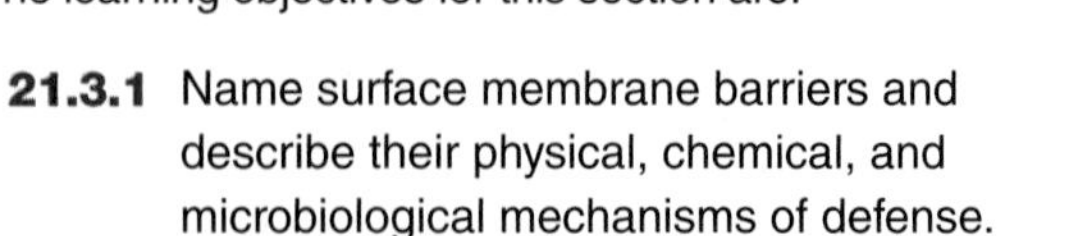

21.3.1 Name surface membrane barriers and describe their physical, chemical, and microbiological mechanisms of defense.

21.3.2 Describe the steps involved in phagocytosis and provide examples of important phagocytic cells in the body.

21.3.3 Describe the functions of natural killer cells.

21.3.4 Explain the role of pattern-recognition receptors in innate defenses.

21.3.5 Explain how complement and interferon function as antimicrobial chemicals.

21.3.6 Define diapedesis, chemotaxis, opsonization, and membrane attack complex, and explain their importance for innate defenses.

21.3.7 Describe the mechanisms that initiate inflammation.

21.3.8 Summarize the cells and chemicals involved in the inflammatory process.

21.3.9 Explain the benefits of inflammation.

21.3.10 Describe the mechanism of fever, including the role of pyrogens.

21.3.11 Explain the benefits of fever.

21.3 Questions

1. Jamaal is in school. People in Jamaal's class have been sneezing and coughing all day long. Explain the role of the mucus in the nose that protects Jamaal from getting sick.

a. The mucus is too dry for bacteria to grow.
b. Mucus constantly leaves the body, taking bacteria with it.
c. The mucus is rich in lysozyme, which breaks down bacteria.
d. Mucus traps bacteria and carries them to the acidic stomach

2. Which of the following are true about macrophages? Please select all that apply.

a. Found in the bloodstream
b. Recruited to the site of infection once a pathogen is detected
c. Engulf debris through cell membrane
d. Kill the bacteria along with themselves

3. What is the function of natural killer cells?

a. To destroy virus-infected cells through apoptosis
b. To destroy pathogens though granules
c. To engulf pathogens to clear the bloodstream

4. Pentraxin is a protein receptor that is embedded within cell membranes. Pentraxin detects a wide variety of pathogens. What is Pentraxin an example of?

a. Interferon
b. Pattern recognition receptor
c. Adaptive immune receptor
d. Prostaglandin

5. You are an epidemiologist and notice that in a new viral infection, the infected cells release cytokines into the extracellular matrix to neighboring cells. What are these cytokines called?

a. Interferons
b. Chemokine
c. Complement proteins
d. Histamine

6. Which of the following is the correct definition of opsonization?

a. Tagging pathogens for phagocytosis by binding them to antibodies
b. Passing of red blood cells through capillaries during inflammation
c. Movement of cells from high concentration to low concentration
d. Engulfing pathogens and breaking down the bacteria

7. Siu is a wrestler for their high school. Siu tried to break their fall and landed on their wrist. They heard a pop and felt a sharp pain. Which of the following are released immediately after tissue cells are injured in order to initiate the inflammatory response? Please select all that apply.

a. Mast cell granules
b. Histamine
c. Interferons

8. What are the functions of prostaglandins? Please select all that apply.

a. To increase the diameter of blood vessels
b. To increase the permeability of capillaries
c. To phagocytose any infiltrated pathogens
d. To mature into macrophages to clean debris

9. Which of the following are benefits of inflammation? Please select all that apply.

a. Encourages entry of clotting factors
b. Facilitates transport of antigens to lymph nodes
c. Prevention of mitosis and resulting cancers
d. Increased tissue elasticity

10. Which of the following is the first step in initiating a fever?

a. Pyrogens are released into the bloodstream
b. The hypothalamus releases prostaglandins
c. Increasing blood flow throughout the body
d. Shivering increases body temperature

11. Pooja woke up one day with a sore throat and a cough. After taking their temperature, Pooja noticed they had a fever. Which of the following are advantages of a fever? Please select all that apply.

a. Slows down enzyme production
b. Increases debris transport to lymph nodes
c. Increases the action of interferons
d. Makes iron less available to bacteria and viruses, which inhibits their replication

21.3 Summary

- Barrier defenses act to prevent pathogens from entering the body, to prevent infections. The main barrier defenses are the skin, mucous membranes, and their various secretions, such as sweat, acid and enzymes in the stomach, and lysozyme in the saliva.
- Other innate immune responses act on pathogens once they have entered the body. The main cells that are responsible for innate immune responses are phagocytes (macrophages and neutrophils), natural killer cells, and granulocytes. The soluble mediators are cytokines and chemokines.
- The inflammatory response is an innate immune response that restricts the spread of infection. It consists of four processes: tissue injury, vasodilation, increased vascular permeability, and recruitment of phagocytes.
- A fever is an innate immune response in which the body temperature is increased in order to make the body uninhabitable for pathogens.

21.4 The Adaptive Immune Response

The learning objectives for this section are:

21.4.1 Define antigen, self-antigen, and antigen receptor.

21.4.2 Compare and contrast antibody-mediated (humoral) and cell-mediated (cellular) immunity.

21.4.3 Describe the immunological memory (anamnestic) response.

21.4.4 Explain the roles of antigen-presenting cells (APCs) and provide examples of cells that function as APCs.

21.4.5 Define major histocompatibility complex (MHC).

21.4.6 Explain the functions of class I and class II major histocompatibility complex (MHC) proteins in adaptive (specific) immunity.

21.4.7 Describe where class I and class II major histocompatibility complex (MHC) proteins are found.

21.4.8 Describe where B and T cells originate, and contrast where they attain their immunocompetence.

21.4.9 Compare and contrast the mechanisms of antigen challenge and the clonal selection processes of B and T cells, including effector cells, helper cells, memory cells, and important cytokines.

21.4.10 Describe the general structure and functions of the various types of lymphocytes (e.g., helper T cells, cytotoxic T cells, regulatory [suppressor] T cells, B cells, plasma cells, memory cells).

21.4.11 Compare and contrast the defense mechanisms and functions of B and T cells.

21.4.12 Interpret a graph of the primary and secondary immune response, in terms of the relative concentrations of different classes of antibodies produced over time.

21.4.13 Describe antibody structure.

21.4.14 Compare and contrast the structure and functions of the classes of antibodies.

21.4.15 Describe mechanisms of antibody action and correlate mechanisms with effector functions.

21.4.16* Compare and contrast the anatomical locations and immune responses against different types of pathogens.

* Objective is not a HAPS Learning Goal.

21.4 Questions

1. Where would you find antigen receptors? Please select all that apply.
   a. B lymphocytes
   b. Antigens
   c. Bacterial cell membranes
   d. T lymphocytes

2. Which of the following are true of cell-mediated or antibody-mediated immune responses? Please select all that apply.
   a. Cell-mediated immune responses are performed by B cells.
   b. In antibody-mediated immunity, antibodies travel to the infection site to attack the pathogens.
   c. Both cell-mediated and antibody-mediated responses are types of adaptive immune responses.
   d. In cell-mediated immune responses, cytokines travel to the site of infection to attack the pathogens.

3. Which of the following explains why children seem to get sick more often than adults?
   a. Children have well-established secondary adaptive responses.
   b. Children are still developing immunological memory.
   c. Children have abnormal primary adaptive responses.
   d. Children lack antibody production until puberty.

4. Which of the following are classified as professional antigen-presenting cells? Please select all that apply.
   a. Macrophages
   b. Somatic cells
   c. T cells
   d. Dendritic cells

5. Which of the following describes the major histocompatibility complex?
   a. Antigens on the surface of specialized cells
   b. Proteins that bring antigenic material to the surface of the cell to display it for certain types of lymphocytes
   c. Cells that secrete cytokines to enhance immune response

6. How are MHC class I and MHC class II molecules similar?
   a. They are both found in all somatic cells.
   b. They are both only found on professional antigen-presenting cells.
   c. Both molecules are monitored by T cells.
   d. Both molecules are monitored by dendritic cells.

7. What type of MHC molecule does a healthy skin cell express?
   a. MHC Class I
   b. MHC Class II
   c. MHC Class III
   d. MHC Class IV

8. B cell lymphoma is a type of cancer in which B cells are over-produced. The treatment is often to prevent mitosis of B cells, sometimes by removing the stem cells that produce them. Which of the following would be the target organ of these treatments?
   a. Blood plasma
   b. Bone marrow
   c. Thymus
   d. Hypothalamus

9. Which of the following is true of the clonal selection in B cells and T cells?
   a. B cells can only bind to antigenic cells for which they contain specific receptors, but T cells can bind to any cells determined to be foreign to the host.
   b. T cells can only bind to antigenic cells for which they contain specific receptors, but B cells can bind to any cells determined to be foreign to the host.
   c. Both B cells and T cells can only bind to antigenic cells for which they contain specific receptors.
   d. Both B cells and T cells can bind to any cells determined to be foreign to the host.

10. Which of the following cells are responsible for suppressing immune responses after the infection has subsided? Please select all that apply.
   a. Helper T cells
   b. Cytotoxic T cells
   c. Regulatory T cells
   d. B cells

11. Which of the following is true about B cell activation?
   a. B cells must differentiate into plasma cells before binding to antigenic cells.
   b. B cells can bind to antigenic cells that fit their receptors, but then require the cytokines from Helper T cells to become fully activated.
   c. B cells must differentiate into memory B cells before binding to antigenic cells.
   d. B cells can only bind to T cells, which then bind to antigenic cells.

12. Jerome was sick with the common cold. After they recovered, they were introduced to the same pathogen again. The following figure portrays the antibody levels when Jerome was sick. Which of the following antibodies is portrayed in purple?

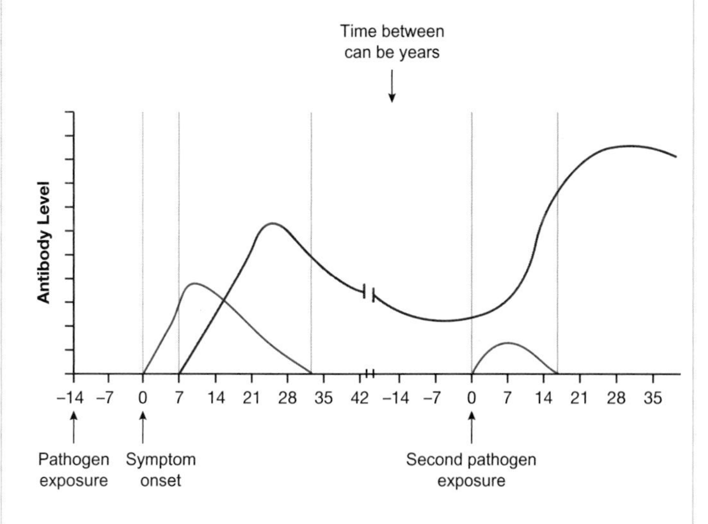

   a. IgG
   b. IgA
   c. IgM
   d. IgE

13. Which is true of antibody structure?

a. The Fc region is comprised of the heavy chains.
b. The base of a Y-shaped antibody is comprised of the light chains.
c. The antigen-binding sites are comprised of the heavy chains.

14. Which is the main class of antibodies of a secondary response in blood?

a. IgA
b. IgE
c. IgG
d. IgM

15. Covid-19 is a respiratory disease in which pathogens are able to permeate the mucosal border and bind to epithelial cells in order to spread. Covid-19 vaccines increase the production of IgA antibodies. Why would increased IgA antibodies help defend against Covid-19?

a. IgA clears pathogens from the blood.
b. IgA neutralizes pathogens within mucus.
c. IgA targets allergic and inflammatory reactions.
d. IgA is the first antibody made during primary response.

16. Tapeworms are long, multicellular parasites that can occupy a host's small intestine if the host eats food contaminated with tapeworm eggs or larvae. Which of the following immune reactions might be effective to combat this parasite?

a. Eosinophils release the contents of their granules.
b. Neutrophils engulf the pathogen, destroying it.
c. NK cells perform apoptosis to destroy pathogen-infected cells.

21.4 Summary

- Adaptive immune responses are slower to respond to the presence of pathogens in the body than the innate immune response, but they are more effective. Adaptive defenses are specific against one type of pathogen, which they must recognize before they can act on them.
- Adaptive immune responses are classified as either (1) cell-mediated immune responses, in which cells travel to the infection site and attack the pathogens, or (2) antibody-mediated immune responses, if antibodies travel to the infection site to act against the pathogens.
- Adaptive immune responses can also be classified as (1) the primary immune response, in the case of the body's first exposure to an antigen, or (2) the secondary immune response, in the case of a subsequent exposure to an antigen. The secondary immune response is faster and more effective than the primary immune response due to immunological memory.
- T cells are responsible for cell-mediated immunity. There are three types of T cells, each with a specific function: (1) helper T cells, (2) cytotoxic T cells, and (3) regulatory T cells.
- B cells are responsible for antibody-mediated immunity, with some help from helper T cells. There are two types of B cells: (1) plasma B cells, which produce antibodies, and (2) memory B cells, which provide long-term immunity.
- Passive immunity is temporary immunity obtained by the transfer of antibodies from one person to another.
- Active immunity is permanent immunity obtained by antigen exposure, the resulting immune response that occurs by a person's own immune system, and the production of memory B cells against the antigen.

22

The Respiratory System

Narcisa/E+/Getty Images

Chapter Introduction

In this chapter we will learn . . .

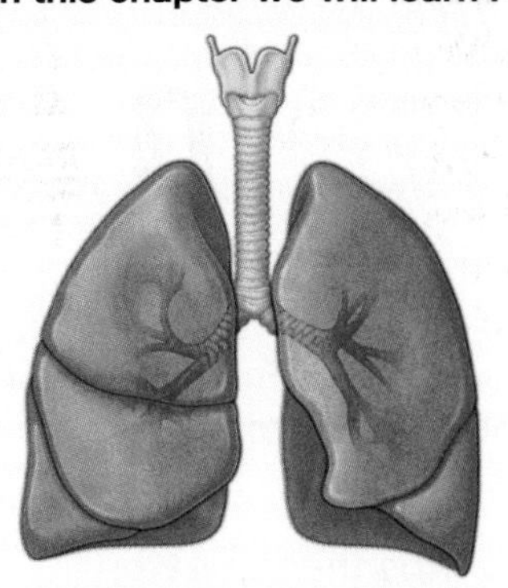

the anatomy of the lungs and how their structure enables the function of oxygenating the blood.

We will also learn more about the physiology of the respiratory gasses in the blood, and how the brain controls our breathing both unconsciously and consciously.

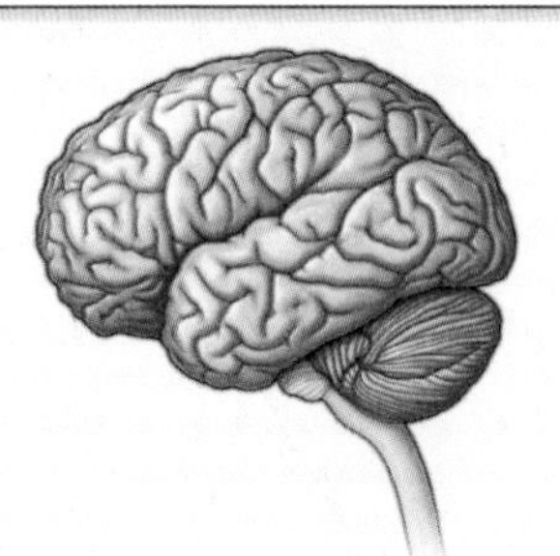

Hold your breath. Really! See how long you can hold your breath as you continue reading. How long can you do it? Chances are you are feeling uncomfortable already. A typical human cannot survive without breathing for more than a few minutes Yet, isn't it interesting that you can control your breathing at all? Compare this to another vital system—try to stop your heart from beating for the next 30 seconds. Human—actually, most vertebrate—breathing muscles are skeletal, which means that they are under voluntary control. Though it can be controlled voluntarily, breathing is regulated by respiratory centers in the brain and the autonomic nervous system to prevent us from avoiding breathing for too long. This is because most cells in the body rely on oxygen to run efficient cellular respiration, the process by which energy is produced in the form of adenosine triphosphate (ATP). Just as critically, cellular respiration, when performed with oxygen, yields carbon dioxide, which must be released to maintain homeostasis. In fact, although oxygen is critical for cell survival, it is actually the buildup of carbon dioxide that drives your need to breathe. Oxygen and carbon dioxide travel through the branched tubes of the respiratory system, traveling to and from the microscopic gas exchange surfaces covered by capillaries. The circulatory and respiratory systems act as a team to fill the blood with oxygen and empty its carbon dioxide waste, ensuring homeostasis for the tissues of the body.

22.1 Functions and Anatomy of the Respiratory System

Learning Objectives: By the end of this section, you will be able to:

22.1.1 Describe the major functions of the respiratory system.

22.1.2 Compare and contrast the general locations and functions of the conducting and respiratory portions (zones) of the respiratory tract.

22.1.3 Identify the anatomical division of the upper versus lower respiratory tract.

22.1.4 List, in order, the respiratory structures that air passes through during inspiration and expiration.

The Human Anatomy and Physiology Society includes more than 1,700 educators who work together to promote excellence in the teaching of this subject area. The HAPS A&P Learning Outcomes measure student mastery of the content typically covered in a two-semester Human A&P curriculum at the undergraduate level. The full Learning Outcomes are available at https://www.hapsweb.org.

22.1.5 Describe the major functions, gross anatomical features, and epithelial lining of the nasal cavity, paranasal sinuses, and pharynx.

22.1.6 Describe the major functions of the larynx.

22.1.7 Describe the anatomical features of the larynx, including the laryngeal cartilages.

22.1.8 Compare and contrast the location, composition, and function of the vestibular folds (false vocal cords) and vocal folds (true vocal cords).

22.1.9 Briefly explain how the vocal folds and the larynx function in phonation.

22.1.10 Describe the major functions of the trachea.

22.1.11 Describe the microscopic anatomy of the trachea, including the significance of the C-shaped hyaline cartilage rings.

22.1.12 Describe the gross anatomical features of the trachea, including its positioning with respect to the esophagus.

22.1.13 Identify and describe the anatomic features of the bronchial tree (e.g., main [primary] bronchi, lobar [secondary] bronchi, segmental [tertiary] bronchi, smaller bronchi, bronchioles, terminal bronchioles, respiratory bronchioles, alveolar ducts, alveolar sacs, and alveoli).

22.1.14 Compare and contrast the main anatomical differences between bronchi and bronchioles.

22.1.15 Describe the histological changes that occur along the bronchial tree from larger to smaller air passageways.

22.1.16 Identify and describe the respiratory membrane, and explain its function.

22.1.17 Compare and contrast the gross anatomic features of the left and right lungs, and explain the reasons for these differences.

22.1.18 Identify and describe the location, structure, and function of the visceral and parietal pleura, serous fluid, and the pleural cavity.

LO 22.1.1

The respiratory system functions in gas exchange as described in the introductory paragraph. But other functions that may not be as obvious to you are speech and immune protection. As illustrated by the coronavirus pandemic, the respiratory system is a major source of immune vulnerability for our human bodies and therefore a site of tremendous immune activity. As a species, humans have the most well-developed and depended upon language skills in all of evolutionary history. Humans are unique in their ability to convey sentences, thoughts, and emotions through language. Much of this ability developed as a side effect of the change in larynx (voice box) anatomy when humans became bipedal and head position changed. A notable side effect to the position of the larynx, which enables our complex speech, is that humans have a risk of food entering the airway, causing choking. This risk is relatively unique to human anatomy, and we will discuss some of the anti-choking mechanisms in place in the airway. Last, but not least in significance, the respiratory system functions in concert with the urinary system to maintain pH homeostasis in the body. Because the respiratory system acts much faster, it is essential in pH homeostasis.

LO 22.1.2

respiratory portion (respiratory zone)

When we consider the anatomy of the respiratory system, we can superficially divide the system into two portions—the **conduction portion** (the tubes of the respiratory tract through air travels) and the **respiratory portion** (the sites of gas exchange between the respiratory system and the blood) (**Figure 22.1**).

Figure 22.1 Overview of the Respiratory System

The respiratory system consists of the lungs, as well as the superior structures that bring air toward them from the external environment.

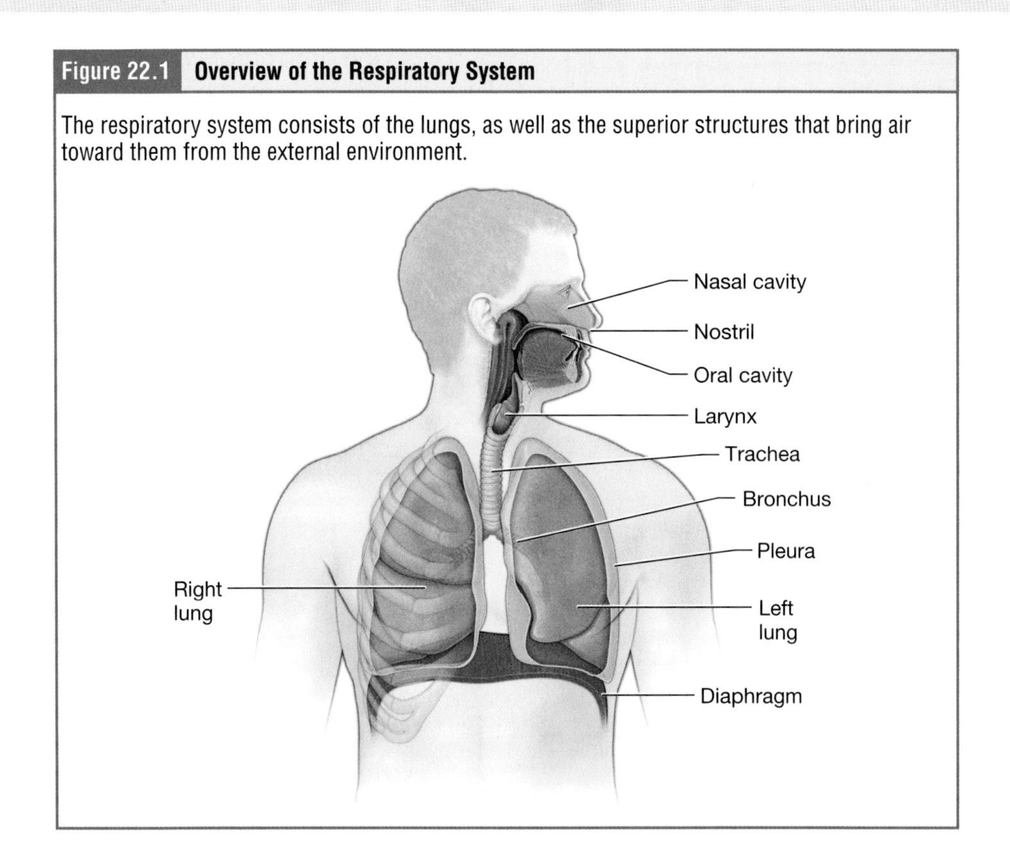

22.1a Conducting Zone

The conducting zone provides a route for incoming and outgoing air, but it also plays a role in removing debris and pathogens from the incoming air, as well as warming and humidifying the air as it travels through. The respiratory tract is divided into the **upper respiratory tract**, consisting of the structures in the head and neck, and the **lower respiratory tract**, consisting of structures in the thorax (**Figure 22.2A**).

LO 22.1.3

The Nose and its Adjacent Structures Air enters the respiratory system through both the nose and the mouth, but the nose serves as the major entrance and exit for the respiratory system.

LO 22.1.4

LO 22.1.5

The **external nose** consists of the surface and skeletal structures that result in the outward appearance of the nose and contribute to its numerous functions (**Figure 22.2B**). The **root** is the region of the nose located between the eyebrows. The **bridge** is the part of the nose that connects the root to the rest of the nose. The **dorsum nasi** is the length of the nose. The **apex** is the tip of the nose. On either side of the apex, the nostrils are formed by the alae (singular = ala). An **ala** is a cartilaginous structure that forms the lateral side of each **naris** (plural = nares), or nostril opening. The **philtrum** is the concave surface that connects the apex of the nose to the upper lip.

Underneath the thin skin of the nose are its skeletal features (**Figure 22.2C**). While the root and bridge of the nose consist of bone, the protruding portion of the nose is composed of cartilage. The **nasal bone** is one of a pair of bones that lies under the root and bridge of the nose. The nasal bone articulates superiorly with the frontal bone and laterally with the maxillary bones. Septal cartilage is flexible hyaline cartilage connected to the nasal bone, forming the dorsum nasi. The **alar cartilage** consists of the apex of the nose; it surrounds each naris.

alar cartilage (major alar cartilage; greater alar cartilage; lower lateral cartilage)

Figure 22.2 The Nose and Upper Respiratory System

(A) The upper respiratory tract consists of the nasal cavity, the pharynx, and the larynx. (B) The structure of the nose consists of cartilage and bone. (C) The external structures of the nose.

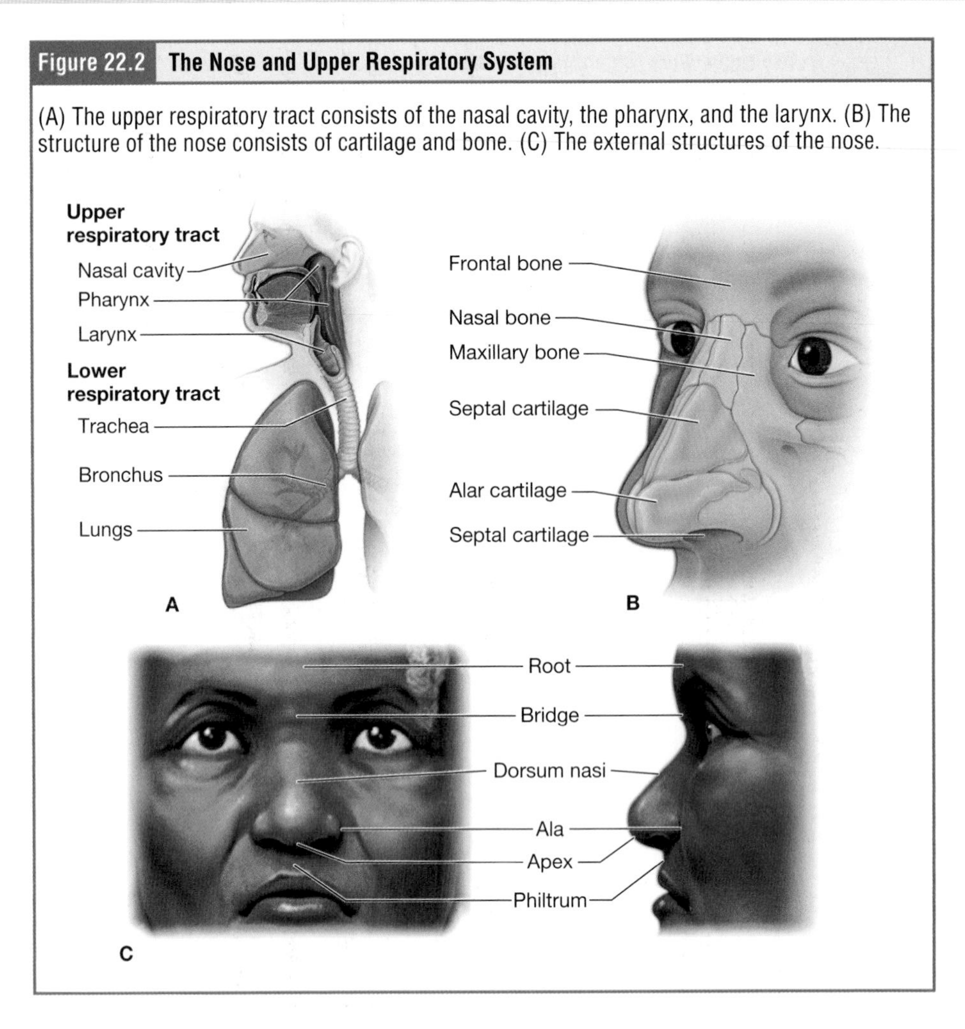

The nares open into the nasal cavity, which is separated into left and right sections by the nasal septum (**Figure 22.3**). The **nasal septum** is formed anteriorly by a portion of the septal cartilage (the flexible portion you can touch with your fingers) and posteriorly by the perpendicular plate of the ethmoid bone (a cranial bone located just posterior to the nasal bones) and the thin vomer bones. Each lateral wall of the nasal cavity has three bony projections, called the *superior*, *middle*, and *inferior nasal conchae*. Whereas the inferior conchae are separate bones, the superior and middle conchae are portions of the ethmoid bone. Conchae are covered with mucous membrane, and as the air is inhaled it swirls throughout the conchae. As it does so, the air is warmed and moistened. The conchae and **meatuses** also conserve water and prevent dehydration of the nasal epithelium by trapping water during exhalation. The floor of the nasal cavity is composed of the palate. The hard palate at the anterior region of the nasal cavity is composed of bone. The soft palate at the posterior portion of the nasal cavity consists of muscle tissue. Air exits the nasal cavities via the internal nares and moves into the pharynx.

Several bones that help form the walls of the nasal cavity have air-containing spaces called the *paranasal sinuses*, which serve to warm and humidify incoming air. Sinuses are lined with a mucosa. Each **paranasal sinus** is named for its associated bone: frontal sinus, maxillary sinus, sphenoidal sinus, and ethmoidal sinus. The sinuses produce mucus, and these air-filled spaces lighten the weight of the skull.

Figure 22.3 The Upper Respiratory System

Air enters and leaves the body through the nose and mouth and flows through the pharynx, which connects the nose and mouth to the lower respiratory system.

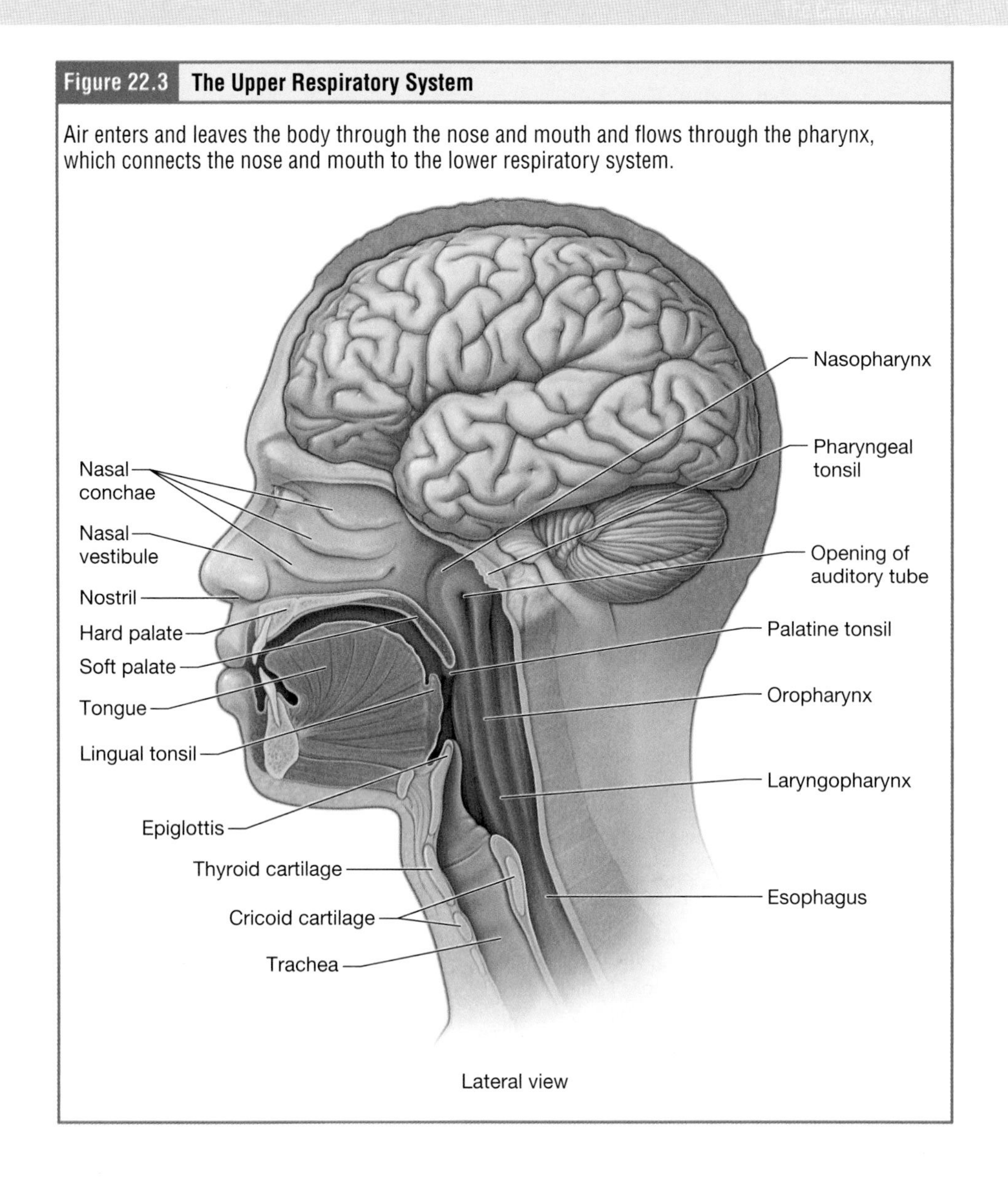

The nares and anterior portion of the nasal cavities are lined with mucous membranes, containing mucous glands and hair follicles that serve to prevent the passage of large debris, such as dirt, through the nasal cavity. An olfactory epithelium used to detect odors is found deeper in the nasal cavity (and discussed in Chapter 15).

The conchae, meatuses, and paranasal sinuses are lined by **respiratory epithelium** composed of pseudostratified ciliated columnar epithelium (**Figure 22.4**). The epithelium contains goblet cells, one of the specialized, columnar epithelial cells that produce mucus to trap debris. The cilia of the respiratory epithelium help remove the mucus and debris from the nasal cavity with a constant beating motion, sweeping materials toward the throat to be swallowed. Interestingly, cold air slows the movement of the cilia, resulting in accumulation of mucus that may in turn lead to a runny nose during cold weather. This moist epithelium functions to warm and humidify incoming air. Capillaries located just beneath the nasal epithelium warm the air by convection. Serous and mucus-producing cells also secrete the lysozyme enzyme and proteins called *defensins*, which have antibacterial properties. Immune cells that patrol the connective tissue deep to the respiratory epithelium provide additional protection.

Learning Connection

Broken Process

If the air-filled sinuses help the head to feel lighter, what would it feel like if these chambers filled with mucus, such as during a sinus infection?

Figure 22.4 Pseudostratified Ciliated Columnar Epithelium

Most of the airway is lined with pseudostratified ciliated columnar epithelium, an epithelium rich with goblet cells, which secrete mucus. The mucus is secreted onto the apical surface of the epithelial cells, where it helps to trap dirt, pathogens, and particles. The cilia sweep the mucus upward, where it can be swallowed into the gastrointestinal tract.

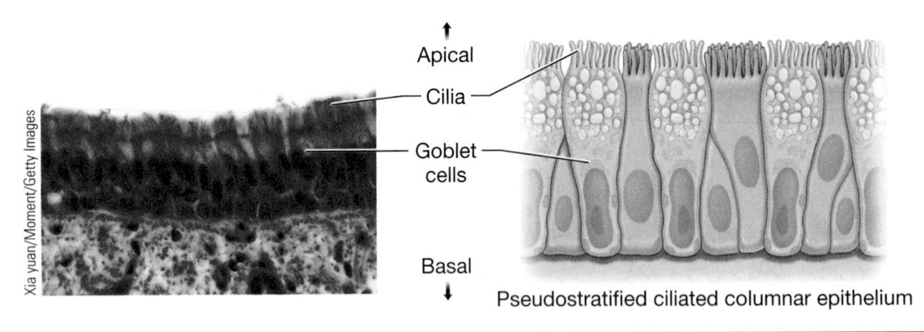

pharynx (throat)

laryngopharynx (hypopharynx)

Pharynx The **pharynx**, or throat, is essentially a tube formed by skeletal muscle and lined by mucous membrane. It begins at the back of the nasal cavities, continues through the back of the mouth, and continues until it splits into the larynx and trachea anteriorly and the esophagus posteriorly. The pharynx has three regions: the **nasopharynx**, which is the pharynx posterior to the nasal cavities, the **oropharynx**, which is the pharynx posterior to the mouth, and the **laryngopharynx** (Figure 22.5).

Figure 22.5 The Parts of the Pharynx

The pharynx, or throat, is a continuous hollow space behind the nose and mouth. The nasopharynx is posterior to, and continuous with, the nasal cavity. The nasopharynx is continuous with the oropharynx, which is posterior to the oral cavity. Inferiorly, the pharynx leads to both the larynx and lower respiratory system and the esophagus and gastrointestinal tract.

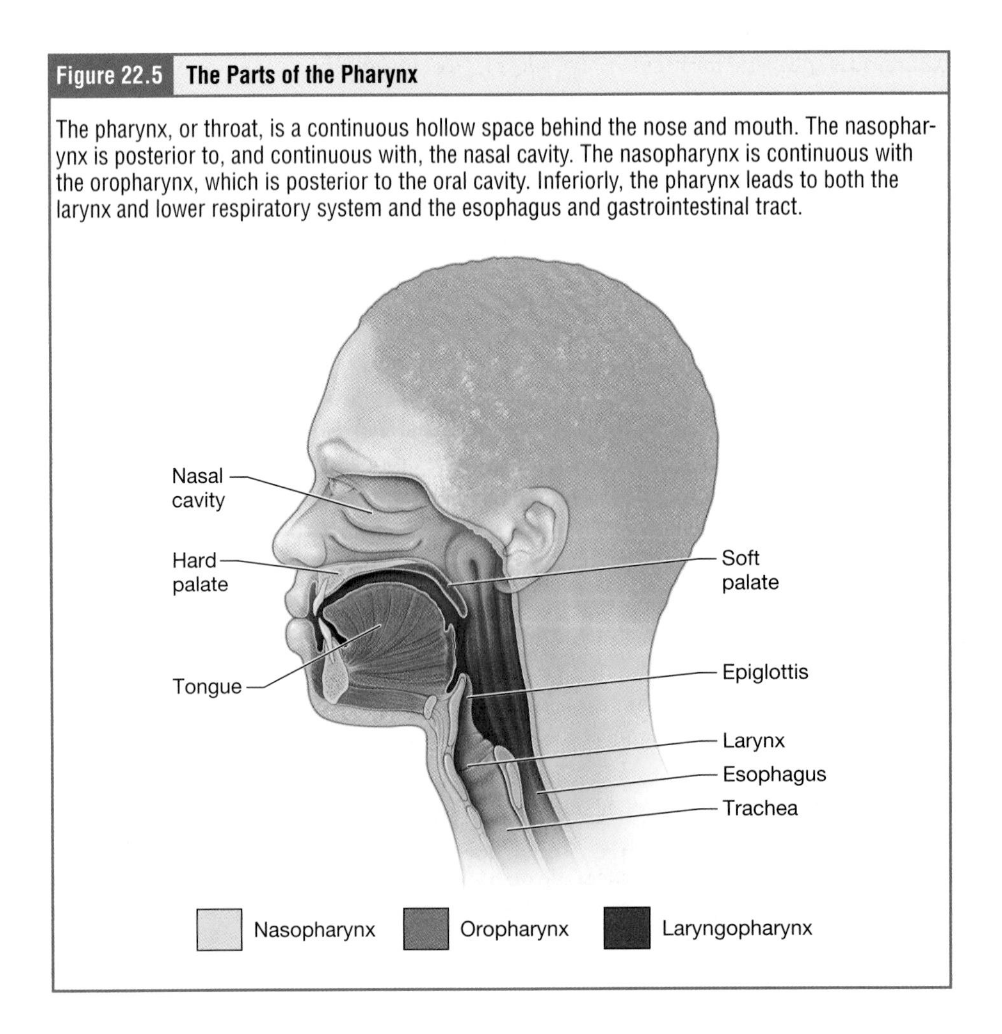

The nasopharynx is structured with the expectation that only air will pass through it; this space is lined by pseudostratified columnar epithelium. The goblet cells of this membrane produce copious amounts of mucus to moisten the apical surface and the air that comes into contact with it. At the top of the nasopharynx are the pharyngeal tonsils. A **pharyngeal tonsil** is an aggregate of lymphoid reticular tissue similar to a lymph node that lies at the superior portion of the nasopharynx. The functions of the pharyngeal tonsils lie both in pathogen destruction and in establishing and maintaining tolerance to the foods we eat. The inferior openings for the auditory tubes are found in the nasopharynx, connecting the air in this region to the middle ear. When an excess of mucus builds up in the nasopharynx during a respiratory infection (cold), this mucus can travel through this connection, fill up the middle ear, and even lead to ear infections. In children, the auditory tubes are more horizontal, making the transport of mucus and bacteria to the middle ear easier. In adults, these tubes are slightly more vertical, making the development of ear infections still possible, but less likely.

pharyngeal tonsil (adenoid)

The **oropharynx** is a passageway for both air and food because it is bordered superiorly by the nasopharynx and anteriorly by the oral cavity. As the nasopharynx becomes the oropharynx, the epithelium changes from pseudostratified ciliated columnar epithelium to stratified squamous epithelium to accommodate the friction of food and drink from our diet. The oropharynx contains two distinct sets of tonsils, the palatine and lingual tonsils. The **palatine tonsils** are located laterally in the oropharynx and the **lingual tonsils** are located at the base of the tongue. The functions they serve are similar to those of the pharyngeal tonsils.

The **laryngopharynx** is inferior to the oropharynx and posterior to the larynx. It continues the route for ingested material and air until its inferior end, where the digestive and respiratory systems diverge. Because both food and air pass through the laryngopharynx, it is lined by stratified squamous epithelium. At its inferior border the laryngopharynx splits into two passages. Food passing through the laryngopharynx will continue into the esophagus, which leads to the gastrointestinal (GI) tract and is found posteriorly, air passing through the laryngopharynx will continue into the larynx, the entrance of the respiratory system.

LO 22.1.6

Larynx The **larynx** is a cartilaginous structure that serves as the entrance to the lower respiratory tract. The larynx connects the pharynx to the trachea and helps regulate the volume of air that enters and leaves the lungs (**Figure 22.6A**). It is also the structure that generates sound for language. Because of this function, the larynx is sometimes called the *voice box*. The structure of the larynx is formed by several pieces of cartilage. Three large cartilage pieces—the thyroid cartilage (anterior), epiglottis (superior), and cricoid cartilage (inferior)—form the major structure of the larynx (**Figures 22.6B** and **22.6C**). The **thyroid cartilage** is the largest piece of cartilage that makes up the larynx. The thyroid cartilage consists of the **laryngeal prominence**, or "Adam's apple," which can be visible in some individuals. The thick **cricoid cartilage** forms a ring, with a wide posterior region and a thinner anterior region. Three smaller, paired cartilages—the **arytenoids**, **corniculates**, and **cuneiforms**—do not contribute to the external structure of the larynx, but function in control over the vocal cords.

larynx (voice box)

LO 22.1.7

laryngeal prominence (Adam's apple)

The **epiglottis**, attached to the thyroid cartilage, is a very flexible piece of elastic cartilage that can close like a lid to cover the opening of the trachea or open to allow air to pass. The act of swallowing causes the pharynx and larynx to lift upward, allowing the pharynx to expand and the epiglottis of the larynx to swing downward, closing the opening to the trachea. These movements produce a larger area for food to pass through, while preventing food and beverages from entering the trachea.

Figure 22.6 The Larynx

(A) The larynx is located at the top of the respiratory tree, just about where the chin and neck meet. You can find your larynx by placing a hand on your throat. (B) The anterior view of the larynx is dominated by the thyroid cartilage, the shape of which is visible from the exterior on some individuals. This large, snowplow-shaped cartilage sits between the hyoid bone and the cricoid cartilage. (C) From the side, we can see the epiglottis, the vestibular and vocal folds, and the smaller cartilages of the larynx, in addition to the thyroid and cricoid cartilages.

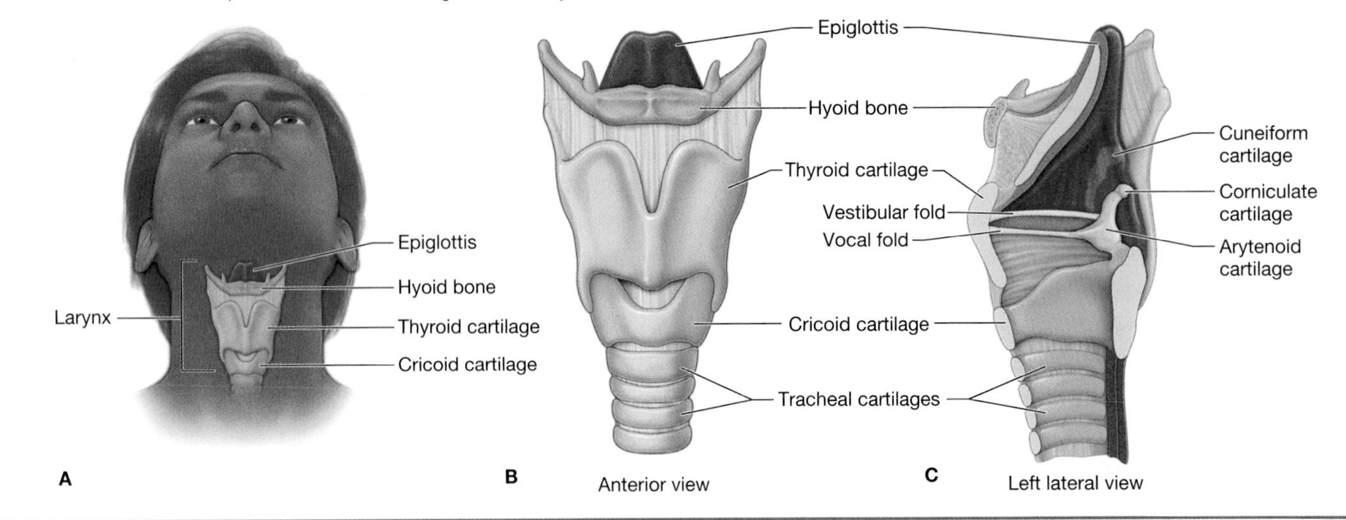

LO 22.1.8

The **glottis** is the space beneath the epiglottis and the soft tissue folds that border it (**Figure 22.7**). The **vestibular folds** are a pair of folded sections of mucous membranes. These folds are one of the most highly innervated and sensitive places in the human body. They function, along with structures in the bronchial tree, to alert the body if material that is dangerous for the airway is entering the respiratory system. You have undoubtedly experienced the result when a crumb or drop of liquid lands on the vestibular folds. A dramatic coughing reaction ensues and the person experiencing this reaction will often excuse themselves by saying that something "went down the wrong pipe." Inferior to the vestibular folds are the vocal folds. The **vocal folds** (sometimes called *vocal cords*) are white, membranous folds attached to the thyroid and arytenoid cartilages. The inner edges of the vocal folds can touch each other. When air is forced through the closed vocal folds, they vibrate against each other and produce sound (**Figure 22.8**). This is not unlike the vibrating strings of instruments such as the guitar

vocal folds (vocal cords, true vocal cords)

LO 22.1.9

Figure 22.7 The Larynx

(A) From the superior view, the larynx is the entrance to the lower respiratory system. The vestibular and vocal folds guard the entrance to the trachea. (B) Laterally, we can see that the vestibular and vocal folds are inferior to the large epiglottis, a membrane-covered shield of elastic cartilage that covers the entrance to the lower respiratory system like a flap during swallowing.

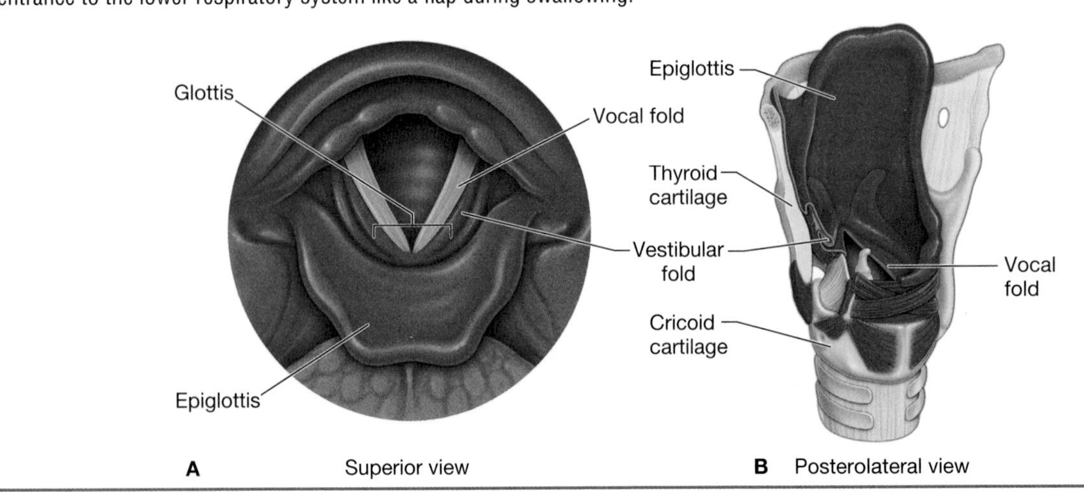

Figure 22.8 The Laryngeal Cartilages

With the membranes removed, the cartilages of the larynx are visible. The vocal folds are stretched between the arytenoid and thyroid cartilages. The arytenoid cartilages rotate atop the cricoid cartilage, pivoting the vocal folds between their open position for breathing or closed position for making sounds (phonation).

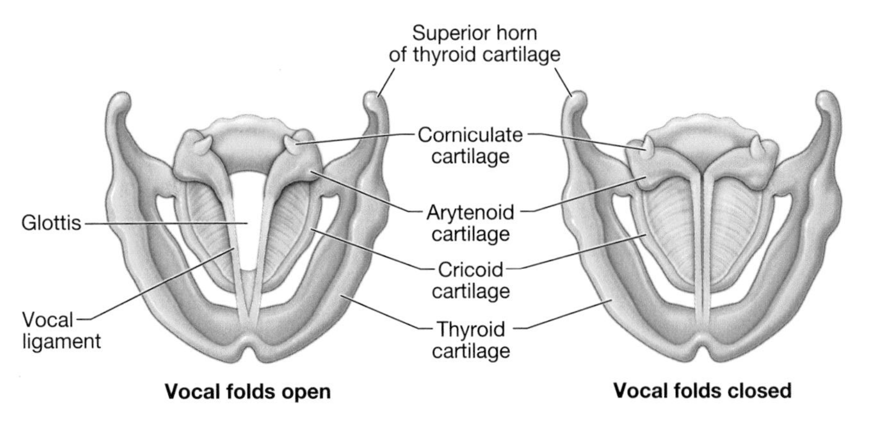

and violin. When air moves through the vocal folds and they are open, not touching, no sound is produced. Our voices do not all sound the same due to anatomical variation in the larynx. If the thyroid cartilage and arytenoid cartilages are further from each other, the vocal folds are longer and produce deeper, lower-pitched sounds (**Figure 22.9**) than if the vocal folds are shorter. This is similar in stringed instruments; the shorter the strings, the higher the pitch of the notes when played.

Figure 22.9 The Length of Vocal Folds Determines Pitch

In stringed instruments (A) as well as vocal folds (B), longer objects vibrate more slowly and therefore produce lower-pitched sounds.

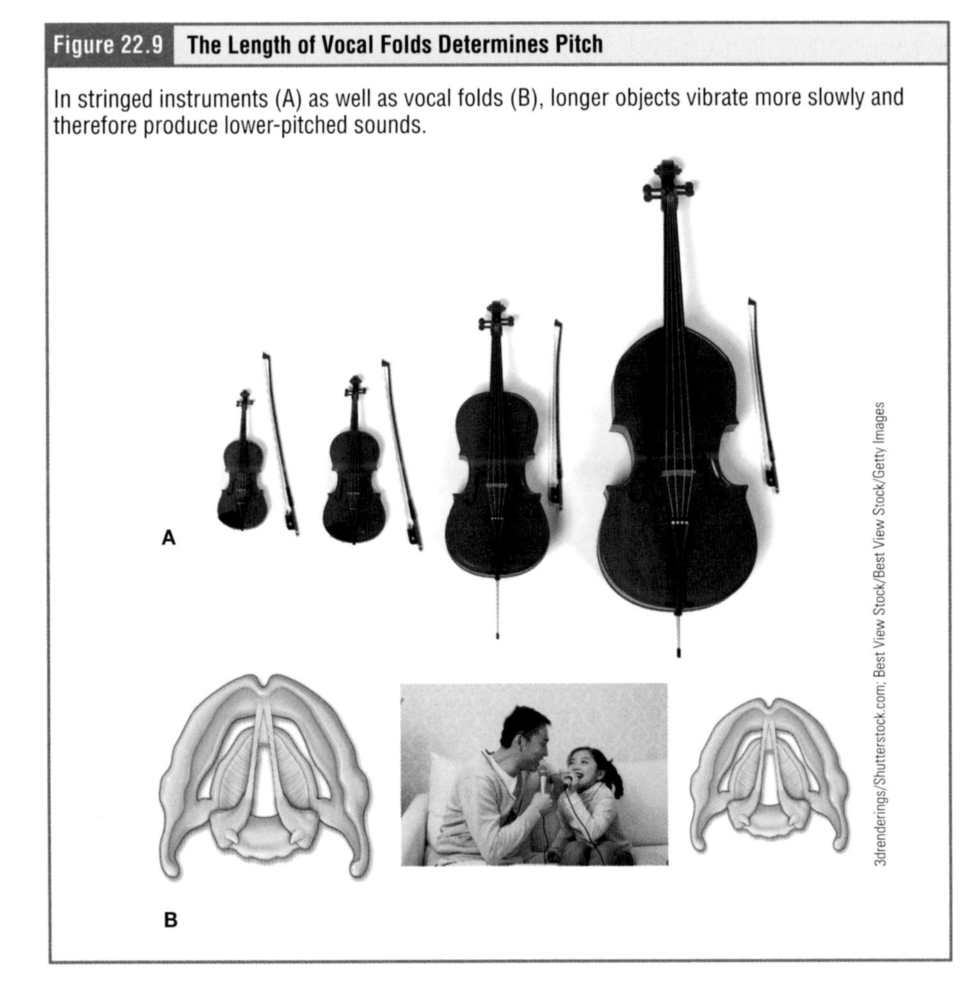

Learning Connection

Broken Process

After hours of loud cheering at an exciting sporting event, you could not cheer; in fact, you could hardly speak at all. You went to the doctor, who said you had developed laryngitis, and that your voice would return in a few days. Look up the meaning of the suffix "-itis." What happens in laryngitis that makes a person unable to speak temporarily? What are other causes of laryngitis besides overuse of the voice?

Continuous with the laryngopharynx, the superior portion of the larynx is lined with stratified squamous epithelium, but this transitions to pseudostratified ciliated columnar epithelium. Similar to the nasal cavity and nasopharynx, this specialized epithelium produces mucus to trap debris and pathogens as they enter the trachea. The cilia beat the mucus upward toward the laryngopharynx, where it can be swallowed down the esophagus.

LO 22.1.10

LO 22.1.11

trachea (windpipe)

LO 22.1.12

Trachea The trachea extends from the larynx toward the lungs, providing a passageway for air into the thorax (**Figure 22.10**). The **trachea** is formed by 16 to 20 stacked, C-shaped pieces of hyaline cartilage that are connected by dense connective tissue. The ends of the hyaline cartilage C rings are joined by the **trachealis muscle**, a flexible muscle that contracts during coughing to narrow the trachea, facilitating the air movement to become more rapid and forceful. The flexible nature of this muscle allows the trachea to stretch and expand slightly during inhalation. In addition, the trachealis muscle allows the esophagus, which is located posteriorly, to expand into the tracheal lumen when swallowing something large (**Figure 22.11**). The trachea is lined with pseudostratified ciliated columnar epithelium, which is continuous with the larynx.

LO 22.1.13

Bronchial Tree At about the level of the sternal angle (the bump that can be felt on your sternum where the manubrium and body meet) the trachea branches into the right and left primary **bronchi** (**Figure 22.12**). These bronchi are also lined by pseudostratified ciliated columnar epithelium containing mucus-producing goblet cells. Rings of cartilage, similar to those of the trachea, support the structure of the bronchi and prevent their collapse. The primary bronchi enter the lungs at the **hilum**, a concave region where blood vessels, lymphatic vessels, and nerves also enter the lungs (**Figure 22.13**). The bronchi continue to branch into a bronchial tree. The **bronchial**

Figure 22.10 Tracheal Rings and the Trachealis Muscle

The length of the trachea and bronchi are lined with C-shaped cartilaginous rings that keep the airway rigidly open. The posterior ends of the cartilaginous rings are joined by the trachealis muscle, which can contract to narrow the airway or relax—especially after swallowing, when the esophagus can distend into the airway.

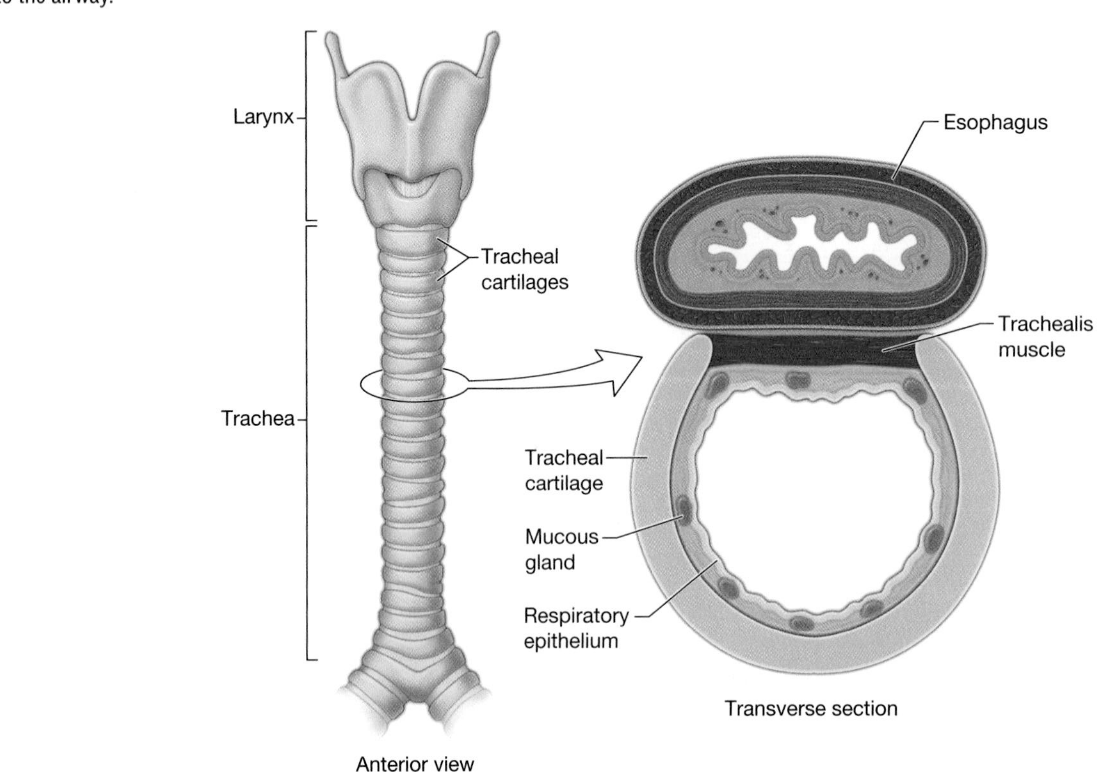

Figure 22.11 The Trachealis Muscle

The esophagus is located directly posterior to the trachea. Between them is the trachealis muscle, which spans the gap between the ends of the cartilaginous rings of the trachea. When the trachealis muscle contracts (middle), it draws the ends of the cartilaginous rings together, narrowing the tracheal lumen. This occurs during coughing, and assists in increasing the velocity and force of air in a cough, which is used to dislodge and expel mucus or other material. When a large chunk, or bolus, of food is swallowed, the trachealis muscle relaxes, allowing the distended walls of the esophagus to push into the space occupied by the trachea.

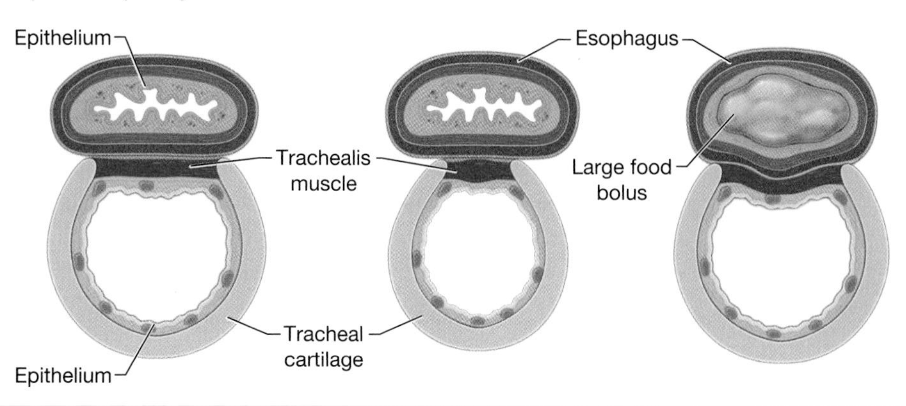

Figure 22.12 The Respiratory System

The upper and lower respiratory tracts are united at the larynx. The lower respiratory tract includes the lungs and the bronchial tree, which begins at the single trachea and splits into right and left bronchi, each of which leads to the bronchial tree within a lung.

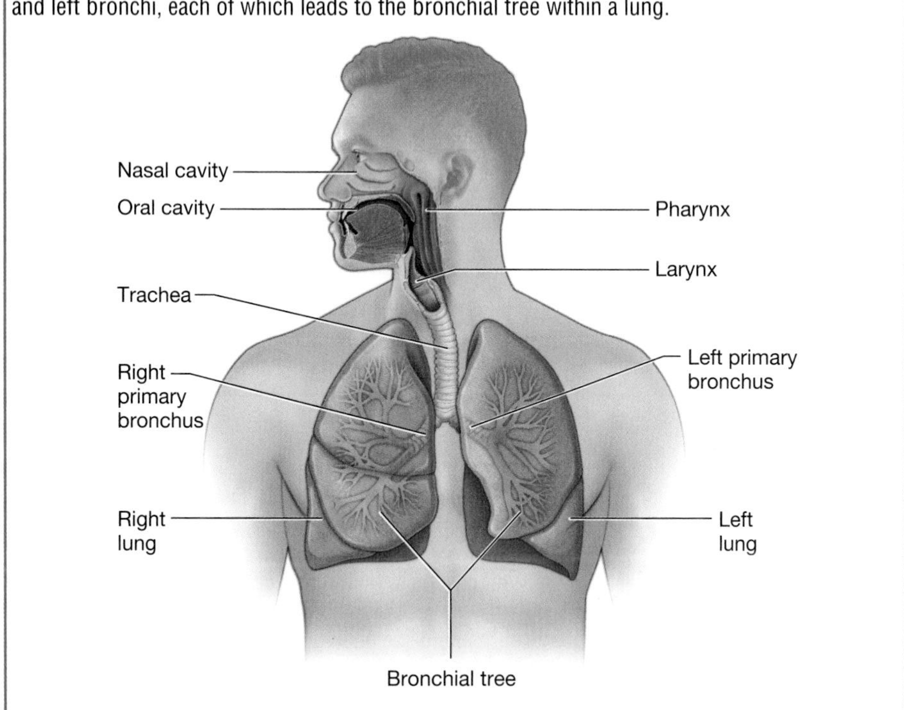

Figure 22.13 The Hilum of the Lung

Each lung has a cluster of vessels that enter/leave at the same location on the medial surface of the lung. This area is called the *hilum*.

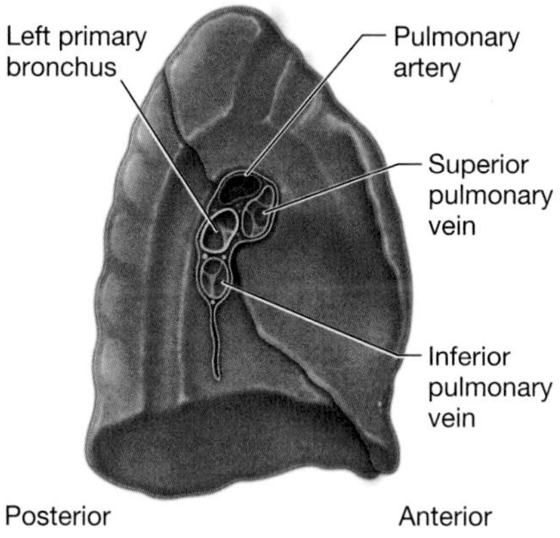

Left primary bronchus
Pulmonary artery
Superior pulmonary vein
Inferior pulmonary vein
Posterior
Anterior

tree is the collective term used for these multiple-branched bronchi. The main function of the bronchi, like other conducting zone structures, is to provide a passageway for air to move into and out of each lung. In addition, the mucus produced by the membrane traps debris and pathogens.

LO 22.1.14

Smaller vessels branching from the smallest bronchi are called **bronchioles**. Bronchioles further branch until they become the tiny terminal bronchioles, which lead to the structures of gas exchange. There are more than 1,000 terminal bronchioles in each lung. Unlike the bronchi, the bronchioles do not contain cartilage, but instead are surrounded by ribbonlike threads of smooth muscle (Figure 22.14). This

LO 22.1.15

Figure 22.14 Alveoli

Clusters of alveoli are found at the ends of the bronchioles. They are covered with a net of capillaries. The pulmonary arteries, which carry deoxygenated blood, transport blood to the pulmonary capillaries. Once oxygenated, the blood leaves in pulmonary veins and returns to the left side of the heart to be pumped to the body.

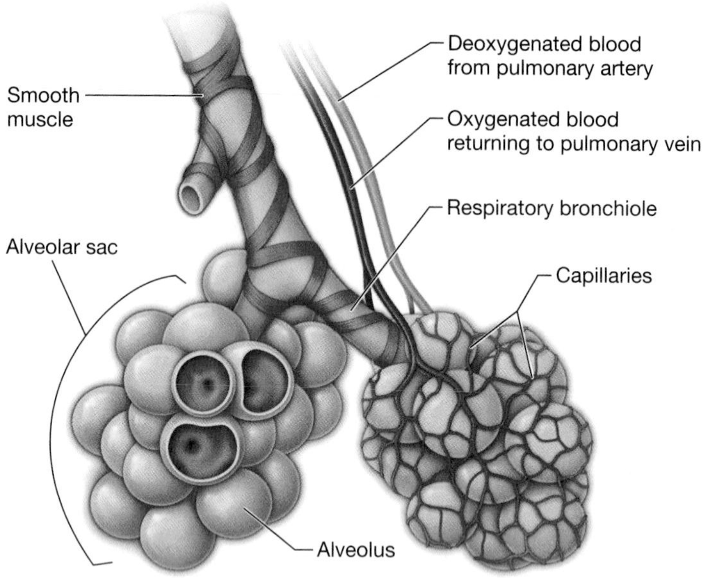

muscular wall can increase the diameter of the bronchioles in **bronchodilation** or decrease the size of the diameter in **bronchoconstriction**. Bronchoconstriction and bronchodilation will be discussed more later on in this chapter, but these passageway changes take place during exercise, quiet breathing, and coughing, among other physiological examples. Whereas the parasympathetic system causes bronchoconstriction, the sympathetic nervous system stimulates bronchodilation.

Cultural Connection

"Just Take a Breath"

How many times have you been upset and been told by someone to "just take a breath" or "take deep breaths." Does it work? What's going on there? Research has been done to study whether breathwork (the conscious control over breathing) changes our physiology. Most studies conclude that there is no significant change to our pulmonary physiology. We can imagine that, as long as the mindful breathing takes place at a pace and cadence that brings in enough oxygen and releases enough carbon dioxide, then a purposeful pattern of breathing might not make much difference to the respiratory system. But what about the other systems? Multiple studies have shown that a regular habit of breathwork or meditation has the ability to lower blood pressure in individuals who are at risk for, or have been diagnosed with, hypertension. Meditation decreases anxiety symptoms in most anxiety sufferers. Because anxiety can increase heart rate and breath rate, using breathwork or meditation as a mechanism for controlling anxiety would have widespread physiological effects. Last, meditation and breathwork have been shown to decrease activity in the amygdala, the fear center in our brains. While scientists do not yet understand how breathwork and meditation are linked to these functions, there is significant emerging data to support the idea that breathwork and meditation can be incorporated into treatment plans for blood pressure and anxiety management and have positive impacts on other brain functions.

22.1b The Respiratory Zone

In contrast to the conducting zone, which functions in air transport, the respiratory zone includes the surfaces across which gasses diffuse and are exchanged between the lungs and the blood. The respiratory zone begins where a terminal bronchiole branches to form a **respiratory bronchiole**, the smallest type of bronchiole (see Figure 22.14), which then leads to an alveolar duct, opening into a cluster of alveoli.

Alveoli An **alveolar duct** is a tube composed of smooth muscle and connective tissue, which opens into a cluster of alveoli. An **alveolus** is one of the many small, grapelike structures that are attached to the alveolar ducts.

An **alveolar sac** is a cluster of many individual alveoli that are responsible for gas exchange. An alveolus has elastic walls that allow the alveolus to stretch during air intake. Alveoli are connected to their neighbors by **alveolar pores**, which help maintain equal air pressure throughout the alveoli and lung (**Figure 22.15**).

alveolar pores (pores of Kohn; interalveolar connections)

Figure 22.15 The Alveolus

(A) Alveoli are hollow spheres; their walls are composed of type I alveolar cells. Scattered within the alveoli are type II alveolar cells, which secrete pulmonary surfactant, and macrophages. Alveoli are clustered together at the end of bronchioles like grapes on a vine. They are connected to their neighbors by small pores. (B) The histology of alveoli reveals how thin their walls are. Type I alveolar cells, which make up 99 percent of the cells in this view, are simple squamous epithelial cells.

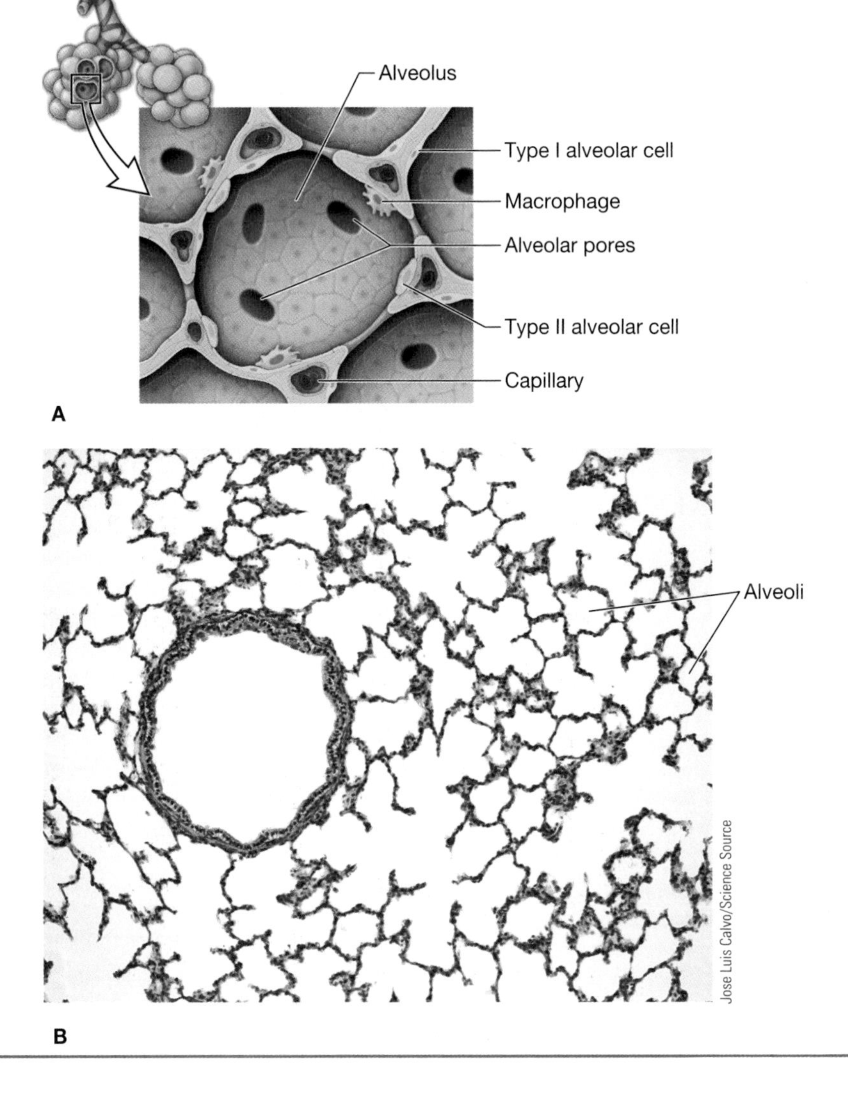

The alveolar wall consists of three major cell types: type I alveolar cells, type II alveolar cells, and alveolar macrophages. A **type I alveolar cell** is a squamous epithelial cell and makes up the walls of the alveoli. These flat cells provide a minimal barrier to the diffusion of the gasses. A **type II alveolar cell** is interspersed among the type I cells and secretes **pulmonary surfactant**, a substance composed of phospholipids and proteins that reduces the surface tension of the alveoli. Roaming around the alveolar wall are **alveolar macrophages**, phagocytic cells of the immune system that remove debris and pathogens.

LO 22.1.16

The simple squamous epithelium formed by type I alveolar cells is attached to a thin, elastic basement membrane (see the "Anatomy of the Respiratory Membrane" feature). This epithelium is extremely thin and borders the endothelial membrane of capillaries. Taken together, the alveoli and capillary membranes form a **respiratory membrane** that allows gasses to cross by simple diffusion, allowing oxygen to be picked up by the blood for transport and CO_2 to be released into the air of the alveoli.

Learning Connection

Image Translation

The terminal bronchioles are part of the conducting zone, while the respiratory bronchioles that branch directly from them are part of the respiratory zone. Look at diagrams or micrographs of terminal and respiratory bronchioles on the Internet. Try to determine what microscopic structures bud off the walls of the respiratory bronchioles but are not present in the walls of the terminal bronchioles. (If you have trouble finding good images, type "Images of terminal and respiratory bronchioles" into your search engine. The images from the websites https://www.quora.com and https://www.differencebetween.com are particularly helpful.)

22.1c The Gross Anatomy of the Lungs

Each **lung** houses structures of both the conducting and respiratory zones. The highly branched structure of the bronchial tree has the end result of creating a very large epithelial surface area, about the size of 5 parking spaces (70 square meters, 750 square feet) that is highly permeable to gasses. Remember that throughout the biological world branching structures are used to maximize surface area to volume relationships, these five parking spaces of surface area are able to fit, highly folded, within your thorax.

Anatomy of...

The Respiratory Membrane

The lumen of the alveolus is filled with air.

Type I alveolar cells make up the walls of the alveolus.

O_2

CO_2

A thin basement membrane unites the thin cells of the alveoli and blood.

The squamous cells of both the alveolar and capillary walls are as thin as possible to enable diffusion of gasses between the alveoli and the blood.

Capillary endothelial cells are simple squamous epithelium.

Red blood cells are packed with hemoglobin, a gas-carrying protein.

The lungs are pyramid-shaped, paired organs that are connected to the trachea by the right and left bronchi; on the inferior surface, the lungs are bordered by the diaphragm. The diaphragm is the flat, dome-shaped muscle located at the base of the lungs and thoracic cavity. The right lung has a bigger volume than the left lung, because of the leftward tilt of the inferior heart, there is less space on the left side of the thorax. The **cardiac notch** is an indentation on the surface of the left lung surrounding the space that the heart occupies (Figure 22.16). The **apex** of each the lung is the superior point that is found beneath each clavicle, while the base is the flattened bottom of each lung that rests on the diaphragm.

LO 22.1.17

Each lung is enclosed within a cavity that is surrounded by the **pleura**. The pleura (plural = pleurae) is a serous membrane that surrounds each lung. The right and left pleurae, which enclose the right and left lungs, respectively, are separated by the mediastinum. The pleurae consist of two layers. The **visceral pleura** is the layer that directly touches the surface of the lungs, and extends into and lines the lung fissures (see Figure 22.16). In contrast, the **parietal pleura** is the outer layer that connects to the thoracic wall, the mediastinum, and the diaphragm. The visceral and parietal pleurae connect to each other at the hilum. The **pleural cavity** is the space between the visceral and parietal layers.

LO 22.1.18

The pleurae perform two major functions: They produce pleural fluid and create cavities that separate the major organs. **Pleural fluid** is secreted by mesothelial cells from both pleural layers and acts to lubricate their surfaces. This lubrication reduces friction between the two layers to prevent trauma during breathing and creates surface tension that helps maintain the position of the lungs against the thoracic wall. This adhesive characteristic of the pleural fluid causes the lungs to enlarge when the thoracic wall expands during ventilation, allowing the lungs to

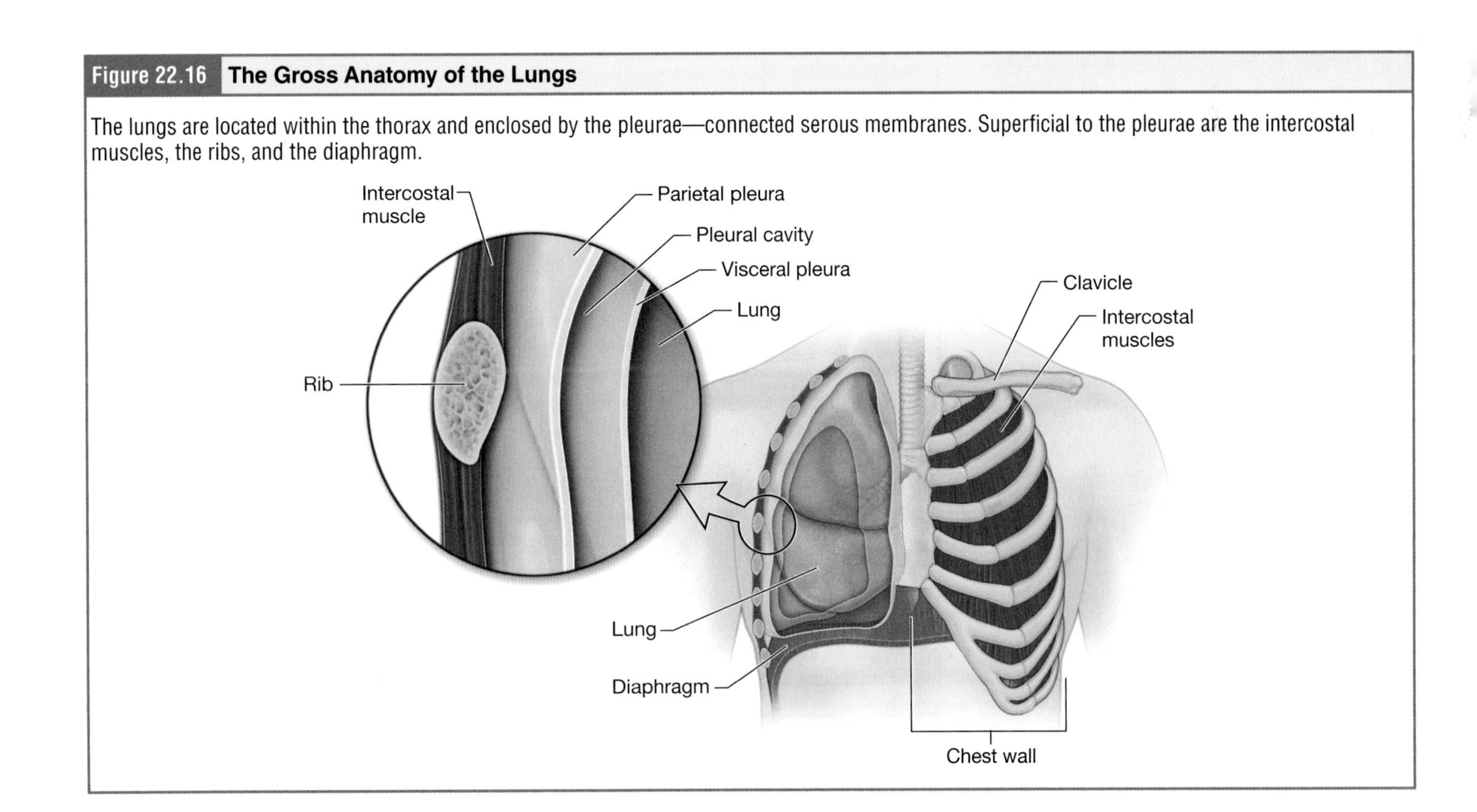

Figure 22.16 The Gross Anatomy of the Lungs

The lungs are located within the thorax and enclosed by the pleurae—connected serous membranes. Superficial to the pleurae are the intercostal muscles, the ribs, and the diaphragm.

superior lobe (upper lobe)

horizontal fissure (minor fissure)

Figure 22.17 The Lobes of the Lungs

Both of the lungs are divided into separate functional lobes. The right lung has three lobes; the left lung, which is smaller due to the presence of the heart, has only two.

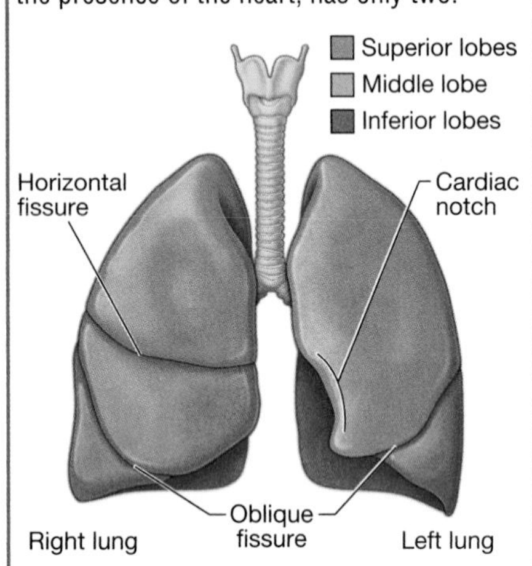

fill with air. The pleurae also create a division between major organs that prevents interference due to the movement of the organs, while preventing the spread of infection.

Each lung is composed of smaller units called *lobes*. Fissures separate these lobes from each other. The right lung consists of three lobes. The **superior lobe** is separated from the **middle lobe** by the **horizontal fissure**. The middle lobe is separated from the **inferior lobe** by the **oblique fissure** (Figure 22.17). The left lung consists of only two lobes, the superior lobe and the inferior lobe, which are separated by an oblique fissure.

Blood Travels Throughout the Lungs The major function of the lungs is to perform gas exchange between the air in the alveoli and the blood within the pulmonary capillaries. Deoxygenated blood travels to the lungs in the left and right **pulmonary arteries**, which are the two branches of the pulmonary trunk. The right pulmonary artery enters the right lung at the hilum (see Figure 22.13) and its many branches carry deoxygenated blood to the alveoli. Capillary beds surround the alveoli (see Figure 22.14) and gas exchange takes place across the respiratory membrane (see the "Anatomy of the Respiratory Membrane" feature). A pulmonary venule drains the capillary bed and carries the now-oxygenated blood toward pulmonary veins, which exit the lungs through the hilum and carry the oxygenated blood to the left side of the heart.

Learning Connection

Chunking

Can you remember other instances in which you have encountered the word "oblique" in anatomy? Do you know what oblique means? If not, based on look of the structures that have the term *oblique* in them, can you guess?

Cultural Connection

Vaping

Traditional or "combustible" cigarettes are full of carcinogens—chemicals that cause cancer. Because of this, about 90 percent of lung cancer deaths are linked to a history of smoking. Vaping, or "e-cigarettes," provide a cleaner alternative, but, as e-cigarette use in the United States soars, scientists are wondering what the long-term effects of this new form of smoking might be. Modern e-cigarettes were invented in China in 2003 by pharmacist Hon Lik, whose father had died from lung cancer as a result of smoking. These e-cigarettes first became available in the United States in 2006, and by 2010, their use had become widespread. Therefore, we still have limited data on long-term use. What is known, however, is that vaping allows the user to take in much higher levels of nicotine—the addictive drug in cigarettes—and so the effects of vaping have more to do with nicotine itself than the other carcinogens present in combustible cigarettes. Nicotine is one of the most addictive substances on the planet, making it very difficult to stop smoking or vaping. The drug itself causes increases in blood pressure in both the pulmonary and systemic circuits, as well as vasoconstriction of the coronary arteries. This combination, in addition to making the heart's job (pumping blood against increased pressure) more difficult, also reduces the delivery of oxygen and nutrients to the heart muscle cells. These effects together can lead to heart changes, including ventricular hypertrophy (enlargement of the ventricular walls), which is often associated with an increased risk of cardiac failure. The effects appear to be especially significant on the right side, and pulmonary hypertension (high blood pressure) appears to be the largest risk of vaping. The intake of nicotine alone appears to increase the risk of several types of cancer. Vaping any substance, whether it be THC (the psychoactive compound in marijuana) or nicotine, appears to cause some inflammation in the lungs. Inflammation here can exacerbate the symptoms and progression of lung disorders, including asthma and COPD (chronic obstructive pulmonary disease).

Learning Check

1. Which of the following is the correct order of the airflow in the upper respiratory tract?
   a. Nasopharynx, laryngopharynx, oropharynx, larynx
   b. Larynx, laryngopharynx, nasopharynx, oropharynx
   c. Nasopharynx, oropharynx, laryngopharynx, larynx
   d. Larynx, nasopharynx, oropharynx, larynx
2. You are eating lunch with a friend when they suddenly have a dramatic, almost choking cough. As they recover, they say to you, "Ugh! Wrong pipe!" and then think for a minute and add, "you're taking anatomy, what is that exactly?" Which of the following is the best explanation?
   a. Their epiglottis was closed over the glottis and food was able to reach the vestibular folds.
   b. Their epiglottis was open and food was able to reach the vestibular folds.
   c. The food entered their trachea and continued down the lower respiratory tract.
   d. The wind from sound production traveled against the bolus of food, suspending it in the esophagus.
3. You are about to take a big exam and are sweating profusely, pacing, your heart is racing, and so on. Are you in a state of sympathetic or parasympathetic dominance? Bronchoconstriction or bronchodilation?
   a. Parasympathetic; bronchoconstriction
   b. Sympathetic; bronchoconstriction
   c. Parasympathetic; bronchodilation
   d. Sympathetic; bronchodilation
4. Which of the following anatomical structures does the left lung have? Please select all that apply.
   a. Superior lobe
   b. Middle lobe
   c. Oblique fissure
   d. Cardiac notch

22.2 The Process of Breathing

Learning Objectives: By the end of this section, you will be able to:

22.2.1 Describe the processes associated with the respiratory system (i.e., ventilation, pulmonary gas exchange [gas exchange between alveoli and blood], transport of gases in blood, tissue gas exchange [gas exchange between blood and body tissues]).

22.2.2 Define pulmonary ventilation, inspiration (inhalation), and expiration (exhalation).

22.2.3 Define atmospheric pressure, intrapulmonary pressure, intrapleural pressure, and transpulmonary pressure.

22.2.4 Identify the muscles used during quiet inspiration, deep inspiration, and forced expiration.

22.2.5 Explain the inverse relationship between gas pressure and volume of the gas (i.e., Boyle's Law) and apply this relationship to explain airflow during inspiration and expiration.

22.2.6 Explain the relationship of intrapleural pressure, transpulmonary pressure, and intrapulmonary pressure relative to atmospheric pressure during ventilation.

22.2.7* Relate the flow equation ($F = \Delta P/R$) to respiration, defining each component.

22.2.8 Explain how pulmonary ventilation is affected by bronchiolar smooth muscle contractions (bronchoconstriction), lung and thoracic wall compliance, and pulmonary surfactant and alveolar surface tension.

22.2.9 Describe the forces that tend to collapse the lungs and those that normally oppose or prevent collapse (e.g., elastic recoil of the lung versus subatmospheric intrapleural pressure).

* Objective is not a HAPS Learning Goal.

Chances are when you hear the word *respiration* you think of moving air in and out of your lungs. However, we also use *respiration* to describe the process of cells using glucose and oxygen to generate ATP. Therefore, we can summarize the interrelated processes of respiration and respiratory system function in four categories:

LO 22.2.1

ventilation (pulmonary ventilation)

- ***Ventilation:*** The movement of air in and out of the lungs
- ***Gas Exchange across the Respiratory Membrane:*** Oxygen and carbon dioxide diffuse across the respiratory membrane according to their concentration gradients, moving from the air-filled alveoli to the blood-filled pulmonary capillaries.
- ***Gas Transport:*** The movement of gasses within the blood around the body.
- ***Gas Exchange between the Blood and the Tissues.***

LO 22.2.2

Pulmonary ventilation is the act of breathing, which can be described as the movement of air into and out of the lungs. Air is a mixture of gasses. Just as with fluids such as blood, air moves by bulk flow, which is driven by pressure gradients (**Figure 22.18**). We discussed these core concepts in our examination of blood flow in Chapter 20. Unlike blood, however, air flows in two directions through the same tubes. Let's examine the pressures involved in breathing. **Atmospheric pressure (P_{atm})** is the pressure of the air in the space around you. **Intrapulmonary pressure (P_{pul})** is the air pressure within the alveoli and **intrapleural pressure (P_{ip})** is the pressure within the pleural cavity, between the two pleural membranes. When atmospheric pressure is greater than intrapulmonary pressure, then air will flow down its pressure gradient from outside toward the lower pressure alveoli—**inhalation**. When atmospheric pressure is lower than intrapulmonary pressure, then air will still flow down its pressure gradient, but this time from inside the lungs to the lower pressure atmosphere—**exhalation** (**Figure 22.19**). In general, two muscle groups are used during normal inspiration: the diaphragm and the external intercostal muscles. Additional muscles can be used if a deeper breath is required. The **diaphragm** is the bell-shaped skeletal

LO 22.2.3

inhalation (inspiration)

exhalation (expiration)

LO 22.2.4

Figure 22.18 Bulk Flow and Pressure Gradients

Gas and fluid mixtures move by bulk flow down pressure gradients. During an inhale, the pressure inside the alveoli (P_{pul}) is lower than the atmospheric pressure (P_{atm}) and so air flows down its pressure gradient into the lungs. During an exhale, P_{pul} is greater than P_{atm}, so air follows its pressure gradient out of the lungs.

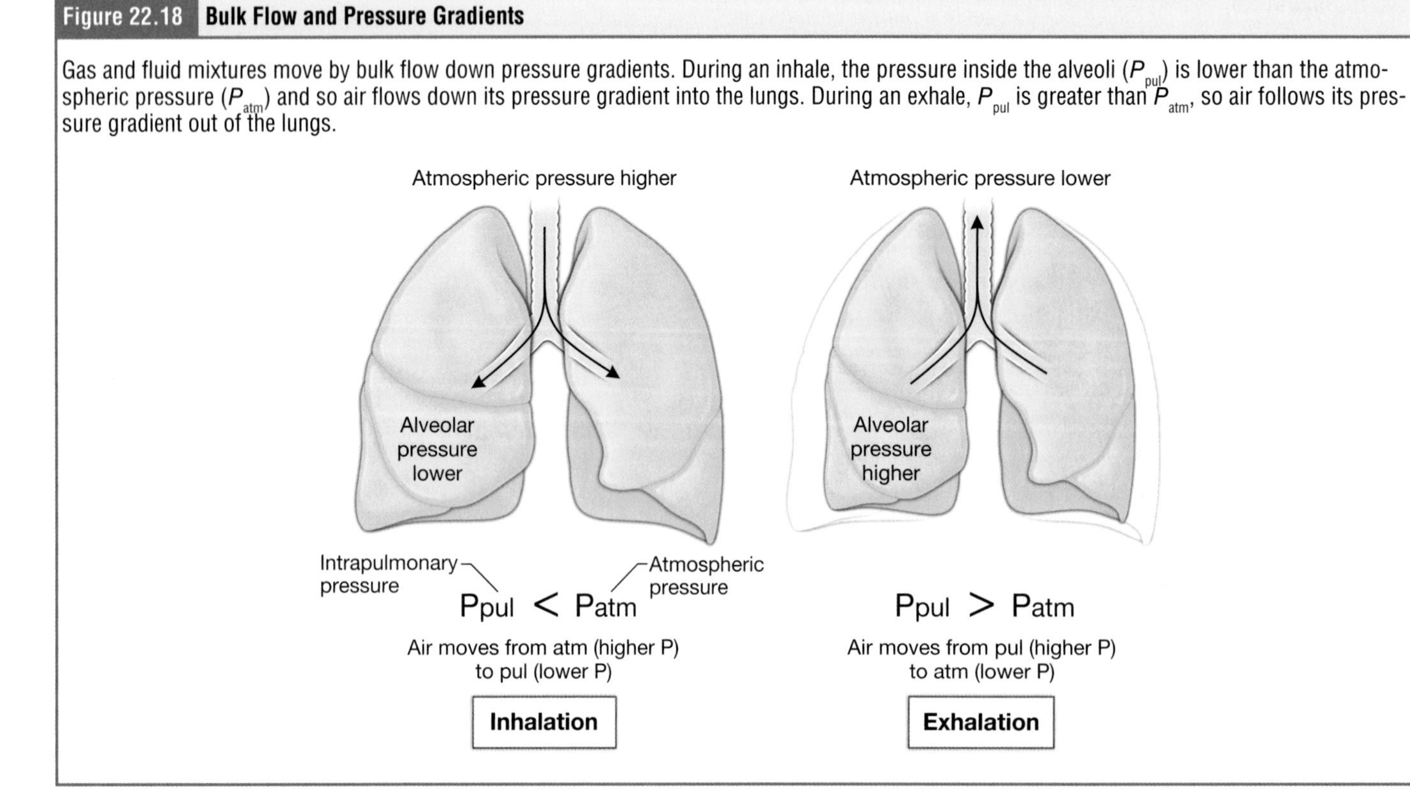

Figure 22.19 Inspiration and Expiration

(A) Inhalation occurs when the external intercostal muscles contract, pulling the ribs upward and laterally. The diaphragm also contracts, moving inferiorly into the abdomen. Together these muscle contractions result in an increase in volume in the thorax, and air moves in. (B) Exhalation is the opposite. The external intercostal muscles relax, and the ribs return to their resting position; the diaphragm also relaxes, moving back up to the thorax. The volume of the thorax decreases, and air flows out into the atmosphere.

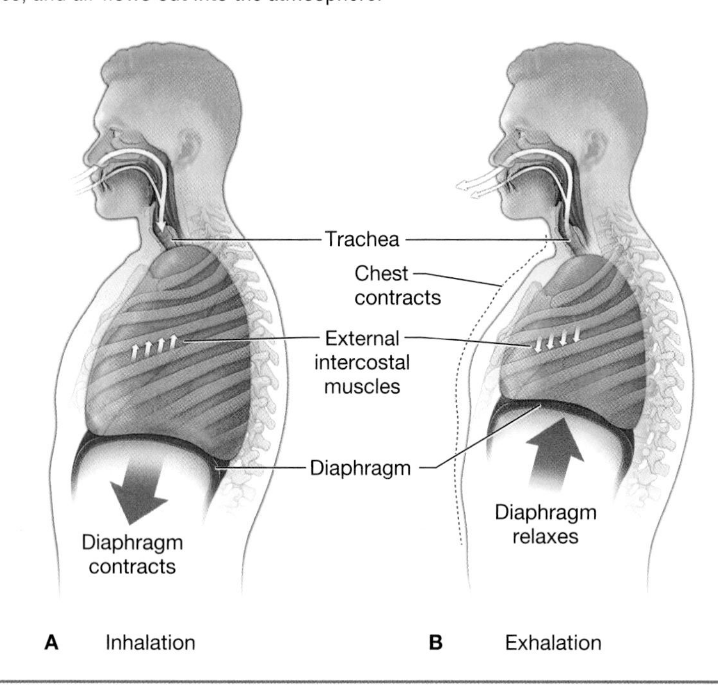

muscle at the bottom of the thorax, separating the thorax from the abdomen. When the diaphragm contracts, it moves inferiorly toward the abdominal cavity, creating a larger thoracic cavity and more space for the lungs. Contraction of the **external intercostal muscles**, which connect each rib to the rib above, moves the ribs upward and outward, causing the rib cage to expand, which increases the volume of the thoracic cavity. Due to the adhesive force of the pleural fluid, the expansion of the thoracic cavity forces the lungs to stretch and expand as well. As a result, a pressure gradient is created that drives air into the lungs.

The process of normal expiration is passive, meaning that energy is not required to push air out of the lungs. Instead, the elasticity of the lung tissue causes the lung to recoil, as the diaphragm and intercostal muscles relax following inspiration. In turn, the thoracic cavity and lungs decrease in volume, causing an increase in intrapulmonary pressure. The intrapulmonary pressure rises above atmospheric pressure, creating a pressure gradient that causes air to leave the lungs.

There are different types, or modes, of breathing that require a slightly different process to allow inspiration and expiration.

- *Quiet breathing* is a mode of breathing that occurs at rest and does not require the cognitive thought of the individual. During quiet breathing, the diaphragm and external intercostals must contract.
- A *deep inhale* requires the diaphragm to contract. As the diaphragm relaxes, air passively leaves the lungs.
- In contrast, a *forced exhale*, also known as *hyperpnea*, is a mode of breathing that can occur during exercise or actions that require the active manipulation of breathing,

Student Study Tip

The pressure that pulls the visceral pleura along with the parietal pleura is the same idea as when you pick up a wet glass from a coaster and the coaster comes off the table along with the glass.

such as singing. During a forced exhale, muscles of the abdomen, including the obliques, contract, forcing abdominal organs upward against the diaphragm. This helps to push the diaphragm further into the thorax, pushing more air out. In addition, the internal intercostals help to compress the rib cage, which also reduces the volume of the thoracic cavity.

Pressure of Gasses In a gas, pressure is a force created by the movement of gas molecules as they collide with the walls of their container. Therefore, a gas that occupies a container with a larger volume, or space within the container, will have fewer collisions with the walls of the container and therefore a lower pressure. Decreasing the volume of the container will result in more collisions, and a higher pressure. Another factor that influences the pressure of a gas is temperature. The temperature of the inside of the lungs is fairly constant, so this factor is not usually considered in physiology. **Boyle's law**, first described by the scientist Robert Boyle, states that if temperature is constant then the pressure of a gas is inversely proportional to its volume: If volume increases, pressure decreases. Likewise, if volume decreases, pressure increases (**Figure 22.20**). Pressure and volume are inversely related ($P = 1/V$).

LO 22.2.5

Because air moves according to pressure gradients, we must consider three types of pressure—atmospheric, intrapulmonary, and intrapleural—when learning about pulmonary ventilation. *Atmospheric pressure* is the amount of force that is exerted by gasses in the air. Atmospheric pressure can be expressed in terms of the unit atmosphere (abbreviated atm) or in millimeters of mercury (mm Hg). One atm is equal to 760 mm Hg, which is the atmospheric pressure at sea level.

Intrapulmonary pressure is the pressure of the air within the alveoli, which changes during the different phases of breathing (**Figure 22.21**). When the intrapulmonary pressure is less than the atmospheric pressure, air will flow from an area of greater pressure (the atmosphere) to the area of lower pressure (the alveoli) and inhalation occurs. To exhale, the intrapulmonary pressure must be greater than the atmospheric pressure.

LO 22.2.6

Intrapleural pressure is the pressure of the fluid within the pleural cavity, between the visceral and parietal pleurae. Similar to intrapulmonary pressure, intrapleural pressure also changes during the different phases of breathing. However, due to certain characteristics of the lungs, the intrapleural pressure is always lower than the intrapulmonary pressure.

Competing forces within the thorax cause the formation of the intrapleural pressure. One of these forces relates to the elasticity of the lungs themselves—elastic tissue pulls the lungs inward, away from the thoracic wall. Surface tension of alveolar fluid, which is mostly water, also creates an inward pull of the lung tissue. This inward tension from the lungs is countered by opposing forces from the pleural fluid and thoracic wall. Surface tension within the pleural cavity pulls the lungs outward. Since the parietal pleura is attached to the thoracic wall, the natural elasticity of the chest wall opposes the inward pull of the lungs. Ultimately, the outward pull is slightly greater than the inward pull. **Transpulmonary pressure** is the difference between the intrapleural and intrapulmonary pressures, and it determines the size of the lungs. A higher transpulmonary pressure corresponds to a larger lung (see Figure 22.21).

Physical Factors Affecting Ventilation The differences in pressures is caused by and dependent upon the contraction and relaxation of muscles. The lungs themselves are passive during breathing, meaning they are not involved in creating the movement that helps inspiration and expiration. But because of the adhesive nature of the pleural fluid, the lungs are pulled outward when the thoracic wall moves during inspiration. The recoil of the thoracic wall during expiration causes a reduction in lung volume.

Figure 22.20 Boyle's Law

Boyle's law states that the pressure of a gas is inversely proportional to the volume of the container that the gas is in. We can see this experimentally when we compress gasses but it is also observable simply by squeezing a balloon and observing the increase in pressure on the balloon's inner surface as the gas inside is compressed. This principle allows us to understand inspiration and expiration. As the size of the thorax increases with inspiration, the gas inside is at a lower pressure. This pressure is lower than the atmospheric pressure, causing air to flow into the thorax from the atmosphere (down its pressure gradient). The opposite occurs during expiration; as the volume of the thorax decreases, air inside the lungs increases in pressure. Eventually, this pressure exceeds the atmospheric pressure and air flows out of the lungs, down its pressure gradient.

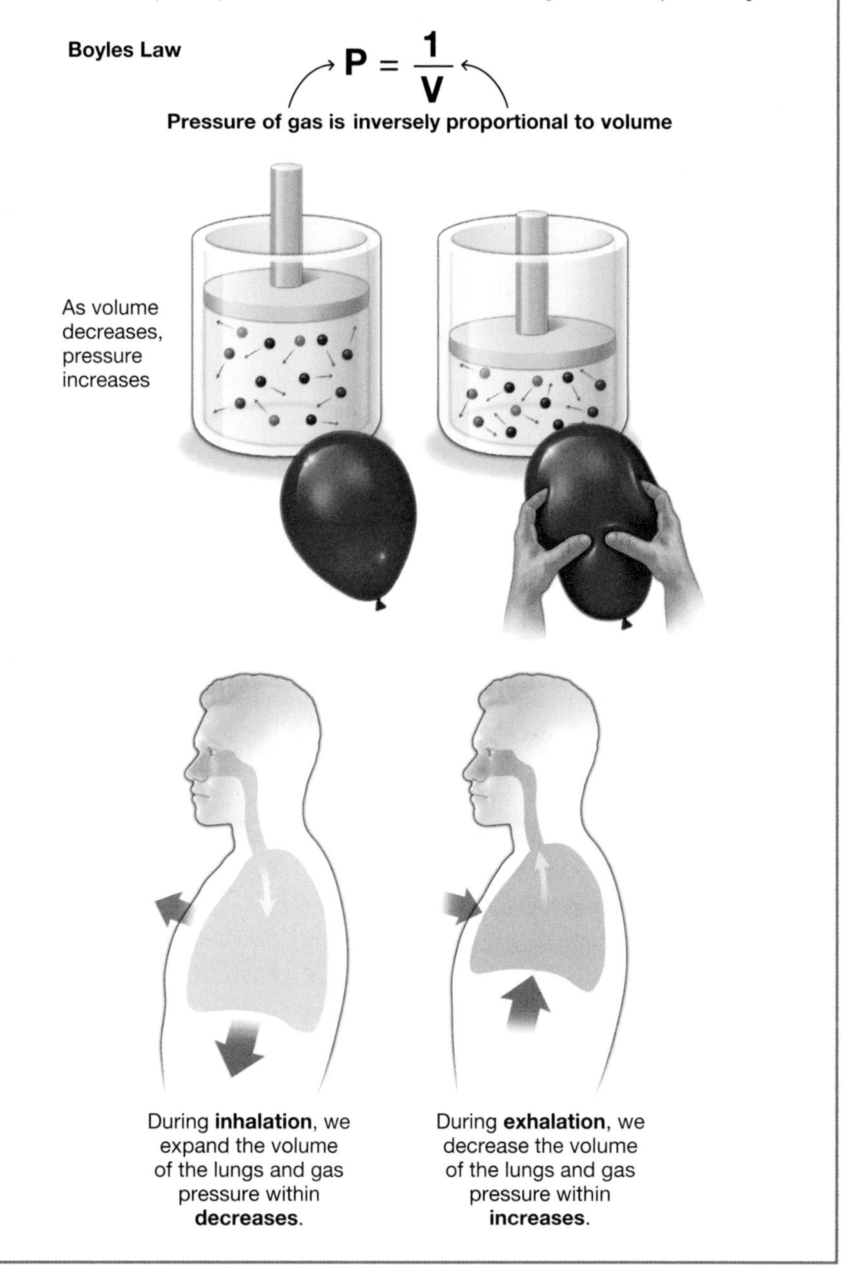
Boyles Law
$P = \frac{1}{V}$
Pressure of gas is inversely proportional to volume
As volume decreases, pressure increases
During **inhalation**, we expand the volume of the lungs and gas pressure within **decreases**.
During **exhalation**, we decrease the volume of the lungs and gas pressure within **increases**.

Other characteristics of the lungs influence the effort that must be expended to ventilate. Resistance is any force that slows flow—in this case, the flow of gasses. We have already discussed the formula to determine flow rate, which is:

LO 22.2.7

$$F = \Delta P/R$$

Figure 22.21 The Pressures of the Lungs

Gasses move according to bulk flow down pressure gradients. There are four pressures in the lungs that are physiologically important. The atmospheric pressure is the pressure of the air outside the body. Transpulmonary pressure is the difference in pressure between the inside of the alveoli and the pleural cavity. The intrapleural pressure is the pressure between the pleurae. The intrapulmonary pressure is the pressure inside the alveoli.

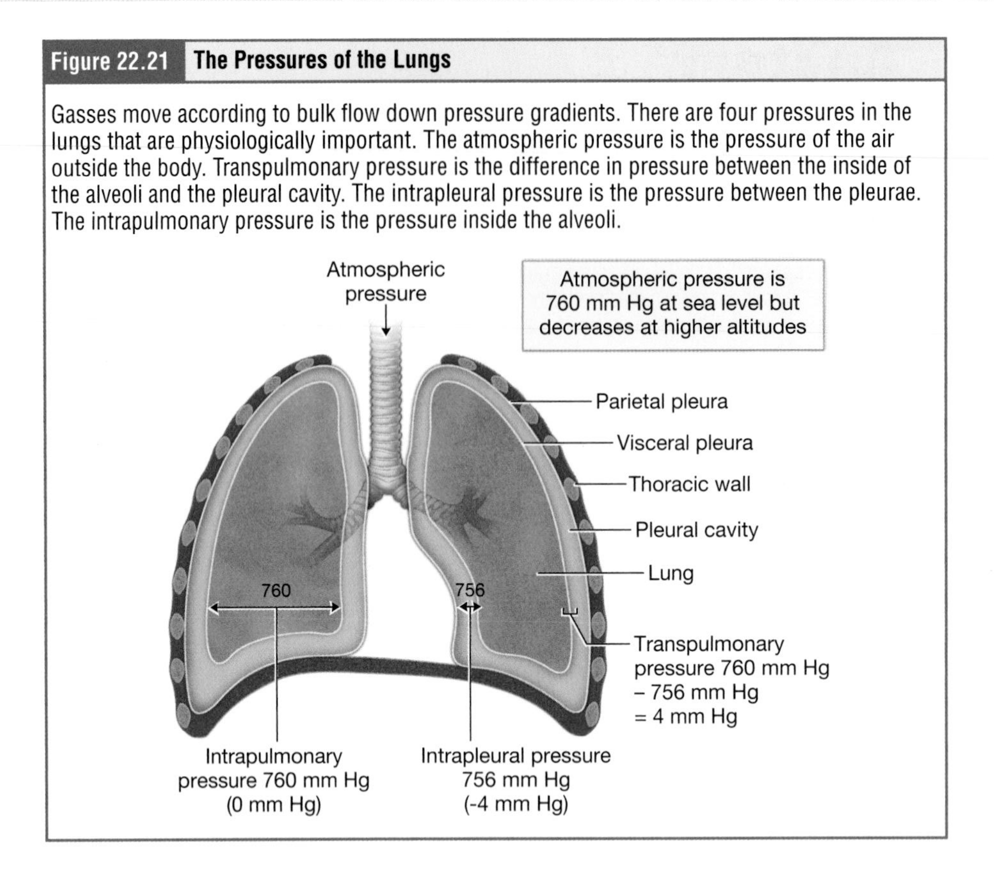

LO 22.2.8

We have already established that air will flow via bulk flow based on differences in pressure between the outside air (P_{atm}) and the air in the alveoli (P_{pul}). The diameter of the airway is the primary factor of resistance. *Bronchoconstriction* and *bronchodilation*—the narrowing and widening of the bronchioles based on smooth muscle contraction—determine the diameter, and are the key factor in resistance. Other resistance factors can be if scar tissue or mucus buildup in the tubes narrows the internal passageway.

Water molecules demonstrate **cohesion**; that is, they like to form hydrogen bonds and stay close to each other. This cohesion can be dangerous for the alveoli, however. Alveolar Type 1 cells have moist surfaces, like most cells of the body. The water molecules on their surfaces are attracted to each other (see the "Anatomy of Surfactant" feature). Because this attraction is so strong, and the alveolar walls are so thin, water cohesion can cause the walls to collapse toward each other. To picture this, imagine that you have two wet tissues that are stuck together. Trying to pull them apart is more likely to rip the tissues than to break the cohesion between the water molecules uniting them. Type II alveolar cells, a second type of cell found in the alveoli, secrete an oily substance called **surfactant** onto the surfaces of the Type I cells, decreasing the water molecule cohesion. Surfactant allows the alveolar walls to get very close together during exhalation but prevents cohesion from allowing them to stick. You might imagine spraying a coat of cooking oil over your wet tissues from earlier. Type II alveolar cells are one of the last cell types to mature during fetal development. For this reason, infants born prematurely may be at increased risk of alveolar collapse due to insufficient surfactant.

LO 22.2.9

Water molecule cohesion may present a risk in the alveoli, but it is a critically important factor in the pleurae. The parietal pleura is adhered tightly to the internal wall of the thorax; the visceral pleura is adhered tightly to the outer surface of the lung. When the thorax expands during inhalation, all of these surfaces (the thorax wall, the parietal pleura, the visceral pleura, and the lung surface) move (**Figure 22.22**).

Anatomy of...

Surfactant

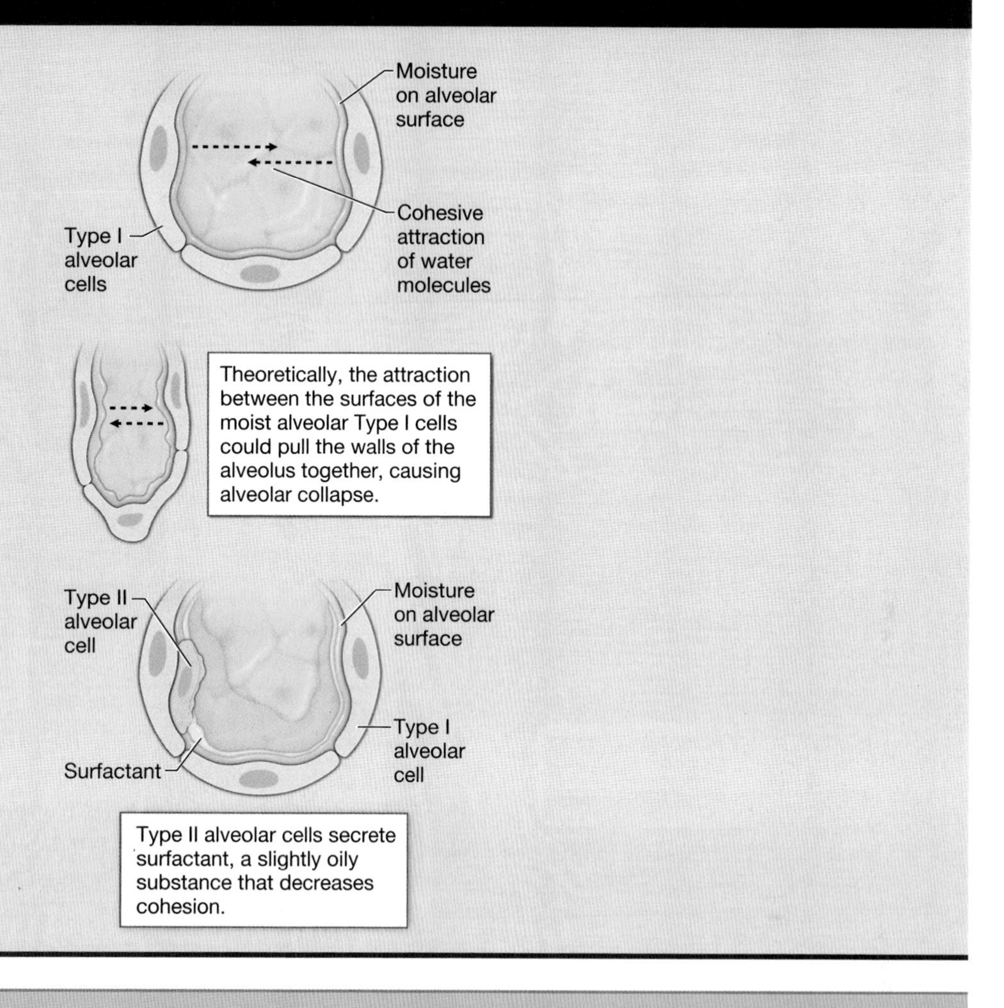
Moisture on alveolar surface
Cohesive attraction of water molecules
Type I alveolar cells
Theoretically, the attraction between the surfaces of the moist alveolar Type I cells could pull the walls of the alveolus together, causing alveolar collapse.
Type II alveolar cell
Moisture on alveolar surface
Type I alveolar cell
Surfactant
Type II alveolar cells secrete surfactant, a slightly oily substance that decreases cohesion.

DIGGING DEEPER:
Respiratory Distress Syndrome

Respiratory distress syndrome (RDS) primarily occurs in infants born prematurely. Up to 50 percent of infants born between 26 and 28 weeks, and fewer than 30 percent of infants born between 30 and 31 weeks, develop RDS. RDS results from insufficient production of pulmonary surfactant, thereby preventing the lungs from properly inflating at birth. A small amount of pulmonary surfactant is produced beginning at around 20 weeks; however, this is not sufficient for inflation of the lungs. As a result, dyspnea occurs, and gas exchange cannot be performed properly. Blood oxygen levels are low, while blood carbon dioxide levels and pH are high.

The primary cause of RDS is premature birth, which may occur due to a variety of known or unknown causes. Other risk factors include gestational diabetes, cesarean delivery, second-born twins, and family history of RDS. The presence of RDS can lead to other serious disorders, such as septicemia (infection of the blood) or pulmonary hemorrhage. Therefore, it is important that RDS is immediately recognized and treated to prevent death and reduce the risk of developing other disorders.

Medical advances have resulted in an improved ability to treat RDS and support the infant until proper lung development can occur. At the time of delivery, treatment may include resuscitation and intubation if the infant does not breathe on their own. These infants would need to be placed on a ventilator to mechanically assist with the breathing process. If spontaneous breathing occurs, application of nasal continuous positive airway pressure (CPAP) may be required. In addition, pulmonary surfactant is typically administered. Death due to RDS has been reduced by 50 percent due to the introduction of pulmonary surfactant therapy. Other therapies may include corticosteroids, supplemental oxygen, and assisted ventilation. Supportive therapies, such as temperature regulation, nutritional support, and antibiotics, may be administered to the premature infant as well.

Figure 22.22 The Pleurae

The parietal pleura and the visceral pleura are held together by the cohesion of water molecules on each of their surfaces. This molecular attraction is responsible for keeping the lungs inflated against the walls of the thorax.

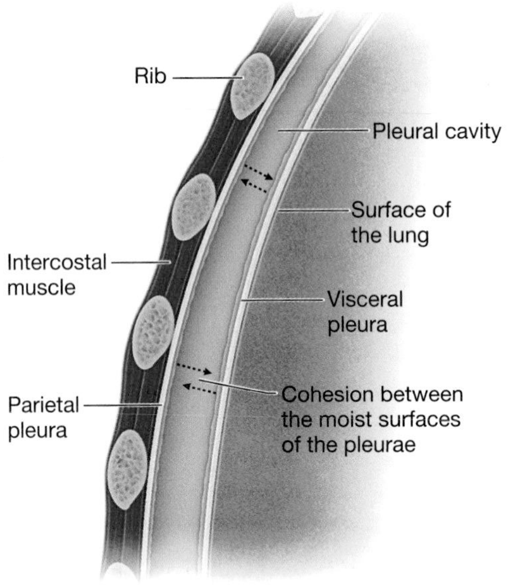

Across most of the wall of the thorax, the parietal and visceral pleurae are not physically connected; they are held together only by water cohesion. For this reason, when air enters the pleural cavity and disrupts the cohesion of water between the two pleura, their movement becomes uncoupled. The parietal pleura continues to move along with the thoracic wall during breathing, but the visceral pleura, and therefore the lung, no longer move with it. This condition, called **pneumothorax**, can result if air enters the pleural cavity from inside the lung (if the lung surface becomes damaged) or from the atmosphere (if the chest wall is damaged) (**Figure 22.23**).

Figure 22.23 Pneumothorax

Cohesion of water molecules is responsible for the adherence of the visceral pleura to the parietal pleura. When air enters the pleural cavity, it can disrupt this cohesion, causing the visceral pleura to pull away from the parietal pleura and the lung to collapse, as it is only held open by the adhesion to the parietal pleura. This condition is called *pneumothorax* and can occur in patients when alveolar damage allows air to escape from the alveoli into this space or when air enters from the outside, which can happen if the chest wall is punctured.

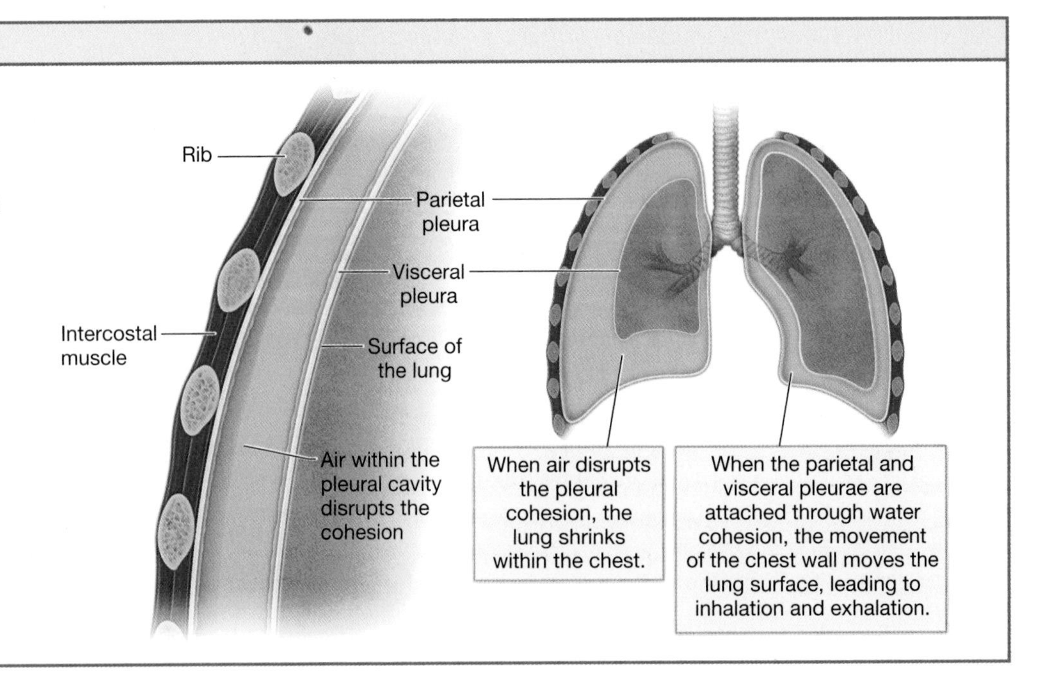

Learning Check

1. Which of the following physiological processes rely on pressure gradients? Please select all that apply.
   a. Pulmonary ventilation
   b. Cardiovascular circulation
   c. Sodium/potassium pump
   d. Osmosis
2. Based on its action, what is the insertion of the external intercostal muscles?
   a. Rib below the muscles
   b. Rib above the muscles
   c. Diaphragm
   d. Bicipital humeral groove
3. What is the name of the law that states that pressure is inversely proportional to volume?
   a. Fick's law
   b. Henry's law
   c. Dalton's law
   d. Boyle's law
4. What happens if intrapleural pressure becomes equal to or greater than atmospheric pressure?
   a. Inhalation
   b. Exhalation
   c. Collapsed lung may result
   d. Excess fluid accumulates in the lungs

22.3 Respiratory Volumes and Capacities

Learning Objectives: By the end of this section, you will be able to:

22.3.1 Define, identify, and determine values for the pulmonary volumes (inspiratory reserve volume [IRV], tidal volume [TV], expiratory reserve volume [ERV], and residual volume [RV]) and the pulmonary capacities (inspiratory capacity [IC], functional residual capacity [FRC], vital capacity [VC], and total lung capacity [TLC]).

22.3.2 Define anatomical dead space.

22.3.3 Explain the effect of anatomical dead space on alveolar ventilation and on the composition of alveolar and expired air.

22.3.4 Define and calculate minute ventilation and alveolar ventilation.

LO 22.3.1

Respiratory volume is the term used for various volumes of air moved by or associated with the lungs at a given point in the respiratory cycle. Most of these volumes are measurable using a **spirometer**, a device that can measure the volume of air you inhale or exhale. Interestingly, spirometers are often programmed with different expected value ranges based on potentially false assumptions about race. While the settings vary only slightly, this is another example where critical thinking and research may improve inclusion in health care. There are four major types of respiratory volumes: tidal, residual, inspiratory reserve, and expiratory reserve (**Figure 22.24**). **Tidal volume (TV)** is the amount of air that normally enters the lungs during quiet breathing. **Expiratory reserve volume (ERV)** is the amount of air you can forcefully exhale past a normal tidal expiration, typically about twice the TV volume. **Inspiratory reserve volume (IRV)** is produced by a deep inhalation, past a tidal inspiration. This is the extra volume that can be brought into the lungs during a forced inspiration. **Residual volume (RV)** is the air left in the lungs if you exhale as much air as possible. You may be wondering why we have RV—why we cannot squeeze out all the air in our lungs—and the answer has to do primarily with the adherence of the lungs to the chest wall. The ribs are bone; they hold the thorax rigidly open, and because the lungs are held to their inner surfaces by the pleurae, the lungs cannot collapse and release all their air under healthy circumstances.

Student Study Tip

The residual volume is air that always *resides* in the lungs.

Figure 22.24 The Lung Volumes and Capacities

Various volumes of the lungs are measured using a spirometer, a machine that measures the volume of air inhaled and exhaled in various circumstances.

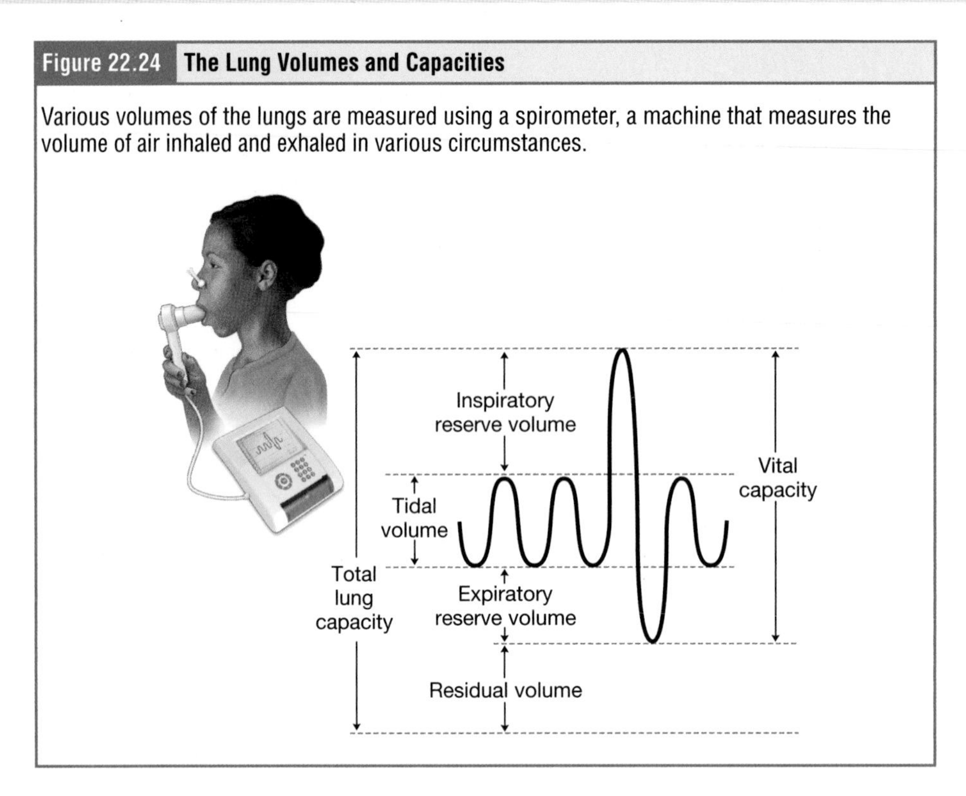

Respiratory capacity is the combination of two or more selected volumes, which further describes the amount of air in the lungs during a given time. For example, **total lung capacity (TLC)** is the sum of all of the lung volumes (TV, ERV, IRV, and RV), which represents the total amount of air a person can hold in the lungs after a deep inhalation. TLC can be as much as six liters, but will average four to five liters for most typically sized adults. **Vital capacity (VC)** is the amount of air a person can move into or out of their lungs; it is the sum of all of the volumes except RV (TV, ERV, and IRV). **Inspiratory capacity (IC)** is the maximum amount of air that can be inhaled past a normal tidal expiration; it is the sum of the TV and IRV. On the other hand, the **functional residual capacity (FRC)** is the amount of air that remains in the lung after a normal tidal expiration; it is the sum of ERV and RV (see Figure 22.24).

Student Study Tip

To practice the relationships among all these volumes and capacities, draw them out and consider how a change in one volume could affect the others.

LO 22.3.2

LO 22.3.3

In addition to the air that creates respiratory volumes, the respiratory system also contains **anatomical dead space**, which is air that is present in the airway but never reaches the alveoli and therefore never participates in gas exchange. **Alveolar dead space** involves air found within alveoli that are unable to function, such as those affected by disease or abnormal blood flow. **Total dead space** is the combination of anatomical dead space and alveolar dead space, and represents all of the air in the respiratory system that is not being used in the gas exchange process.

minute ventilation (minute respiratory volume)

LO 22.3.4

alveolar ventilation (alveolar ventilation rate, AVR)

Minute ventilation is the total volume of air that enters the lungs in one minute; it is the product of the TV of all the breaths taken in that minute and the respiratory rate. **Alveolar ventilation** is a related measurement, but one that subtracts the anatomical dead space and calculates the volume of air that reaches the respiratory membrane in one minute.

Learning Check

1. You take a deep inhale during meditation that lasts longer and brings more air into the lungs than your typical inhale. Which of the following terms can describe the volume of air you have taken in?
   a. Tidal volume
   b. Residual volume
   c. Inspiratory reserve volume
   d. None of the above
2. Which of the following equations accurately compares the three given terms?
   a. Inspiratory capacity = TV + IRV
   b. Inspiratory capacity = TV + IRV + RV
   c. Total lung capacity = TV + IRV
   d. Vital capacity = TV + IRV + RV + ERV
3. What is the difference between minute and alveolar ventilation?
   a. Whether exhalation is taken into account
   b. Whether total dead space is taken into account
   c. Ventilation that occurs within one minute versus one hour, respectively
   d. Whether inhalation is taken into account

22.4 Gas Exchange

Learning Objectives: By the end of this section, you will be able to:

22.4.1 Explain the relationship between the total pressure of gases in a mixture and the partial pressure of an individual gas (i.e., Dalton's Law).

22.4.2 Explain the relationship between the partial pressure of a gas in air, the solubility of that gas in water, and the amount of the gas that can dissolve in water.

22.4.3 Compare and contrast the solubility of oxygen and carbon dioxide in plasma.

22.4.4 Describe oxygen and carbon dioxide concentration gradients and net gas movements between the alveoli and the pulmonary capillaries.

22.4.5 Analyze how oxygen and carbon dioxide movements are affected by changes in partial pressure gradients (e.g., at high altitude), area of the exchange surface, permeability of the exchange surface, and diffusion distance.

22.4.6 Explain the effects of local changes in oxygen and carbon dioxide concentrations on the diameters of pulmonary arterioles and bronchioles.

22.4.7 Describe oxygen and carbon dioxide concentration gradients and net gas movements between systemic capillaries and the body tissues.

22.4.8 Explain the influence of cellular respiration on oxygen and carbon dioxide gradients that govern gas exchange between blood and body tissues.

The purpose of the respiratory system is to perform gas exchange. Pulmonary ventilation provides air to the alveoli for this gas exchange process. At the respiratory membrane, where the alveolar and capillary walls meet, gasses move across the membranes, with oxygen entering the bloodstream and carbon dioxide exiting. It is through

Figure 22.25 Dalton's Law

Gasses can be mixtures of several gasses. Each of the component gasses contributes pressure to the total gas mixture. The pressure of the mixture is the sum of each of the partial pressures.

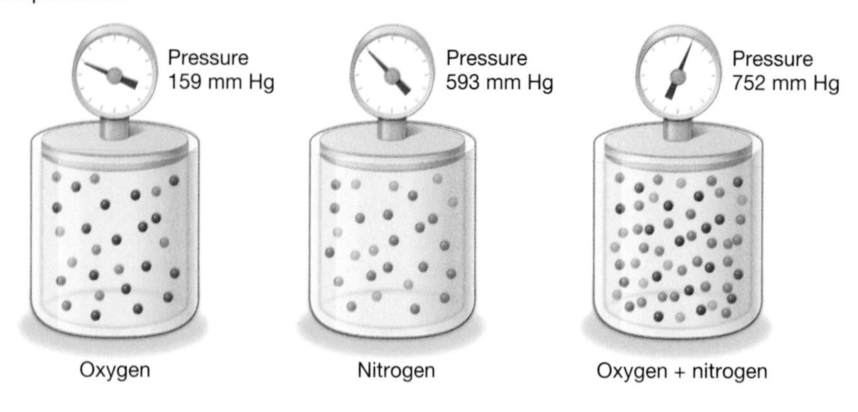

this mechanism that blood is oxygenated and carbon dioxide, the waste product of cellular respiration, is removed from the body.

22.4a Gas Exchange

In order to understand the mechanisms of gas exchange in the lung, it is important to understand the underlying principles of gasses and their behavior. In addition to Boyle's law, several other gas laws help to describe the behavior of gasses.

LO 22.4.1

Dalton's law (Law of partial pressures)

Gas Laws and Air Composition Gas molecules exert force on the surfaces with which they are in contact; this force is called *pressure*. Air is composed of a mixture of gasses including oxygen, nitrogen, and carbon dioxide. The pressure of atmospheric air at sea level is typically 760 mm Hg. But each gas within air contributes to that pressure. Therefore, we can say that each gas contributes a part of the pressure, or *partial pressure*, to the total pressure of the mixture (**Figure 22.25**). **Partial pressure (*Px*)** is the pressure of a single type of gas in a mixture of gasses. **Dalton's law** describes the behavior of gasses in a mixture and states that a specific gas type in a mixture exerts its own pressure; thus, the total pressure exerted by a mixture of gasses is the sum of the partial pressures of the gasses in the mixture. The partial pressure is determined by the composition of the gasses (**Table 22.1**).

Partial pressure is extremely important in predicting the movement of gasses. Recall that gasses tend to equalize their pressure in two regions that are connected. A gas will move from an area where its partial pressure is higher to an area where its

Table 22.1 Partial Pressures of Gasses within Air

Gas	Percentage	Partial Pressure (mm Hg)
N_2	78.6	597
O_2	20.9	159
CO_2	0.04	0.3
H_2O	0.46	3.7
Total Air	100%	760 mm Hg

partial pressure is lower. In this way a partial pressure gradient of a gas is analogous to a concentration gradient. In addition, the greater the partial pressure difference between the two areas, the more rapid the movement of gasses.

Solubility of Gasses in Liquids **Henry's law** describes the behavior of gasses when they come into contact with a liquid. Henry's law states that the concentration of gas in a liquid is directly proportional to the partial pressure of that gas in the air surrounding the liquid. That is to say, a gas will move between a liquid and the gas around it to according to its partial pressure gradient. Let's take an example. If you pour yourself a glass of water before bed, that water has very little dissolved gas in it. As it sits on your nightstand throughout the night, the oxygen within the air is at a much higher concentration than the oxygen gas within the water, and so, over time, the oxygen dissolves in the glass of water (**Figure 22.26**) and in the morning you may notice the small bubbles of dissolved oxygen clinging to the sides of the glass. The ability of a gas to dissolve in a liquid is, of course, also dependent on the solubility of that gas in that liquid. For example, although nitrogen is present in the atmosphere, very little nitrogen dissolves into the blood, because the solubility of nitrogen in blood is very low. Whereas oxygen is minimally soluble in water or plasma, carbon dioxide is much more soluble in water and plasma. Gas molecules establish an equilibrium between those molecules dissolved in liquid and those in air.

LO 22.4.2

LO 22.4.3

Gas exchange occurs at two sites in the body: in the lungs, where oxygen is picked up and carbon dioxide is released at the respiratory membrane, and at the tissues, where oxygen is released and carbon dioxide is picked up. External respiration is the exchange of gasses with the external environment and occurs in the alveoli of the lungs. Internal respiration is the exchange of gasses with the internal environment, and occurs in the tissues. The actual exchange of gasses occurs due to simple diffusion. Energy is not required to move oxygen or carbon dioxide across membranes. Instead, these gasses follow pressure gradients that allow them to diffuse. The anatomy of the lung maximizes the diffusion of gasses: The respiratory membrane is highly permeable to gasses; the respiratory and blood capillary membranes are very thin; and there is a large surface area throughout the lungs.

Figure 22.26 **Henry's Law**

Gasses dissolve in liquids according to their solubility and concentration gradients. This is evident when, over time, oxygen dissolves in a glass of still water.

MirageC/Moment/Getty Images

Figure 22.27 Gas Exchange at the Alveoli

Gasses flow down their partial pressure gradients. Blood entering the pulmonary capillaries has a pO_2 of just 40 mm Hg on average, so the oxygen from the alveoli flows down its partial pressure gradient and enters the blood. Oxygen is poorly soluble in the plasma, so the majority of oxygen binds to hemoglobin molecules in red blood cells. Carbon dioxide is at a higher partial pressure in the blood of the capillaries, so it diffuses into the alveoli, where it can be exhaled.

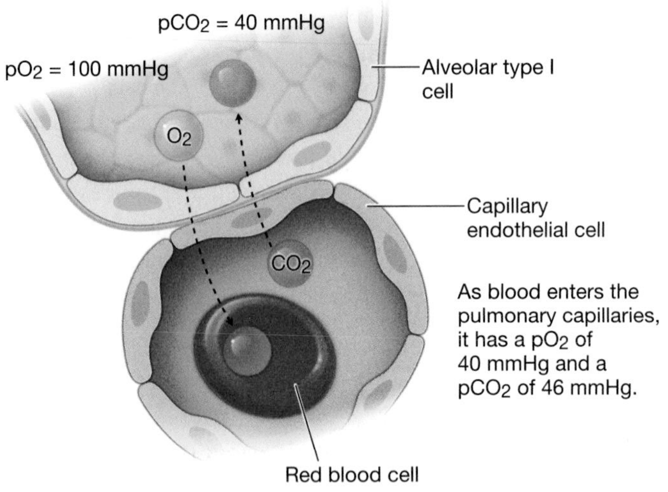

LO 22.4.4

Figure 22.27 illustrates the gradients of oxygen and carbon dioxide between the blood and the alveolus. With every inhalation, the alveoli are refilled with oxygen-rich air. Oxygen diffuses down its partial pressure gradient to the blood plasma, and then to the hemoglobin molecules within the red blood cells. The air from the atmosphere has very little carbon dioxide in it; therefore, the alveolar concentration of carbon dioxide is very low. The majority of carbon dioxide travels within the blood plasma (although carbon dioxide can bind to hemoglobin and be carried within red blood cells), and so it diffuses from the blood plasma, where it is at a very high concentration, to the alveoli, where it is at a very low concentration.

LO 22.4.5

Fick's law tells us that the rate of diffusion is based on a few environmental factors: the available surface area, the concentration gradient, the permeability of the molecule to the membrane, and the membrane thickness.

$$\text{Rate of diffusion} \quad \alpha \quad \frac{\text{Surface area} \times \text{concentration gradient} \times \text{permeability}}{\text{Membrane thickness}}$$

We can apply Fick's law to examine the rate of diffusion of oxygen and carbon dioxide across the respiratory membrane.

- Factors that influence the surface area include the availability of functional alveoli. Alveolar surface area goes down when alveoli are damaged from years of smoking (as in emphysema), as well as if a lung does not inflate properly (such as in pneumothorax).
- The concentration gradient will vary with conditions that alter the exchange of gasses with the atmospheric air. For example, bronchoconstriction will limit gas exchange with the atmosphere, and therefore we can expect the concentration of carbon dioxide in the alveoli to increase and the concentration of oxygen in the alveoli to decrease if insufficient fresh air can be delivered. Additionally, the concentration gradient across the respiratory membrane may change due to changing factors in the tissues (such as an increase in the rate of cellular respiration).

- The permeability of carbon dioxide and oxygen to the cell membranes of the respiratory membrane is not likely to change under any physiological circumstances
- The membrane thickness can change. The plasma membranes themselves will not thicken, but if scar tissue, fibrosis, or mucus are present, the distance across which diffusion must take place will increase, lowering the rate of diffusion.

Ventilation and Perfusion Two important aspects of gas exchange in the lung are ventilation and perfusion. *Ventilation* is the movement of air into and out of the lungs, and **perfusion** is the flow of blood in the pulmonary capillaries. For gas exchange to be efficient, the volumes involved in ventilation and perfusion should be compatible. However, factors such as regional gravitational effects on blood, blocked alveolar ducts, or disease can cause ventilation and perfusion to be imbalanced.

The partial pressure of oxygen in alveolar air is about 100 mm Hg, and the partial pressure of oxygenated blood in pulmonary veins is about 100 mm Hg. When

Learning Connection

Explain a Process

Your friend has a case of pneumonia and is wondering why they are having trouble breathing. Pneumonia causes an accumulation of fluid and pus in the lungs. Which of the previously mentioned factors (alveolar surface area, concentration gradient of oxygen or carbon dioxide, permeability of one of the respiratory gasses, or thickness of the respiratory membrane) is hindered in pneumonia? How would you explain this to your friend in simple terms?

Apply to Pathophysiology

Asthma

Asthma is a common condition that affects the lungs in both adults and children. Approximately 8 percent of adults and 10 percent of children in the United States suffer from asthma. Asthma is the most frequent cause of hospitalization in children.

Asthma is a chronic disease characterized by inflammation and edema of the airway, as well as bronchoconstriction. In some forms of asthma excessive mucus secretion can occur, which further contributes to airway occlusion.

Bronchoconstriction occurs periodically, leading to a reduction in gas exchange between the alveoli and atmospheric air. This event is typically described as an "asthma attack." An attack may be triggered by environmental factors such as dust, pollen, pet hair, or dander, changes in the weather, mold, tobacco smoke, and respiratory infections, or by exercise and stress.

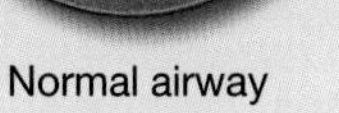

Normal airway

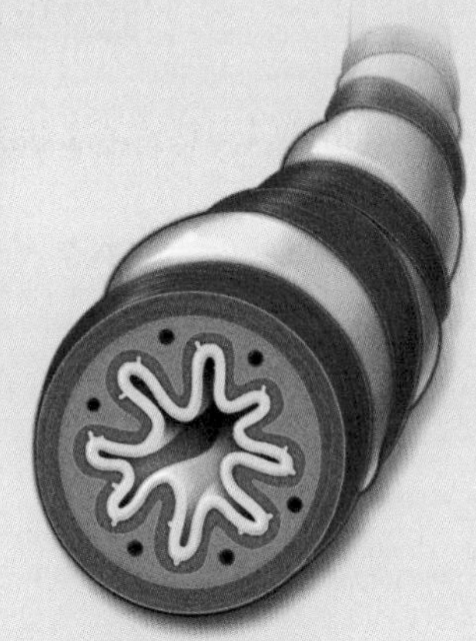

Inflamed airway

Symptoms of an asthma attack involve coughing, shortness of breath, wheezing, and tightness of the chest. Symptoms of a severe asthma attack that requires immediate medical attention would include difficulty breathing that results in blue (cyanotic) lips or face, confusion, drowsiness, a rapid pulse, sweating, and severe anxiety. Asthma severity may increase under the influence of ovarian hormones such as estrogen. Therefore, asthma symptoms may be more noticeable or dangerous between puberty to menopause in individuals with ovaries.

Eliot is a 7-year-old boy experiencing an asthma attack. At this moment Eliot is having bronchoconstriction but no excess mucus production. Predict what changes will occur, if any, to the following physiological variables.

1. Partial pressure of O_2 in the alveoli.
   A) No change
   B) Increase
   C) Decrease
2. Partial pressure of O_2 in the arterial blood.
   A) No change
   B) Increase
   C) Decrease
3. Partial pressure of CO_2 in the systemic arterial blood.
   A) No change
   B) Increase
   C) Decrease
4. pH of the systemic arterial blood.
   A) Increase
   B) No change
   C) Decrease

ventilation is sufficient, air enters the alveoli at a high rate, and the partial pressure of oxygen in the alveoli remains high. In contrast, when ventilation is insufficient, the partial pressure of oxygen in the alveoli drops. Without the large difference in partial pressure between the alveoli and the blood, oxygen does not diffuse efficiently across the respiratory membrane. The body has mechanisms that counteract this problem. In cases in which ventilation is not sufficient for an alveolus, the body redirects blood flow to alveoli that are receiving sufficient ventilation. This is achieved by constricting the pulmonary arterioles that serve the dysfunctional alveolus, which redirects blood to other alveoli that have sufficient ventilation. At the same time, the pulmonary arterioles that serve alveoli receiving sufficient ventilation vasodilate, which brings in greater blood flow. Factors such as carbon dioxide, oxygen, and pH levels can all serve as stimuli for adjusting blood flow in the capillary networks associated with the alveoli.

LO 22.4.6

Student Study Tip

Think of the body's redirecting blood flow as if it were a detour that gets set up following a car accident. Since cars can't get through, traffic is redirected.

Whereas ventilation is regulated by the diameter of the airways, perfusion is regulated by the diameter of the blood vessels. The diameter of the bronchioles is sensitive to the partial pressure of carbon dioxide in the alveoli. A greater partial pressure of carbon dioxide in the alveoli or a lower partial pressure of oxygen in the blood will cause the bronchioles to increase their diameter, allowing carbon dioxide to be exhaled from the body at a greater rate. As mentioned in the previous paragraph, a greater partial pressure of oxygen in the alveoli causes the pulmonary arterioles to dilate, increasing blood flow.

The respiratory membrane is one of two sites of gas exchange across capillary surfaces. The other occurs between the systemic capillaries and the body tissues (**Figure 22.28**). Similar to gas exchange at the respiratory membrane, gas exchange at the tissue membrane also occurs as simple diffusion due to a partial pressure gradient. However, the partial pressure gradients are opposite of those present at the respiratory membrane. The partial pressure of oxygen in tissues is low—about 40 mm Hg—because oxygen is continuously used for cellular respiration. In contrast, the partial pressure of oxygen in the blood is about 100 mm Hg. This creates a pressure gradient that causes

LO 22.4.7

Figure 22.28 Gas Exchange at the Tissues

Gasses flow according to their partial pressure gradients. Oxygen is at its highest partial pressure in a red blood cell, and at its lowest partial pressure in the mitochondria in the tissues; therefore, it flows along its gradient to reach the mitochondria. Carbon dioxide is at its highest partial pressure in the mitochondria and its lowest partial pressure in the blood. It flows along its partial pressure gradient to reach a destination within the red blood cells. There it is either converted to carbonic acid or carried in its CO_2 form.

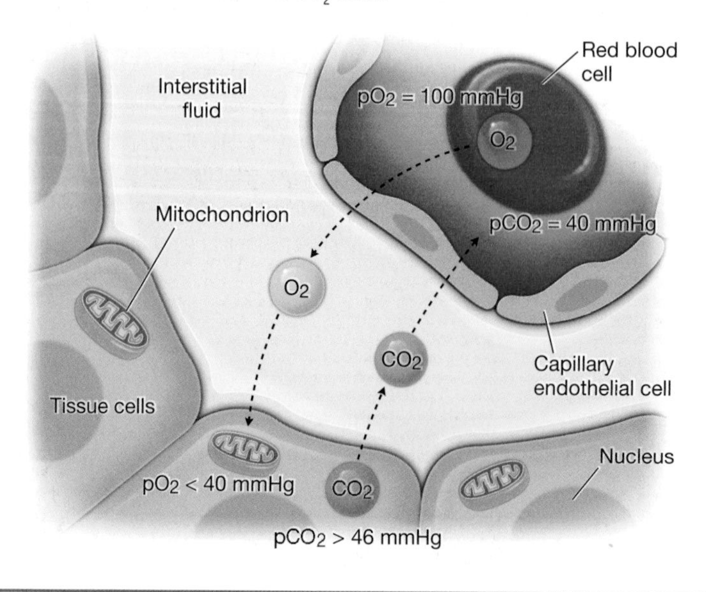

oxygen to dissociate from hemoglobin, diffuse out of the blood, cross the interstitial space, and enter the tissue. Hemoglobin that has little oxygen bound to it loses much of its brightness, so that blood returning to the heart is more burgundy in color.

LO 22.4.8

Considering that cellular respiration continuously produces carbon dioxide, the partial pressure of carbon dioxide is lower in the blood than it is in the tissues, causing carbon dioxide to diffuse out of the tissues, cross the interstitial fluid, and enter the blood. It is then carried back to the lungs either bound to hemoglobin, dissolved in plasma, or as bicarbonate ions. By the time blood returns to the heart, the partial pressure of oxygen is about 40 mm Hg, and the partial pressure of carbon dioxide is about 46 mm Hg. The blood is then pumped back to the lungs to be oxygenated once again.

Apply to Pathophysiology

Altitude

Paolo is a healthy college student from Los Angeles (which is at sea level). Paolo went to Colorado to visit a friend and together they drove to the top of Pike's Peak, a mountain with a peak elevation of approximately 14,000 feet. During the first hour on the peak Paolo walked around taking pictures, but found that they needed to sit down and catch their breath several times. When Paolo did so, they realized they were taking breaths much more frequently than normal. Paolo walked to the weather station to rest and glanced at a barometer (an instrument that measures pressure) hanging overhead. It read 450 mm Hg.

Predict the directional changes that would have taken place in each of the listed parameters by the time Paolo reached the weather station *compared with the values that they would have had at home (sea level).*

1. Partial pressure of O_2 in the alveoli.
   A) Increase
   B) Decrease
   C) No change
2. Hemoglobin O_2 saturation in the systemic arterial blood.
   A) Decrease
   B) Increase
   C) No change
3. Hemoglobin (O_2) saturation in the systemic venous blood.
   A) Increase
   B) Decrease
   C) No change
4. Tidal volume
   A) No change
   B) Decrease
   C) Increase

Learning Check

1. Ventilation transports air using ______, and gas exchange transports respiratory gasses using ______.
   a. Bulk flow; simple diffusion
   b. Bulk flow; bulk flow
   c. Simple diffusion; bulk flow
   d. Simple diffusion; simple diffusion
2. There is insufficient ventilation in alveolus A, but there is sufficient ventilation in alveolus B. Pulmonary arterioles will ______ to alveolus A and ______ to alveolus B.
   a. Vasoconstrict; vasoconstrict
   b. Vasodilate; vasodilate
   c. Vasoconstrict; vasodilate
   d. Vasodilate; vasoconstrict
3. A greater-than-normal partial pressure of carbon dioxide is present in the alveoli. Will the bronchioles constrict or dilate?
   a. Bronchoconstriction will occur.
   b. Bronchodilation will occur.
4. Two 18-year-old males of roughly the same height, weight, and athletic ability both undergo a VO_2 max test, meaning they are testing how much oxygen can be consumed during exercise. Boy 1 smokes, while boy 2 does not. Boy 1 is expected to have a ______ VO_2 max because of the ______ component of Fick's law.
   a. Higher; surface area
   b. Lower; surface area
   c. Higher; concentration gradient
   d. Lower; concentration gradient

22.5 Transport of Gasses

Learning Objectives: By the end of this section, you will be able to:

22.5.1 Describe the ways in which oxygen is transported in blood, and explain the relative importance of each to total oxygen transport.

22.5.2* Correlate hemoglobin saturation to environmental conditions such as the partial pressure of oxygen, temperature, and pH.

22.5.3 Interpret the oxygen–hemoglobin saturation curve at low and high partial pressures of oxygen.

22.5.4 *Describe the oxygen–fetal hemoglobin saturation curve and its impact on oxygen delivery to fetal tissues.

22.5.5 Describe the ways in which carbon dioxide is transported in blood and explain the relative importance of each to total carbon dioxide transport.

22.5.6 Explain the role of each of the following in carbon dioxide transport: carbonic anhydrase, hydrogen ions binding to hemoglobin, the chloride shift, and oxygen–hemoglobin saturation level.

22.5.7 State the reversible chemical equation for the reaction of carbon dioxide and water to carbonic acid and then to hydrogen ion and bicarbonate ion.

22.5.8 Predict how changing the pH or the concentration of bicarbonate ions will affect the partial pressure of carbon dioxide in the plasma.

22.5.9 Predict how changing the partial pressure of carbon dioxide will affect the pH and the concentration of bicarbonate ions in the plasma.

* Objective is not a HAPS Learning Goal.

We have discussed how gasses exchange between the blood and the alveoli and how they exchange between the blood and the tissues, but you may be wondering how gasses are carried within the blood between these locations. Due to their differing solubility in the blood, oxygen and carbon dioxide are carried differently.

22.5a Oxygen Transport in the Blood

LO 22.5.1

Even though oxygen is transported via the blood, you may recall that oxygen is not very soluble in liquids. A small amount of oxygen does dissolve in the blood and is transported in the bloodstream, but it is only about 1.5 percent of the total amount. The majority of oxygen molecules are carried from the lungs to the body's tissues by a specialized transport system, which relies on the erythrocyte—the red blood cell. Erythrocytes contain the protein, hemoglobin, which binds oxygen molecules (**Figure 22.29**). Heme is the portion of hemoglobin that contains iron, and this is where oxygen binds. One hemoglobin molecule contains four iron-containing Heme molecules, and because of this, each hemoglobin molecule is capable of carrying up to four molecules of oxygen. As oxygen diffuses across the respiratory membrane from the alveolus to the capillary, it does not stay dissolved in the plasma for long. Rather, oxygen diffuses into the red blood cell and is bound by hemoglobin, forming **oxyhemoglobin (Hb–O_2)**, or oxygen-bound hemoglobin. Like all proteins, when hemoglobin binds its ligand, it changes shape slightly. This slight shape change of hemoglobin has two notable functional consequences. One is that oxyhemoglobin molecules are colored bright red; when hemoglobin is not bound to oxygen, it is more of a dark brick-red, or brownish red. The second is that each time that oxygen binds hemoglobin, the molecule changes shape slightly in a way that makes the binding of the next molecule more likely. There are multiple factors involved in how readily heme binds to and dissociates from oxygen, which will be discussed in the subsequent sections.

Figure 22.29 Hemoglobin

Hemoglobin is a complex protein that is composed of four subunits. Each subunit carries a heme group which, due to the iron found inside, is able to bind oxygen.

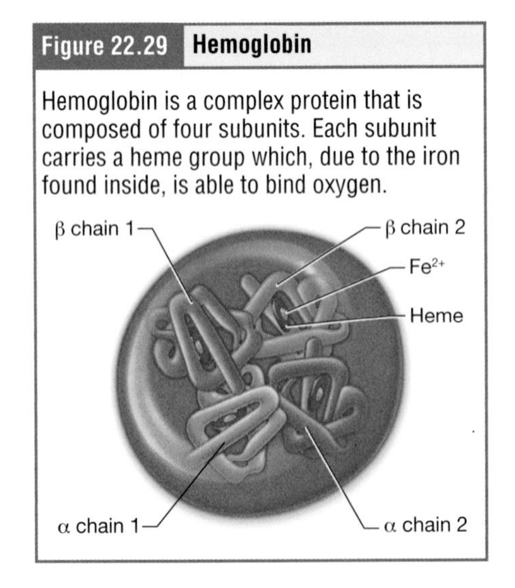

Function of Hemoglobin Hemoglobin is composed of subunits, a protein structure that is referred to as a *quaternary structure*. Each of the four subunits that make up hemoglobin is arranged in a ringlike fashion, with an iron atom bound to the heme in the center of each subunit. When the first oxygen molecule binds, the protein undergoes a small shape change. This new shape actually opens up the hemoglobin molecule slightly, allowing the second molecule of oxygen to bind more easily. As each molecule of oxygen is bound, it further facilitates the binding of the next molecule, until all four heme sites are occupied by oxygen. The opposite occurs as well: after the first oxygen molecule dissociates, the next oxygen molecule dissociates more readily. This progressive loosening of the bound oxygen molecules facilitates tissues to take up as much oxygen as possible, because oxygen is poorly soluble in plasma, so it will readily diffuse to the interstitial fluid and then to the cells of the tissue where there is a low oxygen concentration. When all four heme sites are occupied, the hemoglobin is said to be *saturated*. When one to three heme sites are occupied, the hemoglobin is said to be *partially saturated*. Therefore, when considering the blood as a whole, the percentage of the available heme units that are bound to oxygen at a given time is called **hemoglobin saturation**. Hemoglobin saturation of 100 percent means that every heme unit in all of the erythrocytes of the body is bound to oxygen. In a healthy individual with normal hemoglobin levels, hemoglobin saturation generally ranges from 95 percent to 99 percent. Hemoglobin saturation is measured using a *pulse oximeter*, a small device that can estimate hemoglobin saturation through the skin (**Figure 22.30**).

Student Study Tip

Think of hemoglobin as a series of buses, and each bus has four seats (the heme sites). When all the buses are filled, they are fully saturated.

LO 22.5.2

Oxygen Dissociation from Hemoglobin In addition to saturation, there are other factors that make oxygen more or less likely to bind and unbind from hemoglobin.

Figure 22.30 Pulse Oximeters

Pulse oximeters are small devices that measure the percent of hemoglobin molecules that are saturated with oxygen.

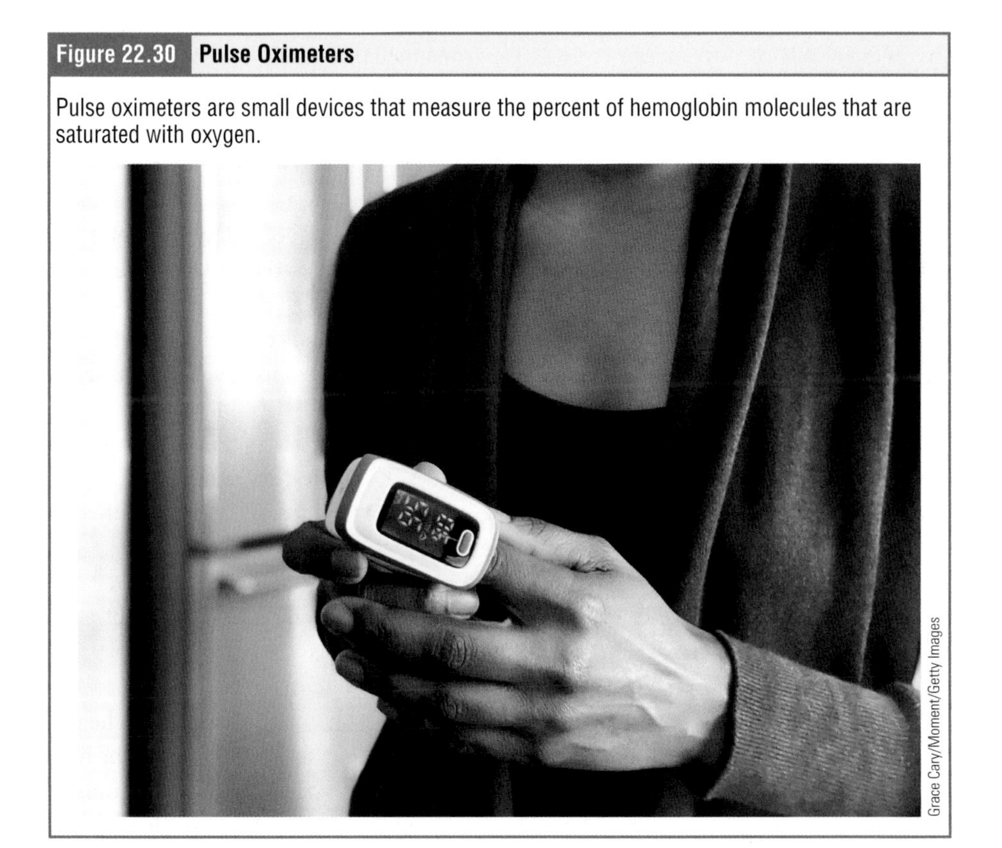

Grace Cary/Moment/Getty Images

We can graph these in order to visualize and predict the binding and unbinding of oxygen from hemoglobin. This type of graph is called an **oxygen–hemoglobin dissociation curve** (see the "Anatomy of an Oxygen–Hemoglobin Dissociation Curve" feature). Partial pressure is an important aspect of the binding of oxygen to, and disassociation from, heme. Remember that gasses travel from an area of higher partial pressure to an area of lower partial pressure. In addition, the affinity of an oxygen molecule for heme increases as more oxygen molecules are bound. **Affinity** can be described as the strength and length of the molecular interaction. You can visualize affinity by thinking of a handshake. A strong affinity would be analogous to a firm and long-lasting handshake; it can be hard to end or pull away from that interaction. Therefore, in the oxygen–hemoglobin saturation curve, as the partial pressure of oxygen increases, a proportionately greater number of oxygen molecules are bound by heme. These oxygen molecules are tightly bound and unlikely to unbind. The oxygen–hemoglobin saturation/dissociation curve also shows that the lower the partial pressure of oxygen, the fewer the oxygen molecules that are bound to heme. At these low partial pressures, the affinity of hemoglobin for oxygen is dramatically lower. That is to say, the "handshake" has become weak and transient, and oxygen molecules unbind from hemoglobin easily. As a result, the partial pressure of oxygen plays a major role in promoting the binding of oxygen to heme in the high-oxygen environment of the respiratory membrane, as well as its dissociation in the low-oxygen environment of the tissues.

The mechanisms behind the oxygen–hemoglobin saturation/dissociation curve also function to distribute oxygen to the tissues that need it most. Some tissues have a higher metabolic rate than others, and highly active tissues, such as muscle, rapidly use oxygen to produce ATP, lowering the partial pressure of oxygen in the tissue to about 20 mm Hg. The partial pressure of oxygen inside capillaries is about 100 mm Hg, so the difference between the two becomes quite high—about 80 mm Hg. As a result, a greater number of oxygen molecules dissociate from hemoglobin and enter the tissues.

Anatomy of...

An Oxygen–Hemoglobin Dissociation Curve

LO 22.5.3

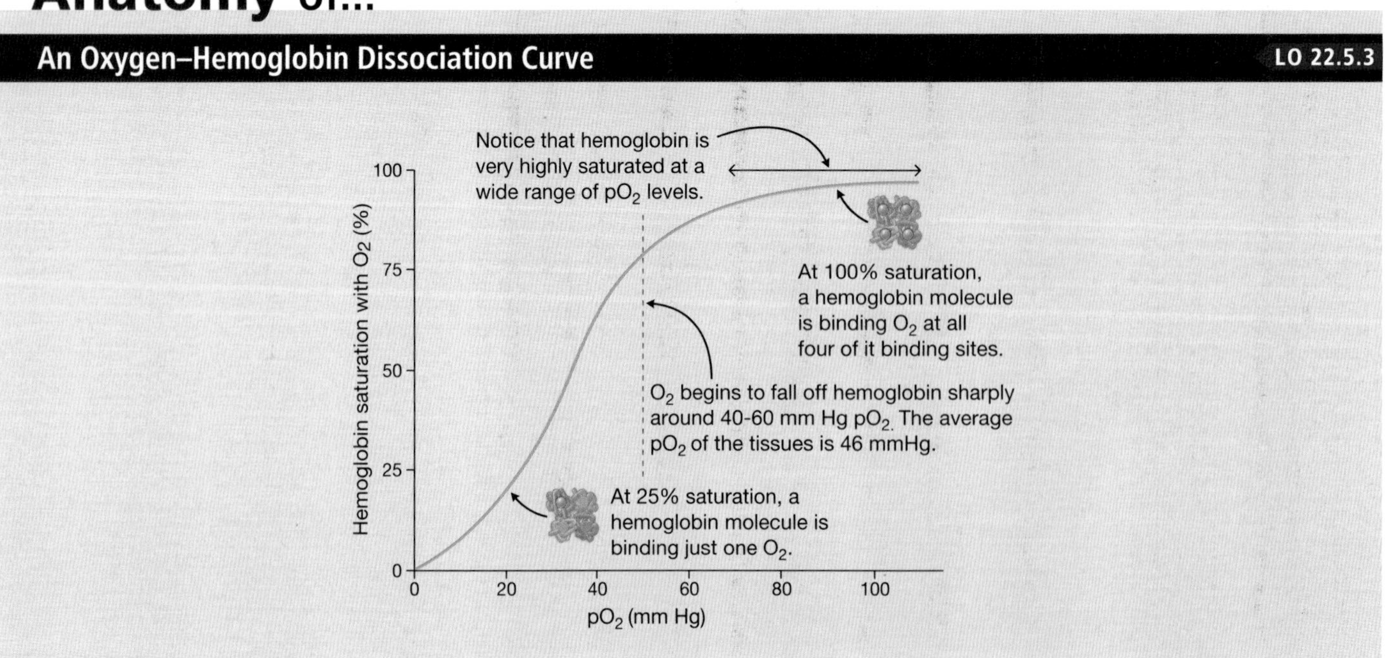

Notice that hemoglobin is very highly saturated at a wide range of pO_2 levels.
At 100% saturation, a hemoglobin molecule is binding O_2 at all four of it binding sites.
O_2 begins to fall off hemoglobin sharply around 40-60 mm Hg pO_2. The average pO_2 of the tissues is 46 mmHg.
At 25% saturation, a hemoglobin molecule is binding just one O_2.
Hemoglobin saturation with O_2 (%)
100
75
50
25
0
0
20
40
60
80
100
pO_2 (mm Hg)

The reverse is true of tissues, such as adipose tissue (body fat), which have lower metabolic rates. Because less oxygen is used by these cells, the partial pressure of oxygen within such tissues remains relatively high, resulting in fewer oxygen molecules dissociating from hemoglobin and entering the tissue interstitial fluid. Although venous blood is said to be deoxygenated, some oxygen is still bound to hemoglobin in its red blood cells. This provides an oxygen reserve that can be used when tissues suddenly demand more oxygen.

Factors other than partial pressure also affect the oxygen–hemoglobin saturation/dissociation curve. These other factors influence hemoglobin's affinity for oxygen and contribute to homeostasis. For example, whereas a higher temperature promotes faster dissociation of hemoglobin and oxygen, a lower temperature inhibits its dissociation (**Figure 22.31A**). You may recall that heat is a byproduct of cellular respiration, and therefore the temperature of a region undergoing an increase in the rate of cellular respiration may be higher than that of an area with a lower rate. At this increased temperature, oxygen is likely to fall off the hemoglobin more readily, making it available for use in the tissues (higher temperatures also increase the rate of

Figure 22.31 Hemoglobin's Affinity for Oxygen Changes with Environmental Conditions

Hemoglobin's affinity for oxygen will change as the (A) temperature and (B) pH of the environment changes. Both of these conditions change due to metabolism, so oxygen is more likely to dissociate from hemoglobin and be available to diffuse into tissues that are warmer or at a lower pH; both are characteristics of increased metabolism and cellular respiration.

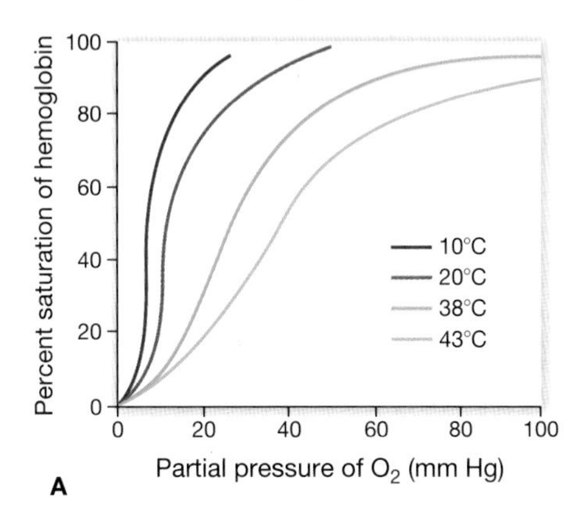

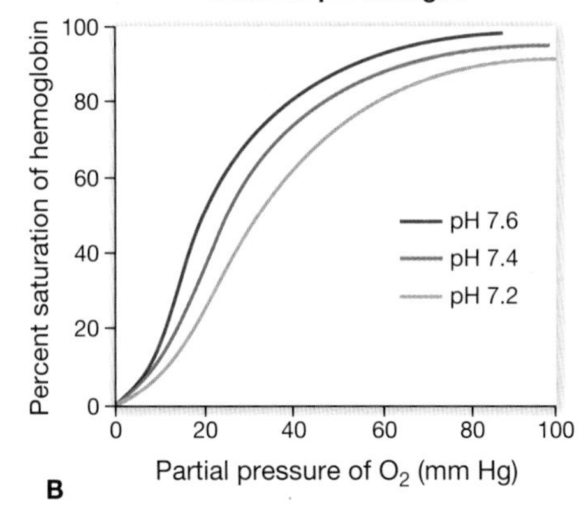

diffusion, so the oxygen will diffuse more rapidly toward the low oxygen environment of the tissue cells).

The pH of the blood is another factor that influences the oxygen–hemoglobin saturation/dissociation curve (**Figure 22.31B**). The relationship between blood pH and hemoglobin's affinity for oxygen was first described in 1904 by Christian Bohr, a Danish physiologist, and is therefore often referred to as the **Bohr effect**. Blood with a lower pH (i.e., more acidic) has hemoglobin with a reduced affinity for oxygen and therefore promotes oxygen dissociation from hemoglobin. In contrast, a higher (i.e., more basic) pH inhibits oxygen dissociation from hemoglobin. The pH of the blood is tightly regulated, but—in ways that we will discuss in a following section—lower-pH environments are common in regions that are undergoing higher metabolic rates. Thus, the decrease in hemoglobin's affinity for oxygen in lower-pH environments is one more way that hemoglobin is tailored to deliver oxygen in the environments where it is most needed.

Fetal Hemoglobin A developing fetus has its own circulation with its own erythrocytes; however, without lungs that are open to the air, the fetus is dependent on the mother's circulation for oxygen. Fetal blood and maternal blood never mix; however, there are capillary beds in the placenta where the fetal blood comes close enough to maternal blood to enable diffusion. Here the fetal blood picks up any gasses and nutrients and is able to get rid of waste products.

LO 22.5.4

The difference in partial pressure of oxygen between maternal and fetal blood is not large. Therefore, not as much oxygen will diffuse readily into fetal plasma. Fetal hemoglobin overcomes this problem by having a greater affinity for oxygen than maternal hemoglobin (**Figure 22.32**). Therefore, as a fetal red blood cell and maternal red blood cell come near each other across a membrane in the placenta, the oxygen on the maternal red blood cell is likely to dissociate and bind to the higher-affinity fetal hemoglobin, but the reverse would never occur.

Figure 22.32 Fetal Hemoglobin

Fetal hemoglobin molecules have a higher affinity for oxygen than maternal hemoglobin molecules. This characteristic enables them to take oxygen from maternal red blood cells when fetal and maternal red blood cells come near each other in separate but adjacent anatomical areas.

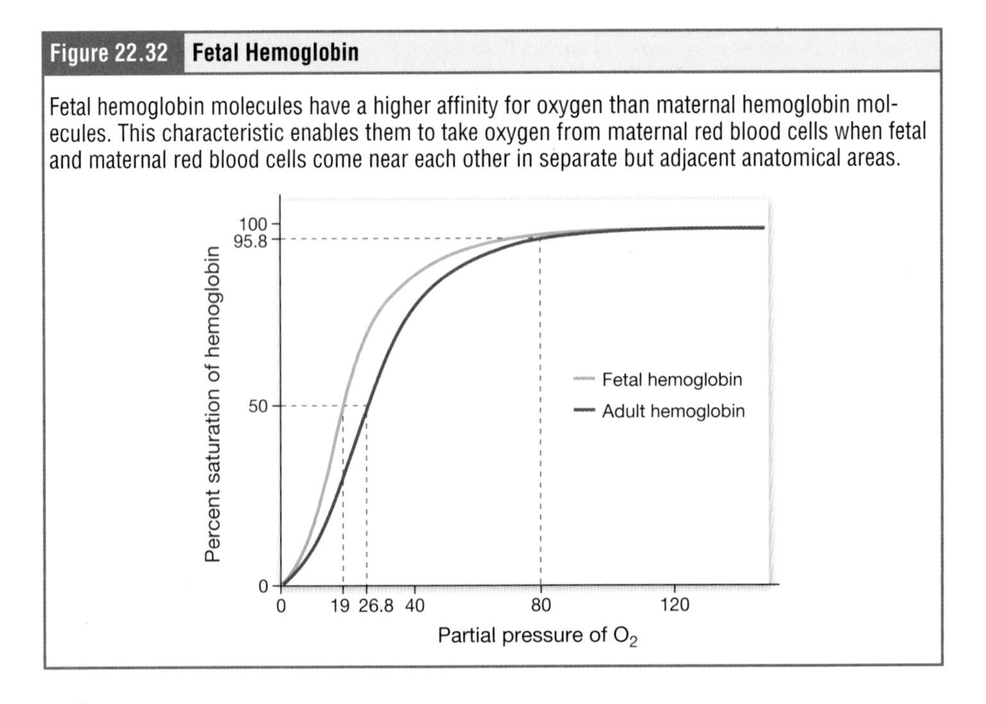

22.5b Carbon Dioxide Transport in the Blood

LO 22.5.5

Carbon dioxide is transported by three major mechanisms. Unlike oxygen, carbon dioxide readily dissolves in the blood plasma, and so some CO_2 molecules will travel in the plasma this way. The vast majority of CO_2 molecules will dissolve in the plasma, but then combine with water in the blood and form bicarbonate (HCO_3^-), which will travel in plasma. The third mechanism of carbon dioxide transport is similar to the transport of oxygen—by binding to hemoglobin and being ferried along in the blood on this molecule (**Figure 22.33**).

Dissolved Carbon Dioxide More soluble than oxygen, carbon dioxide is still not considered to be highly soluble in blood, so only a small fraction—about 7 to 10 percent—of the carbon dioxide that diffuses into the blood from the tissues dissolves in plasma. The dissolved carbon dioxide then travels in the bloodstream and, when the blood reaches the pulmonary capillaries, the dissolved carbon dioxide diffuses across the respiratory membrane into the alveoli, where it is then exhaled during pulmonary ventilation. An additional 20 percent of carbon dioxide is bound by hemoglobin and is transported to the lungs.

LO 22.5.6

Bicarbonate Buffer A large fraction—about 70 percent—of the carbon dioxide molecules that diffuse into the blood is transported to the lungs as bicarbonate. Most bicarbonate is produced in erythrocytes after carbon dioxide diffuses into the capillaries, and subsequently into red blood cells. **Carbonic anhydrase (CA)**, an enzyme carried in red blood cells, combines carbon dioxide and water to form carbonic acid (H_2CO_3), which dissociates into two ions: bicarbonate (HCO_3^-) and hydrogen (H^+). The following formula depicts this reaction:

LO 22.5.7

$$CO_2 + H_2O \overset{CA}{\leftrightarrow} H_2CO_3 \leftrightarrow H^+ + HCO_3^-$$

Bicarbonate tends to build up in the erythrocytes, so that there is a greater concentration of bicarbonate in the erythrocytes than in the surrounding blood plasma. The

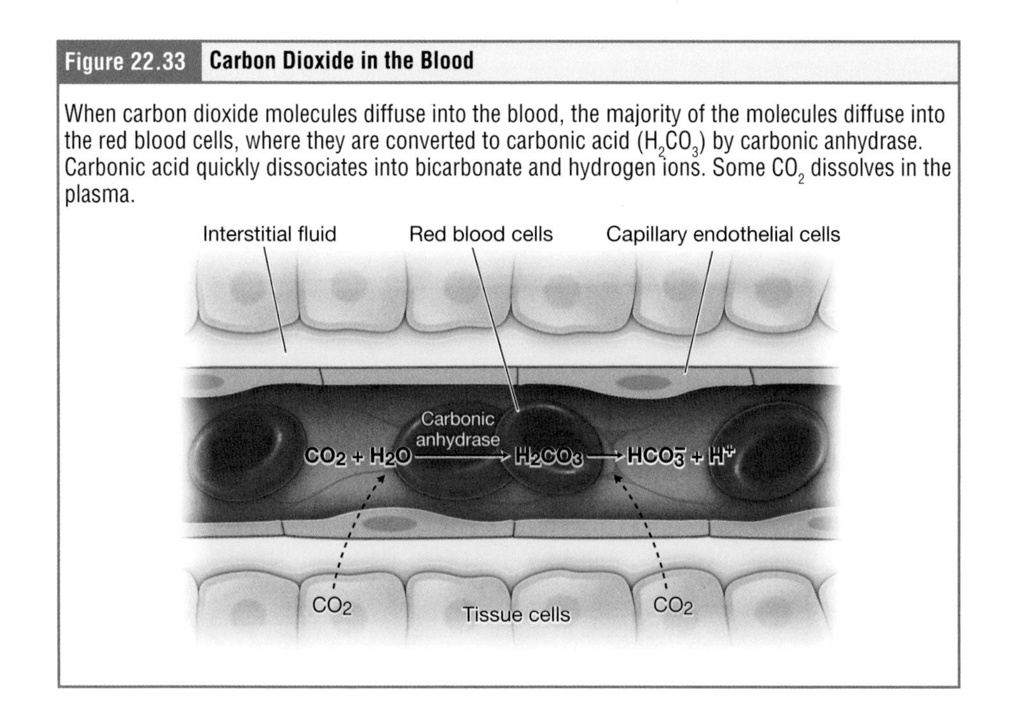

Figure 22.33 Carbon Dioxide in the Blood

When carbon dioxide molecules diffuse into the blood, the majority of the molecules diffuse into the red blood cells, where they are converted to carbonic acid (H_2CO_3) by carbonic anhydrase. Carbonic acid quickly dissociates into bicarbonate and hydrogen ions. Some CO_2 dissolves in the plasma.

H^+ ions may be carried on hemoglobin, or they may leave the erythrocyte and enter the plasma, where they drive down the pH of the blood. This single factor—CO_2—and its ability to acidify the blood is the largest driver of breath rate; getting rid of CO_2 at the lungs is one of the major mechanisms for maintaining CO_2 homeostasis.

At the pulmonary capillaries, the chemical reaction that produced bicarbonate is reversed, and carbon dioxide and water are the products. The reformed carbon dioxide diffuses out of the erythrocytes and into the plasma, where it can further diffuse across the respiratory membrane into the alveoli to be exhaled during pulmonary ventilation.

Similar to the transport of oxygen by heme, the binding and dissociation of carbon dioxide to and from hemoglobin is dependent on the partial pressure of carbon dioxide. Because carbon dioxide is released from the lungs, blood that leaves the lungs and reaches body tissues has a lower partial pressure of carbon dioxide than is found in the tissues. As a result, carbon dioxide leaves the tissues because of its higher partial pressure, enters the blood, and then moves into red blood cells, binding to hemoglobin. In contrast, in the pulmonary capillaries, the partial pressure of carbon dioxide is high compared to that within the alveoli. As a result, carbon dioxide dissociates readily from hemoglobin and diffuses across the respiratory membrane into the air.

Learning Check

LO 22.5.8 LO 22.5.9

1. Picture the process of offloading oxygen from hemoglobin in erythrocytes. Oxygen 1 is released first, followed by oxygen 2, 3, and 4. Which of the oxygen molecules will dissociate most easily?
   a. Oxygen 1
   b. Oxygen 2
   c. Oxygen 3
   d. Oxygen 4
2. Carbon dioxide buildup in the lungs during and after heavy exercise will drive the respiratory rate ______.
   a. Up
   b. Down
   c. No change
   d. Impossible to tell with the given information
3. There is a larger-than-normal concentration of carbon dioxide in the pulmonary capillaries compared to that in the alveoli. What will be the effect on hemoglobin's affinity for carbon dioxide in the pulmonary capillaries?
   a. Increase
   b. Decrease
   c. No change
   d. Impossible to tell with the given information
4. During an event requiring a higher metabolic rate, such as exercise, body temperature ______. This causes a(n) ______ in hemoglobin's affinity for oxygen at the muscle tissues.
   a. Increases; increase
   b. Increases; decrease
   c. Decreases; increase
   d. Decreases; decrease
5. Some students become very nervous when they have to get up in front of the class to present a report, and they begin to hyperventilate. What would you expect to happen to the partial pressure of carbon dioxide in their blood, and how would this affect blood pH?
   a. Hyperventilation would increase the partial pressure of oxygen but would have no effect on the partial pressure of CO_2 or the blood pH.
   b. Hyperventilation would have no effect on the partial pressure of carbon dioxide or pH because the body would balance the effects of hyperventilation by increasing the metabolic rate.
   c. Hyperventilation would decrease the partial pressure of carbon dioxide, which would increase the blood pH.
   d. Hyperventilation would increase the partial pressure of carbon dioxide, which would decrease the blood pH.

22.6 Respiratory Rate and Control of Ventilation

Learning Objectives: By the end of this section, you will be able to:

22.6.1 Describe the locations and functions of the brainstem respiratory centers.

22.6.2 List and describe the major chemical and neural stimuli to the respiratory centers.

22.6.3 Compare and contrast the central and peripheral chemoreceptors.

22.6.4 Define hyperventilation, hypoventilation, panting, eupnea, hyperpnea, and apnea.

Breathing usually occurs without thought, although at times you can consciously control it, such as when you swim under water, sing a song, or blow bubbles. The **respiratory rate** is the total number of breaths that occur each minute. Respiratory rate can be an important indicator of disease, as the rate may increase or decrease during an illness or in a disease condition. The respiratory rate is controlled by the respiratory center located within the medulla oblongata in the brain, which responds primarily to changes in carbon dioxide, oxygen, and pH levels in the blood.

respiratory rate (breathing rate, breath rate)

Respiratory rate decreases across the lifespan as the lungs get bigger in size. A healthy child under 1 year of age has a respiratory rate between 30 and 60 breaths per minute; for an adult, however, 12 to 18 breaths per minute is more typical.

Ventilation Control Centers The control of ventilation is a complex interplay of multiple regions in the brain that signal the muscles used in pulmonary ventilation to contract. The result is typically a rhythmic, consistent ventilation rate that provides the body with sufficient amounts of oxygen, while adequately removing carbon dioxide.

The major brain centers that control pulmonary ventilation are the control centers in the medulla oblongata and the pontine respiratory group (**Figure 22.34**).

LO 22.6.1

The **dorsal respiratory group (DRG)** and the **ventral respiratory group (VRG)** are structures in the medulla oblongata that innervate the muscles of breathing. The two centers control the different sets of muscles, resulting in different breathing functions. The DRG maintains a constant breathing rhythm. Stimulation from the DRG causes these muscles to contract, causing inspiration; a pause in stimulation from the DRG allows the breathing muscles to relax, causing expiration. In contrast, the VRG is involved in deep breathing, as the neurons in the VRG stimulate the accessory muscles of the thorax (such as the serratus anterior and the inner intercostal muscles) resulting in a deep inhale. The VRG also stimulates accessory muscles such as the abdominal muscles to assist with a forced exhale.

The second respiratory center of the brain is located within the pons, called the *pontine respiratory group*, and consists of the apneustic and pneumotaxic centers. The **apneustic center** is a cluster of neuronal cell bodies that stimulate neurons in the DRG, controlling the depth of inspiration, particularly for deep breathing. The **pneumotaxic center** is a network of neurons that inhibits the activity of neurons in the DRG, allowing relaxation after inspiration.

Factors That Affect the Rate and Depth of Respiration The respiratory rate and the depth of inspiration are regulated by the medulla oblongata and the pons; however, these regions of the brain do so in response to systemic stimuli. Multiple systemic factors are involved in stimulating the brain to produce pulmonary ventilation.

Figure 22.34 Regulation of the Breathing Process

The neurological control over breathing is maintained by two locations in the brainstem and is influenced by other brain regions. The pontine respiratory group influences the dorsal respiratory group, which controls regular breathing through its influence on the external intercostals and the diaphragm. The ventral respiratory group influences deep and forced breathing by activating the accessory muscles of breathing, including the internal intercostals and other muscles that insert on the thorax.

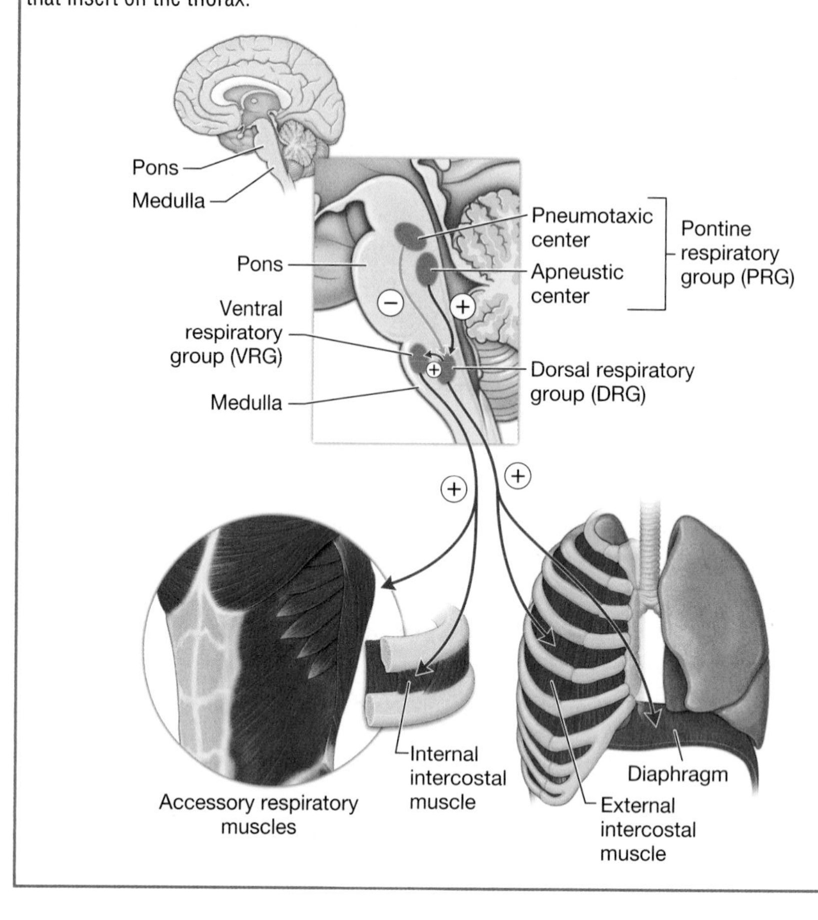

LO 22.6.2

The major factor that stimulates the medulla oblongata and the pons to produce respiration is—surprisingly—not oxygen concentration, but rather the concentration of carbon dioxide in the blood. As discussed in Section 22.5b, carbon dioxide can combine with water to produce carbonic acid, which then dissociates into H^+ ions and bicarbonate ions. While the concentration of CO_2 is not inherently risky to the body, deviations from pH homeostasis are very dangerous to the cells and proteins. Therefore, the CO_2 concentration, because it will lead to a change in pH, must be closely monitored and controlled. Concentrations of chemicals are sensed by chemoreceptors. A **central chemoreceptor** is one of the specialized receptors that are located in the brain and brainstem, while a **peripheral chemoreceptor** is one of the specialized receptors located in the carotid arteries and aortic arch. Concentration changes in both carbon dioxide and hydrogen ions stimulate these receptors, which in turn signal the respiration centers of the brain. The increase in hydrogen ions in the brain triggers the central chemoreceptors to stimulate the respiratory centers to initiate contraction of the diaphragm and intercostal muscles. As a result, the rate and depth of respiration increase, allowing more carbon dioxide to be expelled, which brings more air into and out of the lungs, promoting a reduction in the blood levels of carbon dioxide, and therefore hydrogen ions, in the blood. In contrast, low levels of carbon dioxide in the

LO 22.6.3

blood cause low levels of hydrogen ions in the brain, leading to a decrease in the rate and depth of pulmonary ventilation, producing shallow, slow breathing.

Increased H^+ levels can occur as a side effect of more carbon dioxide being carried in the blood, as well as of other metabolic activities such as lactic acid accumulation after exercise. Peripheral chemoreceptors of the aortic arch and carotid arteries sense arterial levels of hydrogen ions. When peripheral chemoreceptors sense decreasing, or more acidic, pH levels, they stimulate an increase in ventilation to remove carbon dioxide from the blood at a quicker rate. Removal of carbon dioxide from the blood helps to reduce hydrogen ions, thus increasing systemic pH. In **Figure 22.35A**, you can see how quickly the minute ventilation will rise in response to changes in the concentration of H^+ ions in the blood.

Blood levels of oxygen are also an important determinant of respiratory rate. The peripheral chemoreceptors are responsible for sensing large changes in blood oxygen levels. If blood oxygen levels become quite low—about 60 mm Hg or less—then peripheral chemoreceptors stimulate an increase in respiratory activity. The chemoreceptors are only able to sense dissolved oxygen molecules, not the oxygen that is bound to hemoglobin. As you recall, the majority of oxygen is bound by hemoglobin; when dissolved levels of oxygen drop, hemoglobin releases oxygen. Therefore, a large drop in

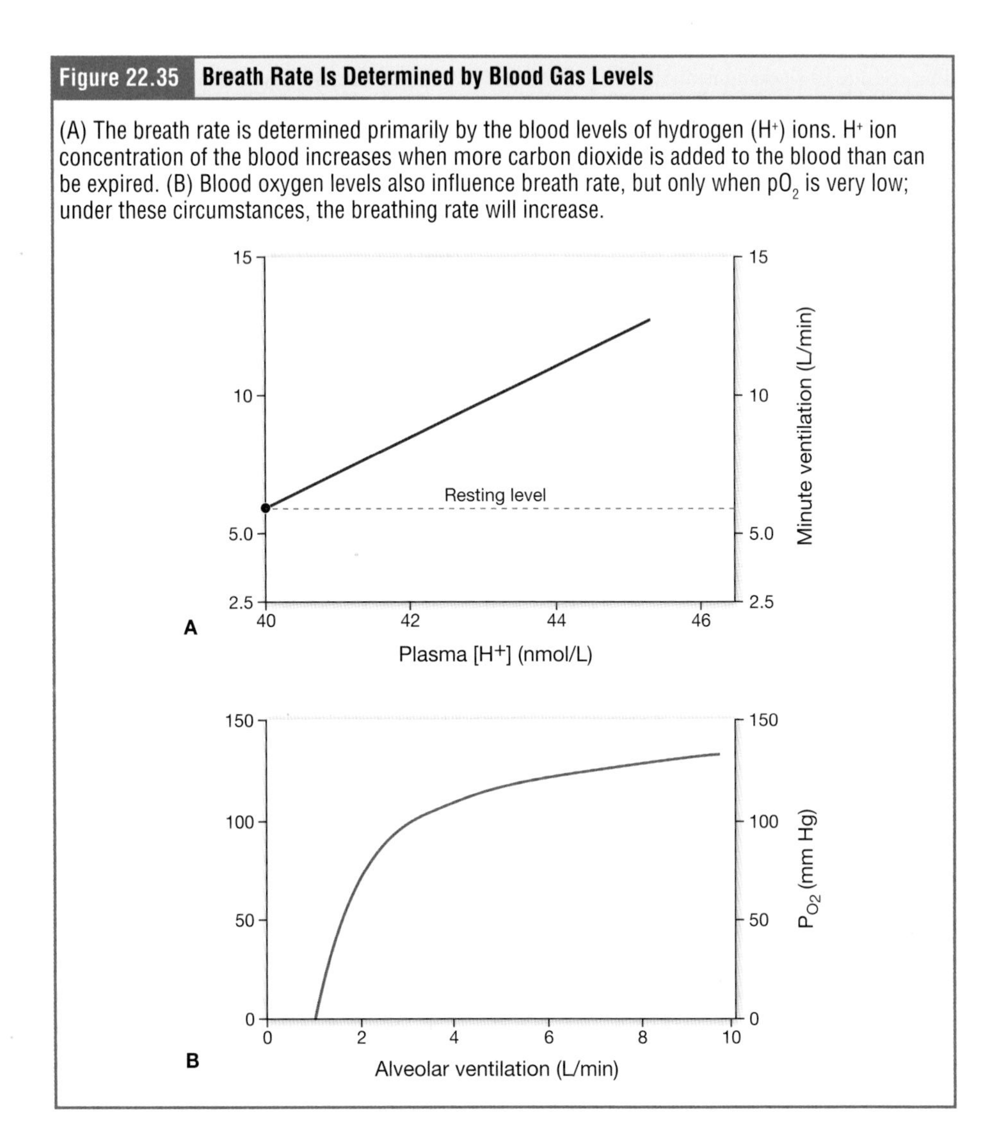

Figure 22.35 Breath Rate Is Determined by Blood Gas Levels

(A) The breath rate is determined primarily by the blood levels of hydrogen (H^+) ions. H^+ ion concentration of the blood increases when more carbon dioxide is added to the blood than can be expired. (B) Blood oxygen levels also influence breath rate, but only when pO_2 is very low; under these circumstances, the breathing rate will increase.

oxygen levels is required to stimulate the chemoreceptors of the aortic arch and carotid arteries. The impact of pO_2 on minute ventilation is graphed in **Figure 22.35B**. Notice that the body is able to tolerate a much larger range of pO_2 levels than it is able to tolerate swings in the H^+ ion concentration of the blood.

The hypothalamus and other brain regions associated with the limbic system also play roles in influencing the regulation of breathing by interacting with the respiratory centers. The hypothalamus and other regions associated with the limbic system are involved in regulating respiration in response to emotions, pain, and temperature. For example, an increase in body temperature causes an increase in respiratory rate. Feeling excited, or the fight-or-flight response, will also result in an increase in respiratory rate.

Because we are able to regulate our breathing for reasons other than gas levels—for example, while swimming—the blood gas levels may deviate from homeostatic range.

LO 22.6.4

Hypoventilation occurs when we do not breath frequently enough to sufficiently maintain blood gas homeostasis. A buildup of CO_2 and a drop in pH occur during hypoventilation. **Hyperventilation** is the opposite; a person hyperventilating is breathing too fast for ideal blood gas homeostasis, and therefore expelling more CO_2 than is ideal. This individual may experience an increase in their blood pH outside the homeostatic range. Hypo- and hyperventilation describe breath rate, but a series of related terms describe breath depth. **Eupnea**, or true breathing, is the depth of breath that is typical of a resting individual. **Hyperpnea** describes breathing that is deeper and more frequent than that of a typical resting individual. **Apnea** describes a spontaneous absence in breathing. Apnea occurs from time to time in sleeping individuals, but when it occurs regularly, sleep apnea can interfere with sleep, preventing the sufferer from feeling rested even after a full night of sleep.

Learning Connection

Micro to Macro

Some people have observed that they can increase their breath-holding time if they hyperventilate just before beginning to hold their breath. Why does this work? Is this related to a change in the blood concentration of oxygen or carbon dioxide?

DIGGING DEEPER:
Sleep Apnea

Sleep apnea is a chronic disorder that can occur in children or adults, and is characterized by the cessation of breathing during sleep. These episodes may last for several seconds or up to several minutes, and may differ in the frequency with which they are experienced. Sleep apnea leads to poor sleep, which is reflected in the symptoms of fatigue, evening napping, irritability, memory problems, and morning headaches. In addition, many individuals with sleep apnea experience a dry throat in the morning after waking from sleep, which may be due to excessive snoring.

There are two types of sleep apnea: *obstructive sleep apnea* and *central sleep apnea*. Obstructive sleep apnea is caused by an obstruction of the airway during sleep, which can occur at different points in the airway, depending on the underlying cause of the obstruction. For example, the tongue and throat muscles of some individuals with obstructive sleep apnea may relax excessively, causing the muscles to push into the airway. Another example is obesity, which is a known risk factor for sleep apnea, as excess adipose tissue in the neck region can push the soft tissues toward the lumen of the airway, causing the trachea to narrow.

In central sleep apnea, the respiratory centers of the brain do not respond properly to rising carbon dioxide levels and therefore do not stimulate the contraction of the diaphragm and intercostal muscles regularly. As a result, inspiration does not occur and breathing stops for a short period. In some cases, the cause of central sleep apnea is unknown. However, some medical conditions, such as stroke and congestive heart failure, may cause damage to the pons or medulla oblongata. In addition, some pharmacologic agents, such as morphine, can affect the respiratory centers, causing a decrease in the respiratory rate. The symptoms of central sleep apnea are similar to those of obstructive sleep apnea.

A diagnosis of sleep apnea is usually done during a sleep study, where the patient is monitored in a sleep laboratory for several nights. The patient's blood oxygen levels, heart rate, respiratory rate, and blood pressure are monitored, as are brain activity and the volume of air that is inhaled and exhaled. Treatment of sleep apnea commonly includes the use of a device called a *continuous positive airway pressure (CPAP) machine* during sleep. The CPAP machine has a mask that covers the nose, or the nose and mouth, and forces air into the airway at regular intervals. This pressurized air can help to gently force the airway to remain open, allowing more normal ventilation to occur. Other treatments include lifestyle changes to decrease weight and eliminate alcohol and other sleep apnea–promoting drugs, as well as changes in sleep position. In addition to these treatments, patients with central sleep apnea may need supplemental oxygen during sleep.

Learning Check

1. Which cluster of neurons controls depth of inspiration and which controls relaxation after inspiration in the dorsal respiratory group?
   a. Apneustic center; pneumotaxic center
   b. Pneumotaxic center; apneustic center
   c. Pneumotaxic center; pneumotaxic center
   d. Apneustic center; apneustic center
2. What is the chief driver of respiratory rate?
   a. Blood oxygen level
   b. Blood nitrogen level
   c. Blood carbon dioxide level
   d. None of the above
3. Say you're having a panic attack and begin hyperventilating. Under these conditions, your body could possibly enter a state of ______.
   a. Acidosis (low blood pH)
   b. Alkalosis (high blood pH)
   c. Neither of the above
   d. Impossible to tell with the given information
4. Which respiratory group is responsible for deep, intentional breathing and sharp, forced exhalation?
   a. Dorsal respiratory group
   b. Ventral respiratory group
   c. Pneumotaxic center
   d. Apneustic center

Chapter Review

22.1 Functions and Anatomy of the Respiratory System

The learning objectives for this section are:

22.1.1 Describe the major functions of the respiratory system.

22.1.2 Compare and contrast the general locations and functions of the conducting and respiratory portions (zones) of the respiratory tract.

22.1.3 Identify the anatomical division of the upper versus lower respiratory tract.

22.1.4 List, in order, the respiratory structures that air passes through during inspiration and expiration.

22.1.5 Describe the major functions, gross anatomical features, and epithelial lining of the nasal cavity, paranasal sinuses, and pharynx.

22.1.6 Describe the major functions of the larynx.

22.1.7 Describe the anatomical features of the larynx, including the laryngeal cartilages.

22.1.8 Compare and contrast the location, composition, and function of the vestibular folds (false vocal cords) and vocal folds (true vocal cords).

22.1.9 Briefly explain how the vocal folds and the larynx function in phonation.

22.1.10 Describe the major functions of the trachea.

22.1.11 Describe the microscopic anatomy of the trachea, including the significance of the C-shaped hyaline cartilage rings.

22.1.12 Describe the gross anatomical features of the trachea, including its positioning with respect to the esophagus.

22.1.13 Identify and describe the anatomic features of the bronchial tree (e.g., main [primary] bronchi, lobar [secondary] bronchi, segmental [tertiary] bronchi, smaller bronchi, bronchioles, terminal bronchioles, respiratory bronchioles, alveolar ducts, alveolar sacs, and alveoli).

22.1.14 Compare and contrast the main anatomical differences between bronchi and bronchioles.

22.1.15 Describe the histological changes that occur along the bronchial tree from larger to smaller air passageways.

22.1.16 Identify and describe the respiratory membrane, and explain its function.

22.1.17 Compare and contrast the gross anatomic features of the left and right lungs, and explain the reasons for these differences.

22.1.18 Identify and describe the location, structure, and function of the visceral and parietal pleura, serous fluid, and the pleural cavity.

22.1 Questions

1. Which of the following are functions of the respiratory system? Please select all that apply.

a. Gas exchange
b. Immune defense
c. Speech production
d. All of the above

2. Which of the following structures are parts of the conducting zone of the respiratory tract? Please select all that apply.

a. Oropharynx
b. Alveoli
c. Trachea
d. Respiratory bronchiole

3. Which of the following structures are included in the lower respiratory tract? Please select all that apply.

a. Lungs
b. The entire conducting zone
c. Respiratory zone
d. Larynx

4. Which is the correct route of air passage through the respiratory tract?

a. Pharynx, nose, larynx, bronchial tree, trachea
b. Larynx, pharynx, nose, trachea, bronchial tree
c. Nose, larynx, pharynx, trachea, bronchial tree
d. Nose, pharynx, larynx, trachea, bronchial tree

5. What type of epithelium makes up the respiratory epithelium of the nasal cavity and pharynx?

a. Pseudostratified ciliated columnar
b. Simple columnar
c. Stratified columnar
d. None of the above

6. Which of the following are functions of the larynx? Please select all that apply.

a. Moisten and warm inspired air
b. Passageway for air and food
c. Sound generation
d. Regulating the volume of air entering and leaving the lungs

7. Sam's thyroid cartilage and arytenoid cartilages are further apart than Alex's. Who will have a higher-pitched voice?

a. Sam
b. Alex

8. You were talking while chewing, so food was entering the pharynx while the epiglottis was open. Which of the following structures will trigger a cough response?

a. Vestibular folds
b. Vocal folds
c. Glottis
d. Epiglottis

9. When sound is being produced, are the true vocal folds closed or open?

a. Closed
b. Open

10. Which responses do the trachealis muscle aid in? Please select all that apply.

a. Vomiting
b. Sneezing
c. Coughing
d. Swallowing

11. Which of the following are true of the microscopic anatomy of the trachea? Please select all that apply.

a. Its cartilaginous C-shaped rings are rigid, so the trachea cannot dilate or constrict.
b. It is lined with pseudostratified ciliated columnar epithelium.
c. It terminates by dividing into the left and right primary bronchi.
d. It lies just superior to the larynx.

12. Where does the trachea lie in relation to the esophagus?

a. In front
b. Behind

c. To the left
d. To the right

13. Which of the following are true of the anatomy of the bronchial tree? Please select all that apply.

a. Some bronchi contain C-rings of hyaline cartilage, like the ones found in the trachea.
b. Heading down the bronchial tree, the first structure that can conduct gas exchange is the terminal bronchioles.
c. Heading down the bronchial tree, the only structure that has a clustered shape is the alveolar sacs.
d. Secondary bronchi branch directly into bronchioles.

14. Which structure do bronchi contain that bronchioles do not?

a. Smooth muscle
b. Cartilage
c. Alveoli
d. Ciliated epithelium

15. All of the following are true about the histology of the bronchial tree *except*:

a. The trachealis muscle is only found in the trachea.
b. The smallest structure of the bronchial tree that contains smooth muscle is an alveolar duct.
c. Cartilage is found in all structures of the bronchial tree, including the alveolar ducts.
d. The alveoli consist of simple squamous epithelium.

16. Which of the following structures make up the respiratory membrane? Please select all that apply.

a. Alveolar membranes
b. Terminal bronchioles
c. Visceral pericardium of the lung
d. Capillary membranes

17. Which of the following anatomical features does the left lobe of the lung contain? Please select all that apply.

a. Superior lobe
b. Middle lobe
c. Horizontal fissure
d. Oblique fissure

18. Which of the following functions do the pleurae serve in the thorax? Please select all that apply.

a. Produce pleural fluid
b. Help expand the lungs in inspiration
c. Separate the major organs
d. All of the above

22.1 Summary

- The functions of the respiratory system are providing oxygen to the body cells, removing carbon dioxide from the body cells, exchanging gas with the blood, maintaining blood pH, creating vocal sounds, and protecting the body against infection.
- The conducting zone of the respiratory system transports gasses (oxygen and carbon dioxide), while the respiratory zone exchanges gasses between the lungs and the blood.
- Oxygen enters the body through the nose, and proceeds to the alveoli in this order: nose, nasal cavity, pharynx, larynx, trachea, (primary, secondary, and tertiary) bronchi, bronchioles, terminal broncioles, respiratory bronchioles, alveolar ducts, alveolar sacs, and alveoli. Oxygen is then transported from the alveoli into the blood for distribution to the body cells. Carbon dioxide proceeds in the opposite direction.
- The nose and nasal cavity are specialized to receive air, filter it, and warm and humidify it as it enters the body.
- The pharynx is the passageway between the nasal cavity and the larynx.
- The larynx contains the vocal cords and acts as an entrance to the trachea.
- The trachea, or windpipe, transports air to the primary bronchi. A primary bronchus supplies air to each lung.
- The bronchail tree begins with the trachea and branches into smaller and smaller tubular air passages. The primary bronchi branch into secondary bronchi, which branch into tertiary bronchi. The tertiary bronchi branch into smaller and smaller bronchioles.
- The only structures that can perform gas exchange are the alveoli and other structures that contain alveoli (respiratory bronchioles, alveolar ducts, and alveolar sacs).
- The respiratory membrane is the membrane over which respiratory gasses are exchanged. It consists of the simple squamous epithelium of an alveolus and pulmonary capillary and the basement membranes between them.

22.2 The Process of Breathing

The learning objectives for this section are:

22.2.1 Describe the processes associated with the respiratory system (i.e., ventilation, pulmonary gas exchange [gas exchange between alveoli and blood], transport of gases in blood, tissue gas exchange [gas exchange between blood and body tissues]).

22.2.2 Define pulmonary ventilation, inspiration (inhalation), and expiration (exhalation).

22.2.3 Define atmospheric pressure, intrapulmonary pressure, intrapleural pressure, and transpulmonary pressure.

22.2.4 Identify the muscles used during quiet inspiration, deep inspiration, and forced expiration.

22.2.5 Explain the inverse relationship between gas pressure and volume of the gas (i.e., Boyle's Law) and apply this relationship to explain airflow during inspiration and expiration.

22.2.6 Explain the relationship of intrapleural pressure, transpulmonary pressure, and intrapulmonary pressure relative to atmospheric pressure during ventilation.

22.2.7* Relate the flow equation ($F = \Delta P/R$) to respiration, defining each component.

22.2.8 Explain how pulmonary ventilation is affected by bronchiolar smooth muscle contractions (bronchoconstriction), lung and thoracic wall compliance, and pulmonary surfactant and alveolar surface tension.

22.2.9 Describe the forces that tend to collapse the lungs and those that normally oppose or prevent collapse (e.g., elastic recoil of the lung versus subatmospheric intrapleural pressure).

* Objective is not a HAPS Learning Goal.

22.2 Questions

1. Between which of the following structures does gas exchange occur? Please select all that apply.
   a. Between the blood and the tissues
   b. Between the bronchi and bronchioles
   c. Between the atmosphere and the nasal cavity
   d. Between the alveoli and the blood (across the respiratory membrane)

2. Which respiratory system function is best described by moving air into and out of the lungs?
   a. Pulmonary ventilation
   b. Gas exchange
   c. Gas transport across the respiratory membrane
   d. Gas exchange between the blood and tissues

3. Which pressure describes the pressure of air in alveoli?
   a. Atmospheric pressure
   b. Intrapulmonary pressure
   c. Intrapleural pressure
   d. Transpulmonary pressure

4. While singing at a concert, is an artist using the internal intercostal muscles to exhale?
   a. Yes
   b. No

5. As Jeff's lungs increase in size, what happens to the pressure of the air inside of them? Which gas law helps you solve this problem?
   a. Increase; Boyle's Law
   b. Decrease; Boyle's Law
   c. Increase; $F = \Delta P/R$
   d. Decrease; $F = \Delta P/R$

6. Choose the correct relationship between atmospheric pressure and intrapulmonary pressure during inhalation.
   a. Atmospheric pressure > intrapulmonary pressure
   b. Intrapulmonary pressure < atmospheric pressure
   c. Intrapulmonary pressure = atmospheric pressure
   d. Impossible to tell with the information given

7. During bronchoconstriction, which variable in the equation $F = \Delta P/R$ changes? Does this change increase or decrease flow rate?
   a. P; Increase
   b. P; Decrease
   c. R; Increase
   d. R; Decrease

8. What effects would a deficiency of surfactant have on the body? Please select all that apply.
   a. It would be more difficult for the lungs to expand.
   b. The cohesion between water molecules in the alveoli would increase.
   c. It would make it difficult to swallow.
   d. It would increase the blood level of oxygen.

9. Which case will result in lung collapse?
   a. Intrapleural pressure > intrapulmonary pressure
   b. Intrapulmonary pressure > intrapleural pressure
   c. Intrapulmonary pressure > atmospheric pressure
   d. Intra-alveolar pressure < atmospheric pressure

22.2 Summary

- Pulmonary ventilation is the process by which the respiratory gasses oxygen and carbon dioxide, are exchanged between the atmosphere and the alveoli. This process consists of inhalation and exhalation.
- Boyle's law states that the volume inside any container is inversely proportional to the pressure. In respiration, the volume of the thoracic cavity is changed purposely, so that the pressure in the thorax will decrease below atmospheric pressure (which causes inhalation) or increases above atmospheric pressure (which causes exhalation).
- Normal, quiet inhalation occurs when the volume of the thorax increases due contraction of the diaphragm and external intercostal muscles. When the intrapulmonary pressure falls below atmospheric pressure, air rushes from the atmosphere into the lungs.
- Normal, quiet exhalation is a passive process, involving elastic recoil of the inspiratory muscles and the thoracic tissues that stretched during inhalation. Exhalation occurs when the volume of the thoracic cavity decreases, increasing the intrapulmonary pressure to a level that is greater than atmospheric pressure.
- Maximal inhalation and exhalation require the contraction of additional muscles.

22.3 Respiratory Volumes and Capacities

The learning objectives for this section are:

22.3.1 Define, identify, and determine values for the pulmonary volumes (inspiratory reserve volume [IRV], tidal volume [TV], expiratory reserve volume [ERV], and residual volume [RV]) and the pulmonary capacities (inspiratory capacity [IC], functional residual capacity [FRC], vital capacity [VC], and total lung capacity [TLC]).

22.3.2 Define anatomical dead space.

22.3.3 Explain the effect of anatomical dead space on alveolar ventilation and on the composition of alveolar and expired air.

22.3.4 Define and calculate minute ventilation and alveolar ventilation.

22.3 Questions

1. In the study of spirometry, what is the functional residual capacity?
   a. The volume of air that is exhaled in a normal breath
   b. The volume of air remaining in the lungs after a maximal exhalation
   c. The volume of air remaining in the lungs after a normal exhalation
   d. The volume of air that could be exhaled after a normal exhalation

2. What is the anatomical dead space?
   a. The volume of inhaled air that never reaches the alveoli and is not exchanged with the blood
   b. The volume of inhaled air that reaches nonfunctional alveoli and is not exchanged with the blood
   c. The total volume of inhaled air that is not exchanged with the blood
   d. The volume of air that always remains in the lungs and cannot be exhaled

3. What effect does the anatomical dead space have on alveolar ventilation?
   a. It increases the alveolar ventilation
   b. It decreases the alveolar ventilation
   c. It has no effect on the alveolar ventilation

4. Kyle measures the volume of air that enters their lab partner's lungs each time they take a breath for one minute, adds the volumes together, and then multiplies the total by the respiratory rate. What is Kyle measuring?
   a. Alveolar ventilation
   b. Minute ventilation
   c. Total lung capacity
   d. Inspiratory capacity

22.3 Summary

- Respiratory volumes and capacities can be measured by a spirometer, using different degrees of effort in the breathing process. These are measured to assess the state of a person's respiratory health or to determine the progression or state of recovery in a person that has a respiratory disorder.
- The four nonoverlapping volumes that comprise total lung capacity are tidal volume, inspiratory reserve volume, expiratory reserve volume, and residual volume.
- The respiratory capacities can be calculated by adding particular respiratory volumes together. These are the inspiratory capacity, vital capacity, functional residual capacity, and total lung capacity.
- Anatomical dead space is the volume of air that never reaches the alveoli and consequently never participates in gas exchange. This air remains in the conducting zone.
- Alveolar dead space is the air present in the alveoli that cannot participate in gas exchange. These alveoli are either damaged or are not associated with functional pulmonary capillaries.
- Minute ventilation is the volume of air inhaled each minute. It can be calculated by multiplying the respiratory rate (RR) by the tidal volume (TV). Alveolar ventilation is a similar measurement, but it takes the total dead space into account. It can be calculated by subtracting the total dead space from the TV, and then multiplying the result by the RR.

22.4 Gas Exchange

The learning objectives for this section are:

22.4.1 Explain the relationship between the total pressure of gases in a mixture and the partial pressure of an individual gas (i.e., Dalton's Law).

22.4.2 Explain the relationship between the partial pressure of a gas in air, the solubility of that gas in water, and the amount of the gas that can dissolve in water.

22.4.3 Compare and contrast the solubility of oxygen and carbon dioxide in plasma.

22.4.4 Describe oxygen and carbon dioxide concentration gradients and net gas movements between the alveoli and the pulmonary capillaries.

22.4.5 Analyze how oxygen and carbon dioxide movements are affected by changes in partial pressure gradients (e.g., at high altitude), area of the exchange surface, permeability of the exchange surface, and diffusion distance.

22.4.6 Explain the effects of local changes in oxygen and carbon dioxide concentrations on the diameters of pulmonary arterioles and bronchioles.

22.4.7 Describe oxygen and carbon dioxide concentration gradients and net gas movements between systemic capillaries and the body tissues.

22.4.8 Explain the influence of cellular respiration on oxygen and carbon dioxide gradients that govern gas exchange between blood and body tissues.

22.4 Questions

1. When you went hiking in the mountains at an elevation of about 10,000 feet, the atmospheric pressure was about 500 mm Hg (compared to 760 mm Hg at sea level). Knowing that air is composed of about 21 Percent oxygen, what is the approximate partial pressure of oxygen at 10,000 feet?

   a. 400 mm Hg
   b. 105 mm Hg
   c. 500 mm Hg
   d. It is not possible to calculate from the given information

2. Which physiological law describes that the concentration of gas in a liquid equilibrates to the air around it?

   a. Dalton's law
   b. Fick's law
   c. Henry's law
   d. Boyle's law

3. Which of the following relationships relates the solubility of oxygen and carbon dioxide in water and blood plasma?

   a. Oxygen $>$ carbon dioxide
   b. Oxygen $<$ carbon dioxide
   c. Oxygen $=$ carbon dioxide

4. Oxygen and carbon dioxide move between the alveoli and the pulmonary capillaries by ______ and ______.

   a. Concentration gradients; diffusion
   b. Pressure gradients; diffusion
   c. Concentration gradients; bulk flow
   d. Pressure gradients; bulk flow

5. Lucy smoked for 10 years, which damages alveolar tissue. Compared to a nonsmoker, one variable in Fick's law is affected. Which variable is affected to the greatest extent?
   a. Concentration gradient
   b. Membrane permeability
   c. Membrane thickness
   d. Surface area

6. In exercise, we want to inhale oxygen and expel carbon dioxide at a higher-than-normal rate. How will the diameter of the bronchioles be affected, in order to help us respond to the increased demands of exercise on the body?
   a. Bronchoconstriction
   b. Bronchodilation
   c. Impossible to tell with the information given

7. Before gas exchange occurs at the tissues, the partial pressure of oxygen in the systemic capillaries is ______ compared to the tissues.
   a. Higher
   b. Lower
   c. The same
   d. Impossible to tell with the information given

8. The blood in the systemic arteries has a(n) ______ concentration of carbon dioxide compared to that of systemic veins.
   a. Higher
   b. Lower
   c. Equal
   d. Impossible to tell with the given information

22.4 Summary

- Dalton's law states that the total pressure in a gas mixture is equal to the sum of the partial pressures exerted by each gas.
- Henry's law states that the concentration of gas in a liquid equilibrates with its concentration in the air around it.
- Gasses move from one location in the body to another based on their partial pressure gradients and their solubilities in the solvent in each location. These factors are particularly important in the movement of oxygen and carbon dioxide across the respiratory membrane and between the blood and the body cells.
- Fick's law states that the rate of diffusion of a gas (or other substance) across a cell membrane is proportional to the surface area for gas exchange and the concentration gradient, and inversely proportional to the thickness of the membrane.
- In order to maintain efficient gas exchange, oxygen delivery to the body cells, and removal of carbon dioxide from the body, the ventilation and perfusion must be balanced in the lungs. Ventilation is the movement of air into and of the lungs. Perfusion is the flow of blood into the pulmonary capillaries. For example, the body responds to poorly ventilated alveoli by vasoconstricting its associated pulmonary capillaries and directing the blood flow to capillaries of functional alveoli.
- Changes in cellular respiration change the oxygen and carbon dioxide gradients in the body, which cause changes in the rate of diffusion between the lungs, body cells, blood and atmosphere. For example, during exercise, the increased generation of carbon dioxide increases the concentration and partial pressure gradients between the active body cells and the blood. This leads to higher rates of diffusion of carbon dioxide into the blood, the alveoli, and the atmosphere.

22.5 Transport of Gasses

The learning objectives for this section are:

HAPS LEARNING GOALS AND OUTCOMES

22.5.1 Describe the ways in which oxygen is transported in blood, and explain the relative importance of each to total oxygen transport.

22.5.2* Correlate hemoglobin saturation to environmental conditions such as the partial pressure of oxygen, temperature, and pH.

22.5.3 Interpret the oxygen–hemoglobin saturation curve at low and high partial pressures of oxygen.

22.5.4 *Describe the oxygen–fetal hemoglobin saturation curve and its impact on oxygen delivery to fetal tissues.

22.5.5 Describe the ways in which carbon dioxide is transported in blood and explain the relative importance of each to total carbon dioxide transport.

22.5.6 Explain the role of each of the following in carbon dioxide transport: carbonic anhydrase, hydrogen ions binding to hemoglobin, the chloride shift, and oxygen–hemoglobin saturation level.

22.5.7 State the reversible chemical equation for the reaction of carbon dioxide and water to carbonic acid and then to hydrogen ion and bicarbonate ion.

22.5.8 Predict how changing the pH or the concentration of bicarbonate ions will affect the partial pressure of carbon dioxide in the plasma.

22.5.9 Predict how changing the partial pressure of carbon dioxide will affect the pH and the concentration of bicarbonate ions in the plasma.

* Objective is not a HAPS Learning Goal.

22.5 Questions

1. Which of the following is the primary oxygen transport method in the blood?

a. Dissolved oxygen in plasma
b. Bound to hemoglobin
c. Dissolved oxygen in buffy coat
d. Bound to white blood cells

2. During a state of lower-than-normal blood pH, hemoglobin's affinity for oxygen will be ______ compared to normal. This is described by ______.

a. Higher; the Bohr effect
b. Lower; the Bohr effect
c. The same; Fick's law
d. Lower; Fick's law

3. According to the oxygen–hemoglobin saturation curve, what is the typical oxygen saturation level in the blood vessels of the lungs?

a. 20 percent
b. 5 percent
c. 75 percent
d. 100 percent

4. Compare fetal hemoglobin's affinity for oxygen to maternal hemoglobin's affinity for oxygen.

a. Higher
b. Lower
c. The same
d. Impossible to tell with the information given

5. How is the majority of CO_2 transported in the blood?

a. Bound to hemoglobin
b. Dissolved in the blood plasma
c. As bicarbonate ions
d. Bound to plasma proteins

6. What is the function of carbonic anhydrase in oxygen or carbon dioxide transport in the blood? Please select all that apply.

a. It is a hormone that increases the binding of oxygen to hemoglobin.
b. It is an enzyme that causes carbon dioxide to bind to water to produce carbonic acid.
c. It is a substance that prevents oxygen molecules from binding to hemoglobin.
d. It is an enzyme that causes carbonic acid to dissociate into carbon dioxide and water.

7. Carbonic acid is formed by carbon dioxide when it comes into contact with water in the blood. What does carbonic acid rapidly dissociate into?

a. H_2 and CO_3
b. H_2CO_3
c. O_2 and H_2C
d. $H^+ + HCO_3^-$

8. After a brief period of hyperventilation, the partial pressure of bicarbonate in your blood is lower than normal. Will the blood be more or less acidic than normal?

a. More
b. Less
c. The same
d. Impossible to tell with the information given

9. During a workout, Sebastian's blood level of carbon dioxide increased. How would this affect the blood pH, compared to that of a resting state?

a. The blood pH would be higher.
b. The blood pH would be lower.
c. The blood pH would be the same.
d. It is impossible to tell from the given information.

22.5 Summary

- Oxygen is transported through the blood by two methods: (1) bound to hemoglobin in red blood cells, and (2) dissolved in the blood plasma. The vast majority of oxygen travels through the blood bound to hemoglobin because it is lipid-soluble, not water-soluble.
- Hemoglobin contains four binding sites for oxygen molecules. Hemoglobin saturation is close to 100 Percent in the lungs and about 75 Percent in venous blood. Oxygen dissociates from hemoglobin in the presence of low partial pressure of oxygen, high body temperature, or low pH. Therefore, oxygen

is delivered to the body cells that require it to the greatest extent.

- Carbon dioxide is transported through the blood by three methods: (1) as part of bicarbonate ions, (2) bound to hemoglobin, and (3) dissolved in the blood plasma. About 70 Percent of carbon dioxide travels through the blood as bicarbonate ions.
- Carbon dioxide binds to water molecules to form carbonic acid, which can then dissociate into hydrogen and bicarbonate ions. This chemical reaction sequence can proceed in either direction, depending on the blood pH, the person's activity level, and the blood level of carbon dioxide.

$$CO_2 + H_2O \leftrightarrow H_2CO_3 \leftrightarrow H^+ + HCO_3^-$$

22.6 Respiratory Rate and Control of Ventilation

The learning objectives for this section are:

22.6.1 Describe the locations and functions of the brainstem respiratory centers.

22.6.2 List and describe the major chemical and neural stimuli to the respiratory centers.

22.6.3 Compare and contrast the central and peripheral chemoreceptors.

22.6.4 Define hyperventilation, hypoventilation, panting, eupnea, hyperpnea, and apnea.

22.6 Questions

 Mini Case 1: Damage to the Respiratory Control Centers

Alex was in a car accident and unfortunately suffered a posterior neck injury. Alex is able to maintain a normal rate and depth of breathing during quiet breathing, but is having trouble with deep breathing. The doctors have said that the neck injury has damaged some of the respiratory centers in Alex's brainstem.

1. Because Alex is having difficulty with deep breathing, which of the respiratory control centers were most likely damaged in the accident? Please select all that apply.
   a. Ventral respiratory group
   b. Dorsal respiratory group
   c. Apneustic center
   d. Pneumotaxic center

2. Under normal circumstances, the factor that stimulates the respiratory centers in the medulla oblongata and pons to increase the rate and depth of breathing is primarily ______.
   a. Oxygen
   b. Nitrogen
   c. Carbon dioxide
   d. Temperature

3. The doctors told Alex that they were lucky that none of their chemoreceptors involved in respiration had been injured in the accident. Which chemoreceptors had the doctors been concerned about, and where are they located?
   a. Peripheral chemoreceptors; aorta and carotid arteries
   b. Central chemoreceptors; aorta and carotid arteries
   c. Periperal chemoreceptors; brain and brainstem
   d. Central chemoreceptors; brain and brainstem

4. Alex's doctors were also relieved that the accident did not cause so much damage to the respiratory centers that Alex was unable to breathe on their own. Which of the following terms refers to an absence of spontaneous breathing?
   a. Eupnea
   b. Apnea
   c. Hypoventilation
   d. Hyperpnea

22.6 Summary

- The processes of respiration are regulated by control centers in the medulla oblongata and pons of the brainstem.
- The dorsal and ventral groups of neurons in the medulla oblongata regulate the respiratory rate and stimulate the muscles of respiration. The apneustic and pneumotaxic centers in the pons regulate the depth of breathing and the duration of each inspiration.
- Central and peripheral chemoreceptors regulate respiration by monitoring the blood levels of carbon dioxide, hydrogen ions, and oxygen, and reporting this information to the respiratory centers. The main factors that increase the respiratory rate are the levels of CO_2 and H^+.
- Hyperventilation occurs when the rate and depth of breathing exceed the body's needs for oxygen inhalation and carbon dioxide exhalation. Hypoventilation occurs when the rate and depth of breathing are insufficient to meet the body's needs for oxygen inhalation and carbon dioxide exhalation.

23

The Digestive System

Prostock-studio/Shutterstock.com

Chapter Introduction

What was the last thing you ate? How long ago was that? While behaviorally we don't eat all of the time, the digestive system is working relatively constantly throughout the day. You may be taking a walk or studying or sleeping, having forgotten all about your last meal or snack, but your stomach and intestines are busy digesting it and absorbing its vitamins and other nutrients. Because humans have evolved over hundreds of thousands of years of food scarcity, our digestive systems are masterful at stripping away all that it can from our food before waste material is excreted. This chapter examines the structure and functions of these organs, and explores the mechanics and chemistry of the digestive processes.

In this chapter we will learn...

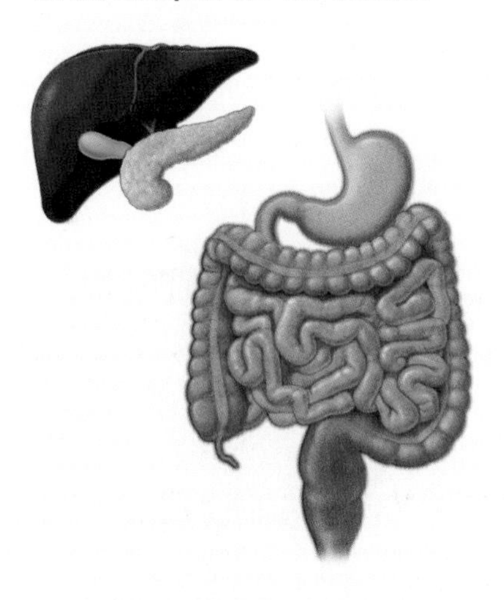

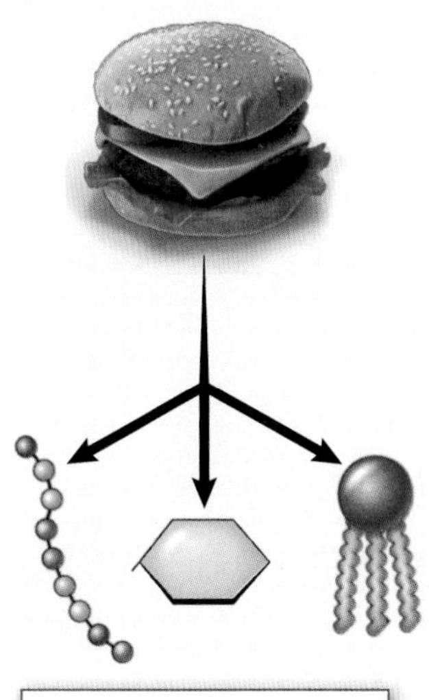

...about the structure and function of the digestive system...

...and how food is broken down into monomers.

23.1 Overview of the Digestive System

LEARNING GOALS AND OUTCOMES

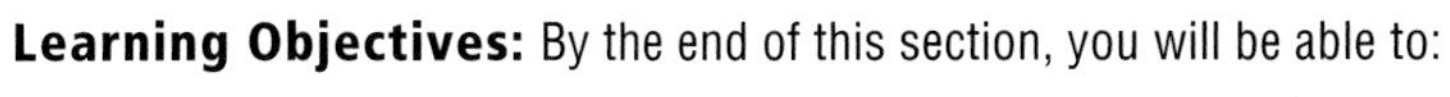

Learning Objectives: By the end of this section, you will be able to:

23.1.1 Describe the major functions of the digestive system.

23.1.2 Explain the differences between the gastrointestinal (GI) tract (alimentary canal) and the accessory digestive organs.

23.1.3 List and identify the organs that compose the gastrointestinal (GI) tract.

23.1.4 Trace the pathway of ingested substances through the gastrointestinal (GI) tract.

LO 23.1.1

LO 23.1.2

The functions of the digestive system are to break down the foods you eat, release their nutrients, and absorb those nutrients into the body as well as to generate, store, and excrete some of our wastes. When you take a bite of food, the food enters a long and complicated tract of connected organs. Each of the organs in the

The Human Anatomy and Physiology Society includes more than 1,700 educators who work together to promote excellence in the teaching of this subject area. The HAPS A&P Learning Outcomes measure student mastery of the content typically covered in a two-semester Human A&P curriculum at the undergraduate level. The full Learning Outcomes are available at https://www.hapsweb.org.

gastrointestinal (GI) tract (alimentary canal)

gastrointestinal (GI) tract has its own function and contributes to the breakdown of your food and absorption of nutrients. There are additional organs outside of the tract that function in digestion, but food does not pass through them. These **accessory digestive organs** are vital to the functioning of the digestive system even though they are not part of the GI tract. The digestive organs are represented in **Figure 23.1**.

As is the case with all body systems, the digestive system does not work in isolation; it functions cooperatively with the other systems of the body. Consider, for example, the interrelationship between the digestive and cardiovascular systems.

Figure 23.1 Organs of the Digestive System

The digestive system is composed of the GI tract and the accessory digestive organs (liver, pancreas, gallbladder, salivary glands).

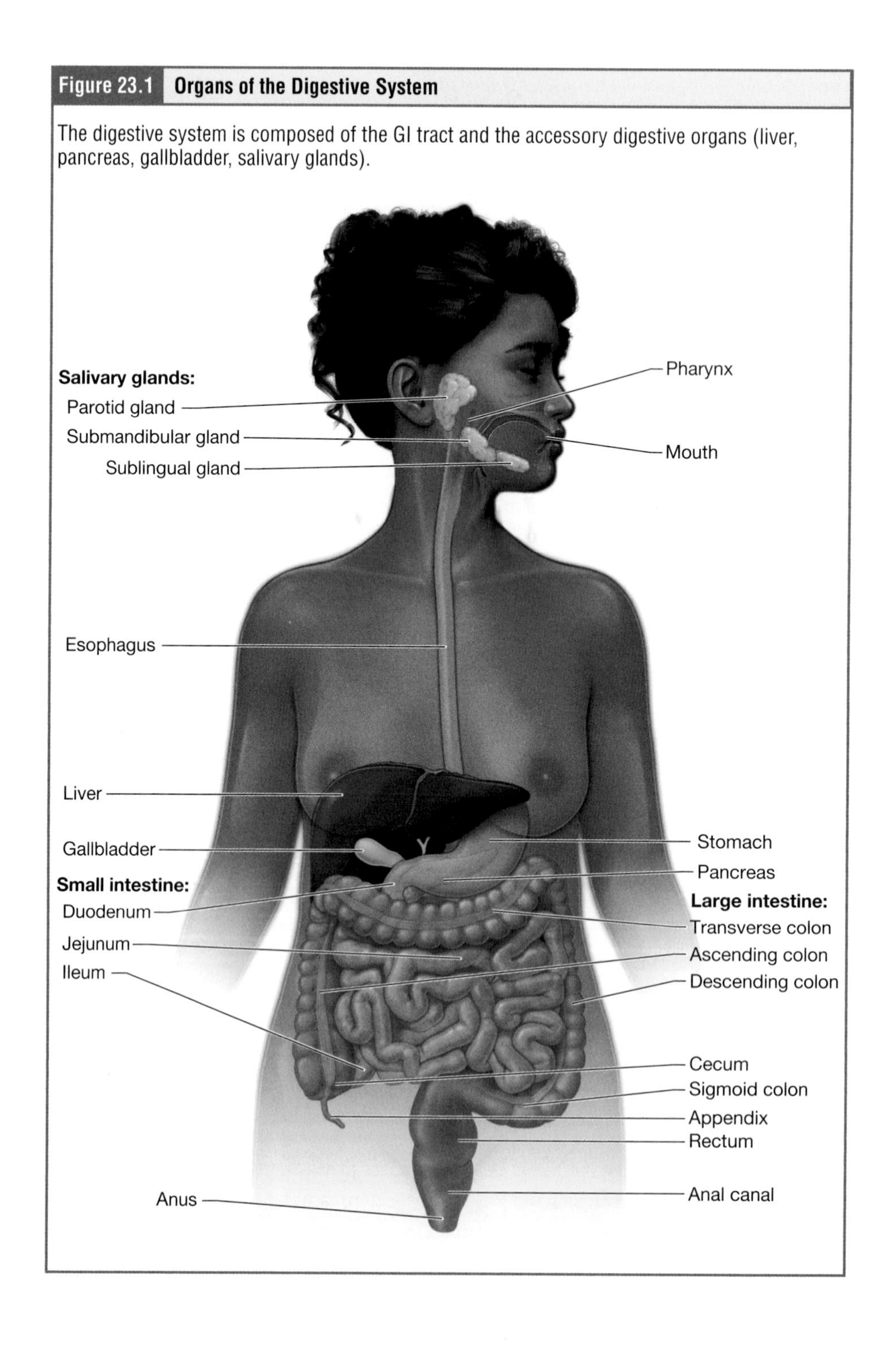

Table 23.1 Connections Among Body Systems

Body	Functional Connections
Cardiovascular	Blood supplies digestive organs with oxygenated blood and removes deoxygenated blood. Blood supply is used to carry absorbed nutrients to the liver for filtering and all over the body for distribution, use, or storage.
Endocrine	Hormones help regulate secretion in digestive glands and accessory organs.
Lymphatic	Mucosa-associated lymphoid tissue and other lymphatic tissue defend against entry of pathogens and maintain immune tolerance to the foods we eat; lymphatic vessels absorb and transport lipids.
Muscular	Skeletal and smooth muscle (depending on location) provide motility within the GI tract.
Nervous	Sensory and motor neurons help regulate secretions and muscle contractions in the digestive tract, provide information to the CNS about "fullness" and "hunger."
Skin	Manufactures Vitamin D in response to sunlight, which enables calcium absorption.
Urinary	Kidneys convert vitamin D into its active form, allowing calcium absorption in the small intestine.

Arteries supply the digestive organs with oxygen and processed nutrients, and veins drain the digestive tract. These intestinal veins, constituting the hepatic portal system, are unique; they do not return blood directly to the heart. Rather, this blood travels to the liver for filtering before the blood completes its circuit back to the heart. The interrelationship of the digestive and endocrine systems is also critical. Hormones secreted by several endocrine glands, as well as endocrine cells of the pancreas, the stomach, and the small intestine, contribute to the control of digestion and nutrient metabolism. Of course, the digestive system is responsible for absorbing the nutrients that fuel all the cells of the body. In turn, the digestive system provides the nutrients to fuel endocrine function. **Table 23.1** gives a quick glimpse at the connections between the digestive system and other body systems.

Organs of the Gastrointestinal Tract The GI tract is a one-way tube about 25 feet in length during life and closer to 35 feet in length when measured after death (the difference has to do with muscle tone, the tract is longer after the muscles relax). Within the organs of the GI tract, our food is broken down into the smallest components possible. Nutrients are absorbed across the wall of the GI tract organs and into the bloodstream or lymph vessels. Any material that cannot be broken down into pieces small enough to absorb, or material that is unusable to us, is compacted into feces and eliminated from the end of the GI tract. Thus, the tube begins at the mouth and terminates at the anus. Between those two points, the food moves through the pharynx, the esophagus, the stomach, the small intestine, and the colon (**Figure 23.2A**). It is progressively broken down all along its route. At both ends (the mouth and anus), the tube is open to the external environment; thus, food and wastes within the GI tract are technically considered to be outside the body. Only through the process of absorption, which occurs across the cells of the wall of the tract, do the nutrients in food enter into and nourish the body's "inside." For this reason, the secretions that are added to the food to help break it down are considered exocrine secretions (secretions to the outside of the body).

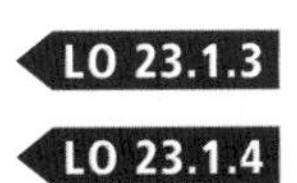

Accessory Structures Each accessory digestive organ aids in the breakdown of food. Within the mouth, the teeth and tongue begin the mechanical breakdown of the food through chewing. The rest of the accessory digestive organs all are secretory in nature, adding water and enzymes to the food. Each contribution is produced in a gland, and contributes to specific components or stages of the breakdown process. These glands are the salivary glands, the gallbladder, the liver, and the pancreas; each is connected to the gut by ducts.

Student Study Tip

Secretory accessory digestive organs are not a part of the main outfit (GI tract), but they ADD to it.

Figure 23.2 The GI Tract and Accessory Organs of the Digestive System

The digestive system includes both (A) a tract of connected organs (the gastrointestinal or GI tract) and (B) organs that contribute to digestion but are found outside the tract (accessory organs).

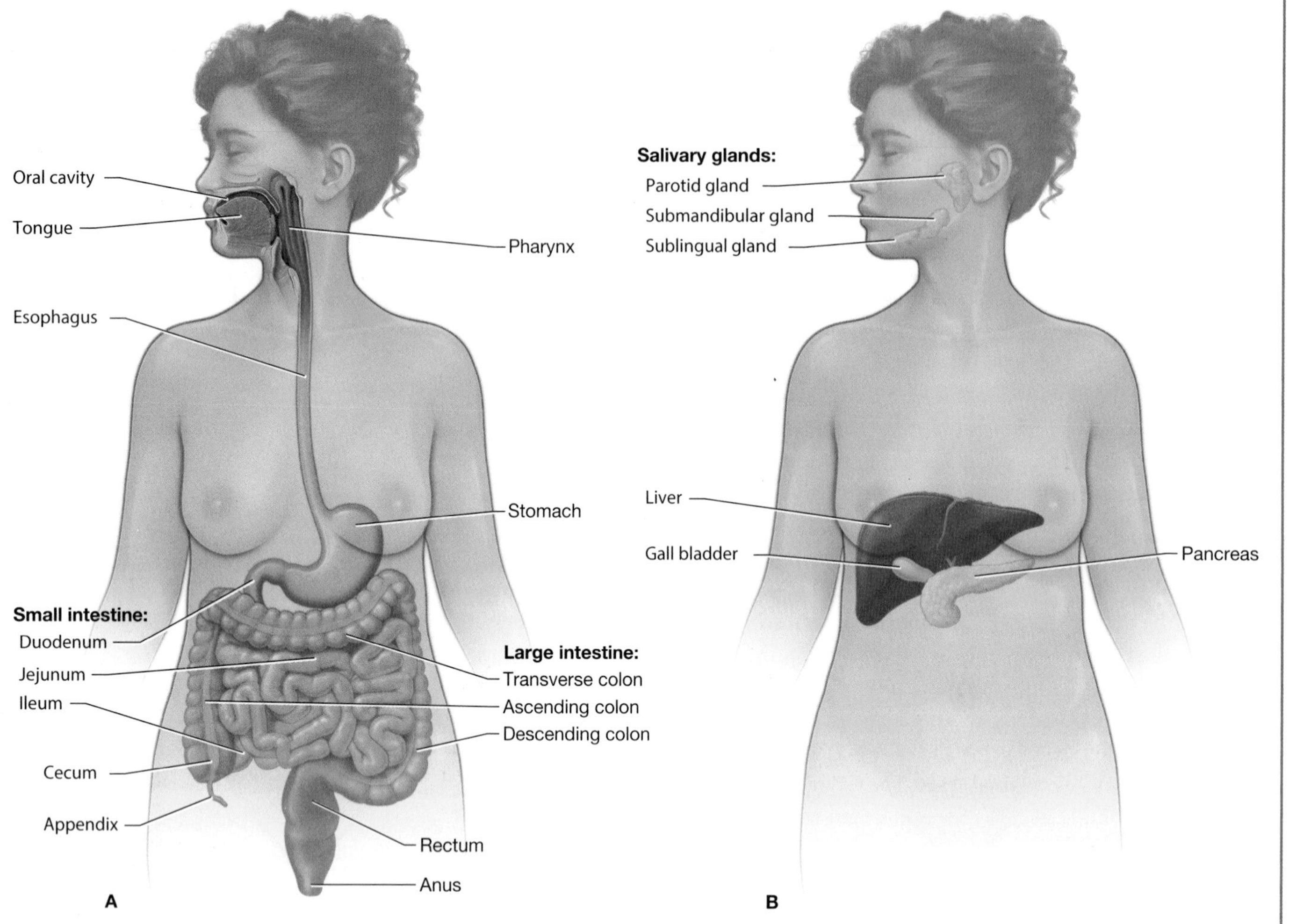

Oral cavity
Tongue
Pharynx
Esophagus
Stomach
Small intestine:
Duodenum
Jejunum
Ileum
Large intestine:
Transverse colon
Ascending colon
Descending colon
Cecum
Appendix
Rectum
Anus
A

Salivary glands:
Parotid gland
Submandibular gland
Sublingual gland
Liver
Gall bladder
Pancreas
B

Learning Check

1. Proteolytic enzymes are secreted into the digestive tract by the pancreas. These enzymes break down protein in the small intestine. Are these secretions considered endocrine (released into the body fluids) or exocrine (released outside the body) secretions?
   a. Endocrine
   b. Exocrine
2. When are nutrients considered to be inside your body?
   a. When they are ingested into the mouth
   b. As soon as they reach the stomach
   c. When they have been absorbed by the cells lining the wall of the GI tract
   d. When they reach the colon

23.2 General Gross and Microscopic Anatomy of the Gastrointestinal (GI) Tract

Learning Objectives: By the end of this section, you will be able to:

23.2.1 Identify and describe the gross anatomic and microscopic structure and function of each of the gastrointestinal (GI) tract tunics (layers): mucosa, submucosa, muscularis (muscularis externa), and serosa or adventitia.

23.2.2 Compare and contrast mechanical digestion and chemical digestion, including where they occur in the digestive system.

23.2.3 Define peristalsis.

23.2.4 Describe the enteric nervous system (ENS) and explain its role in controlling digestive system function.

23.2.5 Identify and describe the location, structure, and function of the visceral and parietal peritoneum, serous fluid, and the peritoneal cavity.

23.2.6 Compare and contrast the locations of the mesenteries (e.g., mesentery proper, mesocolon, lesser omentum, greater omentum).

23.2.7 Define mesentery and explain its function.

23.2.8 Explain the difference between an intraperitoneal and a retroperitoneal organ.

23.2.9 Identify which digestive system organs are intraperitoneal or retroperitoneal.

23.2a Microscopic Structure of the GI Tract

Throughout its length, the GI tract is composed of the same four tissue layers; similar to the tunics of the blood vessels, the composition and structure of each of these layers varies with the function of each organ. Remember that the inside of a hollow body structure (such as the stomach) is called its **lumen**. The four layers of the wall of the GI tract, starting with the innermost layer, are the mucosa, the submucosa, the muscularis, and the serosa (**Figure 23.3**).

LO 23.2.1

The **mucosa** is a layer of mucous membrane, meaning an epithelium enriched with mucus, similar to the epithelial linings of the nose or respiratory system. The epithelium is in direct contact with ingested food, and, like all epithelia, is anchored by a basement membrane. The basement membrane in this case is termed a **lamina propria** because it is more substantial than a typical basement membrane. It is a layer of connective tissue analogous to the dermis of the skin. In addition, the mucosa has a thin smooth muscle layer called the *muscularis mucosa* (not to be confused with the muscularis layer, described in the following text).

- ***Epithelium.*** In the mouth, pharynx, esophagus, and anal canal, the epithelium is primarily a non-keratinized, stratified squamous epithelium. In the stomach and intestines, it is a simple columnar epithelium. Notice that the epithelium is in direct contact with the lumen and the food contents within. The difference between the organs that have a stratified squamous epithelium versus those lined with columnar epithelium is a matter of internal friction. In the mouth, for example, food is sometimes rough as it is not far along in the digestion process, and so the epithelium is stratified to protect the layers beneath from friction. By the time the digested food reaches the stomach and intestines, it is more of a smooth paste and therefore friction is minimal. Moreover, in the small intestine in particular, nutrients must cross the mucosa to be absorbed, so a thicker epithelium is not desirable. Interspersed among its epithelial cells are goblet cells, which secrete mucus and fluid onto the epithelial surface. Enteroendocrine cells

Figure 23.3 Layers of the wall of the GI Tract

Every organ in the GI tract has four layers of different tissues that make up its walls. The hollow center of the organ is the lumen. From the lumen outward, the layers are the mucosa, the submucosa, the muscularis, and the serosa. While all organs have these four layers, their composition and structure vary.

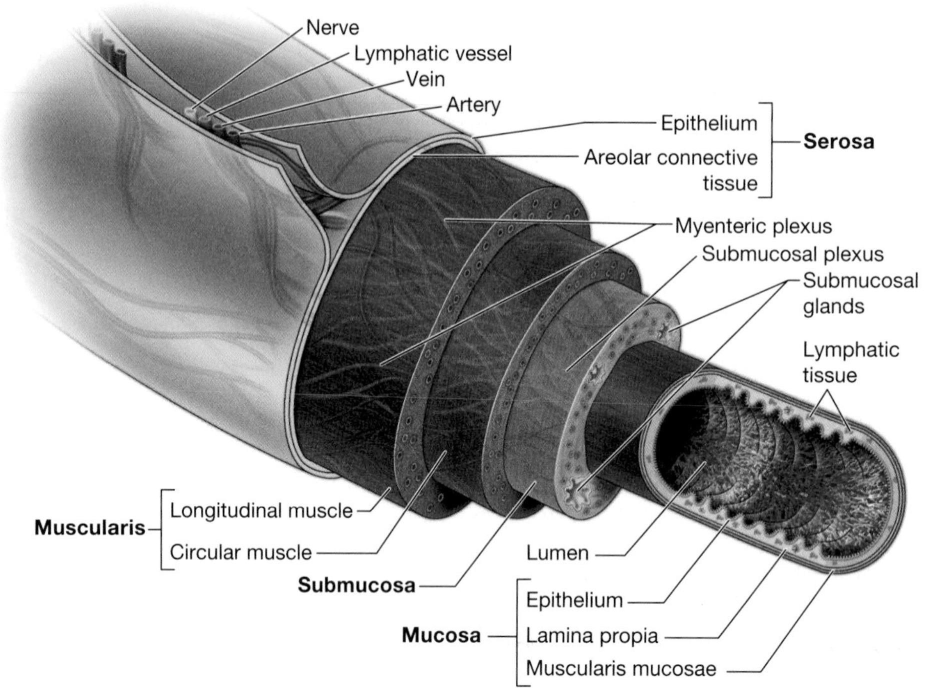

are features of the epithelium of the intestines and function to secrete hormones that regulate some digestive processes. Epithelial cells have a very brief lifespan, averaging only a few days. This process of rapid renewal helps preserve the health of the GI tract and replace cells that are lost to friction.

- ***Lamina propria.*** The lamina propria contains numerous blood and lymphatic vessels that transport nutrients absorbed across the wall of the GI tract organs to other parts of the body. The lamina propria also contains clusters of lymphocytes, making up the mucosa-associated lymphoid tissue (MALT). These lymphocyte clusters are particularly substantial in the small intestine. These patches of MALT serve a dual purpose. These patches of MALT serve a dual purpose: on the one hand, they filter out potentially harmful pathogens from our foods (despite our best efforts, the food we eat is hardly sterile), which contain bacteria, viruses, and fungi; we rarely suffer from foodborne illness because the pathogens stay "outside" the body within the lumen and cannot cross the wall, and because the lymphocytes of the MALT defend us. On the other hand, however, the MALT are also believed to play a role in tolerance—learning about the foods we eat and quelling possible immune reactions to them. Food allergies are a breakdown of this tolerance process.
- ***Muscularis mucosa.*** This thin layer of smooth muscle is in a constant state of tension, pulling the mucosa of the stomach and small intestine into undulating folds. These folds dramatically increase the surface area available for digestion and absorption.

As its name implies, the **submucosa** lies immediately beneath the mucosa. The submucosa is a layer of dense connective tissue that connects the mucosa to the muscularis. Because the mucosal epithelium is avascular, the submucosa is rich with blood and lymphatic vessels and also contains some submucosal glands that release digestive secretions. Additionally, it serves as a conduit for a dense branching network of nerves, the submucosal plexus, which functions as described in the following text.

The third layer of the GI tract is the **muscularis**. The muscularis in most organs is made up of a double layer of smooth muscle: an inner circular layer and an outer longitudinal layer. The contractions of these layers propel food along the tract and also function to mix and churn the digested food. The muscularis is not uniform in each organ, rather, its composition varies from organ-to-organ according to function. In the stomach, there are three layers to the muscularis for extra grinding power. At the ends of the GI tract, including the mouth, pharynx, anterior part of the esophagus, and anus, the muscularis is made up of skeletal muscle instead of smooth muscle, which gives voluntary control over swallowing and defecation.

Student Study Tip

The inner layer is called *Circular*, starting with a "C" just like the word *center*.

LO 23.2.2

The breakdown of food requires the extensive use of enzymes to break the bonds among the food molecules. Once broken, the monomers—the smallest units of the proteins and sugars and fats we eat—can be transported across the membranes of the cells that line the GI tract. This is called **chemical digestion**. However, enzymes cannot act alone. We take food into the body in rather large pieces, and the enzymes can only act on the exposed outer surfaces of these pieces. **Mechanical digestion**, which is initiated by the teeth and continued by the muscularis layer throughout the GI tract, physically grinds the food into smaller and smaller chunks upon which the digestive enzymes can act.

LO 23.2.3

Another mechanical act performed by the muscularis is peristalsis. **Peristalsis** consists of sequential, alternating waves of contraction and relaxation of two muscularis layers, and functions to propel food along the tract (**Figure 23.4**). These waves also play a role in mixing food with digestive juices. Peristalsis is so powerful that foods and liquids you swallow enter your stomach even if you are upside down.

Student Study Tip

Peristalsis in the GI tract performs the same function that your fingers do when they squeeze a toothpaste tube; the muscular contractions force the contents along, and eventually out of, the tube.

Figure 23.4 Peristalsis

Peristalsis is the rhythmic contraction of muscle within the muscularis that propels food along the GI tract.

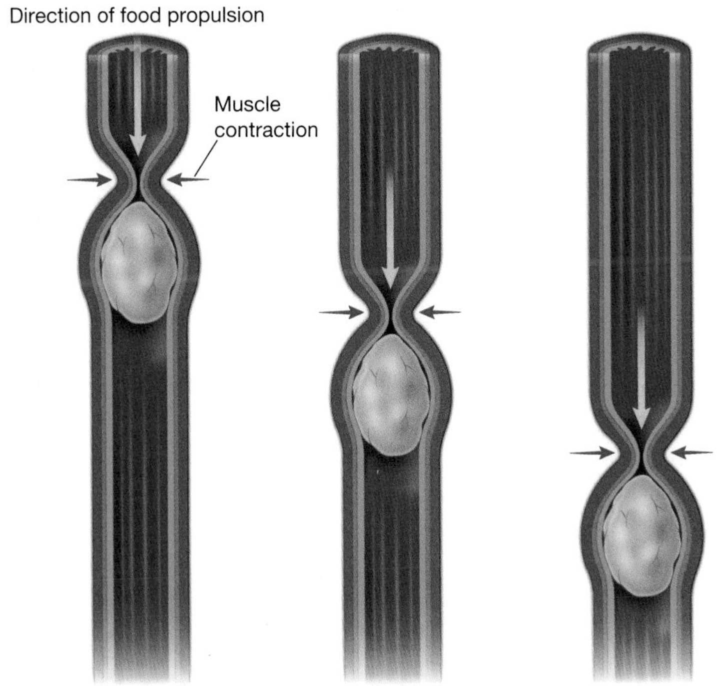

The **serosa** is the portion of the tract which is the furthest from the lumen. It consists of a layer of connective tissue that helps to hold bigger arteries, veins, and nerves to the GI tract wall. All organs of the GI tract have a layer of connective tissue superficial to their muscularis; however, this layer is referred to as the serosa only within the abdominal cavity. The mouth, pharynx, and esophagus have a similar dense sheath of collagen fibers; for these organs, however, this layer is called the **adventitia**. The adventitia serves to hold these organs of the GI tract in place near the vertebral column.

23.2b Nerve Supply

As soon as food enters the mouth, or even when it is just smelled or seen, an orchestrated symphony of events begins throughout the digestive system. The salivary glands and pancreas begin to generate their secretions. Throughout the GI system, coordinated muscle contractions begin to increase motility. All of this activity is coordinated by the nervous system. The sight, smell, feel, or taste of food is detected by receptors that send impulses along the sensory neurons of cranial nerves. With this input, the central nervous system can begin to initiate the process of digestion.

enteric nervous system (intrinsic nervous system)

LO 23.2.4

The central nervous system is not solely or even mainly responsible for GI innervation. There is intrinsic innervation of the GI tract provided by the **enteric nervous system**, a network of nerves that connects the GI tract organs. The enteric nervous system runs from the esophagus to the anus, and contains approximately 100 million neurons—motor and sensory neurons, as well as interneurons. These enteric neurons are grouped into two plexuses. The **myenteric plexus** lies in the muscularis layer and is responsible for **motility**. The **submucosal plexus** lies in the submucosal layer and is responsible for regulating digestive secretions and reacting to the presence of food.

myenteric plexus (plexus of Auerbach)

submucosal plexus (plexus of Meissner)

Extrinsic innervation of the GI tract is provided by the autonomic nervous system, which includes both sympathetic and parasympathetic nerves. In general, sympathetic activation (the fight-or-flight response) decreases GI secretion and motility. In contrast, parasympathetic activation (the rest-and-digest response) increases GI secretion and motility.

23.2c Blood Supply

The blood vessels serving the digestive system have two functions. As it does for all organs of the body, the blood brings oxygen and nutrients to the digestive organs so that the cells can perform cellular respiration and get rid of wastes. Specifically, within the head, neck, and thorax, arteries branching off the aortic arch and thoracic aorta feed the digestive organs. Below the diaphragm the digestive organs are supplied with blood by arteries branching from the abdominal aorta. Whereas the celiac trunk services the liver, stomach, and duodenum, the superior and inferior mesenteric arteries supply blood to the remaining small and large intestines.

The second function of the blood vessels in the digestive system is the transportation of protein and carbohydrate nutrients absorbed across the wall of the GI tract. Lipids are not immediately carried in the blood but are absorbed via *lacteals*, tiny vessels of the lymphatic system.

The veins that collect this nutrient-rich blood empty into the hepatic portal system. This venous network takes the blood into the liver, where the nutrients are either processed or stored for later use. Only then does the blood drained from the GI tract circulate back to the heart. To appreciate just how demanding the digestive process is on the cardiovascular system, consider that while you are "resting and digesting," about one-fourth of the blood in the body is circulating within the intestines.

23.2d The Peritoneum

LO 23.2.5

The abdominal cavity is a fairly large space that holds many organs. These organs are not free to move within the cavity, but rather are held in place by the **peritoneum**, a broad serous membranous sac. The peritoneum is made up of squamous epithelial tissue surrounded by connective tissue. Like the pleura of the lungs and the pericardium of the heart, the peritoneum is composed of two different layers: the **parietal peritoneum**, which lines the abdominal wall, and the **visceral peritoneum**, which envelops the abdominal organs (Figure 23.5). The peritoneal cavity is the space between the visceral and parietal peritoneal surfaces. Within this space, the membranes secrete a watery fluid that acts as a lubricant to minimize friction.

The visceral peritoneum adheres tightly to the surface of the organs and includes multiple large folds that surround various abdominal organs, hugging them to the walls of the abdominal cavity. Within the peritoneum are blood vessels, lymphatic vessels, and nerves. There are four major peritoneal folds. The **lesser omentum** is a vertical sheet that hangs down from the inferior edge of the liver, tying it to the superior edge of the stomach (Figure 23.6). The **greater omentum** hangs down from the inferior surface of the stomach like a curtain, covering the intestines. The **transverse mesocolon** anchors the transverse colon of the large intestine to the posterior wall of the abdominal cavity, and the **mesenteries** perform the same function for the small intestine, suspending each fold from the posterior abdominal wall. Notably, the posterior parietal peritoneum provides the posterior wall of the abdominal cavity, where the mesenteries attach. The posterior parietal peritoneum is directly adjacent to the vertebral column, though there is a small space behind the peritoneum where the kidneys reside; this space is described as **retroperitoneal**. Some digestive organs, such as the stomach, the

lesser omentum (small omentum, gastrohepatic omentum)

greater omentum (great omentum, omentum majus, gastrocolic omentum, epiploon)

LO 23.2.6

LO 23.2.7

LO 23.2.8

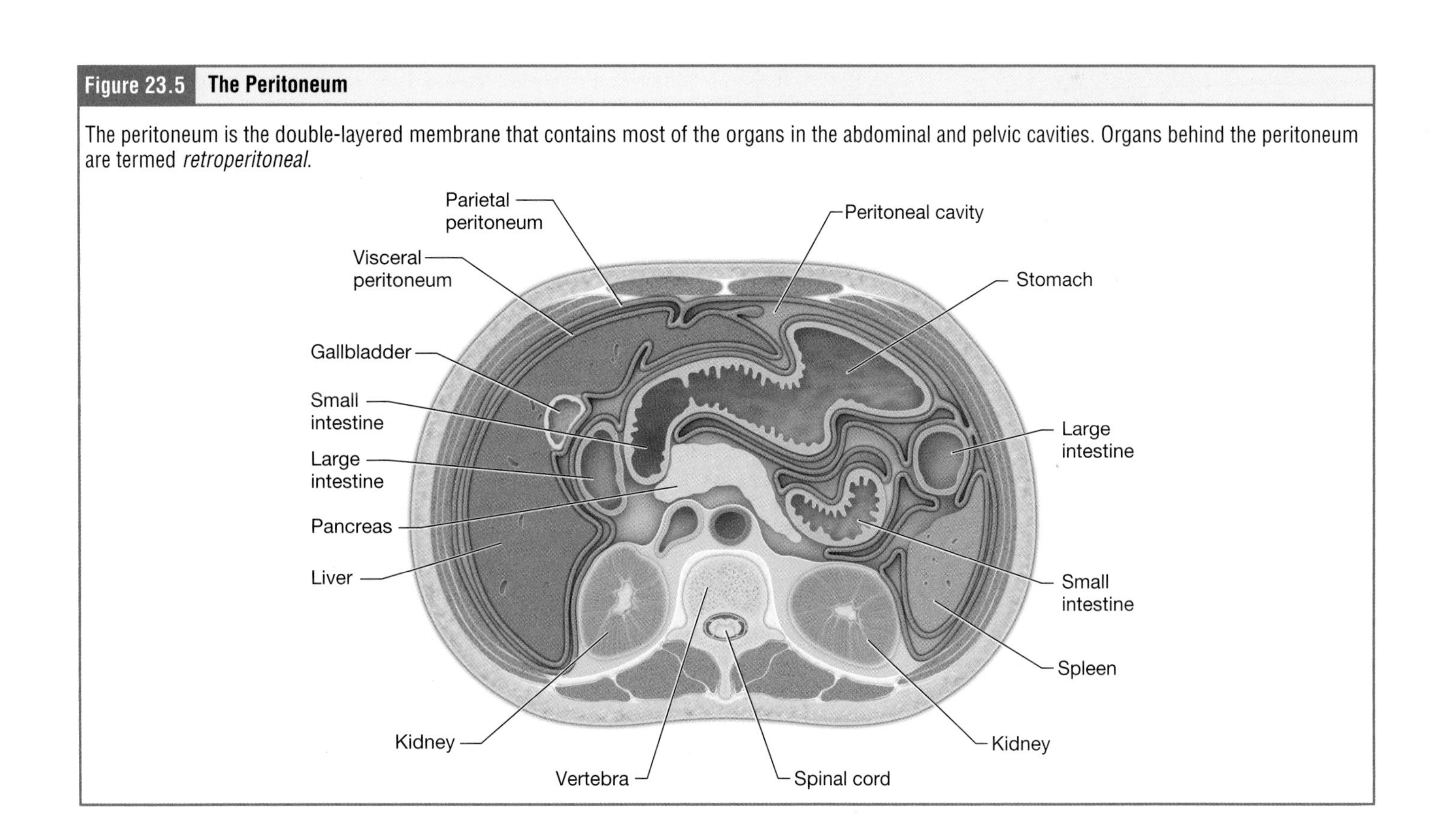

Figure 23.5 The Peritoneum

The peritoneum is the double-layered membrane that contains most of the organs in the abdominal and pelvic cavities. Organs behind the peritoneum are termed *retroperitoneal.*

Figure 23.6 The Mesenteries

(A) Transverse view. The lesser omentum connects the stomach to the liver; the greater omentum hangs down like a curtain over the intestines. The mesentery anchors the small intestine to the posterior body wall. (B) In the cadaver, the omentum hangs in place over the intestines. (C) Once the omentum is removed, the intestines can be seen.

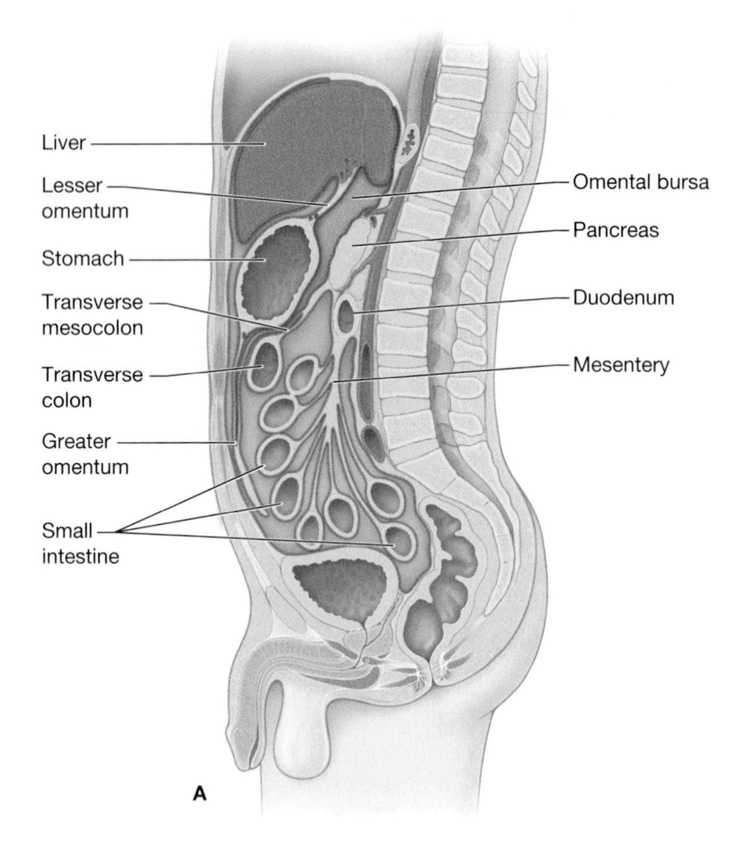

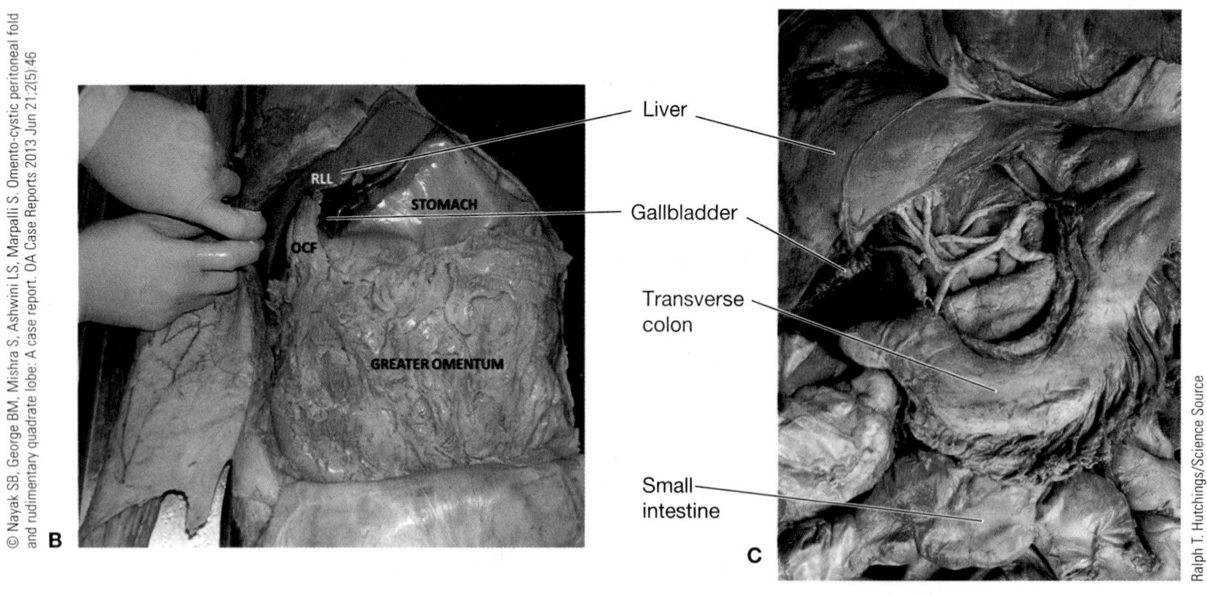

LO 23.2.9

jejunum, and the ileum of the small intestine, and portions of the large intestine, lie completely within the abdominal cavity; they are described as **intraperitoneal**. Other digestive organs, such as the ascending colon, descending colon, and portions of the duodenum, rectum, and pancreas, are retroperitoneal.

Learning Check

1. Which of the following layers of the wall of the GI tract is the most superficial?
   a. Mucosa
   b. Submucosa
   c. Muscularis
   d. Serosa
2. Which type of epithelium would you expect to find in the small intestine?
   a. Simple squamous
   b. Stratified squamous
   c. Simple columnar
   d. Simple cuboidal
3. Which of the following groups of nerves are responsible for motility of the food through the GI tract? Please select all that apply.
   a. Enteric nervous system
   b. Myenteric plexus
   c. Submucosal plexus
   d. Brachial plexus
4. Which of the following hangs down vertically over the small intestines and covers their anterior surface?
   a. Lesser omentum
   b. Greater omentum
   c. Transverse mesocolon
   d. Mesenteries

23.3 The Mouth, Pharynx, and Esophagus

Learning Objectives: By the end of this section, you will be able to:

The Mouth and Pharynx

23.3.1 Identify and describe the boundaries of the oral cavity.

23.3.2 Define mastication.

23.3.3 Compare and contrast the composition and functions of the hard palate, soft palate, and uvula.

23.3.4 Identify and describe the structures (e.g., taste buds, papillae) and the functions of the tongue.

23.3.5 Describe the structure and function of the salivary glands.

23.3.6 Describe the composition and functions of saliva.

23.3.7 Describe the structure and function of teeth.

23.3.8 Identify and describe the different regions of the pharynx with respect to the passage of air and/or food.

23.3.9 List the structures involved in deglutition and explain the process of deglutition, including the changes in position of the glottis and larynx that prevent aspiration.

The Esophagus

23.3.10 Identify and describe the gross anatomy of the esophagus, including its location relative to other body structures.

23.3.11 Describe the general functions of the esophagus.

23.3.12 Describe the anatomic specializations of the esophageal tunics (e.g., composition of the mucosa and muscularis [muscularis externa]) compared to the tunics of the rest of the GI tract.

23.3.13 Relate the anatomic specializations of the esophagus to the organ's functions.

The Stomach

23.3.14 Describe the general functions of the stomach.

23.3.15 Identify and describe the gross anatomy of the stomach, including its location relative to other body structures.

23.3.16 Describe the compositions, locations, and functions of the inferior esophageal (cardiac, lower esophageal) sphincter and the pyloric sphincter.

23.3.17 Identify gastric folds (rugae) and discuss their functional significance.

23.3.18 Describe the anatomic specializations of the stomach tunics compared to the tunics of the rest of the GI tract.

23.3.19 Relate the anatomic specializations of the stomach tunics (e.g., number of layers of muscle in the muscularis [muscularis externa]) to the organ's functions.

23.3.20 Identify and describe the gastric glands, including their cells (e.g., parietal cells, chief cells).

23.3.21 Describe the functions, production, and regulation of secretion of hydrochloric acid (HCl).

23.3.22 Explain the effects of the cephalic phase, gastric phase, and intestinal phase of digestion on various parts of the gastrointestinal (GI) tract.

23.3.23 Explain how volume, chemical composition, and osmolarity of chyme affect motility in the stomach and in the duodenum.

The Small Intestine

23.3.24 Describe the general functions of the small intestine.

23.3.25 Identify the specific segments of the small intestine (i.e., duodenum, jejunum, ileum), including their relative length.

23.3.26 Identify and describe the gross anatomy of the small intestine, including its location relative to other body structures.

23.3.27 Describe the major functions of the biliary apparatus.

23.3.28 Identify and describe the biliary apparatus components (i.e., left and right hepatic ducts, common hepatic duct, cystic duct, common bile duct, main pancreatic duct, hepatopancreatic ampulla [ampulla of Vater], hepatopancreatic sphincter [sphincter of Oddi], major duodenal papilla).

23.3.29 Trace the path of bile and pancreatic juice through the biliary apparatus.

23.3.30 Describe the anatomic specializations of the small intestine tunics (e.g., circular folds [plicae circulares], villi, microvilli) compared to the tunics of the rest of the GI tract.

23.3.31 Relate the anatomic specializations of the small intestine tunics (e.g., circular folds [plicae circulares], villi, microvilli) to the organ's functions.

23.3.32 Identify and describe the function of the following small intestine structures: duodenal glands (Brunner glands), intestinal glands (crypts of Lieberkuhn), and Peyer patches (lymphoid [lymphatic] nodules).

23.3.33 Compare and contrast the following: peristalsis, mixing waves, segmentation, and mass movement.

The Large Intestine, Rectum, and Anal Canal

23.3.34 Describe the general functions of the large intestine, rectum, and anal canal.

23.3.35 Identify and describe the gross anatomy of the large intestine, rectum and anal canal, including their location relative to other body structures.

23.3.36 Identify the specific segments and related flexures of the large intestine.

23.3.37 Compare and contrast the location, composition, and innervation (i.e., somatic versus autonomic) of the internal and external anal sphincters.

23.3.38 Describe the specializations of the large intestine tunics (e.g., composition of the muscularis [muscularis externa]) compared to the tunics of the rest of the GI tract.

23.3.39 Relate the specializations of the large intestine tunics (e.g., composition of the muscularis [muscularis externa]) to the organ's functions.

23.3a The Mouth

oral cavity (buccal cavity)

LO 23.3.1

labia (lips)

The cheeks, tongue, and palate frame the mouth, which is also called the **oral cavity** (or *buccal cavity*). The structures of the mouth are illustrated in **Figure 23.7**.

At the entrance to the mouth are the lips, or **labia** (singular = labium). On their outer surface the labia are composed of skin—keratinized stratified squamous epithelium. On their inner surface the labia are made of a mucous membrane—mucus-covered non-keratinized stratified squamous epithelium. The lips are highly vascular, and their keratin layer is very thin, so the dense vascularization lends a reddish hue to their outer surface. The lips have a huge representative area in the primary

Figure 23.7 The Mouth

The mouth is the entrance to the GI tract. Its structures enable mechanical digestion of ingested food.

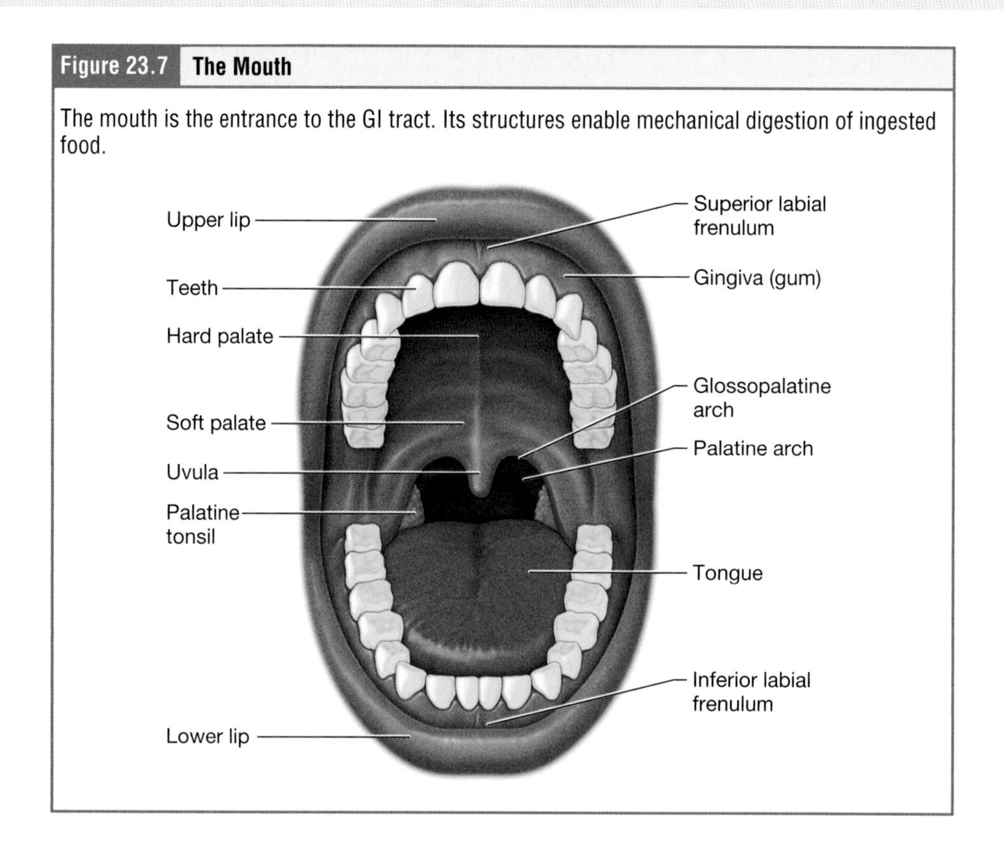

somatosensory cortex, which means that the brain receives a tremendous amount of sensory information from them. This fact probably explains the human fascination with kissing, as well as why babies explore their environments by putting objects in their mouths. The **labial frenulum** is a fold of mucous membrane that attaches the inner surface of each lip to the gum found at the midline of the mouth. The cheeks make up the oral cavity's sidewalls. Like the lips, their outer surface is skin and their inner covering is mucous membrane. Between the skin and mucous membranes of the cheeks and lips are connective tissue and skeletal muscles. These muscles, along with the tongue, enable the movements of chewing, otherwise known as **mastication**.

labial frenulum (lip frenulum)

mastication (chewing)

LO 23.3.2

The oral cavity is bordered by the teeth, gums, and cheeks on the lateral sides. The inferior border is the tongue, and the posterior border contains the opening between the posterior of the oral cavity and throat (oropharynx), called the **fauces**. The superior border, or roof, is called the *palate*. The anterior region of the palate serves as both a barrier between the oral and nasal cavities and a rigid shelf against which the tongue can push food. It is created by the maxillary and palatine bones of the skull and, given its bony structure, is known as the *hard palate*. If you run your tongue along the roof of your mouth, you'll notice that the hard palate ends in the posterior oral cavity, and roof becomes softer. This part of the palate, known as the **soft palate**, is composed mainly of skeletal muscle. You can therefore manipulate, subconsciously, the soft palate—for instance, to yawn, swallow, or sing (see Figure 23.7).

soft palate (velum, palatal velum, muscular palate)

LO 23.3.3

If you examine your posterior soft palate in the mirror, you will notice a flap of soft tissue called the **uvula** that hangs down like an icicle from the center. While it may look strange, it serves an important purpose. When you swallow, the soft palate and uvula move upward, helping to keep foods and liquid from entering the nasal cavity. Unfortunately, it can also contribute to the sound produced by snoring. Two muscular folds extend downward from the soft palate, on either side of the uvula. Toward the front,

uvula (palatine uvula)

Student Study Tip

The Greek suffix *-glossal* means "tongue." This is also seen in the "hypoglossal" cranial nerve.

the **palatoglossal arch** lies next to the base of the tongue; behind it, the **palatopharyngeal arch** forms the superior and lateral margins of the fauces. Between these two arches are the palatine tonsils, clusters of lymphoid tissue that protect the pharynx. The lingual tonsils are located at the base of the tongue.

23.3b The Tongue

LO 23.3.4

palatoglossal arch (glossopalatine arch, anterior pillar of fauces, anterior arch)

palatopharyngeal arch (posterior arch)

The **tongue** is one of the strongest muscles in the body. It is a workhorse, facilitating ingestion, mechanical digestion, sensation (of taste, texture, and temperature of food), swallowing, and language.

The tongue is attached to the mandible, the styloid processes of the temporal bones, and the hyoid bone. The hyoid is unique in that it does not articulate with other bones but rather serves as an anchor to the tongue.

The top and sides of the tongue are studded with **papillae**, small raised bumps of stratified squamous epithelium (**Figure 23.8**) in which taste buds are housed. For more reading on the sense of taste, please see Chapter 15. A fold of mucous membrane on the underside of the tongue, the **lingual frenulum**, tethers the tongue to the floor of the mouth.

23.3c The Salivary Glands

LO 23.3.5

Many small **salivary glands** are housed within the mucous membranes of the mouth and tongue. These minor exocrine glands are constantly secreting **saliva**, a watery substance enriched with several enzymes, either directly into the oral cavity or indirectly through ducts, even while you sleep. In addition to these small contributing glands, there are three major salivary glands. Together, the salivary glands of most adults produce an average of 1 to 1.5 liters of saliva each day. Secretion increases when you see,

Figure 23.8 The Tongue

The tongue is a small but incredibly powerful muscular organ. Its surface is dotted with papillae, which contain taste buds.

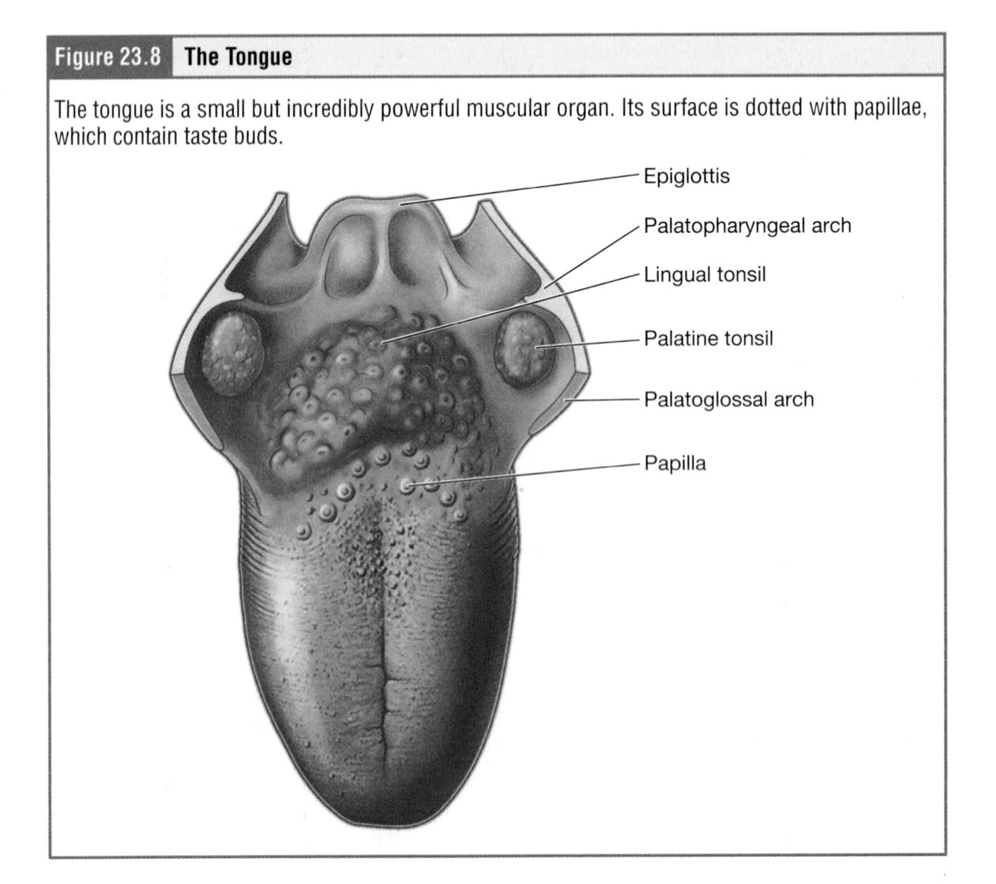

smell, or taste food, because saliva is essential to moisten food and initiate chemical digestion.

The three pairs of major salivary glands, which secrete the majority of saliva into ducts that open into the mouth, are:

- The **submandibular glands**: these are in the floor of the mouth and secrete saliva into the mouth through the submandibular ducts.
- The **sublingual glands**: these lie below the tongue and use the lesser sublingual ducts to secrete saliva into the oral cavity.
- The **parotid glands**: these lie between the skin and the masseter muscle, near the ears. They secrete saliva into the mouth through the parotid duct, which is located on the posterior upper cheek (**Figure 23.9**).

LO 23.3.6

Saliva Saliva is essentially (over 95 percent) water. Dissolved within the water are ions, enzymes, signaling molecules, antibacterial compounds, and waste products. One important ingredient in saliva is the enzyme **salivary amylase**, an enzyme that initiates the breakdown of carbohydrates. While adult salivary glands produce a carbohydrate-digesting enzyme, infant salivary glands actually produce **salivary lipase**, an enzyme that breaks down the fats presents in breast milk. When we are around 2 years of age, the production of enzymes changes to amylase. Many of our favorite foods, including soda, coffee, tea, yogurt, pickles, and many others, are acidic. An overly low pH environment in the mouth can compromise enzymatic function and the integrity of tooth enamel, so bicarbonate and phosphate ions are added to the saliva as buffers,

Figure 23.9 The Salivary Glands

There are three salivary glands on each side of the mouth. The parotid gland is under the skin in the posterior cheek. The submandibular gland sits below the mandible. The sublingual salivary gland is found under the tongue.

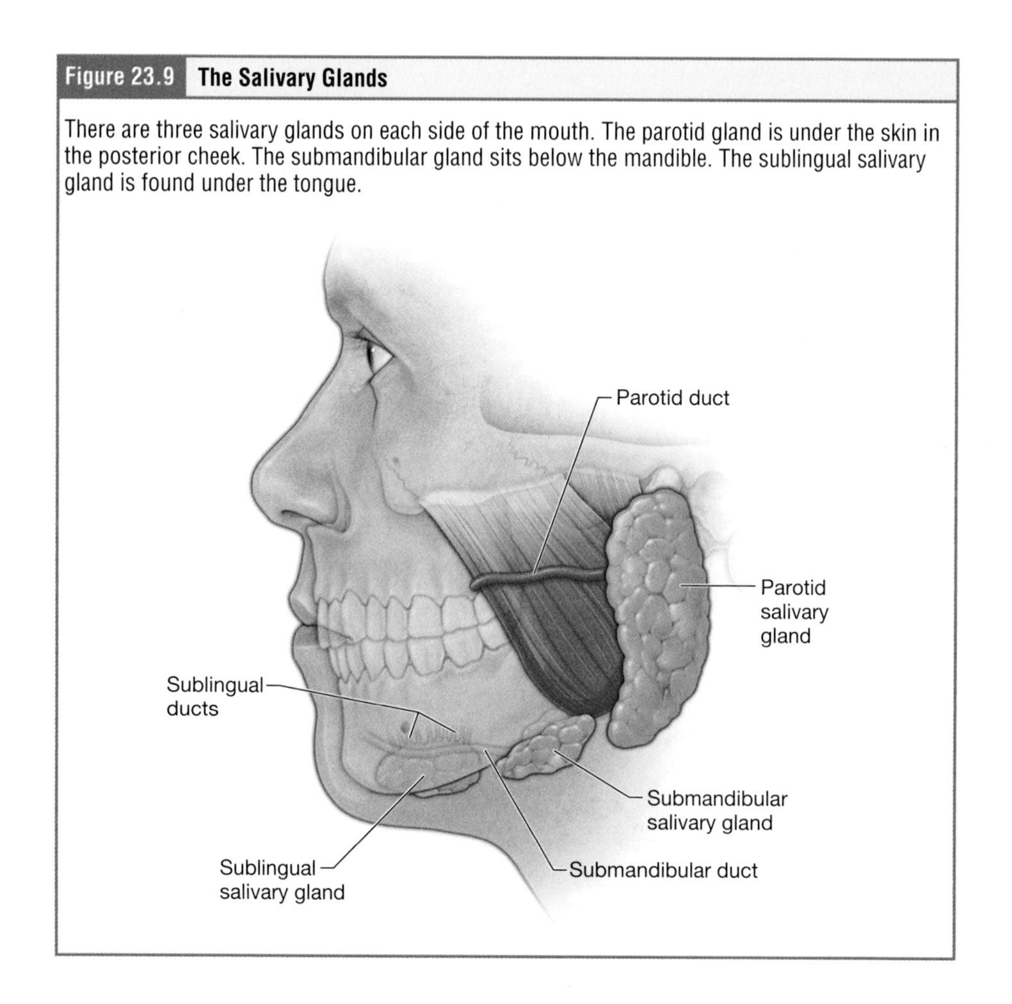

preventing the pH of the mouth from becoming too low. Salivary mucus helps lubricate food; this facilitates movement and swallowing, and enables small taste molecules to reach the taste buds within the papillae. Saliva contains immunoglobulin A—one type of antibody that prevents microbes from penetrating the epithelium—and lysozyme, an antibacterial compound that kills off invading bacteria. Saliva also contains histostatin, which helps to heal wounds quickly. The presence of histostatin, antibodies, and lysozyme in saliva is likely why many animals lick their wounds.

Each of the major salivary glands secretes a unique formulation of saliva. For example, the parotid glands secrete a watery solution that contains salivary amylase. The submandibular glands have cells similar to those of the parotid glands, as well as mucus-secreting cells. The sublingual glands contain mostly mucous cells, and they secrete the thickest saliva with the least amount of salivary amylase. The saliva in our mouths is the end product of the mixing of the individual glands' contributions.

Regulation of Salivation The autonomic nervous system regulates the secretion of saliva. In the absence of food, parasympathetic stimulation keeps saliva flowing at a comfortable level for speaking and breathing. During times of stress, such as before speaking in public, sympathetic stimulation takes over, reducing salivation and producing the symptom of dry mouth often associated with anxiety. When you are dehydrated, salivation is reduced, causing the mouth to feel dry and prompting you to take action to quench your thirst.

Production and secretion of saliva can be stimulated by the sight, smell, and taste of food. Taste receptors on the tongue communicate both with the taste-processing insula and nuclei in the brainstem that regulate the parasympathetic impulses that stimulate salivation. Most saliva is swallowed along with food and is reabsorbed, so that fluid is not lost.

dentes (teeth)

23.3d The Teeth

LO 23.3.7

The teeth, or **dentes** (singular = dens), are organs similar to bones that you use to tear, grind, and otherwise mechanically break down food.

deciduous teeth (baby teeth, milk teeth, primary teeth)

Types of Teeth Teeth begin to erupt from the gums at about 6 months of age. These first 20 **deciduous teeth**, or baby teeth, will be replaced between approximately

DIGGING DEEPER:
Mumps

Paramyxovirus is a virus that spreads through saliva—often spread through coughing, sneezing, sharing utensils, or kissing. Paramyxovirus can infect the nasal passages, the pharynx, and any salivary gland; however, the parotid glands are the usual site of infection. Enlargement and inflammation of the parotid glands is typical, causing a characteristic swelling between the ears and the jaw. Other symptoms include fever and throat pain, which can be severe when swallowing acidic substances such as orange juice.

Mumps was once well-controlled through vaccination, and there were years in the United States in which no or only a few cases of mumps were reported. However, in the past few decades, some individuals and communities have become hesitant to vaccinate their children, and as vaccination rates began to decline, cases of mumps in the United States rose. While vaccination is sufficient to prevent infection in most cases, if the challenge is significant enough, meaning that if the person transmitting the virus to you has a high viral load, then so-called breakthrough infections—infections in a vaccinated person—can occur. Outbreaks of mumps involving a mix of vaccinated and unvaccinated individuals have become frequent on college campuses. Vaccination reduces the chance that you will contract mumps by 88 percent, and breakthrough infections are much less severe. In unvaccinated individuals, however, mumps can be serious or even deadly.

ages 6 and 12 by 32 **permanent teeth**, the ones that last for the rest of the lifetime. Moving from the center of the mouth toward the sides, these are as follows (**Figure 23.10**):

- The eight **incisors**, four top and four bottom. These are the sharp front teeth you use for biting into food or tearing food. For example, biting into a slice of pizza to tear off a piece.
- The four **cuspids** (or canines) flank the incisors and have a pointed edge (cusp) to tear up food. These fang-like teeth are superb for piercing tough or fleshy foods. You might use these to tear off a piece of a chewy candy such as licorice, for example.
- Posterior to the cuspids are the eight **premolars**, which have an overall flatter shape for mashing and grinding.
- The most posterior and largest are the 12 **molars**, which have several points used to crush food so it is ready for swallowing. The third members of each set of three molars, top and bottom, are commonly referred to as the *wisdom teeth*, because their eruption is commonly delayed until early adulthood. It is not uncommon for wisdom teeth to fail to erupt; in these cases, the teeth are typically removed surgically.

permanent teeth (adult teeth, secondary teeth)

cuspids (canines)

premolars (bicuspids)

Anatomy of a Tooth The teeth form a joint with sockets of the maxilla and the mandible. The socket is lined with **gingivae** (commonly called the *gums*), which line the sockets and surround the necks of the teeth. Teeth are held in their sockets by short bands of dense connective tissue called **periodontal ligaments** (**Figure 23.11**).

Figure 23.10 The Teeth

Humans grow an initial set of teeth in infancy and toddlerhood that fall out and are replaced by a second set, called *adult teeth*. The adult teeth come in four varieties, each specialized for a particular function of biting or chewing.

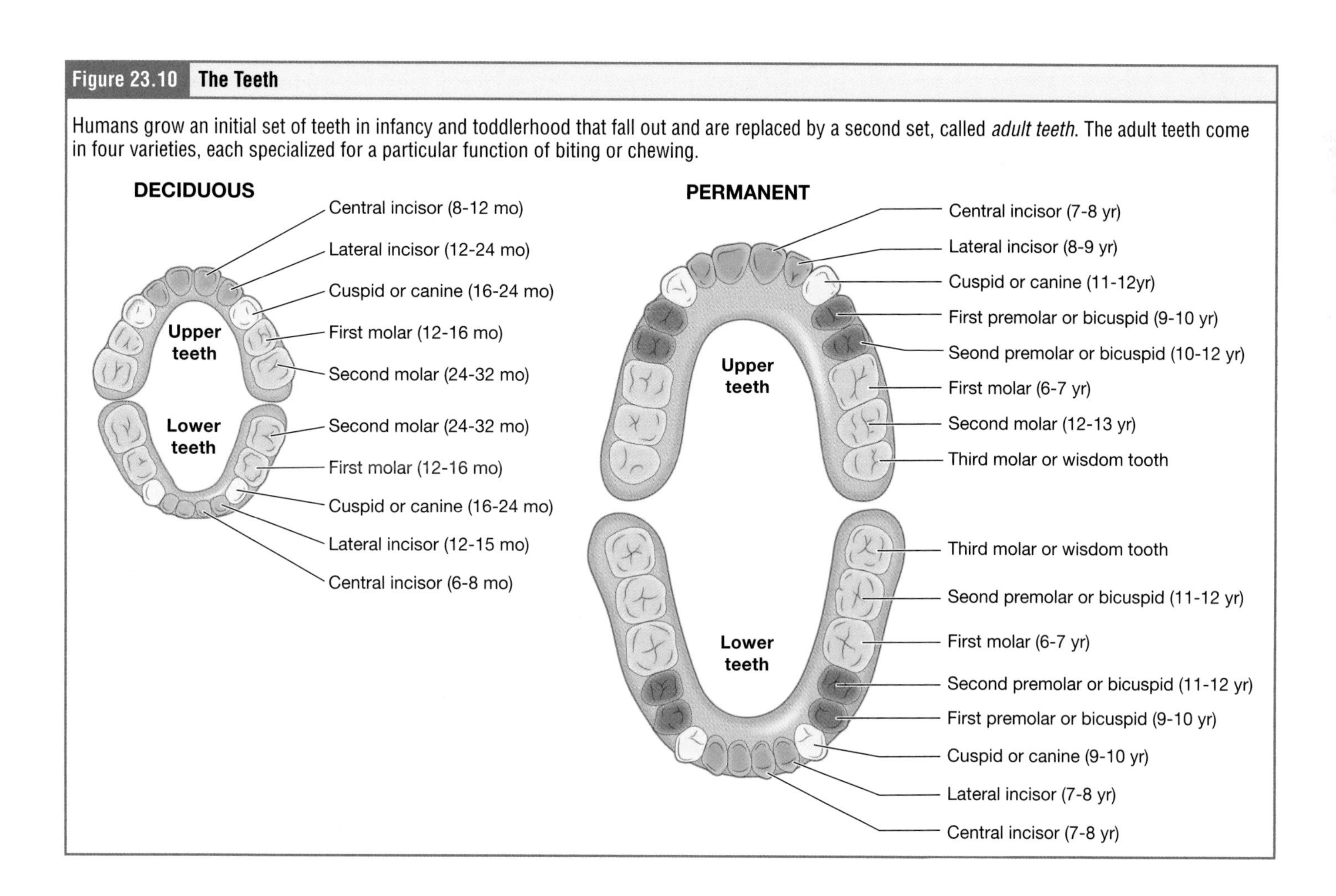

Figure 23.11 Anatomy of a Tooth

An individual tooth has a root, which is housed within a socket in the bone, and through that root runs a canal containing blood vessels and nerves. The top of the tooth, which is visible outside the gums, consists of enamel and dentin; both substances are hard and durable materials.

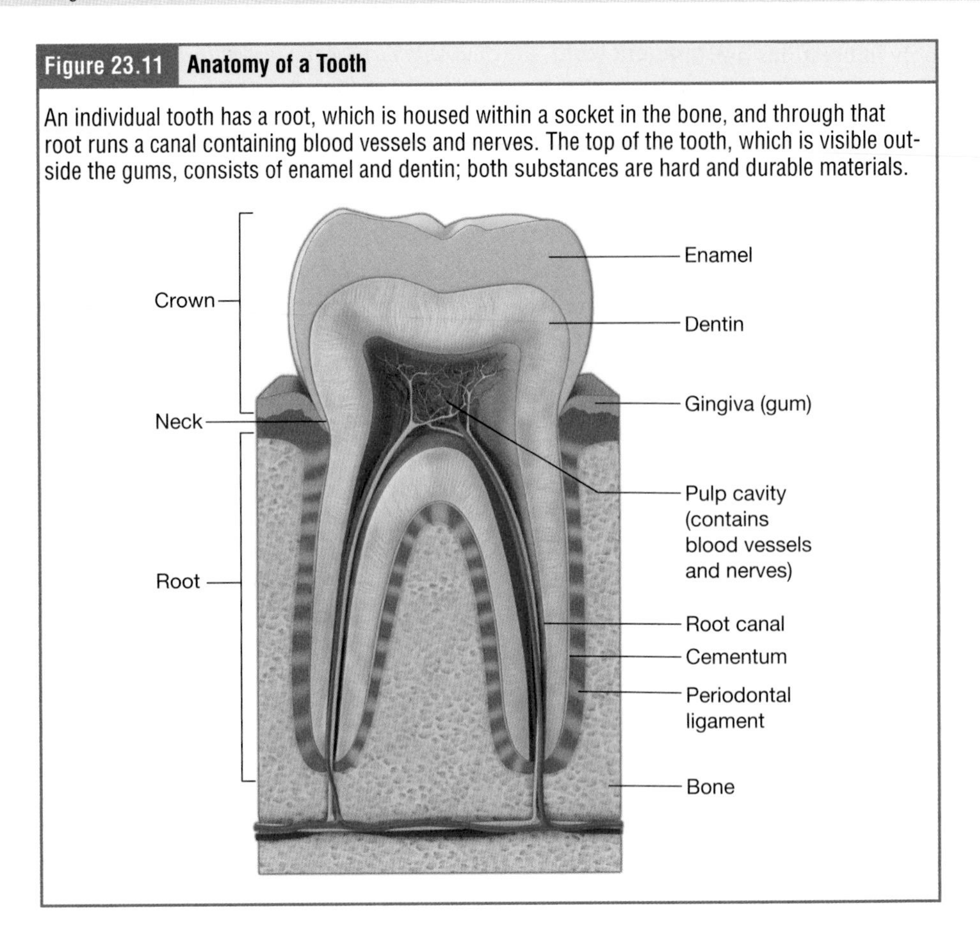

The two main parts of a tooth are the **crown**, which is the portion projecting above the gums, and the **root**, which is embedded within the maxilla and mandible. Both parts contain an inner **pulp cavity**, which contains nerves and blood vessels enmeshed in loose connective tissue. The blood vessels and nerves within the pulp cavity run through the root of the tooth and into the bone (either maxilla or mandible). This passage is called the **root canal**. Surrounding the pulp cavity is **dentin**, a bonelike tissue. In the root of each tooth, the dentin is covered by an even harder bonelike layer called **cementum**. In the crown of each tooth, the dentin is covered by an outer layer of **enamel**, the hardest substance in the body.

Although enamel protects the underlying dentin and pulp cavity, it is still nonetheless susceptible to environmental factors that might cause it to decay. The most common form of damage to the enamel—cavities—can develop when colonies of bacteria feeding on sugars in the mouth release acids that degrade the enamel. Saliva helps to neutralize acids that can damage the enamel.

23.3e The Pharynx

pharynx (throat)

The **pharynx** is a structure involved in both digestion and respiration. It receives food from the mouth, and air from both the mouth and the nasal cavities. When food enters the pharynx, involuntary muscle contractions close off the air passageways.

LO 23.3.8

The pharynx is a short tube of skeletal muscle lined with a mucous membrane that runs from the posterior oral and nasal cavities to the opening of the esophagus and larynx. It has three regions. The most superior, the **nasopharynx**, is involved only in breathing and speech and only air passes through it. The **oropharynx** is the region below the nasopharynx, it is posterior to the mouth and is therefore involved in both breathing and digestion, food, drink, and air all pass through it. The **laryngopharynx** is inferior to the oropharynx and is used for both breathing and digestion (**Figure 23.12**). The laryngopharynx connects to two different inferior structures, like a fork in the road. Anterior and inferior to the laryngopharynx is the larynx, which allows air to flow into the bronchial tree. Posterior and inferior to the laryngopharynx is the esophagus, which allows food and drink to travel to the stomach.

laryngopharynx (hypopharynx)

Histologically, the wall of the oropharynx is similar to that of the oral cavity. The mucosa includes a stratified squamous epithelium that is rich with mucus-producing glands. During swallowing, the skeletal muscles of the pharynx contract, raising and expanding the pharynx to receive the swallowed food. Once received, this set of muscles relaxes and the constrictor muscles of the pharynx contract, forcing the food into the esophagus and initiating peristalsis.

Usually during swallowing, the soft palate and uvula rise reflexively to close off the entrance to the nasopharynx. At the same time, the larynx is pulled superiorly and the cartilaginous epiglottis—its most superior structure—folds down, covering the opening to the larynx, effectively blocking access to the respiratory tree. When the food "goes down the wrong pipe," it goes into the trachea. When food enters the trachea, the reaction is to cough, which usually forces the food up and out of the trachea, and back into the pharynx.

Student Study Tip

When the epiglottis covers the larynx, it functions as a closed garbage can lid. When open, food is able to enter.

Figure 23.12 The Pharynx

The pharynx, or throat, is an open cavity posterior to the nasal cavity and mouth. It has three zones—the nasopharynx, the oropharynx, and the laryngopharynx—all of which are continuous with one another.

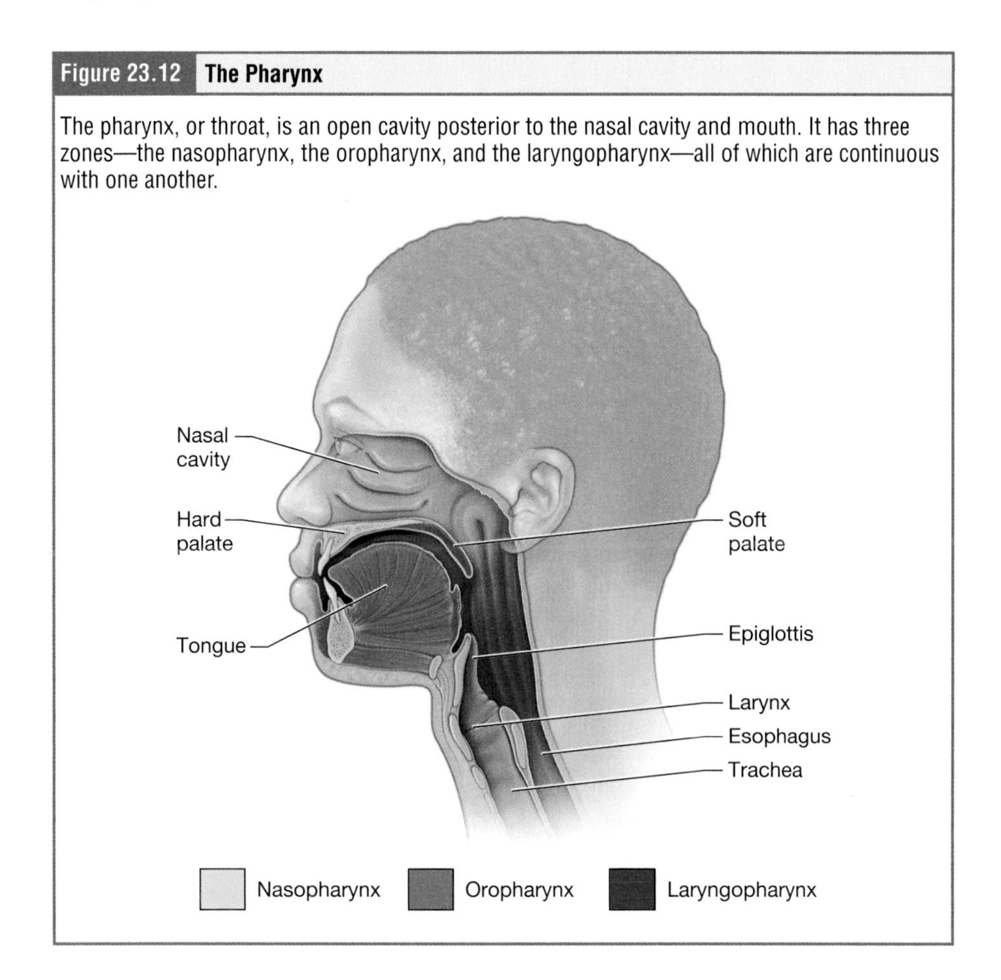

LO 23.3.9

Swallowing, or *deglutition*, begins the movement of food from the mouth to the stomach. The journey from mouth to stomach takes four to eight seconds for solid or semisolid food, and about one second for very soft food and liquids. Although this sounds quick and effortless, swallowing is actually a complex process that involves muscles of the tongue, pharynx, and esophagus. It is aided by the presence of mucus and saliva, and is much tougher in drier environments. There are three stages in swallowing: the voluntary phase, the pharyngeal phase, and the esophageal phase (**Figure 23.13**).

The Voluntary Phase The **voluntary phase** of swallowing is so named because you can control when you swallow food and only skeletal muscles are involved. In this phase, the tongue moves upward and backward against the palate, pushing the chewed food, called a **bolus**, to the back of the oral cavity and into the oropharynx. At this point, the pharyngeal and the esophageal phases—the two involuntary phases of swallowing—begin.

The Pharyngeal Phase The **pharyngeal phase** begins when stimulation of receptors in the oropharynx sends impulses to a nucleus that controls swallowing in the medulla

Figure 23.13 **Swallowing**

Swallowing begins as a voluntary skeletal muscle contraction that moves food from the mouth into the pharynx. As food passes through the upper esophageal sphincter, smooth muscle contractions take over, propelling it downward toward the stomach.

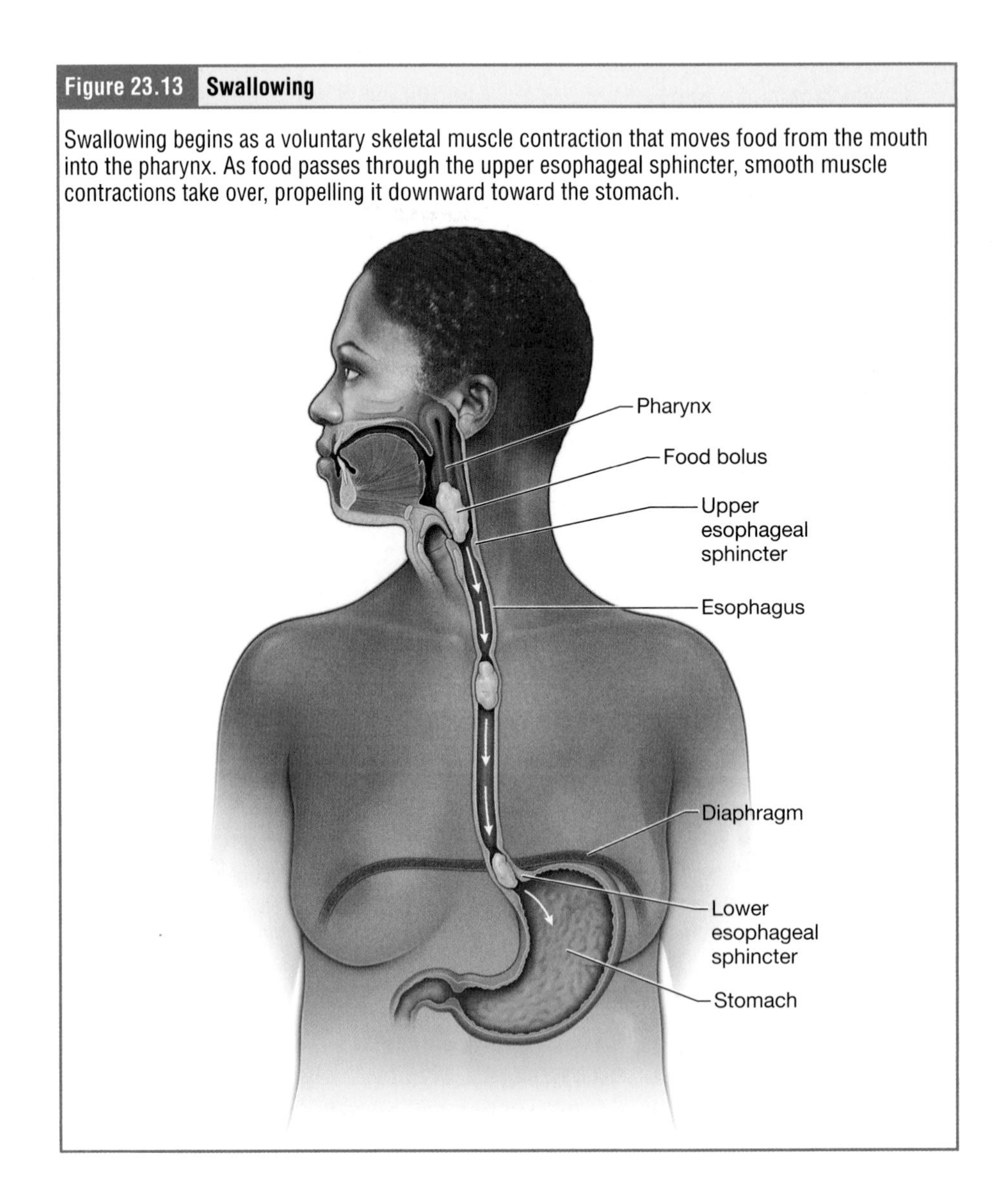

oblongata. Signals are then sent back to the uvula and soft palate, causing them to move upward and close off the nasopharynx. The laryngeal muscles also constrict to prevent aspiration of food into the trachea. At this point, breathing stops for a very brief time.

Contractions of the pharyngeal muscles move the bolus through the oropharynx and laryngopharynx. Relaxation of the upper esophageal sphincter then allows food to enter the esophagus.

The Esophageal Phase The entry of food into the esophagus marks the beginning of the **esophageal phase** of swallowing and the initiation of peristalsis. As in the previous phase, the medulla oblongata controls the muscular contractions. Peristalsis propels the bolus through the esophagus and toward the stomach. When the bolus nears the stomach, the stretching of the esophageal walls initiates a reflexive relaxation of the lower esophageal sphincter that allows the bolus to pass into the stomach. During the esophageal phase, esophageal glands secrete mucus that lubricates the bolus and minimizes friction.

Learning Check

1. Which of the following frames the oral cavity? Please select all that apply.
   a. Cheeks
   b. Palate
   c. Tongue
   d. Labia
2. When you bite into an apple, which of the following is responsible for the sensation of texture?
   a. Papillae
   b. Tongue
   c. Salivary gland
   d. Fauces
3. Which of the following glands would secrete saliva into the floor of the mouth when eating an apple? Please select all that apply.
   a. Submandibular glands
   b. Sublingual glands
   c. Subparotid glands
   d. Parotid glands
4. Cavities form on the most external surface of the tooth. Which of the following structures do cavities form on?
   a. Root
   b. Dentin
   c. Cementum
   d. Enamel
5. Which of the following parts of the pharynx is the most inferior?
   a. Nasopharynx
   b. Oropharynx
   c. Laryngopharynx
   d. Esophagopharynx

23.3f The Esophagus

LO 23.3.10

The **esophagus** is a muscular tube that connects the pharynx to the stomach, **Figure 23.14A**, passing through an opening in the diaphragm, the **esophageal hiatus**, to reach the abdomen. It descends through the entire thorax; therefore, its length varies from human to human, but it is somewhere around 10 inches in length, located posterior to the trachea, and remains in a collapsed form when not engaged in swallowing (**Figure 23.14B**).

LO 23.3.11

Passage of Food through the Esophagus A **sphincter** is a ring of muscle that controls the opening or closing of a passageway, like a valve. The **upper esophageal sphincter**, at the bottom of the pharynx, controls the movement of food into the esophagus. Rhythmic waves of peristalsis, which begin in the upper esophagus, propel

upper esophageal sphincter (inferior pharyngeal sphincter)

Figure 23.14 The Esophagus

(A) The esophagus is a long elastic tube that runs from the mouth to the stomach. It passes through a gap in the diaphragm called the *esophageal hiatus*. (B) A cross-section of the esophagus. Like all organs of the GI tract, the esophagus has a mucosa that lines the lumen, a submucosa, and a muscularis surrounded by a serosa, a sheath of connective tissue.

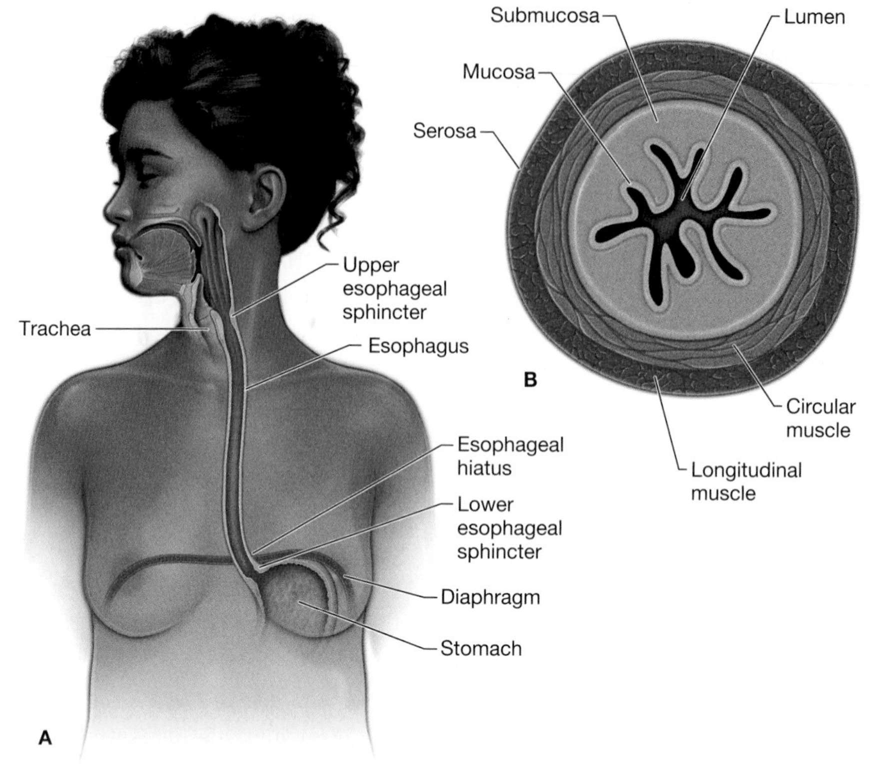

lower esophageal sphincter (cardiac sphincter)

the chewed food toward the stomach. Meanwhile, secretions from the esophageal mucosa lubricate the esophagus and food. Food passes from the esophagus into the stomach through the **lower esophageal sphincter**. The lower esophageal sphincter functions both to regulate food entering the stomach and to prevent stomach acids from escaping into the esophagus. When the lower esophageal sphincter fails to sufficiently hold back stomach acid, this low-pH solution can enter into the esophagus, damaging the esophageal epithelium and causing intense pain known as *gastroesophageal reflux disease (GERD)* or heartburn.

LO 23.3.12

LO 23.3.13

Histology of the Esophagus The mucosa of the esophagus is made up of an epithelial lining that contains non-keratinized, stratified squamous epithelium. This thick epithelium protects the underlying tissue from friction of the undigested food. The mucosa's lamina propria contains mucus-secreting glands. The muscularis layer changes according to location: In the upper third of the esophagus, the muscularis is skeletal muscle. In the middle third, it is both skeletal and smooth muscle. In the lower third, it is entirely smooth muscle. As mentioned previously, the outer layer of the esophagus is called the *adventitia* (not the serosa) because this organ is outside the abdominal cavity.

23.3g The Stomach

LO 23.3.14

The stomach continues the jobs of mechanical and chemical digestion. The stomach also holds the ingested food and releases it slowly into the small intestine, where digestion is completed and absorption begins. The empty stomach is only about the size of your fist, but can stretch to hold as much as 4 liters of food and fluid, or more than 75 times its empty volume, and then return to its resting size when empty. The small intestine, by contrast, cannot distend to hold large quantities of food, so the stretchable nature of the stomach allows us to eat more when it is convenient, and digest over the hours that follow.

As you will see in the following sections, the stomach plays several important roles in digestion, including the initial chemical digestion of proteins. While protein digestion is not completed in the stomach, the stomach does play an essential role in protein digestion. As you remember, proteins are large, highly folded, three-dimensional molecules. Their digestion is complex, and this is detailed in the "Anatomy of Protein Digestion" feature. Little if any nutrient absorption occurs in the stomach, with the exception of the negligible amount of alcohol. The stomach also performs a majority of the mechanical digestion; it has a third layer within the muscularis for extra mixing power.

Student Study Tip

The rugae of the stomach are like a crumpled-up plastic bag that widens as items are placed into it.

LO 23.3.15

Structure There are four main regions in the stomach: the cardia, the fundus, the body, and the pylorus (**Figure 23.15**). The **cardia** is the region just internal to the lower esophageal sphincter; it is the point where food passes into the stomach. Located above

cardia (cardiac region)

Figure 23.15 The Regions of the Stomach

The stomach has five regions, all of which are continuous with each other. The food enters the stomach from the esophagus at the cardia, named for its proximity to the heart. The fundus is an expandable dome-shaped superior region. The body—the largest component of the stomach—comprises the middle of the organ. The antrum is the most inferior, with the pylorus containing the food that is about to pass into the small intestine.

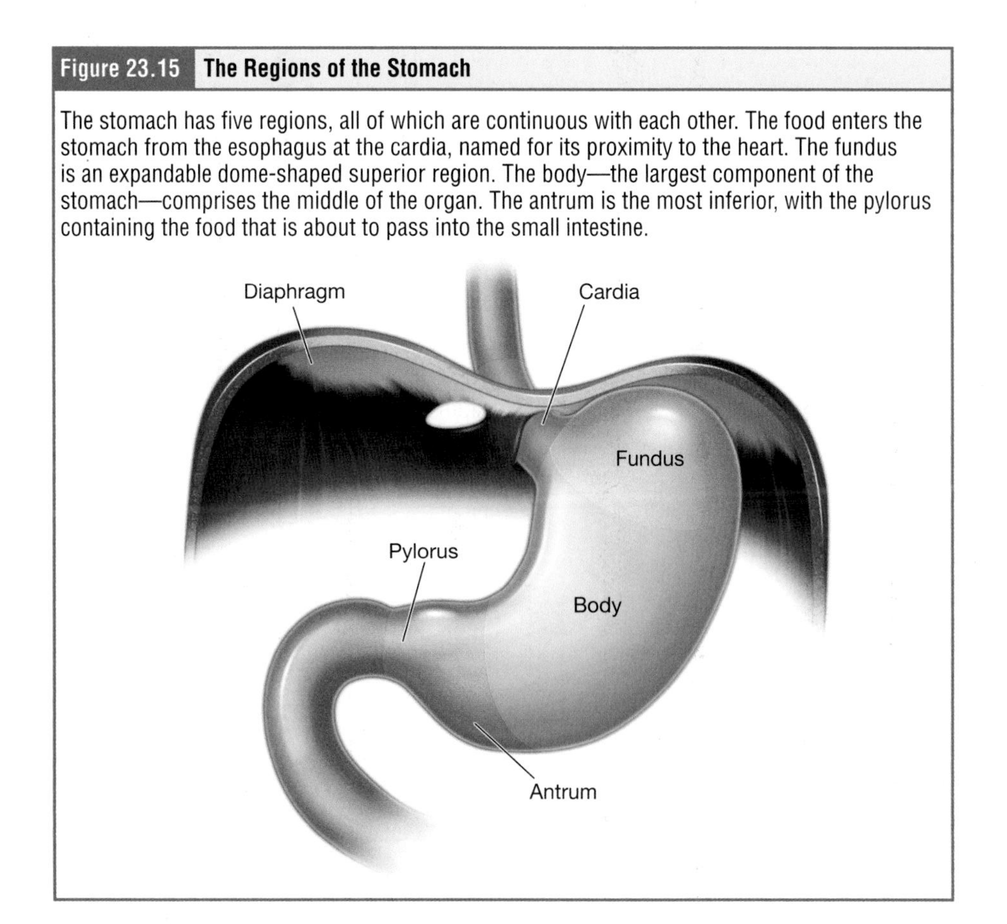

and to the left of the cardia is the dome-shaped **fundus**. Below the fundus is the **body**, the main part of the stomach. The funnel-shaped **pylorus** connects the stomach to the duodenum. The wider end of the funnel, the **pyloric antrum**, connects to the body of the stomach. The narrower end is called the **pyloric canal**, which connects to the small intestine. The smooth muscle **pyloric sphincter** is located at the end of the pyloric canal. This sphincter controls the release of stomach contents from the stomach into the small intestine. The rate at which it opens is highly controlled. The internal lining of the empty stomach is highly folded; each fold is called a **ruga** (plural = rugae). When the stomach is full, the rugae smooth out to permit it to expand in all directions.

LO 23.3.16

LO 23.3.17

The superior curve of the stomach is called the **lesser curvature**; the broader inferior curve is the **greater curvature** (Figure 23.16). The stomach is held in place by the lesser omentum, which extends from the liver to the lesser curvature, and the greater omentum, which runs from the greater curvature to the posterior abdominal wall.

greater curvature (curvatura ventriculi major)

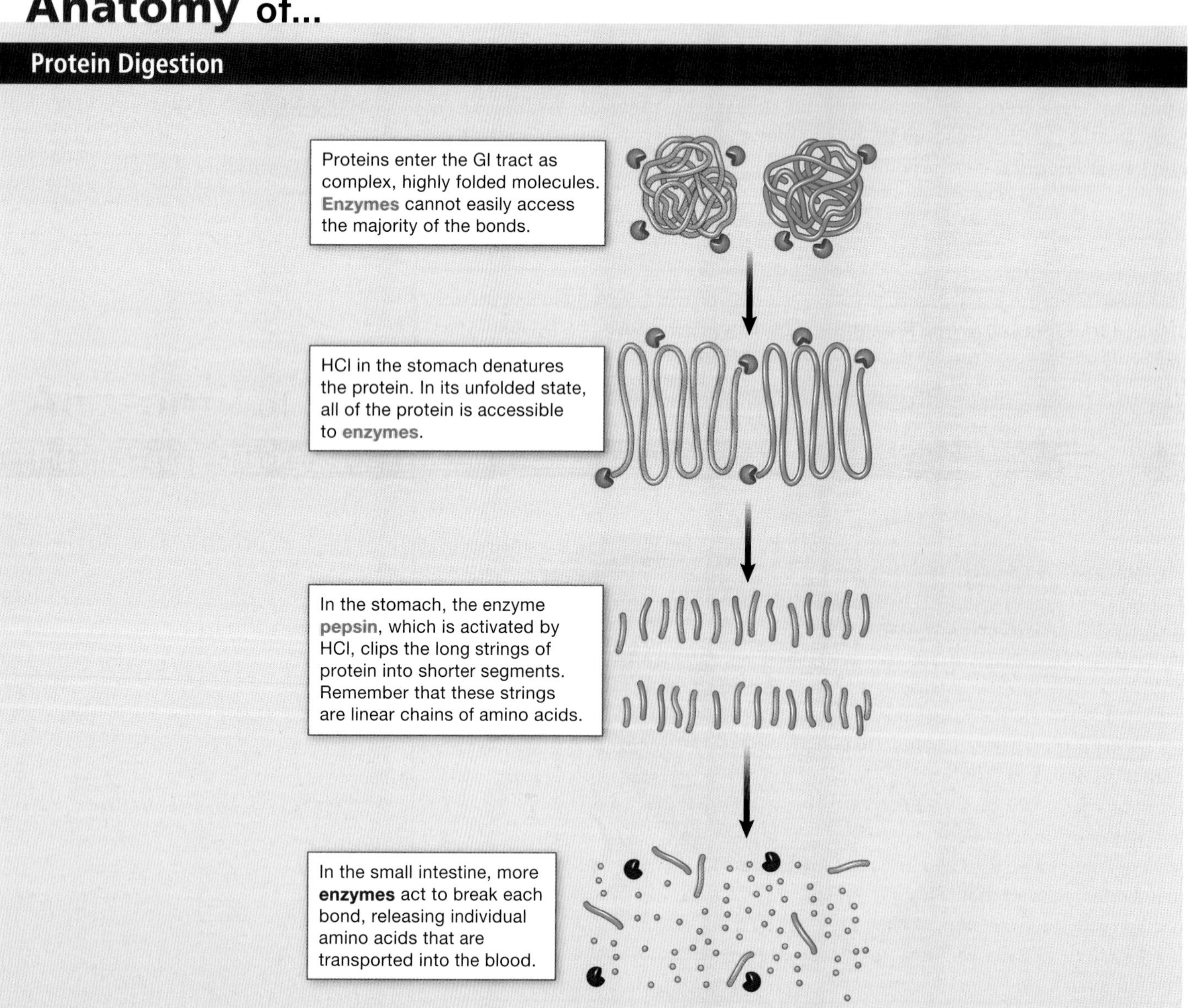

Anatomy of...
Protein Digestion
Proteins enter the GI tract as complex, highly folded molecules. Enzymes cannot easily access the majority of the bonds.
HCl in the stomach denatures the protein. In its unfolded state, all of the protein is accessible to enzymes.
In the stomach, the enzyme pepsin, which is activated by HCl, clips the long strings of protein into shorter segments. Remember that these strings are linear chains of amino acids.
In the small intestine, more enzymes act to break each bond, releasing individual amino acids that are transported into the blood.

Figure 23.16 The Stomach

The stomach connects to the esophagus proximally and the small intestine distally. The stomach wall is composed of three layers of muscle for extra mixing power. The internal surface of the stomach lining is highly folded; the folds are called *rugae*. These features are drawn (A) and shown in a dissected stomach (B).

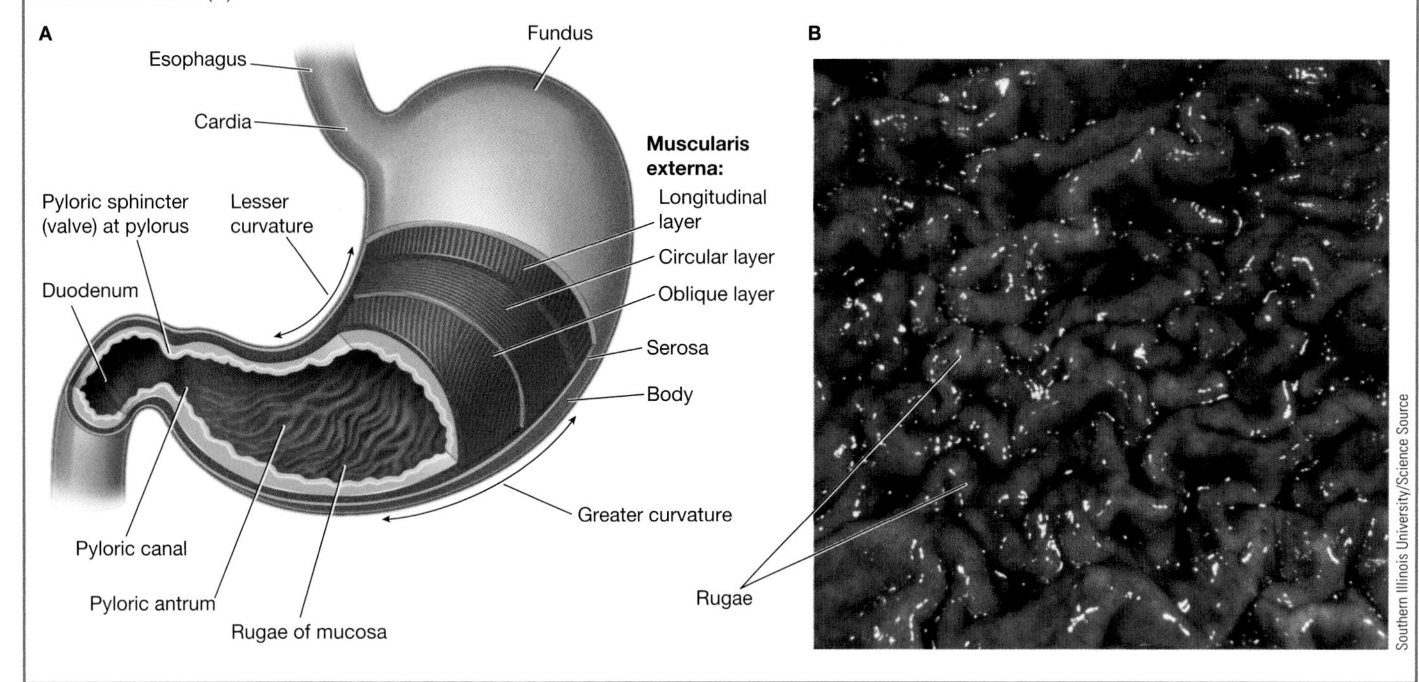

Histology The wall of the stomach is made of the same four layers as most of the rest of the GI tract, but with adaptations to the mucosa and muscularis for the unique functions of this organ. In addition to the typical circular and longitudinal smooth muscle layers, the muscularis has an inner oblique smooth muscle layer (see Figure 23.16 and **Figure 23.17**). As a result, in addition to moving food through the canal, the stomach can vigorously churn food, mechanically breaking it down into smaller particles. Once food is mixed with the gastric secretions, it is called **chyme**.

LO 23.3.18

LO 23.3.19

chyme (chymus)

If you were a piece of food inside the lumen of the stomach, the walls around you would appear full of holes, like a colander. These holes are actually deep pits called **gastric pits**, which are long tunnels leading to gastric glands. Each **gastric gland** secretes a complex digestive fluid referred to as **gastric juice**. Gastric juice is the mixture, but several different types of cells contribute individual components. The surface of the stomach and the shallow regions of the gastric pits are lined with epithelial cells that produce a thick alkaline mucus. This mucus plays an essential role in protecting the epithelial cells from the harsh conditions within the stomach lumen.

LO 23.3.20

Parietal cells—Located primarily in the middle region of the gastric glands are **parietal cells**, which produce both **hydrochloric acid (HCl)** and intrinsic factor. HCl is responsible for the high acidity (pH 1.5 to 3.5) of the stomach contents and is needed to activate the protein-digesting enzyme, pepsin. The acidity also kills much of the bacteria you ingest with food and helps to denature proteins; once the proteins are unfolded, their bonds are more available for enzymatic digestion. **Intrinsic factor** is a glycoprotein necessary for the absorption of vitamin B_{12} in the small intestine.

LO 23.3.21

parietal cells (oxyntic cells)

Chief cells—Deeper within the gastric glands are **chief cells**, which secrete **pepsinogen**, the inactive precursor of pepsin. Pepsinogen is no an active molecule, but once exposed to HCl, it will be converted from pepsinogen to pepsin. **Pepsin** in an enzyme that cleaves the bonds between amino acids and therefore plays an important role in protein digestion.

chief cells (zymogenic cells; peptic cells)

Figure 23.17 The Wall of the Stomach

The inner lining of the stomach is dotted with deep gastric pits. These pits contain parietal cells, chief cells, and enteroendocrine cells, each of which secretes a unique contribution to the process of digestion.

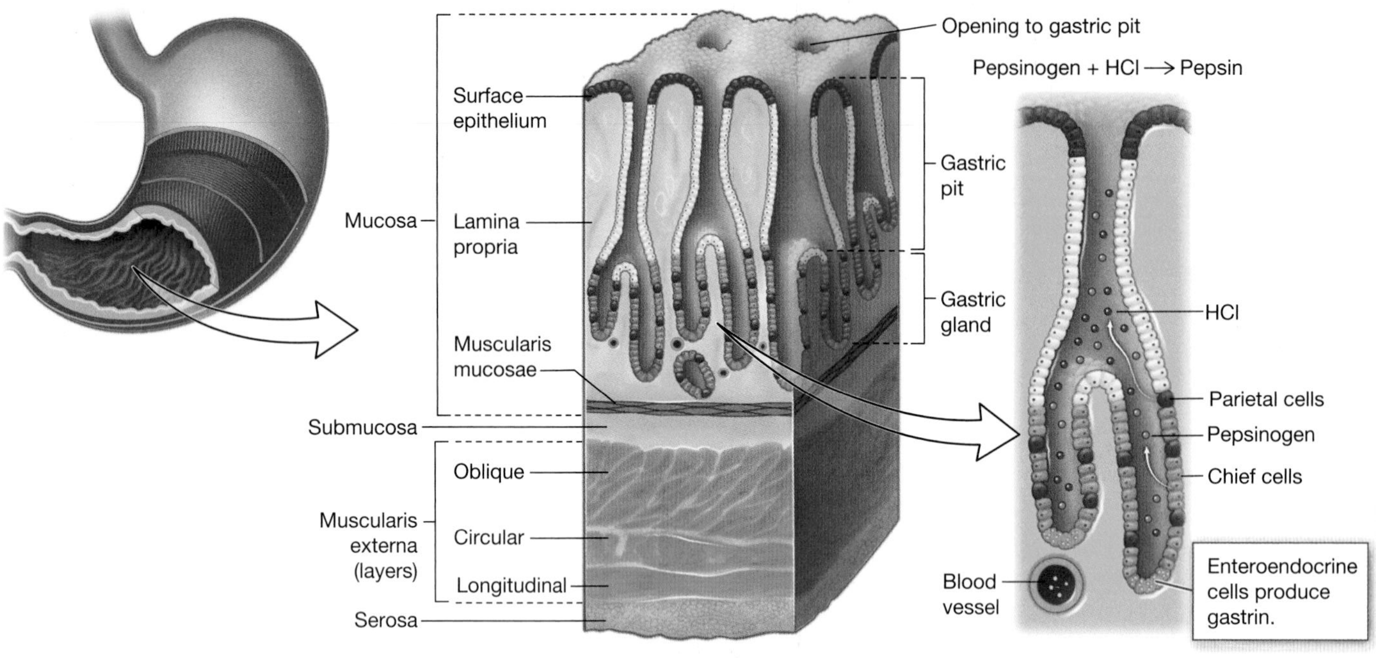

Mucous neck cells—**Mucous neck cells** secrete a thin, acidic mucus that is much different from the mucus secreted by the surface epithelial cells. Scientists have not yet discovered the role of this mucus.

Enteroendocrine cells—Finally, **enteroendocrine cells** found in the gastric glands secrete various hormones into the interstitial fluid of the lamina propria. These include gastrin, which helps regulate stomach motility and secretions. The hormones produced by the enteroendocrine cells regulate digestive functions within the stomach as well as traveling in the bloodstream to other organs to influence their digestive functions. The names and effects of the gastric hormones are illustrated in the "Anatomy of Gastric Hormones" feature.

DIGGING DEEPER: Ulcers—When the Mucosal Barrier Breaks Down

The stomach contains some of the harshest conditions on the planet. The three layers of muscularis make for a lumen experiencing frequent movement and turbulence. The internal pH is the most acidic in the human body. The stomach utilizes a thick, basic mucus to defend the epithelium from these unforgiving conditions. This basic mucosal barrier is effective, but it is not "fail-safe." Furthermore, under some conditions the stomach lining produces less mucus. These conditions include stress, dehydration, excessive caffeine consumption, and the overuse of nonsteroidal anti-inflammatory drugs (NSAIDs) such as Ibuprofen. A bacterium, *Helicobacter pylori*, can also interfere with mucus production and cause or exacerbate the conditions that may lead to mucus breakdown.

When the mucous layer is insufficient, the acidic contents of the stomach can damage the epithelium. The stomach is lined with simple columnar epithelium, so it is ill equipped to withstand harsh conditions that may lead to damage. With damage to the epithelium, underlying structures are exposed and vulnerable.

Some ulcers, if left untreated, become perforated ulcers—open sores in the stomach wall—that can lead to the leakage of stomach contents and inflammation of the structures. These ulcers must be repaired surgically.

Antacids help relieve symptoms of ulcers, such as "burning" pain and indigestion, by increasing the pH of the stomach. When ulcers are caused by NSAID use, switching to other classes of pain relievers allows healing. When caused by *H. pylori* infection, antibiotics are effective.

Anatomy of...

Gastric Hormones

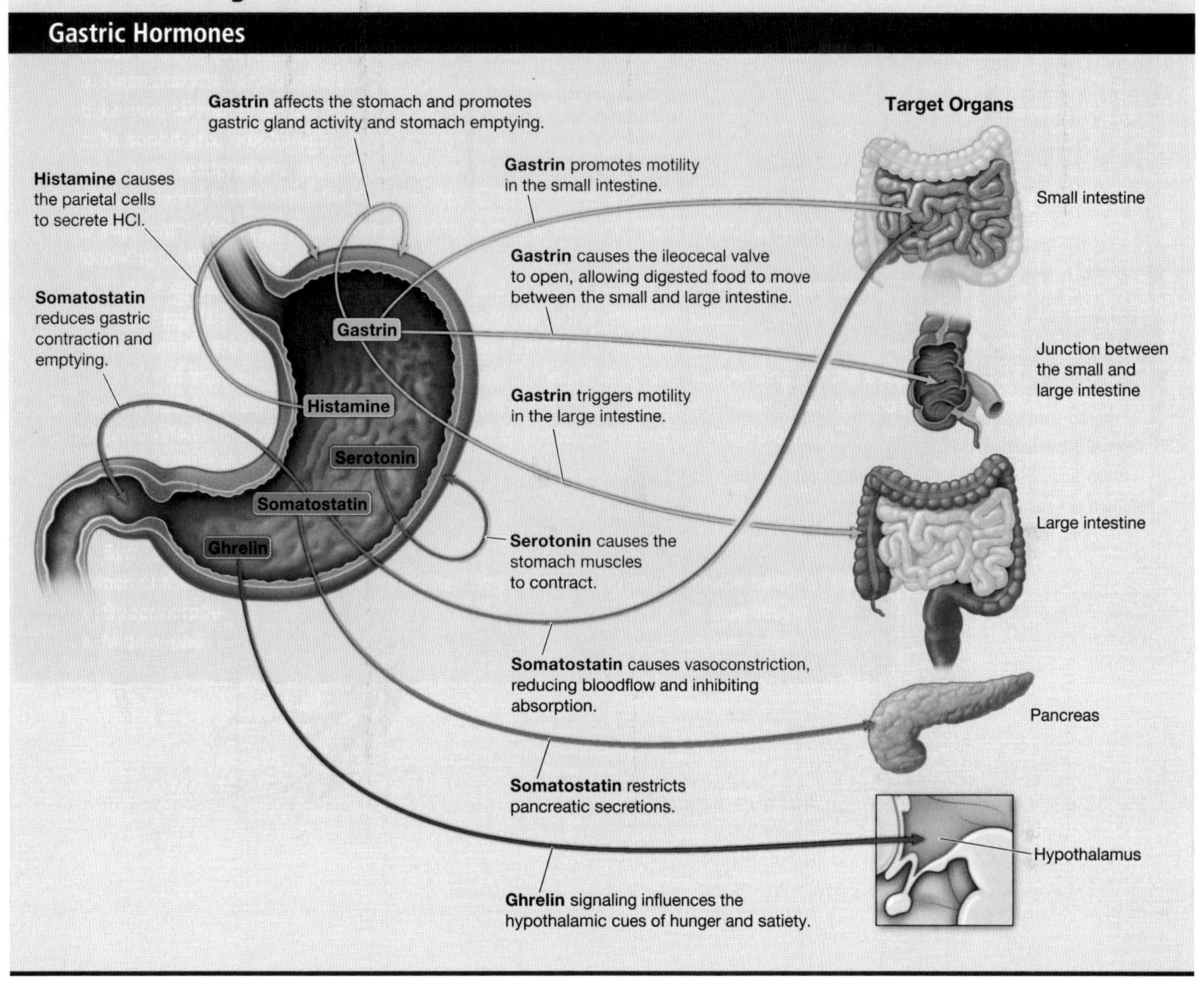

Gastrin affects the stomach and promotes gastric gland activity and stomach emptying.
Histamine causes the parietal cells to secrete HCl.
Somatostatin reduces gastric contraction and emptying.
Gastrin promotes motility in the small intestine.
Gastrin causes the ileocecal valve to open, allowing digested food to move between the small and large intestine.
Gastrin triggers motility in the large intestine.
Serotonin causes the stomach muscles to contract.
Somatostatin causes vasoconstriction, reducing bloodflow and inhibiting absorption.
Somatostatin restricts pancreatic secretions.
Ghrelin signaling influences the hypothalamic cues of hunger and satiety.
Gastrin
Histamine
Serotonin
Somatostatin
Ghrelin
Target Organs
Small intestine
Junction between the small and large intestine
Large intestine
Pancreas
Hypothalamus

23.3h Gastric Secretion

The secretion of gastric juice is controlled by both nerves and hormones. Stimuli in the brain, stomach, and small intestine activate or inhibit gastric juice production. This is why the three phases of gastric secretion are called the *cephalic* (from the head), *gastric* (from the stomach), and *intestinal* phases (**Figure 23.18**). However, once gastric secretion begins, all three phases can occur simultaneously.

The **cephalic phase** of gastric secretion, which is relatively brief, occurs as awareness or experience of food begins. The smell, taste, sight, or thought of food triggers this phase. For example, when you bring a piece of sushi to your lips, impulses from receptors in your taste buds or your nose are relayed to your brain, which sends signals along the vagus nerve toward the stomach that increase gastric secretion. Signals from the brain—for example, seeing something unappealing, a change in mood, or depression—can also suppress appetite.

cephalic phase (reflex phase)

LO 23.3.22

Learning Connection

Broken Process

Some people have to have their stomach removed, due to the presence of a benign or malignant tumor. Make a list of the components of gastric juice and their functions. Which of these components is only produced in the stomach? What effects would you expect gastrectomy to have on processes such as protein digestion and vitamin B_{12} absorption?

Figure 23.18 Regulation of Gastric Secretion

The rate of gastric secretion and motility is regulated through both neural and hormonal inputs. A variety of stimuli from the head, stomach, and intestines regulate these rates.

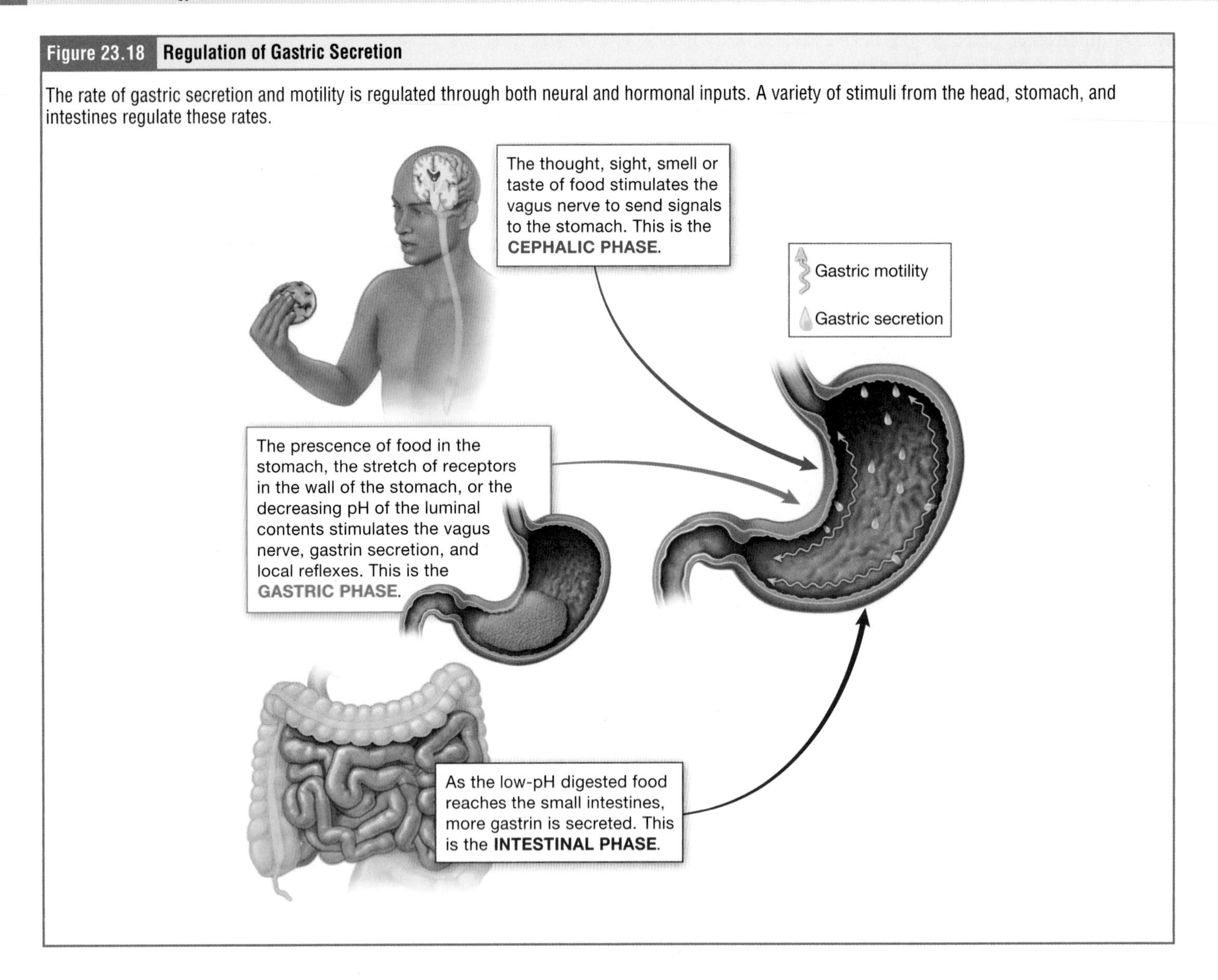

The thought, sight, smell or taste of food stimulates the vagus nerve to send signals to the stomach. This is the CEPHALIC PHASE.
Gastric motility
Gastric secretion
The prescence of food in the stomach, the stretch of receptors in the wall of the stomach, or the decreasing pH of the luminal contents stimulates the vagus nerve, gastrin secretion, and local reflexes. This is the GASTRIC PHASE.
As the low-pH digested food reaches the small intestines, more gastrin is secreted. This is the INTESTINAL PHASE.

LO 23.3.22

The **gastric phase** of secretion lasts three to four hours and is set in motion by local neural and hormonal mechanisms triggered by the entry of food into the stomach. For example, when your stomach fills with food, stretch receptors in the wall of the stomach detect the filling. This stimulates local reflexes that cause the increased secretion of gastric juice. Partially digested proteins, caffeine, and rising pH stimulate the release of gastrin, which in turn induces parietal cells to increase their production of HCl. Additionally, the release of gastrin activates vigorous smooth muscle contractions. The stomach can also inhibit acid secretion. Whenever pH levels drop too low, cells in the stomach react by suspending HCl secretion and increasing mucous secretions. These mechanisms prevent heartburn from developing.

The **intestinal phase** of gastric secretion has both excitatory and inhibitory elements. The proximal portion of the small intestine known as the *duodenum* has a major role in regulating the stomach and its emptying. When partially digested food enters the duodenum, intestinal mucosal cells release gastrin, which further excites gastric juice secretion. This stimulatory activity is brief, however, because when the intestine becomes full of partially digested food, a strong reflex—the *enterogastric* *reflex*—inhibits secretion of gastric juice and closes the pyloric sphincter, blocking additional digested food from entering the duodenum.

23.3i The Mucosal Barrier

The mucosa of the stomach is vulnerable to its constant exposure to the acidity of gastric juice. Gastric enzymes that can digest protein can also digest the stomach itself. The stomach is protected from self-digestion by the **mucosal barrier** (Figure 23.19). This barrier has several components. First, the stomach wall is covered by a thick coating of bicarbonate-rich mucus. This mucus forms a physical barrier, and its bicarbonate ions neutralize acid. Second, the epithelial cells of the stomach's mucosa are joined by tight junctions, which block gastric juice from penetrating the underlying tissue layers. Finally, stem cells located where gastric glands join the gastric pits quickly replace damaged epithelial mucosal cells, when the epithelial cells are shed. In fact, the surface epithelium of the stomach is completely replaced every three to six days.

Gastric Emptying In a process called **gastric emptying**, rhythmic waves of muscle contraction force small amounts of stomach contents, about one teaspoonful at a time, through the pyloric sphincter and into the duodenum. Release of a greater amount of material at one time would overwhelm the capacity of the small intestine to handle it.

LO 23.3.23

Gastric emptying is regulated by both the stomach and the duodenum. The arrival of low-pH stomach contents in the duodenum activates receptors that inhibit gastric secretion and motility. This prevents additional material from being released by the stomach before the duodenum is ready to process it. The contents of the stomach are completely emptied into the duodenum within two to four hours after you eat a meal. Different types of food take different amounts of time to process. Foods heavy in

Figure 23.19 The Gastric Mucosa

The mucosal lining of the stomach is protected by three mechanisms: a thick, basic mucus sits atop the epithelial cells. Tight junctions mechanically unite the cells. Basal stem cells reproduce frequently to replace the epithelial cells when they are damaged.

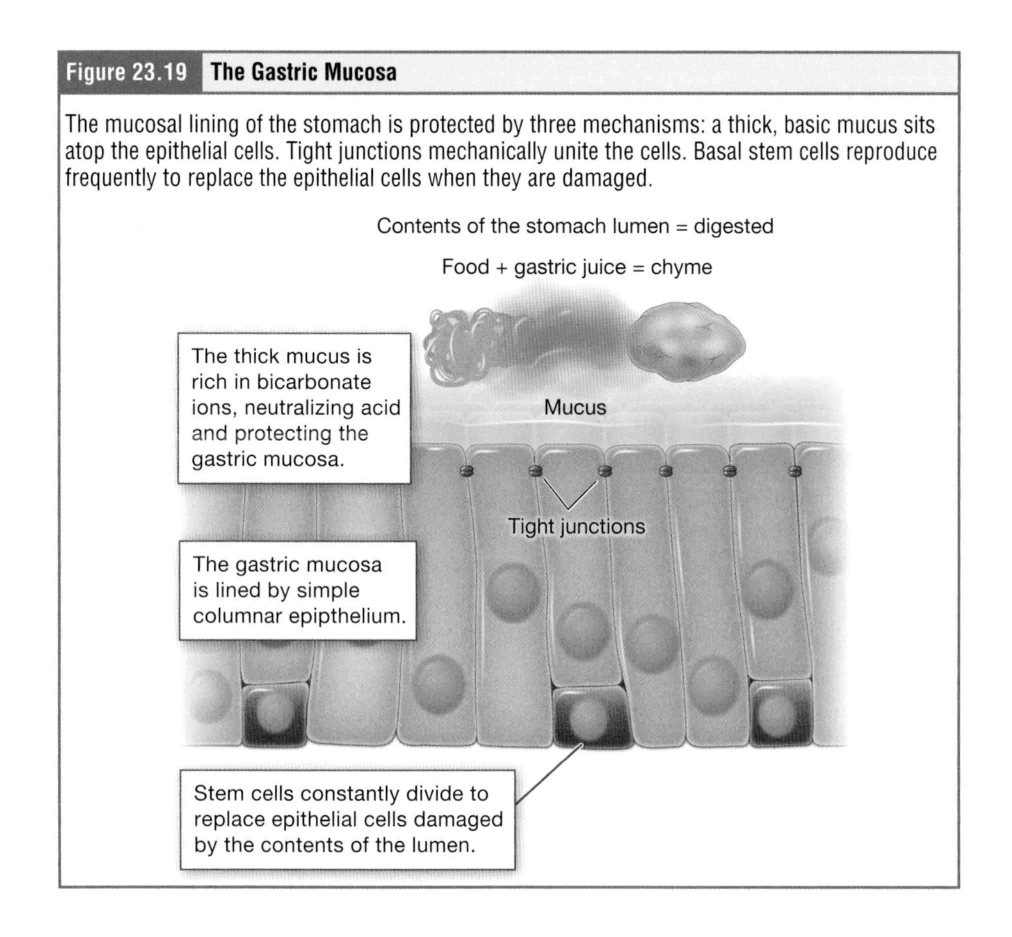

Contents of the stomach lumen = digested
Food + gastric juice = chyme
The thick mucus is rich in bicarbonate ions, neutralizing acid and protecting the gastric mucosa.
Mucus
Tight junctions
The gastric mucosa is lined by simple columnar epipthelium.
Stem cells constantly divide to replace epithelial cells damaged by the contents of the lumen.

carbohydrates empty fastest, followed by high-protein foods. Meals with a high triglyceride content remain in the stomach the longest. Since enzymes in the small intestine digest fats slowly, food can stay in the stomach for six hours or longer when the duodenum is processing fatty chyme. However, note that this is still a fraction of the 24 to 72 hours that full digestion typically takes from start to finish.

Intrinsic Factor Its numerous digestive functions notwithstanding, there is only one stomach function necessary to life: the production of intrinsic factor. The intestinal absorption of vitamin B_{12}, which is necessary for both the production of mature red blood cells and normal neurological functioning, cannot occur without intrinsic factor. Individuals who undergo stomach surgeries that remove some or all of their stomach tissue may require vitamin B_{12} injections to sustain life after the surgery.

Learning Check

1. Which layer allows the stomach to churn food and mechanically break it down into small particles?
   a. Circular muscle layer
   b. Longitudinal muscle layer
   c. Inner oblique muscle layer
2. During gastroesophageal reflux disease (GERD), stomach acid escapes into the esophagus. Damage to which of the following sphincters could be the cause of this condition?
   a. Upper esophageal sphincter
   b. Lower esophageal sphincter
   c. Pyloric sphincter
3. Which of the following cells secrete molecules important for protein digestion? Please select all that apply.
   a. Parietal cells
   b. Chief cells
   c. Mucous neck cells
   d. Enteroendocrine cells
4. Which phase of gastric secretion is triggered by stretch receptors of the stomach?
   a. Cephalic phase
   b. Gastric phase
   c. Intestinal phase

23.3j The Small and Large Intestines

The large and small intestines are two connected organs that together nearly fill the interior of the abdominal cavity. Also referred to as the guts, or the large and small bowels, these organs are extremely long. They have distinct functions and distinct structures. The small intestine is named "small" because of its smaller diameter compared with that of the much wider large intestine. The small intestine, however, is much longer—approximately five times the length of the large intestine. They are connected at the ileocecal valve, through which food passes from the small intestine into the large intestine.

23.3k The Small Intestine

LO 23.3.24

Chyme released from the stomach enters the **small intestine**, which is the primary digestive organ in the body. Not only is this where most breakdown of food molecules occurs, it is also where practically all absorption occurs. Food travels within the small intestine for a long time; in a living person, the small intestine is a long, looping tube

Figure 23.20 Fick's Law

The rate of diffusion is proportional to surface area, concentration gradient, and membrane permeability. It is inversely proportional to membrane thickness.

$$\text{Rate of diffusion} \propto \frac{\text{Surface area} \times \text{Concentration gradient} \times \text{Membrane permeability}}{\text{Membrane thickness}}$$

that can range from 10 feet to as much as 25 feet in length (in a cadaver, due to the loss of muscle tone, it can be nearly twice as long). As we'll see shortly, in addition to its length, the folds and projections of the lining of the small intestine work to give it an enormous surface area, which is approximately 200 square meters, more than 100 times the surface area of your skin. That's roughly the same surface area as half of a basketball court folded and curled and coiled to fit into your abdomen. As Fick's law tells us, surface area is proportional to the rate of diffusion, so this vast amount of surface area enables the absorption of nutrients that occurs within the small intestine (**Figure 23.20**).

Student Study Tip

You can remember the order of the small intestine segments by the mnemonic "Dow Jones Industrial."

LO 23.3.25

Structure The coiled tube of the small intestine is subdivided into three regions. From proximal (at the stomach) to distal, these are the duodenum, the jejunum, and the ileum (**Figure 23.21**).

LO 23.3.26

The shortest region of the small intestine is the **duodenum**, which begins at the pyloric sphincter and stretches for only about 10 inches in most people (but ranges from 8 to 12 inches). Just past the pyloric sphincter, it bends posteriorly behind the peritoneum, becoming retroperitoneal, and then makes a C-shaped curve around the head of the pancreas before ascending anteriorly again to return to the peritoneal cavity and join the jejunum.

Figure 23.21 The Segments of the Small Intestine

From proximal to distal, the three segments of the small intestine are the duodenum, the jejunum, and the ileum.

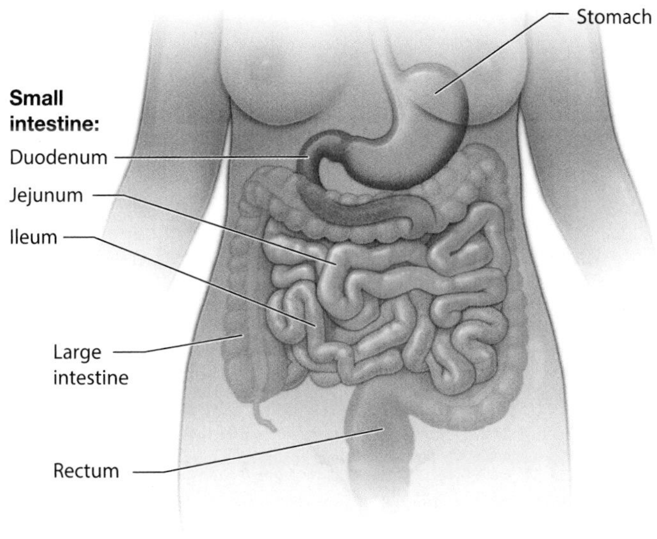

hepatopancreatic ampulla (ampulla of Vater; hepatopancreatic duct)

LO 23.3.27

LO 23.3.28

LO 23.3.29

The duodenum is the shortest of the small intestine segments, but it is the most complex both anatomically and physiologically. The duodenum is the site of the majority of chemical digestion, and this occurs with the secretions of the gallbladder and the pancreas. Therefore, all three of these organs are anatomically connected and their network of connected ducts are collectively referred to as the **biliary apparatus** (**Figure 23.22**). The biliary apparatus meets the duodenum at the **hepatopancreatic ampulla**. Located on the outer surface the duodenal wall, the ampulla is the point at which both the ducts from the gallbladder and pancreas bring their contents into the lumen of the duodenum. The **hepatopancreatic sphincter** (*sphincter of Oddi*) regulates the flow of both bile and pancreatic juice from the ampulla into the duodenum. From within the lumen of the duodenum, there is a small bump that marks the connection to the biliary apparatus; this is the **duodenal papilla**. The contribution to digestion made by the liver and gallbladder is a greenish substance called *bile*. Bile is produced in the liver, where two ducts—the left and right **hepatic ducts**—carry bile inferiorly toward the duodenum. These two ducts converge to form the **common hepatic duct**. The common hepatic duct carries bile to the **common bile duct**, which empties directly into the ampulla and then into the duodenum. Along its length, the common hepatic duct meets the cystic duct, which can carry excess bile to the *gallbladder*. The gallbladder stores and concentrates the bile so that a large quantity can be used when needed. Bile is important in breaking up fats from a meal; we will discuss this process in more length in the next section. The pancreas contributes a mixture of enzymes, called *pancreatic juice*, for digestion. The pancreatic juice is transported to the ampulla in the *main pancreatic duct*.

common bile duct (bile duct)

hepatopancreatic sphincter (sphincter of Oddi; Glisson's sphincter)

Figure 23.22 The Accessory Organs of Digestion

The liver and gallbladder contribute bile to the duodenum. The head of the pancreas is nestled into the curve of the duodenum, and the exocrine secretions of the pancreas are added to the duodenum. Secretions from all three accessory organs enter the duodenum at the ampulla.

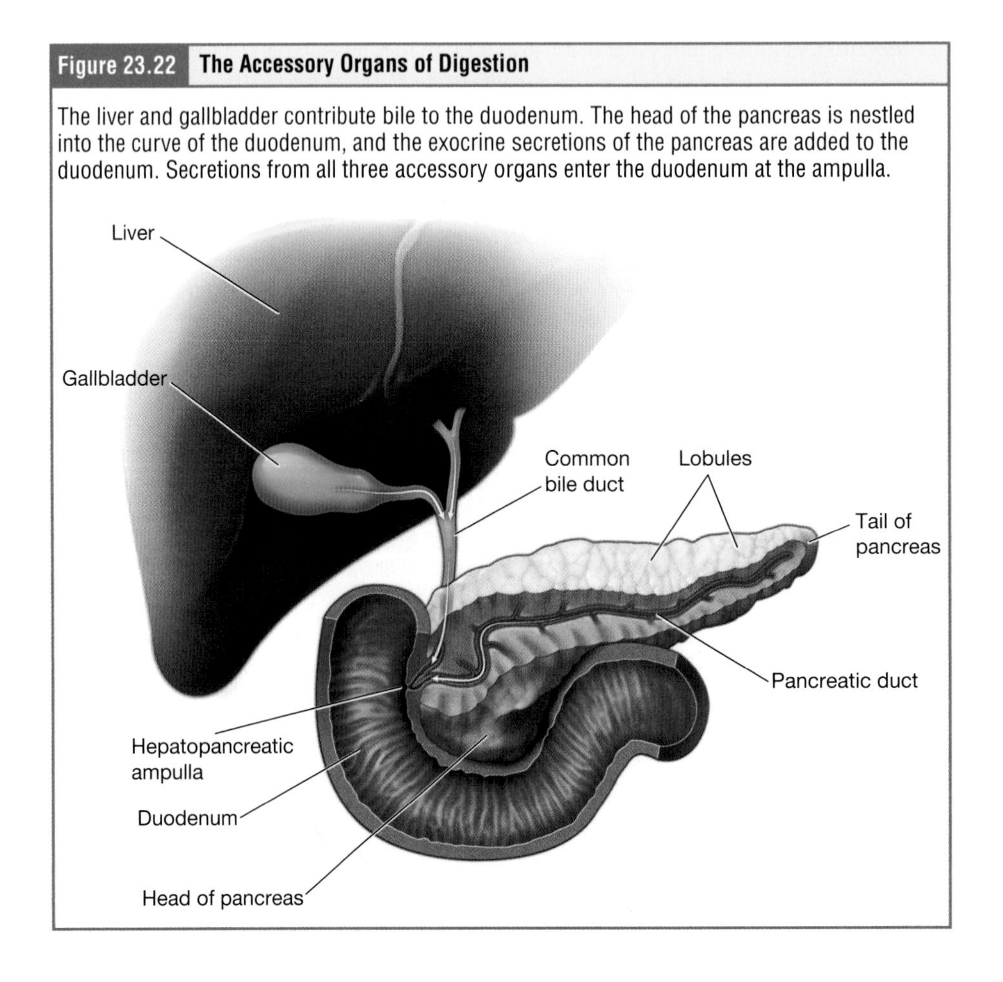

The duodenum ends at its connection to the **jejunum**, the middle segment of the small intestine. The jejunum is approximately eight feet in length in most people, but ranges from four to ten feet. No clear border separates the jejunum from the duodenum or the ileum—the final segment of the small intestine.

The **ileum** is the longest part of the small intestine, measuring about 10 feet in length in most people, but ranging from 6 to 12 feet. It is thicker, more vascular, and has more developed mucosal folds than the jejunum. The ileum joins the cecum—the first portion of the large intestine—at the **ileocecal sphincter** (Figure 23.23). The jejunum and ileum are tethered to the posterior abdominal wall by the mesentery. The large intestine frames these three parts of the small intestine.

ileocecal sphincter (ileocecal valve)

Histology The wall of the small intestine is composed of the same four layers present throughout the organs of the GI tract (the mucosa, submucosa, muscularis, and serosa). However, three features of the mucosa and submucosa—circular folds, villi, and microvilli—are unique (Figure 23.24). Collectively, these features increase the absorptive surface area of the small intestine more than 600-fold. These adaptations are most abundant in the proximal two-thirds of the small intestine, where the majority of absorption occurs.

LO 23.3.30

- ***Circular Folds.*** A **circular fold** is a deep ridge in the mucosa and submucosa. The folds are found in the duodenum and first half of the ileum. They expand the surface area and therefore increase the amount of nutrient absorption possible. Their large size and shape slows the movement of chyme and provides the time needed for nutrients to be fully absorbed.
- ***Villi.*** The surface of the circular folds is not smooth; rather, it is covered in tiny hairlike projections called **villi** (singular = villus) that give the mucosa a furry texture. Each square millimeter of epithelium contains 20 to 40 villi; these greatly

LO 23.3.31

circular fold (plica circularis; valves of Kerckring, valvulae conniventes)

Figure 23.23 The Ileocecal Junction

The ileum—the terminal segment of the small intestine—meets the large intestine at a T-shaped junction called the *ileocecal junction*. Here, a muscular sphincter controls the flow of chyme from the small intestine into the large intestine.

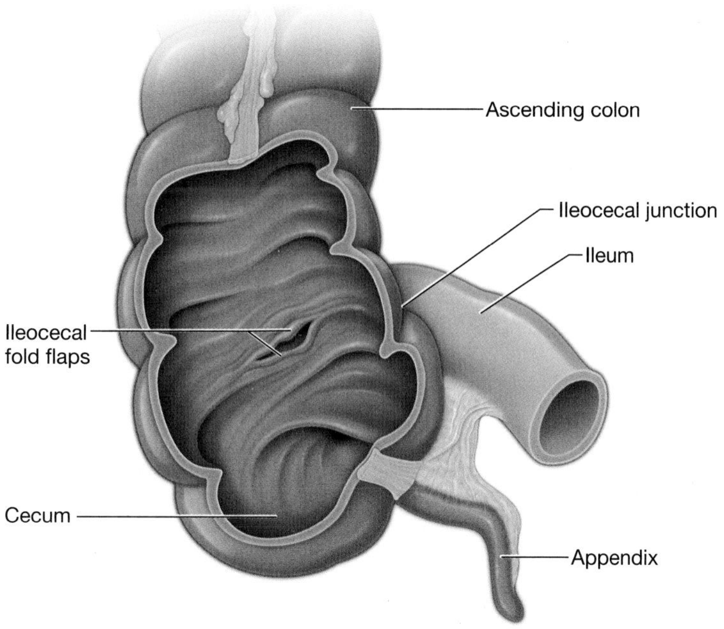

Ascending colon
Ileocecal junction
Ileum
Ileocecal fold flaps
Cecum
Appendix

Figure 23.24 Histology of the Small Intestine

The surface area of the small intestine is 600 times greater than it would be if the internal surface were a smooth tube. Instead, (A) circular folds, (B) villi, and (C) microvilli greatly expand the surface area.

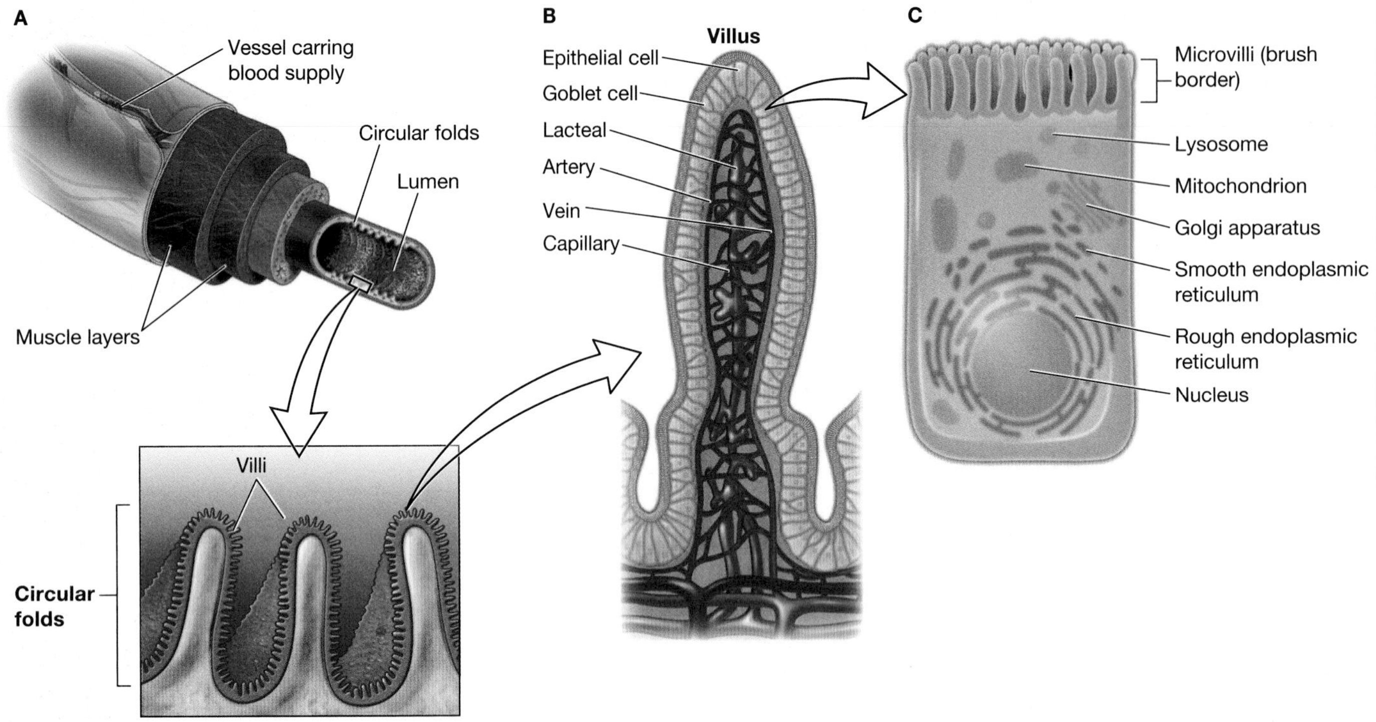

increase the available surface area. Each villus is a projection of both mucosa and submucosa and contains a capillary bed along with its arteriole and venule, as well as a lymphatic capillary called a **lacteal**. As food is digested, the breakdown products are transported across the epithelium of the villi, and then absorbed into the blood or lymph to be transported to the cells of the body. Sugars and amino acids enter the bloodstream directly, but lipid breakdown products are absorbed by the lacteals and transported to the bloodstream via the lymphatic system.

- ***Microvilli.*** As their name suggests, **microvilli** (singular = microvillus) are much smaller than villi. They are fingerlike projections of just the apical membrane of the mucosa's epithelial cells. They are tiny and numerous, like the bristles of a hairbrush, and are collectively termed the **brush border**. Fixed to the surface of the microvilli membranes are enzymes that finish digesting carbohydrates and proteins. Each square millimeter of small intestine contains an estimated 200 million microvilli; this greatly expands the surface area of the plasma membrane, and thus greatly enhances absorption.
- ***Intestinal Glands.*** In addition to the three specialized absorptive features just discussed, the mucosa between the villi is dotted with deep crevices that each lead into an **intestinal gland**, which is formed by cells that line the crevices (see Figure 23.24). These produce **intestinal juice**, a slightly alkaline (pH 7.4–7.8) mixture of water and mucus. Each day, about 0.95–1.9 liters (1–2 quarts) of intestinal juice are secreted to help neutralize the chyme and protect the mucosal cells.

The submucosa of the duodenum is the only site of the complex mucus-secreting **duodenal glands**, which produce an alkaline mucus that buffers the acidic chyme as it enters from the stomach.

brush border (striated border; brush border membrane)

LO 23.3.32

intestinal gland (crypt of Lieberkühn; intestinal crypt)

intestinal juice (succus entericus)

duodenal glands (Brunner's gland)

- ***Intestinal MALT.*** The lamina propria of the small intestine mucosa is studded with quite a bit of MALT. In addition to solitary lymphatic nodules, aggregations of intestinal MALT are concentrated in the distal ileum. These collections of immune cells and tissue serve dual purposes. They stand sentinel to prevent bacteria and other pathogens, which are plentiful in the GI lumen, from entering the bloodstream. They also monitor the digested food molecules, establishing and maintaining immune tolerance to the food we eat. Intestinal MALT nodules are most prominent in young people and become smaller and less numerous as we age, which coincides with the general activity of our immune system.

Mechanical Digestion in the Small Intestine The contractions of intestinal smooth muscles are organized into two motility patterns. Peristalsis in the small intestine, as with other organs, squeezes in a rhythmic pattern to move the digested food distally along the GI tract. **Segmentation** is a back-and-forth mixing pattern, as the rings of smooth muscle repeatedly contract and then relax (**Figure 23.25**). Segmentation does not propel the contents forward along the track but serves to mix the chyme with digestive enzymes.

LO 23.3.33

At the end of the small intestine, the majority of the nutrients have been absorbed and there is far less mass within the lumen. The contents now are a loose mix of unabsorbable or unbroken-down food and water. The ileocecal sphincter is usually in a constricted state, but when motility in the ileum increases, this sphincter relaxes, allowing food residue to enter the first portion of the large intestine, the *cecum*. Relaxation of the ileocecal sphincter is controlled by both nerves and hormones. First, digestive activity in the stomach provokes the **gastroileal reflex**, which increases the force of contractions within the ileum. Second, the stomach releases the hormone gastrin, which enhances ileal motility, thus relaxing the ileocecal sphincter. After chyme passes through, backward pressure helps close the sphincter, preventing backflow into the ileum. Because of this reflex, your lunch is completely emptied from your stomach and small intestine by the time you eat your dinner. It takes about three to five hours for all chyme to leave the small intestine.

Figure 23.25 Segmentation

Segmentation describes contractions of the muscularis that mix chyme together rather than propel it forward along the intestine.

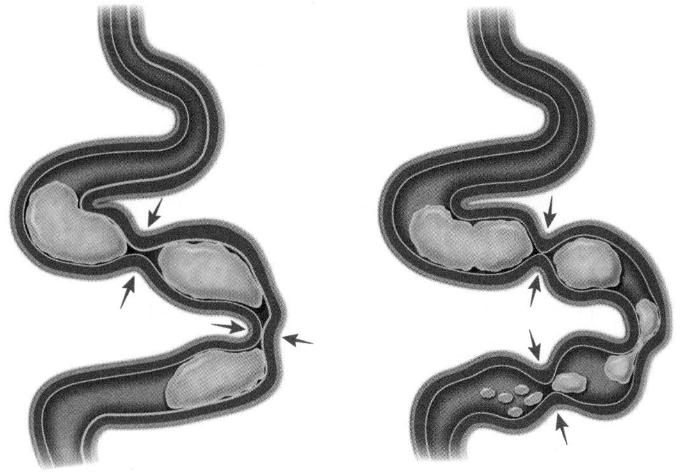

Learning Check

1. Which of the following structures carries excess bile to the gallbladder?
   a. Left hepatic duct
   b. Right hepatic duct
   c. Common hepatic duct
   d. Cystic duct
2. Explain how the villi in the small intestine are different than the rugae in the stomach.
   a. The villi are highly folded while the rugae are smooth.
   b. The villi transport food while the rugae aid in mechanical digestion.
   c. The villi increase surface area while the rugae increase space.
   d. The villi release gastric acid while the rugae absorb nutrients.
3. Which of the following describes the method of mechanical digestion found in the small intestine?
   a. Segmentation
   b. Peristalsis
   c. Propulsion
   d. Haustral contraction

23.3l The Large Intestine

LO 23.3.34

large intestine (colon)

The **large intestine** finishes absorption of nutrients and water, and forms feces. Certain vitamins are synthesized in the large intestine as well. The large intestine of each human body contains trillions of bacteria, representing one of the richest and most diverse bacterial ecosystems on Earth. There are hundreds of different bacterial species, most of which are nonpathogenic, living in balance and competition for the nutritional and space resources of the colon. Many species facilitate chemical digestion and absorption, and some synthesize certain vitamins, including vitamin K. Some are linked to increased immune response. Most bacteria that enter the GI tract are killed by lysozyme, defensins, HCl, or protein-digesting enzymes. The trillions of bacteria that live within the large intestine are referred to as the **microbiota**.

feces (stool)

The small intestine absorbs about 90 percent of the water you ingest. The large intestine absorbs most of the remaining water, a process that converts the liquid chyme residue into semisolid **feces** (also known as *stool*). Feces are composed of undigested food residues, unabsorbed digested substances, millions of bacteria, old epithelial cells from the GI mucosa, inorganic salts, and enough water to let it pass smoothly out of the body. Of every 500 milliliters (17 ounces) of food residue that enters the cecum each day, about 150 milliliters (5 ounces) becomes feces.

LO 23.3.35

Structure The large intestine frames the small intestine on three sides. It is shorter—about five feet in length—than the small intestine, but it is called "large" because it is about three inches wide, more than twice the diameter of the small intestine.

LO 23.3.36

Subdivisions The large intestine is subdivided into four main regions: the cecum, the colon, the rectum, and the anus (**Figure 23.26A**). The colon portion is further divided into the ascending, transverse, descending, and sigmoid colon.

cecum (caecum)

appendix (vermiform appendix)

Student Study Tip

The appendix is like a piñata. When it bursts, bacteria-candy is released.

- *Cecum.* The first part of the large intestine is the **cecum,** a small sac that hangs down (inferiorly) to the ileocecal valve. The cecum is about two to three inches long and ends in the **appendix**, a wormlike ribbon of tissue that is suspended off the end of the cecum. The appendix contains lymphoid tissue and is packed with bacteria. It is thought that the appendix serves as a bacterial reservoir and a small cohort of every bacterial population in the colon is represented here. The bacteria of the colon, often referred to as our *microbiome*, are actually incredibly important to human health. The population of bacteria in the appendix serve to reseed the

Figure 23.26 The Large Intestine

(A) The regions of the large intestine are the cecum, the ascending colon, the transverse colon, the descending colon, the sigmoid colon, the rectum, and the anal canal. (B) The bends in the large intestine are called *flexures.*

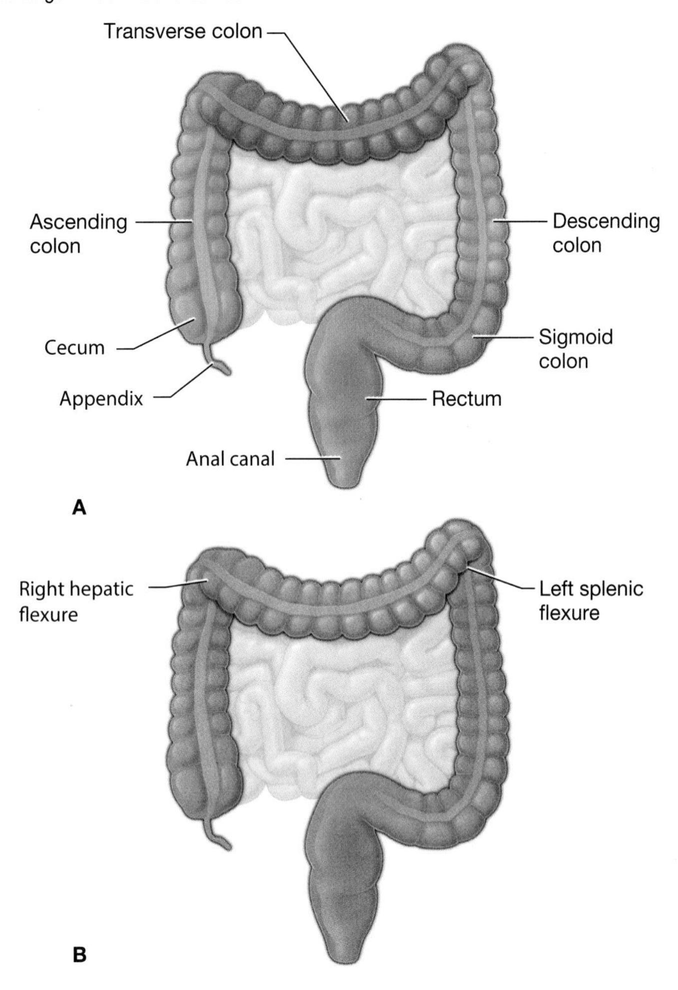

colon if bacteria are lost due to diarrheal illness or other events such as taking an antibiotic medication.

- ***Colon.*** Superiorly, the cecum blends seamlessly with the **colon**. Upon entering the colon, the digested food remnants first travel up the **ascending colon** on the right side of the abdomen. At the inferior surface of the liver, the colon bends to form the **right colic flexure** and becomes the **transverse colon**. The digested material moves through the transverse colon across to the left side of the abdomen, where the colon angles sharply immediately inferior to the spleen at the **left colic flexure**. From there, the **descending colon** runs down the left side of the posterior abdominal wall. Within the pelvis, the colon becomes the s-shaped **sigmoid colon** (see Figure 23.26B). By the time the digested material reaches the end of the sigmoid colon, everything has been absorbed from it and the material is now feces, which is stored in the rectum until it can be eliminated from the body. The ascending colon, the descending colon, and the rectum are retroperitoneal. The transverse and sigmoid colon are tethered to the posterior abdominal wall by the mesocolon.
- ***Rectum.*** The rectum extends anterior to the sacrum and coccyx. The rectum functions to hold feces. Three horizontal folds called the **rectal valves** help separate the feces from gas to prevent the simultaneous passage of feces and gas. The **anal**

colon (large intestine)

right colic flexure (hepatic flexure)

left colic flexure (splenic flexure)

internal anal sphincter (sphincter ani internus)

rectal valves (anal valves)

Figure 23.27 The Rectum and Anus

The rectum functions to hold feces while waiting to excrete them. The anus is gated by two sphincters—an internal smooth muscle sphincter and an external skeletal muscle sphincter.

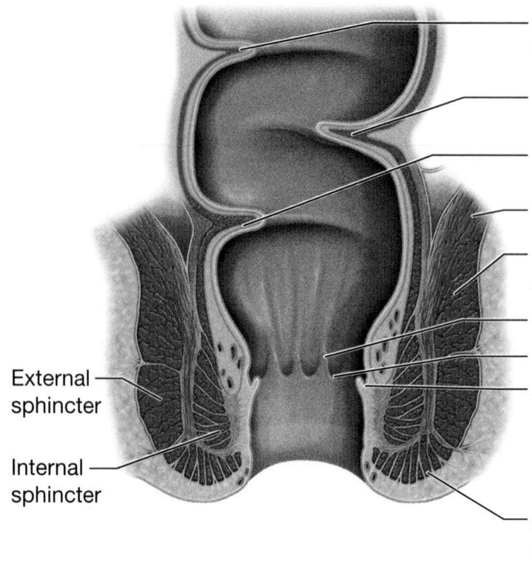

external anal sphincter (sphincter ani externus)

anal sinus (rectal sinus)

canal is lined with stratified squamous epithelium to protect it from the friction of dry fecal material. It is also open to the external environment. The anal canal includes two sphincters (**Figure 23.27**). The **internal anal sphincter** is made of smooth muscle, and its contractions are involuntary. The **external anal sphincter** is made of skeletal muscle, which is under voluntary control. Control over the external anal sphincter is not usually fully developed until 2–3 years of age, which is why infants do not have control over when they defecate. Except when defecating, both usually remain closed.

LO 23.3.37

anal column (Columns of Morgagni; Morgagni's columns)

The anal canal's mucous membrane is organized into longitudinal folds, each called an **anal column**, which house a grid of arteries and veins. Depressions between the anal columns, each called an **anal sinus**, secrete mucus that facilitates defecation.

LO 23.3.38

LO 23.3.39

Histology There are several histological differences between the walls of the large and small intestines (**Figure 23.28**). For example, few enzyme-secreting cells are found in the wall of the large intestine, and there are no circular folds or villi. Remember that these surface-area expanding structures are important for absorption. Since the majority of nutrients are absorbed in the small intestine, they are not present in the large intestine. Other than in the anal canal, the mucosa of both the small and large intestines is simple columnar epithelium made mostly of enterocytes (epithelial cells) and goblet cells. The frequency of goblet cells increases distally; there are many more in the wall of the large intestine. The large intestine also has far more intestinal glands. The glands and goblet cells secrete mucus that eases the movement of feces and protects the intestine from the effects of the acids and gasses produced by bacteria. Water, salts, and vitamins are absorbed across the gut wall.

Learning Connection

Chunking

Mucus is a substance that has many functions. In the respiratory system, for example, it is used to trap and remove pathogens and debris. In the digestive system, it is used for lubrication, to help in the sense of taste, for buffering acid, and to protect organs from the effects of their own secretions. Make a list of the major organs of the GI tract, and add the salivary glands and the pancreas. Then list the function(s) of mucus in each organ.

Anatomy Three features—teniae coli, haustra, and epiploic appendages—are unique to the large intestine (**Figure 23.29**). The **teniae coli** are three bands of smooth muscle that make up the longitudinal muscle layer of the muscularis of the large intestine. Steady contractions of the teniae coli bunch up the colon into a succession of pouches called **haustra** (singular = haustrum), which are responsible for the wrinkled appearance of

Figure 23.28 The Wall of the Large Intestine

The mucosa of the large intestine is smooth with deep pits called *intestinal glands*. The epithelium is heavily studded with goblet cells, which secrete mucus to ease friction.

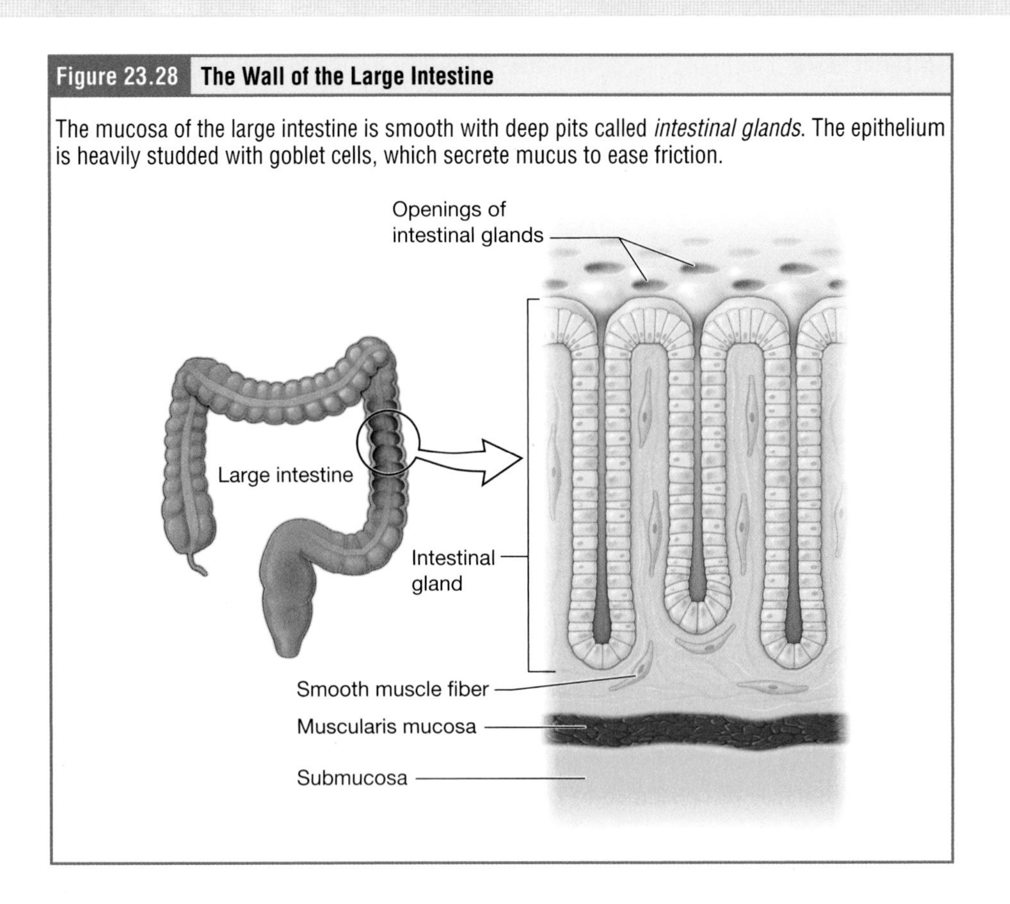

Figure 23.29 Features of the Large Intestine

The external surface of the large intestine is compressed into a series of pouches (haustra), due to the steady contraction of the muscularis muscles within its wall.

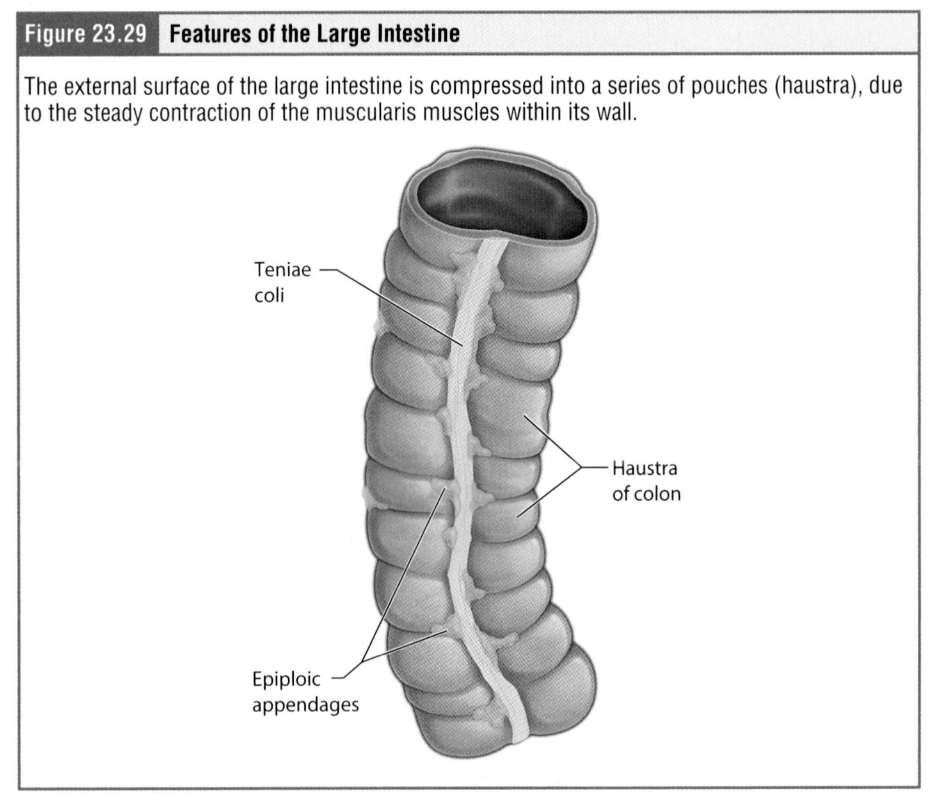

the colon. Attached to the teniae coli are small, fat-filled sacs called **epiploic appendages**. The purpose of these is unknown. Although the rectum and anal canal have neither teniae coli nor haustra, they do have well-developed layers of muscularis that create the strong contractions needed for defecation.

epiploic appendages (appendix epiploica)

Bacterial Flora The human large intestine is home to over 100 trillion bacteria cells. These usually belong to around 300–400 different species. To best imagine this whopping number, picture a small area of a forest. In that area, you may find lots of organisms, from beetles to deer. All of these organisms require both space and nutrients. If the forest changes in some way—for example, if a forest fire eliminates most of the plants—then the beetles and deer and everything in between will be affected. What if just one species changes—for example, one type of fern is wiped out? The types of organisms that depend on that fern might also die out, leaving more space for other organisms to thrive. And so, the relative forest of our large intestines is much the same. The exact mosaic of bacteria present in you is as unique to you as your fingerprint! It is a product of your surroundings and your diet. If you change your diet—for example, become a vegetarian or eliminate processed sugar—the species that depended on those nutrients will die off, making space for other species to thrive. A severe diarrheal illness is analogous to a forest fire; it may wipe out many of the species and provide space for rebuilding a new forest. The bacterial cells of the microbiome are feeding on whatever nutrients are still present in the digested food of the microbiome. Remember that the majority of chemical digestion takes place in the mouth, the stomach, and the duodenum. Therefore, undigested nutrients that make it to the large intestine are largely foods that we do not produce the enzymes to digest. These foods, often collectively called *fiber*, may be broken down by bacteria that do make the enzymes that our human cells do not. The released products of their digestion may be consumed by the bacteria, or they may be transported across the gut wall for our own use. Therefore, because no two humans have the same gut microbiome, no two humans will extract the exact same profile of nutrients from their foods. These breakdown products may also act as signaling molecules, subtly changing our physiological state. Researchers hypothesize that the gut microbiome may influence physiological conditions such as diabetes mellitus, colon cancer, obesity, and even anxiety.

Cultural Connection

Feces as Medicine

One of the advantages of the microbiome is that, since it is filled with bacteria that are competing for the available nutrients, there are rarely leftover nutrients to support additional visiting bacteria. People who have thriving microbiomes are often more resistant to diarrheal illnesses because invading bacteria that may have entered the GI tract along with the meal cannot gain a foothold in the large intestine. There simply isn't enough space or nutrients to support their growth. Not everyone has a robust microbiome, so some individuals are more susceptible than others. Most humans already have a bacterial species living inside their microbiome that is capable of making them very sick. This species, *Clostridioides difficile*, or *C. difficile*, is alive in small numbers in most of us. *C. difficile* cannot compete well with other species and so does not usually grow in large enough numbers to make us sick. When other species are eliminated by diarrheal illness or a course of antibiotics, *C. difficile*, which is resistant to most antibiotics, thrives. *C. difficile* causes severe diarrhea and, because it is resistant to most antibiotics, may be hard to treat. Since competition with other species is typically what keeps *C. difficile* at bay most of the time, one treatment for *C. difficile* infection is to introduce more competition. How? Through a procedure known as *fecal transplant*. In this procedure, donated fecal material is introduced into the patient's colon either through a liquid injection or a capsule. The bacteria from the donated material colonize the colon and—hopefully—outcompete the *C. difficile* for nutrients, limiting *C. difficile* growth and bringing the infection under control. Fecal transplants are incredibly successful for *C. difficile* infection and are being studied for treatment of other diseases, including ulcerative colitis, Crohn's disease, multiple sclerosis, depression, obesity, diabetes, and food allergies.

Motility in the Large Intestine In the large intestine, there are three types of muscle contractions. The first type, peristalsis, is relatively weak here, and because the material is getting less watery as it progresses through the large intestine, more forceful contractions are required. The second type, **haustral contractions**, occur when a haustrum is distended with chyme. The stretch in the wall of that haustrum triggers its muscle to contract, pushing the residue into the next haustrum. These movements also mix the food residue, which helps the large intestine absorb water. The third type is a **mass movement**. These strong waves start midway through the transverse colon and quickly force the contents toward the rectum. Mass movements usually occur three or four times per day, either while you eat or immediately afterward. Distension in the stomach and the breakdown products of digestion in the small intestine provoke the **gastrocolic reflex**, which increases motility, including mass movements, in the colon.

gastrocolic reflex (gastrocolic response)

Feces are eliminated through contractions of the rectal muscles. You help this process by a voluntary procedure called **Valsalva's maneuver**, in which you increase intra-abdominal pressure by contracting your diaphragm and abdominal wall muscles and closing your glottis.

flatus (gas)

DIGGING DEEPER: Gas and Defecation

Although the glands of the large intestine secrete mucus, they do not secrete digestive enzymes. Therefore, chemical digestion in the large intestine occurs exclusively because of bacteria in the lumen of the colon. Through the process of **saccharolytic fermentation**, bacteria break down some of the remaining carbohydrates. This results in the discharge of hydrogen, carbon dioxide, and methane gasses that create **flatus** (gas) in the colon; flatulence is excessive flatus. Each day, up to 1500 milliliters of flatus is produced in the colon. More is produced when you eat foods such as beans, which are rich in otherwise indigestible sugars and complex carbohydrates such as soluble dietary fiber.

The process of defecation begins when mass movements force feces from the colon into the rectum, stretching the rectal wall and provoking the defecation reflex, which eliminates feces from the rectum. This parasympathetic reflex is mediated by the spinal cord. It contracts the sigmoid colon and rectum, relaxes the internal anal sphincter, and initially contracts the external anal sphincter. The presence of feces in the anal canal sends a signal to the brain, which gives you the choice of voluntarily opening the external anal sphincter (defecating) or keeping it temporarily closed. If you decide to delay defecation, it takes a few seconds for the reflex contractions to stop and the rectal walls to relax. The next mass movement will trigger additional defecation reflexes until you defecate.

If defecation is delayed for an extended time, additional water is absorbed, making the feces firmer and potentially leading to constipation. On the other hand, if the waste matter moves too quickly through the intestines, not enough water is absorbed, and diarrhea can result. This can be caused by the ingestion of foodborne pathogens. Diarrhea can also result when the GI tract is unable to fully break down some of the food, changing the osmotic gradient that allows for water absorption.

DIGGING DEEPER: Colorectal Cancer

Each year, approximately 140,000 Americans are diagnosed with colorectal cancer, and another 49,000 die from it, making it one of the deadliest malignancies. People with a family history of colorectal cancer are at increased risk. Smoking, excessive alcohol consumption, and a diet high in animal fat and protein also increase the risk. Despite popular opinion to the contrary, studies support the conclusion that dietary fiber and calcium do not reduce the risk of colorectal cancer.

Colorectal cancer may be signaled by constipation or diarrhea, cramping, abdominal pain, and rectal bleeding. Bleeding from the rectum may be either obvious or occult (hidden in the feces). Since most colon cancers arise from benign mucosal growths called *polyps*, cancer prevention is focused on identifying these polyps. A procedure called a *colonoscopy* is both diagnostic and therapeutic. Colonoscopy allows not only identification of precancerous polyps, but also their removal before they become malignant. Screening for fecal occult blood tests and colonoscopy is recommended for those over 50 years of age.

Learning Check

1. What are the functions of the microbiota in the large intestine? Please select all that apply.
   a. To absorb water from the feces
   b. To help us digest our food
   c. To facilitate mechanical digestion
   d. To synthesize certain vitamins
2. A colonoscopy is a procedure in which a doctor examines the colon using a flexible instrument. It is entered through the anus and travels proximally. If the microscope is currently in the transverse colon, to which of the following regions would the microscope go next?
   a. Left colic flexure
   b. Sigmoid colon
   c. Descending colon
   d. Right colic flexure
3. All of the following methods of motility are utilized in the colon except
   a. Peristalsis
   b. Mass movement
   c. Churning
   d. Haustral contractions

23.4 Accessory Organs in Digestion: The Liver, Pancreas, and Gallbladder

Learning Objectives: By the end of this section, you will be able to:

23.4.1 Describe the general functions of the liver.

23.4.2 Describe the location of the liver relative to other body structures.

23.4.3 Identify and describe the structure of the liver, including the individual lobes, ligaments (e.g., coronary ligament, falciform ligament, round ligament [ligamentum teres]), and the porta hepatis.

23.4.4 Identify and describe the histological components of the classic hepatic lobule.

23.4.5 Define emulsification, and explain how and where bile salts facilitate fat digestion.

23.4.6 Identify and describe the structure and functions of the gallbladder.

23.4.7 Describe the location of the gallbladder relative to other body structures.

23.4.8 Identify and describe the structure and functions of the pancreas.

23.4.9 Describe the location of the pancreas relative to other body structures.

23.4.10 Identify and describe the major histological components of the pancreas (pancreatic acini and pancreatic islets [islets of Langerhans]) and discuss their major functions.

23.4.11 Describe the source, stimuli for release, targets, and actions of gastrointestinal tract hormones (e.g., gastrin, cholecystokinin, secretin).

LO 23.4.1

Chemical digestion in the small intestine relies on the activities of three accessory digestive organs: the liver, the pancreas, and the gallbladder (**Figure 23.30**). The liver is a huge organ, taking up about a quarter of your abdominal cavity and comprising approximately 2 percent of your body weight. It contributes to the function of many body systems. In the digestive system, the liver has two roles. First, the liver produces bile and exports it. Some bile will travel directly to the duodenum; when there is excess bile, the gallbladder can store and concentrate it, releasing the bile later when it is

Figure 23.30 The Accessory Organs of Digestion

The liver, pancreas, and gallbladder are all digestive organs, but they do not belong to the GI tract and food does not pass through them. They contribute to the processes of digestion via their secretions.

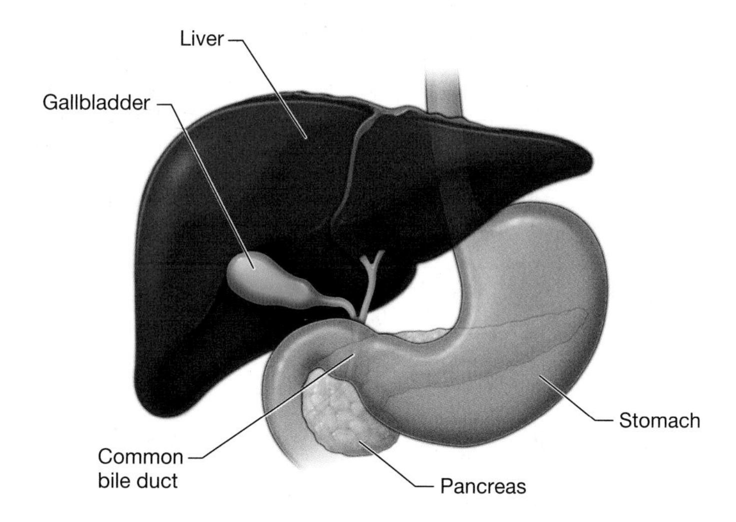

needed in digestion. The liver also filters the blood that comes from the intestines rich with absorbed nutrients. The pancreas produces pancreatic juice, which contains digestive enzymes and bicarbonate ions, and delivers it to the duodenum.

23.4a The Liver

LO 23.4.2

The **liver** plays a number of roles in metabolism and regulation of homeostasis. The liver is also the only organ in the body capable of regeneration. The liver lies inferior to the diaphragm in the right upper quadrant of the abdominal cavity. It is so large that the right kidney is slightly lower than the left to accommodate for the bulk of the liver.

LO 23.4.3

The liver is divided into two primary lobes: a large right lobe and a much smaller left lobe. In the right lobe, two smaller lobes—an inferior quadrate lobe and a posterior caudate lobe (**Figure 23.31**)—can be visualized. The liver is connected to the abdominal wall and diaphragm by five peritoneal folds referred to as ligaments. These are the *falciform ligament*, the *coronary ligament*, two *lateral ligaments*, and the *ligamentum teres hepatis*. The falciform ligament and ligamentum teres hepatis are actually remnants of the umbilical vein, and separate the right and left lobes anteriorly. The lesser omentum tethers the liver to the lesser curvature of the stomach.

Student Study Tip

You can remember which lobe of the liver is larger because it follows the same pattern as the lungs—the left lobe is much smaller.

LO 23.4.4

The **porta hepatis** (translates to "gate to the liver") is the site at which the **hepatic artery** and the **hepatic portal vein** enter the liver. These two vessels, along with the common hepatic duct, enter the liver on the inferior surface (**Figures 23.31C** and **23.31D**). The mass of the liver is organized into lobules, which are the functional units of the liver (**Figure 23.32**). The lobules typically have a hexagonal shape, and each lobule consists of rows of cells, called *hepatocytes*, interspersed with sinusoids, the specialized capillaries of the liver. Branches of the hepatic artery deliver oxygenated blood from the heart to the liver, carrying the oxygen and nutrients that the cells of the liver, **hepatocytes**, need for survival. Branches of the hepatic portal vein carry deoxygenated blood from the stomach, small intestine and large intestine;

porta hepatis (transverse fissure, hilum of the liver)

Figure 23.31 The Liver

(A) In the anterior perspective, the liver is divided into two lobes by the falciform ligament. (B) From the posterior perspective, the smaller caudate and quadrate lobes are visible. (C) The gallbladder can be seen inferiorly.

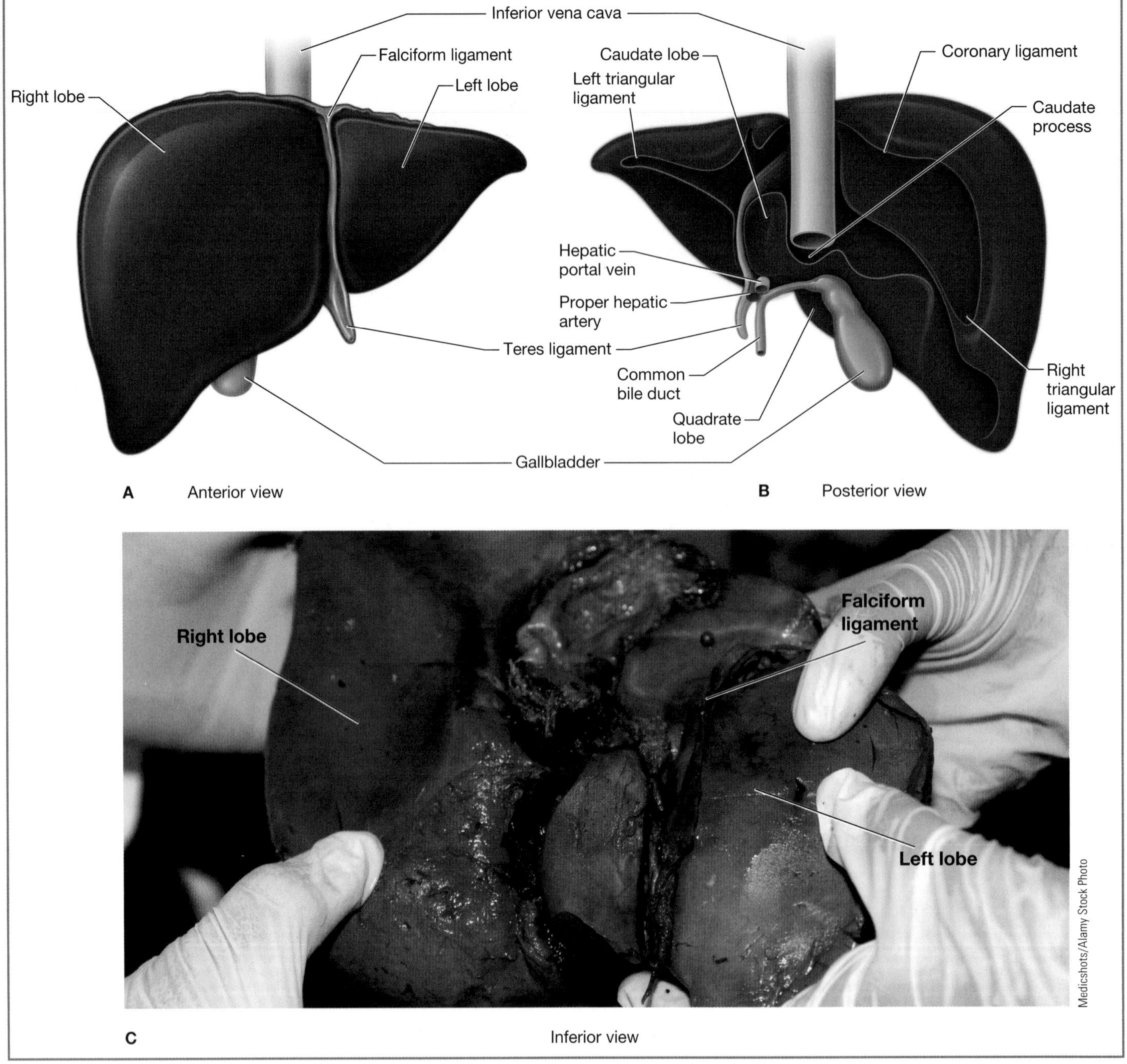

this blood carries nutrients, as well as any toxins absorbed across the gut wall, to the liver for filtration. The blood from these two sources mixes and blood is filtered from the outside of each lobule toward the lobule center. The hepatocytes of the lobule process the bloodborne nutrients and toxins, and release nutrients needed by other cells back into the blood. The blood, now oxygen-poor and filtered, drains into the central vein at the center of the lobule. Central veins converge to form the hepatic vein, which carries this venous blood to the inferior vena cava and on to the right atrium. With this hepatic portal circulation, all blood from the GI tract passes through the liver before joining the general circulation.

Figure 23.32 The Arrangement of Tissues in the Liver

Hepatocytes are organized into lobules. Each lobule filters both oxygenated and deoxygenated blood for red blood cells to recycle as well as toxins and nutrients. The hepatocytes produce bile, which is carried away from the liver lobules in bile ducts.

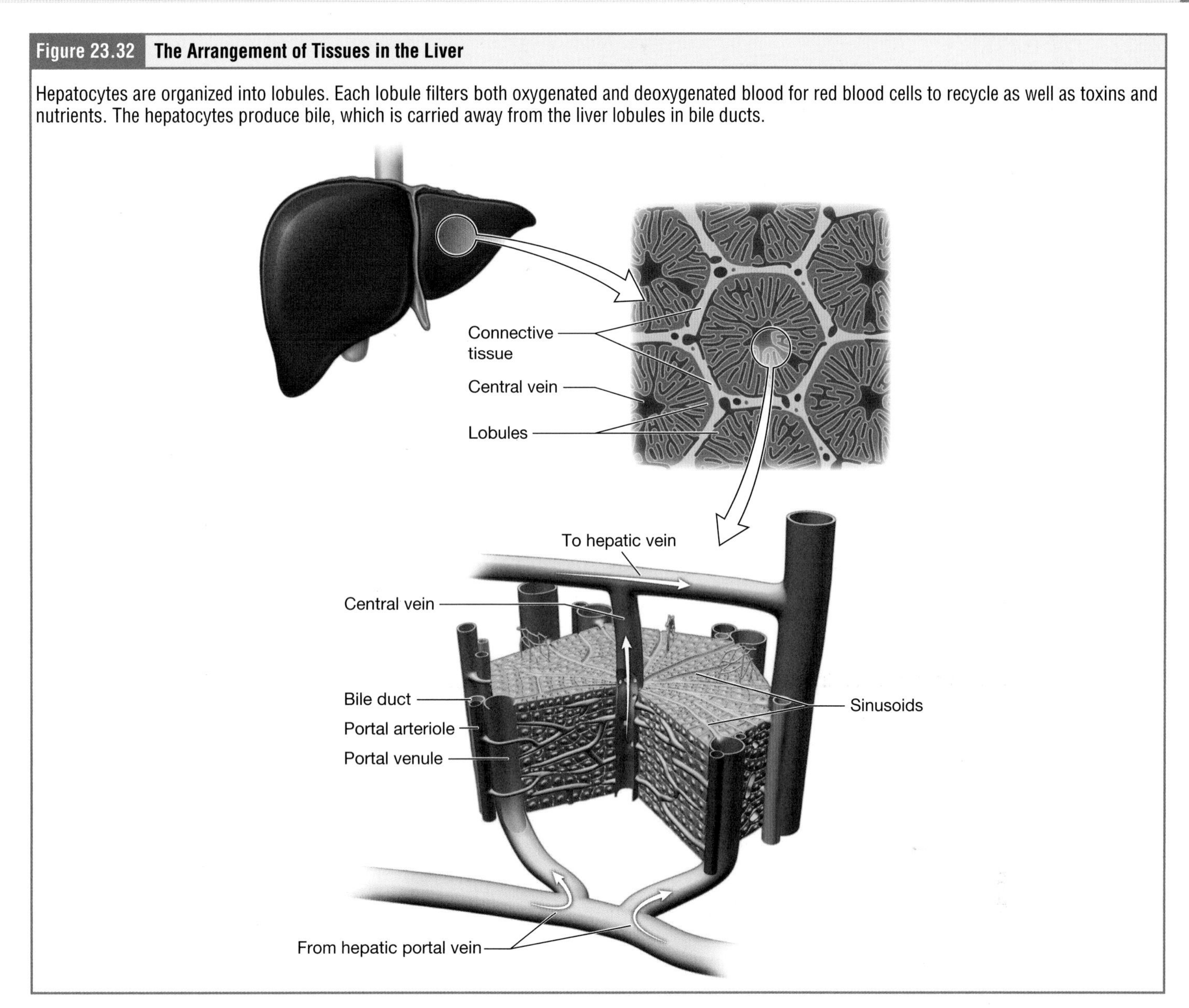

Bile Production and Histology As hepatocytes filter the blood for nutrients, they also remove older or damaged red blood cells from the circulation. As they recycle the red blood cell components, **bilirubin**, a waste product of hemoglobin breakdown, is produced. Bilirubin has a distinctive yellow-brown color. The liver also clears the bilirubin produced by the spleen, which is the other organ that functions to recycle red blood cells. Other breakdown products, including proteins and iron, are also generated and need to be excreted by the liver. Yet, the liver is not open to the outside of the body; getting rid of wastes must occur in a different fashion. The liver collects its waste products and sends them (as part of the bile) to the gallbladder, which empties them into the small intestine, where they become part of the chyme and end up in the feces. The yellow-brown color of bilirubin is actually what gives feces its characteristic brown color. In individuals with liver failure, the feces are white because no bilirubin is being added to them.

But the story of **bile**, the secretory product of the liver, is not a story of simple waste removal. Recall that lipids are hydrophobic; that is, they do not dissolve in water. When fats are ingested in a meal, the lipid molecules tend to aggregate in the watery environment of the stomach and small intestine. Thus, before they can be digested in the watery environment of the small intestine, large lipid globules must be broken down into smaller lipid globules, a process called **emulsification**. Bile acts as an emulsifier; it accomplishes the emulsification of lipids in the small intestine. Once they are emulsified, lipid molecules are accessible to the enzymes for digestion (**Figure 23.33**).

LO 23.4.5

Hepatocytes secrete about one liter of bile each day. The components most critical to emulsification are bile salts and phospholipids, which have a nonpolar (hydrophobic) region as well as a polar (hydrophilic) region. Bile also contains wastes, cholesterol, and ions. Whereas the hydrophobic region interacts with the large lipid molecules, the hydrophilic region interacts with the watery chyme in the intestine. This results in the large lipid globules being pulled apart into many tiny lipid fragments of about one micrometer in diameter. This change dramatically increases the surface area available for lipid-digesting enzyme activity. This is the same way dish soap works on fats mixed with water.

Learning Connection

Explain a Process

Your classmate is having trouble understanding whether the emulsification of fats by bile is a chemical or mechanical process of digestion. What would you tell your classmate, and how would you explain it?

Grooves between the cell membranes of adjacent hepatocytes provide room for each **bile canaliculus** (plural = canaliculi). These small ducts accumulate the bile produced by hepatocytes. Within each lobule are **hepatic sinusoids**, which are large sinusoidal capillaries carrying the blood from nutrient-rich hepatic portal veins and oxygen-rich hepatic arteries. Hepatocytes are tightly packed around the fenestrated endothelium of these spaces, giving them easy access to the blood. The hepatic sinusoids also contain star-shaped **reticuloendothelial cells** (also known as *Kupffer cells*), phagocytes that remove dead red and white blood cells, bacteria, and other foreign material that enters the sinusoids. Here, hepatocytes process the nutrients, toxins, and waste materials carried by the blood. Materials such as bilirubin are processed and excreted

reticuloendothelial cells (Kupffer cells)

Figure 23.33 Fat Emulsification

Because lipids (fats) are hydrophobic, they gather into spheres that exclude water. The function of bile is to interrupt their hydrophobic interactions and force the lipids into smaller particles.

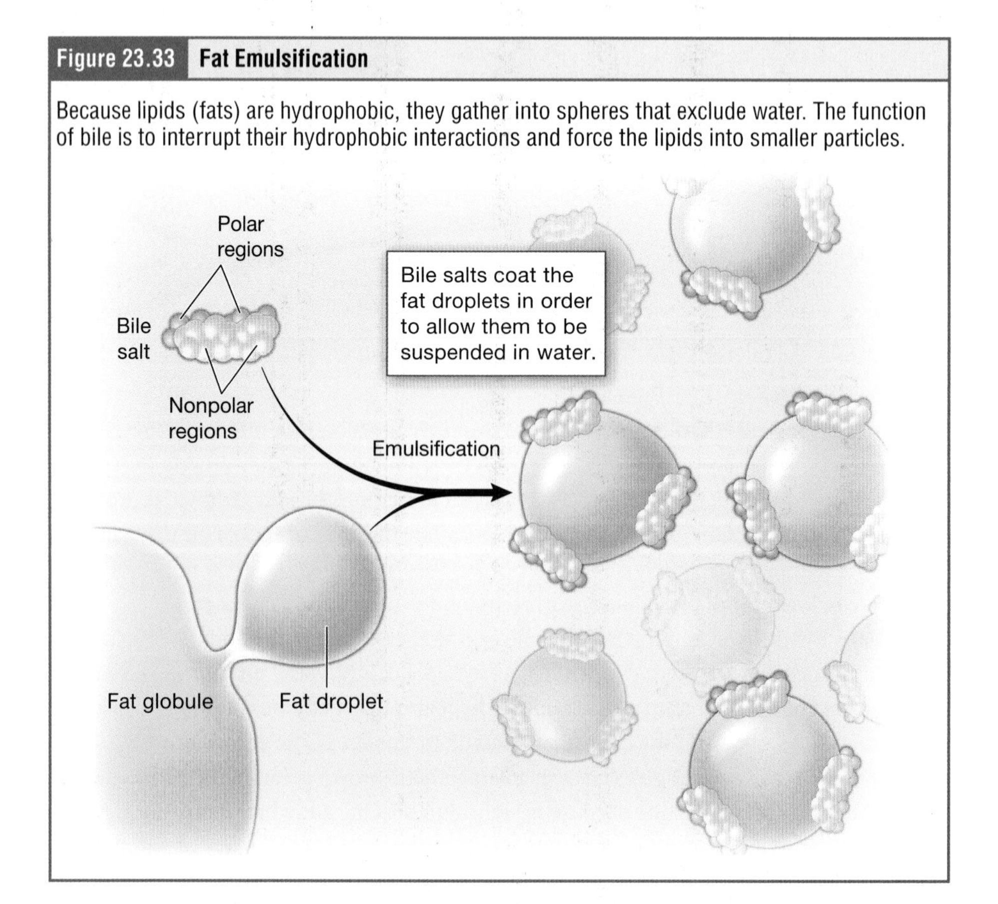

into the bile canaliculi. The bile canaliculi carry bile from within the lobule to the bile ducts on the outer edges of the lobule. The bile ducts unite to form the larger right and left hepatic ducts, which themselves merge and exit the liver as the common hepatic duct. This duct then joins with the cystic duct from the gallbladder to form the common bile duct, through which bile flows into the small intestine. The hepatic sinusoids combine and send blood to a central vein. This means that blood and bile flow in opposite directions to and from the **portal triad**, the group of three structures at the corners of the lobule: a bile duct, a hepatic artery branch, and a hepatic portal vein branch.

Hepatocytes work nonstop, but bile production increases when fatty chyme enters the duodenum and stimulates the secretion of the gut hormone secretin. Between meals, bile is produced but conserved. The valvelike hepatopancreatic ampulla closes, allowing bile to divert to the gallbladder, where it is concentrated and stored until the next meal.

Learning Connection

Broken Process

Sometimes a gallstone that forms in the gallbladder travels down the common bile duct and becomes lodged just before the hepatopancreatic sphincter. What effect would this have on bile release into the duodenum? Do you think this might also have an effect on the release of pancreatic juice?

23.4b The Gallbladder

LO 23.4.6

LO 23.4.7

The **gallbladder** is long and is nested in a shallow area on the posterior aspect of the right lobe of the liver. This muscular sac stores, concentrates, and—when stimulated—propels the bile into the duodenum. It is connected to the common bile duct by the cystic duct (**Figure 23.34**).

23.4c The Pancreas

LO 23.4.8

LO 23.4.9

The **pancreas** lies in the retroperitoneum behind the stomach. Its widest part, the **head of the pancreas**, is nestled into the C-shaped curvature of the duodenum. The body of the pancreas tapers to the left and ends in the tail of the pancreas. The pancreas is unique in that it is both an exocrine and an endocrine organ, comprised of a mix of these two tissue types (**Figure 23.35**).

LO 23.4.10

The exocrine part of the pancreas resembles little grapelike cell clusters, each called an **acinus** (plural = acini), located at the ends of pancreatic ducts. These acinar cells secrete enzyme-rich **pancreatic juice** into the ducts. Similar to venules

Figure 23.34 The Gallbladder

The liver can release bile directly to the duodenum, but bile can also be stored and concentrated in the gallbladder for release during a fatty meal.

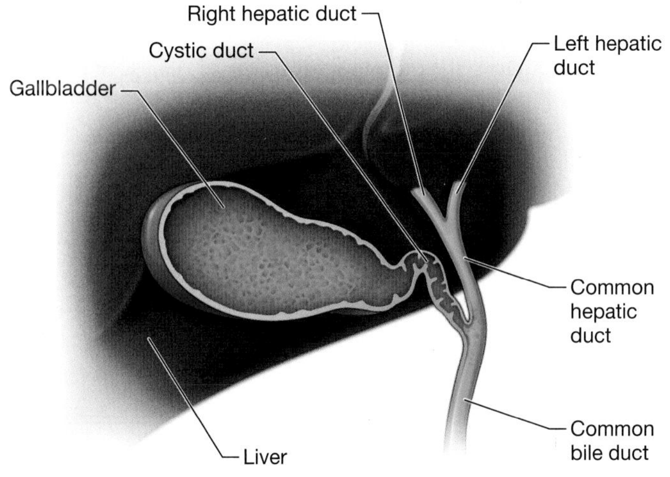

Figure 23.35 The Pancreas

The head of the pancreas is nestled within the C-curve of the duodenum. It produces and delivers pancreatic juice through the pancreatic duct to the duodenum. The pancreas also has endocrine functions in distinct pancreatic islets.

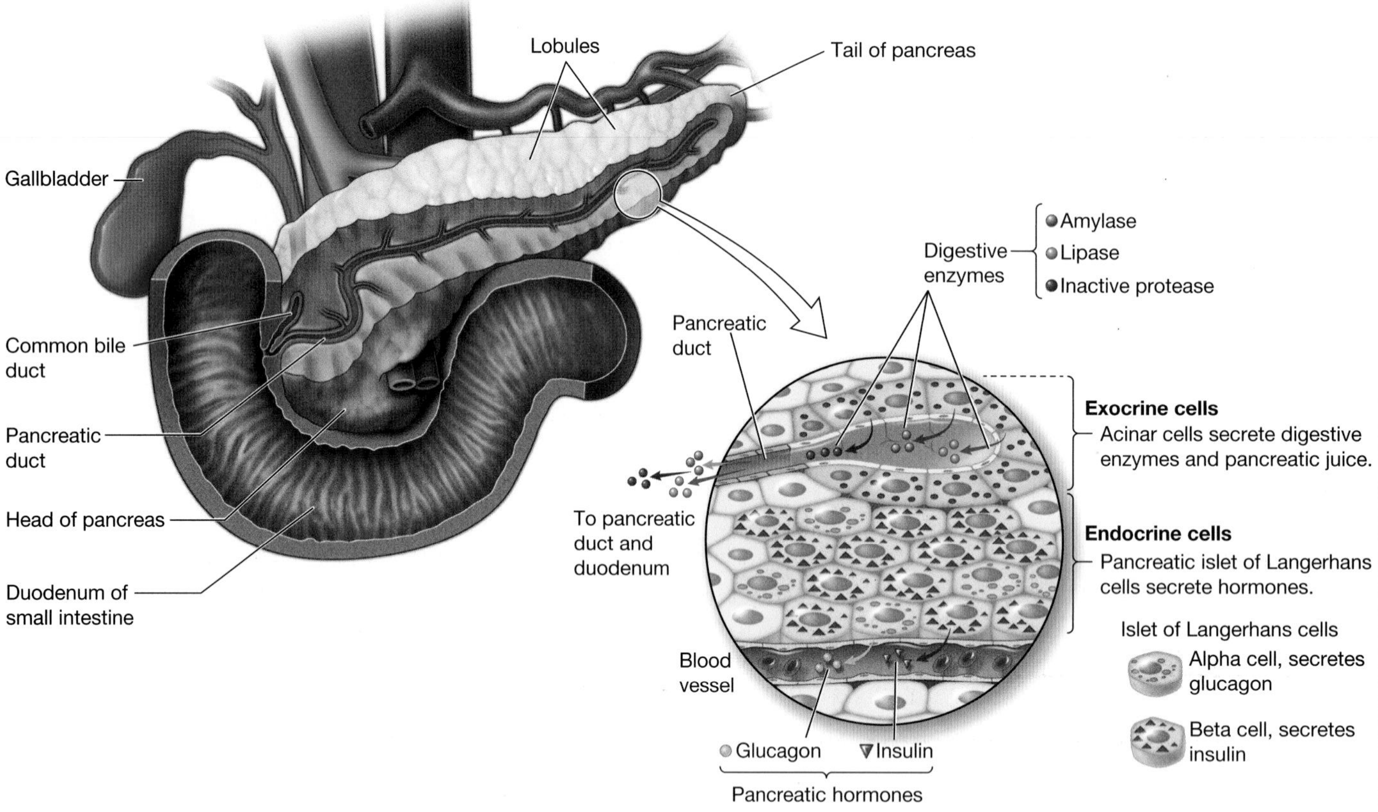

main pancreatic duct (duct of Wirsung)

accessory duct (duct of Santorini; accessory pancreatic duct)

pancreatic islets (islets of Langerhans)

islets of Langerhans (pancreatic islets)

and veins, smaller ducts converge to form larger ducts. The largest duct in the pancreas is the **main pancreatic duct**. It fuses with the common bile duct (carrying bile from the liver and gallbladder) just before entering the duodenum at the hepatopancreatic ampulla. A second and smaller pancreatic duct, the **accessory duct**, runs from the pancreas directly into the duodenum, approximately one inch above the hepatopancreatic ampulla. When present, it is a persistent remnant of pancreatic development.

Scattered through the sea of exocrine acini are small islands of endocrine cells, the **pancreatic islets**, also called **Islets of Langerhans**. These vital cells produce the hormones involved in metabolism and nutrient homeostasis.

Pancreatic Juice The pancreas produces over a liter of pancreatic juice each day. Unlike bile, it is clear and composed mostly of water along with some salts, sodium bicarbonate, and many digestive enzymes. Sodium bicarbonate is responsible for the slight alkalinity of pancreatic juice (pH 7.1–8.2), which serves to buffer the acidic chyme. Once chyme is neutralized, pepsin, which is activated by acidic conditions, is inactivated. Pancreatic enzymes are active in the digestion of sugars, proteins, and fats.

The pancreas produces protein-digesting enzymes in their inactive forms. These enzymes are activated in the duodenum. If produced in an active form, they would digest the pancreas. These enzymes are discussed in detail in the next section.

Apply to Pathophysiology

Ejection Responses (Vomiting and Diarrhea)

When pathogens or toxins are ingested, the body has the ability to eject these harmful substances (along with other ingested foods) through either vomiting or diarrhea. Whereas vomiting gets rid of any material in the stomach or sometimes the proximal portion of the duodenum, diarrhea is effective at eliminating material from the jejunum through the colon.

1. Which of the following statements is accurate?
   A) A low pH means that a solution is acidic and has a high concentration of H^+ ions
   B) A low pH means that a solution is basic and has a high concentration of H^+ ions
   C) A low pH means that a solution is acidic and has a low concentration of H^+ ions
2. Taking your answer from Question 1, which of the ejected solutions would you expect to have a lower pH?
   A) Diarrhea
   B) Vomit
   C) They should have similar or the same pH
3. Keiko has been infected with norovirus, a virus that causes excessive vomiting. They have vomited six times in the last eight hours. What can we assume about Keiko's blood at this time compared to yesterday?
   A) Keiko's blood pH is trending down and homeostatic mechanisms may be required to keep it within homeostatic pH range.
   B) Keiko's blood pH is trending up and homeostatic mechanisms may be required to keep it within homeostatic pH range.
4. In either a person who is sick with an infection, such as Keiko described in Question 3, or in a person experiencing vomiting and diarrhea due to a noninfectious condition, such as a reaction to chemotherapy, which of the ejected solutions would you expect to contain more bacteria?
   A) Diarrhea
   B) Vomit
   C) They would contain similar levels of bacteriav

Pancreatic Secretion Regulation of pancreatic secretion is the job of hormones and the parasympathetic nervous system. The entry of chyme into the duodenum stimulates the release of the hormones **secretin** and **cholecystokinin (CCK)**, which then stimulates the release of pancreatic juice. Parasympathetic regulation occurs mainly during the cephalic and gastric phases of gastric secretion, when vagal stimulation prompts the secretion of pancreatic juice.

Learning Connection

Try Drawing It

Make your own drawing of the GI tract and add the secreting accessory organs to your drawing. Now write down the secretions of each accessory organ and show where they enter the GI tract. This will help you understand what each secretion contributes to the digestion process.

Learning Check

1. Which of the following occur during the filtration process in the liver? Please select all that apply.
   a. Food from the intestine enters into the liver through the hepatic duct.
   b. Toxins from the blood are filtered while the nutrients stay within the blood.
   c. Nutrients are reabsorbed into the blood to travel to cells within the body.
   d. Toxins are removed from the blood and discarded.
2. Which of the following foods would trigger bile production?
   a. A vitamin-rich baby carrot
   b. A carbohydrate-rich slice of bread
   c. A protein-rich chicken wing
   d. A fat-rich hamburger
3. Where is the pancreas located?
   a. Anterior to the liver
   b. Anterior to the kidney
   c. Posterior to the stomach
   d. Posterior to the colon

23.5 Chemical Digestion and Absorption: A Closer Look

Learning Objectives: By the end of this section, you will be able to:

23.5.1 Describe the role of bacteria (microbiome) in digestion.

23.5.2 List the enzymes, their sources, their substrates, and their products of chemical digestion (enzymatic hydrolysis).

23.5.3 Explain the transport processes involved in the absorption of various nutrients.

23.5.4 List the organs and specific structures that facilitate the absorption of nutrients (e.g., monosaccharides, amino acids, fatty acids, monoglycerides).

23.5.5 Describe the absorption of minerals (e.g., calcium, iron) and vitamins (e.g., fat-soluble, water-soluble, B_{12}) in the gastrointestinal (GI) tract.

23.5.6 Describe the process of water absorption in the gastrointestinal (GI) tract.

As you have learned, the process of mechanical digestion is relatively simple. It involves the physical breakdown of food into smaller pieces. Chemical digestion, on the other hand, is a complex process that reduces food into its chemical building blocks, which are then absorbed to nourish the cells of the body (**Figure 23.36**). In this section, you will look more closely at the processes of chemical digestion and absorption.

23.5a Chemical Digestion

You eat hamburgers or burritos or doughnuts, but the cells of your body can only utilize monosaccharides, amino acids, and triglycerides. Only these small subunits are capable of being absorbed by cells that line the GI tract. Therefore, the bonds that hold together the monomers of our foods must be broken; this is accomplished by enzymes through hydrolysis. The many enzymes involved in chemical digestion are summarized in **Table 23.2**. We regularly ingest foods that we do not produce enzymes to digest. There are two possible fates for these foods. The foods that remain completely undigested pass through the GI tract and are excreted in the feces. Foods that can be digested by the bacteria of our microbiome, however, provide nutrients for both the bacteria and our human cells. Therefore, some of our bacterial inhabitants provide us with an unexpected benefit—helping us to digest certain foods and absorb their nutrients. A fantastic example is lactose. Lactose is a disaccharide that requires the enzyme lactase to be digested. Most humans do not produce lactase past toddlerhood. Many of these individuals, however, can still enjoy lactose-containing foods because among the species in their microbiome are bacteria that can produce lactase. Other individuals neither make their own lactase nor have bacteria that produce it.

LO 23.5.1

> **Learning Connection**
>
> ***Chunking***
>
> Compare the digestive enzymes secreted by the stomach with those of the pancreas. Which nutrients can the enzymes in gastric juice help to digest? Which nutrients can pancreatic enzymes digest?

Carbohydrate Digestion Carbohydrates are built from monosaccharides, of which there are three—glucose, fructose, and galactose—that are common in our foods. Sugars, such as table sugar and milk sugar, are disaccharides; they are made from two monosaccharides united by a covalent bond. Disaccharides cannot be transported across cell membranes, but once the bond is broken, monosaccharides can readily be taken up by cells. Starches, such as those found in potatoes or pasta, are complex and large molecules, but they are still built from monosaccharides united by covalent bonds. Each bond requires an enzyme to break it. For example, our

Figure 23.36 Digestion and Absorption

Digestion begins in the mouth and continues as food travels through the small intestine. Most absorption occurs in the small intestine.

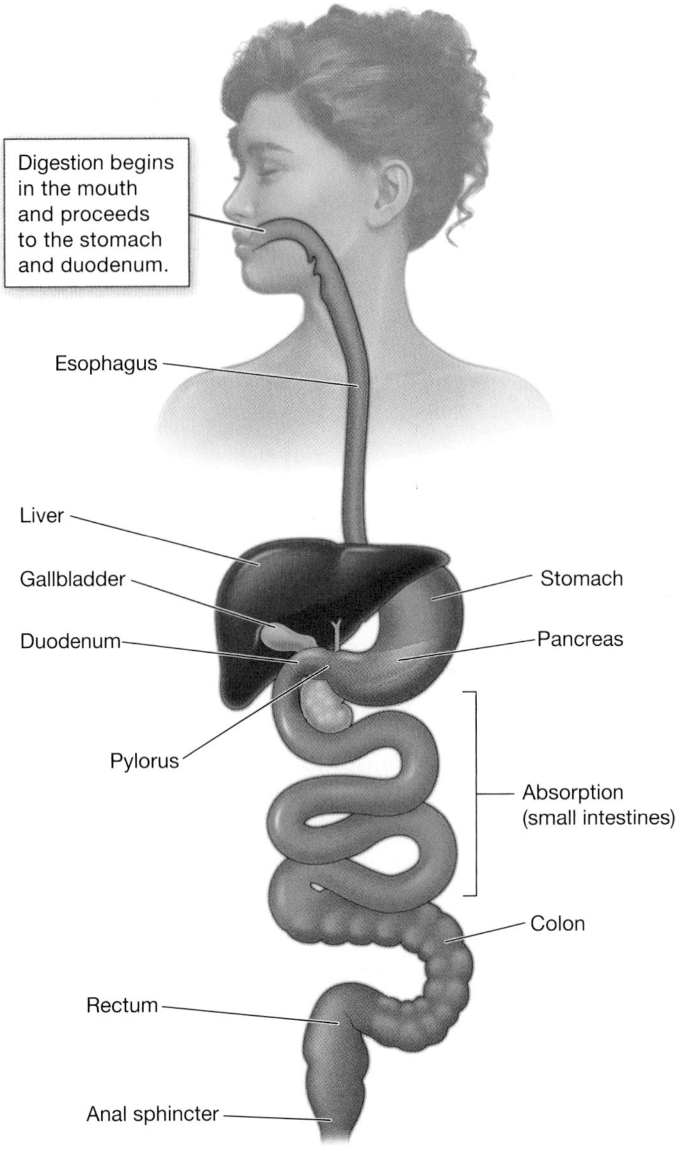

saliva and pancreatic juice contain enzymes, amylases, that can break the bonds among monosaccharides in starches such as pasta, but not the bonds that unite monosaccharides in fibrous polysaccharides, such as cellulose. Therefore, we cannot break down cellulose, and this molecule, often referred to as fiber, comprises our feces.

The chemical digestion of carbohydrates begins in the mouth, by salivary amylase, and is completed in the duodenum by **pancreatic amylase** (**Figure 23.37**). These enzymes primarily function to break complex carbohydrates into smaller pieces. Then brush border enzymes, which are adhered to the surface of the epithelial cells in the duodenum, break off one glucose unit at a time. These brush

Table 23.2 The Digestive Enzymes

Enzyme Category	Enzyme Name	Source	Substrate	Product
Salivary enzymes	Lingual lipase	Lingual glands	Triglycerides	Free fatty acids, and mono- and diglycerides
Salivary enzymes	Salivary amylase	Salivary glands	Polysaccharides	Disaccharides and trisaccharides
Gastric enzymes	Gastric lipase	Chief cells	Triglycerides	Fatty acids and monoglycerides
Gastric enzymes	Pepsin*	Chief cells	Proteins	Peptides
Brush border enzymes	α-Dextrinase	Small intestine	α-Dextrins	Glucose
Brush border enzymes	Enteropeptidase	Small intestine	Trypsinogen	Trypsin
Brush border enzymes	Lactase	Small intestine	Lactose	Glucose and galactose
Brush border enzymes	Maltase	Small intestine	Maltose	Glucose
Brush border enzymes	Nucleosidases and phosphatases	Small intestine	Nucleotides	Phosphates, nitrogenous bases, and pentoses
Brush border enzymes	Peptidases	Small intestine	Aminopeptidase: amino acids at the amino end of peptides Dipeptidase: dipeptides	Aminopeptidase: amino acids and peptides Dipeptidase: amino acids
Brush border enzymes	Sucrase	Small intestine	Sucrose	Glucose and fructose
Pancreatic enzymes	Carboxy-peptidase*	Pancreatic acinar cells	Amino acids at the carboxyl end of peptides	Amino acids and peptides
Pancreatic enzymes	Chymotrypsin*	Pancreatic acinar cells	Proteins	Peptides
Pancreatic enzymes	Elastase*	Pancreatic acinar cells	Proteins	Peptides
Pancreatic enzymes	Nucleases	Pancreatic acinar cells	Ribonuclease: ribonucleic acids Deoxyribonuclease: deoxyribonucleic acids	Nucleotides
Pancreatic enzymes	Pancreatic amylase	Pancreatic acinar cells	Polysaccharides (starches)	α-Dextrins, disaccharides (maltose), trisaccharide (maltotriose)
Pancreatic enzymes	Pancreatic lipase	Pancreatic acinar cells	Triglycerides that have been emulsified by bile salts	Fatty acids and monoglycerides
Pancreatic enzymes	Trypsin*	Pancreatic acinar cells	Proteins	Peptides

*These enzymes have been activated by other substances.

LO 23.5.2

border enzymes are able to hydrolyze the disaccharides sucrose, lactose, and maltose into monosaccharides.

Protein Digestion Proteins are polymers composed of amino acids linked by peptide bonds to form long chains. They are complex, highly folded molecules. Digestion involves first unfolding them, then cleaving the bonds among the amino acids to release these monomers.

The digestion of protein starts in the stomach, where HCl works to denature (or unfold) the proteins and pepsin breaks proteins into smaller polypeptides, which then travel to the small intestine (**Figure 23.38**). Chemical digestion in the small intestine is continued by pancreatic enzymes, including chymotrypsin and trypsin, each of which act on specific bonds in amino acid sequences. Brush border enzymes further break down peptide bonds, releasing amino acids. This results in molecules small enough to traverse the cell membranes and enter the bloodstream.

Figure 23.37 Carbohydrate Digestion

Carbohydrates are first broken down into small chains by amylases, then broken into monomers by brush border enzymes.

Complex carbohydrates enter the GI tract.
Amylases break bonds and yield smaller carbohydrates.
Brush border enzymes are expressed on the surface of epithelial cells in the duodenum.
Brush border enzymes break down carbohydrates and sugars into individual monosaccharides.

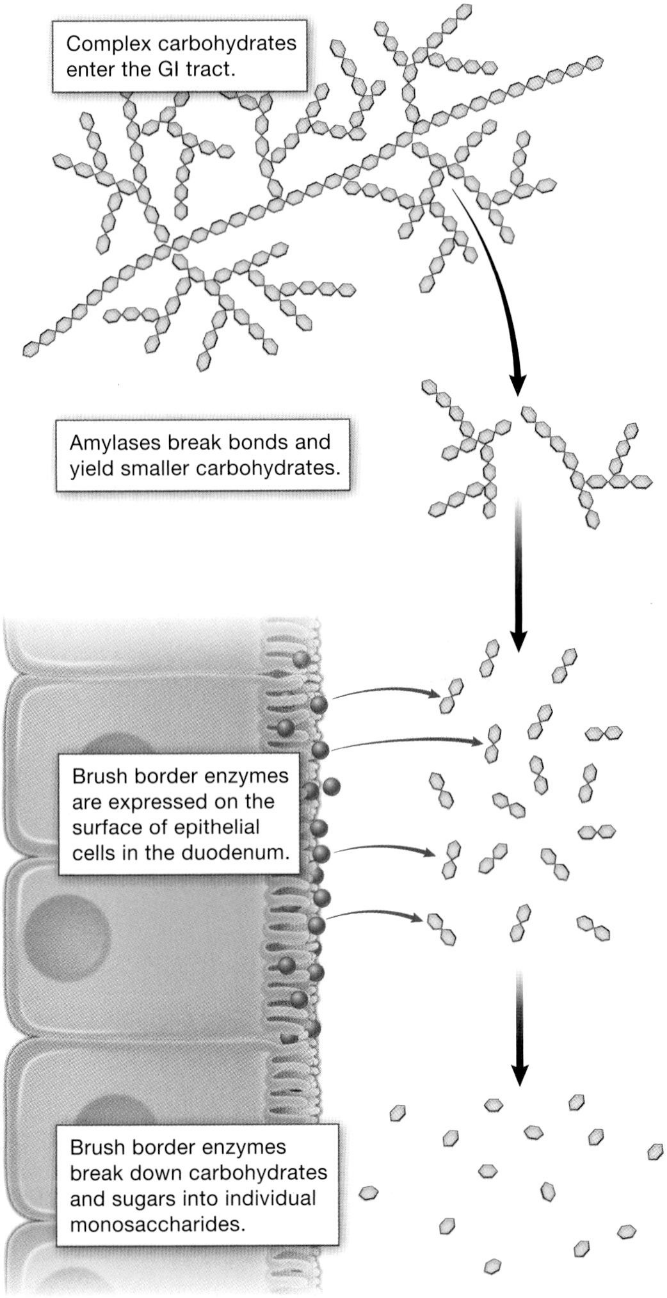

Lipid Digestion The most common dietary lipids are triglycerides, which are made up of a glycerol molecule bound to three fatty acid chains (**Figure 23.39**). Small amounts of dietary cholesterol and phospholipids are also consumed.

Lipases are the enzymes that digest lipids. Lipases are found in saliva and gastric juice in very small quantities, but **pancreatic lipase** does the lion's share of lipid digestion. Pancreatic lipase breaks down each triglyceride into two free fatty acids and a monoglyceride.

Figure 23.38 Protein Digestion

In order for their bonds to be cleaved by enzymes, proteins must first be unfolded/denatured so that the bonds are exposed. Then enzymes can cut the amino acids into shorter segments and finally liberate individual amino acids for absorption across the gut wall.

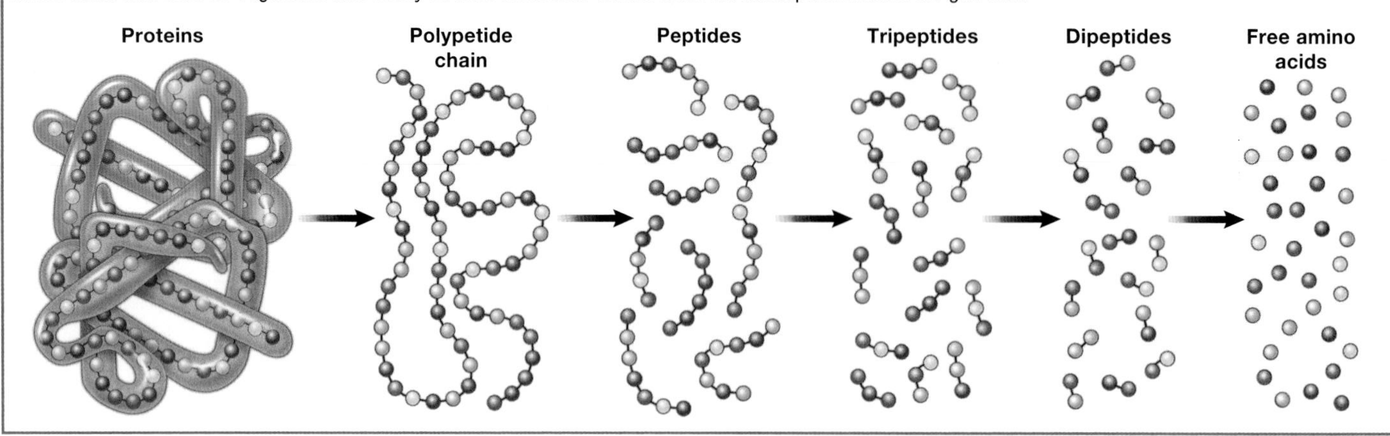

Figure 23.39 Lipid Digestion

The most common form of lipid in the diet is a triglyceride. During digestion, two of the fatty acid chains are freed by lipases, and one fatty acid chain remains bound to a glycerol molecule, producing a monoglyceride.

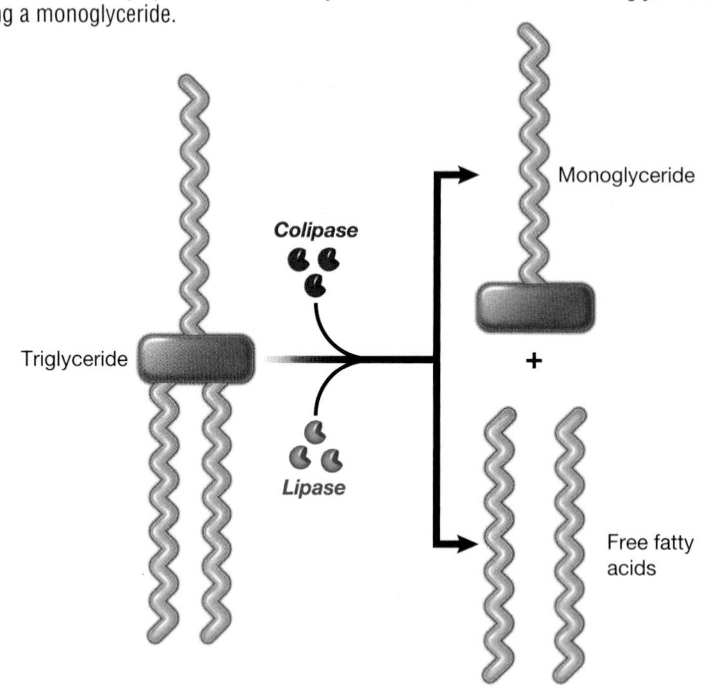

Monoglyceride

Colipase

Triglyceride

\+

Lipase

Free fatty acids

Nucleic Acid Digestion The nucleic acids DNA and RNA are found in most of the foods you eat, since most of the foods you eat come from cellular organisms. Two types of **pancreatic nuclease** are responsible for their digestion: **deoxyribonuclease**, which digests DNA, and **ribonuclease**, which digests RNA. The nucleotides produced by this digestion are further broken down by two intestinal brush border enzymes into monomers that can be absorbed through the cell membranes of the GI tract. The large food molecules that must be broken down into subunits are summarized in **Table 23.3**.

Table 23.3 Absorbable Food Substances

Source	Substance
Carbohydrates	Monosaccharides: glucose, galactose, and fructose
Proteins	Single amino acids, dipeptides, and tripeptides
Triglycerides	Monoglycerides, glycerol, and free fatty acids
Nucleic acids	Pentose sugars, phosphates, and nitrogenous bases

23.5b Absorption

The mechanical and digestive processes have one goal: to convert food into molecules small enough to be absorbed by the epithelial cells of the intestine. The absorptive capacity of these cells is almost endless. Almost all ingested food and about 80–90 percent of the electrolytes and water are absorbed in the small intestine. By the time chyme passes from the small intestine into the large intestine, all that's left is indigestible plant fibers like cellulose, some water, and millions of bacteria (**Figure 23.40**).

LO 23.5.3

LO 23.5.4

Absorption can occur through five mechanisms: (1) active transport, (2) simple diffusion, (3) facilitated diffusion, (4) secondary active transport, and (5) endocytosis. As you will recall from Chapter 4, active transport refers to the movement of a substance across a cell membrane going from an area of lower concentration to an area of higher concentration (against its concentration gradient). In this type of transport, proteins within the cell membrane act as "pumps," using cellular energy (ATP) to move the substance. Simple diffusion refers to the movement of substances from an area of higher concentration to an area of lower concentration across a membrane in which the substance is soluble. Facilitated diffusion refers to the movement of substances from an area of higher to an area of lower concentration using a carrier protein in the cell membrane. Secondary active transport uses the movement of one molecule through the membrane from higher to lower concentration to power the movement of another from lower to higher. Finally, endocytosis is a transportation process in which the cell membrane engulfs material. It requires energy, generally in the form of ATP.

Cultural Connection

Gluten: Friend or Foe?

Gluten is a protein found in wheat and some other grains. When heated, gluten forms long, stretchy fibers that can trap gas, making gluten a critical component to breads that rise as well as lending texture to many foods. About 1 percent of the U.S. population has a gluten-related disease such as celiac disease. For these individuals, ingesting foods with gluten can cause inflammation and damage to the gut epithelium; therefore, gluten must be avoided. But for the other 99 percent of U.S. adults, is gluten part of a healthy diet or something to be avoided? The answer is, it depends. Some individuals with irritable bowel syndrome experience a reduction in symptoms when gluten is removed from their diets. Other individuals, perhaps up to 10–15 percent of adults, have nonceliac gluten sensitivity; these individuals may experience a range of symptoms, such as bloating or gas, that are alleviated when gluten is eliminated from their diets. Some studies have observed benefits of excluding gluten in patients with other diseases such as endometriosis. For the vast majority of U.S. adults, there do not seem to be any physiological or health advantages to a gluten-free diet. In contrast, gluten-free diets tend to carry more heavily processed foods than gluten-containing diets; they also tend to be more expensive on average. One study has even linked gluten-free diets to an increase in coronary artery disease, perhaps because of higher lipid intake when gluten is removed from the diet. A gluten elimination test, and the reintroduction of gluten into the diet, can help any individual understand if they may benefit from a gluten-free diet.

Figure 23.40 Inputs and Outputs of the Digestive System

The absorption of dietary water and nutrients is important, but the absorption of the substances lent to the process of digestion is even more critical. Water, enzymes, and other substances are secreted into the digestive tract every day in saliva, bile, gastric and pancreatic juice, etc., to help process and absorb dietary nutrients; these substances have to be reabsorbed from the digestive tract (mainly the small intestine and colon), so that they will not be excreted from the body in the feces.

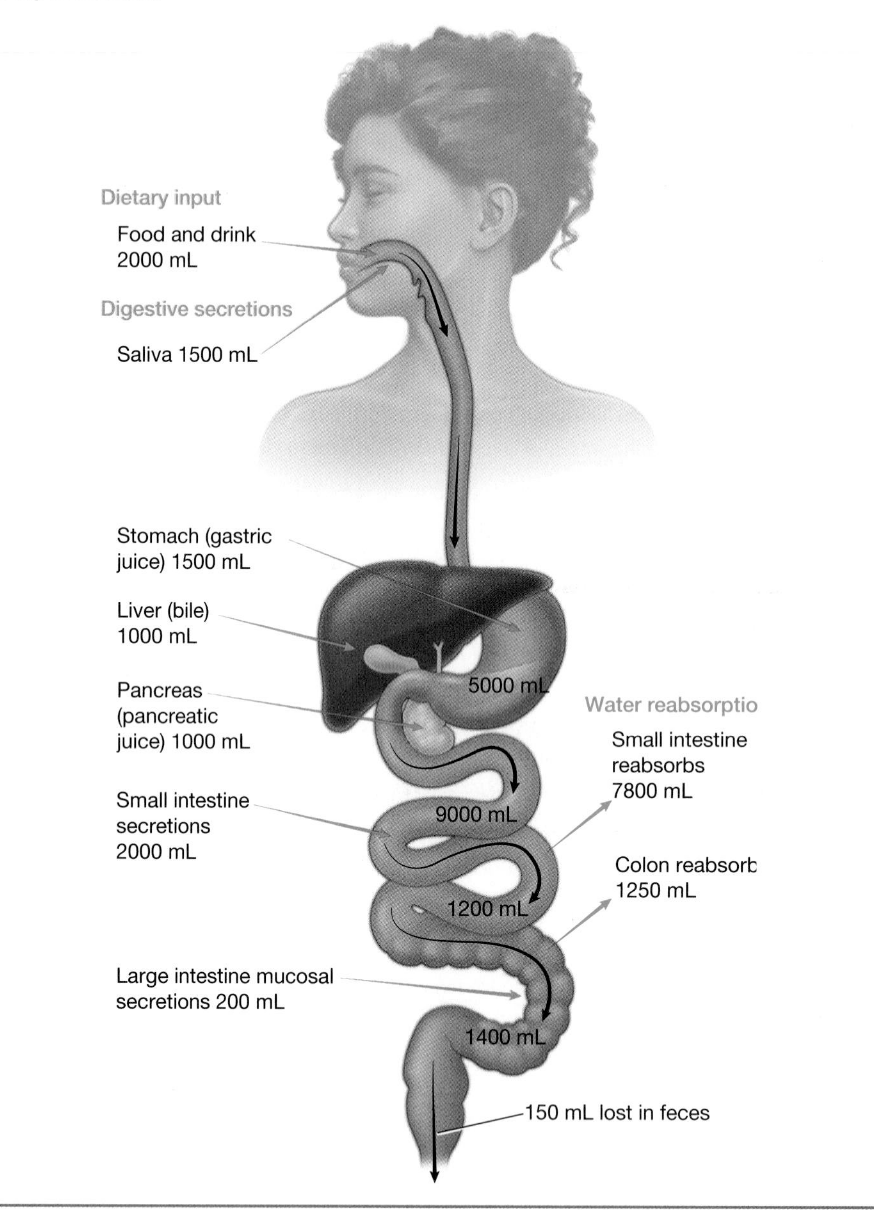

Transport across the membrane is defined by the properties of membrane molecules. The hydrophobic core of the membrane, made up of phospholipid tales, prevents the simple diffusion of water-soluble nutrients. These molecules must use transport proteins embedded in the membrane to enter cells and be absorbed across the gut wall. Transport between the epithelial cells of the intestine is prevented because these cells are bound together by tight junctions. Thus, substances can only enter blood capillaries by passing through the apical surfaces of epithelial cells and into the interstitial fluid. Water-soluble nutrients enter the capillary blood in the villi and travel to the liver via the hepatic portal vein.

Table 23.4 Absorption in the GI Tract

Food	Breakdown Products	Absorption Mechanism	Entry to Bloodstream	Destination
Carbohydrates	Glucose, galactose, or fructose (monosaccharides)	Co-transport with sodium ions (glucose and galactose) or facilitated diffusion (fructose)	Capillary blood in villi	Liver via hepatic portal vein
Protein	Amino acids	Co-transport with sodium ions	Capillary blood in villi	Liver via hepatic portal vein
Lipids	Fatty acids or glycerol	Diffusion into intestinal cells, where they are combined with proteins to create chylomicrons	Lacteals of villi	Systemic circulation via lymph entering thoracic duct
Nucleic Acids	Nucleic acid digestion products	Active transport via membrane carriers	Capillary blood in villi	Liver via hepatic portal vein

In contrast to the water-soluble nutrients, lipid-soluble nutrients can diffuse through the plasma membrane. Once inside the cell, they are packaged for transport at the base of the cell and then enter the lacteals (lymphatic vessels) to be transported to the bloodstream via the lymph. The specific routes of absorption across the epithelial cells are summarized in Table 23.4.

Carbohydrate Absorption All carbohydrates are absorbed in the form of monosaccharides. All digested dietary carbohydrates are absorbed; indigestible fibers are eliminated in the feces. The monosaccharides are transported into the epithelial cells by protein carriers via secondary active transport (i.e., co-transport with sodium ions). The monosaccharides leave these cells via facilitated diffusion and enter the capillaries. The monosaccharide fructose is absorbed and transported by facilitated diffusion alone.

Apply to Pathophysiology

Lactose Intolerance

Lactose intolerance is a condition characterized by indigestion caused by dairy products. All mammals, including humans, are born with the ability to produce lactase, an enzyme that cleaves the bond between the two monosaccharides that make up lactose (milk sugar). In most mammals, lactase production declines over time. In many mammals, it is the decline of lactase production that encourages them to wean, to stop drinking breast milk. In humans, lactase production does generally wane over time, but some human populations, most notably humans descended from northern European populations, are able to maintain the ability to produce lactase as adults.

1. Which compartment of the GI tract would lactase be found in, in an individual who produced it?
   A) Mouth
   B) Stomach
   C) Small or large intestine
2. If lactase were not present, which of the following would occur?
   A) Lactose would be broken apart by HCl instead.
   B) Lactose would remain intact as a disaccharide, never breaking apart.
   C) Lactose would be broken apart by amylases instead.
3. Given your answer to Question 2, would lactose or its monosaccharides be absorbed across the gut wall?
   A) Yes, lactose would be absorbed as a disaccharide.
   B) Lactose can still be broken down without lactase and its monosaccharides will be absorbed across the intestine wall.
   C) Because of its size, the disaccharide lactose would not be absorbed and would be left in the lumen of the intestines.
4. Given your answer to Question 3, would there be an osmotic effect of lactose intolerance?
   A) Yes, individuals without lactase will retain more lactose sugar in the lumen and therefore more water will also remain in the lumen than in an individual without lactose intolerance.
   B) No, individuals without lactase will still be able to absorb lactose and so no retention of water will occur.

Protein Absorption Amino acids, the monomers of proteins, are absorbed through secondary active transport by being linked to sodium. Short chains of two amino acids (dipeptides) or three amino acids (tripeptides) are also transported actively. However, after they enter the absorptive epithelial cells, they are broken down into their amino acids before leaving the cell and entering the capillary blood via diffusion. Infant and toddler gut walls are capable of transporting larger peptides, but this ability is limited in adults.

Lipid Absorption Dietary fats are broken down into long-chain fatty acids, monoglycerides (one fatty acid attached to a glycerol) and glycerol molecules. Short-chain fatty acids are produced by the bacteria in your gut, usually as a byproduct when fiber is digested. Short-chain fatty acids are both small and lipid-soluble, so they are able to enter gut epithelial cells via simple diffusion and then take the same path as monosaccharides and amino acids into the blood capillary of a villus.

The larger long-chain fatty acids and monoglycerides are hydrophobic and tend to clump in the watery intestinal chyme. However, bile salts resolve this issue by enclosing them in a **micelle**, which is a tiny sphere with polar (hydrophilic) ends facing the watery environment and hydrophobic tails turned to the interior (**Figure 23.41**). Dietary cholesterol and fat-soluble vitamins become trapped in the core of the micelle. Micelles squeeze between microvilli and when they come into contact with the cell surface, the lipids are able to cross into the cell via simple diffusion.

Student Study Tip

A micelle is like a hydrophobic vesicle.

Once fatty acids and monoglycerides enter the epithelial cells, they are rebuilt into triglycerides. The triglycerides are mixed with phospholipids and cholesterol and are surrounded with a protein coat. This new complex, called a **chylomicron**, is a water-soluble lipoprotein. After being processed by the Golgi apparatus, chylomicrons are released from the cell (see Figure 23.41). These molecules are too big to pass through the basement membranes of blood capillaries, so instead chylomicrons enter the large pores of lacteals. The lacteals come together to form larger lymphatic vessels. The chylomicrons are transported in the lymphatic vessels and empty through the thoracic duct into the subclavian vein of the circulatory system. Once in the bloodstream, the enzyme **lipoprotein lipase** breaks down the triglycerides of the chylomicrons into free fatty acids and glycerol. These breakdown products then pass through capillary walls to be used for energy by cells or stored in adipose tissue as fat.

chylomicron (ultra low-density lipoprotein)

Nucleic Acid Absorption The products of nucleic acid digestion—pentose sugars, nitrogenous bases, and phosphate ions—are transported by carriers across the villus epithelium via active transport. These products then enter the bloodstream.

LO 23.5.5

Mineral Absorption The electrolytes absorbed by the small intestine are from both GI secretions and ingested foods. Since electrolytes are charged molecules and unable to diffuse across the membrane, most are absorbed via active transport in the small intestine.

In general, all minerals that enter the intestine are absorbed, whether you need them or not. Iron and calcium are exceptions; they are absorbed in the duodenum in amounts that meet the body's current requirements, as follows:

- ***Iron.*** Iron is needed for the production of hemoglobin. Iron is absorbed into epithelial cells via active transport. Once inside mucosal cells, ionic iron binds to the protein ferritin, creating iron-ferritin complexes that store iron until needed. When the body has enough iron, most of the stored iron is lost when worn-out epithelial cells slough off. When the body needs iron because, for example, it is lost during bleeding, there is increased uptake of iron from the intestine and release of iron into the bloodstream.

Figure 23.41 Lipid Absorption

Lipids aggregate as large droplets in chyme because they are hydrophobic. Bile salts emulsify them (break them into smaller clumps) so that lipases are able to separate some of the fatty acids from the glycerol head. Once fatty acids and monoglycerides diffuse into the intestinal epithelial cells, they are reassembled as triglycerides and exported via exocytosis. Triglycerides are too large to be absorbed into the blood capillaries, so they enter the lymphatic lacteals for transportation to the bloodstream.

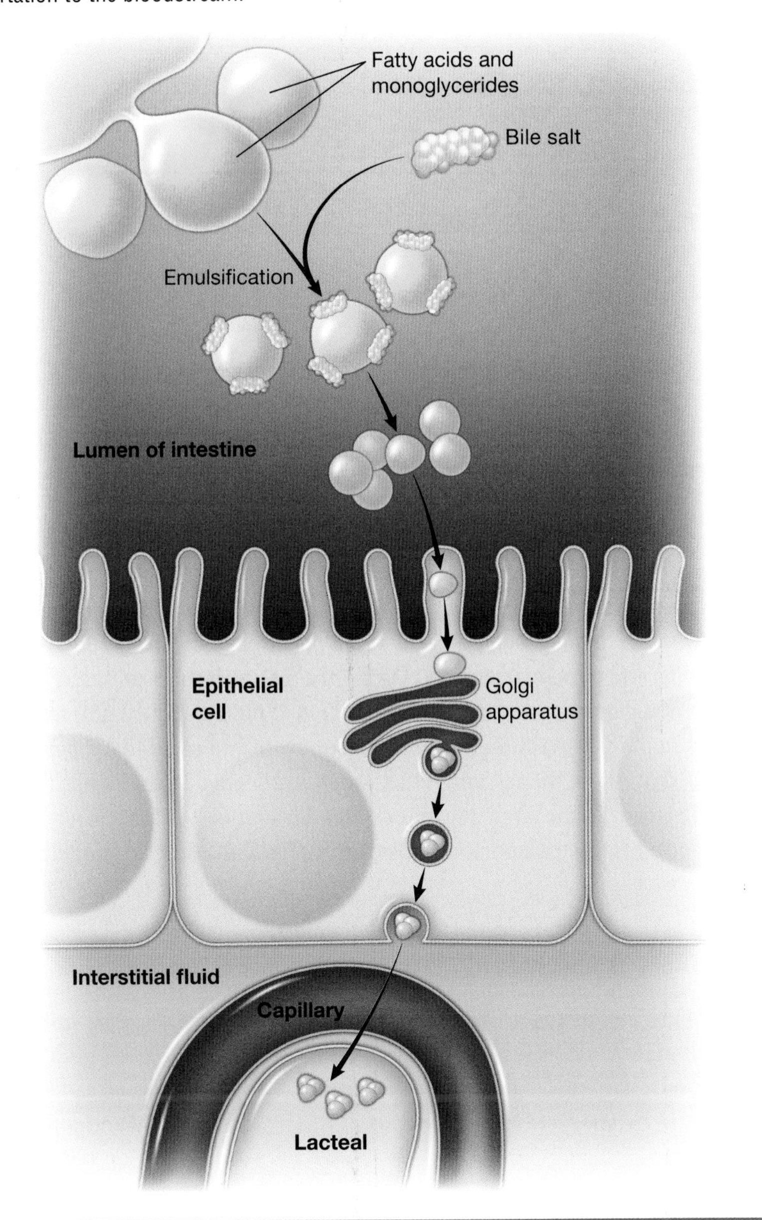

- ***Calcium.*** Blood levels of calcium determine the absorption of dietary calcium. When blood levels of calcium drop, parathyroid hormone (PTH) secreted by the parathyroid glands stimulates the release of calcium ions from bone matrices and increases the conservation of calcium by the kidneys. PTH also upregulates the activation of vitamin D in the kidney, which then facilitates intestinal calcium ion absorption.

Vitamin Absorption The small intestine absorbs the vitamins that occur naturally in food and supplements. Fat-soluble vitamins (A, D, E, and K) are absorbed along with dietary lipids in micelles via simple diffusion. Most water-soluble vitamins (including

DIGGING DEEPER: Aging

Age-related changes in the digestive system begin in the mouth and can affect virtually every aspect of the digestive system. The sense of taste declines, so food isn't as appetizing as it once was. Receptor sensitivity may decline, but taste cell number markedly declines with age. Further, saliva production decreases, making the eating of food less enjoyable; also, without sufficient saliva to solubilize them, it is harder for food molecules to reach the taste cells within the taste buds.

Muscle strength declines throughout the body during aging, which impacts the movement of food through the GI tract. Swallowing can be difficult, and ingested food moves slowly. Because the rate of mitosis in every tissue declines, minor damage to the gut epithelium can be slow or challenging to heal, leading to more significant pathologies such as ulcers or diverticulitis. The slow movement through the GI tract can overdry the feces, leading to constipation. Functional declines in the organs that contribute secretions can lead to insufficient breakdown of nutrients and malabsorption.

most B vitamins and vitamin C) also are absorbed by simple diffusion. An exception is vitamin B_{12}, which is a very large molecule. Intrinsic factor secreted in the stomach binds to vitamin B_{12}, preventing its digestion and creating a complex that binds to mucosal receptors in the small intestine, where it is taken up by endocytosis.

LO 23.5.6

Water Absorption Water absorption is incredibly important. While it varies from person to person and day to day, we take in up to three liters of water through our diet, much of which is absorbed. The critical absorption, however, is the reabsorption of the approximately seven liters of water a day that are required to produce saliva, mucus, gastric juice, bile, and pancreatic juice. If we do not reabsorb this water that is loaned to the process of digestion, we can quickly dehydrate. Some diarrheal illnesses, such as cholera, for example, can result in dehydration so severe and rapid that a person can die within one day. About 90 percent of this water is absorbed in the small intestine the remaining 10 percent in the large intestine. Water absorption occurs entirely by osmosis: as other nutrients are transported across the gut wall, an osmotic gradient is created, and water follows.

Learning Check

1. Which of the following locations are responsible for chemical digestion of carbohydrates? Please select all that apply.
   a. Oral cavity
   b. Stomach
   c. Duodenum
   d. Colon
2. Egg yolks are high in polyunsaturated fats, which are fatty acids containing some double bonds between adjacent carbon atoms. Explain how these fats are absorbed into your body.
   a. They are enclosed in micelles and then cross into the cell through microvilli via simple diffusion.
   b. They directly cross into the epithelial cells through microvilli via simple diffusion.
   c. They are actively transported through the cell walls linked to sodium molecules.
   d. They are transported into the epithelial cells by lipid carriers via secondary active transport.
3. Which minerals are absorbed only when the body needs them? Please select all that apply.
   a. Sodium
   b. Iron
   c. Calcium
   d. Potassium

Chapter Review

23.1 Overview of the Digestive System

The learning objectives for this section are:

23.1.1 Describe the major functions of the digestive system.

23.1.2 Explain the differences between the gastrointestinal (GI) tract (alimentary canal) and the accessory digestive organs.

23.1.3 List and identify the organs that compose the gastrointestinal (GI) tract.

23.1.4 Trace the pathway of ingested substances through the gastrointestinal (GI) tract.

23.1 Questions

1. What are the major functions of the gastrointestinal (GI) tract? Please select all that apply.
   a. To break down food
   b. To absorb nutrients
   c. To secrete enzymes
   d. To move food through the GI tract
2. Is the liver a major organ of the gastrointestinal tract or an accessory digestive organ?
   a. It is an organ of the gastrointestinal tract.
   b. It is an accessory digestive organ.
3. Is the colon a major organ of the gastrointestinal tract or an accessory digestive organ?
   a. It is an organ of the gastrointestinal tract.
   b. It is an accessory digestive organ.
4. You are eating an apple. The parts of the apple are currently in your stomach. Which of the following organs of the GI tract will the apple go to next?
   a. The esophagus
   b. The larynx
   c. The small intestine
   d. The liver

23.1 Summary

- Digestion is accomplished by the organs of the gastrointestinal tract and accessory organs.
- The gastrointestinal (GI) tract is the continuous pathway along which food is propelled as it is digested and absorbe, and the waste is excreted. The organs of the GI tract are the mouth, pharynx, esophagus, stomach, small intestine, large intestine, rectum and anus.
- The accessory digestive organs contribute to the digestion process, but are not part of the food pathway. The accessory organs are the salivary glands, liver, gallbladder, and pancreas.
- The contents of the digestive tract are considered to be outside the body because the two ends of the tract (the mouth and the anus) are open to the outside of the body.

23.2 General Gross and Microscopic Anatomy of the Gastrointestinal (GI) Tract

The learning objectives for this section are:

23.2.1 Identify and describe the gross anatomic and microscopic structure and function of each of the gastrointestinal (GI) tract tunics (layers): mucosa, submucosa, muscularis (muscularis externa), and serosa or adventitia.

23.2.2 Compare and contrast mechanical digestion and chemical digestion, including where they occur in the digestive system.

23.2.3 Define peristalsis.

23.2.4 Describe the enteric nervous system (ENS) and explain its role in controlling digestive system function.

23.2.5 Identify and describe the location, structure, and function of the visceral and parietal peritoneum, serous fluid, and the peritoneal cavity.

23.2.6 Compare and contrast the locations of the mesenteries (e.g., mesentery proper, mesocolon, lesser omentum, greater omentum).

23.2.7 Define mesentery and explain its function.

23.2.8 Explain the difference between an intraperitoneal and a retroperitoneal organ.

23.2.9 Identify which digestive system organs are intraperitoneal or retroperitoneal.

23.2 Questions

1. What are the functions of the layer of the GI tract called the submucosa? Please select all that apply.
   a. It connects the mucosa to the muscularis.
   b. It contains blood and lymphatic vessels and nerves.
   c. It propels food through the GI tract.
   d. It absorbs nutrients across the innermost membrane of the small intestine.

2. What type of digestion do you expect to occur in the mouth? Please select all that apply.
   a. Chemical digestion
   b. Mechanical digestion
   c. Peristalsis
   d. Enzymatic digestion

3. Define peristalsis.
   a. A wave of contraction and relaxation of the two layers of the muscularis
   b. Use of enzymes to break down food molecules
   c. Physically grinding the food into smaller chunks

4. You take a bite of a strawberry. Suddenly, your mouth is filled with saliva, which is breaking down the bites of strawberry into smaller molecules. Which of the following nerve plexuses regulates the secretion of saliva as it reacts to the presence of food?
   a. Myenteric plexus
   b. Submucosal plexus

5. The lesser omentum hangs down from the inferior edge of the liver, tying it to the superior edge of the stomach. Is this an example of a visceral peritoneum or parietal peritoneum?
   a. Visceral peritoneum
   b. Parietal peritoneum

6. Which of the following anchors the transverse colon of the large intestines to the posterior wall of the abdominal cavity?
   a. The lesser omentum
   b. The greater omentum
   c. The transverse mesocolon
   d. The mesenteries

7. What is the function of the mesenteries?
   a. To hang from the stomach and cover the small intestines
   b. To tie the small intestines to the posterior abdominal wall
   c. To hang from the liver and tie it to the stomach
   d. To anchor the large intestines to the posterior abdominal wall

8. The kidneys are located near the posterior abdominal wall. They are situated just behind the posterior parietal peritoneum. Which of the following best describes the space in which the kidneys reside?
   a. Intraperitoneal space
   b. Retroperitoneal space

9. Identify the organs that are considered to be intraperitoneal organs. Please select all that apply.
   a. Esophagus
   b. Kidney
   c. Small intestines
   d. Stomach

23.2 Summary

- There are four major layers of the digestive tract. From deep to superficial, the layers are the mucosa, the submucosa, the muscularis, and the serosa. The mucosa is an epithelial layer that performs protection of underlying cells, secretion and/or absorption. The submucosa is a connective tissue layer that contains the blood vessels that nourish the cells of the mucosa and, in the case of certain organs, absorbs digested nutrients. The muscularis is a layer of muscle that mixes and propels the food as it travels down the GH tract. The serosa is the outermost layer; it protects and lubricates the outer surface of the GI tract.
- Two types of digestion help to break down foods into their building blocks, so the particles are small enough to be absorbed across the mucosa of the small intestine. Mechanical digestion consists of processes that physically break down food into smaller particles, such as mastication in the mouth, churning in the stomach, segmentation in the stomach, and the action of bile in the duodenum. Chemical digestion involves the breakage of chemical bonds in food molecules by enzymes.
- Two types of movements occur along the GI tract: mixing movements and propelling movements. Mixing movements,

such as stomach churning or intestinal segmentation, mix food or chyme with digestive secretions, such as enzymes, hydrochloric acid and buffers. Propelling movements take the form of peristalsis, a wave of muscle contraction that transports food or chyme down the GI tract.

- The intrinsic innervation of the GI tract is provided by the enteric nervous system, which is composed of the myenteric plexus and the submucosal plexus. The myenteric plexus regulates muscle contraction in the GI tract, to control the mixing and propelling movements that occur along the tract. The submucosal plexus regulates glandular secretion in the GI tract, to control the release of various enzymes, buffers and enzymes along the tract.
- The peritoneum holds the organs of the abdominal cavity in place. The peritoneum is composed of two different layers: the parietal peritoneum, which lines the abdominal wall, and the visceral peritoneum, which envelops the abdominal organs. The potential space between the two layers is the peritoneal cavity, which contains lubricating serous fluid.

23.3 The Mouth, Pharynx, and Esophagus

The learning objectives for this section are:

The Mouth and Pharynx

23.3.1 Identify and describe the boundaries of the oral cavity.

23.3.2 Define mastication.

23.3.3 Compare and contrast the composition and functions of the hard palate, soft palate, and uvula.

23.3.4 Identify and describe the structures (e.g., taste buds, papillae) and the functions of the tongue.

23.3.5 Describe the structure and function of the salivary glands.

23.3.6 Describe the composition and functions of saliva.

23.3.7 Describe the structure and function of teeth.

23.3.8 Identify and describe the different regions of the pharynx with respect to the passage of air and/or food.

23.3.9 List the structures involved in deglutition and explain the process of deglutition, including the changes in position of the glottis and larynx that prevent aspiration.

The Esophagus

23.3.10 Identify and describe the gross anatomy of the esophagus, including its location relative to other body structures.

23.3.11 Describe the general functions of the esophagus.

23.3.12 Describe the anatomic specializations of the esophageal tunics (e.g., composition of the mucosa and muscularis [muscularis externa]) compared to the tunics of the rest of the GI tract.

23.3.13 Relate the anatomic specializations of the esophagus to the organ's functions.

The Stomach

23.3.14 Describe the general functions of the stomach.

23.3.15 Identify and describe the gross anatomy of the stomach, including its location relative to other body structures.

23.3.16 Describe the compositions, locations, and functions of the inferior esophageal (cardiac, lower esophageal) sphincter and the pyloric sphincter.

23.3.17 Identify gastric folds (rugae) and discuss their functional significance.

23.3.18 Describe the anatomic specializations of the stomach tunics compared to the tunics of the rest of the GI tract.

23.3.19 Relate the anatomic specializations of the stomach tunics (e.g., number of layers of muscle in the muscularis [muscularis externa]) to the organ's functions.

23.3.20 Identify and describe the gastric glands, including their cells (e.g., parietal cells, chief cells).

23.3.21 Describe the functions, production, and regulation of secretion of hydrochloric acid (HCl).

23.3.22 Explain the effects of the cephalic phase, gastric phase, and intestinal phase of digestion on various parts of the gastrointestinal (GI) tract.

23.3.23 Explain how volume, chemical composition, and osmolarity of chyme affect motility in the stomach and in the duodenum.

The Small Intestine

23.3.24 Describe the general functions of the small intestine.

23.3.25 Identify the specific segments of the small intestine (i.e., duodenum, jejunum, ileum), including their relative length.

23.3.26 Identify and describe the gross anatomy of the small intestine, including its location relative to other body structures.

23.3.27 Describe the major functions of the biliary apparatus.

23.3.28 Identify and describe the biliary apparatus components (i.e., left and right hepatic ducts, common hepatic duct, cystic duct, common bile duct, main pancreatic duct, hepatopancreatic ampulla [ampulla of Vater], hepatopancreatic sphincter [sphincter of Oddi], major duodenal papilla).

23.3.29 Trace the path of bile and pancreatic juice through the biliary apparatus.

23.3.30 Describe the anatomic specializations of the small intestine tunics (e.g., circular folds [plicae circulares], villi, microvilli) compared to the tunics of the rest of the GI tract.

23.3.31 Relate the anatomic specializations of the small intestine tunics (e.g., circular folds [plicae circulares], villi, microvilli) to the organ's functions.

23.3.32 Identify and describe the function of the following small intestine structures: duodenal glands (Brunner glands), intestinal glands (crypts of Lieberkuhn), and Peyer patches (lymphoid [lymphatic] nodules).

23.3.33 Compare and contrast the following: peristalsis, mixing waves, segmentation, and mass movement.

The Large Intestine, Rectum, and Anal Canal

23.3.34 Describe the general functions of the large intestine, rectum, and anal canal.

23.3.35 Identify and describe the gross anatomy of the large intestine, rectum and anal canal, including their location relative to other body structures.

23.3.36 Identify the specific segments and related flexures of the large intestine.

23.3.37 Compare and contrast the location, composition, and innervation (i.e., somatic versus autonomic) of the internal and external anal sphincters.

23.3.38 Describe the specializations of the large intestine tunics (e.g., composition of the muscularis [muscularis externa]) compared to the tunics of the rest of the GI tract.

23.3.39 Relate the specializations of the large intestine tunics (e.g., composition of the muscularis [muscularis externa]) to the organ's functions.

23.3 Questions

1. Which of the following forms the superior boundary of the oral cavity?

a. The tongue
b. The cheek
c. The labia
d. The palate

2. Define mastication.

a. Movements of chewing food
b. Biting or tearing into food
c. Piercing tough or fleshy foods
d. Mashing and grinding food

3. Which of the following is primarily composed of skeletal muscle?

a. The hard palate
b. The soft palate

4. Which of the following structures will allow you to distinguish between sweet and savory-tasting items?

a. The papillae
b. The lingual frenulum
c. The taste buds
d. The labial frenulum

5. If you eat something sour, sometimes you can feel a sharp sensation in the corners of your mouth by your ears. Which of the following salivary glands are responding, causing this sensation?

a. The submandibular glands
b. The sublingual glands
c. The parotid glands
d. The subdural glands

6. Which of the following components of saliva allow you to digest carbohydrates and initiate digestion?
   a. Salivary amylase
   b. Salivary lipase
   c. Salivary chymase
   d. Salivary maltase

7. Which types of teeth allow you to crush chips to prepare them for swallowing? Please select all that apply.
   a. Incisors
   b. Cuspids
   c. Premolars
   d. Molars

8. Which of the following regions of the trachea allows only air to pass through it?
   a. Nasopharynx
   b. Oropharynx
   c. Laryngopharynx

9. Explain the mechanism behind choking.
   a. The epiglottis does not adequately cover the larynx and trachea.
   b. The epiglottis does not adequately cover the esophagus.
   c. The hyoid bone does not adequately cover the trachea.
   d. The hyoid bone does not adequately cover the esophagus.

10. Which of the following describes the anatomical location of the esophagus?
    a. Inferior to the stomach
    b. Superior to the larynx
    c. Inferior to the pharynx
    d. Superior to the nasal cavity

11. Which of the following best describes the role of the esophagus?
    a. To aid in mechanical digestion of food
    b. To transport the bolus to the stomach
    c. To secrete enzymes for chemical digestion
    d. To relay sensory information to the stomach

12. Which type of epithelial tissue would you expect to find in the innermost layer of the esophagus?
    a. Stratified squamous
    b. Simple columnar
    c. Simple cuboidal
    d. Stratified cuboidal

13. Which of the following sphincters allows for food to enter the esophagus?
    a. The upper esophageal sphincter
    b. The lower esophageal sphincter
    c. The pyloric sphincter

14. Which of the following are functions of the stomach? Please select all that apply.
    a. To transport food to the duodenum
    b. To chemically digest protein molecules
    c. To detect chemical concentration changes as food enters the stomach
    d. To interpret stretch input from the esophagus

15. Which of the following structures of the stomach connects it to the duodenum of the small intestine?
    a. Cardia
    b. Fundus
    c. Pylorus
    d. Rugae

16. Which of the following pathologies could be a result of a dysfunctional inferior esophageal sphincter?
    a. Irritable bowel syndrome—excessive contraction of muscles within the colon
    b. Constipation—infrequent or dry bowel movements
    c. Diverticulitis—"outpouchings" within the large intestines
    d. GERD—excess gastric acid escaping into the esophagus

17. What is the functional significance of rugae?
    a. To increase power of contractions during mechanical digestion
    b. To increase surface area for increased secretion of gastric acid
    c. To increase in size in all directions to allow the stomach to expand for storage of food
    d. To increase secretions of pepsin during chemical digestion

18. Which of the following tunics is unique in the stomach compared to the other GI organs?
    a. Inner oblique smooth muscle
    b. Circular smooth muscle
    c. Longitudinal smooth muscle
    d. External oblique smooth muscle

19. Which layer in the wall of the stomach is unique in the GI tract, due to the stomach's churning action?
    a. A thick mucosa containing mucous glands
    b. A thick submucosa containing a layer of elastic connective tissue
    c. A thick muscularis containing oblique muscle, in addition to circular and longitudinal muscle
    d. A thin serosa, which allows for extra expansion of the stomach during churning

20. What could occur if someone lacked parietal cells in their gastric glands? Please select all that apply.

a. Decreased ability to digest proteins
b. Decreased ability to digest carbohydrates
c. Decreased ability to absorb vitamin B_{12}
d. Decreased ability to absorb calcium

21. Which of the following denatures proteins in the food, so that specific enzymes in the stomach can begin protein breakdown?

a. Pancreatic amylase
b. Salivary amylase
c. Pepsin
d. Hydrochloric acid

22. Imagine a large dish of your favorite food. You may notice your mouth starting to water. In which phase of gastric secretion are you?

a. The cephalic phase
b. The gastric phase
c. The intestinal phase

23. Which of the following would decrease the motility in the stomach?

a. Eating a large volume of food
b. Smelling food that you like
c. Emptying chyme into the duodenum
d. Eating alkaline foods

24. Which of the following organs is responsible for absorption of nutrients?

a. The esophagus
b. The oral cavity
c. The stomach
d. The small intestine

25. Which of the following is the longest part of the small intestine?

a. The jejunum
b. The duodenum
c. The ileum
d. The colon

26. Which of the following are true of the gross anatomy of the small intestine? Please select all that apply.

a. The segments of the small intestine, beginning closest to the stomach, are the jejunum, the ileum, and the duodenum.
b. The duodenum is the shortest segment of the small intestine.
c. The jejunum performs most of the final digestion and absorption of nutrients.
d. There is no sphincter muscle running between the small intestine and the large intestine.

27. What is the function of the biliary apparatus?

a. To store bile
b. To produce bile
c. To release bile into the duodenum

28. Into which of the following structures would secretions in the main pancreatic duct proceed next, on their way to the duodenum?

a. The right hepatic duct
b. The common hepatic duct
c. The hepatopancreatic ampulla
d. The cystic duct

29. Which of the following structures regulates the flow of bile and pancreatic juices into the duodenum?

a. The hepatopancreatic sphincter
b. The cystopancreatic sphincter
c. The hepatocystic sphincter
d. The pyloric sphincter

30. Which of the following structures of the GI tract are found only in the small intestine? Please select all that apply.

a. Villi
b. Mucous glands
c. Lacteals
d. Lymphatic nodules

31. Which of the following structures allow(s) for improved absorption of nutrients in the small intestine? Please select all that apply.

a. Microvilli
b. Circular folds
c. Hepatopancreatic ampulla
d. Bile

32. Which of the following produces an alkaline mucus that buffers the acidic chyme as it enters from the stomach?

a. The intestinal glands
b. The duodenal glands
c. Lymphoid nodules
d. The crypts of Lieberkuhn

33. What is the main function of segmentation in the small intestine?

a. To propel chyme from the duodenum to the jejunum
b. To grind and mash the chyme
c. To mix digestive enzymes with chyme
d. To propel chyme into the colon

34. Which of the following are functions of the large intestine? Please select all that apply.

a. Absorption of most dietary nutrients
b. Compacting of indigestible chyme into feces
c. Absorption of some water, electrolytes, and vitamins
d. Production of digestive enzymes

35. All of the following are true of the gross anatomy of the large intestine EXCEPT:

a. Feces moving through the descending colon proceed immediately into the sigmoid colon.
b. The teniae coli create the haustra.
c. The rectum and anal canal contain both teniae coli and haustra.
d. The ascending colon runs up the right side of the body.

Mini Case 1: Inflammatory Bowel Disease

John is a 55-year-old male who arrives at the ER due to stomach cramps, hematochezia (blood in the stool), and diarrhea. During the physical exam, John is found to be very tender in the left, lateral-most part of the abdomen. The medical professionals decide to perform a colonoscopy. In this procedure, a medical professional will insert a small, flexible microscope through the anus, and move it proximally to allow the team to locate any abnormalities.

1. The microscope is currently in the left colic flexure. From the imaging, it seems that there is blood coming from the structure proximal to this region. Which of the following structures could be the source of the problem?

a. Right colic flexure
b. Transverse colon
c. Descending colon
d. Sigmoid colon

2. When the scope was first inserted into the anus, it had to pass through the external and internal anal sphincters. Which of the following are true of these sphincters? Please select all that apply.

a. The internal anal sphincter remains open at all times.
b. The external anal sphincter is composed of skeletal muscle.
c. The internal anal sphincter is under voluntary control.
d. Both sphincters lie distal to the rectum.

When in the impaired region, the medical team found an ulcer in the wall of the colon. This type of ulcer usually indicates the presence of ulcerative colitis, which is one of the two diseases that falls in the category of inflammatory bowel disease (IBD). This ulcer typically forms between the two innermost layers of the colon wall.

3. Between which of the following layers could the ulcer form? Please select all that apply.

a. Mucosa
b. Submucosa
c. Circular muscle
d. Longitudinal muscle

4. As the scope is moving along the innermost layer of the large intestine, a large amount of mucus partially occludes the doctor's view of the inner lining. What are the functions of this mucus in the inner lining of the large intestine?

a. To create a barrier against pathogens
b. To protect the inner lining of the large intestine from its acidic secretions
c. To lubricate the inner lining of the large intestine, so feces can pass through easily
d. To help absorb vitamin C

23.3 Summary

- The oral cavity is the site of food intake, or ingestion. The roof of the oral cavity is formed by the hard and soft palate. The major processes that occur in the oral cavity are mastication, or chewing movements, and the mixing of food with saliva for taste, breakdown of carbohydrates and fats, and swallowing.
- The three major pairs of salivary glands are the parotid, submandibular, and sublingual glands. Each pair secretes saliva of a different composition.
- The teeth in different parts of the mouth are shaped differently, and have different functions. Teeth are classified as incisors, cuspids, premolars or molars.
- The pharynx lies posterior to the nasal and oral cavities. It serves as a passageway for air, food, or both. The three sections of the pharynx are the nasopharynx, oropharynx, and laryngopharynx.
- The esophagus, located inferior to the pharynx, serves as an organ of transport for food to travel from the pharynx to the stomach through a process known as *peristalsis*. It does not secrete any digestive enzymes; it is simply a tubular passageway.
- The stomach consists of four regions: cardia, fundus, body and pyllorus. It receives food from the esophagus via the lower esophageal (cardiac) sphincter. It empties chyme into the duodenum of the small intestine via the pyloric sphincter.

- The stomach performs both mechanical and chemical digestion. The stomach has an extra layer of muscle—the inner oblique smooth muscle layer—that allows for the churning action that accomplishes mechanical digestion. The stomach also secretes a mixture of subtances called gastric juice, which contains hydrochloric acid, pepsinogen, intrinsic factor, mucus, and gastrin. Pepsin begins the digestion of proteins.
- The small intestine consists of three regions: the duodenum, the jejunum, and the ileum. The duodenum is the shortest section, but it is the most involved in digestion because it is where the majority of chemical digestion takes place using digestive enzymes secreted by anatomically connected accessory digestive organs such as the liver and the pancreas. The jejunum performs most of the final nutrient digestion and absorption. The ileum transports the indigestible chyme to the large intestine.
- The small intestine has several structural features that increase surface area for secretion of enzymes and buffers and absorption of nutrients. Circular folds are large folds in the wall of the small intestine. Villi are smaller fingerlike projections of the mucosa and submucosa, and microvilli are microscopic folds on the surface of each cell.
- The biliary apparatus is a group of structures that secretes bile into the duodenum to be mixed with chyme to perform lipid emulsification. It consists of various hepatic ducts from the liver, the cystic duct of the gallbladder, the pancreatic ducts, the hapatopancreatic ampulla and the hepatopancreatic sphincter. The sphincter regulates the release of both bile and pancreatic juice into the duodenum.
- The large intestine frames the small intestine on three sides. It consists of the cecum, colon (ascending, transverse, descending, and sigmoid), rectum and anus.
- The large intestine forms the indigestible chyme into feces, stores it until defecation, and excretes feces to the outside world. It also absorbs vitamins, minerals and water, and houses microbiota that produce vitamins and aid in digestion.

23.4 Accessory Organs in Digestion: The Liver, Pancreas, and Gallbladder

The learning objectives for this section are:

23.4.1 Describe the general functions of the liver.

23.4.2 Describe the location of the liver relative to other body structures.

23.4.3 Identify and describe the structure of the liver, including the individual lobes, ligaments (e.g., coronary ligament, falciform ligament, round ligament [ligamentum teres]), and the porta hepatis.

23.4.4 Identify and describe the histological components of the classic hepatic lobule.

23.4.5 Define emulsification, and explain how and where bile salts facilitate fat digestion.

23.4.6 Identify and describe the structure and functions of the gallbladder.

23.4.7 Describe the location of the gallbladder relative to other body structures.

23.4.8 Identify and describe the structure and functions of the pancreas.

23.4.9 Describe the location of the pancreas relative to other body structures.

23.4.10 Identify and describe the major histological components of the pancreas (pancreatic acini and pancreatic islets [islets of Langerhans]) and discuss their major functions.

23.4.11 Describe the source, stimuli for release, targets, and actions of gastrointestinal tract hormones (e.g., gastrin, cholecystokinin, secretin).

23.4 Questions

1. Which of the following describe(s) the function(s) of the liver? Please select all that apply.
   a. To produce bile
   b. To store bile
   c. To filter waste from blood
   d. To secrete digestive enzymes

2. You are working in the ER when someone comes to you with abdominal pain. You palpate their stomach to localize the tenderness. If you suspect the liver to be the source of the problem, in which of the following regions would you most expect the tenderness?
   a. Right upper quadrant of the abdominal cavity
   b. Left upper quadrant of the abdominal cavity
   c. Right lower quadrant of the abdominal cavity
   d. Left lower quadrant of the abdominal cavity

3. Which of the following binds the liver to the lesser curvature of the stomach?
   a. The ligamentum teres hepatis
   b. The falciform ligament

c. The lesser omentum
d. The coronary ligament

4. Which of the following shapes most accurately represents a hepatic lobule, which is the functional unit of a liver?

a. Circle
b. Square
c. Squamous
d. Hexagonal

5. Which of the following are true of the process of lipid emulsification? Please select all that apply.

a. It is a form of chemical digestion.
b. It helps to break down large lipids into smaller lipids.
c. It is accomplished by bile salts.
d. It takes place in the liver.

6. What path does bile from the gallbladder take to exit into the duodenum?

a. Cystic duct → common bile duct → duodenum
b. Left hepatic duct → right hepatic duct → duodenum
c. Right hepatic duct → cystic duct → duodenum
d. Common bile duct → duodenal duct → duodenum

7. You are a surgeon who is performing surgery on someone with gallstones. To remove the gallstones, you need to access the gallbladder. Where would you find the gallbladder?

a. Anterior to the left lobe of the liver
b. Posterior to the right lobe of the liver
c. Anterior to the left kidney
d. Posterior to the right kidney

8. What is the difference between the main pancreatic duct and the accessory duct?

a. The accessory duct travels from the pancreas to the liver.
b. The main pancreatic duct travels from the pancreas into the duodenum.
c. The accessory duct travels from the stomach into the pancreas.
d. The main pancreatic duct fuses with the common bile duct before emptying into the duodenum.

9. Where is the head of the pancreas located?

a. In the curve of the duodenum
b. On top of the stomach
c. On the underside of the liver
d. Wrapped around the ileum

10. Which of the following is true about pancreatic acini?

a. They produce secretin.
b. They produce cholecystokinin.
c. They produce pancreatic juice.
d. They produce bile.

11. Which organ produces secretin?

a. The stomach
b. The small intestine
c. The salivary gland
d. The pancreas

23.4 Summary

- The liver is the largest internal organ in the human body. It is located in the upper right quadrant of the abdominopelvic cavity.
- The liver plays a tremendous role in many metabolic and homeostatic processes. It has two vital roles in digestion: (1) bile production and (2) filtration of the blood. Bile is an emulsifying agent for lipid breakdown; it helps to prepare the lipids in the chyme for enzymatic breakdown. As blood carrying absorbed nutrients enters the liver through the hepatic portal vein, the liver adjusts nutrient concentrations in the blood and filters toxins out of the blood.
- The gallbladder is a small pear-shaped organ that lies on the posterior side of the right lobe of the liver. The cystic duct is the entry/exit duct for the gallbladder; it merges with the common hepatic duct to form the common bile duct.
- The gallbladder stores and concentrates bile that was produced by the liver. After ingesting a fatty meal, the concentrated bile is released into the common bile duct and then into the duodenum.
- The pancreas is an elongated organ, located in the retroperitoneal area behind the stomach.
- The pancreas is an endocrine organ that secretes insulin and glucagon. It is also an exocrine organ that secretes pancreatic juice through the pancreatic duct into the duodenum. Pancreatic juice contains digestive enzymes that can digest proteins, carbohydrates, fats and nucleic acids; it also contains bicarbonate buffer, which buffers hydrochloric acid in the duodenum.

23.5 Chemical Digestion and Absorption: A Closer Look

The learning objectives for this section are:

23.5.1 Describe the role of bacteria (microbiome) in digestion.

23.5.2 List the enzymes, their sources, their substrates, and their products of chemical digestion (enzymatic hydrolysis).

23.5.3 Explain the transport processes involved in the absorption of various nutrients.

23.5.4 List the organs and specific structures that facilitate the absorption of nutrients (e.g., monosaccharides, amino acids, fatty acids, monoglycerides).

23.5.5 Describe the absorption of minerals (e.g., calcium, iron) and vitamins (e.g., fat-soluble, water-soluble, B_{12}) in the gastrointestinal (GI) tract.

23.5.6 Describe the process of water absorption in the gastrointestinal (GI) tract.

23.5 Questions

1. Yogurt has a high concentration of probiotics that grow healthy and safe bacteria in your gut. What are the advantages of eating yogurt regularly? Please select all that apply.

a. Increased feces production
b. Increased nutrient absorption
c. Increased nutrient digestion
d. Increased water absorption

2. Lactase is an enzyme that serves to breakdown lactose into glucose and galactose. What is the source of lactase?

a. The salivary glands
b. The pancreas
c. The small intestine
d. The stomach

3. How is vitamin C absorbed into the bloodstream?

a. Simple diffusion
b. Facilitated diffusion
c. Active transport
d. Secondary active transport

4. Bread is a food that is rich in carbohydrates. Which of the following organs are responsible for digesting bread? Please select all that apply.

a. The mouth
b. The stomach
c. The small intestine
d. The large intestine

5. How is iron absorbed into the epithelial cells of the small intestine?

a. Osmosis
b. Facilitated diffusion
c. Active transport
d. Secondary active transport

6. How is water absorbed into the bloodstream?

a. Simple diffusion
b. Facilitated diffusion
c. Osmosis
d. Active transport

23.5 Summary

- The microbiome, consisting of the normal bacterial inhabitants of the large intestine, helps with the digestion of some foods. The majority of chemical digestion, however, is accomplished by the digestive enzymes produced in the body.
- Enzymes found in the saliva, gastric juice, pancreatic juice and brush border cells of the small intestine digest the carbohydrates, proteins, lipids, and nucleic acids in our food. Enzymatic breakdown ensures that food particles have been reduced in size to the point that they are able to be absorbed across the cell membranes of the cells lining the inner wall of the small intestine.
- Chemical digestion of carbohydrates starts in the mouth and finishes in the small intestine; chemical digestion of proteins starts in the stomach and finishes in the small intestine; and chemical digestion of lipids starts and finishes in the small intestine.
- Most nutrient absorption occurs in the small intestine. Carbohydrates are broken down into monosaccharides, which are absorbed by secondary active transport. Proteins are broken down into amino acids, which are also absorbed by secondary active transport. Lipids are broken down into fatty acids, monoglycerides and glycerol; these products are absorbed by simple diffusion, sometimes requiring the help of micelles. Nucleic acids are broken down into phosphate, sugar and bases, which are absorbed by active transport.
- Minerals are absorbed by active transport. Vitamins are absorbed by simple diffusion; lipid-soluble vitamins require the help of micelles. Water is absorbed by osmosis.

24

Metabolism and Nutrition

FatCamera/E+/Getty Images

Chapter Introduction

In this chapter we will learn...

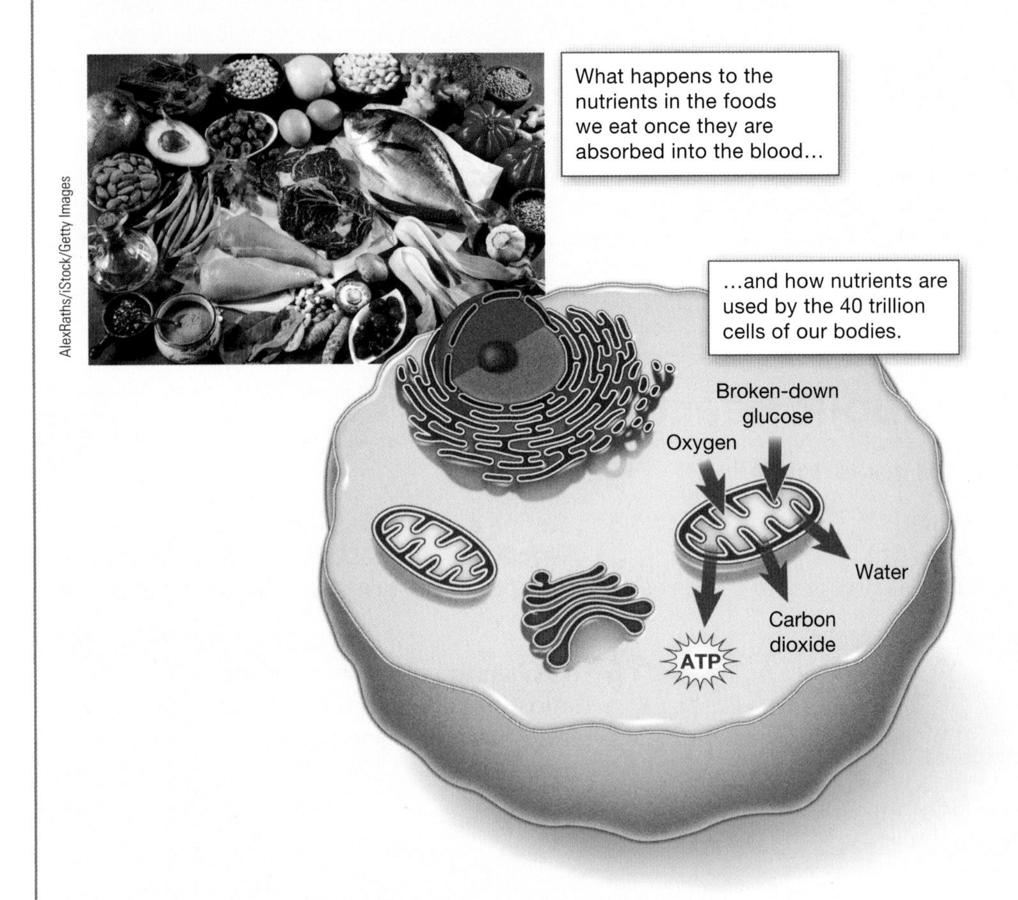
AlexRaths/iStock/Getty Images
What happens to the nutrients in the foods we eat once they are absorbed into the blood...
...and how nutrients are used by the 40 trillion cells of our bodies.
Broken-down glucose
Oxygen
Water
Carbon dioxide
ATP

Eating is not only essential to life; it is also joyful and a way that we often connect with others. The foods we eat may be tied to our culture or identity, or they may serve a perfunctory role of caloric and nutritional input. Other people frequently have opinions about what we should eat, and each of us places values and some care into our food choices. But what does this all mean to your body and the physiological processes it carries out each day? We learned in Chapter 23 about the way that food is broken down in the gastrointestinal (GI) tract so that we can absorb its components into the bloodstream, but what happens to these building blocks once they reach our cells. Cells use the molecules from our diet for metabolic processes that release the energy for the cells to carry out their daily jobs, to manufacture new proteins, cells, and body parts, and to recycle materials in the cell.

This chapter will take you through some of the chemical reactions essential to life, the sum of the chemical reactions in your body is referred to as **metabolism**. The word *metabolism* is sometimes used in everyday conversation to mean the rate at which a person burns calories, but in Anatomy and Physiology the definition is much broader, encompassing not just caloric use, but all types of chemical reactions.

Metabolism varies from one individual to another. Some of the factors that influence metabolism are: age, activity level, fuel consumption, and muscle mass. Your own metabolic rate fluctuates throughout life. Some sources will indicate that biological sex plays a role in metabolism, but it is unclear how much of this impact is based on factors frequently linked to biological sex, such as muscle mass. Estrogen, a. hormone found at much higher concentrations in those with ovaries and during the reproductive years of the lifespan, influences the types and degrees to which adipose tissue is built, and this does influence metabolism. Lastly, your genes play a big role in your metabolism. While there is a tremendous amount of diversity in the processes of metabolism, each person's body engages in the same overall metabolic processes, which we will outline in this chapter.

Cultural Connection

Metabolism and Biological Sex

Many physiology textbooks will state that biological males have higher or more active metabolisms than biological females. A quick internet search will reveal disparate ranges of suggested calories per day for biological men and women. Are these differences due to factors attributable to biological sex, such as hormonal differences, or are these differences due to overgeneralizations often made about male and female bodies? In our current scientific understanding, the answer is both, but mostly the latter. The central component of the statement that biological males require more calories and have faster metabolisms than biological females has to do with the assumption that biological males have more skeletal muscle mass than their biological female counterparts. This assumption is supported by the fact that testosterone promotes muscle anabolism, and testosterone circulates in much higher levels in biological males, but if you compare a highly active individual with less testosterone to a sedentary individual with more testosterone, the active individual is likely to have more muscle mass despite the lack of testosterone. One established effect of estrogen on the metabolism, however, is that estrogen influences lipid production and storage by adipocytes. These pathways likely have a number of effects on the overall metabolism and physiology of individuals with higher circulating estrogen levels, but they may also help to explain why biological males are at a higher risk of cardiovascular disease, due to their higher circulating triglyceride levels.

24.1 Overview of Metabolic Reactions

Learning Objectives: By the end of this section, you will be able to:

24.1.1 Define metabolism, anabolism, and catabolism, and provide examples of anabolic and catabolic reactions.

24.1.2* Define ATP and explain its role in catabolic and anabolic reactions.

24.1.3 Describe common uses in the body for carbohydrates, fats, and proteins.

24.1.4 Compare and contrast the roles of enzymes and cofactors in metabolic processes.

24.1.5* Define a redox reaction, relate redox reactions to metabolism, and explain the component processes of oxidation and reduction.

24.1.6 Explain the roles of coenzyme A, nicotinamide adenine dinucleotide (NAD), and flavin adenine dinucleotide (FAD) in metabolism.

* Objective is not a HAPS Learning Goal.

LO 24.1.1

Your body contains over 200 types of cells, and a total of approximately 40 trillion cells. Each one of those individual cells is capable of some metabolic processes. The different types of cells carry out different metabolic reactions. Metabolism, the sum of all of the chemical reactions in the body, involves both catabolism and anabolism. **Catabolic reactions** are metabolic reactions that break down larger molecules into subunits—for example, when large food molecules are broken down into smaller ones. This allows the body to absorb the nutrients and utilize their energy. Conversely, **anabolic reactions** are those that build larger molecules from smaller ones—for example, when a protein is synthesized from amino acids. Anabolic reactions use the energy produced by catabolic reactions, so both types of reactions are critical to the maintenance of life.

Because catabolic reactions release energy and anabolic reactions use energy, ideally, these types of reactions occur simultaneously and in equal proportions so that energy balance is maintained. If the net energy change is positive (catabolic reactions release more energy than the anabolic reactions use), then the body stores the excess energy

The Human Anatomy and Physiology Society includes more than 1,700 educators who work together to promote excellence in the teaching of this subject area. The HAPS A&P Learning Outcomes measure student mastery of the content typically covered in a two-semester Human A&P curriculum at the undergraduate level. The full Learning Outcomes are available at https://www.hapsweb.org.

by building large molecules that store energy for later use. On the other hand, if the net energy change is negative (catabolic reactions release less energy than anabolic reactions use), the body uses stored energy to compensate for the deficiency of energy released by catabolism. Within the body, energy is held within adenosine triphosphate (ATP), in the process explained in the next section. The body's two storage forms of energy are **glycogen**, which is a polysaccharide that stores glucose molecules, and triglycerides, which are stored in adipose cells (see the "Anatomy of Energy in the Body" feature).

Learning Connection

Micro to Macro

Some athletes, in an effort to increase athletic performance, take anabolic steroids. Do a little research to find out the following information: What are anabolic steroids? What types of molecules are athletes trying to produce in the body? What types of cells are the athletes trying to enlarge? How would you expect the appearance of these athletes to change after they take anabolic steroids?

24.1a Catabolic Reactions

Catabolic reactions break down large organic molecules into smaller subunits, releasing the energy contained in the chemical bonds. Ideally the cells of the body attempt to capture the released energy, but these energy conversions are not 100 percent efficient. As much as is possible, the energy released during catabolic reactions is directly transferred to the high-energy molecule ATP. ATP, the energy currency of cells, can be used immediately to power molecular machines that support cell, tissue, and organ function. This includes building new tissue and repairing damaged tissue. A small amount of ATP can be stored for future use, but cells have a limited capacity for ATP storage. Most stored energy is stored in the form of glycogen or adipose. Energy released from catabolic reactions that is not captured in ATP is given off as heat, which contributes to our body temperature.

Anatomy of...

Energy in the Body

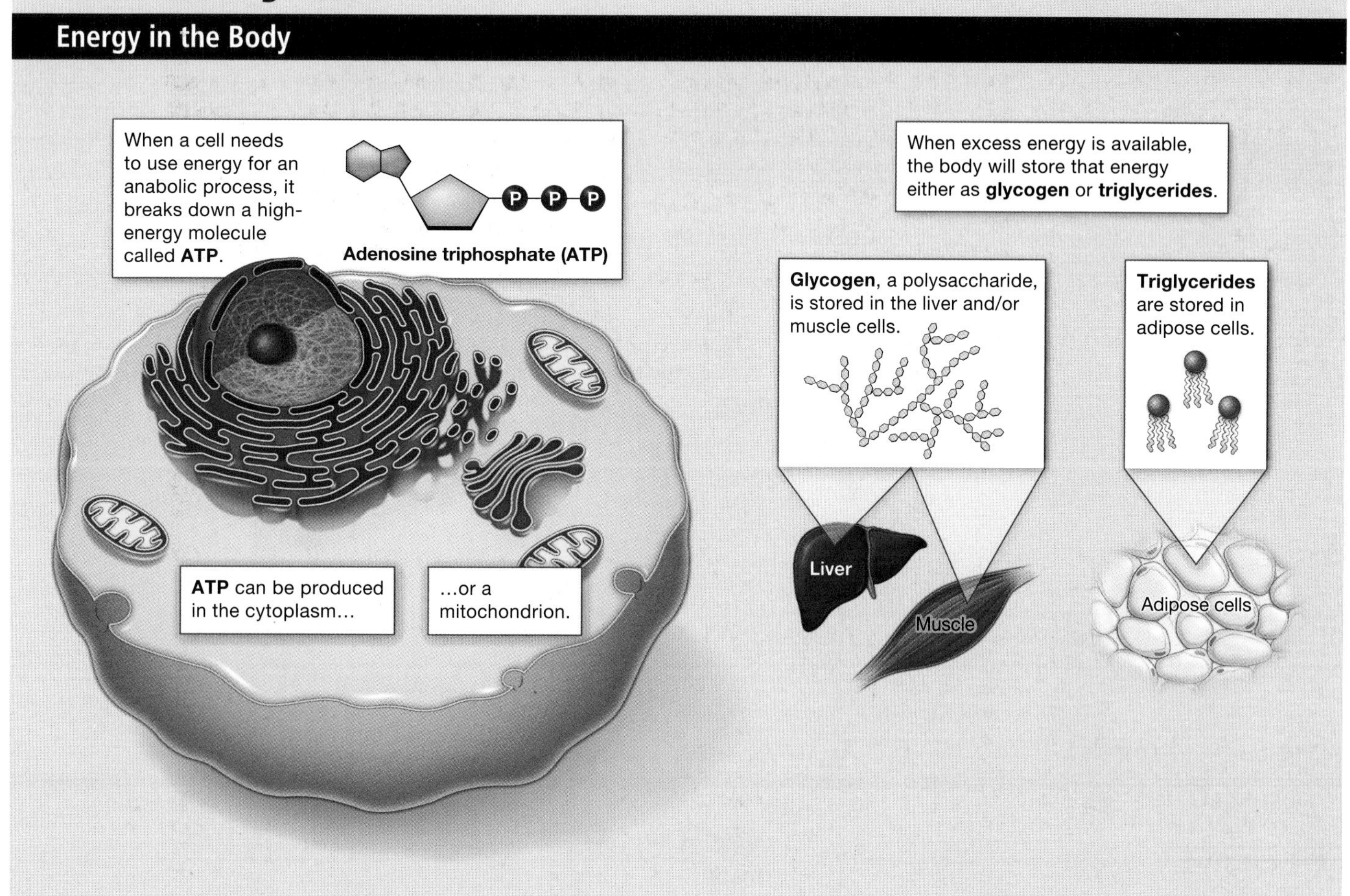
When a cell needs to use energy for an anabolic process, it breaks down a high-energy molecule called ATP.
P P P
Adenosine triphosphate (ATP)
When excess energy is available, the body will store that energy either as glycogen or triglycerides.
Glycogen, a polysaccharide, is stored in the liver and/or muscle cells.
Triglycerides are stored in adipose cells.
Liver
Muscle
Adipose cells
ATP can be produced in the cytoplasm...
...or a mitochondrion.

LO 24.1.2

Structurally, ATP molecules consist of an adenine, a ribose, and three phosphate groups (**Figure 24.1**). The chemical bond between the second and third phosphate groups, termed a *high-energy bond*, represents the greatest source of energy in a cell. It is the first bond that catabolic enzymes break when cells require energy to do work. The products of this reaction are a molecule of adenosine diphosphate (ADP) and a lone phosphate group (Pi). ATP, ADP, and Pi are constantly being cycled through reactions that build ATP to store energy, and reactions that break down ATP to release energy.

LO 24.1.3

Of the four major macromolecular groups (carbohydrates, lipids, proteins, and nucleic acids) that are processed by digestion, carbohydrates are considered the most common source of energy to fuel the body. They take the form of either complex carbohydrates, polysaccharides such as starch and glycogen, or simple sugars (monosaccharides) such as glucose and fructose. Catabolism breaks polysaccharides down into their individual monosaccharides. Among the monosaccharides, glucose is the most common fuel for ATP production in cells, and as such, there are a number of endocrine control mechanisms to regulate glucose concentration in the bloodstream.

Excess glucose is either stored as the complex polymer glycogen as an energy reserve in the liver and skeletal muscles, or it can be converted into fat (triglyceride) in adipose cells (adipocytes).

Dietary lipids can be used immediately to fuel conversion of ADP to ATP, or excess fat is stored in adipose cells in either the subcutaneous tissue under the skin or in other tissues and organs. They are also used to build many of our cellular structures, as well as hormones.

Proteins can be broken down into their monomers, individual amino acids. Amino acids can be used as building blocks of new proteins or broken down further for the production of ATP. Most of the ingested amino acids will be first used in the anabolism

Figure 24.1 Adenosine Triphosphate (ATP)

The currency of cellular energy is ATP. When cells require energy for a metabolic process, they break a phosphate group off of ATP, converting it to adenosine diphosphate (ADP); the energy that is released when the bond is broken fuels cellular work.

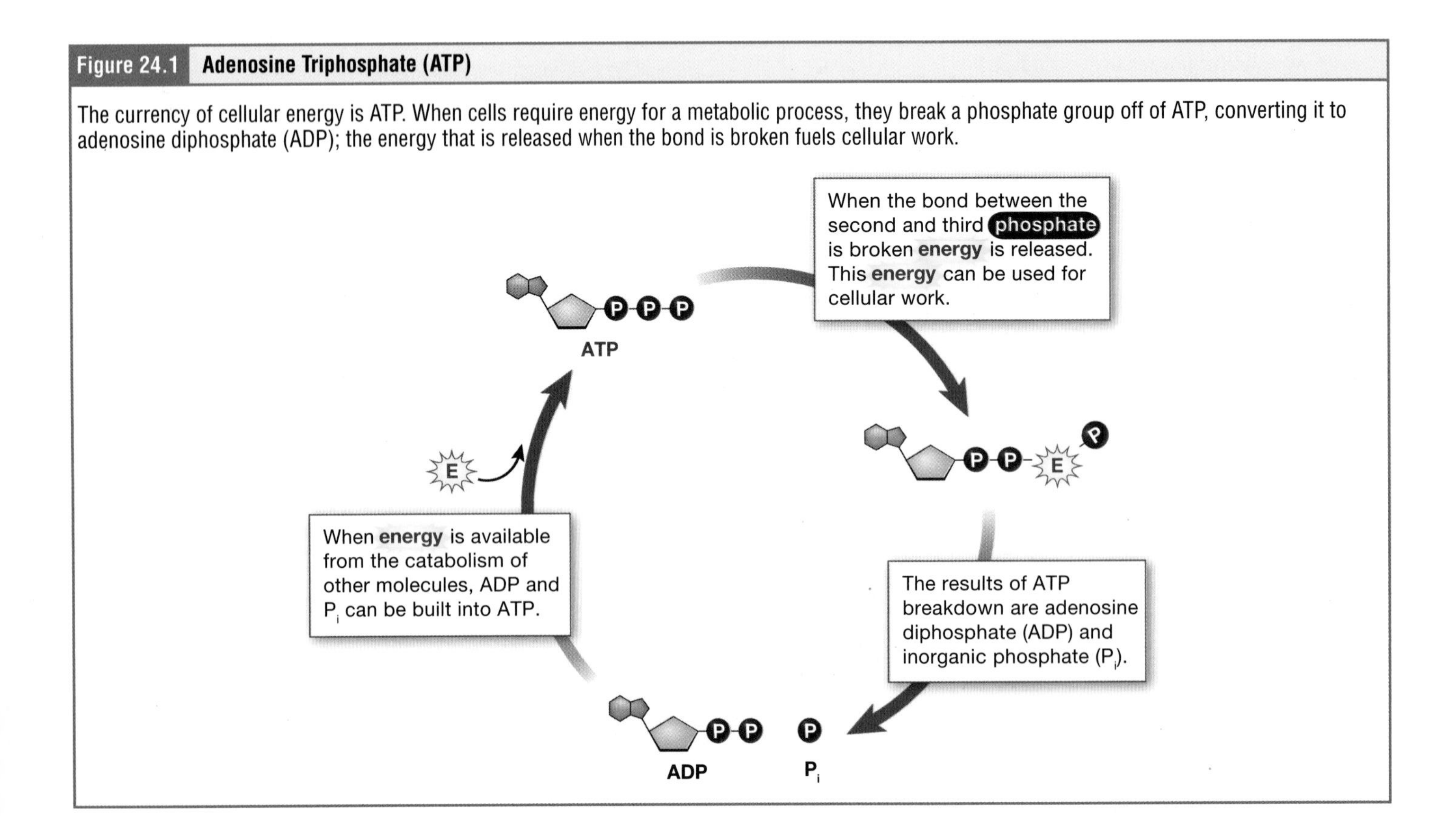

of other proteins; remaining amino acids can be deaminated and used in the conversion of ADP to ATP.

Nucleic acids are present in most of the foods you eat. During digestion, nucleic acids including DNA and various RNAs are broken down into their constituent nucleotides. These nucleotides are readily absorbed and transported throughout the body to be used by individual cells when building new nucleic acid molecules. The catabolism and anabolism of our dietary molecules is summarized in **Figure 24.2**.

24.1b Anabolic Reactions

In contrast to catabolic reactions, *anabolic reactions* involve the joining of smaller molecules into larger ones. Anabolic reactions combine monosaccharides to form polysaccharides, fatty acids to form triglycerides, amino acids to form proteins, and nucleotides to form nucleic acids. These processes require energy in the form of ATP molecules generated by catabolic reactions. Anabolic reactions create new molecules that form new cells and tissues, enable communication among cells, and are vital for the repair and growth of cells, tissues, and organs.

Figure 24.2 Catabolism and Anabolism

Catabolism is the breakdown of molecules; this breakdown releases molecular components and energy. Anabolism is the building of new molecules from smaller molecules and requires energy. The three energy-yielding macromolecules—carbohydrates, proteins, and fats—are each used by the body in different ways.

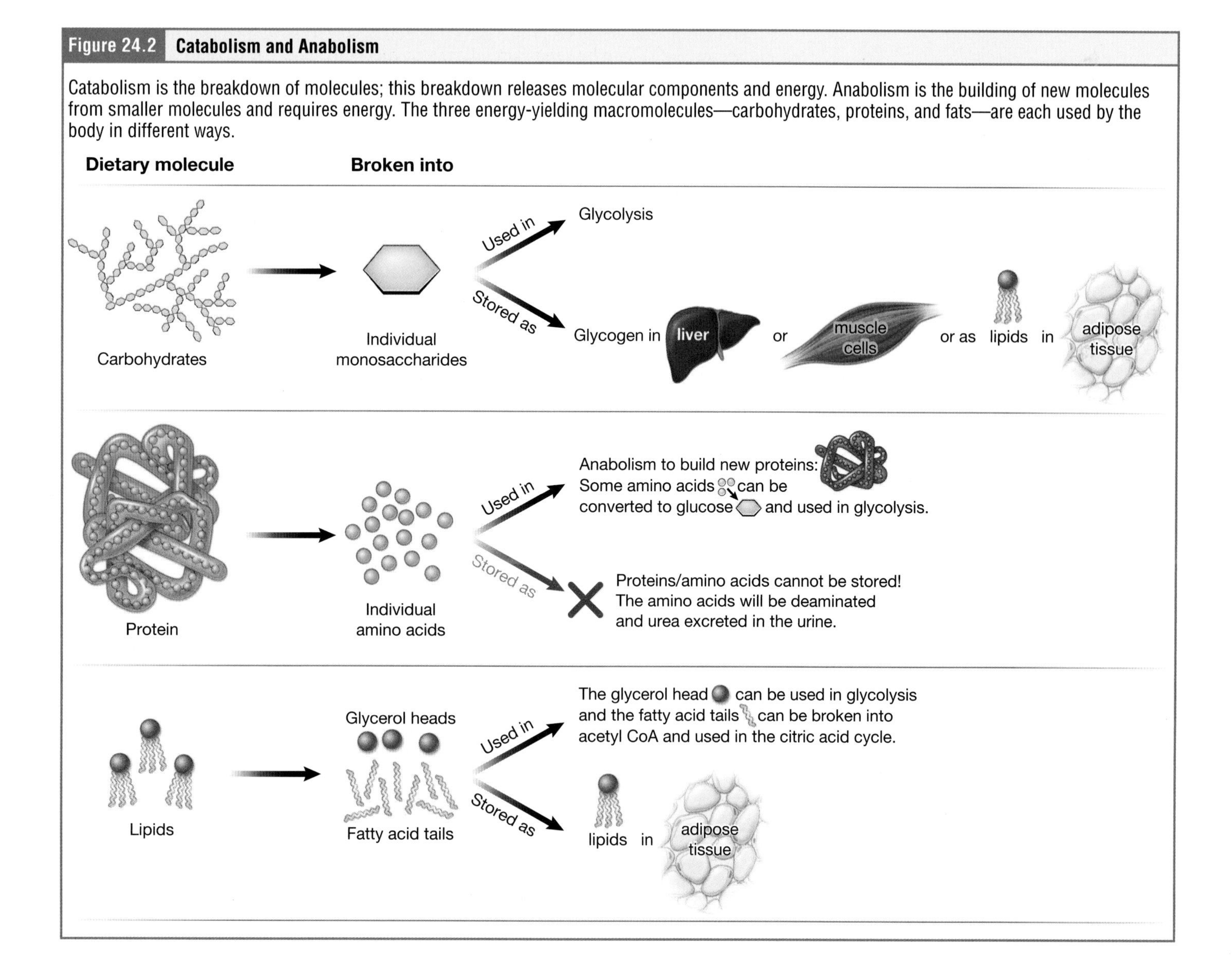

24.1c Metabolic Reactions

LO 24.1.4

Most chemical reactions in the body are accomplished or accelerated by enzymes. **Enzymes** are molecules—usually proteins—that act as a catalyst to a particular chemical reaction. Enzyme activity is tightly regulated in the body in order to maintain homeostasis. Some enzymes require the presence of a cofactor to work. **Cofactors** are substances that bind to enzymes and enable their activity. Cofactors are not usually proteins, but are often small simple molecules; the majority of our vitamins and minerals act as cofactors for enzymes.

oxidation-reduction reaction (redox reaction)

LO 24.1.5

LO 24.1.6

Many of the chemical reactions in the body involve the transfer of electrons from one compound to another. The loss of an electron, or **oxidation**, releases a small amount of energy; both the electron and the energy are then passed to another molecule in the process of **reduction**, or the gaining of an electron (**Figure 24.3A**). These two reactions always happen together in an **oxidation-reduction reaction** (also called a *redox reaction*)—when an electron is passed between molecules, the donor is oxidized and the recipient is reduced (**Figure 24.3B**). The electrons in these reactions commonly come from hydrogen atoms, which consist of an electron and a proton. A molecule gives up a hydrogen atom, in the form of a hydrogen ion (H^+) and an electron, breaking the molecule into smaller parts.

Oxidation-reduction reactions are catalyzed by enzymes that trigger the removal of hydrogen atoms. Like many other enzymes, these enzymes are only active when in the presence of their coenzymes. In these cases, the coenzymes work with enzymes and accept electrons. Once the coenzyme has an extra electron, hydrogen ions are also drawn to the coenzyme. The two most common coenzymes of oxidation-reduction reactions are **nicotinamide adenine dinucleotide (NAD)** and **flavin adenine dinucleotide (FAD)**. Their respective reduced coenzymes are **NADH** and **$FADH_2$**, which are energy-containing molecules used to transfer energy during the production of ATP (**Figure 24.4**). Many food molecules are oxidized in the process of digestion.

Figure 24.3 Redox Reactions

(A) Oxidation reactions are reactions in which an electron is removed from a molecule. A reduction reaction is a reaction in which an electron is added to a molecule. These often occur simultaneously; one molecule is oxidized while the other is reduced. (B) The electrons in these reactions are often obtained from hydrogen, which becomes H^+ once its electron is removed and its proton, which is positively charged, is no longer balanced by the negatively charged electron.

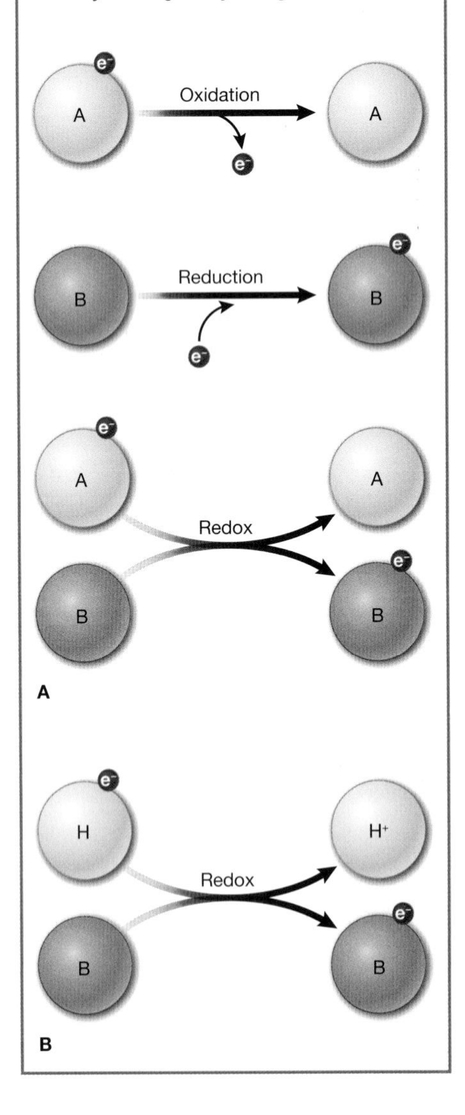

Figure 24.4 Coenzymes of Metabolism

Nicotinamide adenine dinucleotide (NAD) and flavine adenine dinucleotide (FAD) are two common coenzymes in cellular metabolism. These coenzymes accept energy and an electron as nearby molecules are catabolized. The electrons draw H^+ ions toward these molecules.

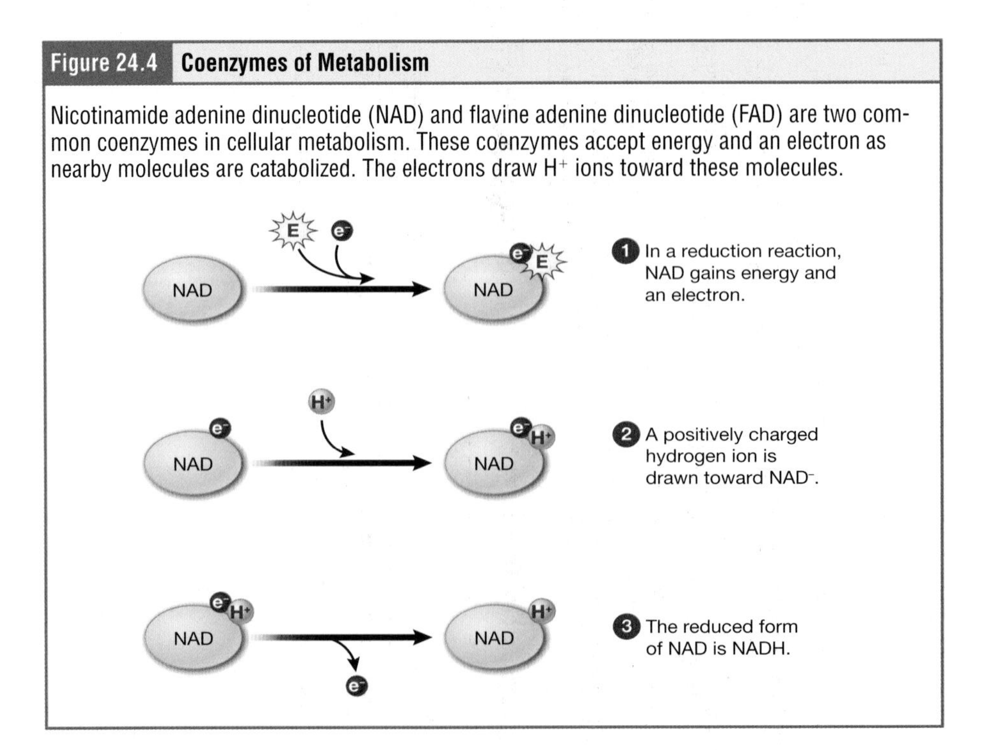

Their energy and electrons are transferred to NAD and FAD in reduction reactions. This energy is harvested in the process of building ATP molecules, which we will discuss in the next section.

NADH (nicotinamide adenine dinucleotide)

$FADH_2$ (Flavin adenine dinucleotide)

Learning Check

1. When you take in fewer calories through food than your body needs (known as *caloric deficit*), your body has to find other means of obtaining energy. This is commonly accomplished by decomposing fat molecules. Which of the following types of reactions best describes this process?
   a. Catabolic reactions
   b. Anabolic reactions
2. Define cofactors.
   a. Molecules that act as catalysts to a particular chemical reaction
   b. Substances that bind to enzymes and enable their activity
   c. Molecules that contain energy and transfer it during reduction reactions
   d. Proteins that serve to maintain homeostasis in the body

24.2 Macronutrients and Metabolism

Learning Objectives: By the end of this section, you will be able to:

24.2.1 Describe the processes of aerobic respiration (e.g., citric acid [Krebs, tricarboxylic acid or TCA] cycle, electron transport chain) in the oxidation of glucose to generate ATP.

24.2.2 Describe the processes of anaerobic respiration (e.g., glycolysis) in the oxidation of carbohydrates to generate ATP.

24.2.3 Describe metabolic pathways that produce or store glucose (e.g., glycogenesis, glycogenolysis, gluconeogenesis).

24.2.4 Describe the anabolic and catabolic processes of fat metabolism (e.g., lipolysis, lipogenesis) and how these processes interact with carbohydrate metabolism.

24.2.5 Describe the anabolic and catabolic processes of protein metabolism (e.g., deamination, transamination) and how these processes interact with carbohydrate metabolism.

24.2.6 Explain the significance of protein intake to nitrogen balance.

As you have read, the goal of digestion is to break down the complex molecules we eat into their monomers. These monomers are then either recycled into the building of new polymers, or further broken down in redox reactions, to fuel the building of ATP, our cells' currency of energy. Of the four types of macromolecules in our food (proteins, lipids, carbohydrates, and nucleic acids), carbohydrates are the main source of energy for ATP building. Carbohydrates are organic molecules composed of carbon, hydrogen, and oxygen atoms. All carbohydrates are sugars or built by adding sugar molecules together. Glucose and fructose are examples of simple sugars; each of these is a monosaccharide, the monomer of carbohydrates. Starch, glycogen, and cellulose are all examples of complex sugars; these are **polysaccharides** and are made of multiple **monosaccharide** molecules. Polysaccharides serve as energy storage (e.g., starch and glycogen) and, in some organisms, polysaccharides are structural components (e.g., cellulose in plants).

Student Study Tip

Think of polysaccharides and monosaccharides as sacks of sugar in storage.

During digestion, carbohydrates are broken down into simple, soluble sugars that can be transported across the intestinal wall and into the circulatory system to be transported throughout the body. Carbohydrate digestion is possible because of the family of amylase enzymes, which break the bonds that join sugars together. Some starches are not digestible to humans because we do not have the enzymes to break their bonds. These carbohydrates are collectively called **fiber**, and they make up a majority of the mass of feces. Those carbohydrates that we can digest are broken into monosaccharides, which are absorbed across the intestinal wall and transported to the tissues. The process of **cellular respiration** breaks apart a monosaccharide, typically glucose, to release the energy stored in its bonds to produce ATP. Cellular respiration can be conceptualized in four linked phases: glycolysis, pyruvate oxidation, citric acid cycle, and electron transport chain (**Figure 24.5**).

LO 24.2.1

24.2a Glycolysis

Cells throughout the body take up circulating glucose (in response to the hormone insulin) and, through a series of reactions called **glycolysis**, transfer some of the energy from within glucose to ADP, forming ATP (**Figure 24.6**). The last step in glycolysis produces the product *pyruvate*.

Glycolysis begins with the phosphorylation of glucose. Phosphorylation is the process of adding a phosphate group to a molecule. Often the phosphate group is the inorganic phosphate (P_i) molecule liberated from the breakdown of ATP into ADP and P_i. In this case, the enzyme hexokinase adds the phosphate group to glucose, forming glucose-6-phosphate. The phosphate group used to form glucose-6-phosphate came from ATP; therefore, this step uses one ATP. Another enzyme, phosphofructokinase, converts glucose-6-phosphate into fructose-6-phosphate. At this point, the sugar is phosphorylated again, and a second ATP is required to donate its phosphate group, forming fructose-1,6-bisphosphate. This six-carbon sugar is split to form two phosphorylated three-carbon molecules—glyceraldehyde-3-phosphate

Figure 24.5 Cellular Respiration of Glucose

Cellular respiration catabolizes glucose molecules through glycolysis, pyruvate oxidation, the citric acid cycle, and the electron transport chain.

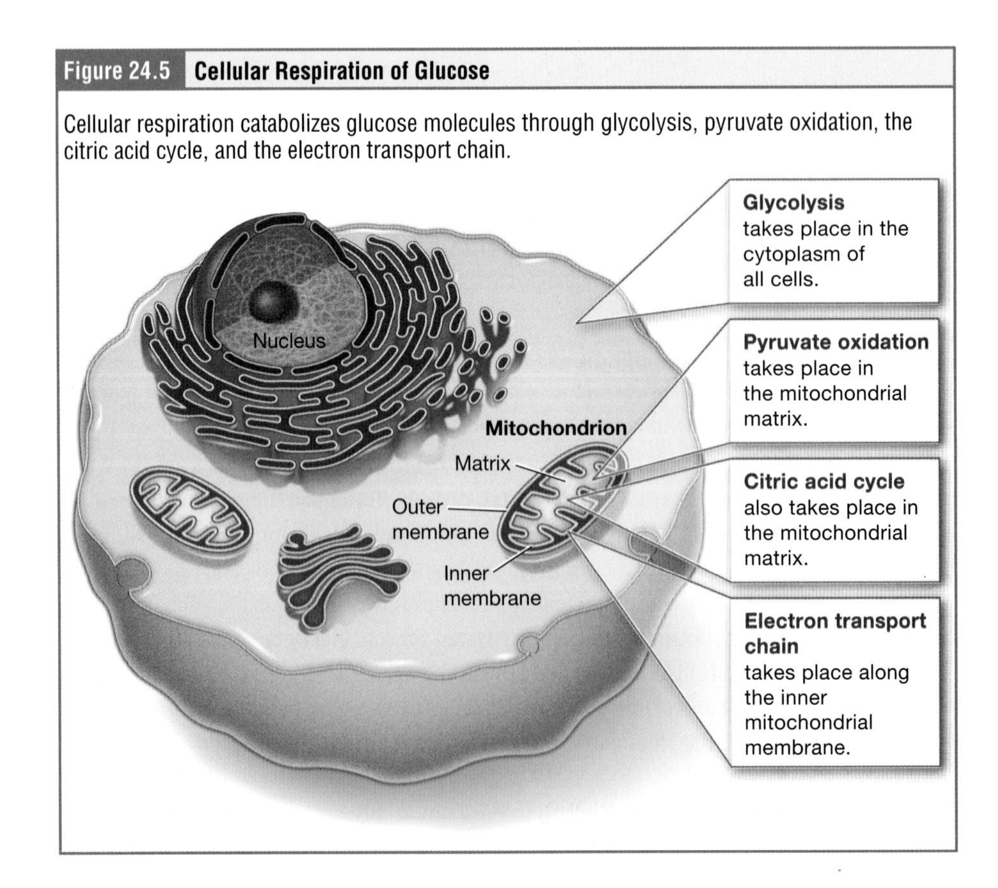
Glycolysis
takes place in the cytoplasm of all cells.
Pyruvate oxidation
takes place in the mitochondrial matrix.
Citric acid cycle
also takes place in the mitochondrial matrix.
Electron transport chain
takes place along the inner mitochondrial membrane.
Nucleus
Mitochondrion
Matrix
Outer membrane
Inner membrane

Figure 24.6 Glycolysis

In glycolysis, a glucose molecule is broken in half, releasing energy and producing two three-carbon molecules of pyruvate.

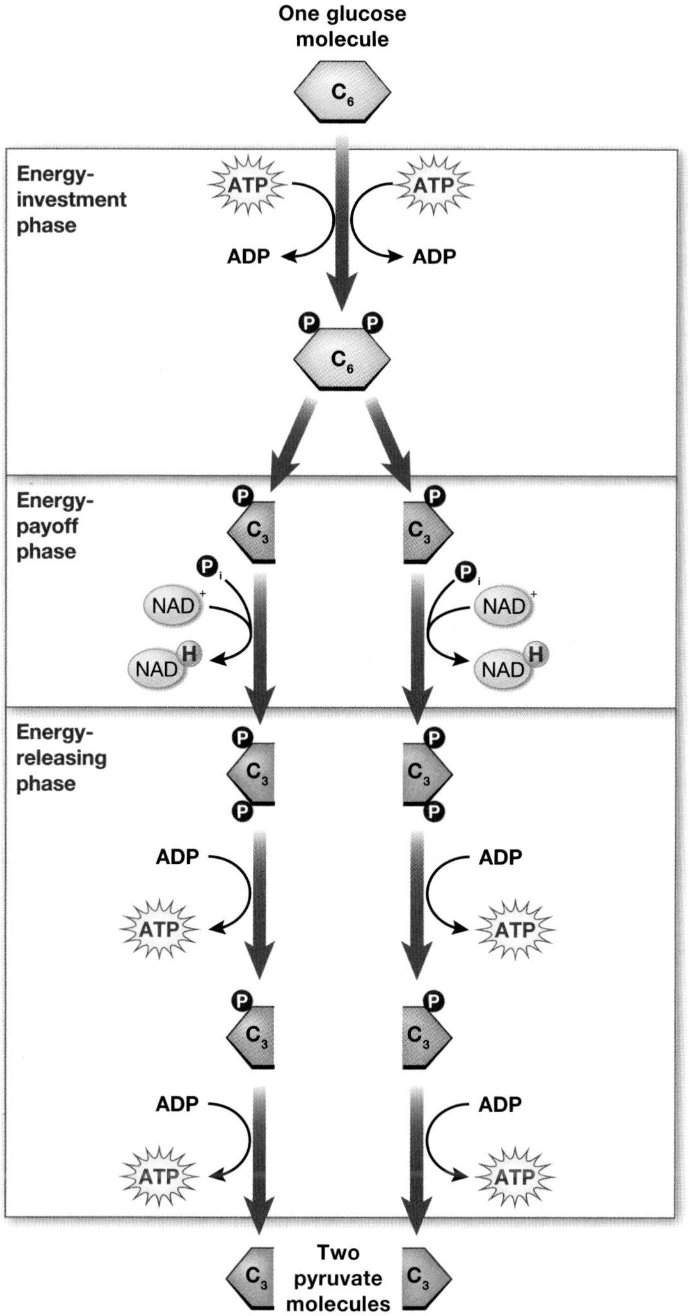

and dihydroxyacetone phosphate—which are both converted into glyceraldehyde-3-phosphate. The glyceraldehyde-3-phosphate is further phosphorylated with groups donated by dihydrogen phosphate present in the cell to form the three-carbon molecule 1,3-bisphosphoglycerate. The energy of this reaction comes from the oxidation of (removal of electrons from) glyceraldehyde-3-phosphate. In a series of reactions leading to pyruvate, the two phosphate groups are then transferred to two ADPs to form two ATPs. Thus, glycolysis uses two ATPs but generates four ATPs, yielding a net

citric acid cycle (Krebs cycle, tricarboxylic acid [TCA] cycle)

gain of two ATPs and two molecules of **pyruvate**. In the presence of oxygen, pyruvate continues on to the **citric acid cycle**, where the additional energy stored in the bonds of the sugar can be extracted.

Glycolysis can be divided into two phases: energy consuming and energy yielding. The first phase is the **energy-consuming phase**, because it requires two ATP molecules to start the reaction. However, the end of the reaction produces four ATPs, resulting in a net gain of two ATP energy molecules.

Glycolysis can be expressed as the following equation:

$$\text{Glucose} + 2\text{ATP} + 2\text{NAD}^+ + 4\text{ADP} + 2\text{P}_i$$
$$\rightarrow 2\text{ Pyruvate} + 4\text{ATP} + 2\text{NADH} + 2\text{H}^+$$

This equation states that glucose, when combined with ATP (the energy source), NAD^+ (a coenzyme that serves as an electron acceptor), and inorganic phosphate, breaks down into two pyruvate molecules, generating four ATP molecules—for a net yield of two ATP—and two energy-containing NADH coenzymes. The NADH that is produced in this process can be used later to produce ATP in the mitochondria if oxygen is present. Importantly, by the end of this process, one glucose molecule generates two pyruvate molecules, two high-energy ATP molecules, and two electron-carrying NADH molecules.

In summary, at the end of glycolysis the cell has a net gain of two ATP molecules, two NADH molecules, and two pyruvate molecules. The cell can use the ATP right away. The NADH and pyruvate molecules can be used for additional energy extraction. The pyruvate molecules have two fates: if oxygen is present, the two pyruvate molecules will go on to the citric acid cycle, where they will be broken down entirely and all of their energy will be eventually converted to ATP. In the absence of sufficient cellular oxygen to undergo the citric acid cycle, the pyruvate molecules will be converted into lactic acid. Note that pyruvates are always converted to lactic acid in humans, but other organisms, notably yeast, are capable of converting pyruvate to other molecules. Collectively, the conversation of pyruvate to other molecules is called *fermentation*.

LO 24.2.2

Anaerobic Respiration In low-oxygen conditions, pyruvate enters an anaerobic pathway in which it is converted into lactic acid (**Figure 24.7A**). Anaerobic respiration occurs in most cells of the body when oxygen is limited. Some cells of the body only use glycolysis; for example, because erythrocytes (red blood cells) lack mitochondria, they must produce their ATP from anaerobic respiration. This is an effective pathway of ATP production for short periods of time, ranging from seconds to a few minutes. The lactic acid produced diffuses into the blood and is carried to the liver, where it is converted back into pyruvate or glucose (**Figure 24.7B**). Sometimes during exercise muscles use ATP faster than oxygen can be delivered to them. In these intense conditions, they supplement their aerobic respiration with additional glycolysis and lactic acid production for rapid ATP production.

Aerobic Respiration In the presence of oxygen, pyruvate can enter the citric acid cycle, where additional energy is extracted. In the citric acid cycle, electrons are transferred from the pyruvate to the receptors NAD^+, GDP, and FAD. The entire original glucose molecule is broken down, its carbon and oxygen atoms becoming the waste product carbon dioxide (CO_2) (**Figure 24.8**). The NADH and $FADH_2$ generated during the citric acid cycle pass electrons on to the electron transport chain, which uses the transferred energy to produce ATP. As the terminal step in the electron transport chain, oxygen is the **terminal electron acceptor** and creates water inside the mitochondria.

Figure 24.7 Anaerobic Respiration

(A) When oxygen is low or mitochondria are unavailable, the pyruvate molecules generated in glycolysis are converted to lactic acid. (B) The lactic acid molecules can travel in the bloodstream to the liver, where they can be converted back to glucose and used in glycolysis again.

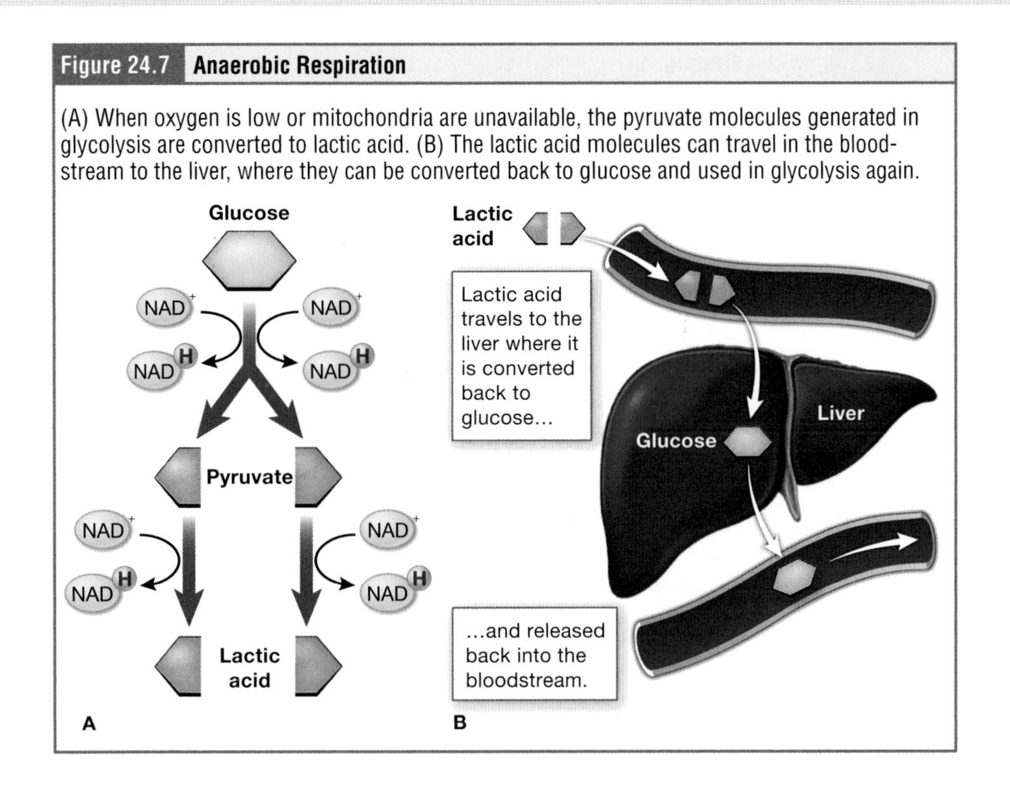

24.2b Citric Acid Cycle

In the citric acid cycle, pyruvate molecules generated during glycolysis are metabolized by enzymes within the mitochondrial matrix (**Figure 24.9**). The energy released by the breakdown of the pyruvate molecules is converted to high-energy molecules, including ATP, NADH, and $FADH_2$. NADH and $FADH_2$ then pass electrons through the electron transport chain in the mitochondria to generate more ATP molecules.

Each of the three-carbon pyruvate molecules that were generated during glycolysis moves from the cytoplasm into the mitochondrial matrix, where each is converted by the enzyme pyruvate dehydrogenase into a two-carbon **acetyl coenzyme A (acetyl CoA)** molecule. This reaction is both an oxidative reaction, meaning that the molecule loses an electron, as well as a decarboxylation reaction, meaning that the molecule loses an atom of carbon. Thus, the three-carbon pyruvate is converted into a two-carbon acetyl CoA molecule, releasing carbon dioxide and transferring two electrons to NAD^+ forming two NADH. Each of the two acetyl CoA molecules enters the citric acid cycle by combining with a four-carbon molecule—oxaloacetate—to form the six-carbon molecule citrate, or citric acid, at the same time releasing the coenzyme A molecule.

Two of the carbons are peeled off the six-carbon citrate molecule systematically, first converting it to a five-carbon molecule and then a four-carbon molecule, ending with oxaloacetate, the beginning of the cycle. It may be helpful to think of oxaloacetate as a bus that drives a circular route; at the beginning of its route it picks up its passengers—the acetyl CoA molecule—and along the way it drops off each of the carbon atoms gained. Oxaloacetate ends the cycle just as it began it, as a four-carbon molecule. Each turn of the cycle—and remember that two pyruvates were produced from glycolysis—will produce one ATP, one $FADH_2$, and three NADH. The $FADH_2$ and NADH will enter the electron transport chain located in the inner mitochondrial membrane.

Figure 24.8 Aerobic Respiration

When both oxygen and mitochondria are available, the pyruvate molecules can be converted to acetyl CoA in the pyruvate oxidation. Acetyl CoA can then be fully broken down in the citric acid cycle. As it is catabolized, the released energy is transferred to NAD and FAD, yielding NADH and $FADH_2$. These molecules then enter the electron transport chain, where their energy can be used to convert ADP to ATP.

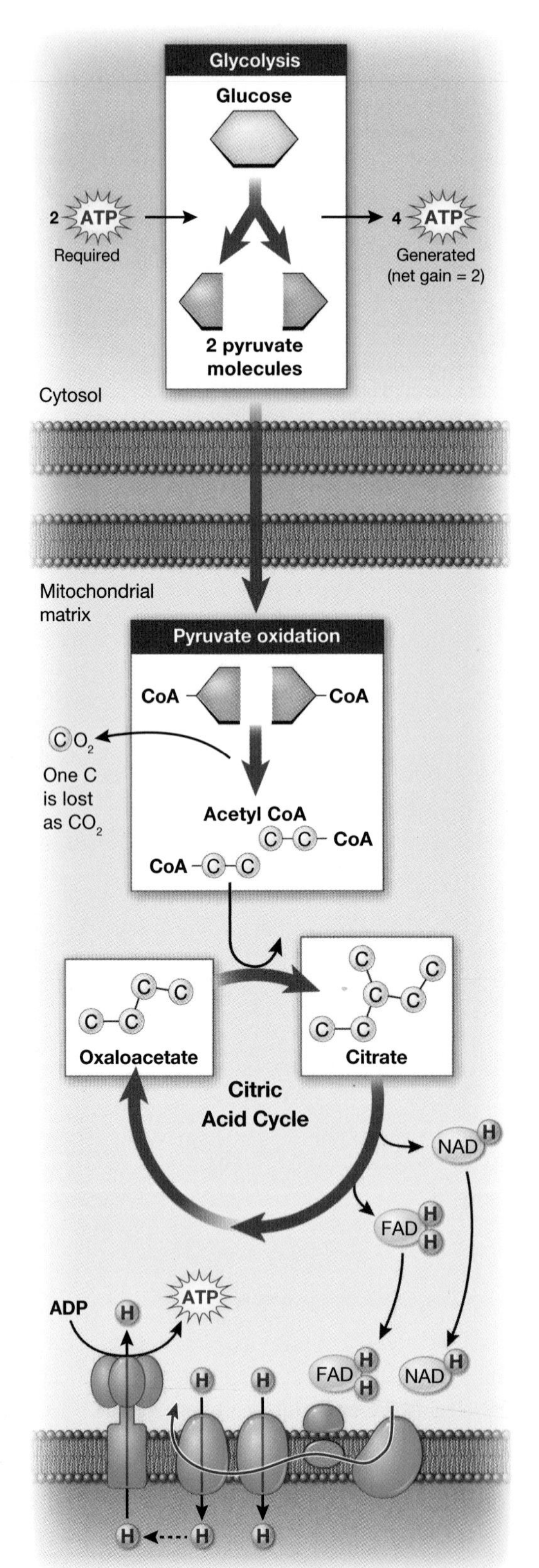

The process of glycolysis takes place in the cytosol of the cell. It splits glucose into two halves, each called *pyruvate*, and yields enough energy for the cell to gain two ATP.

The pyruvate molecules are converted to acetyl CoA in the mitochondrial matrix in the process called *pyruvate oxidation*.

Acetyl CoA is joined to citric acid in the citric acid cycle.

In the electron transport chain, the energy held by NADH and $FADH_2$ is used to chelate an electrochemical gradient that fuels the conversion of ADP to ATP.

Figure 24.9 The Citric Acid Cycle

Each pyruvate molecule that is generated by glycolysis is converted into a two-carbon acetyl CoA molecule in the process of pyruvate oxidation. The acetyl CoA is systematically processed through the citric acid cycle and produces high-energy NADH, $FADH_2$, and ATP molecules.

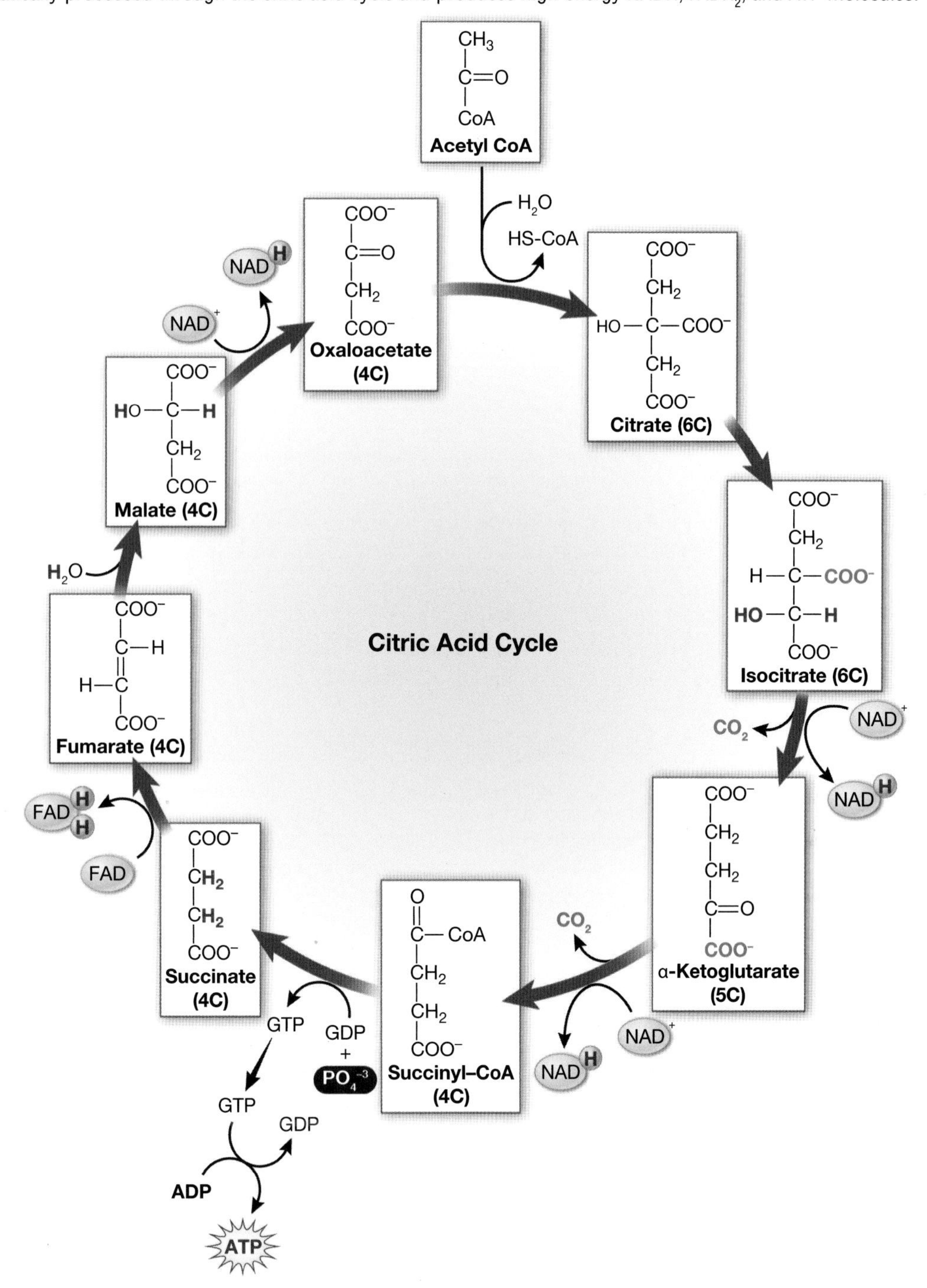

24.2c Oxidative Phosphorylation and the Electron Transport Chain

The **electron transport chain (ETC)** is a system of proteins that can convert NADH and $FADH_2$ back to NAD^+ and FAD and utilize the energy to generate ATP. Electrons from NADH and $FADH_2$ are transferred through protein complexes embedded in the inner mitochondrial membrane by a series of enzymatic reactions. The ETC consists of a series of four enzyme complexes (Complex I–Complex IV) and two

DIGGING DEEPER: Pyruvate Dehydrogenase Complex Deficiency and Phenylketonuria

Pyruvate dehydrogenase complex deficiency (PDCD) and phenylketonuria (PKU) are genetic disorders. Pyruvate dehydrogenase is the enzyme that converts pyruvate into acetyl CoA, the molecule necessary to begin the citric acid cycle to produce ATP. With low levels of the pyruvate dehydrogenase complex (PDC), the rate of cycling through the citric acid cycle is dramatically reduced. This results in a decrease in the total amount of energy that is produced by the cells of the body. PDC deficiency results in a neurodegenerative disease that ranges in severity, depending on the levels of the PDC enzyme. It may cause developmental defects, muscle spasms, and death. Treatments can include diet modification, vitamin supplementation, and gene therapy; however, damage to the central nervous system usually cannot be reversed.

PKU affects about 1 in every 15,000 births in the United States. People afflicted with PKU lack sufficient activity of the enzyme phenylalanine hydroxylase and are therefore unable to break down phenylalanine into tyrosine adequately. Because of this, levels of phenylalanine rise to toxic levels in the body, which results in damage to the central nervous system and brain. Symptoms include delayed neurological development, hyperactivity, mental retardation, seizures, skin rash, tremors, and uncontrolled movements of the arms and legs. Every infant in the United States and Canada is tested at birth to determine whether PKU is present. The earlier a modified diet is begun, the less severe the symptoms will be. The person must closely follow a strict diet that is low in phenylalanine to avoid symptoms and damage.

coenzymes (ubiquinone and cytochrome c), which act as electron carriers and proton pumps used to transfer H^+ ions into the space between the inner and outer mitochondrial membranes (**Figure 24.10**). The ETC couples the transfer of electrons between a donor (such as NADH) and an electron acceptor (such as O_2) with the transfer of protons (H^+ ions) across the inner mitochondrial membrane, enabling the process of **oxidative phosphorylation**. In the presence of oxygen, energy is passed, stepwise, through the electron carriers to gradually collect the energy needed to attach a phosphate to ADP and produce ATP. The role of molecular oxygen, O_2, is as

Figure 24.10 The Electron Transport Chain

The electron transport chain is a series of proteins embedded in the inner mitochondrial membrane that uses the energy from the $FADH_2$ and NADH molecules to pump the hydrogen ions removed from these coenzymes across the membrane, creating an electrochemical gradient. This gradient then drives ATP synthase to convert ADP into ATP.

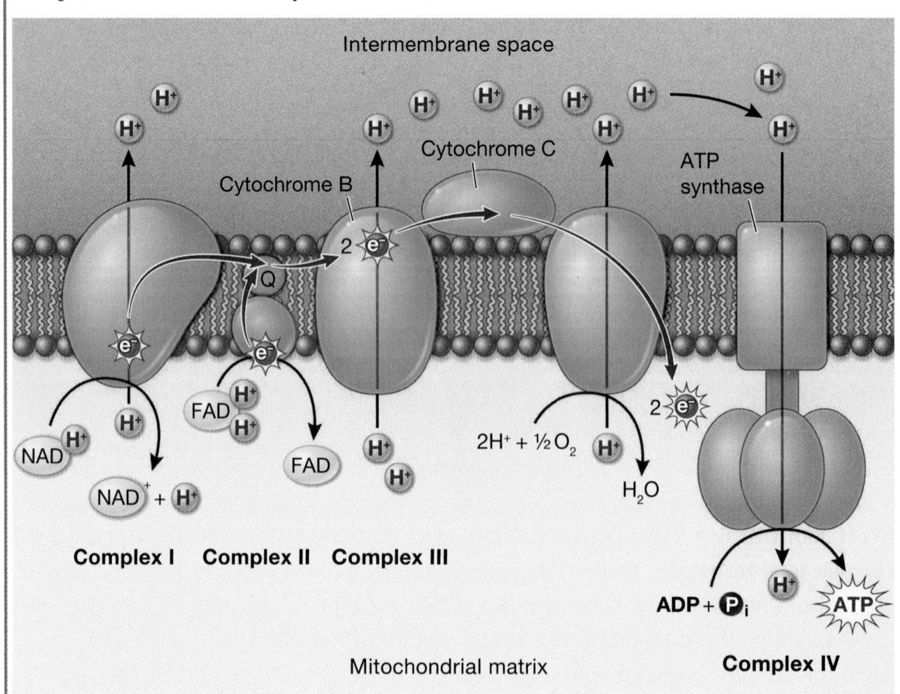

the terminal electron acceptor for the ETC. This means that once the electrons have passed through the entire ETC, giving off energy along the way, they must be passed to a final molecule, which is oxygen. Once on oxygen, the electrons, and their negative charge, attack H^+ ions from the matrix. New water molecules are formed. This is the basis for your need to breathe in oxygen. Without oxygen, electron flow through the ETC ceases.

When NADH and $FADH_2$ reach the ETC, they release their H^+ atoms as well as the electron that holds the H^+ atom to the rest of the molecule. The released electrons are passed along the chain by the protein carriers, such as cytochrome c, which are reduced when they receive the electron and oxidized when passing it on to the next carrier. Each of these reactions releases a small amount of energy, and that energy is used to pump the released H^+ ions across the inner membrane. The accumulation of these H^+ ions in the space between the membranes creates a concentration gradient with more H^+ ions on one side of the membrane than the other.

Also embedded in the inner mitochondrial membrane is an amazing protein complex called **ATP synthase**. It resembles a revolving door and acts effectively as a turbine that is powered by the flow of H^+ ions across the inner membrane. As H^+ ions flow through ATP synthase, down their concentration gradient, the shaft of the complex rotates. The kinetic energy of this rotation enables other portions of ATP synthase to push ADP and Pi together, generating ATP.

In accounting for the total number of ATP produced per glucose molecule through aerobic respiration, it is important to remember the following points:

- A net of two ATP are produced through glycolysis (four produced and two consumed during the energy-consuming stage). However, these two ATP are used for transporting the NADH produced during glycolysis from the cytoplasm into the mitochondria. Therefore, the net production of ATP during glycolysis is zero.
- In all phases after glycolysis, the number of ATP, NADH, and $FADH_2$ produced must be multiplied by two to reflect how each glucose molecule produces two pyruvate molecules.
- In the ETC, about three ATP are produced for every oxidized NADH. However, only about two ATP are produced for every oxidized $FADH_2$. The electrons from $FADH_2$ produce less ATP, because they start at a lower point in the ETC (Complex II) compared to the electrons from NADH (Complex I) (see Figure 24.10).

Therefore, for every glucose molecule that enters aerobic respiration, a net total of 30–36 ATPs are produced; this figure varies based on cell type (**Figure 24.11**).

Other Nutrients and Their Roles in Metabolism So far we have discussed how glucose is utilized in cellular respiration, but of course, our diet is filled with different kinds of molecules. Glucose is the most common starter molecule for cellular respiration. For this reason, our bodies often generate glucose molecules from non-glucose nutrients. **Gluconeogenesis** is the synthesis of new glucose molecules from pyruvate, lactate, glycerol, or the amino acids alanine or glutamine. This process takes place primarily in the liver during periods of low glucose—that is, under conditions of fasting, starvation, and in extreme low/no-carbohydrate diets. So, the following question can be raised: Why would the body create something it has just spent a fair amount of effort to break down? In addition to starting cellular respiration in all cells, some cells can only use glucose as an energy source; therefore, it is essential that the body maintain a minimum blood glucose concentration. Blood glucose levels are tightly regulated to fluctuate only within a limited homeostatic range. When the blood glucose concentration falls below that certain point, gluconeogenesis is required to raise the blood concentration to normal.

Student Study Tip

Think about oxygen as if it were a magnet at the end of a chain, pulling all of the ions toward it. If there were no magnet, they wouldn't move.

Learning Connection

Chunking

Make a chart with the following three columns: glycolysis, citric acid cycle, and oxidative phosphorylation. Now compare these three metabolic processes by writing down the following information about each process: location in which it occurs in the cell, whether oxygen is required, and how many molecules of ATP are generated for each glucose molecule that is being broken down.

Student Study Tip

ATP synthase does exactly what the name suggests. It creates ATP from ADP and P_i.

Learning Connection

Broken Process

There are several types of genetic mitochondrial diseases, including MELAS (Mitochondrial Encephalomyopathy, Lactic Acidosis, and Stroke-like episodes). Since the citric acid cycle and oxidative phosphorylation occur in the mitochondria, it is not surprising that mitochondrial diseases can cause a variety of serious medical issues. Do an internet search to find out which cell types contain the most mitochondria. What types of symptoms would you expect to see in a person with a disease such as MELAS, due to its harmful effects on mitochondrial function?

Figure 24.11 An Overview of Carbohydrate Metabolism

All carbohydrates can be broken down into monomers, which can be fully broken down through the phases of cellular respiration. The energy from these molecules is used to drive the conversion of ADP into ATP.

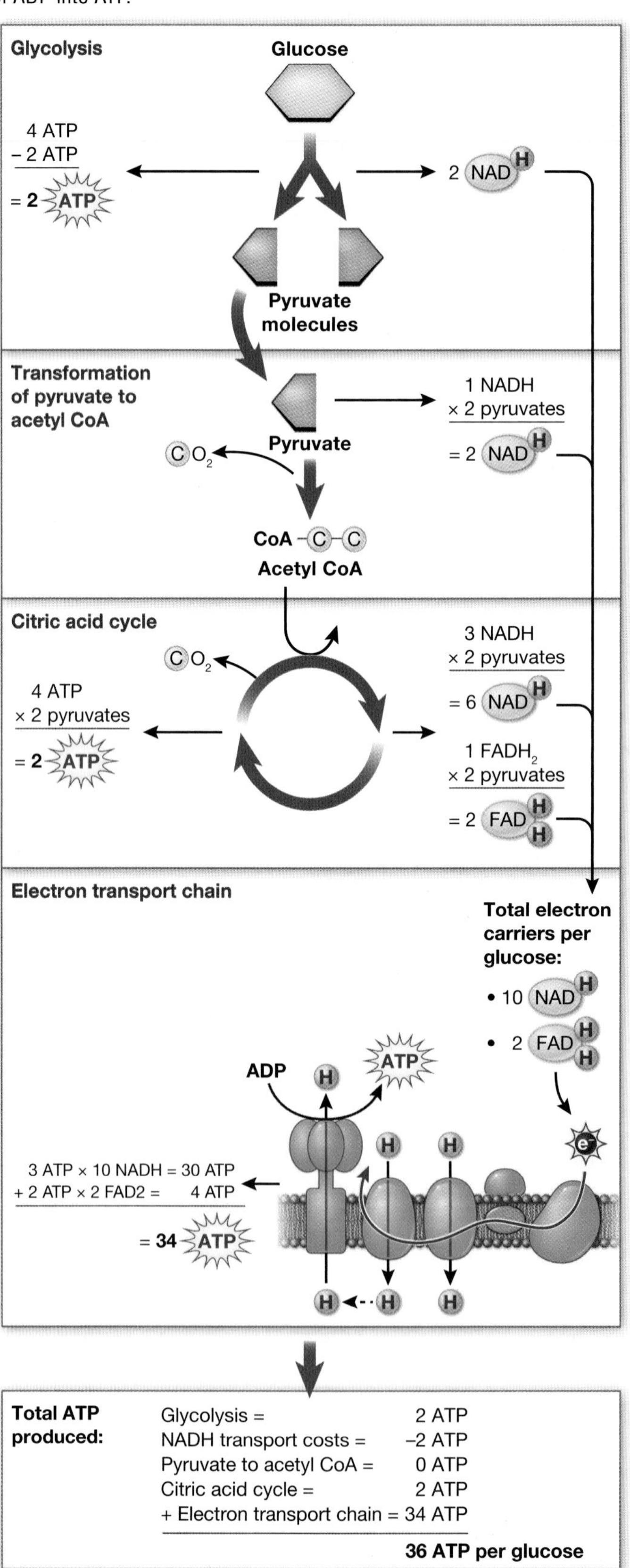

Apply to Pathophysiology

Cyanide Poisoning

Between September 29 and October 5 of 1982, seven individuals died of mysterious causes within a few days of each other. Police were able to determine that all of the murdered individuals lived in suburban Chicago and all had taken Tylenol brand acetaminophen before their deaths. Eventually it was discovered that the murderer had added a toxin to the Tylenol capsules and then put the bottles back on the shelves of Chicago area stores. Excellent detective work, effective use of the media to warn the public, and a nationwide recall of all Tylenol products by the manufacturer within days of the original incident halted the sale and use of Tylenol until all tainted bottles were discovered. The murderer was never identified, and the case remains open to this day.

Prior to their death, all the patients experienced dizziness, confusion, headache, rapid breathing, and vomiting. Most deaths were rapid, occurring within a few hours of the onset of symptoms.

The autopsy reports of all seven victims revealed some similarities:

- All tissues showed damage consistent with death by hypoxia (suffocation or lack of oxygen).
- None of the victims had signs of physical injury.
- Tissue samples (heart, lung, kidney, and liver) all showed necrosis (cellular death).

Lab Values	Average Victim Levels	Healthy Levels
Blood Oxygen level	110 mmHg	75–100 mmHg
Glucose	99 μM	80–100 μM

1. Which cellular organelle(s) might be the source of the problem?
   A) Nucleus
   B) Cytoplasm
   C) Lysosomes
   D) Mitochondria

A blood test for toxins revealed cyanide in each of the victim's bloodstream. Cyanide binds irreversibly to Cytochrome C. When bound, cyanide prevents the transfer of electrons to oxygen.

2. Which stage of cellular respiration is cytochrome C required for?
   A) Glycolysis
   B) Transfer
   C) Citric acid cycle
   D) Electron transport chain

A detailed biochemical analysis can be performed on cells of each victim's organs. Predict whether each metabolite level in the following table will be within the normal range, higher than normal, or lower than normal. For each metabolite listed in the table, describe in your own notes its role in cellular respiration.

Metabolite	Average Victim Levels	Healthy Levels
ATP	#3 A) Higher than normal B) Lower than normal C) Within the normal range	1.6–2.9 nMol/10^6
Glucose	#4 A) Higher than normal B) Lower than normal C) Within the normal range	100 μM
Pyruvate	#5 A) Higher than normal B) Lower than normal C) Within the normal range	25 μM
NAD+	#6 A) Lower than normal B) Higher than normal C) Within the normal range	75 μM
NADH	#7 A) Within the normal range B) Lower than normal C) Higher than normal	50 μM

8. If, as the victims were dying, paramedics had arrived and given them supplemental oxygen, would they have survived?
   A) Yes
   B) No

Pyruvate and lactate—a molecule related to lactic acid—can both be converted to glucose after anaerobic glycolysis. But gluconeogenesis can convert portions of lipids and proteins to glucose as well, as might happen during lipid or protein-rich, but carbohydrate-poor, diets. Triglycerides are the most common form of dietary lipid, and during their digestion they can be broken into glycerol and fatty acid chains. The glycerol can undergo gluconeogenesis to become glucose and begin cellular respiration. There are 20 amino acids commonly found in the proteins of the human diet; most of these are used for anabolism—that is, the building of new proteins. Two of these—alanine and glutamine—can be converted to glucose in the process of gluconeogenesis. Because of this ability, the body can use proteins as a fuel source when blood sugar is low.

LO 24.2.3

In an effort to avoid the need for gluconeogenesis, the body efficiently stores extra glucose as glycogen. This process is called **glycogenesis**. Glycogen is stored both in the liver and in skeletal and cardiac muscle cells. The muscle cells will break down glycogen when they have an urgent need for glucose, but the liver breaks down glycogen and adds the resulting glucose to the bloodstream for any cell in the body to use. The breakdown of glycogen is **glycogenolysis** (**Figure 24.12**). Glycogenolysis can occur any time that there is a need for glucose, such as between meals. The ability to store glucose as glycogen and break it down as needed allows us to space out our meals, or even to fast for an entire day, such as during particular religious observances.

LO 24.2.4

Fats (or triglycerides) within the body are ingested as food or synthesized from carbohydrate precursors. Lipids can be used directly in cellular respiration. The glycerol head of a triglyceride can be used in glycolysis and the fatty acid chains can be broken into small chunks and converted to acetyl CoA for use in the citric acid cycle. When dietary fats are not required for immediate fueling of cellular respiration, they will be stored in adipocytes. **Lipolysis** describes the breakdown of stored fat for release into the bloodstream and use in cellular respiration. When carbohydrates are available in high amounts after a carbohydrate-rich meal, they will first be stored as glycogen, but when glycogen stores are full, carbohydrates can be converted to lipids through the process of **lipogenesis** and stored as fats in adipose tissues. If the production of acetyl CoA from the oxidation of fatty acids outpaces its use in the citric acid cycle, the excess acetyl CoA is converted to ketone bodies. These **ketone bodies** can serve as a fuel source if glucose levels are too low in the body. Ketones serve as fuel in times of prolonged starvation or when patients suffer from uncontrolled diabetes and cannot utilize most of the circulating glucose. In both cases, fat stores are liberated to generate energy through the Krebs cycle and will generate ketone bodies when too much acetyl CoA accumulates.

ketone bodies (ketones)

Student Study Tip

Lipo- means "fat," which we see in these definitions. *Lysis* means "dissolve," and *genesis* means "formation." Use these roots to help remember these new terms.

Figure 24.12 Blood Glucose Homeostasis

The levels of glucose in the blood are kept within a homeostatic range through the actions of glycogenesis—the production of glycogen—and glycogenolysis, the breakdown of glycogen.

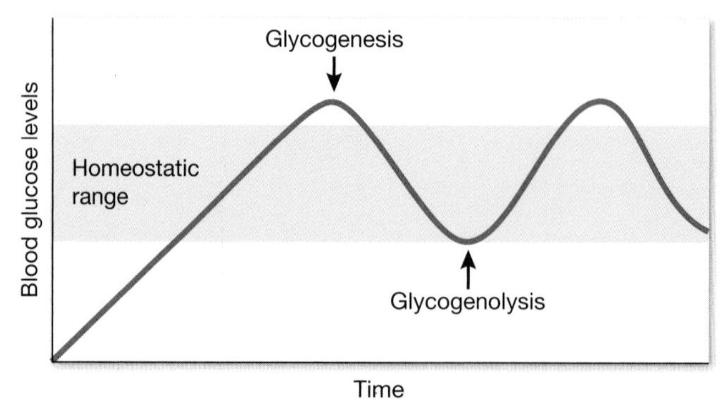

DIGGING DEEPER: Cushing Syndrome and Addison's Disease

As might be expected for a fundamental physiological process like metabolism, errors or malfunctions in metabolic processing lead to a pathophysiology or—if uncorrected—a disease state. Protein or enzyme malfunction can be the consequence of a genetic alteration or mutation. However, normally functioning proteins and enzymes can also have deleterious effects if their availability is not appropriately matched with metabolic need. For example, excessive production of the hormone cortisol gives rise to Cushing syndrome. Clinically, Cushing syndrome is characterized by rapid weight gain (especially in the trunk and face region), depression, and anxiety. Cortisol can be overproduced by the body itself, but exogenous forms of cortisol, such as the drug prednisone, can also cause similar symptoms.

Patients with Cushing syndrome can exhibit high blood glucose levels and show slow growth, accumulation of fat between the shoulders, weak muscles, bone pain (because cortisol causes proteins to be broken down to make glucose via gluconeogenesis), and fatigue. Other symptoms include excessive sweating (hyperhidrosis), capillary dilation, and thinning of the skin, which can lead to easy bruising. The treatments for Cushing syndrome are all focused on reducing excessive cortisol levels.

Insufficient cortisol production is equally problematic. Adrenal insufficiency, or Addison's disease, is characterized by the reduced production of cortisol from the adrenal gland. It can result from malfunction of the adrenal glands—they do not produce enough cortisol—or it can be a consequence of decreased ACTH availability from the pituitary. Patients with Addison's disease may have low blood pressure, paleness, extreme weakness, fatigue, slow or sluggish movements, lightheadedness, and salt cravings due to the loss of sodium and high blood potassium levels (hyperkalemia). Patients also may suffer from loss of appetite, chronic diarrhea, vomiting, mouth lesions, and patchy skin color. Diagnosis typically involves blood tests and imaging tests of the adrenal and pituitary glands. Treatment involves cortisol replacement therapy, which usually must be continued for life.

Organs that have classically been thought to be dependent solely on glucose, such as the brain, can actually use ketones as an alternative energy source. This keeps the brain functioning when glucose is limited. When ketones are produced faster than they can be used, they can be broken down into CO_2 and acetone. The acetone is removed by exhalation. One symptom of ketogenesis is that the patient's breath smells sweet. The carbon dioxide produced can acidify the blood, leading to diabetic ketoacidosis, a dangerous condition in diabetics.

LO 24.2.5

A protein is the most diverse macromolecule in terms of form, size, and shape. Much of the body is made of protein. Cell-signaling receptors, signaling molecules, structural members, enzymes, intracellular trafficking components, extracellular matrix scaffolds, ion pumps, ion channels, oxygen and CO_2 transporters (hemoglobin)—this is not even a complete list! There is protein in bones (collagen), muscles (actin and myosin), and tendons; the hemoglobin that transports oxygen; and enzymes that catalyze all biochemical reactions. Protein is also used for growth and repair.

Most dietary protein is broken down to a collection of amino acids, and these amino acids are used anabolically to build new proteins throughout the body. When other fuel sources are not available, proteins can also fuel cellular respiration. Proteins are not stored for later use, so excess proteins must be converted into glucose or triglycerides to be stored.

When the body has excess sugars, it can store them as glycogen or convert them to lipids. When the body has excess lipids it can store them, but when amino acids exist in excess the body has no capacity or mechanism for their storage; thus, they are either converted into glucose or ketones, or they are broken down. Carbohydrates and fats are made of carbon, hydrogen, and oxygen, but proteins have an additional element, nitrogen. Therefore, amino acid decomposition, has to get rid of the nitrogen as nitrogenous waste. High concentrations of nitrogen are toxic and so the body must have mechanisms to get rid of excess nitrogenous wastes.

LO 24.2.6

The **urea cycle** is a set of biochemical reactions that produces urea from nitrogenous waste in order to prevent a toxic level of nitrogen buildup in the body.

Cultural Connection

Stress and Metabolism

The human species has evolved through hundreds of thousands of years of conditions in which food scarcity has been a very real threat to homeostasis. The body's limitless ability to store energy in the forms of glycogen and triglycerides are the product of evolution under the constraints of caloric limitations. Stress is defined physiologically as any real or perceived threat to our homeostasis. Whether that threat is starvation or predation, stressful circumstances often required early humans to go without food for some period of time. Therefore, our main two stress hormones, cortisol and epinephrine, both enable the body to endure periods of time without food. Epinephrine promotes the breakdown of glycogen, which adds glucose to the bloodstream, increasing blood glucose concentrations. Epinephrine also promotes the breakdown of adipose tissue, adding glycerol and fatty acids to the bloodstream. Therefore, when epinephrine levels are high—which is associated with short-term stress responses and emergency situations—the amount of available energy for fueling cellular respiration soars. Should this individual need to fight off a predator or run to catch a meal, this energy will fuel that process. Cortisol, which is associated with longer term or resigned stress, promotes protein breakdown in the bones and muscles while simultaneously encouraging the liver to increase the rate of gluconeogenesis from amino acid precursors. Cortisol also inhibits growth and reproduction, preserving the blood nutrient supplies for enduring the period of stress. These stress mechanisms work incredibly well to allow the body to survive for periods of time without nutrients or at a higher level of nutrient demand. However, when stress occurs in a chronic fashion, these hormones can have detrimental effects on health and metabolism.

When amino acids are broken down, their amine (nitrogen-containing) groups are converted to ammonia; this process is called **deamination**. In these reactions, an amine group, or ammonium ion, from the amino acid is exchanged with a keto group on another molecule. This **transamination** event creates one molecule that can enter the citric acid cycle and an ammonium ion that enters into the urea cycle to be eliminated.

In the urea cycle, ammonium is combined with CO_2, resulting in urea and water. The urea is eliminated through the kidneys in the urine. Urea is the compound that makes your urine yellow; you must get rid of it regularly, but the volume and color of the urine will vary depending on how much water you have to dilute it.

Learning Check

1. Which of the following molecules converts glucose-6-phosphate into fructose-6-phosphate?
   a. ATP
   b. Phosphofructokinase
   c. Hexokinase
   d. Pyruvate
2. When strength training, muscles are forced to use ATP faster than oxygen can be delivered. What is the product of the reaction that enables muscles to effectively and rapidly produce ATP?
   a. Pyruvate
   b. Glucose
   c. Lactic acid
   d. Carbon dioxide
3. How many ATP molecules will one glucose molecule produce during glycolysis, the citric acid cycle, and the electron transport chain?
   a. 1
   b. 2
   c. 6–10
   d. 30–36
4. Where does gluconeogenesis take place?
   a. Liver
   b. Bone marrow
   c. Stomach
   d. Skeletal muscles

24.3 Metabolic States of the Body

Learning Objectives: By the end of this section, you will be able to:

24.3.1 Compare and contrast carbohydrate, fat, and protein metabolism in the fed (absorptive) and fasted (postabsorptive) states.

Think about all that you accomplish between your meals; you might exercise, study for an exam, write some code or a short story, have a conversation, or accomplish a few errands. Your brain requires a continuous supply of glucose throughout all of these activities, and yet you only eat a few times a day. Humans have evolved elaborate systems for storing and releasing energy so that we can eat infrequently and utilize our other time for activities. If there were no method in place to store excess energy, you would need to eat constantly in order to meet energy demands. Distinct mechanisms are in place to facilitate energy storage, and to make stored energy available during times of fasting and starvation.

LO 24.3.1

24.3a The Absorptive State

absorptive state (fed state)

The **absorptive state**, or the *fed state*, occurs after a meal when your body is digesting the food and absorbing the nutrients. Digestion begins the moment you put food into your mouth, and continues throughout the intestines. The constituent parts of these carbohydrates, fats, and proteins are transported across the intestinal wall and enter the bloodstream (sugars and amino acids) or the lymphatic system (fats). Once in the circulation, the nutrients are taken up by the liver, adipose tissue, or muscle cells that will use or store the molecules.

The absorptive state begins during your meal, and then lasts for a few hours after you have stopped eating. Depending on the amounts and types of nutrients ingested, the absorptive state can last for up to four hours. The fate of the absorbed nutrients depends on the exact physiological needs of the body during the absorptive state. Therefore, because our physiological needs vary, even two people eating the same meal might experience different events during and duration of the absorptive state.

The ingestion of food and the rise of glucose concentrations in the bloodstream stimulate pancreatic beta cells to release **insulin** into the bloodstream. The action of insulin is to permit the movement of glucose from the blood into liver hepatocytes and adipose and muscle cells. Once inside these cells, glucose is immediately converted into glucose-6-phosphate. By doing this, a concentration gradient is established where glucose levels are always higher in the blood than in the cells. This allows for glucose to continue moving from the blood into the cells where it is needed. Insulin also stimulates the storage of glucose as glycogen in the liver and muscle cells where it can be used later. Insulin also promotes the synthesis of protein in muscle. As you will see, muscle protein can be catabolized and used as fuel in times of starvation.

If energy is exerted shortly after eating, the dietary fats and sugars that were just ingested will be processed and used immediately for energy. If not, the excess glucose is stored as glycogen in the liver and muscle cells. Both excess glucose and excess fat can be stored as triglycerides in adipose tissue. Figure 24.13 summarizes the metabolic processes occurring in the body during the absorptive state.

24.3b The Postabsorptive State

postabsorptive state (fasting state)

The **postabsorptive state**, or the *fasting state*, occurs when the food from the last meal has been digested, absorbed, and stored. Most people will enter the postabsorptive state overnight, but skipping or spacing out meals during the day puts your body in the

Figure 24.13 The Absorptive State

The absorptive state occurs while nutrients are being added to the blood throughout food breakdown and nutrient absorption. During this state, food is present in the lumen of the GI tract and insulin is being secreted by the pancreas.

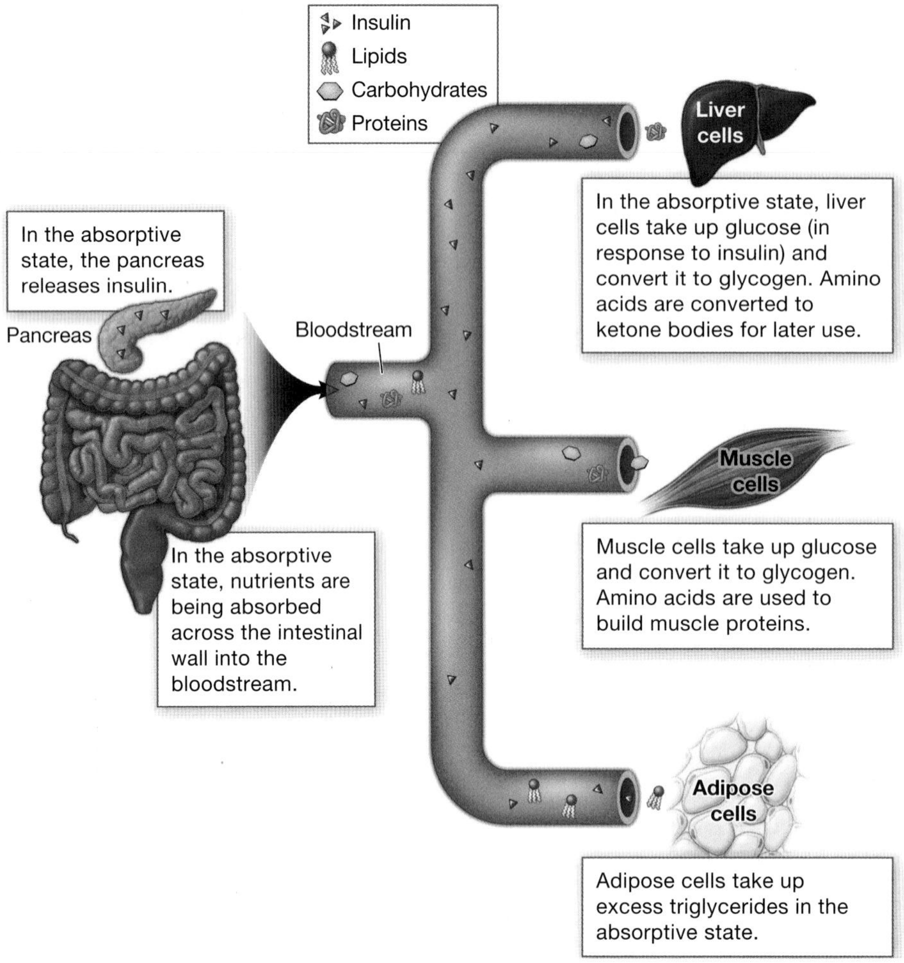

Insulin
Lipids
Carbohydrates
Proteins

In the absorptive state, the pancreas releases insulin.

Pancreas

Bloodstream

In the absorptive state, nutrients are being absorbed across the intestinal wall into the bloodstream.

Liver cells

In the absorptive state, liver cells take up glucose (in response to insulin) and convert it to glycogen. Amino acids are converted to ketone bodies for later use.

Muscle cells

Muscle cells take up glucose and convert it to glycogen. Amino acids are used to build muscle proteins.

Adipose cells

Adipose cells take up excess triglycerides in the absorptive state.

Learning Connection

Chunking

During the absorptive state, excess glucose is stored as glycogen in certain cell types. During the postabsorptive state, stored glycogen is broken down into glucose to use for energy. Using the information in this chapter and in Chapter 17, make a list of hormones that cause glucose to be converted into glycogen and those which cause glycogen to be converted into glucose. Hint: Pay particular attention to insulin, glucagon, cortisol, and growth hormone. After doing this, you will develop a better understanding of the hormonal control of metabolism.

postabsorptive state as well. Glucose levels in the blood begin to drop as the blood glucose is taken up and used by the cells of the body. To preserve blood glucose homeostasis, the body will break down stored glycogen. In response to the decrease in glucose, insulin levels drop and the secretion of another hormone, **glucagon**, increases. In response to the decrease in insulin, glycogen and triglyceride storage slows. However, due to the demands of the tissues and organs, blood glucose levels must be maintained in the normal range of 80–120 mg/ dL. Glucagon acts upon the liver cells, where it inhibits the synthesis of glycogen and stimulates the breakdown of glycogen, releasing glucose. As a result, blood glucose levels begin to rise. Gluconeogenesis will also begin in the liver to replace the glucose that has been used by the peripheral tissues.

After ingestion of food, fats and proteins are processed as described previously; however, the glucose processing changes a bit. The peripheral tissues preferentially absorb glucose. The liver, which normally absorbs and processes glucose, will not do so after a prolonged fast. The gluconeogenesis that has been ongoing in the liver will continue after fasting to replace the glycogen stores that were depleted in the liver. After these stores have been replenished, excess glucose that is absorbed by the liver will be converted into triglycerides and fatty acids for long-term storage. **Figure 24.14** summarizes the metabolic processes occurring in the body during the postabsorptive state.

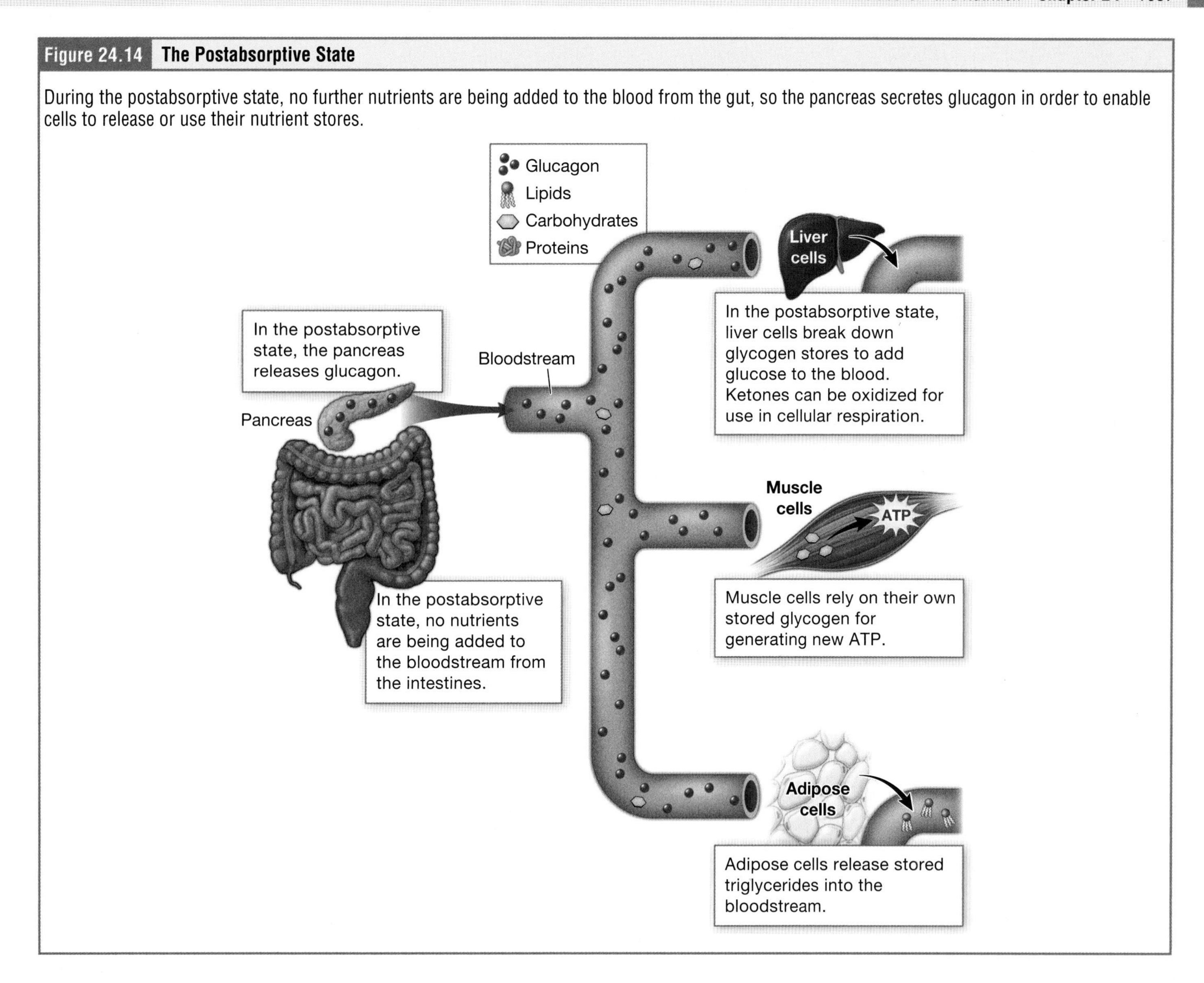

Glucagon
Lipids
Carbohydrates
Proteins
In the postabsorptive state, the pancreas releases glucagon.
Pancreas
Bloodstream
Liver cells
In the postabsorptive state, liver cells break down glycogen stores to add glucose to the blood. Ketones can be oxidized for use in cellular respiration.
Muscle cells
ATP
Muscle cells rely on their own stored glycogen for generating new ATP.
In the postabsorptive state, no nutrients are being added to the bloodstream from the intestines.
Adipose cells
Adipose cells release stored triglycerides into the bloodstream.

Figure 24.14 The Postabsorptive State

During the postabsorptive state, no further nutrients are being added to the blood from the gut, so the pancreas secretes glucagon in order to enable cells to release or use their nutrient stores.

24.3c Starvation

When the body is deprived of nourishment for an extended period of time, it goes into "survival mode." The priority for survival is to provide enough glucose or fuel for the brain. Because glucose levels are very low during starvation, glycolysis will shut off in cells that can use alternative fuels. For example, muscles will switch from using glucose to fatty acids as fuel. Ketones can be used in glucose-dependent cells as gluconeogenesis will begin in the liver to generate glucose from any available pyruvate, lactate, and alanine amino acids. If starvation continues, muscle cells will begin to break down their proteins to liberate amino acids for gluconeogenesis; glycerol from fatty acids can also be liberated and used as a source for gluconeogenesis.

After several days of starvation, ketone bodies become the major source of fuel for the heart and other organs. As starvation continues, fatty acids and triglyceride stores are used to create ketones for the body. This prevents the continued breakdown of proteins that serve as carbon sources for gluconeogenesis. Once these stores are fully depleted, proteins from muscles are released and broken down for glucose synthesis. Overall survival is dependent on the amount of fat and protein stored in the body.

Learning Connection

Micro to Macro

Kwashiorkor is a condition associated with malnutrition in some developing countries. The word *kwashiorkor* is derived from the Ga language of Ghana and refers to a child weaned from its mother's breast by the birth of another baby. The swollen abdomen of a child with this condition is due to a deficiency of protein, not calories. What do you think protein deficiency has to do with swelling of the abdomen? Think about plasma proteins, their role in osmotic balance, and edema. See if you can put this all together and figure out the relationship between protein deficiency and abdominal edema.

Learning Check

1. Which of the following scenarios best describes the postabsorptive state?
   a. Larry was on a day-long hike when they realized they forgot their food.
   b. Sheri was running late and ate a protein bar on their way to work.
   c. Guillie caught the stomach bug and could not eat for several days.
   d. Marlie drinks five smoothies a day as part of their liquid-only diet.
2. Hypoglycemia is a condition that can occur in someone with diabetes. In this case, the person does not have enough glucose in their blood, which can sometimes be life-threatening. In which phase is this most likely to occur?
   a. The absorptive phase
   b. The postabsorptive phase

24.4 Energy and Heat Balance

Learning Objectives: By the end of this section, you will be able to:

24.4.1 Explain the importance of thermoregulation in the body.

24.4.2 Explain how various organ systems and behaviors participate in thermoregulation.

24.4.3 Define metabolic rate and describe the conditions under which basal metabolic rate is measured.

24.4.4 Describe factors that affect metabolic rate.

Imagine you are sitting in an auditorium of 200 students, or standing on a ski slope with 300 skiers. If you ask each person what their body temperature is, they can probably tell you a number that is within a degree or two of being correct, and the two groups will have similar body temperatures despite being in very different environments. But this same crowd would likely be much less precise if asked to guess their heart rate or other physiological variables. Our body temperature is tightly regulated through a process called **thermoregulation**, in which the body can maintain its temperature within certain boundaries, even when the surrounding temperature is very different. The core temperature of the body remains steady at around 36.5–37.5°C (97.7–99.5°F). The process of ATP generation through cellular respiration has a byproduct of heat production; the production of ATP is fairly steady day to day and contributes to the majority of the body's heat. When we have a time when ATP production increases, such as exercise, the body temperature also begins to increase. To maintain homeostasis and prevent too great an increase in temperature, sweating generally occurs to cool the body.

LO 24.4.1

With these examples we can understand that thermoregulation is tightly controlled; but why is its control so important? The answer lies in metabolism. Many of the body's essential molecules, including its enzymes, are proteins. Proteins are large molecules that are held together with a variety of bonds, including hydrogen bonds. These bonds are exquisitely sensitive to changes in temperature and if the body's temperature becomes too high, the proteins may denature and be destroyed. The rate at which chemical reactions can take place requires the constant movement of the molecules of the body. But all molecular movement, including diffusion and binding events, will occur faster at higher temperatures. The body temperature must also be prevented from becoming too low. If the body's temperature falls too low, the chemical reactions similarly cannot occur quickly enough to sustain life.

The hypothalamus in the brain controls body temperature through a mechanism that works similarly to the thermostat in a home (**Figure 24.15**). If the

Figure 24.15 Temperature Homeostasis in the Body

Body temperature is tightly regulated by the hypothalamus.

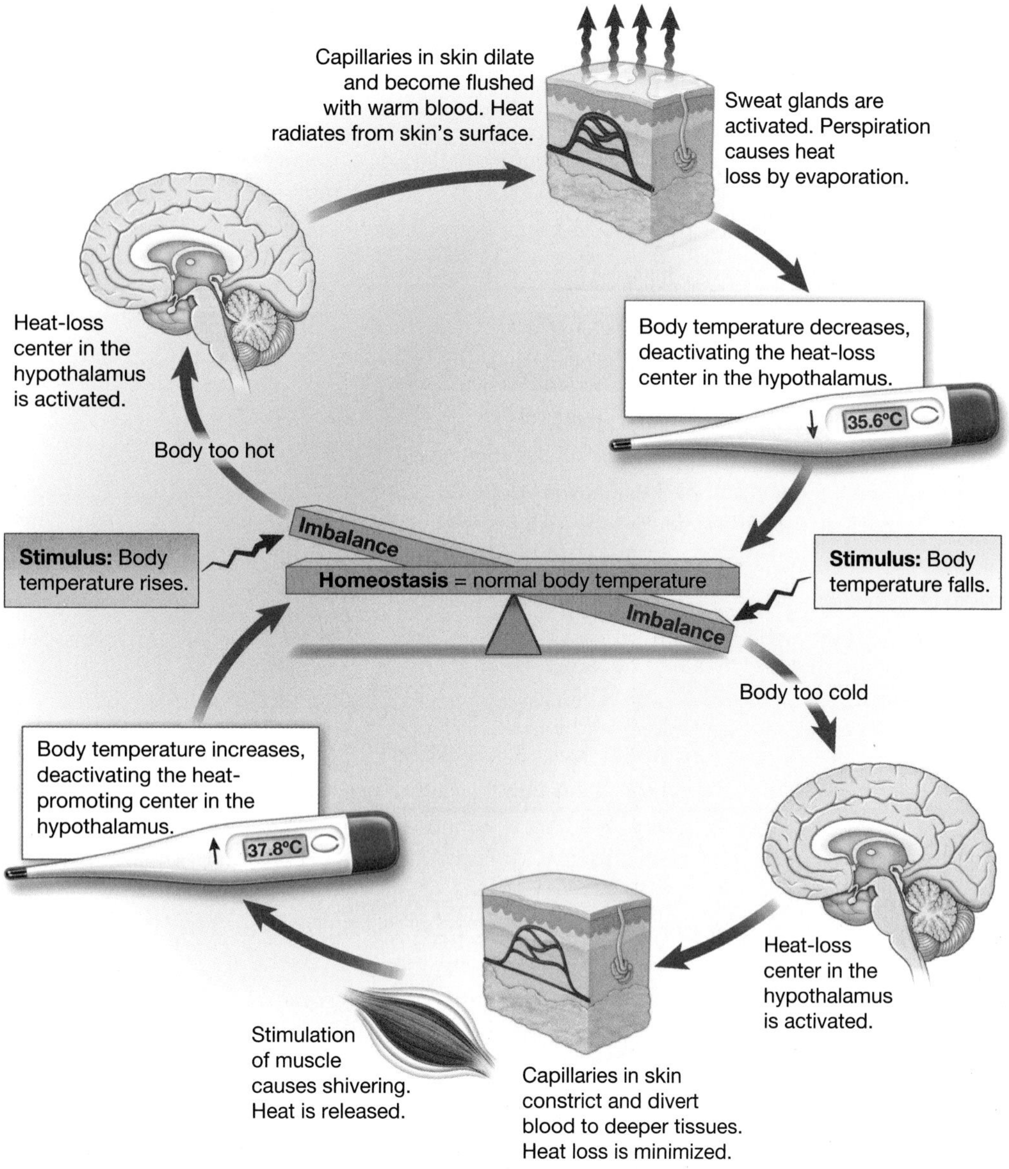

Capillaries in skin dilate and become flushed with warm blood. Heat radiates from skin's surface.

Sweat glands are activated. Perspiration causes heat loss by evaporation.

Heat-loss center in the hypothalamus is activated.

Body temperature decreases, deactivating the heat-loss center in the hypothalamus.

35.6°C

Body too hot

Imbalance

Stimulus: Body temperature rises.

Homeostasis = normal body temperature

Stimulus: Body temperature falls.

Imbalance

Body too cold

Body temperature increases, deactivating the heat-promoting center in the hypothalamus.

37.8°C

Heat-loss center in the hypothalamus is activated.

Stimulation of muscle causes shivering. Heat is released.

Capillaries in skin constrict and divert blood to deeper tissues. Heat loss is minimized.

temperature is too high, the hypothalamus can initiate several processes to lower it. These include increasing the circulation of the blood to the surface of the body to allow for the release of heat off the surface of the skin and initiation of sweating to allow evaporation of water on the skin to cool its surface. Conversely, if the temperature falls below the homeostatic range, the hypothalamus can initiate shivering to generate heat. During shivering, the muscle contractions require an increase in ATP and therefore more heat is generated. In addition, thyroid hormone will stimulate more energy use and heat production by cells throughout the body. Thermoregulation may also trigger behavioral mechanisms such as huddling close to someone, curling the body into a position to minimize heat loss, or pulling on additional layers of clothing or blankets.

LO 24.4.2

24.4a Metabolic Rate

LO 24.4.3

Student Study Tip

If someone asks you to do something you don't want to do, remind them that you can only spend 20 percent of your energy on activity.

LO 24.4.4

The **metabolic rate** is the amount of energy consumed minus the amount of energy expended by the body. The **basal metabolic rate (BMR)** describes the amount of daily energy expended by humans at rest in the postabsorptive state. It measures how much energy the body needs for normal, basic, daily activity. About 70 percent of all daily energy expenditure comes from the basic functions of the organs in the body. Another 20 percent comes from physical activity, and the remaining 10 percent is necessary for body thermoregulation or temperature control. This rate will be higher if a person is more active or has more muscle mass. As you age, the BMR generally decreases as the percentage of muscle mass decreases.

Learning Check

1. Why is having a fever for a long period of time a problem?
   a. Your metabolic rate will increase, and you will be unable to absorb nutrients.
   b. Your enzymes will denature, and you will be unable to carry out chemical reactions.
   c. Your metabolic rate will decrease, and you will be unable to store nutrients.
   d. Your organs will require more energy to carry out their basic functions.
2. Helen works on strength training five times a week while Roberta works on cardiovascular training two times a week. Compare their basal metabolic rates.
   a. Helen's basal metabolic rate is higher than Roberta's.
   b. Roberta's basal metabolic rate is higher than Helen's.
   c. They both have high basal metabolic rates.
   d. They both have low basal metabolic rates.

24.5 Nutrition and Diet

Learning Objectives: By the end of this section, you will be able to:

24.5.1 Define calorie and kilocalorie.

24.5.2 Describe energy yields per gram for carbohydrates, fats, and proteins.

24.5.3 Describe the neural and chemical control of appetite and food intake.

24.5.4 Define nutrient, essential nutrient, and non-essential nutrient.

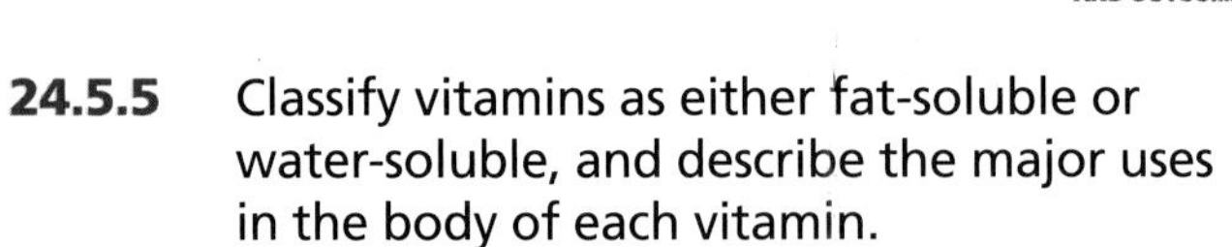

24.5.5 Classify vitamins as either fat-soluble or water-soluble, and describe the major uses in the body of each vitamin.

24.5.6 Name the major minerals (e.g., calcium, sodium, potassium) and trace elements (e.g., iron, iodine, zinc) and their roles within various physiological processes in the body.

The carbohydrates, lipids, and proteins in the foods you eat are used for energy to power molecular, cellular, and organ system activities. Notably, any excess energy is stored primarily as fats. So what happens if you eliminate carbohydrates, or eat primarily proteins and fats? Is there a reason to drink protein shakes as mean replacements? Do we need to take multivitamins? These questions are almost all unique to each individual and in many cases we as scientists are still unraveling the mysteries of nutrients and metabolism. The next few sections will describe, scientifically, some of what we do know about the roles that different nutrients play in our physiology.

24.5a Food and Metabolism

LO 24.5.1

The amount of energy that is needed or ingested per day is measured in calories. The nutritional **Calorie (C)** is the amount of heat it takes to raise 1 kilogram (1,000 grams) of water by 1 degree Celsius. This is different from the calorie (c) used in the physical sciences, which is the amount of heat it takes to raise 1 gram of water by 1 degree Celsius. When we refer to "calorie," we are referring to the nutritional Calorie, which is actually a kilocalorie (or 1,000 calorie units).

Calorie (C) (food calorie, large calorie, kilocalorie)

LO 24.5.2

Our various nutrients range in their calorie density. Whereas carbohydrates and proteins both yield approximately four calories per gram, fats yield nine calories per gram of food.

The amount of calories a person needs per day to sustain (or carry out) daily activities is often listed as between 1,500 and 2,000. However, this number will vary widely among individuals. The total number of calories needed by one person is dependent on their body mass, muscle mass, age, activity level, and the amount they exercise per day. If exercise is regular part of one's day, more calories are required, and individuals who are younger or have more muscle mass will use more calories each day. Weight and body mass index (BMI) are imperfect measures of caloric needs because these measurements cannot differentiate between body weight from adipose tissue and body weight from muscle mass. Elite athletes, such as the tennis player Serena Williams, have been outspoken about the inaccuracy of weight measurements in individuals with very high muscle mass. Sophisticated scales can differentiate and help paint a more accurate picture of BMI and caloric needs.

LO 24.5.3

The relationship between our BMR and our appetite is not direct or linear. Daily, we should take in the number of calories dictated by our BMR plus the amount of calories we use in exercise. Our appetite is influenced by both of these factors along with neural factors and several hormones. There is a satiety center in the central nervous system that play the main role in regulating feeding behavior. Neurons that monitor the degree in the stretch of the wall of the stomach send signals to this center carrying information about the relative fullness or emptiness of the stomach. In addition several hormones influence satiety as well. The hormone **ghrelin** is a hormone that drives our hunger. It is secreted by the parietal cells in the stomach and signals to the brain that the stomach is empty. When we eat, ghrelin production slows and our brains theoretically stop driving feeding behavior. **Leptin** is a hormone produced by the adipocytes. The level of circulating leptin is proportional to the amount of fat stored, and leptin decreases appetite. Theoretically, leptin levels indicate to the brain that the body has sufficient stores of excess calories and therefore appetite is lower. Our understanding of these hormones comes from animal models in which mice can be genetically altered to not express the hormone or its receptor. In humans, the story appears to be much more complex. Leptin is one of many hormones released by adipocytes that affects or alters energy balance and use in the body. The gut microbiome plays a significant role as well. Metabolism research is essential to building a more complete and nuanced scientific understanding of the way that the human body regulates caloric intake and use.

Learning Connection

Explain a Process

Your friend has a young child who will only eat pasta and bread. How would you explain the importance of essential nutrients to your friend? What would you tell your friend about the importance of including fats and proteins, as well as carbohydrates, in their child's diet?

24.5b Vitamins and Nutrients

LO 24.5.4

Nutrients are any substances the body can use in the processes of growth, repair, and metabolism. The body can synthesize many of the wide array of nutrients we need to fuel our daily and life activities. For example, of the 20 amino acids that we use to make the body's proteins, 11 can be synthesized from other precursors; these molecules are **nonessential nutrients**. The molecules the body cannot produce on its own and must therefore obtain from nutritional sources are **essential nutrients**. Iron, for example, is used in the manufacturing of hemoglobin and myoglobin. The body cannot make its own iron and must therefore obtain it from the diet.

LO 24.5.5

Vitamins are organic compounds found in foods and are a necessary part of the biochemical reactions in the body. They are involved in a number of processes, including mineral and bone metabolism, and cell and tissue growth, and they act as cofactors for energy metabolism. The B vitamins, in particular, are important for mitosis and metabolism in cells all over the body; therefore, deficiencies in B vitamins have a wide range of effects (**Tables 24.1** and **24.2**).

You get most of your vitamins through your diet, although some can be formed from the precursors absorbed during digestion. Some vitamins you can obtain either through your diet or by synthesizing the molecule. For example, the body synthesizes vitamin A from the beta-carotene in orange vegetables like carrots and sweet potatoes. Vitamin D is also synthesized in the skin through exposure to sunlight.

Vitamins, like all molecules, are either fat-soluble or water-soluble. Fat-soluble vitamins A, D, E, and K are absorbed through the intestinal tract along with lipids. Because they are carried in lipids, fat-soluble vitamins can accumulate in the lipids stored in the body.

Water-soluble vitamins, including the eight B vitamins and vitamin C, are absorbed with water in the GI tract. These vitamins move easily through bodily fluids, which are water-based, so they are not stored in the body. Excess water-soluble vitamins are excreted in the urine.

24.5c Minerals

LO 24.5.6

Minerals are inorganic compounds that work with other nutrients to ensure the body functions properly. Minerals cannot be made in the body; they all must come from the diet. The amount of minerals in the body is small—only 4 percent of the total body mass—and most of that consists of the minerals that the body requires in larger quantities: potassium, sodium, calcium, phosphorus, magnesium, and chloride. These minerals participate in a variety of physiological processes, such as muscle contraction, that you have read about throughout this text.

Table 24.1 Fat-Soluble Vitamins

Vitamin and Alternative Name	Sources	Recommended Daily Allowance	Function	Problems Associated with Deficiency
A retinal or beta-carotene	Yellow and orange fruits and vegetables, dark green leafy vegetables, eggs, milk, liver	700–900 μg	Eye and bone development, immune function	Night blindness, epithelial changes, immune system deficiency
D cholecalciferol	Dairy products, egg yolks; also synthesized in the skin from exposure to sunlight	5–15 μg	Aids in calcium absorption, promoting bone growth	Rickets, bone pain, muscle weakness, increased risk of death from cardiovascular disease, cognitive impairment, asthma in children, cancer
E tocopherols	Seeds, nuts, vegetable oils, avocados, wheat germ	15 mg	Antioxidant	Anemia
K phylloquinone	Dark green leafy vegetables, broccoli, Brussels sprouts, cabbage	90–120 μg	Blood clotting, bone health	Hemorrhagic disease of newborn in infants; uncommon in adults

Table 24.2 Water-Soluble Vitamins

Vitamin and Alternative Name	Sources	Recommended Daily Allowance	Function	Problems Associated with Deficiency
B_1 thiamine	Whole grains, enriched bread and cereals, milk, meat	1.1–1.2 mg	Carbohydrate metabolism	Beriberi, Wernicke-Korsikoff syndrome
B_2 riboflavin	Brewer's yeast, almonds, milk, organ meats, legumes, enriched breads and cereals, broccoli, asparagus	1.1–1.3 mg	Synthesis of FAD for metabolism, production of red blood cells	Fatigue, slowed growth, digestive problems, light sensitivity, epithelial problems like cracks in the corners of the mouth
B_3 niacin	Meat, fish, poultry, enriched breads and cereals, peanuts	14–16 mg	Synthesis of NAD, nerve function, cholesterol production	Cracked, scaly skin; dementia; diarrhea; also known as *pellagra*
B_5 pantothenic acid	Meat, poultry, potatoes, oats, enriched breads and cereals, tomatoes	5 mg	Synthesis of coenzyme A in fatty acid metabolism	Rare: symptoms may include fatigue, insomnia, depression, irritability
B_6 pyridoxine	Potatoes, bananas, beans, seeds, nuts, meat, poultry, fish, eggs, dark green leafy vegetables, soy, organ meats	1.3–1.5 mg	Sodium and potassium balance, red blood cell synthesis, protein metabolism	Confusion, irritability, depression, mouth and tongue sores
B_7 biotin	Liver, fruits, meats	30 μg	Cell growth, metabolism of fatty acids, production of blood cells	Rare in developed countries; symptoms include dermatitis, hair loss, loss of muscular coordination
B_9 folic acid	Liver, legumes, dark green leafy vegetables, enriched breads and cereals, citrus fruits	400 μg	DNA/protein synthesis	Poor growth, gingivitis, appetite loss, shortness of breath, GI problems, mental deficits
B_{12} cyanocobalamin	Fish, meat, poultry, dairy products, eggs	2.4 μg	Fatty acid oxidation, nerve cell function, red blood cell production	Pernicious anemia, leading to nerve cell damage
C ascorbic acid	Citrus fruits, red berries, peppers, tomatoes, broccoli, dark green leafy vegetables	75–90 mg	Necessary to produce collagen for formation of connective tissue and teeth, and for wound healing	Dry hair, gingivitis, bleeding gums, dry and scaly skin, slow wound healing, easy bruising, compromised immunity; can lead to scurvy

The most common minerals in the body are calcium and phosphorus, both of which are stored in and necessary for the structure of bones. Calcium, of course, is also required for neurons to release neurotransmitters at their synapses and required for all kinds of muscle contraction. Ions of sodium, potassium, and chloride are all utilized in establishing membrane potential in excitable cells such as neurons and muscle cells. Chloride ions are used in osmotic gradients and neuronal function, and iron ions are critical to the formation of hemoglobin. There are additional trace minerals that are still important to the body's functions, but their required quantities are much lower.

Like vitamins, minerals can be consumed in toxic quantities (although it is rare). A healthy diet includes most of the minerals your body requires, so supplements and processed foods can add potentially toxic levels of minerals. **Tables 24.3** and **24.4** provide a summary of minerals and their function in the body.

Table 24.3 Major Minerals

Mineral	Sources	Recommended Daily Allowance	Function	Problems Associated with Deficiency
Potassium	Meats, some fish, fruits, vegetables, legumes, dairy products	4,700 mg	Nerve and muscle function; acts as an electrolyte	Hypokalemia: weakness, fatigue, muscle cramping, GI problems, cardiac problems
Sodium	Table salt, milk, beets, celery, processed foods	2,300 mg	Blood pressure, blood volume, muscle and nerve function	Rare
Calcium	Dairy products, dark green leafy vegetables, blackstrap molasses, nuts, brewer's yeast, some fish	1,000 mg	Bone structure and health; nerve and muscle functions, especially cardiac function	Slow growth, weak and brittle bones
Phosphorous	Meat, milk	700 mg	Bone formation, metabolism, ATP production	Rare
Magnesium	Whole grains, nuts, leafy green vegetables	310–420 mg	Enzyme activation, production of energy, regulation of other nutrients	Agitation, anxiety, sleep problems, nausea and vomiting, abnormal heart rhythms, low blood pressure, muscular problems
Chloride	Most foods, salt, vegetables, especially seaweed, tomatoes, lettuce, celery, olives	2,300 mg	Balance of body fluids, digestion	Loss of appetite, muscle cramps

Table 24.4 Trace Minerals

Mineral	Sources	Recommended Daily Allowance	Function	Problems Associated with Deficiency
Iron	Meat, poultry, fish, shellfish, legumes, nuts, seeds, whole grains, dark leafy green vegetables	8–18 mg	Transport of oxygen in blood, production of ATP	Anemia, weakness, fatigue
Zinc	Meat, fish, poultry, cheese, shellfish	8–11 mg	Immunity, reproduction, growth, blood clotting, insulin and thyroid function	Loss of appetite, poor growth, weight loss, skin problems, hair loss, vision problems, lack of taste or smell
Copper	Seafood, organ meats, nuts, legumes, chocolate, enriched breads and cereals, some fruits and vegetables	900 μg	Red blood cell production, nerve and immune system function, collagen formation, acts as an antioxidant	Anemia, low body temperature, bone fractures, low white blood cell concentration, irregular heartbeat, thyroid problems
Iodine	Fish, shellfish, garlic, lima beans, sesame seeds, soybeans, dark leafy green vegetables	150 μg	Thyroid function	Hypothyroidism: fatigue, weight gain, dry skin, temperature sensitivity
Sulfur	Eggs, meat, poultry, fish, legumes	None	Component of amino acids	Protein deficiency
Fluoride	Fluoridated water	3–4 mg	Maintenance of bone and tooth structure	Increased cavities, weak bones and teeth
Manganese	Nuts, seeds, whole grains, legumes	1.8–2.3 mg	Formation of connective tissue and bones, blood clotting, sex hormone development, metabolism, brain and nerve function	Infertility, bone malformation, weakness, seizures

(*Continued*)

Table 24.4 Trace Minerals (*Continued*)

Mineral	Sources	Recommended Daily Allowance	Function	Problems Associated with Deficiency
Cobalt	Fish, nuts, leafy green vegetables, whole grains	None	Component of vitamin B_{12}	None
Selenium	Brewer's yeast, wheat germ, liver, butter, fish, shellfish, whole grains	55 μg	Antioxidant, thyroid function, immune system function	Muscle pain
Chromium	Whole grains, lean meats, cheese, black pepper, thyme, brewer's yeast	25–35 μg	Insulin function	High blood sugar, triglyceride, and cholesterol levels
Molybdenum	Legumes, whole grains, nuts	45 μg	Cofactor for enzymes	Rare

Apply to Pathophysiology

Hypervitaminosis

Hypervitaminosis—also referred to as a *vitamin overdose*—is a condition in which the body is exposed to a toxic level of a vitamin. Some vitamins are more likely to lead to hypervitaminosis than others when they are ingested in excess.

1. Vitamin D is synthesized in skin cells when exposed to UV radiation, but it can also be obtained from the diet. Vitamin D is synthesized from cholesterol molecules. Given this, do you suspect (or remember) if vitamin D is fat-soluble or water soluble?
   A) Fat soluble
   B) Water soluble
2. Vitamin C is a water-soluble vitamin that cannot be synthesized in the body; it must be obtained from the diet. As such, which of the following would we consider vitamin C to be?
   A) Essential nutrient
   B) Nonessential nutrient
3. Sonia is a college student studying for final exams. Sonia begins to feel a little under-the-weather and suspects they might be getting a cold. Sonia's roommate tells them that vitamin C will lessen their symptoms and gives them vitamin C supplements. Sonia's roommate instructs them to take two times the usual dose. Sonia, desperate to not get sick during finals week, takes four times the usual dose. Since vitamin C is water-soluble, where do we expect the majority of this extra vitamin C to end up?
   A) Built up in the adipose tissue
   B) Filtered into Sonia's urine
   C) Stored in Sonia's muscle cells

Learning Check

1. Which hormone is abundant when you have an adequate amount of fat stored in your body?
   a. Ghrelin
   b. Leptin
2. Salmon is high in vitamin D. Which of the following functions does salmon assist in?
   a. Eye development
   b. Bone growth
   c. Blood clotting
   d. Immune function
3. Which of the following major minerals are important in people who have a history of cardiac disease? Please select all that apply.
   a. Potassium
   b. Calcium
   c. Phosphorous
   d. Sodium

Chapter Review

24.1 Overview of Metabolic Reactions

The learning objectives for this section are:

HAPS LEARNING GOALS AND OUTCOMES

24.1.1 Define metabolism, anabolism, and catabolism, and provide examples of anabolic and catabolic reactions.

24.1.2* Define ATP and explain its role in catabolic and anabolic reactions.

24.1.3 Describe common uses in the body for carbohydrates, fats, and proteins.

24.1.4 Compare and contrast the roles of enzymes and cofactors in metabolic processes.

24.1.5* Define a redox reaction, relate redox reactions to metabolism, and explain the component processes of oxidation and reduction.

24.1.6 Explain the roles of coenzyme A, nicotinamide adenine dinucleotide (NAD), and flavin adenine dinucleotide (FAD) in metabolism.

* Objective is not a HAPS Learning Goal.

24.1 Questions

1. Which of the following is an example of an anabolic reaction?

a. Building glycogen molecules
b. Breaking down glycogen molecules
c. Building and breaking down glycogen molecules

2. Which of the following reactions releases ATP?

a. Anabolic reactions
b. Catabolic reactions

3. Which of the following describes the uses of proteins in the body? Please select all that apply.

a. They rid the body of excess ATP.
b. After breakdown into monomers, the resulting amino acids can be used to synthesize new proteins.
c. Some of their amino acids are used as an energy source in the body.
d. They are used by individual cells to build new nucleic acid molecules

4. Enzymes ________, while cofactors ________.

a. Increase the rate of chemical reactions; decrease the rate of chemical reactions
b. Enable cofactors to increase the rate of chemical reactions; increase the rate of chemical reactions
c. Increase the rate of chemical reactions; enable enzymes to increase the rate of chemical reactions
d. Decrease the rate of chemical reactions; increase the rate of chemical reactions

5. In a redox reaction, which reaction releases energy and an electron from a molecule?

a. Oxidation
b. Reduction

6. Explain the role of NAD during food digestion.

a. As food is digested, energy is transferred to NAD molecules to help produce ATP.
b. To digest food, energy is transferred from NAD molecules to the food.
c. NAD molecules transport nutrients to harvest energy.
d. To digest food, NAD molecules transfer enzymes to the food.

24.1 Summary

- Metabolism is the total of all of the chemical reactions occurring in the body at any time. Metabolic reactions include both catabolic reactions and anabolic reactions.
- Catabolic reactions decompose large molecules to release energy, while anabolic reactions use the energy released during catabolic reactions to synthesize larger molecules.
- Enzymes are proteins that act as catalysts for chemical reactions, while cofactors are substances that bind to enzymes and enable their activity.
- Oxidation reactions release small amounts of energy during the loss of an electron. Reduction reactions require energy in order to gain an electron. These reactions are often coupled together to form an oxidation-reduction reaction.
- Redox reactions are catalyzed by enzymes, which require coenzymes to enable their activity. The coenzymes NAD and FADH are used as electron and hydrogen ion acceptors. Their reduced forms, NADH and $FADH_2$, are

used to transfer energy as food molecules are oxidized in the production of ATP.

- ATP is the molecule that stores energy in its chemical bonds, and releases it to provide energy for processes such as active transport, nerve impulse conduction, muscle contraction, and the synthesis of larger molecules from smaller ones.
- Carbohydrates, lipids and proteins are nutrients that provide the body with energy and building blocks for synthesizing other molecules. Carbohydrates are the body's main source of energy, but the breakdown products of lipids and proteins also provide energy.
- Carbohydrates are broken down to synthesize ATP; excess carbohydrates are stored as glycogen or triglycerides.
- Lipids can be broken down to produce ATP, and are also used as cell membrane components, hormones and vitamins. Excess lipids are stored as triglycerides in the adipose tissue.
- Proteins are broken down into amino acids; most of these amino acids are used to synthesize new proteins. Excess amino acids are deaminated and used to produce ATP. Proteins are not stored in the body.

24.2 Macronutrients and Metabolism

The learning objectives for this section are:

HAPS LEARNING GOALS AND OUTCOMES

24.2.1 Describe the processes of aerobic respiration (e.g., citric acid [Krebs, tricarboxylic acid or TCA] cycle, electron transport chain) in the oxidation of glucose to generate ATP.

24.2.2 Describe the processes of anaerobic respiration (e.g., glycolysis) in the oxidation of carbohydrates to generate ATP.

24.2.3 Describe metabolic pathways that produce or store glucose (e.g., glycogenesis, glycogenolysis, gluconeogenesis).

24.2.4 Describe the anabolic and catabolic processes of fat metabolism (e.g., lipolysis, lipogenesis) and how these processes interact with carbohydrate metabolism.

24.2.5 Describe the anabolic and catabolic processes of protein metabolism (e.g., deamination, transamination) and how these processes interact with carbohydrate metabolism.

24.2.6 Explain the significance of protein intake to nitrogen balance.

24.2 Questions

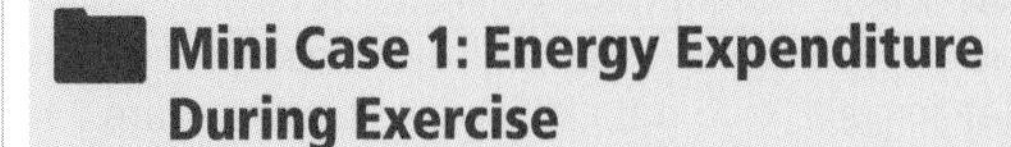

Mini Case 1: Energy Expenditure During Exercise

It is the day after New Year's and you, like millions of people around the world, have made a New Year's resolution to improve your health. This includes improving strength, conditioning, and eating a well-balanced diet. You start off the first day jumping on an exercise bike and working on cardiovascular fitness.

1. Which of the following systems generates the most ATP molecules to help you get through the biking workout?
   a. Glycolysis
   b. Citric acid cycle
   c. Electron transport chain
   d. Anaerobic respiration

The next day you decide to work on strength training. When working out, you want to construct a well-rounded strengthening protocol. You remember that some muscle fibers need oxygen while others do not (Type IIx, also known as *fast-twitch fibers*).

2. If you were training the Type IIx fibers, which process would you expect the generate ATP after a glucose molecule undergoes glycolysis?
   a. Citric acid cycle
   b. Lactic acid cycle
   c. Electron transport chain
   d. Aerobic respiration

After a week into the exercise program, you become more interested in developing a personalized exercise program. In your research, you find that being in a caloric deficit, eating less calories than you burn when working out, will help you lose weight faster.

3. Which of the following do you expect to occur in a single exercise session when there is not enough glucose in the blood to use for energy? Please select all that apply.
   a. New glucose molecules will form from pyruvate and lactate.
   b. Glucose molecules will be stored in the liver as glycogen.
   c. Glycogen molecules will be broken down in the form of glucose.

4. How can fat metabolism contribute to energy production when there is not enough glucose in the blood to supply the energy you need for your exercise session? Please select all that apply.
   a. Glycerol head can be used in glycolysis.
   b. Fatty acid chains can be converted into acetyl CoA.
   c. Glycerol can be converted into acetyl CoA.
   d. Fatty acid chains can be used in the glycolysis.

5. Proteins can also be broken down to supply energy. How is protein metabolism unique? Please select all that apply.
   a. Protein molecules can be converted to carbohydrates and lipids.
   b. Protein molecules cannot be stored so the excess is removed.
   c. Protein molecules are the easiest to convert into energy.
   d. Protein molecules cannot be broken down into subunits.

6. Which of the following would you expect to occur if there was excess protein that could not be used at the time to fuel your exercise session?
   a. The protein could be stored in the form of ketone bodies.
   b. The protein could be stored as muscle fibers.
   c. The amino acids would be removed in the feces via the GI tract.
   d. The amine groups from excess amino acids would be removed in the urine via the urea cycle.

24.2 Summary

- After glucose is ingested, it is first broken down anaerobically, by the process of glycolysis. In this reaction sequence, glucose is broken down into two molecules of pyruvate, and a small amount of ATP is generated. Pyruvate can then enter the citric acid cycle, as long as oxygen is present.
- If there is not enough oxygen present, the pyruvate can be converted into lactic acid. The lactic acid can eventually be converted back into pyruvate or glucose by the liver, providing another opportunity for it to be broken down to produce ATP.
- Carbohydrates, lipids, and proteins can all be broken down aerobically to produce ATP. Aerobic respiration of nutrients requires oxygen, and occurs in the mitochondria in the chemical reaction sequences of the citric acid cycle and the electron transport chain.
- NADH and $FADH_2$ are the reduced forms of the coenzymes NAD and FAD, respectively, which are generated during glycolysis and the citric acid cycle. These coenzymes transport electrons and hydrogen ions to the electron transport chain, during which they are used to produce a large amount of ATP.
- In conditions of glucose deficiency, glucose can be provided by glycogen breakdown (glycogenolysis) or by conversion of other types of molecules into glucose (gluconeogenesis). When glucose is present in excess, it can be stored as glycogen or fat.
- Excess lipids can be stored as triglycerides (lipogenesis). When lipids are needed to provide energy, storage fats can be broken down to provide fatty acids and glycerol (lipolysis).
- Proteins provide amino acids as building blocks for new proteins. Excess amino acids can be deaminated and converted into glucose or broken down to produce ATP.

24.3 Metabolic States of the Body

The learning objective for this section is:

HAPS LEARNING GOALS AND OUTCOMES

24.3.1 Compare and contrast carbohydrate, fat, and protein metabolism in the fed (absorptive) and fasted (postabsorptive) states.

24.3 Question

1. Which of the following occur during the absorptive state? Please select all that apply.
   a. The body is absorbing the ingested nutrients.
   b. The body is storing the nutrients.
   c. There is no food intake.
   d. The glucose levels in the blood drop.

24.3 Summary

- The body experiences different metabolic states involving food storage and release of energy, which depend on the timing of food intake. These states are the absorptive state, the postabsorptive state, and starvation.
- The absorptive (fed) state occurs when food has been eaten recently, and the body is digesting and absorbing the nutrients. Some absorbed nutrients are supplying energy for the body, and excess ingested nutrients are being stored as glycogen or fats in various organs. This state is regulated by insulin, which transports glucose from the blood into the body cells.
- The postabsorptive (fasting) state begins after all ingested nutrients have been digested, absorbed, and used or stored. There are no more nutrients in the gastointestinal tract. This can occur overnight or by waiting several hours between meals. Glucose is obtained by the breakdown of storage molecules that were synthesized during the absorptive state: glycogenolysis, gluconeogenesis and lipolysis. This state is regulated by glucagon.
- In times of extended food deprivation, the body enters a state of starvation, in which its primary goal is survival. Because the brain greatly prefers glucose as its energy source, other types of cells switch to fatty acid and ketone breakdown. Fat stores and muscle proteins can be used for energy if necessary.

24.4 Energy and Heat Balance

The learning objectives for this section are:

24.4.1 Explain the importance of thermoregulation in the body.

24.4.2 Explain how various organ systems and behaviors participate in thermoregulation.

24.4.3 Define metabolic rate and describe the conditions under which basal metabolic rate is measured.

24.4.4 Describe factors that affect metabolic rate.

24.4 Questions

1. What would you expect to happen if your body temperature were too low?
   a. Proteins would be denatured.
   b. The metabolic rate would increase.
   c. The rate of chemical reactions.
   d. Food would not be absorbed

2. You are on the beach on a hot summer day. Which of the following organs are working to maintain thermoregulation? Please select all that apply.
   a. The thyroid
   b. The hypothalamus
   c. The skin
   d. The skeletal muscle

3. Define metabolic rate.
   a. The amount of energy consumed per unit of time
   b. The amount of energy expended per unit of time
   c. The amount of energy expended minus the amount of energy consumed per unit of time
   d. The amount of energy consumed minus the amount of energy expended per unit of time

4. Which of the following factors would affect the metabolic rate? Please select all that apply.
   a. The amount of glycogen stored in the liver
   b. Age
   c. Muscle mass
   d. Your activity level

24.4 Summary

- Thermoregulation, or body temperature regulation, is vital to our survival and good health. The importance of thermoregulation lies in the fact that enzymes, proteins that control chemical reaction rates in the body, function optimally in only a narrow temperature range.
- The hypothalamus contains the thermoregulatory center. It adjusts body temperature through mechanisms such as blood flow distribution changes, sweating or shivering.
- The metabolic rate is the amount of energy expended by the body in a certain amount of time. The basal metabolic rate is the amount of energy expended by the body under basal conditions (resting, awake after overnight fasting, comfortable).
- Factors affecting the metabolic rate include age, muscle mass, activity level, and temperature of the environment.

24.5 Nutrition and Diet

The learning objectives for this section are:

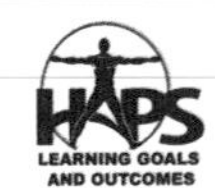

24.5.1 Define calorie and kilocalorie.

24.5.2 Describe energy yields per gram for carbohydrates, fats, and proteins.

24.4.3 Describe the neural and chemical control of appetite and food intake.

24.5.4 Define nutrient, essential nutrient, and non-essential nutrient.

24.5.5 Classify vitamins as either fat-soluble or water-soluble, and describe the major uses in the body of each vitamin.

24.5.6 Name the major minerals (e.g., calcium, sodium, potassium) and trace elements (e.g., iron, iodine, zinc) and their roles within various physiological processes in the body.

24.5 Questions

1. A typical fast-food hamburger has about 500 Calories, 30 grams of fat, 40 grams of carbohydrates, and 25 grams of protein. How many calories—not nutritional Calories—does the burger really have?

a. 5
b. 500
c. 5,000
d. 500,000

2. A typical fast-food hamburger has about 500 Calories, 30 grams of fat, 40 grams of carbohydrates, and 25 grams of protein. How many nutritional Calories come from carbohydrates?

a. 40
b. 160
c. 100
d. 360

3. Explain the role of the hormone ghrelin in the body.

a. Decreases appetite
b. Indicative of fat stored
c. Increases appetite
d. Indicative of protein stored

4. Define *essential nutrient*.

a. Molecules that the body cannot produce on its own
b. Molecules that the body can produce on its own
c. Molecules that should not be ingested by the body
d. Molecules that are only found in foods

5. How is vitamin E absorbed into the body?

a. With lipids
b. With water
c. With proteins
d. With glucose

6. Which major minerals are responsible for nerve function? Please select all that apply.

a. Potassium
b. Phosphorous
c. Magnesium
d. Calcium

24.5 Summary

- Nutritional Calories, or kilocalories, are 1,000 calories. A calorie used in physical science is the amount of heat required to raise the temperature of one gram of water by one degree Celsius. Carbohydrates and proteins yield four nutritional Calories per gram, while fat yields nine nutritional Calories per gram.
- Hunger (appetite) is controlled by neural and hormonal factors. Hunger is increased by the hormone grehlin when the stomach is empty, and decreased by the hormone leptin when fat stores are adequate.
- A nutrient is a substance that the body needs for growth, repair, and metabolism. An essential nutrient is a nutrient that cannot be produced by the body, so it must be obtained through the diet. for example, some amino acids, some fatty acids, and all minerals are essential nutrients. A nonessential nutrient is a nutrient that can be synthesized in the body.
- Vitamins are organic compounds that are absorbed through diet. There are classified as fat-soluble vitamins (A, D, E, and K) or water-soluble vitamins (B vitamins and C). Vitamins are important in growth, mitosis, and metabolism, and many are cofactors in the metabolic processes of food breakdown.
- Minerals are inorganic compounds and cannot be produced in the body. There are six major minerals and eleven trace minerals that are required by the body. In general, they are used in nerve impulse conduction, muscle contraction, bone structure, oxygen transport, and a variety of metabolic processes.

25

The Urinary System

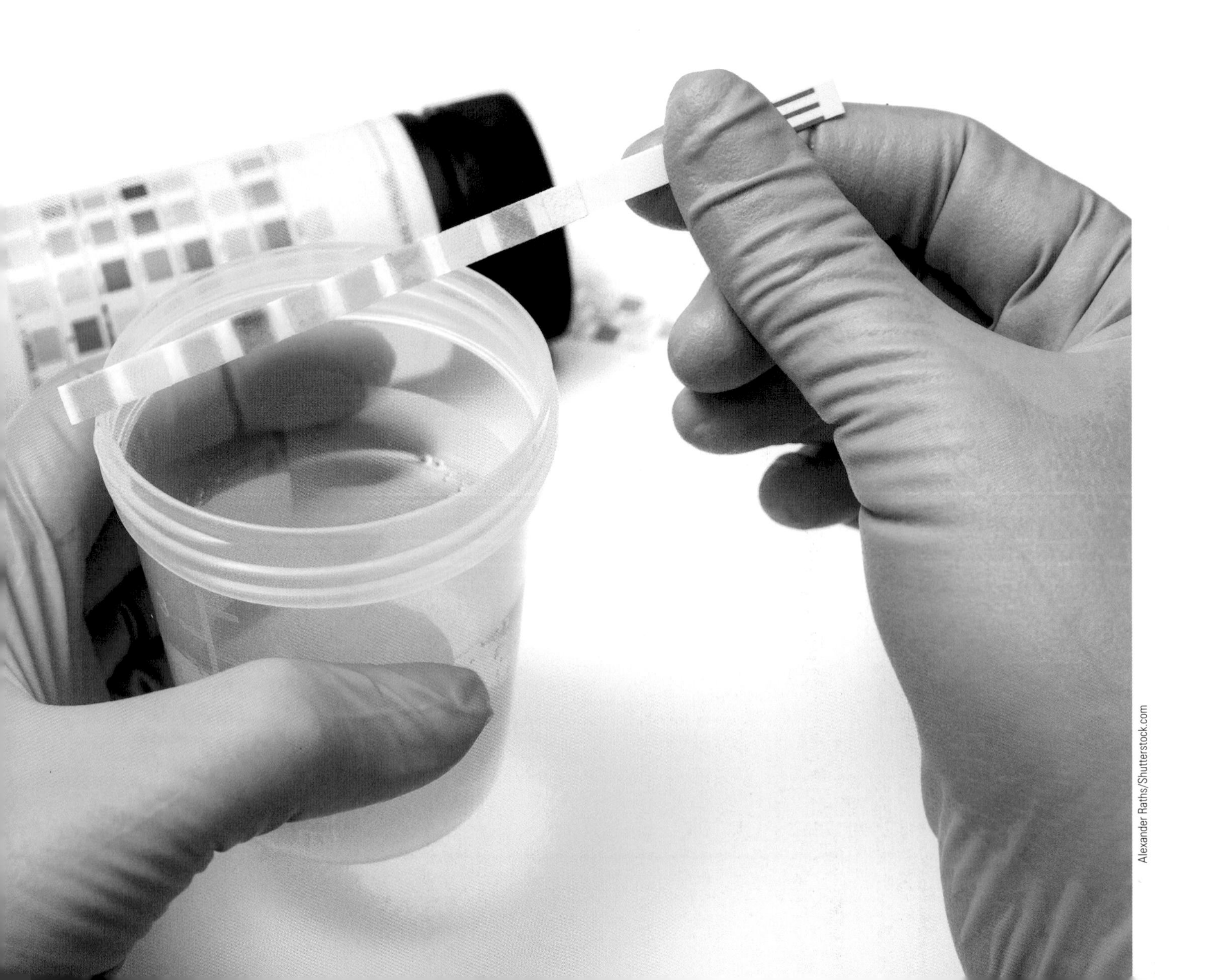

Alexander Raths/Shutterstock.com

Chapter Introduction

In this chapter we will...

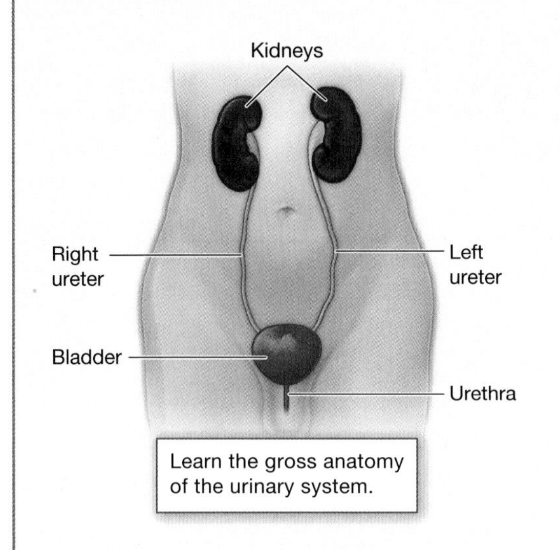

Learn the gross anatomy of the urinary system.

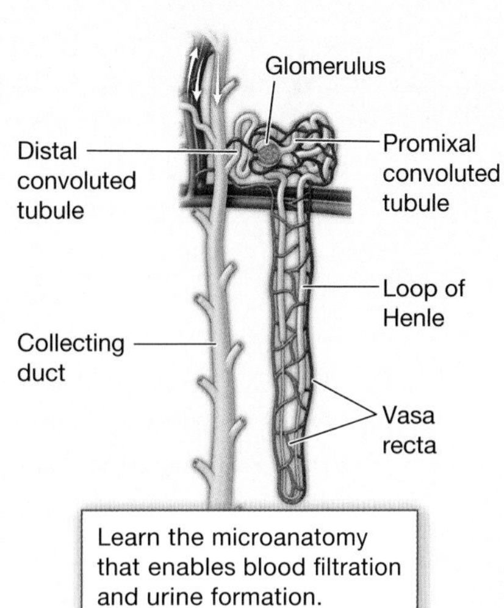

Learn the microanatomy that enables blood filtration and urine formation.

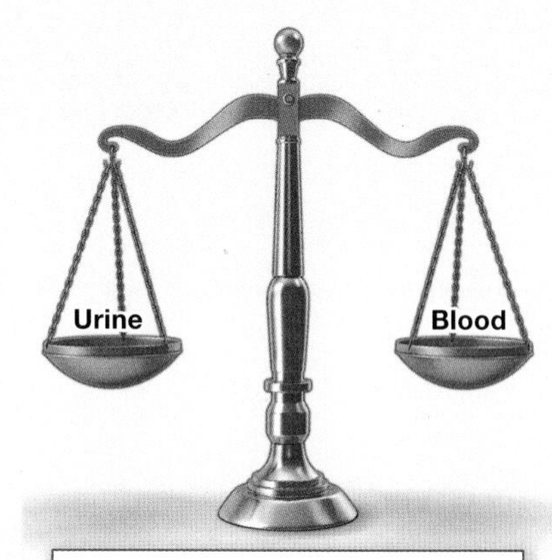

Understand that urine is created to rid the blood of material that it does not need at any given moment. Therefore, urine composition is always changing so that the blood remains in homeostasis.

You've probably already urinated today, perhaps more than once. Likely, you gave the event very little thought other than when to fit it into your schedule. Did you consider that this urine has actually saved your life? Many aspects of your body's homeostasis are critically dependent on the production of urine. Like feces, urine is a combination of material that includes wastes and excess water. The amount of water, solutes, and hydrogen ions in the urine is a critical facet to the maintenance of homeostasis of the body fluids. The urinary system, also called the *renal system*, contributes to homeostasis through several roles, including the production, storage, and excretion of urine.

25.1 Functions of the Urinary System

Learning Objective: By the end of this section, you will be able to:

25.1.1 Describe the major functions of the urinary system and which organs are responsible for those functions.

Most of our organ systems contribute to homeostasis, but perhaps no other organ system has so many homeostatic maintenance jobs as the urinary system. The urinary system:

LO 25.1.1

- ***filters the blood, removing wastes:*** The cells of the body are breaking down and recycling proteins all the time. In this process, nitrogenous wastes, which can be harmful to the body, are generated. These wastes are removed from body fluids and added to the urine.
- ***Regulates the pH of the body:*** While the pH of the body fluids is kept within an incredibly narrow range, the pH of urine varies widely. This is because the kidneys maintain blood pH by adding any excess hydrogen or bicarbonate ions to the urine.

The Human Anatomy and Physiology Society includes more than 1,700 educators who work together to promote excellence in the teaching of this subject area. The HAPS A&P Learning Outcomes measure student mastery of the content typically covered in a two-semester Human A&P curriculum at the undergraduate level. The full Learning Outcomes are available at https://www.hapsweb.org.

- ***Regulates blood pressure:*** Blood volume impacts blood pressure. The higher the volume of blood, the higher the pressure on the walls of the blood vessels. The urinary system will remove excess fluid in the blood when it exists. It also works to conserve blood volume when blood pressure is low by creating lower volume, more concentrated urine.
- ***Regulates the concentration of solutes in the blood:*** What would happen to your heart function if the extracellular K^+ ion concentration became dramatically higher? What cells and systems rely on a dependable extracellular concentration of Na^+ ions? If you can answer those questions, it becomes immediately obvious how critical the regulation of solute concentration is. And yet, today you may have eaten a salty meal and contributed a lot of solutes to the blood. Tomorrow you may be fasting in observance of a religious holiday. Your intake of solutes varies widely, but your cells can only tolerate a limited variation. The kidneys work to excrete or retain solutes in the blood as necessary to maintain body homeostasis.
- ***Influences the concentration of red blood cells through the production of erythropoietin (EPO):*** As blood passes through the kidneys, cells within are monitoring blood for its oxygen-carrying capacity. When the blood is carrying less oxygen than is optimal, the kidneys release EPO, which stimulates red blood cell production in the bone marrow.
- ***Performs the final synthesis step of vitamin D production:*** It converts calcidiol to calcitriol, the active form of vitamin D. Vitamin D is synthesized in the skin in response to UV radiation from the sun, but it is not released in its active form. This final step of active vitamin D production is contributed by the kidneys.

Each of these functions is vital to your well-being and survival. The urinary system, controlled by the nervous system, also stores urine until a convenient time for disposal and then provides the anatomical structures to transport this waste liquid to the outside of the body.

Urine is the byproduct of all this regulation; the materials that must leave the body (i.e., waste products) and the materials that are in excess (i.e., extra fluid) are combined to form urine, keeping the rest of the body fluids in homeostasis. As you read each section, ask yourself this question: "What happens if this does not work?" This question will help you to understand how the urinary system maintains homeostasis and affects all the other systems of the body and the quality of one's life.

Learning Check

1. The homeostasis of which of the following variables are maintained by the kidneys? Please select all that apply.
   a. pH
   b. Heart rate
   c. Body temperature
   d. Blood solute concentration
2. Sam has just undergone a 12-week aerobic training program, in which their VO_{2max}, the maximum amount of oxygen used by the tissues for cellular respiration, increased drastically. Which urinary system hormone contributed to Sam's increased oxygen-carrying capacity?
   a. Insulin
   b. Epinephrine
   c. Erythropoietin
   d. Aldosterone
3. When you are dehydrated, which of the following is true of the urine?
   a. It is concentrated
   b. It is dilute
   c. No urine will be produced
   d. It is impossible to tell with the information given
4. Which of the following best describes the kidneys' contribution to activated vitamin D production and regulation?
   a. They perform every step in activated vitamin D production.
   b. They regulate, but do not play a role in, activated vitamin D production.
   c. They do not play a role in vitamin D production or regulation.
   d. They perform the final synthesis step of activated vitamin D production.

25.2 Gross and Microscopic Anatomy of the Kidney

Learning Objectives: By the end of this section, you will be able to:

25.2.1 Identify and describe the anatomic structure of the kidney, including its coverings.

25.2.2 Distinguish histologically between the renal cortex and the renal medulla.

25.2.3 Trace the path of blood flow through the kidney, from the renal artery to the renal vein.

25.2.4 Identify and describe the vascular elements associated with the nephron (i.e., afferent and efferent arterioles, glomerulus, peritubular capillaries, vasa recta).

25.2.5 Identify and describe the structure of a typical nephron, including the renal corpuscle (i.e., glomerular [Bowman's] capsule, glomerulus) and renal tubule (i.e., proximal convoluted tubule, nephron loop [loop of Henle], distal convoluted tubule).

25.2.6 Describe the filtration structures that lie between the lumen of the glomerular capillaries and the capsular (Bowman) space.

25.2.7 Trace the flow of filtrate from the renal corpuscle through the collecting duct.

25.2.8 Compare and contrast the anatomic structure of the cortical nephrons and juxtamedullary nephrons.

The kidneys lie on either side of the spine within the rib cage. They occupy the retroperitoneal space, meaning that they are posterior to the parietal peritoneum but anterior to the posterior abdominal wall and ribs. Therefore, they are not within the abdominal cavity. The kidneys are well protected by muscle, fat, and ribs. They are probably slightly longer than your fist from end to end and are proportional to the size of your body (i.e. larger, taller individuals have larger, taller kidneys). The kidneys are well vascularized, receiving about 25 percent of the cardiac output at rest.

25.2a External Anatomy

LO 25.2.1

The two kidneys are the same size, but the are not perfectly parallel to each other; the left kidney is slightly lower due to displacement by the very large liver (**Figure 25.1**). Each kidney is covered by a fibrous capsule composed of dense, irregular connective tissue that helps to hold their shape and protect them. This capsule is covered by a shock-absorbing layer of adipose tissue called **perinephric fat**, which in turn is encompassed by a tough renal fascia (**Figure 25.2**). The fascia and, to a lesser extent, the overlying peritoneum serve to firmly anchor the kidneys to the posterior abdominal wall.

Student Study Tip

We've seen that organs tend to fit together like a jigsaw puzzle. This is one example, as are the heart and the lungs.

Each kidney provides a platform for the adrenal gland that sits on its superior surface. The adrenal glands are endocrine glands, producing and secreting a wide array of hormones. These two organs are linked both anatomically and functionally; the adrenal cortex directly influences kidney function through hormone signaling. The adrenal glands are discussed later on in this chapter as well as in Chapter 17.

25.2b Internal Anatomy

medulla (renal medulla)

LO 25.2.2

As with several organs, the tissue within the kidney is organized into an outer region called the **renal cortex** and an inner region called the **medulla** (**Figure 25.3A**). These two layers, analogous to the skin of an orange and its juicy inner sections, are visible in a frontal section through the kidney (**Figure 25.3B**). The **renal columns** are connective tissue extensions that radiate downward from the cortex through the medulla to separate the most characteristic features of the medulla, the **renal pyramids** and **renal papillae**. The papillae are bundles of collecting ducts that transport urine made by nephrons to the **calyces** of the kidney for excretion. At the tip of each renal pyramid is a **minor calyx**;

Figure 25.1 The Kidneys

The kidneys are partially protected by the ribs and are fully encased in fat for protection (not shown).

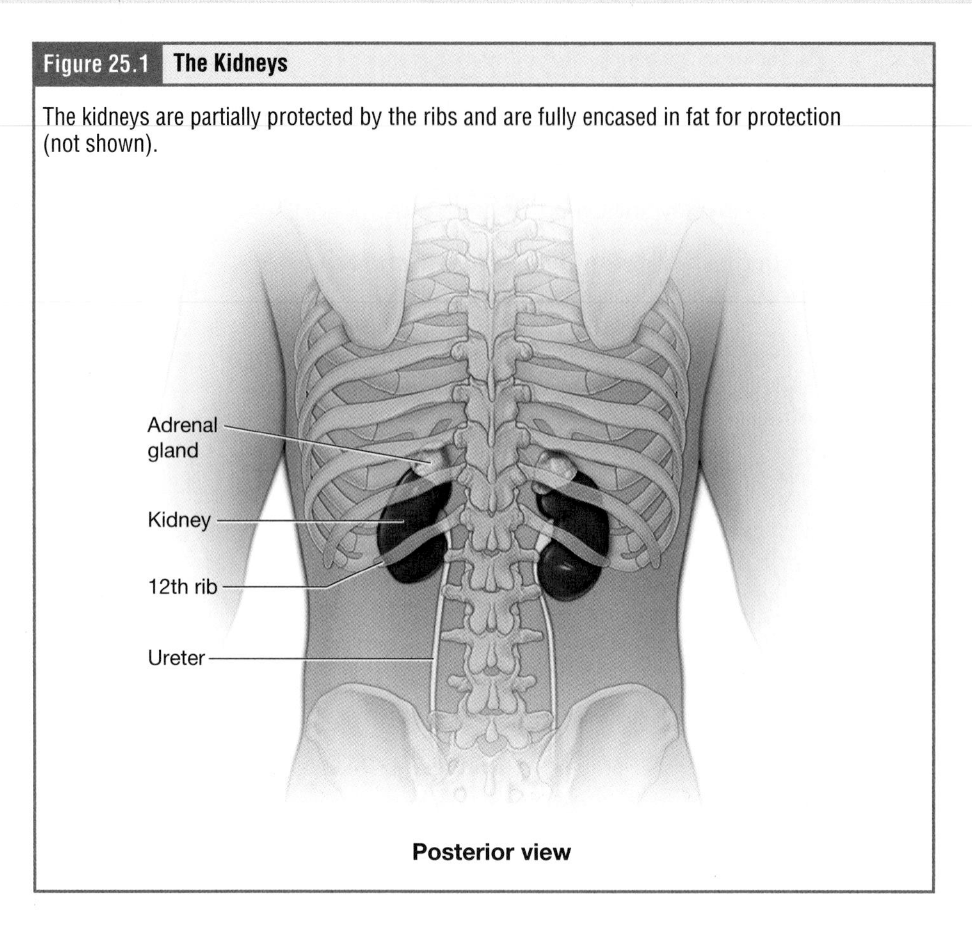

Figure 25.2 External Layers of the Kidneys

Since the kidney is retroperitoneal, it lacks the protection that the peritoneum provides for many abdominal organs. Therefore, the kidney has its own protective coverings; these coverings, from innermost to outermost, are the fibrous capsule, the perinephric fat, and the renal fascia.

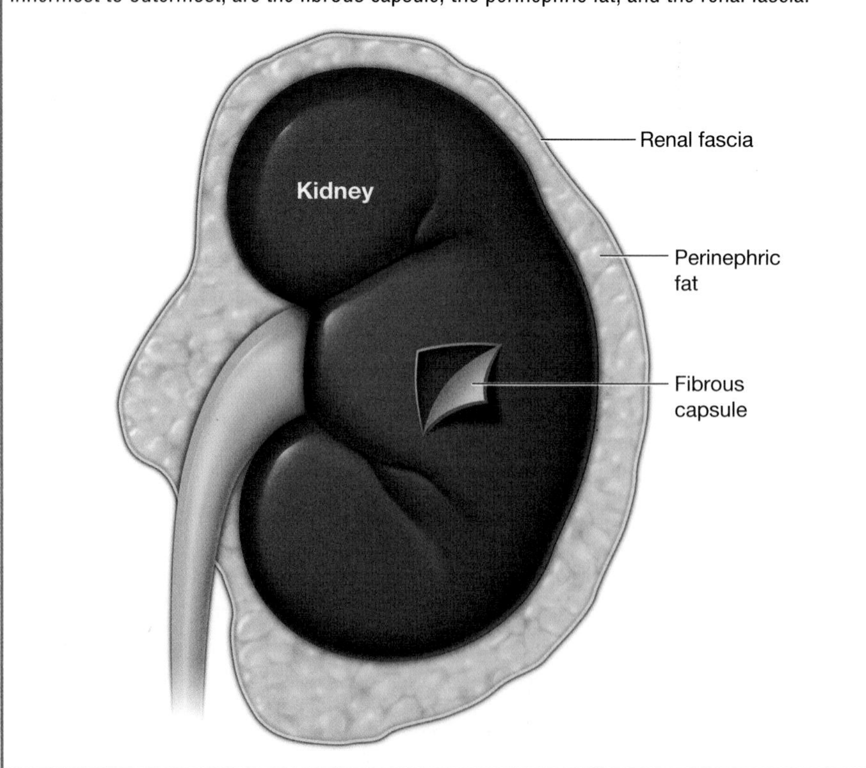

Figure 25.3 The Gross Anatomy of the Kidney

Each kidney has an outer cortex and an inner medulla. The medullary tissue is organized into cones, each of which is referred to as a pyramid. The tips of the pyramids point inward to the center of the kidney, a hollow region known as the *renal pelvis.* Urine is formed in microscopic structures throughout the cortex and medulla and then collects in the renal pelvis before draining into the ureter. (A) Drawing of a kidney and (B) kidney from a cadaver.

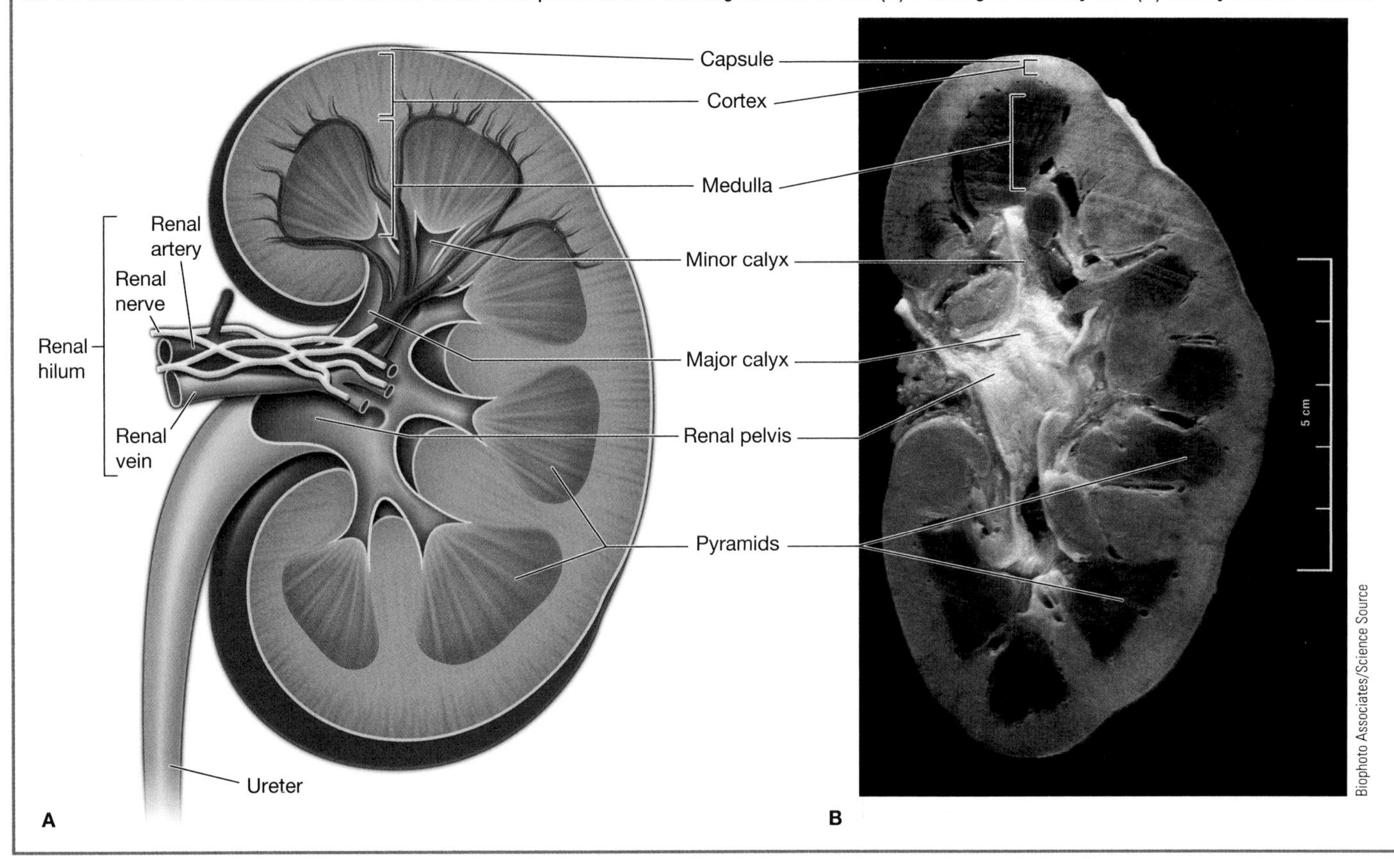

these join together to form a **major calyx**, and eventually all of the major calyces merge to form the *renal pelvis*. The renal columns also serve to divide the kidney into six to eight lobes and provide a supportive framework for vessels that enter and exit the cortex. The pyramids and renal columns taken together constitute the kidney lobes.

Renal Hilum The **renal hilum** is the entry and exit site for structures servicing the kidneys: vessels, nerves, lymphatics, and ureters (**Figure 25.4**). The hilum is tucked into the two curves on the medial side of each kidney. Coming and going from the kidney are two major vessels, the **renal artery**, which brings oxygenated blood to the kidney for filtration, and the **renal vein**, which carries the filtered, deoxygenated blood away from the kidney. Whereas the renal arteries form directly from the descending aorta, the renal veins return cleansed blood directly to the inferior vena cava. The renal nerve supplies innervation to the kidney. The long, hollow tube called the *ureter* carries urine, once produced, away from the kidney and toward the urinary bladder. The renal artery, renal vein, and renal pelvis are arranged in an anterior-to-posterior order.

Internal to the hilum is the **renal pelvis**, a hollow space where urine gathers as it is produced. Urine in the renal pelvis will flow into the ureters to be carried out of the kidney when the smooth muscle in the renal pelvis contracts and funnels the urine into the ureter through peristalsis.

Nephrons and Blood Supply **Nephrons** are the "functional units" of the kidney; they cleanse the blood and balance the constituents of the circulation. There are approximately 1–1.3 million nephrons in each kidney and each needs to be supplied with

Figure 25.4 The Renal Hilum

On the medial side of each kidney is a region containing the vessels that connect the kidneys with the rest of the body. Here, in the hilum, the arterial blood enters in the renal artery while filtered and deoxygenated blood leaves in the renal vein; the urine collects in the ureter and the kidney is innervated by the renal nerve.

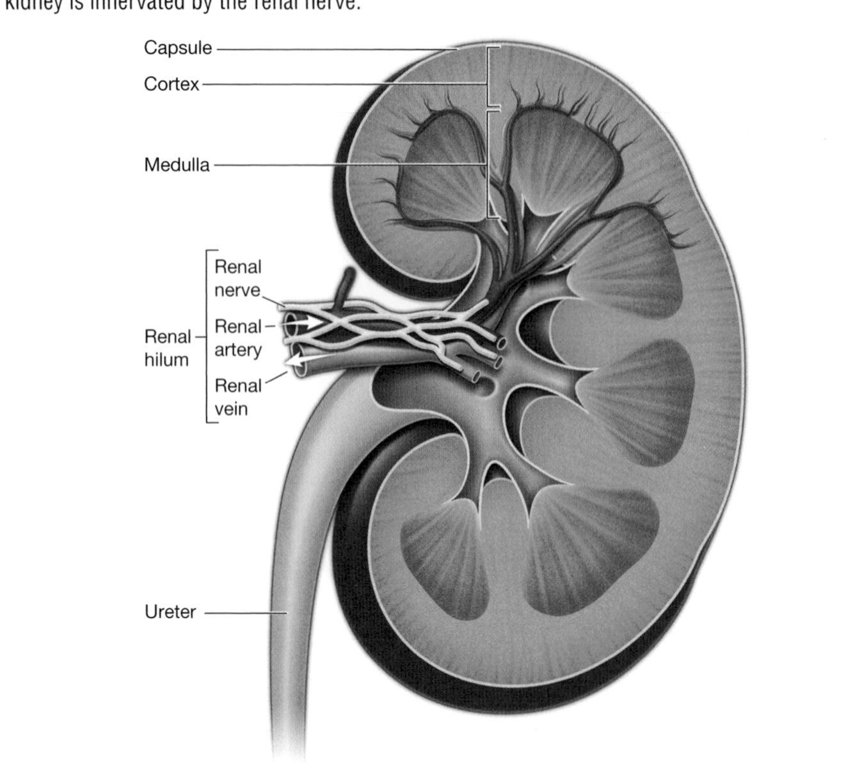

interlobular arteries (cortical radiate arteries)

glomerular capsule (Bowman's capsule)

LO 25.2.3

LO 25.2.4

Bowman's capsule (capsula glomeruli, glomerular capsule)

Student Study Tip

Peritubular capillaries function as a net. If you are net fishing, the fish (valuable material lost out of the glomerulus) are captured by the net (the peritubular capillary bed). We keep these to be reabsorbed into the bloodstream. Unwanted materials, like seaweed, enter the renal tubules by a process called *secretion.*

vasa recta (vasa rectae renis)

oxygenated blood; in addition, the nephrons require access to the blood to filter it. The renal arteries bring oxygenated blood to each kidney. Once inside the kidney at the hilum, the renal artery first divides into **segmental arteries**, followed by further branching to form **interlobar arteries** that pass through the renal columns to reach the cortex (**Figure 25.5**). The interlobar arteries, in turn, branch into **arcuate arteries**, **interlobular arteries**, and then into **afferent arterioles**.

The afferent arterioles form a cluster of high-pressure capillaries called the **glomerulus**. The glomerular capillaries are fenestrated capillaries; that is, their endothelial cells have large gaps—fenestrae—that allow material to pass through the capillary wall. The glomerulus is closely surrounded by the **glomerular capsule**, or **Bowman's capsule**, a cup-shaped enlargement of the kidney tubule. Together, the glomerulus and glomerular capsule form the **renal corpuscle**. The renal corpuscle is the location of blood filtration. Any material in the blood below a certain molecular size will pass through the walls of the glomerulus and enter the glomerular capsule. After passing through the glomerulus and being filtered, the blood enters a second arteriole, the **efferent arteriole** (**Figure 25.6**). The efferent arteriole connects two capillary beds—the glomerulus, which filters the blood, and the **peritubular capillaries**, which surround the renal tubules of the nephron and allow for the exchange of substances between the blood and the renal tubules. Two capillary beds that are connected without returning blood to the heart in between are called a *portal system*. In this portal system, about 20 percent of the volume of the plasma is removed from the blood at the glomerulus. It is filtered through the long tubule connected to the glomerular capsule. Any material selected to return to the blood does so along the peritubular capillaries and the **vasa recta**, the

Figure 25.5 Blood Flow in the Kidney

Oxygenated blood enters the kidney through the renal artery. Within the kidney the arteries branch into smaller and smaller vessels until they give rise to two connected capillary beds. The glomerulus and the peritubular capillaries are a portal system of blood vessels that enable the functions of filtration, reabsorption, and secretion.

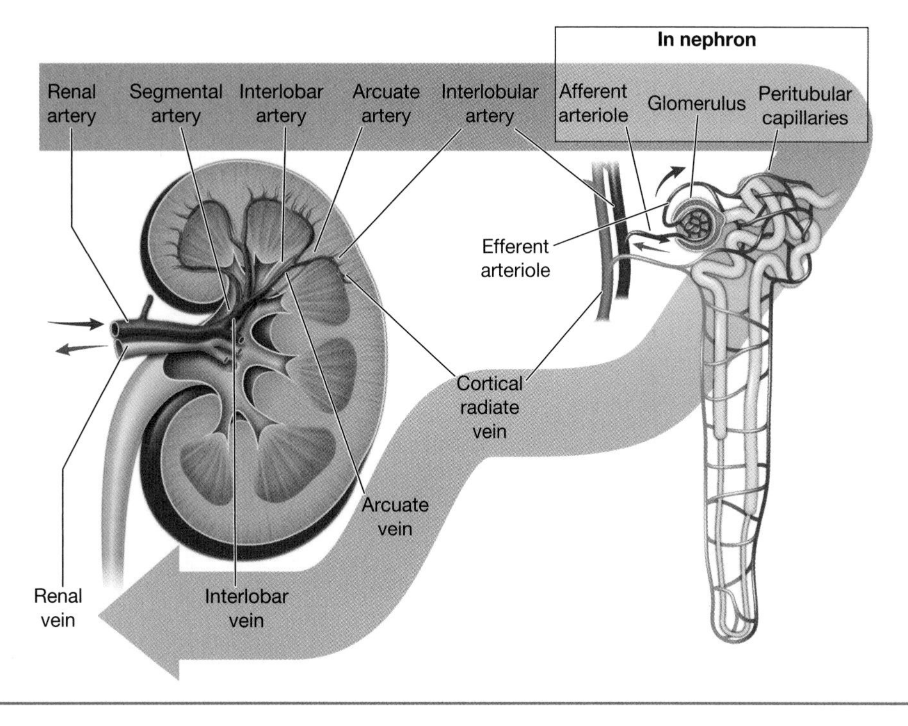

venule that drains the peritubular capillaries. The blood that exits the vasa recta is now filtered and deoxygenated and ready to return to the venous system. The order of veins within the kidney transporting blood back to the renal vein is as follows: **interlobular vein**, the **arcuate vein**, and the **interlobar veins**; these merge to drain into the renal vein (see Figures 25.5 and 25.6). Notice the similarity between interlobular and interlobar veins; although their names differ by only a few letters, these are different structures.

interlobular vein (cortical radiate vein)

25.2c The Structure and Function of the Nephrons

The nephron is, in essence, a blood processing unit. Blood from the renal artery passes through the connected vessels of the nephron. A small portion of the blood plasma is removed from the blood vessels at the renal corpuscle; this plasma, now outside the blood system, is called **filtrate**. As the blood continues along the system of vessels, the filtrate passes through the long, winding nephron **tubule**. The nephron tubule and its peritubular capillary is a site of exchange. Some material is *reabsorbed* back into the blood from the filtrate. Other substances are *secreted* from the blood into the filtrate. At the end of this intertwining exchange system, the blood passes back into the venous circulation and returns to the heart. Any materials left in the tubule become urine.

tubule (renal tubule)

As discussed earlier, the renal corpuscle consists of a nest of capillaries called the *glomerulus* that is largely surrounded by the glomerular capsule. The glomerulus is a high-pressure capillary bed that sits between the afferent and efferent arterioles. The glomerular capsule surrounds the glomerulus to form a lumen, which captures and directs this filtrate to the tubule system. The section of the tubule closest to the glomerulus is called the **proximal convoluted tubule (PCT)**. The PCT eventually narrows and dips into a long loop, the **nephron loop**. In most nephrons, the corpuscle and proximal convoluted tubule are located in the cortex of the kidney; the nephron loop extends

nephron loop (Loop of Henle)

Figure 25.6 Blood Flow in the Nephron

The two linked capillary beds of the portal system of each nephron of the kidney are the glomerulus and the peritubular capillaries. The efferent arteriole is the connecting vessel between the glomerulus, the peritubular capillaries, and the vasa recta.

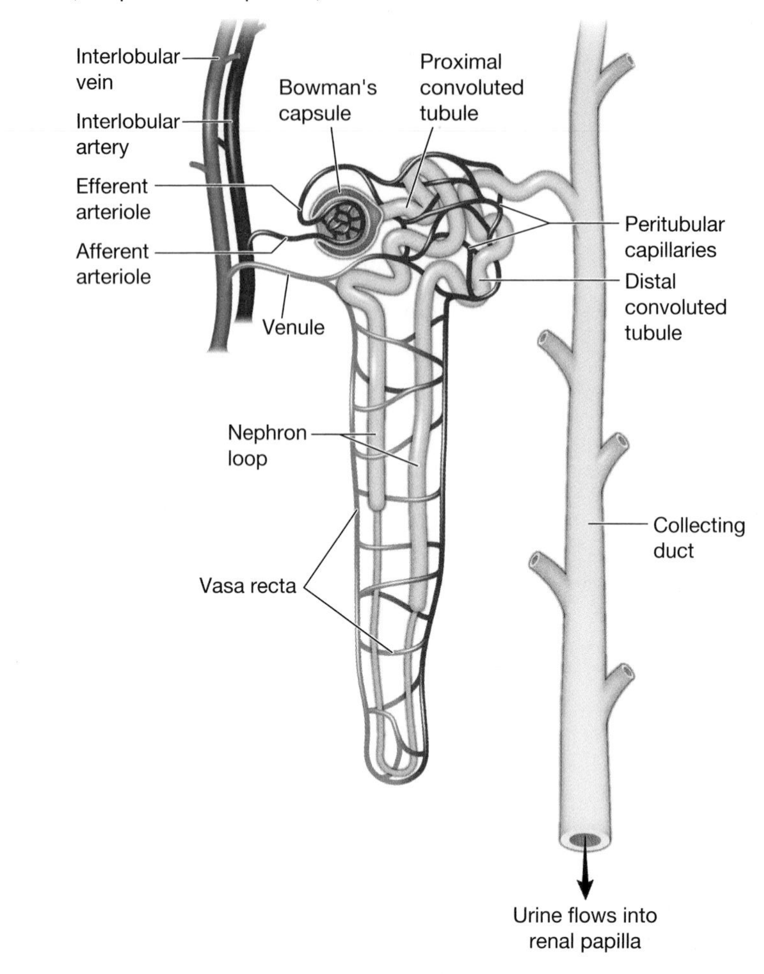

Learning Connection

Explain a Process

Why does the kidney need a portal system? Two sets of linked capillaries? Explain to a friend why the kidney needs to bridge two locations with a capillary set in each, connected by an arteriole. Which set of capillaries is specialized, and which set just performs gas and nutrient exchange? Why do you think the efferent arteriole is called an arteriole instead of a venule?

DIGGING DEEPER:
COVID and Kidney Failure

The disease COVID-19 is caused by infection with the virus SARS-CoV-2. While SARS-CoV-2 is transmitted through respiration, once it is in the body the virus is able to infect a wide range of cells, including kidney tubule epithelial cells. In fact, urine carries a fairly high viral load, indicating that quite a bit of virus ends up in the renal tubules. Individuals without preexisting kidney disease are more likely to develop it after COVID-19, and individuals who had kidney disease before COVID have a much higher chance of developing long-term issues.

Scientists are still trying to unravel the story behind COVID and the kidneys. What, specifically, about COVID disease causes damage to the kidneys? The virus is able to inflict damage on endothelial cells—the cells that line all blood vessels, including glomerular capillaries. Direct damage to the glomeruli might be one cause of kidney damage. In fact, COVID patients are more likely to have protein in their urine and reduced glomerular filtration rate (GFR), both of which are signs of glomerular damage. About 5 percent of COVID patients have a GFR decline of 30 percent or more. Since adult GFR naturally declines by about 1 percent per year, for these patients, surviving COVID infection took the same toll on their renal health as aging 30 years.

The inflammation, circulating clots, and breakdown products from muscle damage and changes in blood pressure and blood perfusion all may play additional roles in initiating or exacerbating kidney disease during and after COVID-19.

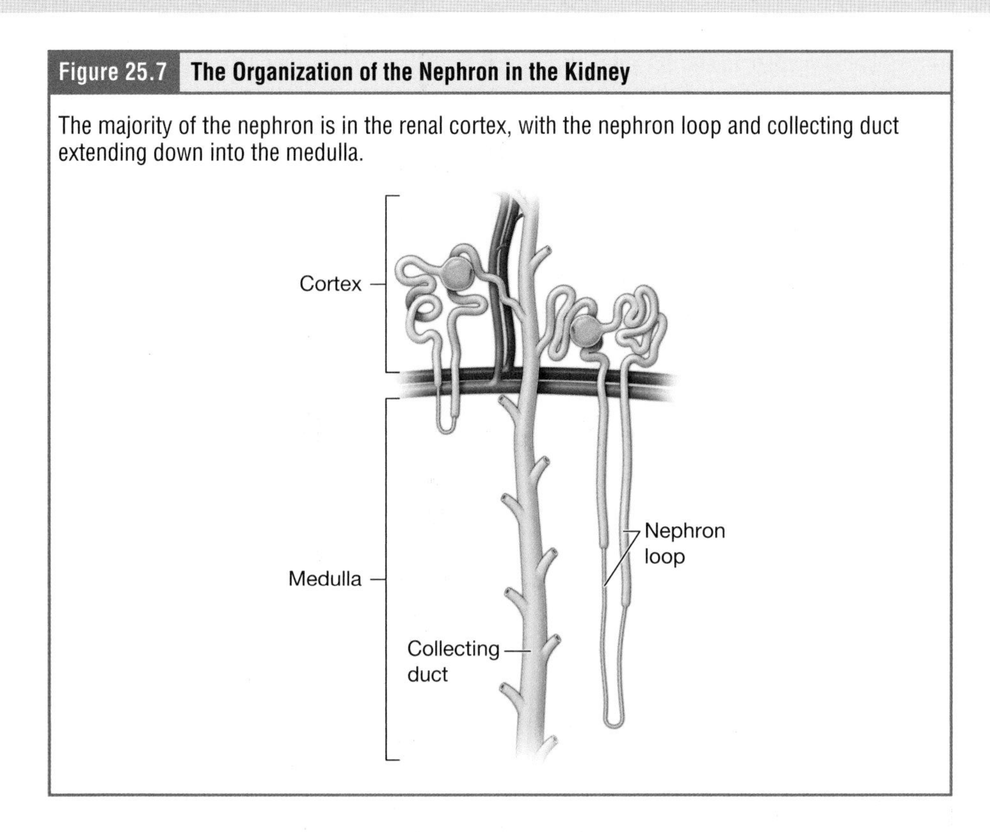

Figure 25.7 The Organization of the Nephron in the Kidney

The majority of the nephron is in the renal cortex, with the nephron loop and collecting duct extending down into the medulla.

down into the medulla (**Figure 25.7**). The nephron loop ascends toward the cortex and widens to form the **distal convoluted tubule (DCT)**.

LO 25.2.5

Most double-walled structures, as we have seen before, use the same structure for naming their layers. The outer layer of the glomerular capsule, is the parietal layer, and it is composed of simple squamous epithelium. The inner layer of the glomerular capsule closely surrounds the surface of the glomerular capillaries and so it is called the *visceral layer of the capsule*. Here, the cells are not squamous, but uniquely shaped cells (**podocytes**) extending fingerlike arms (**pedicels**) to cover the glomerular capillaries (**Figure 25.8**). If you place your hands around a cylinder, such as a mug or water bottle, and alternate your fingers in the grasp, you are forming an enclosure around that water

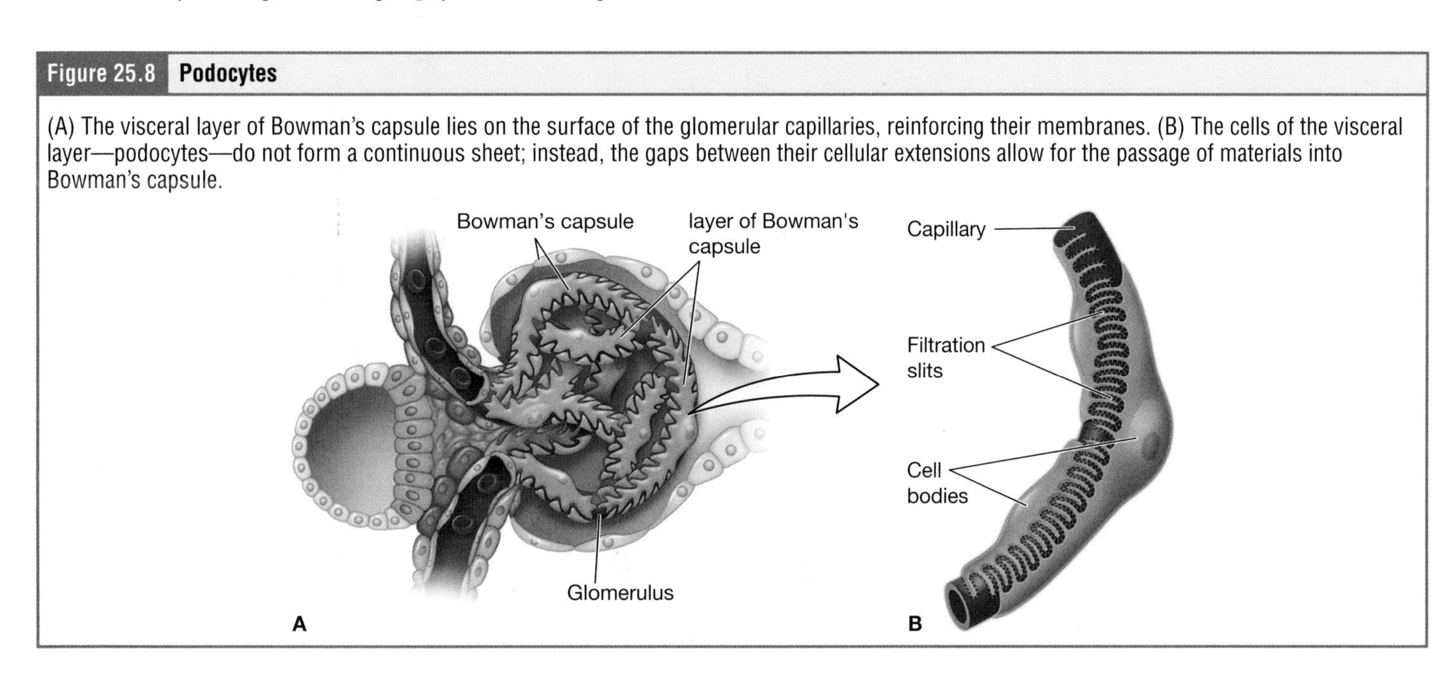

Figure 25.8 Podocytes

(A) The visceral layer of Bowman's capsule lies on the surface of the glomerular capillaries, reinforcing their membranes. (B) The cells of the visceral layer—podocytes—do not form a continuous sheet; instead, the gaps between their cellular extensions allow for the passage of materials into Bowman's capsule.

bottle that is similar to the way the podocytes surround the glomerular capillaries. The glomerular capillary is a fenestrated capillary (**Figure 25.9**). Fenestrated capillaries are one of the three types of capillaries; they have small spaces among the capillary endothelial cells that allow material to leave. The small spaces between the pedicels allow for the passage of material out of the capillary and into the glomerular capsule; these **filtration slits** are small gaps that form a sieve. As blood passes through the glomerulus, 10 to 20 percent of the plasma filters between these sievelike fingers to be captured by the glomerular capsule and funneled to the PCT. Where the fenestrae (gaps) in the glomerular capillaries match the spaces between the podocyte "fingers," the only thing separating the capillary lumen and the lumen of the glomerular capsule is their shared basement membrane (**Figure 25.10**). These three features comprise what is known as the **filtration membrane**. The filtration membrane is not selective; rather, it allows any material below a certain size to pass through between the blood and the lumen of Glomerular capsule. In this way, the filtration membrane can be likened to a kitchen colander; if you think of the blood like a pot of water in which you have cooked some pasta, for example, when you pour the mixture from the pot into the colander everything below a particular size (in this case the water) passes through the holes, while materials above a certain size (i.e., the pasta) are retained.

LO 25.2.6

In the case of the filtration membrane, the holes are large enough to let blood plasma and most dissolved solutes through and into the lumen of the glomerular capsule. Blood cells, platelets, and large proteins, however, are too large for the holes and so these materials are retained within the glomerular capillaries and do not pass into the tubule to become filtrate. An additional factor affecting the ability of substances to cross this barrier is their electric charge. The membrane proteins associated with the filtration membrane are negatively charged, so they tend to repel negatively charged substances and allow positively charged substances to pass more readily. The result is the generation of a filtrate that does not contain cells or large proteins, and has a slight predominance of positively charged substances.

Figure 25.9 Fenestrated Capillary

The glomerulus is composed of fenestrated capillaries. The openings (fenestrae) within the endothelial cells allow any materials smaller than the opening to pass from the glomerular capillaries into Bowman's capsule.

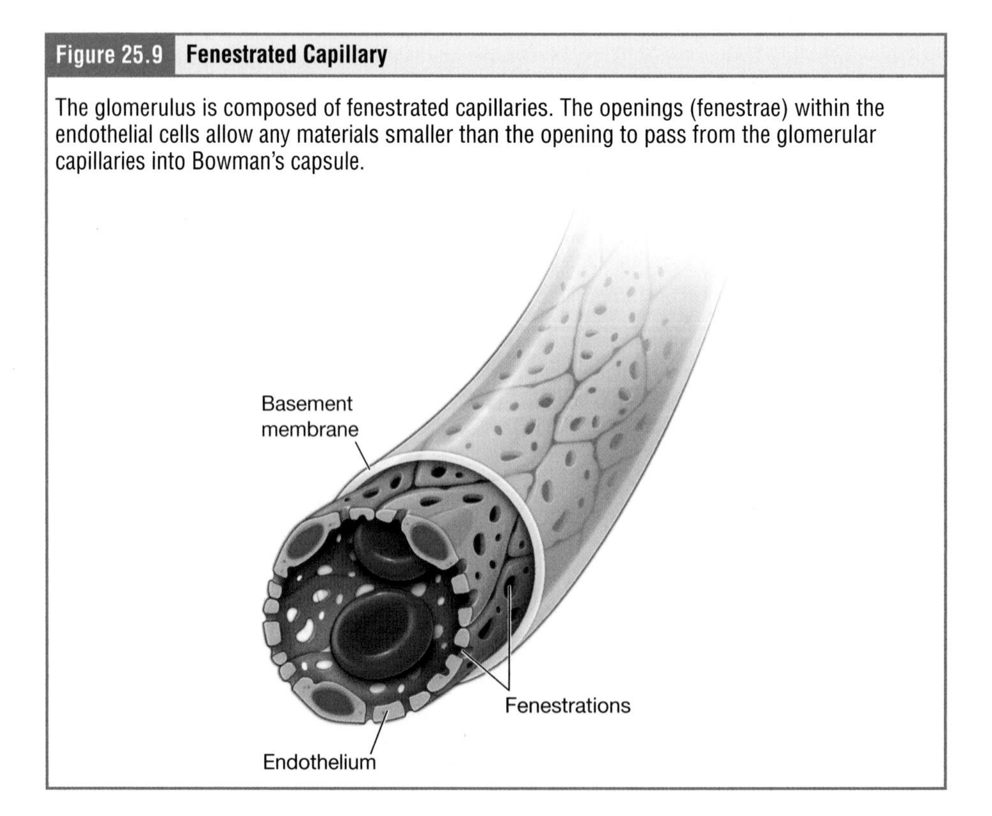

Figure 25.10 The Filtration Slits that Surround the Glomerulus

The filtration slits, made up of the pedicels of the podocytes of the visceral layer of Bowman's capsule, allow all materials below a certain size to pass out of the glomerular capillaries and into the lumen of Bowman's capsule.

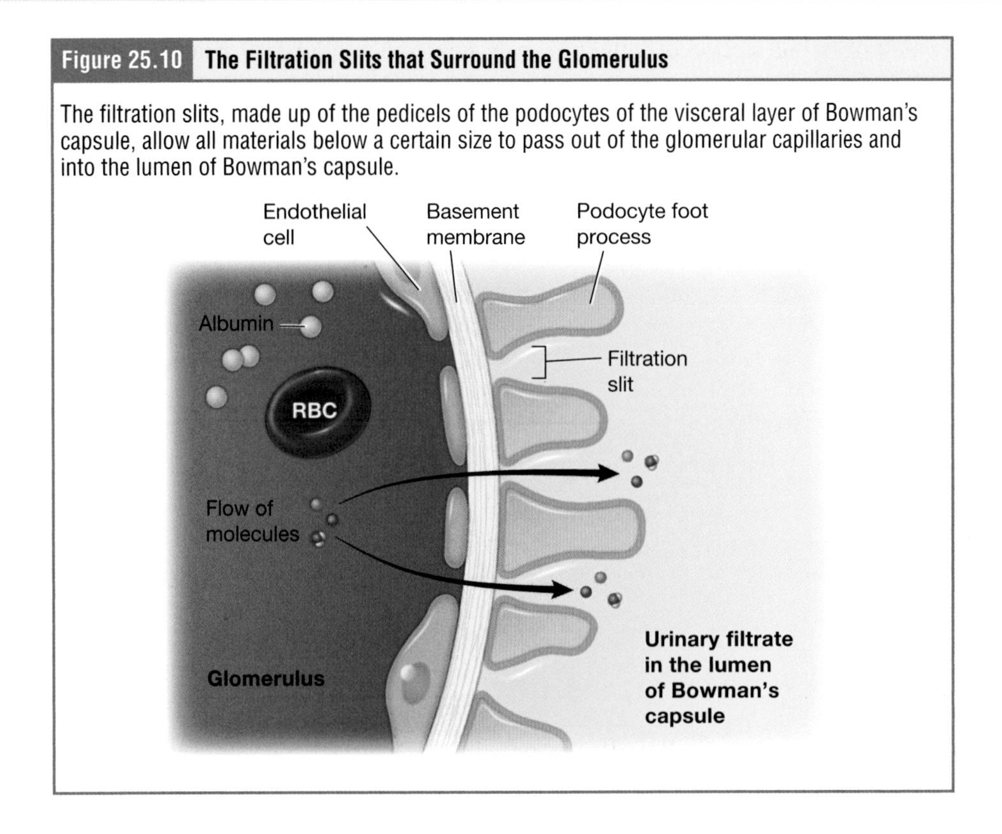

Proximal Convoluted Tubule (PCT) The filtrate within the glomerular capsule drains into the PCT. The inclusion of the word *convoluted* in its name describes its tortuous path, resembling a twisted, looping path of a roller coaster or water slide at a theme park. The wall of the tubule is made of simple cuboidal cells covered with microvilli on their luminal surface, forming a **brush border**. Similar to microvilli we have seen in other places such as the intestinal epithelium, tubule microvilli contribute to the expansion of the surface area in order to maximize the absorption and secretion of solutes (such as Na^+, Cl^-, glucose, and so on). Many solutes are actively transported, so these cells have a significant need for ATP and therefore possess a high concentration of mitochondria.

brush border (striated border; brush border membrane)

Nephron Loop The twisting path of the PCT leads to a long, straight loop of tubule called the *nephron loop* (Figure 25.11). The **descending limb** of the nephron loop is connected to the PCT. Whereas the descending limb consists of an initial short, thick portion and a long, thin portion, the **ascending limb** consists of an initial short, thin portion followed by a long, thick portion. The descending thick portion consists of simple cuboidal epithelium similar to that of the PCT. The descending and ascending thin portions consist of simple squamous epithelium. As you will see later, these are important differences, since different portions of the loop have different permeabilities for solutes and water. The ascending thick portion consists of simple cuboidal epithelium similar to the DCT.

descending limb (nephron loop)

Distal Convoluted Tubule (DCT) The DCT, like the PCT, is very tortuous and formed by simple cuboidal epithelium, but it is shorter than the PCT. Much less reabsorption and secretion of substances is required of the cells in the DCT. Therefore, they have fewer microvilli on their apical surface and fewer mitochondria inside the cell compared to the cells that line the PCT.

Figure 25.11 The Nephron Loop

The fluid that leaves the glomerular capillary, called *filtrate* when it is in the glomerular capsule, flows into a series of tubules. In order, these are the proximal convoluted tubule, the nephron loop, the distal convoluted tubule, and the collecting duct.

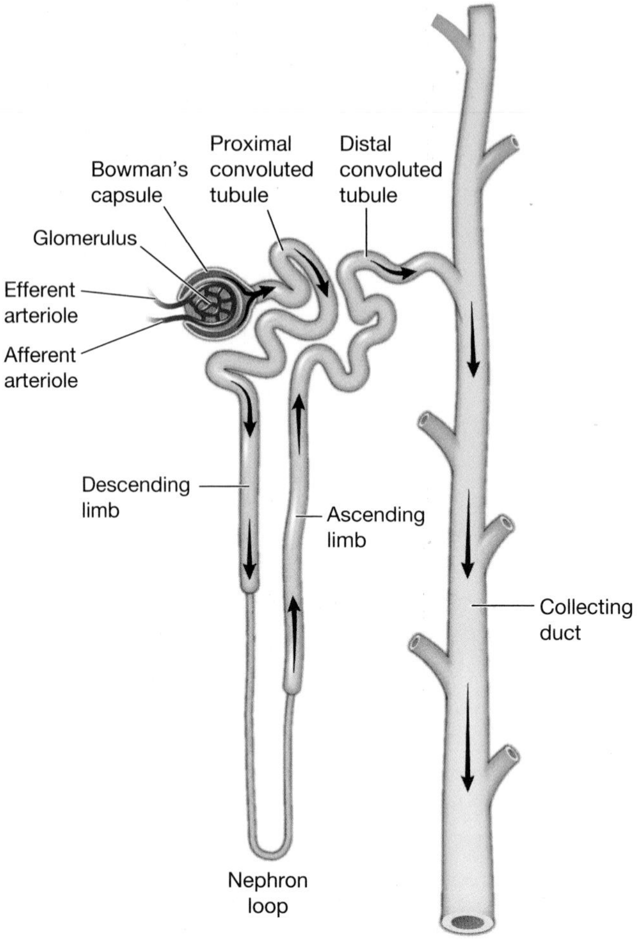

collecting ducts (duct of Bellini)

LO 25.2.7

aquaporin (AQP; water channels)

Collecting Ducts The **collecting ducts** are continuous with the nephron but are not technically part of it. In fact, each duct collects filtrate from several nephrons. The pooled filtrate is modified a final time in the collecting duct. As they descend deeper into the medulla, collecting ducts merge. At their ends, they empty their contents into the minor calyx at the tip of the papilla. They are lined with simple squamous epithelium. Their surfaces express hormone receptors and a physiologically variable number of **aquaporin** channel proteins. Aquaporins are channels that allow water to pass (**Figure 25.12**). The number of aquaporins expressed at any given time changes in response to hormone levels in a process we will learn about later in this chapter. When a high concentration of aquaporins is expressed in the collecting duct, large amounts of water can be recovered from the filtrate and absorbed back into the blood. In this case a lower volume of urine would be produced, but water would remain in the body. This would be desirable in conditions of dehydration or low blood volume. If excess water needs to be removed from the body, fewer aquaporins would be expressed in the collecting duct. More water would stay in the filtrate, producing a high volume of urine.

The cortex contains the majority of nephron structures as well as the afferent and efferent arterioles. Parts of the nephron loops and the collecting ducts are found in the

Figure 25.12 Aquaporins (Water Channels)

Aquaporins are channels embedded within membranes that allow water molecules to pass from one side of the membrane to the other.

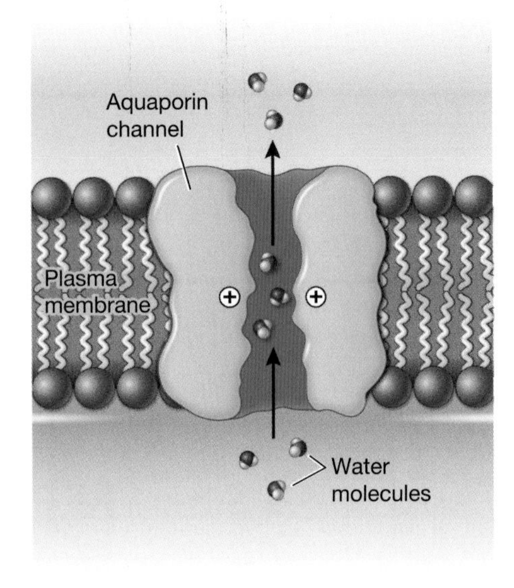

renal medulla (**Figure 25.13**). The majority of nephrons in the kidney—about 85 percent—are classified as **cortical nephrons**. The structures of each cortical nephron are located primarily in the cortex. The tips of their nephron loops reach into the medulla (**Figure 25.14**). In contrast, the remaining 15 percent of nephrons have long nephron loops that extend deep into the medulla and are called **juxtamedullary nephrons**.

LO 25.2.8

Figure 25.13 Renal Cortex versus Renal Medulla

(A) The cortex contains all of the glomeruli and the majority of proximal and distal convoluted tubules. The medulla contains the majority of the nephron loops. (B) Under the microscope, the renal cortex contains notable rounded structures, called *renal corpuscles*, each of which consists of a glomerulus and Bowman's capsule. Increasing the magnification reveals that these are glomeruli and cross sections through the convoluted tubules. The medulla appears to have long stripes or columns; these are the ascending and descending portions of the nephron loops and the collecting ducts.

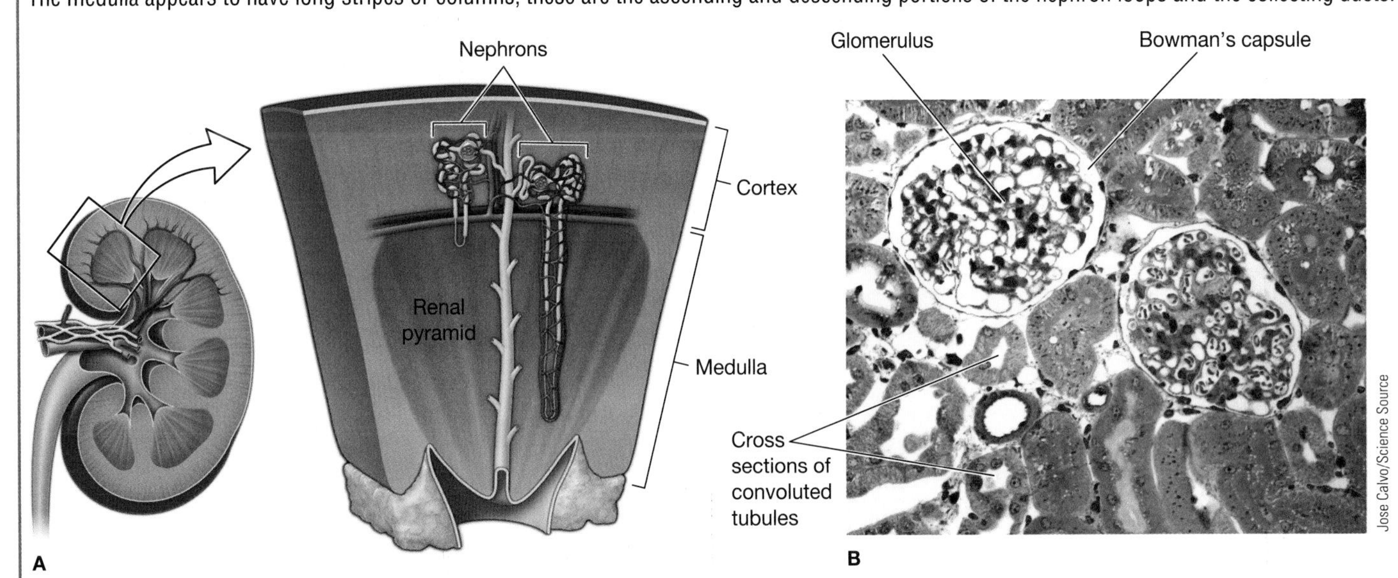

Figure 25.14 Cortical Nephrons versus Juxtamedullary Nephrons

Most of the renal nephrons are cortical nephrons. They have short nephron loops and are located closer to the capsule within the cortex. Juxtamedullary nephrons are farther from the capsule, have long nephron loops, and—notably—have a vasa recta capillary bed that surrounds their nephron loop.

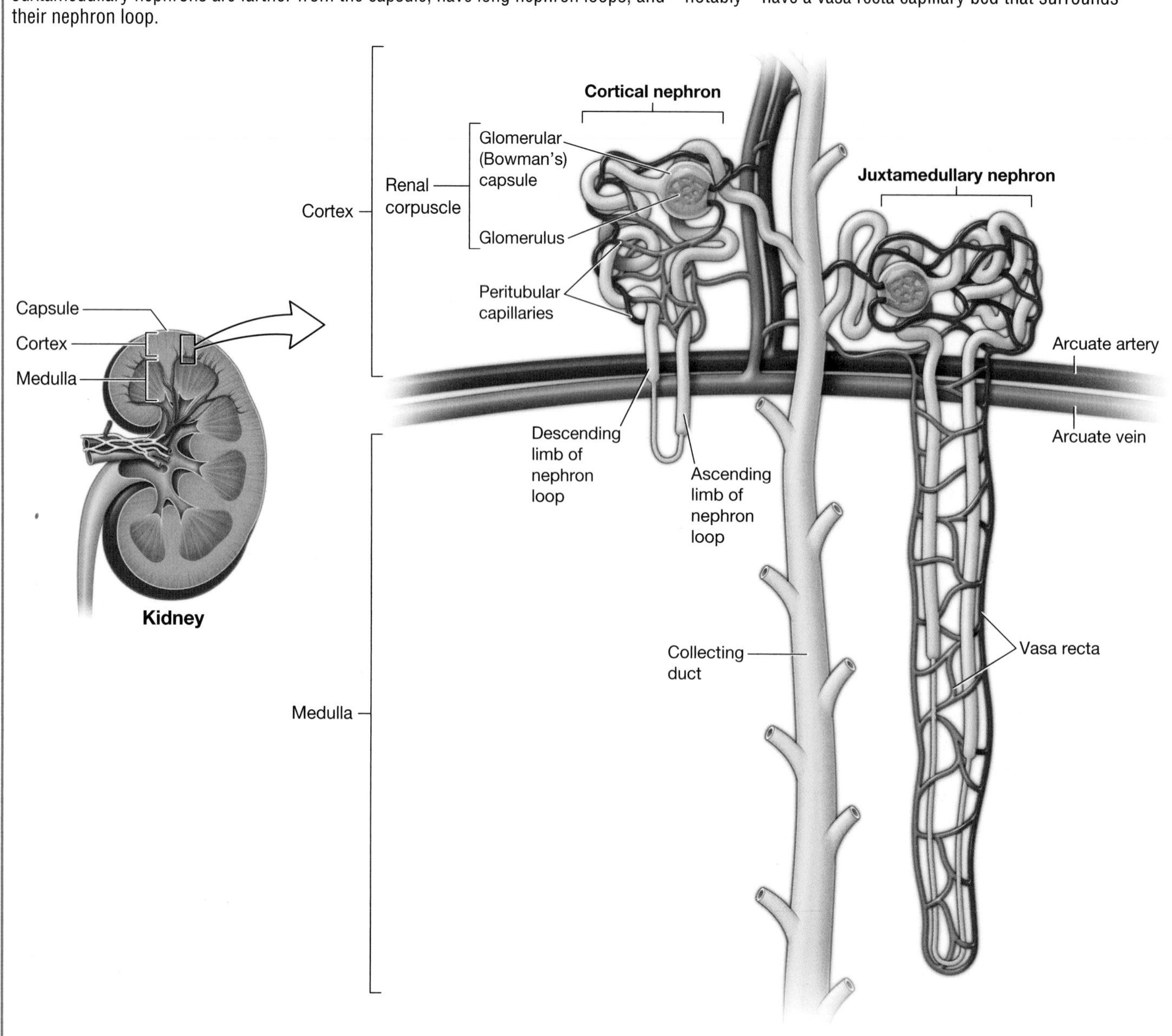

Learning Check

1. Which of the following is the correct order of structures through which fluid flows in the nephron of the kidney?
   a. Proximal convoluted tubule → nephron loop → distal convoluted tubule → collecting duct
   b. Nephron loop → distal convoluted tubule → proximal convoluted tubule → collecting duct
   c. Nephron loop → proximal convoluted tubule → distal convoluted tubule → collecting duct
   d. Proximal convoluted tubule → distal convoluted tubule → nephron loop → collecting duct

2. Which of the following structures runs through the renal hilum and contains filtered blood?
   a. Renal artery
   b. Renal vein
   c. Ureter
   d. Lymph vessels
3. Which type of capillary is the glomerulus?
   a. Continuous
   b. Discontinuous
   c. Fenestrated
   d. Sinusoidal
4. Which of the following is the correct order of branching arteries in the kidney?
   a. Segmental arteries → arcuate arteries → interlobar arteries → interlobular arteries → afferent arterioles
   b. Segmental arteries → interlobar arteries → arcuate arteries → interlobular arteries → afferent arterioles
   c. Afferent arterioles → segmental arteries → interlobar arteries → interlobular arteries → arcuate arteries
   d. Interlobar arteries → segmental arteries → afferent arterioles → interlobular arteries → arcuate arteries

25.3 Physiology of Urine Formation

Learning Objectives: By the end of this section, you will be able to:

25.3.1 Describe the three processes that take place in the nephron (i.e., filtration, reabsorption, and secretion) and explain how the integration of these three processes determines the volume and composition of urine.

25.3.2 Define glomerular filtration rate (GFR) and explain the role of blood pressure, capsule fluid pressure, and colloid osmotic (oncotic) pressure in determining GFR.

25.3.3 Describe factors that can change blood pressure, capsule fluid pressure, and colloid osmotic (oncotic) pressure and thereby change GFR.

25.3.4* Predict the impact on GFR when various physiological factors change.

25.3.5 Compare and contrast tubular reabsorption and secretion with respect to the direction of solute movement and the tubule segments in which each process occurs.

25.3.6* Apply the concepts of transporter specificity, saturation, and competition to explaining variations in tubule reabsorption.

25.3.7 Describe specific mechanisms of transepithelial transport that occur in different parts of the nephron (e.g., active transport, osmosis, facilitated diffusion, electrochemical gradients, receptor-mediated endocytosis, transcytosis).

25.3.8 For the important solutes of the body (e.g., Na^+, K^+, glucose, urea), describe how each segment of the nephron handles the solute and compare the filtration rate of the solute to its excretion rate (i.e., the net handling of the solute by the nephron).

25.3.9 Explain the role of the nephron loop (of Henle), its permeability to water, and the high osmolarity of the interstitial fluid in the renal medulla in the formation of dilute urine.

25.3.10 Trace the changes in filtrate osmolarity as it passes through the segments of the nephron.

25.3.11 Identify the location, structures and cells of the juxtaglomerular apparatus (JGA) and discuss its significance.

25.3.12 Compare and contrast blood plasma, glomerular filtrate, and urine.

25.3.13 For any solute, explain how renal filtration, reabsorption, and secretion determine the excretion rate of that solute.

* Objective is not a HAPS Learning Goal.

The following numbers will vary by individual, but on average, we filter our entire blood volume through the nephrons about once every 30 minutes. During this process, about 20 percent, or one-fifth, of the plasma is removed and becomes filtrate. So if in the span of an hour—the length of a television show or an anatomy and physiology lecture—almost half of your plasma becomes filtrate, how can you possibly survive? Our amazing kidneys have a tremendous filtration capacity that supports homeostasis by quickly ridding our body of excess metabolites and waste products. However, the kidneys must move back a tremendous amount of water and other solutes in order to maintain adequate blood volume and nutrient levels. So far in this chapter we have learned about the process of filtration, but if filtration occurred alone we would quickly die of dehydration. Instead, the kidney uses three processes—filtration, reabsorption, and secretion—in concert to maintain homeostasis through the production of urine.

LO 25.3.1

25.3a Glomerular Filtration Process and Rate

Student Study Tip

Filtration is like unzipping your backpack a little bit and turning it upside down, shaking it and getting rid of anything that falls out. Reabsorption is putting items you need for class back in the bag, and secretion is taking out things that you don't need, that filtration left behind because they were too large to fall out.

Earlier in this chapter we discussed that filtration is a nonspecific process based on size. Filtration occurs as the blood plasma, and all of its solutes below a certain size, move by bulk flow through the filtration slits of the glomerulus and into Bowman's capsule. The volume of filtrate formed per minute by both kidneys is termed the **glomerular filtration rate (GFR)**. When GFR is measured in a laboratory setting, certain physiological conditions are present. The body is at rest, and the heart is pumping the full blood volume (3.5–5.5 liters of blood) per minute. When the body is at rest, approximately 20 percent flows through the kidneys for filtration at any given time. If we consider these average numbers and resting conditions, the GFR ranges between 80 and 140 milliliters of filtrate produced per minute. This amount will equate to a volume of about 150–180 liters per day and 150 liters of filtrate per day is approximately 30 times the total blood volume! Therefore, 99 percent of this filtrate must be returned to the circulation by reabsorption so that only about one to two liters of urine are produced per day.

LO 25.3.2

While these number ranges are stunning, you'll notice that there are wide ranges of values given. Why? For one, we know that no two people are exactly the same. The anatomical and physiological diversity among us means that you could not possibly have the same exact blood volume or internal physiological conditions as the person sitting next to you at any given time. But even within your own body, the amount of filtrate produced within one minute or one day will vary widely based on what you have eaten and drunk that day, how stressed out you are, or what your current metabolic rate is. Let's examine the physiological factors that influence GFR.

LO 25.3.3

GFR is influenced by the hydrostatic pressure and colloid osmotic pressure on either side of the capillary membrane of the glomerulus. In Chapter 20 we discussed the oppositional or "push-pull" forces that regulate the amount of plasma fluid that escapes the capillaries to become interstitial fluid or lymph. These same forces contribute to glomerular filtration, which is similarly based on the bulk flow of plasma out of a capillary (**Figure 25.15**). Recall that whereas fluid mixtures—such as blood—move according to pressure gradients, individual molecules—including water molecules—move according to concentration gradients. Glomerular filtration occurs as hydrostatic pressure forces fluid, along with its solutes, through the semipermeable barrier. **Hydrostatic pressure** is the pressure exerted by a fluid against the walls of its container; in this case we can imagine the hydrostatic pressure exerted by the blood onto the walls of the glomerulus. However, if you have fluid on both sides of a barrier, both fluids exert pressure in opposing directions. Net fluid movement

Figure 25.15 Glomerular Filtration Rate

Glomerular filtration rate is the balance of three factors—the hydrostatic pressure of the glomerulus, the osmotic gradient, and the hydrostatic pressure of Bowman's capsule. These same factors regulate the rate of fluid flow out of capillaries anywhere in the body.

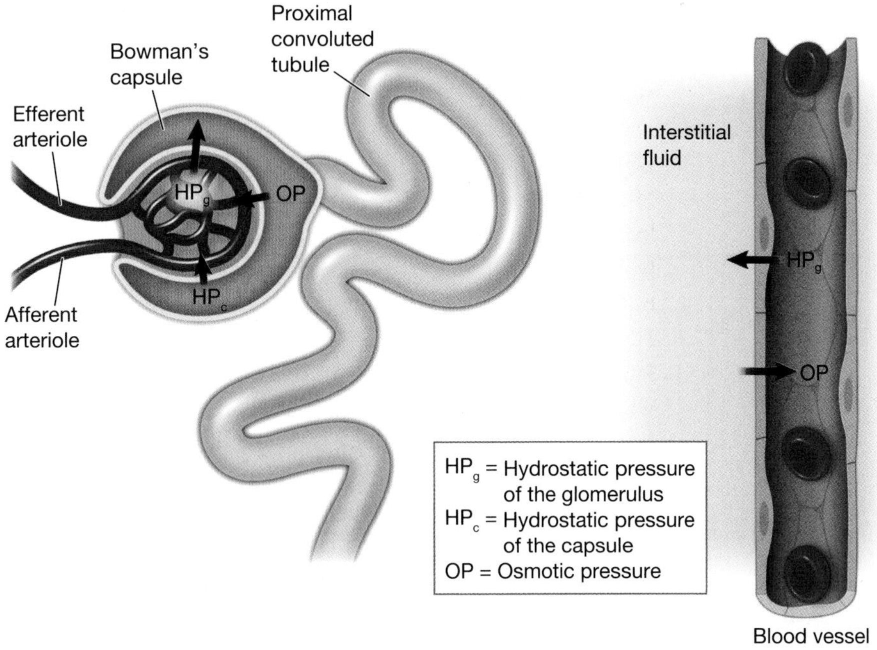

HP_g = Hydrostatic pressure of the glomerulus
HP_c = Hydrostatic pressure of the capsule
OP = Osmotic pressure

will occur in the direction of the lower pressure. Glomerular filtration occurs when glomerular hydrostatic pressure exceeds the luminal hydrostatic pressure of the glomerular capsule (see Figure 25.15). As long as the pressure inside the glomerulus is greater than the pressure of the fluid in the glomerular capsule, there will be net fluid flow from the blood in the glomerulus toward the glomerular capsule. This fluid flow is glomerular filtration. There is an additional pressure to consider: osmotic pressure. Osmosis is the movement of water across a membrane in response to a difference in concentration of water molecules on one side of the membrane compared with the other. Osmotic pressure can be described as the likelihood of water molecules to move across a membrane; in other words, a higher osmotic pressure means that there is a greater concentration gradient difference between one side of the membrane and the other, and water molecules are more likely to move across the membrane. Water molecules move down their concentration gradient, from a region of higher water concentration to a region of lower water concentration. Another way to consider this is that during osmosis, water molecules move from a region of lower osmotic pressure to a region of higher osmotic pressure. As long as the concentration differs, water will move. While most solutes can come to equilibrium across the glomerular membrane, the osmotic pressure is typically higher in the glomerular capillary due to the presence of plasma proteins, which are too large to cross into the glomerular capsule.

To understand why this is so, look more closely at the microenvironment on either side of the filtration membrane. Recall that cells and the medium-to-large proteins cannot pass between the podocyte processes or through the fenestrations of the capillary endothelial cells. This means that red and white blood cells, platelets, albumins, and other proteins too large to pass through the filter remain in the capillary, creating an average colloid osmotic pressure of 30 mm Hg within the capillary. The absence of proteins in the lumen of the glomerular capsule results in an osmotic pressure near zero. Thus, the only pressure moving fluid across the capillary wall into the lumen of glomerular space is the hydrostatic pressure of the glomerulus. The sum of all of the influences, both osmotic and hydrostatic, results in a **net filtration pressure (NFP)** of about 10 mm Hg (Figure 25.16).

25.3b Net Filtration Pressure (NFP)

NFP determines GFR. The three factors of hydrostatic pressure of the glomerulus, hydrostatic pressure of the capsule, and osmotic pressure determine NFP (see Figure 25.16A). You can use this model and change the numbers in order to figure out how glomerular filtration would change in various physiological circumstances. For example, let's consider what would happen in a patient who is experiencing liver failure due to hepatitis infection. The liver produces plasma proteins, so in patients with declining liver function the concentration of plasma proteins decreases as well. Because plasma proteins—and their inability to enter the glomerular capsule—are the main contributors to the osmotic gradient, a decrease in their concentration leads to a corresponding decrease in osmotic pressure. Figure 25.16B illustrates the change in NFP in this physiological condition. You can follow this format to explore two more physiological conditions in the following two "Apply to Pathophysiology" features.

Apply to Pathophysiology

Chronic High Blood Pressure

LO 25.3.4

One of the many consequences of high blood pressure is damage to the delicate, one-cell-thick capillary walls, which may give way under higher pressure. Capillary damage can occur anywhere, including at the glomerular capillaries. In this case study, you are a primary care physician seeing a patient with a history of high blood pressure. At the time of today's office visit, the high blood pressure has begun to damage the glomerular capillaries, though you do not yet know that. Larger openings are now in the filtration membrane and the pressure gradient between the glomerulus and capsular space is exaggerated. Predict changes (increase, decrease, no change) that will be present in the following parameters as a consequence of this change.

1. What will happen to the plasma protein concentration in the blood?
   A) Increase
   B) No change
   C) Decrease
2. What will happen to the osmotic pressure gradient between the glomerular capillary and the filtrate in the glomerular capsule?
   A) Increase
   B) Decrease
   C) No change
3. What will happen to the plasma volume in the system as a whole?
   A) Decrease
   B) No change
   C) Increase
4. What will happen to the glomerular filtration rate in a single nephron?
   A) No change
   B) Decrease
   C) Increase
5. This patient has a history of high blood pressure, but consider that you're seeing the patient after months of glomerular damage. Do you think their blood pressure has decreased, increased, or stayed the same from their previous above-average values?
   A) Decreased
   B) Increased
   C) No change

Apply to Pathophysiology

Tubule Inflammation

Meghan had a sore throat and fever for over a week before their doctor diagnosed them with strep throat, an infection of *streptococcus pyogenes*. The infection spread to the blood and then to the renal tubules, where it caused inflammation. The tubules narrowed due to hardening and inflammation, thereby increasing resistance to the flow of filtrate. Predict the initial changes (increase, decrease, no change) that will occur in the following parameters.

1. What will happen to the glomerular filtration rate in a single nephron?
   A) Decrease
   B) No change
   C) Increase
2. What will happen to the amount of plasma proteins (in the blood)?
   A) Decrease
   B) No change
   C) Increase
3. What will happen to the plasma osmotic pressure?
   A) Increase
   B) Decrease
   C) No change
4. What will happen to the plasma volume?
   A) Increase
   B) Decrease
   C) No change
5. What will happen to this patient's blood pressure?
   A) Increase
   B) Decrease
   C) No change
6. What will happen to the interstitial fluid volume in the legs?
   A) Increase
   B) Decrease
   C) No change

Determination of the GFR is one of the tools used to assess the kidney's function. With up to 180 liters of filtrate passing into the glomerular capsules of the two kidneys each day, it is quite obvious that most of that fluid and its contents must be reabsorbed. That reabsorption occurs in the PCT, nephron loop, DCT, and the collecting ducts (**Figure 25.17** and **Table 25.1**). Filtration is a nonspecific process based on size, but the reabsorption of most materials, including water, is tightly regulated. This control is based on the blood volume and blood pressure and exerted through the action of several hormones. Most water is recovered in the PCT, nephron loop, and DCT. About 10 percent (about 18 liters) reaches the collecting ducts. The collecting ducts can recover almost all of the water passing through them (in cases of dehydration) or almost none of the water (in cases of overhydration). Each molecule of water in the filtrate, which was water in the blood before glomerular filtration, will either be reabsorbed and end

Figure 25.16 Net Filtration Pressure

The net filtration pressure is the pressure that dictates fluid leaving the blood to become filtrate. It is a balance of three forces: the hydrostatic pressure of the glomerulus, the osmotic gradient, and the hydrostatic pressure of the capsule. (A) These are typical values that might be observed under typical health circumstances. (B) These are theoretical values that may occur in an individual experiencing liver failure.

Typical Physiological Circumstances	Liver Failure
HPg = 55 mmHg	HPg = 55 mmHg
OP = −30 mmHg	OP = −15 mmHg
HPc = −15 mmHg	HPc = −15 mmHg
10 mmHg outward pressure	25 mmHg outward pressure
A	B

Figure 25.17 Reabsorption

Different materials are reabsorbed or secreted at different locations along the nephron.

Amino acids, Glucose, Lactate, Protein, Urea, Uric acid, Vitamins, Ca^{2+}, Cl^-, K^+, HCO_3^-, H_2O, Mg^{2+}, Na^+

Cl^-, HCO_3^-, H_2O, Na^+

H^+, K^+, NH_4^+

Creatinine, Urea, Uric acid, Some drugs, H^+, NH_4^+

Cl^-, K^+, Na^+

H_2O, Urea

H_2O

Urea

DIGGING DEEPER:
What is Dialysis?

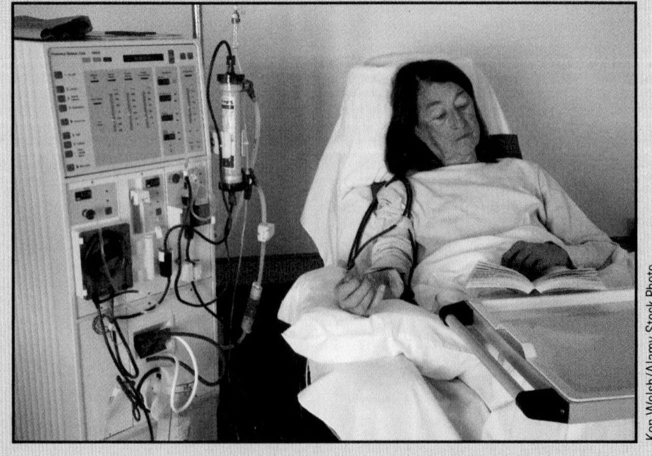

Ken Welsh/Alamy Stock Photo

Dialysis describes medical procedures that fulfill the function of the kidneys by cleansing the blood. Each of our two kidneys has approximately 1.3 million nephrons, which is far more than is needed to maintain homeostasis. Some people can lead very healthy lives after kidney donation. However, damage to the kidneys, when substantial, can threaten homeostasis. In order to perform the functions of the kidney, that is, eliminating excess fluid, toxins, waste products, and ions from the blood, patients suffering from kidney failure can have their blood removed, processed through a machine with filters, and returned to their bodies (see figure). The U.S. Centers for Disease Control estimates that 15 percent of Americans have kidney failure. There are five stages of kidney failure, and patients in Stage 5—the last stage—may require either kidney transplant or dialysis. The risk factors for developing kidney disease are high blood pressure and high blood sugar. Managing blood pressure and blood sugar can help to slow the advancement of kidney disease and prevent kidney failure.

Table 25.1 Substances Secreted or Reabsorbed in the Nephron and Their Locations LO 25.3.5

Substance	PCT	Nephron Loop	DCT	Collecting Ducts
Glucose	Almost 100 percent reabsorbed; secondary active transport with Na^+			
Oligopeptides, proteins, amino acids	Almost 100 percent reabsorbed; symport with Na^+			
Vitamins	Reabsorbed			
Lactate	Reabsorbed			
Creatinine	Secreted			
Urea	50 percent reabsorbed by diffusion; also secreted	Secretion, diffusion in descending limb		Reabsorption in medullary collecting ducts; diffusion
Sodium	65 percent actively reabsorbed	25 percent reabsorbed in thick ascending limb; active transport	5 percent reabsorbed; active	5 percent reabsorbed, stimulated by aldosterone; active
Chloride	Reabsorbed, symport with Na^+, diffusion	Reabsorbed in thin and thick ascending limb; diffusion in ascending limb	Reabsorbed; diffusion	Reabsorbed; symport
Water	67 percent reabsorbed osmotically with solutes	15 percent reabsorbed in descending limb; osmosis	8 percent reabsorbed if ADH; osmosis	Variable amounts reabsorbed, controlled by ADH, osmosis
Bicarbonate	80–90 percent symport reabsorption with Na^+	Reabsorbed, symport with Na^+ and antiport with Cl^-; in ascending limb		Reabsorbed antiport with Cl^-
H^+	Secreted; diffusion		Secreted; active	Secreted; active
NH_4^+	Secreted; diffusion		Secreted; diffusion	Secreted; diffusion
HCO_3^-	Reabsorbed	Reabsorbed; diffusion in ascending limb	Reabsorbed; diffusion	Reabsorbed; antiport with Na^+
Some drugs	Secreted		Secreted; active	Secreted; active
Potassium	65 percent reabsorbed; diffusion	20 percent reabsorbed in thick ascending limb; symport	Secreted; active	Secretion controlled by aldosterone; active
Calcium	Reabsorbed; diffusion	Reabsorbed in thick ascending limb; diffusion		Reabsorbed if parathyroid hormone present; active
Magnesium	Reabsorbed; diffusion	Reabsorbed in thick ascending limb; diffusion	Reabsorbed	
Phosphate	85 percent reabsorbed, inhibited by parathyroid hormone, diffusion		Reabsorbed; diffusion	

up back in the blood, or it will not be reabsorbed, in which case it remains in the kidney tubule, and becomes part of the urine. Therefore, the volume of the urine fluctuates based on how much water needs to be reabsorbed in order to maintain blood pressure homeostasis.

25.3c Mechanisms of Reabsorption and Secretion

The pressure, volume, pH, and osmolarity of the blood are all kept within very tight homeostatic ranges through the action of the kidneys. Any excess materials, plus wastes that the body needs to get rid of, are disposed of together in urine. Urine is the end result of three processes carried out in the nephron: filtration, reabsorption, and secretion. In filtration, as we have just described in the preceding paragraphs, blood plasma, along with any dissolved solutes below a certain size, is removed from the blood and enters the glomerular capsule. As that filtrate passes through the capsule and into the long tubule system attached, it is refined by the processes of reabsorption and secretion. Throughout this process, the fluid, which was called *filtrate* when it left the glomerular capsule, is now called *tubular fluid*. In **reabsorption**, useful materials that were filtered by the glomerulus are moved back from the renal tubules into the blood of the peritubular capillaries. Reabsorbed materials will include water, glucose, amino acids, and many of the solutes such as Na^+, Ca^{2+}, and so on. Materials that are filtered and not reabsorbed will end up in the urine. In **secretion**, substances that were not filtered, but must be excreted in the urine, are transported from the blood of the peritubular capillaries into the fluid of the renal tubules. Molecules that are too large to fit through the fenestrae of the glomerular capillaries are added to the forming urine by secretion. Mechanisms by which substances move across membranes for reabsorption or secretion include: active transport, simple diffusion, facilitated diffusion, secondary active transport, and osmosis. In short, every mechanism for membrane transport, except endocytosis and exocytosis, is used in the kidney tubule for reabsorption and secretion. We discussed these mechanisms thoroughly in Chapter 4, and it might be helpful to review them before proceeding to the rest of this chapter. A brief reminder of transport is provided in the "Anatomy of Membrane Transport" feature.

Anatomy of...

Membrane Transport

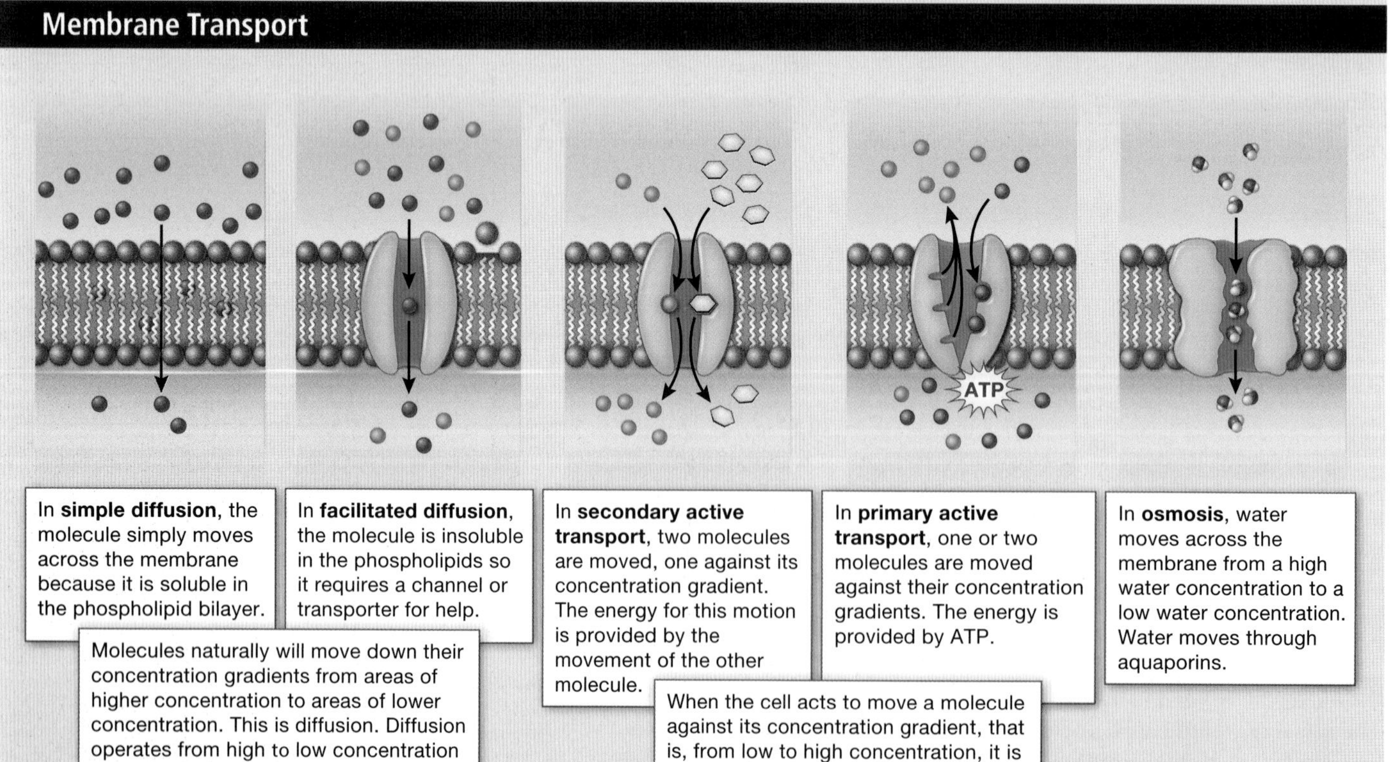

ATP
In **simple diffusion**, the molecule simply moves across the membrane because it is soluble in the phospholipid bilayer.
In **facilitated diffusion**, the molecule is insoluble in the phospholipids so it requires a channel or transporter for help.
Molecules naturally will move down their concentration gradients from areas of higher concentration to areas of lower concentration. This is diffusion. Diffusion operates from high to low concentration and does not require energy.
In **secondary active transport**, two molecules are moved, one against its concentration gradient. The energy for this motion is provided by the movement of the other molecule.
In **primary active transport**, one or two molecules are moved against their concentration gradients. The energy is provided by ATP.
When the cell acts to move a molecule against its concentration gradient, that is, from low to high concentration, it is called *active transport*.
In **osmosis**, water moves across the membrane from a high water concentration to a low water concentration. Water moves through aquaporins.

LO 25.3.6

Remember also that any transporter that has a binding site will be governed by the principles of specificity, saturation and competition.

- ***Specificity:*** A transporter that binds to a molecule is specific for a molecule or a closely related set of molecules. For example, a glucose transporter may be specific to just glucose, or may also be able to bind fructose—a closely related sugar—but it would not be able to bind an amino acid.
- ***Saturation:*** Transporters with binding sites can only bind the number of molecules that they have binding sites for. For example, imagine a transporter that has a single binding site and it takes that transporter one millisecond to complete the movement of its passenger molecule from one side of the membrane to the other. This transporter has a maximum rate of one molecule moved per millisecond. Now imagine that one cell has five of these transporters on its surface. When all five are bound the cell's transport capacity is saturated. The maximum transportation rate for the cell will be five molecules per millisecond. Even if the concentration of the molecule dramatically increases, the rate cannot exceed this saturation level unless more transporters are added.
- ***Competition:*** Two similar molecules may both be able to bind to the same transporter. Using the example provided in the previous paragraph on "Saturation" with a transporter that has a single binding site and is capable of moving one molecule every millisecond, if this transporter were capable of transporting two similar molecules, such as glucose and fructose, then each time it completed a movement it would move one or the other. If there were a 50:50 mixture of glucose and fructose, then we can imagine that over time, such as over a second, the transporter would have moved 1,000 molecules; half of these would be fructose and the other half would be glucose.

Student Study Tip

Two similar molecules are playing a game of musical chairs for one seat.

Examine **Figure 25.18**. When we discuss the process of material being reabsorbed from the nephron tubule back into the blood at the peritubular capillaries, this is

Figure 25.18 Reabsorption and Secretion via Membrane Transport

When considering the mechanisms behind reabsorption and secretion, it is helpful to consider each side of the tubule epithelial cell independently. Transport across the luminal surface is based on the gradients between the intracellular fluid of the cell and the lumen of the tubule. Transport across the basolateral surface is determined by the gradients between the intracellular fluid and the interstitial fluid.

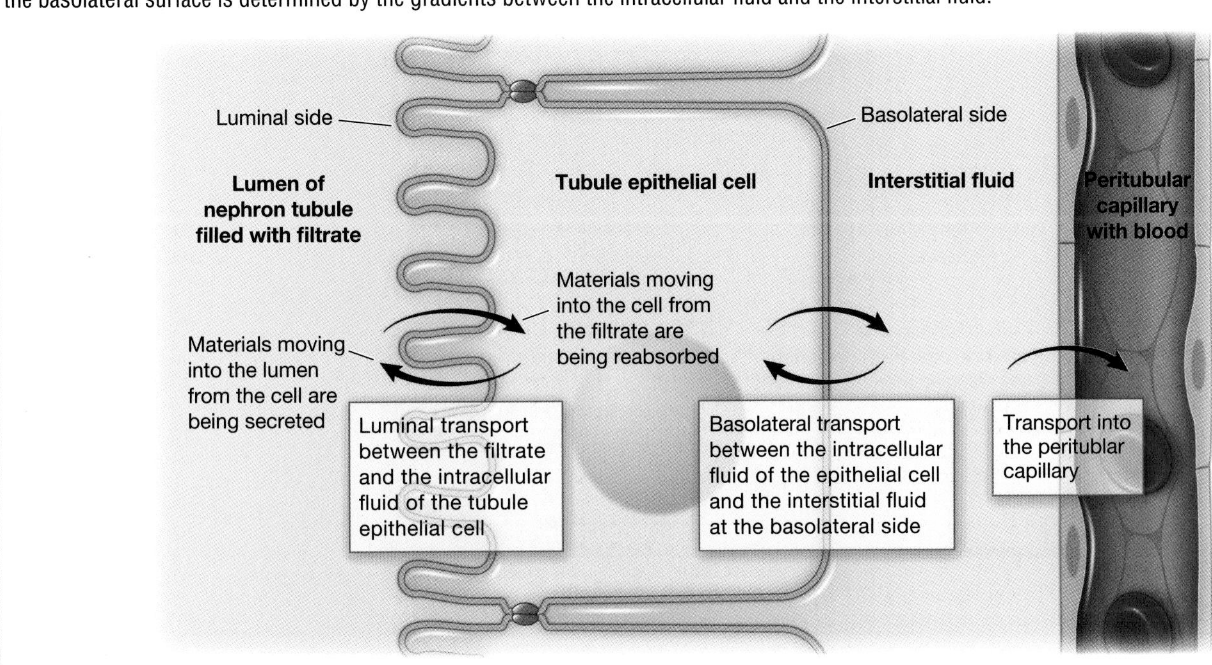
Luminal side
Basolateral side
Lumen of nephron tubule filled with filtrate
Tubule epithelial cell
Interstitial fluid
Peritubular capillary with blood
Materials moving into the cell from the filtrate are being reabsorbed
Materials moving into the lumen from the cell are being secreted
Luminal transport between the filtrate and the intracellular fluid of the tubule epithelial cell
Basolateral transport between the intracellular fluid of the epithelial cell and the interstitial fluid at the basolateral side
Transport into the peritublar capillary

Student Study Tip

Material in the Lumen will be Lost in urine.

transport motion that occurs in three locations. First, the material must cross the **luminal surface** of the tubule epithelial cell, moving from filtrate in the lumen of the tubule to the intracellular fluid of the epithelial cell. Transport across this membrane will be determined by the concentration gradient of the substance between the filtrate and intracellular fluid as well as by what transporters are present and available to facilitate transport. Next, the substance must make its way out of the tubule epithelial cell across the **basolateral surface**. The substance is now in the interstitial fluid and must cross into the peritubular capillaries to complete its reabsorption into the blood.

LO 25.3.7

Each type of reabsorbed material must make its complete journey from tubule lumen to blood. Most substances will use a different form of transport on the luminal side of the epithelial than the type of transport they use on the basolateral side of the epithelial cell. This is because their concentration gradients are different. If the concentration of Na^+ is low in the intracellular fluid of the epithelial cell but high in the interstitial fluid found between the cell and the peritubular capillary, Then the transport across the luminal membrane is from high concentration to low concentration, and would not require energy. The transportation across the basolateral surface, however, would be from low to high concentration; this is not the way the Na^+ ions would tend to flow—it is against their concentration gradient—and so this transport will require energy (**Figure 25.19**).

Na^+/K^+ pumps, also known as *Na^+/K^+ ATPases*, that we met in earlier chapters when we learned about the neuron and the muscle cell, are expressed on the basolateral membrane of the tubular epithelial cells. These transporters constantly pump Na^+ out of the cell, moving the Na^+ ions to the interstitial fluid. Turning our focus to the

Figure 25.19 The Reabsorption of Sodium

The reabsorption of Na^+ on the luminal side is passive through facilitated diffusion. The concentration of Na^+ is higher in the lumen than inside the cell. The intracellular concentration of Na^+ is low because Na^+ leaves the cell via active transport, which pushes Na^+ out of the cell and into the interstitial fluid.

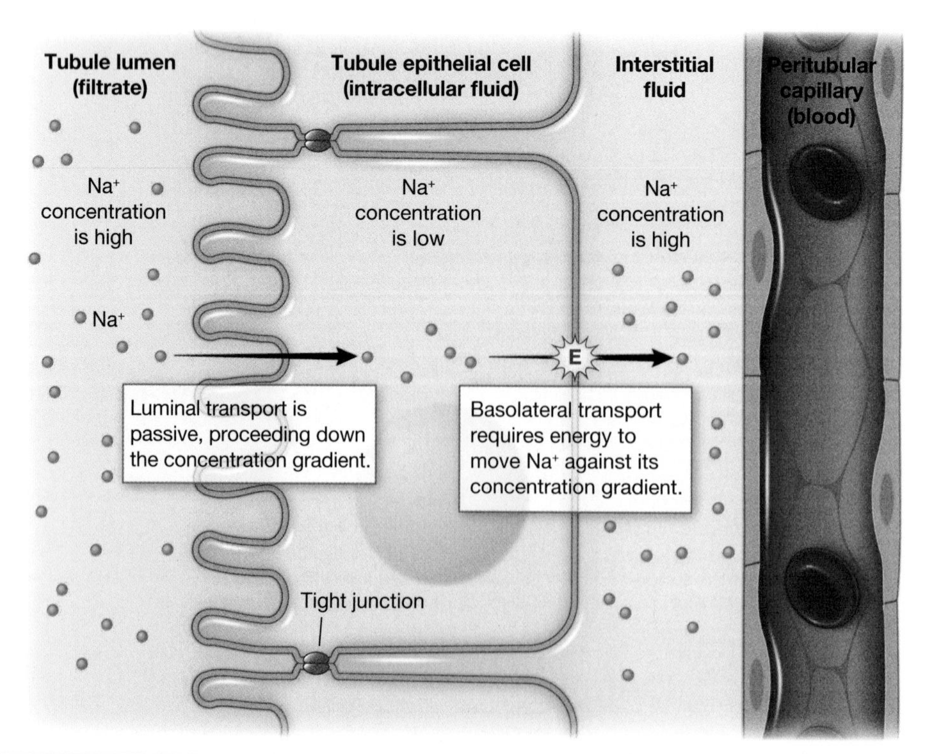
Tubule lumen (filtrate)
Tubule epithelial cell (intracellular fluid)
Interstitial fluid
Peritubular capillary (blood)
Na+ concentration is high
Na+ concentration is low
Na+ concentration is high
Na+
E
Luminal transport is passive, proceeding down the concentration gradient.
Basolateral transport requires energy to move Na+ against its concentration gradient.
Tight junction

luminal membrane, we can see that by steadily pumping Na^+ out of the intracellular fluid there is and will always be a strong electrochemical gradient for Na^+ to move into the cell from the tubular lumen. This setup enables the transport of many substances across the luminal side of the cell.

A Na^+/glucose co-transport protein carries both Na^+ and glucose into the cell; glucose may move against its concentration gradient, but it is carried along by Na^+, which is moving down its concentration gradient. The glucose molecule then diffuses across the basal membrane by facilitated diffusion into the interstitial fluid and from there into peritubular capillaries (**Figure 25.20**).

Most of the Ca^{2+}, Na^+, glucose, and amino acids that were filtered out of the blood at the glomerulus must be reabsorbed and returned to the blood to maintain homeostatic plasma concentrations. Other substances, such as urea, K^+, ammonia (NH_3), creatinine, and some drugs are filtered and additionally secreted into the filtrate as waste products to be excreted from the body in the urine. Acid-base balance is maintained through actions of both the lungs and kidneys: The respiratory system can adjust the frequency of breathing in order to get rid of or conserve H^+, likewise the kidneys secrete or reabsorb H^+ and HCO_3^- as needed to maintain pH homeostasis.

Figure 25.20 The Reabsorption of Glucose and Water is Dependent on the Movement of Na^+

Sodium is actively transported out of the cell on the basolateral side, which keeps the intracellular Na^+ concentration low. On the luminal side, sodium moves in, down its concentration gradient. Glucose is co-transported as Na^+ moves. Glucose, therefore, builds up on the inside of the cell and is able to cross the basolateral side of the cell via facilitated diffusion. As all of these solutes move toward the interstitial fluid, water is drawn there as well based on its osmotic gradient. The only mechanism for water to move across the membrane is through aquaporins.

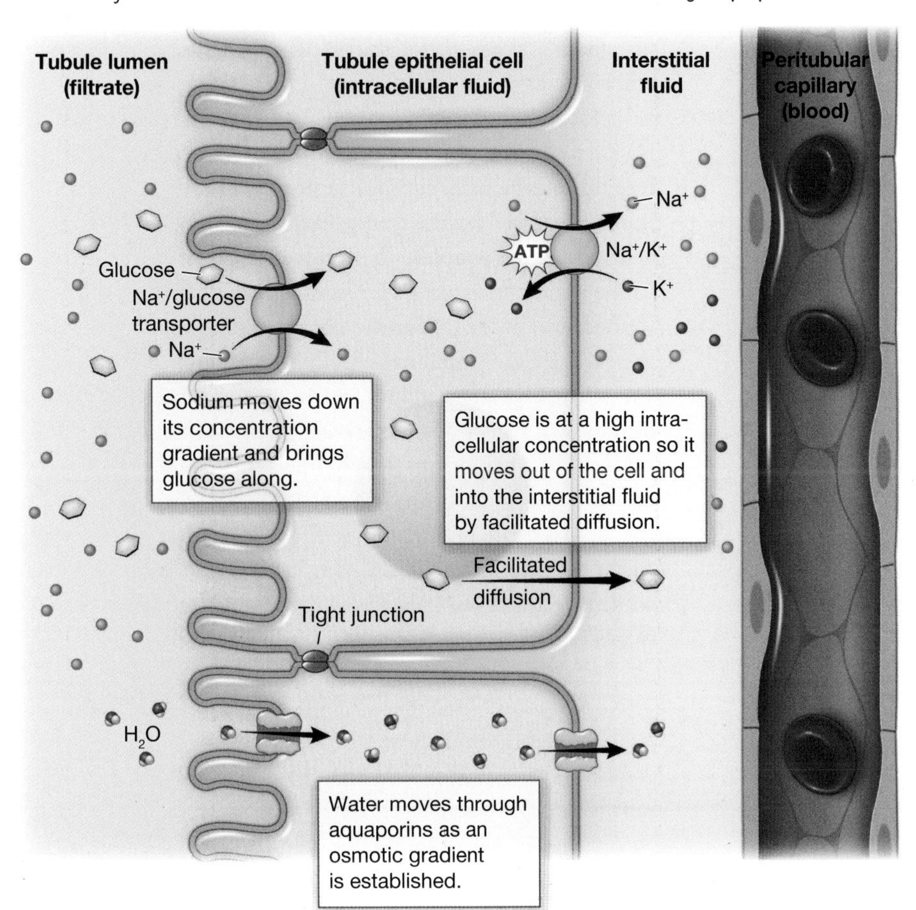

Tubule lumen
(filtrate)
Tubule epithelial cell
(intracellular fluid)
Interstitial
fluid
Peritubular
capillary
(blood)
Na+
ATP
Na+/K+
K+
Glucose
Na+/glucose
transporter
Na+
Sodium moves down its concentration gradient and brings glucose along.
Glucose is at a high intracellular concentration so it moves out of the cell and into the interstitial fluid by facilitated diffusion.
Facilitated
diffusion
Tight junction
H2O
Water moves through aquaporins as an osmotic gradient is established.

Water is reabsorbed passively. The movement of other materials such as Na^+, glucose, and amino acids creates an osmotic gradient in which there is a much higher solute concentration in the interstitial fluid than in the tubule lumen. Remember that the membrane is relatively impermeable to water, so water flows through open aquaporins.

Movement from the interstitial fluid into the peritubular and vasa recta capillaries occurs by bulk flow. It is important to understand the difference between the glomerulus and the peritubular and vasa recta capillaries. The glomerulus has a relatively high pressure inside its capillaries. Because a large proportion of the plasma was removed from the blood during filtration, the blood that flows through the peritubular capillaries and vasa recta has a higher osmolarity and a much lower pressure. Movement of water into the peritubular capillaries and vasa recta will be occur primarily by osmolarity and concentration gradients. Pressure gradients will also guide waters flow into the capillaries. As it moves from the interstitial fluid to the capillaries, many substances move with the water in bulk flow.

LO 25.3.8

More substances move across the membranes of the PCT than any other portion of the nephron. Many of these substances (amino acids and glucose) are moved by cotransport mechanisms along with Na^+. Antiport (a form of cotransport where the molecules move in opposite directions), active transport, diffusion, and facilitated diffusion are additional mechanisms by which substances are moved from one side of a membrane to the other. A few of the substances that are co-transported with Na^+ luminal membrane include Cl^-, Ca^{2+}, amino acids, glucose, and phosphate ions. Like the glucose example showing in Figure 25.20, most of the substances co-transported into the cell with Na^+ exit the cell on the basolateral side by facilitated diffusion.

While the amount of reabsorption of many substances, including water and bicarbonate ions, will vary in the PCT according to the homeostatic needs of the body, just about 100 percent of glucose, amino acids, and other organic substances will be reabsorbed here.

Recovery of bicarbonate (HCO_3^-), which is a very powerful and fast-acting buffer, is vital to the maintenance of acid-base balance. An important enzyme is used to catalyze this mechanism: carbonic anhydrase (CA). This same enzyme and reaction is used in red blood cells in the transportation of CO_2, in the stomach to produce hydrochloric acid, and in the pancreas to produce HCO_3^- to buffer acidic chyme from the stomach. In the kidney, most of the CA is located within the cell, but a small amount is bound to the brush border of the membrane on the apical surface of the cell. In the lumen of the PCT, HCO_3^- combines with hydrogen ions to form carbonic acid (H_2CO_3). This is enzymatically catalyzed into CO and water, which diffuse across the apical membrane into the cell. Water can move osmotically across the lipid bilayer membrane due to the presence of aquaporin water channels. Inside the cell, the reverse reaction occurs to produce bicarbonate ions (HCO_3^-). These bicarbonate ions are co-transported with Na^+ across the basal membrane to the interstitial space around the PCT (**Figure 25.21**). At the same time this is occurring, a Na^+/H^+ antiporter excretes H^+ into the lumen, while it recovers Na^+. Note how the hydrogen ion is recycled so that bicarbonate can be recovered. Also, note that a Na^+ gradient is created by the Na^+/K^+ pump.

$$HCO_3^- + H^+ \leftrightarrow H_2CO_3 \leftrightarrow CO_2 + H_2O$$

The significant recovery of solutes from the PCT lumen to the interstitial space creates an osmotic gradient that promotes water recovery. As noted previously, water moves through channels created by the aquaporin proteins. These proteins are found in all cells in varying amounts and help regulate water movement across membranes and through cells by creating a passageway across the hydrophobic lipid bilayer membrane. Changing the number of aquaporin proteins in membranes of the collecting ducts

Figure 25.21 Reabsorption of Bicarbonate from the PCT

Bicarbonate is reabsorbed by its conversion to carbonic acid (H_2CO_3) and subsequent breakdown into water and carbon dioxide, both of which can diffuse across membranes either by simple or facilitated diffusion.

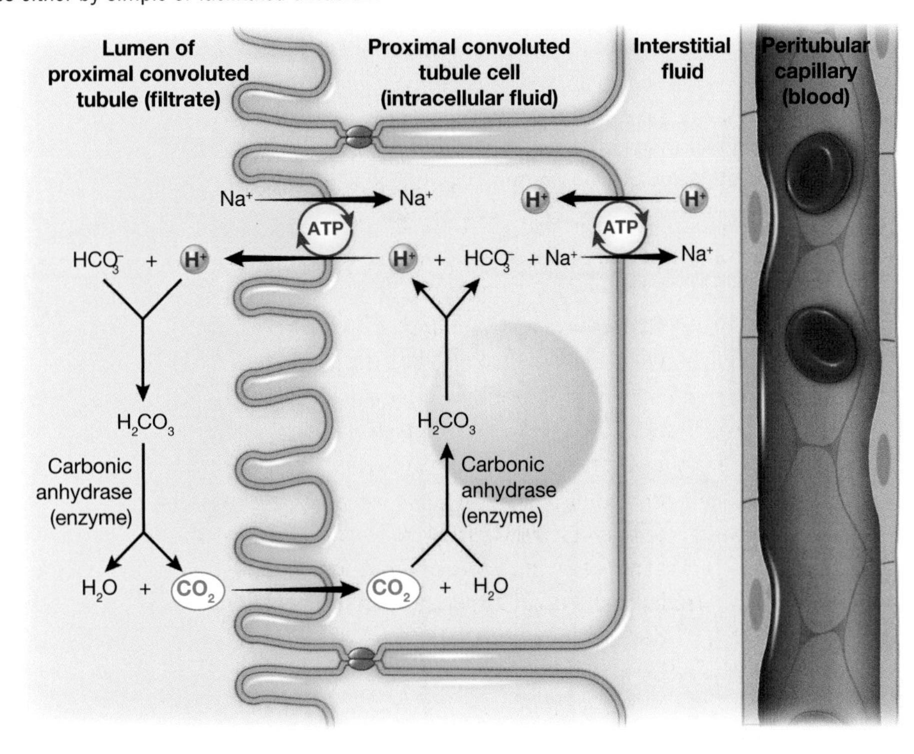

also helps to regulate the osmolarity of the blood. The movement of many positively charged ions also creates an electrochemical gradient. This charge promotes the movement of negative ions toward the interstitial spaces and the movement of positive ions toward the lumen.

Learning Connection

Broken Process

How would you expect urine production and general health to be affected in a person who was born with all cortical nephrons (and no juxtamedullary nephrons)? Remember that cortical nephrons have short nephron loops and juxtamedullary nephrons have very long ones. Hint: Think about the function of the nephron loops in Na^+ and water reabsorption.

LO 25.3.9

25.3d Reabsorption and Secretion in the Nephron Loop

The tubular fluid flows from the PCT and into the nephron loop. The nephron loop is a long, straight, U-shaped tube with two sections: descending and ascending sections (**Figure 25.22**). The descending and ascending portions of the loop are highly specialized to enable recovery of much of the Na^+ and water that were filtered by the glomerulus. As the tubular fluid moves through the loop, the osmolarity of the fluid within the tubule will change. As it flows into the descending loop, the tubular fluid is isosmotic with blood (about 300 milliosmoles per kilogram or mOsmol/kg). During the travel through the descending loop, water leaves the tubular fluid, and the tubular fluid becomes very hypertonic (about 1200 mOsmol/kg). As the filtrate ascends the loop, sodium leaves and the osmolarity of the solution within the tubule decreases; eventually, the solution becomes very hypotonic (100 mOsmol/kg). These changes are accomplished by osmosis in the descending limb and active transport in the ascending limb. Solutes and water recovered from these loops are returned to the circulation by way of the vasa recta. Notice in the panels of Figure 25.22 that the descending and ascending loops each have a thin and thick portion. The thin portions are composed of simple squamous epithelium and the thick portion is composed of simple cuboidal epithelium (**Figure 25.22A**).

Figure 25.22 The Nephron Loop and Countercurrent Multiplier

(A) The parts of the nephron in order from glomerulus to collecting duct. (B) Water leaves the descending loop, steadily increasing the osmlarity of the filtrate. (C) Sodium leaves the filtrate and enters the interstitial fluid surrounding the loop. The resulting increase in osmolarity of the interstitial fluid enables the water to leave in the descending loop.

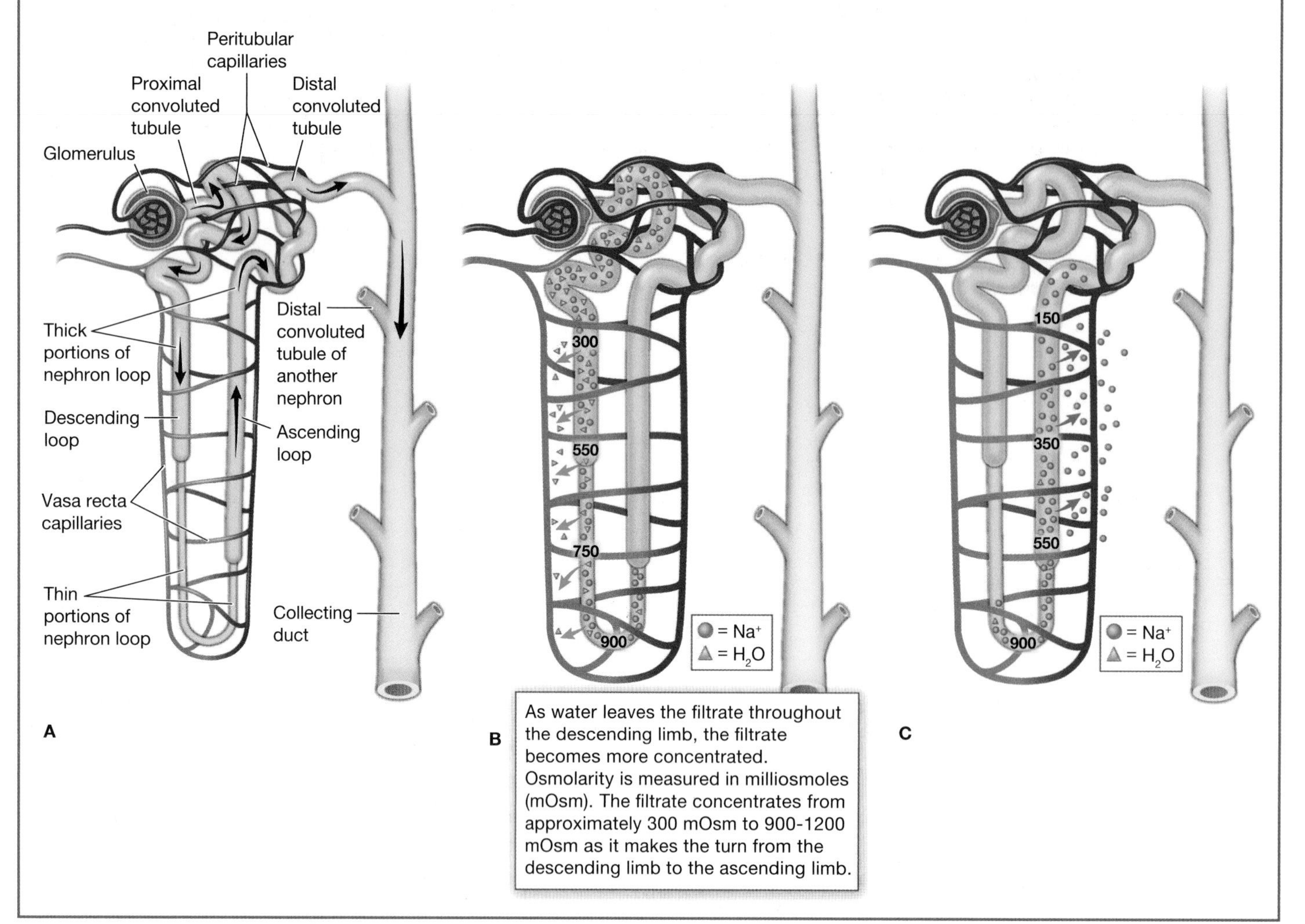

Descending Loop The cell membranes of the thin portion of the descending loop are studded with aquaporin channel proteins. These aquaporin channels are permanently expressed on these cells, they do not change. Therefore, water flows readily out of the descending loop into the surrounding interstitial fluid according to its osmotic gradient. As water leaves, the osmolarity of the fluid within the tubule increases from about 300 mOsmol/kg to 900–1200 mOsmol/kg. While much of the water that was filtered is recovered here, there is still a lot of water left to reabsorb. Modest amounts of urea, Na^+, and other ions are also recovered here (**Figure 25.22B**).

LO 25.3.10

Ascending Loop The ascending loop has a very short thin (and a much longer thick) portion. The cells of the thick portion do not have the brush border for surface area that we have seen in the PCT. These cells are also completely impermeable to water due to the absence of aquaporin proteins, but ions, mainly Na^+ and CL^-, are actively reabsorbed throughout the ascending loop. Because ions are reabsorbed but water cannot follow, the fluid within the loop decreases in osmolarity as the ions move out of the loop and into the interstitial fluid. This has two significant effects: the tubular fluid becomes hypoosmotic as it reaches the DCT; and the interstitial space becomes increasingly hyperosmotic (**Figure 25.22C**).

The Na^+/K^+ ATPase pumps in the basolateral membrane create an electrochemical gradient, allowing reabsorption of Cl^- by Na^+/Cl^- co-transporters in the luminal membrane. Some Cl^- is also able to follow the Na^+ to the interstitial fluid through **leaky tight junctions**. As Cl^- moves through these leaky junctions, Ca^{2+} and Mg^{2+} are drawn through as well by the electrical gradient. These leaky tight junctions are found between cells of the ascending loop, where they allow certain solutes to move according to their electrical and concentration gradient. Most of the K^+ that enters the cell will end up in the lumen by moving down its concentration gradient through channels in the luminal membrane (**Figure 25.23**).

Remember that, while Figure 25.23 illustrates these processes as stepwise, they are actually occurring at the same time. Fluid flow through the nephron is continuous. You might think of the fluid flowing through the tubule as a train, made up of many train cars (**Figure 25.24**). As earlier train cars are moving through the ascending loop, later train cars are moving through the descending loop. As ions are being actively pumped out of the earlier train cars flowing through the ascending loop, a hypertonic interstitial environment is created that impacts the later train cars flowing through the descending loop. The water within the fluid (later train cars) in the descending loop is drawn out of the tubule and into the hyperosmotic interstitial fluid through

Figure 25.23 The Epithelial Cells of the Ascending Loop

In the ascending loop, the transport of ions occurs across the epithelial cells. The concentration of the intracellular fluid is key; it determines the methods of transport (active versus passive) on the luminal and basolateral sides. As more and more solutes are reabsorbed, chloride, calcium, and magnesium are drawn in by their electrical gradients.

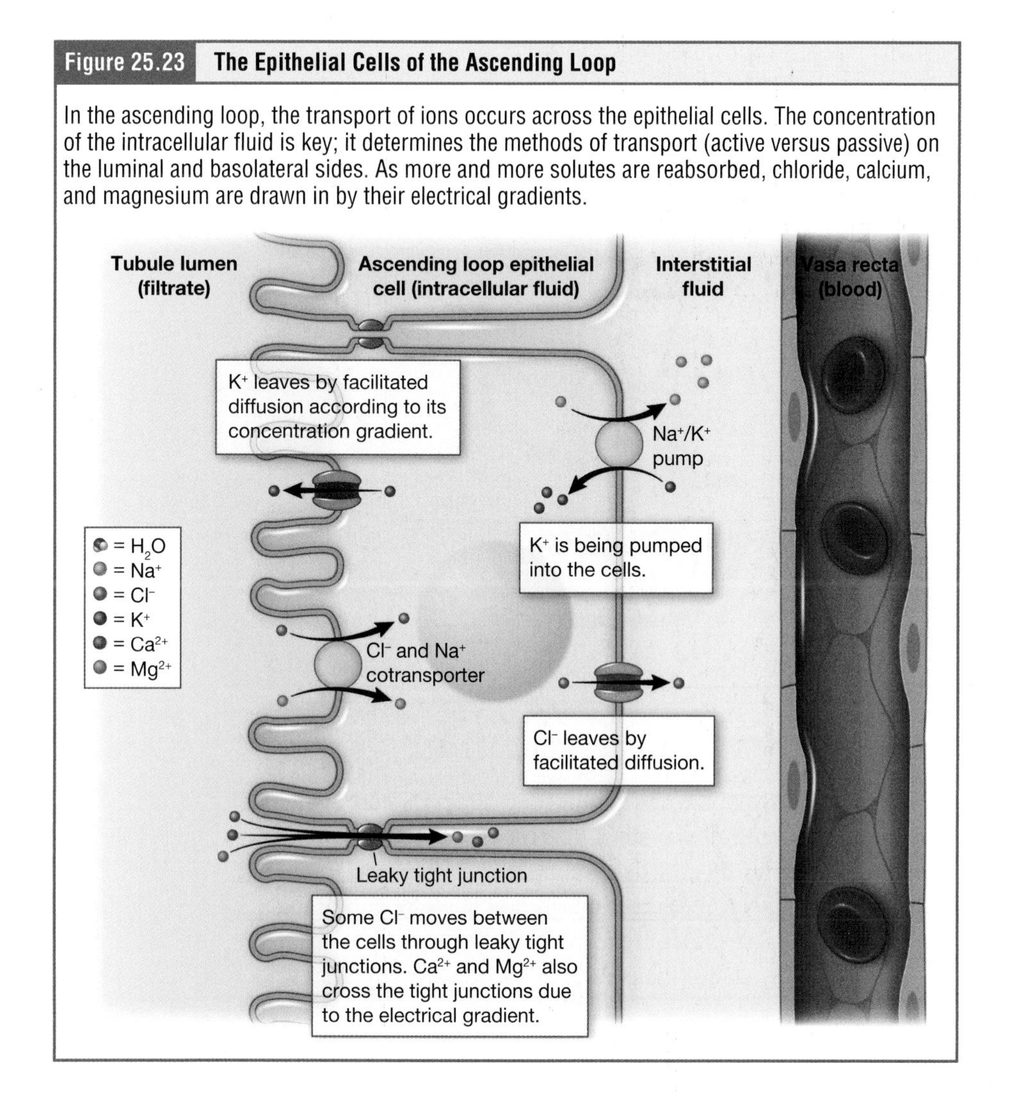
Tubule lumen (filtrate)
Ascending loop epithelial cell (intracellular fluid)
Interstitial fluid
Vasa recta (blood)
K⁺ leaves by facilitated diffusion according to its concentration gradient.
Na⁺/K⁺ pump
K⁺ is being pumped into the cells.
= H₂O
= Na⁺
= Cl⁻
= K⁺
= Ca²⁺
= Mg²⁺
Cl⁻ and Na⁺ cotransporter
Cl⁻ leaves by facilitated diffusion.
Leaky tight junction
Some Cl⁻ moves between the cells through leaky tight junctions. Ca²⁺ and Mg²⁺ also cross the tight junctions due to the electrical gradient.

Figure 25.24 Train Car Analogy for the Countercurrent Multiplier

To visualize how the countercurrent multiplier works, it is helpful to remember that all parts of the nephron are active at all times; therefore, the physiological activity occurring in one location influences the environment at all the other locations of the nephron. In this analogy, you can imagine a long train made of many cars traveling through the nephron. The earlier cars in the train are in the ascending loop; later cars are in the descending loop. As solutes leave the later cars, a hyperosmotic environment is created around the nephron loop, which draws water out of the earlier cars in the train.

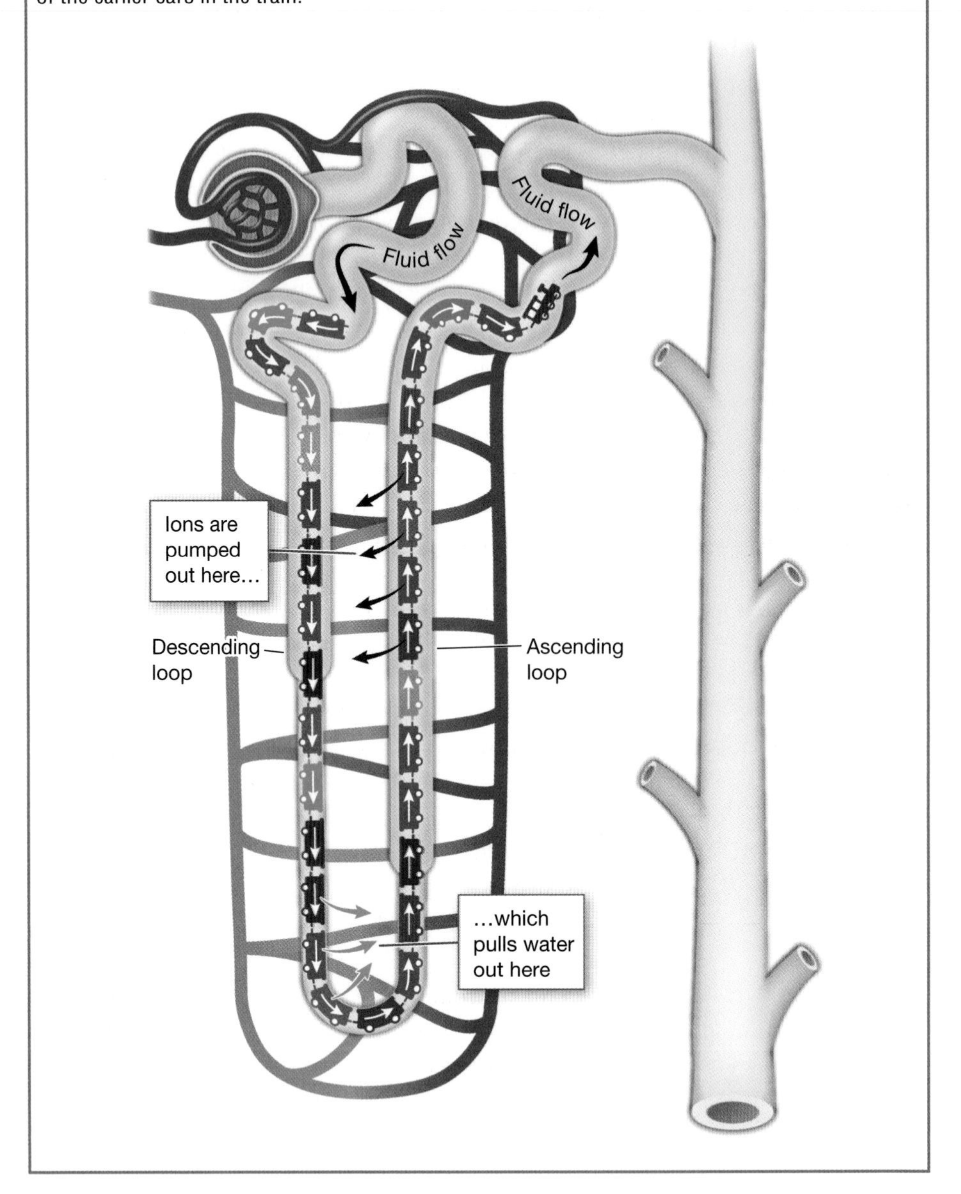

the open aquaporins of the descending loop. This structure is often referred to as the **countercurrent multiplier system**. The *countercurrent* term comes from the fact that fluid flow is downward in the descending loop and upward through the ascending loop. The *multiplier* term describes the action of solute pumps that increase (multiply) the concentrations of solutes in the interstitial space.

Now consider what is happening in the adjacent capillaries, the vasa recta. Both solutes and water are entering due to the reabsorption throughout the nephron loop.

In general, blood flows slowly in capillaries to allow time for exchange of nutrients and wastes. In the vasa recta particularly, this rate of flow is important for an additional reason. The flow must be slow to allow blood cells to lose and regain water without either shrinking or bursting. While some water was already reabsorbed in the peritubular capillaries surrounding the PCT, the blood within the vasa recta is still at a much lower pressure than blood in a typical capillary, contributing to this slow flow.

At the transition from the DCT to the collecting duct, about 20 percent of the original water, and about 10 percent of the sodium, is still present. If no other mechanism for water reabsorption existed, about 20–25 liters of urine would be produced each day and the body would rapidly dehydrate (not to mention that it would be hard to do anything other than urinate for the whole day!). The loss of that much sodium would also be a significant challenge to homeostasis. Thus, we know that further reabsorption must occur in the DCT and collecting duct before urine is excreted from the body.

25.3e Reabsorption and Secretion in the Distal Convoluted Tubule

Approximately 80 percent of filtered water has been recovered by the time the dilute tubular fluid enters the DCT. The DCT will recover another 10–15 percent before the tubular fluid enters the collecting ducts. The amount of solute reabsorption in the DCT will change according to physiological need. When the blood pressure is low and more solute and water needs to be reabsorbed from the urine, hormones (which we will cover later in this chapter) signal to the epithelial cells of the DCT to upregulate their expression of Na^+/K^+ pumps in the basolateral membranes of cells lining the DCT and collecting duct. As we have seen in the other portions of the nephron, the movement of Na^+ out of the lumen and into the interstitial fluid catalyzes a series of events that results in more reabsorption. In particular, the electrical gradient created by Na^+ active transport promotes the movement of Cl^- into the interstitial space by a paracellular route across tight junctions. As these ions move, a solute gradient is established, and water moves by osmosis. Peritubular capillaries receive the solutes and water, returning them to the circulation.

Cells of the DCT also recover Ca^{2+} from the filtrate under the influence of parathyroid hormone (PTH), discussed later.

25.3f The DCT and the Juxtaglomerular Apparatus

LO 25.3.11

The afferent and efferent arterioles emerge from the corpuscle in a V shape. Sandwiched between these arterioles is a section of the DCT as it winds and curves through the kidney cortex. Here, at the juncture where the afferent and efferent arterioles surround this part of the DCT, are a cluster of cells known as the **juxtaglomerular apparatus (JGA)** (Figure 25.25). The wall of the DCT in this location, known as the **macula densa**, forms one part of the JGA. This cluster of cuboidal epithelial cells monitors the fluid composition of filtrate as it flows through the DCT. In response to the concentration of Na^+ in the fluid flowing past them, these cells release paracrine signals. Each cell also has a single cilium that bends as fluid flows past their cell surface and therefore monitors the rate of fluid movement in the tubule.

A second cell type in this apparatus is the **juxtaglomerular cell**. This is a modified, smooth muscle cell lining the afferent arteriole that can contract or relax in response to paracrine signals released by the macula densa. Such contraction and relaxation regulates blood flow to the glomerulus. Blood flow rate influences the GFR. If the osmolarity of the filtrate is too high (hyperosmotic), the juxtaglomerular

Figure 25.25 Juxtaglomerular Apparatus and Glomerulus

(A) The specialized cells of the JGA monitor the composition of the fluid in the DCT and adjust the glomerular filtration rate. (B) A micrograph of the glomerulus and JGA.

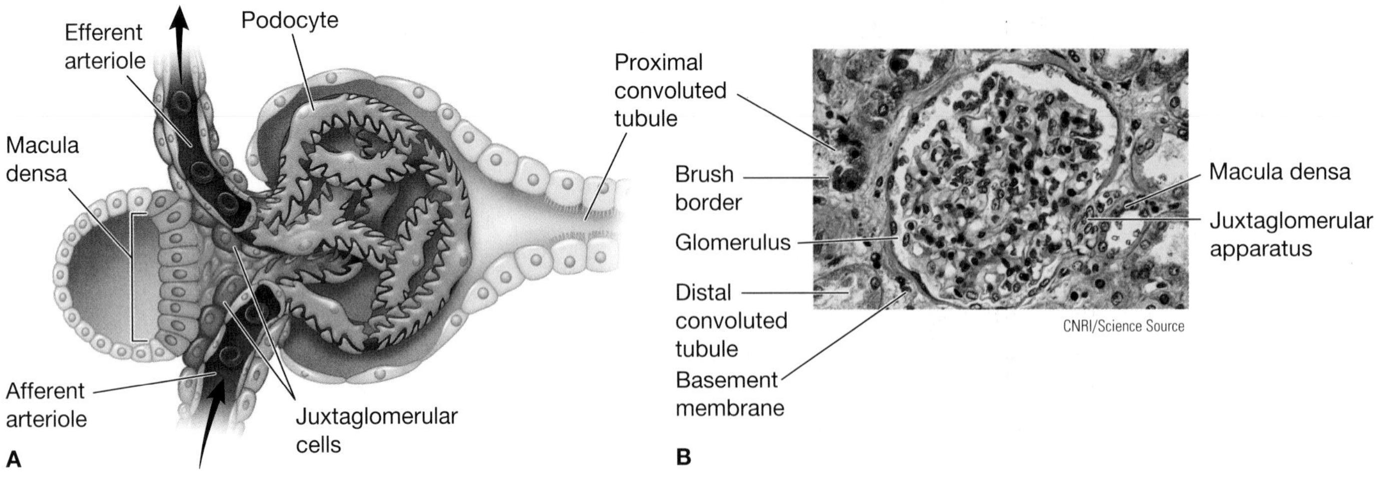

cells will contract, decreasing the GFR so less plasma is filtered, leading to less urine formation and greater fluid retention. This will ultimately decrease blood osmolarity toward the physiologic norm. If the osmolarity of the filtrate is too low, the juxtaglomerular cells will relax, increasing the GFR and enhancing the loss of water to the urine, causing blood osmolarity to rise. In other words, when osmolarity goes up, filtration and urine formation decrease and water is retained. When osmolarity goes down, filtration and urine formation increase and water is lost by way of the urine. The net result of these opposing actions is to keep the rate of filtration relatively constant.

A second function of the macula densa cells is to regulate the release of an enzyme known as *renin* from the juxtaglomerular cells of the afferent arteriole (see Figure 25.14). Renin is a key catalyst of the hormone cascades discussed in the next section.

25.3g Collecting Ducts and Recovery of Water

The walls of the collecting ducts are a mosaic of two different types of cells. **Principal cells** possess channels that permit the movement of Na^+ and K^+. **Intercalated cells** play significant roles in regulating blood pH. Intercalated cells reabsorb K^+ and HCO_3^- while secreting H^+. This function lowers the acidity of the plasma while increasing the acidity of the urine.

Regulation of urine volume and osmolarity are major functions of the collecting ducts. By varying the amount of water that is recovered, the collecting ducts play a major role in maintaining the body's osmolarity. If the blood becomes hyperosmotic, the collecting ducts recover more water to dilute the blood; if the blood becomes hypoosmotic, the collecting ducts recover less of the water, leading to concentration of the blood. This function is regulated by hormones that we will discuss in the next section.

Learning Connection

Explain a Process

Your classmate is having trouble understanding the three processes of urine formation: filtration, reabsorption, and secretion. How would you explain these processes in simple terms to your classmate? What is the purpose of each process? Which of the processes is/are based on molecular size, and which is/are based on the usefulness of the substance?

Cultural Connection

Drug Testing, Athletics, and Beyond

As you have read, urine is formed after blood is filtered and then most materials are reabsorbed. This method is costly in terms of energy! Why go to all the trouble of filtering and reabsorbing? Why not just secrete the materials we want to get rid of into the tubule directly? The answer is a genius trick of evolution. By filtering everything below a certain size out of the blood and then reabsorbing only what the body wants to keep, we are able to make sure that toxins, drugs, and waste products all end up in the urine. Take, for example, the idea that a human is wandering through the woods in a state of starvation; that human might choose to try a mushroom or plant they have never seen before to see if it is a source of nutrition. Let's say that the plant has a toxin in it. If urine were formed simply by selecting the materials we wanted to get rid of, then we would have needed to evolve a transporter capable of secreting that toxin into the urine; but if the plant were new to humans, no such transporter would exist and the toxin would circulate in the blood perpetually. Instead, by forming urine by first filtering everything out and then reabsorbing what we know we can use in the body, toxins and drugs end up cleared from the body relatively quickly in the urine. This also means that we can test urine for the presence of drugs and toxins to evaluate what a person has been exposed to. Competitive athletes frequently need to undergo urinalysis to make sure that they are not consuming drugs that can aid their athletic performance. Other individuals such as pilots, some employees, or some law enforcement agencies may ask or require urinalysis to test for the use of illicit drugs.

Learning Check

1. Which force(s) favor bulk flow into kidney tubules?
   a. Hydrostatic pressure of the glomerulus
   b. Hydrostatic pressure of the capsule
   c. Osmotic pressure in the glomerulus
   d. All of the above
2. Physiological mechanisms favor glucose reabsorption into the bloodstream, but all of the available transporters are in use. What principles are being applied here? Please select all that apply.
   a. Specificity
   b. Saturation
   c. Competition
   d. Sodium potassium pump
3. Which of the following terms refers to the liquid in the glomerular capsule that has been filtered out of the blood? **LO 25.3.12**
   a. Glomerular filtrate
   b. Urine
   c. Blood plasma
   d. All of the above
4. Substance C is removed from the blood via filtration and secretion, while substance B is removed only via filtration. Which substance will be excreted faster? Which is more toxic? **LO 25.3.13**
   a. Substance B; substance B
   b. Substance B; substance C
   c. Substance C; substance C
   d. Substance C; substance B

25.4 Homeostasis and Control over the Formation of Urine

Learning Objectives: By the end of this section, you will be able to:

25.4.1 For the renin-angiotensin system (RAS), describe the factors that initiate renin release, the pathway from angiotensinogen to angiotensin II (ANGII), and the effects of ANGII on various tissues.

25.4.2 Describe the signals that cause release of aldosterone from the adrenal cortex and the effect of aldosterone on the nephron, including the tubule segment involved and the transport mechanisms that are altered by aldosterone.

25.4.3 Describe the effect of vasopressin (ADH, antidiuretic hormone) on the nephron and on the final concentration of urine.

25.4.4 Describe the factors that cause release of natriuretic peptide hormones, their sites of synthesis, and their effects on the nephron.

Several hormones have specific, important roles in regulating kidney function. These hormones have different targets for their actions; some act on the arterioles of the nephron while others regulate the reabsorption of materials. All of the hormones together work to maintain homeostasis in the body.

25.4a Renin–Angiotensin–Aldosterone

Renin is an enzyme that is produced by the juxtaglomerular cells of the afferent arteriole at the JGA (**Figure 25.26**). Before we examine the action of renin, let's take a look at two compounds that are always circulating in the blood. **Angiotensinogen** is a hormone precursor, an inactive form of a hormone (note that the ending *-ogen* is often added to hormone precursors—such as pepsinogen, which we met in the stomach).

Figure 25.26 The Renin-Angiotensin System

(A) Angiotensinogen and angiotensin converting enzyme (ACE) are always present in the bloodstream. (B) Renin is an enzyme added to the blood by the juxtaglomerular apparatus. Renin converts angiotensinogen to angiotensin I, which is an inactive hormone. Angiotensin I is converted to angiotensin II by ACE. Angiotensin II is a powerful amplifier of blood pressure.

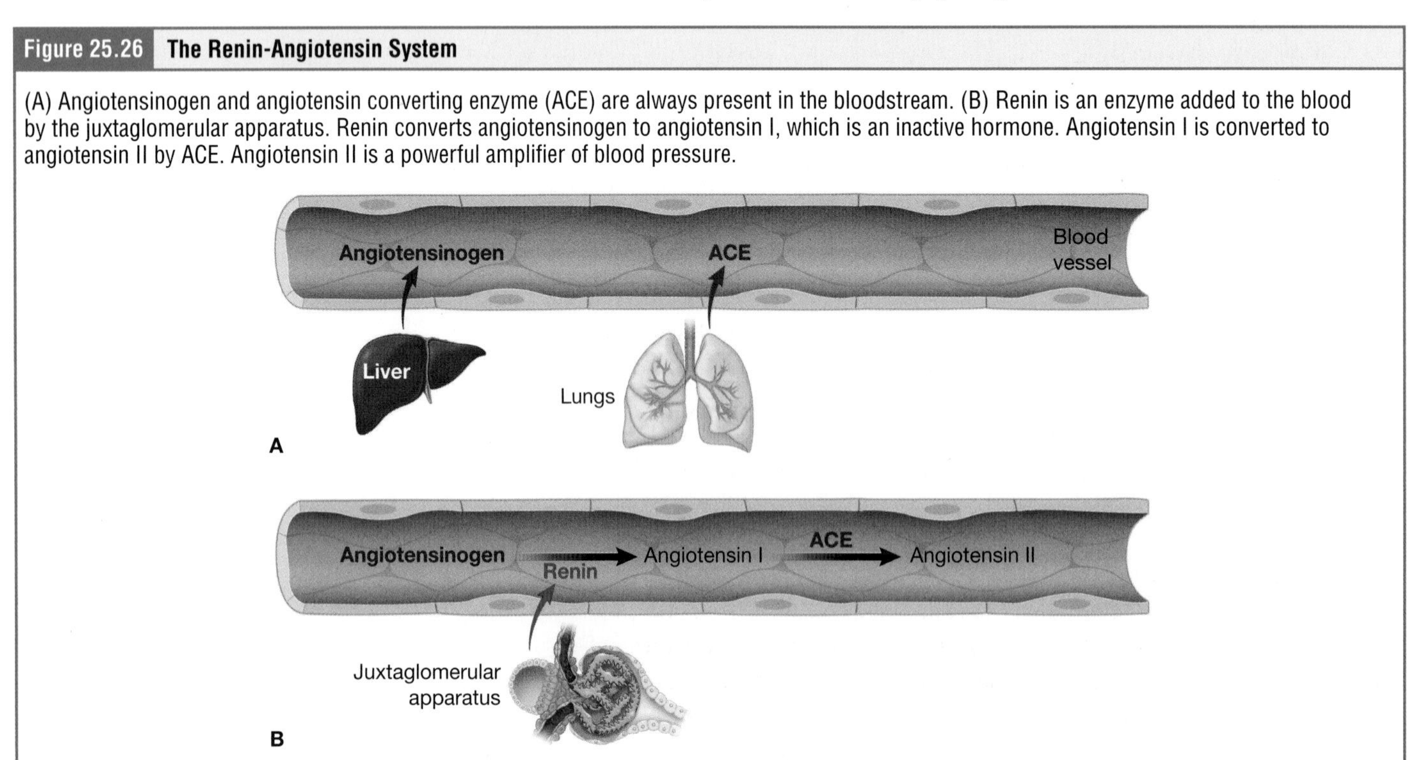

Angiotensin converting enzyme (ACE) is another enzyme, similar to renin, that is capable of converting one hormone to another. Angiotensinogen is made by the liver and added to the blood consistently. ACE is produced in the lungs, and again, added to the blood at a consistent rate over time; therefore, these two molecules are always present in the blood (**Figure 25.26A**).

LO 25.4.1

When blood pressure decreases, the JGA releases renin into the bloodstream. Renin enzymatically converts angiotensinogen into angiotensin I. ACE enzymatically converts inactive angiotensin I into active angiotensin II (**Figure 25.26B**). Angiotensin II is a potent vasoconstrictor that plays an immediate role in the regulation of blood pressure. Remember that the equation that helps us understand blood pressure is the MAP (mean arterial pressure) equation, which states that MAP is the product of heart rate (HR), stroke volume (SV) and TPR (total peripheral resistance):

$$\text{MAP} = \text{HR} \times \text{SV} \times \text{TPR}$$

By increasing vasoconstriction around the body, angiotensin II increases total peripheral resistance (TPR).

In addition to its systemic action, angiotensin II also causes vasoconstriction of both the afferent and efferent arterioles of the glomerulus. This decrease in blood flow to the glomerulus will dramatically decrease GFR and renal blood flow, thereby limiting fluid loss in the urine and preserving blood volume. Angiotensin II also is a stimulator of the sympathetic nervous system, thereby increasing sympathetic output to the heart. This will increase heart rate and stroke volume. Thus, angiotensin II impacts all three variables of the MAP equation and has a profound impact on blood pressure.

Without ACE, angiotensin II cannot be produced. ACE inhibitors are drugs that lower blood pressure by preventing the conversion of angiotensin I to angiotensin II. Approximately one in three Americans are prescribed ACE inhibitors to treat chronic high blood pressure.

LO 25.4.2

Angiotensin II will also stimulate the production of another hormone, aldosterone, which also impacts blood volume and blood pressure through retention of more Na^+ and water.

Aldosterone, often called the *salt-retaining hormone*, is released from the adrenal cortex in response to angiotensin II or directly in response to increased plasma K^+ levels. It promotes Na^+ reabsorption by the nephron, which in turn promotes the retention of water. It is also important in regulating K^+, promoting its excretion. As a result, renin has an immediate effect on blood pressure due to angiotensin II–stimulated vasoconstriction and a prolonged effect through Na^+ recovery due to aldosterone. At the same time that aldosterone causes increased recovery of Na^+, it also causes greater loss of K^+.

Remember that all transporters and receptors are governed by three principles: saturation, competition, and specificity. The activity of the aldosterone receptor can be altered by competition. The hormone progesterone (discussed in greater detail in Chapter 17) is a steroid that is structurally similar to aldosterone. Progesterone and aldosterone can both bind to the aldosterone receptor. Therefore, both hormones can stimulate Na^+ and water reabsorption. Progesterone levels are only high in biological females during some periods of the menstrual cycle. During these times in these individuals, progesterone may contribute to increased retention of water. Physiological differences in the levels of hormones of the renin-angiotensin-aldosterone pathway exist between age-matched biological males and biological females between puberty and menopause. These differences appear to decrease blood pressure in premenopausal biological females and may contribute to the differences in rates of hypertension observed between biological males and biological females.

antidiuretic hormone (ADH) (ADH, vasopressin)

LO 25.4.3

25.4b Antidiuretic Hormone (ADH)

Antidiuretic hormone (ADH) is a peptide hormone released by the posterior pituitary. It promotes the recovery of water and decreases urine volume, thereby increasing blood volume and blood pressure. It does so by promoting the expression of aquaporin proteins into the luminal cell membrane of principal cells of the collecting ducts. The increased expression of aquaporins increases the movement of water from the lumen of the collecting duct into the interstitial space by osmosis. From there, it enters the vasa recta capillaries and the blood. ADH is also called *vasopressin*. Early researchers found that in cases of unusually high secretion of ADH, the hormone caused vasoconstriction, and gave it its name vasopressin. When blood volume or pressure drops, there is a rapid and significant increase in ADH release from the posterior pituitary. Vasoconstriction and increased water reabsorption both increase blood pressure.

natriuretic hormones (natriuretic peptides)

LO 25.4.4

25.4c Natriuretic Hormones

Natriuretic hormones are peptides that stimulate the kidneys to excrete sodium—an effect opposite that of aldosterone. Natriuretic hormones act by inhibiting aldosterone release and therefore inhibiting Na^+ recovery in the collecting ducts. If Na^+ remains in the forming urine, its osmotic force will cause a concurrent loss of water. Natriuretic hormones also inhibit ADH release, which of course will result in less water recovery. Therefore, natriuretic peptides inhibit both Na^+ and water recovery. One example from this family of hormones is atrial natriuretic hormone (ANH), a peptide hormone produced by heart atria in response to overstretching of the atrial wall. The overstretching occurs in persons with increased venous return, which often accompanies elevated blood volume and pressure. It increases GFR through simultaneous vasodilation of the afferent arteriole and vasoconstriction of the efferent arteriole. These events dramatically increase GFR. ANH also decreases sodium reabsorption in the DCT, leading to an increased loss of water and sodium in the forming urine. See **Table 25.2** for information concerning the major hormones that affect GFR and RBF.

Student Study Tip

More space in the afferent arteriole than efferent causes a bottlenecking effect, much like three lanes merging into one on the highway. Traffic (pressure) results.

DIGGING DEEPER:
Drugs that Impact Water Reabsorption

A **diuretic** is a compound that increases urine volume. Diuretics are drugs that can increase water loss by interfering with the reabsorption of solutes and water from the forming urine. They are often prescribed to lower blood pressure. Coffee, tea, and alcoholic beverages are also diuretics. Three familiar drinks contain diuretic compounds: coffee, tea, and alcohol. The caffeine in coffee and tea works by promoting vasodilation in the nephron, which increases GFR. Alcohol increases GFR by inhibiting ADH release from the posterior pituitary, resulting in less water recovery by the collecting duct. In cases of high blood pressure, diuretics may be prescribed to reduce blood volume, and thereby reduce blood pressure. The most frequently prescribed antihypertensive diuretic is hydrochlorothiazide. It inhibits the Na^+/Cl^- symporter in the DCT and collecting duct. The result is a loss of Na^+ with water following passively by osmosis.

Osmotic diuretics promote water loss by osmosis. An example is the indigestible sugar mannitol, which is most often administered to reduce brain swelling after a head injury. However, it is not the only sugar that can produce a diuretic effect. In cases of poorly controlled diabetes mellitus, glucose levels exceed the capacity of the tubular glucose symporters, resulting in glucose in the urine. The unrecovered glucose becomes a powerful osmotic diuretic. Classically, in the days before glucose could be detected in the blood and urine, clinicians identified diabetes mellitus by the three Ps: polyuria (diuresis), polydipsia (increased thirst), and polyphagia (increased hunger).

Table 25.2 Major Hormones That Influence GFR and RBF

	Stimulus	Effect on GFR	Effect on RBF
VASOCONSTRICTORS			
Sympathetic nerves (epinephrine and norepinephrine)	↓ECFV	↓	↓
Angiotensin II	↓ECFV	↓	↓
Endothelin	↑Stretch, bradykinin, angiotensin II, epinephrine ↑ECFV	↓	↓
VASODILATORS			
Prostaglandins (PGE1, PGE2, and PGI2)	↓ECFV ↑shear stress, angiotensin II	No change/↑	↑
Nitric oxide (NO)	↑shear stress, acetylcholine, histamine, bradykinin, ATP, adenosine	↑	↑
Bradykinin	↓Prostaglandins, ACE	↑	↑
Natriuretic peptides (ANP, B-type)	↑ECFV	↑	No change

ACE = angiotensin-converting enzyme; ECFV = extracellular fluid volume; GFR = glomerular filtration rate; RBF = renal blood flow; ANP = atrial natriuretic peptide; B-type = ventricular natriuretic peptide

Learning Connection

Try Drawing It!

After studying the functions of the various parts of the nephron, draw your own nephron. Label the major portions of the nephron. Then add the general functions of each portion to your drawing. Which portions perform functions such as filtration, reabsorption, and secretion? Which portions are affected by aldosterone and ADH? Which portion secretes renin and erythropoietin?

Learning Check

1. Which two hormones have opposite effects? Please select all that apply.
   a. Aldosterone
   b. Antidiuretic hormone
   c. Natriuretic hormones
   d. Angiotensinogen
2. Which of the following scenarios would initiate the renin-angiotensin-aldosterone pathway? Please select all that apply.
   a. Dehydration
   b. Overhydration
   c. Exercise
   d. Body temperature rise
3. Which of the following hormones work together to retain water? Please select all that apply.
   a. Diuretic hormone
   b. Antidiuretic hormone
   c. Aldosterone
   d. Natriuretic hormones
4. After antidiuretic hormone is released, water is reabsorbed. Where does this reabsorbed water go?
   a. Tubule lumen
   b. Renal vein
   c. Glomerulus
   d. Vasa recta

25.5 Additional Endocrine Activities of the Kidney

Learning Objectives: By the end of this section, you will be able to:

25.5.1 Describe the role of the kidney in vitamin D activation.

25.5.2 Describe the role of the kidney in regulating erythropoiesis.

25.5.3 Describe the effect of parathyroid hormone (PTH) on renal handling of calcium and phosphate.

All systems of the body are interrelated. The primary functions of the kidneys are the maintenance of blood pressure, osmolarity and pH homeostasis via filtration, reabsorption, and secretion. The kidneys, however, play several other vital roles in physiology.

LO 25.5.1

25.5a Vitamin D Synthesis

Vitamin D can be obtained through your food, but it also can be synthesized in the skin in response to UV radiation from the sun. The skin synthesizes vitamin D from cholesterol, and the liver converts the vitamin D to calcifediol, or 25-dihydroxycholecalciferol. However, in order for this molecule to become the most active form of vitamin D it must undergo a chemical reaction in the kidney. The kidney completes the conversion to calcitriol, the most active and effective form of vitamin D. These active forms of vitamin D are important for the absorption of Ca^{2+} in the digestive tract, its reabsorption in the kidney, and the maintenance of Ca^{2+} and phosphate homeostasis in the blood. Calcium is vitally important in bone health, muscle contraction, hormone secretion, and neurotransmitter release. Inadequate Ca^{2+} leads to disorders such as osteoporosis and **osteomalacia** in adults and rickets in children. Deficits may also result in problems with cell proliferation, neuromuscular function, blood clotting, and the inflammatory response. Recent research has confirmed that vitamin D receptors are present in most, if not all, cells of the body, reflecting the systemic importance of vitamin D. Many scientists have suggested it be referred to as a hormone (a compound made by the body and released into the bloodstream) rather than a vitamin (a compound obtained in the diet). In many parts of the world vitamin D is just a vitamin in colder winter months when skin exposure to UV radiation from the sun is limited, but a hormone in the warmer months when skin exposure is more common.

LO 25.5.2

25.5b Erythropoiesis

Erythropoietin (EPO) is a hormone that stimulates the formation of red blood cells in the bone marrow. The kidney produces the majority of the body's EPO (the liver produces the remainder). The kidneys secrete EPO in response to low oxygen levels in the blood. For example, if erythrocytes are lost due to severe or prolonged bleeding, or underproduced due to disease or severe malnutrition, the kidneys produce more EPO. The target tissue of EPO is bone marrow; more red blood cells are produced and released into the blood and the oxygen-carrying capacity of the blood increases. If the kidneys are damaged by injury or disease, anemia (which is characterized by inadequate red blood cells in the circulation) can result.

LO 25.5.3

Learning Connection

Chunking

Many hormones and other substances increase or decrease the volume of urine excreted from the body each day. To increase your understanding of this vital regulation, make a list of the following substances: ADH, ANH, Aldosterone, Diuretics, Angiotensin II, Renin, EPO, and Calcitriol. Then list whether each substance increases, decreases, or has no effect on the urine volume.

25.5c Calcium Reabsorption

Receptors for PTH are found in DCT cells. PTH, which was described in Chapter 17, is expressed when blood Ca^{2+} levels are low. When PTH binds to DCT epithelial cells, the cells upregulate their membrane expression of calcium channels on their luminal surface. The channels enhance Ca^{2+} recovery from the forming urine. Calcitriol (1,25 dihydroxyvitamin D, the active form of vitamin D) is also very important for calcium recovery. It induces the production of calcium-binding proteins that transport Ca^{2+} into the cell. These binding proteins are also important for the movement of calcium inside the cell and aid in exocytosis of calcium across the basolateral membrane. Any Ca^{2+} not reabsorbed at this point is lost in the urine.

Learning Check

1. Which of the following hormones are essential for calcium absorption in the intestines? Please select all that apply.
   a. Aldosterone
   b. Activated vitamin D
   c. Parathyroid hormone
   d. Antidiuretic hormone
2. What is the target tissue or organ of erythropoietin?
   a. Kidney
   b. Heart
   c. Skeletal muscle
   d. Bone marrow
3. Which of the following ways can you obtain vitamin D? Please select all that apply.
   a. Food
   b. Kidney production
   c. UV light
   d. Liver production

25.6 Gross and Microscopic Anatomy of the Urinary Tract (Ureters, Urinary Bladder, and Urethra)

Learning Objectives: By the end of this section, you will be able to:

25.6.1 Identify and describe the gross anatomy and location of the ureters, urinary bladder, and urethra.

25.6.2 Identify and describe the microscopic anatomy of the ureters, urinary bladder, and urethra.

25.6.3 Compare and contrast the locations, innervation and functions of the internal urethral sphincter and external urethral sphincter.

25.6.4 Describe voluntary control of micturition.

25.6.5 Describe the micturition reflex and the role of the autonomic nervous system in the reflex.

25.6.6 Trace the path of urine from the collecting duct of the kidney to the external urethral orifice.

As you've read in the preceding sections, the composition of urine is highly variable, because it is composed of whatever material the body needs to get rid of. Typically, urine is acidic and therefore requires specialized structures to remove it from the body safely and efficiently. Blood is filtered continuously because we do not want to have to urinate all the time, urine is stored until it can be eliminated conveniently. All structures involved in the transport and storage of the urine are large enough to be visible to the naked eye. This transport and storage system not only stores the waste, but it protects the tissues from damage due to the wide range of pH and osmolarity of the urine and prevents infection. The urine that collects at the end of the collecting ducts exits the kidneys through the **ureters**, long tubes that transport urine to the urinary bladder. The **urinary bladder** stores the urine until elimination, at which time the urine passes through a single tube, the **urethra**, and into the outside world (Figure 25.27).

25.6a The Ureters

LO 25.6.1

The kidneys and their ureters are completely retroperitoneal. The urinary bladder sits in the pelvic cavity and has a peritoneal covering over its superior surface. As urine is formed, it drains from the collecting ducts into the calyces of the kidney, which merge to form the funnel-shaped renal pelvis in the hilum of each kidney. The renal pelvis

Figure 25.27 The Urinary Tract

After urine is formed by the kidneys, it funnels into the ureters, which transport urine toward the bladder for storage. Urine passes through the urethra as it exits the body.

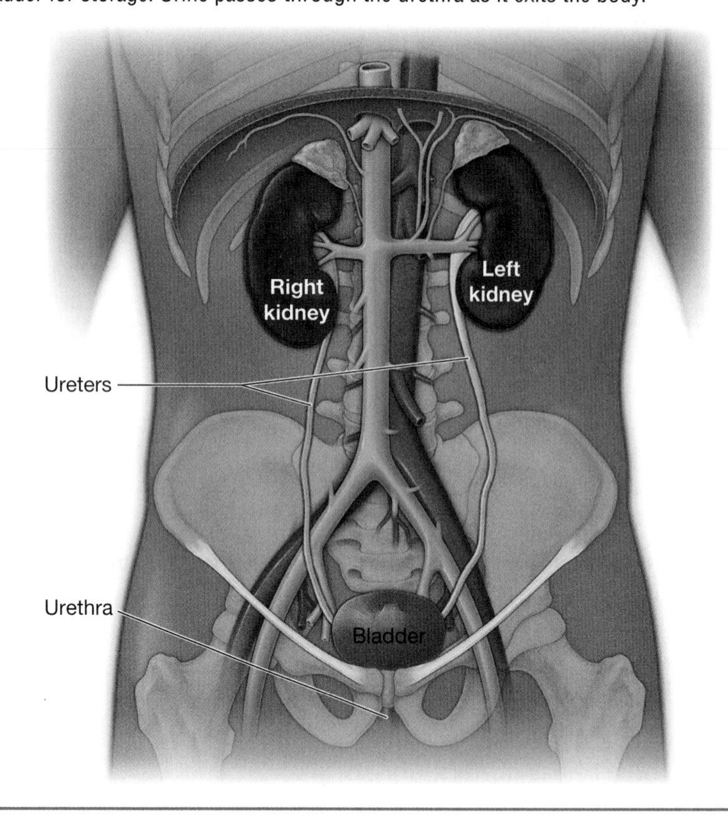

narrows to become the ureter of each kidney. As urine passes through the ureter, it does not passively drain into the bladder but rather is propelled by waves of peristalsis, the same rhythmic muscular contractions that propel food through the GI tract. The ureters connect to the bladder at a diagonal (oblique) angle. This angle is functional; it creates a one-way valve (a **physiological sphincter** rather than an **anatomical sphincter**) that allows urine into the bladder but prevents reflux of urine from the bladder back into the ureter.

physiological sphincter (functional sphincter)

LO 25.6.2

The inner lining of the ureter is lined with transitional epithelium and scattered goblet cells that secrete protective mucus (**Figure 25.28**). Transitional epithelium is known for its elastic qualities, allowing the ureters to distend during the passing of urine. The muscular layer of the ureter consists of longitudinal and circular smooth muscles that create the peristaltic contractions to move the urine into the bladder without the aid of gravity. Finally, a loose connective tissue layer, composed of collagen and fat, anchors the ureters between the parietal peritoneum and the posterior abdominal wall.

25.6b The Bladder

At the end of the ureters, the urine arrives in the urinary bladder (**Figure 25.29**). The bladder is a very elastic organ lined with transitional epithelium. Underneath the transitional epithelium is a muscle woven from crisscrossing bands of smooth muscle called the **detrusor muscle**.

detrusor muscle (detrusor urinae muscle)

In biological females the bladder lies anterior to the uterus, posterior to the pubic bone, and anterior to the rectum (**Figure 25.30**). During late pregnancy, its capacity

Figure 25.28 Ureters

A cross section through a ureter reveals the hollow lumen through which urine passes, a mucosa that is composed of transitional epithelium, two layers of muscle that propel urine via peristalsis, and a connective tissue wrapper called the *adventitia*.

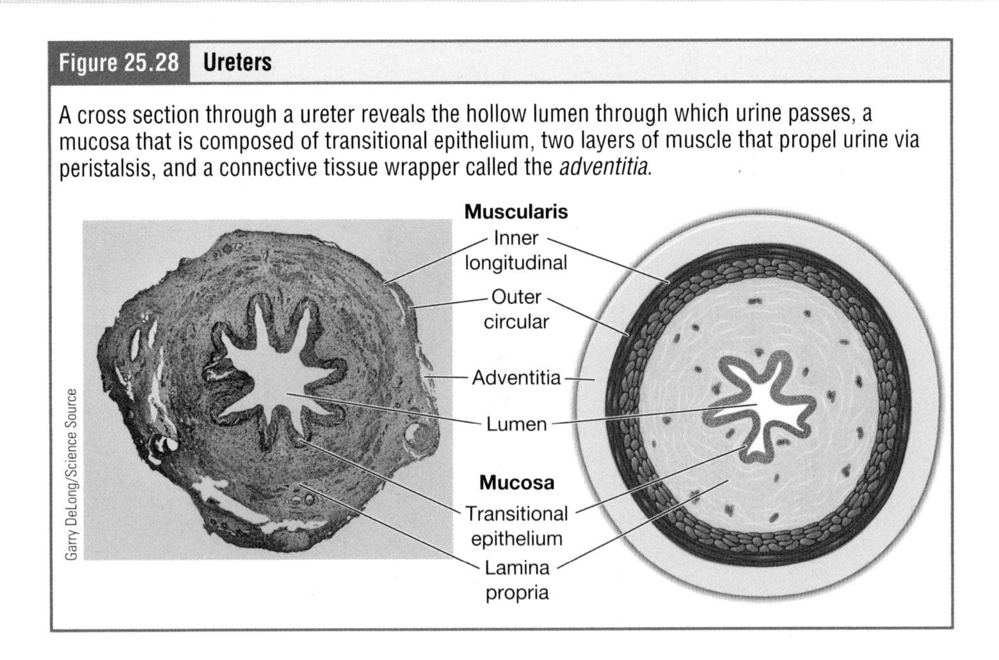

is reduced due to compression by the uterus, resulting in increased frequency of urination. In biological males, the anatomy is similar, minus the uterus, and with the addition of the prostate inferior to the bladder.

Micturition Reflex **Micturition** is a less-often used, but proper term for urination. It results from an interplay of involuntary and voluntary actions by the internal and

micturition (urination)

Figure 25.29 The Urinary Bladder

The urinary bladder is shaped like a funnel, with the ureters entering at oblique (diagonal) angles near the top. The urethra allows the urine to exit the urinary bladder and the body.

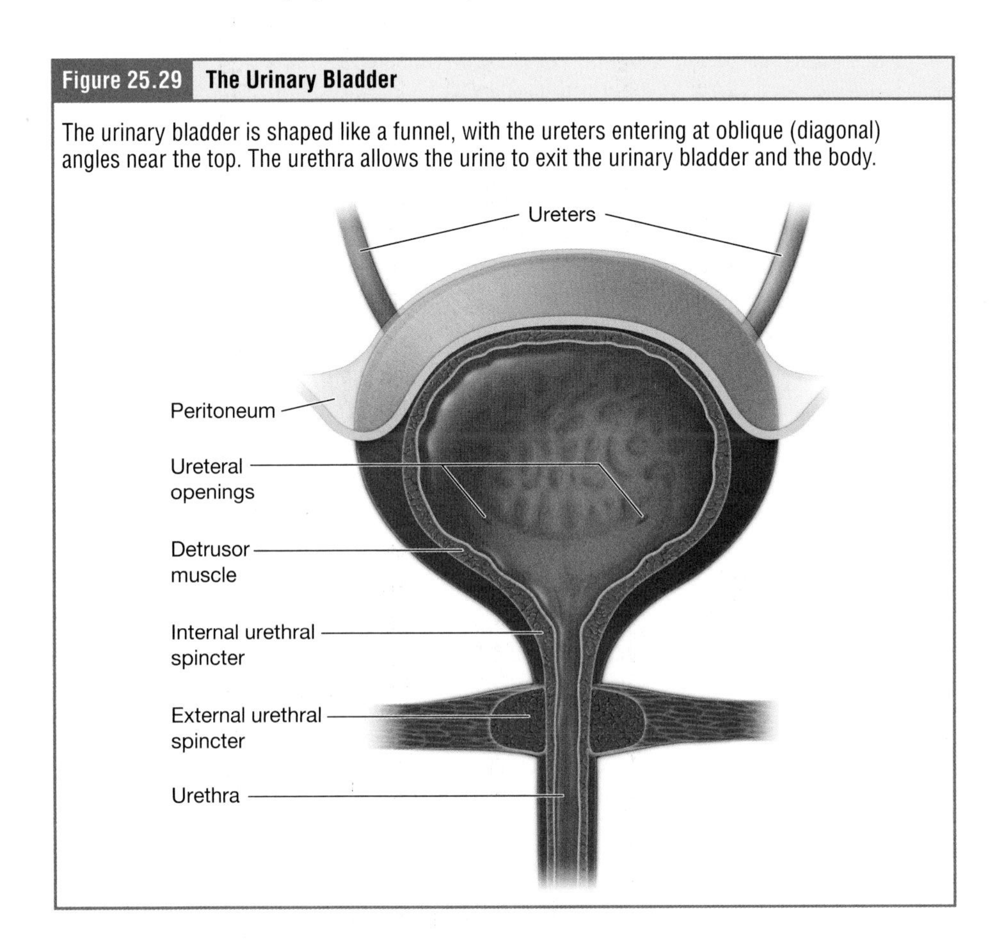

Figure 25.30 Location of the Urinary Bladder and Urethra in the Pelvis

(A) In anatomical females, the urethra is short and the urinary bladder is located anterior and inferior to the uterus. (B) In anatomical males, the urethra is long, tunneling through the prostate and exiting through the penis.

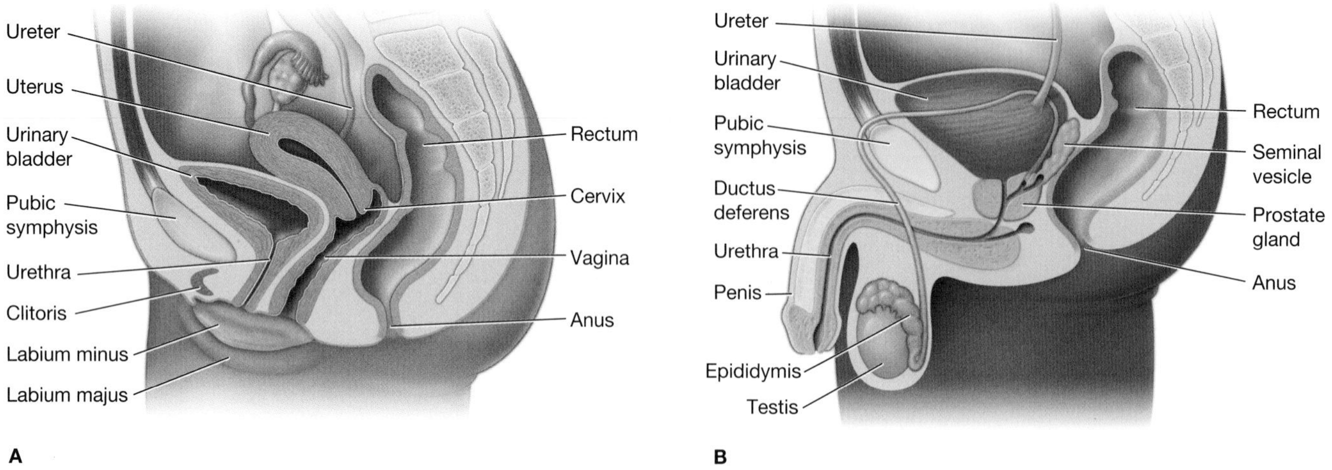

LO 25.6.3

external urethral sphincters. Voluntary control of urination relies on consciously preventing relaxation of the external urethral sphincter to maintain urinary continence. As the bladder fills, subsequent urges become harder to ignore.

Under typical circumstances, micturition is a result of stretch receptors in the bladder wall that transmit nerve impulses to the sacral region of the spinal cord to generate a spinal reflex. The resulting parasympathetic neural outflow causes contraction of the detrusor muscle and relaxation of the involuntary internal urethral sphincter. At the same time, the spinal cord inhibits somatic motor neurons, resulting in the relaxation of the skeletal muscle of the external urethral sphincter. The micturition reflex is active in infants but with maturity, children learn to override the reflex by asserting external sphincter control, thereby delaying urination (i.e., potty training). Because it does not require communication with the brain, this reflex may be preserved even in cases of spinal cord injury. However, relaxation of the external sphincter may not be possible in all cases, and therefore, periodic catheterization may be necessary for bladder emptying.

LO 25.6.4

Nerves involved in the control of urination include the hypogastric, pelvic, and pudendal (**Figure 25.31**). Voluntary micturition requires an intact spinal cord and functional pudendal nerve arising from the **sacral micturition center**. Since the external urinary sphincter is voluntary skeletal muscle, actions by cholinergic neurons maintain contraction (and thereby continence) during filling of the bladder. At the same time, sympathetic nervous activity via the hypogastric nerves suppresses contraction of the detrusor muscle. With further bladder stretch, afferent signals traveling over sacral pelvic nerves activate parasympathetic neurons. This activates efferent neurons to release acetylcholine at the neuromuscular junctions, producing detrusor contraction and bladder emptying.

LO 25.6.5

Learning Connection

Quiz Yourself

After studying the pathways for urine formation and elimination, imagine following a molecule of urea (a waste product of amino acid breakdown) that has been filtered by the glomerulus, but has not been reabsorbed. See if you can list all of the structures through which the urea molecule will travel, from the site of glomerular filtration to its exit from the body.

25.6c The Urethra

The *urethra* transports urine from the bladder to the outside of the body for excretion (**Figure 25.32**).

Whereas the proximal urethra is lined by transitional epithelium, the terminal portion as we get closer to the outside is a nonkeratinized, stratified squamous epithelium.

Figure 25.31 Nerves Involved in Urination

The urinary bladder is innervated by hypogastric and pelvic nerves. The pudendal nerve supplies the external urethral sphincter and therefore the control over urination.

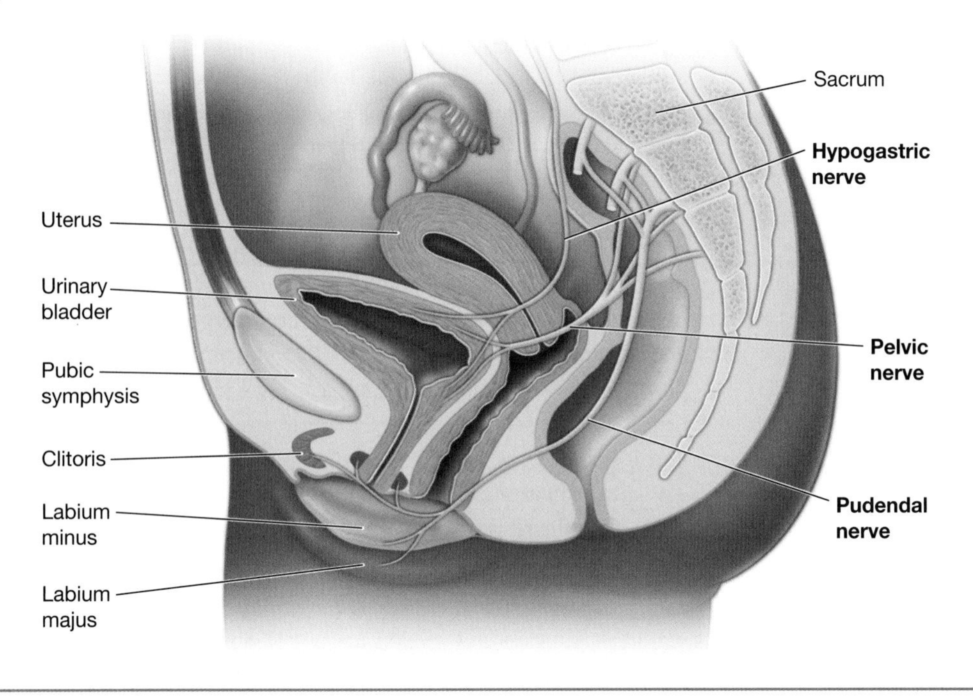

Figure 25.32 The Urethra and the Penis

The urethra is very long in individuals with penises. The first zone, the prostatic zone, tunnels through the prostate gland. The second zone, the membranous zone, is very short but transitions to the penile zone, which travels down the length of the penis.

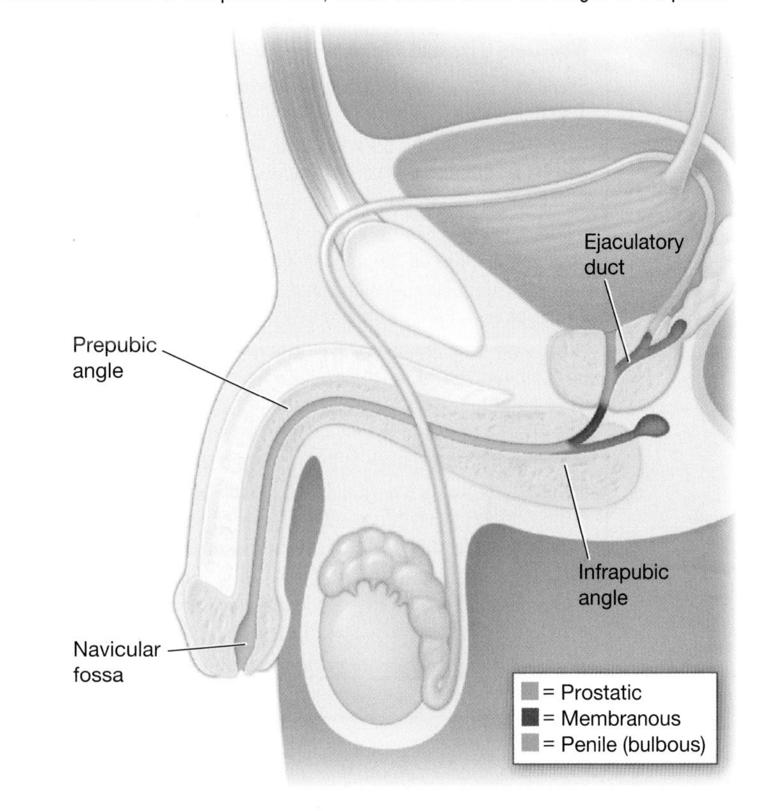

In biological males, the urethra is much longer and a zone of pseudostratified columnar epithelium lines the urethra between these two cell types.

urinary tract infections (UTIs)

The Urethra in Biological Female Bodies The external urethral orifice is embedded in the anterior vaginal wall inferior to the clitoris, superior to the vaginal opening, and medial to the labia minora. Its short length—about four centimeters—is less of a barrier to bacteria than the longer male urethra and therefore the bladder can become infected (**urinary tract infections**) in individuals with these shorter urethras.

membranous urethra (intermediate urethra)

spongy urethra (penile urethra)

The Urethra in Biological Male Bodies The urethra of a biological male is much longer; it tunnels through the **prostate gland**, which is immediately inferior to the bladder, before passing below the pubic symphysis (see Figure 25.30). The urethra is divided into three regions: the **prostatic urethra**, the **membranous urethra**, and the **spongy urethra** (see Figure 25.31). The prostatic urethra passes through the prostate gland. During sexual intercourse, it receives sperm via the ejaculatory ducts and secretions from the seminal vesicles. Paired Cowper's glands (bulbourethral glands) produce and secrete mucus into the urethra to buffer urethral pH during sexual stimulation. The mucus neutralizes the usually acidic environment and lubricates the urethra, decreasing the resistance to ejaculation. The membranous urethra passes through the deep muscles of the perineum, where it is invested by the overlying urethral sphincters. The spongy urethra exits at the tip (external urethral orifice) of the penis after passing through the corpus spongiosum. Mucous glands are found along much of the length of the urethra and protect the urethra from extremes of urine pH. Innervation is the same in both males and females.

Cultural Connection

Who Gets UTIs and Why?

Some individuals suffer from frequent urinary tract infections (UTIs) and other individuals may never have heard of this affliction. Why is there so much diversity in terms of who gets UTIs and how often? UTIs occur when bacteria, typically *E. coli* bacteria, infiltrate the urethra and begin to colonize the bladder. As the bacteria form a film of growth along the inner walls of the bladder, they can cause inflammation and pressure, which triggers the sensation that the bladder is full and urination is needed, even when it isn't. If left untreated, the bacteria can grow and spread up the ureters and into the kidneys, where major, life-threatening structural damage can occur. UTIs are uncomfortable, painful, and can become life-threatening, so why are some people more vulnerable than others? The *E. coli* bacteria that cause UTIs are not contracted from other people; rather, they originate in one's own body. *E. coli* are a common inhabitant of the large intestine and make up a significant amount of the feces. When trace amounts of *E. coli* are left behind after defecation, they can grow along the perineum (the stretch of tissue between anus and urethra) and make their way into the urethra. During urination the bacteria are often flushed out unless they have developed a substantial presence. Therefore, in individuals who have longer spaces between the anus and the urethral orifice, such as individuals who have a penis, the ability of *E. coli* to ever reach the urethra is limited. In individuals with a shorter distance, such as anatomical females with a vagina, it is more likely that significant numbers of *E. coli* bacteria can reach the urethra. In all individuals, frequent flushing with urine is likely to eliminate or reduce the numbers of *E. coli*, so when dehydration causes infrequent urination, UTIs are more likely. In addition to urination, other activities that cause pressure on the urethra—such as sexual intercourse, bike riding, and horseback riding—can all make UTIs more likely.

Learning Check

1. Which of the following is the correct pathway of urine flow out of the body?
   a. Ureters → urinary bladder → urethra
   b. Urinary bladder → ureters → urethra
   c. Urethra → ureters → urinary bladder
   d. Urethra → urinary bladder → ureters
2. What type of epithelium lines the ureters?
   a. Simple squamous
   b. Stratified squamous
   c. Pseudostratified columnar
   d. Transitional
3. Pathology of the external urethral sphincter, causing it to perpetually be relaxed, would result in which of the following conditions?
   a. Inability to urinate
   b. Overproduction of urine
   c. Inability to control urine flow
   d. Underproduction of urine
4. Which of the following urine flow pathways through the kidney is correct? **LO 25.6.6**
   a. Collecting duct → renal pelvis → ureter
   b. Renal pelvis → collecting duct → ureter
   c. Collecting duct → ureter → renal pelvis
   d. Ureter → renal pelvis → collecting duct

25.7 Urine Characteristics and Elimination

Learning Objectives: By the end of this section, you will be able to:

25.7.1* Describe the typical characteristics of urine.

25.7.2* Given a change in the composition of urine (ex: proteinuria) predict the physiological disruption that led to the change.

* Objective is not a HAPS Learning Goal.

The urinary system's ability to filter the blood resides in about two to three million tufts of specialized capillaries—the glomeruli—distributed more or less equally between the two kidneys. Because the glomeruli filter the blood based mostly on particle size, large elements like blood cells, platelets, antibodies, and albumin are excluded. The glomerulus is the first part of the nephron, which then continues as a highly specialized tubular structure responsible for creating the final urine composition. All other solutes, such as ions, amino acids, vitamins, and wastes, are filtered to create a filtrate composition very similar to that of plasma. The glomeruli create about 200 liters (189 quarts) of this filtrate every day, yet you excrete less than 2 liters of waste you call urine.

Characteristics of the urine change, depending on influences such as water intake, exercise, environmental temperature, nutrient intake, and other factors (**Table 25.3**). Some of the characteristics such as color and odor are rough descriptors of your state of hydration. For example, if you exercise or work outside, and sweat a great deal, your urine will turn darker and produce a slight odor, even if you drink plenty of water.

Urinalysis (urine analysis) often provides clues to renal disease. Normally, only traces of protein are found in urine because an intact glomerular membrane will keep

urinalysis (urine analysis)

Table 25.3 Typical Urine Characteristics LO 25.7.1

Characteristic	Normal Values
Color	Pale yellow to deep amber
Odor	Odorless
Volume	750–2000 mL/24 hr
pH	4.5–8.0
Specific gravity	1.003–1.032
Osmolarity	40–1350 mOsmol/kg
Urobilinogen	0.2–1.0 mg/100 mL
White blood cells	0–2 HPF (per high-power field of microscope)
Leukocyte esterase	None
Protein	None or trace
Bilirubin	<0.3 mg/100 mL
Ketones	None
Nitrites	None
Blood	None
Glucose	None

them from entering the filtrate. When higher amounts of protein are found in the urine, damage to the glomeruli is the likely basis. The color of urine is determined mostly by the breakdown products of red blood cell destruction. The "heme" of hemoglobin is converted by the liver into water-soluble forms that can be excreted into the bile and the urine. Blood in the urine typically points to structural damage within the urinary tract such as a kidney stone or a cancer of the urinary system. Dehydration produces darker, concentrated urine that may also possess the slight odor of ammonia. Most of the ammonia produced from protein breakdown is converted into urea by the liver, so ammonia is rarely detected in fresh urine. About one in five people detect a distinctive odor in their urine after consuming asparagus; other foods such as onions, garlic, and fish can impart their own aromas! These food-caused odors are harmless.

Urine volume varies considerably based on the physiological needs of the body. The kidneys must produce a minimum urine volume of about 500 mL/day to rid the body of wastes. Output below this level may be caused by severe dehydration or renal disease. Excessive urine production may be due to diabetes mellitus or diabetes insipidus. In diabetes mellitus, blood glucose levels exceed the number of available sodium-glucose transporters in the kidney, and glucose appears in the urine. The osmotic nature of glucose attracts water, leading to its loss in the urine. In the case of diabetes insipidus, insufficient pituitary ADH release or insufficient numbers of ADH receptors in the collecting ducts means that too few water channels are inserted into the cell membranes that line the collecting ducts of the kidney. Insufficient numbers of water channels (aquaporins) reduce water absorption, resulting in high volumes of very dilute urine.

The pH (hydrogen ion concentration) of the urine can vary more than 1,000-fold, from a normal low of 4.5 to a maximum of 8.0. Diet can influence pH; whereas meats lower the pH, citrus fruits, vegetables, and dairy products raise the pH.

Specific gravity is a measure of the quantity of solutes per unit volume of a solution and is traditionally easier to measure than osmolarity. Urine will always have a specific gravity greater than pure water (water = 1.0) due to the presence of solutes.

DIGGING DEEPER: Monitoring Urine for Signs of Disease

Some glucose may appear in the urine if circulating glucose levels are high enough that all the glucose transporters in the PCT are saturated, so that their capacity to move glucose is exceeded (transport maximum, or Tm). Whereas in biological males, the maximum amount of glucose that can be recovered is about 375 mg/min, in biological females, it is about 300 mg/min. This recovery rate translates to an arterial concentration of about 200 mg/dL. Though an exceptionally high sugar intake might cause sugar to appear briefly in the urine, the appearance of **glycosuria** usually points to type I or II diabetes mellitus. The transport of glucose from the lumen of the PCT to the interstitial space is similar to the way it is absorbed by the small intestine. Glucose and Na^+ bind simultaneously to the same symporters on the apical surface of the cell to be transported in the same direction, toward the interstitial space. Sodium moves down its electrochemical and concentration gradients into the cell and takes glucose with it. Na^+ is then actively pumped out of the cell at the basal surface into the interstitial space. Glucose leaves the cell to enter the interstitial space by facilitated diffusion. The energy to move glucose comes from the Na^+/K^+ ATPase that pumps Na^+ out of the cell on the basal surface. Fifty percent of Cl^- and variable quantities of Ca^{2+}, Mg^{2+}, and HPO_4^{2-} are also recovered in the PCT. Large proteins are rarely found in the urine because they are held back by the glomerular membrane and not filtered. The presence of protein in the urine, called *proteinuria*, indicates glomerular damage, which often results from high blood pressure.

Cells are not normally found in the urine. The presence of leukocytes may indicate an UTI.

Ketones are byproducts of fat metabolism. Finding ketones in the urine suggests that the body is using fat as an energy source in preference to glucose. In diabetes mellitus, when there is not enough insulin (type I diabetes mellitus) or because of insulin resistance (type II diabetes mellitus), there is plenty of glucose, but without the action of insulin, the cells cannot take it up, so it remains in the bloodstream. Instead, the cells are forced to use fat as their energy source, and fat consumed at such a level produces excessive ketones as byproducts. These excess ketones will appear in the urine. Ketones may also appear if there is a severe deficiency of proteins or carbohydrates in the diet.

Nitrates (NO_3^-) occur normally in the urine. Gram-negative bacteria metabolize nitrate into nitrite (NO_2^-), and its presence in the urine is indirect evidence of infection.

Learning Check

1. Sarah's urine is odorless and pale yellow. Is Sarah's urine normal?
   a. Yes
   b. No
   c. Impossible to tell with the information given
2. Does this suggest that Sarah is hydrated or dehydrated?
   a. Hydrated
   b. Dehydrated
   c. Impossible to tell with the information given

   LO 25.7.2
3. Megan's physiology is affected by uncontrolled diabetes mellitus. What will be different in Megan's urine compared with that of someone who is not affected by diabetes?
   a. Protein in the urine
   b. Nitrites in the urine
   c. White blood cells in the urine
   d. Ketones in the urine
4. If proteins cannot be filtered out of the blood during normal glomerular filtration, why do they sometimes appear on urinalysis readings?
   a. They enter the urine by active transport and secretion.
   b. They enter the urine through damaged kidney tubules.
   c. They enter the urine by facilitated diffusion and secretion.
   d. They enter the urine through damaged glomerular membranes.

Chapter Review

25.1 Functions of the Urinary System

The learning objective for this section is:

HAPS LEARNING GOALS AND OUTCOMES

25.1.1 Describe the major functions of the urinary system and which organs are responsible for those functions.

25.1 Question

1. You are at rest and breathing normally, but your blood pH is abnormally low. Which part of the body is most likely malfunctioning under these circumstances?
   a. Stomach
   b. Small intestine
   c. Kidneys
   d. Ureters

25.1 Summary

- The urinary system plays perhaps the largest role in the maintenance of homeostasis.
- The kidney produces, stores, and excretes urine from the body. Urine is derived from the filtration of the blood by the kidney.
- The kidney removes wastes, regulates the pH of the body, regulates blood pressure, regulates the concentration of solutes in the blood, secretes EPO, and performs the final synthesis step of activated vitamin D production.

25.2 Gross and Microscopic Anatomy of the Kidney

The learning objectives for this section are:

HAPS LEARNING GOALS AND OUTCOMES

25.2.1 Identify and describe the anatomic structure of the kidney, including its coverings.

25.2.2 Distinguish histologically between the renal cortex and the renal medulla.

25.2.3 Trace the path of blood flow through the kidney, from the renal artery to the renal vein.

25.2.4 Identify and describe the vascular elements associated with the nephron (i.e., afferent and efferent arterioles, glomerulus, peritubular capillaries, vasa recta).

25.2.5 Identify and describe the structure of a typical nephron, including the renal corpuscle (i.e., glomerular [Bowman's] capsule, glomerulus) and renal tubule (i.e., proximal convoluted tubule, nephron loop [loop of Henle], distal convoluted tubule).

25.2.6 Describe the filtration structures that lie between the lumen of the glomerular capillaries and the capsular (Bowman) space.

25.2.7 Trace the flow of filtrate from the renal corpuscle through the collecting duct.

25.2.8 Compare and contrast the anatomic structure of the cortical nephrons and juxtamedullary nephrons.

25.2 Questions

1. Where are the kidneys situated in the body? Please select all that apply.
   a. Anchored to the anterior abdominal wall
   b. In the peritoneal cavity
   c. Anchored to the posterior abdominal wall
   d. In the retroperitoneal cavity

2. Which of the following structures are within the renal medulla? Please select all that apply.
   a. Renal columns
   b. Renal pyramids
   c. Afferent arterioles
   d. Glomerular capsule

3. Which blood vessels of the kidney travel through the renal columns?

a. Segmental arteries
b. Interlobar arteries
c. Arcuate arteries
d. Interlobular arteries

4. Which of the following is the correct order of branching arteries traveling toward a nephron?

a. Renal artery → segmental arteries → interlobar arteries → arcuate arteries → interlobular arteries → afferent arterioles
b. Efferent arterioles → peritubular capillaries → interlobular veins → arcuate veins → interlobar veins → renal vein
c. Renal artery → interlobar arteries → segmental arteries → interlobular arteries → arcuate arteries → afferent arterioles
d. Peritubular capillaries → arcuate veins → interlobular veins → interlobar veins → renal vein → efferent arterioles

5. On an upper or lower limb, the descriptions of "proximal" and "distal" relate to two points' relative distances from their attachment sites to the body. What do the terms "proximal" and "distal" relate to in terms of the convoluted tubules of the nephron?

a. Distance from the nephron loop
b. Distance from the glomerulus
c. Distance from the renal medulla
d. Distance from the collecting duct

6. Gaps in which of the following structures form filtration slits in the filtration membrane of a nephron?

a. Endothelial cells of the glomerulus
b. Parietal layer of Glomerular capsule
c. Pedicels of Glomerular capsule
d. Visceral layer of the proximal convoluted tubule

7. Which of the following pathways would filtrate follow if not reabsorbed by the renal tubules of the nephron?

a. Distal convoluted tubule → nephron loop → proximal convoluted tubule → collecting duct
b. Proximal convoluted tubule → nephron loop → distal convoluted tubule → collecting duct
c. Collecting duct → proximal convoluted tubule → nephron loop → distal convoluted tubule
d. Collecting duct → proximal convoluted tubule → nephron loop → distal convoluted tubule

8. Compare the lengths of nephron loops of cortical and juxtamedullary nephrons.

a. Cortical nephron loops are longer than juxtamedullary nephron loops.
b. Cortical nephron loops are shorter than juxtamedullary nephron loops.
c. Cortical nephron loops and juxtamedullary nephron loops are equal in length.
d. Either A or C

25.2 Summary

- The kidneys occupy the retroperitoneal space, posterior to the parietal peritoneum.
- Each kidney is covered in a fibrous capsule with a shock-absorbing layer of perinephric fat surrounding the capsule. Fascia anchors the kidney to the posterior abdominal wall. The adrenal glands sit atop the kidneys, secreting a wide array of hormones.
- Inside the kidney, the tissues are organized into the outer renal cortex and inner renal medulla. The nephrons are the functional units of the kidney; they filter the blood, removing wastes and toxins, and balancing the blood concentrations of various solutes and water. Each nephron is an individual site of urine formation.
- The kidney has a unique portal system of blood vessels in the nephrons. It consists of two sets of arterioles and two sets of capillaries. The blood flow through the nephron proceeds in this order: afferent arteriole, glomerular capillaries, efferent arteriole, and peritubular capillaries. The pathway allows the glomerular capillaries (glomerulus) to perform filtration of the blood.
- The glomerulus is surrounded by the glomerular capsule, so filtrate is caught by the urinary system for reabsorption and excretion. The glomerular capsule directs this filtrate to the renal tubules, proceeding in this order: proximal convoluted tubule, nephron loop, and distal convoluted tubule. The distal convoluted tubules of several nephrons empty their urine into the same collecting duct. The urine is then emptied through a renal papilla into a minor calyx.

25.3 Physiology of Urine Formation

The learning objectives for this section are:

HAPS LEARNING GOALS AND OUTCOMES

25.3.1 Describe the three processes that take place in the nephron (i.e., filtration, reabsorption, and secretion) and explain how the integration of these three processes determines the volume and composition of urine.

25.3.2 Define glomerular filtration rate (GFR) and explain the role of blood pressure, capsule fluid pressure, and colloid osmotic (oncotic) pressure in determining GFR.

25.3.3 Describe factors that can change blood pressure, capsule fluid pressure, and colloid osmotic (oncotic) pressure and thereby change GFR.

25.3.4* Predict the impact on GFR when various physiological factors change.

25.3.5 Compare and contrast tubular reabsorption and secretion with respect to the direction of solute movement and the tubule segments in which each process occurs.

25.3.6* Apply the concepts of transporter specificity, saturation, and competition to explaining variations in tubule reabsorption.

25.3.7 Describe specific mechanisms of transepithelial transport that occur in different parts of the nephron (e.g., active transport, osmosis, facilitated diffusion, electrochemical gradients, receptor-mediated endocytosis, transcytosis).

25.3.8 For the important solutes of the body (e.g., Na^+, K^+, glucose, urea), describe how each segment of the nephron handles the solute and compare the filtration rate of the solute to its excretion rate (i.e., the net handling of the solute by the nephron).

25.3.9 Explain the role of the nephron loop (of Henle), its permeability to water, and the high osmolarity of the interstitial fluid in the renal medulla in the formation of dilute urine.

25.3.10 Trace the changes in filtrate osmolarity as it passes through the segments of the nephron.

25.3.11 Identify the location, structures and cells of the juxtaglomerular apparatus (JGA) and discuss its significance.

25.3.12 Compare and contrast blood plasma, glomerular filtrate, and urine.

25.3.13 For any solute, explain how renal filtration, reabsorption, and secretion determine the excretion rate of that solute.

* Objective is not a HAPS Learning Goal.

25.3 Questions

1. Which of the following processes occur(s) at the glomerular capsule?
   a. Filtration
   b. Reabsorption
   c. Secretion
   d. All of the above

2. You are taking an important anatomy exam. Your heart is beating very fast and your blood pressure is high. What is your glomerular filtration rate compared to resting conditions?
   a. Higher
   b. Lower
   c. The same
   d. Impossible to tell with the information given

3. Say you are dehydrated, and therefore your plasma protein concentration is higher. What will happen to your net filtration pressure? Glomerular filtration rate?
   a. Increase; increase
   b. Increase; decrease
   c. Decrease; increase
   d. Decrease; decrease

4. What would happen to glomerular filtration rate if there were damage to the glomerulus, resulting in the formation of wider-than-normal fenestrations?
   a. Increase
   b. Decrease
   c. No change
   d. Impossible to tell with the information given

5. Most substances undergo tubular reabsorption across _____ membranes compared to other points of the nephron.
   a. distal convoluted tubule
   b. proximal convoluted tubule
   c. nephron loop
   d. collecting duct

6. Sally has eaten a large meal with a side of fruit, meaning that they have glucose and fructose in their bloodstream. (The fructose is found in the fruit.) Once these substances are filtered by the glomeruli of the

kidney, one type of transporter reabsorbs both of these molecules back to the bloodstream. After eating this large meal, how would Sally's blood glucose concentration compare to that of their friend, who ate the same large meal, but did not eat the fruit?

a. It would be higher.
b. It would be lower.
c. It would be the same.
d. It is impossible to tell with the information given.

7. Most (80–90 percent of) bicarbonate is reabsorbed at the proximal convoluted tubule. Its reabsorption is coupled to the cotransport with which ion?

a. Potassium
b. Glucose
c. Sodium
d. Calcium

8. As bicarbonate ions are being reabsorbed, a specific type of transporter transports _____ ions into the tubular lumen. Because the two substances are moving in the same direction across the membrane, this transporter is a(n) _____.

a. K^+; cotransporter
b. K^+; antiporter
c. Na^+; cotransporter
d. H^+; antiporter

9. Aquaporins line the descending nephron loop. What effect does the presence of this structure have on the osmolarity inside the descending nephron loop?

a. increase
b. decrease
c. No effect
d. Impossible to tell with the information given

10. Compared to the descending nephron loop, the ascending loop is _____.

a. hypertonic
b. hypotonic
c. isotonic
d. impossible to tell with the information given

11. The juxtaglomerular apparatus impacts glomerular filtration rate. Which physiological condition causes an increase in GFR?

a. High plasma osmolarity
b. Low plasma osmolarity
c. Normal plasma osmolarity
d. Impossible to tell with the information given

12. What is present in glomerular filtrate that may not be present in urine?

a. Toxins
b. Erythrocytes
c. Small substances that are reabsorbed
d. Small substances that are secreted

13. Substance X is very toxic, and your body wants to excrete it as quickly as possible. Which of the following process(es) will take place at the kidney? Please select all that apply.

a. Filtration
b. Reabsorption
c. Secretion

25.3 Summary

- The three basic processes of urine formation are glomerular filtration, reabsoprtion and secretion. Glomerular filtration filters small molecules out of the blood. Most of the useful molecules in the filtrate undergo reabsorption, in which it is transported from the renal tubules back into the blood. Some waste products that were not filtered are added to the forming urine by secretion, in which they are transported from the blood of the peritubular capillaries into the renal tubules.
- On average, we filter our entire blood volume through our kidneys once in every 30 minutes and 20 percent of our blood is lost to filtrate. This varies widely based on factors such as hydration and nutrient concentration in your blood, as nephrons have the capability of changing filtration, reabsorption, and secretion rate.
- Filtration is a nonspecific process based on molecular size. The net filtration pressure, and therefore the glomerular filtration rate, is the product of three factors: (1) the glomerular hydrostatic pressure, which favors filtration, (2) the capsular hydrostatic pressure, which opposes filtration, and (3) the osmotic pressure in the glomerulus, which opposes filtration. The strongest force is typically the glomerular hydrostatic pressure.
- An increase in glomerular filtration rate (GFR) is accomplished by vasodilation of the afferent arteriole or vasoconstriction of the efferent arteriole. This is favorable when there is excess plasma volume.
- The processes of reabsorption and secretion modify the glomerular filtrate into the finished product, urine. These processes occur throughout the renal tubules.
- Various transport methods are used to move substances across the membranes of the renal tubules and peritubular

capillaries, such as facilitated diffusion and active transport. The movement of many substances is coupled to that of sodium ions. The sodium/potassium pump and specific transporters are also utilized in the movement of various substances across the membranes of the cells of the renal tubules and the peritubular capillaries.

- Each portion of the nephron is responsible for different functions involved in modifying the filtrate into urine. For example, the nephron loop concentrates the urine by reabsorbing sodium ions and water.
- The juxtaglomerular apparatus monitors electrolyte concentration in the blood and blood pressure, and secretes renin when necessary to increase the blood pressure.

25.4 Homeostasis and Control over the Formation of Urine

The learning objectives for this section are:

HAPS LEARNING GOALS AND OUTCOMES

25.4.1 For the renin-angiotensin system (RAS), describe the factors that initiate renin release, the pathway from angiotensinogen to angiotensin II (ANGII), and the effects of ANGII on various tissues.

25.4.2 Describe the signals that cause release of aldosterone from the adrenal cortex and the effect of aldosterone on the nephron, including the tubule segment involved and the transport mechanisms that are altered by aldosterone.

25.4.3 Describe the effect of vasopressin (ADH, antidiuretic hormone) on the nephron and on the final concentration of urine.

25.4.4 Describe the factors that cause release of natriuretic peptide hormones, their sites of synthesis, and their effects on the nephron.

25.4 Questions

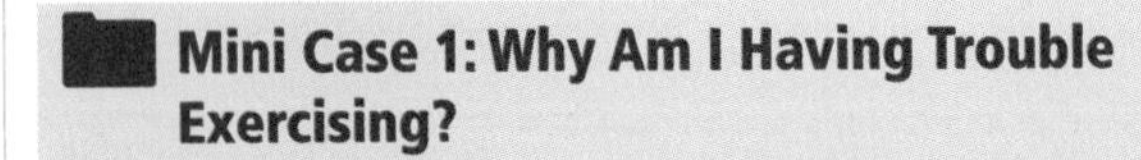

Mini Case 1: Why Am I Having Trouble Exercising?

Jasmyn went to see the doctor thinking they were having heart problems, because exercising had become increasingly difficult over the past few months. Jasmyn could not seem to respond to the increased demands of exercise on their body. The doctor ran some tests on Jasmyn's heart function, and all of the test results were normal. So the doctor started testing hormonal factors that might be causing Jasmyn's problems.

1. Because Jasmyn's blood pressure was not increasing during exercise, the doctor tested for normal functioning of the renin-angiotensin system. Which components of the renin-angiotensin system should always be present in Jasmyn's blood or blood vessels? Please select all that apply.
   a. Angiotensin I
   b. Angiotensin II
   c. Angiotensinogen
   d. Angiotensin converting enzyme

2. The doctor also tested Jasmyn's aldosterone level. All of the following are true of aldosterone except:
   a. Aldosterone secretion can be triggered by the presence of angiotensin II.
   b. Aldosterone typically increases blood pressure.
   c. Aldosterone acts on the nephron to increase reabsorption of sodium ions.
   d. Aldosterone causes increased secretion of water in the kidneys.

3. The level of antidiuretic hormone (ADH) in Jasmyn's blood was also tested. How does ADH affect the nephron, and on which portion of the nephron does it act?
   a. ADH decreases water reabsorption; it acts on the nephron loop.
   b. ADH increases water secretion; it acts on the distal convoluted tubule.
   c. ADH increases water reabsorption; it acts on the collecting ducts.
   d. ADH decreases water secretion; it acts on the proximal convoluted tubule.

4. Finally, the doctor tested the level of Jasmyn's natriuretic peptides. All of the following are true of natriuretic peptides except:
   a. Natriuretic peptides increase blood pressure.
   b. Natriuretic peptides are produced by the heart.
   c. Natriuretic peptides act on the nephron to increase sodium ion secretion.
   d. Natriuretic peptides cause water to be added to the forming urine by osmosis.

The results of Jasmyn's tests showed that the levels of aldosterone, ADH and the natriuretic peptides were all normal, but that they were not producing renin in the nephrons of the kidney. Jasmyn was given medication that mimicked the effects of renin, so they were able to respond once again to the demands of exercise.

25.4 Summary

- The renin-angiotensin-aldosterone system results in aldosterone release, raising blood volume and mean arterial pressure across the cardiovascular system.
- Antidiuretic hormone is a posterior pituitary hormone that works with aldosterone to increase blood volume by increasing aquaporin expression.
- Natriuretic hormones stimulate the kidneys to excrete sodium, decreasing blood volume and pressure by altering the concentration gradient of sodium.

25.5 Additional Endocrine Activities of the Kidney

The learning objectives for this section are:

25.5.1 Describe the role of the kidney in vitamin D activation.

25.5.2 Describe the role of the kidney in regulating erythropoiesis.

25.5.3 Describe the effect of parathyroid hormone (PTH) on renal handling of calcium and phosphate.

25.5 Questions

1. What is the kidney's role in the activation of Vitamin D?
   a. The kidney synthesizes vitamin D in the presence of sunlight.
   b. The kidney converts vitamin A into vitamin D.
   c. The kidney converts cholesterol directly into active vitamin D.
   d. The kidney completes the final step of vitamin D activation, producing calcitriol.
2. The kidney secretes erythropoietin in response to a state of _____ in the blood.
   a. Hypoxia
   b. Hyperoxia
   c. Hyperosmosis
   d. Hypoosmosis
3. Damage to the parathyroid gland during thyroid gland surgery causes which of the following side effects?
   a. Lower than normal blood Ca^{2+} levels
   b. Lower than normal blood pressure
   c. Higher than normal Na^{+} levels
   d. Higher than normal K^{+} levels

25.5 Summary

- The kidney performs the final step of activated vitamin D synthesis, converting calcidiol, which is made in the liver, to active vitamin D, or calcitriol, capable of maintaining calcium and phosphate levels in the body.
- When the kidney detects oxygen deficiency in the blood, it secretes the hormone erythropoietin. This hormone stimulates red blood cell production in the red bone marrow.
- In response to low blood calcium levels, parathyroid hormone (PTH) causes the kidney to convert calcidiol to calcitriol (activated vitamin D); this highly activated form of vitamin D causes increased dietary calcium absorption in the small intestine. PTH also causes calcium reabsorption in the renal tubules.

25.6 Gross and Microscopic Anatomy of the Urinary Tract (Ureters, Urinary Bladder, and Urethra)

The learning objectives for this section are:

25.6.1 Identify and describe the gross anatomy and location of the ureters, urinary bladder, and urethra.

25.6.2 Identify and describe the microscopic anatomy of the ureters, urinary bladder, and urethra.

25.6.3 Compare and contrast the locations, innervation and functions of the internal urethral sphincter and external urethral sphincter.

25.6.4 Describe voluntary control of micturition.

25.6.5 Describe the micturition reflex and the role of the autonomic nervous system in the reflex.

25.6.6 Trace the path of urine from the collecting duct of the kidney to the external urethral orifice.

25.6 Questions

1. The ureters empty urine into which of the following?

a. Urinary bladder
b. Urethra
c. Outside world
d. Prostate or uterus

2. Which of the following types of epithelium line the ureters and urinary bladder? What property does this type exhibit?

a. Transitional; protection
b. Transitional; stretch
c. Stratified squamous; protection
d. Stratified squamous; stretch

3. Which type of muscle is involved in relaxation of the internal urethral sphincter? Is this action voluntary or involuntary?

a. Smooth muscle; involuntary
b. Smooth muscle; voluntary
c. Skeletal muscle; involuntary
d. Skeletal muscle; voluntary

4. Which type of muscle is involved in relaxation of the external urethral sphincter? Is this action voluntary or involuntary?

a. Smooth muscle; involuntary
b. Smooth muscle; voluntary
c. Skeletal muscle; involuntary
d. Skeletal muscle; voluntary

5. A malfunction of parasympathetic control in the sacral region of the spinal cord would prevent which of the following events from happening in the micturition reflex? Please select all that apply.

a. Contraction of detrusor muscle
b. Relaxation of detrusor muscle
c. Relaxation of the internal urethral sphincter
d. Stimulation of somatic motor neurons

6. After urine is filtered through the nephron, trace the correct order of urine flow through excretion to the outside world.

a. Collecting ducts → papillary ducts → minor calyces → major calyces → renal pelvis → ureter → urinary bladder → urethra
b. Urethra → papillary ducts → minor calyces → major calyces → renal pelvis → ureter → urinary bladder → collecting ducts
c. Papillary ducts → minor calyces → major calyces → collecting ducts → renal pelvis → ureter → urinary bladder → urethra
d. Urethra → urinary bladder → ureter → renal pelvis → major calyces → minor calyces → papillary ducts → collecting ducts

25.6 Summary

- Urine exits the kidney in the ureters. Urine is stored in the urinary bladder until a convenient time for urination, or micturition. The urethra, which is much longer in individuals with prostates, carries urine from the bladder to the outside world.
- The ureters, urinary bladder and urethra are lined with transitional epithelium, a unique type of epithelium only found in the urinary system. Transitional epithelium is specialized to allow the organs of the urinary system to stretch without tearing when urine is present and return to their original shape when empty.
- The micturition reflex results from the involuntary control of the smooth muscle of the internal urethral sphincter and the voluntary control of the skeletal muscle of the external urethral sphincter.

25.7 Urine Characteristics and Elimination

The learning objectives for this section are:

25.7.1* Describe the typical characteristics of urine.

25.7.2* Given a change in the composition of urine (ex: proteinuria) predict the physiological disruption that led to the change.

* Objective is not a HAPS Learning Goal.

25.7 Questions

1. Sophie's urine has the following four traits. Three are normal and one is abnormal. Which is abnormal?
   a. No blood
   b. Deep amber color
   c. A high level of proteins
   d. Trace white blood cells

2. Why might there be protein in Sophie's urine?
   a. Damage to the glomerulus
   b. Damage to the renal artery
   c. Damage to the proximal convoluted tubule
   d. Damage to the distal convoluted tubule

25.7 Summary

- Typically, urine is pale yellow to deep amber in color and odorless; a deviation in the color often signals dehydration.
- Urine should have a pH between 4.5 and 8.0, and should contain no ketones, nitrites, blood, glucose and leukocyte esterase.
- Urine also should contain the metabolic wastes urea, uric acid, and creatinine.
- Urinalysis—analysis of the contents and characteristics of urine—often provides clues to renal or other disease conditions if there is a deviation from normal values.

26

Fluid, Electrolyte, and Acid-Base Balance

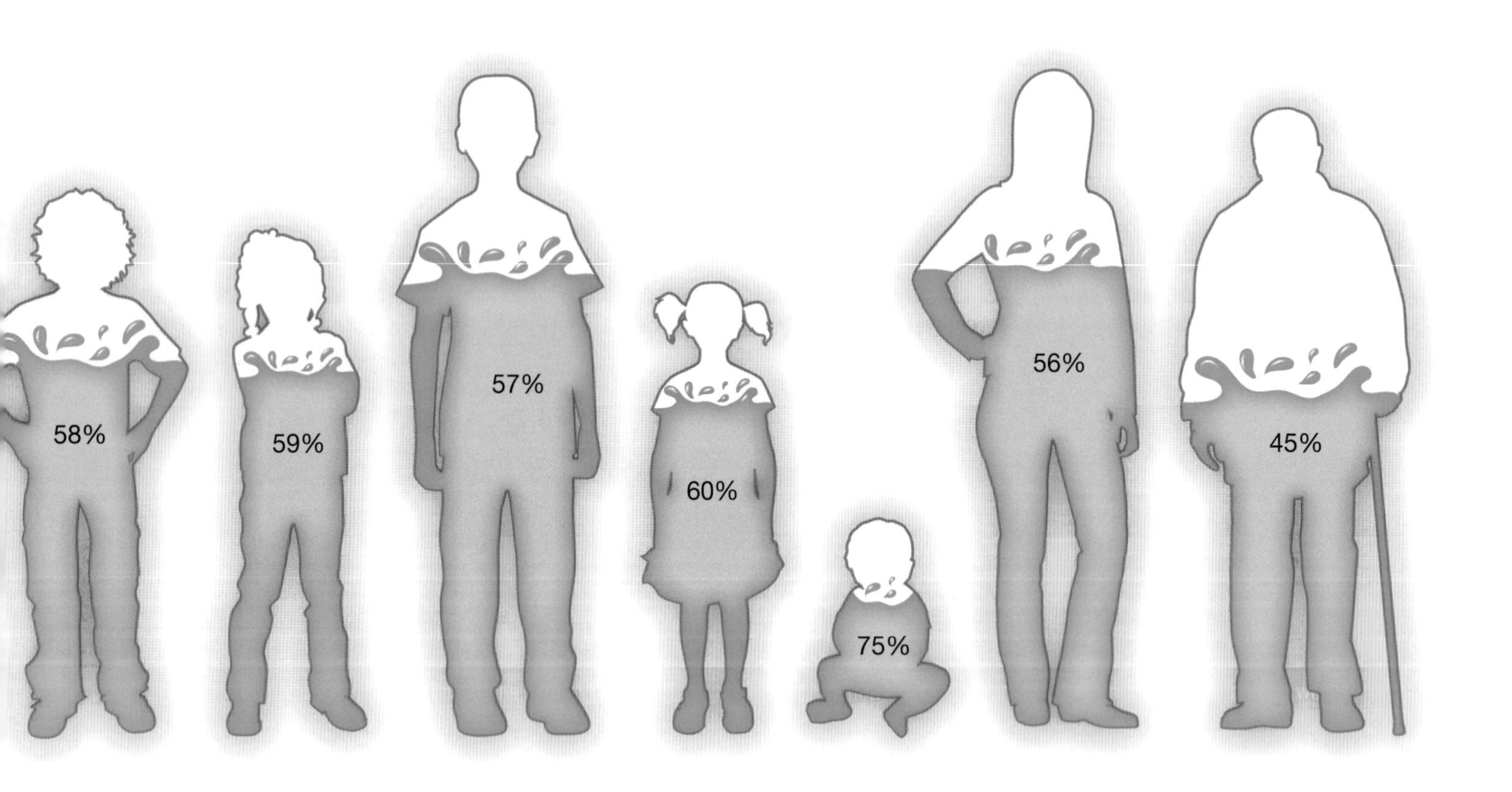

Chapter Introduction

In this chapter we will learn...

how the kidneys, cardiovascular system and lungs work together to maintain water and pH balance in the body.

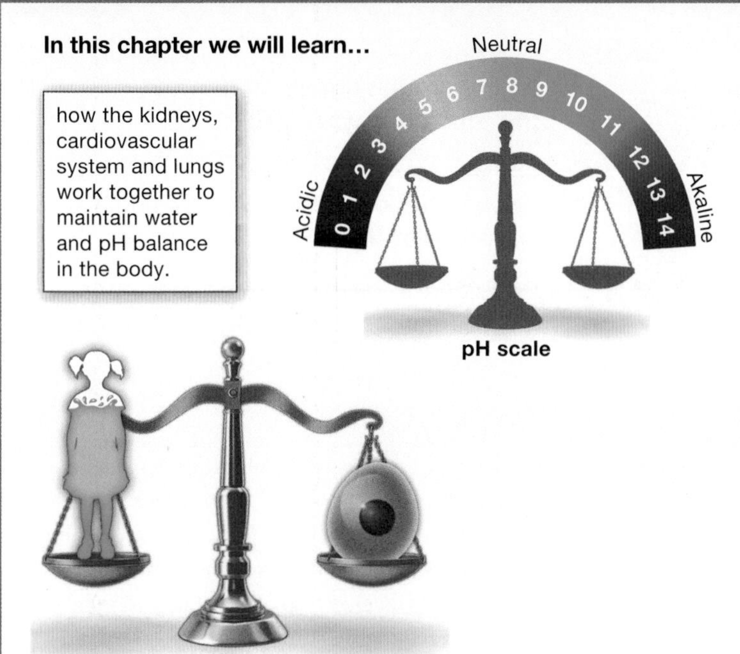

Homeostasis, or the maintenance of stable internal conditions within the body, is a fundamental property of all living things. While homeostasis is maintained in different ways in different organisms, in the human body every homeostatic variable is maintained through the ability to sense a disruption, integrate that sensory input, and enact a response through an effector (**Figure 26.1**). In the human body, the stability of many conditions is maintained through tight homeostatic control. In particular, the substances that participate in chemical reactions must remain within narrow ranges of concentration. Too much or too little of a single substance can disrupt our bodily functions. Because metabolism relies on reactions that are all interconnected, any disruption might affect multiple organs or even organ systems. Water is the most ubiquitous substance in the chemical reactions of life. Water is essential both in terms of body chemistry—water is the universal solvent, and hydrolysis and dehydration reactions are central to all metabolic activity—and the body's physical state (water makes up a majority of our body by weight, and its hydrostatic pressure drives fluid flow and pressure). In this chapter we will explore the mechanisms of water, fluid, and ion homeostasis throughout the body.

Figure 26.1 The Homeostasis Model

Homeostasis is maintained for variables in the body such as temperature, pH, and blood calcium concentration. To maintain homeostasis, the body must have the ability to sense a shift away from the set point, as well as mechanisms to restore the variable to its set point. A control center is in charge of regulating the response.

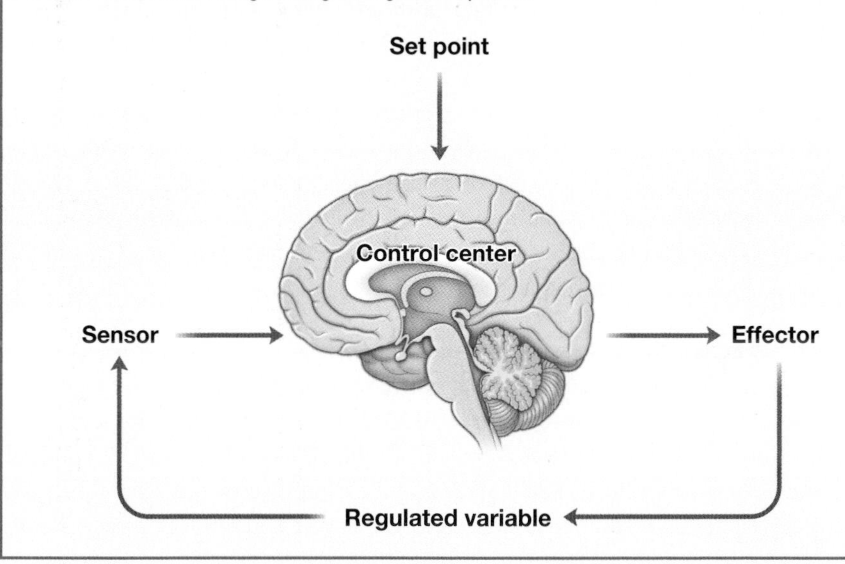

26.1 Body Fluids and Fluid Compartments

Learning Objectives: By the end of this section, you will be able to:

26.1.1 Describe the boundary walls that separate different body fluid compartments and list transport mechanisms by which water and other substances move between compartments.

26.1.2 Explain the subdivision of the extracellular fluid (ECF) compartment into plasma and interstitial fluid (IF), and compare volumes and composition of plasma and IF.

26.1.3* Describe some of the physiological variables (e.g., estrogen signaling) that can influence the total body water (TBW).

26.1.4 Compare and contrast relative volumes and osmolarities of intracellular fluid (ICF) and extracellular fluid (ECF).

26.1.5* Describe the factors that influence the direction and degree of movement of fluid between compartments.

*Objective is not a HAPS Learning Goal

The chemical reactions of life take place in aqueous (watery) solutions. The dissolved substances in a solution are called *solutes* and the solvent is water. In the human body, solutes vary in different parts of the body, but may include proteins—including those that transport lipids, carbohydrates, and (very importantly) electrolytes. In physiology, a mineral that carries an electrical charge (an ion) is called an **electrolyte**. For instance, sodium ions (Na^+) and chloride ions (Cl^-) are often referred to as *electrolytes*. In a store you might find electrolyte drinks; these contain ions (and sometimes quite a bit of sugar!).

In the body, water moves through semipermeable membranes of cells and from one compartment of the body to another by osmosis. **Osmosis** is the diffusion of water from regions of higher water concentration to regions of lower water concentration. We often use the term **osmotic gradient** to describe a difference in water concentration between two locations separated by a membrane. Water will move into and out of cells and tissues via osmosis, according to the osmotic gradient. Therefore, you can imagine that cells would be vulnerable to shifts in the osmotic gradient. If the water concentration outside the cell decreased significantly because of an increase in solutes, water would leave the cell, moving from its higher water concentration inside to its lower water concentration outside. This movement of water would have devastating effects on the cell, causing it to shrink physically and lack water for its chemical processes. Therefore, the balance of solutes and water inside and outside of cells must be maintained to ensure cellular health and function.

LO 26.1.1

Figure 26.2 Water Content of the Body's Organs and Tissues

Water content varies in different body locations, based on structure and function. Whereas teeth and adipose tissue contain just 10 percent water, the brain and heart contain closer to 80 percent.

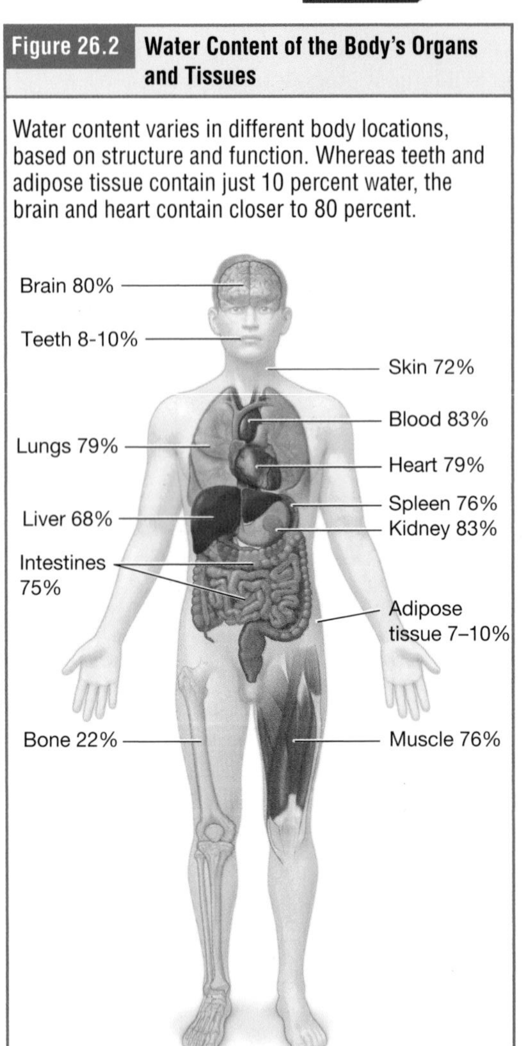

26.1a Body Water Content

Human beings are mostly water, but not all humans have the same proportion of water in their bodies. The body mass of an infant is composed of approximately 75 percent water. Adults have a range that is based on many factors, but typically somewhere between 50 and 60 percent of their body mass consists of water. An elderly adult loses some water mass and carries less water; this is usually around 45 percent of their body mass. The percent of body water changes with development, because the proportions of the body given over to each organ and to muscles, fat, bone, and other tissues change from infancy to adulthood (**Figure 26.2**). Your brain

The Human Anatomy and Physiology Society includes more than 1,700 educators who work together to promote excellence in the teaching of this subject area. The HAPS A&P Learning Outcomes measure student mastery of the content typically covered in a two-semester Human A&P curriculum at the undergraduate level. The full Learning Outcomes are available at https://www.hapsweb.org.

and kidneys contain the highest proportions of water (80–85 percent). In contrast, teeth and adipose tissue have the lowest proportion of water (8–10 percent).

26.1b Fluid Compartments

LO 26.1.2

Bodily fluids can be discussed in terms of their specific **fluid compartment**, a location that is largely separate from another compartment by some form of a physical barrier. The **intracellular fluid (ICF)** compartment is the system that includes all fluid enclosed in cells by their plasma membranes. **Extracellular fluid (ECF)** surrounds all cells in the body. ECF has two primary constituents: the fluid component of the blood (called **plasma**) and the **interstitial fluid (IF)** that surrounds all cells not in the blood (Figure 26.3). Together, the ICF + ECF = **Total body water (TBW).**

Figure 26.3 Fluid Compartments in the Human Body

Fluid is found in the blood and among the tissues. These fluids are both considered extracellular fluids because they are outside of cells. Intracellular fluid is found within cells.

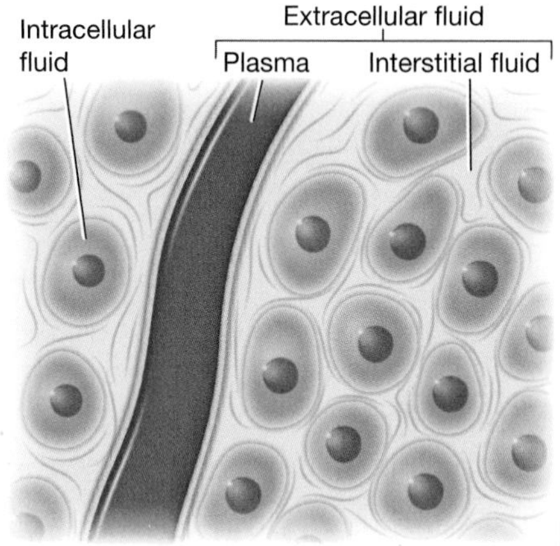

Intracellular Fluid The ICF is the principal component of the cytosol/cytoplasm. The ICF makes up about two-thirds of the total water in the human body (Figure 26.4). The homeostatic regulation of this fluid volume is critical for the health of the cells. If the amount of water inside a cell falls too low, the cytosol becomes too concentrated with solutes to carry on normal cellular activities and the cell physically shrinks; if too much water enters a cell, the cell may burst and be destroyed.

Extracellular Fluid The ECF accounts for the other one-third of the body's water content. Approximately 20 percent of the ECF is found in blood plasma, the other 80 percent is the IF of the tissues. These two compartments are linked, but not continuous. Some material can seamlessly traverse from one compartment to another via diffusion, while other materials are confined in one compartment or the other until transported.

There are a few other body fluids that are contained in compartments. These include the cerebrospinal fluid (CSF) that bathes the brain and spinal cord, lymph in the lymphatic vessels and nodes, the synovial fluid in joints, the pleural fluid in the pleural cavities, the pericardial fluid in the cardiac sac, the peritoneal fluid in the peritoneal cavity, and the aqueous humor of the eye. Because these fluids are outside of cells, these fluids are also considered components of the ECF compartment.

LO 26.1.3

Figure 26.4 Proportion of Total Body Fluid in Each of the Body's Fluid Compartments

Most of the water in the body is intracellular fluid. The second largest volume is the interstitial fluid, which surrounds cells other than blood cells.

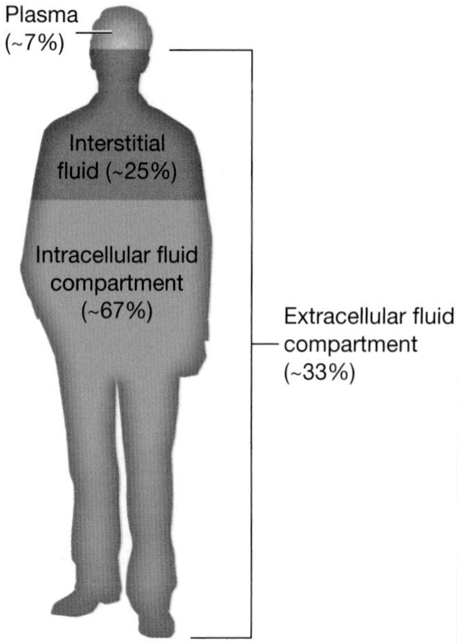

TBW can vary among individuals. One factor is sex hormones. Of the three major sex hormones (estrogen, testosterone, and progesterone), estrogen causes the most water retention. Testosterone can cause water retention, but to a much lower degree, and progesterone promotes the release of water into the urine. Depending on the levels and ratios of these hormones on any given day, the TBW may vary. Adipose tissue, being composed of lipids, has much lower water content than other tissues. Therefore, bodies with more adipose tissue may have a lower TBW overall.

Mild, transient edema of the feet and legs may be caused by sitting or standing in the same position for long periods of time, as in the work of a toll collector or a supermarket cashier. This is because deep veins in the lower limbs rely on skeletal muscle contractions to push on the veins and thus "pump" blood back to the heart. Otherwise, the venous blood pools in the lower limbs and can leak into surrounding tissues.

Medications that can result in edema include vasodilators, calcium channel blockers used to treat hypertension, nonsteroidal anti-inflammatory drugs, estrogen therapies, and some diabetes medications. Underlying medical conditions that can contribute to edema include congestive heart failure, kidney damage and kidney disease, disorders that affect the veins of the legs, and cirrhosis and other liver disorders.

Therapy for edema usually focuses on elimination of the cause. Activities that can reduce the effects of the condition include appropriate exercises to keep the blood and lymph flowing through the affected areas. Other therapies include elevation of the affected part to assist drainage, massage and compression of the areas to move the fluid out of the tissues, and decreased salt intake to decrease sodium and water retention.

DIGGING DEEPER: Edema

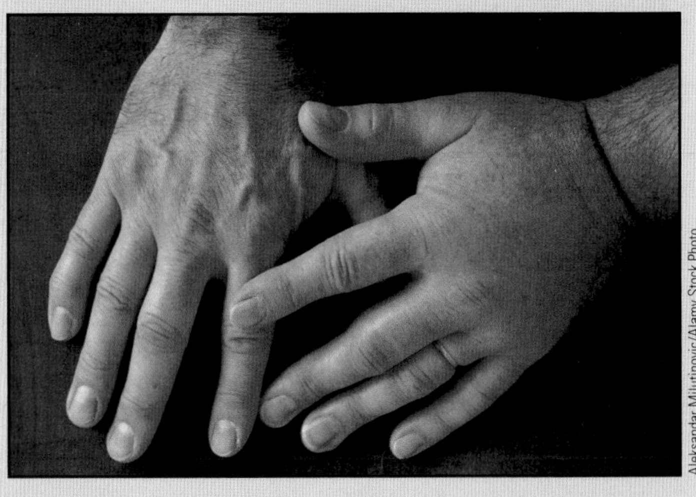
Aleksandar Milutinovic/Alamy Stock Photo

Edema is the accumulation of excess water in the tissues. When interstitial fluid pressure is excessive, sometimes due to high blood pressure or a blockage in the lymph drainage system, notable swelling can result. It is most common in the soft tissues of the extremities.

The physiological causes of edema include water leakage from blood capillaries. Edema is almost always caused by an underlying medical condition, the use of certain therapeutic drugs, pregnancy, localized injury, or an allergic reaction. In the limbs, the symptoms of edema include swelling of the subcutaneous tissues, an increase in the normal size of the limb, and stretched, tight skin. One quick way to check for subcutaneous edema localized in a limb is to press a finger into the suspected area. Edema is likely if the depression persists for several seconds after the finger is removed (which is called *pitting*; see figure).

Pulmonary edema is the presence of excess fluid in the air sacs of the lungs, a common symptom of heart and/or kidney failure. People with pulmonary edema likely will experience difficulty breathing, and they may experience chest pain. Pulmonary edema can be life-threatening, because it compromises gas exchange in the lungs, and anyone having symptoms should immediately seek medical care.

In pulmonary edema resulting from heart failure, excessive leakage of water occurs because fluids get "backed up" in the pulmonary capillaries of the lungs when the left ventricle of the heart is unable to pump sufficient blood into the systemic circulation. Because the left ventricle of the heart is unable to pump out its normal volume of blood, the blood in the pulmonary circulation gets "backed up," into the left atrium, then the pulmonary veins, and finally the pulmonary capillaries. Due to the fact that blood is still being pumped from the pulmonary arteries into the pulmonary capillaries, the hydrostatic pressure within the pulmonary capillaries increases; this results in fluid filtering from the pulmonary capillaries into the lung tissues.

Other causes of edema include damage to blood vessels and/or lymphatic vessels, or a decrease in osmotic pressure in chronic and severe liver disease when the liver is unable to manufacture plasma proteins. A decrease in the normal levels of plasma proteins results in a decrease of colloid osmotic pressure (which counterbalances the hydrostatic pressure) in the capillaries. This process causes loss of water from the blood to the surrounding tissues, resulting in edema.

Learning Check

1. People who have a high-sodium diet have a lot of Na^+ ions in their bloodstream. Which of the following would you expect to see in these people?
   a. Fluid entering the interstitial space from capillaries
   b. Fluid entering capillaries from the interstitial space
   c. Na^+ ions entering the interstitial space from the capillaries
   d. Na^+ ions entering capillaries from the interstitial space
2. Which of the following are part of the intracellular fluid compartment? Please select all that apply.
   a. Cytoplasm within red blood cells
   b. Blood plasma within arteries
   c. Blood plasma within interstitial space
   d. Fluid within muscle cells
3. Identify the correct equation for calculating the total body water (TBW).
   a. ICF + ECF = TBW
   b. IF + ICF = TBW
   c. IF + ECF = TBW
   d. IF + EF = TBW
4. Which of the following could increase the volume of total body water? Please select all that apply.
   a. Low estrogen levels
   b. High testosterone levels
   c. High adipose tissue levels
   d. Low progesterone levels

26.1c Composition of Bodily Fluids

LO 26.1.4

The compositions of the two components of the ECF—plasma and IF—are more similar to each other than either is to the ICF (**Figure 26.5**). Blood plasma has high concentrations of sodium, chloride, bicarbonate, and protein. The IF has high concentrations of sodium, chloride, and bicarbonate, but a relatively lower concentration of protein. In contrast, the ICF has elevated amounts of potassium, phosphate, magnesium, and protein. Overall, whereas the ICF contains high concentrations of potassium and phosphate (HPO^{2-}), both plasma and the ECF contain high concentrations of sodium and chloride.

Most bodily fluids are neutral in charge. Thus, **cations**, or positively charged ions, and **anions**, or negatively charged ions, are balanced in fluids. As seen in the previous graph, whereas sodium (Na^+) ions and chloride (Cl^-) ions are concentrated in the ECF of the body, potassium (K^+) ions are concentrated inside cells. Although sodium and potassium can "leak" through "pores" into and out of cells, respectively, the high levels of potassium and low levels of sodium in the ICF are maintained by sodium-potassium pumps in the cell membranes. These pumps use the energy supplied by ATP to pump sodium out of the cell and potassium into the cell (**Figure 26.6**).

26.1d Fluid Movement between Compartments

LO 26.1.5

Hydrostatic pressure, the force exerted by a fluid against the walls of its container, can cause the movement of fluid out of one compartment and into another. For example, the hydrostatic pressure of blood is the pressure exerted by blood against the walls of the blood vessels. This pressure is a factor of the blood volume (more blood volume = more pressure on the walls) and contributed to by the pumping action of the heart. When the hydrostatic pressure of the blood is higher, plasma will leave the blood and become IF. Capillaries are the only site where exchange of materials between the blood and the IF is possible. Remember from Chapter 20 that the amount of plasma

Figure 26.5 The Composition of Bodily Fluids

The three fluid compartments vary in their volume and composition.

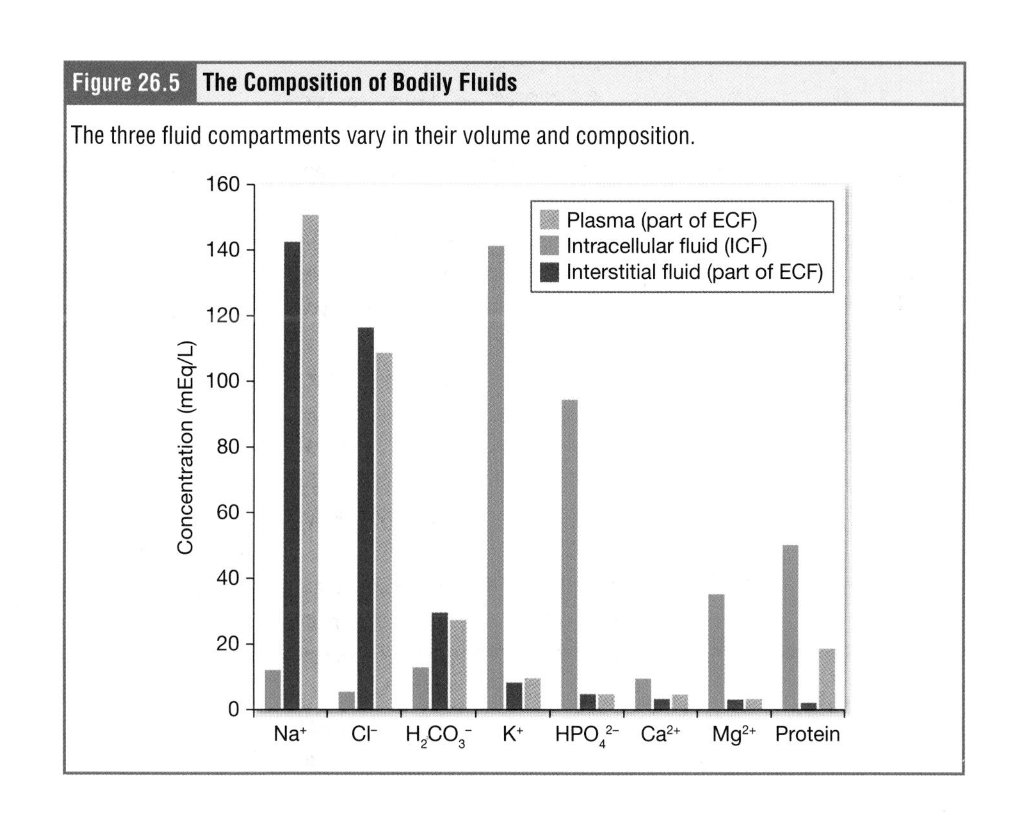

Figure 26.6 The Sodium-Potassium Pump

The sodium-potassium pump transfers sodium out of the cell and into the ECF. The pump also transfers potassium out of the ECF and into the ICF. To build these gradients and transport the ions, the pump requires energy in the form of ATP.

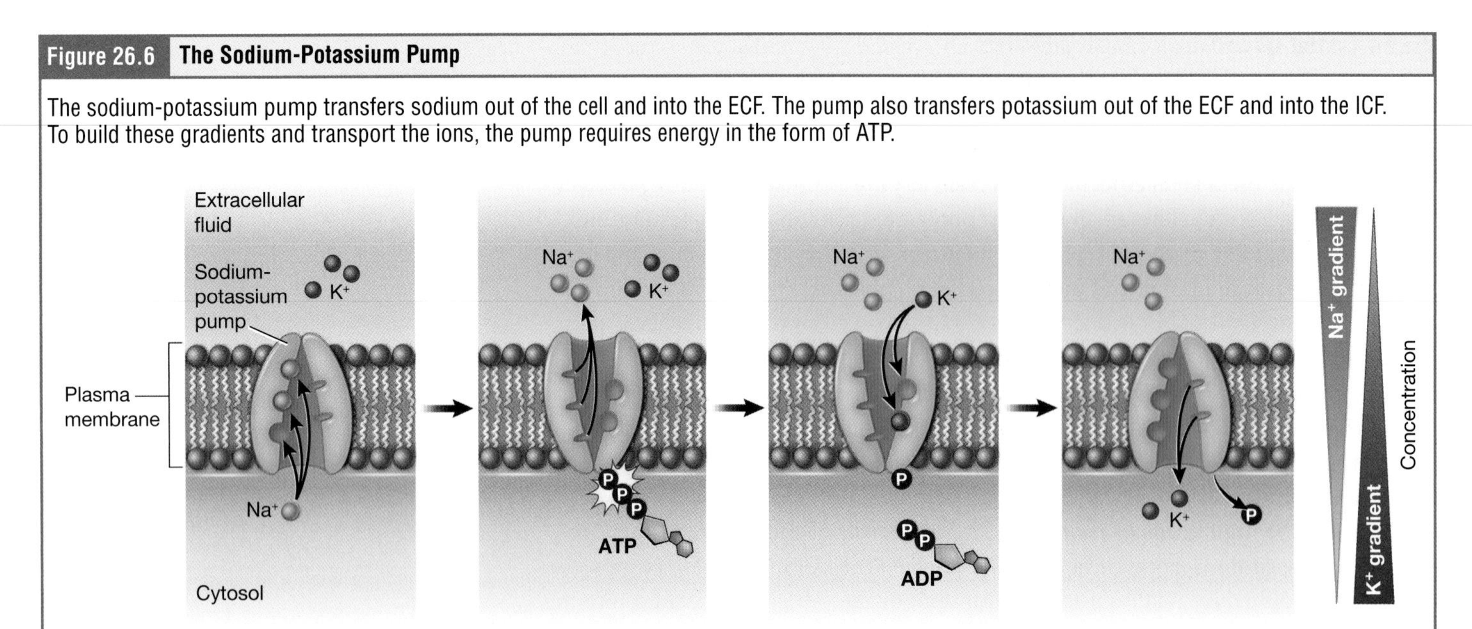

colloid osmotic pressure (oncotic pressure)

that leaves the blood vessels at the capillaries is a balance between hydrostatic pressure and **colloid osmotic pressure**—an osmotic pressure primarily produced by circulating albumin—at the arteriolar end of the capillary (**Figure 26.7**). This pressure forces plasma along with dissolved nutrients and electrolytes out of the capillaries and into the IF of surrounding tissues. This movement of fluid could go either way. The colloid osmotic pressure is relatively constant, so if the hydrostatic pressure is average or higher, more fluid will leave the capillaries and become IF. When blood pressure is very low, fluid may enter the blood from the tissues. Under most typical physiological circumstances, fluid is lost from the blood to the IF and must be returned in order to maintain blood volume. Fluid, along with cellular wastes from the tissues, enters the capillaries at the venule end, where the hydrostatic pressure is lower. Surplus fluid in

Figure 26.7 Capillary Exchange

Two factors regulate the amount of plasma that will exit a blood vessel at a capillary. Those factors are the hydrostatic pressure gradient and the osmotic pressure gradient between the blood vessel and the interstitial fluid.

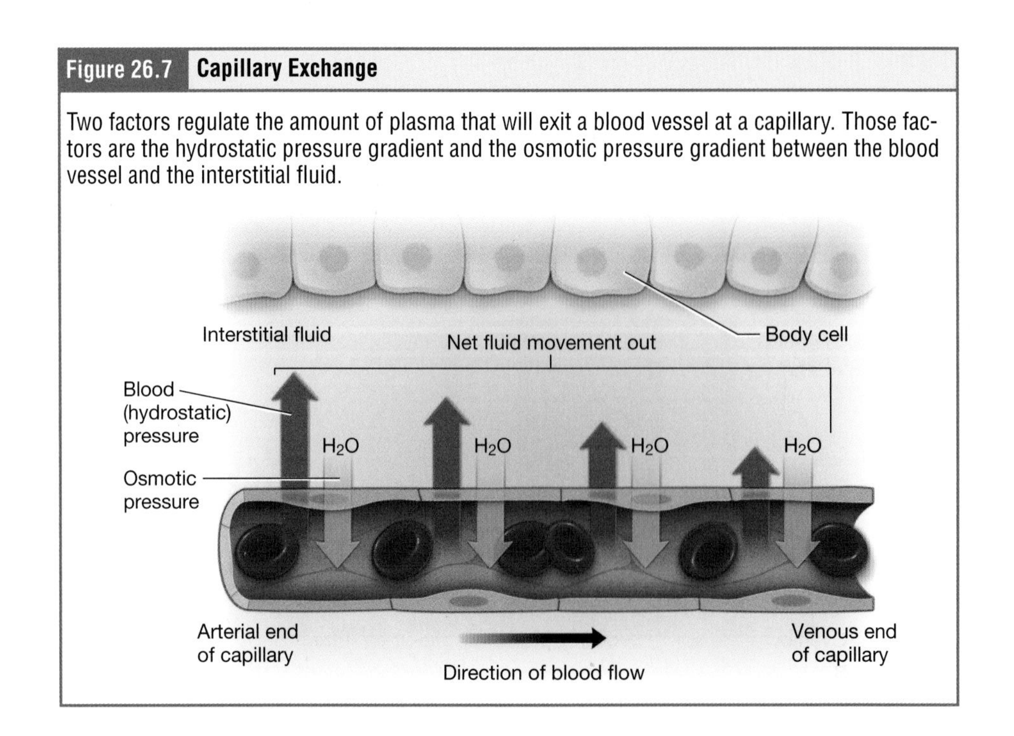

the interstitial space can also be drained from tissues by the lymphatic system, where it reenters the vascular system at the subclavian veins (**Figure 26.8A**). This system, in which fluid is drained and then returned to the circulation, resembles the drain-and-return system of a drinking fountain or a pond (**Figure 26.8B**).

Hydrostatic pressure is especially important in governing the movement of water in the nephrons of the kidneys to ensure proper filtering of the blood to form urine, as you read about in Chapter 25. As hydrostatic pressure in the kidneys increases, the amount of water leaving the capillaries also increases, and more urine filtrate is formed. If hydrostatic pressure in the kidneys drops too low, as can happen in dehydration, the less filtration of the blood will occur. On one hand this will help preserve blood pressure during the period of dehydration, but on the other hand, less nitrogenous waste will be removed from the bloodstream. Extreme dehydration can result in kidney failure as well as blood toxicity.

Fluid also moves between compartments when it follows an osmotic gradient. Recall that an osmotic gradient is produced by the difference in concentration of all solutes on either side of a semipermeable membrane. The magnitude of the osmotic

Learning Connection

Try Drawing It

Make a simple drawing of the following four structures, placing each one in a different corner of a piece of paper: (1) a capillary (containing blood plasma), (2) a body cell (containing intracellular fluid), (3) the interstitial space (containing interstitial fluid), and (4) a lymphatic capillary (containing lymph). Now draw arrows to show which fluids can flow directly from one compartment into another, across a plasma membrane or vessel wall. This exercise will help you see that the interstitial fluid is the only fluid that is in direct contact with all of the other fluids.

Figure 26.8 Plasma, Interstitial Fluid, and Lymph

(A) Fluid is driven out of the blood by hydrostatic pressure; once in the tissues, it is considered interstitial fluid. Most interstitial fluid returns to the capillaries; the rest is drained into lymphatic vessels, from which it is returned to the blood. (B) This system resembles the water drainage system in a fountain or pond. Excess fluid is drained and returned to the fountain pump.

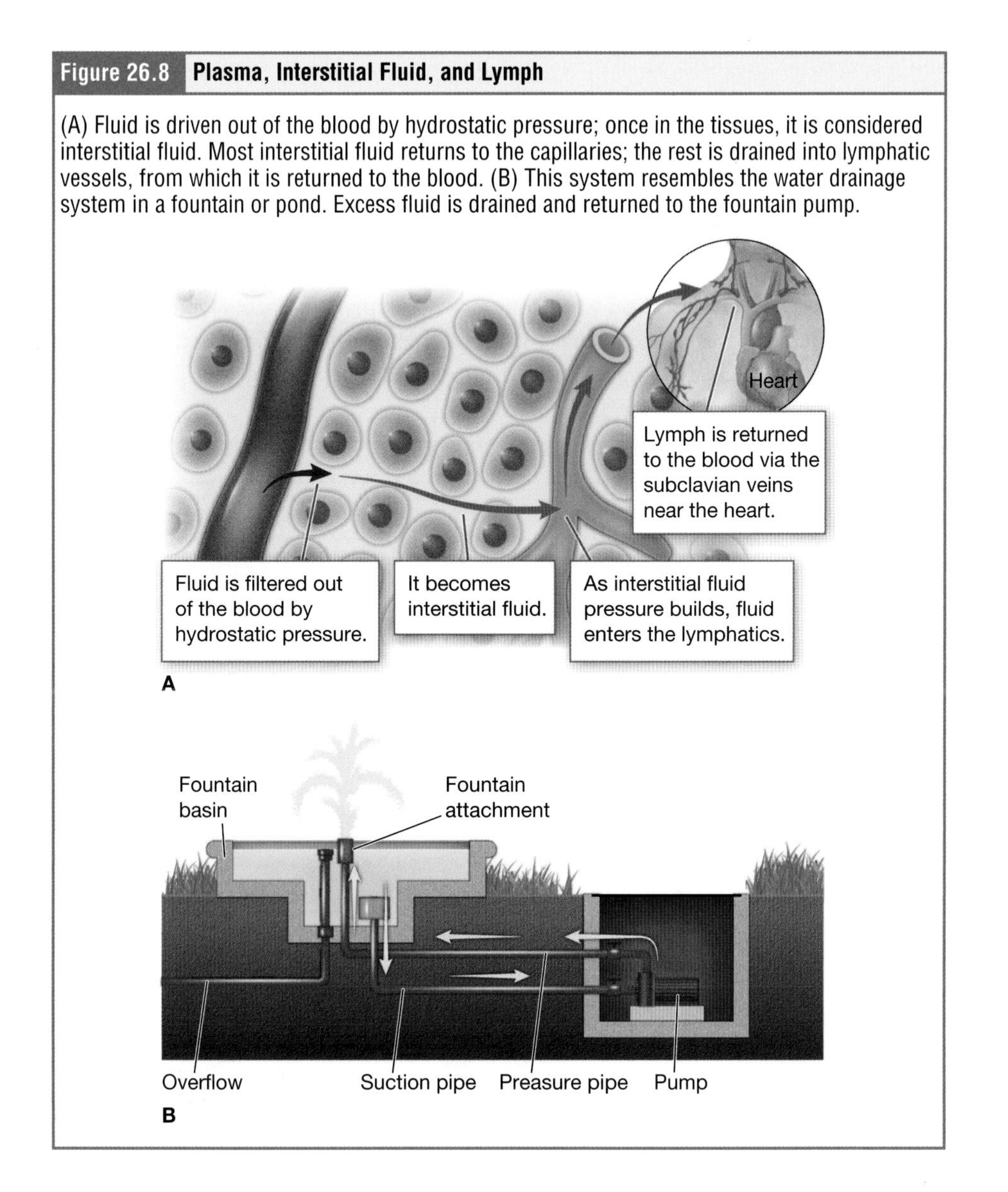

gradient is proportional to the difference in the concentration of solutes on one side of the cell membrane to that on the other side. Water will move by osmosis from the side where its concentration is high (and the concentration of solute is low) to the side of the membrane where its concentration is low (and the concentration of solute is high). In the body, water moves by osmosis from plasma to the IF (and the reverse) and from the IF to the ICF (and the reverse). In the body, water moves constantly into and out of fluid compartments as conditions change in different parts of the body.

For example, if you are sweating, you will lose water through your skin. Sweat is more dilute than body fluids so sweating depletes your tissues of water and increases the solute concentration in those tissues (**Figure 26.9**). As this happens, water diffuses from your blood into sweat glands and surrounding skin tissues that have become dehydrated because of the osmotic gradient. Additionally, as water leaves the blood, it is replaced by the water in other tissues throughout your body that are not dehydrated. If this continues, dehydration spreads throughout the body. When a dehydrated person drinks water and rehydrates, the water is redistributed by the same gradient, but in the opposite direction, replenishing water in all of the tissues.

26.1e Solute Movement between Compartments

Some solutes move between compartments via active transport, which requires energy, and other solutes move via passive transport mechanisms, which do not require energy. All molecules and substances will naturally tend to move along their concentration gradients, from areas of higher concentration to areas of lower concentration. Molecules moving along with their concentration gradient therefore do not need any cellular energy to move; they will do so on their own. Passive transport of a molecule or ion depends on its ability to pass through the membrane. Some molecules, such as gasses and lipids, slip fairly easily across the cell membrane; others, including polar molecules such as glucose, amino acids, and ions, do not. Some of these molecules enter and leave cells using facilitated transport, whereby the molecules move down a concentration gradient through specific protein channels in the membrane. This process does not

Figure 26.9 Sweating

Sweat is hypoosmotic to interstitial fluid. As more and more sweat is produced, for example on a hot day, the interstitial fluid loses more water than solutes. As the interstitial fluid becomes hyperosmotic and lower in volume and pressure, more fluid is drawn into the interstitial fluid from the blood.

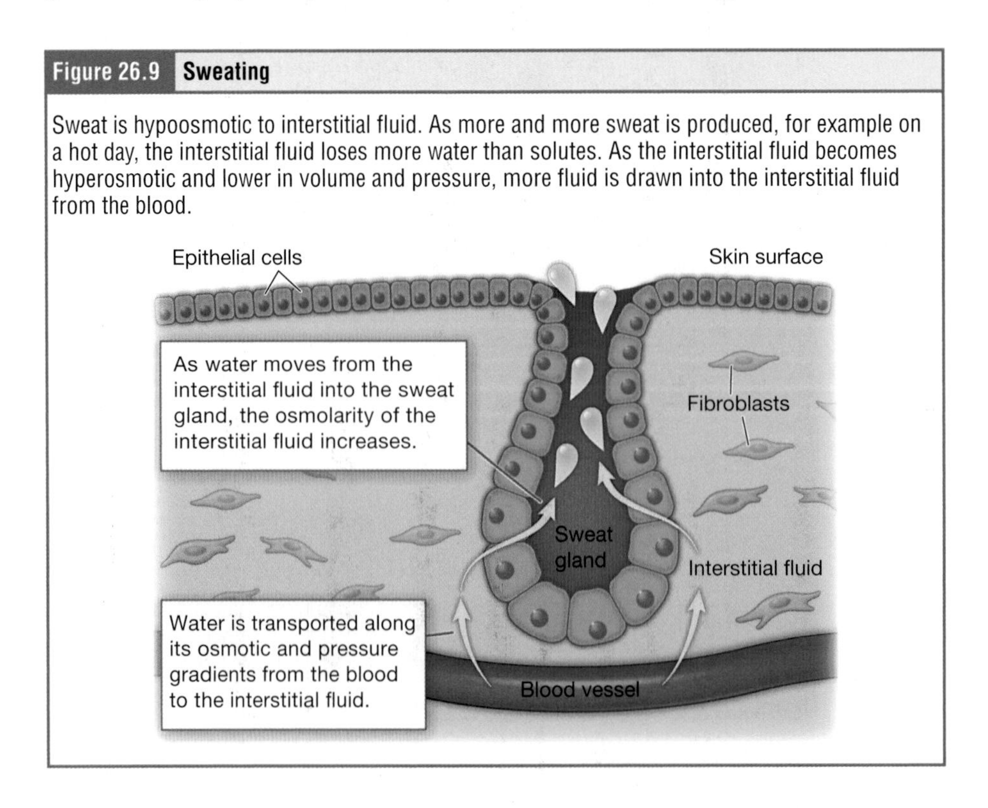
Epithelial cells
Skin surface
As water moves from the interstitial fluid into the sweat gland, the osmolarity of the interstitial fluid increases.
Fibroblasts
Sweat gland
Interstitial fluid
Water is transported along its osmotic and pressure gradients from the blood to the interstitial fluid.
Blood vessel

Figure 26.10 Facilitated Diffusion

Facilitated diffusion allows molecules to cross the cell membrane through a channel or a transporter, but always proceeds down the molecule's concentration gradient, that is to say, from an area of higher concentration to an area of lower concentration.

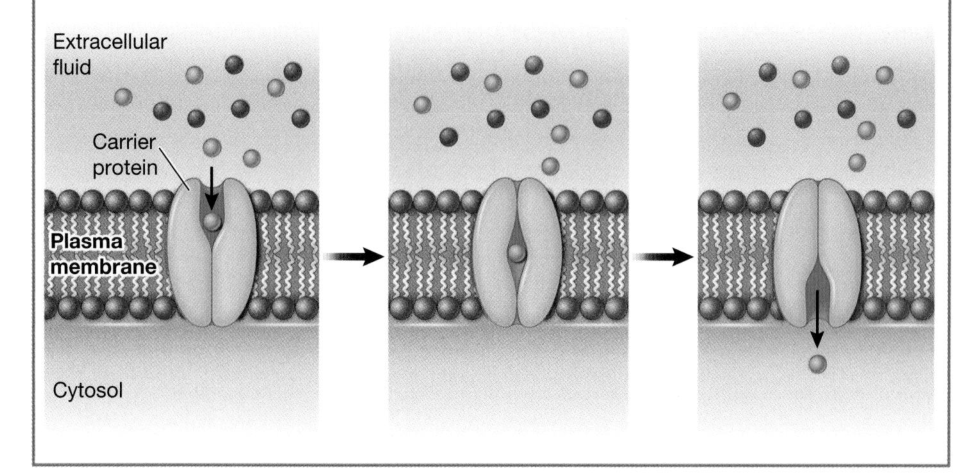

require energy. For example, glucose is transferred into cells by glucose transporters that use facilitated transport (**Figure 26.10**).

Active transport allows cells to move a specific substance against its concentration gradient through a membrane protein; this requires energy in the form of ATP. For example, the sodium-potassium pump employs active transport to pump sodium out of cells and potassium into cells, with both substances moving against their concentration gradients.

Student Study Tip

Think of passive versus active transport in the same way you think of rolling downhill versus rolling uphill. Whereas rolling downhill will happen naturally and does not need any external effort, rolling uphill requires effort and energy.

Learning Connection

Micro to Macro

Why is it absolutely necessary to carefully regulate the movement of electrolytes, such as sodium and potassium, between compartments? If these ions were allowed to pass freely across plasma membranes and the walls of blood vessels, their concentrations would become equal in the intracellular and extracellular fluid compartments. What effects would this have on basic life processes such as nerve impulse conduction and muscle contraction? Could life be maintained under these circumstances?

Cultural Connection

Sugar and Salt in the Western Diet

Dietary sugar (sucrose, glucose, or fructose) and salt (sodium chloride) are prevalent in the Western diet. These molecules all act as solutes in the bloodstream. Sugars are typically cleared from the blood relatively quickly, while salts may take a few hours to decrease in concentration as the kidneys work to rid the blood of excess solute levels. As the molecules are leaving the blood over time, they do change the osmotic gradient, making the plasma more hyperosmotic to the tissues, and drawing water into the bloodstream—increasing both blood volume and pressure.

Most Americans consume more dietary sugar and salt than they need on a daily basis. About 70–75 percent of Americans overconsume salt, which makes sense when you consider that approximately 65 percent of American adults have high blood pressure. When told by their doctors to consume less salt, many adults will decrease the amount of salt they sprinkle on their foods. However, the salt added to the food on our plates makes up just 11 percent of the average American's salt intake. The vast majority of our dietary salt comes from prepared foods. Since it is added to our foods before we even begin to prepare them, reducing salt intake can be hard to accomplish. Americans who have less time for cooking, less access to fresh foods, or are cultured to rely on packaged foods for a variety of reasons find it especially difficult to reduce their salt intake. Similarly, sugar is in a lot of our foods and drinks. While the recommended daily intake of sugar for an adult is 25–35 grams, the average American adult consumes approximately 70 grams! Sugar impacts water balance as well, having a variety of metabolic effects on the body.

Learning Check

1. Identify the anions.
   a. Na^+
   b. Cl^-
   c. K^+
   d. Ca^{2+}
2. In people with high-sodium diets, which of the following symptoms would you expect to see? Please select all that apply.
   a. High blood pressure
   b. Low blood volume
   c. Frequent urination
   d. Decreased heart rate
3. Explain how molecules move between compartments.
   a. Molecules move from low concentrations to high concentrations.
   b. Molecules move from high concentrations to low concentrations.
   c. Molecules remain in their designated compartments.
   d. Molecules remain in equal concentrations in the intracellular and extracellular fluid compartments.

26.2 Water Balance

Learning Objectives: By the end of this section, you will be able to:

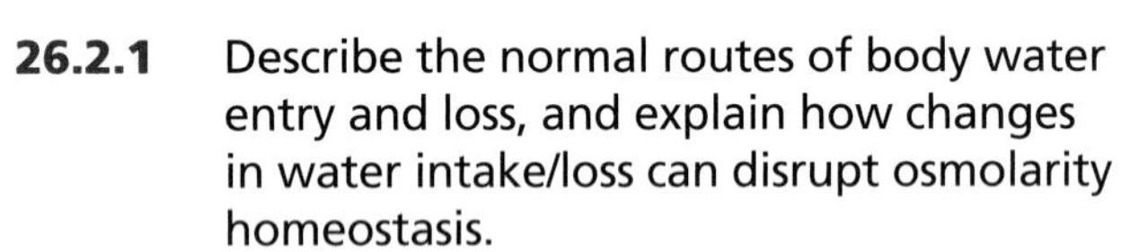

26.2.1 Describe the normal routes of body water entry and loss, and explain how changes in water intake/loss can disrupt osmolarity homeostasis.

26.2.2 Describe behavioral mechanisms that control water intake and loss.

26.2.3 Explain the role of hypothalamic osmoreceptors in regulation of body osmolarity.

26.2.4 Explain how the cardiovascular, endocrine, and urinary systems monitor blood volume and/or blood pressure.

26.2.5 Describe changes to body fluid compartment volumes and osmolarity when a person drinks a large volume of pure water, and then explain the compensatory mechanisms that attempt to restore normal volumes and osmolarity.

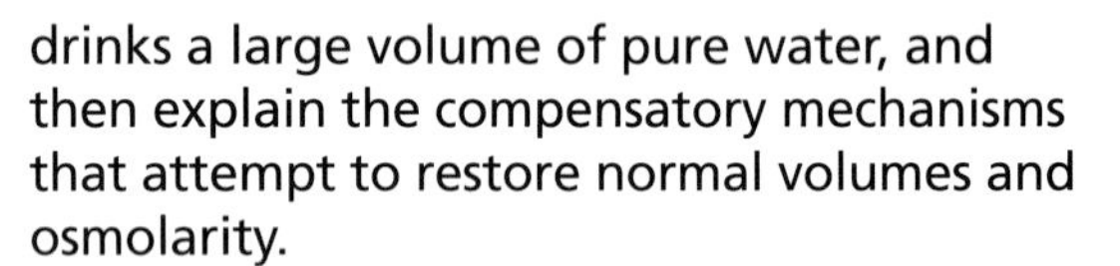

26.2.6 Explain the integrated responses of the cardiovascular, endocrine, and urinary systems to low blood pressure as a result of dehydration.

26.2.7* Compare and contrast dehydration and hemorrhage in terms of blood pressure, blood volume, blood osmolarity, and the compensatory mechanisms used to restore homeostasis.

* Objective is not a HAPS Learning Goal.

LO 26.2.1 Our intake of water or watery fluids can vary widely from day to day. Ideally, we will take in two to three liters of fluid through our diet. Although most of the intake comes through the digestive tract, a small amount per day is generated metabolically, in the last steps of aerobic cellular respiration. Additionally, each day about the same volume (2,500 milliliters) of water leaves the body by different routes; most of this lost water is removed as urine. Thus, it is important to take in a sufficient volume of water to cover these losses. The kidneys also can adjust blood volume though mechanisms that draw water out of the filtrate and urine. The kidneys can regulate water levels in the body; they conserve water if you are dehydrated, and they can make larger volumes of urine to expel excess water if necessary. Water is lost through the skin through evaporation even without sweating, but of course, sweating will increase the amount of water loss. Water is also lost through the respiratory system during breathing, speaking, and singing.

26.2a Regulation of Water Intake

Osmolarity is the ratio of solutes in a solution to a volume of solvent in a solution. *Plasma osmolarity* is thus the ratio of solutes to water in blood plasma. A person's plasma osmolarity value reflects their state of hydration. Plasma osmolarity is maintained within a narrow range by employing several mechanisms that regulate both water intake and output.

LO 26.2.2

Like breathing, drinking water is both voluntary/controllable and driven by homeostatic forces. Consider someone who is experiencing **dehydration**, a net loss of water that results in insufficient water in blood and other tissues. As the blood becomes more concentrated, the thirst response—a sequence of physiological processes—is triggered (Figure 26.11). **Osmoreceptors** are sensory receptors that monitor the

Student Study Tip

If you have ever had a drink that is too concentrated, think about how you could ask for it to be watered down so it would taste better. The osmoreceptors follow the same concept.

Figure 26.11 Thirst

Thirst is an example of a behavioral homeostatic mechanism in that it drives us toward an action.

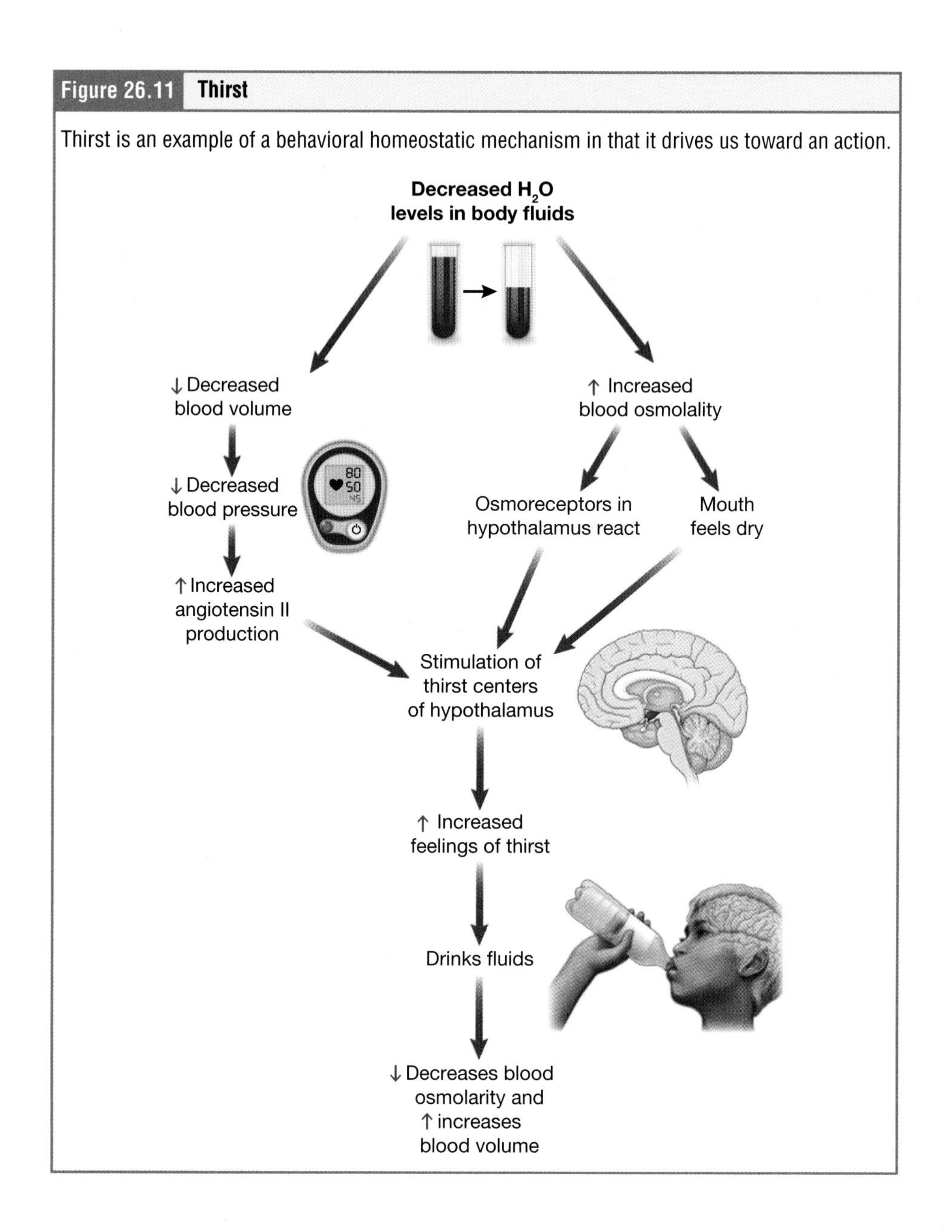

LO 26.2.3 concentration of solutes (osmolarity) of the blood. If blood osmolarity increases above its ideal value, the hypothalamus transmits signals that result in a conscious awareness of thirst. Thirst drives us to drink available beverages, ideally water. The hypothalamus of a dehydrated person also releases antidiuretic hormone (ADH) through the posterior pituitary gland. ADH signals the kidneys to recover water from urine, effectively diluting the blood plasma. To conserve water, the hypothalamus of a dehydrated person also sends signals via the sympathetic nervous system to the salivary glands in the mouth. The signals result in a decrease in watery, serous output (and an increase in stickier, thicker mucus output). These changes in secretions result in a "dry mouth" and the sensation of thirst.

LO 26.2.4 Decreased blood volume resulting from water loss has two additional effects. First, baroreceptors—blood-pressure receptors in the arch of the aorta and the carotid arteries in the neck—detect a decrease in blood pressure that results from decreased blood volume. The heart is ultimately signaled to increase its rate and/or strength of contractions to compensate for the lowered blood pressure. Second, the blood levels of both angiotensin II and aldosterone increase. Aldosterone increases the reabsorption of sodium in the distal tubules of the nephrons in the kidneys, and water follows this reabsorbed sodium back into the blood. Angiotensin II increases the sensation of thirst.

Apply to Pathophysiology

Water Intoxication

LO 26.2.5

In 2007, a 28-year-old individual died during a competition held by a radio station in California. The prize for the winner of the contest was going to be a popular (and hard-to-find) video game system. The contestants were asked to drink as much water as they could without urinating. Let's explore the pathophysiology that led to one contestant's death. As the contestants drank pure water quickly, it entered their GI tract, and was absorbed across their gut walls, entering their bloodstream. (Note: the contestants were drinking gallons of water quickly, not typical amounts).

1. Which of the following words describes the plasma ECF of the contestants compared to the ICF of their red blood cells?
   A) Hyperosmotic
   B) Hypoosmotic
   C) Isosmotic

During the first 20 minutes of the contest we can assume that the absolute amount of sodium in the blood has not changed, but the amount of water has been increasing.

2. Therefore, we can say that the concentration of sodium is ___.
   A) Decreasing
   B) Increasing
   C) No change
3. What is happening to the contestants' blood volume?
   A) Decreasing
   B) Increasing
   C) No change
4. What is happening to the contestants' blood pressure?
   A) Decreasing
   B) Increasing
   C) No change

Other fluid compartments, including the interstitial fluid and cerebrospinal fluid, are physiologically linked to the blood. As blood volume changes, these linked compartments may change in volume as well.

5. Over the first hour, what will happen to the interstitial fluid volume of the tissues?
   A) Decrease
   B) Increase
   C) No change
6. What would happen to the cerebrospinal fluid volume?
   A) Decrease in volume
   B) Increase in volume
   C) No change in volume

If adequate fluids are not consumed, dehydration results and a person's body contains too little water to function correctly. A person who repeatedly vomits or who has diarrhea may become dehydrated, and infants, because their body mass is so low, can become dangerously dehydrated very quickly. Endurance athletes such as distance runners often become dehydrated during long races. Dehydration can be a medical emergency, and a dehydrated person may lose consciousness, become comatose, or die, if their body is not rehydrated quickly.

Recall from Chapter 14 that the brain and CSF are encased in the hard bony shell of the skull. If the IF volume changes, we can sometimes see these changes in edema (swelling) or noticeable sagging of the skin, which is superficial to the IF. Changes in CSF volume are not noticeable from the outside, but increases or decreases in CSF can adversely impact the brain, even leading to death. Under typical circumstances the kidneys will adjust for ingestion of too much fluid. **Diuresis**, which is the production of urine in high volumes, begins about 30 minutes after drinking a large quantity of fluid. In the case of this contest, without being able to shed the excess fluid in the urine, the fluid volume increase became harmful to tissue and organ function and eventually brain function. A **diuretic** is a compound that increases urine output and therefore decreases water conservation by the body. Diuretics are used to treat hypertension, congestive heart failure, and fluid retention associated with menstruation. Alcohol acts as a diuretic by inhibiting the release of ADH. Additionally, caffeine, when consumed in high concentrations, acts as a diuretic.

26.2b Regulation of Water Output

Water loss from the body occurs predominantly through the urine. Urine volume varies widely in response to hydration levels, but there is a minimum volume of urine production required for proper bodily functions. The kidney must get rid of solutes, including excess salts and other water-soluble chemical wastes, most notably creatinine, urea, and uric acid. In order to carry these wastes away, some water must be used to create urine, even during dehydration. If, in cases of extreme dehydration, the body cannot spare enough water to produce the minimum volume of urine, metabolic wastes will build up in the body, resulting in systemic toxicity and compromised organ function.

26.2c Role of ADH

Antidiuretic hormone (ADH), also known as *vasopressin*, controls the availability of aquaporins, and therefore the amount of water reabsorbed from the collecting ducts and tubules in the kidney. This hormone is produced in the hypothalamus and is delivered to the posterior pituitary for storage and release (**Figure 26.12**). When the osmoreceptors in the hypothalamus detect an increase in the solute concentration of blood plasma, the hypothalamus signals for the release of ADH from the posterior pituitary into the blood. When the blood plasma becomes less concentrated and the level of ADH decreases, aquaporins are removed from collecting tubule cell membranes, and the passage of water out of urine and into the blood decreases.

antidiuretic hormone (ADH) (vasopressin)

The control over aquaporin expression and therefore water reabsorption is the major effect of ADH. But ADH also has another effect. It constricts the arterioles in the peripheral circulation, which reduces the flow of blood to the extremities and thereby increases the blood supply to the core of the body.

Figure 26.12 Antidiuretic Hormone (ADH)

ADH is a hormone released by the posterior pituitary that increases water reabsorption in the kidney tubule. ADH is also a vasoconstrictor that increases systemic blood pressure.

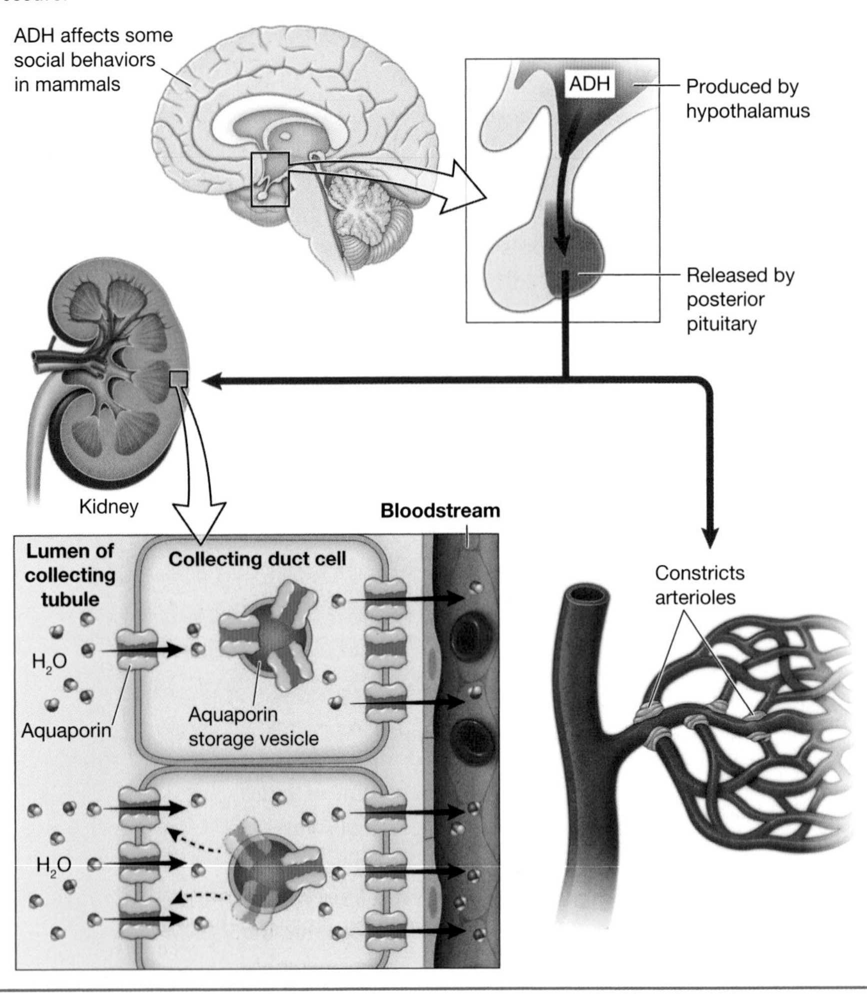

Apply to Pathophysiology

Hemorrhage and Dehydration

LO 26.2.6 LO 26.2.7

Two runners, Casey and Carlie, enroll in a trail running race. On the day of the race, Casey had been out the night before celebrating a friend's recent engagement and drunk more alcohol than they had intended. They are feeling dehydrated and thirsty this morning as the race begins. It's a hot day and Casey is already sweating profusely. Carlie, on the other hand, is well-hydrated and running fast, a personal record in minutes per mile, as the race begins.

Casey: After an hour of intense running in the heat, Casey is rapidly becoming dehydrated.

Carlie: One hour into the race, Carlie slips on a wet rock on the trail. Carlie falls hard and cuts their leg on an exposed ragged stone edge and is bleeding profusely (hemorrhage).

1. What is happening to Casey's blood volume?
   A) It is increasing.
   B) It is decreasing.
   C) There is no change.
2. What is happening to Carlie's blood volume?
   A) It is decreasing.
   B) It is increasing.
   C) There is no change.
3. What is happening to Casey's blood pressure?
   A) It is increasing.
   B) It is decreasing.
   C) There is no change.

4. What is happening to Carlie's blood pressure?
   A) It is decreasing.
   B) It is increasing.
   C) There is no change.
5. What is happening to Casey's blood osmolarity?
   A) It is increasing.
   B) It is decreasing.
   C) There is no change.
6. What is happening to Carlie's blood osmolarity?
   A) It is increasing.
   B) It is decreasing.
   C) There is no change.

The race has EMTs ready and Casey and Carlie both get help. For each runner, the EMTs are trying to figure out what to do to help. Their options are an IV of an isosmotic saline solution, an IV of a slightly hypoosmotic saline solution, or a blood transfusion.

7. Which treatment is best for Casey?
   A) Blood transfusion
   B) Hypoosmotic IV saline
   C) No treatment
8. Which treatment is best for Carlie?
   A) Blood transfusion
   B) Isosmotic IV saline
   C) No treatment

Learning Check

1. Which of the following scenarios could lead to dehydration? Please select all that apply.
   a. High concentrations of sodium in the bloodstream
   b. Low concentrations of electrolytes in the bloodstream
   c. High concentrations of fluid in the bloodstream
   d. Low concentrations of fluid in the bloodstream
2. Define plasma osmolarity.
   a. The difference between solutes and fluid in the blood
   b. The difference in water concentrations between two locations
   c. The ratio of solute concentrations in two different locations
   d. The ratio of solutes to fluid in the blood
3. Explain how alcohol consumption can lead to increased urination.
   a. Alcohol increases the glomerular filtration rate in the nephron.
   b. Alcohol stimulates the release of antidiuretic hormone.
   c. Alcohol decreases solute content in the bloodstream.
   d. Alcohol decreases the rate of water reabsorption in the kidneys.

26.3 Electrolyte Balance

Learning Objectives: By the end of this section, you will be able to:

26.3.1 Explain the importance of maintaining potassium homeostasis with regard to membrane potential, and provide examples of dysfunction that occur when plasma potassium levels are too high or too low.

26.3.2 Explain the importance of maintaining calcium homeostasis, and provide examples of dysfunction that occur when plasma calcium levels are too high or too low.

26.3.3 Describe the integrated responses of the endocrine and urinary systems to disruptions of potassium homeostasis.

26.3.4 Describe the integrated responses of the endocrine, digestive, skeletal, and urinary systems to disruptions of calcium homeostasis.

The body contains a large variety of ions, which perform a variety of functions. Some ions assist in the transmission of electrical impulses along cell membranes in neurons and muscles. Other ions help to stabilize protein structures in enzymes. Still others aid in releasing hormones from endocrine glands. All of the ions in plasma contribute to the osmotic balance that controls the movement of water between cells and their environment.

Electrolytes in living systems include sodium, potassium, chloride, bicarbonate, calcium, phosphate, magnesium, copper, zinc, iron, manganese, molybdenum, copper, and chromium. In terms of body functioning, six electrolytes are most important: sodium, potassium, chloride, bicarbonate, calcium, and phosphate (**Table 26.1**)

26.3a Roles of Electrolytes

These six ions aid in nerve excitability, endocrine secretion, membrane permeability, buffering body fluids, and controlling the movement of fluids between compartments. These ions enter the body through the digestive tract. More than 90 percent of the calcium and phosphate that enters the body is incorporated into bones and teeth, with bone serving as a mineral reserve for these ions. In the event that calcium and phosphate are needed for other functions, bone tissue can be broken down to supply the blood and other tissues with these minerals. Phosphate is a component of nucleic acids; and therefore, blood levels of phosphate will increase whenever nucleic acids are broken down, such as in times of injury or some cancers.

Small amounts of ions are lost from the body in sweat, but remember that sweat is more dilute than body fluids, so the main loss during sweating is water. The primary route of ions loss is through the urine. But when illness strikes and vomiting or diarrhea occur, large amounts of chloride and bicarbonate ions can result.

Table 26.2 lists the range of values for blood plasma, CSF, and urine for the six ions addressed in this section. In a clinical setting, sodium, potassium, and chloride are typically analyzed in a routine urine sample. In contrast, calcium and phosphate analysis requires a collection of urine across a 24-hour period, because the output of these ions can vary considerably over the course of a day. Urine values reflect the rates of excretion of these ions. Bicarbonate is the one ion that is not normally excreted in urine because it is needed for use in the body's buffering systems.

Sodium Sodium is the most common ion in the ECF. Most individuals take in far more sodium through their diet than they need, and excess sodium is cleared from the body in the urine.

Table 26.1 Roles of the Big Six Ions

Name	Symbol	Common in	Role(s)
Sodium	Na^+	ECF	Action potentials, urine formation, blood/body fluid osmolarity, muscle contraction, membrane transport
Potassium	K^+	ICF	Action potentials, muscle contraction
Chloride	Cl^-	ECF	Osmotic balance between ICF and ECF, electrical balance of ECF, function of some neurons
Bicarbonate	HCO^-	Blood/ECF	Buffering acidic changes in body fluids
Calcium	Ca^{2+}	Bones, ECF	Muscle contraction, neurotransmitter release, enzyme activity, clotting
Phosphate	HPO_4^{2+}	Bones, ATP, nucleotides, proteins, phospholipids, ICF	Regulation of protein function, structure

Table 26.2 Reference Values for the Big Six Ions

Name	Chemical Symbol	Plasma	CSF	Urine
Sodium	Na^+	136.00–146.00 (mM)	138.00–150.00 (mM)	40.00–220.00 (mM)
Potassium	K^+	3.50–5.00 (mM)	0.35–3.5 (mM)	25.00–125.00 (mM)
Chloride	Cl^-	98.00–107.00 (mM)	118.00–132.00 (mM)	110.00–250.00 (mM)
Bicarbonate	HCO_3^-	22.00–29.00 (mM)	------	------
Calcium	Ca^{2+}	2.15–2.55 (mmol/day)	------	Up to 7.49 (mmol/day)
Phosphate	HPO^{2-}_4, HPO_4^{2-}	0.81–1.45 (mmol/day)	------	12.90–42.00 (mmol/day)

Hyponatremia or low blood levels of sodium, is usually associated with excess water accumulation in the body, which dilutes the sodium. Other causes of hyponatremia stem from the unusual loss of sodium from the body that would occur during excessive sweating, vomiting, or diarrhea. Excessive production of urine, which can occur in diabetes can also lead to excessive loss of sodium.

Hypernatremia is high blood levels of sodium. It can result from water loss from the blood, resulting in the hemoconcentration of all blood constituents. Hormonal imbalances involving two of the hormones that regulate urine formation—ADH and aldosterone—may also result in higher-than-normal sodium values.

Potassium Potassium is the most common ion found in the ICF. It helps establish the resting membrane potential in neurons and muscle fibers after membrane depolarization and action potentials. The low levels of potassium in blood and CSF are due to the sodium-potassium pumps in cell membranes, which maintain the normal potassium concentration gradients between the ICF and ECF.

LO 26.3.1

Hypokalemia is the term for a low blood level of potassium. Similar to hyponatremia, hypokalemia can occur because of either a reduction of potassium in the body or a change in concentration due to changes in water volume. Loss of potassium can stem from decreased dietary intake, which might happen in starvation. It can also come about from vomiting, diarrhea, or alkalosis.

Hyperkalemia, an elevated blood level of potassium, can impair the function of skeletal muscles, the nervous system, and the heart. Hyperkalemia can result from increased dietary intake of potassium. In such a situation, potassium from the blood ends up in the ECF in abnormally high concentrations.

Chloride Chloride is the most common anion found in the ECF. Chloride is a major contributor to the osmotic pressure gradient between the ICF and ECF, and plays an important role in maintaining proper hydration. Chloride functions to balance cations in the ECF, maintaining the electrical neutrality of this fluid. The paths of secretion and reabsorption of chloride ions in the renal system follow the paths of sodium ions.

Hypochloremia, or low blood chloride levels, can occur because of issues in the kidney nephron. Vomiting, diarrhea, and metabolic acidosis can also lead to hypochloremia. **Hyperchloremia**, or high blood chloride levels, can occur due to dehydration, excessive intake of dietary salt (NaCl) or swallowing of seawater, and cystic fibrosis.

Learning Connection

Broken Process

You have just learned about some of the vital functions of chloride ions in the human body. In Chapter 23, you learned that chloride ions are a component of hydrochloric acid, which is secreted by the stomach. What specific digestive issues would you expect to occur in a person that developed hypochloremia resulting from renal disease?

Bicarbonate Bicarbonate is an anion that is common in the blood. Its principal function is to maintain your body's acid-base balance by being part of buffer systems. This role will be discussed in a different section.

Bicarbonate ions result from a chemical reaction that starts with carbon dioxide (CO_2) and water, two molecules that are produced at the end of cellular respiration. Only a small amount of CO_2 can be dissolved in bodily fluids. Thus, over 90 percent of the CO_2 is converted into bicarbonate ions, HCO_3^-, through the following reactions:

$$CO_2 + H_2O \leftrightarrow HCO_3^- + H^+$$

The bidirectional arrows indicate that the reactions can go in either direction, depending on the concentrations of the reactants and products. Carbon dioxide is produced in large amounts in tissues that have a high metabolic rate, such as skeletal muscles during exercise. Carbon dioxide is converted into bicarbonate in the cytoplasm of red blood cells through the action of an enzyme called **carbonic anhydrase**. Bicarbonate is transported in the blood. Once in the lungs, the reactions reverse direction, and CO_2 is regenerated from bicarbonate to be exhaled.

carbonic anhydrase (carbonate dehydratases)

LO 26.3.2

Calcium Most of the calcium in your body is bound up in bone, contributing to the hardness of the bone. Bones also serve as a mineral reserve for calcium and are able to liberate calcium and add it to the blood when blood levels are low. Teeth also have a high concentration of calcium within them. A little more than one-half of blood calcium is bound to proteins, leaving the rest free floating as calcium ions (Ca^{2+}). Calcium ions are necessary for muscle contraction, enzyme activity, and blood clotting. In addition, calcium helps to stabilize cell membranes and is essential for the release of neurotransmitters from neurons and of hormones from endocrine glands.

Calcium is absorbed through the intestines under the influence of activated vitamin D. A deficiency of vitamin D leads to a decrease in absorbed calcium and, eventually, a depletion of calcium stores from the skeletal system, potentially leading to rickets in children and osteomalacia in adults, contributing to osteoporosis.

Hypocalcemia, or abnormally low blood calcium levels, is seen in hypoparathyroidism, which may follow the removal of the thyroid and parathyroid glands. **Hypercalcemia**, or abnormally high blood calcium levels, is seen in primary hyperparathyroidism. Some forms of cancer may also result in hypercalcemia.

Phosphate Phosphate is present in the body in three ionic forms: $H_2PO_4^-$, HPO_4^{2-}, and PO_4^{3-}. The most common form is HPO_4^{2-}. Bone and teeth bind up 85 percent of the body's phosphate as part of calcium-phosphate salts. Phosphate is found in phospholipids, such as those that make up the cell membrane, as well as in ATP, nucleotides, and buffers.

Hypophosphatemia, or abnormally low blood phosphate levels, occurs with heavy use of antacids, during alcohol withdrawal, and during starvation. **Hyperphosphatemia**, or abnormally increased levels of phosphates in the blood, occurs in kidney failure. Additionally, because phosphate is a major constituent of the ICF, any significant destruction of cells (for example, in injury) can result in dumping of phosphate into the ECF.

LO 26.3.3

26.3b Regulation of Sodium and Potassium

Sodium and potassium levels are kept within homeostatic ranges by the kidneys. If an excess of either of these ions is present in the blood, it can be released into the urine, thereby lowering its level in the blood. When blood levels of these ions are low, the kidneys adjust by reabsorbing as much as possible. Because homeostasis depends on high sodium levels in the blood, most sodium is reabsorbed from the renal filtrate. Likewise, homeostasis depends on low extracellular potassium levels; when potassium levels become excessive, potassium ions are secreted into the filtrate in the renal collecting tubule and exit the body in the urine. The control of this exchange is governed principally by two hormones—aldosterone and angiotensin II—which we also discussed in Chapter 25.

Angiotensin II Angiotensin II is a potent vasoconstrictor and therefore a useful tool to increase systemic blood pressure. This action increases the glomerular filtration rate, resulting in more material filtered out of the glomerular capillaries and into Bowman's capsule. Angiotensin II also signals an increase in the release of aldosterone from the adrenal cortex.

Aldosterone Recall that aldosterone increases the excretion of potassium and the reabsorption of sodium in the distal tubule. Aldosterone is released into the blood when blood levels of potassium increase, if blood levels of sodium severely decrease, or if blood pressure decreases. Its net effect is to conserve and increase water levels in the plasma by reducing the excretion of sodium, and thus water, from the kidneys. In a negative feedback loop, increased osmolality of the ECF (which follows aldosterone-stimulated sodium absorption) inhibits the release of aldosterone (**Figure 26.13**).

In the distal convoluted tubules and collecting ducts of the kidneys, aldosterone stimulates the synthesis and activation of the sodium-potassium pump. Sodium passes from the filtrate, into and through the cells of the tubules and ducts, into the ECF, and then into capillaries. Water follows the sodium due to osmosis. Thus, aldosterone causes an increase in blood sodium levels and blood volume. Aldosterone's effect on potassium is the reverse of that of sodium; under its influence, excess potassium is pumped into the renal filtrate for excretion from the body.

26.3c Regulation of Calcium and Phosphate

Calcium and phosphate are both regulated through the actions of three hormones: parathyroid hormone (PTH), dihydroxyvitamin D (calcitriol), and calcitonin (**Figure 26.14**). All three are released or synthesized in response to the blood levels of calcium.

Figure 26.14 Blood Calcium Homeostasis

Blood calcium levels are regulated by the actions of three hormones: calcitonin, parathyroid hormone, and dihydroxyvitamin D.

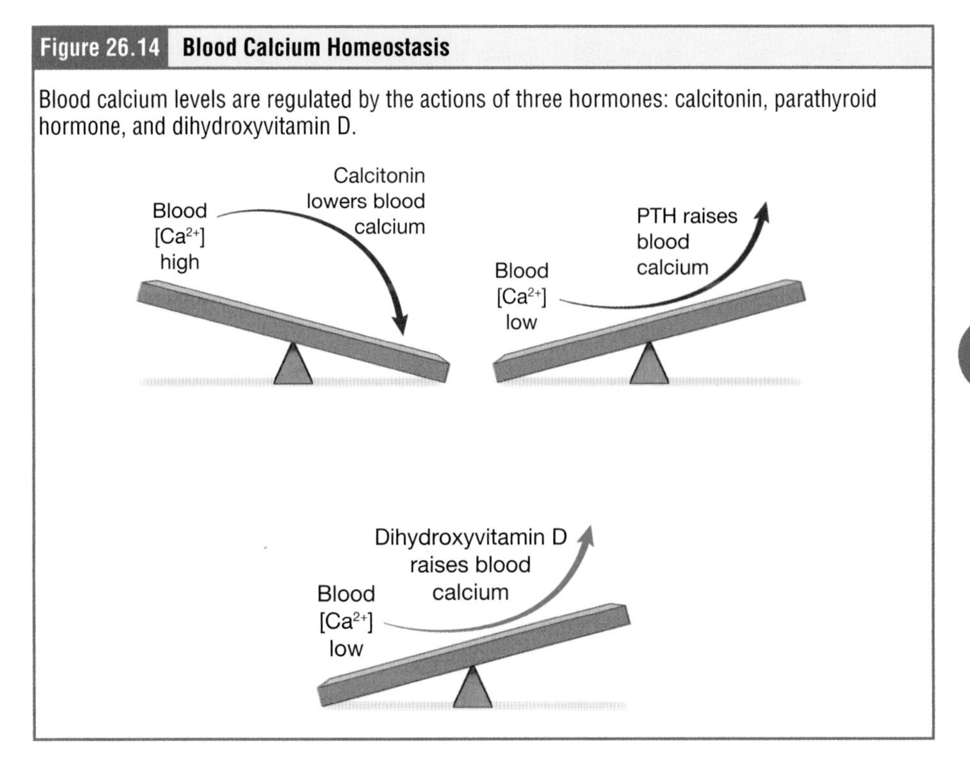

Student Study Tip

You may have heard of blood pressure medications called *ACE inhibitors*. These prevent the formation of angiotensin II, which lowers blood pressure. Think about what you know of these electrolytes to help you understand what would happen if they were not properly regulated.

Figure 26.13 The Aldosterone Feedback Loop

The adrenal glands release aldosterone when stimulated by a change in the concentration of Na^+ or K^+. Aldosterone acts on the kidney tubules, where it increases the reabsorption of Na^+ and thus the reabsorption of water.

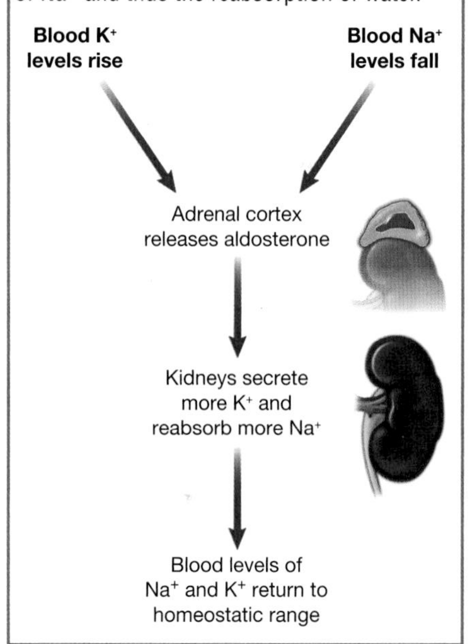

LO 26.3.4

Learning Connection

Quiz Yourself

Hormones play vital roles in water and electrolyte balance. Now that you have studied the effects of antidiuretic hormone, aldosterone, parathyroid hormone, and calcitonin. make a list of these hormones. Then see if you can write down the effects of each hormone on the blood level of the substance(s) that each hormone control(s). Hint: think of water, sodium, potassium, calcium, and phosphate.

PTH is released from the parathyroid gland in response to a decrease in the concentration of blood calcium. The hormone activates osteoclasts to break down bone matrix and release inorganic calcium-phosphate salts. PTH also increases the gastrointestinal absorption of dietary calcium by converting vitamin D into **dihydroxyvitamin D** (calcitriol), an active form of vitamin D that intestinal epithelial cells require to absorb calcium.

dihydroxyvitamin D (calcitriol, 1, 25-dihydroxycholecalciferol)

PTH raises blood calcium levels by inhibiting the loss of calcium through the kidneys. PTH also increases the loss of phosphate through the kidneys.

Calcitonin is released from the thyroid gland in response to elevated blood levels of calcium. The hormone promotes the removal of calcium from the blood and the incorporation of calcium into the bony matrix.

Learning Check

1. Which of the following electrolytes functions to balance cations in the extracellular fluid and maintain electrical neutrality of the fluid?
   a. Sodium
   b. Potassium
   c. Chloride
   d. Bicarbonate
2. Which of the following is a possible cause of hypokalemia?
   a. Vomiting
   b. Weight gain
   c. Blood in urine
   d. Low heart rate
3. Which of the following hormones is responsible for regulating sodium and potassium levels?
   a. Parathyroid hormone
   b. Dihdyroxyvitamin D
   c. Calcitonin
   d. Aldosterone
4. What is the role of calcitonin in the body?
   a. Activation of osteoclasts
   b. Activation of osteoblasts
   c. Absorption of calcium from the small intestine
   d. Reabsorption of sodium by the kidneys

26.4 Acid-Base Balance

Learning Objectives: By the end of this section, you will be able to:

26.4.1* Explain how changes in pH outside the normal range adversely affect body functions and include the normal pH range in arterial blood.

26.4.2 Describe the major buffer systems of the body (e.g., bicarbonate buffer system, protein buffer system) and their locations (e.g., extracellular fluid) in the body.

26.4.3 Explain the relationship between transport of carbon dioxide in the blood and the bicarbonate buffer system in the plasma.

26.4.4 Using the equation $CO_2 + H_2O \leftrightarrow H^+ + HCO_3^-$, explain what happens to pH when arterial blood pCO_2 and HCO_3^- concentrations change.

* Objective is not a HAPS Learning Goal.

Physiological function depends on a very tight balance between the concentrations of acids and bases in the blood. Acid-base balance is measured using the **pH scale**, as shown in **Figure 26.15**. Acids and (to a lesser degree) bases are produced during various metabolic activities. A **buffer** is a solution or compound that resists changes to pH, usually by binding the H^+ ions that can increase acidity or hydroxyl ions that can increase alkalinity. A variety of buffer systems exist in the blood and other bodily fluids and help maintain a narrow pH range in the body. Acid-base homeostasis is so critical for cellular function because our cells are dependent upon their proteins. Membrane proteins, like the Na^+/K^+ pump, contribute to transport and maintenance of membrane potential. Enzymes catalyze metabolic reactions. Contractile proteins allow skeletal, smooth, and cardiac muscle contractions to occur. Structural proteins like collagen form the scaffolding of the tissues of the body. Yet, all of these proteins are vulnerable to pH changes. Proteins will denature—that is to say, their structure will unfold and fall apart—and they will be unable to perform their function when changes away from a neutral pH are sustained in body fluids.

LO 26.4.1

Student Study Tip

Membrane proteins are like very type-A people who need things as specifically controlled as possible.

Remember that carbon dioxide, which is produced during cellular respiration, is converted to bicarbonate and hydrogen ions in water or blood when carbonic anhydrase is present. The formula, also stated earlier in this chapter, looks like this:

$$CO_2 + H_2O \leftrightarrow HCO_3^- + H^+$$

The renal and respiratory systems both are able to rid the body of molecules in order to correct for changes in pH in the blood. In addition, several bodily fluids, such as the blood, have buffering systems in place. Arterial blood maintains a pH of 7.35–7.45. Outside of this range, physiological compensatory mechanisms must kick in to correct the pH back to this range.

LO 26.4.2

Protein Buffers in Blood Plasma and Cells Nearly all proteins can function as buffers. Proteins are made up of amino acids, which contain positively charged amino groups and negatively charged carboxyl groups. The charged regions of these molecules can bind hydrogen and hydroxyl ions, and thus function as buffers. Buffering by proteins accounts for two-thirds of the buffering power of the blood and most of the buffering within cells.

Student Study Tip

Protein buffers are like buffers in a social situation involving two people who probably shouldn't interact with one another, so there is a person in between who is working to prevent a negative conversation.

Hemoglobin as a Buffer Hemoglobin is the principal protein inside of red blood cells and accounts for one-third of the mass of the cell. During the conversion of CO_2 into bicarbonate and hydrogen ions, the hydrogen ions produced can be bound to hemoglobin when the protein is not bound by oxygen. This buffering helps maintain normal pH. The process is reversed in the pulmonary capillaries to reform CO_2, which then can diffuse into the air sacs to be exhaled into the atmosphere. This process is discussed in more detail in Chapter 22.

LO 26.4.3

Bicarbonate-Carbonic Acid Buffer The bicarbonate-carbonic acid buffer works when sodium bicarbonate ($NaHCO_3$) comes into contact with a strong acid, such as HCl; this results in the formation of carbonic acid (H_2CO_3), which is a weak acid, and NaCl. When carbonic acid comes into contact with a strong base, such as NaOH, bicarbonate and water are formed.

$$NaHCO_3 + HCl \rightarrow H_2CO_3 + NaCl$$

(sodium bicarbonate) + (strong acid) → (weak acid) + (salt)

$$H_2CO_3 + NaOH \rightarrow HCO_3^- + H_2O$$

(weak acid) + (strong base) → (bicarbonate) + (water)

Bicarbonate ions and carbonic acid are present in the blood in a 20:1 ratio if the blood pH is within the normal range. With 20 times more bicarbonate than carbonic

Figure 26.15 The pH Scale

The H^+ ion concentration of any solution can be measured and determines the pH value of the solution.

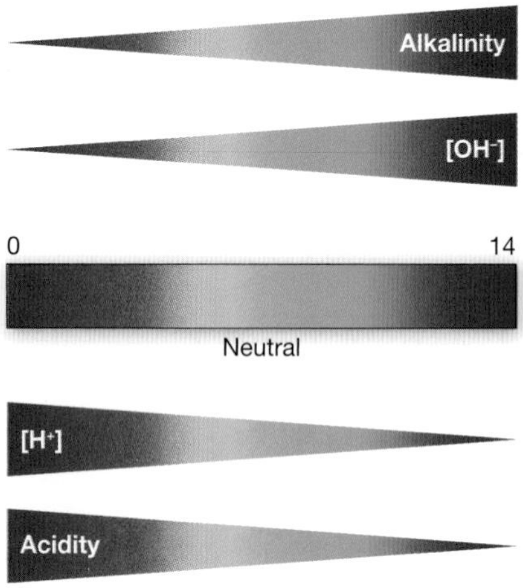

Alkalinity

$[OH^-]$

0

14

Neutral

$[H^+]$

Acidity

acid, this capture system is most efficient at buffering changes that would make the blood more acidic. This is useful because most of the body's metabolic wastes, such as lactic acid and ketones, are acids. Carbonic acid levels in the blood are controlled by the expiration of CO_2 through the lungs. The level of bicarbonate in the blood is controlled through the renal system, where bicarbonate ions in the renal filtrate are conserved and passed back into the blood. However, the bicarbonate buffer is the primary buffering system of the IF surrounding the cells in tissues throughout the body.

26.4a Respiratory Regulation of Acid-Base Balance

The respiratory system contributes to the balance of acids and bases in the body by regulating the blood levels of carbonic acid (**Figure 26.16**). Because CO_2 in the blood reacts with water to form carbonic acid, when the CO_2 level in the blood rises the excess CO_2 reacts with water to form additional carbonic acid, lowering blood pH. Increasing the rate and/or depth of respiration allows you to exhale more CO_2. The loss of CO_2 from the body reduces blood levels of carbonic acid and thereby adjusts the pH upward, toward the normal range. Increases in CO_2 in the blood will occur during increases in cellular respiration, such as during exercise, or when the breath rate is insufficient to rid the body of CO_2, such as when an individual holds their breath. This process also

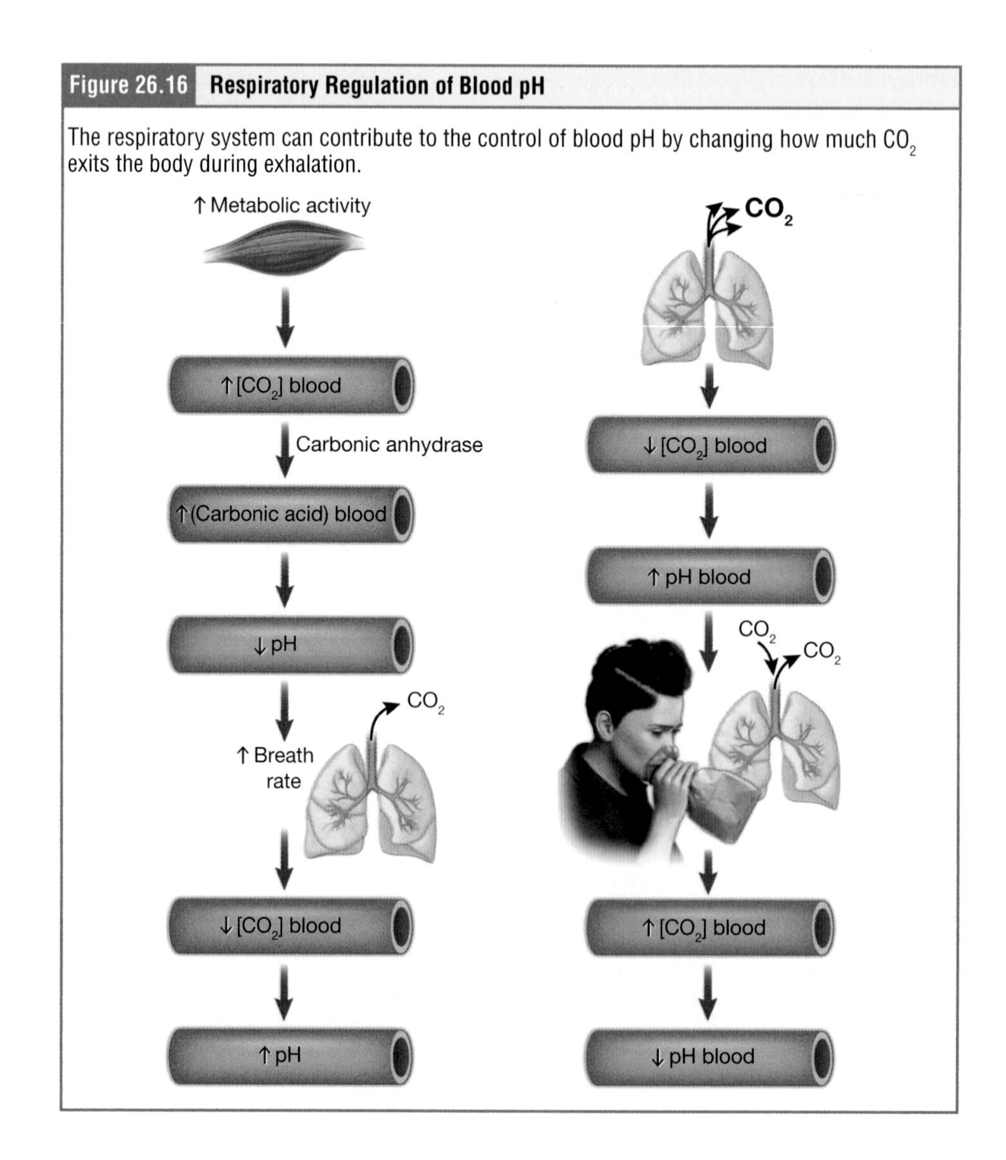

Figure 26.16 Respiratory Regulation of Blood pH

The respiratory system can contribute to the control of blood pH by changing how much CO_2 exits the body during exhalation.

works in the opposite direction. Excessive deep and rapid breathing (as in hyperventilation) gets rid of too much CO_2 and reduces the level of carbonic acid, making the blood too alkaline. Individuals who are hyperventilating are often encouraged to breathe into a paper bag. This "rebreathing" of expelled air rapidly brings blood pH down toward the normal range by increasing the amount of CO_2 within the air in the lungs.

The blood CO_2 concentration is monitored by chemoreceptors found in the walls of the aorta and carotid arteries. These sensors signal the brain to provide immediate adjustments to the respiratory rate if CO_2 levels rise or fall. Additional sensors are found in the brain itself. Changes in the pH of CSF are detected here. Information about deviations in CO_2 or H^+ concentration are sent to the respiratory center in the medulla oblongata, which directly influences the breath rate to bring the pH back into the normal range.

Hypercapnia, or abnormally high blood levels of CO_2, occurs in any situation that impairs respiratory functions, including pneumonia and congestive heart failure. Reduced breathing (hypoventilation) can occur during exposure to opiates (e.g., heroin), or even just by holding one's breath, and will lead to hypercapnia. **Hypocapnia**, or abnormally low blood levels of CO_2, occurs anytime the breath rate exceeds what is required to expel CO_2, such as panic-induced hyperventilation or fever.

hypercapnia (hypercarbia)

hypocapnia (hypocarbia)

26.4b Renal Regulation of Acid-Base Balance

The kidneys are central regulators of the blood pH. They are able to adjust both the H^+ and HCO_3^- levels by getting rid of excessive levels of these ions in the urine.

Both bicarbonate and hydrogen ions are filtered out of the blood at the glomerulus. Under most physiological circumstances, the kidneys will reabsorb much of the bicarbonate, while secreting hydrogen ions, as illustrated in **Figure 26.17** and summarized below:

- Step 1: Sodium ions are reabsorbed from the filtrate in exchange for H^+ by an antiport mechanism in the apical membranes of cells lining the renal tubule.
- Step 2: The cells produce bicarbonate ions that can be released toward peritubular capillaries.
- Step 3: When CO_2 is available, the reaction is driven to the formation of carbonic acid, which dissociates to form a bicarbonate ion and a hydrogen ion.
- Step 4: The bicarbonate ion passes into the peritubular capillaries and returns to the blood. The hydrogen ion is secreted into the filtrate, where it can either become part of new water molecules and be reabsorbed as such, or be removed in the urine.

Cultural Connection

Choosing Your Beverages

It is fair to assume that, for the majority of the millions of years of human evolution, water was the only beverage consumed by humans. Alcoholic beverages began to be made and consumed by humans about 10,000 years ago. Cow's milk has been consumed by humans for about 6,000 years, and tea in its various forms has been a staple of the human diet for at least 5,000 years. Coffee was added to the human beverage menu only about 500 years ago, and carbonated beverages such as soda joined the diet about 250 years ago. These beverages have dramatically changed the human diet. Alcohol, cow's milk, and soda all include sugars that people may not realize they are consuming when they drink. With the exception of milk, all of these drinks are acidic, changing the acid-base balance of the mouth. The acid and sugar combination is particularly detrimental to the environment of our mouths, allowing bacteria to thrive on the sugar while the acidic environment weakens enamel, leading to the risk of cavities.

Figure 26.17 Bicarbonate Regulation in the Kidney

Bicarbonate ions cannot traverse the cells of the renal tubules of the kidney. The action of carbonic anhydrase on both sides of the membrane facilitates bicarbonate balance.

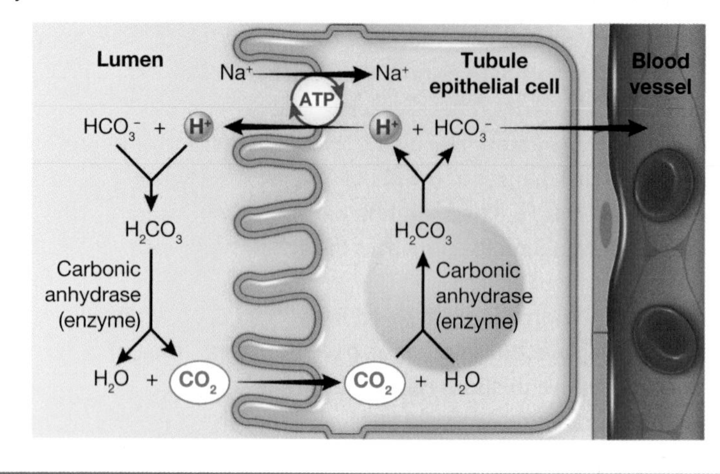

DIGGING DEEPER: Ketoacidosis

Diabetic acidosis, or ketoacidosis, occurs most frequently in people with poorly controlled diabetes mellitus. When certain tissues in the body cannot obtain adequate amounts of glucose, they depend on the breakdown of fatty acids for energy. When acetyl groups break off the fatty acid chains, the acetyl groups non-enzymatically combine to form ketone bodies, acetoacetic acid, beta-hydroxybutyric acid, and acetone, all of which increase the acidity of the blood. In this condition, the brain isn't supplied with enough of its fuel—glucose—to produce all of the ATP it requires to function.

Ketoacidosis can be severe and, if not detected and treated properly, can lead to diabetic coma, which can be fatal. A common early symptom of ketoacidosis is deep, rapid breathing as the body attempts to drive off CO_2 and compensate for the acidosis. Another common symptom is fruity-smelling breath, due to the exhalation of acetone. Other symptoms include dry skin and mouth, a flushed face, nausea, vomiting, and stomach pain. Treatment for diabetic coma is ingestion or injection of sugar; prevention requires the proper daily administration of insulin.

A person who is diabetic and uses insulin can initiate ketoacidosis if a dose of insulin is missed. (Note: ketoacidosis is a form of metabolic acidosis and should not be confused with ketosis, which is a metabolic state in which there are inadequate carbohydrates to burn as fuel for cellular respiration.)

Learning Check

1. Explain how proteins can serve as buffers.
   a. The charged regions can bind to hydrogen and hydroxyl ions.
   b. The proteins attract excess sodium ions.
   c. The amino acids break up sodium bicarbonate into weaker acids.
   d. Hydrogen ion combine with amino acids to form HCl.
2. You are about to give a presentation in front of your colleagues when you suddenly get nervous. You notice your breathing is becoming faster and deeper, and eventually you start hyperventilating. Which of the following explains what is happening in your body?
   a. Increased levels of carbon dioxide
   b. Decreased pH levels
   c. The blood becomes too alkaline
   d. The blood becomes too acidic

3. Which of the following molecules can the kidney regulate to maintain acid-base balance? Please select all that apply.
   a. Hydrogen
   b. Sodium
   c. Bicarbonate
   d. Carbon
4. What would happen to the blood pH if you held your breath as long as you could? Why? LO 26.4.4
   a. The blood pH would decrease due to an increase in CO_2 and H_2CO_3 in the blood.
   b. The blood pH would increase due to a decrease in CO_2 and H_2CO_3 in the blood.
   c. The blood pH would remain the same, because holding your breath has no effect on the CO_2 and H_2CO_3 levels in the blood.

26.5 Acid-Base Homeostasis

HAPS LEARNING GOALS AND OUTCOMES

Learning Objectives: By the end of this section, you will be able to:

26.5.1 Define acidosis and alkalosis.

26.5.2 Compare and contrast metabolic and respiratory causes of pH imbalances.

26.5.3 Explain the relationship between changes in alveolar ventilation (i.e., hypoventilation and hyperventilation), arterial blood pCO_2, arterial blood pH, and arterial blood HCO_3^-.

26.5.4 Describe the concept of compensation in relation to disruption of pH homeostasis.

26.5.5 *Given arterial blood values for pCO_2, pH, and HCO_3^-, determine whether a patient is in acidosis or alkalosis and whether the cause of the pH disturbance is metabolic or respiratory.

LO 26.5.1

Normal arterial blood pH is restricted to a very narrow range of 7.35–7.45. A person who has a blood pH below 7.35 is considered to be in **acidosis**. A sustained blood pH below 7.0 can be fatal. Acidosis has several symptoms, including headache and confusion, and the individual can become lethargic and easily fatigued (**Figure 26.18A**). A person who has a blood pH above 7.45 is considered to be in **alkalosis**, and a pH above 7.8 is fatal. Some symptoms of alkalosis include cognitive impairment (which can progress to unconsciousness), tingling or numbness in the extremities, muscle twitching and spasm, and nausea and vomiting. Both acidosis and alkalosis can be caused by either metabolic or respiratory disorders (**Figure 26.18B**).

As discussed earlier in this chapter, the concentration of carbonic acid is dependent on the balance between how much CO_2 is in the blood compared to how much CO_2 is exhaled through the lungs. Acidosis can occur if more than a typical amount of CO_2 is generated metabolically, or if H^+ is present in excess for other reasons. Acidosis can also occur if the body is not getting rid of sufficient levels of CO_2 via breathing. Alkalosis can occur if CO_2 or H^+ is lost from the body through excessive vomiting or hyperventilation.

26.5a Metabolic Acidosis

Metabolic acidosis occurs when the blood is too acidic (pH below 7.35) due to too little bicarbonate, or excess H^+ ions. The most common cause of metabolic acidosis is the presence of organic acids or excessive ketones in the blood. **Figure 26.19** lists causes of metabolic acidosis and other causes of physiological pH shifts.

Strenuous exercise can cause temporary metabolic acidosis due to the production of lactic acid. Acidosis can also result from the ingestion of specific substances. Metabolic acidosis can also result from the retention of urea and uric acid when the kidneys

Figure 26.18 Symptoms of Acidosis and Alkalosis

Symptoms of acidosis affect several organ systems. Both acidosis and alkalosis can be diagnosed using a blood test.

A Symptoms of acidosis

Central nervous system
- Headache
- Confusion
- Sleepiness
- Loss of consciousness
- Coma

Respiratory system
- Coughing
- Shortness of breath

Heart
- Increased heart rate
- Arrhythmia

Muscular system
- Weakness
- Seizures

Digestive system
- Nausea
- Vomiting
- Diarrhea

Symptoms of alkalosis

Central nervous system
- Light-headedness
- Confusion
- Stupor
- Coma

Peripheral nervous system
- Hand tremor
- Numbness or tingling in the face, hands, or feet

Muscular system
- Twitching
- Prolonged spasms

Digestive system
- Nausea
- Vomiting

Figure 26.19 Causes of Metabolic and Respiratory Acidosis and Alkalosis

LO 26.5.2

Acid-base imbalances are caused by a variety of factors in the body. Acidosis occurs when the blood pH is lower than 7.35, and alkalosis occurs when it is higher than 7.45. Respiratory acidosis is caused by any respiratory condition that results in an excess of CO_2 and H_2CO_3 levels in the blood; respiratory alkalosis results from a deficiency in the CO_2 and H_2CO_3 levels. Metabolic acidosis is caused by a condition (other than that of a respiratory nature) that results in an excess of H^+ ions or a deficiency of HCO_3^- ions in the blood. Metabolic alkalosis is caused by a condition (other than that of a respiratory cause) that results in a deficiency of H^+ ions or an excess of HCO_3^- ions in the blood.

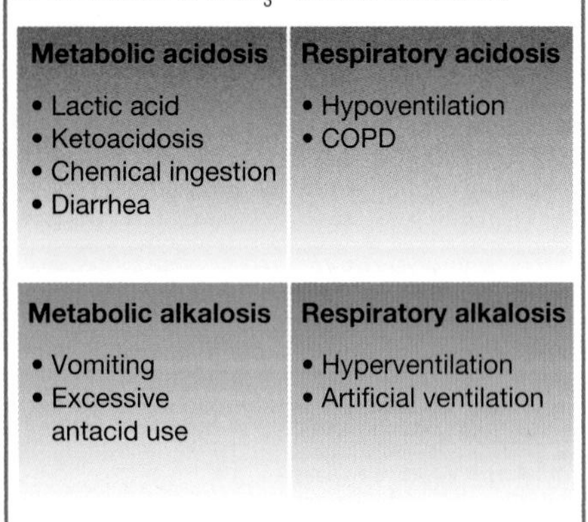

are not functioning properly. Metabolic acidosis can also arise from diabetic ketoacidosis, where an excess of ketones is present in the blood. Another cause of metabolic acidosis is diarrhea. The excessive feces lead to a loss of bicarbonate ions, which are added to the intestines at the duodenum.

Metabolic acidosis
• Lactic acid
• Ketoacidosis
• Chemical ingestion
• Diarrhea
Respiratory acidosis
• Hypoventilation
• COPD
Metabolic alkalosis
• Vomiting
• Excessive antacid use
Respiratory alkalosis
• Hyperventilation
• Artificial ventilation

26.5b Metabolic Alkalosis: Primary Bicarbonate Excess

Metabolic alkalosis is the opposite of metabolic acidosis. It occurs when the blood is too alkaline (pH above 7.45) due to too much bicarbonate.

A transient excess of bicarbonate in the blood can follow ingestion of excessive amounts of antacids for conditions such as stomach acid reflux (also known as *heartburn*). Other causes of metabolic alkalosis include the loss of acid from the stomach through vomiting.

26.5c Respiratory Acidosis: Primary Carbonic Acid/CO_2 Excess

Respiratory acidosis occurs when the blood is overly acidic due to an excess of carbonic acid, resulting from too much CO_2 in the blood. Respiratory acidosis can result from anything that interferes with respiration, such as pneumonia, emphysema, or congestive heart failure.

26.5d Respiratory Alkalosis: Primary Carbonic Acid/CO_2 Deficiency

LO 26.5.3

Respiratory alkalosis occurs when the blood is overly alkaline due to a deficiency in carbonic acid and CO_2 levels in the blood. This condition usually occurs when too much CO_2 is exhaled from the lungs, as occurs in hyperventilation, which is breathing that is deeper or more frequent than normal. An elevated respiratory rate leading to hyperventilation can be due to extreme emotional upset or fear, fever, infections, or hypoxia.

26.5e Compensation Mechanisms

LO 26.5.4

Various compensatory mechanisms exist to maintain blood pH within a narrow range, including buffers, respiration, and renal mechanisms. Although compensatory mechanisms usually work very well, when one of these mechanisms is not working properly (as might occur in kidney failure or respiratory disease), they have their limits. If the pH and bicarbonate-to-carbonic acid ratio are changed too drastically, the body may not be able to compensate. As discussed earlier, extreme changes in pH can denature proteins. Extensive damage to proteins in this way will result in disruption of normal metabolic processes, serious tissue damage, and ultimately death.

Respiratory Compensation Respiratory compensation is an adjustment in breath rate to change the amount of CO_2 exhaled in an attempt to return the blood to homeostatic pH levels. Respiratory compensation can adjust the blood pH within minutes. Respiratory compensation fixes acidosis well, but may not be able to compensate for alkalotic shifts. The respirator response to alkalosis is to increase the amount of CO_2 in the blood by decreasing the respiratory rate to conserve CO_2. There is a limit to the decrease in respiration, due to the need for oxygen. Hence, the respiratory route is less efficient at compensating for metabolic alkalosis than for acidosis.

Metabolic Compensation Metabolic and renal compensation for respiratory diseases that can create acidosis revolves around the conservation of bicarbonate ions. In cases of respiratory acidosis, the kidney increases the conservation of bicarbonate and secretion of H^+. These processes increase the concentration of bicarbonate in the blood, reestablishing the proper relative concentrations of bicarbonate and carbonic acid. In cases of respiratory alkalosis, the kidneys decrease the production of bicarbonate and reabsorb H^+ from the tubular fluid.

Metabolic acidosis is problematic, as lower-than-normal amounts of bicarbonate are present in the blood and the body has a decreased ability to buffer pH changes. The partial pressure of CO_2, or the pCO_2, would be normal at first, but if compensation has occurred, it would decrease as the body reestablishes the proper ratio of bicarbonate and carbonic acid/CO_2.

Respiratory acidosis is problematic, as excess CO_2 is present in the blood. Bicarbonate levels would be normal at first, but if compensation has occurred, they would increase in an attempt to reestablish the proper ratio of bicarbonate and carbonic acid/CO_2.

Student Study Tip

The best way to practice learning these blood gasses is to map out different scenarios. For example: what happens when pCO_2 drops? Will this make the blood pH go up or down? What would the HCO_3^- level have to be to bring pH back to normal?

DIGGING DEEPER:
Diagnosing Acidosis and Alkalosis

LO 26.5.5

While most acid or base shifts will be corrected by compensatory mechanisms, when the body is not able to compensate properly, physiological function will be compromised. If a patient is in alkalosis or acidosis, how would a clinician be able to determine whether the cause was respiratory or metabolic? By pairing tests for pH, and HCO_3^- with tests for CO_2 partial pressure (pCO_2), a clinician can identify acidosis and alkalosis and help to narrow down the possibilities of its cause.

The blood pH value, as shown in Table 26.3, indicates whether the blood is in acidosis, the normal range, or alkalosis. The pCO_2 and total HCO_3^- values aid in determining whether the condition is metabolic or respiratory, and whether the patient has been able to compensate for the problem. Table 26.3 lists the conditions and laboratory results that can be used to classify these conditions. Metabolic acid-base imbalances typically result from kidney disease, and the respiratory system usually compensates by changing the rate and depth of breathing.

Table 26.3 Types of Acid-Base Imbalances

Condition	pH	pCO_2	Total HCO_3^-
Metabolic acidosis	Decreases	Normal then decreases	Decreases
Respiratory acidosis	Decreases	Increases	Normal then increases
Metabolic alkalosis	Increases	Normal then increases	Increases
Respiratory alkalosis	Increases	Decreases	Normal then decreases

Table 26.3 Normal arterial values: pH: 7.35–7.45; pCO_2: biological male: 35–48 mm Hg, biological female: 32–45 mm Hg; Venous bicarbonate levels: 22–29 mM

Learning Connection

Explain a Process

Your neighbor has just learned that they have diabetes mellitus and a complication called *diabetic ketoacidosis.* You have just learned that diabetic ketoacidosis is a type of metabolic acidosis. How would you explain to your neighbor the body's various responses to metabolic acidosis? Which buffers would be helpful in alleviating the acidosis? How would the respiratory system and the kidneys respond to this pH imbalance?

Alkalosis is characterized by a higher-than-normal pH. Metabolic alkalosis is problematic, as elevated pH and excess bicarbonate are present. The pCO_2 would again be normal at first, but if compensation has occurred, it would increase as the body attempts to reestablish the proper ratios of bicarbonate and carbonic acid/CO_2.

Respiratory alkalosis is problematic, as CO_2 deficiency is present in the bloodstream. The bicarbonate concentration would be normal at first. When renal compensation occurs, however, the bicarbonate concentration in blood decreases as the kidneys attempt to reestablish the proper ratios of bicarbonate and carbonic acid/CO_2 by eliminating more bicarbonate to bring the pH into the physiological range.

Learning Check

1. Define metabolic acidosis.
   a. Blood pH falls below 7.35 due to a deficiency of bicarbonate ions.
   b. Blood pH rises above 7.45 due to an excess of bicarbonate ions.
   c. Blood pH falls below 7.35 due to an excess of carbonic acid.
   d. Blood pH rises above 7.45 due to a deficiency of carbonic acid.
2. Identify a possible method of compensation for metabolic acidosis. Please select all that apply.
   a. Increased rate and depth of breathing
   b. Decreased rate and depth of breathing
   c. Decreased bicarbonate production
   d. Decreased reabsorption of H^+ ions

Chapter Review

26.1 Body Fluids and Fluid Compartments

The learning objectives for this section are:

26.1.1 Describe the boundary walls that separate different body fluid compartments and list transport mechanisms by which water and other substances move between compartments.

26.1.2 Explain the subdivision of the extracellular fluid (ECF) compartment into plasma and interstitial fluid (IF), and compare volumes and composition of plasma and IF.

26.1.3* Describe some of the physiological variables (e.g., estrogen signaling) that can influence the total body water (TBW).

26.1.4 Compare and contrast relative volumes and osmolarities of intracellular fluid (ICF) and extracellular fluid (ECF).

26.1.5* Describe the factors that influence the direction and degree of movement of fluid between compartments.

* Objective is not a HAPS Learning Goal.

26.1 Questions

1. Which of the following separates the intercellular compartment from the extracellular compartment?
   a. Cell membrane
   b. Arterial wall
   c. Mitochondrial membrane
   d. Muscles and ligaments

2. Identify the scenarios that would cause blood plasma to flow into the interstitial space. Please select all that apply.
   a. There is a higher concentrations of calcium ions in the capillaries.
   b. There is a higher concentration of fluid in the interstitial space.
   c. There is a higher concentration of fluid in the capillaries.
   d. There is a higher concentration of sodium in the interstitial space.

3. During pregnancy, females have a steadily increasing amount of estrogen and progesterone flowing through their blood. Which of the following best explains the total body water levels?
   a. Increases
   b. Decreases
   c. Remains at the same or almost the same level

4. Which of the following are true of the intracellular (ICF) and extracellular (ECF) fluid compartments? Please select all that apply.
   a. The ICF contains about one-third of the total body water.
   b. The ECF contains a higher concentration of sodium ions than the ICF.
   c. The ECF is composed of about 80% interstitial fluid and 20% blood plasma.
   d. The ICF contains a higher concentration of chloride ions than the ECF.

5. Identify the scenario that would cause sodium ions water to flow into a blood cell from the blood plasma. Please select all that apply.
   a. There is a higher concentration of sodium ions in the blood cell.
   b. There is a higher concentration of fluid in the blood plasma.
   c. There is a higher concentration of sodium in the blood plasma.
   d. There is a higher concentration of fluid in the blood cell.

26.1 Summary

- Fluid and electrolytes occupy specific compartments throughout the body. These compartments are separated by cell membranes.
- Total body water (TBW) consists of the intracellular fluid (ICF) compartment and the extracellular fluid (ECF) compartment. The ICF consists of the fluid within the cells; the ECF is composed mainly of the interstitial fluid and the blood plasma.
- Electrolytes are charged atoms, or ions, that are capable of carrying an electrical charge. Although most body fluids are electrically neutral, they vary greatly in their concentrations of various electrolytes. For example, the ECF contains a higher concentration of sodium ions than the ICF, but the ICF contains a higher concentration of potassium ions than the ECF.
- The movement of electrolytes across cell membranes is carefully controlled, so water acts to equilibrate osmotic concentrations between adjacent fluid compartments.
- The movement of water from one compartment to another occurs due to the osmotic gradient between the compartments. Water diffuses across cell membranes from regions of higher water concentration to regions of lower water concentration; this is called *osmosis*. Water also diffuses across cell membranes from regions of lower solute concentration to regions of higher solute concentration.
- Water movement between compartments can also be explained by the opposing forces of hydrostatic and osmotic pressure. Water moves from a region of higher to lower hydrostatic (or fluid) pressure. Water also moves from a region of lower to higher osmotic pressure; this pressure is higher in regions that contain a high concentration of albumins, a type of plasma protein. The balance between these pressures will determine the direction in which water flows between compartments.

26.2 Water Balance

The learning objectives for this section are:

HAPS LEARNING GOALS AND OUTCOMES

26.2.1 Describe the normal routes of body water entry and loss, and explain how changes in water intake/loss can disrupt osmolarity homeostasis.

26.2.2 Describe behavioral mechanisms that control water intake and loss.

26.2.3 Explain the role of hypothalamic osmoreceptors in regulation of body osmolarity.

26.2.4 Explain how the cardiovascular, endocrine, and urinary systems monitor blood volume and/or blood pressure.

26.2.5 Describe changes to body fluid compartment volumes and osmolarity when a person drinks a large volume of pure water, and then explain the compensatory mechanisms that attempt to restore normal volumes and osmolarity.

26.2.6 Explain the integrated responses of the cardiovascular, endocrine, and urinary systems to low blood pressure as a result of dehydration.

26.2.7* Compare and contrast dehydration and hemorrhage in terms of blood pressure, blood volume, blood osmolarity, and the compensatory mechanisms used to restore homeostasis.

* Objective is not a HAPS Learning Goal.

26.2 Questions

1. Which of the following are ways that water enters into the body fluids? Please select all that apply.
   a. Urination
   b. Diet
   c. Aerobic respiration
   d. Anaerobic respiration

2. Which of the following decreases water loss? Please select all that apply.
   a. Decreased alcohol consumption
   b. Increased caffeine intake
   c. Prescribed diuretic medication
   d. Decreased fluid intake

3. Which of the following are true of the hypothalamic osmoreceptors? Please select all that apply.
   a. Osmoreceptors monitor solute concentration of the blood.
   b. The osmoreceptors stimulate ADH secretion in an overhydrated person.
   c. The main role of osmoreceptors is to monitor oxygen concentration in the blood.
   d. An increase in blood osmolarity stimulates the osmoreceptros to trigger the sensation of thirst.

4. Which of the following body systems regulate blood volume and blood pressure? Please select all that apply.
   a. Cardiovascular
   b. Endocrine
   c. Neurological
   d. Musculoskeletal

5. After a long day of soccer practice, you drink a large bottle of water. Which of the following changes in plasma osmolarity would you expect to see?
   a. Increased
   b. Decreased

6. Which of the following compensations would occur if there were decreased blood volume? Please select all that apply.
   a. Increased heart rate
   b. Decreased angiotensin I
   c. Increased aldosterone
   d. Decreased angiotensin II

7. Which of the following symptoms would you expect to see in a person that was dehydrated? Please select all that apply.
   a. Increased blood pressure
   b. Increased heart rate
   c. Increased sweating
   d. Increased blood osmolarity

26.2 Summary

- Water enters the internal environment of the body fluids mainly through the diet, but a small amount of water is also produced in metabolic reactions. Most water leaves the body through the urine, but small amounts are lost in the feces, sweat, and breathing, and between the skin cells.
- Water balance is vital to the maintenance of homeostasis. Osmolarity, the ratio of solutes to the volume of solvent in the body fluids, is important for properly functioning bodily processes such as nerve impulse conduction, muscle contraction, and enzyme-controlled metabolic reactions.
- Dehydration and overhydration (water intoxication) are both dangerous to the homeostatic balance of the internal environment. Dehydration concentrates the solutes, such as electrolytes, while overhydration diluted the solutes. In both situations, the hypothalamic osmoreceptors, the cardiovascular baroreceptors, and the renal receptors relay changes to various brain centers that respond to correct the deviation from normal blood osmolarity.
- Osmoreceptors in the hypothalamus monitor the osmolarity of the blood plasma and respond to dehydration or overhydration by stimulating or inhibiting the secretion of antidiuretic hormone (ADH) respectively.

- Baroreceptors monitor blood volume and blood pressure and adjust the heart rate and force of contraction as needed to respond to deviations from normal values. For example, the renin-angiotensin system is stimulated by decreased blood volume or pressure; its function is to increase the blood volume and pressure.

26.3 Electrolyte Balance

The learning objectives for this section are:

HAPS LEARNING GOALS AND OUTCOMES

26.3.1 Explain the importance of maintaining potassium homeostasis with regard to membrane potential, and provide examples of dysfunction that occur when plasma potassium levels are too high or too low.

26.3.2 Explain the importance of maintaining calcium homeostasis, and provide examples of dysfunction that occur when plasma calcium levels are too high or too low.

26.3.3 Describe the integrated responses of the endocrine and urinary systems to disruptions of potassium homeostasis.

26.3.4 Describe the integrated responses of the endocrine, digestive, skeletal, and urinary systems to disruptions of calcium homeostasis.

26.3 Questions

1. Which of the following electrolytes functions to restore resting membrane potential after the depolarization phase of an action potential?
   a. Sodium
   b. Chloride
   c. Potassium
   d. Phosphate

2. Which of the following leads to hypercalcemia? Please select all that apply.
   a. Some types of cancer
   b. Cystic fibrosis
   c. Increased ADH
   d. Hyperparathyroidism

3. Explain how angiotensin II maintains potassium homeostasis.
   a. Signals an increase in the release of aldosterone
   b. Increases reabsorption of sodium and water
   c. Activates osteoclasts to break down bone matrix
   d. Triggers the thalamus to increase fluid intake

4. Jakob does not regularly intake calcium in their diet. Jakob's blood test shows that they have the typical amount of calcium in their body. Which of the following explains the normal findings of Jakob's blood test?
   a. Angiotensin II reabsorbed calcium into the bloodstream.
   b. Aldosterone activates osteoclasts to increase bone density.
   c. Calcitonin activates osteoblasts to deposit calcium into the bone matrix.
   d. Parathyroid hormone increases osteoclast activity to break down bone matrix.

26.3 Summary

- The most important electrolytes in the human body are sodium, potassium, chloride, bicarbonate, calcium, and phosphate ions. It is vital to normal functioning of the body that these ions remain in homeostatic balance. When any electrolyte is in a state of imbalance, it adversely affects the processes in which the electrolyte participates.
- The electrolytes function in nerve impulse conduction, muscle contraction, bone and tooth health, endocrine secretion, membrane permeability, the buffering of body fluids, and the regulation of water between fluid compartments.
- Electrolytes are obtained through the diet, and lost in the urine, feces and sweat.
- Sodium imbalances, such as hypernatremia or hyponatremia, have several causes, and results in disruption to nerve impulse conduction, muscle contraction, and water balance.
- Potassium imbalances, such as hyperkalemia and hypokalemia, also have a variety of causes, and results in a disruption to nerve impulse conduction, muscle contraction, heart function, and water balance.
- Calcium imbalances, such as hypercalcemia and hypocalcemia, cause problems with bone and tooth strength, muscle contraction, enzyme activity and blood clotting.
- The endocrine system is active in regulating electrolyte balance. Sodium and potassium balance are regulated by angiotensin II and aldosterone. Calcium and phosphate balance are regulated by parathyroid hormone (PTH) and calcitonin.

26.4 Acid-Base Balance

The learning objectives for this section are:

26.4.1* Explain how changes in pH outside the normal range adversely affect body functions and include the normal pH range in arterial blood.

26.4.2 Describe the major buffer systems of the body (e.g., bicarbonate buffer system, protein buffer system) and their locations (e.g., extracellular fluid) in the body.

26.4.3 Explain the relationship between transport of carbon dioxide in the blood and the bicarbonate buffer system in the plasma.

26.4.4 Using the equation $CO_2 + H_2O \leftrightarrow H^+ + HCO_3^-$, explain what happens to pH when arterial blood pCO_2 and HCO_3^- concentrations change.

* Objective is not a HAPS Learning Goal.

26.4 Questions

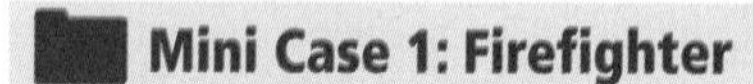

Mini Case 1: Firefighter

Jose is a firefighter for their community. While fighting a fire, Jose had to run inside to rescue a child. Inside the house, Jose noticed that their ventilated mask was defective and carbon dioxide was getting in. Jose started running out, but it was too late; they started having a coughing attack and their team rushed them to the hospital.

1. You are a doctor on staff. What are you most concerned about with Jose and their inhalation of carbon dioxide from the burning building?
   a. Jose may have hypocapnia.
   b. Jose's proteins may denature.
   c. Jose's liver may shut down.
   d. Jose's lungs may develop pneumonia.

2. Which of the following molecules can serve as a buffer to help Jose reach the normal pH level? Please select all that apply.
   a. Hemoglobin
   b. Hydrochloric acid
   c. Sodium bicarbonate
   d. Fatty acids

3. How would sodium bicarbonate help Jose regain pH balance?
   a. It would break down the strong acid into a weak base and water.
   b. It would break down the strong acid into a weak acid and salt.
   c. It would break down the strong base into a weak acid and salt.
   d. It would break down the strong base into a weak base and water.

4. Using the equation $CO_2 + H_2O \leftrightarrow H^+ + HCO_3^-$, explain what happened to Jose.
   a. Jose's H_2O content increased.
   b. Jose's H^+ concentration increased.
   c. Jose's HCO_3^- concentration decreased.
   d. Jose's CO_2 concentration decreased.

26.4 Summary

- The hydrogen ion concentration, or the pH, of the body fluids is vital for the normal functioning of proteins, such as enzymes, protein hormones, cell membrane proteins (such as pumps), structural proteins, and contractile proteins.
- A pH, or acid-base, imbalance denatures proteins, and stops them from performing their functions, such as catalyzing metabolic reactions (enzymes), exerting endocrine control (hormones), maintaining the membrane potential (the sodium-potassium pump), maintaining healthy structural support (collagen), and contracting our muscles (contractile proteins).
- The normal pH of arterial blood is 7.4, with a normal range of 7.35 to 7.45. When the blood pH is out of balance, a variety of buffers in the body resist the pH change and act to bring the pH back into its normal range.
- There are several buffer systems in the body fluids, such as the protein and bicarbonate buffer systems.
- The respiratory system acts as a buffer by increasing or decreasing the rate and depth of breathing. Increasing the rate and depth of breathing eliminates carbon dioxide, carbonic acid, and hydrogen ions from the body, which results in the removal of excess acidity from the body.
- The kidney acts as a buffer by increasing or decreasing the excretion of hydrogen ions or bicarbonate ions in the urine. Increasing the excretion of hydrogen ions in the urine removes excess acidity from the body.

26.5 Acid-Base Homeostasis

The learning objectives for this section are:

26.5.1 Define acidosis and alkalosis.

26.5.2 Compare and contrast metabolic and respiratory causes of pH imbalances.

26.5.3 Explain the relationship between changes in alveolar ventilation (i.e., hypoventilation and hyperventilation), arterial blood pCO_2, arterial blood pH, and arterial blood HCO_3^-.

26.5.4 Describe the concept of compensation in relation to disruption of pH homeostasis.

26.5.5 *Given arterial blood values for pCO_2, pH, and HCO_3^-, determine whether a patient is in acidosis or alkalosis and whether the cause of the pH disturbance is metabolic or respiratory.

26.5 Questions

1. Define alkalosis.

a. The blood pH falls below 7.35
b. The blood pH rises above 7.45.

2. Which of the following would lead to metabolic alkalosis?

a. An excess of hydrogen ions in the blood
b. An excess of bicarbonate ions in the blood
c. An excess of carbonic acid in the blood
d. An excess of potassium ions in the blood

3. You are walking through an amusement park and notice the person in front of you hyperventilating. Which of the following is happening in their blood? Please select all that apply.

a. Increased pH level
b. Increased level of bicarbonate
c. Decreased level of carbon dioxide
d. Decreased level of hydrogen ions

4. What is a possible metabolic compensation for respiratory acidosis?

a. Kidneys increase conservation of bicarbonate ions.
b. Kidneys decrease secretion of H^+ ions.
c. Kidneys reabsorb H^+ ions from the tubular fluid.
d. Kidneys increase production of hydrochloric acid.

5. You are a nurse practitioner in the emergency room when you just receive a patient's bloodwork. The lab values read: pH: 7.55, pCO_2: 25 mm Hg, HCO_3^-: 24 mEq/L. Normal lab values should read: pH: 7.35–7.45, pCO_2: 32–48 mm Hg, HCO_3^-: 22–29 mEq/L. Which of the following is most likely the patient's diagnosis?

a. Respiratory acidosis
b. Respiratory alkalosis
c. Metabolic acidosis
d. Metabolic alkalosis

26.5 Summary

- When the pH of the body fluids drops below 7.35, the body is in a state of acidosis. When the pH rises above 7.45, the body is in a state of alkalosis.
- Metabolic acidosis occurs when the blood becomes too acidic due to a deficiency of bicarbonate ions. Metabolic alkalosis occurs when the blood becomes too alkaline due to an excess of bicarbonate ions.
- Respiratory acidosis occurs when the blood becomes too acidic due to an excess of carbonic acid, derived from an excess of carbon dioxide in the blood. Respiratory alkalosis occurs when the blood becomes too alkaline due to a deficiency of carbonic acid, derived from a deficiency of carbon dioxide.
- Compensation mechanisms work to counteract the effects of a pH imbalance. Compensation methods include chemical buffers, changes to the process of respiration, and changes to the kidney's excretion of specific ions in the urine.
- Respiratory compensation for metabolic pH imbalances involves changes in the rate and depth of breathing. Metabolic (or renal) compensation for respiratory pH imbalances involves changes in the level of excretion of hydrogen or bicarbonate ions in the urine.

27

The Reproductive Systems

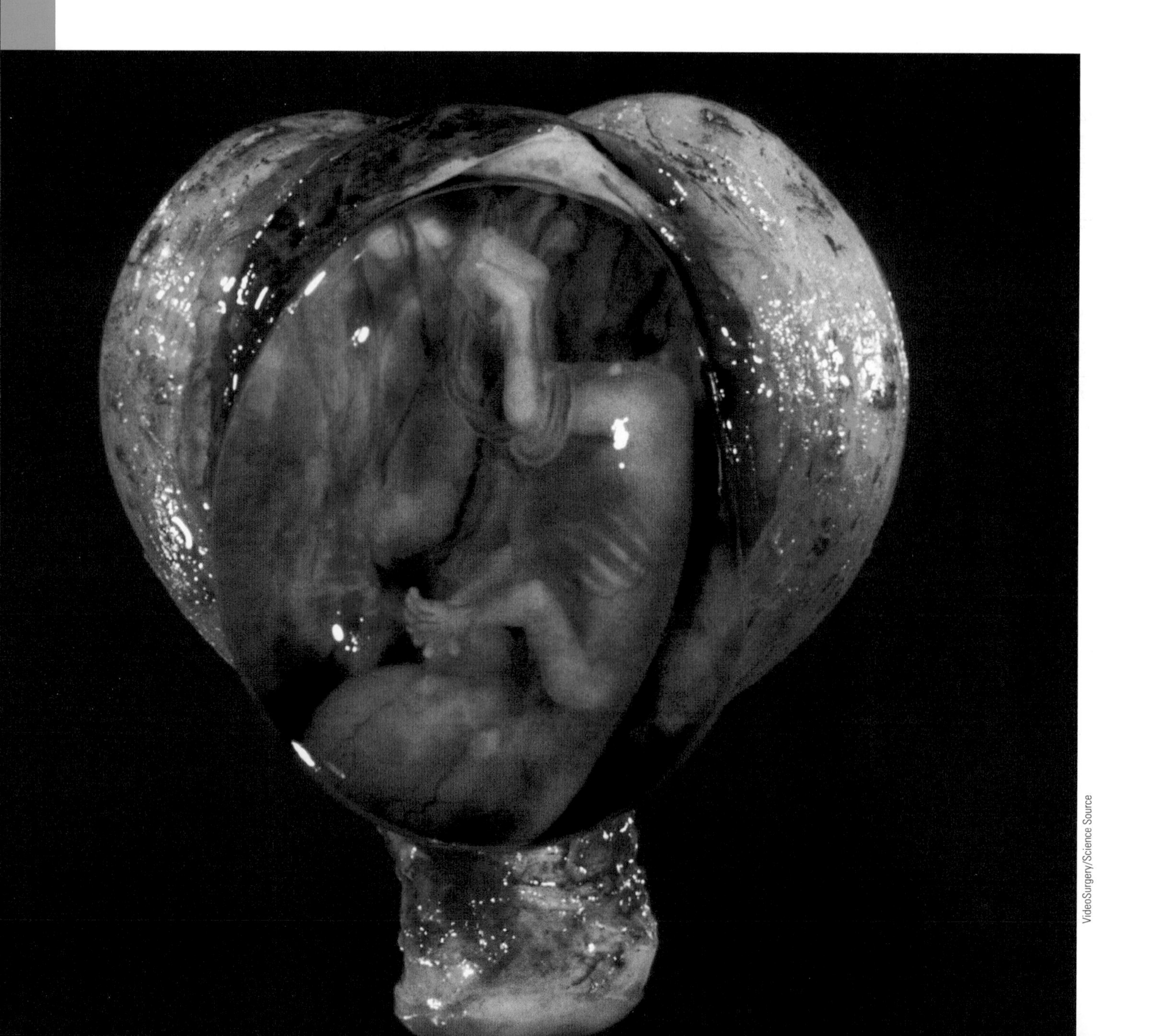
VideoSurgery/Science Source

Chapter Introduction

In this chapter we will learn...

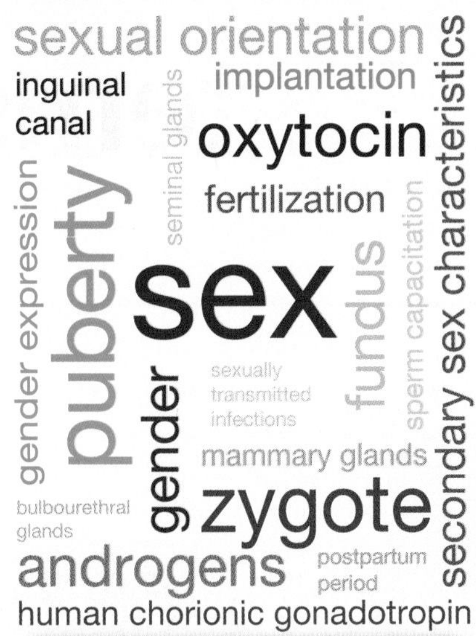

The words and their correct uses associated with sex and reproduction

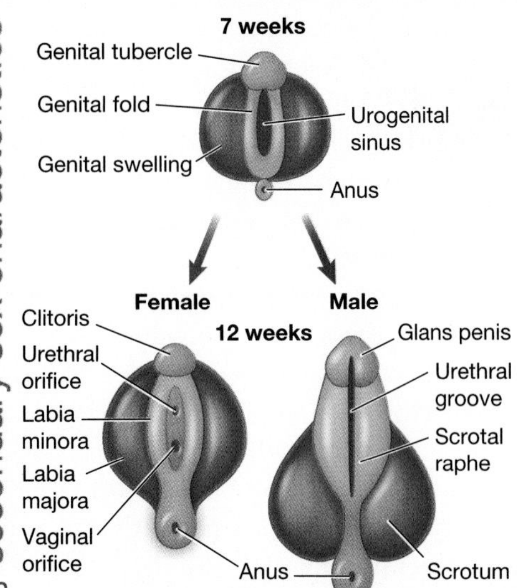

The development and adult anatomy of the structures of reproduction

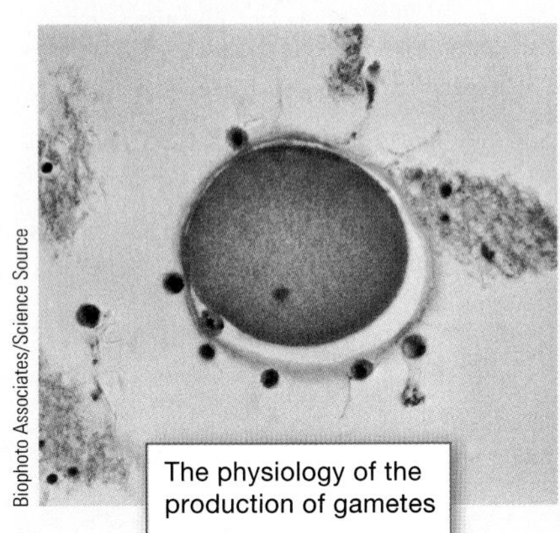

Biophoto Associates/Science Source

The physiology of the production of gametes

David Isaacson/Alamy Stock Photo

And, when pregnancy does occur, the events and physiology of pregnancy

In each of the other systems of the human body presented in this book we have discussed that system's contribution to homeostasis. Disruption of the functions of these systems can lead to a deviation from homeostasis and often a resulting disease state. The reproductive systems, however, do not contribute to the homeostasis of the body. Their functions, the production of **gametes** (germ cells such as eggs and sperm), pleasure, and sometimes the production of a new human, do not contribute to the physiological well-being of the body. Production of gametes, pregnancy, birth, and lactation are all incredibly energy-intensive activities. The processes of the reproductive systems undoubtedly contribute to the human species and often to the psychological well-being of an individual.

In this chapter we will rely on some precursor knowledge. If these are rusty spots in your physiological understand you may want to revisit these topics in earlier chapters before embarking on this one.

- Meiosis (Chapter 4)
- Cell differentiation (Chapter 4)
- Tissue types (Chapter 5)
- Hormone types, functions and tropic hormone relationships (Chapter 17)

27.1 Overview of Human Reproductive Systems

HAPS LEARNING GOALS AND OUTCOMES

Learning Objectives: By the end of this section, you will be able to:

27.1.1* Define and contrast the terms *sex*, *gender*, *gender expression*, and *sexual orientation*.

27.1.2 Describe the functions of the hormones involved in the regulation of the reproductive processes (e.g., gonadotropin releasing hormone [GnRH], follicle stimulating hormone [FSH], luteinizing hormone [LH], androgens, inhibin, estrogens, progesterone).

* Objective is not a HAPS Learning Goal.

The Human Anatomy and Physiology Society includes more than 1,700 educators who work together to promote excellence in the teaching of this subject area. The HAPS A&P Learning Outcomes measure student mastery of the content typically covered in a two-semester Human A&P curriculum at the undergraduate level. The full Learning Outcomes are available at https://www.hapsweb.org.

LO 27.1.1

Sex (anatomical sex, biological sex)

Open up most texts or websites that talk about reproduction and two words—*female* and *male*—appear more than any others. These terms describe the sex of an individual. **Sex** can include considerations of a person's anatomy, chromosomes, or hormones. The International Olympic Committee (IOC) is highly invested in trying to define criteria for two sexes for the purposes of sorting athletes into competition brackets (e.g., men's basketball and women's basketball) but this committee has found over time that sex can often be very difficult to define in absolute terms or values. Throughout the last 100 years the IOC has attempted different definitions of sex from anatomy to chromosomes, but the committee is still redefining criteria today. In this text we use modifications of these terms—*biological female* and *biological male*—to refer to individuals with prototypical anatomy and hormone levels within ranges that are typical for males and females; however, we acknowledge that many individuals may not fit neatly into these two categories while still being completely healthy. Biological sex is very different from gender or gender expression. A person's **gender** (sometimes called *gender identity*) reflects who a person knows themselves to be. Gender may or may not match biological sex, and it also may not fit into our two biological categories. **Gender expression** is the sum of the characteristics that an individual shows the world, things such as the clothes they wear or their demeanor. None of these terms discussed so far are related to **sexual orientation**, a term that describes who an individual is sexually attracted to.

In most cases, sex is assigned at birth by the attending medical team, based entirely on external genitalia. As you will read about in the next section, the development of these anatomical structures, both internal and external, is based largely on hormone signaling; that is, both the levels of pertinent hormones as well as the expression of their receptors. The physiology of these structures, how they function, is governed by hormones as well. Interestingly, all healthy humans have all of the hormones we will describe in this chapter; they just have different amounts of them. Since they are all present in each human body, let's get to know the hormones first (**Figure 27.1**).

LO 27.1.2

- *Gonadotropin Releasing Hormone (GnRH):* GnRH is a tropic hormone, secreted by the hypothalamus. It regulates the secretion of FSH and LH by the anterior pituitary.
- *Follicle Stimulating Hormone (FSH):* FSH is a hormone released by the anterior pituitary that regulates the development of gametes.
- *Luteinizing Hormone (LH):* LH is a hormone released by the anterior pituitary that affects the ovaries and testes differently. In ovaries, LH promotes the release of the eggs in a process called *ovulation*. In the testes, LH promotes the synthesis of the hormone testosterone.
- *Androgens:* A group of hormones, including testosterone, made both in the gonads and the adrenal cortex; they influence gamete production, sex drive, and tissue function all over the body. Androgens are derived from cholesterol (**Figure 27.1B**).
- *Inhibin:* A hormone made in the gonads as part of the feedback system to regulate the anterior pituitary.
- *Estrogens:* A group of hormones made in the gonads, adipose tissue, and the adrenal cortex that influence gamete production, enhance sex drive, and influence tissue function all over the body. Estrogens are derived from cholesterol (see Figure 27.1B).
- *Progesterone:* A hormone made by the corpus luteum in the ovaries (when these are present) and the placenta (when this is present). Progesterone is also made by

Figure 27.1 **Hormones of Reproduction**

(A) The hormones that control gametogenesis and reproduction are secreted by the gonads in response to tropic hormones produced by the hypothalamus and anterior pituitary. (B) Steroid hormones are derived from cholesterol. Testosterone and estrogen have similar structures.

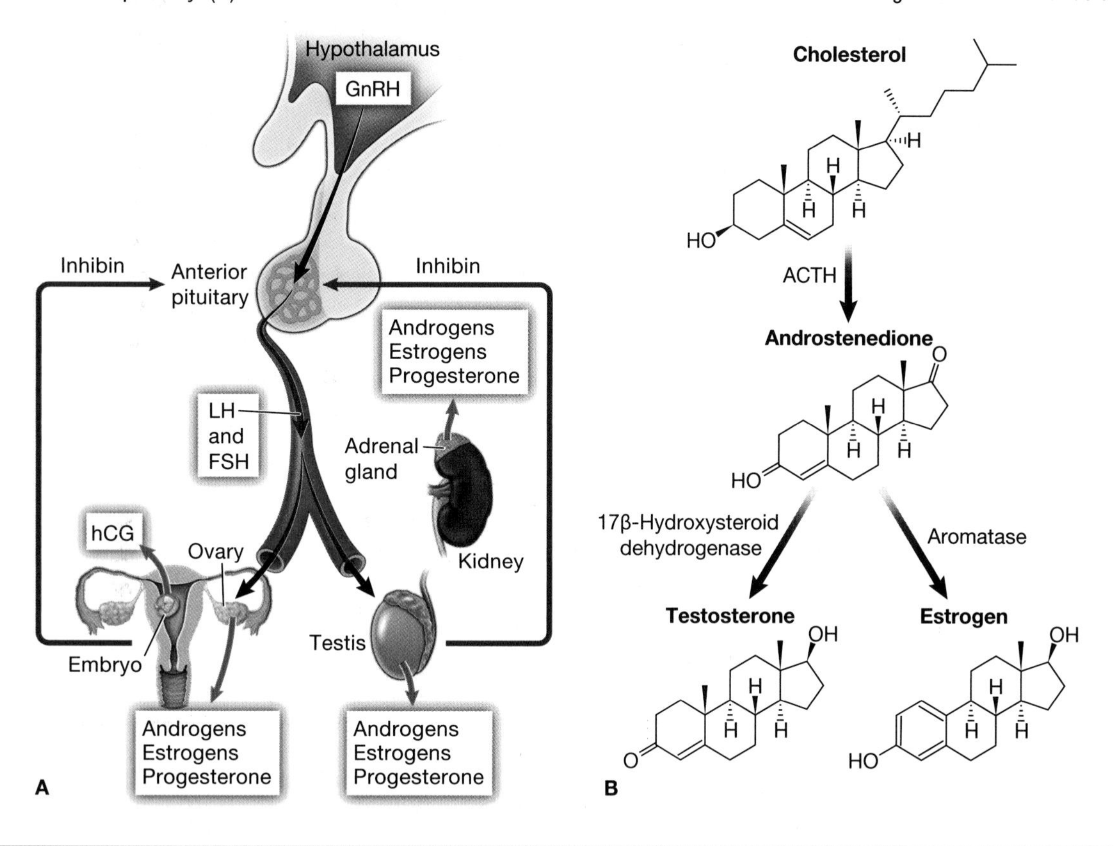

the adrenal cortex and the testes. Progesterone's primary target is the lining of the uterus, though it has a few other effects throughout the body.

- *Human chorionic gonadotropin (hCG):* A hormone made by an embryo that regulates some aspects of development; it also feeds back to the ovaries to encourage continued signaling of estrogen and progesterone.

Learning Check

1. Which of the following terms describes the genitalia, hormones, and chromosomes a person is born with?
   a. Sex
   b. Gender
   c. Gender expression
   d. Sexual orientation
2. How does inhibin's action relate to those of luteinizing hormone (LH) and follicle stimulating hormone (FSH)?
   a. FSH regulates inhibin and LH.
   b. Inhibin regulates LH and FSH.
   c. LH regulates FSH and inhibin.
   d. Inhibin regulates FSH but not LH.

3. Where are estrogens made? Please select all that apply.
   a. Gonads
   b. Anterior pituitary
   c. Adipose tissue
   d. Adrenal cortex
4. How are human chorionic gonadotropin (hCG), estrogen, and progesterone related?
   a. hCG stimulates estrogen and progesterone secretion.
   b. hCG inhibits estrogen and progesterone secretion.
   c. hCG inhibits estrogen secretion and stimulates progesterone secretion.
   d. hCG stimulates estrogen secretion and inhibits progesterone secretion.

27.2 Development of Reproductive Structures

Learning Objectives: By the end of this section, you will be able to:

27.2.1* Identify homologues of reproductive system structures.

27.2.2* Describe the physiological and anatomical changes that occur during puberty.

27.2.3* Identify the structures of human mammary glands and describe their functions.

* Objective is not a HAPS Learning Goal.

Although we have built our understanding of the body one system at a time throughout this book, the embryo develops its body systems concurrently, building the anatomy for each system nearly simultaneously. This development begins soon after fertilization of the egg, the event that begins pregnancy. When an egg (which carries an X chromone as its twenty-third chromosome) is fertilized by a sperm (which might carry either an X or Y chromosome as its twenty-third chromosome), they merge their genetic information, producing a diploid embryo that either has two X chromosomes or one X and one Y chromosome. The genes on these chromosomes guide development of the reproductive system. When we observe pregnancy, we count its duration beginning with the first day of the last menstrual cycle, since that is the last observable event. Within the body, fertilization can occur two to three weeks after that date and embryogenesis begins soon after.

27.2a Development of the Sexual Organs in the Embryo and Fetus

LO 27.2.1

The first seven weeks of embryogenesis are called the *indifferent stage* because, while the sexual organs are developing, the development is inside the embryo rather than external, and there is no difference among any embryos, regardless of which chromosomes they carry inside their cells. Development of reproductive structures is so interesting because if you compare them to the development of other cells, for example cardiac muscle cells, these cells proceed along a developmental path until they are fully differentiated, functional cells. The cells that will develop into the reproductive structures, however, proceed forward until they come to a developmental fork in the road. These cells and structures will proceed along one path if a certain set of signals is present, or along a different path if different signals are present. The same embryonic structures could develop one way to become testes or a different way to become ovaries. Another structure could develop into a penis or develop into a clitoris; it all depends on the signals that are present at precise moments.

The big kickoff event in reproductive development occurs around weeks 4–5, when primordial germ cells migrate into the embryo. These cells, which began on the yolk sac, will become sperm or eggs (oocytes). They migrate into the embryo and onto a slim ridge of tissue at the posterior of the body, called the **genital ridge**, analogous to the posterior of the abdominal/pelvic cavities (**Figure 27.2**). The surface of these genital ridges begins to differentiate into a network of tubes of the reproductive system. These tubes will be used later to transport either sperm or eggs. Once the primordial germ cells have arrived, they send the signals that drive gonadal development.

Week 7 is the transition from the indifferent stage to a differentiated stage. In the indifferent stage all structures develop the same way, regardless of the chromosomes present. At week 7 we start to see differences in the way the development proceeds based on the chromosomes the embryo has. The Y chromosome is very small, but it does have a few genes of interest. There is a sex determining region of the Y chromosome, typically referred to as **SRY**, which is a gene region that spans two genes—one for an enzyme that catalyzes the production of testosterone and the other for a hormone called *anti-Mullerian hormone (AMH)*, a hormone produced only in the XY embryo. In an XY embryo, production of these hormones—testosterone and AMH—begins in week 7. In an XX embryo, these hormones will not be produced at this time.

SRY (sex-determining region of the Y chromosome)

By the end of week 7 there are two sets of ducts in the embryo—the **mesonephric ducts** and the **paramesonephric ducts** (**Figure 27.3**). All embryos will be secreting hCG during this early phase of development, but if testosterone signaling is also present, hCG and testosterone work together to drive the development of interstitial cells. These cells, once they begin developing, will secrete testosterone. Thus, testosterone production in the early embryo follows a positive feedback pattern. If testosterone signaling is present, interstitial cells will develop; if interstitial cells develop, they secrete more testosterone. Receptors for testosterone are expressed on both sets of ducts in the embryo. Testosterone acts as a "go" signal for the differentiation of the mesonephric ducts (**Figure 27.4A**). The SRY region also contains the gene for AMH. This hormone provides a "stop" signal for the differentiation of paramesonephric ducts. Testosterone and AMH also provide "stop" signals for mitosis of germ cells. Thus, in individuals who had testosterone and AMH signaling during the embryonic period, germ cell mitosis will not begin until puberty.

mesonephric ducts (Wolffian ducts)

paramesonephric ducts (Mullerian ducts)

Figure 27.2 Migration of Primordial Germ Cells

Primordial germ cells develop in the yolk sac and migrate into the embryo, taking up residence on a structure known as the *genital ridge*. The genial ridge will give rise to the gonads.

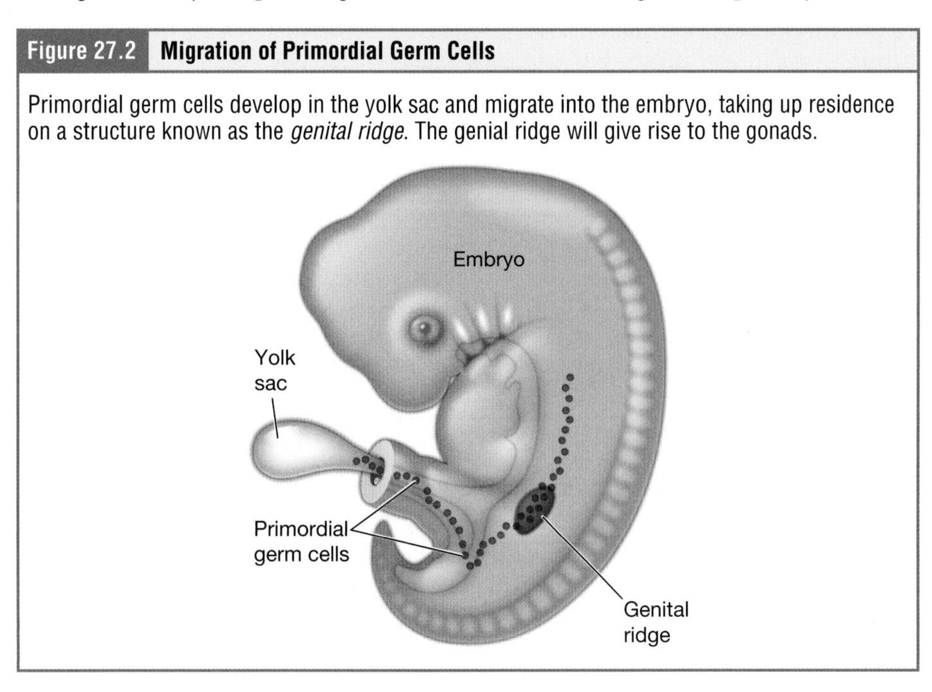

Figure 27.3 Development of the Internal Ducts of the Reproductive System

Early in embryogenesis, two sets of ducts—mesonephric and paramesonephric—develop. One set of ducts will develop into internal reproductive ducts and organs; the other will be destroyed.

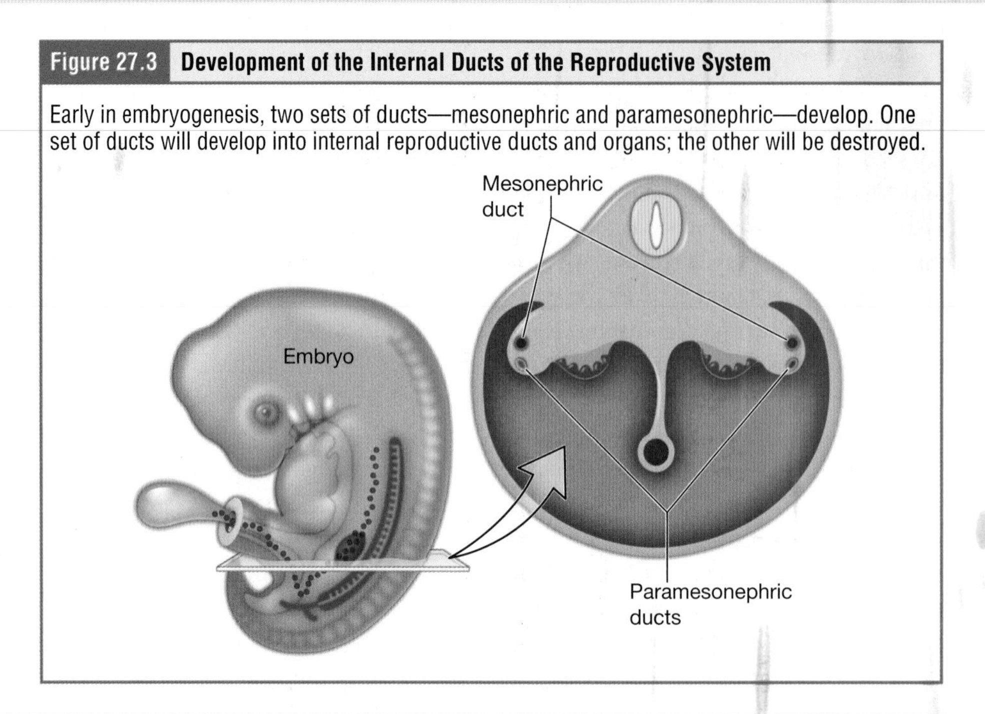
Mesonephric duct
Embryo
Paramesonephric ducts

Figure 27.4 Mesonephric Ducts and Paramesonephric Ducts

The mesonephric ducts and paramesonephric ducts are the primordial structures that can become the adult gonads. Early in embryogenesis, both sets of ducts develop. Depending on which signals are present during embryonic development, only one will persist and develop into the internal ducts of the reproductive system. (A) In the presence of testosterone, the interstitial cells develop and secrete more testosterone, which drives the development of mesonephric ducts into several internal male reproductive structures, such as the ductus deferens and the seminal vesicles. (B) In the absence of testosterone, the interstitial cells and mesonephric ducts do not continue developing; they shrink and disappear over time. The paramesonephric ducts will thrive and develop into female reproductive structures such as the uterine tubes and uterus if AMH is absent.

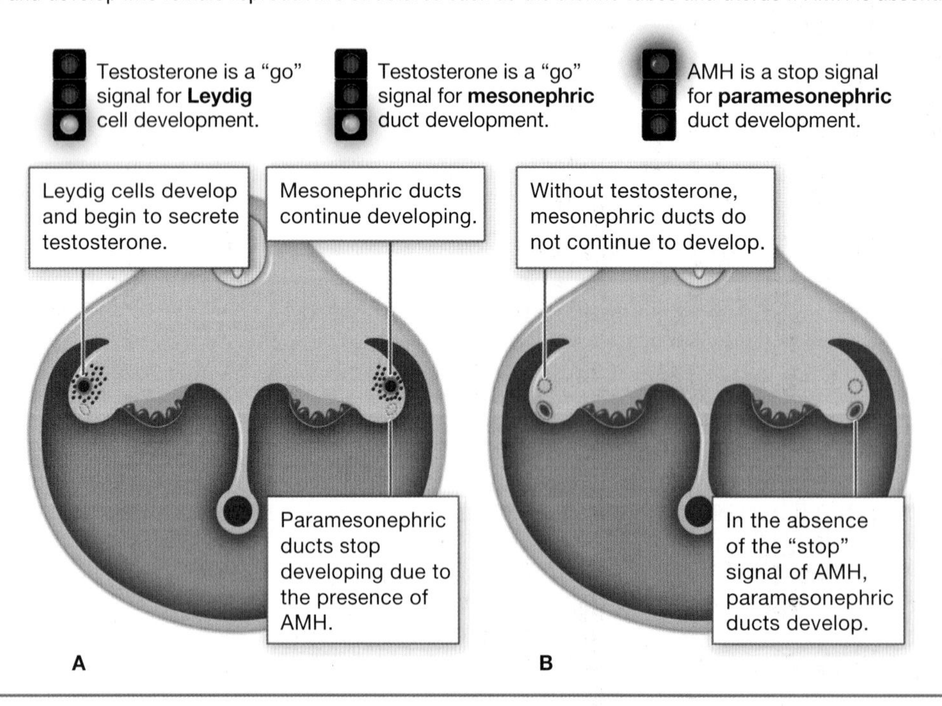
Testosterone is a "go" signal for **Leydig** cell development.
Testosterone is a "go" signal for **mesonephric** duct development.
AMH is a stop signal for **paramesonephric** duct development.
Leydig cells develop and begin to secrete testosterone.
Mesonephric ducts continue developing.
Without testosterone, mesonephric ducts do not continue to develop.
Paramesonephric ducts stop developing due to the presence of AMH.
In the absence of the "stop" signal of AMH, paramesonephric ducts develop.
A
B

Mesonephric duct development requires testosterone, and paramesonephric duct development is suppressed in the presence of testosterone. So, when SRY and testosterone are absent, paramesonephric ducts develop and mesonephric ducts do not (**Figure 27.4B**). Additionally, there are a number of signaling factors that must be present and active for functional paramesonephric duct development. Since testosterone also provides a stop signal for germ cell mitosis, in its absence mitosis and even early stages of meiosis proceed forward. Therefore, in individuals without active testosterone signaling in early embryonic development, germ cells proliferate and begin meiosis, seeding the population of gametes that will be used later. These individuals are born with all of the gametes—oocytes in this case—that they will ever have in their lifetimes. Paramesonephric ducts will grow and develop, becoming uterine tubes. At their inferior ends they will fuse and form the primordial uterus and upper one-third of the vagina (see **Figure 27.5**, comparing embryonic and adult structures).

During weeks 12–20, testosterone continues to drive the development of the mesonephric ducts into seminiferous tubules. These structures are solid during early life, but will hollow out at puberty to become the location for the development of sperm.

On the external surface, an epidermis covers the entire embryo. The epidermis between the forming embryonic legs has folds—the **genital folds**—surrounding a small opening, the **urogenital sinus**. Under the influence of estrogen, the urogenital sinus invaginates inward to become the lower two-thirds of the vagina. As this new lower half invaginates to meet the upper half, the two halves eventually fuse to become the full vagina, sometimes leaving a thin sheet of membrane between these structures, which is sometimes called the **hymen**. This thin membrane usually disappears either right before or right after birth. The ovaries are developing too; they also require the expression of estrogen. The primordial germ cells, encased in small rings of cells, are

Figure 27.5 Fetal Development of External Genitalia

At 7 weeks of development, the external genitalia are bipotential; they can become the genital structures associated with either sex. We can consider the adult structures such as the labia and the scrotum, or the clitoris and the penis, to be homologues, as they consist of the same tissues and develop from the same embryonic structures.

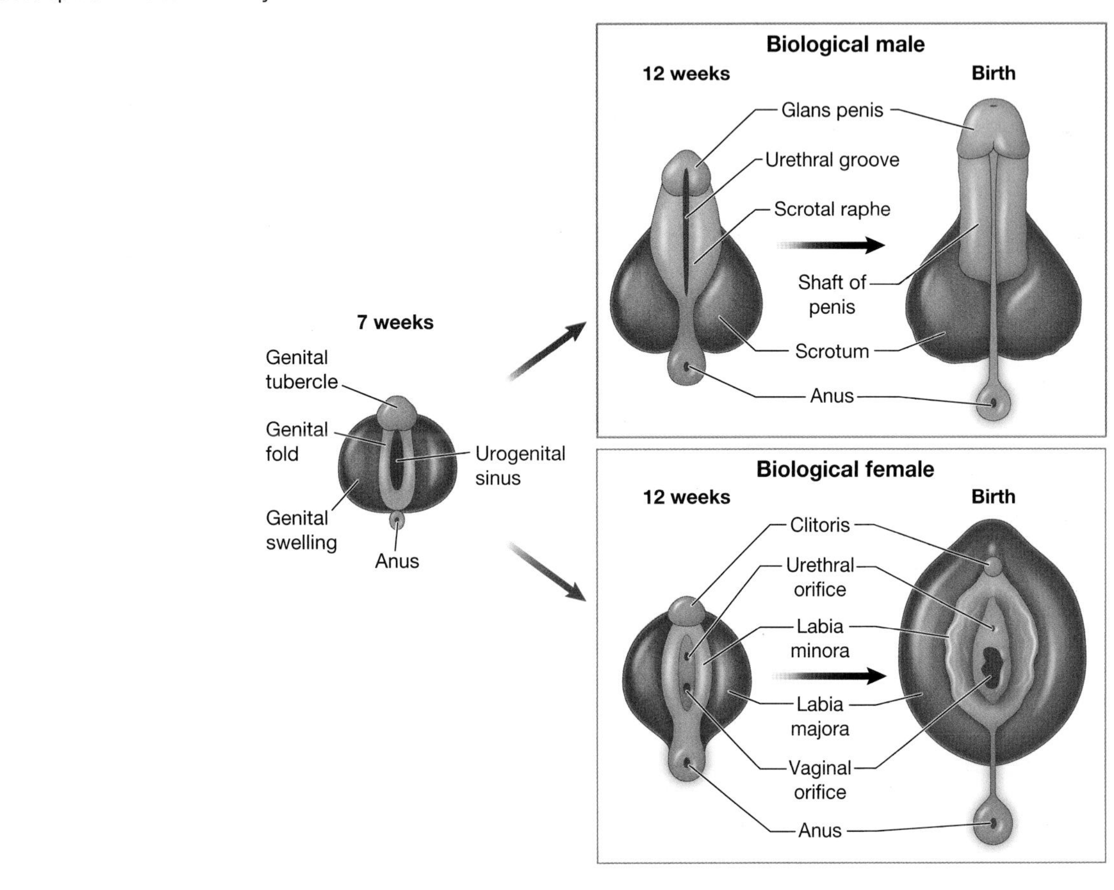

within the developing ovaries; by the time the fetus is finished with development, they will have approximately 2 million of these primordial eggs. This supply will be all that the individual has for life, for in humans, further mitosis of the primordial germ cells does not happen (it does in some mammals, but not in humans).

The genital folds, and a small knob of tissue above them called the *genital tubercle*, are described as being *bipotential*, meaning that they can become external genitalia of either biological males or biological females. Each of these is a path of development, and it is important to note that beginning a path and developing along that path requires the continued presence of both hormonal signals and receptors. Small differences in the use of these signaling pathways can alter development, and there is a wide variety in the development of our external genitalia. In broad strokes, the genital folds can enlarge and remain separate, in which case they become the external structures known as *labia* (see Figure 27.5). Or the folds may meet and fuse at the midline of the body, becoming a single pocket of tissue known in the adult as the *scrotum*. Likewise, the genital tubercle, a bulb of tissue at the anterior portion of the genital folds, is comprised of a spongy erectile tissue. This tissue is soft most of the time, but during arousal can fill with blood to become hard and erect. Under one set of signals, the majority of the genital tubercle structure protrudes from the body and is called the *penis*. Under a different set of signals, the majority of the genital tubercle structure is hidden on the inside of the body and is called the *clitoris*. Even within one developmental tract there can be a wide degree of diversity. For example, genital folds can become labia, and labia can develop naturally into many sizes, colors, and shapes, all of which are normal.

27.2b Further Sexual Development Occurs at Puberty

LO 27.2.2

Puberty is the stage of development at which individuals become sexually mature and ready for reproduction, should they so choose. The hormonal control of puberty is very similar in all individuals. While the physiological control and sequence of events of puberty does not vary much from one individual to another, the timing of these events can vary widely. As shown in **Figure 27.6**, a concerted release of hormones from the hypothalamus (GnRH), the anterior pituitary (LH and FSH), and the gonads is responsible for the maturation of the reproductive systems and the development of **secondary sex characteristics**, which are physical changes outside the reproductive organs, such as the growth of pubic hair, that accompany puberty.

The first changes begin in late childhood, sometimes between ages 8 and 10, when the production of LH becomes detectable. The release of LH occurs primarily at night during sleep and precedes the physical changes of puberty by several years. In prepubescent children, the sensitivity of the negative feedback system in the hypothalamus and pituitary is very high. This means that very low concentrations of androgens or estrogens will negatively feed back onto the hypothalamus and pituitary, keeping the production of GnRH, LH, and FSH low.

As an individual approaches puberty, two changes in sensitivity occur. The first is a decrease of sensitivity in the hypothalamus and pituitary to negative feedback, meaning that it takes increasingly larger concentrations of hormones to stop the production of LH and FSH. The second change in sensitivity is an increase in sensitivity of the gonads to the FSH and LH signals, meaning the gonads of adults are more responsive to gonadotropins than are the gonads of children. As a result of these two changes, the levels of LH and FSH slowly increase and lead to the enlargement and maturation of the gonads, which in turn leads to secretion of higher levels of sex hormones and the initiation of spermatogenesis and folliculogenesis.

In addition to age, multiple factors, including genetics, environment, and psychological stress, can affect the age of onset of puberty. One of the more important influences may be

Figure 27.6 Puberty Causes a Change in Hormone Release

During puberty, the release of LH and FSH from the anterior pituitary stimulates the gonads to increase their production of sex hormones, resulting in changes to the external genitalia and other tissues throughout the body.

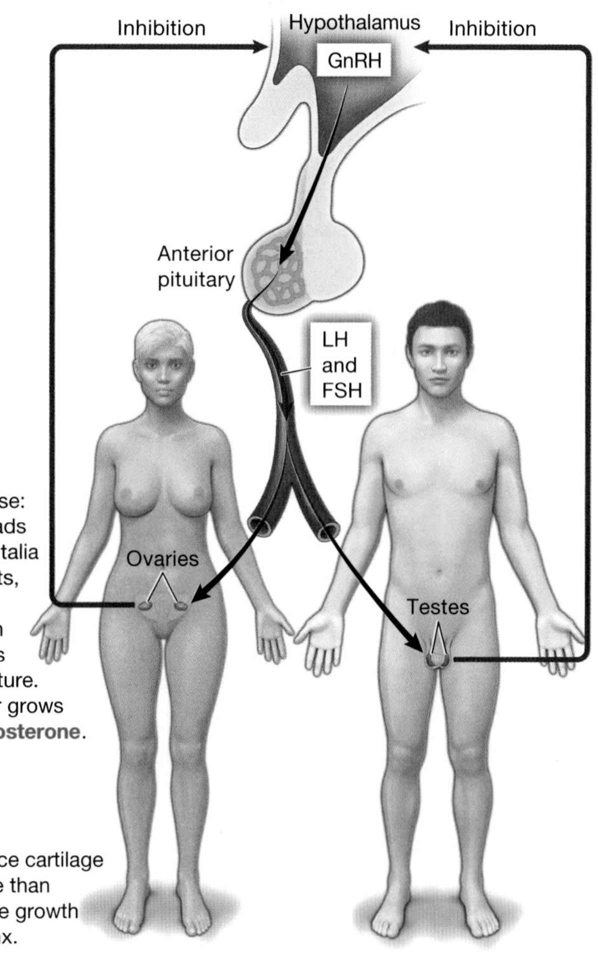

1. At the beginning of puberty, often around ages 8 or 9, the hypothalamus begins to increase its production of a tropic hormone called *gonadotropin-releasing hormone (GnRH).*
2. GnRH impacts the anterior pituitary, causing it to release follicle situlating hormone (FSH) and luteinizing hormone (LH).
3. In response to FSH and LH, the gonads secrete more **estrogen** and **testosterone**, which has impacts all over the body.
4. Effects of sex hormone release:
   - Gametogenesis in the gonads
   - Growth of the external genitalia
   - **Estrogen** impacts ligaments, causing hips to widen
   - **Estrogen** causes mitosis in breast tissue, which causes breasts to develop and mature.
   - Pubic, facial, and body hair grows under the influence of **testosterone**.
   - **Testosterone** impacts muscle development.
   - **Estrogen** impacts bone matrix development.
   - Both sex hormones influence cartilage growth, **testosterone** more than **estrogen**, which causes the growth of long bones and the larynx.
5. **Estrogen** and **testosterone** inhibit the hypothalamus from releasing GnRH.

nutrition; historical data demonstrate that as the human diet changed to include better and more consistent nutrition, the age of first menstruation (a signal that puberty is well underway) decreased from an average age of approximately 17 years of age in 1860 to the current age of approximately 12.75 years in 1960, in the United States. The accumulation of adipose tissue also impacts the onset of puberty, indicating that sexual maturity and readiness for reproduction are linked to availability of calories. Reproduction is incredibly calorie- and nutrient-intensive, so linking sexual maturity to nutrient availability makes a lot of sense.

27.2c Breasts

LO 27.2.3

Mammary glands, or *breasts*, are present in all individuals. They increase in size under the influence of estrogens, which promote mitosis in mammary tissue. Therefore, biological females tend to have larger breasts than biological males, who have lower levels of circulating estrogens. Due to the effects of estrogen, breasts begin their growth during puberty and may change in size throughout life. Breast size can also vary slightly along with hormone shifts during the menstrual cycle. In all bodies, adipose tissue is

Figure 27.7 Breast Anatomy

All adults have mammary glands, or breasts. Estrogen promotes mitosis in breast tissue, making breasts larger in individuals with higher circulating estrogen levels. Mammary glands are capable of producing milk, which exits through the nipple during lactation.

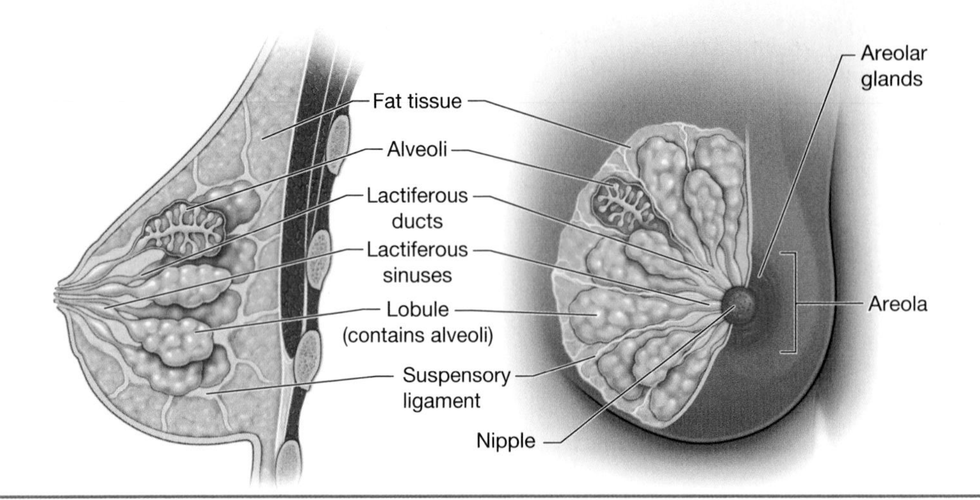

stored in the mammary region, among the glands. The amount of adipose tissue stored is the main factor that determines how large an individual's breasts are.

Mammary glands are modified sweat glands. Under the influence of prolactin, an anterior pituitary hormone, these glands are capable of generating milk. Breast milk is a mixture, of proteins, fats, sugars, water, and antibodies unique to each individual milk producer. The milk itself exits the breast through the **nipple**, a cylindrical projection on the anterior of the breast. Each nipple is connected to the glands via 15 to 20 **lactiferous ducts**. These lactiferous ducts each extend to a **lactiferous sinus** that connects to a glandular lobe within the breast itself that contains groups of milk-secreting cells in clusters called **alveoli** (**Figure 27.7**). The clusters can change in size depending on the amount of milk in the alveolar lumen. Once milk is made in the alveoli, it will be ejected through the ducts in response to secretion of the posterior pituitary hormone oxytocin.

Apply to Pathophysiology

XY Androgen Insensitivity

Individuals with XY androgen insensitivity have one X chromosome and one Y chromosome, but they lack the expression of androgen receptors due to a genetic mutation.

1. Will individuals with XY androgen insensitivity have an SRY region?
   A) Yes, these individuals will have the SRY region.
   B) No, these individuals will not have the SRY region.
2. During embryonic development, will they develop and differentiate the mesonephric ducts?
   A) Yes, because they carry a Y chromosome, they will develop and differentiate mesonephric ducts.
   B) While the mesonephric ducts will initially develop, they will not differentiate.
   C) They will never have mesonephric ducts develop or differentiate.
3. Will an adult with XY androgen insensitivity have mammary glands?
   A) An adult with XY androgen insensitivity will entirely lack mammary glands.
   B) An adult with XY androgen insensitivity will have mammary glands.
4. Will an adult with XY androgen insensitivity grow facial hair?
   A) Yes, facial hair will be present due to the presence of the Y chromosome.
   B) No, facial hair will not be present because growth of hair on the face requires testosterone signaling.

Learning Check

1. The gametes in the biological male reproductive system are called *sperm*. What is the homologue in biological females?
   a. Fallopian tubes
   b. Ovaries
   c. Follicles
   d. Oocytes
2. Sadie has a lower circulating level of estrogen than Sam. Who is expected to have larger mammary glands?
   a. Sadie
   b. Sam
   c. Impossible to tell with the information given
3. True/False: The control and sequence of events in puberty significantly varies among individuals.
   a. True
   b. False
4. The groups of milk-secreting cells in the glandular lobules of the breasts are called ______.
   a. Nipples
   b. Alveoli
   c. Mammary glands
   d. Lactiferous sinuses

27.3 Anatomy of Biological Females

Learning Objectives: By the end of this section, you will be able to:

27.3.1 Identify and describe the structure and functions of the external genitalia (e.g., mons pubis, labia majora, labia minora, clitoris, greater vestibular glands).

27.3.2 Identify and describe the structure and function of the vagina.

27.3.3 Identify and describe the gross anatomy, microscopic anatomy, and functions of the ovaries.

27.3.4 Identify and describe the ligaments of the female reproductive system (e.g., broad ligament, ovarian ligament, suspensory ligament of the ovary, round ligament of the uterus).

27.3.5 Describe the pathway of the oocyte from the ovary to the uterus.

27.3.6 Identify and describe the gross anatomy, microscopic anatomy, and functions of the uterus and uterine (fallopian) tubes.

The reproductive systems of all humans can be grouped by function. For both biological sexes we will discuss the external genitalia, which have functions in intercourse and pleasure as well as protection from infection. The **gonads** are the structures that produce gametes and reproductive hormones. In all mammals, one sex carries the additional reproductive tasks of housing the developing fetus(es) and delivering them to the outside world, and the option of feeding offspring through lactation. Therefore, the anatomy of biological females includes the structures involved in these separate but linked tasks.

27.3a External Genitalia

LO 27.3.1

vulva (pudendum)

The term **vulva** (Figure 27.8) encompasses many of the genital structures. The **mons pubis** is a pad of fat that is located at the anterior, over the pubic bone. At the onset of puberty, it becomes covered in pubic hair. There are two folds of epidermis, both called *labia* (which is Latin for lips), that comprise the majority of the visible genitals. The larger and more external **labia majora** are hair-covered skin folds that begin just posterior to the mons pubis. They cover the hairless inner **labia minora**. Both sets of

Figure 27.8 The Vulva

The term *vulva* encompasses all of the external female genital structures: the clitoris, the labia majora, and the labia minora. (A) Superficial view of the external female reproductive organs. (B) Enlarged view of some of the external female reproductive structures.

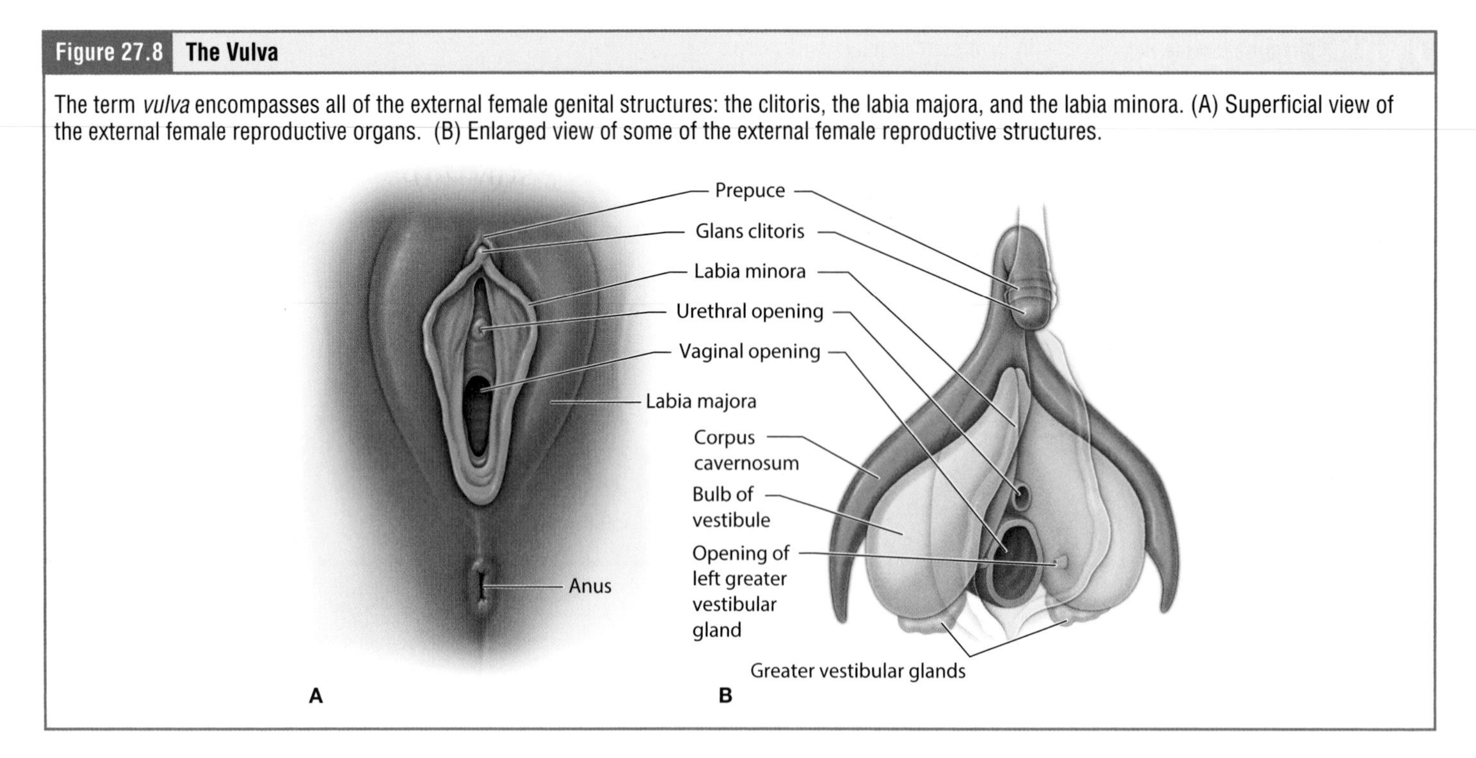

labia, but the labia minora in particular, naturally vary in shape and size from person to person. The labia minora serve to protect the urethra and vagina.

clitoris (glans clitoris)

bulbs of the vestibule (vestibular bulbs)

greater vestibular glands (Bartholin's glands)

The superior, anterior portions of the labia minora come together to encircle the **clitoris** (or *glans clitoris*), an organ that originates from the same cells as the glans penis and has abundant nerves that make it important in sexual sensation and orgasm. The vast majority of the clitoris tissue is internal; it is a horseshoe-shaped structure of erectile tissue, homologous to the erectile tissue of the penis, that straddles the internal anterior walls of the vagina. The clitoris flanks two erectile tissue bulbs called the **bulbs of the vestibule**. The vaginal opening is located between the opening of the urethra and the anus. It is flanked by openings to the **greater vestibular glands** (Bartholin's glands).

27.3b Vagina

LO 27.3.2

The **vagina** is a muscular canal that serves as the entrance and exit of the reproductive tract. At its superior end the vagina is gated by the cervix, the doorway to the uterus (**Figure 27.9**). The walls of the vagina are lined with an outer, fibrous adventitia; a middle layer of smooth muscle; and an inner mucous membrane with transverse folds called **rugae**. Together, the relaxation of the smooth muscle and stretch of the rugae allow the expansion of the vagina to accommodate intercourse and childbirth. The vagina and labia are both lubricated by a unique mucus produced by the vestibular glands.

The vagina is open to the outside world and is therefore vulnerable to microbes. During the onset of puberty, the pH of the vagina decreases from neutral (around a pH of 7) to acidic (below 4.5). At menopause, when there is a dramatic decrease in hormone production in the gonads, the pH of the vagina will return to the neutral range. In all phases of life, the vagina is home to a population of microorganisms. The specific microbe populations will shift with the pH changes across the lifespan. The microbes that call the vagina home actually help to protect against infection by pathogenic bacteria, yeast, or other organisms by utilizing the space and nutrients while causing no harm to their host. The most predominant bacteria in the vagina are from

Figure 27.9 Reproductive Structures of Biological Females

The organs involved in gametogenesis and reproduction are all housed within the pelvic cavity.

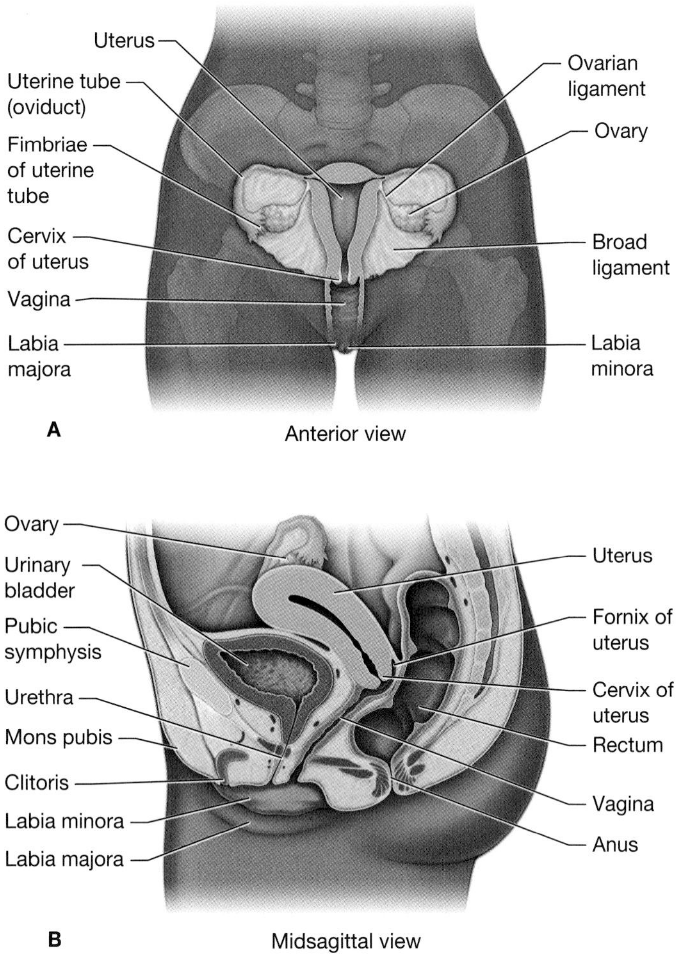

Uterus
Uterine tube (oviduct)
Fimbriae of uterine tube
Cervix of uterus
Vagina
Labia majora
Ovarian ligament
Ovary
Broad ligament
Labia minora

A Anterior view

Ovary
Urinary bladder
Pubic symphysis
Urethra
Mons pubis
Clitoris
Labia minora
Labia majora
Uterus
Fornix of uterus
Cervix of uterus
Rectum
Vagina
Anus

B Midsagittal view

the genus *Lactobacillus*. This family of beneficial bacterial flora secretes lactic acid, and thus protects the vagina by maintaining an acidic pH. Potential pathogens are less likely to survive in these acidic conditions. Lactic acid, in combination with other vaginal secretions, makes the vagina a self-cleansing organ. The vagina is also home to a yeast, *Candida albicans*, which lives in the mouth of all humans as well. While the presence of these two populations is normal and healthy, disturbances in their numbers can cause their host discomfort. For example, if the bacterial population dwindles, such as during a course of antibiotics taken for an infection elsewhere, the yeast population can grow to inhabit the space and consume the nutrients usually taken by the bacteria. In these cases, called *yeast infections*, there is no new microbe infection just an overgrowth of a population that was already present.

27.3c Ovaries

LO 27.3.3

The **ovaries** are the gonads that produce oocytes (eggs). There are typically two ovaries, each oval-shaped and approximately the size of almonds, though their size will change throughout the course of each menstrual cycle as well as during pregnancy (see Figure 27.9). The ovaries are located within the pelvic cavity and are suspended by a net of ligaments and connective tissue. The **mesovarium**, an extension of the peritoneum, connects the ovaries to the **broad ligament**, a blanket of connective tissue that covers and connects the internal reproductive structures. Extending from the mesovarium itself is the **suspensory ligament** that contains the ovarian blood and lymph vessels. Finally, the ovary itself is attached to the uterus via the **ovarian ligament** (**Figure 27.10A**).

LO 27.3.4

The ovary itself is clothed in an outer covering of cuboidal epithelium called the **ovarian surface epithelium**, just under it a dense connective tissue layer called the *tunica albuginea* can be found. Beneath the tunica albuginea, each ovary has an outer **ovarian cortex** and an inner **ovarian medulla**. The cortex is filled with *ovarian follicles*, the cellular structures that include eggs. The growth and development of ovarian follicles will be described shortly. The ovarian medulla is the site of blood vessels, lymph vessels, and the nerves of the ovary.

27.3d The Uterine Tubes

LO 27.3.5

uterine tube (fallopian tube or oviduct)

Student Study Tip

Tubes function in Transportation.

While the ovaries are anchored to the uterus by the mesovarium and broad ligament, these are solid structures and do not permit the passage of eggs toward the uterus. Instead, at their lateral edges, each ovary abuts a **uterine tube**. These serve as the conduit of the egg from the ovary to the uterus (**Figure 27.10B**). Each of the two uterine tubes is close to, but not directly connected to, the ovary. The uterine tubes are continuous tunnels that transmit eggs to the uterus, but functionally we consider them in sections. As an egg leaves the ovary it will first pass into the **infundibulum**, a wide portion of the tube that flares out with slender, fingerlike projections called **fimbriae**. The middle region of the tube, called the **ampulla**, is where fertilization often occurs. The last stretch, the **isthmus**, is the narrow medial end of each uterine tube that is connected to the uterus. The uterine tubes also have three layers: an outer serosa, a middle smooth muscle layer, and an inner mucosal layer. In addition to its mucus-secreting cells, the inner mucosa contains ciliated cells that beat in the direction of the uterus, producing a current that will be critical to move the egg.

LO 27.3.6

Figure 27.10 The Uterus and Ovaries

(A) The uterus and ovaries are tethered into place in the pelvic cavity by a series of ligaments. (B) Some of the ligaments are connected directly to the uterine tubes, which provide a pathway for eggs to travel from the ovary to the uterus.

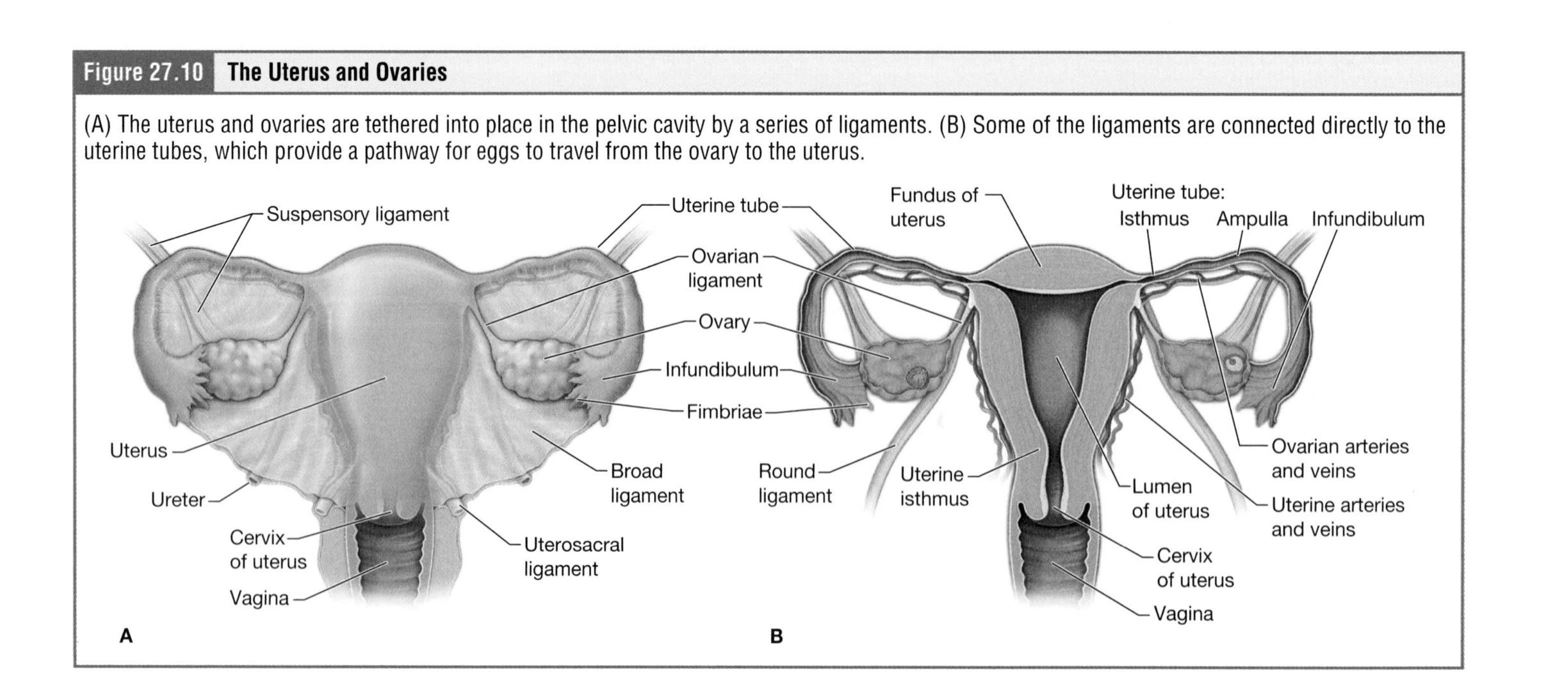

In the process of ovulation, an egg is released from the ovary into the peritoneal cavity. The nearby uterine tube receives the oocyte. Unlike sperm, oocytes lack flagella, and therefore cannot move on their own. So how do they travel into the uterine tube and toward the uterus? The high concentrations of estrogen at the time of ovulation induce contractions of the smooth muscle along the length of the uterine tube. These contractions occur every four to eight seconds, and the result is a coordinated movement that sweeps the surface of the ovary and the pelvic cavity. Current flowing toward the uterus is generated by coordinated beating of the cilia that line the lumen along the length of the uterine tube. These cilia beat more strongly in response to the high estrogen concentrations that occur around the time of ovulation. As a result of these mechanisms, the egg is pulled into the interior of the tube. Once inside, the muscular contractions and beating cilia move the egg slowly toward the uterus. If sperm are present in the reproductive tract, they typically meet the egg while it is still moving through the ampulla.

Student Study Tip

High estrogen concentrations signal that a follicle has developed enough for ovulation to occur. Lower estrogen concentrations signal immaturity.

If the oocyte is fertilized, the resulting **zygote** will begin to divide into two cells, then four, and so on, as it makes its way through the uterine tube and into the uterus. There, it will implant and continue to grow. If the egg is not fertilized, it will simply degrade—either in the uterine tube or in the uterus, where it may be shed with the next menstrual period.

zygote (fertilized egg)

27.3e The Uterus and Cervix

The **uterus** is a muscular organ, roughly the size of a small pear, located in the pelvic cavity posterior to the bladder (see Figure 27.9). Inferiorly, it connects to the vagina, and superior-laterally it connects to the uterine tubes. If pregnancy does occur, the growing embryo implants in the uterus and grows there for the duration of the pregnancy.

Several ligaments maintain the position of the uterus within the abdominopelvic cavity. The broad ligament, mentioned earlier, is a fold of peritoneum that serves as a primary support for the uterus, extending laterally from both sides of the uterus and attaching it to the pelvic wall. The **round ligament** attaches to the uterus near the uterine tubes and extends to the labia majora. Finally, the uterosacral ligament stabilizes the uterus posteriorly by its connection from the cervix to the pelvic wall (**Figure 27.11**).

Figure 27.11 Round Ligament Uterosacral Ligament

The round ligament tethers the uterus anteriorly.

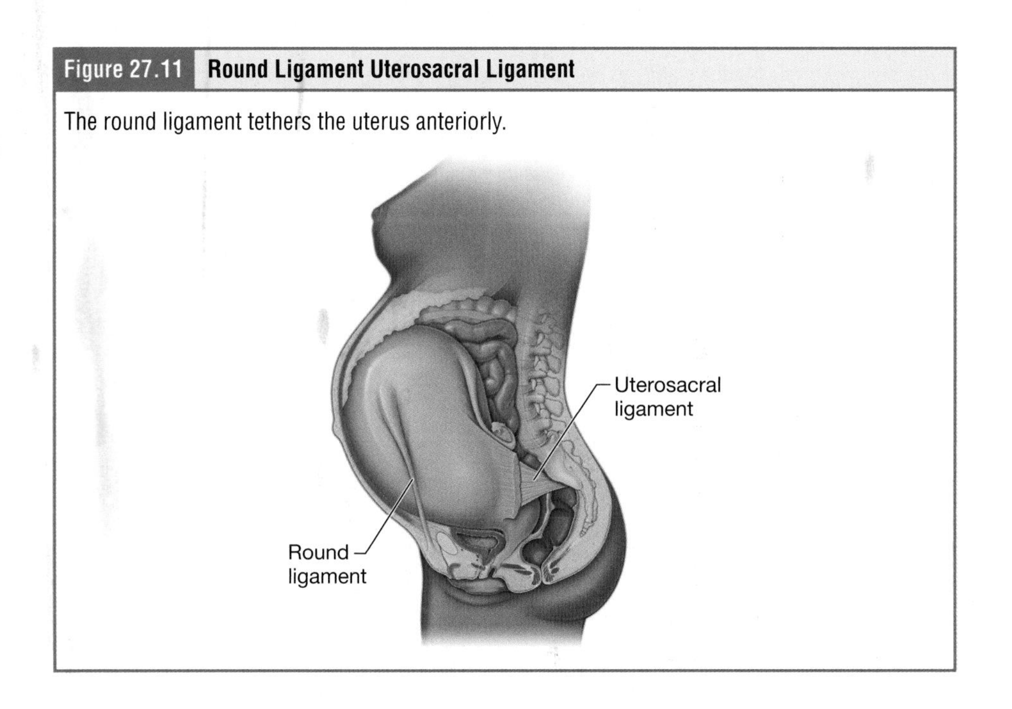

Figure 27.12 The Regions of the Uterus

The uterus is divided into three functional regions: the fundus, the body, and the cervix.

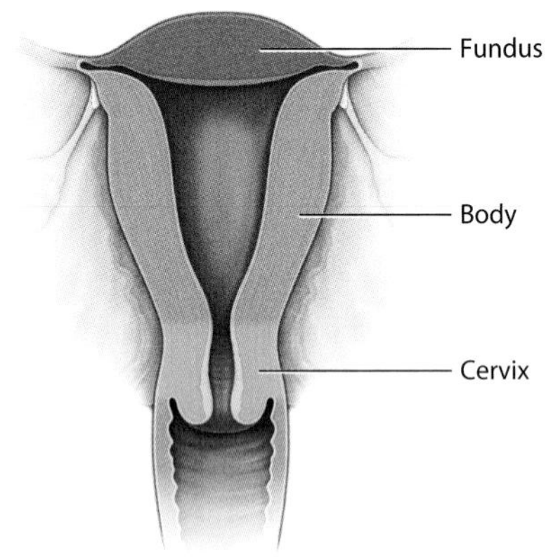

Fundus
Body
Cervix

There are three regions to the uterus and three layers. The region of the uterus superior to the opening of the uterine tubes is called the **fundus**. The middle section of the uterus is called the **body of the uterus**. The **cervix** is the narrow inferior portion of the uterus that projects into the vagina (**Figure 27.12**). The cervix is a narrow, almost closed channel through which material can pass from the uterus to the vagina or from vagina to uterus. The cervix produces mucus that obstructs the path through this channel. The consistency of the mucus changes throughout the menstrual cycle to allow easier passage at particular times in the cycle.

The wall of the uterus is made up of three layers. The most superficial layer is the serous membrane, or **perimetrium**, which consists of epithelial tissue that covers the exterior portion of the uterus. The middle layer, or **myometrium**, is a thick layer of smooth muscle responsible for uterine contractions (**Figure 27.13**). Most of the uterus is myometrial tissue, and the muscle fibers run horizontally, vertically, and diagonally, allowing the powerful contractions that can expel the contents of the uterus, whether they be a fetus being born at the end of pregnancy or menstrual tissue during menstruation.

The innermost layer of the uterus is called the **endometrium**. The endometrium contains a connective tissue lining, the **lamina propria**, which is covered by simple columnar epithelial tissue that lines the lumen. Deep mucus glands span the epithelium and extend deep into the lamina propria. Also, within the lamina propria, unique spiral arteries are found. These arteries supply the endometrium with oxygenated blood and nutrients. The cells of the lamina propria proliferate in response to estrogen, and the endometrium becomes thicker in response. At the end of a menstrual cycle in which fertilization did not occur, the spiral arteries constrict, cutting the supply of nutrients to the endometrium. The majority of this tissue layer dies off and the myometrial contractions will help it separate from the wall of the uterus and be expelled through the cervix and the vagina. The spiral arteries also play acritical role in pregnancy, so we will revisit them later in the chapter.

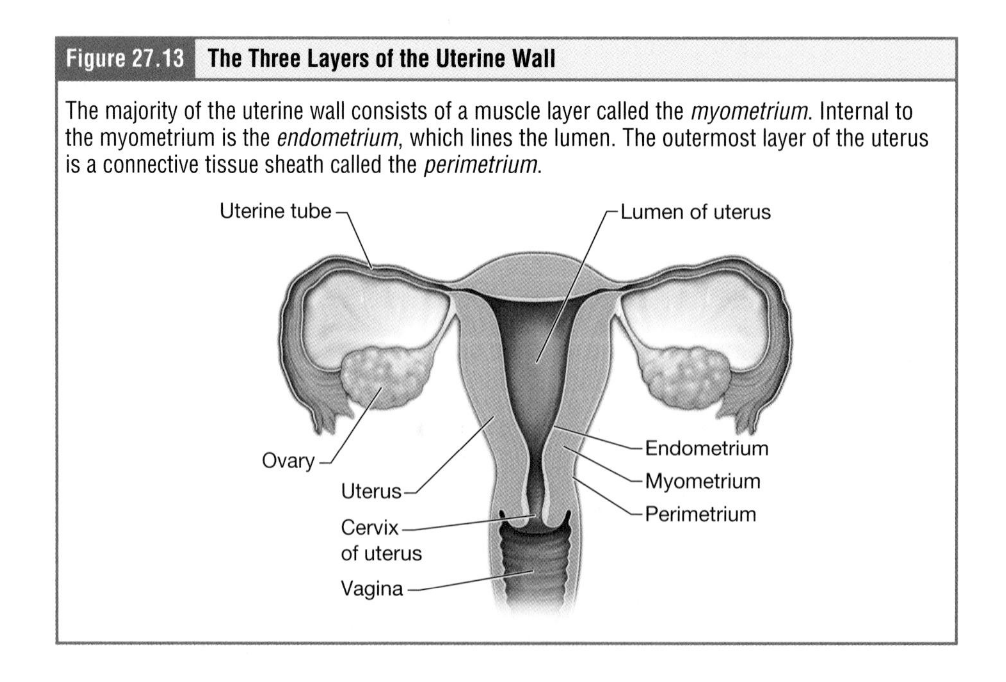

Figure 27.13 The Three Layers of the Uterine Wall

The majority of the uterine wall consists of a muscle layer called the *myometrium.* Internal to the myometrium is the *endometrium,* which lines the lumen. The outermost layer of the uterus is a connective tissue sheath called the *perimetrium.*

Learning Check

1. Which ligament attaches the ovary to the uterus?
   a. Mesovarium
   b. Broad ligament
   c. Suspensory ligament
   d. Ovarian ligament
2. What is the function of the labia minora?
   a. Protect the labia majora
   b. Protect the testes
   c. Protect the urethra and vagina
3. At puberty, the pH of the vagina (increases/decreases) and remains this way until menopause.
   a. Increases
   b. Decreases
   c. Impossible to tell with the information given
4. Which of the following is the correct pathway traveled by the oocyte on its way to the uterus after ovulation?
   a. Isthmus, ampulla, infundibulum, uterus
   b. Isthmus, infundibulum, ampulla, uterus
   c. Infundibulum, ampulla, isthmus, uterus
   d. Ampulla, infundibulum, isthmus, uterus

27.4 The Ovarian and Uterine Cycles

Learning Objectives: By the end of this section, you will be able to:

27.4.1 Define the process of oogenesis (oocyte development).

27.4.2 Describe the stages of folliculogenesis (ovarian follicle development).

27.4.3 Describe endocrine regulation of oogenesis, folliculogenesis, and the ovarian cycle.

27.4.4 Describe a typical ovarian cycle and explain how the process of folliculogenesis spans multiple ovarian cycles.

27.4.5 Define ovulation, and explain the role of luteinizing hormone (LH) in ovulation.

27.4.6 Name the phases of the uterine (menstrual) cycle, and describe the anatomical changes in the uterine wall that occur during each phase.

27.4.7 Describe the correlation between the uterine and ovarian cycles.

Human embryos form from the fusion of two gametes: egg and sperm. Each of these brings half of the chromosomes required for life. Once the fertilized egg has a full set of chromosomes, this single cell will divide through the process of mitosis. After a few days of mitosis, the ball of identical cells will begin to change; cells will differentiate—that is, take on unique characteristics. Cells will migrate and the embryo will begin to take a shape that somewhat resembles a human. As development progresses, the embryo will form internal organs. Once the body plan is established and growth (instead of formation) is the primary objective, we will use the term *fetus*. Human (and all mammalian) fetuses grow and develop inside the uterus, with continual access to nutrients by way of exchange between maternal and fetal blood that occurs across a fetal structure called the *placenta*. Therefore, when we consider the reproductive functions of humans, we consider that, in addition to preparing gametes, biological females will as also be preparing the uterus for the possibility of housing a developing embryo/fetus. The preparation of these two sites occurs in a cyclical fashion. A cycle takes 21–35 days for most biological women. There is a huge degree of diversity, but we will use 28 days here. We will examine the ovarian and uterine cycles separately, but keep in mind that the two cycles occur concurrently and pregnancy can only happen if they are coordinated. Biological males, who will be considered in Sections 27.5 and 27.6, only prepare gametes and therefore do not require coordination or have a cyclical nature to their reproductive function.

27.4a The Ovarian Cycle

Eggs are housed and readied within the ovaries. The ovaries will go through a set of predictable, hormone-driven events to mature and release a single (typically) egg each cycle. The ovarian cycle includes two interrelated processes: **oogenesis** (the production of eggs) and **folliculogenesis** (the growth and development of ovarian follicles).

LO 27.4.1

Oogenesis Oogenesis begins with the ovarian stem cells, or **oogonia** (see Figure 27.11). Oogonia are formed during fetal development and divide via mitosis. During the fetal period, prior to birth, oogonia begin the process of meiosis, forming **primary oocytes**. These primary oocytes are then stopped in this stage of meiosis I and eggs in this arrested state can be found in the ovaries throughout childhood. The ovarian cycle, which occurs an average of once every 28 days beginning at puberty and continuing until menopause (the cessation of a reproductive function in biological women), will stimulate one of these primary oocytes to mature and complete meiosis I, to prepare it for possible fertilization. The ovarian cycle can only begin with these arrested primary oocytes; therefore, the number of primary oocytes present in the ovaries at birth represents the total number of oocytes available for the individual's entire lifetime. This number declines from 2 to 1 million in an infant, to approximately 400,000 at puberty, to 0 by the end of menopause.

Once a primary oocyte has progressed through the stages of meiosis and maturity, it is released from the ovary in the process of **ovulation**. Just prior to ovulation, a surge of LH triggers the resumption of meiosis in a primary oocyte. This initiates the transition from primary to secondary oocyte. However, as you can see in **Figure 27.14**, this cell division does not result in two identical cells. Instead, the cytoplasm is divided unequally, and one daughter cell is much larger than the other. This larger cell, the **secondary oocyte**, eventually leaves the ovary during ovulation. The smaller cell, called the first **polar body**, contains half of the chromosomes; it eventually disintegrates.

Ovulation will occur if the secondary oocyte is mature and ready for fertilization, but even at this point the secondary oocyte has not yet completed meiosis II. Meiosis II is only completed *if* a sperm fertilizes the secondary oocyte. Meiosis then resumes and completes and the nucleus of the egg is ready to fuse with the nucleus of the sperm, producing a new diploid cell that is a unique combination of the genomes of the egg and sperm. As meiosis is completed, the oocyte releases one or two second polar bodies containing unneeded chromosomes. Therefore, even though oogenesis produces up to four cells, only one survives to become a fertilizable egg. Upon fertilization, the fertilized egg is called a **zygote** (**Figure 27.15**).

Sperm are streamlined for travel; they are little more than traveling nuclei. Therefore, the majority of the cytoplasm and organelles of the zygote come from the egg. Nutrients contained in the cytoplasm of the egg supply the developing zygote during the period between fertilization and implantation into the uterus. The mitochondria

Figure 27.14 Oogenesis

The mitotic events that generate eggs are completed before birth and all the eggs are arrested in development. At the onset of puberty, meiosis begins again and is only completed in eggs that are fertilized.

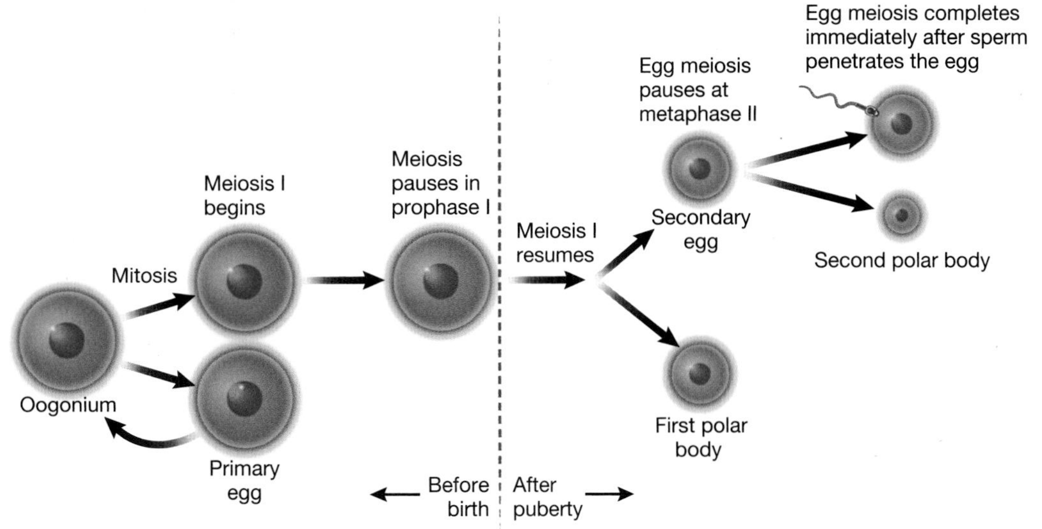

Oogonium
Mitosis
Primary egg
Meiosis I begins
Meiosis pauses in prophase I
Before birth
After puberty
Meiosis I resumes
Secondary egg
First polar body
Egg meiosis pauses at metaphase II
Egg meiosis completes immediately after sperm penetrates the egg
Second polar body

of the egg become all of the mitochondria of the zygote/embryo/fetus/offspring, which has interesting implications because mitochondria contain their own DNA. Therefore, all mitochondrial DNA is maternally inherited, meaning that you can trace mitochondrial DNA throughout all of your female ancestors.

Folliculogenesis During their time in the ovaries, eggs are enveloped within a nest of cells called an **ovarian follicle**. Primary oocytes are encased in **primordial follicles**, so during childhood all follicles in the ovary will be primordial follicles (**Figure 27.16**). Primordial follicles have only a single flat layer of support cells, called **granulosa cells**, that surround the oocyte. In an adult ovary, almost all of the follicles will be primordial follicles, but the follicles that surround maturing eggs during the ovarian cycle are changing too. The follicles are growing and developing in a process called *folliculogenesis*.

Folliculogenesis begins with primordial follicles in a resting state. After puberty, a few primordial follicles will respond to a recruitment signal each day and will join a pool of immature growing follicles called **primary follicles**. Primary follicles start with a single layer of granulosa cells, but the granulosa cells then become active and transition from a flat or squamous shape to a rounded, cuboidal shape as they increase in size and proliferate. As the granulosa cells divide, the follicles—now called **secondary follicles** (see Figure 27.16)—increase in diameter, adding a new outer layer of connective tissue, blood vessels, and **thecal cells**—cells that work with the granulosa cells to produce estrogens. With each ovarian cycle, 15–20 primordial follicles begin folliculogenesis.

Within the growing secondary follicle, the primary oocyte now secretes a thin jelly-like coat called the *zona pellucida* that will play a critical role in fertilization. A thick fluid, called *follicular fluid*, that has formed between the granulosa cells also begins to collect into one large pool, or **antrum**. Follicles in which the antrum has become large and fully formed are considered **tertiary follicles** (or *antral follicles*). Several follicles reach the tertiary stage at the same time, and one will continue to grow and develop until ovulation; the rest will undergo a process called *atresia*. **Atresia** is the death of ovarian follicle; this can occur at any point during follicular development. This process, in which many follicles are readied but only one (typically) is ovulated, acts as a quality control mechanism to ovulate

LO 27.4.2

Figure 27.15 Fertilization

When the nucleus from the sperm enters the cytoplasm of the egg, the egg is fertilized and becomes diploid. The fertilized egg is called a *zygote*.

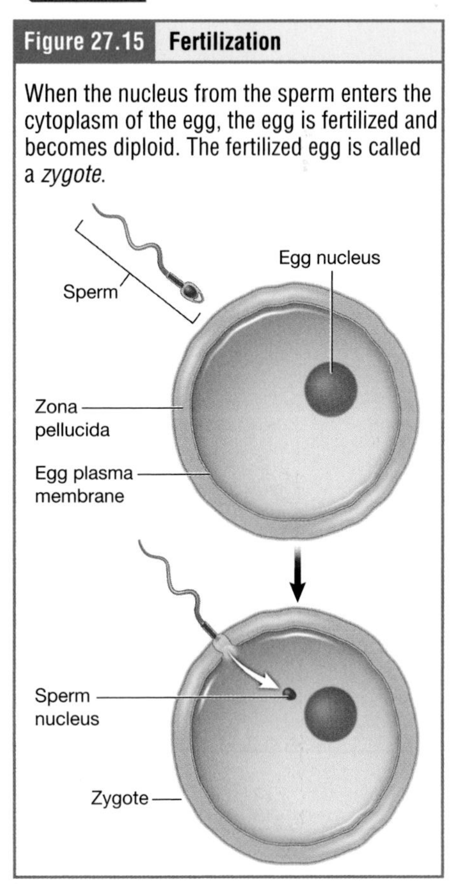

tertiary follicles (antral follicles, mature follicles, mature antral follicles, Graafian follicles)

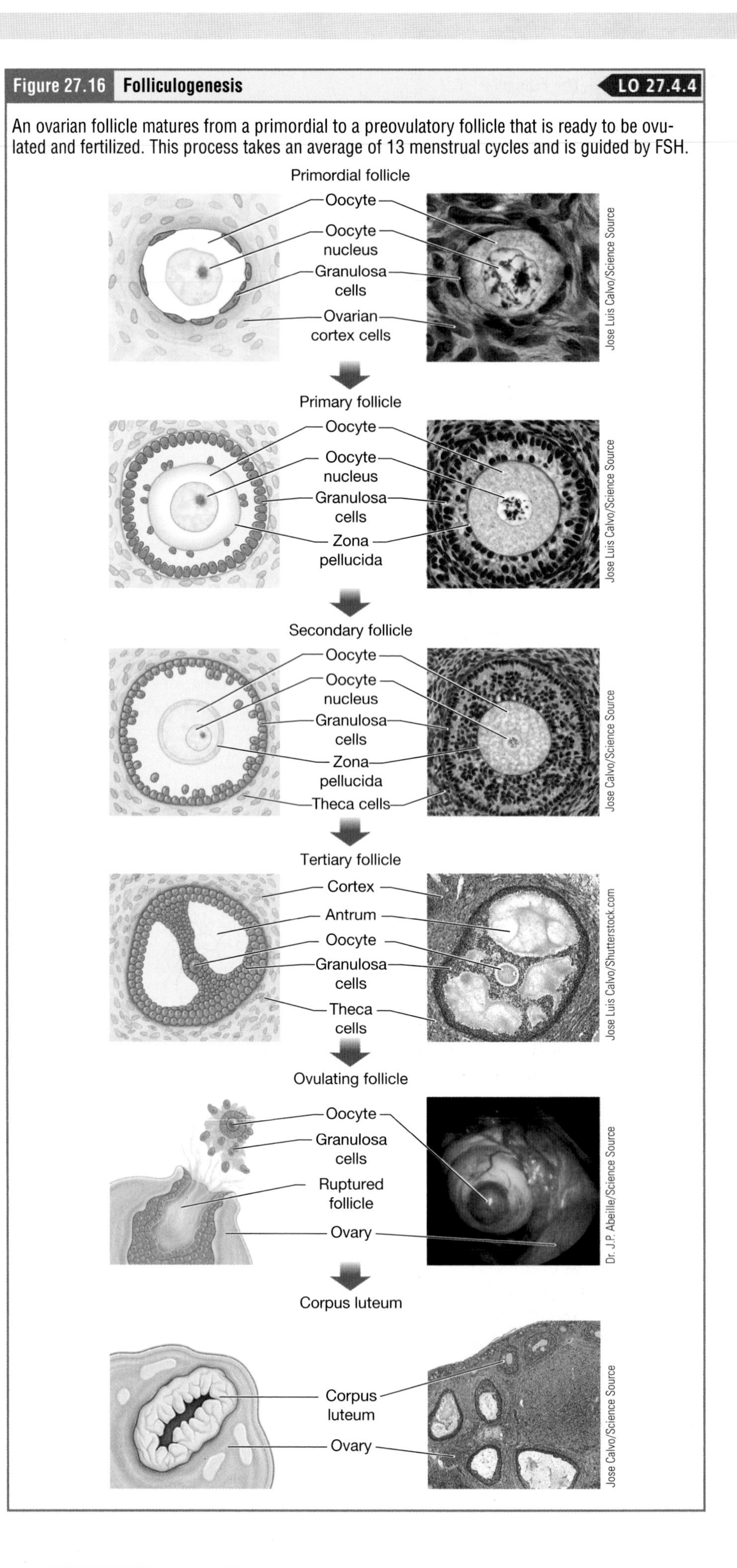

Figure 27.16 Folliculogenesis LO 27.4.4

An ovarian follicle matures from a primordial to a preovulatory follicle that is ready to be ovulated and fertilized. This process takes an average of 13 menstrual cycles and is guided by FSH.

Figure 27.17 The Follicular Phase

FSH drives the development of follicles in the ovary. As they mature, they contribute more and more estradiol to the bloodstream.

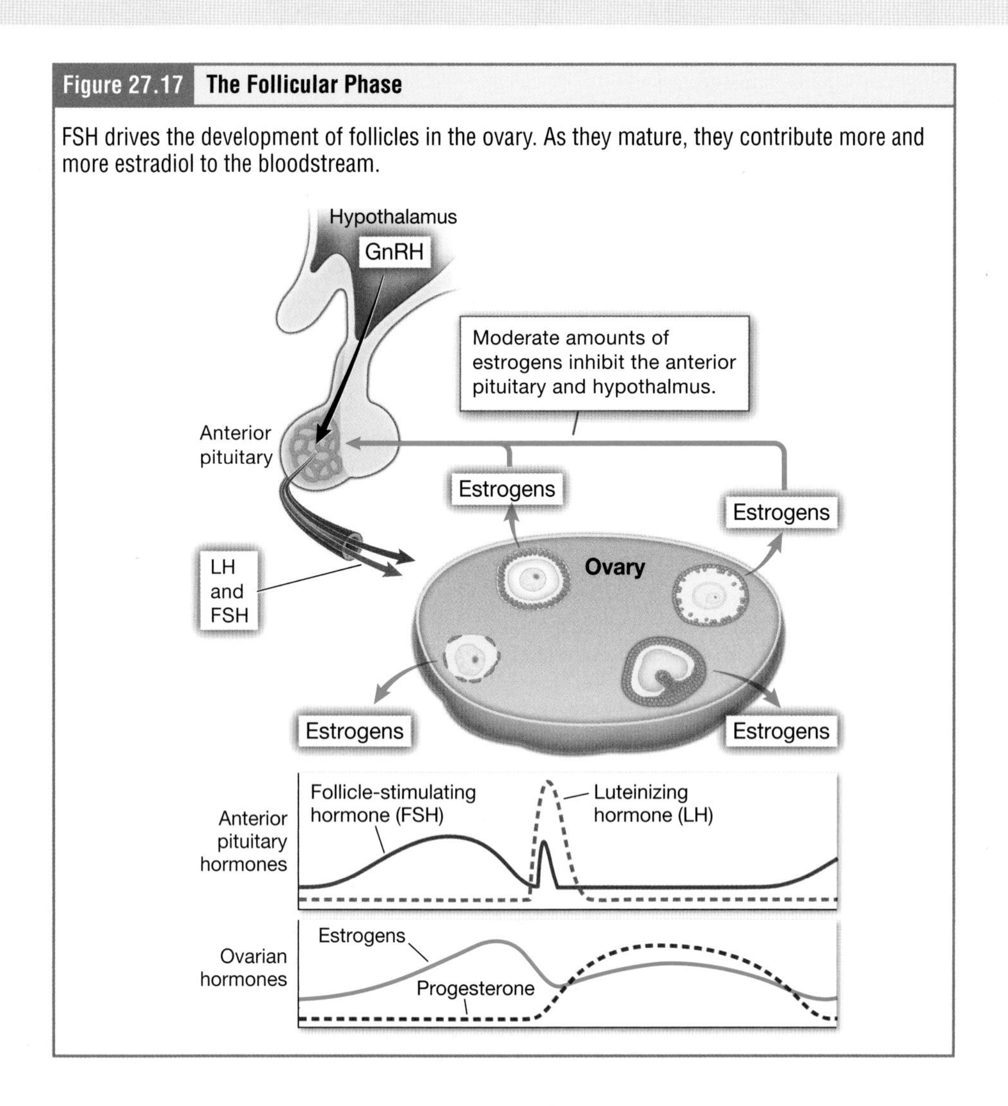

only the most viable egg of those that begin oogenesis. Roughly 99 percent of the follicles in the ovary will undergo atresia, which can occur at any stage of folliculogenesis.

Hormonal Control of the Ovarian Cycle The process of development that we have just described, from primordial follicle to early tertiary follicle, takes approximately two months in humans. The final stages of development occur over a course of approximately 28 days. These changes are regulated by several hormones, including GnRH, LH, and FSH.

LO 27.4.3

The hypothalamus produces GnRH, a hormone that signals the anterior pituitary gland to produce the hormones FSH and LH (**Figure 27.17**). These hormones leave the pituitary and travel through the bloodstream to the ovaries, where they bind to receptors on the granulosa cells and thecal cells of the follicles. FSH stimulates the follicles to grow (hence its name of *follicle stimulating hormone*), and the handful of tertiary follicles expand in size. The release of LH also stimulates the granulosa cells and thecal cells of the follicles to produce the sex steroid hormone *estradiol*, a type of estrogen. This phase of the ovarian cycle, when the tertiary follicles are growing and secreting estrogen, is known as the *follicular phase*.

The more granulosa cells and thecal cells a follicle has (that is, the larger and more developed it is), the more estrogen it will produce in response to LH stimulation. As a result of these large follicles producing large amounts of estrogen, blood levels of estrogen increase. This moderately higher concentration of estrogen will inhibit the hypothalamus and anterior pituitary, thereby reducing the production of GnRH, LH, and FSH. Because the large tertiary follicles require FSH to grow and survive, this decline

Cultural Connection

Twins

You may have heard that twins run in families. This is only partly true. There are two types of twins, commonly known as *identical* and *fraternal*. Identical twins do not run in families; rather, they form from a completely random event that is just as likely to occur in one pregnancy as in another. Out of every 1,000 pregnancies, 1 will result in identical twins. In identical twin pregnancies, one egg is fertilized by one sperm, forming a single zygote. The zygote goes through an initial round of mitosis, resulting in a two-cell embryo. In the next round of mitosis the embryo grows to four cells, and in the next eight, and so on. At the two- , four- , or eight-cell stage, the embryo might spontaneously separate into two clusters of cells (see figure). Each cluster of cells is identical to the other, carrying 100 percent of the same genes. If both embryos implant, the resulting pregnancy is termed **monozygotic** because it came from the same zygote. Fraternal, or nonidentical, twins, on the other hand, occur when two eggs are ovulated; once in the uterine tubes, both eggs are fertilized by separate sperm (see figure). Because these embryos resulted from two different eggs and two different sperm, the twins are no more alike or different than any other set of siblings, sharing on average about 50 percent of their genome. These twin pregnancies are called **dizygotic** because they result from two different zygotes. Whereas monozygotic twins are random, dizygotic twin pregnancies do run in families. The tendency to ovulate more than one egg is called *hyperovulation*, and about 4 percent of biological females are prone to hyperovulation. There are two genes that contribute to the likelihood of hyperovulation. Both genes play a role in FSH signaling. The first gene, FSHB, causes this individual to secrete more FSH. More FSH in the blood stimulates more follicles to grow and increases the chance of concurrent ovulation of more than one egg. The second gene, SMAD3, increases responsiveness to FSH. While individuals with alterations in SMAD3 do not have more FSH in their blood, they respond more to typical levels of FSH. When it comes to hormones, we can always remember that it takes two to tango—either an increase in the signal or an increase in its receptor will result in more signaling.

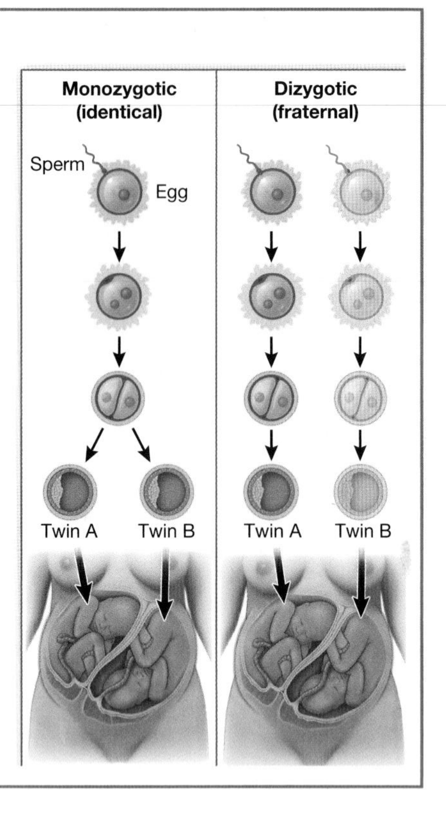

in FSH caused by negative feedback leads most of them to die (atresia). Typically, only one follicle will survive this reduction in FSH, and this follicle will be the one that releases an oocyte. Which follicle is the one to survive while the others die off? There are several factors—including size, the number of granulosa cells, and the number of FSH receptors on those granulosa cells—that predict follicular success and contribute to a follicle becoming the one surviving dominant follicle.

After the process of atresia, only the one dominant follicle remains in the ovary, but it is still secreting estrogen. As it continues to grow, it produces more estrogen than all of the developing follicles did together. It produces so much estrogen that the normal negative feedback in the hypothalamus does not occur. Instead, these extremely high concentrations of systemic plasma estrogen trigger a regulatory switch in the anterior pituitary, which responds by secreting a surge of LH and FSH into the bloodstream (**Figure 27.18**). This positive relationship, in which a high amount of estrogen triggers release of more LH and FSH, only occurs at this point in the cycle. The moderate amounts of estrogen produced earlier in the cycle inhibit FSH and LH.

LO 27.4.5

Student Study Tip

This pressure creates an antral fluid jet: internal pressure is just like a hose being turned on with a ball covering the nozzle. The pressure from the antrum shoots the ball (the oocyte) from the hose (antrum) and into the external environment (peritoneal cavity).

It is this large burst of LH (called the *LH surge*) that leads to ovulation of the dominant follicle. The LH surge also triggers the release of enzymes that break down structural proteins in the ovary wall on the surface of the bulging dominant follicle (**Figure 27.19**). This degradation of the wall, combined with pressure from the large, fluid-filled antrum, results in the expulsion of the oocyte surrounded by granulosa cells into the peritoneal cavity. This release is ovulation.

In the next section, you will follow the ovulated oocyte as it travels toward the uterus, but there is one more important event that occurs in the ovarian cycle. The surge of LH also stimulates a change in the granulosa and thecal cells that remain in the follicle after the oocyte has been ovulated. This change is called *luteinization* (recall that LH stands for "luteinizing hormone"), and it transforms the leftover follicle into an endocrine structure called the **corpus luteum**, a term meaning "yellowish body" (**Figure 27.20**). The corpus

Figure 27.18 **Triggering Ovulation**

The large amounts of estrogen released by the preovulatory follicle into the bloodstream trigger the anterior pituitary to release a surge of LH. This LH surge causes the follicle to release the egg during ovulation.

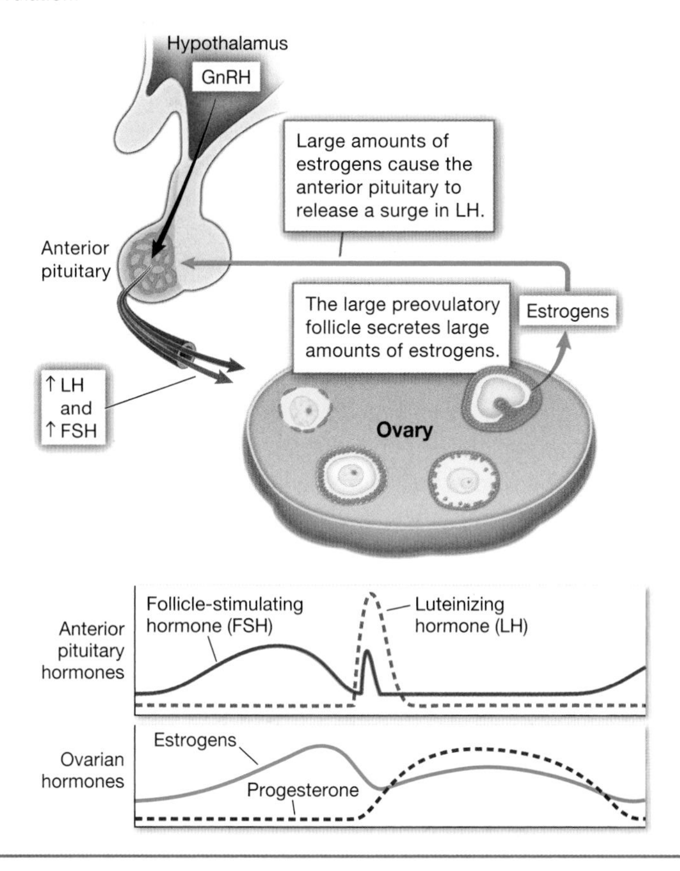

Figure 27.19 **Ovulation**

The preovulatory follicle produces a large, raised bump on the surface of the ovary. Once LH is received by the follicle, the ovarian wall on the follicle rips open and the egg, along with a stream of antral fluid, is released.

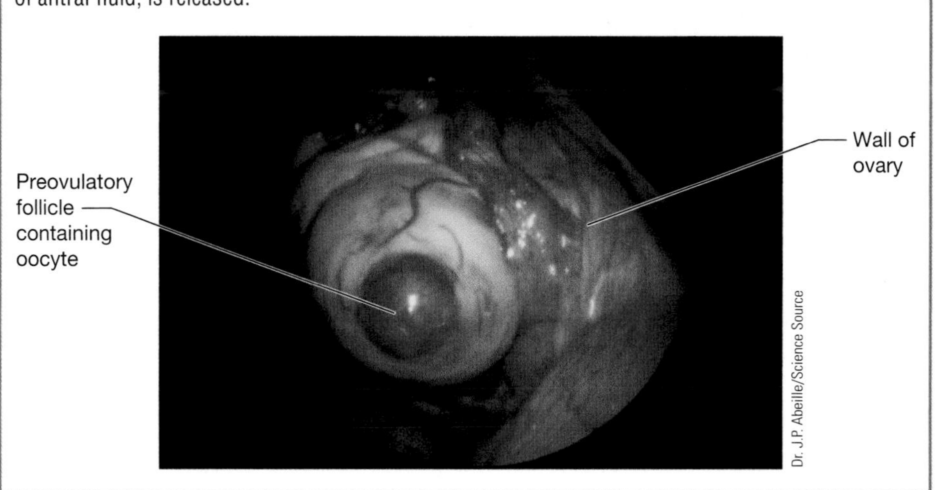

Figure 27.20 The Corpus Luteum

The corpus luteum forms from the leftover follicle after ovulation. It engages in active secretion of two hormones—estrogen and progesterone—and will last for 10–12 days, unless another hormone (hCG) is present, in which case the corpus luteum will persist and release hormones for the next 10–12 weeks.

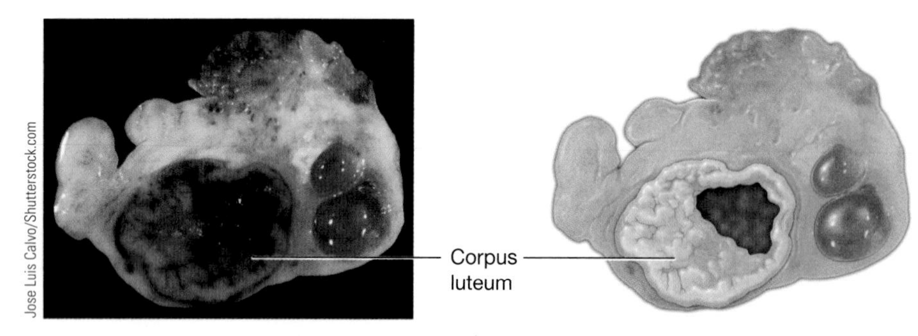

luteum is capable of producing both estrogen and the steroid hormone progesterone. Progesterone levels increase in the blood. Progesterone plays two key roles at this time: it affects the uterus, as we will discuss in the next section, but it also affects the hypothalamus and anterior pituitary, inhibiting their secretion of GnRH, LH, and FSH. At this point in the cycle, and for as long as the corpus luteum stays active in the ovary, blood levels of GnRH, LH, and FSH will remain low and no new dominant follicles will develop.

The postovulatory phase is known as the *luteal phase* of the ovarian cycle. The corpus luteum will endure within the ovary, secreting hormones into the bloodstream, for approximately 10 to 12 days. If the ovulated egg is fertilized, the embryo will secrete a signal—hCG, which will cause the corpus luteum to persist and continue its production of hormones. If no hCG signal is received, the corpus luteum will begin to die. It slowly degrades into the **corpus albicans**, a nonfunctional structure that will disintegrate in the ovary over a period of several months. As progesterone levels decline, FSH and LH inhibition reduces, and these hormones are once again secreted by the anterior pituitary. A new follicular phase begins, with a new cohort of early tertiary follicles beginning to grow and secrete estrogen (**Figure 27.21**).

27.4b The Uterine Cycle

Now that we have discussed the maturation of the follicles in the ovary, and the hormones that cycle throughout the 28-day (average) ovarian cycle, we can examine the impact of these hormones on the uterus. The endometrium—the lining of the uterus—is

Figure 27.21 Blood Levels of the Ovarian and Anterior Pituitary Hormones

The release of the ovarian hormones estrogen and progesterone is influenced by tropic hormones from the anterior pituitary.

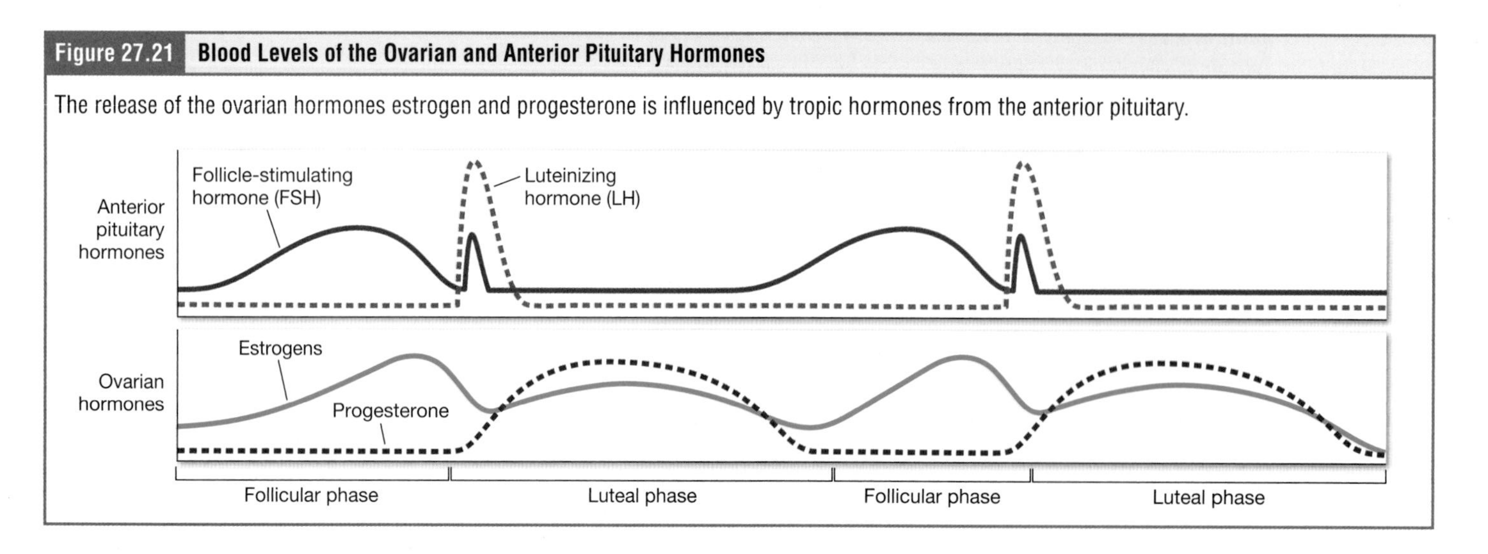

highly responsive to estrogens. If we examine the blood levels of estrogen across the cycle, we can see that estrogen blood levels ramp up during the follicular phase and are moderate throughout the luteal phase. In addition, progesterone is secreted only during the luteal phase. Let's discuss the impact of each of these hormones on the endometrium.

- Estrogens: estrogens promote mitosis, in that they trigger the proliferation of endometrial cells and therefore the expansion of the endometrial layer. The first half of the cycle is marked by the rapid proliferation called the *proliferative phase*. Estrogen also increases the expression of progesterone receptors on endometrial cells.
- Progesterone: progesterone causes endometrial secretory cells to make and secrete their products. Progesterone signaling is only meaningfully present in this second half of the uterine cycle; this half is usually called the *secretory phase* for this reason. Progesterone also inhibits myometrial contractions, thereby preventing the shedding of the endometrial lining.

Because the blood levels of the hormones estrogen and progesterone fluctuate in a predictable pattern over the course of ovarian cycle, we can describe the concurrent uterine events as the **uterine cycle** (menstrual cycle)—the series of changes in which the uterine lining is shed, rebuilds, and prepares for implantation.

uterine cycle (menstrual cycle)

The shedding of the uterine lining results in the discharge of blood and tissue through the cervix and the vagina. In A&P, we typically refer to this event using the word **menses**, but in English it is most commonly called the **menstrual period** or just *period* (of course, many people in many cultures have other names for this event). Day 1 of the ovarian and uterine cycles is counted as the first day of a noticeable menstrual period. While there as there are two phases of the ovarian cycle, there are three phases of the menstrual cycle—the menses phase, the proliferative phase, and the secretory phase.

menstrual period (period)

Menses Phase The **menses phase** of the menstrual cycle is the phase during which the lining is shed—that is, the days that the menstrual period occurs. Although it averages approximately five days, the menses phase can last from two to seven days, or longer. As shown in Figure 27.22, the menses phase occurs during the early days of the follicular phase of the ovarian cycle, when progesterone, FSH, and LH levels are low. Recall that progesterone concentrations decline as a result of the degradation of the corpus luteum, marking the end of the luteal phase. This decline in progesterone triggers the shedding of the endometrium.

menses phase (menstrual period, period, menstruation)

LO 27.4.6

Proliferative Phase A few days into the cycle, the tertiary follicles begin to secrete estradiol into the bloodstream. At this point menstrual flow has usually slowed or stopped and the endometrium begins to proliferate in response to the increasing blood levels of estrogen. This is the **proliferative phase** of the menstrual cycle (see Figure 27.22). Estrogen levels will increase steadily until ovulation, and so more and more proliferation within the endometrium occurs during this time.

Secretory Phase In addition to prompting the LH surge, high estrogen levels increase the uterine tube contractions that help propel the ovulated oocyte on its journey toward the uterus. High estrogen levels also slightly decrease the acidity of the vagina, making it more hospitable to sperm. In the ovary, the leftover follicle is becoming the corpus luteum, which secretes progesterone. Progesterone inhibits uterine muscle contractions, preventing the contractions that trigger the loss of the endometrium. As long as progesterone levels in the blood remain moderate or high, the endometrium stays in place and is ready for implantation of an embryo, if fertilization and embryogenesis occur. At this time, progesterone also stimulates the endometrial glands to secrete a nutrient-rich mucus that will serve as nourishment

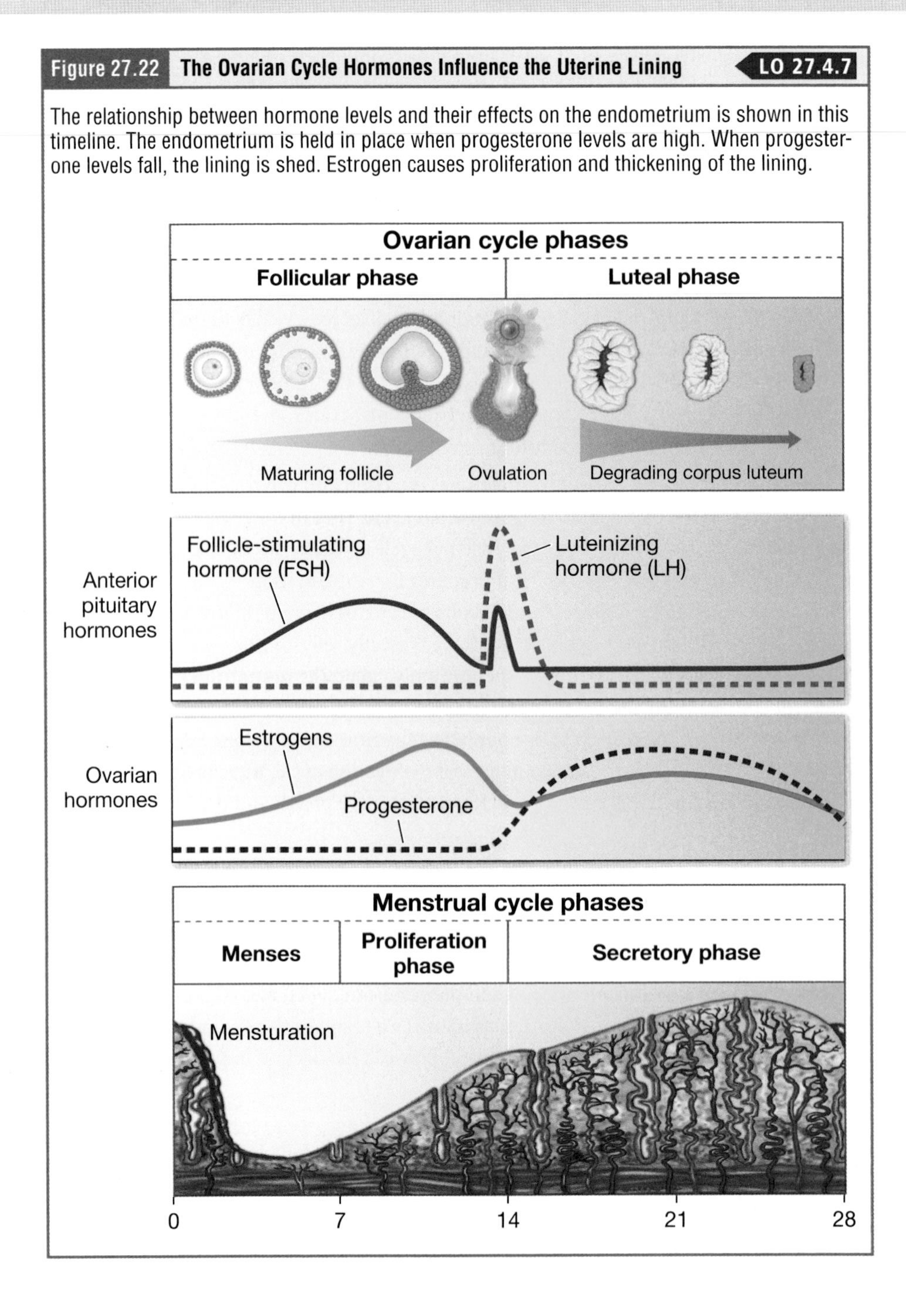

Figure 27.22 The Ovarian Cycle Hormones Influence the Uterine Lining LO 27.4.7

The relationship between hormone levels and their effects on the endometrium is shown in this timeline. The endometrium is held in place when progesterone levels are high. When progesterone levels fall, the lining is shed. Estrogen causes proliferation and thickening of the lining.

for the embryo if it implants; this is called the **secretory phase** of the menstrual cycle. The spiral arteries are elongating and widening to deliver increased blood flow to this developing endometrium.

If no fertilization of the egg occurs within approximately 10 to 12 days, the corpus luteum will begin to degrade into the corpus albicans. Levels of both estrogen and progesterone will fall, and the endometrium will thin. The spiral arteries will constrict, reducing oxygen supply, and the cells of the endometrial tissue will die. Without sufficient progesterone to inhibit uterine muscle contractions, the myometrium will contract, causing the shedding of the dead endometrial layer. The tissue and blood will pull away from the uterine wall into the lumen of the uterus and be discarded through the cervix and vagina as menses.

Learning Check

1. In which menstrual cycle phase is the corpus luteum present?
   a. Menses phase
   b. Proliferative phase
   c. Secretory phase
   d. Both A and C
2. Which of the following processes is about equal to one menstrual cycle in length?
   a. Oogenesis
   b. Folliculogenesis
   c. Both A and B
   d. Neither A nor B
3. Secretion of which of the following hormones is directly stimulated by GnRH?
   a. FSH
   b. Estrogen
   c. FSH and LH
   d. Progesterone
4. Which gamete(s), if any, has/have mitochondrial DNA?
   a. Sperm
   b. Egg
   c. Both
   d. Neither

27.5 Anatomy of Biological Males

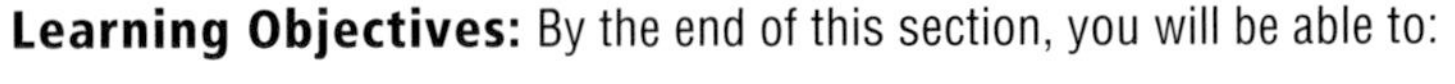
Learning Objectives: By the end of this section, you will be able to:

27.5.1 Identify and describe the structure and functions of the male external genitalia (e.g., scrotum, penis).

27.5.2 Identify and describe the gross anatomy, microscopic anatomy, and functions of the testes.

27.5.3 Identify and describe the gross anatomy, microscopic anatomy, and functions of the epididymis.

27.5.4 Identify and describe the structure and functions of the spermatic cord and male reproductive ducts (e.g., ductus [vas] deferens, ejaculatory duct, urethra).

27.5.5 Describe the production, composition, and functions of semen.

27.5.6 Identify and describe the structure and functions of accessory glands (i.e., seminal glands [seminal vesicles], prostate gland, bulbourethral [Cowper] glands).

27.5.7 Describe the pathway of sperm from the seminiferous tubules to the external urethral orifice of the penis.

Just as we organized the anatomy in the discussion of biological females in Section 27.3, we will group the anatomical structures by function. As with biological females, the external genitalia have functions in intercourse and pleasure. The gonads are the structures that produce gametes and reproductive hormones. During reproduction, biological males deliver gametes; they do not offer a site for fertilization or house a developing fetus. Therefore, our discussion of the structures involved in the reproductive system of biological males is somewhat shorter.

LO 27.5.1

Figure 27.23 The External Genitalia of a Biological Male

From the external perspective, the penis and scrotum are visible.

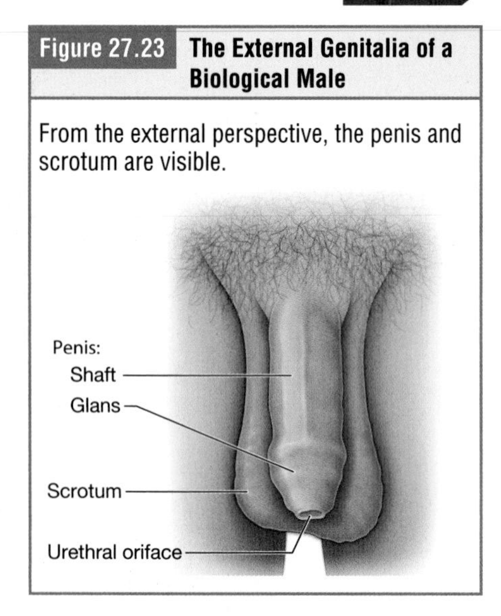

glans (glans penis)

27.5a External Genitalia

The external structures of biological males include the penis and the scrotum (Figure 27.23). The **penis** is a rodlike external organ through which the urethra passes to allow either urine or semen to exit the body. *Semen* is a mixture of sperm and other fluids and enzymes that help the sperm to travel safely within the vagina and uterus. The scrotum is a saclike structure found posterior to the penis that holds the testes, which are the gonads.

27.5b The Penis

The penis is the male homologue of the clitoris; both structures are composed of erectile tissue that is capable of filling with blood and becoming more firm and rigid during arousal.

The penis is considered an external structure, but, like the clitoris, a substantial amount of the tissue of the penis is inside the body. The **root** is internal and the shaft and glans are the portions of the penis visible externally. The **shaft** is the rodlike column of the penis and the **glans** is the expanded head at the distal end. The glans is covered in foreskin (prepuce) and contains the **external urethral orifice**, the opening of the urethra at the tip of the penis. The urethra traverses the entire length of the penis (Figures 27.24 and 27.25). The shaft is composed of three columnlike chambers of erectile tissue that span the length of the shaft. Each of the two larger lateral chambers is called a **corpus cavernosum** (plural = corpora cavernosa). Together, these make up the bulk of the penis. The **corpus spongiosum**, which can be felt as a raised ridge on the erect penis, is a smaller chamber that surrounds the urethra. Remember that these structures are homologues of the clitoris, discussed earlier. The glans and the foreskin contain a dense concentration of nerve endings, as well as sebaceous glands that secrete a waxy fluid that lubricates and protects the sensitive skin of the glans penis. A surgical procedure called *circumcision*, often performed for religious or social reasons, removes the prepuce; see the Cultural Connection "Circumcision" feature.

Cultural Connection

Circumcision

Circumcision is a procedure that removes the foreskin from the penis. Most often the procedure is performed on babies shortly after their birth, sometimes as part of a religious or cultural ceremony. It is the world's oldest planned surgery and dates back at least 15,000 years. Circumcision is commonly practiced in the Jewish, Muslim, and Druze faiths, but became more common outside of religious practices in the eighteenth and nineteenth centuries in many cultures. There are a few minor medical benefits to circumcision: it is associated with slightly lower rates of human papilloma virus (HPV) infection, a virus that is preventable with vaccination and still infects approximately 45 percent of sexually active biological males. Circumcision also decreases the already very low rates of urinary tract infections in biological males. The procedure carries a risk of infection, pain, and potentially reduced sexual sensation. Circumcision is very common in the United States but less common worldwide. One controversy around circumcision is that it is generally performed on babies before they can consent.

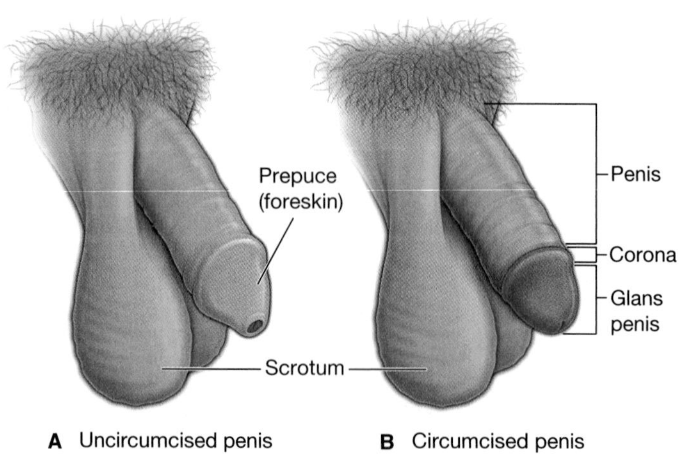

Figure 27.24 The Reproductive Anatomy of a Biological Male

The testes are the gonads of a biological male; they are housed within the scrotum. The gametes (sperm) are released from the testes and travel within the ductus deferens to exit the body at the tip of the penis. (A) Midsagittal view of male reproductive system. (B) Anterior view of external male reproductive organs. (C) Cross-section through the shaft of the penis.

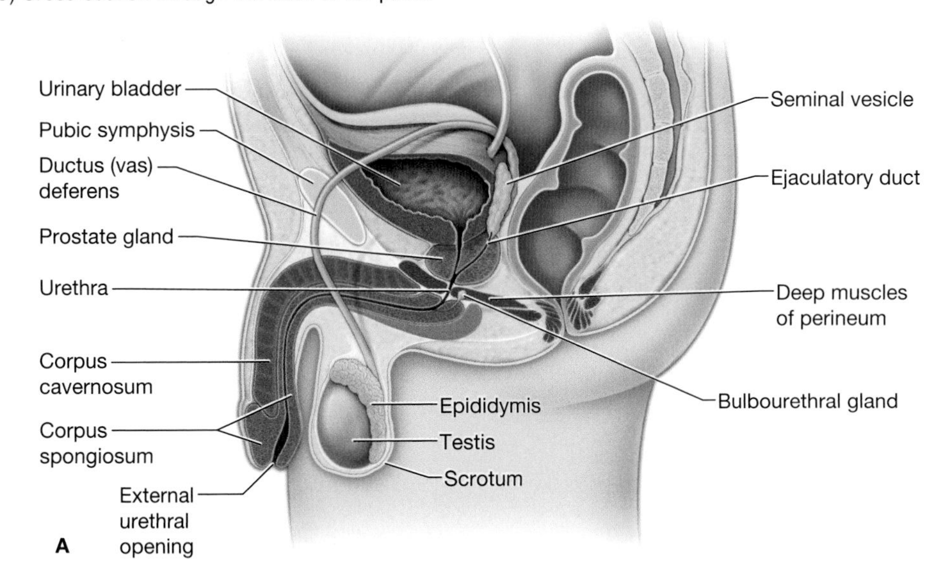

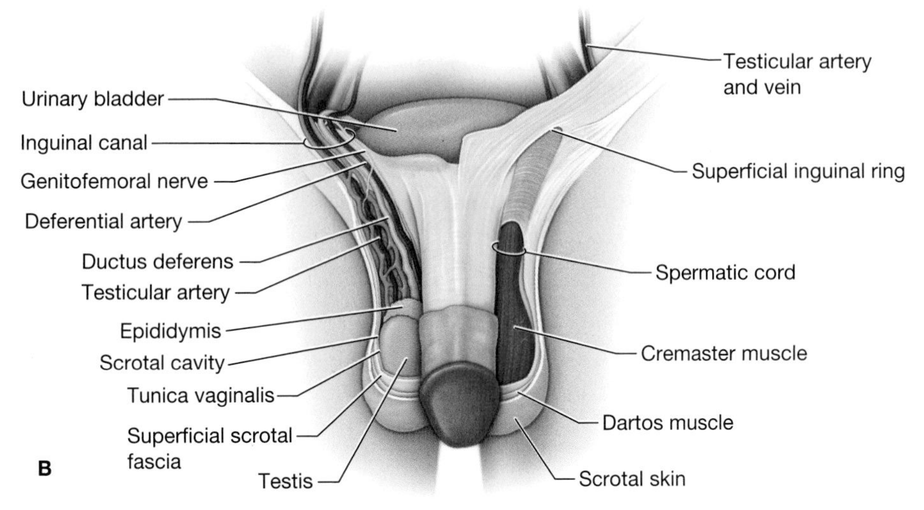

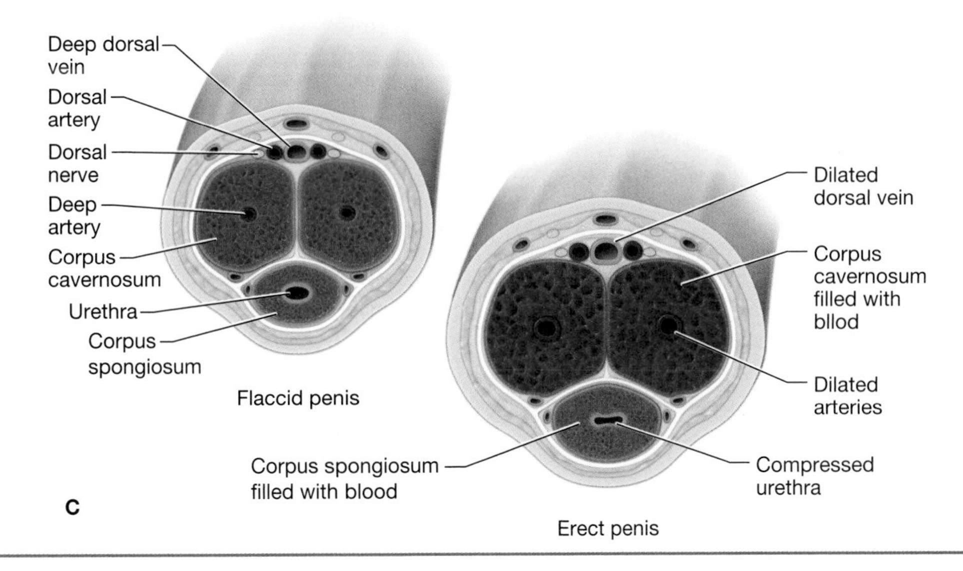

Figure 27.25 The Scrotum and Testes

(A) The testes are housed within the scrotum due to the fact that sperm are produced at an ideal temperature that is lower than body temperature. (B) Two muscles, the cremaster and the dartos muscles, raise or lower the testes to bring them either closer to, or farther away from, the body to regulate testicular temperature. (C) The spermatic cord contains the testicular blood vessels, lymphatic vessels, nerves, and the ductus deferens.

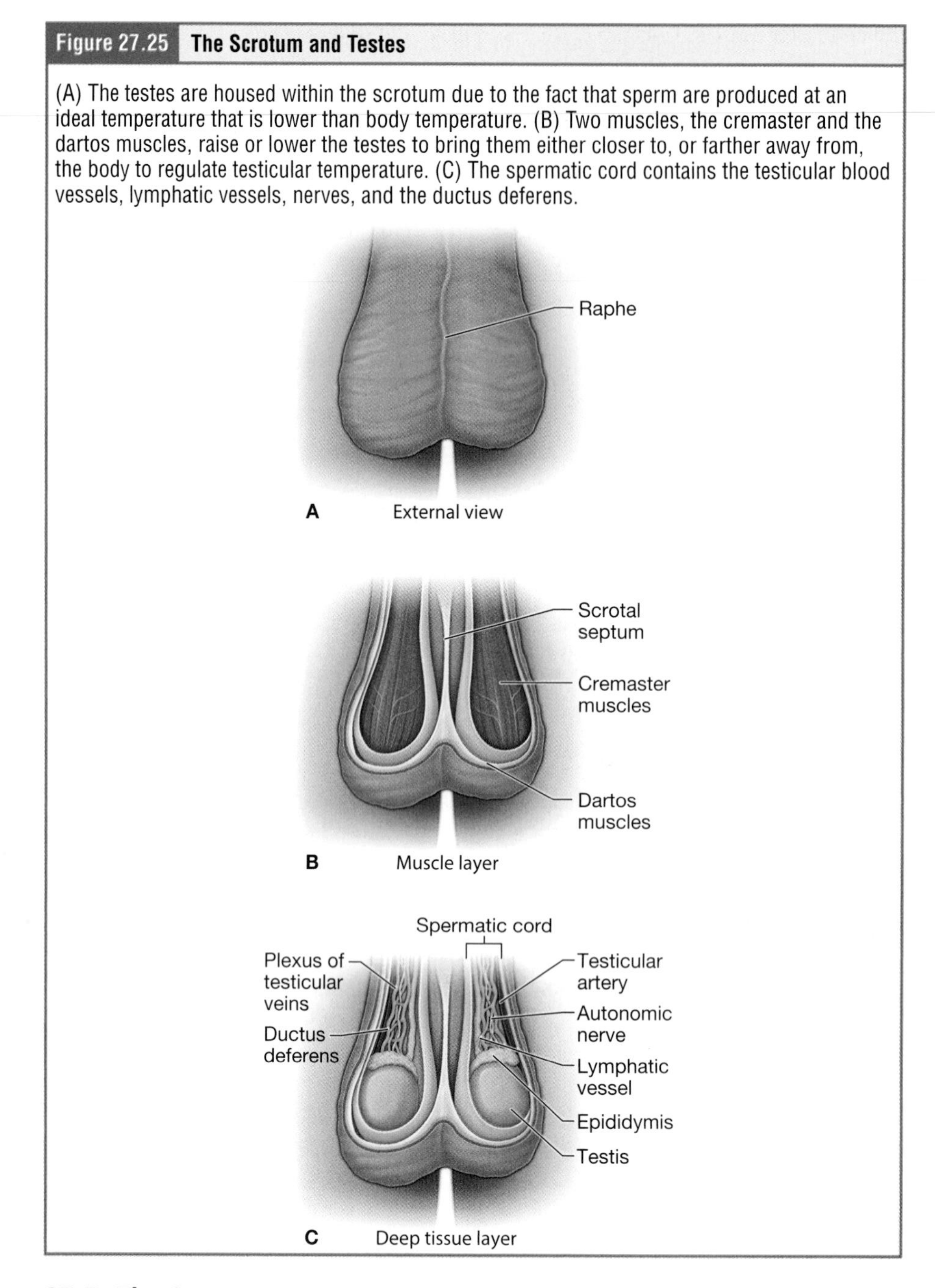

27.5c The Scrotum

The testes are housed within a skin-covered, highly pigmented, muscular sac called the **scrotum**, which extends from the body behind the penis (see Figure 27.24). This location is important in sperm production. The ideal temperature for making sperm, which occurs within the testes, is actually at 93.2° F or 35° C. Since body temperature is typically maintained at a slightly warmer temperature, the testes are housed outside of the body and can be lifted or dropped as necessary to achieve an optimal temperature for sperm production.

The **dartos muscle** makes up the subcutaneous muscle layer of the scrotum (see Figure 27.25). It continues internally to make up the scrotal septum, a wall that divides the scrotum into two compartments, each housing one testis. Descending from the internal oblique muscle of the abdominal wall are the two **cremaster muscles**, which cover each testis like a muscular net. By contracting simultaneously, the dartos and cremaster muscles can elevate the testes in cold weather (or cold water), moving the testes closer to the body and decreasing the surface area of the scrotum to retain heat. Alternatively, as the environmental temperature increases, the muscles within the scrotum relax, moving the testes farther from the body core and increasing scrotal surface area, which promotes heat loss. The scrotum is the male homologue of the labia majora. Both labia and the scrotum form from the genital folds. In biological males, if certain signals are present during fetal development, the genital folds fuse, leaving a noticeable raised medial thickening along the surface of the scrotum called the **raphe**.

Learning Connection

Broken Process

During fetal development in some biological males, the testes do not descend from the abdominopelvic cavity into the scrotum before birth; this is called *cryptorchidism* and can occur with one or both testes. How would you expect this to affect sperm production and fertility?

27.5d The Testes

LO 27.5.2

The **testes** (singular = testis) are the male gonads; like the ovaries, they produce both gametes—in this case sperm—and hormones. The group of hormones made by the testes are the **androgens**, the most common of which is testosterone.

The paired testes are oval in shape and approximately four to five centimeters in length. Within the scrotum, they are surrounded by two distinct layers of protective connective tissue (**Figure 27.26**). The outer **tunica vaginalis** is a serous membrane that has both a parietal and a visceral layer. Beneath the tunica vaginalis is the **tunica albuginea**, a tough, white, dense connective tissue layer covering the testis itself. Not only does the tunica albuginea cover the outside of the testis, it also invaginates to form **septa** that divide the testis into 300 to 400 structures called **lobules**. Each lobule contains a long, coiled tube called the *seminiferous tubule*.

The tightly coiled **seminiferous tubules** form the bulk of each testis. They are composed of developing sperm cells surrounding a lumen—the hollow center of the tubule, where formed sperm are released into a system of ducts called the **rete testes**. Sperm leave the rete testes, and the testis itself, through the 15 to 20 efferent

Figure 27.26 Anatomy of the Testis

A single testis is shown here in sagittal view. The testis is subdivided into lobules by invaginations of the tunica albuginea. Coiled within each lobule are the seminiferous tubules, the site of sperm production. Fully-formed sperm move to the epididymis, where they mature. They leave the epididymis during ejaculation and enter the ductus deferens.

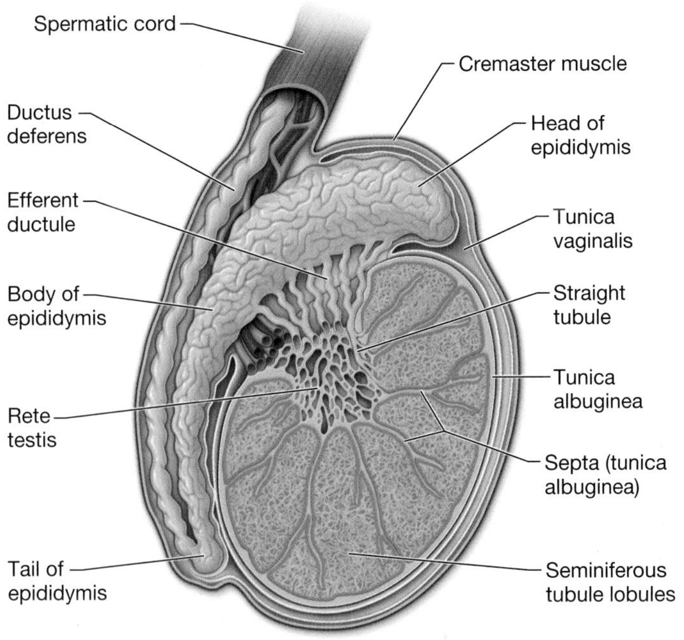

LO 27.5.3

ductus deferens (vas deferens)

ductules that cross the tunica albuginea. Hugging the posterior wall of each testis is an **epididymis**, where sperm mature. Once ready to be ejaculated, the sperm move into the **ductus deferens**—a very long tube that carries sperm up into the pelvic cavity and merges with the urethra, which carries sperm the final length of their journey out through the penis.

Student Study Tip

The epididymis acts as a sperm nursery. Nurseries help children, and sperm, to mature before they are ready for their next adventure.

The ductus deferens is bundled together with blood vessels and nerves into a structure called the **spermatic cord** (see Figure 27.26). Each spermatic cord extends superiorly into the abdominal cavity through the **inguinal canal** in the abdominal wall. From here, each ductus deferens curves around the sides of the bladder and into the *prostate gland*, a small gland approximately the size of a walnut located beneath the urinary bladder. Within the prostate, each ductus deferens forms a junction with the singular urethra.

LO 27.5.4

LO 27.5.5

Sperm make up only 5 percent of the final volume of **semen**, the thick, milky fluid that is ejaculated. The bulk of semen is produced by three critical accessory glands of the male reproductive system—the seminal vesicles, the prostate, and the bulbourethral glands.

seminal vesicles (seminal glands)

Seminal Vesicles The paired **seminal vesicles** are glands that contribute approximately 60 percent of the semen volume. They are found superior to the prostate, along the posterior border of the urinary bladder. Seminal vesicle fluid contains large amounts of fructose, which is used by the sperm mitochondria to generate ATP to allow movement through the female reproductive tract. Seminal fluid is basic, and this high pH functions to neutralize the acidic environment of the vagina.

LO 27.5.6

The fluid—now containing both sperm and seminal vesicle secretions—next moves into the junction between the ductus deferens and the urethra within the prostate gland.

Prostate Gland The single, centrally located **prostate gland** sits anterior to the rectum at the base of the bladder surrounding the prostatic urethra (the portion of the urethra that runs within the prostate). The prostate is formed of both muscular and glandular

DIGGING DEEPER:
The Aging Prostate Gland

The prostate gland normally doubles in size during puberty. At approximately age 25, it gradually begins to enlarge again. This enlargement does not usually cause problems; however, abnormal growth of the prostate, or benign prostatic hyperplasia (BPH), can cause constriction of the urethra as it passes through the middle of the prostate gland, leading to a number of lower urinary tract symptoms, such as a frequent and intense urge to urinate, a weak stream, and a sensation that the bladder has not emptied completely. By age 60, approximately 40 percent of biological men have some degree of BPH. By age 80, the number of affected individuals has jumped to as many as 80 percent. Treatments for BPH attempt to relieve the pressure on the urethra so that urine can flow more normally. Mild to moderate symptoms are treated with medication, whereas severe enlargement of the prostate is treated by surgery in which a portion of the prostate tissue is removed.

Another common disorder involving the prostate is prostate cancer. According to the U.S. Centers for Disease Control and Prevention (CDC), prostate cancer is the second most common type of cancer in men. However, some forms of prostate cancer grow very slowly and thus may not ever require treatment. Aggressive forms of prostate cancer, in contrast, involve metastasis to vulnerable organs like the lungs and brain. There is no link between BPH and prostate cancer, but the symptoms are similar. Prostate cancer is detected by a medical history, a blood test, and/or a rectal exam that allows physicians to palpate the prostate and check for unusual masses. If a mass is detected, the cancer diagnosis is confirmed by biopsy of the cells.

tissues. It excretes an alkaline, milky fluid to the passing semen that is critical to coagulate the semen following ejaculation.

bulbourethral glands (Cowper's glands)

Bulbourethral Glands The final addition to semen is made by two **bulbourethral glands** (or *Cowper's glands*) that release a thick, salty fluid that lubricates the end of the urethra and the vagina, and helps to clean urine residues from the penile urethra. The fluid from these accessory glands is released during sexual arousal, and shortly before the release of the semen. It is therefore sometimes called *pre-ejaculate*. It is important to note that, in addition to the lubricating proteins, it is possible for bulbourethral fluid to pick up sperm already present in the urethra, and therefore it alone may be able to cause pregnancy.

Learning Connection

Quiz Yourself

You have learned that many reproductive organs and glands in biological males and females form from the same embryonic structures. For each of the following structures in a biological female—ovary, clitoris, labia, vestibular glands, and vaginal orifice—see if you can remember the analogous structure in the biological male.

Learning Check

1. Which of the following structures does not contribute cells or secretions to the semen?
   a. Seminal vesicles
   b. Testes
   c. Tunica albuginea
   d. Prostate gland
2. Which of the following is the correct pathway of sperm as they mature and leave the body during ejaculation? LO 27.5.7
   a. Testis, ductus deferens, epididymis, urethra, ejaculatory duct
   b. Testis, epididymis, ductus deferens, ejaculatory duct, urethra
   c. Ductus deferens, epididymis, urethra, seminiferous tubules, ejaculatory duct
   d. Epididymis, ductus deferens, ejaculatory duct, seminiferous tubules, urethra
3. Sean, a biological male, is enjoying a nice hot day at the beach. Will his dartos muscle and his two cremaster muscles be contracted?
   a. Yes
   b. No
   c. Impossible to tell with the information given
4. The tunica albuginea is renamed ______ when it invaginates the testes to divide the larger structure into lobules.
   a. Tunica vaginalis
   b. Septa
   c. Raphe
   d. Seminiferous tubule

27.6 Spermatogenesis and Spermiogenesis

Learning Objectives: By the end of this section, you will be able to:

27.6.1 Define the processes of spermatogenesis and spermiogenesis.

27.6.2 Describe the stages of spermatogenesis in the seminiferous tubule, including the roles of the nurse (sustentacular, Sertoli) cells and interstitial cells (of Leydig).

27.6.3 Describe endocrine regulation of spermatogenesis.

As mentioned at the beginning of Section 27.5, because biological males do not house the fetus as it develops, the physiology of biological male reproduction is limited to the production of gametes and hormones. These gametes—sperm—are the product of two linked processes: spermatogenesis and spermiogenesis. **Spermatogenesis** is the generation of haploid sperm cells through mitosis and then meiosis. These haploid cells, though, are not yet ready to fertilize an egg. First, they must mature and differentiate to become cells capable of movement and digestion of the egg's zona pellucida. **Spermiogenesis** describes this maturation and differentiation to produce functional sperm; it is the last stage of spermatogenesis, in which spermatids develop into sperm.

LO 27.6.1

Two cell types in the testes collaborate in the making of sperm. Interstitial cells, found outside the seminiferous tubules, are responsible for generating androgens—especially testosterone, which drives spermatogenesis and spermiogenesis forward. Inside the seminiferous tubules we can find nurse cells, which directly support the process of making sperm. Inside the seminiferous tubules we can also find five types of developing sperm cells. As these cells develop, they progress from the basement membrane—at the perimeter of the tubule—toward the lumen (**Figure 27.27**). Let's look more closely at these cell types.

27.6a Nurse Cells

nurse cells (sustentacular cells, Sertoli cells)

Nurse cells make up the wall of the seminiferous tubule and surround the developing sperm through all stages of their development. Nurse cells secrete signaling molecules that promote sperm production and can control whether sperm cells live or die. Tight junctions between these nurse cells create the **blood-testis barrier**, which keeps bloodborne substances, as well as antibodies and immune cells, from reaching the developing sperm.

LO 27.6.2

27.6b Gametes

All cells originate from a stem cell population, and the stem cells that give rise to sperm are **spermatogonia** (singular = spermatogonium). They are found lining the basement membrane inside the tubule (see Figure 27.27).

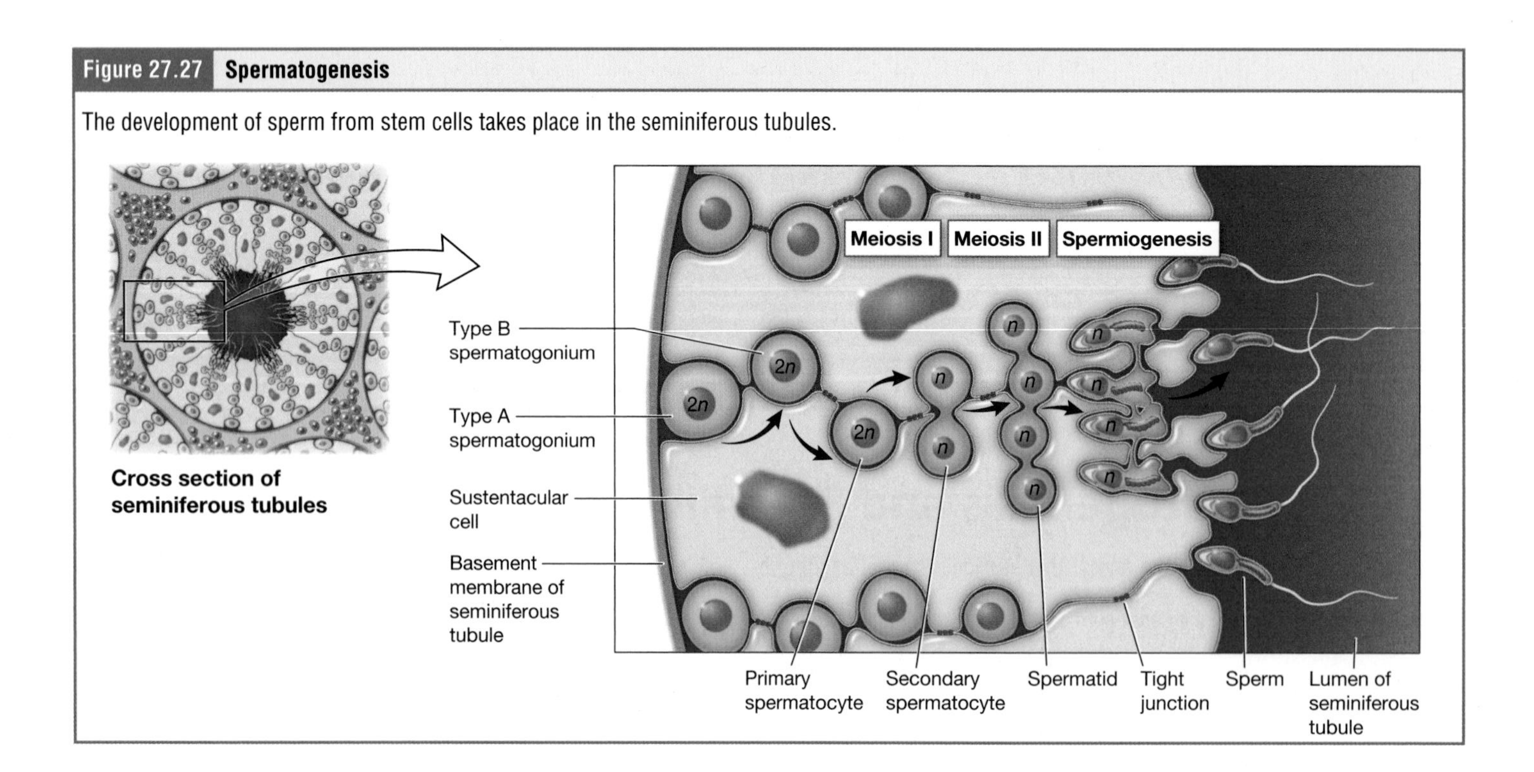

Figure 27.27 Spermatogenesis

The development of sperm from stem cells takes place in the seminiferous tubules.

27.6c Spermatogenesis

The process of spermatogenesis begins at puberty and continues relatively constantly throughout the rest of the lifespan. In biological females, the processes of gametogenesis and readying the uterus are synced, such that typically one gamete is readied every 28 days on average and the uterus prepares for the possibility that that singular gamete will be fertilized and development into a fetus would need to be supported. In contrast, each seminiferous tubule undergoes its own cycle, but the cycle of one seminiferous tubule is not linked or coordinated with any other. Therefore, the production of sperm within the entire human is relatively constant.

Following a single cell, the process of generating sperm from a spermatogonium takes approximately 64 days. The process of spermatogenesis begins with mitosis of the diploid spermatogonia (see Figure 27.27). Because these cells are diploid ($2n$), they each have a complete copy of genetic material, or 46 chromosomes. However, mature gametes are haploid ($1n$), containing 23 chromosomes—meaning that daughter cells of spermatogonia must undergo a second cellular division through the process of meiosis.

Two identical diploid cells result from spermatogonia mitosis. One of these cells remains a spermatogonium, meaning that it remains a stem cell for future use. The other becomes a primary **spermatocyte**, which undergoes meiosis to yield haploid gametes. The resulting cells (secondary spermatocytes) undergo meiosis II, resulting in four haploid cells, each called a **spermatid**. Although they have reduced their chromosome number, early spermatids look very similar to cells in the earlier stages of spermatogenesis, with a round shape, central nucleus, and large amount of cytoplasm. Spermiogenesis transforms these early spermatids into sperm by reducing the amount of cytoplasm and elongating the flagellum in preparation for swimming. At the end of this process the formed sperm can be found within the wall of the seminiferous tubule nearest the lumen. Eventually, the sperm are released into the lumen and are moved along a series of ducts in the testis toward the epididymis and ductus deferens.

Learning Connection

Quiz Yourself

After studying the processes of oogenesis and spermatogenesis, compare and contrast the two processes. How are they alike? How are they different? How many chromosomes are found in mature sperm and mature oocytes? How many functional oocytes are produced for every oogonium that undergoes meiosis? How many functional sperm are produced for every spermatogonium that undergoes meiosis?

27.6d Spermiogenesis

Sperm are smaller than most cells in the body; in fact, it would take 85,000 sperm cells to equal the volume of one egg released from the ovary. Approximately 100 to 300 million sperm are produced each day. As is true for all cells in the body, the structure of sperm cells speaks to their function. Sperm are almost entirely made up of their nucleus; they are literally traveling missiles of DNA. Attached to their nuclear head, they have a mid-piece and tail region (**Figure 27.28**). The head contains the extremely compact nucleus and very little cytoplasm. The head is covered by the **acrosome**, a "cap" that is filled with lysosomal enzymes important for digesting the egg's coat, the zona pellucida. Tightly

Figure 27.28 Sperm

Sperm are streamlined for speed during travel. Their haploid nucleus is surrounded by an *acrosome*, a membrane containing enzymes that can digest the protective layers around the oocyte at the time of fertilization. Their flagellum is powered by ATP produced by mitochondria in the mid-piece.

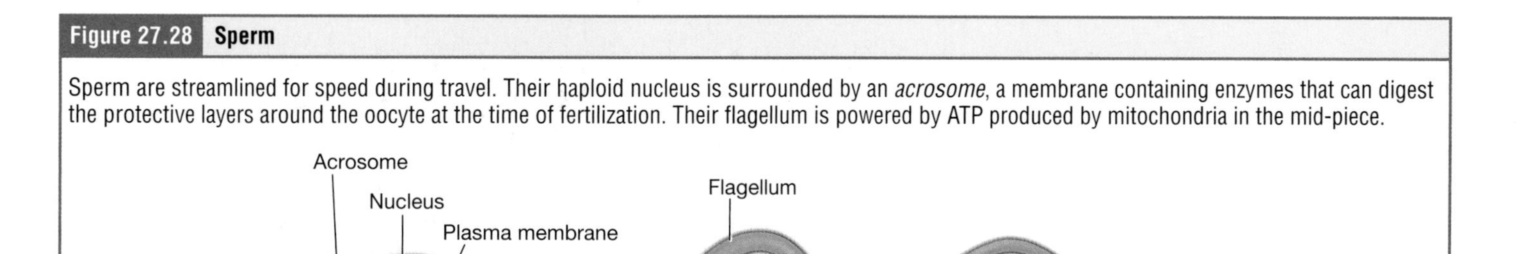

Learning Connection

Try Drawing It

Sperm follow an interesting pathway during ejaculation. Starting with the site of production in the testes, draw the pathway that sperm take as they leave the body. Label each organ that the sperm pass through, and show where each gland contributes fluid to the semen.

packed mitochondria fill the mid-piece of the sperm. ATP produced by these mitochondria will power the long flagellum, which enables movement.

Once fully formed, sperm are released into the lumen of the seminiferous tubules, surrounded by testicular fluid, and moved to the *epididymis* (plural = epididymides), a coiled tube attached to the testis where newly formed sperm continue to mature (see Figure 27.26). Though the epididymis does not take up much room in its tightly coiled state, it would be approximately 20 feet long if straightened. It takes an average of 12 days for sperm to move through the coils. As they are moved by smooth muscle contractions, the sperm further mature and develop the ability to move their flagella and therefore power their own movement. The mature sperm are stored in the tail of the epididymis until ejaculation occurs. During ejaculation, sperm exit the tail of the epididymis and are pushed by smooth muscle contraction to the ductus deferens. Should they end up in the vagina, it is there that sperm first swim.

27.6e Testosterone

interstitial cells (cells of Leydig)

In addition to spermatogenesis and spermiogenesis, the testes function as endocrine organs. Testosterone, an androgen, is a steroid hormone produced by **interstitial cells**. The term *interstitial* means "within the tissue"; it is the same adjective we use throughout this book to describe interstitial fluid. Interstitial cells are so named because they are located *between* the seminiferous tubules rather than within them (**Figure 27.29**). In embryos with the SRY gene region, testosterone is secreted by interstitial cells by the seventh week of development, with peak concentrations reached in the second

Figure 27.29 Interstitial Cells

There are two types of cells that are central to the functions of the testes. Interstitial cells, which lie between the seminiferous tubules, produce testosterone. Nurse (sustentacular) cells line the walls of the seminiferous tubules and support the maturation of sperm.

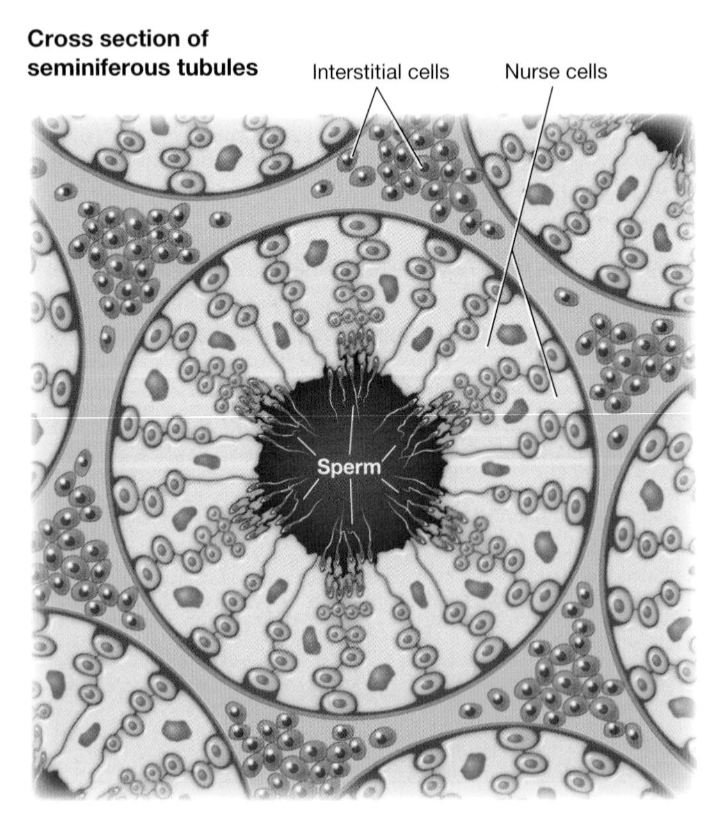

trimester. This early release of testosterone drives the development of the sex organs. In childhood, testosterone concentrations are low. They increase during puberty, activating characteristic physical changes and initiating spermatogenesis.

LO 27.6.3

Functions of Testosterone **Testosterone** is a steroid hormone derived from cholesterol. Testosterone produced by interstitial cells, in addition to acting locally on nurse cells in the nearby seminiferous tubules, is also released into the systemic circulation. Like all hormones, testosterone will act on any tissues that express testosterone receptors. Testosterone and other androgens play an important role in muscle development, bone growth, skin, and hair characteristics, behavior, and maintaining libido (sex drive). While estrogen and testosterone play a small role, testosterone is the prime driver of libido in all humans. Every adult human body has some testosterone signaling. Testosterone is produced in both ovaries and testes, as well as in every adult's adrenal glands.

Control of Testosterone The intricate interplay between the endocrine system and the reproductive system is shown in **Figure 27.30**.

Figure 27.30 Regulation of Testosterone Production

Testosterone is produced in response to tropic hormones from the hypothalamus and the anterior pituitary.

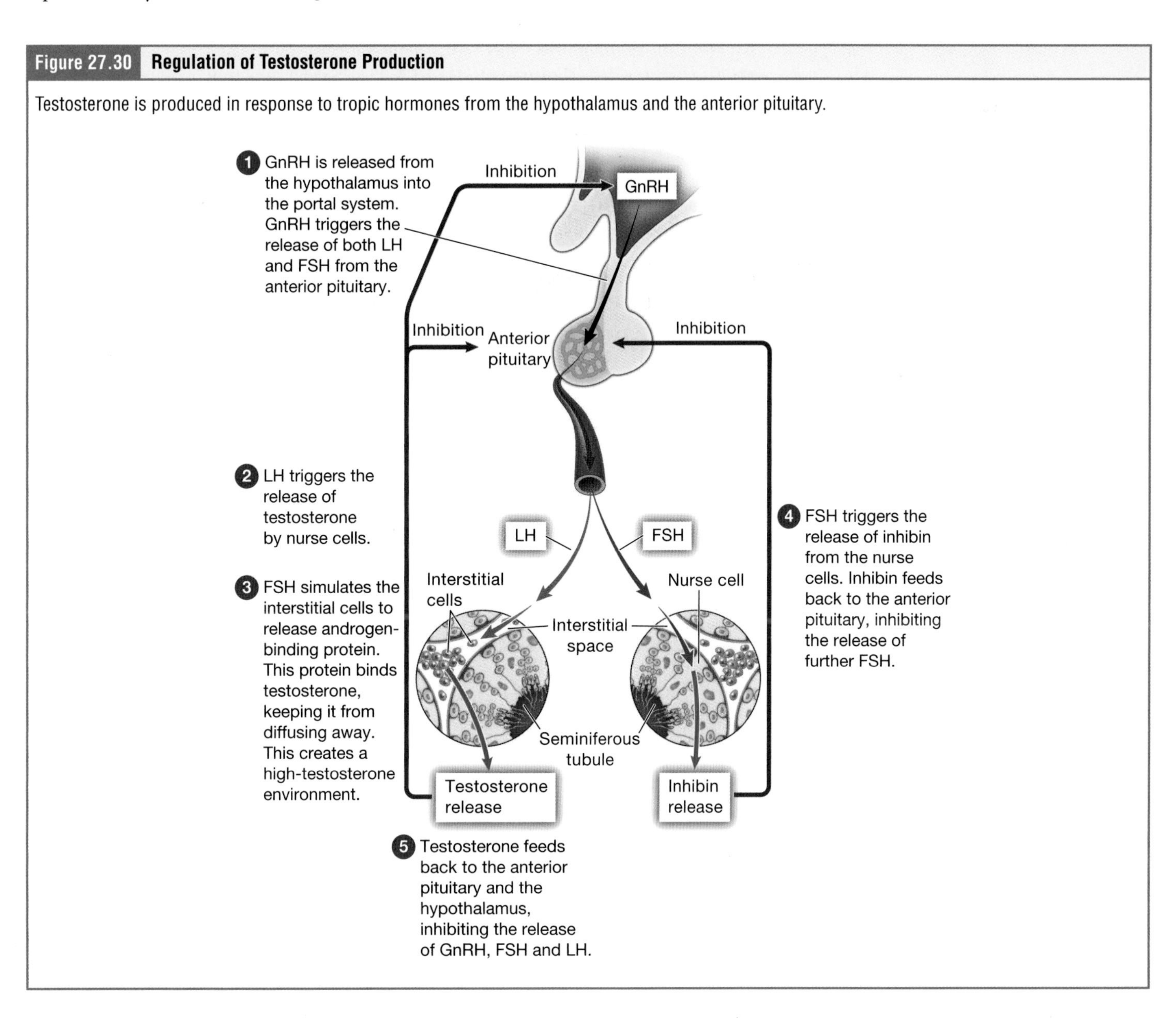
1 GnRH is released from the hypothalamus into the portal system. GnRH triggers the release of both LH and FSH from the anterior pituitary.
Inhibition
GnRH
Inhibition
Anterior pituitary
Inhibition
2 LH triggers the release of testosterone by nurse cells.
LH
FSH
4 FSH triggers the release of inhibin from the nurse cells. Inhibin feeds back to the anterior pituitary, inhibiting the release of further FSH.
3 FSH simulates the interstitial cells to release androgen-binding protein. This protein binds testosterone, keeping it from diffusing away. This creates a high-testosterone environment.
Interstitial cells
Nurse cell
Interstitial space
Seminiferous tubule
Testosterone release
Inhibin release
5 Testosterone feeds back to the anterior pituitary and the hypothalamus, inhibiting the release of GnRH, FSH and LH.

Testosterone production by the interstitial cells is regulated by the hypothalamus and the anterior pituitary gland. The regulation begins in the hypothalamus. Pulsatile release of GnRH from the hypothalamus stimulates the endocrine release of hormones from the anterior pituitary gland. Binding of GnRH to its receptors on the anterior pituitary gland stimulates release of LH and FSH. FSH binds predominantly to the nurse cells within the seminiferous tubules and promotes spermatogenesis. FSH also stimulates the nurse cells to produce hormones called *inhibins*, which function to inhibit FSH release from the pituitary, thus reducing testosterone secretion. LH binds to receptors on interstitial cells, causing them to increase their production of testosterone.

A negative feedback loop predominantly controls the synthesis and secretion of both FSH and LH. Low blood concentrations of testosterone stimulate the hypothalamic release of GnRH. GnRH then stimulates the anterior pituitary to secrete LH into the bloodstream. LH binds to LH receptors on interstitial cells and stimulates the release of testosterone. When concentrations of testosterone in the blood reach a critical threshold, testosterone will bind to androgen receptors on both the hypothalamus and the anterior pituitary, inhibiting the synthesis and secretion of GnRH and LH, respectively. When the blood concentrations of testosterone once again decline, testosterone no longer interacts with the receptors to the same degree and GnRH and LH are once again secreted, stimulating more testosterone production. This same process occurs with FSH and inhibin to control spermatogenesis.

Learning Connection

Explain a Process

Explain to a classmate the similarities and differences between the hormonal pathways for reproduction in biological males and females. Be sure to discuss GnRH, FSH, LH, estrogen, and testosterone. Where is each hormone produced? What are the basic functions of each hormone?

Learning Check

1. Which of the following types of cells forms the blood-testis barrier?
   a. Leydig cells
   b. Interstitial cells
   c. Nurse cells
2. What are gametes produced by biological males called? Are these gametes haploid or diploid?
   a. Sperm; diploid
   b. Sperm; haploid
   c. Spermatids; diploid
   d. Spermatids; haploid
3. In the biological male body, are testosterone levels homeostatically regulated? If so, are they regulated by a positive or negative feedback mechanism?
   a. No
   b. Yes; positive feedback
   c. Yes; negative feedback
   d. Impossible to tell with the information given
4. During the process of spermatogenesis, which developing gamete is the first one to be haploid?
   a. Spermatogonium
   b. Spermatid
   c. Primary spermatocyte
   d. Secondary spermatocyte

27.7 Sex (Coitus)

Learning Objectives: By the end of this section, you will be able to:

27.7.1* List and describe the physiology and stages of sexual responses.

27.7.2* Compare and contrast STIs in terms of their causative agents, symptoms, and diseases.

27.7.3* Explain how different examples of birth control methods work.

* Objective is not a HAPS Learning Goal.

While all humans and all sexual responses look and feel different from each other, we can broadly group the physiological changes that occur during sex into three stages: excitement/plateau, orgasm, and resolution.

LO 27.7.1

- **Excitement and plateau stages:** During these phases the body prepares for and/or engages in sex (which is often called *coitus* in biological/anatomical/physiological texts). The excitement phase may begin with any number of stimuli, from a thought or dream to physical touch. Excitement is characterized by **systemic myotonia**, which means that muscle tension increases all over the body. There are increases in heart rate, breathing rate, and blood pressure. Parasympathetic neurons fire signals to structures in the genitals, causing the release of nitric oxide, a potent vasodilator. Vasodilation of arteries and simultaneous vasoconstriction of veins occurs in response. The result is that more blood enters the genital region than can leave, and the spongy corpora tissues flood with blood and become swollen, enlarged, and rigid. This is called **vasocongestion** and occurs in each sex in their homologous structures. In biological females, the labia minora and clitoris both experience vasocongestion; in biological males, the homologues of these structures are the scrotum and penis.

  Other excitement changes include closing off the urethra to prevent urination. This occurs in biological males with the contraction of the urethral sphincter, in biological females, the swollen internal structures of the clitoris close off the urethra. The uterus, which normally leans anteriorly over the urinary bladder, pulls up, becoming almost vertical in the pelvic cavity and elongating the vagina.
- **Orgasm stage**: **Orgasm** is a brief but intense response ranging in duration from one to fifteen seconds that may occur at the end of the excitement and plateau phase. During orgasm, the sympathetic nervous system is dominant, driving the heart rate up as high as 180 beats per minute (near the maximum heart rate of an individual), increasing the breath rate up to 40 breaths per minute, and increasing blood pressure. Orgasm is characterized by two physiological events: a surge in blood oxytocin levels and pelvic muscle contractions. **Oxytocin**, a posterior pituitary hormone, has a relaxing effect on the body, promoting the activation of the parasympathetic nervous system. Oxytocin also promotes social bonding and feelings of affection. Pelvic muscle contractions occur at intervals of 0.8 second in all individuals. In biological females, it is hypothesized (though in debate) that these contractions may help propel sperm, if present, toward the egg for fertilization. Pelvic muscle contraction in biological males may expel semen from the urethra. However, **ejaculation**, the expulsion of semen from the penis, may not always occur simultaneously with orgasm.
- **Resolution stage:** The resolution stage follows orgasm, should orgasm occur. For individuals who do not experience orgasm, the resolution stage does not occur either. In the resolution stage, the parasympathetic nervous system becomes dominant. Heart rate, breathing rate, and blood pressure all decrease back to resting levels. Oxytocin surges throughout the resolution stage. In biological males, there is an extended refractory period in the parasympathetic neurons, preventing these neurons from firing signals and therefore preventing erection. In biological females, the refractory period is short, allowing for further signaling in these neurons, and multiple cycles through the excitement/plateau and orgasm phases.

Learning Connection

Chunking

Make a list of the common STIs discussed here. Then classify them in terms of whether they are bacterial or viral, and whether they are completely curable or not.

DIGGING DEEPER: STIs

LO 27.7.2

sexually transmitted infections (STIs; sexually transmitted diseases, STDs)

In addition to sharing pleasure and the possible result of pregnancy, coitus also presents a significant risk of sharing illness. In this brief section we will discuss infections that can be shared during sex, but keep in mind that other infections of the reproductive organs, for example yeast infections or urinary tract infections, are not due to pathogens that are transmitted during intercourse. The United States is experiencing a dramatic rise in **sexually transmitted infections (STIs)**. While some may argue that the rise in cases is due to an increase in the average number of partners an individual has in their lifetimes, comparing the United States to Europe provides clarity. While the average number of STIs has declined in European countries after they implemented routine screening (see Figure A), the United States, which has no similar screening program, has experienced a corresponding increase in the incidence of some types of STIs. Exploring the pathogens in STIs will provide some insight as to why routine screening is so important.

Though there are fungal and parasitic infections that can be transmitted sexually as well, we will discuss viral and bacterial pathogens here. The major viruses that can be passed sexually include human immunodeficiency virus (HIV), human papilloma virus (HPV), and herpes simplex virus (HSV).

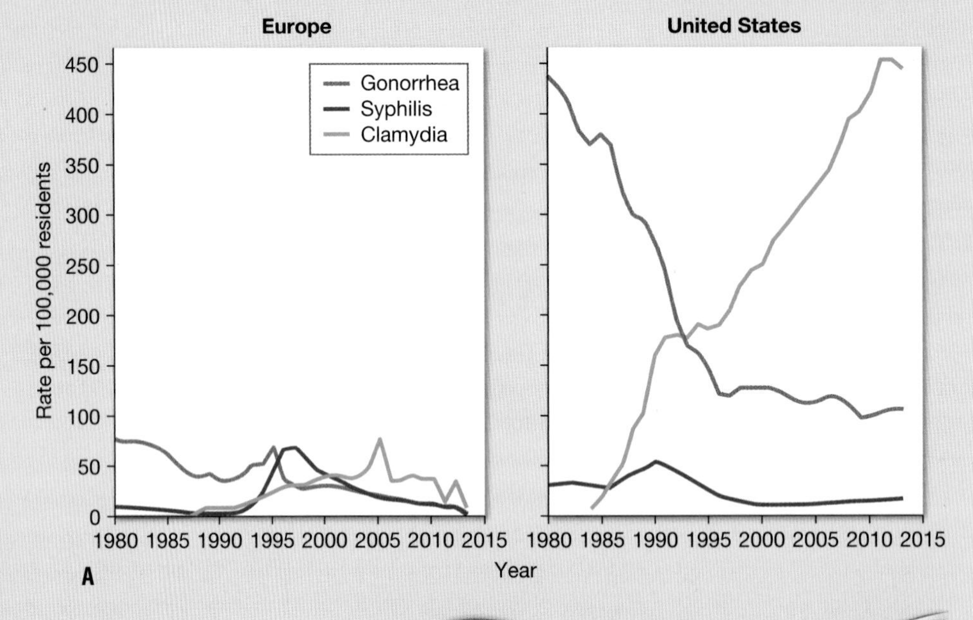

A

Of these three infections, only one of the viruses tends to cause immediate symptoms after infection. Contracting HIV usually is followed by an immediate viremia—an increase in viral load in the blood. The result is a systemic response, an acute illness that is similar to having a case of the flu. After the acute viremia, HIV infection is asymptomatic (does not produce symptoms) for a long period of months to years. During this time, viral levels in the blood increase slowly. HIV infects cells of the immune system, slowly killing these cells and disabling immune responses. HIV is a lifelong infection that can lead to death. Only two individuals in the world have ever been cured, but progress of the viral infection can be stalled with antiretroviral medication.

HPV is a family of viruses with many strains that commonly cause small warts on the genitals. These warts are not painful, and the virus causes very few other issues. In rare cases, however, HPV can cause cervical cancer, which can be deadly. HPV is the most common STI in the United States; it is also preventable with a vaccine.

Most HSV infections are asymptomatic; the only symptom is occasional sores on the lips or genitals. HSV is a lifelong infection with no cure; however, antiviral medication can help control outbreaks. There are several strains of HSV. Cold sores are a common manifestation of herpes virus infections. A different herpes virus, the varicella zoster virus, causes chickenpox.

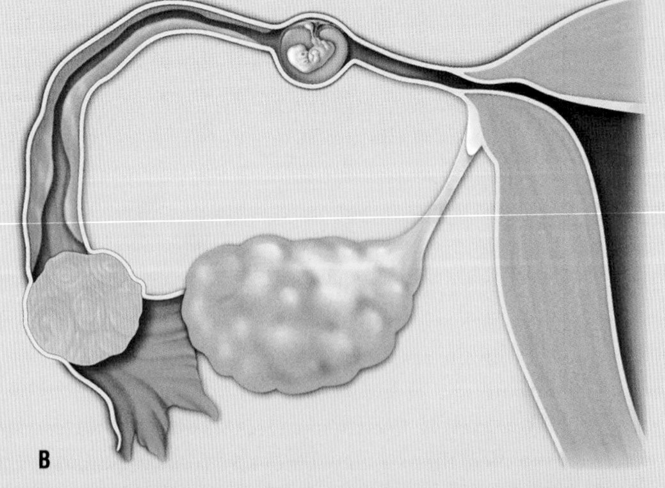
B

The three common bacterial STIs are chlamydia, gonorrhea, and syphilis. All three infections are curable with antibiotics once the infection is detected. Some strains of gonorrhea, however, are antibiotic-resistant, making these infections both uncurable and untreatable. Syphilis, if left untreated, can be deadly. The majority of cases of bacterial STIs are asymptomatic, which enables spread if infections are not detected by routine screening. Any bacterial infection of the reproductive tract can cause a condition called *pelvic inflammatory disease* (see Figure B) in which inflammation occurs in the uterine tubes. The resulting tissue changes and scar tissue can be permanent, forever narrowing the lumen of the uterine tubes. Pelvic inflammatory disease increases the chance of fertility challenges if the fertilized eggs struggle to pass through the narrowed tubes. Damage to the cilia can make endometriosis more likely to occur and increases the risk of ectopic pregnancy.

Apply to Pathophysiology

Birth Control

LO 27.7.3

While a more appropriate term might be *fertilization control* or *implantation control*, the term *birth control* encompasses a wide variety of methods to allow sex and ejaculation to take place without a risk of unwanted pregnancy. Some methods of birth control are currently in use, while some are in clinical trials and others are theoretical.

1. Vasectomy is a procedure in which the ductus deferens is cut prior to its junction with the urethra. Which of the following is true in an individual who has had a vasectomy?
   A) He will still ejaculate, but will release only immature sperm.
   B) He will still ejaculate, but no sperm will be released.
   C) He will not ejaculate.
2. Would a birth control pill that supplied a consistent, high level of inhibin every day decrease gametogenesis?
   A) It would decrease oogenesis but not spermatogenesis.
   B) It would decrease spermatogenesis but not oogenesis.
   C) It would decrease both spermatogenesis and oogenesis.
3. Would a birth control pill that supplied a consistent, moderate level of estrogen every day prevent pregnancy?
   A) Yes, follicles will mature but ovulation will not occur.
   B) Yes, follicles will not mature or ovulate.
   C) No, both folliculogenesis and ovulation will still occur.
4. Would a gel that lowers the pH of the vagina, applied just before sex, prevent fertilization?
   A) No, the oocyte and sperm would both be available, fertilization would still occur.
   B) Yes, although the oocyte would still be available, sperm would be unlikely to make it past the cervix.
   C) Yes, the lowered pH would cause deterioration of the oocyte, preventing fertilization.

Barrier methods, including condoms, diaphragms, and sponges, are the most common and least invasive forms of birth control. These methods simply offer a physical barrier that prevents sperm from reaching the egg. Condoms—though not diaphragms or sponges—also prevent transmission of most STDs and are therefore the most protective option for engaging in sex. Intrauterine devices (IUDs) are a very common, semipermanent form of birth control. These devices, if inserted properly by a healthcare provider, can offer protection from pregnancy for years. Once removed, the effects of the IUD are reversible.

Learning Check

1. Oxytocin surges during which stage(s) of coitus? Please select all that apply.
   a. Excitation and plateau stages
   b. Orgasm stage
   c. Resolution stage
2. During the excitement stage, arteries ______ and veins ______ to achieve vasocongestion in the genital region.
   a. Vasoconstrict; vasoconstrict
   b. Vasoconstrict; vasodilate
   c. Vasodilate; vasodilate
   d. Vasodilate; vasoconstrict
3. Which of the following stimuli are able to trigger the excitement phase of coitus?
   a. Thoughts and memories
   b. Physical touch
   c. Media
   d. All of the above
4. Some birth control pills contain high concentrations of synthetic forms of estrogen and progesterone, which suppress ovulation. Which phase of the ovarian cycle is being mimicked by this combination of hormones?
   a. The luteal phase
   b. The follicular phase
   c. The menses phase

27.8 Pregnancy, Birth, and Lactation

LEARNING GOALS AND OUTCOMES

Learning Objectives: By the end of this section, you will be able to:

27.8.1 Define fertilization.

27.8.2 Describe the processes that facilitate fertilization (e.g., sperm capacitation, acrosomal reaction).

27.8.3 Describe the three phases of fertilization (i.e., corona radiata penetration, zona pellucida penetration, fusion of the oocyte and sperm plasma membranes).

27.8.4 Describe the formation and function of the placenta and extraembryonic membranes.

27.8.5 Describe the hormones associated with pregnancy and the effects of these hormones.

27.8.6 Describe physiological changes that permit pregnancy (e.g., change in blood volume) and their reversal during the postpartum period.

27.8.7 Define parturition (labor).

27.8.8 Describe the three stages of labor.

27.8.9 Define the postpartum period.

27.8.10 Describe the hormonal regulation of lactation.

In anatomy and physiology, we refer to these structures as reproductive, while acknowledging that many humans never reproduce. In cases where reproduction—that is, giving birth to offspring—does occur, the defining events are: fertilization, implantation of the embryo in the uterine wall, development of the fetus during pregnancy, and birth.

LO 27.8.1

27.8a Fertilization

Fertilization is the event in which a sperm encounters and penetrates an egg, resulting in the combining of their genetic material. Sperm may reach the egg in as little as 30 minutes after being deposited in the vagina, but it will take them another 10 hours to become capable of penetrating the egg. This process, called **sperm capacitation**, occurs only in sperm that have entered the vagina and uterus. The fluids of the reproductive tract of biological females contain enzymes that alter the cell membrane of sperm, making them more likely to be able to fuse their membrane with the egg's membrane. Recall in Figure 27.28 that sperm carry a cap above their nucleus called the *acrosome*. The membrane superficial to the acrosome is hardened by cholesterol, preventing premature rupture of the acrosome. Enzymes within the fluids of the vagina and uterus remove this cholesterol, making the membrane and the acrosome below it vulnerable to rupture. When the sperm reaches the ovulated egg—which is coated with the jellylike zona pellucida and a crown of granulosa cells—its now fragile membrane ruptures and it releases the enzymes of the acrosome (**Figure 27.31**). The first few sperm to reach the egg are probably not going to be able to fertilize it. Instead, each sperm undergoes an **acrosomal reaction**, in which the acrosome ruptures and releases its enzymes. This paves the way for other sperm by removing granulosa cells and chipping away at the zona pellucida. Eventually, one sperm merges its membrane with the membrane of the egg, pushing its nucleus into the egg cytoplasm, fertilizing it.

LO 27.8.2

LO 27.8.3

27.8b The Preembryonic Stage

The zygote now has a novel set of 46 chromosomes and begins to expand its cell number by repeated mitotic divisions. The zygote is not getting bigger during this stage, but just repeatedly replicating the nuclei and dividing cytoplasm (**Figure 27.32**). By approximately 16 days after fertilization, the first cell differentiation occurs. Cells along the

Figure 27.31 Acrosomal Reactions

It takes several sperm and their acrosomal reactions to clear the path through the granulosa cells for the fertilizing sperm to be able to penetrate the zona pellucida and merge its membrane with that of the egg.

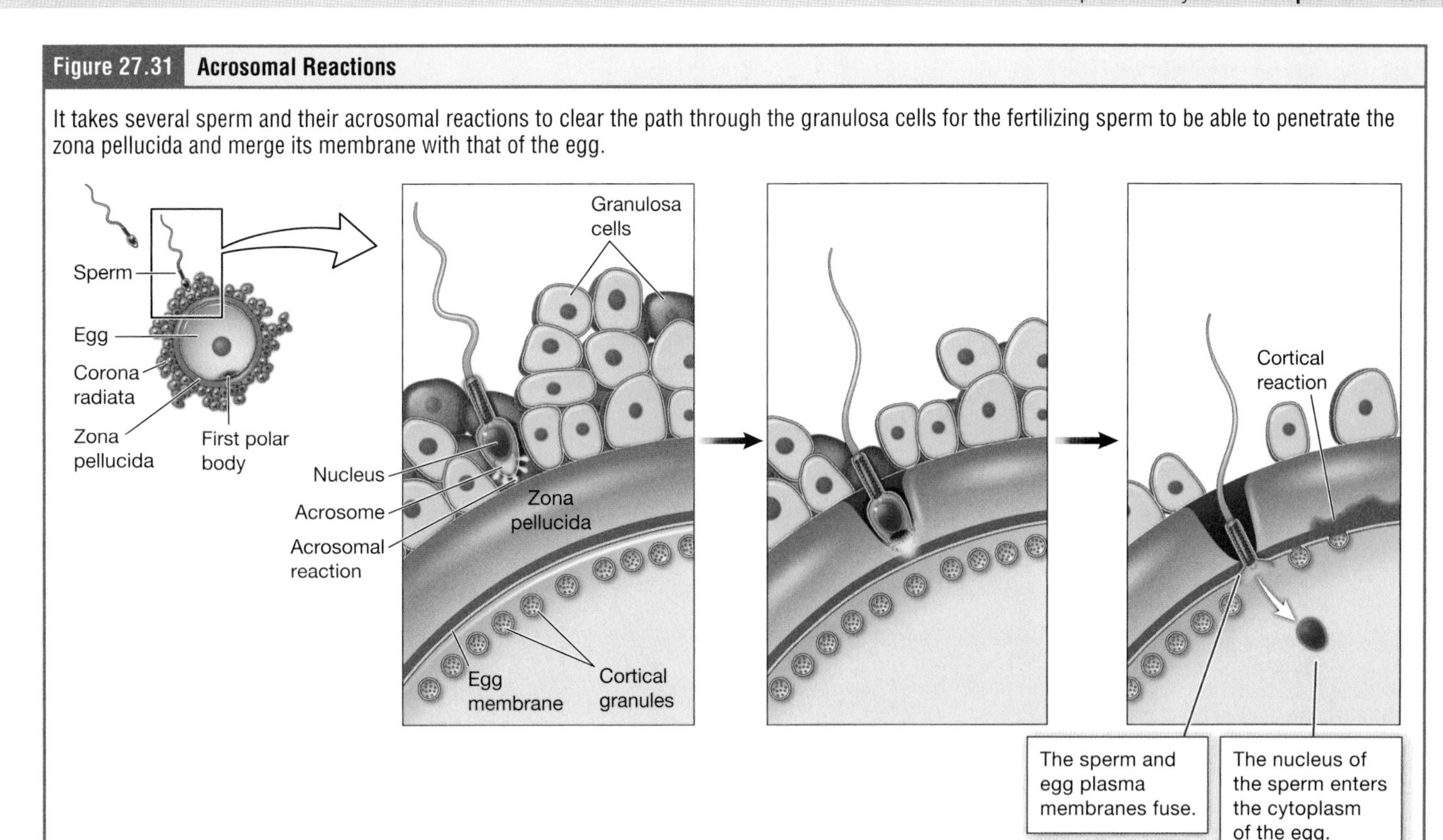

DIGGING DEEPER: Mitochondrial DNA

When we talk about human DNA, we are usually referring to nuclear DNA; that is, the DNA coiled into chromosomal bundles in the nuclei of our cells. We inherit half of our nuclear DNA from the egg and half from the sperm. However, mitochondrial DNA (mtDNA) comes only from the mitochondria in the cytoplasm of the egg. The egg carries mtDNA from the biological female who generated it, and so on. Each of our cells contains approximately 1700 mitochondria, with each mitochondrion packed with mtDNA containing approximately 37 genes.

Mutations (changes) in mtDNA occur spontaneously in a somewhat organized pattern at regular intervals in human history. By analyzing these mutational relationships, researchers have been able to determine that we can all trace our ancestry back to one woman who lived in Africa about 200,000 years ago. Scientists have given this woman the biblical name Eve, although she is not, of course, the first *Homo sapiens* female. More precisely, she is our most recent common ancestor through matrilineal descent.

This does not mean that everyone's mtDNA today looks exactly like that of our ancestral Eve. Because of the spontaneous mutations in mtDNA that have occurred over the centuries, researchers can map different "branches" off of the "main trunk" of our mtDNA family tree. Your mtDNA might have a pattern of mutations that aligns more closely with one branch, and your neighbor's may align with another branch. Still, all branches eventually lead back to Eve.

But what happened to the mtDNA of all of the other *Homo sapiens* females who were living at the time of Eve? Researchers explain that, over the centuries, their female descendants died childless or with only male children, and thus, their maternal line—and its mtDNA—ended.

outer rim of the preembryo differentiate to become the **trophoblast**, a structure that will develop into the placenta and other structures required for pregnancy that are not part of the fetus. Once the trophoblast has developed, the preembryo can implant in the uterine wall. The implanted structure, now called an *embryo*, begins to develop.

Figure 27.32 Events after Fertilization

Fertilization usually occurs in the uterine tubes in cases of successful implantation. After fertilization, the preembryo divides over several days while moving along the uterine tube. By the time it reaches the lumen of the uterus, the cells of the preembryo have differentiated to allow for implantation in the endometrium.

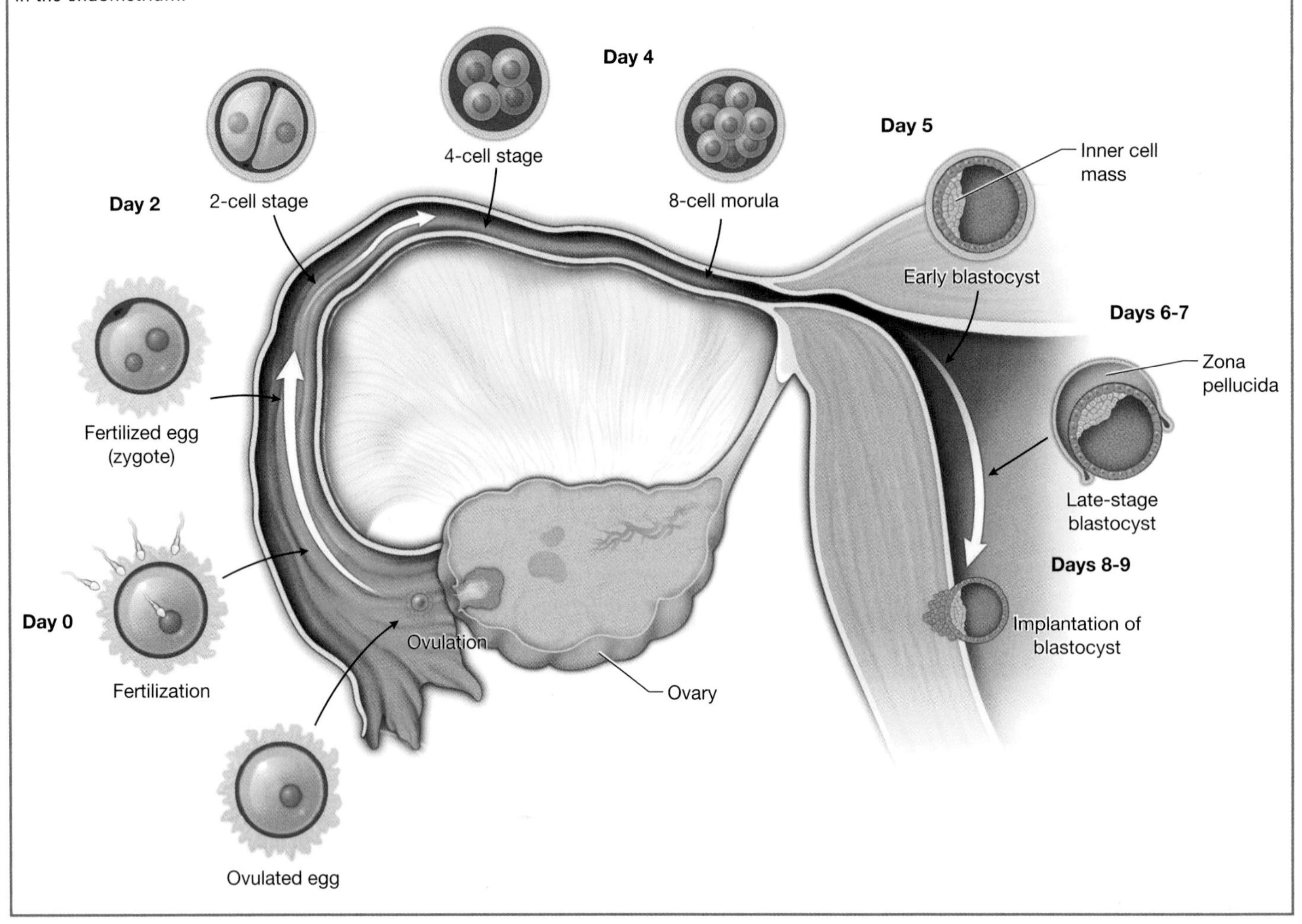

Implantation, the process in which the preembryo attaches and embeds in the endometrium, can only occur when the preembryo has developed a trophoblast and the endometrium is ready. The preembryo and the endometrium each express complimentary adhesive molecules, which can allow the embryo to stick onto the uterine wall like Velcro. If the uterus is not prepared and has not upregulated its adhesive molecules, the preembryo will continue through the uterus and pass through the cervix. Likewise, if the egg is fertilized late and does not sufficiently develop its trophoblast and upregulate its adhesive molecules, it will pass through the cervix before it is capable of implantation.

27.8c Development of the Placenta

LO 27.8.4

The embryo continues to grow through mitosis and differentiate its cells in preparation for developing organs and external structures. Its growth is fueled by the nutrients in the endometrium, secreted there during the secretory phase. The embryo will eventually run out of local nutrients, and so a central event during early pregnancy is the establishment and growth of the placenta. The cells of the trophoblast,

Figure 27.33 The Placenta

(A) The placenta is a fetal structure. (B) It resembles a tree of fetal blood vessels covered with a membrane. (C) The cells of the placenta are called *cytotrophoblasts*. Some of these cells cover the fetal blood vessels and others invade into the uterine wall to renovate the uterine arteries into blood jets.

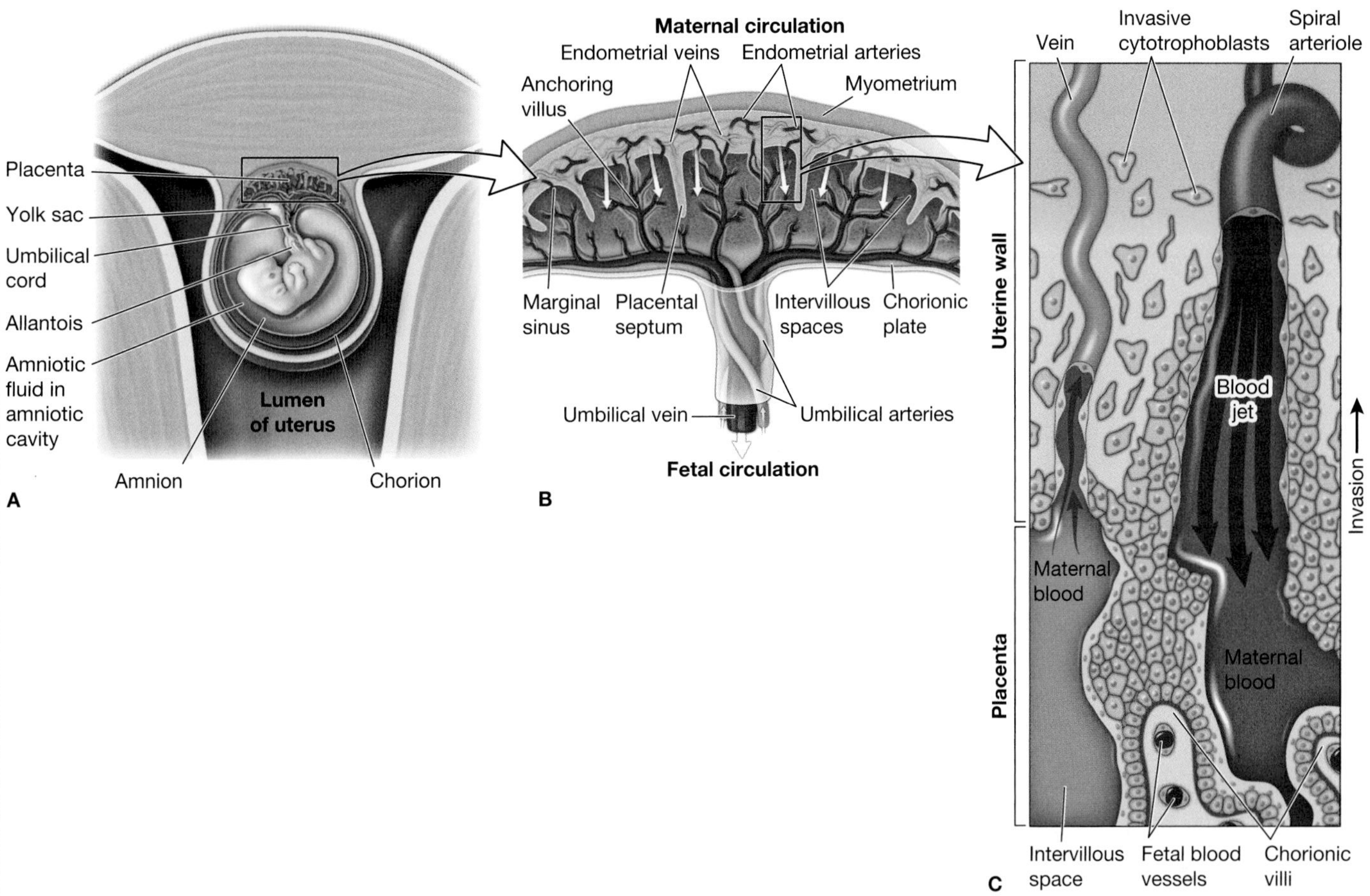

each called a **cytotrophoblast**, migrate through the endometrium in search of the uterine spiral arteries. Once they reach the spiral arteries, they invade these structures, replacing the endothelium and assuming control over the artery diameter. The maternal arteries widen, delivering more blood to the uterus (**Figure 27.33A**). Meanwhile, other cytotrophoblast cells are developing a branching structure—the placenta—in which fetal capillaries will grow (**Figure 27.33B**). These placental capillaries are connected to the blood vessels of the fetus by the umbilical cord. Only fetal blood travels within the fetus, umbilical cord, and placenta. The placenta is a fantastic example of the anatomical theme of branching structure in order to maximize surface area.

As the placenta grows, the uterine spiral arteries must dilate more and more to keep the placental branches coated with maternal blood. The maternal blood supply must increase by 30 percent in order to meet the increased circulatory demands of a full-term placenta. That means that the mother carries up to two extra liters of blood at the end of the pregnancy than she did before becoming pregnant. To generate the new red blood cells for this blood, dietary iron is important for hematopoiesis. The maternal blood delivers oxygen and nutrients to the **placenta** and removes waste. These materials all diffuse or are transported across the cytotrophoblast surface so that

Student Study Tip

To understand placental function, think of where else you have seen this theme, and how different structures (such as the lungs) make use of the same concepts.

exchange takes place between the fetal and maternal blood, but the two bloods never mix (**Figure 27.33C**). With access to maternal nutrients, embryonic development can continue and the primitive versions of each of the organ systems will all be present by the end of the embryonic period, which is defined as the first eight weeks of pregnancy.

27.8d Hormones of Pregnancy

LO 27.8.5

Student Study Tip

hCG holds the corpus luteum intact.

Once the preembryo develops the trophoblast, it begins to secrete **human chorionic gonadotrophin (hCG)**. In earlier parts of this chapter, we discussed that hCG serves as a signal from the developing embryo to the corpus luteum to encourage the corpus luteum to stay, grow, and continue its secretion of progesterone. Once the embryo has implanted, the placenta continues hCG secretion; this continues until around week 10–12 of pregnancy (**Figure 27.34**), at which point the placenta is substantial enough to take over the role of hormone secretion and the corpus luteum can degrade.

LO 27.8.6

By the end of pregnancy, the circulating levels of estrogens are 30 times higher than pre-pregnancy highs. Remember that estrogens are hormones that promote mitosis, and during pregnancy estrogens encourage tissue growth in both mother and fetus. The mother's uterus must expand in size to continue to house the growing fetus, and the mother's breasts increase in size in preparation for lactation. Estrogen also affects ligaments, causing them to loosen. Ligaments all over the body loosen, but in particular the ligaments that bind the pelvis loosen, allowing it to widen in preparation for birth. Estrogens are produced by the corpus luteum for the first 10–12 weeks of pregnancy, at which point the placenta takes over estrogen production.

Recall that progesterone acts to prevent the uterine contractions that would expel the endometrium. Therefore, in order for pregnancy to occur, progesterone must be continually secreted from the secretory phase through the end of pregnancy (see Figure 27.34). Progesterone is secreted by the corpus luteum for the first 10–12 weeks of pregnancy, at which point the placenta takes over progesterone production. This change in location of progesterone production is not always seamless; indeed, one of the several understood causes of miscarriage is unexpected drops in progesterone production.

Figure 27.34 Hormone Levels during Pregnancy

The hormone hCG is a signal from the fertilized egg/embryo to the corpus luteum instructing it to remain viable until the placenta takes over hormone synthesis. Estrogens and progesterone are produced first by the corpus luteum and then by the placenta.

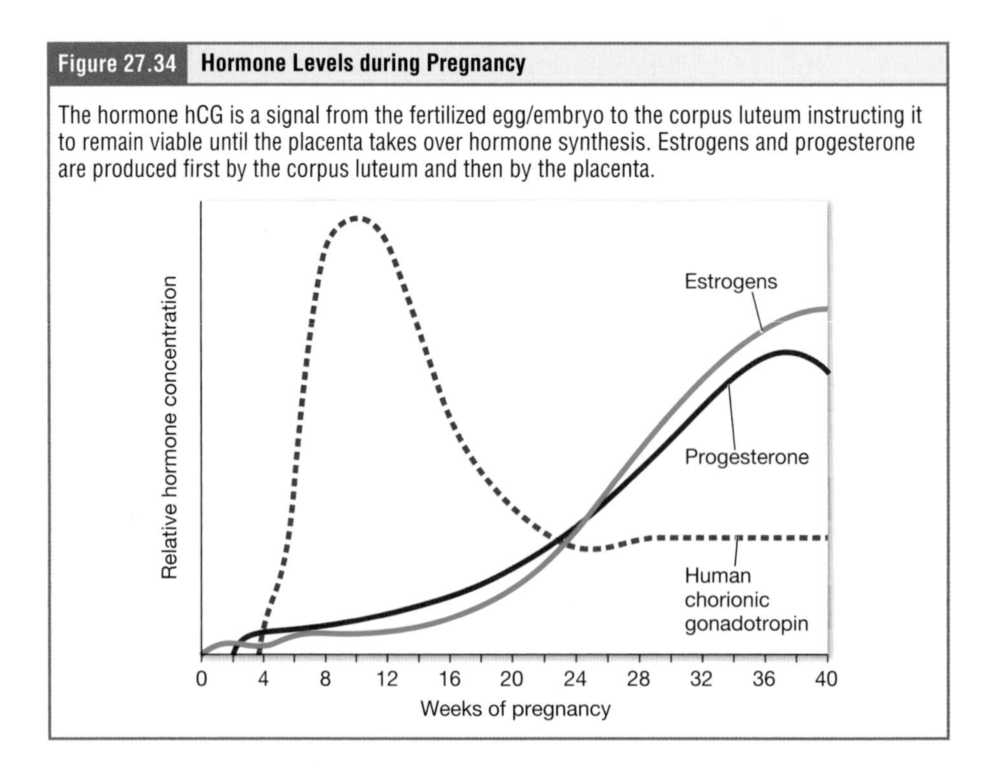

High circulating levels of progesterone and estrogen inhibit the hypothalamus and the anterior pituitary from secreting GnRH, LH, and FSH, preventing the development of follicles until after the pregnancy has ended.

27.8e Labor and Childbirth

LO 27.8.7

The smooth muscle of the myometrium contracts randomly throughout pregnancy. These contractions, known as *Braxton-Hicks contractions*, are not the organized, powerful contractions of labor, but are nonetheless a nuisance. Smooth muscle has the property of stretch contractility—that is, stretch of smooth muscle initiates contraction. This alone could explain why Braxton-Hicks contractions increase in strength as pregnancy (and fetal growth) progresses. It also possible that estrogen may play a role in this contractility. However, as pregnancy nears full term (average is 40 weeks since the first day of the last menstrual period), the myometrium increases its expression of oxytocin receptors. Secretion of oxytocin from the posterior pituitary may also increase. This interplay sets the stage for a dramatic surge in oxytocin secretion that drives the uterine contractions of childbirth. Oxytocin drives labor forward in a positive feedback loop. As the fetus grows, the weight of the fetal head on the cervix slowly stretches it. Stretch receptors in the cervix feed back to the posterior pituitary, driving the release of oxytocin. Oxytocin causes uterine contractions, which push the fetal head onto the cervix, stretching it further. Labor proceeds in three stages:

LO 27.8.8

Dilation Stage During dilation, the cervix opens and thins in response to the weight and/or pressure of the fetal head against its internal side. This process, as described previously, is driven by oxytocin and the resulting uterine contractions. Dilation is typically the longest stage of labor, but its duration can vary widely from up to 24 hours to as little as a few minutes.

Expulsion Stage This stage lasts anywhere from 1 to 60 minutes and involves the expulsion of the fetus, now called the *baby*, from the mother's body. Expulsion begins as the baby's head passes the cervix and stretches the walls of the vagina and ends when the baby is fully out of the mother's body. The umbilical cord must be cut for the baby to move away from the mother's body, as the placenta is still within the uterus.

Placental Stage Oxytocin levels are still very high and the uterine contractions continue after the baby is out of the body. The oxytocin contributes to maternal bonding and the continued contractions have two purposes. The contractions cause the placenta to collapse, pulling away from the uterine wall and into the lumen of the uterus, then through the cervix and out the vagina. The uterine contractions also help to constrict the spiral arteries, limiting the loss of maternal blood.

27.8f Postpartum

LO 27.8.9

The first six weeks after birth, the **postpartum period**, are a time of intense physiological changes. The uterus must return to its pre-pregnancy size, no small feat for an organ that rapidly grew from about the size of a pear to about the size of a watermelon in the 40 weeks preceding birth. Uterine cells digest themselves with their own lysosomal enzymes, a process called *autolysis*. This self-digestion eliminates many of the cells of the uterus, and the uterus slowly returns to its original size.

27.8g Lactation

LO 27.8.10

The hormones of pregnancy prepare the breast for lactation. **Lactation**, or milk production, is governed by positive feedback loops. Prolactin, the anterior pituitary hormone that controls milk production, is secreted throughout pregnancy. After birth, prolactin levels drop back to their pre-pregnancy levels and will not rise again unless breastfeeding occurs. Therefore, a mother who does not breastfeed will not produce milk beyond what is in the breasts at the time of childbirth. In mothers who do breastfeed, prolactin secretion increases after each feeding, driving the production of milk for the next feeding. Milk synthesis also requires other hormones, including cortisol, growth hormone, parathyroid hormone, and insulin in order to mobilize the nutrients necessary for milk production. The nutrients in breast milk may come from the lactating mother's diet or they may be harvested from her nutrient stores, including her own tissues as necessary.

Milk production—driven by prolactin—occurs on a positive feedback loop, and milk ejection, which is driven by oxytocin, operates on a feedback loop as well. The baby suckling at the breast stimulates oxytocin secretion, which causes milk ejection, and milk coming out of the nipple further increases the likelihood that the baby will continue to suckle at the breast. Oxytocin also increases parasympathetic nervous system activity, allowing a feeling of relaxation in the lactating mother, and promotes bonding between mother and infant.

Lactation suppresses the secretion of estrogen, which impacts maternal physiology in a number of ways. Early after birth, lactation facilitates the shrinking of the uterus through estrogen suppression (estrogen provides signals for mitosis of uterine cells). The duration of lactation, and the number of children that the mother breastfed, are both negatively correlated with breast cancer incidence. In other words, more breastfeeding, and breastfeeding for longer, helps to decrease the chance of breast cancer later in life. It is likely that the break from estrogen synthesis during lactation decreases breast cancer rates because estrogen promotes mitosis in breast tissue. The suppression of estrogen may also provide a break from fertility, stalling the resumption of follicle development and ovulation. In each period of lactation, the length of time that estrogen is suppressed varies; therefore, lactation alone is not a sufficient birth control method if a subsequent pregnancy is not desired.

Learning Check

1. Sally and Shannon are both biological females and new mothers. Sally has decided to breastfeed her child, while Shannon has chosen formula-feeding. What will Sally's prolactin levels be in comparison to Shannon's?
   a. Sally will have higher circulating prolactin.
   b. Shannon will have higher circulating prolactin.
   c. Sally and Shannon will have about equal levels of circulating prolactin.
   d. Impossible to tell with the information given
2. During pregnancy, which of the following hormones prevents uterine contractions, so that the endometrium is not shed?
   a. Progesterone
   b. FSH
   c. Estrogen
   d. Inhibin
3. During the placental stage of labor, is there pressure on the cervix? What do oxytocin levels look like in the body?
   a. Yes; high
   b. Yes; low
   c. No; high
   d. No; low
4. Emily is a biological female. Her lysosomal enzyme activity is low in the postpartum period. Which of the following processes will be slow while Emily's body returns to its pre-birth state?
   a. Ligament tightening
   b. Blood volume decreasing
   c. Respiratory rate decrease
   d. Uterus shrinking

Chapter Review

27.1 Overview of Human Reproductive Systems

The learning objectives for this section are:

27.1.1* Define and contrast the terms *sex*, *gender*, *gender expression*, and *sexual orientation*.

27.1.2 Describe the functions of the hormones involved in the regulation of the reproductive processes (e.g., gonadotropin releasing hormone [GnRH], follicle stimulating hormone [FSH], luteinizing hormone [LH], androgens, inhibin, estrogens, progesterone).

* Objective is not a HAPS Learning Goal.

27.1 Questions

1. Which of the following statements is true of the term *gender*?

a. It refers to whom a person is sexually attracted to.

b. It is also called *gender identity*.

c. It refers to a person's anatomy, chromosomes, and hormones present at birth.

d. It must match biological sex.

2. Which of the following hormones is/are produced by the hypothalamus and target(s) the anterior pituitary gland?

a. GnRH

b. LH/FSH

c. Androgens

d. Progesterone

27.1 Summary

- The terms *sex*, *gender*, *gender expression*, and *sexual orientation* are all distinct. *Sex* refers to that characteristics that a person is born with in terms of anatomy, chromosomes, and hormones. *Gender* may or may not match sex, nor fit neatly into two distinct categories of male or female. *Gender expression* relates to a person's demeanor, and *sexual orientation* relates to whom the individual is sexually attracted to.
- All humans have the following hormones, but produce different levels of each one: gonadotropin releasing hormone (GnRH), follicle stimulating hormone (FSH), luteinizing hormone (LH), androgens, inhibin, estrogens, and progesterone. Each of these has a distinct function.
- GnRH stimulates the secretion of FSH and LH from the anterior pituitary gland. FSH stimulates ovarian follicle development and spermatogenesis. LH stimulates ovulation in biological females and sex hormone production in both sexes. Testosterone causes development of the sex organs, maintenance of the secondary sex characteristics, and spermatogenesis in biological males. Estrogens cause sex organ development, oogenesis, and maintenance of the secondary sex characteristics, and control the uterine cycle in biological females. Progesterone mainly acts on the uterus during the secretory phase of the uterine cycle. Inhibin exerts negative feedback on FSH to prevent its oversecretion.

27.2 Development of Reproductive Systems

The learning objectives for this section are:

27.2.1* Identify homologues of reproductive system structures.

27.2.2* Describe the physiological and anatomical changes that occur during puberty.

27.2.3* Identify the structures of human mammary glands and describe their functions.

* Objective is not a HAPS Learning Goal.

27.2 Questions

1. The genital folds surrounding the urogenital sinus can become external genitalia of biological males or females. If the genital folds remain separate, which is the developing fetus's biological sex?

a. Male

b. Female

c. Impossible to tell with the information given

2. Which of the following hormonal changes occur during puberty?
   a. Increased LH and FSH production
   b. Increased estrogen and testosterone production
   c. Decreased negative feedback on the hypothalamus and anterior pituitary
   d. All of the above

3. All of the following are true of the structure of the mammary glands except:
   a. Mammary glands are actually modified sweat glands.
   b. Both biological males and females deposit adipose tissue in the breasts.
   c. Breast growth at puberty in biological females is regulated by the hormone progesterone.
   d. Lactiferous ducts drain milk directly into the nipple.

27.2 Summary

- Development of reproductive structures begins at approximately weeks 4–5 of embryogenesis, when primordial germ cells migrate into the embryo to become sperm or eggs.
- Week 7 marks the transition from the indifferent stage to a differentiated stage, due to the presence or absence of the SRY gene region on the Y chromosome.
- All embryos develop two internal reproductive duct systems; the mesonephric duct system develops in biological males and the paramesonephric duct system develops in biological females. The duct systems eventually develop into the internal reproductive structures.
- The genital folds and genital tubercles develop into homologous structures of the external genitalia in both biological sexes.
- Puberty is the stage of development in which individuals become ready for reproduction. Changes in hormone levels trigger the development of secondary sex characteristics, the growth of the reproductive organs, and the beginning of the uterine cycles in biological females.
- Breasts are present in all individuals; they are larger in those with a higher level of circulating estrogen (biological females). The mammary glands function in lactation or milk production, to nourish a newborn baby. Milk production is under the control of prolactin, and milk ejection is under the control of oxytocin.

27.3 Anatomy of Biological Females

The learning objectives for this section are:

HAPS LEARNING GOALS AND OUTCOMES

27.3.1 Identify and describe the structure and functions of the external genitalia (e.g., mons pubis, labia majora, labia minora, clitoris, greater vestibular glands).

27.3.2 Identify and describe the structure and function of the vagina.

27.3.3 Identify and describe the gross anatomy, microscopic anatomy, and functions of the ovaries.

27.3.4 Identify and describe the ligaments of the female reproductive system (e.g., broad ligament, ovarian ligament, suspensory ligament of the ovary, round ligament of the uterus).

27.3.5 Describe the pathway of the oocyte from the ovary to the uterus.

27.3.6 Identify and describe the gross anatomy, microscopic anatomy, and functions of the uterus and uterine (fallopian) tubes.

27.3 Questions

1. Which of the following structures make up the vulva? Please select all that apply.
   a. Vagina
   b. Labia majora
   c. Labia minora
   d. Clitoris

2. All of the following are true of the vagina except:
   a. The cervix forms a superior gate into the vagina.
   b. The inner lining of the vagina contains folds called *rugae*, just like the stomach.
   c. All throughout the lifespan, the vagina houses a large population of microorganisms.
   d. The vagina has a pH of about 10 all throughout the lifespan.

3. In which part of the ovary does folliculogenesis occur?
   a. Tunica albuginea
   b. Cortex
   c. Medulla
   d. Mesovarium

4. The ovarian vein is contained within which ligament of the biological female reproductive anatomy?
   a. Broad ligament
   b. Suspensory ligament
   c. Mesovarium
   d. Ovarian ligament

5. Which is the correct path of the oocyte through the female reproductive tract, including the sites of fertilization and implantation?
   a. Uterus, infundibulum, ampulla, isthmus, ovary
   b. Ovary, ampulla, infundibulum, isthmus, uterus
   c. Ovary, infundibulum, ampulla, isthmus, uterus
   d. Uterus, infundibulum, isthmus, ampulla, ovary

6. Which of the following is the correct order of the regions of the uterus from top to bottom?
   a. Fundus, body, cervix
   b. Body, cervix, fundus
   c. Cervix, fundus, body
   d. Body, fundus, cervix

27.3 Summary

- The external genitalia, or vulva, of the biological female function in protection of internal structures (labia majora, labia minora, and mons pubis), lubrication for sexual intercourse (vestibular glands), and orgasm (clitoris).
- The vagina is the tubular organ that lies between the uterus and the external environment. It receives sperm for fertilization, excretes uterine substances during the menses, and serves as part of the birth canal. The vagina maintains a low pH during the reproductive years to be inhospitable to microbes.
- The ovaries are the gonads of biological females; they are analogous to the testes in biological males. The ovaries function in oogenesis, folliculogenesis, and ovulation.
- The uterine tubes connect the ovaries to the uterus and are the site of fertilization. They are composed of three sections: the infundibulum, ampulla, and isthmus. Ovulated eggs are directed into the uterine tubes by the fimbriae of the infundibulum.
- The uterus is the muscular organ in the reproductive system of a biological female. It connects to the vagina anteriorly and sits posteriorly to the bladder. The uterus houses the fetus during pregnancy. Its strong muscle contractions propel the fetus to the outside world during childbirth.

27.4 The Ovarian and Uterine Cycles

The learning objectives for this section are:

27.4.1 Define the process of oogenesis (oocyte development).

27.4.2 Describe the stages of folliculogenesis (ovarian follicle development).

27.4.3 Describe endocrine regulation of oogenesis, folliculogenesis, and the ovarian cycle.

27.4.4 Describe a typical ovarian cycle and explain how the process of folliculogenesis spans multiple ovarian cycles.

27.4.5 Define ovulation, and explain the role of luteinizing hormone (LH) in ovulation.

27.4.6 Name the phases of the uterine (menstrual) cycle, and describe the anatomical changes in the uterine wall that occur during each phase.

27.4.7 Describe the correlation between the uterine and ovarian cycles.

27.4 Questions

1. What stages of oocyte development occur in the ovaries of fetuses before birth? Please select all that apply.
   a. Secondary oocytes
   b. Polar body
   c. Oogonia
   d. Primary oocytes

2. An ovarian follicle is currently undergoing atresia during folliculogenesis. In which stage(s) of development could this follicle be?
   a. A primary follicle
   b. A secondary follicle
   c. A tertiary follicle
   d. All of the above could be correct

3. In the hormonal control of the biological female reproductive system, a hormone that once caused negative feedback now turns positive, effecting change in the follicle. ______ turns from exerting negative feedback on LH to exerting positive feedback, triggering ______.

a. Progesterone; ovulation
b. Progesterone; oogenesis
c. Estrogen; ovulation
d. Estrogen; oogenesis

4. Which of the following processes requires about 13 menstrual cycles to complete?

a. Oogenesis
b. Folliculogenesis
c. Corpus luteum formation
d. Ovulation

5. Which of the following events occurs at ovulation? Please select all that apply.

a. At ovulation, the oocyte is released both from its follicle and from the wall of the ovary.
b. As soon as the oocyte is ovulated, the follicle degenerates.
c. During most menstrual cycles, oocytes from several dominant follicles are ovulated.
d. A few layers of granulosa cells are ovulated along with the oocyte.

6. Which of the following correctly pairs the phase of the uterine cycle with the events occurring in the uterus at that time?

a. Menses; the uterine wall is thickening
b. Secretory; endometrial glands are producing nutrients for a possible implanted embryo
c. Proliferative; granulosa cells are reproducing in the dominant follicle

7. When the uterine cycle is in the proliferative phase, in which phase is the ovarian cycle?

a. Menses
b. Luteal
c. Secretory
d. Follicular

27.4 Summary

- In order to have successful fertilization, the sperm and egg must fuse. The biological female prepares eggs for fertilization through the processes of the ovarian cycle and the uterine cycle. The ovarian cycle consists of oogenesis (egg production and maturation) and folliculogenesis (production and maturation of ovarian follicles).
- Oogenesis, part of the ovarian cycle, begins with ovarian stem cells called *oogonia*. They develop into primary oocytes, encased in primordial follicles, and then into secondary oocytes and polar bodies. During each month of the uterine (menstrual) cycle, a secondary oocyte is released from a tertiary follicle during ovulation.
- The ovarian cycle is controlled by the hormones GnRH, FSH, LH, estrogen, and progesterone. GrRH stimulates the release of FSH and LH from the anterior pituitary gland. FSH stimulates follicles to grow in folliculogenesis. LH stimulates the thecal and granulosa cells of the follicle to release estrogen, which controls several aspects of the ovarian and uterine cycles.
- Ovulation is triggered by a surge of LH from the anterior pituitary gland. A secondary oocyte, along with its surrounding layers, is released from the ovary and swept into the uterine tube. If sperm are present in the uterine tube, fertilization may occur.
- The uterine (menstrual) cycle is the series of events in which the uterine lining is shed, rebuilds, and prepares for possible implantation of an embryo under the influence of estrogen and progesterone. It consists of three phases: the menses phase, the proliferative phase, and the secretory phase.
- In menses, low progesterone levels trigger the shedding of the endometrium in the absence of fertilization. The proliferative phase is correlated with the follicular phase in the ovary; as the endometrium is thickening, an oocyte is maturing and preparing for ovulation. At the end of this phase, ovulation occurs. During the secretory phase, endometrial glands are secreting nutrients for a possible implanted embryo. At the same time, the corpus luteum, which formed from the remnants of the follicle after ovulation, is secreting estrogen, progesterone, and inhibin.

27.5 Anatomy of Biological Males

The learning objectives for this section are:

HAPS LEARNING GOALS AND OUTCOMES

27.5.1 Identify and describe the structure and functions of the male external genitalia (e.g., scrotum, penis).

27.5.2 Identify and describe the gross anatomy, microscopic anatomy, and functions of the testes.

27.5.3 Identify and describe the gross anatomy, microscopic anatomy, and functions of the epididymis.

27.5.4 Identify and describe the structure and functions of the spermatic cord and male reproductive ducts (e.g., ductus [vas] deferens, ejaculatory duct, urethra).

27.5.5 Describe the production, composition, and functions of semen.

27.5.6 Identify and describe the structure and functions of accessory glands (i.e., seminal glands [seminal vesicles], prostate gland, bulbourethral [Cowper] glands).

27.5.7 Describe the pathway of sperm from the seminiferous tubules to the external urethral orifice of the penis.

27.5 Questions

1. Though the penis is considered an external structure, a substantial amount of the penis is inside the body. What is the portion of the penis that lies inside the abdominopelvic cavity called?
   a. Shaft
   b. Glans
   c. Root
   d. Corpus cavernosum

2. The scrotum in biological males houses the reproductive structure that has which function?
   a. Erection
   b. Sperm production
   c. Androgen production
   d. Both B and C

3. Where is the epididymis located in relation to the testes?
   a. Along the anterior wall
   b. In the center of the testes
   c. Along the posterior wall
   d. Along the interior border

4. An issue with torsion of the spermatic cord creates twisting in which of the following structures of the biological male reproductive system? Please select all that apply.
   a. Epididymis
   b. Nerve supply to the testes
   c. Blood supply to the testes
   d. Ductus deferens

Mini Case 1: Seminal Vesiculitis

Jim, a biological male, and his partner, Susan, a biological female, were having problems having a baby. Susan was thoroughly examined and found to have a healthy reproductive system. When Jim was examined, it was found that they had a condition called *seminal vesiculitis*, which was determined to be the cause of the couple's infertility.

1. One of Jim's symptoms was that their ejaculate contained an abnormally low semen volume. The bulk of the semen volume consists of secretions of the accessory glands of the biological male reproductive system. What functions do these secretions serve? Please select all that apply.
   a. To provide fuel to the sperm for movement
   b. To coagulate the semen, thickening it
   c. To liquefy the semen, making it thinner
   d. To clean urine residues from the penile urethra

2. Seminal vesiculitis is an inflammation of the seminal vesicles, sometimes caused by an infection and often due to an unknown cause. Jim's tests did not reveal any infection of the seminal vesicles. Why would seminal vesiculitis cause Jim's semen volume to be abnormally low?
   a. When the seminal vesicles are not functioning properly, their cells are unable to contribute their typical 60 percent of the volume of the semen.
   b. When the seminal vesicles are inflamed, the entire urethra is swollen to the point that the semen cannot pass through it.
   c. When the seminal vesicles are inflamed, it prevents the testes from producing sperm (which make up 50 percent of the volume of the semen).
   d. When the seminal vesicles are inflamed, the prostate gland and bulbourethral glands stop functioning too, so very little fluid is secreted into the semen during ejaculation.

3. Although Jim's semen volume was abnormally low, they were still able to ejaculate. Which of the following is the correct pathway that sperm follow while traveling through the reproductive tract of the biological male during ejaculation?
   a. Seminiferous tubules, epididymis, rete testes, ductus deferens, external urethral orifice, urethra
   b. Seminiferous tubules, rete testes, epididymis, ductus deferens, urethra, external urethral orifice
   c. External urethral orifice, ductus deferens, epididymis, rete testes, urethra, seminiferous tubules
   d. Urethra, external urethral orifice, epididymis, ductus deferens, rete testes, seminiferous tubules

27.5 Summary

- The external genitalia of the biological male consists of the penis and the scrotum. The penis is homologous to the female clitoris, and the scrotum is analogous to the labia.
- The penis functions to deliver sperm to the vagina during sexual intercourse. It contains erectile tissue, which allows for this function. The urethra, which runs down the entire length of the penis, is the organ that transports both urine and semen out of the body.
- The scrotum houses the testes outside of the abdominopelvic cavity, to maintain the temperature required for optimum sperm production.
- The testes are the gonads of the biological male; they produce gametes (sperm) and hormones. The gametes are completely formed in the seminiferous tubules of the testes; they then mature in the epididymis, moving into the ductus deferens when matured.
- There are three types of glands in the male reproductive system: seminal vesicles contribute the majority of the semen volume; the prostate gland excretes an alkaline fluid into the semen; and the bulbourethral glands release pre-ejaculate, which functions in cleaning urine residue from the urethra, and also contributes a small amount of fluid to the semen. Sperm contribute only about 5% of the volume of the semen.

27.6 Spermatogenesis and Spermiogenesis

The learning objectives for this section are:

27.6.1 Define the processes of spermatogenesis and spermiogenesis.

27.6.2 Describe the stages of spermatogenesis in the seminiferous tubule, including the roles of the nurse (sustentacular, Sertoli) cells and interstitial cells (of Leydig).

27.6.3 Describe endocrine regulation of spermatogenesis.

27.6 Questions

1. Which of the following processes describes the maturation and differentiation of spermatids to produce functional sperm?
   a. Spermatogenesis
   b. Spermiogenesis
   c. Oogenesis
   d. Folliculogenesis

2. Spermatogenesis begins with ______ and ends with ______.
   a. Spermatogonia; secondary spermatocytes
   b. Spermatids; spermatogonia
   c. Primary spermatocytes; spermatids
   d. Spermatogonia; sperm

3. Before the onset of puberty in biological male bodies, sperm are not yet produced. This can be due to low levels of which of the following hormones? Please select all that apply.
   a. LH
   b. FSH
   c. Progesterone
   d. Testosterone

27.6 Summary

- Spermatogenesis describes the formation of haploid gametes—sperm—through mitosis and then meiosis. These haploid cells are not yet ready to fertilize an egg.
- Spermiogenesis is the last stage of spermatogenesis, in which spermatids mature into sperm.
- In the testes, two cell types interact in the formation of sperm. Interstitial cells produce testosterone, and nurse cells promote spermatogenesis and produce inhibin, which exerts negative feedback on the secretion of FSH, to maintain control of spermatogenesis.
- Spermatogenesis begins with primitive cells called spermatogonia. These cells develop into primary spermatocytes, which undergo meiosis I to become secondary spermatocytes. These then undergo meiosis II to become spermatids, which mature into sperm. Sperm are the haploid gametes that can participate in fertilization.

27.7 Sex (Coitus)

The learning objectives for this section are:

27.7.1* List and describe the physiology and stages of sexual responses.

27.7.2* Compare and contrast STIs in terms of their causative agents, symptoms, and diseases.

27.7.3* Explain how different examples of birth control methods work.

* Objective is not a HAPS Learning Goal.

27.7 Questions

1. Which of the following stages of sexual responses is categorized in part by the development of muscular tension all over the body?
   a. Excitement and plateau stages
   b. Orgasm stage
   c. Resolution stage
   d. Both A and B

2. Which of the following common sexually transmitted infections is viral and cause immediate symptoms following infection?
   a. Herpes
   b. HIV
   c. Chlamydia
   d. Syphilis

3. Aromatase inhibitors are a type of medication used for the treatment of breast cancer and endometriosis, but could potentially be used as a method of birth control. Aromatase converts androgens into estrogens. Under the influence of this inhibitor, will ovulation occur in biological female bodies?
   a. Yes
   b. No
   c. Impossible to tell with the information given

27.7 Summary

- The sexual response occurs in a series of stages: (1) the excitement and plateau stage, (2) the orgasm stage, and (3) the resolution stage.
- The excitement stage of coitus is the phase in which the body prepares for and/or engages in sex; it is characterized by systemic myotonia, vasocongestion, and closing off of the urethra in the biological male to prevent urination during ejaculation.
- The orgasm stage of coitus is an intense response of 1–15 seconds during which the sympathetic nervous system is dominant, increasing breathing rate and blood pressure. Oxytocin surges and pelvic floor muscles contract.
- The resolution stage follows orgasm (if orgasm occurs). Parasympathetic nerves become dominant, returning breathing rate and blood pressure to a normal level though oxytocin levels continue to surge.
- Refractory periods from parasympathetic neurons prevent erection after one cycle of coitus has occurred. The biological male refractory period is longer than that of the female.
- Sexually transmitted infections, or STIs, are infections that can be passed from one person to another through sexual contact. Most of the STI pathogens are bacterial or viral. Many are treatable or curable.
- Birth control is the voluntary regulation of fertilization or implantation. There are many birth control methods, and each has a different level of effectiveness. Mechanisms include barrier methods, intrauterine devices, and various hormonal contraceptives that disrupt ovulation.

27.8 Pregnancy, Birth, and Lactation

The learning objectives for this section are:

27.8.1 Define fertilization.

27.8.2 Describe the processes that facilitate fertilization (e.g., sperm capacitation, acrosomal reaction).

27.8.3 Describe the three phases of fertilization (i.e., corona radiata penetration, zona pellucida penetration, fusion of the oocyte and sperm plasma membranes).

27.8.4 Describe the formation and function of the placenta and extraembryonic membranes.

27.8.5 Describe the hormones associated with pregnancy and the effects of these hormones.

27.8.6 Describe physiological changes that permit pregnancy (e.g., change in blood volume) and their reversal during the postpartum period.

27.8.7 Define parturition (labor).

27.8.8 Describe the three stages of labor.

27.8.9 Define the postpartum period.

27.8.10 Describe the hormonal regulation of lactation.

27.8 Questions

1. Mixing of biological male and female gametes and genetic information is accomplished during which of the following stages of reproduction?
   a. Fertilization
   b. Implantation
   c. Preembryonic stage

2. Which of the following processes must occur for fertilization to take place? Please select all that apply.
   a. Sperm capacitation
   b. A surge of testosterone
   c. hGC release
   d. Acrosomal reaction

3. Which phase of fertilization involves the rupture of the layer covering the sperm nucleus, which releases enzymes that can break down the corona radiata and the zona pellucida?
   a. Sperm capacitation
   b. Acrosomal reaction
   c. Spermiogenesis
   d. Atresia

4. Which of the following are true of the formation of the placenta? Please select all that apply.
   a. The placenta forms from structures of both the mother and the fetus.
   b. Maternal and fetal blood come in direct contact within the placenta.
   c. Once the umbilical cord forms, the placenta degenerates.
   d. The placenta connects placental capillaries to fetal blood vessels.

5. Which of the following hormones is secreted only during pregnancy, and functions to keep the corpus luteum viable until the placenta is established?
   a. Estrogen
   b. Human placental lactogen
   c. Progesterone
   d. Human chorionic gonadotropin

6. All of the following changes occur in the mother's body during pregnancy except:
   a. Ligament loosening all over the body
   b. A 30-fold increase in the level of estrogen
   c. An increase in uterine contractions, caused by progesterone
   d. A cessation of ovulation

7. Which of the following events occur during the process of labor? Please select all the apply.
   a. Oxytocin causes the strong uterine contractions of childbirth.
   b. Oxytocin secretion during labor operates in a negative feedback loop.
   c. The last stage of labor involves the birth of the baby.
   d. The placenta is expelled from the mother's body after the birth of the baby.

8. In which stage(s) of labor do uterine contractions occur?
   a. Dilation
   b. Expulsion
   c. Placental
   d. All of the above

9. Which of the following are true of the postpartum period? Please select all that apply.
   a. It lasts for 6 months after childbirth.
   b. It involves a decrease in the size of the uterus.
   c. The uterus never returns to its original size.
   d. Many uterine cells are destroyed by their own lysosomal enzymes.

10. Milk production is regulated by the hormone _________, while milk ejection is regulated by the hormone _________.
    a. Estrogen; progesterone
    b. Prolactin; oxytocin
    c. Insulin; glucagon
    d. Oxytocin; prolactin

27.8 Summary

- The defining event in reproduction is fertilization, in which the acrosomal enzymes of the sperm penetrate the corona radiata and the zona pellucida of the egg, which enables the combining their genetic material in a process that will take about 10 hours after the sperm has reached the egg.
- Implantation of the preembryo in the endometrium occurs several days after fertilization; implantation becomes possible only after the first differentiation event of the preembryo to become a trophoblast.
- After implantation, the embryo is nourished by the endometrial glands until the placenta is established. The placenta is produced from structures from both the embryo and the mother. The placenta will provide nutrients and oxygen to the fetus, and will remove wastes, throughout the rest of the pregnancy.
- The labor of childbirth occurs in three stages: (1) the dilation stage, in which the cervix is dilated to accommodate the passage of the fetus's head, (2) the expulsion stage, in which the fetus is propelled out of the mother's body, and (3) the placental stage, in which the placenta is expelled from the mother's body. Oxytocin plays a vital role in causing the uterine contractions of childbirth.
- The postpartum period, which lasts for six weeks after childbirth, is characterized by changes in the mother's body, such as by uterine shrinkage and possible lactation. Prolactin promotes lactation and is stimulated by breastfeeding; levels of prolactin are much lower in people who do not breastfeed.

28

Anatomical and Physiological Response to Disease

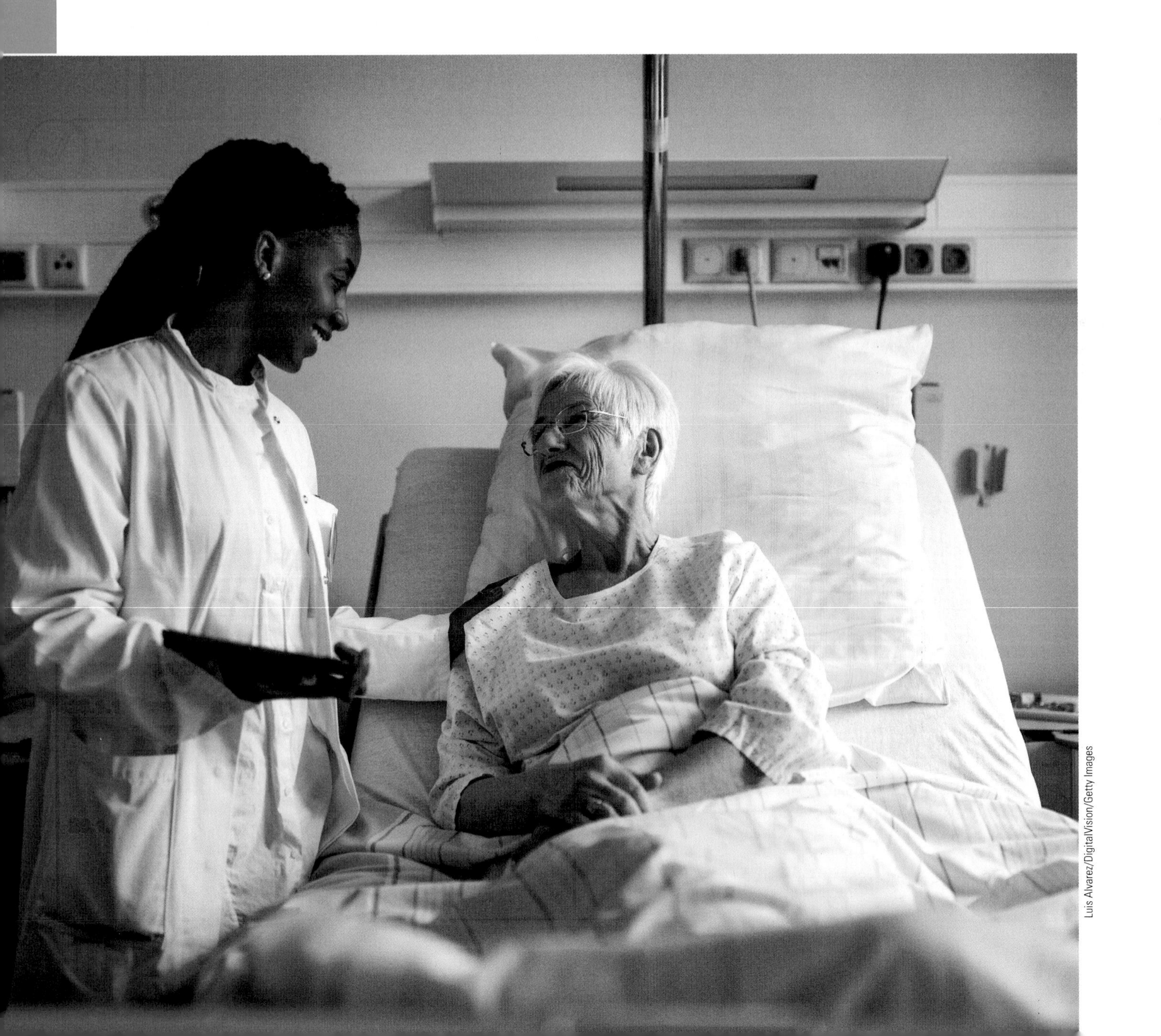
Luis Alvarez/DigitalVision/Getty Images

Chapter Introduction

In this chapter you will...

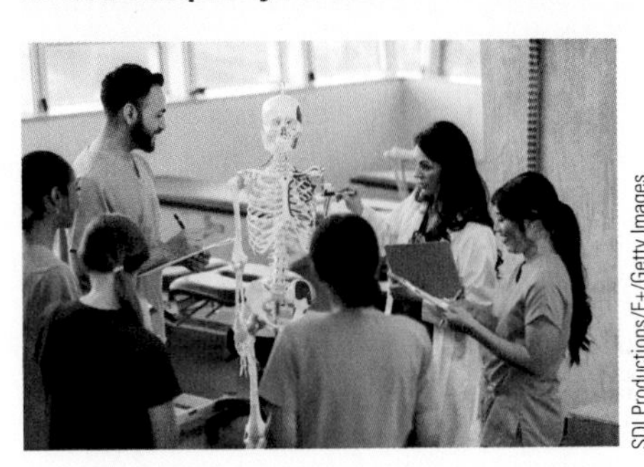
SDI Productions/E+/Getty Images

IvanRiver/Shutterstock.com

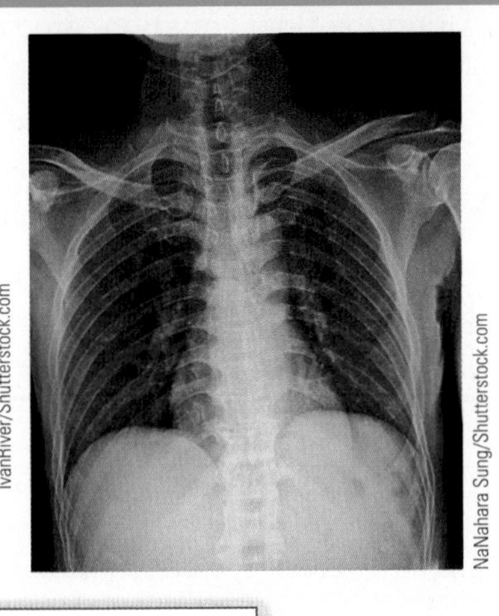
NaNahara Sung/Shutterstock.com

apply what you've learned in anatomy and physiology so far. . .

to interpret test results in clinical case studies.

Throughout this text we have built an understanding of anatomy and physiology by exploring one system at a time. In this section we will examine the interconnectedness of the human body by looking at disease states. During disease, there is often one precipitating event—one failing organ, one misfiring pathway, one type of cell, or one gene mutation. From that singular issue, however, malfunction cascades through the body because no single organ, gene, or cell operates in isolation.

This chapter is a collection of case studies. Each tells the story of one individual with one disease. You may choose to tackle the cases individually or work your way through the entire chapter at once. Each case provides some narrative but calls on your brain to tell the whole story. Each case is accompanied by review questions, which ask you to recall information you have learned in A&P that is pertinent to each case, and analysis questions, which ask you to think critically with the information you have to predict changes that might occur in the course of the disease.

Note: None of the Learning Objectives in this chapter are from HAPS. They are all created by the author for this unique chapter of case studies.

28.1 Hypertension Case Study

Learning Objectives: By the end of this case study you will be able to:

28.1.1 Use the principles of diffusion and osmosis to predict various changes in blood pressure, fluid osmolarity, sweat production, or other physiological functions (Sections 4.1 and 20.2).

28.1.2 List the physiological components of mean arterial pressure (Section 20.2).

28.1.3 Describe the physiological factors that influence each component of mean arterial pressure (Section 20.2).

28.1.4 Describe the relationship between blood volume and blood pressure (Section 20.2).

28.1.5 Describe the control over hematopoiesis (Section 18.2).

28.1.6 Predict how changes in diet or lifestyle may impact mean arterial pressure (Section 20.2).

28.1.7 Identify the layers of blood vessels and their histological components (Section 20.1).

28.1.8 Compare and contrast the characteristics of arteries and veins (Section 19.1).

28.1.9 Describe, compare, and contrast different types of molecules found in the human body (Section 3.5).

28.1.10 List and describe the components of the flow equation (Section 22.2).

28.1.11 Using the flow equation, predict how different changes to physiological conditions will impact flow rate (Section 22.2).

28.1.12 Describe the hormonal control over blood pressure (Section 26.2).

28.1.13 Predict the impact on blood pressure in various physiological conditions (Sections 20.2, 20.3, and 20.4).

Hypertension, or chronic high blood pressure, is a disease that affects 47 percent of American adults. We learned in Chapter 20 that blood pressure is monitored by baroreceptors in the aorta and carotid arteries. These receptors detect changes in blood pressure and send this information to a nucleus in the medulla oblongata. The medulla integrates this information and sends efferent signals to the heart to adjust its rate.

1. The hormone aldosterone is secreted when the blood level of potassium is too high. Besides increasing the excretion of potassium ions into the urine, aldosterone also causes the reabsorption of sodium ions. How would this be expected to affect the blood pressure? Please select all that apply.
   a. It would decrease the blood pressure due to a decrease in the blood level of potassium.
   b. It would decrease the blood pressure due to increased water loss in the urine.
   c. It would increase the blood pressure due to an increased sodium level in the blood.
   d. It would increase the blood pressure due to an increase in the blood volume.

Remember that mean arterial pressure (MAP) is a homeostatically regulated variable. There are three factors that contribute to MAP: heart rate (HR), stroke volume (SV), and total peripheral resistance (TPR). We consider MAP using the following equation:

$$\text{MAP} = \text{HR} \times \text{SV} \times \text{TPR}$$

2. Using the given equation, if the baroreceptors in the carotid bodies detected an increase in blood pressure, they would send signals to the heart to adjust its rate. Those signals would ______.
   a. increase heart rate
   b. decrease heart rate

If the body has a system in place to detect and adjust for changes in blood pressure, how is it possible that an individual, let alone such vast numbers of people, develop chronic high blood pressure? Most hypertension is defined as idiopathic, meaning that we don't know what causes it (the prefix *idio-* also gives rise to the word *idiot*). But what we do know is that in individuals who develop hypertension, the baroreceptors acclimate to higher sustained MAP and the response system no longer functions correctly.

3. Which of the following factors directly impact blood pressure? Please select all that apply.
   a. Blood volume
   b. Vasoconstriction
   c. The number of red blood cells in the blood
   d. The number of white blood cells in the blood

4. Which of the following changes would lead to a change in blood volume?
   a. A change in blood osmolarity
   b. A change in hydration

c. A change in kidney function and urine production
d. All of the above

5. Concerning vasoconstriction, which of the following are true? Please select all that apply.
   a. It has a lower impact on resistance to flow than other resistance factors.
   b. The stress hormone, cortisol, triggers systemic vasoconstriction.
   c. Vasoconstriction occurs due to activation of the parasympathetic nervous system.
   d. Due to anatomical differences, vasoconstriction occurs more in arteries and arterioles than veins and venules.
6. How does an increase in blood volume affect the blood pressure?
   a. It decreases the blood pressure by vasodilating the arteries.
   b. It increases the blood pressure by increasing the cardiac output.
   c. It increases the blood pressure by vasoconstricting the veins.
   d. It has no effect on the blood pressure, since blood pressure is regulated by the brain.
7. Concerning the number of red blood cells in the blood, which of the following would lead to a change in red blood cell count?
   a. Blood loss from donation, trauma, or menstruation
   b. An increased blood cell creation rate in the bone marrow
   c. Injection of the hormone erythropoietin
   d. All of the above
8. Concerning the number of white blood cells in the blood, which of the following would lead to a change in white blood cell count?
   a. Infection with a viral pathogen
   b. Increase in blood osmolarity
   c. Decrease in blood glucose level between meals
9. Which of the following physiological changes would cause a temporary rise in blood pressure in an otherwise healthy individual? Please select all that apply.
   a. Rise in blood cortisol levels due to stress
   b. Decrease in blood glucose between meals
   c. Increase in blood sodium levels following a salty snack
   d. Decrease in heart rate when lying down

In many individuals with hypertension, scientists hypothesize that one or more of the factors that lead to temporary increases in blood pressure from Question 7 occurs frequently over the years. Over time, the baroreceptors desensitize and stop responding to the constant elevated blood pressure.

10. If you are a doctor or a nurse practitioner seeing a patient for only the first or second time and their blood pressure is higher than the healthy range, which of the following recommendations do you make?
    a. Come back soon for a repeat blood pressure measurement
    b. Stay away from salty foods
    c. Eat more frequent meals
    d. Take blood pressure medication
    e. Get more sleep each night

Juan is a 47-year-old man who has been moving around for work a lot over the last decade. Because Juan rarely lives in a new city for more than a year, they have not formed a long-lasting relationship with a primary care doctor and has only been seen when they are ill. Today Juan is visiting a doctor that they have seen twice before, and the doctor notes that Juan has had a high blood pressure reading at each visit. The doctor asks about Juan's lifestyle and notes that Juan has a stressful job and rarely makes time for exercise. The doctor asks if Juan's blood pressure has been high at other visits at other doctors' offices, but Juan doesn't remember. Let's take a look at Juan's blood vessels.

Figure 28.1 **Cross Section through a Blood Vessel Wall**

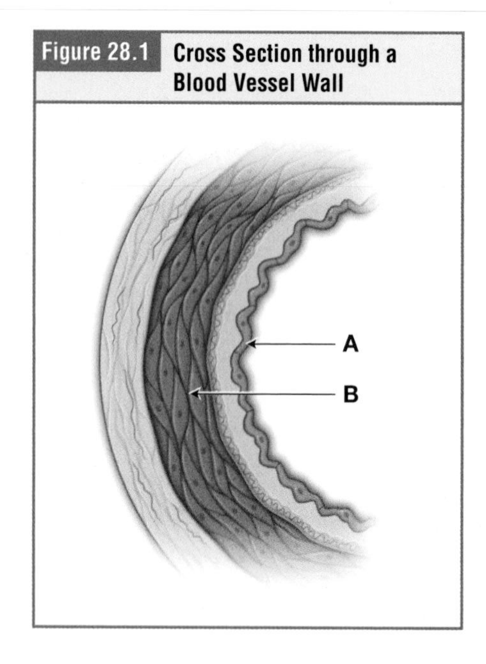

11. Examine **Figure 28.1**. What is the correct term for the layer labeled A?
    a. Endothelium
    b. Endovesolium
    c. Epithelium
    d. Epidermis
12. In Figure 28.1, what type of tissue is layer A composed of?
    a. Connective tissue
    b. Epithelium
    c. Smooth muscle
13. In Figure 28.1, what type of tissue is layer B composed of?
    a. Connective tissue
    b. Epithelium
    c. Smooth muscle
14. Veins have a much thinner layer B (see Figure 28.1) than arteries do. When exposed to higher blood pressure, which do you think will be true of veins, compared to arteries?
    a. Vein walls will be more rigid and therefore expand less in response to pressure.
    b. Artery walls will be more rigid and therefore expand less in response to pressure.
    c. Layer B will not influence the vessels' response to pressure.
15. Given your answer to Question 14, let's take a look at Juan's vessel walls. They have been exposed to high blood pressure over a period of years. Which type of vessel do you suspect has endured more damage due to Juan's hypertension?
    a. Arteries and arterioles
    b. Veins and Venules
    c. Both sets of vessels have endured the same level of damage.

Damage to blood vessel walls is healed through a complex interaction of cells and molecules. Because the lumen of the vessel is filled with a liquid, one component of the healing process is a hydrophobic patch.

16. Which of the following types of molecules is the most hydrophobic and could be used to patch the vessel lining during the healing process?
    a. Proteins
    b. Carbohydrates
    c. Cholesterol molecules

On the inside of the vessel wall, underneath the hydrophobic patch, an inflammatory process is at work healing the damage. Because inflammation often involves swelling, the lumen of the vessel is substantially narrowed (**Figure 28.2**). If this patch persists, it is often called an **atherosclerotic plaque**.

atherosclerotic plaque (atherosclerosis, plaque)

The flow of any fluid, including blood, can be summarized in the formula:

$$\text{Flow} \propto \text{pressure gradient}/\text{resistance}$$

17. As the inflammation develops and the plaque is pushed into the lumen of the blood vessel, which aspect of the flow equation changes?
    a. Resistance
    b. Pressure gradient
    c. They both change equally
    d. Neither changes

18. Given what you answered in Question 17, what do you predict happens to blood flow through the affected vessel?
    a. Blood flow increases
    b. Blood flow decreases
    c. No change to blood flow

In order to protect their vessels from damage and try to prevent atherosclerosis, patients who have chronic hypertension are often prescribed medicines. One common class of medicine prescribed to hypertensive patients is ACE inhibitors. These medications work by inhibiting the production of angiotensin-converting enzyme (ACE).

19. Where is this enzyme found?
    a. The blood
    b. The adrenal glands
    c. The kidneys
20. What is the function of ACE in the body?
    a. It converts renin to angiotensin I.
    b. It converts angiotensin I to angiotensin II.
    c. It converts angiotensin II to aldosterone.
21. If you inhibit the step you chose in Question 20, which of the following statements is true about the balance between blood volume and urine volume?
    a. Blood volume decreases and urine volume increases
    b. Blood volume increases and urine volume decreases
    c. No impact on blood and urine volume
22. Consider your choices for Questions 18 and 19. Which part of the MAP equation do you think changes with the use of an ACE inhibitor? Please select all that apply.
    a. HR (heart rate)
    b. SV (stroke volume)
    c. TPR (total peripheral resistance)

Figure 28.2 An Atherosclerotic Plaque

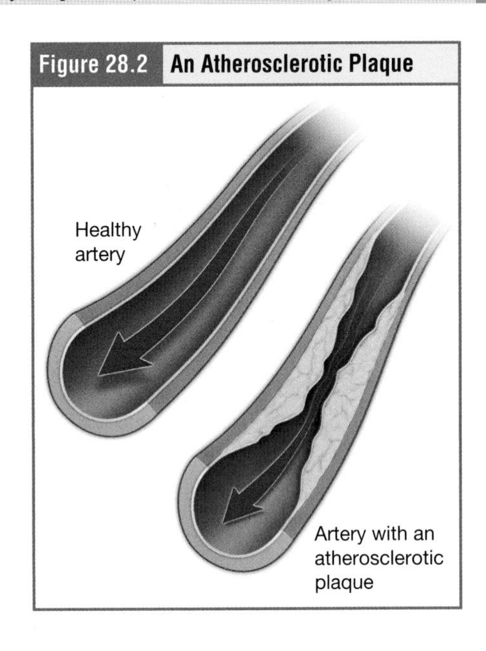

28.2 Cystic Fibrosis Case Study

Learning Objectives: By the end of this case study you will be able to:

28.2.1 Describe and identify the typical components of a cell membrane (Section 4.1).

28.2.2 Describe the types of transport across the cell membrane (Section 4.1).

28.2.3 Compare and contrast the structure and function of exocrine glands (Section 5.2).

28.2.4 Describe the components of barrier immunity in the lungs including the cilial elevator, mucus, and tight junctions (Section 21.1).

28.2.5 List the components of Fick's law (Section 22.4).

28.2.6 Apply Fick's law to perturbations in physiological circumstances to predict changes in diffusion (Section 22.4).

28.2.7 Connect the CO_2 concentration of a fluid to its pH using the carbonic acid equation (Section 22.5).

28.2.8 Describe the functions of the organ systems of the human body (Section 2.6).

28.2.9 Use the principles of diffusion and osmosis to predict various changes in blood pressure, fluid osmolarity, sweat production, or other physiological functions (Sections 4.1 and 20.2).

28.2.10 Define the terms *hypoosmotic, isosmotic, hyperosmotic, hypotonic, isotonic,* and *hypertonic* (Section 4.1).

28.2.11 Describe the functions of the organs of the gastrointestinal tract and the accessory organs of digestion (Section 23.1).

Cystic fibrosis (CF) is a genetic disease in which a mutation occurs in the gene for the cystic fibrosis transmembrane conductance regulator (CFTR). CFTR is a membrane channel that permits substances to cross from one side of the plasma membrane to the other. It is expressed in exocrine glands. Eliot is a 1-year-old child with frequent hospitalizations due to respiratory infections. Eliot's doctor is starting to suspect that CF may be the cause.

1. Membrane channels consist of ______.
   a. proteins
   b. cholesterol
   c. carbohydrates
2. Membrane channels permit substances to pass from one side of the membrane to another from a region of higher concentration to a region of lower concentration; this is an example of ______.
   a. primary active transport
   b. secondary active transport
   c. simple diffusion
   d. facilitated diffusion
3. CFTR is expressed in the membranes of cells in exocrine glands. Which of the following organs or structures function as exocrine glands (they might or might not have other functions)? Please select all that apply.
   a. Sweat glands
   b. Pancreas
   c. Thyroid gland
   d. Liver
   e. Anterior pituitary gland

CFTR permits the transport of Cl^- anions, and as they move, they change the gradient, drawing water through aquaporins to the side with the Cl^- ions (**Figure 28.3**). In CF, the mutation in the gene for the channel results in a nonfunctional channel and no anions can pass through it; therefore, water does not travel either.

28.2a Lungs

The inner lining of the lungs, like the inside of all hollow organs in the body, is lined with an epithelium. The lung epithelium is coated with a layer of mucus that protects the epithelium by trapping pathogens and particles. The apical surface of the lung epithelial cells is covered with cilia, which beat the mucus upward toward the top of the trachea in order to get rid of the mucus and any particles or pathogens trapped within (**Figure 28.4**). The production of the lung mucus involves the active transport of Na^+ into the lumen of the lungs; Cl^- follows through the CFTR and water follows by osmosis through aquaporins. Other substances, such as lipids and proteins, are added as well. In individuals with CF, Cl^- cannot cross the membrane and so less water is added to the mucus that sits on top of the lung epithelium, resulting in much thicker mucus.

4. How would the thicker mucus of a CF patient cause the frequent lung infections seen in this disease?
   a. The thicker mucus attracts more pathogens to move in from the outside air.
   b. The thicker mucus clogs the cilial elevator, resulting in pathogens persisting in the lungs that would otherwise be swept upward, out of the lungs.
   c. The thicker mucus has more ions in it, which are important nutrients for pathogens, allowing them to live longer in the lungs.
5. Which law helps us to understand the rate of diffusion in various physiological circumstances?
   a. Boyle's law
   b. Dalton's law

Figure 28.3 **The CFTR Protein in Healthy Individuals (top) and Individuals with Cystic Fibrosis (bottom)**

As Cl^- moves through CFTR and crosses the membrane,...
...it draws H_2O by creating a gradient.
CFTR
H_2O
Aquaporin channel
Plasma membrane
Cl^-
H_2O
CFTR
Aquaporin channel
Plasma membrane
Cl^-
H_2O
When the CFTR has a mutation and no longer functions to allow Cl^- to move through the membrane,...
...no gradient is created, so no H_2O movement occurs.

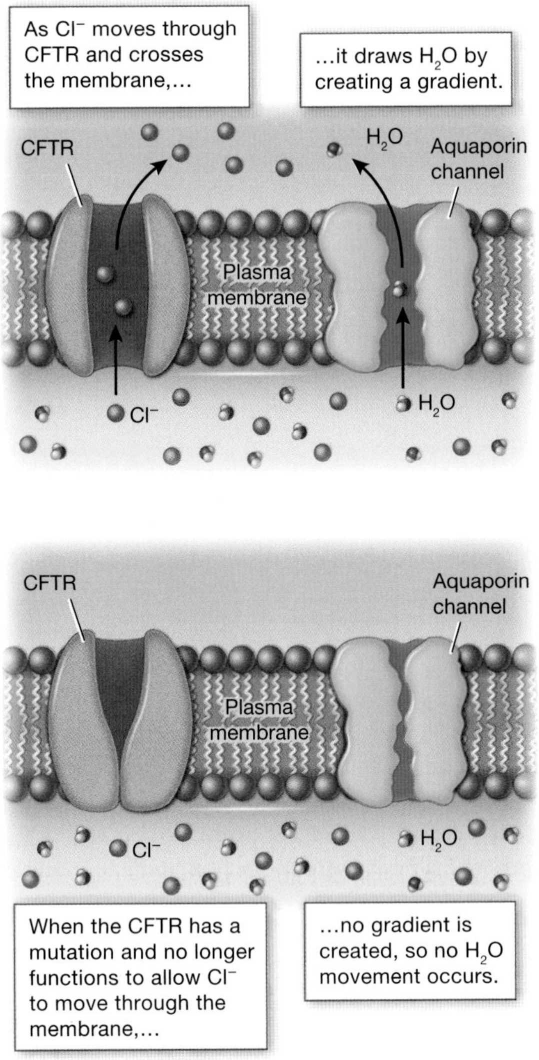

c. Fick's law
d. Starling's law

6. If Eliot has thicker mucus that is harder to clear from the lungs, what effect would this mucus have on gas diffusion across the alveolar membrane?
   a. Increase the rate of diffusion
   b. Decrease the rate of diffusion
   c. No change to the rate of diffusion.
7. Given your answer to Question 6, what do you expect might be true of the concentration of CO_2 in Eliot's arterial blood?
   a. The amount of CO_2 would be higher than the normal level.
   b. The amount of CO_2 would be lower than the normal level.
   c. There is no change to the amount of CO_2 in Eliot's arterial blood.
8. Given your answers to Questions 6 and 7, what would happen to the pH of Eliot's blood?
   a. pH would increase (would become alkaline).
   b. pH would decrease (would become acidic).
   c. There would be no changes to the pH of the blood.

Figure 28.4 **Lung Epithelia in Healthy Individuals (top) and Individuals with Cystic Fibrosis (bottom)**

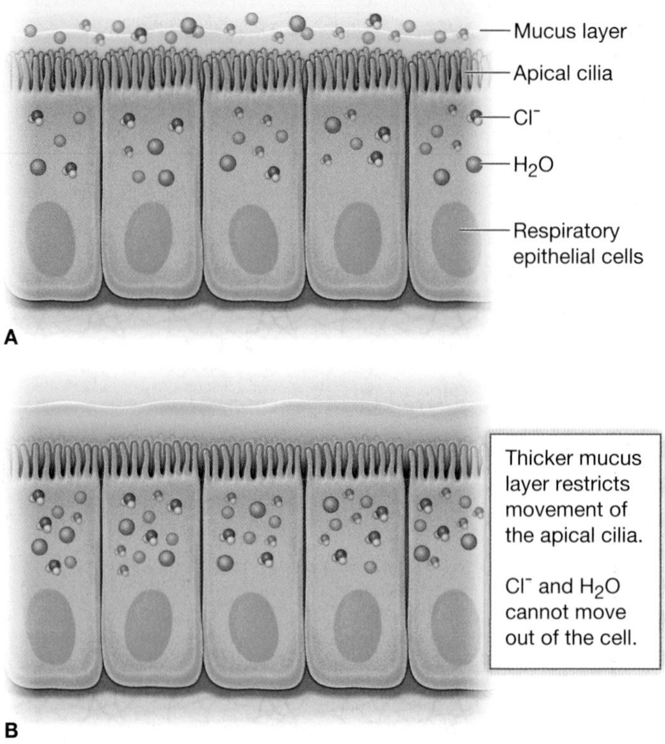

9. If a physiological factor of the lungs is limiting their ability to regulate the pH of the blood, which body system can contribute to maintaining pH homeostasis?
   a. Digestive
   b. Urinary
   c. Muscular
10. One of the therapies for individuals with CF is inhaled hypertonic saline. This saline, which is 7.0 percent NaCl instead of the more common 0.9 percent NaCl, helps to thin the mucus of a CF patient. How does adding hypertonic saline thin the mucus? Where does the water that thins the mucus come from?
    a. The saline brings water from the atmospheric air with it as its inhaled.
    b. By increasing the osmolarity of the mucus, more water will be drawn out through the lung epithelial cells.

Eliot's pediatrician recommends that they be tested for CF. The doctor explains that a common test is a "sweat test." During this test, a small collection container is adhered to the patient's skin to collect their sweat (**Figure 28.5**).

Figure 28.6 illustrates the production of sweat under typical circumstances. Since Na^+ and Cl^- ions are actively pumped into the sweat gland, water follows. Cl^- moves back to the interstitial fluid through the CFTR channel but water does not move back due to restrictions in aquaporins.

11. Under the typical circumstances outlined in Figure 28.6, how would sweat compare to interstitial fluid?
    a. It would be hypoosmotic.

Figure 28.5 **The Sweat Test Collects a Sample of the Patient's Sweat to Test for Cystic Fibrosis**

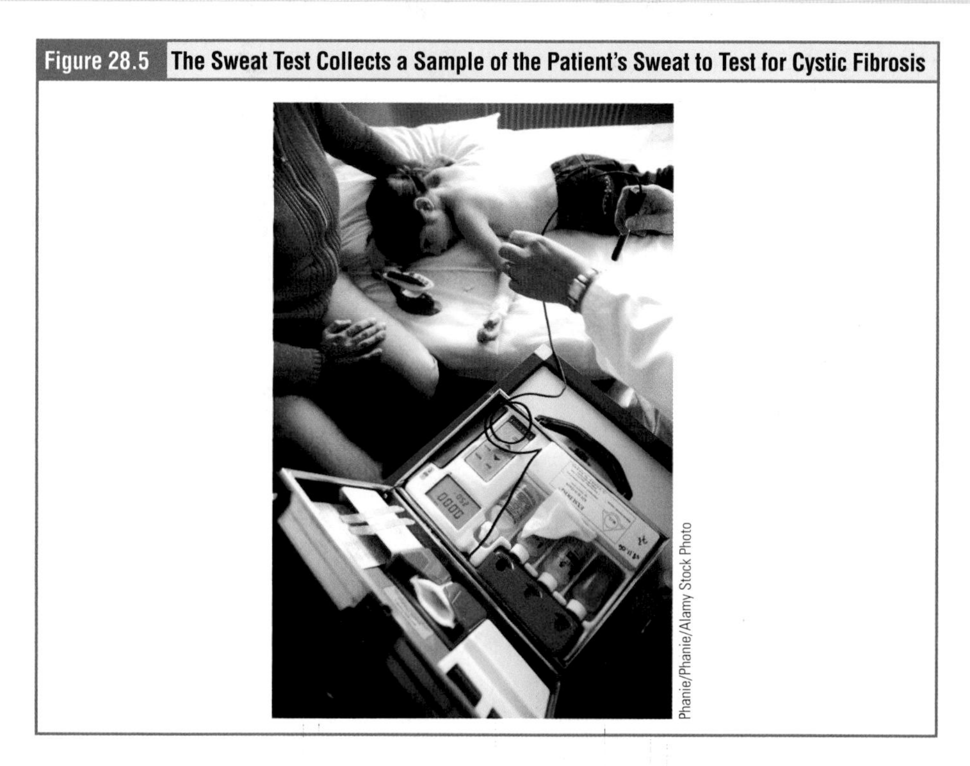

b. It would be hyperosmotic.
c. It would be isosmotic.

12. Now imagine the sweat gland of a person with CF. If Eliot has CF, how would Eliot's sweat compare to that of an individual without CF?
    a. Eliot's sweat would have fewer water molecules.
    b. Eliot's sweat would have more Cl^- ions.
    c. Eliot's sweat would have fewer Cl^- ions.

Figure 28.6 **Sweat Production in an Individual with Cystic Fibrosis**

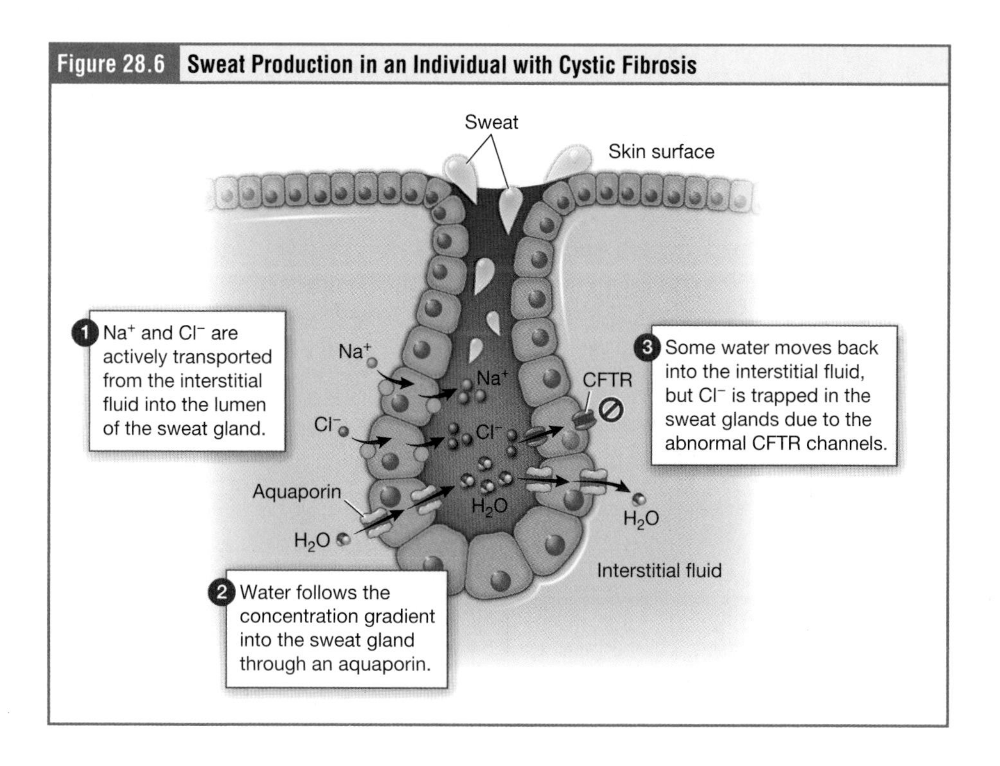

Sweat
Skin surface
1 Na⁺ and Cl⁻ are actively transported from the interstitial fluid into the lumen of the sweat gland.
Na⁺
Na⁺
Cl⁻
Cl⁻
CFTR
3 Some water moves back into the interstitial fluid, but Cl⁻ is trapped in the sweat glands due to the abnormal CFTR channels.
Aquaporin
H₂O
H₂O
H₂O
Interstitial fluid
2 Water follows the concentration gradient into the sweat gland through an aquaporin.

13. One of the symptoms of CF is fatty stool (meaning that the patients observe globs or streaks of fat in their feces). CFTR is expressed in all exocrine glands. The altered function of which of the following glands could explain this symptom?
    a. Pancreas
    b. Sweat
    c. Thyroid

28.3 Pulmonary Embolism Case Study

Learning Objectives: By the end of this case study you will be able to:

28.3.1 List and identify the anatomical regions of the human body (Section 2.7).

28.3.2 Describe the path of blood flow from the left ventricle to the left atrium (Section 19.1).

28.3.3 Using the flow equation, predict how different changes to physiological conditions will impact flow rate (Section 22.2).

28.3.4 Describe the physiological factors that influence each component of mean arterial pressure (Section 20.2).

28.3.5 Using the MAP equation, predict physiological changes that would occur in order to maintain MAP (Section 20.2).

28.3.6 Describe preload and afterload and physiological impact of changes to these factors (Section 19.4).

David, a 59-year-old male, collapses in the baggage claim area of an airport after getting off a long flight. EMTs transport him to the hospital. David feels like he can't get an adequate breath and complains of dizziness. Upon arrival in the ER an echocardiogram (ultrasound imaging of the heart) is performed and it is noted that his right ventricle appears slightly dilated. Dilation of a heart chamber typically appears when that chamber is working harder than usual. The ER doctors perform blood tests for proteins typically found in the blood during heart attacks and found that these tests were negative, indicating that David is not having a heart attack.

A CT scan with contrast dye of his chest is performed and bilateral pulmonary embolisms are found. Embolisms are blood clots or other substances that have become lodged in blood vessels and have stopped blood flow. Embolisms typically result when **emboli**, blood clots or pieces of material flowing through blood vessels, lodge in a blood vessel through which they cannot pass, due to its diameter.

To figure out how to treat the embolisms, the doctors perform further tests to investigate their identity and origin. Ultrasounds of the David's legs reveal three **deep vein thromboses (DVTs)** in the patient's right lower extremity. **Thromboses** are blood clots. When the clots or pieces of the clots come loose, they travel in the blood vessels and are now called *emboli*. If they get stuck in a smaller vessel, they are then known as *embolisms*. Pulmonary embolisms often result from DVTs, commonly called *blood clots*, which break loose and move to the lungs (**Figure 28.7**).

thromboses (blood clots)

1. What region is the likely origin of the embolisms?
    a. Brachial
    b. Thoracic
    c. Cervical
    d. Femoral
2. Will the *pulmonary embolisms* be in pulmonary arteries or veins?
    a. Pulmonary veins
    b. Pulmonary arteries
    c. Could be either

Figure 28.7 The Major Veins of the Body

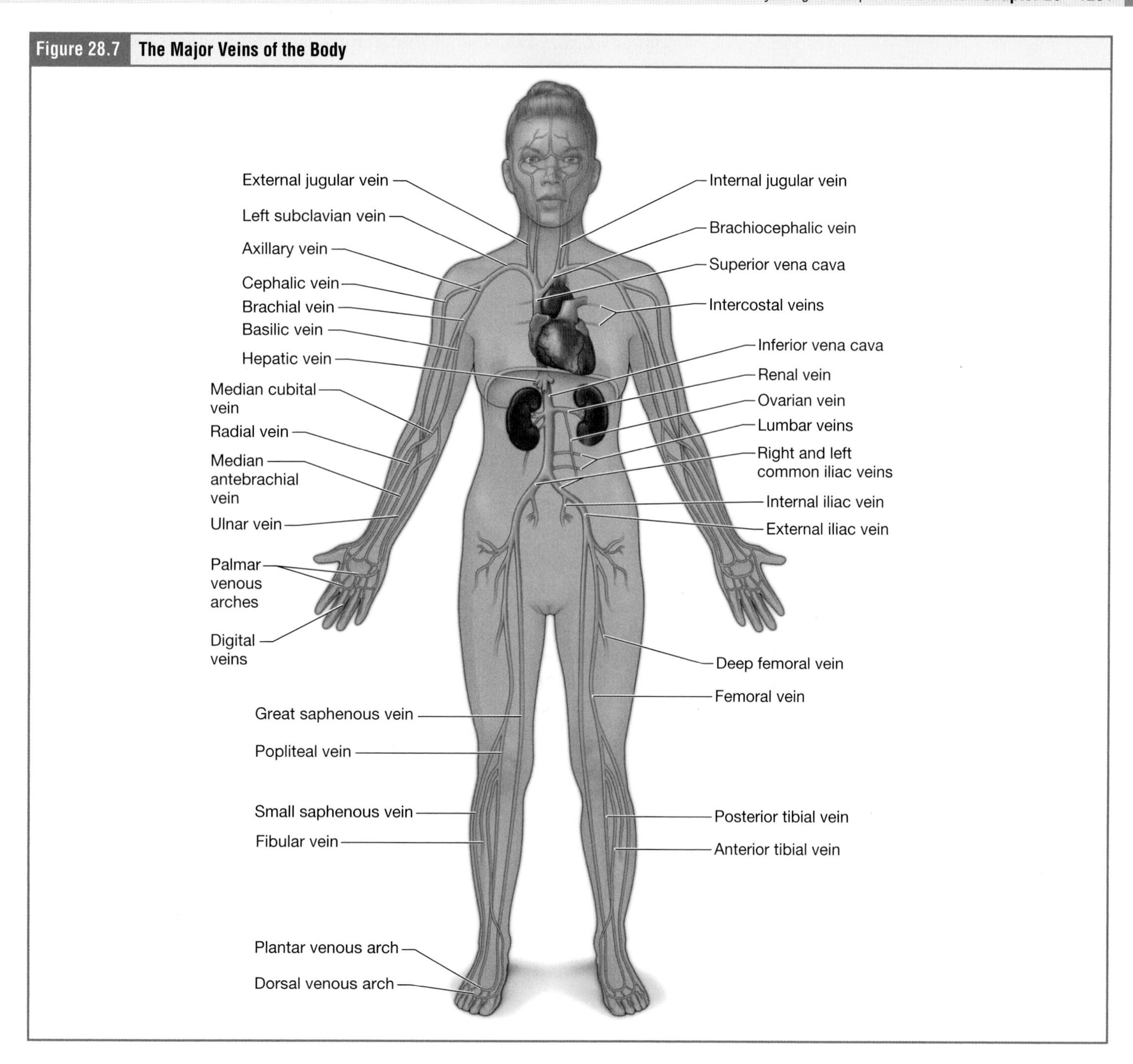

3. Why did a blood clot from the leg end up in the lungs? Why did it not get stuck earlier? To answer this question, select the correct path that the emboli took as they traveled from the legs to the lungs.
   a. Femoral vein, external iliac vein, inferior vena cava, right atrium, right ventricle, pulmonary trunk, pulmonary artery
   b. Femoral vein, saphenous vein, inferior vena cava, right atrium, right ventricle, pulmonary vein
   c. Saphenous vein, internal iliac vein, superior vena cava, right atrium, right ventricle, pulmonary trunk, pulmonary artery
4. With the pulmonary emboli in place, what happens to the volume of blood returning to the left atrium?
   a. A higher volume of blood would be returning to the left ventricle than under normal circumstances.

b. A lower volume of blood would be returning to the left ventricle than under normal circumstances.
c. There would be no change to the volume of blood returning to the left ventricle during this time.

Remember that mean arterial pressure (MAP) is a homeostatically regulated variable. There are three factors that contribute to MAP: heart rate (HR), stroke volume (SV), and total peripheral resistance (TPR). We consider MAP using the following equation:

$$\text{MAP} = \text{HR} \times \text{SV} \times \text{TPR}$$

5. If David is suffering from pulmonary embolisms, which MAP factor is *directly* impacted?
   a. Heart rate
   b. Stroke volume
   c. Total peripheral resistance
6. To compensate, the body will alter the other factors. Which of the following symptoms is it likely that David noticed before he collapsed?
   a. Numbness in his left leg
   b. Racing heart
   c. Abdominal pain
7. The echocardiogram showed a dilated right ventricle. What might this indicate?
   a. Decrease in pulmonary blood pressure
   b. Increase in pulmonary blood pressure
   c. Emboli are stuck in coronary vessels, decreasing oxygen delivery to the heart muscle
8. Will the dissolving clots pass through any heart structures as they dissolve away? If so, which ones?
   a. No, the dissolving clots will not pass through the heart.
   b. Yes, the dissolving clots will pass through the right atrium and right ventricle.
   c. Yes, the dissolving clots will pass through the left atrium and left ventricle.

DVTs are linked to genetics and certain lifestyle factors. Of all patients who have experienced a DVT, 30 percent will get one again. To avoid further PEs, the doctors implant a filter into the bloodstream to catch any emboli moving from the extremities or pelvis *before reaching the heart.*

9. Would the doctors be able to place *one* filter that would catch emboli originating in both the arms and the legs?
   a. Yes, one filter will catch emboli from both locations.
   b. No, more than one filter would be needed.
10. To catch any emboli from the legs or pelvis, which vein would be the best location for the filter?
    a. Superior vena cava
    b. Hepatic portal vein
    c. Inferior vena cava
    d. Femoral vein
11. To place a filter that would catch emboli originating in the arm, the doctor inserts a catheter with the filter into the right internal jugular vein and passes it through various structures until it reaches the desired location for filter placement. Which structures does the filter pass through?
    a. Right brachiocephalic vein and superior vena cava
    b. Right iliac vein and right subclavian vein
    c. Right ventricle and right pulmonary trunk
    d. Right pulmonary artery and right pulmonary vein

28.4 Femoroacetabular Impingement Case Study

Learning Objectives: By the end of this case study you will be able to:

28.4.1 Identify the anatomical components (muscles, cartilages, tendons, ligaments, and bones) of the lower limb (Section 12.1).

28.4.2 Identify the arches of the foot and describe the components that contribute to each arch (Section 12.4).

28.4.3 List the actions of the muscles of the body (Section 12.1).

28.4.4 Describe movements or structures on the body using directional anatomical terms (Section 2.7).

28.4.5 Predict functional changes to joint, structures and movements by applying your knowledge of anatomical components (Sections 10.5 and 10.6).

Femoroacetabular impingement is a hip disorder that can be caused by an issue in the foot, far away from the hip. It starts with a weak muscle, and because the hip, knee, and ankle are all functionally connected, it has radiating consequences throughout the leg. In the following set of questions, consider Alice, a 24-year-old athlete who will be diagnosed with femoroacetabular impingement by the end of this case study.

1. Which of the images in **Figure 28.8** is correctly labeled?
   a. A
   b. B
2. Which of these muscles wraps around the medial side of the foot and inserts on its sole?
   a. Tibialis anterior
   b. Peroneus longus
   c. Peroneus brevis
   d. Gastrocnemius
3. What is the name of the sheet of connective tissue that spans the sole of the foot?
   a. Plantar aponeurosis
   b. Dorsal aponeurosis
   c. Flexor retinaculum
   d. Deltoid ligament
4. If the tibialis anterior were weak and its tendon were slack instead of taut, which foot arch would fall?
   a. Medial longitudinal arch
   b. Lateral longitudinal arch
   c. Transverse arch

Figure 28.8 The Knee Joint

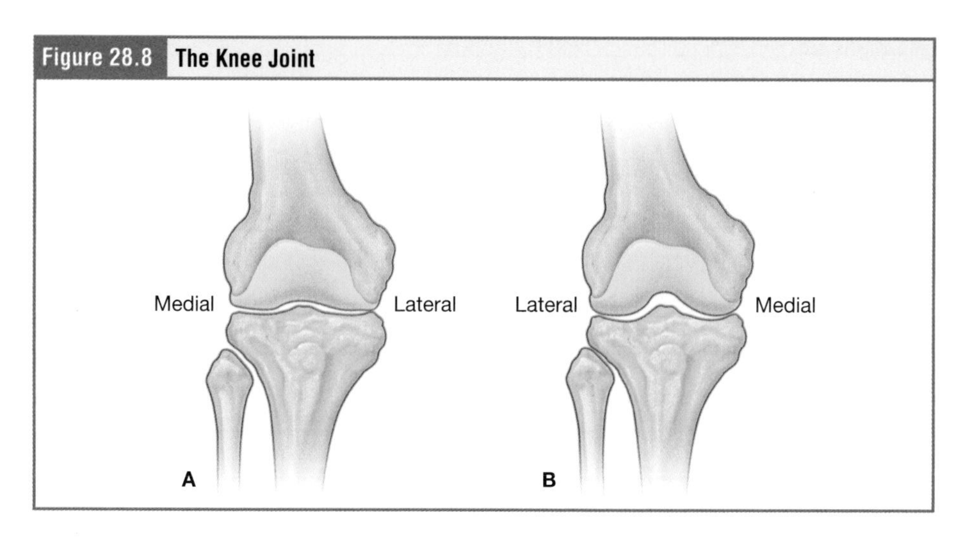

Figure 28.9 **The Ankle Joint**

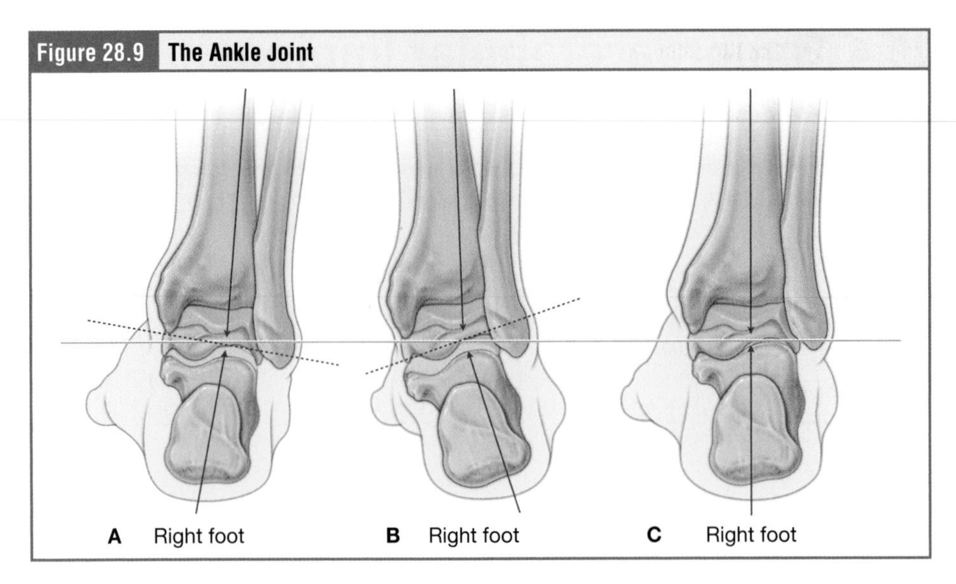

5. Which of the following exercises would strengthen the tibialis anterior?
   a. Dorsiflex the foot (pull the toes up toward the front of the shin)
   b. Plantarflex the foot (point the toes like a ballet dancer)
   c. Flex (bend) the knee
   d. Evert the foot (show the sole of the foot medially)
6. A weak tibialis anterior allows for pronation, a condition in which the ankle rolls or tilts medially. Which of the images in **Figure 28.9** shows pronation?
   a. A
   b. B
   c. C
7. The talocrural joint, which includes the talus, tibia, and fibula, is one of two joints involved in pronation. During excessive pronation, we see that the tibia is slightly rotated, affecting the knee. In which direction would you expect that rotation?
   a. External rotation
   b. Internal rotation
8. Eventually the knee is impacted, and a condition called *genu valgum* develops. With what you already know about the ankle and the tibia, which of the images in **Figure 28.10** represents the knees of this patient?
   a. A
   b. B
   c. C

Figure 28.10 **Genu Valgum Is a Condition of the Knee**

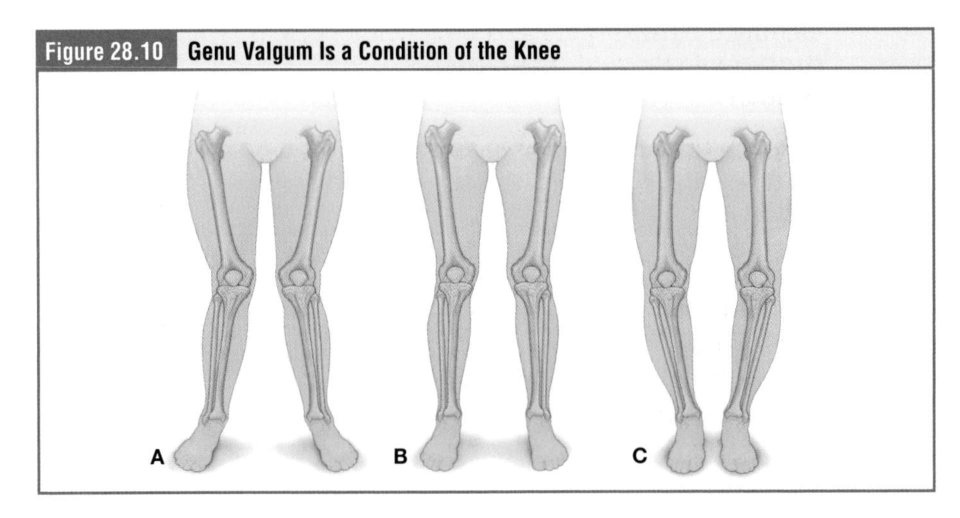

Figure 28.11 The Hip Joint

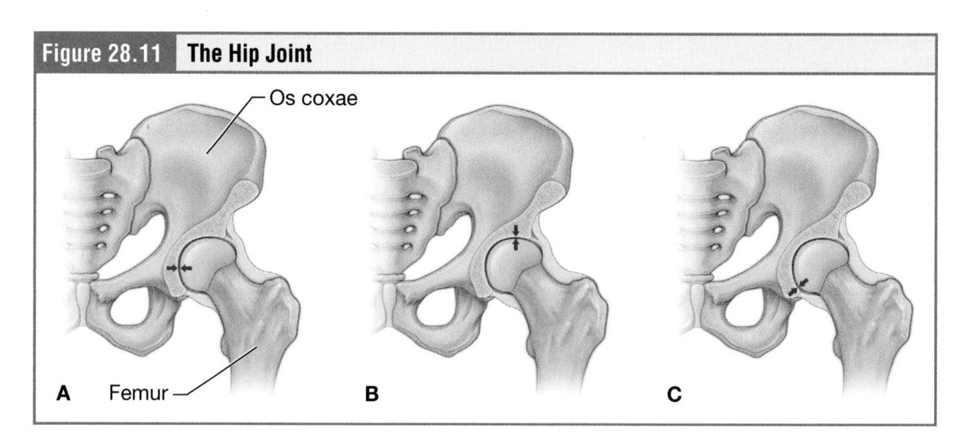

9. If we examine the knee in genu valgum, we can see that its asymmetry is exaggerated. Which side is experiencing excessive compression?
   a. Medial
   b. Lateral
10. Which side is experiencing excessive stretch?
    a. Medial
    b. Lateral
11. Taking the information you have gathered about the angle of the femur, where do you think the stress will be in the hip joint? Examine the images in Figure 28.11 and choose the most likely site of impingement.
    a. A
    b. B
    c. C
12. FAI can cause tears in the fibrocartilage of the hip. What is this structure called?
    a. Bursa
    b. Labrum
    c. Meniscus

28.5 Arthritis Case Study

Learning Objectives: By the end of this case study you will be able to:

28.5.1 Identify the tissue types that compose various structures in the body and their functions (Section 7.1).

28.5.2 Describe the functions of structures, features, and fluids of the body (Sections 4.1 and 21.1).

28.5.3 Interpret images of anatomical structures (Section 2.7).

28.5.4 Identify and describe the functions of the cells and their cellular components that compose the tissues of the body (Sections 5.1 and 5.2).

Osteoarthritis is the most common form of arthritis, affecting approximately 10 percent of adult Americans. It is sometimes called *degenerative joint disease* because it is a disease in which portions of the joints can deteriorate over time. Osteoarthritis is progressive, and while it cannot be reversed, treatment can slow its progress. Husna is a 63-year-old teacher. Through the pandemic, Husna taught via Zoom and began sitting in chairs more than they ever had before. After two years, they find that they are having trouble walking up stairs or taking longer walks in their neighborhood. Their knees, in particular, feel painful. They arrive at their doctor to try to figure out what is going on.

1. In Husna's knee joints, which type of tissue covers their femurs and tibiae?
   a. Hyaline cartilage
   b. Elastic cartilage
   c. Fibrous connective tissue
   d. Epithelium
2. What are the functions of synovial fluid? Please select all that apply.
   a. To supply the cells of the bones with nutrients
   b. To nourish the articular cartilage
   c. To lubricate the joint
   d. To provide the collagen fibers in the ligaments with oxygen

Husna's doctor orders MRIs and X-rays of their knees.

3. Correctly identify the medial and lateral sides of the knee in **Figure 28.12**. Which side is medial?
   a. A
   b. B
4. One of the images shown in **Figure 28.13** is of a knee without arthritis. The other is of Husna's knee, which has arthritis. In the progression of arthritis, the cartilage at the joint wears down over time, resulting in a joint where the bones are much closer together. Which image is of Husna's knee?
   a. A
   b. B
5. Without the protection of a thick layer of articular cartilage, the bones experience much more friction with movement. People with arthritis describe their movements as painful. From what structure(s) is the pain originating?
   a. Tendons and ligaments at work to stabilize the joint
   b. Bones
   c. Cartilage

Husna's doctor diagnoses them with osteoarthritis. The doctor notes that Husna's body mass index (BMI) value is in the obese range and recommends that they lose weight.

Figure 28.12 An MRI of the Knee Joint

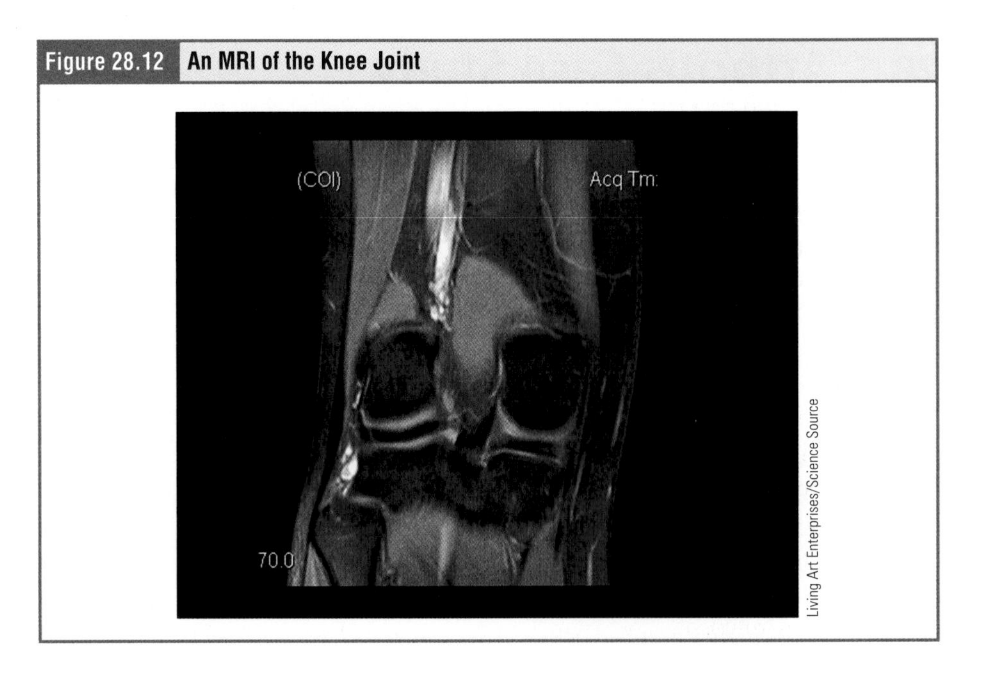

Living Art Enterprises/Science Source

Figure 28.13 **Two X-rays of the Knee Joint**

A healthy knee (on the left) and a knee with osteoarthritis (on the right).

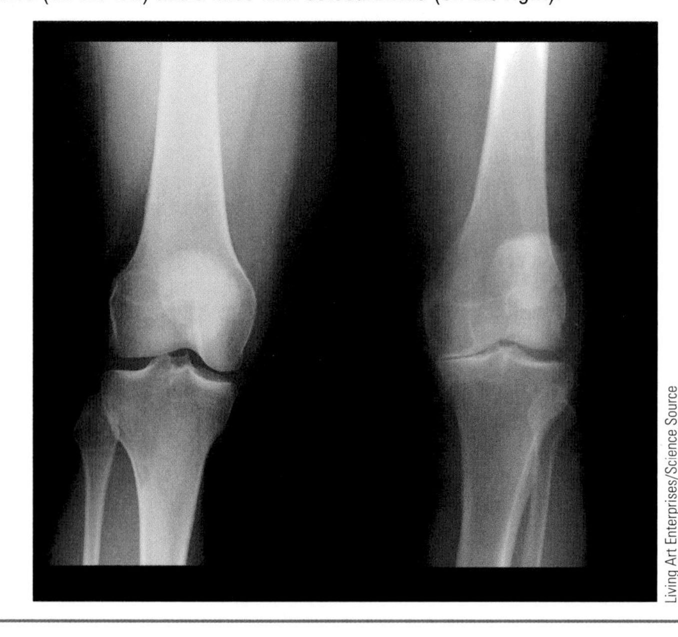

Living Art Enterprises/Science Source

The doctor says that carrying more body weight will put stress on Husna's knees. The doctor's final recommendation is that Husna become more active and convert some of the time they spend sitting to standing or walking activities. These two pieces of advice seem contradictory to Husna, if their weight is an issue, won't standing cause more stress on their knees? As Husna; leaves the doctor's office, they call their friend, a physiology professor, with their questions.

6. Sedentary lifestyles are a risk factor for osteoarthritis. Why is movement of a synovial joint important in maintaining its health?
   a. Movement and compression in a synovial joint help to circulate the synovial fluid through the articular cartilage.
   b. Movement and compression of bones helps to drive blood flow through them.
   c. Movement and compression of joints strengthens the tendons that hold the joint together.

Husna's friend explains that osteoarthritis is a disease that occurs when cartilage breakdown exceeds cartilage synthesis and draws Husna a picture (**Figure 28.14**).

7. Which cells are responsible for the synthesis of cartilage matrix?
   a. Osteocytes
   b. Osteoblasts
   c. Chondrocytes
8. Husna's friend explains that obesity is a risk factor for arthritis but that is not really about weight putting mechanical stress on the knees. Athletes who have heavier bodies due to musculature are not at risk of osteoarthritis. Instead, obesity is a risk factor because it contributes to inflammation. What is the link between inflammation and obesity?
   a. Adipose tissue can release adipokines, cytokines that promote inflammation.
   b. Extra nutrients allow the immune system to create more inflammatory compounds.
   c. Extra nutrients allow more bacteria to thrive in the body, creating an inflammatory environment.

Figure 28.14 **Cartilage Synthesis Should Balance Cartilage Breakdown, but Breakdown Exceeds Synthesis in Cases of Osteoarthritis**

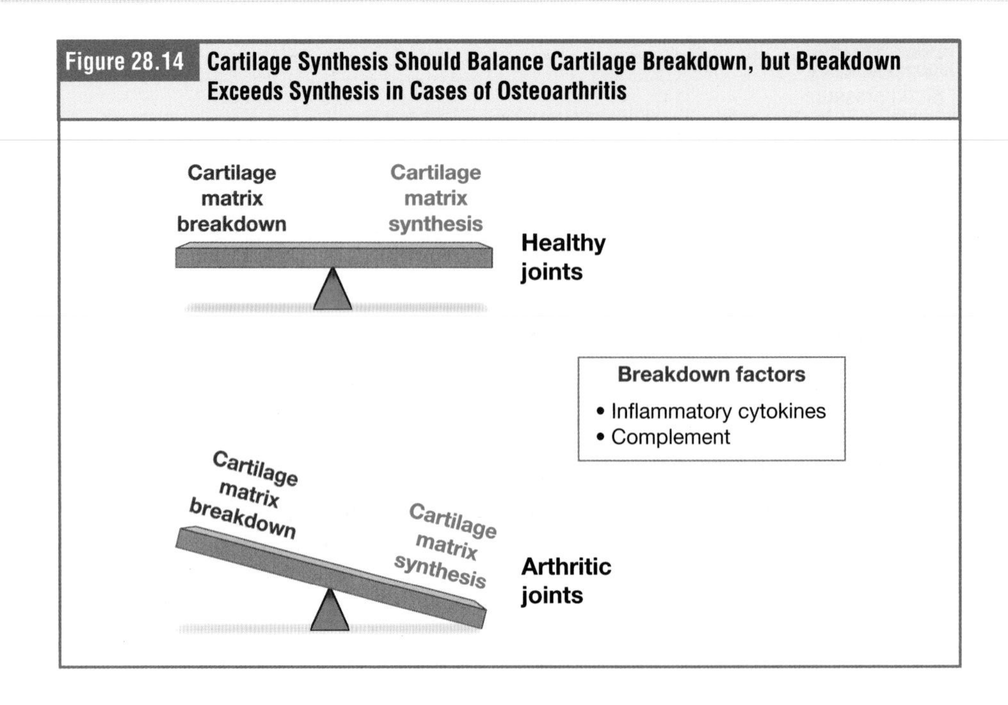

28.6 Cardiac Failure Case Study

Learning Objectives: By the end of this case study you will be able to:

28.6.1 Connect the functions of the systems of the body to predict how they contribute or compensate to maintain homeostasis (Sections 7.3, 7.6, and 18.1).

28.6.2 Interpret images of anatomical structures (Section 2.7).

28.6.3 List the components of Fick's law (Section 22.4).

28.6.4 Apply Fick's law to perturbations in physiological circumstances to predict changes in diffusion (Section 22.4).

28.6.5 Predict changes to blood flow during perturbations of cardiac function (Sections 19.1 and 20.2).

28.6.6 Predict fluid flow based on Starling's forces (Section 19.4).

28.6.7 Identify and describe the functions of the cells and their cellular components that compose the tissues of the body (Sections 5.1 and 5.2).

28.6.8 Evaluate potential treatments or therapies using an understanding of the pathophysiology of a disease (Section 20.2).

Angelo, a 68-year-old-male with a history of asthma and high blood pressure, arrives in the ER with difficulty breathing. Angelo is in relatively poor health with very little daily exercise and consumes a diet that includes daily intake of red meat, vegetables, fruits, and processed foods. Angelo reports that he has been coughing for several days and has taken his asthma inhaler (albuterol) many times with seemingly no relief from the shortness of breath. His breathing is now visibly and audibly labored. You are the emergency room physician assigned to Angelo's case. You order a number of tests, including a chest X-ray that reveals fluid buildup in his lungs. Additional test results are summarized below.

	Patient's Values	Normal Values
Blood pressure	145/95 mmHg	120/80 mmHg
Pulse	110 beats/min	70 beats/min
Respiratory rate	30 breaths/min	12–20 breaths/min
Troponin	25 μg/L	Less than 10 μg/L
Hematocrit	50%	42–54%

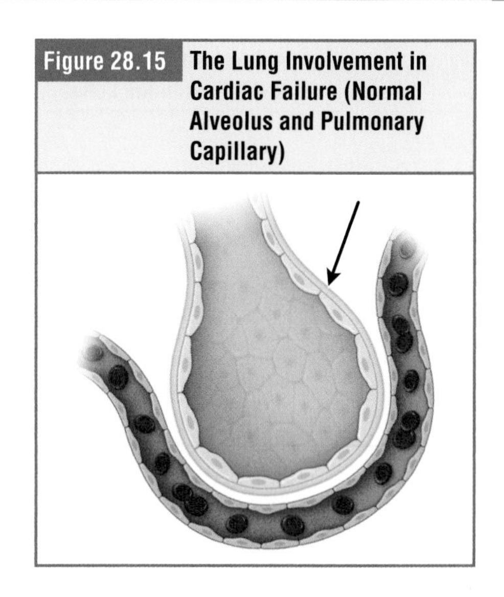

Figure 28.15 The Lung Involvement in Cardiac Failure (Normal Alveolus and Pulmonary Capillary)

1. What is the structure indicated by the arrow in **Figure 28.15**?
   a. An alveolus
   b. A lung
   c. A heart chamber
2. Which of the images in **Figure 28.16** has a double-headed arrow that accurately shows the diffusion distance (the distance across which oxygen and carbon dioxide must diffuse to be exchanged) from the alveoli to the blood?
   a. A
   b. B
   c. C
3. What do we use Fick's Law to determine?
   a. The relationship between pressure and volume of gasses
   b. The relationship between different gasses in a mixture
   c. The rate of diffusion
   d. The rate of flow

Examine **Figure 28.17**. This illustrates the situation in Angelo's lungs at the time he arrives at the ER.

4. According to Fick's Law, the presence of fluid in the lungs would have what effect on the rate of diffusion?
   a. No effect
   b. Increase
   c. Decrease
5. What component of Fick's Law is affected here?
   a. Surface area
   b. Membrane thickness
   c. Membrane permeability
   d. Concentration gradient
   e. None, no effect

Back in the ER, your next test is a pulse oximetry test, a measure of the amount of oxygen circulating on red blood cells. Angelo's pulse oximeter level registers slightly lower than normal.

6. Given the fluid in the lungs, why do you think Angelo's pulse oximeter readings were not far lower than normal?
   a. There is oxygen in water, so the presence of fluid in the lungs is not harmful.
   b. Angelo's body was compensating by elevating the breathing rate.
   c. Oxygen diffuses readily in water, so the added distance for diffusion is not an issue.

Based on the data, you and your colleague begin to discuss a possible diagnosis. Your colleague suggests pneumonia, based on the fluid in the lungs. You, however, disagree because there is no fever. Your alternative hypothesis is cardiac failure—a condition in which the heart cannot eject enough blood into the systemic circulation, resulting in a blood backup.

Figure 28.16 Diffusion Distance within the Lungs

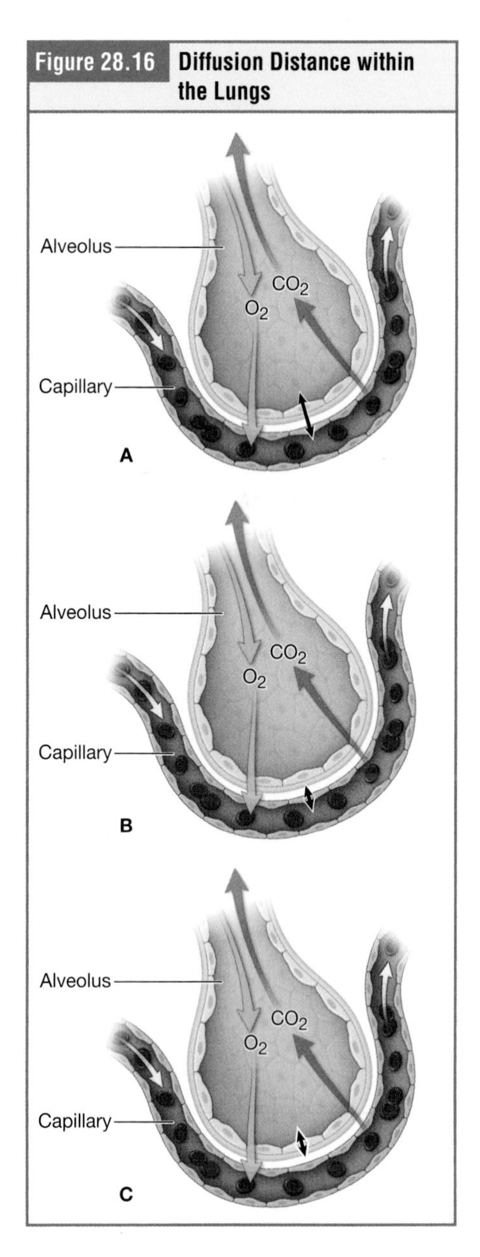

Figure 28.17 **Lung Involvement in Cardiac Failure (Fluid in the Lungs)**

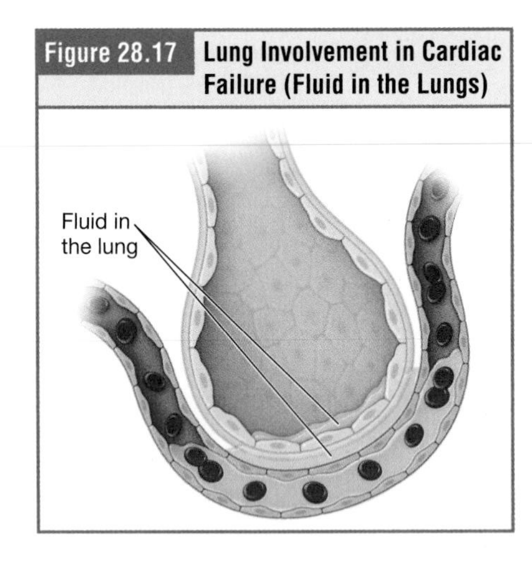

7. If the heart is not able to eject enough blood into the systemic circulation, which side of the heart is failing to eject sufficient blood?
   a. The right side
   b. The left side
8. Where does the blood come from that fills the left side of the heart?
   a. The systemic circulation
   b. The pulmonary circulation
9. If you are correct that Angelo is experiencing cardiac failure, where did the fluid in his lungs come from?
   a. Blood plasma diffusing into the lung space
   b. Inflammation and swelling in the lung tissue from irritation
   c. A puncture in the lung due to coughing and labored breathing
   d. The fluid descended from the upper airways
   e. It is normal mucus due to a severe allergic response
10. To test your hypothesis of cardiac failure, you schedule the patient for an echocardiogram. If you are correct in your diagnosis, what do you expect the echocardiogram to reveal?
   a. The wall of the right ventricle will be enlarged.
   b. All walls of the heart will be noticeably thinner.
   c. The volume of blood in all chambers of the heart will be lower.
   d. The ejection fraction will be lower than normal.
   e. The ejection fraction will be higher than normal.
11. While you are waiting for Angelo's echocardiogram results to come back, you look back over his chart for other evidence to support your hypothesis of cardiac failure. You notice his elevated troponin levels. Where do we find the protein troponin in the body?
   a. Inside cardiac and skeletal muscle cells
   b. Inside bone matrix
   c. Inside the tissues of the kidneys
12. Which of the following could explain the elevated troponin levels in Angelo's bloodstream?
   a. Starved of oxygen, cardiac muscle cells are beginning to die.
   b. Starved of oxygen, neurons in the brain are beginning to die.
   c. Damaged by excessive contracting, cardiac muscle cells are beginning to die.
   d. Damaged by excessive contracting, cells of the lungs are beginning to die.

The echocardiogram results come back and confirm your suspicion of cardiac failure. In this condition, the heart cannot meet its demands. It can precipitate either because the cardiac muscle has weakened or the afterload, the existing pressure in the systemic circulation, is too high. In order to eject blood, the ventricle must generate greater pressure than that of the aorta, because blood (and all fluids) can only flow from areas of greater pressure to areas of lesser pressure. If the pressure in the aorta increases, the ventricle must increase the amount of pressure it generates during contraction to exceed the aortic pressure. You can think of it like lifting a dumbbell with your arm. If you are used to lifting 8-pound dumbbells but today you try to lift 15-pound dumbbells, your muscle needs to be strong enough to generate more power. Often times in cardiac failure patients, both factors have changed at the same time. The systemic blood pressure is higher that it has ever been and the cardiac muscle is weakening due to issues such as atherosclerotic plaque in the coronary vessels. Using our analogy, it is as if your arm muscles were weakened due to injury at the same time you swapped your 8-pound dumbbell for a 15-pound dumbbell.

13. Which of the following treatments might help this patient?
    a. Medication that lowers the heart rate
    b. Antibiotics
    c. Bronchodilators
    d. Medication that lowers blood pressure
    e. Cough suppressants

14. Based on the information about this patient, do you think that surgical techniques to widen his coronary arteries would be effective?
    a. Yes, this would be an effective treatment for this patient.
    b. No, this treatment is not called for at this time.

28.7 Stroke Case Study

Learning Objectives: By the end of this case study you will be able to:

28.7.1 Identify the regions of the brain and list their functions (Section 14.2).

28.7.2 Explain the relationship between sides of the body and sides of the brain (Section 14.2).

28.7.3 Identify the blood vessels of the brain (Section 20.1).

28.7.4 Explain the process of blood coagulation (Section 18.3).

28.7.5 Identify and describe the functions of the cells and their cellular components that compose the tissues of the body (Sections 5.1 and 5.2).

A stroke is a temporary reduction or interruption of blood flow within the brain. Strokes usually occur either when a clot, or embolus, travels within the blood vessels and then gets stuck in a smaller vessel or when a blood vessel bursts. The symptoms of a stroke are:

- Sudden confusion, trouble speaking or understanding speech
- Paralysis or numbness in one location—usually one side of the face, one arm, or one leg
- Trouble seeing in one or both eyes
- Trouble walking
- Sudden severe headache

1. If a person experiencing a stroke is having trouble speaking, which part of the brain is most likely affected?
   a. Occipital lobe
   b. Insula
   c. Wernicke's area
   d. Corpus callosum
2. If a person experiencing a stroke is having trouble seeing, which part of the brain is most likely affected?
   a. Occipital lobe
   b. Insula
   c. Wernicke's area
   d. Parietal lobe

A common teaching tip is that in a stroke you need to act F.A.S.T. This stands for:

Face—ask the patient to smile; does one side of their face droop?
Arms—ask the patient to raise both arms; does one arm drift downwards?

Figure 28.18 The Arteries of the Brain

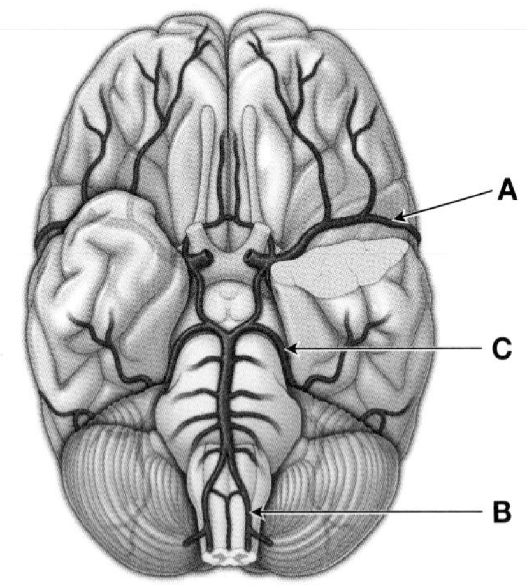

Anterior

A

C

B

Posterior

Speech—ask the patient to talk; is their speech slurred?
Time—if one or more of these symptoms are present, act fast by calling 911.

3. If you ask a patient who you suspect might be having a stroke to smile and the left side of their mouth droops, on which side of the brain is the stroke occurring?
   a. The left side
   b. The right side
   c. It could be either side

Let's examine a specific case. Tony has experienced a stroke after an embolus got stuck and occluded their right middle cerebral artery.

4. Which of the labeled arteries in **Figure 28.18** is the middle cerebral artery?
   a. A
   b. B
   c. C
5. Given the location of the middle cerebral artery, do you think Tony will experience issues with vision? That is to say, do you think that the areas of Tony's brain that process visual information will be affected?
   a. The areas of the brain that process visual information might be affected, as they are served by the middle cerebral artery.
   b. The areas of the brain that process visual information are far from the middle cerebral artery and are served by other arteries, so are unlikely to be affected.
6. Given the location of the middle cerebral artery, do you think Tony will experience issues with walking or other motor activities? That is to say, do you think that the areas of Tony's brain that control skeletal muscles will be affected?
   a. The areas of the brain that control the muscles might be affected, as they are close to the middle cerebral artery.
   b. The areas of the brain that control the muscles are far from the middle cerebral artery and are served by other arteries, so are unlikely to be affected.

Tony's husband calls 911 and Tony is swiftly taken to the ER in an ambulance. The doctors in the ER give Tony a drug called *tissue plasminogen activating factor*.

7. Plasminogen is a zymogen, a pre-enzyme that is made by the cells that line blood vessels. These cells are called ______.
   a. Endovascular cells
   b. Endocardial cells
   c. Endothelial cells
8. Tissue plasminogen activating factor works by converting plasminogen to plasmin, an active enzyme that breaks up blood clots. Which of the following are plausible mechanisms for the action of plasmin?
   a. Breaks bonds within fibrin molecules
   b. Breaks bonds within water molecules
   c. Interferes with the association between phospholipids

In addition to its function as explored in Question 8, plasmin also breaks bonds in laminin—a protein, similar to collagen in function, that makes up the architecture of some tissues. Through its action on laminin, plasmin is an important component of ovulation, helping to create a break in the wall of the ovary through which the egg can escape.

Back at the hospital, one of the doctors comes out to talk to Tony's husband. The doctor says that the drug has helped to break up the clot in Tony's middle cerebral artery, and that Tony will survive this stroke. Though Tony is young, they will have long-lasting effects from the stroke.

9. Why will Tony be unable to recover fully?
   a. Full recovery from a stroke is often impossible because neurons in adults are nonmitotic, so the brain cannot regrow damaged areas.

b. Full recovery from a stroke is only possible when certain regions of the brain, such as the insula lobe, are the ones damaged, since Tony's damage was to the lateral portions of their brain, they will not be able to recover.
c. Brain function requires connections between the brain and the effector organs, such as muscles. These locations may never fully reconnect, and therefore the effects of the stroke may last a long time.

28.8 Type 2 Diabetes Mellitus Case Study

Learning Objectives: By the end of this case study you will be able to:

28.8.1 Define the terms *hypoosmotic, isosmotic, hyperosmotic, hypotonic, isotonic,* and *hypertonic* (Section 4.1).

28.8.2 Identify and describe the functions of the cells and their cellular components that compose the tissues of the body (Sections 5.1 and 5.2).

28.8.3 List the hormones of the body, their functions and the glands that produce them (Section 17.2).

28.8.4 Explain the roles of the various components of a signaling pathway (e.g., the signal, the receptor, the target cell, the response) (Sections 13.1 and 17.1).

28.8.5 Predict the consequences of changes or perturbations in a signaling pathway (Section 17.1).

28.8.6 List the locations (tissue types or organs) and mechanisms for energy and nutrient storage in the body (Section 24.5).

28.8.7 Identify, compare, and contrast the different components of innate and adaptive immune responses (Section 21.2).

28.8.8 List, compare, and contrast the different types of pathogens that can infect the human body (Section 21.4).

Betty is a 62-year-old female who works at an insurance company. She works at a desk, drives to work, and on weekends most of her time is spent driving all over the state to watch her granddaughter's hockey games. While she is busy, Betty is not active, and she is sedentary, meaning she spends most of her time sitting. Today Betty is at her annual physical exam at her doctor's office. The medical assistant who measures her vitals, height, and weight tells Betty that her BMI is 31, which is technically obese, although BMI can be a poor predictor of the amount of adipose tissue or overall health of an individual. The doctor orders a panel of blood tests and tests her blood sugar level. The test reveals that Betty's blood sugar is far outside the typical range. Betty's doctor shows her how her blood sugar level compares to typical values on the chart in **Figure 28.19**.

1. What is the difference between the terms *tonicity* and *osmolarity*?
   a. Whereas *osmolarity* is used to describe body fluids in disease states, *tonicity* is used to describe body fluids from healthy individuals.
   b. Whereas *osmolarity* is used to compare two solutions, *tonicity* is used to compare a fluid to the cell that it surrounds.
   c. Whereas *osmolarity* is used to describe the interstitial fluid, *tonicity* is used to describe the blood.
2. Betty's doctor has healthy blood sugar regulation and has not eaten in the last four hours. If we compared Betty's blood to her doctor's blood at this moment, how could we correctly describe them?
   a. Betty's doctor's blood is hypotonic to Betty's blood.
   b. Betty's blood is hyperosmotic to her doctor's blood.
   c. Betty's blood is isotonic to her doctor's blood.

Betty's doctor orders more tests. A week later, Betty's doctor calls her to let her know that they are diagnosing Betty with Type 2 Diabetes Mellitus (T2DM), a chronic metabolic

Figure 28.19 **A Graph of Blood Sugar Over Time After a Meal**

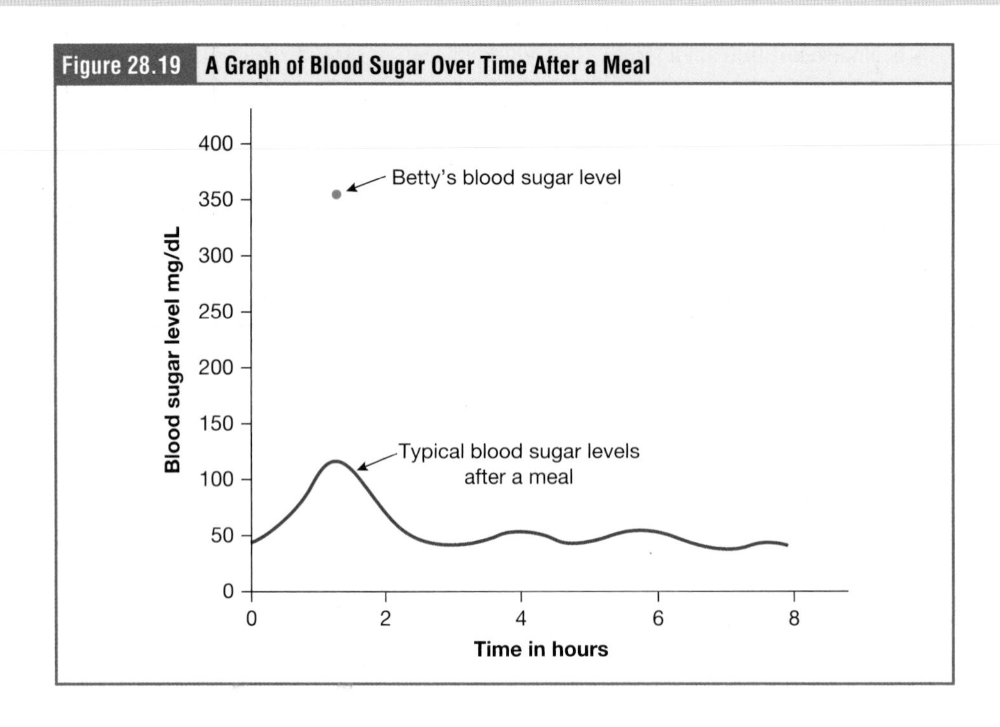

disease with a primary symptom of elevated blood glucose levels. Betty's doctor explains that the greatest risk factor for T2DM is obesity, and the second greatest risk factor is a sedentary lifestyle. Betty has both of these. Betty's doctor explains that the reason that obesity is a risk factor is that adipose tissue contributes to inflammatory conditions.

3. What factors are released by adipose tissue that contribute to inflammation?
   a. Adipokines
   b. Complement proteins
   c. Antibodies

Betty has a cousin who also has diabetes. Betty calls her cousin to compare therapies and her cousin explains that they do not have the same disease. Betty's cousin has Type 1 Diabetes Mellitus (T1DM), which has a different pathophysiology. In both types of diabetes, insulin signaling isn't working as it should.

4. Which organ secretes insulin?
   a. The liver
   b. The pancreas
   c. The thyroid
   d. The anterior pituitary

The cells responsible for making insulin are called *beta cells* or *β-cells*. In T1DM, β-cells die through immune attack. In T2DM, β-cells fall into dysfunction due to inflammation and metabolic stress. The factors that drive β-cell dysfunction in T2DM are a proinflammatory state and excess nutrients. In T2DM, however, another factor is contributing to the disease. Not only is less insulin being made by the β-cells, but insulin receptors have become less sensitive, a condition called **insulin resistance**. In these cases, even when insulin is present in the bloodstream it may not have an effect on its target cells.

5. In T1DM, β-cells will never again be able to produce insulin because the immune system has killed them. Scientists are currently trying to develop a treatment to replace lost β-cells with new ones. In the case of someone with early T2DM, would this treatment be effective? Why or why not?
   a. Yes, if the β-cells were replaced, then more insulin might be produced and the symptoms of diabetes would resolve.

b. No, even if the β-cells were replaced, the immune system might just kill them off again.
c. No, even if the β-cells were replaced and insulin levels were higher, the insulin resistance would still be a factor in the disease.

6. What is the link between the hormone insulin and the blood sugar test taken in the doctor's office?
   a. Insulin functions to release stored sugars from the liver.
   b. Insulin functions to allow the brain to store sugar between meals.
   c. Insulin functions to help cells all over the body take sugar out of the blood.

Let's dive into the function of insulin a bit further. Specifically, when insulin binds to the insulin receptor on cell types all over the body, the activated insulin receptor causes the upregulation of glucose transporters. These glucose transporters are called *GLUTs*. This process is illustrated in **Figure 28.20**.

7. What does the term *upregulation* mean?
   a. To express, or increase expression of, a molecule on the surface of the cell
   b. To increase the number of copies of a gene that an individual has through gene therapy techniques
   c. To make more of a particular protein through transcription and translation

Figure 28.20 **Insulin Regulates the Expression of Glucose Transporters on the Cell Surface**

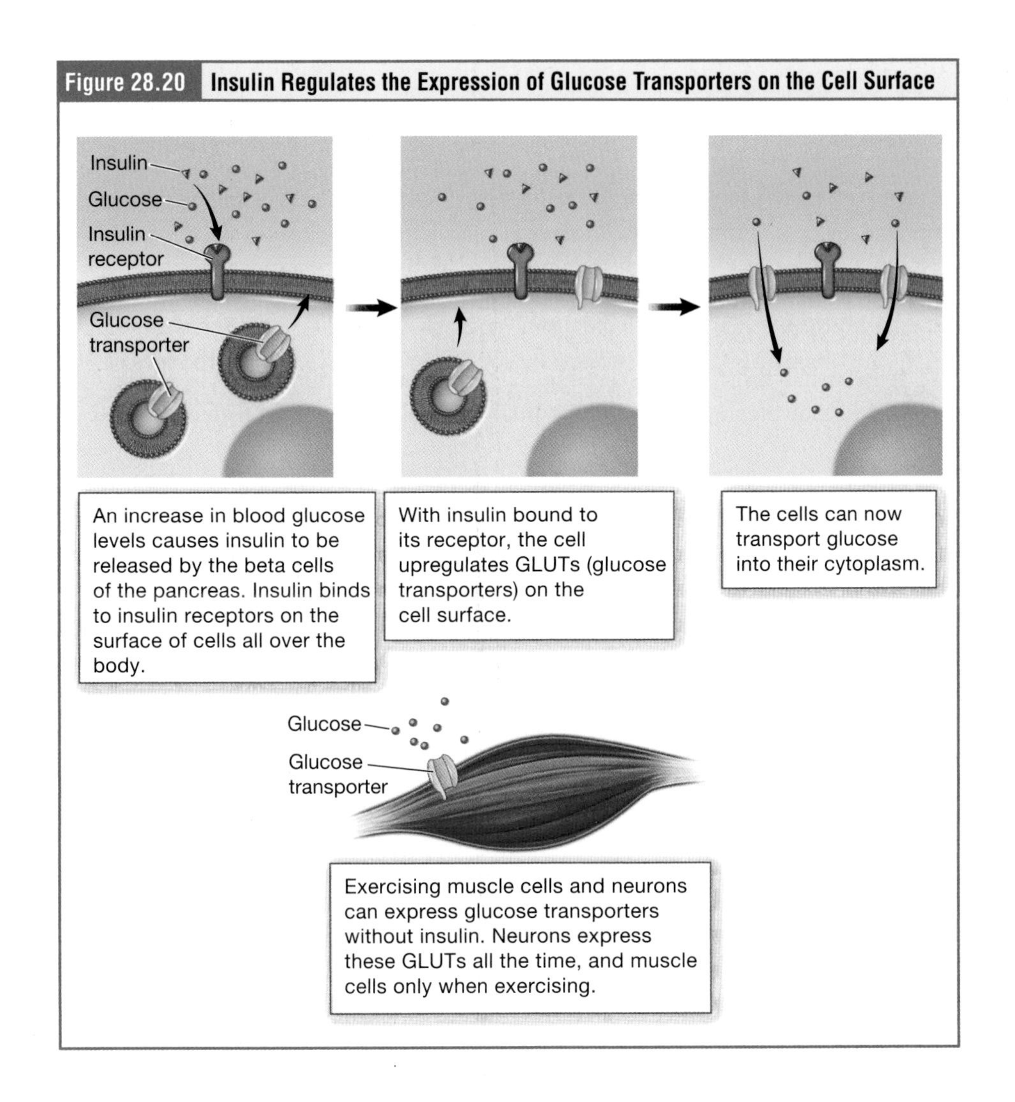

Insulin
Glucose
Insulin receptor
Glucose transporter
An increase in blood glucose levels causes insulin to be released by the beta cells of the pancreas. Insulin binds to insulin receptors on the surface of cells all over the body.
With insulin bound to its receptor, the cell upregulates GLUTs (glucose transporters) on the cell surface.
The cells can now transport glucose into their cytoplasm.
Glucose
Glucose transporter
Exercising muscle cells and neurons can express glucose transporters without insulin. Neurons express these GLUTs all the time, and muscle cells only when exercising.

There are 14 versions of GLUTs in the human body; not all cells express all of these molecules. Skeletal muscle cells and neurons express GLUT-4, a version of glucose transporters that is not regulated by insulin. Neurons have GLUT-4 on the surface of their cells all the time, but skeletal muscle cells only express GLUT-4 on their surfaces when they are actively exercising.

Betty's doctor prescribes metformin, a drug commonly used to treat T2DM, and recommends that Betty begin a regular exercise program. Betty's doctor explains that exercise has both a preventative and therapeutic effect.

8. What effect will exercise have on Betty's blood sugar levels?
   a. Exercise will cause Betty's blood sugar level to rise.
   b. Exercise will cause Betty's blood sugar level to fall.
   c. Exercise will have no effect on Betty's blood sugar level.

Metformin also increases the expression of GLUT4, but in addition it has several other effects including preventing the breakdown of glycogen into glucose.

9. What organ stores glycogen and is able to break it down to add glucose to the blood when needed?
   a. The liver
   b. The pancreas
   c. Bone
   d. The stomach
10. How would we classify glycogen?
   a. Glycogen is a triglyceride.
   b. Glycogen is a nucleic acid.
   c. Glycogen is a polysaccharide.

By reducing the circulating blood sugar levels over time through treatments such as increased exercise and the use of metformin, often the patient's insulin resistance improves. Scientists are trying to uncover what exactly causes insulin resistance in the first place. Some theories include inflammation, high circulating nutrient levels over time, and autoimmune action. An autoimmune disease is one in which the immune system attacks a component of the body, mistaking it for a pathogen.

11. If the insulin resistance observed in T2DM had an autoimmune cause, which of the following might the scientists find in the blood of T2DM patients such as Betty?
   a. Antibodies that bind to the insulin receptor
   b. Antibodies that bind to glucose
   c. Antibodies that bind to surface features of β-cells

A few weeks later Betty sits down with a diabetes educator. The educator talks to Betty about her lifestyle and some simple habits and diet adjustments that might help. Of course, they talk a lot about exercise. Toward the end of the meeting, the diabetes educator also mentions that Betty and all individuals with T2DM are more prone to certain infections such as sinus infections and yeast infections. These infections are caused by bacteria and fungi. Betty has been hearing a lot in the news about viruses lately and asks the diabetes educator if she is more likely to catch a virus. The diabetes counselor clarifies: individuals with T2DM are not more likely to catch any infections. However, once in the body, bacteria and fungi, not viruses, are more likely to thrive.

12. Why are bacteria and fungi, not viruses, more likely to thrive in an individual with T2DM?
   a. Betty's immune system is altered against bacteria and fungi.

b. Betty's fluids have more sugar in them, providing more food for bacteria and fungi.
c. Because bacteria and fungi are larger than viruses, they are harder for the immune system to find.

During the COVID-19 pandemic, obesity and diabetes were both listed as risk factors for more severe disease caused by the SARS CoV-2 virus. It is thought that most conditions that indicate chronic inflammation were risk factors for more severe outcomes. While scientists are still working to understand the pathophysiology of COVID-19 disease, it is clear that underlying inflammation was a significant risk factor.

28.9 Chronic Obstructive Pulmonary Disease Case Study

Learning Objectives: By the end of this case study you will be able to:

28.9.1 Identify and describe the relationship among the structures of the lungs (Section 22.1).

28.9.2 List and describe the components of the flow equation (Section 22.2).

28.9.3 Using the flow equation, predict how different changes to physiological conditions will impact flow rate (Section 22.2).

28.9.4 Identify and describe the functions of the cells and their cellular components that compose the tissues of the body (Sections 5.1 and 5.2).

28.9.5 Compare and contrast the functions, structures and characteristics of the four major macromolecules (Section 3.5).

28.9.6 Predict changes to pO_2 and pCO_2 with various changes or perturbations to metabolism or lung function (Sections 26.4 and 26.5).

28.9.7 Connect the CO_2 concentration of a fluid to its pH using the carbonic acid equation (Section 22.5).

28.9.8 Define, compare, and contrast the terms *hypercapnia* and *hypoxia* (Section 26.5).

28.9.9 Define, compare, and contrast the terms *acidosis* and *alkalosis* (Section 26.5).

28.9.10 Describe a hematocrit and the physiological information can be gained from this test (Section 18.1).

28.9.11 List the hormones of the body, their functions and the glands that produce them (Section 17.2).

28.9.12 List the products and reactants of cellular respiration (Section 22.4).

Chronic obstructive pulmonary disease (COPD) is a disorder of chronic inflammation of the lungs that results from smoking. Smoking (any substance), vaping, or breathing heavily polluted air all cause inflammation in the lungs. Individuals who develop COPD have an exaggerated inflammatory response to these stimuli. The inflammation is particularly severe in the small airways of the lungs.

1. Identify the correctly labeled diagram of the lungs in **Figure 28.21**.
   a. Diagram A
   b. Diagram B
   c. Diagram C
2. Are the smaller and smallest airways found closer to or farther from the bronchi?
   a. Closer to
   b. Farther from

Figure 28.21 **Anatomy of the Lungs**

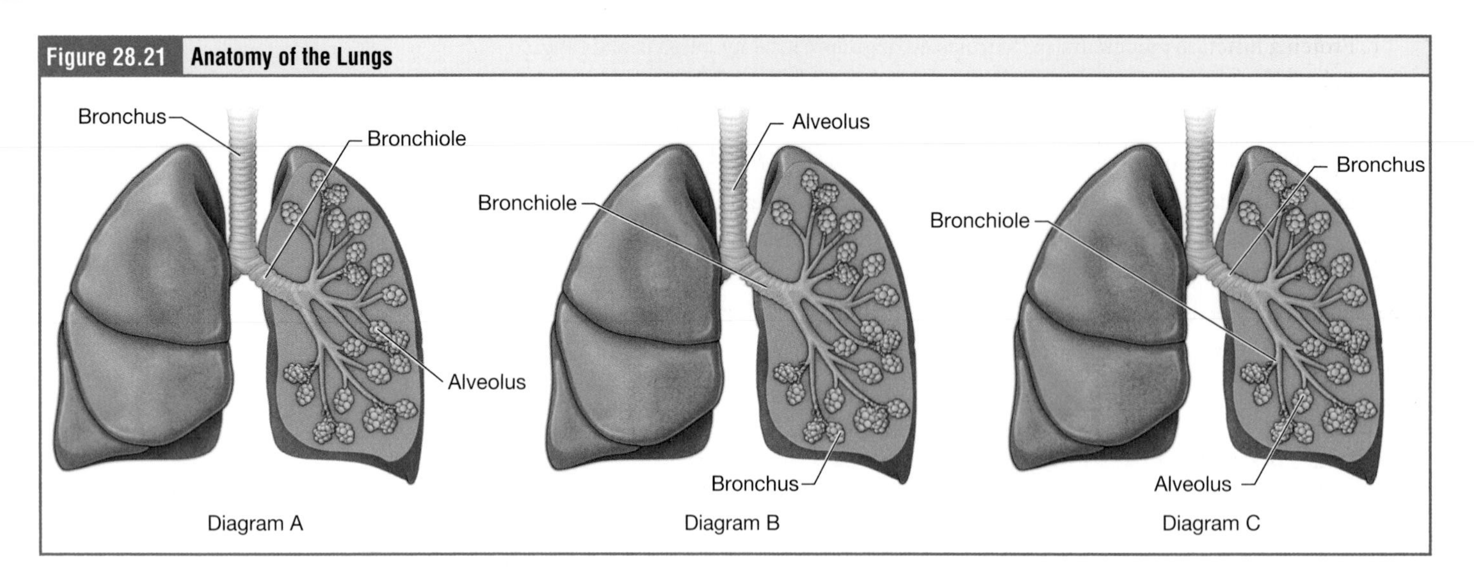

We use an equation, often called the *flow rate equation*, to examine factors that influence flow. Flow can be air flow, fluid flow, or even can be modified to represent the flow of molecules. The equation is written as follows:

$$F = \frac{\Delta P}{R}$$

3. F in the equation stands for flow; what are the other two components?
   a. ΔP indicates an electrical charge and R indicates resistance.
   b. ΔP indicates a pressure gradient and R indicates resistance.
   c. ΔP indicates a pressure gradient and R indicates a reservoir or holding tank.
4. Which of the following components of the flow equation is directly affected by the increased inflammation?
   a. ΔP
   b. R
   c. Neither factor is impacted
5. What happens to flow (F) with increased inflammation?
   a. Flow increases
   b. Flow decreases
   c. There are no changes to flow caused by the inflammation
6. One of the changes characteristic of COPD is an increase in the number of goblet cells. Which of the following symptoms of COPD would be most directly related to this change?
   a. Chronic wet cough
   b. Increased breathing rate
   c. Increased red blood cell volume

The inflammation seen in COPD can be measured by the increased presence of immune cells. COPD patients have much higher levels of neutrophils, macrophages, and killer T cells present in their lungs. These cells cause a variety of issues, including the secretion of proteases, enzymes that degrade proteins.

7. Which of the following are roles of proteins in the body?
   a. Proteins function as enzymes, catalyzing chemical reactions in cells.
   b. Proteins lend strength and stability to the architecture of the extracellular matrix.

c. Proteins function as signaling molecules and receptors for signaling molecules.
d. All of the above are functions of proteins.

Mike is a 71-year-old former smoker with COPD. Mike's doctor has recently retired and Mike is meeting with a new doctor and describing their medical history.

8. Which of the lungs in the **Figure 28.22** do you think is Mike's?
   a. A
   b. B

The new doctor observes that Mike is breathing a little faster than the doctor would expect for a 71-year-old. The doctor orders a test that measures arterial blood gas levels and pH.

9. What do you expect of Mike's arterial oxygen level (pO_2) compared with that of an individual without COPD?
   a. Mike's pO_2 will be higher than that of an individual without COPD.
   b. Mike's pO_2 will be lower than that of an individual without COPD.
   c. Mike's pO_2 will be similar to that of an individual without COPD.
10. What do you expect of Mike's arterial carbon dioxide level (pCO_2) compared with that of an individual without COPD?
   a. Mike's pCO_2 will be higher than that of an individual without COPD.
   b. Mike's pCO_2 will be lower than that of an individual without COPD.
   c. Mike's pCO_2 will be similar to that of an individual without COPD.

Remember that changes in CO_2 level in the blood lead to changes in pH level. The connection is through an enzyme called *carbonic anhydrase*. The action of this enzyme can be summarized in **Figure 28.23**.

11. What do you expect of Mike's arterial pH compared with that of an individual without COPD?
   a. Mike's pH will be higher than that of an individual without COPD.
   b. Mike's pH will be lower than that of an individual without COPD.
   c. Mike's pH will be similar to that of an individual without COPD.
12. Which of the following terms might be used to describe Mike's current condition?
   a. Hypercapnia
   b. Hypoxia
   c. Acidosis
   d. All of these terms could be applied to Mike

Figure 28.22 The Lungs of a COPD Patient and of a Healthy Individual

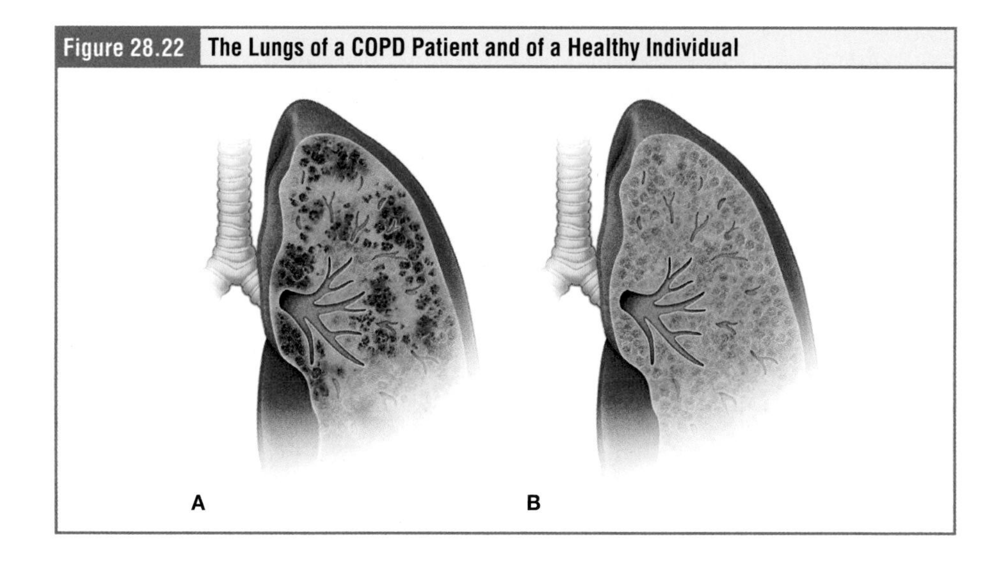

Figure 28.23 The Carbonic Anhydrase Equation

$$CO_2 + H_2O \xrightleftharpoons{\text{carbonic anhydrase}} H_2CO_3 \rightleftharpoons HCO_3^- + H^+$$

carbon dioxide + water — *carbonic acid* — *bicarbonate + hydrogen ion*

Considering that Mike's COPD has been a chronic condition over a period of years, there have been other changes to Mike's physiology in response. The doctor looks at Mike's blood work and notices that Mike has a very high hematocrit.

13. What does a hematocrit measure?
    a. The percentage of blood volume that is composed of red blood cells
    b. The relative amount of ions in the blood
    c. The level of nutrients available in the bloodstream
14. As the condition(s) described in Question 12 precipitated over several years, Mike's body responded by releasing a hormone that resulted in the increased hematocrit. Which of the following hormones could cause an increased hematocrit?
    a. Parathyroid hormone
    b. Estrogen
    c. Prolactin
    d. Erythropoietin

Mike returns to the doctor's office a few weeks later with a fever. It turns out that Mike's fever is resulting from an infected cut on their leg. Their lungs are unaffected by the infection, but nonetheless they are breathing hard and fast. Let's dig through the physiology of fever in order to understand the impact on Mike's lungs. Remember that most of the heat that contributes to and maintains body temperature is a byproduct of cellular respiration. In **Figure 28.24**, the equation for cellular respiration is partially drawn. The following two questions ask you to complete this equation.

15. Which molecule belongs in the first blank of the equation?
    a. O_2
    b. CO_2
    c. PO_4^-
16. Which molecule belongs in the second blank?
    a. O_2
    b. CO_2
    c. PO_4^-
17. What do we expect to happen to Mike's pCO_2 and pH levels in their blood while they have a fever?
    a. pCO_2 increases and pH decreases farther from Mike's baseline.

Figure 28.24 The Cellular Respiration Equation

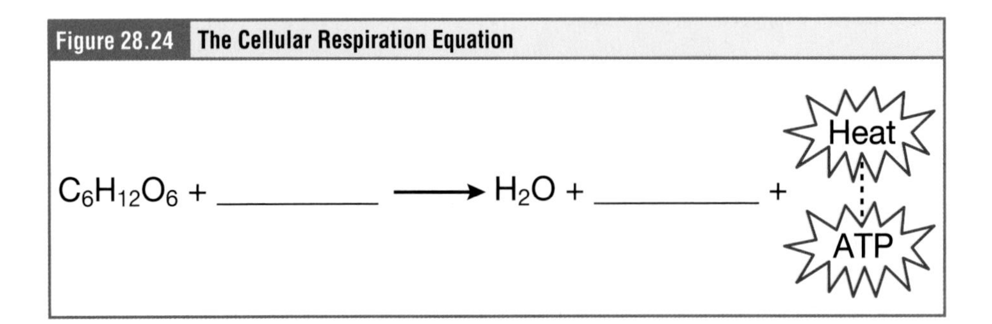

b. pCO_2 decreases and pH increases farther from Mike's baseline.
c. pCO_2 increases, but pH stays the same.
d. pCO_2 stays the same, but pH increases.

18. Increased body temperature or metabolism can be very difficult for COPD patients to endure. In addition to fever, which of the other following factors would cause the same changes to Mike's pCO_2 and pH levels? Please select all that apply.
    a. An increase in thyroid hormone
    b. Traveling to a higher elevation/altitude
    c. Deep breathing such as in yoga or meditation
    d. Exercise

28.10 Pregnancy Case Study

Learning Objectives: By the end of this case study you will be able to:

28.10.1 Identify the regions of the spinal column (Section 8.3).

28.10.2 Identify the structures of a vertebra (Section 8.3).

28.10.3 Identify the structures of the pelvis (Section 9.3).

28.10.4 Identify the ligaments of the spinal column (Section 8.3).

28.10.5 Interpret a physiological graph (Section 27.8).

28.10.6 Identify the layers of blood vessels and their histological components (Section 20.1).

28.10.7 Using the flow equation, predict how different changes to physiological conditions will impact flow rate (Section 22.2).

28.10.8 Describe the relationship between blood volume and blood pressure (Section 20.2).

28.10.9 List the locations (tissue types or organs) and mechanisms for energy and nutrient storage in the body (Section 24.5).

28.10.10 List the hormones of the body, their functions and the glands that produce them (Section 17.2).

28.10.11 Identify, compare, and contrast the different components of innate and adaptive immune responses (Section 21.2).

28.10.12 Describe the types of transport across the cell membrane (Section 4.1).

28.10.13 Describe, compare, and contrast the cellular processes of mitosis, apoptosis, and autolysis (Section 4.5).

28.10.14 Identify the muscles of the pelvic floor and abdomen (Section 2.7).

Brittany is a 28-year-old nulliparous female who is currently 36 weeks pregnant. **Nulliparous** means that Brittany has had no prior births. At a checkup with her obstetrician (a doctor who specializes in pregnancy and birth), Brittany describes significant lower back pain. As Brittany stands in the office, she resembles the individual in **Figure 28.25**. Brittany's doctor writes the term *lordosis* in Brittany's medical file. *Lordosis* describes an exaggeration in the curve of the lower back.

1. Which region of the spinal column is excessively curved?
   a. Cervical
   b. Thoracic
   c. Lumbar
2. Compared to a nonpregnant individual, which of the following is true of the affected vertebrae?
   a. The vertebral bodies will be closer together and the spinous processes will be further apart.

Figure 28.25 **Lordosis Is a Condition That May Develop at the End of Pregnancy**

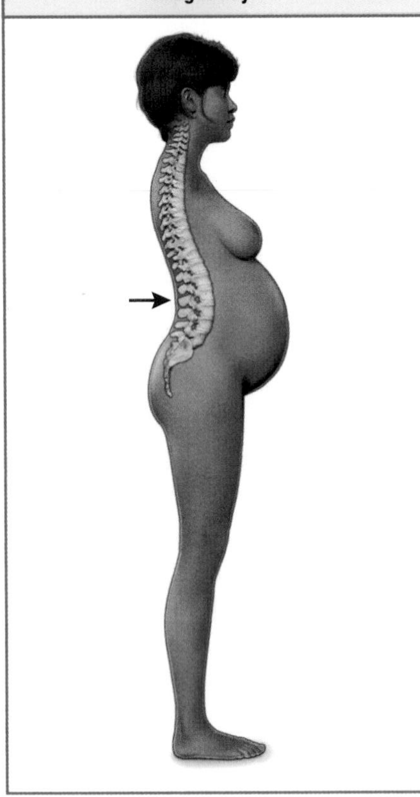

b. The vertebral bodies will be further apart and the spinous processes will be forced closer together.
c. The vertebral bodies and spinous processes will be the same between a pregnant and a nonpregnant individual, but the transverse processes will be forced closer together.
d. With more anterior and posterior space present between the vertebrae, the intervertebral discs will expand and take up more space.

In high amounts, estrogen can cause collagen degradation. Tendons and ligaments, being very collagen-rich, weaken and loosen in these conditions. Weakening of the anterior longitudinal ligament and ligaments of the sacroiliac joint are likely contributing to Brittany's lower back pain.

3. In **Figure 28.26**, which of the labels indicates the sacroiliac joint?
   a. A
   b. B
   c. C
   d. D
4. In Figure 28.26, which of the labels indicates the anterior longitudinal ligament?
   a. A
   b. B
   c. C
   d. D
5. Examine the graph of the hormones of pregnancy in **Figure 28.27**. If we assume that estrogen-mediated ligament changes are contributing to Brittany's lower back pain, what can the obstetrician tell Brittany to expect in the final four weeks of her pregnancy?
   a. Brittany's estrogen levels will decrease, and ligaments will tighten, alleviating the pain that she is experiencing now.
   b. Brittany's ligaments will become twice as loose as they are now because there will be a twofold increase in circulating estrogen levels in the final weeks of the pregnancy.

Figure 28.26 **The Pelvis and Its Ligaments**

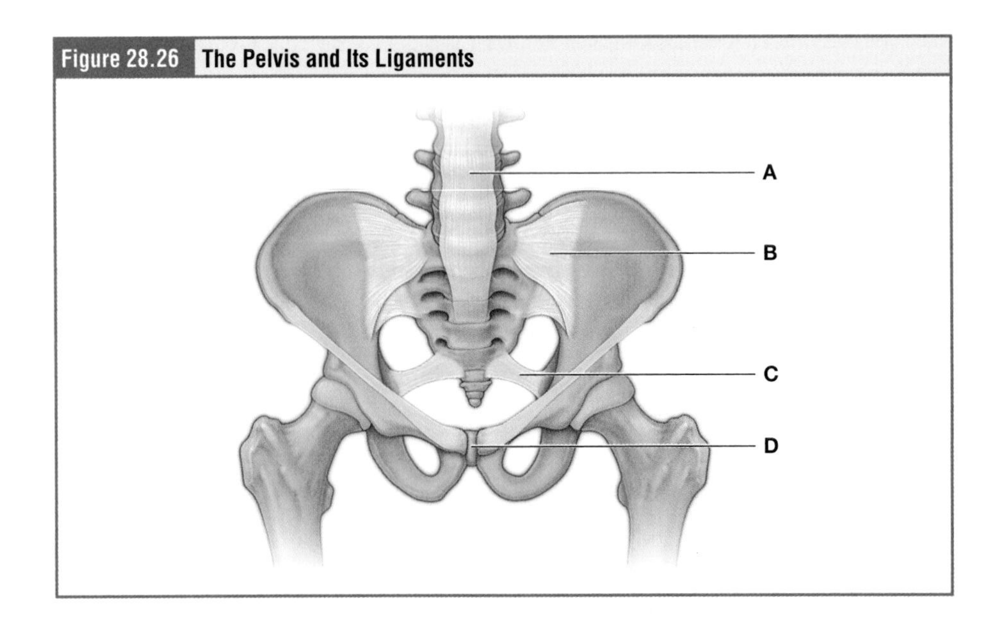

Figure 28.27 The Hormones of Pregnancy

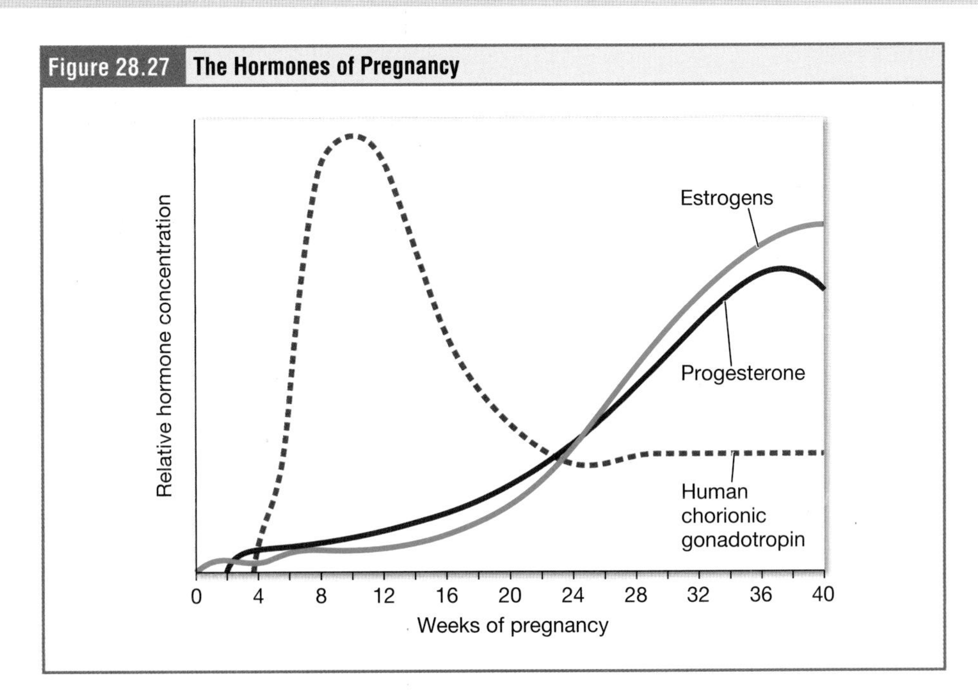

c. Brittany's estrogen level and ligament loosening will not get significantly worse in the final weeks of her pregnancy.

During the last 10 weeks of pregnancy, the pelvic brim does increase in diameter by about 10–15 percent.

6. Which of the labels in **Figure 28.28** correctly indicates the pelvic brim?
   a. A
   b. B
   c. C
7. When the obstetrician steps out of the exam room to collect some test results, the medical student being trained by them asks Brittany how she has been feeling in general. Which of these is likely to be a central complaint? Use **Figure 28.29** for assistance.
   a. "I feel an odd pain on my right side when I have to defecate, that was never there before."
   b. "I'm so much hungrier at every meal, I used to eat small meals, but now I can really put it away, I've been going to a lot of all-you-can-eat buffets"
   c. "I've noticed that I have a lot more trouble digesting fatty meals than I did before. I tried to each nachos, and they really didn't agree with me."
   d. "I just have to urinate all the time, it's like my bladder has shrunk in half."

At almost full-term, the placenta of Brittney's fetus is nearly nine inches in diameter and weighs about a pound and a half. The placenta, an entirely fetal structure, is disc-shaped and packed with treelike branches. Each branch contains fetal capillaries filled with fetal blood. The cells of the placenta also invade into the uterine wall, where they migrate until they find access to maternal arteries. Within these arteries, placental cells replace the innermost lining of the artery in order to facilitate the changes to the artery necessary for pregnancy.

8. What is the innermost lining of the artery called?
   a. The tunica intima
   b. The tunica media
   c. The tunica endothelia

Figure 28.28 The Osteology of the Pelvis

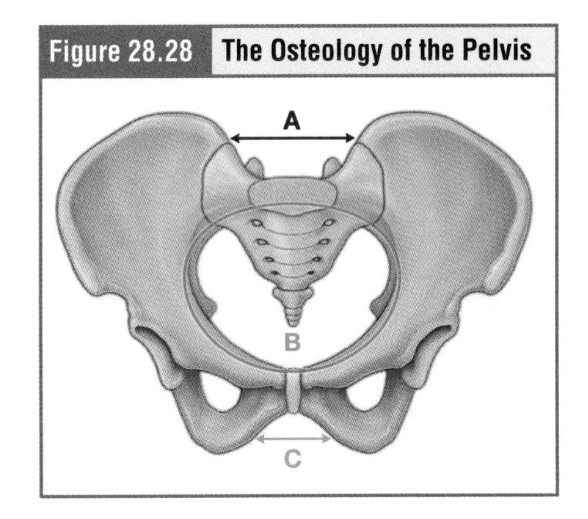

Figure 28.29 **A Sagittal Section of an Abdomen and Pelvic Cavity during Pregnancy**

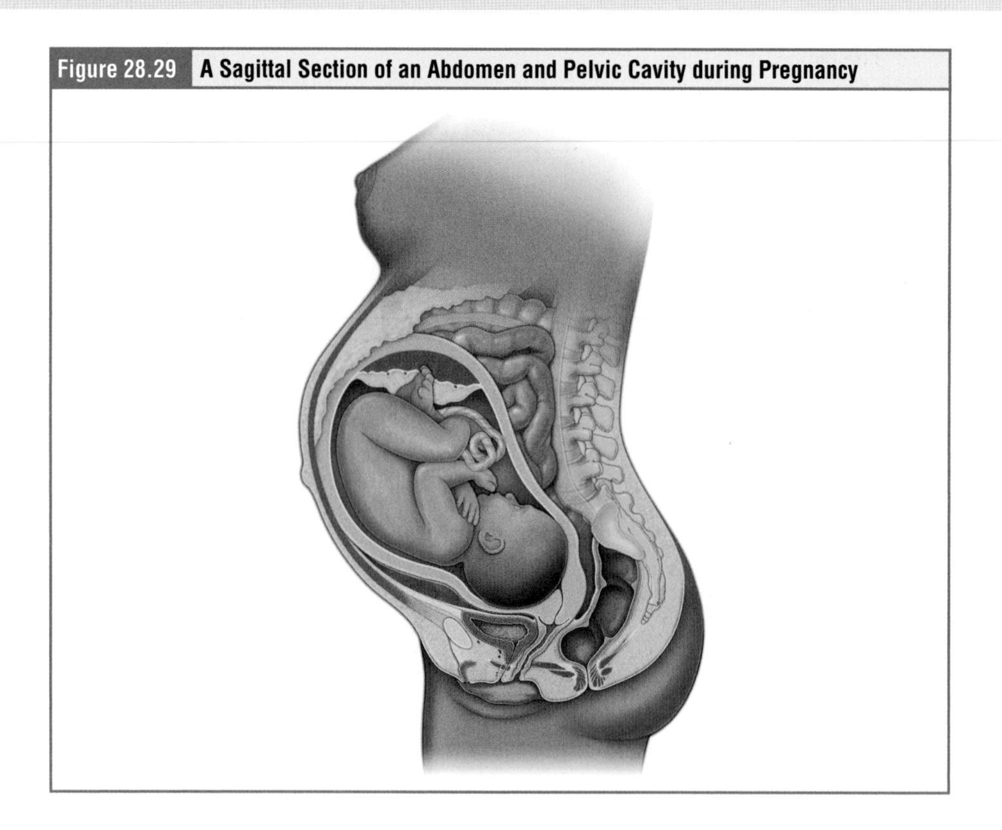

9. Once the placental cells replace the innermost lining of the arteries, they are able to widen the arteries. This is one example of vasodilation. What will happen to blood flow through these vessels?
   a. Blood flow will increase.
   b. Blood flow will decrease.
   c. There will be no change to blood flow based on this alone.
10. Once the uterine arteries have been remodeled by the placental cells, they are responsible for perfusing not just the uterine wall, but also for covering the entire surface of the placenta. This represents a significant increase to the surface area of the mother's cardiovascular system. Which of the following statements is true?
    a. Blood pressure should not be affected, so maternal blood volume will only adjust as needed to meet the oxygen and nutrient demands of the fetus.
    b. To maintain blood pressure, maternal blood vessels in other locations will die off and be regrown after the pregnancy ends.
    c. To maintain blood pressure, maternal blood volume must increase proportionally to the increased surface area.
11. The growing fetus grows a skeleton out of hyaline cartilage, then slowly ossifies it by adding calcium and phosphorous to the cartilage matrix. These minerals must be obtained from maternal blood. If there is an increased need for mineral but not an increased supply of these minerals from the diet, where in the maternal body can the minerals come from?
    a. The liver
    b. The bones

c. The pancreas
d. The cerebrospinal fluid

12. Assuming that minerals are being liberated from the maternal source you chose in Question 11, which hormone is probably elevated in the maternal bloodstream to facilitate this?
    a. Thyroid hormone
    b. Growth hormone
    c. Calcitonin
    d. Parathyroid hormone

At this doctor visit, the obstetrician recommends that Brittany get a booster vaccination for COVID-19. The doctor tells Brittany that her immune system will produce antibodies against the virus and these antibodies will cross the placenta and circulate in the baby for months after birth.

13. Which of the following best describes an antibody?
    a. A protein that is capable of binding to and possibly disabling a pathogen
    b. A sugar that is capable of coating a pathogen to disable it from binding to cells
    c. An enzyme the degrades a pathogen, killing it
14. Which type of Brittany's cells are capable of making antibodies?
    a. Macrophages
    b. T lymphocytes
    c. B lymphocytes
    d. Natural Killer (NK) cells
15. The placenta has a layer of cells that create a barricade between the maternal blood and the fetal capillaries. How would an antibody cross the placenta?
    a. Transcytosis (endocytosis on one side, vesicle transport across the cell, exocytosis on the other side)
    b. Facilitated diffusion through a protein channel or transporter
    c. Simple diffusion
16. The uterus grows during pregnancy; by the end of a full-term pregnancy, it is 500 percent larger than its pre-pregnancy size. Afterward, it shrinks back down to its original size. What are the names for the cellular processes responsible?
    a. Mitosis during the pregnancy and autolysis after the pregnancy
    b. Endocytosis during the pregnancy and exocytosis after the pregnancy
    c. Cytosis during the pregnancy and apoptosis after the pregnancy

Three and a half weeks later, Brittany gives birth to a beautiful baby. She and her partner are excited and happy about their new child. However, after a month has passed, Brittany returns to her obstetrician. She notes that, occasionally, she thinks her underwear is slightly wet, and suspects she might be leaking urine. She has noticed this especially after laughing. The doctor explains that tears to pelvic floor muscles during childbirth can cause this **incontinence** (urine leakage).

17. Which muscle is most likely torn during childbirth?
    a. Levator ani
    b. Psoas major
    c. Illiacus

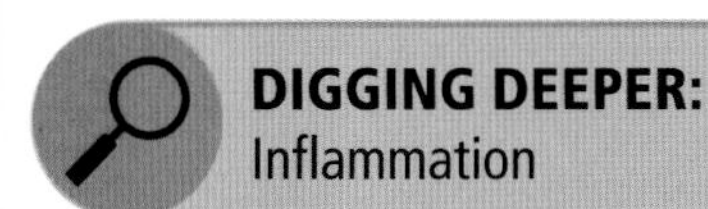

DIGGING DEEPER: Inflammation

Throughout this chapter we often described inflammation as a root cause of many diseases. But inflammation, at its core, is the body's natural response to injury and infection. Without inflammation we would have a very difficult time responding to pathogens or healing a sprained ankle. So what is inflammation, and how can it be so important in some circumstances and so harmful in others?

You may have heard the phrase "all good things must come to an end," and that is certainly true of our inflammatory response. Inflammation can be **acute**, meaning severe but having a short duration, or it can be **chronic**, meaning persisting over time or recurring often. There are many parts of our immune system that turn on or exacerbate inflammation, but there are also several types of immune system cells that function to turn off an immune response, including inflammation once it is over. In a healthy body, the chemicals and cells that turn on or exacerbate inflammation are at very low levels most of the time. They activate and their levels increase when needed and they are turned off and return to a low resting level between infections or injuries.

Some individuals are more pro-inflammatory. There are genetic components and also, as we have seen in some of the case studies in this chapter, there are physiological components as well. One component, as we saw often in this chapter, are adipokines. Adipokines are a set of signaling chemicals and hormones released from adipose tissue. Adipokines are pro-inflammatory and contribute significantly to chronic inflammation.

Smoking, high nutrient levels, and age are all factors that are linked to chronic inflammation as well. Frequent injury is a risk factor for chronic inflammation, and this likely makes sense because the more times we enter into an acute inflammation phase, the more active these components of our immune system are.

Exercise helps to clear inflammatory components, especially a signaling molecule called *interleukin-1* (IL-1) from the bloodstream. Therefore, even if you have other risk factors for chronic inflammation, such as a family history, being older, or frequent injury, staying active can help clear these chemicals from the bloodstream and keep chronic inflammation under control.

Anatomy & Physiology: Answers Appendix

Answers to Learning Checks and Chapter Review Questions

Chapter 2

Learning Check Questions, Section 2.3:

1. a
2. b
3. b

Learning Check Questions, Section 2.6:

1. c
2. b
3. c

Learning Check Questions, Section 2.7c:

1. a
2. d
3. a
4. b, d

Learning Check Questions, Section 2.7d:

1. a
2. b
3. b
4. a
5. b

Learning Check Questions, Section 2.7e:

1. a, c
2. a
3. c

Chapter Review Questions, Section 2.1:

1. a
2. b

Chapter Review Questions, Section 2.2:

1. a
2. a

Chapter Review Questions, Section 2.3:

1. c
2. d

Chapter Review Questions, Section 2.4:

1. b

Chapter Review Questions, Section 2.4 (Mini Case 1: Flow Rate):

1. a
2. b

Chapter Review Questions, Section 2.5:

1. b
2. a
3. c
4. b
5. b

Chapter Review Questions, Section 2.6:

1. c
2. b
3. b, c
4. b

Chapter Review Questions, Section 2.7:

1. a, c
2. b
3. a
4. d
5. a, d
6. b
7. a
8. a, b, c
9. c
10. a
11. a

Chapter 3

Learning Check Questions, Section 3.1b:

1. c
2. a, c
3. a, b
4. a

Learning Check Questions, Section 3.1c:

1. a
2. b
3. c
4. a, c

Learning Check Questions, Section 3.2:

1. a
2. b
3. a

Learning Check Questions, Section 3.3:

1. c
2. a
3. d
4. c

Learning Check #1 Questions, Section 3.4c:

1. a
2. d
3. c

Learning Check #2 Questions, Section 3.4c:

1. d
2. c
3. a, c
4. b

Learning Check Questions, Section 3.5a:

1. b
2. d
3. a

Learning Check Questions, Section 3.5c:

1. c. d
2. b
3. b

Learning Check Questions, Section 3.5d:

1. a, b, c
2. a
3. b
4. a, b

Learning Check Questions, Section 3.5e:

1. a
2. d
3. a

Chapter Review Questions, Section 3.1:

1. a, b
2. c
3. c
4. b

Chapter Review Questions, Section 3.1 (Mini Case 1: Exercise Stress Test):

1. a
2. b
3. b

Chapter Review Questions, Section 3.2:

1. b
2. b
3. a

Chapter Review Questions, Section 3.3:

1. a, b
2. a, b, c
3. b, c, d
4. c
5. b

Chapter Review Questions, Section 3.4:

1. c
2. d
3. a
4. c
5. a

Chapter Review Questions, Section 3.5:

1. a
2. b
3. c
4. b
5. b
6. a, b
7. a

Chapter 4

Learning Check Questions, Section 4.1b:

1. a, b
2. a, d
3. c

Learning Check Questions, Section 4.1c:

1. b
2. b
3. d
4. d
5. a, b

Learning Check Questions, Section 4.2c:

1. d
2. a
3. a
4. a

Learning Check Questions, Section 4.3:

1. c
2. a
3. a

Learning Check Questions, Section 4.4b:

1. b
2. a, b
3. c
4. a

Learning Check Questions #1, Section 4.5b:

1. d
2. a
3. b
4. c

Learning Check Questions #2, Section 4.5b:

1. a, b
2. a
3. a

Chapter Review Questions, Section 4.1:

1. a, d
2. c
3. d
4. b

Chapter Review Questions, Section 4.1 (Mini Case 1: Peripheral Edema):

1. c
2. a
3. c
4. a

Chapter Review Questions, Section 4.2:

1. b
2. d
3. c
4. a
5. a, b

Chapter Review Questions, Section 4.3:

1. a
2. d

Chapter Review Questions, Section 4.4:

1. c
2. b
3. a

Chapter Review Questions, Section 4.5:

1. a
2. a
3. a
4. c
5. c

Chapter Review Questions, Section 4.6:

1. a, d
2. a, b

Chapter 5

Learning Check Questions, Section 5.1c:

1. a, b
2. a
3. b, d

Learning Check Questions, Section 5.2a:

1. b
2. a
3. d
4. b

Learning Check Questions, Section 5.2c:

1. a
2. d
3. a
4. a
5. d

Learning Check Questions, Section 5.3b:

1. b
2. a, b
3. a

Learning Check Questions, Section 5.3e:

1. a
2. a
3. d
4. b, c

Learning Check Questions, Section 5.4:

1. a, b
2. a
3. c

Learning Check Questions, Section 5.5:

1. d
2. a
3. a

Learning Check Questions, Section 5.6:

1. b
2. c

Learning Check Questions, Section 5.7c:

1. a
2. a, b, c
3. a, b

Chapter Review Questions, Section 5.1:

1. b
2. c
3. a
4. b
5. d

Chapter Review Questions, Section 5.2:

1. a, b
2. c
3. b
4. d
5. a
6. a

Chapter Review Questions, Section 5.3:

1. d
2. a
3. b

Chapter Review Questions, Section 5.4:

1. d
2. b
3. b

Chapter Review Questions, Section 5.5 (Mini Case 1: Multiple Sclerosis):

1. a
2. c

Chapter Review Questions, Section 5.6:

1. d
2. c, d

Chapter Review Questions, Section 5.7:

1. a
2. a
3. a, b
4. a

Chapter 6

Learning Check #1 Questions, Section 6.1a:

1. a
2. b
3. a
4. d

Learning Check #2 Questions, Section 6.1a:

1. b
2. d
3. a, c

Learning Check Questions, Section 6.1c:

1. a
2. c
3. b

Learning Check Questions, Section 6.2d:

1. a, b
2. b
3. d
4. b
5. a, d

Learning Check Questions, Section 6.4a:

1. b
2. b
3. b
4. b

Chapter Review Questions, Section 6.1:

1. d
2. c
3. a
4. a
5. c
6. c, d
7. a
8. b
9. a
10. b, d

Chapter Review Questions, Section 6.1(Mini Case 1: Why is grandma always cold?):

1. c, d
2. b, c
3. b

Chapter Review Questions, Section 6.2:

1. d
2. c
3. a
4. a, d
5. a, d

Chapter Review Questions, Section 6.3:

1. b
2. b

Chapter Review Questions, Section 6.4:

1. b
2. a

Chapter 7

Learning Check Questions, Section 7.1b:

1. a
2. d
3. b

Learning Check Questions, Section 7.2e:

1. a
2. b
3. d

Learning Check Questions, Section 7.2f:

1. a, b
2. a, b, d
3. c
4. c

Learning Check Questions, Section 7.3c:

1. c
2. a, d
3. d
4. a

Learning Check Questions, Section 7.4b:

1. a
2. d
3. b

Learning Check Questions, Section 7.5f:

1. b
2. a
3. b

Learning Check Questions, Section 7.6b:

1. b
2. b

Chapter Review Questions, Section 7.1:

1. a
2. d
3. b
4. a, d

Chapter Review Questions, Section 7.2:

1. d
2. c
3. d

Chapter Review Questions, Section 7.3:

1. c
2. b, d
3. b
4. a

Chapter Review Questions, Section 7.4:

1. a
2. c

Chapter Review Questions, Section 7.5:

1. a
2. d

Chapter Review Questions, Section 7.5 (in Mini Case 1: Fractures of the Epiphyseal Plate):

1. a, b
2. c
3. c
4. b

Chapter Review Questions, Section 7.6:

1. b, d
2. a
3. b, d
4. b

Chapter 8

Learning Check Questions, Section 8.1b:

1. a, b
2. a
3. b

Learning Check Questions, Section 8.2a:

1. a
2. c
3. b
4. d
5. b

Learning Check Questions, Section 8.2g:

1. d
2. c
3. a
4. a, c

Learning Check Questions, Section 8.4b:

1. a
2. a
3. c
4. d
5. d

Chapter Review Questions, Section 8.1:

1. c, d

Chapter Review Questions, Section 8.2:

1. a, b
2. a, d
3. c
4. b

Chapter Review Questions, Section 8.3:

1. b, c

Chapter Review Questions, Section 8.3 (Mini Case 1: Low Back Pain):

1. a
2. c, d
3. a, d

Chapter Review Questions, Section 8.4:

1. b
2. b, c
3. c

Chapter 9

Learning Check Questions, Section 9.1b:

1. a, b
2. c
3. b

Learning Check Questions, Section 9.2c:

1. a
2. c
3. b

Learning Check Questions, Section 9.3b:

1. b
2. d
3. b
4. b

Learning Check Questions, Section 9.4d:

1. d
2. b
3. b
4. a, b

Chapter Review Questions, Section 9.1:

1. a
2. a

Chapter Review Questions, Section 9.2:

1. a
2. b

Chapter Review Questions, Section 9.3 (Mini Case 1: Pregnancy):

1. b
2. c
3. b, d

Chapter Review Questions, Section 9.4:

1. b
2. b, c

Chapter 10

Learning Check Questions, Section 10.1b:

1. a, d
2. c
3. b

Learning Check Questions, Section 10.2c:

1. c
2. a

Learning Check Questions, Section 10.3b:

1. a, b
2. b

Learning Check Questions, Section 10.4c:

1. b
2. c
3. a
4. b

Learning Check Questions, Section 10.5:

1. a, b
2. a, c
3. a, b

Learning Check Questions, Section 10.6:

1. a
2. c

Chapter Review Questions, Section 10.1:

1. c
2. a
3. b

Chapter Review Questions, Section 10.2:

1. a
2. a
3. a

Chapter Review Questions, Section 10.3:

1. a, b
2. b
3. a, b

Chapter Review Questions, Section 10.4 (Mini Case 1: Shoulder Dislocation):

1. a
2. b, c
3. b

Chapter Review Questions, Section 10.5:

1. b
2. b, d

Chapter Review Questions, Section 10.6:

1. c
2. b

Chapter 11

Learning Check Questions, Section 11.1:

1. d
2. c
3. a
4. c, d

Learning Check Questions, Section 11.2c:

1. b
2. c
3. b
4. a, b

Learning Check Questions, Section 11.3c:

1. a, b, d
2. a
3. a
4. d

Learning Check Questions, Section 11.4a:

1. b
2. a, b
3. a, c
4. b

Learning Check Questions, Section 11.5d:

1. b, c
2. c, d
3. b
4. a

Learning Check Questions, Section 11.6:

1. a, c
2. b, c, d

Learning Check Questions, Section 11.7:

1. c, d
2. a, b
3. d

Chapter Review Questions, Section 11.1:

1. b
2. d
3. d

Chapter Review Questions, Section 11.2:

1. b
2. c
3. c
4. a
5. b, d
6. a

Chapter Review Question, Section 11.3

1. a

Chapter Review Questions, Section 11.3 (Mini Case 1: Myasthenia Gravis):

1. a, d
2. b, d
3. a, b, d

Chapter Review Questions, Section 11.4:

1. b
2. b
3. b
4. a

Chapter Review Questions, Section 11.5:

1. d
2. a
3. b
4. b
5. b
6. b
7. d

Chapter Review Questions, Section 11.6:

1. b
2. b, d

Chapter Review Questions, Section 11.7:

1. b, c
2. a

Chapter 12

Learning Check Questions, Section 12.1b:

1. a, d
2. b
3. d
4. a

Learning Check Questions, Section 12.2:

1. a
2. b
3. a, b

Learning Check Questions, Section 12.3f:

1. a, b
2. a
3. a, d
4. a, b

Learning Check Questions, Section 12.3h:

1. c
2. b
3. c
4. a

Learning Check Questions, Section 12.3i:

1. b
2. b
3. a, b

Learning Check Questions, Section 12.4d:

1. a
2. a, c, d
3. b, c
4. a

Learning Check Questions, Section 12.4f:

1. a
2. a, b
3. b, d
4. a

Chapter Review Questions, Section 12.1:

1. a, b, d
2. d

Chapter Review Questions, Section 12.2:

1. b, c

Chapter Review Questions, Section 12.3:

1. a
2. a, b

Chapter Review Questions, Section 12.4 (Mini Case 1: Spinal Cord Injury):

1. b
2. a

Chapter 13

Learning Check Questions, Section 13.1b:

1. b
2. a
3. a
4. b

Learning Check Questions, Section 13.2b:

1. b
2. a
3. a, d
4. b, d

Learning Check #1 Questions, Section 13.3a:

1. a
2. a, c
3. a
4. c

Learning Check #2 Questions, Section 13.3a:

1. c
2. b
3. a
4. a

Learning Check Questions, Section 13.4a:

1. a
2. c
3. a

Chapter Review Questions, Section 13.1:

1. a, b, c
2. c, d
3. d
4. c
5. a
6. b

Chapter Review Questions, Section 13.2:

1. c
2. a
3. a
4. b
5. b, d

Chapter Review Questions, Section 13.3:

1. a
2. c
3. b
4. b, d
5. a, b
6. a
7. a, c
8. c
9. c
10. a
11. b
12. a, c

Chapter Review Questions, Section 13.4:

1. b
2. b, d

Chapter Review Questions, Section 13.4 (Mini Case 1: Parkinson's Disease):

1. a
2. a, b, c
3. d
4. b

Chapter 14

Learning Check Questions, Section 14.1b:

1. a
2. d
3. a
4. a
5. b

Learning Check Questions, Section 14.2b:

1. a, d
2. b
3. a
4. a, d

Learning Check 1 Questions, Section 14.2d:

1. c
2. c
3. a, d
4. a, b

Learning Check 2 Questions, Section 14.2h:

1. b
2. b
3. a

Learning Check Questions, Section 14.3b:

1. c
2. b, c
3. c
4. a

Chapter Review Questions, Section 14.1:

1. c, d
2. a
3. a
4. c
5. a, b
6. a
7. a
8. d
9. c
10. a, c
11. c
12. a
13. c

Chapter Review Questions, Section 14.2:

1. a, c
2. c
3. b
4. c
5. a
6. a, d
7. b, c, d
8. a
9. a
10. a
11. c
12. a, b
13. d
14. a
15. a
16. a, b
17. d
18. a
19. b
20. b, c
21. a
22. b
23. a, b

Chapter Review Questions, Section 14.2 (Mini Case 1: Spinal Nerve Compression):

1. a
2. b
3. d
4. c

Chapter Review Questions, Section 14.3:

1. c
2. c
3. a, b
4. c
5. a
6. b
7. c, d

Chapter 15

Learning Check Questions, Section 15.1a:

1. a
2. c
3. a, c
4. a, c

Learning Check Questions, Section 15.2:

1. b
2. a
3. b
4. a

Learning Check Questions, Section 15.3a:

1. a, b
2. c
3. b

Learning Check Questions, Section 15.4a:

1. b
2. d
3. a
4. b

Learning Check Questions, Section 15.4b:

1. b, c
2. b
3. c, d

Learning Check Questions, Section 15.4e:

1. d
2. a
3. a, c

Chapter Review Questions, Section 15.1:

1. b
2. b
3. c
4. b, d
5. b
6. a, d
7. a

Chapter Review Questions, Section 15.2:

1. b
2. b
3. a
4. a, b
5. c

Chapter Review Questions, Section 15.3 (Mini Case 1: Diabetes):

1. b
2. b
3. b

Chapter Review Questions, Section 15.4:

1. a
2. a
3. b
4. b
5. b, d
6. a
7. a, d
8. c
9. b
10. a
11. a, c
12. a
13. a, c
14. a, c
15. a
16. b, c
17. a
18. b
19. b
20. b
21. b
22. a
23. d
24. c
25. b
26. b
27. a
28. a
29. a
30. c
31. a, b, d

Chapter 16

Learning Check Questions, Section 16.1b:

1. a, c
2. c
3. c, d
4. d

Learning Check Questions, Section 16.2:

1. c, d
2. b
3. c

Learning Check Questions, Section 16.3a:

1. a, b
2. c, d
3. d

Learning Check Questions, Section 16.4a:

1. b, d
2. b
3. a, b, c
4. a, b

Chapter Review Questions, Section 16.1 (Mini Case 1: Spinal Cord Injury):

1. a, c
2. a
3. b, c
4. a, c, d

Chapter Review Questions, Section 16.2:

1. a
2. b

Chapter Review Questions, Section 16.3:

1. b, d

Chapter Review Questions, Section 16.4:

1. b
2. a
3. a
4. a

Chapter 17

Learning Check Questions, Section 17.1b:

1. a
2. a
3. b

Learning Check Questions, Section 17.2d:

1. d
2. a
3. b, d
4. b

Learning Check Questions, Section 17.3b:

1. a
2. b
3. a
4. b

Learning Check Questions, Section 17.4b:

1. a, d
2. a, b
3. b
4. a
5. a

Learning Check Questions, Section 17.4m:

1. b
2. b, d
3. b
4. a

Chapter Review Questions, Section 17.1:

1. a, b, d
2. a
3. b

Chapter Review Questions, Section 17.2:

1. b

Chapter Review Questions, Section 17.2 (Mini Case 1: Running Away from a Thunderstorm):

1. b, d
2. c
3. a
4. b, c

Chapter Review Questions, Section 17.3:

1. a
2. b
3. b
4. b, d
5. c

Chapter Review Questions, Section 17.4:

1. b, d
2. a
3. a
4. b, c
5. d
6. d
7. a
8. a, b, c

Chapter 18

Learning Check Questions, Section 18.1h:

1. b
2. c
3. a
4. d

Learning Check Questions, Section 18.2c:

1. a
2. a, b
3. c

Learning Check Questions, Section 18.3e:

1. a
2. a
3. a
4. a, c

Learning Check Questions, Section 18.4a:

1. b
2. d
3. b
4. a, b, c, d

Chapter Review Questions, Section 18.1:

1. c
2. b
3. c
4. d
5. a
6. d
7. b, c
8. b, c
9. d
10. c

Chapter Review Questions, Section 18.2:

1. d
2. b
3. c
4. b, c, d

Chapter Review Questions, Section 18.3:

1. c
2. c, d
3. a
4. d
5. a, c
6. a, d
7. b, d

Chapter Review Questions, Section 18.4:

1. a

Chapter Review Questions, Section 18.4 (Mini Case 1: Blood Typing and Transfusions):

1. c
2. d
3. a, c
4. b

Chapter 19

Learning Check Questions, Section 19.1d:

1. a
2. d
3. c
4. a
5. c

Learning Check Questions, Section 19.1f:

1. a
2. c
3. b
4. b
5. a, b

Learning Check Questions, Section 19.2c:

1. c
2. b
3. a
4. a

Learning Check Questions, Section 19.2e:

1. a
2. b
3. b, c
4. a

Learning Check Questions, Section 19.3c:

1. a
2. b
3. a

Learning Check Questions, Section 19.4d:

1. a
2. d
3. c
4. a

Chapter Review Questions, Section 19.1:

1. b
2. b, c
3. a, d
4. c, d
5. c
6. c
7. d
8. d
9. a
10. c, d

Chapter Review Questions, Section 19.2:

1. d
2. b, d
3. c
4. a, c
5. a
6. b, c
7. c
8. d
9. c
10. b

Chapter Review Questions, Section 19.2 (Mini Case 1: Atrial Flutter):

1. c
2. b
3. a
4. c

Chapter Review Questions, Section 19.3:

1. b
2. b, c
3. b
4. a, c
5. c
6. a, d
7. c
8. d
9. c
10. b

Chapter Review Questions, Section 19.4:

1. c
2. a
3. a
4. b, c
5. a
6. c, d
7. a, c
8. b
9. a, c
10. c
11. d
12. a, b

Chapter 20

Learning Check Questions, Section 20.1i:

1. a
2. b
3. b
4. c
5. b

Learning Check Questions, Section 20.2b:

1. a
2. d
3. a

Learning Check Questions, Section 20.2f:

1. a
2. a
3. b

Learning Check Questions, Section 20.3a:

1. a
2. b

Learning Check Questions, Section 20.4c:

1. a
2. b
3. b, c
4. a

Learning Check Questions, Section 20.5e:

1. a
2. c
3. a
4. d

Learning Check Questions, Section 20.5f:

1. a, d
2. c
3. a
4. b

Learning Check Questions, Section 20.6:

1. a, d
2. a

Chapter Review Questions, Section 20.1:

1. c
2. a
3. a
4. a
5. b
6. c
7. b
8. b

Chapter Review Questions, Section 20.2:

1. a, d
2. b
3. b, d
4. c, d
5. b

Chapter Review Questions, Section 20.2 (Mini Case 1: Peripheral Vascular Disease):

1. b
2. a, b
3. b

Chapter Review Questions, Section 20.3:

1. b
2. b
3. b

Chapter Review Questions, Section 20.4:

1. b, d
2. b
3. a, c
4. d
5. a
6. a
7. c

Chapter Review Questions, Section 20.5:

1. b
2. b, c
3. b
4. b

Chapter Review Questions, Section 20.6:

1. b
2. a
3. b

Chapter 21

Learning Check Questions, Section 21.1b:

1. b
2. b
3. a
4. a

Learning Check Questions, Section 21.2:

1. a
2. b

Learning Check Questions, Section 21.3b:

1. c
2. a, b
3. b, c

Learning Check Questions, Section 21.3e:

1. c
2. b
3. b, c
4. b, c

Learning Check Questions, Section 21.4g:

1. b
2. a
3. c

Learning Check Questions, Section 21.4l:

1. c
2. a
3. a

Chapter Review Questions, Section 21.1:

1. b
2. c

Chapter Review Questions, Section 21.1 (Mini Case 1: Lymphedema):

1. c, d
2. b
3. a
4. b, c

Chapter Review Questions, Section 21.2:

1. a, b
2. b
3. a
4. b

Chapter Review Questions, Section 21.3:

1. c, d
2. b, c
3. a
4. b
5. a
6. a
7. a, b
8. a, b
9. a, b
10. a
11. c, d

Chapter Review Questions, Section 21.4:

1. a, d
2. b, c
3. b
4. a, d
5. b
6. c
7. a
8. b
9. c
10. a, c
11. b
12. a
13. a
14. c
15. b
16. a

Chapter 22

Learning Check Questions, Section 22.1c:

1. c
2. b
3. d
4. a, c, d

Learning Check Questions, Section 22.2:

1. a, b
2. a
3. d
4. c

Learning Check Questions, Section 22.3:

1. c
2. a
3. b

Learning Check Questions, Section 22.4a:

1. a
2. c
3. b
4. b

Learning Check Questions, Section 22.5b:

1. d
2. a

3. b
4. b
5. c

Learning Check Questions, Section 22.6:

1. a
2. c
3. b
4. b

Chapter Review Questions, Section 22.1:

1. d
2. a, c
3. a, c
4. d
5. a
6. c, d
7. b
8. a
9. a
10. c, d
11. b, c
12. a
13. a, c
14. b
15. c
16. a, d
17. a, d
18. d

Chapter Review Questions, Section 22.2:

1. a, d
2. a
3. b
4. a
5. b
6. a
7. d
8. a, b
9. a

Chapter Review Questions, Section 22.3:

1. c
2. a
3. b
4. b

Chapter Review Questions, Section 22.4:

1. b
2. c
3. b
4. b
5. d
6. b
7. a
8. b

Chapter Review Questions, Section 22.5:

1. b
2. b
3. d
4. a
5. c
6. b, d
7. d
8. b
9. b

Chapter Review Questions, Section 22.6 (Mini Case 1: Damage to the Respiratory Control Centers):

1. a, c
2. c
3. d
4. b

Chapter 23

Learning Check Questions, Section 23.1:

1. b
2. c

Learning Check Questions, Section 23.2d:

1. d
2. c
3. a, b
4. b

Learning Check Questions, Section 23.3e:

1. a, b, c
2. b
3. a, b
4. d
5. c

Learning Check Questions, Section 23.3i:

1. c
2. b
3. a, b
4. b

Learning Check Questions, Section 23.3k:

1. d
2. c
3. a

Learning Check Questions, Section 23.3l:

1. b, d
2. d
3. c

Learning Check Questions, Section 23.4c:

1. b, c, d
2. d
3. c

Learning Check Questions, Section 23.5b:

1. a, c
2. a
3. b, c

Chapter Review Questions, Section 23.1:

1. a, b, c, d
2. b
3. a
4. c

Chapter Review Questions, Section 23.2:

1. a, b
2. a, b, d
3. a
4. b
5. a
6. c
7. b
8. b
9. c, d

Chapter Review Questions, Section 23.3:

1. d
2. a
3. b
4. c
5. c
6. a
7. c, d
8. a
9. a
10. c
11. b
12. a
13. a
14. a, b, c
15. c
16. d
17. c
18. a
19. c
20. a, c
21. d
22. a
23. c
24. d
25. c
26. b, c
27. c
28. c
29. a
30. a, c
31. a, b
32. b
33. c
34. b, c
35. c

Chapter Review Questions, Section 23.3 (Mini Case 1: Inflammatory Bowel Disease):

1. b
2. b, d
3. a, b
4. c

Chapter Review Questions, Section 23.4:

1. a, c
2. a
3. c
4. d
5. b, c
6. a
7. b
8. d
9. a
10. c
11. b

Chapter Review Questions, Section 23.5:

1. b, c
2. c
3. a
4. a, c
5. c
6. c

Chapter 24

Learning Check Questions, Section 24.1c:

1. a
2. b

Learning Check Questions, Section 24.2c:

1. b
2. c
3. d
4. a

Learning Check Questions, Section 24.3c:

1. a
2. b

Learning Check Questions, Section 24.4a:

1. b
2. a

Learning Check Questions, Section 24.5c:

1. b
2. b
3. a, b, d

Chapter Review Questions, Section 24.1:

1. a
2. b
3. b, c
4. c
5. a
6. c

Chapter Review Questions, Section 24.2 (Mini Case 1: Energy Expenditure During Exercise):

1. c
2. b
3. a, c
4. a, b, c
5. a, b
6. d

Chapter Review Questions, Section 24.3:

1. a, b

Chapter Review Questions, Section 24.4:

1. c
2. b, c
3. d
4. b, c, d

Chapter Review Questions, Section 24.5:

1. d
2. b
3. c
4. a
5. a
6. a, d

Chapter 25

Learning Check Questions, Section 25.1:

1. a, d
2. c
3. a
4. d

Learning Check Questions, Section 25.2c:

1. a
2. b
3. c
4. b

Learning Check Questions, Section 25.3g:

1. a
2. a, b
3. a
4. c

Learning Check Questions, Section 25.4c:

1. a, c
2. a, c, d
3. b, c
4. d

Learning Check Questions, Section 25.5c:

1. b, c
2. d
3. a, c

Learning Check Questions, Section 25.6c:

1. a
2. d
3. c
4. a

Learning Check Questions, Section 25.7:

1. a
2. a
3. d
4. d

Chapter Review Questions, Section 25.1:

1. c

Chapter Review Questions, Section 25.2:

1. c, d
2. a, b
3. b
4. a
5. b
6. c
7. b
8. b

Chapter Review Questions, Section 25.3:

1. a
2. a
3. d
4. a
5. b
6. b
7. c
8. c
9. a
10. b
11. b
12. c
13. a, c

Chapter Review Questions, Section 25.4 (Mini Case 1: Why Am I Having Trouble Exercising?):

1. c, d
2. d
3. c
4. a

Chapter Review Questions, Section 25.5:

1. d
2. a
3. a

Chapter Review Questions, Section 25.6:

1. a
2. b
3. a
4. d
5. a, c
6. a, c

Chapter Review Questions, Section 25.7:

1. c
2. a

Chapter 26

Learning Check Questions, Section 26.1b:

1. b
2. a, d
3. a
4. b, d

Learning Check Questions, Section 26.1e:

1. b
2. a, c
3. b

Learning Check Questions, Section 26.2c:

1. b, d
2. d
3. d

Learning Check Questions, Section 26.3b:

1. c
2. a
3. d
4. b

Learning Check Questions, Section 26.4b:

1. a
2. c
3. a, c
4. a

Learning Check Questions, Section 26.5e:

1. a
2. a, d

Chapter Review Questions, Section 26.1:

1. a
2. c, d
3. c
4. b, c
5. a, b

Chapter Review Questions, Section 26.2:

1. b, c
2. a, d
3. a, d
4. a, b
5. b
6. a, c
7. b, d

Chapter Review Questions, Section 26.3:

1. c
2. a, d
3. a
4. d

Chapter Review Questions, Section 26.4 (Mini Case 1: Firefighter):

1. b
2. a, c
3. b
4. b

Chapter Review Questions, Section 26.5:

1. b
2. b
3. a, c
4. a
5. b

Chapter 27

Learning Check Questions, Section 27.1:

1. a
2. d
3. a, c, d
4. a

Learning Check Questions, Section 27.2c:

1. d
2. b
3. b
4. b

Learning Check Questions, Section 27.3e:

1. d
2. c
3. b
4. c

Learning Check Questions, Section 27.4b:

1. c
2. d
3. c
4. b

Learning Check Questions, Section 27.5d:

1. c
2. b
3. b
4. b

Learning Check Questions, Section 27.6e:

1. c
2. b
3. c
4. d

Learning Check Questions, Section 27.7:

1. b, c
2. d
3. d
4. a

Learning Check Questions, Section 27.8g:

1. a
2. a
3. c
4. d

Chapter Review Questions, Section 27.1:

1. b
2. a

Chapter Review Questions, Section 27.2:

1. b
2. d
3. c

Chapter Review Questions, Section 27.3:

1. b, c, d
2. d
3. b
4. b
5. c
6. a

Chapter Review Questions, Section 27.4:

1. c, d
2. d
3. c
4. b
5. a, d
6. b
7. d

Chapter Review Questions, Section 27.5:

1. c
2. d
3. c
4. a, b, c, d

Chapter Review Questions, Section 27.5 (Mini Case 1: Seminal Vesiculitis):

1. a, b, d
2. a
3. b

Chapter Review Questions, Section 27.6:

1. b
2. d
3. a, b, d

Chapter Review Questions, Section 27.7:

1. a
2. b
3. b

Chapter Review Questions, Section 27.8:

1. a
2. a, d
3. b
4. a, d
5. d
6. c
7. a, d
8. d
9. b, d
10. b

Chapter 28

Case Study Questions, Hypertension, Section 28.1:

1. c, d
2. b
3. a, b, c, d
4. d
5. b, d
6. b
7. d
8. a
9. a, c
10. a, b
11. a
12. b
13. c
14. a
15. a
16. c
17. a
18. b
19. a
20. b
21. a
22. b, c

Case Study Questions, Cystic Fibrosis, Section 28.2:

1. a
2. d
3. a, b, d
4. b
5. c
6. b
7. a
8. b
9. b
10. b
11. a
12. b
13. a

Case Study Questions, Pulmonary Embolism, Section 28.3:

1. d
2. b
3. a
4. b
5. b
6. b
7. b
8. c
9. b
10. c
11. a

Case Study Questions, Femoroacetabular Impingement, Section 28.4:

1. b
2. a
3. a
4. a
5. a
6. b
7. b
8. a
9. b
10. a
11. c
12. b

Case Study Questions, Arthritis, Section 28.5:

1. a
2. b, c
3. a
4. b
5. b
6. a
7. c
8. a

Case Study Questions, Cardiac Failure, Section 28.6:

1. a
2. a
3. c
4. c
5. b
6. b
7. b
8. b
9. a
10. d
11. a
12. a
13. d
14. a

Case Study Questions, Stroke, Section 28.7:

1. c
2. a
3. b
4. a
5. b
6. a
7. c
8. a
9. a

Case Study Questions, Type 2 Diabetes Mellitus, Section 28.8:

1. b
2. b
3. a
4. b
5. c
6. c
7. a
8. b
9. a
10. c
11. a
12. b

Case Study Questions, Chronic Obstructive Pulmonary Disease, Section 28.9:

1. c
2. b
3. b
4. b
5. b
6. a
7. d
8. b
9. b
10. a
11. b
12. d
13. a
14. d
15. a
16. b
17. a
18. a, d

Case Study Questions, Pregnancy, Section 28.10:

1. c
2. b
3. b
4. a
5. c
6. b
7. d
8. a
9. a
10. c
11. b
12. d
13. a
14. c
15. a
16. a
17. a

Glossary

A

A band the portion of the sarcomere that contains thick filaments; in some portions, the thick filaments overlap with the thin filaments; appears as dark band under a light microscope.
abdominal aortic plexus (Latin: *plexus* - braid) an autonomic nerve network that contains the celiac, superior mesenteric, and inferior mesenteric plexuses; innervates much of the abdominal viscera.
abducens nerve (CN VI) (Latin: *abducens* - abducting) the sixth cranial nerve; contributes to eye movement via control of specific extraocular muscles.
abduction (Latin: *abductio* - movement away from center) motion in the coronal plane that pulls a structure away from the midline of the body.
abductor digiti minimi hyothenar muscle; abducts the little finger.
abductor pollicis brevis (Latin: *brevis* - short) thenar muscle; abducts the thumb.
abductor pollicis longus (Latin: *longus* - long) deep posterior muscle of the forearm that inserts into the first metacarpal.
ABO blood group a classification system used in blood typing, based on whether "A" and/or "B" antigens are present on the cell membranes of a person's erythrocytes.
absolute refractory period (Latin: *refractarius* - to break into pieces) phase of the refractory period during which no action potential can start, regardless of stimulus strength.
absorptive state (*aka* fed state) the period of time, up to about four hours during and after ingesting food, during which nutrients are present in the GI tract and are being digested and absorbed.
accessory digestive organs organs of the digestive system that food does not pass through, but that contribute to the digestion process; consist of the teeth, tongue, salivary glands, liver, gallbladder, and pancreas.
accessory duct (*aka* duct of Santorini; accessory pancreatic duct) a small duct, which runs from the main pancreatic duct into the duodenum, without going through the hepatopancreatic sphincter; not present in all people.
acetabular labrum fibrocartilage lip surrounding the outer margin of the acetabulum.
acetabulum (Latin: *acetabulum* - a shallow cup) deep, cup-shaped cavity on the lateral side of the hip bone; formed by the convergence of the ilium, pubis, and ischium.
acetyl coenzyme A (acetyl CoA) (Latin: *acetum* - vinegar) a two-carbon molecule produced from pyruvate, amino acids, and fatty acids, which enters the citric acid cycle.
acetylcholine (ACh) a neurotransmitter used in the central nervous system and in neuromuscular junctions; binds to receptors on the motor end-plate to trigger depolarization.
acetylcholinesterase (AChE) the enzyme that decomposes acetylcholine (Ach) in the synaptic cleft after Ach has transmitted its message to the muscle cell or postsynaptic neuron.
acid (Latin: *acidus:* sour) a compound that releases hydrogen ions (H^+) into a solution.
acidosis (Latin: *acidus* - sour; *osis* - a state) a condition in which the blood pH is below 7.35.
acromegaly (Greek: *akron* - tip; *megas* – large) disorder that results in bone growth in the hands, feet, and face in response to excessive levels of GH during adulthood (after the ossification of the long bones has occurred).
acromial end of the clavicle (*aka* lateral end of the clavicle) (Greek: *akron* - tip; ōmos – shoulder) lateral end of the clavicle; articulates with the acromion of the scapula.
acromioclavicular joint (*aka* AC joint) (Greek: *akron* - tip; ōmos - shoulder; Latin: *clavicula* - a small key) point of articulation between the acromion of the scapula and the lateral end of the clavicle.
acromion (Greek: *akron* - tip; *ōmos* - shoulder) broad, flat projection that extends laterally from the scapular spine to form the tip of the scapula.
acrosomal reaction (Greek: *akro* - at the end; *soma* - the body) the release of enzymes when the acrosome ruptures.
acrosome (Greek: *akro* - at the end; *soma* - the body) an enzyme filled cap that covers the extremely compact nucleus of the nuclear head.
actin protein subunit that makes up much of the thin myofilaments in a sarcomere.
action potential (*pl:* action potentials) (Latin: *actio* - to do; *potentia* - power) electrical signal unique to neurons and muscle fibers; caused by a change in voltage across a cell membrane in response to a stimulus. These electrochemical impulses allow the neurons to communicate throughout the body.
activation energy (Latin: *actus* - to do; Greek: *energeia* - in work) amount of energy greater than the energy contained in the reactants, which must be overcome for a reaction to proceed.
active immunity immunity obtained via antigen exposure, which evokes an immune response by the person's own immune system; this results in the production of antibodies and immunological memory against the antigen.
active transport (Latin: *actio* - to do; *transporto* - to carry over) form of transport across the cell membrane that requires energy from the cell.
acuity (French: *acutus* - sharpness) a term used to describe how accurately our brain can understand a stimulus; often applied to the sense of vision.
acute (Latin: *acutus* – sharp) short-term; a term describing an illness with rapid onset and severe symptoms.
adaptation (Latin: *adapto* - to adjust) a process by which a sensory neuron stops or slows its rate of neurotransmitter release even though it is still receiving graded potentials.
adaptive immune response (Latin: *adaptare* - adjust) the relatively slow, specific immune response, directed against a particular type of pathogenic organism; performed by B and T lymphocytes and antigen-presenting cells.
adduction (Latin: *adduco* - to bring toward) motion in the coronal plane that brings a structure toward or past the midline of the body.
adductor brevis (Latin: *brevis* - short) muscle that laterally and medially rotates the thigh.
adductor longus (Latin: *longus* – long) muscle that laterally and medially rotates, and flexes the thigh.
adductor magnus (Latin: *magnus* - large) muscle that laterally and medially rotates and flexes the thigh.
adductor pollicis (*aka* thenar muscle) muscle that adducts the thumb.
adenylyl cyclase enzyme activated by a G protein which converts ATP to cAMP.
adipocytes (*sing*: adipocyte) (*aka* fat cells) (Latin: *adip* - fat; Greek: *kytos* - cell) specialized lipid storage cells.
adipokines (Greek: *lipo* - fat; *kinēsis* - movement) hormones produced and secreted by adipose tissue, which have roles in the regulation of appetite, metabolism, inflammation, blood pressure, and the stoppage of bleeding.
adipose tissue (Latin: *adip* - fat; French: *tissu* - woven) specialized areolar tissue largely composed of adipocytes.
adrenal artery (Latin: *ad* - to; *ren* - kidney) an artery that branches off of the abdominal aorta, and provides blood to the adrenal (suprarenal) glands.
adrenal cortex (*aka* suprarenal cortex) (Latin: *ad* - to; *ren* – kidney) the outer layer of the adrenal gland, which secretes several hormones called corticoids; these steroid hormones regulate glucose balance, electrolyte balance, and some reproductive function.
adrenal gland (*aka* suprarenal gland) (Latin: *ad* - to; *ren* - kidney; *glans* - acorn) a gland that sits on top of each kidney; the medulla secretes epinephrine and norepinephrine as sympathetic hormones, and the cortex secretes steroid hormones that regulate glucose and electrolyte balance and help to supplement certain functions of the sex hormones.
adrenal glands (*aka* suprarenal glands) (Latin: *ad* - to; *ren* - kidney; *glans* – acorn) endocrine glands composed of the adrenal cortex and adrenal medulla; located on top of the kidneys;

cortex secretes steroid hormones and medulla secretes epinephrine and norepinephrine.
adrenal medulla (*aka* suprarenal medulla) (Latin: *ad* - to; *ren* - kidney; *medulla* – marrow) the internal portion of the adrenal gland; secretes the hormones epinephrine and norepinephrine.
adrenal medulla (*aka* suprarenal medulla) (Latin: *ad* - to; *ren* - kidney; *medulla* - marrow) the inner layer of the adrenal glands; functions as a component of the sympathetic nervous system response; secretes epinephrine and norepinephrine.
adrenal vein (Latin: *ad* - to; *ren* - kidney) a vein that drains blood from the adrenal (suprarenal) glands; the right adrenal vein drains directly into the inferior vena cava, and the left adrenal vein drains into the left renal vein.
adrenergic (Latin: *ad* - to; *ren* - kidney; Greek: *ergon* - work) a type of synapse at which norepinephrine (noradrenaline) or epinephrine (adrenaline) is released, or a type of receptor that binds to norepinephrine or epinephrine.
adrenocorticotropic hormone (ACTH) (*aka* corticotropin) (Latin: *ad* - to; *ren* - kidney; Greek: *trophē* - nurturing) a hormone produced by the anterior pituitary gland; stimulates the adrenal cortex to secrete corticosteroid hormones, such as cortisol.
adventitia the outermost layer of tubular organs of the GI tract that lie outside of the abdominopelvic cavity, such as the pharynx and esophagus; consists of dense connective tissue.
aerobic respiration (*aka* oxidative phosphorylation) (Greek: *aēr* - air; Latin: *respiratio* – breathe) breakdown of pyruvate in the presence of oxygen; results in ATP formation.
afferent (Latin: *afferens* - to bring to) nerves that bring input toward the central nervous system.
afferent arterioles (Latin: *arteriola* - an artery) arterioles that transport oxygen-rich blood from the cortical radiate (intralobular) arteries to the glomerulus in each nephron.
afferent branch (Latin: *afferens* - to bring to) the component of a reflex arc that transports sensory information into the central nervous system for integration.
afferent lymphatic vessels (Latin: *lympha* - water) lymphatic vessels that drain lymph into a lymph node to be filtered.
affinity (Latin: *affinitatem* – neighborhood) the strength with which a molecule binds to another molecule or to its binding site; often used to describe the strength with which hemoglobin binds to oxygen or an enzyme binds to its substrate.
afterload the force the ventricles must develop to effectively pump blood against the resistance in the vessels.
agglutination (Latin: *agglutinare* - to glue) the clumping of erythrocytes containing certain antigens by antibodies that specifically bind to those antigens
agonist (*aka* prime mover) (Greek: *agōn* – contest) prime mover whose action is responsible for a particular movement.
agranular leukocytes (Greek: *leuko* - white; *kytos* - cell) leukocytes (monocytes and lymphocytes) that contain few cytoplasmic granules.
ala (*pl*: alae) (Latin: *alaris* - wing) small, rounded, cartilaginous structure that forms the lateral wall of a nostril.
alar cartilage (*aka* major alar cartilage; greater alar cartilage; lower lateral cartilage) (Latin: *alaris* - wing; *cartilaginem* - gristle) a mass of cartilage that helps to support the apex of the nose and provide the shape of the nares.
albumins (Latin: *albumen* - white of egg) a group of plasma proteins that maintains the osmotic pressure in the blood; makes up the majority of plasma proteins.
aldosterone a mineralocorticoid hormone produced by the zona glomerulosa of the adrenal cortex; involved in regulation of sodium and potassium levels in the blood and of blood pressure.
alkalosis (Latin: *alkali* - soda; *osis* - a state) a condition in which the blood pH is above 7.45.
alpha (α)-adrenergic receptor (Latin: *ad* - to; *ren* - kidney; Greek: *ergon* - work) one class of receptors of the adrenergic system; binds epinephrine and norepinephrine; consists of three subtypes: α1, α2, and α3.
alpha cell a pancreatic islet cell type that produces the hormone glucagon.
alveolar dead space (Latin: *alveolus* - basin) the volume of air in nonfunctional alveoli, which does not participate in gas exchange.
alveolar duct (Latin: *alveolus* - basin) a small tubular organ that branches off of a respiratory bronchiole and opens into a cluster of alveoli called an alveolar sac.
alveolar macrophages phagocytic cells found in the alveoli of the lungs, which engulf and destroy pathogens and debris.
alveolar pores (*aka* pores of Kohn; interalveolar connections) (Latin: *alveolus* - basin) openings in the walls of the alveoli that allow air to flow between adjacent alveoli.
alveolar process of the maxilla (Latin: *alveus* - cavity; *processus* - advance; *maxilla* - jawbone) curved, inferior margin of the maxillary bone; forms the upper jaw and contains the upper teeth.
alveolar sac (Latin: *alveolus* - basin) a cluster of alveoli that branches from an alveolar duct.
alveolar ventilation (*aka* alveolar ventilation rate, AVR) (Latin: *alveolus* - basin) the volume of air that reaches the respiratory membrane for gas exchange per unit of time; often expressed in units of mL/min.
alveoli (Latin: *alveolus* - basin) rounded sacs found in clusters within the mammary lobules of the breast; composed of cells that secrete milk to nourish a baby; (Note: the term *alveoli* also refers to saclike structures in the lungs).
alveolus (*pl*: **alveoli**) (Latin: *alveolus* – basin) a small, rounded, saclike structure that performs gas exchange in the lungs; consists of simple squamous epithelium for rapid gas exchange with the blood of the pulmonary capillaries.
amino acid (Latin: *acidus* - sour) building block of proteins; characterized by an amino and carboxyl functional groups and a variable side-chain.
amphiarthrosis (*aka* amphiarthrotic joint) (Greek: *amphi* - both sides; *arthrōsis* - joint) joint at which slight motion is possible.
amphipathic (Greek: *amphi* - on both sides; *pathos* - feeling) describes a molecule that has two sides with different polarity, and therefore two sides with differences in water solubility.
ampulla (*pl*: ampullae) () an expanded structure at the base of each semicircular canal containing the crista ampullaris, the organ that responds to rotational movement of the head 000; the middle section of a uterine tube, which lies between the infundibulum and isthmus; common site of fertilization.
amygdala (*pl*: amygdalae) (Greek: *amygdalē* – almond) a midbrain nucleus located within the temporal lobe of the cerebrum; involved in memory and emotional behavior.
anabolic reactions (Greek: *anabole* - building up) chemical reactions in the body that chemically combine smaller molecules into larger ones.
anabolism (Greek: *anabolē* - to build up) process of building new molecules.
anal canal (Latin: *anus* – ring) the short portion of the large intestine between the rectum and anus.
anal column (*aka* Columns of Morgagni; Morgagni's columns) (Latin: *anus* - ring) a ridge in the mucosa of the anal canal, where the anal canal meets the skin around the anus.
anal sinus (*aka* rectal sinus) (Latin: *anus* - ring; *sinus* - cavity) an indentation between anal columns, which produces mucus for lubrication of the feces as they exit the body.
anal triangle (Latin: *tri* - three; *angulus* - angles) posterior triangle of the perineum; contains the anus.
anaphase (Greek: *ana* - up; *phasis* - appearance) third stage of mitosis (and meiosis), during which sister chromatids separate and move toward the poles of the dividing cell.
anatomical dead space (Greek: *ana* – up; *tomia* - cutting) the volume of air occupying the airway that does not reach the alveoli, and does not participate in gas exchange.
anatomical neck (Greek: *ana* – up; *tomia* - cutting) region of the humerus where the smooth head meets the rest of the bone.
anatomical position (Greek: *ana* - up; *tomia* - cutting) standard position to reference when describing locations and directions on the human body.
anatomical sphincter (Latin: *sphincter* - contractile muscle) a ring of smooth or skeletal muscle that encircles the lumen of a vessel or hollow organ; regulates the flow of a substance via contraction and relaxation.

anatomy (Greek: *ana* - up; *tomia* – cutting) science that studies the form and composition of the structures of the body.
anchoring junction (*pl*: anchoring junctions) mechanically attaches adjacent cells to each other or to the extracellular matrix.
anconeus small muscle that acts as an extensor of the forearm.
androgens (Greek: *andro* - male; *genes* - produce) a group of hormones made from cholesterol.
anemia (Greek: *an* - lack of; *haima* - blood) a deficiency in the oxygen-carrying capacity of the blood, due to a lack of sufficient red blood cells or hemoglobin.
angiogenesis (Greek: *angeion* - blood vessel; *genesis* - production) the formation of new blood vessels from other blood vessels in the body.
angiotensinogen (Greek: *angeion* - vessel of the body; *genus* - produced) the inactive precursor for angiotensin I, which is synthesized by the liver and released into the blood; converted into active angiotensin I by the enzymes renin; an intermediate in the renin-angiotensin system, which functions to increase blood pressure.
angle of the mandible (Latin: *angulus* - angle; *mando* - to chew) corner of the mandible, where body and ramus meet.
angle of the rib (Latin: *angulus* - angle) portion of the rib body with the greatest curvature.
anion (*pl*: anions) (Greek: *anion* - going up) negatively charged ion.
ankle joint joint formed inferiorly by the articulation between the talus bone of the foot, and superiorly by the distal end of the tibia, medial malleolus of the tibia, and lateral malleolus of the fibula.
annular ligament (Latin: *anulus* – ring) ligament that encircles and supports the head of the radius at the proximal radioulnar joint.
antagonist muscle whose action is the opposite of an agonist.
antebrachium (Latin: *ante* - before; *brachium* - arm) the portion of the upper limb that is between the elbow and wrist joint.
anterior (*aka* ventral) (Latin: *ante* - in front) describes the front or direction toward the front of the body.
anterior arch (Latin: *anterior* - earlier; *arcus* - bow) anterior portion of the ring-shaped C1 vertebra.
anterior cardiac veins (Latin: *anterior* - before; Greek: *cardiac* - *kardia* meaning heart) cardiac veins found on the anterior surface of the heart.
anterior cavity (*aka* ventral cavity) (Latin: *cavus* - hollow) large body cavity located anterior to the spine; includes the pleural cavities for the lungs, pericardial cavity for the heart, and peritoneal cavity for the abdominal and pelvic organs.
anterior cerebral artery (Latin: *ante* - before; *cerebro* - brain) an artery that branches from the internal carotid artery; supplies blood to portions of the frontal and parietal lobes of the cerebrum.
anterior columns (*aka* ventral columns, anterior funiculi) (Latin: *ante* - front) white matter located between the anterior horns of the spinal cord; composed of groups of axons of both ascending and descending tracts.
anterior communicating artery (Latin: *ante* - before) an artery that unites the anterior cerebral arteries in the cerebral arterial circle; supplies blood to portions of the frontal lobe of the cerebrum.
anterior compartment of the thigh area that includes muscles that flex the thigh and extend the knee.
anterior cranial fossa (*sing*: fossum) (Latin: *anterior* - earlier; *fossa* - trench; Greek: *kranion* - skull) most anterior and shallowest cranial fossa making up the base of the skull; extends from the frontal bone to the lesser wing of the sphenoid bone.
anterior cruciate ligament (ACL) (Latin: *ante* - in front; *cruciatus* - resembling cross; *ligamentum* - a band) intracapsular ligament that extends from anterior, superior surface of the tibia to the inner aspect of the lateral condyle of the femur.
anterior horn (*aka* ventral horn) (Latin: *ante* - front) a region of gray matter in the spinal cord, which contains multipolar motor neurons.
anterior inferior iliac spine (*aka* AIIS) (Latin: *ante* - before; *inferior* - lower; *spina* – backbone) small protuberance on the anterior margin of the ilium, inferior to the anterior superior iliac spine.
anterior interventricular artery (*aka* left anterior descending artery, LAD) (Latin: *anterior* - before; Greek: *cardiac* - *kardia* meaning heart) major branch of the left coronary+D4vessels that parallel the small cardiac arteries and drain the anterior surface of the right ventricle; bypass the coronary sinus and drain directly into the right atrium.
anterior interventricular sulcus (Latin: *anterior* - before; Greek: *cardiac* - *kardia* meaning heart) the groove in the heart wall in which the anterior interventricular artery sits.
anterior median fissure (Latin: *ante* - front; *medianus* - middle; *fissura* - cleft) a deep midline groove in the anterior spinal cord, which marks the separation between the right and left sides of the cord.
anterior root (*aka* ventral root) (Latin: *ante* - front) a bundle of axons of multipolar motor neurons that transmits nerve impulses from the anterior horn of the spinal cord to the skeletal muscles; briefly runs through a spinal nerve along with a posterior root.
anterior superior iliac spine (*aka* ASIS) (Latin: *ante* - before; *superior* - upper; *spina* - backbone) rough, rounded projection on the anterior end of the iliac crest.
anterior thalamic nuclei (Latin: *ante* - front; *nucleus* - a little nut) groups of neuron cell bodies in the thalamus that contribute to the function of the limbic system.
anterior tibial artery (Latin: *ante* - before; *tibia* - shinbone) an artery that branches off of the popliteal artery; provides blood to the muscles and skin of the anterior and lateral areas of the leg, and continues as the dorsalis pedis artery in the foot.
anterior tibial vein the anterior of the two tibial veins, which drains blood from the dorsal venous arch in the foot and empties it into the popliteal vein.
antibodies (*aka* immunoglobulins, gamma globulins) (Greek: *anti* - opposite) proteins produced by B lymphocytes that defend the body from infection and disease, by acting against specific antigens on pathogens such as bacteria, viruses, and fungi via direct attack, inflammation, or complement activation.
antibody-mediated responses (Greek: *anti* – against) a set of actions of the adaptive immune response that involves antibodies traveling to an infection site and organizing an immune attack on pathogenic organisms.
anticoagulant (Greek: *anti* - opposite; Latin: *coagulum* - to condense) a substance, such as heparin, that inhibits blood coagulation in healthy endothelial cells.
anticodon (*pl*: anticodons) (Greek: *anti* - against) sequence of three nucleotides on a tRNA molecule that is complementary to three nucleotides (a codon) on an mRNA molecule.
antidiuretic hormone (ADH) (*aka* ADH, vasopressin) (Greek: *anti* - against; *diouretikos* - promoting urine) a peptide hormone, produced by the hypothalamus and released by the posterior pituitary gland, that promotes the reabsorption of water in the distal convoluted tubules and collecting ducts of the nephrons; this decreases the urine volume and increases the blood volume and blood pressure.
antigen (Greek: *anti* - opposite; *genes* - produced) a molecule that evokes an immune response in the body, since it is recognized as being "non-self" or foreign by the immune system.
antigen-binding site (Greek: *anti* - against) the portion of a receptor on the cell membrane of a T cell, B cell, or antibody, which is specific for binding to a particular antigen.
antigenic determinant (*aka* epitope) (Greek: *anti* - against) the portion of an antigen that an antibody or lymphocyte receptor recognizes and binds to; forms a complex with a major histocompatibility complex (MHC) protein for antigen presentation to T lymphocytes.
antigen-presenting cells (APCs) (*aka* accessory cells) (Greek: *anti* - against) cells that display antigens on their own cell membranes, bound to major histocompatibility complex proteins, to help T cells to recognize their antigens.
antigens (Greek: *anti* - against) molecules that evoke an immune response in the body, since they are recognized as being foreign by B and T lymphocytes.
antiporters secondary active transporters that move two substances across the membrane in opposite directions.

antithrombin (Greek: *anti* - opposite) a plasma protein that acts as an anticoagulant by inactivating thrombin and other clotting factors.
antrum (Greek: *antron* - a cave) a cavity inside a mature (tertiary) ovarian follicle that is filled with fluid.
anulus fibrosus (Latin: *anulus* - ring; *fibra* - fiber) tough, fibrous outer layer of the intervertebral disk; firmly anchored to the outer margins of the adjacent vertebrae.
aorta the largest artery of the systemic circuit and the body, which transports blood from the left ventricle to all body tissues except the respiratory tissues of the lungs.
aortic arch the curved portion of the aorta, which lies between the ascending aorta and the descending aorta.
aortic baroreceptors pressure receptors found in the ascending aorta, which detect changes in blood pressure and relay the information to the cardiovascular center in the brain in order to maintain blood pressure within homeostatic range.
aortic hiatus (Latin: *hiatus* - an opening) an opening in the diaphragm through which the aorta passes from the thorax into the abdomen.
aortic semilunar valve (*aka* aortic valve) (Latin: *semi* - half; *lunaris* - moon) valve in the heart located between the left ventricle and the aorta; prevents backflow of blood from the aorta into the left ventricle while the heart is relaxing.
apex (Latin: *apex* - summit or tip) the "point" of the heart at the inferior aspect 000; the tip of the external nose, which protrudes more anteriorly than any other portion of the nose 000; the pointed region at the superior tip of each lung.
apical (Latin: *apex* - summit or tip) generally, the part of a cell or tissue facing an external space or internal environment.
apnea a temporary stoppage or absence of breathing.
apneustic center one of the control centers for respiration in the pons, which activates the neurons of the dorsal respiratory group, to regulate the depth of inhalation.
apocrine secretion (Greek: *apo-krinō* - to separate; Latin: *secerno* - to separate) type of nonvesicular secretion which results in the discharge of a portion of the cytoplasm.
aponeurosis broad, tendonlike sheet of connective tissue that may fuse with the epimysium, connecting to another skeletal muscle or to a bone.
apoptosis (*aka* programmed cell death) (Greek: *apo* - off; *ptosis* - a falling) purposeful cell death.
appendicular (Latin: *appendo* - to hang something onto something) referring to the arms and legs.
appendicular skeleton (Latin: *appendo* - to hang something onto something; Greek: *skeletos* - skeleton) the 126 bones of the upper and lower limbs and the girdle bones that attach the limbs to the skeleton.
appendix (*aka* vermiform appendix) (Latin: *appendo* - to hang something onto something) a tubular lymphatic organ attached to the cecum of the large intestine.
appositional bone growth (Latin: *appono* - to place at) bone growth that increases the bone's diameter.
appositional cartilage growth (Latin: *appono* - to place at) cartilage growth where the cartilage grows wider as mesenchymal cells in the perichondrium differentiate into chondroblasts and then secrete matrix around themselves.
aquaporin (*aka* AQP, water channels) (Latin: *aqua* - water; *porus* - a pore) a channel that transports water through the plasma membrane of a cell; activated in the collecting ducts of the kidney by ADH.
aqueous (Latin: *aqua* - water) watery.
aqueous humor (Latin: *aqua* - water; *umor* - liquid) watery fluid that fills the anterior cavity of the eye.
arachnoid mater (Greek: *arachnē* - cobweb; *eidos* - resemblance; Latin: *mater* - mother) the middle layer of the three meninges; a thin, weblike membrane that forms a loose sac around the brain and spinal cord.
arbor vitae (Latin: *arbor* - tree; *vita* - life) the inner white matter tracts that form a treelike shape in the cerebellum.
arcuate arteries (Latin: *arcuatus* - bent like a bow) arteries that transport blood from the interlobar arteries to the cortical radiate (intralobular) arteries in the kidney; form arches around the bases of the renal pyramids.
arcuate fibers (Latin: *arcuatus* - bent like a bow) short, U-shaped nerve tracts that connect association areas within a single lobe of the cerebrum.
arcuate vein (Latin: *arcuatus* - bent like a bow) a vein that transports oxygen-poor blood from the cortical radiate (intralobular) veins to the interlobar veins in the kidney; arches around the base of each renal pyramid.
areolar tissue (*aka* loose connective tissue) (Latin: *area* - space; French: *tissu* - woven) a type of connective tissue characterized by widely spaced fibers and interstitial fluid.
arterial anastomosis (Greek: *anastomosis* - opening) a blood flow pattern in which multiple arteries converge to supply their blood to a common capillary bed.
arteries (*sing*: **artery**) (Greek: *arteria* - pipe) blood vessels that carry blood away from the heart.
arteriole (*aka* resistance vessel) a blood vessel with a small diameter that transports blood from an artery to a capillary.
arteriovenous anastomosis (Greek: *anastomosis* - opening) a blood vessel that directly connects a small artery to a small vein in the hands and feet, in order to bypass a capillary bed; in open position, used to maintain core body temperature when a person becomes too cold.
articular capsule (Latin: *articulo* - to articulate; *capsula* - box) structure made of fibrous connective tissue that forms the walls of the joint cavity of a synovial joint.
articular cartilage (Latin: *articulo* - to articulate; *cartilago* - gristle) thin layer of hyaline cartilage covering the epiphyses, or articulating surfaces of the bones, at a synovial joint; reduces friction and absorbs shocks.
articular disc (*aka* meniscus) (Latin: *articulo* - to articulate) fibrocartilage structure located between the bones of some synovial joints that strongly unites the bones; provides padding or smooths motion between the bones.
articular tubercle (Latin: *articulatio* - a forming of vines; *tuberculum* - a knob) smooth ridge located immediately anterior to the mandibular fossa; makes up part of the temporomandibular joint.
articulation (Latin: *articulo* - to articulate) a joint where two bone surfaces come together.
artificial pacemaker medical device that transmits electrical signals to the heart to ensure that it contracts and pumps blood to the body.
arytenoids small, paired cartilages of the larynx that help to control the tension of the vocal cords.
ascending aorta (Latin: *ascendere* - to climb up) the first portion of the aorta, which extends upward out of the left ventricle; transports blood between the left ventricle and the aortic arch.
ascending colon (Latin: *ascendere* - to climb up) the region of the colon between the cecum and the transverse colon, which runs upward in the right side of the abdominopelvic cavity.
ascending limb (Latin: *ascendere* - to climb up) the portion of a nephron loop (loop of Henle) that transports filtrate from the descending limb of the nephron loop to the distal convoluted tubule (DCT) in each nephron.
ascending tracts (Latin: *ascendere* - to climb up) axons within the spinal cord that transmit sensory information toward the brain.
association areas (Latin: *associo* - to join to) regions of the brain that allow for specific recognition.
association tracts (Latin: *associo* - to join to) small tracts of white matter that run between association areas in the same cerebral hemisphere.
astrocytes (Greek: *astron* - star; *kytos* - cell) star-shaped glial cells that maintain concentration of chemicals in the extracellular space, remove excess signaling molecules, and react to tissue damage.
atherosclerotic plaque (*aka* atherosclerosis, plaque) (Greek: *atheroma* - porridge; *sklerosis* - hardening; Dutch: *placken* - to plaster) a mass of fatty substances (including cholesterol) and calcium, that accumulates in the inner walls of arteries; leads to a thickening or hardening of the arteries.
atlas (Greek: *Atlas* - the mythic titan who supported the heavens on his shoulders) first cervical vertebra.
atmospheric pressure (Patm) the force exerted by the gas mixture in the air on all objects with which it is in contact.
atom (Greek: *atomos* - indivisible) the smallest unit of a chemical element.

atomic number (Greek: *atomos* - indivisible) number of protons in the nucleus of an atom.
atomic weight (*aka* atomic mass) (Greek: *atomos* – indivisible) weight of an atom, measured in daltons (*aka* atomic mass units or amu); roughly equal to the number of protons and neutrons in an atom's nucleus.
ATP synthase (Greek: *syn* - together) an enzyme in the inner membrane of mitochondria that adds a phosphate group to ADP to generate ATP, using the energy of a hydrogen ion gradient across the membrane.
atresia (Greek: *atretos* - not perforated) the death of ovarian follicle and the egg it contains that can occur at any point during follicular development.
atrial natriuretic peptide (ANP) a peptide hormone produced by cells in the walls of the atria in response to increased blood volume or pressure; signals the kidneys to increase sodium excretion in the urine.
atrioventricular (AV) node (Latin: *atrium* – entry hall) clump of myocardial cells located in the inferior portion of the right atrium within the atrioventricular septum; receives the impulse from the SA node, pauses, and then transmits it into specialized conducting cells within the interventricular septum.
atrioventricular bundle (*aka* bundle of His) (Latin: *atrium* - entry hall) group of specialized myocardial conductile cells that transmit the impulse from the AV node through the interventricular septum; form the left and right atrioventricular bundle branches.
atrioventricular bundle branches (*aka* bundle branches) (Latin: *atrium* - entry hall) specialized myocardial conductile cells that arise from the bifurcation of the atrioventricular bundle and pass through the interventricular septum; lead to the Purkinje fibers and also to the right papillary muscle via the moderator band.
atrioventricular valves (Latin: *atrium* - entry hall) one-way valves located between the atria and ventricles; the valve on the right is called the tricuspid valve, and the one on the left is the mitral or bicuspid valve.
atrium (*pl*: atria) (Latin: *atrium* - entry hall) upper or receiving chamber of the heart that pumps blood into the lower chambers just prior to their contraction; the right atrium receives blood from the systemic circuit that flows into the right ventricle; the left atrium receives blood from the pulmonary circuit that flows into the left ventricle.
auditory canal (*aka* external auditory canal, external auditory meatus, external acoustic meatus) (Latin: *audio* - to hear) a structure in the ear that funnels sound from the auricle into the tympanic membrane.
auricle (*aka* auricula) (Latin: *auris* - an ear) (1) fleshy portion of the external ear; located on the lateral aspect of the head; (2) extension of an atrium visible on the superior surface of the heart.
auricular surface of the ilium (Latin: *auris* - an ear) rough area on the posterior, medial side of the ilium of the os coxae,
autocrine signaling (Greek: *auto* - self; *krinō* - to separate) a method of chemical communication in which a chemical signal affects the cell that releases it.
autoimmune diseases diseases that result from a person producing antibodies against self-antigens and attacking their own cells or tissues.
autolysis (Greek: *autos:* self; *lysis:* dissolution) breakdown of cells by their own enzymatic action.
autonomic nervous system (ANS) functional division of the nervous system responsible for involuntary control of the body, usually to maintain homeostasis.
autonomic plexuses (Latin: *plexus* - braid) networks of sympathetic and parasympathetic axons that innervate organs of the thorax, abdomen, and pelvis.
autonomic tone a low level of sympathetic or parasympathetic activity that keeps certain organs functioning at their homeostatic ranges in the resting state.
autorhythmicity ability of cardiac muscle to initiate its own electrical impulse that triggers the mechanical contraction that pumps blood at a fixed pace without nervous or endocrine control.
avascular without blood supply.
axial referring to the trunk and head.
axial skeleton (Greek: *skeletos* - skeleton) the bones of the head, neck, torso, and back.
axillary artery (Latin: *axilla* - armpit) a continuation of the subclavian artery as it passes behind the clavicle; runs between the subclavian artery and the brachial artery, and supplies blood to the shoulder, upper arm, and chest.
axillary nerve (Latin: *axilla* - armpit) a nerve of the brachial plexus that innervates the skin and muscles of the shoulder; located in the armpit region.
axillary vein (Latin: *axilla* - armpit) a vein in the axillary region that drains blood from the upper limb; arises from the merging of the brachial and basilic veins and continues to become the subclavian vein.
axis the second cervical vertebra.
axon (*pl*: axons) (Greek: *axōn* - axis) part of the neuron that extends from the cell body to another cell.
axon hillock (Greek: *axon* - axis) tapered region of the cell body of the neuron where the axon emerges; location where signals from different dendrites converge to become a single signal.
axon terminals (*aka* synaptic boutons, terminal boutons, end feet) (Greek: *axon* - axis; Latin: *terminus* - a limit) the end of the axon; usually includes several branches that extend toward the target cell.
azygos vein a vein that arises in the posterior wall of the abdominal cavity, travels upward through the diaphragm into the thoracic cavity just to the right of the vertebral column, and empties into the superior vena cava; drains blood from many of the thoracic and abdominal muscles.

B

B cell receptors (*aka* BCR) receptors on the cell membrane of B cells that are specific for a particular type of antigen; when a B cell encounters an antigen that matches its receptors, the B cell binds to the antigen.
B lymphocytes (*aka* B cells) (Latin: *lympha* - water; Greek: *kytos* – cell) a diverse group of lymphocytes that provides antibody-mediated (humoral) immunity, by acting against specific pathogens; some produce antibodies and some remember previously encountered pathogens to prevent a subsequent infection from those pathogens.
ball-and-socket joint synovial joint formed by the articulation between the rounded head of one bone and the bowl-shaped socket of the adjacent bone.
baroreceptors (*aka* pressoreceptors) (Greek: *baros* - weight; Latin: *recipere* - to receive) special mechanoreceptors that detect changes in pressure, such as blood pressure, by detecting the degree of stretch of blood vessel walls.
basal generally, the part of a cell or tissue facing the extracellular matrix.
basal cell carcinoma (Latin: *basis* - foundation; Greek: *kytos* - cell; *karkinōma* - cancerous tumor) cancer that originates from stratum basale of the epidermis.
basal epithelial cells (*aka* basal cells) stem cells inferior to the taste buds that can reproduce to replace the taste receptor cells if they age or are damaged.
basal forebrain region of the cerebrum containing nuclei related to learning and memory.
basal metabolic rate (BMR) the amount of energy used by the body in a certain amount of time when at rest, to maintain organ function; can be measured in Calories/hour.
basal nuclei (*aka* basal ganglia) (Latin: *nucleus* - a little nut) nuclei of the cerebrum (with a few components in the upper brainstem and diencephalon) responsible for cognitive processing, planning and initiation of movements.
base (Latin: *basis* - bottom) (1) compound that accepts hydrogen ions (H+) in solution 00; (2) bottom of a structure 000; (3) the flat portion at the posterior of the heart formed by the merging of the pulmonary veins with the left atrium.
base of the skull (Latin: *basis* - bottom; Middle English: *skulle* - a bowl) bottom of the skull.
basement membrane (Latin: *membrana* - a skin or membrane that covers a part of your body) a thin layer of fibrous material that supports and anchors the basal epithelium; made up of the lamina lucida and the lamina densa.
basilar artery an artery formed from the union of the vertebral arteries at the base of the brain; supplies blood to several regions of the brain.

basilar membrane a membrane that forms the floor of the cochlear duct and the roof of the scala tympani.
basilic vein a superficial vein that courses up the medial side of the arm; originates from the palmar and dorsal venous arches and merges with the brachial vein in the upper arm; empties into the axillary vein.
basolateral surface (Latin: *basis* - foundation; *lateralis* - belonging to the side) the surface of an epithelial cell of a renal tubule that borders on the interstitial fluid (and lies farther from the lumen of the tubule).
basophils (Latin: *basis* - base; *philus* - loving) granulocytes that secrete heparin and histamine to enhance inflammation; make up the smallest percentage of leukocytes, and are stained by basic (alkaline) dyes.
bedsores (*aka* decubitus ulcers) sores on the skin caused by long-term pressure necrosis in the epidermis; most common in patients with reduced mobility.
belly (*aka* muscle belly) the plump center of a muscle.
benign (Latin: *benignus* - kind) a tumor that does not break off and spread to other areas in the body.
beta (β)-adrenergic receptor (Latin: *ad* - to; *ren* - kidney; Greek: *ergon* - work) one class of receptors of the adrenergic system; binds epinephrine and norepinephrine; consists of three subtypes: β_1, β_2, and β_3.
beta cell a pancreatic islet cell type that produces the hormone insulin.
biaxial joint (Latin: *bi*- two; Latin: *axis* - axle) joint that allows for motion within two planes.
biceps brachii (Latin: *bi* - two; *caput* - head; *brachium* - arm) two-headed muscle that crosses the shoulder and elbow joints to flex the forearm; assists in supinating the forearm at the radioulnar joints and flexing the arm at the shoulder.
biceps femoris (Latin: *bi* - two; *caput* - head; *femur* - thigh) one of three hamstring muscles; functions in flexion of the leg at the knee joint, rotation of the leg, and extension of the hip.
bicipital groove (*aka* intertubercular groove, sulcus intertubercularis) (Latin: *bi* - two; *caput* - head) narrow groove between the greater and lesser tubercles of the humerus.
bile (Latin: *bilis* - fluid secreted by the liver) a green, alkaline mixture of substances produced by the liver, which functions in lipid emulsification and excretion of wastes, such as bile pigments, from the body.
bile canaliculus (*pl*: canaliculi) (Latin: *bilis* - fluid secreted by the liver) a small duct that collects bile from the hepatocytes, and transports it to the bile ductules.
biliary apparatus (Latin: *bilis* - fluid secreted by the liver; *apparatus* - equipment) a network of interconnected ducts between the pancreas, liver, gallbladder, and duodenum, which functions in the secretion and excretion of bile.
bilirubin (Latin: *bilis* - bile (fluid secreted by the liver); *ruber* - red) a yellow bile pigment produced in the decomposition of heme groups from hemoglobin molecules during erythrocyte breakdown; excreted in the bile as a waste product. Produces the brown color of feces.
biliverdin (Latin: *bilis* - bile; *verde* - green) a green pigment formed in the decomposition of heme groups from hemoglobin molecules during erythrocyte breakdown; converted into bilirubin by the liver and excreted in the bile as a waste product.
biogenic amine (Greek: *bios* - life; *genesis* - origin) class of neurotransmitters derived from amino acids (includes serotonin, dopamine, and norepinephrine).
bipedalism (Latin: *pes* - foot) the ability to walk on two feet.
bipennate (Latin: *bi* - two; *penna* - feather) a pennate muscle with fascicles on both sides of the tendon.
bipolar neuron shape with two processes extending from the neuron cell body, the axon and one dendrite.
bipolar cells (Latin: *bi* - two) cell type in the retina anterior to the photoreceptor cells; connect the photoreceptors to the ganglion cells.
blood portion of the cardiovascular system that transports gases, nutrients, wastes, and other substances around the body; consists of plasma (the liquid matrix) and formed elements (erythrocytes, leukocytes, and platelets).
blood flow the continuous circulation of blood through the blood vessels and organs; typically expressed in units of volume/time, such as L/min.
blood hydrostatic pressure (Greek: *hydro* - water; *statos* - stationary) the force or fluid pressure exerted by the blood against the walls of blood vessels.
blood pressure force or hydrostatic pressure applied by the blood against the wall of a blood vessel; expressed in units of mm Hg.
blood-brain barrier (BBB) physiological barrier separating the circulatory system and the central nervous system; restricts the flow of substances into the CNS.
blood-testis barrier (Latin: *testis* - testicle) a physical wall composed of the tight junctions between nurse (Sertoli, sustentacular) cells, which stops substances, hormones, and pathogens from coming in contact with the developing sperm during spermatogenesis.
Bloom's Taxonomy well-accepted framework in education for considering the skills involved in a question or task; named for Benjamin Bloom, an American educational psychologist.
body the main and middle region of the stomach.
body of the mandible (Latin: *mando* - to chew) horizontal portion of the L-shaped mandible.
body of the rib the angle and anterior surface of the rib, excluding the head, neck, and tubercle.
body of the uterus (Latin: *uterus* - womb) the main and middle region of the uterus.
Bohr effect the concept that a decrease in blood pH or an increase in the partial pressure of carbon dioxide leads to the dissociation of oxygen from hemoglobin; first described in 1904 by Christian Bohr, a Danish physiologist.
bolus (Greek: *bōlos* - lump) a mass of food that has been mixed with saliva and is ready for swallowing.
bond electrical force linking atoms.
bone hard, dense connective tissue that provides rigidity and strength to the skeleton.
Bowman's capsule (*aka* capsula glomeruli, glomerular capsule) the cup-shaped beginning of a renal tubule; along with the glomerulus, makes up the renal corpuscle; receives the filtrate from the glomerulus during glomerular filtration and transports it to the proximal convoluted tubule.
Boyle's law a law stating that the volume and pressure inside a contained area are inversely proportional; for example, the higher the volume inside the lungs, the lower the pressure; formula: $P1V1 = P2V2$. Named for Robert Boyle, an Anglo-Irish scientist.
brachial artery (Latin: *brachium* - arm) a continuation of the axillary artery as it passes into the upper arm; provides blood to the upper arm, and terminates in the elbow region by dividing into the radial and ulnar arteries.
brachial plexus (Latin: *brachium* - arm; *plexus* - braid) a group of axons of spinal nerves, associated with spinal nerves C4 through T1.
brachial vein (Latin: *brachium* - arm) a deep vein of the arm that arises from the merging of the radial and ulnar veins in the forearm; empties into the axillary vein.
brachialis (Latin: *brachium* - arm) muscle deep to the biceps brachii; a major flexor of the forearm.
brachiocephalic vein (*aka* innominate vein) (Latin: *brachium* - arm; *cephalicus* - pertaining to head) a vein formed by the merging of the internal jugular vein with the subclavian vein; the two brachiocephalic veins then merge to form the superior vena cava; drains blood from the upper limb, head, neck, and a portion of the thorax.
brachioradialis (Latin: *brachium* - arm; *radius* - ray) muscle that can quickly flex the forearm or help lift a load.
brachium (Latin: *brachium* - arm) the portion of the upper limb located between the shoulder and elbow joints.
bradycardia (Greek, *bradys* - slow; *cardiac* - *kardia* meaning heart) slow heart rate, typically under 60bpm for most individuals.
brain organ located in the skull made up of nervous tissue; component of the central nervous system.
bridge the region of the external nose that lies between the root and the lower portion of the nose; formed by the nasal bones.
broad ligament a large fold of the peritoneum that connects the uterus to the wall of the pelvis; also supports the ovary and uterine tube.
Broca's area (*aka* motor speech area) a region of the frontal lobe of the cerebrum, responsible for the production of language or controlling

movements responsible for speech; located on the left side in most people.
bronchi (Greek: *bronkhos* - wind pipe) tubular organs of the respiratory system; the primary bronchi branch directly off the trachea, the secondary bronchi branch off the primary bronchi, and the tertiary bronchi branch off the secondary bronchi.
bronchial artery (Greek: *bronkhos* - windpipe) an artery that branches off of the thoracic aorta, and is part of the systemic circuit; supplies blood to the tissues of the bronchi, lungs, and pleural membranes.
bronchial tree (Greek: *bronkhos* - windpipe) the name for the upside-down branching structure of the respiratory system; the trachea represents the tree trunk, the bronchi represent the tree limbs, and the bronchioles represent the smaller branches of the tree.
bronchial vein (Greek: *bronkhos* - windpipe) a vein of the systemic circuit that drains blood from the lungs, bronchi, and pleural membranes; empties into the azygos vein.
bronchioles (Greek: *bronkhos* - windpipe) tiny tubular organs, 0.5 to 1 mm in diameter, that branch off small bronchi, and branch successively into smaller and smaller tubes; some, such as terminal bronchioles, are conduction structures, while others, such as respiratory bronchioles, are gas exchange structures.
bronchoconstriction (Greek: *bronkhos* - windpipe; Latin: *constrictionem* - drawing together) a decrease in the diameter of the bronchi and bronchioles, due to contraction of the smooth muscle in their walls.
bronchodilation (Greek: *bronkhos* - windpipe; Latin: *dilatare* - enlarge) an increase in the diameter of the bronchi and bronchioles, due to relaxation of the smooth muscle in their walls.
bronchus-associated lymphoid tissue (BALT) (Greek: *bronkhos* - wind pipe; Latin: *lympha* - water) lymphoid nodules found in the mucosa of the respiratory tract, especially the bronchi; a type of MALT (mucosa-associated lymphoid tissue), which protects against infection by pathogenic microorganisms entering the body in inhaled air.
brush border (*aka* striated border; brush border membrane) the luminal surface of the proximal convoluted tubules in the kidney, formed by microvilli on the surface of the simple cuboidal epithelial cells; greatly increases surface area for reabsorption and secretion of various substances.
buccinator (Latin: *bucca* - cheek) large facial muscle that compresses the cheek.
buffer solution or compound containing a weak acid or a weak base that opposes wide fluctuations in the pH of body fluids.
bulbourethral glands (*aka* Cowper's glands) (Latin: *bulbus* - bulb-shaped; *urethra* - passage for urine) small glands of the reproductive system of a biological male that secrete a mucus-containing fluid that lubricates the penis during sexual intercourse and becomes part of the semen during ejaculation.
bulbous corpuscles (*aka* Ruffini ending, Ruffini corpuscle) (Latin: *corpus* - body) stretch receptors that transduce stretching of the skin.
bulbs of the vestibule (*aka* vestibular bulbs) masses of erectile tissue located on both sides of the vaginal orifice.
bulbus cordis (Greek: *bulbus* - a swelling; Latin: *cord* - a string) portion of the primitive heart tube that will eventually develop into the right ventricle.
bulk flow the movement of a relatively large amount of a substance or mixture from a region of higher pressure to a region of lower pressure.
burn death of skin cells resulting from exposure to intense heat, radiation, electricity, or chemicals.
bursa (*pl*: bursae) (Latin: *bursa* - purse) small sac of synovial fluid surrounded by connective tissue; prevents friction between adjacent structures by separating structures and providing cushioning.

C

calcaneal tendon (*aka* Achilles tendon) (Latin: *calcaneum* - heel) strong tendon that inserts into the calcaneal bone of the ankle; site of insertion of the superficial muscles in the posterior compartment of the lower leg.
calcaneus (*aka* heel bone) (Latin: *calcaneum* – heel) large posterior, inferior tarsal bone; forms the heel of the foot.
calcitonin a peptide hormone produced by the parafollicular cells of the thyroid gland; functions to decrease blood calcium levels when they are too high.
calcitriol active form of vitamin D that allows for absorption of calcium into the bone.
callus (Latin: *callus* - hard skin) a fibrocartilaginous template for later mineralization formed during bone growth.
calmodulin regulatory protein that controls cross-bridge formation and facilitates contraction in smooth muscles; activated by the binding of calcium ions.
Calorie (C) (*aka* food calorie, large calorie, kilocalorie) (Latin: *calor* - heat) the amount of heat required to increase the temperature of 1 kg (1000 g) of water by 1°C; written with a capital "C."
calvaria (*sing*: calvarium) (*aka* skullcap) (Latin: *calvaria* - skull) rounded top, or "cap," of the skull.
calyces (*sing*: calyx) (Greek: *kalyx* - seed pod) cuplike drainage tubes that receive urine from the collecting ducts and transport it to the renal pelvis of the kidney.
canaliculi (sing: canaliculus) (Latin: *canaliculus* - a small channel) channels within the bone matrix.
capillary (Latin: *capillaris* - hairlike) the smallest of the blood vessels, which exchange oxygen, carbon dioxide, nutrients, and wastes between the blood and the cells.
capillary bed (Latin: *capillaris* - hairlike) an interwoven network of capillaries that transports blood between an arteriole and a venule.
capitate (Latin: *caput* - head) from the lateral side, the head-shaped third of the four distal carpal bones; articulates with the scaphoid and lunate proximally, the trapezoid laterally, the hamate medially, and primarily with the third metacarpal distally.
capitulum (*aka* lateral condyle of the humerus) (Latin: *caput* - head) knoblike bony structure of the lateral, distal end of the humerus; located lateral to the trochlea.
capsaicin molecule that binds to ion channels on heat-sensing nociceptors; the basis for "hot" sensations caused by spicy food.
carbohydrate class of organic compounds built from sugars, molecules containing carbon, hydrogen, and oxygen.
carbonic anhydrase (CA) (*aka* carbonate dehydratases) the enzyme that catalyzes the chemical reaction between carbon dioxide and water to form the compound carbonic acid (H_2CO_3).
cardia (*aka* cardiac region) (Greek: *cardia* – heart) the region of the stomach that encircles the opening through which food enters the stomach from the esophagus; named for its proximity to the heart.
cardiac cycle (Greek: *cardiac* - *kardia* meaning heart) period of time between the onset of atrial contraction (atrial systole) and ventricular relaxation (ventricular diastole).
cardiac muscle (Greek: *cardiac* - *kardia* meaning heart) striated muscle that forms the contractile walls of the heart; each muscle cell contains a single nucleus and contracts autonomously; cells are connected to one another via gap junctions.
cardiac notch (Greek: *cardiac* - *kardia* meaning heart) depression in the medial surface of the inferior lobe of the left lung where the apex of the heart is located.
cardiac output (CO) (Greek: *cardiac* - *kardia* meaning heart) the volume of blood that the left ventricle ejects in one minute.
cardiac plexus (Latin: *cardiacus* - heart; *plexus* - braid) a network of sympathetic and parasympathetic axons in the thorax, which innervates the heart, increasing or decreasing the heart rate in response to changing conditions.
cardiac reserve (Greek: *cardiac* - *kardia* meaning heart) difference between maximum and resting CO.
cardiomyocytes (*sing*: **cardiomyocyte**) (Greek: *kardia* - heart; *mys* - muscle; *kytos* - cell) cells that make up the heart muscles.
carotid baroreceptors (Greek: *karotides* - great arteries of the neck) pressure receptors found in the carotid sinuses, which detect changes in blood pressure and relay the information to the cardiovascular center in the brain in order to maintain blood pressure within homeostatic range.
carotid body (Greek: *karotides* - great arteries of the neck) chemoreceptors that detect changes

in oxygen concentration; located in the carotid arteries, at the point of division into the internal and external carotid arteries; relay information to the cardiac and respiratory control centers in the medulla oblongata of the brain to help maintain blood oxygen level in its homeostatic range.
carotid canal (Greek: *karotides* - great arteries of the neck; Latin: *canal* - channel) the zig-zag passageway through which the internal carotid artery enters the skull; enters on the inferior aspect, anteromedial to the styloid process and opening to the middle cranial cavity near the posterior-lateral base of the sella turcica.
carpal bone (Greek: *karpus* - wrist) one of the eight bones that form the wrist and base of the hand.
cartilage (Latin: *cartilago* - gristle) connective tissue that is less rigid than bone that provides flexibility and smooth surfaces to the skeleton.
cartilaginous joint (Latin: *cartilago* - gristle) joint at which the adjacent bones are united by hyaline cartilage or fibrocartilage.
catabolic reactions (Greek: *kataballein* - to throw down) chemical reactions in the body that decompose larger molecules into smaller ones.
catabolism (Greek: *katabolē* - a casting down) The process of breaking molecules down.
catalyst a substance that increases the rate of a chemical reaction without itself being changed in the process.
cation (*pl*: **cations**) (Greek: *katiōn* - going down) an atom with a positive charge.
cauda equina (Latin: *cauda*- tail; *equina* - horse) a long bundle of spinal nerve roots descending below the first lumbar vertebra, encased within the inferior portion of the spine.
caudate (*aka* caudate nucleus) (Latin: *cauda*- tail) a long nucleus located in the cerebrum; along with the putamen, makes up the striatum of the basal nuclei.
caval opening (Latin: *cavus* - hollow) opening in the diaphragm through which the inferior vena cava passes.
caveolae small indentations in the cell membranes of smooth muscle cells, which help in intercellular communication, endocytosis, and exocytosis.
cecum (*aka* caecum) (Latin: *caecum* - gut) a pouchlike structure that forms the beginning of the large intestine; lies between the ileum of the small intestine and the ascending colon of the large intestine.
celiac ganglion (*pl*: celiac ganglia) (Greek: *koilia* - belly; *ganglion* - knot) one of the three collateral (prevertebral) ganglia of the sympathetic system; projects to the digestive system.
celiac plexus (*aka* solar plexus) (Greek: *koilia* - belly; *plexus* - braid) one of the networks of sympathetic and parasympathetic nerve fibers that make up the abdominal aortic plexus; transmits pain impulses from some of the visceral organs and stimulates peristalsis.
celiac trunk (*aka* celiac artery) (Latin: *coeliacus* - pertaining to bowels) a single artery that arises from the abdominal aorta, and branches into the left gastric artery, splenic artery, and common hepatic artery.
cell (Latin: *cella* - a storeroom, a chamber) smallest independently functioning unit of all organisms; in animals, a cell contains cytoplasm, composed of fluid and organelles.
cell body (*pl*: cell bodies) (*aka* soma) (Greek: *kytos* - cell) part of the neuron that contains most of the cytoplasm, organelles, and the nucleus.
cell cycle (*pl*: cell cycles) (Latin: *cella* - chamber; Greek: *kyklos* - circles) life cycle of a single cell, until its division into two new daughter cells.
cell membrane (*pl*: cell membranes) (aka plasma membrane) (Latin: *cella* - chamber; *membrana* - a skin that covers part of the body) membrane surrounding all cells, composed primarily of phospholipids.
cellular respiration a series of metabolic processes by which food molecules are broken down to produce ATP in the cell; consists of glycolysis (when glucose is being broken down), the citric acid cycle, and the electron transport chain.
cellular responses (*aka* cell-mediated immunity) a set of actions of the adaptive immune response that involves T cells traveling to an infection site and directing an attack on infected body cells, cancerous cells, or donor graft cells.
cementum a calcified type of connective tissue, similar to bone, that helps to hold each tooth in place; connects the root of the tooth to its periodontal ligament.
central canal (*aka* Haversian canal) (Latin: *canal* - a small channel) 1) longitudinal channel in the center of the osteon that contains blood vessels, nerves, and lymphatic vessels; 2) a long tubular channel running down the center of the spinal cord; circulates cerebrospinal fluid.
central chemoreceptor one of the receptors in the medulla oblongata of the brainstem that detects changes in the concentration of hydrogen ions (the pH) in the extracellular fluid of the brain.
central nervous system (CNS) anatomical division of the nervous system; includes the brain and spinal cord.
central sulcus (*aka* sulcus of Rolando) (Latin: *sulcus* - a ditch) a deep groove of the cerebral cortex, which divides the frontal lobe from the parietal lobes.
centriole (*pl*: centrioles) (Greek: *kentron* - center) small organelle is the origin of the microtubule spindle that moves chromosomes during cell replication.
centromere (*pl*: centromeres) (Greek: *kentron:* center; *meros:* part) region of attachment for two sister chromatids.
cephalic phase (*aka* reflex phase) (Greek: *kephalikos* - pertaining to the head) the first phase of gastric secretion that occurs before food reaches the stomach; occurs when a person thinks about, sees, smells, or tastes food.
cephalic vein (Latin: *cephalicus*- pertaining to head) a superficial vein that courses up the lateral side of the arm; originates from the dorsal venous arch and empties into the axillary vein.
cerebellum (Latin: *cerebellum* - little brain) a region of the adult brain that develops from the metencephalon; responsible for balance, coordination, and complex motor functions.
cerebral arterial circle (*aka* circle of Willis) (Latin: *cerebrum* - brain) a circular arterial anastomosis located at the base of the brain, composed of arteries that supply blood to the brain; consists of many branches entering and exiting the circle; resembles a traffic circle 000 20 (27)
cerebral aqueduct (*aka* aqueduct of Sylvius) (Latin: *cerebrum* - brain; *aquaeductus* - canal) a tubular channel that transports cerebrospinal fluid between the third and fourth ventricles; located in the midbrain.
cerebral cortex (Latin: *cerebrum* - brain) the outer layer of the cerebrum; consists of gray matter and is marked by ridges and indentations known as gyri and sulci.
cerebral hemisphere (Latin: *cerebrum* - brain; Greek: *hemi* - half; *sphaira* - ball) one of the two halves of the cerebrum; left and right hemispheres are connected by the corpus callosum and other commissural tracts.
cerebral peduncles large bulges on the anterior side of the midbrain, which contain the major descending tracts that transport motor impulses to the spinal cord.
cerebrospinal fluid (CSF) (Latin: *cerebrum* - brain; *spinalis* - spine) fluid within the subarachnoid space that provides a liquid cushion and nutrition to the brain and spinal cord.
cerebrum (Latin: *cerebrum* - brain) a large region of the adult brain that develops from the telencephalon; responsible for higher neurological functions including memory, language, and consciousness.
cervical curve (Latin: *cervix* – neck) posteriorly concave curvature of the neck region.
cervical plexus (Latin: *cervix* - neck; *plexus* - braid) nerve plexus composed of the axons from spinal nerves C1 through C5.
cervix (Latin: *cervix* - neck) the inferior portion of the uterus, that connects to the vagina.
channel protein (*pl*: channel proteins) (*aka* transmembrane protein) (Latin: *canalis* - canal; Greek: *protos* - first) membrane-spanning protein that has an inner tunnel for molecules or ions to pass from one side of the membrane to the other.
checkpoint (*pl*: checkpoints) a pause in the cell cycle at which certain conditions must be met in order for the cell to proceed to the next phase.
chemical digestion the breakdown of food via the action of enzymes on specific chemical bonds within the food molecules.
chemical energy (Greek: *chēmeia* - alchemy; *energeia* - in work) form of energy that is absorbed as chemical bonds form, stored as they are maintained, and released as they are broken.
chemical synapse (Greek: *syn* - with; *hapto* - to clasp) connection between a neuron and

a target cell where neurotransmitters diffuse across a short distance.
chemically gated ion channel (*aka* ligand-gated ion channel) a protein channel in the plasma (cell) membrane that regulates the passage of specific ions through the membrane; for example, a chemically-gated sodium channel opens in skeletal muscle cells in response to the binding of acetylcholine.
chemokine (Greek: *kinetikos* - moving) a chemical messenger, secreted by a damaged cell, that attracts leukocytes to infection or injury sites.
chemoreceptors (Greek: *chēmeia* - alchemy; Latin: *recipere* - to receive) sensory receptor cells that detect changes in the concentration of specific chemical stimuli.
chief cells (*aka* zymogenic cells, peptic cells) cells of the parathyroid gland that produce and secrete parathyroid hormone 000; glandular cells of the stomach that secrete pepsinogen, the inactive precursor of the enzyme pepsin.
cholecystokinin (CCK) (Greek: *khole* - gall; *kystis* - bladder) a hormone released by the duodenum when chyme enters the duodenum from the stomach; stimulates release of pancreatic juice and bile, and opens the hepatopancreatic sphincter.
cholinergic (Greek: *ergon* - work) a type of synapse at which acetylcholine is released, or a type of receptor that binds to acetylcholine.
chondroblasts (Greek: *chondrion* - cartilage; *blastos* - germ) cells capable of generating cartilage matrix.
chondrocytes (Greek: *chondrion* - cartilage; *kytos* - cell) cartilage cells; cells that are differentiated from chondroblasts and occupy lacunae.
chordae tendineae (*aka* tendinous cords) (Latin: *chord* - string or rope; *tendo* - stretch) stringlike extensions of tough connective tissue that extend from the flaps of the atrioventricular valves to the papillary muscles.
choroid (*aka* choroida, choroid coat) (Greek: *chorioeidēs* - like a membrane) highly vascular loose connective tissue in the wall of the eye that provides blood supply to the outer retina; forms the middle layer of the wall of the eye.
choroid plexus (Greek: *choroeidēs* - membrane; Latin: *plexus* - braid) a structure within the ventricles of the brain containing blood capillaries lined by ependymal cells that filter blood to produce CSF.
chromaffin (Greek: *chrōma* - color; Latin: *affinis* - affinity) neuroendocrine cells that make up the adrenal medulla; produce epinephrine and norepinephrine.
chromatin (Greek: *chrōma* - color) loosely wound DNA and histone proteins.
chromosome (*pl*: chromosomes) (Greek: *chrōma* - color; *sōma* - body) condensed version of chromatin.
chronic (Greek: *chronos* - time) long-term; a term describing an illness that develops gradually and lasts for a long period of time.
chylomicron (*aka* ultra-low-density lipoproteins) a mixture of lipids and proteins, which transports lipids around the body and delivers them to various tissues and organs.
chyme (*aka* chymus) (Latin: *khylos* - juice) a pasty mixture of food and gastric juice, produced by the churning action of the stomach.
cilia (*sing*: cilum) (Latin: *cilium* - eyelid) small appendages on some cells that include microtubules; capable of a sweeping motion across the cell surface.
ciliary body (*aka* ciliary bodies) (Latin: *ciliaris* - relating to the eye) a mass of smooth muscle attached to the lens of the eye by suspensory ligaments; controls lens shape through the zonule fibers.
ciliary ganglion (Latin: *cilium* - eyelid; Greek: *ganglion* – knot) a group of neuron cell bodies of the parasympathetic nervous system; located in the posterior eye orbit; contains axons of the oculomotor nerve that control constriction of the pupil and shape of the lens.
cingulate gyrus (Greek: *gyros* - circle) a component of the limbic function, which functions in the expression of emotion.
circadian rhythm 24-hour cycles that correspond to changes in bodily processes over the course of a day; influenced by the amount of light entering the retina.
circular (*aka* sphincter) (Latin: *circulus* - circle) muscles whose fascicles are arranged in a circle.
circular fold (*aka* plica circulare, valves of Kerckring, valvulae conniventes) a pleat or convolution in the mucosa and submucosa of the small intestine, which increases the surface area for secretion and absorption.
circumduction (Latin: *circum* - around; *ductus* - to draw) circular movement; involves a sequential combination of flexion, adduction, extension, and abduction.
circumflex artery (Latin: *circum* - all around; Greek: *arteria* - pipe) branch of the left coronary artery that follows the coronary sulcus.
cisterna chyli (*aka* receptaculum chyli) (Latin: *cisterna* - underground water reservoir) a saclike vessel that forms the beginning of the thoracic duct; formed from the merging of the two lumbar trunks and the intestinal trunk.
citric acid cycle (*aka* Krebs cycle, tricarboxylic acid (TCA) cycle) a metabolic cycle that produces ATP, NADH, and $FADH_2$ in the mitochondria from the breakdown of carbohydrates, lipids or amino acids; also produces CO_2 from pyruvate in the case of glucose breakdown.
class switching the process by which an activated B cell shifts its production of IgM antibodies to a different class (IgA, IgE, or IgG) while retaining antigen specificity.
clavicle (*aka* collarbone) an S-shaped bone that connects the sternum to the scapula.
clavicular notch shallow depressions located on the superior-lateral sides of the sternal manubrium.
cleavage furrow a contractile band made up of microfilaments (actin fibers) that forms around the midline of the cell during cytokinesis, squeezing the two cells apart.
clitoris (*aka* glans clitoris) (Greek: *kleis* - a key) a small mass of erectile tissue that lies anterior to the vestibule in biological females; contains many nerves and is sensitive to sexual stimulation.
clonal expansion the reproduction of a particular type of lymphocytes, which results from the binding of an antigen to a specific lymphocyte receptor.
clonal selection the binding of an antigen only to lymphocytes that contain specific receptors for that antigen.
clotting factors (*aka* coagulation factors) (Middle English: *clott* - lump) a group of 12 substances that participate in the reaction cascade of blood coagulation.
coagulation (Latin: *coagulum* - to condense) the final and most effective step of hemostasis, in which the liquid portion of the blood is converted into a gelatinous clot via a cascade of chemical reactions, and blood cells and platelets are embedded in a mass of fibrin threads.
coccyx (*aka* tailbone) (Greek: *kokkyx* - the coccyx) bone located at inferior end of the adult vertebral column; formed by the fusion of four coccygeal vertebrae during adulthood.
cochlea (Latin: *cochlea* - snail) fluid-filled tube located in the inner ear; contains structures to detect and transduce sound stimuli.
cochlear duct (*aka* scala media) (Latin: *cochlea* - snail) space within inner ear containing the spiral organ (organ of Corti); adjacent to the scala tympani and scala vestibuli.
cochlear nucleus (*pl*: cochlear nuclei) (Latin: *cochlea* - snail; *nucleus* - a little nut) a structure in the medulla oblongata that is responsible for transmitting sound impulses.
cofactors substances that bind to enzymes and enable their activity; typically take the form of metal ions or vitamins.
cohesion (Latin: *cohaesionem* - sticking together) sticking together; in the respiratory system, the tendency of water molecules to form hydrogen bonds with adjacent water molecules in the alveoli and the pleural cavity, holding them in close proximity by charge attraction.
collagen (Greek: *kolia* - glue; *-gen* - producing) proteins which form flexible, tough fibers that give connective tissue tensile strength.
collagen fibers (Greek: *kolia* - glue; *-gen* - producing; Latin: *fibra* - fiber) flexible fibrous proteins that give connective tissue tensile strength.
collateral ganglia (*aka* prevertebral ganglia) (Latin: *prae* - before; *vertebral* - joint) sympathetic ganglia, which lie anterior to the vertebral column; consist of the celiac, superior mesenteric, and inferior mesenteric ganglia; run between the sympathetic chain ganglia and abdominal organs that they regulate.
collecting ducts (*aka* duct of Bellini) the most distal portions of the renal tubules, which collect filtrate from the distal convoluted tubules (DCTs) of several nephrons.
colloid (Greek: *kolla* - glue; *eidos* - appearance) 1) liquid mixture in which the solute particles consist of clumps of molecules large enough to

scatter light; 2) a viscous fluid filling the central cavity of thyroid follicles; precursor to thyroid hormones.

colloid osmotic pressure (*aka* oncotic pressure) (Greek: *kolla* - glue; *osmos* - pushing fluid) osmotic pressure, caused mainly by albumins, which draws plasma from the interstitial fluid into the capillaries; opposes the hydrostatic pressure in the capillaries.

colon (*aka* large intestine) the main part of the large intestine, which lies between the cecum and rectum; consists of ascending, transverse, descending, and sigmoid portions.

commissural tracts (*aka* commissural fibers) tracts that connect corresponding regions of gray matter in the two cerebral hemispheres.

common bile duct (*aka* bile duct) a duct formed by the convergence of the common hepatic duct and the cystic duct; transports bile from the liver and gallbladder to the duodenum via the hepatopancreatic sphincter.

common carotid artery (Greek: *kerotides* - great arteries of the neck) an artery that supplies blood to the head, neck, and brain; the right common carotid artery is a branch of the brachiocephalic artery, and the left common carotid artery is a branch of the aortic arch.

common hepatic artery (Latin: *hepaticus* - pertaining to the liver) an artery that branches off of the celiac trunk, and supplies blood to the liver, the gallbladder, portions of the stomach, and the pancreas.

common hepatic duct (Latin: *hepaticus* - pertaining to the liver) a large duct formed by the merging of several smaller hepatic ducts in the liver; merges with the cystic duct from the gallbladder to form the common bile duct.

common iliac artery (Latin: *iliacus* - pertaining to colic) a terminal branch of the abdominal aorta that divides into the internal and external iliac arteries; supplies blood to the pelvis, the lower abdominal area, and the lower limbs.

common iliac vein (Latin: *iliacus* - pertaining to colic) a vein that arises from the merging of the internal and external iliac veins in the pelvic region; drains blood from the pelvic organs and the legs; the two common iliac veins merge to form the inferior vena cava.

common pathway the final reaction sequence of the blood coagulation pathway, which results in the formation of a blood clot; can be initiated by the intrinsic or extrinsic pathway.

compact bone dense, strong bone type; located under the periosteum, provides support and protection.

complement system (*aka* complement cascade) (Latin: *complementum* - that which completes) a group of several proteins that circulate in the blood plasma, which can be activated either by antibodies or an innate immune response; once activated, it works in a cascade of reactions, with each reaction product activating the next protein in the sequence; the various proteins function in opsonization, enhancement of phagocytosis, or lysis (rupture) of the antigenic cell.

compliance the ability of a blood vessel to stretch or expand; the opposite of rigidity.

compound (Latin: *compono* - to place together) substance composed of two or more different elements joined by chemical bonds.

computed tomography (CT) (CT scans are sometimes called "cat scans") (Greek: *tomos* - cutting) medical imaging technique in which a computer-enhanced cross-sectional X-ray image is obtained.

concentration (Latin: *con* - together; *centrum* - center) number of particles within a given space.

concentration gradient (*pl:* concentration gradients) (Latin: *con* - together; *centrum* - center) difference in the concentration of an ion or molecule between two locations.

concentric contraction (Latin: *contractus* - to draw together) shortening of muscle to move a load.

conduction portion (Latin: *conductio* - bring together) the tubular organs of the respiratory tract which transport air into and out of the lungs, without participating in gas exchange.

condylar process of the mandible (Latin: *processus* - advance; *mando* - to chew) posterior projection of the ramus, topped by the condyle.

condyle oval-shaped structure located at the top of the condylar process of the mandible; forms a joint with the temporal bone.

condyloid joint (*aka* ellipsoid joint) (Greek: *kondylos* - resembling knuckles) synovial joint formed by the articulation of a shallow depression at the end of one bone and a rounded end from a second bone.

cone photoreceptors (Greek: *phōt* - light; Latin: *recipere* - to receive) type of retinal receptor cell specialized for color vision; contain opsin proteins.

congenital heart defects defects in the heart structure that are present at or before birth.

conjunctiva (Latin: *conjugo* - to join together) a layer of stratified columnar epithelium which forms the inner lining of the eyelid, and folds back to form a covering over the sclera.

connective tissue (French: *tissu* - woven) type of tissue that serves to hold in place, connect, and integrate the body's organs and systems.

connective tissue proper (French: *tissu* - woven) connective tissue containing a viscous matrix, fibers, and cells.

contact inhibition (Latin: *con* - together; *tactus* - to touch; *inhibitus* - to keep back) a form of cell replication control that prevents cells from replicating if there are cells in all around it.

continuous capillaries capillaries in which the endothelial cells are joined by tight junctions, but also contain intercellular clefts that allow the movement of small molecules across the capillary walls; the most abundant type of capillary.

continuous conduction (Latin: *conductio* - to lead) propagation of an action potential along an unmyelinated axon.

contractility (Latin: *contractus* - to draw together) the ability to shorten with force by pulling on attachment points.

contraction phase (Latin: *contractus* - to draw together) phase of muscle contraction during which calcium ions have bound to troponin, and sarcomeres are shortening.

contralateral reflex (Latin: *latus* - side; *reflexus* - a bending back) reflexes that begin and end on opposite sides of the body.

convergent (Latin: *convergere* - to incline together) a muscle with a widespread origin over a single area whose fascicles come to a single, common insertion point.

coracobrachialis (Greek: *korakōdēs* - like a crow's beak; Latin: *brachium* - arm) muscle that adducts and flexes the arm.

coracohumeral ligament (Greek: *korakodes* - resembling a crow; Latin: *humerus* - arm; *ligamentum* - a band) ligament that runs from the coracoid process of the scapula to the anterior humerus.

coracoid process (Greek: *korakōdēs* - like a raven's beak) lateral, hook-shaped anterior projection from the superior margin of the scapula.

cornea (Latin: *corneus* - horny) clear portion of the fibrous layer of the eye; lies anterior to the iris and pupil and helps to refract light rays entering the eye.

corniculates small, paired cartilages of the larynx that help control the tension of the vocal cords.

coronal suture (Latin: *corona* - crown) joint that joins the frontal bone parietal bones across the top of the skull.

coronary arteries (Latin: *corona* - crown; Greek: *arteria* - pipe) branches of the ascending aorta that supply blood to the heart.

coronary sinus (Latin: *corona* - crown) large, thin-walled vein on the posterior surface of the heart that lies within the atrioventricular sulcus and drains the heart myocardium directly into the right atrium.

coronary sulcus (Latin: *corona* - crown) sulcus that marks the boundary between the atria and ventricles.

coronary veins (Latin: *corona* - crown) veins that drain blood from the heart wall and return this deoxygenated blood to the right atrium.

coronary vessels (Latin: *corona* - crown) the general term for both arteries and veins that carry blood to and from the heart wall.

coronoid fossa (Greek: *korōnē* - a crow; *eidos* - resembling; Latin: *fossa* - a trench) depression located on the anterior surface of the humerus superior to the trochlea; accommodates the coronoid process of the ulna when the elbow is maximally flexed.

coronoid process of the mandible (Greek: *korōnē* - a crow; *eidos* - resembling; Latin: *processus* - advance; *mando* - to chew) anterior projection from the ramus; site of attachment for one of the biting muscles.

coronoid process of the ulna (Greek: *korōnē* - a crow; Greek: *ōlenē* - ulna) projecting bony lip on the anterior, proximal ulna; forms the anterior edge of the trochlear notch.

corpus albicans (Latin: *corpus* - body) a remnant of the corpus luteum that settles in the

ovary after degeneration of the corpus luteum; name means "white body."
corpus callosum (Latin: *corpus* - body;) large white matter commissural tract between the two cerebral hemispheres; the major pathway for communication between the hemispheres.
corpus cavernosum (*pl*: corpora cavernosa) (Latin: *corpus* - body; *cavus* - hollow) one of two columns of erectile tissue that becomes engorged with blood during the erection process; found in both the penis and the clitoris.
corpus luteum (Latin: *corpus* - body; *luteus* - yellow) an endocrine structure formed by the remnants of the ovarian follicle after ovulation; secretes estrogen, progesterone, and inhibin.
corpus spongiosum (Latin: *corpus* - body) a single column of erectile tissue in the penis that surrounds the urethra and becomes engorged with blood during sexual stimulation.
cortex (*pl*: cortices) (Latin: *cortex* - bark of a tree) in hair, the second or middle layer of compressed keratinocytes 000; the outer region of a lymph node, which consists of follicles containing dividing B cells.
cortical nephrons (Latin: *cortex* - bark of a tree; Greek: *nephros* - kidney) nephrons that reside mainly in the renal cortex, whose nephron loops (loops of Henle) extend a small distance into the renal medulla.
corticoids a group of steroid hormones released by the adrenal cortex that regulate glucose and electrolyte balance and supplement the sex hormones.
cortisol a glucocorticoid produced by the zona fasciculata of the adrenal cortex; promotes the breakdown of glucose and other stored nutrients in response to stress.
costal cartilages (Latin: *costa* - ribs; *cartilago* - gristle) strips of hyaline cartilage that attach the ribs to the sternum.
costal groove (Latin: *costa* - ribs) shallow groove along the inferior margin of the body of a rib that allows blood vessels and a nerve to pass from the spinal cord to the thoracic wall.
countercurrent multiplier system (Latin: *contra* - opposite; *currere* - to flow) a method used by the kidney to concentrate the urine, by reabsorbing a large amount of sodium ions and water from the filtrate; occurs in the descending and ascending limbs of the nephron loop, which flow in opposite directions; involves reabsorption of water in the descending limb in response to active reabsorption of sodium ions in the descending limb.
covalent bond chemical bond in which two atoms share electrons, thereby completing their valence shells.
cranial cavity (Greek: *kranion* - skull; Latin: *cavus* - hollow) division of the posterior body cavity that houses the brain.
cranial dural septa (*sing*: septum) (Latin: *saeptum* - wall) folds in the dura mater that separate structures to stabilize the brain within the cranium; consists of the falx cerebri, falx cerebelli, tentorium cerebelli, and diaphragma sellae.
cranial nerves twelve pairs of nerves that are directly connected to the brain; responsible for the sensory and motor functions of the head, neck, thoracic and abdominal organs.
cranial reflexes (Greek: *kranion* - skull; Latin: *reflexus* - a bending back) reflexes that involve the brain.
craniosacral system (*aka* parasympathetic system) the parasympathetic division of the autonomic nervous system; axons emerge from the central nervous system through the brainstem and sacral portion of the spinal cord; associated with maintaining homeostasis while a person is in a "rest and digest" or relaxed state.
cranium (Greek: *kranion* - skull) the rounded portion of the skull that surrounds the brain; excludes the facial bones.
creatine phosphate molecule that regenerates ATP from ADP in muscle cells, by donating a phosphate group to ADP.
cremaster muscles muscles that descend from the internal oblique muscle of the abdominal wall and cover each testis like a muscular net that is responsible for elevating the testes.
cribriform plates (Latin: *cribrum* - a sieve; *forma* - form; Greek: *platys* - flat) small, flattened areas with many tiny holes located on the sides of the crista galli on the ethmoid bone.
cricoid cartilage (Latin: *cartilago* - gristle) a ring-shaped mass of hyaline cartilage that lies between the thyroid cartilage and the trachea in the larynx.
crista galli (Latin: *crista* - crest; *galli* - rooster) small bony projection of the ethmoid bone located at the midline of the cranial floor; provides an anterior attachment point for a covering layer of the brain.
cristae (*sing*: crista) (Latin: *crista* - crest) the folds of the inner membrane of mitochondria.
crossed-extensor reflexes (Latin: *reflexus* - a bending back) contralateral reflexes that activate the opposite side of the body, usually due to painful stimulus that needs to be avoided.
crown the region of a tooth that protrudes externally from the gums (gingiva).
cuboid bone (Greek: *kybos* - cube; *eidos* – resembling) tarsal bone that articulates posteriorly with the calcaneus, medially with the lateral cuneiform bone, and anteriorly with the fourth and fifth metatarsal bones.
cuneiforms (Latin: *cuneus* - wedge-shaped) small, paired cartilages of the larynx that help control the tension of the vocal cords.
cupula (*pl*: cupulae) (Latin: *cupa* - a tub) jelly-like cap over the hair cells in the crista ampullaris of the semicircular canals that bends the stereocilia of hair cells when the head rotates.
cuspids (*aka* canines) (Latin: *cuspis* - point) long, pointed teeth, that lie between the incisors and bicuspids (premolars), which are used for tearing food.
cutaneous membrane (*aka* skin) (Latin: *cutis* - skin; *membra* - a skin or membrane that covers a part of your body) epithelial tissue made up of stratified squamous epithelial cells that cover the outside of the body.
cuticle (Latin: *cuticula* - skin) in hair, the outermost layer of hard keratinocytes.
cyclic adenosine monophosphate (cAMP) the most common second messenger molecule in the cells; triggers a phosphorylation cascade following hormone-receptor binding.
cystic artery (Greek: *kystis* - bladder) an artery that branches off of the common hepatic artery, and provides blood to the gallbladder.
cytokine (*pl*: **cytokines**) (Greek: *kytos* - cell; *kineto* - to move) a chemical messenger, secreted by a macrophage, lymphocyte (such as a helper T cell), neutrophil, mast cell, or other cell, that delivers information to nearby cells; used in various immune responses, such as inflammation, and cell differentiation.
cytokines (*sing*: **cytokine**) (Greek: *kytos* - cell; *kinēsis* - movement) a group of chemical messengers, secreted by certain cells of the immune system, that affect growth and development of other cells and immune responses via autocrine, paracrine, or endocrine signaling; examples are interferons and interleukins.
cytokinesis (Greek: *kytos* - cell; *kinesis* - movement) final stage in cell division, where the cytoplasm divides to form two separate daughter cells.
cytology (Greek: *kytos* - cell; *logos* - study) the study of cells.
cytoplasm (Greek: *kytos* - a hollow cell; *plasma* - thing formed) internal material of the cell including the water-based fluid called cytosol, and all the other organelles and solutes.
cytoskeleton (Greek: *kytos:* a hollow cell; *skeletos* - skeleton) "skeleton" of a cell; formed by fibrous proteins that support the cell's shape and enable movement.
cytosol (Greek: *kytos* - a hollow cell; *sol* - soluble) clear, semifluid medium of the cytoplasm, made up mostly of water.
cytotoxic T cells (CD8 T cells) (*aka* T_c cells) (Greek: *kytos* - cell) T lymphocytes (T cells) that destroy cells that contain foreign antigens, such as virus-infected and tumor cells.
cytotrophoblast (Greek: *kytos* - a hollow cell; *trophe* - nourishment; *blast* - to build) cells of trophoblast.

D

Dalton's law (*aka* Law of partial pressures) a law stating that the total pressure exerted by a gas mixture (such as the air) equals the sum of the partial pressures of all of the gases that are present in the mixture.
dark adaptation (Latin: *adapto* - to adjust) a process that uses the stored rhodopsin within rods to allow us to see in dark spaces.
dartos muscle the muscles that make up the subcutaneous muscle layer of the scrotum.
deamination the process in which an amine (nitrogen-containing) group is removed from an amino acid; the amine group is converted into ammonia and then urea for excretion.
deciduous teeth (*aka* baby teeth, milk teeth, primary teeth) (Latin: *deciduus* - that which

falls down) the first set of 20 teeth to erupt in the mouth; eventually are replaced by the 32 permanent teeth.
decomposition reaction (Latin: *re-* - again; *actio* – action) type of catabolic reaction in which one or more bonds within a larger molecule are broken, resulting in the release of smaller molecules or atoms.
decussates (Latin: *decusso* - to form an X) crosses the midline; often refers to the crossover of nerves that allows the left side of the brain to control muscles on the right side of the body.
deep describes a position farther from the surface of the body.
deep brachial arteries (Latin: *brachio* - arm) arteries that provide blood to the posterior surface of the arm.
deep femoral artery (*aka* deep artery of the thigh) (Latin: *femur* - thigh) an artery that branches off of the femoral artery; supplies blood to the thigh muscles and branches to form the lateral and medial circumflex arteries.
deep femoral vein (*aka* profunda femoris vein) (Latin: *femur* - thigh) a deep vein that drains the deep regions of the thigh, and empties to the femoral vein.
deep vein thromboses (DVTs) (Greek: *thrombosis* - clumping) blood clots that form in one of the deep veins of the body, and decrease blood flow; often occur in a vein of the leg or pelvis.
dehydration (Latin: *de* - down; Greek: *hydro* – water) a condition in which the body does not contain enough water to perform basic bodily functions properly; occurs when water output exceeds water intake on a daily basis.
dehydration synthesis a reaction where one reactant gives up a hydrogen atom, and the other reactant gives up a hydroxyl group, to form a water molecule as a byproduct.
delta cell a pancreatic islet cell type; secretes the hormone somatostatin.
deltoid (Greek: *deltoeidēs* - shaped like a delta) thick muscle that caps the lateral shoulder; major abductor of the arm; facilitates flexing and medial rotation of the arm.
deltoid tuberosity (Greek: *deltoeidēs* - resembling the shape of a delta; Latin: *tuberosus* - full of lumps) roughened protrusion on the lateral mid-shaft of the humerus.
denaturation (Latin: *de* - from; Greek: *hydr* - water; *syn* - together; *thesis* - arranging) change in the structure of a molecule most often a protein.
dendrites (*sing*: dendrite) (*aka* branches) (Greek: *dendritēs* - related to a tree) part of the neuron that receives signals from neighboring cells.
dendrons highly branched processes extending from the neuron cell body; receive information from other neurons and sensory cells.
dendritic cells (Greek: *dendrites* - pertaining to a tree) phagocytic cells similar to macrophages in function, which transport antigens to lymph nodes and function as antigen-presenting cells for T cells.
dens bony projection that extends upward from the body on the anterior side of the second cervical vertebra.
dense body smooth muscle structure that attaches the thin filaments to the sarcolemma; shortens the muscle fiber as thin filaments slide past thick filaments.
dentes (*sing*: dens) (aka teeth) the teeth, which lie in the oral cavity, and are used to bite, tear, and grind food.
dentin (Latin: *dens* - tooth) a hard bonelike material that lies just under the enamel of a tooth crown; makes up a large majority of the tooth.
deoxygenated (*aka* oxygen-poor) (Latin: *de* - out of) a state in which the hemoglobin molecules in erythrocytes are not completely saturated with oxygen molecules (70–80 percent).
deoxyribonuclease an enzyme, produced by the pancreas, that digests deoxyribonucleic acid (DNA).
deoxyribonucleic acid (DNA) (Latin: *de* - away; Greek: *oxys* - keen; Latin: *nux* - nut; *acidus* - sour) deoxyribose-containing nucleotide that stores genetic information.
depolarize to make the membrane potential of a cell less negative.
depression (Latin: *depressio* - to press down) inferior movement of the mandible or scapula.
dermal papillae (*pl*: dermal papillae) (Greek: *derma* - skin) projections of the papillary layer of the dermis that increase surface contact between the epidermis and dermis.
dermcidin an antimicrobial chemical found in sweat.
dermis (Greek: *derma* - skin) layer of tissue between the epidermis and hypodermis; composed mainly of connective tissue; contains blood vessels, lymph vessels, nerves, and other structures.
descending aorta (Latin: *descendere* - come down) the portion of the aorta that extends inferiorly from the aortic arch to the termination point, where it splits into the common iliac arteries; divided into thoracic and abdominal segments.
descending colon the portion of the colon that runs down the left side of the body; connects the transverse colon to the sigmoid colon.
descending limb (*aka* nephron loop) (Latin: *descendere* - come down) the portion of a nephron loop (loop of Henle) that transports filtrate from the proximal convoluted tubule (PCT) to the ascending limb of the nephron loop in each nephron.
descending tracts nerve pathways that carry motor commands from the brain to the muscles via the spinal cord.
desmosomes (*sing*: desmosome) (Greek: *desmos* - a band; *sōma* - body) a type of anchoring junction that unites two neighboring cells structurally.
detrusor muscle (*aka* detrusor urinae muscle) (Latin: *detrudere* - to thrust down) the smooth muscle that makes up the wall of the bladder; contains muscle fibers running in all directions, to decrease the size of the bladder and maintain its rounded shape while emptying urine.
diabetes mellitus a condition caused by dysfunction or autoimmune destruction of the beta cells of the pancreas (Type 1) or cellular resistance to insulin (Type 2); results in abnormally high blood glucose levels.
diacylglycerol (DAG) a molecule produced following the cleavage of a membrane-bound phospholipid by phospholipase C; activates protein kinases, initiating various responses in the target cells.
diapedesis (*aka* emigration) (Greek: *diapedan* - to ooze through) a process by which leukocytes move from a capillary into the tissues by squeezing between adjacent cells of the capillary wall.
diaphragm (Greek: *diaphragma* - barrier) a flat, dome-shaped muscular organ that separates the thoracic cavity from the abdominopelvic cavity; one of the main muscles of inspiration.
diaphragma sella (Greek: *diaphragma* - a partition wall) one of the cranial dural septa; a fold of dura mater that contains an opening for the pituitary stalk.
diaphysis (Greek: *diaphysis* - a growing between) tubular shaft that runs between two epiphyses.
diarthrosis (*pl*: diarthroses) (Greek: *di-* two; *arthron* - joint) joint at which a lot of motion is possible; freely mobile joint.
diastole (Greek: *diastole* - dilation/relaxation) relaxation of the heart wall, also the term for the phases of the cardiac cycle in which relaxation occurs.
diastolic pressure (Greek: *diastole* - dilation) the minimum blood pressure in the large arteries that occurs during ventricular relaxation (diastole); expressed in units of mm Hg.
dicrotic notch (Greek: *dicro* - forked) dip in the aortic pressure graph that occurs at the moment of isovolumic ventricular diastole.
diencephalon (Greek: *dia* - through; *enkephalos* - brain) a major region of the adult brain that includes the thalamus and hypothalamus; retains its name from embryonic development.
differentiation (Latin: *differ* - differ) the process in which a stem cell becomes a specialized cell in the human body.
diffusion (Latin: difusus - to pour in different directions) the movement of molecules in response to concentration gradients.
diffusion (Latin: *diffuses* - to pour in different directions) movement of a substance from an area of higher concentration to one of lower concentration.
digastric (Greek: *di* - two; *gastēr* - belly) one of the suprahyoid muscles; has anterior and posterior bellies and elevates the hyoid bone and larynx during swallowing.
digital arteries (Latin: *digitus* - finger or toe) arteries that arise from the superficial and deep palmar arches; provide blood to the digits (fingers).
digital veins (Latin: *digitus* - finger or toe) veins that drain the digits (fingers or toes) and

empty blood into the palmar arches (hand) or dorsal venous arch (foot).
dihydroxyvitamin D (*aka* calcitriol, 1, 25-dihydroxycholecalciferol) (Greek: *di* - two; *hydro* - water) a hormone derived from vitamin D; considered to be the active form of vitamin D; causes calcium absorption in the small intestine.
diploid (*aka* 2n) (Greek: *diploos* - double; *eidos* - resemblance) condition of a cell having two sets of chromosomes, one set inherited from the sperm and one inherited from the egg.
disaccharides (Greek: *dis* - two; *sakcharon* - sugar) pair of carbohydrate monomers bonded by dehydration synthesis via a glycosidic bond.
distal (Latin: *distalis* - far away from center) describes the position of a structure on a limb farther from the point of attachment or the trunk of the body.
distal convoluted tubule (DCT) (Latin: *distantem* - remote; *convolutus* - to roll together) the part of the nephron distal to the nephron loop, which transports filtrate between the nephron loop and the collecting ducts; conducts reabsorption and secretion of several substances.
distal radioulnar joint (Latin: *distalis* - far away from center; *radius* - ray; Greek: *ōlenē* - ulna) articulation between the ulnar notch of the radius and the head of the ulna.
distal tibiofibular joint (Latin: *distalis* - far away from center; *tibia* - shin bone; *fibula* - brooch) articulation between the fibular notch of the tibia and the distal fibula.
diuresis (Greek: *diourein* - to urinate) excess loss of water in the urine, leading to the excretion of a large volume of urine.
diuretic (Greek: *diouretikos* - promoting urine) a substance that increases the volume of urine excreted by the body.
diuretic (Greek: *diouretikos* - prompting urine) a substance that increases urine output and therefore decreases water conservation by the body.
dizygotic (Greek: *di* - two; *zygotos* - yoked; *ikos* - pertaining to) twins that originate from two different eggs and two different sperm.
DNA polymerase (*pl*: polymerases) (Greek: *poly* - many; *meros* - part) enzymes that add nucleotides to a growing new strand of DNA during DNA replication.
DNA replication (Greek: *replico* - to fold back) process of duplicating the DNA of the genome.
dorsal arch (*aka* arcuate arch, arcuate artery) (Latin: *dorsalis* - back) an arterial arch formed by the merging of the dorsalis pedis artery with the medial and plantar arteries; provides blood to the metatarsal region of the foot and the digits (toes).
dorsal group (Latin: *dorsalis* - back) area of the foot that includes the extensor digitorum brevis.
dorsal interossei (Latin: *dorsalis* - back; *inter* - between; *os* - bone) intermediate muscles of the hand; abduct and flex the three middle fingers at the metacarpophalangeal joints and extend them at the interphalangeal joints.
dorsal (posterior) nerve root (Latin: *dorsalis* - back) sensory root of the spinal cord; enters at the posterior side of the spinal cord.
dorsal respiratory group (DRG) (Latin: *dorsalis* - of the back) one of the control centers for respiration in the medulla oblongata, which causes contraction of the diaphragm and intercostal muscles, in order to initiate inhalation.
dorsal venous arch (Latin: *dorsalis* - back) a curved superficial vein that arises from the dorsal metatarsal and dorsal digital veins, and drains into the great and small saphenous veins and the tibial veins.
dorsalis pedis artery (Latin: *dorsalis* - *back*; *pedalis* - foot) an artery that branches off of the anterior tibial artery; provides blood to the ankle and dorsal portion of the foot.
dorsiflexion (Latin: *dorsalis* - back; *flecto* - to bend) motion in the ankle that lifts the top of the foot toward the anterior leg.
dorsum nasi (Latin: *dorsalis* - of the back) the long, middle region of the external nose that lies between the bridge and the apex.
downregulation a reduction in the number of active receptors in/on a target cell for a hormone that is present in high concentration; this decreases the response of the cell to the hormone.
dual innervation (Latin: *duo* - two) the regulation of organ function by both the sympathetic and parasympathetic divisions of the autonomic nervous system.
ductus arteriosus (Latin: *ductus* - pipe) a short vessel that transports blood from the pulmonary trunk to the aortic arch in the fetal circulation, bypassing the fetal lungs, which are nonfunctional; after birth, closes to become the ligamentum arteriosum.
ductus deferens (*aka* vas deferens) a tube that carries sperm from the epididymis to the ejaculatory duct.
duodenal glands (*aka* Brunner's glands) (Latin: *duodecim* - twelve) glands found in the submucosa of the duodenum; secrete mucus.
duodenal papilla (Latin: *duodecim* - twelve; *papilla* - nipple) a small, raised area in the duodenum that marks the point at which the common bile duct and main pancreatic duct converge to release their secretions into the duodenum.
duodenum (Latin: *duodecim* – twelve) the first portion of the small intestine, which runs between the pyloric sphincter and the jejunum; receives chyme from the stomach and mixes it with bile and pancreatic juice.
ductus venosus (Latin: *ductus* - pipe) a short vessel that transports oxygen-rich blood from the umbilical vein to the inferior vena cava in the fetus, bypassing the fetal liver.
dura mater (Latin: *mater* - mother) the outermost layer of the three meninges; a thick, tough layer that surrounds and protects the brain and spinal cord.
dural venous sinuses venous channels between the two layers of the dura mater, which collect venous blood from smaller veins of the CNS and drain into the internal jugular veins for return to the heart.
dwarfism disorder in children, whose epiphyseal plates of the long bones are still active, caused by an abnormally low level of GH; results in growth impairment, especially in the long bones of the upper and lower limbs.
dynamic equilibrium (Greek: *dynamis* - power; Latin: *aequilibrium* - an equal balance) our sense of head or body position while the head or body is in motion.

E

eccentric contraction (Greek: *ek* - out; *kentron* - center) contraction characterized by a decrease in muscle tension as the muscle stretches to a length greater than its resting length.
eccrine sweat glands (Greek: *ekkrino* - to secrete; *glans* - acorn) sweat glands that produce hypotonic sweat for thermoregulation.
echocardiogram (Greek: *echo* - a returned sound; *cardiac* - *kardia* meaning heart) an ultrasound of the heart that allows blood flow to be visualized.
edema (Latin: *oidema* - a swelling tumor) an accumulation of excess fluid in the interstitial spaces, resulting in swelling; results from blockage of lymphatic drainage or an imbalance between capillary outflow and lymph collection.
effector (Latin: *effector* - producer) an organ or muscle that responds to a signal.
efferent (Latin: *effere* - to bring out) a neuron that brings input away from the central nervous system.
efferent arteriole (Latin: *effere* - to carry away; *arteriola* - an artery) an arteriole that transports blood from the glomerulus to the peritubular capillaries in each nephron.
efferent branch (Latin: *effere* - to bring out) component of a reflex arc that conveys information from the central nervous system to the target or effector organ.
efferent lymphatic vessels (Latin: *effere* - to bring out; *lympha* - water) lymphatic vessels that transport filtered lymph out of a lymph node.
ejaculation (Latin: *ejaculari* - to shoot out) the explosion of semen from the penis.
ejection fraction portion of the blood that is pumped or ejected from the heart with each contraction; mathematically represented by SV divided by EDV.
elastic artery (*aka* conducting artery) a large artery that contains many elastic fibers in the tunica media; transports blood to smaller (muscular) arteries.
elastic cartilage (Greek: elastreō - drive; Latin: cartilago - gristle) cartilage that contains elastic proteins; found in places that need to stretch or change shape and then return, such as the external ears, tip of the nose, and the epiglottis.
elastic fibers (Greek: *elastreō* - drive; Latin: *fibra* – fiber) in connective tissue, fibrous protein containing a high percentage of elastin, which allows the fibers to stretch and compress, and return to original size.

elasticity (Greek: *elastrēo* - drive) the ability to recoil back to original length due to the presence of elastic fibers.
elastin protein that allows the structure to return to its original shape after being stretched, compressed, or twisted.
elbow joint joint formed by the articulations between the trochlea of the humerus and the trochlear notch of the ulna, and the capitulum of the humerus and the head of the radius.
elbow joint hinge joint formed by the articulation of the humerus, radius, and ulna.
electrical gradient (*pl*: gradients) (Greek: *electron* - amber on which static electricity can be generated by friction; Latin: *gradus* - step) difference in the electrical charge (potential) between two regions.
electrical synapse (Greek: *syn* - with; *hapto* - to clasp) connection between two electrically active cells where ions flow directly through channels spanning adjacent cell membranes.
electrocardiogram (ECG) (*aka* EKG) (Greek: *echo* - a returned sound; *cardiac* - *kardia* meaning heart) surface recording of the electrical activity of the heart that can be used for diagnosis of irregular heart function.
electrolyte (Greek: *elektro* - electrical; *lytos* - loosed) a mineral that carries an electrical charge, such as an ion.
electrolytes (Greek: *ēlektron* - electricity generated by friction; *lysos* - soluble) particles capable of conducting electrical currents in a solution.
electron (Greek: *ēlektron* - electricity generated by friction) subatomic particle having a negative charge and nearly no mass; found orbiting the atom's nucleus.
electron shell (Greek: *ēlektron* - electricity generated by friction) area of space a given distance from an atom's nucleus in which electrons are grouped.
electron transport chain (ETC) a metabolic reaction sequence of food breakdown that produces ATP by transporting electrons along a series of electron carriers; uses oxygen, electrons, and hydrogen ions to produce water.
eleidin clear protein rich in lipids, located in the stratum lucidum; derived from keratohyalin and provides a barrier to water.
element (Latin: *elementum* - a rudiment) substance that cannot be created or broken down by ordinary chemical means.
elevation (Latin: *elevo* - to lift up) superior movement of the mandible or scapula.
emboli (*sing*: embolus) (Greek: *embolismos* - intercalation) blood clots or substances that are traveling in a blood vessel; if an embolus lodges in a blood vessel and stops blood flow, it is called an embolism.
embolism (Greek: *embolismos* - intercalation) the blockage of a blood vessel by a lodged embolus, which is a blood clot that is traveling in the circulating blood; common site of blockage is in a pulmonary artery, where it is called a pulmonary embolism.
embolus (*pl*: emboli) (Greek: *embolismos* - intercalation) a thrombus (blood clot) or mass of fatty plaque that has broken away from its site of formation and is traveling through the bloodstream.
emulsification (Latin: *emuls* - to milk out) a process by which specific bile components help to break down large lipid globules into smaller ones; occurs as bile salts and phospholipids in the bile surround small lipid globules to prevent them from reaggregating into large globules.
enamel the hard bonelike covering over the crown of a tooth; overlies the dentin.
encapsulated receptors (*aka* encapsulated nerve endings) (Latin: *encapsa* - in a box; *recipere* - to receive) anatomically complex structures where dendrites of sensory neurons are wrapped in such a way that enables their function.
end-diastolic volume (EDV) (*aka* preload) the amount of blood in the ventricles at the end of atrial systole just prior to ventricular contraction.
end-systolic volume (ESV) amount of blood remaining in each ventricle following systole.
endergonic reaction (Greek: *endo* - within; *ergon* - work; Latin: *re* - again; *actio* - action) reactions that absorb energy.
endocardium (Greek: *endo* - within, inside) innermost layer of the heart lining the heart chambers and heart valves; composed of endothelium reinforced with a thin layer of connective tissue that binds to the myocardium.
endochondral ossification (Greek: *endo* - within; *chondrion* - cartilage; *ossificatio* - to make bone) the process of building a bone from a template of hyaline cartilage.
endocrine gland (*pl*: endocrine glands) (Greek: *endo* - within; *krinōs* - to separate; Latin: *glans* - acorn) structure that synthesizes and releases chemical signals into the interstitial fluid to be delivered to target organs by the bloodstream.
endocrine glands (Greek: *endo* - within; *krinō* - to separate; Latin: *glans* – acorn) tissues or organs that secrete hormones into the surrounding body fluid, which are carried throughout the body to their target cells by the blood.
endocrine signaling (Greek: *endo* - within; *krinō* - to separate) chemical signals secreted by endocrine glands into the bloodstream, which induce responses in their target cells.
endocrine system (Greek: *endo* - within; *krinō* - to separate) cells, tissues, and organs of the body that secrete hormones into the extracellular fluid as a primary or secondary function.
endocytosis (Greek: *endo* - within; *kytos* - a hollow cell; *osis* - condition) import of material into the cell within a membrane-bound vesicle.
endogenous opioids (*aka* endorphins) (Greek: *endo* - within; *gen* - production) neurotransmitters that are released by inhibitory neurons; they prevent pain from being fully felt.
endolymph (*aka* Scarpa fluid) (Greek: *endo* - within; Latin: *lympha* - a clear fluid) a fluid that occupies the cochlear duct of the inner ear; derived from blood plasma but with a slightly different composition.
endometrium (Greek: *endo* - within; *meter* - mother) the inner layer of the wall of the uterus; thickens during the proliferative and secretory phases of the uterine cycle and is shed during the menstrual phase.
endomysium (Greek: *endo* - within; *mys* - muscle) thin layer of collagen and reticular fibers that surrounds each muscle fiber within the fascicle; contains extracellular fluid and nutrients.
endoneurium (Greek: *endo* - within; *neuron* - nerve) a layer of loose connective tissue that surrounds individual axons in a fascicle.
endoplasmic reticulum (ER) (*pl*: endoplasmic reticula) (Greek: *endo:* within; *plastos:* formed; Latin: *reticulum* - a small net) cellular organelle that consists of connected membranous sacs. When associated with ribosomes, it is called rough ER, when not associated with ribosomes, called smooth ER.
endothelial cells (Greek: *endo* - within; *thele* - nipple; *kytos* - cell) the simple squamous epithelial cells that line the lumen of all types of blood vessels.
endosteum (Greek: *endo* - within; *osteon* - bone) the membranous lining of the medullary cavity of a bone.
endothelins powerful vasoconstrictors secreted by the endothelial cells of blood vessels, which contract the smooth muscle in the walls of the blood vessels in order to increase blood pressure or decrease local blood flow.
endothelium (Greek: *endo* - within; *thēlē* - nipple) layer of smooth, simple squamous epithelium that lines the endocardium and blood vessels.
energy (Greek: *energeia* - in work) the capacity to do work.
energy-consuming phase the first of two stages in the reaction sequence of glycolysis, in which two molecules of ATP are used up in order to begin the reaction.
enteric nervous system (Greek: *enterikos* - intestinal) a portion of the PNS, which regulates the organs and processes of the digestive system.
enteric nervous system (*aka* intrinsic nervous system) (Greek: *enterikos* - intestinal) a network of nerves that lies within the wall of the GI tract; makes up most of the submucosal plexus, which controls glandular secretion, and the myenteric plexus, which controls motility (movement).
enteroendocrine cells (Greek: *enteron* - intestine; *endo* - within; *krinein* - distinguish) glandular cells of the stomach that secrete hormones.
enzyme (Greek: *en* - in; Latin: *zyme* - leaven) protein or RNA that catalyzes chemical reactions.
enzymes protein molecules that act as catalysts in particular chemical reactions; speed up the reaction rate, but do not cause reactions to occur that would not occur in nature.
eosinophils (Greek: *eos* - dawn; Latin: *philus* - loving) granulocytes that secrete anti-inflammatory chemicals and substances that destroy parasitic worms; make up a small percentage of leukocytes, and are stained by the dye, eosin.

ependymal cells (Greek: *ependyma* - an upper garment) glial cell type in the CNS that filters blood to produce cerebrospinal fluid.
epicardium (Greek: *epi* - upon) the outermost layer of the heart wall.
epicranial aponeurosis (*aka* galea aponeurosis) broad tendon connecting the frontalis and occipitalis.
epidermis (Greek: *epi* -upon; *derma* - skin) outermost tissue layer of the skin; composed of keratinized squamous epithelium.
epididymis (*pl*: epididymides) (Greek: *epi* - on; *didymos* - testicle) a tubular organ of the reproductive system of a biological male, in which sperm reside and mature for about two weeks after leaving the testis.
epidural space (Greek: *epi* - upon) a potential space between the dura mater and the cranial bones, and between the meningeal dura mater and the vertebral foramina.
epiglottis (Greek: *epi* - upon; *glossa* - tongue) an oval mass of elastic cartilage in the larynx; remains open to allow air to flow into the larynx and trachea and closes during swallowing to prevent food from entering these organs of the airway.
epimysium (Greek: *epi* - upon; *mys* - muscle) sheath of dense, irregular connective tissue that surrounds an entire skeletal muscle.
epinephrine (*aka* adrenaline) (Greek: *epi* - upon; *nephros* - kidney) signaling molecule (hormone) secreted by the adrenal medulla as part of the sympathetic response.
epinephrine (*aka* adrenaline) (Greek: *epi* - upon; *nephros* - kidney) an anime hormone secreted by the adrenal medulla in response to short-term stress; prepares the body for fight-or-flight responses, such as an increase in heart rate, breathing rate, and blood pressure, and dilation of the pupils and airways.
epineurium (Greek: *epi* - upon; *neuron* - nerve) a layer of fibrous connective tissue surrounding a nerve.
epiphyseal line (Greek: *epi* - upon; *physis* - growth) remnant of the epiphyseal plate following ossification.
epiphyseal plate (*aka* growth plate) (Greek: *epi* - upon; *physis* - growth) layer of hyaline cartilage within a growing bone; replaced by bone tissue during growth.
epiphyses (Greek: *epi* - upon; *physis* – growth) wide sections at each end of a long bone.
epiploic appendages (*aka* appendix epiploica) small fatty pouches that are suspended from the peritoneum of the large intestine; associated with the teniae coli.
epithalamus (Greek: *epi* - upon; Latin: *thalamus* - bedroom) a portion of the diencephalon that contains the pineal gland.
epithelial tissue (Greek: *epi* - upon, *thēlē*- nipple; French: *tissu* - woven) highly cellular tissue that serves primarily as a covering or lining of body parts.
epithelium (*pl*: epithelia) (Greek: *epi* - upon, *thēlē*- nipple) the lining of the tissue that is connected to the "outside" world.
equilibrium (*pl*: equilibria) (Latin: *aequus* - equal) a dynamic state at which molecules or ions move randomly, but there is no net movement of molecules or ions in either direction of the cell membrane.
equilibrium (Latin: *aequilibrium* - an equal balance) the sense of balance, derived from balance organs in the inner ear; consists of the senses of static and dynamic equilibrium.
erector spinae group muscle mass that forms the primary extensor of the vertebral column; consists of the iliocostalis, longissimus, and spinalis muscle groups.
erythrocytes (*sing*: erythrocyte) (*aka* red blood cells) (Greek: *erythros* - red; *kytos* - cell) cells that transport oxygen and other gases.
erythrocytes (*aka* red blood cells, RBCs) (Greek: *erythro* - red; *kytos* - cell) blood cells that transport oxygen and some carbon dioxide through the blood; make up the vast majority of the formed elements of the blood.
erythropoiesis (Greek: *erythro* - red; *poiesis* - a making) the production of new erythrocytes in the red bone marrow.
erythropoietin (EPO) (Greek: *erythro* - red; *poiesis* - a making) a hormone produced by the kidney, which causes the red bone marrow to produce more erythrocytes in low-oxygen conditions.
esophageal artery (Greek: *oisophagos* - passage for food) an artery that branches off of the thoracic aorta; provides blood to a portion of the esophagus.
esophageal hiatus (Greek: *oisophagos* - passage for food; Latin: *hiare* - gape) an opening in the diaphragm through which the esophagus passes from the thoracic cavity into the abdominopelvic cavity.
esophageal phase the third and final phase of swallowing, in which food is passing through the esophagus on its way to the stomach; peristaltic waves help the food pass through the esophagus.
esophageal plexus (Greek: *oisophagos* - passage for food; Latin: *plexus* - braid) an autonomic network that consists of both sympathetic and parasympathetic nerve fibers; located in the thoracic cavity, and regulates muscle contraction in the esophagus.
esophageal vein (Greek: *oisophagos* - passage for food) a vein that drains blood from the inferior part of the esophagus, and empties into the azygos vein.
esophagus (Greek: *oisophagos* - passage for food) a tubular organ of the digestive system, which transports food from the pharynx to the stomach via peristaltic waves.
essential nutrients molecules that the body cannot produce at all, or cannot produce in adequate quantity, and therefore must obtain from nutritional sources.
ethmoid bone (Greek: *ēthmos* - sieve) unpaired bone located between the two eye orbits; forms a barrier between the nose and the brain.
ethmoid sinus (Greek: *ēthmos* - sieve; Latin: *sinus* – cavity) small spaces separated by very thin bony walls located in the lateral aspects of the ethmoid bone.
eupnea true breathing; the normal rate and depth of breathing.
eversion (Latin: *everto* - to overturn) the turning of the foot to angle the bottom of the foot away from the midline of the body.
exchange reaction (Latin: *re* - again; *actio* - action) type of chemical reaction in which bonds are both formed and broken, resulting in the transfer of components.
excitability (Latin: *excito* - arouse) in muscles, the ability to contract or relax based on electrical properties of the plasma membrane.
excitation-contraction coupling (Latin: *excito* - arouse; *contractus* - to draw together) the union of events beginning with motor neuron signaling to a skeletal muscle fiber and ending with contraction of the fiber's sarcomeres.
excitatory postsynaptic potential (EPSP) graded potential in the postsynaptic membrane caused by depolarization; makes an action potential more likely.
excitement and plateau stages the phase where the body prepares for and/or engages in sex.
excursion (Latin: *ex* - out; *currere* - to run) sideways movement of the mandible (to the left or right).
exergonic reaction (Greek: *exo* - outside; *ergon* - work; Latin: *re* - again; *actio* – action) reactions that release energy.
exhalation (*aka* expiration) (Latin: *exhalare* - to breathe out) the process of moving air out of the lungs into the atmosphere.
exocrine gland (*pl*: exocrine glands) (Greek: *exo* - outside; *krinōs* - to separate; Latin: *glans* - acorn) structure that synthesizes and releases substances through ducts onto body surfaces.
exocytosis (Greek: *exō* - outside; *kytos* - a hollow cell; *osis* - condition) export of a substance out of a cell when a membrane-bound vesicle fuses with the cell membrane.
expiratory reserve volume (ERV) (Latin: *externus* - outside) the additional amount of air that can be exhaled after a normal exhalation.
extensibility (Latin: *extensus* - stretched out) the ability to stretch or extend.
extension (Latin: *extendo* - to stretch out) movement in the sagittal plane that increases the angle of a joint.
extensor carpi radialis brevis (Latin: *carpi* - wrist; *brevis* - short) superficial posterior muscle of the forearm that extends and abducts the hand at the wrist.
extensor carpi ulnaris (Latin: *carpi* - wrist) superficial posterior muscle of the forearm that extends and adducts the hand.
extensor digiti minimi superficial posterior muscle of the forearm that extends the little finger.
extensor digitorum superficial posterior muscle of the forearm that extends the hand at the wrist and the phalanges.
extensor digitorum brevis dorsal group muscle that extends the toes.

extensor digitorum longus muscle that is lateral to the tibialis anterior; raises the front of the foot.
extensor hallucis longus (Latin: *hallex* - great toe) muscle that is deep to the tibialis anterior; raises the front of the foot.
extensor indicis deep posterior muscle of the forearm that inserts onto the tendon of the extensor digitorum of the index finger.
extensor pollicis brevis deep posterior muscle of the forearm that inserts onto the base of the proximal phalanx of the thumb.
extensor pollicis longus deep posterior muscle of the forearm that inserts onto the base of the distal phalanx of the thumb.
extensor radialis longus superficial posterior muscle of the forearm that extends and abducts the hand at the wrist.
extensor retinaculum (Latin: *retinere* - to hold back) retinaculum that extends over the dorsal surface of the hand.
external acoustic meatus (Latin: *externus* - outside; *meatus* - to pass; Greek: *akoustikos* - sound) a canal in the center of the temporal bone; tunnel through which sound reaches the inner ear.
external anal sphincter (*aka* sphincter ani externus) (Latin: *sphincter* - contractile muscle) a ring of skeletal muscle in the anal canal, which provides voluntary control over defecation.
external carotid artery (Latin: *externus* - outwards; Greek: *karotides* - great arteries of the neck) an artery that branches from the common carotid artery and supplies blood to various structures around the face, scalp, lower jaw, neck, esophagus, and larynx.
external ear (Latin: *externus* - outside) the auricle, ear canal, and tympanic membrane.
external elastic membrane (Latin: *externus* - outside) a layer of elastic fibers that lies between the tunica media and the tunica externa in large arteries.
external iliac artery (Latin: *externus* - outside; *iliacus* - pertaining to colic) an artery that branches off of the common iliac artery and provides blood to the lower limbs and a few structures of the anterior wall of the abdominal cavity.
external iliac vein (Latin: *externus* - outside; *iliacus* - pertaining to colic) a continuation of the femoral vein as it enters the trunk of the body; drains blood from the legs and empties into the common iliac vein.
external intercostal (Latin: *inter* - between; *costa* - rib) one of 11 pairs of superior intercostal muscles; raise the rib cage, resulting in expansion.
external intercostal muscles (Latin: *externus* - outside; *inter* - between; *costa* – rib) short muscles between the ribs, which expand the thoracic cavity for inhalation.
external jugular vein (Latin: *externus* - outside; *iugulum* - collarbone) a vein found in the superficial portion of the neck; empties blood from the superficial regions of the head and scalp; drains into the subclavian vein.
external nose (Latin: *externus* - outside) the portion of the nose that is visible from the outside; the outer portion is composed of skin, and the inner part is composed of cartilage and parts of the frontal, nasal, and maxillary bones.
external oblique (Latin: *externus* - outside; *obliquus* - slanted) superficial abdominal muscle composed of fibers that extend inferiorly and medially.
external occipital protuberance (Latin: *externus* - outside) small bump at the midline on the posterior skull on the occipital bone; attachment site for a ligament of the posterior neck.
external urethral orifice (Latin: *urethra* - passage for urine; *orificium* - an opening) The opening of the urethra.
exteroceptor (*aka* outer ear) (Latin: *externus* - outside; *recipere* - to receive) sensory receptor that reacts to stimuli in the external environment.
extracellular fluid (ECF) (Latin: *extra* - outside; Greek: *kytos* - a hollow cell; Latin: *fluo* - to flow) fluid exterior to cells; includes the interstitial fluid, blood plasma, and fluid found in other reservoirs in the body.
extracellular fluid (ECF) (Latin: *extra* - outside; Greek: *kytos* - cell) all fluid outside of the cells; consists of interstitial fluid, blood plasma, lymph, and specialized fluids such as synovial fluid in the joints, cerebrospinal fluid, and aqueous and vitreous humors in the eye.
extracellular matrix (ECM) a network of proteins and polysaccharides, outside of cells, that supports and guides the cells developmentally and physically.
extraocular muscles (Latin: *extra* - outside; *oculus* - eye) six skeletal muscles that lie outside the eyeball that control eye movement.
extrinsic (Latin: *extrinsecus* - from without) muscle that originates outside of the area in which it acts.
extrinsic eye muscles (Latin: *extrinsecus* - from without) muscles that move the eyeball; originate outside the eye and insert onto the outer surface of the white of the eye.
extrinsic ligament (Latin: *extrinsecus* - from without) ligament located on the outside of the articular capsule.
extrinsic pathway (Latin: *extrinsecus* - from without) one of the two pathways that initiate blood coagulation; once activated by a factor found outside of the blood, it leads to the activation of the common pathway of coagulation.

F

facet flattened area on a bone for that serves as a site of articulation or connection to a muscle.
facial bones fourteen bones that form the structures of the face, upper and lower jaws, and the hard palate.
facial nerve (CN VII) seventh cranial nerve; regulates the muscles of facial expressions, part of the sense of taste and the production of saliva.
facilitated diffusion (*aka* mediated diffusion) (Latin: *facilis* - easy; *diffuses* - to pour in different directions) process by which a substance moves from areas where it is more concentrated to less concentrated, but requires a membrane protein to move.
$FADH_2$ (*aka* Flavin adenine dinucleotide) the reduced form of the coenzyme FAD; transports electrons and hydrogen ions from the citric acid cycle to the electron transport chain.
false ribs ribs 8–12; these ribs do not attach directly to the sternum.
falx cerebelli (Latin: *falx* - sickle) one of the cranial dural septa that form a partition between the two halves of the cerebellum.
falx cerebri (Latin: *falx* - sickle) the largest of the cranial dural septa; forms a partition between the right and left cerebral hemispheres.
fascia (Latin: *fascia* - a band) a sheet of connective tissue.
fascia (Latin: *fascia* - a band or fillet) a layer of dense (fibrous) connective tissue that closely surrounds an entire skeletal muscle and lies external to the epimysium.
fascicle (Latin: *fascis* - bundle) 1) a bundled group of skeletal muscle fibers wrapped by the perimysium; 2) bundles of axons within a nerve, surrounded by a layer of connective tissue called the perineurium.
fast glycolytic (FG) (*aka* fast-twitch muscle fibers, Type IIb muscle fibers; Type Iix fibers) (Greek: *glykys* - sweet; *lysis* - a loosening) muscle fibers that contract relatively quickly and use glycolysis to produce ATP.
fast oxidative (FO) (*aka* Type IIa muscle fibers, intermediate fibers) muscle fibers that contract relatively quickly and use a mixture of aerobic respiration and glycolysis to produce ATP.
fat pads small masses of adipose tissue found at many joints, which cushion the bones of the joint.
fauces (Latin: *fauces* – gullet) the passageway from the posterior portion of the oral cavity to the pharynx.
Fc region the "tail" portion of a Y-shaped antibody molecule; called the "fragment crystallizable" region, it is the area in which the two heavy chains of the antibody lie side by side; binds to some complement proteins and some plasma membrane proteins of various cells of the immune system.
feces (*aka* stool) (Latin: *faeces* - sediment) the waste product of the digestion process; consists of water, undigested materials, bacteria, and organic wastes.
femoral artery (Latin: *femur* - thigh) a continuation of the external iliac artery as it enters the thigh; supplies blood to the structures of the thigh, and is renamed to the popliteal artery as it runs down the posterior side of the knee.
femoral circumflex vein (Latin: *femur* - thigh; *circumflexus* - bent around) a vein that encircles the neck of the femur; drains the head and neck of the femur, and empties into the femoral vein.
femoral nerve (Latin: *femur* - thigh) a peripheral nerve of the anterior leg, which arises from the lumbar plexus; innervates the anterior portion of the thigh.

femoral triangle (Latin: *femur* – thigh) region at the junction between the hip and the leg; includes the pectineus, femoral artery, femoral nerve, femoral vein, and deep inguinal lymph nodes.
femoral vein (Latin: *femur* - thigh) a continuation of the popliteal vein in the thigh, which empties into the external iliac vein in the trunk of the body; drains blood from the thigh; tributaries include the great saphenous, the deep femoral, and the femoral circumflex veins.
femur (*aka* thigh bone) (Latin: *femur* - thigh) long, single bone of the thigh.
fenestrated capillary (Latin: *fenestrare* - window) capillaries that contain fenestrations (pores) in the endothelial cells that allow for the movement of larger molecules across the capillary walls; less common than continuous capillaries but more common than sinusoid capillaries.
ferritin an iron-storage protein found in the liver, spleen, and bone marrow, and sometimes in the blood plasma.
fertilization (Latin: *fertilis* - fruitful) The event in which a sperm encounters and penetrates an egg, resulting in the combining of their genetic material.
fever an increase of more than 1°C (1.8°F) above the set point for body temperature; a mechanism used by the immune system to create a hostile environment for the replication of pathogens and to increase the reaction rate of the immune response.
fiber carbohydrates, such as cellulose, that are not digestible by human enzymes because the enzymes cannot break the chemical bonds within them.
fibrillation quivering or irregular contraction.
fibrin (Latin: *fibra* – fiber) an insoluble protein produced by the activation of the plasma protein fibrinogen, which forms a threadlike mesh that traps platelets and blood cells in a blood clot.
fibrinogen (Latin: *fibra* - fiber; Greek: *genes* - to produce) a plasma protein that is converted into fibrin during the process of blood clotting; produced by the liver.
fibrinolysis (Latin: *fibra* - fiber; Greek: *lysis* - to loosen) the decomposition and removal of a blood clot after healing has occurred, by dissolving of the fibrin mesh by the enzyme plasmin.
fibroblast (*pl*: fibroblasts) (Latin: *fibra* - fiber) the most abundant cell type in connective tissue, secrete protein fibers that form connective tissue.
fibrocartilage (Latin: *fibra* - fiber; *cartilago* - gristle) thick, tough bundles of collagen fibers embedded in ground substance.
fibrous joint (Latin: *fibra* - fiber) joint at which the adjacent bones are directly connected by fibrous connective tissue.
fibrous layer (Latin: *fibra* - fiber) the outer layer of the eye, which consists of dense connective tissue and provides a strong framework for the eye.
fibrous skeleton of the heart the collective term for the valves of the heart, which are fibrous and nonconductive in composition.
fibula (Latin: *fibula* - brooch) thin, lateral bone of the lower leg.
fibular collateral ligament (Latin: *fibula* - brooch) extrinsic ligament on the lateral side of the knee joint that runs from the lateral epicondyle of the femur to the head of the fibula.
fibular nerve (*aka* common fibular nerve, common peroneal nerve, external popliteal nerve, lateral popliteal nerve) (Latin: *fibula* - brooch) a peripheral nerve of the posterior leg, which branches from the sciatic nerve and innervates the skin and muscles of the lateral portion of the leg.
fibular notch (Latin: *fibula* - brooch) wide groove located on the lateral side of the distal tibia; articulates with the distal end of the fibula.
fibular vein (Latin: *fibula* - brooch) a vein that drains blood from the muscles and skin in the region of the fibula, and empties into the popliteal vein.
fibularis brevis (*aka* peroneus brevis) muscle of the lateral compartment of the lower leg; plantar flexes the foot and ankle.
fibularis longus (*aka* peroneus longus) muscle of the lateral compartment of the lower leg; plantar flexes the foot and ankle.
fibularis tertius (*aka* peroneus tertius) small muscle that originates on the anterior surface of the fibula; associated with and sometimes fused to the extensor digitorum longus.
filling time in the heart, the duration of ventricular diastole, during which filling occurs.
filtrate the fluid that occupies the renal tubules of the nephron; produced from the small molecules of the blood plasma that are filtered out of the glomerulus in the renal corpuscle.
filtration membrane the filtration apparatus that allows small molecules to pass from the blood of the glomerulus into the lumen of the glomerular (Bowman's) capsule in each renal corpuscle; consists of endothelium of the glomerulus (which is fenestrated), the podocytes of the glomerular capsule, and the basement membranes between them.
filtration slits long, narrow gaps in the pedicels (foot processes) of podocytes, the cells that surround the glomeruli; allow substances filtered by the glomeruli to enter the glomerular (Bowman's) capsule.
fimbriae (Latin: *fimbria* - fringe) the fingerlike extensions from the infundibulum of the uterine tubes; sweep over the edge of the ovary to help direct an ovulated oocyte into the uterine tube.
first-degree burn superficial burn that affects only the epidermis.
first messenger an extracellular substance that binds to a cell surface receptor, triggering intracellular activity via a second messenger.
fixator synergist that works by stabilizing the insertion of the prime mover.
flagellum (*pl*: flagella) (Latin: *flagrum* - whip) appendage on certain cells formed by microtubules and modified for movement.
flat bone thin and often curved bone; provides protection for internal organs and points of attachment for muscles.
flatus (*aka* gas) (Latin: *flatus* - a blowing) gas in the stomach or the intestine; derived from swallowing air and from fermentation of materials in the digestive tract by bacterial inhabitants.
flavin adenine dinucleotide (FAD) (Latin: *flavus* - yellow) a coenzyme that transports electrons and hydrogen ions from the citric acid cycle to the electron transport chain.
flexion (Latin: *flecto* - to bend) movement in the sagittal plane that decreases the angle of a joint.
flexor carpi radialis (Latin: *carpi* - wrist) the most lateral anterior superficial muscle of the forearm; flexes and abducts the hand at the wrist.
flexor carpi ulnaris (Latin: *carpi* - wrist) anterior superficial muscle of the forearm; flexes and adducts the hand at the wrist.
flexor digiti minimi brevis hypothenar muscle; flexes the little finger.
flexor digitorum longus deep muscle in the posterior compartment of the leg; flexes the four small toes.
flexor digitorum profundus (Latin: *profundus* - deep) deep anterior muscle of the forearm; flexes the phalanges of the fingers and the hand at the wrist.
flexor digitorum superficialis the most medial anterior superficial muscle of the forearm that flexes the hand and the digits.
flexor hallucis longus deep muscle in the posterior compartment of the leg; flexes the big toe.
flexor pollicis brevis thenar muscle; flexes the thumb.
flexor pollicis longus deep anterior muscle of the forearm that flexes the distal phalanx of the thumb.
flexor retinaculum (*aka* transverse carpal ligament) (Latin: *retinere* - to hold back) retinaculum that extends over the palmar surface of the hand.
floating ribs ribs 11–12; these ribs do not attach to the sternum or to the costal cartilage of another rib.
fluid compartment fluid-containing regions of the body, which are separated from other regions by barriers such as plasma membranes.
fluid connective tissue (Latin: *fluidus* - to flow; French: *tissu* - woven) blood and lymph; cells that circulate in a liquid extracellular matrix containing nutrients, salts, and dissolved proteins.
follicle-stimulating hormone (FSH) (Latin: *folliculus* - a small bag) an anterior pituitary hormone, which stimulates the production and maturation of sex cells in the ovary or testis.
folliculogenesis (Latin: *folliculus* - a small bag; Greek: *genesis* - origin) the process of development of ovarian follicles and their maturation from primordial to tertiary follicles.
fontanelle (*pl*: **fontanelles**) (*aka* "soft spot") (French: *fontaine* - spring) area of dense connective tissue that separates the bones of the skull

prior to birth and during the first year after birth; disappear by age 2.
foramen lacerum (Latin: *foramen* - opening; *lacero* - to tear to pieces) irregular opening at the base of the skull, immediately inferior to the exit of carotid canal.
foramen magnum (Latin: *foramen* - opening; *magnus* - large) large opening in the occipital bone of the skull through which the spinal cord emerges and the vertebral arteries enter the cranium.
foramen ovale (Latin: *foramen* – opening; *ovum* - egg) 1) large opening in the floor of the middle cranial fossa; allows for the passage of the spinal cord through the skull; 2) an opening between the right and left atria in the fetal heart, which transports oxygen-rich blood from the right atrium directly into the left atrium; this bypasses the fetal lungs, as they are nonfunctional.
foramen rotundum (Latin: *foramen* - opening) round opening in the floor of the middle cranial fossa, inferior to the superior orbital fissure; exit point for sensory nerve supplying the cheek, nose, and upper teeth.
foramen spinosum (Latin: *foramen* - opening; *spina* - thornlike projection) small opening in the floor of the middle cranial fossa, located posterior-lateral to the foramen ovale; entry point for artery supplying covering layers that surround the brain.
forebrain (*aka* prosencephalon) anterior region of the adult brain that includes the cerebrum and diencephalon.
formed elements the cells and fragments of the blood; consist of erythrocytes, all five types of leukocytes, and platelets; make up about 45 percent of the blood volume.
fornix (Latin: *fornix* – arch) a bundle of nerves (white matter) that lies inferior to the corpus callosum; a component of the limbic system, which connects other limbic system structures to each other; important in memory and emotion.
fossa (*pl*: fossae) (*aka* cavity) (Latin: *fossa* – trench) shallow depression on a bone's surface.
fossa ovalis (Latin: *fossa* - ditch) oval-shaped depression in the interatrial septum that marks the former location of the foramen ovale.
fourth ventricle (Latin: *ventriculus* - belly) one of the chambers in the brain that produces and circulates cerebrospinal fluid; lies between the cerebellum and the brainstem; opens into the subarachnoid space through the median and lateral apertures.
fovea capitis (Latin: *fossa* - trench; *caput* - head) indentation on the head of the femur; serves as the site of attachment for a ligament.
fovea centralis (Latin: *fovea* - a pit; *centralis* - central) an indented area in the center of the macula lutea that contains the highest density of cones and almost no rods.
fracture hematoma (Latin: *fractura* - to break; Greek: *haima* - blood; *-oma* - tumor) blood clot that forms at the site of a broken bone to prevent further blood loss.
Frank-Starling mechanism relationship between ventricular stretch and contraction in which the force of heart contraction is directly proportional to the initial length of the muscle fiber.
frontal eye fields regions of the frontal lobe of the cerebrum, associated with voluntary movement of the eyes.
frontal bone unpaired bone that forms forehead, roof of orbit, and floor of anterior cranial fossa.
frontal lobe a large region of the cerebrum that lies directly beneath the frontal bone; responsible for planning and initiating muscle contraction, including contractions for speech and eye movements, and higher-order cognitive behaviors.
frontal plane (*aka* coronal plane) two-dimensional, vertical plane that divides a structure into anterior and posterior halves.
frontal sinus (Latin: *sinus* - cavity) air-filled space within the frontal bone; just above the eyebrows.
frontalis anterior region of the occipitofrontalis muscle.
functional group (Latin: *functus* - to perform) group of atoms linked by strong covalent bonds that tends to behave as a distinct unit in chemical reactions with other atoms.
functional residual capacity (FRC) the amount of air left in the lungs after a normal exhalation; can be calculated by taking the sum of the ERV and RV.
fundus (Latin: *fundus* - bottom) the rounded portion of the stomach, located above the cardiac region; often traps swallowed air 000; the rounded region of the uterus that extends superiorly to the points of entry of the uterine tubes.
fusiform (Latin: *fusus* - spindle; *forma* - form) spindle-shaped parallel muscles with a belly.

G

G_1 phase (*aka* gap one phase) first phase of the cell cycle, after a new cell is born.
G_2 phase (*aka* gap two phase) third phase of the cell cycle, after the DNA synthesis phase.
G protein (*aka* guanine nucleotide-binding protein) an intracellular protein associated with a cell membrane hormone receptor; upon hormone-receptor binding, it activates an ion channel or enzyme to carry out the effects of the hormone.
gallbladder (Greek: *gaster* - stomach) an accessory organ of the digestive system, which stores and concentrates bile, and secretes it into the common bile duct via the cystic duct.
gametes (Greek: *gamete* - a wife) sex cells (sperm or egg) that contain the genetic material to produce offspring; haploid cells.
ganglion (Greek: *ganglion* - knot) a group of neuron cell bodies in the peripheral nervous system.
ganglion cells (Greek: *ganglion* - a swelling) cells that form the anterior layer of the retina, which synapse with the bipolar cells and leave the eye to become the optic nerve.
gap junction (*pl*: gap junctions) intercellular passageway that allows small molecules and ions to move between the cytoplasm of adjacent cells.
gas transport the movement of respiratory gases (oxygen and carbon dioxide) throughout the body within the blood.
gastric emptying (Greek: *gaster* - stomach) the process of gradually propelling chyme from the stomach into the duodenum; may take up to four hours after a large meal.
gastric gland (Greek: *gaster* - stomach) a tubular gland in the mucosa of the stomach, containing cells that produce the components of gastric juice, such as pepsinogen, hydrochloric acid, intrinsic factor, mucus, and hormones.
gastric juice (Greek: *gaster* - stomach) a mixture of the secretory products of the stomach; consists of hydrochloric acid, pepsinogen, pepsin, intrinsic factor, mucus, and hormones such as gastrin.
gastric phase (Greek: *gaster* - stomach) the middle stage of gastric secretion that starts when food enters the stomach from the esophagus.
gastric pits (Greek: *gaster* - stomach) tiny depressions in the mucosa of the stomach, which lead into the gastric glands.
gastrin a peptide hormone secreted by the stomach and duodenum in response to stomach distention or the presence of food in the stomach; stimulates the release of gastric juice and gastric motility (mixing and propelling movements).
gastrocnemius (Greek: *gastroknēmia* - calf of the leg) most superficial and visible muscle of the calf.
gastrocolic reflex (*aka* gastrocolic response) (Greek: *gaster* - stomach; *kolikos* - lower intestine) a strong movement of compacting feces down the colon, stimulated when food enters the stomach, which makes room for the recently ingested food in the GI tract; often causes the urge for defecation just after a meal.
gastroileal reflex (Greek: *gaster* - stomach; Latin: *ileum* - groin) an increase in segmentation contractions in the ileum of the small intestine that occurs when food enters the stomach; propels chyme toward the large intestine to make room for recently ingested food in the small intestine.
gastrointestinal (GI) tract (*aka* alimentary canal) (Greek: *gaster* - stomach; Latin: *intestinum* - a gut) a long passageway consisting of several tubular organs, that helps to break down food, absorb the nutrients into the body, and generate, store, and excrete the waste; consists of the mouth, pharynx, esophagus, stomach, small intestine, large intestine, rectum, and anus.
gender reflects who a person knows themselves to be.
gender expression the sum of the characteristics that an individual shows the world, things like the clothes that they wear or their demeanor.

gene (*pl*: genes) (Greek: *genos* – birth) functional stretch of DNA that provides the genetic information necessary to build a protein.
genicular artery (Latin: *geniculatus* - having knots) an artery that branches off of the femoral artery; provides blood to the knee area.
genioglossus (Greek: *geneion* - chin; *glōssa* - tongue) muscle originating on the mandible; allows the tongue to move downward and to the front.
geniohyoid [muscle] (Greek: *geneion* - chin; *hyoeidēs* - y-shaped) one of the suprahyoid muscles; depresses the mandible and raises and pulls the hyoid bone anteriorly.
genital folds (Latin: *genitalis* - fruitful) the folds on the epidermis between the forming legs of the embryo.
genital ridge (Latin: *genitalis* - fruitful) the precursor of gonads during embryonic development.
genome (*pl*: genomes) (Greek: *genos* - birth; *ome* - fixed system) entire complement of an organism's DNA; found within virtually every cell.
germ cells (*sing*: germ cell) (Latin: *germen* - bud; Greek: *kytos* - a hollow cell) cells that produce eggs and sperm.
ghrelin a hormone produced by the stomach when it is empty, which is responsible for stimulating the appetite.
gigantism (Greek: *gigas* - giant) a disorder in children, whose epiphyseal plates in the long bones have not yet ossified, by the secretion of abnormally large amounts of GH, resulting in excessive growth.
gingivae (*sing*: gingiva) (Latin: *gingivae* - gums) the gums that surround the neck of a tooth.
glabella (Latin: *glabellus* - hairless) slight depression at the anterior midline of the frontal bone.
glans (*aka* glans penis) (Latin: *glans* - acorn) the most distal portion, or head, of the penis or the clitoris.
glenohumeral joint (*aka* shoulder joint; GH joint) (Greek: *glēnoeidēs* - resembling a socket; Latin: *humerus* - arm) the shoulder joint; formed by the articulation between the glenoid cavity of the scapula and the head of the humerus; allows for flexion/extension, abduction/adduction, circumduction, and medial/ lateral rotation of the humerus.
glenohumeral ligament (Greek: *glēnoeidēs* - resembling a socket; Latin: *humerus* - arm) one of three ligaments that strengthen the anterior and superior sides of the glenohumeral joint.
glenoid cavity (*aka* glenoid fossa) (Greek: *glēnē* - socket of joint; Latin: *cavus* - hollow) shallow depression on the lateral scapula, forms the corner between the superior and lateral borders; articulates with the head of the humerus.
glenoid labrum (Greek: *glēnoeidēs* - resembling a socket) slightly raised fibrocartilage lip at the perimeter of the glenoid cavity.
glial cell (*aka* neuroglial cells) (Greek: *glia* - glue) various cell types that make up neural tissue excluding neurons; cells are responsible for maintenance and support of neurons.
glial cells (*sing*: glial cell) (Greek: *glia* - glue; *kytos* - cell) nervous tissue cells that play a supportive role to neurons and increase their function.
globin (Latin: *globus* - ball) a protein that makes up a portion of a hemoglobin molecule; binds to a heme group to form hemoglobin.
globulins (Latin: *globus* - ball) a group of plasma proteins that functions as transport proteins for lipids and some ions or as antibodies.
globus pallidus (Latin: *globus* - globe) a layered nucleus just medial to the putamen in the basal nuclei.
glomerular capsule (*aka* Bowman's capsule) the cup-shaped beginning of a renal tubule; along with the glomerulus, makes up the renal corpuscle; receives the filtrate from the glomerulus during glomerular filtration and transports it to the proximal convoluted tubule.
glomerular filtration rate (GFR) (Latin: *glomerationem* - ball) the amount of filtrate produced by the glomeruli in the kidneys each minute; usually stated in mL/min.
glomerulus (*pl*: glomeruli) (Latin: *glomerationem* - ball) a cluster of capillaries encased in a glomerular (Bowman's) capsule; a renal corpuscle consists of a glomerulus and a glomerular capsule; filters small molecules out of the blood.
glossopharyngeal nerve (CN IX) ninth cranial nerve; responsible for controlling muscles in the oral cavity and upper throat, as well as part of the sense of taste and the production of saliva.
glottis (Greek: *glossa* – tongue) the opening between the vocal folds (vocal cords) through which air passes for sound production.
glucagon (Greek: *glyco* - sweet; *agon* - push forward) a hormone secreted by the pancreas that acts upon the liver cells to inhibit glycogen synthesis and stimulate glycogen breakdown, releasing glucose into the blood.
glucocorticoids hormones produced by the zona fasciculata of the adrenal cortex, which regulate glucose metabolism.
gluconeogenesis (Greek: *glyco* - sweet; *neo* - new; *genesis* - origin) the formation of new glucose molecules from non-carbohydrates, such as pyruvate, lactate, glycerol, and certain amino acids; occurs when the blood glucose level is too low.
gluteal tuberosity (Greek: *gloutos* - buttock; Latin: *tuberosus* - full of lumps) roughened area of the proximal femur; extends inferiorly from the base of the greater trochanter.
gluteus maximus (Greek: *gloutos* - buttock) the largest muscle in the buttocks.
gluteus medius (Greek: *gloutos* - buttock) the second largest muscle in the buttocks.
gluteus minimus (Greek: *gloutos* - buttock) the smallest muscle in the buttocks.
glycocalyx (*pl*: glycocalyces) (Greek: *glykys* - sweet; *kalyx* - shell) coating of sugar molecules that surrounds the cell membrane.
glycogen (Greek: *glykys* - sweet) a polysaccharide consisting of a long, branching chain of glucose molecules, which serves as the storage form of excess glucose.
glycogenesis (Greek: *glykys* - sweet; *genesis* - origin) a process in which the body synthesizes glycogen, a molecule that stores excess glucose for later use.
glycogenolysis (Greek: *glykys* - sweet; *lysis* - loosening) the process in which stored glycogen is broken down into glucose, in order to increase the blood glucose level when it is too low.
glycolysis (Greek: *glykys* - sweet; *lysis* – a loosening) an anaerobic sequence of metabolic reactions that decomposes glucose into two molecules of pyruvate and results in a net gain of two molecules of ATP.
glycoprotein (*pl*: glycoproteins) (Greek: *glykys* - sweet; *protos* - first) protein that has one or more carbohydrates attached.
glycosuria a state in which glucose is found in the urine; often results from high blood glucose due to untreated diabetes mellitus.
goblet cells (*sing*: goblet cell) (Greek: *kytos* - cell) mucous-secreting, unicellular exocrine glands found in columnar epithelium of mucous membranes.
goiter an externally visible enlargement of the thyroid gland, which results from overstimulation and growth of the thyroid gland by thyroid stimulating hormone (TSH); occurs in response to iodine deficiency or hyperthyroidism.
Golgi apparatus (*pl*: apparati) (Latin: *apparatus:* preparation) cellular organelle formed by a series of flattened, membrane-bound sacs that functions in protein modification, tagging, packaging, and transport.
gomphosis (Greek: *gomphos* - bolt) fibrous joint that anchors the root of a tooth into its bony jaw socket via strong periodontal ligaments.
gonadal artery (Greek: *gonos* - child) an artery that branches off of the abdominal aorta, and supplies blood to the gonads; in females, supplies blood to the ovary and uterine tube, and is also called the ovarian artery; in males, supplies blood to the testis, and is also called the testicular artery.
gonadal veins (Greek: *gonos* - child) a vein that drains the gonads; drains the ovary in the female and is also called the ovarian vein; in the male, drains the testis and is also called the testicular vein.
gonads (Greek: *gonos* - child) the primary reproductive organs that produce sex cells (sperm or eggs) and sex hormones.
gracilis (Latin: *gracilis* - slender) long, slender muscle running down the medial side of the thigh, from the pubis to the tibia; flexes the knee and adducts the thigh.
graded potential change in membrane potential that varies in size.
granular cells cells such as neutrophils, eosinophils, basophils, and mast cells, which contain granules consisting of chemicals that are toxic to pathogens.
granular leukocytes (Greek: *leuko* - white; *kytos* - cell) leukocytes (neutrophils, eosinophils,

and basophils) that contain many large, visible cytoplasmic granules filled with chemicals that help in the immune response against pathogens.
granulation tissue (Latin: *granulum* - grain; French - *tissu* - woven) a collagen bridge that is formed across the epidermis to promote growth during tissue healing.
granules vesicles in the cytoplasm of leukocytes that contain chemicals used in the immune response; used to destroy pathogenic microorganisms or to regulate the inflammatory response.
granulosa cells (Latin: *granulum* - granule) the cells that surround and nourish the oocyte in an ovarian follicle.
gray matter a region in the brain or spinal cord that consists mainly of neuron cell bodies, dendrites, and unmyelinated axons.
gray rami communicantes (*sing*: ramus communicans) (*aka* gray communicating branch, gray communicating ramus) (Latin: *ramus* - branch) unmyelinated axons that provide short connections between a sympathetic chain ganglion and the spinal nerve that contains the postganglionic sympathetic fiber.
great cardiac vein (Greek: *cardiac* - *kardia* meaning heart) vessel that follows the interventricular sulcus on the anterior surface of the heart and flows along the coronary sulcus into the coronary sinus on the posterior surface; parallels the anterior interventricular artery and drains the areas supplied by this vessel.
great saphenous vein a large superficial vein that travels up the medial side of the leg and thigh; arises from the dorsal venous arch of the foot, and empties into the femoral vein; drains blood from the superficial regions of the leg and thigh; longest vein in the body.
greater curvature (*aka* curvatura ventriculi major) the broad inferior convex curve of the stomach.
greater omentum (*aka* great omentum, omentum majus, gastrocolic omentum, epiploon) a large mesenterial sheet that hangs down from the inferior surface of the stomach like a curtain covering the small intestine.
greater pelvis (*aka* greater pelvic cavity, false pelvis) (Latin: *pelvis* - basin) space superior the pelvic brim.
greater sciatic notch (Latin: *sciaticus* - corruption of; Greek: *ischiadikos* - hip joint) inverted U-shaped indentation on the posterior margin of the ilium, superior to the ischial spine.
greater splanchnic nerve (Greek: *splankhna* - inward parts) a sympathetic nerve arising in the T5-T9 level of the thoracic spinal cord and ending in the celiac or superior mesenteric ganglion; regulates function of the organs of the gastrointestinal tract.
greater trochanter (Greek: *trochantēr* - a runner) on the lateral side of the neck of the femur; bony expansion of the femur that projects superiorly from the base of the femoral neck.
greater tubercle (Latin: *tuberosus* - full of lumps) large prominence on the lateral side of the proximal humerus.
greater vestibular glands (*aka* Bartholin's glands) glands in the vestibule of biological females that produce mucus that lubricates the vulva.
greater wings of the sphenoid bone (Latin: *sphenoides* - resembling a wedge) lateral projections of the sphenoid bone that extend from the sella turcica, forming the base of the skull.
gross anatomy (*aka* macroscopic anatomy) (Latin: *grossus* - thick; Greek: *ana* - up; *tomē* - cutting) study of the larger structures of the body that can be seen with naked eyes.
ground substance fluid or semifluid portion of the extracellular matrix of connective tissue.
growth factors (*sing*: growth factor) hormonal signaling in females that triggers estrogen to promote growth of breast tissue to prepare it for lactate production.
growth hormone (GH) a protein hormone produced by the anterior pituitary, which promotes cell growth and tissue building and influences nutrient metabolism.
gyrus (Greek: *guros* - a ring) a raised ridge on the surface of the cerebrum.

H

H zone the area in the center of the A band of a sarcomere, in which only thick filaments are found.
habenular nuclei (Latin: *habenula* - strap; *nucleus* - a little nut) small groups of neuron cell bodies in the epithalamus of the diencephalon, which are part of the limbic system; involved in dopamine secretion.
hair bulb deepest part of the hair follicle; surrounds the dermal papilla.
hair cells mechanoreceptor cells of the inner ear that transduce sound stimuli.
hair follicle (Latin: *folliculus* - a small bag) cavity or sac from which hair originates; originates in the epidermis but partially located in the dermis.
hair matrix (*pl*: hair matrices) (Latin: *matrix* - womb) basal epithelial cells from which a strand of hair.
hair root portion of hair follicle that is between the hair bulb and hair shaft, anchored to the follicle.
hair shaft externally visible portion of hair; composed of keratin.
hallux (*aka* big toe, great toe) (Latin: *hallex* - great toe) digit 1 of the foot; the big toe.
hamate (Latin: *hamatus* - a hook) from the lateral side, the hook-shaped fourth of the four distal carpal bones; articulates with the lunate and triquetrum proximally, the fourth and fifth metacarpals distally, and the capitate laterally.
hamstring group three long muscles on the back of the knee; extend the thigh.
hand region of the upper limb distal to the wrist.
haploid (Greek: *haplous* - single; *eidos* - appearance) homologous pairs of chromosomes, usually 23.
haustra (*sing*: haustrum) a rounded pouch in the colon, formed by the muscle tone of the teniae coli.
haustral contractions the segmentation contractions that occur in the large intestine, and last for about one minute; stimulated by the entry of chyme into the colon.
head of the femur (Latin: *femur* - thigh) rounded, proximal end of the femur; articulates with the acetabulum of the hip bone.
head of the fibula (Latin: *fibula* - brooch) knob-shaped, proximal end of the fibula.
head of the humerus (Latin: *humerus* - shoulder) smooth, rounded region located on medial side of the proximal humerus; forms the glenohumeral (shoulder) joint with the glenoid fossa of the scapula.
head of the pancreas (Greek: *pankreas* - sweetbread) the widest part of the pancreas, which lies in the C-shaped curvature of the duodenum.
head of the radius (Latin: *radius* - ray) disc-shaped proximal end of the radius; articulates with the capitulum at the elbow joint, and with the radial notch of the ulna at the proximal radioulnar joint.
head of the rib posterior end of a rib that articulates with the costal facet of the thoracic vertebra.
head of the ulna (Greek: *ōlenē* - ulna) distal end of the ulna; articulates with the ulnar notch of the distal radius.
heart block interruption in the normal conduction pathway.
heart rate (HR) number of times the heart contracts (beats) per minute.
heart sounds sounds heard via auscultation with a stethoscope of the closing of the atrioventricular valves ("lub") and semilunar valves ("dub").
heavy chain the larger chain of proteins found in an antibody molecule; antibodies contain two heavy chains and two light chains.
helper T cells (CD4+ T cells) (*aka* T_h cells) T lymphocytes (T cells) that activate B cells, cytotoxic T cells, macrophages, and other immune cells, and regulate the immune response, by secreting cytokines.
hematocrit (*aka* packed cell volume) (Greek: *haima* - blood; *krino* - to separate) the percentage of a blood sample that consists of erythrocytes; can be determined by centrifugation.
hematopoeisis (*aka* hemopoiesis) (Greek: *haima* - blood; *poiesis* - a making) the production of all types of blood cells (erythrocytes, leukocytes, and platelets) in the red bone marrow.
hematopoietic stem cells (*aka* **hemopoietic stem cells**) (Greek: *haima* - blood; *poiesis* - a making) hemocytoblasts; pluripotent stem cells of the red bone marrow, from which all of the formed elements of the blood (erythrocytes, leukocytes, and platelets) arise.
heme (Greek: *haima* - blood) the iron-containing portion of a hemoglobin molecule; binds to oxygen molecules in erythrocytes and gives blood its red color.

hemiazygos vein (Latin: *hemi* - half) a vein that arises in the posterior wall of the abdominal cavity, runs upward just to the left of the vertebral column and parallel to the azygos vein; end empties into the azygos vein in the thorax; drains blood from part of the esophagus and other structures of the mediastinum.
hemidesmosomes (*sing*: hemidesmosome) (Greek: *hemi* - half; *desmos* - a band; *sōma* – body) a type of anchoring junction that unites a cell to the extracellular matrix.
hemoglobin (Greek: *haima* - blood; Latin: *globus* - ball) a protein that makes up a portion of erythrocytes; binds to and transports oxygen and some carbon dioxide molecules through the blood.
hemoglobin saturation (Greek: *haima* - blood; *globule* - ball) the percentage of oxygen-binding sites on hemoglobin that are occupied with oxygen molecules.
hemolysis (Greek: *haima* - blood; Latin: *lysis* - to loosen) the rupture and destruction of erythrocytes, and hemoglobin release into the blood.
hemolytic disease of the newborn (HDN) (*aka* erythroblastosis fetalis) (Greek: *haima* - blood) a disorder that occurs in an Rh+ fetus or baby of an Rh- mother, in which the Rh+ erythrocytes are agglutinated and destroyed by anti-Rh antibodies from the mother.
hemorrhage (Greek: *haima* - blood; *rrhagia* - a bursting forth) loss of an excessive amount of blood from the body.
hemosiderin (Greek: *haima* - blood; *sideros* - iron) an iron-storage complex found in the bone marrow, liver, and spleen; found within cells, but not in the blood plasma.
hemostasis (Greek: *haima* - blood; *stasis* - standing) the process by which bleeding is stopped in the body; typically involves vascular spasm, platelet plug formation, and blood coagulation.
Henry's law a law stating that the amount of a specific gas that dissolves in a liquid it is in contact with is directly proportional to the partial pressure of that gas.
heparin (Latin: *hepar* - liver) an anticoagulant found in mast cells and basophils, and on the surface of healthy endothelial cells, which prevents clot formation from extending into healthy tissue.
hepatic artery (Latin: *hepaticus* - pertaining to the liver) the artery that delivers oxygen-rich blood to the liver.
hepatic artery proper (Latin: *hepaticus* - pertaining to the liver) an artery that branches off of the common hepatic artery, and provides blood to the liver.
hepatic ducts (Latin: *hepaticus* – pertaining to the liver) two ducts, one on the left and one on the right side of the liver, that converge to form the common hepatic duct; form part of the transportation pathway through which bile travels from the liver to the duodenum.
hepatic portal system (Latin: *hepaticus* - pertaining to the liver) a system of blood vessels that transports blood from the digestive organs directly to the liver for the adjustment of nutrient concentration and the removal of toxins, before it enters the inferior vena cava and the systemic circuit.
hepatic portal vein (Latin: *hepaticus* - pertaining to the liver) the vein that transports oxygen-poor blood, containing absorbed nutrients, to the liver.
hepatic sinusoids (Latin: *hepaticus* - pertaining to the liver; *sinus* - cavity) specialized capillaries between plates of hepatocytes, which transport a mixture of blood from branches of the hepatic portal vein and hepatic artery through the hepatic lobules.
hepatic vein (Latin: *hepaticus* - pertaining to the liver) a vein that drains blood from the liver and drains into the inferior vena cava.
hepatocytes (Latin: *hepaticus* - pertaining to the liver; Greek: *kytos* - cell) liver cells, which produce bile and adjust nutrient concentrations in the blood as it flows through the liver.
hepatopancreatic ampulla (*aka* ampulla of Vater, hepatopancreatic duct) (Latin: *hepaticus* - pertaining to the liver; *ampulla* - flask) a rounded structure in the wall of the duodenum, located at the site at which the main pancreatic duct and the common bile duct merge; transports bile and pancreatic juice through the major duodenal papilla into the duodenum.
hepatopancreatic sphincter (*aka* sphincter of Oddi, Glisson's sphincter) (Latin: *hepaticus* - pertaining to the liver; *sphincter* - contractile muscle) a ring of smooth muscle in the wall of the duodenum that controls the flow of bile and pancreatic juice into the duodenum.
hilum the indented entry/exit point on the medial surface of each lung, through which nerves, blood and lymphatic vessels, and a primary bronchus enters/exits the lung.
hindbrain (*aka* rhombencephalon) region of the embryonic brain that develops into the metencephalon and myelencephalon, which eventually develop into the pons, cerebellum, and medulla oblongata.
hinge joint synovial joint formed by the articulation of the convex surface of one bone with the concave surface of a second bone.
hippocampus (Greek: *hippocampos* – seahorse) midbrain nucleus involved in long-term memory formation and emotional responses.
histamine (Greek: *histos* - web) a chemical secreted by basophils and mast cells; produces many of the effects of inflammation and allergy, such as vasodilation and increased permeability of blood vessels, bronchoconstriction of the airways, increased mucus production, and decreased blood pressure.
histology (Greek: *histos* - tissue; *logos* - study) microscopic study of tissue appearance, organization, and function.
histone (*pl*: histones) a protein that associates with DNA in the nucleus to form chromatin.
hole opening or groove in the bone; allows blood vessels or nerves to pass through.
holocrine secretion (Greek: *holos* - whole; *krinōs* - to separate; Latin: *secerno* - to separate) the release of compounds via the rupture and destruction of a gland cell.
homeostasis (Greek: *homoisis* - similar; *stasis* - standing) dynamic state of stable internal conditions within the body systems that living organisms maintain.
homologous (Greek: *homo* - the same; *logos* - relation) describes two copies of the same chromosome (not identical), one inherited from each parent.
hook of the hamate bone (Latin: *hamatus* - a hook) bony extension from the anterior side of the hamate bone.
horizontal fissure (*aka* minor fissure) a double fold of the visceral pleura that separates the superior lobe from the middle lobe of the right lung.
horizontal plate extension of the palatine bone that joins with a paired horizontal plate to form the posterior hard palate.
hormone receptor (Latin: *recipio* - receiver) a protein inside a cell or within the cell membrane that binds to a specific hormone and initiates a cellular response.
hormones chemical signals secreted by endocrine glands to induce a response in target cells or tissues.
human chorionic gonadotropin (hCG) (Greek: *khorion* - membrane closing the fetus; *gonos* - child; *tropikos* - pertaining to a change) a hormone that is secreted as a signal from the developing embryo.
humerus (Latin: *humerus* - shoulder) single long bone of the brachium.
hyaline cartilage (Greek: *hyalos* - glass; Latin: *cartilago* - gristle) most common type of cartilage, smooth and made of short, dispersed collagen fibers with a high proportion of proteoglycans; the most abundant type of cartilage in the skeletal system, found at the ends of bones at sites of articulation.
hydrochloric acid (HCl) an acid secreted by the parietal cells of the stomach, which converts pepsinogen into active pepsin; helps to destroy microorganisms that enter the body in food, and breaks down plant cell walls to help in digestion.
hydrogen bond (Greek: *hydōr* - water; *gen* - producing) dipole-dipole bond in which a hydrogen atom covalently bonded to an electronegative atom is weakly attracted to a second electronegative atom.
hydrolysis reaction (Greek: *hydōr* - water; *lysis* - dissolution) A reaction where a water molecule splits another molecule so that one product binds with the hydrogen ion, and the other product binds with the hydroxyl group.
hydrophilic (Greek: *hydōr* - water; *philos* - loving) describes a substance or structure attracted to water.
hydrophobic (Greek: *hydōr* - water; *phobos* - fear) describes a substance or structure repelled by water.
hydrostatic pressure the pressure exerted by a fluid against the inner walls of its container; in the nephrons, this type of pressure is exerted by blood against the walls of the glomerular

capillaries and by filtrate against the walls of the glomerular capsule.
hydrostatic pressure (Greek: *hydōr* - water; *statikos* - stand) the pressure exerted by a liquid against the wall that contains it; in the body, caused by the force of gravity and the pumping action of the heart.
hymen (Greek: *hymen* - membrane) a crescent-shaped or ring-shaped mucosal membrane that lies just inside the vaginal orifice.
hyoglossus (Greek: *hyoeidēs* - shaped like U; *glōssa* - tongue) muscle originating on the hyoid bone; flattens the tongue and moves it to the back.
hyoid bone (Greek: *hyoeidēs* - shaped like U) small, independent, U-shaped bone located in upper neck; serves as an anchor for some muscles of the tongue; does not contact any other bone.
hypercalcemia (Greek: *hyper* – over) the condition of having an abnormally high concentration of calcium in the blood.
hypercapnia (*aka* hypercarbia) (Greek: *hyper* - over) an abnormally high concentration of carbon dioxide in the blood.
hyperchloremia (Greek: *hyper* - over) an abnormally high concentration of chloride in the blood.
hyperextend (Greek: *hyper* - over) to increase the joint angle beyond 180 degrees.
hyperextension (Greek: *hyper* - over; Latin: *extendo* - to stretch out) an increase in the joint angle beyond 180 degrees.
hyperkalemia (Greek: *hyper* – over; *kalio* - potassium) an abnormally high concentration of potassium in the blood.
hypernatremia (Greek: *hyper* – over; *natrio* - sodium) an abnormally high concentration of sodium in the blood.
hyperparathyroidism (Greek: *hyper* - over) a disorder caused by overproduction of parathyroid hormone (PTH), resulting in excessive calcium resorption from bone and decreased calcium excretion in the urine.
hyperphosphatemia (Greek: *hyper* – over; *fosforiko* - phosphate) an abnormally high concentration of phosphate in the blood.
hyperpnea (Greek: *hyper* - over; *apnoia* - absence of respiration) an increase in the depth of breathing, sometimes accompanied by an increase in breathing rate, due to an increased need for oxygen (such as during exercise).
hypertension (*aka* high blood pressure) (Greek: *hyper* - over) a state in which the blood pressure is chronically elevated to a level greater than 140/90.
hyperthyroidism a disorder caused by overproduction of thyroid hormone (TH), resulting in increased metabolic rate, increased appetite, weight loss, heat sensitivity, and anxiety.
hypertonic (Greek: *hyper* - over; *tonos* - tension) describes a solution concentration that is higher than a reference concentration.
hypertrophies (Greek: *hyper* - beyond, over, more) an increase in growth.
hyperventilation (Greek: *hyper* - over; Latin: *ventus* - wind) an increase in the rate and depth of ventilation, which results in excess excretion of carbon dioxide, a decrease in the blood level of carbon dioxide, and a high blood pH.
hypervolemia (Greek: *hyper* - over) an abnormally high volume of extracellular fluid or blood in the body.
hypocalcemia (Greek: *hypo* - under) the condition of having an abnormally low concentration of calcium in the blood.
hypocapnia (*aka* hypocarbia) (Greek: *hypo* - under) an abnormally low concentration of carbon dioxide in the blood.
hypochloremia (Greek: *hypo* - under) an abnormally low concentration of chloride in the blood.
hypodermis (*aka* subcutaneous layer) (Greek: *hypo* - under; *derma* - skin) connective tissue connecting the skin to the underlying fascia.
hypogastric plexus (Greek: *hypo* - under; Latin: *plexus* - braid) a nerve network that lies anterior to the L5 and S1 vertebrae; contains both sympathetic and parasympathetic nerve fibers and innervates several pelvic organs.
hypoglossal canal (Greek: *hypo* - under; *glossa* - tongue) paired passages along the anterior-lateral edge of the foramen magnum; allow for the passage of a nerve to the tongue.
hypoglossal nerve (CN XII) (Greek: *hypo* - under; *glossa* - tongue) twelfth cranial nerve; is responsible for controlling the muscles of the lower throat and tongue.
hypokalemia (Greek: *hypo* – under; *kalio* - potassium) an abnormally low concentration of potassium in the blood.
hyponatremia (Greek: *hypo* – under; *natrio* - sodium) an abnormally low concentration of sodium in the blood.
hypoparathyroidism (Greek: *hypo* - under) a disorder caused by underproduction of parathyroid hormone (PTH), resulting in an abnormally low level of blood calcium.
hypophosphatemia (Greek: *hypo* – under; *fosforiko* - phosphate) an abnormally low concentration of phosphate in the blood.
hypothalamic-hypophyseal portal system (*aka* hypophyseal portal system) (Greek: *hypo* - under) a specialized system of blood vessels, consisting of capillary beds in the hypothalamus and the anterior pituitary gland, connected by the hypophyseal portal veins; transports releasing and inhibiting (tropic) hormones from the hypothalamus to the anterior pituitary gland.
hypothalamus (Greek: *hypo* - under) a structure of the diencephalon inferior and anterior to the thalamus; has both neural and endocrine functions, and is responsible for monitoring many homeostatic functions, such as hunger, body weight, thirst, body temperature, heart rate, and pituitary gland activity.
hypothenar (Greek: *hypo* - under; *thenar* - the palm) muscles located on the medial aspect of the palm.
hypothyroidism (Greek: *hypo* - under) a condition resulting from deficient TH production; characterized by low metabolic rate, fatigue, lethargy, cold sensitivity, and decreased appetite.
hypotonic (Greek: *hypo* - under; *tonos* - tension) describes a solution concentration that is lower than a reference concentration.
hypoventilation (Greek: *hypo* - under; Latin: *ventus* - wind) a decrease in the rate or depth of breathing, which results in insufficient oxygen being inhaled and insufficient carbon dioxide being exhaled.
hypovolemia (Greek: *hypo* - under) an abnormally low volume of extracellular fluid or blood in the body.
hypoxemia (Greek: *hypo* - under; *haima* - blood) abnormally low oxygen level in the blood, in which oxygen saturation is lower than 95 percent.
hypoxia (Greek: *hypo* - under) an oxygen deficiency in a particular portion of the body.

I

I band the portion of the sarcomere in which only thin filaments are found; lies between adjacent A bands, and contains the Z disks.
IgA a class of antibodies found in the secretions of exocrine glands, such as milk, saliva, sweat, and gastric and intestinal fluid; very effective against pathogens of the digestive and respiratory systems.
IgD a class of antibodies found mainly on the cell membranes of B cells, but also found in small quantities in the blood plasma; functions as an antigen receptor for B cells.
IgE a class of antibodies found in the secretions of exocrine glands; binds to mast cells and and basophils to promote inflammatory and allergic responses in the body.
IgG the most abundant class of antibodies, comprising about 80 percent of all antibodies; found in blood plasma and interstitial fluid, and can cross the placenta.
IgM the first class of antibodies to be secreted during the initial response to an antigen; circulates mainly in the blood plasma and lymph for a short time, and levels decline as IgG antibodies increase.
ileocecal sphincter (*aka* ileocecal valve) (Latin: *ileum* - intestines; *sphincter* - contractile muscle) a ring of muscle that regulates the flow of chyme from the ileum of the small intestine into the cecum of the large intestine.
ileum (Latin: *ileum* - intestines) the final and longest segment of the small intestine, which runs between the jejunum and the cecum of the large intestine.
iliac crest (Latin: *ilium* - groin; *crista* - crest) arched, superior ridge of the ilium.
iliac fossa (Latin: *ilium* - groin; *fossa* - trench) sloping depression found on the anterior and medial surfaces of the upper ilium.
iliacus (Latin: *ilium* - groin) muscle that makes up part of the posterior abdominal wall.

iliococcygeus (Latin: *iliacus* - hip bone) one of the levator ani muscles; supports pelvic organs and aids in urination and defecation.
iliocostalis (Latin: *ilium* - groin; *costa* – rib) the lateral group of muscles of the erector spinae group; helps to extend the vertebral column.
iliofemoral ligament (Latin: *ilium* - groin; *femur* - thigh; *ligamentum* - a band) thick ligament that runs from the ilium to the femur; located on the superior-anterior aspect of the hip joint.
iliopsoas group (Latin: *ilium* - groin) muscle group made up of the psoas major and iliacus muscles.
iliotibial tract (*aka* Iliotibial band, IT band) long, flat tendon that runs along the lateral thigh and inserts at the knee.
ilium (Latin: *ilium* - groin) large, superior region of the hip bone.
immune system a body system consisting of organs, cells, and chemical substances that protects the body against infection by pathogenic organisms, by destroying the invading organisms and preventing their proliferation.
immunoglobulins (*aka* antibodies, gamma globulins) proteins produced by B lymphocytes that defend the body from infection and disease by acting against specific antigens on pathogens such as bacteria, viruses, and fungi.
immunological memory the ability of the immune system to remember and recognize previously encountered antigens, so it can launch a strong, rapid immune response upon subsequent exposure.
implantation (French: *implanter* - implant) The process in which the preembryo attaches and embeds in the endometrium; can only occur when the preembryo has developed a trophoblast and endometrium is ready.
incisors sharp, anterior teeth, which are used for biting and cutting into food.
incontinence (Latin: *incontinentia* - inability to contain) the inability to control urination or defecation.
incus (*aka* anvil) (Latin: *incus* - anvil) ossicle that connects the malleus to the stapes.
inferior (*aka* caudal) describes a position below or lower than another part of the body proper.
inferior angle of the scapula (Latin: *inferior* - below; *scapulae* - shoulder blades) corner of the scapula located between the medial and lateral borders.
inferior articular process (Latin: *articulo* - to articulate) flat, downward-facing bony process that extends from the vertebral arch of a vertebra to the superior articular process of the next lower vertebra.
inferior colliculi (Latin: *inferior* - below) two of the four structures of the corpora quadrigemina, which lie along the anterior surface of the midbrain; part of the brainstem auditory pathway.
inferior extensor retinaculum (Latin: *retinere* - to hold back) band of connective tissue holding the tendons of the muscles of the anterior compartment of the lower leg.
inferior gemellus (Latin: *gemellus* - twin) muscle deep to the gluteus maximus on the lateral surface of the thigh; laterally rotates the femur at the hip.
inferior lobe (Latin: *inferior* – lower) the bottom lobe of each lung, which rests on the superior surface of the diaphragm.
inferior mesenteric artery (Latin: *inferior* - lower; *mesenterie* - middle of intestines) an artery that branches off of the abdominal aorta, and provides blood to the portion of the large intestine between the transverse colon and the sigmoid colon, and the rectum.
inferior mesenteric ganglion (Latin: *inferior* - lower; Greek: *ganglion* - knot) one of the three collateral (prevertebral) ganglia of the sympathetic system; projects to the digestive system.
inferior mesenteric plexus (Latin: *inferior* - low; *plexus* - braid) one of the three nerve networks of the abdominal aortic plexus; innervates the distal portion of the large intestine and the rectum.
inferior nasal conchae (Latin: *nasus* - nose; *concha* – shell) one of two paired bones that project from the lateral walls of the nasal cavity.
inferior phrenic artery (Latin: *inferior* - lower; *phrenicus* - diaphragm) an artery that branches off of the abdominal aorta, and provides blood to the inferior portion of the diaphragm.
inferior pubic ramus (Latin: *inferior* - lower; *pubes* - groin) narrow segment of bone that extends laterally from the pubic body.
inferior sagittal sinus (Latin: *inferior* - lower; *sagitta* – an arrow; *sinus* – curve) one of the cranial dural septa, which is located between the two cerebral hemispheres; lies within the falx cerebri.
inferior vena cava (Latin: *inferior* - lower; *vena* - vein; *cava* - hollow) one of the largest veins in the systemic circuit; formed by the merging of the common iliac veins; drains blood from most body regions inferior to the diaphragm, such as the lower limbs and the inferior portion of the trunk; empties into the right atrium.
inflammation (Latin: *inflammare* - to set fire) an innate or adaptive immune response that serves to wall off a site of injury or infection, to prevent it from spreading to the rest of the body; also attracts immune cells to the area to destroy the pathogenic organisms and remove the dead cells and debris; properties include redness, swelling, heat, and pain.
infrahyoid muscles (Latin: *infra* - below; *hyoeidēs* - shaped like U) anterior neck muscles inferior to the hyoid bone.
infraorbital foramen (Latin: *infra* - below; *orbis* - circle; *foramen* - opening) opening located on anterior skull, below the orbit.
infraspinatus (Latin: *infra* - below; *spina* - spine) muscle located inferior to the spine of the scapula that laterally rotates the arm.
infraspinous fossa (Latin: *infra* - below; *spina* - backbone; *fossa* - trench) broad depression on the posterior scapula, inferior to the spine.
infundibulum (*aka* pituitary stalk) (Latin: *infundibulum* – funnel) 1) a long, narrow stem of tissue connecting the hypothalamus to the posterior pituitary; contains the axons of neurons that secrete oxytocin and antidiuretic hormone (ADH) 2) the widest part of the uterine tube, which lies closest to the ovary; contains the fimbriae which surround the end of the ovary that releases oocytes.
inguinal canal (Latin: *inguinalis* - groin) an opening in the wall of the abdominal cavity through which either the spermatic cord or round ligament runs.
inhalation (*aka* inspiration) (Latin: *inhalare* - inhale) the process of moving air into the lungs from the atmosphere.
inhibitory postsynaptic potential (IPSP) graded potential in the postsynaptic membrane caused by hyperpolarization; makes an action potential less likely.
innate immune response (Latin: *innatus* - inborn) the rapid, nonspecific immune response against foreign antigens associated with pathogenic organisms; examples include barriers such as the skin, inflammation, phagocytosis, and the action of neutrophils and natural killer (NK) cells and various chemical substances, such as enzymes and interferon.
inner ear structure within the bony labyrinth of the temporal bone; composed of the cochlea, the vestibule, and the semicircular canals.
innermost intercostal (Latin: *inter* - between; *costa* - rib) deepest intercostal muscles; act as synergists for the internal intercostals.
inorganic compound (Latin: *compono* - to place together) substance that does not contain both carbon and hydrogen.
inositol triphosphate (IP_3) a molecule produced when then enzyme phospholipase C (PLC) cleaves a membrane-bound phospholipid; causes calcium ions to be released from storage, so they can acts as second messengers for certain hormones.
insertion (Latin: *insertio* - to plant in) moveable end of a muscle; attaches to the structure being pulled during muscle contraction.
inspiratory capacity (IC) (Latin: *inspirare* - breathe upon) the maximal amount of air that can be inhaled after a normal exhalation; can be calculated by taking the sum of the TV and IRV.
inspiratory reserve volume (IRV) (Latin: *inspirare* - breathe upon) the additional amount of air that can be inhaled after a normal inhalation.
insulin (Latin: *insula* - island) a pancreatic hormone produced by beta cells that enhances the cellular uptake and utilization of glucose; its systemic effect is to decrease the blood glucose level.
insulin resistance (Latin: *insula* - island; *resistentia* - resistance) a condition in which insulin receptors have become less sensitive to insulin; this leads to reduced uptake of glucose into liver, muscle, and adipose cells and an increased blood glucose level.
insulin-like growth factors (IGFs) proteins that enhance cellular proliferation, inhibit apoptosis, and stimulate the cellular uptake of amino

acids; produced by the liver and other tissues when stimulated by GH.
integumentary system (Latin: *tegumentum* - a covering) the skin, hair, nails, and exocrine glands.
interatrial septum cardiac septum located between the two atria; contains the fossa ovalis after birth.
intercalated cells (*aka* IC) (Latin: *intercalatus* - interposed) cells found in the distal convoluted tubules and collecting ducts of the kidney, which help to regulate blood pH by secreting hydrogen ions or reabsorbing bicarbonate or potassium ions.
intercalated disc (Latin: *intercalatus* - interposed) physical junction between adjacent cardiac muscle cells; consisting of desmosomes, specialized linking proteoglycans, and gap junctions that allow passage of ions between the two cells.
intercondylar eminence (Latin: *eminentia* - raised surface) irregular, elevated area of the superior end of the tibia, located between the articulating surfaces of the medial and lateral condyles.
intercondylar fossa (Latin: *fossa* - trench) deep depression located on the posterior side of the distal femur; separates the medial and lateral condyles.
intercostal arteries (Latin: *interi* - between; *costa* - rib) branches of the thoracic aorta, which supply blood to the structures of the vertebral column, segments of the spinal cord, and the skin and muscles of the back.
intercostal artery (Latin: *interi* - between; *costa* - rib) an artery that branches off of the thoracic aorta, which supplies blood to the structures of the vertebral column, segments of the spinal cord, and the skin and muscles of the back.
intercostal muscles (Latin: *inter* - between; *costa* - rib) three sets of muscles that span the spaces between the ribs.
intercostal nerves (Latin: *inter* - between; *costa* - rib) nerves originating from the T2 through T11 spinal nerves; located between the ribs.
intercostal vein (Latin: *interi* - between; *costa* - rib) a vein that drains the muscles of the wall of the thorax, and empties into the azygos vein.
interferons a type of cytokine, secreted by virus-infected cells, which travels to nearby cells and causes them to produce antiviral proteins; this prevents viral replication in these cells.
interlobar arteries (Latin: *inter* - between) arteries that transport oxygen-rich blood from the segmental arteries to the arcuate arteries in the kidney; run through the renal columns between the renal pyramids.
interlobar veins (Latin: *inter* - between) veins that transport oxygen-poor blood from the arcuate veins to the renal vein in the kidney; run through the renal columns between the renal pyramids.
interlobular arteries (*aka* cortical radiate arteries) arteries that transport oxygen-rich blood from the arcuate arteries to the afferent arterioles of the renal corpuscles; also supply blood to the tissue of the renal cortex.
interlobular vein (*aka* cortical radiate vein) (Latin: *inter* - between) a vein that transports oxygen-poor blood from the peritubular capillaries of the nephrons to the arcuate vein in the kidney.
intermediate (Latin: *inter* - between; *medius* - middle) muscles located intermediate to the thenar and hypothenar muscles.
intermediate cuneiform (Latin: *inter* - between; *medius* - middle; *cuneus* - wedge) center of the three cuneiform tarsal bones; articulates posteriorly with the navicular bone, medially with the medial cuneiform bone, laterally with the lateral cuneiform bone, and anteriorly with the second metatarsal bone.
intermediate filament (*pl*: filaments) (Latin: *inter* - between; *medius* - middle; *filamentum* - thread) type of cytoskeletal filament made of keratin, characterized by an intermediate thickness, and playing a role in resisting cellular tension.
internal acoustic meatus opening in the temporal bone; provides passage for the nerve from the hearing and equilibrium organs of the inner ear.
internal anal sphincter (*aka* sphincter ani internus) (Latin: *sphincter* - contractile muscle) a ring of smooth muscle in the anal canal, which is involuntary and helps to regulate the defecation process.
internal carotid artery (Latin: *internus* - internal; Greek: *karotides* - great arteries of the neck) an artery that branches from the common carotid artery, and joins with the vertebral artery at the base of the brain to form the cerebral arterial circle; provides the main blood supply to the brain.
internal elastic membrane (*aka* internal elastic lamina) (Latin: *internus* - internal) a layer of elastic fibers that lies between the tunica intima and the tunica media in large arteries.
internal iliac artery (Latin: *internus* - internal; *iliacus* - pertaining to colic) an artery that branches off of the common iliac artery; provides blood to pelvic organs, such as the urinary bladder, rectum, uterus (females) and ductus deferens (males), and to the external genitalia.
internal iliac vein (Latin: *internus* - internal; *iliacus* - pertaining to colic) a vein that drains blood from the pelvic viscera and skin; empties into the common iliac vein.
internal intercostal (Latin: *internus* - internal; *inter* - between; *costa* - rib) one of 11 pairs of intermediate intercostal muscles; draw the ribs together, resulting in the contraction of the rib cage.
internal jugular vein (Latin: *internus* - internal; *iugula* - collarbone) a vein that arises from the dural venous sinuses of the brain, travels through the jugular foramen of the skull, and empties blood into the subclavian vein; drains blood from a large portion of the brain; considered to be the counterpart of the carotid artery.
internal oblique (Latin: *internus* - internal; *obliquus* - slanted) intermediate abdominal muscle; composed of fibers that run perpendicular to those of the external oblique.
internal thoracic artery (*aka* mammary artery) (Latin: *internus* - internal; *thorax* - chest) an artery that branches off of the subclavian artery; supplies blood to the thymus, pericardium, and the anterior wall of the thorax.
internal thoracic vein (*aka* internal mammary vein) (Latin: *internus* - internal; *thorax* - chest) a vein that drains the anterior region of the thorax, and empties into the brachiocephalic vein.
interneurons (*aka* association neurons) neurons that integrate and process sensory input or relay the signal within the central nervous system.
internodal pathway specialized conductile cells within the atria that transmit the impulse from the SA node throughout the myocardial cells of the atrium and to the AV node.
interoceptor (Latin: *inter* - between; *capere* - to take) sensory receptor that interprets stimuli from internal organs and tissues.
interosseous membrane (*aka* interosseous ligament) (Latin: *inter* - between; *os* - bone; *membrana* - skin) broad sheet of fibrous connective tissue; may fill in wide gaps formed by syndesmoses.
interosseous membrane of the forearm (Latin: *inter* - between; *os* - bone; *membrana* - skin) sheet of thin, dense connective tissue; unites the radius and ulna.
interosseous membrane of the leg (Latin: *inter* - between; *os* - bone; *membrana* - skin) sheet of dense connective tissue that joins the tibia and fibula.
interphase (Latin: *inter* - between) entire life cycle of a cell, excluding mitosis.
interstitial bone growth (Latin: *interstitium* - to stand) growth during which a bone grows in length.
interstitial cells (aka cells of Leydig) (Latin: *interstitium* - interval) cells that produce testosterone and are located between the coiled seminiferous tubules rather than within them.
interstitial fluid (IF) (Latin: *interstitium* - space between) the fluid that lies in the tissue spaces and surrounds the cells; exchanges substances with the cells and the blood plasma.
interthalamic adhesion (*aka* intermediate mass) (Latin: *inter* - between; *adhaesiōn* - sticking; Greek: *thalamos* - inner room) a connection between the two oval masses that make up the thalamus; found in most people.
intertrochanteric crest (Latin: *inter* - between; Greek: *trochanter* - a runner) ridge on the posterior side of the proximal femur; runs between the greater and lesser trochanters.
intertrochanteric line (Latin: *inter* - between; Greek: *trochanter* - a runner) rough ridge running between the greater and lesser trochanters on the anterior side of the femur.
interventricular foramen (*aka* foramen of Monro) (Latin: *inter* - between; *ventriculus* - diminutive; *foramen* - an opening) a small

opening between each of the lateral ventricles and the third ventricle in the brain; allows for the passage of CSF.
interventricular septum (Latin: *inter* - between; *ventriculus* - diminutive; *septum* - partition) the wall between the right and left ventricles of the heart.
intervertebral disc (Latin: *inter* - between; *vertebra* - joint) fibrocartilaginous structure located between adjacent vertebrae that joins the vertebrae.
intervertebral foramen (Latin: *inter* - between; *vertebra* - joint; *foramen* - opening) the opening through which spinal nerves exit the vertebral column; located between adjacent vertebrae.
intestinal gland (*aka* crypt of Lieberkühn, intestinal crypt) a tubular depression between the villi of the small intestine, which secretes intestinal juice, a watery fluid that serves as a transport medium for absorbing nutrients.
intestinal juice (*aka* succus entericus) a watery secretion of the cells in the intestinal glands, which helps absorb nutrients from the lumen of the small intestine.
intestinal phase the third and final phase of gastric secretion, which starts when chyme enters the intestine; accounts for a small percentage of gastric secretion.
intracapsular ligament (Latin: *intra* - within; *capsula* - box; *ligamentum* - a band) ligament located within the articular capsule.
intracellular fluid (ICF) (Latin: *intra* - within; Greek: *kytos* - cell) the fluid within the cells.
intramembranous ossification (Latin: *intra* - within; Greek: *ossificatio* - to make bone) the process of building a bone within a sheet of connective tissue.
intramural ganglia (Latin: *intra* - within; *muralis* - pertaining to a wall; Greek: *ganglion* - knot) terminal ganglia of the parasympathetic system; located within the walls of the target or effector organs.
intraperitoneal (Latin: *intra* - within; *peritonaeum* - abdominal membrane) lying within the peritoneal cavity; examples of intraperitoneal digestive organs are the stomach and most of the small intestine.
intrapleural pressure (Pip) (Latin: *intra* - within; Greek: *pleuron* - ribs) the pressure of the air inside the pleural cavity, between the visceral pleura and the parietal pleura.
intrapulmonary pressure (Ppul) (Latin: *intra* - within; *pulmonis* - lung) the pressure within the alveoli.
intrinsic (Latin: *intrinsecus* - on the inside) muscle that originates and is located inside the area in which it acts.
intrinsic factor a glycoprotein secreted by the parietal cells of the stomach, which is necessary for the absorption of vitamin B_{12} in the small intestine.
intrinsic ligament (Latin: *intrinsecus* - on the inside; *ligamentum* - a band) ligament that is fused to or incorporated into the wall of the articular capsule.
intrinsic muscles of the hand (Latin: *intrinsecus* - on the inside) muscles that originate and insert within the hand; allow for precise movements of the fingers.
intrinsic pathway (Latin: *intrinsecus* - on the inside) one of the two pathways that initiate blood coagulation; once activated by factors found inside the blood, it leads to the activation of the common pathway of blood coagulation.
intrinsic reflex (Latin: *intrinsecus* - on the inside; *reflexus* - a bending back) reflex that is developed in the fetal period and is present at birth.
inversion (Latin: *inverto* - to turn upside down) the turning of the foot to angle the bottom of the foot toward the midline of the body.
ion (Greek: *ion* - going) atom with an overall positive or negative charge.
ionic bond (Greek: *ion* - going) attraction between an anion and a cation.
ipsilateral reflex (Latin: *ipse* - same; *latus* - side; *reflexus* - a bending back) reflex that begins and ends on the same side of the body.
iris (Latin: *iris* - rainbow) colored portion of the anterior eye; composed of smooth muscle that surrounds the pupil and regulates the amount of light entering the eye.
irregular bone a bone that does not have an easily characterized shape; shapes tend to be complex.
ischemia (Latin: *ischaemus* - stopping blood) a deficiency in blood flow to a particular portion of the body, which often results in inadequate oxygen delivery to the cells, causing cellular death.
ischial ramus (Greek: *ischion* - hip joint) narrow anterior, superior projection from the ischial tuberosity.
ischial spine (Greek: *ischion* - hip joint; Latin: *spina* - backbone) projection of the posterior margin of the ischium; separates the greater sciatic notch and lesser sciatic notch.
ischial tuberosity (*aka* “sit bone”) (Greek: *ischion* - hip joint; Latin: *tuberosus* - full of lumps) large, roughened area of the inferior ischium.
ischiococcygeus (*aka* coccygeus) (Greek: *ischion* - hip joint) muscle that assists the pubococcygeus; pulls the coccyx anteriorly.
ischiofemoral ligament (Greek: *ischion* - hip joint; Latin: *femur* - thigh; *ligamentum* - a band) thick ligament that runs from the ischium to the femur; located on the posterior aspect of the hip joint.
ischium (Greek: *ischion* - hip joint) posteroinferior region of the hip bone.
islets of Langerhans (*aka* pancreatic islets) small groups of endocrine cells in the pancreas that produce the hormones insulin and glucagon.
isometric contraction (Greek: *isos* - equal; *metron* - measure) muscle contraction that does not move a load or result in a change in muscle length.
isosmotic (Greek: *isos* - equal; *ōsmos* - impulsion) two solutions that have the same concentration of solutes.
isotonic (Greek: *isos* - equal; *tonos*- tension) a solution outside a cell with the same concentration of solutes as the intracellular fluid.
isotonic contraction (Greek: *isos* - equal; *tonos* - tension) muscle contraction that moves a load or involves a change in muscle length.
isotope (Greek: *isos* - equal; *topos* - place, space) one of the variations of an element in which the number of neutrons differ from each other.
isovolumic contraction (Greek: *isos* - equal) (*aka* isovolumetric contraction) initial phase of ventricular contraction in which tension and pressure in the ventricle increase, but no blood is pumped or ejected from the heart.
isovolumic ventricular relaxation phase (Greek: *isos* - equal) initial phase of the ventricular diastole when pressure in the ventricles drops below pressure in the two major arteries, the pulmonary trunk, and the aorta, and blood attempts to flow back into the ventricles, producing the dicrotic notch of the ECG and closing the two semilunar valves.
isthmus (Greek: *isthmos* - narrow passage) the narrowest region of a uterine tube, that lies closest to the uterus.

J

jejunum (Latin: *ieiunus* - hungry) the middle portion of the small intestine, which lies between the duodenum and the ileum; performs most of the final digestion and absorption of dietary nutrients.
joint a structure in which adjacent bones and/or cartilage come together.
joint cavity (*aka* synovial cavity) (Latin: *cavus* - hollow) fluid-filled space formed by the articular capsule of a synovial joint; houses the articulating surfaces of adjacent bones.
jugular foramen (Latin: *jugulum* - throat; *foramen* - an opening) irregularly shaped opening located in the lateral floor of the posterior cranial cavity.
juxtaglomerular apparatus (JGA) (Latin: *iuxta* - beside; *glomerationem* - ball; *apparatus* - tools) a structure that lies at the point at which the distal convoluted tubule passes between the afferent and efferent arterioles of the nephron; functions in the regulation of blood pressure, electrolyte balance, glomerular filtration rate, and blood flow to the kidney.
juxtaglomerular cell (Latin: *iuxta* - beside; *glomerationem* - ball) a specialized smooth muscle cell found in the wall of the afferent ateriole; secretes the enzyme renin in response to a decrease in blood pressure.
juxtamedullary nephrons (Latin: *iuxta* - beside; *medulla* - marrow; Greek: *nephros* - kidney) nephrons whose renal corpuscles lie close to the border of the renal cortex and medulla, which contain very long nephron loops that extend deep into the renal medulla.

K

keloid (Greek: *kēlē* - tumor; *eidos* - appearance) type of scar that is raised compared to the surrounding tissue.
keratin (Greek: *keras* - horn) an intracellular fibrous protein that is found in the dead cells on the apical surface; gives skin, hair, and nails their hard, water-resistant properties.
keratinocyte (Greek: *keras* - horn; *kytos* - cell) epidermal cell that produces keratin.
keratohyalin (Greek: *keras* - horn; *hyalos* - glass) keratin precursor, found in some granules in the stratum granulosum.
ketone bodies (*aka* ketones) small organic molecules produced by the breakdown of acetyl coenzyme A, which can be used as an energy source when glucose is not present in sufficient quantity; can cause ketoacidosis when present in excess, such as during uncontrolled diabetes mellitus.
kinetic energy (Greek: *kinētos* - moving; *energeia* - in work) energy that matter possesses because of its motion.
knee joint joint formed by the articulations between the medial and lateral condyles of the femur and the medial and lateral condyles of the tibia.
kyphosis (*aka* humpback, hunchback) (Greek: *kyphōs* - crooked; Latin: *osis* - a state) disorder characterized by posterior curvature of the vertebral column in the thoracic region.

L

labia (*sing*: labium) (*aka* lips) (Latin: *labia* - lips) the lips, which help in eating, facial expression, and closing the mouth to keep out objects.
labia majora (Latin: *labia* - lips; *magnus* - large) larger, outer folds of skin that protect the other organs of the vulva in biological females; lie posterior to the mons pubis.
labia minora (Latin: *labia* - lips; *minor* - small) smaller, inner folds of skin that lie inside the labia majora; surround the vestibule in biological females.
labial frenulum (*aka* lip frenulum) (Latin: *labia* - lips) a small fold of tissue that connects the inner lip to the gums; lies along the midline of the mouth.
lacrimal bone (Latin: *lacrima* - tear) small, rectangular paired bone that contribute to the anterior-medial wall of each orbit; contain a tunnel between the inner corner of the eye and the nasal cavity.
lacrimal duct (Latin: *lacrima* - tear; *duco* - to lead) a duct located in the medial corner of the eye; drains tears into the nasal cavity.
lacrimal glands (Latin: *lacrima* - tear; *glans* - an acorn) glands located just above the lateral side of the upper eyelid; produce tears.
lactation (Latin: *lactationem* - suckling) the production of milk by a biological female to nourish a newborn baby; begins two to three days after childbirth.
lacteals (Latin: *lacteus* - milky) lymphatic capillaries in the villi of the small intestine that absorb and transport dietary lipids and lipid-soluble vitamins to the bloodstream.
lactic acid (Latin: *lac* - milk) product of anaerobic breakdown of pyruvate.
lactiferous ducts ducts that transport milk from the individual mammary glands to the nipple.
lactiferous sinus expanded areas of the lactiferous ducts that collect the milk and transport it to the nipple.
lacunae (*sing*: lacuna) (Latin: *lacus* - a hollow lake) small spaces in bone or cartilage tissue occupied by osteocytes or chondrocytes, respectively.
lacunae (*sing*: lacuna) (Latin: *lacus* - a hollow lake) small holes in cartilage that house chondrocytes.
lambdoid suture (Latin: *lambdoid* - resembling Greek letter lambda (λ) in form) V-shaped juncture that joins the occipital bone and the right and left parietal bones on the posterior skull.
lamellae (Latin: *lamina* - leaf) concentric rings of calcified matrix that support osteons.
lamellated corpuscles (*aka* Pacinian corpuscle) (Latin: *lamina* - leaf; *corpus* - body) encapsulated receptors found deep in the dermis or subcutaneous tissues that transduce deep pressure and vibration.
lamina densa supportive membrane that is denser and more structurally woven of tough collagen fibers.
lamina lucida (Latin: *lucidus* - clear) supportive membrane that is a mixture of glycoproteins and collagen, which provide the immediate attachment site for the epithelium.
lamina propria (Latin: *lamina* - thin plate; *proprius* - proper) a thickened basement membrane found in the deep portion of the mucosa of many tubular organs; consists of connective tissue.
laminae (*sing*: = lamina) (Latin: *lamina* - layer) portion of the vertebral arch, which extends between the transverse and spinous process.
Langerhans cell specialized dendritic cell found in the stratum spinosum that engulfs foreign bodies and damaged cells.
large intestine (*aka* colon) the portion of the of the GI tract, or alimentary canal, that lies between the small intestine and the rectum; absorbs water, vitamins, and electrolytes, and produces the feces.
laryngeal prominence (*aka* Adam's apple) (Greek: *larynx* - upper windpipe; Latin: *prominentia* - projection) the point of fusion of the two hyaline cartilage plates of the thyroid cartilage; this protruding area, which can be seen from the outside, is called the "Adam's apple."
laryngopharynx (*aka* hypopharynx) (Greek: *larynx* - upper windpipe; *pharynx* - throat) the most inferior part of the pharynx. Lies posteriorly to the larynx, and connects the oropharynx to the esophagus and trachea; transports respiratory gases to and from the larynx and transports food to the esophagus.
–**larynx** (*aka* voice box) (Greek: *larynx* - upper windpipe) an oval cartilaginous structure that lies anterior to the esophagus and between the pharynx and trachea; contains the vocal cords, produces vocal sounds, protects the trachea from food entry, and controls the amount of air entering or leaving the lungs.
late ventricular relaxation a phase of the cardiac cycle where the ventricles completely relax, becoming a point of low pressure in the circulatory system.
latent period (Latin: *lateo* - to lie hidden) phase of muscle contraction during which the action potential is propagated along the sarcolemma and calcium ions are released.
lateral (Latin: *latus* - side) describes the side or direction toward the side of the body.
lateral border of the scapula (Latin: *latus* - side; *scapulae* - shoulder blades) lateral margin of the scapula.
lateral circumflex artery (Latin: *lateralis* - side; *circumflex* - bent around) an artery that branches off of the deep femoral artery; surrounds the neck of the femur, and provides blood to the some of the deep thigh muscles and nearby skin.
lateral columns (*aka* lateral funiculi) (Latin: *lateralis* – side) masses of white matter on both sides of the spinal cord between the posterior horn and the anterior horn; composed of many different groups of axons of both ascending and descending tracts.
lateral condyle of the femur (Latin: *latus* - side; *femur* - thigh; Greek: *kondylōma* - resembling knuckle) smooth, articulating surface of the lateral expansion of the distal femur.
lateral condyle of the tibia (Latin: *latus* - side; *tibia* - shin bone; Greek: *kondylōma* - resembling knuckle) smooth, flat, proximal surface on the lateral side of the tibia.
lateral cuneiform (Latin: *latus* - side; *cuneus* - wedge) most lateral of the cuneiform tarsal bones; articulates posteriorly with the navicular bone, medially with the intermediate cuneiform bone, laterally with the cuboid bone, and anteriorly with the third metatarsal bone.
lateral epicondyle of the humerus (Latin: *latus* - side; *epi* - on; *kondyloma* - resembling knuckle; *humerus* - shoulder) small bony projection on the lateral side of the distal humerus.
lateral excursion (Latin: *latus* - side) side-to-side motion of the mandible away from the midline, to the right or left side.
lateral flexion (Latin: *latus* - side; *flecto* - to bend) bending of the body or neck toward the left or right.
lateral geniculate nucleus (Latin: *latus* - side; *genu* - knee; *nucleus* - a little nut) thalamic target of the axons of the visual system that projects to the visual cortex.
lateral horns (Latin: *lateralis* - side) masses of gray matter in the thoracic and upper lumbar regions of the spinal cord; contain neurons of the autonomic nervous system.
lateral malleolus (Latin: *latus* - side; *malleus* - mallet) expanded bump formed by the distal end of the fibula.

lateral meniscus (Latin: *latus* - side; Greek: *mēniskos* - crescent) c-shaped articular disc located between the lateral condyle of the femur and the lateral condyle of the tibia.
lateral plantar artery (Latin: *lateralis* - side; *planta* - sole of the foot) an artery formed by the splitting of the posterior tibial artery in the foot; provides blood to the lateral plantar region of the foot.
lateral pterygoid plates (Latin: *latus* - side; Greek: *pteryx* - wing; *eidos* - resemblance) paired, downward bony projections of the sphenoid bone located on the inferior skull; sites of attachment for chewing muscles.
lateral rotation (*aka* external rotation) (Latin: *latus* - side; *rotatio* - rotation) movement that rotates the limb so the anterior surface moves away from the midline of the body.
lateral sacral crest (Latin: *latus* - side; *sacrum* - sacred) paired ridges along the lateral sides of the posterior sacrum; formed by a fusion of the transverse processes of the sacral vertebrae.
lateral sulcus (Latin: *lateralis* - side; *sulcus* - a ditch) a deep groove on the surface of the cerebral cortex that separates the temporal lobe and the frontal and parietal lobes.
lateral supracondylar ridge (Latin: *latus* - side; *supra* - above; Greek: *kondylōma* - resembling knuckles) roughened bony ridge that runs along the lateral side of the distal humerus, superior to the lateral epicondyle.
lateral ventricles (Latin: *lateralis* - side; *ventriculus* - belly) paired superior chambers housed within the cerebral hemispheres, which produce and circulate CSF.
lateralis (Latin: *latus* - side) located laterally to.
latissimus dorsi (Latin: *latus* - side; *dorsalis* - back) broad, triangular muscle located on the inferior area of the back.
leak channel (*aka* non-gated channel) ion channel that opens randomly.
leaky tight junctions modified tight junctions found specifically in the ascending limb of the nephron loop, through which certain solutes, such as chloride, calcium, and magnesium ions, are able to pass; contain fewer proteins than other types of tight junctions.
learned reflex (Latin: *reflexus* - a bending back) a reflex that is developed after birth.
left atrioventricular valve (*aka* bicuspid valve, mitral valve) (Latin: *atrium* - entry hall) valve located between the left atrium and ventricle; consists of two flaps of tissue.
left colic flexure (*aka* splenic flexure) (Greek: *kolikos* - lower intestine; Latin: *flextura* - bend) the bend in the colon between the transverse and descending colon; located close to the spleen.
left coronary artery (Latin: *corona* - crown) one of two arteries that branch off the aorta to bring oxygenated blood to the heart wall; distributes blood to the left atrium, the left ventricle, and the intraventricular septum.
left gastric artery (Greek: *gaster* - stomach) an artery that branches off of the celiac trunk, and provides blood to the stomach and a portion of the esophagus.
lens clear disc through which light enters the eye; helps to focus light on the fovea centralis.
leptin (Greek: *leptos* - thin) a protein hormone secreted by adipose cells in response to food consumption; promotes satiety by informing the brain when lipid stores are sufficient in the body.
lesser curvature the superior concave curve of the stomach.
lesser omentum (*aka* small omentum, gastrohepatic omentum) a mesenterial sheet that hangs down from the inferior edge of the liver, connecting it to the superior edge of the stomach.
lesser pelvis (*aka* lesser pelvic cavity, true pelvis) (Latin: *pelvis* - basin) narrow, rounded space located within the pelvis, defined superiorly by the pelvic brim and inferiorly by the pelvic outlet.
lesser sciatic notch (Latin: *sciaticus* - corruption of; Greek: *ischiadikos* - hip joint) slightly curved posterior margin of the ischium, superior to the ischial tuberosity.
lesser splanchnic nerve (Greek: *splankhna* - inward parts) a sympathetic nerve arising in the T10–T11 level of the thoracic spinal cord and ending in the superior mesenteric ganglion; regulates function of certain organs of the gastrointestinal tract.
lesser trochanter (Greek: *trochantēr* - a runner) small, bony prominence on the medial aspect of the femur, at the base of the femoral neck.
lesser tubercle (Latin: *tuberosus* - full of lumps) small prominence on anterior side of the proximal humerus.
lesser wings of the sphenoid bone (Greek: *sphēnoeidēs* - resembling a wedge) paired lateral extensions of the sphenoid bone; form the ridge separating the anterior and middle cranial fossae.
leukocytes (*aka* white blood cells, WBCs) (Greek: *leukos* - white; *kytos* - cell) blood cells that defend the body from infection and disease; consisting of neutrophils, eosinophils, basophils, monocytes, and lymphocytes, they make up a small percentage of the formed elements of the blood.
levator ani (Latin: *levo* - to lift) pelvic muscles that support the pelvic viscera; resists pressure produced by contractions of the abdominal muscles.
ligamentum teres (*aka* ligament of the head of the femur) (Latin: *ligamentum* - a band; *femur* - thigh) intracapsular ligament that spans between the acetabulum and the femoral head.
ligand (*pl*: ligands) (Latin: *ligos:* to bind) a molecule that binds with specificity to, and activates, a specific receptor molecule.
ligand-gated channel (*aka* ionotropic receptors) (Latin: *ligo* - to bind) ion channel that opens in response to the binding of a molecule to its extracellular region.
light adaptation (Latin: *adapto* - to adjust) a process where the irises constrict immediately to reduce the amount of light entering the eyes, and the rods turn off so the cones can adjust their signaling.
light chain the smaller chain of proteins found in an antibody molecule; antibodies contain two heavy chains and two light chains.
limbic system a collection of cerebral structures involved in memory, emotion, and behavior.
linea alba (Latin: *albus* - white) white, fibrous band made of the bilateral rectus sheaths that join at the body's anterior midline.
linea aspera (*aka* rough line) (Latin: *linea* - line) roughened ridge that runs longitudinally along the posterior mid-femur.
lingual frenulum (Latin: *lingua* - tongue) a small fold of tissue that connects the tongue to the floor of the oral cavity; helps to hold the tongue in place and aids in speech and eating.
lingual tonsils (Latin: *linguae* - of the tongue) a mass of lymphoid tissue that lies on the posterior side of the tongue; helps remove and destroy pathogens in incoming air and food.
lipid (Greek: *lipose* - fat) class of nonpolar organic compounds built from hydrocarbons and distinguished by the fact that they are not soluble in water.
lipogenesis (Greek: *lipos* - fat; *genesis* - origin) the process of fatty acid or triglyceride synthesis, which occurs in the liver or adipose tissue.
lipolysis (Greek: *lipos* - fat; *lysis* - loosening) the breakdown of storage fats (triglycerides) in the adipose tissue; this adds glycerol and fatty acids to the blood, to be used by several tissues as an energy source.
lipoprotein lipase (Greek: *lipos* - fat) an enzyme found in the capillaries of adipose tissue and the liver, which decomposes the triglycerides in chylomicrons into fatty acids, monoglycerides, and glycerol.
liver an accessory organ of the digestive system; secretes bile and adjusts nutrient concentrations in the blood; the largest internal organ of the human body; also accomplishes many types of blood and metabolic regulation.
lobules substructures within testis that are divided by septa.
long bone bone that is longer than it is wide, as well as cylindrical in shape; long bones function as levers moved by muscle contraction.
longissimus the middle group of muscles of the erector spinae group; lies between the iliocostalis and spinalis muscles; helps to extend the vertebral column.
longitudinal fasciculi (Latin: *fascis* - bundle) long association tracts that connect functional areas between one lobe and another within the same cerebral hemisphere.
longitudinal fissure (Latin: *fissura* - fissure) large, deep groove that separates the cerebral hemispheres along the midline.
lordosis (*aka* swayback) (Greek: *lordōs* - bent backwards) condition characterized by excessive

vertebral column curvature in the anterior lumbar region.
lower esophageal sphincter (*aka* cardiac sphincter) (Greek: *oisophagos* - passage for food; Latin: *sphincter* - contractile muscle) a ring of smooth muscle that regulates the passage of food from the esophagus into the stomach.
lower leg the region of the lower limb that is made up of the two bones that run parallel to each other and sit between the knee joint and the ankle joint.
lower respiratory tract the portion of the respiratory tract that lies within the thorax; consists of the trachea, bronchi, bronchioles, alveolar ducts, alveolar sacs, and alveoli.
lumbar arteries (Latin: *lumbaris* – loin) four pairs of arteries that branch off of the abdominal aorta; provide blood to the muscles and skin of the lumbar region, spinal cord, lumbar vertebrae, and the posterior portion of the abdominal wall.
lumbar curve (Latin: *lumbus* - a loin) posteriorly concave curvature of the lumbar region of the vertebral column.
lumbar plexus (Latin: *lumbus* - a loin; *plexus* - braid) a nerve plexus arising from the lumbar spinal nerves giving rise to the nerves of the pelvic region and anterior leg.
lumbar veins (Latin: *lumbaris* - loin) veins that drain blood from the lumbar region of the abdominal wall and a portion of the spinal cord; empty into either the hemiazygos or the azygos vein, and eventually into the superior vena cava.
lumbar vertebrae (Latin: *lumbus* - a loin) five vertebrae (L1–L5) located in the lumbar region of the vertebral column; characterized by large, thick vertebral bodies.
lumbrical (Latin: *lumbricus* - earthworm) intermediate muscle of the hand; flexes each finger at the metacarpophalangeal joints and extend each finger at the interphalangeal joints.
lumen (Latin: *lumen* - light) 1) an enclosed space, lined with epithelia; 2) the interior of a tubular organ or blood vessel through which a substance is transported;3) the inside of a tubular or hollow body structure.
luminal surface (Latin: *lumen* - light) the surface of an epithelial cell of a renal tubule that borders on the lumen of the renal tubule (and lies farther from the interstitial fluid).
lunate (Latin: *luna* - moon) from the lateral side, the moon-shaped second of the four proximal carpal bones; articulates with the radius proximally, the capitate and hamate distally, the scaphoid laterally, and the triquetrum medially.
lung a major organ of the respiratory system that performs air movement and gas exchange.
lunula (*aka* lunulum) (Latin: *lunula* - little moon) basal area of the nail body that consists of a crescent-shaped layer of epithelium.
luteinizing hormone (LH) a hormone secreted by the anterior pituitary gland; triggers ovulation and formation of the corpus luteum in females; stimulates the production of ovarian hormones in females and testosterone in males.
lymph (Latin: *lympha* - water) the fluid flowing through the lymphatic system; derived from interstitial fluid that has entered a lymphatic capillary.
lymph node (Latin: *lympha* - water) a small, bean-shaped organ of the lymphatic system, which lies along a lymphatic vessel; filters lymph, removing pathogenic organisms and engaging in antigen recognition and attack.
lymphatic capillaries (Latin: *lympha* - water) the smallest vessels of the lymphatic system; pick up interstitial fluid from the tissue spaces and transport it through the lymphatic system, to eventually return it to the bloodstream.
lymphatic system (Latin: *lympha* - water) a body system consisting of lymphatic vessels, cells, lymph nodes, and other organs; transports fluid from the interstitial spaces back to the bloodstream.
lymphatic trunks (Latin: *lympha* - water) large lymphatic vessels that receive lymph from smaller lymphatic vessels and empty it into lymphatic ducts.
lymphocytes (Latin: *lympha* - water; Greek: *kytos* - cell) agranular leukocytes (white blood cells), which function in adaptive (specific) immunity; types of lymphocytes include B-lymphocytes, T-lymphocytes, and NK (natural killer) cells; comprise approximately 25 to 33 percent of the leukocyte population.
lymphoid nodules (Latin: *lympha* - water) unencapsulated masses of lymphoid tissue found mainly in the walls of organs of the respiratory and digestive tracts; protect the body from infection by pathogens entering the body in the air or food.
lymphoid stem cells (Latin: *lympha* - water; Greek: *eidos* - appearance) partially differentiated hemopoietic stem cells that arise from hemocytoblasts and give rise to lymphocytes.
lysosome (*pl*: lysosomes) (Greek: *lysis* - loosening: *soma* - body) membrane-bound cellular organelle originating from the Golgi apparatus and containing digestive enzymes.
lysozyme (Greek: *lysis* - loosen; *soma* - body) an enzyme found in saliva, mucus, tears, and milk, which destroys bacteria by digesting their cell walls.

M

M line A line in the center of the sarcomere, consisting of proteins that anchor the centers of the thick filaments in place.
macromolecule (Greek: *makros* - large; Latin: *molecula* - mass) large molecule formed by covalent bonding.
macrophage (*pl*: **macrophages**) (Greek: *makros* - large; *phago* - to eat) large phagocytic cell, found in various tissues of the body, which arises from the differentiation of monocytes after they leave the bloodstream.
macula (*pl*: maculae) (Latin: *macula* - a spot) flat membrane on which hair cells sit, in the organs of static equilibrium in the vestibule of the inner ear.
macula densa (Latin: *macula* - spot) a group of cells found in the portion of the distal convoluted tubule that makes up the juxtaglomerular apparatus; detects changes in Na^+ concentration in the filtrate, as a measure of electrolyte concentration in the blood.
macula lutea (Latin: *macula* - a spot; *luteus* - saffron-yellow) a small region of the posterior retina with a high proportion of cones; the focal point for light rays entering the eye.
magnetic resonance imaging (MRI) medical imaging technique in which a device generates a magnetic field to obtain detailed sectional images of the internal structures of the body.
main pancreatic duct (*aka* duct of Wirsung) the duct that transports pancreatic juice from the pancreas into the duodenum.
major calyx (Greek: *kalyx* - seed pod) a large tubular structure that transports urine from several minor calyces to the renal pelvis of the kidney.
major histocompatibility complex (MHC) a group of genes that codes for MHC proteins, which form complexes with antigenic determinants for antigen presentation to T cells.
malignant (*aka* metastatic tumor) (Latin: *maligno* - to do anything maliciously).
malleus (*aka* hammer) (Latin: *malleus* - hammer) ossicle that is attached to the tympanic membrane that articulates with the incus.
mammary glands (Latin: *mamma* - breast) modified sweat glands, located in the breast, that secrete milk to nourish a baby; present in both biological males and females, but develop in biological females under the direction of estrogens, human placental lactogen, and prolactin.
mammillary bodies paired structures of the limbic system that function in the transmission of olfactory information.
mandible (Latin: *mando* - to chew) single bone that forms the lower jaw; the only moveable bone of the skull.
mandibular foramen (Latin: *mando* - to chew; *foramen* - opening) opening on the medial side of the mandibular ramus; sensory nerve and blood vessels supplying the lower teeth enter through this opening.
mandibular fossa (Latin: *mando* - to chew; *fossa* - trench) deep, oval-shaped depression located on the external base of the skull; site of joint between mandible and skull.
mandibular notch (Latin: *mando* - to chew) broad U-shaped curve located between the condylar process and coronoid process of the mandible.
manubrium (Latin: *manubrium* - handle) wider, superior portion of the sternum.
marginal arteries (Greek: *arteria* - pipe) branches of the right coronary artery that supply blood to the superficial portions of the right ventricle.
mass amount of matter contained within an object.
mass movement a strong, slow, peristaltic wave that occurs in the transverse colon a few

times each day, often after meals; propels a large amount of feces down the colon at a time.
masseter (Latin: *masticare* - to chew) primary muscle involved in chewing; elevates the mandible to close the mouth.
mast cell (*aka* mastocyte; labrocyte) a cell found near blood and lymphatic vessels, under the skin, and in various connective tissues, that secretes heparin (which prevents blood clotting), histamine (which promotes the inflammatory response), and prostaglandins; also participates in certain allergic responses.
mastication (Latin: *masticare* - to chew) the act of chewing food, which reduces the size of the food particles and prepares the food for swallowing.
mastoid process (Greek: *mastos* - the breast; *eidos* - resemblance) large bony prominence of the temporal bone; located on the inferior, lateral skull, just behind the earlobe.
matter (Latin: *materies* - substance) physical substance; that which occupies space and has mass.
maxillary bone (*aka* maxilla) (Latin: *maxilla* - jawbone) paired bone that forms the upper jaw, part of the eye orbit, the lateral base of the nose, and anterior portion of the hard palate.
maxillary sinuses (Latin: *maxilla* - jawbone; *sinus* - cavity) large, air-filled space located within the maxillary bone; largest of the paranasal sinuses.
maxillary vein (Latin: *maxilla* - jawbone) a vein that drains blood from structures of the maxillary region and drains into the external jugular vein.
mean arterial pressure (MAP) the average blood pressure driving blood to the cells; estimated by adding one-third of the pulse pressure to the diastolic pressure.
meatuses a group of three narrow air passages (superior, middle, and inferior) in the nasal cavity, formed by the conchae, which increase the surface area to warm, filter, and humidify incoming air.
mechanical digestion the process of breaking down food into smaller particles to prepare it for enzymatic digestion; includes chewing, stomach churning, and intestinal segmentation.
mechanically gated channel (Greek: *mechanikos* - mechanical) ion channel that opens when a distortion of the cell membrane affects the structure of the channel.
mechanoreceptors (Greek: *mēchanē* - machine; Latin: *recipere* - to receive) receptor cells that detect physical stimuli, such as pressure and vibration.
medial (Latin: *medialis* - middle) describes the middle or direction toward the midline of the body.
medial border of the scapula (Latin: *medialis* - middle; *scapulae* - shoulder blades) medial margin of the scapula.
medial compartment of the thigh (Latin: *medialis* - middle) area that includes the adductor brevis, adductor longus, adductor magnus, pectineus, gracilis, and their associated blood vessels and nerves.
medial condyle of the femur (Latin: *medialis* - middle; *femur* - thigh; Greek: *kondylōma* - resembling knuckle) smooth, articulating surface of the medial expansion of the distal femur.
medial condyle of the tibia (Latin: *medialis* - middle; *tibia* - shin bone; Greek: *kondylōma* - resembling knuckle) smooth, flat, proximal surface on the medial side of the tibia.
medial cuneiform (Latin: *medialis* - middle; *cuneus* - wedge) most medial of the cuneiform tarsal bones; articulates posteriorly with the navicular bone, laterally with the intermediate cuneiform bone, and anteriorly with the first and second metatarsal bones.
medial epicondyle of the femur (Latin: *medialis* - middle; *epi* - on; *kondyloma* - resembling knuckle; *femur* - thigh) rough area on the medial side of the medial condyle of the distal femur.
medial epicondyle of the humerus (Latin: *medialis* - middle; *epi* - on; *kondyloma* - resembling knuckle; *humerus* - shoulder) prominent bony projection on the medial side of the distal humerus.
medial excursion (Latin: *medialis* - middle) side-to-side motion of the mandible toward the midline.
medial malleolus (Latin: *medialis* - middle; *malleus* - mallet) distal medial expansion of the tibia.
medial meniscus (Latin: *medialis* - middle; Greek: *mēniskos* - crescent) C-shaped articular disc located between the medial condyle of the femur and medial condyle of the tibia.
medial plantar artery (Latin: *medialis* - middle; *planta* - sole of the foot) an artery formed by the splitting of the posterior tibial artery in the foot; provides blood to the medial plantar region of the foot.
medial pterygoid processes (Latin: *medialis* - middle; Greek: *pteryx* - wing; *eidos* - resemblance) paired projections of the sphenoid bone located on the inferior skull; forms the posterior portion of the nasal cavity.
medial rotation (*aka* internal rotation) (Latin: *medialis* - middle; *rotatio* - rotation) movement that brings the anterior surface of a limb towards the midline of the body.
medialis (Latin: *medialis* - middle) located medially to.
median antebrachial vein (Latin: *medianus* - middle; *ante* - before; *brachium* - arm) a vein that runs parallel to the ulnar vein, and lies between the radial ulnar veins; arises from the palmar venous arches and empties into the basilic vein.
median cubital vein (Latin: *medianus* - middle; *cubitus* - elbow) a superficial vein of the antecubital region, that connects the cephalic vein to the basilic vein; often used for drawing blood.
median nerve (Latin: *medianus* - middle) a peripheral nerve of the arm, which innervates the skin and muscles of the forearm, hand, and fingers.
median sacral artery (Latin: *medianus* - middle; *sacrum* - sacred) a single artery that branches off of the abdominal aorta just as it terminates by dividing into the common iliac arteries; supplies blood to the sacrum and coccyx.
median sacral crest (Latin: *medianus* - middle; *sacral* - sacred) bumpy ridge extending down the midline of the posterior sacrum; remnant of the fused spinous processes.
mediastinal artery an artery that branches off of the thoracic aorta; provides blood supply to the posterior portion to the mediastinum, such as lymph nodes and a portion of the esophagus.
mediastinum (Latin: *medius* - middle) the subdivision in the center of the thoracic cavity; contains the heart and lungs.
medulla (Latin: *medulla* - marrow) in hair, the innermost layer of keratinocytes and the spaces between them 000; the inner portion of the kidney, which consists of the renal pyramids; *aka* renal medulla.
medullary cavity (Latin: *medulla* - marrow; *cavus* - hollow) long, hollow region within the diaphysis; filled with bone marrow.
megakaryocytes (Greek: *mega* - great; *karyon* - nut; *kytos* - cell) very large cells of the bone marrow, that release small cytoplasmic fragments that become platelets.
meiosis the process by which gametes (sperm and egg) are produced.
melanin (Greek: *melan* - black) one of the pigments found in the skin; determines the color of hair and skin and protects cells from UV radiation damage.
melanocyte (Greek: *melan* - black; *kytos* - cell) cell found in the stratum basale that produces the pigment melanin.
melanoma (Greek: *melan* - black; *-oma* - tumor) type of skin cancer that is characterized by the uncontrolled growth of melanocytes in the epidermis.
membrane potential (Latin: *membrana* - skin) The electrical charge difference between the inside and the outside of a cell membrane.
membranous urethra (*aka* intermediate urethra) (Greek: *ourethra* - passage for urine) the middle portion of the urethra, that passes through the external urethral sphincter and runs between the prostatic urethra and the spongy (penile) urethra.
memory T cells T lymphocytes (T cells) that remain dormant after their first exposure to an antigen, but remember the antigen and remain in the body fluids for a long period of time to respond to subsequent exposure to that antigen.
meninges (Greek: *mēninx* - membrane) the three membranes that surround and protect the brain and spinal cord; consist of the dura mater, the arachnoid mater, and the pia mater.
meningitis (Greek: *mēninx* - membrane; *itis* - inflammation) inflammation of the meninges, the protective membranes covering the brain and spinal cord; typically results from a viral or bacterial infection.
meniscus (Greek: *mene* - moon; *mēniskos* - crescent) c-shaped articular disc found between

articulating surfaces in some joints; located between the femoral and tibial condyles in the knee joint; composed of fibrocartilage.
menses (Latin: *menses* - month) the material that is sloughed off of the inner part of the endometrium and new blood vessels of the uterine wall though the vagina; shedding occurs during each uterine cycle between menarche and menopause; also used to describe the process of shedding of blood and endometrial tissue that occurs each month.
menses phase (*aka* menstrual period, period, menstruation) (Latin: *menses* - month) the portion of the uterine cycle during which the lining of the endometrium and the new blood vessels that arose to nourish it are shed from the uterus through the vagina; occurs each month in which fertilization has not occurred.
menstrual period (*aka* period) (Latin: *menses* - month) another term for menses, this is the material that is shed from the uterus and exits the body through the vagina.
Merkel cell (*pl*: **Merkel cells**) (*aka* Merkel-Ranvier cell, tactile epithelial cell) sensory receptor cell of the stratum basale connected to sensory nerves; responds to touch.
merocrine secretion (Greek: *mēros* - share; *krinōs* - to share; Latin: *secerno* - to separate) the release of a substance from a gland via exocytosis; the most common type of exocrine secretion.
mesenchymal cells (Greek: *mesos* - middle; *enkyma* - fusion) embryonic cells that gather together and begin to differentiate into specialized cells.
mesenteries (Latin: *mesenterium* - middle of the intestine) double-layered sheets of the peritoneum that suspend the digestive organs from the posterior wall of the abdominal cavity; contain blood and lymphatic vessels and nerves.
mesonephric ducts (*aka* Wolffian ducts) (Greek: *mesos* - middle; *nephros* - kidney) ducts that develop in embryos at week 7 which require testosterone for development.
mesovarium (Greek: *mesos* - middle) an extension of the peritoneum.
messenger RNA (mRNA) nucleotide molecule that serves as an intermediate in the genetic code between DNA and protein.
metabolic acidosis (Latin: *acidus* - sour; *osis* - a state) a condition in which the pH of the blood is below 7.35 (too acidic), due to a nonrespiratory cause; causes include lactic acidosis and ketoacidosis; characterized by an excess of hydrogen ions or a deficiency of bicarbonate ions in the blood plasma.
metabolic alkalosis (Latin: *alkali* - soda; *osis* - a state) a condition in which the pH of the blood is above 7.45 (too alkaline or basic), due to a nonrespiratory cause; causes include vomiting of the stomach contents or taking too many antiacids; characterized by a deficiency of hydrogen ions or an excess of bicarbonate ions in the blood plasma.
metabolic rate the amount of energy the body expends per unit of time; can be measured in Calories/hour.
metabolism (Greek: *metabole* - change) sum of all chemical reactions in the body, including both anabolic and catabolic reactions.
metacarpal bone (Greek: *meta* - between; *karpos* - wrist) one of the five bones that form the palm of the hand.
metaphase (Greek: *meta* - between; *phasis* - an appearance) the stage of mitosis (and meiosis), characterized by the alignment of chromosomes in the center of the cell.
metaphysis (Greek: *meta* - between; *physis* - growth) a narrow bone region where the diaphysis meets the epiphysis; contains the epiphyseal plate.
metarteriole a blood vessel that transports blood between a terminal arteriole and a capillary bed.
metastasis (Greek: *metastasis* - removal or change) the process by which cancer cells mobilize and establish tumors in other parts of the body.
metatarsal bone (Greek: *meta* - between; *tarsos* - tarsus) one of the five bones that form the anterior region of the foot.
metencephalon (Greek: *meta* - after; *enkephalos* - brain) a primary vesicle of the embryonic brain; develops into the pons and cerebellum in the adult brain.
MHC class I proteins found on all cells of the body except erythrocytes, which form complexes with antigenic determinants to present the antigens to the CD8 molecules on cytotoxic T cells.
MHC class II proteins found only on macrophages, dendritic cells, and B cells, which form complexes with antigenic determinants to present the antigens to the CD4 molecules on helper T cells.
micelle a loose aggregate of lipids that lies close to the mucosa of the small intestine, which regulates the absorption of lipids by intestinal epithelial cells.
microbiota (Greek: *mikros* - small) the trillions of bacteria that live within the large intestine.
microfilament (*pl*: microfilaments) (Greek: *mikros* - small; Latin: *filamentum* - thread) the thinnest of the cytoskeletal filaments; composed of actin subunits that function in muscle contraction and cellular structural support.
microglia (Greek: *mikros* - small; *glia* - glue) glial cell type in the CNS that serves as the resident immune cells.
microscopic anatomy (Greek: *mikros* - small; *skopeō* - to view; *ana* - up; *tome* - cutting) study of very small structures of the body using magnification.
microtubule (*pl*: microtubules) (Greek: *mikros* - small; Latin: *tubus* - tube) the thickest of the cytoskeletal filaments, composed of tubulin subunits that function in cellular movement and structural support.
microvilli (*sing*: microvillus) (Greek: *mikros* - small; Latin: *villi* - shaggy hair) small projections on the surface of cells that function to increase surface area.
microvilli (*sing*: microvillus) (Greek: *mikros* - small; Latin: *villus* - shaggy hair) tiny fingerlike projections of the plasma membrane of the epithelial cells lining the small intestine, which provide increased surface area for secretion and absorption.
micturition (*aka* urination) (Latin: *micturitium* - to desire to urinate) urination or emptying of the urinary bladder.
midbrain (*aka* mesencephalon) the middle region of the adult brain.
middle cardiac vein (Greek: *cardiac* - *kardia* meaning heart) vessel that parallels and drains the areas supplied by the posterior interventricular artery; drains into the great cardiac vein.
middle cerebral artery (Latin: *cerebro* - brain) an artery that branches from the internal carotid artery; supplies blood to portions of the frontal, temporal, and parietal lobes of the cerebrum.
middle cranial fossa (Latin: *medialis* - middle; Greek: *kranion* - skull) centrally located cranial fossa; extends from the lesser wings of the sphenoid bone to the petrous ridge.
middle ear space spanned by the ossicles.
middle lobe the lobe of the right lung that lies between the superior and inferior lobes.
middle nasal concha (Latin: *medialis* - middle; *nasus* - nose; *concha* - shell) thin, curved projection of the ethmoid bone; located between the superior and inferior conchae within the nasal cavity.
middle sacral vein (*aka* median sacral vein) a vein that drains blood from a portion of the sacrum into the common iliac vein.
mineralocorticoids hormones produced by the zona glomerulosa of the adrenal cortex, such as aldosterone; regulate fluid and electrolyte balance.
minerals inorganic compounds which are not used as energy sources, but are needed in the body for a variety of functions (nerve impulse conduction, muscle contraction, adding strength to structures such as bones, being incorporated into specific hormones, etc.).
minor calyx (Greek: *kalyx* - seed pod) a small tubular structure that transports urine from the collecting ducts in the renal papillae (the tips of the renal pyramids) to a major calyx
minute ventilation (*aka* minute respiratory volume) the total volume of inhaled air that enters or exits the lungs in one minute; can be calculated by multiplying the number of breaths per minute by the tidal volume.
mitochondrion (*pl*: mitochondria) (Greek: *mitos* - thread; *chondros* - granule) one of the cellular organelles bound by a double lipid bilayer that function primarily in the production of cellular energy (ATP).
mitosis (Greek: *mitos* - thread) division of genetic material, during which the cell nucleus breaks down and two new, fully functional, nuclei are formed.
mitotic phase (*pl*: phases) (Greek: *mitos* - thread) phase of the cell cycle in which a cell undergoes mitosis.
mitotic spindle (*pl*: spindles) (Greek: *mitos:* thread) network of microtubules, originating

from centrioles, that arranges and pulls apart chromosomes during mitosis.
moderator band band of myocardium covered by endocardium that arises from the inferior portion of the interventricular septum in the right ventricle and crosses to the anterior papillary muscle; contains conductile fibers that carry electrical signals followed by contraction of the heart.
molarity (Latin: *moles* - mass) moles of the molecule per liter.
molars the broad, flattened teeth that lie in the posterior portion of the mouth; used for grinding food down to smaller particles.
molecule (Latin: *molecula* - mass) two or more atoms covalently bonded together.
monocytes (Greek: *mono* - one; *kytos* - cell) agranular leukocytes that conduct phagocytosis of pathogens; once they leave the bloodstream, they differentiate into macrophages.
monomer (Greek: *mono* – single; *meros* - part) individual units that make up a larger molecule.
monosaccharide (*aka* simple sugar) (Greek: *mono* - single; Latin: *saccharum* - sugar) monomer of carbohydrate; a single sugar molecule, containing between three and seven carbon atoms; examples are glucose, fructose, and galactose.
monosynaptic reflex (Greek: *monos* - single; *syn* - together; Latin: *reflexus* - a bending back) a reflex that only contains one synapse, which is found between the sensory and motor neurons.
monozygotic (Greek: *monos* - one; *zygon* - yoke) identical twins resulting from the same zygote.
mons pubis (Latin: *mons* - mountain; *pubes* - groin) a skin-covered mound of adipose tissue that covers and protects the symphysis pubis in biological females; lies anterior to the labia majora.
motility (Latin: *movere* - to move) movement; in the case of the digestive system, the movement of food in the GI tract, such as stomach churning (mixing) or peristalsis (propulsion down the GI tract).
motor end plate (Latin: *moveo* - to move) region of sarcolemma located at the neuromuscular junction of a skeletal muscle cell.
motor neurons (*aka* efferent neurons) (Latin: *motor* - to move; Greek: *neuro* – nerve) nerve fibers that carry signals that emanate from the central nervous system and connect to effectors.
motor unit (Latin: *moveo* - to move) group of muscle fibers innervated by a single motor neuron.
mucosa (Latin: *mucosus* – mucous) the innermost layer of the GI tract; consists of an epithelial layer, a lamina propria (connective tissue) and a muscularis mucosae (smooth muscle).
mucosa-associated lymphoid tissue (MALT) (Latin: *mucosus* - mucous; *lympha* – water) lymphoid nodules found in the mucosa of the digestive, respiratory, genital, and urinary tracts, as well as the breast, eye, and skin, which protect against infection by pathogenic organisms entering the body from the external environment.
mucosal barrier (Latin: *mucosus* - mucous) a layer that separates the epithelial cells of the mucosa of the stomach and intestines from the contents of the lumen; protects the mucosal epithelial cells from being destroyed by their own secretions; consists of a layer of alkaline mucus and tight junctions between adjacent cells.
mucous gland (*pl*: mucous glands) (Latin: *mucosus* - mucous; *glans* - acorn) group of cells that secrete mucous, watery to viscous fluid rich in mucin.
mucous membranes (*sing*: mucous membrane) (Latin: *mucosus* - mucous; *membrana* - a skin or membrane that covers a part of your body) epithelial and connective tissues that line the body cavities and passageways open to the external environment, which are kept moist by exocrine mucous secretions.
mucous neck cells (Latin: *mucosus* - mucous) cells in the upper portion of the gastric glands, which produce acidic mucus; the function of this mucus is unknown.
multiaxial joint (Latin: *multus* – much) joint that allows for motion within several planes.
multipennate (Latin: *multus* - much; *penna* - feather) a pennate muscle in which the tendon branches within the muscle.
Multiple Sclerosis (MS) (Greek: *sklērōsis* - hardness) autoimmune disease characterized by the loss of myelin at one or several places in the nervous system.
multipolar neuron shape with multiple processes extending from the cell body, the axon and two or more dendrites.
murmur unusual heart sound detected by auscultation; typically related to septal or valve defects.
muscarinic receptor (*pl*: **muscarinic receptors**) one type of acetylcholine receptor protein; also binds to the mushroom poison muscarine.
muscle spindle bundle of muscle cells that are wrapped with sensory neuron dendrites.
muscle tension the force generated by the contraction of a muscle.
muscle tissue (Latin: *musculus* - muscle) contractile tissue capable of generating tension in response to stimulation; produces movement.
muscle tone (Greek: *tonos* - tension) baseline level of muscle contraction; occurs when the muscle is not producing movement.
muscular artery (*aka* distributing artery) a medium-sized artery that contains many smooth muscle fibers in the tunica media; receives blood from elastic arteries and branches into small arteries and arterioles.
muscularis the muscle layer of the wall of the GI tract; consists of smooth muscle in most areas, but skeletal muscle in a few areas; lies between the submucosa and the serosa.
mutation (*pl*: mutations) (Latin: *muto* - to change) an error made with DNA replication.
myelencephalon a secondary vesicle of the embryonic brain; develops into the medulla oblongata in the adult brain.
myelin insulating layer of lipids that surround the axons of some neurons, formed by glial cells.
myelin sheath fatty layer of insulation that surrounds some axons; facilitates the transmission of electrical signals down the axon.
myeloid stem cells (Greek: *myelos* - marrow; *eidos* - appearance) partially differentiated hemopoietic stem cells that arise from hemocytoblasts and give rise to erythrocytes, neutrophils, eosinophils, basophils, monocytes, and platelets.
myenteric plexus (*aka* plexus of Auerbach) a nerve network in the muscularis of the wall of the GI tract, which controls motility; makes up a portion of the enteric nervous system.
mylohyoid (Greek: *mylē* - molar teeth; *hyoeidēs* - U-shaped) one of the suprahyoid muscles; lifts the larynx and presses the tongue to the top of the mouth.
myocardial conducting cells (Greek: *cardiac* - *kardia* meaning heart) specialized cells that transmit electrical impulses throughout the heart and trigger contraction by the myocardial contractile cells.
myocardial contractile cells (Greek: *cardiac* - *kardia* meaning heart) bulk of the cardiac muscle cells in the atria and ventricles that conduct impulses and contract to propel blood.
myocardium (Greek: *cardiac* - *kardia* meaning heart) thickest layer of the heart composed of cardiac muscle cells built upon a framework of primarily collagenous fibers and blood vessels that supply it and the nervous fibers that help to regulate it.
myofibril (Greek: *mys* - muscle) the contractile machinery of muscle cells; long cylinders of contractile proteins that shorten during muscle contraction.
myofilaments (Greek: *mys* - muscle) protein subunits that make up the myofibrils in muscle cells; composed mainly of actin or myosin.
myogenic response (Greek: *mys* - muscle; *genesisi* - to produce) a change in the diameter of arterioles, which occurs in response to a change in blood flow; increased blood flow stretches the smooth muscle cells in the walls of the arterioles, leading to constriction and decreased blood flow; decreased blood flow has the opposite effect.
myogram (Greek: *mys* - muscle; *gramma* - a drawing) a graph that represents muscle tension produced over time.
myometrium (Greek: *mys* - muscle) the thick muscle layer of the wall of the uterus; contractions of this muscle layer excrete the menses each month, and also expel the fetus during childbirth.
myosin (Greek: *mys* - muscle) protein subunit that makes up much of the thick myofilaments within a sarcomere.

N

NADH (*aka* nicotinamide adenine dinucleotide) the reduced form of the coenzme NAD; transports electrons and hydrogen ions from

glycolysis and the citric acid cycle to the electron transport chain.
nail bed epidermal structure that produces the nail body.
nail body (*pl*: nail bodies) hard, bladelike keratinous plate that forms the nail.
nail cuticle (*aka* eponychium) (Latin: *cuticula* - skin) fold of epithelium that extends over the nail bed at the meeting of the proximal nail fold and nail body.
nail root protected region at the proximal side of the nail bed from which the nail body grows.
naris (*pl*: nares) (Latin: *naris* - nostril) the opening into one of the nostrils.
nasal bone (Latin: *nasus* – nose) a rectangular bone that supports the root and bridge of the nose.
nasal bones (Latin: *nasus* - nose) small, paired bones that articulate to form the bridge of the nose.
nasal conchae (Latin: *nasus* - nose; *concha* - shell) mucous membrane-covered bony plates that project from the lateral walls of the nasal cavity; function to subdivide the nasal cavity.
nasal septum (Latin: *nasus* - nose; *saeptum* - a partition) flat structure that divides the nasal cavity into halves; formed by the perpendicular plate of the ethmoid bone and the vomer bone.
nasal septum (Latin: *nasus* - nose; *saeptum* - a fence) a partition that divides the nasal cavity into left and right portions; composed of septal cartilage and portions of the ethmoid and vomer bones.
nasopharynx (Latin: *nasus* - nose; Greek: *pharynx* - throat) the most superior part of the pharynx, which lies posteriorly to the nasal cavity and transports air from the nasal cavity to the oropharynx.
nasopharynx (Latin: *nasus* - nose; Greek: *pharynx* - throat) the most superior region of the pharynx; lies posterior to the nasal cavity, transports only air and is involved in the breathing process.
natriuretic hormones (*aka* natriuretic peptides) peptide hormones, produced mainly in the heart and central nervous system, that cause increased sodium excretion in the urine; increased water excretion occurs along with the sodium excretion; considered antagonists of aldosterone and antidiuretic hormone.
natural killer (NK) cells a small population of lymphocytes, which acts nonspecifically to defend the body against virus-infected and tumor cells.
navicular (Latin: *navis* - ship) last bone in the proximal row of tarsal bones; articulates posteriorly with the talus bone, laterally with the cuboid bone, and anteriorly with the medial, intermediate, and lateral cuneiform bones.
neck of the femur (Latin: *femur* - thigh) narrowed region of the femur; immediately inferior to the head of the femur.
neck of the radius (Latin: *radius* - ray) narrow region immediately distal to the head of the radius.
neck of the rib narrowed region of a rib; located next to the rib head.
negative feedback homeostatic mechanism that tends to stabilize an upset in the body's physiological condition by preventing an excessive response to a stimulus.
negative inotropic factors factors that negatively impact or lower heart contractility.
nephron loop (*aka* Loop of Henle) (Greek: *nephros* - kidney) the portion of the nephron that extends into the renal medulla of the kidney; transports filtrate between the proximal convoluted tubule and the distal convoluted tubule; performs reabsorption of NaCl and water.
nephrons (Greek: *nephros* - kidney) the structural and functional units of the kidney; conduct the filtration, reabsorption, and secretion processes of urine formation; consist of the renal corpuscles and renal tubules; empty urine into the collecting ducts.
nerve a bundle of axons in the peripheral nervous system.
nerve plexuses (Latin: *plexus* - braid) branching networks of interlacing spinal nerves.
nervi vasorum nerve fibers found in the walls of blood vessels, that stimulate contraction of the smooth muscle tissue in the walls of the vessels.
nervous tissue (French: *tissu* - woven) tissue that is capable of sending and receiving impulses throughout the body using electrochemical communication.
net filtration pressure (NFP) 1) the force that determines whether blood flows out of or into the capillaries; calculated by subtracting the blood colloidal osmotic pressure from the capillary hydrostatic pressure. 2) the net pressure that determines whether glomerular filtration will occur; can be calculated by subtracting the capsular hydrostatic pressure and the glomerular osmotic pressure from the glomerular hydrostatic pressure.
neural crest cells (Greek: *neuro* - nerve) scattered cells surrounding the developing neural tube; differentiate into peripheral nerves.
neural layer (*aka* retina) (Greek: *neuro* - nerve) the innermost layer of the eye; contains the nervous tissue responsible for photoreception.
neural tube a hollow tube extending along the posterior side of the developing embryo; precursor to the central nervous system.
neurilemma cell (*aka* Schwann cell) (Greek: *neuro* - nerve; *lemma* - husk) a type of glial cell that insulates axons with myelin in the peripheral nervous system.
neurofibril node (*aka* nodes of Ranvier) each gap along the axon that contributes to saltatory conduction.
neuromuscular junction (NMJ) (Greek: *neuro* - nerve; *mys* - muscle) point of contact between the skeletal muscle and the controlling neuron; site of muscle stimulation.
neurons (*sing*: neuron) (Greek: *neuro* - a nerve) neural cells that respond to stimuli and send impulses throughout the body using electrochemical signals.
neurons (*sing*: neuron) (Greek: *neuro* – nerve) neural cells responsible for communication; generate and propagate electrical signals into, within, and out of the nervous system.
neurotransmitter (Greek: *neuro* - nerve; Latin: *transmitto* - to send across) chemical messenger released by nerve terminals; binds to and activates receptors on target cells.
neurotransmitter (Greek: *neuro* - nerve; Latin: *transmitto* - to send across) chemical signal released from a neuron at a synapse in order to cause a change in the target cell.
neurotransmitters chemicals released by neurons in synapses to communicate with other neurons, muscle cells or gland cells.
neutral (Latin: *neuter* - neither) particles that carry no charge.
neutralization (Latin: *neuter* - neither) the binding of an antibody to a virus or bacterial toxin, rendering the pathogen unable to bind to body cell receptors and therefore harmless.
neutron (Latin: *neuter* - neither) heavy subatomic particle having no electrical charge and found in the atom's nucleus.
neutrophils (Latin: *neutra* - neutral; Greek: *philos* - fond) granulocytes that are especially effective in the phagocytosis of bacteria, make up the largest percentage of the leukocytes, and are stained by neutral dyes.
neutrophils (Latin: *neuter* - neither) the most abundant of the five types of leukocytes (white blood cells) in the body; strong phagocytic cells that engulf mainly bacteria; circulate in the blood, but also leave the bloodstream to migrate to infection sites via chemotaxis.
nicotinamide adenine dinucleotide (NAD) a coenzyme that transports electrons and hydrogen ions from glycolysis and the citric acid cycle to the electron transport chain.
nicotinic receptor (*pl*: **nicotinic receptors**) one type of chemically gated ion channel acetylcholine receptor protein; characterized by also binding to nicotine.
nipple a cylindrical projection on the anterior of the breast.
nociceptors (Latin: *nocio* - pain; *capere* - to take) nerve fibers specific to communicating pain stimuli to the brain and spinal cord.
nonessential nutrients nutrients that can be produced in adequate amounts by the body, so they do not have to be obtained through the diet; includes 11 amino acids that can be synthesized from other precursors and certain fatty acids.
nonpolar molecule (Latin: *polus* - pole; *molecula* - mass) molecules that do not have a slightly positive or slightly negative charge.
norepinephrine signaling molecule released as a neurotransmitter by most postganglionic sympathetic fibers; also secreted as an amine hormone by the chromaffin cells of the adrenal medulla in response to short-term stress; prepares the body for fight-or-flight responses, such as an increase in heart rate, breathing rate, and

blood pressure, and dilation of the pupils and airways
nuclear pore (*pl*: pores) (Latin: *nucleus* - a little nut; Greek: *poros* – passageway) the small, protein-lined openings found scattered throughout the nuclear envelope.
nucleolus (*pl*: nucleoli) (Latin: *nucleus* - a little nut) small region of the nucleus that functions in ribosome synthesis.
nucleosome (*pl*: nucleosomes) (Latin: *nucleus* - a little nut; Greek: *soma* - body) unit of chromatin consisting of a DNA strand wrapped around histone proteins.
nucleotide class of organic compounds composed of one or more phosphate groups, a pentose sugar, and a base.
nucleus (*pl*: nuclei) (Latin: *nucleus* - a little nut) 1) cell's central organelle; contains the cell's DNA; 2) a mass of neuron cell bodies in the central nervous system.
nucleus pulposus (Latin: *nucleus* - a little nut; *pulpa* - flesh) gel-like center region of an intervertebral disc.
nulliparous (Latin: *nullus* - no; *para* - to produce) a term describing a biological female who has never given birth to a baby.
nurse cells (*aka* sustentacular cells, Sertoli cells) cells that make up the wall of the seminiferous tubule and surround the developing sperm through all stages of their development.
nutrients substances that the body uses in the process of growth, repair, and metabolism.

O

oblique (Latin: *obliquus* - slanted) referring to the direction of muscle fibers; at an angle.
oblique fissure (Latin: *obliquus* - slanting; *fissura* - a cleft) a double fold of the visceral pleura that separates the superior and inferior lobes of the left lung and separates the middle and inferior lobes of the right lung.
obturator externus (Latin: *obturo* - to occlude; *externus* - external) muscle deep to the gluteus maximus on the lateral surface of the thigh; laterally rotates the femur at the hip.
obturator foramen (Latin: *obturo* - to occlude; *foramen* - an opening) large opening in the anterior hip bone formed at the junction of the rami of the pubis and ischium.
obturator internus (Latin: *obturo* - to occlude) muscle deep to the gluteus maximus on the lateral surface of the thigh; laterally rotates the femur at the hip.
occipital bone single bone that forms the posterior skull and the posterior base of the cranial cavity.
occipital condyle paired, oval-shaped bony knobs located on the inferior skull, on either side of the foramen magnum; form joints with the first cervical vertebra.
occipital lobe a region of the cerebrum beneath the occipital bone of the cranium; responsible for processing visual stimuli
occipital sinus (*sinus* - curve) a small dural venous sinus within the falx cerebelli.
occipitalis posterior region of the occipitofrontalis muscle.
occipitofrontalis muscle that moves the scalp and eyebrows; consists of the frontal belly and occipital belly.
oculomotor nerve (CN III) third cranial nerve; responsible for eye movements by controlling four of the extraocular muscles.
odorant molecules molecules that bind to receptor proteins in olfactory neurons, stimulating olfaction
olecranon fossa (Greek: ōlenē + *kranion* - head of the ulna; Latin: *fossa* - trench) large depression on the posterior side of the distal humerus; accommodates the olecranon process of the ulna when the elbow is fully extended.
olecranon process (Greek: ōlenē + *kranion* - head of the ulna) curved extension of the ulna, formed by the posterior and superior portions of the proximal ulna; fits into the olecranon fossa of the humerus when the elbow is extended.
olfaction (Latin: *olfacio* - to smell) the sense of smell.
olfaction (Latin: *olfactus* - to smell) the sense of smell.
olfactory bulb (Latin: *olfactus* - to smell) central target of the first cranial nerve; located on the inferior surface of the frontal lobe.
olfactory bulbs (Latin: *olfacio* - to smell) structures that lie at the anterior inferior portion of the brain, that transmit nerve impulses from the olfactory receptor cells to the brain; participate in the sense of smell and in limbic system function.
olfactory epithelium (Latin: *olfactus* - to smell) region within the superior nasal cavity where olfactory neurons are located.
olfactory foramina (Latin: *olfactio* - to smell; *foramen* - opening) small holes in the cribriform plates that allow for olfactory neurons to pass through.
olfactory nerve (CN I) (Latin: *olfacio* - to smell) first cranial nerve; responsible for olfaction.
olfactory sensory neuron (Latin: *olfactus* - to smell; Greek: *neuro* - nerve) receptor cell of the olfactory system, the axons of which compose the first cranial nerve with dendrites extending into the mucous lining of the nasal cavity.
oligodendrocyte (Greek: *oligos* - few; *dendron* - tree; *kytos* - cell) glial cell type in the CNS that insulates axons in myelin.
omohyoid [muscle] (Greek: *omōs* - shoulder; *hyoeidēs* - U-shaped) one of the infrahyoid muscles; has anterior and posterior bellies and depresses the hyoid bone.
oogenesis (Greek: *oon* – egg; Latin: *genesis* - origin) the process of oocyte production and development in the biological female; includes the production and mitosis of oogonia and development of primary oocytes by meiosis during fetal development; further development into secondary oocytes occurs only upon fertilization.
oogonia (Greek: *oon* - egg) the most primitive form of the gametes produced by biological females; divide by mitosis to increase the supply and begin Meiosis I during fetal development of the biological female to form primary oocytes.
ophthalmic artery (Latin: *ophthalmicus* - pertaining to eyes) an artery that branches from the internal carotid artery; supplies blood to the eyes, forehead, and nose.
opponens digiti minimi (Latin: *oppono* - oppose) hypothenar muscle; brings the little finger across the palm to meet the thumb.
opponens pollicis (Latin: *oppono* - oppose) thenar muscle; moves the thumb across the palm to meet another finger.
opposition (Latin: *oppono* - oppose) movement of the thumb that brings the tip into contact with the tip of a finger.
opsins photopigments specialized to detect certain wavelengths of light located in cone photoreceptors.
opsonization the binding of a pathogenic organism to a phagocyte by an antibody or complement protein; this aids in the process of phagocytosis
optic canal (Greek: *optikos* - eyes; *canal* – channel) opening located at the anterior lateral corner of the sella turcica; provides passage of the optic nerve to the orbit.
optic chiasm (Greek: *optikos* - eye; *chiasma* - crossing) the point of decussation (crossover) of the optic nerves.
optic nerve (CN II) (Greek: *optikos* - eye; *neuro* – nerve) second cranial nerve; transmits sensory nerve impulses from the ganglion cells of the retina.
oral cavity (*aka* buccal cavity) the mouth; bordered by the lips, fauces, cheeks, palate, and floor of the mouth; occupied mainly by the tongue and teeth.
orbicularis oculi (Latin: *orbiculus* - a small circle; *oculi* - of the eye) circular muscle responsible for closing the eye.
orbicularis oris (Latin: *orbiculus* - a small circle; *oris* - mouth) circular muscle surrounding the mouth.
orbit a protective space for the eyeball and muscles that move the eyeball and upper eyelid.
organ (Greek: *organon* - a tool) functionally distinct structure composed of two or more types of tissues.
organ system (Greek: *organon* - a tool; *systema* - an organized whole) group of organs that work together to contribute to a particular function.
organelle (Greek: *organon* - a tool) tiny functioning units within a cell.
organelle (*pl*: organelles) (Greek: *organon* - a tool) any of several different types of membrane-enclosed specialized structures in the cell that perform specific functions for the cell.
organic compound (Latin: *compono* - to place together) substance that contains both carbon and hydrogen.
organism living being composed of one or more cells that can independently perform all physiologic functions necessary for life
orgasm (Greek: *orgasmos* - excitement) a brief but intense response ranging from 1–15 seconds

that may occur at the end of the excitement and plateau phase.
orgasm stage (Greek: *orgasmos* - excitement) a stage where an orgasm occurs, usually after the excitement phase.
origin (Latin: *origio* - source) fixed (unmovable) end of a muscle; generally attached to a bone.
oropharynx (Latin: *oris* - mouth; Greek: *pharynx* - throat) the middle part of the pharynx; lies posteriorly to the oral cavity and runs between the nasopharynx and the laryngopharynx; a passageway for air and food.
os coxae (*aka* Coxal bone, hip bone) (Latin: *os* - bone; *coxae* - hip) one of the two hip bones; considered part of the appendicular skeleton, and is attached to the axial skeleton at its joint with the sacrum of the vertebral column.
osmolarity (Greek: *osmos* - pushing fluid) the concentration of solutes in a solution, such as the total number of solutes per liter of solution.
osmoreceptors (Greek: *osmos* - pushing fluid) specialized hypothalamic receptor cells that detect changes in the solute concentration in the blood.
osmosis (Greek: *osmos* - pushing fluid) the diffusion of water from regions of higher water concentration to regions of lower water concentration across a selectively permeable membrane.
osmotic gradient (Greek: *osmos* - pushing fluid) a difference in water concentration between two regions separated by a membrane.
ossicles (Latin: *ossiculum* - bone) three tiny bones of the middle ear, called the malleus, incus, and stapes, which transmit sound vibrations from the tympanic membrane to the fluid of the inner ear.
ossification (*aka* osteogenesis) (Greek: *ossificatio* - to make bone) the development of bone.
ossification center (Greek: *osteon* - bone) cluster of early osteoblasts present during intramembranous ossification.
osteoarthritis (Greek: *osteon* - bone; *arthrosi* - joint; *-itis* - inflammation) a painful disorder that occurs when there is a reduced amount of hyaline cartilage present in the joints.
osteoblasts (Greek: *osteon* - bone; *blastos* - germ) cells that form new bone matrix by synthesizing and secreting collagen fibers and calcium salts.
osteoclast (Greek: *osteon* - bone; *klastos* - breaker) multinucleate cells that break down bone.
osteocytes (Greek: *osteon* - bone; *kytos* - cells) most abundant cell in mature bone; maintain the mineral concentration in the matrix via secretion of enzymes.
osteogenic cells (Greek: *osteon* - bone; Latin: *genesis* - production) bone stem cells; the only bone cells capable of dividing.
osteoid (Greek: *osteon* - bone; *eidos* - resemblance) soft matrix that hardens when calcium is deposited on it; secreted by osteoblasts.
osteomalacia (Greek: *osteon* - bone; *malakia* - softness) a condition in adults that leads to softening of the bones due to a deficiency of vitamin D or calcium; corresponds to rickets in children.
osteon (*aka* Haversian system) (Greek: *ostéon* - bone) microscopic structural unit of compact bone; composed of concentric layers of lamellae.
osteoporosis (Greek: *osteon* - bone; *poros* - pore; *-osis* - condition) condition characterized by a decrease in bone mass.
otic nucleus (*aka* otic ganglion) (Greek: *otikos* - ear; Latin: *nucleus* - a little nut) a parasympathetic network of neuron cell bodies within the brainstem, associated with the glossopharyngeal nerve; regulates production of saliva by the parotid salivary gland and helps to control the chewing process.
ovarian cortex (Latin: *cortex* - bark of a tree) the region of the ovary that is filled with ovarian follicles.
ovarian follicle (Latin: *folliculus* - a little bag) a structure in the ovary consisting of an oocyte and the granulosa cells that surround and nourish it. 27 (13)
ovarian ligament a ligament that attaches the ovary to the uterus.
ovarian medulla (Latin: *medulla* - marrow) the site of the blood vessels, lymph vessels, and the nerves of the ovary.
ovarian surface epithelium an outer covering of cuboidal epithelium that surrounds the ovary.
ovaries (*sing*: **ovary**) (Latin: *ovum* - egg) the gonads of biological females; produce sex hormones and gametes (eggs).
ovulation (Latin: *ovulum* - formation of ovules) the release of an egg, along with the surrounding zona pellucida and corona radiata cells, from the mature follicle and the wall of the ovary.
oxidation a chemical reaction in which an electron is lost by a molecule.
oxidation-reduction reaction (*aka* redox reaction) a chemical reaction in which an electron is transferred from one molecule to a different molecule; in the process, one molecule is oxidized and the other is reduced.
oxidative phosphorylation the process of adding phosphate groups to ADP to form ATP, using oxygen and energy from the electron transport chain.
oxygen–hemoglobin dissociation curve a graph that plots the percentage of oxygen saturation of hemoglobin against the partial pressure of oxygen
oxygenated (*aka* oxygen-rich) 1) a state in which the vast majority of hemoglobin molecules in erythrocytes are almost completely saturated with oxygen molecules (close to 100 percent); 2) blood that has a higher oxygen content, generally 95 percent saturation or above.
oxyhemoglobin (Hb–O_2) hemoglobin to which oxygen is bound.
oxyphil cells (Greek: *oxys* - sour; *philos* - fond) a type of cells found in the parathyroid glands whose function is not yet known.
oxytocin (Greek: *oxus* - swift; *tokos* - childbirth) a peptide hormone produced by the hypothalamus and stored in the posterior pituitary gland; stimulates uterine contractions during labor, milk ejection during breastfeeding, and feelings of attachment.
oxytocin a posterior pituitary hormone that has many systemic effects including social bonding, uterine contractions, and activation of the parasympathetic nervous system

P

P wave component of the electrocardiogram that represents the depolarization of the atria.
pacemaker cluster of specialized myocardial cells known as the SA node that initiates the sinus rhythm.
palatine bones paired, irregularly shaped bones form small parts of the nasal cavity and the medial wall of the orbit, and the posterior hard palate.
palatine tonsils (Latin: *palatum* - roof of mouth) paired lymphoid structures in the posterior portion of the oropharynx, that remove and destroy pathogens in incoming air and food.
palatoglossal arch (*aka* glossopalatine arch, anterior pillar of fauces, anterior arch) (Latin: *palatum* - roof of mouth; Greek: *glossa* - tongue) one of the anterior muscular ridges that connects the tongue to the soft palate; forms the border between the oral cavity and the oropharynx.
palatoglossus (Latin: *palatum* - palate; *glōssa* - tongue) muscle originating on the soft palate; elevates the back of the tongue.
palatopharyngeal arch (*aka* posterior arch) one of the posterior muscular ridges that connects the soft palate to the oropharynx.
palmar arches (Latin: *palma* - palm of hands) arterial arches formed by the merging of the radial and ulnar arteries; provide blood to the hand and give off branches that supply the fingers; consist of superficial and deep arches.
palmar interossei (Latin: *palma* - palm *inter* - between; *os* - bone) intermediate muscles of the hand; abduct and flex each finger at the metacarpophalangeal joints and extend each finger at the interphalangeal joints.
palmar venous arches (Latin: *palma* - palm of hands) curved veins that drain blood from the hands and fingers, and empty into the radial and ulnar veins.
palmaris longus (Latin: *palma* - palm) anterior superficial muscle of the forearm that provides weak flexion of the hand at the wrist.
pancreas (Greek: *pankreas* - sweetbread) an organ with both endocrine and exocrine functions; secretes hormones that regulate blood glucose and appetite and produces digestive enzymes and buffers; located posterior to the stomach.
pancreas (Greek: *pankreas* - sweetbread) one of the accessory digestive organs, which lies posterior to the stomach; the exocrine portion secretes pancreatic juice, and the endocrine portion secretes insulin and glucagon.
pancreatic amylase (Greek: *pankreas* - sweetbread) an enzyme secreted by the pancreas that

breaks down complex carbohydrates, such as starch and glycogen, into maltose in the small intestine.
pancreatic islets (Greek: *pankreas* - sweetbread) specialized groups of pancreatic cells with an endocrine function; secrete insulin, glucagon, and somatostatin.
pancreatic islets (*aka* islets of Langerhans) (Greek: *pankreas* - sweetbread) small groups of endocrine cells in the pancreas that produce the hormones insulin and glucagon.
pancreatic juice (Greek: *pankreas* - sweetbread) a mixture of substances secreted by the pancreas, which is delivered to the duodenum via the main pancreatic duct; contains several digestive enzymes and bicarbonate ions.
pancreatic lipase (Greek: *pankreas* - sweetbread) an enzyme produced by the pancreas that breaks down lipids in the small intestine.
pancreatic nuclease (Greek: *pankreas* - sweetbread) an enzyme produced by the pancreas that breaks down nucleic acids in the small intestine.
papillae (*sing*: papilla) (Latin: *papilla* - nipple) small, fingerlike projections of body structures; lingual papillae house taste buds on the tongue; the duodenal papilla connects the common bile duct and main pancreatic duct to the duodenum.
papillary layer (Latin: *papilla* - nipple) superficial layer of the dermis, made of areolar connective tissue; contains fibroblasts and small blood vessels, along with other structures.
papillary muscle (Latin: *papilla* - nipple) extension of the myocardium in the ventricles to which the chordae tendineae attach.
paracrine signaling (Greek: *para* - near; *krinō* - to separate) local intercellular communication in which a chemical is released by one cell and induces a response in a neighboring cell.
parahippocampal gyrus (Greek: *para* - near; *hippocampos* - seahorse; *gyros* - circles) a region of gray matter surrounding the hippocampus in the limbic system, which functions in memory retrieval.
parallel (Greek: *para* - alongside; *allēlōn* - of one another) muscles whose fibers are arranged in the same direction as the fascicles.
paramesonephric ducts (*aka* Mullerian ducts) (Latin: *parare* - to make ready; Greek: *mesos* - middle; *nephros* - kidney) a duct that develops in an embryo at the end of week 7 in the absence of testosterone and presence of other signaling factors.
paranasal sinus (*pl*: **paranasal sinuses**) (Greek: *para* - near; Latin: *nasus* - nose; *sinus* - fold) one of the chambers within a skull bone (frontal, maxillary, sphenoid, and ethmoid bones) that opens into the nasal cavity; serves to resonate the voice and decrease the weight of the skull.
parasympathetic division (Greek: *para* - near; *sympathētikos* - to feel with) division of the autonomic nervous system responsible for restful and digestive functions; dominant between emergency situations.
parasympathetic dominance (Greek: *para* - near; *sympathētikos* - to feel with) a state in which the parasympathetic nervous system is overactive, causing various symptoms such as lethargy, lack of motivation, and frequent hunger.
parathyroid glands (Greek: *para* - near; *thyreoeidēs* - shield-shaped; *glans* - acorn) small, oval glands embedded in the posterior surface of the thyroid gland; secrete parathyroid hormone.
parathyroid hormone (PTH) (Greek: *para* - near; *thyreoeidēs* - shield-shaped) a peptide hormone produced by the parathyroid glands, which increases the blood calcium level.
parietal bones (Latin: *parietalis* – wall) paired bones that form the superior lateral sides of the skull.
parietal cells (*aka* oxyntic cells) (Latin: *parietalis* - wall) cells of the gastric glands that secrete hydrochloric acid (to convert pepsinogen into pepsin) and intrinsic factor (to help in the absorption of vitamin B_{12}).
parietal lobe (Latin: *parietalis* - wall) a region of the cerebrum that lies superior to the lateral sulcus, directly beneath the parietal bone of the cranium; associated with somatosensation.
parietal peritoneum (Latin: *parietalis* - wall; *peritoneum* - abdominal membrane) the outer layer of the double-layered membrane (the peritoneum) that lines the wall of the abdominopelvic cavity.
parietal pleura (Latin: *parietalis* - of walls; Greek: *pleuron* - a rib) the outer layer of the double-layered pleura, which lines the inside of the thoracic wall, mediastinum, and superior surface of the diaphragm.
parotid glands (Greek: *para* - beside; *ot* - ear) the largest pair of salivary glands in the body, found inferior and anterior to the ears; secrete serous saliva containing amylase into the roof of the mouth.
partial pressure (P_x) the force exerted by a specific gas in a mixture, such as the pressure exerted by oxygen in the air.
passive immunity (Latin: *passivus* - to endure; *immunis* - exempt) temporary immunity obtained by the transfer of antibodies to a person, either via the injection of antibodies (gammaglobulins) or passed through the placenta to a fetus or through the milk to a baby.
passive transport (Latin: *passivus* - to endure; *transporto* - to carry over) form of transport across the cell membrane that does not require input of cellular energy.
patella (*aka* kneecap) (Latin: *patella* - small, shallow dish) sesamoid bone of the knee; articulates with the distal femur.
patellar ligament (Latin: *patella* - small, shallow dish; *ligamentum* - a band) inferior end of the quadriceps tendon, continuation of the patellar tendon; located between the patella and the tibia, just below the knee.
patellar surface (Latin: *patella* - small, shallow dish) wide groove on the anterior side of the distal femur, at the meeting of the medial and lateral condyles; site of articulation for the patella.
pathogen (Greek: *pathos* - disease; French: *genique* - producing) a microorganism that causes infection or disease in the body, such as a virus or bacterium.
pathology (*pl*: pathologies) (Greek: *pathos* – disease; *logos* - study) the study of changes that occur during the course of a disease.
pattern recognition receptor (PRR) a protein receptor found in the cell membranes of certain cells, such as macrophages and dendritic cells, that recognizes and binds to specific foreign molecules commonly found on the surface of bacteria and viruses; utilized in the innate immune response.
pectinate muscles (Latin: *pectinatus* - resembling a comb) muscular ridges seen on the anterior surface of the right atrium.
pectineal line (Latin: *pectineus* - comblike) narrow ridge that runs along the superior margin of the superior pubic ramus.
pectineus (Latin: *pectineus* - comblike) muscle that adducts and flexes the femur at the hip.
pectoralis major (Latin: *pectoralis* - breastplate) large, fan-shaped muscle that covers much of the superior portion of the anterior thorax.
pectoralis minor (Latin: *pectoralis* - breastplate) anterior shoulder muscle; moves the scapula and assists in inhalation.
pedicels (Latin: *pedicellus* - footstalk) projections from the podocytes that closely surround the glomerular capillaries; contain the filtration slits though which small molecules are filter from the glomerulus into the glomerular capsule.
pedicle (Latin: *pediculus* - footstalk) one of the lateral sides of the vertebral arch, attached to the vertebral body.
pelvic brim (*aka* pelvic inlet) (Latin: *pelvis* - basin) the dividing line between the greater and lesser pelvic regions; formed anteriorly by the superior margin of the pubic symphysis and posteriorly by the pectineal lines of each pubis, the arcuate lines of each ilium, and the sacral promontory.
pelvic diaphragm (Latin: *pelvis* - basin) muscular sheet spanning from the pubis to the coccyx.
pelvic inlet (Latin: *pelvis* - basin) the roof of the lesser pelvis.
pelvic outlet (Latin: *pelvis* - basin) inferior floor of the lesser pelvis; formed by the inferior margin of the pubic symphysis, right and left ischiopubic rami and sacrotuberous ligaments, and the tip of the coccyx.
pelvic splanchnic nerves (Greek: *splankhna* - inward parts) parasympathetic nerves that arise from the S2-S4 spinal nerves; innervate organs such as the distal portion of the large intestine, the urinary bladder, and some of the reproductive organs.
pelvis (Latin: *pelvis* - basin) the right and left hip bones, sacrum, and coccyx.
penis (Latin: *penis* - penis) an organ of the external genitalia in biological males;

participates in sexual intercourse and houses the urethra through which both semen and urine are transported to the outside of the body through the urethra.

pennate (Latin: *penna* - feather) a muscle whose fibers blend into a tendon that runs through the central region of the muscle.

pepsin (Greek: *pepsis* - digestion) an enzyme secreted by the stomach that breaks down proteins in food by cleaving the bonds between amino acids.

pepsinogen (Greek: *pepsis* - digestion; *genesis* - origin) the inactive form of the enzyme pepsin, secreted by the chief cells of the gastric mucosa; converted into pepsin by hydrochloric acid.

peptide bond covalent bond formed by dehydration synthesis between two amino acids.

perforating canals (*aka* Volkmann's canal) (Latin: *perforo* - to bore through; *canal* - a channel) channel connecting to the central canal; contains vessels and nerves connect osteon to osteon, and osteons to the endosteum and periosteum.

perforating fibers (Latin: *perforo* - to bore through; *fibra* - fibers) fibers that make tiny holes in the outer surface of bones.

perfusion (Latin: *perfusionem* - a pouring over) 1) delivery of blood into the capillaries in order to supply blood to the cells of the body; 2) the flow of blood in the pulmonary capillaries; must be balanced with ventilation in order to provide effective gas exchange in the lungs.

pericardial artery (Greek: *peri* - around; Latin: *cardiacus* - heart) an artery that branches off of the thoracic aorta; provides blood to the pericardium, the serous membrane that surrounds the heart.

pericardial cavity (Greek: *peri* - around; *cardiac* - *kardia* meaning heart) cavity surrounding the heart filled with a lubricating serous fluid that reduces friction as the heart contracts.

pericardium (*aka* pericardial sac) (Greek: *peri* - around; *cardiac* - *kardia* meaning heart) membrane that separates the heart from other mediastinal structures; consists of two distinct, fused sublayers: the fibrous pericardium and the parietal pericardium.

perichondrium (Greek: *peri* - around; *chondros* - cartilage) layer of vascularized dense, irregular connective tissue that encapsulates cartilage.

perimetrium Greek: *peri* - around;) the outermost layer of the wall of the uterus.

perimysium (Greek: *peri* - around; *mys* - muscle) layer of connective tissue that closely surrounds a fascicle.

perinephric fat (Greek: *peri* - around; *nephros* - kidney) a shock-absorbing layer of adipose tissue that surrounds the kidney; lies between the fibrous capsule of the kidney and the renal fascia.

perineum diamond-shaped space between the pubic symphysis, the coccyx and the ischial tuberosities, anterior to the pelvic diaphragm.

perineurium (Greek: peri - *around; neuron* - nerve) a layer of fibrous connective tissue surrounding the fascicles of a nerve.

periodontal ligament (*pl*: **periodontal ligaments**) (Greek: *peri* - around; *odous* - tooth; Latin: *ligamentum* - a band) one of the short bands of dense connective tissue that anchors the root of a tooth into its socket in the maxilla or mandible.

periosteum (Greek: *peri* - around; *osteon* - bone) membrane covering the outer surface of bone.

peripheral chemoreceptor (Greek: *peripheria* - outer surface) one of the specialized chemoreceptors located in the carotid arteries and aortic arch; detects changes in the H^+ or CO_2 concentration in the blood and transmits this information to the respiratory centers in the brainstem.

peripheral nervous system (PNS) anatomical division of the nervous system including all parts except the brain and spinal cord; largely located outside of the brain and vertebral column.

peripheral protein (*pl*: proteins) (Greek: *peri* - around; *pherō* - to carry; *protos* - first) membrane-associated protein that does not span the width of the lipid bilayer, but is attached peripherally to integral proteins, membrane lipids, or other components of the membrane.

peristalsis (Greek: *peri* - around; *stalsis* - constriction) a wave of muscle contraction that propels food down the GI tract.

peritoneum (Greek: *peri* - around; *tonos* - stretched) a double-layered serous membrane that surrounds the abdominopelvic organs and lines the abdominopelvic cavity.

peritubular capillaries (Greek: *peri* - around) the second capillary system of the nephrons; surround the renal tubules, and participate in the reabsorption and secretion process of urine formation.

permanent teeth (aka adult teeth, secondary teeth) the 32 adult teeth, which replace the baby teeth.

peroxisome (*pl*: peroxisomes) (Greek: *sōma* - body) membrane-bound organelle that contains enzymes primarily responsible for detoxifying harmful substances.

perpendicular plate of the ethmoid bone (Greek: *ēthmos* - sieve) the largest extension of the ethmoid bone; a downward, midline extension that forms the superior portion of the nasal septum.

petrous portion of the temporal bones (Greek: *petro* - rock) bumpy posterior portion of the temporal bone.

pH (*aka* Potential for Hydrogen) negative logarithm of the hydrogen ion (H+) concentration of a solution.

pH scale (*aka* Potential for Hydrogen scale) measures the concentration of hydrogen ions in a solution.

pH scale a scale that represents the hydrogen ion concentration of a solution; runs from 0 to 14 (a pH of 7 is neutral, a pH lower than 7 is acidic, and a pH higher than 7 is alkaline or basic).

phagocytic cells (*aka* phagocytes) (Greek: *phago* - eating) cells that can engulf pathogens or other cellular debris and digest them with their lysosomal enzymes.

phagocytosis (Greek: *phagō* - to eat; *kytos* - a hollow cell; *osis* - condition) endocytosis of large particles.

phagocytosis (Greek: *phago* – eating; *kytos* - cell) the engulfment and destruction of pathogenic organisms and debris by specific types of cells such as macrophages and neutrophils.

phalanx bone of the foot (*pl*: phalanges) (Greek: *phalanx* - bone between two joints of the fingers and toes) one of the 14 bones that form the toes.

phalanx bone of the hand (*pl*: phalanges) (Greek: *phalanx* - bone between two joints of the fingers and toes) one of the 14 bones that form the thumb and fingers.

phantom pain (Greek: *phantasma* - an appearance) the continued perception of sensation from the missing limb after it has been lost or removed.

pharyngeal phase the second of three phases of the swallowing process; during this phase, the soft palate blocks the nasal cavity and the larynx blocks the trachea, preventing food from entering the airways as it passes through the pharynx.

pharyngeal tonsil (*aka* adenoid) (Greek: *pharynx* - throat) a lymphoid structure in the posterior wall of the nasopharynx, which removes and destroys pathogens in inhaled air.

pharynx (*aka* throat) (Greek: *pharynx* - throat) a tubular passageway in the conducting zone that lies posteriorly to the nasal and oral cavities and the larynx; lined mainly with pseudostratified ciliated columnar epithelium; transports air toward the larynx and food toward the esophagus; also called the throat.

phasic receptors (Latin: *recipere* - to receive) the sensory receptors that adapt quickly to repeated or prolonged stimuli.

pheromones (Greek: *pherō* - to carry; *hormaō* - stimulate) a type of chemical signal that organisms can use to communicate with each other.

philtrum (Greek: *philtron* - love-charm) the indented region of the face that lies between the nose and the upper lip.

phosphodiesterase (PDE) a cytosolic enzyme that deactivates cAMP, limiting the duration of cellular responses.

phospholipid a lipid compound in which a phosphate group is combined with a diglyceride.

phosphorylation cascade a signaling event in which many protein kinases phosphorylate numerous and various cellular proteins, including other enzymes, amplifying the cellular response to a small number of hormone molecules.

photoisomerization (Greek: *phōt* - light; *isos* - equal; *meros* - part) change in the conformation of a chemical bond from *cis* to *trans* as a result of interaction with a photon.

photon (Greek: *phōt* - light) a single unit of light.

photoreceptors (Greek: *phōt* - light; Latin: *recipere* - to receive) receptor cells found in the eye that respond to photons.

phrenic nerve (Greek: *phrenicus* - diaphragm, mind) a peripheral nerve of the cervical plexus, which innervates the diaphragm.
phrenic vein (Greek: *phrenicus* – diaphragm) the vein that drains venous blood from the diaphragm.
physiological sphincter (*aka* functional sphincter) (Latin: *sphincter* - contractile muscle) a relatively weak smooth muscle sphincter, which cannot be distinguished from nearby muscle tissue by appearance or feel; able to perform as a sphincter via muscle contraction rather than muscle thickness (as in an anatomical sphincter).
physiology (Greek: *physis* - nature) science that studies the functioning of the body structures including their relevant chemistry, biochemistry, and physics.
pia mater (Latin: *pia* - tender; *mater* - mother) the thin, transparent, innermost membrane of the meninges, which directly covers the brain and spinal cord.
pineal gland (Latin: *pineus* - relating to pine; *glans* - acorn) a small gland within the epithalamus that helps to regulate wake-sleep cycles, via the secretion of the hormone melatonin.
pinocytosis (Greek: *pineō* - to drink; *kytos* - a hollow cell; *osis* - condition) endocytosis of fluid.
piriformis (Latin: *pirum* - pear; *forma* - form) muscle deep to the gluteus maximus on the lateral surface of the thigh; laterally rotates the femur at the hip.
pisiform (Latin: *pisum* - pea; *forma* - appearance) from the lateral side, the pea-shaped fourth of the four proximal carpal bones; articulates with the anterior surface of the triquetrum.
pituitary fossa (Latin: *pituita* - phlegm; *fossa* - trench) rounded depression in the floor of the sella turcica; houses and protects the pituitary gland.
pituitary glands (Latin: *pituita* - phlegm; *glans* - acorn) bean-sized organ suspended from the hypothalamus that produces hormones in response to hypothalamic stimulation; composed of the anterior and posterior pituitary glands.
pivot joint (*aka* trochoid joint) synovial joint at which the rounded portion of a bone is enclosed within a ring formed by a ligament and an articulating bone; the bone rotates within the ring.
placenta (Latin: *placenta* - a cake) a fetal structure that provides an exchange site between the maternal and fetal circulation.
plane imaginary two-dimensional surface that passes through the body.
plane joint (*aka* gliding joint) synovial joint formed between articulation of the flat surfaces of adjacent bones.
plantar aponeurosis (Latin: *planta* - sole of the foot) thick sheet of connective tissue running from the calcaneus bone to the toes; supports the sole of the foot.
plantar arch (Latin: *planta* – sole of the foot) an arterial arch formed by the merging of the dorsalis pedis artery with the medial and plantar arteries; provides blood to the distal portion of the plantar surface of the foot and the digits (toes).
plantar flexion (Latin: *planta* - sole of the foot; *flecto* - to bend) motion in the ankle that lifts the heel of the foot from the ground; pointing the toes downward.
plantar group (Latin: *planta* - sole of the foot) group of intrinsic foot muscles containing four layers.
plantar veins (Latin: *planta* - sole of the foot) veins that drain blood from the plantar (inferior) region of the foot, and empty into the plantar venous arch and eventually into the posterior tibial vein.
plantar venous arch (Latin: *planta* - sole of the foot) a curved deep vein that arises from the plantar veins, and drains into the anterior and posterior tibial veins.
plantaris (Latin: *planta* - sole of the foot) muscle that runs diagonally between the gastrocnemius and the soleus.
plasma (Greek: *plasma* - something formed) liquid extracellular matrix in fluid connective tissue.
plasma (Greek: *plasma* - something formed) the liquid portion of the blood.
plasma (Greek: *plasma* - something formed) the fluid component of the blood; considered to be the matrix of the blood.
plasma cells a group of differentiated B cells that secretes antibodies against specific antigens.
plasmin (Greek: *plasma* - something formed) an enzyme in the blood that dissolves fibrin and removes a blood clot after healing has occurred.
platelet plug (Greek: *platys* - flat) a mass of platelets that accumulates in damaged blood vessel walls to help stop bleeding; formation of a platelet plug is the second step of hemostasis, between vascular spasm and coagulation.
platelets (*aka* thrombocytes) (Greek: *platys* - flat) small cytoplasmic fragments of megakaryocytes that function in hemostasis (the stoppage of bleeding); make up a small percentage of the formed elements of the blood.
platysma (Greek: *platysma* - a flat plate) an anterior neck muscle that tenses the neck and draws the corners of the mouth downward, as in pouting.
pleura (Greek: *pleura* - a rib) serous membrane that lines the pleural cavity and covers the lungs.
pleura (Greek: *pleuron* - a rib) the double-layered serous membrane that surrounds each lung.
pleural cavity (Greek: *pleuron* - a rib) the potential space between the visceral and parietal layers of the pleura.
pleural fluid (Greek: *pleuron* - a rib) the serous fluid that occupies the pleural cavity; serves as a lubricant between the layers of the pleura, to prevent friction during breathing movements.
pneumotaxic center (Greek: *pneumonas* – lung) one of the control centers for respiration in the pons, which inhibits neural activity in the dorsal respiratory group, to terminate each inhalation and help regulate the rate and depth of breathing.
pneumothorax an abnormal accumulation of air in the pleural cavity; can lead to lung collapse.
podocytes (Greek: *pod* – foot; *kytos* - cells) specialized cells of the inner layer of the glomerular (Bowman's) capsule, which closely surround the glomerular capillaries; the filtration slits in the pedicels of these cells permit small molecules to enter the glomerular capsule.
polar body (Latin: *polus* - pole) a tiny cell produced during the process of meiosis in oocytes, which rids the developing oocyte of extra chromosomal material; contains very little cytoplasm and is programmed to degenerate.
polar molecule (Latin: *polus* - pole; *molecula* - mass) molecule with regions that have opposite charges resulting from uneven numbers of electrons in the nuclei of the atoms participating in the covalent bond.
polarized (Latin: *polus* - pole) a state in which a charge difference exists between the inside and outside of a cell membrane.
pollex (*aka* thumb) (Latin: *pollex* - thumb) digit 1 of the hand; the thumb.
polycythemia (Greek: *poly* – many; *kytos* – cell; *haima* - blood) a condition in which an overabundance of erythrocytes causes impaired oxygen delivery to the body cells; can be caused by dehydration or excess erythrocyte production.
polymer (Greek: *polys* – many) larger molecule made up of smaller monomers.
polypeptide (*pl*: polypeptides) (Greek: *polys* - many) chain of amino acids linked by peptide bonds.
polysaccharide (Greek: *polys* - many; Latin: *saccharum* - sugar) compound consisting of more than two carbohydrate monomers bonded by dehydration synthesis via glycosidic bonds.
polysaccharides (Latin: *saccharum* - sugar) complex carbohydrates consisting of chains of monosaccharides, mainly glucose; examples are starch, glycogen, and cellulose.
polysynaptic reflex (Greek: *polys* - many; *syn* - together; Latin: *reflexus* - a bending back) a reflex that contains more than one synapse and therefore involves interneurons in addition to sensory and motor neurons.
pons (Latin: *pons* - bridge) the bulging middle portion of the brainstem, which connects the cerebellum to the brainstem and the spinal cord to some of the higher brain centers; functions in information relay and in the control of respiration.
popliteal artery (Latin: *popliteus* - back of knee joint) a continuation of the femoral artery as it passes posteriorly to the knee joint; divides into the anterior and posterior tibial arteries.
popliteal fossa (Latin: *popliteus* - the ham of the knees; *fossa* - a trench) diamond-shaped space at the back of the knee; formed by the tendons of the hamstring muscles.
popliteal vein (Latin: *popliteus* - back of knee joint) a vein that arises from the merging of the anterior and posterior tibial veins in the

posterior knee region; drains blood from the lower leg and posterior knee areas, and continues as the femoral vein.
popliteus (Latin: *popliteus* - the ham of the knees) deep muscle in the posterior compartment of the leg; flexes the leg at the knee and creates the floor of the popliteal fossa.
porta hepatis (*aka* transverse fissure, hilum of the liver) (Latin: *hepaticus* - liver) the region at which the hepatic artery and hepatic portal vein transport blood into the liver and the common hepatic duct transports bile out of the liver.
portal system (Latin: *portalis* - pertaining to gate) a system of blood vessels that connects two capillary beds, and transports blood from one tissue to the other without the blood circulating elsewhere or returning to the heart.
portal system a specialized blood flow pattern that contains two capillary beds instead of one; for example, in the hepatic portal system, blood passes through a capillary bed in organs such as the stomach and small intestine, and another capillary bed in the liver; performs a specific function in the body.
portal triad a group of three vessels that runs through each corner of the hepatic lobules; consists of a branch of the bile duct, the hepatic artery, and the hepatic portal vein.
positive chemotaxis (Greek: *chemo* - chemical; *taxis* - orderly arrangement) process by which a cell is attracted to a specific chemical stimulus and moves toward it; occurs when a leukocyte is attracted to an infection or injury site by a particular chemical.
positive feedback mechanism that intensifies a change in the body's physiological condition in response to a stimulus.
positive inotropic factors (Greek: *is* - sinew, tendon, force, strength; *tropos* - turn, direction, way) factors that positively impact or increase heart contractility.
positron emission tomography (PET) (Greek: *tomos* - cutting) medical imaging technique in which radioisotopes are traced to reveal metabolic and physiological functions in tissues.
postabsorptive state (*aka* fasting state) the period of time during which there are no nutrients in the GI tract to be absorbed, and stored nutrients are being broken down for energy.
postcentral gyrus (Greek: *gyrus* - circle) a raised ridge in the cerebral cortex, just posterior to the central sulcus; site of initial somatosensory processing.
posterior (*aka* dorsal) (Latin: *posterus* - following) describes direction toward the back of the body.
posterior arch posterior portion of the ring-shaped C1 vertebra
posterior cardiac vein (Greek: *cardiac* - *kardia* meaning heart) vessel that parallels and drains the areas supplied by the marginal artery branch of the circumflex artery; drains into the great cardiac vein.
posterior cavity (*aka* dorsal cavity) (Latin: *cavus* - hollow) body cavity that houses the brain and spinal cord.
posterior cerebral artery (Latin: *cerebro* - brain) an artery that branches off of the basilar artery, and forms part of the cerebral arterial circle; supplies blood to portions of the occipital and temporal lobes of the cerebrum.
posterior columns (*aka* posterior funiculi) (Latin: *posterus* - following) masses of white matter of the spinal cord located between the posterior horns; composed of ascending tracts that carry sensory information to the brain.
posterior communicating artery (Latin: *posterus* - following) an artery that connects the posterior cerebral artery to the middle cerebral artery, and forms part of the cerebral arterial circle; supplies blood to the posterior portions of the cerebrum.
posterior compartment of the thigh area that includes muscles that flex the knee and extend the thigh.
posterior cranial fossa (Greek: *kranion* - skull; Latin: *fossa* - trench) deepest cranial fossa; extends from the petrous ridge to the occipital bone.
posterior cruciate ligament (PCL) (Latin: *cruciatus* - resembling cross; *ligamentum* - a band) intracapsular ligament that extends from the posterior, superior surface of the tibia to the inner aspect of the medial condyle of the femur.
posterior horn (Latin: *posterus* - following) a gray matter region of the spinal cord containing the axons of sensory neurons.
posterior inferior iliac spine (*aka* PIIS) projection on the inferior margin of the auricular surface on the posterior ilium.
posterior interventricular artery (*aka* posterior descending artery) branch of the right coronary artery that runs along the posterior portion of the interventricular sulcus toward the apex of the heart and gives rise to branches that supply the interventricular septum and portions of both ventricles.
posterior interventricular sulcus sulcus located between the left and right ventricles on the anterior surface of the heart.
posterior median sulcus (Latin: *posterus* - following) a midline groove on the posterior side of the spinal cord; separates the right and left sides of the spinal cord.
posterior root ganglion (*aka* dorsal root ganglion) (Latin: *posterus* - following; Greek: *ganglion* - knot) a group of cell bodies of sensory neurons that extend into the spinal cord through the posterior root.
posterior superior iliac spine (*aka* PSIS) rounded area on the posterior end of the iliac crest.
posterior tibial artery (Latin: *posterior* - behind; *tibia* - shinbone) an artery that branches off of the popliteal artery; provides blood to the lateral and posterior areas of the leg and the bottom of the foot, and branches to form the fibular (peroneal) artery and the medial and lateral plantar arteries.
posterior tibial vein (Latin: *posterior* – behind; *tibia* - shinbone) a vein that arises from the dorsal venous arch and empties into the popliteal vein; drains blood from the posterior tibial region.
posterolateral sulcus (Latin: *posterus* - following; *lateralis* - side; *sulcus* - furrow) a groove on the posterior spinal cord, found at the point of entry of the posterior nerve roots into the spinal cord.
postganglionic neuron (Greek: *ganglion* - knot) a neuron that transmits neural information from an autonomic ganglion to an effector organ.
postpartum period the interval between childbirth and six weeks of age in a newborn baby; characterized by major changes in both the mother's body and the baby.
postsynaptic potential (PSP) graded potential in the dendrites of a neuron receiving signals caused by the binding of neurotransmitter to protein receptors.
potential energy (Latin: *potential* - power; Greek: *energeia* - in work) stored energy matter possesses because of the positioning or structure of its components.
power stroke a change in shape of the myosin head (cross-bridge) due to its binding to actin.
PP cell (*aka* pancreatic polypeptide cells, gamma cells, F cells) a cell type in the pancreatic islets, which secretes pancreatic polypeptide; this hormone regulates appetite, metabolism, and motility (movement) in the gastrointestinal (GI) tract.
precapillary sphincters (Latin: *pre* - before; *capillaris* - hair-like; Greek: *sphincter* - contractile muscle) rings of smooth muscle that encircle the arteriolar ends of capillaries, and control the blood flow to the capillaries.
precentral gyrus (Latin: *prae* - before; Greek: *gyros* - circle) a raised ridge in the frontal lobe of the cerebral cortex, just anterior to the central sulcus, which contains the primary motor cortex.
prefrontal lobe (Latin: *prae* - before; Greek: *lobos* - lobe) an anterior region of the frontal lobe responsible for the cognitive basis of personality, short-term memory and consciousness.
preganglionic neuron (Latin: *prae* - before; Greek: *ganglion* - knot) a neuron that transmits neural information from the spinal cord to an autonomic ganglion.
preload (*aka* end diastolic volume) amount of blood in the ventricles at the end of atrial systole just prior to ventricular contraction.
premolars (*aka* bicuspid) teeth that lie between the canines and molars; used for grinding food.
premotor area (Latin: *prae* - before; *moveo* - mover) a region of the frontal lobe, that is responsible for planning movement.
pressure gradient the difference in pressure between two points or regions; in the blood, this refers to the difference in blood pressure between two portions of a blood vessel or two regions of the vascular system.
primary follicles (Latin: *folliculus* - a little bag) ovarian follicles that develop from primordial ovarian follicles, as the primary oocyte enlarges

and the layer of surrounding granulosa cells develops from squamous to cuboidal cells.
primary motor cortex (Latin: *primarius* - first; *moveo* - mover) a functional region of the cerebral cortex that lies in the precentral gyrus; transmits nerve impulses to the spinal cord, which then direct movements of the skeletal muscles.
primary oocytes (Greek: *oon* - egg; *kytos* - cell) immature eggs formed by meiosis.
primary response (*aka* primary immune response) the immune response of the body during the first exposure to a pathogen.
primary somatosensory cortex (Latin: *primarius* - first) a functional area of the cerebral cortex that lies in the postcentral gyrus; processes sensory information such as that of touch, pressure, tickle, pain, itch, and vibration.
primary vesicles (Latin: *primarius* - first; *vesicula* - a blister) three chambers that form from the development of the anterior end of the neural tube in the embryonic brain; consist of the forebrain, midbrain, and hindbrain.
prime mover (*aka* agonist) principal muscle responsible for a movement.
primordial follicles (Latin: *folliculus* - a little bag) the most primitive of the ovarian follicles; composed only of an oocyte and one layer of surrounding squamous epithelial (granulosa) cells.
principal cells cells of the collecting ducts which contain channels for sodium reabsorption and potassium secretion; regulated by the hormone aldosterone.
product (Latin: *productus* - to lead forth) one or more substances produced by a chemical reaction.
projection (Latin: *projectio* - to throw before) an area of the bone that projects above the bone surface; serve as points for ligaments and tendons to attach.
projection tracts (*aka* projection fibers) (Latin: *projectio* - to throw before) the large white matter bundles of motor and sensory axons that transmit nerve impulses between the brain and spinal cord.
Prolactin (PRL) a hormone produced by the anterior pituitary gland, which promotes development of the mammary glands and the production of breast milk in lactating individuals.
proliferative phase (French: *proliferation* - producing offspring) the portion of the uterine cycle in which the endometrium of the uterus is thickening under the direction of estrogen.
proliferative zone (Latin: *proles* - offspring; *fero* - to bear) region of the epiphyseal plate made up of dividing or recently divided chondrocytes.
pronated position (Latin: *pronus* - bent forward) forearm position in which the radius and ulna are crossed and the palm faces posteriorly.
pronation (Latin: *pronus* - bent forward) positioning the forearm so that the palm of the hand faces posteriorly; if the arm is outstretched, positioning of the forearm so that the palm faces inferiorly.
pronation (Latin: *pronus* - bent forward) motion that moves the forearm from supinated to pronated position.
pronator quadratus (Latin: *pronus* - bent forward; *quadratus* - square) muscle that pronates the forearm; originates on the ulna and inserts on the radius. 12 (p. 15)
pronator teres (Latin: *pronus* - bent forward; *teres* - round) muscle that pronates the forearm; originates on the humerus and inserts on the radius. 12 (p. 15)
prone (Latin: *pronus* - bent forward) anterior side of the body down so that the posterior is viewable/up.
prophase (Greek: *pro* - before; *phasis:* appearance) first stage of mitosis (and meiosis), characterized by breakdown of the nuclear envelope and condensing of the chromatin to form chromosomes.
proprioception (*aka* kinesthesia) (Latin: *proprius* - one's own; *capere* - to take) the sense or awareness of body position, movement, and location in space.
proprioceptor (Latin: *proprius* - one's own; *capere* - to take) receptor that detects stimuli from a moving part of the body, such as joints muscles and tendons.
prostaglandins (*sing*: **prostaglandin**) a group of signaling molecules with a wide range of effects, such as vasodilation of blood vessels during the inflammatory response, stimulating pain receptors during inflammation, and causing muscle contraction during childbirth.
prostate gland a ring-shaped gland at the base of the urinary bladder in biological males; surrounds the prostatic urethra and secretes an alkaline fluid that becomes part of the semen during ejaculation.
prostatic urethra (Greek: *ourethra* - passage for urine) the portion of the urethra that passes through the prostate gland; runs between the urinary bladder and the membranous urethra.
protein (Greek: *protos* - first) class of organic compounds that are composed of many amino acids linked together by peptide bonds.
protein kinase (Greek: *protos* - first) an enzyme that activates other proteins via phosphorylation (covalently adding a phosphate group).
proteoglycans (*sing*: proteoglycan) large, negatively charged molecules that are a combination of carbohydrates and proteins.
proteome (Greek: *protos* - first) full complement of proteins produced by a cell (determined by the cell's specific gene expression).
proton (Greek: *protos* - first) heavy subatomic particle having a positive charge and found in the atom's nucleus.
protraction (Latin: *protractus* - to draw forth) anterior movement of a portion of the body, such as forward thrusting of the mandible or scapula.
proximal (Latin: *proximus* - nearest) describes a position or structure on a limb nearer to the point of attachment or the trunk of the body.
proximal convoluted tubule (PCT) (Latin: *convolutus* - to roll together) a coiled tubule of the nephron, which transports filtrate from the glomerular capsule to the nephron loop; performs a significant portion of reabsorption and secretion for the nephron.
proximal radioulnar joint (Latin: *proximus* - nearest; *radius* - ray; Greek: *ōlenē* - ulna) articulation formed by the radial notch of the ulna and the head of the radius.
proximal radioulnar joint (Latin: *proximus* - nearest; *radius* - ray; Greek: *olene* - elbow) uniaxial pivot joint located between head of radius and radial notch of ulna; allows for forearm movements.
proximal tibiofibular joint (Latin: *proximus* - nearest; *tibia* - shin bone; *fibula* - brooch) articulation between the inferior aspect of the lateral condyle of the tibia and the head of the fibula.
pseudostratified columnar epithelium (Greek: *pseudo* - fake; *epi* - upon; *thēlē*- nipple; Latin: *stratum* - layer; *columna* - cylindrical) tissue that consists of a single layer of irregularly shaped and sized cells that give the appearance of multiple layers when sectioned.
psoas major (Greek: *psoa* - the muscles of the loins) muscle that makes up part of the posterior abdominal wall.
pterygoid muscles (Greek: *pterygoeides* - shaped like a wing) muscles found in the skull, which are responsible for elevation, depression, protraction, and side-to-side movement of the mandible. 12 (p. 7)
pterygopalatine ganglion (*aka* sphenopalatine ganglion, nasal ganglion) (Greek: *pteryx* - wing; *ganglion* - knot) a mass of neuron cell bodies of the parasympathetic nervous system, which regulates the lacrimal glands, pharynx, and blood vessels of the eye cerebrum.
puberty a stage of life in which the reproductive organs mature and a person becomes able to reproduce
pubic body (Latin: *pubes* - groin) enlarged, medial portion of the pubis.
pubic symphysis (Latin: *pubes* - groin; Greek: *symphysis* - growing together) articulation between the pubic bodies of the two ossa coxae.
pubic tubercle (Latin: *pubes* - groin; *tuberosus* - full of lumps) small bump on the superior aspect of the pubic body.
pubis (Latin: *pubes* - groin) anterior region of the hip bone.
pubococcygeus one muscle of the levator ani
pubofemoral ligament (Latin: *pubes* - groin; *femur* - thigh; *ligamentus* - a band) thick ligament that runs from the pubis to the femur; located on the anterior- inferior aspect of the hip joint.
pulmonary arteries (Latin: *pulmo* - lung; Greek: *arteria* - pipe) left and right branches of the pulmonary trunk that carry deoxygenated blood from the heart to each of the lungs.
pulmonary arteries (Latin: *pulmonarius* – lungs) branches of the pulmonary trunk, which transport deoxygenated blood toward the lungs;

branch into pulmonary arterioles, which then branch into pulmonary capillaries.
pulmonary capillaries (Latin: *pulmonarius* - lungs) capillaries surrounding the alveoli of the lungs where gas exchange occurs: carbon dioxide exits the blood and oxygen enters.
pulmonary circuit (Latin: *pulmonarius* – lungs) the pathway that blood follows as it travels from the heart to the lungs, and then back to the heart; functions in gas exchange between the blood and the lungs.
pulmonary plexus (Latin: *pulmonarius* - lungs; *plexus* - braid) a mass of sympathetic and parasympathetic axons located in the thorax, which innervates the lungs, regulating bronchodilation and bronchoconstriction.
pulmonary semilunar valve (*aka* pulmonary valve, pulmonic valve, right semilunar valve) (Latin: *pulmonarius* - lungs; Latin: *semi* - half; *luna* - moon) valve at the base of the pulmonary trunk that prevents backflow of blood into the right ventricle; consists of three flaps.
pulmonary surfactant (Latin: *pulmonarius* - lungs) a mixture of phospholipids and proteins, secreted by the type II alveolar cells, which decreases the surface tension between water molecules in the alveoli.
pulmonary trunk (Latin: *pulmonarius* - lungs) a short, large-diameter arterial blood vessel that transports blood ejected from the right ventricle, and soon divides into the right and left pulmonary arteries.
pulmonary veins (Latin: *pulmonarius* - lungs) veins that carry highly oxygenated blood into the left atrium, which pumps the blood into the left ventricle, which in turn pumps oxygenated blood into the aorta and to the many branches of the systemic circuit.
pulmonary ventilation the breathing process, consisting of inhalation and exhalation.
pulp cavity the inner part of the crown of a tooth, which contains nerves and blood vessels; continues into the root of the tooth as the root canal.
pulse a throbbing in an artery that can be felt through the skin, due to the alternating dilation and recoil that occurs as blood moves through it; used as a measure of heart rate
pulse pressure the difference between the systolic and diastolic blood pressures; calculated by subtracting the diastolic pressure from the systolic pressure.
pupil an opening in the iris, located at the center of the eye, that allows light to enter.
purine nitrogen-containing base with a double ring structure; adenine and guanine
Purkinje fibers specialized myocardial conduction fibers that arise from the bundle branches and spread the impulse to the myocardial contraction fibers of the ventricles.
putamen (Latin: *putamen* - to prune) a structure of the basal nuclei, located within the anterior regions of the frontal and parietal lobes; forms the striatum with the caudate nucleus.
pyloric sphincter (Greek: *pyloros* - gate; Latin: *sphincter* - contractile muscle) a ring of smooth muscle that lies between the stomach and the duodenum, which regulates the rate of gastric emptying into the duodenum.
pyloric antrum (Greek: *pyloros* - gate; Latin: *antrum* - cave) the wider portion of the pyloric region of the stomach, which lies between the body of the stomach and the pyloric canal.
pyloric canal (Greek: *pyloros* - gate) the narrower portion of the pyloric region of the stomach, which lies between pyloric antrum and the pyloric sphincter.
pylorus (Greek: *pyloros* - gate) the J-shaped curve at the distal end of the stomach, which lies between the body of the stomach and the duodenum.
pyramidal tracts (*aka* corticospinal tracts) nerve pathways that transmit motor information from the precentral gyri of the cerebral cortex to the brainstem or spinal cord; cross over on the medulla oblongata.
pyrimidine nitrogen-containing molecule with a single ring structure; cytosine, thiamine, and uracil.
pyruvate a three-carbon molecule produced when glucose is broken down during glycolysis; converted into acetyl coenzyme A, a molecule that enters the citric acid cycle

Q

quadratus femoris (Latin: *quadratus* - square; *femoris* - thigh) muscle deep to the gluteus maximus on the lateral surface of the thigh; laterally rotates the femur at the hip.
quadratus lumborum (Latin: *quadratus* - square; *lumbus* - a loin) a muscle of the posterior abdominal wall, which aids in extension of the lumbar portion of the vertebral column, lateral flexion of the vertebral column, and maintenance of upright posture.
quadriceps femoris group (Latin: *quattuor* - four; *caput* - head; *femur* - thigh) four muscles in the anterior compartment of the thigh; extend and stabilize the knee.
quadriceps tendon (*aka* patellar tendon) (Latin: *quattuor* - four; *caput* – head) tendon common to all four quadriceps muscles; inserts on the tibial tuberosity.
QRS complex component of the electrocardiogram that represents the depolarization of the ventricles and includes, as a component, the repolarization of the atria.

R

radial artery one of the terminal branches of the brachial artery; runs along the radius on the lateral side of the forearm; in the hand, merges with the ulnar artery to form the palmar arches; provides blood to the forearm and hand.
radial collateral ligament (Latin: *radius* - ray; *ligamentum* - a band) ligament strengthening the articular capsule of the elbow; located on the lateral side.
radial fossa (Latin: *radius* - ray; *fossa* - trench) small depression on the anterior humerus superior to the capitulum; provides space for the head of the radius when the elbow is maximally flexed.
radial nerve (Latin: *radialis* - ray) a large peripheral nerve that runs through the arm from the brachial plexus, and innervates portions of the upper limb, including the hand and fingers.
radial notch of the ulna (Latin: *radius* - ray; Greek: *ōlenē* - ulna) lateral side of the proximal ulna; articulates with the head of the radius at the proximal radioulnar joint.
radial tuberosity (Latin: *radius* - ray; *tuberosus* - full of lumps) rough protuberance on the medial side of the proximal radius; serves as a point of attachment for muscles.
radial vein a vein that originates from the palmar venous arches and runs parallel to the radius on the lateral side of the forearm; empties into the brachial vein; drains blood from the hand and forearm.
radioactive isotope (*aka* radioisotopes) (Latin: *radius* - ray; Greek: *isos* - equal; *topos* – part) unstable, heavy isotope that gives off subatomic particles, or electromagnetic energy, as it decays.
radiocarpal joint (Latin: *radius* – ray; Greek: *karpos* - wrist) articulation formed by the distal end of the radius and the fibrocartilaginous disc that unites the distal radius and ulna bone, and two carpal bones.
radius (Latin: *radius* - ray) bone located on the lateral side of the antebrachium of each upper limb.
ramus of the mandible (Latin: *mando* - to chew) posterior, vertically oriented portion of the mandible.
raphe (Greek: *raphe* - seam) a noticeable raised thickening along the surface of the scrotum where the genital folds fused.
reabsorption the process in which substances that were initially filtered by the glomerulus are transported back into the blood; reabsorbed substances are transported from the renal tubules into the peritubular capillaries.
reactant (Latin: *re* - again; *action* – action) one or more substances that enter into the reaction.
receptive field (*aka* sensory space) (Latin: *recipere* - to receive) a region of tissue space within which a receptor receives information.
receptors (*sing*: **receptor**) (Latin: *recipere* – to receive) protein molecules that detect and bind to specific stimuli, triggering some type of chemical reaction or change in the cell.
receptor-mediated endocytosis (Latin: *recipere* - to receive; *mediates* - to divide in the middle; Greek: *endo* - within; *kytos* - a hollow cell; *osis* - condition) endocytosis of ligands attached to membrane-bound receptors.
recruitment an increase in muscle contraction as more motor units are employed.
rectal valves (*aka* anal valves) a group of three folds in the rectum, which keep feces apart from gas; this helps prevent feces and gas from being eliminated from the body simultaneously.
rectus (Latin: *rectus* - straight) referring to the direction of muscle fibers; straight.
rectus abdominis (Latin: *rectus* - straight) paired long, linear abdominal muscle; originates

at the pubic crest and symphysis and extends the length of the trunk.
rectus femoris (Latin: *rectus* - straight; *femur* - thigh) one of four quadricep muscles; located on the anterior aspect of the thigh.
rectus sheaths (Latin: *rectus* - straight) tissue that composes the linea alba.
red marrow connective tissue in the interior cavity of a bone; site of hematopoiesis.
red pulp an area in the spleen that consists mainly of red blood cells; removes damaged or worn-out red blood cells from the circulation.
reduction (Latin: *reductio* - to lead back) 1) the process of aligning the bones for proper healing; 2) a chemical reaction in which an electron is gained by a molecule.
referred pain somatic pain that is caused by a visceral stimulus
reflexes (Latin: *reflexus* - a bending back) connections between the sensory and motor neurons that do not include the higher brain centers or include conscious or voluntary aspects of movement.
refractory period (*pl*: **refractory periods**) (Latin: *refractarius* - to break into pieces) the period of time after the start of an action potential before the next action potential can begin.
regional anatomy (Greek: *ana* - up; *tomē* – cutting) an approach to studying the structures of the body that considers all of the structures found near each other at the same time.
regulatory T cells (*aka* suppressor T cells, T_{reg} cells) a special group of helper T cells that reduces the immune response by secreting inhibitory cytokines to lessen the activity of other T cells.
relative refractory period (Latin: *refractarius* - to break into pieces) phase of the refractory period when a new action potential can only be initiated by a stronger stimulus than the current action potential.
relaxation phase phase of muscle contraction during which contraction stops, and the sarcomeres return to their resting length.
REM (rapid eye movement) sleep the phase of sleep in which the eyes a re moving behind closed eyelids; dreaming occurs during this phase.
remodeling the process of adding or removing minerals to bone depending on the homeostatic needs of the body
renal artery (Latin: *renes* - kidneys) the major artery that transports oxygen-rich blood to the kidney; branches directly off the abdominal aorta.
renal columns (Latin: *renes* - kidneys) connective tissue bands running between the renal pyramids, that transport blood vessels between the renal cortex and the renal medulla.
renal corpuscle (Latin: *renes* - kidneys; *corpuscle* - small particle) the beginning of a nephron, which performs the process of filtration during urine formation; consists of a glomerulus and a glomerular capsule.
renal cortex (Latin: *renes* - kidneys; *cortex* - bark of a tree) the outer portion of the kidney, which contains all of the renal corpuscles of the nephrons.
renal hilum (Latin: *renes* - kidneys; *hilum* - little thing) the medial indentation in the kidney, which serves as the entry/exit point for the renal artery, renal vein, and renal nerves.
renal papillae (Latin: *renes* - kidneys; *papilla* - nipple) the pointed tips of the renal pyramids that face the renal pelvis; site at which urine is drained from the collecting ducts into the minor calyces.
renal pelvis (Latin: *renes* - kidneys) the funnel-shaped region of the kidney that collects all of the urine for each kidney; formed by the merging of the major calyces; after leaving the kidney at the hilum, narrows to become the ureter.
renal pyramids (Latin: *renes* - kidneys) cone-shaped units of the renal medulla, which consist of the collecting ducts and the long nephron loops of juxtamedullary nephrons.
renal vein (Latin: *renes* - kidneys) the major vein that transports the filtered, oxygen-poor blood out of the kidney; drains into the inferior vena cava.
renin (Latin: *renes* - kidneys) an enzyme secreted by juxtaglomerular cells of the juxtaglomerular apparatus, when blood pressure is reduced; converts the plasma protein angiotensinogen into angiotensin I; a participant in the renin-angiotensin system.
repolarization phase return of the membrane potential to its steady-state voltage following an action potential.
repolarizes (Latin: *polaris* - polar) increases the negativity inside an excitable cell, after the depolarization phase of an action potential; returns the cell to its resting potential.
reposition movement that returns the thumb from opposition to anatomical position.
reserve zone (Latin: *reservo* - to keep back) region of the epiphyseal plate closest to the epiphyseal end, made up of chondrocytes in a cartilage matrix.
residual volume (RV) the amount of air that always remains in the lungs, even after forced exhalation.
resolution stage a refractory period that occurs after orgasm.
respiratory acidosis (Latin: *respirare* - breathe; *acidus* - sour; *osis* - a state) a condition in which the pH of the blood is below 7.35 (too acidic), due to a respiratory cause; causes include pneumonia, asthma, and obstruction of the airways; characterized by an excess of carbon dioxide in the blood plasma, which results in an excess of carbonic acid.
respiratory alkalosis (Latin: *respirare* - breathe;–*alkali* - soda; *osis* - a state) a condition in which the pH of the blood is above 7.45 (too alkaline or basic), due to a respiratory cause; causes include hyperventilation due to anxiety, fever or trauma; characterized by a deficiency of carbon dioxide in the blood plasma, which results in a deficiency of carbonic acid.
respiratory bronchiole (Latin: *respirare* - breathe; Greek: *bronkhos* - windpipe) a bronchiole that branches off of a terminal bronchiole and leads into an alveolar duct; the first structure of the bronchial tree to be able to perform gas exchange, due to the presence of alveoli budding from its walls.
respiratory epithelium (Latin: *respirare* - breathe; Greek: *epi* - upon; *thele* - nipple) the pseudostratified ciliated columnar epithelium that lines the nasal cavity, parts of the pharynx, trachea, conchae, and sinuses; removes pathogens by trapping them in the mucus produced by goblet cells and sweeping them out of the respiratory tract with cilia.
respiratory membrane (Latin: *respirare* - breathe) a membrane composed of the simple squamous epithelium of alveoli and their adjacent pulmonary capillaries and their fused basement membranes; rapid gas exchange occurs across this membrane.
respiratory portion (Latin: *respirare* - breathe) (*aka* respiratory zone) the location within the respiratory system where gas exchange with the blood can occur; consists of the respiratory bronchioles, alveolar ducts, alveolar sacs, and alveoli.
respiratory pump (Latin: *respirare* - breathe) the changes in thoracic volume and pressure during respiration, that help venous blood from the legs and trunk return to the heart; during inhalation, decreased thoracic pressure promotes the movement of venous blood into the thoracic cavity; then during exhalation, increased thoracic pressure helps to move blood into the heart.
respiratory rate (*aka* breathing rate) (Latin: *respirare* - breathe) the number of breaths a person takes during one minute.
respiratory volume (Latin: *respirare* - breathe) an amount of air that can be moved into or out of the lungs, or that occupies the lungs, at various phases of the respiratory cycle.
response one function of the nervous system; causes target tissue to produce an event as a result of stimulation
resting membrane potential (Greek: *membrana* - skin) the electrical charge difference between the inside and outside of a cell membrane when the cell is at rest (not conducting action potentials).
resting potential (*aka* resting membrane potential) the difference in voltage across a cell membrane under steady-state conditions; typically -70 mV.
rete testes (Latin: *rete* - net; *testis* - testicle) a system of ducts into which sperm is released.
reticular fibers (Latin: *reticulum* - a small net) within connective tissue; narrow, branching fibers made up of the same monomers found in collagen, which cross-link to form supporting "nets."
reticular formation (Latin: *reticulum* – a small net) diffuse region of connected gray matter throughout the brainstem and parts of the diencephalon; helps to regulate sleep and wakefulness, cardiovascular function, moods, and various cognitive functions.

reticular layer (Latin: *reticulum* - a small net) layer of connective tissue underlying the papillary layer; reticulated appearance due to the abundant collagen and elastin fibers.
reticular tissue tissue composed of reticular fibers, loosely woven together to form a mesh-like supportive framework for the cells of some organs such as lymphatic tissue, the spleen, and the liver.
reticuloendothelial cells (*aka* Kupffer cells) (Latin: *reticulum* – a small net; Greek: *endo* - within; *thele* - nipple) a diverse group of phagocytic cells in the reticular connective tissue of the immune system as well as the liver; in the hepatic sinusoids, Kupffer cells serve to filter out pathogens from the blood and break down damaged red blood cells.
retina (Latin: *rete* - net) the innermost layer of the wall of the eyeball; contains the rods and cones (the photoreceptor cells for the sense of vision).
retinacula (*sing*: retinaculum) (Latin: *retinere* - to retain) fibrous bands that surround the tendons at the wrist, knee, and ankle.
retinal (Latin: *retina* - retina) cofactor contained in opsin proteins that undergoes a biochemical change when struck by a photon.
retract (Latin: *retraho* - to draw back) backward movement of a portion of the body, such as posterior movement of the mandible.
retraction (Latin: *retraho* - to draw back) posterior movement of the mandible or scapula.
retroperitoneal (Latin: *retro* - behind; *peritonaeum* - abdominal membrane) posterior to the peritoneum; organs such as the kidneys, duodenum, portions of the colon, and the rectum reside in this region.
reuptake a process in which cells can take the released neurotransmitter back into the presynaptic cell.
Rh blood group a classification system used in blood typing, based on whether "D"(or "Rh") antigens are present on the cell membranes of a person's erythrocytes.
rhodopsin photopigment located in the rod photoreceptors.
rhomboid major posterior shoulder muscle; attaches the vertebral border of the scapula to the spinous process of the thoracic vertebrae.
rhomboid minor posterior shoulder muscle; attaches the vertebral border of the scapula to the spinous process of the thoracic vertebrae.
rib flattened, curved bones of the chest wall; articulated posteriorly with the thoracic vertebrae.
ribonuclease a pancreatic enzyme that breaks down ribonucleic acid (RNA).
ribonucleic acid (RNA) (Latin: a*cidus* - sour) ribose-containing nucleotide that helps manifest the genetic code as protein.
ribosomal RNA (rRNA) RNA that makes up the subunits of a ribosome.
rickets disease caused by vitamin D deficiency during bone development, leading to weak and soft bones.
right atrioventricular valve (*aka* tricuspid valve) (Latin: *atrium* - entry hall; *ventriculum* - belly) valve located between the right atrium and ventricle; consists of three flaps of tissue.
right colic flexure (*aka* hepatic flexure) (Greek: *kolikos* - lower intestine; Latin: *flextura* – bend) the bend in the colon between the ascending and transverse colon; located close to the liver.
right coronary artery (Latin: *corona* - crown) one of two arteries that branch off the aorta to bring oxygenated blood to the heart wall; distributes blood to the right atrium, portions of the left and right ventricles, and the heart conduction system.
right gastric artery an artery that branches off of the common hepatic artery, and supplies blood to a portion of the stomach.
right and left coronary arteries (Latin: *corona* - crown) the two arteries that branch off the aorta to bring oxygenated blood t the heart wall.
right lymphatic duct (Latin: *lympha* - water) the smaller of the two lymphatic collecting ducts; drains lymph from the right side of the head, upper limb, and chest into the right subclavian vein.
RNA polymerase (*pl*: polymerases) (Greek: *poly* - many; *meros* - part) enzyme that adds nucleotides to a growing strand of new RNA for the eventual goal of protein synthesis.
rod photoreceptors (Greek: *phōt* - light; Latin: *recipere* - to receive) type of retinal receptor cell specialized for low-light vision; contain the pigment rhodopsin.
root 1) the upper portion of the external nose, which runs between the eyebrows; 2) the region of a tooth which lies below the gum line, and anchors the tooth to the alveolar process of the maxilla or mandible; 3) the internal part of the penis.
root canal the passageway through which blood vessels and nerves within the pulp cavity of a tooth run through the root into the maxilla or mandible.
rotation (Latin: *rotare* - to rotate) twisting movement of the body or a limb around the long axis.
rotator cuff (*aka* musculotendinous cuff) (Latin: *rotare* - to rotate) four muscles that cross the shoulder joint, joining the scapula to the humerus; provides the primary structural support for the shoulder joint; composed of the supraspinatus, infraspinatus, teres minor, and subscapularis muscles.
rough ER (RER) an organelle in a cell that functions to synthesize and modify proteins.
round ligament a ligament that attaches to the uterus near the uterine tubes.
ruga (*pl*: **rugae**) (Latin: *ruga* - wrinkle) a fold in the inner lining (mucosa and submucosa) of certain organs to allow for expansion; rugae are visible in the stomach when it is empty but disappear when it is full following a meal; similarly, rugae in the vagina permit it to stretch during childbirth and sexual intercourse.

S

S phase stage of the cell cycle during which DNA replication occurs.
saccharolytic fermentation (Latin: *saccharum* - sugar) the breakdown of carbohydrates by the bacterial inhabitants of the large intestine.
sacral canal (Latin: *sacral* - sacred; *canalis* - channel) bony tunnel that passes inferiorly through the sacrum.
sacral foramina (Latin: *sacral* - sacred; *foramen* - an opening) series of paired openings through which nerves exit the anterior (ventral) and posterior (dorsal) aspects of the sacrum.
sacral hiatus (Latin: *sacral* - sacred; *hiatus* - an opening) the inferior opening of the sacral canal; located near the inferior tip of the sacrum.
sacral micturition center (Latin: *sacrum* - sacred; *micturire* - to desire to urinate) a bundle of neurons in the S2–S4 region of the sacral spinal cord, which helps to control the process of voluntary urination.
sacral plexus (Latin: *plexus* - braid) a nerve plexus associated with the lower lumbar (L4 and L5) and sacral (S1 to S4) spinal nerves; innervates structures in the lower limb, and the gluteal and pelvic regions.
sacral promontory (Latin: *sacral* - sacred; *promontorium* - projection) anterior lip of the body of the first sacral vertebra.
sacrococcygeal curve (Latin: *sacral* - sacred; Greek: *kokkyx* - the coccyx) a primary curvature of the vertebral column retained in adults; anteriorly concave curvature formed by the sacrum and coccyx.
sacroiliac joint (*aka* SI joint) (Latin: *sacrum* - sacred; *ilium* - groin) largely immobile articulation between the auricular surfaces of the sacrum and ilium.
sacrum (Latin: *sacrum* - sacred) single bone forming part of the posterior portion of the pelvis; located near the inferior end of the adult vertebral column; formed by the fusion of the sacral vertebrae.
saddle joint (*aka* sellar joint) synovial joint formed by the articulation of the convex and concave surfaces of two saddle-shaped bones.
sagittal plane (Latin: *sagitta* - an arrow) two-dimensional, vertical plane that divides the body or organ into right and left portions.
sagittal suture (Latin: *sagitta* - an arrow; *sutura* - seam) joint that unites the parietal bones at the superior midline along the top of the skull.
saliva the secretory product of the salivary glands, which is released into the mouth to help with carbohydrate breakdown, buffering of acids, and the binding of food particles together for swallowing.
salivary amylase a digestive enzyme found in saliva that breaks down starch and glycogen into the disaccharide, maltose.
salivary glands exocrine glands that secrete the fluid, saliva, into the mouth; the main pairs are the parotid, submandibular, and sublingual glands.

salivary lipase an enzyme present only in the saliva of infants; breaks down the fat in breast milk.
salt a substance that dissolves into ions other than H^+ and OH^- when dissolved in water.
saltatory conduction (Latin: *saltio* - to jump; *conductio* - to lead) propagation of the action potential along a myelinated axon.
saphenous nerve a peripheral sensory nerve of the lower limb, which branches from the femoral nerve and innervates the skin of the medial side of the lower leg, ankle, and great toe.
sarcolemma (Greek: *sarx* - flesh; *lemma* - husk) the plasma membrane of a skeletal muscle cell.
sarcomere (Greek: *sarx* - flesh; *meros* - part) repeating contractile unit of skeletal muscle; consists of overlapping thick and thin myofilaments.
sarcoplasm (Greek: *sarx* - flesh; *plasma* - a thing formed) the cytoplasm of a skeletal muscle cell.
sarcoplasmic reticulum (SR) (Greek: *sarx* - flesh; *plasma* - a thing formed; Latin: *rete* - net) specialized smooth endoplasmic reticulum of a skeletal muscle cell, which stores and releases calcium ions for muscle contraction. 11 (mp. 4)
sartorius (*aka* tailor's muscle) (Latin: *sartor* - a tailor) band-like muscle that extends from the anterior superior iliac spine to the medial side of the proximal tibia; flexes, abducts, and laterally rotates the leg at the hip and flexes the knee.
satellite cell (Latin: *satelles* - attendant) glial cell type in the PNS that surrounds and supports the cell bodies of sensory neurons.
saturated fats (Latin: *saturo* - to fill) lipids whose carbon molecules are bound to as many hydrogens as possible.
scala tympani (*aka* tympanic duct, tympanic ramp) (Latin: *scala* - a stairway; Greek: *tympano* - drum) portion of the cochlea extending from the apex to the round window.
scala vestibuli (*aka* vestibular duct, vestibular ramp) (Latin: *scala* - a stairway; *vestibulum* - cavity) portion of the cochlea extending from the oval window to the apex.
scalenes (Greek: *scalēnos* - uneven) synergists to the sternocleidomastoid; involved in neck flexion.
scaphoid (Greek: *skaphē* - boat; *eidos* - resemblance) from the lateral side, the boat-shaped first of the four proximal carpal bones; articulates with the radius proximally, the trapezoid, trapezium, and capitate distally, and the lunate medially.
scapula (*pl*: scapulae) (*aka* shoulder blade) (Latin: *scapulae* - shoulder blades) flat, triangular bone located on the posterior side of the shoulder.
scar (Greek: *eschara* - scab) collagen-rich tissue formed after the process of wound healing that is different from normal skin; does not contain accessory structures.
sciatic nerve (Latin: *sciaticus* - corruption of hip joint) a large peripheral mixed nerve from the sacral plexus, that is composed of the tibial nerve and fibular nerve; extends across the hip joint, and provides sensory and motor control of portions of the thigh, lower leg, and foot.
sciatica (Latin: *sciaticus* - corruption of hip joint) a painful condition resulting from irritation or compression of the sciatic nerve or any of the spinal nerves that give rise to it.
sclera (Greek: *sklēros* - hard) white portion of the fibrous layer of the eye.
scoliosis (Greek: *skoliōsis* - a crookedness) abnormal lateral curvature and twisting of the vertebral column.
scrotum (Latin: *scrotum* - a skin) a sac of skin and muscle tissue that contains the testes and helps with temperature regulation of the developing sperm.
sebaceous gland (Latin: *sebum* - oily; *glans* - acorn) oil gland found in the dermis, usually associated with a hair follicle; secretes sebum to lubricate and waterproof skin and hair.
sebum (Latin: *sebum* - oily) oily mixture of lipids that lubricates the skin, hair, and stratum corneum.
second messenger a molecule inside a cell, which acts as a part of a signaling cascade in response to a hormone binding to a receptor and subsequent activation of a G protein.
second-degree burn a burn that injures the epidermis and a portion of the dermis.
secondary active transport (Latin: *actio* - to do; *transporto* - to carry over) a phase during interphase where a cell undergoes DNA replication.
secondary adaptive response an immune response that occurs upon the second or subsequent exposures to a pathogen, which occurs more quickly and produces a higher concentration of antibodies than the primary response
secondary follicles (Latin: *folliculus* - a little bag) ovarian follicles that develop from primary ovarian follicles; consist of a primary oocyte and several layers of surrounding granulosa cells.
secondary oocyte (Greek: *oon* - egg; *kytos* - cell) the larger of the cells after the cell is divided which eventually leaves the ovary during ovulation.
secondary sex characteristics physical traits that appear at puberty which do not directly participate in reproduction; for example, breast development or changes in the pattern of hair growth.
secondary vesicles (Latin: *vesicula* - a blister) the five vesicles that develop from primary vesicles in the embryonic brain; consist of the telencephalon, diencephalon, midbrain (mesencephalon), metencephalon, and myelencephalon.
secretin a hormone released by the duodenum when chyme enters the duodenum; stimulates the release of pancreatic juice containing bicarbonate buffer into the duodenum.
secretion the process in which substances that were not filtered by the glomerulus can be added to the forming urine; involves the movement of wastes from the blood of the peritubular capillaries into the fluid of the renal tubules
secretory phase the portion of the uterine cycle in which the glands of the endometrium are secreting nutrients, such as glycogen and lipids, to nourish an embryo, if one should implant in the lining of the endometrium.
section in anatomy, a single flat surface that results when a three-dimensional structure has been cut.
segmental arteries (Latin: *segmentum* - a piece cut off) arteries that branch from the renal artery as it enters the kidney at the hilum; branch into the interlobar arteries.
segmentation contraction and relaxation cycles of small sections of the intestine, which alternate to break down and mix food, placing it in close proximity with digestive juices and the inner wall of the intestine
selectively permeable (Latin: *seligo* - to select; *permeao* - to pass through) feature of any barrier that allows certain substances to cross but excludes others.
sella turcica (*aka* Turkish saddle) (Latin: *sella* - saddle; *turcica* - Turkish) bony region of the sphenoid bone located at midline of the middle cranial fossa.
semen (Latin: *semen* - seed) the reproductive fluid of biological males, consisting of sperm and nutritive fluids, buffers and antibiotics from the seminal vesicles, prostate gland and bulbourethral glands.
semilunar valves (Latin: *semi* - half; *Luna* - moon) valves located at the base of the pulmonary trunk and at the base of the aorta.
semimembranosus (Latin: *semi* - half) one of three hamstring muscles.
seminal vesicles (*aka* seminal glands) (Latin: *seminalis* - seed) glands that contribute approximately 60 percent of the semen volume.
seminiferous tubules (Latin: *seminalis* - seed) long, coiled, tubular structures that make up the majority of the testes; sites of spermatogenesis.
semitendinosus (Latin: *semi* - half; *tendo* - stretch) one of three hamstring muscles.
sensation (Latin: *sentire* - to perceive) 1) one function of the nervous system; receives information from the environment and translates it into the electrical signals that are transmitted via nervous tissue; 2) the conscious awareness of stimuli.
sensory neurons (*aka* afferent neurons) (Latin: *sensatio* - feeling; Greek: *neuro* - nerve) neurons that collect information about the internal and external environment and send it toward the central nervous system.
sensory receptor (Latin: *sensatio* - feeling; *recipere* - to receive) cells or structures that detect sensation.
septa (Latin: *saeptum* - a partition) connective tissue that divides lobules within the testis.
septal cartilage (Latin: *saeptum* - a partition; *cartilago* - gristle) cartilaginous structure that forms the anterior nasal septum.
septal nuclei (Latin: *saeptum* - wall) groups of neuron cell bodies in the limbic system, close to the corpus callosum; function in the experiences of pleasure and reinforcement.

septum (*pl*: septa) (Latin: *septum* - fence, partition) a wall or partition that divides the heart into chambers.
septum primum (Latin: *septum* - fence, partition) flap of tissue in the fetus that covers the foramen ovale within a few seconds after birth.
serosa the outermost layer of the wall of the GI tract; continuous with the visceral peritoneum of organs of the abdominopelvic cavity.
serous gland (*pl*: serous glands) (Latin: *glans* - acorn) gland that produces enzyme-rich secretions derived from blood plasma.
serous membranes (*sing*: **serous membrane**) (*aka* serosa) (Latin: *membrana* - membrane) epithelial membranes lining body cavities that do not open to the outside; lining both the cavity wall and organ surface; 2) membrane that covers organs and releases a fluid to reduce friction.
serratus anterior (Latin: *serratus* - a saw) anterior shoulder muscle; originates on the ribs and inserts onto the scapula.
sesamoid bone (Greek: *sēsamoeidēs* - resembling a sesame seed) round bone suspended within a tendon or ligament; functions to protect the tendon from compressive forces.
setpoint ideal value or narrow range for a physiological parameter
sex considerations of a person's anatomy, chromosomes, or hormones.
sexual orientation describes who an individual is sexually attracted to.
sexually transmitted infections (*aka* STIs; sexually transmitted diseases, STDs) infections that can be contagious through sexual intercourse.
shaft the long cylindrical portion of the penis; visible externally.
shaft of the femur (Latin: *femur* - thigh) cylinder-shaped central portion of the femur.
shaft of the humerus (Latin: *humerus* - shoulder) cylindrical central region of the humerus.
shaft of the tibia (Latin: *tibia* - shin bone) central portion of the tibia.
short bone cube-shaped bone that is approximately equal in length, width, and thickness; provides limited motion.
shoulder girdle (*aka* pectoral girdle) the region formed by the scapula and clavicle, which attaches each upper limb to the axial skeleton.
sickle cell disease (*aka* sickle cell anemia) hereditary blood disorder, involving abnormal hemoglobin structure; results in production of sickle-shaped erythrocytes that do not transport oxygen well and break down easily.
sigmoid colon (Greek: *sigmoeidēs* - S-shaped) the final region of the colon, which connects the descending colon to the rectum.
simple columnar epithelium (Latin: *columna* - cylindrical; Greek: *epi* - upon; *thēlē* - nipple) tissue that consists of a single layer of column-like cells; typically characterized by elongated, basal nuclei.
simple cuboidal epithelium (Greek: *kybos* - cube; *epi* - upon; *thēlē* - nipple) tissue that consists of a single layer of cube-shaped cells; typically characterized by round, central nuclei.
simple squamous epithelium (Latin: *squamosus* - scaly; Greek: *epi* - upon; *thēlē* - nipple) tissue that consists of a single layer of flat, scale-like cells; typically characterized by flat, horizontal nuclei.
sinoatrial (SA) node known as the pacemaker, a specialized clump of myocardial conducting cells located in the superior portion of the right atrium that has the highest inherent rate of depolarization that then spreads throughout the heart.
sinus rhythm normal contractile pattern of the heart.
sinus venosus develops into the posterior portion of the right atrium, the SA node, and the coronary sinus.
sinusoid capillaries (*aka* sinusoidal capillaries, sinusoids) capillaries that contain large clefts between the endothelial cells and openings in the basement membranes, which allow for the movement of large molecules and cells across the capillary walls; uncommon in the body.
sister chromatid (*pl*: chromatids) (Greek: *chrōma:* color) one of a pair of identical chromosomes, formed during DNA replication.
skeletal muscle striated muscle tissue made up of long, multinucleated cells (muscle fibers); often attached to bone; under voluntary or involuntary control; cells contain multiple nuclei; striated appearance; arranged in bundles surrounded by connective tissue.
skeletal muscle pump the compression of deep veins by the contracting skeletal muscles surrounding them, which helps venous blood, especially in the legs, to return to the heart.
skull (Middle English: *skulle* - a bowl) complex bony structure consisting of 22 bones that forms the head, face, and jaws.
sleep apnea a serious condition in which a person frequently stops and starts breathing while asleep.
slow oxidative (SO) (*aka* slow twitch muscle fibers, Type I muscle fibers, fatigue-resistant muscle fibers) muscle fibers that contract relatively slowly and use aerobic respiration to produce ATP.
small cardiac vein (Greek: *cardiac* - *kardia* meaning heart) parallels the right coronary artery and drains blood from the posterior surfaces of the right atrium and ventricle; drains into the great cardiac vein.
small intestine a major organ of the GI tract, which accomplishes most of the digestion and absorption of nutrients
small saphenous vein a vein that arises from the dorsal veinous arch and dorsal vein of the little toe, and empties into the popliteal vein; travels up the lateral side of the leg, and drains the superficial areas of the foot and lower leg.
smooth ER (SER) an organelle that functions to synthesize and modify carbohydrates as well as serve some detoxification functions.
smooth muscle nonstriated muscle tissue, under involuntary control; associated with hair follicles, internal organ walls, and blood vessels, made up of single-nucleated cells that are football-shaped and to not appear striated.
sodium/potassium (Na^+/K^+) pump a specialized ion channel in the plasma membrane that uses energy from ATP to actively transport three Na^+ ions out of the cell and two K^+ ions into the cell; maintains the ion gradients which maintain the resting potential of the cell.
soft palate (*aka* velum, palatal velum, muscular palate) (Latin: *palatum* - roof of mouth) the muscular portion of the palate, which separates the oral cavity from the nasal cavity; lies posterior to the hard (bony) palate, and contains the uvula, which closes the entrance to the nasal cavity during swallowing.
solar plexus (*aka* celiac plexus) (Latin: *plexus* - braid) one of the networks of sympathetic and parasympathetic nerve fibers that makes up the abdominal aortic plexus; transmits pain impulses from some of the visceral organs and stimulates peristalsis.
soleus (Latin: *solea* - sole of the foot) flat, wide muscle deep to the gastrocnemius.
solute (Latin: *solutus* - dissolved) substance that is put into a solution to dissolve.
solution (Latin: *solutus* - dissolve) homogeneous liquid mixture in which a solute is dissolved into molecules within a solvent.
solvent (Latin: *solvere* - to loosen) a substance in a solution that dissolves a solute.
somatic cell (*pl*: cells) (Greek: *soma* - body) all cells of the body excluding gamete cells.
somatic nervous system (SNS) (Greek: *sōma* – body) functional division of the nervous system concerned with voluntary motor responses and conscious perception.
somatic reflex (Greek: *sōma* - body; Latin: *reflexus* - a bending back) reflexes where the motor response is carried out by skeletal muscles.
somatosensation general sensations associated with the body
spatial summation (Latin: *summatio* - to sum up) the summation of graded potentials from several nearby synapses within a close period of time.
sperm capacitation (Latin: *capacitas* - make capable) the process in which the sperm penetrates the outer layers of the egg.
spermatic cord (Greek: *sperma* - seed) a tubular structure that consists of the ductus deferens, nerves, and blood and lymphatic vessels that supply the testes; travels through the inguinal canal.
spermatid (Greek: *sperma* - seed) a cell that arises from a secondary spermatocyte during meiosis I, and will eventually mature into a sperm.
spermatocyte (Greek: *sperma* - seed) a cell that arises from the mitosis of a spermatogonium; later participates in two meiotic divisions to become a spermatid and then a sperm.
spermatogenesis (Greek: *sperma* - seed; *genesis* - origin) the process of sperm production, which occurs in the seminiferous tubules of the testes.

spermatogonia (*sing*: spermatogonium) (Greek: *sperma* - seed) the most primitive form of spermatogenic cells; eventually develop into primary and secondary spermatocytes, spermatids, and finally sperm.
spermiogenesis (Greek: *sperma* - seed) the last stage of spermatogenesis, in which spermatids mature into sperm (or spermatozoa).
sphenoid bone (Latin: *sphenoides* - resembling a wedge) single bone that forms the central base of skull and articulates with almost every other bone in the skull.
sphenoid sinus (Latin: *sphenoides* - resembling a wedge; *sinus* - cavity) air-filled space located within the body of the sphenoid bone; most posterior of the paranasal sinuses.
sphincter (Greek: *sphinktēr* - a band) a ring-shaped muscle that encircles a passageway in the body; it constricts or relaxes to regulate the flow of a substance (such as urine or ingested food) through the passageway.
sphygmomanometer an instrument used to measure the blood pressure, consisting of an inflatable cuff, bulb (for inflating the cuff), and a pressure gauge.
spinal accessory nerve (CN XI) (*aka* accessory nerve) eleventh cranial nerve; responsible for control of certain muscles of the neck and back.
spinal cavity (*aka* vertebral cavity) (Latin: *cavus* - hollow) portion of the dorsal cavity containing the spinal cord.
spinal cord a column of nervous tissue located within the vertebral cavity; component of the central nervous system.
spinal nerves the 31 pairs of mixed nerves attached to the spinal cord.
spinal reflex (Latin: *reflexus* - a bending back) reflexes that involve the spinal cord.
spinalis the medial group of muscles of the erector spinae group; located medially to the iliocostalis and longissimus muscles; helps with extension of the vertebral column.
spine of the scapula (*aka* scapular spine) (Latin: *spina* - backbone; *scapulae* - shoulder blades) posterior ridge passing mediolaterally along the scapular surface.
spinous process single bony process that projects posteriorly from the midline of the back; functions as a site of muscle attachment
spiral organs (Greek: *organum* - instrument) small structures within the cochlear duct that transduce the wave motions of the vestibuli and the scala tympani.
spirometer (Latin: *spira* - a coil; Greek: *metron* - a measure) a device that measures the volume of air that a person inhales or exhales, as a result of different degrees of effort in the breathing process; used to test various respiratory volumes in patients with certain respiratory conditions.
spleen one of the secondary lymphoid organs, and the largest lymphatic organ in the body; removes pathogenic organisms from the blood, produces B lymphocytes (white pulp) for the adaptive immune response, and removes worn-out or damaged red blood cells and platelets from the bloodstream (red pulp)
splenic artery an artery that branches off of the celiac trunk, provides blood to the spleen and pancreas.
splenius (Greek: *splēnion* - a bandage) posterior neck muscles; originate at the midline and run laterally and superiorly to their insertions; includes the splenius capitis and the splenius cervicis.
splenius capitis (Greek: *splēnion* - a bandage; Latin: *caput* - head) posterior neck muscle that inserts into the head region.
splenius cervicis (Greek: *splēnion* - a bandage; Latin: *cervix* - neck) posterior neck muscle that inserts into the cervical region.
splicing the process of modifying a pre-mRNA transcript by removing certain, typically non-coding, regions
spongy bone (*aka* cancellous bone) bone containing lacunae and osteocytes arranged in trabeculae.
spongy urethra (*aka* penile urethra) (Greek: *ourethra* - passage for urine) the longest portion of the male urethra; runs through the entire length of the penis and is surrounded by the corpus spongiosum.
spontaneous depolarization (*aka* prepotential depolarization) the mechanism that accounts for the autorhythmic property of cardiac muscle; the membrane potential increases as sodium ions diffuse through the always-open sodium ion channels and causes the electrical potential to rise.
squamous cell carcinoma (Latin: *squamosus* - scaly; *karkinos* - crab (i.e., Cancer, the crab)) type of skin cancer that originates from the keratinocytes of the stratum spinosum of the epidermis.
squamous portion of the temporal bone (Latin: *squamosus* - scaly; *tempus* – temple) flattened, upper portion of the temporal bone.
squamous suture (Latin: *squamosus* - scaly; *sutura* – seam) joint that unites the parietal bone to the squamous portion of the temporal bone.
SRY (*aka* sex determining region of the Y chromosome) region of the Y chromosome that contains genes important in sexual development.
stapes (*aka* stirrup) ossicle that is attached to the oval window of the inner ear.
static equilibrium (Latin: *aequilibrium* - an equal balance) our sense of head or body position while the head or body is not in motion.
stem cell (*pl*: cells) cell that is has the ability to produce additional stem cells rather than becoming further specialized.
sternal angle (Greek: *sterno* - chest) junction between the manubrium and body of the sternum; site of attachment of the second rib to the sternum.
sternal end of the clavicle (*aka* medial end of the clavicle) (Greek: *sterno* - chest) triangle-shaped medial end of the clavicle; articulates with the manubrium of the sternum.
sternoclavicular joint (*aka* SC joint) (Greek: *sterno* - chest) juncture between the lateral end of the sternum and the sternal end of the clavicle.
sternocleidomastoid (Greek: *sterno* - chest) prime mover that laterally flexes the head.
sternohyoid (Greek: *sterno* - chest; *hyoeidēs* - U-shaped) one of the infrahyoid muscles; depresses the hyoid bone.
sternothyroid (Greek: *sterno* - chest; *thyreoeidēs* - shield-shaped) one of the infrahyoid muscles; depresses the thyroid cartilage of the larynx.
sternum (Greek: *sterno* - chest) elongated, flat bony structure located at the center of the anterior chest.
steroid (*aka* sterol) lipid compound composed of four hydrocarbon rings bonded to a variety of other atoms and molecules.
straight sinus a vein that drains blood from the great cerebral vein and the inferior sagittal sinus in the brain; eventually empties into the internal jugular vein via the transverse sinus; one of the dural venous sinuses; runs vertically along the deepest border of the tentorium cerebelli.
stratified columnar epithelium (Latin: *stratum* - layer; *columna* - cylindrical; Greek: *epi* - upon; *thēlē*- nipple) uncommon human tissue consisting of two or more layers of column-like cells, located in some ducts.
stratified cuboidal epithelium (Latin: *stratum* - layer; Greek: *epi* - upon; *thēlē* - nipple; *kybos* - cube) uncommon human tissue consisting of two or more layers of cube-shaped cells, located in some ducts.
stratified squamous epithelium (Latin: *stratum* - layer; *squamosus* - scaly; Greek: *epi* - upon; *thēlē* - nipple) tissue consisting of several layers of cells, the most apical of which are flat; protects body surfaces.
stratum basale (Latin: *stratum* - layer– *basis* - foundation) the deepest layer of the epidermis, composed of epidermal stem cells; connects the epidermis to the basement membrane.
stratum corneum (Latin: *stratum* - layer; *corneus* - horn) most superficial layer of the epidermis; composed of layers of dead cells.
stratum granulosum (Latin: *stratum* - layer; *granulum* - grain) layer of the epidermis superficial to the stratum spinosum; characterized by the presence of intracellular granules.
stratum lucidum (Latin: *stratum* - layer; *lucidus* – clear) layer of the epidermis between the stratum granulosum and stratum corneum, found in the skin of the palms and soles of the feet.
stratum spinosum (Latin: *stratum* - layer; *spinosus* - thorny) layer of the epidermis superficial to the stratum basale; composed of football-shaped keratinocytes, joined together by desmosomes.
stress a physical, emotional, or psychological strain on a person; derived from a real or imagined threat to their homeostasis; activates the sympathetic division of the autonomic nervous

system, which responds with changes such as increased heart and breathing rate, bronchodilation, and dilation of the pupils.
stress-relaxation response contraction and relaxation of smooth muscle cells stimulated by the stretching of the muscle.
stretch mark mark formed on the skin due to expansion of the dermis beyond its elastic limits, resulting in the breakage of collagen fibers.
stretch reflex (Latin: *reflexus* - a bending back) type of reflex which responds to the activation of muscle spindle stretch receptors; results in contraction of the muscle to maintain a constant length.
striation (*pl*: striations) (Latin: *striatus* – furrowed) in muscle tissue; the appearance of a banded pattern formed by the alignment of parallel actin and myosin filaments.
striatum (Latin: *striatus* - furrowed) part of the basal nuclei, composed of the caudate nucleus and putamen; receives input from the cerebral cortex.
stroke (*aka* cerebrovascular accident, CVA) a disruption of blood flow to the brain, resulting from blockage or bursting of an artery; results in oxygen deprivation in the brain and death of brain cells.
stroke volume (SV) amount of blood pumped by each ventricle per contraction; also, the difference between EDV and ESV.
styloglossus (Greek: *stylos* - pillar; *glõssa* - tongue) muscle originating on the styloid bone; allows the tongue to move upward and to the back.
stylohyoid (Greek: *stylos* - pillar; *hyoeidēs* - U-shaped) one of the suprahyoid muscles; moves the hyoid bone posteriorly and elevates the larynx.
styloid process (Greek: *stylos* - pillar; *eidos* - resemblance) elongated, downward bony projection, located posterior to the mandibular fossa on the external base of the skull.
styloid process of the radius (Greek: *stylos* - pillar; *eidos* - resemblance; Latin: *radius* - ray) delicate inferior projection on the lateral end of the distal radius.
styloid process of the ulna (Greek: *stylos* - pillar; *eidos* - resemblance; *ōlenē* – ulna) delicate inferior projection located on the medial end of the distal ulna.
stylomastoid foramen (Greek: *stylos* - pillar; Greek: *mastós* - breast; *eidos* - resemblance; *foramen* - opening) small opening between the styloid process and mastoid process; point of exit for the cranial nerve that supplies facial muscles.
subarachnoid space (Greek: *arachne* - cobweb) the space between the arachnoid mater and pia mater, which contains CSF.
subclavian artery arteries that supply blood to the arms, and give off branches that supply the brain and spinal cord, shoulder area, chest, and back; the left subclavian branches directly from the aortic arch, while the right subclavian branches from the brachiocephalic artery.
subclavian vein a continuation of the axillary vein as it passes behind the clavicle; drains blood from the upper limb and scapular region; drains into the brachiocephalic vein.
subclavius (Latin: *sub* - under) anterior shoulder muscle; stabilizes the clavicle during movement.
subdural space the area between the dura mater and arachnoid mater; occupied by a small amount of serous fluid.
sublingual glands (Latin: *sub* - under; *lingualis* - tongue) one of the major pairs of salivary glands; lies under the tongue; secretes mucous saliva into the floor of the mouth.
submandibular ganglion (Latin: *mandibula* - jaw; Greek: *ganglion* - knot) a group of cell bodies of parasympathetic nerves, which receives axons of the facial nerve; regulates secretion of saliva from the submandibular and sublingual salivary glands.
submandibular glands (Latin: *sub* - under; *mandibula* – jaw) one of the major pairs of salivary glands, which lies deep to the mandible; secretes mixed saliva (part serous and part mucous) into the floor of the mouth.
submucosa (Latin: *sub* - under) the layer of areolar connective tissue in the wall of the GI tract that is located between the mucosa and the muscularis.
submucosal plexus (*aka* plexus of Meissner) a network of nerves in the submucosa of the wall of the GI tract; regulates glandular secretion.
subpubic angle (Latin: *sub* - under; *pubes* – groin) inverted V-shape formed by at the pubic symphysis.
subscapular fossa (Latin: *sub* - under; *scapulae* - shoulder blades; *fossa* - trenches) broad depression on the anterior surface of the scapula.
subcapsular sinus (Latin: *sub* - under; *sinus* - a fold) the region of a lymph node through which lymph enters; lies just inside of the capsule; filters the lymph, bringing pathogens and debris into contact with phagocytic cells that can engulf and destroy them.
subscapular vein a vein that drains the subscapular area, and drains into the axillary vein.
subscapularis (Latin: *sub* - under) muscle that medially rotates the arm; originates on the anterior scapula.
substance P (*aka* SP) the neurotransmitter used by nociceptors to communicate to the secondary neuron in the pain pathway.
substantia nigra (Latin: *substantia* - material; *nigra* - black) a midbrain nucleus that releases dopamine to the basal nuclei; functions in the inhibition of excess movement and the production of smooth movements.
substrate (Latin: *substerno* - to spread under) reactant in an enzymatic reaction.
subthalamus (Latin: *-thalamus* - bedroom) structure within the diencephalon containing the subthalamic nuclei, which help the basal nuclei in the control of movement.
sulci (*sing*: **sulcus**) (Latin: *sulcus* - furrow, wrinkle) grooves or indentations on the surface of the cerebrum; lie between gyri (raised ridges).
sulcus (*pl*: **sulci**) (Latin: *sulcus* - groove) fat-filled groove visible on the surface of the heart; coronary vessels are also located in these areas.
summate (*aka* sum) (Latin: *summatio* - to sum up) to add together or combine effects.
summation (Latin: *summatio* - to sum up) an increase in muscle tension, due to the additive effects of several action potentials.
superficial (Latin: *superficialis* - surface) describes a position nearer to the surface of the body.
superior (*aka* cranial) (Latin: *superiorem* - higher) describes a position above or closer to the head.
superior angle of the scapula (Latin: *scapulae* - shoulder blades) corner of the scapula located between the superior and medial borders.
superior articular process flat projection extending upward from the vertebral arch of a vertebra; articulates with the inferior articular process of the adjacent vertebra.
superior articular processes of the sacrum (Latin: *sacral* - sacred) paired projections from the base of the sacrum that extend upward and articulate with the inferior articular processes of the L5 vertebra.
superior border of the scapula (Latin: *scapulae* - shoulder blades) superior margin of the scapula.
superior cervical ganglion (Latin: *superiorem* - higher; *cervix* - neck; Greek: *ganglion* - knot) a mass of neuron cell bodies of the sympathetic nervous system, which innervates the pineal gland, the eye, blood vessels of the head, the thyroid gland, and the salivary glands.
superior colliculi (Latin: *superiorem* - higher) two of the four structures of the corpora quadrigemina of the midbrain; responsible for combining sensory information with motor output.
superior extensor retinaculum (Latin: *retinere* - to retain) band of connective tissue holding the tendons of the muscles of the anterior compartment of the lower leg.
superior gemellus (Latin: *gemellus* - twin) muscle deep to the gluteus maximus on the lateral surface of the thigh; laterally rotates the femur at the hip.
superior lobe (*aka* upper lobe) (Latin: *superiorem* - higher) the uppermost (most superior) lobe of each lung.
superior mesenteric artery an artery that branches off of the abdominal aorta, and provides blood to the small intestine, the pancreas, and the proximal portion of the large intestine.
superior mesenteric ganglion (Latin: *superiorem* - higher; Greek: *ganglion* – knot) one of the collateral (prevertebral) ganglia of the sympathetic nervous system; projects to the digestive system.
superior mesenteric plexus (Latin: *superiorem* - higher; *plexus* - braid) one of the three groups of axons that make up the abdominal aortic plexus; innervates the small intestine, the

proximal portions of the large intestine, and the pancreas.
superior nasal concha (*pl*: conchae) (Latin: *nasus* - nose; *concha* - shell) thin, curved projection of the ethmoid bone; smallest and most superiorly located of the nasal conchae.
superior nuchal line (Latin: *nucha* - nape) paired bony lines on the posterior skull that extend laterally from the external occipital protuberance; sites of attachment for muscles of the neck
superior olivary nucleus (Latin: *superiorem* - higher; *oliva* - olive; *nucleus* - a little nut) a group of cell bodies within the pons that transmit sensory information about sound.
superior orbital fissure (Latin: *superiorem* - higher; *fissura* - cleft) irregularly shaped opening located laterally to the optic canal; provides passage for the artery supplying the eyeball.
superior phrenic artery an artery that branches off of the thoracic aorta and provides the blood supply to the superior portion of the diaphragm.
superior pubic ramus (Latin: *superiorem* - higher; *pubes* - groin) segment of bone that passes laterally from the pubic body to the ilium.
superior sagittal sinus (Latin: *superiorem* - higher) a large vein that lies between the layers of the dura mater in the falx cerebri; drains cerebrospinal fluid from the arachnoid villi and blood from the superior cerebrum; eventually empties into the internal jugular vein via the transverse sinuses; one of the dural venous sinuses.
superior vena cava one of the largest veins of the systemic circuit; formed by the merging of the left and right brachiocephalic veins; empties blood from the head, neck, chest, and upper limbs; drains blood into the right atrium of the heart.
supinated position (Latin: *supinare* - to bend backwards) forearm position in which the radius and ulna are parallel and the palms face forward.
supination (Latin: *supinare* - to bend backwards) positioning the forearm so that the palm of the hand faces anteriorly; if the arm is outstretched, positioning the forearm so that the palm faces superiorly.
supinator (Latin: *supinare* - to bend backwards) prime supinator muscle of the forearm.
supine (Latin: *supinare* - to bend backwards) position of the body resting on the posterior so that anterior structures are visible and accessible.
supportive connective tissue (Latin: *supporto* - to carry; French: *tissu* - woven) bone and cartilage; structural connective tissue.
suprachiasmatic nucleus (Latin: *superiorem* - higher; *nucleus* - a little nut; Greek: *chiasma* - cross) hypothalamic target of a small number of axons projecting from the optic chiasm; helps establish the circadian rhythm.
suprahyoid muscles (Greek: *hyoeidēs* - U-shaped) anterior neck muscles superior to the hyoid bone.
supraorbital foramen (Latin: *supra* - above; *foramen* - opening) small opening located at the superior margin of the orbit; provides passage for the sensory nerve of the forehead.
supraorbital margin (Latin: *supra* - above) feature of the frontal bone, superior rim of the orbit.
supraspinatus muscle located superior to the spine of the scapula that abducts the arm
supraspinous fossa (Latin: *supra* - above; *spina* - spine; *fossa* – trench) narrow depression on the posterior scapula, superior to the spine.
suprasternal notch (Latin: *supra* - above; Greek: *sterno* - chest) a shallow U-shaped border on the top of the manubrium.
surfactant a substance composed of a mixture of phospholipids and proteins, which reduces the surface tenson between the water molecules lining the alveolar walls.
surgical neck region of the humerus where the proximal end meets the narrower shaft; often the site of fractures
suspension (Latin: *suspendo* - to hang up) liquid mixture in which particles distributed in the liquid settle out over time.
suspensory ligament (*pl*: **suspensory ligaments**) (Latin: *suspendo* - to hang up) 1) a ligament that connects the ovary to the wall of the pelvis; also, one of the ligaments that supports the breast by connecting it to the chest wall and the dermis; 2) a set of ligaments that connect the ciliary body to the lens of the eye; regulate tension on the lens, so that it can change shape to focus on near or far objects.
suture (Latin: *sutura* - seam) immobile joint at which adjacent bones of the skull are connected by fibrous connective tissue.
swallowing a complex process, involving a sequence of muscle contractions, that begins the movement of food from the mouth to the stomach
sympathetic chain ganglia (*aka* paravertebral ganglia) (Greek: *sympathētikos* - to feel with; *ganglion* - knot) long strings of ganglia adjacent to the vertebral column on both sides; transport nerve fibers of the sympathetic nervous system.
sympathetic division (Greek: *sympathētikos* - to feel with) portion of the autonomic nervous system associated with preservation of life in emergency situations.
sympathetic dominance (Greek: *sympathētikos* - to feel with) a state that exists in the body when the sympathetic nervous system is overactive; associated with symptoms such as difficulty sleeping and relaxing, tightness in the neck and shoulders, and digestive upset.
symphysis (Greek: *symphysis* - growing together) cartilaginous joint in which bones are joined by a fibrocartilage pad; amphiarthrotic.
symporters (*sing*: symporter) (Greek: *syn* - with; Latin: *porto* - to carry) secondary active transporters that move two substances in the same direction.
synapse (Greek: *syn* - with; *hapto* - to clasp) narrow junction between two neurons across which a chemical signal passes.
synaptic cleft (Greek: *syn* - together; *hapto* - to clasp) 1) small space between a nerve terminal and the motor end plate of a muscle cell; 2) the small gap between two cells at a chemical synapse.
synarthrosis (Greek: *syn* - together; *arthrosis* - joints) joint at which no motion is possible.
synchondrosis (Greek: *syn* - together; *chondros* - cartilage) cartilaginous joint in which bones are joined together by hyaline cartilage, or where bone is joined to hyaline cartilage.
syndesmosis (Greek: *syndesmos* - a fastening) type of fibrous joint in which two parallel bones are united by fibrous connective tissue.
synergist (Greek: *synergia* - together) muscle that contributes to the movement of a prime mover; often by flexing a joint or stabilizing the insertion.
synovial fluid (Greek: *syn* - together; Latin: *ovum* - egg; *fluidus* - to flow) thick, slimy fluid that is released into a synovial joint; lubricates the joint and nourishes the cells of cartilage within the joint.
synovial joint joint at which the articulating surfaces of the bones are not directly connected, but are located within a cavity formed by an articular capsule.
synovial membrane (Greek: *syn* - together; Latin: *ovum* - egg; *membrana* - membrane) connective tissue membrane that lines the cavities of a freely movable joint; composed of cells that produce synovial fluid.
synovial membrane thin layer that lines the inner surface of the articular capsule.
synthesis reaction (Greek: *syn* - together; *thesis* - arranging; Latin: *re* - again; *actio* - action) type of anabolic reaction in which two or more atoms or molecules bond, resulting in a larger molecule.
systemic anatomy (Greek: *systema* - an organized whole; Greek: *ana* - up; *tomē* - cutting) an approach to studying the structures of the body that considers all of the structures within an organ system at one time.
systemic circuit the pathway that blood follows as it travels from the heart to the body cells, and then back to the heart; functions in gas and nutrient exchange between the blood and the cells of the body.
systemic myotonia (Greek: *mys* - muscle) a temporary condition in which muscles all over the body contract and maintain tension.
systole (Greek: *systole* - contraction) period of time when the heart muscle is contracting.
systolic pressure the maximum blood pressure in the large arteries that occurs during ventricular contraction (systole); expressed in units of mm Hg.

T

T cell tolerance a lack of responsiveness of T cells against self-antigens, due to the destruction of T cells that recognize these antigens.
T lymphocytes (*aka* T cells) (Latin: *lympa* - water; *kytos* - cell) a diverse group of

lymphocytes that provides cell-mediated immunity by physically attacking specific pathogens.
T wave component of the electrocardiogram that represents the repolarization of the ventricles.
tactile corpuscles (*aka* Meissner's corpuscles) (Latin: *tactus* - to touch; *corpus* – body) encapsulated receptors found within the dermal papillae that transduce light touch.
talus (Latin: *talus* - ankle) most superior tarsal bone; articulates superiorly with the tibia and fibula at the ankle joint, inferiorly with the calcaneus bone and anteriorly with the navicular bone.
target cells cells that contain receptors for a particular hormone; the only cells which a hormone can affect.
tarsal bone (Greek: *tarsos* - ship) one of the seven bones that make up the posterior foot and ankle.
taste buds structures within the papilla of the tongue that contain gustatory receptor cells.
taste receptor cells (*aka* gustatory cells) (Latin: *recipere* - to receive) specialized epithelial cells that are the receptors and transducers of taste stimuli.
telencephalon one of the secondary vesicles of the embryonic brain; develops into the cerebrum of the adult brain.
telophase (Greek: *telos* - end; *phasis* - appearance) final stage of mitosis (and meiosis), preceding cytokinesis, characterized by the formation of two new daughter nuclei.
temporal bone (Latin: *tempus* - temple) one of a pair of bones that forms the lateral, inferior portion of the skull; subdivided into squamous, mastoid, and petrous regions.
temporal fossa (Latin: *tempus* - temple; *fossa* - trench) shallow space on the lateral side of the brain case.
temporal lobe (Latin: *temporalis* - temples) a region of the cerebral cortex directly beneath the temporal bone of the cranium; functions in the understanding of language, memory, processing of auditory information, and some visual perception.
temporal process of the zygomatic bone (Latin: *temporalis* - temple; Greek: *zygon* - yoke) short projection of the zygomatic bone; forms the anterior portion of the zygomatic arch.
temporal summation (Latin: *summatio* - to sum up) the summation of graded potentials from a single synapse within a close period of time.
temporal vein (*aka* superficial temporal vein) a vein that lies on the lateral side of the head; drains blood from the temporal area, outer ear, and superficial portions of the head, and empties into the external jugular vein.
temporalis (Latin: *temporalis* - temple) muscle that retracts the mandible; assists in chewing.
temporomandibular joint (TMJ) (Latin: *temporalis* - temple; *mando* - to chew) articulation formed by the condyle of the mandible and the mandibular fossa and articular tubercle of the temporal bone; allows for depression/elevation, protraction/retraction, and excursion of the mandible.
tendon (Latin: *tendo* - to stretch) structure made of dense connective tissue that attaches muscle to bone.
tendon reflexes (Latin: *reflexus* - a bending back) a reflex that serves to protect the muscles during contraction by stretching a tendon during muscle contraction.
tendon sheath (Latin: *tendo* - to stretch) elongated structure made of connective tissue that surrounds a tendon where the tendon crosses a joint; contains synovial fluid to reduce friction and allow for smooth movement of the tendon.
teniae coli three longitudinal bands of smooth muscle that run along most of the length of the large intestine
tensor fasciae latae (Latin: *tendo* - to stretch; *fascia* - a band) a rectangular muscle on the lateral side of the upper thigh; flexes, abducts, and rotates the thigh medially.
tentorium cerebelli (Latin: *tentorium* - tent; *cerebrum* - brain) one of the four cranial dural septa; provides a roof over the cerebellum, separating the cerebellum from the rest of the brain.
teres major (Latin: *teres* – round) thick, flat muscle located inferior to the teres minor; extends the arm and assists with adduction and medial rotation of the arm.
teres minor (Latin: *teres* - round) long muscle that laterally rotates and extends the arm.
terminal electron acceptor an oxygen atom, which binds to electrons and hydrogen ions at the end of the electron transport chain, and forms water molecules in the process of ATP synthesis.
terminal ganglia (Greek: *ganglion* – knot) ganglia of the parasympathetic division of the autonomic system; located close to the effector organ.
tertiary follicles (*aka* antral follicles, mature follicles, mature antral follicles, Graafian follicles) (Latin: *tertiarius* - third; *folliculus* - little bag) ovarian follicles consisting of a primary or secondary oocyte, a zona pellucida, a single fluid-filled antrum, and several layers of granulosa cells; ready for ovulation.
testes (*sing*: testis) (Latin: *testis* - testicle) the gonads that produce sperm and sex hormones including testosterone in biological males.
tetanus (Greek: *tetanos* - convulsive tension) continuous muscle contraction.
thalamus (Latin: *thalamus* - bedroom) a major region of the diencephalon; a collection of nuclei that relays sensory and motor information between the cerebral cortex and the periphery, spinal cord, or brainstem.
thalassemia (Greek: *thalassa* - the sea; *haima* - blood) hereditary blood disorder in which the body produces abnormal hemoglobin molecules; results in production of abnormal erythrocytes, short erythrocyte life span, and poor oxygen-carrying capacity.
theca cells cells that surround developing ovarian follicles in the ovary; cells of the theca interna produce androgens, and cells of the theca externa from a connective tissue sheath around the follicles.
thenar (Greek: *thenar* - the palm) muscles located on the lateral (radial/thumb) aspect of the palm.
thermoreceptors receptors that detect and respond to moderate temperature changes; exist as free nerve endings on sensory neurons.
thermoregulation the control of body temperature, in order to keep enzyme activity and chemical reaction rates at their homeostatic levels
thick filament myosin strand made up of multiple heads; projects from the center of the sarcomere toward the Z disk
thigh area of the lower limb between the hip and knee joints
thin filament strand of actin and a troponin-tropomyosin complex; projects from the Z disc toward the center of the sarcomere.
third ventricle (Latin: *ventriculus* - belly) one of the four chambers in the brain that produces and circulates cerebrospinal fluid; located in the diencephalon.
third-degree burn a burn that fully penetrates and destroys the affected epidermis and dermis; affects sensory function.
thoracic aorta the part of the descending aorta that lies superior to the diaphragm.
thoracic cage (Greek: *thōrax* - chest) forms the chest; consists of the sternum, 12 pairs of ribs and their costal cartilages; protects the heart and lungs.
thoracic cavity (Greek: *thōrax* - chest; Latin: *cavus* - hollow) division of the anterior cavity that contains the heart, lungs, esophagus, and trachea.
thoracic curve (Greek: *thōrax* - chest) in adults, anterior curvature of the thoracic vertebral column region.
thoracic duct (Latin: *thorax* - chest) the larger of the two lymphatic collecting ducts; drains lymph from the lower limbs, abdomen, left upper limb, and the left side of the head and chest, and empties it into the left subclavian vein.
thoracic vertebrae (Greek: *thōrax* - chest; Latin: *vertebra* - joint of the spine) twelve vertebrae, located in the thoracic region of the vertebral column.
thoracolumbar system (*aka* sympathetic system) (Greek: *thōrax* - chest; Latin: *lumbus* - loin) alternate name for the sympathetic division of the autonomic nervous system.
thoroughfare channel a blood vessel that transports blood between capillaries and a venule; when precapillary sphincters are closed, it allows blood to flow from a metarteriole into a venule.
threshold voltage minimum voltage across a membrane at which an action potential is initiated.
thrombin (Greek: *thrombos* - clot) the active enzyme in the blood coagulation pathway, which converts the plasma protein fibrinogen

into active fibrin; this is the last step in the formation of a blood clot.
thrombopoietin (Greek: *thrombos* - clot; *poiesis* - a making) a hormone produced by the liver and kidneys that causes megakaryocytes in the red bone marrow to release platelets (thrombocytes).
thromboses (*sing*: thrombosis) (*aka* blood clots) (Greek: *thrombosis* - clumping) blood clots that form in veins and decrease blood flow.
thrombus a blood clot that forms in a blood vessel, especially a vein.
thymosins (Greek: *thumos* - warty excrescence) hormones produced and secreted by the thymus; contribute to the development and differentiation of T cells.
thymus (Greek: *thumos* - a warty excrescence) one of the primary lymphoid organs; responsible for the maturation and proliferation of T lymphocytes received from the red bone marrow.
thyrocervical artery an artery that branches from the subclavian artery, and supplies blood to the thyroid gland, and structures of the neck, upper back, and shoulder.
thyrohyoid (Greek: *hyoeidēs* - U-shaped) one of the infrahyoid muscles; depresses the hyoid bone and elevates the thyroid cartilage of the larynx
thyroid cartilage (Greek: *thyreoiedes* - shield-shaped) the largest cartilage of the larynx; consists of two plates of hyaline cartilage that fuse in the middle to form the laryngeal prominence or "Adam's apple."
thyroid follicles (Greek: *thyreoiedes* - shield-shaped; Latin: *folliculus* - a small bag) spherical structures composed of a ring of thyroid epithelial cells surrounding a central cavity filled with gelatinous colloid; the cells produce thyroid hormones and the colloid stores them.
thyroid gland (Greek: *thyreoiedes* - shield-shaped; *glans* – acorn) a large, butterfly-shaped endocrine gland located over the anterior surface of the trachea, inferior to the larynx; responsible for the synthesis of thyroid hormones and calcitonin.
thyroid-stimulating hormone (TSH) (*aka* thyrotropin) (Greek: *thyreoiedes* - shield-shaped) an anterior pituitary hormone that stimulates the secretion of thyroid hormones.
thyroxine (*aka* T4 hormone, tetraiodothyronine) an amino acid–derived thyroid hormone that contains four iodine atoms; one of the two hormones comprising thyroid hormone.
tibia (*aka* shin bone) the larger and more medial of the two bones of each lower leg; articulates with the femur to form the knee joint; important weight-bearing bone.
tibial collateral ligament (Latin: *tibia* - shin bone; *ligamentum* - a band) extrinsic ligament on the medial knee that runs from the medial epicondyle of the femur to the medial tibia.
tibial nerve (Latin: *tibia* – shin bone) a peripheral nerve of the posterior leg; branches from the sciatic nerve and innervates the skin and muscles of the leg, lower leg, and foot.
tibial tuberosity (Latin: *tibia* - shin bone; *tuberosus* - full of lumps) prominent elevated area located on the anterior surface of the proximal tibia.
tibialis anterior (Latin: *tibia* - shin bone) long, thick muscle located on the lateral surface of the tibia; raises the front of the foot.
tibialis posterior (*aka* posterior tibialis) (Latin: *tibia* - shin bone) deep muscle in the posterior compartment of the leg; plantar flexes and inverts the foot.
tidal volume (TV) the amount of air that enters or leaves the lungs during one normal breath
tight junction (*pl*: tight junctions) impermeable membrane fusion between adjacent cells.
tissue (French: *tissu* - woven) an organized unit of a type of cells that have a similar structure and that work together to perform a specific function.
tissue membrane (French: *tissu* - woven; Latin: *membrana* - membrane). thin layer or sheet of cells that covers the outside of the body, organs, and internal cavities.
tongue one of the accessory organs of the digestive system, which occupies a large portion of the oral cavity; composed of skeletal muscle; active in the chewing and swallowing processes.
tonic receptors (Greek: *tonikos* - tone; Latin: *recipere* - to receive) receptors that do not adapt, or adapt very slowly, even when the stimulus remains constant.
tonicity the concentration of solute in the extracellular fluid compared to the concentration of solute in the intracellular fluid.
tonsillar crypts deep indentations of the epithelium of a tonsil, which accumulate materials entering the body through eating and breathing; contain macrophages and leukocytes, which interact with incoming materials to destroy pathogens.
tonsils masses of lymphoid nodules associated with the pharynx; consist of the pharyngeal tonsil (adenoid), the palatine tonsils, and the lingual tonsil
total body water (TBW) the total volume of water contained in the intracellular and extracellular fluid compartments of the body.
total dead space the total volume of inhaled air that does not undergo gas exchange; can be calculated by taking the sum of the anatomical and alveolar dead spaces.
total lung capacity (TLC) the maximal amount of air that the lungs can hold; can be calculated by taking the sum of the TV, ERV, IRV, and RV
trabeculae (*sing*: trabecula) (Latin: *trabs* - a beam) lattice-like beams making up the matrix of spongy bone.
trabeculae carneae (Latin: *trabecula* - a small beam) ridges of muscle covered by endocardium located in the ventricles.
trachea (*aka* windpipe) (Greek: *trakheia* – windpipe) a tubular organ that runs vertically from the larynx to the primary bronchi; contains cartilaginous C-shaped incomplete rings that keep it open at all times; provides the air passageway between the larynx and the lungs.
trachealis muscle (Greek: *trakheia* – windpipe) smooth muscle that connects the ends of the C-shaped rings of cartilage in the posterior wall of the trachea; contracts to constrict the trachea and relaxes to dilate the trachea.
tract a bundle of myelinated axons in the central nervous system.
trans-fatty acids (Latin: *trans* - through; *acidus* - sour) unsaturated fats that are chemically treated to add hydrogen to bring them closer to saturated fats.
transamination the transfer of an amine group from an amino acid to an alpha keto acid, to produce a new amino acid and a new alpha keto acid.
transcription (Latin: *transcribere* - to copy) process of producing an mRNA molecule that is complementary to a particular gene of DNA.
transcription factor (*pl*: factors) (Latin: *transcribere* - to copy) one of the proteins that regulate the transcription of genes.
transducers (Latin: *trans* - through; *ductus* - to lead across) receptor cells that are able to convert one type of stimulus into electrical signals.
transfer RNA (tRNA) (Latin: *transfero* - to bear across) molecules of RNA that serve to bring amino acids to a growing polypeptide strand and properly place them into the sequence.
transferrin (Latin: *trans* - through; *ferrum* - iron) a plasma protein that transports iron through the body; important for delivering iron from recycled erythrocytes to the red bone marrow for new erythrocyte production.
transitional epithelium (Latin: *transito* - to go across; Greek: *epi* - upon; *thēlē*- nipple) specialized stratified epithelium located in the urinary tract; characterized by cells that change shape in response to the presence of urine; cells are dome-shaped when unstretched.
translation (Latin: *translatio* - a transferring) process of producing a protein from the nucleotide sequence code of an mRNA transcript.
transmembrane protein (*pl*: proteins) (Latin: *trans* - through; *membrana* - a skin or membrane that covers part of the body; *prōtos:* first) a protein that spans the membrane from its intracellular side to its extracellular side.
transpulmonary pressure (Latin: *trans* - through; *pulmo* - lungs) difference between the intrapleural pressure (pressure in the pleural cavity) and intra-alveolar pressure (pressure inside the alveoli).
transverse colon the portion of the colon that lies between the ascending colon and the descending colon; runs across the upper portion of the abdominopelvic cavity from right to left.
transverse foramen (Latin: *transversus* - crosswise; *foramen* - opening) opening located in the transverse processes of the cervical vertebrae.
transverse mesocolon a mesentery that connects the transverse colon to the posterior abdominal wall.
transverse plane (Latin: *transversus* - crosswise) two-dimensional, horizontal slice that divides the body or organ into superior and inferior portions.

transverse processes (Latin: *transversus* - crosswise) paired bony processes that extend laterally from the junction between the pedicle and lamina of the vertebra.
transverse sinuses (*aka* lateral sinuses) (Latin: *transversus* - crosswise; *sinus* - cavity) dural venous sinuses in the tentorium cerebelli that drain blood from the posterior portion of the head.
transversospinalis muscles (Latin: *transversus* - crosswise) muscles originating at the transverse processes which insert at the spinous processes of the vertebrae.
transversus abdominis (Latin: *transversus* - crosswise) deepest abdominal muscle; composed of fibers arranged transversely around the abdomen.
trapezium (Greek: *trapezion* - table) from the lateral side, the table-shaped first of the four distal carpal bones; articulates with the scaphoid proximally, the first and second metacarpals distally, and the trapezoid medially.
trapezius (Greek: *trapezion* - table) posterior shoulder muscle; stabilizes the upper part of the back.
trapezoid (Greek: *trapeza* - a table; *eidos* - resembling) from the lateral side, the table-shaped second of the four distal carpal bones; articulates with the scaphoid proximally, the second metacarpal distally, the trapezium laterally, and the capitate medially.
triceps brachii (Latin: *tri* - three; *caput* - head; *brachium* - arm) extensor muscle of the forearm.
trigeminal nerve (CN V) (Latin: *trigeminus* - triplet) fifth cranial nerve; responsible for cutaneous sensations of the face and controlling some of the chewing muscles.
triglyceride (Latin: *tri-*: three) lipid compound composed of a glycerol molecule bonded with three fatty acid chains.
triiodothyronine (*aka* T3) an amino acid–derived thyroid hormone that contains three iodine atoms; one of the two hormones comprising thyroid hormone.
triquetrum (Latin: *triquetrus* - three-cornered) from the lateral side, the three-cornered third of the four proximal carpal bones; articulates with the lunate laterally, the hamate distally, and has a facet for the pisiform.
trochlea (Greek: *trochileia* - a pulley) pulley-shaped region of the distal end of the humerus; articulates at the elbow with the trochlear notch of the ulna.
trochlear nerve (CN IV) (Latin: *trochlea* - a pulley) fourth cranial nerve; contributes to eye movement through control of certain extraocular muscles.
trochlear notch (Latin: *trochlea* - a pulley) C-shaped depression on the anterior side of the proximal ulna; articulates with the trochlea of the humerus.
trophoblast (Greek: *trophe* – nourishment) a structure that will yield the placenta and other structures required for pregnancy that are not part of the fetus.
tropic hormones (Greek: *trope* - turning) hormones that act on a specific endocrine gland to trigger the release of another hormone from that gland.
tropomyosin (Greek: *trope* - turning; *mys* - muscle) regulatory protein that makes up part of the thin myofilament; blocks myosin-binding sites to prevent actin from binding to myosin when a muscle cell is relaxed.
troponin (Greek: *trope* - turning) regulatory protein that makes up part of the thin myofilament; binds to actin, tropomyosin, and calcium; changes the shape of the tropomyosin molecule to allow myosin to bind to actin during muscle contraction.
true ribs ribs 1–7; attach directly to the sternum via the costal cartilage.
truncus arteriosus portion of the primitive heart that will eventually divide and give rise to the ascending aorta and pulmonary trunk.
trunk a short, large-diameter blood vessel that splits into smaller blood vessels
T-tubules deep invaginations in the sarcolemma that transmit electrical signals into the sarcoplasmic reticulum of muscle cells
tubercle of the rib (Latin: *tuberculum* - a knob) small bump located at the posterior end of a rib; articulates with the transverse process of a thoracic vertebra.
tubule (*aka* renal tubule) a duct within the kidney that transports glomerular filtrate (tubular fluid) toward a collecting duct.
tunica albuginea (Latin: *tunica* - coat; *albugo* - whiteness) a dense connective tissue layer under the ovarian surface epithelium.
tunica externa (*aka* tunica adventitia) (Latin: *tunica* - coat; *externa* - coming from outside) the outermost layer of the wall of a vessel; not present in capillaries.
tunica intima (*aka* tunica interna) (Latin: *tunica* - coat; *intima* - innermost) the innermost layer of the wall of a blood vessel.
tunica media (Latin: *tunica* - coat; *media* - middle) the middle layer of the wall of a blood vessel; not present in capillaries.
tunica vaginalis (Latin: *tunica* - coat; *vaginalis* - sheath) a serous membrane that has both a parietal and visceral layer.
twitch the contraction of a single muscle fiber caused by a single action potential.
tympanic membrane (*aka* eardrum) structure at the internal end of the auditory canal; vibrates when struck by sound waves, and transmits the sound waves to the ossicles.
type I alveolar cell (Latin: *alveolus* - basin) a simple squamous epithelial cell that makes up the alveolar walls; performs rapid diffusion of respiratory gases with the pulmonary capillaries, due to its flattened shape and single-layered structure.
type II alveolar cell (Latin: *alveolus* - basin) a cuboidal epithelial cell found in the walls of the alveoli; secretes pulmonary surfactant to decrease the surface tension between water molecules in the alveoli, allowing them to expand easily during inhalation.

U

ulna (Greek: *ōlenē* - ulna) bone located on the medial side of the antebrachium of each upper limb.
ulnar artery one of the terminal branches of the brachial artery; runs along the ulna on the medial side of the forearm; in the hand, merges with the radial artery to form the palmar arches; supplies blood to the forearm and hand.
ulnar collateral arteries arteries that supply blood to structures in the the elbow area; arise from the brachial artery.
ulnar collateral ligament (Greek: *ōlenē* - ulna; *ligamentum* - a band) triangle-shaped ligament strengthening the articular capsule of the elbow; located on the medial side.
ulnar nerve (Greek: *ōlenē* - ulna) a peripheral nerve of the arm that runs parallel to the radial nerve, close to the ulna; innervates certain muscles of the forearm and hand.
ulnar notch of the radius (Greek: *ōlenē* - ulna; Latin: *radius* - ray) smooth area on the medial side of the distal radius; articulates with the head of the ulna to form the distal radioulnar joint.
ulnar vein (Greek: *ōlenē* - ulna) a vein that originates from the palmar venous arches and runs parallel to the ulna on the medial side of the forearm; empties into the brachial vein; drains blood from the hand and forearm.
ultrasonography (Latin: *ultra-* beyond; *sonus* - sound; Greek: *graphō* - to write) application of ultrasonic waves to visualize body structures such as tendons and organs.
umbilical arteries two arteries that transport oxygen-poor blood and wastes from the fetus to the placenta; arise from the internal iliac arteries of the fetus, travel through the umbilical cord, and empty into the placenta.
umbilical vein a single vein that transports blood containing oxygen and nutrients from the placenta to the fetus; blood travels from the placenta into the umbilical cord, then into the fetal liver or ductus venosus, then into the inferior vena cava, and finally into the right atrium of the fetal heart.
unencapsulated receptors (*aka* unencapsulated nerve endings) (Medieval English: *un* - not; Latin: *encapsa* - in a box; *recipere* - to receive) dendrites of the sensory neuron enmeshed in the surrounding tissue.
uniaxial joint (Latin: *unus* - one) joint that allows for motion within a single plane.
unipennate (Latin: *unus* - one; *penna* - feather) a pennate muscle whose fasicles are located on one side of the tendon.
unipolar neuron shape with only one process extending from the cell body which splits into the axon and the dendrites.
unsaturated fats (Latin: *saturo* - to fill) lipids with double bonds and a zigzag fatty acid chain that cannot get close together.
upper esophageal sphincter (*aka* inferior pharyngeal sphincter) (Latin: *sphincter* - contractile muscle) a ring of skeletal muscle that surrounds the upper portion of the esophagus,

and regulates the passage of food from the pharynx to the esophagus.
upper respiratory tract the portion of the respiratory tract that lies in the head and neck; consists of the nasal cavity, pharynx, and larynx.
upregulation an increase in the number of active receptors in/on a target cell for a hormone that is present in low concentration; this increases the response of the cell to the hormone.
urea cycle (Greek: *ouron* – urine) a reaction sequence that rids the body of excess nitrogen and ammonia, by converting amine groups from excess amino acids into ammonia and combining the ammonia with carbon dioxide to form urea; the urea is then excreted from the body in the urine.
ureter (Greek: *oureter* - urinary duct of kidney) a long, hollow tubular organ that transports urine from the kidney to the urinary bladder; begins as a narrowing of the renal pelvis as it leaves the kidney.
urethra (Greek: *ourethra* - passage for urine) a tubular organ that transports urine from the urinary bladder to the outside of the body.
urinalysis (*aka* urine analysis) (Greek: *ouron* - urine; *analysis* - breaking up) a group of tests conducted on a urine sample to monitor general health and diagnose various disease conditions.
urinary bladder (Greek: *ouron* - urine) a saclike organ that stores urine until elimination; receives urine from the ureters and empties into the urethra.
urinary tract infections (*aka* UTIs) (Greek: *ouron* - urine) bacterial infections of any of the structures of the urinary tract; most begin in the urethra, but may spread into the urinary bladder, ureters, and kidneys.
urogenital sinus (Greek: *ouron* - urine; Latin: *genitalis* - fruitful) a small opening between the genital folds.
urogenital triangle (Greek: *ouron* - urine; *genitalis* - fruitful) anterior triangle of the perineum; includes the external genitals.
uterine cycle (*aka* menstrual cycle) a monthly reproductive cycle in the biological female between puberty and menopause, that is characterized by recurring changes in the thickness of the uterine wall.
uterine tube (*aka* fallopian tube, oviduct) (Latin: *uterinus* - pertaining to the womb) tubular organ that transports ovulated oocytes from the ovary to the uterus; the site of fertilization of an oocyte by a sperm.
uterus (Latin: *uterus* - womb) the highly muscular organ which houses a developing embryo and fetus, and propels the fetus to the outside world during childbirth.
uvula (*aka* palatine uvula) (Latin: *uvula* - bunch of grapes) a fingerlike projection of soft tissue that hangs down from the center of the soft palate into the throat; prevents food from entering the nasal cavity during swallowing.

V

vagina (Latin: *vagina* - sheath) a tubular organ in biological females that connects the uterus to the outside of the body; receives semen for fertilization, forms part of the birth canal, and excretes the menstrual flow.
vagus nerve (CN X) (Latin: *vagus* - wandering) tenth cranial nerve; responsible for contributing to homeostatic control of the organs of the thoracic and upper abdominal cavities.
valence shell (*aka* outermost shell) (Latin: *valentia* - strength) outermost electron shell of an atom.
Valsalva's maneuver a voluntary process involving closure of the glottis and contraction of the muscles of the abdominal wall to increase the pressure in the abdominal cavity; used to help expel feces and to lift heavy objects.
valve in the cardiovascular system, a specialized structure located within the heart or vessels that ensures one-way flow of blood.
varicosity (Latin: *varix* - a dilated vein; Greek: *osis* - condition) structure in some autonomic neurons resembling a string of swellings along the length of the fiber; releases neurotransmitter molecules to exert control over various target organs
vasa recta (*aka* vasa rectae renis) (Latin: *vas* - vessel; *recta* - right) long straight capillaries that run parallel to the nephron loops of the juxtamedullary nephrons; function in urine concentration.
vasa vasorum (Latin: *vas* - vessel; *vasorum* - of vessels) blood vessels present in the walls of blood vessels; supply nutrients and oxygen to, and remove wastes and carbon dioxide from, the cells of the blood vessels.
vascular layer (Latin: *vasculum* - a small vessel) the middle layer of the eye that is primarily connective tissue but is rich with blood vessels and intrinsic muscles.
vascular spasm (Latin: *vasculum* - a small vessel; Greek: *spasmos* - to pull) the first step of hemostasis, in which the walls of the damaged blood vessel constrict, due to contraction of their smooth muscle.
vascular tone (Latin: *vasculum* – small vessel; a state of slight contraction that exists in the smooth muscle tissue of a blood vessel.
vasocongestion more blood enters the genital region than can leave and the spongy corpora tissues flood with blood and become swollen, enlarged, and rigid.
vasoconstriction contraction of the smooth muscle tissue in the wall of a blood vessel, resulting in a decrease in the diameter of the vessel.
vasodilation (Latin: *vas* - a vessel; *dilato* - to spread out) relaxation of the smooth muscle tissue in the wall of a blood vessel, resulting in the widening of blood vessels.
vastus intermedius one of four quadricep muscles; located between the vastus lateralis and vastus medialis and is deep to the rectus femoris.
vastus lateralis (Latin: *latus* - side) one of four quadricep muscles; located on the lateral aspect of the thigh.
vastus medialis (Latin: *medialis* - to middle) one of four quadricep muscles; located on the medial aspect of the thigh.
vein (*pl*: **veins**) (Latin: *vena* - vein) a blood vessel that transports blood toward the heart.
venae cavae the two largest veins in the human body.
venous anastomosis a blood flow pattern in which venules divide and drain blood into several veins.
venous return the volume of blood that returns to the atria from the veins in one cardiac cycle.
ventilation (*aka* pulmonary ventilation) (Latin: *ventilatus* - a breeze) the process of breathing, or moving of air into and out of the lungs; consists of inhalation and exhalation.
ventral (anterior) nerve root motor root of the spinal nerve; consists of axons emerging from the anterior side of the spinal cord.
ventral respiratory group (VRG) (Latin: *ventralis* - pertaining to belly) one of the control centers for respiration in the medulla oblongata, which causes contraction of the accessory muscles involved in forced inhalation and exhalation.
ventricle (*pl*: **ventricles**) (Latin: *ventriculus* - belly) (1) one of the primary pumping chambers of the heart located in the lower portion of the heart; the left ventricle is the major pumping chamber on the lower left side of the heart that ejects blood into the systemic circuit via the aorta and receives blood from the left atrium; the right ventricle is the major pumping chamber on the lower right side of the heart that ejects blood into the pulmonary circuit via the pulmonary trunk and receives blood from the right atrium. (2) space within the brain where cerebrospinal fluid is produced.
ventricular ejection phase second phase of ventricular systole during which blood is pumped from the ventricle.
venule a small vessel that transports blood from a capillary into a vein.
vertebrae (Latin: *vertebrum* - joint) irregular bone located in the neck and back regions of the vertebral column; typically consist of a body, vertebral arch, and processes.
vertebral arch (Latin: *vertebrum* - joint) posterior portion of the vertebra surrounding the spinal cord; consists of two pedicles and two laminae.
vertebral artery (Latin: *vertebrum* - joint) an artery that branches off of the subclavian artery and passes through the transverse foramina of the cervical vertebrae, continuing to the brain; joins the internal carotid artery to form the cerebral arterial circle; supplies blood to the central nervous system.
vertebral column (Latin: vertebrum - joint) all of the vertebrae and intervertebral disks that support the head, neck and body and protect the spinal cord.

vertebral foramen (Latin: vertebrum - joint; *foramen* - opening) opening within a vertebra, enclosed by the vertebral arch; provides passage for the spinal cord.
vertebral vein (Latin: *vertebrum* - joint) a vein that drains blood from the spinal cord, cervical vertebrae, and neck; runs through the intervertebral foramina of the cervical vertebrae, and empties into the brachiocephalic vein.
vesicle (*pl*: vesicles) (Latin: *vesicula* - a blister) membrane-bound pocket that contains and moves or stores materials.
vestibular folds (Latin: *vestibulum* - entrance) folded regions of mucous membrane that lie superior to the vocal folds in the larynx; do not produce sound, but help support the epiglottis during the swallowing process and to clear debris from the larynx by triggering the cough reflex.
vestibulocochlear nerve (CN VIII) (*aka* auditory nerve, acoustic nerve) eighth cranial nerve; responsible for the senses of hearing and equilibrium (balance).
villi (*sing*: villus) (Latin: *villus* - shaggy hair) fingerlike projections of the mucosa of the small intestine, which increase the surface area for secretion and absorption.
visceral peritoneum (Latin: *viscera* - internal organs; *peritonaeum* - abdominal membrane) the inner layer of the double-layered membrane (the peritoneum) that lines the wall of the abdominopelvic cavity; continuous with the outermost layer of the abdominopelvic organs.
visceral reflex (Latin: *viscera* - internal organs; *reflexus* - a bending back) a reflex where the motor response involves smooth or cardiac muscle or a gland.
visceral pleura (Latin: *viscera* - internal organs) the inner layer of the double-layered pleura, which is in direct contact with the outer surface of the lung; also forms partitions between the lung fissures.
vision (Latin: *visio* - to see) sense of sight based on the transduction of light stimuli by photoreceptors in the eye.
vital capacity (VC) the amount of air that can be forcefully exhaled after a forced inhalation; can be calculated by taking the sum of the TV, ERV, and IRV.
vitamin (*pl*: **vitamins**) (Latin: *vita* - life) any organic substance obtained in the diet that is essential to human metabolism.
vitamin D (Latin: *vita* - life) fat-soluble compound that aids absorption of calcium and phosphates in the intestine; precursor is synthesized in the skin when exposed to UV radiation.
vitamins (*sing*: **vitamin**) (Latin: *vita* - life) organic compounds which are not used as energy sources in the body, but instead are necessary for the utilization of nutrients such as carbohydrates, lipids, and proteins; required in small amounts, and cannot be synthesized in the body.
vitreous humor (Latin: *vitreus* - glassy) jellylike fluid that fills the posterior cavity of the eye.
vocal folds (*aka* vocal cords, true vocal cords) white, membranous folds in the larynx, which produce sound when air is forced through them in their closed position, causing them to vibrate.
voltage-gated channel (*pl*: **voltage-gated channels**) ion channel in the cell membrane of a neuron or muscle that opens because of a change in the electrical properties of the membrane (membrane potential) where it is located.
voltage-gated potassium channels ion channels in the cell membrane of a neuron or muscle cell that open to allow K^+ ions to leave the cell in response to a change in the membrane potential; open during the repolarization phase of an action potential.
voltage-gated sodium channels ion channels in the cell membrane of a neuron or muscle cell that open to allow Na^+ ions to enter the cell in response to a change in the membrane potential; open during the depolarization phase of an action potential.
voluntary phase the first of three phases of the swallowing process, in which a mass (bolus) of food and saliva is propelled from the mouth into the oropharynx
vomer bone single bone that forms the inferior and posterior parts of the nasal septum.
vulva (*aka* pudendum) (Latin: *vulva* - womb) a group of external reproductive organs in biological females that includes the labia majora, labia minora, clitoris, mons pubis, vestibule, and vestibular glands.

W

white blood cells (*aka* leukocytes, leucocytes) cells that defend the body against infection; consist of phagocytic cells (such as monocytes), lymphocytes, and granular cells (such as neutrophils, eosinophils, and basophils).
white matter a mass of myelinated axons in the central nervous system.
white pulp the portion of the spleen that contains B cells and T cells, and macrophages; filter pathogens from the blood and engage in antigen recognition and attack.
white rami communicantes (*aka* white communicating branch, white communicating ramus) (Latin: *ramus* - branch) myelinated preganglionic sympathetic fibers that provide a connection between a sympathetic chain ganglion and the spinal nerve.
withdrawal reflex (Latin: *reflexus* - a bending back) a reflex in which the nociceptors that are activated by a painful stimulus activate the motor neurons responsible for contraction of a muscle to withdraw the body part from the painful stimulus.

X

X-ray form of high-energy electromagnetic radiation capable of penetrating solids; used in medicine as a diagnostic aid to visualize body structures such as bones.
xiphoid process (Greek: *xiphos* - sword) the inferior tip of the sternum; cartilaginous early in life, but ossifies beginning in middle age.

Y

yellow marrow adipose tissue in the cavity of bones where fat is stored for energy.

Z

Z disc (*aka* Z line) a disc of proteins that forms the outer boundary of a sarcomere; anchors the thin filaments in place, and indirectly anchors the thick filaments in place via titins.
zona fasciculata the middle region of the adrenal cortex, which produces glucocorticoid hormones such as cortisol.
zona glomerulosa the outermost region of the adrenal cortex, which produces mineralocorticoid hormones such as aldosterone.
zona reticularis (*aka* reticulate zone) the deepest or innermost region of the adrenal cortex, which produces adrenal androgens.
zone of calcified matrix region of the epiphyseal plate closest to the diaphysis; connects the epiphyseal plate to the diaphysis.
zone of mature cartilage the layer of cartilage in which the chondrocytes are older and larger than those in the proliferative zone.
zygomatic arch (Greek: *zygon* - yoke) arch formed by the temporal process of the zygomatic bone and the temporal bone; located on the lateral skull.
zygomatic bones (Greek: *zygon* - yoke) paired bones that contribute to the lateral wall of the orbit and the anterior zygomatic arch.
zygomatic process of the temporal bone (Greek: *zygon* - yoke; Latin: *temporalis* - temple) anterior projection of the temporal bone; forms the posterior portion of the zygomatic arch.
zygomaticus major (Greek: *zygon* - yoke) the larger of a pair of muscles that draws the corners of the mouth upward, as in smiling; runs from the zygomatic bone to the corners of the mouth.
zygomaticus minor (Greek: *zygon* - yoke) the smaller of a pair of muscles that draws the corners of the mouth upward, as in smiling; runs from the zygomatic bone to the corners of the mouth.
zygote (Greek: *zygōtós* - yoked) a fertilized egg.

Index

Page numbers followed by *f* denote figures; those followed by *t* denote tables, and those followed by *b* denote boxes.

B

C

D

G

H

J

K

L

M

N

Q

R

S

U

V

W

X

Y

Z